Powerful, easy-to-use gradebook offers…

Grades-at-a-glance—Optional color-coding identifies vulnerable students and challenging assignments.

Detailed individual student work—View student answers submitted, hints opened, and time spent on each part.

Collective student work—Assignment Summary shows relative difficulty, time spent, even the most common wrong answers to any exercise. Compare your students' performance with the system average.

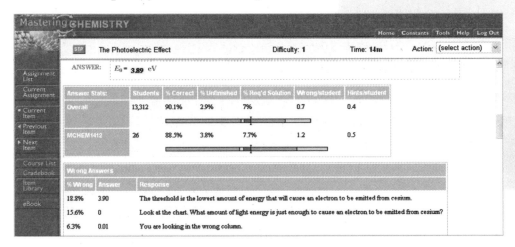

Diagnostics-at-a-glance—With one click, charts summarize the most difficult problems, vulnerable students, grade distribution, even score improvement over time.

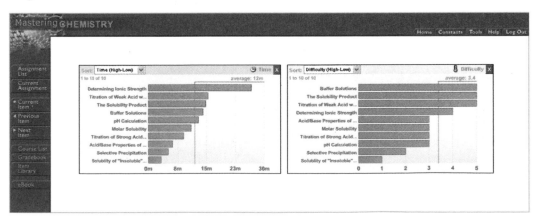

http://www.masteringchemistry.com/

CHEMISTRY
THE CENTRAL SCIENCE

ELEVENTH EDITION

CHEMISTRY
THE CENTRAL SCIENCE

Theodore L. Brown
University of Illinois at Urbana-Champaign

H. Eugene LeMay, Jr.
University of Nevada, Reno

Bruce E. Bursten
University of Tennessee, Knoxville

Catherine J. Murphy
University of South Carolina

With Contributions from
Patrick Woodward
The Ohio State University

PEARSON

Prentice
Hall

Pearson Education International

Editor in Chief, Science: Nicole Folchetti
Acquisitions Editor: Andrew Gilfillan
Associate Editor: Jennifer Hart
Project Manager: Donna Young
Editor in Chief, Development: Ray Mullaney
Development Editor: Karen Nein
Marketing Manager: Elizabeth Averbeck
Senior Operations Supervisor: Alan Fischer
Art Director: Maureen Eide
Assistant Art Director: Suzanne Behnke
Interior Design: John Christiana
Cover Design: Maureen Eide, John Christiana
Composition: Preparé, Inc.
Copy Editor: Marcia Youngman
Proofreader: Karen Bosch
Art Project Manager: Connie Long
Art Studio: Production Solutions
Production Manager/Illustration Art Director: Sean Hogan
Assistant Manager, Art Production: Ronda Whitson
Illustrations: Royce Copenheaver, Daniel Knopsnyder, Mark Landis
Editorial Assistant: Jonathan Colon
Production Assistant: Asha Rohra
Marketing Assistant: Che Flowers
Assessment Content Developer: Kristin Mayo
Lead Media Project Manager: Richard Barnes
Media Project Manager: Nicole Jackson
Assistant Managing Editor, Science: Gina M. Cheselka
Project Manager, Print Supplements: Ashley M. Booth
Manager, Visual Research: Beth Brenzel
Image Permission Coordinator: Debbie Hewitson
Photo Researcher: Truitt & Marshall

© 2009, 2006, 2003, 2000, 1997, 1994, 1991, 1988, 1985, 1981, 1977 by Pearson Education, Inc.
Pearson Education, Inc.
Upper Saddle River, NJ 07458

ISBN 0-13-235848-4
ISBN 978-0-13-235-848-4

Pearson Education Ltd., *London*
Pearson Education Australia Pty. Ltd., *Sydney*
Pearson Education Singapore, Pte. Ltd.
Pearson Education North Asia Ltd., *Hong Kong*
Pearson Education Canada, Inc., *Toronto*
Pearson Educacíon de Mexico, S.A. de C.V.
Pearson Education—Japan, *Tokyo*
Pearson Education Malaysia, Pte. Ltd
Pearson Education, Upper Saddle River, *New Jersey*

To our students, whose enthusiasm and curiosity have often inspired us, and whose questions and suggestions have sometimes taught us.

Brief Contents

Preface xxiv

About the Authors xxxiii

1 Introduction: Matter and Measurement 1

2 Atoms, Molecules, and Ions 36

3 Stoichiometry: Calculations with Chemical Formulas and Equations 78

4 Aqueous Reactions and Solution Stoichiometry 118

5 Thermochemistry 164

6 Electronic Structure of Atoms 210

7 Periodic Properties of the Elements 254

8 Basic Concepts of Chemical Bonding 296

9 Molecular Geometry and Bonding Theories 340

10 Gases 392

11 Intermolecular Forces, Liquids, and Solids 436

12 Modern Materials 480

13 Properties of Solutions 526

14 Chemical Kinetics 572

15 Chemical Equilibrium 626

16 Acid–Base Equilibria 666

17 Additional Aspects of Aqueous Equilibria 718

18 Chemistry of the Environment 766

19 Chemical Thermodynamics 800

20 Electrochemistry 842

21 Nuclear Chemistry 892

22 Chemistry of the Nonmetals 930

23 Metals and Metallurgy 980

24 Chemistry of Coordination Compounds 1012

25 The Chemistry of Life: Organic and Biological Chemistry 1050

Appendices

A Mathematical Operations 1104

B Properties of Water 1111

C Thermodynamic Quantities for Selected Substances at 298.15 K (25 °C) 1112

D Aqueous Equilibrium Constants 1115

E Standard Reduction Potentials at 25 °C 1117

Answers to Selected Exercises A-1

Answers to "Give It Some Thought" A-36

Glossary G-1

Photo/Art Credits P-1

Index I-1

Contents

Preface xxiv

About the Authors xxxiii

1 Introduction: Matter and Measurement 1

1.1 THE STUDY OF CHEMISTRY 2
The Atomic and Molecular Perspective of Chemistry 2   Why Study
Chemistry? 3

1.2 CLASSIFICATIONS OF MATTER 4
States of Matter 4   Pure Substances 5   Elements 6   Compounds 6
Mixtures 8

1.3 PROPERTIES OF MATTER 9
Physical and Chemical Changes 10   Separation of Mixtures 11

1.4 UNITS OF MEASUREMENT 13
SI Units 13   Length and Mass 14   Temperature 15   Derived SI Units 16
Volume 16   Density 17

1.5 UNCERTAINTY IN MEASUREMENT 20
Precision and Accuracy 20   Significant Figures 21   Significant Figures in
Calculations 23

1.6 DIMENSIONAL ANALYSIS 24
Using Two or More Conversion Factors 26   Conversions Involving Volume 27

SUMMARY AND KEY TERMS 29   KEY SKILLS 30
KEY EQUATIONS 30   VISUALIZING CONCEPTS 30
EXERCISES 31   ADDITIONAL EXERCISES 33

Chemistry Put to Work Chemistry and the Chemical Industry 4
A Closer Look The Scientific Method 13
Chemistry Put to Work Chemistry in the News 18
Strategies in Chemistry Estimating Answers 26
Strategies in Chemistry The Importance of Practice 28

2 Atoms, Molecules, and Ions 36

2.1 THE ATOMIC THEORY OF MATTER 38

2.2 THE DISCOVERY OF ATOMIC STRUCTURE 39
Cathode Rays and Electrons 39   Radioactivity 41
The Nuclear Atom 41

2.3 THE MODERN VIEW OF ATOMIC STRUCTURE 43
Atomic Numbers, Mass Numbers, and Isotopes 44

2.4 ATOMIC WEIGHTS 46
The Atomic Mass Scale 46   Average Atomic Masses 47

2.5 THE PERIODIC TABLE 48

2.6 MOLECULES AND MOLECULAR COMPOUNDS 51
Molecules and Chemical Formulas 52   Molecular and Empirical Formulas 52
Picturing Molecules 53

2.7 IONS AND IONIC COMPOUNDS 54
Predicting Ionic Charges 55   Ionic Compounds 56

2.8 NAMING INORGANIC COMPOUNDS 59
Names and Formulas of Ionic Compounds 59   Names and Formulas of Acids 64
Names and Formulas of Binary Molecular Compounds 65

2.9 SOME SIMPLE ORGANIC COMPOUNDS 66
Alkanes 66   Some Derivatives of Alkanes 66

**SUMMARY AND KEY TERMS 68   KEY SKILLS 69
VISUALIZING CONCEPTS 69   EXERCISES 70
ADDITIONAL EXERCISES 75**

A Closer Look Basic Forces 45
A Closer Look The Mass Spectrometer 48
A Closer Look Glenn Seaborg and Seaborgium 51
Chemistry and Life Elements Required by Living Organisms 57
Strategies in Chemistry Pattern Recognition 58

3 Stoichiometry: Calculations with Chemical Formulas and Equations 78

3.1 CHEMICAL EQUATIONS 80
Balancing Equations 80   Indicating the States of Reactants and Products 83

3.2 SOME SIMPLE PATTERNS OF CHEMICAL REACTIVITY 84
Combination and Decomposition Reactions 84   Combustion in Air 86

3.3 FORMULA WEIGHTS 87
Formula and Molecular Weights 87   Percentage Composition from Formulas 88

3.4 AVOGADRO'S NUMBER AND THE MOLE 89
Molar Mass 91   Interconverting Masses and Moles 93   Interconverting Masses and Numbers of Particles 94

3.5 EMPIRICAL FORMULAS FROM ANALYSES 95
Molecular Formula from Empirical Formula 96   Combustion Analysis 97

3.6 QUANTITATIVE INFORMATION FROM BALANCED EQUATIONS 98

3.7 LIMITING REACTANTS 102
Theoretical Yields 105

**SUMMARY AND KEY TERMS 107   KEY SKILLS 108
KEY EQUATIONS 108   VISUALIZING CONCEPTS 108
EXERCISES 109   ADDITIONAL EXERCISES 115
INTEGRATIVE EXERCISES 116**

Strategies in Chemistry Problem Solving 89
Chemistry and Life Glucose Monitoring 102
Strategies in Chemistry How to Take a Test 106

4 Aqueous Reactions and Solution Stoichiometry 118

4.1 GENERAL PROPERTIES OF AQUEOUS SOLUTIONS 120
Electrolytic Properties 120   Ionic Compounds in Water 120   Molecular Compounds in Water 122   Strong and Weak Electrolytes 122

4.2 PRECIPITATION REACTIONS 124
Solubility Guidelines for Ionic Compounds 124   Exchange (Metathesis) Reactions 126   Ionic Equations 127

4.3 ACID–BASE REACTIONS 128
Acids 129   Bases 129   Strong and Weak Acids and Bases 130
Identifying Strong and Weak Electrolytes 131   Neutralization Reactions and Salts 132   Acid–Base Reactions with Gas Formation 134

4.4 OXIDATION-REDUCTION REACTIONS 135
Oxidation and Reduction 136   Oxidation Numbers 137
Oxidation of Metals by Acids and Salts 138   The Activity Series 140

4.5   CONCENTRATIONS OF SOLUTIONS 142

Molarity 144   Expressing the Concentration of an Electrolyte 145
Interconverting Molarity, Moles, and Volume 145   Dilution 146

4.6   SOLUTION STOICHIOMETRY AND CHEMICAL ANALYSIS 149

Titrations 150

**SUMMARY AND KEY TERMS 154   KEY SKILLS 155
KEY EQUATIONS 155   VISUALIZING CONCEPTS 156
EXERCISES 156   ADDITIONAL EXERCISES 161
INTEGRATIVE EXERCISES 162**

Chemistry Put to Work Antacids 135
A Closer Look The Aura of Gold 143
Strategies in Chemistry Analyzing Chemical Reactions 143
Chemistry and Life Drinking Too Much Water Can Kill You 147

5 Thermochemistry   164

5.1   THE NATURE OF ENERGY 166

Kinetic Energy and Potential Energy 166   Units of Energy 167   System and
Surroundings 168   Transferring Energy: Work and Heat 168

5.2   THE FIRST LAW OF THERMODYNAMICS 170

Internal Energy 170   Relating ΔE to Heat and Work 171   Endothermic and
Exothermic Processes 172   State Functions 172

5.3   ENTHALPY 174

5.4   ENTHALPIES OF REACTION 177

5.5   CALORIMETRY 179

Heat Capacity and Specific Heat 179   Constant-Pressure Calorimetry 182
Bomb Calorimetry (Constant-Volume Calorimetry) 183

5.6   HESS'S LAW 184

5.7   ENTHALPIES OF FORMATION 188

Using Enthalpies of Formation to Calculate Enthalpies of Reaction 190

5.8   FOODS AND FUELS 193

Foods 193   Fuels 195   Other Energy Sources 197

**SUMMARY AND KEY TERMS 199   KEY SKILLS 200
KEY EQUATIONS 201   VISUALIZING CONCEPTS 201
EXERCISES 202   ADDITIONAL EXERCISES 207
INTEGRATIVE EXERCISES 209**

A Closer Look Energy, Enthalpy, and P-V Work 176
Strategies in Chemistry Using Enthalpy as a Guide 180
Chemistry and Life The Regulation of Human Body Temperature 185
Chemistry Put to Work The Hybrid Car 196

6 Electronic Structure of Atoms   210

6.1   THE WAVE NATURE OF LIGHT 212

6.2   QUANTIZED ENERGY AND PHOTONS 215

Hot Objects and the Quantization of Energy 215   The Photoelectric Effect
and Photons 216

6.3   LINE SPECTRA AND THE BOHR MODEL 218

Line Spectra 218   Bohr's Model 219   The Energy States of the Hydrogen
Atom 220   Limitations of the Bohr Model 222

6.4   THE WAVE BEHAVIOR OF MATTER 222

The Uncertainty Principle 223

6.5   QUANTUM MECHANICS AND ATOMIC ORBITALS 224

Orbitals and Quantum Numbers 226

6.6   REPRESENTATIONS OF ORBITALS 228
The *s* Orbitals 228   The *p* Orbitals 231   The *d* and *f* Orbitals 232

6.7   MANY-ELECTRON ATOMS 232
Orbitals and Their Energies 232   Electron Spin and the Pauli Exclusion Principle 233

6.8   ELECTRON CONFIGURATIONS 234
Hund's Rule 235   Condensed Electron Configurations 237
Transition Metals 238   The Lanthanides and Actinides 239

6.9   ELECTRON CONFIGURATIONS AND THE PERIODIC TABLE 240
Anomalous Electron Configurations 243

SUMMARY AND KEY TERMS 244   KEY SKILLS 245
KEY EQUATIONS 246   VISUALIZING CONCEPTS 246
EXERCISES 247   ADDITIONAL EXERCISES 251
INTEGRATIVE EXERCISES 253

A Closer Look The Speed of Light 214
A Closer Look Measurement and the Uncertainty Principle 225
A Closer Look Probability Density and Radial Probability Functions 230
A Closer Look Experimental Evidence for Electron Spin 234
Chemistry and Life Nuclear Spin and Magnetic Resonance Imaging 236

7   Periodic Properties of the Elements   254

7.1   DEVELOPMENT OF THE PERIODIC TABLE 256

7.2   EFFECTIVE NUCLEAR CHARGE 257

7.3   SIZES OF ATOMS AND IONS 259
Periodic Trends in Atomic Radii 262   Periodic Trends in Ionic Radii 263

7.4   IONIZATION ENERGY 264
Variations in Successive Ionization Energies 266   Periodic Trends in First Ionization Energies 267   Electron Configurations of Ions 268

7.5   ELECTRON AFFINITIES 270

7.6   METALS, NONMETALS, AND METALLOIDS 271
Metals 272   Nonmetals 274   Metalloids 276

7.7   GROUP TRENDS FOR THE ACTIVE METALS 276
Group 1A: The Alkali Metals 276   Group 2A: The Alkaline Earth Metals 279

7.8   GROUP TRENDS FOR SELECTED NONMETALS 281
Hydrogen 281   Group 6A: The Oxygen Group 282   Group 7A:
The Halogens 283   Group 8A: The Noble Gases 284

SUMMARY AND KEY TERMS 286   KEY SKILLS 287
KEY EQUATIONS 287   VISUALIZING CONCEPTS 288
EXERCISES 289   ADDITIONAL EXERCISES 292
INTEGRATIVE EXERCISES 294

A Closer Look Effective Nuclear Charge 260
Chemistry and Life Ionic Size Makes a Big Difference 265
Chemistry and Life The Improbable Development of Lithium Drugs 280

8   Basic Concepts of Chemical Bonding   296

8.1   CHEMICAL BONDS, LEWIS SYMBOLS, AND THE OCTET RULE 298
Lewis Symbols 298   The Octet Rule 299

8.2   IONIC BONDING 299
Energetics of Ionic Bond Formation 301   Electron Configurations of Ions of the *s*- and *p*-Block Elements 302   Transition-Metal Ions 303

8.3 COVALENT BONDING 305
Lewis Structures 305   Multiple Bonds 307

8.4 BOND POLARITY AND ELECTRONEGATIVITY 307
Electronegativity 308   Electronegativity and Bond Polarity 308
Dipole Moments 310   Differentiating Ionic and Covalent Bonding 312

8.5 DRAWING LEWIS STRUCTURES 314
Formal Charge 316

8.6 RESONANCE STRUCTURES 319
Resonance in Benzene 321

8.7 EXCEPTIONS TO THE OCTET RULE 322
Odd Number of Electrons 322   Less than an Octet of Valence Electrons 322
More than an Octet of Valence Electrons 323

8.8 STRENGTHS OF COVALENT BONDS 325
Bond Enthalpies and the Enthalpies of Reactions 326
Bond Enthalpy and Bond Length 329

SUMMARY AND KEY TERMS 331   KEY SKILLS 332
KEY EQUATIONS 332   VISUALIZING CONCEPTS 332
EXERCISES 333   ADDITIONAL EXERCISES 336
INTEGRATIVE EXERCISES 338

A Closer Look Calculation of Lattice Energies: The Born–Haber Cycle 304
A Closer Look Oxidation Numbers, Formal Charges, and Actual Partial Charges 318
Chemistry Put to Work Explosives and Alfred Nobel 328

9 Molecular Geometry and Bonding Theories   340

9.1 MOLECULAR SHAPES 342

9.2 THE VSEPR MODEL 344
The Effect of Nonbonding Electrons and Multiple Bonds on
Bond Angles 348   Molecules with Expanded Valence Shells 349
Shapes of Larger Molecules 352

9.3 MOLECULAR SHAPE AND MOLECULAR POLARITY 353

9.4 COVALENT BONDING AND ORBITAL OVERLAP 355

9.5 HYBRID ORBITALS 357
sp Hybrid Orbitals 357 sp^2 and sp^3 Hybrid Orbitals 358
Hybridization Involving d Orbitals 360   Hybrid Orbital Summary 360

9.6 MULTIPLE BONDS 362
Resonance Structures, Delocalization, and π Bonding 365
General Conclusions 368

9.7 MOLECULAR ORBITALS 368
The Hydrogen Molecule 368   Bond Order 370

9.8 SECOND-ROW DIATOMIC MOLECULES 371
Molecular Orbitals for Li_2 and Be_2 371   Molecular Orbitals from
$2p$ Atomic Orbitals 372   Electron Configurations for B_2 Through Ne_2 375
Electron Configurations and Molecular Properties 376
Heteronuclear Diatomic Molecules 379

SUMMARY AND KEY TERMS 381   KEY SKILLS 383
KEY EQUATIONS 383   VISUALIZING CONCEPTS 383
EXERCISES 384   ADDITIONAL EXERCISES 388
INTEGRATIVE EXERCISES 390

Chemistry and Life The Chemistry of Vision 367
A Closer Look Phases in Atomic and Molecular Orbitals 373
Chemistry Put to Work Orbitals and Energy 380

10 Gases   392

10.1 CHARACTERISTICS OF GASES 394

10.2 PRESSURE 395
Atmospheric Pressure and the Barometer 395

10.3 THE GAS LAWS 398
The Pressure-Volume Relationship: Boyle's Law 399   The Temperature-Volume Relationship: Charles's Law 400   The Quantity-Volume Relationship: Avogadro's Law 400

10.4 THE IDEAL-GAS EQUATION 402
Relating the Ideal-Gas Equation and the Gas Laws 404

10.5 FURTHER APPLICATIONS OF THE IDEAL-GAS EQUATION 406
Gas Densities and Molar Mass 406   Volumes of Gases in Chemical Reactions 408

10.6 GAS MIXTURES AND PARTIAL PRESSURES 410
Partial Pressures and Mole Fractions 411   Collecting Gases over Water 412

10.7 KINETIC-MOLECULAR THEORY 414
Distributions of Molecular Speed 414   Application to the Gas Laws 415

10.8 MOLECULAR EFFUSION AND DIFFUSION 417
Graham's Law of Effusion 418   Diffusion and Mean Free Path 419

10.9 REAL GASES: DEVIATIONS FROM IDEAL BEHAVIOR 420
The van der Waals Equation 422

**SUMMARY AND KEY TERMS 425   KEY SKILLS 426
KEY EQUATIONS 426   VISUALIZING CONCEPTS 426
EXERCISES 427   ADDITIONAL EXERCISES 432
INTEGRATIVE EXERCISES 434**

Chemistry and Life Blood Pressure 398
Strategies in Chemistry Calculations Involving Many Variables 404
Chemistry Put to Work Gas Pipelines 409
A Closer Look The Ideal-Gas Equation 416
Chemistry Put to Work Gas Separations 420

11 Intermolecular Forces, Liquids, and Solids   436

11.1 A MOLECULAR COMPARISON OF GASES, LIQUIDS, AND SOLIDS 438

11.2 INTERMOLECULAR FORCES 439
Ion–Dipole Forces 440   Dipole–Dipole Forces 440   London Dispersion Forces 441   Hydrogen Bonding 443   Comparing Intermolecular Forces 446

11.3 SOME PROPERTIES OF LIQUIDS 447
Viscosity 447   Surface Tension 448

11.4 PHASE CHANGES 449
Energy Changes Accompanying Phase Changes 449   Heating Curves 450   Critical Temperature and Pressure 452

11.5 VAPOR PRESSURE 453
Explaining Vapor Pressure on the Molecular Level 454   Volatility, Vapor Pressure, and Temperature 454   Vapor Pressure and Boiling Point 455

11.6 PHASE DIAGRAMS 456
The Phase Diagrams of H_2O and CO_2 457

11.7 STRUCTURES OF SOLIDS 458
Unit Cells 460   The Crystal Structure of Sodium Chloride 461
Close Packing of Spheres 463

11.8 BONDING IN SOLIDS 464
Molecular Solids 464   Covalent-Network Solids 466   Ionic Solids 467
Metallic Solids 467

SUMMARY AND KEY TERMS 469   KEY SKILLS 470
VISUALIZING CONCEPTS 471   EXERCISES 472
ADDITIONAL EXERCISES 477   INTEGRATIVE EXERCISES 478

Chemistry Put to Work Supercritical Fluid Extraction 453
A Closer Look The Clausius–Clapeyron Equation 456
A Closer Look X-Ray Diffraction by Crystals 465
A Closer Look The Third Form of Carbon 468

12 Modern Materials   480

12.1 CLASSES OF MATERIALS 482

12.2 ELECTRONIC STRUCTURE OF MATERIALS 482

12.3 SEMICONDUCTORS 484
Semiconductor Doping 488   The Silicon Chip 489   Solar Energy
Conversion 489   Semiconductor Light-Emitting Diodes 492

12.4 CERAMICS 493
Applications of Ceramics 494   Making Ceramics 494

12.5 SUPERCONDUCTORS 495
Ceramic Superconductors 497

12.6 POLYMERS AND PLASTICS 499
Making Polymers 499   Structure and Physical Properties of Polymers 502

12.7 BIOMATERIALS 505
Characteristics of Biomaterials 507   Polymeric Biomaterials 507
Heart Repair 508   Vascular Grafts 509   Artificial Tissue 509

12.8 LIQUID CRYSTALS 510
Types of Liquid Crystalline Phases 511

12.9 NANOMATERIALS 513
Semiconductors on the Nanoscale 513   Metals on the Nanoscale 515
Carbon Nanotubes 515

SUMMARY AND KEY TERMS 517   KEY SKILLS 519
VISUALIZING CONCEPTS 519   EXERCISES 520
ADDITIONAL EXERCISES 524   INTEGRATIVE EXERCISES 525

A Closer Look The Transistor 490
Chemistry Put to Work Cell Phone Tower Range 498
Chemistry Put to Work Recycling Plastics 501
Chemistry Put to Work Toward the Plastic Car 506
Chemistry Put to Work Liquid Crystal Displays 513

13 Properties of Solutions   526

13.1 THE SOLUTION PROCESS 528
The Effect of Intermolecular Forces 528   Energy Changes and Solution
Formation 529   Solution Formation, Spontaneity, and Entropy 531
Solution Formation and Chemical Reactions 533

13.2 SATURATED SOLUTIONS AND SOLUBILITY 534

13.3 FACTORS AFFECTING SOLUBILITY 535
Solute–Solvent Interactions 536   Pressure Effects 539
Temperature Effects 541

13.4 WAYS OF EXPRESSING CONCENTRATION 542
Mass Percentage, ppm, and ppb 542   Mole Fraction, Molarity, and Molality 543   Conversion of Concentration Units 544

13.5 COLLIGATIVE PROPERTIES 546
Lowering the Vapor Pressure 546   Boiling-Point Elevation 549
Freezing-Point Depression 550   Osmosis 551   Determination of Molar Mass 555

13.6 COLLOIDS 556
Hydrophilic and Hydrophobic Colloids 557   Removal of Colloidal Particles 560

**SUMMARY AND KEY TERMS 561   KEY SKILLS 562
KEY EQUATIONS 563   VISUALIZING CONCEPTS 563
EXERCISES 565   ADDITIONAL EXERCISES 569
INTEGRATIVE EXERCISES 570**

A Closer Look Hydrates 533
Chemistry and Life Fat- and Water-Soluble Vitamins 538
Chemistry and Life Blood Gases and Deep-Sea Diving 540
A Closer Look Ideal Solutions with Two or More Volatile Components 548
A Closer Look Colligative Properties of Electrolyte Solutions 554
Chemistry and Life Sickle-Cell Anemia 559

14 Chemical Kinetics   572

14.1 FACTORS THAT AFFECT REACTION RATES 574

14.2 REACTION RATES 575
Change of Rate with Time 577   Instantaneous Rate 577   Reaction Rates and Stoichiometry 578

14.3 THE RATE LAW: THE EFFECT OF CONCENTRATION ON RATE 580
Reaction Orders: The Exponents in the Rate Law 581   Units of Rate Constants 583   Using Initial Rates to Determine Rate Laws 584

14.4 THE CHANGE OF CONCENTRATION WITH TIME 585
First-Order Reactions 586   Second-Order Reactions 588
Half-life 589

14.5 TEMPERATURE AND RATE 591
The Collision Model 592   The Orientation Factor 592
Activation Energy 592   The Arrhenius Equation 594   Determining the Activation Energy 595

14.6 REACTION MECHANISMS 597
Elementary Reactions 597   Multistep Mechanisms 598   Rate Laws for Elementary Reactions 599   The Rate-Determining Step for a Multistep Mechanism 600   Mechanisms with a Slow Initial Step 601   Mechanisms with a Fast Initial Step 602

14.7 CATALYSIS 605
Homogeneous Catalysis 605   Heterogeneous Catalysis 606
Enzymes 608

**SUMMARY AND KEY TERMS 613   KEY SKILLS 614
KEY EQUATIONS 614   VISUALIZING CONCEPTS 615
EXERCISES 616   ADDITIONAL EXERCISES 622
INTEGRATIVE EXERCISES 624**

A Closer Look Using Spectroscopic Methods to Measure Reaction Rates 580
Chemistry Put to Work Methyl Bromide in the Atmosphere 590
Chemistry Put to Work Catalytic Converters 608
Chemistry and Life Nitrogen Fixation and Nitrogenase 610

15 Chemical Equilibrium   626

15.1   THE CONCEPT OF EQUILIBRIUM 628

15.2   THE EQUILIBRIUM CONSTANT 630
Evaluating K_c 632   Equilibrium Constants in Terms of Pressure, K_p 633
Equilibrium Constants and Units 635

**15.3   INTERPRETING AND WORKING WITH EQUILIBRIUM
CONSTANTS 635**
The Magnitude of Equilibrium Constants 635   The Direction of the
Chemical Equation and K 636   Relating Chemical Equations and
Equilibrium Constants 637

15.4   HETEROGENEOUS EQUILIBRIA 639

15.5   CALCULATING EQUILIBRIUM CONSTANTS 641

15.6   APPLICATIONS OF EQUILIBRIUM CONSTANTS 644
Predicting the Direction of Reaction 644   Calculating Equilibrium
Concentrations 645

15.7   LE CHÂTELIER'S PRINCIPLE 648
Change in Reactant or Product Concentrations 649   Effects of Volume and
Pressure Changes 649   Effect of Temperature Changes 651   The Effect of
Catalysts 654

**SUMMARY AND KEY TERMS 657   KEY SKILLS 657
KEY EQUATIONS 658   VISUALIZING CONCEPTS 658
EXERCISES 660   ADDITIONAL EXERCISES 662
INTEGRATIVE EXERCISES 664**

Chemistry Put to Work Haber Process 631
Chemistry Put to Work Controlling Nitric Oxide Emissions 656

16 Acid–Base Equilibria   666

16.1   ACIDS AND BASES: A BRIEF REVIEW 668

16.2   BRØNSTED–LOWRY ACIDS AND BASES 668
The H^+ Ion in Water 669   Proton-Transfer Reactions 669   Conjugate
Acid–Base Pairs 670   Relative Strengths of Acids and Bases 672

16.3   THE AUTOIONIZATION OF WATER 673
The Ion Product of Water 674

16.4   THE pH SCALE 675
pOH and Other "p" Scales 678   Measuring pH 678

16.5   STRONG ACIDS AND BASES 679
Strong Acids 679   Strong Bases 680

16.6   WEAK ACIDS 681
Calculating K_a from pH 682   Percent Ionization 683   Using K_a to Calculate
pH 684   Polyprotic Acids 688

16.7   WEAK BASES 690
Types of Weak Bases 692

16.8   RELATIONSHIP BETWEEN K_a AND K_b 693

16.9   ACID–BASE PROPERTIES OF SALT SOLUTIONS 695
An Anion's Ability to React with Water 696   A Cation's Ability to React with
Water 696   Combined Effect of Cation and Anion in Solution 697

16.10   ACID–BASE BEHAVIOR AND CHEMICAL STRUCTURE 699
Factors That Affect Acid Strength 699   Binary Acids 699
Oxyacids 700   Carboxylic Acids 702

16.11   LEWIS ACIDS AND BASES 704
Hydrolysis of Metal Ions 705

**SUMMARY AND KEY TERMS 707   KEY SKILLS 708
KEY EQUATIONS 708   VISUALIZING CONCEPTS 709
EXERCISES 710   ADDITIONAL EXERCISES 715
INTEGRATIVE EXERCISES 716**

Chemistry Put to Work Amines and Amine Hydrochlorides 694
Chemistry and Life The Amphiprotic Behavior of Amino Acids 703

17 Additional Aspects of Aqueous Equilibria   718

17.1   THE COMMON-ION EFFECT 720

17.2   BUFFERED SOLUTIONS 723
Composition and Action of Buffered Solutions 723   Calculating the pH of a Buffer 724   Buffer Capacity and pH Range 726   Addition of Strong Acids or Bases to Buffers 727

17.3   ACID–BASE TITRATIONS 730
Strong Acid–Strong Base Titrations 730   Weak Acid–Strong Base Titrations 733   Titrations of Polyprotic Acids 737

17.4   SOLUBILITY EQUILIBRIA 737
The Solubility-Product Constant, K_{sp} 738   Solubility and K_{sp} 738

17.5   FACTORS THAT AFFECT SOLUBILITY 741
Common-Ion Effect 741   Solubility and pH 742   Formation of Complex Ions 745   Amphoterism 748

17.6   PRECIPITATION AND SEPARATION OF IONS 750
Selective Precipitation of Ions 751

17.7   QUALITATIVE ANALYSIS FOR METALLIC ELEMENTS 753
**SUMMARY AND KEY TERMS 756   KEY SKILLS 757
KEY EQUATIONS 757   VISUALIZING CONCEPTS 757
EXERCISES 759   ADDITIONAL EXERCISES 763
INTEGRATIVE EXERCISES 764**

Chemistry and Life Blood as a Buffered Solution 729
A Closer Look Limitations of Solubility Products 741
Chemistry and Life Sinkholes 744
Chemistry and Life Tooth Decay and Fluoridation 747

18 Chemistry of the Environment   766

18.1   EARTH'S ATMOSPHERE 768
Composition of the Atmosphere 769

18.2   OUTER REGIONS OF THE ATMOSPHERE 770
Photodissociation 770   Photoionization 772

18.3   OZONE IN THE UPPER ATMOSPHERE 772
Depletion of the Ozone Layer 774

18.4   CHEMISTRY OF THE TROPOSPHERE 775
Sulfur Compounds and Acid Rain 776   Carbon Monoxide 778   Nitrogen Oxides and Photochemical Smog 779   Water Vapor, Carbon Dioxide, and Climate 780

18.5   THE WORLD OCEAN 783
Seawater 783   Desalination 784

18.6   FRESHWATER 785
Dissolved Oxygen and Water Quality 786   Treatment of Municipal Water Supplies 786

18.7   GREEN CHEMISTRY 788
Solvents and Reagents 789   Other Processes 790   The Challenges of Water
Purification 790
SUMMARY AND KEY TERMS 792   KEY SKILLS 793
VISUALIZING CONCEPTS 794   EXERCISES 794
ADDITIONAL EXERCISES 797   INTEGRATIVE EXERCISES 798

A Closer Look Other Greenhouse Gases 782
A Closer Look Water Softening 788

19 Chemical Thermodynamics   800

19.1   SPONTANEOUS PROCESSES 802
Seeking a Criterion for Spontaneity 804   Reversible and
Irreversible Processes 804

**19.2   ENTROPY AND THE SECOND LAW
OF THERMODYNAMICS 806**
Entropy Change 806 ΔS for Phase Changes 807   The Second Law of
Thermodynamics 808

19.3   THE MOLECULAR INTERPRETATION OF ENTROPY 809
Molecular Motions and Energy 809   Boltzmann's Equation
and Microstates 810   Making Qualitative Predictions About ΔS 813
The Third Law of Thermodynamics 816

19.4   ENTROPY CHANGES IN CHEMICAL REACTIONS 817
Entropy Changes in the Surroundings 818

19.5   GIBBS FREE ENERGY 819
Standard Free-Energy of Formation 822

19.6   FREE ENERGY AND TEMPERATURE 824

19.7   FREE ENERGY AND THE EQUILIBRIUM CONSTANT 826
SUMMARY AND KEY TERMS 832   KEY SKILLS 832
KEY EQUATIONS 833   VISUALIZING CONCEPTS 833
EXERCISES 834   ADDITIONAL EXERCISES 839
INTEGRATIVE EXERCISES 840

A Closer Look The Entropy Change when a Gas Expands Isothermally 808
A Closer Look Entropy and Probability 812
Chemistry and Life Entropy and Life 815
A Closer Look What's "Free" about Free Energy? 822
Chemistry and Life Driving Nonspontaneous Reactions 830

20 Electrochemistry   842

**20.1   OXIDATION STATES AND OXIDATION-REDUCTION
REACTIONS 844**

20.2   BALANCING OXIDATION-REDUCTION EQUATIONS 846
Half-Reactions 846   Balancing Equations by the Method
of Half-Reactions 846   Balancing Equations for Reactions Occurring
in Basic Solution 849

20.3   VOLTAIC CELLS 851
A Molecular View of Electrode Processes 854

20.4   CELL EMF UNDER STANDARD CONDITIONS 855
Standard Reduction (Half-Cell) Potentials 856   Strengths of Oxidizing and
Reducing Agents 860

20.5   FREE ENERGY AND REDOX REACTIONS 862
EMF and ΔG 863

20.6   CELL EMF UNDER NONSTANDARD CONDITIONS 865
The Nernst Equation 865   Concentration Cells 867

20.7   BATTERIES AND FUEL CELLS 870
Lead-Acid Battery 871   Alkaline Battery 872   Nickel-Cadmium, Nickel-Metal-Hydride, and Lithium-Ion Batteries 872   Hydrogen Fuel Cells 873

20.8   CORROSION 874
Corrosion of Iron 874   Preventing the Corrosion of Iron 875

20.9   ELECTROLYSIS 876
Quantitative Aspects of Electrolysis 878   Electrical Work 879

**SUMMARY AND KEY TERMS 881   KEY SKILLS 882
KEY EQUATIONS 883   VISUALIZING CONCEPTS 883
EXERCISES 884   ADDITIONAL EXERCISES 889
INTEGRATIVE EXERCISES 890**

Chemistry and Life Heartbeats and Electrocardiography 868
Chemistry Put to Work Direct Methanol Fuel Cells 874

21 Nuclear Chemistry   892

21.1   RADIOACTIVITY 894
Nuclear Equations 895   Types of Radioactive Decay 896

21.2   PATTERNS OF NUCLEAR STABILITY 898
Neutron-to-Proton Ratio 898   Radioactive Series 900
Further Observations 901

21.3   NUCLEAR TRANSMUTATIONS 901
Accelerating Charged Particles 902   Using Neutrons 903
Transuranium Elements 903

21.4   RATES OF RADIOACTIVE DECAY 903
Radiometric Dating 905   Calculations Based on Half-life 905

21.5   DETECTION OF RADIOACTIVITY 908
Radiotracers 910

21.6   ENERGY CHANGES IN NUCLEAR REACTIONS 911
Nuclear Binding Energies 912

21.7   NUCLEAR POWER: FISSION 913
Nuclear Reactors 915

21.8   NUCLEAR POWER: FUSION 918

21.9   RADIATION IN THE ENVIRONMENT AND LIVING SYSTEMS 919
Radiation Doses 920   Radon 921

**SUMMARY AND KEY TERMS 924   KEY SKILLS 925
KEY EQUATIONS 925   VISUALIZING CONCEPTS 925
EXERCISES 926   ADDITIONAL EXERCISES 928
INTEGRATIVE EXERCISES 929**

Chemistry and Life Medical Applications of Radiotracers 910
A Closer Look The Dawning of the Nuclear Age 915
Chemistry Put to Work Environmental Applications of Radioisotopes 921
Chemistry and Life Radiation Therapy 922

22 Chemistry of the Nonmetals   930

**22.1   GENERAL CONCEPTS: PERIODIC TRENDS
AND CHEMICAL REACTIONS 932**
Chemical Reactions 933

22.2   HYDROGEN 935
Isotopes of Hydrogen 935   Properties of Hydrogen 936   Preparation of Hydrogen 936   Uses of Hydrogen 937   Binary Hydrogen Compounds 937

22.3   GROUP 8A: THE NOBLE GASES 938
Noble-Gas Compounds 939

22.4   GROUP 7A: THE HALOGENS 940
Properties and Preparation of the Halogens 940   Uses of the Halogens 941
The Hydrogen Halides 942   Interhalogen Compounds 943   Oxyacids and
Oxyanions 943

22.5   OXYGEN 944
Properties of Oxygen 944   Preparation of Oxygen 944   Uses of Oxygen
945   Ozone 945   Oxides 946   Peroxides and Superoxides 948

22.6   THE OTHER GROUP 6A ELEMENTS: S, Se, Te, AND Po 948
General Characteristics of the Group 6A Elements 949   Occurrence and
Preparation of S, Se, and Te 949   Properties and Uses of Sulfur,
Selenium, and Tellurium 949   Sulfides 950   Oxides, Oxyacids, and
Oxyanions of Sulfur 950

22.7   NITROGEN 952
Properties of Nitrogen 952   Preparation and Uses of Nitrogen 952
Hydrogen Compounds of Nitrogen 953   Oxides and Oxyacids
of Nitrogen 954

22.8   THE OTHER GROUP 5A ELEMENTS: P, As, Sb, AND Bi 956
General Characteristics of the Group 5A Elements 956   Occurrence, Isolation,
and Properties of Phosphorus 957   Phosphorus Halides 957
Oxy Compounds of Phosphorus 958

22.9   CARBON 960
Elemental Forms of Carbon 961   Oxides of Carbon 962   Carbonic Acid
and Carbonates 963   Carbides 964   Other Inorganic Compounds of
Carbon 965

22.10   THE OTHER GROUP 4A ELEMENTS: Si, Ge, Sn, AND Pb 965
General Characteristics of the Group 4A Elements 965   Occurrence and
Preparation of Silicon 966   Silicates 966   Glass 968   Silicones 969

22.11   BORON 969
SUMMARY AND KEY TERMS 971   KEY SKILLS 972
VISUALIZING CONCEPTS 973   EXERCISES 974
ADDITIONAL EXERCISES 977   INTEGRATIVE EXERCISES 978
A Closer Look The Hydrogen Economy 937
Chemistry and Life How Much Perchlorate Is Too Much? 944
Chemistry and Life Nitroglycerin and Heart Disease 956
Chemistry and Life Arsenic in Drinking Water 960
Chemistry Put to Work Carbon Fibers and Composites 962

23 Metals and Metallurgy   980

23.1   OCCURRENCE AND DISTRIBUTION OF METALS 982
Minerals 982   Metallurgy 983

23.2   PYROMETALLURGY 984
The Pyrometallurgy of Iron 985   Formation of Steel 986

23.3   HYDROMETALLURGY 987
Hydrometallurgy of Gold 987   Hydrometallurgy of Aluminum 988

23.4   ELECTROMETALLURGY 988
Electrometallurgy of Sodium 988   Electrometallurgy of Aluminum 989
Electrorefining of Copper 990

23.5   METALLIC BONDING 991
Physical Properties of Metals 992   Electron-Sea Model
for Metallic Bonding 992   Molecular-Orbital Model for Metals 994

23.6   ALLOYS 995
Intermetallic Compounds 997

23.7   TRANSITION METALS 998

Physical Properties 999   Electron Configurations and Oxidation States 1000
Magnetism 1001

23.8   CHEMISTRY OF SELECTED TRANSITION METALS 1002

Chromium 1002   Iron 1003   Copper 1004

SUMMARY AND KEY TERMS 1006   KEY SKILLS 1007
VISUALIZING CONCEPTS 1007   EXERCISES 1008
ADDITIONAL EXERCISES 1010   INTEGRATIVE EXERCISES 1011

A Closer Look Charles M. Hall 990
A Closer Look Shape-Memory Alloys 998

24 Chemistry of Coordination Compounds   1012

24.1   METAL COMPLEXES 1014

The Development of Coordination Chemistry: Werner's Theory 1014
The Metal–Ligand Bond 1016   Charges, Coordination Numbers, and
Geometries 1016

24.2   LIGANDS WITH MORE THAN ONE DONOR ATOM 1019

Metals and Chelates in Living Systems 1021

24.3   NOMENCLATURE OF COORDINATION CHEMISTRY 1025

24.4   ISOMERISM 1026

Structural Isomerism 1027   Stereoisomerism 1027

24.5   COLOR AND MAGNETISM 1031

Color 1031   Magnetism 1033

24.6   CRYSTAL-FIELD THEORY 1033

Electron Configurations in Octahedral Complexes 1037   Tetrahedral and
Square-Planar Complexes 1038

SUMMARY AND KEY TERMS 1042   KEY SKILLS 1043
VISUALIZING CONCEPTS 1043   EXERCISES 1044
ADDITIONAL EXERCISES 1046   INTEGRATIVE EXERCISES 1048

A Closer Look Entropy and the Chelate Effect 1021
Chemistry and Life The Battle for Iron in Living Systems 1024
A Closer Look Charge-Transfer Color 1040

25 The Chemistry of Life: Organic and Biological Chemistry   1050

25.1   SOME GENERAL CHARACTERISTICS OF ORGANIC MOLECULES 1052

The Structures of Organic Molecules 1052   The Stabilities of Organic
Substances 1053   Solubility and Acid–Base Properties of Organic
Substances 1054

25.2   INTRODUCTION TO HYDROCARBONS 1054

25.3   ALKANES, ALKENES, AND ALKYNES 1055

Structures of Alkanes 1056   Structural Isomers 1057   Nomenclature of
Alkanes 1058   Cycloalkanes 1060   Reactions of Alkanes 1061   Alkenes
1062   Alkynes 1064   Addition Reactions of Alkenes and Alkynes 1065
Mechanism of Addition Reactions 1066   Aromatic Hydrocarbons 1068

25.4   ORGANIC FUNCTIONAL GROUPS 1070

Alcohols 1072   Ethers 1073   Aldehydes and Ketones 1073   Carboxylic
Acids and Esters 1074   Amines and Amides 1077

25.5   **CHIRALITY IN ORGANIC CHEMISTRY** 1078

25.6   **INTRODUCTION TO BIOCHEMISTRY** 1080

25.7   **PROTEINS** 1080

Amino Acids 1080   Polypeptides and Proteins 1082

Protein Structure 1084

25.8   **CARBOHYDRATES** 1086

Disaccharides 1087   Polysaccharides 1088

25.9   **LIPIDS** 1089

25.10 **NUCLEIC ACIDS** 1090

SUMMARY AND KEY TERMS 1095   **KEY SKILLS** 1096
VISUALIZING CONCEPTS 1096   **EXERCISES** 1097
ADDITIONAL EXERCISES 1102   **INTEGRATIVE EXERCISES** 1103

Chemistry Put to Work Gasoline 1061
Chemistry and Life Polycyclic Aromatic Hydrocarbons 1070
Chemistry Put to Work Portrait of an Organic Chemical 1076
Chemistry and Life The Origins of Chirality in Living Systems 1085
Strategies in Chemistry What Now? 1094

Appendices

A   Mathematical Operations 1104

B   Properties of Water 1111

C   Thermodynamic Quantities for Selected Substances
at 298.15 K (25 °C) 1112

D   Aqueous Equilibrium Constants 1115

E   Standard Reduction Potentials at 25 °C 1117

Answers to Selected Exercises A-1

Answers to "Give It Some Thought" A-36

Glossary G-1

Photo/Art Credits P-1

Index I-1

Chemical Applications and Essays

Strategies in Chemistry

Estimating Answers 26
The Importance of Practice 28
Pattern Recognition 58
Problem Solving 89
How to Take a Test 106
Analyzing Chemical Reactions 143
Using Enthalpy as a Guide 180
Calculations Involving Many Variables 404
What Now? 1094

Chemistry and Life

Elements Required by Living Organisms 57
Glucose Monitoring 102
Drinking Too Much Water Can Kill You 147
The Regulation of Human Body Temperature 185
Nuclear Spin and Magnetic Resonance Imaging 236
Ionic Size Makes a Big Difference 265
The Improbable Development of Lithium Drugs 280
The Chemistry of Vision 367
Blood Pressure 398
Fat- and Water-Soluble Vitamins 538
Blood Gases and Deep-Sea Diving 540
Sickle-Cell Anemia 559
Nitrogen Fixation and Nitrogenase 610
The Amphiprotic Behavior of Amino Acids 703
Blood as a Buffered Solution 729
Sinkholes 744
Tooth Decay and Fluoridation 747
Entropy and Life 815
Driving Nonspontaneous Reactions 830
Heartbeats and Electrocardiography 868
Medical Applications of Radiotracers 910
Radiation Therapy 922
How Much Perchlorate Is Too Much? 944
Nitroglycerin and Heart Disease 956
Arsenic in Drinking Water 960

The Battle for Iron in Living Systems 1024
Polycyclic Aromatic Hydrocarbons 1070
The Origins of Chirality in Living Systems 1085

Chemistry Put to Work

Chemistry and the Chemical Industry 4
Chemistry in the News 18
Antacids 135
The Hybrid Car 196
Explosives and Alfred Nobel 328
Orbitals and Energy 380
Gas Pipelines 409
Gas Separations 420
Supercritical Fluid Extraction 453
Cell Phone Tower Range 498
Recycling Plastics 501
Toward the Plastic Car 506
Liquid Crystal Displays 513
Methyl Bromide in the Atmosphere 590
Catalytic Converters 608
The Haber Process 631
Controlling Nitric Oxide Emissions 656
Amines and Amine Hydrochlorides 694
Direct Methanol Fuel Cells 874
Environmental Applications of Radioisotopes 921
Carbon Fibers and Composites 962
Gasoline 1061
Portrait of an Organic Chemical 1076

A Closer Look

The Scientific Method 13
Basic Forces 45
The Mass Spectrometer 48
Glenn Seaborg and Seaborgium 51
The Aura of Gold 143
Energy, Enthalpy, and P-V Work 176

The Speed of Light 214

Measurement and the Uncertainty Principle 225

Probability Density and Radial Probability Functions 230

Experimental Evidence for Electron Spin 234

Effective Nuclear Charge 260

Calculation of Lattice Energies: The Born–Haber Cycle 304

Oxidation Numbers, Formal Charges, and Actual Partial Charges 318

Phases in Atomic and Molecular Orbitals 373

The Ideal-Gas Equation 416

The Clausius–Clapeyron Equation 456

X-Ray Diffraction by Crystals 465

The Third Form of Carbon 468

The Transistor 490

Hydrates 533

Ideal Solutions with Two or More Volatile Components 548

Colligative Properties of Electrolyte Solutions 554

Using Spectroscopic Methods to Measure Reaction Rates 580

Limitations of Solubility Products 741

Other Greenhouse Gases 782

Water Softening 788

The Entropy Change when a Gas Expands Isothermally 808

Entropy and Probability 812

What's "Free" About Free Energy? 822

The Dawning of the Nuclear Age 915

The Hydrogen Economy 937

Charles M. Hall 990

Shape-Memory Alloys 998

Entropy and the Chelate Effect 1021

Charge-Transfer Color 1040

Preface

TO THE INSTRUCTOR

Philosophy

This is the eleventh edition of a text that has enjoyed unprecedented success over its many editions. It is fair to ask why there needs to be yet another edition. The answer in part lies in the nature of chemistry itself, a dynamic science in a process of continual discovery. New research leads to new applications of chemistry in other fields of science and in technology. In addition, environmental and economic concerns bring about changes in the place of chemistry in society. We want our textbook to reflect that dynamic, changing character. We also want it to convey the excitement that scientists experience in making new discoveries and contributing to our understanding of the physical world.

In addition, new ideas about how to present chemistry are being offered by teachers of chemistry, and many of these new ideas are reflected in how the textbook is organized and the ways in which individual topics are presented. New technologies and new devices to assist students in learning lead to new ways of presenting learning materials: the Internet, computer-based classroom projection tools, and more effective means of testing, to name just a few. All of these factors impact on how the text and the accompanying supplementary materials are modified from one edition to the next.

Our aim in revising the text has been to ensure that it remains a central, indispensable learning tool for the student. It is the one device that can be carried everywhere and used at any time, and as such, it is a one-stop source of all the information that the student is likely to need for learning, skill development, reference, and test preparation.

We believe that students are more enthusiastic about learning chemistry when they see its importance to their own goals and interests. With this in mind, we have highlighted many important applications of chemistry in everyday life. At the same time, the text provides the background in modern chemistry that students need to serve their professional interests and, as appropriate, to prepare for more advanced chemistry courses.

If the text is to support your role as teacher effectively, it must be addressed to the students. We have done our best to keep our writing clear and interesting and the book attractive and well-illustrated. Furthermore, we have provided numerous in-text study aids for students, including carefully placed descriptions of problem-solving strategies. Together, we have logged many years of teaching experience. We hope this is evident in our pacing and choice of examples.

A textbook is only as useful to students as the instructor permits it to be. This book is loaded with many features that can help students learn and that can guide them as they acquire both conceptual understanding and problem-solving skills. But the text and all the supplementary materials provided to support its use must work in concert with the instructor. There is a great deal for the students to use here, too much for all of it to be absorbed by any one student. You, the instructor, are the guide to a proper use of the book. Only with your active help will the students be able to fully utilize all that the text and its supplements offer. Students care about grades, of course, but with encouragement, they can also care about learning just because the subject matter is interesting. Please consider emphasizing features of the book that can materially enhance student appreciation of chemistry, such as the *Chemistry Put to Work* and *Chemistry and Life* boxes that show how chemistry impacts modern life and its relationship to health and life processes. Learn to use, and urge students to use, the rich Internet resources available. Emphasize conceptual understanding, and place less emphasis on simple manipulative, algorithmic problem solving. Spending less time on solving a variety of gas law problems, for example, can open up opportunities to talk about chemistry and the environment.

Organization and Contents

The first five chapters give a largely macroscopic, phenomenological view of chemistry. The basic concepts introduced—such as nomenclature, stoichiometry, and thermochemistry—provide necessary background for many of the laboratory experiments usually performed in general chemistry. We believe that an early introduction to thermochemistry is desirable because so much of our understanding of chemical processes is based on considerations of energy change. Thermochemistry is also important when we come to a discussion of bond enthalpies. We believe we have produced an effective, balanced approach to teaching thermodynamics in general chemistry. It is no easy matter to walk the narrow pathway between—on the one hand—trying to teach too much at too high a level and—on the other—resorting to oversimplifications. As with the book as a whole, the emphasis has been on imparting *conceptual* understanding, as opposed to presenting equations into which students are supposed to plug numbers.

The next four chapters (Chapters 6–9) deal with electronic structure and bonding. We have largely retained our presentation of atomic orbitals. For more advanced students, *Closer Look* boxes deal with radial probability functions and the nature of antibonding orbitals. In Chapter 7 we have improved our discussion of atomic and ionic radii. The focus of the text then changes to the next level of the organization of matter: the states of matter (Chapters 10 and 11) and solutions (Chapter 13). Also included in this section is an applications chapter on the chemistry of modern materials (Chapter 12), which builds on the student's understanding of chemical bonding and intermolecular interactions. This chapter has again received substantial revision, in keeping with the rapid pace of change in technology. It has been reorganized to emphasize a classification of materials based on their electronic bonding characteristics. This chapter provides an opportunity to show how the sometimes abstract concept of chemical bonding impacts real world applications. The modular organization of the chapter allows you to tailor your coverage to focus on those materials (semiconductors, polymers, biomaterials, nanotechnology, etc.) that are most relevant to your students.

The next several chapters examine the factors that determine the speed and extent of chemical reactions: kinetics (Chapter 14), equilibria (Chapters 15–17), thermodynamics (Chapter 19), and electrochemistry (Chapter 20). Also in this section is a chapter on environmental chemistry (Chapter 18), in which the concepts developed in preceding chapters are applied to a discussion of the atmosphere and hydrosphere.

After a discussion of nuclear chemistry (Chapter 21), the final chapters survey the chemistry of nonmetals, metals, organic chemistry, and biochemistry (Chapters 22–25). Chapter 22 has been shortened slightly. Chapter 23 now contains a modern treatment of band structure and bonding in metals. A brief discussion of lipids has been added to Chapter 25. These final chapters are developed in a parallel fashion and can be treated in any order.

Our chapter sequence provides a fairly standard organization, but we recognize that not everyone teaches all the topics in exactly the order we have chosen. We have therefore made sure that instructors can make common changes in teaching sequence with no loss in student comprehension. In particular, many instructors prefer to introduce gases (Chapter 10) after stoichiometry or after thermochemistry rather than with states of matter. The chapter on gases has been written to permit this change with *no* disruption in the flow of material. It is also possible to treat the balancing of redox equations (Sections 20.1 and 20.2) earlier, after the introduction of redox reactions in Section 4.4. Finally, some instructors like to cover organic chemistry (Chapter 25) right after bonding (Chapter 9). This, too, is a largely seamless move.

We have introduced students to descriptive organic and inorganic chemistry by integrating examples throughout the text. You will find pertinent and relevant examples of "real" chemistry woven into all the chapters as a means to illustrate principles and applications. Some chapters, of course, more directly address the properties of elements and their compounds, especially Chapters 4, 7, 12, 18, and 22–25. We also incorporate descriptive organic and inorganic chemistry in the end-of-chapter exercises.

Changes in This Edition

Some of the changes in the eleventh edition made in individual chapters have already been mentioned. More broadly, we have introduced a number of new features that are general throughout the text. *Chemistry: The Central Science* has traditionally been valued for its clarity of writing, its scientific accuracy and currency, its strong end-of-chapter exercises, and its consistency in level of coverage. In making changes, we have made sure not to compromise these characteristics. At the same time, we have responded to feedback received from the faculty and students who used previous editions. To make the text easier for students to use, we have continued to employ an open, clean design in the layout of the book. Illustrations that lend themselves to a more schematic, bolder presentation of the underlying principles have been introduced or revised from earlier versions. The art program in general has been strengthened, to better convey the beauty, excitement, and concepts of chemistry to students. The chapter-opening photos have been integrated into the introduction to each chapter, and thus made more relevant to the chapter's contents.

We have continued to use the What's Ahead overview at the opening of each chapter, introduced in the ninth edition, but we have changed the format to make the materials more useful to students. *Concept links* (∞) continue to provide easy-to-see cross-references to pertinent material covered earlier in the text. The essays titled *Strategies in Chemistry*, which provide advice to students on problem solving and "thinking like a chemist," continue to be an important feature. The *Give It Some Thought* exercises that we introduced in the tenth edition have proven to be very popular, and we have increased their number. These are informal, rather sharply focused questions that give students opportunities to test whether they are actually "getting it" as they read along. We have continued to emphasize conceptual exercises in the end-of-chapter exercise materials. The *Visualizing Concepts* category of exercise has been continued in this edition. These exercises are designed to facilitate concept understanding through use of models, graphs, and other visual materials. They precede the regular end-of-chapter exercises and are identified in each case with the relevant chapter section number. We continue to use multi-focus graphics to depict topics in macroscopic, microscopic, symbolic, and conceptual representation so students learn to see chemistry the way scientists do, from a variety of perspectives. The *Integrative Exercises*, which give students the opportunity to solve more challenging problems that integrate concepts from the present chapter with those of previous chapters, have also been increased in number.

New essays in our well-received *Chemistry Put to Work* and *Chemistry and Life* series emphasize world events, scientific discoveries, and medical breakthroughs that have occurred since publication of the tenth edition. We maintain our focus on the positive aspects of chemistry, without neglecting the problems that can arise in an increasingly technological world. Our goal is to help students appreciate the real-world perspective of chemistry and the ways in which chemistry affects their lives.

A minor change that you will see throughout the text is the use of condensed structural formulas for simple carboxylic acids. For example, we now write CH_3COOH for acetic acid instead of $HC_2H_3O_2$.

You'll also find that we've

- Revised or replaced some of the end-of-chapter Exercises, with particular focus on the black-numbered exercises (those not answered in the Appendix).

- Integrated more conceptual questions into the end-of-chapter material. For the convenience of instructors, these are identified by the CQ annotation in the Annotated Instructor's Edition, but not in the student edition of the text.

- Carried the stepwise Analyze, Plan, Solve, Check problem-solving strategy into nearly all of the Sample Exercises of the book to provide additional guidance in problem solving.

- Expanded the use of dual-column problem-solving strategies in many Sample Exercises to more clearly outline the process underlying mathematical calculations, thereby helping students to better perform mathematical calculations.

- Added both Key Skills and Key Equations sections to the end of chapter material to help students focus their study.

TO THE STUDENT

Chemistry: The Central Science, Eleventh Edition, has been written to introduce you to modern chemistry. As authors, we have, in effect, been engaged by your instructor to help you learn chemistry. Based on the comments of students and instructors who have used this book in its previous editions, we believe that we have done that job well. Of course, we expect the text to continue to evolve through future editions. We invite you to write to us to tell us what you like about the book so that we will know where we have helped you most. Also, we would like to learn of any shortcomings, so that we might further improve the book in subsequent editions. Our addresses are given at the end of the Preface.

Advice for Learning and Studying Chemistry

Learning chemistry requires both the assimilation of many new concepts and the development of analytical skills. In this text we have provided you with numerous tools to help you succeed in both. If you are going to succeed in your course in chemistry, you will have to develop good study habits. Science courses, and chemistry in particular, make different demands on your learning skills than other types of courses. We offer the following tips for success in your study of chemistry:

Don't fall behind! As your chemistry course moves along, new topics will build on material already presented. If you don't keep up in your reading and problem solving, you will find it much harder to follow the lectures and discussions on current topics. "Cramming" just before an exam has been shown to be an ineffective way to study any subject, chemistry included.

Focus your study. The amount of information you will be expected to learn can sometimes seem overwhelming. It is essential to recognize those concepts and skills that are particularly important. Pay attention to what your instructor is emphasizing. As you work through the Sample Exercises and homework assignments, try to see what general principles and skills they deal with. Use the **What's Ahead** feature at the beginning of each chapter to help orient you to what is important in each chapter. A single reading of a chapter will simply not be enough for successful learning of chapter concepts and problem-solving skills. You will need to go over assigned materials more than once. Don't skip the **Give It Some Thought** features, **Sample Exercises**, and **Practice Exercises**.

They are your guides to whether you are actually learning the material. The **Key Skills** and **Key Equations** at the end of each chapter should help you focus your study.

Keep good lecture notes. Your lecture notes will provide you with a clear and concise record of what your instructor regards as the most important material to learn. Use your lecture notes in conjunction with this text; that's your best way to determine which material to study.

Skim topics in the text before they are covered in lecture. Reviewing a topic before lecture will make it easier for you to take good notes. First read the introduction and Summary, then quickly read through the chapter, skipping Sample Exercises and supplemental sections. Pay attention to the titles of sections and subsections, which give you a feeling for the scope of topics. Try to avoid thinking that you must learn and understand everything right away.

After lecture, carefully read the topics covered in class. As you read, pay attention to the concepts presented and to the application of these concepts in the Sample Exercises. Once you think you understand a Sample Exercise, test your understanding by working the accompanying Practice Exercise.

Learn the language of chemistry. As you study chemistry, you will encounter many new words. It is important to pay attention to these words and to know their meanings or the entities to which they refer. Knowing how to identify chemical substances from their names is an important skill; it can help you avoid painful mistakes on examinations. For example, "chlorine" and "chloride" refer to very different things.

Attempt the assigned end-of-chapter exercises. Working the exercises that have been selected by your instructor provides necessary practice in recalling and using the essential ideas of the chapter. You cannot learn merely by observing; you must be a participant. In particular, try to resist checking the *Student Solutions Manual* (if you have one) until you have made a sincere effort to solve the exercise yourself. If you really get stuck on an exercise, however, get help from your instructor, your teaching assistant, or another student. Spending more than 20 minutes on a single exercise is rarely effective unless you know that it is particularly challenging.

Make use of the online resources. Some things are more easily learned by discovery, and others are best shown in three dimensions. If your instructor has included Mastering Chemistry with your book, take advantage of the unique tools it provides to get the most out of your time in chemistry.

The bottom line is to work hard, study effectively, and use the tools that are available to you, including this textbook. We want to help you learn more about the world of chemistry and why chemistry is the central science. If you learn chemistry well, you can be the life of the party, impress your friends and parents, and ... well, also pass the course with a good grade.

ACKNOWLEDGMENTS

The production of a textbook is a team effort requiring the involvement of many people besides the authors. Many people contributed hard work and talent to bring this edition to life. Although their names don't appear on the cover of the book, their creativity, time, and support has been instrumental in all stages of its development and production.

Each of us has benefited greatly from discussions with colleagues and from correspondence with both instructors and students both here and abroad. Colleagues have also helped immensely by reviewing our materials, sharing their insights, and providing suggestions for improvements. On this edition we were particularly blessed with an exceptional group of accuracy checkers who read through our materials looking for both technical inaccuracies and typographical errors.

Eleventh Edition Reviewers

Patricia Amateis	Virginia Polytechnic Institute and State University	Pamela Marks	Arizona State University
Todd L. Austell	University of North Carolina, Chapel Hill	Shelley Minteer	Saint Louis University
Melita Balch	University of Illinois at Chicago	Jessica Orvis	Georgia Southern University
David L. Cedeño	Illinois State University	Jason Overby	College of Charleston
Elzbieta Cook	Louisiana State University	Amy L. Rogers	College of Charleston
Enriqueta Cortez	South Texas College	Mark Schraf	West Virginia University
Stephen Drucker	University of Wisconsin-Eau Claire	Susan M. Shih	College of DuPage
James M. Farrar	University of Rochester	David Soriano	University of Pittsburgh-Bradford
Gregory M. Ferrence	Illinois State University	Domenic J. Tiani	University of North Carolina, Chapel Hill
Paul A. Flowers	University of North Carolina at Pembroke	John B. Vincent	University of Alabama
Cheryl B. Frech	University of Central Oklahoma	Michael J. Van Stipdonk	Wichita State University
Kenneth A. French	Blinn College	Thomas R. Webb	Auburn University
Eric Goll	Brookdale Community College	Paul G. Wenthold	Purdue University
Carl A. Hoeger	University of California, San Diego	Wayne Wesolowski	University Of Arizona
Michael O. Hurst	Georgia Southern University	Charles A. Wilkie	Marquette University
Milton D. Johnston, Jr.	University of South Florida	Troy Wood	SUNY Buffalo
Robley J. Light	Florida State University	Dr. Susan M. Zirpoli	Slippery Rock University

Eleventh Edition Accuracy Checkers

Margaret Asirvatham	University of Colorado	Barbara Mowery	York College
Louis J. Kirschenbaum	University of Rhode Island	Richard Perkins	University of Southwest Louisiana
Richard Langley	Stephen F. Austin State University	Kathy Thrush Shaginaw	Villanova University
Albert Martin	Moravian College	Maria Vogt	Bloomfield College
Stephen Mezyk	California State University, Long Beach		

MasteringChemistry Summit Participants

Phil Bennett	Santa Fe Community College	Bob Pribush	Butler University
Jo Blackburn	Richland College	Al Rives	Wake Forest University
John Bookstaver	St. Charles Community College	Joel Russell	Oakland University
David Carter	Angelo State University	Greg Szulczewski	University of Alabama, Tuscaloosa
Doug Cody	Nassau Community College	Matt Tarr	University of New Orleans
Tom Dowd	Harper College	Dennis Taylor	Clemson University
Palmer Graves	Florida International University	Harold Trimm	Broome Community College
Margie Haak	Oregon State University	Emanuel Waddell	University of Alabama, Huntsville
Brad Herrick	Colorado School of Mines	Kurt Winklemann	Florida Institute of Technology
Jeff Jenson	University of Findlay	Klaus Woelk	University of Missouri, Rolla
Jeff McVey	Texas State University	Steve Wood	Brigham Young University
Gary Michels	Creighton University		

Reviewers of Previous Editions of *Chemistry: The Central Science*

S.K. Airee	University of Tennessee	Leon Borowski	Diablo Valley College
John J. Alexander	University of Cincinnati	Simon Bott	University of Houston
Robert Allendoerfer	SUNY Buffalo	Kevin L. Bray	Washington State University
Patricia Amateis	Virginia Polytechnic Institute and State University	Daeg Scott Brenner	Clark University
		Gregory Alan Brewer	Catholic University of America
Sandra Anderson	University of Wisconsin	Karen Brewer	Virginia Polytechnic Institute and State University
John Arnold	University of California		
Socorro Arteaga	El Paso Community College	Edward Brown	Lee University
Rosemary Bartoszek-Loza	The Ohio State University	Gary Buckley	Cameron University
Boyd Beck	Snow College	Carmela Byrnes	Texas A&M University
Amy Beilstein	Centre College	B. Edward Cain	Rochester Institute of Technology
Victor Berner	New Mexico Junior College	Kim Calvo	University of Akron
Narayan Bhat	University of Texas, Pan American	Donald L. Campbell	University of Wisconsin
Merrill Blackman	United States Military Academy	Gene O. Carlisle	Texas A&M University
Salah M. Blaih	Kent State University	Elaine Carter	Los Angeles City College
James A. Boiani	SUNY Geneseo	Robert Carter	University of Massachusetts

Ann Cartwright — San Jacinto Central College

Dana Chatellier — University of Delaware

Stanton Ching — Connecticut College

Paul Chirik — Cornell University

William Cleaver — University of Vermont

Beverly Clement — Blinn College

Robert D. Cloney — Fordham University

John Collins — Broward Community College

Edward Werner Cook — Tunxis Community Technical College

Elzbieta Cool — University of Calgary

Thomas Edgar Crumm — Indiana University of Pennsylvania

Dwaine Davis — Forsyth Tech Community College

Ramón López de la Vega — Florida International University

Nancy De Luca — University of Massachusetts, Lowell North Campus

Angel de Dios — Georgetown University

John M. DeKorte — Glendale Community College

Daniel Domin — Tennessee State University

James Donaldson — University of Toronto

Bill Donovan — University of Akron

Ronald Duchovic — Indiana University-Purdue University at Fort Wayne

David Easter — Southwest Texas State University

Joseph Ellison — United States Military Academy

George O. Evans II — East Carolina University

Clark L. Fields — University of Northern Colorado

Jennifer Firestine — Lindenwood University

Jan M. Fleischner — College of New Jersey

Michelle Fossum — Laney College

Roger Frampton — Tidewater Community College

Joe Franek — University of Minnesota

David Frank — California State University

Ewa Fredette — Moraine Valley College

Karen Frindell — Santa Rosa Junior College

John I. Gelder — Oklahoma State University

Paul Gilletti — Mesa Community College

Peter Gold — Pennsylvania State University

James Gordon — Central Methodist College

Thomas J. Greenbowe — Iowa State University

Michael Greenlief — University of Missouri

Eric P. Grimsrud — Montana State University

John Hagadorn — University of Colorado

Randy Hall — Louisiana State University

John M. Halpin — New York University

Marie Hankins — University of Southern Indiana

Robert M. Hanson — St. Olaf College

Daniel Haworth — Marquette University

Inna Hefley — Blinn College

David Henderson — Trinity College

Paul Higgs — Barry University

Gary G. Hoffman — Florida International University

Deborah Hokien — Marywood University

Robin Horner — Fayetteville Tech Community College

Roger K. House — Moraine Valley College

William Jensen — South Dakota State University

Janet Johannessen — County College of Morris

Andrew Jones — Southern Alberta Institute of Technology

Booker Juma — Fayetteville State University

Ismail Kady — East Tennessee State University

Siam Kahmis — University of Pittsburgh

Steven Keller — University of Missouri

John W. Kenney — Eastern New Mexico University

Neil Kestner — Louisiana State University

Leslie Kinsland — University of Louisiana

Donald Kleinfelter — University of Tennessee, Knoxville

David Kort — George Mason University

George P. Kreishman — University of Cincinnati

Paul Kreiss — Anne Arundel Community College

Manickham Krishnamurthy — Howard University

Brian D. Kybett — University of Regina

William R. Lammela — Nazareth College

John T. Landrum — Florida International University

Richard Langley — Stephen F. Austin State University

N. Dale Ledford — University of South Alabama

Ernestine Lee — Utah State University

David Lehmpuhl — University of Southern Colorado

Donald E. Linn, Jr. — Indiana University-Purdue University Indianapolis

David Lippmann — Southwest Texas State

Patrick Lloyd — Kingsborough Community College

Encarnacion Lopez — Miami Dade College, Wolfson

Arthur Low — Tarleton State University

Gary L. Lyon — Louisiana State University

Preston J. MacDougall — Middle Tennessee State University

Jeffrey Madura — Duquesne University

Larry Manno — Triton College

Asoka Marasinghe — Moorhead State University

Earl L. Mark — ITT Technical Institute

Albert H. Martin — Moravian College

Przemyslaw Maslak — Pennsylvania State University

Hilary L. Maybaum — ThinkQuest, Inc.

Armin Mayr — El Paso Community College

Marcus T. McEllistrem — University of Wisconsin

Craig McLauchlan — Illinois State University

William A. Meena — Valley College

Joseph Merola — Virginia Polytechnic Institute and State University

Stephen Mezyk — California State University

Eric Miller — San Juan College

Gordon Miller — Iowa State University

Massoud (Matt) Miri — Rochester Institute of Technology

Mohammad Moharerrzadeh — Bowie State University

Tracy Morkin — Emory University

Barbara Mowery — Yorktown, VA

Kathleen E. Murphy — Daemen College

Kathy Nabona — Austin Community College

Robert Nelson — Georgia Southern University

Al Nichols — Jacksonville State University

Ross Nord — Eastern Michigan University

Mark Ott — Jackson Community College

Robert H. Paine — Rochester Institute of Technology

Robert T. Paine — University of New Mexico

Sandra Patrick — Malaspina University College

Mary Jane Patterson — Brazosport College

Tammi Pavelec — Lindenwood University

Albert Payton — Broward Community College

Christopher J. Peeples — University of Tulsa

Kim Percell — Cape Fear Community College

Gita Perkins	Estrella Mountain Community College	Vince Sollimo	Burlington Community College
Richard Perkins	University of Louisiana	Eugene Stevens	Binghamton University
Nancy Peterson	North Central College	James Symes	Cosumnes River College
Robert C. Pfaff	Saint Joseph's College	Iwao Teraoka	Polytechnic University
John Pfeffer	Highline Community College	Kathy Thrush	Villanova University
Lou Pignolet	University of Minnesota	Edmund Tisko	University of Nebraska at Omaha
Bernard Powell	University of Texas	Richard S. Treptow	Chicago State University
Jeffrey A. Rahn	Eastern Washington University	Michael Tubergen	Kent State University
Steve Rathbone	Blinn College	Claudia Turro	The Ohio State University
Scott Reeve	Arkansas State University	James Tyrell	Southern Illinois University
John Reissner	University of North Carolina	Philip Verhalen	Panola College
Helen Richter	University of Akron	Ann Verner	University of Toronto at Scarborough
Thomas Ridgway	University of Cincinnati	Edward Vickner	Gloucester County Community College
Mark G. Rockley	Oklahoma State University		
Lenore Rodicio	Miami Dade College	John Vincent	University of Alabama
Jimmy R. Rogers	University of Texas at Arlington	Maria Vogt	Bloomfield College
Steven Rowley	Middlesex Community College	Tony Wallner	Barry University
James E. Russo	Whitman College	Lichang Wang	Southern Illinois University
Theodore Sakano	Rockland Community College	Karen Weichelman	University of Louisiana-Lafayette
Michael J. Sanger	University of Northern Iowa	Laurence Werbelow	New Mexico Institute of Mining and Technology
Jerry L. Sarquis	Miami University		
James P. Schneider	Portland Community College	Sarah West	University of Notre Dame
Mark Schraf	West Virginia University	Linda M. Wilkes	University at Southern Colorado
Gray Scrimgeour	University of Toronto	Darren L. Williams	West Texas A&M University
Paula Secondo	Western Connecticut State University	Thao Yang	University of Wisconsin
David Shinn	University of Hawaii at Hilo	David Zax	Cornell University

We would also like to express our gratitude to our many team members at Prentice Hall whose hard work, imagination, and commitment have contributed so greatly to the final form of this edition: Nicole Folchetti, our Editor in Chief, who has brought energy and imagination not only to this edition but to earlier ones as well; Andrew Gilfillan, our Chemistry Editor, for his enthusiasm and support; Jonathan Colon, Editorial Assistant, who coordinated all reviews and accuracy checking; Ray Mullaney, our Developmental Editor in Chief, who has been with us since the very first edition of this textbook, helping guide its evolution, infusing it with a sense of style and quality; Karen Nein, our Development Editor, whose diligence and careful attention to detail were invaluable to this revision; Marcia Youngman, our copy editor, for her keen eye; Maureen Eide, our Art Director, who managed the complex task of bringing our sketches into final form; Jennifer Hart, who coordinated all the supplements that accompany the text; Donna Young, our Production Editor, who always seemed to be available for questions any time of the day or night, and who managed the complex responsibilities of bringing the design, photos, artwork, and writing together with efficiency and good cheer. The Prentice Hall team is a first-class operation.

There are many others who also deserve special recognition including the following: Jerry Marshall (Truitt and Marshall), our photo researcher, who was so effective in finding photos to bring chemistry to life for students; Linda Brunauer (Santa Clara University) for her marvelous efforts in preparing the Annotated Instructor's Edition of the text; Roxy Wilson (University of Illinois) for performing the difficult job of working out solutions to the end-of-chapter exercises.

Finally, we wish to thank our families and friends for their love, support, encouragement, and patience as we brought this eleventh edition to completion.

Theodore L. Brown
School of Chemical Sciences
University of Illinois,
  Urbana-Champaign
Urbana, IL 61801
tlbrown@uiuc.edu

H. Eugene LeMay, Jr.
Department of Chemistry
University of Nevada
Reno, NV 89557
lemay@unr.edu

Bruce E. Bursten
College of Arts and Sciences
University of Tennessee
Knoxville, TN 37996
bbursten@utk.edu

Catherine J. Murphy
Department of Chemistry
  and Biochemistry
University of South Carolina
Columbia, SC 29208
murphy@mail.chem.sc.edu

Patrick M. Woodward
Department of Chemistry
The Ohio State University
Columbus, OH 43210
woodward@chemistry.
ohio-state.edu

List of Resources

For Students

MasteringChemistry: MasteringChemistry is the first adaptive-learning online homework system. It provides selected end-of-chapter problems from the text, as well as hundreds of tutorials with automatic grading, immediate answer-specific feedback, and simpler questions on request. Based on extensive research of precise concepts students struggle with, MasteringChemistry uniquely responds to your immediate learning needs, thereby optimizing your study time.

Solutions to Red Exercises (0-13-600287-0) Prepared by Roxy Wilson of the University of Illinois–Urbana-Champaign. Full solutions to all of the red-numbered exercises in the text are provided. (Short answers to red exercises are found in the appendix of the text).

Solutions to Black Exercises (0-13-600324-9) Prepared by Roxy Wilson of the University of Illinois–Urbana-Champaign. Full solutions to all of the black-numbered exercises in the text are provided.

Student's Guide (0-13-600264-1) Prepared by James C. Hill of California State University. This book assists students through the text material with chapter overviews, learning objectives, a review of key terms, as well as self-tests with answers and explanations. This edition also features the addition of MCAT practice questions.

Laboratory Experiments (0-13-600285-4) Prepared by John H. Nelson and Kenneth C. Kemp, both of the University of Nevada. This manual contains 43 finely tuned experiments chosen to introduce students to basic lab techniques and to illustrate core chemical principles. This new edition has been revised to correlate more tightly with the text and now features a guide on how to keep a lab report notebook. You can also customize these labs through Catalyst, our custom database program. For more information, visit www.prenhall.com/catalyst.

Virtual ChemLab: An easy to use simulation of five different general chemistry laboratories that can be used to supplement a wet lab, for pre-laboratory and post-laboratory activities, for homework or quiz assignments, or for classroom demonstrations.

For Instructors

Annotated Instructor's Edition (0-13-601250-7) Prepared by Linda Brunauer of Santa Clara University. Provides marginal notes and information for instructors and TAs, including transparency icons, suggested lecture demonstrations, teaching tips, and background references from the chemical education literature.

Full Solutions Manual (0-13-600325-7) Prepared by Roxy Wilson of the University of Illinois–Urbana-Champaign. This manual contains all end-of-chapter exercises in the text. With an instructor's permission, this manual may be made available to students.

Instructor's Resource Center on CD-DVD (0-13-600281-1) This resource provides an integrated collection of resources to help you make efficient and effective use of your time. This CD/DVD features most art from the text, including figures and tables in PDF format for high-resolution printing, as well as four pre-built PowerPoint™ presentations. The first presentation contains the images embedded within PowerPoint slides. The second includes a complete lecture outline that is modifiable by the user. The final two presentations contain worked "in chapter" sample exercises and questions to be used with Classroom Response Systems. This CD/DVD also contains movies, animations, and electronic files of the Instructor's Resource Manual.

Test Item File (0-13-601251-5) Prepared by Joseph P. Laurino and Donald Cannon, both of the University of Tampa. The Test Item File now provides a selection of more than 4000 test questions.

Instructor's Resource Manual (0-13-600237-4) Prepared by Linda Brunauer of Santa Clara University and Elizabeth Cook of Louisiana State University. Organized by chapter, this manual offers detailed lecture outlines and complete descriptions of all available lecture demonstrations, the interactive media assets, common student misconceptions, and more.

Transparencies (0-13-601256-6) Approximately 300 full-color transparencies put principles into visual perspective and save you time when preparing lectures.

Annotated Instructor's Edition to *Laboratory Experiments* (0-13-6002862) Prepared by John H. Nelson and Kenneth C. Kemp, both of the University of Nevada. This AIE combines the full student lab manual with appendices covering the proper disposal of chemical waste, safety instructions for the lab, descriptions of standard lab equipment, answers to questions, and more.

BlackBoard and WebCT: Practice and assessment materials are available upon request in these course management platforms.

About the Authors

THEODORE L. BROWN received his Ph.D. from Michigan State University in 1956. Since then, he has been a member of the faculty of the University of Illinois, Urbana-Champaign, where he is now Professor of Chemistry, Emeritus. He served as Vice Chancellor for Research, and Dean, The Graduate College, from 1980 to 1986, and as Founding Director of the Arnold and Mabel Beckman Institute for Advanced Science and Technology from 1987 to 1993. Professor Brown has been an Alfred P. Sloan Foundation Research Fellow and has been awarded a Guggenheim Fellowship. In 1972 he was awarded the American Chemical Society Award for Research in Inorganic Chemistry, and received the American Chemical Society Award for Distinguished Service in the Advancement of Inorganic Chemistry in 1993. He has been elected a Fellow of both the American Association for the Advancement of Science and the American Academy of Arts and Sciences.

H. EUGENE LeMAY, JR., received his B.S. degree in Chemistry from Pacific Lutheran University (Washington) and his Ph.D. in Chemistry in 1966 from the University of Illinois (Urbana). He then joined the faculty of the University of Nevada, Reno, where he is currently Professor of Chemistry, Emeritus. He has enjoyed Visiting Professorships at the University of North Carolina at Chapel Hill, at the University College of Wales in Great Britain, and at the University of California, Los Angeles. Professor LeMay is a popular and effective teacher, who has taught thousands of students during more than 35 years of university teaching. Known for the clarity of his lectures and his sense of humor, he has received several teaching awards, including the University Distinguished Teacher of the Year Award (1991) and the first Regents' Teaching Award given by the State of Nevada Board of Regents (1997).

BRUCE E. BURSTEN received his Ph.D. in Chemistry from the University of Wisconsin in 1978. After two years as a National Science Foundation Postdoctoral Fellow at Texas A&M University, he joined the faculty of The Ohio State University, where he rose to the rank of Distinguished University Professor. In 2005, he moved to his present position at the University of Tennessee, Knoxville as Distinguished Professor of Chemistry and Dean of the College of Arts and Sciences. Professor Bursten has been a Camille and Henry Dreyfus Foundation Teacher-Scholar and an Alfred P. Sloan Foundation Research Fellow, and he has been elected a Fellow of the American Association for the Advancement of Science. At Ohio State he has received the University Distinguished Teaching Award in 1982 and 1996, the Arts and Sciences Student Council Outstanding Teaching Award in 1984, and the University Distinguished Scholar Award in 1990. He received the Spiers Memorial Prize and Medal of the Royal Society of Chemistry in 2003, and the Morley Medal of the Cleveland Section of the American Chemical Society in 2005. He was elected President of the American Chemical Society for 2008. In addition to his teaching and service activities, Professor Bursten's research program focuses on compounds of the transition-metal and actinide elements.

CATHERINE J. MURPHY received two B.S. degrees, one in Chemistry and one in Biochemistry, from the University of Illinois, Urbana-Champaign, in 1986. She received her Ph.D. in Chemistry from the University of Wisconsin in 1990. She was a National Science Foundation and National Institutes of Health Postdoctoral Fellow at the California Institute of Technology from 1990 to 1993. In 1993, she joined the faculty of the University of South Carolina, Columbia, where she is currently the Guy F. Lipscomb Professor of Chemistry. Professor Murphy has been honored for both research and teaching as a Camille Dreyfus Teacher-Scholar, an Alfred P. Sloan Foundation Research Fellow, a Cottrell Scholar of the Research Corporation, a National Science Foundation CAREER Award winner, and a subsequent NSF Award for Special Creativity. She has also received a USC Mortar Board Excellence in Teaching Award, the USC Golden Key Faculty Award for Creative Integration of Research and Undergraduate Teaching, the USC Michael J. Mungo Undergraduate Teaching Award, and the USC Outstanding Undergraduate Research Mentor Award. Since 2006, Professor Murphy has served as a Senior Editor to the *Journal of Physical Chemistry*. Professor Murphy's research program focuses on the synthesis and optical properties of inorganic nanomaterials, and on the local structure and dynamics of the DNA double helix.

Contributing Author

PATRICK M. WOODWARD received B.S. degrees in both Chemistry and Engineering from Idaho State University in 1991. He received a M.S. degree in Materials Science and a Ph.D. in Chemistry from Oregon State University in 1996. He spent two years as a postdoctoral researcher in the Department of Physics at Brookhaven National Laboratory. In 1998, he joined the faculty of the Chemistry Department at The Ohio State University where he currently holds the rank of Associate Professor. He has enjoyed visiting professorships at the University of Bordeaux, in France, and the University of Sydney, in Australia. Professor Woodward has been an Alfred P. Sloan Foundation Research Fellow and a National Science Foundation CAREER Award winner. He currently serves as an Associate Editor to the *Journal of Solid State Chemistry* and as the director of the Ohio REEL program, an NSF funded center that works to bring authentic research experiments into the laboratories of first and second year chemistry classes in 15 colleges and universities across the state of Ohio. Professor Woodward's research program focuses on understanding the links between bonding, structure, and properties of solid state inorganic functional materials.

INTRODUCTION: MATTER AND MEASUREMENT

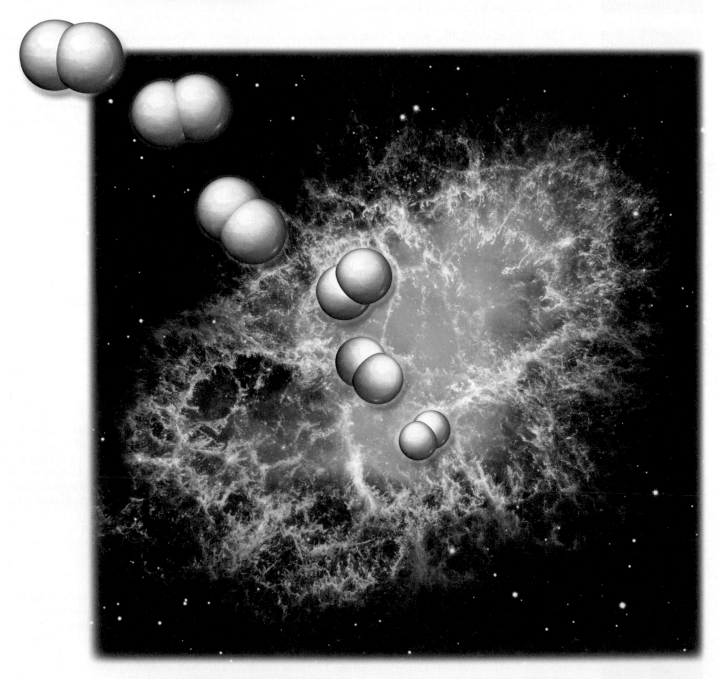

HUBBLE SPACE TELESCOPE IMAGE of the Crab Nebula, a 6-light-year-wide expanding remnant of a star's supernova explosion. The orange filaments are the tattered remains of the star and consist mostly of hydrogen, the simplest and most plentiful element in the universe. Hydrogen occurs as molecules in cool regions, as atoms in hotter regions, and as ions in the hottest regions. The processes that occur within stars are responsible for creating other chemical elements from hydrogen.

WHAT'S AHEAD

1.1 The Study of Chemistry
We begin by providing a brief perspective of what chemistry is about and why it is useful to learn chemistry.

1.2 Classifications of Matter
Next, we examine some fundamental ways to classify materials, distinguishing between *pure substances* and *mixtures* and noting that there are two fundamentally different kinds of pure substances: *elements* and *compounds*.

1.3 Properties of Matter
We then consider some of the different kinds of characteristics, or *properties*, that we use to characterize, identify, and separate substances.

1.4 Units of Measurement
We observe that many properties rely on quantitative measurements, involving both numbers and units. The units of measurement used throughout science are those of the *metric system*, a decimal system of measurement.

1.5 Uncertainty in Measurement
We also observe that the uncertainties inherent in all measured quantities are expressed by the number of *significant figures* used to report the number. Significant figures are also used to express the uncertainties associated with calculations involving measured quantities.

1.6 Dimensional Analysis
We recognize that units as well as numbers are carried through calculations and that obtaining correct units for the result of a calculation is an important way to check whether the calculation is correct.

HAVE YOU EVER WONDERED why ice melts and water evaporates? Why do leaves turn colors in the fall, and how does a battery generate electricity? Why does keeping foods cold slow their spoilage, and how do our bodies use food to maintain life?

Chemistry answers these questions and countless others like them. **Chemistry** is the study of materials and the changes that materials undergo. One of the joys of learning chemistry is seeing how chemical principles operate in all aspects of our lives, from everyday activities like lighting a match to more far-reaching matters like the development of drugs to cure cancer. Chemical principles also operate in the far reaches of our galaxy (chapter-opening photo) as well as within and around us.

This first chapter lays a foundation for our study of chemistry by providing an overview of what chemistry is about and dealing with some fundamental concepts of matter and scientific measurements. The list above, entitled "What's Ahead," gives a brief overview of the organization of this chapter and some of the ideas that we will consider. As you study, keep in mind that the chemical facts and concepts you are asked to learn are not ends in themselves; they are tools to help you better understand the world around you.

1.1 THE STUDY OF CHEMISTRY

Before traveling to an unfamiliar city, you might look at a map to get some sense of where you are heading. Because chemistry may be unfamiliar to you, it's useful to get a general idea of what lies ahead before you embark on your journey. In fact, you might even ask why you are taking the trip.

The Atomic and Molecular Perspective of Chemistry

Chemistry is the study of the properties and behavior of matter. **Matter** is the physical material of the universe; it is anything that has mass and occupies space. A **property** is any characteristic that allows us to recognize a particular type of matter and to distinguish it from other types. This book, your body, the clothes you are wearing, and the air you are breathing are all samples of matter. Not all forms of matter are so common or so familiar. Countless experiments have shown that the tremendous variety of matter in our world is due to combinations of only about 100 very basic, or elementary, substances called **elements**. As we proceed through this text, we will seek to relate the properties of matter to its composition, that is, to the particular elements it contains.

Chemistry also provides a background to understanding the properties of matter in terms of **atoms**, the almost infinitesimally small building blocks of matter. Each element is composed of a unique kind of atom. We will see that the properties of matter relate to both the kinds of atoms the matter contains (*composition*) and to the arrangements of these atoms (*structure*).

Atoms can combine to form **molecules** in which two or more atoms are joined together in specific shapes. Throughout this text you will see molecules represented using colored spheres to show how their component atoms connect to each other (Figure 1.1▼). The color provides a convenient and easy way to distinguish between the atoms of different elements. For examples, compare the molecules of ethanol and ethylene glycol in Figure 1.1. Notice that these molecules have different compositions and structures. Ethanol contains only one oxygen atom, which is depicted by one red sphere. In contrast, ethylene glycol has two atoms of oxygen.

Even apparently minor differences in the composition or structure of molecules can cause profound differences in their properties. Ethanol, also called grain alcohol, is the alcohol in beverages such as beer and wine. Ethylene glycol, on the other hand, is a viscous liquid used as automobile antifreeze. The properties of these two substances differ in many ways, including the temperatures at which they freeze and boil. The biological activities of the two molecules are also quite different. Ethanol is consumed throughout the world, but you should *never* consume ethylene glycol because it is highly toxic. One of the challenges that chemists undertake is to alter the composition or structure of molecules in a controlled way, creating new substances with different properties.

Every change in the observable world—from boiling water to the changes that occur as our bodies combat invading viruses—has its basis in the world of atoms and molecules. Thus, as we proceed with our study of chemistry, we will find ourselves thinking in two realms: the *macroscopic* realm

▼ **Figure 1.1 Molecular models.** The white, dark gray, and red spheres represent atoms of hydrogen, carbon, and oxygen, respectively.

(a) Oxygen

(b) Water

(c) Carbon dioxide

(d) Ethanol

(e) Ethylene glycol

(f) Aspirin

of ordinary-sized objects (*macro* = large) and the *submicroscopic* realm of atoms and molecules. We make our observations in the macroscopic world—in the laboratory and in our everyday surroundings. To understand that world, however, we must visualize how atoms and molecules behave at the submicroscopic level. Chemistry is the science that seeks to understand the properties and behavior of matter by studying the properties and behavior of atoms and molecules.

GIVE IT SOME THOUGHT

(a) In round numbers, about how many elements are there? **(b)** What submicroscopic particles are the building blocks of matter?

Why Study Chemistry?

Chemistry provides important understanding of our world and how it works. It is an extremely practical science that greatly impacts our daily lives. Indeed, chemistry lies near the heart of many matters of public concern: improvement of health care; conservation of natural resources; protection of the environment; and provision of our everyday needs for food, clothing, and shelter. Using chemistry, we have discovered pharmaceutical chemicals that enhance our health and prolong our lives. We have increased food production through the use of fertilizers and pesticides, and we have developed plastics and other materials that are used in almost every facet of our lives. Unfortunately, some chemicals also have the potential to harm our health or the environment. As educated citizens and consumers, it is in our best interest to understand the profound effects, both positive and negative, that chemicals have on our lives and to strike an informed balance about their uses.

Most of you are studying chemistry, however, not merely to satisfy your curiosity or to become more informed consumers or citizens, but because it is an essential part of your curriculum. Your major might be biology, engineering, pharmacy, agriculture, geology, or some other field. Why do so many diverse subjects share an essential tie to chemistry? The answer is that chemistry, by its very nature, is the *central science*, central to a fundamental understanding of other sciences and technologies. For example, our interactions with the material world raise basic questions about the materials around us. What are their compositions and properties? How do they interact with us and our environment? How, why, and when do they undergo change? These questions are important whether the material is part of high-tech computer chips, a pigment used by a Renaissance painter, or the DNA that transmits genetic information in our bodies (Figure 1.2▼).

By studying chemistry, you will learn to use the powerful language and ideas that have evolved to describe and enhance our understanding of matter. The language of chemistry is a universal scientific language that is widely used

▼ Figure 1.2 **Chemistry helps us better understand materials.** (a) A microscopic view of an EPROM (Erasable Programmable Read-Only Memory) silicon microchip. (b) A Renaissance painting, *Young Girl Reading*, by Vittore Carpaccio (1472–1526). (c) A long strand of DNA that has spilled out of the damaged cell wall of a bacterium.

(a)

(b)

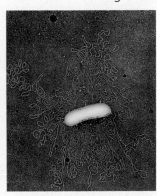
(c)

Chemistry Put to Work   CHEMISTRY AND THE CHEMICAL INDUSTRY

Many people are familiar with common household chemicals such as those shown in Figure 1.3▶, but few realize the size and importance of the chemical industry. Worldwide sales of chemicals and related products manufactured in the United States total approximately $550 billion annually. The chemical industry employs more than 10% of all scientists and engineers and is a major contributor to the US economy.

Vast amounts of chemicals are produced each year and serve as raw materials for a variety of uses, including the manufacture of metals, plastics, fertilizers, pharmaceuticals, fuels, paints, adhesives, pesticides, synthetic fibers, microprocessor chips, and numerous other products. Table 1.1▼ lists the top eight chemicals produced in the United States. We will discuss many of these substances and their uses as the course progresses.

People who have degrees in chemistry hold a variety of positions in industry, government, and academia. Those who work in the chemical industry find positions as laboratory chemists, carrying out experiments to develop new products (research and development), analyzing materials (quality control), or assisting customers in using products (sales and service). Those with more experience or training may work as managers or company directors. A chemistry degree also can prepare you for alternate careers in teaching, medicine, biomedical research, information science, environmental work, technical sales, work with government regulatory agencies, and patent law.

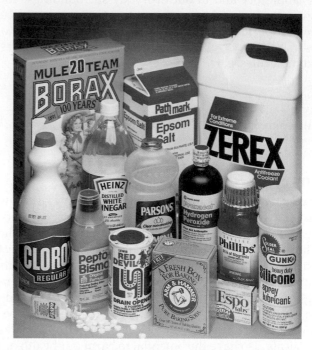

▲ **Figure 1.3 Household chemicals.** Many common supermarket products have very simple chemical compositions.

TABLE 1.1 ■ The Top Eight Chemicals Produced by the Chemical Industry in 2006[a]				
Rank	Chemical	Formula	2006 Production (billions of pounds)	Principal End Uses
1	Sulfuric acid	H_2SO_4	79	Fertilizers, chemical manufacturing
2	Ethylene	C_2H_4	55	Plastics, antifreeze
3	Lime	CaO	45	Paper, cement, steel
4	Propylene	C_3H_6	35	Plastics
5	Phosphoric acid	H_3PO_4	24	Fertilizers
6	Ammonia	NH_3	23	Fertilizers
7	Chlorine	Cl_2	23	Bleaches, plastics, water purification
8	Sodium hydroxide	$NaOH$	18	Aluminum production, soap

[a]Most data from *Chemical and Engineering News*, July 2, 2007, pp. 57, 60.

in other disciplines. Furthermore, an understanding of the behavior of atoms and molecules provides powerful insights into other areas of modern science, technology, and engineering.

1.2 CLASSIFICATIONS OF MATTER

Let's begin our study of chemistry by examining some fundamental ways in which matter is classified and described. Two principal ways of classifying matter are according to its physical state (as a gas, liquid, or solid) and according to its composition (as an element, compound, or mixture).

States of Matter

A sample of matter can be a gas, a liquid, or a solid. These three forms of matter are called the **states of matter**. The states of matter differ in some of their simple observable properties. A **gas** (also known as *vapor*) has no fixed volume or shape; rather, it conforms to the volume and shape of its container. A gas can be compressed to occupy a smaller volume, or it can expand to occupy a larger one. A **liquid** has a distinct volume independent of its container but has no specific shape. A liquid assumes the shape of the portion of the container that it occupies. A **solid** has both a definite shape and a definite volume. Neither liquids nor solids can be compressed to any appreciable extent.

The properties of the states of matter can be understood on the molecular level (Figure 1.4▶). In a gas the molecules are far apart and are moving at high speeds, colliding repeatedly with each other and with the walls of the container. Compressing a gas decreases the amount of space between molecules, increases the frequency of collisions between molecules, but does not alter the size or shape of the molecules. In a liquid the molecules are packed closely together but still move rapidly. The rapid movement allows the molecules to slide over each other; thus, a liquid pours easily. In a solid the molecules are held tightly together, usually in definite arrangements in which the molecules can wiggle only slightly in their otherwise fixed positions.

Gas	Liquid	Solid

▲ Figure 1.4 **The three physical states of water—water vapor, liquid water, and ice.** In this photo we see both the liquid and solid states of water. We cannot see water vapor. What we see when we look at steam or clouds is tiny droplets of liquid water dispersed in the atmosphere. The molecular views show that the molecules in the gas are much further apart than those in the liquid or solid. The molecules in the liquid do not have the orderly arrangement seen in the solid.

Pure Substances

Most forms of matter that we encounter—for example, the air we breathe (a gas), gasoline for cars (a liquid), and the sidewalk on which we walk (a solid)—are not chemically pure. We can, however, resolve, or separate, these forms of matter into different pure substances. A **pure substance** (usually referred to simply as a *substance*) is matter that has distinct properties and a composition that does not vary from sample to sample. Water and ordinary table salt (sodium chloride), the primary components of seawater, are examples of pure substances.

All substances are either elements or compounds. **Elements** cannot be decomposed into simpler substances. On the molecular level, each element is composed of only one kind of atom [Figure 1.5(a and b)▼]. **Compounds** are

(a) Atoms of an element   (b) Molecules of an element   (c) Molecules of a compound   (d) Mixture of elements and a compound

▲ Figure 1.5 **Molecular comparison of element, compounds, and mixtures.** Each element contains a unique kind of atom. Elements might consist of individual atoms, as in (a), or molecules, as in (b). Compounds contain two or more different atoms chemically joined together, as in (c). A mixture contains the individual units of its components, shown in (d) as both atoms and molecules.

▶ Figure 1.6 **Relative abundances of elements.** Elements in percent by mass in (a) Earth's crust (including oceans and atmosphere) and (b) the human body.

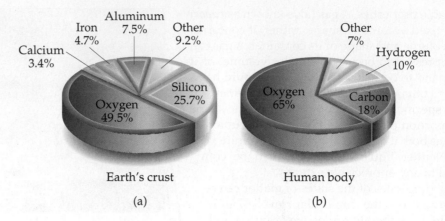

Earth's crust

(a)

Human body

(b)

substances composed of two or more elements; they contain two or more kinds of atoms [Figure 1.5(c)]. Water, for example, is a compound composed of two elements: hydrogen and oxygen. Figure 1.5(d) shows a mixture of substances. **Mixtures** are combinations of two or more substances in which each substance retains its own chemical identity.

Elements

Currently, 117 elements are known. These elements vary widely in their abundance, as shown in Figure 1.6▲. For example, only five elements—oxygen, silicon, aluminum, iron, and calcium—account for over 90% of Earth's crust (including oceans and atmosphere). Similarly, just three elements—oxygen, carbon, and hydrogen—account for over 90% of the mass of the human body.

Some of the more common elements are listed in Table 1.2▼, along with the chemical abbreviations, or chemical *symbols*, used to denote them. The symbol for each element consists of one or two letters, with the first letter capitalized. These symbols are mostly derived from the English name for the element, but sometimes they are derived from a foreign name instead (last column in Table 1.2). You will need to know these symbols and learn others as we encounter them in the text.

All the known elements and their symbols are listed on the front inside cover of this text. The table in which the symbol for each element is enclosed in a box is called the *periodic table*. In the periodic table the elements are arranged in vertical columns so that closely related elements are grouped together. We describe the periodic table in more detail in Section 2.5.

GIVE IT SOME THOUGHT

Which element is most abundant in both Earth's crust and in the human body? What is the symbol for this element?

Compounds

Most elements can interact with other elements to form compounds. For example, consider the fact that when hydrogen gas burns in oxygen gas, the

TABLE 1.2 ■ Some Common Elements and Their Symbols					
Carbon	C	Aluminum	Al	Copper	Cu (from *cuprum*)
Fluorine	F	Bromine	Br	Iron	Fe (from *ferrum*)
Hydrogen	H	Calcium	Ca	Lead	Pb (from *plumbum*)
Iodine	I	Chlorine	Cl	Mercury	Hg (from *hydrargyrum*)
Nitrogen	N	Helium	He	Potassium	K (from *kalium*)
Oxygen	O	Lithium	Li	Silver	Ag (from *argentum*)
Phosphorus	P	Magnesium	Mg	Sodium	Na (from *natrium*)
Sulfur	S	Silicon	Si	Tin	Sn (from *stannum*)

Oxygen gas, O_2

Water, H_2O

Hydrogen gas, H_2

▲ **Figure 1.7 Electrolysis of water.** Water decomposes into its component elements, hydrogen and oxygen, when a direct electrical current is passed through it. The volume of hydrogen, which is collected in the right tube of the apparatus, is twice the volume of oxygen, which is collected in the left tube.

elements hydrogen and oxygen combine to form the compound water. Conversely, water can be decomposed into its component elements by passing an electrical current through it, as shown in Figure 1.7▲. Pure water, regardless of its source, consists of 11% hydrogen and 89% oxygen by mass. This macroscopic composition corresponds to the molecular composition, which consists of two hydrogen atoms combined with one oxygen atom:

Hydrogen atom   Oxygen atom   Water molecule

(written H)   (written O)   (written H_2O)

The elements hydrogen and oxygen themselves exist naturally as diatomic (two-atom) molecules:

Oxygen molecule   (written O_2)

Hydrogen molecule   (written H_2)

As seen in Table 1.3▼, the properties of water bear no resemblance to the properties of its component elements. Hydrogen, oxygen, and water are each a unique substance, a consequence of the uniqueness of their respective molecules.

TABLE 1.3 ■ Comparison of Water, Hydrogen, and Oxygen			
	Water	**Hydrogen**	**Oxygen**
State[a]	Liquid	Gas	Gas
Normal boiling point	100 °C	−253 °C	−183 °C
Density[a]	1000 g/L	0.084 g/L	1.33 g/L
Flammable	No	Yes	No

[a]At room temperature and atmospheric pressure. (See Section 10.2.)

The observation that the elemental composition of a pure compound is always the same is known as the **law of constant composition** (or the **law of definite proportions**). French chemist Joseph Louis Proust (1754–1826) first put forth the law in about 1800. Although this law has been known for 200 years, the general belief persists among some people that a fundamental difference exists between compounds prepared in the laboratory and the corresponding compounds found in nature. However, a pure compound has the same composition and properties regardless of its source. Both chemists and nature must use the same elements and operate under the same natural laws. When two materials differ in composition and properties, we know that they are composed of different compounds or that they differ in purity.

▲ GIVE IT SOME THOUGHT

Hydrogen, oxygen, and water are all composed of molecules. What is it about a molecule of water that makes it a compound, whereas hydrogen and oxygen are elements?

Mixtures

Most of the matter we encounter consists of mixtures of different substances. Each substance in a mixture retains its own chemical identity and its own properties. In contrast to a pure substance that has a fixed composition, the composition of a mixture can vary. A cup of sweetened coffee, for example, can contain either a little sugar or a lot. The substances making up a mixture (such as sugar and water) are called *components* of the mixture.

Some mixtures do not have the same composition, properties, and appearance throughout. Both rocks and wood, for example, vary in texture and appearance throughout any typical sample. Such mixtures are *heterogeneous* [Figure 1.8(a) ▼]. Mixtures that are uniform throughout are *homogeneous*. Air is a homogeneous mixture of the gaseous substances nitrogen, oxygen, and smaller amounts of other substances. The nitrogen in air has all the properties that pure nitrogen does because both the pure substance and the mixture contain the same nitrogen molecules. Salt, sugar, and many other substances dissolve in water to form homogeneous mixtures [Figure 1.8(b)]. Homogeneous mixtures are also called **solutions**. Although the term solution conjures an image of a liquid in a beaker or flask, solutions can be solids, liquids, or gases. Figure 1.9 ▶ summarizes the classification of matter into elements, compounds, and mixtures.

▶ **Figure 1.8 Mixtures.** (a) Many common materials, including rocks, are heterogeneous. This close-up photo is of malachite, a copper mineral. (b) Homogeneous mixtures are called solutions. Many substances, including the blue solid shown in this photo (copper sulfate), dissolve in water to form solutions.

(a)   (b)

◄ **Figure 1.9 Classification of matter.** At the chemical level all matter is classified ultimately as either elements or compounds.

■ **SAMPLE EXERCISE 1.1** | Distinguishing Among Elements, Compounds, and Mixtures

"White gold," used in jewelry, contains gold and another "white" metal such as palladium. Two different samples of white gold differ in the relative amounts of gold and palladium that they contain. Both samples are uniform in composition throughout. Without knowing any more about the materials, use Figure 1.9 to classify white gold.

SOLUTION

Because the material is uniform throughout, it is homogeneous. Because its composition differs for the two samples, it cannot be a compound. Instead, it must be a homogeneous mixture.

■ **PRACTICE EXERCISE**

Aspirin is composed of 60.0% carbon, 4.5% hydrogen, and 35.5% oxygen by mass, regardless of its source. Use Figure 1.9 to characterize and classify aspirin.
Answer: It is a compound because it has constant composition and can be separated into several elements.

1.3 PROPERTIES OF MATTER

Every substance has a unique set of properties. For example, the properties listed in Table 1.3 allow us to distinguish hydrogen, oxygen, and water from one another. The properties of matter can be categorized as physical or chemical. **Physical properties** can be observed without changing the identity and composition of the substance. These properties include color, odor, density, melting point, boiling point, and hardness. **Chemical properties** describe the way a substance may change, or *react*, to form other substances. A common chemical property is flammability, the ability of a substance to burn in the presence of oxygen.

Some properties, such as temperature, melting point, and density, are called **intensive properties**. They do not depend on the amount of the sample being examined and are particularly useful in chemistry because many of these properties can be used to *identify* substances. **Extensive properties** of substances depend on the quantity of the sample, with two examples being mass and volume. Extensive properties relate to the *amount* of substance present.

Physical and Chemical Changes

As with the properties of a substance, the changes that substances undergo can be classified as either physical or chemical. During a **physical change** a substance changes its physical appearance but not its composition. (That is, it is the same substance before and after the change.) The evaporation of water is a physical change. When water evaporates, it changes from the liquid state to the gas state, but it is still composed of water molecules, as depicted earlier in Figure 1.4. All **changes of state** (for example, from liquid to gas or from liquid to solid) are physical changes.

In a **chemical change** (also called a **chemical reaction**) a substance is transformed into a chemically different substance. When hydrogen burns in air, for example, it undergoes a chemical change because it combines with oxygen to form water. The molecular-level view of this process is depicted in Figure 1.10▼.

Chemical changes can be dramatic. In the account that follows, Ira Remsen, author of a popular chemistry text published in 1901, describes his first

CHEMICAL REACTION (CHEMICAL CHANGE)
In a chemical reaction, the chemical identities of substances change.
Here hydrogen and oxygen undergo a chemical change to form water.

Burn

Hydrogen and oxygen gases.

When hydrogen burns, it undergoes a chemical change.

To form water, which is produced in the flame.

▲ Figure 1.10 **A chemical reaction.**

(a)   (b)   (c)

▲ **Figure 1.11 The chemical reaction between a copper penny and nitric acid.** The dissolved copper produces the blue-green solution; the reddish brown gas produced is nitrogen dioxide.

experiences with chemical reactions. The chemical reaction that he observed is shown in Figure 1.11 ▲.

> While reading a textbook of chemistry, I came upon the statement "nitric acid acts upon copper," and I determined to see what this meant. Having located some nitric acid, I had only to learn what the words "act upon" meant. In the interest of knowledge I was even willing to sacrifice one of the few copper cents then in my possession. I put one of them on the table, opened a bottle labeled "nitric acid," poured some of the liquid on the copper, and prepared to make an observation. But what was this wonderful thing which I beheld? The cent was already changed, and it was no small change either. A greenish-blue liquid foamed and fumed over the cent and over the table. The air became colored dark red. How could I stop this? I tried by picking the cent up and throwing it out the window. I learned another fact: nitric acid acts upon fingers. The pain led to another unpremeditated experiment. I drew my fingers across my trousers and discovered nitric acid acts upon trousers. That was the most impressive experiment I have ever performed. I tell of it even now with interest. It was a revelation to me. Plainly the only way to learn about such remarkable kinds of action is to see the results, to experiment, to work in the laboratory.

GIVE IT SOME THOUGHT

Which of the following is a physical change, and which is a chemical change? Explain. **(a)** Plants use carbon dioxide and water to make sugar. **(b)** Water vapor in the air on a cold day forms frost.

Separation of Mixtures

Because each component of a mixture retains its own properties, we can separate a mixture into its components by taking advantage of the differences in their properties. For example, a heterogeneous mixture of iron filings and gold filings could be sorted individually by color into iron and gold. A less tedious approach would be to use a magnet to attract the iron filings, leaving the gold ones behind. We can also take advantage of an important chemical difference between these two metals: Many acids dissolve iron but not gold. Thus, if we put our mixture into an appropriate acid, the acid would dissolve the iron and the gold would be left behind. The two could then be separated by *filtration*, a

(a)   (b)

▲ Figure 1.12 **Separation by filtration.** A mixture of a solid and a liquid is poured through a porous medium, in this case filter paper. The liquid passes through the paper while the solid remains on the paper.

procedure illustrated in Figure 1.12 ◄. We would have to use other chemical reactions, which we will learn about later, to transform the dissolved iron back into metal.

An important method of separating the components of a homogeneous mixture is *distillation*, a process that depends on the different abilities of substances to form gases. For example, if we boil a solution of salt and water, the water evaporates, forming a gas, and the salt is left behind. The gaseous water can be converted back to a liquid on the walls of a condenser, as shown in the apparatus depicted in Figure 1.13 ▼.

The differing abilities of substances to adhere to the surfaces of various solids such as paper and starch can also be used to separate mixtures. This ability is the basis of *chromatography* (literally "the writing of colors"), a technique that can give beautiful and dramatic results. An example of the chromatographic separation of ink is shown in Figure 1.14 ▼.

▶ Figure 1.13 **Distillation.** A simple apparatus for the separation of a sodium chloride solution (salt water) into its components. Boiling the solution vaporizes the water, which is condensed, then collected in the receiving flask. After all the water has boiled away, pure sodium chloride remains in the boiling flask.

Salt water

Boiling flask

Cold water out

Condenser

Cold water in

Clamp

Receiving flask

Burner

Pure water

▶ Figure 1.14 **Separation of ink into components by paper chromatography.** (a) Water begins to move up the paper. (b) Water moves past the ink spot, dissolving different components of the ink at different rates. (c) The ink has separated into its several different components.

(a)

(b)

(c)

▲ **A Closer Look**   THE SCIENTIFIC METHOD

Although two scientists rarely approach the same problem in exactly the same way, they use guidelines for the practice of science that are known as the **scientific method**. These guidelines are outlined in Figure 1.15▼. We begin our study by collecting information, or *data*, by observation and experiment. The collection of information, however, is not the ultimate goal. The goal is to find a pattern or sense of order in our observations and to understand the origin of this order.

As we perform our experiments, we may begin to see patterns that lead us to a *tentative explanation*, or **hypothesis**, that guides us in planning further experiments. Eventually, we may be able to tie together a great number of observations in a single statement or equation called a scientific law. A **scientific law** *is a concise verbal statement or a mathematical equation that summarizes a broad variety of observations and experiences*. We tend to think of the laws of nature as the basic rules under which nature operates. However, it is not so much that matter obeys the laws of nature, but rather that the laws of nature describe the behavior of matter.

At many stages of our studies we may propose explanations of why nature behaves in a particular way. If a hypothesis is sufficiently general and is continually effective in predicting facts yet to be observed, it is called a theory. A **theory** *is an explanation of the general causes of certain phenomena, with considerable*

evidence or facts to support it. For example, Einstein's theory of relativity was a revolutionary new way of thinking about space and time. It was more than just a simple hypothesis, however, because it could be used to make predictions that could be tested experimentally. When these experiments were conducted, the results were generally in agreement with the predictions and were not explainable by earlier theories. Thus, the theory of relativity was supported, but not proven. Indeed, theories can never be proven to be absolutely correct.

As we proceed through this text, we will rarely have the opportunity to discuss the doubts, conflicts, clashes of personalities, and revolutions of perception that have led to our present ideas. We need to be aware that just because we can spell out the results of science so concisely and neatly in textbooks, it does not mean that scientific progress is smooth, certain, and predictable. Some of the ideas we present in this text took centuries to develop and involved many scientists. We gain our view of the natural world by standing on the shoulders of the scientists who came before us. Take advantage of this view. As you study, exercise your imagination. Don't be afraid to ask daring questions when they occur to you. You may be fascinated by what you discover!

Related Exercise: 1.57

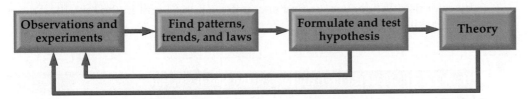

▲ **Figure 1.15 The scientific method.** The scientific method is a general approach to solving problems that involves making observations, seeking patterns in the observations, formulating hypotheses to explain the observations, and testing these hypotheses by further experiments. Those hypotheses that withstand such tests and prove themselves useful in explaining and predicting behavior become known as theories.

1.4 UNITS OF MEASUREMENT

Many properties of matter are *quantitative*; that is, they are associated with numbers. When a number represents a measured quantity, the units of that quantity must always be specified. To say that the length of a pencil is 17.5 is meaningless. Expressing the number with its units, 17.5 centimeters (cm), properly specifies the length. The units used for scientific measurements are those of the **metric system**.

The metric system, which was first developed in France during the late eighteenth century, is used as the system of measurement in most countries throughout the world. The United States has traditionally used the English system, although use of the metric system has become more common. For example, the contents of most canned goods and soft drinks in grocery stores are now given in metric as well as in English units, as shown in Figure 1.16▶.

SI Units

In 1960 an international agreement was reached specifying a particular choice of metric units for use in scientific measurements. These preferred units are

▲ **Figure 1.16 Metric units.** Metric measurements are increasingly common in the United States, as exemplified by the volume printed on this soda can.

TABLE 1.4 ■ SI Base Units		
Physical Quantity	Name of Unit	Abbreviation
Mass	Kilogram	kg
Length	Meter	m
Time	Second	s[a]
Temperature	Kelvin	K
Amount of substance	Mole	mol
Electric current	Ampere	A
Luminous intensity	Candela	cd

[a]The abbreviation sec is frequently used.

called **SI units**, after the French *Système International d'Unités*. This system has seven *base units* from which all other units are derived. Table 1.4▲ lists these base units and their symbols. In this chapter we will consider the base units for length, mass, and temperature.

In the metric system, prefixes are used to indicate decimal fractions or multiples of various units. For example, the prefix *milli-* represents a 10^{-3} fraction of a unit: A milligram (mg) is 10^{-3} gram (g), a millimeter (mm) is 10^{-3} meter (m), and so forth. Table 1.5▼ presents the prefixes commonly encountered in chemistry. In using SI units and in working problems throughout this text, you must be comfortable using exponential notation. If you are unfamiliar with exponential notation or want to review it, refer to Appendix A.1.

Although non-SI units are being phased out, some are still commonly used by scientists. Whenever we first encounter a non-SI unit in the text, the proper SI unit will also be given.

TABLE 1.5 ■ Selected Prefixes Used in the Metric System			
Prefix	Abbreviation	Meaning	Example
Giga	G	10^9	1 gigameter (Gm) = 1×10^9 m
Mega	M	10^6	1 megameter (Mm) = 1×10^6 m
Kilo	k	10^3	1 kilometer (km) = 1×10^3 m
Deci	d	10^{-1}	1 decimeter (dm) = 0.1 m
Centi	c	10^{-2}	1 centimeter (cm) = 0.01 m
Milli	m	10^{-3}	1 millimeter (mm) = 0.001 m
Micro	μ[a]	10^{-6}	1 micrometer (μm) = 1×10^{-6} m
Nano	n	10^{-9}	1 nanometer (nm) = 1×10^{-9} m
Pico	p	10^{-12}	1 picometer (pm) = 1×10^{-12} m
Femto	f	10^{-15}	1 femtometer (fm) = 1×10^{-15} m

[a]This is the Greek letter mu (pronounced "mew").

GIVE IT SOME THOUGHT

Which of the following quantities is the smallest: 1 mg, 1μg, or 1 pg?

Length and Mass

The SI base unit of *length* is the meter (m), a distance only slightly longer than a yard. The relations between the English and metric system units that we will use most frequently in this text appear on the back inside cover. We will discuss how to convert English units into metric units, and vice versa, in Section 1.6.

Mass* is a measure of the amount of material in an object. The SI base unit of mass is the kilogram (kg), which is equal to about 2.2 pounds (lb). This base unit is unusual because it uses a prefix, *kilo-*, instead of the word *gram* alone. We obtain other units for mass by adding prefixes to the word *gram*.

■■■ **SAMPLE EXERCISE 1.2** | **Using Metric Prefixes**

What is the name given to the unit that equals **(a)** 10^{-9} gram, **(b)** 10^{-6} second, **(c)** 10^{-3} meter?

SOLUTION

In each case we can refer to Table 1.5, finding the prefix related to each of the decimal fractions: **(a)** nanogram, ng, **(b)** microsecond, μs, **(c)** millimeter, mm.

■■■ **PRACTICE EXERCISE**

(a) What decimal fraction of a second is a picosecond, ps? **(b)** Express the measurement 6.0×10^3 m using a prefix to replace the power of ten. **(c)** Use exponential notation to express 3.76 mg in grams.
Answers: **(a)** 10^{-12} second, **(b)** 6.0 km, **(c)** 3.76×10^{-3} g

---

Temperature

Temperature is a measure of the hotness or coldness of an object. Indeed, temperature is a physical property that determines the direction of heat flow. Heat always flows spontaneously from a substance at higher temperature to one at lower temperature. Thus, we feel the influx of heat when we touch a hot object, and we know that the object is at a higher temperature than our hand.

The temperature scales commonly employed in scientific studies are the Celsius and Kelvin scales. The **Celsius scale** is also the everyday scale of temperature in most countries (Figure 1.17▶). It was originally based on the assignment of 0 °C to the freezing point of water and 100 °C to its boiling point at sea level (Figure 1.18▼).

▲ **Figure 1.17 Australian stamp.** Many countries employ the Celsius temperature scale in everyday use, as illustrated by this stamp.

373 K	100 °C	212 °F	Water boils
310 K	37.0 °C	98.6 °F	Normal body temperature
273 K	0 °C	32 °F	Water freezes

Kelvin scale   Celsius scale   Fahrenheit scale

◀ **Figure 1.18 Comparison of the Kelvin, Celsius, and Fahrenheit temperature scales.** The freezing point and boiling point of water as well as normal human body temperature is indicated on each of the scales.

---

*Mass and weight are not interchangeable terms but are often incorrectly thought to be the same. The weight of an object is the force that its mass exerts due to gravity. In space, where gravitational forces are very weak, an astronaut can be weightless, but he or she cannot be massless. In fact, the astronaut's mass in space is the same as it is on Earth.

The **Kelvin scale** is the SI temperature scale, and the SI unit of temperature is the kelvin (K). Historically, the Kelvin scale was based on the properties of gases; its origins will be considered in Chapter 10. Zero on this scale is the lowest attainable temperature, -273.15 °C, a temperature referred to as *absolute zero*. Both the Celsius and Kelvin scales have equal-sized units—that is, a kelvin is the same size as a degree Celsius. Thus, the Kelvin and Celsius scales are related as follows:

$$K = °C + 273.15 \qquad\qquad [1.1]$$

The freezing point of water, 0 °C, is 273.15 K (Figure 1.18). Notice that we do not use a degree sign (°) with temperatures on the Kelvin scale.

The common temperature scale in the United States is the *Fahrenheit scale*, which is not generally used in scientific studies. On the Fahrenheit scale, water freezes at 32 °F and boils at 212 °F. The Fahrenheit and Celsius scales are related as follows:

$$°C = \frac{5}{9}(°F - 32) \quad \text{or} \quad °F = \frac{9}{5}(°C) + 32 \qquad [1.2]$$

■ **SAMPLE EXERCISE 1.3** | Converting Units of Temperature

If a weather forecaster predicts that the temperature for the day will reach 31 °C, what is the predicted temperature **(a)** in K, **(b)** in °F?

SOLUTION

(a) Using Equation 1.1, we have K = 31 + 273 = 304 K

(b) Using Equation 1.2, we have °F = $\frac{9}{5}$ (31) + 32 = 56 + 32 = 88 °F

■ **PRACTICE EXERCISE**

Ethylene glycol, the major ingredient in antifreeze, freezes at -11.5 °C. What is the freezing point in **(a)** K, **(b)** °F?
Answers: **(a)** 261.7 K, **(b)** 11.3 °F

Derived SI Units

The SI base units in Table 1.4 are used to derive the units of other quantities. To do so, we use the defining equation for the quantity, substituting the appropriate base units. For example, speed is defined as the ratio of distance traveled to elapsed time. Thus, the SI unit for speed—m/s, which we read as "meters per second"—is the SI unit for distance (length), m, divided by the SI unit for time, s. We will encounter many derived units, such as those for force, pressure, and energy, later in this text. In this chapter we examine the derived units for volume and density.

Volume

The *volume* of a cube is given by its length cubed, (length)^3. Thus, the SI unit of volume is the SI unit of length, m, raised to the third power. The cubic meter, or m^3, is the volume of a cube that is 1 m on each edge. Smaller units, such as cubic centimeters, cm^3 (sometimes written as cc), are frequently used in chemistry. Another unit of volume commonly used in chemistry is the *liter* (L), which equals a cubic decimeter, dm^3, and is slightly larger than a quart. The liter is the first metric unit we have encountered that is *not* an SI unit. There are 1000 milliliters (mL) in a liter (Figure 1.19◄), and each milliliter is the same volume as a cubic centimeter: 1 mL = 1 cm^3. The terms *milliliter* and *cubic centimeter* are used interchangeably in expressing volume.

The devices used most frequently in chemistry to measure volume are illustrated in Figure 1.20▶. Syringes, burets, and pipets deliver liquids with more precision than graduated cylinders. Volumetric flasks are used to contain specific volumes of liquid.

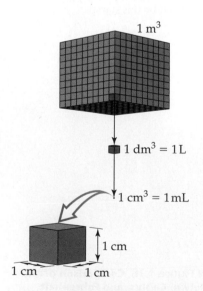

1 m^3

1 dm^3 = 1 L

1 cm^3 = 1 mL

1 cm

1 cm   1 cm

▲ **Figure 1.19 Volume relationships.** The volume occupied by a cube that is 1 m on each edge is a cubic meter, 1 m^3 (top). Each cubic meter contains 1000 dm^3 (middle). A liter is the same volume as a cubic decimeter, 1 L = 1 dm^3. Each cubic decimeter contains 1000 cubic centimeters, 1 dm^3 = 1000 cm^3. Each cubic centimeter equals 1 milliliter, 1 cm^3 = 1 mL (bottom).

Stopcock, a valve to control the liquid flow

Graduated cylinder   Syringe   Buret   Pipet   Volumetric flask

GIVE IT SOME THOUGHT

Which of the following quantities represents a volume measurement: 15 m^2; $2.5 \times 10^2 \text{ m}^3$; 5.77 L/s? How do you know?

Density

Density is a property of matter that is widely used to characterize a substance. Density is defined as the amount of mass in a unit volume of the substance:

$$\text{Density} = \frac{\text{mass}}{\text{volume}} \qquad [1.3]$$

The densities of solids and liquids are commonly expressed in units of grams per cubic centimeter (g/cm^3) or grams per milliliter (g/mL). The densities of some common substances are listed in Table 1.6▼. It is no coincidence that the density of water is 1.00 g/mL; the gram was originally defined as the mass of 1 mL of water at a specific temperature. Because most substances change volume when they are heated or cooled, densities are temperature dependent. When reporting densities, the temperature should be specified. If no temperature is reported, we usually assume that the temperature is 25 °C, close to normal room temperature.

The terms *density* and *weight* are sometimes confused. A person who says that iron weighs more than air generally means that iron has a higher density than air—1 kg of air has the same mass as 1 kg of iron, but the iron occupies a smaller volume, thereby giving it a higher density. If we combine two liquids that do not mix, the less dense liquid will float on the denser liquid.

TABLE 1.6 ■ Densities of Some Selected Substances at 25 °C	
Substance	**Density (g/cm³)**
Air	0.001
Balsa wood	0.16
Ethanol	0.79
Water	1.00
Ethylene glycol	1.09
Table sugar	1.59
Table salt	2.16
Iron	7.9
Gold	19.32

Chemistry is a very lively, active field of science. Because chemistry is so central to our lives, reports on matters of chemical significance appear in the news nearly every day. Some reports tell of recent breakthroughs in the development of new pharmaceuticals, materials, and processes. Others deal with environmental and public safety issues. As you study chemistry, we hope you will develop the skills to better understand the importance of chemistry in your life. By way of examples, here are summaries of a few recent stories in which chemistry plays a role.

Biofuels Reality Check

With the Energy Policy Act of 2005, the United States Congress has given a big push to fuels derived from biomass as a renewable, homegrown alternative to gasoline. The law requires that 4 billion gallons of the so-called renewable fuel be mixed with gasoline in 2007, increasing to 7.5 billion gallons by 2012. The United States currently consumes about 140 billion gallons of gasoline per year.

Although the Act does not dictate which renewable fuels to use, ethanol derived from corn currently dominates the alternatives with 40% of all gasoline now containing some ethanol. A blend of 10% ethanol and 90% gasoline, called E10, is the most common blend because it can be used in virtually all vehicles. Blends of 85% ethanol and 15% gasoline, called E85, are also available but can be used only with specially modified engines in what are called flexible-fuel vehicles (FFVs) (Figure 1.21 ▼).

▲ **Figure 1.21 A gasoline pump that dispenses E85 ethanol.**

When it comes to ethanol's pros and cons, there is no shortage of disagreement. In 2006, researchers at the University of Minnesota calculated that "Even dedicating all U.S. corn and soybean production to biofuels would meet only 12% of gasoline and 6% of diesel demand." The conversion of a much wider range of plant material, making use of a much greater fraction of the available plant matter, into fuels will be necessary to improve these numbers substantially. Because most cellulose of which plants are formed does not readily convert to ethanol, a great deal of research will be needed to solve this challenging problem. Meanwhile, it is worth reflecting that a 3% improvement in vehicle efficiency of fuel use would displace more gasoline use than the entire 2006 US ethanol production.

New Element Created

A new entry has been made to the list of elements. The production of the newest and heaviest element—element 118—was announced in October 2006. The synthesis of element 118 resulted from studies performed from 2002 to 2006 at the Joint Institute for Nuclear Research (JINR) in Dubna, Russia. JINR scientists and their collaborators from Lawrence Livermore National Laboratory in California announced that they had produced three atoms of the new element, one atom in 2002 and two more in 2005.

The new element was formed by striking a target of californium atoms (element 98) with a highly energetic beam consisting of the nuclei of calcium atoms (element 20) in a device called a particle accelerator. Occasionally, the nuclei from the atoms of the two different elements fused to form the new, superheavy element 118. The 2002 experiment took four months and used a beam of 2.5×10^{19} calcium atoms to produce the single atom of element 118.

The three atoms of element 118 created during these experiments came and went in a literal flash. On the average, the atoms survived for just 0.9 milliseconds before decomposing.

These experimental results were met with praise but also caution from other scientists in the field, particularly given the difficult history of element 118. Another California lab, the Lawrence Berkeley National Laboratory, announced that it discovered element 118 in 1999 but retracted the claim two years later after an investigation found that one of the researchers had fabricated data.

This discovery brings the total number of elements created by the Livermore-Dubna collaboration to five: elements 113, 114, 115, 116, and 118. As of this writing, element 118 has not yet been named.

Important Antibiotic Modified to Combat Bacterial Resistance

Vancomycin is an antibiotic of last resort—used only when other antibacterial agents are ineffective. Some bacteria have developed a resistance to vancomycin, causing researchers to modify the molecular structure of the substance to make it

(a)   (b)

◄ Figure 1.22 **Comparing CO and CH₂ groups.** Two molecules, one containing the CO group (left) and one containing the CH₂ group (right), are shown. The subtle difference between these two molecules is like that produced when the structure of the much more complex vancomycin molecule was modified.

more effective in killing bacteria. This approach was based on the knowledge that vancomycin works by binding to a particular protein, called a glycoprotein, that is essential to forming the walls of bacterial cells. Researchers have now synthesized an analog of vancomycin in which a CO group in the molecule has been converted to a CH₂ group (Figure 1.22▲). This molecular modification increases the compound's binding affinity with the glycoprotein in the cell walls of vancomycin-resistant bacteria. The analog is 100 times more active than vancomycin against vancomycin-resistant bacteria.

The Hole Story

Ozone in the upper atmosphere protects life on Earth by blocking harmful ultraviolet rays coming from the sun. The "ozone hole" is a severe depletion of the ozone layer high above Antarctica. Human-produced compounds that release chlorine and bromine into the stratosphere are the primary cause of the ozone hole.

The production of ozone depleting chemicals has been banned since 1996, although emissions of previously produced and stored amounts of those chemicals that are not destroyed or recycled will continue. Scientists had predicted that the ozone hole would disappear by 2050 because of the ban. The 2006 World Meteorological Organization/United Nations Environment Programme Scientific Assessment of Ozone De-

pletion, however, recently issued its report changing this estimate. Based on a combination of new ozone measurements, computer models, and revised estimates of the existing stores of ozone-depleting chemicals, scientists now estimate the date for full Antarctic ozone recovery to be 2065.

Replacing the Lightbulb through Chemistry

If you want to save the world from global warming, you can start by replacing incandescent lightbulbs, which waste about 90% of the energy supplied to them by producing heat. A promising place to look for replacement bulbs is in the field of light-emitting diodes (LEDs). Red LEDs and those emitting other colors are found everywhere these days: in flashlights, traffic lights, car taillights, and a host of electronics applications (Figure 1.23▼). But to really make it big in the world, LEDs need to be capable of producing white light at a reasonable cost.

Progress is being made in forming high-efficiency LEDs based on organic films that emit white light. In these devices a light-emitting material is sandwiched between two electrical connectors. When electricity passes through the organic film, oppositely charged particles combine and give off light. White organic LEDs have been steadily improving, and are now about as efficient as fluorescent tubes. More work needs to be done before these devices can replace the lightbulb, but progress has been rapid.

▶ Figure 1.23 **Sign made from LEDs.**

■ **SAMPLE EXERCISE 1.4** | Determining Density and Using Density to Determine Volume or Mass

(a) Calculate the density of mercury if 1.00×10^2 g occupies a volume of 7.36 cm³.
(b) Calculate the volume of 65.0 g of the liquid methanol (wood alcohol) if its density is 0.791 g/mL.
(c) What is the mass in grams of a cube of gold (density = 19.32 g/cm³) if the length of the cube is 2.00 cm?

SOLUTION

(a) We are given mass and volume, so Equation 1.3 yields

$$\text{Density} = \frac{\text{mass}}{\text{volume}} = \frac{1.00 \times 10^2 \text{ g}}{7.36 \text{ cm}^3} = 13.6 \text{ g/cm}^3$$

(b) Solving Equation 1.3 for volume and then using the given mass and density gives

$$\text{Volume} = \frac{\text{mass}}{\text{density}} = \frac{65.0 \text{ g}}{0.791 \text{ g/mL}} = 82.2 \text{ mL}$$

(c) We can calculate the mass from the volume of the cube and its density. The volume of a cube is given by its length cubed:

$$\text{Volume} = (2.00 \text{ cm})^3 = (2.00)^3 \text{ cm}^3 = 8.00 \text{ cm}^3$$

Solving Equation 1.3 for mass and substituting the volume and density of the cube, we have

$$\text{Mass} = \text{volume} \times \text{density} = (8.00 \text{ cm}^3)(19.32 \text{ g/cm}^3) = 155 \text{ g}$$

■ **PRACTICE EXERCISE**

(a) Calculate the density of a 374.5-g sample of copper if it has a volume of 41.8 cm³. **(b)** A student needs 15.0 g of ethanol for an experiment. If the density of ethanol is 0.789 g/mL, how many milliliters of ethanol are needed? **(c)** What is the mass, in grams, of 25.0 mL of mercury (density = 13.6 g/mL)?
Answers: **(a)** 8.96 g/cm³, **(b)** 19.0 mL, **(c)** 340 g

1.5 UNCERTAINTY IN MEASUREMENT

Two kinds of numbers are encountered in scientific work: *exact numbers* (those whose values are known exactly) and *inexact numbers* (those whose values have some uncertainty). Most of the exact numbers that we will encounter in this course have defined values. For example, there are exactly 12 eggs in a dozen, exactly 1000 g in a kilogram, and exactly 2.54 cm in an inch. The number 1 in any conversion factor between units, as in 1 m = 100 cm or 1 kg = 2.2046 lb, is also an exact number. Exact numbers can also result from counting numbers of objects. For example, we can count the exact number of marbles in a jar or the exact number of people in a classroom.

Numbers obtained by measurement are always *inexact*. The equipment used to measure quantities always has inherent limitations (equipment errors), and there are differences in how different people make the same measurement (human errors). Suppose that ten students with ten balances are given the same dime and told to determine its mass. The ten measurements will probably vary slightly from one another for various reasons. The balances might be calibrated slightly differently, and there might be differences in how each student reads the mass from the balance. Remember: *Uncertainties always exist in measured quantities.* Counting very large numbers of objects usually has some associated error as well. Consider, for example, how difficult it is to obtain accurate census information for a city or vote counts for an election.

▲ **GIVE IT SOME THOUGHT**

Which of the following is an inexact quantity: **(a)** the number of people in your chemistry class, **(b)** the mass of a penny, **(c)** the number of grams in a kilogram?

Precision and Accuracy

The terms precision and accuracy are often used in discussing the uncertainties of measured values. **Precision** is a measure of how closely individual measurements agree with one another. **Accuracy** refers to how closely individual measurements agree with the correct, or "true," value. The analogy of darts stuck in a dartboard pictured in Figure 1.24▶ illustrates the difference between these two concepts.

In the laboratory we often perform several different "trials" of the same experiment and average the results. The precision of the measurements is often expressed in terms of what is called the *standard deviation*, which reflects how much the individual measurements differ from the average, as described in Appendix A. We gain confidence in our measurements if we obtain nearly the same value each time—that is, the standard deviation is small. Figure 1.24 should remind us, however, that precise measurements could be inaccurate. For example, if a very sensitive balance is poorly calibrated, the masses we measure will be consistently either high or low. They will be inaccurate even if they are precise.

Significant Figures

Suppose you determine the mass of a dime on a balance capable of measuring to the nearest 0.0001 g. You could report the mass as 2.2405 ± 0.0001 g. The ± notation (read "plus or minus") expresses the magnitude of the uncertainty of your measurement. In much scientific work we drop the ± notation with the understanding that there is always some uncertainty in the last digit of the measured quantity. That is, *measured quantities are generally reported in such a way that only the last digit is uncertain.*

Figure 1.25▼ shows a thermometer with its liquid column between the scale marks. We can read the certain digits from the scale and estimate the uncertain one. From the scale marks on the thermometer, we see that the liquid is between the 25 °C and 30 °C marks. We might estimate the temperature to be 27 °C, being somewhat uncertain of the second digit of our measurement.

All digits of a measured quantity, including the uncertain one, are called **significant figures.** A measured mass reported as 2.2 g has two significant figures, whereas one reported as 2.2405 g has five significant figures. The greater the number of significant figures, the greater is the certainty implied for the measurement. When multiple measurements are made of a quantity, the results can be averaged, and the number of significant figures estimated by using statistical methods.

Good accuracy
Good precision

Poor accuracy
Good precision

Poor accuracy
Poor precision

▲ Figure 1.24 **Precision and accuracy.** The distribution of darts on a target illustrates the difference between accuracy and precision.

◀ Figure 1.25 **Significant figures in measurements.** The thermometer has markings every 5 °C. The temperature is between 25 °C and 30 °C and is approximately 27 °C. The two significant figures in the measurement include the second digit, which is estimated by reading between the scale marks.

■ **SAMPLE EXERCISE 1.5** | Relating Significant Figures to the Uncertainty of a Measurement

What difference exists between the measured values 4.0 g and 4.00 g?

SOLUTION

Many people would say there is no difference, but a scientist would note the difference in the number of significant figures in the two measurements. The value 4.0 has two significant figures, while 4.00 has three. This difference implies that the first measurement has more uncertainty. A mass of 4.0 g indicates that the uncertainty is in the first decimal place of the measurement. Thus, the mass might be anything between 3.9 and 4.1 g, which we can represent as 4.0 ± 0.1 g. A measurement of 4.00 g implies that the uncertainty is in the second decimal place. Thus, the mass might be anything between 3.99 and 4.01 g, which we can represent as 4.00 ± 0.01 g. Without further information, we cannot be sure whether the difference in uncertainties of the two measurements reflects the precision or accuracy of the measurement.

■ **PRACTICE EXERCISE**

A balance has a precision of ±0.001 g. A sample that has a mass of about 25 g is placed on this balance. How many significant figures should be reported for this measurement?

Answer: five, as in the measurement 24.995 g, the uncertainty being in the third decimal place

"MY GOODNESS, IT'S 12:15:0936420175! TIME FOR LUNCH."

ScienceCartoonPlus.com

To determine the number of significant figures in a reported measurement, read the number from left to right, counting the digits starting with the first digit that is not zero. *In any measurement that is properly reported, all nonzero digits are significant.* Zeros, however, can be used either as part of the measured value or merely to locate the decimal point. Thus, zeros may or may not be significant, depending on how they appear in the number. The following guidelines describe the different situations involving zeros:

1. Zeros *between* nonzero digits are always significant—1005 kg (four significant figures); 1.03 cm (three significant figures).

2. Zeros *at the beginning* of a number are never significant; they merely indicate the position of the decimal point—0.02 g (one significant figure); 0.0026 cm (two significant figures).

3. Zeros *at the end* of a number are significant if the number contains a decimal point—0.0200 g (three significant figures); 3.0 cm (two significant figures).

A problem arises when a number ends with zeros but contains no decimal point. In such cases, it is normally assumed that the zeros are not significant. Exponential notation (Appendix A) can be used to clearly indicate whether zeros at the end of a number are significant. For example, a mass of 10,300 g can be written in exponential notation showing three, four, or five significant figures depending on how the measurement is obtained:

$$1.03 \times 10^4 \text{ g} \qquad \text{(three significant figures)}$$
$$1.030 \times 10^4 \text{ g} \qquad \text{(four significant figures)}$$
$$1.0300 \times 10^4 \text{ g} \qquad \text{(five significant figures)}$$

In these numbers all the zeros to the right of the decimal point are significant (rules 1 and 3). (The exponential term does not add to the number of significant figures.)

■ **SAMPLE EXERCISE 1.6** | Determining the Number of Significant Figures in a Measurement

How many significant figures are in each of the following numbers (assume that each number is a measured quantity): **(a)** 4.003, **(b)** 6.023×10^{23}, **(c)** 5000?

SOLUTION

(a) Four; the zeros are significant figures. **(b)** Four; the exponential term does not add to the number of significant figures. **(c)** One. We assume that the zeros are not significant when there is no decimal point shown. If the number has more significant figures, a decimal point should be employed or the number written in exponential notation. Thus, 5000. has four significant figures, whereas 5.00×10^3 has three.

■ **PRACTICE EXERCISE**
How many significant figures are in each of the following measurements: **(a)** 3.549 g, **(b)** 2.3×10^4 cm, **(c)** 0.00134 m^3?
Answers: **(a)** four, **(b)** two, **(c)** three

Significant Figures in Calculations

When carrying measured quantities through calculations, *the least certain measurement limits the certainty of the calculated quantity and thereby determines the number of significant figures in the final answer.* The final answer should be reported with only one uncertain digit. To keep track of significant figures in calculations, we will make frequent use of two rules, one for addition and subtraction, and another for multiplication and division.

1. *For addition and subtraction*, the result has the same number of decimal places as the measurement with the fewest decimal places. When the result contains more than the correct number of significant figures, it must be rounded off. Consider the following example in which the uncertain digits appear in color:

This number limits	20.42	← two decimal places
the number of significant	1.322	← three decimal places
figures in the result →	83.1	← one decimal place
	104.842	← round off to one decimal place (104.8)

We report the result as 104.8 because 83.1 has only one decimal place.

2. *For multiplication and division*, the result contains the same number of significant figures as the measurement with the fewest significant figures. When the result contains more than the correct number of significant figures, it must be rounded off. For example, the area of a rectangle whose measured edge lengths are 6.221 cm and 5.2 cm should be reported as 32 cm^2 even though a calculator shows the product of 6.221 and 5.2 to have more digits:

Area = (6.221 cm) (5.2 cm) = 32.3492 cm^2 ⟹ round off to 32 cm^2

We round off to two significant figures because the least precise number—5.2 cm—has only two significant figures.

Notice that for addition and subtraction, decimal places are counted; whereas for multiplication and division, significant figures are counted.

In determining the final answer for a calculated quantity, *exact numbers* can be treated as if they have an infinite number of significant figures. This rule applies to many definitions between units. Thus, when we say, "There are 12 inches in 1 foot," the number 12 is exact, and we need not worry about the number of significant figures in it.

In *rounding off numbers*, look at the leftmost digit to be removed:

- If the leftmost digit removed is less than 5, the preceding number is left unchanged. Thus, rounding 7.248 to two significant figures gives 7.2.
- If the leftmost digit removed is 5 or greater, the preceding number is increased by 1. Rounding 4.735 to three significant figures gives 4.74, and rounding 2.376 to two significant figures gives 2.4.*

■ **SAMPLE EXERCISE 1.7** | **Determining the Number of Significant Figures in a Calculated Quantity**

The width, length, and height of a small box are 15.5 cm, 27.3 cm, and 5.4 cm, respectively. Calculate the volume of the box, using the correct number of significant figures in your answer.

Your instructor may want you to use a slight variation on the rule when the leftmost digit to be removed is exactly 5, with no following digits or only zeros. One common practice is to round up to the next higher number if that number will be even, and down to the next lower number otherwise. Thus, 4.7350 would be rounded to 4.74, and 4.7450 would also be rounded to 4.74.

SOLUTION

The product of the width, length, and height determines the volume of a box. In reporting the product, we can show only as many significant figures as given in the dimension with the fewest significant figures, that for the height (two significant figures):

$$\text{Volume} = \text{width} \times \text{length} \times \text{height}$$

$$= (15.5 \text{ cm}) (27.3 \text{ cm})(5.4 \text{ cm}) = 2285.01 \text{ cm}^3 \Rightarrow 2.3 \times 10^3 \text{ cm}^3$$

When we use a calculator to do this calculation, the display shows 2285.01, which we must round off to two significant figures. Because the resulting number is 2300, it is best reported in exponential notation, 2.3×10^3, to clearly indicate two significant figures.

■ **PRACTICE EXERCISE**

It takes 10.5 s for a sprinter to run 100.00 m. Calculate the average speed of the sprinter in meters per second, and express the result to the correct number of significant figures.
Answer: 9.52 m/s (three significant figures)

■ **SAMPLE EXERCISE 1.8** | **Determining the Number of Significant Figures in a Calculated Quantity**

A gas at 25 °C fills a container whose volume is $1.05 \times 10^3 \text{ cm}^3$. The container plus gas have a mass of 837.6 g. The container, when emptied of all gas, has a mass of 836.2 g. What is the density of the gas at 25 °C?

SOLUTION

To calculate the density, we must know both the mass and the volume of the gas. The mass of the gas is just the difference in the masses of the full and empty container:

$$(837.6 - 836.2) \text{ g} = 1.4 \text{ g}$$

In subtracting numbers, we determine the number of significant figures in our result by counting decimal places in each quantity. In this case each quantity has one decimal place. Thus, the mass of the gas, 1.4 g, has one decimal place.

Using the volume given in the question, $1.05 \times 10^3 \text{ cm}^3$, and the definition of density, we have

$$\text{Density} = \frac{\text{mass}}{\text{volume}} = \frac{1.4 \text{ g}}{1.05 \times 10^3 \text{ cm}^3}$$

$$= 1.3 \times 10^{-3} \text{ g/cm}^3 = 0.0013 \text{ g/cm}^3$$

In dividing numbers, we determine the number of significant figures in our result by counting the number of significant figures in each quantity. There are two significant figures in our answer, corresponding to the smaller number of significant figures in the two numbers that form the ratio. Notice that in this example, following the rules for determining significant figures gives an answer containing only two significant figures, even though each of the measured quantities contained at least three significant figures.

■ **PRACTICE EXERCISE**

To how many significant figures should the mass of the container be measured (with and without the gas) in Sample Exercise 1.8 for the density to be calculated to three significant figures?
Answer: five (For the difference in the two masses to have three significant figures, there must be two decimal places in the masses of the filled and empty containers. Therefore, each mass must be measured to five significant figures.)

When a calculation involves two or more steps and you write down answers for intermediate steps, retain at least one additional digit—past the number of significant figures—for the intermediate answers. This procedure ensures that small errors from rounding at each step do not combine to affect the final result. When using a calculator, you may enter the numbers one after another, rounding only the final answer. Accumulated rounding-off errors may account for small differences among results you obtain and answers given in the text for numerical problems.

1.6 DIMENSIONAL ANALYSIS

Throughout the text we use an approach called **dimensional analysis** as an aid in problem solving. In dimensional analysis we carry units through all calculations. Units are multiplied together, divided into each other, or "canceled."

Using dimensional analysis helps ensure that the solutions to problems yield the proper units. Moreover, it provides a systematic way of solving many numerical problems and of checking our solutions for possible errors.

The key to using dimensional analysis is the correct use of conversion factors to change one unit into another. A **conversion factor** is a fraction whose numerator and denominator are the same quantity expressed in different units. For example, 2.54 cm and 1 in. are the same length, 2.54 cm = 1 in. This relationship allows us to write two conversion factors:

$$\frac{2.54 \text{ cm}}{1 \text{ in.}} \quad \text{and} \quad \frac{1 \text{ in.}}{2.54 \text{ cm}}$$

We use the first of these factors to convert inches to centimeters. For example, the length in centimeters of an object that is 8.50 in. long is given by

$$\text{Number of centimeters} = (8.50 \text{ in.}) \overset{\text{Desired unit}}{\underset{\text{Given unit}}{\frac{2.54 \text{ cm}}{1 \text{ in.}}}} = 21.6 \text{ cm}$$

The unit inches in the denominator of the conversion factor cancels the unit inches in the given data (8.50 *inches*). The unit centimeters in the numerator of the conversion factor becomes the unit of the final answer. Because the numerator and denominator of a conversion factor are equal, multiplying any quantity by a conversion factor is equivalent to multiplying by the number 1 and so does not change the intrinsic value of the quantity. The length 8.50 in. is the same as the length 21.6 cm.

In general, we begin any conversion by examining the units of the given data and the units we desire. We then ask ourselves what conversion factors we have available to take us from the units of the given quantity to those of the desired one. When we multiply a quantity by a conversion factor, the units multiply and divide as follows:

$$\text{Given unit} \times \frac{\text{desired unit}}{\text{given unit}} = \text{desired unit}$$

If the desired units are not obtained in a calculation, then an error must have been made somewhere. Careful inspection of units often reveals the source of the error.

■ SAMPLE EXERCISE 1.9 | Converting Units

If a woman has a mass of 115 lb, what is her mass in grams? (Use the relationships between units given on the back inside cover of the text.)

SOLUTION

Because we want to change from lb to g, we look for a relationship between these units of mass. From the back inside cover we have 1 lb = 453.6 g. To cancel pounds and leave grams, we write the conversion factor with grams in the numerator and pounds in the denominator:

$$\text{Mass in grams} = (115 \text{ lb})\left(\frac{453.6 \text{ g}}{1 \text{ lb}}\right) = 5.22 \times 10^4 \text{ g}$$

The answer can be given to only three significant figures, the number of significant figures in 115 lb. The process we have used is diagrammed in the margin.

■ PRACTICE EXERCISE

By using a conversion factor from the back inside cover, determine the length in kilometers of a 500.0-mi automobile race.
Answer: 804.7 km

Strategies in Chemistry   ESTIMATING ANSWERS

A friend once remarked cynically that calculators let you get the wrong answer more quickly. What he was implying by that remark was that unless you have the correct strategy for solving a problem and have punched in the correct numbers, the answer will be incorrect. If you learn to *estimate* answers, however, you will be able to check whether the answers to your calculations are reasonable.

The idea is to make a rough calculation using numbers that are rounded off in such a way that the arithmetic can be easily performed without a calculator. This approach is often referred to as making a "ballpark" estimate, meaning that while it does not give an exact answer, it gives one that is roughly the right size. By working with units using dimensional analysis and by estimating answers, we can readily check the reasonableness of our answers to calculations.

GIVE IT SOME THOUGHT

How do we determine how many digits to use in conversion factors, such as the one between pounds and grams in Sample Exercise 1.9?

Using Two or More Conversion Factors

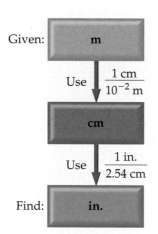

It is often necessary to use several conversion factors in solving a problem. As an example, let's convert the length of an 8.00-m rod to inches. The table on the back inside cover does not give the relationship between meters and inches. It *does*, however, give the relationship between centimeters and inches. (1 in. = 2.54 cm). From our knowledge of metric prefixes, we know that 1 cm = 10^{-2} m. Thus, we can convert step by step, first from meters to centimeters, and then from centimeters to inches as diagrammed in the margin.

Combining the given quantity (8.00 m) and the two conversion factors, we have

$$\text{Number of inches} = (8.00 \text{ m})\left(\frac{1 \text{ cm}}{10^{-2} \text{ m}}\right)\left(\frac{1 \text{ in.}}{2.54 \text{ cm}}\right) = 315 \text{ in.}$$

The first conversion factor is applied to cancel meters and convert the length to centimeters. Thus, meters are written in the denominator and centimeters in the numerator. The second conversion factor is written to cancel centimeters, so it has centimeters in the denominator and inches, the desired unit, in the numerator.

■ SAMPLE EXERCISE 1.10 | Converting Units Using Two or More Conversion Factors

The average speed of a nitrogen molecule in air at 25 °C is 515 m/s. Convert this speed to miles per hour.

SOLUTION

To go from the given units, m/s, to the desired units, mi/hr, we must convert meters to miles and seconds to hours. From our knowledge of metric prefixes we know that 1 km = 10^3 m. From the relationships given on the back inside cover of the book, we find that 1 mi = 1.6093 km. Thus, we can convert m to km and then convert km to mi. From our knowledge of time we know that 60 s = 1 min and 60 min = 1 hr. Thus, we can convert s to min and then convert min to hr. The overall process is diagrammed in the margin.

Applying first the conversions for distance and then those for time, we can set up one long equation in which unwanted units are canceled:

$$\text{Speed in mi/hr} = \left(515\frac{\text{m}}{\text{s}}\right)\left(\frac{1 \text{ km}}{10^3 \text{ m}}\right)\left(\frac{1 \text{ mi}}{1.6093 \text{ km}}\right)\left(\frac{60 \text{ s}}{1 \text{ min}}\right)\left(\frac{60 \text{ min}}{1 \text{ hr}}\right)$$

$$= 1.15 \times 10^3 \text{ mi/hr}$$

Our answer has the desired units. We can check our calculation, using the estimating procedure described in the previous "Strategies" box. The given speed is about

500 m/s. Dividing by 1000 converts m to km, giving 0.5 km/s. Because 1 mi is about 1.6 km, this speed corresponds to 0.5/1.6 = 0.3 mi/s. Multiplying by 60 gives about 0.3 × 60 = 20 mi/min. Multiplying again by 60 gives 20 × 60 = 1200 mi/hr. The approximate solution (about 1200 mi/hr) and the detailed solution (1150 mi/hr) are reasonably close. The answer to the detailed solution has three significant figures, corresponding to the number of significant figures in the given speed in m/s.

■ PRACTICE EXERCISE

A car travels 28 mi per gallon of gasoline. How many kilometers per liter will it go?
Answer: 12 km/L

Conversions Involving Volume

The conversion factors previously noted convert from one unit of a given measure to another unit of the same measure, such as from length to length. We also have conversion factors that convert from one measure to a different one. The density of a substance, for example, can be treated as a conversion factor between mass and volume. Suppose that we want to know the mass in grams of two cubic inches (2.00 in.^{3}) of gold, which has a density of 19.3 g/cm^3. The density gives us the following factors:

$$\frac{19.3\ \text{g}}{1\ \text{cm}^3} \quad \text{and} \quad \frac{1\ \text{cm}^3}{19.3\ \text{g}}$$

Because the answer we want is a mass in grams, we can see that we will use the first of these factors, which has mass in grams in the numerator. To use this factor, however, we must first convert cubic inches to cubic centimeters. The relationship between in.^{3} and cm^3 is not given on the back inside cover, but the relationship between inches and centimeters is given: 1 in. = 2.54 cm (exactly). Cubing both sides of this equation gives (1 in.)^3 = (2.54 cm)^3, from which we write the desired conversion factor:

$$\frac{(2.54\ \text{cm})^3}{(1\ \text{in.})^3} = \frac{(2.54)^3\ \text{cm}^3}{(1)^3\ \text{in.}^3} = \frac{16.39\ \text{cm}^3}{1\ \text{in.}^3}$$

Notice that both the numbers and the units are cubed. Also, because 2.54 is an exact number, we can retain as many digits of (2.54)^3 as we need. We have used four, one more than the number of digits in the density (19.3 g/cm^3). Applying our conversion factors, we can now solve the problem:

$$\text{Mass in grams} = (2.00\ \cancel{\text{in.}^3})\left(\frac{16.39\ \cancel{\text{cm}^3}}{1\ \cancel{\text{in.}^3}}\right)\left(\frac{19.3\ \text{g}}{1\ \cancel{\text{cm}^3}}\right) = 633\ \text{g}$$

The procedure is diagrammed below. The final answer is reported to three significant figures, the same number of significant figures as in 2.00 in.^{3} and 19.3 g.

Given: Find:

| in.^{3} | → Use → | cm^3 | → Use → | g |

$$\left(\frac{2.54\ \text{cm}}{1\ \text{in.}}\right)^3 \qquad \frac{19.3\ \text{g}}{1\ \text{cm}^3}$$

■ SAMPLE EXERCISE 1.11 | Converting Volume Units

Earth's oceans contain approximately 1.36 × 10^9 km^3 of water. Calculate the volume in liters.

SOLUTION

This problem involves conversion of km^3 to L. From the back inside cover of the text we find 1 L = 10^{-3} m^3, but there is no relationship listed involving km^3. From our knowledge of metric prefixes, however, we have 1 km = 10^3 m and we can use this relationship between lengths to write the desired conversion factor between volumes:

$$\left(\frac{10^3 \text{ m}}{1 \text{ km}}\right)^3 = \frac{10^9 \text{ m}^3}{1 \text{ km}^3}$$

Thus, converting from km^3 to m^3 to L, we have

$$\text{Volume in liters} = (1.36 \times 10^9 \text{ km}^3)\left(\frac{10^9 \text{ m}^3}{1 \text{ km}^3}\right)\left(\frac{1 \text{ L}}{10^{-3} \text{ m}^3}\right) = 1.36 \times 10^{21} \text{ L}$$

■ PRACTICE EXERCISE

If the volume of an object is reported as 5.0 ft^3, what is the volume in cubic meters?
Answer: 0.14 m^3

Strategies in Chemistry   THE IMPORTANCE OF PRACTICE

If you have ever played a musical instrument or participated in athletics, you know that the keys to success are practice and discipline. You cannot learn to play a piano merely by listening to music, and you cannot learn how to play basketball merely by watching games on television. Likewise, you cannot learn chemistry by merely watching your instructor do it. Simply reading this book, listening to lectures, or reviewing notes will not usually be sufficient when exam time comes around. Your task is not merely to understand how someone else uses chemistry, but to be able to do it yourself. That takes practice on a regular basis, and anything that you have to do on a regular basis requires self-discipline until it becomes a habit.

Throughout the book, we have provided sample exercises in which the solutions are shown in detail. A practice exercise, for which only the answer is given, accompanies each sample exercise. It is important that you use these exercises as learn-

ing aids. End-of-chapter exercises provide additional questions to help you understand the material in the chapter. Red numbers indicate exercises for which answers are given at the back of the book. A review of basic mathematics is given in Appendix A.

The practice exercises in this text and the homework assignments given by your instructor provide the minimal practice that you will need to succeed in your chemistry course. Only by working all the assigned problems will you face the full range of difficulty and coverage that your instructor expects you to master for exams. There is no substitute for a determined and perhaps lengthy effort to work problems on your own. If you are stuck on a problem, however, ask for help from your instructor, a teaching assistant, a tutor, or a fellow student. Spending an inordinate amount of time on a single exercise is rarely effective unless you know that it is particularly challenging and requires extensive thought and effort.

■ SAMPLE EXERCISE 1.12 | Conversions Involving Density

What is the mass in grams of 1.00 gal of water? The density of water is 1.00 g/mL.

SOLUTION

Before we begin solving this exercise, we note the following:

1. We are given 1.00 gal of water (the known, or given, quantity) and asked to calculate its mass in grams (the unknown).
2. We have the following conversion factors either given, commonly known, or available on the back inside cover of the text:

$$\frac{1.00 \text{ g water}}{1 \text{ mL water}} \quad \frac{1 \text{ L}}{1000 \text{ mL}} \quad \frac{1 \text{ L}}{1.057 \text{ qt}} \quad \frac{1 \text{ gal}}{4 \text{ qt}}$$

The first of these conversion factors must be used as written (with grams in the numerator) to give the desired result, whereas the last conversion factor must be inverted in order to cancel gallons:

$$\text{Mass in grams} = (1.00 \text{ gal})\left(\frac{4 \text{ qt}}{1 \text{ gal}}\right)\left(\frac{1 \text{ L}}{1.057 \text{ qt}}\right)\left(\frac{1000 \text{ mL}}{1 \text{ L}}\right)\left(\frac{1.00 \text{ g}}{1 \text{ mL}}\right)$$

$$= 3.78 \times 10^3 \text{ g water}$$

The units of our final answer are appropriate, and we've also taken care of our significant figures. We can further check our calculation by the estimation procedure. We can round 1.057 off to 1. Focusing on the numbers that do not equal 1 then gives merely 4 × 1000 = 4000 g, in agreement with the detailed calculation.

In cases such as this you may also be able to use common sense to assess the reasonableness of your answer. In this case we know that most people can lift a gallon of milk with one hand, although it would be tiring to carry it around all day. Milk is mostly water and will have a density that is not too different than water. Therefore, we might estimate that in familiar units a gallon of water would have mass that was more than 5 lbs but less than 50 lbs. The mass we have calculated is 3.78 kg $\times$ 2.2 lb/kg = 8.3 lbs—an answer that is reasonable at least as an order of magnitude estimate.

■ PRACTICE EXERCISE

The density of benzene is 0.879 g/mL. Calculate the mass in grams of 1.00 qt of benzene.
Answer: 832 g

C H A P T E R   R E V I E W

Following each chapter you will find a summary that highlights important content of the chapter. The summary contains all the key terms from the chapter in their contexts. A list of key skills and key equations follows the summary. These review materials are important tools to help you prepare for exams.

SUMMARY AND KEY TERMS

Introduction and Section 1.1 **Chemistry** is the study of the composition, structure, properties, and changes of **matter**. The composition of matter relates to the kinds of **elements** it contains. The structure of matter relates to the ways the **atoms** of these elements are arranged. A **property** is any characteristic that gives a sample of matter its unique identity. A **molecule** is an entity composed of two or more atoms with the atoms attached to one another in a specific way.

Section 1.2 Matter exists in three physical states, **gas, liquid,** and **solid,** which are known as the **states of matter**. There are two kinds of **pure substances: elements** and **compounds**. Each element has a single kind of atom and is represented by a chemical symbol consisting of one or two letters, with the first letter capitalized. Compounds are composed of two or more elements joined chemically. The **law of constant composition**, also called the **law of definite proportions**, states that the elemental composition of a pure compound is always the same. Most matter consists of a mixture of substances. **Mixtures** have variable compositions and can be either homogeneous or heterogeneous; homogeneous mixtures are called **solutions**.

Section 1.3 Each substance has a unique set of **physical properties** and **chemical properties** that can be used to identify it. During a **physical change**, matter does not change its composition. **Changes of state** are physical changes. In a **chemical change (chemical reaction)** a substance is transformed into a chemically different substance. **Intensive properties** are independent of the amount of matter examined and are used to identify substances. **Extensive properties** relate to the amount of substance present. Differences in physical and chemical properties are used to separate substances.

The **scientific method** is a dynamic process used to answer questions about our physical world. Observations and experiments lead to **scientific laws**, general rules that summarize how nature behaves. Observations also lead to tentative explanations or **hypotheses**. As a hypothesis is tested and refined, a **theory** may be developed.

Section 1.4 Measurements in chemistry are made using the **metric system**. Special emphasis is placed on a particular set of metric units called **SI units**, which are based on the meter, the kilogram, and the second as the basic units of length, **mass**, and time, respectively. The metric system employs a set of prefixes to indicate decimal fractions or multiples of the base units. The SI temperature scale is the **Kelvin scale**, although the **Celsius scale** is frequently used as well. **Density** is an important property that equals mass divided by volume.

Section 1.5 All measured quantities are inexact to some extent. The **precision** of a measurement indicates how closely different measurements of a quantity agree with one another. The **accuracy** of a measurement indicates how well a measurement agrees with the accepted or "true" value. The **significant figures** in a measured quantity include one estimated digit, the last digit of the measurement. The significant figures indicate the extent of the uncertainty of the measurement. Certain rules must be followed so that a calculation involving measured quantities is reported with the appropriate number of significant figures.

Section 1.6 In the **dimensional analysis** approach to problem solving, we keep track of units as we carry measurements through calculations. The units are multiplied together, divided into each other, or canceled like algebraic quantities. Obtaining the proper units for the final result is an important means of checking the method of calculation. When converting units and when carrying out several other types of problems, **conversion factors** can be used. These factors are ratios constructed from valid relations between equivalent quantities.

KEY SKILLS

- Distinguish among elements, compounds, and mixtures.
- Memorize symbols of common elements and common metric prefixes.
- Use significant figures, scientific notation, metric units, and dimensional analysis in calculations.

KEY EQUATIONS

- $K = °C + 273.15$ [1.1]   Interconverting between Celsius (°C) and Kelvin (K) temperatures scales

- $°C = \dfrac{5}{9}(°F - 32)$ or $°F = \dfrac{9}{5}(°C) + 32$ [1.2]   Interconverting between Celsius (°C) and Fahrenheit (°F) temperature scales

- $\text{Density} = \dfrac{\text{mass}}{\text{volume}}$ [1.3]   Definition of density

VISUALIZING CONCEPTS

The exercises in this section are intended to probe your understanding of key concepts rather than your ability to utilize formulas and perform calculations. Those exercises with red exercise numbers have answers in the back of the book.

1.1 Which of the following figures represents (a) a pure element, (b) a mixture of two elements, (c) a pure compound, (d) a mixture of an element and a compound? (More than one picture might fit each description.) [Section 1.2]

(i)   (ii)   (iii)

(iv)   (v)   (vi)

1.2 Does the following diagram represent a chemical or physical change? How do you know? [Section 1.3]

1.3 Identify each of the following as measurements of length, area, volume, mass, density, time, or temperature: (a) 5 ns, (b) 5.5 kg/m^3, (c) 0.88 pm, (d) 540 km^2, (e) 173 K, (f) 2 mm^3, (g) 23 °C. [Section 1.4]

1.4 Three spheres of equal size are composed of aluminum (density = 2.70 g/cm^3), silver (density = 10.49 g/cm^3), and nickel (density = 8.90 g/cm^3). List the spheres from lightest to heaviest.

1.5 The following dartboards illustrate the types of errors often seen when one measurement is repeated several times. The bull's-eye represents the "true value," and the darts represent the experimental measurements. Which board best represents each of the following scenarios: (a) measurements both accurate and precise, (b) measurements precise but inaccurate, (c) measurements imprecise but yield an accurate average? [Section 1.5]

(i)   (ii)   (iii)

1.6 (a) What is the length of the pencil in the following figure if the scale reads in centimeters? How many significant figures are there in this measurement? (b) An oven thermometer with a circular scale reading degrees Fahrenheit is shown. What temperature does the scale indicate? How many significant figures are in the measurement? [Section 1.5]

1.7 What is wrong with the following statement? Twenty years ago an ancient artifact was determined to be 1900 years old. It must now be 1920 years old. [Section 1.5]

1.8 **(a)** How many significant figures should be reported for the volume of the metal bar shown below? **(b)** If the mass of the bar is 104.7 g, how many significant figures should be reported when its density is calculated using the calculated volume? [Section 1.5]

2.5 cm

1.25 cm

5.30 cm

1.9 When you convert units, how do you decide which part of the conversion factor is in the numerator and which is in the denominator? [Section 1.6]

1.10 Draw a logic map indicating the steps you would take to convert miles per hour to kilometers per second. Write down the conversion factor for each step, as done in the diagram on page 26. [Section 1.6]

EXERCISES

Classification and Properties of Matter

The following exercises are divided into sections that deal with specific topics in the chapter. These exercises are grouped in pairs, with the answer given in the back of the book to the odd-numbered exercise, as indicated by the red exercise number. Those exercises whose number appears in brackets are more challenging than the nonbracketed exercises.

1.11 Classify each of the following as a pure substance or a mixture. If a mixture, indicate whether it is homogeneous or heterogeneous: **(a)** rice pudding, **(b)** seawater, **(c)** magnesium, **(d)** gasoline.

1.12 Classify each of the following as a pure substance or a mixture. If a mixture, indicate whether it is homogeneous or heterogeneous: **(a)** air, **(b)** tomato juice, **(c)** iodine crystals, **(d)** sand.

1.13 Give the chemical symbol or name for the following elements, as appropriate: **(a)** sulfur, **(b)** magnesium, **(c)** potassium, **(d)** chlorine, **(e)** copper, **(f)** F, **(g)** Ni, **(h)** Na, **(i)** Al, **(j)** Si.

1.14 Give the chemical symbol or name for each of the following elements, as appropriate: **(a)** carbon, **(b)** nitrogen, **(c)** bromine, **(d)** zinc, **(e)** iron, **(f)** P, **(g)** Ca, **(h)** He, **(i)** Pb, **(j)** Ag.

1.15 A solid white substance A is heated strongly in the absence of air. It decomposes to form a new white substance B and a gas C. The gas has exactly the same properties as the product obtained when carbon is burned in an excess of oxygen. Based on these observations, can we determine whether solids A and B and the gas C are elements or compounds? Explain your conclusions for each substance.

1.16 In 1807 the English chemist Humphry Davy passed an electric current through molten potassium hydroxide and isolated a bright, shiny reactive substance. He claimed the discovery of a new element, which he named potassium. In those days, before the advent of modern instruments, what was the basis on which one could claim that a substance was an element?

1.17 In the process of attempting to characterize a substance, a chemist makes the following observations: The substance is a silvery white, lustrous metal. It melts at 649 °C and boils at 1105 °C. Its density at 20 °C is 1.738 g/cm^3. The substance burns in air, producing an intense white light. It reacts with chlorine to give a brittle white solid. The substance can be pounded into thin sheets or drawn into wires. It is a good conductor of electricity. Which of these characteristics are physical properties, and which are chemical properties?

1.18 Read the following description of the element zinc, and indicate which are physical properties and which are chemical properties. Zinc is a silver–gray–colored metal that melts at 420 °C. When zinc granules are added to dilute sulfuric acid, hydrogen is given off and the metal dissolves. Zinc has a hardness on the Mohs scale of 2.5 and a density of 7.13 g/cm^3 at 25 °C. It reacts slowly with oxygen gas at elevated temperatures to form zinc oxide, ZnO.

1.19 Label each of the following as either a physical process or a chemical process: **(a)** corrosion of aluminum metal, **(b)** melting of ice, **(c)** pulverizing an aspirin, **(d)** digesting a candy bar, **(e)** explosion of nitroglycerin.

1.20 A match is lit and held under a cold piece of metal. The following observations are made: **(a)** The match burns. **(b)** The metal gets warmer. **(c)** Water condenses on the metal. **(d)** Soot (carbon) is deposited on the metal. Which of these occurrences are due to physical changes, and which are due to chemical changes?

1.21 Suggest a method of separating each of the following mixtures into two components: **(a)** sugar and sand, **(b)** iron and sulfur.

1.22 A beaker contains a clear, colorless liquid. If it is water, how could you determine whether it contained dissolved table salt? Do *not* taste it!

Units and Measurement

1.23 What exponential notation do the following abbreviations represent: (a) d, (b) c, (c) f, (d) μ, (e) M, (f) k, (g) n, (h) m, (i) p?

1.24 Use appropriate metric prefixes to write the following measurements without use of exponents: (a) 6.35×10^{-2} L, (b) 6.5×10^{-6} s, (c) 9.5×10^{-4} m, (d) 4.23×10^{-9} m^3, (e) 12.5×10^{-8} kg, (f) 3.5×10^{-10} g, (g) 6.54×10^{9} fs.

1.25 Make the following conversions: (a) 62 °F to °C, (b) 216.7 °C to °F, (c) 233 °C to K, (d) 315 K to °F, (e) 2500 °F to K.

1.26 (a) The temperature on a warm summer day is 87 °F. What is the temperature in °C? (b) Many scientific data are reported at 25 °C. What is this temperature in kelvins and in degrees Fahrenheit? (c) Suppose that a recipe calls for an oven temperature of 175 °F. Convert this temperature to degrees Celsius and to kelvins. (d) The melting point of sodium bromide (a salt) is 755 °C. Calculate this temperature in °F and in kelvins. (e) Neon, a gaseous element at room temperature, is used to make electronic signs. Neon has a melting point of −248.6 °C and a boiling point of −246.1 °C. Convert these temperatures to kelvins.

1.27 (a) A sample of carbon tetrachloride, a liquid once used in dry cleaning, has a mass of 39.73 g and a volume of 25.0 mL at 25 °C. What is its density at this temperature? Will carbon tetrachloride float on water? (Materials that are less dense than water will float.) (b) The density of platinum is 21.45 g/cm^3 at 20 °C. Calculate the mass of 75.00 cm^3 of platinum at this temperature. (c) The density of magnesium is 1.738 g/cm^3 at 20 °C. What is the volume of 87.50 g of this metal at this temperature?

1.28 (a) A cube of osmium metal 1.500 cm on a side has a mass of 76.31 g at 25 °C. What is its density in g/cm^3 at this temperature? (b) The density of titanium metal is 4.51 g/cm^3 at 25 °C. What mass of titanium displaces 125.0 mL of water at 25 °C? (c) The density of benzene at 15 °C is 0.8787 g/mL. Calculate the mass of 0.1500 L of benzene at this temperature.

1.29 (a) To identify a liquid substance, a student determined its density. Using a graduated cylinder, she measured out a 45-mL sample of the substance. She then measured the mass of the sample, finding that it weighed 38.5 g. She knew that the substance had to be either isopropyl alcohol (density 0.785 g/mL) or toluene (density 0.866 /mL). What are the calculated density and the probable identity of the substance? (b) An experiment requires 45.0 g of ethylene glycol, a liquid whose density is 1.114 g/mL. Rather than weigh the sample on a balance, a chemist chooses to dispense the liquid using a graduated cylinder. What volume of the liquid should he use? (c) A cubic piece of metal measures 5.00 cm on each edge. If the metal is nickel, whose density is 8.90 g/cm^3, what is the mass of the cube?

1.30 (a) After the label fell off a bottle containing a clear liquid believed to be benzene, a chemist measured the density of the liquid to verify its identity. A 25.0-mL portion of the liquid had a mass of 21.95 g. A chemistry handbook lists the density of benzene at 15 °C as 0.8787 g/mL. Is the calculated density in agreement with the tabulated value? (b) An experiment requires 15.0 g of cyclohexane, whose density at 25 °C is 0.7781 g/mL. What volume of cyclohexane should be used? (c) A spherical ball of lead has a diameter of 5.0 cm. What is the mass of the sphere if lead has a density of 11.34 g/cm^3? (The volume of a sphere is $(\frac{4}{3})\pi r^3$ where r is the radius.)

1.31 Gold can be hammered into extremely thin sheets called gold leaf. If a 200-mg piece of gold (density $= 19.32$ g/cm^3) is hammered into a sheet measuring 2.4×1.0 ft, what is the average thickness of the sheet in meters? How might the thickness be expressed without exponential notation, using an appropriate metric prefix?

1.32 A cylindrical rod formed from silicon is 16.8 cm long and has a mass of 2.17 kg. The density of silicon is 2.33 g/cm^3. What is the diameter of the cylinder? (The volume of a cylinder is given by $\pi r^2 h$, where r is the radius, and h is its length.)

Uncertainty in Measurement

1.33 Indicate which of the following are exact numbers: (a) the mass of a paper clip, (b) the surface area of a dime, (c) the number of inches in a mile, (d) the number of ounces in a pound, (e) the number of microseconds in a week, (f) the number of pages in this book.

1.34 Indicate which of the following are exact numbers: (a) the mass of a 32-oz can of coffee, (b) the number of students in your chemistry class, (c) the temperature of the surface of the sun, (d) the mass of a postage stamp, (e) the number of milliliters in a cubic meter of water, (f) the average height of students in your school.

1.35 What is the number of significant figures in each of the following measured quantities? (a) 358 kg, (b) 0.054 s, (c) 6.3050 cm, (d) 0.0105 L, (e) 7.0500×10^{-3} m^3.

1.36 Indicate the number of significant figures in each of the following measured quantities: (a) 3.774 km, (b) 205 m^2, (c) 1.700 cm, (d) 350.00 K, (e) 307.080 g.

1.37 Round each of the following numbers to four significant figures, and express the result in standard exponential notation: (a) 102.53070, (b) 656,980, (c) 0.008543210, (d) 0.000257870, (e) −0.0357202.

1.38 (a) The diameter of Earth at the equator is 7926.381 mi. Round this number to three significant figures, and express it in standard exponential notation. (b) The circumference of Earth through the poles is 40,008 km. Round this number to four significant figures, and express it in standard exponential notation.

1.39 Carry out the following operations, and express the answers with the appropriate number of significant figures.
(a) $12.0550 + 9.05$
(b) $257.2 - 19.789$
(c) $(6.21 \times 10^3)(0.1050)$
(d) $0.0577/0.753$

1.40 Carry out the following operations, and express the answer with the appropriate number of significant figures.
(a) $320.5 - (6104.5/2.3)$
(b) $[(285.3 \times 10^5) - (1.200 \times 10^3)] \times 2.8954$
(c) $(0.0045 \times 20,000.0) + (2813 \times 12)$
(d) $863 \times [1255 - (3.45 \times 108)]$

Dimensional Analysis

1.41 Using your knowledge of metric units, English units, and the information on the back inside cover, write down the conversion factors needed to convert (a) mm to nm, (b) mg to kg, (c) km to ft, (d) in.^{3} to cm^3.

1.42 Using your knowledge of metric units, English units, and the information on the back inside cover, write down the conversion factors needed to convert (a) μm to mm, (b) ms to ns, (c) mi to km, (d) ft^3 to L.

1.43 Perform the following conversions: (a) 0.076 L to mL, (b) 5.0×10^{-8} m to nm, (c) 6.88×10^5 ns to s, (d) 0.50 lb to g, (e) 1.55 kg/m^3 to g/L, (f) 5.850 gal/hr to L/s.

1.44 (a) The speed of light in a vacuum is 2.998×10^8 m/s. Calculate its speed in km/hr. (b) The Sears Tower in Chicago is 1454 ft tall. Calculate its height in meters. (c) The Vehicle Assembly Building at the Kennedy Space Center in Florida has a volume of 3,666,500 m^3. Convert this volume to liters, and express the result in standard exponential notation. (d) An individual suffering from a high cholesterol level in her blood has 232 mg of cholesterol per 100 mL of blood. If the total blood volume of the individual is 5.2 L, how many grams of total blood cholesterol does the individual's body contain?

1.45 Perform the following conversions: (a) 5.00 days to s, (b) 0.0550 mi to m, (c) $1.89/gal to dollars per liter, (d) 0.510 in./ms to km/hr, (e) 22.50 gal/min to L/s, (f) 0.02500 ft^3 to cm^3.

1.46 Carry out the following conversions: (a) 0.105 in. to mm, (b) 0.650 qt to mL, (c) 8.75 μm/s to km/hr, (d) 1.955 m^3 to yd^3, (e) $3.99/lb to dollars per kg, (f) 8.75 lb/ft^3 to g/mL.

1.47 (a) How many liters of wine can be held in a wine barrel whose capacity is 31 gal? (b) The recommended adult dose of Elixophyllin®, a drug used to treat asthma, is 6 mg/kg of body mass. Calculate the dose in milligrams for a 150-lb person. (c) If an automobile is able to travel 254 mi on 11.2 gal of gasoline, what is the gas mileage in km/L? (d) A pound of coffee beans yields 50 cups of coffee (4 cups = 1 qt). How many milliliters of coffee can be obtained from 1 g of coffee beans?

1.48 (a) If an electric car is capable of going 225 km on a single charge, how many charges will it need to travel from Boston, Massachusetts, to Miami, Florida, a distance of 1486 mi, assuming that the trip begins with a full charge? (b) If a migrating loon flies at an average speed of 14 m/s, what is its average speed in mi/hr? (c) What is the engine piston displacement in liters of an engine whose displacement is listed as 450 in.^{3}? (d) In March 1989 the *Exxon Valdez* ran aground and spilled 240,000 barrels of crude petroleum off the coast of Alaska. One barrel of petroleum is equal to 42 gal. How many liters of petroleum were spilled?

1.49 The density of air at ordinary atmospheric pressure and 25 °C is 1.19 g/L. What is the mass, in kilograms, of the air in a room that measures $12.5 \times 15.5 \times 8.0$ ft?

1.50 The concentration of carbon monoxide in an urban apartment is 48 μg/m^3. What mass of carbon monoxide in grams is present in a room measuring $9.0 \times 14.5 \times 18.8$ ft?

1.51 By using estimation techniques, arrange these items in order from shortest to longest: a 57-cm length of string, a 14-in. long shoe, and a 1.1-m length of pipe.

1.52 By using estimation techniques, determine which of the following is the heaviest and which is the lightest: a 5-lb bag of potatoes, a 5-kg bag of sugar, or 1 gal of water (density = 1.0 g/mL).

1.53 The Morgan silver dollar has a mass of 26.73 g. By law, it was required to contain 90% silver, with the remainder being copper. (a) When the coin was minted in the late 1800s, silver was worth $1.18 per troy ounce (31.1 g). At this price, what is the value of the silver in the silver dollar? (b) Today, silver sells for about $13.25 per troy ounce. How many Morgan silver dollars are required to obtain $25.00 worth of pure silver?

1.54 A copper refinery produces a copper ingot weighing 150 lb. If the copper is drawn into wire whose diameter is 8.25 mm, how many feet of copper can be obtained from the ingot? The density of copper is 8.94 g/cm^3 (Assume that the wire is a cylinder whose volume is $V = \pi r^2 h$, where r is its radius and h is its height or length.)

ADDITIONAL EXERCISES

The exercises in this section are not divided by category, although they are roughly in the order of the topics in the chapter. They are not paired.

1.55 What is meant by the terms composition and structure when referring to matter?

1.56 (a) Classify each of the following as a pure substance, a solution, or a heterogeneous mixture: a gold coin, a cup of coffee, a wood plank. (b) What ambiguities are there in answering part (a) from the descriptions given?

1.57 (a) What is the difference between a hypothesis and a theory? (b) Explain the difference between a theory and a scientific law. Which addresses how matter behaves, and which addresses why it behaves that way?

1.58 A sample of ascorbic acid (vitamin C) is synthesized in the laboratory. It contains 1.50 g of carbon and 2.00 g of oxygen. Another sample of ascorbic acid isolated from citrus fruits contains 6.35 g of carbon. How many grams of oxygen does it contain? Which law are you assuming in answering this question?

1.59 Two students determine the percentage of lead in a sample as a laboratory exercise. The true percentage is 22.52%. The students' results for three determinations are as follows:
   1. 22.52, 22.48, 22.54
   2. 22.64, 22.58, 22.62
   (a) Calculate the average percentage for each set of data, and tell which set is the more accurate based on the average. **(b)** Precision can be judged by examining the average of the deviations from the average value for that data set. (Calculate the average value for each data set, then calculate the average value of the absolute deviations of each measurement from the average.) Which set is more precise?

1.60 Is the use of significant figures in each of the following statements appropriate? Why or why not? **(a)** The 2005 circulation of *National Geographic* was 7,812,564. **(b)** On July 1, 2005, the population of Cook County, Illinois, was 5,303,683. **(c)** In the United States, 0.621% of the population has the surname Brown.

1.61 What type of quantity (for example, length, volume, density) do the following units indicate: **(a)** mL, **(b)** cm^2, **(c)** mm^3, **(d)** mg/L, **(e)** ps, **(f)** nm, **(g)** K?

1.62 Give the derived SI units for each of the following quantities in base SI units: **(a)** acceleration = distance/time^2; **(b)** force = mass × acceleration; **(c)** work = force × distance; **(d)** pressure = force/area; **(e)** power = work/time.

1.63 The distance from Earth to the Moon is approximately 240,000 mi. **(a)** What is this distance in meters? **(b)** The peregrine falcon has been measured as traveling up to 350 km/hr in a dive. If this falcon could fly to the Moon at this speed, how many seconds would it take?

1.64 The US quarter has a mass of 5.67 g and is approximately 1.55 mm thick. **(a)** How many quarters would have to be stacked to reach 575 ft, the height of the Washington Monument? **(b)** How much would this stack weigh? **(c)** How much money would this stack contain? **(d)** At the beginning of 2007, the national debt was $8.7 trillion. How many stacks like the one described would be necessary to pay off this debt?

1.65 In the United States, water used for irrigation is measured in acre-feet. An acre-foot of water covers an acre to a depth of exactly 1 ft. An acre is 4840 yd^2. An acre-foot is enough water to supply two typical households for 1.00 yr. **(a)** If desalinated water costs $1950 per acre-foot, how much does desalinated water cost per liter? **(b)** How much would it cost one household per day if it were the only source of water?

1.66 Suppose you decide to define your own temperature scale using the freezing point (−11.5 °C) and boiling point (197.6 °C) of ethylene glycol. If you set the freezing point as 0 °G and the boiling point as 100 °G, what is the freezing point of water on this new scale?

1.67 The liquid substances mercury (density = 13.5 g/mL), water (1.00 g/mL), and cyclohexane (0.778 g/mL) do not form a solution when mixed, but separate in distinct layers. Sketch how the liquids would position themselves in a test tube.

1.68 Small spheres of equal mass are made of lead (density = 11.3 g/cm^3), silver (10.5 g/cm^3), and aluminum (2.70 g/cm^3). Without doing a calculation, list the spheres in order from the smallest to the largest.

1.69 Water has a density of 0.997 g/cm^3 at 25 °C; ice has a density of 0.917 g/cm^3 at −10 °C. **(a)** If a soft-drink bottle whose volume is 1.50 L is completely filled with water and then frozen to −10 °C, what volume does the ice occupy? **(b)** Can the ice be contained within the bottle?

1.70 A 32.65-g sample of a solid is placed in a flask. Toluene, in which the solid is insoluble, is added to the flask so that the total volume of solid and liquid together is 50.00 mL. The solid and toluene together weigh 58.58 g. The density of toluene at the temperature of the experiment is 0.864 g/mL. What is the density of the solid?

1.71 **(a)** You are given a bottle that contains 4.59 cm^3 of a metallic solid. The total mass of the bottle and solid is 35.66 g. The empty bottle weighs 14.23 g. What is the density of the solid? **(b)** Mercury is traded by the "flask," a unit that has a mass of 34.5 kg. What is the volume of a flask of mercury if the density of mercury is 13.5 g/mL? **(c)** A thief plans to steal a gold sphere with a radius of 28.9 cm from a museum. If the gold has a density of 19.3 g/cm^3 what is the mass of the sphere? [The volume of a sphere is $V = (4/3)\pi r^3$.] Is he likely to be able to walk off with it unassisted?

1.72 Automobile batteries contain sulfuric acid, which is commonly referred to as "battery acid." Calculate the number of grams of sulfuric acid in 0.500 L of battery acid if the solution has a density of 1.28 g/mL and is 38.1% sulfuric acid by mass.

1.73 A 40-lb container of peat moss measures 14 × 20 × 30 in. A 40-lb container of topsoil has a volume of 1.9 gal. **(a)** Calculate the average densities of peat moss and topsoil in units of g/cm^3. Would it be correct to say that peat moss is "lighter" than topsoil? Explain. **(b)** How many bags of the peat moss are needed to cover an area measuring 10. ft by 20. ft to a depth of 2.0 in.?

1.74 A coin dealer offers to sell you an ancient gold coin that is 2.2 cm in diameter and 3.0 mm in thickness. **(a)** The density of gold is 19.3 g/cm^3. How much should the coin weigh if it is pure gold? **(b)** If gold sells for $640 per troy ounce, how much is the gold content worth? (1 troy ounce = 31.1 g).

1.75 A package of aluminum foil contains 50 ft^2 of foil, which weighs approximately 8.0 oz. Aluminum has a density of 2.70 g/cm^3. What is the approximate thickness of the foil in millimeters?

1.76 A 15.0-cm long cylindrical glass tube, sealed at one end, is filled with ethanol. The mass of ethanol needed to fill the tube is found to be 11.86 g. The density of ethanol is 0.789 g/mL. Calculate the inner diameter of the tube in centimeters.

1.77 Gold is alloyed (mixed) with other metals to increase its hardness in making jewelry. **(a)** Consider a piece of gold jewelry that weighs 9.85 g and has a volume of

0.675 cm^3. The jewelry contains only gold and silver, which have densities of 19.3 g/cm^3 and 10.5 g/cm^3, respectively. If the total volume of the jewelry is the sum of the volumes of the gold and silver that it contains, calculate the percentage of gold (by mass) in the jewelry. **(b)** The relative amount of gold in an alloy is commonly expressed in units of karats. Pure gold is 24-karat, and the percentage of gold in an alloy is given as a percentage of this value. For example, an alloy that is 50% gold is 12-karat. State the purity of the gold jewelry in karats.

1.78 Suppose you are given a sample of a homogeneous liquid. What would you do to determine whether it is a solution or a pure substance?

1.79 Chromatography (Figure 1.14) is a simple, but reliable, method for separating a mixture into its constituent substances. Suppose you are using chromatography to separate a mixture of two substances. How would you know whether the separation is successful? Can you propose a means of quantifying how good or how poor the separation is?

1.80 You are assigned the task of separating a desired granular material, with a density of 3.62 g/cm^3, from an undesired granular material that has a density of 2.04 g/cm^3. You want to do this by shaking the mixture in a liquid in which the heavier material will fall to the bottom and the lighter material will float. A solid will float on any liquid that is more dense. Using the internet or a handbook of chemistry, find the densities of the following substances: carbon tetrachloride, hexane, benzene, and methylene iodide. Which of these liquids will serve your purpose, assuming no chemical interaction between the liquid and the solids?

1.81 In 2006, Professor Galen Suppes, from the University of Missouri-Columbia, was awarded a Presidential Green Challenge Award for his system of converting glycerin, $C_3H_5(OH)_3$, a by-product of biodiesel production, to propylene glycol, $C_3H_6(OH)_2$. Propylene glycol pro-

duced in this way will be cheap enough to replace the more toxic ethylene glycol that is the primary ingredient in automobile antifreeze. **(a)** If 50.0 mL of propylene glycol has a mass of 51.80 g, what is its density? **(b)** To obtain the same antifreeze protection requires 76 g of propylene glycol to replace each 62 g of ethylene glycol. Calculate the mass of propylene glycol required to replace 1.00 gal of ethylene glycol. The density of ethylene glycol is 1.12 g/mL. **(c)** Calculate the volume of propylene glycol, in gallons, needed to produce the same antifreeze protection as 1.00 gallon of ethylene glycol.

1.82 The concepts of accuracy and precision are not always easy to grasp. Here are two sets of studies: **(a)** The mass of a secondary weight standard is determined by weighing it on a very precise balance under carefully controlled laboratory conditions. The average of 18 different weight measurements is taken as the weight of the standard. **(b)** A group of 10,000 males between the ages of 50 and 55 is surveyed to ascertain a relationship between calorie intake and blood cholesterol level. The survey questionnaire is quite detailed, asking the respondents about what they eat, smoking and drinking habits, and so on. The results are reported as showing that for men of comparable lifestyles, there is a 40% chance of the blood cholesterol level being above 230 for those who consume more than 40 calories per gram of body weight per day, as compared with those who consume fewer than 30 calories per gram of body weight per day.

Discuss and compare these two studies in terms of the precision and accuracy of the result in each case. How do the two studies differ in nature in ways that affect the accuracy and precision of the results? What makes for high precision and accuracy in any given study? In each of these studies, what factors might not be controlled that could affect the accuracy and precision? What steps can be taken generally to attain higher precision and accuracy?

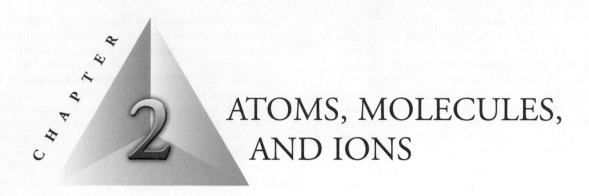

CHAPTER 2

ATOMS, MOLECULES, AND IONS

Sickered

A CIRCLE OF INIDIVUAL IRON ATOMS on a copper surface, as viewed by a technique known as scanning tunneling microscopy (STM). The image is artificially colored to enhance it. The shapes of the iron atoms in the STM image are distorted, and the atoms of the copper surface are not revealed.

WHAT'S AHEAD

2.1 The Atomic Theory of Matter
We begin by giving a brief history of the notion of *atoms*—the smallest pieces of matter.

2.2 The Discovery of Atomic Structure
We then look at some of the key experiments that led to the discovery of *electrons* and to the *nuclear model* of the atom.

2.3 The Modern View of Atomic Structure
We explore the modern theory of atomic structure, including the ideas of *atomic numbers, mass numbers,* and *isotopes.*

2.4 Atomic Weights
We introduce the concept of *atomic weights* and how they relate to the masses of individual atoms.

2.5 The Periodic Table
We examine the organization of the elements into the *periodic table,* in which elements are put in order of increasing atomic number and grouped by chemical similarity.

2.6 Molecules and Molecular Compounds
We discuss the assemblies of atoms called *molecules* and how their compositions are represented by *empirical* and *molecular formulas.*

2.7 Ions and Ionic Compounds
We learn that atoms can gain or lose electrons to form *ions.* We also look at how to use the periodic table to predict the charges on ions and the empirical formulas of *ionic compounds.*

2.8 Naming Inorganic Compounds
We consider the systematic way in which substances are named, called *nomenclature,* and how this nomenclature is applied to inorganic compounds.

2.9 Some Simple Organic Compounds
We introduce some basic ideas of *organic chemistry,* which is the chemistry of the element carbon.

LOOK AROUND YOU. Notice the great variety of colors, textures, and other properties in the materials that surround you—the colors in a garden scene, the texture of the fabric in your clothes, the solubility of sugar in a cup of coffee, the

transparency of a window. The materials in our world exhibit a striking and seemingly infinite variety.

We can classify properties in different ways, but how do we understand and explain them? What makes diamonds transparent and hard, while table salt is brittle and dissolves in water? Why does paper burn, and why does water quench fires? The structure and behavior of atoms are key to understanding both the physical and chemical properties of matter.

Remarkably, the diversity of these properties we see around us results from only about 100 different elements and therefore about 100 chemically different kinds of atoms. In a sense, the atoms are like the 26 letters of the English alphabet that join in different combinations to form the immense number of words in our language. But how do atoms combine with one another? What rules govern the ways in which atoms can combine? How do the properties of a substance relate to the kinds of atoms it contains? Indeed, what is an atom like, and what makes the atoms of one element different from those of another?

The chapter-opening photograph is an image of a circle of 48 iron atoms arranged on a copper metal surface. The diameter of the circle is about 1/20,000 the diameter of a human hair. Atoms are indeed very tiny entities.

This very striking image reveals the power of modern experimental methods to identify individual atoms, but it does not reveal the structures of the atoms themselves. Fortunately, we can use a variety of experimental techniques to probe the atom to gain a clearer understanding of what it is like. In this chapter we begin to explore the fascinating world of atoms that we discover by such experiments. We will examine the basic structure of the atom and briefly discuss the formation of molecules and ions, thereby providing a foundation for exploring chemistry more deeply in later chapters.

2.1 THE ATOMIC THEORY OF MATTER

Philosophers from the earliest times have speculated about the nature of the fundamental "stuff" from which the world is made. Democritus (460–370 BC) and other early Greek philosophers thought that the material world must be made up of tiny indivisible particles that they called *atomos*, meaning "indivisible or uncuttable." Later, Plato and Aristotle formulated the notion that there can be no ultimately indivisible particles. The "atomic" view of matter faded for many centuries during which Aristotelean philosophy dominated Western culture.

The notion of atoms reemerged in Europe during the seventeenth century, when scientists tried to explain the properties of gases. Air is composed of something invisible and in constant motion; we can feel the motion of the wind against us, for example. It is natural to think of tiny invisible particles as giving rise to these familiar effects. Isaac Newton (1642–1727), the most famous scientist of his time, favored the idea of atoms. But thinking of atoms as invisible particles in air is very different from thinking of atoms as the fundamental building blocks of elements.

As chemists learned to measure the amounts of elements that reacted with one another to form new substances, the ground was laid for an atomic theory that linked the idea of elements with the idea of atoms. That theory came from the work of an English schoolteacher, John Dalton (Figure 2.1 ◄), during the period from 1803–1807. Dalton's atomic theory involved the following postulates:

1. Each element is composed of extremely small particles called atoms.

2. All atoms of a given element are identical to one another in mass and other properties, but the atoms of one element are different from the atoms of all other elements.

3. The atoms of one element cannot be changed into atoms of a different element by chemical reactions; atoms are neither created nor destroyed in chemical reactions.

4. Compounds are formed when atoms of more than one element combine; a given compound always has the same relative number and kind of atoms.

According to Dalton's atomic theory, **atoms** are the smallest particles of an element that retain the chemical identity of the element. ∞ (Section 1.1) As noted in the postulates of Dalton's theory, an element is composed of only one kind of atom. A compound, in contrast, contains atoms of two or more elements.

Dalton's theory explains several simple laws of chemical combination that were known during his time. One of these laws was the *law of constant composition*: In a given compound, the relative numbers and kinds of atoms are constant. ∞ (Section 1.2) This law is the basis of Dalton's Postulate 4. Another fundamental chemical law was the *law of conservation of mass* (also known as the *law of conservation of matter*): The total mass of materials present after a chemical reaction is the same as the total mass present before the reaction. This law is the basis for Postulate 3. Dalton proposed that atoms always retain their identities and that atoms taking part in a chemical reaction rearrange to give new chemical combinations.

▲ Figure 2.1 **John Dalton (1766–1844).** Dalton was the son of a poor English weaver. He began teaching at the age of 12. He spent most of his years in Manchester, where he taught both grammar school and college. His lifelong interest in meteorology led him to study gases, then chemistry, and eventually atomic theory.

A good theory should explain the known facts and predict new ones. Dalton used his theory to deduce the *law of multiple proportions*: If two elements A and B combine to form more than one compound, the masses of B that can combine with a given mass of A are in the ratio of small whole numbers. We can illustrate this law by considering the substances water and hydrogen peroxide, both of which consist of the elements hydrogen and oxygen. In forming water, 8.0 g of oxygen combine with 1.0 g of hydrogen. In forming hydrogen peroxide, 16.0 g of oxygen combine with 1.0 g of hydrogen. In other words, the ratio of the mass of oxygen per gram of hydrogen in the two compounds is 2:1. Using the atomic theory, we can conclude that hydrogen peroxide contains twice as many atoms of oxygen per hydrogen atom as does water.

GIVE IT SOME THOUGHT

One compound of carbon and oxygen contains 1.333 g of oxygen per gram of carbon, whereas a second compound contains 2.666 g of oxygen per gram of carbon. **(a)** What chemical law do these data illustrate? **(b)** If the first compound has an equal number of oxygen and carbon atoms, what can we conclude about the composition of the second compound?

2.2 THE DISCOVERY OF ATOMIC STRUCTURE

Dalton reached his conclusion about atoms based on chemical observations in the macroscopic world of the laboratory. Neither he nor those who followed him during the century after his work was published had direct evidence for the existence of atoms. Today, however, we can use powerful instruments to measure the properties of individual atoms and even provide images of them (Figure 2.2▶).

As scientists began to develop methods for more detailed probing of the nature of matter, the atom, which was supposed to be indivisible, began to show signs of a more complex structure: We now know that the atom is composed of still smaller **subatomic particles**. Before we summarize the current model of atomic structure, we will briefly consider a few of the landmark discoveries that led to that model. We will see that the atom is composed in part of electrically charged particles, some with a positive (+) charge and some with a negative (−) charge. As we discuss the development of our current model of the atom, keep in mind a simple statement of the behavior of charged particles: *Particles with the same charge repel one another, whereas particles with unlike charges attract one another.*

▲ **Figure 2.2 An image of the surface of the semiconductor GaAs (gallium arsenide).** This image was obtained by a technique called scanning tunneling microscopy. The color was added to the image by computer to distinguish the gallium atoms (blue spheres) from the arsenic atoms (red spheres).

▼ **Figure 2.3 Cathode-ray tube.** (a) In a cathode-ray tube, electrons move from the negative electrode (cathode) to the positive electrode (anode). (b) A photo of a cathode-ray tube containing a fluorescent screen to show the path of the cathode rays. (c) The path of the cathode rays is deflected by the presence of a magnet.

Cathode Rays and Electrons

During the mid-1800s, scientists began to study electrical discharge through partially evacuated tubes (tubes that had been pumped almost empty of air), such as those shown in Figure 2.3▼. When a high voltage was applied to the

Partially evacuated glass vessel

(−)   (+)

High voltage

(a)

(b)

(c)

▶ Figure 2.4 **Cathode-ray tube with perpendicular magnetic and electric fields.** The cathode rays (electrons) originate from the negative plate on the left and are accelerated toward the positive plate on the right, which has a hole in its center. A narrow beam of electrons passes through the hole and is then deflected by the magnetic and electric fields. The three paths result from different strengths of the magnetic and electric fields. The charge-to-mass ratio of the electron can be determined by measuring the effects that the magnetic and electric fields have on the direction of the beam.

electrodes in the tube, radiation was produced. This radiation, called **cathode rays**, originated from the negative electrode, or cathode. Although the rays themselves could not be seen, their movement was detected because the rays cause certain materials, including glass, to *fluoresce*, or to give off light.

Scientists held conflicting views about the nature of the cathode rays. It was not initially clear whether the rays were an invisible stream of particles or a new form of radiation. Experiments showed that cathode rays are deflected by electric or magnetic fields in a way consistent with their being a stream of negative electrical charge [Figure 2.3(c)]. The British scientist J. J. Thomson observed many properties of the cathode rays, including the fact that they are the same regardless of the identity of the cathode material. In a paper published in 1897, Thomson summarized his observations and concluded that cathode rays are streams of negatively charged particles. Thomson's paper is generally accepted as the "discovery" of what later became known as the *electron*.

Thomson constructed a cathode-ray tube having a fluorescent screen at one end, such as that shown in Figure 2.4▲, so that he could quantitatively measure the effects of electric and magnetic fields on the thin stream of electrons passing through a hole in the positively charged electrode. These measurements made it possible to calculate a value of 1.76×10^8 coulombs per gram for the ratio of the electron's electrical charge to its mass.*

Once the charge-to-mass ratio of the electron was known, measuring either the charge or the mass of an electron would yield the value of the other quantity. In 1909, Robert Millikan (1868–1953) of the University of Chicago succeeded in measuring the charge of an electron by performing a series of experiments described in Figure 2.5▼. He then calculated the mass of the electron by using

▶ Figure 2.5 **Millikan's oil-drop experiment.** A representation of the apparatus Millikan used to measure the charge of the electron. Small drops of oil, which had picked up extra electrons, were allowed to fall between two electrically charged plates. Millikan monitored the drops, measuring how the voltage on the plates affected their rate of fall. From these data he calculated the charges on the drops. His experiment showed that the charges were always integral multiples of 1.602×10^{-19} C, which he deduced was the charge of a single electron.

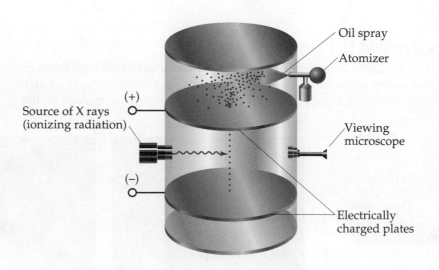

The coulomb (C) is the SI unit for electrical charge.

his experimental value for the charge, 1.602×10^{-19} C, and Thomson's charge-to-mass ratio, 1.76×10^8 C/g:

$$\text{Electron mass} = \frac{1.602 \times 10^{-19} \text{ C}}{1.76 \times 10^8 \text{ C/g}} = 9.10 \times 10^{-28} \text{ g}$$

This result agrees well with the presently accepted value for the mass of the electron, 9.10938×10^{-28} g. This mass is about 2000 times smaller than that of hydrogen, the lightest atom.

Radioactivity

In 1896 the French scientist Henri Becquerel (1852–1908) was studying a uranium compound when he discovered that it spontaneously emits high-energy radiation. This spontaneous emission of radiation is called **radioactivity**. At Becquerel's suggestion Marie Curie (Figure 2.6▶) and her husband, Pierre, began experiments to isolate the radioactive components of the compound.

Further study of the nature of radioactivity, principally by the British scientist Ernest Rutherford (Figure 2.7▶), revealed three types of radiation: alpha (α), beta (β), and gamma (γ) radiation. Each type differs in its response to an electric field, as shown in Figure 2.8▼. The paths of both α and β radiation are bent by the electric field, although in opposite directions; γ radiation is unaffected.

Rutherford showed that both α and β rays consist of fast-moving particles, which were called α and β particles. In fact, β particles are high-speed electrons and can be considered the radioactive equivalent of cathode rays. They are attracted to a positively charged plate. The α particles have a positive charge and are attracted toward a negative plate. In units of the charge of the electron, β particles have a charge of $1-$ and α particles a charge of $2+$. Each α particle has a mass about 7400 times that of an electron. Gamma radiation is high-energy radiation similar to X-rays; it does not consist of particles and carries no charge. We will discuss radioactivity in greater detail in Chapter 21.

The Nuclear Atom

With the growing evidence that the atom is composed of smaller particles, attention was given to how the particles fit together. During the early 1900s Thomson reasoned that because electrons contribute only a very small fraction of the mass of an atom, they probably were responsible for an equally small fraction of the atom's size. He proposed that the atom consisted of a uniform positive sphere of matter in which the electrons were embedded, as shown in

▲ **Figure 2.6 Marie Sklodowska Curie (1867–1934).** When M. Curie presented her doctoral thesis, it was described as the greatest single contribution of any doctoral thesis in the history of science. Among other things, Curie discovered two new elements, polonium and radium. In 1903 Henri Becquerel, M. Curie, and her husband, Pierre, were jointly awarded the Nobel Prize in physics. In 1911 M. Curie won a second Nobel Prize, this time in chemistry.

▲ **Figure 2.7 Ernest Rutherford (1871–1937).** Rutherford, whom Einstein called "the second Newton," was born and educated in New Zealand. In 1895 he was the first overseas student ever to be awarded a position at the Cavendish Laboratory at Cambridge University in England, where he worked with J. J. Thomson. In 1898 he joined the faculty of McGill University in Montreal. While at McGill, Rutherford did his research on radioactivity that led to his being awarded the 1908 Nobel Prize in chemistry. In 1907 Rutherford moved back to England to be a faculty member at Manchester University, where in 1910 he performed his famous α-particle scattering experiments that led to the nuclear model of the atom. In 1992 his native New Zealand honored Rutherford by putting his likeness, along with his Nobel Prize medal, on their $100 currency note.

▲ **Figure 2.8 Behavior of alpha (α), beta (β), and gamma (γ) rays in an electric field.** The α rays consist of positively charged particles and are therefore attracted to the negatively charged plate. The β rays consist of negatively charged particles and are attracted to the positively charged plate. The γ rays, which carry no charge, are unaffected by the electric field.

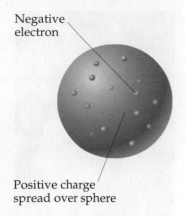

Negative electron

Positive charge spread over sphere

▲ Figure 2.9 J. J. Thomson's "plum-pudding" model of the atom. Thomson pictured the small electrons to be embedded in the atom much like raisins in a pudding or seeds in a watermelon. Ernest Rutherford proved this model wrong.

Figure 2.9 ◄. This so-called "plum-pudding" model, named after a traditional English dessert, was very short lived.

In 1910, Rutherford and his coworkers performed an experiment that disproved Thomson's model. Rutherford was studying the angles at which α particles were deflected, or *scattered*, as they passed through a thin gold foil only a few thousand atoms thick (Figure 2.10▼). Rutherford and his coworkers discovered that almost all the α particles passed directly through the foil without deflection. A few particles were found to be deflected by approximately 1 degree, consistent with Thomson's plum-pudding model. Just for the sake of completeness, Rutherford suggested that Ernest Marsden, an undergraduate student working in the laboratory, look for evidence of scattering at large angles. To everyone's surprise, a small amount of scattering was observed at large angles. Some particles were even scattered back in the direction from which they had come. The explanation for these results was not immediately obvious, but they were clearly inconsistent with Thomson's plum-pudding model.

By 1911, Rutherford was able to explain these observations. He postulated that most of the mass of each gold atom in his foil and all of its positive charge reside in a very small, extremely dense region, which he called the **nucleus**. He postulated further that most of the total volume of an atom is empty space in which electrons move around the nucleus. In the α-scattering experiment, most α particles passed directly through the foil because they did not encounter the minute nucleus of any gold atom; they merely passed through the empty space making up the greatest part of all the atoms in the foil. Occasionally, however, an α particle came close to a gold nucleus. The repulsion between the highly charged gold nucleus and the α particle was strong enough to deflect the less massive α particle, as shown in Figure 2.11►.

Subsequent experimental studies led to the discovery of both positive particles (*protons*) and neutral particles (*neutrons*) in the nucleus. Protons were discovered in 1919 by Rutherford. In 1932 British scientist James Chadwick (1891–1972) discovered neutrons. We examine these particles more closely in Section 2.3.

► Figure 2.10 Rutherford's experiment on the scattering of α particles. The red lines represent the paths of the α particles. When the incoming beam strikes the gold foil, most particles pass straight through the foil, but some are scattered.

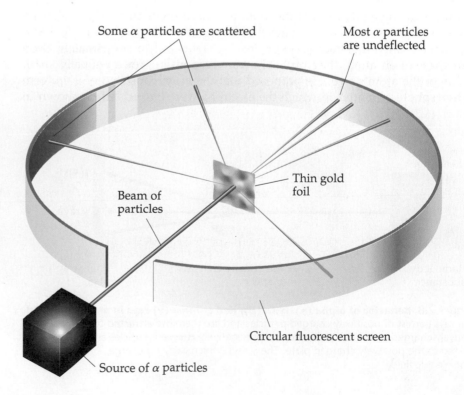

Some α particles are scattered

Most α particles are undeflected

Thin gold foil

Beam of particles

Circular fluorescent screen

Source of α particles

> GIVE IT SOME THOUGHT

What happens to most of the α particles that strike the gold foil in Rutherford's experiment? Why do they behave that way?

2.3 THE MODERN VIEW OF ATOMIC STRUCTURE

Since the time of Rutherford, physicists have learned much about the detailed composition of atomic nuclei. In the course of these discoveries, the list of particles that make up nuclei has grown long and continues to increase. As chemists, however, we can take a very simple view of the atom because only three subatomic particles—the **proton, neutron,** and **electron**—have a bearing on chemical behavior.

The charge of an electron is -1.602×10^{-19} C, and that of a proton is $+1.602 \times 10^{-19}$ C. The quantity 1.602×10^{-19} C is called the **electronic charge**. For convenience, the charges of atomic and subatomic particles are usually expressed as multiples of this charge rather than in coulombs. Thus, the charge of the electron is $1-$, and that of the proton is $1+$. Neutrons are uncharged and are therefore electrically neutral (which is how they received their name). *Every atom has an equal number of electrons and protons, so atoms have no net electrical charge.*

Protons and neutrons reside together in the nucleus of the atom, which, as Rutherford proposed, is extremely small. The vast majority of an atom's volume is the space in which the electrons reside. The electrons are attracted to the protons in the nucleus by the electrostatic force that exists between particles of opposite electrical charge. In later chapters we will see that the strength of the attractive forces between electrons and nuclei can be used to explain many of the differences between different elements.

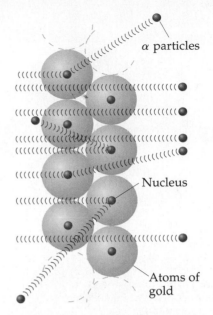

▲ **Figure 2.11 Rutherford's model explaining the scattering of α particles.** The gold foil is several thousand atoms thick. Because most of the volume of each atom is empty space, most α particles pass through the foil without deflection. When an α particle passes very close to a gold nucleus, however, it is repelled, causing its path to be altered.

> GIVE IT SOME THOUGHT

(a) If an atom has 15 protons, how many electrons does it have? **(b)** Where do the protons reside in an atom?

Atoms have extremely small masses. The mass of the heaviest known atom, for example, is approximately 4×10^{-22} g. Because it would be cumbersome to express such small masses in grams, we use instead the **atomic mass unit**, or amu.* One amu equals 1.66054×10^{-24} g. The masses of the proton and neutron are very nearly equal, and both are much greater than that of the electron: A proton has a mass of 1.0073 amu, a neutron 1.0087 amu, and an electron 5.486×10^{-4} amu. Because it would take 1836 electrons to equal the mass of 1 proton, the nucleus contains most of the mass of an atom. Table 2.1 ▼ summarizes the charges and masses of the subatomic particles. We will have more to say about atomic masses in Section 2.4.

Atoms are also extremely small. Most atoms have diameters between 1×10^{-10} m and 5×10^{-10} m, or 100–500 pm. A convenient, although non-SI, unit of length used to express atomic dimensions is the **angstrom** (Å). One

TABLE 2.1 ■ Comparison of the Proton, Neutron, and Electron		
Particle	**Charge**	**Mass (amu)**
Proton	Positive (1+)	1.0073
Neutron	None (neutral)	1.0087
Electron	Negative (1−)	5.486×10^{-4}

The SI abbreviation for the atomic mass unit is u. We will use the more common abbreviation amu.

angstrom equals 10^{-10} m. Thus, atoms have diameters of approximately 1–5 Å. The diameter of a chlorine atom, for example, is 200 pm, or 2.0 Å. Both picometers and angstroms are commonly used to express the dimensions of atoms and molecules.

■ SAMPLE EXERCISE 2.1 | Illustrating the Size of an Atom

The diameter of a US penny is 19 mm. The diameter of a silver atom, by comparison, is only 2.88 Å. How many silver atoms could be arranged side by side in a straight line across the diameter of a penny?

SOLUTION

The unknown is the number of silver (Ag) atoms. We use the relationship 1 Ag atom = 2.88 Å as a conversion factor relating the number of atoms and distance. Thus, we can start with the diameter of the penny, first converting this distance into angstroms and then using the diameter of the Ag atom to convert distance to the number of Ag atoms:

$$\text{Ag atoms} = (19 \text{ mm})\left(\frac{10^{-3} \text{ m}}{1 \text{ mm}}\right)\left(\frac{1 \text{ Å}}{10^{-10} \text{ m}}\right)\left(\frac{1 \text{ Ag atom}}{2.88 \text{ Å}}\right) = 6.6 \times 10^7 \text{ Ag atoms}$$

That is, 66 million silver atoms could sit side by side across a penny!

■ PRACTICE EXERCISE

The diameter of a carbon atom is 1.54 Å. **(a)** Express this diameter in picometers. **(b)** How many carbon atoms could be aligned side by side in a straight line across the width of a pencil line that is 0.20 mm wide?
Answers: **(a)** 154 pm, **(b)** 1.3×10^6 C atoms

The diameters of atomic nuclei are approximately 10^{-4} Å, only a small fraction of the diameter of the atom as a whole. You can appreciate the relative sizes of the atom and its nucleus by imagining that if the hydrogen atom were as large as a football stadium, the nucleus would be the size of a small marble. Because the tiny nucleus carries most of the mass of the atom in such a small volume, it has an incredible density—on the order of 10^{13}–10^{14} g/cm^3. A matchbox full of material of such density would weigh over 2.5 billion tons! Astrophysicists have suggested that the interior of a collapsed star may approach this density.

An illustration of the atom that incorporates the features we have just discussed is shown in Figure 2.12▼. The electrons, which take up most of the volume of the atom, play the major role in chemical reactions. The significance of representing the region containing the electrons as an indistinct cloud will become clear in later chapters when we consider the energies and spatial arrangements of the electrons.

Atomic Numbers, Mass Numbers, and Isotopes

What makes an atom of one element different from an atom of another element? For example, how does an atom of carbon differ from an atom of oxygen? The significant difference is in their subatomic compositions. The atoms of each element have a characteristic number of protons. Indeed, the number

▼ Figure 2.12 **The structure of the atom.** Neon gas is composed of atoms. The nucleus, which contains protons and neutrons, is the location of virtually all the mass of the atom. The rest of the atom is the space in which the light, negatively charged electrons reside.

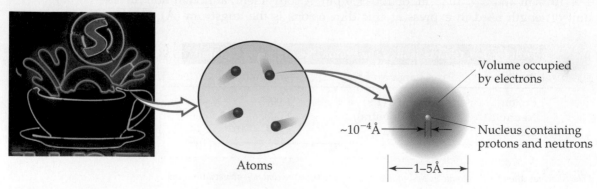

Atoms $\sim10^{-4}$Å   Volume occupied by electrons   Nucleus containing protons and neutrons   ⟵1–5Å⟶

There are four basic forces known in nature: (1) gravitational, (2) electromagnetic, (3) strong nuclear, and (4) weak nuclear. *Gravitational forces* are attractive forces that act between all objects in proportion to their masses. Gravitational forces between atoms or between subatomic particles are so small that they are of no chemical significance.

Electromagnetic forces are attractive or repulsive forces that act between either electrically charged or magnetic objects. Electric and magnetic forces are intimately related. Electric forces are of fundamental importance in understanding the chemical behavior of atoms. The magnitude of the electric force between two charged particles is given by *Coulomb's law*: $F = kQ_1Q_2/d^2$, where Q_1 and Q_2 are the magnitudes of the charges on the two particles, d is the distance between their centers, and k is a constant determined by the units for Q and d. A negative value for the force indicates attraction, whereas a positive value indicates repulsion.

All nuclei except those of hydrogen atoms contain two or more protons. Because like charges repel, electrical repulsion would cause the protons to fly apart if a stronger attractive force, called the *strong nuclear force*, did not keep them together. This force acts between subatomic particles, as in the nucleus. At this distance, the strong nuclear force is stronger than the electric force and holds the nucleus together.

The *weak nuclear force* is weaker than the electric force but stronger than the gravitational force. We are aware of its existence only because it shows itself in certain types of radioactivity. ***Related Exercises: 2.83(b) and 2.89***

of protons in the nucleus of an atom of any particular element is called that element's **atomic number**. Because an atom has no net electrical charge, the number of electrons it contains must equal the number of protons. All atoms of carbon, for example, have six protons and six electrons, whereas all atoms of oxygen have eight protons and eight electrons. Thus, carbon has atomic number 6, whereas oxygen has atomic number 8. The atomic number of each element is listed with the name and symbol of the element on the inside front cover of the text.

Atoms of a given element can differ in the number of neutrons they contain and consequently in mass. For example, most atoms of carbon have six neutrons, although some have more and some have less. The symbol $^{12}_{6}C$ (read "carbon twelve," carbon-12) represents the carbon atom containing six protons and six neutrons. The atomic number is shown by the subscript, and the superscript, called the **mass number**, is the total number of protons plus neutrons in the atom:

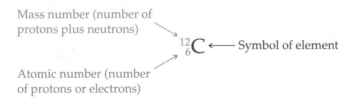

Mass number (number of protons plus neutrons)

$^{12}_{6}C$ ← Symbol of element

Atomic number (number of protons or electrons)

Because all atoms of a given element have the same atomic number, the subscript is redundant and is often omitted. Thus, the symbol for carbon-12 can be represented simply as ^{12}C. As one more example of this notation, atoms that contain six protons and eight neutrons have a mass number of 14 and are represented as $^{14}_{6}C$ or ^{14}C and referred to as carbon-14.

Atoms with identical atomic numbers but different mass numbers (that is, same number of protons but different numbers of neutrons) are called **isotopes** of one another. Several isotopes of carbon are listed in Table 2.2▼. We will

TABLE 2.2 ■ Some Isotopes of Carbon*			
Symbol	Number of Protons	Number of Electrons	Number of Neutrons
^{11}C	6	6	5
^{12}C	6	6	6
^{13}C	6	6	7
^{14}C	6	6	8

*Almost 99% of the carbon found in nature is ^{12}C.

generally use the notation with superscripts only when referring to a particular isotope of an element.

■ **SAMPLE EXERCISE 2.2** | **Determining the Number of Subatomic Particles in Atoms**

How many protons, neutrons, and electrons are in (a) an atom of ^{197}Au; (b) an atom of strontium-90?

SOLUTION

(a) The superscript 197 is the mass number, the sum of the number of protons plus the number of neutrons. According to the list of elements given inside the front cover, gold has an atomic number of 79. Consequently, an atom of ^{197}Au has 79 protons, 79 electrons, and $197 - 79 = 118$ neutrons. (b) The atomic number of strontium (listed inside the front cover) is 38. Thus, all atoms of this element have 38 protons and 38 electrons. The strontium-90 isotope has $90 - 38 = 52$ neutrons.

■ **PRACTICE EXERCISE**

How many protons, neutrons, and electrons are in (a) a ^{138}Ba atom, (b) an atom of phosphorus-31?
Answer: (a) 56 protons, 56 electrons, and 82 neutrons; (b) 15 protons, 15 electrons, and 16 neutrons.

■ **SAMPLE EXERCISE 2.3** | **Writing Symbols for Atoms**

Magnesium has three isotopes, with mass numbers 24, 25, and 26. (a) Write the complete chemical symbol (superscript and subscript) for each of them. (b) How many neutrons are in an atom of each isotope?

SOLUTION

(a) Magnesium has atomic number 12, so all atoms of magnesium contain 12 protons and 12 electrons. The three isotopes are therefore represented by $^{24}_{12}$Mg, $^{25}_{12}$Mg, and $^{26}_{12}$Mg. (b) The number of neutrons in each isotope is the mass number minus the number of protons. The numbers of neutrons in an atom of each isotope are therefore 12, 13, and 14, respectively.

■ **PRACTICE EXERCISE**

Give the complete chemical symbol for the atom that contains 82 protons, 82 electrons, and 126 neutrons.
Answer: $^{208}_{82}$Pb

2.4 ATOMIC WEIGHTS

Atoms are small pieces of matter, so they have mass. In this section we will discuss the mass scale used for atoms and introduce the concept of *atomic weights*. In Section 3.3 we will extend these concepts to show how atomic masses are used to determine the masses of compounds and *molecular weights*.

The Atomic Mass Scale

Although scientists of the nineteenth century knew nothing about subatomic particles, they were aware that atoms of different elements have different masses. They found, for example, that each 100.0 g of water contains 11.1 g of hydrogen and 88.9 g of oxygen. Thus, water contains $88.9/11.1 = 8$ times as much oxygen, by mass, as hydrogen. Once scientists understood that water contains two hydrogen atoms for each oxygen atom, they concluded that an oxygen atom must have $2 \times 8 = 16$ times as much mass as a hydrogen atom. Hydrogen, the lightest atom, was arbitrarily assigned a relative mass of 1 (no units). Atomic masses of other elements were at first determined relative to this value. Thus, oxygen was assigned an atomic mass of 16.

Today we can determine the masses of individual atoms with a high degree of accuracy. For example, we know that the ^{1}H atom has a mass of 1.6735×10^{-24} g and the ^{16}O atom has a mass of 2.6560×10^{-23} g. As we noted in Section 2.3, it is convenient to use the *atomic mass unit* (amu) when dealing with these extremely small masses:

$$1 \text{ amu} = 1.66054 \times 10^{-24} \text{ g and } 1 \text{ g} = 6.02214 \times 10^{23} \text{ amu}$$

The atomic mass unit is presently defined by assigning a mass of exactly 12 amu to an atom of the ^{12}C isotope of carbon. In these units, an ^{1}H atom has a mass of 1.0078 amu and an ^{16}O atom has a mass of 15.9949 amu.

Average Atomic Masses

Most elements occur in nature as mixtures of isotopes. We can determine the *average atomic mass* of an element by using the masses of its various isotopes and their relative abundances. Naturally occurring carbon, for example, is composed of 98.93% ^{12}C and 1.07% ^{13}C. The masses of these isotopes are 12 amu (exactly) and 13.00335 amu, respectively. We calculate the average atomic mass of carbon from the fractional abundance of each isotope and the mass of that isotope:

$$(0.9893)(12 \text{ amu}) + (0.0107)(13.00335 \text{ amu}) = 12.01 \text{ amu}$$

The average atomic mass of each element (expressed in atomic mass units) is also known as its **atomic weight**. Although the term *average atomic mass* is more proper, the term *atomic weight* is more common. The atomic weights of the elements are listed in both the periodic table and the table of elements inside the front cover of this text.

GIVE IT SOME THOUGHT

A particular atom of chromium has a mass of 52.94 amu, whereas the atomic weight of chromium is 51.99 amu. Explain the difference in the two masses.

SAMPLE EXERCISE 2.4 | **Calculating the Atomic Weight of an Element from Isotopic Abundances**

Naturally occurring chlorine is 75.78% ^{35}Cl, which has an atomic mass of 34.969 amu, and 24.22% ^{37}Cl, which has an atomic mass of 36.966 amu. Calculate the average atomic mass (that is, the atomic weight) of chlorine.

SOLUTION

We can calculate the average atomic mass by multiplying the abundance of each isotope by its atomic mass and summing these products. Because 75.78% = 0.7578 and 24.22% = 0.2422, we have

$$\text{Average atomic mass} = (0.7578)(34.969 \text{ amu}) + (0.2422)(36.966 \text{ amu})$$
$$= 26.50 \text{ amu} + 8.953 \text{ amu}$$
$$= 35.45 \text{ amu}$$

This answer makes sense: The average atomic mass of Cl is between the masses of the two isotopes and is closer to the value of ^{35}Cl, which is the more abundant isotope.

PRACTICE EXERCISE

Three isotopes of silicon occur in nature: ^{28}Si (92.23%), which has an atomic mass of 27.97693 amu; ^{29}Si (4.68%), which has an atomic mass of 28.97649 amu; and ^{30}Si (3.09%), which has an atomic mass of 29.97377 amu. Calculate the atomic weight of silicon.
Answer: 28.09 amu

A Closer Look   THE MASS SPECTROMETER

The most direct and accurate means for determining atomic and molecular weights is provided by the **mass spectrometer** (Figure 2.13 ▼). A gaseous sample is introduced at *A* and bombarded by a stream of high-energy electrons at *B*. Collisions between the electrons and the atoms or molecules of the gas produce positively charged particles, mostly with a 1+ charge. These charged particles are accelerated toward a negatively charged wire grid (*C*). After the particles pass through the grid, they encounter two slits that allow only a narrow beam of particles to pass. This beam then passes between the poles of a magnet, which deflects the particles into a curved path, much as electrons are deflected by a magnetic field (Figure 2.4). For charged particles with the same charge, the extent of deflection depends on mass—the more massive the particle, the less the deflection. The particles are thereby separated according to their masses. By changing the strength of the magnetic field or the accelerating voltage on the negatively charged grid, charged particles of various masses can be selected to enter the detector at the end of the instrument.

A graph of the intensity of the detector signal versus particle atomic mass is called a *mass spectrum*. The mass spectrum of chlorine atoms, shown in Figure 2.14 ▼, reveals the presence of two isotopes. Analysis of a mass spectrum gives both the masses of the charged particles reaching the detector and their relative abundances. The abundances are obtained from the signal intensities. Knowing the atomic mass and the abundance of each isotope allows us to calculate the atomic weight of an element, as shown in Sample Exercise 2.4.

Mass spectrometers are used extensively today to identify chemical compounds and analyze mixtures of substances. Any molecule that loses electrons can fall apart, forming an array of positively charged fragments. The mass spectrometer measures the masses of these fragments, producing a chemical "fingerprint" of the molecule and providing clues about how the atoms were connected in the original molecule. Thus, a chemist might use this technique to determine the molecular structure of a newly synthesized compound or to identify a pollutant in the environment.

Related Exercises: 2.33, 2.34, 2.35(b), 2.36, 2.93, and 2.94

▲ **Figure 2.13 A mass spectrometer.** Cl atoms are introduced on the left side of the spectrometer and are ionized to form Cl^+ ions, which are then directed through a magnetic field. The paths of the ions of the two isotopes of Cl diverge as they pass through the magnetic field. As drawn, the spectrometer is tuned to detect $^{35}Cl^+$ ions. The heavier $^{37}Cl^+$ ions are not deflected enough for them to reach the detector.

▲ **Figure 2.14 Mass spectrum of atomic chlorine.** The fractional abundances of the ^{35}Cl and ^{37}Cl isotopes of chlorine are indicated by the relative signal intensities of the beams reaching the detector of the mass spectrometer.

2.5 THE PERIODIC TABLE

Dalton's atomic theory set the stage for a vigorous growth in chemical experimentation during the early 1800s. As the body of chemical observations grew and the list of known elements expanded, attempts were made to find regular patterns in chemical behavior. These efforts culminated in the development of the periodic table in 1869. We will have much to say about the periodic table in later chapters, but it is so important and useful that you should become acquainted with it now. You will quickly learn that *the periodic table is the most significant tool that chemists use for organizing and remembering chemical facts*.

Many elements show very strong similarities to one another. The elements lithium (Li), sodium (Na), and potassium (K) are all soft, very reactive metals,

◀ Figure 2.15 **Arranging the elements by atomic number reveals a periodic pattern of properties.** This periodic pattern is the basis of the periodic table.

for example. The elements helium (He), neon (Ne), and argon (Ar) are all very nonreactive gases. If the elements are arranged in order of increasing atomic number, their chemical and physical properties show a repeating, or periodic, pattern. For example, each of the soft, reactive metals—lithium, sodium, and potassium—comes immediately after one of the nonreactive gases—helium, neon, and argon—as shown in Figure 2.15▲.

The arrangement of elements in order of increasing atomic number, with elements having similar properties placed in vertical columns, is known as the **periodic table**. The periodic table is shown in Figure 2.16▼ and is also given on the front inside cover of the text. For each element in the table, the atomic number and atomic symbol are given. The atomic weight is often given as well, as in the following typical entry for potassium:

19 ← — atomic number
K ← — atomic symbol
39.0983 ← — atomic weight

You may notice slight variations in periodic tables from one book to another or between those in the lecture hall and in the text. These are simply matters of style, or they might concern the particular information included. There are no fundamental differences.

1A 1																	8A 18
1 **H**	2A 2											3A 13	4A 14	5A 15	6A 16	7A 17	2 **He**
3 **Li**	4 **Be**											5 **B**	6 **C**	7 **N**	8 **O**	9 **F**	10 **Ne**
11 **Na**	12 **Mg**	3B 3	4B 4	5B 5	6B 6	7B 7	8	8B 9	10	1B 11	2B 12	13 **Al**	14 **Si**	15 **P**	16 **S**	17 **Cl**	18 **Ar**
19 **K**	20 **Ca**	21 **Sc**	22 **Ti**	23 **V**	24 **Cr**	25 **Mn**	26 **Fe**	27 **Co**	28 **Ni**	29 **Cu**	30 **Zn**	31 **Ga**	32 **Ge**	33 **As**	34 **Se**	35 **Br**	36 **Kr**
37 **Rb**	38 **Sr**	39 **Y**	40 **Zr**	41 **Nb**	42 **Mo**	43 **Tc**	44 **Ru**	45 **Rh**	46 **Pd**	47 **Ag**	48 **Cd**	49 **In**	50 **Sn**	51 **Sb**	52 **Te**	53 **I**	54 **Xe**
55 **Cs**	56 **Ba**	71 **Lu**	72 **Hf**	73 **Ta**	74 **W**	75 **Re**	76 **Os**	77 **Ir**	78 **Pt**	79 **Au**	80 **Hg**	81 **Tl**	82 **Pb**	83 **Bi**	84 **Po**	85 **At**	86 **Rn**
87 **Fr**	88 **Ra**	103 **Lr**	104 **Rf**	105 **Db**	106 **Sg**	107 **Bh**	108 **Hs**	109 **Mt**	110 **Ds**	111 **Rg**	112	113	114	115	116		118

	Metals	57 **La**	58 **Ce**	59 **Pr**	60 **Nd**	61 **Pm**	62 **Sm**	63 **Eu**	64 **Gd**	65 **Tb**	66 **Dy**	67 **Ho**	68 **Er**	69 **Tm**	70 **Yb**
	Metalloids	89 **Ac**	90 **Th**	91 **Pa**	92 **U**	93 **Np**	94 **Pu**	95 **Am**	96 **Cm**	97 **Bk**	98 **Cf**	99 **Es**	100 **Fm**	101 **Md**	102 **No**
	Nonmetals														

▲ Figure 2.16 **Periodic table of the elements.** Different colors are used to show the division of the elements into metals, metalloids, and nonmetals.

The horizontal rows of the periodic table are called **periods**. The first period consists of only two elements, hydrogen (H) and helium (He). The second and third periods, which begin with lithium (Li) and sodium (Na), respectively, consist of eight elements each. The fourth and fifth periods contain 18 elements. The sixth period has 32 elements, but for it to fit on a page, 14 of these elements (those with atomic numbers 57–70) appear at the bottom of the table. The seventh and last period is incomplete, but it also has 14 of its members placed in a row at the bottom of the table.

The vertical columns of the periodic table are called **groups**. The way in which the groups are labeled is somewhat arbitrary. Three labeling schemes are in common use, two of which are shown in Figure 2.16. The top set of labels, which have A and B designations, is widely used in North America. Roman numerals, rather than Arabic ones, are often employed in this scheme. Group 7A, for example, is often labeled VIIA. Europeans use a similar convention that numbers the columns from 1A through 8A and then from 1B through 8B, thereby giving the label 7B (or VIIB) instead of 7A to the group headed by fluorine (F). In an effort to eliminate this confusion, the International Union of Pure and Applied Chemistry (IUPAC) has proposed a convention that numbers the groups from 1 through 18 with no A or B designations, as shown in the lower set of labels at the top of the table in Figure 2.16. We will use the traditional North American convention with Arabic numerals.

Elements that belong to the same group often exhibit similarities in physical and chemical properties. For example, the "coinage metals"—copper (Cu), silver (Ag), and gold (Au)— belong to group 1B. As their name suggests, the coinage metals are used throughout the world to make coins. Many other groups in the periodic table also have names, as listed in Table 2.3▼.

We will learn in Chapters 6 and 7 that the elements in a group of the periodic table have similar properties because they have the same arrangement of electrons at the periphery of their atoms. However, we need not wait until then to make good use of the periodic table; after all, chemists who knew nothing about electrons developed the table! We can use the table, as they intended, to correlate the behaviors of elements and to aid in remembering many facts. You will find it helpful to refer to the periodic table frequently when studying the remainder of this chapter.

Except for hydrogen, all the elements on the left side and in the middle of the periodic table are **metallic elements**, or **metals**. The majority of elements are metallic; they all share characteristic properties, such as luster and high electrical and heat conductivity. All metals, with the exception of mercury (Hg), are solids at room temperature. The metals are separated from the **nonmetallic elements**, or **nonmetals**, by a diagonal steplike line that runs from boron (B) to astatine (At), as shown in Figure 2.16. Hydrogen, although on the left side of the periodic table, is a nonmetal. At room temperature some of the nonmetals are gaseous, some are solid, and one is liquid. Nonmetals generally differ from the metals in appearance (Figure 2.17 ◄) and in other physical properties. Many of the elements that lie along the line that separates metals from nonmetals, such as antimony (Sb), have properties that fall between those of metals and those of nonmetals. These elements are often referred to as **metalloids**.

▲ Figure 2.17 **Some familiar examples of metals and nonmetals.** The nonmetals (from bottom left) are sulfur (yellow powder), iodine (dark, shiny crystals), bromine (reddish brown liquid and vapor in glass vial), and three samples of carbon (black charcoal powder, diamond, and graphite in the pencil lead). The metals are in the form of an aluminum wrench, copper pipe, lead shot, silver coins, and gold nuggets.

TABLE 2.3 ■ Names of Some Groups in the Periodic Table		
Group	Name	Elements
1A	Alkali metals	Li, Na, K, Rb, Cs, Fr
2A	Alkaline earth metals	Be, Mg, Ca, Sr, Ba, Ra
6A	Chalcogens	O, S, Se, Te, Po
7A	Halogens	F, Cl, Br, I, At
8A	Noble gases (or rare gases)	He, Ne, Ar, Kr, Xe, Rn

A Closer Look   GLENN SEABORG AND SEABORGIUM

Prior to 1940 the periodic table ended at uranium, element number 92. Since that time, no scientist has had a greater effect on the periodic table than Glenn Seaborg. Seaborg (Figure 2.18▶) became a faculty member in the chemistry department at the University of California, Berkeley in 1937. In 1940 he and his colleagues Edwin McMillan, Arthur Wahl, and Joseph Kennedy succeeded in isolating plutonium (Pu) as a product of the reaction between uranium and neutrons. We will talk about reactions of this type, called *nuclear reactions*, in Chapter 21.

During the period 1944 through 1958, Seaborg and his coworkers also identified various products of nuclear reactions as being the elements having atomic numbers 95 through 102. All these elements are radioactive and are not found in nature; they can be synthesized only via nuclear reactions. For their efforts in identifying the elements beyond uranium (the *transuranium* elements), McMillan and Seaborg shared the 1951 Nobel Prize in chemistry.

From 1961 to 1971, Seaborg served as the chairman of the U.S. Atomic Energy Commission (now the Department of Energy). In this position he had an important role in establishing international treaties to limit the testing of nuclear weapons. Upon his return to Berkeley, he was part of the team that in 1974 first identified element number 106. Another team at Berkeley corroborated that discovery in 1993. In 1994, to honor Seaborg's many contributions to the discovery of new elements, the American Chemical Society proposed that element

◀ Figure 2.18 **Glenn Seaborg (1912–1999).** The photograph shows Seaborg at Berkeley in 1941 using a Geiger counter to try to detect radiation produced by plutonium. Geiger counters will be discussed in Section 21.5.

number 106 he named "seaborgium," with a proposed symbol of Sg. After several years of controversy about whether an element should be named after a living person, the IUPAC officially adopted the name seaborgium in 1997. Seaborg became the first person to have an element named after him while he was still alive.

Related Exercise: 2.96

GIVE IT SOME THOUGHT

Chlorine is a halogen. Locate this element in the periodic table. **(a)** What is its symbol? **(b)** In what period and in what group is the element located? **(c)** What is its atomic number? **(d)** Is chlorine a metal or nonmetal?

■ **SAMPLE EXERCISE 2.5** | **Using the Periodic Table**

Which two of the following elements would you expect to show the greatest similarity in chemical and physical properties: B, Ca, F, He, Mg, P?

SOLUTION

Elements that are in the same group of the periodic table are most likely to exhibit similar chemical and physical properties. We therefore expect that Ca and Mg should be most alike because they are in the same group (2A, the alkaline earth metals).

■ **PRACTICE EXERCISE**

Locate Na (sodium) and Br (bromine) on the periodic table. Give the atomic number of each, and label each a metal, metalloid, or nonmetal.
Answer: Na, atomic number 11, is a metal; Br, atomic number 35, is a nonmetal.

2.6 MOLECULES AND MOLECULAR COMPOUNDS

Even though the atom is the smallest representative sample of an element, only the noble-gas elements are normally found in nature as isolated atoms. Most matter is composed of molecules or ions, both of which are formed from atoms. We examine molecules here and ions in Section 2.7.

▶ **Figure 2.19 Diatomic molecules.** Seven common elements exist as diatomic molecules at room temperature.

A **molecule** is an assembly of two or more atoms tightly bound together. The resultant "package" of atoms behaves in many ways as a single, distinct object, just as a cell phone composed of many parts can be recognized as a single object. We will discuss the forces that hold the atoms together (the chemical bonds) in Chapters 8 and 9.

Molecules and Chemical Formulas

Many elements are found in nature in molecular form; that is, two or more of the same type of atom are bound together. For example, the oxygen normally found in air consists of molecules that contain two oxygen atoms. We represent this molecular form of oxygen by the **chemical formula** O_2 (read "oh two"). The subscript in the formula tells us that two oxygen atoms are present in each molecule. A molecule that is made up of two atoms is called a **diatomic molecule**. Oxygen also exists in another molecular form known as *ozone*. Molecules of ozone consist of three oxygen atoms, making the chemical formula for this substance O_3. Even though "normal" oxygen (O_2) and ozone (O_3) are both composed only of oxygen atoms, they exhibit very different chemical and physical properties. For example, O_2 is essential for life, but O_3 is toxic; O_2 is odorless, whereas O_3 has a sharp, pungent smell.

The elements that normally occur as diatomic molecules are hydrogen, oxygen, nitrogen, and the halogens. Their locations in the periodic table are shown in Figure 2.19▲. When we speak of the substance hydrogen, we mean H_2 unless we explicitly indicate otherwise. Likewise, when we speak of oxygen, nitrogen, or any of the halogens, we are referring to O_2, N_2, F_2, Cl_2, Br_2, or I_2. Thus, the properties of oxygen and hydrogen listed in Table 1.3 are those of O_2 and H_2. Other, less common forms of these elements behave much differently.

Compounds that are composed of molecules contain more than one type of atom and are called **molecular compounds**. A molecule of water, for example, consists of two hydrogen atoms and one oxygen atom and is therefore represented by the chemical formula H_2O. Lack of a subscript on the O indicates one atom of O per water molecule. Another compound composed of these same elements (in different relative proportions) is hydrogen peroxide, H_2O_2. The properties of hydrogen peroxide are very different from the properties of water.

Several common molecules are shown in Figure 2.20◀. Notice how the composition of each compound is given by its chemical formula. Notice also that these substances are composed only of nonmetallic elements. *Most molecular substances that we will encounter contain only nonmetals.*

Hydrogen, H_2 (a)

Oxygen, O_2 (b)

Water, H_2O (c)

Hydrogen peroxide, H_2O_2 (d)

Carbon monoxide, CO (e)

Carbon dioxide, CO_2 (f)

Methane, CH_4 (g)

Ethylene, C_2H_4 (h)

▲ **Figure 2.20 Molecular models of some simple molecules.** Notice how the chemical formulas of these substances correspond to their compositions.

Molecular and Empirical Formulas

Chemical formulas that indicate the actual numbers and types of atoms in a molecule are called **molecular formulas**. (The formulas in Figure 2.20 are molecular formulas.) Chemical formulas that give only the relative number of atoms of each type in a molecule are called **empirical formulas**. The subscripts in an empirical formula are always the smallest possible whole-number ratios. The molecular formula for hydrogen peroxide is H_2O_2, for example, whereas its empirical formula is HO. The molecular formula for ethylene is C_2H_4, and its empirical formula is CH_2. For many substances, the molecular formula and the empirical formula are identical, as in the case of water, H_2O.

Molecular formulas provide more information about molecules than do empirical formulas. Whenever we know the molecular formula of a compound, we can determine its empirical formula. The converse is not true, however. If we know the empirical formula of a substance, we cannot determine its molecular formula unless we have more information. So why do chemists bother with empirical formulas? As we will see in Chapter 3, certain common methods of analyzing substances lead to the empirical formula only. Once the empirical formula is known, additional experiments can give the information needed to convert the empirical formula to the molecular one. In addition, there are substances, such as the most common forms of elemental carbon, that do not exist as isolated molecules. For these substances, we must rely on empirical formulas. Thus, all the common forms of elemental carbon are represented by the element's chemical symbol, C, which is the empirical formula for all the forms.

■ SAMPLE EXERCISE 2.6 | Relating Empirical and Molecular Formulas

Write the empirical formulas for the following molecules: **(a)** glucose, a substance also known as either blood sugar or dextrose, whose molecular formula is $C_6H_{12}O_6$; **(b)** nitrous oxide, a substance used as an anesthetic and commonly called laughing gas, whose molecular formula is N_2O.

SOLUTION

(a) The subscripts of an empirical formula are the smallest whole-number ratios. The smallest ratios are obtained by dividing each subscript by the largest common factor, in this case 6. The resultant empirical formula for glucose is CH_2O.
(b) Because the subscripts in N_2O are already the lowest integral numbers, the empirical formula for nitrous oxide is the same as its molecular formula, N_2O.

■ PRACTICE EXERCISE

Give the empirical formula for the substance called *diborane*, whose molecular formula is B_2H_6.
Answer: BH_3

Picturing Molecules

The molecular formula of a substance summarizes the composition of the substance but does not show how the atoms come together to form the molecule. The **structural formula** of a substance shows which atoms are attached to which within the molecule. For example, the structural formulas for water, hydrogen peroxide, and methane (CH_4) can be written as follows:

$$H{-}\underset{H}{\overset{}{O}} \qquad H{-}O{-}\underset{}{\overset{H}{O}} \qquad H{-}\underset{\underset{H}{|}}{\overset{\overset{H}{|}}{C}}{-}H$$

Water   Hydrogen peroxide   Methane

The atoms are represented by their chemical symbols, and lines are used to represent the bonds that hold the atoms together.

A structural formula usually does not depict the actual geometry of the molecule, that is, the actual angles at which atoms are joined together. A structural formula can be written as a *perspective drawing*, however, to give some sense of three-dimensional shape, as shown in Figure 2.21▶.

Scientists also rely on various models to help visualize molecules. *Ball-and-stick models* show atoms as spheres and bonds as sticks. This type of model has the advantage of accurately representing the angles at which the atoms are attached to one another within the molecule (Figure 2.21). In a ball-and-stick

$$H{-}\underset{\underset{H}{|}}{\overset{\overset{H}{|}}{C}}{-}H$$

Structural formula

Perspective drawing

Ball-and-stick model

Space-filling model

▲ **Figure 2.21 Different representations of the methane (CH₄) molecule.** Structural formulas, perspective drawings, ball-and-stick models, and space-filling models each help us visualize the ways atoms are attached to each other in molecules. In the perspective drawing, solid lines represent bonds in the plane of the paper, the solid wedge represents a bond that extends out from the plane of the paper, and dashed lines represent bonds behind the paper.

model, balls of the same size may represent all atoms, or the relative sizes of the balls may reflect the relative sizes of the atoms. Sometimes the chemical symbols of the elements are superimposed on the balls, but often the atoms are identified simply by color.

A *space-filling model* depicts what the molecule would look like if the atoms were scaled up in size (Figure 2.21). These models show the relative sizes of the atoms, but the angles between atoms, which help define their molecular geometry, are often more difficult to see than in ball-and-stick models. As in ball-and-stick models, the identities of the atoms are indicated by their colors, but they may also be labeled with the element's symbol.

GIVE IT SOME THOUGHT

The structural formula for the substance ethane is shown here:

$$
\begin{array}{c}
\text{H} \quad \text{H} \\
| \quad \; | \\
\text{H}-\text{C}-\text{C}-\text{H} \\
| \quad \; | \\
\text{H} \quad \text{H}
\end{array}
$$

(a) What is the molecular formula for ethane? **(b)** What is its empirical formula? **(c)** Which kind of molecular model would most clearly show the angles between atoms?

2.7 IONS AND IONIC COMPOUNDS

The nucleus of an atom is unchanged by chemical processes, but some atoms can readily gain or lose electrons. If electrons are removed from or added to a neutral atom, a charged particle called an **ion** is formed. An ion with a positive charge is called a **cation** (pronounced CAT-ion); a negatively charged ion is called an **anion** (AN-ion).

To see how ions form, consider the sodium atom, which has 11 protons and 11 electrons. This atom easily loses one electron. The resulting cation has 11 protons and 10 electrons, which means it has a net charge of $1+$.

The net charge on an ion is represented by a superscript. The superscripts $+$, $2+$, and $3+$, for instance, mean a net charge resulting from the *loss* of one, two, and three electrons, respectively. The superscripts $-$, $2-$, and $3-$ represent net charges resulting from the *gain* of one, two, and three electrons, respectively. Chlorine, with 17 protons and 17 electrons, for example, can gain an electron in chemical reactions, producing the Cl^- ion:

In general, metal atoms tend to lose electrons to form cations, whereas nonmetal atoms tend to gain electrons to form anions.

■ **SAMPLE EXERCISE 2.7** | **Writing Chemical Symbols for Ions**

Give the chemical symbol, including mass number, for each of the following ions:
(a) The ion with 22 protons, 26 neutrons, and 19 electrons; **(b)** the ion of sulfur that has 16 neutrons and 18 electrons.

SOLUTION

(a) The number of protons (22) is the atomic number of the element. By referring to a periodic table or list of elements, we see that the element with atomic number 22 is titanium (Ti). The mass number of this isotope of titanium is $22 + 26 = 48$ (the sum of the protons and neutrons). Because the ion has three more protons than electrons, it has a net charge of 3+. Thus, the symbol for the ion is $^{48}Ti^{3+}$.
(b) By referring to a periodic table or a table of elements, we see that sulfur (S) has an atomic number of 16. Thus, each atom or ion of sulfur must contain 16 protons. We are told that the ion also has 16 neutrons, meaning the mass number of the ion is $16 + 16 = 32$. Because the ion has 16 protons and 18 electrons, its net charge is 2−. Thus, the symbol for the ion is $^{32}S^{2-}$.

In general, we will focus on the net charges of ions and ignore their mass numbers unless the circumstances dictate that we specify a certain isotope.

■ **PRACTICE EXERCISE**

How many protons, neutrons, and electrons does the $^{79}Se^{2-}$ ion possess?
Answer: 34 protons, 45 neutrons, and 36 electrons

In addition to simple ions, such as Na^+ and Cl^-, there are **polyatomic ions**, such as NH_4^+ (ammonium ion) and SO_4^{2-} (sulfate ion). These latter ions consist of atoms joined as in a molecule, but they have a net positive or negative charge. We will consider further examples of polyatomic ions in Section 2.8.

It is important to realize that the chemical properties of ions are very different from the chemical properties of the atoms from which the ions are derived. The difference is like the change from Dr. Jekyll to Mr. Hyde: Although a given atom and its ion may be essentially the same (plus or minus a few electrons), the behavior of the ion is very different from that of the atom.

Predicting Ionic Charges

Many atoms gain or lose electrons to end up with the same number of electrons as the noble gas closest to them in the periodic table. The members of the noble-gas family are chemically very nonreactive and form very few compounds. We might deduce that this is because their electron arrangements are very stable. Nearby elements can obtain these same stable arrangements by losing or gaining electrons. For example, the loss of one electron from an atom of sodium leaves it with the same number of electrons as the neutral neon atom (atomic number 10). Similarly, when chlorine gains an electron, it ends up with 18, the same number of electrons as in argon (atomic number 18). We will use this simple observation to explain the formation of ions until Chapter 8, where we discuss chemical bonding.

■ **SAMPLE EXERCISE 2.8** | **Predicting the Charges of Ions**

Predict the charge expected for the most stable ion of barium and for the most stable ion of oxygen.

SOLUTION

We will assume that these elements form ions that have the same number of electrons as the nearest noble-gas atom. From the periodic table, we see that barium has atomic number 56. The nearest noble gas is xenon, atomic number 54. Barium can attain a stable arrangement of 54 electrons by losing two of its electrons, forming the Ba^{2+} cation.

Oxygen has atomic number 8. The nearest noble gas is neon, atomic number 10. Oxygen can attain this stable electron arrangement by gaining two electrons, thereby forming the O^{2-} anion.

■ **PRACTICE EXERCISE**

Predict the charge expected for the most stable ion of **(a)** aluminum and **(b)** fluorine.
Answer: **(a)** 3+; **(b)** 1−

1A																7A	8A
H^+																H^-	
	2A											3A	4A	5A	6A		N
Li^+														N^{3-}	O^{2-}	F^-	O
Na^+	Mg^{2+}			Transition metals								Al^{3+}			S^{2-}	Cl^-	B L E
K^+	Ca^{2+}														Se^{2-}	Br^-	G A S
Rb^+	Sr^{2+}														Te^{2-}	I^-	E S
Cs^+	Ba^{2+}																

▲ **Figure 2.22 Charges of some common ions.** Notice that the steplike line that divides metals from nonmetals also separates cations from anions. Hydrogen forms both 1+ and 1− ions.

The periodic table is very useful for remembering the charges of ions, especially those of the elements on the left and right sides of the table. As Figure 2.22 ▲ shows, the charges of these ions relate in a simple way to their positions in the table. On the left side of the table, for example, the group 1A elements (the alkali metals) form 1+ ions, and the group 2A elements (the alkaline earths) form 2+ ions. On the other side of the table the group 7A elements (the halogens) form 1− ions, and the group 6A elements form 2− ions. As we will see later in the text, many of the other groups do not lend themselves to such simple rules.

Ionic Compounds

A great deal of chemical activity involves the transfer of electrons from one substance to another. As we just saw, ions form when one or more electrons transfer from one neutral atom to another. Figure 2.23 ▼ shows that when elemental sodium is allowed to react with elemental chlorine, an electron transfers from a neutral sodium atom to a neutral chlorine atom. We are left with a Na^+ ion and a Cl^- ion. Because objects of opposite charge attract, the Na^+ and the Cl^- ions bind together to form the compound sodium chloride (NaCl). Sodium chloride, which we know better as common table salt, is an example of an **ionic compound**, a compound that contains both positively and negatively charged ions.

We can often tell whether a compound is ionic (consisting of ions) or molecular (consisting of molecules) from its composition. In general, cations are metal ions, whereas anions are nonmetal ions. Consequently, *ionic compounds*

▼ **Figure 2.23 The formation of an ionic compound.** (a) The transfer of an electron from a neutral Na atom to a neutral Cl atom leads to the formation of a Na^+ ion and a Cl^- ion. (b) Arrangement of these ions in solid sodium chloride (NaCl). (c) A sample of sodium chloride crystals.

(a)   (b)   (c)

are generally combinations of metals and nonmetals, as in NaCl. In contrast, molecular compounds are generally composed of nonmetals only, as in H₂O.

■ **SAMPLE EXERCISE 2.9** | **Identifying Ionic and Molecular Compounds**

Which of the following compounds would you expect to be ionic: N_2O, Na_2O, $CaCl_2$, SF_4?

SOLUTION

We would predict that Na_2O and $CaCl_2$ are ionic compounds because they are composed of a metal combined with a nonmetal. The other two compounds, composed entirely of nonmetals, are predicted (correctly) to be molecular compounds.

■ **PRACTICE EXERCISE**

Which of the following compounds are molecular: CBr_4, FeS, P_4O_6, PbF_2?
Answer: CBr_4 and P_4O_6

The ions in ionic compounds are arranged in three-dimensional structures. The arrangement of Na^+ and Cl^- ions in NaCl is shown in Figure 2.23. Because there is no discrete molecule of NaCl, we are able to write only an empirical formula for this substance. In fact, only empirical formulas can be written for most ionic compounds.

Chemistry and Life ELEMENTS REQUIRED BY LIVING ORGANISMS

Figure 2.24 ▼ shows the elements that are essential for life. More than 97% of the mass of most organisms comprises just six elements—oxygen, carbon, hydrogen, nitrogen, phosphorus, and sulfur. Water (H_2O) is the most common compound in living organisms, accounting for at least 70% of the mass of most cells. Carbon is the most prevalent element (by mass) in the solid components of cells. Carbon atoms are found in a vast variety of organic molecules, in which the carbon atoms are bonded to other carbon atoms or to atoms of other elements, principally H, O, N, P, and S. All proteins, for example, contain the following group of atoms that occurs repeatedly within the molecules:

$$-\overset{\displaystyle O}{\underset{\displaystyle R}{\overset{\displaystyle \|}{N}-C-}}$$

(R is either an H atom or a combination of atoms such as CH_3.)

In addition, 23 more elements have been found in various living organisms. Five are ions that are required by all organisms: Ca^{2+}, Cl^-, Mg^{2+}, K^+, and Na^+. Calcium ions, for example, are necessary for the formation of bone and for the transmission of signals in the nervous system, such as those that trigger the contraction of cardiac muscles, causing the heart to beat. Many other elements are needed in only very small quantities and consequently are called *trace* elements. For example, trace quantities of copper are required in the diet of humans to aid in the synthesis of hemoglobin.

▼ **Figure 2.24 Biologically essential elements.** The elements that are essential for life are indicated by colors. Red denotes the six most abundant elements in living systems (hydrogen, carbon, nitrogen, oxygen, phosphorus, and sulfur). Blue indicates the five next most abundant elements. Green indicates the elements needed in only trace amounts.

1A																	8A
H																	He
Li	Be											B	C	N	O	F	Ne
Na	Mg	3B	4B	5B	6B	7B	8	9	10	1B	2B	Al	Si	P	S	Cl	Ar
K	Ca	Sc	Ti	V	Cr	Mn	Fe	Co	Ni	Cu	Zn	Ga	Ge	As	Se	Br	Kr
Rb	Sr	Y	Zr	Nb	Mo	Tc	Ru	Rh	Pd	Ag	Cd	In	Sn	Sb	Te	I	Xe
Cs	Ba	La	Hf	Ta	W	Re	Os	Ir	Pt	Au	Hg	Tl	Pb	Bi	Po	At	Rn

Someone once said that drinking at the fountain of knowledge in a chemistry course is like drinking from a fire hydrant. Indeed, the pace can sometimes seem brisk. More to the point, however, we can drown in the facts if we do not see the general patterns. The value of recognizing patterns and learning rules and generalizations is that the patterns, rules, and generalizations free us from learning (or trying to memorize) many individual facts. The patterns, rules, and generalizations tie ideas together so that we do not get lost in the details.

Many students struggle with chemistry because they do not see how the topics relate to one another, how ideas connect together. They therefore treat every idea and problem as being unique instead of as an example or application of a general rule, procedure, or relationship. You can avoid this pitfall by remembering the following: Begin to notice the structure of the topic you are studying. Pay attention to the trends and rules given to summarize a large body of information. Notice, for example, how atomic structure helps us understand the existence of isotopes (as seen in Table 2.2) and how the periodic table aids us in remembering the charges of ions (as seen in Figure 2.22). You may surprise yourself by observing patterns that are not even explicitly spelled out yet. Perhaps you have even noticed certain trends in chemical formulas. Moving across the periodic table from element 11 (Na), we find that the elements form compounds with F having the following compositions: NaF, MgF_2, and AlF_3. Does this trend continue? Do SiF_4, PF_5, and SF_6 exist? Indeed they do. If you have noticed trends like this from the scraps of information you have seen so far, then you are ahead of the game and you have already prepared yourself for some topics we will address in later chapters.

We can readily write the empirical formula for an ionic compound if we know the charges of the ions of which the compound is composed. Chemical compounds are always electrically neutral. Consequently, the ions in an ionic compound always occur in such a ratio that the total positive charge equals the total negative charge. Thus, there is one Na^+ to one Cl^- (giving $NaCl$), one Ba^{2+} to two Cl^- (giving $BaCl_2$), and so forth.

As you consider these and other examples, you will see that if the charges on the cation and anion are equal, the subscript on each ion will be 1. If the charges are not equal, the charge on one ion (without its sign) will become the subscript on the other ion. For example, the ionic compound formed from Mg (which forms Mg^{2+} ions) and N (which forms N^{3-} ions) is Mg_3N_2:

$$Mg^{②+} \bowtie N^{③-} \longrightarrow Mg_3N_2$$

GIVE IT SOME THOUGHT

Why don't we write the formula for the compound formed by Ca^{2+} and O^{2-} as Ca_2O_2?

■ **SAMPLE EXERCISE 2.10** | **Using Ionic Charge to Write Empirical Formulas for Ionic Compounds**

What are the empirical formulas of the compounds formed by **(a)** Al^{3+} and Cl^- ions, **(b)** Al^{3+} and O^{2-} ions, **(c)** Mg^{2+} and NO_3^- ions?

SOLUTION

(a) Three Cl^- ions are required to balance the charge of one Al^{3+} ion. Thus, the formula is $AlCl_3$.
(b) Two Al^{3+} ions are required to balance the charge of three O^{2-} ions (that is, the total positive charge is $6+$, and the total negative charge is $6-$). Thus, the formula is Al_2O_3.
(c) Two NO_3^- ions are needed to balance the charge of one Mg^{2+}. Thus, the formula is $Mg(NO_3)_2$. In this case the formula for the entire polyatomic ion NO_3^- must be enclosed in parentheses so that it is clear that the subscript 2 applies to all the atoms of that ion.

■ **PRACTICE EXERCISE**

Write the empirical formulas for the compounds formed by the following ions:
(a) Na^+ and PO_4^{3-}, **(b)** Zn^{2+} and SO_4^{2-}, **(c)** Fe^{3+} and CO_3^{2-}.
Answers: **(a)** Na_3PO_4, **(b)** $ZnSO_4$, **(c)** $Fe_2(CO_3)_3$

2.8   NAMING INORGANIC COMPOUNDS

To obtain information about a particular substance, you must know its chemical formula and name. The names and formulas of compounds are essential vocabulary in chemistry. The system used in naming substances is called **chemical nomenclature** from the Latin words *nomen* (name) and *calare* (to call).

There are now more than 19 million known chemical substances. Naming them all would be a hopelessly complicated task if each had a special name independent of all others. Many important substances that have been known for a long time, such as water (H_2O) and ammonia (NH_3), do have individual, traditional names (so-called "common" names). For most substances, however, we rely on a systematic set of rules that leads to an informative and unique name for each substance, a name based on the composition of the substance.

The rules for chemical nomenclature are based on the division of substances into categories. The major division is between organic and inorganic compounds. *Organic compounds* contain carbon, usually in combination with hydrogen, oxygen, nitrogen, or sulfur. All others are *inorganic compounds*. Early chemists associated organic compounds with plants and animals, and they associated inorganic compounds with the nonliving portion of our world. Although this distinction between living and nonliving matter is no longer pertinent, the classification between organic and inorganic compounds continues to be useful. In this section we consider the basic rules for naming inorganic compounds, and in Section 2.9 we will introduce the names of some simple organic compounds. Among inorganic compounds, we will consider three categories: ionic compounds, molecular compounds, and acids.

Names and Formulas of Ionic Compounds

Recall from Section 2.7 that ionic compounds usually consist of metal ions combined with nonmetal ions. The metals form the positive ions, and the nonmetals form the negative ions. Let's examine the naming of positive ions, then the naming of negative ones. After that, we will consider how to put the names of the ions together to identify the complete ionic compound.

1. Positive Ions (Cations)

   (a) *Cations formed from metal atoms have the same name as the metal:*

   Na^+ sodium ion Zn^{2+} zinc ion Al^{3+} aluminum ion

   Ions formed from a single atom are called *monatomic ions*.

   (b) *If a metal can form different cations, the positive charge is indicated by a Roman numeral in parentheses following the name of the metal:*

   Fe^{2+} iron(II) ion Cu^+ copper(I) ion
   Fe^{3+} iron(III) ion Cu^{2+} copper(II) ion

   Ions of the same element that have different charges exhibit different properties, such as different colors (Figure 2.25▶).

   Most of the metals that can form more than one cation are *transition metals*, elements that occur in the middle block of elements, from group 3B to group 2B in the periodic table. The charges of these ions are indicated by Roman numerals. The metals that form only one cation are those of group 1A (Na^+, K^+, and Rb^+) and group 2A (Mg^{2+}, Ca^{2+}, Sr^{2+}, and Ba^{2+}), as well as Al^{3+} (group 3A) and two transition-metal ions: Ag^+ (group 1B) and Zn^{2+} (group 2B). Charges are not expressed explicitly when naming these ions. However, if there is any doubt in your mind whether a metal forms more than one cation, use a Roman numeral to indicate the charge. It is never wrong to do so, even though it may be unnecessary.

   An older method still widely used for distinguishing between two differently charged ions of a metal is to apply the ending *-ous* or *-ic*.

▲ **Figure 2.25   Ions of the same element with different charges exhibit different properties.** Compounds containing ions of the same element but with different charge can be very different in appearance and properties. Both substances shown are complex compounds of iron that also contain K^+ and CN^- ions. The substance on the left is potassium ferrocyanide, which contains Fe(II) bound to CN^- ions. The substance on the right is potassium ferricyanide, which contains Fe(III) bound to CN^- ions. Both substances are used extensively in blueprinting and other dyeing processes.

These endings represent the lower and higher charged ions, respectively. They are added to the root of the element's Latin name:

Fe^{2+}	ferrous ion	Cu^+	cuprous ion
Fe^{3+}	ferric ion	Cu^{2+}	cupric ion

Although we will only rarely use these older names in this text, you might encounter them elsewhere.

(c) *Cations formed from nonmetal atoms have names that end in* -ium:

$$NH_4^+ \quad \text{ammonium ion} \qquad H_3O^+ \quad \text{hydronium ion}$$

These two ions are the only ions of this kind that we will encounter frequently in the text. They are both polyatomic. The vast majority of cations are monatomic metal ions.

The names and formulas of some common cations are shown in Table 2.4▼; they are also included in a table of common ions in the back inside cover of the text. The ions listed on the left side in Table 2.4 are the monatomic ions that do not have variable charges. Those listed on the right side are either polyatomic cations or cations with variable charges. The Hg_2^{2+} ion is unusual because this metal ion is not monatomic. It is called the mercury(I) ion because it can be thought of as two Hg^+ ions bound together. The cations that you will encounter most frequently are shown in boldface. You should learn these cations first.

▲ **GIVE IT SOME THOUGHT**

Why is CrO named using a Roman numeral, chromium(II) oxide, whereas CaO is named without a Roman numeral in the name, calcium oxide?

2. Negative Ions (Anions)

(a) *The names of monatomic anions are formed by replacing the ending of the name of the element with* -ide:

$$H^- \quad \text{hydride ion} \qquad O^{2-} \quad \text{oxide ion} \qquad N^{3-} \quad \text{nitride ion}$$

TABLE 2.4 ■ Common Cations*

Charge	Formula	Name	Formula	Name
1+	**H^+**	**Hydrogen ion**	NH_4^+	**Ammonium ion**
	Li^+	Lithium ion	Cu^+	Copper(I) or cuprous ion
	Na^+	**Sodium ion**		
	K^+	**Potassium ion**		
	Cs^+	Cesium ion		
	Ag^+	**Silver ion**		
2+	**Mg^{2+}**	**Magnesium ion**	Co^{2+}	Cobalt(II) or cobaltous ion
	Ca^{2+}	**Calcium ion**	**Cu^{2+}**	**Copper(II)** or cupric ion
	Sr^{2+}	Strontium ion	**Fe^{2+}**	**Iron(II)** or ferrous ion
	Ba^{2+}	Barium ion	Mn^{2+}	Manganese(II) or manganous ion
	Zn^{2+}	**Zinc ion**	Hg_2^{2+}	Mercury(I) or mercurous ion
	Cd^{2+}	Cadmium ion	**Hg^{2+}**	**Mercury(II)** or mercuric ion
			Ni^{2+}	Nickel(II) or nickelous ion
			Pb^{2+}	**Lead(II)** or plumbous ion
			Sn^{2+}	Tin(II) or stannous ion
3+	**Al^{3+}**	**Aluminum ion**	Cr^{3+}	Chromium(III) or chromic ion
			Fe^{3+}	**Iron(III)** or ferric ion

*The most common ions are in boldface.

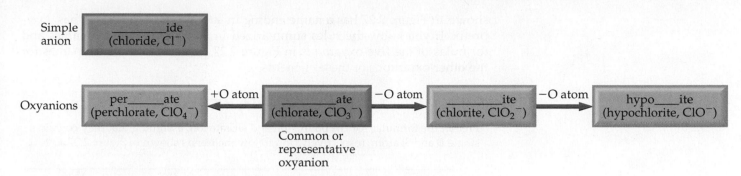

▲ **Figure 2.26 Summary of the procedure for naming anions.** The root of the name (such as "chlor" for chlorine) goes in the blank.

A few simple polyatomic anions also have names ending in *-ide*:

OH^- hydroxide ion CN^- cyanide ion O_2^{2-} peroxide ion

(b) *Polyatomic anions containing oxygen have names ending in -ate or -ite.* These anions are called **oxyanions**. The ending *-ate* is used for the most common oxyanion of an element. The ending *-ite* is used for an oxyanion that has the same charge but one O atom fewer:

NO_3^- nitrate ion SO_4^{2-} sulfate ion
NO_2^- nitrite ion SO_3^2 sulfite ion

Prefixes are used when the series of oxyanions of an element extends to four members, as with the halogens. The prefix *per-* indicates one more O atom than the oxyanion ending in *-ate*; the prefix *hypo-* indicates one O atom fewer than the oxyanion ending in *-ite*:

ClO_4^- perchlorate ion (one more O atom than chlorate)
ClO_3^-   **chlorate ion**
ClO_2^- chlorite ion (one O atom fewer than chlorate)
ClO^- hypochlorite ion (one O atom fewer than chlorite)

These rules are summarized in Figure 2.26 ▲.

GIVE IT SOME THOUGHT

What information is conveyed by the endings *-ide*, *-ate*, and *-ite* in the name of an anion?

Students often have a hard time remembering the number of oxygen atoms in the various oxyanions and the charges of these ions. Figure 2.27▼ lists the oxyanions of C, N, P, S, and Cl that contain the maximum number of O atoms. The periodic pattern seen in these formulas can help you remember them. Notice that C and N, which are in the second period of the periodic table, have only three O atoms each, whereas P, S, and Cl, which are in the third period, have four O atoms each. If we begin at the lower right side of the figure, with Cl, we see that the charges increase from right to left, from 1− for Cl (ClO_4^-) to 3− for P (PO_4^{3-}). In the second period the charges also increase from right to left, from 1− for N (NO_3^-) to 2− for C (CO_3^{2-}). Each anion

	4A	5A	6A	7A
2	CO_3^{2-} **Carbonate ion**	NO_3^- **Nitrate ion**		
3		PO_4^{3-} **Phosphate ion**	SO_4^{2-} **Sulfate ion**	ClO_4^- **Perchlorate ion**

◀ **Figure 2.27 Common oxyanions.** The composition and charges of common oxyanions are related to their location in the periodic table.

shown in Figure 2.27 has a name ending in -*ate*. The ClO_4^- ion also has a *per-* prefix. If you know the rules summarized in Figure 2.26 and the names and formulas of the five oxyanions in Figure 2.27, you can deduce the names for the other oxyanions of these elements.

▲ **GIVE IT SOME THOUGHT**

Predict the formulas for the borate ion and silicate ion, assuming that they contain a single B and Si atom, respectively, and follow the trends shown in Figure 2.27▲.

■■ **SAMPLE EXERCISE 2.11** | **Determining the Formula of an Oxyanion from Its Name**

Based on the formula for the sulfate ion, predict the formula for **(a)** the selenate ion and **(b)** the selenite ion. (Sulfur and selenium are both members of group 6A and form analogous oxyanions.)

SOLUTION

(a) The sulfate ion is SO_4^{2-} The analogous selenate ion is therefore SeO_4^{2-}.
(b) The ending -*ite* indicates an oxyanion with the same charge but one O atom fewer than the corresponding oxyanion that ends in -*ate*. Thus, the formula for the selenite ion is SeO_3^{2-}.

■■ **PRACTICE EXERCISE**

The formula for the bromate ion is analogous to that for the chlorate ion. Write the formula for the hypobromite and perbromate ions.
Answer: BrO^- and BrO_4^-

(c) *Anions derived by adding H^+ to an oxyanion are named by adding as a prefix the word* hydrogen *or* dihydrogen, *as appropriate:*

CO_3^{2-}	carbonate ion	PO_4^{3-}	phosphate ion
HCO_3^-	hydrogen carbonate ion	$H_2PO_4^-$	dihydrogen phosphate ion

Notice that each H^+ reduces the negative charge of the parent anion by one. An older method for naming some of these ions is to use the prefix *bi-*. Thus, the HCO_3^- ion is commonly called the bicarbonate ion, and HSO_4^- is sometimes called the bisulfate ion.

The names and formulas of the common anions are listed in Table 2.5▼ and on the back inside cover of the text. Those anions whose names end in -*ide* are listed on the left portion of Table 2.5, and those

TABLE 2.5 ■ Common Anions*

Charge	Formula	Name	Formula	Name
1−	H^-	Hydride ion	CH_3COO^- (or $C_2H_3O_2^-$)	**Acetate ion**
	F^-	**Fluoride ion**	ClO_3^-	Chlorate ion
	Cl^-	**Chloride ion**	ClO_4^-	**Perchlorate ion**
	Br^-	**Bromide ion**	NO_3^-	**Nitrate ion**
	I^-	**Iodide ion**	MnO_4^-	Permanganate ion
	CN^-	Cyanide ion		
	OH^-	**Hydroxide ion**		
2−	O^{2-}	**Oxide ion**	CO_3^{2-}	**Carbonate ion**
	O_2^{2-}	Peroxide ion	CrO_4^{2-}	Chromate ion
	S^{2-}	**Sulfide ion**	$Cr_2O_7^{2-}$	Dichromate ion
			SO_4^{2-}	**Sulfate ion**
3−	N^{3-}	Nitride ion	PO_4^{3-}	**Phosphate ion**

* The most common ions are in boldface.

whose names end in -*ate* are listed on the right. The most common of these ions are shown in boldface. You should learn names and formulas of these anions first. The formulas of the ions whose names end with -*ite* can be derived from those ending in -*ate* by removing an O atom. Notice the location of the monatomic ions in the periodic table. Those of group 7A always have a 1− charge (F^-, Cl^-, Br^-, and I^-), and those of group 6A have a 2− charge (O^{2-} and S^{2-}).

3. Ionic Compounds

Names of ionic compounds consist of the cation name followed by the anion name:

$CaCl_2$	calcium chloride
$Al(NO_3)_3$	aluminum nitrate
$Cu(ClO_4)_2$	copper(II) perchlorate (or cupric perchlorate)

In the chemical formulas for aluminum nitrate and copper(II) perchlorate, parentheses followed by the appropriate subscript are used because the compounds contain two or more polyatomic ions.

■ **SAMPLE EXERCISE 2.12** | Determining the Names of Ionic Compounds from Their Formulas

Name the following compounds: **(a)** K_2SO_4, **(b)** $Ba(OH)_2$, **(c)** $FeCl_3$.

SOLUTION

Each compound is ionic and is named using the guidelines we have already discussed. In naming ionic compounds, it is important to recognize polyatomic ions and to determine the charge of cations with variable charge.

(a) The cation in this compound is K^+, and the anion is SO_4^{2-}. (If you thought the compound contained S^{2-} and O^{2-} ions, you failed to recognize the polyatomic sulfate ion.) Putting together the names of the ions, we have the name of the compound, potassium sulfate.
(b) In this case the compound is composed of Ba^{2+} and OH^- ions. Ba^{2+} is the barium ion and OH^- is the hydroxide ion. Thus, the compound is called barium hydroxide.
(c) You must determine the charge of Fe in this compound because an iron atom can form more than one cation. Because the compound contains three Cl^- ions, the cation must be Fe^{3+} which is the iron(III), or ferric, ion. The Cl^- ion is the chloride ion. Thus, the compound is iron(III) chloride or ferric chloride.

■ **PRACTICE EXERCISE**

Name the following compounds: **(a)** NH_4Br, **(b)** Cr_2O_3, **(c)** $Co(NO_3)_2$.
Answers: **(a)** ammonium bromide, **(b)** chromium(III) oxide, **(c)** cobalt(II) nitrate

---

■ **SAMPLE EXERCISE 2.13** | Determining the Formulas of Ionic Compounds from Their Names

Write the chemical formulas for the following compounds: **(a)** potassium sulfide, **(b)** calcium hydrogen carbonate, **(c)** nickel(II) perchlorate.

SOLUTION

In going from the name of an ionic compound to its chemical formula, you must know the charges of the ions to determine the subscripts.

(a) The potassium ion is K^+, and the sulfide ion is S^{2-}. Because ionic compounds are electrically neutral, two K^+ ions are required to balance the charge of one S^{2-} ion, giving the empirical formula of the compound, K_2S.
(b) The calcium ion is Ca^{2+}. The carbonate ion is CO_3^{2-}, so the hydrogen carbonate ion is HCO_3^-. Two HCO_3^- ions are needed to balance the positive charge of Ca^{2+}, giving $Ca(HCO_3)_2$.
(c) The nickel(II) ion is Ni^{2+}. The perchlorate ion is ClO_4^-. Two ClO_4^- ions are required to balance the charge on one Ni^{2+} ion, giving $Ni(ClO_4)_2$.

■ **PRACTICE EXERCISE**

Give the chemical formula for **(a)** magnesium sulfate, **(b)** silver sulfide, **(c)** lead(II) nitrate.
Answers: **(a)** $MgSO_4$, **(b)** Ag_2S, **(c)** $Pb(NO_3)_2$

▶ Figure 2.28 **Relating names of anions and acids.** Summary of the way in which anion names and acid names are related. The prefixes *per-* and *hypo-* are retained in going from the anion to the acid.

Names and Formulas of Acids

Acids are an important class of hydrogen-containing compounds, and they are named in a special way. For our present purposes, an *acid* is a substance whose molecules yield hydrogen ions (H^+) when dissolved in water. When we encounter the chemical formula for an acid at this stage of the course, it will be written with H as the first element, as in HCl and H_2SO_4.

An acid is composed of an anion connected to enough H^+ ions to neutralize, or balance, the anion's charge. Thus, the SO_4^{2-} ion requires two H^+ ions, forming H_2SO_4. The name of an acid is related to the name of its anion, as summarized in Figure 2.28 ▲.

1. *Acids containing anions whose names end in* -ide *are named by changing the* -ide *ending to* -ic, *adding the prefix* hydro- *to this anion name, and then following with the word* acid, *as in the following examples:*

Anion	Corresponding Acid
Cl^- (chloride)	HCl (hydrochloric acid)
S^{2-} (sulfide)	H_2S (hydrosulfuric acid)

2. *Acids containing anions whose names end in* -ate *or* -ite *are named by changing* -ate *to* -ic *and* -ite *to* -ous, *and then adding the word* acid. *Prefixes in the anion name are retained in the name of the acid. These rules are illustrated by the oxyacids of chlorine:*

Anion	Corresponding Acid
ClO_4^- (perchloric)	$HClO_4$ (perchloric acid)
ClO_3^- (chlorate)	$HClO_3$ (chloric acid)
ClO_2^- (chlorite)	$HClO_2$ (chlorous acid)
ClO^- (hypochlorite)	HClO (hypochlorous acid)

■ **SAMPLE EXERCISE 2.14** | **Relating the Names and Formulas of Acids**

Name the following acids: **(a)** HCN, **(b)** HNO_3, **(c)** H_2SO_4, **(d)** H_2SO_3.

SOLUTION

(a) The anion from which this acid is derived is CN^-, the cyanide ion. Because this ion has an *-ide* ending, the acid is given a *hydro-* prefix and an *-ic* ending: hydrocyanic acid. Only water solutions of HCN are referred to as hydrocyanic acid: The pure compound, which is a gas under normal conditions, is called hydrogen cyanide. Both hydrocyanic acid and hydrogen cyanide are *extremely* toxic.

(b) Because NO_3^- is the nitrate ion, HNO_3 is called nitric acid (the *-ate* ending of the anion is replaced with an *-ic* ending in naming the acid).
(c) Because SO_4^{2-} is the sulfate ion, H_2SO_4 is called sulfuric acid.
(d) Because SO_3^{2-} is the sulfite ion, H_2SO_3 is sulfurous acid (the *-ite* ending of the anion is replaced with an *-ous* ending).

■ **PRACTICE EXERCISE**
Give the chemical formulas for **(a)** hydrobromic acid, **(b)** carbonic acid.
Answers: **(a)** HBr, **(b)** H_2CO_3

Names and Formulas of Binary Molecular Compounds

The procedures used for naming *binary* (two-element) molecular compounds are similar to those used for naming ionic compounds:

1. *The name of the element farther to the left in the periodic table is usually written first.* An exception to this rule occurs in the case of compounds that contain oxygen. Oxygen is always written last except when combined with fluorine.

2. *If both elements are in the same group in the periodic table, the one having the higher atomic number is named first.*

3. *The name of the second element is given an -ide ending.*

4. *Greek prefixes (Table 2.6▶) are used to indicate the number of atoms of each element.* The prefix *mono-* is never used with the first element. When the prefix ends in *a* or *o* and the name of the second element begins with a vowel (such as *oxide*), the *a* or *o* of the prefix is often dropped.

The following examples illustrate these rules:

Cl_2O	dichlorine monoxide	NF_3	nitrogen trifluoride
N_2O_4	dinitrogen tetroxide	P_4S_{10}	tetraphosphorus decasulfide

It is important to realize that you cannot predict the formulas of most molecular substances in the same way that you predict the formulas of ionic compounds. For this reason, we name molecular compounds using prefixes that explicitly indicate their composition. Molecular compounds that contain hydrogen and one other element are an important exception, however. These compounds can be treated as if they were neutral substances containing H^+ ions and anions. Thus, you can predict that the substance named hydrogen chloride has the formula HCl, containing one H^+ to balance the charge of one Cl^-. (The name hydrogen chloride is used only for the pure compound; water solutions of HCl are called hydrochloric acid.) Similarly, the formula for hydrogen sulfide is H_2S because two H^+ are needed to balance the charge on S^{2-}.

TABLE 2.6 ■ **Prefixes Used in Naming Binary Compounds Formed between Nonmetals**

Prefix	Meaning
Mono-	1
Di-	2
Tri-	3
Tetra-	4
Penta-	5
Hexa-	6
Hepta-	7
Octa-	8
Nona-	9
Deca-	10

■ **SAMPLE EXERCISE 2.15** | **Relating the Names and Formulas of Binary Molecular Compounds**

Name the following compounds: **(a)** SO_2, **(b)** PCl_5, **(c)** N_2O_3.

SOLUTION

The compounds consist entirely of nonmetals, so they are molecular rather than ionic. Using the prefixes in Table 2.6, we have **(a)** sulfur dioxide, **(b)** phosphorus pentachloride, and **(c)** dinitrogen trioxide.

■ **PRACTICE EXERCISE**
Give the chemical formula for **(a)** silicon tetrabromide, **(b)** disulfur dichloride.
Answers: **(a)** $SiBr_4$, **(b)** S_2Cl_2

2.9 SOME SIMPLE ORGANIC COMPOUNDS

The study of compounds of carbon is called **organic chemistry**, and as noted earlier in the chapter, compounds that contain carbon and hydrogen, often in combination with oxygen, nitrogen, or other elements, are called *organic compounds*. We will examine organic compounds and organic chemistry in some detail in Chapter 25. You will see a number of organic compounds throughout this text; many of them have practical applications or are relevant to the chemistry of biological systems. Here we present a very brief introduction to some of the simplest organic compounds to provide you with a sense of what these molecules look like and how they are named.

Alkanes

Compounds that contain only carbon and hydrogen are called **hydrocarbons**. In the most basic class of hydrocarbons, each carbon atom is bonded to four other atoms. These compounds are called **alkanes**. The three simplest alkanes, which contain one, two, and three carbon atoms, respectively, are methane (CH_4), ethane (C_2H_6), and propane (C_3H_8). The structural formulas of these three alkanes are as follows:

We can make longer alkanes by adding additional carbon atoms to the "skeleton" of the molecule.

Although the hydrocarbons are binary molecular compounds, they are not named like the binary inorganic compounds discussed in Section 2.8. Instead, each alkane has a name that ends in *-ane*. The alkane with four carbon atoms is called butane. For alkanes with five or more carbon atoms, the names are derived from prefixes such as those in Table 2.6. An alkane with eight carbon atoms, for example, is called *octane* (C_8H_{18}), where the *octa-* prefix for eight is combined with the *-ane* ending for an alkane. Gasoline consists primarily of octanes, as will be discussed in Chapter 25.

Some Derivatives of Alkanes

Other classes of organic compounds are obtained when hydrogen atoms of alkanes are replaced with *functional groups*, which are specific groups of atoms. An **alcohol**, for example, is obtained by replacing an H atom of an alkane with an —OH group. The name of the alcohol is derived from that of the alkane by adding an *-ol* ending:

Alcohols have properties that are very different from the properties of the alkanes from which the alcohols are obtained. For example, methane, ethane, and propane are all colorless gases under normal conditions, whereas methanol, ethanol, and propanol are colorless liquids. We will discuss the reasons for these differences in properties in Chapter 11.

The prefix "1" in the name 1-propanol indicates that the replacement of H with OH has occurred at one of the "outer" carbon atoms rather than the "middle" carbon atom. A different compound called 2-propanol (also known as isopropyl alcohol) is obtained if the OH functional group is attached to the middle carbon atom:

$$
\begin{array}{ccc}
 & \text{H H H} & \\
 & | \ \ | \ \ | & \\
\text{H}-\text{C}-\text{C}-\text{C}-\text{OH} \\
 & | \ \ | \ \ | & \\
 & \text{H H H} &
\end{array}
\qquad
\begin{array}{ccc}
 & \text{H H H} & \\
 & | \ \ | \ \ | & \\
\text{H}-\text{C}-\text{C}-\text{C}-\text{H} \\
 & | \ \ | \ \ | & \\
 & \text{H O H} & \\
 & | & \\
 & \text{H} &
\end{array}
$$

1-Propanol   2-Propanol

Ball-and-stick models of these two molecules are presented in Figure 2.29▶.

Much of the richness of organic chemistry is possible because organic compounds can form long chains of carbon–carbon bonds. The series of alkanes that begins with methane, ethane, and propane and the series of alcohols that begins with methanol, ethanol, and propanol can both be extended for as long as we desire, in principle. The properties of alkanes and alcohols change as the chains get longer. Octanes, which are alkanes with eight carbon atoms, are liquids under normal conditions. If the alkane series is extended to tens of thousands of carbon atoms, we obtain *polyethylene*, a solid substance that is used to make thousands of plastic products, such as plastic bags, food containers, and laboratory equipment.

▲ **Figure 2.29 The two forms of propanol (C_3H_7OH).** (a) 1-Propanol, in which the OH group is attached to one of the end carbon atoms, and (b) 2-propanol, in which the OH group is attached to the middle carbon atom.

■ **SAMPLE EXERCISE 2.16** | **Writing Structural and Molecular Formulas for Hydrocarbons**

Consider the alkane called *pentane*. **(a)** Assuming that the carbon atoms are in a straight line, write a structural formula for pentane. **(b)** What is the molecular formula for pentane?

SOLUTION

(a) Alkanes contain only carbon and hydrogen, and each carbon atom is attached to four other atoms. Because the name pentane contains the prefix *penta-* for five (Table 2.6), we can assume that pentane contains five carbon atoms bonded in a chain. If we then add enough hydrogen atoms to make four bonds to each carbon atom, we obtain the following structural formula:

$$
\begin{array}{ccccc}
\text{H} & \text{H} & \text{H} & \text{H} & \text{H} \\
| & | & | & | & | \\
\text{H}-\text{C}-\text{C}-\text{C}-\text{C}-\text{C}-\text{H} \\
| & | & | & | & | \\
\text{H} & \text{H} & \text{H} & \text{H} & \text{H}
\end{array}
$$

This form of pentane is often called *n*-pentane, where the *n*- stands for "normal" because all five carbon atoms are in one line in the structural formula.
(b) Once the structural formula is written, we can determine the molecular formula by counting the atoms present. Thus, *n*-pentane has the formula C_5H_{12}.

■ **PRACTICE EXERCISE**

Butane is the alkane with four carbon atoms. **(a)** What is the molecular formula of butane? **(b)** What are the name and molecular formula of an alcohol derived from butane?
Answers: **(a)** C_4H_{10}, **(b)** butanol, $C_4H_{10}O$ or C_4H_9OH

CHAPTER REVIEW

SUMMARY AND KEY TERMS

Sections 2.1 and 2.2 **Atoms** are the basic building blocks of matter. They are the smallest units of an element that can combine with other elements. Atoms are composed of even smaller particles, called **subatomic particles**. Some of these subatomic particles are charged and follow the usual behavior of charged particles: Particles with the same charge repel one another, whereas particles with unlike charges are attracted to one another. We considered some of the important experiments that led to the discovery and characterization of subatomic particles. Thomson's experiments on the behavior of **cathode rays** in magnetic and electric fields led to the discovery of the electron and allowed its charge-to-mass ratio to be measured. Millikan's oil-drop experiment determined the charge of the electron. Becquerel's discovery of **radioactivity**, the spontaneous emission of radiation by atoms, gave further evidence that the atom has a substructure. Rutherford's studies of how thin metal foils scatter α particles showed that the atom has a dense, positively charged **nucleus**.

Section 2.3 Atoms have a nucleus that contains **protons** and **neutrons**; **electrons** move in the space around the nucleus. The magnitude of the charge of the electron, 1.602×10^{-19} C, is called the **electronic charge**. The charges of particles are usually represented as multiples of this charge—an electron has a $1-$ charge, and a proton has a $1+$ charge. The masses of atoms are usually expressed in terms of **atomic mass units** (1 amu $= 1.66054 \times 10^{-24}$ g). The dimensions of atoms are often expressed in units of **angstroms** (1 Å $= 10^{-10}$ m).

Elements can be classified by **atomic number**, the number of protons in the nucleus of an atom. All atoms of a given element have the same atomic number. The **mass number** of an atom is the sum of the numbers of protons and neutrons. Atoms of the same element that differ in mass number are known as **isotopes**.

Section 2.4 The atomic mass scale is defined by assigning a mass of exactly 12 amu to a ^{12}C atom. The **atomic weight** (average atomic mass) of an element can be calculated from the relative abundances and masses of that element's isotopes. The **mass spectrometer** provides the most direct and accurate means of experimentally measuring atomic (and molecular) weights.

Section 2.5 The **periodic table** is an arrangement of the elements in order of increasing atomic number. Elements with similar properties are placed in vertical columns. The elements in a column are known as a periodic **group**. The elements in a horizontal row are known as a **period**. The **metallic elements (metals)**, which comprise the majority of the elements, dominate the left side and the middle of the table; the **nonmetallic elements (nonmetals)**

are located on the upper right side. Many of the elements that lie along the line that separates metals from nonmetals are **metalloids**.

Section 2.6 Atoms can combine to form **molecules**. Compounds composed of molecules (**molecular compounds**) usually contain only nonmetallic elements. A molecule that contains two atoms is called a **diatomic molecule**. The composition of a substance is given by its **chemical formula**. A molecular substance can be represented by its **empirical formula**, which gives the relative numbers of atoms of each kind. It is usually represented by its **molecular formula**, however, which gives the actual numbers of each type of atom in a molecule. **Structural formulas** show the order in which the atoms in a molecule are connected. Ball-and-stick models and space-filling models are often used to represent molecules.

Section 2.7 Atoms can either gain or lose electrons, forming charged particles called **ions**. Metals tend to lose electrons, becoming positively charged ions (**cations**). Nonmetals tend to gain electrons, forming negatively charged ions (**anions**). Because **ionic compounds** are electrically neutral, containing both cations and anions, they usually contain both metallic and nonmetallic elements. Atoms that are joined together, as in a molecule, but carry a net charge are called **polyatomic ions**. The chemical formulas used for ionic compounds are empirical formulas, which can be written readily if the charges of the ions are known. The total positive charge of the cations in an ionic compound equals the total negative charge of the anions.

Section 2.8 The set of rules for naming chemical compounds is called **chemical nomenclature**. We studied the systematic rules used for naming three classes of inorganic substances: ionic compounds, acids, and binary molecular compounds. In naming an ionic compound, the cation is named first and then the anion. Cations formed from metal atoms have the same name as the metal. If the metal can form cations of differing charges, the charge is given using Roman numerals. Monatomic anions have names ending in *-ide*. Polyatomic anions containing oxygen and another element (**oxyanions**) have names ending in *-ate* or *-ite*.

Section 2.9 **Organic chemistry** is the study of compounds that contain carbon. The simplest class of organic molecules is the **hydrocarbons**, which contain only carbon and hydrogen. Hydrocarbons in which each carbon atom is attached to four other atoms are called **alkanes**. Alkanes have names that end in *-ane*, such as methane and ethane. Other organic compounds are formed when an H atom of a hydrocarbon is replaced with a functional group. An **alcohol**, for example, is a compound in which an H atom of a hydrocarbon is replaced by an OH func- tional group. Alcohols have names that end in -*ol*, such as methanol and ethanol.

KEY SKILLS

- Describe the basic postulates of Dalton's atomic theory.
- Describe the key experiments that led to the discovery of electrons and to the nuclear model of the atom.
- Describe the structure of the atom in terms of protons, neutrons, and electrons.
- Describe the electric charge and relative masses of protons, neutrons, and electrons.
- Use chemical symbols together with atomic number and mass number to express the subatomic composition of isotopes.
- Understand atomic weights and how they relate to the masses of individual atoms and to their natural abundances.
- Describe how elements are organized in the periodic table by atomic number and by similarities in chemical behav- ior, giving rise to periods and groups.
- Describe the locations of metals and nonmetals in the periodic table.
- Distinguish between molecular substances and ionic substances in terms of their composition.
- Distinguish between empirical formulas and molecular formulas.
- Describe how molecular formulas and structural formulas are used to represent the compositions of molecules.
- Explain how ions are formed by the gain or loss of electrons and be able to use the periodic table to predict the charges of ions.
- Write the empirical formulas of ionic compounds, given their charges.
- Write the name of an ionic compound given its chemical formula, or write the chemical formula given its name.
- Name or write chemical formulas for binary inorganic compounds and for acids.
- Identify organic compounds and name simple alkanes and alcohols.

VISUALIZING CONCEPTS

2.1 A charged particle is caused to move between two elec- trically charged plates, as shown below.

(a) Why does the path of the charged particle bend? **(b)** What is the sign of the electrical charge on the parti- cle? **(c)** As the charge on the plates is increased, would you expect the bending to increase, decrease, or stay the same? **(d)** As the mass of the particle is increased while the speed of the particles remains the same, would you expect the bending to increase, decrease, or stay the same? [Section 2.2]

2.2 The following diagram is a representation of 20 atoms of a fictitious element, which we will call nevadium (Nv). The red spheres are ^{293}Nv, and the blue spheres are ^{295}Nv. **(a)** Assuming that this sample is a statistically representative sample of the element, calculate the per- cent abundance of each element. **(b)** If the mass of ^{293}Nv is 293.15 amu and that of ^{295}Nv is 295.15 amu, what is the atomic weight of Nv? [Section 2.4]

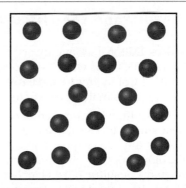

2.3 Four of the boxes in the following periodic table are col- ored. Which of these are metals and which are non- metals? Which one is an alkaline earth metal? Which one is a noble gas? [Section 2.5]

2.4 Does the following drawing represent a neutral atom or an ion? Write its complete chemical symbol including mass number, atomic number, and net charge (if any). [Sections 2.3 and 2.7]

16 protons + 16 neutrons

18 electrons

2.5 Which of the following diagrams most likely represents an ionic compound, and which represents a molecular one? Explain your choice. [Sections 2.6 and 2.7]

   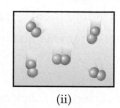

(i)   (ii)

2.6 Write the chemical formula for the following compound. Is the compound ionic or molecular? Name the compound. [Sections 2.6 and 2.8]

2.7 Five of the boxes in the following periodic table are colored. Predict the charge on the ion associated with each of these elements. [Section 2.7]

2.8 The following diagram represents an ionic compound in which the red spheres represent cations and blue spheres represent anions. Which of the following formulas is consistent with the drawing: KBr, K_2SO_4, $Ca(NO_3)_2$, $Fe_2(SO_4)_3$? Name the compound. [Sections 2.7 and 2.8]

EXERCISES

Atomic Theory and the Discovery of Atomic Structure

2.9 How does Dalton's atomic theory account for the fact that when 1.000 g of water is decomposed into its elements, 0.111 g of hydrogen and 0.889 g of oxygen are obtained regardless of the source of the water?

2.10 Hydrogen sulfide is composed of two elements: hydrogen and sulfur. In an experiment, 6.500 g of hydrogen sulfide is fully decomposed into its elements. **(a)** If 0.384 g of hydrogen is obtained in this experiment, how many grams of sulfur must be obtained? **(b)** What fundamental law does this experiment demonstrate? **(c)** How is this law explained by Dalton's atomic theory?

2.11 A chemist finds that 30.82 g of nitrogen will react with 17.60 g, 35.20 g, 70.40 g, or 88.00 g of oxygen to form four different compounds. **(a)** Calculate the mass of oxygen per gram of nitrogen in each compound. **(b)** How do the numbers in part (a) support Dalton's atomic theory?

2.12 In a series of experiments, a chemist prepared three different compounds that contain only iodine and fluorine and determined the mass of each element in each compound:

Compound	Mass of Iodine (g)	Mass of Fluorine (g)
1	4.75	3.56
2	7.64	3.43
3	9.41	9.86

(a) Calculate the mass of fluorine per gram of iodine in each compound. **(b)** How do the numbers in part (a) support the atomic theory?

2.13 Summarize the evidence used by J. J. Thomson to argue that cathode rays consist of negatively charged particles.

2.14 An unknown particle is caused to move between two electrically charged plates, as illustrated in Figure 2.8. Its path is deflected by a smaller magnitude in the opposite direction from that of a beta particle. What can you conclude about the charge and mass of this unknown particle?

2.15 **(a)** Figure 2.5 shows the apparatus used in the Millikan oil-drop experiment with the positively charged plate above the negatively charged plate. What do you think would be the effect on the rate of oil drops descending if the charges on the plates were reversed (negative above positive)? **(b)** In his original series of experiments, Millikan measured the charge on 58 separate oil drops. Why do you suppose he chose so many drops before reaching his final conclusions?

2.16 Millikan determined the charge on the electron by studying the static charges on oil drops falling in an electric field. A student carried out this experiment using several oil drops for her measurements and calcu-

lated the charges on the drops. She obtained the following data:

Droplet	Calculated Charge (C)
A	1.60×10^{-19}
B	3.15×10^{-19}
C	4.81×10^{-19}
D	6.31×10^{-19}

(a) What is the significance of the fact that the droplets carried different charges? **(b)** What conclusion can the student draw from these data regarding the charge of the electron? **(c)** What value (and to how many significant figures) should she report for the electronic charge?

Modern View of Atomic Structure; Atomic Weights

2.17 The radius of an atom of krypton (Kr) is about 1.9 Å. **(a)** Express this distance in nanometers (nm) and in picometers (pm). **(b)** How many krypton atoms would have to be lined up to span 1.0 mm? **(c)** If the atom is assumed to be a sphere, what is the volume in cm^3 of a single Kr atom?

2.18 An atom of tin (Sn) has a diameter of about 2.8×10^{-8} cm. **(a)** What is the radius of a tin atom in angstroms (Å) and in meters (m)? **(b)** How many Sn atoms would have to be placed side by side to span a distance of 6.0 μm? **(c)** If you assume that the tin atom is a sphere, what is the volume in m^3 of a single atom?

2.19 Answer the following questions without referring to Table 2.1: **(a)** What are the main subatomic particles that make up the atom? **(b)** What is the relative charge (in multiples of the electronic charge) of each of the particles? **(c)** Which of the particles is the most massive? **(d)** Which is the least massive?

2.20 Determine whether each of the following statements is true or false. If false, correct the statement to make it true: **(a)** The nucleus has most of the mass and comprises most of the volume of an atom; **(b)** every atom of a given element has the same number of protons; **(c)** the number of electrons in an atom equals the number of neutrons in the atom; **(d)** the protons in the nucleus of the helium atom are held together by a force called the strong nuclear force.

2.21 **(a)** Define atomic number and mass number. **(b)** Which of these can vary without changing the identity of the element?

2.22 **(a)** Which two of the following are isotopes of the same element: $^{31}_{16}X$, $^{31}_{15}X$, $^{32}_{16}X$? **(b)** What is the identity of the element whose isotopes you have selected?

2.23 How many protons, neutrons, and electrons are in the following atoms: **(a)** ^{40}Ar, **(b)** ^{65}Zn, **(c)** ^{70}Ga, **(d)** ^{80}Br, **(e)** ^{184}W, **(f)** ^{243}Am?

2.24 Each of the following isotopes is used in medicine. Indicate the number of protons and neutrons in each isotope: **(a)** phosphorus-32, **(b)** chromium-51, **(c)** cobalt-60, **(d)** technetium-99, **(e)** iodine-131; **(f)** thallium-201.

2.25 Fill in the gaps in the following table, assuming each column represents a neutral atom:

Symbol	^{52}Cr				
Protons		25			82
Neutrons		30	64		
Electrons			48	86	
Mass no.				222	207

2.26 Fill in the gaps in the following table, assuming each column represents a neutral atom:

Symbol	^{65}Zn				
Protons		44			92
Neutrons		57	49		
Electrons			38	47	
Mass no.				108	235

2.27 Write the correct symbol, with both superscript and subscript, for each of the following. Use the list of elements inside the front cover as needed: **(a)** the isotope of platinum that contains 118 neutrons, **(b)** the isotope of krypton with mass number 84, **(c)** the isotope of arsenic with mass number 75, **(d)** the isotope of magnesium that has an equal number of protons and neutrons.

2.28 One way in which Earth's evolution as a planet can be understood is by measuring the amounts of certain isotopes in rocks. One quantity recently measured is the ratio of ^{129}Xe to ^{130}Xe in some minerals. In what way do these two isotopes differ from one another? In what respects are they the same?

2.29 **(a)** What isotope is used as the standard in establishing the atomic mass scale? **(b)** The atomic weight of boron is reported as 10.81, yet no atom of boron has the mass of 10.81 amu. Explain.

2.30 **(a)** What is the mass in amu of a carbon-12 atom? **(b)** Why is the atomic weight of carbon reported as 12.011 in the table of elements and the periodic table in the front inside cover of this text?

2.31 Only two isotopes of copper occur naturally, ^{63}Cu (atomic mass = 62.9296 amu; abundance 69.17%) and ^{65}Cu (atomic mass = 64.9278 amu; abundance 30.83%). Calculate the atomic weight (average atomic mass) of copper.

2.32 Rubidium has two naturally occurring isotopes, rubidium-85 (atomic mass = 84.9118 amu; abundance = 72.15%) and rubidium-87 (atomic mass = 86.9092 amu; abundance = 27.85%). Calculate the atomic weight of rubidium.

2.33 **(a)** In what fundamental way is mass spectrometry related to Thomson's cathode-ray experiments (Figure 2.4)? **(b)** What are the labels on the axes of a mass spectrum? **(c)** To measure the mass spectrum of an atom, the atom must first lose one or more electrons. Why is this so?

2.34 **(a)** The mass spectrometer in Figure 2.13 has a magnet as one of its components. What is the purpose of the magnet? **(b)** The atomic weight of Cl is 35.5 amu. However, the mass spectrum of Cl (Figure 2.14) does not show a peak at this mass. Explain. **(c)** A mass spectrum of phosphorus (P) atoms shows only a single peak at a mass of 31. What can you conclude from this observation?

2.35 Naturally occurring magnesium has the following isotopic abundances:

Isotope	Abundance	Atomic mass (amu)
^{24}Mg	78.99%	23.98504
^{25}Mg	10.00%	24.98584
^{26}Mg	11.01%	25.98259

(a) What is the average atomic mass of Mg? **(b)** Sketch the mass spectrum of Mg.

2.36 Mass spectrometry is more often applied to molecules than to atoms. We will see in Chapter 3 that the *molecular weight* of a molecule is the sum of the atomic weights of the atoms in the molecule. The mass spectrum of H_2 is taken under conditions that prevent decomposition into H atoms. The two naturally occurring isotopes of hydrogen are ^1H (atomic mass = 1.00783 amu; abundance 99.9885%) and ^2H (atomic mass = 2.01410 amu; abundance 0.0115%). **(a)** How many peaks will the mass spectrum have? **(b)** Give the relative atomic masses of each of these peaks. **(c)** Which peak will be the largest, and which the smallest?

The Periodic Table; Molecules and Ions

2.37 For each of the following elements, write its chemical symbol, locate it in the periodic table, and indicate whether it is a metal, metalloid, or nonmetal: **(a)** chromium, **(b)** helium, **(c)** phosphorus, **(d)** zinc, **(e)** magnesium, **(f)** bromine, **(g)** arsenic.

2.38 Locate each of the following elements in the periodic table; indicate whether it is a metal, metalloid, or nonmetal; and give the name of the element: **(a)** Ca, **(b)** Ti, **(c)** Ga, **(d)** Th, **(e)** Pt, **(f)** Se, **(g)** Kr.

2.39 For each of the following elements, write its chemical symbol, determine the name of the group to which it belongs (Table 2.3), and indicate whether it is a metal, metalloid, or nonmetal: **(a)** potassium, **(b)** iodine, **(c)** magnesium, **(d)** argon, **(e)** sulfur.

2.40 The elements of group 4A show an interesting change in properties moving down the group. Give the name and chemical symbol of each element in the group, and label it as a nonmetal, metalloid, or metal.

2.41 What can we tell about a compound when we know the empirical formula? What additional information is conveyed by the molecular formula? By the structural formula? Explain in each case.

2.42 Two compounds have the same empirical formula. One substance is a gas, the other is a viscous liquid. How is it possible for two substances with the same empirical formula to have markedly different properties?

2.43 Write the empirical formula corresponding to each of the following molecular formulas: **(a)** Al_2Br_6, **(b)** C_8H_{10}, **(c)** $C_4H_8O_2$, **(d)** P_4O_{10}, **(e)** $C_6H_4Cl_2$, **(f)** $B_3N_3H_6$.

2.44 Determine the molecular and empirical formulas of the following: **(a)** The organic solvent *benzene*, which has six carbon atoms and six hydrogen atoms; **(b)** the compound *silicon tetrachloride*, which has a silicon atom and four chlorine atoms and is used in the manufacture of computer chips; **(c)** the reactive substance *diborane*, which has two boron atoms and six hydrogen atoms; **(d)** the sugar called *glucose*, which has six carbon atoms, twelve hydrogen atoms, and six oxygen atoms.

2.45 How many hydrogen atoms are in each of the following: **(a)** C_2H_5OH, **(b)** $Ca(CH_3COO)_2$, **(c)** $(NH_4)_3PO_4$?

2.46 How many of the indicated atoms are represented by each chemical formula: **(a)** carbon atoms in $C_2H_5COOCH_3$, **(b)** oxygen atoms in $Ca(ClO_4)_2$, **(c)** hydrogen atoms in $(NH_4)_2HPO_4$?

2.47 Write the molecular and structural formulas for the compounds represented by the following molecular models:

(a)   (b)

(c)   (d)

2.48 Write the molecular and structural formulas for the compounds represented by the following models:

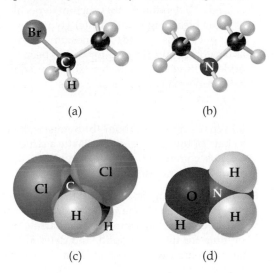

(a)   (b)

(c)   (d)

2.49 Fill in the gaps in the following table:

Symbol	$^{59}Co^{3+}$			
Protons		34	76	80
Neutrons		46	116	120
Electrons		36		78
Net charge			2+	

2.50 Fill in the gaps in the following table:

Symbol	$^{31}P^{3-}$			
Protons		35	49	
Neutrons		45	66	118
Electrons			46	76
Net charge		1−		3+

2.51 Each of the following elements is capable of forming an ion in chemical reactions. By referring to the periodic table, predict the charge of the most stable ion of each: **(a)** Mg, **(b)** Al, **(c)** K, **(d)** S, **(e)** F.

2.52 Using the periodic table, predict the charges of the ions of the following elements: **(a)** Ga, **(b)** Sr, **(c)** As, **(d)** Br, **(e)** Se.

2.53 Using the periodic table to guide you, predict the chemical formula and name of the compound formed by the following elements: **(a)** Ga and F, **(b)** Li and H, **(c)** Al and I, **(d)** K and S.

2.54 The most common charge associated with silver in its compounds is 1+. Indicate the chemical formulas you would expect for compounds formed between Ag and **(a)** iodine, **(b)** sulfur, **(c)** fluorine.

2.55 Predict the chemical formula for the ionic compound formed by **(a)** Ca^{2+} and Br^-, **(b)** K^+ and CO_3^{2-}, **(c)** Al^{3+} and CH_3COO^-, **(d)** NH_4^+ and SO_4^{2-}, **(e)** Mg^{2+} and PO_4^{3-}.

2.56 Predict the chemical formulas of the compounds formed by the following pairs of ions: **(a)** Cu^{2+} and Br^-, **(b)** Fe^{3+} and O^{2-}, **(c)** Hg_2^{2+} and CO_3^{2-}, **(d)** Ca^{2+} and AsO_4^{3-}, **(e)** NH_4^+ and CO_3^{2-}.

2.57 Complete the table by filling in the formula for the ionic compound formed by each pair of cations and anions, as shown for the first pair.

Ion	K^+	NH_4^+	Mg^{2+}	Fe^{3+}
Cl^-	KCl			
OH^-				
CO_3^{2-}				
PO_4^{3-}				

2.58 Complete the table by filling in the formula for the ionic compound formed by each pair of cations and anions, as shown for the first pair.

Ion	Na^+	Ca^{2+}	Fe^{2+}	Al^{3+}
O^{2-}	Na_2O			
NO_3^-				
SO_4^{2-}				
AsO_4^{3-}				

2.59 Predict whether each of the following compounds is molecular or ionic: **(a)** B_2H_6, **(b)** CH_3OH, **(c)** $LiNO_3$, **(d)** Sc_2O_3, **(e)** CsBr, **(f)** NOCl, **(g)** NF_3, **(h)** Ag_2SO_4.

2.60 Which of the following are ionic, and which are molecular? **(a)** PF_5, **(b)** NaI, **(c)** SCl_2, **(d)** $Ca(NO_3)_2$, **(e)** $FeCl_3$, **(f)** LaP, **(g)** $CoCO_3$, **(h)** N_2O_4.

Naming Inorganic Compounds; Organic Molecules

2.61 Give the chemical formula for **(a)** chlorite ion, **(b)** chloride ion, **(c)** chlorate ion, **(d)** perchlorate ion, **(e)** hypochlorite ion.

2.62 Selenium, an element required nutritionally in trace quantities, forms compounds analogous to sulfur. Name the following ions: **(a)** SeO_4^{2-}, **(b)** Se^{2-}, **(c)** HSe^-, **(d)** $HSeO_3^-$.

2.63 Give the names and charges of the cation and anion in each of the following compounds: **(a)** CaO, **(b)** Na_2SO_4, **(c)** $KClO_4$, **(d)** $Fe(NO_3)_2$, **(e)** $Cr(OH)_3$.

2.64 Give the names and charges of the cation and anion in each of the following compounds: **(a)** CuS, **(b)** Ag_2SO_4, **(c)** $Al(ClO_3)_3$, **(d)** $Co(OH)_2$, **(e)** $PbCO_3$.

2.65 Name the following ionic compounds: **(a)** MgO, **(b)** $AlCl_3$, **(c)** Li_3PO_4, **(d)** $Ba(ClO_4)_2$, **(e)** $Cu(NO_3)_2$, **(f)** $Fe(OH)_2$, **(g)** $Ca(C_2H_3O_2)_2$, **(h)** $Cr_2(CO_3)_3$, **(i)** K_2CrO_4, **(j)** $(NH_4)_2SO_4$.

2.66 Name the following ionic compounds: **(a)** K_2O, **(b)** $NaClO_2$, **(c)** $Sr(CN)_2$, **(d)** $Co(OH)_2$, **(e)** $Fe_2(CO_3)_3$, **(f)** $Cr(NO_3)_3$, **(g)** $(NH_4)_2SO_3$, **(h)** NaH_2PO_4, **(i)** $KMnO_4$, **(j)** $Ag_2Cr_2O_7$.

2.67 Write the chemical formulas for the following compounds: **(a)** aluminum hydroxide, **(b)** potassium sulfate, **(c)** copper(I) oxide, **(d)** zinc nitrate, **(e)** mercury(II) bromide, **(f)** iron(III) carbonate, **(g)** sodium hypobromite.

2.68 Give the chemical formula for each of the following ionic compounds: **(a)** sodium phosphate, **(b)** zinc nitrate, **(c)** barium bromate, **(d)** iron(II) perchlorate, **(e)** cobalt(II) hydrogen carbonate, **(f)** chromium(III) acetate, **(g)** potassium dichromate.

2.69 Give the name or chemical formula, as appropriate, for each of the following acids: **(a)** $HBrO_3$, **(b)** HBr, **(c)** H_3PO_4, **(d)** hypochlorous acid, **(e)** iodic acid, **(f)** sulfurous acid.

2.70 Provide the name or chemical formula, as appropriate, for each of the following acids: **(a)** hydrobromic acid, **(b)** hydrosulfuric acid, **(c)** nitrous acid, **(d)** H_2CO_3, **(e)** $HClO_3$, **(f)** $HC_2H_3O_2$.

2.71 Give the name or chemical formula, as appropriate, for each of the following binary molecular substances: **(a)** SF_6, **(b)** IF_5, **(c)** XeO_3, **(d)** dinitrogen tetroxide, **(e)** hydrogen cyanide, **(f)** tetraphosphorus hexasulfide.

2.72 The oxides of nitrogen are very important components in urban air pollution. Name each of the following compounds: **(a)** N_2O, **(b)** NO, **(c)** NO_2, **(d)** N_2O_5, **(e)** N_2O_4.

2.73 Write the chemical formula for each substance mentioned in the following word descriptions (use the front inside cover to find the symbols for the elements you don't know). **(a)** Zinc carbonate can be heated to form zinc oxide and carbon dioxide. **(b)** On treatment with hydrofluoric acid, silicon dioxide forms silicon tetrafluoride and water. **(c)** Sulfur dioxide reacts with water to form sulfurous acid. **(d)** The substance phosphorus trihydride, commonly called phosphine, is a toxic gas. **(e)** Perchloric acid reacts with cadmium to form cadmium(II) perchlorate. **(f)** Vanadium(III) bromide is a colored solid.

2.74 Assume that you encounter the following sentences in your reading. What is the chemical formula for each substance mentioned? **(a)** Sodium hydrogen carbonate is used as a deodorant. **(b)** Calcium hypochlorite is used in some bleaching solutions. **(c)** Hydrogen cyanide is a very poisonous gas. **(d)** Magnesium hydroxide is used as a cathartic. **(e)** Tin(II) fluoride has been used as a fluoride additive in toothpastes. **(f)** When cadmium sulfide is treated with sulfuric acid, fumes of hydrogen sulfide are given off.

2.75 **(a)** What is a hydrocarbon? **(b)** Butane is the alkane with a chain of four carbon atoms. Write a structural formula for this compound, and determine its molecular and empirical formulas.

2.76 **(a)** What ending is used for the names of alkanes? **(b)** Hexane is an alkane whose structural formula has all its carbon atoms in a straight chain. Draw the structural formula for this compound, and determine its molecular and empirical formulas. (*Hint:* You might need to refer to Table 2.6.)

2.77 **(a)** What is a functional group? **(b)** What functional group characterizes an alcohol? **(c)** With reference to Exercise 2.75, write a structural formula for 1-butanol, the alcohol derived from butane, by making a substitution on one of the end carbon atoms.

2.78 **(a)** What do ethane and ethanol have in common? **(b)** How does 1-propanol differ from propane?

ADDITIONAL EXERCISES

2.79 Describe a major contribution to science made by each of the following scientists: **(a)** Dalton, **(b)** Thomson, **(c)** Millikan, **(d)** Rutherford.

2.80 How did Rutherford interpret the following observations made during his α-particle scattering experiments? **(a)** Most α particles were not appreciably deflected as they passed through the gold foil. **(b)** A few α particles were deflected at very large angles. **(c)** What differences would you expect if beryllium foil were used instead of gold foil in the α-particle scattering experiment?

2.81 Suppose a scientist repeats the Millikan oil-drop experiment, but reports the charges on the drops using an unusual (and imaginary) unit called the *warmomb* (wa). He obtains the following data for four of the drops:

Droplet	Calculated Charge (wa)
A	3.84×10^{-8}
B	4.80×10^{-8}
C	2.88×10^{-8}
D	8.64×10^{-8}

(a) If all the droplets were the same size, which would fall most slowly through the apparatus? **(b)** From these data, what is the best choice for the charge of the electron in warmombs? **(c)** Based on your answer to part (b), how many electrons are there on each of the droplets? **(d)** What is the conversion factor between warmombs and coulombs?

2.82 The natural abundance of ^3He is 0.000137%. **(a)** How many protons, neutrons, and electrons are in an atom of ^3He? **(b)** Based on the sum of the masses of their subatomic particles, which is expected to be more massive, an atom of ^3He or an atom of ^3H (which is also called *tritium*)? **(c)** Based on your answer for part (b), what would need to be the precision of a mass spectrometer that is able to differentiate between peaks that are due to $^3\text{He}^+$ and $^3\text{H}^+$?

2.83 An α particle is the nucleus of an ^4He atom. **(a)** How many protons and neutrons are in an α particle? **(b)** What force holds the protons and neutrons together in the α particle? **(c)** What is the charge on an α particle in units of electronic charge? **(d)** The charge-to-mass ratio of an α particle is 4.8224×10^4 C/g. Based on the charge on the particle, calculate its mass in grams and in amu. **(e)** By using the data in Table 2.1, compare your answer for part (d) with the sum of the masses of the individual subatomic particles. Can you explain the difference in mass? (If not, we will discuss such mass differences further in Chapter 21.)

2.84 A cube of gold that is 1.00 cm on a side has a mass of 19.3 g. A single gold atom has a mass of 197.0 amu. **(a)** How many gold atoms are in the cube? **(b)** From the information given, estimate the diameter in Å of a single gold atom. **(c)** What assumptions did you make in arriving at your answer for part (b)?

2.85 The diameter of a rubidium atom is 4.95 Å. We will consider two different ways of placing the atoms on a surface. In arrangement A, all the atoms are lined up with one another. Arrangement B is called a *close-packed* arrangement because the atoms sit in the "depressions" formed by the previous row of atoms:

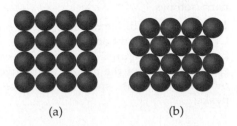

(a)   (b)

(a) Using arrangement A, how many Rb atoms could be placed on a square surface that is 1.0 cm on a side? **(b)** How many Rb atoms could be placed on a square surface that is 1.0 cm on a side, using arrangement B? **(c)** By what factor has the number of atoms on the surface increased in going to arrangement B from arrangement A? If extended to three dimensions, which arrangement would lead to a greater density for Rb metal?

2.86 **(a)** Assuming the dimensions of the nucleus and atom shown in Figure 2.12, what fraction of the *volume* of the atom is taken up by the nucleus? **(b)** Using the mass of the proton from Table 2.1 and assuming its diameter is 1.0×10^{-15} m, calculate the density of a proton in g/cm^3.

2.87 Identify the element represented by each of the following symbols and give the number of protons and neutrons in each: **(a)** $^{74}_{33}\text{X}$, **(b)** $^{127}_{53}\text{X}$, **(c)** $^{152}_{63}\text{X}$, **(d)** $^{209}_{83}\text{X}$.

2.88 The element oxygen has three naturally occurring isotopes, with 8, 9, and 10 neutrons in the nucleus, respectively. **(a)** Write the full chemical symbols for these three isotopes. **(b)** Describe the similarities and differences between the three kinds of atoms of oxygen.

2.89 Use Coulomb's law, $F = kQ_1Q_2/d^2$, to calculate the electric force on an electron ($Q = -1.6 \times 10^{-19}$ C) exerted by a single proton if the particles are 0.53×10^{-10} m apart. The constant k in Coulomb's law is 9.0×10^9 N·m^2/C^2. (The unit abbreviated N is the Newton, the SI unit of force.)

2.90 The element lead (Pb) consists of four naturally occurring isotopes with atomic masses 203.97302, 205.97444, 206.97587, and 207.97663 amu. The relative abundances of these four isotopes are 1.4, 24.1, 22.1, and 52.4%, respectively. From these data, calculate the atomic weight of lead.

2.91 Gallium (Ga) consists of two naturally occurring isotopes with masses of 68.926 and 70.925 amu. **(a)** How many protons and neutrons are in the nucleus of each isotope? Write the complete atomic symbol for each, showing the atomic number and mass number. **(b)** The average atomic mass of Ga is 69.72 amu. Calculate the abundance of each isotope.

2.92 Using a suitable reference such as the *CRC Handbook of Chemistry and Physics* or http://www.webelements.com, look up the following information for nickel: **(a)** the number of known isotopes, **(b)** the atomic masses (in amu) and the natural abundance of the five most abundant isotopes.

2.93 There are two different isotopes of bromine atoms. Under normal conditions, elemental bromine consists of Br_2 molecules (Figure 2.19), and the mass of a Br_2 molecule is the sum of the masses of the two atoms in the molecule. The mass spectrum of Br_2 consists of three peaks:

Mass (amu)	Relative Size
157.836	0.2569
159.834	0.4999
161.832	0.2431

(a) What is the origin of each peak (of what isotopes does each consist)? **(b)** What is the mass of each isotope? **(c)** Determine the average molecular mass of a Br_2 molecule. **(d)** Determine the average atomic mass of a bromine atom. **(e)** Calculate the abundances of the two isotopes.

2.94 It is common in mass spectrometry to assume that the mass of a cation is the same as that of its parent atom. **(a)** Using data in Table 2.1, determine the number of significant figures that must be reported before the difference in mass of ^1H and $^1H^+$ is significant. **(b)** What percentage of the mass of an ^1H atom does the electron represent?

2.95 From the following list of elements—Ar, H, Ga, Al, Ca, Br, Ge, K, O—pick the one that best fits each description. Use each element only once: **(a)** an alkali metal, **(b)** an alkaline earth metal, **(c)** a noble gas, **(d)** a halogen, **(e)** a metalloid, **(f)** a nonmetal listed in group 1A, **(g)** a metal that forms a 3+ ion, **(h)** a nonmetal that forms a 2− ion, **(i)** an element that resembles aluminum.

2.96 The first atoms of seaborgium (Sg) were identified in 1974. The longest-lived isotope of Sg has a mass number of 266. **(a)** How many protons, electrons, and neutrons are in an ^{266}Sg atom? **(b)** Atoms of Sg are very unstable, and it is therefore difficult to study this element's properties. Based on the position of Sg in the periodic table, what element should it most closely resemble in its chemical properties?

2.97 From the molecular structures shown here, identify the one that corresponds to each of the following species:

(a) chlorine gas; **(b)** propane, **(c)** nitrate ion; **(d)** sulfur trioxide; **(e)** methyl chloride, CH_3Cl.

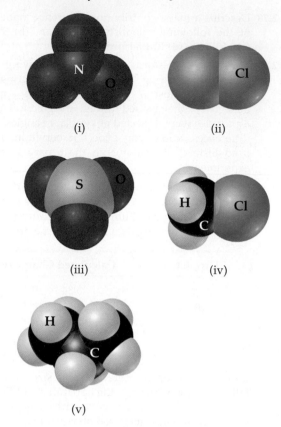

2.98 Name each of the following oxides. Assuming that the compounds are ionic, what charge is associated with the metallic element in each case? **(a)** NiO, **(b)** MnO_2, **(c)** Cr_2O_3, **(d)** MoO_3.

2.99 Iodic acid has the molecular formula HIO_3. Write the formulas for the following: **(a)** the iodate anion, **(b)** the periodate anion, **(c)** the hypoiodite anion, **(d)** hypoiodous acid, **(e)** periodic acid.

2.100 Elements in the same group of the periodic table often form oxyanions with the same general formula. The anions are also named in a similar fashion. Based on these observations, suggest a chemical formula or name, as appropriate, for each of the following ions: **(a)** BrO_4^-, **(b)** SeO_3^{2-}, **(c)** arsenate ion, **(d)** hydrogen tellurate ion.

2.101 Carbonic acid occurs in carbonated beverages. When allowed to react with lithium hydroxide it produces lithium carbonate. Lithium carbonate is used to treat depression and bipolar disorder. Write chemical formulas for carbonic acid, lithium hydroxide, and lithium carbonate.

2.102 Give the chemical names of each of the following familiar compounds: **(a)** NaCl (table salt) , **(b)** $NaHCO_3$ (baking soda), **(c)** NaOCl (in many bleaches), **(d)** NaOH (caustic soda), **(e)** $(NH_4)_2CO_3$ (smelling salts), **(f)** $CaSO_4$ (plaster of Paris).

2.103 Many familiar substances have common, unsystematic names. For each of the following, give the correct systematic name: **(a)** saltpeter, KNO_3; **(b)** soda ash, Na_2CO_3; **(c)** lime, CaO; **(d)** muriatic acid, HCl; **(e)** Epsom salts, $MgSO_4$; **(f)** milk of magnesia, $Mg(OH)_2$.

2.104 Many ions and compounds have very similar names, and there is great potential for confusing them. Write the correct chemical formulas to distinguish between **(a)** calcium sulfide and calcium hydrogen sulfide, **(b)** hydrobromic acid and bromic acid, **(c)** aluminum nitride and aluminum nitrite, **(d)** iron(II) oxide and iron(III) oxide, **(e)** ammonia and ammonium ion, **(f)** potassium sulfite and potassium bisulfite, **(g)** mercurous chloride and mercuric chloride, **(h)** chloric acid and perchloric acid.

2.105 The compound *cyclohexane* is an alkane in which six carbon atoms form a ring. The partial structural formula of the compound is as follows:

(a) Complete the structural formula for cyclohexane. **(b)** Is the molecular formula for cyclohexane the same as that for *n*-hexane, in which the carbon atoms are in a straight line? If possible, comment on the source of any differences. **(c)** Propose a structural formula for *cyclohexanol*, the alcohol derived from cyclohexane.

2.106 The periodic table helps organize the chemical behaviors of the elements. As a class discussion or as a short essay, describe how the table is organized, and mention as many ways as you can think of in which the position of an element in the table relates to the chemical and physical properties of the element.

STOICHIOMETRY: CALCULATIONS WITH CHEMICAL FORMULAS AND EQUATIONS

SUGAR CARAMELIZING. Major changes in the appearance of compounds are indications of chemical reactions. Here, prolonged heating of sucrose, common table sugar, produces caramel.

WHAT'S AHEAD

3.1 Chemical Equations
We begin by considering how we can use chemical formulas to write equations that represent chemical reactions.

3.2 Some Simple Patterns of Chemical Reactivity
We then examine some simple chemical reactions: *combination reactions, decomposition reactions,* and *combustion reactions.*

3.3 Formula Weights
We can obtain quantitative information from chemical formulas by using their *formula weights.*

3.4 Avogadro's Number and the Mole
We use chemical formulas to relate the masses of substances to the numbers of atoms, molecules, or ions contained in the substances, a relationship that leads to the crucially important concept of a *mole.* A *mole* is 6.022×10^{23} objects (atoms, molecules, ions, etc.).

3.5 Empirical Formulas from Analyses
We apply the mole concept to determine chemical formulas from the masses of each element in a given quantity of a compound.

3.6 Quantitative Information from Balanced Equations
We use the quantitative information inherent in chemical formulas and equations together with the mole concept to predict the amounts of substances consumed or produced in chemical reactions.

3.7 Limiting Reactants
We recognize that one of the reactants may be used up before the others in a chemical reaction. This is the *limiting reactant.* The reaction therefore stops, leaving some of the excess starting material unreacted.

YOU POUR VINEGAR INTO A glass of water containing baking soda, and bubbles form. You strike a match and use the flame to light a candle. You heat sugar in a pan, and it turns brown (caramelizes). The bubbles, the flame, and the color change are visual evidence that something is happening.

To an experienced eye, these visual changes indicate a chemical change, or chemical reaction. The study of chemical changes is at the heart of chemistry. Some chemical changes are simple; others are complex. Some are dramatic; some are very subtle. Even as you sit reading this chapter, chemical changes are occurring within your body. Chemical changes that occur in your eyes and brain, for example, allow you to see these words and think about them. Although such chemical changes are not as obvious as some, they are nevertheless remarkable for how they allow us to function.

In this chapter we begin to explore some important aspects of chemical change. Our focus will be both on the use of chemical formulas to represent re-actions and on the quantitative information we can obtain about the amounts of substances involved in reactions. **Stoichiometry** (pronounced stoy-key-OM-uh-tree) is the area of study that examines the quantities of substances consumed and produced in chemical reactions. The name is derived from the Greek *stoicheion* ("element") and *metron* ("measure"). This study of stoichiome-try provides an essential set of tools that is widely used in chemistry. Aspects of stoichiometry include such diverse problems as measuring the concentra-tion of ozone in the atmosphere, determining the potential yield of gold from an ore, and assessing different processes for converting coal into gaseous fuels.

Stoichiometry is built on an understanding of atomic masses ∞ (Section 2.4), chemical formulas, and the law of conservation of mass. ∞ (Section 2.1)

▲ **Figure 3.1 Antoine Lavoisier (1734–1794).** Lavoisier conducted many important studies on combustion reactions. Unfortunately, the French Revolution cut his career short. He was a member of the French nobility and a tax collector. He was guillotined in 1794 during the final months of the Reign of Terror. He is now generally considered to be the father of modern chemistry because he conducted carefully controlled experiments and used quantitative measurements.

The French nobleman and scientist Antoine Lavoisier (Figure 3.1 ◄) discovered this important chemical law during the late 1700s. In a chemistry text published in 1789, Lavoisier stated the law in this eloquent way: "We may lay it down as an incontestable axiom that, in all the operations of art and nature, nothing is created; an equal quantity of matter exists both before and after the experiment. Upon this principle, the whole art of performing chemical experiments depends." With the advent of Dalton's atomic theory, chemists understood the basis for this law: *Atoms are neither created nor destroyed during any chemical reaction*. The changes that occur during any reaction merely rearrange the atoms. The same collection of atoms is present both before and after the reaction.

3.1 CHEMICAL EQUATIONS

Chemical reactions are represented in a concise way by **chemical equations**. When the gas hydrogen (H_2) burns, for example, it reacts with oxygen (O_2) in the air to form water (H_2O). We write the chemical equation for this reaction as follows:

$$2\,H_2 + O_2 \longrightarrow 2\,H_2O \qquad\qquad [3.1]$$

We read the + sign as "reacts with" and the arrow as "produces." The chemical formulas to the left of the arrow represent the starting substances, called **reactants**. The chemical formulas to the right of the arrow represent substances produced in the reaction, called **products**. The numbers in front of the formulas are *coefficients*. (As in algebraic equations, the numeral 1 is usually not written.) The coefficients indicate the relative numbers of molecules of each kind involved in the reaction.

Because atoms are neither created nor destroyed in any reaction, a chemical equation must have an equal number of atoms of each element on each side of the arrow. When this condition is met, the equation is said to be *balanced*. On the right side of Equation 3.1, for example, there are two molecules of H_2O, each composed of two atoms of hydrogen and one atom of oxygen. Thus, $2\,H_2O$ (read "two molecules of water") contains $2 \times 2 = 4$ H atoms and $2 \times 1 = 2$ O atoms. Notice that the number of atoms is obtained by multiplying the coefficient and the subscripts in the chemical formula. Because there are four H atoms and two O atoms on each side of the equation, the equation is balanced. We can represent the balanced equation by the following molecular models, which illustrate that the number of atoms of each kind is the same on both sides of the arrow:

▲ **GIVE IT SOME THOUGHT**

How many atoms of Mg, O, and H are represented by $3\,Mg(OH)_2$?

Balancing Equations

Once we know the formulas of the reactants and products in a reaction, we can write an unbalanced equation. We then balance the equation by determining the coefficients that provide equal numbers of each type of atom on each side of the equation. For most purposes, a balanced equation should contain the smallest possible whole-number coefficients.

In balancing an equation, you need to understand the difference between a coefficient in front of a formula and a subscript within a formula. Refer to Figure 3.2 ►. Notice that changing a subscript in a formula—from H_2O to H_2O_2,

Chemical symbol	Meaning		Composition
H_2O	One molecule of water:		Two H atoms and one O atom
$2\,H_2O$	Two molecules of water:		Four H atoms and two O atoms
H_2O_2	One molecule of hydrogen peroxide:		Two H atoms and two O atoms

◀ **Figure 3.2 The difference between a subscript and a coefficient.** Notice how adding the coefficient 2 in front of the formula (line 2) has a different effect on the implied composition than adding the subscript 2 to the formula (in line 3). The number of atoms of each type (listed under composition) is obtained by multiplying the coefficient and the subscript associated with each element in the formula.

for example—changes the identity of the chemical. The substance H_2O_2, hydrogen peroxide, is quite different from the substance H_2O, water. *You should never change subscripts when balancing an equation.* In contrast, placing a coefficient in front of a formula changes only the *amount* of the substance and not its *identity*. Thus, $2\,H_2O$ means two molecules of water, $3\,H_2O$ means three molecules of water, and so forth.

To illustrate the process of balancing an equation, consider the reaction that occurs when methane (CH_4), the principal component of natural gas, burns in air to produce carbon dioxide gas (CO_2) and water vapor (H_2O) (Figure 3.3▼). Both of these products contain oxygen atoms that come from O_2 in the air. Thus, O_2 is a reactant, and the unbalanced equation is

$$CH_4 + O_2 \longrightarrow CO_2 + H_2O \quad \text{(unbalanced)} \qquad [3.2]$$

It is usually best to balance first those elements that occur in the fewest chemical formulas on each side of the equation. In our example both C and H appear in only one reactant and, separately, in one product each. So we begin by focusing on CH_4. Let's consider first carbon and then hydrogen.

One molecule of the reactant CH_4 contains the same number of C atoms (one) as one molecule of the product CO_2. The coefficients for these substances *must* be the same, therefore, we start the balancing process by choosing the coefficient one for each. However, one molecule of CH_4 contains more H atoms (four) than one molecule of the product H_2O (two). If we place a coefficient 2 in front of H_2O, there will be four H atoms on each side of the equation:

$$CH_4 + O_2 \longrightarrow CO_2 + 2\,H_2O \quad \text{(unbalanced)} \qquad [3.3]$$

Products

Reactants

O_2

CO_2 and H_2O

CH_4

◀ **Figure 3.3 Methane reacts with oxygen to produce the flame in a Bunsen burner.** The methane (CH_4) in natural gas and oxygen (O_2) from the air are the reactants in the reaction, while carbon dioxide (CO_2) and water vapor (H_2O) are the products.

At this stage the products have more O atoms (four—two from CO_2 and two from $2 H_2O$) than the reactants (two). If we place the coefficient 2 in front of the reactant O_2, we balance the equation by making the number of O atoms equal on both sides of the equation:

$$CH_4 + 2 O_2 \longrightarrow CO_2 + 2 H_2O \quad \text{(balanced)} \qquad [3.4]$$

The molecular view of the balanced equation is shown in Figure 3.4 ◄. We see one C, four H, and four O atoms on each side of the arrow, indicating that the equation is balanced.

The approach we have taken in arriving at balanced Equation 3.4 is largely trial and error. We balance each kind of atom in succession, adjusting coefficients as necessary. This approach works for most chemical equations.

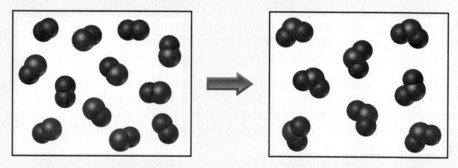

$$CH_4 \quad + \quad 2 O_2 \quad \longrightarrow \quad CO_2 \quad + \quad 2 H_2O$$

$$\begin{pmatrix} 1\ C \\ 4\ H \end{pmatrix} \qquad (4\ O) \qquad\qquad \begin{pmatrix} 1\ C \\ 2\ O \end{pmatrix} \quad \begin{pmatrix} 2\ O \\ 4\ H \end{pmatrix}$$

▲ **Figure 3.4 Balanced chemical equation for the combustion of CH_4.** The drawings of the molecules involved call attention to the conservation of atoms through the reaction.

SAMPLE EXERCISE 3.1 | Interpreting and Balancing Chemical Equations

The following diagram represents a chemical reaction in which the red spheres are oxygen atoms and the blue spheres are nitrogen atoms. (a) Write the chemical formulas for the reactants and products. (b) Write a balanced equation for the reaction. (c) Is the diagram consistent with the law of conservation of mass?

SOLUTION

(a) The left box, which represents the reactants, contains two kinds of molecules, those composed of two oxygen atoms (O_2) and those composed of one nitrogen atom and one oxygen atom (NO). The right box, which represents the products, contains only molecules composed of one nitrogen atom and two oxygen atoms (NO_2).

(b) The unbalanced chemical equation is

$$O_2 + NO \longrightarrow NO_2 \quad \text{(unbalanced)}$$

This equation has three O atoms on the left side of the arrow and two O atoms on the right side. We can increase the number of O atoms by placing a coefficient 2 on the product side:

$$O_2 + NO \longrightarrow 2 NO_2 \quad \text{(unbalanced)}$$

Now there are two N atoms and four O atoms on the right. Placing the coefficient 2 in front of NO balances both the number of N atoms and O atoms:

$$O_2 + 2 NO \longrightarrow 2 NO_2 \quad \text{(balanced)}$$

(c) The left box (reactants) contains four O_2 molecules and eight NO molecules. Thus, the molecular ratio is one O_2 for each two NO as required by the balanced equation. The right box (products) contains eight NO_2 molecules. The number of NO_2 molecules on the right equals the number of NO molecules on the left as the balanced equation requires. Counting the atoms, we find eight N atoms in the eight NO molecules in the box on the left. There are also $4 \times 2 = 8$ O atoms in the O_2 molecules and eight O atoms in the NO molecules, giving a total of 16 O atoms. In the box on the right, we find eight N atoms and $8 \times 2 = 16$ O atoms in the eight NO_2 molecules. Because there are equal numbers of both N and O atoms in the two boxes, the drawing is consistent with the law of conservation of mass.

■ **PRACTICE EXERCISE**

In the following diagram, the white spheres represent hydrogen atoms, and the blue spheres represent nitrogen atoms. To be consistent with the law of conservation of mass, how many NH_3 molecules should be shown in the right box?

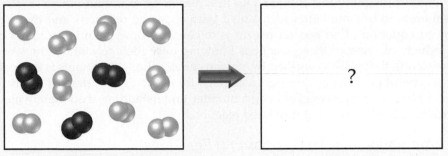

Answer: Six NH_3 molecules

Indicating the States of Reactants and Products

Additional information is often added to the formulas in balanced equations to indicate the physical state of each reactant and product. We use the symbols (*g*), (*l*), (*s*), and (*aq*) for gas, liquid, solid, and aqueous (water) solution, respectively. Thus, Equation 3.4 can be written

$$CH_4(g) + 2\,O_2(g) \longrightarrow CO_2(g) + 2\,H_2O(g) \qquad\qquad [3.5]$$

Sometimes the conditions (such as temperature or pressure) under which the reaction proceeds appear above or below the reaction arrow. The symbol Δ (the Greek uppercase letter delta) is often placed above the arrow to indicate the addition of heat.

■ **SAMPLE EXERCISE 3.2** | **Balancing Chemical Equations**

Balance this equation:

$$Na(s) + H_2O(l) \longrightarrow NaOH(aq) + H_2(g)$$

SOLUTION

Begin by counting each kind of atom on both sides of the arrow. The Na and O atoms are balanced—one Na and one O on each side—but there are two H atoms on the left and three H atoms on the right. Thus, we need to increase the number of H atoms on the left. To begin balancing H, let's try placing the coefficient 2 in front of H_2O:

$$Na(s) + 2\,H_2O(l) \longrightarrow NaOH(aq) + H_2(g)$$

Beginning this way does not balance H but does increase the number of H atoms among the reactants, which we need to do. Adding the coefficient 2 causes O to be unbalanced; we will take care of that after we balance H. Now that we have 2 H_2O on the left, we can balance H by putting the coefficient 2 in front of NaOH on the right:

$$Na(s) + 2\,H_2O(l) \longrightarrow 2\,NaOH(aq) + H_2(g)$$

Balancing H in this way fortuitously brings O into balance. But notice that Na is now unbalanced, with one Na on the left and two on the right. To rebalance Na, we put the coefficient 2 in front of the reactant:

$$2\,Na(s) + 2\,H_2O(l) \longrightarrow 2\,NaOH(aq) + H_2(g)$$

Finally, we check the number of atoms of each element and find that we have two Na atoms, four H atoms, and two O atoms on each side of the equation. The equation is balanced.

Comment Notice that in balancing this equation, we moved back and forth placing a coefficient in front of H_2O, then NaOH, and finally Na. In balancing equations, we often find ourselves following this pattern of moving back and forth from one side of the arrow to the other, placing coefficients first in front of a formula on one side and then in front of a formula on the other side until the equation is balanced. You can always tell if you have balanced your equation correctly, no matter how you did it, by checking that the number of atoms of each element is the same on both sides of the arrow.

■ **PRACTICE EXERCISE**

Balance the following equations by providing the missing coefficients:
(a) __Fe(s) + __$O_2(g)$ $\longrightarrow$ __$Fe_2O_3(s)$
(b) __$C_2H_4(g)$ + __$O_2(g)$ $\longrightarrow$ __$CO_2(g)$ + __$H_2O(g)$
(c) __Al(s) + __HCl(aq) $\longrightarrow$ __$AlCl_3(aq)$ + __$H_2(g)$
Answers: **(a)** 4, 3, 2; **(b)** 1, 3, 2, 2; **(c)** 2, 6, 2, 3

3.2 SOME SIMPLE PATTERNS OF CHEMICAL REACTIVITY

In this section we examine three simple kinds of reactions that we will see frequently throughout this chapter. Our first reason for examining these reactions is merely to become better acquainted with chemical reactions and their balanced equations. Our second reason is to consider how we might predict the products of some of these reactions knowing only their reactants. The key to predicting the products formed by a given combination of reactants is recognizing general patterns of chemical reactivity. Recognizing a pattern of reactivity for a class of substances gives you a broader understanding than merely memorizing a large number of unrelated reactions.

Combination and Decomposition Reactions

Table 3.1 ▼ summarizes two simple types of reactions: combination and decomposition reactions. In **combination reactions** two or more substances react to form one product. There are many examples of combination reactions, especially those in which elements combine to form compounds. For example, magnesium metal burns in air with a dazzling brilliance to produce magnesium oxide, as shown in Figure 3.5 ▶:

$$2\,Mg(s) + O_2(g) \longrightarrow 2\,MgO(s) \qquad [3.6]$$

This reaction is used to produce the bright flame generated by flares and some fireworks.

When a combination reaction occurs between a metal and a nonmetal, as in Equation 3.6, the product is an ionic solid. Recall that the formula of an ionic compound can be determined from the charges of the ions involved. ∞∞ (Section 2.7) When magnesium reacts with oxygen, for example, the magnesium loses electrons and forms the magnesium ion, Mg^{2+}. The oxygen gains electrons and forms the oxide ion, O^{2-}. Thus, the reaction product is MgO. You should be able to recognize when a reaction is a combination reaction and to predict the products of a combination reaction in which the reactants are a metal and a nonmetal.

▲ GIVE IT SOME THOUGHT

When Na and S undergo a combination reaction, what is the chemical formula of the product?

In a **decomposition reaction** one substance undergoes a reaction to produce two or more other substances. Many compounds undergo decomposition reactions when heated. For example, many metal carbonates decompose to form metal oxides and carbon dioxide when heated:

$$CaCO_3(s) \overset{\Delta}{\longrightarrow} CaO(s) + CO_2(g) \qquad [3.7]$$

TABLE 3.1 ■ Combination and Decomposition Reactions

Combination Reactions

$A + B \longrightarrow C$ $C(s) + O_2(g) \longrightarrow CO_2(g)$ $N_2(g) + 3\,H_2(g) \longrightarrow 2\,NH_3(g)$ $CaO(s) + H_2O(l) \longrightarrow Ca(OH)_2(s)$	Two reactants combine to form a single product. Many elements react with one another in this fashion to form compounds.

Decomposition Reactions

$C \longrightarrow A + B$ $2\,KClO_3(s) \longrightarrow 2\,KCl(s) + 3\,O_2(g)$ $PbCO_3(s) \longrightarrow PbO(s) + CO_2(g)$ $Cu(OH)_2(s) \longrightarrow CuO(s) + H_2O(l)$	A single reactant breaks apart to form two or more substances. Many compounds react this way when heated.

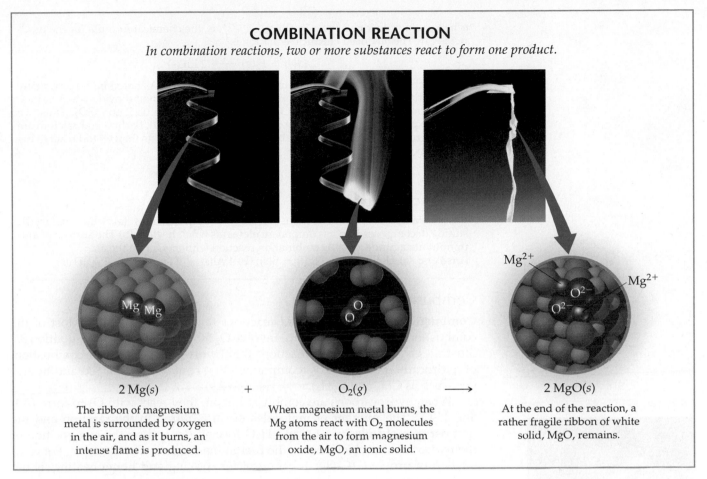

COMBINATION REACTION

In combination reactions, two or more substances react to form one product.

| $2 Mg(s)$ | $+$ | $O_2(g)$ | $\longrightarrow$ | $2 MgO(s)$ |

The ribbon of magnesium metal is surrounded by oxygen in the air, and as it burns, an intense flame is produced.

When magnesium metal burns, the Mg atoms react with O_2 molecules from the air to form magnesium oxide, MgO, an ionic solid.

At the end of the reaction, a rather fragile ribbon of white solid, MgO, remains.

▲ Figure 3.5 **Combustion of magnesium metal in air.**

The decomposition of $CaCO_3$ is an important commercial process. Limestone or seashells, which are both primarily $CaCO_3$, are heated to prepare CaO, which is known as lime or quicklime. About 2×10^{10} kg (20 million tons) of CaO is used in the United States each year, principally in making glass, in obtaining iron from its ores, and in making mortar to bind bricks.

The decomposition of sodium azide (NaN_3) rapidly releases $N_2(g)$, so this reaction is used to inflate safety air bags in automobiles (Figure 3.6 ▶):

$$2 NaN_3(s) \longrightarrow 2 Na(s) + 3 N_2(g) \qquad [3.8]$$

The system is designed so that an impact ignites a detonator cap, which in turn causes NaN_3 to decompose explosively. A small quantity of NaN_3 (about 100 g) forms a large quantity of gas (about 50 L). We will consider the volumes of gases produced in chemical reactions in Section 10.5.

▲ Figure 3.6 **An automobile air bag.** The decomposition of sodium azide, $NaN_3(s)$, is used to inflate automobile air bags. When properly ignited, the NaN_3 decomposes rapidly, forming nitrogen gas, $N_2(g)$, which expands the air bag.

■ **SAMPLE EXERCISE 3.3** | **Writing Balanced Equations for Combination and Decomposition Reactions**

Write balanced equations for the following reactions: **(a)** The combination reaction that occurs when lithium metal and fluorine gas react. **(b)** The decomposition reaction that occurs when solid barium carbonate is heated. (Two products form: a solid and a gas.)

SOLUTION

(a) The symbol for lithium is Li. With the exception of mercury, all metals are solids at room temperature. Fluorine occurs as a diatomic molecule (see Figure 2.19). Thus, the reactants are Li(s) and $F_2(g)$. The product will be composed of a metal and a nonmetal, so we expect it to be an ionic solid. Lithium ions have a 1+ charge, Li^+, whereas fluoride ions have a 1− charge, F⁻. Thus, the chemical formula for the product is LiF. The balanced chemical equation is

$$2\,Li(s) + F_2(g) \longrightarrow 2\,LiF(s)$$

(b) The chemical formula for barium carbonate is $BaCO_3$. As noted in the text, many metal carbonates decompose to form metal oxides and carbon dioxide when heated. In Equation 3.7, for example, $CaCO_3$ decomposes to form CaO and CO_2. Thus, we would expect that $BaCO_3$ decomposes to form BaO and CO_2. Barium and calcium are both in group 2A in the periodic table, which further suggests they would react in the same way:

$$BaCO_3(s) \longrightarrow BaO(s) + CO_2(g)$$

▮ PRACTICE EXERCISE

Write balanced chemical equations for the following reactions: **(a)** Solid mercury(II) sulfide decomposes into its component elements when heated. **(b)** The surface of aluminum metal undergoes a combination reaction with oxygen in the air.
Answers: **(a)** $HgS(s) \longrightarrow Hg(l) + S(s)$; **(b)** $4\,Al(s) + 3\,O_2(g) \longrightarrow 2\,Al_2O_3(s)$

Combustion in Air

Combustion reactions are rapid reactions that produce a flame. Most of the combustion reactions we observe involve O_2 from air as a reactant. Equation 3.5 illustrates a general class of reactions that involve the burning, or combustion, of hydrocarbon compounds (compounds that contain only carbon and hydrogen, such as CH_4 and C_2H_4). ∞ (Section 2.9)

When hydrocarbons are combusted in air, they react with O_2 to form CO_2 and H_2O.* The number of molecules of O_2 required in the reaction and the number of molecules of CO_2 and H_2O formed depend on the composition of the hydrocarbon, which acts as the fuel in the reaction. For example, the combustion of propane (C_3H_8), a gas used for cooking and home heating, is described by the following equation:

$$C_3H_8(g) + 5\,O_2(g) \longrightarrow 3\,CO_2(g) + 4\,H_2O(g) \qquad [3.9]$$

The state of the water, $H_2O(g)$ or $H_2O(l)$, depends on the conditions of the reaction. Water vapor, $H_2O(g)$, is formed at high temperature in an open container. The blue flame produced when propane burns is shown in Figure 3.7 ◄.

Combustion of oxygen-containing derivatives of hydrocarbons, such as CH_3OH, also produces CO_2 and H_2O. The simple rule that hydrocarbons and related oxygen-containing derivatives of hydrocarbons form CO_2 and H_2O when they burn in air summarizes the behavior of about 3 million compounds. Many substances that our bodies use as energy sources, such as the sugar glucose ($C_6H_{12}O_6$), similarly react in our bodies with O_2 to form CO_2 and H_2O. In our bodies, however, the reactions take place in a series of intermediate steps that occur at body temperature. These reactions that involve intermediate steps are described as *oxidation reactions* instead of combustion reactions.

▲ **Figure 3.7 Propane burning in air.** The liquid propane, C_3H_8, vaporizes and mixes with air as it escapes through the nozzle. The combustion reaction of C_3H_8 and O_2 produces a blue flame.

▮ SAMPLE EXERCISE 3.4 | Writing Balanced Equations for Combustion Reactions

Write the balanced equation for the reaction that occurs when methanol, $CH_3OH(l)$, is burned in air.

SOLUTION

When any compound containing C, H, and O is combusted, it reacts with the $O_2(g)$ in air to produce $CO_2(g)$ and $H_2O(g)$. Thus, the unbalanced equation is

$$CH_3OH(l) + O_2(g) \longrightarrow CO_2(g) + H_2O(g)$$

When there is an insufficient quantity of O_2 present, carbon monoxide (CO) will be produced along with the CO_2; this is called incomplete combustion. If the amount of O_2 is severely restricted, fine particles of carbon that we call soot will be produced. Complete combustion produces only CO_2 and H_2O. Unless specifically stated to the contrary, we will always take combustion to mean complete combustion.

In this equation the C atoms are balanced with one carbon on each side of the arrow. Because CH₃OH has four H atoms, we place the coefficient 2 in front of H₂O to balance the H atoms:

$$CH_3OH(l) + O_2(g) \longrightarrow CO_2(g) + 2\,H_2O(g)$$

Adding the coefficient balances H but gives four O atoms in the products. Because there are only three O atoms in the reactants (one in CH_3OH and two in O_2), we are not finished yet. We can place the fractional coefficient $\frac{3}{2}$ in front of O_2 to give a total of four O atoms in the reactants (there are $\frac{3}{2} \times 2 = 3$ O atoms in $\frac{3}{2}\,O_2$):

$$CH_3OH(l) + \tfrac{3}{2}\,O_2(g) \longrightarrow CO_2(g) + 2\,H_2O(g)$$

Although the equation is now balanced, it is not in its most conventional form because it contains a fractional coefficient. If we multiply each side of the equation by 2, we will remove the fraction and achieve the following balanced equation:

$$2\,CH_3OH(l) + 3\,O_2(g) \longrightarrow 2\,CO_2(g) + 4\,H_2O(g)$$

■ **PRACTICE EXERCISE**

Write the balanced equation for the reaction that occurs when ethanol, $C_2H_5OH(l)$, is burned in air.
Answer: $C_2H_5OH(l) + 3\,O_2(g) \longrightarrow 2\,CO_2(g) + 3\,H_2O(g)$

3.3 FORMULA WEIGHTS

Chemical formulas and chemical equations both have a *quantitative* significance; the subscripts in formulas and the coefficients in equations represent precise quantities. The formula H_2O indicates that a molecule of this substance (water) contains exactly two atoms of hydrogen and one atom of oxygen. Similarly, the coefficients in a balanced chemical equation indicate the relative quantities of reactants and products. But how do we relate the numbers of atoms or molecules to the amounts we measure in the laboratory? Although we cannot directly count atoms or molecules, we can indirectly determine their numbers if we know their masses. Therefore, before we can pursue the quantitative aspects of chemical formulas or equations, we must examine the masses of atoms and molecules, which we do in this section and the next.

Formula and Molecular Weights

The **formula weight** of a substance is the sum of the atomic weights of each atom in its chemical formula. Using atomic masses from a periodic table, we find, for example, that the formula weight of sulfuric acid (H_2SO_4) is 98.1 amu:*

$$\begin{aligned} \text{FW of } H_2SO_4 &= 2(\text{AW of H}) + (\text{AW of S}) + 4(\text{AW of O}) \\ &= 2(1.0\text{ amu}) + 32.1\text{ amu} + 4(16.0\text{ amu}) \\ &= 98.1\text{ amu} \end{aligned}$$

For convenience, we have rounded off all the atomic weights to one place beyond the decimal point. We will round off the atomic weights in this way for most problems.

If the chemical formula is merely the chemical symbol of an element, such as Na, then the formula weight equals the atomic weight of the element, in this case 23.0 amu. If the chemical formula is that of a molecule, then the formula weight is also called the **molecular weight**. The molecular weight of glucose ($C_6H_{12}O_6$), for example, is

$$\text{MW of } C_6H_{12}O_6 = 6(12.0\text{ amu}) + 12(1.0\text{ amu}) + 6(16.0\text{ amu}) = 180.0\text{ amu}$$

Because ionic substances, such as NaCl, exist as three-dimensional arrays of ions (Figure 2.23), it is inappropriate to speak of molecules of NaCl. Instead,

The abbreviation AW is used for atomic weight, FW for formula weight, and MW for molecular weight.

we speak of *formula units*, represented by the chemical formula of the substance. The formula unit of NaCl consists of one Na^+ ion and one Cl^- ion. Thus, the formula weight of NaCl is the mass of one formula unit:

$$FW \text{ of NaCl} = 23.0 \text{ amu} + 35.5 \text{ amu} = 58.5 \text{ amu}$$

■ SAMPLE EXERCISE 3.5 | Calculating Formula Weights

Calculate the formula weight of **(a)** sucrose, $C_{12}H_{22}O_{11}$ (table sugar), and **(b)** calcium nitrate, $Ca(NO_3)_2$.

SOLUTION

(a) By adding the atomic weights of the atoms in sucrose, we find the formula weight to be 342.0 amu:

$$\begin{aligned}
12 \text{ C atoms} &= 12(12.0 \text{ amu}) = 144.0 \text{ amu} \\
22 \text{ H atoms} &= 22(1.0 \text{ amu}) = \underline{22.0 \text{ amu}} \\
11 \text{ O atoms} &= 11(16.0 \text{ amu}) = \underline{176.0 \text{ amu}} \\
&\phantom{= 11(16.0 \text{ amu}) =} 342.0 \text{ amu}
\end{aligned}$$

(b) If a chemical formula has parentheses, the subscript outside the parentheses is a multiplier for all atoms inside. Thus, for $Ca(NO_3)_2$, we have

$$\begin{aligned}
1 \text{ Ca atom} &= 1(40.1 \text{ amu}) = 40.1 \text{ amu} \\
2 \text{ N atoms} &= 2(14.0 \text{ amu}) = 28.0 \text{ amu} \\
6 \text{ O atoms} &= 6(16.0 \text{ amu}) = \underline{96.0 \text{ amu}} \\
&\phantom{= 6(16.0 \text{ amu}) =} 164.1 \text{ amu}
\end{aligned}$$

■ PRACTICE EXERCISE

Calculate the formula weight of **(a)** $Al(OH)_3$ and **(b)** CH_3OH.
Answers: **(a)** 78.0 amu, **(b)** 32.0 amu

Percentage Composition from Formulas

Occasionally we must calculate the *percentage composition* of a compound—that is, the percentage by mass contributed by each element in the substance. For example, to verify the purity of a compound, we can compare the calculated percentage composition of the substance with that found experimentally. Forensic chemists, for example, will measure the percentage composition of an unknown white powder and compare it to the percentage compositions for sugar, salt, or cocaine to identify the powder. Calculating percentage composition is a straightforward matter if the chemical formula is known. The calculation depends on the formula weight of the substance, the atomic weight of the element of interest, and the number of atoms of that element in the chemical formula:

$$\% \text{ element} = \frac{\left(\begin{array}{c}\text{number of atoms} \\ \text{of that element}\end{array}\right)\left(\begin{array}{c}\text{atomic weight} \\ \text{of element}\end{array}\right)}{\text{formula weight of compound}} \times 100\% \quad [3.10]$$

■ SAMPLE EXERCISE 3.6 | Calculating Percentage Composition

Calculate the percentage of carbon, hydrogen, and oxygen (by mass) in $C_{12}H_{22}O_{11}$.

SOLUTION

Let's examine this question using the problem-solving steps in the "Strategies in Chemistry: Problem Solving" essay that appears on the next page.

Analyze We are given a chemical formula, $C_{12}H_{22}O_{11}$, and asked to calculate the percentage by mass of its component elements (C, H, and O).

Plan We can use Equation 3.10, relying on a periodic table to obtain the atomic weight of each component element. The atomic weights are first used to determine the formula weight of the compound. (The formula weight of $C_{12}H_{22}O_{11}$, 342.0 amu, was calculated in Sample Exercise 3.5.) We must then do three calculations, one for each element.

Solve Using Equation 3.10, we have

$$\%C = \frac{(12)(12.0 \text{ amu})}{342.0 \text{ amu}} \times 100\% = 42.1\%$$

$$\%H = \frac{(22)(1.0 \text{ amu})}{342.0 \text{ amu}} \times 100\% = 6.4\%$$

$$\%O = \frac{(11)(16.0 \text{ amu})}{342.0 \text{ amu}} \times 100\% = 51.5\%$$

Check The percentages of the individual elements must add up to 100%, which they do in this case. We could have used more significant figures for our atomic weights, giving more significant figures for our percentage composition, but we have adhered to our suggested guideline of rounding atomic weights to one digit beyond the decimal point.

■ PRACTICE EXERCISE

Calculate the percentage of nitrogen, by mass, in $Ca(NO_3)_2$.
Answer: 17.1%

3.4 AVOGADRO'S NUMBER AND THE MOLE

Even the smallest samples that we deal with in the laboratory contain enormous numbers of atoms, ions, or molecules. For example, a teaspoon of water (about 5 mL) contains 2×10^{23} water molecules, a number so large that it almost defies comprehension. Chemists, therefore, have devised a special counting unit for describing such large numbers of atoms or molecules.

In everyday life we use counting units such as a dozen (12 objects) and a gross (144 objects) to deal with modestly large quantities. In chemistry the unit for dealing with the number of atoms, ions, or molecules in a common-sized sample is the **mole**, abbreviated mol.* A mole is the amount of matter that contains as many objects (atoms, molecules, or whatever objects we are considering) as the number of atoms in exactly 12 g of isotopically pure ^{12}C. From experiments, scientists have determined this number to be 6.0221421×10^{23}. Scientists call this number **Avogadro's number**, in honor of the Italian scientist Amedeo Avogadro (1776–1856). Avogadro's number has the symbol N_A, which we will usually round to $6.02 \times 10^{23} \text{ mol}^{-1}$. The unit mol^{-1} ("inverse mole" or "per mole") reminds us that there are 6.02×10^{23} objects per one mole. A mole of atoms, a mole of molecules, or a mole of anything else all contain Avogadro's number of these objects:

$$1 \text{ mol } ^{12}C \text{ atoms} = 6.02 \times 10^{23} \, ^{12}C \text{ atoms}$$

$$1 \text{ mol } H_2O \text{ molecules} = 6.02 \times 10^{23} H_2O \text{ molecules}$$

$$1 \text{ mol } NO_3^- \text{ ions} = 6.02 \times 10^{23} NO_3^- \text{ ions}$$

Strategies in Chemistry   PROBLEM SOLVING

Practice is the key to success in problem solving. As you practice, you can improve your skills by following these steps:

Step 1: Analyze the problem. Read the problem carefully for understanding. What does it say? Draw any picture or diagram that will help you to visualize the problem. Write down both the data you are given and the quantity that you need to obtain (the unknown).

Step 2: Develop a plan for solving the problem. Consider the possible paths between the given information and the unknown. What principles or equations relate the known data to the unknown? Recognize that some data may not be given explicitly in the problem; you may be expected to know certain quantities (such as Avogadro's number) or look them up in tables (such as

atomic weights). Recognize also that your plan may involve either a single step or a series of steps with intermediate answers.

Step 3: Solve the problem. Use the known information and suitable equations or relationships to solve for the unknown. Dimensional analysis (Section 1.6) is a very useful tool for solving a great number of problems. Be careful with significant figures, signs, and units.

Step 4: Check the solution. Read the problem again to make sure you have found all the solutions asked for in the problem. Does your answer make sense? That is, is the answer outrageously large or small, or is it in the ballpark? Finally, are the units and significant figures correct?

---

*The term mole comes from the Latin word moles, meaning "a mass." The term molecule is the diminutive form of this word and means "a small mass."

Avogadro's number is so large that it is difficult to imagine. Spreading 6.02×10^{23} marbles over the entire surface of Earth would produce a continuous layer about 3 miles thick. If Avogadro's number of pennies were placed side by side in a straight line, they would encircle Earth 300 trillion (3×10^{14}) times.

■■ **SAMPLE EXERCISE 3.7** | **Estimating Numbers of Atoms**

Without using a calculator, arrange the following samples in order of increasing numbers of carbon atoms: 12 g ^{12}C, 1 mol C_2H_2, 9×10^{23} molecules of CO_2.

SOLUTION

Analyze We are given amounts of different substances expressed in grams, moles, and number of molecules and asked to arrange the samples in order of increasing numbers of C atoms.

Plan To determine the number of C atoms in each sample, we must convert g ^{12}C, 1 mol C_2H_2, and 9×10^{23} molecules CO_2 all to numbers of C atoms. To make these conversions, we use the definition of mole and Avogadro's number.

Solve A mole is defined as the amount of matter that contains as many units of the matter as there are C atoms in exactly 12 g of ^{12}C. Thus, 12 g of ^{12}C contains 1 mol of C atoms (that is, 6.02×10^{23} C atoms). One mol of C_2H_2 contains 6×10^{23} C_2H_2 molecules. Because there are two C atoms in each C_2H_2 molecule, this sample contains 12×10^{23} C atoms. Because each CO_2 molecule contains one C atom, the sample of CO_2 contains 9×10^{23} C atoms. Hence, the order is 12 g ^{12}C (6×10^{23} C atoms) < 9×10^{23} CO_2 molecules (9×10^{23} C atoms) < 1 mol C_2H_2 (12×10^{23} C atoms).

Check We can check our results by comparing the number of moles of C atoms in each sample because the number of moles is proportional to the number of atoms. Thus, 12 g of ^{12}C is 1 mol C; 1 mol of C_2H_2 contains 2 mol C, and 9×10^{23} molecules of CO_2 contain 1.5 mol C, giving the same order as above: 12 g ^{12}C (1 mol C) < 9×10^{23} CO_2 molecules (1.5 mol C) < 1 mol C_2H_2 (2 mol C).

■■ **PRACTICE EXERCISE**

Without using a calculator, arrange the following samples in order of increasing number of O atoms: 1 mol H_2O, 1 mol CO_2, 3×10^{23} molecules O_3.
Answer: 1 mol H_2O (6×10^{23} O atoms) < 3×10^{23} molecules O_3 (9×10^{23} O atoms) < 1 mol CO_2 (12×10^{23} O atoms)

■■ **SAMPLE EXERCISE 3.8** | **Converting Moles to Number of Atoms**

Calculate the number of H atoms in 0.350 mol of $C_6H_{12}O_6$.

SOLUTION

Analyze We are given both the amount of a substance (0.350 mol) and its chemical formula ($C_6H_{12}O_6$). The unknown is the number of H atoms in the sample.

Plan Avogadro's number provides the conversion factor between the number of moles of $C_6H_{12}O_6$ and the number of molecules of $C_6H_{12}O_6$. Once we know the number of molecules of $C_6H_{12}O_6$, we can use the chemical formula, which tells us that each molecule of $C_6H_{12}O_6$ contains 12 H atoms. Thus, we convert moles of $C_6H_{12}O_6$ to molecules of $C_6H_{12}O_6$ and then determine the number of atoms of H from the number of molecules of $C_6H_{12}O_6$:

$$\text{Moles } C_6H_{12}O_6 \longrightarrow \text{molecules } C_6H_{12}O_6 \longrightarrow \text{atoms H}$$

Solve

$$\text{H atoms} = (0.350 \text{ mol } C_6H_{12}O_6)\left(\frac{6.02 \times 10^{23} \text{ molecules } C_6H_{12}O_6}{1 \text{ mol } C_6H_{12}O_6}\right)\left(\frac{12 \text{ H atoms}}{1 \text{ molecule } C_6H_{12}O_6}\right)$$

$$= 2.53 \times 10^{24} \text{ H atoms}$$

Check The magnitude of our answer is reasonable. It is a large number about the magnitude of Avogadro's number. We can also make the following ballpark calculation: Multiplying $0.35 \times 6 \times 10^{23}$ gives about 2×10^{23} molecules. Multiplying this result by 12 gives $24 \times 10^{23} = 2.4 \times 10^{24}$ H atoms, which agrees with the previous, more detailed calculation. Because we were asked for the number of H atoms, the units of our answer are correct. The given data had three significant figures, so our answer has three significant figures.

■ **PRACTICE EXERCISE**

How many oxygen atoms are in **(a)** 0.25 mol $Ca(NO_3)_2$ and **(b)** 1.50 mol of sodium carbonate?
Answers: **(a)** 9.0×10^{23}, **(b)** 2.71×10^{24}

Molar Mass

A dozen (12) is the same number whether we have a dozen eggs or a dozen elephants. Clearly, however, a dozen eggs does not have the same mass as a dozen elephants. Similarly, a mole is always the *same number* (6.02×10^{23}), but 1-mole samples of different substances will have *different masses*. Compare, for example, 1 mol of ^{12}C and 1 mol of ^{24}Mg. A single ^{12}C atom has a mass of 12 amu, whereas a single ^{24}Mg atom is twice as massive, 24 amu (to two significant figures). Because a mole always has the same number of particles, a mole of ^{24}Mg must be twice as massive as a mole of ^{12}C. Because a mole of ^{12}C has a mass of 12 g (by definition), then a mole of ^{24}Mg must have a mass of 24 g. This example illustrates a general rule relating the mass of an atom to the mass of Avogadro's number (1 mol) of these atoms: *The mass of a single atom of an element (in amu) is numerically equal to the mass (in grams) of 1 mol of that element.* This statement is true regardless of the element:

1 atom of ^{12}C has a mass of 12 amu $\Rightarrow$ 1 mol ^{12}C has a mass of 12 g

1 atom of Cl has an atomic weight of 35.5 amu $\Rightarrow$ 1 mol Cl has a mass of 35.5 g

1 atom of Au has an atomic weight of 197 amu $\Rightarrow$ 1 mol Au has a mass of 197 g

Notice that when we are dealing with a particular isotope of an element, we use the mass of that isotope; otherwise we use the atomic weight (the average atomic mass) of the element. ∞ (Section 2.4)

For other kinds of substances, the same numerical relationship exists between the formula weight (in amu) and the mass (in grams) of one mole of that substance:

1 H_2O molecule has a mass of 18.0 amu $\Rightarrow$ 1 mol H_2O has a mass of 18.0 g

1 NO_3^- ion has a mass of 62.0 amu $\Rightarrow$ 1 mol NO_3^- has a mass of 62.0 g

1 NaCl unit has a mass of 58.5 amu $\Rightarrow$ 1 mol NaCl has a mass of 58.5 g

Figure 3.8 ▼ illustrates the relationship between the mass of a single molecule of H_2O and that of a mole of H_2O.

▲ **GIVE IT SOME THOUGHT**

(a) Which has more mass, a mole of water (H_2O) or a mole of glucose ($C_6H_{12}O_6$)?
(b) Which contains more molecules, a mole of water or a mole of glucose?

The mass in grams of one mole of a substance (that is, the mass in grams per mol) is called the **molar mass** of the substance. *The molar mass (in g/mol) of any substance is always numerically equal to its formula weight (in amu).* The substance NaCl, for example, has a formula weight of 58.5 amu and a molar mass

▶ **Figure 3.8 Comparing the mass of 1 molecule H_2O and 1 mol H_2O.**
Notice that both masses have the same number but have different units (18.0 amu compared to 18.0 g) representing the huge difference in mass.

Single molecule

1 molecule H_2O
(18.0 amu)

Avogadro's number of molecules
(6.02×10^{23})

Laboratory-size sample

1 mol H_2O
(18.0 g)

TABLE 3.2 ▪ Mole Relationships

Name of Substance	Formula	Formula Weight (amu)	Molar Mass (g/mol)	Number and Kind of Particles in One Mole
Atomic nitrogen	N	14.0	14.0	6.02×10^{23} N atoms
Molecular nitrogen	N_2	28.0	28.0	$\begin{cases} 6.02 \times 10^{23} \ N_2 \ \text{molecules} \\ 2(6.02 \times 10^{23}) \ \text{N atoms} \end{cases}$
Silver	Ag	107.9	107.9	6.02×10^{23} Ag atoms
Silver ions	Ag^+	107.9[a]	107.9	6.02×10^{23} Ag^+ ions
Barium chloride	$BaCl_2$	208.2	208.2	$\begin{cases} 6.02 \times 10^{23} \ BaCl_2 \ \text{units} \\ 6.02 \times 10^{23} \ Ba^{2+} \ \text{ions} \\ 2(6.02 \times 10^{23}) \ Cl^- \ \text{ions} \end{cases}$

[a] Recall that the electron has negligible mass; thus, ions and atoms have essentially the same mass.

▲ **Figure 3.9 One mole each of a solid, a liquid, and a gas.** One mole of NaCl, the solid, has a mass of 58.45 g. One mole of H_2O, the liquid, has a mass of 18.0 g and occupies a volume of 18.0 mL. One mole of O_2, the gas, has a mass of 32.0 g and occupies a balloon whose diameter is 35 cm.

of 58.5 g/mol. Further examples of mole relationships are shown in Table 3.2▲. Figure 3.9◄ shows 1-mole quantities of several common substances.

The entries in Table 3.2 for N and N_2 point out the importance of stating the exact chemical form of a substance when we use the mole concept. Suppose you read that 1 mol of nitrogen is produced in a particular reaction. You might interpret this statement to mean 1 mol of nitrogen atoms (14.0 g). Unless otherwise stated, however, what is probably meant is 1 mol of nitrogen molecules, N_2 (28.0 g), because N_2 is the most common chemical form of the element. To avoid ambiguity, it is important to state explicitly the chemical form being discussed. Using the chemical formula N_2 avoids ambiguity.

▪ SAMPLE EXERCISE 3.9 | Calculating Molar Mass

What is the mass in grams of 1.000 mol of glucose, $C_6H_{12}O_6$?

SOLUTION

Analyze We are given a chemical formula and asked to determine its molar mass.

Plan The molar mass of a substance is found by adding the atomic weights of its component atoms.

Solve

$$\begin{aligned} 6 \text{ C atoms} &= 6(12.0 \text{ amu}) = \quad 72.0 \text{ amu} \\ 12 \text{ H atoms} &= 12(1.0 \text{ amu}) = \quad 12.0 \text{ amu} \\ 6 \text{ O atoms} &= 6(16.0 \text{ amu}) = \underline{\quad 96.0 \text{ amu}} \\ & \qquad\qquad\qquad\qquad\quad 180.0 \text{ amu} \end{aligned}$$

Because glucose has a formula weight of 180.0 amu, one mole of this substance has a mass of 180.0 g. In other words, $C_6H_{12}O_6$ has a molar mass of 180.0 g/mol.

Check The magnitude of our answer seems reasonable, and g/mol is the appropriate unit for the molar mass.

Comment Glucose is sometimes called dextrose. Also known as blood sugar, glucose is found widely in nature, occurring in honey and fruits. Other types of sugars used as food are converted into glucose in the stomach or liver before the body uses them as energy sources. Because glucose requires no conversion, it is often given intravenously to patients who need immediate nourishment. People who have diabetes must carefully monitor the amount of glucose in their blood (See "Chemistry and Life" box in Section 3.6).

▪ PRACTICE EXERCISE

Calculate the molar mass of $Ca(NO_3)_2$.
Answer: 164.1 g/mol

Interconverting Masses and Moles

Conversions of mass to moles and of moles to mass are frequently encountered in calculations using the mole concept. These calculations are simplified using dimensional analysis, as shown in Sample Exercises 3.10 and 3.11.

■ **SAMPLE EXERCISE 3.10** | **Converting Grams to Moles**

Calculate the number of moles of glucose ($C_6H_{12}O_6$) in 5.380 g of $C_6H_{12}O_6$.

SOLUTION

Analyze We are given the number of grams of a substance and its chemical formula and asked to calculate the number of moles.

Plan The molar mass of a substance provides the factor for converting grams to moles. The molar mass of $C_6H_{12}O_6$ is 180.0 g/mol (Sample Exercise 3.9).

Solve Using 1 mol $C_6H_{12}O_6$ = 180.0 g $C_6H_{12}O_6$ to write the appropriate conversion factor, we have

$$\text{Moles } C_6H_{12}O_6 = (5.380 \text{ g } C_6H_{12}O_6)\left(\frac{1 \text{ mol } C_6H_{12}O_6}{180.0 \text{ g } C_6H_{12}O_6}\right) = 0.02989 \text{ mol } C_6H_{12}O_6$$

Check Because 5.380 g is less than the molar mass, a reasonable answer is less than one mole. The units of our answer (mol) are appropriate. The original data had four significant figures, so our answer has four significant figures.

■ **PRACTICE EXERCISE**

How many moles of sodium bicarbonate ($NaHCO_3$) are in 508 g of $NaHCO_3$?
Answer: 6.05 mol $NaHCO_3$

■ **SAMPLE EXERCISE 3.11** | **Converting Moles to Grams**

Calculate the mass, in grams, of 0.433 mol of calcium nitrate.

SOLUTION

Analyze We are given the number of moles and the name of a substance and asked to calculate the number of grams in the sample.

Plan To convert moles to grams, we need the molar mass, which we can calculate using the chemical formula and atomic weights.

Solve Because the calcium ion is Ca^{2+} and the nitrate ion is NO_3^-, calcium nitrate is $Ca(NO_3)_2$. Adding the atomic weights of the elements in the compound gives a formula weight of 164.1 amu. Using 1 mol $Ca(NO_3)_2$ = 164.1 g $Ca(NO_3)_2$ to write the appropriate conversion factor, we have

$$\text{Grams } Ca(NO_3)_2 = (0.433 \text{ mol } Ca(NO_3)_2)\left(\frac{164.1 \text{ g } Ca(NO_3)_2}{1 \text{ mol } Ca(NO_3)_2}\right) = 71.1 \text{ g } Ca(NO_3)_2$$

Check The number of moles is less than 1, so the number of grams must be less than the molar mass, 164.1 g. Using rounded numbers to estimate, we have 0.5 × 150 = 75 g. The magnitude of our answer is reasonable. Both the units (g) and the number of significant figures (3) are correct.

■ **PRACTICE EXERCISE**

What is the mass, in grams, of **(a)** 6.33 mol of $NaHCO_3$ and **(b)** 3.0×10^{-5} mol of sulfuric acid?
Answers: **(a)** 532 g, **(b)** 2.9×10^{-3} g

▶ Figure 3.10 **Procedure for interconverting the mass and the number of formula units of a substance.** The number of moles of the substance is central to the calculation; thus, the mole concept can be thought of as the bridge between the mass of a substance in grams and the number of formula units.

Interconverting Masses and Numbers of Particles

The mole concept provides the bridge between mass and the number of particles. To illustrate how we can interconvert mass and numbers of particles, let's calculate the number of copper atoms in an old copper penny. Such a penny weighs about 3 g, and we will assume that it is 100% copper:

$$\text{Cu atoms} = (3 \text{ g Cu})\left(\frac{1 \text{ mol Cu}}{63.5 \text{ g Cu}}\right)\left(\frac{6.02 \times 10^{23} \text{ Cu atoms}}{1 \text{ mol Cu}}\right)$$
$$= 3 \times 10^{22} \text{ Cu atoms}$$

We have rounded our answer to one significant figure, since we used only one significant figure for the mass of the penny. Notice how dimensional analysis ∞ (Section 1.6) provides a straightforward route from grams to numbers of atoms. The molar mass and Avogadro's number are used as conversion factors to convert grams → moles → atoms. Notice also that our answer is a very large number. Any time you calculate the number of atoms, molecules, or ions in an ordinary sample of matter, you can expect the answer to be very large. In contrast, the number of moles in a sample will usually be much smaller, often less than 1. The general procedure for interconverting mass and number of formula units (atoms, molecules, ions, or whatever is represented by the chemical formula) of a substance is summarized in Figure 3.10▲.

■ **SAMPLE EXERCISE 3.12** | Calculating the Number of Molecules
                                                     and Number of Atoms from Mass

(a) How many glucose molecules are in 5.23 g of $C_6H_{12}O_6$? **(b)** How many oxygen atoms are in this sample?

SOLUTION

Analyze We are given the number of grams and the chemical formula and asked to calculate (a) the number of molecules and (b) the number of O atoms in the sample.

(a) Plan The strategy for determining the number of molecules in a given quantity of a substance is summarized in Figure 3.10. We must convert 5.23 g $C_6H_{12}O_6$ to moles $C_6H_{12}O_6$, which can then be converted to molecules $C_6H_{12}O_6$. The first conversion uses the molar mass of $C_6H_{12}O_6$: 1 mol $C_6H_{12}O_6$ = 180.0 g $C_6H_{12}O_6$. The second conversion uses Avogadro's number.

Solve
Molecules $C_6H_{12}O_6$
$$= (5.23 \text{ g } C_6H_{12}O_6)\left(\frac{1 \text{ mol } C_6H_{12}O_6}{180.0 \text{ g } C_6H_{12}O_6}\right)\left(\frac{6.02 \times 10^{23} \text{ molecules } C_6H_{12}O_6}{1 \text{ mol } C_6H_{12}O_6}\right)$$
$$= 1.75 \times 10^{22} \text{ molecules } C_6H_{12}O_6$$

Check The magnitude of the answer is reasonable. Because the mass we began with is less than a mole, there should be fewer than 6.02×10^{23} molecules. We can make a ballpark estimate of the answer: $5/200 = 2.5 \times 10^{-2}$ mol; $2.5 \times 10^{-2} \times 6 \times 10^{23} = 15 \times 10^{21} = 1.5 \times 10^{22}$ molecules. The units (molecules) and significant figures (three) are appropriate.

(b) Plan To determine the number of O atoms, we use the fact that there are six O atoms in each molecule of $C_6H_{12}O_6$. Thus, multiplying the number of molecules $C_6H_{12}O_6$ by the factor (6 atoms O/1 molecule $C_6H_{12}O_6$) gives the number of O atoms.

Solve
$$\text{Atoms O} = (1.75 \times 10^{22} \text{ molecules } C_6H_{12}O_6)\left(\frac{6 \text{ atoms O}}{1 \text{ molecule } C_6H_{12}O_6}\right)$$
$$= 1.05 \times 10^{23} \text{ atoms O}$$

Check The answer is simply 6 times as large as the answer to part (a). The number of significant figures (three) and the units (atoms O) are correct.

3.5 EMPIRICAL FORMULAS FROM ANALYSES

As we learned in Section 2.6, the empirical formula for a substance tells us the relative number of atoms of each element it contains. The empirical formula H_2O shows that water contains two H atoms for each O atom. This ratio also applies on the molar level: 1 mol of H_2O contains 2 mol of H atoms and 1 mol of O atoms. Conversely, *the ratio of the number of moles of each element in a compound gives the subscripts in a compound's empirical formula.* In this way, the mole concept provides a way of calculating the empirical formulas of chemical substances, as shown in the following examples.

Mercury and chlorine combine to form a compound that is 73.9% mercury and 26.1% chlorine by mass. This means that if we had a 100.0-g sample of the solid, it would contain 73.9 g of mercury (Hg) and 26.1 g of chlorine (Cl). (Any size sample can be used in problems of this type, but we will generally use 100.0 g to simplify the calculation of mass from percentage.) Using the atomic weights of the elements to give us molar masses, we can calculate the number of moles of each element in the sample:

$$(73.9 \text{ g Hg})\left(\frac{1 \text{ mol Hg}}{200.6 \text{ g Hg}}\right) = 0.368 \text{ mol Hg}$$

$$(26.1 \text{ g Cl})\left(\frac{1 \text{ mol Cl}}{35.5 \text{ g Cl}}\right) = 0.735 \text{ mol Cl}$$

We then divide the larger number of moles (0.735) by the smaller (0.368) to obtain a Cl: Hg mole ratio of 1.99:1:

$$\frac{\text{moles of Cl}}{\text{moles of Hg}} = \frac{0.735 \text{ mol Cl}}{0.368 \text{ mol Hg}} = \frac{1.99 \text{ mol Cl}}{1 \text{ mol Hg}}$$

Because of experimental errors, the results may not lead to exact integers for the ratios of moles. The number 1.99 is very close to 2, so we can confidently conclude that the empirical formula for the compound is $HgCl_2$. The empirical formula is correct because its subscripts are the smallest integers that express the *ratios* of atoms present in the compound. ∞ (Section 2.6) The general procedure for determining empirical formulas is outlined in Figure 3.11 ▼.

Given:

Find:

▲ Figure 3.11 **Procedure for calculating an empirical formula from percentage composition.** The central part of the calculation is determining the number of moles of each element in the compound. The procedure is also summarized as "percent to mass, mass to mole, divide by small, multiply 'til whole."

■ **SAMPLE EXERCISE 3.13** | Calculating an Empirical Formula

Ascorbic acid (vitamin C) contains 40.92% C, 4.58% H, and 54.50% O by mass. What is the empirical formula of ascorbic acid?

SOLUTION

Analyze We are to determine an empirical formula of a compound from the mass percentages of its elements.

Plan The strategy for determining the empirical formula involves the three steps given in Figure 3.11.

Solve We *first* assume, for simplicity, that we have exactly 100 g of material (although any mass can be used). In 100 g of ascorbic acid, therefore, we have

40.92 g C, 4.58 g H, and 54.50 g O.

Second, we calculate the number of moles of each element:

$$\text{Moles C} = (40.92 \text{ g C})\left(\frac{1 \text{ mol C}}{12.01 \text{ g C}}\right) = 3.407 \text{ mol C}$$

$$\text{Moles H} = (4.58 \text{ g H})\left(\frac{1 \text{ mol H}}{1.008 \text{ g H}}\right) = 4.54 \text{ mol H}$$

$$\text{Moles O} = (54.50 \text{ g O})\left(\frac{1 \text{ mol O}}{16.00 \text{ g O}}\right) = 3.406 \text{ mol O}$$

Third, we determine the simplest whole-number ratio of moles by dividing each number of moles by the smallest number of moles, 3.406:

$$\text{C:} \frac{3.407}{3.406} = 1.000 \qquad \text{H:} \frac{4.54}{3.406} = 1.33 \qquad \text{O:} \frac{3.406}{3.406} = 1.000$$

The ratio for H is too far from 1 to attribute the difference to experimental error; in fact, it is quite close to $1\frac{1}{3}$. This suggests that if we multiply the ratio by 3, we will obtain whole numbers:

$$\text{C:H:O} = 3(1:1.33:1) = 3:4:3$$

The whole-number mole ratio gives us the subscripts for the empirical formula:

$$C_3H_4O_3$$

Check It is reassuring that the subscripts are moderately sized whole numbers. Otherwise, we have little by which to judge the reasonableness of our answer.

■ **PRACTICE EXERCISE**

A 5.325-g sample of methyl benzoate, a compound used in the manufacture of perfumes, contains 3.758 g of carbon, 0.316 g of hydrogen, and 1.251 g of oxygen. What is the empirical formula of this substance?
Answer: C_4H_4O

Molecular Formula from Empirical Formula

For any compound, the formula obtained from percentage compositions is always the empirical formula. We can obtain the molecular formula from the empirical formula if we are given the molecular weight or molar mass of the compound. *The subscripts in the molecular formula of a substance are always a whole-number multiple of the corresponding subscripts in its empirical formula.* ∞ (Section 2.6) This multiple can be found by comparing the empirical formula weight with the molecular weight:

$$\text{Whole-number multiple} = \frac{\text{molecular weight}}{\text{empirical formula weight}} \qquad [3.11]$$

In Sample Exercise 3.13, for example, the empirical formula of ascorbic acid was determined to be $C_3H_4O_3$, giving an empirical formula weight of 3(12.0 amu) + 4(1.0 amu) + 3(16.0 amu) = 88.0 amu. The experimentally determined molecular weight is 176 amu. Thus, the molecular weight is 2 times the empirical formula weight (176/88.0 = 2.00), and the molecular formula must therefore have twice as many of each kind of atom as the empirical formula. Consequently, we multiply the subscripts in the empirical formula by 2 to obtain the molecular formula: $C_6H_8O_6$.

■ **SAMPLE EXERCISE 3.14** | Determining a Molecular Formula

Mesitylene, a hydrocarbon that occurs in small amounts in crude oil, has an empirical formula of C_3H_4. The experimentally determined molecular weight of this substance is 121 amu. What is the molecular formula of mesitylene?

SOLUTION

Analyze We are given an empirical formula and a molecular weight and asked to determine a molecular formula.

Plan The subscripts in the molecular formula of a compound are whole-number multiples of the subscripts in its empirical formula. To find the appropriate multiple, we must compare the molecular weight with the formula weight of the empirical formula.

Solve First, we calculate the formula weight of the empirical formula, C_3H_4:

$$3(12.0 \text{ amu}) + 4(1.0 \text{ amu}) = 40.0 \text{ amu}$$

Next, we divide the molecular weight by the empirical formula weight to obtain the multiple used to multiply the subscripts in C_3H_4:

$$\frac{\text{Molecular weight}}{\text{Empirical formula weight}} = \frac{121}{40.0} = 3.02$$

Only whole-number ratios make physical sense because we must be dealing with whole atoms. The 3.02 in this case could result from a small experimental error in the molecular weight. We therefore multiply each subscript in the empirical formula by 3 to give the molecular formula: C_9H_{12}.

Check We can have confidence in the result because dividing the molecular weight by the formula weight yields nearly a whole number.

■ PRACTICE EXERCISE

Ethylene glycol, the substance used in automobile antifreeze, is composed of 38.7% C, 9.7% H, and 51.6% O by mass. Its molar mass is 62.1 g/mol. **(a)** What is the empirical formula of ethylene glycol? **(b)** What is its molecular formula?
Answers: **(a)** CH_3O, **(b)** $C_2H_6O_2$

Combustion Analysis

The empirical formula of a compound is based on experiments that give the number of moles of each element in a sample of the compound. The word "empirical" means "based on observation and experiment." Chemists have devised a number of experimental techniques to determine empirical formulas. One technique is combustion analysis, which is commonly used for compounds containing principally carbon and hydrogen as their component elements.

When a compound containing carbon and hydrogen is completely combusted in an apparatus such as that shown in Figure 3.12▶, the carbon in the compound is converted to CO_2, and the hydrogen is converted to H_2O. ∞ (Section 3.2) The amounts of CO_2 and H_2O produced are determined by measuring the mass increase in the CO_2 and H_2O absorbers. From the masses of CO_2 and H_2O we can calculate the number of moles of C and H in the original compound and thereby the empirical formula. If a third element is present in the compound, its mass can be determined by subtracting the masses of C and H from the compound's original mass. Sample Exercise 3.15 shows how to determine the empirical formula of a compound containing C, H, and O.

▲ Figure 3.12 **Apparatus to determine percentages of carbon and hydrogen in a compound.** The compound is combusted to form CO_2 and H_2O. Copper oxide helps to oxidize traces of carbon and carbon monoxide to carbon dioxide and to oxidize hydrogen to water.

■ SAMPLE EXERCISE 3.15 | Determining Empirical Formula by Combustion Analysis

Isopropyl alcohol, a substance sold as rubbing alcohol, is composed of C, H, and O. Combustion of 0.255 g of isopropyl alcohol produces 0.561 g of CO_2 and 0.306 g of H_2O. Determine the empirical formula of isopropyl alcohol.

SOLUTION

Analyze We are told that isopropyl alcohol contains C, H, and O atoms and given the quantities of CO_2 and H_2O produced when a given quantity of the alcohol is combusted. We must use this information to determine the empirical formula for isopropyl alcohol, a task that requires us to calculate the number of moles of C, H, and O in the sample.

Plan We can use the mole concept to calculate the number of grams of C present in the CO_2 and the number of grams of H present in the H_2O. These amounts are the quantities of C and H present in the isopropyl alcohol before combustion. The number of grams of O in the compound equals the mass of the isopropyl alcohol minus the sum of the C and H masses. Once we have the number of grams of C, H, and O in the sample, we can then proceed as in Sample Exercise 3.13. We can calculate the number of moles of each element, and determine the mole ratio, which gives the subscripts in the empirical formula.

Solve To calculate the number of grams of C, we first use the molar mass of CO_2, 1 mol CO_2 = 44.0 g CO_2, to convert grams of CO_2 to moles of CO_2. Because each CO_2 molecule has only 1 C atom, there is 1 mol of C atoms per mole of CO_2 molecules. This fact allows us to convert the moles of CO_2 to moles of C. Finally, we use the molar mass of C, 1 mol C = 12.0 g C, to convert moles of C to grams of C. Combining the three conversion factors, we have

$$\text{Grams C} = (0.561 \text{ g } CO_2)\left(\frac{1 \text{ mol } CO_2}{44.0 \text{ g } CO_2}\right)\left(\frac{1 \text{ mol C}}{1 \text{ mol } CO_2}\right)\left(\frac{12.0 \text{ g C}}{1 \text{ mol C}}\right) = 0.153 \text{ g C}$$

The calculation of the number of grams of H from the grams of H_2O is similar, although we must remember that there are 2 mol of H atoms per 1 mol of H_2O molecules:

$$\text{Grams H} = (0.306 \text{ g } H_2O)\left(\frac{1 \text{ mol } H_2O}{18.0 \text{ g } H_2O}\right)\left(\frac{2 \text{ mol H}}{1 \text{ mol } H_2O}\right)\left(\frac{1.01 \text{ g H}}{1 \text{ mol H}}\right) = 0.0343 \text{ g H}$$

The total mass of the sample, 0.255 g, is the sum of the masses of the C, H, and O. Thus, we can calculate the mass of O as follows:

$$\begin{aligned}\text{Mass of O} &= \text{mass of sample} - (\text{mass of C} + \text{mass of H}) \\ &= 0.255 \text{ g} - (0.153 \text{ g} + 0.0343 \text{ g}) = 0.068 \text{ g O}\end{aligned}$$

We then calculate the number of moles of C, H, and O in the sample:

$$\text{Moles C} = (0.153 \text{ g C})\left(\frac{1 \text{ mol C}}{12.0 \text{ g C}}\right) = 0.0128 \text{ mol C}$$

$$\text{Moles H} = (0.0343 \text{ g H})\left(\frac{1 \text{ mol H}}{1.01 \text{ g H}}\right) = 0.0340 \text{ mol H}$$

$$\text{Moles O} = (0.068 \text{ g O})\left(\frac{1 \text{ mol O}}{16.0 \text{ g O}}\right) = 0.0043 \text{ mol O}$$

To find the empirical formula, we must compare the relative number of moles of each element in the sample. The relative number of moles of each element is found by dividing each number by the smallest number, 0.0043. The mole ratio of C:H:O so obtained is 2.98:7.91:1.00. The first two numbers are very close to the whole numbers 3 and 8, giving the empirical formula C_3H_8O.

Check The subscripts work out to be moderately sized whole numbers, as expected.

■ **PRACTICE EXERCISE**

(a) Caproic acid, which is responsible for the foul odor of dirty socks, is composed of C, H, and O atoms. Combustion of a 0.225-g sample of this compound produces 0.512 g CO_2 and 0.209 g H_2O. What is the empirical formula of caproic acid? **(b)** Caproic acid has a molar mass of 116 g/mol. What is its molecular formula?
Answers: **(a)** C_3H_6O, **(b)** $C_6H_{12}O_2$

▲ **GIVE IT SOME THOUGHT**

In Sample Exercise 3.15, how do you explain the fact that the ratios C:H:O are 2.98:7.91:1.00, rather than exact integers 3:8:1?

3.6 QUANTITATIVE INFORMATION FROM BALANCED EQUATIONS

The coefficients in a chemical equation represent the relative numbers of molecules in a reaction. The mole concept allows us to convert this information to the masses of the substances. Consider the following balanced equation:

$$2 H_2(g) + O_2(g) \longrightarrow 2 H_2O(l) \qquad [3.12]$$

The coefficients indicate that two molecules of H_2 react with each molecule of O_2 to form two molecules of H_2O. It follows that the relative numbers of moles are identical to the relative numbers of molecules:

$2 H_2(g)$	+	$O_2(g)$	$\longrightarrow$	$2 H_2O(l)$
2 molecules		1 molecule		2 molecules
2(6.02 × 10²³ molecules)		1(6.02 × 10²³ molecules)		2(6.02 × 10²³ molecules)
2 mol		1 mol		2 mol

We can generalize this observation for all balanced chemical equations: *The coefficients in a balanced chemical equation indicate both the relative numbers of molecules (or formula units) in the reaction and the relative numbers of moles.* Table 3.3 ▼ further summarizes this result and shows how it corresponds to the law of conservation of mass. Notice that the total mass of the reactants (4.0 g + 32.0 g) equals the total mass of the products (36.0 g).

TABLE 3.3 ■ Information from a Balanced Equation				
Equation:	**2 H$_2$(g)**	**+ O$_2$(g)**	**⟶**	**2 H$_2$O(l)**
Molecules:	2 molecules H$_2$	+ 1 molecule O$_2$	⟶	2 molecules H$_2$O
Mass (amu):	4.0 amu H$_2$	+ 32.0 amu O$_2$	⟶	36.0 amu H$_2$O
Amount (mol):	2 mol H$_2$	+ 1 mol O$_2$	⟶	2 mol H$_2$O
Mass (g):	4.0 g H$_2$	+ 32.0 g O$_2$	⟶	36.0 g H$_2$O

The quantities 2 mol H$_2$, 1 mol O$_2$, and 2 mol H$_2$O, which are given by the coefficients in Equation 3.12, are called *stoichiometrically equivalent quantities.* The relationship between these quantities can be represented as

$$2 \text{ mol } H_2 \simeq 1 \text{ mol } O_2 \simeq 2 \text{ mol } H_2O$$

where the $\simeq$ symbol means "is stoichiometrically equivalent to." In other words, Equation 3.12 shows 2 mol of H$_2$ and 1 mol of O$_2$ forming 2 mol of H$_2$O. These stoichiometric relations can be used to convert between quantities of reactants and products in a chemical reaction. For example, the number of moles of H$_2$O produced from 1.57 mol of O$_2$ can be calculated as follows:

$$\text{Moles } H_2O = (1.57 \text{ mol } O_2)\left(\frac{2 \text{ mol } H_2O}{1 \text{ mol } O_2}\right) = 3.14 \text{ mol } H_2O$$

GIVE IT SOME THOUGHT

When 1.57 mol O$_2$ reacts with H$_2$ to form H$_2$O, how many moles of H$_2$ are consumed in the process?

As an additional example, consider the combustion of butane (C$_4$H$_{10}$), the fuel in disposable cigarette lighters:

$$2 \text{ C}_4\text{H}_{10}(l) + 13 \text{ O}_2(g) \longrightarrow 8 \text{ CO}_2(g) + 10 \text{ H}_2\text{O}(g) \qquad [3.13]$$

Let's calculate the mass of CO$_2$ produced when 1.00 g of C$_4$H$_{10}$ is burned. The coefficients in Equation 3.13 tell how the amount of C$_4$H$_{10}$ consumed is related to the amount of CO$_2$ produced: 2 mol C$_4$H$_{10}$ $\simeq$ 8 mol CO$_2$. To use this relationship we must use the molar mass of C$_4$H$_{10}$ to convert grams of C$_4$H$_{10}$ to moles of C$_4$H$_{10}$. Because 1 mol C$_4$H$_{10}$ = 58.0 g C$_4$H$_{10}$, we have

$$\text{Moles } C_4H_{10} = (1.00 \text{ g } C_4H_{10})\left(\frac{1 \text{ mol } C_4H_{10}}{58.0 \text{ g } C_4H_{10}}\right)$$

$$= 1.72 \times 10^{-2} \text{ mol } C_4H_{10}$$

We can then use the stoichiometric factor from the balanced equation, 2 mol C$_4$H$_{10}$ $\simeq$ 8 mol CO$_2$, to calculate moles of CO$_2$:

$$\text{Moles } CO_2 = (1.72 \times 10^{-2} \text{ mol } C_4H_{10})\left(\frac{8 \text{ mol } CO_2}{2 \text{ mol } C_4H_{10}}\right)$$

$$= 6.88 \times 10^{-2} \text{ mol } CO_2$$

Finally, we can calculate the mass of the CO_2, in grams, using the molar mass of CO_2 (1 mol CO_2 = 44.0 g CO_2):

$$\text{Grams } CO_2 = (6.88 \times 10^{-2} \text{ mol } CO_2)\left(\frac{44.0 \text{ g } CO_2}{1 \text{ mol } CO_2}\right)$$

$$= 3.03 \text{ g } CO_2$$

Thus, the conversion sequence is

These steps can be combined in a single sequence of factors:

$$\text{Grams } CO_2 = (1.00 \text{ g } C_4H_{10})\left(\frac{1 \text{ mol } C_4H_{10}}{58.0 \text{ g } C_4H_{10}}\right)\left(\frac{8 \text{ mol } CO_2}{2 \text{ mol } C_4H_{10}}\right)\left(\frac{44.0 \text{ g } CO_2}{1 \text{ mol } CO_2}\right)$$

$$= 3.03 \text{ g } CO_2$$

Similarly, we can calculate the amount of O_2 consumed or H_2O produced in this reaction. For example, to calculate the amount of O_2 consumed, we again rely on the coefficients in the balanced equation to give us the appropriate stoichiometric factor: 2 mol $C_4H_{10} \simeq 13$ mol O_2:

$$\text{Grams } O_2 = (1.00 \text{ g } C_4H_{10})\left(\frac{1 \text{ mol } C_4H_{10}}{58.0 \text{ g } C_4H_{10}}\right)\left(\frac{13 \text{ mol } O_2}{2 \text{ mol } C_4H_{10}}\right)\left(\frac{32.0 \text{ g } O_2}{1 \text{ mol } O_2}\right)$$

$$= 3.59 \text{ g } O_2$$

Figure 3.13▼ summarizes the general approach used to calculate the quantities of substances consumed or produced in chemical reactions. The balanced chemical equation provides the relative numbers of moles of reactants and products in the reaction.

▶ Figure 3.13 **The procedure for calculating amounts of reactants or products in a reaction.** The number of grams of a reactant consumed or of a product formed in a reaction can be calculated, starting with the number of grams of one of the other reactants or products. Notice how molar masses and the coefficients in the balanced equation are used.

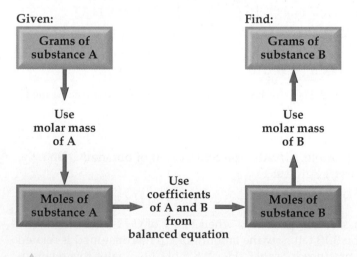

GIVE IT SOME THOUGHT

If 20.00 g of a compound reacts completely with 30.00 g of another compound in a combination reaction, how many grams of product were formed?

■ **SAMPLE EXERCISE 3.16** | Calculating Amounts of Reactants and Products

How many grams of water are produced in the oxidation of 1.00 g of glucose, $C_6H_{12}O_6$?

$$C_6H_{12}O_6(s) + 6\,O_2(g) \longrightarrow 6\,CO_2(g) + 6\,H_2O(l)$$

SOLUTION

Analyze We are given the mass of a reactant and are asked to determine the mass of a product in the given equation.

Plan The general strategy, as outlined in Figure 3.13, requires three steps. First, the amount of $C_6H_{12}O_6$ must be converted from grams to moles. Second, we can use the balanced equation, which relates the moles of $C_6H_{12}O_6$ to the moles of H_2O: 1 mol $C_6H_{12}O_6 \simeq$ 6 mol H_2O. Third, we must convert the moles of H_2O to grams.

Solve First, use the molar mass of $C_6H_{12}O_6$ to convert from grams $C_6H_{12}O_6$ to moles $C_6H_{12}O_6$:

$$\text{Moles } C_6H_{12}O_6 = (1.00 \text{ g } C_6H_{12}O_6)\left(\frac{1 \text{ mol } C_6H_{12}O_6}{180.0 \text{ g } C_6H_{12}O_6}\right)$$

Second, use the balanced equation to convert moles of $C_6H_{12}O_6$ to moles of H_2O:

$$\text{Moles } H_2O = (1.00 \text{ g } C_6H_{12}O_6)\left(\frac{1 \text{ mol } C_6H_{12}O_6}{180.0 \text{ g } C_6H_{12}O_6}\right)\left(\frac{6 \text{ mol } H_2O}{1 \text{ mol } C_6H_{12}O_6}\right)$$

Third, use the molar mass of H_2O to convert from moles of H_2O to grams of H_2O:

$$\text{Grams } H_2O = (1.00 \text{ g } C_6H_{12}O_6)\left(\frac{1 \text{ mol } C_6H_{12}O_6}{180.0 \text{ g } C_6H_{12}O_6}\right)\left(\frac{6 \text{ mol } H_2O}{1 \text{ mol } C_6H_{12}O_6}\right)\left(\frac{18.0 \text{ g } H_2O}{1 \text{ mol } H_2O}\right)$$
$$= 0.600 \text{ g } H_2O$$

The steps can be summarized in a diagram like that in Figure 3.13:

Check An estimate of the magnitude of our answer, 18/180 = 0.1 and 0.1 × 6 = 0.6, agrees with the exact calculation. The units, grams H_2O, are correct. The initial data had three significant figures, so three significant figures for the answer is correct.

Comment An average person ingests 2 L of water daily and eliminates 2.4 L. The difference between 2 L and 2.4 L is produced in the metabolism of foodstuffs, such as in the oxidation of glucose. (*Metabolism* is a general term used to describe all the chemical processes of a living animal or plant.) The desert rat (kangaroo rat), on the other hand, apparently never drinks water. It survives on its metabolic water.

■ **PRACTICE EXERCISE**

The decomposition of $KClO_3$ is commonly used to prepare small amounts of O_2 in the laboratory: $2 KClO_3(s) \longrightarrow 2 KCl(s) + 3 O_2(g)$. How many grams of O_2 can be prepared from 4.50 g of $KClO_3$?
Answer: 1.77 g

■ **SAMPLE EXERCISE 3.17** | Calculating Amounts of Reactants and Products

Solid lithium hydroxide is used in space vehicles to remove the carbon dioxide exhaled by astronauts. The lithium hydroxide reacts with gaseous carbon dioxide to form solid lithium carbonate and liquid water. How many grams of carbon dioxide can be absorbed by 1.00 g of lithium hydroxide?

SOLUTION

Analyze We are given a verbal description of a reaction and asked to calculate the number of grams of one reactant that reacts with 1.00 g of another.

Plan The verbal description of the reaction can be used to write a balanced equation:

$$2 LiOH(s) + CO_2(g) \longrightarrow Li_2CO_3(s) + H_2O(l)$$

We are given the grams of LiOH and asked to calculate grams of CO_2. We can accomplish this task by using the following sequence of conversions:

$$\text{Grams LiOH} \longrightarrow \text{moles LiOH} \longrightarrow \text{moles } CO_2 \longrightarrow \text{grams } CO_2$$

The conversion from grams of LiOH to moles of LiOH requires the molar mass of LiOH (6.94 + 16.00 + 1.01 = 23.95 g/mol). The conversion of moles of LiOH to moles of CO_2 is based on the balanced chemical equation: 2 mol LiOH $\simeq$ 1 mol CO_2. To convert the number of moles of CO_2 to grams, we must use the molar mass of CO_2: 12.01 + 2(16.00) = 44.01 g/mol.

Solve

$$(1.00 \text{ g LiOH})\left(\frac{1 \text{ mol LiOH}}{23.95 \text{ g LiOH}}\right)\left(\frac{1 \text{ mol CO}_2}{2 \text{ mol LiOH}}\right)\left(\frac{44.01 \text{ g CO}_2}{1 \text{ mol CO}_2}\right) = 0.919 \text{ g CO}_2$$

Check Notice that $23.95 \approx 24$, $24 \times 2 = 48$, and $44/48$ is slightly less than 1. The magnitude of the answer is reasonable based on the amount of starting LiOH; the significant figures and units are appropriate, too.

■ PRACTICE EXERCISE

Propane, C_3H_8, is a common fuel used for cooking and home heating. What mass of O_2 is consumed in the combustion of 1.00 g of propane?
Answer: 3.64 g

Chemistry and Life   GLUCOSE MONITORING

Over 20 million Americans have diabetes. In the world, the number approaches 172 million. Diabetes is a disorder of metabolism in which the body cannot produce or properly use the hormone insulin. One signal that a person is diabetic is that the concentration of glucose in her or his blood is higher than normal. Therefore, people who are diabetic need to measure their blood glucose concentrations regularly. Untreated diabetes can cause severe complications such as blindness and loss of limbs.

How does insulin relate to glucose? The body converts most of the food we eat into glucose. After digestion, glucose is delivered to cells via the bloodstream; cells need glucose to live. Insulin must be present for glucose to enter the cells. Normally, the body adjusts the concentration of insulin automatically, in concert with the glucose concentration after eating. However, in a diabetic person, little or no insulin is produced (Type 1 diabetes), or the cells cannot take up insulin properly (Type 2 diabetes). The result is that the blood glucose concentration is too high. People normally have a range of 70–120 mg glucose per deciliter of blood (about 4–6 mmol glucose per liter of blood). If a person has not eaten for 8 hours or more, he or she would be diagnosed as diabetic if the glucose levels were 126 mg/dL or higher. In the United States, diabetics monitor their blood glucose concentrations in mg/dL; in Europe, they use different units—millimoles glucose per liter of blood.

Glucose monitors work by the introduction of blood from a person, usually by a prick of the finger, onto a small strip of paper that contains numerous chemicals that react specifically with glucose. Insertion of the strip into a small battery-operated reader gives the glucose concentration (Figure 3.14▼). The actual mechanism of the readout varies for different devices—it may be a small electrical current or a measure of light produced in a chemical reaction. Depending on the result, a diabetic person may need to receive an injection of insulin or simply stop eating sweets for a while.

▲ **Figure 3.14 Glucose meter.** This is an example of a commercial glucose meter and its readout.

3.7 LIMITING REACTANTS

Suppose you wish to make several sandwiches using one slice of cheese and two slices of bread for each sandwich. Using Bd = bread, Ch = cheese, and Bd_2Ch = sandwich, the recipe for making a sandwich can be represented like a chemical equation:

$$2 \text{ Bd} + \text{Ch} \longrightarrow Bd_2Ch$$

If you have 10 slices of bread and 7 slices of cheese, you will be able to make only five sandwiches before you run out of bread. You will have 2 slices of cheese left over. The amount of available bread limits the number of sandwiches.

An analogous situation occurs in chemical reactions when one of the reactants is used up before the others. The reaction stops as soon as any one of the reactants is totally consumed, leaving the excess reactants as leftovers.

Suppose, for example, that we have a mixture of 10 mol H_2 and 7 mol O_2, which react to form water:

$$2 H_2(g) + O_2(g) \longrightarrow 2 H_2O(g)$$

Because 2 mol $H_2 \doteq 1$ mol O_2, the number of moles of O_2 needed to react with all the H_2 is

$$\text{Moles } O_2 = (10 \text{ mol } H_2)\left(\frac{1 \text{ mol } O_2}{2 \text{ mol } H_2}\right) = 5 \text{ mol } O_2$$

Because 7 mol O_2 was available at the start of the reaction, 7 mol $O_2 - 5$ mol O_2 = 2 mol O_2 will still be present when all the H_2 is consumed. The example we have considered is depicted on a molecular level in Figure 3.15▶.

The reactant that is completely consumed in a reaction is called either the **limiting reactant** or *limiting reagent* because it determines, or limits, the amount of product formed. The other reactants are sometimes called either *excess reactants* or *excess reagents*. In our example, H_2 is the limiting reactant, which means that once all the H_2 has been consumed, the reaction stops. O_2 is the excess reactant, and some is left over when the reaction stops.

There are no restrictions on the starting amounts of the reactants in any reaction. Indeed, many reactions are carried out using an excess of one reagent. The quantities of reactants consumed and the quantities of products formed, however, are restricted by the quantity of the limiting reactant. When a combustion reaction takes place in the open air, oxygen is plentiful and is therefore the excess reactant. You may have had the unfortunate experience of running out of gasoline while driving. The car stops because you have run out of the limiting reactant in the combustion reaction, the fuel.

Before we leave our present example, let's summarize the data in a tabular form:

Before reaction

10 H_2 and 7 O_2

After reaction

10 H_2O and 2 O_2

▲ **Figure 3.15 Example illustrating a limiting reactant.** Because the H_2 is completely consumed, it is the limiting reagent in this case. Because there is a stoichiometric excess of O_2, some is left over at the end of the reaction. The amount of H_2O formed is related directly to the amount of H_2 consumed.

	2 $H_2(g)$	+	$O_2(g)$	$\longrightarrow$	2 $H_2O(g)$
Initial quantities:	10 mol		7 mol		0 mol
Change (reaction):	−10 mol		−5 mol		+10 mol
Final quantities:	0 mol		2 mol		10 mol

The initial amounts of the reactants are what we started with (10 mol H_2 and 7 mol O_2). The second line in the table (change) summarizes the amounts of the reactants consumed and the amount of the product formed in the reaction. These quantities are restricted by the quantity of the limiting reactant and depend on the coefficients in the balanced equation. The mole ratio H_2: O_2: H_2O = 10:5:10 conforms to the ratio of the coefficients in the balanced equation, 2:1:2. The changes are negative for the reactants because they are consumed during the reaction and positive for the product because it is formed during the reaction. Finally, the quantities in the third line of the table (final quantities) depend on the initial quantities and their changes, and these entries are found by adding the entries for the initial quantity and change for each column. No amount of the limiting reactant (H_2) remains at the end of the reaction. All that remains is 2 mol O_2 and 10 mol H_2O.

■■■ **SAMPLE EXERCISE 3.18** | Calculating the Amount of Product Formed from a Limiting Reactant

The most important commercial process for converting N_2 from the air into nitrogen-containing compounds is based on the reaction of N_2 and H_2 to form ammonia (NH_3):

$$N_2(g) + 3 H_2(g) \longrightarrow 2 NH_3(g)$$

How many moles of NH_3 can be formed from 3.0 mol of N_2 and 6.0 mol of H_2?

SOLUTION

Analyze We are asked to calculate the number of moles of product, NH_3, given the quantities of each reactant, N_2 and H_2, available in a reaction. Thus, this is a limiting reactant problem.

Plan If we assume that one reactant is completely consumed, we can calculate how much of the second reactant is needed in the reaction. By comparing the calculated quantity with the available amount, we can determine which reactant is limiting. We then proceed with the calculation, using the quantity of the limiting reactant.

Solve The number of moles of H_2 needed for complete consumption of 3.0 mol of N_2 is:

$$\text{Moles } H_2 = (3.0 \text{ mol } N_2)\left(\frac{3 \text{ mol } H_2}{1 \text{ mol } N_2}\right) = 9.0 \text{ mol } H_2$$

Because only 6.0 mol H_2 is available, we will run out of H_2 before the N_2 is gone, and H_2 will be the limiting reactant. We use the quantity of the limiting reactant, H_2, to calculate the quantity of NH_3 produced:

$$\text{Moles } NH_3 = (6.0 \text{ mol } H_2)\left(\frac{2 \text{ mol } NH_3}{3 \text{ mol } H_2}\right) = 4.0 \text{ mol } NH_3$$

Comment The table on the right summarizes this example:

	$N_2(g)$	+	$3 H_2(g)$	$\longrightarrow$	$2 NH_3(g)$
Initial quantities:	3.0 mol		6.0 mol		0 mol
Change (reaction):	−2.0 mol		−6.0 mol		+4.0 mol
Final quantities:	1.0 mol		0 mol		4.0 mol

Notice that we can calculate not only the number of moles of NH_3 formed but also the number of moles of each of the reactants remaining after the reaction. Notice also that although the number of moles of H_2 present at the beginning of the reaction is greater than the number of moles of N_2 present, the H_2 is nevertheless the limiting reactant because of its larger coefficient in the balanced equation.

Check The summarizing table shows that the mole ratio of reactants used and product formed conforms to the coefficients in the balanced equation, 1:3:2. Also, because H_2 is the limiting reactant, it is completely consumed in the reaction, leaving 0 mol at the end. Because 6.0 mol H_2 has two significant figures, our answer has two significant figures.

■■■ **PRACTICE EXERCISE**

Consider the reaction $2 Al(s) + 3 Cl_2(g) \longrightarrow 2 AlCl_3(s)$. A mixture of 1.50 mol of Al and 3.00 mol of Cl_2 is allowed to react. **(a)** Which is the limiting reactant? **(b)** How many moles of $AlCl_3$ are formed? **(c)** How many moles of the excess reactant remain at the end of the reaction?
Answers: **(a)** Al, **(b)** 1.50 mol, **(c)** 0.75 mol Cl_2

■■■ **SAMPLE EXERCISE 3.19** | Calculating the Amount of Product Formed from a Limiting Reactant

Consider the following reaction that occurs in a fuel cell:

$$2 H_2(g) + O_2(g) \longrightarrow 2 H_2O(g)$$

This reaction, properly done, produces energy in the form of electricity and water. Suppose a fuel cell is set up with 150 g of hydrogen gas and 1500 grams of oxygen gas (each measurement is given with two significant figures). How many grams of water can be formed?

SOLUTION

Analyze We are asked to calculate the amount of a product, given the amounts of two reactants, so this is a limiting reactant problem.

Plan We must first identify the limiting reagent. To do so, we can calculate the number of moles of each reactant and compare their ratio with that required by the balanced equation. We then use the quantity of the limiting reagent to calculate the mass of water that forms.

Solve From the balanced equation, we have the following stoichiometric relations:

$$2 \text{ mol } H_2 \simeq 1 \text{ mol } O_2 \simeq 2 \text{ mol } H_2O$$

Using the molar mass of each substance, we can calculate the number of moles of each reactant:

$$\text{Moles } H_2 = (150 \text{ g } H_2)\left(\frac{1 \text{ mol } H_2}{2.00 \text{ g } H_2}\right) = 75 \text{ mol } H_2$$

$$\text{Moles } O_2 = (1500 \text{ g } O_2)\left(\frac{1 \text{ mol } O_2}{32.0 \text{ g } O_2}\right) = 47 \text{ mol } O_2$$

Thus, there are more moles of H_2 than O_2. The coefficients in the balanced equation indicate, however, that the reaction requires 2 moles of H_2 for every 1 mole of O_2. Therefore, to completely react all the O_2, we would need $2 \times 47 = 94$ moles of H_2. Since there are only 75 moles of H_2, H_2 is the limiting reagent. We therefore use the quantity of H_2 to calculate the quantity of product formed. We can begin this calculation with the grams of H_2, but we can save a step by starting with the moles of H_2 that were calculated previously in the exercise:

$$\text{Grams } H_2O = (75 \text{ moles } H_2)\left(\frac{2 \text{ mol } H_2O}{2 \text{ mol } H_2}\right)\left(\frac{18.0 \text{ g } H_2O}{1 \text{ mol } H_2O}\right)$$
$$= 1400 \text{ g } H_2O \text{ (to two significant figures)}$$

Check The magnitude of the answer seems reasonable. The units are correct, and the number of significant figures (two) corresponds to those in the numbers of grams of the starting materials.

Comment The quantity of the limiting reagent, H_2, can also be used to determine the quantity of O_2 used (37.5 mol = 1200 g). The number of grams of the excess oxygen remaining at the end of the reaction equals the starting amount minus the amount consumed in the reaction, $1500 \text{ g} - 1200 \text{ g} = 300 \text{ g}$.

■ **PRACTICE EXERCISE**

A strip of zinc metal with a mass of 2.00 g is placed in an aqueous solution containing 2.50 g of silver nitrate, causing the following reaction to occur:

$$Zn(s) + 2\,AgNO_3(aq) \longrightarrow 2\,Ag(s) + Zn(NO_3)_2(aq)$$

(a) Which reactant is limiting? **(b)** How many grams of Ag will form? **(c)** How many grams of $Zn(NO_3)_2$ will form? **(d)** How many grams of the excess reactant will be left at the end of the reaction?
Answers: **(a)** $AgNO_3$, **(b)** 1.59 g, **(c)** 1.39 g, **(d)** 1.52 g Zn

Theoretical Yields

The quantity of product that is calculated to form when all of the limiting reactant reacts is called the **theoretical yield**. The amount of product actually obtained in a reaction is called the *actual yield*. The actual yield is almost always less than (and can never be greater than) the theoretical yield. There are many reasons for this difference. Part of the reactants may not react, for example, or they may react in a way different from that desired (side reactions). In addition, it is not always possible to recover all of the product from the reaction mixture. The **percent yield** of a reaction relates the actual yield to the theoretical (calculated) yield:

$$\text{Percent yield} = \frac{\text{actual yield}}{\text{theoretical yield}} \times 100\% \qquad \text{[3.14]}$$

Strategies in Chemistry   HOW TO TAKE A TEST

At about this time in your study of chemistry, you are likely to face your first hour-long examination. The best way to prepare for the exam is to study and do homework diligently and to make sure you get help from the instructor on any material that is unclear or confusing. (See the advice for learning and studying chemistry presented in the preface of the book.) We present here some general guidelines for taking tests.

Depending on the nature of your course, the exam could consist of a variety of different types of questions. Let's consider some of the more common types and how they can best be addressed.

1. *Multiple-choice questions* In large-enrollment courses, the most common kind of testing device is the multiple-choice question. You are given the problem and usually presented with four or five possible answers from which you must select the correct one. The first thing to realize is that the instructor has written the question so that all of the answer choices appear at first glance to be correct. (There would be little point in offering choices you could

tell were wrong even without knowing much about the concept being tested.) Thus, you should not jump to the conclusion that because one of the choices looks correct, it must be so.

If a multiple-choice question involves a calculation, perform the calculation, quickly double-check your work, and *only then* compare your answer with the choices. If you find a match, you have probably found the correct answer. Keep in mind, though, that your instructor has anticipated the most common errors one can make in solving a given problem and has probably listed the incorrect answers resulting from those errors. Always double-check your reasoning and make sure to use dimensional analysis to arrive at the correct answer, with the correct units.

In multiple-choice questions that do not involve calculations, if you are not sure of the correct choice, eliminate all the choices you know for sure to be incorrect. Additionally, the reasoning you used in eliminating incorrect choices will help you in reasoning about which choice is correct.

■■ SAMPLE EXERCISE 3.20 | **Calculating the Theoretical Yield and Percent Yield for a Reaction**

Adipic acid, $H_2C_6H_8O_4$, is used to produce nylon. The acid is made commercially by a controlled reaction between cyclohexane (C_6H_{12}) and O_2:

$$2 C_6H_{12}(l) + 5 O_2(g) \longrightarrow 2 H_2C_6H_8O_4(l) + 2 H_2O(g)$$

(a) Assume that you carry out this reaction starting with 25.0 g of cyclohexane and that cyclohexane is the limiting reactant. What is the theoretical yield of adipic acid?
(b) If you obtain 33.5 g of adipic acid from your reaction, what is the percent yield of adipic acid?

SOLUTION

Analyze We are given a chemical equation and the quantity of the limiting reactant (25.0 g of C_6H_{12}). We are asked first to calculate the theoretical yield of a product ($H_2C_6H_8O_4$) and then to calculate its percent yield if only 33.5 g of the substance is actually obtained.

Plan (a) The theoretical yield, which is the calculated quantity of adipic acid formed in the reaction, can be calculated using the following sequence of conversions:

$$g\ C_6H_{12} \longrightarrow mol\ C_6H_{12} \longrightarrow mol\ H_2C_6H_8O_4 \longrightarrow g\ H_2C_6H_8O_4$$

(b) The percent yield is calculated by comparing the actual yield (33.5 g) to the theoretical yield using Equation 3.14.

Solve

(a) Grams $H_2C_6H_8O_4 = (25.0\ \text{g}\ C_6H_{12})\left(\dfrac{1\ \text{mol}\ C_6H_{12}}{84.0\ \text{g}\ C_6H_{12}}\right)\left(\dfrac{2\ \text{mol}\ H_2C_6H_8O_4}{2\ \text{mol}\ C_6H_{12}}\right)\left(\dfrac{146.0\ \text{g}\ H_2C_6H_8O_4}{1\ \text{mol}\ H_2C_6H_8O_4}\right)$

$= 43.5\ \text{g}\ H_2C_6H_8O_4$

(b) Percent yield $= \dfrac{\text{actual yield}}{\text{theoretical yield}} \times 100\% = \dfrac{33.5\ \text{g}}{43.5\ \text{g}} \times 100\% = 77.0\%$

Check Our answer in (a) has the appropriate magnitude, units, and significant figures. In (b) the answer is less than 100% as necessary.

2. *Calculations in which you must show your work* Your instructor may present you with a numerical problem in which you are to show your work in arriving at a solution. In questions of this kind, you may receive partial credit even if you do not arrive at the correct answer, depending on whether the instructor can follow your line of reasoning. It is important, therefore, to be as neat and organized as you can be, given the pressures of exam taking. It is helpful in approaching such questions to take a few moments to think about the direction you are going to take in solving the problem. You may even want to write a few words or a diagram on the test paper to indicate your approach. Then write out your calculations as neatly as you can. Show the units for every number you write down, and use dimensional analysis as much as you can, showing how units cancel.

3. *Questions requiring drawings* Sometimes a test question will require you to draw a chemical structure, a diagram related to chemical bonding, or a figure showing some kind of chemical process. Questions of this kind will come later in the course, but it is useful to talk about them here. (You should review this box before each exam you take, to remind yourself of good exam-taking practices.) Be sure to label your drawing as completely as possible.

4. *Other types of questions* Other exam questions you might encounter include true-false questions and ones in which you are given a list and asked to indicate which members of the list match some criterion given in the question. Often students answer such questions incorrectly because, in their haste, they misunderstand the nature of the question. Whatever the form of the question, ask yourself this: What is the instructor testing here? What material am I supposed to know that this question covers?

Finally, if you find that you simply do not understand how to arrive at a reasoned response to a question, do not linger over the question. Put a check next to it and go on to the next one. If time permits, you can come back to the unanswered questions, but lingering over a question when nothing is coming to mind is wasting time you may need to finish the exam.

◼ PRACTICE EXERCISE

Imagine that you are working on ways to improve the process by which iron ore containing Fe_2O_3 is converted into iron. In your tests you carry out the following reaction on a small scale:

$$Fe_2O_3(s) + 3\ CO(g) \longrightarrow 2\ Fe(s) + 3\ CO_2(g)$$

(a) If you start with 150 g of Fe_2O_3 as the limiting reagent, what is the theoretical yield of Fe? **(b)** If the actual yield of Fe in your test was 87.9 g, what was the percent yield?
Answers: **(a)** 105 g Fe, **(b)** 83.7%

CHAPTER REVIEW

SUMMARY AND KEY TERMS

Introduction and Section 3.1 The study of the quantitative relationships between chemical formulas and chemical equations is known as **stoichiometry**. One of the important concepts of stoichiometry is the law of conservation of mass, which states that the total mass of the products of a chemical reaction is the same as the total mass of the reactants. The same numbers of atoms of each type are present before and after a chemical reaction. A balanced **chemical equation** shows equal numbers of atoms of each element on each side of the equation. Equations are balanced by placing coefficients in front of the chemical formulas for the **reactants** and **products** of a reaction, *not* by changing the subscripts in chemical formulas.

Section 3.2 Among the reaction types described in this chapter are (1) **combination reactions**, in which two reactants combine to form one product; (2) **decomposition reactions**, in which a single reactant forms two or more products; and (3) **combustion reactions** in oxygen, in which a hydrocarbon or related compound reacts with O_2 to form CO_2 and H_2O.

Section 3.3 Much quantitative information can be determined from chemical formulas and balanced chemical equations by using atomic weights. The **formula weight** of a compound equals the sum of the atomic weights of the atoms in its formula. If the formula is a molecular formula, the formula weight is also called the **molecular weight**. Atomic weights and formula weights can be used to determine the elemental composition of a compound.

Section 3.4 A **mole** of any substance is **Avogadro's number** (6.02×10^{23}) of formula units of that substance. The mass of a mole of atoms, molecules, or ions (the **molar mass**) equals the formula weight of that material expressed in grams. The mass of one molecule of H_2O, for example, is 18 amu, so the mass of 1 mol of H_2O is 18 g. That is, the molar mass of H_2O is 18 g/mol.

Section 3.5 The empirical formula of any substance can be determined from its percent composition by calculating the relative number of moles of each atom in 100 g of the substance. If the substance is molecular in nature, its molecular formula can be determined from the empirical formula if the molecular weight is also known.

Sections 3.6 and 3.7 The mole concept can be used to calculate the relative quantities of reactants and products in chemical reactions. The coefficients in a balanced equation give the relative numbers of moles of the reactants and products. To calculate the number of grams of a product from the number of grams of a reactant, first convert grams of reactant to moles of reactant. Then use the coefficients in the balanced equation to convert the number of moles of reactant to moles of product. Finally, convert moles of product to grams of product.

A **limiting reactant** is completely consumed in a reaction. When it is used up, the reaction stops, thus limiting the quantities of products formed. The **theoretical yield** of a reaction is the quantity of product calculated to form when all of the limiting reagent reacts. The actual yield of a reaction is always less than the theoretical yield. The **percent yield** compares the actual and theoretical yields.

KEY SKILLS

- Balance chemical equations.
- Calculate molecular weights.
- Convert grams to moles and moles to grams using molar masses.
- Convert number of molecules to moles and moles to number of molecules using Avogadro's number.
- Calculate the empirical and molecular formula of a compound from percentage composition and molecular weight.
- Calculate amounts, in grams or moles, of reactants and products for a reaction.
- Calculate the percent yield of a reaction.

KEY EQUATIONS

- $$\% \text{ element} = \frac{\left(\begin{array}{c}\text{number of atoms}\\\text{of that element}\end{array}\right)\left(\begin{array}{c}\text{atomic weight}\\\text{of element}\end{array}\right)}{\text{formula weight of compound}} \times 100\% \qquad [3.10]$$

  This is the formula to calculate the mass percentage of each element in a compound. The sum of all the percentages of all the elements in a compound should add up to 100%.

- $$\% \text{ yield} = \frac{(\text{actual yield})}{(\text{theoretical yield})} \times 100\% \qquad [3.14]$$

  This is the formula to calculate the percent yield of a reaction. The percent yield can never be more than 100%.

VISUALIZING CONCEPTS

3.1 The reaction between reactant A (blue spheres) and reactant B (red spheres) is shown in the following diagram:

Based on this diagram, which equation best describes the reaction? [Section 3.1]

(a) $A_2 + B \longrightarrow A_2B$

(b) $A_2 + 4 B \longrightarrow 2 AB_2$

(c) $2 A + B_4 \longrightarrow 2 AB_2$

(d) $A + B_2 \longrightarrow AB_2$

3.2 Under appropriate experimental conditions, H_2 and CO undergo a combination reaction to form CH_3OH. The drawing below represents a sample of H_2. Make a corresponding drawing of the CO needed to react completely with the H_2. How did you arrive at the number of CO molecules in your drawing? [Section 3.2]

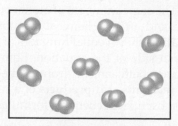

3.3 The following diagram represents the collection of elements formed by a decomposition reaction. (a) If the blue spheres represent N atoms and the red ones represent O atoms, what was the empirical formula of the original compound? (b) Could you draw a diagram representing the molecules of the compound that had been decomposed? Why or why not? [Section 3.2]

3.4 The following diagram represents the collection of CO_2 and H_2O molecules formed by complete combustion of a hydrocarbon. What is the empirical formula of the hydrocarbon? [Section 3.2]

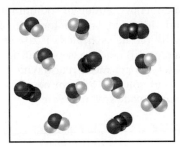

3.5 Glycine, an amino acid used by organisms to make proteins, is represented by the molecular model below.
  (a) Write its molecular formula.
  (b) Determine its molar mass.
  (c) Calculate the mass of 3 moles of glycine.
  (d) Calculate the percent nitrogen by mass in glycine. [Sections 3.3 and 3.5]

3.6 The following diagram represents a high-temperature reaction between CH_4 and H_2O. Based on this reaction,

how many moles of each product can be obtained starting with 4.0 mol CH_4? [Section 3.6]

3.7 Nitrogen (N_2) and hydrogen (H_2) react to form ammonia (NH_3). Consider the mixture of N_2 and H_2 shown in the accompanying diagram. The blue spheres represent N, and the white ones represent H. Draw a representation of the product mixture, assuming that the reaction goes to completion. How did you arrive at your representation? What is the limiting reactant in this case? [Section 3.7]

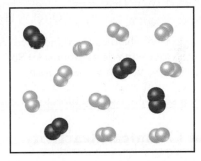

3.8 Nitrogen monoxide and oxygen react to form nitrogen dioxide. Consider the mixture of NO and O_2 shown in the accompanying diagram. The blue spheres represent N, and the red ones represent O. **(a)** Draw a representation of the product mixture, assuming that the reaction goes to completion. What is the limiting reactant in this case? **(b)** How many NO_2 molecules would you draw as products if the reaction had a percent yield of 75%? [Section 3.7]

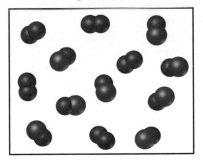

EXERCISES

Balancing Chemical Equations

3.9 **(a)** What scientific principle or law is used in the process of balancing chemical equations? **(b)** In balancing equations, why should you not change subscripts in chemical formulas? **(c)** How would one write out liquid water, water vapor, aqueous sodium chloride, and solid sodium chloride in chemical equations?

3.10 **(a)** What is the difference between adding a subscript 2 to the end of the formula for CO to give CO_2 and adding a coefficient in front of the formula to give 2 CO?

(b) Is the following chemical equation, as written, consistent with the law of conservation of mass?

$$3\,Mg(OH)_2\,(s) + 2\,H_3PO_4(aq) \longrightarrow Mg_3(PO_4)_2(s) + 6\,H_2O(l)$$

Why or why not?

3.11 Balance the following equations:
  (a) $CO(g) + O_2(g) \longrightarrow CO_2(g)$
  (b) $N_2O_5(g) + H_2O(l) \longrightarrow HNO_3(aq)$
  (c) $CH_4(g) + Cl_2(g) \longrightarrow CCl_4(l) + HCl(g)$

(d) $Al_4C_3(s) + H_2O(l) \longrightarrow Al(OH)_3(s) + CH_4(g)$

(e) $C_5H_{10}O_2(l) + O_2(g) \longrightarrow CO_2(g) + H_2O(g)$

(f) $Fe(OH)_3(s) + H_2SO_4(aq) \longrightarrow$
$$Fe_2(SO_4)_3(aq) + H_2O(l)$$

(g) $Mg_3N_2(s) + H_2SO_4(aq) \longrightarrow$
$$MgSO_4(aq) + (NH_4)_2SO_4(aq)$$

3.12 Balance the following equations:

(a) $Li(s) + N_2(g) \longrightarrow Li_3N(s)$

(b) $La_2O_3(s) + H_2O(l) \longrightarrow La(OH)_3(aq)$

(c) $NH_4NO_3(s) \longrightarrow N_2(g) + O_2(g) + H_2O(g)$

(d) $Ca_3P_2(s) + H_2O(l) \longrightarrow Ca(OH)_2(aq) + PH_3(g)$

(e) $Ca(OH)_2(aq) + H_3PO_4(aq) \longrightarrow$
$$Ca_3(PO_4)_2(s) + H_2O(l)$$

(f) $AgNO_3(aq) + Na_2SO_4(aq) \longrightarrow$
$$Ag_2SO_4(s) + NaNO_3(aq)$$

(g) $CH_3NH_2(g) + O_2(g) \longrightarrow$
$$CO_2(g) + H_2O(g) + N_2(g)$$

3.13 Write balanced chemical equations to correspond to each of the following descriptions: **(a)** Solid calcium carbide, CaC_2, reacts with water to form an aqueous solution of calcium hydroxide and acetylene gas, C_2H_2. **(b)** When solid potassium chlorate is heated, it decomposes to form solid potassium chloride and oxygen gas. **(c)** Solid zinc metal reacts with sulfuric acid to form hydrogen gas and an aqueous solution of zinc sulfate. **(d)** When liquid phosphorus trichloride is added to water, it reacts to form aqueous phosphorous acid, $H_3PO_3(aq)$, and aqueous hydrochloric acid. **(e)** When hydrogen sulfide gas is passed over solid hot iron(III) hydroxide, the resultant reaction produces solid iron(III) sulfide and gaseous water.

3.14 Write balanced chemical equations to correspond to each of the following descriptions: **(a)** When sulfur trioxide gas reacts with water, a solution of sulfuric acid forms. **(b)** Boron sulfide, $B_2S_3(s)$, reacts violently with water to form dissolved boric acid, H_3BO_3, and hydrogen sulfide gas. **(c)** When an aqueous solution of lead(II) nitrate is mixed with an aqueous solution of sodium iodide, an aqueous solution of sodium nitrate and a yellow solid, lead iodide, are formed. **(d)** When solid mercury(II) nitrate is heated, it decomposes to form solid mercury(II) oxide, gaseous nitrogen dioxide, and oxygen. **(e)** Copper metal reacts with hot concentrated sulfuric acid solution to form aqueous copper(II) sulfate, sulfur dioxide gas, and water.

Patterns of Chemical Reactivity

3.15 **(a)** When the metallic element sodium combines with the nonmetallic element bromine, $Br_2(l)$, how can you determine the chemical formula of the product? How do you know whether the product is a solid, liquid, or gas at room temperature? Write the balanced chemical equation for the reaction. **(b)** When a hydrocarbon burns in air, what reactant besides the hydrocarbon is involved in the reaction? What products are formed? Write a balanced chemical equation for the combustion of benzene, $C_6H_6(l)$, in air.

3.16 **(a)** Determine the chemical formula of the product formed when the metallic element calcium combines with the nonmetallic element oxygen, O_2. Write the balanced chemical equation for the reaction. **(b)** What products form when a compound containing C, H, and O is completely combusted in air? Write a balanced chemical equation for the combustion of acetone, $C_3H_6O(l)$, in air.

3.17 Write a balanced chemical equation for the reaction that occurs when **(a)** $Mg(s)$ reacts with $Cl_2(g)$; **(b)** barium carbonate decomposes into barium oxide and carbon dioxide gas when heated; **(c)** the hydrocarbon styrene, $C_8H_8(l)$, is combusted in air; **(d)** dimethylether, $CH_3OCH_3(g)$, is combusted in air.

3.18 Write a balanced chemical equation for the reaction that occurs when **(a)** aluminum metal undergoes a combination reaction with $O_2(g)$; **(b)** copper(II) hydroxide decomposes into copper(II) oxide and water when heated; **(c)** heptane, $C_7H_{16}(l)$, burns in air; **(d)** the gasoline additive MTBE (methyl tert-butyl ether), $C_5H_{12}O(l)$, burns in air.

3.19 Balance the following equations, and indicate whether they are combination, decomposition, or combustion reactions:

(a) $Al(s) + Cl_2(g) \longrightarrow AlCl_3(s)$

(b) $C_2H_4(g) + O_2(g) \longrightarrow CO_2(g) + H_2O(g)$

(c) $Li(s) + N_2(g) \longrightarrow Li_3N(s)$

(d) $PbCO_3(s) \longrightarrow PbO(s) + CO_2(g)$

(e) $C_7H_8O_2(l) + O_2(g) \longrightarrow CO_2(g) + H_2O(g)$

3.20 Balance the following equations, and indicate whether they are combination, decomposition, or combustion reactions:

(a) $C_3H_6(g) + O_2(g) \longrightarrow CO_2(g) + H_2O(g)$

(b) $NH_4NO_3(s) \longrightarrow N_2O(g) + H_2O(g)$

(c) $C_5H_6O(l) + O_2(g) \longrightarrow CO_2(g) + H_2O(g)$

(d) $N_2(g) + H_2(g) \longrightarrow NH_3(g)$

(e) $K_2O(s) + H_2O(l) \longrightarrow KOH(aq)$

Formula Weights

3.21 Determine the formula weights of each of the following compounds: **(a)** nitric acid, HNO_3; **(b)** $KMnO_4$; **(c)** $Ca_3(PO_4)_2$; **(d)** quartz, SiO_2; **(e)** gallium sulfide; **(f)** chromium(III) sulfate, **(g)** phosphorus trichloride.

3.22 Determine the formula weights of each of the following compounds: **(a)** nitrous oxide, N_2O, known as laughing gas and used as an anesthetic in dentistry; **(b)** benzoic acid, $HC_7H_5O_2$, a substance used as a food preservative; **(c)** $Mg(OH)_2$, the active ingredient in milk of magnesia; **(d)** urea, $(NH_2)_2CO$, a compound used as a nitrogen fertilizer; **(e)** isopentyl acetate, $CH_3CO_2C_5H_{11}$, responsible for the odor of bananas.

3.23 Calculate the percentage by mass of oxygen in the following compounds: **(a)** morphine, $C_{17}H_{19}NO_3$; **(b)** codeine, $C_{18}H_{21}NO_3$; **(c)** cocaine, $C_{17}H_{21}NO_4$; **(d)** tetracycline, $C_{22}H_{24}N_2O_8$; **(e)** digitoxin, $C_{41}H_{64}O_{13}$; **(f)** vancomycin, $C_{66}H_{75}Cl_2N_9O_{24}$.

3.24 Calculate the percentage by mass of the indicated element in the following compounds: **(a)** carbon in acetylene, C_2H_2, a gas used in welding; **(b)** hydrogen in ascorbic acid, $HC_6H_7O_6$, also known as vitamin C; **(c)** hydrogen in ammonium sulfate, $(NH_4)_2SO_4$, a substance used as a nitrogen fertilizer; **(d)** platinum in $PtCl_2(NH_3)_2$, a chemotherapy agent called cisplatin; **(e)** oxygen in the female sex hormone estradiol, $C_{18}H_{24}O_2$; **(f)** carbon in capsaicin, $C_{18}H_{27}NO_3$, the compound that gives the hot taste to chili peppers.

3.25 Based on the following structural formulas, calculate the percentage of carbon by mass present in each compound:

(a) Benzaldehyde (almond fragrance)

(b) Vanillin (vanilla flavor)

(c) Isopentyl acetate (banana flavor)

3.26 Calculate the percentage of carbon by mass in each of the compounds represented by the following models:

(a)

(b)

(c)

(d)

Avogadro's Number and the Mole

3.27 **(a)** What is Avogadro's number, and how is it related to the mole? **(b)** What is the relationship between the formula weight of a substance and its molar mass?

3.28 **(a)** What is the mass, in grams, of a mole of ^{12}C? **(b)** How many carbon atoms are present in a mole of ^{12}C?

3.29 Without doing any detailed calculations (but using a periodic table to give atomic weights), rank the following samples in order of increasing number of atoms: 0.50 mol H_2O, 23 g Na, 6.0×10^{23} N_2 molecules.

3.30 Without doing any detailed calculations (but using a periodic table to give atomic weights), rank the following samples in order of increasing number of atoms: 3.0×10^{23} molecules of H_2O_2, 2.0 mol CH_4, 32 g O_2.

3.31 What is the mass, in kilograms, of an Avogadro's number of people, if the average mass of a person is 160 lb? How does this compare with the mass of Earth, 5.98×10^{24} kg?

3.32 If Avogadro's number of pennies is divided equally among the 300 million men, women, and children in the United States, how many dollars would each receive? How does this compare with the gross domestic product of the United States, which was $13.5 trillion in 2006? (The GDP is the total market value of the nation's goods and services.)

3.33 Calculate the following quantities:
(a) mass, in grams, of 0.105 moles sucrose ($C_{12}H_{22}O_{11}$)
(b) moles of $Zn(NO_3)_2$ in 143.50 g of this substance
(c) number of molecules in 1.0×10^{-6} mol CH_3CH_2OH
(d) number of N atoms in 0.410 mol NH_3

3.34 Calculate the following quantities
(a) mass, in grams, of 5.76×10^{-3} mol of CdS
(b) number of moles of NH_4Cl in 112.6 g of this substance
(c) number of molecules in 1.305×10^{-2} mol C_6H_6
(d) number of O atoms in 4.88×10^{-3} mol $Al(NO_3)_3$

3.35 **(a)** What is the mass, in grams, of 2.50×10^{-3} mol of ammonium phosphate?
(b) How many moles of chloride ions are in 0.2550 g of aluminum chloride?
(c) What is the mass, in grams, of 7.70×10^{20} molecules of caffeine, $C_8H_{10}N_4O_2$?
(d) What is the molar mass of cholesterol if 0.00105 mol weighs 0.406 g?

3.36 **(a)** What is the mass, in grams, of 0.0714 mol of iron(III) sulfate?
(b) How many moles of ammonium ions are in 8.776 g of ammonium carbonate?

(c) What is the mass, in grams, of 6.52×10^{21} molecules of aspirin, $C_9H_8O_4$?

(d) What is the molar mass of diazepam (Valium®) if 0.05570 mol weighs 15.86 g?

3.37 The molecular formula of allicin, the compound responsible for the characteristic smell of garlic, is $C_6H_{10}OS_2$. (a) What is the molar mass of allicin? (b) How many moles of allicin are present in 5.00 mg of this substance? (c) How many molecules of allicin are in 5.00 mg of this substance? (d) How many S atoms are present in 5.00 mg of allicin?

3.38 The molecular formula of aspartame, the artificial sweetener marketed as NutraSweet®, is $C_{14}H_{18}N_2O_5$. (a) What is the molar mass of aspartame? (b) How many moles of aspartame are present in 1.00 mg of aspartame? (c) How many molecules of aspartame are present in 1.00 mg of aspartame? (d) How many hydrogen atoms are present in 1.00 mg of aspartame?

3.39 A sample of glucose, $C_6H_{12}O_6$, contains 1.250×10^{21} carbon atoms. (a) How many atoms of hydrogen does it contain? (b) How many molecules of glucose does it contain? (c) How many moles of glucose does it contain? (d) What is the mass of this sample in grams?

3.40 A sample of the male sex hormone testosterone, $C_{19}H_{28}O_2$, contains 7.08×10^{20} hydrogen atoms. (a) How many atoms of carbon does it contain? (b) How many molecules of testosterone does it contain? (c) How many moles of testosterone does it contain? (d) What is the mass of this sample in grams?

3.41 The allowable concentration level of vinyl chloride, C_2H_3Cl, in the atmosphere in a chemical plant is 2.0×10^{-6} g/L. How many moles of vinyl chloride in each liter does this represent? How many molecules per liter?

3.42 At least 25 μg of tetrahydrocannabinol (THC), the active ingredient in marijuana, is required to produce intoxication. The molecular formula of THC is $C_{21}H_{30}O_2$. How many moles of THC does this 25 μg represent? How many molecules?

Empirical Formulas

3.43 Give the empirical formula of each of the following compounds if a sample contains (a) 0.0130 mol C, 0.0390 mol H, and 0.0065 mol O; (b) 11.66 g iron and 5.01 g oxygen; (c) 40.0% C, 6.7% H, and 53.3% O by mass.

3.44 Determine the empirical formula of each of the following compounds if a sample contains (a) 0.104 mol K, 0.052 mol C, and 0.156 mol O; (b) 5.28 g Sn and 3.37 g F; (c) 87.5% N and 12.5% H by mass.

3.45 Determine the empirical formulas of the compounds with the following compositions by mass:
(a) 10.4% C, 27.8% S, and 61.7% Cl
(b) 21.7% C, 9.6% O, and 68.7% F
(c) 32.79% Na, 13.02% Al, and 54.19% F

3.46 Determine the empirical formulas of the compounds with the following compositions by mass:
(a) 55.3% K, 14.6% P, and 30.1% O
(b) 24.5% Na, 14.9% Si, and 60.6% F
(c) 62.1% C, 5.21% H, 12.1% N, and 20.7% O

3.47 What is the molecular formula of each of the following compounds?
(a) empirical formula CH_2, molar mass = 84 g/mol
(b) empirical formula NH_2Cl, molar mass = 51.5 g/mol

3.48 What is the molecular formula of each of the following compounds?
(a) empirical formula HCO_2, molar mass = 90.0 g/mol
(b) empirical formula C_2H_4O, molar mass = 88 g/mol

3.49 Determine the empirical and molecular formulas of each of the following substances:
(a) Styrene, a compound substance used to make Styrofoam® cups and insulation, contains 92.3% C and 7.7% H by mass and has a molar mass of 104 g/mol.
(b) Caffeine, a stimulant found in coffee, contains 49.5% C, 5.15% H, 28.9% N, and 16.5% O by mass and has a molar mass of 195 g/mol.
(c) Monosodium glutamate (MSG), a flavor enhancer in certain foods, contains 35.51% C, 4.77% H, 37.85%

O, 8.29% N, and 13.60% Na, and has a molar mass of 169 g/mol.

3.50 Determine the empirical and molecular formulas of each of the following substances:
(a) Ibuprofen, a headache remedy, contains 75.69% C, 8.80% H, and 15.51% O by mass, and has a molar mass of 206 g/mol.
(b) Cadaverine, a foul smelling substance produced by the action of bacteria on meat, contains 58.55% C, 13.81% H, and 27.40% N by mass; its molar mass is 102.2 g/mol.
(c) Epinephrine (adrenaline), a hormone secreted into the bloodstream in times of danger or stress, contains 59.0% C, 7.1% H, 26.2% O, and 7.7% N by mass; its MW is about 180 amu.

3.51 (a) Combustion analysis of toluene, a common organic solvent, gives 5.86 mg of CO_2 and 1.37 mg of H_2O. If the compound contains only carbon and hydrogen, what is its empirical formula? (b) Menthol, the substance we can smell in mentholated cough drops, is composed of C, H, and O. A 0.1005-g sample of menthol is combusted, producing 0.2829 g of CO_2 and 0.1159 g of H_2O. What is the empirical formula for menthol? If menthol has a molar mass of 156 g/mol, what is its molecular formula?

3.52 (a) The characteristic odor of pineapple is due to ethyl butyrate, a compound containing carbon, hydrogen, and oxygen. Combustion of 2.78 mg of ethyl butyrate produces 6.32 mg of CO_2 and 2.58 mg of H_2O. What is the empirical formula of the compound? (b) Nicotine, a component of tobacco, is composed of C, H, and N. A 5.250-mg sample of nicotine was combusted, producing 14.242 mg of CO_2 and 4.083 mg of H_2O. What is the empirical formula for nicotine? If nicotine has a molar mass of 160 ± 5 g/mol, what is its molecular formula?

3.53 Washing soda, a compound used to prepare hard water for washing laundry, is a hydrate, which means that a certain number of water molecules are included in the solid structure. Its formula can be written as $Na_2CO_3 \cdot xH_2O$, where x is the number of moles of H_2O per mole of Na_2CO_3. When a 2.558-g sample of washing soda is heated at 25 °C, all the water of hydration is lost, leaving 0.948 g of Na_2CO_3. What is the value of x?

Calculations Based on Chemical Equations

3.55 Why is it essential to use balanced chemical equations when determining the quantity of a product formed from a given quantity of a reactant?

3.56 What parts of balanced chemical equations give information about the relative numbers of moles of reactants and products involved in a reaction?

3.57 Hydrofluoric acid, HF(aq), cannot be stored in glass bottles because compounds called silicates in the glass are attacked by the HF(aq). Sodium silicate (Na_2SiO_3), for example, reacts as follows:

$$Na_2SiO_3(s) + 8\ HF(aq) \longrightarrow$$
$$H_2SiF_6(aq) + 2\ NaF(aq) + 3\ H_2O(l)$$

(a) How many moles of HF are needed to react with 0.300 mol of Na_2SiO_3?
(b) How many grams of NaF form when 0.500 mol of HF reacts with excess Na_2SiO_3?
(c) How many grams of Na_2SiO_3 can react with 0.800 g of HF?

3.58 The fermentation of glucose ($C_6H_{12}O_6$) produces ethyl alcohol (C_2H_5OH) and CO_2:

$$C_6H_{12}O_6(aq) \longrightarrow 2\ C_2H_5OH(aq) + 2\ CO_2(g)$$

(a) How many moles of CO_2 are produced when 0.400 mol of $C_6H_{12}O_6$ reacts in this fashion?
(b) How many grams of $C_6H_{12}O_6$ are needed to form 7.50 g of C_2H_5OH?
(c) How many grams of CO_2 form when 7.50 g of C_2H_5OH are produced?

3.59 Several brands of antacids use $Al(OH)_3$ to react with stomach acid, which contains primarily HCl:

$$Al(OH)_3(s) + HCl(aq) \longrightarrow AlCl_3(aq) + H_2O(l)$$

(a) Balance this equation.
(b) Calculate the number of grams of HCl that can react with 0.500 g of $Al(OH)_3$.
(c) Calculate the number of grams of $AlCl_3$ and the number of grams of H_2O formed when 0.500 g of $Al(OH)_3$ reacts.
(d) Show that your calculations in parts (b) and (c) are consistent with the law of conservation of mass.

3.60 An iron ore sample contains Fe_2O_3 together with other substances. Reaction of the ore with CO produces iron metal:

$$Fe_2O_3(s) + CO(g) \longrightarrow Fe(s) + CO_2(g)$$

(a) Balance this equation.
(b) Calculate the number of grams of CO that can react with 0.150 kg of Fe_2O_3.
(c) Calculate the number of grams of Fe and the number of grams of CO_2 formed when 0.150 kg of Fe_2O_3 reacts.

3.54 Epsom salts, a strong laxative used in veterinary medicine, is a hydrate, which means that a certain number of water molecules are included in the solid structure. The formula for Epsom salts can be written as $MgSO_4 \cdot xH_2O$, where x indicates the number of moles of H_2O per mole of $MgSO_4$. When 5.061 g of this hydrate is heated to 250 °C, all the water of hydration is lost, leaving 2.472 g of $MgSO_4$. What is the value of x?

(d) Show that your calculations in parts (b) and (c) are consistent with the law of conservation of mass.

3.61 Aluminum sulfide reacts with water to form aluminum hydroxide and hydrogen sulfide. (a) Write the balanced chemical equation for this reaction. (b) How many grams of aluminum hydroxide are obtained from 14.2 g of aluminum sulfide?

3.62 Calcium hydride reacts with water to form calcium hydroxide and hydrogen gas. (a) Write a balanced chemical equation for the reaction. (b) How many grams of calcium hydride are needed to form 8.500 g of hydrogen?

3.63 Automotive air bags inflate when sodium azide, NaN_3, rapidly decomposes to its component elements:

$$2\ NaN_3(s) \longrightarrow 2\ Na(s) + 3\ N_2(g)$$

(a) How many moles of N_2 are produced by the decomposition of 1.50 mol of NaN_3?
(b) How many grams of NaN_3 are required to form 10.0 g of nitrogen gas?
(c) How many grams of NaN_3 are required to produce 10.0 ft^3 of nitrogen gas, about the size of an automotive air bag, if the gas has a density of 1.25 g/L?

3.64 The complete combustion of octane, C_8H_{18}, the main component of gasoline, proceeds as follows:

$$2\ C_8H_{18}(l) + 25\ O_2(g) \longrightarrow 16\ CO_2(g) + 18\ H_2O(g)$$

(a) How many moles of O_2 are needed to burn 1.25 mol of C_8H_{18}?
(b) How many grams of O_2 are needed to burn 10.0 g of C_8H_{18}?
(c) Octane has a density of 0.692 g/mL at 20 °C. How many grams of O_2 are required to burn 1.00 gal of C_8H_{18}?

3.65 A piece of aluminum foil 1.00 cm square and 0.550 mm thick is allowed to react with bromine to form aluminum bromide as shown in the accompanying photo.

(a) How many moles of aluminum were used? (The density of aluminum is 2.699 g/cm^3.) **(b)** How many grams of aluminum bromide form, assuming the aluminum reacts completely?

3.66 Detonation of nitroglycerin proceeds as follows:

$$4\,C_3H_5N_3O_9(l) \longrightarrow$$
$$12\,CO_2(g) + 6\,N_2(g) + O_2(g) + 10\,H_2O(g)$$

(a) If a sample containing 2.00 mL of nitroglycerin (density = 1.592 g/mL) is detonated, how many total moles of gas are produced? **(b)** If each mole of gas occupies 55 L under the conditions of the explosion, how many liters of gas are produced? **(c)** How many grams of N$_2$ are produced in the detonation?

Limiting Reactants; Theoretical Yields

3.67 **(a)** Define the terms *limiting reactant* and *excess reactant*. **(b)** Why are the amounts of products formed in a reaction determined only by the amount of the limiting reactant? **(c)** Why should you base your choice of what compound is the limiting reactant on its number of initial moles, not on its initial mass in grams?

3.68 **(a)** Define the terms *theoretical yield*, *actual yield*, and *percent yield*. **(b)** Why is the actual yield in a reaction almost always less than the theoretical yield? **(c)** Can a reaction ever have 110% actual yield?

3.69 A manufacturer of bicycles has 4815 wheels, 2305 frames, and 2255 handlebars. **(a)** How many bicycles can be manufactured using these parts? **(b)** How many parts of each kind are left over? **(c)** Which part limits the production of bicycles?

3.70 A bottling plant has 121,515 bottles with a capacity of 355 mL, 122,500 caps, and 40,875 L of beverage. **(a)** How many bottles can be filled and capped? **(b)** How much of each item is left over? **(c)** Which component limits the production?

3.71 Sodium hydroxide reacts with carbon dioxide as follows:

$$2\,NaOH(s) + CO_2(g) \longrightarrow Na_2CO_3(s) + H_2O(l)$$

Which reagent is the limiting reactant when 1.85 mol NaOH and 1.00 mol CO$_2$ are allowed to react? How many moles of Na$_2$CO$_3$ can be produced? How many moles of the excess reactant remain after the completion of the reaction?

3.72 Aluminum hydroxide reacts with sulfuric acid as follows:

$$2\,Al(OH)_3(s) + 3\,H_2SO_4(aq) \longrightarrow$$
$$Al_2(SO_4)_3(aq) + 6\,H_2O(l)$$

Which reagent is the limiting reactant when 0.500 mol Al(OH)$_3$ and 0.500 mol H$_2$SO$_4$ are allowed to react? How many moles of Al$_2$(SO$_4$)$_3$ can form under these conditions? How many moles of the excess reactant remain after the completion of the reaction?

3.73 The fizz produced when an Alka-Seltzer® tablet is dissolved in water is due to the reaction between sodium bicarbonate (NaHCO$_3$) and citric acid (H$_3$C$_6$H$_5$O$_7$):

$$3\,NaHCO_3(aq) + H_3C_6H_5O_7(aq) \longrightarrow$$
$$3\,CO_2(g) + 3\,H_2O(l) + Na_3C_6H_5O_7(aq)$$

In a certain experiment 1.00 g of sodium bicarbonate and 1.00 g of citric acid are allowed to react. **(a)** Which is the limiting reactant? **(b)** How many grams of carbon dioxide form? **(c)** How many grams of the excess reac-

tant remain after the limiting reactant is completely consumed?

3.74 One of the steps in the commercial process for converting ammonia to nitric acid is the conversion of NH$_3$ to NO:

$$4\,NH_3(g) + 5\,O_2(g) \longrightarrow 4\,NO(g) + 6\,H_2O(g)$$

In a certain experiment, 1.50 g of NH$_3$ reacts with 2.75 g of O$_2$ **(a)** Which is the limiting reactant? **(b)** How many grams of NO and of H$_2$O form? **(c)** How many grams of the excess reactant remain after the limiting reactant is completely consumed? **(d)** Show that your calculations in parts (b) and (c) are consistent with the law of conservation of mass.

3.75 Solutions of sodium carbonate and silver nitrate react to form solid silver carbonate and a solution of sodium nitrate. A solution containing 3.50 g of sodium carbonate is mixed with one containing 5.00 g of silver nitrate. How many grams of sodium carbonate, silver nitrate, silver carbonate, and sodium nitrate are present after the reaction is complete?

3.76 Solutions of sulfuric acid and lead(II) acetate react to form solid lead(II) sulfate and a solution of acetic acid. If 7.50 g of sulfuric acid and 7.50 g of lead(II) acetate are mixed, calculate the number of grams of sulfuric acid, lead(II) acetate, lead(II) sulfate, and acetic acid present in the mixture after the reaction is complete.

3.77 When benzene (C$_6$H$_6$) reacts with bromine (Br$_2$), bromobenzene (C$_6$H$_5$Br) is obtained:

$$C_6H_6 + Br_2 \longrightarrow C_6H_5Br + HBr$$

(a) What is the theoretical yield of bromobenzene in this reaction when 30.0 g of benzene reacts with 65.0 g of bromine? **(b)** If the actual yield of bromobenzene was 42.3 g, what was the percentage yield?

3.78 When ethane (C$_2$H$_6$) reacts with chlorine (Cl$_2$), the main product is C$_2$H$_5$Cl; but other products containing Cl, such as C$_2$H$_4$Cl$_2$, are also obtained in small quantities. The formation of these other products reduces the yield of C_2H_5Cl. **(a)** Calculate the theoretical yield of C_2H_5Cl when 125 g of C_2H_6 reacts with 255 g of Cl_2, assuming that C_2H_6 and Cl_2 react only to form C_2H_5Cl and HCl. **(b)** Calculate the percent yield of C_2H_5Cl if the reaction produces 206 g of C_2H_5Cl.

3.79 Hydrogen sulfide is an impurity in natural gas that must be removed. One common removal method is called the Claus process, which relies on the reaction:

$$8 H_2S(g) + 4 O_2(g) \longrightarrow S_8(l) + 8 H_2O(g)$$

Under optimal conditions the Claus process gives 98% yield of S_8 from H_2S. If you started with 30.0 grams of H_2S and 50.0 grams of O_2, how many grams of S_8 would be produced, assuming 98% yield?

3.80 When hydrogen sulfide gas is bubbled into a solution of sodium hydroxide, the reaction forms sodium sulfide and water. How many grams of sodium sulfide are formed if 1.50 g of hydrogen sulfide is bubbled into a solution containing 2.00 g of sodium hydroxide, assuming that the sodium sulfide is made in 92.0% yield?

ADDITIONAL EXERCISES

3.81 Write the balanced chemical equations for **(a)** the complete combustion of acetic acid (CH_3COOH), the main active ingredient in vinegar; **(b)** the decomposition of solid calcium hydroxide into solid calcium(II) oxide (lime) and water vapor; **(c)** the combination reaction between nickel metal and chlorine gas.

3.82 The effectiveness of nitrogen fertilizers depends on both their ability to deliver nitrogen to plants and the amount of nitrogen they can deliver. Four common nitrogen-containing fertilizers are ammonia, ammonium nitrate, ammonium sulfate, and urea [$(NH_2)_2CO$]. Rank these fertilizers in terms of the mass percentage nitrogen they contain.

3.83 **(a)** Diamond is a natural form of pure carbon. How many moles of carbon are in a 1.25-carat diamond (1 carat = 0.200 g)? How many atoms are in this diamond? **(b)** The molecular formula of acetylsalicylic acid (aspirin), one of the most common pain relievers, is $C_9H_8O_4$. How many moles of $C_9H_8O_4$ are in a 0.500-g tablet of aspirin? How many molecules of $C_9H_8O_4$ are in this tablet?

3.84 **(a)** One molecule of the antibiotic known as penicillin G has a mass of 5.342×10^{-21} g. What is the molar mass of penicillin G? **(b)** Hemoglobin, the oxygen-carrying protein in red blood cells, has four iron atoms per molecule and contains 0.340% iron by mass. Calculate the molar mass of hemoglobin.

3.85 Very small crystals composed of 1000 to 100,000 atoms, called quantum dots, are being investigated for use in electronic devices.
 (a) A quantum dot was made of solid silicon in the shape of a sphere, with a diameter of 4 nm. Calculate the mass of the quantum dot, using the density of silicon (2.3 g/cm^3).
 (b) How many silicon atoms are in the quantum dot?
 (c) The density of germanium is 5.325 g/cm^3. If you made a 4 nm quantum dot of germanium, how many Ge atoms would it contain? Assume the dot is spherical.

3.86 Serotonin is a compound that conducts nerve impulses in the brain. It contains 68.2 mass percent C, 6.86 mass percent H, 15.9 mass percent N, and 9.08 mass percent O. Its molar mass is 176 g/mol. Determine its molecular formula.

3.87 The koala dines exclusively on eucalyptus leaves. Its digestive system detoxifies the eucalyptus oil, a poison to other animals. The chief constituent in eucalyptus oil is a substance called eucalyptol, which contains 77.87% C,

11.76% H, and the remainder O. **(a)** What is the empirical formula for this substance? **(b)** A mass spectrum of eucalyptol shows a peak at about 154 amu. What is the molecular formula of the substance?

3.88 Vanillin, the dominant flavoring in vanilla, contains C, H, and O. When 1.05 g of this substance is completely combusted, 2.43 g of CO_2 and 0.50 g of H_2O are produced. What is the empirical formula of vanillin?

[3.89] An organic compound was found to contain only C, H, and Cl. When a 1.50-g sample of the compound was completely combusted in air, 3.52 g of CO_2 was formed. In a separate experiment the chlorine in a 1.00-g sample of the compound was converted to 1.27 g of AgCl. Determine the empirical formula of the compound.

[3.90] An oxybromate compound, $KBrO_x$, where x is unknown, is analyzed and found to contain 52.92% Br. What is the value of x?

[3.91] An element X forms an iodide (XI_3) and a chloride (XCl_3). The iodide is quantitatively converted to the chloride when it is heated in a stream of chlorine:

$$2 XI_3 + 3 Cl_2 \longrightarrow 2 XCl_3 + 3 I_2$$

If 0.5000 g of XI_3 is treated, 0.2360 g of XCl_3 is obtained. **(a)** Calculate the atomic weight of the element X. **(b)** Identify the element X.

3.92 If 1.5 mol of each of the following compounds is completely combusted in oxygen, which one will produce the largest number of moles of H_2O? Which will produce the least? Explain. C_2H_5OH, C_3H_8, $CH_3CH_2COCH_3$.

3.93 A method used by the U.S. Environmental Protection Agency (EPA) for determining the concentration of ozone in air is to pass the air sample through a "bubbler" containing sodium iodide, which removes the ozone according to the following equation:

$$O_3(g) + 2 NaI(aq) + H_2O(l) \longrightarrow$$
$$O_2(g) + I_2(s) + 2 NaOH(aq)$$

(a) How many moles of sodium iodide are needed to remove 5.95×10^{-6} mol of O_3? **(b)** How many grams of sodium iodide are needed to remove 1.3 mg of O_3?

3.94 A chemical plant uses electrical energy to decompose aqueous solutions of NaCl to give Cl_2, H_2, and NaOH:

$$2 NaCl(aq) + 2 H_2O(l) \longrightarrow$$
$$2 NaOH(aq) + H_2(g) + Cl_2(g)$$

If the plant produces 1.5×10^6 kg (1500 metric tons) of Cl_2 daily, estimate the quantities of H_2 and NaOH produced.

3.95 The fat stored in the hump of a camel is a source of both energy and water. Calculate the mass of H_2O produced by metabolism of 1.0 kg of fat, assuming the fat consists entirely of tristearin ($C_{57}H_{110}O_6$), a typical animal fat, and assuming that during metabolism, tristearin reacts with O_2 to form only CO_2 and H_2O.

[3.96] When hydrocarbons are burned in a limited amount of air, both CO and CO_2 form. When 0.450 g of a particular hydrocarbon was burned in air, 0.467 g of CO, 0.733 g of CO_2, and 0.450 g of H_2O were formed. **(a)** What is the empirical formula of the compound? **(b)** How many grams of O_2 were used in the reaction? **(c)** How many grams would have been required for complete combustion?

3.97 A mixture of $N_2(g)$ and $H_2(g)$ reacts in a closed container to form ammonia, $NH_3(g)$. The reaction ceases before either reactant has been totally consumed. At this stage 3.0 mol N_2, 3.0 mol H_2, and 3.0 mol NH_3 are present. How many moles of N_2 and H_2 were present originally?

[3.98] A mixture containing $KClO_3$, K_2CO_3, $KHCO_3$, and KCl was heated, producing CO_2, O_2, and H_2O gases according to the following equations:

$$2 \, KClO_3(s) \longrightarrow 2 \, KCl(s) + 3 \, O_2(g)$$
$$2 \, KHCO_3(s) \longrightarrow K_2O(s) + H_2O(g) + 2 \, CO_2(g)$$
$$K_2CO_3(s) \longrightarrow K_2O(s) + CO_2(g)$$

The KCl does not react under the conditions of the reaction. If 100.0 g of the mixture produces 1.80 g of H_2O, 13.20 g of CO_2, and 4.00 g of O_2, what was the composition of the original mixture? (Assume complete decomposition of the mixture.)

3.99 When a mixture of 10.0 g of acetylene (C_2H_2) and 10.0 g of oxygen (O_2) is ignited, the resultant combustion reaction produces CO_2 and H_2O. **(a)** Write the balanced chemical equation for this reaction. **(b)** Which is the limiting reactant? **(c)** How many grams of C_2H_2, O_2, CO_2, and H_2O are present after the reaction is complete?

3.100 Aspirin ($C_9H_8O_4$) is produced from salicylic acid ($C_7H_6O_3$) and acetic anhydride ($C_4H_6O_3$):

$$C_7H_6O_3 + C_4H_6O_3 \longrightarrow C_9H_8O_4 + HC_2H_3O_2$$

(a) How much salicylic acid is required to produce 1.5×10^2 kg of aspirin, assuming that all of the salicylic acid is converted to aspirin? **(b)** How much salicylic acid would be required if only 80% of the salicylic acid is converted to aspirin? **(c)** What is the theoretical yield of aspirin if 185 kg of salicylic acid is allowed to react with 125 kg of acetic anhydride? **(d)** If the situation described in part (c) produces 182 kg of aspirin, what is the percentage yield?

INTEGRATIVE EXERCISES

(These exercises require skills from earlier chapters as well as skills from the present chapter.)

3.101 Consider a sample of calcium carbonate in the form of a cube measuring 2.005 in. on each edge. If the sample has a density of 2.71 g/cm^3, how many oxygen atoms does it contain?

3.102 **(a)** You are given a cube of silver metal that measures 1.000 cm on each edge. The density of silver is 10.5 g/cm^3. How many atoms are in this cube? **(b)** Because atoms are spherical, they cannot occupy all of the space of the cube. The silver atoms pack in the solid in such a way that 74% of the volume of the solid is actually filled with the silver atoms. Calculate the volume of a single silver atom. **(c)** Using the volume of a silver atom and the formula for the volume of a sphere, calculate the radius in angstroms of a silver atom.

3.103 **(a)** If an automobile travels 225 mi with a gas mileage of 20.5 mi/gal, how many kilograms of CO_2 are produced? Assume that the gasoline is composed of octane, $C_8H_{18}(l)$, whose density is 0.69 g/mL. **(b)** Repeat the calculation for a truck that has a gas mileage of 5 mi/gal.

3.104 In 1865 a chemist reported that he had reacted a weighed amount of pure silver with nitric acid and had recovered all the silver as pure silver nitrate. The mass ratio of silver to silver nitrate was found to be 0.634985. Using only this ratio and the presently accepted values for the atomic weights of silver and oxygen, calculate the atomic weight of nitrogen. Compare this calculated atomic weight with the currently accepted value.

3.105 A particular coal contains 2.5% sulfur by mass. When this coal is burned at a power plant, the sulfur is con-

verted into sulfur dioxide gas, which is a pollutant. To reduce sulfur dioxide emissions, calcium oxide (lime) is used. The sulfur dioxide reacts with calcium oxide to form solid calcium sulfite. **(a)** Write the balanced chemical equation for the reaction. **(b)** If the coal is burned in a power plant that uses 2000 tons of coal per day, what mass of calcium oxide is required daily to eliminate the sulfur dioxide? **(c)** How many grams of calcium sulfite are produced daily by this power plant?

3.106 Copper is an excellent electrical conductor widely used in making electric circuits. In producing a printed circuit board for the electronics industry, a layer of copper is laminated on a plastic board. A circuit pattern is then printed on the board using a chemically resistant polymer. The board is then exposed to a chemical bath that reacts with the exposed copper, leaving the desired copper circuit, which has been protected by the overlaying polymer. Finally, a solvent removes the polymer. One reaction used to remove the exposed copper from the circuit board is

$$Cu(s) + Cu(NH_3)_4Cl_2(aq) + 4 \, NH_3(aq) \longrightarrow$$
$$2 \, Cu(NH_3)_4Cl(aq)$$

A plant needs to produce 5000 circuit boards, each with a surface area measuring 2.0 in. $\times$ 3.0 in. The boards are covered with a 0.65-mm layer of copper. In subsequent processing, 85% of the copper is removed. Copper has a density of 8.96 g/cm^3. Calculate the masses of $Cu(NH_3)_4$ Cl_2 and NH_3 needed to produce the circuit boards, assuming that the reaction used gives a 97% yield.

3.107 Hydrogen cyanide, HCN, is a poisonous gas. The lethal dose is approximately 300 mg HCN per kilogram of air when inhaled. **(a)** Calculate the amount of HCN that gives the lethal dose in a small laboratory room measuring $12 \times 15 \times 8.0$ ft. The density of air at 26 °C is 0.00118 g/cm^3. **(b)** If the HCN is formed by reaction of NaCN with an acid such as H_2SO_4, what mass of NaCN gives the lethal dose in the room?

$$2\,NaCN(s) + H_2SO_4(aq) \longrightarrow Na_2SO_4(aq) + 2\,HCN(g)$$

(c) HCN forms when synthetic fibers containing Orlon® or Acrilan® burn. Acrilan® has an empirical formula of CH_2CHCN, so HCN is 50.9% of the formula by mass. A rug measures 12×15 ft and contains 30 oz of Acrilan® fibers per square yard of carpet. If the rug burns, will a lethal dose of HCN be generated in the room? Assume that the yield of HCN from the fibers is 20% and that the carpet is 50% consumed.

3.108 The source of oxygen that drives the internal combustion engine in an automobile is air. Air is a mixture of gases, which are principally N_2 (~79%) and O_2 (~20%).

In the cylinder of an automobile engine, nitrogen can react with oxygen to produce nitric oxide gas, NO. As NO is emitted from the tailpipe of the car, it can react with more oxygen to produce nitrogen dioxide gas. **(a)** Write balanced chemical equations for both reactions. **(b)** Both nitric oxide and nitrogen dioxide are pollutants that can lead to acid rain and global warming; collectively, they are called "NO$_x$" gases. In 2004, the United States emitted an estimated 19 million tons of nitrogen dioxide into the atmosphere. How many grams of nitrogen dioxide is this? **(c)** The production of NO$_x$ gases is an unwanted side reaction of the main engine combustion process that turns octane, C_8H_{18}, into CO_2 and water. If 85% of the oxygen in an engine is used to combust octane, and the remainder used to produce nitrogen dioxide, calculate how many grams of nitrogen dioxide would be produced during the combustion of 500 grams of octane.

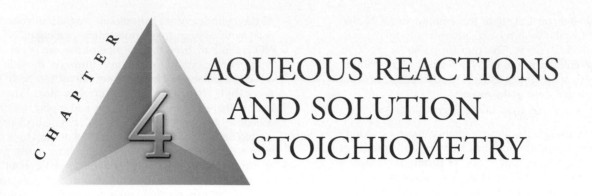

CHAPTER 4

AQUEOUS REACTIONS AND SOLUTION STOICHIOMETRY

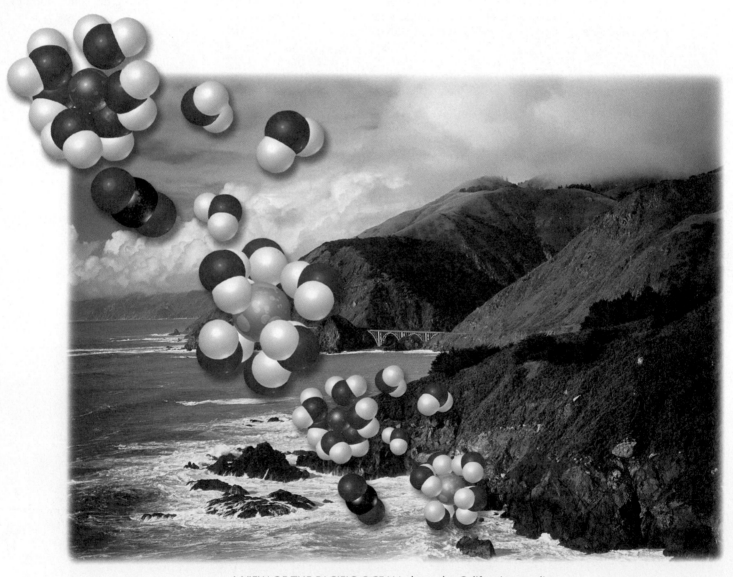

A VIEW OF THE PACIFIC OCEAN along the California coastline.

WHAT'S AHEAD

4.1 General Properties of Aqueous Solutions
We begin by examining the nature of the substances dissolved in water, whether they exist in water as ions, molecules, or some mixture of the two. This information is necessary to understand the nature of reactants in aqueous solutions.

4.2 Precipitation Reactions
We will identify reactions in which soluble reactants yield an insoluble product.

4.3 Acid–Base Reactions
We will explore reactions in which protons, H^+ ions, are transferred between reactants.

4.4 Oxidation-Reduction Reactions
We will examine reactions in which electrons are transferred between reactants.

4.5 Concentrations of Solutions
We will recognize that the amount of a compound in a given volume of a solution is called its *concentration*. Concentration can be expressed in a number of ways, the most usual of which is moles of compound per liter of solution (*molarity*).

4.6 Solution Stoichiometry and Chemical Analysis
We will use what we have learned about stoichiometry and concentration to calculate the amounts or concentrations of substances in solution.

THE WATERS OF THE PACIFIC OCEAN, seen in this chapter-opening photograph of the California coast, are part of the World Ocean that covers almost two-thirds of our planet. Water has been the key to much of Earth's evolutionary history. Life itself almost certainly originated in water, and the

need for water by all forms of life has helped determine diverse biological structures. Your own body is about 60% water by mass. We will see repeatedly throughout this text that water possesses many unusual properties essential to supporting life on Earth.

The waters of the World Ocean may not appear to be any different from those of Lake Tahoe or the water that flows from your kitchen faucet, but a taste of seawater is all it takes to demonstrate that there is an important difference. Water has an exceptional ability to dissolve a wide variety of substances. Water on Earth—whether it is drinking water from the tap, water from a clear mountain stream, or seawater—invariably contains a variety of dissolved substances. A solution in which water is the dissolving medium is called an **aqueous solution**. Seawater is different from what we call "freshwater" because it has a much higher total concentration of dissolved ionic substances.

Water is the medium for most of the chemical reactions that take place within us and around us. Nutrients dissolved in blood are carried to our cells, where they enter into reactions that help keep us alive. Automobile parts rust

▲ **Figure 4.1 Limestone cave.** When CO_2 dissolves in water, the resulting solution is slightly acidic. Limestone caves are formed by the dissolving action of this acidic solution acting on $CaCO_3$ in the limestone.

when they come into frequent contact with aqueous solutions that contain various dissolved substances. Spectacular limestone caves (Figure 4.1 ◄) are formed by the dissolving action of underground water containing carbon dioxide, $CO_2(aq)$:

$$CaCO_3(s) + H_2O(l) + CO_2(aq) \longrightarrow Ca(HCO_3)_2(aq) \qquad [4.1]$$

In Chapter 3 we saw a few simple types of chemical reactions and their descriptions. In this chapter we continue to examine chemical reactions by focusing on aqueous solutions. A great deal of important chemistry occurs in aqueous solutions, so we need to learn the vocabulary and concepts used to describe and understand this chemistry. In addition, we will extend the concepts of stoichiometry that we learned in Chapter 3 by considering how solution concentrations are expressed and used.

4.1 GENERAL PROPERTIES OF AQUEOUS SOLUTIONS

Recall that a *solution* is a homogeneous mixture of two or more substances. ∞ (Section 1.2) The substance present in the greatest quantity is usually called the **solvent**. The other substances in the solution are called the **solutes**; they are said to be dissolved in the solvent. When a small amount of sodium chloride (NaCl) is dissolved in a large quantity of water, for example, the water is the solvent and the sodium chloride is the solute.

Electrolytic Properties

Imagine preparing two aqueous solutions—one by dissolving a teaspoon of table salt (sodium chloride) in a cup of water and the other by dissolving a teaspoon of table sugar (sucrose) in a cup of water. Both solutions are clear and colorless. How do they differ? One way, which might not be immediately obvious, is in their electrical conductivity: The salt solution is a good conductor of electricity; the sugar solution is not.

Whether a solution conducts electricity can be determined by using a device such as that shown in Figure 4.2 ▶. To light the bulb, an electric current must flow between the two electrodes that are immersed in the solution. Although water itself is a poor conductor of electricity, the presence of ions causes aqueous solutions to become good conductors. Ions carry electrical charge from one electrode to the other, completing the electrical circuit. Thus, the conductivity of NaCl solutions indicates the presence of ions in the solution. The lack of conductivity of sucrose solutions indicates the absence of ions. When NaCl dissolves in water, the solution contains Na^+ and Cl^- ions, each surrounded by water molecules. When sucrose ($C_{12}H_{22}O_{11}$) dissolves in water, the solution contains only neutral sucrose molecules surrounded by water molecules.

A substance (such as NaCl) whose aqueous solutions contain ions is called an **electrolyte**. A substance (such as $C_{12}H_{22}O_{11}$) that does not form ions in solution is called a **nonelectrolyte**. The difference between NaCl and $C_{12}H_{22}O_{11}$ arises largely because NaCl is ionic, whereas $C_{12}H_{22}O_{11}$ is molecular.

Ionic Compounds in Water

Recall from Section 2.7 and especially Figure 2.23 that solid NaCl consists of an orderly arrangement of Na^+ and Cl^- ions. When NaCl dissolves in water, each ion separates from the solid structure and disperses throughout the solution, as shown in Figure 4.3(a) ▶. The ionic solid *dissociates* into its component ions as it dissolves.

ELECTROLYTIC PROPERTIES

One way to differentiate two aqueous solutions is to employ a device that measures their electrical conductivities. The ability of a solution to conduct electricity depends on the number of ions it contains. An electrolyte solution contains ions that serve as charge carriers, causing the bulb to light.

No ions

A nonelectrolyte solution does not contain ions, and the bulb does not light.

Few ions

If the solution contains a small number of ions, the bulb will be only dimly lit.

Many ions

If the solution contains a large number of ions, the bulb will be brightly lit.

▲ Figure 4.2 **Measuring ion concentrations using conductivity.**

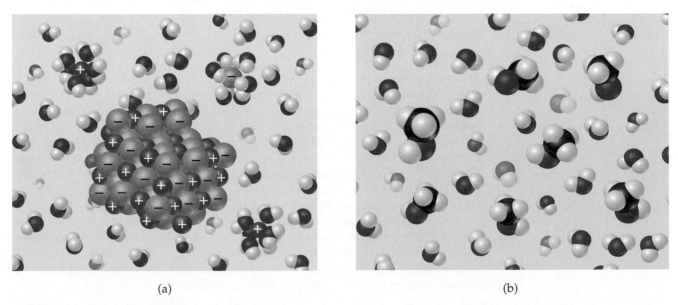

(a)   (b)

▲ Figure 4.3 **Dissolution in water.** (a) When an ionic compound dissolves in water, H_2O molecules separate, surround, and disperse the ions into the liquid. (b) Methanol, CH_3OH, a molecular compound, dissolves without forming ions. The methanol molecules contain black spheres, which represent carbon atoms. In both parts (a) and (b) the water molecules have been moved apart so that the solute particles can be seen more clearly.

Water is a very effective solvent for ionic compounds. Although water is an electrically neutral molecule, one end of the molecule (the O atom) is rich in electrons and has a partial negative charge, denoted by $\delta-$. The other end (the H atoms) has a partial positive charge, denoted by $\delta+$ as shown in the margin. Positive ions (cations) are attracted by the negative end of H_2O, and negative ions (anions) are attracted by the positive end.

As an ionic compound dissolves, the ions become surrounded by H_2O molecules, as shown in Figure 4.3(a). The ions are said to be solvated. We denote these ions in chemical equations by writing them as $Na^+(aq)$ and $Cl^-(aq)$, where "aq" is an abbreviation for "aqueous." ∞ (Section 3.1) The **solvation** process helps stabilize the ions in solution and prevents cations and anions from recombining. Furthermore, because the ions and their shells of surrounding water molecules are free to move about, the ions become dispersed uniformly throughout the solution.

We can usually predict the nature of the ions present in a solution of an ionic compound from the chemical name of the substance. Sodium sulfate (Na_2SO_4), for example, dissociates into sodium ions (Na^+) and sulfate ions (SO_4^{2-}). You must remember the formulas and charges of common ions (Tables 2.4 and 2.5) to understand the forms in which ionic compounds exist in aqueous solution.

GIVE IT SOME THOUGHT

What dissolved species are present in a solution of **(a)** KCN, **(b)** $NaClO_4$?

Molecular Compounds in Water

When a molecular compound dissolves in water, the solution usually consists of intact molecules dispersed throughout the solution. Consequently, most molecular compounds are nonelectrolytes. As we have seen, table sugar (sucrose) is an example of a nonelectrolyte. As another example, a solution of methanol (CH_3OH) in water consists entirely of CH_3OH molecules dispersed throughout the water [Figure 4.3(b)].

A few molecular substances, however, have aqueous solutions that contain ions. Acids are the most important of these solutions. For example, when $HCl(g)$ dissolves in water to form hydrochloric acid, $HCl(aq)$, it *ionizes*; that is, it dissociates into $H^+(aq)$ and $Cl^-(aq)$ ions.

Strong and Weak Electrolytes

Two categories of electrolytes—strong and weak—differ in the extent to which they conduct electricity. **Strong electrolytes** are those solutes that exist in solution completely or nearly completely as ions. Essentially all soluble ionic compounds (such as NaCl) and a few molecular compounds (such as HCl) are strong electrolytes. **Weak electrolytes** are those solutes that exist in solution mostly in the form of molecules with only a small fraction in the form of ions. For example, in a solution of acetic acid (CH_3COOH) most of the solute is present as $CH_3COOH(aq)$ molecules. Only a small fraction (about 1%) of the CH_3COOH is present as $H^+(aq)$ and $CH_3COO^-(aq)$ ions.*

We must be careful not to confuse the extent to which an electrolyte dissolves with whether it is strong or weak. For example, CH_3COOH is extremely soluble in water but is a weak electrolyte. $Ba(OH)_2$, on the other hand, is not very soluble, but the amount of the substance that does dissolve dissociates almost completely. Thus, $Ba(OH)_2$ is a strong electrolyte.

---

*The chemical formula of acetic acid is sometimes written as $HC_2H_3O_2$ with the acidic H written in front of the chemical formula, so the formula looks like that of other common acids such as HCl. The formula CH_3COOH conforms to the molecular structure of acetic acid, with the acidic H on the O atom at the end of the formula.

When a weak electrolyte such as acetic acid ionizes in solution, we write the reaction in the following manner:

$$CH_3COOH(aq) \rightleftharpoons CH_3COO^-(aq) + H^+(aq) \qquad [4.2]$$

The half-arrows in both directions mean that the reaction is significant in both directions. At any given moment some CH_3COOH molecules are ionizing to form H^+ and CH_3COO^- ions. At the same time, H^+ and CH_3COO^- ions are recombining to form CH_3COOH. The balance between these opposing processes determines the relative numbers of ions and neutral molecules. This balance produces a state of **chemical equilibrium** in which the relative numbers of each type of ion or molecule in the reaction are constant over time. This equilibrium condition varies from one weak electrolyte to another. Chemical equilibria are extremely important. We will devote Chapters 15–17 to examining them in detail.

Chemists use half-arrows in both directions to represent the ionization of weak electrolytes and a single arrow to represent the ionization of strong electrolytes. Because HCl is a strong electrolyte, we write the equation for the ionization of HCl as follows:

$$HCl(aq) \longrightarrow H^+(aq) + Cl^-(aq) \qquad [4.3]$$

The absence of a reverse arrow indicates that the H^+ and Cl^- ions have no tendency to recombine in water to form HCl molecules.

In the following sections we will look more closely at how we can use the composition of a compound to predict whether it is a strong electrolyte, weak electrolyte, or nonelectrolyte. For the moment, you need only to remember that *soluble ionic compounds are strong electrolytes.* We identify ionic compounds as those composed of metals and nonmetals—such as NaCl, $FeSO_4$, and $Al(NO_3)_3$—or compounds containing the ammonium ion, NH_4^+—such as NH_4Br and $(NH_4)_2CO_3$.

▲ GIVE IT SOME THOUGHT

Which solute will cause the lightbulb in the experiment shown in Figure 4.2 to glow more brightly, CH_3OH or $MgBr_2$?

■ **SAMPLE EXERCISE 4.1** | **Relating Relative Numbers of Anions and Cations to Chemical Formulas**

The diagram on the right represents an aqueous solution of one of the following compounds: $MgCl_2$, KCl, or K_2SO_4. Which solution does the drawing best represent?

SOLUTION

Analyze: We are asked to associate the charged spheres in the diagram with ions present in a solution of an ionic substance.

Plan: We examine the ionic substances given in the problem to determine the relative numbers and charges of the ions that each contains. We then correlate these charged ionic species with the ones shown in the diagram.

Solve: The diagram shows twice as many cations as anions, consistent with the formulation K_2SO_4.

Check: Notice that the total net charge in the diagram is zero, as it must be if it is to represent an ionic substance.

■ **PRACTICE EXERCISE**

If you were to draw diagrams (such as that shown on the right) representing aqueous solutions of each of the following ionic compounds, how many anions would you show if the diagram contained six cations? **(a)** $NiSO_4$, **(b)** $Ca(NO_3)_2$, **(c)** Na_3PO_4, **(d)** $Al_2(SO_4)_3$
Answers: **(a)** 6, **(b)** 12, **(c)** 2, **(d)** 9

4.2 PRECIPITATION REACTIONS

Figure 4.4▼ shows two clear solutions being mixed. One solution contains lead nitrate, $Pb(NO_3)_2$, and the other contains potassium iodide (KI). The reaction between these two solutes produces an insoluble yellow product. Reactions that result in the formation of an insoluble product are called **precipitation reactions**. A **precipitate** is an insoluble solid formed by a reaction in solution. In Figure 4.4 the precipitate is lead iodide (PbI_2), a compound that has a very low solubility in water:

$$Pb(NO_3)_2(aq) + 2\ KI(aq) \longrightarrow PbI_2(s) + 2\ KNO_3(aq) \qquad [4.4]$$

The other product of this reaction, potassium nitrate (KNO_3), remains in solution.

Precipitation reactions occur when certain pairs of oppositely charged ions attract each other so strongly that they form an insoluble ionic solid. To predict whether certain combinations of ions form insoluble compounds, we must consider some guidelines concerning the solubilities of common ionic compounds.

Solubility Guidelines for Ionic Compounds

The **solubility** of a substance at a given temperature is the amount of the substance that can be dissolved in a given quantity of solvent at the given temperature.

▼ Figure 4.4 **A precipitation reaction.**

PRECIPITATION REACTION

Reactions that result in the formation of an insoluble product are known as precipitation reactions.

2 KI(aq)   +   Pb(NO$_3$)$_2$(aq) $\longrightarrow$ PbI$_2$(s) + 2 KNO$_3$(aq)

The addition of a colorless solution of potassium iodide (KI)

to a colorless solution of lead nitrate

produces a yellow precipitate of lead iodide (PbI$_2$) that slowly settles to the bottom of the beaker.

TABLE 4.1 ■ Solubility Guidelines for Common Ionic Compounds in Water		
Soluble Ionic Compounds		**Important Exceptions**
Compounds containing	NO_3^-	None
	CH_3COO^-	None
	Cl^-	Compounds of Ag^+, Hg_2^{2+}, and Pb^{2+}
	Br^-	Compounds of Ag^+, Hg_2^{2+}, and Pb^{2+}
	I^-	Compounds of Ag^+, Hg_2^{2+}, and Pb^{2+}
	SO_4^{2-}	Compounds of Sr^{2+}, Ba^{2+}, Hg_2^{2+}, and Pb^{2+}
Insoluble Ionic Compounds		**Important Exceptions**
Compounds containing	S^{2-}	Compounds of NH_4^+, the alkali metal cations, and Ca^{2+}, Sr^{2+}, and Ba^{2+}
	CO_3^{2-}	Compounds of NH_4^+ and the alkali metal cations
	PO_4^{3-}	Compounds of NH_4^+ and the alkali metal cations
	OH^-	Compounds of the alkali metal cations, and NH_4^+, Ca^{2+}, Sr^{2+}, and Ba^{2+}

For instance, only 1.2×10^{-3} mol of PbI_2 dissolves in a liter of water at 25 °C. In our discussions, any substance with a solubility less than 0.01 mol/L will be referred to as *insoluble*. In those cases the attraction between the oppositely charged ions in the solid is too great for the water molecules to separate the ions to any significant extent; the substance remains largely undissolved.

Unfortunately, there are no rules based on simple physical properties such as ionic charge to guide us in predicting whether a particular ionic compound will be soluble. Experimental observations, however, have led to guidelines for predicting solubility for ionic compounds. For example, experiments show that all common ionic compounds that contain the nitrate anion, NO_3^-, are soluble in water. Table 4.1▲ summarizes the solubility guidelines for common ionic compounds. The table is organized according to the anion in the compound, but it also reveals many important facts about cations. Note that *all common ionic compounds of the alkali metal ions (group 1A of the periodic table) and of the ammonium ion (NH_4^+) are soluble in water.*

■ SAMPLE EXERCISE 4.2 | Using Solubility Rules

Classify the following ionic compounds as soluble or insoluble in water: **(a)** sodium carbonate (Na_2CO_3), **(b)** lead sulfate ($PbSO_4$).

SOLUTION

Analyze: We are given the names and formulas of two ionic compounds and asked to predict whether they are soluble or insoluble in water.

Plan: We can use Table 4.1 to answer the question. Thus, we need to focus on the anion in each compound because the table is organized by anions.

Solve:
(a) According to Table 4.1, most carbonates are insoluble. But carbonates of the alkali metal cations (such as sodium ion) are an exception to this rule and are soluble. Thus, Na_2CO_3 is soluble in water.
(b) Table 4.1 indicates that although most sulfates are water soluble, the sulfate of Pb^{2+} is an exception. Thus, $PbSO_4$ is insoluble in water.

■ PRACTICE EXERCISE

Classify the following compounds as soluble or insoluble in water: **(a)** cobalt(II) hydroxide, **(b)** barium nitrate, **(c)** ammonium phosphate.
Answers: **(a)** insoluble, **(b)** soluble, **(c)** soluble

To predict whether a precipitate forms when we mix aqueous solutions of two strong electrolytes, we must (1) note the ions present in the reactants, (2) consider the possible combinations of the cations and anions, and (3) use Table 4.1 to determine if any of these combinations is insoluble. For example, will a precipitate form when solutions of $Mg(NO_3)_2$ and $NaOH$ are mixed? Both $Mg(NO_3)_2$ and $NaOH$ are soluble ionic compounds and strong electrolytes. Mixing $Mg(NO_3)_2(aq)$ and $NaOH(aq)$ first produces a solution containing Mg^{2+}, NO_3^-, Na^+, and OH^- ions. Will either of the cations interact with either of the anions to form an insoluble compound? In addition to the reactants, the other possible interactions are Mg^{2+} with OH^- and Na^+ with NO_3^-. From Table 4.1 we see that hydroxides are generally insoluble. Because Mg^{2+} is not an exception, $Mg(OH)_2$ is insoluble and will thus form a precipitate. $NaNO_3$, however, is soluble, so Na^+ and NO_3^- will remain in solution. The balanced equation for the precipitation reaction is

$$Mg(NO_3)_2(aq) + 2\,NaOH(aq) \longrightarrow Mg(OH)_2(s) + 2\,NaNO_3(aq) \quad [4.5]$$

Exchange (Metathesis) Reactions

Notice in Equation 4.5 that the cations in the two reactants exchange anions—Mg^{2+} ends up with OH^-, and Na^+ ends up with NO_3^-. The chemical formulas of the products are based on the charges of the ions—two OH^- ions are needed to give a neutral compound with Mg^{2+}, and one NO_3^- ion is needed to give a neutral compound with Na^+. ∞ (Section 2.7) The equation can be balanced only after the chemical formulas of the products have been determined.

Reactions in which positive ions and negative ions appear to exchange partners conform to the following general equation:

$$AX + BY \longrightarrow AY + BX \quad [4.6]$$

Example: $AgNO_3(aq) + KCl(aq) \longrightarrow AgCl(s) + KNO_3(aq)$

Such reactions are called **exchange reactions**, or **metathesis reactions** (meh-TATH-eh-sis, which is the Greek word for "to transpose"). Precipitation reactions conform to this pattern, as do many acid–base reactions, as we will see in Section 4.3.

To complete and balance a metathesis equation, we follow these steps:

1. Use the chemical formulas of the reactants to determine the ions that are present.
2. Write the chemical formulas of the products by combining the cation from one reactant with the anion of the other. (Use the charges of the ions to determine the subscripts in the chemical formulas.)
3. Finally, balance the equation.

■■ SAMPLE EXERCISE 4.3 | Predicting a Metathesis Reaction

(a) Predict the identity of the precipitate that forms when solutions of $BaCl_2$ and K_2SO_4 are mixed.
(b) Write the balanced chemical equation for the reaction.

SOLUTION

Analyze: We are given two ionic reactants and asked to predict the insoluble product that they form.

Plan: We need to write down the ions present in the reactants and to exchange the anions between the two cations. Once we have written the chemical formulas for these products, we can use Table 4.1 to determine which is insoluble in water. Knowing the products also allows us to write the equation for the reaction.

Solve:
(a) The reactants contain Ba^{2+}, Cl^-, K^+, and SO_4^{2-} ions. If we exchange the anions, we will have $BaSO_4$ and KCl. According to Table 4.1, most compounds of SO_4^{2-} are soluble but those of Ba^{2+} are not. Thus, $BaSO_4$ is insoluble and will precipitate from solution. KCl, on the other hand, is soluble.

(b) From part (a) we know the chemical formulas of the products, $BaSO_4$ and KCl. The balanced equation with phase labels shown is

$$BaCl_2(aq) + K_2SO_4(aq) \longrightarrow BaSO_4(s) + 2\ KCl(aq)$$

▪ PRACTICE EXERCISE

(a) What compound precipitates when solutions of $Fe_2(SO_4)_3$ and $LiOH$ are mixed? **(b)** Write a balanced equation for the reaction. **(c)** Will a precipitate form when solutions of $Ba(NO_3)_2$ and KOH are mixed?
Answers: **(a)** $Fe(OH)_3$; **(b)** $Fe_2(SO_4)_3(aq) + 6\ LiOH(aq) \longrightarrow$
$2\ Fe(OH)_3(s) + 3\ Li_2SO_4(aq)$; **(c)** no (both possible products are water soluble)

Ionic Equations

In writing chemical equations for reactions in aqueous solution, it is often useful to indicate explicitly whether the dissolved substances are present predominantly as ions or as molecules. Let's reconsider the precipitation reaction between $Pb(NO_3)_2$ and $2\ KI$, shown previously in Figure 4.4:

$$Pb(NO_3)_2(aq) + 2\ KI(aq) \longrightarrow PbI_2(s) + 2\ KNO_3(aq)$$

An equation written in this fashion, showing the complete chemical formulas of the reactants and products, is called a **molecular equation** because it shows the chemical formulas of the reactants and products without indicating their ionic character. Because $Pb(NO_3)_2$, KI, and KNO_3 are all soluble ionic compounds and therefore strong electrolytes, we can write the chemical equation to indicate explicitly the ions that are in the solution:

$$Pb^{2+}(aq) + 2\ NO_3^-(aq) + 2\ K^+(aq) + 2\ I^-(aq) \longrightarrow$$
$$PbI_2(s) + 2\ K^+(aq) + 2\ NO_3^-(aq) \qquad [4.7]$$

An equation written in this form, with all soluble strong electrolytes shown as ions, is called a **complete ionic equation**.

Notice that $K^+(aq)$ and $NO_3^-(aq)$ appear on both sides of Equation 4.7. Ions that appear in identical forms among both the reactants and products of a complete ionic equation are called **spectator ions**. These ions are present but play no direct role in the reaction. When spectator ions are omitted from the equation (they cancel out like algebraic quantities), we are left with the **net ionic equation**:

$$Pb^{2+}(aq) + 2\ I^-(aq) \longrightarrow PbI_2(s) \qquad [4.8]$$

A net ionic equation includes only the ions and molecules directly involved in the reaction. Charge is conserved in reactions, so the sum of the charges of the ions must be the same on both sides of a balanced net ionic equation. In this case the 2+ charge of the cation and the two 1− charges of the anions add to give zero, the charge of the electrically neutral product. *If every ion in a complete ionic equation is a spectator, then no reaction occurs.*

◢ GIVE IT SOME THOUGHT

Are any spectator ions shown in the following chemical equation?
$$Ag^+(aq) + Na^+(aq) + Cl^-(aq) \longrightarrow AgCl(s) + Na^+(aq)$$

Net ionic equations are widely used to illustrate the similarities between large numbers of reactions involving electrolytes. For example, Equation 4.8 expresses the essential feature of the precipitation reaction between any strong electrolyte containing Pb^{2+} and any strong electrolyte containing I^-: The $Pb^{2+}(aq)$ and $I^-(aq)$ ions combine to form a precipitate of PbI_2. Thus, a net ionic equation demonstrates that more than one set of reactants can lead to the same net reaction. For example, aqueous solutions of KI and MgI_2 share many chemical similarities because both contain I^- ions. The complete equation, on the other hand, identifies the actual reactants that participate in a reaction.

The following steps summarize the procedure for writing net ionic equations:

1. Write a balanced molecular equation for the reaction.
2. Rewrite the equation to show the ions that form in solution when each soluble strong electrolyte dissociates into its component ions. *Only strong electrolytes dissolved in aqueous solution are written in ionic form.*
3. Identify and cancel **spectator ions**.

■ SAMPLE EXERCISE 4.4 | Writing a Net Ionic Equation

Write the net ionic equation for the precipitation reaction that occurs when solutions of calcium chloride and sodium carbonate are mixed.

SOLUTION

Analyze: Our task is to write a net ionic equation for a precipitation reaction, given the names of the reactants present in solution.

Plan: We first need to write the chemical formulas of the reactants and products and then determine which product is insoluble. We then write and balance the molecular equation. Next, we write each soluble strong electrolyte as separated ions to obtain the complete ionic equation. Finally, we eliminate the spectator ions to obtain the net ionic equation.

Solve: Calcium chloride is composed of calcium ions, Ca^{2+}, and chloride ions, Cl^-; hence an aqueous solution of the substance is $CaCl_2(aq)$. Sodium carbonate is composed of Na^+ ions and CO_3^{2-} ions; hence an aqueous solution of the compound is $Na_2CO_3(aq)$. In the molecular equations for precipitation reactions, the anions and cations appear to exchange partners. Thus, we put Ca^{2+} and CO_3^{2-} together to give $CaCO_3$ and Na^+ and Cl^- together to give NaCl. According to the solubility guidelines in Table 4.1, $CaCO_3$ is insoluble and NaCl is soluble. The balanced molecular equation is

$$CaCl_2(aq) + Na_2CO_3(aq) \longrightarrow CaCO_3(s) + 2\,NaCl(aq)$$

In a complete ionic equation, *only* dissolved strong electrolytes (such as soluble ionic compounds) are written as separate ions. As the (*aq*) designations remind us, $CaCl_2$, Na_2CO_3, and NaCl are all dissolved in the solution. Furthermore, they are all strong electrolytes. $CaCO_3$ is an ionic compound, but it is not soluble. We do not write the formula of any insoluble compound as its component ions. Thus, the complete ionic equation is

$$Ca^{2+}(aq) + 2\,Cl^-(aq) + 2\,Na^+(aq) + CO_3^{2-}(aq) \longrightarrow$$
$$CaCO_3(s) + 2\,Na^+(aq) + 2\,Cl^-(aq)$$

Cl^- and Na^+ are spectator ions. Canceling them gives the following net ionic equation:

$$Ca^{2+}(aq) + CO_3^{2-}(aq) \longrightarrow CaCO_3(s)$$

Check: We can check our result by confirming that both the elements and the electric charge are balanced. Each side has one Ca, one C, and three O, and the net charge on each side equals 0.

Comment: If none of the ions in an ionic equation is removed from solution or changed in some way, then they all are spectator ions and a reaction does not occur.

■ PRACTICE EXERCISE

Write the net ionic equation for the precipitation reaction that occurs when aqueous solutions of silver nitrate and potassium phosphate are mixed.
Answers: $3\,Ag^+(aq) + PO_4^{3-}(aq) \longrightarrow Ag_3PO_4(s)$

▲ Figure 4.5 **Some common household acids (left) and bases (right).**

4.3 ACID–BASE REACTIONS

Many acids and bases are industrial and household substances (Figure 4.5 ◀), and some are important components of biological fluids. Hydrochloric acid, for example, is an important industrial chemical and the main constituent of gastric juice in your stomach. Acids and bases are also common electrolytes.

Acids

Acids are substances that ionize in aqueous solutions to form hydrogen ions, thereby increasing the concentration of $H^+(aq)$ ions. Because a hydrogen atom consists of a proton and an electron, H^+ is simply a proton. Thus, acids are often called proton donors. Molecular models of three common acids, HCl, HNO_3 and CH_3COOH, are shown in the margin.

Just as cations are surrounded and bound by water molecules (see Figure 4.3[a]), the proton is also solvated by water molecules. The nature of the proton in water is discussed in detail in Section 16.2. In writing chemical equations involving the proton in water, we represent it simply as $H^+(aq)$.

Molecules of different acids can ionize to form different numbers of H^+ ions. Both HCl and HNO_3 are *monoprotic* acids, which yield one H^+ per molecule of acid. Sulfuric acid, H_2SO_4, is a *diprotic* acid, one that yields two H^+ per molecule of acid. The ionization of H_2SO_4 and other diprotic acids occurs in two steps:

$$H_2SO_4(aq) \longrightarrow H^+(aq) + HSO_4^-(aq) \qquad [4.9]$$

$$HSO_4^-(aq) \rightleftharpoons H^+(aq) + SO_4^{2-}(aq) \qquad [4.10]$$

Although H_2SO_4 is a strong electrolyte, only the first ionization is complete. Thus, aqueous solutions of sulfuric acid contain a mixture of $H^+(aq)$, $HSO_4^-(aq)$, and $SO_4^{2-}(aq)$.

The molecule CH_3COOH (acetic acid) that we have mentioned frequently is the primary component in vinegar. Acetic acid has four hydrogens, but only one of them is capable of being ionized in water. Only the hydrogen that is bound to oxygen in the COOH group will ionize in water; the other hydrogens are bound to carbon and do not break their C—H bonds in water. We will discuss acids much more in Chapter 16.

HCl

HNO₃

CH₃COOH

GIVE IT SOME THOUGHT

The structural formula of citric acid, a main component of citrus fruits, is shown here:

$$
\begin{array}{c}
\text{H} \\
| \\
\text{H}-\text{C}-\text{COOH} \\
| \\
\text{HO}-\text{C}-\text{COOH} \\
| \\
\text{H}-\text{C}-\text{COOH} \\
| \\
\text{H}
\end{array}
$$

How many $H^+(aq)$ can be generated by each citric acid molecule when citric acid is dissolved in water?

Bases

Bases are substances that accept (react with) H^+ ions. Bases produce hydroxide ions (OH^-) when they dissolve in water. Ionic hydroxide compounds such as NaOH, KOH, and $Ca(OH)_2$ are among the most common bases. When dissolved in water, they dissociate into their component ions, introducing OH^- ions into the solution.

Compounds that do not contain OH^- ions can also be bases. For example, ammonia (NH_3) is a common base. When added to water, it accepts an H^+ ion from the water molecule and thereby produces an OH^- ion (Figure 4.6▶):

$$NH_3(aq) + H_2O(l) \rightleftharpoons NH_4^+(aq) + OH^-(aq) \qquad [4.11]$$

Ammonia is a weak electrolyte because only a small fraction of the NH_3 (about 1%) forms NH_4^+ and OH^- ions.

H_2O OH^-

NH_3 NH_4^+

▲ **Figure 4.6 Hydrogen ion transfer.** An H_2O molecule acts as a proton donor (acid), and NH_3 as a proton acceptor (base). Only a fraction of the NH_3 reacts with H_2O; NH_3 is a weak electrolyte.

TABLE 4.2 ■ Common Strong Acids and Bases	
Strong Acids	**Strong Bases**
Hydrochloric, HCl	Group 1A metal hydroxides (LiOH, NaOH, KOH, RbOH, CsOH)
Hydrobromic, HBr	Heavy group 2A metal hydroxides [Ca(OH)$_2$, Sr(OH)$_2$, Ba(OH)$_2$]
Hydroiodic, HI	
Chloric, HClO$_3$	
Perchloric, HClO$_4$	
Nitric, HNO$_3$	
Sulfuric, H$_2$SO$_4$	

Strong and Weak Acids and Bases

Acids and bases that are strong electrolytes (completely ionized in solution) are called **strong acids** and **strong bases**. Those that are weak electrolytes (partly ionized) are called **weak acids** and **weak bases**. Strong acids are more reactive than weak acids when the reactivity depends only on the concentration of $H^+(aq)$. The reactivity of an acid, however, can depend on the anion as well as on $H^+(aq)$. For example, hydrofluoric acid (HF) is a weak acid (only partly ionized in aqueous solution), but it is very reactive and vigorously attacks many substances, including glass. This reactivity is due to the combined action of $H^+(aq)$ and $F^-(aq)$.

Table 4.2▲ lists the common strong acids and bases. You should commit these to memory. As you examine this table, notice that some of the most common acids, such as HCl, HNO$_3$, and H$_2$SO$_4$, are strong. (For H$_2$SO$_4$, as we noted earlier, only the first proton completely ionizes.) Three of the strong acids are the hydrogen compounds of the halogen family. (HF, however, is a weak acid.) The list of strong acids is very short. Most acids are weak. The only common strong bases are the hydroxides of Li$^+$, Na$^+$, K$^+$, Rb$^+$, and Cs$^+$ (the alkali metals, group 1A) and the hydroxides of Ca^{2+}, Sr^{2+}, and Ba^{2+} (the heavy alkaline earths, group 2A). These are the common soluble metal hydroxides. Most other metal hydroxides are insoluble in water. The most common weak base is NH$_3$, which reacts with water to form OH$^-$ ions (Equation 4.11).

GIVE IT SOME THOUGHT

Which of the following is a strong acid: H$_2$SO$_3$, HBr, CH$_3$COOH?

■ SAMPLE EXERCISE 4.5 | Comparing Acid Strengths

The following diagrams represent aqueous solutions of three acids (HX, HY, and HZ) with water molecules omitted for clarity. Rank them from strongest to weakest.

HX   HY   HZ

SOLUTION

Analyze: We are asked to rank three acids from strongest to weakest, based on schematic drawings of their solutions.

Plan: We can examine the drawings to determine the relative numbers of uncharged molecular species present. The strongest acid is the one with the most H^+ ions and fewest undissociated acid molecules in solution. The weakest acid is the one with the largest number of undissociated molecules.

Solve: The order is HY > HZ > HX. HY is a strong acid because it is totally ionized (no HY molecules in solution), whereas both HX and HZ are weak acids, whose solutions consist of a mixture of molecules and ions. Because HZ contains more H^+ ions and fewer molecules than HX, it is a stronger acid.

■ PRACTICE EXERCISE

Imagine a diagram showing 10 Na^+ ions and 10 OH^- ions. If this solution were mixed with the one pictured on the previous page for HY, what would the diagram look like that represents the solution after any possible reaction? (H^+ ions will react with OH^- ions to form H_2O.)
Answer: The final diagram would show 10 Na^+ ions, 2 OH^- ions, 8 Y^- ions, and 8 H_2O molecules.

Identifying Strong and Weak Electrolytes

If we remember the common strong acids and bases (Table 4.2) and also remember that NH_3 is a weak base, we can make reasonable predictions about the electrolytic strength of a great number of water-soluble substances. Table 4.3▼ summarizes our observations about electrolytes. To classify a soluble substance as a strong electrolyte, weak electrolyte, or nonelectrolyte, we simply work our way down and across this table. We first ask ourselves whether the substance is ionic or molecular. If it is ionic, it is a strong electrolyte. The second column of Table 4.3 tells us that all ionic compounds are strong electrolytes. If the substance we want to classify is molecular, we ask whether it is an acid or a base. (Does it have H first in the chemical formula or contain a COOH group?) If it is an acid, we rely on the memorized list from Table 4.2 to determine whether it is a strong or weak electrolyte: All strong acids are strong electrolytes, and all weak acids are weak electrolytes. If an acid is not listed in Table 4.2, it is probably a weak acid and therefore a weak electrolyte. For example, H_3PO_4, H_2SO_3, and $HC_7H_5O_2$ are not listed in Table 4.2 and are weak acids. If the substance we want to classify is a base, we again turn to Table 4.2 to determine whether it is one of the listed strong bases. NH_3 is the only molecular base that we consider in this chapter, and Table 4.3 tells us it is a weak electrolyte. (There are compounds called amines that are related to NH_3 and are also molecular bases, but we will not consider them until Chapter 16.) Finally, any molecular substance that we encounter in this chapter that is not an acid or NH_3 is probably a nonelectrolyte.

TABLE 4.3 ■ Summary of the Electrolytic Behavior of Common Soluble Ionic and Molecular Compounds		
	Strong Electrolyte	**Weak Electrolyte**	**Nonelectrolyte**
Ionic	All	None	None
Molecular	Strong acids (see Table 4.2)	Weak acids Weak bases	All other compounds

■ **SAMPLE EXERCISE 4.6** | Identifying Strong, Weak, and Nonelectrolytes

Classify each of the following dissolved substances as a strong electrolyte, weak electrolyte, or nonelectrolyte: $CaCl_2$, HNO_3, C_2H_5OH (ethanol), $HCOOH$ (formic acid), KOH.

SOLUTION

Analyze: We are given several chemical formulas and asked to classify each substance as a strong electrolyte, weak electrolyte, or nonelectrolyte.

Plan: The approach we take is outlined in Table 4.3. We can predict whether a substance is ionic or molecular based on its composition. As we saw in Section 2.7, most ionic compounds we encounter in this text are composed of a metal and a nonmetal, whereas most molecular compounds are composed only of nonmetals.

Solve: Two compounds fit the criteria for ionic compounds: $CaCl_2$ and KOH. Because Table 4.3 tells us that all ionic compounds are strong electrolytes, that is how we classify these two substances. The three remaining compounds are molecular. Two, HNO_3 and $HCOOH$, are acids. Nitric acid, HNO_3, is a common strong acid, as shown in Table 4.2, and therefore is a strong electrolyte. Because most acids are weak acids, our best guess would be that $HCOOH$ is a weak acid (weak electrolyte). This is correct. The remaining molecular compound, C_2H_5OH, is neither an acid nor a base, so it is a nonelectrolyte.

Comment: Although C_2H_5OH has an OH group, it is not a metal hydroxide; thus, it is not a base. Rather, it is a member of a class of organic compounds that have C—OH bonds, which are known as alcohols. ∞ (Section 2.9). The COOH group is called the "carboxylic acid group" (Chapter 16). Molecules that have this group are all weak acids.

■ **PRACTICE EXERCISE**

Consider solutions in which 0.1 mol of each of the following compounds is dissolved in 1 L of water: $Ca(NO_3)_2$ (calcium nitrate), $C_6H_{12}O_6$ (glucose), CH_3COONa (sodium acetate), and CH_3COOH (acetic acid). Rank the solutions in order of increasing electrical conductivity, based on the fact that the greater the number of ions in solution, the greater the conductivity.
Answers: $C_6H_{12}O_6$ (nonelectrolyte) < CH_3COOH (weak electrolyte, existing mainly in the form of molecules with few ions) < CH_3COONa (strong electrolyte that provides two ions, Na^+ and CH_3COO^-) < $Ca(NO_3)_2$ (strong electrolyte that provides three ions, Ca^{2+} and $2\ NO_3^-$)

Neutralization Reactions and Salts

The properties of acidic solutions are quite different from those of basic solutions. Acids have a sour taste, whereas bases have a bitter taste.* Acids can change the colors of certain dyes in a specific way that differs from the effect of a base (Figure 4.7 ◄). The dye known as litmus, for example, is changed from blue to red by an acid and from red to blue by a base. In addition, acidic and basic solutions differ in chemical properties in several important ways that we will explore in this chapter and in later chapters.

When a solution of an acid and a solution of a base are mixed, a **neutralization reaction** occurs. The products of the reaction have none of the characteristic properties of either the acidic solution or the basic solution. For example, when hydrochloric acid is mixed with a solution of sodium hydroxide, the following reaction occurs:

▲ Figure 4.7 **The acid–base indicator bromthymol blue.** The indicator is blue in basic solution and yellow in acidic solution. The left flask shows the indicator in the presence of a base, aqueous ammonia (labeled as ammonium hydroxide). The right flask shows the indicator in the presence of hydrochloric acid, HCl.

$$\underset{\text{(acid)}}{HCl(aq)} + \underset{\text{(base)}}{NaOH(aq)} \longrightarrow \underset{\text{(water)}}{H_2O(l)} + \underset{\text{(salt)}}{NaCl(aq)} \qquad [4.12]$$

Water and table salt, NaCl, are the products of the reaction. By analogy to this reaction, the term **salt** has come to mean any ionic compound whose cation comes from a base (for example, Na^+ from NaOH) and whose anion comes from an acid (for example, Cl^- from HCl). In general, *a neutralization reaction between an acid and a metal hydroxide produces water and a salt.*

Tasting chemical solutions is not a good practice. However, we have all had acids such as ascorbic acid (vitamin C), acetylsalicylic acid (aspirin), and citric acid (in citrus fruits) in our mouths, and we are familiar with their characteristic sour taste. Soaps, which are basic, have the characteristic bitter taste of bases.

(a)   (b)   (c)

Because HCl, NaOH, and NaCl are all soluble strong electrolytes, the complete ionic equation associated with Equation 4.12 is

$$H^+(aq) + Cl^-(aq) + Na^+(aq) + OH^-(aq) \longrightarrow$$
$$H_2O(l) + Na^+(aq) + Cl^-(aq) \qquad [4.13]$$

Therefore, the net ionic equation is

$$H^+(aq) + OH^-(aq) \longrightarrow H_2O(l) \qquad [4.14]$$

Equation 4.14 summarizes the essential feature of the neutralization reaction between any strong acid and any strong base: $H^+(aq)$ and $OH^-(aq)$ ions combine to form H_2O.

Figure 4.8▲ shows the reaction between hydrochloric acid and the base $Mg(OH)_2$, which is insoluble in water. A milky white suspension of $Mg(OH)_2$ called milk of magnesia is seen dissolving as the neutralization reaction occurs:

Molecular equation:

$$Mg(OH)_2(s) + 2\,HCl(aq) \longrightarrow MgCl_2(aq) + 2\,H_2O(l) \qquad [4.15]$$

Net ionic equation:

$$Mg(OH)_2(s) + 2\,H^+(aq) \longrightarrow Mg^{2+}(aq) + 2\,H_2O(l) \qquad [4.16]$$

Notice that the OH^- ions (this time in a solid reactant) and H^+ ions combine to form H_2O. Because the ions exchange partners, neutralization reactions between acids and metal hydroxides are also metathesis reactions.

▲ Figure 4.8 **Reaction of Mg(OH)₂(s) with acid.** (a) Milk of magnesia is a suspension of magnesium hydroxide, $Mg(OH)_2(s)$, in water. (b) The magnesium hydroxide dissolves upon the addition of hydrochloric acid, $HCl(aq)$. (c) The final clear solution contains soluble $MgCl_2(aq)$, as shown in Equation 4.15.

▪ SAMPLE EXERCISE 4.7 | Writing Chemical Equations for a Neutralization Reaction

(a) Write a balanced molecular equation for the reaction between aqueous solutions of acetic acid (CH_3COOH) and barium hydroxide, $Ba(OH)_2$. **(b)** Write the net ionic equation for this reaction.

SOLUTION

Analyze: We are given the chemical formulas for an acid and a base and asked to write a balanced molecular equation and then a net ionic equation for their neutralization reaction.

Plan: As Equation 4.12 and the italicized statement that follows it indicate, neutralization reactions form two products, H_2O and a salt. We examine the cation of the base and the anion of the acid to determine the composition of the salt.

Solve:

(a) The salt will contain the cation of the base (Ba^{2+}) and the anion of the acid (CH_3COO^-). Thus, the formula of the salt is $Ba(CH_3COO)_2$. According to the solubility guidelines in Table 4.1, this compound is soluble. The unbalanced molecular equation for the neutralization reaction is

$$CH_3COOH(aq) + Ba(OH)_2(aq) \longrightarrow H_2O(l) + Ba(CH_3COO^-)_2(aq)$$

To balance this molecular equation, we must provide two molecules of CH_3COOH to furnish the two CH_3COO^- ions and to supply the two H^+ ions needed to combine with the two OH^- ions of the base. The balanced molecular equation is

$$2\,CH_3COOH(aq) + Ba(OH)_2(aq) \longrightarrow 2\,H_2O(l) + Ba(CH_3COO)_2(aq)$$

(b) To write the net ionic equation, we must determine whether each compound in aqueous solution is a strong electrolyte. CH_3COOH is a weak electrolyte (weak acid), $Ba(OH)_2$ is a strong electrolyte, and $Ba(CH_3COO)_2$ is also a strong electrolyte (ionic compound). Thus, the complete ionic equation is

$$2\,CH_3COOH(aq) + Ba^{2+}(aq) + 2\,OH^-(aq) \longrightarrow$$
$$2\,H_2O(l) + Ba^{2+}(aq) + 2\,CH_3COO^-(aq)$$

Eliminating the spectator ions gives

$$2\,CH_3COOH(aq) + 2\,OH^-(aq) \longrightarrow 2\,H_2O(l) + 2\,CH_3COO^-(aq)$$

Simplifying the coefficients gives the net ionic equation:

$$CH_3COOH(aq) + OH^-(aq) \longrightarrow H_2O(l) + CH_3COO^-(aq)$$

Check: We can determine whether the molecular equation is correctly balanced by counting the number of atoms of each kind on both sides of the arrow. (There are 10 H, 6 O, 4 C, and 1 Ba on each side.) However, it is often easier to check equations by counting groups: There are 2 CH_3COO groups, as well as 1 Ba, and 4 additional H atoms and 2 additional O atoms on each side of the equation. The net ionic equation checks out because the numbers of each kind of element and the net charge are the same on both sides of the equation.

■ **PRACTICE EXERCISE**

(a) Write a balanced molecular equation for the reaction of carbonic acid (H_2CO_3) and potassium hydroxide (KOH). **(b)** Write the net ionic equation for this reaction.
Answers: **(a)** $H_2CO_3(aq) + 2\,KOH(aq) \longrightarrow 2\,H_2O(l) + K_2CO_3(aq)$; **(b)** $H_2CO_3(aq) + 2\,OH^-(aq) \longrightarrow 2\,H_2O(l) + CO_3^{2-}(aq)$. ($H_2CO_3$ is a weak acid and therefore a weak electrolyte, whereas KOH, a strong base, and K_2CO_3, an ionic compound, are strong electrolytes.)

Acid–Base Reactions with Gas Formation

Many bases besides OH^- react with H^+ to form molecular compounds. Two of these that you might encounter in the laboratory are the sulfide ion and the carbonate ion. Both of these anions react with acids to form gases that have low solubilities in water. Hydrogen sulfide (H_2S), the substance that gives rotten eggs their foul odor, forms when an acid such as HCl(aq) reacts with a metal sulfide such as Na_2S:

Molecular equation: $2\,HCl(aq) + Na_2S(aq) \longrightarrow H_2S(g) + 2\,NaCl(aq)$ [4.17]

Net ionic equation: $2\,H^+(aq) + S^{2-}(aq) \longrightarrow H_2S(g)$ [4.18]

Carbonates and bicarbonates react with acids to form CO_2 gas. Reaction of CO_3^{2-} or HCO_3^- with an acid first gives carbonic acid (H_2CO_3). For example, when hydrochloric acid is added to sodium bicarbonate, the following reaction occurs:

$$HCl(aq) + NaHCO_3(aq) \longrightarrow NaCl(aq) + H_2CO_3(aq) \qquad [4.19]$$

Carbonic acid is unstable. If carbonic acid is present in solution in sufficient concentrations, it decomposes to form H_2O and CO_2, which escapes from the solution as a gas.

$$H_2CO_3(aq) \longrightarrow H_2O(l) + CO_2(g) \qquad [4.20]$$

The decomposition of H_2CO_3 produces bubbles of CO_2 gas, as shown in Figure 4.9 ◀. The overall reaction is summarized by the following equations:

▲ **Figure 4.9 Carbonates react with acids to form carbon dioxide gas.** Here $NaHCO_3$ (white solid) reacts with hydrochloric acid. The bubbles contain CO_2.

Molecular equation:

$$HCl(aq) + NaHCO_3(aq) \longrightarrow NaCl(aq) + H_2O(l) + CO_2(g) \qquad [4.21]$$

Net ionic equation: $H^+(aq) + HCO_3^-(aq) \longrightarrow H_2O(l) + CO_2(g)$ [4.22]

Chemistry Put to Work   ANTACIDS

Your stomach secretes acids to help digest foods. These acids, which include hydrochloric acid, contain about 0.1 mol of H^+ per liter of solution. The stomach and digestive tract are normally protected from the corrosive effects of stomach acid by a mucosal lining. Holes can develop in this lining, however, allowing the acid to attack the underlying tissue, causing painful damage. These holes, known as ulcers, can be caused by the secretion of excess acids or by a weakness in the digestive lining. Studies indicate, however, that many ulcers are caused by bacterial infection. Between 10 and 20% of Americans suffer from ulcers at some point in their lives. Many others experience occasional indigestion or heartburn that is due to digestive acids entering the esophagus.

We can address the problem of excess stomach acid in two simple ways: (1) removing the excess acid, or (2) decreasing the production of acid. Those substances that remove excess acid are called *antacids*, whereas those that decrease the production of acid are called *acid inhibitors*. Figure 4.10▼ shows several common, over-the-counter antacids.

Antacids are simple bases that neutralize digestive acids. They are able to neutralize acids because they contain hydroxide, carbonate, or bicarbonate ions. Table 4.4◄ lists the active ingredients in some antacids.

The newer generation of antiulcer drugs, such as Tagamet® and Zantac®, are acid inhibitors. They act on acid-producing cells in the lining of the stomach. Formulations that control acid in this way are now available as over-the-counter drugs.

Related Exercise: 4.95

TABLE 4.4 ■ Some Common Antacids

Commercial Name	Acid-Neutralizing Agents
Alka-Seltzer®	$NaHCO_3$
Amphojel®	$Al(OH)_3$
Di-Gel®	$Mg(OH)_2$ and $CaCO_3$
Milk of Magnesia	$Mg(OH)_2$
Maalox®	$Mg(OH)_2$ and $Al(OH)_3$
Mylanta®	$Mg(OH)_2$ and $Al(OH)_3$
Rolaids®	$NaAl(OH)_2CO_3$
Tums®	$CaCO_3$

▲ Figure 4.10 **Antacids.** These products all serve as acid-neutralizing agents in the stomach.

Both $NaHCO_3$ and Na_2CO_3 are used as acid neutralizers in acid spills. The bicarbonate or carbonate salt is added until the fizzing due to the formation of $CO_2(g)$ stops. Sometimes sodium bicarbonate is used as an antacid to soothe an upset stomach. In that case the HCO_3^- reacts with stomach acid to form $CO_2(g)$. The fizz when Alka-Seltzer® tablets are added to water arises from the reaction of sodium bicarbonate and citric acid.

GIVE IT SOME THOUGHT

By analogy to examples already given in the text, predict what gas forms when $Na_2SO_3(s)$ is treated with $HCl(aq)$.

4.4 OXIDATION-REDUCTION REACTIONS

In precipitation reactions, cations and anions come together to form an insoluble ionic compound. In neutralization reactions H^+ ions and OH^- ions come together to form H_2O molecules. Now let's consider a third important kind of reaction, one in which electrons are transferred between reactants. Such reactions are called **oxidation-reduction**, or *redox*, **reactions**.

Oxidation and Reduction

The corrosion of iron (rusting) and of other metals, such as the corrosion of the terminals of an automobile battery, are familiar processes. What we call *corrosion* is the conversion of a metal into a metal compound by a reaction between the metal and some substance in its environment. Rusting, as shown in Figure 4.11 ▲, involves the reaction of oxygen with iron in the presence of water.

When a metal corrodes, it loses electrons and forms cations. Calcium, for example, is vigorously attacked by acids to form calcium ions:

$$Ca(s) + 2\,H^+(aq) \longrightarrow Ca^{2+}(aq) + H_2(g) \qquad [4.23]$$

When an atom, ion, or molecule has become more positively charged (that is, when it has lost electrons), we say that it has been oxidized. *Loss of electrons by a substance is called* **oxidation.** Thus, Ca, which has no net charge, is *oxidized* (undergoes oxidation) in Equation 4.23, forming Ca^{2+}.

The term oxidation is used because the first reactions of this sort to be studied thoroughly were reactions with oxygen. Many metals react directly with O_2 in air to form metal oxides. In these reactions the metal loses electrons to oxygen, forming an ionic compound of the metal ion and oxide ion. For example, when calcium metal is exposed to air, the bright metallic surface of the metal tarnishes as CaO forms:

$$2\,Ca(s) + O_2(g) \longrightarrow 2\,CaO(s) \qquad [4.24]$$

As Ca is oxidized in Equation 4.24, oxygen is transformed from neutral O_2 to two O^{2-} ions (Figure 4.12 ▼). When an atom, ion, or molecule has become more negatively charged (gained electrons), we say that it is *reduced.*

$2\,Ca(s) + O_2(g)$ $2\,CaO(s)$

▲ Figure 4.12 **Oxidation of calcium metal by molecular oxygen.** The oxidation involves transfer of electrons from the metal to O_2, eventually leading to formation of CaO.

The gain of electrons by a substance is called **reduction**. When one reactant loses electrons (that is, when it is *oxidized*), another reactant must gain them. The oxidation of one substance is always accompanied by the reduction of another as electrons are transferred between them.

Substance **oxidized** (loses electron)

Substance **reduced** (gains electron)

Oxidation Numbers

Before we can properly identify an oxidation-reduction reaction, we must have a kind of bookkeeping system—a way of keeping track of the electrons gained by the substance reduced and those lost by the substance oxidized. The concept of oxidation numbers (also called *oxidation states*) was devised as a way of doing this. Each atom in a neutral molecule or charged species is assigned an **oxidation number,** which is the actual charge for a monatomic ion. Otherwise, the oxidation number is the hypothetical charge assigned to the atom, assuming that the electrons are *completely* held by one atom or the other. The oxidation numbers of certain atoms change in an oxidation-reduction reaction. Oxidation occurs when the oxidation number increases, whereas reduction occurs when the oxidation number decreases.

We use the following rules for assigning oxidation numbers:

1. *For an atom in its* **elemental form***, the oxidation number is always zero.*
   Thus, each H atom in the H_2 molecule has an oxidation number of 0, and each P atom in the P_4 molecule has an oxidation number of 0.

2. *For any* **monatomic ion** *the oxidation number equals the charge on the ion.*
   Thus, K^+ has an oxidation number of +1, S^{2-} has an oxidation number of −2, and so forth. The alkali metal ions (group 1A) always have a 1+ charge, and therefore the alkali metals always have an oxidation number of +1 in their compounds. Similarly, the alkaline earth metals (group 2A) are always +2, and aluminum (group 3A) is always +3 in its compounds. (In writing oxidation numbers, we will write the sign before the number to distinguish them from the actual electronic charges, which we write with the number first.)

3. *Nonmetals* usually have negative oxidation numbers, although they can sometimes be positive:
   (a) *The oxidation number of* **oxygen** *is usually* −2 in both ionic and molecular compounds. The major exception is in compounds called peroxides, which contain the O_2^{2-} ion, giving each oxygen an oxidation number of −1.
   (b) *The oxidation number of* **hydrogen** *is usually* +1 *when bonded to nonmetals and* −1 *when bonded to metals.*
   (c) *The oxidation number of* **fluorine** *is* −1 *in all compounds.* The other **halogens** have an oxidation number of −1 in most binary compounds. When combined with oxygen, as in oxyanions, however, they have positive oxidation states.

4. **The sum of the oxidation numbers** *of all atoms in a neutral compound is zero. The sum of the oxidation numbers in a polyatomic ion equals the charge of the ion.* For example, in the hydronium ion, H_3O^+, the oxidation number of each hydrogen is +1 and that of oxygen is −2. Thus, the sum of the oxidation numbers is $3(+1) + (-2) = +1$, which equals the net charge of the ion. This rule is very useful in obtaining the oxidation number of one atom in a compound or ion if you know the oxidation numbers of the other atoms, as illustrated in Sample Exercise 4.8.

▲ GIVE IT SOME THOUGHT

(a) What noble gas element has the same number of electrons as the fluoride ion?
(b) What is the oxidation number of that noble gas?

■■ **SAMPLE EXERCISE 4.8** | Determining Oxidation Numbers

Determine the oxidation number of sulfur in each of the following: **(a)** H_2S, **(b)** S_8, **(c)** SCl_2, **(d)** Na_2SO_3, **(e)** SO_4^{2-}.

SOLUTION

Analyze: We are asked to determine the oxidation number of sulfur in two molecular species, in the elemental form, and in two ionic substances.

Plan: In each species the sum of oxidation numbers of all the atoms must equal the charge on the species. We will use the rules outlined above to assign oxidation numbers.

Solve:

(a) When bonded to a nonmetal, hydrogen has an oxidation number of +1 (rule 3b). Because the H_2S molecule is neutral, the sum of the oxidation numbers must equal zero (rule 4). Letting x equal the oxidation number of S, we have $2(+1) + x = 0$. Thus, S has an oxidation number of −2.
(b) Because this is an elemental form of sulfur, the oxidation number of S is 0 (rule 1).
(c) Because this is a binary compound, we expect chlorine to have an oxidation number of −1 (rule 3c). The sum of the oxidation numbers must equal zero (rule 4). Letting x equal the oxidation number of S, we have $x + 2(-1) = 0$. Consequently, the oxidation number of S must be +2.
(d) Sodium, an alkali metal, always has an oxidation number of +1 in its compounds (rule 2). Oxygen has a common oxidation state of −2 (rule 3a). Letting x equal the oxidation number of S, we have $2(+1) + x + 3(-2) = 0$. Therefore, the oxidation number of S in this compound is +4.
(e) The oxidation state of O is −2 (rule 3a). The sum of the oxidation numbers equals −2, the net charge of the SO_4^{2-} ion (rule 4). Thus, we have $x + 4(-2) = -2$. From this relation we conclude that the oxidation number of S in this ion is +6.

Comment: These examples illustrate that the oxidation number of a given element depends on the compound in which it occurs. The oxidation numbers of sulfur, as seen in these examples, range from −2 to +6.

■■ **PRACTICE EXERCISE**

What is the oxidation state of the boldfaced element in each of the following: **(a)** $\mathbf{P}_2O_5$, **(b)** $Na\mathbf{H}$, **(c)** $\mathbf{Cr}_2O_7^{2-}$, **(d)** $\mathbf{Sn}Br_4$, **(e)** $Ba\mathbf{O}_2$?
Answers: **(a)** +5, **(b)** −1, **(c)** +6, **(d)** +4, **(e)** −1

Oxidation of Metals by Acids and Salts

There are many kinds of redox reactions. For example, combustion reactions are redox reactions because elemental oxygen is converted to compounds of oxygen. ∞ (Section 3.2) In this chapter we consider the redox reactions between metals and either acids or salts. In Chapter 20 we will examine more complex kinds of redox reactions.

The reaction of a metal with either an acid or a metal salt conforms to the following general pattern:

$$A + BX \longrightarrow AX + B \qquad [4.25]$$

Examples:
$$Zn(s) + 2\,HBr(aq) \longrightarrow ZnBr_2(aq) + H_2(g)$$
$$Mn(s) + Pb(NO_3)_2(aq) \longrightarrow Mn(NO_3)_2(aq) + Pb(s)$$

These reactions are called **displacement reactions** because the ion in solution is displaced or replaced through oxidation of an element.

Many metals undergo displacement reactions with acids, producing salts and hydrogen gas. For example, magnesium metal reacts with hydrochloric acid to form magnesium chloride and hydrogen gas (Figure 4.13 ▶).

◀ Figure 4.13 **Reaction of magnesium with acid.** The bubbles are due to hydrogen gas, H_2.

To show that oxidation and reduction have occurred, the oxidation number for each atom is shown below the chemical equation for this reaction:

$$Mg(s) + 2\ HCl(aq) \longrightarrow MgCl_2(aq) + H_2(g) \qquad [4.26]$$

$$\begin{array}{ccccc} 0 & +1 & -1 & +2 & -1 & 0 \end{array}$$

Notice that the oxidation number of Mg changes from 0 to +2. The increase in the oxidation number indicates that the atom has lost electrons and has therefore been oxidized. The H^+ ion of the acid decreases in oxidation number from +1 to 0, indicating that this ion has gained electrons and has therefore been reduced. The oxidation number of the Cl^- ion remains −1, and it is a spectator ion in the reaction. The net ionic equation is as follows:

$$Mg(s) + 2\ H^+(aq) \longrightarrow Mg^{2+}(aq) + H_2(g) \qquad [4.27]$$

Metals can also be oxidized by aqueous solutions of various salts. Iron metal, for example, is oxidized to Fe^{2+} by aqueous solutions of Ni^{2+} such as $Ni(NO_3)_2(aq)$:

Molecular equation: $Fe(s) + Ni(NO_3)_2(aq) \longrightarrow Fe(NO_3)_2(aq) + Ni(s)$ [4.28]

Net ionic equation: $Fe(s) + Ni^{2+}(aq) \longrightarrow Fe^{2+}(aq) + Ni(s)$ [4.29]

The oxidation of Fe to form Fe^{2+} in this reaction is accompanied by the reduction of Ni^{2+} to Ni. Remember: *Whenever one substance is oxidized, some other substance must be reduced.*

■ SAMPLE EXERCISE 4.9 | Writing Molecular and Net Ionic Equations for Oxidation-Reduction Reactions

Write the balanced molecular and net ionic equations for the reaction of aluminum with hydrobromic acid.

SOLUTION

Analyze: We must write two equations—molecular and net ionic—for the redox reaction between a metal and an acid.

Plan: Metals react with acids to form salts and H_2 gas. To write the balanced equations, we must write the chemical formulas for the two reactants and then determine the formula of the salt. The salt is composed of the cation formed by the metal and the anion of the acid.

Solve: The formulas of the given reactants are Al and HBr. The cation formed by Al is Al^{3+}, and the anion from hydrobromic acid is Br^-. Thus, the salt formed in the reaction is $AlBr_3$. Writing the reactants and products and then balancing the equation gives this molecular equation:

$$2\,Al(s) + 6\,HBr(aq) \longrightarrow 2\,AlBr_3(aq) + 3\,H_2(g)$$

Both HBr and $AlBr_3$ are soluble strong electrolytes. Thus, the complete ionic equation is

$$2\,Al(s) + 6\,H^+(aq) + 6\,Br^-(aq) \longrightarrow 2\,Al^{3+}(aq) + 6\,Br^-(aq) + 3\,H_2(g)$$

Because Br^- is a spectator ion, the net ionic equation is

$$2\,Al(s) + 6\,H^+(aq) \longrightarrow 2\,Al^{3+}(aq) + 3\,H_2(g)$$

Comment: The substance oxidized is the aluminum metal because its oxidation state changes from 0 in the metal to +3 in the cation, thereby increasing in oxidation number. The H^+ is reduced because its oxidation state changes from +1 in the acid to 0 in H_2.

■ PRACTICE EXERCISE

(a) Write the balanced molecular and net ionic equations for the reaction between magnesium and cobalt(II) sulfate. **(b)** What is oxidized and what is reduced in the reaction?

Answers: **(a)** $Mg(s) + CoSO_4(aq) \longrightarrow MgSO_4(aq) + Co(s)$;
$Mg(s) + Co^{2+}(aq) \longrightarrow Mg^{2+}(aq) + Co(s)$; **(b)** Mg is oxidized and Co^{2+} is reduced.

The Activity Series

Can we predict whether a certain metal will be oxidized either by an acid or by a particular salt? This question is of practical importance as well as chemical interest. According to Equation 4.28, for example, it would be unwise to store a solution of nickel nitrate in an iron container because the solution would dissolve the container. When a metal is oxidized, it reacts to form various compounds. Extensive oxidation can lead to the failure of metal machinery parts or the deterioration of metal structures.

Different metals vary in the ease with which they are oxidized. Zn is oxidized by aqueous solutions of Cu^{2+}, for example, but Ag is not. Zn, therefore, loses electrons more readily than Ag; that is, Zn is easier to oxidize than Ag.

A list of metals arranged in order of decreasing ease of oxidation is called an **activity series**. Table 4.5 ▶ gives the activity series in aqueous solution for many of the most common metals. Hydrogen is also included in the table. The metals at the top of the table, such as the alkali metals and the alkaline earth metals, are most easily oxidized; that is, they react most readily to form compounds. They are called the *active metals*. The metals at the bottom of the activity series, such as the transition elements from groups 8B and 1B, are very stable and form compounds less readily. These metals, which are used to make coins and jewelry, are called *noble metals* because of their low reactivity.

The activity series can be used to predict the outcome of reactions between metals and either metal salts or acids. *Any metal on the list can be oxidized by the ions of elements below it.* For example, copper is above silver in the series.

TABLE 4.5 ■ Activity Series of Metals in Aqueous Solution

Metal	Oxidation Reaction
Lithium	$Li(s) \longrightarrow Li^+(aq) + e^-$
Potassium	$K(s) \longrightarrow K^+(aq) + e^-$
Barium	$Ba(s) \longrightarrow Ba^{2+}(aq) + 2e^-$
Calcium	$Ca(s) \longrightarrow Ca^{2+}(aq) + 2e^-$
Sodium	$Na(s) \longrightarrow Na^+(aq) + e^-$
Magnesium	$Mg(s) \longrightarrow Mg^{2+}(aq) + 2e^-$
Aluminum	$Al(s) \longrightarrow Al^{3+}(aq) + 3e^-$
Manganese	$Mn(s) \longrightarrow Mn^{2+}(aq) + 2e^-$
Zinc	$Zn(s) \longrightarrow Zn^{2+}(aq) + 2e^-$
Chromium	$Cr(s) \longrightarrow Cr^{3+}(aq) + 3e^-$
Iron	$Fe(s) \longrightarrow Fe^{2+}(aq) + 2e^-$
Cobalt	$Co(s) \longrightarrow Co^{2+}(aq) + 2e^-$
Nickel	$Ni(s) \longrightarrow Ni^{2+}(aq) + 2e^-$
Tin	$Sn(s) \longrightarrow Sn^{2+}(aq) + 2e^-$
Lead	$Pb(s) \longrightarrow Pb^{2+}(aq) + 2e^-$
Hydrogen	$H_2(g) \longrightarrow 2\,H^+(aq) + 2e^-$
Copper	$Cu(s) \longrightarrow Cu^{2+}(aq) + 2e^-$
Silver	$Ag(s) \longrightarrow Ag^+(aq) + e^-$
Mercury	$Hg(l) \longrightarrow Hg^{2+}(aq) + 2e^-$
Platinum	$Pt(s) \longrightarrow Pt^{2+}(aq) + 2e^-$
Gold	$Au(s) \longrightarrow Au^{3+}(aq) + 3e^-$

Ease of oxidation increases →

Thus, copper metal will be oxidized by silver ions, as pictured in Figure 4.14 ▼:

$$Cu(s) + 2\,Ag^+(aq) \longrightarrow Cu^{2+}(aq) + 2\,Ag(s) \qquad [4.30]$$

The oxidation of copper to copper ions is accompanied by the reduction of silver ions to silver metal. The silver metal is evident on the surface of the copper wires in Figure 4.14(b) and 4.14(c). The copper(II) nitrate produces a blue color in the solution, which is most evident in part (c).

$Cu(s)$ + $2\,AgNO_3(aq)$ → $2\,Ag(s)$ + $Cu(NO_3)_2(aq)$

◀ Figure 4.14 **Reaction of copper with silver ion.** When copper metal is placed in a solution of silver nitrate, a redox reaction occurs, forming silver metal and a blue solution of copper(II) nitrate.

Which is the more easily reduced, $Mg^{2+}(aq)$ or $Ni^{2+}(aq)$?

Only those metals above hydrogen in the activity series are able to react with acids to form H_2. For example, Ni reacts with $HCl(aq)$ to form H_2:

$$Ni(s) + 2\,HCl(aq) \longrightarrow NiCl_2(aq) + H_2(g) \qquad [4.31]$$

Because elements below hydrogen in the activity series are not oxidized by H^+, Cu does not react with $HCl(aq)$. Interestingly, copper does react with nitric acid, as shown previously in Figure 1.11. This reaction, however, is not a simple oxidation of Cu by the H^+ ions of the acid. Instead, the metal is oxidized to Cu^{2+} by the nitrate ion of the acid, accompanied by the formation of brown nitrogen dioxide, $NO_2(g)$:

$$Cu(s) + 4\,HNO_3(aq) \longrightarrow Cu(NO_3)_2(aq) + 2\,H_2O(l) + 2\,NO_2(g) \qquad [4.32]$$

What substance is reduced as copper is oxidized in Equation 4.32? In this case the NO_2 results from the reduction of NO_3^-. We will examine reactions of this type in more detail in Chapter 20.

■ **SAMPLE EXERCISE 4.10** | **Determining When an Oxidation-Reduction Reaction Can Occur**

Will an aqueous solution of iron(II) chloride oxidize magnesium metal? If so, write the balanced molecular and net ionic equations for the reaction.

SOLUTION

Analyze: We are given two substances—an aqueous salt, $FeCl_2$, and a metal, Mg—and asked if they react with each other.

Plan: A reaction will occur if Mg is above Fe in the activity series, Table 4.5. If the reaction occurs, the Fe^{2+} ion in $FeCl_2$ will be reduced to Fe, and the elemental Mg will be oxidized to Mg^{2+}.

Solve: Because Mg is above Fe in the table, the reaction will occur. To write the formula for the salt that is produced in the reaction, we must remember the charges on common ions. Magnesium is always present in compounds as Mg^{2+}; the chloride ion is Cl^-. The magnesium salt formed in the reaction is $MgCl_2$, meaning the balanced molecular equation is

$$Mg(s) + FeCl_2(aq) \longrightarrow MgCl_2(aq) + Fe(s)$$

Both $FeCl_2$ and $MgCl_2$ are soluble strong electrolytes and can be written in ionic form. Cl^- then, is a spectator ion in the reaction. The net ionic equation is

$$Mg(s) + Fe^{2+}(aq) \longrightarrow Mg^{2+}(aq) + Fe(s)$$

The net ionic equation shows that Mg is oxidized and Fe^{2+} is reduced in this reaction.

Check: Note that the net ionic equation is balanced with respect to both charge and mass.

■ **PRACTICE EXERCISE**

Which of the following metals will be oxidized by $Pb(NO_3)_2$: Zn, Cu, Fe?
Answer: Zn and Fe

4.5 CONCENTRATIONS OF SOLUTIONS

The behavior of solutions often depends on the nature of the solutes and their concentrations. Scientists use the term **concentration** to designate the amount of solute dissolved in a given quantity of solvent or quantity of solution. The concept of concentration is intuitive: The greater the amount of solute dissolved in a certain amount of solvent, the more concentrated the resulting solution. In chemistry we often need to express the concentrations of solutions quantitatively.

A Closer Look   THE AURA OF GOLD

Gold has been known since the earliest records of human existence. Throughout history people have cherished gold, have fought for it, and have died for it.

The physical and chemical properties of gold serve to make it a special metal. First, its intrinsic beauty and rarity make it precious. Second, gold is soft and can be easily formed into artistic objects, jewelry, and coins (Figure 4.15 ▶). Third, gold is one of the least active metals (Table 4.5). It is not oxidized in air and does not react with water. It is unreactive toward basic solutions and nearly all acidic solutions. As a result, gold can be found in nature as a pure element rather than combined with oxygen or other elements, which accounts for its early discovery.

Many of the early studies of the reactions of gold arose from the practice of alchemy, in which people attempted to turn cheap metals, such as lead, into gold. Alchemists discovered that gold can be dissolved in a 3:1 mixture of concentrated hydrochloric and nitric acids, known as aqua regia ("royal water"). The action of nitric acid on gold is similar to that on copper (Equation 4.32) in that the nitrate ion, rather than H^+, oxidizes the metal to Au^{3+}. The Cl^- ions interact with Au^{3+} to form highly stable $AuCl_4^-$ ions. The net ionic equation for the reaction of gold with aqua regia is

$$Au(s) + NO_3^-(aq) + 4\,H^+(aq) + 4\,Cl^-(aq) \longrightarrow$$
$$AuCl_4^-(aq) + 2\,H_2O(l) + NO(g)$$

All the gold ever mined would easily fit in a cube 19 m on a side and weighing about 1.1×10^8 kg (125,000 tons). More than 90% of this amount has been produced since the beginning of the California gold rush of 1848. Each year, worldwide production of gold amounts to about 1.8×10^6 kg (2000 tons). By contrast, over 1.5×10^{10} kg (16 million tons) of aluminum are produced annually. Gold is used mainly in jewelry (73%), coins (10%), and electronics (9%). Its use in electronics relies on its excellent conductivity and its corrosion resistance. Gold is used, for example, to plate contacts in electrical switches, relays, and connections. A typical touch-tone telephone contains 33 gold-plated contacts. Gold is also used in computers and other microelectronic devices where fine gold wire is used to link components.

Because of its resistance to corrosion by acids and other substances found in saliva, gold is an ideal metal for dental crowns and caps, which accounts for about 3% of the annual use of the element. The pure metal is too soft to use in dentistry, so it is combined with other metals to form alloys.
Related Exercise: 4.97

◀ Figure 4.15 **Portrait of Pharaoh Tutankhamun (1346–1337 BC) made of gold and precious stones.** This highly prized article is from the inner coffin of the tomb of Tutankhamun.

Strategies in Chemistry   ANALYZING CHEMICAL REACTIONS

In this chapter you have been introduced to a great number of chemical reactions. A major difficulty that students face in trying to master material of this sort is gaining a "feel" for what happens when chemicals are allowed to react. In fact, you might marvel at the ease with which your professor or teaching assistant can figure out the results of a chemical reaction. One of our goals in this textbook is to help you become more adept at predicting the outcomes of reactions. The key to gaining this "chemical intuition" is understanding how to categorize reactions.

Attempting to memorize the many individual reactions in chemistry would be a futile task. It is far more fruitful to try to recognize patterns to determine the general category of a reaction, such as metathesis or oxidation-reduction. Thus, when you are faced with the challenge of predicting the outcome of a chemical reaction, ask yourself the following pertinent questions:

- What are the reactants in the reaction?
- Are they electrolytes or nonelectrolytes?
- Are they acids and bases?
- If the reactants are electrolytes, will metathesis produce a precipitate? Water? A gas?

- If metathesis cannot occur, can the reactants possibly engage in an oxidation-reduction reaction? This requires that there be both a reactant that can be oxidized and one that can be reduced.

By asking questions such as these, you should be able to predict what might happen during the reaction. You might not always be entirely correct, but if you keep your wits about you, you will not be far off. As you gain experience with chemical reactions, you will begin to look for reactants that might not be immediately obvious, such as water from the solution or oxygen from the atmosphere.

One of the greatest tools available to us in chemistry is experimentation. If you perform an experiment in which two solutions are mixed, you can make observations that help you understand what is happening. For example, using the information in Table 4.1 to predict whether a precipitate will form is not nearly as exciting as actually seeing the precipitate form, as in Figure 4.4. Careful observations in the laboratory portion of the course will make your lecture material easier to master.

| (a) | (b) | (c) | (d) |

▲ Figure 4.16 **Procedure for preparation of 0.250 L of 1.00 _M_ solution of CuSO₄.** (a) Weigh out 0.250 mol (39.9 g) of $CuSO_4$ (formula weight = 159.6 amu). (b) Put the $CuSO_4$ (solute) into a 250-mL volumetric flask, and add a small quantity of water. (c) Dissolve the solute by swirling the flask. (d) Add more water until the solution just reaches the calibration mark etched on the neck of the flask. Shake the stoppered flask to ensure complete mixing.

Molarity

Molarity (symbol M) expresses the concentration of a solution as the number of moles of solute in a liter of solution (soln):

$$\text{Molarity} = \frac{\text{moles solute}}{\text{volume of solution in liters}} \qquad [4.33]$$

A 1.00 molar solution (written 1.00 M) contains 1.00 mol of solute in every liter of solution. Figure 4.16▲ shows the preparation of 250.0 mL of a 1.00 M solution of $CuSO_4$ by using a volumetric flask that is calibrated to hold exactly 250.0 mL. First, 0.250 mol of $CuSO_4$ (39.9 g) is weighed out and placed in the volumetric flask. Water is added to dissolve the salt, and the resultant solution is diluted to a total volume of 250.0 mL. The molarity of the solution is (0.250 mol $CuSO_4$)/(0.250 L soln) = 1.00 M.

> ### GIVE IT SOME THOUGHT
>
> Which is more concentrated, a $1.00 \times 10^{-2}\ M$ solution of sucrose or a $1.00 \times 10^{-4}\ M$ solution of sucrose?

■■■ **SAMPLE EXERCISE 4.11** | Calculating Molarity

Calculate the molarity of a solution made by dissolving 23.4 g of sodium sulfate (Na_2SO_4) in enough water to form 125 mL of solution.

SOLUTION

Analyze: We are given the number of grams of solute (23.4 g), its chemical formula (Na_2SO_4), and the volume of the solution (125 ml). We are asked to calculate the molarity of the solution.

Plan: We can calculate molarity using Equation 4.33. To do so, we must convert the number of grams of solute to moles and the volume of the solution from milliliters to liters.

Solve: The number of moles of Na_2SO_4 is obtained by using its molar mass:

$$\text{Moles } Na_2SO_4 = (23.4\ \text{g } Na_2SO_4)\left(\frac{1\ \text{mol } Na_2SO_4}{142\ \text{g } Na_2SO_4}\right) = 0.165\ \text{mol } Na_2SO_4$$

Converting the volume of the solution to liters:

$$\text{Liters soln} = (125\ \text{mL})\left(\frac{1\ \text{L}}{1000\ \text{mL}}\right) = 0.125\ \text{L}$$

Thus, the molarity is

$$\text{Molarity} = \frac{0.165\ \text{mol } Na_2SO_4}{0.125\ \text{L soln}} = 1.32\ \frac{\text{mol } Na_2SO_4}{\text{L soln}} = 1.32\ M$$

Check: Because the numerator is only slightly larger than the denominator, it is reasonable for the answer to be a little over 1 M. The units (mol/L) are appropriate for molarity, and three significant figures are appropriate for the answer because each of the initial pieces of data had three significant figures.

■ **PRACTICE EXERCISE**

Calculate the molarity of a solution made by dissolving 5.00 g of glucose ($C_6H_{12}O_6$) in sufficient water to form exactly 100 mL of solution.
Answer: 0.278 *M*

Expressing the Concentration of an Electrolyte

When an ionic compound dissolves, the relative concentrations of the ions introduced into the solution depend on the chemical formula of the compound. For example, a 1.0 *M* solution of NaCl is 1.0 *M* in Na^+ ions and 1.0 *M* in Cl^- ions. Similarly, a 1.0 *M* solution of Na_2SO_4 is 2.0 *M* in Na^+ ions and 1.0 *M* in SO_4^{2-} ions. Thus, the concentration of an electrolyte solution can be specified either in terms of the compound used to make the solution (1.0 *M* Na_2SO_4) or in terms of the ions that the solution contains (2.0 *M* Na^+ and 1.0 *M* SO_4^{2-}).

■ **SAMPLE EXERCISE 4.12** | **Calculating Molar Concentrations of Ions**

What are the molar concentrations of each of the ions present in a 0.025 *M* aqueous solution of calcium nitrate?

SOLUTION

Analyze: We are given the concentration of the ionic compound used to make the solution and asked to determine the concentrations of the ions in the solution.

Plan: We can use the subscripts in the chemical formula of the compound to determine the relative concentrations of the ions.

Solve: Calcium nitrate is composed of calcium ions (Ca^{2+}) and nitrate ions (NO_3^-), so its chemical formula is $Ca(NO_3)_2$. Because there are two NO_3^- ions for each Ca^{2+} ion in the compound, each mole of $Ca(NO_3)_2$ that dissolves dissociates into 1 mol of Ca^{2+} and 2 mol of NO_3^-. Thus, a solution that is 0.025 *M* in $Ca(NO_3)_2$ is 0.025 *M* in Ca^{2+} and $2 \times 0.025\ M = 0.050\ M$ in NO_3^-:

$$\frac{\text{mol } NO_3^-}{L} = \left(\frac{0.025\ \text{mol } Ca(NO_3)_2}{L}\right)\left(\frac{2\ \text{mol } NO_3}{1\ \text{mol } Ca(NO_3)_2}\right) = 0.050\ M$$

Check: The concentration of NO_3^- ions is twice that of Ca^{2+} ions, as the subscript 2 after the NO_3^- in the chemical formula $Ca(NO_3)_2$ suggests it should be.

■ **PRACTICE EXERCISE**

What is the molar concentration of K^+ ions in a 0.015 *M* solution of potassium carbonate?
Answer: 0.030 *M* K^+

Interconverting Molarity, Moles, and Volume

The definition of molarity (Equation 4.33) contains three quantities—molarity, moles solute, and liters of solution. If we know any two of these, we can calculate the third. For example, if we know the molarity of a solution, we can calculate the number of moles of solute in a given volume. Molarity, therefore, is a conversion factor between volume of solution and moles of solute. Calculation of the number of moles of HNO_3 in 2.0 L of 0.200 *M* HNO_3 solution illustrates the conversion of volume to moles:

$$\text{moles } HNO_3 = (2.0\ \text{L soln})\left(\frac{0.200\ \text{mol } HNO_3}{1\ \text{L soln}}\right)$$
$$= 0.40\ \text{mol } HNO_3$$

Dimensional analysis can be used in this conversion if we express molarity as moles/liter soln. To obtain moles, therefore, we multiply liters and molarity:
moles = liters × molarity = liters × moles/liter.

To illustrate the conversion of moles to volume, let's calculate the volume of 0.30 M HNO_3 solution required to supply 2.0 mol of HNO_3:

$$\text{Liters soln} = (2.0 \text{ mol } HNO_3)\left(\frac{1 \text{ L soln}}{0.30 \text{ mol } HNO_3}\right) = 6.7 \text{ L soln}$$

In this case we must use the reciprocal of molarity in the conversion: liters = moles $\times$ $1/M$ = moles $\times$ liters/mole.

■ SAMPLE EXERCISE 4.13 | Using Molarity to Calculate Grams of Solute

How many grams of Na_2SO_4 are required to make 0.350 L of 0.500 M Na_2SO_4?

SOLUTION

Analyze: We are given the volume of the solution (0.350 L), its concentration (0.500 M), and the identity of the solute Na_2SO_4 and asked to calculate the number of grams of the solute in the solution.

Plan: We can use the definition of molarity (Equation 4.33) to determine the number of moles of solute, and then convert moles to grams using the molar mass of the solute.

$$M_{Na_2SO_4} = \frac{\text{moles } Na_2SO_4}{\text{liters soln}}$$

Solve: Calculating the moles of Na_2SO_4 using the molarity and volume of solution gives

$$M_{Na_2SO_4} = \frac{\text{moles } Na_2SO_4}{\text{liters soln}}$$

$$\text{moles } Na_2SO_4 = \text{liters soln} \times M_{Na_2SO_4}$$

$$= (0.350 \text{ L soln})\left(\frac{0.500 \text{ mol } Na_2SO_4}{1 \text{ L soln}}\right)$$

$$= 0.175 \text{ mol } Na_2SO_4$$

Because each mole of Na_2SO_4 weighs 142 g, the required number of grams of Na_2SO_4 is

$$\text{grams } Na_2SO_4 = (0.175 \text{ mol } Na_2SO_4)\left(\frac{142 \text{ g } Na_2SO_4}{1 \text{ mol } Na_2SO_4}\right) = 24.9 \text{ g } Na_2SO_4$$

Check: The magnitude of the answer, the units, and the number of significant figures are all appropriate.

■ PRACTICE EXERCISE

(a) How many grams of Na_2SO_4 are there in 15 mL of 0.50 M Na_2SO_4? **(b)** How many milliliters of 0.50 M Na_2SO_4 solution are needed to provide 0.038 mol of this salt?
Answers: **(a)** 1.1 g, **(b)** 76 mL

Dilution

Solutions that are used routinely in the laboratory are often purchased or prepared in concentrated form (called *stock solutions*). Hydrochloric acid, for example, is purchased as a 12 M solution (concentrated HCl). Solutions of lower concentrations can then be obtained by adding water, a process called **dilution.***

Let's look at how to prepare a dilute solution from a concentrated one. Suppose we wanted to prepare 250.0 mL (that is, 0.2500 L) of 0.100 M $CuSO_4$ solution by diluting a stock solution containing 1.00 M $CuSO_4$. When solvent is added to dilute a solution, the number of moles of solute remains unchanged.

$$\text{Moles solute before dilution} = \text{moles solute after dilution} \qquad [4.34]$$

*In diluting a concentrated acid or base, the acid or base should be added to water and then further diluted by adding more water. Adding water directly to concentrated acid or base can cause spattering because of the intense heat generated.

Chemistry and Life   DRINKING TOO MUCH WATER CAN KILL YOU

For a long time it was thought that dehydration was a potential danger for people engaged in extended vigorous activity. Thus, athletes were encouraged to drink lots of water while engaged in active sport. The trend toward extensive hydration has spread throughout society; many people carry water bottles everywhere and dutifully keep well hydrated.

It turns out, though, that in some circumstances, drinking too much water is a greater danger than not drinking enough. Excess water consumption can lead to *hyponatremia*, a condition in which the concentration of sodium ion in the blood is too low. In the past decade at least four marathon runners have died from hyponatremia-related trauma, and dozens more have become seriously ill. For example, a first-time marathoner named Hillary Bellamy, running in the Marine Corps marathon in 2003, collapsed near mile 22 and died the next day. One physician who treated her said that she died from hyponatremia-induced brain swelling, the result of drinking too much water before and during the race.

The normal blood sodium level is 135 to 145 m*M (millimolar)*. When that level drops to as low as 125 m*M*, dizziness and confusion set in. A concentration below 120 m*M* can be critical. Low sodium level in the blood causes brain tissue to swell. Dangerously low levels can occur in a marathon runner or other active athlete who is sweating out salt at the same time that excessive salt-free water is being drunk to compensate for water loss. The condition affects women more than men because of their generally different body composition and patterns of metabolism. Drinking a sport drink, such as Gatorade, which contains some electrolytes, helps to prevent hyponatremia (Figure 4.17▶).

Contrary to popular belief, dehydration is not as likely as overhydration to present a life-threatening situation, though it can contribute to heat stroke when the temperature is high. Athletes frequently lose several pounds in the course of extreme workouts, all in the form of water loss, with no lasting adverse effects. When, for instance, Amby Burfoot ran in the Boston Marathon in 1968, his body weight went from 138 to 129 pounds during the race. He lost 6.5% of his body weight while winning the men's competition that year. Weight losses of this magnitude are typical of elite marathon runners, who produce tremendous amounts of heat and sweat and cannot afford to slow down for much drinking.
Related Exercises: 4.63, 4.64

▲ **Figure 4.17 Water stations.** To help prevent overhydration, the number of water stations such as this one has been reduced in many marathon events.

Because we know both the volume and concentration of the dilute solution, we can calculate the number of moles of $CuSO_4$ it contains.

$$\text{moles } CuSO_4 \text{ in dil soln} = (0.2500 \; \text{L soln})\left(0.100 \; \frac{\text{mol } CuSO_4}{\text{L soln}} \right)$$

$$= 0.0250 \text{ mol } CuSO_4$$

Now we can calculate the volume of the concentrated solution needed to provide 0.0250 mol $CuSO_4$:

$$\text{Liters of conc soln} = (0.0250 \; \text{mol } CuSO_4)\left(\frac{1 \text{ L soln}}{1.00 \; \text{mol } CuSO_4} \right) = 0.0250 \text{ L}$$

Thus, this dilution is achieved by withdrawing 0.0250 L (that is, 25.0 mL) of the 1.00 *M* solution using a pipet, adding it to a 250-mL volumetric flask, and then diluting it to a final volume of 250.0 mL, as shown in Figure 4.18▼. Notice that the diluted solution is less intensely colored than the concentrated one.

In laboratory situations, calculations of this sort are often made very quickly with a simple equation that can be derived by remembering that the number of moles of solute is the same in both the concentrated and dilute solutions and that moles = molarity × liters:

$$\text{moles solute in conc soln} = \text{moles solute in dil soln}$$

$$M_{\text{conc}} \times V_{\text{conc}} = M_{\text{dil}} \times V_{\text{dil}}$$

[4.35]

▶ Figure 4.18 **Procedure for preparing 250 mL of 0.100 M CuSO$_4$ by dilution of 1.00 M CuSO$_4$.** (a) Draw 25.0 mL of the 1.00 M solution into a pipet. (b) Add this to a 250-mL volumetric flask. (c) Add water to dilute the solution to a total volume of 250 mL.

(a)   (b)   (c)

GIVE IT SOME THOUGHT

How is the molarity of a 0.50 M KBr solution changed when water is added to double its volume?

The molarity of the more concentrated stock solution (M_{conc}) is always larger than the molarity of the dilute solution (M_{dil}). Because the volume of the solution increases upon dilution, V_{dil} is always larger than V_{conc}. Although Equation 4.35 is derived in terms of liters, any volume unit can be used as long as that same unit is used on both sides of the equation. For example, in the calculation we did for the CuSO$_4$ solution, we have

$$(1.00\ M)(V_{conc}) = (0.100\ M)(250.\ \text{mL})$$

Solving for V_{conc} gives $V_{conc} = 25.0$ mL as before.

■ SAMPLE EXERCISE 4.14 | Preparing a Solution by Dilution

How many milliliters of 3.0 M H$_2$SO$_4$ are needed to make 450 mL of 0.10 M H$_2$SO$_4$?

SOLUTION

Analyze: We need to dilute a concentrated solution. We are given the molarity of a more concentrated solution (3.0 M) and the volume and molarity of a more dilute one containing the same solute (450 mL of 0.10 M solution). We must calculate the volume of the concentrated solution needed to prepare the dilute solution.

Plan: We can calculate the number of moles of solute, H$_2$SO$_4$, in the dilute solution and then calculate the volume of the concentrated solution needed to supply this amount of solute. Alternatively, we can directly apply Equation 4.35. Let's compare the two methods.

Solve: Calculating the moles of H$_2$SO$_4$ in the dilute solution:

$$\text{moles H}_2\text{SO}_4 \text{ in dilute solution} = (0.450\ \text{L soln})\left(\frac{0.10\ \text{mol H}_2\text{SO}_4}{1\ \text{L soln}}\right)$$

$$= 0.045\ \text{mol H}_2\text{SO}_4$$

Calculating the volume of the concentrated solution that contains 0.045 mol H$_2$SO$_4$:

$$\text{L conc soln} = (0.045\ \text{mol H}_2\text{SO}_4)\left(\frac{1\ \text{L soln}}{3.0\ \text{mol H}_2\text{SO}_4}\right) = 0.015\ \text{L soln}$$

Converting liters to milliliters gives 15 mL.

If we apply Equation 4.35, we get the same result:

$$(3.0 \ M)(V_{\text{conc}}) = (0.10 \ M)(450 \ \text{mL})$$

$$(V_{\text{conc}}) = \frac{(0.10 \ M)(450 \ \text{mL})}{3.0 \ M} = 15 \ \text{mL}$$

Either way, we see that if we start with 15 mL of 3.0 M H$_2$SO$_4$ and dilute it to a total volume of 450 mL, the desired 0.10 M solution will be obtained.

Check: The calculated volume seems reasonable because a small volume of concentrated solution is used to prepare a large volume of dilute solution.

■■ **PRACTICE EXERCISE**

(a) What volume of 2.50 M lead(II) nitrate solution contains 0.0500 mol of Pb^{2+}?
(b) How many milliliters of 5.0 M K$_2$Cr$_2$O$_7$ solution must be diluted to prepare 250 mL of 0.10 M solution? **(c)** If 10.0 mL of a 10.0 M stock solution of NaOH is diluted to 250 mL, what is the concentration of the resulting stock solution?
Answers: **(a)** 0.0200 L = 20.0 mL, **(b)** 5.0 mL, **(c)** 0.40 M

4.6 SOLUTION STOICHIOMETRY
AND CHEMICAL ANALYSIS

Imagine that you have to determine the concentrations of several ions in a sample of lake water. Although many instrumental methods have been developed for such analyses, chemical reactions such as those discussed in this chapter continue to be used. In Chapter 3 we learned that if you know the chemical equation and the amount of one reactant consumed in the reaction, you can calculate the quantities of other reactants and products. In this section we briefly explore such analyses of solutions.

Recall that the coefficients in a balanced equation give the relative number of moles of reactants and products. ∞ (Section 3.6) To use this information, we must convert the quantities of substances involved in a reaction into moles. When we are dealing with grams of substances, as we were in Chapter 3, we use the molar mass to achieve this conversion. When we are working with solutions of known molarity, however, we use molarity and volume to determine the number of moles (moles solute = $M \times$ L). Figure 4.19▼ summarizes this approach to using stoichiometry.

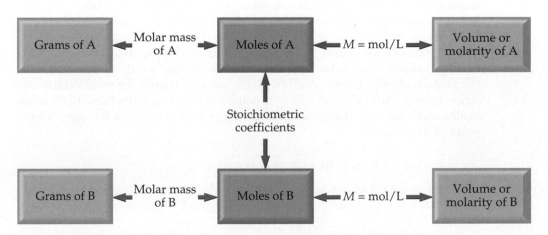

▲ **Figure 4.19 Problem-solving procedure.** Outline of the procedure used to solve stoichiometry problems that involve measured (laboratory) units of mass, solution concentration (molarity), or volume.

■ SAMPLE EXERCISE 4.15 | Using Mass Relations in a Neutralization Reaction

How many grams of $Ca(OH)_2$ are needed to neutralize 25.0 mL of 0.100 M HNO_3?

SOLUTION

Analyze: The reactants are an acid, HNO_3, and a base, $Ca(OH)_2$. The volume and molarity of HNO_3 are given, and we are asked how many grams of $Ca(OH)_2$ are needed to neutralize this quantity of HNO_3.

Plan: We can use the molarity and volume of the HNO_3 solution to calculate the number of moles of HNO_3. We then use the balanced equation to relate the moles of HNO_3 to moles of $Ca(OH)_2$. Finally, we can convert moles of $Ca(OH)_2$ to grams. These steps can be summarized as follows:

$$L_{HNO_3} \times M_{HNO_3} \Rightarrow mol\ HNO_3 \Rightarrow mol\ Ca(OH)_2 \Rightarrow g\ Ca(OH)_2$$

Solve: The product of the molar concentration of a solution and its volume in liters gives the number of moles of solute:

$$\text{moles } HNO_3 = L_{HNO_3} \times M_{HNO_3} = (0.0250\ \text{L})\left(0.100\ \frac{\text{mol } HNO_3}{\text{L}}\right)$$

$$= 2.50 \times 10^{-3}\ \text{mol } HNO_3$$

Because this is an acid–base neutralization reaction, HNO_3 and $Ca(OH)_2$ react to form H_2O and the salt containing Ca^{2+} and NO_3^-:

$$2\ HNO_3(aq) + Ca(OH)_2(s) \longrightarrow 2\ H_2O(l) + Ca(NO_3)_2(aq)$$

Thus, 2 mol $HNO_3 \approxeq 1$ mol $Ca(OH)_2$. Therefore,

$$\text{grams } Ca(OH)_2 = (2.50 \times 10^{-3}\ \text{mol } HNO_3)\left(\frac{1\ \text{mol } Ca(OH)_2}{2\ \text{mol } HNO_3}\right)\left(\frac{74.1\ g\ Ca(OH)_2}{1\ \text{mol } Ca(OH)_2}\right)$$

$$= 0.0926\ g\ Ca(OH)_2$$

Check: The size of the answer is reasonable. A small volume of dilute acid will require only a small amount of base to neutralize it.

■ PRACTICE EXERCISE

(a) How many grams of NaOH are needed to neutralize 20.0 mL of 0.150 M H_2SO_4 solution? **(b)** How many liters of 0.500 M HCl(aq) are needed to react completely with 0.100 mol of $Pb(NO_3)_2(aq)$, forming a precipitate of $PbCl_2(s)$?
Answers: **(a)** 0.240 g, **(b)** 0.400 L

Titrations

To determine the concentration of a particular solute in a solution, chemists often carry out a **titration**, which involves combining a sample of the solution with a reagent solution of known concentration, called a **standard solution**. Titrations can be conducted using acid–base, precipitation, or oxidation-reduction reactions. Suppose we have an HCl solution of unknown concentration and an NaOH solution we know to be 0.100 M. To determine the concentration of the HCl solution, we take a specific volume of that solution, say 20.00 mL. We then slowly add the standard NaOH solution to it until the neutralization reaction between the HCl and NaOH is complete. The point at which stoichiometrically equivalent quantities are brought together is known as the **equivalence point** of the titration.

▲ GIVE IT SOME THOUGHT

25.00 mL of a 0.100 M HBr solution is titrated with a 0.200 M NaOH solution. How many mL of the NaOH solution are required to reach the equivalence point?

To titrate an unknown with a standard solution, there must be some way to determine when the equivalence point of the titration has been reached. In acid–base titrations, dyes known as acid–base **indicators** are used for this purpose.

(a)   (b)   (c)

▲ **Figure 4.20 Change in appearance of a solution containing phenolphthalein indicator as base is added.**
Before the end point, the solution is colorless (a). As the end point is approached, a pale pink color forms where the base is added (b). At the end point, this pale pink color extends throughout the solution after mixing. As even more base is added, the intensity of the pink color increases (c).

For example, the dye known as phenolphthalein is colorless in acidic solution but is pink in basic solution. If we add phenolphthalein to an unknown solution of acid, the solution will be colorless, as seen in Figure 4.20(a)▲. We can then add standard base from a buret until the solution barely turns from colorless to pink, as seen in Figure 4.20(b). This color change indicates that the acid has been neutralized and the drop of base that caused the solution to become colored has no acid to react with. The solution therefore becomes basic, and the dye turns pink. The color change signals the *end point* of the titration, which usually coincides very nearly with the equivalence point. Care must be taken to choose indicators whose end points correspond to the equivalence point of the titration. We will consider this matter in Chapter 17. The titration procedure is summarized in Figure 4.21▼.

20.0 mL of acid solution

Pipet

Initial volume reading

Buret

Standard NaOH solution

20.0 mL of acid solution

Final volume reading

Neutralized solution (indicator has changed color)

(a)   (b)   (c)

◀ **Figure 4.21 Procedure for titrating an acid against a standardized solution of NaOH.**
(a) A known quantity of acid is added to a flask. (b) An acid–base indicator is added, and standardized NaOH is added from a buret. (c) The equivalence point is signaled by a color change in the indicator.

SAMPLE EXERCISE 4.16 | Determining the Quantity of Solute by Titration

The quantity of Cl^- in a municipal water supply is determined by titrating the sample with Ag^+. The reaction taking place during the titration is

$$Ag^+(aq) + Cl^-(aq) \longrightarrow AgCl(s)$$

The end point in this type of titration is marked by a change in color of a special type of indicator. **(a)** How many grams of chloride ion are in a sample of the water if 20.2 mL of 0.100 M Ag^+ is needed to react with all the chloride in the sample? **(b)** If the sample has a mass of 10.0 g, what percent Cl^- does it contain?

SOLUTION

Analyze: We are given the volume (20.2 mL) and molarity (0.100 M) of a solution of Ag^+ and the chemical equation for reaction of this ion with Cl^-. We are asked first to calculate the number of grams of Cl^- in the sample and, second, to calculate the mass percent of Cl^- in the sample.

(a) Plan: We begin by using the volume and molarity of Ag^+ to calculate the number of moles of Ag^+ used in the titration. We can then use the balanced equation to determine the moles of Cl^- in the sample and from that the grams of Cl^-.

Solve:

$$\text{moles } Ag^+ = (20.2 \text{ mL soln})\left(\frac{1 \text{ L soln}}{1000 \text{ mL soln}}\right)\left(0.100 \frac{\text{mol } Ag^+}{\text{L soln}}\right)$$

$$= 2.02 \times 10^{-3} \text{ mol } Ag^+$$

From the balanced equation we see that 1 mol $Ag^+ \hat{=}$ 1 mol Cl^-. Using this information and the molar mass of Cl, we have

$$\text{grams } Cl^- = (2.02 \times 10^{-3} \text{ mol } Ag^+)\left(\frac{1 \text{ mol } Cl^-}{1 \text{ mol } Ag^+}\right)\left(\frac{35.5 \text{ g } Cl^-}{1 \text{ mol } Cl^-}\right)$$

$$= 7.17 \times 10^{-2} \text{ g } Cl^-$$

(b) Plan: To calculate the percentage of Cl^- in the sample, we compare the number of grams of Cl^- in the sample, 7.17×10^{-2} g, with the original mass of the sample, 10.0 g.

Solve: Percent $Cl^- = \dfrac{7.17 \times 10^{-2} \text{ g}}{10.0 \text{ g}} \times 100\% = 0.717\% \text{ } Cl^-$

Comment: Chloride ion is one of the most common ions in water and sewage. Ocean water contains 1.92% Cl^-. Whether water containing Cl^- tastes salty depends on the other ions present. If the only accompanying ions are Na^+, a salty taste may be detected with as little as 0.03% Cl^-.

PRACTICE EXERCISE

A sample of an iron ore is dissolved in acid, and the iron is converted to Fe^{2+}. The sample is then titrated with 47.20 mL of 0.02240 M MnO_4^- solution. The oxidation-reduction reaction that occurs during titration is as follows:

$$MnO_4^-(aq) + 5 Fe^{2+}(aq) + 8 H^+(aq) \longrightarrow Mn^{2+}(aq) + 5 Fe^{3+}(aq) + 4 H_2O(l)$$

(a) How many moles of MnO_4^- were added to the solution? **(b)** How many moles of Fe^{2+} were in the sample? **(c)** How many grams of iron were in the sample? **(d)** If the sample had a mass of 0.8890 g, what is the percentage of iron in the sample?
Answers: **(a)** 1.057×10^{-3} mol MnO_4^- **(b)** 5.286×10^{-3} mol Fe^{2+}, **(c)** 0.2952 g, **(d)** 33.21%

SAMPLE EXERCISE 4.17 | Determining Solution Concentration Via an Acid–Base Titration

One commercial method used to peel potatoes is to soak them in a solution of NaOH for a short time, remove them from the NaOH, and spray off the peel. The concentration of NaOH is normally in the range of 3 to 6 M. The NaOH is analyzed periodically. In one such analysis, 45.7 mL of 0.500 M H_2SO_4 is required to neutralize a 20.0-mL sample of NaOH solution. What is the concentration of the NaOH solution?

SOLUTION

Analyze: We are given the volume (45.7 mL) and molarity (0.500 M) of an H_2SO_4 solution that reacts completely with a 20.0-mL sample of NaOH. We are asked to calculate the molarity of the NaOH solution.

Plan: We can use the volume and molarity of the H_2SO_4 to calculate the number of moles of this substance. Then, we can use this quantity and the balanced equation for the reaction to calculate the number of moles of NaOH. Finally, we can use the moles of NaOH and the volume of this solution to calculate molarity.

Solve: The number of moles of H_2SO_4 is given by the product of the volume and molarity of this solution:

$$\text{moles } H_2SO_4 = (45.7 \text{ mL soln})\left(\frac{1 \text{ L soln}}{1000 \text{ mL soln}}\right)\left(0.500 \frac{\text{mol } H_2SO_4}{\text{L soln}}\right)$$

$$= 2.28 \times 10^{-2} \text{ mol } H_2SO_4$$

Acids react with metal hydroxides to form water and a salt. Thus, the balanced equation for the neutralization reaction is

$$H_2SO_4(aq) + 2 \text{ NaOH}(aq) \longrightarrow 2 H_2O(l) + Na_2SO_4(aq)$$

According to the balanced equation, 1 mol $H_2SO_4 \backsimeq 2$ mol NaOH. Therefore,

$$\text{moles NaOH} = (2.28 \times 10^{-2} \text{ mol } H_2SO_4)\left(\frac{2 \text{ mol NaOH}}{1 \text{ mol } H_2SO_4}\right)$$

$$= 4.56 \times 10^{-2} \text{ mol NaOH}$$

Knowing the number of moles of NaOH present in 20.0 mL of solution allows us to calculate the molarity of this solution:

$$\text{Molarity NaOH} = \frac{\text{mol NaOH}}{\text{L soln}} = \left(\frac{4.56 \times 10^{-2} \text{ mol NaOH}}{20.0 \text{ mL soln}}\right)\left(\frac{1000 \text{ mL soln}}{1 \text{ L soln}}\right)$$

$$= 2.28 \frac{\text{mol NaOH}}{\text{L soln}} = 2.28 \ M$$

■ PRACTICE EXERCISE

What is the molarity of an NaOH solution if 48.0 mL is needed to neutralize 35.0 mL of 0.144 M H_2SO_4?
Answers: 0.210 M

■ SAMPLE INTEGRATIVE EXERCISE | Putting Concepts Together

Note: Integrative exercises require skills from earlier chapters as well as ones from the present chapter.

A sample of 70.5 mg of potassium phosphate is added to 15.0 mL of 0.050 M silver nitrate, resulting in the formation of a precipitate. **(a)** Write the molecular equation for the reaction. **(b)** What is the limiting reactant in the reaction? **(c)** Calculate the theoretical yield, in grams, of the precipitate that forms.

SOLUTION

(a) Potassium phosphate and silver nitrate are both ionic compounds. Potassium phosphate contains K^+ and PO_4^{3-} ions, so its chemical formula is K_3PO_4. Silver nitrate contains Ag^+ and NO_3^- ions, so its chemical formula is $AgNO_3$. Because both reactants are strong electrolytes, the solution contains K^+, PO_4^{3-}, Ag^+, and NO_3^- ions before the reaction occurs. According to the solubility guidelines in Table 4.1, Ag^+ and PO_4^{3-} form an insoluble compound, so Ag_3PO_4 will precipitate from the solution. In contrast, K^+ and NO_3^- will remain in solution because KNO_3 is water soluble. Thus, the balanced molecular equation for the reaction is

$$K_3PO_4(aq) + 3 \text{ AgNO}_3(aq) \longrightarrow Ag_3PO_4(s) + 3 KNO_3(aq)$$

(b) To determine the limiting reactant, we must examine the number of moles of each reactant. ∞(Section 3.7) The number of moles of K_3PO_4 is calculated from the mass of the sample using the molar mass as a conversion factor. ∞(Section 3.4)

The molar mass of K_3PO_4 is $3(39.1) + 31.0 + 4(16.0) = 212.3$ g/mol. Converting milligrams to grams and then to moles, we have

$$(70.5 \text{ mg } K_3PO_4)\left(\frac{10^{-3} \text{ g } K_3PO_4}{1 \text{ mg } K_3PO_4}\right)\left(\frac{1 \text{ mol } K_3PO_4}{212.3 \text{ g } K_3PO_4}\right) = 3.32 \times 10^{-4} \text{ mol } K_3PO_4$$

We determine the number of moles of $AgNO_3$ from the volume and molarity of the solution. ∞ (Section 4.5) Converting milliliters to liters and then to moles, we have

$$(15.0 \text{ mL})\left(\frac{10^{-3} \text{ L}}{1 \text{ mL}}\right)\left(\frac{0.050 \text{ mol } AgNO_3}{\text{L}}\right) = 7.5 \times 10^{-4} \text{ mol } AgNO_3$$

Comparing the amounts of the two reactants, we find that there are $(7.5 \times 10^{-4})/(3.32 \times 10^{-4}) = 2.3$ times as many moles of $AgNO_3$ as there are moles of K_3PO_4. According to the balanced equation, however, 1 mol K_3PO_4 requires 3 mol of $AgNO_3$. Thus, there is insufficient $AgNO_3$ to consume the K_3PO_4, and $AgNO_3$ is the limiting reactant.

(c) The precipitate is Ag_3PO_4, whose molar mass is $3(107.9) + 31.0 + 4(16.0) = 418.7$ g/mol. To calculate the number of grams of Ag_3PO_4 that could be produced in this reaction (the theoretical yield), we use the number of moles of the limiting reactant, converting mol $AgNO_3 \Rightarrow$ mol $Ag_3PO_4 \Rightarrow$ g Ag_3PO_4. We use the coefficients in the balanced equation to convert moles of $AgNO_3$ to moles Ag_3PO_4, and we use the molar mass of Ag_3PO_4 to convert the number of moles of this substance to grams.

$$(7.5 \times 10^{-4} \text{ mol } AgNO_3)\left(\frac{1 \text{ mol } Ag_3PO_4}{3 \text{ mol } AgNO_3}\right)\left(\frac{418.7 \text{ g } Ag_3PO_4}{1 \text{ mol } Ag_3PO_4}\right) = 0.10 \text{ g } Ag_3PO_4$$

The answer has only two significant figures because the quantity of $AgNO_3$ is given to only two significant figures.

CHAPTER REVIEW

SUMMARY AND KEY TERMS

Introduction and Section 4.1 Solutions in which water is the dissolving medium are called **aqueous solutions**. The component of the solution that is in the greater quantity is the **solvent**. The other components are **solutes**.

Any substance whose aqueous solution contains ions is called an **electrolyte**. Any substance that forms a solution containing no ions is a **nonelectrolyte**. Electrolytes that are present in solution entirely as ions are **strong electrolytes**, whereas those that are present partly as ions and partly as molecules are **weak electrolytes**. Ionic compounds dissociate into ions when they dissolve, and they are strong electrolytes. The solubility of ionic substances is made possible by **solvation**, the interaction of ions with polar solvent molecules. Most molecular compounds are nonelectrolytes, although some are weak electrolytes, and a few are strong electrolytes. When representing the ionization of a weak electrolyte in solution, half-arrows in both directions are used, indicating that the forward and reverse reactions can achieve a chemical balance called a **chemical equilibrium**.

Section 4.2 Precipitation reactions are those in which an insoluble product, called a **precipitate**, forms. Solubility guidelines help determine whether or not an ionic compound will be soluble in water. (The **solubility** of a substance is the amount that dissolves in a given quantity of solvent.) Reactions such as precipitation reactions, in which cations and anions appear to exchange partners, are called **exchange reactions**, or **metathesis reactions**.

Chemical equations can be written to show whether dissolved substances are present in solution predominantly as ions or molecules. When the complete chemical formulas of all reactants and products are used, the equation is called a **molecular equation**. A **complete ionic equation** shows all dissolved strong electrolytes as their component ions. In a **net ionic equation**, those ions that go through the reaction unchanged (**spectator ions**) are omitted.

Section 4.3 Acids and bases are important electrolytes. **Acids** are proton donors; they increase the concentration of $H^+(aq)$ in aqueous solutions to which they are added.

Bases are proton acceptors; they increase the concentration of $OH^-(aq)$ in aqueous solutions. Those acids and bases that are strong electrolytes are called **strong acids** and **strong bases**, respectively. Those that are weak electrolytes are **weak acids** and **weak bases**. When solutions of acids and bases are mixed, a **neutralization reaction** results. The neutralization reaction between an acid and a metal hydroxide produces water and a **salt**. Gases can also be formed as a result of acid–base reactions. The reaction of a sulfide with an acid forms $H_2S(g)$; the reaction between a carbonate and an acid forms $CO_2(g)$.

Section 4.4 **Oxidation** is the loss of electrons by a substance, whereas **reduction** is the gain of electrons by a substance. **Oxidation numbers** keep track of electrons during chemical reactions and are assigned to atoms using specific rules. The oxidation of an element results in an increase in its oxidation number, whereas reduction is accompanied by a decrease in oxidation number. Oxidation is always accompanied by reduction, giving **oxidation-reduction**, or redox, **reactions**.

Many metals are oxidized by O_2, acids, and salts. The redox reactions between metals and acids and between metals and salts are called **displacement reactions**. The products of these displacement reactions are always an element (H_2 or a metal) and a salt. Comparing such reactions allows us to rank metals according to their ease of oxidation. A list of metals arranged in order of decreasing ease of oxidation is called an **activity series**. Any metal on the list can be oxidized by ions of metals (or H^+) below it in the series.

Section 4.5 The composition of a solution expresses the relative quantities of solvent and solutes that it contains. One of the common ways to express the **concentration** of a solute in a solution is in terms of molarity. The **molarity** of a solution is the number of moles of solute per liter of solution. Molarity makes it possible to interconvert solution volume and number of moles of solute. Solutions of known molarity can be formed either by weighing out the solute and diluting it to a known volume or by the **dilution** of a more concentrated solution of known concentration (a stock solution). Adding solvent to the solution (the process of dilution) decreases the concentration of the solute without changing the number of moles of solute in the solution ($M_{conc} \times V_{conc} = M_{dil} \times V_{dil}$).

Section 4.6 In the process called **titration**, we combine a solution of known concentration (a **standard solution**) with a solution of unknown concentration to determine the unknown concentration or the quantity of solute in the unknown. The point in the titration at which stoichiometrically equivalent quantities of reactants are brought together is called the **equivalence point**. An **indicator** can be used to show the end point of the titration, which coincides closely with the equivalence point.

KEY SKILLS

- Recognize compounds as acids or bases, and as strong, weak, or nonelectrolytes.
- Recognize reactions as acid–base, precipitation, metathesis, or redox.
- Be able to calculate moles or grams of substances in solution using molarity.
- Understand how to carry out a dilution to achieve a desired solution concentration.
- Understand how to perform and interpret the results of a titration.

KEY EQUATIONS

- $$\text{Molarity} = \frac{\text{moles solute}}{\text{volume of solution in liters}} \qquad [4.33]$$

  Molarity is the most commonly used unit of concentration in chemistry.

- $$M_{conc} \times V_{conc} = M_{dil} \times V_{dil} \qquad [4.35]$$

  When adding solvent to a concentrated solution to make a dilute solution, molarities and volumes of both concentrated and dilute solutions can be calculated if three of the quantities are known.

VISUALIZING CONCEPTS

4.1 Which of the following schematic drawings best describes a solution of Li$_2$SO$_4$ in water (water molecules not shown for simplicity)? [Section 4.1]

(a)   (b)   (c)

4.2 Methanol, CH$_3$OH, and hydrogen chloride, HCl, are both molecular substances, yet an aqueous solution of methanol does not conduct an electrical current, whereas a solution of HCl does conduct. Account for this difference. [Section 4.1]

4.3 Aqueous solutions of three different substances, AX, AY, and AZ, are represented by the three diagrams below. Identify each substance as a strong electrolyte, weak electrolyte, or nonelectrolyte. [Section 4.1]

AX   AY   AZ

(a)   (b)   (c)

4.4 A 0.1 M solution of acetic acid, CH$_3$COOH, causes the lightbulb in the apparatus of Figure 4.2 to glow about as brightly as a 0.001 M solution of HBr. How do you account for this fact? [Section 4.1]

4.5 You are presented with three white solids, A, B, and C, which are glucose (a sugar substance), NaOH, and AgBr. Solid A dissolves in water to form a conducting solution. B is not soluble in water. C dissolves in water to form a nonconducting solution. Identify A, B, and C. [Section 4.2]

4.6 We have seen that ions in aqueous solution are stabilized by the attractions between the ions and the water molecules. Why then do some pairs of ions in solution form precipitates? [Section 4.2]

4.7 Which of the following ions will *always* be a spectator ion in a precipitation reaction? **(a)** Cl$^-$, **(b)** NO$_3^-$, **(c)** NH$_4^+$, **(d)** S^{2-}, **(e)** SO$_4^{2-}$. Explain briefly. [Section 4.2]

4.8 The labels have fallen off two bottles, one containing Mg(NO$_3$)$_2$ and the other containing Pb(NO$_3$)$_2$. You have a bottle of dilute H$_2$SO$_4$. How could you use it to test a portion of each solution to identify which solution is which? [Section 4.2]

4.9 Explain how a redox reaction involves electrons in the same way that an acid–base reaction involves protons. [Sections 4.3 and 4.4]

4.10 If you want to double the concentration of a solution, how could you do it? [Section 4.5]

EXERCISES

Electrolytes

4.11 When asked what causes electrolyte solutions to conduct electricity, a student responds that it is due to the movement of electrons through the solution. Is the student correct? If not, what is the correct response?

4.12 When methanol, CH$_3$OH, is dissolved in water, a nonconducting solution results. When acetic acid, CH$_3$COOH, dissolves in water, the solution is weakly conducting and acidic in nature. Describe what happens upon dissolution in the two cases, and account for the different results.

4.13 We have learned in this chapter that many ionic solids dissolve in water as strong electrolytes, that is, as separated ions in solution. What properties of water facilitate this process?

4.14 What does it mean to say that ions are hydrated when an ionic substance dissolves in water?

4.15 Specify what ions are present in solution upon dissolving each of the following substances in water: **(a)** ZnCl$_2$, **(b)** HNO$_3$, **(c)** (NH$_4$)$_2$SO$_4$, **(d)** Ca(OH)$_2$.

4.16 Specify what ions are present upon dissolving each of the following substances in water: **(a)** MgI$_2$, **(b)** Al(NO$_3$)$_3$, **(c)** HClO$_4$, **(d)** NaCH$_3$COO.

4.17 Formic acid, HCOOH, is a weak electrolyte. What solute particles are present in an aqueous solution of this compound? Write the chemical equation for the ionization of HCOOH.

4.18 Acetone, CH$_3$COCH$_3$, is a nonelectrolyte; hypochlorous acid, HClO, is a weak electrolyte; and ammonium chloride, NH$_4$Cl, is a strong electrolyte. **(a)** What are the solute particles present in aqueous solutions of each compound? **(b)** If 0.1 mol of each compound is dissolved in solution, which one contains 0.2 mol of solute particles, which contains 0.1 mol of solute particles, and which contains somewhere between 0.1 and 0.2 mol of solute particles?

Precipitation Reactions and Net Ionic Equations

4.19 Using solubility guidelines, predict whether each of the following compounds is soluble or insoluble in water: **(a)** $NiCl_2$, **(b)** Ag_2S, **(c)** Cs_3PO_4, **(d)** $SrCO_3$, **(e)** $PbSO_4$.

4.20 Predict whether each of the following compounds is soluble in water: **(a)** $Ni(OH)_2$, **(b)** $PbBr_2$, **(c)** $Ba(NO_3)_2$, **(d)** $AlPO_4$, **(e)** $AgCH_3COO$.

4.21 Will precipitation occur when the following solutions are mixed? If so, write a balanced chemical equation for the reaction. **(a)** Na_2CO_3 and $AgNO_3$, **(b)** $NaNO_3$ and $NiSO_4$, **(c)** $FeSO_4$ and $Pb(NO_3)_2$.

4.22 Identify the precipitate (if any) that forms when the following solutions are mixed, and write a balanced equation for each reaction. **(a)** $Ni(NO_3)_2$ and $NaOH$, **(b)** $NaOH$ and K_2SO_4, **(c)** Na_2S and $Cu(CH_3COO)_2$.

4.23 Name the spectator ions in any reactions that may be involved when each of the following pairs of solutions are mixed.
(a) $Na_2CO_3(aq)$ and $MgSO_4(aq)$
(b) $Pb(NO_3)_2(aq)$ and $Na_2S(aq)$
(c) $(NH_4)_3PO_4(aq)$ and $CaCl_2(aq)$

4.24 Write balanced net ionic equations for the reactions that occur in each of the following cases. Identify the spectator ion or ions in each reaction.

(a) $Cr_2(SO_4)_3(aq) + (NH_4)_2CO_3(aq) \longrightarrow$
(b) $Ba(NO_3)_2(aq) + K_2SO_4(aq) \longrightarrow$
(c) $Fe(NO_3)_2(aq) + KOH(aq) \longrightarrow$

4.25 Separate samples of a solution of an unknown salt are treated with dilute solutions of HBr, H_2SO_4, and NaOH. A precipitate forms in all three cases. Which of the following cations could the solution contain: K^+; Pb^{2+}; Ba^{2+}?

4.26 Separate samples of a solution of an unknown ionic compound are treated with dilute $AgNO_3$, $Pb(NO_3)_2$, and $BaCl_2$. Precipitates form in all three cases. Which of the following could be the anion of the unknown salt: Br^-; CO_3^{2-}; NO_3^-?

4.27 You know that an unlabeled bottle contains a solution of one of the following: $AgNO_3$, $CaCl_2$, or $Al_2(SO_4)_3$. A friend suggests that you test a portion of the solution with $Ba(NO_3)_2$ and then with NaCl solutions. Explain how these two tests together would be sufficient to determine which salt is present in the solution.

4.28 Three solutions are mixed together to form a single solution. One contains 0.2 mol $Pb(CH_3COO)_2$, the second contains 0.1 mol Na_2S, and the third contains 0.1 mol $CaCl_2$. **(a)** Write the net ionic equations for the precipitation reaction or reactions that occur. **(b)** What are the spectator ions in the solution?

Acid–Base Reactions

4.29 Which of the following solutions has the largest concentration of solvated protons: **(a)** 0.2 M LiOH, **(b)** 0.2 M HI, **(c)** 1.0 M methyl alcohol (CH_3OH)? Explain.

4.30 Which of the following solutions is the most basic? **(a)** 0.6 M NH_3, **(b)** 0.150 M KOH, **(c)** 0.100 M $Ba(OH)_2$. Explain.

4.31 What is the difference between **(a)** a monoprotic acid and a diprotic acid, **(b)** a weak acid and a strong acid, **(c)** an acid and a base?

4.32 Explain the following observations: **(a)** NH_3 contains no OH^- ions, and yet its aqueous solutions are basic; **(b)** HF is called a weak acid, and yet it is very reactive; **(c)** although sulfuric acid is a strong electrolyte, an aqueous solution of H_2SO_4 contains more HSO_4^- ions than SO_4^{2-} ions.

4.33 HCl, HBr, and HI are strong acids, yet HF is a weak acid. What does this mean in terms of the extent to which these substances are ionized in solution?

4.34 What is the relationship between the solubility rules in Table 4.1 and the list of strong bases in Table 4.2? Another way of asking this question is, why is $Cd(OH)_2$, for example, not listed as a strong base in Table 4.2?

4.35 Label each of the following substances as an acid, base, salt, or none of the above. Indicate whether the substance exists in aqueous solution entirely in molecular form, entirely as ions, or as a mixture of molecules and ions. **(a)** HF; **(b)** acetonitrile, CH_3CN; **(c)** $NaClO_4$; **(d)** $Ba(OH)_2$.

4.36 An aqueous solution of an unknown solute is tested with litmus paper and found to be acidic. The solution is weakly conducting compared with a solution of NaCl of the same concentration. Which of the following substances could the unknown be: KOH, NH_3, HNO_3, $KClO_2$, H_3PO_3, CH_3COCH_3 (acetone)?

4.37 Classify each of the following substances as a nonelectrolyte, weak electrolyte, or strong electrolyte in water: **(a)** H_2SO_3, **(b)** C_2H_5OH (ethanol), **(c)** NH_3, **(d)** $KClO_3$, **(e)** $Cu(NO_3)_2$.

4.38 Classify each of the following aqueous solutions as a nonelectrolyte, weak electrolyte, or strong electrolyte: **(a)** $HClO_4$, **(b)** HNO_3, **(c)** NH_4Cl, **(d)** CH_3COCH_3 (acetone), **(e)** $CoSO_4$, **(f)** $C_{12}H_{22}O_{11}$ (sucrose).

4.39 Complete and balance the following molecular equations, and then write the net ionic equation for each:
(a) $HBr(aq) + Ca(OH)_2(aq) \longrightarrow$
(b) $Cu(OH)_2(s) + HClO_4(aq) \longrightarrow$
(c) $Al(OH)_3(s) + HNO_3(aq) \longrightarrow$

4.40 Write the balanced molecular and net ionic equations for each of the following neutralization reactions:
(a) Aqueous acetic acid is neutralized by aqueous potassium hydroxide.
(b) Solid chromium(III) hydroxide reacts with nitric acid.
(c) Aqueous hypochlorous acid and aqueous calcium hydroxide react.

4.41 Write balanced molecular and net ionic equations for the following reactions, and identify the gas formed in each: **(a)** solid cadmium sulfide reacts with an aqueous solution of sulfuric acid; **(b)** solid magnesium carbonate reacts with an aqueous solution of perchloric acid.

4.42 Because the oxide ion is basic, metal oxides react readily with acids. **(a)** Write the net ionic equation for the following reaction:

$$FeO(s) + 2\,HClO_4(aq) \longrightarrow Fe(ClO_4)_2(aq) + H_2O(l)$$

(b) Based on the equation in part (a), write the net ionic equation for the reaction that occurs between $NiO(s)$ and an aqueous solution of nitric acid.

4.43 Write a balanced molecular equation and a net ionic equation for the reaction that occurs when **(a)** solid $CaCO_3$ reacts with an aqueous solution of nitric acid; **(b)** solid iron(II) sulfide reacts with an aqueous solution of hydrobromic acid.

4.44 As K_2O dissolves in water, the oxide ion reacts with water molecules to form hydroxide ions. Write the molecular and net ionic equations for this reaction. Based on the definitions of acid and base, what ion is the base in this reaction? What is the acid? What is the spectator ion in the reaction?

Oxidation-Reduction Reactions

4.45 Define oxidation and reduction in terms of **(a)** electron transfer and **(b)** oxidation numbers.

4.46 Can oxidation occur without accompanying reduction? Explain.

4.47 Which circled region of the periodic table shown here contains the most readily oxidized elements? Which contains the least readily oxidized?

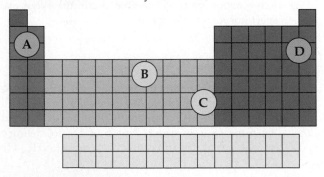

4.48 From the elements listed in Table 4.5, select an element that lies in region A of the periodic table shown above and an element that lies in region C. Write a balanced oxidation-reduction equation that shows the oxidation of one metal and reduction of an ion of the other. You will need to decide which element is oxidized and which is reduced.

4.49 Determine the oxidation number for the indicated element in each of the following substances: **(a)** S in SO_2, **(b)** C in $COCl_2$, **(c)** Mn in MnO_4^-, **(d)** Br in HBrO, **(e)** As in As_4, **(f)** O in K_2O_2.

4.50 Determine the oxidation number for the indicated element in each of the following compounds: **(a)** Ti in TiO_2, **(b)** Sn in $SnCl_3^-$, **(c)** C in $C_2O_4^{2-}$, **(d)** N in N_2H_4, **(e)** N in HNO_2, **(f)** Cr in $Cr_2O_7^{2-}$.

4.51 Which element is oxidized and which is reduced in the following reactions?

(a) $N_2(g) + 3\,H_2(g) \longrightarrow 2\,NH_3(g)$

(b) $3\,Fe(NO_3)_2(aq) + 2\,Al(s) \longrightarrow$
$$3\,Fe(s) + 2\,Al(NO_3)_3(aq)$$

(c) $Cl_2(aq) + 2\,NaI(aq) \longrightarrow I_2(aq) + 2\,NaCl(aq)$

(d) $PbS(s) + 4\,H_2O_2(aq) \longrightarrow PbSO_4(s) + 4\,H_2O(l)$

4.52 Which of the following are redox reactions? For those that are, indicate which element is oxidized and which is reduced. For those that are not, indicate whether they are precipitation or acid–base reactions.

(a) $Cu(OH)_2(s) + 2\,HNO_3(aq) \longrightarrow$
$$Cu(NO_3)_2(aq) + 2\,H_2O(l)$$

(b) $Fe_2O_3(s) + 3\,CO(g) \longrightarrow 2\,Fe(s) + 3\,CO_2(g)$

(c) $Sr(NO_3)_2(aq) + H_2SO_4(aq) \longrightarrow$
$$SrSO_4(s) + 2\,HNO_3(aq)$$

(d) $4\,Zn(s) + 10\,H^+(aq) + 2\,NO_3^-(aq) \longrightarrow$
$$4\,Zn^{2+}(aq) + N_2O(g) + 5\,H_2O(l)$$

4.53 Write balanced molecular and net ionic equations for the reactions of **(a)** manganese with dilute sulfuric acid; **(b)** chromium with hydrobromic acid; **(c)** tin with hydrochloric acid; **(d)** aluminum with formic acid, HCOOH.

4.54 Write balanced molecular and net ionic equations for the reactions of **(a)** hydrochloric acid with nickel; **(b)** dilute sulfuric acid with iron; **(c)** hydrobromic acid with magnesium; **(d)** acetic acid, CH_3COOH, with zinc.

4.55 Using the activity series (Table 4.5), write balanced chemical equations for the following reactions. If no reaction occurs, simply write NR. **(a)** Iron metal is added to a solution of copper(II) nitrate; **(b)** zinc metal is added to a solution of magnesium sulfate; **(c)** hydrobromic acid is added to tin metal; **(d)** hydrogen gas is bubbled through an aqueous solution of nickel(II) chloride; **(e)** aluminum metal is added to a solution of cobalt(II) sulfate.

4.56 Based on the activity series (Table 4.5), what is the outcome (if any) of each of the following reactions?

(a) $Mn(s) + NiCl_2(aq) \longrightarrow$

(b) $Cu(s) + Cr(CH_3COO)_3(aq) \longrightarrow$

(c) $Cr(s) + NiSO_4(aq) \longrightarrow$

(d) $Pt(s) + HBr(aq) \longrightarrow$

(e) $H_2(g) + CuCl_2(aq) \longrightarrow$

4.57 The metal cadmium tends to form Cd^{2+} ions. The following observations are made: (i) When a strip of zinc metal is placed in $CdCl_2(aq)$, cadmium metal is deposited on the strip. (ii) When a strip of cadmium metal is placed in $Ni(NO_3)_2(aq)$, nickel metal is deposited on the strip. **(a)** Write net ionic equations to explain each of the observations made above. **(b)** What can you conclude about the position of cadmium in the activity series? **(c)** What experiments would you need to perform to locate more precisely the position of cadmium in the activity series?

4.58 (a) Use the following reactions to prepare an activity series for the halogens:

$$Br_2(aq) + 2\,NaI(aq) \longrightarrow 2\,NaBr(aq) + I_2(aq)$$
$$Cl_2(aq) + 2\,NaBr(aq) \longrightarrow 2\,NaCl(aq) + Br_2(aq)$$

(b) Relate the positions of the halogens in the periodic table with their locations in this activity series. **(c)** Predict whether a reaction occurs when the following reagents are mixed: $Cl_2(aq)$ and $KI(aq)$; $Br_2(aq)$ and $LiCl(aq)$.

Solution Composition; Molarity

4.59 (a) Is the concentration of a solution an intensive or an extensive property? **(b)** What is the difference between 0.50 mol HCl and 0.50 M HCl?

4.60 (a) Suppose you prepare 500 mL of a 0.10 M solution of some salt and then spill some of it. What happens to the concentration of the solution left in the container? **(b)** Suppose you prepare 500 mL of a 0.10 M aqueous solution of some salt and let it sit out, uncovered, for a long time, and some water evaporates. What happens to the concentration of the solution left in the container? **(c)** A certain volume of a 0.50 M solution contains 4.5 g of a salt. What mass of the salt is present in the same volume of a 2.50 M solution?

4.61 (a) Calculate the molarity of a solution that contains 0.0250 mol NH_4Cl in exactly 500 mL of solution. **(b)** How many moles of HNO_3 are present in 50.0 mL of a 2.50 M solution of nitric acid? **(c)** How many milliliters of 1.50 M KOH solution are needed to provide 0.275 mol of KOH?

4.62 (a) Calculate the molarity of a solution made by dissolving 0.750 grams of Na_2SO_4 in enough water to form exactly 850 mL of solution. **(b)** How many moles of $KMnO_4$ are present in 250 mL of a 0.0475 M solution? **(c)** How many milliliters of 11.6 M HCl solution are needed to obtain 0.250 mol of HCl?

4.63 The average adult human male has a total blood volume of 5.0 L. If the concentration of sodium ion in this average individual is 0.135 M, what is the mass of sodium ion circulating in the blood?

4.64 A person suffering from hyponatremia has a sodium ion concentration in the blood of 0.118 M and a total blood volume of 4.6 L. What mass of sodium chloride would need to be added to the blood to bring the sodium ion concentration up to 0.138 M, assuming no change in blood volume?

4.65 The concentration of alcohol (CH_3CH_2OH) in blood, called the "blood alcohol concentration" or BAC, is given in units of grams of alcohol per 100 mL of blood. The legal definition of intoxication, in many states of the United States, is that the BAC is 0.08 or higher. What is the concentration of alcohol, in terms of molarity, in blood if the BAC is 0.08?

4.66 The average adult male has a total blood volume of 5.0 L. After drinking a few beers, he has a BAC of 0.10 (see Exercise 4.65). What mass of alcohol is circulating in his blood?

4.67 Calculate **(a)** the number of grams of solute in 0.250 L of 0.175 M KBr, **(b)** the molar concentration of a solution containing 14.75 g of $Ca(NO_3)_2$ in 1.375 L, **(c)** the volume of 1.50 M Na_3PO_4 in milliliters that contains 2.50 g of solute.

4.68 (a) How many grams of solute are present in 50.0 mL of 0.488 M $K_2Cr_2O_7$? **(b)** If 4.00 g of $(NH_4)_2SO_4$ is dissolved in enough water to form 400 mL of solution, what is the molarity of the solution? **(c)** How many milliliters of 0.0250 M $CuSO_4$ contain 1.75 g of solute?

4.69 (a) Which will have the highest concentration of potassium ion: 0.20 M KCl, 0.15 M K_2CrO_4, or 0.080 M K_3PO_4? **(b)** Which will contain the greater number of moles of potassium ion: 30.0 mL of 0.15 M K_2CrO_4 or 25.0 mL of 0.080 M K_3PO_4?

4.70 In each of the following pairs, indicate which has the higher concentration of Cl^- ion: **(a)** 0.10 M $CaCl_2$ or 0.15 M KCl solution, **(b)** 100 mL of 0.10 M KCl solution or 400 mL of 0.080 M LiCl solution, **(c)** 0.050 M HCl solution or 0.020 M $CdCl_2$ solution.

4.71 Indicate the concentration of each ion or molecule present in the following solutions: **(a)** 0.25 M $NaNO_3$, **(b)** 1.3×10^{-2} M $MgSO_4$, **(c)** 0.0150 M $C_6H_{12}O_6$, **(d)** a mixture of 45.0 mL of 0.272 M NaCl and 65.0 mL of 0.0247 M $(NH_4)_2CO_3$. Assume that the volumes are additive.

4.72 Indicate the concentration of each ion present in the solution formed by mixing **(a)** 42.0 mL of 0.170 M NaOH and 37.6 mL of 0.400 M NaOH, **(b)** 44.0 mL of 0.100 M Na_2SO_4 and 25.0 mL of 0.150 M KCl, **(c)** 3.60 g KCl in 75.0 mL of 0.250 M $CaCl_2$ solution. Assume that the volumes are additive.

4.73 (a) You have a stock solution of 14.8 M NH_3. How many milliliters of this solution should you dilute to make 1000.0 mL of 0.250 M NH_3? **(b)** If you take a 10.0-mL portion of the stock solution and dilute it to a total volume of 0.500 L, what will be the concentration of the final solution?

4.74 (a) How many milliliters of a stock solution of 10.0 M HNO_3 would you have to use to prepare 0.450 L of 0.500 M HNO_3? **(b)** If you dilute 25.0 mL of the stock solution to a final volume of 0.500 L, what will be the concentration of the diluted solution?

4.75 (a) Starting with solid sucrose, $C_{12}H_{22}O_{11}$, describe how you would prepare 250 mL of a 0.250 M sucrose solution. **(b)** Describe how you would prepare 350.0 mL of 0.100 M $C_{12}H_{22}O_{11}$ starting with 3.00 L of 1.50 M $C_{12}H_{22}O_{11}$.

4.76 **(a)** How would you prepare 175.0 mL of 0.150 M AgNO$_3$ solution starting with pure solute? **(b)** An experiment calls for you to use 100 mL of 0.50 M HNO$_3$ solution. All you have available is a bottle of 3.6 M HNO$_3$. How would you prepare the desired solution?

[4.77] Pure acetic acid, known as glacial acetic acid, is a liquid with a density of 1.049 g/mL at 25 °C. Calculate the molarity of a solution of acetic acid made by dissolving

20.00 mL of glacial acetic acid at 25 °C in enough water to make 250.0 mL of solution.

[4.78] Glycerol, $C_3H_8O_3$, is a substance used extensively in the manufacture of cosmetics, foodstuffs, antifreeze, and plastics. Glycerol is a water-soluble liquid with a density of 1.2656 g/mL at 15 °C. Calculate the molarity of a solution of glycerol made by dissolving 50.000 mL glycerol at 15 °C in enough water to make 250.00 mL of solution.

Solution Stoichiometry; Titrations

4.79 What mass of KCl is needed to precipitate the silver ions from 15.0 mL of 0.200 M AgNO$_3$ solution?

4.80 What mass of NaOH is needed to precipitate the Cd^{2+} ions from 35.0 mL of 0.500 M Cd(NO$_3$)$_2$ solution?

4.81 **(a)** What volume of 0.115 M HClO$_4$ solution is needed to neutralize 50.00 mL of 0.0875 M NaOH? **(b)** What volume of 0.128 M HCl is needed to neutralize 2.87 g of Mg(OH)$_2$? **(c)** If 25.8 mL of AgNO$_3$ is needed to precipitate all the Cl$^-$ ions in a 785-mg sample of KCl (forming AgCl), what is the molarity of the AgNO$_3$ solution? **(d)** If 45.3 mL of 0.108 M HCl solution is needed to neutralize a solution of KOH, how many grams of KOH must be present in the solution?

4.82 **(a)** How many milliliters of 0.120 M HCl are needed to completely neutralize 50.0 mL of 0.101 M Ba(OH)$_2$ solution? **(b)** How many milliliters of 0.125 M H$_2$SO$_4$ are needed to neutralize 0.200 g of NaOH? **(c)** If 55.8 mL of BaCl$_2$ solution is needed to precipitate all the sulfate ion in a 752-mg sample of Na$_2$SO$_4$, what is the molarity of the solution? **(d)** If 42.7 mL of 0.208 M HCl solution is needed to neutralize a solution of Ca(OH)$_2$, how many grams of Ca(OH)$_2$ must be in the solution?

4.83 Some sulfuric acid is spilled on a lab bench. You can neutralize the acid by sprinkling sodium bicarbonate on it and then mopping up the resultant solution. The sodium bicarbonate reacts with sulfuric acid as follows:

2 NaHCO$_3$(s) + H$_2$SO$_4$(aq) $\longrightarrow$
$$Na_2SO_4(aq) + 2\,H_2O(l) + 2\,CO_2(g)$$

Sodium bicarbonate is added until the fizzing due to the formation of CO$_2$(g) stops. If 27 mL of 6.0 M H$_2$SO$_4$ was spilled, what is the minimum mass of NaHCO$_3$ that must be added to the spill to neutralize the acid?

4.84 The distinctive odor of vinegar is due to acetic acid, CH$_3$COOH, which reacts with sodium hydroxide in the following fashion:

CH$_3$COOH(aq) + NaOH(aq) $\longrightarrow$
$$H_2O(l) + NaC_2H_3O_2(aq)$$

If 3.45 mL of vinegar needs 42.5 mL of 0.115 M NaOH to reach the equivalence point in a titration, how many grams of acetic acid are in a 1.00-qt sample of this vinegar?

4.85 A sample of solid Ca(OH)$_2$ is stirred in water at 30 °C until the solution contains as much dissolved Ca(OH)$_2$ as it can hold. A 100-mL sample of this solution is withdrawn and titrated with 5.00×10^{-2} M HBr. It requires 48.8 mL of the acid solution for neutralization. What is the molarity of the Ca(OH)$_2$ solution? What is the solubility of Ca(OH)$_2$ in water, at 30 °C, in grams of Ca(OH)$_2$ per 100 mL of solution?

4.86 In a laboratory, 6.82 g of Sr(NO$_3$)$_2$ is dissolved in enough water to form 0.500 L of solution. A 0.100-L sample is withdrawn from this stock solution and titrated with a 0.0245 M solution of Na$_2$CrO$_4$. What volume of Na$_2$CrO$_4$ solution is needed to precipitate all the Sr^{2+}(aq) as SrCrO$_4$?

4.87 A solution of 100.0 mL of 0.200 M KOH is mixed with a solution of 200.0 mL of 0.150 M NiSO$_4$. **(a)** Write the balanced chemical equation for the reaction that occurs. **(b)** What precipitate forms? **(c)** What is the limiting reactant? **(d)** How many grams of this precipitate form? **(e)** What is the concentration of each ion that remains in solution?

4.88 A solution is made by mixing 12.0 g of NaOH and 75.0 mL of 0.200 M HNO$_3$. **(a)** Write a balanced equation for the reaction that occurs between the solutes. **(b)** Calculate the concentration of each ion remaining in solution. **(c)** Is the resultant solution acidic or basic?

[4.89] A 0.5895-g sample of impure magnesium hydroxide is dissolved in 100.0 mL of 0.2050 M HCl solution. The excess acid then needs 19.85 mL of 0.1020 M NaOH for neutralization. Calculate the percent by mass of magnesium hydroxide in the sample, assuming that it is the only substance reacting with the HCl solution.

[4.90] A 1.248-g sample of limestone rock is pulverized and then treated with 30.00 mL of 1.035 M HCl solution. The excess acid then requires 11.56 mL of 1.010 M NaOH for neutralization. Calculate the percent by mass of calcium carbonate in the rock, assuming that it is the only substance reacting with the HCl solution.

ADDITIONAL EXERCISES

4.91 Explain why a titration experiment is a good way to measure the unknown concentration of a compound in solution.

4.92 The accompanying photo shows the reaction between a solution of $Cd(NO_3)_2$ and one of Na_2S. What is the identity of the precipitate? What ions remain in solution? Write the net ionic equation for the reaction.

4.93 Suppose you have a solution that might contain any or all of the following cations: Ni^{2+}, Ag^+, Sr^{2+}, and Mn^{2+}. Addition of HCl solution causes a precipitate to form. After filtering off the precipitate, H_2SO_4 solution is added to the resultant solution and another precipitate forms. This is filtered off, and a solution of NaOH is added to the resulting solution. No precipitate is observed. Which ions are present in each of the precipitates? Which of the four ions listed above must be absent from the original solution?

4.94 You choose to investigate some of the solubility guidelines for two ions not listed in Table 4.1, the chromate ion (CrO_4^{2-}) and the oxalate ion ($C_2O_4^{2-}$). You are given 0.01 M solutions (A, B, C, D) of four water-soluble salts:

Solution	Solute	Color of Solution
A	Na_2CrO_4	Yellow
B	$(NH_4)_2C_2O_4$	Colorless
C	$AgNO_3$	Colorless
D	$CaCl_2$	Colorless

When these solutions are mixed, the following observations are made:

Expt Number	Solutions Mixed	Result
1	A + B	No precipitate, yellow solution
2	A + C	Red precipitate forms
3	A + D	No precipitate, yellow solution
4	B + C	White precipitate forms
5	B + D	White precipitate forms
6	C + D	White precipitate forms

(a) Write a net ionic equation for the reaction that occurs in each of the experiments. **(b)** Identify the precipitate formed, if any, in each of the experiments. **(c)** Based on these limited observations, which ion tends to form the more soluble salts, chromate or oxalate?

4.95 Antacids are often used to relieve pain and promote healing in the treatment of mild ulcers. Write balanced net ionic equations for the reactions between the HCl(aq) in the stomach and each of the following substances used in various antacids: **(a)** $Al(OH)_3(s)$, **(b)** $Mg(OH)_2(s)$, **(c)** $MgCO_3(s)$, **(d)** $NaAl(CO_3)(OH)_2(s)$, **(e)** $CaCO_3(s)$.

[4.96] Salts of the sulfite ion, SO_3^{2-}, react with acids in a way similar to that of carbonates. **(a)** Predict the chemical formula, and name the weak acid that forms when the sulfite ion reacts with acids. **(b)** The acid formed in part (a) decomposes to form water and a gas. Predict the molecular formula, and name the gas formed. **(c)** Use a source book such as the *CRC Handbook of Chemistry and Physics* to confirm that the substance in part (b) is a gas under normal room-temperature conditions. **(d)** Write balanced net ionic equations of the reaction of HCl(aq) with (i) $Na_2SO_3(aq)$, (ii) $Ag_2SO_3(s)$, (iii) $KHSO_3(s)$, and (iv) $ZnSO_3(aq)$.

[4.97] The commercial production of nitric acid involves the following chemical reactions:

$$4\,NH_3(g) + 5\,O_2(g) \longrightarrow 4\,NO(g) + 6\,H_2O(g)$$
$$2\,NO(g) + O_2(g) \longrightarrow 2\,NO_2(g)$$
$$3\,NO_2(g) + H_2O(l) \longrightarrow 2\,HNO_3(aq) + NO(g)$$

(a) Which of these reactions are redox reactions? **(b)** In each redox reaction identify the element undergoing oxidation and the element undergoing reduction.

4.98 Use Table 4.5 to predict which of the following ions can be reduced to their metal forms by reacting with zinc: **(a)** $Na^+(aq)$, **(b)** $Pb^{2+}(aq)$, **(c)** $Mg^{2+}(aq)$, **(d)** $Fe^{2+}(aq)$, **(e)** $Cu^{2+}(aq)$, **(f)** $Al^{3+}(aq)$. Write the balanced net ionic equation for each reaction that occurs.

[4.99] Lanthanum metal forms cations with a charge of 3+. Consider the following observations about the chemistry of lanthanum: When lanthanum metal is exposed to air, a white solid (compound A) is formed that contains lanthanum and one other element. When lanthanum metal is added to water, gas bubbles are observed and a different white solid (compound B) is formed. Both A and B dissolve in hydrochloric acid to give a clear solution. When either of these solutions is evaporated, a soluble white solid (compound C) remains. If compound C is dissolved in water and sulfuric acid is added, a white precipitate (compound D) forms. **(a)** Propose identities for the substances A, B, C, and D. **(b)** Write net ionic equations for all the reactions described. **(c)** Based on the preceding observations, what can be said about the position of lanthanum in the activity series (Table 4.5)?

4.100 A 35.0-mL sample of 1.00 M KBr and a 60.0-mL sample of 0.600 M KBr are mixed. The solution is then heated to evaporate water until the total volume is 50.0 mL. What is the molarity of the KBr in the final solution?

4.101 Using modern analytical techniques, it is possible to detect sodium ions in concentrations as low as 50 pg/mL. What is this detection limit expressed in **(a)** molarity of Na^+, **(b)** Na^+ ions per cubic centimeter?

4.102 Hard water contains Ca^{2+}, Mg^{2+}, and Fe^{2+}, which interfere with the action of soap and leave an insoluble coating on the insides of containers and pipes when heated. Water softeners replace these ions with Na^+. If 1500 L of hard water contains 0.020 M Ca^{2+} and 0.0040 M Mg^{2+}, how many moles of Na^+ are needed to replace these ions?

4.103 Tartaric acid, $H_2C_4H_4O_6$, has two acidic hydrogens. The acid is often present in wines and precipitates from solution as the wine ages. A solution containing an unknown concentration of the acid is titrated with NaOH. It requires 24.65 mL of 0.2500 M NaOH solution to titrate both acidic protons in 50.00 mL of the tartaric acid solution. Write a balanced net ionic equation for the neutralization reaction, and calculate the molarity of the tartaric acid solution.

4.104 The concentration of hydrogen peroxide in a solution is determined by titrating a 10.0-mL sample of the solution with permanganate ion.

$$2\,MnO_4^-(aq) + 5\,H_2O_2(aq) + 6\,H^+(aq) \longrightarrow$$
$$2\,Mn^{2+}(aq) + 5\,O_2(g) + 8\,H_2O(l)$$

If it takes 14.8 mL of 0.134 M MnO_4^- solution to reach the equivalence point, what is the molarity of the hydrogen peroxide solution?

[4.105] A solid sample of $Zn(OH)_2$ is added to 0.350 L of 0.500 M aqueous HBr. The solution that remains is still acidic. It is then titrated with 0.500 M NaOH solution, and it takes 88.5 mL of the NaOH solution to reach the equivalence point. What mass of $Zn(OH)_2$ was added to the HBr solution?

INTEGRATIVE EXERCISES

4.106 **(a)** By titration, 15.0 mL of 0.1008 M sodium hydroxide is needed to neutralize a 0.2053-g sample of an organic acid. What is the molar mass of the acid if it is monoprotic? **(b)** An elemental analysis of the acid indicates that it is composed of 5.89% H, 70.6% C, and 23.5% O by mass. What is its molecular formula?

4.107 A 3.455-g sample of a mixture was analyzed for barium ion by adding a small excess of sulfuric acid to an aqueous solution of the sample. The resultant reaction produced a precipitate of barium sulfate, which was collected by filtration, washed, dried, and weighed. If 0.2815 g of barium sulfate was obtained, what was the mass percentage of barium in the sample?

[4.108] A tanker truck carrying 5.0×10^3 kg of concentrated sulfuric acid solution tips over and spills its load. If the sulfuric acid is 95.0% H_2SO_4 by mass and has a density of 1.84 g/mL, how many kilograms of sodium carbonate must be added to neutralize the acid?

4.109 A sample of 5.53 g of $Mg(OH)_2$ is added to 25.0 mL of 0.200 M HNO_3. **(a)** Write the chemical equation for the reaction that occurs. **(b)** Which is the limiting reactant in the reaction? **(c)** How many moles of $Mg(OH)_2$, HNO_3, and $Mg(NO_3)_2$ are present after the reaction is complete?

4.110 A sample of 1.50 g of lead(II) nitrate is mixed with 125 mL of 0.100 M sodium sulfate solution. **(a)** Write the chemical equation for the reaction that occurs. **(b)** Which is the limiting reactant in the reaction? **(c)** What are the concentrations of all ions that remain in solution after the reaction is complete?

4.111 A mixture contains 76.5% NaCl, 6.5% $MgCl_2$, and 17.0% Na_2SO_4 by mass. What is the molarity of Cl^- ions in a solution formed by dissolving 7.50 g of the mixture in enough water to form 500.0 mL of solution?

[4.112] The average concentration of bromide ion in seawater is 65 mg of bromide ion per kg of seawater. What is the molarity of the bromide ion if the density of the seawater is 1.025 g/mL?

[4.113] The mass percentage of chloride ion in a 25.00-mL sample of seawater was determined by titrating the sample with silver nitrate, precipitating silver chloride. It took 42.58 mL of 0.2997 M silver nitrate solution to reach the equivalence point in the titration. What is the mass percentage of chloride ion in the seawater if its density is 1.025 g/mL?

4.114 The arsenic in a 1.22-g sample of a pesticide was converted to AsO_4^{3-} by suitable chemical treatment. It was then titrated using Ag^+ to form Ag_3AsO_4 as a precipitate. **(a)** What is the oxidation state of As in AsO_4^{3-}? **(b)** Name Ag_3AsO_4 by analogy to the corresponding compound containing phosphorus in place of arsenic. **(c)** If it took 25.0 mL of 0.102 M Ag^+ to reach the equivalence point in this titration, what is the mass percentage of arsenic in the pesticide?

[4.115] The newest U.S. standard for arsenate in drinking water, mandated by the Safe Drinking Water Act, required that by January 2006, public water supplies must contain no greater than 10 parts per billion (ppb) arsenic. If this arsenic is present as arsenate, AsO_4^{3-}, what mass of sodium arsenate would be present in a 1.00-L sample of drinking water that just meets the standard?

[4.116] The safe drinking water standard for arsenic (which is usually found as arsenate, see 4.115) is 50 parts per billion (ppb) in most developing countries. **(a)** How many grams of sodium arsenate are in 55 gallons of water, if the concentration of arsenate is 50 ppb? **(b)** In 1993, naturally occurring arsenic was discovered as a major contaminant in the drinking water across the country of Bangladesh. Approximately 12 million people in Bangladesh still drink water from wells that have higher concentrations of arsenic than the standard. Recently, a chemistry professor from George Mason University was awarded a $1 million Grainger Challenge Prize for Sustainability for his development of a simple, inexpensive system for filtering naturally occuring arsenic from drinking water. The system uses buckets of sand, cast iron, activated carbon, and wood chips for trapping arsenic-containing minerals. Assuming the efficiency of such a bucket system is 90% (meaning, 90% of the arsenic that comes in is retained in the bucket and 10% passes out of the bucket), how many times should water that is 500 ppb in arsenic be passed through to meet the 50 ppb standard?

[4.117] Federal regulations set an upper limit of 50 parts per million (ppm) of NH_3 in the air in a work environment [that is, 50 molecules of $NH_3(g)$ for every million molecules in the air]. Air from a manufacturing operation was drawn through a solution containing 1.00×10^2 mL of 0.0105 M HCl. The NH_3 reacts with HCl as follows:

$$NH_3(aq) + HCl(aq) \longrightarrow NH_4Cl(aq)$$

After drawing air through the acid solution for 10.0 min at a rate of 10.0 L/min, the acid was titrated. The remaining acid needed 13.1 mL of 0.0588 M NaOH to reach the equivalence point. **(a)** How many grams of NH_3 were drawn into the acid solution? **(b)** How many ppm of NH_3 were in the air? (Air has a density of 1.20 g/L and an average molar mass of 29.0 g/mol under the conditions of the experiment.) **(c)** Is this manufacturer in compliance with regulations?

5 THERMOCHEMISTRY

WILDFLOWERS. Plants use solar energy to carry out photosynthesis of carbohydrates, which provide the energy needed for plant growth. Among the molecules produced by plants are O_2 and ethylene, C_2H_4.

WHAT'S AHEAD

5.1 The Nature of Energy
We begin by considering the nature of *energy* and the forms it takes, notably *kinetic energy* and *potential energy*. We will also discuss the units used in measuring energy and the fact that energy can be used to accomplish *work* or to transfer *heat*. To study energy changes, we focus on a particular part of the universe, which we call the *system*. Everything else is called the *surroundings*.

5.2 The First Law of Thermodynamics
We will then explore the *first law of thermodynamics*: Energy cannot be created or destroyed, but it can be transformed from one form to another or transferred between systems and surroundings. The energy possessed by a system is called its *internal energy*. Internal energy, *E*, is a *state function*, a quantity whose value depends only on the state, or condition, of a system now, not on how the system came to be in that state.

5.3 Enthalpy
Next, we will encounter a state function called *enthalpy*, *H*, that relates to energy. This function is useful because the change in enthalpy, ΔH, measures the quantity of heat energy gained or lost by a system in a process occurring under a constant pressure.

5.4 Enthalpies of Reaction
We will see that the enthalpy change associated with a chemical reaction (ΔH_{rxn}) is given by the enthalpies of the products minus the enthalpies of the reactants. This quantity is directly proportional to the amount of reactant consumed in the process.

5.5 Calorimetry
We will next examine *calorimetry*, which is an experimental technique that we use to measure heat changes in chemical processes.

5.6 Hess's Law
We will observe that the enthalpy change for a reaction can be calculated using appropriate enthalpy changes for other reactions. To do so, we apply *Hess's law*, which uses the fact that enthalpy, like energy, is a state function.

5.7 Enthalpies of Formation
Then we will discuss how to establish standard values for enthalpy changes in chemical reactions and how to use them to calculate enthalpy changes for reactions (ΔH_{rxn}).

5.8 Foods and Fuels
Finally, we will examine foods and fuels as sources of energy and discuss some related health and social issues.

MODERN SOCIETY DEPENDS ON ENERGY for its existence. Energy is used to drive our machinery and appliances, to power our transportation vehicles, and to keep us warm in the winter and cool in the summer. It is not just modern society, however, that depends on energy. Energy is

necessary for all life. Plants, such as those in the chapter-opening photograph, use solar energy to carry out photosynthesis, allowing the plants to grow. The plants, in turn, provide food, from which we humans derive the energy we need to move, to maintain body temperature, and to carry out bodily functions. What exactly is energy, though, and what principles are involved in its many transactions and transformations, such as those from the sun to plants to animals?

In this chapter we begin to explore energy and its changes. We are motivated partly by the fact that energy changes invariably accompany chemical reactions. Indeed, sometimes we use a chemical reaction specifically to obtain energy, as when we burn fuels. Thus, energy is very much a chemical topic. Nearly all of the energy on which we depend is derived from chemical reactions, whether those reactions are associated with the combustion of fuels, the discharge of a battery, or the metabolism of our foods. If we are to properly understand chemistry, we must also understand the energy changes that accompany chemical reactions.

The study of energy and its transformations is known as **thermodynamics** (Greek: *thérme-*, "heat"; *dy'namis*, "power"). In this chapter we will examine an aspect of thermodynamics that involves the relationships between chemical reactions and energy changes involving heat. This portion of thermodynamics is called **thermochemistry**. We will discuss additional aspects of thermodynamics in Chapter 19.

5.1 THE NATURE OF ENERGY

Although the idea of energy is a familiar one, it is a bit challenging to deal with the concept in a precise way. **Energy** is commonly defined as *the capacity to do work or to transfer heat.* This definition requires us to understand the concepts of work and heat. We can think of **work** as *the energy used to cause an object with mass to move against a force* and **heat** as *the energy used to cause the temperature of an object to increase* (Figure 5.1 ◀). We will consider each of these concepts more closely to give them fuller meaning. Let's begin by examining the ways in which matter can possess energy and how that energy can be transferred from one piece of matter to another.

Kinetic Energy and Potential Energy

Objects, whether they are tennis balls or molecules, can possess **kinetic energy**, the energy of *motion*. The magnitude of the kinetic energy, E_k, of an object depends on its mass, m, and speed, v:

$$E_k = \tfrac{1}{2} mv^2 \qquad [5.1]$$

Equation 5.1 shows that the kinetic energy increases as the speed of an object increases. For example, a car moving at 55 miles per hour (mph) has greater kinetic energy than it does at 40 mph. For a given speed the kinetic energy increases with increasing mass. Thus, a large sport-utility vehicle traveling at 55 mph has greater kinetic energy than a small sedan traveling at the same speed because the SUV has greater mass than the sedan. Atoms and molecules have mass and are in motion. They therefore possess kinetic energy.

All other kinds of energy—the energy stored in chemical bonds, the energy of attraction of north and south poles of magnets, for example—are potential energy. An object can possess **potential energy** by virtue of its *position* relative to other objects. Potential energy arises when a force operates on an object. A **force** is any kind of push or pull exerted on an object. The most familiar force is the pull of gravity. Think of a cyclist poised at the top of a hill, as illustrated in Figure 5.2 ▼. Gravity acts upon her and her bicycle, exerting a force directed toward the center of Earth. At the top of the hill the cyclist and her bicycle possess a certain potential energy by virtue of their elevation. The potential energy, E_p,

(a)

(b)

▲ **Figure 5.1 Work and heat.** Energy can be used to achieve two basic types of tasks: (a) Work is energy used to cause an object with mass to move. (b) Heat is energy used to cause the temperature of an object to increase.

▶ **Figure 5.2 Potential energy and kinetic energy.** (a) A bicycle at the top of a hill has a high potential energy relative to the bottom of the hill. (b) As the bicycle proceeds down the hill, the potential energy is converted into kinetic energy.

(a)   (b)

is given by the equation $E_p = mgh$, where m is the mass of the object in question (in this case the cyclist and her bicycle), h is the height of the object relative to some reference height, and g is the gravitational constant, 9.8 m/s^2. Once in motion, without any further effort on her part, the cyclist gains speed as the bicycle rolls down the hill. Her potential energy decreases as she moves downward, but the energy does not simply disappear. It is converted to other forms of energy, principally kinetic energy, the energy of motion. This example illustrates that forms of energy are interconvertible.

Gravity is an important force for large objects, such as the cyclist and Earth. Chemistry, however, deals mostly with extremely small objects—atoms and molecules. Gravitational forces play a negligible role in the ways these submicroscopic objects interact with one another. Forces that arise from electrical charges are more important when dealing with atoms and molecules.

One of the most important forms of potential energy in chemistry is *electrostatic potential energy*, which arises from the interactions between charged particles. The electrostatic potential energy, E_{el}, is proportional to the electrical charges on the two interacting objects, Q_1 and Q_2, and is inversely proportional to the distance separating them:

$$E_{el} = \frac{\kappa Q_1 Q_2}{d} \qquad [5.2]$$

Here κ is simply a constant of proportionality, $8.99 \times 10^9 \text{ J-m/C}^2$. (C is the coulomb, a unit of electrical charge ∞ (Section 2.2), and J is the joule, a unit of energy that we will soon discuss.) When dealing with molecular-level objects, the electrical charges Q_1 and Q_2 are typically on the order of magnitude of the charge of the electron (1.60×10^{-19} C). When Q_1 and Q_2 have the same sign (for example, both are positive), the two charges repel one another, pushing them apart; E_{el} is positive. When they have opposite signs, they attract one another, pulling them toward each other; E_{el} is negative. The lower the energy of a system, the more stable it is. Thus, the more strongly opposite charges interact, the more stable the system.

One of our goals in chemistry is to relate the energy changes that we see in our macroscopic world to the kinetic or potential energy of substances at the atomic or molecular level. Many substances—fuels, for example—release energy when they react. The *chemical energy* of these substances is due to the potential energy stored in the arrangements of their atoms. Likewise, we will see that the energy a substance possesses because of its temperature (its *thermal energy*) is associated with the kinetic energy of the molecules in the substance.

GIVE IT SOME THOUGHT

What are the terms for the energy an object possesses **(a)** because of its motion, **(b)** because of its position? What terms are used to describe changes of energy associated with **(c)** temperature changes, **(d)** moving an object against a force?

Units of Energy

The SI unit for energy is the **joule** (pronounced "jool"), J, in honor of James Joule (1818–1889), a British scientist who investigated work and heat: $1 \text{ J} = 1 \text{ kg-m}^2/\text{s}^2$. A mass of 2 kg moving at a speed of 1 m/s possesses a kinetic energy of 1 J:

$$E_k = \tfrac{1}{2} mv^2 = \tfrac{1}{2}(2 \text{ kg})(1 \text{ m/s})^2 = 1 \text{ kg-m}^2/\text{s}^2 = 1 \text{ J}$$

A joule is not a large amount of energy, and we will often use *kilojoules* (kJ) in discussing the energies associated with chemical reactions.

Traditionally, energy changes accompanying chemical reactions have been expressed in calories, a non-SI unit still widely used in chemistry, biology, and biochemistry. A **calorie** (cal) was originally defined as the amount of energy required to raise the temperature of 1 g of water from 14.5 °C to 15.5 °C. A calorie is now defined in terms of the joule:

$$1 \text{ cal} = 4.184 \text{ J (exactly)}$$

A related energy unit used in nutrition is the nutritional *Calorie* (note that this unit is capitalized): 1 Cal = 1000 cal = 1 kcal.

System and Surroundings

When we analyze energy changes, we need to focus our attention on a limited and well-defined part of the universe to keep track of the energy changes that occur. The portion we single out for study is called the **system**; everything else is called the **surroundings**. When we study the energy change that accompanies a chemical reaction in the laboratory, the reactants and products constitute the system. The container and everything beyond it are considered the surroundings.

Systems may be open, closed, or isolated. An *open* system is one in which matter and energy can be exchanged with the surroundings. A boiling pot of water on a stove, without its lid, is an open system: heat comes into the system from the stove, and water is released to the surroundings as steam.

The systems we can most readily study in thermochemistry are called *closed systems*. A closed system can exchange energy but not matter with its surroundings. For example, consider a mixture of hydrogen gas, H_2, and oxygen gas, O_2, in a cylinder, as illustrated in Figure 5.3 ◄. The system in this case is just the hydrogen and oxygen; the cylinder, piston, and everything beyond them (including us) are the surroundings. If the hydrogen and oxygen react to form water, energy is liberated:

$$2 H_2(g) + O_2(g) \longrightarrow 2 H_2O(g) + \text{energy}$$

Although the chemical form of the hydrogen and oxygen atoms in the system is changed by this reaction, the system has not lost or gained mass; it undergoes no exchange of matter with its surroundings. However, it can exchange energy with its surroundings in the form of *work* and *heat*.

An *isolated* system is one in which neither energy nor matter can be exchanged with the surroundings. An insulated thermos containing hot coffee approximates an isolated system. We know, however, that the coffee eventually cools, so it is not perfectly isolated.

▲ **Figure 5.3 A closed system and its surroundings.** Hydrogen and oxygen gases are confined in a cylinder with a movable piston. If we are interested only in the properties of these gases, the gases are the system and the cylinder and piston are part of the surroundings. Because the system can exchange energy (in the form of heat and work) but not matter with its surroundings, it is a closed system.

▲ GIVE IT SOME THOUGHT

Is a human being an isolated, closed, or open system? Explain your choice.

Transferring Energy: Work and Heat

Figure 5.1 illustrates the two ways that we experience energy changes in our everyday lives—in the form of work or heat. In Figure 5.1(a) energy is transferred from the tennis racquet to the ball, changing the direction and speed of the ball's movement. In Figure 5.1(b) energy is transferred in the form of heat. Indeed, energy is transferred between systems and surroundings in two general ways, as work or heat.

Energy used to cause an object to move against a force is called *work*. Thus, we can define work, w, as the energy transferred when a force moves an object. The magnitude of this work equals the product of the force, F, and the distance, d, that the object is moved:

$$w = F \times d \qquad [5.3]$$

We perform work, for example, when we lift an object against the force of gravity or when we bring two like charges closer together. If we define the object as the system, then we—as part of the surroundings—are performing work on that system, transferring energy to it.

The other way in which energy is transferred is as heat. *Heat* is the energy transferred from a hotter object to a colder one. Or stating this idea in a slightly more abstract but nevertheless useful way, heat is the energy transferred between a system and its surroundings because of their difference in temperature. A combustion reaction, such as the burning of natural gas illustrated in Figure 5.1(b), releases the chemical energy stored in the molecules of the fuel. ∞ (Section 3.2) If we define the substances involved in the reaction as the system and everything else as the surroundings, we find that the released energy causes the temperature of the system to increase. Energy in the form of heat is then transferred from the hotter system to the cooler surroundings.

■ SAMPLE EXERCISE 5.1 | Describing and Calculating Energy Changes

A bowler lifts a 5.4-kg (12-lb) bowling ball from ground level to a height of 1.6 m (5.2 feet) and then drops the ball back to the ground. **(a)** What happens to the potential energy of the bowling ball as it is raised from the ground? **(b)** What quantity of work, in J, is used to raise the ball? **(c)** After the ball is dropped, it gains kinetic energy. If we assume that all of the work done in part (b) has been converted to kinetic energy by the time the ball strikes the ground, what is the speed of the ball at the instant just before it hits the ground? (Note: The force due to gravity is $F = m \times g$, where m is the mass of the object and g is the gravitational constant; $g = 9.8$ m/s^2.)

SOLUTION

Analyze: We need to relate the potential energy of the bowling ball to its position relative to the ground. We then need to establish the relationship between work and the change in potential energy of the ball. Finally, we need to connect the change in potential energy when the ball is dropped with the kinetic energy attained by the ball.

Plan: We can calculate the work done in lifting the ball by using Equation 5.3: $w = F \times d$. The kinetic energy of the ball at the moment of impact equals its initial potential energy. We can use the kinetic energy and Equation 5.1 to calculate the speed, v, at impact.

Solve:

(a) Because the bowling ball is raised to a greater height above the ground, its potential energy increases.

(b) The ball has a mass of 5.4 kg, and it is lifted a distance of 1.6 m. To calculate the work performed to raise the ball, we use both Equation 5.3 and $F = m \times g$ for the force that is due to gravity:

$$w = F \times d = m \times g \times d = (5.4 \text{ kg})(9.8 \text{ m/s}^2)(1.6 \text{ m}) = 85 \text{ kg-m}^2/\text{s}^2 = 85 \text{ J}$$

Thus, the bowler has done 85 J of work to lift the ball to a height of 1.6 m.

(c) When the ball is dropped, its potential energy is converted to kinetic energy. At the instant just before the ball hits the ground, we assume that the kinetic energy is equal to the work done in part (b), 85 J:

$$E_k = \tfrac{1}{2} mv^2 = 85 \text{ J} = 85 \text{ kg-m}^2/\text{s}^2$$

We can now solve this equation for v:

$$v^2 = \left(\frac{2E_k}{m}\right) = \left(\frac{2(85 \text{ kg-m}^2/\text{s}^2)}{5.4 \text{ kg}}\right) = 31.5 \text{ m}^2/\text{s}^2$$

$$v = \sqrt{31.5 \text{ m}^2/\text{s}^2} = 5.6 \text{ m/s}$$

Check: Work must be done in part (b) to increase the potential energy of the ball, which is in accord with our experience. The units are appropriate in both parts (b) and (c). The work is in units of J and the speed in units of m/s. In part (c) we have carried an additional digit in the intermediate calculation involving the square root, but we report the final value to only two significant figures, as appropriate.

Comment: A speed of 1 m/s is roughly 2 mph, so the bowling ball has a speed greater than 10 mph upon impact.

■ PRACTICE EXERCISE

What is the kinetic energy, in J, of **(a)** an Ar atom moving with a speed of 650 m/s, **(b)** a mole of Ar atoms moving with a speed of 650 m/s (Hint: 1 amu = 1.66×10^{-27} kg)
Answers: **(a)** 1.4×10^{-20} J, **(b)** 8.4×10^3 J

5.2 THE FIRST LAW OF THERMODYNAMICS

We have seen that the potential energy of a system can be converted into kinetic energy, and vice versa. We have also seen that energy can be transferred back and forth between a system and its surroundings in the forms of work and heat. In general, energy can be converted from one form to another, and it can be transferred from one place to another. All of these transactions proceed in accord with one of the most important observations in science—that energy can be neither created nor destroyed. A simple statement known as the **first law of thermodynamics** summarizes this simple truth: *Energy is conserved.* Any energy that is lost by the system must be gained by the surroundings, and vice versa. To apply the first law of thermodynamics quantitatively, we must first define the energy of a system more precisely.

Internal Energy

The **internal energy** of a system is the sum of *all* the kinetic and potential energies of all its components. For the system in Figure 5.3, for example, the internal energy includes the motions of the H_2 and O_2 molecules through space, their rotations, and their internal vibrations. It also includes the energies of the nuclei of each atom and of the component electrons. We represent the internal energy with the symbol E. (Some texts, particularly more advanced ones, use the symbol U.) We generally do not know the actual numerical value of E. What we can hope to know, however, is ΔE (read "delta E"),* the change in E that accompanies a change in the system.

Imagine that we start with a system with an initial internal energy, $E_{initial}$. The system then undergoes a change, which might involve work being done or heat being transferred. After the change, the final internal energy of the system is E_{final}. We define the *change* in internal energy, ΔE, as the difference between E_{final} and $E_{initial}$:

$$\Delta E = E_{final} - E_{initial} \qquad [5.4]$$

To apply the first law of thermodynamics, we need only the value of ΔE. We do not really need to know the actual values of E_{final} or $E_{initial}$ for the system.

Thermodynamic quantities such as ΔE have three parts: (1) a number and (2) a unit, that together give the magnitude of the change, and (3) a sign that gives the direction. A *positive* value of ΔE results when $E_{final} > E_{initial}$, indicating the system has gained energy from its surroundings. A *negative* value of ΔE is obtained when $E_{final} < E_{initial}$, indicating the system has lost energy to its surroundings. Notice that we are taking the point of view of the system rather than that of the surroundings in discussing the energy changes. We need to remember, however, that any change in the energy of the system is accompanied by an opposite change in the energy of the surroundings. These features of energy changes are summarized in Figure 5.4▼.

▼ **Figure 5.4 Changes in internal energy.** (a) When a system loses energy, that energy is released to the surroundings. The loss of energy is represented by an arrow that points downward between the initial and final states of the system. In this case, the energy change of the system, $\Delta E = E_{final} - E_{initial}$, is negative. (b) When a system gains energy, that energy is gained from the surroundings. In this case, the gain of energy is represented by an arrow that points upward between the initial and final states of the system, and the energy change of the system is positive. Notice in both (a) and (b) that the vertical arrow originates at the initial state and has its head at the final state.

The symbol Δ is commonly used to denote change. For example, a change in height, h, can be represented by Δh.

In a chemical reaction, the initial state of the system refers to the reactants, and the final state refers to the products. When hydrogen and oxygen form water at a given temperature, the system loses energy to the surroundings. Because energy is lost from the system, the internal energy of the products (final state) is less than that of the reactants (initial state), and ΔE for the process is negative. Thus, the *energy diagram* in Figure 5.5▶ shows that the internal energy of the mixture of H_2 and O_2 is greater than that of H_2O.

▲ **Figure 5.5   Energy diagram for the interconversion of $H_2(g)$, $O_2(g)$, and $H_2O(l)$.** A system composed of $H_2(g)$ and $O_2(g)$ has a greater internal energy than one composed of $H_2O(l)$. The system loses energy ($\Delta E < 0$) when H_2 and O_2 are converted to H_2O. It gains energy ($\Delta E > 0$) when H_2O is decomposed into H_2 and O_2.

> ### GIVE IT SOME THOUGHT
>
> The internal energy for $Mg(s)$ and $Cl_2(g)$ is greater than that of $MgCl_2(s)$. Sketch an energy diagram that represents the reaction $MgCl_2(s) \longrightarrow Mg(s) + Cl_2(g)$.

Relating ΔE to Heat and Work

As we noted in Section 5.1, a system may exchange energy with its surroundings as heat or as work. The internal energy of a system changes in magnitude as heat is added to or removed from the system or as work is done on it or by it. If we think of internal energy as the system's bank account of energy, we see that deposits or withdrawals can be made either in terms of heat or in terms of work. Deposits increase the energy of the system (positive ΔE), whereas withdrawals decrease the energy of the system (negative ΔE).

We can use these ideas to write a very useful algebraic expression of the first law of thermodynamics. When a system undergoes any chemical or physical change, the magnitude and sign of the accompanying change in internal energy, ΔE, is given by the heat added to or liberated from the system, q, plus the work done on or by the system, w:

$$\Delta E = q + w \qquad [5.5]$$

When heat is added to a system or work is done on a system, its internal energy increases. Therefore, when heat is transferred to the system from the surroundings, q has a positive value. Adding heat to the system is like making a deposit to the energy account—the total amount of energy goes up. Likewise, when work is done on the system by the surroundings, w has a positive value (Figure 5.6▶). Work also is a deposit, increasing the internal energy of the system. Conversely, both the heat lost by the system to the surroundings and the work done by the system on the surroundings have negative values; that is, they lower the internal energy of the system. They are energy withdrawals and as a result, lower the total amount of energy in the energy account. The sign conventions for q, w, and ΔE are summarized in Table 5.1▼. Notice that any energy entering the system as either heat or work carries a positive sign.

▲ **Figure 5.6   Sign conventions for heat and work.** Heat, q, gained by a system and work, w, done on a system are both positive quantities. Both increase the internal energy, E, of the system, causing ΔE to be a positive quantity.

TABLE 5.1 ▪ Sign Conventions for q, w, and ΔE

For q	+ means system *gains* heat	− means system *loses* heat
For w	+ means work done *on* system	− means work done *by* system
For ΔE	+ means *net gain* of energy by system	− means *net loss* of energy by system

---

■ **SAMPLE EXERCISE 5.2** | **Relating Heat and Work to Changes of Internal Energy**

Two gases, $A(g)$ and $B(g)$, are confined in a cylinder-and-piston arrangement like that in Figure 5.3. Substances A and B react to form a solid product: $A(g) + B(g) \rightarrow C(s)$. As the reaction occurs, the system loses 1150 J of heat to the surroundings. The piston moves downward as the gases react to form a solid. As the volume of the gas decreases under the constant pressure of the atmosphere, the surroundings do 480 J of work on the system. What is the change in the internal energy of the system?

SOLUTION

Analyze: The question asks us to determine ΔE, given information about q and w.

Plan: We first determine the signs of q and w (Table 5.1) and then use Equation 5.5, $\Delta E = q + w$, to calculate ΔE.

Solve: Heat is transferred from the system to the surroundings, and work is done on the system by the surroundings, so q is negative and w is positive: $q = -1150$ J and $w = 480$ kJ. Thus,

$$\Delta E = q + w = (-1150 \text{ J}) + (480 \text{ J}) = -670 \text{ J}$$

The negative value of ΔE tells us that a net quantity of 670 J of energy has been transferred from the system to the surroundings.

Comment: You can think of this change as a decrease of 670 J in the net value of the system's energy bank account (hence the negative sign); 1150 J is withdrawn in the form of heat, while 480 J is deposited in the form of work. Notice that as the volume of the gases decreases, work is being done on the system by the surroundings, resulting in a deposit of energy.

■ PRACTICE EXERCISE

Calculate the change in the internal energy of the system for a process in which the system absorbs 140 J of heat from the surroundings and does 85 J of work on the surroundings.
Answer: +55 J

(a)

(b)

▲ **Figure 5.7 Examples of endothermic and exothermic reactions.** (a) When ammonium thiocyanate and barium hydroxide octahydrate are mixed at room temperature, an endothermic reaction occurs: $2 \text{ NH}_4\text{SCN}(s) + \text{Ba(OH)}_2 \cdot 8 \text{ H}_2\text{O}(s) \longrightarrow \text{Ba(SCN)}_2(aq) + 2 \text{ NH}_3(aq) + 10 \text{ H}_2\text{O}(l)$. As a result, the temperature of the system drops from about 20 °C to −9 °C. (b) The reaction of powdered aluminum with Fe_2O_3 (the thermite reaction) is highly exothermic. The reaction proceeds vigorously to form Al_2O_3 and molten iron: $2 \text{ Al}(s) + \text{Fe}_2\text{O}_3(s) \longrightarrow \text{Al}_2\text{O}_3(s) + 2 \text{ Fe}(l)$.

Endothermic and Exothermic Processes

When a process occurs in which the system absorbs heat, the process is called **endothermic**. (*Endo-* is a prefix meaning "into.") During an endothermic process, such as the melting of ice, heat flows *into* the system from its surroundings. If we, as part of the surroundings, touch a container in which ice is melting, it feels cold to us because heat has passed from our hands to the container.

A process in which the system loses heat is called **exothermic**. (*Exo-* is a prefix meaning "out of.") During an exothermic process, such as the combustion of gasoline, heat *exits* or flows *out* of the system and into the surroundings. Figure 5.7 ◄ shows two examples of chemical reactions: one endothermic and the other highly exothermic. In the endothermic process shown in Figure 5.7(a), the temperature in the beaker decreases. In this example the system consists of the chemical reactants and products. The solvent in which they are dissolved is part of the surroundings. Heat flows from the solvent, as part of the surroundings, into the system as reactants are converted to products. Thus, the temperature of the solution drops.

▲ **GIVE IT SOME THOUGHT**

Using Figure 5.5 as a reference, indicate whether the reaction $2 \text{ H}_2\text{O}(l) \longrightarrow 2 \text{ H}_2(g) + \text{O}_2(g)$ is exothermic or endothermic. What feature(s) of the figure indicate whether the reaction is exothermic or endothermic?

State Functions

Although we usually have no way of knowing the precise value of the internal energy of a system, E, it does have a fixed value for a given set of conditions. The conditions that influence internal energy include the temperature and pressure. Furthermore, the total internal energy of a system is proportional to the total quantity of matter in the system because energy is an extensive property. ∞ (Section 1.3)

Suppose we define our system as 50 g of water at 25 °C, as in Figure 5.8 ►. The system could have arrived at this state by cooling 50 g of water from 100 °C or by melting 50 g of ice and subsequently warming the water to 25 °C.

50 g
H_2O (*l*)
100 °C

Cooling →

50 g
H_2O (*l*)
25 °C

← Heating

50 g
H_2O (*s*)
0 °C

◀ **Figure 5.8 Internal energy, *E*, a state function.** *E* depends only on the present state of the system and not on the path by which it arrived at that state. The internal energy of 50 g of water at 25 °C is the same whether the water is cooled from a higher temperature to 25 °C or warmed from a lower temperature to 25 °C.

The internal energy of the water at 25 °C is the same in either case. Internal energy is an example of a **state function**, a property of a system that is determined by specifying the system's condition, or state (in terms of temperature, pressure, and so forth). *The value of a state function depends only on the present state of the system, not on the path the system took to reach that state.* Because *E* is a state function, Δ*E* depends only on the initial and final states of the system, not on how the change occurs.

An analogy may help to explain the difference between quantities that are state functions and those that are not. Suppose you are traveling between Chicago and Denver. Chicago is 596 ft above sea level; Denver is 5280 ft above sea level. No matter what route you take, the altitude change will be 4684 ft. The distance you travel, however, will depend on your route. Altitude is analogous to a state function because the change in altitude is independent of the path taken. Distance traveled is not a state function.

Some thermodynamic quantities, such as *E*, are state functions. Other quantities, such as *q* and *w*, are not. Although Δ*E* = *q* + *w* does not depend on how the change occurs, the specific amounts of heat and work produced depend on the way in which the change is carried out, analogous to the choice of travel route between Chicago and Denver. Nevertheless, if changing the path by which a system goes from an initial state to a final state increases the value of *q*, that path change will also decrease the value of *w* by exactly the same amount. The result is that the value for Δ*E* for the two paths will be the same.

We can illustrate this principle with the example of a flashlight battery as our system. In Figure 5.9 ▶, we consider two possible ways of discharging the battery at constant temperature. If a coil of wire shorts out the battery, no work is accomplished because nothing is moved against a force. All the energy is lost from the battery in the form of heat. (The wire coil will get warmer and release heat to the surrounding air.) On the other hand, if the battery is used to make a small motor turn, the discharge of the battery produces work. Some heat will be released as well, although not as much as when the battery is shorted out. The magnitudes of *q* and *w* are different for these two cases. If the initial and final states of the battery are identical in both cases, however, then Δ*E* = *q* + *w* must be the same in both cases because Δ*E* is a state function. Thus, Δ*E* depends only on the initial and final states of the system, regardless of how the transfers of energy occur in terms of heat and work.

(a)

(b)

Charged battery

Heat

Heat

Work

Energy
Δ*E* lost by
battery

Discharged battery

▲ **Figure 5.9 Internal energy is a state function, but heat and work are not.** The amounts of heat and work transferred between the system and the surroundings depend on the way in which the system goes from one state to another. (a) A battery shorted out by a wire loses energy to the surroundings only as heat; no work is performed by the system. (b) A battery discharged through a motor loses energy as work (to make the fan turn) and also loses energy as heat. Now, however, the amount of heat lost is much less than in (a). The value of Δ*E* is the same for both processes even though the values of *q* and *w* in (a) are different from the values of *q* and *w* in (b).

◣ **GIVE IT SOME THOUGHT**

In what ways is the balance in your checkbook a state function?

5.3 ENTHALPY

The chemical and physical changes that occur around us, such as photosynthesis in the leaves of a plant, the evaporation of water from a lake, or a reaction in an open beaker in a laboratory, occur at essentially constant atmospheric pressure. The changes can result in the release or absorption of heat or can be accompanied by work that is done by or on the system. The heat flow is the easiest change to measure, so we will begin to focus on that aspect of reactions. Nevertheless, we still need to account for any work that accompanies the process.

Most commonly, the only kind of work produced by chemical or physical changes open to the atmosphere is the mechanical work associated with a change in the volume of the system. Consider, for example, the reaction of zinc metal with hydrochloric acid solution:

$$Zn(s) + 2\,H^+(aq) \longrightarrow Zn^{2+}(aq) + H_2(g) \qquad [5.6]$$

If we carry out this reaction in the laboratory hood in an open beaker, we can see the evolution of hydrogen gas, but it may not be so obvious that work is being done. Still, the hydrogen gas that is being produced must expand against the existing atmosphere, which requires the system to do work. We can see this better by conducting the reaction in a closed vessel at constant pressure, as illustrated in Figure 5.10▼. In this apparatus the piston moves up or down to maintain a constant pressure in the reaction vessel. If we assume for simplicity that the piston has no mass, the pressure in the apparatus is the same as the atmospheric pressure outside the apparatus. As the reaction proceeds, H_2 gas forms, and the piston rises. The gas within the flask is thus doing work on the surroundings by lifting the piston against the force of atmospheric pressure that presses down on it.

The work involved in the expansion or compression of gases is called **pressure-volume work** (or P-V work). When the pressure is constant, as in our example, the sign and magnitude of the pressure-volume work is given by

$$w = -P\,\Delta V \qquad [5.7]$$

where P is pressure and ΔV is the change in volume of the system ($\Delta V = V_{final} - V_{initial}$). The negative sign in Equation 5.7 is necessary to conform to the sign conventions given in Table 5.1. Thus, when the volume expands, ΔV is a positive quantity and w is a negative quantity. That is, energy leaves the system as work, indicating that work is done *by* the system *on* the surroundings. On the other hand, when a gas is compressed, ΔV is a negative quantity

▶ Figure 5.10 **A system that does work on its surroundings.** (a) An apparatus for studying the reaction of zinc metal with hydrochloric acid at constant pressure. The piston is free to move up and down in its cylinder to maintain a constant pressure equal to atmospheric pressure inside the apparatus. Notice the pellets of zinc in the L-shaped arm on the left. When this arm is rotated, the pellets will fall into the main container and the reaction will begin. (b) When zinc is added to the acid solution, hydrogen gas is evolved. The hydrogen gas does work on the surroundings, raising the piston against atmospheric pressure to maintain constant pressure inside the reaction vessel.

Piston

H_2 gas plus original atmosphere

Zn

HCl solution

Zn

HCl solution

(a)   (b)

(the volume decreases), which makes w a positive quantity. That is, energy enters the system as work, indicating that work is done on the system by the surroundings. The "A Closer Look" box discusses pressure-volume work in more detail, but all you really need to keep in mind for now is Equation 5.7, which applies to processes occurring at constant pressure. We will take up the properties of gases in more detail in Chapter 10.

GIVE IT SOME THOUGHT

If a system does not change its volume during the course of a process, does it do pressure-volume work?

A thermodynamic function called **enthalpy** (from the Greek word *enthalpein*, meaning "to warm") accounts for heat flow in processes occurring at constant pressure when no forms of work are performed other than P-V work. Enthalpy, which we denote by the symbol H, equals the internal energy plus the product of the pressure and volume of the system:

$$H = E + PV \qquad [5.8]$$

Enthalpy is a state function because internal energy, pressure, and volume are all state functions.

When a change occurs at constant pressure, the change in enthalpy, ΔH, is given by the following relationship:

$$\Delta H = \Delta(E + PV)$$
$$= \Delta E + P\,\Delta V \qquad [5.9]$$

That is, the change in enthalpy equals the change in internal energy plus the product of the constant pressure times the change in volume.

We can gain further insight into enthalpy change by recalling that $\Delta E = q + w$ (Equation 5.5) and that the work involved in the expansion or compression of gases is $w = -P\,\Delta V$. If we substitute $-w$ for $P\,\Delta V$ and $q + w$ for ΔE into Equation 5.9, we have

$$\Delta H = \Delta E + P\,\Delta V = (q_P + w) - w = q_P \qquad [5.10]$$

where the subscript P on the heat, q, emphasizes changes at constant pressure. Thus, *the change in enthalpy equals the heat gained or lost at constant pressure.* Because q_P is something we can either measure or readily calculate and because so many physical and chemical changes of interest to us occur at constant pressure, enthalpy is a more useful function than internal energy. For most reactions the difference in ΔH and ΔE is small because $P\,\Delta V$ is small.

When ΔH is positive (that is, when q_P is positive), the system has gained heat from the surroundings (Table 5.1), which is an endothermic process. When ΔH is negative, the system has released heat to the surroundings, which is an exothermic process. These cases are diagrammed in Figure 5.11▶.

Because H is a state function, ΔH (which equals q_P) depends only on the initial and final states of the system, not on how the change occurs. At first glance this statement might seem to contradict our earlier discussion in Section 5.2, in which we said that q is *not* a state function. There is no contradiction, however, because the relationship between ΔH and heat (q_P) has the special limitations that only P-V work is involved and the pressure is constant.

▲ **Figure 5.11 Endothermic and exothermic processes.** (a) If the system absorbs heat (endothermic process), ΔH will be positive ($\Delta H > 0$). (b) If the system loses heat (exothermic process), ΔH will be negative ($\Delta H < 0$).

GIVE IT SOME THOUGHT

What is the advantage of using enthalpy rather than internal energy to describe energy changes in reactions?

A Closer Look   ENERGY, ENTHALPY, AND *P-V* WORK

In chemistry we are interested mainly in two types of work: electrical work and mechanical work done by expanding gases. We will focus here only on the latter, called pressure-volume, or *P-V*, work. Expanding gases in the cylinder of an automobile engine do *P-V* work on the piston; this work eventually turns the wheels. Expanding gases from an open reaction vessel do *P-V* work on the atmosphere. This work accomplishes nothing in a practical sense, but we must keep track of all work, useful or not, when monitoring the energy changes of a system.

Consider a gas confined to a cylinder with a movable piston of cross-sectional area *A* (Figure 5.12▼). A downward force, *F*, acts on the piston. The *pressure*, *P*, on the gas is the force per area: $P = F/A$. We will assume that the piston is weightless and that the only pressure acting on it is the *atmospheric pressure* that is due to the weight of Earth's atmosphere, which we will assume to be constant.

Suppose the gas in the cylinder expands and the piston moves a distance, Δh. From Equation 5.3, the magnitude of the work done by the system equals the distance moved times the force acting on the piston:

$$\text{Magnitude of work} = \text{force} \times \text{distance} = F \times \Delta h \quad [5.11]$$

We can rearrange the definition of pressure, $P = F/A$, to $F = P \times A$. In addition, the volume change, ΔV, resulting from the movement of the piston, is the product of the cross-sectional area of the piston and the distance it moves: $\Delta V = A \times \Delta h$. Substituting into Equation 5.11,

$$\text{Magnitude of work} = F \times \Delta h = P \times A \times \Delta h$$
$$= P \times \Delta V$$

Because the system (the confined gas) is doing work on the surroundings, the work is a negative quantity:

$$w = -P\,\Delta V \quad [5.12]$$

Now, if *P-V* work is the only work that can be done, we can substitute Equation 5.12 into Equation 5.5 to give

$$\Delta E = q + w = q - P\,\Delta V \quad [5.13]$$

When a reaction is carried out in a constant-volume container ($\Delta V = 0$), the heat transferred equals the change in internal energy:

$$\Delta E = q_V \quad \text{(constant volume)} \quad [5.14]$$

The subscript *V* indicates that the volume is constant.

Most reactions are run under constant-pressure conditions. In this case Equation 5.13 becomes

$$\Delta E = q_P - P\,\Delta V \text{ or}$$
$$q_P = \Delta E + P\,\Delta V \quad \text{(constant pressure)} \quad [5.15]$$

But we see from Equation 5.9 that the right-hand side of Equation 5.15 is just the enthalpy change under constant-pressure conditions. Thus, $\Delta H = q_P$, as we saw earlier in Equation 5.10.

In summary, the change in internal energy is equal to the heat gained or lost at constant volume; the change in enthalpy is equal to the heat gained or lost at constant pressure. The difference between ΔE and ΔH is the amount of *P-V* work done by the system when the process occurs at constant pressure, $-P\,\Delta V$. The volume change accompanying many reactions is close to zero, which makes $P\,\Delta V$, and therefore the difference between ΔE and ΔH, small. It is generally satisfactory to use ΔH as the measure of energy changes during most chemical processes.

Related Exercises: 5.33, 5.34, 5.35, 5.36

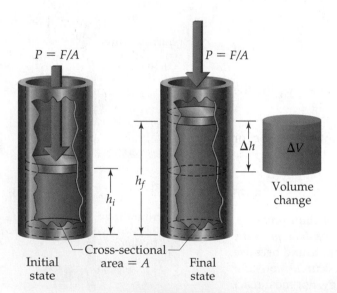

▲ **Figure 5.12 Pressure-volume work.** A piston moving upward, expanding the volume of the system against an external pressure, *P*, does work on the surroundings. The amount of work done by the system on the surroundings is $w = -P\,\Delta V$.

■ **SAMPLE EXERCISE 5.3** | Determining the Sign of ΔH

Indicate the sign of the enthalpy change, ΔH, in each of the following processes carried out under atmospheric pressure, and indicate whether the process is endothermic or exothermic: **(a)** An ice cube melts; **(b)** 1 g of butane (C_4H_{10}) is combusted in sufficient oxygen to give complete combustion to CO_2 and H_2O.

SOLUTION

Analyze: Our goal is to determine whether ΔH is positive or negative for each process. Because each process appears to occur at constant pressure, the enthalpy change of each one equals the amount of heat absorbed or released, $\Delta H = q_P$.

Plan: We must predict whether heat is absorbed or released by the system in each process. Processes in which heat is absorbed are endothermic and have a positive sign for ΔH; those in which heat is released are exothermic and have a negative sign for ΔH.

Solve: In (a) the water that makes up the ice cube is the system. The ice cube absorbs heat from the surroundings as it melts, so ΔH is positive and the process is endothermic. In (b) the system is the 1 g of butane and the oxygen required to combust it. The combustion of butane in oxygen gives off heat, so ΔH is negative and the process is exothermic.

■ PRACTICE EXERCISE

Suppose we confine 1 g of butane and sufficient oxygen to completely combust it in a cylinder like that in Figure 5.12. The cylinder is perfectly insulating, so no heat can escape to the surroundings. A spark initiates combustion of the butane, which forms carbon dioxide and water vapor. If we used this apparatus to measure the enthalpy change in the reaction, would the piston rise, fall, or stay the same?
Answer: The piston must move to maintain a constant pressure in the cylinder. The products contain more molecules of gas than the reactants, as shown by the balanced equation

$$2\,C_4H_{10}(g) + 13\,O_2(g) \longrightarrow 8\,CO_2(g) + 10\,H_2O(g)$$

As a result, the piston would rise to make room for the additional molecules of gas. Heat is given off, so the piston would also rise an additional amount to accommodate the expansion of the gases because of the temperature increase.

5.4 ENTHALPIES OF REACTION

Because $\Delta H = H_{final} - H_{initial}$, the enthalpy change for a chemical reaction is given by the enthalpy of the products minus the enthalpy of the reactants:

$$\Delta H = H_{products} - H_{reactants} \qquad [5.16]$$

The enthalpy change that accompanies a reaction is called the **enthalpy of reaction**, or merely the *heat of reaction*, and is sometimes written ΔH_{rxn}, where "rxn" is a commonly used abbreviation for "reaction."

The combustion of hydrogen is shown in Figure 5.13▼. When the reaction is controlled so that 2 mol $H_2(g)$ burn to form 2 mol $H_2O(g)$ at a constant pressure, the system releases 483.6 kJ of heat. We can summarize this information as

$$2\,H_2(g) + O_2(g) \longrightarrow 2\,H_2O(g) \qquad \Delta H = -483.6\ kJ \qquad [5.17]$$

(a)   (b)   (c)

▲ **Figure 5.13 Exothermic reaction of hydrogen with oxygen.** (a) A candle is held near a balloon filled with hydrogen gas and oxygen gas. (b) The $H_2(g)$ ignites, reacting with $O_2(g)$ to form $H_2O(g)$. The resultant explosion produces a ball of flame. The system gives off heat to its surroundings. (c) The enthalpy diagram for this reaction, showing its exothermic character.

ΔH is negative, so this reaction is exothermic. Notice that ΔH is reported at the end of the balanced equation, without explicitly mentioning the amounts of chemicals involved. In such cases the coefficients in the balanced equation represent the number of moles of reactants and products producing the associated enthalpy change. Balanced chemical equations that show the associated enthalpy change in this way are called *thermochemical equations*.

▲ GIVE IT SOME THOUGHT

What information is summarized by the coefficients in a *thermochemical equation*?

▲ **Figure 5.14 The burning of the hydrogen-filled airship *Hindenburg*.** This photograph was taken only 22 seconds after the first explosion occurred. This tragedy, which occurred in Lakehurst, New Jersey, on May 6, 1937, led to the discontinuation of hydrogen as a buoyant gas in such craft. Modern-day blimps are filled with helium, which is not as buoyant as hydrogen but is not flammable.

The enthalpy change accompanying a reaction may also be represented in an *enthalpy diagram* such as that shown in Figure 5.13(c). Because the combustion of $H_2(g)$ is exothermic, the enthalpy of the products in the reaction is lower than the enthalpy of the reactants. The enthalpy of the system is lower after the reaction because energy has been lost in the form of heat released to the surroundings.

The reaction of hydrogen with oxygen is highly exothermic (ΔH is negative and has a large magnitude), and it occurs rapidly once it starts. It can occur with explosive violence, too, as demonstrated by the disastrous explosions of the German airship *Hindenburg* in 1937 (Figure 5.14 ◄) and the space shuttle *Challenger* in 1986.

The following guidelines are helpful when using thermochemical equations and enthalpy diagrams:

1. *Enthalpy is an extensive property.* The magnitude of ΔH, therefore, is directly proportional to the amount of reactant consumed in the process. For the combustion of methane to form carbon dioxide and liquid water, for example, 890 kJ of heat is produced when 1 mol of CH_4 is burned in a constant-pressure system:

$$CH_4(g) + 2\,O_2(g) \longrightarrow CO_2(g) + 2\,H_2O(l) \qquad \Delta H = -890 \text{ kJ} \qquad [5.18]$$

Because the combustion of 1 mol of CH_4 with 2 mol of O_2 releases 890 kJ of heat, the combustion of 2 mol of CH_4 with 4 mol of O_2 releases twice as much heat, 1780 kJ.

2. *The enthalpy change for a reaction is equal in magnitude, but opposite in sign, to ΔH for the reverse reaction.* For example, if we could reverse Equation 5.18 so that $CH_4(g)$ and $O_2(g)$ formed from $CO_2(g)$ and $H_2O(l)$, ΔH for the process would be +890 kJ:

$$CO_2(g) + 2\,H_2O(l) \longrightarrow CH_4(g) + 2\,O_2(g) \qquad \Delta H = 890 \text{ kJ} \qquad [5.19]$$

When we reverse a reaction, we reverse the roles of the products and the reactants. As a result, the reactants in a reaction become the products of the reverse reaction, and so forth. From Equation 5.16, we can see that reversing the products and reactants leads to the same magnitude, but a change in sign for ΔH_{rxn}. This relationship is diagrammed for Equations 5.18 and 5.19 in Figure 5.15 ◄.

3. *The enthalpy change for a reaction depends on the state of the reactants and products.* If the product in the combustion of methane (Equation 5.18) were gaseous H_2O instead of liquid H_2O, ΔH_{rxn} would be −802 kJ instead of

Enthalpy →

$CH_4(g) + 2\,O_2(g)$

$\Delta H_1 = -890 \text{ kJ}$

$\Delta H_2 = 890 \text{ kJ}$

$CO_2(g) + 2\,H_2O(l)$

▲ **Figure 5.15 ΔH for a reverse reaction.** Reversing a reaction changes the sign but not the magnitude of the enthalpy change: $\Delta H_2 = -\Delta H_1$.

−890 kJ. Less heat would be available for transfer to the surroundings because the enthalpy of $H_2O(g)$ is greater than that of $H_2O(l)$. One way to see this is to imagine that the product is initially liquid water. The liquid water must be converted to water vapor, and the conversion of 2 mol $H_2O(l)$ to 2 mol $H_2O(g)$ is an endothermic process that absorbs 88 kJ:

$$2\,H_2O(l) \longrightarrow 2\,H_2O(g) \qquad \Delta H = +88\ kJ \qquad [5.20]$$

Thus, it is important to specify the states of the reactants and products in thermochemical equations. In addition, we will generally assume that the reactants and products are both at the same temperature, 25 °C, unless otherwise indicated.

■ **SAMPLE EXERCISE 5.4** | **Relating ΔH to Quantities of Reactants and Products**

How much heat is released when 4.50 g of methane gas is burned in a constant-pressure system? (Use the information given in Equation 5.18.)

SOLUTION

Analyze: Our goal is to use a thermochemical equation to calculate the heat produced when a specific amount of methane gas is combusted. According to Equation 5.18, 890 kJ is released by the system when 1 mol CH_4 is burned at constant pressure ($\Delta H = -890$ kJ).

Plan: Equation 5.18 provides us with a stoichiometric conversion factor: 1 mol $CH_4 \simeq -890$ kJ. Thus, we can convert moles of CH_4 to kJ of energy. First, however, we must convert grams of CH_4 to moles of CH_4. Thus, the conversion sequence is grams CH_4 (given) → moles CH_4 → kJ (unknown to be found).

Solve: By adding the atomic weights of C and 4 H, we have 1 mol CH_4 = 16.0 g CH_4. We can use the appropriate conversion factors to convert grams of CH_4 to moles of CH_4 to kilojoules:

$$Heat = (4.50\ g\ CH_4)\left(\frac{1\ mol\ CH_4}{16.0\ g\ CH_4}\right)\left(\frac{-890\ kJ}{1\ mol\ CH_4}\right) = -250\ kJ$$

The negative sign indicates that the system released 250 kJ into the surroundings.

■ **PRACTICE EXERCISE**

Hydrogen peroxide can decompose to water and oxygen by the following reaction:

$$2\,H_2O_2(l) \longrightarrow 2\,H_2O(l) + O_2(g) \qquad \Delta H = -196\ kJ$$

Calculate the value of q when 5.00 g of $H_2O_2(l)$ decomposes at constant pressure.
Answer: −14.4 kJ

In many situations it is valuable to know the enthalpy change associated with a given chemical process. As we will see in the following sections, ΔH can be determined directly by experiment or calculated from the known enthalpy changes of other reactions by invoking the first law of thermodynamics.

5.5 CALORIMETRY

The value of ΔH can be determined experimentally by measuring the heat flow accompanying a reaction at constant pressure. Typically, we can determine the magnitude of the heat flow by measuring the magnitude of the temperature change the heat flow produces. The measurement of heat flow is **calorimetry**; a device used to measure heat flow is a **calorimeter**.

Heat Capacity and Specific Heat

The more heat an object gains, the hotter it gets. All substances change temperature when they are heated, but the magnitude of the temperature change produced by a given quantity of heat varies from substance to substance.

Strategies in Chemistry   USING ENTHALPY AS A GUIDE

If you hold a brick in the air and let it go, it will fall as the force of gravity pulls it toward Earth. A process that is thermodynamically favored to happen, such as a falling brick, is called a *spontaneous* process. A spontaneous process can be either fast or slow. Speed is not the issue in thermodynamics.

Many chemical processes are thermodynamically favored, or spontaneous, too. By "spontaneous," we do not mean that the reaction will form products without any intervention. That can be the case, but often some energy must be imparted to get the process started. The enthalpy change in a reaction gives one indication as to whether the reaction is likely to be spontaneous. The combustion of $H_2(g)$ and $O_2(g)$, for example, is a highly exothermic process:

$$H_2(g) + \tfrac{1}{2}O_2(g) \longrightarrow H_2O(g) \qquad \Delta H = -242 \text{ kJ}$$

Hydrogen gas and oxygen gas can exist together in a volume indefinitely without noticeable reaction occurring, as in Figure 5.13(a). Once initiated, however, energy is rapidly transferred from the system (the reactants) to the surroundings. As the reaction proceeds, large amounts of heat are released, which greatly increases the temperature of the reactants and the products. The system then loses enthalpy by transferring the heat to the surroundings. (Recall that the first law of thermodynamics indicates that the total energy of the system plus the surroundings will not change; energy is conserved.)

Enthalpy change is not the only consideration in the spontaneity of reactions, however, nor is it a foolproof guide. For example, the melting of ice is an endothermic process:

$$H_2O(s) \longrightarrow H_2O(l) \qquad \Delta H = +6.01 \text{ kJ}$$

Even though this process is endothermic, it is spontaneous at temperatures above the freezing point of water (0 °C). The reverse process, the freezing of water to ice, is spontaneous at temperatures below 0 °C. Thus, we know that ice at room temperature will melt and that water put into a freezer at −20 °C will turn into ice. Both of these processes are spontaneous under different conditions even though they are the reverse of one another. In Chapter 19 we will address the spontaneity of processes more fully. We will see why a process can be spontaneous at one temperature, but not at another, as is the case for the conversion of water to ice.

Despite these complicating factors, however, you should pay attention to the enthalpy changes in reactions. As a general observation, when the enthalpy change is large, it is the dominant factor in determining spontaneity. Thus, reactions for which ΔH is *large* and *negative* tend to be spontaneous. Reactions for which ΔH is *large* and *positive* tend to be spontaneous only in the reverse direction. The enthalpy of a reaction can be estimated in a number of ways. From these estimates, the likelihood of the reaction being thermodynamically favorable can be predicted.
Related Exercises: 5.45, 5.46

▲ **Figure 5.16 Specific heat of water.** Specific heat indicates the amount of heat that must be added to one gram of a substance to raise its temperature by 1 K (or 1 °C). Specific heats can vary slightly with temperature, so for precise measurements the temperature is specified. The specific heat of $H_2O(l)$ at 14.5 °C is 4.184 J/g-K; the addition of 4.184 J of heat to 1 g of liquid water at this temperature raises the temperature to 15.5 °C. This amount of energy defines the calorie: 1 cal = 4.184 J.

The temperature change experienced by an object when it absorbs a certain amount of heat is determined by its **heat capacity**, C. The heat capacity of an object is the amount of heat required to raise its temperature by 1 K (or 1 °C). The greater the heat capacity, the greater the heat required to produce a given increase in temperature.

For pure substances the heat capacity is usually given for a specified amount of the substance. The heat capacity of one mole of a substance is called its **molar heat capacity**, C_m. The heat capacity of one gram of a substance is called its *specific heat capacity*, or merely its **specific heat** (Figure 5.16 ◄). The specific heat, C_s, of a substance can be determined experimentally by measuring the temperature change, ΔT, that a known mass, m, of the substance undergoes when it gains or loses a specific quantity of heat, q:

$$\text{Specific heat} = \frac{(\text{quantity of heat transferred})}{(\text{grams of substance}) \times (\text{temperature change})}$$

$$C_s = \frac{q}{m \times \Delta T} \qquad [5.21]$$

For example, 209 J is required to increase the temperature of 50.0 g of water by 1.00 K. Thus, the specific heat of water is

$$C_s = \frac{209 \text{ J}}{(50.0 \text{ g})(1.00 \text{ K})} = 4.18 \frac{\text{J}}{\text{g-K}}$$

A temperature change in kelvins is equal in magnitude to the temperature change in degrees Celsius: ΔT in K = ΔT in °C. ∞(Section 1.4) When the

| TABLE 5.2 ■ Specific Heats of Some Substances at 298 K | | | |
| Elements | | Compounds | |
Substance	Specific Heat (J/g-K)	Substance	Specific Heat (J/g-K)
$N_2(g)$	1.04	$H_2O(l)$	4.18
$Al(s)$	0.90	$CH_4(g)$	2.20
$Fe(s)$	0.45	$CO_2(g)$	0.84
$Hg(l)$	0.14	$CaCO_3(s)$	0.82

sample gains heat (positive q), the temperature of the sample increases (positive ΔT). Rearranging Equation 5.21, we get

$$q = C_s \times m \times \Delta T \qquad [5.22]$$

Thus, we can calculate the quantity of heat that a substance has gained or lost by using its specific heat together with its measured mass and temperature change.

The specific heats of several substances are listed in Table 5.2▲. Notice that the specific heat of liquid water is higher than those of the other substances listed. For example, it is about five times as great as that of aluminum metal. The high specific heat of water affects Earth's climate because it makes the temperatures of the oceans relatively resistant to change. It also is very important in maintaining a constant temperature in our bodies, as we will discuss in the "Chemistry and Life" box later in this chapter.

▲ **GIVE IT SOME THOUGHT**

Which substance in Table 5.2 will undergo the greatest temperature change when the same mass of each substance absorbs the same quantity of heat?

■ **SAMPLE EXERCISE 5.5** | Relating Heat, Temperature Change, and Heat Capacity

(a) How much heat is needed to warm 250 g of water (about 1 cup) from 22 °C (about room temperature) to near its boiling point, 98 °C? The specific heat of water is 4.18 J/g-K. **(b)** What is the molar heat capacity of water?

SOLUTION

Analyze: In part **(a)** we must find the quantity of heat (q) needed to warm the water, given the mass of water (m), its temperature change (ΔT), and its specific heat (C_s). In part **(b)** we must calculate the molar heat capacity (heat capacity per mole, C_m) of water from its specific heat (heat capacity per gram).

Plan: **(a)** Given C_s, m, and ΔT, we can calculate the quantity of heat, q, using Equation 5.22. **(b)** We can use the molar mass of water and dimensional analysis to convert from heat capacity per gram to heat capacity per mole.

Solve:

(a) The water undergoes a temperature change of

$$\Delta T = 98\ °C - 22\ °C = 76\ °C = 76\ K$$

Using Equation 5.22, we have

$$q = C_s \times m \times \Delta T$$
$$= (4.18\ \text{J/g-K})(250\ \text{g})(76\ \text{K}) = 7.9 \times 10^4\ \text{J}$$

(b) The molar heat capacity is the heat capacity of one mole of substance. Using the atomic weights of hydrogen and oxygen, we have

$$1\ \text{mol } H_2O = 18.0\ \text{g } H_2O$$

From the specific heat given in part (a), we have

$$C_m = \left(4.18\ \frac{\text{J}}{\text{g-K}}\right)\left(\frac{18.0\ \text{g}}{1\ \text{mol}}\right) = 75.2\ \text{J/mol-K}$$

■ **PRACTICE EXERCISE**

(a) Large beds of rocks are used in some solar-heated homes to store heat. Assume that the specific heat of the rocks is 0.82 J/g-K. Calculate the quantity of heat absorbed by 50.0 kg of rocks if their temperature increases by 12.0 °C. **(b)** What temperature change would these rocks undergo if they emitted 450 kJ of heat?
Answers: **(a)** 4.9×10^5 J, **(b)** 11 K decrease = 11 °C decrease

Thermometer

Glass stirrer

Cork stopper

Two Styrofoam®
cups nested
together containing
reactants in solution

▲ Figure 5.17 Coffee-cup
calorimeter. This simple apparatus is
used to measure heat-accompanying
reactions at constant pressure.

Constant-Pressure Calorimetry

The techniques and equipment employed in calorimetry depend on the nature of the process being studied. For many reactions, such as those occurring in solution, it is easy to control pressure so that ΔH is measured directly. (Recall that $\Delta H = q_P$.) Although the calorimeters used for highly accurate work are precision instruments, a very simple "coffee-cup" calorimeter, as shown in Figure 5.17 ◄, is often used in general chemistry labs to illustrate the principles of calorimetry. Because the calorimeter is not sealed, the reaction occurs under the essentially constant pressure of the atmosphere.

In this case there is no physical boundary between the system and the surroundings. The reactants and products of the reaction are the system, and the water in which they dissolve as well as the calorimeter are part of the surroundings. If we assume that the calorimeter perfectly prevents the gain or loss of heat from the solution, the heat gained by the solution must be produced from the chemical reaction under study. In other words, the heat produced by the reaction, q_{rxn}, is entirely absorbed by the solution; it does not escape the calorimeter. (We also assume that the calorimeter itself does not absorb heat. In the case of the coffee-cup calorimeter, this is a reasonable approximation because the calorimeter has a very low thermal conductivity and heat capacity.) For an exothermic reaction, heat is "lost" by the reaction and "gained" by the solution, so the temperature of the solution rises. The opposite occurs for an endothermic reaction. The heat gained by the solution, q_{soln}, is therefore equal in magnitude to q_{rxn} but opposite in sign: $q_{soln} = -q_{rxn}$. The value of q_{soln} is readily calculated from the mass of the solution, its specific heat, and the temperature change:

$$q_{soln} = (\text{specific heat of solution}) \times (\text{grams of solution}) \times \Delta T = -q_{rxn} \quad [5.23]$$

For dilute aqueous solutions the specific heat of the solution will be approximately the same as that of water, 4.18 J/g-K.

Equation 5.23 makes it possible to calculate q_{rxn} from the temperature change of the solution in which the reaction occurs. A temperature increase ($\Delta T > 0$) means the reaction is exothermic ($q_{rxn} < 0$).

GIVE IT SOME THOUGHT

(a) How are the energy changes of a system and its surroundings related? (b) How is the heat gained or lost by a system related to the heat gained or lost by its surroundings?

■ SAMPLE EXERCISE 5.6 | Measuring ΔH Using a Coffee-Cup Calorimeter

When a student mixes 50 mL of 1.0 M HCl and 50 mL of 1.0 M NaOH in a coffee-cup calorimeter, the temperature of the resultant solution increases from 21.0 °C to 27.5 °C. Calculate the enthalpy change for the reaction in kJ/mol HCl, assuming that the calorimeter loses only a negligible quantity of heat, that the total volume of the solution is 100 mL, that its density is 1.0 g/mL, and that its specific heat is 4.18 J/g-K.

SOLUTION

Analyze: Mixing solutions of HCl and NaOH results in an acid–base reaction:

$$HCl(aq) + NaOH(aq) \longrightarrow H_2O(l) + NaCl(aq)$$

We need to calculate the heat produced per mole of HCl, given the temperature increase of the solution, the number of moles of HCl and NaOH involved, and the density and specific heat of the solution.

Plan: The total heat produced can be calculated using Equation 5.23. The number of moles of HCl consumed in the reaction must be calculated from the volume and molarity of this substance, and this amount then used to determine the heat produced per mol HCl.

Solve:

Because the total volume of the solution is 100 mL, its mass is	$(100 \text{ mL})(1.0 \text{ g/mL}) = 100 \text{ g}$
The temperature change is	$\Delta T = 27.5 \,°C - 21.0 \,°C = 6.5 \,°C = 6.5 \text{ K}$
Using Equation 5.23, we have	$q_{rxn} = -C_s \times m \times \Delta T$
	$= -(4.18 \text{ J/g-K})(100 \text{ g})(6.5 \text{ K}) = -2.7 \times 10^3 \text{ J} = -2.7 \text{ kJ}$
Because the process occurs at constant pressure,	$\Delta H = q_P = -2.7 \text{ kJ}$

To express the enthalpy change on a molar basis, we use the fact that the number of moles of HCl is given by the product of the respective solution volumes (50 mL = 0.050 L) and concentrations (1.0 M = 1.0 mol/L):

$(0.050 \text{ L})(1.0 \text{ mol/L}) = 0.050 \text{ mol}$

Thus, the enthalpy change per mole of HCl is

$\Delta H = -2.7 \text{ kJ/0.050 mol} = -54 \text{ kJ/mol}$

Check: ΔH is negative (exothermic), which is expected for the reaction of an acid with a base and evidenced by the fact that the reaction causes the temperature of the solution to increase. The molar magnitude of the heat produced seems reasonable.

■ PRACTICE EXERCISE

When 50.0 mL of 0.100 M AgNO$_3$ and 50.0 mL of 0.100 M HCl are mixed in a constant-pressure calorimeter, the temperature of the mixture increases from 22.30 °C to 23.11 °C. The temperature increase is caused by the following reaction:

$$AgNO_3(aq) + HCl(aq) \longrightarrow AgCl(s) + HNO_3(aq)$$

Calculate ΔH for this reaction in kJ/mol AgNO$_3$, assuming that the combined solution has a mass of 100.0 g and a specific heat of 4.18 J/g °C.
Answer: $-68,000 \text{ J/mol} = -68 \text{ kJ/mol}$

Bomb Calorimetry (Constant-Volume Calorimetry)

One of the most important types of reactions studied using calorimetry is combustion, in which a compound (usually an organic compound) reacts completely with excess oxygen. ∞ (Section 3.2) Combustion reactions are most conveniently studied using a **bomb calorimeter**, a device shown schematically in Figure 5.18▶. The substance to be studied is placed in a small cup within a sealed vessel called a *bomb*. The bomb, which is designed to withstand high pressures, has an inlet valve for adding oxygen and electrical contacts to initiate the combustion. After the sample has been placed in the bomb, the bomb is sealed and pressurized with oxygen. It is then placed in the calorimeter, which is essentially an insulated container, and covered with an accurately measured quantity of water. When all the components within the calorimeter have come to the same temperature, the combustion reaction is initiated by passing an electrical current through a fine wire that is in contact with the sample. When the wire becomes sufficiently hot, the sample ignites.

Heat is released when combustion occurs. This heat is absorbed by the calorimeter contents, causing a rise in the temperature of the water. The temperature of the water is very carefully measured before reaction and then after reaction when the contents of the calorimeter have again arrived at a common temperature.

To calculate the heat of combustion from the measured temperature increase in the bomb calorimeter, we must know the total heat capacity of the calorimeter, C_{cal}. This quantity is determined by combusting a sample that releases a known quantity of heat and measuring the resulting temperature change. For example, the combustion of exactly 1 g of benzoic acid, C$_6$H$_5$COOH, in a bomb calorimeter produces 26.38 kJ of heat. Suppose 1.000 g of benzoic acid is combusted in a calorimeter, and it increases the temperature by 4.857 °C. The heat capacity of the calorimeter is then given by C_{cal} = 26.38 kJ/4.857 °C = 5.431 kJ/°C. Once we know the value of C_{cal}, we can measure temperature changes produced by other reactions, and from these we can calculate the heat evolved in the reaction, q_{rxn}:

$$q_{rxn} = -C_{cal} \times \Delta T \qquad [5.24]$$

▲ **Figure 5.18 Bomb calorimeter.** This device is used to measure heat accompanying combustion reactions at constant volume.

Labels (top to bottom): Motorized stirrer; Electrical leads for igniting sample; Thermometer; Insulated container; O$_2$ inlet; Bomb (reaction chamber); Fine wire in contact with sample; Cup holding sample; Water

■ SAMPLE EXERCISE 5.7 | Measuring q_{rxn} Using a Bomb Calorimeter

Methylhydrazine (CH_6N_2) is used as a liquid rocket fuel. The combustion of methylhydrazine with oxygen produces $N_2(g)$, $CO_2(g)$, and $H_2O(l)$:

$$2\,CH_6N_2(l) + 5\,O_2(g) \longrightarrow 2\,N_2(g) + 2\,CO_2(g) + 6\,H_2O(l)$$

When 4.00 g of methylhydrazine is combusted in a bomb calorimeter, the temperature of the calorimeter increases from 25.00 °C to 39.50 °C. In a separate experiment the heat capacity of the calorimeter is measured to be 7.794 kJ/°C. Calculate the heat of reaction for the combustion of a mole of CH_6N_2.

SOLUTION

Analyze: We are given a temperature change and the total heat capacity of the calorimeter. We are also given the amount of reactant combusted. Our goal is to calculate the enthalpy change per mole for combustion of the reactant.

Plan: We will first calculate the heat evolved for the combustion of the 4.00-g sample. We will then convert this heat to a molar quantity.

Solve:

For combustion of the 4.00-g sample of methylhydrazine, the temperature change of the calorimeter is

$$\Delta T = (39.50\ °C - 25.00\ °C) = 14.50\ °C$$

We can use ΔT and the value for C_{cal} to calculate the heat of reaction (Equation 5.24):

$$q_{rxn} = -C_{cal} \times \Delta T = -(7.794\ kJ/°C)(14.50\ °C) = -113.0\ kJ$$

We can readily convert this value to the heat of reaction for a mole of CH_6N_2:

$$\left(\frac{-113.0\ kJ}{4.00\ g\ CH_6N_2}\right) \times \left(\frac{46.1\ g\ CH_6N_2}{1\ mol\ CH_6N_2}\right) = -1.30 \times 10^3\ kJ/mol\ CH_6N_2$$

Check: The units cancel properly, and the sign of the answer is negative as it should be for an exothermic reaction.

■ PRACTICE EXERCISE

A 0.5865-g sample of lactic acid ($HC_3H_5O_3$) is burned in a calorimeter whose heat capacity is 4.812 kJ/°C. The temperature increases from 23.10 °C to 24.95 °C. Calculate the heat of combustion of lactic acid (a) per gram and (b) per mole.
Answers: (a) −15.2 kJ/g, (b) −1370 kJ/mol

Because the reactions in a bomb calorimeter are carried out under constant-volume conditions, the heat transferred corresponds to the change in internal energy, ΔE, rather than the change in enthalpy, ΔH (Equation 5.14). For most reactions, however, the difference between ΔE and ΔH is very small. For the reaction discussed in Sample Exercise 5.7, for example, the difference between ΔE and ΔH is only about 1 kJ/mol—a difference of less than 0.1%. It is possible to correct the measured heat changes to obtain ΔH values, and these form the basis of the tables of enthalpy change that we will see in the following sections. However, we need not concern ourselves with how these small corrections are made.

5.6 HESS'S LAW

Many enthalpies of reaction have been measured and tabulated. In this section and the next we will see that it is often possible to calculate the ΔH for a reaction from the tabulated ΔH values of other reactions. Thus, it is not necessary to make calorimetric measurements for all reactions.

Because enthalpy is a state function, the enthalpy change, ΔH, associated with any chemical process depends only on the amount of matter that undergoes change and on the nature of the initial state of the reactants and the final state of the products. This means that whether a particular reaction is carried out in one step or in a series of steps, the sum of the enthalpy changes associated with the individual steps must be the same as the enthalpy change associated with the one-step process. As an example, the combustion of methane gas, $CH_4(g)$, to

Chemistry and Life   THE REGULATION OF HUMAN BODY TEMPERATURE

For most of us, the question "Are you running a fever?" was one of our first introductions to medical diagnosis. Indeed, a deviation in body temperature of only a few degrees indicates that something is amiss. In the laboratory you may have tried to maintain a solution at a constant temperature, only to find how difficult it can be. Yet our bodies manage to maintain a near-constant temperature in spite of widely varying weather, levels of physical activity, and periods of high metabolic activity (such as after a meal). How does the human body manage this task, and how does it relate to some of the topics we have discussed in this chapter?

Maintaining a near-constant temperature is one of the primary physiological functions of the human body. Normal body temperature generally ranges from 35.8 °C to 37.2 °C (96.5 °F–99 °F). This very narrow temperature range is essential to proper muscle function and to the control of the rates of the biochemical reactions in the body. You will learn more about the effects of temperature on reaction rates in Chapter 14.

A portion of the human brain stem called the *hypothalamus* regulates the temperature. The hypothalamus acts as a thermostat for body temperature. When body temperature rises above the high end of the normal range, the hypothalamus triggers mechanisms to lower the temperature. It likewise triggers mechanisms to increase the temperature if body temperature drops too low.

To understand qualitatively how the body's heating and cooling mechanisms operate, we can view the body as a thermodynamic system. The body increases its internal energy content by ingesting foods from the surroundings. ∞ (Section 5.8) The foods, such as glucose ($C_6H_{12}O_6$), are metabolized—a process that is essentially controlled oxidation to CO_2 and H_2O:

$$C_6H_{12}O_6(s) + 6 O_2(g) \longrightarrow 6 CO_2(g) + 6 H_2O(l)$$
$$\Delta H = -2803 \text{ kJ}$$

Roughly 40% of the energy produced is ultimately used to do work in the form of muscle contractions and nerve cell activities. The remainder of the energy is released as heat, part of which is used to maintain body temperature. When the body produces too much heat, as in times of heavy physical exertion, it dissipates the excess to the surroundings.

Heat is transferred from the body to its surroundings primarily by *radiation, convection,* and *evaporation.* Radiation is the direct loss of heat from the body to cooler surroundings, much as a hot stovetop radiates heat to its surroundings. Convection is heat loss by virtue of heating air that is in contact with the body. The heated air rises and is replaced with cooler air, and the process continues. Warm clothing, which usually consists of insulating layers of material with "dead air" in between, decreases convective heat loss in cold weather. Evaporative cooling occurs when perspiration is generated at the skin surface by the sweat glands. Heat is removed from the body as the perspiration evaporates into the surroundings. Perspiration is predominantly water, so the process involved is the endothermic conversion of liquid water into water vapor:

$$H_2O(l) \longrightarrow H_2O(g) \qquad \Delta H = +44.0 \text{ kJ}$$

The speed with which evaporative cooling occurs decreases as the atmospheric humidity increases, which is why people seem to be more sweaty and uncomfortable on hot and humid days.

When the hypothalamus senses that the body temperature has risen too high, it increases heat loss from the body in two principal ways. First, it increases the flow of blood near the surface of the skin, which allows for increased radiational and convective cooling. The reddish "flushed" appearance of a hot individual is the result of this increased subsurface blood flow. Second, the hypothalamus stimulates the secretion of perspiration from the sweat glands, which increases evaporative cooling. During periods of extreme activity, the amount of liquid secreted as perspiration can be as high as 2 to 4 liters per hour. As a result, water must be replenished to the body during these periods (Figure 5.19▼). If the body loses too much fluid through perspiration, it will no longer be able to cool itself and blood volume decreases, which can lead to *heat exhaustion* or the more serious and potentially fatal *heat stroke*, during which the body temperature can rise to as high as 41 °C to 45 °C (106 °F–113 °F). On the other hand, replenishing water without replenishing the electrolytes that are lost during perspiration can also lead to serious problems as pointed out in the "Chemistry in Life" box in Section 4.5.

When body temperature drops too low, the hypothalamus decreases the blood flow to the surface of the skin, thereby decreasing heat loss. It also triggers small involuntary contractions of the muscles; the biochemical reactions that generate the energy to do this work also generate more heat for the body. When these contractions get large enough—as when the body feels a chill—a *shiver* results. If the body is unable to maintain a temperature above 35 °C (95 °F), the very dangerous condition called *hypothermia* can result.

▲ **Figure 5.19 Marathon runner drinking water.** Runners must constantly replenish the water in their bodies that is lost through perspiration.

form $CO_2(g)$ and liquid water can be thought of as occurring in two steps: (1) the combustion of $CH_4(g)$ to form $CO_2(g)$ and gaseous water, $H_2O(g)$ and (2) the condensation of gaseous water to form liquid water, $H_2O(l)$. The enthalpy change for the overall process is simply the sum of the enthalpy changes for these two steps:

$$CH_4(g) + 2\,O_2(g) \longrightarrow CO_2(g) + 2\,H_2O(g) \qquad \Delta H = -802 \text{ kJ}$$

(Add) $\qquad 2\,H_2O(g) \longrightarrow 2\,H_2O(l) \qquad\qquad\qquad\quad \Delta H = -\ 88 \text{ kJ}$

$$CH_4(g) + 2\,O_2(g) + 2\,H_2O(g) \longrightarrow CO_2(g) + 2\,H_2O(l) + 2\,H_2O(g) \quad \Delta H = -890 \text{ kJ}$$

The net equation is

$$CH_4(g) + 2\,O_2(g) \longrightarrow CO_2(g) + 2\,H_2O(l) \qquad \Delta H = -890 \text{ kJ}$$

To obtain the net equation, the sum of the reactants of the two equations is placed on one side of the arrow and the sum of the products on the other side. Because $2\,H_2O(g)$ occurs on both sides, it can be canceled like an algebraic quantity that appears on both sides of an equal sign. Figure 5.20◄ compares the two-step reaction with the direct one.

Hess's law states that *if a reaction is carried out in a series of steps, ΔH for the overall reaction will equal the sum of the enthalpy changes for the individual steps.* The overall enthalpy change for the process is independent of the number of steps or the particular nature of the path by which the reaction is carried out. This principle is a consequence of the fact that enthalpy is a state function, and ΔH is therefore independent of the path between the initial and final states. We can therefore calculate ΔH for any process, as long as we find a route for which ΔH is known for each step. This means that a relatively small number of experimental measurements can be used to calculate ΔH for a vast number of different reactions.

Hess's law provides a useful means of calculating energy changes that are difficult to measure directly. For instance, it is impossible to measure directly the enthalpy for the combustion of carbon to form carbon monoxide. Combustion of 1 mol of carbon with 0.5 mol of O_2 produces both CO and CO_2, leaving some carbon unreacted. However, solid carbon and carbon monoxide can both be completely burned in O_2 to produce CO_2. We can use the enthalpy changes of these reactions to calculate the heat of combustion of C to CO, as shown in Sample Exercise 5.8.

▲ **Figure 5.20 An enthalpy diagram comparing a one-step and a two-step process for a reaction.** The enthalpy change of the direct reaction on the left equals the sum of the two steps on the right. That is, ΔH for the overall reaction equals the sum of the ΔH values for the two steps shown.

GIVE IT SOME THOUGHT

What effect do the following changes have on ΔH for a reaction: **(a)** reversing a reaction, **(b)** multiplying coefficients by 2?

■■■ **SAMPLE EXERCISE 5.8** | Using Hess's Law to Calculate ΔH

The enthalpy of reaction for the combustion of C to CO_2 is -393.5 kJ/mol C, and the enthalpy for the combustion of CO to CO_2 is -283.0 kJ/mol CO:

(1) $\qquad\qquad C(s) + O_2(g) \longrightarrow CO_2(g) \qquad \Delta H_1 = -393.5 \text{ kJ}$

(2) $\qquad\qquad CO(g) + \tfrac{1}{2}O_2(g) \longrightarrow CO_2(g) \qquad \Delta H_2 = -283.0 \text{ kJ}$

Using these data, calculate the enthalpy for the combustion of C to CO:

(3) $\qquad\qquad C(s) + \tfrac{1}{2}O_2(g) \longrightarrow CO(g) \qquad \Delta H_3 = ?$

SOLUTION

Analyze: We are given two thermochemical equations, and our goal is to combine them in such a way as to obtain the third equation and its enthalpy change.

Plan: We will use Hess's law. In doing so, we first note the numbers of moles of substances among the reactants and products in the target equation, (3). We then manipulate equations (1) and (2) to give the same number of moles of these substances, so that when the resulting equations are added, we obtain the target equation. At the same time, we keep track of the enthalpy changes, which we add.

Solve: To use equations (1) and (2), we arrange them so that C(*s*) is on the reactant side and CO(*g*) is on the product side of the arrow, as in the target reaction, equation (3). Because equation (1) has C(*s*) as a reactant, we can use that equation just as it is. We need to turn equation (2) around, however, so that CO(*g*) is a product. Remember that when reactions are turned around, the sign of ΔH is reversed. We arrange the two equations so that they can be added to give the desired equation:

$$C(s) + O_2(g) \longrightarrow CO_2(g) \qquad \Delta H_1 = -393.5 \text{ kJ}$$
$$CO_2(g) \longrightarrow CO(g) + \tfrac{1}{2}O_2(g) \qquad -\Delta H_2 = 283.0 \text{ kJ}$$
$$\overline{C(s) + \tfrac{1}{2}O_2(g) \longrightarrow CO(g) \qquad \Delta H_3 = -110.5 \text{ kJ}}$$

When we add the two equations, $CO_2(g)$ appears on both sides of the arrow and therefore cancels out. Likewise, $\tfrac{1}{2}O_2(g)$ is eliminated from each side.

Comment: It is sometimes useful to add subscripts to the enthalpy changes, as we have done here, to keep track of the associations between the chemical reactions and their ΔH values.

■ PRACTICE EXERCISE

Carbon occurs in two forms, graphite and diamond. The enthalpy of the combustion of graphite is -393.5 kJ/mol and that of diamond is -395.4 kJ/mol:

$$C(graphite) + O_2(g) \longrightarrow CO_2(g) \qquad \Delta H_1 = -393.5 \text{ kJ}$$
$$C(diamond) + O_2(g) \longrightarrow CO_2(g) \qquad \Delta H_2 = -395.4 \text{ kJ}$$

Calculate ΔH for the conversion of graphite to diamond:

$$C(graphite) \longrightarrow C(diamond) \qquad \Delta H_3 = ?$$

Answer: $\Delta H_3 = +1.9$ kJ

■ SAMPLE EXERCISE 5.9 | Using Three Equations with Hess's Law to Calculate ΔH

Calculate ΔH for the reaction

$$2\,C(s) + H_2(g) \longrightarrow C_2H_2(g)$$

given the following chemical equations and their respective enthalpy changes:

$$C_2H_2(g) + \tfrac{5}{2}O_2(g) \longrightarrow 2\,CO_2(g) + H_2O(l) \qquad \Delta H = +1299.6 \text{ kJ}$$
$$2\,C(s) + O_2(g) \longrightarrow CO_2(g) \qquad \Delta H = -393.5 \text{ kJ}$$
$$H_2(g) + \tfrac{1}{2}O_2(g) \longrightarrow H_2O(l) \qquad \Delta H = -285.8 \text{ kJ}$$

SOLUTION

Analyze: We are given a chemical equation and asked to calculate its ΔH using three chemical equations and their associated enthalpy changes.

Plan: We will use Hess's law, summing the three equations or their reverses and multiplying each by an appropriate coefficient so that they add to give the net equation for the reaction of interest. At the same time, we keep track of the ΔH values, reversing their signs if the reactions are reversed and multiplying them by whatever coefficient is employed in the equation.

Solve: Because the target equation has C_2H_2 as a product, we turn the first equation around; the sign of ΔH is therefore changed. The desired equation has 2 C(*s*) as a reactant, so we multiply the second equation and its ΔH by 2. Because the target equation has H_2 as a reactant, we keep the third equation as it is. We then add the three equations and their enthalpy changes in accordance with Hess's law:

$$2\,CO_2(g) + H_2O(l) \longrightarrow C_2H_2(g) + \tfrac{5}{2}O_2(g) \qquad \Delta H = 1299.6 \text{ kJ}$$
$$2\,C(s) + 2\,O_2(g) \longrightarrow 2\,CO_2(g) \qquad \Delta H = -787.0 \text{ kJ}$$
$$H_2(g) + \tfrac{1}{2}O_2(g) \longrightarrow H_2O(l) \qquad \Delta H = -285.8 \text{ kJ}$$
$$\overline{2\,C(s) + H_2(g) \longrightarrow C_2H_2(g) \qquad \Delta H = 226.8 \text{ kJ}}$$

When the equations are added, there are 2 CO_2, $\tfrac{5}{2}O_2$, and H_2O on both sides of the arrow. These are canceled in writing the net equation.

Check: The procedure must be correct because we obtained the correct net equation. In cases like this you should go back over the numerical manipulations of the ΔH values to ensure that you did not make an inadvertent error with signs.

■ PRACTICE EXERCISE

Calculate ΔH for the reaction

$$NO(g) + O(g) \longrightarrow NO_2(g)$$

given the following information:

$$NO(g) + O_3(g) \longrightarrow NO_2(g) + O_2(g) \qquad \Delta H = -198.9 \text{ kJ}$$
$$O_3(g) \longrightarrow \tfrac{3}{2} O_2(g) \qquad \Delta H = -142.3 \text{ kJ}$$
$$O_2(g) \longrightarrow 2\, O(g) \qquad \Delta H = 495.0 \text{ kJ}$$

Answer: -304.1 kJ

▲ **Figure 5.21 An enthalpy diagram illustrating Hess's law.** Because H is a state function, the enthalpy change for the combustion of 1 mol CH_4 is independent of whether the reaction takes place in one or more steps: $\Delta H_1 = \Delta H_2 + \Delta H_3$.

The key point of these examples is that H is a state function, so *for a particular set of reactants and products, ΔH is the same whether the reaction takes place in one step or in a series of steps.* For example, consider the reaction of methane (CH_4) and oxygen (O_2) to form CO_2 and H_2O. We can envision the reaction forming CO_2 directly or with the initial formation of CO, which is then combusted to CO_2. These two paths are compared in Figure 5.21 ◄. Because H is a state function, both paths *must* produce the same value of ΔH. In the enthalpy diagram, that means $\Delta H_1 = \Delta H_2 + \Delta H_3$.

5.7 ENTHALPIES OF FORMATION

By using the methods we have just discussed, we can calculate the enthalpy changes for a great many reactions from tabulated ΔH values. Many experimental data are tabulated according to the type of process. For example, extensive tables exist of *enthalpies of vaporization* (ΔH for converting liquids to gases), *enthalpies of fusion* (ΔH for melting solids), *enthalpies of combustion* (ΔH for combusting a substance in oxygen), and so forth. A particularly important process used for tabulating thermochemical data is the formation of a compound from its constituent elements. The enthalpy change associated with this process is called the **enthalpy of formation** (or *heat of formation*) and is labeled ΔH_f, where the subscript *f* indicates that the substance has been *formed* from its component elements.

The magnitude of any enthalpy change depends on the conditions of temperature, pressure, and state (gas, liquid, or solid crystalline form) of the reactants and products. To compare the enthalpies of different reactions, we must define a set of conditions, called a *standard state*, at which most enthalpies are tabulated. The standard state of a substance is its pure form at atmospheric pressure (1 atm; ∞Section 10.2) and the temperature of interest, which we usually choose to be 298 K (25 °C).* The **standard enthalpy change** of a reaction is defined as the enthalpy change when all reactants and products are in their standard states. We denote a standard enthalpy change as $\Delta H°$, where the superscript ° indicates standard-state conditions.

---

*The definition of the standard state for gases has been changed to 1 bar (1 atm = 1.013 bar), a slightly lower pressure than the value of 1 atm that is used for the data in this text. For most purposes, this change makes very little difference in the standard changes.

The **standard enthalpy of formation** of a compound, ΔH_f°, is the change in enthalpy for the reaction that forms one mole of the compound from its elements, with all substances in their standard states:

elements (in standard states) $\longrightarrow$ compound (in standard state) $\quad \Delta H_f^\circ$

We usually report ΔH_f° values at 298 K. If an element exists in more than one form under standard conditions, the most stable form of the element is usually used for the formation reaction. For example, the standard enthalpy of formation for ethanol, C_2H_5OH, is the enthalpy change for the following reaction:

$$2\ C(graphite) + 3\ H_2(g) + \tfrac{1}{2}O_2(g) \longrightarrow C_2H_5OH(l) \quad \Delta H_f^\circ = -277.7\ kJ \quad [5.25]$$

The elemental source of oxygen is O_2, not O or O_3, because O_2 is the stable form of oxygen at 298 K and standard atmospheric pressure. Similarly, the elemental source of carbon is graphite and not diamond, because graphite is more stable (lower energy) at 298 K and standard atmospheric pressure (see Practice Exercise 5.8). Likewise, the most stable form of hydrogen under standard conditions is $H_2(g)$, so this is used as the source of hydrogen in Equation 5.25.

The stoichiometry of formation reactions always indicates that one mole of the desired substance is produced, as in Equation 5.25. As a result, enthalpies of formation are reported in kJ/mol of the substance being formed. Several standard enthalpies of formation are given in Table 5.3▼. A more complete table is provided in Appendix C.

By definition, *the standard enthalpy of formation of the most stable form of any element is zero* because there is no formation reaction needed when the element is already in its standard state. Thus, the values of ΔH_f° for C(graphite), $H_2(g)$, $O_2(g)$, and the standard states of other elements are zero by definition.

GIVE IT SOME THOUGHT

In Table 5.3, the standard enthalpy of formation of $C_2H_2(g)$ is listed as 226.7 kJ/mol. Write the thermochemical equation associated with ΔH_f° for this substance.

TABLE 5.3 ■ Standard Enthalpies of Formation, ΔH_f°, at 298 K

Substance	Formula	ΔH_f° (kJ/mol)	Substance	Formula	ΔH_f° (kJ/mol)
Acetylene	$C_2H_2(g)$	226.7	Hydrogen chloride	$HCl(g)$	−92.30
Ammonia	$NH_3(g)$	−46.19	Hydrogen fluoride	$HF(g)$	−268.60
Benzene	$C_6H_6(l)$	49.0	Hydrogen iodide	$HI(g)$	25.9
Calcium carbonate	$CaCO_3(s)$	−1207.1	Methane	$CH_4(g)$	−74.80
Calcium oxide	$CaO(s)$	−635.5	Methanol	$CH_3OH(l)$	−238.6
Carbon dioxide	$CO_2(g)$	−393.5	Propane	$C_3H_8(g)$	−103.85
Carbon monoxide	$CO(g)$	−110.5	Silver chloride	$AgCl(s)$	−127.0
Diamond	$C(s)$	1.88	Sodium bicarbonate	$NaHCO_3(s)$	−947.7
Ethane	$C_2H_6(g)$	−84.68	Sodium carbonate	$Na_2CO_3(s)$	−1130.9
Ethanol	$C_2H_5OH(l)$	−277.7	Sodium chloride	$NaCl(s)$	−410.9
Ethylene	$C_2H_4(g)$	52.30	Sucrose	$C_{12}H_{22}O_{11}(s)$	−2221
Glucose	$C_6H_{12}O_6(s)$	−1273	Water	$H_2O(l)$	−285.8
Hydrogen bromide	$HBr(g)$	−36.23	Water vapor	$H_2O(g)$	−241.8

■ **SAMPLE EXERCISE 5.10** | Identifying Equations Associated with Enthalpies of Formation

For which of the following reactions at 25 °C would the enthalpy change represent a standard enthalpy of formation? For each that does not, what changes are needed to make it an equation whose ΔH is an enthalpy of formation?

(a) $2\,Na(s) + \frac{1}{2}O_2(g) \longrightarrow Na_2O(s)$

(b) $2\,K(l) + Cl_2(g) \longrightarrow 2\,KCl(s)$

(c) $C_6H_{12}O_6(s) \longrightarrow 6\,C(diamond) + 6\,H_2(g) + 3\,O_2(g)$

SOLUTION

Analyze: The standard enthalpy of formation is represented by a reaction in which each reactant is an element in its standard state and the product is one mole of the compound.

Plan: We need to examine each equation to determine, first, whether the reaction is one in which one mole of substance is formed from the elements. Next, we need to determine whether the reactant elements are in their standard states.

Solve: In (a) 1 mol Na_2O is formed from the elements sodium and oxygen in their proper states, solid Na and O_2 gas, respectively. Therefore, the enthalpy change for reaction (a) corresponds to a standard enthalpy of formation.

In (b) potassium is given as a liquid. It must be changed to the solid form, its standard state at room temperature. Furthermore, two moles of product are formed, so the enthalpy change for the reaction as written is twice the standard enthalpy of formation of $KCl(s)$. The equation for the formation reaction of 1 mol of $KCl(s)$ is

$$K(s) + \frac{1}{2}Cl(g) \longrightarrow KCl(s)$$

Reaction (c) does not form a substance from its elements. Instead, a substance decomposes to its elements, so this reaction must be reversed. Next, the element carbon is given as diamond, whereas graphite is the standard state of carbon at room temperature and 1 atm pressure. The equation that correctly represents the enthalpy of formation of glucose from its elements is

$$6\,C(graphite) + 6\,H_2(g) + 3\,O_2(g) \longrightarrow C_6H_{12}O_6(s)$$

■ **PRACTICE EXERCISE**

Write the equation corresponding to the standard enthalpy of formation of liquid carbon tetrachloride (CCl_4).
Answer: $C(graphite) + 2\,Cl_2(g) \longrightarrow CCl_4(l)$

Using Enthalpies of Formation to Calculate Enthalpies of Reaction

Tabulations of ΔH_f°, such as those in Table 5.3 and Appendix C, have many important uses. As we will see in this section, we can use Hess's law to calculate the standard enthalpy change for any reaction for which we know the ΔH_f° values for all reactants and products. For example, consider the combustion of propane gas, $C_3H_8(g)$, with oxygen to form $CO_2(g)$ and $H_2O(l)$ under standard conditions:

$$C_3H_8(g) + 5\,O_2(g) \longrightarrow 3\,CO_2(g) + 4\,H_2O(l)$$

We can write this equation as the sum of three formation reactions:

$$C_3H_8(g) \longrightarrow 3\,C(s) + 4\,H_2(g) \qquad \Delta H_1 = -\Delta H_f^\circ[C_3H_8(g)] \qquad [5.26]$$

$$3\,C(s) + 3\,O_2(g) \longrightarrow 3\,CO_2(g) \qquad \Delta H_2 = 3\Delta H_f^\circ[CO_2(g)] \qquad [5.27]$$

$$4\,H_2(g) + 2\,O_2(g) \longrightarrow 4\,H_2O(l) \qquad \Delta H_3 = 4\Delta H_f^\circ[H_2O(l)] \qquad [5.28]$$

$$\overline{C_3H_8(g) + 5\,O_2(g) \longrightarrow 3\,CO_2(g) + 4\,H_2O(l)} \quad \Delta H_{rxn}^\circ = \Delta H_1 + \Delta H_2 + \Delta H_3 \quad [5.29]$$

From Hess's law we can write the standard enthalpy change for the overall reaction, Equation 5.29, as the sum of the enthalpy changes for the processes in

Equations 5.26 through 5.28. We can then use values from Table 5.3 to compute a numerical value for $\Delta H°$ for the overall reaction:

$$\Delta H°_{rxn} = \Delta H_1 + \Delta H_2 + \Delta H_3$$

$$= -\Delta H°_f[C_3H_8(g)] + 3\Delta H°_f[CO_2(g)] + 4\Delta H°_f[H_2O(l)] \qquad [5.30]$$

$$= -(-103.85 \text{ kJ}) + 3(-393.5 \text{ kJ}) + 4(-285.8 \text{ kJ}) = -2219.9 \text{ kJ}$$

Several aspects of this calculation depend on the guidelines we discussed in Section 5.4.

1. Equation 5.26 is the reverse of the formation reaction for $C_3H_8(g)$, so the enthalpy change for this reaction is $-\Delta H°_f[C_3H_8(g)]$.

2. Equation 5.27 is the formation reaction for 3 mol of $CO_2(g)$. Because enthalpy is an extensive property, the enthalpy change for this step is $3\Delta H°_f[CO_2(g)]$. Similarly, the enthalpy change for Equation 5.28 is $4\Delta H°_f[H_2O(l)]$. The reaction specifies that $H_2O(l)$, was produced, so be careful to use the value of $\Delta H°_f$ for $H_2O(l)$, not $H_2O(g)$.

3. We assume that the stoichiometric coefficients in the balanced equation represent moles. For Equation 5.29, therefore, the value $\Delta H°_{rxn} = -2220$ kJ represents the enthalpy change for the reaction of 1 mol C_3H_8 and 5 mol O_2 to form 3 mol CO_2 and 4 mol H_2O. The product of the number of moles and the enthalpy change in kJ/mol has the units kJ: (number of moles) $\times$ ($\Delta H°_f$ in kJ/mol) $-$ kJ. We therefore report $\Delta H°_{rxn}$ in kJ.

Figure 5.22▼ presents an enthalpy diagram for Equation 5.29, showing how it can be broken into steps involving formation reactions.

We can break down any reaction into formation reactions as we have done here. When we do, we obtain the general result that the standard enthalpy change of a reaction is the sum of the standard enthalpies of formation of the products minus the standard enthalpies of formation of the reactants:

$$\Delta H°_{rxn} = \Sigma n\Delta H°_f(\text{products}) - \Sigma m\Delta H°_f(\text{reactants}) \qquad [5.31]$$

The symbol Σ (sigma) means "the sum of," and n and m are the stoichiometric coefficients of the chemical equation. The first term in Equation 5.31 represents the formation reactions of the products, which are written in the "forward" direction, that is, elements reacting to form products. This term is analogous to Equations 5.27 and 5.28 in the previous example. The second term represents the reverse of the formation reactions of the reactants, as in Equation 5.26, which is why the $\Delta H°_f$ values have a minus sign in front of them.

◀ Figure 5.22 **An enthalpy diagram relating the enthalpy change for a reaction to enthalpies of formation.** For the combustion of propane gas, $C_3H_8(g)$, the reaction is $C_3H_8(g) + 5 O_2(g) \longrightarrow 3 CO_2(g) + 4 H_2O(l)$. We can imagine this reaction as occurring in three steps. First, $C_3H_8(g)$ is decomposed to its elements, so $\Delta H_1 = \Delta H°_f[C_3H_8(g)]$. Second, 3 mol $CO_2(g)$ are formed, so $\Delta H_2 = 3\Delta H°_f[CO_2(g)]$. Finally, 4 mol $H_2O(l)$ are formed, so $\Delta H_3 = 4\Delta H°_f[H_2O(l)]$. Hess's law tells us that $\Delta H°_{rxn} = \Delta H_1 + \Delta H_2 + \Delta H_3$. This same result is given by Equation 5.31 because $\Delta H°_f[O_2(g)] = 0$.

■ SAMPLE EXERCISE 5.11 | Calculating an Enthalpy of Reaction from Enthalpies of Formation

(a) Calculate the standard enthalpy change for the combustion of 1 mol of benzene, $C_6H_6(l)$, to form $CO_2(g)$ and $H_2O(l)$. (b) Compare the quantity of heat produced by combustion of 1.00 g propane to that produced by 1.00 g benzene.

SOLUTION

Analyze: (a) We are given a reaction [combustion of $C_6H_6(l)$ to form $CO_2(g)$ and $H_2O(l)$] and asked to calculate its standard enthalpy change, $\Delta H°$. (b) We then need to compare the quantity of heat produced by combustion of 1.00 g C_6H_6 with that produced by 1.00 g C_3H_8, whose combustion was treated above in the text. (See Equations 5.29 and 5.30.)

Plan: (a) We need to write the balanced equation for the combustion of C_6H_6. We then look up $\Delta H_f°$ values in Appendix C or in Table 5.3 and apply Equation 5.31 to calculate the enthalpy change for the reaction. (b) We use the molar mass of C_6H_6 to change the enthalpy change per mole to that per gram. We similarly use the molar mass of C_3H_8 and the enthalpy change per mole calculated in the text above to calculate the enthalpy change per gram of that substance.

Solve:

(a) We know that a combustion reaction involves $O_2(g)$ as a reactant. Thus, the balanced equation for the combustion reaction of 1 mol $C_6H_6(l)$ is

$$C_6H_6(l) + \tfrac{15}{2}O_2(g) \longrightarrow 6\,CO_2(g) + 3\,H_2O(l)$$

We can calculate $\Delta H°$ for this reaction by using Equation 5.31 and data in Table 5.3. Remember to multiply the $\Delta H_f°$ value for each substance in the reaction by that substance's stoichiometric coefficient. Recall also that $\Delta H_f° = 0$ for any element in its most stable form under standard conditions, so $\Delta H_f°[O_2(g)] = 0$

$$\begin{aligned}\Delta H_{rxn}° &= [6\Delta H_f°(CO_2) + 3\Delta H_f°(H_2O)] - [\Delta H_f°(C_6H_6) + \tfrac{15}{2}\Delta H_f°(O_2)] \\ &= [6(-393.5\text{ kJ}) + 3(-285.8\text{ kJ})] - \left[(49.0\text{ kJ}) + \tfrac{15}{2}(0\text{ kJ})\right] \\ &= (-2361 - 857.4 - 49.0)\text{ kJ} \\ &= -3267\text{ kJ}\end{aligned}$$

(b) From the example worked in the text, $\Delta H° = -2220$ kJ for the combustion of 1 mol of propane. In part (a) of this exercise we determined that $\Delta H° = -3267$ kJ for the combustion of 1 mol benzene. To determine the heat of combustion per gram of each substance, we use the molar masses to convert moles to grams:

$C_3H_8(g)$: $(-2220\text{ kJ/mol})(1\text{ mol}/44.1\text{ g}) = -50.3\text{ kJ/g}$
$C_6H_6(l)$: $(-3267\text{ kJ/mol})(1\text{ mol}/78.1\text{ g}) = -41.8\text{ kJ/g}$

Comment: Both propane and benzene are hydrocarbons. As a rule, the energy obtained from the combustion of a gram of hydrocarbon is between 40 and 50 kJ.

■ PRACTICE EXERCISE

Using the standard enthalpies of formation listed in Table 5.3, calculate the enthalpy change for the combustion of 1 mol of ethanol:

$$C_2H_5OH(l) + 3\,O_2(g) \longrightarrow 2\,CO_2(g) + 3\,H_2O(l)$$

Answer: −1367 kJ

■ SAMPLE EXERCISE 5.12 | Calculating an Enthalpy of Formation Using an Enthalpy of Reaction

The standard enthalpy change for the reaction

$$CaCO_3(s) \longrightarrow CaO(s) + CO_2(g)$$

is 178.1 kJ. From the values for the standard enthalpies of formation of $CaO(s)$ and $CO_2(g)$ given in Table 5.3, calculate the standard enthalpy of formation of $CaCO_3(s)$.

SOLUTION

Analyze: We need to obtain $\Delta H_f°\,(CaCO_3)$.

Plan: We begin by writing the expression for the standard enthalpy change for the reaction:

$$\Delta H_{rxn}° = [\Delta H_f°(CaO) + \Delta H_f°(CO_2)] - \Delta H_f°(CaCO_3)$$

Solve: Inserting the given $\Delta H_{rxn}°$ and the known $\Delta H_f°$ values from Table 5.3 or Appendix C, we have

$$178.1\text{ kJ} = -635.5\text{ kJ} - 393.5\text{ kJ} - \Delta H_f°(CaCO_3)$$

Solving for $\Delta H_f°\,(CaCO_3)$ gives

$$\Delta H_f°\,(CaCO_3) = -1207.1\text{ kJ/mol}$$

Check: We expect the enthalpy of formation of a stable solid such as calcium carbonate to be negative, as obtained.

■ **PRACTICE EXERCISE**
Given the following standard enthalpy change, use the standard enthalpies of formation in Table 5.3 to calculate the standard enthalpy of formation of $CuO(s)$:

$$CuO(s) + H_2(g) \longrightarrow Cu(s) + H_2O(l) \qquad \Delta H° = -129.7 \text{ kJ}$$

Answer: -156.1 kJ/mol

5.8 FOODS AND FUELS

Most chemical reactions used for the production of heat are combustion reactions. The energy released when one gram of a material is combusted is often called its **fuel value**. Although fuel values represent the heat *released* in a combustion reaction, fuel values are reported as positive numbers. The fuel value of any food or fuel can be measured by calorimetry.

Foods

Most of the energy our bodies need comes from carbohydrates and fats. The forms of carbohydrate known as starch are decomposed in the intestines into glucose, $C_6H_{12}O_6$. Glucose is soluble in blood, and in the human body it is known as blood sugar. It is transported by the blood to cells, where it reacts with O_2 in a series of steps, eventually producing $CO_2(g)$, $H_2O(l)$, and energy:

$$C_6H_{12}O_6(s) + 6\,O_2(g) \longrightarrow 6\,CO_2(g) + 6\,H_2O(l) \qquad \Delta H° = -2803 \text{ kJ}$$

The breakdown of carbohydrates is rapid, so their energy is quickly supplied to the body. However, the body stores only a very small amount of carbohydrates. The average fuel value of carbohydrates is 17 kJ/g (4 kcal/g).

Like carbohydrates, fats produce CO_2 and H_2O when metabolized and when subjected to combustion in a bomb calorimeter. The reaction of tristearin, $C_{57}H_{110}O_6$, a typical fat, is as follows:

$$2\,C_{57}H_{110}O_6(s) + 163\,O_2(g) \longrightarrow 114\,CO_2(g) + 110\,H_2O(l)$$
$$\Delta H° = -75{,}520 \text{ kJ}$$

The body uses the chemical energy from foods to maintain body temperature (see the "Chemistry and Life" box in Section 5.6), to contract muscles, and to construct and repair tissues. Any excess energy is stored as fats. Fats are well suited to serve as the body's energy reserve for at least two reasons: (1) They are insoluble in water, which facilitates storage in the body; and (2) they produce more energy per gram than either proteins or carbohydrates, which makes them efficient energy sources on a mass basis. The average fuel value of fats is 38 kJ/g (9 kcal/g).

The metabolism of proteins in the body produces less energy than combustion in a calorimeter because the products are different. Proteins contain nitrogen, which is released in the bomb calorimeter as N_2. In the body this nitrogen ends up mainly as urea, $(NH_2)_2CO$. Proteins are used by the body mainly as building materials for organ walls, skin, hair, muscle, and so forth. On average, the metabolism of proteins produces 17 kJ/g (4 kcal/g), the same as for carbohydrates.

The fuel values for a variety of common foods are shown in Table 5.4▼. Labels on packaged foods show the amounts of carbohydrate, fat, and protein contained in an average serving, as well as the amount of energy supplied by a serving (Figure 5.23►). The amount of energy our bodies require varies considerably depending on such factors as weight, age, and muscular activity. About 100 kJ per kilogram of body weight per day is required to keep the body functioning at a minimal level. An average 70-kg (154-lb) person expends about 800 kJ/hr when doing light work, such as slow walking or light gardening. Strenuous activity, such as running, often requires 2000 kJ/hr or more. When the energy value, or caloric content, of our food exceeds the energy we expend, our body stores the surplus as fat.

▲ Figure 5.23 **Labels of processed foods showing nutritional information.** Such labels give information about the quantities of different nutrients and the energy value (caloric value) in an average serving.

TABLE 5.4 ■ Compositions and Fuel Values of Some Common Foods

	Approximate Composition (% by mass)			Fuel Value	
	Carbohydrate	Fat	Protein	kJ/g	kcal/g (Cal/g)
Carbohydrate	100	–	–	17	4
Fat	–	100	–	38	9
Protein	–	–	100	17	4
Apples	13	0.5	0.4	2.5	0.59
Beer*	1.2	–	0.3	1.8	0.42
Bread	52	3	9	12	2.8
Cheese	4	37	28	20	4.7
Eggs	0.7	10	13	6.0	1.4
Fudge	81	11	2	18	4.4
Green beans	7.0	–	1.9	1.5	0.38
Hamburger	–	30	22	15	3.6
Milk (whole)	5.0	4.0	3.3	3.0	0.74
Peanuts	22	39	26	23	5.5

*Beers typically contain 3.5% ethanol, which has fuel value.

GIVE IT SOME THOUGHT

Which releases the greatest amount of energy per gram upon metabolism, carbohydrates, proteins, or fats?

■ SAMPLE EXERCISE 5.13 | Comparing Fuel Values

A plant such as celery contains carbohydrates in the form of starch and cellulose. These two kinds of carbohydrates have essentially the same fuel values when combusted in a bomb calorimeter. When we consume celery, however, our bodies receive fuel value from the starch only. What can we conclude about the difference between starch and cellulose as foods?

SOLUTION

If cellulose does not provide fuel value, we must conclude that it is not converted in the body into CO_2 and H_2O, as starch is. A slight, but critical, difference in the structures of starch and cellulose explains why only starch is broken down into glucose in the body. Cellulose passes through without undergoing significant chemical change. It serves as fiber, or roughage, in the diet, but provides no caloric value.

■ PRACTICE EXERCISE

The nutritional label on a bottle of canola oil indicates that 10 g of the oil has an energy value of 86 kcal. A similar label on a bottle of pancake syrup indicates that 60 mL (about 60 g) has an energy value of 200 kcal. Account for the difference.
Answer: The oil has a fuel value of 8.6 kcal/g, whereas the syrup has a fuel value of about 3.3 kcal/g. The higher fuel value for the canola oil arises because the oil is essentially pure fat, whereas the syrup is a solution of sugars (carbohydrates) in water. The oil has a higher fuel value per gram; in addition, the syrup is diluted by water.

■ SAMPLE EXERCISE 5.14 | Estimating the Fuel Value of a Food from Its Composition

(a) A 28-g (1-oz) serving of a popular breakfast cereal served with 120 mL of skim milk provides 8 g protein, 26 g carbohydrates, and 2 g fat. Using the average fuel values of these kinds of substances, estimate the energy value (caloric content) of this serving.
(b) A person of average weight uses about 100 Cal/mi when running or jogging. How many servings of this cereal provide the energy value requirements for running 3 mi?

SOLUTION

(a) **Analyze:** The energy value of the serving will be the sum of the energy values of the protein, carbohydrates, and fat.

Plan: We are given the masses of the protein, carbohydrates, and fat contained in the combined . We can use the data in Table 5.4 to convert these masses to their energy values, which we can sum to get the total energy value.

Solve: $(8 \text{ g protein})\left(\dfrac{17 \text{ kJ}}{1 \text{ g protein}}\right) + (26 \text{ g carbohydrate})\left(\dfrac{17 \text{ kJ}}{1 \text{ g carbohydrate}}\right) +$

$$(2 \text{ g fat})\left(\dfrac{38 \text{ kJ}}{1 \text{ g fat}}\right) = 650 \text{ kJ (to two significant figures)}$$

This corresponds to 160 kcal:

$$(650 \text{ kJ})\left(\dfrac{1 \text{ kcal}}{4.18 \text{ kJ}}\right) = 160 \text{ kcal}$$

Recall that the dietary Calorie is equivalent to 1 kcal. Thus, the serving provides 160 Cal.

(b) Analyze: Here we are faced with the reverse problem, calculating the quantity of food that provides a specific energy value.

Plan: The problem statement provides a conversion factor between Calories and miles. The answer to part (a) provides us with a conversion factor between servings and Calories.

Solve: We can use these factors in a straightforward dimensional analysis to determine the number of servings needed, rounded to the nearest whole number:

$$\text{Servings} = (3 \text{ mi})\left(\dfrac{100 \text{ Cal}}{1 \text{ mi}}\right)\left(\dfrac{1 \text{ serving}}{160 \text{ Cal}}\right) = 2 \text{ servings}$$

■ PRACTICE EXERCISE

(a) Dry red beans contain 62% carbohydrate, 22% protein, and 1.5% fat. Estimate the fuel value of these beans. **(b)** During a very light activity, such as reading or watching television, the average adult uses about 7 kJ/min. How many minutes of such activity can be sustained by the energy provided by a serving of chicken noodle soup containing 13 g protein, 15 g carbohydrate, and 5 g fat?
Answers: **(a)** 15 kJ/g, **(b)** 95 min

Fuels

The elemental compositions and fuel values of several common fuels are compared in Table 5.5▼. During the complete combustion of fuels, carbon is converted to CO_2 and hydrogen is converted to H_2O, both of which have large negative enthalpies of formation. Consequently, the greater the percentage of carbon and hydrogen in a fuel, the higher its fuel value. Compare, for example, the compositions and fuel values of bituminous coal and wood. The coal has a higher fuel value because of its greater carbon content.

In 2005 the United States consumed 1.05×10^{17} kJ of energy. This value corresponds to an average daily energy consumption per person of 9.6×10^5 kJ which is roughly 100 times greater than the per capita food-energy needs.

TABLE 5.5 ■ Fuel Values and Compositions of Some Common Fuels

	Approximate Elemental Composition (mass %)			
	C	H	O	Fuel Value (kJ/g)
Wood (pine)	50	6	44	18
Anthracite coal (Pennsylvania)	82	1	2	31
Bituminous coal (Pennsylvania)	77	5	7	32
Charcoal	100	0	0	34
Crude oil (Texas)	85	12	0	45
Gasoline	85	15	0	48
Natural gas	70	23	0	49
Hydrogen	0	100	0	142

▲ Figure 5.24 Sources of energy consumed in the United States. In 2005 the United States consumed a total of 1.05×10^{17} kJ of energy.

Nuclear (8.1%)
Renewable energy (6.0%)
Coal (22.9%)
Petroleum (40.4%)
Natural gas (22.6%)

Although the population of the United States is only about 4.5% of the world's population, the U.S. accounts for nearly one-fourth of the world's total energy consumption. Figure 5.24 ◄ illustrates the sources of this energy.

Coal, petroleum, and natural gas, which are the world's major sources of energy, are known as **fossil fuels**. All have formed over millions of years from the decomposition of plants and animals and are being depleted far more rapidly than they are being formed. **Natural gas** consists of gaseous hydrocarbons, compounds of hydrogen and carbon. It contains primarily methane (CH_4), with small amounts of ethane (C_2H_6), propane (C_3H_8), and butane (C_4H_{10}). We determined the fuel value of propane in Sample Exercise 5.11. **Petroleum** is a liquid composed of hundreds of compounds, most of which are hydrocarbons, with the remainder being chiefly organic compounds containing sulfur, nitrogen, or oxygen. **Coal**, which is solid, contains hydrocarbons of high molecular weight as well as compounds containing sulfur, oxygen, or nitrogen. Coal is the most abundant fossil fuel; it constitutes 80% of the fossil fuel reserves of the United States and 90% of those of the world. However, the use of coal presents a number of problems. Coal is a complex mixture of substances, and it contains components that cause air pollution. When coal is combusted, the sulfur it contains is converted mainly to sulfur dioxide, SO_2, a very troublesome air pollutant. Because coal is a solid, recovery from its underground deposits is expensive and often dangerous. Furthermore, coal deposits are not always close to locations of high-energy use, so there are often substantial shipping costs.

One promising way to utilize coal reserves is to use them to produce a mixture of gaseous hydrocarbons called *syngas* (for "*syn*thesis *gas*"). In this process, called *coal gasification*, the coal typically is pulverized and treated with superheated steam. Sulfur-containing compounds, water, and carbon dioxide can be removed from the products, leading to a gaseous mixture of CH_4, H_2, and CO, all of which have high fuel values:

$$\text{Coal + steam} \xrightarrow{\text{conversion}} \text{complex mixture} \xrightarrow{\text{purification}}$$

$$\text{mixture of } CH_4, H_2, CO \text{ (syngas)}$$

Chemistry Put to Work   THE HYBRID CAR

The hybrid cars now entering the automobile marketplace nicely illustrate the convertibility of energy from one form to another. Hybrid cars run on either gasoline or electricity. The so-called "full hybrids" are cars capable of running at low speeds on a battery-powered electrical engine alone (Figure 5.25 ►). The "mild hybrid" cars are best described as electrically assisted gasoline engines.

Full hybrid cars are more efficient than the mild hybrid designs but are more costly to produce and require more technological advances than the mild hybrid versions. The mild hybrids are likely to be more widely produced and sold within the next several years. Let's consider how they operate and some of the interesting thermodynamic considerations they incorporate.

Figure 5.26 ► shows a schematic diagram of the power system for a mild hybrid car. In addition to the 12-volt battery that is standard on conventional autos, the hybrid car typically carries a 48-volt battery pack. The electrical energy from this battery pack is not employed to move the car directly; an electrical engine capable of doing that, as in the full hybrids, requires from 150 to 300 volts. (The popular Toyota Prius has a battery pack consisting of 228 cells of 1.2 V each, thereby generating a nominal voltage of 270 V.)

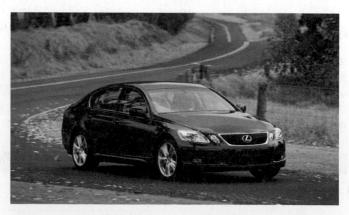

▲ Figure 5.25 A hybrid car. The Lexus GS450h is the first rear-drive full hybrid luxury car, evidence of the widespread movement of hybrid cars into the marketplace.

In the mild hybrid cars the added electrical source is employed to run various auxiliary devices that would otherwise be run off the gasoline engine, such as water pump, power steering, and air systems. To save on energy, when the hybrid

Because it is gaseous, syngas can be easily transported in pipelines. Additionally, because much of the sulfur in coal is removed during the gasification process, combustion of syngas causes less air pollution than burning coal. For these reasons, the economical conversion of coal and petroleum into "cleaner" fuels such as syngas and hydrogen is a very active area of current research in chemistry and engineering.

Other Energy Sources

Nuclear energy is energy that is released in either the splitting or the fusion (combining) of the nuclei of atoms. Nuclear power is currently used to produce about 22% of the electric power in the United States and comprises about 8% of the total U.S. energy production (Figure 5.24). Nuclear energy is, in principle, free of the polluting emissions that are a major problem in the generation of energy from fossil fuels. However, nuclear power plants produce radioactive waste products, and their use has therefore been fraught with controversy. We will discuss issues related to the production of nuclear energy in Chapter 21.

Fossil fuel and nuclear energy are *nonrenewable* sources of energy; they are limited resources that we are consuming at a much greater rate than they are being regenerated. Eventually these fuels will be expended, although estimates vary greatly as to when this will occur. Because nonrenewable sources of energy will eventually be used up, a great deal of research is being conducted into sources of **renewable energy**, energy sources that are essentially inexhaustible. Renewable energy sources include *solar energy* from the Sun, *wind energy* harnessed by windmills, *geothermal energy* from the heat stored in the mass of Earth, *hydroelectric energy* from flowing rivers, and *biomass energy* from crops, such as trees and corn, and from biological waste matter. Currently, renewable sources provide about 6.0% of the U.S. annual energy consumption, with hydroelectric and biomass sources as the major contributors.

Providing our future energy needs will most certainly depend on developing the technology to harness solar energy with greater efficiency. Solar energy is the world's largest energy source. On a clear day about 1 kJ of solar energy

car comes to a stop, the engine shuts off. It restarts automatically when the driver presses the accelerator. This feature saves fuel that would otherwise be used to keep the engine idling at traffic lights and other stopping situations.

▲ **Figure 5.26 Schematic diagram of a mild hybrid car.**
The 48-volt battery pack provides energy for operating several auxiliary functions. It is recharged from the engine and through the braking system.

The idea is that the added electrical system will improve overall fuel efficiency of the car. The added battery, moreover, is not supposed to need recharging from an external power source. Where, then, can the improved fuel efficiency come from? Clearly, if the battery pack is to continue to operate auxiliary devices such as the water pump, it must be recharged. We can think of it this way: The source of the voltage that the battery develops is a chemical reaction. Recharging the battery thus represents a conversion of mechanical energy into chemical potential energy. The recharging occurs in part through the agency of an alternator, which runs off the engine and provides a recharging voltage. In the mild hybrid car, the braking system serves as an additional source of mechanical energy for recharging. When the brakes are applied in a conventional car, the car's kinetic energy is converted through the brake pads in the wheels into heat, so no useful work is done. In the hybrid car, some of the car's kinetic energy is used to recharge the battery when the brakes are applied. Thus, kinetic energy that would otherwise be dissipated as heat is partially converted into useful work. Overall, the mild hybrid cars are expected to yield 10–20% improvements in fuel economy as compared with similar conventional cars.

reaches each square meter of Earth's surface every second. The average solar energy that falls on only 0.1% of U.S. land area is equivalent to all the energy that this nation currently uses. Harnessing this energy is difficult because it is dilute (it is distributed over a wide area) and it varies with time of day and weather conditions. The effective use of solar energy will depend on the development of some means of storing the collected energy for use at a later time. Any practical means for doing this will almost certainly involve use of an endothermic chemical process that can be later reversed to release heat. One such reaction is the following:

$$CH_4(g) + H_2O(g) + heat \longleftrightarrow CO(g) + 3 H_2(g)$$

This reaction proceeds in the forward direction at high temperatures, which can be obtained in a solar furnace. The CO and H_2 formed in the reaction could then be stored and allowed to react later, with the heat released being put to useful work.

A survey taken about 25 years ago at Walt Disney's EPCOT Center revealed that nearly 30% of the visitors expected that solar energy would be the principal source of energy in the United States in the year 2000. The future of solar energy has proven to be a lot like the Sun itself: big and bright but farther away than it seems. Nevertheless, important progress has been made in recent years. Perhaps the most direct way to make use of the Sun's energy is to convert it directly into electricity by use of photovoltaic devices, sometimes called *solar cells*. The efficiencies of solar energy conversion by use of such devices have increased dramatically during the past few years because of intensive research efforts. Photovoltaics are vital to the generation of power for the space station. More significant for our Earth-bound concerns, the unit costs of solar panels have been steadily declining, even as their efficiencies have improved dramatically.

In 2006 construction was started in southern Portugal on what the builders claim will be the world's biggest solar energy power station. The first module of the station is planned to cover about 150 acres and is capable of generating 11 MW (megawatts) of electrical power—enough for 8000 homes. When fully constructed, the plant is projected to cover 620 acres and supply over 100 MW of power. Several other large solar plants with capacities over 100 MW have also been announced in Australia, Israel, and China.

■ SAMPLE INTEGRATIVE EXERCISE | Putting Concepts Together

Trinitroglycerin, $C_3H_5N_3O_9$ (usually referred to simply as nitroglycerin), has been widely used as an explosive. Alfred Nobel used it to make dynamite in 1866. Rather surprisingly, it also is used as a medication, to relieve angina (chest pains resulting from partially blocked arteries to the heart) by dilating the blood vessels. The enthalpy of decomposition at 1 atm pressure of trinitroglycerin to form nitrogen gas, carbon dioxide gas, liquid water, and oxygen gas at 25 °C is -1541.4 kJ/mol. **(a)** Write a balanced chemical equation for the decomposition of trinitroglycerin. **(b)** Calculate the standard heat of formation of trinitroglycerin. **(c)** A standard dose of trinitroglycerin for relief of angina is 0.60 mg. If the sample is eventually oxidized in the body (not explosively, though!) to nitrogen gas, carbon dioxide gas, and liquid water, what number of calories is released? **(d)** One common form of trinitroglycerin melts at about 3 °C. From this information and the formula for the substance, would you expect it to be a molecular or ionic compound? Explain. **(e)** Describe the various conversions of forms of energy when trinitroglycerin is used as an explosive to break rockfaces in highway construction.

SOLUTION

(a) The general form of the equation we must balance is

$$C_3H_5N_3O_9(l) \longrightarrow N_2(g) + CO_2(g) + H_2O(l) + O_2(g)$$

We go about balancing in the usual way. To obtain an even number of nitrogen atoms on the left, we multiply the formula for $C_3H_5N_3O_9$ by 2. This then gives us 3 mol of N_2, 6 mol of CO_2 and 5 mol of H_2O. Everything is balanced except for oxygen. We have an odd number of oxygen atoms on the right. We can balance the oxygen by adding $\frac{1}{2}$ mol of O_2 on the right:

$$2 C_3H_5N_3O_9(l) \longrightarrow 3 N_2(g) + 6 CO_2(g) + 5 H_2O(l) + \tfrac{1}{2} O_2(g)$$

We multiply through by 2 to convert all coefficients to whole numbers:

$$4\,C_3H_5N_3O_9(l) \longrightarrow 6\,N_2(g) + 12\,CO_2(g) + 10\,H_2O(l) + O_2(g)$$

(At the temperature of the explosion, water is a gas. The rapid expansion of the gaseous products creates the force of an explosion.)

(b) The heat of formation is the enthalpy change in the balanced chemical equation:

$$3\,C(s) + \tfrac{3}{2}N_2(g) + \tfrac{5}{2}H_2(g) + \tfrac{9}{2}O_2(g) \longrightarrow C_3H_5N_3O_9(l) \qquad \Delta H_f^\circ = ?$$

We can obtain the value of ΔH_f° by using the equation for the heat of decomposition of trinitroglycerin:

$$4\,C_3H_5N_3O_9(l) \longrightarrow 6\,N_2(g) + 12\,CO_2(g) + 10\,H_2O(l) + O_2(g)$$

The enthalpy change in this reaction is $4(-1541.4 \text{ kJ}) = -6165.6$ kJ. [We need to multiply by 4 because there are 4 mol of $C_3H_5N_3O_9(l)$ in the balanced equation.] This enthalpy change is given by the sum of the heats of formation of the products minus the heats of formation of the reactants, each multiplied by its coefficient in the balanced equation:

$$-6165.6 \text{ kJ} = \{6\Delta H_f^\circ[N_2(g)] + 12\Delta H_f^\circ[CO_2(g)] + 10\Delta H_f^\circ[H_2O(l)] + \Delta H_f^\circ[O_2(g)]\} \\ - 4\Delta H_f^\circ(C_3H_5N_3O_9(l))$$

The ΔH_f° values for $N_2(g)$ and $O_2(g)$ are zero, by definition. We look up the values for $H_2O(l)$ and $CO_2(g)$ from Table 5.3 and find that

$$-6165.6 \text{ kJ} = 12(-393.5 \text{ kJ}) + 10(-285.8 \text{ kJ}) - 4\Delta H_f^\circ(C_3H_5N_3O_9(l))$$

$$\Delta H_f^\circ(C_3H_5N_3O_9(l)) = -353.6 \text{ kJ/mol}$$

(c) We know that on oxidation 1 mol of $C_3H_5N_3O_9(l)$ yields 1541.4 kJ. We need to calculate the number of moles of in $C_3H_5N_3O_9(l)$ in 0.60 mg:

$$0.60 \times 10^{-3} \text{ g } C_3H_5N_3O_9\left(\frac{1 \text{ mol } C_3H_5N_3O_9}{227 \text{ g } C_3H_5N_3O_9}\right)\left(\frac{1541.4 \text{ kJ}}{1 \text{ mol } C_3H_5N_3O_9}\right) = 4.1 \times 10^{-3} \text{ kJ}$$

$$= 4.1 \text{ J}$$

(d) Because trinitroglycerin melts below room temperature, we expect that it is a molecular compound. With few exceptions, ionic substances are generally hard, crystalline materials that melt at high temperatures. ∞ (Sections 2.5 and 2.6) Also, the molecular formula suggests that it is likely to be a molecular substance. All the elements of which it is composed are nonmetals.

(e) The energy stored in trinitroglycerin is chemical potential energy. When the substance reacts explosively, it forms substances such as carbon dioxide, water, and nitrogen gas, which are of lower potential energy. In the course of the chemical transformation, energy is released in the form of heat; the gaseous reaction products are very hot. This very high heat energy is transferred to the surroundings; the gases expand against the surroundings, which may be solid materials. Work is done in moving the solid materials and imparting kinetic energy to them. For example, a chunk of rock might be impelled upward. It has been given kinetic energy by transfer of energy from the hot, expanding gases. As the rock rises, its kinetic energy is transformed into potential energy. Eventually, it again acquires kinetic energy as it falls to Earth. When it strikes Earth, its kinetic energy is converted largely to thermal energy, though some work may be done on the surroundings as well.

CHAPTER REVIEW

SUMMARY AND KEY TERMS

Introduction and Section 5.1 **Thermodynamics** is the study of energy and its transformations. In this chapter we have focused on **thermochemistry**, the transformations of energy—especially heat—during chemical reactions. An object can possess energy in two forms: (1) **kinetic energy** is the energy due to the motion of the object; and (2) **potential energy** is the energy that an object possesses by virtue of its position relative to other objects. An electron in motion near a proton, for example, has kinetic energy because of its motion and potential energy because of its electrostatic attraction to the proton. The SI unit of energy is the **joule** (J): $1 \text{ J} = 1 \text{ kg-m}^2/\text{s}^2$. Another common energy unit is the **calorie** (cal), which was originally defined as the quantity of energy necessary to increase the temperature of 1 g of water by 1 °C: $1 \text{ cal} = 4.184 \text{ J}$. When we study thermodynamic properties, we define a specific amount of matter as the **system**. Everything outside the system is the **surroundings**. When we study a chemical reaction, the system is generally the reactants and products. A closed system can exchange energy, but not matter, with the surroundings. Energy can be transferred between the system and the surroundings as work or heat. **Work** is the energy expended to move an object against a **force**. **Heat** is the energy that is transferred from a hotter object to a colder one. **Energy** is the capacity to do work or to transfer heat.

Section 5.2 The **internal energy** of a system is the sum of all the kinetic and potential energies of its component parts. The internal energy of a system can change because of energy transferred between the system and the surroundings. According to the **first law of thermodynamics**, the change in the internal energy of a system, ΔE, is the sum of the heat, q, transferred into or out of the system and the work, w, done on or by the system: $\Delta E = q + w$. Both q and w have a sign that indicates the direction of energy transfer. When heat is transferred from the surroundings to the system, $q > 0$. Likewise, when the surroundings do work on the system, $w > 0$. In an **endothermic** process the system absorbs heat from the surroundings; in an **exothermic** process the system releases heat to the surroundings. The internal energy, E, is a **state function**. The value of any state function depends only on the state or condition of the system and not on the details of how it came to be in that state. The heat, q, and the work, w, are not state functions; their values depend on the particular way in which a system changes its state.

Sections 5.3 and 5.4 When a gas is produced or consumed in a chemical reaction occurring at constant pressure, the system may perform **pressure-volume (P-V) work** against the prevailing pressure. For this reason, we define a new state function called **enthalpy**, H, which is related to energy: $H = E + PV$. In systems where only pressure-volume work that is due to gases is involved, the change in the enthalpy of a system, ΔH, equals the heat gained or lost by the system at constant pressure, $\Delta H = q_P$. For an endothermic process, $\Delta H > 0$; for an exothermic process, $\Delta H < 0$. Every substance has a characteristic enthalpy. In a chemical process, the **enthalpy of reaction** is the enthalpy of the products minus the enthalpy of the reactants: $\Delta H_{rxn} = H \text{ (products)} - H \text{ (reactants)}$. Enthalpies of reaction follow some simple rules: (1) The enthalpy of reaction is proportional to the amount of reactant that reacts. (2) Reversing a reaction changes the sign of ΔH. (3) The enthalpy of reaction depends on the physical states of the reactants and products.

Section 5.5 The amount of heat transferred between the system and the surroundings is measured experimentally by **calorimetry**. A **calorimeter** measures the temperature change accompanying a process. The temperature change of a calorimeter depends on its **heat capacity**, the amount of heat required to raise its temperature by 1 K. The heat capacity for one mole of a pure substance is called its **molar heat capacity**; for one gram of the substance, we use the term

specific heat. Water has a very high specific heat, 4.18 J/g-K. The amount of heat, q, absorbed by a substance is the product of its specific heat (C_s), its mass, and its temperature change: $q = C_s \times m \times \Delta T$. If a calorimetry experiment is carried out under a constant pressure, the heat transferred provides a direct measure of the enthalpy change of the reaction. Constant-volume calorimetry is carried out in a vessel of fixed volume called a **bomb calorimeter**. Bomb calorimeters are used to measure the heat evolved in combustion reactions. The heat transferred under constant-volume conditions is equal to ΔE. However, corrections can be applied to ΔE values to yield enthalpies of combustion.

Section 5.6 Because enthalpy is a state function, ΔH depends only on the initial and final states of the system. Thus, the enthalpy change of a process is the same whether the process is carried out in one step or in a series of steps. **Hess's law** states that if a reaction is carried out in a series of steps, ΔH for the reaction will be equal to the sum of the enthalpy changes for the steps. We can therefore calculate ΔH for any process, as long as we can write the process as a series of steps for which ΔH is known.

Section 5.7 The **enthalpy of formation**, ΔH_f, of a substance is the enthalpy change for the reaction in which the substance is formed from its constituent elements. The **standard enthalpy change** of a reaction, $\Delta H°$, is the enthalpy change when all reactants and products are at 1 atm pressure and a specific temperature, usually 298 K (25 °C). Combining these ideas, the **standard enthalpy of formation**, $\Delta H_f°$, of a substance is the change in enthalpy for the reaction that forms one mole of the substance from its elements in their most stable form with all reactants and products at 1 atm pressure and usually 298 K. For any element in its most stable state at 298 K and 1 atm pressure, $\Delta H_f° = 0$. The standard enthalpy change for any reaction can be readily calculated from the standard enthalpies of formation of the reactants and products in the reaction:

$$\Delta H°_{rxn} = \Sigma n \Delta H_f° \text{(products)} - \Sigma m \Delta H_f° \text{(reactants)}$$

Section 5.8 The **fuel value** of a substance is the heat released when one gram of the substance is combusted. Different types of foods have different fuel values and differing abilities to be stored in the body. The most common fuels are hydrocarbons that are found as **fossil fuels**, such as **natural gas**, **petroleum**, and **coal**. Coal is the most abundant fossil fuel, but the sulfur present in most coals causes air pollution. Coal gasification is one possible way to use existing resources as sources of cleaner energy in the future. Sources of **renewable energy** include solar energy, wind energy, biomass, and hydroelectric energy. Nuclear power does not utilize fossil fuels but does create serious waste-disposal problems.

KEY SKILLS

- Interconvert energy units.
- Express the relationships among the quantities q, w, ΔE, and ΔH. Understand their sign conventions, including how the signs of q and ΔH relate to whether a process is exothermic or endothermic.

- State the first law of thermodynamics.
- Understand the concept of a state function and be able to give examples.
- Use thermochemical equations to relate the amount of heat energy transferred in reactions at constant pressure (ΔH) to the amount of substance involved in the reaction.
- Calculate the heat transferred in a process from temperature measurements together with heat capacities or specific heats (calorimetry).
- Use Hess's law to determine enthalpy changes for reactions.
- Use standard enthalpies of formation to calculate ΔH° for reactions.

KEY EQUATIONS

- $E_k = \frac{1}{2}mv^2$ [5.1]   Kinetic energy

- $\Delta E = E_{final} - E_{initial}$ [5.4]   The change in internal energy

- $\Delta E = q + w$ [5.5]   Relates the change in internal energy to heat and work (the first law of thermodynamics)

- $w = -P\,\Delta V$ [5.7]   The work done by an expanding gas at constant pressure

- $\Delta H = \Delta E + P\,\Delta V = q_P$ [5.10]   Enthalpy change at constant pressure

- $q = C_s \times m \times \Delta T$ [5.22]   Heat gained or lost based on specific heat, mass, and temperature change.

- $\Delta H^\circ_{rxn} = \Sigma n \Delta H^\circ_f(\text{products}) - \Sigma m \Delta H^\circ_f(\text{reactants})$ [5.31]   Standard enthalpy change of a reaction

VISUALIZING CONCEPTS

5.1 Imagine a book that is falling from a shelf. At a particular moment during its fall, the book has a kinetic energy of 13 J and a potential energy with respect to the floor of 72 J. How does the book's kinetic energy and its potential energy change as it continues to fall? What is its total kinetic energy at the instant just before it strikes the floor? [Section 5.1]

5.2 Consider the accompanying energy diagram.
(a) Does this diagram represent an increase or decrease in the internal energy of the system?
(b) What sign is given to ΔE for this process? **(c)** If there is no work associated with the process, is it exothermic or endothermic? [Section 5.2]

5.3 The contents of the closed box in each of the following illustrations represent a system, and the arrows show the changes to the system during some process. The lengths of the arrows represent the relative magnitudes of q and w **(a)** Which of these processes is endothermic? **(b)** For which of these processes, if any, is $\Delta E < 0$? **(c)** For which process, if any, is there a net gain in internal energy? [Section 5.2]

5.4 Imagine that you are climbing a mountain. **(a)** Is the distance you travel to the top a state function? Why or why not? **(b)** Is the change in elevation between your base camp and the peak a state function? Why or why not? [Section 5.2]

5.5 In the cylinder diagrammed below, a chemical process occurs at constant temperature and pressure. **(a)** Is the sign of w indicated by this change positive or negative? **(b)** If the process is endothermic, does the internal energy of the system within the cylinder increase or decrease during the change and is ΔE positive or negative? [Sections 5.2 and 5.3]

5.6 Imagine a container placed in a tub of water, as depicted in the accompanying diagram. **(a)** If the contents of the container are the system and heat is able to flow through the container walls, what qualitative changes will occur in the temperatures of the system and in its surroundings? What is the sign of q associated with each change? From the system's perspective, is the process exothermic or endothermic? **(b)** If neither the volume nor the pressure of the system changes during the process, how is the change in internal energy related to the change in enthalpy? [Sections 5.2 and 5.3]

350 K   290 K

5.7 Which will release more heat as it cools from 50 °C to 25 °C, 1 kg of water or 1 kg of aluminum? How do you know? [Section 5.5]

5.8 A gas-phase reaction was run in an apparatus designed to maintain a constant pressure. **(a)** Write a balanced chemical equation for the reaction depicted, and predict whether w is positive, negative, or zero. **(b)** Using data from Appendix C, determine ΔH for the formation of one mole of the product. Why is this enthalpy change called the enthalpy of formation of the involved product? [Sections 5.3 and 5.7]

P   P

● O
● N

5.9 Consider the two diagrams below. **(a)** Based on (*i*), write an equation showing how ΔH_A is related to ΔH_B and ΔH_C. How do both diagram (*i*) and your equation relate to the fact that enthalpy is a state function? **(b)** Based on (*ii*), write an equation relating ΔH_Z to the other enthalpy changes in the diagram. **(c)** How do these diagrams relate to Hess's law? [Section 5.6]

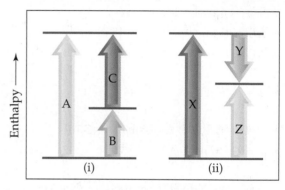

(i)   (ii)

5.10 Does ΔH_{rxn} for the reaction represented by the following equation equal the standard enthalpy of formation for $CH_3OH(l)$? Why or why not? [Section 5.7]

$$C(graphite) + 4\,H(g) + O(g) \longrightarrow CH_3OH(l)$$

EXERCISES

The Nature of Energy

5.11 In what two ways can an object possess energy? How do these two ways differ from one another?

5.12 Suppose you toss a tennis ball upward. **(a)** Does the kinetic energy of the ball increase or decrease as it moves higher? **(b)** What happens to the potential energy of the ball as it moves higher? **(c)** If the same amount of energy were imparted to a ball the same size as a tennis ball, but of twice the mass, how high would it go in comparison to the tennis ball? Explain your answers.

5.13 **(a)** Calculate the kinetic energy in joules of a 45-g golf ball moving at 61 m/s. **(b)** Convert this energy to calories. **(c)** What happens to this energy when the ball lands in a sand trap?

5.14 **(a)** What is the kinetic energy in joules of an 850-lb motorcycle moving at 66 mph? **(b)** By what factor will the kinetic energy change if the speed of the motorcycle is decreased to 33 mph? **(c)** Where does the kinetic energy of the motorcycle go when the rider brakes to a stop?

5.15 The use of the British thermal unit (Btu) is common in much engineering work. A Btu is the amount of heat required to raise the temperature of 1 lb of water by 1 °F. Calculate the number of joules in a Btu.

5.16 A watt is a measure of power (the rate of energy change) equal to 1 J/s. **(a)** Calculate the number of joules in a kilowatt-hour. **(b)** An adult person radiates heat to the surroundings at about the same rate as a 100-watt electric incandescent lightbulb. What is the total amount of energy in kcal radiated to the surroundings by an adult in 24 hours?

5.17 **(a)** What is meant by the term *system* in thermodynamics? **(b)** What is a *closed system*?

5.18 In a thermodynamic study a scientist focuses on the properties of a solution in an apparatus as illustrated. A solution is continuously flowing into the apparatus at the top and out at the bottom, such that the amount of solution in the apparatus is constant with time. **(a)** Is the solution in the apparatus a closed system, open system, or isolated system? Explain your choice. **(b)** If it is not a closed system, what could be done to make it a closed system?

In

Out

5.19 (a) What is work? (b) How do we determine the amount of work done, given the force associated with the work?

5.20 (a) What is heat? (b) Under what conditions is heat transferred from one object to another?

5.21 Identify the force present, and explain whether work is being performed in the following cases: (a) You lift a pencil off the top of a desk. (b) A spring is compressed to half its normal length.

5.22 Identify the force present, and explain whether work is done when (a) a positively charged particle moves in a circle at a fixed distance from a negatively charged particle; (b) an iron nail is pulled off a magnet.

The First Law of Thermodynamics

5.23 (a) State the first law of thermodynamics. (b) What is meant by the *internal energy* of a system? (c) By what means can the internal energy of a closed system increase?

5.24 (a) Write an equation that expresses the first law of thermodynamics in terms of heat and work. (b) Under what conditions will the quantities q and w be negative numbers?

5.25 Calculate ΔE, and determine whether the process is endothermic or exothermic for the following cases: (a) A system absorbs 105 kJ of heat from its surroundings while doing 29 kJ of work on the surroundings; (b) $q = 1.50$ kJ and $w = -657$ J; (c) the system releases 57.5 kJ of heat while doing 22.5 kJ of work on the surroundings.

5.26 For the following processes, calculate the change in internal energy of the system and determine whether the process is endothermic or exothermic: (a) A balloon is heated by adding 850 J of heat. It expands, doing 382 J of work on the atmosphere. (b) A 50-g sample of water is cooled from 30 °C to 15 °C, thereby losing approximately 3140 J of heat. (c) A chemical reaction releases 6.47 kJ of heat and does no work on the surroundings.

5.27 A gas is confined to a cylinder fitted with a piston and an electrical heater, as shown in the accompanying illustration. Suppose that current is supplied to the heater so that 100 J of energy is added. Consider two different situations. In case (1) the piston is allowed to move as the energy is added. In case (2) the piston is fixed so that it cannot move. (a) In which case does the gas have the higher temperature after addition of the electrical energy? Explain.

(b) What can you say about the values of q and w in each of these cases? (c) What can you say about the relative values of ΔE for the system (the gas in the cylinder) in the two cases?

5.28 Consider a system consisting of two oppositely charged spheres hanging by strings and separated by a distance r_1, as shown in the accompanying illustration. Suppose they are separated to a larger distance r_2, by moving them apart along a track. (a) What change, if any, has occurred in the potential energy of the system? (b) What effect, if any, does this process have on the value of ΔE? (c) What can you say about q and w for this process?

5.29 (a) What is meant by the term *state function?* (b) Give an example of a quantity that is a state function and one that is not. (c) Is work a state function? Why or why not?

5.30 Indicate which of the following is independent of the path by which a change occurs: (a) the change in potential energy when a book is transferred from table to shelf, (b) the heat evolved when a cube of sugar is oxidized to $CO_2(g)$ and $H_2O(g)$, (c) the work accomplished in burning a gallon of gasoline.

Enthalpy

5.31 (a) Why is the change in enthalpy usually easier to measure than the change in internal energy? (b) For a given process at constant pressure, ΔH is negative. Is the process endothermic or exothermic?

5.32 (a) Under what condition will the enthalpy change of a process equal the amount of heat transferred into or out of the system? (b) During a constant-pressure process the system absorbs heat from the surroundings. Does the enthalpy of the system increase or decrease during the process?

5.33 You are given ΔH for a process that occurs at constant pressure. What additional information do you need to determine ΔE for the process?

5.34 Suppose that the gas-phase reaction $2\,NO(g) + O_2(g) \longrightarrow 2\,NO_2(g)$ were carried out in a constant-volume container at constant temperature. Would the measured heat change represent ΔH or ΔE? If there is a difference, which quantity is larger for this reaction? Explain.

5.35 A gas is confined to a cylinder under constant atmospheric pressure, as illustrated in Figure 5.3. When the gas undergoes a particular chemical reaction, it releases 79 kJ of heat to its surroundings and does 18 kJ of P-V work on its surroundings. What are the values of ΔH and ΔE for this process?

5.36 A gas is confined to a cylinder under constant atmospheric pressure, as illustrated in Figure 5.3. When 378 J of heat is added to the gas, it expands and does 56 J of work on the surroundings. What are the values of ΔH and ΔE for this process?

5.37 The complete combustion of acetic acid, $CH_3COOH(l)$, to form $H_2O(l)$ and $CO_2(g)$ at constant pressure releases 871.7 kJ of heat per mole of CH_3COOH. **(a)** Write a balanced thermochemical equation for this reaction. **(b)** Draw an enthalpy diagram for the reaction.

5.38 The decomposition of zinc carbonate, $ZnCO_3(s)$, into zinc oxide, $ZnO(s)$, and $CO_2(g)$ at constant pressure requires the addition of 71.5 kJ of heat per mole of $ZnCO_3$. **(a)** Write a balanced thermochemical equation for the reaction. **(b)** Draw an enthalpy diagram for the reaction.

5.39 Consider the following reaction, which occurs at room temperature and pressure:

$$2\, Cl(g) \longrightarrow Cl_2(g) \qquad \Delta H = -243.4 \text{ kJ}$$

Which has the higher enthalpy under these conditions, $2\, Cl(g)$ or $Cl_2(g)$?

5.40 Without referring to tables, predict which of the following has the higher enthalpy in each case: **(a)** 1 mol $CO_2(s)$ or 1 mol $CO_2(g)$ at the same temperature, **(b)** 2 mol of hydrogen atoms or 1 mol of H_2, **(c)** 1 mol $H_2(g)$ and 0.5 mol $O_2(g)$ at 25 °C or 1 mol $H_2O(g)$ at 25 °C, **(d)** 1 mol $N_2(g)$ at 100 °C or 1 mol $N_2(g)$ at 300 °C.

5.41 Consider the following reaction:

$$2\, Mg(s) + O_2(g) \longrightarrow 2\, MgO(s) \qquad \Delta H = -1204 \text{ kJ}$$

(a) Is this reaction exothermic or endothermic? **(b)** Calculate the amount of heat transferred when 2.4 g of $Mg(s)$ reacts at constant pressure. **(c)** How many grams of MgO are produced during an enthalpy change of −96.0 kJ? **(d)** How many kilojoules of heat are absorbed when 7.50 g of $MgO(s)$ is decomposed into $Mg(s)$ and $O_2(g)$ at constant pressure?

5.42 Consider the following reaction:

$$CH_3OH(g) \longrightarrow CO(g) + 2\, H_2(g) \qquad \Delta H = +90.7 \text{ kJ}$$

(a) Is heat absorbed or released in the course of this reaction? **(b)** Calculate the amount of heat transferred when 45.0 g of $CH_3OH(g)$ is decomposed by this reaction at constant pressure. **(c)** For a given sample of CH_3OH, the enthalpy change on reaction is 25.8 kJ. How many grams of hydrogen gas are produced? What is the value of ΔH for the reverse of the previous reaction? **(d)** How

many kilojoules of heat are released when 50.9 g of $CO(g)$ reacts completely with $H_2(g)$ to form $CH_3OH(g)$ at constant pressure?

5.43 When solutions containing silver ions and chloride ions are mixed, silver chloride precipitates:

$$Ag^+(aq) + Cl^-(aq) \longrightarrow AgCl(s) \qquad \Delta H = -65.5 \text{ kJ}$$

(a) Calculate ΔH for production of 0.200 mol of AgCl by this reaction. **(b)** Calculate ΔH for the production of 2.50 g of AgCl. **(c)** Calculate ΔH when 0.150 mmol of AgCl dissolves in water.

5.44 At one time, a common means of forming small quantities of oxygen gas in the laboratory was to heat $KClO_3$:

$$2\, KClO_3(s) \longrightarrow 2\, KCl(s) + 3\, O_2(g) \qquad \Delta H = -89.4 \text{ kJ}$$

For this reaction, calculate ΔH for the formation of **(a)** 0.632 mol of O_2 and **(b)** 8.57 g of KCl. **(c)** The decomposition of $KClO_3$ proceeds spontaneously when it is heated. Do you think that the reverse reaction, the formation of $KClO_3$ from KCl and O_2, is likely to be feasible under ordinary conditions? Explain your answer.

5.45 Consider the combustion of liquid methanol, $CH_3OH(l)$:

$$CH_3OH(l) + \tfrac{3}{2} O_2(g) \longrightarrow CO_2(g) + 2\, H_2O(l)$$
$$\Delta H = -726.5 \text{ kJ}$$

(a) What is the enthalpy change for the reverse reaction? **(b)** Balance the forward reaction with whole-number coefficients. What is ΔH for the reaction represented by this equation? **(c)** Which is more likely to be thermodynamically favored, the forward reaction or the reverse reaction? **(d)** If the reaction were written to produce $H_2O(g)$ instead of $H_2O(l)$, would you expect the magnitude of ΔH to increase, decrease, or stay the same? Explain.

5.46 Consider the decomposition of liquid benzene, $C_6H_6(l)$, to gaseous acetylene, $C_2H_2(g)$:

$$C_6H_6(l) \longrightarrow 3\, C_2H_2(g) \qquad \Delta H = +630 \text{ kJ}$$

(a) What is the enthalpy change for the reverse reaction? **(b)** What is ΔH for the formation of 1 mol of acetylene? **(c)** Which is more likely to be thermodynamically favored, the forward reaction or the reverse reaction? **(d)** If $C_6H_6(g)$ were consumed instead of $C_6H_6(l)$, would you expect the magnitude of ΔH to increase, decrease, or stay the same? Explain.

Calorimetry

5.47 **(a)** What are the units of molar heat capacity? **(b)** What are the units of specific heat? **(c)** If you know the specific heat of copper, what additional information do you need to calculate the heat capacity of a particular piece of copper pipe?

5.48 Two solid objects, A and B, are placed in boiling water and allowed to come to temperature there. Each is then lifted out and placed in separate beakers containing 1000 g water at 10.0 °C. Object A increases the water temperature by 3.50 °C; B increases the water temperature by 2.60 °C. **(a)** Which object has the larger heat capacity? **(b)** What can you say about the specific heats of A and B?

5.49 **(a)** What is the specific heat of liquid water? **(b)** What is the molar heat capacity of liquid water? **(c)** What is the heat capacity of 185 g of liquid water? **(d)** How many kJ of heat are needed to raise the temperature of 10.00 kg of liquid water from 24.6 °C to 46.2 °C?

5.50 **(a)** Which substance in Table 5.2 requires the smallest amount of energy to increase the temperature of 50.0 g of that substance by 10 K? **(b)** Calculate the energy needed for this temperature change.

5.51 The specific heat of iron metal is 0.450 J/g-K. How many J of heat are necessary to raise the temperature of a 1.05-kg block of iron from 25.0 °C to 88.5 °C?

5.52 The specific heat of ethylene glycol is 2.42 J/g-K. How many J of heat are needed to raise the temperature of 62.0 g of ethylene glycol from 13.1 °C to 40.5 °C?

5.53 When a 9.55-g sample of solid sodium hydroxide dissolves in 100.0 g of water in a coffee-cup calorimeter (Figure 5.17), the temperature rises from 23.6 °C to 47.4 °C. Calculate ΔH (in kJ/mol NaOH) for the solution process

$$NaOH(s) \longrightarrow Na^+(aq) + OH^-(aq)$$

Assume that the specific heat of the solution is the same as that of pure water.

5.54 **(a)** When a 3.88-g sample of solid ammonium nitrate dissolves in 60.0 g of water in a coffee-cup calorimeter (Figure 5.17), the temperature drops from 23.0 °C to 18.4 °C. Calculate ΔH (in kJ/mol NH_4NO_3) for the solution process

$$NH_4NO_3(s) \longrightarrow NH_4^+(aq) + NO_3^-(aq)$$

Assume that the specific heat of the solution is the same as that of pure water. **(b)** Is this process endothermic or exothermic?

5.55 A 2.200-g sample of quinone ($C_6H_4O_2$) is burned in a bomb calorimeter whose total heat capacity is 7.854 kJ/°C. The temperature of the calorimeter increases from 23.44 °C to 30.57 °C. What is the heat of combustion per gram of quinone? Per mole of quinone?

5.56 A 1.800-g sample of phenol (C_6H_5OH) was burned in a bomb calorimeter whose total heat capacity is 11.66 kJ/°C. The temperature of the calorimeter plus contents increased from 21.36 °C to 26.37 °C. **(a)** Write a balanced chemical equation for the bomb calorimeter reaction. **(b)** What is the heat of combustion per gram of phenol? Per mole of phenol?

5.57 Under constant-volume conditions the heat of combustion of glucose ($C_6H_{12}O_6$) is 15.57 kJ/g. A 2.500-g sample of glucose is burned in a bomb calorimeter. The temperature of the calorimeter increased from 20.55 °C to 23.25 °C. **(a)** What is the total heat capacity of the calorimeter? **(b)** If the size of the glucose sample had been exactly twice as large, what would the temperature change of the calorimeter have been?

5.58 Under constant-volume conditions the heat of combustion of benzoic acid (C_6H_5COOH) is 26.38 kJ/g. A 1.640-g sample of benzoic acid is burned in a bomb calorimeter. The temperature of the calorimeter increases from 22.25 °C to 27.20 °C. **(a)** What is the total heat capacity of the calorimeter? **(b)** A 1.320-g sample of a new organic substance is combusted in the same calorimeter. The temperature of the calorimeter increases from 22.14 °C to 26.82 °C. What is the heat of combustion per gram of the new substance? **(c)** Suppose that in changing samples, a portion of the water in the calorimeter were lost. In what way, if any, would this change the heat capacity of the calorimeter?

Hess's Law

5.59 What is the connection between Hess's law and the fact that H is a state function?

5.60 Consider the following hypothetical reactions:

$$A \longrightarrow B \quad \Delta H = +30 \text{ kJ}$$
$$B \longrightarrow C \quad \Delta H = +60 \text{ kJ}$$

(a) Use Hess's law to calculate the enthalpy change for the reaction $A \longrightarrow C$. **(b)** Construct an enthalpy diagram for substances A, B, and C, and show how Hess's law applies.

5.61 Calculate the enthalpy change for the reaction

$$P_4O_6(s) + 2 O_2(g) \longrightarrow P_4O_{10}(s)$$

given the following enthalpies of reaction:

$$P_4(s) + 3 O_2(g) \longrightarrow P_4O_6(s) \quad \Delta H = -1640.1 \text{ kJ}$$
$$P_4(s) + 5 O_2(g) \longrightarrow P_4O_{10}(s) \quad \Delta H = -2940.1 \text{ kJ}$$

5.62 From the enthalpies of reaction

$$2 H_2(g) + O_2(g) \longrightarrow 2 H_2O(g) \quad \Delta H = -483.6 \text{ kJ}$$
$$3 O_2(g) \longrightarrow 2 O_3(g) \quad \Delta H = +284.6 \text{ kJ}$$

calculate the heat of the reaction

$$3 H_2(g) + O_3(g) \longrightarrow 3 H_2O(g)$$

5.63 From the enthalpies of reaction

$$H_2(g) + F_2(g) \longrightarrow 2 HF(g) \quad \Delta H = -537 \text{ kJ}$$
$$C(s) + 2 F_2(g) \longrightarrow CF_4(g) \quad \Delta H = -680 \text{ kJ}$$
$$2 C(s) + 2 H_2(g) \longrightarrow C_2H_4(g) \quad \Delta H = +52.3 \text{ kJ}$$

calculate ΔH for the reaction of ethylene with F_2:

$$C_2H_4(g) + 6 F_2(g) \longrightarrow 2 CF_4(g) + 4 HF(g)$$

5.64 Given the data

$$N_2(g) + O_2(g) \longrightarrow 2 NO(g) \quad \Delta H = +180.7 \text{ kJ}$$
$$2 NO(g) + O_2(g) \longrightarrow 2 NO_2(g) \quad \Delta H = -113.1 \text{ kJ}$$
$$2 N_2O(g) \longrightarrow 2 N_2(g) + O_2(g) \quad \Delta H = -163.2 \text{ kJ}$$

use Hess's law to calculate ΔH for the reaction

$$N_2O(g) + NO_2(g) \longrightarrow 3 NO(g)$$

Enthalpies of Formation

5.65 **(a)** What is meant by the term *standard conditions*, with reference to enthalpy changes? **(b)** What is meant by the term *enthalpy of formation*? **(c)** What is meant by the term *standard enthalpy of formation*?

5.66 **(a)** Why are tables of standard enthalpies of formation so useful? **(b)** What is the value of the standard enthalpy of formation of an element in its most stable form? **(c)** Write the chemical equation for the reaction whose enthalpy change is the standard enthalpy of formation of glucose, $C_6H_{12}O_6(s)$, $\Delta H_f^\circ[C_6H_{12}O_6]$.

5.67 For each of the following compounds, write a balanced thermochemical equation depicting the formation of one mole of the compound from its elements in their standard states and use Appendix C to obtain the value of ΔH_f°: **(a)** $NH_3(g)$, **(b)** $SO_2(g)$, **(c)** $RbClO_3(s)$, **(d)** $NH_4NO_3(s)$.

5.68 Write balanced equations that describe the formation of the following compounds from elements in their standard states, and use Appendix C to obtain the values of their standard enthalpies of formation: **(a)** $HBr(g)$, **(b)** $AgNO_3(s)$, **(c)** $Fe_2O_3(s)$, **(d)** $CH_3COOH(l)$.

5.69 The following is known as the thermite reaction [Figure 5.7(b)]:

$$2\,Al(s) + Fe_2O_3(s) \longrightarrow Al_2O_3(s) + 2\,Fe(s)$$

This highly exothermic reaction is used for welding massive units, such as propellers for large ships. Using standard enthalpies of formation in Appendix C, calculate ΔH° for this reaction.

5.70 Many cigarette lighters contain liquid butane, $C_4H_{10}(l)$. Using standard enthalpies of formation, calculate the quantity of heat produced when 5.00 g of butane is completely combusted in air under standard conditions.

5.71 Using values from Appendix C, calculate the standard enthalpy change for each of the following reactions:
(a) $2\,SO_2(g) + O_2(g) \longrightarrow 2\,SO_3(g)$
(b) $Mg(OH)_2(s) \longrightarrow MgO(s) + H_2O(l)$
(c) $N_2O_4(g) + 4\,H_2(g) \longrightarrow N_2(g) + 4\,H_2O(g)$
(d) $SiCl_4(l) + 2\,H_2O(l) \longrightarrow SiO_2(s) + 4\,HCl(g)$

5.72 Using values from Appendix C, calculate the value of ΔH° for each of the following reactions:
(a) $4\,HBr(g) + O_2(g) \longrightarrow 2\,H_2O(l) + 2\,Br_2(l)$
(b) $2\,Na(OH)(s) + SO_3(g) \longrightarrow Na_2SO_4(s) + H_2O(g)$
(c) $CH_4(g) + 4\,Cl_2(g) \longrightarrow CCl_4(l) + 4\,HCl(g)$
(d) $Fe_2O_3(s) + 6\,HCl(g) \longrightarrow 2\,FeCl_3(s) + 3\,H_2O(g)$

5.73 Complete combustion of 1 mol of acetone (C_3H_6O) liberates 1790 kJ:

$$C_3H_6O(l) + 4\,O_2(g) \longrightarrow 3\,CO_2(g) + 3\,H_2O(l)$$
$$\Delta H^\circ = -1790\ kJ$$

Using this information together with data from Appendix C, calculate the enthalpy of formation of acetone.

5.74 Calcium carbide (CaC_2) reacts with water to form acetylene (C_2H_2) and $Ca(OH)_2$. From the following enthalpy of reaction data and data in Appendix C, calculate ΔH_f° for $CaC_2(s)$:

$$CaC_2(s) + 2\,H_2O(l) \longrightarrow Ca(OH)_2(s) + C_2H_2(g)$$
$$\Delta H^\circ = -127.2\ kJ$$

5.75 Gasoline is composed primarily of hydrocarbons, including many with eight carbon atoms, called *octanes*. One of the cleanest-burning octanes is a compound called 2,3,4-trimethylpentane, which has the following structural formula:

$$\begin{array}{ccccc} & CH_3 & CH_3 & CH_3 & \\ & | & | & | & \\ H_3C- & CH- & CH- & CH- & CH_3 \end{array}$$

The complete combustion of one mole of this compound to $CO_2(g)$ and $H_2O(g)$ leads to $\Delta H^\circ = -5064.9$ kJ/mol. **(a)** Write a balanced equation for the combustion of 1 mol of $C_8H_{18}(l)$. **(b)** Write a balanced equation for the formation of $C_8H_{18}(l)$ from its elements. **(c)** By using the information in this problem and data in Table 5.3, calculate ΔH_f° for 2,3,4-trimethylpentane.

5.76 Naphthalene ($C_{10}H_8$) is a solid aromatic compound often sold as mothballs. The complete combustion of this substance to yield $CO_2(g)$ and $H_2O(l)$ at 25 °C yields 5154 kJ/mol. **(a)** Write balanced equations for the formation of naphthalene from the elements and for its combustion. **(b)** Calculate the standard enthalpy of formation of naphthalene.

5.77 Ethanol (C_2H_5OH) is currently blended with gasoline as an automobile fuel. **(a)** Write a balanced equation for the combustion of liquid ethanol in air. **(b)** Calculate the standard enthalpy change for the reaction, assuming $H_2O(g)$ as a product. **(c)** Calculate the heat produced per liter of ethanol by combustion of ethanol under constant pressure. Ethanol has a density of 0.789 g/mL. **(d)** Calculate the mass of CO_2 produced per kJ of heat emitted.

5.78 Methanol (CH_3OH) is used as a fuel in race cars. **(a)** Write a balanced equation for the combustion of liquid methanol in air. **(b)** Calculate the standard enthalpy change for the reaction, assuming $H_2O(g)$ as a product. **(c)** Calculate the heat produced by combustion per liter of methanol. Methanol has a density of 0.791 g/mL. **(d)** Calculate the mass of CO_2 produced per kJ of heat emitted.

Foods and Fuels

5.79 **(a)** What is meant by the term *fuel value*? **(b)** Which is a greater source of energy as food, 5 g of fat or 9 g of carbohydrate?

5.80 **(a)** Why are fats well suited for energy storage in the human body? **(b)** A particular chip snack food is composed of 12% protein, 14% fat, and the rest carbohydrate. What percentage of the calorie content of this food is fat? **(c)** How many grams of protein provide the same fuel value as 25 g of fat?

5.81 A serving of condensed cream of mushroom soup contains 7 g fat, 9 g carbohydrate, and 1 g protein. Estimate the number of Calories in a serving.

5.82 A pound of plain M&M® candies contains 96 g fat, 320 g carbohydrate, and 21 g protein. What is the fuel value in kJ in a 42-g (about 1.5 oz) serving? How many Calories does it provide?

5.83 The heat of combustion of fructose, $C_6H_{12}O_6$, is −2812 kJ/mol. If a fresh golden delicious apple weighing 4.23 oz (120 g) contains 16.0 g of fructose, what caloric content does the fructose contribute to the apple?

5.84 The heat of combustion of ethanol, $C_2H_5OH(l)$, is −1367 kJ/mol. A batch of Sauvignon Blanc wine contains 10.6% ethanol by mass. Assuming the density of the wine to be 1.0 g/mL, what caloric content does the alcohol (ethanol) in a 6-oz glass of wine (177 mL) have?

5.85 The standard enthalpies of formation of gaseous propyne (C_3H_4), propylene (C_3H_6), and propane (C_3H_8) are +185.4, +20.4, and −103.8 kJ/mol, respectively. **(a)** Calculate the heat evolved per mole on combustion of each substance to yield $CO_2(g)$ and $H_2O(g)$. **(b)** Calculate the heat evolved on combustion of 1 kg of each substance.

ADDITIONAL EXERCISES

5.87 At 20 °C (approximately room temperature) the average velocity of N_2 molecules in air is 1050 mph. **(a)** What is the average speed in m/s? **(b)** What is the kinetic energy (in J) of an N_2 molecule moving at this speed? **(c)** What is the total kinetic energy of 1 mol of N_2 molecules moving at this speed?

5.88 Suppose an Olympic diver who weighs 52.0 kg executes a straight dive from a 10-m platform. At the apex of the dive, the diver is 10.8 m above the surface of the water. **(a)** What is the potential energy of the diver at the apex of the dive, relative to the surface of the water? **(b)** Assuming that all the potential energy of the diver is converted into kinetic energy at the surface of the water, at what speed in m/s will the diver enter the water? **(c)** Does the diver do work on entering the water? Explain.

5.89 When a mole of dry ice, $CO_2(s)$, is converted to $CO_2(g)$ at atmospheric pressure and −78 °C, the heat absorbed by the system exceeds the increase in internal energy of the CO_2. Why is this so? What happens to the remaining energy?

5.90 The air bags that provide protection in autos in the event of an accident expand because of a rapid chemical reaction. From the viewpoint of the chemical reactants as the system, what do you expect for the signs of q and w in this process?

[5.91] An aluminum can of a soft drink is placed in a freezer. Later, you find that the can is split open and its contents frozen. Work was done on the can in splitting it open. Where did the energy for this work come from?

[5.92] A sample of gas is contained in a cylinder-and-piston arrangement. It undergoes the change in state shown in the drawing. **(a)** Assume first that the cylinder and piston are perfect thermal insulators that do not allow heat to be transferred. What is the value of q for the state

change? What is the sign of w for the state change? What can be said about ΔE for the state change? **(b)** Now assume that the cylinder and piston are made up of a thermal conductor such as a metal. During the state change, the cylinder gets warmer to the touch. What is the sign of q for the state change in this case? Describe the difference in the state of the system at the end of the process in the two cases. What can you say about the relative values of ΔE?

[5.93] Limestone stalactites and stalagmites are formed in caves by the following reaction:

$$Ca^{2+}(aq) + 2\,HCO_3^-(aq) \longrightarrow$$
$$CaCO_3(s) + CO_2(g) + H_2O(l)$$

If 1 mol of $CaCO_3$ forms at 298 K under 1 atm pressure, the reaction performs 2.47 kJ of P-V work, pushing back the atmosphere as the gaseous CO_2 forms. At the same time, 38.95 kJ of heat is absorbed from the environment. What are the values of ΔH and of ΔE for this reaction?

[5.94] Consider the systems shown in Figure 5.9. In one case the battery becomes completely discharged by running the current through a heater, and in the other by running a fan. Both processes occur at constant pressure. In both cases the change in state of the system is the same: The battery goes from being fully charged to being fully discharged. Yet in one case the heat evolved is large, and in the other it is small. Is the enthalpy change the same in the two cases? If not, how can enthalpy be considered a state function? If it is, what can you say about the relationship between enthalpy change and q in this case, as compared with others that we have considered?

5.95 The enthalpy change for melting ice at 0 °C and constant atmospheric pressure is 6.01 kJ/mol. Calculate the quantity of energy required to melt a moderately large iceberg with a mass of 1.25 million metric tons. (A metric ton is 1000 kg.)

5.96 Comparing the energy associated with the rainstorm and that of a conventional explosive gives some idea of the immense amount of energy associated with a storm. **(a)** The heat of vaporization of water is 44.0 kJ/mol. Calculate the quantity of energy released when enough water vapor condenses to form 0.50 inches of rain over an area of one square mile. **(b)** The energy released when one ton of dynamite explodes is 4.2×10^6 kJ. Calculate the number of tons of dynamite needed to provide the energy of the storm in part (a).

5.97 A house is designed to have passive solar energy features. Brickwork incorporated into the interior of the house acts as a heat absorber. Each brick weighs

(c) Which is the most efficient fuel in terms of heat evolved per unit mass?

5.86 It is interesting to compare the "fuel value" of a hydrocarbon in a world where fluorine rather than oxygen is the combustion agent. The enthalpy of formation of $CF_4(g)$ is −679.9 kJ/mol. Which of the following two reactions is the more exothermic?

$$CH_4(g) + 2\,O_2(g) \longrightarrow CO_2(g) + 2\,H_2O(g)$$
$$CH_4(g) + 4\,F_2(g) \longrightarrow CF_4(g) + 4\,HF(g)$$

approximately 1.8 kg. The specific heat of the brick is 0.85 J/g-K. How many bricks must be incorporated into the interior of the house to provide the same total heat capacity as 1.7×10^3 gal of water?

[5.98] A coffee-cup calorimeter of the type shown in Figure 5.17 contains 150.0 g of water at 25.1 °C. A 121.0-g block of copper metal is heated to 100.4 °C by putting it in a beaker of boiling water. The specific heat of Cu(s) is 0.385 J/g-K. The Cu is added to the calorimeter, and after a time the contents of the cup reach a constant temperature of 30.1°C. (a) Determine the amount of heat, in J, lost by the copper block. (b) Determine the amount of heat gained by the water. The specific heat of water is 4.18 J/g-K. (c) The difference between your answers for (a) and (b) is due to heat loss through the Styrofoam® cups and the heat necessary to raise the temperature of the inner wall of the apparatus. The heat capacity of the calorimeter is the amount of heat necessary to raise the temperature of the apparatus (the cups and the stopper) by 1 K. Calculate the heat capacity of the calorimeter in J/K. (d) What would be the final temperature of the system if all the heat lost by the copper block were absorbed by the water in the calorimeter?

[5.99] (a) When a 0.235-g sample of benzoic acid is combusted in a bomb calorimeter, the temperature rises 1.642 °C. When a 0.265-g sample of caffeine, $C_8H_{10}O_2N_4$, is burned, the temperature rises 1.525 °C. Using the value 26.38 kJ/g for the heat of combustion of benzoic acid, calculate the heat of combustion per mole of caffeine at constant volume. (b) Assuming that there is an uncertainty of 0.002 °C in each temperature reading and that the masses of samples are measured to 0.001 g, what is the estimated uncertainty in the value calculated for the heat of combustion per mole of caffeine?

5.100 How many grams of methane [$CH_4(g)$] must be combusted to heat 1.00 kg of water from 25.0 °C to 90.0 °C, assuming $H_2O(l)$ as a product and 100% efficiency in heat transfer?

5.101 Meals-ready-to-eat (MREs) are military meals that can be heated on a flameless heater. The heat is produced by the following reaction: $Mg(s) + 2 H_2O(l) \longrightarrow$ $Mg(OH)_2(s) + H_2(g)$. (a) Calculate the standard enthalpy change for this reaction. (b) Calculate the number of grams of Mg needed for this reaction to release enough energy to increase the temperature of 25 mL of water from 15 °C to 85 °C.

5.102 Burning methane in oxygen can produce three different carbon-containing products: soot (very fine particles of graphite), CO(g), and $CO_2(g)$. (a) Write three balanced equations for the reaction of methane gas with oxygen to produce these three products. In each case assume that $H_2O(l)$ is the only other product. (b) Determine the standard enthalpies for the reactions in part (a). (c) Why, when the oxygen supply is adequate, is $CO_2(g)$ the predominant carbon-containing product of the combustion of methane?

5.103 (a) Calculate the standard enthalpy of formation of gaseous diborane (B_2H_6) using the following thermochemical information:

$$4 B(s) + 3 O_2(g) \longrightarrow 2 B_2O_3(s) \qquad \Delta H° = -2509.1 \text{ kJ}$$
$$2 H_2(g) + O_2(g) \longrightarrow 2 H_2O(l) \qquad \Delta H° = -571.7 \text{ kJ}$$
$$B_2H_6(g) + 3 O_2(g) \longrightarrow B_2O_3(s) + 3 H_2O(l) \quad \Delta H° = -2147.5 \text{ kJ}$$

(b) Pentaborane (B_5H_9) is another boron hydride. What experiment or experiments would you need to perform to yield the data necessary to calculate the heat of formation of $B_5H_9(l)$? Explain by writing out and summing any applicable chemical reactions.

5.104 From the following data for three prospective fuels, calculate which could provide the most energy per unit volume:

Fuel	Density at 20 °C (g/cm³)	Molar Enthalpy of Combustion kJ/mol
Nitroethane, $C_2H_5NO_2(l)$	1.052	−1368
Ethanol, $C_2H_5OH(l)$	0.789	−1367
Methylhydrazine, $CH_6N_2(l)$	0.874	−1305

5.105 The hydrocarbons acetylene (C_2H_2) and benzene (C_6H_6) have the same empirical formula. Benzene is an "aromatic" hydrocarbon, one that is unusually stable because of its structure. (a) By using the data in Appendix C, determine the standard enthalpy change for the reaction $3 C_2H_2(g) \longrightarrow C_6H_6(l)$. (b) Which has greater enthalpy, 3 mol of acetylene gas or 1 mol of liquid benzene? (c) Determine the fuel value in kJ/g for acetylene and benzene.

[5.106] Ammonia (NH_3) boils at −33 °C; at this temperature it has a density of 0.81 g/cm³. The enthalpy of formation of $NH_3(g)$ is −46.2 kJ/mol, and the enthalpy of vaporization of $NH_3(l)$ is 23.2 kJ/mol. Calculate the enthalpy change when 1 L of liquid NH_3 is burned in air to give $N_2(g)$ and $H_2O(g)$. How does this compare with ΔH for the complete combustion of 1 L of liquid methanol, $CH_3OH(l)$? For $CH_3OH(l)$, the density at 25 °C is 0.792 g/cm³, and $\Delta H_f°$ equals −239 kJ/mol.

[5.107] Three common hydrocarbons that contain four carbons are listed here, along with their standard enthalpies of formation:

Hydrocarbon	Formula	$\Delta H_f°$ (kJ/mol)
1,3-Butadiene	$C_4H_6(g)$	111.9
1-Butene	$C_4H_8(g)$	1.2
n-Butane	$C_4H_{10}(g)$	−124.7

(a) For each of these substances, calculate the molar enthalpy of combustion to $CO_2(g)$ and $H_2O(l)$. (b) Calculate the fuel value in kJ/g for each of these compounds. (c) For each hydrocarbon, determine the percentage of hydrogen by mass. (d) By comparing your answers for parts (b) and (c), propose a relationship between hydrogen content and fuel value in hydrocarbons.

5.108 The two common sugars, glucose ($C_6H_{12}O_6$) and sucrose ($C_{12}H_{22}O_{11}$), are both carbohydrates. Their standard enthalpies of formation are given in Table 5.3. Using these data, (a) calculate the molar enthalpy of combustion to $CO_2(g)$ and $H_2O(l)$ for the two sugars; (b) calculate the enthalpy of combustion per gram of each sugar; (c) determine how your answers to part (b) compare to the average fuel value of carbohydrates discussed in Section 5.8.

5.109 A 200-lb man decides to add to his exercise routine by walking up three flights of stairs (45 ft) 20 times per day. He figures that the work required to increase his potential energy in this way will permit him to eat an extra order of French fries, at 245 Cal, without adding to his weight. Is he correct in this assumption?

5.110 The sun supplies about 1.0 kilowatt of energy for each square meter of surface area (1.0 kW/m², where a watt = 1 J/s). Plants produce the equivalent of about 0.20 g of sucrose ($C_{12}H_{22}O_{11}$) per hour per square meter. Assuming that the sucrose is produced as follows, calculate the percentage of sunlight used to produce sucrose.

$$12\ CO_2(g) + 11\ H_2O(l) \longrightarrow C_{12}H_{22}O_{11} + 12\ O_2(g)$$
$$\Delta H = 5645\ kJ$$

[5.111] It is estimated that the net amount of carbon dioxide fixed by photosynthesis on the landmass of Earth is 5.5×10^{16} g/yr of CO_2. Assume that all this carbon is converted into glucose. **(a)** Calculate the energy stored by photosynthesis on land per year in kJ. **(b)** Calculate the average rate of conversion of solar energy into plant energy in MW (1W = 1 J/s). A large nuclear power plant produces about 10^3 MW. The energy of how many such nuclear power plants is equivalent to the solar energy conversion?

INTEGRATIVE EXERCISES

5.112 Consider the combustion of a single molecule of $CH_4(g)$ forming $H_2O(l)$ as a product. **(a)** How much energy, in J, is produced during this reaction? **(b)** A typical X-ray photon has an energy of 8 keV. How does the energy of combustion compare to the energy of the X-ray photon?

5.113 Consider the dissolving of NaCl in water, illustrated in Figure 4.3. Assume the system consists of 0.1 mol NaCl and 1 L of water. Considering that the NaCl readily dissolves in the water and that the ions are strongly stabilized by the water molecules, as shown in the figure, is it safe to conclude that the dissolution of NaCl in water results in a lower enthalpy for the system? Explain your response. What experimental evidence would you examine to test this question?

5.114 Consider the following unbalanced oxidation-reduction reactions in aqueous solution:

$$Ag^+(aq) + Li(s) \longrightarrow Ag(s) + Li^+(aq)$$
$$Fe(s) + Na^+(aq) \longrightarrow Fe^{2+}(aq) + Na(s)$$
$$K(s) + H_2O(l) \longrightarrow KOH(aq) + H_2(g)$$

(a) Balance each of the reactions. **(b)** By using data in Appendix C, calculate $\Delta H°$ for each of the reactions. **(c)** Based on the values you obtain for $\Delta H°$, which of the reactions would you expect to be thermodynamically favored? (That is, which would you expect to be spontaneous?) **(d)** Use the activity series to predict which of these reactions should occur. ∞ (Section 4.4) Are these results in accord with your conclusion in part (c) of this problem?

[5.115] Consider the following acid-neutralization reactions involving the strong base NaOH(aq):

$$HNO_3(aq) + NaOH(aq) \longrightarrow NaNO_3(aq) + H_2O(l)$$
$$HCl(aq) + NaOH(aq) \longrightarrow NaCl(aq) + H_2O(l)$$
$$NH_4^+(aq) + NaOH(aq) \longrightarrow NH_3(aq) + Na^+(aq) + H_2O(l)$$

(a) By using data in Appendix C, calculate $\Delta H°$ for each of the reactions. **(b)** As we saw in Section 4.3, nitric acid and hydrochloric acid are strong acids. Write net ionic equations for the neutralization of these acids. **(c)** Compare the values of $\Delta H°$ for the first two reactions. What can you conclude? **(d)** In the third equation $NH_4^+(aq)$ is acting as an acid. Based on the value of $\Delta H°$ for this reaction, do you think it is a strong or a weak acid? Explain.

5.116 Consider two solutions, the first being 50.0 mL of 1.00 M $CuSO_4$ and the second 50.0 mL of 2.00 M KOH. When the two solutions are mixed in a constant-pressure calorimeter, a precipitate forms and the temperature of the mixture rises from 21.5 °C to 27.7 °C. **(a)** Before mixing, how many grams of Cu are present in the solution of $CuSO_4$? **(b)** Predict the identity of the precipitate in the reaction. **(c)** Write complete and net ionic equations for the reaction that occurs when the two solutions are mixed. **(d)** From the calorimetric data, calculate ΔH for the reaction that occurs on mixing. Assume that the calorimeter absorbs only a negligible quantity of heat, that the total volume of the solution is 100.0 mL, and that the specific heat and density of the solution after mixing are the same as that of pure water.

5.117 The precipitation reaction between $AgNO_3(aq)$ and NaCl(aq) proceeds as follows:

$$AgNO_3(aq) + NaCl(aq) \longrightarrow NaNO_3(aq) + AgCl(s)$$

(a) By using Appendix C, calculate $\Delta H°$ for the net ionic equation of this reaction. **(b)** What would you expect for the value of $\Delta H°$ of the overall molecular equation compared to that for the net ionic equation? Explain. **(c)** Use the results from (a) and (b) along with data in Appendix C to determine the value of $\Delta H_f^°$ for $AgNO_3(aq)$.

[5.118] A sample of a hydrocarbon is combusted completely in $O_2(g)$ to produce 21.83 g $CO_2(g)$, 4.47 g $H_2O(g)$, and 311 kJ of heat. **(a)** What is the mass of the hydrocarbon sample that was combusted? **(b)** What is the empirical formula of the hydrocarbon? **(c)** Calculate the value of $\Delta H_f^°$ per empirical-formula unit of the hydrocarbon. **(d)** Do you think that the hydrocarbon is one of those listed in Appendix C? Explain your answer.

5.119 The methane molecule, CH_4, has the geometry shown in Figure 2.21. Imagine a hypothetical process in which the methane molecule is "expanded," by simultaneously extending all four C—H bonds to infinity. We then have the process

$$CH_4(g) \longrightarrow C(g) + 4\ H(g)$$

(a) Compare this process with the reverse of the reaction that represents the standard enthalpy of formation. **(b)** Calculate the enthalpy change in each case. Which is the more endothermic process? What accounts for the difference in $\Delta H°$ values? **(c)** Suppose that 3.45 g $CH_4(g)$ is reacted with 1.22 g $F_2(g)$, forming $CF_4(g)$ and HF(g) as sole products. What is the limiting reagent in this reaction? If the reaction occurs at constant pressure, what amount of heat is evolved?

ELECTRONIC STRUCTURE OF ATOMS

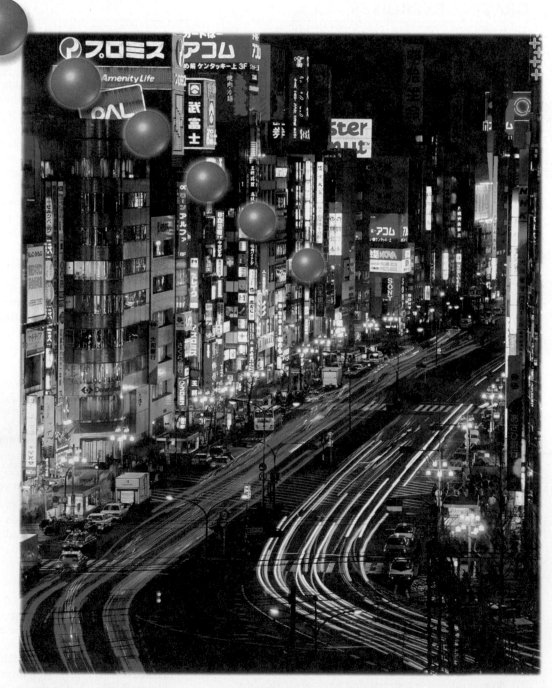

THE GLASS TUBES OF NEON LIGHTS contain various gases that can be excited by electricity. Light is produced when electrically excited atoms return to their lowest-energy states.

WHAT'S AHEAD

6.1 The Wave Nature of Light
We will learn that light (radiant energy, or *electromagnetic radiation*) has wavelike properties and so is characterized by *wavelength, frequency,* and *speed.*

6.2 Quantized Energy and Photons
We will recognize that many different types of experiments indicate that electromagnetic radiation also has particle-like properties and can be described in terms of *photons,* "particles" of light.

6.3 Line Spectra and the Bohr Model
We will explore the fact that atoms give off characteristic colors of light (*line spectra*), when appropriately stimulated. Line spectra provide clues about how electrons are arranged in atoms. Experiments show that electrons exist only at certain energy levels around a nucleus and that energy is involved in moving an electron from one level to another. The Bohr model of the atom pictures the atom as a miniature solar system, with the nucleus of an atom as the "Sun" about which electrons, like planets, orbit.

6.4 The Wave Behavior of Matter
We recognize that matter also has wavelike properties that are manifested at the atomic scale. Because of the fundamental particle-wave duality of matter, it is impossible to determine simultaneously the exact position and the exact motion of an electron in an atom (*Heisenberg's Uncertainty Principle*).

6.5 Quantum Mechanics and Atomic Orbitals
We will discover that classical mechanics treats matter as particles, while quantum mechanics treats matter as waves. We will see that we can describe how an electron exists in atoms by thinking about the electron as a standing wave surrounding the nucleus. The *wave functions* that mathematically describe the electron's position and energy in an atom are called *atomic orbitals*. The orbitals can be described in a shorthand notation using *quantum numbers*.

6.6 Representations of Orbitals
We will see how to draw pictures of orbitals and interpret graphs of electron density.

6.7 Many-Electron Atoms
We will recognize that the energy levels for one electron in an atom are altered when multiple electrons are in an atom. The electron itself has a quantum-mechanical property called *spin*. The *Pauli exclusion principle* states that no two electrons in an atom can have the same four quantum numbers (three for the orbital and one for the spin). Therefore, an orbital can hold a maximum of two electrons.

6.8 Electron Configurations
We will realize that knowing the energies of orbitals as well as some fundamental characteristics of electrons described by *Hund's rule* allows us to determine how electrons are distributed among various orbitals in an atom (*electron configurations*).

6.9 Electron Configurations and the Periodic Table
We will observe that the electron configuration of an atom is related to the location of the element in the periodic table.

WHAT HAPPENS WHEN someone switches on a neon light? The electrons in the neon atoms, which are excited to a higher energy by electricity, emit light when they drop back down to a lower energy. The pleasing glow that results is explained by one of the most revolutionary

discoveries of the twentieth century, namely, the *quantum theory*. This theory explains much of the behavior of electrons in atoms. We will see that the behavior of electrons in an atom is quite unlike anything we see in our macroscopic world.

In this chapter we will explore the quantum theory and its importance in chemistry. We will begin by looking more closely at the nature of light and how our description of light was changed by the quantum theory. We will explore some of the tools used in *quantum mechanics*, the "new" physics that had to be developed to describe atoms correctly. We will then use the quantum

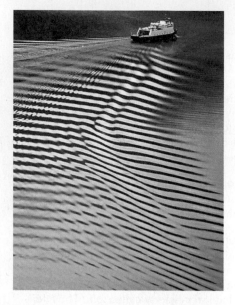

▲ **Figure 6.1 Water waves.** The movement of the boat through the water forms waves. The regular variation of the peaks and troughs enables us to sense the motion, or *propagation*, of the waves.

(a)

(b)

▲ **Figure 6.2 Characteristics of water waves.** (a) The distance between corresponding points on each wave is called the *wavelength*. In this drawing, the two corresponding points are two peaks, but they could be any other two corresponding points, such as two adjacent troughs. (b) The number of times per second that the cork bobs up and down is called the *frequency* of the wave.

theory to describe the arrangements of electrons in atoms—what we call the **electronic structure** of atoms. The electronic structure of an atom refers to the number of electrons in an atom as well as the distribution of the electrons around the nucleus and their energies. We will see that the quantum description of the electronic structure of atoms helps us to understand the elegant arrangement of the elements in the periodic table—why, for example, helium and neon are both unreactive gases, whereas sodium and potassium are both soft, reactive metals. In the chapters that follow, we will see how the concepts of quantum theory are used to explain trends in the periodic table and the formation of bonds between atoms.

6.1 THE WAVE NATURE OF LIGHT

Much of our present understanding of the electronic structure of atoms has come from analysis of the light either emitted or absorbed by substances. To understand electronic structure, therefore, we must first learn more about light. The light that we can see with our eyes, *visible light*, is an example of **electromagnetic radiation.** Because electromagnetic radiation carries energy through space, it is also known as *radiant energy*. There are many types of electromagnetic radiation in addition to visible light. These different forms—such as the radio waves that carry music to our radios, the infrared radiation (heat) from a glowing fireplace, and the X-rays used by a dentist—may seem very different from one another, but they all share certain fundamental characteristics.

All types of electromagnetic radiation move through a vacuum at a speed of 3.00×10^8 m/s, the *speed of light*. All have wavelike characteristics similar to those of waves that move through water. Water waves are the result of energy imparted to the water, perhaps by the dropping of a stone or the movement of a boat on the water surface (Figure 6.1◄). This energy is expressed as the up-and-down movements of the water.

A cross section of a water wave (Figure 6.2◄) shows that it is periodic, which means that the pattern of peaks and troughs repeats itself at regular intervals. The distance between two adjacent peaks (or between two adjacent troughs) is called the **wavelength.** The number of complete wavelengths, or *cycles*, that pass a given point each second is the **frequency** of the wave. We can measure the frequency of a water wave by counting the number of times per second that a cork bobbing on the water moves through a complete cycle of upward and downward motion.

Just as with water waves, we can assign a frequency and wavelength to electromagnetic waves, as illustrated in Figure 6.3▼. These and all other wave

▼ **Figure 6.3 Characteristics of electromagnetic waves.** Radiant energy has wave characteristics; it consists of electromagnetic waves. Notice that the shorter the wavelength, λ, the higher the frequency, ν. The wavelength in (b) is half as long as that in (a), and the frequency of the wave in (b) is therefore twice as great as the frequency in (a). The *amplitude* of the wave relates to the intensity of the radiation, which is the maximum extent of the oscillation of the wave. In these diagrams amplitude is measured as the vertical distance from the midline of the wave to its peak. The waves in (a) and (b) have the same amplitude. The wave in (c) has the same frequency as that in (b), but its amplitude is lower.

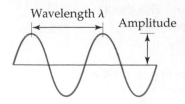

(a) Two complete cycles of wavelength λ

(b) Wavelength half of that in (a); frequency twice as great as in (a)

(c) Same frequency as (b), smaller amplitude

characteristics of electromagnetic radiation are due to the periodic oscillations in the intensities of the electric and magnetic fields associated with the radiation.

The speed of water waves can vary depending on how they are created—for example, the waves produced by a speedboat travel faster than those produced by a rowboat. In contrast, all electromagnetic radiation moves at the same speed, namely the speed of light. As a result, the wavelength and frequency of electromagnetic radiation are always related in a straightforward way. If the wavelength is long, there will be fewer cycles of the wave passing a point per second, and the frequency will be low. Conversely, for a wave to have a high frequency, the wave must have a short wavelength—that is, the distance between the peaks of the waves is small. This inverse relationship between the frequency and the wavelength of electromagnetic radiation can be expressed by the equation

$$c = \lambda \nu \qquad [6.1]$$

where c is the speed of light, λ (lambda) is the wavelength, and ν (nu) is the frequency.

Why do different forms of electromagnetic radiation have different properties? Their differences are due to their different wavelengths, which are expressed in units of length. Figure 6.4▲ shows the various types of electromagnetic radiation arranged in order of increasing wavelength, a display called the *electromagnetic spectrum*. Notice that the wavelengths span an enormous range. The wavelengths of gamma rays are similar to the diameters of atomic nuclei, whereas the wavelengths of radio waves can be longer than a football field. Notice also that visible light, which corresponds to wavelengths of about 400 to 700 nm (4×10^{-7} m to 7×10^{-7} m), is an extremely small portion of the electromagnetic spectrum. We can see visible light because of the chemical reactions it triggers in our eyes. The unit of length normally chosen to express wavelength depends on the type of radiation, as shown in Table 6.1▼.

Frequency is expressed in cycles per second, a unit also called a *hertz* (Hz). Because it is understood that cycles are involved, the units of frequency are normally given simply as "per second," which is denoted by s^{-1} or /s. For example, a frequency of 820 kilohertz (kHz), a typical frequency for an AM radio station, could be written as 820 kHz, 820,000 Hz, 820,000 s^{-1}, or 820,000/s.

▲ **Figure 6.4 The electromagnetic spectrum.** Wavelengths in the spectrum range from very short gamma rays to very long radio waves. Notice that the color of visible light can be expressed quantitatively by wavelength.

TABLE 6.1 ■ Common Wavelength Units for Electromagnetic Radiation

Unit	Symbol	Length (m)	Type of Radiation
Angstrom	Å	10^{-10}	X-ray
Nanometer	nm	10^{-9}	Ultraviolet, visible
Micrometer	μm	10^{-6}	Infrared
Millimeter	mm	10^{-3}	Microwave
Centimeter	cm	10^{-2}	Microwave
Meter	m	1	TV, radio
Kilometer	km	1000	Radio

A Closer Look | THE SPEED OF LIGHT

How do we know that light has a speed and does not move infinitely fast?

During the late 1600s, the Danish astonomer Ole Rømer (1644–1710) measured the orbits of several of Jupiter's moons. These moons move much faster than our own—they have orbits of 1–7 days and are eclipsed by Jupiter's shadow at every revolution. Over many months, Rømer measured discrepancies of up to 10 minutes in the times of these orbits. He reasoned that one possible explanation for the discrepancies was that Jupiter was farther from Earth at different times of the year. Thus, the light from the Sun, which reflected off Jupiter and ultimately to his telescope, had farther to travel at different times of the year. Rømer's data led to an estimate of 3.5×10^8 m/s for the speed of light. In 1704, Isaac Newton (1643–1727) used estimates of the distance from the Sun to Earth and the time it takes for the light to travel that distance to calculate the speed of light as $2.4 - 2.7 \times 10^8$ m/s.

In 1927, A. A. Michelson (1852–1931) performed a famous experiment to determine the speed of light. Michelson set up a rotating mirror system at the top of Mount Wilson in southern California (Figure 6.5►). The mirror system bounced light from the top of Mount Wilson to the top of Mount San Antonio, 22 miles away, where another mirror system bounced the light back to Mount Wilson. If the speed of light was instantaneous, or an exact multiple of the turn speed of the rotating mirror, the reflected spot of light would appear exactly super-

imposed on the original spot. Michelson was able to change the speed of the rotating mirror and measure small displacements in the position of the reflected spot. The value for the speed of light (in air) based on this experiment is $2.9980 \pm 0.0002 \times 10^8$ m/s. The main source of error is the distance between the mirrors at the tops of the two mountains, which was measured within a fifth of an inch in 22 miles.

▲ Figure 6.5 **View of Mount San Antonio from the top of Mount Wilson.** The mountains are 22 miles apart.

■■ SAMPLE EXERCISE 6.1 | Concepts of Wavelength and Frequency

Two electromagnetic waves are represented in the margin. **(a)** Which wave has the higher frequency? **(b)** If one wave represents visible light and the other represents infrared radiation, which wave is which?

SOLUTION

(a) The lower wave has a longer wavelength (greater distance between peaks). The longer the wavelength, the lower the frequency ($\nu = c/\lambda$). Thus, the lower wave has the lower frequency, and the upper wave has the higher frequency.
(b) The electromagnetic spectrum (Figure 6.4) indicates that infrared radiation has a longer wavelength than visible light. Thus, the lower wave would be the infrared radiation.

■■ PRACTICE EXERCISE

If one of the waves in the margin represents blue light and the other red light, which is which?
Answer: The expanded visible-light portion of Figure 6.4 tells you that red light has a longer wavelength than blue light. The lower wave has the longer wavelength (lower frequency) and would be the red light.

■■ SAMPLE EXERCISE 6.2 | Calculating Frequency from Wavelength

The yellow light given off by a sodium vapor lamp used for public lighting has a wavelength of 589 nm. What is the frequency of this radiation?

SOLUTION

Analyze: We are given the wavelength, λ, of the radiation and asked to calculate its frequency, ν.

Plan: The relationship between the wavelength (which is given) and the frequency (which is the unknown) is given by Equation 6.1. We can solve this equation for ν and then use the values of λ and c to obtain a numerical answer. (The speed of light, c, is a fundamental constant whose value is 3.00×10^8 m/s.)

Solve: Solving Equation 6.1 for frequency gives $\nu = c/\lambda$. When we insert the values for c and λ, we note that the units of length in these two quantities are different. We can convert the wavelength from nanometers to meters, so the units cancel:

$$\nu = \frac{c}{\lambda} = \left(\frac{3.00 \times 10^8 \text{ m/s}}{589 \text{ nm}}\right)\left(\frac{1 \text{ nm}}{10^{-9} \text{ m}}\right) = 5.09 \times 10^{14} \text{ s}^{-1}$$

Check: The high frequency is reasonable because of the short wavelength. The units are proper because frequency has units of "per second," or s^{-1}.

■ **PRACTICE EXERCISE**

(a) A laser used in eye surgery to fuse detached retinas produces radiation with a wavelength of 640.0 nm. Calculate the frequency of this radiation. (b) An FM radio station broadcasts electromagnetic radiation at a frequency of 103.4 MHz (megahertz; MHz $= 10^6$ s^{-1}). Calculate the wavelength of this radiation. The speed of light is 2.998×10^8 m/s to four significant digits.
Answers: (a) 4.688×10^{14} s^{-1}, (b) 2.901 m

▲ **GIVE IT SOME THOUGHT**

Our bodies are penetrated by X-rays but not by visible light. Is this because X-rays travel faster than visible light?

6.2 QUANTIZED ENERGY AND PHOTONS

Although the wave model of light explains many aspects of its behavior, this model cannot explain several phenomena. Three of these are particularly pertinent to our understanding of how electromagnetic radiation and atoms interact: (1) the emission of light from hot objects (referred to as *blackbody radiation* because the objects studied appear black before heating), (2) the emission of electrons from metal surfaces on which light shines (the *photoelectric effect*), and (3) the emission of light from electronically excited gas atoms (*emission spectra*). We examine the first two phenomena here and the third in Section 6.3.

Hot Objects and the Quantization of Energy

When solids are heated, they emit radiation, as seen in the red glow of an electric stove burner and the bright white light of a tungsten lightbulb. The wavelength distribution of the radiation depends on temperature; a red-hot object is cooler than a white-hot one (Figure 6.6▶). During the late 1800s, a number of physicists were studying this phenomenon, trying to understand the relationship between the temperature and the intensity and wavelengths of the emitted radiation. The prevailing laws of physics could not account for the observations.

In 1900 a German physicist named Max Planck (1858–1947) solved the problem by assuming that energy can be either released or absorbed by atoms only in discrete "chunks" of some minimum size. Planck gave the name **quantum** (meaning "fixed amount") to the smallest quantity of energy that can be emitted or absorbed as electromagnetic radiation. He proposed that the energy, E, of a single quantum equals a constant times the frequency of the radiation:

$$E = h\nu \qquad [6.2]$$

The constant h is called **Planck's constant** and has a value of 6.626×10^{-34} joule-second (J-s). According to Planck's theory, matter is allowed to emit and absorb energy only in whole-number multiples of $h\nu$, such as $h\nu$, $2h\nu$, $3h\nu$, and so forth. If the quantity of energy emitted by an atom is $3h\nu$, for example, we say that three quanta of energy have been emitted (quanta

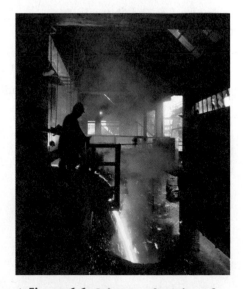

▲ **Figure 6.6 Color as a function of temperature.** The color and intensity of the light emitted by a hot object depend on the temperature of the object. The temperature is highest at the center of this pour of molten steel. As a result, the light emitted from the center is most intense and of shortest wavelength.

▲ **Figure 6.7 A model for quantized energy.** The potential energy of a person walking up a ramp (a) increases in a uniform, continuous manner, whereas that of a person walking up steps (b) increases in a stepwise, quantized manner.

being the plural of quantum). Because the energy can be released only in specific amounts, we say that the allowed energies are *quantized*—their values are restricted to certain quantities. Planck's revolutionary proposal that energy is quantized was proved correct, and he was awarded the 1918 Nobel Prize in Physics for his work on the quantum theory.

If the notion of quantized energies seems strange, it might be helpful to draw an analogy by comparing a ramp and a staircase (Figure 6.7 ◄). As you walk up a ramp, your potential energy increases in a uniform, continuous manner. When you climb a staircase, you can step only *on* individual stairs, not *between* them, so that your potential energy is restricted to certain values and is therefore quantized.

If Planck's quantum theory is correct, why aren't its effects more obvious in our daily lives? Why do energy changes seem continuous rather than quantized, or "jagged"? Notice that Planck's constant is an extremely small number. Thus, a quantum of energy, $h\nu$, is an extremely small amount. Planck's rules regarding the gain or loss of energy are always the same, whether we are concerned with objects on the size scale of our ordinary experience or with microscopic objects. With everyday macroscopic objects, however, the gain or loss of a single quantum of energy goes completely unnoticed. In contrast, when dealing with matter at the atomic level, the impact of quantized energies is far more significant.

▲ **GIVE IT SOME THOUGHT**

The temperature of stars is gauged by their colors. For example, red stars have a lower temperature than blue-white stars. How is this temperature scale consistent with Planck's assumption?

The Photoelectric Effect and Photons

A few years after Planck presented his theory, scientists began to see its applicability to a great many experimental observations. They recognized that Planck's theory had within it the seeds of a revolution in the way we view the

▼ **Figure 6.8 The photoelectric effect.** When photons of sufficiently high energy strike a metal surface, electrons are emitted from the metal, as in (a). The photoelectric effect is the basis of the photocell shown in (b). The emitted electrons are drawn toward the positive terminal. As a result, current flows in the circuit. Photocells are used in photographic light meters as well as in numerous other electronic devices.

physical world. In 1905, Albert Einstein (1879–1955) used Planck's quantum theory to explain the **photoelectric effect,** which is illustrated in Figure 6.8 ◄. Experiments had shown that light shining on a clean metal surface causes the surface to emit electrons. Each metal has a minimum frequency of light below which no electrons are emitted. For example, light with a frequency of $4.60 \times 10^{14}\,\text{s}^{-1}$ or greater will cause cesium metal to emit electrons, but light of lower frequency has no effect.

To explain the photoelectric effect, Einstein assumed that the radiant energy striking the metal surface does not behave like a wave but rather as if it were a stream of tiny energy packets. Each energy packet, called a **photon,** behaves like a tiny particle. Extending Planck's quantum theory,

Einstein deduced that each photon must have an energy equal to Planck's constant times the frequency of the light:

$$\text{Energy of photon} = E = h\nu \qquad [6.3]$$

Thus, radiant energy itself is quantized.

Under the right conditions, a photon can strike a metal surface and be absorbed. When this happens, the photon can transfer its energy to an electron in the metal. A certain amount of energy—called the *work function*—is required for an electron to overcome the attractive forces that hold it in the metal. If the photons of the radiation impinging on the metal have less energy than the work function, electrons do not acquire sufficient energy to escape from the metal surface, even if the light beam is intense. If the photons of radiation have sufficient energy, electrons are emitted from the metal. If the photons have more than the minimum energy required to free electrons, the excess energy appears as the kinetic energy of the emitted electrons. Einstein won the Nobel Prize in Physics in 1921 for his explanation of the photoelectric effect.

To better understand what a photon is, imagine that you have a light source that produces radiation with a single wavelength. Further suppose that you could switch the light on and off faster and faster to provide ever-smaller bursts of energy. Einstein's photon theory tells us that you would eventually come to the smallest energy burst, given by $E = h\nu$. This smallest burst of energy consists of a single photon of light.

■ SAMPLE EXERCISE 6.3 | Energy of a Photon

Calculate the energy of one photon of yellow light with a wavelength of 589 nm.

SOLUTION

Analyze: Our task is to calculate the energy, E, of a photon, given $\lambda = 589$ nm.

Plan: We can use Equation 6.1 to convert the wavelength to frequency:

$$\nu = c/\lambda$$

We can then use Equation 6.3 to calculate energy:

$$E = h\nu$$

Solve: The frequency, ν, is calculated from the given wavelength, as shown in Sample Exercise 6.2:

$$\nu = c/\lambda = 5.09 \times 10^{14}\,\text{s}^{-1}$$

The value of Planck's constant, h, is given both in the text and in the table of physical constants on the inside front cover of the text, and so we can easily calculate E:

$$E = (6.626 \times 10^{-34}\,\text{J-s})(5.09 \times 10^{14}\,\text{s}^{-1}) = 3.37 \times 10^{-19}\,\text{J}$$

Comment: If one photon of radiant energy supplies 3.37×10^{-19} J, then one mole of these photons will supply

$$(6.02 \times 10^{23}\,\text{photons/mol})(3.37 \times 10^{-19}\,\text{J/photon}) = 2.03 \times 10^5\,\text{J/mol}$$

This is the magnitude of enthalpies of reactions (Section 5.4), so radiation can break chemical bonds, producing what are called *photochemical reactions.*

■ PRACTICE EXERCISE

(a) A laser emits light with a frequency of $4.69 \times 10^{14}\,\text{s}^{-1}$. What is the energy of one photon of the radiation from this laser? **(b)** If the laser emits a pulse of energy containing 5.0×10^{17} photons of this radiation, what is the total energy of that pulse? **(c)** If the laser emits 1.3×10^{-2} J of energy during a pulse, how many photons are emitted during the pulse?
Answers: **(a)** 3.11×10^{-19} J, **(b)** 0.16 J, **(c)** 4.2×10^{16} photons

---

The idea that the energy of light depends on its frequency helps us understand the diverse effects that different kinds of electromagnetic radiation have on matter. For example, the high frequency (short wavelength) of X-rays

▲ **Figure 6.9 Quantum giants.** Niels Bohr (right) with Albert Einstein. Bohr (1885–1962) made major contributions to the quantum theory. From 1911 to 1913 Bohr studied in England, working first with J. J. Thomson at Cambridge University and then with Ernest Rutherford at the University of Manchester. He published his quantum theory of the atom in 1914 and was awarded the Nobel Prize in Physics in 1922.

▲ **Figure 6.10 Monochromatic radiation.** Lasers produce light of a single wavelength, which we call *monochromatic light*. Different lasers produce light of different wavelengths. The photo shows beams from a variety of lasers that produce visible light of different colors. Other lasers produce light that is not visible, including infrared and ultraviolet light.

▶ **Figure 6.11 Creating a spectrum.** A continuous visible spectrum is produced when a narrow beam of white light is passed through a prism. The white light could be sunlight or light from an incandescent lamp.

(Figure 6.4) causes X-ray photons to have energy high enough to cause tissue damage and even cancer. Thus, signs are normally posted around X-ray equipment warning us of high-energy radiation.

Although Einstein's theory of light as a stream of particles rather than a wave explains the photoelectric effect and a great many other observations, it also poses a dilemma. Is light a wave, or does it consist of particles? The only way to resolve this dilemma is to adopt what might seem to be a bizarre position: We must consider that light possesses both wavelike and particle-like characteristics and, depending on the situation, will behave more like a wave or more like particles. We will soon see that this dual nature of light is also characteristic of matter.

GIVE IT SOME THOUGHT

What has more energy, a photon of infrared light or a photon of ultraviolet light?

6.3 LINE SPECTRA AND THE BOHR MODEL

The work of Planck and Einstein paved the way for understanding how electrons are arranged in atoms. In 1913, the Danish physicist Niels Bohr (Figure 6.9 ▲) offered a theoretical explanation of *line spectra*, another phenomenon that had puzzled scientists during the nineteenth century. We will examine this phenomenon and then consider how Bohr used the ideas of Planck and Einstein.

Line Spectra

A particular source of radiant energy may emit a single wavelength, as in the light from a laser (Figure 6.10 ◀). Radiation composed of a single wavelength is said to be *monochromatic*. However, most common radiation sources, including lightbulbs and stars, produce radiation containing many different wavelengths. A **spectrum** is produced when radiation from such sources is separated into its different wavelength components. Figure 6.11▼ shows how a prism spreads light from a light source into its component wavelengths. The resulting

Screen

Prism

Slit

Light source

spectrum consists of a continuous range of colors—violet merges into blue, blue into green, and so forth, with no blank spots. This rainbow of colors, containing light of all wavelengths, is called a **continuous spectrum.** The most familiar example of a continuous spectrum is the rainbow produced when raindrops or mist acts as a prism for sunlight.

Not all radiation sources produce a continuous spectrum. When a high voltage is applied to tubes that contain different gases under reduced pressure, the gases emit different colors of light (Figure 6.12▶). The light emitted by neon gas is the familiar red-orange glow of many "neon" lights, whereas sodium vapor emits the yellow light characteristic of some modern streetlights. When light coming from such tubes is passed through a prism, only a few wavelengths are present in the resultant spectra, as shown in Figure 6.13▼. Each wavelength is represented by a colored line in one of these spectra. A spectrum containing radiation of only specific wavelengths is called a **line spectrum.**

▲ **Figure 6.12 Atomic emission.** Different gases emit light of different characteristic colors upon excitation by an electrical discharge: (a) hydrogen, (b) neon.

When scientists first detected the line spectrum of hydrogen in the mid-1800s, they were fascinated by its simplicity. At that time, only the four lines in the visible portion of the spectrum were observed, as shown in Figure 6.13. These lines correspond to wavelengths of 410 nm (violet), 434 nm (blue), 486 nm (blue-green), and 656 nm (red). In 1885, a Swiss schoolteacher named Johann Balmer showed that the wavelengths of these four visible lines of hydrogen fit an intriguingly simple formula that related the wavelengths of the visible line spectrum to integers. Later, additional lines were found to occur in the ultraviolet and infrared regions of the hydrogen spectrum. Soon Balmer's equation was extended to a more general one, called the *Rydberg equation*, which allowed the calculation of the wavelengths of all the spectral lines of hydrogen:

$$\frac{1}{\lambda} = (R_H)\left(\frac{1}{n_1^2} - \frac{1}{n_2^2}\right) \qquad [6.4]$$

In this formula λ is the wavelength of a spectral line, R_H is the *Rydberg constant* ($1.096776 \times 10^7 \text{ m}^{-1}$), and n_1 and n_2 are positive integers, with n_2 being larger than n_1. How could the remarkable simplicity of this equation be explained? It took nearly 30 more years to answer this question.

(a)

(b)

◀ **Figure 6.13 Line spectra.** Spectra obtained from the electrical discharge from (a) Na, (b) H. Light of only a few specific wavelengths is produced, as shown by colored lines in the spectra.

Bohr's Model

Rutherford's discovery of the nuclear nature of the atom ∞ (Section 2.2) suggests that the atom can be thought of as a "microscopic solar system" in which electrons orbit the nucleus. To explain the line spectrum of hydrogen, Bohr assumed that electrons move in circular orbits around the nucleus. According to classical physics, however, an electrically charged particle (such as an electron) that moves in a circular path should continuously lose energy by emitting electromagnetic radiation. As the electron loses energy, it should spiral into the positively charged nucleus. This spiraling obviously does not happen since hydrogen atoms are stable. So how can we explain this apparent violation of the laws of physics? Bohr approached this problem in much the same way that Planck had approached the problem of the nature of the radiation emitted by hot objects: Bohr assumed that the prevailing laws of physics were inadequate to describe all aspects of atoms. Furthermore, Bohr adopted Planck's idea that energies are quantized.

Bohr based his model on three postulates:

1. Only orbits of certain radii, corresponding to certain definite energies, are permitted for the electron in a hydrogen atom.
2. An electron in a permitted orbit has a specific energy and is in an "allowed" energy state. An electron in an allowed energy state will not radiate energy and therefore will not spiral into the nucleus.
3. Energy is emitted or absorbed by the electron only as the electron changes from one allowed energy state to another. This energy is emitted or absorbed as a photon, $E = h\nu$.

▲ GIVE IT SOME THOUGHT

Before reading further about the details of Bohr's model, speculate as to how they explain the fact that hydrogen gas emits a line spectrum (Figure 6.13) rather than a continuous spectrum.

The Energy States of the Hydrogen Atom

Starting with his three postulates and using classical equations for motion and for interacting electrical charges, Bohr calculated the energies corresponding to each allowed orbit for the electron in the hydrogen atom. Ultimately, the energies that Bohr calculated fit the formula

$$E = (-hcR_H)\left(\frac{1}{n^2}\right) = (-2.18 \times 10^{-18} \text{ J})\left(\frac{1}{n^2}\right) \qquad [6.5]$$

In this equation, h, c, and R_H are Planck's constant, the speed of light, and the Rydberg constant, respectively. The product of these three constants equals 2.18×10^{-18} J. The integer n, which can have whole number values of 1, 2, 3, ... to infinity (∞), is called the *principal quantum number*. Each orbit corresponds to a different value of n, and the radius of the orbit gets larger as n increases. Thus, the first allowed orbit (the one closest to the nucleus) has $n = 1$, the next allowed orbit (the one second closest to the nucleus) has $n = 2$, and so forth. The electron in the hydrogen atom can be in any allowed orbit. Equation 6.5 tells us the energy that the electron will have, depending on which orbit it is in.

The energies of the electron of a hydrogen atom given by Equation 6.5 are negative for all values of n. The lower (more negative) the energy is, the more stable the atom will be. The energy is lowest (most negative) for $n = 1$. As n gets larger, the energy becomes successively less negative and therefore increases. We can liken the situation to a ladder in which the rungs are numbered from the bottom rung on up. The higher one climbs the ladder (the greater the value of n), the higher the energy. The lowest energy state ($n = 1$, analogous to the bottom rung) is called the **ground state** of the atom. When the electron is in a higher senergy (less negative) orbit—$n = 2$ or higher—the atom is said to be in an **excited state**. Figure 6.14 ◀ shows the energy of the electron in a hydrogen atom for several values of n.

What happens to the orbit radius and the energy as n becomes infinitely large? The radius increases as n^2, so we reach a point at which the electron is completely separated from the nucleus. When $n = \infty$, the energy is zero:

$$E = (-2.18 \times 10^{-18} \text{ J})\left(\frac{1}{\infty^2}\right) = 0$$

▲ **Figure 6.14 Energy levels in the hydrogen atom from the Bohr model.** The arrows refer to the transitions of the electron from one allowed energy state to another. The states shown are those for which $n = 1$ through $n = 6$ and the state for $n = \infty$ for which the energy, E, equals zero.

Thus, the state in which the electron is removed from the nucleus is the reference, or zero-energy, state of the hydrogen atom. This zero-energy state is *higher* in energy than the states with negative energies.

In his third postulate, Bohr assumed that the electron could "jump" from one allowed energy state to another by either absorbing or emitting photons whose radiant energy corresponds exactly to the energy difference between the two states. Energy must be absorbed for an electron to move to a higher energy state (one with a higher value of n). Conversely, radiant energy is emitted when the electron jumps to a lower energy state (one with a lower value of n). Thus, if the electron jumps from an initial state that has energy E_i to a final state of energy E_f, the change in energy is

$$\Delta E = E_f - E_i = E_{\text{photon}} = h\nu \qquad [6.6]$$

Bohr's model of the hydrogen atom states, therefore, that only the specific frequencies of light that satisfy Equation 6.6 can be absorbed or emitted by the atom.

Substituting the energy expression in Equation 6.5 into Equation 6.6 and recalling that $\nu = c/\lambda$, we have

$$\Delta E = h\nu = \frac{hc}{\lambda} = (-2.18 \times 10^{-18}\,\text{J})\left(\frac{1}{n_f^2} - \frac{1}{n_i^2}\right) \qquad [6.7]$$

In this equation n_i and n_f are the principal quantum numbers of the initial and final states of the atom, respectively. If n_f is smaller than n_i, the electron moves closer to the nucleus and ΔE is a negative number, indicating that the atom releases energy. For example, if the electron moves from $n_i = 3$ to $n_f = 1$, we have

$$\Delta E = (-2.18 \times 10^{-18}\,\text{J})\left(\frac{1}{1^2} - \frac{1}{3^2}\right) = (-2.18 \times 10^{-18}\,\text{J})\left(\frac{8}{9}\right) = -1.94 \times 10^{-18}\,\text{J}$$

Knowing the energy for the emitted photon, we can calculate either its frequency or its wavelength. For the wavelength, we have

$$\lambda = \frac{c}{\nu} = \frac{hc}{\Delta E} = \frac{(6.626 \times 10^{-34}\,\text{J-s})(3.00 \times 10^{8}\,\text{m/s})}{1.94 \times 10^{-18}\,\text{J}} = 1.03 \times 10^{-7}\,\text{m}$$

We have not included the negative sign of the energy in this calculation because wavelength and frequency are always reported as positive quantities. The direction of energy flow is indicated by saying that a photon of wavelength 1.03×10^{-7} m has been *emitted*.

If we solve Equation 6.7 for $1/\lambda$, we find that this equation derived from Bohr's theory corresponds to the Rydberg equation, Equation 6.4, which was obtained using experimental data:

$$\frac{1}{\lambda} = \frac{-hcR_H}{hc}\left(\frac{1}{n_f^2} - \frac{1}{n_i^2}\right) = R_H\left(\frac{1}{n_i^2} - \frac{1}{n_f^2}\right)$$

Thus, the existence of discrete spectral lines can be attributed to the quantized jumps of electrons between energy levels.

GIVE IT SOME THOUGHT

As the electron in a hydrogen atom jumps from the $n = 3$ orbit to the $n = 7$ orbit, does it absorb energy or emit energy?

■ **SAMPLE EXERCISE 6.4** | **Electronic Transitions in the Hydrogen Atom**

Using Figure 6.14, predict which of the following electronic transitions produces the spectral line having the longest wavelength: $n = 2$ to $n = 1$, $n = 3$ to $n = 2$, or $n = 4$ to $n = 3$.

SOLUTION

The wavelength increases as frequency decreases ($\lambda = c/\nu$). Hence the longest wavelength will be associated with the lowest frequency. According to Planck's equation, $E = h\nu$, the lowest frequency is associated with the lowest energy. In Figure 6.14 the shortest vertical line represents the smallest energy change. Thus, the $n = 4$ to $n = 3$ transition produces the longest wavelength (lowest frequency) line.

■ **PRACTICE EXERCISE**

Indicate whether each of the following electronic transitions emits energy or requires the absorption of energy: **(a)** $n = 3$ to $n = 1$; **(b)** $n = 2$ to $n = 4$.
Answers: **(a)** emits energy, **(b)** requires absorption of energy

Limitations of the Bohr Model

While the Bohr model explains the line spectrum of the hydrogen atom, it cannot explain the spectra of other atoms, except in a rather crude way. Bohr also avoided the problem of why the negatively charged electron would not just fall into the positively charged nucleus by simply assuming it would not happen. Therefore, there is a problem with describing an electron merely as a small particle circling about the nucleus. As we will see in Section 6.4, the electron exhibits wavelike properties, a fact that any acceptable model of electronic structure must accommodate. As it turns out, the Bohr model was only an important step along the way toward the development of a more comprehensive model. What is most significant about Bohr's model is that it introduces two important ideas that are also incorporated into our current model: (1) Electrons exist only in certain discrete energy levels, which are described by quantum numbers, and (2) energy is involved in moving an electron from one level to another. We will now start to develop the successor to the Bohr model, which requires that we take a closer look at the behavior of matter.

6.4 THE WAVE BEHAVIOR OF MATTER

In the years following the development of Bohr's model for the hydrogen atom, the dual nature of radiant energy became a familiar concept. Depending on the experimental circumstances, radiation appears to have either a wavelike or a particle-like (photon) character. Louis de Broglie (1892–1987), who was working on his Ph.D. thesis in physics at the Sorbonne in Paris, boldly extended this idea. If radiant energy could, under appropriate conditions, behave as though it were a stream of particles, could matter, under appropriate conditions, possibly show the properties of a wave? Suppose that the electron orbiting the nucleus of a hydrogen atom could be thought of as a wave, with a characteristic wavelength, rather than as a particle. De Broglie suggested that as the electron moves about the nucleus, it is associated with a particular wavelength. He went on to propose that the characteristic wavelength of the electron, or of any other particle, depends on its mass, m, and on its velocity, v, (where h is Planck's constant):

$$\lambda = \frac{h}{mv} \qquad [6.8]$$

The quantity mv for any object is called its **momentum.** De Broglie used the term **matter waves** to describe the wave characteristics of material particles.

Because de Broglie's hypothesis is applicable to all matter, any object of mass m and velocity v would give rise to a characteristic matter wave. However, Equation 6.8 indicates that the wavelength associated with an object of ordinary size, such as a golf ball, is so tiny as to be completely out of the range of any possible observation. This is not so for an electron because its mass is so small, as we see in Sample Exercise 6.5.

SAMPLE EXERCISE 6.5 | Matter Waves

What is the wavelength of an electron moving with a speed of 5.97×10^6 m/s? The mass of the electron is 9.11×10^{-31} kg.

SOLUTION

Analyze: We are given the mass, m, and velocity, v, of the electron, and we must calculate its de Broglie wavelength, λ.

Plan: The wavelength of a moving particle is given by Equation 6.8, so λ is calculated by inserting the known quantities h, m, and v. In doing so, however, we must pay attention to units.

Solve: Using the value of Planck's constant, $h = 6.626 \times 10^{-34}$ J-s

and recalling that $1 \text{ J} = 1 \text{ kg-m}^2/\text{s}^2$

we have the following:

$$\lambda = \frac{h}{mv}$$

$$= \frac{(6.626 \times 10^{-34} \text{ J-s})}{(9.11 \times 10^{-31} \text{ kg})(5.97 \times 10^6 \text{ m/s})} \left(\frac{1 \text{ kg} \cdot \text{m}^2/\text{s}^2}{1 \text{ J}} \right)$$

$$= 1.22 \times 10^{-10} \text{ m} = 0.122 \text{ nm} = 1.22 \text{ Å}$$

Comment: By comparing this value with the wavelengths of electromagnetic radiation shown in Figure 6.4, we see that the wavelength of this electron is about the same as that of X-rays.

PRACTICE EXERCISE

Calculate the velocity of a neutron whose de Broglie wavelength is 500 pm. The mass of a neutron is given in the table inside the back cover of the text.
Answer: 7.92×10^2 m/s

Within a few years after de Broglie published his theory, the wave properties of the electron were demonstrated experimentally. As electrons passed through a crystal, they were diffracted by the crystal, just as X-rays are diffracted. Thus, a stream of moving electrons exhibits the same kinds of wave behavior as electromagnetic radiation.

The technique of electron diffraction has been highly developed. In the electron microscope, for instance, the wave characteristics of electrons are used to obtain images at the atomic scale. This microscope is an important tool for studying surface phenomena at very high magnifications. Electron microscopes can magnify objects by 3,000,000 times (x), far more than can be done with visible light (1000x), because the wavelength of the electrons is so small compared to visible light. Figure 6.15▶ is a photograph of an electron microscope image.

▲ Figure 6.15 **Electrons as waves.** The dots you see in this transmission electron micrograph are columns of atoms. Their regular spacing at the atomic level proves that this material is crystalline. Because this crystal is only about 15 nm in diameter, it is a nanocrystal, which has unusual properties that we will discuss in Chapter 12.

GIVE IT SOME THOUGHT

A baseball pitcher throws a fastball that moves at 95 miles per hour. Does that moving baseball generate matter waves? If so, can we observe them?

The Uncertainty Principle

The discovery of the wave properties of matter raised some new and interesting questions about classical physics. Consider, for example, a ball rolling down a ramp. Using the equations of classical physics, we can calculate the ball's position, direction of motion, and speed at any time, with great accuracy.

▲ **Figure 6.16 Werner Heisenberg (1901–1976).** During his postdoctoral assistantship with Niels Bohr, Heisenberg formulated his famous uncertainty principle. At the age of 25, he became the chair in theoretical physics at the University of Leipzig. At 32 he was one of the youngest scientists to receive the Nobel Prize.

Can we do the same for an electron, which exhibits wave properties? A wave extends in space, and therefore its location is not precisely defined. We might therefore anticipate that it is impossible to determine exactly where an electron is located at a specific time.

The German physicist Werner Heisenberg (Figure 6.16 ◄) proposed that the dual nature of matter places a fundamental limitation on how precisely we can know both the location and the momentum of any object. The limitation becomes important only when we deal with matter at the subatomic level (that is, with masses as small as that of an electron). Heisenberg's principle is called the **uncertainty principle.** When applied to the electrons in an atom, this principle states that it is inherently impossible for us to know simultaneously both the exact momentum of the electron and its exact location in space.

Heisenberg mathematically related the uncertainty of the position (Δx) and the uncertainty in momentum $\Delta(mv)$ to a quantity involving Planck's constant:

$$\Delta x \cdot \Delta(mv) \geq \frac{h}{4\pi} \qquad [6.9]$$

A brief calculation illustrates the dramatic implications of the uncertainty principle. The electron has a mass of 9.11×10^{-31} kg and moves at an average speed of about 5×10^6 m/s in a hydrogen atom. Let's assume that we know the speed to an uncertainty of 1% (that is, an uncertainty of $(0.01)(5 \times 10^6 \text{ m/s}) = 5 \times 10^4$ m/s) and that this is the only important source of uncertainty in the momentum, so that $\Delta(mv) = m\Delta v$. We can then use Equation 6.9 to calculate the uncertainty in the position of the electron:

$$\Delta x \geq \frac{h}{4\pi m \Delta v} = \frac{(6.626 \times 10^{-34} \text{ J-s})}{4\pi(9.11 \times 10^{-31} \text{ kg})(5 \times 10^4 \text{ m/s})} = 1 \times 10^{-9} \text{ m}$$

Because the diameter of a hydrogen atom is only about 1×10^{-10} m, the uncertainty is an order of magnitude greater than the size of the atom. Thus, we have essentially no idea of where the electron is located within the atom. On the other hand, if we were to repeat the calculation with an object of ordinary mass such as a tennis ball, the uncertainty would be so small that it would be inconsequential. In that case, m is large and Δx is out of the realm of measurement and therefore of no practical consequence.

De Broglie's hypothesis and Heisenberg's uncertainty principle set the stage for a new and more broadly applicable theory of atomic structure. In this new approach, any attempt to define precisely the instantaneous location and momentum of the electron is abandoned. The wave nature of the electron is recognized, and its behavior is described in terms appropriate to waves. The result is a model that precisely describes the energy of the electron while describing its location not precisely, but in terms of probabilities.

▲ **GIVE IT SOME THOUGHT**

What is the principal reason that the uncertainty principle should be considered when discussing electrons and other subatomic particles, but is not so necessary when discussing our macroscopic world?

6.5 QUANTUM MECHANICS AND ATOMIC ORBITALS

In 1926 the Austrian physicist Erwin Schrödinger (1887–1961) proposed an equation, now known as Schrödinger's wave equation, that incorporates both the wavelike behavior and the particle-like behavior of the electron. His work opened a new way of dealing with subatomic particles, known as either *quantum mechanics* or *wave mechanics*. The application of Schrödinger's equation

Whenever any measurement is made, some uncertainty exists. Our experience with objects of ordinary dimensions, such as balls or trains or laboratory equipment, indicates that using more precise instruments can decrease the uncertainty of a measurement. In fact, we might expect that the uncertainty in a measurement can be made indefinitely small. However, the uncertainty principle states that there is an actual limit to the accuracy of measurements. This limit is not a restriction on how well instruments can be made; rather, it is inherent in nature. This limit has no practical consequences when dealing with ordinary-sized objects, but its implications are enormous when dealing with subatomic particles, such as electrons.

To measure an object, we must disturb it, at least a little, with our measuring device. Imagine using a flashlight to locate a large rubber ball in a dark room. You see the ball when the light from the flashlight bounces off the ball and strikes your eyes. When a beam of photons strikes an object of this size, it does not alter its position or momentum to any practical extent. Imagine, however, that you wish to locate an electron by similarly bouncing light off it into some detector. Objects can be located to an accuracy no greater than the wavelength of the radiation used. Thus, if we want an accurate position measurement for an electron, we must use a

short wavelength. This means that photons of high energy must be employed. The more energy the photons have, the more momentum they impart to the electron when they strike it, which changes the electron's motion in an unpredictable way. The attempt to measure accurately the electron's position introduces considerable uncertainty in its momentum; the act of measuring the electron's position at one moment makes our knowledge of its future position inaccurate.

Suppose, then, that we use photons of longer wavelength. Because these photons have lower energy, the momentum of the electron is not so appreciably changed during measurement, but its position will be correspondingly less accurately known. This is the essence of the uncertainty principle: *There is an uncertainty in simultaneously knowing either the position or the momentum of the electron that cannot be reduced beyond a certain minimum level.* The more accurately one is known, the less accurately the other is known. Although we can never know the exact position and momentum of the electron, we can talk about the probability of its being at certain locations in space. In Section 6.5 we introduce a model of the atom that provides the probability of finding electrons of specific energies at certain positions in atoms. *Related Exercises: 6.45 and 6.46*

requires advanced calculus, and we will not be concerned with the details of his approach. We will, however, qualitatively consider the results he obtained, because they give us a powerful new way to view electronic structure. Let's begin by examining the electronic structure of the simplest atom, hydrogen.

In the same way that a plucked guitar string vibrates as a standing wave, Schrödinger treated the electron as a standing circular wave around the nucleus. Just as the plucked guitar string produces a fundamental frequency and higher overtones (harmonics), there is a lowest-energy standing wave, and higher-energy ones, for an electron in an atom. Solving Schrödinger's equation leads to a series of mathematical functions called **wave functions** that describe the electron in an atom. These wave functions are usually represented by the symbol ψ (the lowercase Greek letter *psi*). Although the wave function itself has no direct physical meaning, the square of the wave function, ψ^2, provides information about an electron's location when the electron is in an allowed energy state.

For the hydrogen atom, the allowed energies are the same as those predicted by the Bohr model. However, the Bohr model assumes that the electron is in a circular orbit of some particular radius about the nucleus. In the quantum mechanical model, the electron's location cannot be described so simply. According to the uncertainty principle, if we know the momentum of the electron with high accuracy, our simultaneous knowledge of its location is very uncertain. Thus, we cannot hope to specify the exact location of an individual electron around the nucleus. Rather, we must be content with a kind of statistical knowledge. In the quantum mechanical model, we therefore speak of the *probability* that the electron will be in a certain region of space at a given instant. As it turns out, the square of the wave function, ψ^2, at a given point in space represents the probability that the electron will be found at that location. For this reason, ψ^2 is called either the **probability density** or the **electron density.**

One way of representing the probability of finding the electron in various regions of an atom is shown in Figure 6.17▶. In this figure the density of the dots represents the probability of finding the electron. The regions with a high

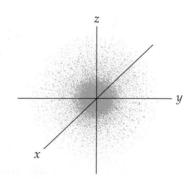

▲ **Figure 6.17 Electron-density distribution.** This rendering represents the probability of where in the space surrounding the nucleus the electron is to be found in a hydrogen atom in its ground state.

density of dots correspond to relatively large values for ψ^2 and are therefore regions where there is a high probability of finding the electron. In Section 6.6 we will say more about the ways in which we can represent electron density.

▲ **GIVE IT SOME THOUGHT**

Is there a difference between stating, "The electron is located at a particular point in space" and "There is a high probability that the electron is located at a particular point in space"?

Orbitals and Quantum Numbers

The solution to Schrödinger's equation for the hydrogen atom yields a set of wave functions and corresponding energies. These wave functions are called **orbitals.** Each orbital describes a specific distribution of electron density in space, as given by the orbital's probability density. Each orbital, therefore, has a characteristic energy and shape. For example, the lowest-energy orbital in the hydrogen atom has an energy of -2.18×10^{-18} J and the shape illustrated in Figure 6.17. Note that an *orbital* (quantum mechanical model) is not the same as an *orbit* (Bohr model). The quantum mechanical model does not refer to orbits, because the motion of the electron in an atom cannot be precisely measured or tracked (Heisenberg uncertainty principle).

The Bohr model introduced a single quantum number, n, to describe an orbit. The quantum mechanical model uses three quantum numbers, n, l, and m_l, which result naturally from the mathematics used, to describe an orbital. Let's consider what information we obtain from each of these quantum numbers and how they are interrelated.

1. The *principal quantum number, n,* can have positive integral values of 1, 2, 3, and so forth. As n increases, the orbital becomes larger, and the electron spends more time farther from the nucleus. An increase in n also means that the electron has a higher energy and is therefore less tightly bound to the nucleus. For the hydrogen atom, $E_n = -(2.18 \times 10^{-18} \text{ J})(1/n^2)$, as in the Bohr model.

2. The second quantum number—the *angular momentum quantum number, l*— can have integral values from 0 to $(n - 1)$ for each value of n. This quantum number defines the shape of the orbital. (We will consider these shapes in Section 6.6.) The value of l for a particular orbital is generally designated by the letters s, p, d, and f,* corresponding to l values of 0, 1, 2, and 3, respectively, as summarized here:

Value of l	0	1	2	3
Letter used	s	p	d	f

3. The *magnetic quantum number, m_l,* can have integral values between $-l$ and l, including zero. This quantum number describes the orientation of the orbital in space, as we will discuss in Section 6.6.

Notice that because the value of n can be any positive integer, an infinite number of orbitals for the hydrogen atom is possible. The electron in a hydrogen atom is described by only one of these orbitals at any given time—we say that the electron *occupies* a certain orbital. The remaining orbitals are *unoccupied* for that particular state of the hydrogen atom. We will see that we are mainly interested in the orbitals of the hydrogen atom with small values of n.

---

*The letters s, p, d, and f come from the words sharp, principal, diffuse, and fundamental, which were used to describe certain features of spectra before quantum mechanics was developed.

TABLE 6.2 ▪ Relationship among Values of n, l, and m_l through $n = 4$					
n	Possible Values of l	Subshell Designation	Possible Values of m_l	Number of Orbitals in Subshell	Total Number of Orbitals in Shell
1	0	1s	0	1	1
2	0	2s	0	1	
	1	2p	1, 0, −1	3	4
3	0	3s	0	1	
	1	3p	1, 0, −1	3	
	2	3d	2, 1, 0, −1, −2	5	9
4	0	4s	0	1	
	1	4p	1, 0, −1	3	
	2	4d	2, 1, 0, −1, −2	5	
	3	4f	3, 2, 1, 0, −1, −2, −3	7	16

GIVE IT SOME THOUGHT

What is the difference between an *orbit* (Bohr model) and an *orbital* (quantum mechanical model)?

The collection of orbitals with the same value of n is called an **electron shell.** All the orbitals that have $n = 3$, for example, are said to be in the third shell. Further, the set of orbitals that have the same n and l values is called a **subshell.** Each subshell is designated by a number (the value of n) and a letter (s, p, d, or f, corresponding to the value of l). For example, the orbitals that have $n = 3$ and $l = 2$ are called 3d orbitals and are in the 3d subshell.

Table 6.2▲ summarizes the possible values of the quantum numbers l and m_l for values of n through $n = 4$. The restrictions on the possible values of the quantum numbers give rise to the following very important observations:

1. The shell with principal quantum number n will consist of exactly n subshells. Each subshell corresponds to a different allowed value of l from 0 to $(n − 1)$. Thus, the first shell ($n = 1$) consists of only one subshell, the 1s ($l = 0$); the second shell ($n = 2$) consists of two subshells, the 2s ($l = 0$) and 2p ($l = 1$); the third shell consists of three subshells, 3s, 3p, and 3d, and so forth.

2. Each subshell consists of a specific number of orbitals. Each orbital corresponds to a different allowed value of m_l. For a given value of l, there are $(2l + 1)$ allowed values of m_l, ranging from $−l$ to $+l$. Thus, each s ($l = 0$) subshell consists of one orbital, each p ($l = 1$) subshell consists of three orbitals, each d ($l = 2$) subshell consists of five orbitals, and so forth.

3. The total number of orbitals in a shell is n^2, where n is the principal quantum number of the shell. The resulting number of orbitals for the shells—1, 4, 9, 16—is related to a pattern seen in the periodic table: We see that the number of elements in the rows of the periodic table—2, 8, 18, and 32—equals twice these numbers. We will discuss this relationship further in Section 6.9.

Figure 6.18▶ shows the relative energies of the hydrogen atom orbitals through $n = 3$. Each box represents an orbital; orbitals of the same subshell, such as the 2p, are grouped together. When the electron occupies the lowest-energy orbital (1s), the hydrogen atom is said to be in its *ground state.* When the

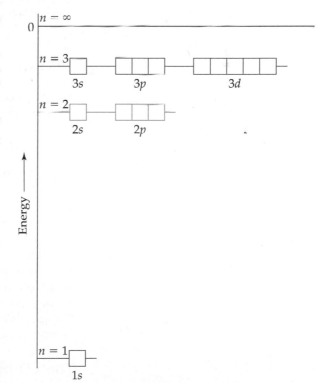

▲ **Figure 6.18 Orbital energy levels in the hydrogen atom.** Each box represents an orbital. Note that all orbitals with the same value for the principal quantum number, n, have the same energy. This is true only in one-electron systems, such as the hydrogen atom.

electron occupies any other orbital, the atom is in an *excited state*. At ordinary temperatures, essentially all hydrogen atoms are in the ground state. The electron can be excited to a higher-energy orbital by absorption of a photon of appropriate energy.

GIVE IT SOME THOUGHT

In Figure 6.18, why is the energy difference between the $n = 1$ and $n = 2$ levels so much greater than the energy difference between the $n = 2$ and $n = 3$ levels?

■ SAMPLE EXERCISE 6.6 | Subshells of the Hydrogen Atom

(a) Without referring to Table 6.2, predict the number of subshells in the fourth shell, that is, for $n = 4$. (b) Give the label for each of these subshells. (c) How many orbitals are in each of these subshells?

Analyze and Plan: We are given the value of the principal quantum number, n. We need to determine the allowed values of l and m_l for this given value of n and then count the number of orbitals in each subshell.

SOLUTION

There are four subshells in the fourth shell, corresponding to the four possible values of l (0, 1, 2, and 3).

These subshells are labeled 4s, 4p, 4d, and 4f. The number given in the designation of a subshell is the principal quantum number, n; the letter designates the value of the angular momentum quantum number, l: for $l = 0$, s; for $l = 1$, p; for $l = 2$, d; for $l = 3$, f.

There is one 4s orbital (when $l = 0$, there is only one possible value of m_l: 0). There are three 4p orbitals (when $l = 1$, there are three possible values of m_l: 1, 0, and −1). There are five 4d orbitals (when $l = 2$, there are five allowed values of m_l: 2, 1, 0, −1, −2). There are seven 4f orbitals (when $l = 3$, there are seven permitted values of m_l: 3, 2, 1, 0, −1, −2, −3).

■ PRACTICE EXERCISE

(a) What is the designation for the subshell with $n = 5$ and $l = 1$? (b) How many orbitals are in this subshell? (c) Indicate the values of m_l for each of these orbitals.
Answers: (a) 5p; (b) 3; (c) 1, 0, −1

6.6 REPRESENTATIONS OF ORBITALS

In our discussion of orbitals so far, we have emphasized their energies. But the wave function also provides information about the electron's location in space when it occupies an orbital. Let's examine the ways that we can picture the orbitals. In doing so, we will examine some important aspects of the electron-density distributions of the orbitals. First, we will look at the three-dimensional shape of the orbital—is it spherical, for example, or does it have directionality? Second, we will examine how the probability density changes as we move on a straight line farther and farther from the nucleus. Finally, we will look at the typical three-dimensional sketches that chemists use in describing the orbitals.

The s Orbitals

One representation of the lowest-energy orbital of the hydrogen atom, the 1s, is shown in Figure 6.17. This type of drawing, which shows the distribution of electron density around the nucleus, is one of the several ways we use to help us visualize orbitals. The first thing that we notice about the electron density for the 1s orbital is that it is *spherically symmetric*—in other words, the electron density at a given distance from the nucleus is the same regardless of the direction in which we proceed from the nucleus. All of the other s orbitals (2s, 3s, 4s, and so forth) are spherically symmetric as well. Recall that the l quantum number

(a)   (b)   (c)

▲ **Figure 6.19 Radial probability functions for the 1s, 2s, and 3s orbitals.** These plots show the probability of finding the electron as a function of distance from the nucleus. As n increases, the most likely distance at which to find the electron moves farther from the nucleus, similar to the Bohr model. In the 2s and 3s orbitals the radial probability function drops to zero at certain distances from the nucleus but then rises again. The points at which the probability is zero are called *nodes*.

for the s orbitals is 0; therefore the m_l quantum number must be 0. Therefore, for each value of n, there is only one s orbital.

So what is different about the s orbitals having different n quantum numbers? For example, how does the electron-density distribution of the hydrogen atom change when the electron is excited from the 1s orbital to the 2s orbital? To address questions like this, we must look at the *radial probability density*, that is, the probability that we will find the electron at a specific distance from the nucleus. In Figure 6.19 ▲ we have plotted the radial probability density for the 1s orbital as a function of r, the distance from the nucleus. The resulting curve is the **radial probability function** for the 1s orbital. (Radial probability functions are described more fully in the "A Closer Look" box in this section.) We see that the probability of finding the electron rises rapidly as we move away from the nucleus, maximizing at a distance of 0.529 Å from the nucleus, and then falls off rapidly. Thus, when the electron occupies the 1s orbital, it is *most likely* to be found 0.529 Å from the nucleus*. We still use the probabilistic description, consistent with the uncertainty principle. Notice also that the probability of finding the electron at a distance greater than 3 Å from the nucleus is essentially zero.

Figure 6.19(b) shows the radial probability function for the 2s orbital of the hydrogen atom. We can see three significant differences between this plot and that for the 1s orbital: (1) There are two separate maxima in the radial probability function for the 2s orbital, namely a small peak at about $r = 0.5$ Å and a much larger peak at about $r = 3$ Å; (2) Between these two peaks is a point at which the function goes to zero (at about $r = 1$ Å). An intermediate point at which a probability function goes to zero is called a **node.** There is a zero probability of finding the electron at a distance corresponding to a node, even though the electron might be found at shorter or longer distances; (3) The radial probability function for the 2s orbital is significantly broader (more spread out) than that for the 1s orbital. Thus, for the 2s orbital, there is a larger range of distances from the

*In the quantum mechanical model, the most probable distance at which to find the electron in the 1s orbital—0.529 Å—is identical to the radius of the orbit predicted by Bohr for $n = 1$. The distance 0.529 Å is often called the Bohr radius.

1s

2s

3s

▲ **Figure 6.20 Contour representations of the 1s, 2s, and 3s orbitals.** The relative radii of the spheres correspond to a 90% probability of finding the electron within each sphere.

nucleus at which there is a high probability of finding the electron than for the 1s orbital. This trend continues for the 3s orbital, as shown in Figure 6.19(c). The radial probability function for the 3s orbital has three peaks of increasing size, with the largest peak maximizing even farther from the nucleus (at about $r = 7$ Å) at which it has two nodes and is even more spread out.

The radial probability functions in Figure 6.19 tell us that as n increases, there is also an increase in the most likely distance from the nucleus to find the electron. In other words, the size of the orbital increases with increasing n, just as it did in the Bohr model.

One widely used method of representing orbitals is to display a boundary surface that encloses some substantial portion, say 90%, of the total electron density for the orbital. For the s orbitals, these contour representations are spheres. The contour representations of the 1s, 2s, and 3s orbitals are shown in Figure 6.20◄. All the orbitals have the same shape, but they differ in size. Although the details of how the electron density varies within the contour representation are lost in these representations, this is not a serious disadvantage. For more qualitative discussions, the most important features of orbitals are their relative sizes and their shapes, which are adequately displayed by contour representations.

GIVE IT SOME THOUGHT

How many maxima would you expected to find in the radial probability function for the 4s orbital of the hydrogen atom? How many nodes would you expect in the 4s radial probability function?

A Closer Look   PROBABILITY DENSITY AND RADIAL PROBABILITY FUNCTIONS

The quantum mechanical description of the hydrogen atom requires that we talk about the position of the electron in the atom in a somewhat unfamiliar way. In classical physics, we can exactly pinpoint the position and velocity of an orbiting object, such as a planet orbiting a star. Under quantum mechanics, however, we must describe the position of the electron in the hydrogen atom in terms of probabilities rather than an exact location—an exact answer would violate the uncertainty principle, which becomes important when considering subatomic particles. The information we need about the probability of finding the electron is contained in the wave functions, ψ, that are obtained when Schrödinger's equation is solved. Remember that there are an infinite number of wave functions (orbitals) for the hydrogen atom, but the electron can occupy only one of them at any given time. Here we will discuss briefly how we can use the orbitals to obtain radial probability functions, such as those in Figure 6.19.

In Section 6.5 we stated that the square of the wave function, ψ^2, gives the probability that the electron is at any one given point in space. Recall that this quantity is called the *probability density* for the point. For a spherically symmetric s orbital, the value of ψ depends only on the distance from the nucleus, r. Let's consider a straight line outward from the nucleus, as shown in Figure 6.21▶. The probability of finding the electron at distance r from the nucleus along that line is $[\psi(r)]^2$, where $\psi(r)$ is the value of ψ at distance r. Figure 6.23▶ shows plots of $[\psi(r)]^2$ as a function of r for the 1s, 2s, and 3s orbitals of the hydrogen atom.

You will notice that the plots in Figure 6.23 look distinctly different from the radial probability functions plotted in Figure 6.19. These two types of plots for the s orbitals are very closely related, but they provide somewhat different information. The probability density, $[\psi(r)]^2$, tells us the probability of finding the electron at *a specific* point in space that is at distance r from the nucleus. The radial probability function, which we will denote $P(r)$, tells us the probability of finding the electron at *any* point that is distance r from the nucleus. In other words, to get $P(r)$ we need to "add up" the probabilities of finding the electron over all the points at distance r from the nucleus. The difference between these descriptions may seem rather subtle, but mathematics provides us with a precise way to connect them.

A sphere of radius r around the nucleus

A point at distance r from the nucleus

▲ **Figure 6.21 Probability at a point.** The probability density, $\psi(r)^2$ gives the probability that the electron will be found at *a specific point* at distance r from the nucleus. The radial probability function, $4\psi r^2 \psi(r)^2$, gives the probability that the electron will be found at *any* point distance r from the nucleus—in other words, at any point on the sphere of radius r.

The *p* Orbitals

The distribution of electron density for a 2*p* orbital is shown in Figure 6.22(a) ▼. As we can see from this figure, the electron density is not distributed in a spherically symmetric fashion as in an *s* orbital. Instead, the electron density is concentrated in two regions on either side of the nucleus, separated by a node at the nucleus. We say that this dumbbell-shaped orbital has two *lobes*. Recall that we are making no statement of how the electron is moving within the orbital. The only thing Figure 6.22(a) portrays is the *averaged* distribution of the electron density in a 2*p* orbital.

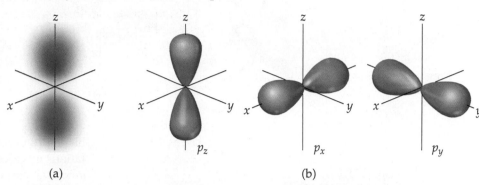

(a)

(b)

Beginning with the $n = 2$ shell, each shell has three *p* orbitals. Recall that the *l* quantum number for *p* orbitals is 1. Therefore, the magnetic quantum number m_l can have three possible values: $-1, 0,$ and $+1$. Thus, there are three 2*p* orbitals, three 3*p* orbitals, and so forth, corresponding to the three possible values of m_l. Each set of *p* orbitals has the dumbbell shapes shown in Figure 6.22(a) for the 2*p* orbitals. For each value of *n*, the three *p* orbitals have the same size and shape but differ from one another in spatial orientation. We usually represent *p* orbitals by drawing the shape and orientation of their wave functions, as shown in Figure 6.22(b). It is convenient to label these as the p_x, p_y and p_z orbitals. The letter subscript indicates the Cartesian axis along which the

▲ **Figure 6.22 The *p* orbitals.** (a) Electron-density distribution of a 2*p* orbital. (b) Contour representations of the three *p* orbitals. Note that the subscript on the orbital label indicates the axis along which the orbital lies.

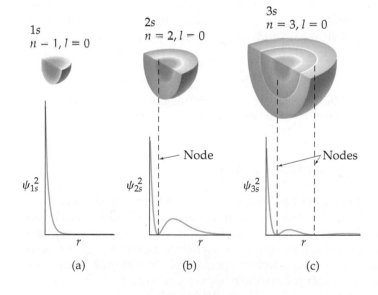

(a)   (b)   (c)

◀ **Figure 6.23 Probability density distribution in 1*s*, 2*s*, and 3*s* orbitals.** The lower part of the figure shows how the probability density, $\psi(r)^2$, varies as a function of distance *r* from the nucleus. The upper part of the figure shows a cutaway of the spherical electron density in each of the *s* orbitals.

to describe, however. The radial probability function at distance *r*, $P(r)$, is simply the probability density at distance *r*, $[\psi(r)]^2$ multiplied by the surface area of the sphere, which is given by the formula $4\pi r^2$:

$$P(r) = 4\pi r^2 [\psi(r)]^2$$

Thus, the plots of $P(r)$ in Figure 6.19 are equal to the plots of $[\psi(r)]^2$ in Figure 6.23 multiplied by $4\pi r^2$. The fact that $4\pi r^2$ increases rapidly as we move away from the nucleus makes the two sets of plots look very different. For example, the plot of $[\psi(r)]^2$ for the 3*s* orbital (Figure 6.23) shows that the function generally gets smaller the farther we go from the nucleus. But when we multiply by $4\pi r^2$, we see peaks that get larger and larger as we move away from the nucleus (Figure 6.19). We will see that the radial probability functions in Figure 6.19 provide us with the more useful information because they tell us the probability for finding the electron at *all* points distance *r* from the nucleus, not just one particular point.

Related Exercises: 6.48, 6.57, 6.58, and 6.91

As shown in Figure 6.21, the collection of points at distance *r* from the nucleus is simply a sphere of radius *r*. The probability density at every point on that sphere is $[\psi(r)]^2$. To add up all of the individual probability densities requires the use of calculus and is beyond the scope of this text (in the language of calculus "we integrate the probability density over the surface of the sphere"). The result we obtain is easy

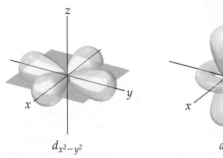

▲ Figure 6.24 Contour representations of the five *d* orbitals.

orbital is oriented.* Like *s* orbitals, *p* orbitals increase in size as we move from $2p$ to $3p$ to $4p$, and so forth.

The *d* and *f* Orbitals

When *n* is 3 or greater, we encounter the *d* orbitals (for which $l = 2$). There are five $3d$ orbitals, five $4d$ orbitals, and so forth because in each shell there are five possible values for the m_l quantum number: $-2, -1, 0, 1,$ and 2. The different *d* orbitals in a given shell have different shapes and orientations in space, as shown in Figure 6.24 ◀. Four of the *d*-orbital contour representations have a "four-leaf clover" shape, and each lies primarily in a plane. The d_{xy}, d_{xz}, and d_{yz} lie in the *xy*, *xz*, and *yz* planes, respectively, with the lobes oriented *between* the axes. The lobes of the $d_{x^2-y^2}$ orbital also lie in the *xy* plane, but the lobes lie *along* the *x* and *y* axes. The d_{z^2} orbital looks very different from the other four: It has two lobes along the *z* axis and a "doughnut" in the *xy* plane. Even though the d_{z^2} orbital looks different from the other *d* orbitals, it has the same energy as the other four *d* orbitals. The representations in Figure 6.24 are commonly used for all *d* orbitals, regardless of principal quantum number.

When *n* is 4 or greater, there are seven equivalent *f* orbitals (for which $l = 3$). The shapes of the *f* orbitals are even more complicated than those of the *d* orbitals and are not presented here. As you will see in the next section, however, you must be aware of *f* orbitals as we consider the electronic structure of atoms in the lower part of the periodic table.

In many instances later in the text you will find that knowing the number and shapes of atomic orbitals will help you understand chemistry at the molecular level. You will therefore find it useful to memorize the shapes of the orbitals shown in Figures 6.20, 6.23, and 6.24.

▲ **GIVE IT SOME THOUGHT**

Note in Figure 6.22(a) that the color is deep pink in the interior of each lobe but fades to pale pink at the edges. What does this change in color represent?

6.7 MANY-ELECTRON ATOMS

One of our goals in this chapter has been to determine the electronic structures of atoms. So far, we have seen that quantum mechanics leads to a very elegant description of the hydrogen atom. This atom, however, has only one electron. How must our description of the electronic structure of atoms change when we consider atoms with two or more electrons (a *many-electron* atom)? To describe these atoms, we must consider the nature of orbitals and their relative energies as well as how the electrons populate the available orbitals.

Orbitals and Their Energies

The quantum mechanical model would not be very useful if we could not extend what we have learned about hydrogen to other atoms. Fortunately, we can describe the electronic structure of a many-electron atom in terms of orbitals

We cannot make a simple correspondence between the subscripts (x, y, and z) and the allowed m_l values (1, 0, and −1). To explain why this is so is beyond the scope of an introductory text.

like those of the hydrogen atom. Thus, we can continue to designate orbitals as 1s, $2p_x$, and so forth. Further, these orbitals have the same general shapes as the corresponding hydrogen orbitals.

Although the shapes of the orbitals for many-electron atoms are the same as those for hydrogen, the presence of more than one electron greatly changes the energies of the orbitals. In hydrogen the energy of an orbital depends only on its principal quantum number, n (Figure 6.18); the 3s, 3p, and 3d subshells all have the same energy, for instance. In a many-electron atom, however, the electron–electron repulsions cause the different subshells to be at different energies, as shown in Figure 6.25▶. To understand why this is so, we must consider the forces between the electrons and how these forces are affected by the shapes of the orbitals. We will, however, forgo this analysis until Chapter 7.

The important idea is this: *In a many-electron atom, for a given value of n, the energy of an orbital increases with increasing value of l.* You can see this illustrated in Figure 6.25. Notice, for example, that the $n = 3$ orbitals (red) increase in energy in the order 3s < 3p < 3d. Figure 6.25 is a *qualitative* energy-level diagram; the exact energies of the orbitals and their spacings differ from one atom to another. Notice that all orbitals of a given subshell (such as the five 3d orbitals) still have the same energy as one another, just as they do in the hydrogen atom. Orbitals with the same energy are said to be **degenerate.**

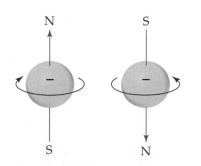

▲ **Figure 6.25 Orbital energy levels in many-electron atoms.** In a many-electron atom, the energies of the subshells in each shell follow the order $ns < np < nd < nf$. As in Figure 6.18, each box represents an orbital.

GIVE IT SOME THOUGHT

For a many-electron atom, can we predict unambiguously whether the 4s orbital is lower in energy or higher in energy than the 3d orbitals?

Electron Spin and the Pauli Exclusion Principle

We have now seen that we can use hydrogen-like orbitals to describe many-electron atoms. What, however, determines which orbitals the electrons reside in? That is, how do the electrons of a many-electron atom populate the available orbitals? To answer this question, we must consider an additional property of the electron.

When scientists studied the line spectra of many-electron atoms in great detail, they noticed a very puzzling feature: Lines that were originally thought to be single were actually closely spaced pairs. This meant, in essence, that there were twice as many energy levels as there were "supposed" to be. In 1925 the Dutch physicists George Uhlenbeck and Samuel Goudsmit proposed a solution to this dilemma. They postulated that electrons have an intrinsic property, called **electron spin,** that causes each electron to behave as if it were a tiny sphere spinning on its own axis.

By now it probably does not surprise you to learn that electron spin is quantized. This observation led to the assignment of a new quantum number for the electron, in addition to n, l, and m_l, which we have already discussed. This new quantum number, the **spin magnetic quantum number,** is denoted m_s (the subscript s stands for *spin*). Two possible values are allowed for m_s, $+\frac{1}{2}$ or $-\frac{1}{2}$, which was first interpreted as indicating the two opposite directions in which the electron can spin. A spinning charge produces a magnetic field. The two opposite directions of spin therefore produce oppositely directed magnetic fields, as shown in Figure 6.26▶.* These two opposite magnetic fields lead to the splitting of spectral lines into closely spaced pairs.

Electron spin is crucial for understanding the electronic structures of atoms. In 1925 the Austrian-born physicist Wolfgang Pauli (1900–1958) discovered the principle that governs the arrangements of electrons in many-electron atoms. The **Pauli exclusion principle** states that *no two electrons in an atom can have the same*

▲ **Figure 6.26 Electron spin.** The electron behaves as if it were spinning about an axis, thereby generating a magnetic field whose direction depends on the direction of spin. The two directions for the magnetic field correspond to the two possible values for the spin quantum number, m_s.

As we discussed earlier, the electron has both particle-like and wavelike properties. Thus, the picture of an electron as a spinning charged sphere is, strictly speaking, just a useful pictorial representation that helps us understand the two directions of magnetic field that an electron can possess.

Even before electron spin had been proposed, there was experimental evidence that electrons had an additional property that needed explanation. In 1921, Otto Stern and Walter Gerlach succeeded in separating a beam of neutral atoms into two groups by passing them through a nonhomogeneous magnetic field. Their experiment is diagrammed in Figure 6.27▶. Let's assume that they used a beam of hydrogen atoms (in actuality, they used silver atoms, which contain just one unpaired electron). We would normally expect that neutral atoms would not be affected by a magnetic field. However, the magnetic field arising from the electron's spin interacts with the magnet's field, deflecting the atom from its straight-line path. As shown in Figure 6.27, the magnetic field splits the beam in two, suggesting that there are two (and only two) equivalent values for the electron's own magnetic field. The Stern–Gerlach experiment could be readily interpreted once it was realized that there are exactly two values for the spin of the electron. These values will produce equal magnetic fields that are opposite in direction.

Related Exercise: 6.94

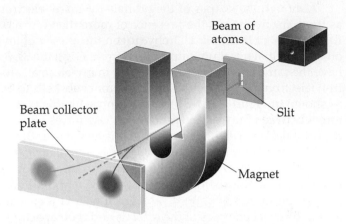

▲ **Figure 6.27 The Stern–Gerlach experiment.** Atoms in which the electron spin quantum number (m_s) of the unpaired electron is $+\frac{1}{2}$ are deflected in one direction, and those in which m_s is $-\frac{1}{2}$ are deflected in the other.

set of four quantum numbers n, l, m_l, and m_s. For a given orbital ($1s$, $2p_z$, and so forth), the values of n, l, and m_l are fixed. Thus, if we want to put more than one electron in an orbital *and* satisfy the Pauli exclusion principle, our only choice is to assign different m_s values to the electrons. Because there are only two such values, we conclude that *an orbital can hold a maximum of two electrons and they must have opposite spins*. This restriction allows us to index the electrons in an atom, giving their quantum numbers and thereby defining the region in space where each electron is most likely to be found. It also provides the key to one of the great problems in chemistry—understanding the structure of the periodic table of the elements. We will discuss these issues in the next two sections.

6.8 ELECTRON CONFIGURATIONS

Armed with knowledge of the relative energies of orbitals and the Pauli exclusion principle, we are now in a position to consider the arrangements of electrons in atoms. The way in which the electrons are distributed among the various orbitals of an atom is called the **electron configuration** of the atom. The most stable electron configuration of an atom—the ground state—is that in which the electrons are in the lowest possible energy states. If there were no restrictions on the possible values for the quantum numbers of the electrons, all the electrons would crowd into the $1s$ orbital because it is the lowest in energy (Figure 6.25). The Pauli exclusion principle tells us, however, that there can be at most two electrons in any single orbital. Thus, *the orbitals are filled in order of increasing energy, with no more than two electrons per orbital.* For example, consider the lithium atom, which has three electrons. (Recall that the number of electrons in a neutral atom equals its atomic number.) The $1s$ orbital can accommodate two of the electrons. The third one goes into the next lowest energy orbital, the $2s$.

We can represent any electron configuration by writing the symbol for the occupied subshell and adding a superscript to indicate the number of electrons in that subshell. For example, for lithium we write $1s^2 2s^1$ (read "$1s$ two, $2s$ one"). We can also show the arrangement of the electrons as

Li   ⏹⏫   ⏹↑

$1s$ $2s$

In this kind of representation, which we call an *orbital diagram*, each orbital is denoted by a box and each electron by a half arrow. A half arrow pointing up (↑) represents an electron with a positive spin magnetic quantum number ($m_s = +\frac{1}{2}$) and a half arrow pointing down (↓) represents an electron with a negative spin magnetic quantum number ($m_s = -\frac{1}{2}$). This pictorial representation of electron spin is quite convenient. In fact, chemists and physicists often refer to electrons as "spin-up" and "spin-down" rather than specifying the value for m_s.

Electrons having opposite spins are said to be *paired* when they are in the same orbital (↑↓). An *unpaired electron* is one not accompanied by a partner of opposite spin. In the lithium atom the two electrons in the 1s orbital are paired and the electron in the 2s orbital is unpaired.

Hund's Rule

Consider now how the electron configurations of the elements change as we move from element to element across the periodic table. Hydrogen has one electron, which occupies the 1s orbital in its ground state.

$$\text{H} \quad \boxed{\uparrow} \quad : 1s^1$$
$$1s$$

The choice of a spin-up electron here is arbitrary; we could equally well show the ground state with one spin-down electron in the 1s orbital. It is customary, however, to show unpaired electrons with their spins up.

The next element, helium, has two electrons. Because two electrons with opposite spins can occupy an orbital, both of helium's electrons are in the 1s orbital.

$$\text{He} \quad \boxed{\uparrow\downarrow} \quad : 1s^2$$
$$1s$$

The two electrons present in helium complete the filling of the first shell. This arrangement represents a very stable configuration, as is evidenced by the chemical inertness of helium.

The electron configurations of lithium and several elements that follow it in the periodic table are shown in Table 6.3 ▼. For the third electron of lithium, the change in principal quantum number represents a large jump in energy and a corresponding jump in the average distance of the electron from the nucleus.

TABLE 6.3 ■ Electron Configurations of Several Lighter Elements

Element	Total Electrons	Orbital Diagram					Electron Configuration
		1s	2s	2p		3s	
Li	3	↑↓	↑				$1s^2 2s^1$
Be	4	↑↓	↑↓				$1s^2 2s^2$
B	5	↑↓	↑↓	↑			$1s^2 2s^2 2p^1$
C	6	↑↓	↑↓	↑ ↑			$1s^2 2s^2 2p^2$
N	7	↑↓	↑↓	↑ ↑ ↑			$1s^2 2s^2 2p^3$
Ne	10	↑↓	↑↓	↑↓ ↑↓ ↑↓			$1s^2 2s^2 2p^6$
Na	11	↑↓	↑↓	↑↓ ↑↓ ↑↓		↑	$1s^2 2s^2 2p^6 3s^1$

Chemistry and Life   NUCLEAR SPIN AND MAGNETIC RESONANCE IMAGING

A major challenge facing medical diagnosis is seeing inside the human body from the outside. Until recently, this was accomplished primarily by using X-rays to image human bones, muscles, and organs. However, there are several drawbacks to using X-rays for medical imaging. First, X-rays do not give well-resolved images of overlapping physiological structures. Moreover, because damaged or diseased tissue often yields the same image as healthy tissue, X-rays frequently fail to detect illness or injuries. Finally, X-rays are high-energy radiation that can cause physiological harm, even in low doses.

During the 1980s, a new technique called *magnetic resonance imaging* (MRI) moved to the forefront of medical imaging technology. The foundation of MRI is a phenomenon called nuclear magnetic resonance (NMR), which was discovered in the mid-1940s. Today NMR has become one of the most important spectroscopic methods used in chemistry. It is based on the observation that, like electrons, the nuclei of many elements possess an intrinsic spin. Like electron spin, nuclear spin is quantized. For example, the nucleus of ^1H (a proton) has two possible magnetic nuclear spin quantum numbers, $+\frac{1}{2}$ and $-\frac{1}{2}$. The hydrogen nucleus is the most common one studied by NMR.

A spinning hydrogen nucleus acts like a tiny magnet. In the absence of external effects, the two spin states have the same energy. However, when the nuclei are placed in an external magnetic field, they can align either parallel or opposed (antiparallel) to the field, depending on their spin. The parallel alignment is lower in energy than the antiparallel one by a certain amount, ΔE (Figure 6.28◄). If the nuclei are irradiated with photons with energy equal to ΔE, the spin of the nuclei can be "flipped," that is, excited from the parallel to the antiparallel alignment. Detection of the flipping of nuclei between the two spin states leads to an NMR spectrum. The radiation used in an NMR experiment is in the radiofrequency range, typically 100 to 900 MHz, which is far less energetic per photon than X-rays.

Because hydrogen is a major constituent of aqueous body fluids and fatty tissue, the hydrogen nucleus is the most convenient one for study by MRI. In MRI a person's body is placed in a strong magnetic field. By irradiating the body with pulses of radiofrequency radiation and using sophisticated detection techniques, tissue can be imaged at specific depths within the body, giving pictures with spectacular detail (Figure 6.29▼). The ability to sample at different depths allows medical technicians to construct a three-dimensional picture of the body.

MRI has none of the disadvantages of X-rays. Diseased tissue appears very different from healthy tissue, resolving overlapping structures at different depths in the body is much easier, and the radiofrequency radiation is not harmful to humans in the doses used. The technique has had such a profound influence on the modern practice of medicine that Paul Lauterbur, a chemist, and Peter Mansfield, a physicist, were awarded the 2003 Nobel Prize in Physiology or Medicine for their discoveries concerning MRI. The major drawback of this technique is expense: The current cost of a new MRI instrument for clinical applications is over $1.5 million.
Related Exercises: 6.94 and 6.95

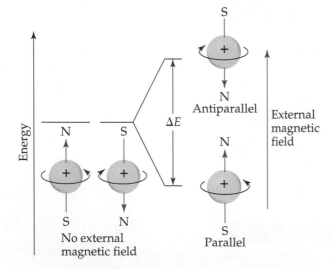

▲ **Figure 6.28 Nuclear spin.** Like electron spin, nuclear spin generates a small magnetic field and has two allowed values. In the absence of an external magnetic field (left), the two spin states have the same energy. If an external magnetic field is applied (right), the parallel alignment of the nuclear magnetic field is lower in energy than the antiparallel alignment. The energy difference, ΔE, is in the radio frequency portion of the electromagnetic spectrum.

◄ **Figure 6.29 MRI image.** This image of a human head, obtained using MRI, shows the structures of a normal brain, airways, and facial tissues.

It represents the start of a new shell occupied with electrons. As you can see by examining the periodic table, lithium starts a new row of the table. It is the first member of the alkali metals (group 1A).

The element that follows lithium is beryllium; its electron configuration is $1s^2 2s^2$ (Table 6.3). Boron, atomic number 5, has the electron configuration $1s^2 2s^2 2p^1$. The fifth electron must be placed in a $2p$ orbital because the $2s$ orbital is filled. Because all the three $2p$ orbitals are of equal energy, it does not matter which $2p$ orbital is occupied.

With the next element, carbon, we encounter a new situation. We know that the sixth electron must go into a $2p$ orbital. However, does this new electron go into the $2p$ orbital that already has one electron, or into one of the other two $2p$ orbitals? This question is answered by **Hund's rule**, which states that *for degenerate orbitals, the lowest energy is attained when the number of electrons with the same spin is maximized.* This means that electrons will occupy orbitals singly to the maximum extent possible and that these single electrons in a given subshell will all have the same spin magnetic quantum number. Electrons arranged in this way are said to have *parallel spins*. For a carbon atom to achieve its lowest energy, therefore, the two $2p$ electrons will have the same spin. For this to happen, the electrons must be in different $2p$ orbitals, as shown in Table 6.3. Thus, a carbon atom in its ground state has two unpaired electrons. Similarly, for nitrogen in its ground state, Hund's rule requires that the three $2p$ electrons singly occupy each of the three $2p$ orbitals. This is the only way that all three electrons can have the same spin. For oxygen and fluorine, we place four and five electrons, respectively, in the $2p$ orbitals. To achieve this, we pair up electrons in the $2p$ orbitals, as we will see in Sample Exercise 6.7.

Hund's rule is based in part on the fact that electrons repel one another. By occupying different orbitals, the electrons remain as far as possible from one another, thus minimizing electron–electron repulsions.

■ **SAMPLE EXERCISE 6.7** | Orbital Diagrams and Electron Configurations

Draw the orbital diagram for the electron configuration of oxygen, atomic number 8. How many unpaired electrons does an oxygen atom possess?

SOLUTION

Analyze and Plan: Because oxygen has an atomic number of 8, each oxygen atom has 8 electrons. Figure 6.25 shows the ordering of orbitals. The electrons (represented as arrows) are placed in the orbitals (represented as boxes) beginning with the lowest-energy orbital, the $1s$. Each orbital can hold a maximum of two electrons (the Pauli exclusion principle). Because the $2p$ orbitals are degenerate, we place one electron in each of these orbitals (spin-up) before pairing any electrons (Hund's rule).

Solve: Two electrons each go into the $1s$ and $2s$ orbitals with their spins paired. This leaves four electrons for the three degenerate $2p$ orbitals. Following Hund's rule, we put one electron into each $2p$ orbital until all three orbitals have one electron each. The fourth electron is then paired up with one of the three electrons already in a $2p$ orbital, so that the representation is

$$\boxed{\uparrow\downarrow}\quad\boxed{\uparrow\downarrow}\quad\boxed{\uparrow\downarrow}\,\boxed{\uparrow}\,\boxed{\uparrow}$$

$$1s\qquad 2s\qquad\quad 2p$$

The corresponding electron configuration is written $1s^2 2s^2 2p^4$. The atom has two unpaired electrons.

■ **PRACTICE EXERCISE**

(a) Write the electron configuration for phosphorus, element 15. **(b)** How many unpaired electrons does a phosphorus atom possess?
Answers: **(a)** $1s^2 2s^2 2p^6 3s^2 3p^3$, **(b)** three

Condensed Electron Configurations

The filling of the $2p$ subshell is complete at neon (Table 6.3), which has a stable configuration with eight electrons (an *octet*) in the outermost occupied shell. The next element, sodium, atomic number 11, marks the beginning of a new row of the periodic table. Sodium has a single $3s$ electron beyond the stable configuration of neon. We can therefore abbreviate the electron configuration of sodium as

$$\text{Na:}\quad [\text{Ne}]3s^1$$

The symbol [Ne] represents the electron configuration of the ten electrons of neon, $1s^2 2s^2 2p^6$. Writing the electron configuration as $[Ne]3s^1$ helps focus attention on the outermost electrons of the atom, which are the ones largely responsible for the chemical behavior of an element.

We can generalize what we have just done for the electron configuration of sodium. In writing the *condensed electron configuration* of an element, the electron configuration of the nearest noble-gas element of lower atomic number is represented by its chemical symbol in brackets. For example, we can write the electron configuration of lithium as

$$\text{Li:} \quad [He]2s^1$$

We refer to the electrons represented by the symbol for a noble gas as the *noble-gas core* of the atom. More usually, these inner-shell electrons are referred to as the **core electrons.** The electrons given after the noble-gas core are called the *outer-shell electrons.* The outer-shell electrons include the electrons involved in chemical bonding, which are called the **valence electrons.** For lighter elements (those with atomic number of 30 or less), all of the outer-shell electrons are valence electrons. As we will discuss later, many of the heavier elements have completely filled subshells that are not involved in bonding and are therefore not considered valence electrons.

By comparing the condensed electron configuration of lithium with that of sodium, we can appreciate why these two elements are so similar chemically. They have the same type of electron configuration in the outermost occupied shell. Indeed, all the members of the alkali metal group (1A) have a single *s* valence electron beyond a noble-gas configuration.

Transition Metals

The noble-gas element argon marks the end of the row started by sodium. The configuration for argon is $1s^2 2s^2 2p^6 3s^2 3p^6$. The element following argon in the periodic table is potassium (K), atomic number 19. In all its chemical properties, potassium is clearly a member of the alkali metal group. The experimental facts about the properties of potassium leave no doubt that the outermost electron of this element occupies an *s* orbital. But this means that the electron with the highest energy has *not* gone into a 3*d* orbital, which we might have expected it to do. Here the ordering of energy levels is such that the 4*s* orbital is lower in energy than the 3*d* orbital (Figure 6.25). Hence, the condensed electron configuration of potassium is

$$\text{K:} \quad [Ar]4s^1$$

Following the complete filling of the 4*s* orbital (this occurs in the calcium atom), the next set of orbitals to be filled is the 3*d* (You will find it helpful as we go along to refer often to the periodic table on the front inside cover.) Beginning with scandium and extending through zinc, electrons are added to the five 3*d* orbitals until they are completely filled. Thus, the fourth row of the periodic table is ten elements wider than the two previous rows. These ten elements are known as either **transition elements** or **transition metals.** Note the position of these elements in the periodic table.

In deriving the electron configurations of the transition elements, the orbitals are filled in accordance with Hund's rule—electrons are added to the 3*d* orbitals singly until all five orbitals have one electron each. Additional electrons are then placed in the 3*d* orbitals with spin pairing until the shell is completely filled.

The condensed electron configurations and the corresponding orbital diagram representations of two transition elements are as follows:

Once all the 3d orbitals have been filled with two electrons each, the 4p orbitals begin to be occupied until the completed octet of outer electrons ($4s^2 4p^6$) is reached with krypton (Kr), atomic number 36, another of the noble gases. Rubidium (Rb) marks the beginning of the fifth row. Refer again to the periodic table on the front inside cover. Notice that this row is in every respect like the preceding one, except that the value for n is greater by 1.

▲ GIVE IT SOME THOUGHT

Based on the structure of the periodic table, which becomes occupied first, the 6s orbital or the 5d orbitals?

The Lanthanides and Actinides

The sixth row of the periodic table begins similarly to the preceding one: one electron in the 6s orbital of cesium (Cs) and two electrons in the 6s orbital of barium (Ba). Notice, however, that the periodic table then has a break, and the subsequent set of elements (elements 57–70) is placed below the main portion of the table. This place is where we begin to encounter a new set of orbitals, the 4f.

There are seven degenerate 4f orbitals, corresponding to the seven allowed values of m_l, ranging from 3 to −3. Thus, it takes 14 electrons to fill the 4f orbitals completely. The 14 elements corresponding to the filling of the 4f orbitals are known as either the **lanthanide elements** or the **rare earth elements**. These elements are set below the other elements to avoid making the periodic table unduly wide. The properties of the lanthanide elements are all quite similar, and these elements occur together in nature. For many years it was virtually impossible to separate them from one another.

Because the energies of the 4f and 5d orbitals are very close to each other, the electron configurations of some of the lanthanides involve 5d electrons. For example, the elements lanthanum (La), cerium (Ce), and praseodymium (Pr) have the following electron configurations:

$$[Xe]6s^2\, 5d^1 \qquad [Xe]6s^2\, 5d^1\, 4f^1 \qquad [Xe]6s^2\, 4f^3$$
$$\text{Lanthanum} \qquad \text{Cerium} \qquad \text{Praseodymium}$$

Because La has a single 5d electron, it is sometimes placed below yttrium (Y) as the first member of the third series of transition elements; Ce is then placed as the first member of the lanthanides. Based on their chemistry, however, La can be considered the first element in the lanthanide series. Arranged this way, there are fewer apparent exceptions to the regular filling of the 4f orbitals among the subsequent members of the series.

After the lanthanide series, the third transition element series is completed by the filling of the 5d orbitals, followed by the filling of the 6p orbitals. This brings us to radon (Rn), heaviest of the known noble-gas elements.

The final row of the periodic table begins by filling the 7s orbitals. The **actinide elements,** of which uranium (U, element 92) and plutonium (Pu, element 94) are the best known, are then built up by completing the 5f orbitals. The actinide elements are radioactive, and most of them are not found in nature.

6.9 ELECTRON CONFIGURATIONS AND THE PERIODIC TABLE

Our rather brief survey of electron configurations of the elements has taken us through the periodic table. We have seen that the electron configurations of the elements are related to their locations in the periodic table. The periodic table is structured so that elements with the same pattern of outer-shell (valence) electron configuration are arranged in columns. For example, the electron configurations for the elements in groups 2A and 3A are given in Table 6.4 ◄. We see that all the 2A elements have ns^2 outer configurations, while the all the 3A elements have ns^2np^1 configurations.

Earlier, in Table 6.2, we saw that the total number of orbitals in each shell is equal to n^2: 1, 4, 9, or 16. Because each orbital can hold two electrons, each shell can accommodate up to $2n^2$ electrons: 2, 8, 18, or 32. The structure of the periodic table reflects this orbital structure. The first row has two elements, the second and third rows have eight elements, the fourth and fifth rows have 18 elements, and the sixth row has 32 elements (including the lanthanide metals). Some of the numbers repeat because we reach the end of a row of the periodic table before a shell completely fills. For example, the third row has eight elements, which corresponds to filling the 3s and 3p orbitals. The remaining orbitals of the third shell, the 3d orbitals, do not begin to fill until the fourth row of the periodic table (and after the 4s orbital is filled). Likewise, the 4d orbitals do not begin to fill until the fifth row of the table, and the 4f orbitals don't begin filling until the sixth row.

All these observations are evident in the structure of the periodic table. For this reason, we will emphasize that *the periodic table is your best guide to the order in which orbitals are filled.* You can easily write the electron configuration of an element based on its location in the periodic table. The pattern is summarized in Figure 6.30 ◄. Notice that the elements can be grouped by the *type* of orbital into which the electrons are placed. On the left are *two* columns of elements, depicted in blue. These elements, known as the alkali metals (group 1A) and alkaline earth metals (group 2A), are those in which the valence s orbitals are being filled. On the right is a pink block of *six* columns. These are the elements in which the valence p orbitals are being filled. The s block and the p block of the periodic table together are the **representative elements,** which are sometimes called the **main-group elements.**

In the middle of Figure 6.30 is a gold block of *ten* columns containing the transition metals. These are the elements in which the valence d orbitals are being filled. Below the main portion of the table are two tan rows containing 14 columns. These elements are often referred to as the *f***-block metals,** because they are the ones in which the valence f orbitals are being filled. Recall that the numbers 2, 6, 10, and 14 are precisely the number of electrons that can fill the s, p, d, and f subshells, respectively. Recall also that the 1s subshell is the first s subshell, the 2p is the first p subshell, the 3d is the first d subshell, and the 4f is the first f subshell.

TABLE 6.4 ■ Electron Configurations of the Group 2A and 3A Elements	
Group 2A	
Be	$[He]2s^2$
Mg	$[Ne]3s^2$
Ca	$[Ar]4s^2$
Sr	$[Kr]5s^2$
Ba	$[Xe]6s^2$
Ra	$[Rn]7s^2$
Group 3A	
B	$[He]2s^22p^1$
Al	$[Ne]3s^23p^1$
Ga	$[Ar]3d^{10}4s^24p^1$
In	$[Kr]4d^{10}5s^25p^1$
Tl	$[Xe]4f^{14}5d^{10}6s^26p^1$

Representative s-block elements

d-Block metals (transition metals)

Representative p-block elements

f-Block metals

▲ **Figure 6.30 Regions of the periodic table.** This block diagram of the periodic table shows the order in which electrons are added to orbitals as we move through the table from beginning to end.

■ SAMPLE EXERCISE 6.8 | Electron Configurations for a Group

What is the characteristic valence electron configuration of the group 7A elements, the halogens?

SOLUTION

Analyze and Plan: We first locate the halogens in the periodic table, write the electron configurations for the first two elements, and then determine the general similarity between them.

Solve: The first member of the halogen group is fluorine, atomic number 9. The condensed electron configuration for fluorine is

$$F: \quad [He]2s^2 2p^5$$

Similarly, that for chlorine, the second halogen, is

$$Cl: \quad [Ne]3s^2 3p^5$$

From these two examples, we see that the characteristic valence electron configuration of a halogen is $ns^2 np^5$, where n ranges from 2 in the case of fluorine to 6 in the case of astatine.

■ PRACTICE EXERCISE

Which family of elements is characterized by an $ns^2 np^2$ electron configuration in the outermost occupied shell?
Answer: group 4A

■ SAMPLE EXERCISE 6.9 | Electron Configurations from the Periodic Table

(a) Write the electron configuration for bismuth, element number 83. **(b)** Write the condensed electron configuration for this element. **(c)** How many unpaired electrons does each atom of bismuth possess?

SOLUTION

(a) We write the electron configuration by moving across the periodic table one row at a time and writing the occupancies of the orbital corresponding to each row (refer to Figure 6.29).

First row $1s^2$
Second row $2s^2 2p^6$
Third row $3s^2 3p^6$
Fourth row $4s^2 3d^{10} 4p^6$
Fifth row $5s^2 4d^{10} 5p^6$
Sixth row $6s^2 4f^{14} 5d^{10} 6p^3$
Total: $1s^2 2s^2 2p^6 3s^2 3p^6 3d^{10} 4s^2 4p^6 4d^{10} 4f^{14} 5s^2 5p^6 5d^{10} 6s^2 6p^3$

Note that 3 is the lowest possible value that n may have for a d orbital and that 4 is the lowest possible value of n for an f orbital.

The total of the superscripted numbers should equal the atomic number of bismuth, 83. The electrons may be listed, as shown above in the "Total" row, in the order of increasing principal quantum number. However, it is equally correct to list the orbitals in the order in which they are read from Figure 6.30: $1s^2 2s^2 2p^6 3s^2 3p^6 4s^2 3d^{10} 4p^6 5s^2 4d^{10} 5p^6 6s^2 4f^{14} 5d^{10} 6p^3$.

(b) We write the condensed electron configuration by locating bismuth on the periodic table and then moving *backward* to the nearest noble gas, which is Xe, element 54. Thus, the noble-gas core is [Xe]. The outer electrons are then read from the periodic table as before. Moving from Xe to Cs, element 55, we find ourselves in the sixth row. Moving across this row to Bi gives us the outer electrons. Thus, the abbreviated electron configuration is $[Xe]6s^2 4f^{14} 5d^{10} 6p^3$ or $[Xe]4f^{14} 5d^{10} 6s^2 6p^3$.

(c) We can see from the abbreviated electron configuration that the only partially occupied subshell is the $6p$. The orbital diagram representation for this subshell is

In accordance with Hund's rule, the three $6p$ electrons occupy the three $6p$ orbitals singly, with their spins parallel. Thus, there are three unpaired electrons in each atom of bismuth.

■ PRACTICE EXERCISE

Use the periodic table to write the condensed electron configurations for **(a)** Co (atomic number 27), **(b)** Te (atomic number 52).
Answers: **(a)** $[Ar]4s^2 3d^7$ or $[Ar]3d^7 4s^2$, **(b)** $[Kr]5s^2 4d^{10} 5p^4$ or $[Kr]4d^{10} 5s^2 5p^4$

Main group and transition elements (Core configuration indicated at left):

Core	1A / 1	2A / 2	3B / 3	4B / 4	5B / 5	6B / 6	7B / 7	8B / 8	8B / 9	8B / 10	1B / 11	2B / 12	3A / 13	4A / 14	5A / 15	6A / 16	7A / 17	8A / 18
	1 H $1s^1$																	2 He $1s^2$
[He]	3 Li $2s^1$	4 Be $2s^2$											5 B $2s^2 2p^1$	6 C $2s^2 2p^2$	7 N $2s^2 2p^3$	8 O $2s^2 2p^4$	9 F $2s^2 2p^5$	10 Ne $2s^2 2p^6$
[Ne]	11 Na $3s^1$	12 Mg $3s^2$											13 Al $3s^2 3p^1$	14 Si $3s^2 3p^2$	15 P $3s^2 3p^3$	16 S $3s^2 3p^4$	17 Cl $3s^2 3p^5$	18 Ar $3s^2 3p^6$
[Ar]	19 K $4s^1$	20 Ca $4s^2$	21 Sc $3d^1 4s^2$	22 Ti $3d^2 4s^2$	23 V $3d^3 4s^2$	24 Cr $3d^5 4s^1$	25 Mn $3d^5 4s^2$	26 Fe $3d^6 4s^2$	27 Co $3d^7 4s^2$	28 Ni $3d^8 4s^2$	29 Cu $3d^{10} 4s^1$	30 Zn $3d^{10} 4s^2$	31 Ga $3d^{10} 4s^2 4p^1$	32 Ge $3d^{10} 4s^2 4p^2$	33 As $3d^{10} 4s^2 4p^3$	34 Se $3d^{10} 4s^2 4p^4$	35 Br $3d^{10} 4s^2 4p^5$	36 Kr $3d^{10} 4s^2 4p^6$
[Kr]	37 Rb $5s^1$	38 Sr $5s^2$	39 Y $4d^1 5s^2$	40 Zr $4d^2 5s^2$	41 Nb $4d^3 5s^2$	42 Mo $4d^5 5s^1$	43 Tc $4d^5 5s^2$	44 Ru $4d^7 5s^1$	45 Rh $4d^8 5s^1$	46 Pd $4d^{10}$	47 Ag $4d^{10} 5s^1$	48 Cd $4d^{10} 5s^2$	49 In $4d^{10} 5s^2 5p^1$	50 Sn $4d^{10} 5s^2 5p^2$	51 Sb $4d^{10} 5s^2 5p^3$	52 Te $4d^{10} 5s^2 5p^4$	53 I $4d^{10} 5s^2 5p^5$	54 Xe $4d^{10} 5s^2 5p^6$
[Xe]	55 Cs $6s^1$	56 Ba $6s^2$	71 Lu $4f^{14} 5d^1 6s^2$	72 Hf $4f^{14} 5d^2 6s^2$	73 Ta $4f^{14} 5d^3 6s^2$	74 W $4f^{14} 5d^4 6s^2$	75 Re $4f^{14} 5d^5 6s^2$	76 Os $4f^{14} 5d^6 6s^2$	77 Ir $4f^{14} 5d^7 6s^2$	78 Pt $4f^{14} 5d^9 6s^1$	79 Au $4f^{14} 5d^{10} 6s^1$	80 Hg $4f^{14} 5d^{10} 6s^2$	81 Tl $4f^{14} 5d^{10} 6s^2 6p^1$	82 Pb $4f^{14} 5d^{10} 6s^2 6p^2$	83 Bi $4f^{14} 5d^{10} 6s^2 6p^3$	84 Po $4f^{14} 5d^{10} 6s^2 6p^4$	85 At $4f^{14} 5d^{10} 6s^2 6p^5$	86 Rn $4f^{14} 5d^{10} 6s^2 6p^6$
[Rn]	87 Fr $7s^1$	88 Ra $7s^2$	103 Lr $5f^{14} 6d^1 7s^2$	104 Rf $5f^{14} 6d^2 7s^2$	105 Db $5f^{14} 6d^3 7s^2$	106 Sg $5f^{14} 6d^4 7s^2$	107 Bh $5f^{14} 6d^5 7s^2$	108 Hs $5f^{14} 6d^6 7s^2$	109 Mt $5f^{14} 6d^7 7s^2$	110 Ds	111 Rg	112	113	114	115	116		118

Lanthanide series [Xe]:

57 La $5d^1 6s^2$	58 Ce $4f^1 5d^1 6s^2$	59 Pr $4f^3 6s^2$	60 Nd $4f^4 6s^2$	61 Pm $4f^5 6s^2$	62 Sm $4f^6 6s^2$	63 Eu $4f^7 6s^2$	64 Gd $4f^7 5d^1 6s^2$	65 Tb $4f^9 6s^2$	66 Dy $4f^{10} 6s^2$	67 Ho $4f^{11} 6s^2$	68 Er $4f^{12} 6s^2$	69 Tm $4f^{13} 6s^2$	70 Yb $4f^{14} 6s^2$

Actinide series [Rn]:

89 Ac $6d^1 7s^2$	90 Th $6d^2 7s^2$	91 Pa $5f^2 6d^1 7s^2$	92 U $5f^3 6d^1 7s^2$	93 Np $5f^4 6d^1 7s^2$	94 Pu $5f^6 7s^2$	95 Am $5f^7 7s^2$	96 Cm $5f^7 6d^1 7s^2$	97 Bk $5f^9 7s^2$	98 Cf $5f^{10} 7s^2$	99 Es $5f^{11} 7s^2$	100 Fm $5f^{12} 7s^2$	101 Md $5f^{13} 7s^2$	102 No $5f^{14} 7s^2$

Legend: ☐ Metals   ☐ Metalloids   ☐ Nonmetals

▲ Figure 6.31 **Valence electron configurations of the elements.**

Figure 6.31 ▲ gives the valence ground-state electron configurations for all the elements. You can use this figure to check your answers as you practice writing electron configurations. We have written these configurations with orbitals listed in order of increasing principal quantum number. As we saw in Sample Exercise 6.9, the orbitals can also be listed in order of filling, as they would be read off the periodic table.

The electron configurations in Figure 6.31 allow us to reexamine the concept of *valence electrons*. Notice, for example, that as we proceed from Cl ([Ne]$3s^2 3p^5$) to Br ([Ar]$3d^{10} 4s^2 4p^5$) we have added a complete subshell of $3d$ electrons to the outer-shell electrons beyond the noble-gas core of Ar. Although the $3d$ electrons are outer-shell electrons, they are not involved in chemical bonding and are therefore not considered valence electrons. Thus, we consider only the $4s$ and $4p$ electrons of Br to be valence electrons. Similarly, if we compare the electron configuration of Ag and Au, Au has a completely full $4f^{14}$ subshell beyond its noble-gas core, but those $4f$ electrons are not involved in bonding. In general, *for representative elements we do not consider completely full* d *or* f *subshells to be among the valence electrons*, and *for transition elements we likewise do not consider a completely full* f *subshell to be among the valence electrons*.

Anomalous Electron Configurations

If you inspect Figure 6.31 closely, you will see that the electron configurations of certain elements appear to violate the rules we have just discussed. For example, the electron configuration of chromium is $[Ar]3d^54s^1$ rather than the $[Ar]3d^44s^2$ configuration we might have expected. Similarly, the configuration of copper is $[Ar]3d^{10}4s^1$ instead of $[Ar]3d^94s^2$. This anomalous behavior is largely a consequence of the closeness of the $3d$ and $4s$ orbital energies. It frequently occurs when there are enough electrons to lead to precisely half-filled sets of degenerate orbitals (as in chromium) or to completely fill a d subshell (as in copper). There are a few similar cases among the heavier transition metals (those with partially filled $4d$ or $5d$ orbitals) and among the f-block metals. Although these minor departures from the expected are interesting, they are not of great chemical significance.

GIVE IT SOME THOUGHT

The elements Ni, Pd, and Pt are all in the same group. By examining the electron configurations for these elements in Figure 6.31, what can you conclude about the relative energies of the nd and $(n + 1)$ s orbitals for this group?

■ SAMPLE INTEGRATIVE EXERCISE | Putting Concepts Together

Boron, atomic number 5, occurs naturally as two isotopes, ^{10}B and ^{11}B, with natural abundances of 19.9% and 80.1%, respectively. **(a)** In what ways do the two isotopes differ from each other? Does the electronic configuration of ^{10}B differ from that of ^{11}B? **(b)** Draw the orbital diagram for an atom of ^{11}B. Which electrons are the valence electrons? **(c)** Indicate three major ways in which the $1s$ electrons in boron differ from its $2s$ electrons. **(d)** Elemental boron reacts with fluorine to form BF_3, a gas. Write a balanced chemical equation for the reaction of solid boron with fluorine gas. **(e)** ΔH_f° for $BF_3(g)$ is -1135.6 kJ mol^{-1}. Calculate the standard enthalpy change in the reaction of boron with fluorine. **(f)** When BCl_3, also a gas at room temperature, comes into contact with water, the two react to form hydrochloric acid and boric acid, H_3BO_3, a very weak acid in water. Write a balanced net ionic equation for this reaction.

SOLUTION

(a) The two isotopes of boron differ in the number of neutrons in the nucleus. ∞∞ (Sections 2.3 and 2.4) Each of the isotopes contains five protons, but ^{10}B contains five neutrons, whereas ^{11}B contains six neutrons. The two isotopes of boron have identical electron configurations, $1s^22s^22p^1$, because each has five electrons.
(b) The complete orbital diagram is

$$1s \qquad 2s \qquad\quad 2p$$

The valence electrons are the ones in the outermost occupied shell, the $2s^2$ and $2p^1$ electrons. The $1s^2$ electrons constitute the core electrons, which we represent as [He] when we write the condensed electron configuration, $[He]2s^22p^1$.
(c) The $1s$ and $2s$ orbitals are both spherical, but they differ in three important respects: First, the $1s$ orbital is lower in energy than the $2s$ orbital. Second, the average distance of the $2s$ electrons from the nucleus is greater than that of the $1s$ electrons, so the $1s$ orbital is smaller than the $2s$. Third, the $2s$ orbital has one node, whereas the $1s$ orbital has no nodes (Figure 6.19).
(d) The balanced chemical equation is

$$2\ B(s) + 3\ F_2(g) \longrightarrow 2\ BF_3(g)$$

(e) $\Delta H^\circ = 2(-1135.6) - [0 + 0] = -2271.2$ kJ. The reaction is strongly exothermic.
(f) $BCl_3(g) + 3\ H_2O(l) \longrightarrow H_3BO_3(aq) + 3\ H^+(aq) + 3\ Cl^-(aq)$. Note that because H_3BO_3 is a very weak acid, its chemical formula is written in molecular form, as discussed in Section 4.3.

CHAPTER REVIEW

SUMMARY AND KEY TERMS

Introduction and Section 6.1 The **electronic structure** of an atom describes the energies and arrangement of electrons around the atom. Much of what is known about the electronic structure of atoms was obtained by observing the interaction of light with matter. Visible light and other forms of **electromagnetic radiation** (also known as radiant energy) move through a vacuum at the speed of light, $c = 3.00 \times 10^8$ m/s. Electromagnetic radiation has both electric and magnetic components that vary periodically in wavelike fashion. The wave characteristics of radiant energy allow it to be described in terms of **wavelength,** λ and **frequency,** ν, which are interrelated: $c = \lambda\nu$.

Section 6.2 Planck proposed that the minimum amount of radiant energy that an object can gain or lose is related to the frequency of the radiation: $E = h\nu$. This smallest quantity is called a **quantum** of energy. The constant h is called **Planck's constant:** $h = 6.626 \times 10^{-34}$ J-s. In the quantum theory, energy is quantized, meaning that it can have only certain allowed values. Einstein used the quantum theory to explain the **photoelectric effect,** the emission of electrons from metal surfaces by light. He proposed that light behaves as if it consists of quantized energy packets called **photons.** Each photon carries energy, $E = h\nu$.

Section 6.3 Dispersion of radiation into its component wavelengths produces a **spectrum.** If the spectrum contains all wavelengths, it is called a **continuous spectrum**; if it contains only certain specific wavelengths, the spectrum is called a **line spectrum.** The radiation emitted by excited hydrogen atoms forms a line spectrum; the frequencies observed in the spectrum follow a simple mathematical relationship that involves small integers.

Bohr proposed a model of the hydrogen atom that explains its line spectrum. In this model the energy of the electron in the hydrogen atom depends on the value of a number n, called the quantum number. The value of n must be a positive integer (1, 2, 3, ...), and each value of n corresponds to a different specific energy, E_n. The energy of the atom increases as n increases. The lowest energy is achieved for $n = 1$; this is called the **ground state** of the hydrogen atom. Other values of n correspond to **excited states** of the atom. Light is emitted when the electron drops from a higher energy state to a lower energy state; light must be absorbed to excite the electron from a lower energy state to a higher one. The frequency of light emitted or absorbed must be such that $h\nu$ equals the difference in energy between two allowed states of the atom.

Section 6.4 De Broglie proposed that matter, such as electrons, should exhibit wavelike properties. This hypothesis of **matter waves** was proved experimentally by observing the diffraction of electrons. An object has a characteristic wavelength that depends on its **momentum,** mv: $\lambda = h/mv$. Discovery of the wave properties of the electron led to Heisenberg's **uncertainty principle,** which states that there is an inherent limit to the accuracy with which the position and momentum of a particle can be measured simultaneously.

Section 6.5 In the quantum mechanical model of the hydrogen atom, the behavior of the electron is described by mathematical functions called **wave functions,** denoted with the Greek letter ψ. Each allowed wave function has a precisely known energy, but the location of the electron cannot be determined exactly; rather, the probability of it being at a particular point in space is given by the **probability density,** ψ^2. The **electron density** distribution is a map of the probability of finding the electron at all points in space.

The allowed wave functions of the hydrogen atom are called **orbitals.** An orbital is described by a combination of an integer and a letter, corresponding to values of three quantum numbers for the orbital. The principal quantum number, n, is indicated by the integers 1,2,3, ... This quantum number relates most directly to the size and energy of the orbital. The angular momentum quantum number, l, is indicated by the letters s, p, d, f, and so on, corresponding to the values of 0, 1,2,3, ... The l quantum number defines the shape of the orbital. For a given value of n, l can have integer values ranging from 0 to $(n - 1)$. The magnetic quantum number, m_l, relates to the orientation of the orbital in space. For a given value of l, m_l can have integral values ranging from $-l$ to l, including 0. Cartesian labels can be used to label the orientations of the orbitals. For example, the three $3p$ orbitals are designated $3p_x$, $3p_y$, and $3p_z$, with the subscripts indicating the axis along which the orbital is oriented.

An **electron shell** is the set of all orbitals with the same value of n, such as $3s$, $3p$, and $3d$. In the hydrogen atom all the orbitals in an electron shell have the same energy. A **subshell** is the set of one or more orbitals with the same n and l values; for example, $3s$, $3p$, and $3d$ are each subshells of the $n = 3$ shell. There is one orbital in an s subshell, three in a p subshell, five in a d subshell, and seven in an f subshell.

Section 6.6 Contour representations are useful for visualizing the spatial characteristics (shapes) of the orbitals. Represented this way, s orbitals appear as spheres that increase in size as n increases. The **radial probability function** tells us the probability that the electron will be found at a certain distance from the nucleus. The wave function for each p orbital has two lobes on opposite sides of the nucleus. They are oriented along the x-, y-, and z-axes. Four of the d orbitals appear as shapes with four lobes around the nucleus; the fifth one, the d_{z^2} orbital, is represented as two lobes along the z-axis and a "doughnut" in the xy plane. Regions in which the wave function is zero are called **nodes.** There is zero probability that the electron will be found at a node.

Section 6.7 In many-electron atoms, different subshells of the same electron shell have different energies. For a given value of n, the energy of the subshells increases as the value of l increases: $ns < np < nd < nf$. Orbitals within the same subshell are **degenerate,** meaning they have the same energy.

Electrons have an intrinsic property called **electron spin,** which is quantized. The **spin magnetic quantum number,** m_s, can have two possible values, $+\frac{1}{2}$ and $-\frac{1}{2}$, which can be envisioned as the two directions of an electron spinning about an axis. The **Pauli exclusion principle** states that no two electrons in an atom can have the same values for n, l, m_l, and m_s. This principle places a limit of two on the number of electrons that can occupy any one atomic orbital. These two electrons differ in their value of m_s.

Sections 6.8 and 6.9 The **electron configuration** of an atom describes how the electrons are distributed among the orbitals of the atom. The ground-state electron configurations are generally obtained by placing the electrons in the atomic orbitals of lowest possible energy with the restriction that each orbital can hold no more than two electrons. When electrons occupy a subshell with more than one degenerate orbital, such as the $2p$ subshell, **Hund's rule** states that the lowest energy is attained by maximizing the number of electrons with the same electron spin.

For example, in the ground-state electron configuration of carbon, the two $2p$ electrons have the same spin and must occupy two different $2p$ orbitals.

Elements in any given group in the periodic table have the same type of electron arrangements in their outermost shells. For example, the electron configurations of the halogens fluorine and chlorine are $[\text{He}]2s^2 2p^5$ and $[\text{Ne}]3s^2 3p^5$, respectively. The outer-shell electrons are those that lie outside the orbitals occupied in the next lowest noble-gas element. The outer-shell electrons that are involved in chemical bonding are the **valence electrons** of an atom; for the elements with atomic number 30 or less, all the outer-shell electrons are valence electrons. The electrons that are not valence electrons are called **core electrons.**

The periodic table is partitioned into different types of elements, based on their electron configurations. Those elements in which the outermost subshell is an s or p subshell are called the **representative (or main-group) elements.** The alkali metals (group 1A), halogens (group 7A), and noble gases (group 8A) are representative elements. Those elements in which a d subshell is being filled are called the **transition elements (or transition metals).** The elements in which the $4f$ subshell is being filled are called the **lanthanide (or rare earth) elements.** The **actinide elements** are those in which the $5f$ subshell is being filled. The lanthanide and actinide elements are collectively referred to as the **f-block metals.** These elements are shown as two rows of 14 elements below the main part of the periodic table. The structure of the periodic table, summarized in Figure 6.30, allows us to write the electron configuration of an element from its position in the periodic table.

KEY SKILLS

- Be able to calculate the wavelength of electromagnetic radiation given its frequency or its frequency given its wavelength.
- Be able to order the common kinds of radiation in the electromagnetic spectrum according to their wavelengths or energy.
- Understand the concept of photons, and be able to calculate their energies given either their frequency or wavelength.
- Be able to explain how line spectra of the elements relate to the idea of quantized energy states of electrons in atoms.
- Be familiar with the wavelike properties of matter.
- Understand how the uncertainty principle limits how precisely we can specify the position and the momentum of subatomic particles such as electrons.
- Know how the quantum numbers relate to the number and type of orbitals, and recognize the different orbital shapes.
- Interpret radial probability function graphs for the orbitals.
- Be able to draw an energy-level diagram for the orbitals in a many-electron atom, and describe how electrons populate the orbitals in the ground-state of an atom, using the Pauli Exclusion Principle and Hund's rule.
- Be able to use the periodic table to write abbreviated electron configurations and determine the number of unpaired electrons in an atom.

KEY EQUATIONS

- $c = \lambda\nu$ [6.1] light as a wave: c = speed of light (3.00×10^8 m/s), λ = wavelength in meters, ν = frequency in s^{-1}

- $E = h\nu$ [6.2] light as a particle (photon): E = energy of photon in Joules, h = Planck's constant (6.626×10^{-34} J-s), ν = frequency in s^{-1} (same frequency as previous formula)

- $\lambda = h/mv$ [6.8] matter as a wave: λ = wavelength, h = Planck's constant, m = mass of object in kg, v = speed of object in m/s

- $\Delta x \cdot \Delta(mv) \geq \dfrac{h}{4\pi}$ [6.9] Heisenberg's uncertainty principle. The uncertainty in position (Δx) and momentum ($\Delta(mv)$) of an object cannot be zero; the smallest value of their product is $h/4\pi$

VISUALIZING CONCEPTS

6.1 Consider the water wave shown here. **(a)** How could you measure the speed of this wave? **(b)** How would you determine the wavelength of the wave? **(c)** Given the speed and wavelength of the wave, how could you determine the frequency of the wave? **(d)** Suggest an independent experiment to determine the frequency of the wave. [Section 6.1]

6.2 A popular kitchen appliance produces electromagnetic radiation with a frequency of 2450 MHz. With reference to Figure 6.4, answer the following: **(a)** Estimate the wavelength of this radiation. **(b)** Would the radiation produced by the appliance be visible to the human eye? **(c)** If the radiation is not visible, do photons of this radiation have more or less energy than photons of visible light? **(d)** Propose the identity of the kitchen appliance. [Section 6.1]

6.3 As shown in the accompanying photograph, an electric stove burner on its highest setting exhibits an orange glow. **(a)** When the burner setting is changed to low, the burner continues to produce heat but the orange glow disappears. How can this observation be explained with reference to one of the fundamental observations that led to the notion of quanta? **(b)** Suppose that the energy provided to the burner could be increased beyond the highest setting of the stove. What would we expect to observe with regard to visible light emitted by the burner? [Section 6.2]

6.4 The familiar phenomenon of a rainbow results from the diffraction of sunlight through raindrops. **(a)** Does the wavelength of light increase or decrease as we proceed outward from the innermost band of the rainbow? **(b)** Does the frequency of light increase or decrease as we proceed outward? **(c)** Suppose that instead of sunlight, the visible light from a hydrogen discharge tube (Figure 6.12) was used as the light source. What do you think the resulting "hydrogen discharge rainbow" would look like? [Section 6.3]

6.5 A certain quantum mechanical system has the energy levels shown in the diagram below. The energy levels are indexed by a single quantum number n that is an integer. **(a)** As drawn, which quantum numbers are involved in the transition that requires the most energy? **(b)** Which quantum numbers are involved in the transition that requires the least energy? **(c)** Based on the drawing, put the following in order of increasing wavelength of the light absorbed or emitted during the transition: (*i*) $n = 1$ to $n = 2$; (*ii*) $n = 3$ to $n = 2$; (*iii*) $n = 2$ to $n = 4$; (*iv*) $n = 3$ to $n = 1$. [Section 6.3]

6.6 Consider a fictitious one-dimensional system with one electron. The wave function for the electron, drawn at the top of the next column, is $\psi(x) = \sin x$ from $x = 0$ to $x = 2\pi$. **(a)** Sketch the probability density, $\psi^2(x)$, from $x = 0$ to $x = 2\pi$ **(b)** At what value or values of x will there be the greatest probability of finding the electron? **(c)** What is the probability that the electron will be found at $x = \pi$? What is such a point in a wave function called? [Section 6.5]

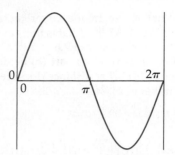

6.7 The contour representation of one of the orbitals for the $n = 3$ shell of a hydrogen atom is shown below. **(a)** What is the quantum number l for this orbital? **(b)** How do we label this orbital? **(c)** How would you modify this sketch to show the analogous orbital for the $n = 4$ shell? [Section 6.6]

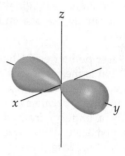

6.8 The drawing below shows part of the orbital diagram for an element. **(a)** As drawn, the drawing is *incorrect*. Why? **(b)** How would you correct the drawing without changing the number of electrons? **(c)** To which group in the periodic table does the element belong? [Section 6.8]

EXERCISES

The Wave Nature of Light

6.9 What are the basic SI units for **(a)** the wavelength of light, **(b)** the frequency of light, **(c)** the speed of light?

6.10 **(a)** What is the relationship between the wavelength and the frequency of radiant energy? **(b)** Ozone in the upper atmosphere absorbs energy in the 210–230-nm range of the spectrum. In what region of the electromagnetic spectrum does this radiation occur?

6.11 Label each of the following statements as true or false. For those that are false, correct the statement. **(a)** Visible light is a form of electromagnetic radiation. **(b)** The frequency of radiation increases as the wavelength increases. **(c)** Ultraviolet light has longer wavelengths than visible light. **(d)** X-rays travel faster than microwaves. **(e)** Electromagnetic radiation and sound waves travel at the same speed.

6.12 Determine which of the following statements are false, and correct them. **(a)** Electromagnetic radiation is incapable of passing through water. **(b)** Electromagnetic radiation travels through a vacuum at a constant speed, regardless of wavelength. **(c)** Infrared light has higher frequencies than visible light. **(d)** The glow from a fireplace, the energy within a microwave oven, and a foghorn blast are all forms of electromagnetic radiation.

6.13 Arrange the following kinds of electromagnetic radiation in order of increasing wavelength: infrared, green light, red light, radio waves, X-rays, ultraviolet light.

6.14 List the following types of electromagnetic radiation in order of increasing wavelength: **(a)** the gamma rays produced by a radioactive nuclide used in medical imaging; **(b)** radiation from an FM radio station at 93.1 MHz on the dial; **(c)** a radio signal from an AM radio station at 680 kHz on the dial; **(d)** the yellow light from sodium vapor streetlights; **(e)** the red light of a light-emitting diode, such as in a calculator display.

6.15 **(a)** What is the frequency of radiation that has a wavelength of 10 μm, about the size of a bacterium? **(b)** What is the wavelength of radiation that has a frequency of 5.50×10^{14} s^{-1}? **(c)** Would the radiations in part (a) or part (b) be visible to the human eye? **(d)** What distance does electromagnetic radiation travel in 50.0 μs?

6.16 **(a)** What is the frequency of radiation whose wavelength is 10.0 Å? **(b)** What is the wavelength of radiation that has a frequency of 7.6×10^{10} s^{-1}? **(c)** Would the radiations in part (a) or part (b) be detected by an X-ray detector? **(d)** What distance does electromagnetic radiation travel in 25.5 fs?

6.17 An argon ion laser emits light at 532 nm. What is the frequency of this radiation? Using Figure 6.4, predict the color associated with this wavelength.

6.18 It is possible to convert radiant energy into electrical energy using photovoltaic cells. Assuming equal efficiency of conversion, would infrared or ultraviolet radiation yield more electrical energy on a per-photon basis?

Quantized Energy and Photons

6.19 If human height were quantized in one-foot increments, what would happen to the height of a child as she grows up?

6.20 Einstein's 1905 paper on the photoelectric effect was the first important application of Planck's quantum hypothesis. Describe Planck's original hypothesis, and explain how Einstein made use of it in his theory of the photoelectric effect.

6.21 **(a)** A red laser pointer emits light with a wavelength of 650 nm. What is the frequency of this light? **(b)** What is the energy of 1 mole of these photons? **(c)** The laser pointer emits light because electrons in the material are excited (by a battery) from their ground state to an upper excited state. When the electrons return to the ground state they lose the excess energy in the form of 650 nm photons. What is the energy gap between the ground state and excited state in the laser material?

6.22 If you put 120 volts of electricity through a pickle, the pickle will smoke and start glowing an orange-yellow color. The light is emitted because the sodium ions in the pickle become excited; their return to the ground state results in light emission (see Figure 6.13b and Sample Exercise 6.3). **(a)** The wavelength of this emitted light is 589 nm. Calculate its frequency. **(b)** What is the energy of 0.10 mole of these photons? **(c)** Calculate the energy gap between the excited and ground states for the sodium ion. **(d)** If you soaked the pickle for a long time in a different salt solution, such as strontium chloride, would you still observe 589 nm light emission? Why or why not?

6.23 **(a)** Calculate and compare the energy of a photon of wavelength 3.3 μm with that of wavelength 0.154 nm. **(b)** Use Figure 6.4 to identify the region of the electromagnetic spectrum to which each belongs.

6.24 An AM radio station broadcasts at 1010 kHz, and its FM partner broadcasts at 98.3 MHz. Calculate and compare the energy of the photons emitted by these two radio stations.

6.25 One type of sunburn occurs on exposure to UV light of wavelength in the vicinity of 325 nm. **(a)** What is the energy of a photon of this wavelength? **(b)** What is the

energy of a mole of these photons? **(c)** How many photons are in a 1.00 mJ burst of this radiation? **(d)** These UV photons can break chemical bonds in your skin to cause sunburn—a form of radiation damage. If the 325-nm radiation provides exactly the energy to break an average chemical bond in the skin, estimate the average energy of these bonds in kJ/mol.

6.26 The energy from radiation can be used to cause the rupture of chemical bonds. A minimum energy of 941 kJ/mol is required to break the nitrogen–nitrogen bond in N$_2$. What is the longest wavelength of radiation that possesses the necessary energy to break the bond? What type of electromagnetic radiation is this?

6.27 A diode laser emits at a wavelength of 987 nm. **(a)** In what portion of the electromagnetic spectrum is this radiation found? **(b)** All of its output energy is absorbed in a detector that measures a total energy of 0.52 J over a period of 32 s. How many photons per second are being emitted by the laser?

6.28 A stellar object is emitting radiation at 3.55 mm. **(a)** What type of electromagnetic spectrum is this radiation? **(b)** If the detector is capturing 3.2×10^8 photons per second at this wavelength, what is the total energy of the photons detected in one hour?

6.29 Molybdenum metal must absorb radiation with a minimum frequency of 1.09×10^{15} s^{-1} before it can eject an electron from its surface via the photoelectric effect. **(a)** What is the minimum energy needed to eject an electron? **(b)** What wavelength of radiation will provide a photon of this energy? **(c)** If molybdenum is irradiated with light of wavelength of 120 nm, what is the maximum possible kinetic energy of the emitted electrons?

6.30 Sodium metal requires a photon with a minimum energy of 4.41×10^{-19} J to emit electrons. **(a)** What is the minimum frequency of light necessary to emit electrons from sodium via the photoelectric effect? **(b)** What is the wavelength of this light? **(c)** If sodium is irradiated with light of 439 nm, what is the maximum possible kinetic energy of the emitted electrons? **(d)** What is the maximum number of electrons that can be freed by a burst of light whose total energy is 1.00 μJ?

Bohr's Model; Matter Waves

6.31 Explain how the existence of line spectra is consistent with Bohr's theory of quantized energies for the electron in the hydrogen atom.

6.32 **(a)** In terms of the Bohr theory of the hydrogen atom, what process is occurring when excited hydrogen atoms

emit radiant energy of certain wavelengths and only those wavelengths? **(b)** Does a hydrogen atom "expand" or "contract" as it moves from its ground state to an excited state?

6.33 Is energy emitted or absorbed when the following electronic transitions occur in hydrogen: (a) from $n = 4$ to $n = 2$, (b) from an orbit of radius 2.12 Å to one of radius 8.46 Å, (c) an electron adds to the H^+ ion and ends up in the $n = 3$ shell?

6.34 Indicate whether energy is emitted or absorbed when the following electronic transitions occur in hydrogen: (a) from $n = 2$ to $n = 6$, (b) from an orbit of radius 4.76 Å to one of radius 0.529 Å, (c) from the $n = 6$ to the $n = 9$ state.

6.35 (a) Using Equation 6.5, calculate the energy of an electron in the hydrogen atom when $n = 2$ and when $n = 6$. Calculate the wavelength of the radiation released when an electron moves from $n = 6$ to $n = 2$. Is this line in the visible region of the electromagnetic spectrum? If so, what color is it? (b) Calculate the energies of an electron in the hydrogen atom for $n = 1$ and for $n = (\infty)$. How much energy does it require to move the electron out of the atom completely (from $n = 1$ to $n = \infty$), according to Bohr? Put your answer in kJ/mol. (c) The energy for the process $H + \text{energy} \rightarrow H^+ + e^-$ is called the ionization energy of hydrogen. The experimentally determined value for the ionization energy of hydrogen is 1310 kJ/mol. How does this compare to your calculation?

6.36 For each of the following electronic transitions in the hydrogen atom, calculate the energy, frequency, and wavelength of the associated radiation, and determine whether the radiation is emitted or absorbed during the transition: (a) from $n = 4$ to $n = 1$, (b) from $n = 5$ to $n = 2$, (c) from $n = 3$ to $n = 6$. Does any of these transitions emit or absorb visible light?

6.37 The visible emission lines observed by Balmer all involved $n_f = 2$. (a) Explain why only the lines with $n_f = 2$ were observed in the visible region of the electromagnetic spectrum. (b) Calculate the wavelengths of the first three lines in the Balmer series—those for which $n_i = 3, 4$, and 5—and identify these lines in the emission spectrum shown in Figure 6.13.

6.38 The Lyman series of emission lines of the hydrogen atom are those for which $n_f = 1$. (a) Determine the region of the electromagnetic spectrum in which the lines of the Lyman series are observed. (b) Calculate the wavelengths of the first three lines in the Lyman series—those for which $n_i = 2, 3$, and 4.

6.39 One of the emission lines of the hydrogen atom has a wavelength of 93.8 nm. (a) In what region of the electromagnetic spectrum is this emission found? (b) Determine the initial and final values of n associated with this emission.

6.40 The hydrogen atom can absorb light of wavelength 2626 nm. (a) In what region of the electromagnetic spectrum is this absorption found? (b) Determine the initial and final values of n associated with this absorption.

6.41 Use the de Broglie relationship to determine the wavelengths of the following objects: (a) an 85-kg person skiing at 50 km/hr, (b) a 10.0-g bullet fired at 250 m/s, (c) a lithium atom moving at 2.5×10^5 m/s, (d) an ozone (O_3) molecule in the upper atmosphere moving at 550 m/s.

6.42 Among the elementary subatomic particles of physics is the muon, which decays within a few nanoseconds after formation. The muon has a rest mass 206.8 times that of an electron. Calculate the de Broglie wavelength associated with a muon traveling at a velocity of 8.85×10^5 cm/s.

6.43 Neutron diffraction is an important technique for determining the structures of molecules. Calculate the velocity of a neutron needed to achieve a wavelength of 0.955 Å. (Refer to the inside cover for the mass of the neutron).

6.44 The electron microscope has been widely used to obtain highly magnified images of biological and other types of materials. When an electron is accelerated through a particular potential field, it attains a speed of 9.38×10^6 m/s. What is the characteristic wavelength of this electron? Is the wavelength comparable to the size of atoms?

6.45 Using Heisenberg's uncertainty principle, calculate the uncertainty in the position of (a) a 1.50-mg mosquito moving at a speed of 1.40 m/s if the speed is known to within ± 0.01 m/s; (b) a proton moving at a speed of $(5.00 \pm 0.01) \times 10^4$ m/s. (The mass of a proton is given in the table of fundamental constants in the inside cover of the text.)

6.46 Calculate the uncertainty in the position of (a) an electron moving at a speed of $(3.00 \pm 0.01) \times 10^5$ m/s, (b) a neutron moving at this same speed. (The masses of an electron and a neutron are given in the table of fundamental constants in the inside cover of the text.) (c) What are the implications of these calculations to our model of the atom?

Quantum Mechanics and Atomic Orbitals

6.47 (a) Why does the Bohr model of the hydrogen atom violate the uncertainty principle? (b) In what way is the description of the electron using a wave function consistent with de Broglie's hypothesis? (c) What is meant by the term *probability density*? Given the wave function, how do we find the probability density at a certain point in space?

6.48 (a) According to the Bohr model, an electron in the ground state of a hydrogen atom orbits the nucleus at a specific radius of 0.53 Å. In the quantum mechanical description of the hydrogen atom, the most probable distance of the electron from the nucleus is 0.53 Å. Why are these two statements different? (b) Why is the use of Schrödinger's wave equation to describe the location of a particle very different from the description obtained from classical physics? (c) In the quantum mechanical description of an electron, what is the physical significance of the square of the wave function, ψ^2?

6.49 (a) For $n = 4$, what are the possible values of l? (b) For $l = 2$, what are the possible values of m_l? (c) If m_l is 2, what are the possible values for l?

6.50 How many possible values for l and m_l are there when (a) $n = 3$; (b) $n = 5$?

6.51 Give the numerical values of n and l corresponding to each of the following orbital designations: (a) $3p$, (b) $2s$, (c) $4f$, (d) $5d$.

6.52 Give the values for n, l, and m_l for (a) each orbital in the $2p$ subshell, (b) each orbital in the $5d$ subshell.

6.53 Which of the following represent impossible combinations of n and l: (a) $1p$, (b) $4s$, (c) $5f$, (d) $2d$?

6.54 For the table below, write which orbital goes with the quantum numbers. Don't worry about x, y, z subscripts. If the quantum numbers are not allowed, write "not allowed."

n	l	m_l	Orbital
2	1	−1	$2p$ (example)
1	0	0	
3	−3	2	
3	2	−2	
2	0	−1	
0	0	0	
4	2	1	
5	3	0	

6.55 Sketch the shape and orientation of the following types of orbitals: (a) s, (b) p_z, (c) d_{xy}.

6.56 Sketch the shape and orientation of the following types of orbitals: (a) p_x, (b) d_{z^2}, (c) $d_{x^2-y^2}$.

6.57 (a) What are the similarities and differences between the $1s$ and $2s$ orbitals of the hydrogen atom? (b) In what sense does a $2p$ orbital have directional character? Compare the "directional" characteristics of the p_x and $d_{x^2-y^2}$ orbitals (that is, in what direction or region of space is the electron density concentrated?). (c) What can you say about the average distance from the nucleus of an electron in a $2s$ orbital as compared with a $3s$ orbital? (d) For the hydrogen atom, list the following orbitals in order of increasing energy (that is, most stable ones first): $4f$, $6s$, $3d$, $1s$, $2p$.

6.58 (a) With reference to Figure 6.19, what is the relationship between the number of nodes in an s orbital and the value of the principal quantum number? (b) Identify the number of nodes; that is, identify places where the electron density is zero, in the $2p_x$ orbital; in the $3s$ orbital. (c) What information is obtained from the radial probability functions in Figure 6.19? (d) For the hydrogen atom, list the following orbitals in order of increasing energy: $3s$, $2s$, $2p$, $5s$, $4d$.

Many-Electron Atoms and Electron Configurations

6.59 For a given value of the principal quantum number, n, how do the energies of the s, p, d, and f subshells vary for (a) hydrogen, (b) a many-electron atom?

6.60 (a) The average distance from the nucleus of a $3s$ electron in a chlorine atom is smaller than that for a $3p$ electron. In light of this fact, which orbital is higher in energy? (b) Would you expect it to require more or less energy to remove a $3s$ electron from the chlorine atom, as compared with a $2p$ electron? Explain.

6.61 (a) What experimental evidence is there for the electron having a "spin"? (b) Draw an energy-level diagram that shows the relative energetic positions of a $1s$ orbital and a $2s$ orbital. Put two electrons in the $1s$ orbital. (c) Draw an arrow showing the excitation of an electron from the $1s$ to the $2s$ orbital.

6.62 (a) State the Pauli exclusion principle in your own words. (b) The Pauli exclusion principle is, in an important sense, the key to understanding the periodic table. Explain why.

6.63 What is the maximum number of electrons that can occupy each of the following subshells: (a) $3p$, (b) $5d$, (c) $2s$, (d) $4f$?

6.64 What is the maximum number of electrons in an atom that can have the following quantum numbers: (a) $n = 2$, $m_s = -\frac{1}{2}$, (b) $n = 5$, $l = 3$; (c) $n = 4$, $l = 3$, $m_l = -3$; (d) $n = 4$, $l = 1$, $m_l = 1$?

6.65 (a) What are "valence electrons"? (b) What are "core electrons"? (c) What does each box in an orbital diagram represent? (d) What quantity is represented by the direction (up or down) of the half-arrows in an orbital diagram?

6.66 For each element, count the number of valence electrons, core electrons, and unpaired electrons in the ground state: (a) carbon, (b) phosphorus, (c) neon.

6.67 Write the condensed electron configurations for the following atoms, using the appropriate noble-gas core abbreviations: (a) Cs, (b) Ni, (c) Se, (d) Cd, (e) U, (f) Pb.

6.68 Write the condensed electron configurations for the following atoms, and indicate how many unpaired electrons each has: (a) Ga, (b) Ca, (c) V, (d) I, (e) Y, (f) Pt, (g) Lu.

6.69 Ions also have electron configurations (Section 7.4). Cations have fewer valence electrons, and anions have more valence electrons, respectively, than their parent atoms. For example, chloride, Cl^-, has an electron configuration of $1s^2 2s^2 2p^6 3s^2 3p^6$, for a total of 18 electrons, compared to 17 for neutral chlorine, the element. Na has an electron configuration of $1s^2 2s^2 2p^6 3s^1$, but Na^+ has an electron configuration of $1s^2 2s^2 2p^6$. Write out the electron configurations for (a) F^-, (b) I^-, (c) O^{2-}, (d) K^+, (e) Mg^{2+}, (f) Al^{3+}.

6.70 In the transition metals (the *d*-block), the electron configuration of cations is different than what you might expect. Instead of the d electrons being lost first, s electrons are lost first. For example, the electron configuration of iron, Fe, is $[Ar]4s^23d^6$; but the electron configuration of Fe^{2+} is $[Ar]3d^6$; the 4s electrons are eliminated to make the cation. Write out the electron configurations of **(a)** Zn^{2+} **(b)** Pt^{2+} **(c)** Cr^{3+} **(d)** Ti^{4+}.

6.71 Identify the specific element that corresponds to each of the following electron configurations: **(a)** $1s^22s^2$, **(b)** $1s^22s^22p^4$, **(c)** $[Ar]4s^13d^5$, **(d)** $[Kr]5s^24d^{10}5p^4$, **(e)** $1s^1$.

6.72 Identify the group of elements that corresponds to each of the following generalized electron configurations:
(a) [noble gas] ns^2np^5
(b) [noble gas] $ns^2(n-1)d^2$
(c) [noble gas] $ns^2(n-1)d^{10}np^1$
(d) [noble gas] $ns^2(n-2)f^6$

6.73 What is wrong with the following electron configurations for atoms in their ground states? **(a)** $1s^22s^23s^1$, **(b)** $[Ne]2s^22p^3$, **(c)** $[Ne]3s^23d^5$.

6.74 The following electron configurations represent excited states. Identify the element, and write its ground-state condensed electron configuration. **(a)** $1s^22s^23p^24p^1$, **(b)** $[Ar]3d^{10}4s^14p^45s^1$, **(c)** $[Kr]4d^65s^25p^1$.

ADDITIONAL EXERCISES

6.75 Consider the two waves shown here, which we will consider to represent two electromagnetic radiations:
(a) What is the wavelength of wave A? Of wave B?
(b) What is the frequency of wave A? Of wave B?
(c) Identify the regions of the electromagnetic spectrum to which waves A and B belong.

6.76 Certain elements emit light of a specific wavelength when they are burned. Historically, chemists used such emission wavelengths to determine whether specific elements were present in a sample. Some characteristic wavelengths for some of the elements are

Ag	328.1 nm	Fe	372.0 nm
Au	267.6 nm	K	404.7 nm
Ba	455.4 nm	Mg	285.2 nm
Ca	422.7 nm	Na	589.6 nm
Cu	324.8 nm	Ni	341.5 nm

(a) Determine which elements emit radiation in the visible part of the spectrum. **(b)** Which element emits photons of highest energy? Of lowest energy? **(c)** When burned, a sample of an unknown substance is found to emit light of frequency $6.59 \times 10^{14} \, s^{-1}$. Which of these elements is probably in the sample?

6.77 In June 2004, the Cassini–Huygens spacecraft began orbiting Saturn and transmitting images to Earth. The closest distance between Saturn and Earth is 746 million miles. What is the minimum amount of time it takes for the transmitted signals to travel from the spacecraft to Earth?

6.78 The rays of the Sun that cause tanning and burning are in the ultraviolet portion of the electromagnetic spectrum. These rays are categorized by wavelength. So-called UV-A radiation has wavelengths in the range of 320–380 nm, whereas UV-B radiation has wavelengths in the range of 290–320 nm. **(a)** Calculate the frequency of light that has a wavelength of 320 nm. **(b)** Calculate the energy of a mole of 320-nm photons. **(c)** Which are more energetic, photons of UV-A radiation or photons of UV-B radiation? **(d)** The UV-B radiation from the Sun is considered a greater cause of sunburn in humans than is UV-A radiation. Is this observation consistent with your answer to part (c)?

6.79 The watt is the derived SI unit of power, the measure of energy per unit time: $1 \, W = 1 \, J/s$. A semiconductor laser in a CD player has an output wavelength of 780 nm and a power level of 0.10 mW. How many photons strike the CD surface during the playing of a CD 69 minutes in length?

6.80 The color wheel (Figure 24.24) is a convenient way to relate what colors of light are absorbed by a sample and the visible appearance of the sample. If all visible colors are absorbed by a sample, the sample appears black. If no colors are absorbed by the sample, the sample appears white. If the sample absorbs red, then what we see is green; such *complementary colors* are across the wheel from each other. Carrots appear orange because they contain a compound called carotene. Based on the color wheel, what is the possible wavelength range for the light absorbed by carotene?

6.81 A photocell, such as the one illustrated in Figure 6.8(b), is a device used to measure the intensity of light. In a certain experiment, when light of wavelength 630 nm is directed onto the photocell, electrons are emitted at the rate of 2.6×10^{-12} Coulombs/sec. Assume that each photon that impinges on the photocell emits one electron. How many photons per second are striking the photocell? How much energy per second is the photocell absorbing?

6.82 The light-sensitive substance in black-and-white photographic film is AgBr. Photons provide the energy necessary to transfer an electron from Br^- to Ag^+ to produce elemental Ag and Br and thereby darken the film. **(a)** If a minimum energy of 2.00×10^5 J/mol is needed for this process, what is the minimum energy needed from each photon? **(b)** Calculate the wavelength of the light necessary to provide photons of this energy. **(c)** Explain why this film can be handled in a darkroom under red light.

6.83 In an experiment to study the photoelectric effect, a scientist measures the kinetic energy of ejected electrons as a function of the frequency of radiation hitting a metal surface. She obtains the following plot: The point labeled "ν_0" corresponds to light with a wavelength of 680 nm.

(a) What is the value of ν_0 in s^{-1}? **(b)** What is the value of the work function of the metal in units of kJ/mol of ejected electrons? **(c)** What happens when the metal is irradiated with light of frequency less than ν_0? **(d)** Note that when the frequency of the light is greater than ν_0, the plot shows a straight line with a nonzero slope. Why is this the case? **(e)** Can you determine the slope of the line segment discussed in part (d)? Explain.

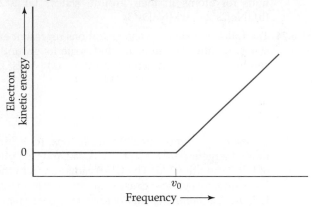

6.84 The series of emission lines of the hydrogen atom for which $n_f = 3$ is called the *Paschen series*. **(a)** Determine the region of the electromagnetic spectrum in which the lines of the Paschen series are observed. **(b)** Calculate the wavelengths of the first three lines in the Paschen series—those for which $n_i = 4, 5,$ and 6.

6.85 When the spectrum of light from the Sun is examined in high resolution in an experiment similar to that illustrated in Figure 6.11, dark lines are evident. These are called Fraunhofer lines, after the scientist who studied them extensively in the early nineteenth century. Altogether, about 25,000 lines have been identified in the solar spectrum between 2950 Å and 10,000 Å. The Fraunhofer lines are attributed to absorption of certain wavelengths of the Sun's "white" light by gaseous elements in the Sun's atmosphere. **(a)** Describe the process that causes absorption of specific wavelengths of light from the solar spectrum. **(b)** If a scientist wanted to know which Fraunhofer lines belonged to a given element, say neon, what experiments could she conduct here on Earth to provide data?

[6.86] Bohr's model can be used for hydrogen-like ions—ions that have only one electron, such as He^+ and Li^{2+}. **(a)** Why is the Bohr model applicable to He^+ ions but not to neutral He atoms? **(b)** The ground-state energies of H, He^+, and Li^{2+} are tabulated as follows:

Atom or ion	H	He^+	Li^{2+}
Ground-state energy	-2.18×10^{-18} J	-8.72×10^{-18} J	-1.96×10^{-17} J

By examining these numbers, propose a relationship between the ground-state energy of hydrogen-like systems and the nuclear charge, Z. **(c)** Use the relationship you derive in part (b) to predict the ground-state energy of the C^{5+} ion.

6.87 Under appropriate conditions, molybdenum emits X-rays that have a characteristic wavelength of 0.711 Å. These X-rays are used in diffraction experiments to determine the structures of molecules. **(a)** Why are X-rays, and not visible light, suitable for the determination of structure at the atomic level? **(b)** How fast would an electron have to be moving to have the same wavelength as these X-rays?

[6.88] An electron is accelerated through an electric potential to a kinetic energy of 18.6 keV. What is its characteristic wavelength? [*Hint:* Recall that the kinetic energy of a moving object is $E = \frac{1}{2}mv^2$, where m is the mass of the object and v is the speed of the object.]

6.89 In the television series *Star Trek*, the transporter beam is a device used to "beam down" people from the *Starship Enterprise* to another location, such as the surface of a planet. The writers of the show put a "Heisenberg compensator" into the transporter beam mechanism. Explain why such a compensator would be necessary to get around Heisenberg's uncertainty principle.

6.90 Which of the quantum numbers governs **(a)** the shape of an orbital, **(b)** the energy of an orbital, **(c)** the spin properties of the electron, **(d)** the spatial orientation of the orbital?

[6.91] Consider the discussion of radial probability functions in the "A Closer Look" box in Section 6.6. **(a)** What is the difference between the probability density as a function of r and the radial probability function as a function of r? **(b)** What is the significance of the term $4\pi r^2$ in the radial probability functions for the s orbitals? **(c)** Based on Figures 6.19 and 6.23, make sketches of what you think the probability density as a function of r and the radial probability function would look like for the $4s$ orbital of the hydrogen atom.

6.92 The "magic numbers" in the periodic table are the atomic numbers of elements with high stability (the noble gases): 2, 10, 18, 36, 54, and 86. In terms of allowed values of orbitals and spin quantum numbers, explain why these electron arrangements correspond to special stability.

[6.93] For non-spherically symmetric orbitals, the contour representations (as in Figures 6.23 and 6.24) suggest where nodal planes exist (that is, where the electron density is zero). For example, the p_x orbital has a node wherever $x = 0$. This equation is satisfied by all points on the yz plane, so this plane is called a nodal plane of the p_x orbital. **(a)** Determine the nodal plane of the p_z orbital. **(b)** What are the two nodal planes of the d_{xy} orbital? **(c)** What are the two nodal planes of the $d_{x^2-y^2}$ orbital?

[6.94] As noted in Figure 6.26, the spin of an electron generates a magnetic field, with spin-up and spin-down electrons having opposite fields. In the absence of a magnetic field, a spin-up and a spin-down electron have the same energy. **(a)** Why do you think that the use of a magnet was important in the discovery of electron spin (see the "A Closer Look" box in Section 6.8)? **(b)** Imagine that the two spinning electrons in Figure 6.26 were placed between the poles of a horseshoe magnet, with the north pole of the magnet at the top of the figure. Based on what you know about magnets, would you expect the left or right electron in the figure to have the lower energy? **(c)** A phenomenon called *electron spin resonance* (ESR) is closely related to nuclear magnetic resonance. In ESR a compound with an unpaired electron is placed in a magnetic field, which causes the unpaired electron to have two different energy states analogous to Figure 6.28.

ESR uses microwave radiation to excite the unpaired electron from one state to the other. Based on your reading of the "Chemistry and Life" box in Section 6.8, does an ESR experiment require photons of greater or lesser energy than an NMR experiment?

[6.95] The "Chemistry and Life" box in Section 6.8 described the techniques called NMR and MRI. (a) Instruments for obtaining MRI data are typically labeled with a frequency, such as 600 MHz. Why do you suppose this label is relevant to the experiment? (b) What is the value of ΔE in Figure 6.28 that would correspond to the absorption of a photon of radiation with frequency 450 MHz? (c) In general, the stronger the magnetic field, the greater the information obtained from an NMR or MRI experiment. Why do you suppose this is the case?

6.96 Suppose that the spin quantum number, m_s, could have *three* allowed values instead of two. How would this affect the number of elements in the first four rows of the periodic table?

6.97 Using only a periodic table as a guide, write the condensed electron configurations for the following atoms: (a) Se, (b) Rh, (c) Si, (d) Hg, (e) Hf.

6.98 Scientists have speculated that element 126 might have a moderate stability, allowing it to be synthesized and characterized. Predict what the condensed electron configuration of this element might be.

INTEGRATIVE EXERCISES

6.99 Microwave ovens use microwave radiation to heat food. The energy of the microwaves is absorbed by water molecules in food, then transferred to other components of the food. (a) Suppose that the microwave radiation has a wavelength of 11.2 cm. How many photons are required to heat 200 mL of coffee from 23 °C to 60 °C? (b) Suppose the microwave's power is 900 W (1 Watt = 1 Joule/sec). How long would you have to heat the coffee in part (a)?

6.100 The stratospheric ozone (O_3) layer helps to protect us from harmful ultraviolet radiation. It does so by absorbing ultraviolet light and falling apart into an O_2 molecule and an oxygen atom, a process known as photodissociation.

$$O_3(g) \longrightarrow O_2(g) + O(g)$$

Use the data in Appendix C to calculate the enthalpy change for this reaction. What is the maximum wavelength a photon can have if it is to possess sufficient energy to cause this dissociation? In what portion of the spectrum does this wavelength occur?

6.101 The discovery of hafnium, element number 72, provided a controversial episode in chemistry. G. Urbain, a French chemist, claimed in 1911 to have isolated an element number 72 from a sample of rare earth (elements 58–71) compounds. However, Niels Bohr believed that hafnium was more likely to be found along with zirconium than with the rare earths. D. Coster and G. von Hevesy, working in Bohr's laboratory in Copenhagen, showed in 1922 that element 72 was present in a sample of Norwegian zircon, an ore of zirconium. (The name hafnium comes from the Latin name for Copenhagen, *Hafnia*). (a) How would you use electron configuration arguments to justify Bohr's prediction? (b) Zirconium, hafnium's neighbor in group 4B, can be produced as a metal by reduction of solid $ZrCl_4$ with molten sodium metal. Write a balanced chemical equation for the reaction. Is this an oxidation-reduction reaction? If yes, what is reduced and what is oxidized? (c) Solid zirconium dioxide, ZrO_2, is reacted with chlorine gas in the presence of carbon. The products of the reaction are $ZrCl_4$ and two gases, CO_2 and CO in the ratio 1 : 2 Write a balanced chemical equation for the reaction. Starting with a 55.4-g sample of ZrO_2, calculate the mass of $ZrCl_4$ formed, assuming that ZrO_2 is the limiting reagent and assuming 100% yield. (d) Using their electron configurations, account for the fact that Zr and Hf form chlorides MCl_4 and oxides MO_2.

6.102 (a) Account for formation of the following series of oxides in terms of the electron configurations of the elements and the discussion of ionic compounds in Section 2.7: K_2O, CaO, Sc_2O_3, TiO_2, V_2O_5, CrO_3. (b) Name these oxides. (c) Consider the metal oxides whose enthalpies of formation (in kJ mol^{-1}) are listed here.

Oxide	$K_2O(s)$	$CaO(s)$	$TiO_2(s)$	$V_2O_5(s)$
ΔH_f°	−363.2	−635.1	−938.7	−1550.6

Calculate the enthalpy changes in the following general reaction for each case:

$$M_nO_m(s) + H_2(g) \longrightarrow nM(s) + mH_2O(g)$$

(You will need to write the balanced equation for each case, then compute ΔH°.) (d) Based on the data given, estimate a value of ΔH_f° for $Sc_2O_3(s)$.

6.103 The first 25 years of the twentieth century were momentous for the rapid pace of change in scientists' understanding of the nature of matter. (a) How did Rutherford's experiments on the scattering of α particles by a gold foil set the stage for Bohr's theory of the hydrogen atom? (b) In what ways is de Broglie's hypothesis, as it applies to electrons, consistent with J. J. Thomson's conclusion that the electron has mass? In what sense is it consistent with proposals that preceded Thomson's work, that the cathode rays are a wave phenomenon?

[6.104] The two most common isotopes of uranium are ^{235}U and ^{238}U. (a) Compare the number of protons, the number of electrons, and the number of neutrons in atoms of these two isotopes. (b) Using the periodic table in the front inside cover, write the electron configuration for a U atom. (c) Compare your answer to part (b) to the electron configuration given in Figure 6.31. How can you explain any differences between these two electron configurations? (d) ^{238}U undergoes radioactive decay to ^{234}Th. How many protons, electrons, and neutrons are gained or lost by the ^{238}U atom during this process? (e) Examine the electron configuration for Th in Figure 6.31. Are you surprised by what you find? Explain.

6.105 Imagine sunlight falling on three square areas. One is an inert black material. The second is a photovoltaic cell surface, which converts radiant energy into electricity. The third is an area on a green tree leaf. Draw diagrams that show the energy conversions in each case, using Figure 5.9 as a model. How are these three examples related to the idea of sustainable energy sources?

CHAPTER 7

PERIODIC PROPERTIES OF THE ELEMENTS

OIL PAINTS CONTAIN PIGMENTS, which are usually highly colored salts, suspended in an organic carrier composed of a variety of heavy hydrocarbon molecules. This painting, by the famous French Impressionist Claude Monet (1840–1926), is entitled *La rue Montorgueil, fête du 30 juin 1878*. Photo credit: Claude Monet (1840–1926) "Rue Montorgueil in Paris, Festival of 30 June 1878," 1878. Herve Lewandowski/Reunion des Muses Nationaux/Art Resource, NY.

WHAT'S AHEAD

7.1 Development of the Periodic Table
We begin our discussion with a brief history of the periodic table.

7.2 Effective Nuclear Charge
We next explore the many properties of atoms that depend both on the net attraction of the outer electrons to the nucleus and on the average distance of those electrons from the nucleus. The net positive charge of the nucleus experienced by the outer electrons is called the *effective nuclear charge*.

7.3 Sizes of Atoms and Ions
We explore the relative sizes (or radii) of atoms and ions, both of which follow trends that are related to their placement in the periodic table.

7.4 Ionization Energy
We next encounter the energy—called ionization energy—required to remove one or more electrons from an atom. The periodic trends in ionization energy depend upon variations in effective nuclear charge and atomic radii.

7.5 Electron Affinities
Next we will examine periodic trends in the energy released when an electron is added to an atom.

7.6 Metals, Nonmetals, and Metalloids
We recognize that the physical and chemical properties of metals and nonmetals are distinctly different and can be understood from the fundamental characteristics of atoms discussed earlier in the chapter, particularly ionization energy. Metalloids display properties that are intermediate between metals and nonmetals.

7.7 Group Trends for the Active Metals
We examine some periodic trends in the chemistry of the active metals (groups 1A and 2A).

7.8 Group Trends for Selected Nonmetals
We examine some periodic trends in the chemistry of hydrogen, as well as groups 6A, 7A, and 8A.

THE BEAUTY OF AN IMPRESSIONIST OIL PAINTING, such as the Monet masterpiece shown here, begins with chemistry. Colorful inorganic salts are suspended in various organic media that contain hydrocarbons and other molecular substances. Indeed, the great painters

had a gift for using compounds of elements that span nearly the entire periodic table.

Today the periodic table is still the most significant tool chemists have for organizing and remembering chemical facts. As we saw in Chapter 6, the periodic nature of the table arises from the repeating patterns in the electron configurations of the elements. Elements in the same column of the table contain the same number of electrons in their **valence orbitals**, the occupied orbitals that hold the electrons involved in bonding. For example, O ([He]$2s^2 2p^4$) and S ([Ne]$3s^2 3p^4$) are both members of group 6A. The similarity of the electron distribution in their valence s and p orbitals leads to similarities in the properties of these two elements.

When we compare O and S, however, it is apparent that they exhibit differences as well, not the least of which is that oxygen is a colorless gas at room

▲ Figure 7.1 Oxygen and sulfur.
Because they are both group 6A
elements, oxygen and sulfur have many
chemical similarities. They also have
many differences, however, including the
forms they take at room temperature.
Oxygen consists of O_2 molecules that
appear as a colorless gas (shown here
enclosed in a glass container on the left).
In contrast, sulfur consists of S_8
molecules that form a yellow solid.

temperature, whereas sulfur is a yellow solid (Figure 7.1 ◀)! One of the major
differences between atoms of these two elements is their electron configurations: the outermost electrons of O are in the second shell, whereas those of S
are in the third shell. We will see that electron configurations can be used to
explain differences as well as similarities in the properties of elements.

In this chapter we explore how some of the important properties of elements change as we move across a row or down a column of the periodic
table. In many cases the trends within a row or column allow us to make predictions about the physical and chemical properties of the elements.

7.1 DEVELOPMENT OF THE PERIODIC TABLE

The discovery of the chemical elements has been an ongoing process since ancient times (Figure 7.2 ▼). Certain elements, such as gold, appear in nature in elemental form and were thus discovered thousands of years ago. In contrast,
some elements, such as technetium, are radioactive and intrinsically unstable.
We know about them only because of technology developed during the twentieth century.

The majority of the elements, although stable, readily form compounds.
Consequently, they are not found in nature in their elemental form. For centuries, therefore, scientists were unaware of their existence. During the early
nineteenth century, advances in chemistry made it easier to isolate elements
from their compounds. As a result, the number of known elements more than
doubled from 31 in 1800 to 63 by 1865.

As the number of known elements increased, scientists began to investigate
the possibilities of classifying them in useful ways. In 1869, Dmitri Mendeleev
in Russia and Lothar Meyer in Germany published nearly identical classification schemes. Both scientists noted that similar chemical and physical properties recur periodically when the elements are arranged in order of increasing
atomic weight. Scientists at that time had no knowledge of atomic numbers.
Atomic weights, however, generally increase with increasing atomic number,
so both Mendeleev and Meyer fortuitously arranged the elements in proper sequence. The tables of elements advanced by Mendeleev and Meyer were the
forerunners of the modern periodic table.

Although Mendeleev and Meyer came to essentially the same conclusion
about the periodicity of elemental properties, Mendeleev is given credit for advancing his ideas more vigorously and stimulating much new work in chemistry.

**► Figure 7.2 Discovering the
elements.** Periodic table showing the
dates of discovery of the elements.

H																	He
Li	Be											B	C	N	O	F	Ne
Na	Mg											Al	Si	P	S	Cl	Ar
K	Ca	Sc	Ti	V	Cr	Mn	Fe	Co	Ni	Cu	Zn	Ga	Ge	As	Se	Br	Kr
Rb	Sr	Y	Zr	Nb	Mo	Tc	Ru	Rh	Pd	Ag	Cd	In	Sn	Sb	Te	I	Xe
Cs	Ba	Lu	Hf	Ta	W	Re	Os	Ir	Pt	Au	Hg	Tl	Pb	Bi	Po	At	Rn
Fr	Ra	Lr	Rf	Db	Sg	Bh	Hs	Mt	Ds	Rg							

La	Ce	Pr	Nd	Pm	Sm	Eu	Gd	Tb	Dy	Ho	Er	Tm	Yb
Ac	Th	Pa	U	Np	Pu	Am	Cm	Bk	Cf	Es	Fm	Md	No

Ancient Times   1735–1843   1894–1918

Middle Ages–1700   1843–1886   1923–1961   1965–

TABLE 7.1 ■ Comparison of the Properties of Eka-Silicon Predicted by Mendeleev with the Observed Properties of Germanium

Property	Mendeleev's Predictions for Eka-Silicon (made in 1871)	Observed Properties of Germanium (discovered in 1886)
Atomic weight	72	72.59
Density (g/cm³)	5.5	5.35
Specific heat (J/g-K)	0.305	0.309
Melting point (°C)	High	947
Color	Dark gray	Grayish white
Formula of oxide	XO_2	GeO_2
Density of oxide (g/cm³)	4.7	4.70
Formula of chloride	XCl_4	$GeCl_4$
Boiling point of chloride (°C)	A little under 100	84

His insistence that elements with similar characteristics be listed in the same family forced him to leave several blank spaces in his table. For example, both gallium (Ga) and germanium (Ge) were unknown at that time. Mendeleev boldly predicted their existence and properties, referring to them as *eka-aluminum* ("under" aluminum) and *eka-silicon* ("under" silicon), respectively, after the elements under which they appear in the periodic table. When these elements were discovered, their properties closely matched those predicted by Mendeleev, as shown in Table 7.1 ▲.

In 1913, two years after Rutherford proposed the nuclear model of the atom ∞ (Section 2.2), an English physicist named Henry Moseley (1887–1915) developed the concept of atomic numbers. Moseley determined the frequencies of X-rays emitted as different elements were bombarded with high-energy electrons. He found that each element produces X-rays of a unique frequency; furthermore, he found that the frequency generally increased as the atomic mass increased. He arranged the X-ray frequencies in order by assigning a unique whole number, called an *atomic number*, to each element. Moseley correctly identified the atomic number as the number of protons in the nucleus of the atom. ∞ (Section 2.3)

The concept of atomic number clarified some problems in the periodic table of Moseley's day, which was based on atomic weights. For example, the atomic weight of Ar (atomic number 18) is greater than that of K (atomic number 19), yet the chemical and physical properties of Ar are much more like that of Ne and Kr than they are like Na and Rb. However, when the elements are arranged in order of increasing atomic number, rather than increasing atomic weight, Ar and K appear in their correct places in the table. Moseley's studies also made it possible to identify "holes" in the periodic table, which led to the discovery of other previously unknown elements.

GIVE IT SOME THOUGHT

Arranging the elements by atomic weight leads to a slightly different order than arranging them by atomic number. Why does this happen? Can you find an example, other than the case of Ar and K discussed above, where the order of the elements would be different if the elements were arranged in order of increasing atomic weight?

7.2 EFFECTIVE NUCLEAR CHARGE

Because electrons are negatively charged, they are attracted to nuclei, which are positively charged. Many of the properties of atoms depend on their electron configurations and on how strongly their outer electrons are attracted to the nucleus. Coulomb's law tells us that the strength of the interaction between two electrical charges depends on the magnitudes of the charges and on the distance between them. ∞(Section 2.3) Thus, the force of attraction between an electron and the nucleus depends on the magnitude of the net nuclear charge acting on the electron and on the average distance between the nucleus and the electron. The force of attraction increases as the nuclear charge increases, and it decreases as the electron moves farther from the nucleus.

In a many-electron atom, each electron is simultaneously attracted to the nucleus and repelled by the other electrons. In general, there are so many electron–electron repulsions that we cannot analyze the situation exactly. We can, however, estimate the net attraction of each electron to the nucleus by considering how it interacts with the *average* environment created by the nucleus and the other electrons in the atom. This approach allows us to treat each electron individually as though it were moving in the net electric field created by the nucleus and the electron density of the other electrons. We can view this net electric field as if it results from a single positive charge located at the nucleus, called the **effective nuclear charge**, Z_{eff}.

It is important to realize that the effective nuclear charge acting on an electron in an atom is smaller than the *actual* nuclear charge because the effective nuclear charge also accounts for the repulsion of the electron by the other electrons in the atom—in other words, $Z_{eff} < Z$. Let's consider how we can get a sense of the magnitude of Z_{eff} for an electron in an atom.

A valence electron in an atom is attracted to the nucleus of the atom and is repelled by the other electrons in the atom. In particular, the electron density that is due to the inner (core) electrons is particularly effective at partially canceling the attraction of the valence electron to the nucleus. We say that the inner electrons partially *shield* or *screen* the outer electrons from the attraction of the nucleus. We can therefore write a simple relationship between the effective nuclear charge, Z_{eff}, and the number of protons in the nucleus, Z:

$$Z_{eff} = Z - S \qquad [7.1]$$

The factor S is a positive number called the *screening constant*. It represents the portion of the nuclear charge that is screened from the valence electron by the other electrons in the atom. Because the core electrons are most effective at screening a valence electron from the nucleus, *the value of S is usually close to the number of core electrons in an atom.* Electrons in the same valence shell do not screen one another very effectively, but they do affect the value of S slightly (see "A Closer Look" on Effective Nuclear Charge). Let's take a look at a Na atom to see what we would expect for the magnitude of Z_{eff}. Sodium (atomic number 11) has a condensed electron configuration of [Ne]$3s^1$. The nuclear charge of the atom is 11+, and the Ne inner core consists of ten electrons ($1s^2 2s^2 2p^6$). Very roughly then, we would expect the $3s$ valence electron of the Na atom to experience an effective nuclear charge of about $11 - 10 = 1+$, as pictured in a simplified way in Figure 7.3(a) ◀. The situation is a bit more complicated because of the electron distributions of atomic orbitals. ∞(Section 6.6) Recall that a $3s$ electron has a small probability of being found close to the nucleus and inside the core electrons, as shown in Figure 7.3(b). Thus, there is a probability that the $3s$ electron will experience a greater attraction than our simple model suggests, which will increase the value of Z_{eff} somewhat. Indeed, the value of Z_{eff} obtained from detailed calculations indicate that the effective nuclear charge acting on the $3s$ electron in Na is 2.5+.

▼ **Figure 7.3 Effective nuclear charge.** (a) The effective nuclear charge experienced by the valence electron in sodium depends mostly on the 11+ charge of the nucleus and the 10− charge of the neon core. If the neon core were totally effective in shielding the valence electron from the nucleus, then the valence electron would experience an effective nuclear charge of 1+. (b) The $3s$ electron has some probability of being inside the Ne core. Because of this "penetration," the core is not completely effective in screening the $3s$ electron from the nucleus. Thus, the effective nuclear charge experienced by the $3s$ electron is somewhat greater than 1+.

(a)

(b)

The notion of effective nuclear charge also explains an important effect we noted in Section 6.7, namely, that for a many-electron atom the energies of orbitals with the same n value increase with increasing l value. For example, consider a carbon atom, for which the electron configuration is $1s^2 2s^2 2p^2$. The energy of the $2p$ orbital ($l = 1$) is somewhat higher than that of the $2s$ orbital ($l = 0$) even though both of these orbitals are in the $n - 2$ shell (Figure 6.25). The reason that these orbitals have different energy in a many-electron atom is due to the radial probability functions for the orbitals, shown in Figure 7.4▶. Notice that the $2s$ probability function has a small peak fairly close to the nucleus, whereas the $2p$ probability function does not. As a result, an electron in the $2s$ orbital is less effectively screened by the core orbitals than is an electron in the $2p$ orbital. In other words, the electron in the $2s$ orbital experiences a higher effective nuclear charge than one in the $2p$ orbital. The greater attraction between the 2s electron and the nucleus leads to a lower energy for the $2s$ orbital than for the $2p$ orbital. The same reasoning explains the general trend in orbital energies ($ns < np < nd$) in many-electron atoms.

Finally, let's examine the trends in Z_{eff} for valence electrons as we move from one element to another in the periodic table. *The effective nuclear charge increases as we move across any row (period) of the table.* Although the number of core electrons stays the same as we move across the row, the actual nuclear charge increases. The valence electrons added to counterbalance the increasing nuclear charge shield one another very ineffectively. Thus, the effective nuclear charge, Z_{eff}, increases steadily. For example, the $1s^2$ core electrons of lithium ($1s^2 2s^1$) shield the 2s valence electron from the 3+ nucleus fairly efficiently. Consequently, the outer electron experiences an effective nuclear charge of roughly $3 - 2 = 1+$. For beryllium ($1s^2 2s^2$) the effective nuclear charge experienced by each 2s valence electron is larger; in this case, the inner 1s electrons are shielding a 4+ nucleus, and each 2s electron only partially shields the other from the nucleus. Consequently, the effective nuclear charge experienced by each 2s electron is about $4 - 2 = 2+$.

Going down a column, the effective nuclear charge experienced by valence electrons changes far less than it does across a row. For example, we would expect the effective nuclear charge for the outer electrons in lithium and sodium to be about the same, roughly $3 - 2 = 1+$ for lithium and $11 - 10 = 1+$ for sodium. In fact, however, the effective nuclear charge increases slightly as we go down a family because larger electron cores are less able to screen the outer electrons from the nuclear charge. In the case of the alkali metals, the value of Z_{eff} increases from 1.3+ for lithium, to 2.5+ for sodium, to 3.5+ for potassium. Nevertheless, the increase in effective nuclear charge that occurs moving down a column is smaller than the change that occurs when moving horizontally from one side of the periodic table to the other.

▲ **Figure 7.4 2s and 2p radial functions.** The radial probability function for the 2s orbital of the hydrogen atom (red curve) shows a "bump" of probability close to the nucleus, whereas that for the 2p orbital (blue curve) does not. As a result, an electron in the 2s orbital for a many-electron atom "sees" more of the nuclear charge than does an electron in the 2p orbital. Hence, the effective nuclear charge experienced by the 2s electron is greater than that for the 2p electron.

▲ **GIVE IT SOME THOUGHT**

Which would you expect to experience a greater effective nuclear charge, a 2p electron of a Ne atom or a 3s electron of a Na atom?

7.3 SIZES OF ATOMS AND IONS

Size is one of the important properties of an atom or an ion. We often think of atoms and ions as hard, spherical objects. According to the quantum mechanical model, however, atoms and ions do not have sharply defined boundaries at which the electron distribution becomes zero. ∞ (Section 6.5) Nevertheless,

A Closer Look | EFFECTIVE NUCLEAR CHARGE

To get a better sense of how the effective nuclear charge varies as both the charge of the nucleus and the number of electrons increase, consider Figure 7.5▶. This figure shows the variation in Z_{eff} for elements in the second (Li – Ne) and third (Na – Ar) periods. The effective nuclear charge is plotted for electrons in the $1s$ subshell (in red) and for the outermost valence electrons (in blue). These values are considered the most accurate estimate of Z_{eff}. Although the details of how they are calculated are beyond the scope of our discussion, the trends are instructive. The effective nuclear charge as felt by electrons in the $1s$ subshell closely track the increasing charge of the nucleus, Z, (in black) because the other electrons do little to shield these innermost electrons from the charge of the nucleus.

Compare the values of Z_{eff} experienced by the inner core electron (in red) with those experienced by the outermost electrons (in blue). The values of Z_{eff} as felt by the outermost electrons are smaller because of screening by the inner electrons. In addition, the effective nuclear charge felt by the outermost electrons does not increase as steeply with increasing atomic number, because the valence electrons make a small, but non-negligible contribution to the screening constant, S. However, the most striking feature associated with the Z_{eff} for the outermost electrons is the sharp drop between the last element of the second period (Ne) and the first element of the third period (Na). This drop reflects the fact that the core electrons are much more effective than the valence electrons at screening the charge of the nucleus.

Effective nuclear charge is a tool that can be used to understand many physically measurable quantities, such as ionization energy, atomic radii, and electron affinity. Therefore, it is desirable to have a simple method for estimating Z_{eff} without resorting to sophisticated calculations or memorization. The charge of the nucleus, Z, is known exactly, so the challenge boils down to estimating accurately the value of the screening constant, S. The approach outlined in the text was to estimate S by assuming that the core electrons contribute a value of 1.00 to the screening constant, and the outer electrons contribute nothing to S. A more accurate approach is to use an empirical set of rules developed by John Slater. If we limit ourselves to elements that do not have electrons in d or f subshells (the rules become somewhat more complicated for such elements), Slater's rules are straightforward to apply. Electrons with larger values of n, the principal quantum number, than the electron of interest contribute 0 to the value of the screening constant, S. Electrons with the same value of n, as the electron of interest contribute 0.35 to the value of S. Electrons with a

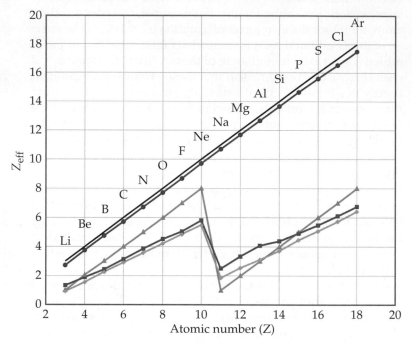

▲ **Figure 7.5 Variations in effective nuclear charge.**
This figure shows the variation in effective nuclear charge, Z_{eff}, as felt by the innermost electrons, those in the $1s$ subshell (red circles), which closely tracks the increase in nuclear charge, Z (black line). The Z_{eff} felt by the outermost valence electrons (blue squares) not only is significantly smaller than Z, it does not evolve linearly with increasing atomic number. It is also possible to estimate Z_{eff} using certain rules. The values shown as green triangles were obtained by assuming the core electrons are completely effective at screening and the valence electrons do not screen the nuclear charge at all. The values shown as gray diamonds were estimated using Slater's rules.

principal quantum number one less than the electron of interest contribute 0.85 to the value of S, while those with even smaller values of n contribute 1.00 to the value of S. Values of Z_{eff} as estimated using the simple method outlined in the text (in green), as well as those estimated with Slater's rules (in grey), are plotted in Figure 7.5. While neither of these methods exactly replicate the values of Z_{eff} obtained from more sophisticated calculations, we see that both methods effectively capture the periodic variation in Z_{eff}. While Slater's approach is more accurate, the method outlined in the text does a reasonably good job of estimating the effective nuclear charge, despite its simplicity. For our purposes, we can assume that the screening constant, S, in Equation 7.1 is roughly equal to the number of core electrons.

Related Exercises: 7.13, 7.14, 7.33, 7.34, 7.85, 7.87

we can define atomic size in several ways, based on the distances between atoms in various situations.

Imagine a collection of argon atoms in the gas phase. When two atoms collide with each other in the course of their motions, they ricochet apart—somewhat like billiard balls. This movement happens because the electron clouds of the colliding atoms cannot penetrate each other to any significant extent. The closest distances separating the nuclei during such collisions determine the *apparent* radii of the argon atoms. We might call this radius the *nonbonding atomic radius* of an atom.

When two atoms are chemically bonded to each other, as in the Cl_2 molecule, an attractive interaction exists between the two atoms leading to a chemical bond. We will discuss the nature of such bonding in Chapter 8. For now, the only thing we need to realize is that this attractive interaction brings the two atoms closer together than they would be in a nonbonding collision. We can define an atomic radius based on the distances separating the nuclei of atoms when they are chemically bonded to each other. This distance, called the **bonding atomic radius**, is shorter than the nonbonding atomic radius, as illustrated in Figure 7.6▶. Unless otherwise noted, we will refer to the bonding atomic radius when we speak of the size of an atom.

Scientists have developed a variety of experimental techniques for measuring the distances separating nuclei in molecules. From observations of these distances in many molecules, each element can be assigned a bonding atomic radius. For example, in the I_2 molecule, the distance separating the iodine nuclei is observed to be 2.66 Å.* We can define the bonding atomic radius of iodine on this basis to be one-half of the bond distance, namely 1.33 Å. Similarly, the distance separating two adjacent carbon nuclei in diamond, which is a three-dimensional solid network, is 1.54 Å; thus, the bonding atomic radius of carbon is assigned the value 0.77 Å. The radii of other elements can be similarly defined (Figure 7.7▶). (For helium and neon, the bonding radii must be estimated because there are no known compounds of these elements.)

Knowing atomic radii allows us to estimate the bond lengths between different elements in molecules. For example, the $Cl—Cl$ bond length in Cl_2 is 1.99 Å, so a radius of 0.99 Å is assigned to Cl. In the compound CCl_4 the measured length of the $C—Cl$ bond is 1.77 Å, very close to the sum $(0.77 + 0.99 Å)$ of the atomic radii of C and Cl.

If we consider the radius of an atom as one-half the distance between the nuclei of atoms that are held together by a bond, you may wonder why the spheres representing atoms seem to overlap in some drawings, such as those in Figures 1.1 and 2.20. These depictions of molecules are called space-filling models, and they use nonbonding atomic radii (also called *van der Waals radii*) to represent the sizes of different elements. Even in these representations the distance between the centers of two adjacent atoms in the molecule is determined from bonding atomic radii (also called *covalent radii*).

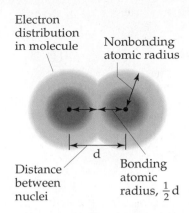

▲ **Figure 7.6 Distinction between nonbonding and bonding atomic radii.** The nonbonding atomic radius is the effective radius of an atom when it is not involved in bonding to another atom. Values of bonding atomic radius are obtained from measurements of interatomic distances in chemical compounds.

▲ **Figure 7.7 Trends in atomic radii.** Bonding atomic radii for the first 54 elements of the periodic table. The height of the bar for each element is proportional to its radius, giving a "relief map" view of the radii.

*Remember: The angstrom $(1 Å = 10^{-10} m)$ is a convenient metric unit for atomic measurements of length. The angstrom is not an SI unit. The most commonly used SI unit for such measurements is the picometer $(1 pm = 10^{-12} m; 1 Å = 100 pm)$.

■ SAMPLE EXERCISE 7.1 | Bond Lengths in a Molecule

Natural gas used in home heating and cooking is odorless. Because natural gas leaks pose the danger of explosion or suffocation, various smelly substances are added to the gas to allow detection of a leak. One such substance is methyl mercaptan, CH_3SH, whose structure is shown in the margin. Use Figure 7.7 to predict the lengths of the C—S, C—H, and S—H bonds in this molecule.

SOLUTION

Analyze and Plan: We are given three bonds and the list of bonding atomic radii. We will assume that each bond length is the sum of the radii of the two atoms involved.

Solve: Using radii for C, S, and H from Figure 7.7, we predict

$$C—S \text{ bond length} = \text{radius of C} + \text{radius of S}$$

$$= 0.77 \text{ Å} + 1.02 \text{ Å} = 1.79 \text{ Å}$$

$$C—H \text{ bond length} = 0.77 \text{ Å} + 0.37 \text{ Å} = 1.14 \text{ Å}$$

$$S—H \text{ bond length} = 1.02 \text{ Å} + 0.37 \text{ Å} = 1.39 \text{ Å}$$

Check: The experimentally determined bond lengths in methyl mercaptan (taken from the chemical literature) are C—S = 1.82 Å, C—H = 1.10 Å, and S—H = 1.33 Å. (In general, the lengths of bonds involving hydrogen show larger deviations from the values predicted by the sum of the atomic radii than do those bonds involving larger atoms.)

Comment: Notice that the estimated bond lengths using bonding atomic radii are close, but not exact matches, to the experimental bond lengths. Atomic radii must be used with some caution in estimating bond lengths. In Chapter 8 we will examine some of the average lengths of common types of bonds.

■ PRACTICE EXERCISE

Using Figure 7.7, predict which will be greater, the P—Br bond length in PBr_3 or the As—Cl bond length in $AsCl_3$.
Answer: P—Br

Periodic Trends in Atomic Radii

If we examine the "relief map" of atomic radii shown in Figure 7.7, we observe two interesting trends in the data:

1. Within each column (group), atomic radius tends to increase from top to bottom. This trend results primarily from the increase in the principal quantum number (n) of the outer electrons. As we go down a column, the outer electrons have a greater probability of being farther from the nucleus, causing the atom to increase in size.

2. Within each row (period), atomic radius tends to decrease from left to right. The major factor influencing this trend is the increase in the effective nuclear charge (Z_{eff}) as we move across a row. The increasing effective nuclear charge steadily draws the valence electrons closer to the nucleus, causing the atomic radius to decrease.

▲ GIVE IT SOME THOUGHT

In section 7.2 we said that the effective nuclear charge generally increases when you move down a column of the periodic table, whereas in Chapter 6 we saw that the "size" of an orbital increases as the principal quantum number increases. With respect to atomic radii, do these trends work together or against each other? Which effect is larger?

■ SAMPLE EXERCISE 7.2 | Atomic Radii

Referring to a periodic table, arrange (as much as possible) the following atoms in order of increasing size: $_{15}P$, $_{16}S$, $_{33}As$, $_{34}Se$. (Atomic numbers are given for the elements to help you locate them quickly in the periodic table.)

SOLUTION

Analyze and Plan: We are given the chemical symbols for four elements. We can use their relative positions in the periodic table and the two periodic trends just described to predict the relative order of their atomic radii.

Solve: Notice that P and S are in the same row of the periodic table, with S to the right of P. Therefore, we expect the radius of S to be smaller than that of P. (Radii decrease as we move from left to right.) Likewise, the radius of Se is expected to be smaller than that of As. We also notice that As is directly below P and that Se is directly below S. We expect, therefore, that the radius of As is greater than that of P and the radius of Se is greater than that of S. From these observations, we predict S < P, P < As, S < Se, and Se < As. We can therefore conclude that S has the smallest radius of the four elements and that As has the largest radius.

Using just the two trends described above, we cannot determine whether P or Se has the larger radius. To go from P to Se in the periodic table, we must move down (radius tends to increase) and to the right (radius tends to decrease). In Figure 7.7 we see that the radius of Se (1.16 Å) is greater than that of P (1.06 Å). If you examine the figure carefully, you will discover that for the s- and p-block elements the increase in radius moving down a column tends to be the greater effect. There are exceptions, however.

Check: From Figure 7.7, we have S (1.02 Å) < P (1.06 Å) < Se (1.16 Å) < As (1.19 Å).

Comment: Note that the trends we have just discussed are for the s- and p-block elements. You will see in Figure 7.7 that the transition elements do not show a regular decrease upon moving from left to right across a row.

■ PRACTICE EXERCISE

Arrange the following atoms in order of increasing atomic radius: $_{11}Na$, $_4Be$, $_{12}Mg$.
Answer: Be < Mg < Na

Periodic Trends in Ionic Radii

Just as bonding atomic radii can be determined from interatomic distances in elements, ionic radii can be determined from interatomic distances in ionic compounds. Like the size of an atom, the size of an ion depends on its nuclear charge, the number of electrons it possesses, and the orbitals in which the valence electrons reside. The formation of a cation vacates the most spatially extended occupied orbitals in an atom and decreases the number of electron–electron repulsions. Therefore, *cations are smaller than their parent atoms*, as illustrated in Figure 7.8▶. The opposite is true of anions. When electrons are added

▶ **Figure 7.8 Cation and anion size.** Comparisons of the radii, in Å, of neutral atoms and ions for several of the groups of representative elements. Neutral atoms are shown in gray, cations in red, and anions in blue.

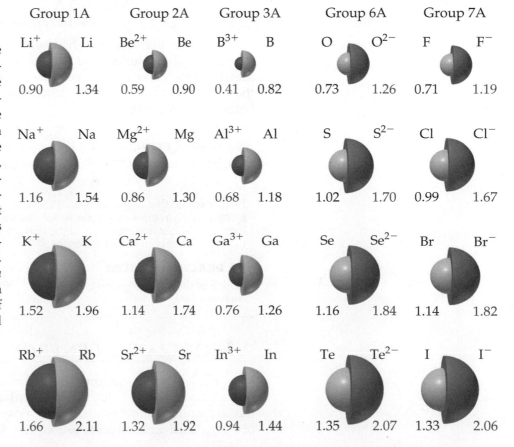

Group 1A		Group 2A		Group 3A		Group 6A		Group 7A	
Li^+	Li	Be^{2+}	Be	B^{3+}	B	O	O^{2-}	F	F^-
0.90	1.34	0.59	0.90	0.41	0.82	0.73	1.26	0.71	1.19
Na^+	Na	Mg^{2+}	Mg	Al^{3+}	Al	S	S^{2-}	Cl	Cl^-
1.16	1.54	0.86	1.30	0.68	1.18	1.02	1.70	0.99	1.67
K^+	K	Ca^{2+}	Ca	Ga^{3+}	Ga	Se	Se^{2-}	Br	Br^-
1.52	1.96	1.14	1.74	0.76	1.26	1.16	1.84	1.14	1.82
Rb^+	Rb	Sr^{2+}	Sr	In^{3+}	In	Te	Te^{2-}	I	I^-
1.66	2.11	1.32	1.92	0.94	1.44	1.35	2.07	1.33	2.06

to an atom to form an anion, the increased electron-electron repulsions cause the electrons to spread out more in space. Thus, *anions are larger than their parent atoms.*

For ions carrying the same charge, size increases as we move down a column in the periodic table. This trend is also seen in Figure 7.8. As the principal quantum number of the outermost occupied orbital of an ion increases, the radius of the ion increases.

■ SAMPLE EXERCISE 7.3 | Atomic and Ionic Radii

Arrange these atoms and ions in order of decreasing size: Mg^{2+}, Ca^{2+}, and Ca.

SOLUTION

Cations are smaller than their parent atoms, and so the Ca^{2+} ion is smaller than the Ca atom. Because Ca is below Mg in group 2A of the periodic table, Ca^{2+} is larger than Mg^{2+}. Consequently, $Ca > Ca^{2+} > Mg^{2+}$.

■ PRACTICE EXERCISE

Which of the following atoms and ions is largest: S^{2-}, S, O^{2-}?
Answer: S^{2-}

An **isoelectronic series** is a group of ions all containing the same number of electrons. For example, each ion in the isoelectronic series O^{2-}, F^-, Na^+, Mg^{2+}, Al^{3+} has 10 electrons. In any isoelectronic series we can list the members in order of increasing atomic number; therefore, nuclear charge increases as we move through the series. (Recall that the charge on the nucleus of an atom or monatomic ion is given by the atomic number of the element.) Because the number of electrons remains constant, the radius of the ion decreases with increasing nuclear charge, as the electrons are more strongly attracted to the nucleus:

$$\longrightarrow \text{Increasing nuclear charge} \longrightarrow$$

O^{2-}	F^-	Na^+	Mg^{2+}	Al^{3+}
1.26 Å	1.19 Å	1.16 Å	0.86 Å	0.68 Å

$$\longleftarrow \text{Decreasing ionic radius} \longrightarrow$$

Notice the positions and the atomic numbers of these elements in the periodic table. The nonmetal anions precede the noble gas Ne in the table. The metal cations follow Ne. Oxygen, the largest ion in this isoelectronic series, has the lowest atomic number, 8. Aluminum, the smallest of these ions, has the highest atomic number, 13.

■ SAMPLE EXERCISE 7.4 | Ionic Radii in an Isoelectronic Series

Arrange the ions K^+, Cl^-, Ca^{2+}, and S^{2-} in order of decreasing size.

SOLUTION

First, we note that this is an isoelectronic series of ions, with all ions having 18 electrons. In such a series, size decreases as the nuclear charge (atomic number) of the ion increases. The atomic numbers of the ions are S (16), Cl (17), K (19), and Ca (20). Thus, the ions decrease in size in the order $S^{2-} > Cl^- > K^+ > Ca^{2+}$.

■ PRACTICE EXERCISE

Which of the following ions is largest, Rb^+, Sr^{2+}, or Y^{3+}?
Answer: Rb^+

7.4 IONIZATION ENERGY

The ease with which electrons can be removed from an atom or ion has a major impact on chemical behavior. The **ionization energy** of an atom or ion is the minimum energy required to remove an electron from the ground state of the isolated gaseous atom or ion. The *first ionization energy*, I_1, is the energy

I onic size plays a major role in determining the properties of ions in solution. For example, a small difference in ionic size is often sufficient for one metal ion to be biologically important and another not to be. To illustrate, let's examine some of the biological chemistry of the zinc ion (Zn^{2+}) and compare it with the cadmium ion (Cd^{2+}).

Recall from the "Chemistry and Life" box in Section 2.7 that zinc is needed in our diets in trace amounts. Zinc is an essential part of several enzymes—the proteins that facilitate or regulate the speeds of key biological reactions. For example, one of the most important zinc-containing enzymes is *carbonic anhydrase*. This enzyme is found in red blood cells. Its job is to facilitate the reaction of carbon dioxide (CO_2) with water to form the bicarbonate ion (HCO_3^-):

$$CO_2(aq) + H_2O(l) \longrightarrow HCO_3^-(aq) + H^+(aq) \qquad [7.2]$$

You might be surprised to know that our bodies need an enzyme for such a simple reaction. In the absence of carbonic anhydrase, however, the CO_2 produced in cells when they are oxidizing glucose or other fuels in vigorous exercise would be cleared out much too slowly. About 20% of the CO_2 produced by cell metabolism binds to hemoglobin and is carried to the lungs, where it is expelled. About 70% of the CO_2 produced is converted to bicarbonate ion through the action of carbonic anhydrase. When the CO_2 has been converted into bicarbonate ion, it diffuses into the blood plasma and eventually is passed into the lungs in the reverse of Equation 7.2. These processes are illustrated in Figure 7.9 ▶. In the absence of zinc, the carbonic anhydrase would be inactive, and serious imbalances would result in the amount of CO_2 present in blood.

Zinc is also found in several other enzymes, including some found in the liver and kidneys. It is obviously essential to life. By contrast, cadmium, zinc's neighbor in group 2B, is extremely toxic to humans. But why are two elements so dif-

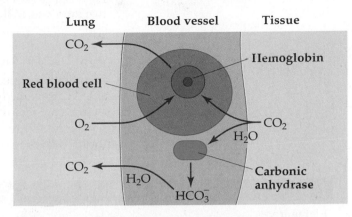

▲ **Figure 7.9 Ridding the body of carbon dioxide.** Illustration of the flow of CO_2 from tissues into blood vessels and eventually into the lungs. About 20% of the CO_2 binds to hemoglobin and is released in the lungs. About 70% is converted by carbonic anhydrase into HCO_3^- ion, which remains in the blood plasma until the reverse reaction releases CO_2 into the lungs. Small amounts of CO_2 simply dissolve in the blood plasma and are released in the lungs.

ferent? Both occur as 2+ ions, but Zn^{2+} is smaller than Cd^{2+}. The radius of Zn^{2+} is 0.88 Å; that of Cd^{2+} is 1.09 Å. Can this difference be the cause of such a dramatic reversal of biological properties? The answer is that while size is not the only factor, it is very important. In the carbonic anhydrase enzyme the Zn^{2+} ion is found electrostatically bonded to atoms on the protein, as shown in Figure 7.10 ▼. It turns out that Cd^{2+} binds in this same place preferentially over Zn^{2+}, thus displacing it. When Cd^{2+} is present instead of Zn^{2+}, however, the reaction of CO_2 with water is not facilitated. More seriously, Cd^{2+} inhibits reactions that are essential to the kidneys' functioning. *Related Exercises: 7.30, 7.91, and 7.92*

◄ **Figure 7.10 A zinc-containing enzyme.** The enzyme called carbonic anhydrase (left) catalyzes the reaction between CO_2 and water to form HCO_3^-. The ribbon represents the folding of the protein chain. The "active site" of the enzyme (represented by the ball-and-stick model) is where the reaction occurs. (H atoms have been excluded from this model for clarity.) The red sphere represents the oxygen of a water molecule that is bound to the zinc ion (gold sphere) at the center of the active site. The water molecule is replaced by CO_2 in the reaction. The bonds coming off the five-member rings attach the active site to the protein (nitrogen and carbon atoms are represented by blue and black spheres, respectively).

needed to remove the first electron from a neutral atom. For example, the first ionization energy for the sodium atom is the energy required for the process

$$Na(g) \longrightarrow Na^+(g) + e^- \qquad [7.3]$$

The *second ionization energy*, I_2, is the energy needed to remove the second electron, and so forth, for successive removals of additional electrons. Thus, I_2 for the sodium atom is the energy associated with the process

$$Na^+(g) \longrightarrow Na^{2+}(g) + e^- \qquad [7.4]$$

The greater the ionization energy, the more difficult it is to remove an electron.

▲ GIVE IT SOME THOUGHT

Light can be used to ionize atoms and ions, as in Equations 7.3 and 7.4. Which of the two processes, [7.3] or [7.4], would require shorter wavelength radiation?

Variations in Successive Ionization Energies

Ionization energies for the elements sodium through argon are listed in Table 7.2▼. Notice that the values for a given element increase as successive electrons are removed: $I_1 < I_2 < I_3$, and so forth. This trend exists because with each successive removal, an electron is being pulled away from an increasingly more positive ion, requiring increasingly more energy.

TABLE 7.2 ■ Successive Values of Ionization Energies, I, for the Elements Sodium through Argon (kJ/mol)

Element	I_1	I_2	I_3	I_4	I_5	I_6	I_7
Na	496	4562	(inner-shell electrons)				
Mg	738	1451	7733				
Al	578	1817	2745	11,577			
Si	786	1577	3232	4356	16,091		
P	1012	1907	2914	4964	6274	21,267	
S	1000	2252	3357	4556	7004	8496	27,107
Cl	1251	2298	3822	5159	6542	9362	11,018
Ar	1521	2666	3931	5771	7238	8781	11,995

A second important feature shown in Table 7.2 is the sharp increase in ionization energy that occurs when an inner-shell electron is removed. For example, consider silicon, whose electron configuration is $1s^2 2s^2 2p^6 3s^2 3p^2$ or [Ne]$3s^2 3p^2$. The ionization energies increase steadily from 786 kJ/mol to 4356 kJ/mol for the loss of the four electrons in the outer $3s$ and $3p$ subshells. Removal of the fifth electron, which comes from the $2p$ subshell, requires a great deal more energy: 16,091 kJ/mol. The large increase occurs because the $2p$ electron is much more likely to be found close to the nucleus than are the four $n = 3$ electrons, and therefore the $2p$ electron experiences a much greater effective nuclear charge than do the $3s$ and $3p$ electrons.

▲ GIVE IT SOME THOUGHT

Which would you expect to be greater, I_1 for a boron atom or I_2 for a carbon atom?

Every element exhibits a large increase in ionization energy when electrons are removed from its noble-gas core. This observation supports the idea that only the outermost electrons, those beyond the noble-gas core, are involved in the sharing and transfer of electrons that give rise to chemical bonding and reactions. The inner electrons are too tightly bound to the nucleus to be lost from the atom or even shared with another atom.

■ SAMPLE EXERCISE 7.5 | Trends in Ionization Energy

Three elements are indicated in the periodic table in the margin. Based on their locations, predict the one with the largest second ionization energy.

SOLUTION

Analyze and Plan: The locations of the elements in the periodic table allow us to predict the electron configurations. The greatest ionization energies involve removal of core electrons. Thus, we should look first for an element with only one electron in the outermost occupied shell.

Solve: The element in group 1A (Na), indicated by the red box, has only one valence electron. The second ionization energy of this element is associated, therefore, with the removal of a core electron. The other elements indicated, S (green box) and Ca (blue box), have two or more valence electrons. Thus, Na should have the largest second ionization energy.

Check: If we consult a chemistry handbook, we find the following values for the second ionization energies (I_2) of the respective elements: Ca (1,145 kJ/mol) < S (2,252 kJ/mol) < Na (4,562 kJ/mol).

■ PRACTICE EXERCISE

Which will have the greater third ionization energy, Ca or S?
Answer: Ca

Periodic Trends in First Ionization Energies

We have seen that the ionization energy for a given element increases as we remove successive electrons. What trends do we observe in ionization energy as we move from one element to another in the periodic table? Figure 7.11 ▶ shows a graph of I_1 versus atomic number for the first 54 elements. The important trends are as follows:

1. Within each row (period) of the table, I_1 generally increases with increasing atomic number. The alkali metals show the lowest ionization energy in each row, and the noble gases show the highest. There are slight irregularities in this trend that we will discuss shortly.

2. Within each column (group) of the table, the ionization energy generally decreases with increasing atomic number. For example, the ionization energies of the noble gases follow the order He > Ne > Ar > Kr > Xe.

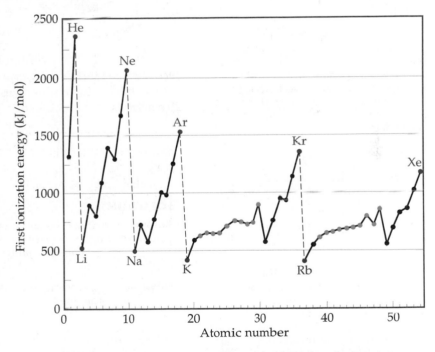

▲ **Figure 7.11 First ionization energy versus atomic number.** The red dots mark the beginning of a period (alkali metals), the blue dots mark the end of a period (noble gases), and the black dots indicate *s*- and *p*-block elements, while green dots are used to represent the transition metals.

3. The *s*- and *p*-block elements show a larger range of values of I_1 than do the transition-metal elements. Generally, the ionization energies of the transition metals increase slowly as we proceed from left to right in a period. The *f*-block metals, which are not shown in Figure 7.11, also show only a small variation in the values of I_1.

The periodic trends in the first ionization energies of the *s*- and *p*-block elements are further illustrated in Figure 7.12▼.

In general, smaller atoms have higher ionization energies. The same factors that influence atomic size also influence ionization energies. The energy needed to remove an electron from the outermost occupied shell depends on both the effective nuclear charge and the average distance of the electron from the nucleus. Either increasing the effective nuclear charge or decreasing the distance from the nucleus increases the attraction between the electron and the nucleus.

As this attraction increases, it becomes more difficult to remove the electron and, thus, the ionization energy increases. As we move across a period, there is both an increase in effective nuclear charge and a decrease in atomic radius, causing the ionization energy to increase. As we move down a column, however, the atomic radius increases, while the effective nuclear charge increases rather gradually. Thus, the attraction between the nucleus and the electron decreases, causing the ionization energy to decrease.

The irregularities within a given row are somewhat subtler but still readily explained. For example, the decrease in ionization energy from beryllium ($[He]2s^2$) to boron ($[He]2s^22p^1$), shown in Figures 7.11 and 7.12, occurs because the third valence electron of B must occupy the $2p$ subshell, which is empty for Be. Recall that, as we discussed earlier, the $2p$ subshell is at a higher energy than the $2s$ subshell (Figure 6.23). The decrease in ionization energy when moving from nitrogen ($[He]2s^22p^3$) to oxygen ($[He]2s^22p^4$) is because of the repulsion of paired electrons in the p^4 configuration, as shown in Figure 7.13 ◄. Remember that according to Hund's rule, each electron in the p^3 configuration resides in a different p orbital, which minimizes the electron–electron repulsion among the three $2p$ electrons. ∞ (Section 6.8)

▲ **Figure 7.12 Trends in first ionization energy.** First ionization energies for the s- and p-block elements in the first six periods. The ionization energy generally increases from left to right and decreases from top to bottom. The ionization energy of astatine has not been determined.

2p | 1↓ | 1 | 1 |
Oxygen

2p | 1 | 1 | 1 |
Nitrogen

▲ **Figure 7.13 2p orbital filling in nitrogen and oxygen.** The presence of a fourth electron in the 2p orbitals of oxygen leads to an extra repulsion associated with putting two electrons in a single orbital. This repulsion is responsible for the lower first ionization energy of oxygen.

■ **SAMPLE EXERCISE 7.6** | Periodic Trends in Ionization Energy

Referring to a periodic table, arrange the following atoms in order of increasing first ionization energy: Ne, Na, P, Ar, K.

SOLUTION

Analyze and Plan: We are given the chemical symbols for five elements. To rank them according to increasing first ionization energy, we need to locate each element in the periodic table. We can then use their relative positions and the trends in first ionization energies to predict their order.

Solve: Ionization energy increases as we move left to right across a row. It decreases as we move from the top of a group to the bottom. Because Na, P, and Ar are in the same row of the periodic table, we expect I_1 to vary in the order Na < P < Ar.

Because Ne is above Ar in group 8A, we expect Ne to have the greater first ionization energy: Ar < Ne. Similarly, K is the alkali metal directly below Na in group 1A, and so we expect I_1 for K to be less than that of Na: K < Na.

From these observations, we conclude that the ionization energies follow the order

K < Na < P < Ar < Ne

Check: The values shown in Figure 7.12 confirm this prediction.

■ **PRACTICE EXERCISE**

Which has the lowest first ionization energy, B, Al, C, or Si? Which has the highest first ionization energy?
Answer: Al lowest, C highest

Electron Configurations of Ions

When electrons are removed from an atom to form a cation, they are always removed first from the occupied orbitals having the largest principal quantum number, n. For example, when one electron is removed from a lithium atom ($1s^2 2s^1$), it is the $2s^1$ electron that is removed:

$$\text{Li } (1s^2 2s^1) \Rightarrow \text{Li}^+ (1s^2) + e^-$$

Likewise, when two electrons are removed from Fe ($[\text{Ar}]3d^6 4s^2$), the $4s^2$ electrons are the ones removed:

$$\text{Fe } ([\text{Ar}]3d^6 4s^2) \Rightarrow \text{Fe}^{2+} ([\text{Ar}]3d^6) + 2e^-$$

If an additional electron is removed, forming Fe^{3+}, it now comes from a $3d$ orbital because all the orbitals with $n = 4$ are empty:

$$\text{Fe}^{2+} ([\text{Ar}]3d^6) \Rightarrow \text{Fe}^{3+} ([\text{Ar}]3d^5) + e^-$$

It may seem odd that the $4s$ electrons are removed before the $3d$ electrons in forming transition-metal cations. After all, in writing electron configurations, we added the $4s$ electrons before the $3d$ ones. In writing electron configurations for atoms, however, we are going through an imaginary process in which we move through the periodic table from one element to another. In doing so, we are adding both an electron to an orbital and a proton to the nucleus to change the identity of the element. In ionization, we do not reverse this process because no protons are being removed.

If there is more than one occupied subshell for a given value of n the electrons are first removed from the orbital with the highest value of l. For example a tin atom loses its $5p$ electrons before it loses its $5s$ electrons:

$$\text{Sn } ([\text{Kr}]4d^{10}5s^2 5p^2) \Rightarrow \text{Sn}^{2+} ([\text{Kr}]4d^{10}5s^2) + 2e^- \Rightarrow \text{Sn}^{4+} ([\text{Kr}]4d^{10}) + 4e^-$$

When electrons are added to an atom to form an anion, they are added to the empty or partially filled orbital having the lowest value of n. For example, when an electron is added to a fluorine atom to form the F^- ion, the electron goes into the one remaining vacancy in the $2p$ subshell:

$$\text{F } (1s^2 2s^2 2p^5) + e^- \Rightarrow \text{F}^- (1s^2 2s^2 2p^6)$$

▲ **GIVE IT SOME THOUGHT**

Would Cr^{3+} and V^{2+} have the same or different electron configurations?

■■ **SAMPLE EXERCISE 7.7** | **Electron Configurations of Ions**

Write the electron configuration for **(a)** Ca^{2+} **(b)** Co^{3+}, and **(c)** S^{2-}.

SOLUTION

Analyze and Plan: We are asked to write electron configurations for three ions. To do so, we first write the electron configuration of the parent atom. We then remove electrons to form cations or add electrons to form anions. Electrons are first removed from the orbitals having the highest value of n. They are added to the empty or partially filled orbitals having the lowest value of n.

Solve:

(a) Calcium (atomic number 20) has the electron configuration

$$\text{Ca: } [\text{Ar}]4s^2$$

To form a 2+ ion, the two outer electrons must be removed, giving an ion that is isoelectronic with Ar:

$$\text{Ca}^{2+}: [\text{Ar}]$$

(b) Cobalt (atomic number 27) has the electron configuration

$$\text{Co: } [\text{Ar}]3d^7 4s^2$$

To form a 3+ ion, three electrons must be removed. As discussed in the text preceding this Sample Exercise, the $4s$ electrons are removed before the $3d$ electrons. Consequently, the electron configuration for Co^{3+} is

$$\text{Co}^{3+}: [\text{Ar}]3d^6$$

(c) Sulfur (atomic number 16) has the electron configuration

$$S: [Ne]3s^2 3p^4$$

To form a 2− ion, two electrons must be added. There is room for two additional electrons in the $3p$ orbitals. Thus, the S^{2-} electron configuration is

$$S^{2-}: [Ne]3s^2 3p^6 = [Ar]$$

Comment: Remember that many of the common ions of the *s*- and *p*-block elements, such as Ca^{2+} and S^{2-}, have the same number of electrons as the closest noble gas. ∞ (Section 2.7)

■ **PRACTICE EXERCISE**

Write the electron configuration for **(a)** Ga^{3+}, **(b)** Cr^{3+}, and **(c)** Br^-.
Answers: **(a)** $[Ar]3d^{10}$, **(b)** $[Ar]3d^3$, **(c)** $[Ar]3d^{10}4s^24p^6 = [Kr]$

7.5 ELECTRON AFFINITIES

The first ionization energy of an atom is a measure of the energy change associated with removing an electron from the atom to form a positively charged ion. For example, the first ionization energy of $Cl(g)$, 1251 kJ/mol, is the energy change associated with the process

$$\textit{Ionization energy:}\quad Cl(g) \longrightarrow Cl^+(g) + e^- \qquad \Delta E = 1251 \text{ kJ/mol} \qquad [7.5]$$
$$\qquad\qquad\qquad [Ne]3s^23p^5 \qquad [Ne]3s^23p^4$$

The positive value of the ionization energy means that energy must be put into the atom to remove the electron.

In addition, most atoms can gain electrons to form negatively charged ions. The energy change that occurs when an electron is added to a gaseous atom is called the **electron affinity** because it measures the attraction, or *affinity*, of the atom for the added electron. For most atoms, energy is released when an electron is added. For example, the addition of an electron to a chlorine atom is accompanied by an energy change of −349 kJ/mol, the negative sign indicating that energy is released during the process. We therefore say that the electron affinity of Cl is −349 kJ/mol:*

$$\textit{Electron affinity:}\quad Cl(g) + e^- \longrightarrow Cl^-(g) \qquad \Delta E = -349 \text{ kJ/mol} \qquad [7.6]$$
$$\qquad\qquad\qquad [Ne]3s^23p^5 \qquad\quad [Ne]3s^23p^6$$

It is important to understand the difference between ionization energy and electron affinity: Ionization energy measures the ease with which an atom *loses* an electron, whereas electron affinity measures the ease with which an atom *gains* an electron.

The greater the attraction between a given atom and an added electron, the more negative the atom's electron affinity will be. For some elements, such as the noble gases, the electron affinity has a positive value, meaning that the anion is higher in energy than are the separated atom and electron:

$$Ar(g) + e^- \longrightarrow Ar^-(g)^- \qquad \Delta E > 0 \qquad\qquad [7.7]$$
$$[Ne]3s^23p^6 \qquad\quad [Ne]3s^23p^64s^1$$

The fact that the electron affinity is a positive number means that an electron will not attach itself to an Ar atom; the Ar^- ion is unstable and does not form.

Two sign conventions are used for electron affinity. In most introductory texts, including this one, the thermodynamic sign convention is used: a negative sign indicates that the addition of an electron is an exothermic process, as in the electron affinity given for chlorine, −349 kJ/mol. Historically, however, electron affinity has been defined as the energy released when an electron is added to a gaseous atom or ion. Because 349 kJ/mol is released when an electron is added to Cl(g), the electron affinity by this convention would be +349 kJ/mol.

Figure 7.14▶ shows the electron affinities for the *s*- and *p*-block elements in the first five rows of the periodic table. Notice that the trends in electron affinity as we proceed through the periodic table are not as evident as they were for ionization energy. The halogens, which are one electron shy of a filled *p* subshell, have the most-negative electron affinities. By gaining an electron, a halogen atom forms a stable negative ion that has a noble-gas configuration (Equation 7.6). The addition of an electron to a noble gas, however, requires that the electron reside in a higher-energy subshell that is empty in the neutral atom (Equation 7.7). Because occupying a higher-energy subshell is energetically very unfavorable, the electron affinity is highly positive. The electron affinities of Be and Mg are positive for the same reason; the added electron would reside in a previously empty *p* subshell that is higher in energy.

The electron affinities of the group 5A elements (N, P, As, Sb) are also interesting. Because these elements have half-filled *p* subshells, the added electron must be put in an orbital that is already occupied, resulting in larger electron–electron repulsions. Consequently, these elements have electron affinities that are either positive (N) or less negative than their neighbors to the left (P, As, Sb). Recall that we saw a discontinuity in the regular periodic trends for first ionization energy in Section 7.4 for the same reason.

Electron affinities do not change greatly as we move down a group. For example, consider the electron affinities of the halogens (Figure 7.14). For F, the added electron goes into a 2*p* orbital, for Cl a 3*p* orbital, for Br a 4*p* orbital, and so forth. As we proceed from F to I, therefore, the average distance between the added electron and the nucleus steadily increases, causing the electron-nucleus attraction to decrease. However, the orbital that holds the outermost electron is increasingly spread out, so that as we proceed from F to I, the electron-electron repulsions are also reduced. As a result, the reduction in the electron-nucleus attraction is counterbalanced by the reduction in electron–electron repulsions.

H −73							He >0
Li −60	Be >0	B −27	C −122	N >0	O −141	F −328	Ne >0
Na −53	Mg >0	Al −43	Si −134	P −72	S −200	Cl −349	Ar >0
K −48	Ca −2	Ga −30	Ge −119	As −78	Se −195	Br −325	Kr >0
Rb −47	Sr −5	In −30	Sn −107	Sb −103	Te −190	I −295	Xe >0
1A	2A	3A	4A	5A	6A	7A	8A

▲ **Figure 7.14 Electron affinity.** Electron affinities in kJ/mol for the *s*- and *p*-block elements in the first five rows of the periodic table. The more negative the electron affinity, the greater the attraction of the atom for an electron. An electron affinity > 0 indicates that the negative ion is higher in energy than the separated atom and electron.

GIVE IT SOME THOUGHT

Suppose you were asked for a value for the first ionization energy of a Cl⁻(*g*) ion. What is the relationship between this quantity and the electron affinity of Cl(*g*)?

7.6 METALS, NONMETALS, AND METALLOIDS

Atomic radii, ionization energies, and electron affinities are properties of individual atoms. With the exception of the noble gases, however, none of the elements exists in nature as individual atoms. To get a broader understanding of the properties of elements, we must also examine periodic trends in properties that involve large collections of atoms.

The elements can be broadly grouped into the categories of metals, nonmetals, and metalloids. ∞ (Section 2.5) This classification is shown in Figure 7.15▶. Roughly three-quarters of the elements are metals, situated in the left and middle portions of the table. The nonmetals are located

▼ **Figure 7.15 Metals, metalloids, and nonmetals.** The majority of elements are metals. Metallic character increases from right to left across a period and also increases from top to bottom in a group.

Increasing metallic character →

Increasing metallic character ↓

1A 1												3A 13	4A 14	5A 15	6A 16	7A 17	8A 18
1 H	2A 2																2 He
3 Li	4 Be											5 B	6 C	7 N	8 O	9 F	10 Ne
11 Na	12 Mg	3B 3	4B 4	5B 5	6B 6	7B 7	8	8B 9	10	1B 11	2B 12	13 Al	14 Si	15 P	16 S	17 Cl	18 Ar
19 K	20 Ca	21 Sc	22 Ti	23 V	24 Cr	25 Mn	26 Fe	27 Co	28 Ni	29 Cu	30 Zn	31 Ga	32 Ge	33 As	34 Se	35 Br	36 Kr
37 Rb	38 Sr	39 Y	40 Zr	41 Nb	42 Mo	43 Tc	44 Ru	45 Rh	46 Pd	47 Ag	48 Cd	49 In	50 Sn	51 Sb	52 Te	53 I	54 Xe
55 Cs	56 Ba	71 Lu	72 Hf	73 Ta	74 W	75 Re	76 Os	77 Ir	78 Pt	79 Au	80 Hg	81 Tl	82 Pb	83 Bi	84 Po	85 At	86 Rn
87 Fr	88 Ra	103 Lr	104 Rf	105 Db	106 Sg	107 Bh	108 Hs	109 Mt	110 Ds	111 Rg	112	113	114	115	116		118

	57 La	58 Ce	59 Pr	60 Nd	61 Pm	62 Sm	63 Eu	64 Gd	65 Tb	66 Dy	67 Ho	68 Er	69 Tm	70 Yb
	89 Ac	90 Th	91 Pa	92 U	93 Np	94 Pu	95 Am	96 Cm	97 Bk	98 Cf	99 Es	100 Fm	101 Md	102 No

Metals
Metalloids
Nonmetals

TABLE 7.3 ■ Characteristic Properties of Metals and Nonmetals	
Metals	**Nonmetals**
Have a shiny luster; various colors, although most are silvery	Do not have a luster; various colors
Solids are malleable and ductile	Solids are usually brittle; some are hard, some are soft
Good conductors of heat and electricity	Poor conductors of heat and electricity
Most metal oxides are ionic solids that are basic	Most nonmetal oxides are molecular substances that form acidic solutions
Tend to form cations in aqueous solution	Tend to form anions or oxyanions in aqueous solution

▲ **Figure 7.16 The luster of metals.** Metallic objects are readily recognized by their characteristic shiny luster.

at the top right corner, and the metalloids lie between the metals and nonmetals. Hydrogen, which is located at the top left corner, is a nonmetal. This is why we set off hydrogen from the remaining group 1A elements in Figure 7.15 by inserting a space between the H box and the Li box. Some of the distinguishing properties of metals and nonmetals are summarized in Table 7.3▲.

The more an element exhibits the physical and chemical properties of metals, the greater its **metallic character**. As indicated in Figure 7.15, metallic character generally increases as we proceed down a column of the periodic table and increases as we proceed from right to left in a row. Let's now examine the close relationships that exist between electron configurations and the properties of metals, nonmetals, and metalloids.

Metals

Most metallic elements exhibit the shiny luster that we associate with metals (Figure 7.16◄). Metals conduct heat and electricity. In general they are malleable (can be pounded into thin sheets) and ductile (can be drawn into wires). All are solids at room temperature except mercury (melting point = −39 °C), which is a liquid. Two metals melt at slightly above room temperature, cesium at 28.4 °C and gallium at 29.8 °C. At the other extreme, many metals melt at very high temperatures. For example, chromium melts at 1900 °C.

Metals tend to have low ionization energies and therefore tend to form positive ions relatively easily. As a result, metals are oxidized (lose electrons) when they undergo chemical reactions. This fact is illustrated in Figure 7.17▼, where the first ionization energies of the metals, nonmetals, and metalloids are compared.

Among the fundamental atomic properties (radius, electron configuration, electron affinity, and so forth), the first ionization energy is the best indicator of whether an element will behave as a metal or a nonmetal. The relative ease of oxidation of common metals was discussed earlier, in Section 4.4. As we noted there, many metals are oxidized by a variety of common substances, including O_2 and acids.

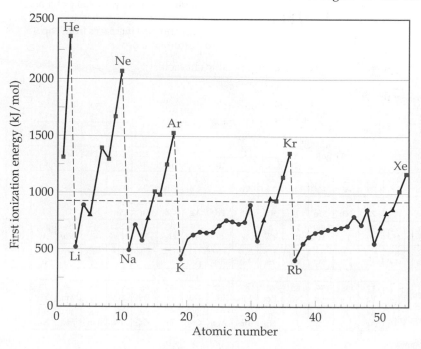

◄ **Figure 7.17 A comparison of the first ionization energies of metals vs. nonmetals.** Values of first ionization energy for metals are markedly lower than those of nonmetals. The red circles correspond to metallic elements, the blue squares to nonmetals, and the black triangles to metalloids. The dashed line at 925 kJ/mol separates the metals from the nonmetals.

1A	2A											3A	4A	5A	6A	7A	8A
H^+																H^-	N
Li^+														N^{3-}	O^{2-}	F^-	O
Na^+	Mg^{2+}				Transition metals							Al^{3+}		P^{3-}	S^{2-}	Cl^-	B L E
K^I	Ca^{2+}	Sc^{3+}	Ti^{4+}	V^{5+} V^{4+}	Cr^{3+}	Mn^{2+} Mn^{4+} Fe^{2+} Fe^{3+}	Co^{2+} Co^{3+}	Ni^{2+}	Cu^+ Cu^{2+}	Zn^{2+}					Se^{2-}	Br^-	G A S E S
Rb^+	Sr^{2+}							Pd^{2+}	Ag^+	Cd^{2+}		Sn^{2+} Sn^{4+}	Sb^{3+} Sb^{5+}	Te^{2-}	I^-		
Cs^+	Ba^{2+}							Pt^{2+}	Au^+ Au^{3+}	Hg_2^{2+} Hg^{2+}		Pb^{2+} Pb^{4+}	Bi^{3+} Bi^{5+}				

▲ Figure 7.18 **Representative oxidation states.** Oxidation states found in ionic compounds, including some examples of higher oxidation states adopted by *p*-block metals. Notice that the steplike line that divides metals from nonmetals also separates cations from anions.

Figure 7.18▲ shows the oxidation states of some representative ions of both metals and nonmetals. As we noted in Section 2.7, the charge on any alkali metal ion is always 1+, and that on any alkaline earth metal is always 2+ in their compounds. For atoms belonging to either of these groups, the outer *s* electrons are easily lost, yielding a noble-gas electron configuration. For metals belonging to groups with partially occupied *p* orbitals (Groups 3A–7A), the observed cations are formed either by losing only the outer *p* electrons (such as Sn^{2+}) or the outer *s* and *p* electrons (such as Sn^{4+}). The charge on transition-metal ions does not follow an obvious pattern. One of the characteristic features of the transition metals is their ability to form more than one positive ion. For example, iron may be 2+ in some compounds and 3+ in others.

GIVE IT SOME THOUGHT

Based on periodic trends discussed in this chapter, can you see a general relationship between the trends in metallic character and those for ionization energy?

Compounds of metals with nonmetals tend to be ionic substances. For example, most metal oxides and halides are ionic solids. To illustrate, the reaction between nickel metal and oxygen produces nickel oxide, an ionic solid containing Ni^{2+} and O^{2-} ions:

$$2\,Ni(s) + O_2(g) \longrightarrow 2\,NiO(s) \qquad [7.8]$$

The oxides are particularly important because of the great abundance of oxygen in our environment.

Most metal oxides are basic. Those that dissolve in water react to form metal hydroxides, as in the following examples:

$$\text{Metal oxide} + \text{water} \longrightarrow \text{metal hydroxide}$$

$$Na_2O(s) + H_2O(l) \longrightarrow 2\,NaOH(aq) \qquad [7.9]$$

$$CaO(s) + H_2O(l) \longrightarrow Ca(OH)_2(aq) \qquad [7.10]$$

The basicity of metal oxides is due to the oxide ion, which reacts with water according to the net ionic equation

$$O^{2-}(aq) + H_2O(l) \longrightarrow 2\,OH^-(aq) \qquad [7.11]$$

▶ Figure 7.19 **Metal oxides react with acids.** (a) Nickel oxide (NiO), nitric acid (HNO$_3$), and water. (b) NiO is insoluble in water, but reacts with HNO$_3$ to give a green solution of the salt Ni(NO$_3$)$_2$.

NiO

(a)   (b)

Metal oxides also demonstrate their basicity by reacting with acids to form a salt plus water, as illustrated in Figure 7.19▲:

$$\text{Metal oxide} + \text{acid} \longrightarrow \text{salt} + \text{water}$$

$$NiO(s) + 2\,HNO_3(aq) \longrightarrow Ni(NO_3)_2(aq) + H_2O(l) \qquad [7.12]$$

In contrast, we will soon see that nonmetal oxides are acidic, dissolving in water to form acidic solutions and reacting with bases to form salts.

■ **SAMPLE EXERCISE 7.8** | Metal Oxides

(a) Would you expect scandium oxide to be a solid, liquid, or gas at room temperature? **(b)** Write the balanced chemical equation for the reaction of scandium oxide with nitric acid.

SOLUTION

Analyze and Plan: We are asked about one physical property of scandium oxide—its state at room temperature—and one chemical property—how it reacts with nitric acid.

Solve:

(a) Because scandium oxide is the oxide of a metal, we would expect it to be an ionic solid. Indeed it is, with the very high melting point of 2485 °C.

(b) In its compounds, scandium has a 3+ charge, Sc^{3+}; the oxide ion is O^{2-}. Consequently, the formula of scandium oxide is Sc$_2$O$_3$. Metal oxides tend to be basic and therefore to react with acids to form a salt plus water. In this case the salt is scandium nitrate, Sc(NO$_3$)$_3$. The balanced chemical equation is

$$Sc_2O_3(s) + 6\,HNO_3(aq) \longrightarrow 2\,Sc(NO_3)_3(aq) + 3\,H_2O(l)$$

■ **PRACTICE EXERCISE**

Write the balanced chemical equation for the reaction between copper(II) oxide and sulfuric acid.

Answer: CuO(s) + H$_2$SO$_4$(aq) $\longrightarrow$ CuSO$_4$(aq) + H$_2$O(l)

▼ Figure 7.20 **The diversity of nonmetals.** Nonmetallic elements are diverse in their appearances. Shown here are (clockwise from top left) sulfur, bromine, phosphorus, iodine, and carbon.

Nonmetals

Nonmetals vary greatly in appearance (Figure 7.20 ◀). They are not lustrous and generally are poor conductors of heat and electricity. Their melting points are generally lower than those of metals (although diamond, a form of carbon, melts at 3570 °C). Under ordinary conditions, seven nonmetals exist as diatomic molecules. Five of these are gases (H$_2$, N$_2$, O$_2$, F$_2$, and Cl$_2$), one is a liquid (Br$_2$), and one is a volatile solid (I$_2$). Excluding the noble gases, the remaining nonmetals are solids that can be either hard, such as diamond, or soft, such as sulfur.

Because of their electron affinities, nonmetals tend to gain electrons when they react with metals. For example, the reaction of aluminum with bromine produces aluminum bromide, an ionic compound containing the aluminum ion, Al^{3+}, and the bromide ion, Br^-:

$$2\,Al(s) + 3\,Br_2(l) \longrightarrow 2\,AlBr_3(s) \qquad [7.13]$$

A nonmetal typically will gain enough electrons to fill its outermost occupied *p* subshell, giving a noble-gas electron configuration. For example, the bromine atom gains one electron to fill its 4*p* subshell:

$$Br\ ([Ar]4s^2 3d^{10} 4p^5) + e^- \Rightarrow Br^-\ ([Ar]4s^2 3d^{10} 4p^6)$$

Compounds composed entirely of nonmetals are typically molecular substances. For example, the oxides, halides, and hydrides of the nonmetals are molecular substances that tend to be gases, liquids, or low-melting solids at room temperature.

Most nonmetal oxides are acidic; those that dissolve in water react to form acids, as in the following examples:

$$Nonmetal\ oxide + water \longrightarrow acid$$

$$CO_2(g) + H_2O(l) \longrightarrow H_2CO_3(aq) \qquad [7.14]$$

$$P_4O_{10}(s) + 6\,H_2O(l) \longrightarrow 4\,H_3PO_4(aq) \qquad [7.15]$$

The reaction of carbon dioxide with water (Figure 7.21 ▼) accounts for the acidity of carbonated water and, to some extent, rainwater. Because sulfur is present in oil and coal, combustion of these common fuels produces sulfur dioxide and sulfur trioxide. These substances dissolve in water to produce *acid rain*, a major pollution problem in many parts of the world. Like acids, most nonmetal oxides dissolve in basic solutions to form a salt plus water:

$$Nonmetal\ oxide + base \longrightarrow salt + water$$

$$CO_2(g) + 2\,NaOH(aq) \longrightarrow Na_2CO_3(aq) + H_2O(l) \qquad [7.16]$$

GIVE IT SOME THOUGHT

A compound ACl_3 (A is an element) has a melting point of $-112\ ^\circ C$. Would you expect the compound to be a molecular or ionic substance? If you were told that element A was either scandium (Sc) or phosphorus (P), which do you think would be a more likely choice?

(a)   (b)

◀ Figure 7.21 **The reaction of CO₂ with water.** (a) The water has been made slightly basic and contains a few drops of bromthymol blue, an acid–base indicator that is blue in basic solution. (b) Upon the addition of a piece of solid carbon dioxide, $CO_2(s)$, the color changes to yellow, indicating an acidic solution. The mist is due to water droplets condensed from the air by the cold CO_2 gas.

■ SAMPLE EXERCISE 7.9 | Nonmetal Oxides

Write the balanced chemical equations for the reactions of solid selenium dioxide with **(a)** water, **(b)** aqueous sodium hydroxide.

SOLUTION

Analyze and Plan: We first note that selenium (Se) is a nonmetal. We therefore need to write chemical equations for the reaction of a nonmetal oxide, first with water and then with a base, NaOH. Nonmetal oxides are acidic, reacting with water to form an acid and with bases to form a salt and water.

Solve:

(a) Selenium dioxide is SeO_2. Its reaction with water is like that of carbon dioxide (Equation 7.14):

$$SeO_2(s) + H_2O(l) \longrightarrow H_2SeO_3(aq)$$

(It does not matter that SeO_2 is a solid and CO_2 is a gas under ambient conditions; the point is that both are water-soluble nonmetal oxides.)
(b) The reaction with sodium hydroxide is like the reaction summarized by Equation 7.16:

$$SeO_2(s) + 2\,NaOH(aq) \longrightarrow Na_2SeO_3(aq) + H_2O(l)$$

■ PRACTICE EXERCISE

Write the balanced chemical equation for the reaction of solid tetraphosphorus hexoxide with water.
Answer: $P_4O_6(s) + 6\,H_2O(l) \longrightarrow 4\,H_3PO_3(aq)$

Metalloids

Metalloids have properties intermediate between those of metals and those of nonmetals. They may have *some* characteristic metallic properties but lack others. For example, silicon *looks* like a metal (Figure 7.22 ◄), but it is brittle rather than malleable and is a much poorer conductor of heat and electricity than are metals. Compounds of metalloids can have characteristics of the compounds of metals or nonmetals, depending on the specific compound.

Several of the metalloids, most notably silicon, are electrical semiconductors and are the principal elements used in the manufacture of integrated circuits and computer chips. One of the reasons metalloids such as silicon can be used for integrated circuits is the fact that their electrical conductivity is intermediate between that of metals and nonmetals. Very pure silicon is an electrical insulator, but its conductivity can be dramatically increased with the addition of specific impurities (dopants). This modification provides a mechanism for controlling the electrical conductivity by controlling the chemical composition. We will return to this point in Chapter 12.

▲ **Figure 7.22 Elemental silicon.** Silicon is an example of a metalloid. Although it looks metallic, silicon is brittle and is a poor thermal and electrical conductor as compared to metals. Large crystals of silicon are sliced into thin wafers for use in integrated circuits.

▼ **Figure 7.23 Alkali metals.** Sodium and the other alkali metals are soft enough to be cut with a knife. The shiny metallic surface quickly tarnishes as the metal reacts with oxygen in the air.

7.7 GROUP TRENDS FOR THE ACTIVE METALS

Our discussion of atomic radius, ionization energy, electron affinity, and metallic character gives some idea of the way the periodic table can be used to organize and remember facts. As we have seen, elements in a group possess general similarities. However, trends also exist as we move through a group. In this section we will use the periodic table and our knowledge of electron configurations to examine the chemistry of the **alkali metals** (group 1A) and the **alkaline earth metals** (group 2A).

Group 1A: The Alkali Metals

The alkali metals are soft metallic solids (Figure 7.23 ◄). All have characteristic metallic properties such as a silvery, metallic luster and high thermal and electrical conductivities. The name *alkali* comes from an Arabic word meaning "ashes." Many compounds of sodium and potassium, two alkali metals, were isolated from wood ashes by early chemists.

TABLE 7.4 ■ Some Properties of the Alkali Metals

Element	Electron Configuration	Melting Point (°C)	Density (g/cm³)	Atomic Radius (Å)	I_1 (kJ/mol)
Lithium	[He]$2s^1$	181	0.53	1.34	520
Sodium	[Ne]$3s^1$	98	0.97	1.54	496
Potassium	[Ar]$4s^1$	63	0.86	1.96	419
Rubidium	[Kr]$5s^1$	39	1.53	2.11	403
Cesium	[Xe]$6s^1$	28	1.88	2.25	376

Sodium and potassium are relatively abundant in Earth's crust, in seawater, and in biological systems. We all have sodium ions in our bodies. However, ingesting too much sodium can raise our blood pressure. Potassium is also prevalent in our bodies; a 140-pound person contains about 130 g of potassium, as K^+ ions in intracellular fluids. Plants require potassium for growth and development (Figure 7.24▶).

Some of the physical and chemical properties of the alkali metals are given in Table 7.4▲. The elements have low densities and melting points, and these properties vary in a fairly regular way with increasing atomic number. We can also see some of the usual trends as we move down the group, such as increasing atomic radius and decreasing first ionization energy. For each row of the periodic table, the alkali metal has the lowest I_1 value (Figure 7.11), which reflects the relative ease with which its outer s electron can be removed. As a result, the alkali metals are all very reactive, readily losing one electron to form ions carrying a 1+ charge. ∞ (Section 4.4)

The alkali metals exist in nature only as compounds. The metals combine directly with most nonmetals. For example, they react with hydrogen to form hydrides and with sulfur to form sulfides:

$$2 M(s) + H_2(g) \longrightarrow 2 MH(s) \qquad [7.17]$$

$$2 M(s) + S(s) \longrightarrow M_2S(s) \qquad [7.18]$$

(The symbol M in Equations 7.17 and 7.18 represents any one of the alkali metals.) In hydrides of the alkali metals (LiH, NaH, and so forth), hydrogen is present as H^-, called the **hydride ion**. The hydride ion, which is a hydrogen atom that has *gained* an electron, is distinct from the hydrogen ion, H^+, formed when a hydrogen atom *loses* its electron.

The alkali metals react vigorously with water, producing hydrogen gas and a solution of an alkali metal hydroxide:

$$2 M(s) + 2 H_2O(l) \longrightarrow 2 MOH(aq) + H_2(g) \qquad [7.19]$$

These reactions are very exothermic. In many cases enough heat is generated to ignite the H_2, producing a fire or sometimes even an explosion (Figure 7.25▼).

▲ Figure 7.24 **Elements in fertilizers.** Fertilizers contain large quantities of potassium, phosphorus, and nitrogen to meet the needs of growing plants.

◀ Figure 7.25 **The alkali metals react vigorously with water.** (a) The reaction of lithium is evidenced by the bubbling of escaping hydrogen gas. (b) The reaction of sodium is more rapid and is so exothermic that the hydrogen gas produced burns in air. (c) Potassium reacts almost explosively.

(a)   (b)   (c)

▶ **Figure 7.26 Flame tests.** (a) Li (crimson red), (b) Na (yellow), and (c) K (lilac).

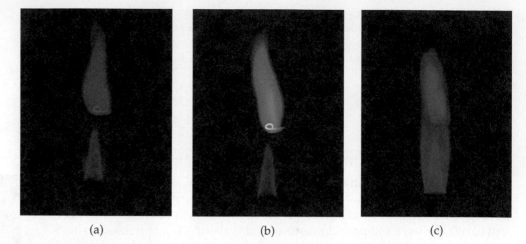

(a)   (b)   (c)

This reaction is most violent for the heavier members of the group, in keeping with their weaker hold on the single valence electron.

The reactions between the alkali metals and oxygen are more complex. When oxygen reacts with metals, metal oxides, which contain the O^{2-} ion, are usually formed. Indeed, lithium shows this reactivity:

$$4\,Li(s) + O_2(g) \longrightarrow 2\,Li_2O(s)$$
$$\text{lithium oxide} \qquad\qquad [7.20]$$

When dissolved in water, Li_2O and other soluble metal oxides react with water to form hydroxide ions from the reaction of O^{2-} ions with H_2O (Equation 7.11).

In contrast, the other alkali metals all react with oxygen to form metal *peroxides*, which contain the O_2^{2-} ion. For example, sodium forms sodium peroxide, Na_2O_2.

$$2\,Na(s) + O_2(g) \longrightarrow Na_2O_2(s)$$
$$\text{sodium peroxide} \qquad\qquad [7.21]$$

Potassium, rubidium, and cesium also form compounds that contain the O_2^- ion, which we call the *superoxide ion*. For example, potassium forms potassium superoxide, KO_2:

$$K(s) + O_2(g) \longrightarrow KO_2(s)$$
$$\text{potassium superoxide} \qquad\qquad [7.22]$$

You should be aware that the reactions shown in Equations 7.21 and 7.22 are somewhat surprising; in most cases, the reaction of oxygen with a metal forms the metal oxide.

As is evident from Equations 7.19 through 7.22, the alkali metals are extremely reactive toward water and oxygen. Because of this, the metals are usually stored submerged in a liquid hydrocarbon, such as mineral oil or kerosene.

Although alkali metal ions are colorless, each emits a characteristic color when placed in a flame (Figure 7.26▲). The ions are reduced to gaseous metal atoms in the central region of the flame. The high temperature of the flame then excites the valence electron to a higher-energy orbital, causing the atom to be in an excited state. The atom then emits energy in the form of visible light as the electron falls back into the lower-energy orbital and the atom returns to its ground state. Sodium, for instance, gives a yellow flame because of emission at 589 nm. This wavelength is produced when the excited valence electron drops from the $3p$ subshell to the lower-energy $3s$ subshell. The characteristic yellow emission of sodium is the basis for sodium vapor lamps (Figure 7.27◀).

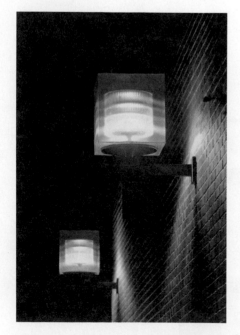

▲ **Figure 7.27 Light from sodium.** Sodium vapor lamps, which are used for commercial and highway lighting, have a yellow glow because of the emission from excited sodium atoms.

⬛ **GIVE IT SOME THOUGHT**

Cesium metal tends to be the most reactive of the stable alkali metals (francium, element number 87, is radioactive and has not been extensively studied). What *atomic* property of Cs is most responsible for its high reactivity?

■■ SAMPLE EXERCISE 7.10 | Reactions of an Alkali Metal

Write a balanced equation that predicts the reaction of cesium metal with (a) $Cl_2(g)$, (b) $H_2O(l)$, (c) $H_2(g)$.

SOLUTION

Analyze and Plan: Cesium is an alkali metal (atomic number 55). We therefore expect that its chemistry will be dominated by oxidation of the metal to Cs^+ ions. Further, we recognize that Cs is far down the periodic table, which means it will be among the most active of all metals and will probably react with all three of the substances listed.

Solve: The reaction between Cs and Cl_2 is a simple combination reaction between two elements, one a metal and the other a nonmetal, forming the ionic compound CsCl:

$$2\,Cs(s) + Cl_2(g) \longrightarrow 2\,CsCl(s)$$

By analogy to Equations 7.19 and 7.17, respectively, we predict the reactions of cesium with water and hydrogen to proceed as follows:

$$2\,Cs(s) + 2\,H_2O(l) \longrightarrow 2\,CsOH(aq) + H_2(g)$$
$$2\,Cs(s) + H_2(g) \longrightarrow 2\,CsH(s)$$

All three of these reactions are redox reactions where cesium forms a Cs^+ ion in the product. The chloride (Cl^-), hydroxide (OH^-), and hydride (H^-) ions are all 1− ions, which means the final products have 1:1 stoichiometry with Cs^+.

■■ PRACTICE EXERCISE

Write a balanced equation for the reaction between potassium metal and elemental sulfur.
Answer: $2\,K(s) + S(s) \longrightarrow K_2S(s)$

Group 2A: The Alkaline Earth Metals

Like the alkali metals, the group 2A elements are all solids at room temperature and have typical metallic properties, some of which are listed in Table 7.5▼. Compared with the alkali metals, the alkaline earth metals are harder and more dense, and they melt at higher temperatures.

The first ionization energies of the alkaline earth elements are low, but they are not as low as those of the alkali metals. Consequently, the alkaline earth metals are less reactive than their alkali metal neighbors. As we noted in Section 7.4, the ease with which the elements lose electrons decreases as we move across the periodic table from left to right and increases as we move down a group. Thus, beryllium and magnesium, the lightest members of the alkaline earth metals, are the least reactive.

The trend of increasing reactivity within the group is shown by the way the alkaline earth metals behave in the presence of water. Beryllium does not react with water or steam, even when heated red-hot. Magnesium reacts slowly with liquid water and more readily with steam to form magnesium oxide and hydrogen:

$$Mg(s) + H_2O(g) \longrightarrow MgO(s) + H_2(g) \qquad [7.23]$$

Calcium and the elements below it react readily with water at room temperature (although more slowly than the alkali metals adjacent to them in the

TABLE 7.5 ■ Some Properties of the Alkaline Earth Metals

Element	Electron Configuration	Melting Point (°C)	Density (g/cm³)	Atomic Radius (Å)	I_1 (kJ/mol)
Beryllium	$[He]2s^2$	1287	1.85	0.90	899
Magnesium	$[Ne]3s^2$	650	1.74	1.30	738
Calcium	$[Ar]4s^2$	842	1.55	1.74	590
Strontium	$[Kr]5s^2$	777	2.63	1.92	549
Barium	$[Xe]6s^2$	727	3.51	1.98	503

Chemistry and Life   THE IMPROBABLE DEVELOPMENT OF LITHIUM DRUGS

The alkali metal ions tend to play a rather unexciting role in most chemical reactions in general chemistry. As noted in Section 4.2, all salts of the alkali metal ions are soluble, and the ions are spectators in most aqueous reactions (except for those involving the alkali metals in their elemental form, such as in Equations 7.17 through 7.22). However, the alkali metal ions play an important role in human physiology. Sodium and potassium ions, for example, are major components of blood plasma and intracellular fluid, respectively, with average concentrations of 0.1 M. These electrolytes serve as vital charge carriers in normal cellular function. We will see in Chapter 20 that these ions are also two of the principal ions involved in regulation of the heart.

In contrast, the lithium ion (Li^+) has no known function in normal human physiology. Since the discovery of lithium in 1817, however, people thought salts of the element possessed almost mystical healing powers. There were even claims that lithium ions were an ingredient in ancient "fountain of youth" formulas. In 1927, Mr. C. L. Grigg began marketing a soft drink that contained lithium. The original unwieldy name of the beverage was "Bib-Label Lithiated Lemon-Lime Soda," which was soon changed to the simpler and more familiar name Seven Up® (Figure 7.28 ▶).

Because of concerns of the Food and Drug Administration, lithium was removed from Seven Up® during the early 1950s. At nearly the same time, psychiatrists discovered that the lithium ion has a remarkable therapeutic effect on the mental disorder called *bipolar affective disorder*, or *manic-depressive illness*. Over 1 million Americans suffer from this psychosis, undergoing severe mood swings from deep depression to a manic euphoria. The lithium ion smooths these mood swings, allowing the bipolar patient to function more effectively in daily life.

The antipsychotic action of Li^+ was discovered by accident during the late 1940s by Australian psychiatrist John Cade as he was researching the use of uric acid—a component of urine—to treat manic-depressive illness. He administered the acid to manic laboratory animals in the form of its most soluble salt, lithium urate, and found that many of the manic symptoms seemed to disappear. Later studies showed that uric acid has no role in the therapeutic effects observed; rather, the seemingly innocuous Li^+ ions were responsible. Because lithium overdose can cause severe side effects in humans, including kidney failure and death, lithium salts were not approved as antipsychotic drugs for humans until 1970. Today Li^+ is usually administered orally in the form of Li_2CO_3, which is the active ingredient in prescription drugs such as Eskalith®. Lithium drugs are effective for about 70% of the bipolar patients who take it.

In this age of sophisticated drug design and biotechnology, the simple lithium ion is still the most effective treatment of this destructive psychological disorder. Remarkably, in spite of intensive research, scientists still do not fully understand the biochemical action of lithium that leads to its therapeutic effects. Because of its similarity to the Na^+ ion, the Li^+ ion is incorporated into blood plasma, where it can affect the behavior of both nerve cells and muscle cells. The Li^+ ion has a smaller radius than the Na^+ ion (Figure 7.8), so its interaction with molecules in cells is somewhat different. Other studies indicate that Li^+ alters the function of certain neurotransmitters, which might lead to its effectiveness as an antipsychotic drug.

▲ **Figure 7.28 Lithium no more.** The soft drink Seven Up® originally contained lithium citrate, the lithium salt of citric acid. The lithium was claimed to give the beverage healthful benefits, including "an abundance of energy, enthusiasm, a clear complexion, lustrous hair, and shining eyes!" The lithium was removed from the beverage in the early 1950s, about the same time that the antipsychotic action of Li^+ was discovered.

▲ **Figure 7.29 Elemental calcium solution.** Calcium metal reacts with water to form hydrogen gas and aqueous calcium hydroxide, $Ca(OH)_2(aq)$.

periodic table), as shown in Figure 7.29 ◀. The reaction between calcium and water, for example, is

$$Ca(s) + 2\,H_2O(l) \longrightarrow Ca(OH)_2(aq) + H_2(g) \qquad [7.24]$$

The reactions represented in Equations 7.23 and 7.24 illustrate the dominant pattern in the reactivity of the alkaline earth elements: they tend to lose their two outer *s* electrons and form 2+ ions. For example, magnesium reacts with chlorine at room temperature to form $MgCl_2$ and burns with dazzling brilliance in air to give MgO (Figure 3.5):

$$Mg(s) + Cl_2(g) \longrightarrow MgCl_2(s) \qquad [7.25]$$

$$2\,Mg(s) + O_2(g) \longrightarrow 2\,MgO(s) \qquad [7.26]$$

In the presence of O_2, magnesium metal is protected from many chemicals by a thin surface coating of water-insoluble MgO. Thus, even though Mg is high in the activity series ∞ (Section 4.4), it can be incorporated into lightweight structural alloys used in, for example, automobile wheels. The heavier alkaline earth metals (Ca, Sr, and Ba) are even more reactive toward nonmetals than is magnesium.

The heavier alkaline earth ions give off characteristic colors when strongly heated in a flame. The colored flame produced by calcium is brick red, that of strontium is crimson red, and that of barium is green. Strontium salts produce the brilliant red color in fireworks, and barium salts produce the green color (Figure 7.30▶).

Both magnesium and calcium are essential for living organisms (Figure 2.24). Calcium is particularly important for growth and maintenance of bones and teeth (Figure 7.31▶). In humans 99% of the calcium is found in the skeletal system.

▲ **Figure 7.30 Fireworks display.** The colors in a fireworks display originate from the characteristic emissions of elements, including the alkaline earths. In this display the crimson color comes from strontium, while the green color comes from barium.

▲ GIVE IT SOME THOUGHT

Calcium carbonate, $CaCO_3$, is often used as a dietary calcium supplement for bone health. Although $CaCO_3(s)$ is insoluble in water (Table 4.1), it can be taken orally to allow for the delivery of $Ca^{2+}(aq)$ ions to the musculoskeletal system. Why is this the case? [*Hint:* Recall the reactions of metal carbonates that were discussed in Section 4.3.]

▲ **Figure 7.31 Calcium in the body.** This X-ray photograph shows the bone structure of the human hand. The primary mineral in bone and teeth is hydroxyapatite, $Ca_5(PO_4)_3OH$, in which calcium is present as Ca^{2+}.

7.8 GROUP TRENDS FOR SELECTED NONMETALS

Hydrogen

Hydrogen, the first element in the periodic table, has a $1s^1$ electron configuration, and for this reason its usual position in the table is above the alkali metals. However, hydrogen does not truly belong to any particular group. Unlike the alkali metals, hydrogen is a nonmetal that occurs as a colorless diatomic gas, $H_2(g)$, under most conditions. Nevertheless, hydrogen can be metallic under tremendous pressures. The interiors of the planets Jupiter and Saturn, for example, are believed to consist of a dense core of rock and ice surrounded by a thick shell of metallic hydrogen. The metallic hydrogen is in turn surrounded by a layer of liquid molecular hydrogen that gradually transforms to a mixture of gaseous hydrogen and helium upon moving closer to the surface. The Cassini–Huygens satellite mission, launched in 1997, has provided a wealth of new information regarding Saturn and its largest moon, Titan, including their chemical compositions. The spacecraft successfully entered orbit around Saturn in June 2004, and in January 2005 the Huygens probe descended through Titan's atmosphere (Figure 7.32▶).

Owing to the complete absence of nuclear shielding of its sole electron, the ionization energy of hydrogen, 1312 kJ/mol, is more than double that of any of the alkali metals (Figure 7.11). In fact, hydrogen's ionization energy is comparable to the I_1 values of other nonmetals, such as oxygen and chlorine. As a result, hydrogen does not lose its valence electron as easily as do the alkali metals. Whereas the alkali metals readily lose their valence electron to nonmetals to form ionic compounds, hydrogen shares its electron with nonmetals and thereby forms molecular compounds. The reactions between hydrogen and nonmetals can be quite exothermic, as evidenced by the combustion reaction between hydrogen and oxygen to form water (Figure 5.13):

▲ **Figure 7.32 Exploring Saturn.** This panoramic view of Saturn was taken from the Cassini spacecraft. From this perspective Saturn shelters the spacecraft from the Sun's blinding glare.

$$2\,H_2(g) + O_2(g) \longrightarrow 2\,H_2O(l) \qquad \Delta H° = -571.7 \text{ kJ} \qquad [7.27]$$

We have also seen (Equation 7.17) that hydrogen reacts with active metals to form solid metal hydrides that contain the hydride ion, H^-. The fact that hydrogen can gain an electron further illustrates that it is not truly a member of the alkali metal family. In fact, in terms of chemical reactivity, hydrogen has more in common with the halogens (Group 7A) than with the alkali metals.

In addition to its ability to form covalent bonds and metal hydrides, probably the most important characteristic of hydrogen is its ability to lose its electron to form a cation. Indeed, the aqueous chemistry of hydrogen is dominated by the $H^+(aq)$ ion, which we first encountered in Chapter 4. ∞ (Section 4.1) We will study this important ion in greater detail in Chapter 16.

Group 6A: The Oxygen Group

As we proceed down group 6A, there is a change from nonmetallic to metallic character (Figure 7.15). Oxygen, sulfur, and selenium are typical nonmetals. Tellurium has some metallic properties and is classified as a metalloid. Polonium, which is radioactive and quite rare, is a metal.

Oxygen is a colorless gas at room temperature; all of the other members of group 6A are solids. Some of the physical properties of the group 6A elements are given in Table 7.6 ▼.

As we saw in Section 2.6, oxygen is encountered in two molecular forms, O_2 and O_3. The O_2 form is the more common one. People generally mean O_2 when they say "oxygen," although the name *dioxygen* is more descriptive. The O_3 form is called **ozone**. The two forms of oxygen are examples of *allotropes*. Allotropes are different forms of the same element in the same state. (In this case both forms are gases.) About 21% of dry air consists of O_2 molecules. Ozone is present in very small amounts in the upper atmosphere and in polluted air. It is also formed from O_2 in electrical discharges, such as in lightning storms:

$$3\,O_2(g) \longrightarrow 2\,O_3(g) \qquad \Delta H° = 284.6 \text{ kJ} \qquad [7.28]$$

This reaction is strongly endothermic, telling us that O_3 is less stable than O_2.

It is interesting to consider the differences in physical and chemical properties of the allotropes of oxygen. Although both O_2 and O_3 are colorless and therefore do not absorb visible light, O_3 absorbs certain wavelengths of ultraviolet light that O_2 does not. Because of this difference, the presence of ozone in the upper atmosphere is beneficial, filtering out harmful UV light ∞ (Section 18.3). Ozone and dioxygen also have different chemical properties. Ozone, which has a pungent odor, is a powerful oxidizing agent. Because of this property ozone is sometimes added to water to kill bacteria or used in low levels to help to purify air. However, the reactivity of ozone also makes its presence in polluted air near Earth's surface detrimental to human health.

Oxygen has a great tendency to attract electrons from other elements (to *oxidize* them). Oxygen in combination with a metal is almost always present as the oxide ion, O^{2-}. This ion has a noble-gas configuration and is particularly stable. As shown in Equation 7.27, the formation of nonmetal oxides is also often very exothermic and thus energetically favorable.

TABLE 7.6 ■ Some Properties of the Group 6A Elements

Element	Electron Configuration	Melting Point (°C)	Density	Atomic Radius (Å)	I_1 (kJ/mol)
Oxygen	$[He]2s^2 2p^4$	−218	1.43 g/L	0.73	1314
Sulfur	$[Ne]3s^2 3p^4$	115	1.96 g/cm^3	1.02	1000
Selenium	$[Ar]3d^{10}\,4s^2\,4p^4$	221	4.82 g/cm^3	1.16	941
Tellurium	$[Kr]4d^{10}\,5s^2\,5p^4$	450	6.24 g/cm^3	1.35	869
Polonium	$[Xe]4f^{14}\,5d^{10}\,6s^2\,6p^4$	254	9.20 g/cm^3	—	812

In our discussion of the alkali metals, we noted two less common oxygen anions—the peroxide (O_2^{2-}) ion and the superoxide (O_2^-) ion. Compounds of these ions often react with themselves to produce an oxide and O_2. For example, aqueous hydrogen peroxide, H_2O_2, slowly decomposes into water and O_2 at room temperature:

$$2 \, H_2O_2(aq) \longrightarrow 2 \, H_2O(l) + O_2(g) \qquad \Delta H° = -196.1 \text{ kJ} \qquad [7.29]$$

For this reason, bottles of aqueous hydrogen peroxide are topped with caps that are able to release the $O_2(g)$ produced before the pressure inside becomes too great (Figure 7.33▶).

After oxygen, the most important member of group 6A is sulfur. This element also exists in several allotropic forms, the most common and stable of which is the yellow solid having the molecular formula S_8. This molecule consists of an eight-membered ring of sulfur atoms, as shown in Figure 7.34▼. Even though solid sulfur consists of S_8 rings, we usually write it simply as $S(s)$ in chemical equations to simplify the stoichiometric coefficients.

Like oxygen, sulfur has a tendency to gain electrons from other elements to form sulfides, which contain the S^{2-} ion. In fact, most sulfur in nature is present as metal sulfides. Sulfur is below oxygen in the periodic table, and the tendency of sulfur to form sulfide anions is not as great as that of oxygen to form oxide ions. As a result, the chemistry of sulfur is more complex than that of oxygen. In fact, sulfur and its compounds (including those in coal and petroleum) can be burned in oxygen. The main product is sulfur dioxide, a major pollutant:

$$S(s) + O_2(g) \longrightarrow SO_2(g) \qquad [7.30]$$

We will discuss the environmental aspects of sulfur oxide chemistry in greater depth in Chapter 18.

▲ **Figure 7.33 Hydrogen peroxide solution.** Bottles of this common antiseptic are topped with a cap that allows any excess pressure created by $O_2(g)$ to be released from the bottle. Hydrogen peroxide is often stored in dark-colored or opaque bottles to minimize exposure to light, which accelerates its decomposition.

◀ **Figure 7.34 Elemental sulfur.** At room temperature, the most common allotropic form of sulfur is an eight-member ring, S_8.

Group 7A: The Halogens

The group 7A elements are known as the **halogens**, after the Greek words *halos* and *gennao*, meaning "salt formers." Some of the properties of these elements are given in Table 7.7▼. Astatine, which is both extremely rare and radioactive, is omitted because many of its properties are not yet known.

Unlike the group 6A elements, all the halogens are typical nonmetals. Their melting and boiling points increase with increasing atomic number. Fluorine and chlorine are gases at room temperature, bromine is a liquid, and iodine is a solid. Each element consists of diatomic molecules: F_2, Cl_2, Br_2, and I_2. Fluorine gas is pale yellow; chlorine gas is yellow-green; bromine liquid is reddish

TABLE 7.7 ■ Some Properties of the Halogens					
Element	Electron Configuration	Melting Point (°C)	Density	Atomic Radius (Å)	I_1 (kJ/mol)
Fluorine	$[He]2s^22p^5$	−220	1.69 g/L	0.71	1681
Chlorine	$[Ne]3s^23p^5$	−102	3.12 g/L	0.99	1251
Bromine	$[Ar]3d^{10}4s^24p^5$	−7.3	3.12 g/cm^3	1.14	1140
Iodine	$[Kr]4d^{10}5s^25p^5$	114	4.94 g/cm^3	1.33	1008

▲ Figure 7.35 Elemental halogens.
All three of these elements—from left to
right, iodine (I_2), bromine (Br_2), and
chlorine (Cl_2)—exist as diatomic
molecules.

brown and readily forms a reddish brown vapor; and solid iodine is grayish
black and readily forms a violet vapor (Figure 7.35 ◄).

The halogens have highly negative electron affinities (Figure 7.14). Thus, it
is not surprising that the chemistry of the halogens is dominated by their tendency to gain electrons from other elements to form halide ions, X^-. (In many
equations X is used to indicate any one of the halogen elements.) Fluorine and
chlorine are more reactive than bromine and iodine. In fact, fluorine removes
electrons from almost any substance with which it comes into contact, including
water, and usually does so very exothermically, as in the following examples:

$$2\,H_2O(l) + 2\,F_2(g) \longrightarrow 4\,HF(aq) + O_2(g) \qquad \Delta H = -758.9\ \text{kJ} \qquad [7.31]$$

$$SiO_2(s) + 2\,F_2(g) \longrightarrow SiF_4(g) + O_2(g) \qquad \Delta H = -704.0\ \text{kJ} \qquad [7.32]$$

As a result, fluorine gas is difficult and dangerous to use in the laboratory,
requiring specialized equipment.

Chlorine is the most industrially useful of the halogens. In 2005, total production was 22 billion pounds, making it the eighth most produced chemical in
the United States. Unlike fluorine, chlorine reacts slowly with water to form relatively stable aqueous solutions of HCl and HOCl (hypochlorous acid):

$$Cl_2(g) + H_2O(l) \longrightarrow HCl(aq) + HOCl(aq) \qquad [7.33]$$

Chlorine is often added to drinking water and swimming pools, where the
HOCl(aq) that is generated serves as a disinfectant.

The halogens react directly with most metals to form ionic halides. The halogens also react with hydrogen to form gaseous hydrogen halide compounds:

$$H_2(g) + X_2 \longrightarrow 2\,HX(g) \qquad [7.34]$$

These compounds are all very soluble in water and dissolve to form the hydrohalic acids. As we discussed in Section 4.3, HCl(aq), HBr(aq), and HI(aq) are
strong acids, whereas HF(aq) is a weak acid.

▲ GIVE IT SOME THOUGHT

Can you use data in Table 7.7 to provide estimates for the atomic radius and first ionization energy of an astatine atom?

Group 8A: The Noble Gases

The group 8A elements, known as the **noble gases**, are all nonmetals that are
gases at room temperature. They are all *monatomic* (that is, they consist of single
atoms rather than molecules). Some physical properties of the noble-gas elements
are listed in Table 7.8 ▼. The high radioactivity of radon (Rn, atomic number 86)
has limited the study of its reaction chemistry and some of its properties.

The noble gases have completely filled s and p subshells. All elements of
group 8A have large first ionization energies, and we see the expected decrease
as we move down the column. Because the noble gases possess such stable electron configurations, they are exceptionally unreactive. In fact, until the early

TABLE 7.8 ■ Some Properties of the Noble Gases

Element	Electron Configuration	Boiling Point (K)	Density (g/L)	Atomic Radius* (Å)	I_1 (kJ/mol)
Helium	$1s^2$	4.2	0.18	0.32	2372
Neon	$[He]2s^22p^6$	27.1	0.90	0.69	2081
Argon	$[Ne]3s^23p^6$	87.3	1.78	0.97	1521
Krypton	$[Ar]3d^{10}4s^24p^6$	120	3.75	1.10	1351
Xenon	$[Kr]4d^{10}5s^25p^6$	165	5.90	1.30	1170
Radon	$[Xe]4f^{14}5d^{10}6s^26p^6$	211	9.73	1.45	1037

*Only the heaviest of the noble-gas elements form chemical compounds. Thus, the atomic radii for the lighter noble-gas elements are estimated values.

1960s the elements were called the *inert gases* because they were thought to be incapable of forming chemical compounds. In 1962, Neil Bartlett at the University of British Columbia reasoned that the ionization energy of Xe might be low enough to allow it to form compounds. In order for this to happen, Xe would have to react with a substance with an extremely high ability to remove electrons from other substances, such as fluorine. Bartlett synthesized the first noble-gas compound by combining Xe with the fluorine-containing compound PtF_6. Xenon also reacts directly with $F_2(g)$ to form the molecular compounds XeF_2, XeF_4, and XeF_6 (Figure 7.36▶). Krypton has a higher I_1 value than xenon and is therefore less reactive. In fact, only a single stable compound of krypton is known, KrF_2. In 2000, Finnish scientists reported the first neutral molecule that contains argon, the HArF molecule, which is stable only at low temperatures.

▲ **Figure 7.36 A compound of xenon.** Crystals of XeF_4, which is one of the very few compounds that contain a group 8A element.

■ **SAMPLE INTEGRATIVE EXERCISE** | **Putting Concepts Together**

The element bismuth (Bi, atomic number 83) is the heaviest member of group 5A. A salt of the element, bismuth subsalicylate, is the active ingredient in Pepto-Bismol®, an over-the-counter medication for gastric distress.

(a) The covalent atomic radii of thallium (Tl) and lead (Pb) are 1.48 Å and 1.47 Å, respectively. Using these values and those in Figure 7.7, predict the covalent atomic radius of the element bismuth (Bi). Explain your answer.
(b) What accounts for the general increase in atomic radius going down the group 5A elements?
(c) Another major use of bismuth has been as an ingredient in low-melting metal alloys, such as those used in fire sprinkler systems and in typesetting. The element itself is a brittle white crystalline solid. How do these characteristics fit with the fact that bismuth is in the same periodic group with such nonmetallic elements as nitrogen and phosphorus?
(d) Bi_2O_3 is a basic oxide. Write a balanced chemical equation for its reaction with dilute nitric acid. If 6.77 g of Bi_2O_3 is dissolved in dilute acidic solution to make 0.500 L of solution, what is the molarity of the solution of Bi^{3+} ion?
(e) ^{209}Bi is the heaviest stable isotope of any element. How many protons and neutrons are present in this nucleus?
(f) The density of Bi at 25 °C is 9.808 g/cm^3. How many Bi atoms are present in a cube of the element that is 5.00 cm on each edge? How many moles of the element are present?

SOLUTION

(a) Note that there is a gradual decrease in radius of the elements in Groups 3A–5A as we proceed across the fifth period, that is, in the series In–Sn–Sb. Therefore, it is reasonable to expect a decrease of about 0.02 Å as we move from Pb to Bi, leading to an estimate of 1.45 Å. The tabulated value is 1.46 Å.
(b) The general increase in radius with increasing atomic number in the group 5A elements occurs because additional shells of electrons are being added, with corresponding increases in nuclear charge. The core electrons in each case largely shield the outermost electrons from the nucleus, so the effective nuclear charge does not vary greatly as we go to higher atomic numbers. However, the principal quantum number, n, of the outermost electrons steadily increases, with a corresponding increase in orbital radius.
(c) The contrast between the properties of bismuth and those of nitrogen and phosphorus illustrates the general rule that there is a trend toward increased metallic character as we move down in a given group. Bismuth, in fact, is a metal. The increased metallic character occurs because the outermost electrons are more readily lost in bonding, a trend that is consistent with its lower ionization energy.
(d) Following the procedures described in Section 4.2 for writing molecular and net ionic equations, we have the following:

Molecular equation: $Bi_2O_3(s) + 6\,HNO_3(aq) \longrightarrow 2\,Bi(NO_3)_3(aq) + 3\,H_2O(l)$

Net ionic equation: $Bi_2O_3(s) + 6\,H^+(aq) \longrightarrow 2\,Bi^{3+}(aq) + 3\,H_2O(l)$

In the net ionic equation, nitric acid is a strong acid and $Bi(NO_3)_3$ is a soluble salt, so we need only show the reaction of the solid with the hydrogen ion forming the $Bi^{3+}(aq)$ ion and water.
To calculate the concentration of the solution, we proceed as follows (Section 4.5):

$$\frac{6.77\ \text{g}\ Bi_2O_3}{0.500\ \text{L soln}} \times \frac{1\ \text{mol}\ Bi_2O_3}{466.0\ \text{g}\ Bi_2O_3} \times \frac{2\ \text{mol}\ Bi^{3+}}{1\ \text{mol}\ Bi_2O_3} = \frac{0.0581\ \text{mol}\ Bi^{3+}}{\text{L soln}} = 0.0581\ M$$

(e) We can proceed as in Section 2.3. Bismuth is element 83; there are therefore 83 protons in the nucleus. Because the atomic mass number is 209, there are $209 - 83 = 126$ neutrons in the nucleus.

(f) We proceed as in Sections 1.4 and 3.4: the volume of the cube is $(5.00)^3$ cm^3 = 125 cm^3. Then we have

$$125 \text{ cm}^3 \text{ Bi} \times \frac{9.808 \text{ g Bi}}{1 \text{ cm}^3} \times \frac{1 \text{ mol Bi}}{209.0 \text{ g Bi}} = 5.87 \text{ mol Bi}$$

$$5.87 \text{ mol Bi} \times \frac{6.022 \times 10^{23} \text{ atom Bi}}{1 \text{ mol Bi}} = 3.54 \times 10^{24} \text{ atoms Bi}$$

CHAPTER REVIEW

SUMMARY AND KEY TERMS

Introduction and Section 7.1 The periodic table was first developed by Mendeleev and Meyer on the basis of the similarity in chemical and physical properties exhibited by certain elements. Moseley established that each element has a unique atomic number, which added more order to the periodic table. We now recognize that elements in the same column of the periodic table have the same number of electrons in their **valence orbitals**. This similarity in valence electronic structure leads to the similarities among elements in the same group. The differences among elements in the same group arise because their valence orbitals are in different shells.

Section 7.2 Many properties of atoms are due to the average distance of the outer electrons from the nucleus and to the **effective nuclear charge** experienced by these electrons. The core electrons are very effective in screening the outer electrons from the full charge of the nucleus, whereas electrons in the same shell do not screen each other effectively. As a result, the effective nuclear charge experienced by valence electrons increases as we move left to right across a period.

Section 7.3 The size of an atom can be gauged by its **bonding atomic radius**, based on measurements of the distances separating atoms in their chemical compounds. In general, atomic radii increase as we go down a column in the periodic table and decrease as we proceed left to right across a row.

Cations are smaller than their parent atoms; anions are larger than their parent atoms. For ions of the same charge, size increases going down a column of the periodic table. An **isoelectronic series** is a series of ions that has the same number of electrons. For such a series, size decreases with increasing nuclear charge as the electrons are attracted more strongly to the nucleus.

Section 7.4 The first **ionization energy** of an atom is the minimum energy needed to remove an electron from the atom in the gas phase, forming a cation. The second ionization energy is the energy needed to remove a second electron, and so forth. Ionization energies show a sharp increase after all the valence electrons have been removed, because of the much higher effective nuclear charge experienced by the core electrons. The first ionization energies of the elements show periodic trends that are opposite those seen for atomic radii, with smaller atoms having higher first ionization energies. Thus, first ionization energies decrease as we go down a column and increase as we proceed left to right across a row.

We can write electron configurations for ions by first writing the electron configuration of the neutral atom and then removing or adding the appropriate number of electrons. Electrons are removed first from the orbitals with the largest value of n. If there are two valence orbitals with the same value of n (such as $4s$ and $4p$), then the electrons are lost first from the orbital with a higher value of l (in this case, $4p$). Electrons are added to orbitals in the reverse order.

Section 7.5 The **electron affinity** of an element is the energy change upon adding an electron to an atom in the gas phase, forming an anion. A negative electron affinity means that the anion is stable; a positive electron affinity means that the anion is not stable relative to the separated atom and electron, in which case its exact value cannot be measured. In general, electron affinities become more negative as we proceed from left to right across the periodic table. The halogens have the most-negative electron affinities. The electron affinities of the noble gases are positive because the added electron would have to occupy a new, higher-energy subshell.

Section 7.6 The elements can be categorized as metals, nonmetals, and metalloids. Most elements are metals; they occupy the left side and the middle of the periodic table. Nonmetals appear in the upper-right section of the table. Metalloids occupy a narrow band between the metals and nonmetals. The tendency of an element to exhibit the properties of metals, called the **metallic character**, increases as we proceed down a column and decreases as we proceed from left to right across a row.

Metals have a characteristic luster, and they are good conductors of heat and electricity. When metals react with nonmetals, the metal atoms are oxidized to cations and ionic substances are generally formed. Most metal oxides are basic; they react with acids to form salts and water.

Nonmetals lack metallic luster and are generally poor conductors of heat and electricity. Several are gases at room temperature. Compounds composed entirely of nonmetals are generally molecular. Nonmetals usually form anions in their reactions with metals. Nonmetal oxides are acidic; they react with bases to form salts and water. Metalloids have properties that are intermediate between those of metals and nonmetals.

Section 7.7 The periodic properties of the elements can help us understand the properties of groups of the representative elements. The **alkali metals** (group 1A) are soft metals with low densities and low melting points. They have the lowest ionization energies of the elements. As a result, they are very reactive toward nonmetals, easily losing their outer s electron to form 1+ ions. The **alkaline earth metals** (group 2A) are harder and more dense and have higher melting points than the alkali metals. They are also very reactive toward nonmetals, although not as reactive as the alkali metals. The alkaline earth metals readily lose their two outer s electrons to form 2+ ions. Both alkali and alkaline earth metals react with hydrogen to form ionic substances that contain the **hydride ion**, H^-.

Section 7.8 Hydrogen is a nonmetal with properties that are distinct from any of the groups of the periodic table. It forms molecular compounds with other nonmetals, such as oxygen and the halogens.

Oxygen and sulfur are the most important elements in group 6A. Oxygen is usually found as a diatomic molecule, O_2. **Ozone**, O_3, is an important allotrope of oxygen. Oxygen has a strong tendency to gain electrons from other elements, thus oxidizing them. In combination with metals, oxygen is usually found as the oxide ion, O^{2-}, although salts of the peroxide ion, O_2^{2-}, and superoxide ion, O_2^-, are sometimes formed. Elemental sulfur is most commonly found as S_8 molecules. In combination with metals, it is most often found as the sulfide ion, S^{2-}.

The **halogens** (group 7A) are nonmetals that exist as diatomic molecules. The halogens have the most negative electron affinities of the elements. Thus their chemistry is dominated by a tendency to form 1− ions, especially in reactions with metals.

The **noble gases** (group 8A) are nonmetals that exist as monatomic gases. They are very unreactive because they have completely filled s and p subshells. Only the heaviest noble gases are known to form compounds, and they do so only with very active nonmetals, such as fluorine.

KEY SKILLS

- Understand the meaning of effective nuclear charge, Z_{eff}, and how Z_{eff} depends upon nuclear charge and electron configuration.
- Use the periodic table to predict the trends in atomic radii, ionic radii, ionization energy, and electron affinity.
- Understand how the radius of an atom changes upon losing electrons to form a cation or gaining electrons to form an anion.
- Understand how the ionization energy changes as we remove successive electrons. Recognize the jump in ionization energy that occurs when the ionization corresponds to removing a core electron.
- Be able to write the electron configurations of ions.
- Understand how irregularities in the periodic trends for electron affinity can be related to electron configuration.
- Recognize the differences in chemical and physical properties of metals and nonmetals, including the basicity of metal oxides and the acidity of nonmetal oxides.
- Understand how the atomic properties, such as ionization energy and electron configuration, are related to the chemical reactivity and physical properties of the alkali and alkaline earth metals (groups 1A and 2A).
- Be able to write balanced equations for the reactions of the group 1A and 2A metals with water, oxygen, hydrogen, and the halogens.
- Understand and recognize the unique characteristics of hydrogen.
- Understand how the atomic properties (such as ionization energy, electron configuration, and electron affinity) of group 6A, 7A, and 8A elements are related to their chemical reactivity and physical properties.

KEY EQUATIONS

- $Z_{eff} = Z - S$ [7.1]   Estimating effective nuclear charge

VISUALIZING CONCEPTS

7.1 We can draw an analogy between the attraction of an electron to a nucleus and seeing a lightbulb—in essence, the more nuclear charge the electron "sees," the greater the attraction. **(a)** Within this analogy, discuss how the shielding by core electrons is analogous to putting a frosted-glass lampshade between the lightbulb and your eyes, as shown in the illustration. **(b)** Explain how we could mimic moving to the right in a row of the periodic table by changing the wattage of the lightbulb. **(c)** How would you change the wattage of the bulb and/or the frosted glass to mimic the effect of moving down a column of the periodic table? [Section 7.2]

Observer

Light bulb   Frosted glass

7.2 Fluorine has atomic number 9. If we represent the radius of a fluorine atom with the billiard ball illustrated here, would the analogy be more appropriate for the bonding or nonbonding atomic radius? If we used the same billiard ball to illustrate the concept of fluorine's bonding atomic radius, would we overestimate or underestimate the bonding atomic radius? Explain. [Section 7.3]

7.3 Consider the A_2X_4 molecule depicted below, where A and X are elements. The A—A bond length in this molecule is d_1, and the four A—X bond lengths are each d_2. **(a)** In terms of d_1 and d_2, how could you define the bonding atomic radii of atoms A and X? **(b)** In terms of d_1 and d_2, what would you predict for the X—X bond length of an X_2 molecule? [Section 7.3]

7.4 Make a simple sketch of the shape of the main part of the periodic table, as shown. **(a)** Ignoring H and He, write a single straight arrow from the element with the smallest bonding atomic radius to the element with the largest. **(b)** Ignoring H and He, write a single straight arrow from the element with the smallest first ionization energy to the element with the largest. **(c)** What significant observation can you make from the arrows you drew in parts (a) and (b)? [Sections 7.3 and 7.4]

7.5 In the chemical process called *electron transfer*, an electron is transferred from one atom or molecule to another (We will talk about electron transfer extensively in Chapter 20.) A simple electron transfer reaction is

$$A(g) + A(g) \longrightarrow A^+(g) + A^-(g)$$

In terms of the ionization energy and electron affinity of atom A, what is the energy change for this reaction? For a representative nonmetal such as chlorine, is this process exothermic? For a representative metal such as sodium, is this process exothermic? [Sections 7.4 and 7.5]

7.6 An element X reacts with $F_2(g)$ to form the molecular product shown below. **(a)** Write a balanced equation for this reaction (do not worry about the phases for X and the product). **(b)** Do you think that X is a metal or nonmetal? Explain. [Section 7.6]

EXERCISES

Periodic Table; Effective Nuclear Charge

7.7 Why did Mendeleev leave blanks in his early version of the periodic table? How did he predict the properties of the elements that belonged in those blanks?

7.8 The prefix *eka-* comes from the Sanskrit word for one. Mendeleev used this prefix to indicate that the unknown element was one place away from the known element that followed the prefix. For example, *eka-silicon*, which we now call germanium, is one element below silicon. Mendeleev also predicted the existence of *eka-manganese*, which was not experimentally confirmed until 1937 because this element is radioactive and does not occur in nature. Based on the periodic table shown in Figure 7.2, what do we now call the element Mendeleev called *eka-manganese*?

7.9 In Chapter 1 we learned that silicon is the second most abundant element in Earth's crust, accounting for more than one-fourth of the mass of the crust (Figure 1.6). Yet we see that silicon is not among the elements that have been known since ancient times (Figure 7.2), whereas iron, which accounts for less than 5% of Earth's crust, has been known since prehistoric times. Given silicon's abundance how do you account for its relatively late discovery?

7.10 (a) During the period from about 1800 to about 1865, the atomic weights of many elements were accurately measured. Why was this important to Mendeleev's formulation of the periodic table? (b) What property of the atom did Moseley associate with the wavelength of X-rays emitted from an element in his experiments? (c) Why are chemical and physical properties of the elements more closely related to atomic number than they are to atomic weight?

7.11 (a) What is meant by the term *effective nuclear charge*? (b) How does the effective nuclear charge experienced by the valence electrons of an atom vary going from left to right across a period of the periodic table?

7.12 (a) How is the concept of effective nuclear charge used to simplify the numerous electron-electron repulsions in a many-electron atom? (b) Which experiences a greater effective nuclear charge in a Be atom, the $1s$ electrons or the $2s$ electrons? Explain.

7.13 Detailed calculations show that the value of Z_{eff} for Na and K atoms is 2.51+ and 3.49+, respectively. (a) What value do you estimate for Z_{eff} experienced by the outermost electron in both Na and K by assuming core electrons contribute 1.00 and valence electrons contribute 0.00 to the screening constant? (b) What values do you estimate for Z_{eff} using Slater's rules? (c) Which approach gives a more accurate estimate of Z_{eff}? (d) Does either method of approximation account for the gradual increase in Z_{eff} that occurs upon moving down a group?

7.14 Detailed calculations show that the value of Z_{eff} for Si and Cl atoms is 4.29+ and 6.12+, respectively. (a) What value do you estimate for Z_{eff} experienced by the outermost electron in both Si and Cl by assuming core electrons contribute 1.00 and valence electrons contribute 0.00 to the screening constant? (b) What values do you estimate for Z_{eff} using Slater's rules? (c) Which approach gives a more accurate estimate of Z_{eff}? (d) Which method of approximation more accurately accounts for the steady increase in Z_{eff} that occurs upon moving left to right across a period?

7.15 Which will experience the greater effective nuclear charge, the electrons in the $n = 3$ shell in Ar or the $n = 3$ shell in Kr? Which will be closer to the nucleus? Explain.

7.16 Arrange the following atoms in order of increasing effective nuclear charge experienced by the electrons in the $n = 3$ electron shell: K, Mg, P, Rh, and Ti. Explain the basis for your order.

Atomic and Ionic Radii

7.17 (a) Because an exact outer boundary cannot be measured or even calculated for an atom, how are atomic radii determined? (b) What is the difference between a bonding radius and a nonbonding radius? (c) For a given element, which one is larger?

7.18 (a) Why does the quantum mechanical description of many-electron atoms make it difficult to define a precise atomic radius? (b) When nonbonded atoms come up against one another, what determines how closely the nuclear centers can approach?

7.19 The distance between W atoms in tungsten metal is 2.74 Å. What is the atomic radius of a tungsten atom in this environment? (This radius is called the *metallic radius*.)

7.20 Based on the radii presented in Figure 7.7, predict the distance between Si atoms in solid silicon.

7.21 Estimate the As—I bond length from the data in Figure 7.7, and compare your value to the experimental As—I bond length in arsenic triiodide, AsI_3, 2.55 Å.

7.22 The experimental Bi—I bond length in bismuth triiodide, BiI_3, is 2.81 Å. Based on this value and data in Figure 7.7, predict the atomic radius of Bi.

7.23 How do the sizes of atoms change as we move (a) from left to right across a row in the periodic table, (b) from top to bottom in a group in the periodic table? (c) Arrange the following atoms in order of increasing atomic radius: F, P, S, As.

7.24 **(a)** Among the nonmetallic elements, the change in atomic radius in moving one place left or right in a row is smaller than the change in moving one row up or down. Explain these observations. **(b)** Arrange the following atoms in order of increasing atomic radius: Si, Al, Ge, Ga.

7.25 Using only the periodic table, arrange each set of atoms in order of increasing radius: **(a)** Ca, Mg, Be; **(b)** Ga, Br, Ge; **(c)** Al, Tl, Si.

7.26 Using only the periodic table, arrange each set of atoms in order of increasing radius: **(a)** Ba, Ca, Na; **(b)** Sn, Sb, As; **(c)** Al, Be, Si.

7.27 **(a)** Why are monatomic cations smaller than their corresponding neutral atoms? **(b)** Why are monatomic anions larger than their corresponding neutral atoms? **(c)** Why does the size of ions increase as one proceeds down a column in the periodic table?

7.28 Explain the following variations in atomic or ionic radii: **(a)** $I^- > I > I^+$, **(b)** $Ca^{2+} > Mg^{2+} > Be^{2+}$, **(c)** $Fe > Fe^{2+} > Fe^{3+}$.

7.29 Consider a reaction represented by the following spheres:

Reactants   Products

Which sphere represents a metal and which a nonmetal? Explain.

7.30 Consider the following spheres:

Which one represents Ca, which Ca^{2+}, and which Mg^{2+}?

7.31 **(a)** What is an isoelectronic series? **(b)** Which neutral atom is isoelectronic with each of the following ions: Al^{3+}, Ti^{4+}, Br^-, Sn^{2+}?

7.32 Some ions do not have a corresponding neutral atom that has the same electron configuration. For each of the following ions identify the neutral atom that has the same number of electrons and determine if this atom has the same electron configuration. If such an atom does not exist explain why: **(a)** Cl^-, **(b)** Sc^{3+}, **(c)** Fe^{2+}, **(d)** Zn^{2+}, **(e)** Sn^{4+}.

7.33 Consider the isoelectronic ions F^- and Na^+. **(a)** Which ion is smaller? **(b)** Using Equation 7.1 and assuming that core electrons contribute 1.00 and valence electrons contribute 0.00 to the screening constant, S, calculate Z_{eff} for the 2p electrons in both ions. **(c)** Repeat this calculation using Slater's rules to estimate the screening constant, S. **(d)** For isoelectronic ions, how are effective nuclear charge and ionic radius related?

7.34 Consider the isoelectronic ions Cl^- and K^+. **(a)** Which ion is smaller? **(b)** Use Equation 7.1 and assuming that core electrons contribute 1.00 and valence electrons contribute nothing to the screening constant, S, calculate Z_{eff} for these two ions. **(c)** Repeat this calculation using Slater's rules to estimate the screening constant, S. **(d)** For isoelectronic ions how are effective nuclear charge and ionic radius related?

7.35 Consider S, Cl, and K and their most common ions. **(a)** List the atoms in order of increasing size. **(b)** List the ions in order of increasing size. **(c)** Explain any differences in the orders of the atomic and ionic sizes.

7.36 For each of the following sets of atoms and ions, arrange the members in order of increasing size: **(a)** Se^{2-}, Te^{2-}, Se; **(b)** Co^{3+}, Fe^{2+}, Fe^{3+}; **(c)** Ca, Ti^{4+}, Sc^{3+}; **(d)** Be^{2+}, Na^+, Ne.

7.37 For each of the following statements, provide an explanation: **(a)** O^{2-} is larger than O; **(b)** S^{2-} is larger than O^{2-}; **(c)** S^{2-} is larger than K^+; **(d)** K^+ is larger than Ca^{2+}.

7.38 In the ionic compounds LiF, NaCl, KBr, and RbI, the measured cation–anion distances are 2.01 Å (Li–F), 2.82 Å (Na–Cl), 3.30 Å (K–Br), and 3.67 Å (Rb–I), respectively. **(a)** Predict the cation–anion distance using the values of ionic radii given in Figure 7.8. **(b)** Is the agreement between the prediction and the experiment perfect? If not, why not? **(c)** What estimates of the cation–anion distance would you obtain for these four compounds using *bonding atomic radii*? Are these estimates as accurate as the estimates using ionic radii?

Ionization Energies; Electron Affinities

7.39 Write equations that show the processes that describe the first, second, and third ionization energies of a boron atom.

7.40 Write equations that show the process for **(a)** the first two ionization energies of tin and **(b)** the fourth ionization energy of titanium.

7.41 **(a)** Why are ionization energies always positive quantities? **(b)** Why does F have a larger first ionization energy than O? **(c)** Why is the second ionization energy of an atom always greater than its first ionization energy?

7.42 **(a)** Why does Li have a larger first ionization energy than Na? **(b)** The difference between the third and fourth ionization energies of scandium is much larger than the difference between the third and fourth ionization energies of titanium. Why? **(c)** Why does Li have a much larger second ionization energy than Be?

7.43 **(a)** What is the general relationship between the size of an atom and its first ionization energy? **(b)** Which element in the periodic table has the largest ionization energy? Which has the smallest?

7.44 **(a)** What is the trend in first ionization energies as one proceeds down the group 7A elements? Explain how this trend relates to the variation in atomic radii. **(b)** What is the trend in first ionization energies as one moves across the fourth period from K to Kr? How does this trend compare with the trend in atomic radii?

7.45 Based on their positions in the periodic table, predict which atom of the following pairs will have the larger first ionization energy: **(a)** Cl, Ar; **(b)** Be, Ca; **(c)** K, Co; **(d)** S, Ge; **(e)** Sn, Te.

7.46 For each of the following pairs, indicate which element has the larger first ionization energy: **(a)** Ti, Ba; **(b)** Ag, Cu; **(c)** Ge, Cl; **(d)** Pb, Sb. (In each case use electron configuration and effective nuclear charge to explain your answer.)

7.47 Write the electron configurations for the following ions: **(a)** In^{3+}, **(b)** Sb^{3+}, **(c)** Te^{2-}, **(d)** Te^{6+}, **(e)** Hg^{2+}, **(f)** Rh^{3+}.

7.48 Write electron configurations for the following ions, and determine which have noble-gas configurations: **(a)** Cr^{3+}, **(b)** N^{3-}, **(c)** Sc^{3+}, **(d)** Cu^{2+}, **(e)** Tl^+, **(f)** Au^+.

7.49 Write the electron configuration for **(a)** the Ni^{2+} ion and **(b)** the Sn^{2+} ion. How many unpaired electrons does each contain?

7.50 Identify the element whose ions have the following electron configurations: **(a)** a 2+ ion with $[Ar]3d^9$, **(b)** a 1+ ion with $[Xe]4f^{14}5d^{10}6s^2$. How many unpaired electrons does each ion contain?

7.51 The first ionization energy of Ar and the electron affinity of Ar are both positive values. What is the significance of the positive value in each case?

7.52 The electron affinity of lithium is a negative value, whereas the electron affinity of beryllium is a positive value. Use electron configurations to account for this observation.

7.53 While the electron affinity of bromine is a negative quantity, it is positive for Kr. Use the electron configurations of the two elements to explain the difference.

7.54 What is the relationship between the ionization energy of an anion with a 1− charge such as F^- and the electron affinity of the neutral atom, F?

7.55 Consider the first ionization energy of neon and the electron affinity of fluorine. **(a)** Write equations, including electron configurations, for each process. **(b)** These two quantities will have opposite signs. Which will be positive, and which will be negative? **(c)** Would you expect the **magnitudes** of these two quantities to be equal? If not, which one would you expect to be larger? Explain your answer.

7.56 Write an equation for the process that corresponds to the electron affinity of the Mg^+ ion. Also write the electron configurations of the species involved. What is the magnitude of the energy change in the process? [*Hint:* The answer is in Table 7.2.]

Properties of Metals and Nonmetals

7.57 How are metallic character and first ionization energy related?

7.58 Arrange the following pure solid elements in order of increasing electrical conductivity: Ge, Ca, S, and Si. Explain the reasoning you used.

7.59 If we look at groups 3A through 5A, we see two metalloids for groups 4A (Si, Ge) and 5A (As, Sb), but only one metalloid in group 3A (B). To maintain a regular geometric pattern one might expect that aluminum would also be a metalloid, giving group 3A two metalloids. What can you say about the metallic character of aluminum with respect to its neighbors based on its first ionization energy?

7.60 For each of the following pairs, which element will have the greater metallic character: **(a)** Li or Be, **(b)** Li or Na, **(c)** Sn or P, **(d)** Al or B?

7.61 Predict whether each of the following oxides is ionic or molecular: SO_2, MgO, Li_2O, P_2O_5, Y_2O_3, N_2O, and XeO_3. Explain the reasons for your choices.

7.62 Some metal oxides, such as Sc_2O_3, do not react with pure water, but they do react when the solution becomes either acidic or basic. Do you expect Sc_2O_3 to react when the solution becomes acidic or when it becomes basic? Write a balanced chemical equation to support your answer.

7.63 **(a)** What is meant by the terms acidic oxide and basic oxide? **(b)** How can we predict whether an oxide will be acidic or basic, based on its composition?

7.64 Arrange the following oxides in order of increasing acidity: CO_2, CaO, Al_2O_3, SO_3, SiO_2, and P_2O_5.

7.65 Chlorine reacts with oxygen to form Cl_2O_7. **(a)** What is the name of this product (see Table 2.6)? **(b)** Write a balanced equation for the formation of $Cl_2O_7(l)$ from the elements. **(c)** Under usual conditions, Cl_2O_7 is a colorless liquid with a boiling point of 81 °C. Is this boiling point expected or surprising? **(d)** Would you expect Cl_2O_7 to be more reactive toward $H^+(aq)$ or $OH^-(aq)$? Explain.

[7.66] An element X reacts with oxygen to form XO_2 and with chlorine to form XCl_4. XO_2 is a white solid that melts at high temperatures (above 1000 °C). Under usual conditions, XCl_4 is a colorless liquid with a boiling point of 58 °C. **(a)** XCl_4 reacts with water to form XO_2 and another product. What is the likely identity of the other product? **(b)** Do you think that element X is a metal, nonmetal, or metalloid? Explain. **(c)** By using a sourcebook such as the *CRC Handbook of Chemistry and Physics*, try to determine the identity of element X.

7.67 Write balanced equations for the following reactions: **(a)** barium oxide with water, **(b)** iron(II) oxide with perchloric acid, **(c)** sulfur trioxide with water, **(d)** carbon dioxide with aqueous sodium hydroxide.

7.68 Write balanced equations for the following reactions: **(a)** potassium oxide with water, **(b)** diphosphorus trioxide with water, **(c)** chromium(III) oxide with dilute hydrochloric acid, **(d)** selenium dioxide with aqueous potassium hydroxide.

Group Trends in Metals and Nonmetals

7.69 Compare the elements sodium and magnesium with respect to the following properties: **(a)** electron configuration, **(b)** most common ionic charge, **(c)** first ionization energy, **(d)** reactivity toward water, **(e)** atomic radius. Account for the differences between the two elements.

7.70 (a) Compare the electron configurations and atomic radii (see Figure 7.7) of rubidium and silver. In what respects are their electronic configurations similar? Account for the difference in radii of the two elements. **(b)** As with rubidium, silver is most commonly found as the 1+ ion, Ag^+. However, silver is far less reactive. Explain these observations.

7.71 (a) Why is calcium generally more reactive than magnesium? **(b)** Why is calcium generally less reactive than potassium?

7.72 (a) Why is cesium more reactive toward water than is lithium? **(b)** One of the alkali metals reacts with oxygen to form a solid white substance. When this substance is dissolved in water, the solution gives a positive test for hydrogen peroxide, H_2O_2. When the solution is tested in a burner flame, a lilac-purple flame is produced. What is the likely identity of the metal? **(c)** Write a balanced chemical equation for reaction of the white substance with water.

7.73 Write a balanced equation for the reaction that occurs in each of the following cases: **(a)** Potassium metal burns in an atmosphere of chlorine gas. **(b)** Strontium oxide is added to water. **(c)** A fresh surface of lithium metal is exposed to oxygen gas. **(d)** Sodium metal is reacted with molten sulfur.

7.74 Write a balanced equation for the reaction that occurs in each of the following cases: **(a)** Cesium is added to water. **(b)** Stontium is added to water. **(c)** Sodium reacts with oxygen. **(d)** Calcium reacts with iodine.

7.75 (a) If we arrange the elements of the second period (Li–Ne) in order of increasing first ionization energy, where would hydrogen fit into this series? **(b)** If we now arrange the elements of the third period (Na–Ar) in order of increasing first ionization energy, where would lithium fit into this series? **(c)** Are these series consistent with the assignment of hydrogen as a nonmetal and lithium as a metal?

7.76 (a) As described in Section 7.7, the alkali metals react with hydrogen to form hydrides and react with halogens—for example, fluorine—to form halides. Compare the roles of hydrogen and the halogen in these reactions. In what sense are the forms of hydrogen and halogen in the products alike? **(b)** Write balanced equations for the reaction of fluorine with calcium and for the reaction of hydrogen with calcium. What are the similarities among the products of these reactions?

7.77 Compare the elements fluorine and chlorine with respect to the following properties: **(a)** electron configuration, **(b)** most common ionic charge, **(c)** first ionization energy, **(d)** reactivity toward water, **(e)** electron affinity, **(f)** atomic radius. Account for the differences between the two elements.

7.78 Little is known about the properties of astatine, At, because of its rarity and high radioactivity. Nevertheless, it is possible for us to make many predictions about its properties. **(a)** Do you expect the element to be a gas, liquid, or solid at room temperature? Explain. **(b)** What is the chemical formula of the compound it forms with Na?

7.79 Until the early 1960s the group 8A elements were called the inert gases; before that they were called the rare gases. The term rare gases was dropped after it was discovered that argon accounts for roughly 1% of Earth's atmosphere. **(a)** Why was the term inert gases dropped? **(b)** What discovery triggered this change in name? **(c)** What name is applied to the group now?

7.80 Why does xenon react with fluorine, whereas neon does not?

7.81 Write a balanced equation for the reaction that occurs in each of the following cases: **(a)** Ozone decomposes to dioxygen. **(b)** Xenon reacts with fluorine. (Write three different equations.) **(c)** Sulfur reacts with hydrogen gas. **(d)** Fluorine reacts with water.

7.82 Write a balanced equation for the reaction that occurs in each of the following cases: **(a)** Chlorine reacts with water. **(b)** Barium metal is heated in an atmosphere of hydrogen gas. **(c)** Lithium reacts with sulfur. **(d)** Fluorine reacts with magnesium metal.

ADDITIONAL EXERCISES

7.83 Consider the stable elements through lead ($Z = 82$). In how many instances are the atomic weights of the elements in the reverse order relative to the atomic numbers of the elements? What is the explanation for these cases?

7.84 (a) Which will have the lower energy, a $4s$ or a $4p$ electron in an As atom? **(b)** How can we use the concept of effective nuclear charge to explain your answer to part (a)?

7.85 (a) If the core electrons were totally effective at shielding the valence electrons and the valence electrons provided no shielding for each other, what would be the effective nuclear charge acting on the $3s$ and $3p$ valence electrons in P? **(b)** Repeat these calculations using Slater's rules. **(c)** Detailed calculations indicate that the effective nuclear charge is 5.6+ for the $3s$ electrons and 4.9+ for the $3p$ electrons. Why are the values for the $3s$ and $3p$ electrons different? **(d)** If you remove a single electron from a P atom, which orbital will it come from? Explain.

7.86 Nearly all the mass of an atom is in the nucleus, which has a very small radius. When atoms bond together (for example, two fluorine atoms in F_2), why is the distance separating the nuclei so much larger than the radii of the nuclei?

[7.87] Consider the change in effective nuclear charge experienced by a $2p$ electron as we proceed from C to N. **(a)** Based on a simple model in which core electrons screen the valence electrons completely and valence electrons do not screen other valence electrons, what do you predict for the change in Z_{eff} from C to N? **(b)** What change do you predict using Slater's rules? **(c)** The actual change in Z_{eff} from C to N is 0.70+. Which approach to estimating Z_{eff} is more accurate? **(d)** The change in Z_{eff} from N to O is smaller than that from C to N. Can you provide an explanation for this observation?

7.88 As we move across a period of the periodic table, why do the sizes of the transition elements change more gradually than those of the representative elements?

7.89 In the series of group 5A hydrides, of general formula MH_3, the measured bond distances are P—H, 1.419 Å; As—H, 1.519 Å; Sb—H, 1.707 Å. **(a)** Compare these values with those estimated by use of the atomic radii in Figure 7.7. **(b)** Explain the steady increase in M—H bond distance in this series in terms of the electronic configurations of the M atoms.

7.90 It is possible to produce compounds of the form $GeClH_3$, $GeCl_2H_2$, and $GeCl_3H$. What values do you predict for the Ge—H and Ge—Cl bond lengths in these compounds?

7.91 Note from the following table that the increase in atomic radius in moving from Zr to Hf is smaller than in moving from Y to La. Suggest an explanation for this effect.

Atomic Radii (Å)

Sc	1.44	Ti	1.36
Y	1.62	Zr	1.48
La	1.69	Hf	1.50

7.92 The "Chemistry and Life" box on ionic size in Section 7.3 compares the ionic radii of Zn^{2+} and Cd^{2+} **(a)** The 2+ ion of which other element seems the most obvious one to compare to Zn^{2+} and Cd^{2+}? **(b)** With reference to Figure 2.24, is the element in part (a) essential for life? **(c)** Estimate the ionic radius of the 2+ ion of the element in part (a). Explain any assumptions you have made. **(d)** Would you expect the 2+ ion of the element in part (a) to be physiologically more similar to Zn^{2+} or to Cd^{2+}? **(e)** Use a sourcebook or a Web search to determine whether the element in part (a) is toxic to humans.

[7.93] The ionic substance strontium oxide, SrO, forms from the direct reaction of strontium metal with molecular oxygen. The arrangement of the ions in solid SrO is analogous to that in solid NaCl (see Figure 2.23) and is shown here. **(a)** Write a balanced equation for the formation of SrO(s) from the elements. **(b)** Based on the ionic radii in Figure 7.8, predict the length of the side of the cube in the figure (the distance from the center of an atom at one corner to the center of an atom at a neighboring corner). **(c)** The experimental density of SrO is 5.10 g/cm^3. Given your answer to part (b), what is the number of formula units of SrO that are contained in the cube in the figure? (We will examine structures like those in the figure more closely in Chapter 11.)

7.94 Explain the variation in ionization energies of carbon, as displayed in the following graph:

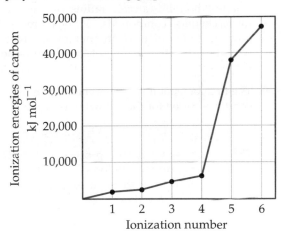

7.95 Do you agree with the following statement? "A negative value for the electron affinity of an atom occurs when the outermost electrons incompletely shield one another from the nucleus." If not, change it to make it more nearly correct in your view. Apply either the statement as given or your revised statement to explain why the electron affinity of bromine is -325 kJ/mol and that for its neighbor Kr is > 0.

7.96 Use orbital diagrams to illustrate what happens when an oxygen atom gains two electrons. Why is it extremely difficult to add a third electron to the atom?

[7.97] Use electron configurations to explain the following observations: **(a)** The first ionization energy of phosphorus is greater than that of sulfur. **(b)** The electron affinity of nitrogen is lower (less negative) than those of both carbon and oxygen. **(c)** The second ionization energy of oxygen is greater than the first ionization energy of fluorine. **(d)** The third ionization energy of manganese is greater than those of both chromium and iron.

7.98 The following table gives the electron affinities, in kJ/mol, for the group 1B and group 2B metals: **(a)** Why are the electron affinities of the group 2B elements greater than zero? **(b)** Why do the electron affinities of the group 1B elements become more negative as we move down the group? [*Hint:* Examine the trends in the electron affinity of other groups as we proceed down the periodic table.]

Cu	Zn
−119	> 0
Ag	Cd
−126	> 0
Au	Hg
−223	> 0

7.99 Hydrogen is an unusual element because it behaves in some ways like the alkali metal elements and in other ways like a nonmetal. Its properties can be explained in part by its electron configuration and by the values for its ionization energy and electron affinity. **(a)** Explain why the electron affinity of hydrogen is much closer to the values for the alkali elements than for the halogens. **(b)** Is the following statement true? "Hydrogen has the smallest bonding atomic radius of any element that forms chemical compounds." If not, correct it. If it is, explain in terms of electron configurations. **(c)** Explain why the ionization energy of hydrogen is closer to the values for the halogens than for the alkali metals.

[7.100] The first ionization energy of the oxygen molecule is the energy required for the following process:

$$O_2(g) \longrightarrow O_2^+(g) + e^-$$

The energy needed for this process is 1175 kJ/mol, very similar to the first ionization energy of Xe. Would you expect O_2 to react with F_2? If so, suggest a product or products of this reaction.

7.101 Based on your reading of this chapter, arrange the following in order of increasing melting point: K, Br_2, Mg, and O_2. Explain the factors that determine the order.

7.102 The element strontium is used in a variety of industrial processes. It is not an extremely hazardous substance, but low levels of strontium ingestion could affect the health of children. Radioactive strontium is very hazardous; it was a by-product of nuclear weapons testing and was found widely distributed following nuclear tests. Calcium is quite common in the environment, including food products, and is frequently present in drinking water. Discuss the similarities and differences between calcium and strontium, and indicate how and why strontium might be expected to accompany calcium in water supplies, uptake by plants, and so on.

[7.103] There are certain similarities in properties that exist between the first member of any periodic family and the element located below it and to the right in the periodic table. For example, in some ways Li resembles Mg, Be resembles Al, and so forth. This observation is called the diagonal relationship. Using what we have learned in this chapter, offer a possible explanation for this relationship.

[7.104] A historian discovers a nineteenth-century notebook in which some observations, dated 1822, on a substance thought to be a new element, were recorded. Here are some of the data recorded in the notebook: Ductile, silver-white, metallic looking. Softer than lead. Unaffected by water. Stable in air. Melting point: 153 °C Density: 7.3 g/cm³. Electrical conductivity: 20% that of copper. Hardness: About 1% as hard as iron. When 4.20 g of the unknown is heated in an excess of oxygen, 5.08 g of a white solid is formed. The solid could be sublimed by heating to over 800 °C. **(a)** Using information in the text and a handbook of chemistry, and making allowances for possible variations in numbers from current values, identify the element reported. **(b)** Write a balanced chemical equation for the reaction with oxygen. **(c)** Judging from Figure 7.2, might this nineteenth-century investigator have been the first to discover a new element?

INTEGRATIVE EXERCISES

[7.105] Moseley established the concept of atomic number by studying X-rays emitted by the elements. The X-rays emitted by some of the elements have the following wavelengths:

Element	Wavelength (Å)
Ne	14.610
Ca	3.358
Zn	1.435
Zr	0.786
Sn	0.491

(a) Calculate the frequency, ν, of the X-rays emitted by each of the elements, in Hz. **(b)** Using graph paper (or suitable computer software), plot the square root of ν versus the atomic number of the element. What do you observe about the plot? **(c)** Explain how the plot in part (b) allowed Moseley to predict the existence of undiscovered elements. **(d)** Use the result from part (b) to predict the X-ray wavelength emitted by iron. **(e)** A particular element emits X-rays with a wavelength of 0.980 Å. What element do you think it is?

[7.106] **(a)** Write the electron configuration for Li, and estimate the effective nuclear charge experienced by the valence electron. **(b)** The energy of an electron in a one-electron atom or ion equals $(-2.18 \times 10^{-18} \text{ J})\left(\dfrac{Z^2}{n^2}\right)$ where Z is the nuclear charge and n is the principal quantum number of the electron. Estimate the first ionization energy of Li. **(c)** Compare the result of your calculation with the value reported in table 7.4, and explain the difference. **(d)** What value of the effective nuclear charge gives the proper value for the ionization energy? Does this agree with your explanation in (c)?

[7.107] One way to measure ionization energies is photoelectron spectroscopy (PES), a technique based on the photoelectric effect. ∞ (Section 6.2) In PES, monochromatic light is directed onto a sample, causing electrons to be emitted. The kinetic energy of the emitted electrons is measured. The difference between the energy of the photons and the kinetic energy of the electrons corresponds to the energy needed to remove the electrons (that is, the ionization energy). Suppose that a PES experiment is performed in which mercury vapor is irradiated with ultraviolet light of wavelength 58.4 nm. **(a)** What is the energy of a photon of this light, in eV? **(b)** Write an equation that shows the process corresponding to the first ionization energy of Hg. **(c)** The kinetic energy of the emitted electrons is measured to be 10.75 eV. What is the first ionization energy of Hg, in kJ/mol? **(d)** With reference to Figure 7.11, determine which of the halogen elements has a first ionization energy closest to that of mercury.

7.108 Consider the gas-phase transfer of an electron from a sodium atom to a chlorine atom:

$$Na(g) + Cl(g) \longrightarrow Na^+(g) + Cl^-(g)$$

(a) Write this reaction as the sum of two reactions, one that relates to an ionization energy and one that relates to an electron affinity. **(b)** Use the result from part (a), data in this chapter, and Hess's law to calculate the enthalpy of the above reaction. Is the reaction exothermic or endothermic? **(c)** The reaction between sodium metal and chlorine gas is highly exothermic and produces NaCl(s), whose structure was discussed in Section 2.7. Comment on this observation relative to the calculated enthalpy for the aforementioned gas-phase reaction.

[7.109] When magnesium metal is burned in air (Figure 3.5), two products are produced. One is magnesium oxide, MgO. The other is the product of the reaction of Mg with molecular nitrogen, magnesium nitride. When water is added to magnesium nitride, it reacts to form magnesium oxide and ammonia gas. **(a)** Based on the charge of the nitride ion (Table 2.5), predict the formula of magnesium nitride. **(b)** Write a balanced equation for the reaction of magnesium nitride with water. What is the driving force for this reaction? **(c)** In an experiment a piece of magnesium ribbon is burned in air in a crucible.

The mass of the mixture of MgO and magnesium nitride after burning is 0.470 g. Water is added to the crucible, further reaction occurs, and the crucible is heated to dryness until the final product is 0.486 g of MgO. What was the mass percentage of magnesium nitride in the mixture obtained after the initial burning? **(d)** Magnesium nitride can also be formed by reaction of the metal with ammonia at high temperature. Write a balanced equation for this reaction. If a 6.3-g Mg ribbon reacts with 2.57 g $NH_3(g)$ and the reaction goes to completion, which component is the limiting reactant? What mass of $H_2(g)$ is formed in the reaction? **(e)** The standard enthalpy of formation of solid magnesium nitride is −461.08 kJ/mol. Calculate the standard enthalpy change for the reaction between magnesium metal and ammonia gas.

7.110 **(a)** The experimental Bi—Br bond length in bismuth tribromide, $BiBr_3$, is 2.63 Å. Based on this value and the data in Figure 7.7, predict the atomic radius of Bi. **(b)** Bismuth tribromide is soluble in acidic solution. It is formed by treating solid bismuth(III) oxide with aqueous hydrobromic acid. Write a balanced chemical equation for this reaction. **(c)** While bismuth(III) oxide is soluble in acidic solutions, it is insoluble in basic solutions such as NaOH(aq). Based on these properties, is bismuth characterized as a metallic, metalloid, or nonmetallic element? **(d)** Treating bismuth with fluorine gas forms BiF_5. Use the electron configuration of Bi to explain the formation of a compound with this formulation. **(e)** While it is possible to form BiF_5 in the manner just described, pentahalides of bismuth are not known for the other halogens. Explain why the pentahalide might form with fluorine, but not with the other halogens. How does the behavior of bismuth relate to the fact that xenon reacts with fluorine to form compounds, but not with the other halogens?

7.111 Potassium superoxide, KO_2, is often used in oxygen masks (such as those used by firefighters) because KO_2 reacts with CO_2 to release molecular oxygen. Experiments indicate that 2 mol of $KO_2(s)$ react with each mole of $CO_2(g)$. **(a)** The products of the reaction are $K_2CO_3(s)$ and $O_2(g)$. Write a balanced equation for the reaction between $KO_2(s)$ and $CO_2(g)$. **(b)** Indicate the oxidation number for each atom involved in the reaction in part (a). What elements are being oxidized and reduced? **(c)** What mass of $KO_2(s)$ is needed to consume 18.0 g $CO_2(g)$? What mass of $O_2(g)$ is produced during this reaction?

8 BASIC CONCEPTS OF CHEMICAL BONDING

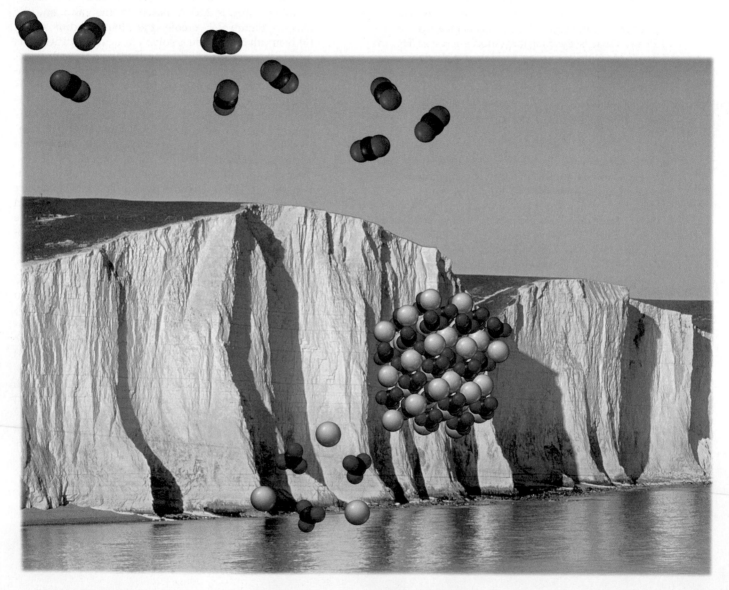

THE WHITE CLIFFS OF DOVER in southeastern England are made up largely of chalk, a porous form of limestone. The mineral calcite, with a composition of $CaCO_3$, is the predominant chemical substance in both chalk and limestone. Much of Earth's calcite is produced by marine organisms, which combine Ca^{2+} and CO_3^{2-} ions to form shells of $CaCO_3$. The presence of CO_3^{2-} ions in oceans can be traced to dissolved CO_2 from the atmosphere.

8.1 Chemical Bonds, Lewis Symbols, and the Octet Rule
We can broadly characterize chemical bonds into three types: *ionic, covalent,* and *metallic.* In evaluating bonding, *Lewis symbols* provide a useful shorthand notation for keeping track of the valence electrons in atoms and ions.

8.2 Ionic Bonding
We will observe that in ionic substances the atoms are held together by the electrostatic attractions between ions of opposite charge. We will study the energetics of formation of ionic substances and describe the *lattice energy* of these substances.

8.3 Covalent Bonding
We also recognize that the atoms in molecular substances are held together by the sharing of one or more electron pairs between atoms. In general the electrons are shared in such a way that each atom attains an *octet* of electrons.

8.4 Bond Polarity and Electronegativity
We define *electronegativity* as the ability of an atom in a compound to attract electrons to itself. In general, electron pairs will be shared unequally between atoms of differing electronegativity, leading to *polar covalent bonds.*

8.5 Drawing Lewis Structures
We will see that *Lewis structures* are a simple yet powerful way of predicting the covalent bonding patterns within molecules. In addition to the octet rule, we will see that the concept of *formal charge* can be used to identify the most favorable Lewis structure.

8.6 Resonance Structures
We observe that in some cases more than one equivalent Lewis structure can be drawn for a molecule or polyatomic ion. The actual structure in such cases is a blend of two or more contributing Lewis structures, called *resonance structures.*

8.7 Exceptions to the Octet Rule
We recognize that the octet rule is more of a guideline than an inviolate rule. Exceptions to the octet rule include molecules with an odd number of electrons; molecules where large differences in electronegativity prevent an atom from completing its octet; and, most commonly, molecules where an element from the third period or lower attains more than an octet of electrons.

8.8 Strengths of Covalent Bonds
We observe that bond strengths vary with the number of shared electron pairs as well as other factors. We can use *average bond enthalpy* values to estimate the enthalpies of reactions in cases where thermodynamic data such as heats of formation are unavailable.

CALCIUM CARBONATE, CaCO$_3$, IS ONE OF THE MOST INTERESTING AND VERSATILE COMPOUNDS on the planet. It accounts for roughly 4% of Earth's crust and is the major component of rocks such as limestone and marble. The white cliffs of Dover in southeastern England are one of the most

famous natural formations of $CaCO_3$. The cliffs consist almost entirely of a porous form of limestone called chalk.

Unlike most inorganic substances, $CaCO_3$ is widely used by living organisms and is found in objects such as seashells, coral, eggshells, and pearls. Calcium carbonate plays a role in the complex chemistry associated with the greenhouse effect because it is formed in the oceans through the reaction between calcium ions and dissolved carbon dioxide. When limestone is heated to elevated temperatures, $CaCO_3$ decomposes into solid CaO (the important industrial chemical called quicklime) and gaseous CO_2. Calcium carbonate undergoes the reactions that it does because the atoms in $CaCO_3$ are held together by a combination of two different types of bonds. The carbon and oxygen atoms that make up the carbonate ion share valence electrons, resulting in the formation of *covalent bonds.* The oppositely charged Ca^{2+} and CO_3^{2-} ions are held together by electrostatic attractions, which are called *ionic bonds.*

The properties of substances are determined in large part by the *chemical bonds* that hold their atoms together. What determines the type of bonding in each substance? How do the characteristics of these bonds give rise to different physical and chemical properties? The keys to answering the first question are found in the electronic structures of the atoms involved, which we discussed in Chapters 6 and 7. In this chapter and the next, we will examine the relationship between the electronic structures of atoms and the chemical bonds they form. We will also see how the properties of ionic and covalent substances arise from the distributions of electronic charge within atoms, ions, and molecules.

8.1 CHEMICAL BONDS, LEWIS SYMBOLS, AND THE OCTET RULE

Whenever two atoms or ions are strongly attached to each other, we say there is a **chemical bond** between them. There are three general types of chemical bonds: ionic, covalent, and metallic. Figure 8.1 ◄ shows examples of substances in which we find each of these types of attractive forces.

The term **ionic bond** refers to electrostatic forces that exist between ions of opposite charge. Ions may be formed from atoms by the transfer of one or more electrons from one atom to another. Ionic substances generally result from the interaction of metals on the left side of the periodic table with nonmetals on the right side (excluding the noble gases, group 8A). Ionic bonding will be discussed in Section 8.2.

A **covalent bond** results from the sharing of electrons between two atoms. The most familiar examples of covalent bonding are seen in the interactions of nonmetallic elements with one another. We devote much of this chapter and the next to describing and understanding covalent bonds.

Metallic bonds are found in metals, such as copper, iron, and aluminum. Each atom in a metal is bonded to several neighboring atoms. The bonding electrons are relatively free to move throughout the three-dimensional structure of the metal. Metallic bonds give rise to such typical metallic properties as high electrical conductivity and luster. We will examine these bonds in Chapter 23.

Lewis Symbols

The electrons involved in chemical bonding are the *valence electrons,* which, for most atoms, are those residing in the outermost occupied shell of an atom. ∞ (Section 6.8) The American chemist G. N. Lewis (1875–1946) suggested a simple way of showing the valence electrons in an atom and tracking them in the course of bond formation, using what are now known as *Lewis electron-dot symbols,* or merely Lewis symbols.

The **Lewis symbol** for an element consists of the chemical symbol for the element plus a dot for each valence electron. Sulfur, for example, has the electron configuration $[Ne]3s^23p^4$; its Lewis symbol therefore shows six valence electrons:

The dots are placed on the four sides of the atomic symbol: the top, the bottom, and the left and right sides. Each side can accommodate up to two electrons. All four sides of the symbol are equivalent, which means that the choice of which sides accommodate the fifth and sixth electrons is arbitrary.

▼ **Figure 8.1 Chemical bonds.** Examples of substances in which (a) ionic, (b) covalent, and (c) metallic bonds are found.

Magnesium oxide

Potassium dichromate   Nickel(II) oxide

(a)

Sulfur

Bromine   Sucrose

(b)

Magnesium

Gold   Copper

(c)

TABLE 8.1 ■ Lewis Symbols

Element	Electron Configuration	Lewis Symbol	Element	Electron Configuration	Lewis Symbol
Li	$[He]2s^1$	Li·	Na	$[Ne]3s^1$	Na·
Be	$[He]2s^2$	·Be·	Mg	$[Ne]3s^2$	·Mg·
B	$[He]2s^2 2p^1$	·Ḃ·	Al	$[Ne]3s^2 3p^1$	·Ȧl·
C	$[He]2s^2 2p^2$	·Ċ·	Si	$[Ne]3s^2 3p^2$	·Ṡi·
N	$[He]2s^2 2p^3$	·Ṅ:	P	$[Ne]3s^2 3p^3$	·Ṗ:
O	$[He]2s^2 2p^4$	:Ö:	S	$[Ne]3s^2 3p^4$	:Ṡ:
F	$[He]2s^2 2p^5$	·F̈:	Cl	$[Ne]3s^2 3p^5$	·C̈l:
Ne	$[He]2s^2 2p^6$	:N̈e:	Ar	$[Ne]3s^2 3p^6$	:Är:

The electron configurations and Lewis symbols for the representative elements of the second and third rows of the periodic table are shown in Table 8.1 ▲. Notice that the number of valence electrons in any representative element is the same as the group number of the element. For example, the Lewis symbols for oxygen and sulfur, members of group 6A, both show six dots.

▲ **GIVE IT SOME THOUGHT**

Are all three of the following Lewis symbols for Cl correct?

:C̈l·   :C̈l: :C̈l·

The Octet Rule

Atoms often gain, lose, or share electrons to achieve the same number of electrons as the noble gas closest to them in the periodic table. The noble gases have very stable electron arrangements, as evidenced by their high ionization energies, low affinity for additional electrons, and general lack of chemical reactivity. ∞(Section 7.8) Because all noble gases (except He) have eight valence electrons, many atoms undergoing reactions also end up with eight valence electrons. This observation has led to a guideline known as the **octet rule:** *Atoms tend to gain, lose, or share electrons until they are surrounded by eight valence electrons.*

An octet of electrons consists of full *s* and *p* subshells in an atom. In terms of Lewis symbols, an octet can be thought of as four pairs of valence electrons arranged around the atom, as in the Lewis symbol for Ne in Table 8.1. There are many exceptions to the octet rule, but it provides a useful framework for introducing many important concepts of bonding.

8.2 IONIC BONDING

When sodium metal, Na(*s*), is brought into contact with chlorine gas, $Cl_2(g)$, a violent reaction ensues (Figure 8.2 ▼). The product of this very exothermic reaction is sodium chloride, NaCl(*s*):

$$Na(s) + \tfrac{1}{2}Cl_2(g) \longrightarrow NaCl(s) \qquad \Delta H_f^\circ = -410.9 \text{ kJ} \qquad [8.1]$$

Sodium chloride is composed of Na^+ and Cl^- ions, which are arranged in a regular three-dimensional array, as shown in Figure 8.3 ▼.

The formation of Na^+ from Na and Cl^- from Cl_2 indicates that an electron has been lost by a sodium atom and gained by a chlorine atom—we can envision an *electron transfer* from the Na atom to the Cl atom. Two of the atomic properties we discussed in Chapter 7 give us an indication of how readily electron transfer

FORMATION OF SODIUM CHLORIDE

When metals and nonmetals react, electrons are transferred from the metal atoms to the nonmetal atoms, forming ions. The principal reason that ionic compounds are stable is the attraction between ions of unlike charge, which draws them together, releasing energy and causing the ions to form a solid array or lattice.

A container of chlorine gas and a container of sodium metal.

Formation of NaCl begins as sodium is added to the chlorine.

The reaction a few minutes later, strongly exothermic, giving off both heat and light.

▲ Figure 8.2 **Formation of sodium chloride.**

▲ Figure 8.3 **The crystal structure of sodium chloride.** In this three-dimensional array of ions, each Na^+ ion is surrounded by six Cl^- ions, and each Cl^- ion is surrounded by six Na^+ ions.

occurs: the ionization energy, which indicates how easily an electron can be removed from an atom, and the electron affinity, which measures how much an atom wants to gain an electron. ∞ (Sections 7.4 and 7.5) Electron transfer to form oppositely charged ions occurs when one of the atoms readily gives up an electron (low ionization energy) and the other atom readily gains an electron (high electron affinity). Thus, NaCl is a typical ionic compound because it consists of a metal of low ionization energy and a nonmetal of high electron affinity. Using Lewis electron-dot symbols (and showing a chlorine atom rather than the Cl_2 molecule), we can represent this reaction as follows:

$$Na\cdot + \cdot\ddot{\underset{..}{Cl}}: \longrightarrow Na^+ + [:\ddot{\underset{..}{Cl}}:]^- \qquad [8.2]$$

The arrow indicates the transfer of an electron from the Na atom to the Cl atom. Each ion has an octet of electrons, the octet on Na^+ being the $2s^22p^6$ electrons that lie below the single $3s$ valence electron of the Na atom. We have put a bracket around the chloride ion to emphasize that all eight electrons are located on the Cl^- ion.

GIVE IT SOME THOUGHT

Describe the electron transfers that occur in the formation of calcium oxide from elemental calcium and oxygen.

Energetics of Ionic Bond Formation

As seen in Figure 8.2, the reaction of sodium with chlorine is *very* exothermic. In fact, Equation 8.1 is the reaction for the formation of NaCl(*s*) from its elements, so that the enthalpy change for the reaction is ΔH_f° for NaCl(*s*). In Appendix C we see that the heat of formation of other ionic substances is also quite negative. What factors make the formation of ionic compounds so exothermic?

In Equation 8.2 we represented the formation of NaCl as the transfer of an electron from Na to Cl. Recall from our discussion of ionization energies, however, that the loss of electrons from an atom is always an endothermic process. ∞ (Section 7.4) Removing an electron from Na(*g*) to form $Na^+(g)$, for instance, requires 496 kJ/mol. Conversely, when a nonmetal gains an electron, the process is generally exothermic, as seen from the negative electron affinities of the elements. ∞ (Section 7.5) Adding an electron to Cl(*g*), for example, releases 349 kJ/mol. If the transfer of an electron from one atom to another were the only factor in forming an ionic bond, the overall process would not be exothermic. For example, removing an electron from Na(*g*) and adding it to Cl(*g*) is an endothermic process that requires 496 − 349 = 147 kJ/mol. This endothermic process corresponds to the formation of sodium and chloride ions that are infinitely far apart—in other words, the positive energy change assumes that the ions are not interacting with one another, which is quite different from the situation in ionic solids.

GIVE IT SOME THOUGHT

Consider the ionization energies of the alkali metals and the electron affinities of the halogens given in Chapter 7. Can you find any pair where the transfer of an electron from the alkali metal to the halogen would be an exothermic process?

The principal reason that ionic compounds are stable is the attraction between ions of unlike charge. This attraction draws the ions together, releasing energy and causing the ions to form a solid array, or lattice, such as that shown for NaCl in Figure 8.3. A measure of just how much stabilization results from arranging oppositely charged ions in an ionic solid is given by the **lattice energy**, which is *the energy required to completely separate a mole of a solid ionic compound into its gaseous ions.*

To get a picture of this process for NaCl, imagine that the structure shown in Figure 8.3 expands from within, so that the distances between the ions increase until the ions are very far apart. This process requires 788 kJ/mol, which is the value of the lattice energy:

$$NaCl(s) \longrightarrow Na^+(g) + Cl^-(g)$$

$$\Delta H_{lattice} = +788 \text{ kJ/mol} \quad [8.3]$$

Notice that this process is highly endothermic. The reverse process—the coming together of $Na(g)^+$ and $Cl(g)^-$ to form NaCl(*s*)—is therefore highly exothermic $\Delta H = -788$ kJ/mol.

Table 8.2▶ lists the lattice energies of NaCl and other ionic compounds. All are large positive values, indicating that the ions are strongly attracted to one another in these solids. The energy released by the

TABLE 8.2 ■ Lattice Energies for Some Ionic Compounds

Compound	Lattice Energy (kJ/mol)	Compound	Lattice Energy (kJ/mol)
LiF	1030	MgCl$_2$	2326
LiCl	834	SrCl$_2$	2127
LiI	730		
NaF	910	MgO	3795
NaCl	788	CaO	3414
NaBr	732	SrO	3217
NaI	682		
KF	808	ScN	7547
KCl	701		
KBr	671		
CsCl	657		
CsI	600		

attraction between ions of unlike charge more than makes up for the endother-mic nature of ionization energies, making the formation of ionic compounds an exothermic process. The strong attractions also cause most ionic materials to be hard and brittle, with high melting points—for example, NaCl melts at 801 °C.

The magnitude of the lattice energy of a solid depends on the charges of the ions, their sizes, and their arrangement in the solid. We saw in Chapter 5 that the potential energy of two interacting charged particles is given by

$$E_{el} = \frac{\kappa Q_1 Q_2}{d} \qquad [8.4]$$

In this equation Q_1 and Q_2 are the charges on the particles, d is the distance be-tween their centers, and κ is a constant, 8.99×10^9 J-m/C^2. ∞ (Section 5.1) Equation 8.4 indicates that the attractive interaction between two oppositely charged ions increases as the magnitudes of their charges increase and as the distance between their centers decreases. Thus, *for a given arrangement of ions, the lattice energy increases as the charges on the ions increase and as their radii decrease.* The magnitude of lattice energies depends predominantly on the ionic charges because ionic radii vary over only a limited range.

■ SAMPLE EXERCISE 8.1 | Magnitudes of Lattice Energies

Without consulting Table 8.2, arrange the following ionic compounds in order of in-creasing lattice energy: NaF, CsI, and CaO.

SOLUTION

Analyze: From the formulas for three ionic compounds, we must determine their relative lattice energies.

Plan: We need to determine the charges and relative sizes of the ions in the com-pounds. We can then use Equation 8.4 qualitatively to determine the relative ener-gies, knowing that the larger the ionic charges, the greater the energy and the farther apart the ions are, the lower the energy.

Solve: NaF consists of Na^+ and F^- ions, CsI of Cs^+ and I^- ions, and CaO of Ca^{2+} and O^{2-} ions. Because the product of the charges, Q_1Q_2, appears in the numerator of Equation 8.4, the lattice energy will increase dramatically when the charges of the ions increase. Thus, we expect the lattice energy of CaO, which has 2+ and 2− ions, to be the greatest of the three.

The ionic charges in NaF and CsI are the same. As a result, the difference in their lattice energies will depend on the difference in the distance between the centers of the ions in their lattice. Because ionic size increases as we go down a group in the period-ic table (Section 7.3), we know that Cs^+ is larger than Na^+ and I^- is larger than F^-. Therefore the distance between the Na^+ and F^- ions in NaF will be less than the dis-tance between the Cs^+ and I^- ions in CsI. As a result, the lattice energy of NaF should be greater than that of CsI. In order of increasing energy, therefore, we have CsI < NaF < CaO.

Check: Table 8.2 confirms our predicted order is correct.

■ PRACTICE EXERCISE

Which substance would you expect to have the greatest lattice energy, MgF_2, CaF_2, or ZrO_2?
Answer: ZrO_2

Electron Configurations of Ions of the *s*- and *p*-Block Elements

We began considering the electron configurations of ions in Section 7.4. In light of our examination of ionic bonding, we will continue with that discussion here. The energetics of ionic bond formation helps explain why many ions tend to have noble-gas electron configurations. For example, sodium readily loses one electron to form Na^+, which has the same electron configuration as Ne:

$$\begin{aligned} Na \qquad & 1s^2 2s^2 2p^6 3s^1 = [\text{Ne}]3s^1 \\ Na^+ \qquad & 1s^2 2s^2 2p^6 \;\; = [\text{Ne}] \end{aligned}$$

Even though lattice energy increases with increasing ionic charge, we never find ionic compounds that contain Na^{2+} ions. The second electron removed would have to come from an inner shell of the sodium atom, and removing electrons from an inner shell requires a very large amount of energy. ∞ (Section 7.4) The increase in lattice energy is not enough to compensate for the energy needed to remove an inner-shell electron. Thus, sodium and the other group 1A metals are found in ionic substances only as 1+ ions.

Similarly, the addition of electrons to nonmetals is either exothermic or only slightly endothermic as long as the electrons are being added to the valence shell. Thus, a Cl atom easily adds an electron to form Cl^-, which has the same electron configuration as Ar:

$$Cl \qquad 1s^2 2s^2 2p^6 3s^2 3p^5 = [Ne]3s^2 3p^5$$
$$Cl^- \qquad 1s^2 2s^2 2p^6 3s^2 3p^6 = [Ne]3s^2 3p^6 = [Ar]$$

To form a Cl^{2-} ion, the second electron would have to be added to the next higher shell of the Cl atom, which is energetically very unfavorable. Therefore, we never observe Cl^{2-} ions in ionic compounds. Based on these concepts, we expect that ionic compounds of the representative metals from groups 1A, 2A, and 3A will contain cations with charges of 1+, 2+, and 3+, respectively. Likewise, ionic compounds of the representative nonmetals of groups 5A, 6A, and 7A usually contain anions of charge 3−, 2−, and 1−, respectively.

■■■ **SAMPLE EXERCISE 8.2** | **Charges on Ions**

Predict the ion generally formed by **(a)** Sr, **(b)** S, **(c)** Al.

SOLUTION

Analyze: We must decide how many electrons are most likely to be gained or lost by atoms of Sr, S, and Al.

Plan: In each case we can use the element's position in the periodic table to predict whether it will form a cation or an anion. We can then use its electron configuration to determine the ion that is likely to be formed.

Solve: (a) Strontium is a metal in group 2A and will therefore form a cation. Its electron configuration is $[Kr]5s^2$, and so we expect that the two valence electrons can be lost easily to give an Sr^{2+} ion. **(b)** Sulfur is a nonmetal in group 6A and will thus tend to be found as an anion. Its electron configuration ($[Ne]3s^2 3p^4$) is two electrons short of a noble-gas configuration. Thus, we expect that sulfur will form S^{2-} ions. **(c)** Aluminum is a metal in group 3A. We therefore expect it to form Al^{3+} ions.

Check: The ionic charges we predict here are confirmed in Tables 2.4 and 2.5.

■■■ **PRACTICE EXERCISE**

Predict the charges on the ions formed when magnesium reacts with nitrogen.
Answer: Mg^{2+} and N^{3-}

Transition-Metal Ions

Because ionization energies increase rapidly for each successive electron removed, the lattice energies of ionic compounds are generally large enough to compensate for the loss of up to only three electrons from atoms. Thus, we find cations with charges of 1+, 2+, or 3+ in ionic compounds. Most transition metals, however, have more than three electrons beyond a noble-gas core. Silver, for example, has a $[Kr]4d^{10}5s^1$ electron configuration. Metals of group 1B (Cu, Ag, Au) often occur as 1+ ions (as in CuBr and AgCl). In forming Ag^+, the 5s electron is lost, leaving a completely filled 4d subshell. As in this example, transition metals generally do not form ions that have a noble-gas configuration. The octet rule, although useful, is clearly limited in scope.

Recall from our discussion in Section 7.4 that when a positive ion is formed from an atom, electrons are always lost first from the subshell having the

A Closer Look   CALCULATION OF LATTICE ENERGIES: THE BORN–HABER CYCLE

Lattice energy is a useful concept because it relates directly to the stability of an ionic solid. Unfortunately, the lattice energy cannot be determined directly by experiment. It can, however, be calculated by envisioning the formation of an ionic compound as occurring in a series of well-defined steps. We can then use Hess's law (Section 5.6) to put these steps together in a way that gives us the lattice energy for the compound. By so doing, we construct a **Born–Haber cycle**, a thermochemical cycle named after the German scientists Max Born (1882–1970) and Fritz Haber (1868–1934), who introduced it to analyze the factors contributing to the stability of ionic compounds.

In the Born–Haber cycle for NaCl, we consider the formation of $NaCl(s)$ from the elements $Na(s)$ and $Cl_2(g)$ by two different routes, as shown in Figure 8.4▶. The enthalpy change for the direct route (red arrow) is the heat of formation of $NaCl(s)$:

$$Na(s) + \tfrac{1}{2}Cl_2(g) \longrightarrow NaCl(s)$$
$$\Delta H_f^\circ[NaCl(s)] = -411 \text{ kJ} \quad [8.5]$$

The indirect route consists of five steps, shown by the green arrows in Figure 8.4. First, we generate gaseous atoms of sodium by vaporizing sodium metal. Then we form gaseous atoms of chlorine by breaking the bonds in the Cl_2 molecules. The enthalpy changes for these processes are available to us as enthalpies of formation (Appendix C):

$$Na(s) \longrightarrow Na(g) \quad \Delta H_f^\circ[Na(g)] = 108 \text{ kJ} \quad [8.6]$$
$$\tfrac{1}{2}Cl_2(g) \longrightarrow Cl(g) \quad \Delta H_f^\circ[Cl(g)] = 122 \text{ kJ} \quad [8.7]$$

Both of these processes are endothermic; energy is required to generate gaseous sodium and chlorine atoms.

In the next two steps we remove the electron from $Na(g)$ to form $Na^+(g)$ and then add the electron to $Cl(g)$ to form $Cl^-(g)$. The enthalpy changes for these processes equal the first ionization energy of Na, $I_1(Na)$, and the electron affinity of Cl, denoted $E(Cl)$, respectively: ∞ (Sections 7.4, 7.5)

$$Na(g) \longrightarrow Na^+(g) + e^- \quad \Delta H = I_1(Na) = 496 \text{ kJ} \quad [8.8]$$
$$Cl(g) + e^- \longrightarrow Cl^-(g) \quad \Delta H = E(Cl) = -349 \text{ kJ} \quad [8.9]$$

Finally, we combine the gaseous sodium and chloride ions to form solid sodium chloride. Because this process is just the reverse of the lattice energy (breaking a solid into gaseous ions), the enthalpy change is the negative of the lattice energy, the quantity that we want to determine:

$$Na^+(g) + Cl^-(g) \longrightarrow NaCl(s)$$
$$\Delta H = -\Delta H_{lattice} = ? \quad [8.10]$$

The sum of the five steps in the indirect path (green arrows) gives us $NaCl(s)$ from $Na(s)$ and $\tfrac{1}{2}Cl_2(g)$. Thus, from Hess's law we know that the sum of the enthalpy changes for these five steps equals that for the direct path, indicated by the red arrow, Equation 8.5:

$$\Delta H_f^\circ[NaCl(s)] = \Delta H_f^\circ[Na(g)] + \Delta H_f^\circ[Cl(g)]$$
$$+ I_1(Na) + E(Cl) - \Delta H_{lattice}$$

$$-411 \text{ kJ} = 108 \text{ kJ} + 122 \text{ kJ} + 496 \text{ kJ} - 349 \text{ kJ} - \Delta H_{lattice}$$

Solving for $\Delta H_{lattice}$:

$$\Delta H_{lattice} = 108 \text{ kJ} + 122 \text{ kJ} + 496 \text{ kJ} - 349 \text{ kJ} + 411 \text{ kJ}$$
$$= 788 \text{ kJ}$$

Thus, the lattice energy of NaCl is 788 kJ/mol.

Related Exercises: 8.26, 8.27, and 8.28

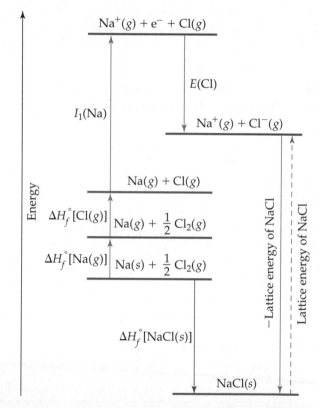

▲ **Figure 8.4 The Born–Haber cycle.** This representation shows the energetic relationships in the formation of ionic solids from the elements. By Hess's law, the enthalpy of formation of $NaCl(s)$ from elemental sodium and chlorine (Equation 8.5) is equal to the sum of the energies of several individual steps (Equations 8.6 through 8.10).

largest value of n. Thus, *in forming ions, transition metals lose the valence-shell s electrons first, then as many d electrons as are required to reach the charge of the ion.* Let's consider Fe, which has the electron configuration $[Ar]3d^64s^2$. In forming the Fe^{2+} ion, the two $4s$ electrons are lost, leading to an $[Ar]3d^6$ configuration. Removal of an additional electron gives the Fe^{3+} ion, whose electron configuration is $[Ar]3d^5$.

GIVE IT SOME THOUGHT

Which element forms a 2+ ion that has the electron configuration $[Kr]4d^8$?

8.3 COVALENT BONDING

Ionic substances possess several characteristic properties. They are usually brittle substances with high melting points. They are usually crystalline. Furthermore, ionic crystals often can be cleaved; that is, they break apart along smooth, flat surfaces. These characteristics result from electrostatic forces that maintain the ions in a rigid, well-defined, three-dimensional arrangement such as that shown in Figure 8.3.

The vast majority of chemical substances do not have the characteristics of ionic materials. Most of the substances with which we come into daily contact—such as water—tend to be gases, liquids, or solids with low melting points. Many, such as gasoline, vaporize readily. Many are pliable in their solid forms—for example, plastic bags and paraffin.

For the very large class of substances that do not behave like ionic substances, we need a different model for the bonding between atoms. G. N. Lewis reasoned that atoms might acquire a noble-gas electron configuration by sharing electrons with other atoms. As we noted in Section 8.1, a chemical bond formed by sharing a pair of electrons is called a *covalent bond*.

The hydrogen molecule, H_2, provides the simplest example of a covalent bond. When two hydrogen atoms are close to each other, electrostatic interactions occur between them. The two positively charged nuclei repel each other, the two negatively charged electrons repel each other, and the nuclei and electrons attract each other, as shown in Figure 8.5▶. Because the H_2 molecule exists as a stable entity, the attractive forces must exceed the repulsive ones. Let's take a closer look at the attractive forces that hold this molecule together.

By using quantum mechanical methods analogous to those employed for atoms ∞ (Section 6.5), it is possible to calculate the distribution of electron density in molecules. Such a calculation for H_2 shows that the attractions between the nuclei and the electrons cause electron density to concentrate between the nuclei, as shown in Figure 8.5(b). As a result, the overall electrostatic interactions are attractive. Thus, the atoms in H_2 are held together principally because the two nuclei are electrostatically attracted to the concentration of negative charge between them. In essence, the shared pair of electrons in any covalent bond acts as a kind of "glue" to bind atoms together.

(a)

(b)

▲ **Figure 8.5 The covalent bond in H_2.** (a) The attractions and repulsions among electrons and nuclei in the hydrogen molecule. (b) Electron distribution in the H_2 molecule. The concentration of electron density between the nuclei leads to a net attractive force that constitutes the covalent bond holding the molecule together.

GIVE IT SOME THOUGHT

If a H_2 molecule is ionized to form H_2^+, it will change the bond strength. Based on the simple description of covalent bonding given above, would you expect the H—H bond in H_2^+ to be weaker or stronger than the H—H bond in H_2?

Lewis Structures

The formation of covalent bonds can be represented using Lewis symbols for the constituent atoms. The formation of the H_2 molecule from two H atoms, for example, can be represented as

$$\text{H·} + \text{·H} \longrightarrow \text{H:H}$$

In this way, each hydrogen atom acquires a second electron, achieving the stable, two-electron, noble-gas electron configuration of helium.

The formation of a bond between two Cl atoms to give a Cl_2 molecule can be represented in a similar way:

$$\text{:}\overset{..}{\text{Cl}}\text{·} + \text{·}\overset{..}{\text{Cl}}\text{:} \longrightarrow \text{:}\overset{..}{\text{Cl}}\text{:}\overset{..}{\text{Cl}}\text{:}$$

By sharing the bonding electron pair, each chlorine atom has eight electrons (an octet) in its valence shell. It thus achieves the noble-gas electron configuration of argon.

The structures shown here for H_2 and Cl_2 are called **Lewis structures** (or Lewis electron-dot structures). In writing Lewis structures, we usually show each electron pair shared between atoms as a line and the unshared electron pairs as dots. Written this way, the Lewis structures for H_2 and Cl_2 are

$$H\!-\!H \qquad :\!\ddot{\underset{..}{C}}l\!-\!\ddot{\underset{..}{C}}l\!:$$

For the nonmetals, the number of valence electrons in a neutral atom is the same as the group number. Therefore, one might predict that 7A elements, such as F, would form one covalent bond to achieve an octet; 6A elements, such as O, would form two covalent bonds; 5A elements, such as N, would form three covalent bonds; and 4A elements, such as C, would form four covalent bonds. These predictions are borne out in many compounds. For example, consider the simple hydrogen compounds of the nonmetals of the second row of the periodic table:

$$H\!-\!\ddot{\underset{..}{F}}\!: \qquad H\!-\!\ddot{\underset{..}{O}}\!: \qquad H\!-\!\overset{..}{\underset{|}{N}}\!-\!H \qquad H\!-\!\overset{\displaystyle H}{\underset{\displaystyle H}{\overset{|}{\underset{|}{C}}}}\!-\!H$$
$$ H \qquad\quad H$$

Thus, the Lewis model succeeds in accounting for the compositions of many compounds of nonmetals, in which covalent bonding predominates.

■■ **SAMPLE EXERCISE 8.3** | Lewis Structure of a Compound

Given the Lewis symbols for the elements nitrogen and fluorine shown in Table 8.1, predict the formula of the stable binary compound (a compound composed of two elements) formed when nitrogen reacts with fluorine, and draw its Lewis structure.

SOLUTION

Analyze: The Lewis symbols for nitrogen and fluorine reveal that nitrogen has five valence electrons and fluorine has seven.

Plan: We need to find a combination of the two elements that results in an octet of electrons around each atom in the compound. Nitrogen requires three additional electrons to complete its octet, whereas fluorine requires only one. Sharing a pair of electrons between one N atom and one F atom will result in an octet of electrons for fluorine but not for nitrogen. We therefore need to figure out a way to get two more electrons for the N atom.

Solve: Nitrogen must share a pair of electrons with three fluorine atoms to complete its octet. Thus, the Lewis structure for the resulting compound, NF_3, is

$$\cdot\ddot{N}\cdot \; + \; 3\cdot\ddot{\underset{..}{F}}\!: \quad\longrightarrow\quad :\ddot{\underset{..}{F}}\!:\!\ddot{N}\!:\!\ddot{\underset{..}{F}}\!: \quad\longrightarrow\quad :\ddot{\underset{..}{F}}\!-\!\overset{..}{\underset{|}{N}}\!-\!\ddot{\underset{..}{F}}\!:$$
$$ \ddot{\underset{..}{F}} :\ddot{\underset{..}{F}}\!:$$

Check: The Lewis structure in the center shows that each atom is surrounded by an octet of electrons. Once you are accustomed to thinking of each line in a Lewis structure as representing *two* electrons, you can just as easily use the structure on the right to check for octets.

■■ **PRACTICE EXERCISE**

Compare the Lewis symbol for neon with the Lewis structure for methane, CH_4. In what important way are the electron arrangements about neon and carbon alike? In what important respect are they different?

Answer: Both atoms have an octet of electrons about them. However, the electrons about neon are unshared electron pairs, whereas those about carbon are shared with four hydrogen atoms.

Multiple Bonds

The sharing of a pair of electrons constitutes a single covalent bond, generally referred to simply as a **single bond**. In many molecules, atoms attain complete octets by sharing more than one pair of electrons. When two electron pairs are shared, two lines are drawn, representing a **double bond**. In carbon dioxide, for example, bonding occurs between carbon, with four valence electrons, and oxygen, with six:

$$:\ddot{O}: + \cdot\dot{C}\cdot + :\ddot{O}: \longrightarrow \ddot{O}::C::\ddot{O} \quad (or\ \ddot{O}=C=\ddot{O})$$

As the diagram shows, each oxygen acquires an octet of electrons by sharing two electron pairs with carbon. Carbon, on the other hand, acquires an octet of electrons by sharing two electron pairs in each of the two bonds it forms with oxygen.

A **triple bond** corresponds to the sharing of three pairs of electrons, such as in the N_2 molecule:

$$:\dot{N}\cdot + \cdot\dot{N}: \longrightarrow :N:::N: \quad (or\ :N\equiv N:)$$

Because each nitrogen atom possesses five electrons in its valence shell, three electron pairs must be shared to achieve the octet configuration.

The properties of N_2 are in complete accord with its Lewis structure. Nitrogen is a diatomic gas with exceptionally low reactivity that results from the very stable nitrogen–nitrogen bond. Study of the structure of N_2 reveals that the nitrogen atoms are separated by only 1.10 Å. The short N—N bond distance is a result of the triple bond between the atoms. From structure studies of many different substances in which nitrogen atoms share one or two electron pairs, we have learned that the average distance between bonded nitrogen atoms varies with the number of shared electron pairs:

N—N	N=N	N≡N
1.47 Å	1.24 Å	1.10 Å

As a general rule, the distance between bonded atoms decreases as the number of shared electron pairs increases. The distance between the nuclei of the atoms involved in a bond is called the **bond length** for the bond. We first encountered bond lengths in Section 7.3 in our discussion of atomic radii, and we will discuss them further in Section 8.8.

GIVE IT SOME THOUGHT

The C—O bond length in carbon monoxide, CO, is 1.13 Å, whereas the C—O bond length in CO_2 is 1.24 Å. Without drawing a Lewis structure, do you think that carbon monoxide has a single, double, or triple bond between the C and O atoms?

8.4 BOND POLARITY AND ELECTRONEGATIVITY

When two identical atoms bond, as in Cl_2 or H_2, the electron pairs must be shared equally. In highly ionic compounds, on the other hand, such as NaCl, there is relatively little sharing of electrons, which means that NaCl is best described as composed of Na^+ and Cl^- ions. The $3s$ electron of the Na atom is, in effect, transferred completely to chlorine. The bonds that are found in most substances fall somewhere between these extremes.

The concept of **bond polarity** helps describe the sharing of electrons between atoms. A **nonpolar covalent bond** is one in which the electrons are shared equally between two atoms, as in the Cl_2 and N_2 examples we just cited.

In a **polar covalent bond**, one of the atoms exerts a greater attraction for the bonding electrons than the other. If the difference in relative ability to attract electrons is large enough, an ionic bond is formed.

Electronegativity

We use a quantity called electronegativity to estimate whether a given bond will be nonpolar covalent, polar covalent, or ionic. **Electronegativity** is defined as the ability of an atom *in a molecule* to attract electrons to itself. The greater an atom's electronegativity, the greater is its ability to attract electrons to itself. The electronegativity of an atom in a molecule is related to its ionization energy and electron affinity, which are properties of isolated atoms. The *ionization energy* measures how strongly a gaseous atom holds on to its electrons, ∞ (Section 7.4) while the *electron affinity* is a measure of how strongly an atom attracts additional electrons. ∞ (Section 7.5) An atom with a very negative electron affinity and high ionization energy will both attract electrons from other atoms and resist having its electrons attracted away; it will be highly electronegative.

Numerical estimates of electronegativity can be based on a variety of properties, not just ionization energy and electron affinity. The American chemist Linus Pauling (1901–1994) developed the first and most widely used electronegativity scale; he based his scale on thermochemical data. Figure 8.6 ◀ shows Pauling's electronegativity values for many of the elements. The values are unitless. Fluorine, the most electronegative element, has an electronegativity of 4.0. The least electronegative element, cesium, has an electronegativity of 0.7. The values for all other elements lie between these two extremes.

Within each period there is generally a steady increase in electronegativity from left to right; that is, from the most metallic to the most nonmetallic elements. With some exceptions (especially within the transition metals), electronegativity decreases with increasing atomic number in any one group. This is what we might expect because we know that ionization energies tend to decrease with increasing atomic number in a group and electron affinities do not change very much. You do not need to memorize numerical values for electronegativity. Instead, you should know the periodic trends so that you can predict which of two elements is more electronegative.

▼ **Figure 8.6 Electronegativities of the elements.** Electronegativity generally increases from left to right across a period and decreases from top to bottom down a group.

GIVE IT SOME THOUGHT

How does the *electronegativity* of an element differ from its *electron affinity*?

Electronegativity and Bond Polarity

We can use the difference in electronegativity between two atoms to gauge the polarity of the bonding between them. Consider these three fluorine-containing compounds:

Compound	F_2	HF	LiF
Electronegativity difference	$4.0 - 4.0 = 0$	$4.0 - 2.1 = 1.9$	$4.0 - 1.0 = 3.0$
Type of bond	Nonpolar covalent	Polar covalent	Ionic

F_2 HF   LiF

▲ **Figure 8.7 Electron density distribution.** This computer-generated rendering shows the calculated electron-density distribution on the surface of the F_2, HF, and LiF molecules. The regions of relatively low electron density (net positive charge) appear blue, those of relatively high electron density (net negative charge) appear red, and regions that are close to electrically neutral appear green.

In F_2 the electrons are shared equally between the fluorine atoms, and thus the covalent bond is *nonpolar*. In general, a nonpolar covalent bond results when the electronegativities of the bonded atoms are equal.

In HF the fluorine atom has a greater electronegativity than the hydrogen atom, with the result that the sharing of electrons is unequal—the bond is polar. In general, a polar covalent bond results when the atoms differ in electronegativity. In HF the more electronegative fluorine atom attracts electron density away from the less electronegative hydrogen atom, leaving a partial positive charge on the hydrogen atom and a partial negative charge on the fluorine atom. We can represent this charge distribution as

$$\overset{\delta+}{H} - \overset{\delta-}{F}$$

The $\delta+$ and $\delta-$ (read "delta plus" and "delta minus") symbolize the partial positive and negative charges, respectively.

In LiF the electronegativity difference is very large, meaning that the electron density is shifted far toward F. The resultant bond is therefore most accurately described as *ionic*. This shift of electron density toward the more electronegative atom can be seen in the results of calculations of electron density distributions. For the three species in our example, the calculated electron density distributions are shown in Figure 8.7▲. The regions of space that have relatively higher electron density are shown in red, and those with a relatively lower electron density are shown in blue. You can see that in F_2 the distribution is symmetrical, in HF it is clearly shifted toward fluorine, and in LiF the shift is even greater. These examples illustrate, therefore, that *the greater the difference in electronegativity between two atoms, the more polar their bond.*

▲ **GIVE IT SOME THOUGHT**

Based on differences in electronegativity, how would you characterize the bonding in silicon nitride, Si_3N_4? Would you expect the bonds between Si and N to be nonpolar, polar covalent, or ionic?

■ **SAMPLE EXERCISE 8.4** | Bond Polarity

In each case, which bond is more polar: **(a)** B—Cl or C—Cl, **(b)** P—F or P—Cl? Indicate in each case which atom has the partial negative charge.

SOLUTION

Analyze: We are asked to determine relative bond polarities, given nothing but the atoms involved in the bonds.

Plan: Because we are not asked for quantitative answers, we can use the periodic table and our knowledge of electronegativity trends to answer the question.

Solve:

(a) The chlorine atom is common to both bonds. Therefore, the analysis reduces to a comparison of the electronegativities of B and C. Because boron is to the left of carbon in the periodic table, we predict that boron has the lower electronegativity. Chlorine, being on the right side of the table, has a higher electronegativity. The more polar bond will be the one between the atoms having the lowest electronegativity (boron) and the highest electronegativity (chlorine). Consequently, the B—Cl bond is more polar; the chlorine atom carries the partial negative charge because it has a higher electronegativity.

(b) In this example phosphorus is common to both bonds, and the analysis reduces to a comparison of the electronegativities of F and Cl. Because fluorine is above chlorine in the periodic table, it should be more electronegative and will form the more polar bond with P. The higher electronegativity of fluorine means that it will carry the partial negative charge.

Check:

(a) Using Figure 8.6: The difference in the electronegativities of chlorine and boron is $3.0 - 2.0 = 1.0$; the difference between chlorine and carbon is $3.0 - 2.5 = 0.5$. Hence the B—Cl bond is more polar, as we had predicted.

(b) Using Figure 8.6: The difference in the electronegativities of chlorine and phosphorus is $3.0 - 2.1 = 0.9$; the difference between fluorine and phosphorus is $4.0 - 2.1 = 1.9$. Hence the P—F bond is more polar, as we had predicted.

■ PRACTICE EXERCISE

Which of the following bonds is most polar: S—Cl, S—Br, Se—Cl, or Se—Br?
Answer: Se—Cl

Dipole Moments

The difference in electronegativity between H and F leads to a polar covalent bond in the HF molecule. As a consequence, there is a concentration of negative charge on the more electronegative F atom, leaving the less electronegative H atom at the positive end of the molecule. A molecule such as HF, in which the centers of positive and negative charge do not coincide, is said to be a **polar molecule**. Thus, we describe both bonds and entire molecules as being polar and nonpolar.

We can indicate the polarity of the HF molecule in two ways:

$$\overset{\delta+}{\text{H}}-\overset{\delta-}{\text{F}} \quad or \quad \overset{\longleftrightarrow}{\text{H}-\text{F}}$$

Recall from the preceding subsection that "$\delta+$" and "$\delta-$" indicate the partial positive and negative charges on the H and F atoms. In the notation on the right, the arrow denotes the shift in electron density toward the fluorine atom. The crossed end of the arrow can be thought of as a plus sign that designates the positive end of the molecule.

Polarity helps determine many of the properties of substances that we observe at the macroscopic level, in the laboratory and in everyday life. Polar molecules align themselves with respect to one another, with the negative end of one molecule and the positive end of another attracting each other. Polar molecules are likewise attracted to ions. The negative end of a polar molecule is attracted to a positive ion, and the positive end is attracted to a negative ion. These interactions account for many properties of liquids, solids, and solutions, as you will see in Chapters 11, 12, and 13.

How can we quantify the polarity of a molecule? Whenever a distance separates two electrical charges of equal magnitude but opposite sign, a **dipole** is established. The quantitative measure of the magnitude of a dipole is called its **dipole moment**, denoted μ. If a distance r separates two equal and opposite charges $Q+$ and $Q-$, the magnitude of the dipole moment is the product of Q and r (Figure 8.8 ◄):

$$\mu = Qr \qquad [8.11]$$

▲ **Figure 8.8 Dipole and dipole moment.** When charges of equal magnitude and opposite sign $Q+$ and $Q-$ are separated by a distance r, a dipole is produced. The size of the dipole is given by the dipole moment, μ, which is the product of the charge separated and the distance of separation between the charge centers: $\mu = Qr$.

The dipole moment increases as the magnitude of charge that is separated increases and as the distance between the charges increases. For a nonpolar molecule, such as F_2, the dipole moment is zero because there is no charge separation.

GIVE IT SOME THOUGHT

The molecules chlorine monofluoride, ClF, and iodine monofluoride, IF, are examples of *interhalogen* compounds—compounds that contain bonds between different halogen elements. Which of these molecules will have the larger dipole moment?

Dipole moments are usually reported in *debyes* (D), a unit that equals 3.34×10^{-30} coulomb-meters (C-m). For molecules, we usually measure charge in units of the electronic charge e, 1.60×10^{-19} C, and distance in units of angstroms. Suppose that two charges $1+$ and $1-$ (in units of e) are separated by a distance of 1.00 Å. The dipole moment produced is

$$\mu = Qr = (1.60 \times 10^{-19}\,\text{C})(1.00\,\text{Å})\left(\frac{10^{-10}\,\text{m}}{1\,\text{Å}}\right)\left(\frac{1\,\text{D}}{3.34 \times 10^{-30}\,\text{C-m}}\right) = 4.79\,\text{D}$$

Measurement of the dipole moments can provide us with valuable information about the charge distributions in molecules, as illustrated in Sample Exercise 8.5.

■ SAMPLE EXERCISE 8.5 | Dipole Moments of Diatomic Molecules

The bond length in the HCl molecule is 1.27 Å. **(a)** Calculate the dipole moment, in debyes, that would result if the charges on the H and Cl atoms were $1+$ and $1-$, respectively. **(b)** The experimentally measured dipole moment of $HCl(g)$ is 1.08 D. What magnitude of charge, in units of e, on the H and Cl atoms would lead to this dipole moment?

SOLUTION

Analyze and Plan: We are asked in (a) to calculate the dipole moment of HCl that would result if there were a full charge transferred from H to Cl. We can use Equation 8.11 to obtain this result. In (b), we are given the actual dipole moment for the molecule and will use that value to calculate the actual partial charges on the H and Cl atoms.

Solve:

(a) The charge on each atom is the electronic charge, $e = 1.60 \times 10^{-19}$ C. The separation is 1.27 Å. The dipole moment is therefore

$$\mu = Qr = (1.60 \times 10^{-19}\,\text{C})(1.27\,\text{Å})\left(\frac{10^{-10}\,\text{m}}{1\,\text{Å}}\right)\left(\frac{1\,\text{D}}{3.34 \times 10^{-30}\,\text{C-m}}\right) = 6.08\,\text{D}$$

(b) We know the value of μ, 1.08 D and the value of r, 1.27 Å. We want to calculate the value of Q:

$$Q = \frac{\mu}{r} = \frac{(1.08\,\text{D})\left(\dfrac{3.34 \times 10^{-30}\,\text{C-m}}{1\,\text{D}}\right)}{(1.27\,\text{Å})\left(\dfrac{10^{-10}\,\text{m}}{1\,\text{Å}}\right)} = 2.84 \times 10^{-20}\,\text{C}$$

We can readily convert this charge to units of e:

$$\text{Charge in } e = (2.84 \times 10^{-20}\,\text{C})\left(\frac{1\,e}{1.60 \times 10^{-19}\,\text{C}}\right) = 0.178e$$

Thus, the experimental dipole moment indicates that the charge separation in the HCl molecule is

$$\overset{0.178+}{\text{H}} - \overset{0.178-}{\text{Cl}}$$

Because the experimental dipole moment is less than that calculated in part (a), the charges on the atoms are much less than a full electronic charge. We could have anticipated this because the H—Cl bond is polar covalent rather than ionic.

■ PRACTICE EXERCISE

The dipole moment of chlorine monofluoride, $ClF(g)$, is 0.88 D. The bond length of the molecule is 1.63 Å. **(a)** Which atom is expected to have the partial negative charge? **(b)** What is the charge on that atom, in units of e?
Answers: **(a)** F, **(b)** 0.11−

TABLE 8.3 ■ Bond Lengths, Electronegativity Differences, and Dipole Moments of the Hydrogen Halides

Compound	Bond Length (Å)	Electronegativity Difference	Dipole Moment (D)
HF	0.92	1.9	1.82
HCl	1.27	0.9	1.08
HBr	1.41	0.7	0.82
HI	1.61	0.4	0.44

Table 8.3 ◄ presents the bond lengths and dipole moments of the hydrogen halides. Notice that as we proceed from HF to HI, the electronegativity difference decreases and the bond length increases. The first effect decreases the amount of charge separated and causes the dipole moment to decrease from HF to HI, even though the bond length is increasing. Calculations identical to those used in Sample Exercise 8.5 show that the actual charges on the atoms decrease from 0.41+/0.41− in HF to 0.057+/0.057− in HI. We can "visualize" the varying degree of electronic charge shift in these substances from computer-generated renderings based on calculations of electron distribution, as shown in Figure 8.9▼. For these molecules, the change in the electronegativity difference has a greater effect on the dipole moment than does the change in bond length.

▲ **GIVE IT SOME THOUGHT**

The bond between carbon and hydrogen is one of the most important types of bonds in chemistry. The length of a H—C bond is approximately 1.1 Å. Based on this distance and differences in electronegativity, would you predict the dipole moment of an individual H—C bond to be larger or smaller than the dipole moment of the H—I bond?

Before leaving this section let's return to the LiF molecule pictured in Figure 8.7. Under standard conditions LiF exists as an extended ionic solid with an arrangement of atoms analogous to the sodium chloride structure shown in Figure 8.3. However, it is possible to generate LiF molecules by vaporizing the solid at high temperature. The molecules have a dipole moment of 6.28 D and a bond distance of 1.53 Å. From these values we can calculate the charge on lithium and fluorine to be 0.857+ and 0.857−, respectively. This bond is extremely polar, and the presence of such large charges strongly favors the formation of an extended ionic lattice whereby each lithium ion is surrounded by fluoride ions and vice versa.

Differentiating Ionic and Covalent Bonding

To understand the interactions responsible for chemical bonding, it is advantageous to treat ionic and covalent bonding separately. That is the approach taken in this chapter, as well as in most undergraduate-level chemistry texts. The partitioning of bonding into ionic and covalent extremes is considered when we name chemical substances. We saw in Section 2.8 that there are two general approaches

HF

HCl

HBr

HI

▲ **Figure 8.9 Charge separation in the hydrogen halides.** Blue represents regions of lowest electron density, red regions of highest electron density. In HF the strongly electronegative F pulls much of the electron density away from H. In HI the I, being much less electronegative than F, does not attract the shared electrons as strongly, and consequently there is far less polarization of the bond.

to naming binary compounds: one used for ionic compounds and the other for molecular ones. However, in reality there is a continuum between the extremes of ionic and covalent bonding. This lack of a well-defined separation between the two types of bonding may seem unsettling or confusing at first.

Fortunately, the simple models of ionic and covalent bonding presented in this chapter go quite a long way toward understanding and predicting the structures and properties of chemical compounds. When covalent bonding is dominant, more often than not we expect compounds to exist as molecules*, with all of the properties we associate with molecular substances, such as relatively low melting and boiling points and non-electrolyte behavior when dissolved in water. Furthermore, we will see later that the polarity of covalent bonds has important consequence for the properties of substances. On the other hand, when ionic bonding is dominant, we expect the compounds to possess very different properties. Ionic compounds tend to be brittle, high-melting solids with extended lattice structures and they exhibit strong electrolyte behavior when dissolved in water.

There are of course exceptions to these general stereotypes, some of which we will examine later in the book. Nonetheless, the ability to quickly categorize the predominant bonding interactions in a substance as covalent or ionic imparts considerable insight into the properties of that substance. The question then becomes what is the best system for recognizing which type of bonding will be dominant?

The simplest approach one can take is to assume that the interaction between a metal and a nonmetal will be ionic, while the interaction between two nonmetals will be covalent. While this classification scheme is reasonably predictive, there are far too many exceptions to use it blindly. For example, tin is a metal and chlorine is a nonmetal, but $SnCl_4$ is a molecular substance that exists as a colorless liquid at room temperature. It freezes at $-33\,°C$ and boils at $114\,°C$. Clearly the bonding in this substance is better described as polar covalent than ionic. A more sophisticated approach, as outlined in the preceding discussion, is to use the difference in electronegativity as the main criterion for determining whether ionic or covalent bonding will be dominant. This approach correctly predicts the bonding in $SnCl_4$ to be polar covalent based on an electronegativity difference of 1.2 and at the same time correctly predicts the bonding in NaCl to be predominantly ionic based on an electronegativity difference of 2.1.

Evaluating the bonding based on electronegativity difference is a useful system, but it has one shortcoming. Our electronegativity scale does not explicitly take into account changes in bonding that accompany changes in the oxidation state of the metal. For example, the electronegativity difference between manganese and oxygen is $3.5 - 1.5 = 2.0$, which falls in the range where the bonding is normally considered to be ionic (the electronegativity difference for NaCl was 2.1). Therefore, it is not surprising to learn that manganese (II) oxide, MnO, is a green solid that melts at $1842\,°C$ and has the same crystal structure as NaCl.

However, it would be incorrect to assume that the bonding between manganese and oxygen is always ionic. Manganese (VII) oxide, which has the formula Mn_2O_7, is a green liquid that freezes at $5.9\,°C$, signaling that covalent rather than ionic bonding is dominant. The change in the oxidation state of manganese is responsible for the change in bonding type. As a general principle, whenever the oxidation state of the metal increases, it will lead to an increase in the degree of covalent character in the bonding. When the oxidation state of the metal becomes highly positive (roughly speaking +4 or larger), we should expect a significant degree of covalency in bonds it forms with nonmetals. In such instances you should not be surprised if a compound or polyatomic ion (such as MnO_4^- or CrO_4^{2-}) exhibits the general properties of molecular, rather than ionic compounds.

*There are some obvious exceptions to this rule, such as the network solids, including diamond, silicon, and germanium, where an extended structure is formed even though the bonding is clearly covalent. These examples are discussed in more detail in Section 11.8.

▲ GIVE IT SOME THOUGHT

You encounter two substances: one is a yellow solid that melts at 41 °C and boils at 131 °C. The other is a green solid that melts at 2320 °C. If you are told that one of the compounds is Cr_2O_3 and the other is OsO_4, which one would you expect to be the yellow solid?

8.5 DRAWING LEWIS STRUCTURES

Lewis structures can help us understand the bonding in many compounds and are frequently used when discussing the properties of molecules. For this reason, drawing Lewis structures is an important skill that you should practice. To do so, you should follow a regular procedure. First we will outline the procedure, and then we will go through several examples.

1. *Sum the valence electrons from all atoms.* (Use the periodic table as necessary to help you determine the number of valence electrons in each atom.) For an anion, add one electron to the total for each negative charge. For a cation, subtract one electron from the total for each positive charge. Do not worry about keeping track of which electrons come from which atoms. Only the total number is important.

2. *Write the symbols for the atoms to show which atoms are attached to which, and connect them with a single bond* (a dash, representing *two* electrons). Chemical formulas are often written in the order in which the atoms are connected in the molecule or ion. The formula HCN, for example, tells you that the carbon atom is bonded to the H and to the N. In many polyatomic molecules and ions, the central atom is usually written first, as in $CO_3{}^{2-}$ and SF_4. Remember that the central atom is generally less electronegative than the atoms surrounding it. In other cases, you may need more information before you can draw the Lewis structure.

3. *Complete the octets around all the atoms bonded to the central atom.* Remember, however, that you use only a single pair of electrons around hydrogen.

4. *Place any leftover electrons on the central atom,* even if doing so results in more than an octet of electrons around the atom. In Section 8.7 we will discuss molecules that do not adhere to the octet rule.

5. *If there are not enough electrons to give the central atom an octet, try multiple bonds.* Use one or more of the unshared pairs of electrons on the atoms bonded to the central atom to form double or triple bonds.

■ **SAMPLE EXERCISE 8.6** | Drawing Lewis Structures

Draw the Lewis structure for phosphorus trichloride, PCl_3.

SOLUTION

Analyze and Plan: We are asked to draw a Lewis structure from a molecular formula. Our plan is to follow the five-step procedure just described.

Solve:

First, we sum the valence electrons. Phosphorus (group 5A) has five valence electrons, and each chlorine (group 7A) has seven. The total number of valence electrons is therefore

$$5 + (3 \times 7) = 26$$

Second, we arrange the atoms to show which atom is connected to which, and we draw a single bond between them. There are various ways the atoms might be arranged. In binary (two-element) compounds, however, the first element listed in the chemical formula is generally surrounded by the remaining atoms. Thus, we begin with a skeleton structure that shows a single bond between the phosphorus atom and each chlorine atom:

Cl—P—Cl
     |
    Cl

(It is not crucial to place the atoms in exactly this arrangement.)

Third, we complete the octets on the atoms bonded to the central atom. Placing octets around each Cl atom accounts for 24 electrons (remember, each line in our structure represents *two* electrons):

$$:\ddot{\underset{..}{Cl}}—P—\ddot{\underset{..}{Cl}}:$$
$$|$$
$$:\ddot{\underset{..}{Cl}}:$$

Fourth, we place the remaining two electrons on the central atom, completing the octet around it:

$$:\ddot{\underset{..}{Cl}}—\ddot{P}—\ddot{\underset{..}{Cl}}:$$
$$|$$
$$:\ddot{\underset{..}{Cl}}:$$

This structure gives each atom an octet, so we stop at this point. (Remember that in achieving an octet, the bonding electrons are counted for both atoms.)

■ **PRACTICE EXERCISE**

(a) How many valence electrons should appear in the Lewis structure for CH_2Cl_2?
(b) Draw the Lewis structure.

Answers: **(a)** 20, **(b)**

$$H$$
$$|$$
$$:\ddot{\underset{..}{Cl}}—C—\ddot{\underset{..}{Cl}}:$$
$$|$$
$$H$$

■ **SAMPLE EXERCISE 8.7** | **Lewis Structures with Multiple Bonds**

Draw the Lewis structure for HCN.

SOLUTION

Hydrogen has one valence electron, carbon (group 4A) has four, and nitrogen (group 5A) has five. The total number of valence electrons is therefore $1 + 4 + 5 = 10$. In principle, there are different ways in which we might choose to arrange the atoms. Because hydrogen can accommodate only one electron pair, it always has only one single bond associated with it in any compound. Therefore, C—H—N is an impossible arrangement. The remaining two possibilities are H—C—N and H—N—C. The first is the arrangement found experimentally. You might have guessed this to be the atomic arrangement because **(a)** the formula is written with the atoms in this order, and **(b)** carbon is less electronegative than nitrogen. Thus, we begin with a skeleton structure that shows single bonds between hydrogen, carbon, and nitrogen:

$$H—C—N$$

These two bonds account for four electrons. If we then place the remaining six electrons around N to give it an octet, we do not achieve an octet on C:

$$H—C—\ddot{\underset{..}{N}}:$$

We therefore try a double bond between C and N, using one of the unshared pairs of electrons we placed on N. Again, there are fewer than eight electrons on C, and so we next try a triple bond. This structure gives an octet around both C and N:

$$H—C \overset{\frown}{\underset{\smile}{N}}: \quad \longrightarrow \quad H—C≡N:$$

We see that the octet rule is satisfied for the C and N atoms, and the H atom has two electrons around it. This appears to be a correct Lewis structure.

■ **PRACTICE EXERCISE**

Draw the Lewis structure for **(a)** NO^+ ion, **(b)** C_2H_4.

Answers: **(a)** $[:N≡O:]^+$, **(b)**

$$\begin{array}{cc} H\diagdown & \diagup H \\ & C=C \\ H\diagup & \diagdown H \end{array}$$

■ **SAMPLE EXERCISE 8.8** | Lewis Structure for a Polyatomic Ion

Draw the Lewis structure for the BrO_3^- ion.

SOLUTION

Bromine (group 7A) has seven valence electrons, and oxygen (group 6A) has six. We must now add one more electron to our sum to account for the 1− charge of the ion. The total number of valence electrons is therefore $7 + (3 \times 6) + 1 = 26$. For oxyanions—$BrO_3^-$, SO_4^{2-}, NO_3^-, CO_3^{2-}, and so forth—the oxygen atoms surround the central nonmetal atoms. After following this format and then putting in the single bonds and distributing the unshared electron pairs, we have

$$\left[:\ddot{O}-\overset{..}{Br}-\ddot{O}: \atop :\ddot{O}: \right]^-$$

Notice here and elsewhere that the Lewis structure for an ion is written in brackets with the charge shown outside the brackets at the upper right.

■ **PRACTICE EXERCISE**

Draw the Lewis structure for **(a)** ClO_2^- ion, **(b)** PO_4^{3-} ion.

Answers: **(a)** $\left[:\ddot{O}-\ddot{Cl}-\ddot{O}: \right]^-$ **(b)** $\left[:\ddot{O}-\overset{\textstyle :\ddot{O}:}{\underset{\textstyle :\ddot{O}:}{P}}-\ddot{O}: \right]^{3-}$

Formal Charge

When we draw a Lewis structure, we are describing how the electrons are distributed in a molecule (or polyatomic ion). In some instances we can draw several different Lewis structures that all obey the octet rule. How do we decide which one is the most reasonable? One approach is to do some "bookkeeping" of the valence electrons to determine the formal charge of each atom in each Lewis structure. The **formal charge** of any atom in a molecule is the charge the atom would have if all the atoms in the molecule had the same electronegativity (that is, if each bonding electron pair in the molecule were shared equally between its two atoms).

To calculate the formal charge on any atom in a Lewis structure, we assign the electrons to the atom as follows:

1. *All* unshared (nonbonding) electrons are assigned to the atom on which they are found.

2. For any bond—single, double, or triple—*half* of the bonding electrons are assigned to each atom in the bond.

The formal charge of each atom is then calculated *by subtracting the number of electrons assigned to the atom from the number of valence electrons in the isolated atom.*

Let's illustrate this procedure by calculating the formal charges on the C and N atoms in the cyanide ion, CN^-, which has the Lewis structure

$$[:C\equiv N:]^-$$

For the C atom, there are 2 nonbonding electrons and 3 electrons from the 6 in the triple bond ($\frac{1}{2} \times 6 = 3$) for a total of 5. The number of valence electrons on a neutral C atom is 4. Thus, the formal charge on C is $4 - 5 = -1$. For N, there are 2 nonbonding electrons and 3 electrons from the triple bond. Because the number of valence electrons on a neutral N atom is 5, its formal charge is $5 - 5 = 0$. Thus, the formal charges on the atoms in the Lewis structure of CN^- are

$$[:\overset{-1}{C}\equiv\overset{0}{N}:]^-$$

Notice that the sum of the formal charges equals the overall charge on the ion, 1−. The formal charges on a neutral molecule must add to zero, whereas those on an ion add to give the overall charge on the ion.

The concept of formal charge can help us choose between alternative Lewis structures. We will consider the CO_2 molecule to see how this is done. As shown in Section 8.3, CO_2 is represented as having two double bonds. However, we can also satisfy the octet rule by drawing a Lewis structure having one single bond and one triple bond. Calculating the formal charge for each atom in these structures, we have

	Ö=C=Ö			:Ö—C≡O:		
Valence electrons:	6	4	6	6	4	6
−(Electrons assigned to atom):	6	4	6	7	4	5
Formal charge:	0	0	0	−1	0	+1

Note that in both cases the formal charges add up to zero, as they must because CO_2 is a neutral molecule. So, which is the correct structure? As a general rule, when several Lewis structures are possible, we will use the following guidelines to choose the most correct one:

1. We generally choose the Lewis structure in which the atoms bear formal charges closest to zero.

2. We generally choose the Lewis structure in which any negative charges reside on the more electronegative atoms.

Thus, the first Lewis structure of CO_2 is preferred because the atoms carry no formal charges and so satisfy the first guideline.

Although the concept of formal charge helps us choose between alternative Lewis structures, it is very important that you remember that *formal charges do not represent real charges on atoms*. These charges are just a bookkeeping convention. The actual charge distributions in molecules and ions are determined not by formal charges but by a number of factors, including the electronegativity differences between atoms.

GIVE IT SOME THOUGHT

Suppose that a Lewis structure for a neutral fluorine-containing molecule results in a formal charge on the fluorine atom of +1. What conclusion would you draw?

■ SAMPLE EXERCISE 8.9 | Lewis Structures and Formal Charges

The following are three possible Lewis structures for the thiocyanate ion, NCS^-:

$$[:\ddot{N}-C\equiv S:]^-   \quad   [\ddot{N}=C=\ddot{S}]^-   \quad   [:N\equiv C-\ddot{S}:]^-$$

(a) Determine the formal charges of the atoms in each structure. (b) Which Lewis structure is the preferred one?

SOLUTION

(a) Neutral N, C, and S atoms have five, four, and six valence electrons, respectively. We can determine the following formal charges in the three structures by using the rules we just discussed:

$$\begin{array}{ccc} -2 \quad 0 \quad +1 & -1 \quad 0 \quad 0 & 0 \quad 0 \quad -1 \\ [:\ddot{N}-C\equiv S:]^- & [\ddot{N}=C=\ddot{S}]^- & [:N\equiv C-\ddot{S}:]^- \end{array}$$

As they must, the formal charges in all three structures sum to 1−, the overall charge of the ion.

(b) We will use the guidelines for the best Lewis structure to determine which of the three structures is likely the most correct. As discussed in Section 8.4, N is more electronegative than C or S. Therefore, we expect that any negative formal charge will reside on the N atom (guideline 2). Further, we usually choose the Lewis structure that produces the formal charges of smallest magnitude (guideline 1). For these two reasons, the middle structure is the preferred Lewis structure of the NCS⁻ ion.

■ PRACTICE EXERCISE

The cyanate ion (NCO⁻), like the thiocyanate ion, has three possible Lewis structures. **(a)** Draw these three Lewis structures, and assign formal charges to the atoms in each structure. **(b)** Which Lewis structure is the preferred one?

$$
\begin{array}{ccc}
-2 \quad 0 \quad +1 & -1 \quad 0 \quad 0 & 0 \quad 0 \quad -1 \\
\end{array}
$$

Answers: **(a)** $[:\overset{..}{\underset{..}{N}}-C\equiv O:]^-$ $[\overset{..}{N}=C=\overset{..}{O}:]^-$ $[:N\equiv C-\overset{..}{\underset{..}{O}}:]^-$

     (i)   (ii)   (iii)

(b) Structure (iii), which places a negative charge on oxygen, the most electronegative of the three elements, is the preferred Lewis structure.

A Closer Look | OXIDATION NUMBERS, FORMAL CHARGES, AND ACTUAL PARTIAL CHARGES

In Chapter 4 we introduced the rules for assigning *oxidation numbers* to atoms. The concept of electronegativity is the basis of these numbers. The oxidation number of an atom is the charge it would have if its bonds were completely ionic. That is, in determining the oxidation number, all shared electrons are counted with the more electronegative atom. For example, consider the Lewis structure of HCl shown in Figure 8.10(a)▼. To assign oxidation numbers, the pair of electrons in the covalent bond between the atoms is assigned to the more electronegative Cl atom. This procedure gives Cl eight valence-shell electrons, one more than the neutral atom. Thus, its oxidation number is −1. Hydrogen has no valence electrons when they are counted this way, giving it an oxidation number of +1.

In this section we have just considered another way of counting electrons that gives rise to *formal charges*. The formal charge is assigned by ignoring electronegativity and assigning equally the electrons in bonds between the bonded atoms. Consider again the HCl molecule, but this time divide the bonding pair of electrons equally between H and Cl as shown in Figure 8.10(b). In this case Cl has seven assigned electrons, the same as that of the neutral Cl atom. Thus, the formal charge of Cl in this compound is 0. Likewise, the formal charge of H is also 0.

Neither the oxidation number nor the formal charge gives an accurate depiction of the actual charges on atoms. Oxidation numbers overstate the role of electronegativity, and formal charges ignore it completely. It seems reasonable that electrons in covalent bonds should be apportioned according to the relative electronegativities of the bonded atoms. From Figure 8.6 we see that Cl has an electronegativity of 3.0, while that of H is 2.1. The more electronegative Cl atom might therefore be expected to have roughly $3.0/(3.0 + 2.1) = 0.59$ of the electrical charge in the bonding pair, whereas the H atom has $2.1/(3.0 + 2.1) = 0.41$ of the charge. Because the bond consists of two electrons, the Cl atom's share is $0.59 \times 2e = 1.18e$, or $0.18e$ more than the neutral Cl atom. This gives rise to a partial charge of 0.18− on Cl and 0.18+ on H (notice again that we place the + and − signs before the magnitude when speaking about oxidation numbers and formal charges, but after the magnitude when talking about actual charges).

The dipole moment of HCl gives an experimental measure of the partial charges on each atom. In Sample Exercise 8.5 we saw that the dipole moment of HCl indicates a charge separation with a partial charge of 0.178+ on H and 0.178− on Cl, in remarkably good agreement with our simple approximation based on electronegativities. Although that type of calculation provides "ballpark" numbers for the magnitude of charge on atoms, the relationship between electronegativities and charge separation is generally more complicated. As we have already seen, computer programs employing quantum mechanical principles have been developed to obtain more accurate estimates of the partial charges on atoms, even in complex molecules. Figure 8.10(c) shows a graphical representation of the charge distribution in HCl.

Related Exercises: 8.6, 8.47, 8.48, 8.49, 8.50, 8.88, and 8.89

◀ **Figure 8.10 Oxidation number and formal charge.** (a) The oxidation number for any atom in a molecule is determined by assigning all shared electrons to the more electronegative atom (in this case Cl). (b) Formal charges are derived by dividing all shared electron pairs equally between the bonded atoms. (c) The calculated distribution of electron density on an HCl molecule. Regions of relatively more negative charge are red; those of more positive charge are blue. Negative charge is clearly localized on the chlorine atom.

(a)   (b)   (c)

8.6 RESONANCE STRUCTURES

We sometimes encounter molecules and ions in which the experimentally determined arrangement of atoms is not adequately described by a single Lewis structure. Consider a molecule of ozone, O_3, which is a bent molecule with two equal O—O bond lengths (Figure 8.11▶). Because each oxygen atom contributes 6 valence electrons, the ozone molecule has 18 valence electrons. In writing the Lewis structure, we find that we must have one O—O single bond and one O—O double bond to attain an octet of electrons about each atom:

▲ **Figure 8.11 Ozone.** Molecular structure (top) and electron-distribution diagram (bottom) for the ozone molecule, O_3.

However, this structure cannot by itself be correct because it requires that one O—O bond be different from the other, contrary to the observed structure—we would expect the O=O double bond to be shorter than the O—O single bond. ∞(Section 8.3) In drawing the Lewis structure, however, we could just as easily have put the O=O bond on the left:

The placement of the atoms in these two alternative but completely equivalent Lewis structures for ozone is the same, but the placement of the electrons is different. Lewis structures of this sort are called **resonance structures.** To describe the structure of ozone properly, we write both Lewis structures and use a double-headed arrow to indicate that the real molecule is described by an average of the two resonance structures:

To understand why certain molecules require more than one resonance structure, we can draw an analogy to the mixing of paint (Figure 8.12▶). Blue and yellow are both primary colors of paint pigment. An equal blend of blue and yellow pigments produces green pigment. We cannot describe green paint in terms of a single primary color, yet it still has its own identity. Green paint does not oscillate between its two primary colors: It is not blue part of the time and yellow the rest of the time. Similarly, molecules such as ozone cannot be described as oscillating between the two individual Lewis structures shown above.

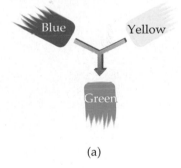

The true arrangement of the electrons in molecules such as O_3 must be considered as a blend of two (or more) Lewis structures. By analogy to the green paint, the molecule has its own identity separate from the individual resonance structures. For example, the ozone molecule always has two equivalent O—O bonds whose lengths are intermediate between the lengths of an oxygen–oxygen single bond and an oxygen–oxygen double bond. Another way of looking at it is to say that the rules for drawing Lewis structures do not allow us to have a single structure that adequately represents the ozone molecule. For example, there are no rules for drawing half-bonds. We can get around this limitation by drawing two equivalent Lewis structures that, when averaged, amount to something very much like what is observed experimentally.

▲ **Figure 8.12 Resonance.** Describing a molecule as a blend of different resonance structures is similar to describing a paint color as a blend of primary colors. (a) Green paint is a blend of blue and yellow. We cannot describe green as a single primary color. (b) The ozone molecule is a blend of two resonance structures. We cannot describe the ozone molecule in terms of a single Lewis structure.

▲ **GIVE IT SOME THOUGHT**

The O—O bonds in ozone are often described as "one-and-a-half" bonds. Is this description consistent with the idea of resonance?

As an additional example of resonance structures, consider the nitrate ion, NO_3^-, for which three equivalent Lewis structures can be drawn:

Notice that the arrangement of atoms is the same in each structure—only the placement of electrons differs. In writing resonance structures, the same atoms must be bonded to each other in all structures, so that the only differences are in the arrangements of electrons. All three Lewis structures taken together adequately describe the nitrate ion, in which all three N—O bond lengths are the same.

▲ **GIVE IT SOME THOUGHT**

In the same sense that we describe the O—O bonds in O_3 as "one-and-a-half" bonds, how would you describe the N—O bonds in NO_3^-?

In some instances all the possible Lewis structures for a species may not be equivalent to one another. Instead, one or more may represent a more stable arrangement than other possibilities. We will encounter examples of this as we proceed.

■ **SAMPLE EXERCISE 8.10** | **Resonance Structures**

Which is predicted to have the shorter sulfur–oxygen bonds, SO_3 or SO_3^{2-}?

SOLUTION

The sulfur atom has six valence electrons, as does oxygen. Thus, SO_3 contains 24 valence electrons. In writing the Lewis structure, we see that three equivalent resonance structures can be drawn:

As was the case for NO_3^-, the actual structure of SO_3 is an equal blend of all three. Thus, each S—O bond distance should be about one-third of the way between that of a single and that of a double bond (see the Give It Some Thought exercise above). That is, they should be shorter than single bonds but not as short as double bonds.

The SO_3^{2-} ion has 26 electrons, which leads to a Lewis structure in which all the S—O bonds are single bonds:

There are no other reasonable Lewis structures for this ion. It can be described quite well by a single Lewis structure rather than by multiple resonance structures.

Our analysis of the Lewis structures leads us to conclude that SO_3 should have the shorter S—O bonds and SO_3^{2-} the longer ones. This conclusion is correct: The experimentally measured S—O bond lengths are 1.42 Å in SO_3 and 1.51 Å in SO_3^{2-}.

■ **PRACTICE EXERCISE**

Draw two equivalent resonance structures for the formate ion, HCO_2^-.

Answer:

Resonance in Benzene

Resonance is an extremely important concept in describing the bonding in organic molecules, particularly in the ones called *aromatic* molecules. Aromatic organic molecules include the hydrocarbon called *benzene*, which has the molecular formula C_6H_6 (Figure 8.13▶). The six C atoms of benzene are bonded in a hexagonal ring, and one H atom is bonded to each C atom.

We can write two equivalent Lewis structures for benzene, each of which satisfies the octet rule. These two structures are in resonance:

Each resonance structure shows three C—C single bonds and three C=C double bonds, but the double bonds are in different places in the two structures. The experimental structure of benzene shows that all six C—C bonds are of equal length, 1.40 Å, intermediate between the typical bond lengths for a C—C single bond (1.54 Å) and a C=C double bond (1.34 Å).

Benzene is commonly represented by omitting the hydrogen atoms attached to carbon and showing only the carbon-carbon framework with the vertices unlabeled. In this convention, the resonance in the benzene molecule is represented either by two structures separated by a double-headed arrow, as with our other examples, or by a shorthand notation in which we draw a hexagon with a circle in it:

The shorthand notation on the right reminds us that benzene is a blend of two resonance structures—it emphasizes that the C=C double bonds cannot be assigned to specific edges of the hexagon. Chemists use both representations of benzene interchangeably.

The bonding arrangement in benzene confers special stability to the molecule. As a result, literally millions of organic compounds contain the six-membered rings characteristic of benzene. Many of these compounds are important in biochemistry, in pharmaceuticals, and in the production of modern materials. We will say more about the bonding in benzene in Chapter 9 and about its unusual stability in Chapter 25.

(a)

(b)

▲ **Figure 8.13 Benzene, an "aromatic" organic compound.**
(a) Benzene is obtained from the distillation of fossil fuels. More than 4 billion pounds of benzene is produced annually in the United States. Because benzene is a carcinogen, its use is closely regulated. (b) The benzene molecule is a regular hexagon of carbon atoms with a hydrogen atom bonded to each one.

◢ **GIVE IT SOME THOUGHT**

Each Lewis structure of benzene has three C=C double bonds. Another hydrocarbon containing three C=C double bonds is *hexatriene*, C_6H_8. A Lewis structure of hexatriene is

Would you expect hexatriene to have multiple resonance structures like benzene? If not, why is this molecule different from benzene with respect to resonance?

8.7 EXCEPTIONS TO THE OCTET RULE

The octet rule is so simple and useful in introducing the basic concepts of bonding that you might assume it is always obeyed. In Section 8.2, however, we noted its limitation in dealing with ionic compounds of the transition metals. The octet rule also fails in many situations involving covalent bonding. These exceptions to the octet rule are of three main types:

1. Molecules and polyatomic ions containing an odd number of electrons
2. Molecules and polyatomic ions in which an atom has fewer than an octet of valence electrons
3. Molecules and polyatomic ions in which an atom has more than an octet of valence electrons

Odd Number of Electrons

In the vast majority of molecules and polyatomic ions, the total number of valence electrons is even, and complete pairing of electrons occurs. However, in a few molecules and polyatomic ions, such as ClO_2, NO, NO_2, and O_2^-, the number of valence electrons is odd. Complete pairing of these electrons is impossible, and an octet around each atom cannot be achieved. For example, NO contains $5 + 6 = 11$ valence electrons. The two most important Lewis structures for this molecule are

$$\ddot{N}{=}\ddot{O} \quad \text{and} \quad \dot{N}{=}\ddot{O}$$

> ### GIVE IT SOME THOUGHT
>
> Which of the Lewis structures shown above for NO would be preferred based on analysis of the formal charges?

Less than an Octet of Valence Electrons

A second type of exception occurs when there are fewer than eight valence electrons around an atom in a molecule or polyatomic ion. This situation is also relatively rare (with the exception of hydrogen and helium as we have already discussed), most often encountered in compounds of boron and beryllium. As an example, let's consider boron trifluoride, BF_3. If we follow the first four steps of the procedure at the beginning of Section 8.5 for drawing Lewis structures, we obtain the structure

There are only six electrons around the boron atom. In this Lewis structure the formal charges on both the B and the F atoms are zero. We could complete the octet around boron by forming a double bond (step 5). In so doing, we see that there are three equivalent resonance structures (the formal charges on each atom are shown in red):

These Lewis structures force a fluorine atom to share additional electrons with the boron atom, which is inconsistent with the high electronegativity of fluorine. In fact, the formal charges tell us that this is an unfavorable situation.

In each of the Lewis structures, the F atom involved in the B=F double bond has a formal charge of +1, while the less electronegative B atom has a formal charge of −1. Thus, the Lewis structures in which there is a B=F double bond are less important than the one in which there are fewer than an octet of valence electrons around boron:

Most important   Less important

We usually represent BF_3 solely by the leftmost resonance structure, in which there are only six valence electrons around boron. The chemical behavior of BF_3 is consistent with this representation. In particular, BF_3 reacts very energetically with molecules having an unshared pair of electrons that can be used to form a bond with boron. For example, BF_3 reacts with ammonia, NH_3, to form the compound NH_3BF_3:

$$
\begin{array}{ccc}
\text{H} & & \text{F} \\
| & & | \\
\text{H}-\text{N:} & + & \text{B}-\text{F} \\
| & & | \\
\text{H} & & \text{F}
\end{array}
\longrightarrow
\begin{array}{cc}
\text{H} & \text{F} \\
| & | \\
\text{H}-\text{N}-\text{B}-\text{F} \\
| & | \\
\text{H} & \text{F}
\end{array}
$$

In this stable compound, boron has an octet of valence electrons. We will consider reactions of this type in more detail in Chapter 16 when we study Lewis acids and bases. ∞ (Section 16.11)

More than an Octet of Valence Electrons

The third and largest class of exceptions consists of molecules or polyatomic ions in which there are more than eight electrons in the valence shell of an atom. When we draw the Lewis structure for PCl_5, for example, we are forced to "expand" the valence shell and place ten electrons around the central phosphorus atom:

Other examples of molecules and ions with "expanded" valence shells are SF_4, AsF_6^-, and ICl_4^-. The corresponding molecules with a second-period atom as the central atom, such as NCl_5 and OF_4, do *not* exist. Let's take a look at why expanded valence shells are observed only for elements in period 3 and beyond in the periodic table.

Elements of the second period have only the 2s and 2p valence orbitals available for bonding. Because these orbitals can hold a maximum of eight electrons, we never find more than an octet of electrons around elements from the second period. Elements from the third period and beyond, however, have *ns*, *np*, and unfilled *nd* orbitals that can be used in bonding. For example, the orbital diagram for the valence shell of a phosphorus atom is

1↓	1	1	1					

3s   3p   3d

Although third-period elements often satisfy the octet rule, as in PCl_3, they also often exceed an octet by seeming to use their empty d orbitals to accommodate additional electrons.*

Size also plays an important role in determining whether an atom in a molecule or polyatomic ion can accommodate more than eight electrons in its valence shell. The larger the central atom is, the larger the number of atoms that can surround it. The number of molecules and ions with expanded valence shells therefore increases with increasing size of the central atom. The size of the surrounding atoms is also important. Expanded valence shells occur most often when the central atom is bonded to the smallest and most electronegative atoms, such as F, Cl, and O.

■ **SAMPLE EXERCISE 8.11** | Lewis Structure for an Ion with an Expanded Valence Shell

Draw the Lewis structure for ICl_4^-.

SOLUTION

Iodine (group 7A) has seven valence electrons. Each chlorine (group 7A) also has seven. An extra electron is added to account for the 1− charge of the ion. Therefore, the total number of valence electrons is

$$7 + 4(7) + 1 = 36$$

The I atom is the central atom in the ion. Putting eight electrons around each Cl atom (including a pair of electrons between I and each Cl to represent the single bond between these atoms) requires $8 \times 4 = 32$ electrons.

We are thus left with $36 - 32 = 4$ electrons to be placed on the larger iodine:

$$\left[\begin{array}{ccc} :\ddot{C}l: & & :\ddot{C}l: \\ & I & \\ :\ddot{C}l: & & :\ddot{C}l: \end{array} \right]^-$$

Iodine has 12 valence electrons around it, four more than needed for an octet.

■ PRACTICE EXERCISE

(a) Which of the following atoms is never found with more than an octet of valence electrons around it: S, C, P, Br? **(b)** Draw the Lewis structure for XeF_2.

Answers: **(a)** C, **(b)** $:\ddot{F}—\ddot{X}e—\ddot{F}:$

At times you may see Lewis structures written with an expanded valence shell even though structures can be written with an octet. For example, consider the following Lewis structures for the phosphate ion, PO_4^{3-}:

$$\left[\begin{array}{c} :\ddot{O}:^{-1} \\ | \\ :\ddot{O}—\overset{+1}{P}—\ddot{O}:^{-1} \\ {}^{-1} | ^{-1} \\ :\ddot{O}: \end{array} \right]^{3-} \qquad \left[\begin{array}{c} :\ddot{O}:^{-1} \\ || \\ {}^{0}\ddot{O}=\overset{0}{P}—\ddot{O}:^{-1} \\ | \\ :\ddot{O}:^{-1} \end{array} \right]^{3-}$$

The formal charges on the atoms are shown in red. In the Lewis structure shown on the left, the P atom obeys the octet rule. In the Lewis structure shown on the right, the P atom has an expanded valence shell of five electron pairs[†] leading to smaller formal charges on the atoms. Which Lewis structure is a better representation of the bonding in PO_4^{3} ? Theoretical calculations based on quantum

---

Based on theoretical calculations, some chemists have questioned whether valence d orbitals are actually used in the bonding of molecules and ions with expanded valence shells. Nevertheless, the presence of valence d orbitals in period 3 and beyond provides the simplest explanation of this phenomenon, especially within the scope of a general chemistry textbook.

†*The Lewis structure shown on the right has four equivalent resonance forms. Only one is shown for clarity.*

mechanics suggest that the structure on the left is the best single Lewis structure for the phosphate ion. In general, when choosing between alternative Lewis structures, if it is possible to draw a Lewis structure where the octet rule is satisfied without using multiple bonds, that structure will be preferred.

8.8 STRENGTHS OF COVALENT BONDS

The stability of a molecule is related to the strengths of the covalent bonds it contains. The strength of a covalent bond between two atoms is determined by the energy required to break that bond. It is easiest to relate bond strength to the enthalpy change in reactions in which bonds are broken. ∞∞ (Section 5.4) The **bond enthalpy** is the enthalpy change, ΔH, for the breaking of a particular bond in one mole of a gaseous substance. For example, the bond enthalpy for the bond between chlorine atoms in the Cl_2 molecule is the enthalpy change when 1 mol of Cl_2 is dissociated into chlorine atoms:

$$:\ddot{C}l-\ddot{C}l:(g) \longrightarrow 2\ :\ddot{C}l\cdot(g)$$

We use the designation D(bond type) to represent bond enthalpies.

It is relatively simple to assign bond enthalpies to bonds that are found in diatomic molecules, such as the Cl—Cl bond in Cl_2, D(Cl—Cl), or the H—Br bond in HBr, D(H—Br). The bond enthalpy is just the energy required to break the diatomic molecule into its component atoms. Many important bonds, such as the C—H bond, exist only in polyatomic molecules. For these types of bonds, we usually use *average* bond enthalpies. For example, the enthalpy change for the following process in which a methane molecule is decomposed to its five atoms (a process called *atomization*) can be used to define an average bond enthalpy for the C—H bond, D (C—H):

$$H-\underset{\underset{\displaystyle H}{\vert}}{\overset{\overset{\displaystyle H}{\vert}}{C}}-H(g) \longrightarrow \cdot\dot{C}\cdot(g) + 4\,H\cdot(g) \qquad \Delta H = 1660 \text{ kJ}$$

Because there are four equivalent C—H bonds in methane, the heat of atomization is equal to the sum of the bond enthalpies of the four C—H bonds. Therefore, the average C—H bond enthalpy for CH_4 is D(C—H) = (1660/4) kJ/mol = 415 kJ/mol.

The bond enthalpy for a given set of atoms, say C—H, depends on the rest of the molecule of which the atom pair is a part. However, the variation from one molecule to another is generally small, which supports the idea that bonding electron pairs are localized between atoms. If we consider C—H bond enthalpies in many different compounds, we find that the average bond enthalpy is 413 kJ/mol, which compares closely with the 415 kJ/mol value calculated from CH_4.

▲ GIVE IT SOME THOUGHT

The hydrocarbon *ethane*, C_2H_6, was first introduced in Section 2.9. How could you use the enthalpy of atomization of $C_2H_6(g)$ along with the value of D(C—H) to provide an estimate for D(C—C)?

Table 8.4▼ lists several average bond enthalpies. *The bond enthalpy is always a positive quantity*; energy is always required to break chemical bonds. Conversely, energy is always released when a bond forms between two gaseous atoms or molecular fragments. The greater the bond enthalpy is, the stronger the bond.

TABLE 8.4 ■ Average Bond Enthalpies (kJ/mol)							
Single Bonds							
C—H	413	N—H	391	O—H	463	F—F	155
C—C	348	N—N	163	O—O	146		
C—N	293	N—O	201	O—F	190	Cl—F	253
C—O	358	N—F	272	O—Cl	203	Cl—Cl	242
C—F	485	N—Cl	200	O—I	234		
C—Cl	328	N—Br	243			Br—F	237
C—Br	276			S—H	339	Br—Cl	218
C—I	240	H—H	436	S—F	327	Br—Br	193
C—S	259	H—F	567	S—Cl	253		
		H—Cl	431	S—Br	218	I—Cl	208
Si—H	323	H—Br	366	S—S	266	I—Br	175
Si—Si	226	H—I	299			I—I	151
Si—C	301						
Si—O	368						
Si—Cl	464						
Multiple Bonds							
C=C	614	N=N	418	O_2	495		
C≡C	839	N≡N	941				
C=N	615	N=O	607	S=O	523		
C≡N	891			S=S	418		
C=O	799						
C≡O	1072						

A molecule with strong chemical bonds generally has less tendency to undergo chemical change than does one with weak bonds. This relationship between strong bonding and chemical stability helps explain the chemical form in which many elements are found in nature. For example, Si—O bonds are among the strongest ones that silicon forms. It should not be surprising, therefore, that SiO_2 and other substances containing Si—O bonds (silicates) are so common; it is estimated that over 90% of Earth's crust is composed of SiO_2 and silicates.

Bond Enthalpies and the Enthalpies of Reactions

We can use average bond enthalpies to estimate the enthalpies of reactions in which bonds are broken and new bonds are formed. This procedure allows us to estimate quickly whether a given reaction will be endothermic ($\Delta H > 0$) or exothermic ($\Delta H < 0$) even if we do not know ΔH_f° for all the chemical species involved.

Our strategy for estimating reaction enthalpies is a straightforward application of Hess's law. ∞ (Section 5.6) We use the fact that breaking bonds is always an endothermic process, and bond formation is always exothermic. We therefore imagine that the reaction occurs in two steps: (1) We supply enough energy to break those bonds in the reactants that are not present in the products. In this step the enthalpy of the system is increased by the sum of the bond enthalpies of the bonds that are broken. (2) We form the bonds in the products that were not present in the reactants. This step releases energy and therefore lowers the enthalpy of the system by the sum of the bond enthalpies of the bonds that are formed. The enthalpy of the reaction, ΔH_{rxn}, is estimated as the sum of the bond enthalpies of the bonds broken minus the sum of the bond enthalpies of the bonds formed:

$$\Delta H_{rxn} = \Sigma(\text{bond enthalpies of bonds broken}) -$$

$$\Sigma(\text{bond enthalpies of bonds formed}) \qquad [8.12]$$

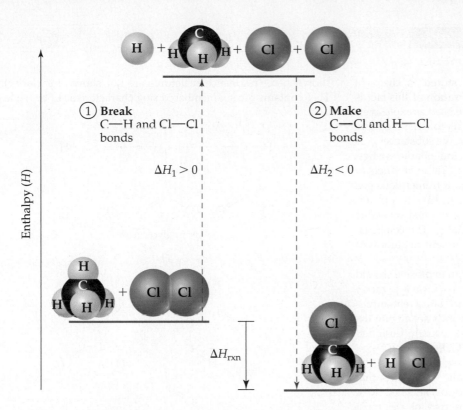

▲ Figure 8.14 **Using bond enthalpies to calculate ΔH_{rxn}.** Average bond enthalpies are used to estimate ΔH_{rxn} for the reaction in Equation 8.13. Breaking the C—H and Cl—Cl bonds produces a positive enthalpy change (ΔH_1), whereas making the C—Cl and H—Cl bonds causes a negative enthalpy change (ΔH_2). The values of ΔH_1 and ΔH_2 are estimated from the values in Table 8.4. From Hess's law, $\Delta H_{rxn} = \Delta H_1 + \Delta H_2$.

Consider, for example, the gas-phase reaction between methane, CH_4, and chlorine to produce methyl chloride, CH_3Cl, and hydrogen chloride, HCl:

$$H—CH_3(g) + Cl—Cl(g) \longrightarrow Cl—CH_3(g) + H—Cl(g) \quad \Delta H_{rxn} = ? \quad [8.13]$$

Our two-step procedure is outlined in Figure 8.14▲. We note that in the course of this reaction the following bonds are broken and made:

> *Bonds broken:* 1 mol C—H, 1 mol Cl—Cl
>
> *Bonds made:* 1 mol C—Cl, 1 mol H—Cl

We first supply enough energy to break the C—H and Cl—Cl bonds, which will raise the enthalpy of the system. We then form the C—Cl and H—Cl bonds, which will release energy and lower the enthalpy of the system. By using Equation 8.12 and the data in Table 8.4, we estimate the enthalpy of the reaction as

$$\Delta H_{rxn} = [D(C—H) + D(Cl—Cl) - [D(C—Cl) + D(H—Cl)]$$
$$= (413 \text{ kJ} + 242 \text{ kJ}) - (328 \text{ kJ} + 431 \text{ kJ}) = -104 \text{ kJ}$$

The reaction is exothermic because the bonds in the products (especially the H—Cl bond) are stronger than the bonds in the reactants (especially the Cl—Cl bond).

We usually use bond enthalpies to estimate ΔH_{rxn} only if we do not have the needed ΔH_f° values readily at hand. For the above reaction, we cannot calculate ΔH_{rxn} from ΔH_f° values and Hess's law because the value of ΔH_f° for $CH_3Cl(g)$ is not given in Appendix C. If we obtain the value of ΔH_f° for $CH_3Cl(g)$ from another source (such as the *CRC Handbook of Chemistry and Physics*) and use Equation 5.31, we find that $\Delta H_{rxn} = -99.8$ kJ for the reaction in Equation 8.13. Thus, the use of average bond enthalpies provides a reasonably accurate estimate of the actual reaction enthalpy change.

It is important to remember that bond enthalpies are derived for *gaseous* molecules and that they are often *averaged* values. Nonetheless, average bond enthalpies are useful for estimating reaction enthalpies quickly, especially for gas-phase reactions.

Chemistry Put to Work   EXPLOSIVES AND ALFRED NOBEL

Enormous amounts of energy can be stored in chemical bonds. Perhaps the most graphic illustration of this fact is seen in certain molecular substances that are used as explosives. Our discussion of bond enthalpies allows us to examine more closely some of the properties of such explosive substances.

An explosive must have the following characteristics: (1) It must decompose very exothermically; (2) the products of its decomposition must be gaseous, so that a tremendous gas pressure accompanies the decomposition; (3) its decomposition must occur very rapidly; and (4) it must be stable enough so that it can be detonated predictably. The combination of the first three effects leads to the violent evolution of heat and gases.

To give the most exothermic reaction, an explosive should have weak chemical bonds and should decompose into molecules with very strong bonds. Looking at bond enthalpies (Table 8.4), the $N\equiv N$, $C\equiv O$, and $C=O$ bonds are among the strongest. Not surprisingly, explosives are usually designed to produce the gaseous products $N_2(g)$, $CO(g)$, and $CO_2(g)$. Water vapor is nearly always produced as well.

Many common explosives are organic molecules that contain nitro (NO_2) or nitrate (NO_3) groups attached to a carbon skeleton. The Lewis structures of two of the most familiar explosives, nitroglycerin and trinitrotoluene (TNT), are shown here (resonance structures are not shown for clarity). TNT contains the six-membered ring characteristic of benzene.

Nitroglycerin

Trinitrotoluene (TNT)

SAMPLE EXERCISE 8.12 | Using Average Bond Enthalpies

Using Table 8.4, estimate ΔH for the following reaction (where we explicitly show the bonds involved in the reactants and products):

SOLUTION

Analyze: We are asked to estimate the enthalpy change for a chemical process by using average bond enthalpies for the bonds that are broken in the reactants and formed in the products.

Plan: Among the reactants, we must break six C—H bonds and a C—C bond in C_2H_6; we also break $\frac{7}{2}O_2$ bonds. Among the products, we form four C=O bonds (two in each CO_2) and six O—H bonds (two in each H_2O).

Solve: Using Equation 8.12 and data from Table 8.4, we have

$$\Delta H = 6D(C-H) + D(C-C) + \tfrac{7}{2}D(O_2) - 4D(C=O) - 6D(O-H)$$
$$= 6(413\text{ kJ}) + 348\text{ kJ} + \tfrac{7}{2}(495\text{ kJ}) - (4(799\text{ kJ}) + 6(463\text{ kJ}))$$
$$= 4558\text{ kJ} - 5974\text{ kJ}$$
$$= -1416\text{ kJ}$$

Check: This estimate can be compared with the value of -1428 kJ calculated from more accurate thermochemical data; the agreement is good.

PRACTICE EXERCISE

Using Table 8.4, estimate ΔH for the reaction

$$H-N-N-H(g) \longrightarrow N\equiv N(g) + 2\,H-H(g)$$

Answer: -86 kJ

Nitroglycerin is a pale yellow, oily liquid. It is highly *shock-sensitive*: Merely shaking the liquid can cause its explosive decomposition into nitrogen, carbon dioxide, water, and oxygen gases:

$$4\ C_3H_5N_3O_9(l) \longrightarrow$$
$$6\ N_2(g) + 12\ CO_2(g) + 10\ H_2O(g) + O_2(g)$$

The large bond enthalpies of the N_2 molecules (941 kJ/mol), CO_2 molecules (2 × 799 kJ/mol), and water molecules (2 × 463 kJ/mol) make this reaction enormously exothermic. Nitroglycerin is an exceptionally unstable explosive because it is in nearly perfect *explosive balance*: With the exception of a small amount of $O_2(g)$ produced, the only products are N_2, CO_2, and H_2O. Note also that, unlike combustion reactions (Section 3.2), explosions are entirely self-contained. No other reagent, such as $O_2(g)$, is needed for the explosive decomposition.

Because nitroglycerin is so unstable, it is difficult to use as a controllable explosive. The Swedish inventor Alfred Nobel (Figure 8.15▶) found that mixing nitroglycerin with an absorbent solid material such as diatomaceous earth or cellulose gives a solid explosive (*dynamite*) that is much safer than liquid nitroglycerin.

Related Exercises: 8.93 and 8.94

▲ **Figure 8.15 Alfred Nobel (1833–1896), the Swedish inventor of dynamite.** By many accounts Nobel's discovery that nitroglycerin could be made more stable by absorbing it onto cellulose was an accident. This discovery made Nobel a very wealthy man. He was also a complex and lonely man, however, who never married, was frequently ill, and suffered from chronic depression. He had invented the most powerful military explosive to date, but he strongly supported international peace movements. His will stated that his fortune be used to establish prizes awarding those who "have conferred the greatest benefit on mankind," including the promotion of peace and "fraternity between nations." The Nobel Prize is probably the most coveted award that a scientist, writer, or peace advocate can receive.

Bond Enthalpy and Bond Length

Just as we can define an average bond enthalpy, we can also define an average bond length for a number of common bond types. Some of these are listed in Table 8.5▼. Of particular interest is the relationship among bond enthalpy, bond length, and the number of bonds between the atoms. For example, we can use data in Tables 8.4 and 8.5 to compare the bond lengths and bond enthalpies of carbon–carbon single, double, and triple bonds:

C—C	C=C	C≡C
1.54 Å	1.34 Å	1.20 Å
348 kJ/mol	614 kJ/mol	839 kJ/mol

As the number of bonds between the carbon atoms increases, the bond enthalpy increases and the bond length decreases; that is, the carbon atoms are held more closely and more tightly together. In general, *as the number of bonds between two atoms increases, the bond grows shorter and stronger.*

TABLE 8.5 ■ Average Bond Lengths for Some Single, Double, and Triple Bonds

Bond	Bond Length (Å)	Bond	Bond Length (Å)
C—C	1.54	N—N	1.47
C=C	1.34	N=N	1.24
C≡C	1.20	N≡N	1.10
C—N	1.43	N—O	1.36
C=N	1.38	N=O	1.22
C≡N	1.16		
		O—O	1.48
C—O	1.43	O=O	1.21
C=O	1.23		
C≡O	1.13		

■ **SAMPLE INTEGRATIVE EXERCISE** | **Putting Concepts Together**

Phosgene, a substance used in poisonous gas warfare duirng World War I, is so named because it was first prepared by the action of sunlight on a mixture of carbon monoxide and chlorine gases. Its name comes from the Greek words *phos* (light) and *genes* (born of). Phosgene has the following elemental composition: 12.14% C, 16.17% O, and 71.69% Cl by mass. Its molar mass is 98.9 g/mol. **(a)** Determine the molecular formula of this compound. **(b)** Draw three Lewis structures for the molecule that satisfy the octet rule for each atom. (The Cl and O atoms bond to C.) **(c)** Using formal charges, determine which Lewis structure is the most important one. **(d)** Using average bond enthalpies, estimate ΔH for the formation of gaseous phosgene from $CO(g)$ and $Cl_2(g)$.

SOLUTION

(a) The empirical formula of phosgene can be determined from its elemental composition. ∞ (Section 3.5) Assuming 100 g of the compound and calculating the number of moles of C, O, and Cl in this sample, we have

$$(12.14 \text{ g C})\left(\frac{1 \text{ mol C}}{12.01 \text{ g C}}\right) = 1.011 \text{ mol C}$$

$$(16.17 \text{ g O})\left(\frac{1 \text{ mol O}}{16.00 \text{ g O}}\right) = 1.011 \text{ mol O}$$

$$(71.69 \text{ g Cl})\left(\frac{1 \text{ mol Cl}}{35.45 \text{ g Cl}}\right) = 2.022 \text{ mol Cl}$$

The ratio of the number of moles of each element, obtained by dividing each number of moles by the smallest quantity, indicates that there is one C and one O for each two Cl in the empirical formula, $COCl_2$.

The molar mass of the empirical formula is $12.01 + 16.00 + 2(35.45) = 98.91$ g/mol, the same as the molar mass of the molecule. Thus, $COCl_2$ is the molecular formula.

(b) Carbon has four valence electrons, oxygen has six, and chlorine has seven, giving $4 + 6 + 2(7) = 24$ electrons for the Lewis structures. Drawing a Lewis structure with all single bonds does not give the central carbon atom an octet. Using multiple bonds, three structures satisfy the octet rule:

(c) Calculating the formal charges on each atom gives

The first structure is expected to be the most important one because it has the lowest formal charges on each atom. Indeed, the molecule is usually represented by this Lewis structure.

(d) Writing the chemical equation in terms of the Lewis structures of the molecules, we have

Thus, the reaction involves breaking a $C\equiv O$ bond and a $Cl-Cl$ bond and forming a $C=O$ bond and two $C-Cl$ bonds. Using bond enthalpies from Table 8.4, we have

$$\Delta H = D(C\equiv O) + D(Cl-Cl) - (D(C=O) + 2D(C-Cl))$$
$$= 1072 \text{ kJ} + 242 \text{ kJ} - (799 \text{ kJ} + 2(328 \text{ kJ})) = -141 \text{ kJ}$$

CHAPTER REVIEW

SUMMARY AND KEY TERMS

Introduction and Section 8.1 In this chapter we have focused on the interactions that lead to the formation of **chemical bonds**. We classify these bonds into three broad groups: **ionic bonds**, which result from the electrostatic forces that exist between ions of opposite charge; **covalent bonds**, which result from the sharing of electrons by two atoms; and **metallic bonds**, which result from a delocalized sharing of electrons in metals. The formation of bonds involves interactions of the outermost electrons of atoms, their valence electrons. The valence electrons of an atom can be represented by electron-dot symbols, called **Lewis symbols**. The tendencies of atoms to gain, lose, or share their valence electrons often follow the **octet rule**, which can be viewed as an attempt by atoms to achieve a noble-gas electron configuration.

Section 8.2 Ionic bonding results from the transfer of electrons from one atom to another, leading to the formation of a three-dimensional lattice of charged particles. The stabilities of ionic substances result from the strong electrostatic attractions between an ion and the surrounding ions of opposite charge. The magnitude of these interactions is measured by the **lattice energy**, which is the energy needed to separate an ionic lattice into gaseous ions. Lattice energy increases with increasing charge on the ions and with decreasing distance between the ions. The **Born–Haber cycle** is a useful thermochemical cycle in which we use Hess's law to calculate the lattice energy as the sum of several steps in the formation of an ionic compound.

Section 8.3 A covalent bond results from the sharing of electrons. We can represent the electron distribution in molecules by means of **Lewis structures**, which indicate how many valence electrons are involved in forming bonds and how many remain as unshared electron pairs. The octet rule helps determine how many bonds will be formed between two atoms. The sharing of one pair of electrons produces a **single bond**; the sharing of two or three pairs of electrons between two atoms produces **double** or **triple bonds**, respectively. Double and triple bonds are examples of multiple bonding between atoms. The **bond length** between two bonded atoms is the distance between the two nuclei. The bond length decreases as the number of bonds between the atoms increases.

Section 8.4 In covalent bonds, the electrons may not necessarily be shared equally between two atoms. **Bond polarity** helps describe unequal sharing of electrons in a bond. In a **nonpolar covalent bond** the electrons in the bond are shared equally by the two atoms; in a **polar covalent bond** one of the atoms exerts a greater attraction for the electrons than the other.

Electronegativity is a numerical measure of the ability of an atom to compete with other atoms for the electrons shared between them. Fluorine is the most electronegative element, meaning it has the greatest ability to attract electrons from other atoms. Electronegativity values range from 0.7 for Cs to 4.0 for F. Electronegativity generally increases from left to right in a row of the periodic table, and decreases going down a column. The difference in the electronegativities of bonded atoms can be used to determine the polarity of a bond. The greater the electronegativity difference the more polar the bond.

A **polar molecule** is one whose centers of positive and negative charge do not coincide. Thus, a polar molecule has a positive side and a negative side. This separation of charge produces a **dipole**, the magnitude of which is given by the **dipole moment**, which is measured in debyes (D). Dipole moments increase with increasing amount of charge separated and increasing distance of separation. Any diatomic molecule $X-Y$ in which X and Y have different electronegativities is a polar molecule.

Most bonding interactions lie between the extremes of covalent and ionic bonding. While it is generally true that the bonding between a metal and a nonmetal is predominantly ionic, exceptions to this guideline are not uncommon when the difference in electronegativity of the atoms is relatively small or when the oxidation state of the metal becomes large.

Sections 8.5 and 8.6 If we know which atoms are connected to one another, we can draw Lewis structures for molecules and ions by a simple procedure. Once we do so, we can determine the **formal charge** of each atom in a Lewis structure, which is the charge that the atom would have if all atoms had the same electronegativity. Most acceptable Lewis structures will have low formal charges with any negative formal charges residing on more electronegative atoms.

Sometimes a single Lewis structure is inadequate to represent a particular molecule (or ion). In such situations, we describe the molecule by using two or more **resonance structures** for the molecule. The molecule is envisioned as a blend of these multiple resonance structures. Resonance structures are important in describing the bonding in molecules such as ozone, O_3, and the organic molecule benzene, C_6H_6.

Section 8.7 The octet rule is not obeyed in all cases. Exceptions occur when (a) a molecule has an odd number of electrons, (b) it is not possible to complete an octet around an atom without forcing an unfavorable distribution of electrons, or (c) a large atom is surrounded by a sufficiently large number of small electronegative atoms that it has more than an octet of electrons around it. In this last case we envision using the unfilled d orbitals of the large atom to "expand" the valence shell of the atom. Expanded valence shells are observed for atoms in the third row and beyond in the periodic table, for which low-energy d orbitals are available.

Section 8.8 The strength of a covalent bond is measured by its **bond enthalpy**, which is the molar enthalpy change upon breaking a particular bond. Average bond enthalpies can be determined for a wide variety of covalent bonds. The strengths of covalent bonds increase with the number of electron pairs shared between two atoms.

We can use bond enthalpies to estimate the enthalpy change during chemical reactions in which bonds are broken and new bonds formed. The average bond length between two atoms decreases as the number of bonds between the atoms increases, consistent with the bond being stronger as the number of bonds increases.

KEY SKILLS

- Write Lewis symbols for atoms and ions.
- Understand lattice energy and be able to arrange compounds in order of increasing lattice energy based on the charges and sizes of the ions involved.
- Use atomic electron configurations and the octet rule to write Lewis structures for molecules to determine their electron distribution.
- Use electronegativity differences to identify nonpolar covalent, polar covalent, and ionic bonds.
- Calculate charge separation in diatomic molecules based on the experimentally measured dipole moment and bond distance.
- Calculate formal charges from Lewis structures and use those formal charges to identify the most favorable Lewis structures.
- Recognize molecules where resonance structures are needed to describe the bonding.
- Recognize exceptions to the octet rule and draw accurate Lewis structures even when the octet rule is not obeyed.
- Understand the relationship between bond type (single, double, and triple), bond strength (or enthalpy) and bond length.
- Use bond enthalpies to estimate enthalpy changes for reactions involving gas phase reactants and products.

KEY EQUATIONS

- $E_{el} = \dfrac{\kappa Q_1 Q_2}{d}$ [8.4]

  The potential energy of two interacting charges

- $\mu = Qr$ [8.11]

  The dipole moment of two charges of equal magnitude but opposite sign, separated by a distance r

- $\Delta H_{rxn} = \Sigma(\text{bond enthalpies of bonds broken}) -$ [8.12]
  $\Sigma(\text{bond enthalpies of bonds formed})$

  The enthalpy change as a function of bond enthalpies for reactions involving gas phase molecules.

VISUALIZING CONCEPTS

8.1 For each of these Lewis symbols, indicate the group in the periodic table in which the element X belongs:
(a) $\cdot\overset{\cdot}{X}\cdot$ (b) $\cdot X\cdot$ (c) $:\overset{\cdot}{X}\cdot$

8.2 Illustrated at right are four ions—A, B, X, and Y— showing their relative ionic radii. The ions shown in red carry positive charges: a 2+ charge for A and a 1+ charge for B. Ions shown in blue carry negative charges: a 1– charge for X and a 2– charge for Y. (a) Which combinations of these ions produce ionic compounds where there is a 1:1 ratio of cations and anions? (b) Among those compounds, which combination of ions leads to the ionic compound having the largest lattice energy? (c) Which combination of ions leads to the ionic compound having the smallest lattice energy? [Section 8.2]

8.3 The orbital diagram below shows the valence electrons for a 2+ ion of an element. (a) What is the element? (b) What is the electron configuration of an atom of this element? [Section 8.2]

3d

8.4 In the Lewis structure shown below, A, D, E, Q, X, and Z represent elements in the first two rows of the periodic table (H—Ne). Identify all six elements so that the formal charges of all atoms are zero. [Section 8.3]

$$\begin{array}{c} \ddot{\textrm{:E:}} \quad \textrm{X} \\ \| \quad | \\ \ddot{\textrm{:A}}\!-\!\textrm{D}\!-\!\ddot{\textrm{Q}}\!-\!\textrm{Z} \\ \ddot{} \end{array}$$

8.5 The partial Lewis structure below is for a hydrocarbon molecule. In the full Lewis structure, each carbon atom satisfies the octet rule, and there are no unshared electron pairs in the molecule. The carbon–carbon bonds are labeled 1, 2, and 3. **(a)** Determine where the hydrogen atoms are in the molecule. **(b)** Rank the carbon–carbon bonds in order of increasing bond length. **(c)** Rank the carbon–carbon bonds in order of increasing bond enthalpy. [Section 8.3 and 8.8]

$$\textrm{C}\overset{1}{=}\textrm{C}\overset{2}{-}\textrm{C}\overset{3}{\equiv}\textrm{C}$$

8.6 Consider the Lewis structure for the polyatomic oxyanion shown below, where X is an element from the 3rd period (Na—Ar). By changing the overall charge, n, from 1− to 2− to 3− we get three different polyatomic ions. For each of these ions **(a)** Identify the central atom, X. **(b)** Determine the formal charge of the central atom, X. **(c)** Draw a Lewis structure that makes the formal charge on the central atom equal to zero. **(d)** If the Lewis structure you drew in part (c) differs from the Lewis structure shown below, which one do you think is the best one (if you were able to choose only a single Lewis structure)? [Sections 8.5, 8.6, and 8.7]

$$\left[\begin{array}{c} \ddot{\textrm{:O:}} \\ | \\ \ddot{\textrm{:O}}\!-\!\textrm{X}\!-\!\ddot{\textrm{O:}} \\ | \\ \ddot{\textrm{:O:}} \end{array} \right]^{n-}$$

EXERCISES

Lewis Symbols

8.7 **(a)** What are valence electrons? **(b)** How many valence electrons does a nitrogen atom possess? **(c)** An atom has the electron configuration $1s^2 2s^2 2p^6 3s^2 3p^2$. How many valence electrons does the atom have?

8.8 **(a)** What is the octet rule? **(b)** How many electrons must a sulfur atom gain to achieve an octet in its valence shell? **(c)** If an atom has the electron configuration $1s^2 2s^2 2p^3$, how many electrons must it gain to achieve an octet?

8.9 Write the electron configuration for phosphorus. Identify the valence electrons in this configuration and the nonvalence electrons. From the standpoint of chemical reactivity, what is the important difference between them?

8.10 **(a)** Write the electron configuration for the element titanium, Ti. How many valence electrons does this atom possess? **(b)** Hafnium, Hf, is also found in group 4B. Write the electron configuration for Hf. **(c)** Both Ti and Hf behave as though they possess the same number of valence electrons. Which of the subshells in the electron configuration of Hf behave as valence orbitals? Which behave as core orbitals?

8.11 Write the Lewis symbol for atoms of each of the following elements: **(a)** Al, **(b)** Br, **(c)** Ar, **(d)** Sr.

8.12 What is the Lewis symbol for each of the following atoms or ions: **(a)** Ca, **(b)** P, **(c)** Mg^{2+}, **(d)** S^{2-}?

Ionic Bonding

8.13 Using Lewis symbols, diagram the reaction between magnesium and oxygen atoms to give the ionic substance MgO.

8.14 Use Lewis symbols to represent the reaction that occurs between Ca and F atoms.

8.15 Predict the chemical formula of the ionic compound formed between the following pairs of elements: **(a)** Al and F, **(b)** K and S, **(c)** Y and O, **(d)** Mg and N.

8.16 Which ionic compound is expected to form from combining the following pairs of elements: **(a)** barium and fluorine, **(b)** cesium and chlorine, **(c)** lithium and nitrogen, **(d)** aluminum and oxygen?

8.17 Write the electron configuration for each of the following ions, and determine which ones possess noble-gas configurations: **(a)** Sr^{2+}, **(b)** Ti^{2+}, **(c)** Se^{2-}, **(d)** Ni^{2+}, **(e)** Br^-, **(f)** Mn^{3+}.

8.18 Write electron configurations for the following ions, and determine which have noble-gas configurations: **(a)** Zn^{2+}, **(b)** Te^{2-}, **(c)** Sc^{3+}, **(d)** Rh^{3+}, **(e)** Tl^+, **(f)** Bi^{3+}.

8.19 **(a)** Define the term *lattice energy*. **(b)** Which factors govern the magnitude of the lattice energy of an ionic compound?

8.20 NaCl and KF have the same crystal structure. The only difference between the two is the distance that separates cations and anions. **(a)** The lattice energies of NaCl and KF are given in Table 8.2. Based on the lattice energies, would you expect the Na–Cl or the K–F distance to be longer? **(b)** Use the ionic radii given in Figure 7.8 to estimate the Na–Cl and K–F distances. Does this estimate agree with the prediction you made based upon the lattice energies?

8.21 The ionic substances KF, CaO, and ScN are isoelectronic (they have the same number of electrons). Examine the lattice energies for these substances in Table 8.2, and account for the trends you observe.

8.22 **(a)** Does the lattice energy of an ionic solid increase or decrease (i) as the charges of the ions increase, (ii) as the sizes of the ions increase? **(b)** Using a periodic table, arrange the following substances according to their expected lattice energies, listing them from lowest lattice energy to the highest: ScN, KBr, MgO, NaF. Compare your list with the data in Table 8.2.

8.23 The lattice energies of KBr and CsCl are nearly equal (Table 8.2). What can you conclude from this observation?

8.24 Explain the following trends in lattice energy: (a) $CaF_2 > BaF_2$; (b) $NaCl > RbBr > CsBr$; (c) $BaO > KF$.

8.25 Energy is required to remove two electrons from Ca to form Ca^{2+} and is required to add two electrons to O to form O^{2-}. Why, then, is CaO stable relative to the free elements?

8.26 Construct a Born–Haber cycle for the formation of the hypothetical compound $NaCl_2$, where the sodium ion has a 2+ charge (the 2nd ionization energy for sodium is given in Table 7.2). (a) How large would the lattice energy need to be for the formation of $NaCl_2$ to be exothermic? (b) If we were to estimate the lattice energy of $NaCl_2$ to be roughly equal to that of $MgCl_2$ (2326 kJ/mol from Table 8.2), what value would you obtain for the standard enthalpy of formation, ΔH_f°, of $NaCl_2$?

8.27 Use data from Appendix C, Figure 7.12, and Figure 7.14 to calculate the lattice energy of RbCl. Is this value greater than or less than the lattice energy of NaCl? Explain.

8.28 (a) Based on the lattice energies of $MgCl_2$ and $SrCl_2$ given in Table 8.2, what is the range of values that you would expect for the lattice energy of $CaCl_2$? (b) Using data from Appendix C, Figure 7.12, and Figure 7.14 and the value of the second ionization energy for Ca, 1145 kJ/mol, calculate the lattice energy of $CaCl_2$.

Covalent Bonding, Electronegativity, and Bond Polarity

8.29 (a) What is meant by the term *covalent bond*? (b) Give three examples of covalent bonding. (c) A substance XY, formed from two different elements, boils at −33 °C. Is XY likely to be a covalent or an ionic substance? Explain.

8.30 Which of these elements is unlikely to form covalent bonds: S, H, K, Ar, Si? Explain your choices.

8.31 Using Lewis symbols and Lewis structures, diagram the formation of $SiCl_4$ from Si and Cl atoms.

8.32 Use Lewis symbols and Lewis structures to diagram the formation of PF_3 from P and F atoms.

8.33 (a) Construct a Lewis structure for O_2 in which each atom achieves an octet of electrons. (b) Explain why it is necessary to form a double bond in the Lewis structure. (c) The bond in O_2 is shorter than the O—O bond in compounds that contain an O—O single bond. Explain this observation.

8.34 (a) Construct a Lewis structure for hydrogen peroxide, H_2O_2, in which each atom achieves an octet of electrons. (b) Do you expect the O—O bond in H_2O_2 to be longer or shorter than the O—O bond in O_2?

8.35 (a) What is meant by the term electronegativity? (b) On the Pauling scale what is the range of electronegativity values for the elements? (c) Which element has the greatest electronegativity? (d) Which element has the smallest electronegativity?

8.36 (a) What is the trend in electronegativity going from left to right in a row of the periodic table? (b) How do electronegativity values generally vary going down a column in the periodic table? (c) How do periodic trends in electronegativity relate to those for ionization energy and electron affinity?

8.37 Using only the periodic table as your guide, select the most electronegative atom in each of the following sets: (a) Se, Rb, O, In; (b) Al, Ca, C, Si; (c) Ge, As, P, Sn; (d) Li, Rb, Be, Sr.

8.38 By referring only to the periodic table, select (a) the most electronegative element in group 6A; (b) the least elec-tronegative element in the group Al, Si, P; (c) the most electronegative element in the group Ga, P, Cl, Na; (d) the element in the group K, C, Zn, F, that is most likely to form an ionic compound with Ba.

8.39 Which of the following bonds are polar: (a) B—F, (b) Cl—Cl, (c) Se—O, (d) H—I? Which is the more electronegative atom in each polar bond?

8.40 Arrange the bonds in each of the following sets in order of increasing polarity: (a) C—F, O—F, Be—F; (b) O—Cl, S—Br, C—P; (c) C—S, B—F, N—O.

8.41 The dipole moment and bond distance measured for the highly reactive gas phase OH molecule are 1.78 D and 0.98 Å, respectively. (a) Given these values calculate the effective charges on the H and O atoms of the OH molecule in units of the electronic charges e. (b) Is this bond more or less polar than the H—Cl bond in an HCl molecule? (c) Is that what you would have expected based on electronegativities?

8.42 The iodine monobromide molecule, IBr, has a bond length of 2.49 Å and a dipole moment of 1.21 D. (a) Which atom of the molecule is expected to have a negative charge? Explain. (b) Calculate the effective charges on the I and Br atoms in IBr, in units of the electronic charge e.

8.43 In the following pairs of binary compounds determine which one is a molecular substance and which one is an ionic substance. Use the appropriate naming convention (for ionic or molecular substances) to assign a name to each compound: (a) SiF_4 and LaF_3, (b) $FeCl_2$ and $ReCl_6$, (c) $PbCl_4$ and RbCl.

8.44 In the following pairs of binary compounds determine which one is a molecular substance and which one is an ionic substance. Use the appropriate naming convention (for ionic or molecular substances) to assign a name to each compound: (a) $TiCl_4$ and CaF_2, (b) ClF_3 and VF_3, (c) $SbCl_5$ and AlF_3.

Lewis Structures; Resonance Structures

8.45 Draw Lewis structures for the following: **(a)** SiH_4, **(b)** CO, **(c)** SF_2, **(d)** H_2SO_4 (H is bonded to O), **(e)** ClO_2^- **(f)** NH_2OH.

8.46 Write Lewis structures for the following: **(a)** H_2CO (both H atoms are bonded to C), **(b)** H_2O_2, **(c)** C_2F_6 (contains a C—C bond), **(d)** AsO_3^{3-}, **(e)** H_2SO_3 (H is bonded to O), **(f)** C_2H_2.

8.47 **(a)** When talking about atoms in a Lewis structure, what is meant by the term *formal charge*? **(b)** Does the formal charge of an atom represent the actual charge on that atom? Explain. **(c)** How does the formal charge of an atom in a Lewis structure differ from the oxidation number of the atom?

8.48 **(a)** Write a Lewis structure for the phosphorus trifluoride molecule, PF_3. Is the octet rule satisfied for all the atoms in your structure? **(b)** Determine the oxidation numbers of the P and F atoms. **(c)** Determine the formal charges of the P and F atoms. **(d)** Is the oxidation number for the P atom the same as its formal charge? Explain why or why not.

8.49 Write Lewis structures that obey the octet rule for each of the following, and assign oxidation numbers and formal charges to each atom: **(a)** NO^+, **(b)** $POCl_3$ (P is bonded to the three Cl atoms and to the O), **(c)** ClO_4^-, **(d)** $HClO_3$ (H is bonded to O).

8.50 For each of the following molecules or ions of sulfur and oxygen, write a single Lewis structure that obeys the octet rule, and calculate the oxidation numbers and formal charges on all the atoms: **(a)** SO_2, **(b)** SO_3, **(c)** SO_3^{2-}. **(d)** Arrange these molecules/ions in order of increasing S—O bond distance.

8.51 **(a)** Write one or more appropriate Lewis structures for the nitrite ion, NO_2^-. **(b)** With what allotrope of oxygen is it isoelectronic? **(c)** What would you predict for the lengths of the bonds in NO_2^- relative to N—O single bonds?

8.52 Consider the nitryl cation, NO_2^+. **(a)** Write one or more appropriate Lewis structures for this ion. **(b)** Are resonance structures needed to describe the structure? **(c)** With what familiar molecule is it isoelectronic?

8.53 Predict the ordering of the C—O bond lengths in CO, CO_2, and CO_3^{2-}.

8.54 Based on Lewis structures, predict the ordering of N—O bond lengths in NO^+, NO_2^-, and NO_3^-.

8.55 **(a)** Use the concept of resonance to explain why all six C—C bonds in benzene are equal in length. **(b)** Are the C—C bond lengths in benzene shorter than C—C single bonds? Are they shorter than C=C double bonds?

8.56 Mothballs are composed of naphthalene, $C_{10}H_8$, a molecule of which consists of two six-membered rings of carbon fused along an edge, as shown in this incomplete Lewis structure:

(a) Write two complete Lewis structures for naphthalene. **(b)** The observed C—C bond lengths in the molecule are intermediate between C—C single and C=C double bonds. Explain. **(c)** Represent the resonance in naphthalene in a way analogous to that used to represent it in benzene.

Exceptions to the Octet Rule

8.57 **(a)** State the octet rule. **(b)** Does the octet rule apply to ionic as well as to covalent compounds? Explain, using examples as appropriate.

8.58 Considering the nonmetals, what is the relationship between the group number for an element (carbon, for example, belongs to group 4A; see the periodic table on the inside front cover) and the number of single covalent bonds that element needs to form to conform to the octet rule?

8.59 What is the most common exception to the octet rule? Give two examples.

8.60 For elements in the third row of the periodic table and beyond, the octet rule is often not obeyed. What factors are usually cited to explain this fact?

8.61 Draw the Lewis structures for each of the following ions or molecules. Identify those that do not obey the octet rule, and explain why they do not. **(a)** SO_3^{2-}, **(b)** AlH_3, **(c)** N_3^-, **(d)** CH_2Cl_2, **(e)** SbF_5.

8.62 Draw the Lewis structures for each of the following molecules or ions. Which do not obey the octet rule? **(a)** NH_4^+, **(b)** SCN^-, **(c)** PCl_3, **(d)** TeF_4, **(e)** XeF_2.

8.63 In the vapor phase, $BeCl_2$ exists as a discrete molecule. **(a)** Draw the Lewis structure of this molecule, using only single bonds. Does this Lewis structure satisfy the octet rule? **(b)** What other resonance forms are possible that satisfy the octet rule? **(c)** Using formal charges, select the resonance form from among all the Lewis structures that is most important in describing $BeCl_2$.

8.64 **(a)** Describe the molecule chlorine dioxide, ClO_2, using three possible resonance structures. **(b)** Do any of these resonance structures satisfy the octet rule for every atom in the molecule? Why or why not? **(c)** Using formal charges, select the resonance structure(s) that is (are) most important.

Bond Enthalpies

8.65 Using the bond enthalpies tabulated in Table 8.4, estimate ΔH for each of the following gas-phase reactions:

(a)

$$\underset{H}{\overset{H}{>}}C{=}C\underset{H}{\overset{H}{<}} + H{-}O{-}O{-}H \longrightarrow$$

$$H{-}O{-}\underset{\underset{H}{|}}{\overset{\overset{H}{|}}{C}}{-}\underset{\underset{H}{|}}{\overset{\overset{H}{|}}{C}}{-}O{-}H$$

(b)

$$\underset{H}{\overset{H}{>}}C{=}C\underset{H}{\overset{H}{<}} + H{-}C{\equiv}N \longrightarrow$$

$$H{-}\underset{\underset{H}{|}}{\overset{\overset{H}{|}}{C}}{-}\underset{\underset{H}{|}}{\overset{\overset{H}{|}}{C}}{-}C{\equiv}N$$

(c) $2\,Cl{-}\underset{\underset{Cl}{|}}{N}{-}Cl \longrightarrow N{\equiv}N + 3\,Cl{-}Cl$

8.66 Using bond enthalpies (Table 8.4), estimate ΔH for the following gas-phase reactions:

(a) $Br{-}\underset{\underset{Br}{|}}{\overset{\overset{Br}{|}}{C}}{-}H + Cl{-}Cl \longrightarrow Br{-}\underset{\underset{Br}{|}}{\overset{\overset{Br}{|}}{C}}{-}Cl + H{-}Cl$

(b) $H{-}S{-}\underset{\underset{H}{|}}{\overset{\overset{H}{|}}{C}}{-}\underset{\underset{H}{|}}{\overset{\overset{H}{|}}{C}}{-}S{-}H + 2\,H{-}Br \longrightarrow$

$$Br{-}\underset{\underset{H}{|}}{\overset{\overset{H}{|}}{C}}{-}\underset{\underset{H}{|}}{\overset{\overset{H}{|}}{C}}{-}Br + 2\,H{-}S{-}H$$

(c) $H{-}\underset{\underset{H}{|}}{N}{-}\underset{\underset{H}{|}}{N}{-}H + Cl{-}Cl \longrightarrow 2\,H{-}\underset{\underset{Cl}{|}}{N}{-}Cl$

8.67 Using bond enthalpies (Table 8.4), estimate ΔH for each of the following reactions:

(a) $2\,CH_4(g) + O_2(g) \longrightarrow 2\,CH_3OH(g)$
(b) $H_2(g) + Br_2(g) \longrightarrow 2\,HBr(g)$
(c) $2\,H_2O_2(g) \longrightarrow 2\,H_2O(g) + O_2(g)$

[8.68] Use bond enthalpies (Table 8.4) to estimate the enthalpy change for each of the following reactions:
(a) $C_3H_8(g) + 5\,O_2(g) \longrightarrow 3\,CO_2(g) + 4\,H_2O(g)$
(b) $C_2H_5OH(g) + 3\,O_2(g) \longrightarrow 2\,CO_2(g) + 3\,H_2O(g)$
(c) $8\,H_2S(g) \longrightarrow 8\,H_2(g) + S_8(s)$

8.69 Ammonia is produced directly from nitrogen and hydrogen by using the Haber process. The chemical reaction is

$$N_2(g) + 3\,H_2(g) \longrightarrow 2\,NH_3(g)$$

(a) Use bond enthalpies (Table 8.4) to estimate the enthalpy change for the reaction, and tell whether this reaction is exothermic or endothermic. **(b)** Compare the enthalpy change you calculate in (a) to the true enthalpy change as obtained using ΔH_f° values.

8.70 (a) Use bond enthalpies to estimate the enthalpy change for the reaction of hydrogen with ethene:

$$H_2(g) + C_2H_4(g) \longrightarrow C_2H_6(g)$$

(b) Calculate the standard enthalpy change for this reaction, using heats of formation. Why does this value differ from that calculated in (a)?

8.71 Given the following bond-dissociation energies, calculate the average bond enthalpy for the Ti—Cl bond.

	ΔH (kJ/mol)
$TiCl_4(g) \longrightarrow TiCl_3(g) + Cl(g)$	335
$TiCl_3(g) \longrightarrow TiCl_2(g) + Cl(g)$	423
$TiCl_2(g) \longrightarrow TiCl(g) + Cl(g)$	444
$TiCl(g) \longrightarrow Ti(g) + Cl(g)$	519

[8.72] (a) Using average bond enthalpies, predict which of the following reactions will be most exothermic:
(i) $C(g) + 2\,F_2(g) \longrightarrow CF_4(g)$
(ii) $CO(g) + 3\,F_2 \longrightarrow CF_4(g) + OF_2(g)$
(iii) $CO_2(g) + 4\,F_2 \longrightarrow CF_4(g) + 2\,OF_2(g)$
(b) Explain the trend, if any, that exists between reaction exothermicity and the extent to which the carbon atom is bonded to oxygen.

ADDITIONAL EXERCISES

8.73 How many elements in the periodic table are represented by a Lewis symbol with a single dot? Are all these elements in the same group? Explain.

8.74 (a) Explain the following trend in lattice energy: BeH_2, 3205 kJ/mol; MgH_2, 2791 kJ/mol; CaH_2, 2410 kJ/mol; SrH_2, 2250 kJ/mol; BaH_2, 2121 kJ/mol. **(b)** The lattice energy of ZnH_2 is 2870 kJ/mol. Based on the data given in part (a), the radius of the Zn^{2+} ion is expected to be closest to that of which group 2A element?

8.75 Based on data in Table 8.2, estimate (within 30 kJ/mol) the lattice energy for **(a)** LiBr, **(b)** CsBr, **(c)** $CaCl_2$.

8.76 Would you expect AlN to have a lattice energy that is larger or smaller than ScN? Explain.

[8.77] From the ionic radii given in Figure 7.8, calculate the potential energy of a Ca^{2+} and O^{2-} ion pair that are just touching (the magnitude of the electronic charge is given on the back inside cover). Calculate the energy of a mole of such pairs. How does this value compare with the lattice energy of CaO (Table 8.2)? Explain the difference.

[8.78] From Equation 8.4 and the ionic radii given in Figure 7.8, calculate the potential energy of the following pairs of ions. Assume that the ions are separated by a distance equal to the sum of their ionic radii: (a) Na^+, Br^-; (b) Rb^+, Br^-; (c) Sr^{2+}, S^{2-}.

8.79 (a) How does a polar molecule differ from a nonpolar one? (b) Atoms X and Y have different electronegativities. Will the diatomic molecule X—Y necessarily be polar? Explain. (c) What factors affect the size of the dipole moment of a diatomic molecule?

8.80 Which of the following molecules or ions contain polar bonds: (a) P_4, (b) H_2S, (c) NO_2^-, (d) S_2^{2-}?

8.81 To address energy and environmental issues, there is great interest in powering vehicles with hydrogen rather than gasoline. One of the most attractive aspects of the "hydrogen economy" is the fact that in principle the only emission would be water. However, two daunting obstacles must be overcome before this vision can become a reality. First, an economical method of producing hydrogen must be found. Second, a safe, lightweight, and compact way of storing hydrogen must be found. The hydrides of light metals are attractive for hydrogen storage because they can store a high weight percentage of hydrogen in a small volume. One of the most attractive hydrides is $NaAlH_4$, which can release 5.6% of its mass as H_2 upon decomposing to $NaH(s)$, $Al(s)$, and $H_2(g)$. $NaAlH_4$ possesses both covalent bonds, which hold polyatomic anions together, and ionic bonds. (a) Write a balanced equation for the decomposition of $NaAlH_4$. (b) Which element in $NaAlH_4$ is the most electronegative? Which one is the least electronegative? (c) Based on electronegativity differences, what do you think is the identity of the polyatomic anion? Draw a Lewis structure for this ion.

8.82 For the following collection of nonmetallic elements, O, P, Te, I, B, (a) which two would form the most polar single bond? (b) Which two would form the longest single bond? (c) Which two would be likely to form a compound of formula XY_2? (d) Which combinations of elements would likely yield a compound of empirical formula X_2Y_3? In each case explain your answer.

8.83 You and a partner are asked to complete a lab entitled "Oxides of Ruthenium" that is scheduled to extend over two lab periods. The first lab, which is to be completed by your partner, is devoted to carrying out compositional analysis. In the second lab, you are to determine melting points. Upon going to lab you find two unlabeled vials, one containing a soft yellow substance and the other a black powder. You also find the following notes in your partner's notebook—Compound 1: 76.0% Ru and 24.0% O (by mass), Compound 2: 61.2% Ru and 38.8% O (by mass). (a) What is the empirical formula for Compound 1? (b) What is the empirical formula for Compound 2? (c) Upon determining the melting points of these two compounds, you find that the yellow compound melts at 25 °C, while the black powder does not melt up to the maximum temperature of your apparatus, 1200 °C. What is the identity of the yellow compound? What is the identity of the black compound? Be sure to use the appropriate naming convention depending upon whether the compound is better described as a molecular or ionic compound.

8.84 You and a partner are asked to complete a lab entitled "Fluorides of Group 6B Metals" that is scheduled to extend over two lab periods. The first lab, which is to be completed by your partner, is devoted to carrying out compositional analysis. In the second lab, you are to determine melting points. Upon going to lab you find two unlabeled vials, one containing a colorless liquid and the other a green powder. You also find the following notes in your partner's notebook—Compound 1: 47.7% Cr and 52.3% F (by mass), Compound 2: 45.7% Mo and 54.3% F (by mass). (a) What is the empirical formula for Compound 1? (b) What is the empirical formula for Compound 2? (c) Upon determining the melting points of these two compounds you find that the colorless liquid solidifies at 18 °C, while the green powder does not melt up to the maximum temperature of your apparatus, 1200 °C. What is the identity of the colorless liquid? What is the identity of the green powder? Be sure to use the appropriate naming convention depending upon whether the compound is better described as a molecular or ionic compound.

[8.85] (a) Triazine, $C_3H_3N_3$, is like benzene except that in triazine every other C—H group is replaced by a nitrogen atom. Draw the Lewis structure(s) for the triazine molecule. (b) Estimate the carbon–nitrogen bond distances in the ring.

[8.86] Using the electronegativities of Br and Cl, estimate the partial charges on the atoms in the Br—Cl molecule. Using these partial charges and the atomic radii given in Figure 7.7, estimate the dipole moment of the molecule. The measured dipole moment is 0.57 D.

8.87 Although I_3^- is known, F_3^- is not. Using Lewis structures, explain why F_3^- does not form.

8.88 Calculate the formal charge on the indicated atom in each of the following molecules or ions: (a) the central oxygen atom in O_3, (b) phosphorus in PF_6^-, (c) nitrogen in NO_2, (d) iodine in ICl_3, (e) chlorine in $HClO_4$ (hydrogen is bonded to O).

8.89 (a) Determine the formal charge on the chlorine atom in the hypochlorite ion, ClO^-, and the perchlorate ion, ClO_4^-, using resonance structures where the Cl atom has an octet. (b) What are the oxidation numbers of chlorine in ClO^- and in ClO_4^-? (c) Is it uncommon for the formal charge and the oxidation state to be different? Explain. (d) Perchlorate is a much stronger oxidizing agent than hypochlorite. Would you expect there to be any relationship between the oxidizing power of the oxyanion and either the oxidation state or the formal charge of chlorine?

8.90 The following three Lewis structures can be drawn for N_2O:

$$:N{\equiv}N{-}\ddot{O}: \longleftrightarrow :\ddot{N}{-}N{\equiv}O: \longleftrightarrow :\ddot{N}{=}N{=}\ddot{O}:$$

(a) Using formal charges, which of these three resonance forms is likely to be the most important? (b) The N—N bond length in N_2O is 1.12 Å, slightly longer than a typical $N{\equiv}N$ bond; and the N—O bond length is 1.19 Å, slightly shorter than a typical $N{=}O$ bond. (See Table 8.5.) Rationalize these observations in terms of the resonance structures shown previously and your conclusion for (a).

8.91 An important reaction for the conversion of natural gas to other useful hydrocarbons is the conversion of methane to ethane.

$$2 \, CH_4(g) \longrightarrow C_2H_6(g) + H_2(g)$$

In practice, this reaction is carried out in the presence of oxygen, which converts the hydrogen produced to water.

$$2 \, CH_4(g) + \tfrac{1}{2} O_2(g) \longrightarrow C_2H_6(g) + H_2O(g)$$

Use bond enthalpies (Table 8.4) to estimate ΔH for these two reactions. Why is the conversion of methane to ethane more favorable when oxygen is used?

8.92 Two compounds are isomers if they have the same chemical formula but a different arrangement of atoms. Use bond enthalpies (Table 8.4) to estimate ΔH for each of the following gas-phase isomerization reactions, and indicate which isomer has the lower enthalpy:

(a)

Ethanol   Dimethyl ether

(b)

Ethylene oxide   Acetaldehyde

(c)

Cyclopentene   Pentadiene

(d)

Methyl isocyanide   Acetonitrile

[8.93] With reference to the "Chemistry Put to Work" box on explosives, **(a)** use bond enthalpies to estimate the enthalpy change for the explosion of 1.00 g of nitroglycerin. **(b)** Write a balanced equation for the decomposition of TNT.

Assume that, upon explosion, TNT decomposes into $N_2(g)$, $CO_2(g)$, $H_2O(g)$, and $C(s)$.

[8.94] The "plastic" explosive C-4, often used in action movies, contains the molecule *cyclotrimethylenetrinitramine*, which is often called RDX (for Royal Demolition eXplosive):

Cyclotrimethylenetrinitramine (RDX)

(a) Complete the Lewis structure for the molecule by adding unshared electron pairs where they are needed. **(b)** Does the Lewis structure you drew in part (a) have any resonance structures? If so, how many? **(c)** The molecule causes an explosion by decomposing into $CO(g)$, $N_2(g)$, and $H_2O(g)$. Write a balanced equation for the decomposition reaction. **(d)** With reference to Table 8.4, which is the weakest type of bond in the molecule? **(e)** Use average bond enthalpies to estimate the enthalpy change when 5.0 g of RDX decomposes.

8.95 The bond lengths of carbon–carbon, carbon–nitrogen, carbon–oxygen, and nitrogen–nitrogen single, double, and triple bonds are listed in Table 8.5. Plot bond enthalpy (Table 8.4) versus bond length for these bonds. What do you conclude about the relationship between bond length and bond enthalpy? What do you conclude about the relative strengths of C—C, C—N, C—O, and N—N bonds?

8.96 Use the data in Table 8.5 and the following data: S—S distance in $S_8 = 2.05$ Å; S—O distance in $SO_2 = 1.43$ Å to answer the following questions: **(a)** Predict the bond length in a S—N single bond. **(b)** Predict the bond length in a S—O single bond. **(c)** Why is the S—O bond length in SO_2 considerably shorter than your predicted value for the S—O single bond? **(d)** When elemental sulfur, S_8, is carefully oxidized, a compound S_8O is formed, in which one of the sulfur atoms in the S_8 ring is bonded to an oxygen atom. The S—O bond length in this compound is 1.48 Å. In light of this information, write Lewis structures that can account for the observed S—O bond length. Does the sulfur bearing the oxygen in this compound obey the octet rule?

INTEGRATIVE EXERCISES

8.97 The Ti^{2+} ion is isoelectronic with the Ca atom. **(a)** Are there any differences in the electron configurations of Ti^{2+} and Ca? **(b)** With reference to Figure 6.25, comment on the changes in the ordering of the $4s$ and $3d$ subshells in Ca and Ti^{2+}. **(c)** Will Ca and Ti^{2+} have the same number of unpaired electrons? Explain.

[8.98] **(a)** Write the chemical equations that are used in calculating the lattice energy of $SrCl_2(s)$ via a Born–Haber cycle. **(b)** The second ionization energy of $Sr(g)$ is 1064 kJ/mol. Use this fact along with data in Appendix C, Figure 7.12, Figure 7.14, and Table 8.2 to calculate ΔH_f° for $SrCl_2(s)$.

[8.99] The electron affinity of oxygen is -141 kJ/mol, corresponding to the reaction

$$O(g) + e^- \longrightarrow O^-(g)$$

The lattice energy of $K_2O(s)$ is 2238 kJ/mol. Use these data along with data in Appendix C and Figure 7.12 to calculate the "second electron affinity" of oxygen, corresponding to the reaction

$$O^-(g) + e^- \longrightarrow O^{2-}(g)$$

8.100 The reaction of indium, In, with sulfur leads to three binary compounds, which we will assume to be purely ionic. The three compounds have the following properties:

Compound	Mass % In	Melting Point (°C)
A	87.7	653
B	78.2	692
C	70.5	1050

(a) Determine the empirical formulas of compounds A, B, and C. **(b)** Give the oxidation state of In in each of the three compounds. **(c)** Write the electron configuration for the In ion in each compound. Do any of these configurations correspond to a noble-gas configuration? **(d)** In which compound is the ionic radius of In expected to be smallest? Explain. **(e)** The melting point of ionic compounds often correlates with the lattice energy. Explain the trends in the melting points of compounds A, B, and C in these terms.

[8.101] One scale for electronegativity is based on the concept that the electronegativity of any atom is proportional to the ionization energy of the atom minus its electron affinity: electronegativity = $k(IE - EA)$, where k is a proportionality constant. **(a)** How does this definition explain why the electronegativity of F is greater than that of Cl even though Cl has the greater electron affinity? **(b)** Why are both ionization energy and electron affinity relevant to the notion of electronegativity? **(c)** By using data in Chapter 7, determine the value of k that would lead to an electronegativity of 4.0 for F under this definition. **(d)** Use your result from part (c) to determine the electronegativities of Cl and O using this scale. Do these values follow the trend shown in Figure 8.6?

8.102 The compound chloral hydrate, known in detective stories as knockout drops, is composed of 14.52% C, 1.83% H, 64.30% Cl, and 19.35% O by mass and has a molar mass of 165.4 g/mol. **(a)** What is the empirical formula of this substance? **(b)** What is the molecular formula of this substance? **(c)** Draw the Lewis structure of the molecule, assuming that the Cl atoms bond to a single C atom and that there are a C—C bond and two C—O bonds in the compound.

8.103 Barium azide is 62.04% Ba and 37.96% N. Each azide ion has a net charge of 1−. **(a)** Determine the chemical formula of the azide ion. **(b)** Write three resonance structures for the azide ion. **(c)** Which structure is most important? **(d)** Predict the bond lengths in the ion.

8.104 Acetylene (C_2H_2) and nitrogen (N_2) both contain a triple bond, but they differ greatly in their chemical properties. **(a)** Write the Lewis structures for the two substances. **(b)** By referring to Appendix C, look up the enthalpies of formation of acetylene and nitrogen and compare their reactivities. **(c)** Write balanced chemical equations for the complete oxidation of N_2 to form $N_2O_5(g)$ and of acetylene to form $CO_2(g)$ and $H_2O(g)$. **(d)** Calculate the enthalpy of oxidation per mole of N_2 and C_2H_2 (the enthalpy of formation of $N_2O_5(g)$ is 11.30 kJ/mol). How do these comparative values relate to your response to part (b)? Both N_2 and C_2H_2 possess triple bonds with quite high bond enthalpies (Table 8.4). What aspect of chemical bonding in these molecules or

in the oxidation products seems to account for the difference in chemical reactivities?

[8.105] Under special conditions, sulfur reacts with anhydrous liquid ammonia to form a binary compound of sulfur and nitrogen. The compound is found to consist of 69.6% S and 30.4% N. Measurements of its molecular mass yield a value of 184.3 g mol^{-1}. The compound occasionally detonates on being struck or when heated rapidly. The sulfur and nitrogen atoms of the molecule are joined in a ring. All the bonds in the ring are of the same length. **(a)** Calculate the empirical and molecular formulas for the substance. **(b)** Write Lewis structures for the molecule, based on the information you are given. (*Hint:* You should find a relatively small number of dominant Lewis structures.) **(c)** Predict the bond distances between the atoms in the ring. (*Note:* The S—S distance in the S_8 ring is 2.05Å.) **(d)** The enthalpy of formation of the compound is estimated to be 480 kJ mol^{-1}. ΔH_f° of S(g) is 222.8 kJ mol^{-1}. Estimate the average bond enthalpy in the compound.

[8.106] A common form of elemental phosphorus is the tetrahedral P_4 molecule, where all four phosphorus atoms are equivalent:

At room temperature phosphorus is a solid. **(a)** Do you think there are any unshared pairs of electrons in the P_4 molecule? **(b)** How many P—P bonds are there in the molecule? **(c)** Can you draw a Lewis structure for a linear P_4 molecule that satisfies the octet rule? **(d)** Using formal charges, what can you say about the stability of the linear molecule vs. that of the tetrahedral molecule?

[8.107] Consider benzene (C_6H_6) in the gas phase. **(a)** Write the reaction for breaking all the bonds in $C_6H_6(g)$, and use data in Appendix C to determine the enthalpy change for this reaction. **(b)** Write a reaction that corresponds to breaking all the carbon–carbon bonds in $C_6H_6(g)$. **(c)** By combining your answers to parts (a) and (b) and using the average bond enthalpy for C—H from Table 8.4, calculate the average bond enthalpy for the carbon–carbon bonds in $C_6H_6(g)$. **(d)** Comment on your answer from part (c) as compared to the values for C—C single bonds and C=C double bonds in Table 8.4.

8.108 Average bond enthalpies are generally defined for gas-phase molecules. Many substances are liquids in their standard state. ∞ (Section 5.7) By using appropriate thermochemical data from Appendix C, calculate average bond enthalpies in the liquid state for the following bonds, and compare these values to the gas-phase values given in Table 8.4: **(a)** Br—Br, from $Br_2(l)$; **(b)** C—Cl, from $CCl_4(l)$; **(c)** O—O, from $H_2O_2(l)$ (assume that the O—H bond enthalpy is the same as in the gas phase). **(d)** What can you conclude about the process of breaking bonds in the liquid as compared to the gas phase? Explain the difference in the ΔH values between the two phases.

CHAPTER 9

MOLECULAR GEOMETRY AND BONDING THEORIES

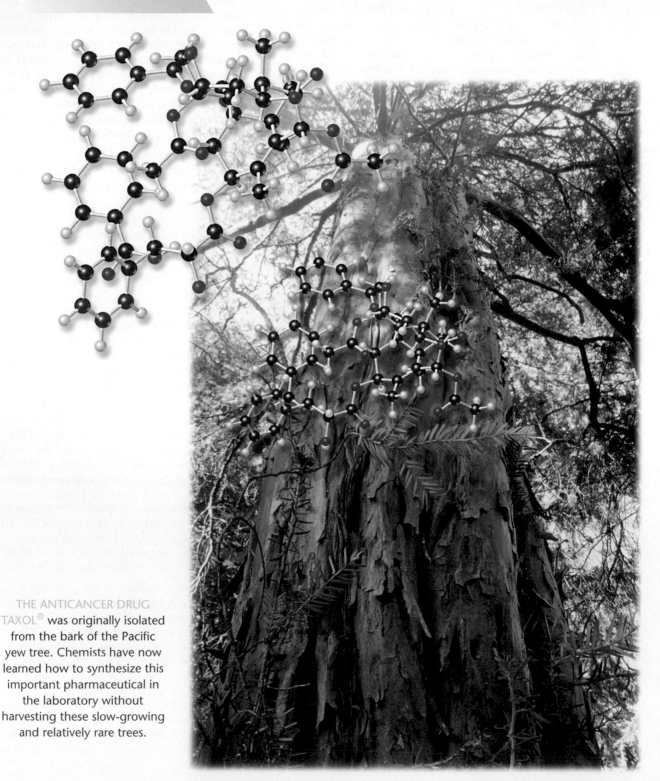

THE ANTICANCER DRUG TAXOL® was originally isolated from the bark of the Pacific yew tree. Chemists have now learned how to synthesize this important pharmaceutical in the laboratory without harvesting these slow-growing and relatively rare trees.

9.1 Molecular Shapes
We begin by discussing how *molecular geometries* are described and by examining some common geometries exhibited by molecules.

9.2 The VSEPR Model
We next consider how molecular geometries can be predicted using a simple model (the *valence-shell electron-pair repulsion, or VSEPR,* model), which is based on Lewis structures and the repulsions between regions of high electron density.

9.3 Molecular Shape and Molecular Polarity
Once we know the type of bonds in a molecule and the geometry of the molecule, we can determine whether the molecule is *polar* or *nonpolar*.

9.4 Covalent Bonding and Orbital Overlap
We recognize that electrons are shared between atoms in a covalent bond. In *valence-bond theory,* the bonding electrons are visualized as originating in atomic orbitals on two different atoms. The covalent bond between the atoms arises when these orbitals overlap in space.

9.5 Hybrid Orbitals
To account for molecular shape, we consider how the orbitals of one atom mix with one another, or *hybridize,* to create *hybrid orbitals* suitable for bonding within molecules.

9.6 Multiple Bonds
We consider that overlapping atomic orbitals that underlie covalent bonding in a molecule can overlap in multiple ways, based on geometry, to produce *sigma* (σ) and *pi* (π) bonds between atoms. Single bonds generally consist of one sigma bond between the bonded atoms; multiple bonds involve one sigma and one or more pi bonds.

9.7 Molecular Orbitals
We evaluate a more sophisticated treatment of bonding that uses *molecular orbital theory,* which introduces the concepts of *bonding and antibonding molecular orbitals* arising from atomic orbitals.

9.8 Second-Row Diatomic Molecules
We consider how molecular orbital theory is used to construct *energy level diagrams* for second-row diatomic molecules.

WE SAW IN CHAPTER 8 THAT LEWIS STRUCTURES help us understand the compositions of molecules and their covalent bonds. However, Lewis structures do not show one of the most important aspects of molecules—their overall shapes. Molecules have shapes and sizes that are defined by the

angles and distances between the nuclei of their component atoms. Indeed, chemists often refer to molecular *architecture* when describing the distinctive shapes and sizes of molecules.

The shape and size of a molecule of a particular substance, together with the strength and polarity of its bonds, largely determine the properties of that substance. Some of the most dramatic examples of the important roles of molecular shape and size are seen in biochemical reactions and in substances produced by living species. For example, the chapter-opening photograph shows a Pacific yew tree, a species that grows along the Pacific coast of the northwestern United States and Canada. In 1967 two chemists isolated from the bark of the Pacific yew small amounts of a molecule that was found to be among the most effective treatments for breast and ovarian cancer. This molecule, now known as the pharmaceutical Taxol®, has a complex molecular architecture that leads to its powerful therapeutic effectiveness. Even a small modification to the shape and size of the molecule decreases its effectiveness and can lead to the formation of a substance toxic to humans. Chemists now know how to synthesize the drug in the laboratory, which has made it more available and has saved the slow-growing Pacific yew tree from possible extinction. Before the drug was synthesized, six trees had to be harvested to provide the Taxol® necessary to treat one cancer patient.

Our first goal in this chapter is to learn the relationship between two-dimensional Lewis structures and three-dimensional molecular shapes. Armed with this knowledge, we can then examine more closely the nature of covalent bonds. The lines that are used to depict bonds in Lewis structures provide important clues about the orbitals that molecules use in bonding. By examining these orbitals, we can gain a greater understanding of the behavior of molecules. You will find that the material in this chapter will help you in later discussions of the physical and chemical properties of substances.

9.1 MOLECULAR SHAPES

In Chapter 8 we used Lewis structures to account for the formulas of covalent compounds. ∞ (Section 8.5) Lewis structures, however, do not indicate the shapes of molecules; they simply show the number and types of bonds between atoms. For example, the Lewis structure of CCl_4 tells us only that four Cl atoms are bonded to a central C atom:

$$:\ddot{C}l - C - \ddot{C}l:$$

The Lewis structure is drawn with the atoms all in the same plane. As shown in Figure 9.1 ◄, however, the actual three-dimensional arrangement of the atoms shows the Cl atoms at the corners of a *tetrahedron*, a geometric object with four corners and four faces, each of which is an equilateral triangle.

The overall shape of a molecule is determined by its **bond angles**, the angles made by the lines joining the nuclei of the atoms in the molecule. The bond angles of a molecule, together with the bond lengths ∞ (Section 8.8), accurately define the shape and size of the molecule. In CCl_4 the bond angles are defined as the angles between the C—Cl bonds. You should be able to see that there are six Cl—C—Cl angles in CCl_4, and they all have the same value— 109.5°, which is characteristic of a tetrahedron. In addition, all four C—Cl bonds are the same length (1.78 Å). Thus, the shape and size of CCl_4 are completely described by stating that the molecule is tetrahedral with C—Cl bonds of length 1.78 Å.

In our discussion of the shapes of molecules, we will begin with molecules (and ions) that, like CCl_4, have a single central atom bonded to two or more atoms of the same type. Such molecules conform to the general formula AB_n, in which the central atom A is bonded to n B atoms. Both CO_2 and H_2O are AB_2 molecules, for example, whereas SO_3 and NH_3 are AB_3 molecules, and so on.

The possible shapes of AB_n molecules depend on the value of n. We observe only a few general shapes for a given value of n. Those commonly found for AB_2 and AB_3 molecules are shown in Figure 9.2▶. Therefore, an AB_2 molecule must be either linear (bond angle = 180°) or bent (bond angle ≠ 180°). For example, CO_2 is linear, and SO_2 is bent. For AB_3 molecules, the two most common

(a)

Bond distance, 1.78Å

Cl

109.5°

(b)

Cl

C

(c)

▲ **Figure 9.1 Tetrahedral geometry.** (a) A tetrahedron is an object with four faces and four vertices. Each face is an equilateral triangle. (b) The geometry of the CCl_4 molecule. Each C—Cl bond in the molecule points toward a vertex of a tetrahedron. All of the C—Cl bonds are the same length, and all of the Cl—C—Cl bond angles are the same. This type of drawing of CCl_4 is called a ball-and-stick model. (c) A representation of CCl_4, called a space-filling model. It shows the relative sizes of the atoms, but the geometry is somewhat harder to see.

◀ **Figure 9.2 Shapes of AB₂ and AB₃ molecules.** Top: AB_2 molecules can be either linear or bent. Bottom: Three possible shapes for AB_3 molecules.

shapes place the B atoms at the corners of an equilateral triangle. If the A atom lies in the same plane as the B atoms, the shape is called *trigonal planar*. If the A atom lies above the plane of the B atoms, the shape is called *trigonal pyramidal* (a pyramid with an equilateral triangle as its base). For example, SO_3 is trigonal planar, and NF_3 is trigonal pyramidal. Some AB_3 molecules, such as ClF_3, exhibit the more unusual *T shape* shown in Figure 9.2.

The shape of any particular AB_n molecule can usually be derived from one of the five basic geometric structures shown in Figure 9.3▶. Starting with a tetrahedron, for example, we can remove atoms successively from the corners as shown in Figure 9.4▼. When an atom is removed from one corner of the tetrahedron, the remaining fragment has a trigonal-pyramidal geometry, such as that found for NF_3. When two atoms are removed, a bent geometry results.

Why do so many AB_n molecules have shapes related to the basic structures in Figure 9.3, and can we predict these shapes? When A is a representative element (one of the elements from the *s* block or *p* block of the periodic table), we can answer these questions by using the **valence-shell electron-pair repulsion (VSEPR) model**. Although the name is rather imposing, the model is quite simple. It has useful predictive capabilities, as we will see in Section 9.2.

▼ **Figure 9.3 Shapes of AB$_n$ molecules.** For molecules whose formula is of the general form AB_n, there are five fundamental shapes.

Linear   Trigonal planar   Tetrahedral

Trigonal bipyramidal   Octahedral

Tetrahedral   Trigonal pyramidal   Bent

◀ **Figure 9.4 Derivatives from the AB$_n$ geometries.** Additional molecular shapes can be obtained by removing corner atoms from the basic shapes shown in Figure 9.3. Here we begin with a tetrahedron and successively remove corners, producing first a trigonal-pyramidal geometry and then a bent geometry, each having ideal bond angles of 109.5°. Molecular shape is meaningful only when there are at least three atoms. If there are only two, they must be arranged next to each other, and no special name is given to describe the molecule.

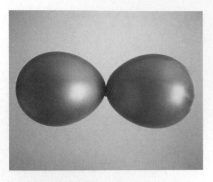

(a) Two balloons adopt a
linear arrangement.

(b) Three balloons adopt a
trigonal-planar arrangement.

(c) Four balloons adopt a
tetrahedral arrangement.

▲ Figure 9.5 **A balloon analogy for
electron domains.** Balloons tied
together at their ends naturally adopt
their lowest-energy arrangement.

▲ **GIVE IT SOME THOUGHT**

One of the common shapes for AB_4 molecules is *square planar*: All five atoms lie in the same plane, the B atoms lie at the corners of a square, and the A atom is at the center of the square. Which of the shapes in Figure 9.3 could lead to a square-planar geometry upon the removal of one or more atoms?

9.2 THE VSEPR MODEL

Imagine tying two identical balloons together at their ends. As shown in Figure 9.5(a) ◄, the balloons naturally orient themselves to point away from each other; that is, they try to "get out of each other's way" as much as possible. If we add a third balloon, the balloons orient themselves toward the vertices of an equilateral triangle as shown in Figure 9.5(b). If we add a fourth balloon, they adopt a tetrahedral shape [Figure 9.5(c)]. We see that an optimum geometry exists for each number of balloons.

In some ways the electrons in molecules behave like the balloons shown in Figure 9.5. We have seen that a single covalent bond is formed between two atoms when a pair of electrons occupies the space between the atoms. ∞ (Section 8.3) A **bonding pair** of electrons thus defines a region in which the electrons will most likely be found. We will refer to such a region as an **electron domain**. Likewise, a **nonbonding pair** (or **lone pair**) of electrons defines an electron domain that is located principally on one atom. For example, the Lewis structure of NH_3 has four electron domains around the central nitrogen atom (three bonding pairs and one nonbonding pair):

Each multiple bond in a molecule also constitutes a single electron domain. Thus, the following resonance structure for O_3 has three electron domains around the central oxygen atom (a single bond, a double bond, and a nonbonding pair of electrons):

$$\ddot{\ddot{O}}\!-\!\ddot{O}\!=\!\ddot{\ddot{O}}$$

In general, *each nonbonding pair, single bond, or multiple bond produces an electron domain around the central atom.*

▲ **GIVE IT SOME THOUGHT**

An AB_3 molecule has the resonance structure

$$\overset{:\ddot{B}:}{\underset{:\ddot{B}-A-\ddot{B}:}{\|}}$$

Does this Lewis structure follow the octet rule? How many electron domains are there around the A atom?

The VSEPR model is based on the idea that electron domains are negatively charged and therefore repel one another. Like the balloons in Figure 9.5, electron domains try to stay out of one another's way. *The best arrangement of a given number of electron domains is the one that minimizes the repulsions among them.* In fact, the analogy between electron domains and balloons is so close that the same preferred geometries are found in both cases. Like the balloons in

TABLE 9.1 ■ Electron-Domain Geometries as a Function of the Number of Electron Domains

Number of Electron Domains	Arrangement of Electron Domains	Electron-Domain Geometry	Predicted Bond Angles
2	180°	Linear	180°
3	120°	Trigonal planar	120°
4	109.5°	Tetrahedral	109.5°
5	90° 120°	Trigonal bipyramidal	120° 90°
6	90° 90°	Octahedral	90°

Figure 9.5, two electron domains are arranged *linearly*, three domains are arranged in a *trigonal-planar* fashion, and four are arranged *tetrahedrally*. These arrangements, together with those for five electron domains (*trigonal bipyramidal*) and six electron domains (*octahedral*), are summarized in Table 9.1▲. If you compare the geometries in Table 9.1 with those in Figure 9.3, you will see that they are the same. *The shapes of different AB$_n$ molecules or ions depend on the number of electron domains surrounding the central A atom.*

The arrangement of electron domains about the central atom of an AB$_n$ molecule or ion is called its **electron-domain geometry**. In contrast, the **molecular geometry** is the arrangement of *only the atoms* in a molecule or ion—any nonbonding pairs are not part of the description of the molecular geometry. In the VSEPR model, we predict the electron-domain geometry. From knowing how many domains are due to nonbonding pairs, we can then predict the molecular geometry of a molecule or ion from its electron-domain geometry.

$$NH_3 \Rightarrow \text{(Lewis structure)} \Rightarrow \text{(Electron-domain geometry, tetrahedral)} \Rightarrow \text{(Molecular geometry, trigonal pyramidal)}$$

Lewis structure

Electron-domain geometry (tetrahedral)

Molecular geometry (trigonal pyramidal)

▲ Figure 9.6 **The molecular geometry of NH₃.** The geometry is predicted by first drawing the Lewis structure, then using the VSEPR model to determine the electron-domain geometry, and finally focusing on the atoms themselves to describe the molecular geometry.

When all the electron domains in a molecule arise from bonds, the molecular geometry is identical to the electron-domain geometry. When, however, one or more of the domains involve nonbonding pairs of electrons, we must remember to ignore those domains when predicting molecular shape. Consider the NH_3 molecule, which has four electron domains around the nitrogen atom (Figure 9.6 ▲). We know from Table 9.1 that the repulsions among four electron domains are minimized when the domains point toward the vertices of a tetrahedron, so the electron-domain geometry of NH_3 is tetrahedral. We know from the Lewis structure of NH_3 that one of the electron domains is due to a nonbonding pair of electrons, which will occupy one of the four vertices of the tetrahedron. Hence the molecular geometry of NH_3 is trigonal pyramidal, as shown in Figure 9.6. Notice that the tetrahedral arrangement of the four electron domains leads us to predict the trigonal-pyramidal molecular geometry.

We can generalize the steps we follow in using the VSEPR model to predict the shapes of molecules or ions:

1. Draw the *Lewis structure* of the molecule or ion, and count the total number of electron domains around the central atom. Each nonbonding electron pair, each single bond, each double bond, and each triple bond counts as an electron domain.

2. Determine the *electron-domain geometry* by arranging the electron domains about the central atom so that the repulsions among them are minimized, as shown in Table 9.1.

3. Use the arrangement of the bonded atoms to determine the *molecular geometry*.

Figure 9.6 shows how these steps are applied to predict the geometry of the NH_3 molecule. Because the trigonal-pyramidal molecular geometry is based on tetrahedral electron-domain geometry, the *ideal bond angles* are 109.5°. As we will soon see, bond angles deviate from the ideal angles when the surrounding atoms and electron domains are not identical.

Let's apply these steps to determine the shape of the CO_2 molecule. We first draw its Lewis structure, which reveals two electron domains (two double bonds) around the central carbon:

$$\ddot{O}{=}C{=}\ddot{O}$$

Two electron domains will arrange themselves to give a linear electron-domain geometry (Table 9.1). Because neither domain is a nonbonding pair of electrons, the molecular geometry is also linear, and the $O{-}C{-}O$ bond angle is 180°.

Table 9.2 ▶ summarizes the possible molecular geometries when an AB_n molecule has four or fewer electron domains about A. These geometries are important because they include all the commonly occurring shapes found for molecules or ions that obey the octet rule.

TABLE 9.2 ■ Electron-Domain Geometries and Molecular Shapes for Molecules with Two, Three, and Four Electron Domains around the Central Atom

Number of Electron Domains	Electron-Domain Geometry	Bonding Domains	Nonbonding Domains	Molecular Geometry	Example
2	Linear	2	0	Linear	$\ddot{O}=C=\ddot{O}$
3	Trigonal planar	3	0	Trigonal planar	
		2	1	Bent	$\left[\ddot{O} \diagdown \overset{..}{N} \diagup \ddot{O} \right]^{-}$
4	Tetrahedral	4	0	Tetrahedral	
		3	1	Trigonal pyramidal	
		2	2	Bent	

■ **SAMPLE EXERCISE 9.1** | Using the VSEPR Model

Use the VSEPR model to predict the molecular geometry of **(a)** O_3, **(b)** $SnCl_3^-$.

SOLUTION

Analyze: We are given the molecular formulas of a molecule and a polyatomic ion, both conforming to the general formula AB_n and both having a central atom from the p block of the periodic table.

Plan: To predict the molecular geometries of these species, we first draw their Lewis structures and then count the number of electron domains around the central atom. The number of electron domains gives the electron-domain geometry. We then obtain the molecular geometry from the arrangement of the domains that are due to bonds.

Solve:

(a) We can draw two resonance structures for O_3:

$$:\ddot{O}-\ddot{O}=\ddot{O} \longleftrightarrow \ddot{O}=\ddot{O}-\ddot{O}:$$

Because of resonance, the bonds between the central O atom and the outer O atoms are of equal length. In both resonance structures the central O atom is bonded to the two outer O atoms and has one nonbonding pair. Thus, there are three electron domains about the central O atoms. (Remember that a double bond counts as a single electron domain.) The arrangement of three electron domains is trigonal planar (Table 9.1). Two of the domains are from bonds, and one is due to a nonbonding pair. So, the molecule has a bent shape with an ideal bond angle of 120° (Table 9.2).

As this example illustrates, when a molecule exhibits resonance, any one of the resonance structures can be used to predict the molecular geometry.

(b) The Lewis structure for the $SnCl_3^-$ ion is

$$\left[:\ddot{\underset{..}{Cl}}-\overset{}{Sn}-\ddot{\underset{..}{Cl}}: \right]^-$$
$$:\ddot{\underset{..}{Cl}}:$$

The central Sn atom is bonded to the three Cl atoms and has one nonbonding pair. Therefore, the Sn atom has four electron domains around it. The resulting electron-domain geometry is tetrahedral (Table 9.1) with one of the corners occupied by a nonbonding pair of electrons. The molecular geometry is therefore trigonal pyramidal (Table 9.2), like that of NH_3.

■ **PRACTICE EXERCISE**

Predict the electron-domain geometry and the molecular geometry for **(a)** $SeCl_2$, **(b)** CO_3^{2-}.
Answers: **(a)** tetrahedral, bent; **(b)** trigonal planar, trigonal planar

---

Bonding electron pair

Nuclei

Nonbonding pair

Nucleus

▲ **Figure 9.7 Relative "sizes" of bonding and nonbonding electron domains.**

The Effect of Nonbonding Electrons and Multiple Bonds on Bond Angles

We can refine the VSEPR model to predict and explain slight distortions of molecules from the ideal geometries summarized in Table 9.2. For example, consider methane (CH_4), ammonia (NH_3), and water (H_2O). All three have tetrahedral electron-domain geometries, but their bond angles differ slightly:

Notice that the bond angles decrease as the number of nonbonding electron pairs increases. A bonding pair of electrons is attracted by both nuclei of the bonded atoms. By contrast, a nonbonding pair is attracted primarily by only one nucleus. Because a nonbonding pair experiences less nuclear attraction, its electron domain is spread out more in space than is the electron domain for a bonding pair, as shown in Figure 9.7 ◄. Nonbonding pairs of electrons, therefore, take up more space than bonding pairs. As a result, *the electron domains for nonbonding electron pairs exert greater repulsive forces on adjacent electron domains and tend to compress the bond angles.* Using the analogy in Figure 9.5, we can envision the domains for nonbonding electron pairs as represented by balloons that are slightly larger and slightly fatter than those for bonding pairs.

Because multiple bonds contain a higher electronic-charge density than single bonds, multiple bonds also represent larger electron domains ("fatter balloons"). Consider the Lewis structure of phosgene, $COCl_2$:

$$\begin{array}{c} :\!\ddot{C}l \\ | \\ C\!=\!\ddot{O} \\ | \\ :\!\ddot{C}l \end{array}$$

Because three electron domains surround the central carbon atom, we might expect a trigonal-planar geometry with 120° bond angles. The double bond, however, seems to act much like a nonbonding pair of electrons, reducing the Cl—C—Cl bond angle from the ideal angle of 120° to an actual angle of 111.4°:

$$\begin{array}{c} Cl \; \leftarrow 124.3° \\ 111.4°\, C\!=\!O \\ Cl \; \leftarrow 124.3° \end{array}$$

In general, *electron domains for multiple bonds exert a greater repulsive force on adjacent electron domains than do electron domains for single bonds.*

GIVE IT SOME THOUGHT

One of the resonance structures of the nitrate ion, NO_3^-, is

$$\left[\begin{array}{c} :\!O\!: \\ \| \\ :\!\ddot{O}\!-\!N\!-\!\ddot{O}\!: \end{array} \right]^-$$

The bond angles in this ion are exactly 120°. Is this observation consistent with the above discussion of the effect of multiple bonds on bond angles?

Molecules with Expanded Valence Shells

So far, our discussion of the VSEPR model has considered molecules with no more than an octet of electrons around the central atom. Recall, however, that when the central atom of a molecule is from the third period of the periodic table and beyond, that atom may have more than four electron pairs around it. ∞ (Section 8.7) Molecules with five or six electron domains around the central atom display a variety of molecular geometries based on the *trigonal-bipyramidal* (five electron domains) or the *octahedral* (six electron domains) electron-domain geometries, as shown in Table 9.3▼.

The most stable electron-domain geometry for five electron domains is the trigonal bipyramid (two trigonal pyramids sharing a base). Unlike the arrangements we have seen to this point, the electron domains in a trigonal bipyramid can point toward two geometrically distinct types of positions. Two of the five domains point toward *axial positions*, and the remaining three domains point toward *equatorial positions* (Figure 9.8▶). Each axial domain makes a 90° angle with any equatorial domain. Each equatorial domain makes a 120° angle with either of the other two equatorial domains and a 90° angle with either axial domain.

Suppose a molecule has five electron domains, one or more of which originates from a nonbonding pair. Will the electron domains from the nonbonding pairs occupy axial or equatorial positions? To answer this question, we must determine which location minimizes the repulsions between the electron domains. Repulsions between domains are much greater when they are situated 90° from each other than when they are at 120°. An equatorial domain is 90° from only two other domains (the two axial domains). By contrast, an axial domain is situated 90° from *three* other domains (the three equatorial domains). Hence, an equatorial domain experiences less repulsion than an axial domain. Because the domains from nonbonding pairs exert larger repulsions than those from bonding pairs, they always occupy the equatorial positions in a trigonal bipyramid.

Axial position

Equatorial positions

Axial position

▲ **Figure 9.8 Trigonal-bipyramidal geometry.** Five electron domains arrange themselves around a central atom as a trigonal bipyramid. The three *equatorial* electron domains define an equilateral triangle. The two *axial* domains lie above and below the plane of the triangle. If a molecule has nonbonding electron domains, they will occupy the equatorial positions.

TABLE 9.3 ■ Electron-Domain Geometries and Molecular Shapes for Molecules with Five and Six Electron Domains around the Central Atom

Total Electron Domains	Electron-Domain Geometry	Bonding Domains	Nonbonding Domains	Molecular Geometry	Example
5	Trigonal bipyramidal	5	0	Trigonal bipyramidal	PCl_5
		4	1	Seesaw	SF_4
		3	2	T-shaped	ClF_3
		2	3	Linear	XeF_2
6	Octahedral	6	0	Octahedral	SF_6
		5	1	Square pyramidal	BrF_5
		4	2	Square planar	XeF_4

GIVE IT SOME THOUGHT

It might seem that a square planar geometry of four electron domains around a central atom would be more favorable than a tetrahedron. Can you rationalize why the tetrahedron is preferred, based on angles between electron domains?

The most stable electron-domain geometry for six electron domains is the *octahedron*. As shown in Figure 9.9 ▶, an octahedron is a polyhedron with eight faces and six vertices, each of which is an equilateral triangle. If an atom has six electron domains around it, that atom can be visualized as being at the center of the octahedron with the electron domains pointing toward the six vertices. All the bond angles in an octahedron are 90°, and all six vertices are equivalent. Therefore, if an atom has five bonding electron domains and one nonbonding domain, we can put the nonbonding domain at any of the six vertices of the octahedron. The result is always a *square-pyramidal* molecular geometry. When there are two nonbonding electron domains, however, their repulsions are minimized by pointing them toward opposite sides of the octahedron, producing a *square-planar* molecular geometry, as shown in Table 9.3.

▲ Figure 9.9 **An octahedron.** The octahedron is an object with eight faces and six vertices. Each face is an equilateral triangle.

■ **SAMPLE EXERCISE 9.2** | Molecular Geometries of Molecules with Expanded Valence Shells

Use the VSEPR model to predict the molecular geometry of **(a)** SF_4, **(b)** IF_5.

SOLUTION

Analyze: The molecules are of the AB_n type with a central atom from the *p* block of the periodic table.

Plan: We can predict their structures by first drawing Lewis structures and then using the VSEPR model to determine the electron-domain geometry and molecular geometry.

Solve:

(a) The Lewis structure for SF_4 is

The sulfur has five electron domains around it: four from the S—F bonds and one from the nonbonding pair. Each domain points toward a vertex of a trigonal bipyramid. The domain from the nonbonding pair will point toward an equatorial position. The four bonds point toward the remaining four positions, resulting in a molecular geometry that is described as seesaw-shaped:

Comment: The experimentally observed structure is shown on the right. We can infer that the nonbonding electron domain occupies an equatorial position, as predicted. The axial and equatorial S—F bonds are slightly bent back away from the nonbonding domain, suggesting that the bonding domains are "pushed" by the nonbonding domain, which is larger and has greater repulsion (Figure 9.7).

(b) The Lewis structure of IF_5 is

The iodine has six electron domains around it, one of which is from a nonbonding pair. The electron-domain geometry is therefore octahedral, with one position occupied by the nonbonding pair. The resulting molecular geometry is therefore *square pyramidal* (Table 9.3):

Comment: Because the domain for the nonbonding pair is larger than the other domains, the four F atoms in the base of the pyramid are tipped up slightly toward the F atom on top. Experimentally, we find that the angle between the base and top F atoms is 82°, smaller than the ideal 90° angle of an octahedron.

■ **PRACTICE EXERCISE**

Predict the electron-domain geometry and molecular geometry of **(a)** ClF_3, **(b)** ICl_4^-.
Answers: **(a)** trigonal bipyramidal, T-shaped; **(b)** octahedral, square planar

Shapes of Larger Molecules

Although the structures of the molecules and ions we have already considered contain only a single central atom, the VSEPR model can be extended to more complex molecules. Consider the acetic acid molecule, with the following Lewis structure:

Acetic acid has three interior atoms: the left C atom, the central C atom, and the right-most O atom. We can use the VSEPR model to predict the geometry about each of these atoms individually:

	H | H—C— | H	:O: || —C—	.. O—H ..
Number of electron domains	4	3	4
Electron-domain geometry	Tetrahedral	Trigonal planar	Tetrahedral
Predicted bond angles	109.5°	120°	109.5°

The leftmost C has four electron domains (all from bonding pairs), and so the geometry around that atom is tetrahedral. The central C has three electron domains (counting the double bond as one domain). Thus, the geometry around that atom is trigonal planar. The O atom has four electron domains (two from bonding pairs and two from nonbonding pairs), so its electron-domain geometry is tetrahedral, and the molecular geometry around O is bent. The bond angles about the central C atom and the O atom are expected to deviate slightly from the ideal values of 120° and 109.5° because of the spatial demands of multiple bonds and nonbonding electron pairs. The structure of the acetic acid molecule is shown in Figure 9.10 ◄.

▲ Figure 9.10 **Ball-and-stick (top) and space-filling (bottom) representations of acetic acid, CH₃COOH.**

■ **SAMPLE EXERCISE 9.3** | **Predicting Bond Angles**

Eyedrops for dry eyes usually contain a water-soluble polymer called *poly(vinyl alcohol)*, which is based on the unstable organic molecule called *vinyl alcohol*:

$$H-\overset{..}{\underset{..}{O}}-\overset{\overset{H}{|}}{C}=\overset{\overset{H}{|}}{C}-H$$

Predict the approximate values for the H—O—C and O—C—C bond angles in vinyl alcohol.

SOLUTION

Analyze: We are given a molecular structure and asked to determine two bond angles in the structure.

Plan: To predict a particular bond angle, we consider the middle atom of the angle and determine the number of electron domains surrounding that atom. The ideal angle corresponds to the electron-domain geometry around the atom. The angle will be compressed somewhat by nonbonding electrons or multiple bonds.

Solve: For the H—O—C bond angle, the middle O atom has four electron domains (two bonding and two nonbonding). The electron-domain geometry around O is therefore tetrahedral, which gives an ideal angle of 109.5°. The H—O—C angle will be compressed somewhat by the nonbonding pairs, so we expect this angle to be slightly less than 109.5°.

To predict the O—C—C bond angle, we must examine the leftmost C atom, which is the central atom for this angle. There are three atoms bonded to this C atom and no nonbonding pairs, and so it has three electron domains about it. The predicted electron-domain geometry is trigonal planar, resulting in an ideal bond angle of 120°. Because of the larger size of the C=C domain, however, the O—C—C bond angle should be slightly greater than 120°.

■ PRACTICE EXERCISE

Predict the H—C—H and C—C—C bond angles in the following molecule, called *propyne*:

$$H-\underset{\underset{H}{|}}{\overset{\overset{H}{|}}{C}}-C\equiv C-H$$

Answers: 109.5°, 180°

9.3 MOLECULAR SHAPE AND MOLECULAR POLARITY

We now have a sense of the shapes that molecules adopt and why they do so. We will spend the rest of this chapter looking more closely at the ways in which electrons are shared to form the bonds between atoms in molecules. We will begin by returning to a topic that we first discussed in Section 8.4, namely *bond polarity* and *dipole moments*.

Recall that bond polarity is a measure of how equally the electrons in a bond are shared between the two atoms of the bond: As the difference in electronegativity between the two atoms increases, so does the bond polarity. ∞ (Section 8.4) We saw that the dipole moment of a diatomic molecule is a quantitative measure of the amount of charge separation in the molecule. The charge separation in molecules has a significant effect on physical and chemical properties. We will see in Chapter 11, for example, how molecular polarity affects boiling points, melting points, and other physical properties.

For a molecule that consists of more than two atoms, *the dipole moment depends on both the polarities of the individual bonds and the geometry of the molecule.* For each bond in the molecule, we can consider the **bond dipole**, which is the dipole moment that is due only to the two atoms in that bond. Consider the linear CO_2 molecule, for example. As shown in Figure 9.11(a) ▶, each C=O bond is polar, and because the C=O bonds are identical, the bond dipoles are equal in magnitude. A plot of the electron density of the CO_2 molecule, shown in Figure 9.11(b), clearly shows that the bonds are polar: On the oxygen atoms, regions of high electron density (red) are at the ends of the molecule. On the carbon atom, regions of low electron density (blue) are in the center. But what can we say about the *overall* dipole moment of the CO_2 molecule?

Bond dipoles and dipole moments are *vector* quantities; that is, they have both a magnitude and a direction. The *overall* dipole moment of a polyatomic molecule is the vector sum of its bond dipoles. Both the magnitudes *and* the directions of the bond dipoles must be considered when summing these vectors. The two bond dipoles in CO_2, although equal in magnitude, are exactly opposite in direction. Adding them together is the same as adding two numbers that are equal in magnitude but opposite in sign, such as 100 + (−100). The bond dipoles, like the numbers, "cancel" each other. Therefore, the overall dipole moment of CO_2 is zero, even though the individual bonds are polar. Thus, the geometry of the molecule dictates that the overall dipole moment be zero, making CO_2 a *nonpolar* molecule.

Bond dipoles

Overall dipole moment = 0

(a)

(b)

▲ Figure 9.11 **CO₂, a nonpolar molecule.** (a) The overall dipole moment of a molecule is the sum of its bond dipoles. In CO_2 the bond dipoles are equal in magnitude, but exactly oppose each other. The overall dipole moment is zero, therefore, making the molecule nonpolar. (b) The electron-density model shows that the regions of higher electron density (red) are at the ends of the molecule while the region of lower electron density (blue) is at the center.

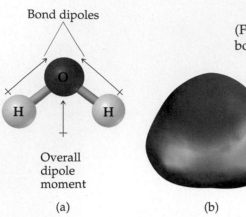

Bond dipoles

Overall dipole moment

(a)   (b)

▲ **Figure 9.12 The dipole moment of a bent molecule.** (a) In H_2O the bond dipoles are equal in magnitude, but do not exactly oppose each other. The molecule has a nonzero dipole moment overall, making the molecule polar. (b) The electron-density model shows that one end of the molecule has more electron density (the oxygen end) while the other end has less electron density (the hydrogens).

Now let's consider H_2O, which is a bent molecule with two polar bonds (Figure 9.12 ◀). Again, the two bonds in the molecule are identical, and the bond dipoles are equal in magnitude. Because the molecule is bent, however, the bond dipoles do not directly oppose each other and therefore do not cancel each other. Hence, the H_2O molecule has an overall nonzero dipole moment ($\mu = 1.85$ D). Because H_2O has a nonzero dipole moment, it is a *polar* molecule. The oxygen atom carries a partial negative charge, and the hydrogen atoms each have a partial positive charge, as shown by the electron-density model in Figure 9.12(b).

▲ **GIVE IT SOME THOUGHT**

The molecule $O=C=S$ has a Lewis structure analogous to that of CO_2 and is a linear molecule. Will it necessarily have a zero dipole moment like that of CO_2?

Figure 9.13▼ shows examples of polar and nonpolar molecules, all of which have polar bonds. The molecules in which the central atom is symmetrically surrounded by identical atoms (BF_3 and CCl_4) are nonpolar. For AB_n molecules in which all the B atoms are the same, certain symmetrical shapes—linear (AB_2), trigonal planar (AB_3), tetrahedral and square planar (AB_4), trigonal bipyramidal (AB_5), and octahedral (AB_6)—must lead to nonpolar molecules even though the individual bonds might be polar.

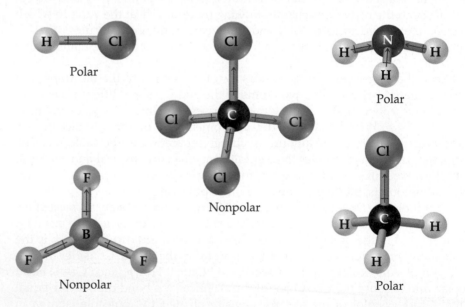

▶ **Figure 9.13 Molecules containing polar bonds.** Two of these molecules have a zero dipole moment because their bond dipoles cancel one another, while the other molecules are polar.

■ **SAMPLE EXERCISE 9.4** | **Polarity of Molecules**

Predict whether the following molecules are polar or nonpolar: **(a)** BrCl, **(b)** SO_2, **(c)** SF_6.

SOLUTION

Analyze: We are given the molecular formulas of several substances and asked to predict whether the molecules are polar.

Plan: If the molecule contains only two atoms, it will be polar if the atoms differ in electronegativity. If the molecule contains three or more atoms, its polarity depends on both its molecular geometry and the polarity of its bonds. Thus, we must draw a Lewis structure for each molecule containing three or more atoms and determine its molecular geometry. We then use the relative electronegativities of the atoms in each bond to determine the direction of the bond dipoles. Finally, we see if the bond dipoles cancel each other to give a nonpolar molecule or reinforce each other to give a polar one.

Solve:

(a) Chlorine is more electronegative than bromine. All diatomic molecules with polar bonds are polar molecules. Consequently, BrCl will be polar, with chlorine carrying the partial negative charge:

$$\overset{\longleftarrow \quad \longrightarrow}{Br-Cl}$$

The actual dipole moment of BrCl, as determined by experimental measurement, is $\mu = 0.57$ D.

(b) Because oxygen is more electronegative than sulfur, SO_2 has polar bonds. Three resonance forms can be written for SO_2:

$$:\overset{..}{O}-\overset{..}{S}=\overset{..}{O}: \longleftrightarrow :\overset{..}{O}=\overset{..}{S}-\overset{..}{O}: \longleftrightarrow :\overset{..}{O}=\overset{..}{S}=\overset{..}{O}:$$

For each of these, the VSEPR model predicts a bent geometry. Because the molecule is bent, the bond dipoles do not cancel, and the molecule is polar:

Experimentally, the dipole moment of SO_2 is $\mu = 1.63$ D.

(c) Fluorine is more electronegative than sulfur, so the bond dipoles point toward fluorine. The six S—F bonds are arranged octahedrally around the central sulfur:

Because the octahedral geometry is symmetrical, the bond dipoles cancel, and the molecule is nonpolar, meaning that $\mu = 0$.

■ **PRACTICE EXERCISE**

Determine whether the following molecules are polar or nonpolar: **(a)** NF_3, **(b)** BCl_3.
Answers: **(a)** polar because polar bonds are arranged in a trigonal-pyramidal geometry, **(b)** nonpolar because polar bonds are arranged in a trigonal-planar geometry

9.4 COVALENT BONDING AND ORBITAL OVERLAP

The VSEPR model provides a simple means for predicting the shapes of molecules. However, it does not explain why bonds exist between atoms. In developing theories of covalent bonding, chemists have approached the problem from another direction, using quantum mechanics. How can we use atomic orbitals to explain bonding and to account for the geometries of molecules? The marriage of Lewis's notion of electron-pair bonds and the idea of atomic orbitals leads to a model of chemical bonding called **valence-bond theory**. By extending this approach to include the ways in which atomic orbitals can mix with one another, we obtain a picture that corresponds nicely to the VSEPR model.

In the Lewis theory, covalent bonding occurs when atoms share electrons, which concentrates electron density between the nuclei. In the valence-bond theory, we visualize the buildup of electron density between two nuclei as occurring when a valence atomic orbital of one atom shares space, or *overlaps*, with that of another atom. The overlap of orbitals allows two electrons of opposite spin to share the common space between the nuclei, forming a covalent bond.

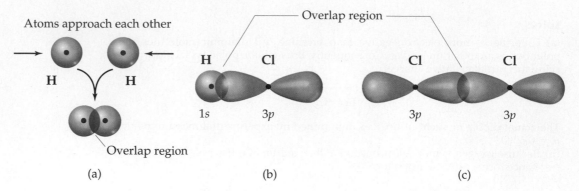

Atoms approach each other

H   H

Overlap region

(a)

Overlap region

H   Cl

1s   3p

Cl   Cl

3p   3p

(b)

(c)

▲ **Figure 9.14 The overlap of orbitals to form covalent bonds.** (a) The bond in H_2 results from the overlap of two $1s$ orbitals from two H atoms. (b) The bond in HCl results from the overlap of a $1s$ orbital of H and one of the lobes of a $3p$ orbital of Cl. (c) The bond in Cl_2 results from the overlap of two $3p$ orbitals from two Cl atoms.

The coming together of two H atoms to form H_2 is depicted in Figure 9.14(a) ▲. Each atom has a single electron in a $1s$ orbital. As the orbitals overlap, electron density is concentrated between the nuclei. Because the electrons in the overlap region are simultaneously attracted to both nuclei, they hold the atoms together, forming a covalent bond.

The idea of orbital overlap producing a covalent bond applies equally well to other molecules. In HCl, for example, chlorine has the electron configuration $[Ne]3s^23p^5$. All the valence orbitals of chlorine are full except one $3p$ orbital, which contains a single electron. This electron pairs with the single electron of H to form a covalent bond. Figure 9.14(b) shows the overlap of the $3p$ orbital of Cl with the $1s$ orbital of H. Likewise, we can explain the covalent bond in the Cl_2 molecule in terms of the overlap of the $3p$ orbital of one atom with the $3p$ orbital of another, as shown in Figure 9.14(c).

There is always an optimum distance between the two bonded nuclei in any covalent bond. Figure 9.15▼ shows how the potential energy of the system changes as two H atoms come together to form an H_2 molecule. At infinite distance, the atoms do not "feel" each other and so the energy approaches zero. As the distance between the atoms decreases, the overlap between their $1s$ orbitals increases. Because of the resultant increase in electron density between the nuclei, the potential energy of the system decreases. That is, the strength of the bond increases, as shown by the decrease in the energy on the curve. However, the curve also shows that as the atoms come very close together, the energy increases sharply. This increase, which becomes significant at short internuclear distances, is due mainly to the electrostatic repulsion between the nuclei. The internuclear distance, or bond length, is the distance that corresponds to the minimum of the potential-energy curve. The potential energy at this minimum corresponds to the bond strength. Thus, the observed bond length is the distance at which the attractive forces between unlike charges (electrons and nuclei) are balanced by the repulsive forces between like charges (electron–electron and nucleus–nucleus).

▶ **Figure 9.15 Formation of the H_2 molecule.** Plot of the change in potential energy as two hydrogen atoms come together to form the H_2 molecule. The minimum in the energy, at 0.74 Å, represents the equilibrium bond distance. The energy at that point, −436 kJ/mol, corresponds to the energy change for formation of the H—H bond.

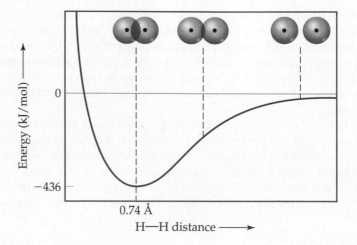

Energy (kJ/mol)

0

−436

0.74 Å

H—H distance

◢ GIVE IT SOME THOUGHT

If you could put pressure on the hydrogen molecule so that its bond length decreased, would its bond strength increase or decrease? (Refer to Figure 9.15).

9.5 HYBRID ORBITALS

The VSEPR model, simple as it is, does a surprisingly good job at predicting molecular shape, despite the fact that it has no obvious relationship to the filling and shapes of atomic orbitals. For example, based on the shapes and orientations of the 2s and 2p orbitals on a carbon atom, it is not obvious why a CH_4 molecule should have a tetrahedral geometry. How can we reconcile the notion that covalent bonds are formed from the overlap of atomic orbitals with the molecular geometries that come from the VSEPR model?

To explain geometries, we assume that the atomic orbitals on an atom (usually the central atom) mix to form new orbitals called **hybrid orbitals**. The shape of any hybrid orbital is different from the shapes of the original atomic orbitals. The process of mixing atomic orbitals is called **hybridization**. The total number of atomic orbitals on an atom remains constant, however, and so the number of hybrid orbitals on an atom equals the number of atomic orbitals that are mixed.

Let's examine the common types of hybridization. As we do so, notice the connection between the type of hybridization and the five basic electron-domain geometries predicted by the VSEPR model: linear, trigonal planar, tetrahedral, trigonal bipyramidal, and octahedral.

sp Hybrid Orbitals

To illustrate the process of hybridization, consider the BeF_2 molecule, which is generated when solid BeF_2 is heated to high temperatures. The Lewis structure of BeF_2 is

$$:\ddot{F}—Be—\ddot{F}:$$

The VSEPR model correctly predicts that BeF_2 is linear with two identical Be—F bonds. How can we use valence-bond theory to describe the bonding? The electron configuration of F ($1s^22s^22p^5$) indicates there is an unpaired electron in a 2p orbital. This 2p electron can be paired with an unpaired electron from the Be atom to form a polar covalent bond. Which orbitals on the Be atom, however, overlap with those on the F atoms to form the Be—F bonds?

The orbital diagram for a ground-state Be atom is

Because it has no unpaired electrons, the Be atom in its ground state is incapable of forming bonds with the fluorine atoms. The Be atom could form two bonds, however, by "promoting" one of the 2s electrons to a 2p orbital:

The Be atom now has two unpaired electrons and can therefore form two polar covalent bonds with the F atoms. The two bonds would not be identical, however, because a Be 2s orbital would be used to form one of the bonds and a 2p orbital would be used for the other. Therefore, although the promotion of an electron allows two Be—F bonds to form, we still have not explained the structure of BeF_2.

▲ **Figure 9.16 Formation of *sp* hybrid orbitals.** One *s* orbital and one *p* orbital can hybridize to form two equivalent *sp* hybrid orbitals. The two hybrid orbitals have their large lobes pointing in opposite directions, 180° apart.

We can solve this dilemma by "mixing" the 2*s* orbital with one of the 2*p* orbitals to generate two new orbitals, as shown in Figure 9.16▲. Like *p* orbitals, each of the new orbitals has two lobes. Unlike *p* orbitals, however, one lobe is much larger than the other. The two new orbitals are identical in shape, but their large lobes point in opposite directions. These two new orbitals are hybrid orbitals. Because we have hybridized one *s* and one *p* orbital, we call each hybrid an *sp* hybrid orbital. *According to the valence-bond model, a linear arrangement of electron domains implies* sp *hybridization.*

For the Be atom of BeF$_2$, we write the orbital diagram for the formation of two *sp* hybrid orbitals as follows:

$$1s \qquad sp \qquad 2p$$

The electrons in the *sp* hybrid orbitals can form two-electron bonds with the two fluorine atoms (Figure 9.17◄). Because the *sp* hybrid orbitals are equivalent but point in opposite directions, BeF$_2$ has two identical bonds and a linear geometry. The remaining two 2*p* orbitals remain unhybridized.

Large lobe of *sp* hybrid orbital

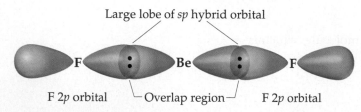

F 2*p* orbital — Overlap region — F 2*p* orbital

▲ **Figure 9.17 Formation of two equivalent Be—F bonds in BeF$_2$.** Each *sp* hybrid orbital on Be overlaps with a 2*p* orbital on F to form a bond. The two bonds are equivalent to each other and form an angle of 180°.

▲ **GIVE IT SOME THOUGHT**

Suppose that the two unhybridized 2*p* orbitals on Be were used to make the Be—F bonds in BeF$_2$. Would the two bonds be equivalent to each other? What would be the expected F—Be—F bond angle?

sp^2 and *sp*^{3} Hybrid Orbitals

Whenever we mix a certain number of atomic orbitals, we get the same number of hybrid orbitals. Each of these hybrid orbitals is equivalent to the others but points in a different direction. Thus, mixing one 2*s* and one 2*p* orbital yields two equivalent *sp* hybrid orbitals that point in opposite directions (Figure 9.16). Other combinations of atomic orbitals can be hybridized to obtain different geometries. In BF$_3$, for example, mixing the 2*s* and two of the 2*p* orbitals yields three equivalent *sp*^{2} (pronounced "s-p-two") hybrid orbitals (Figure 9.18▶).

The three *sp*^{2} hybrid orbitals lie in the same plane, 120° apart from one another (Figure 9.18). They are used to make three equivalent bonds with the three fluorine atoms, leading to the trigonal-planar geometry of BF$_3$. Notice that an unfilled 2*p* orbital remains unhybridized. This unhybridized orbital will be important when we discuss double bonds in Section 9.6.

▲ **GIVE IT SOME THOUGHT**

In an *sp*^{2} hybridized atom, what is the orientation of the unhybridized *p* orbital relative to the three *sp*^{2} hybrid orbitals?

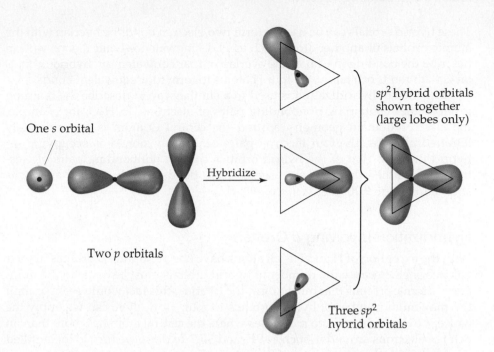

◀ Figure 9.18 **Formation of sp^2 hybrid orbitals.** One s orbital and two p orbitals can hybridize to form three equivalent sp^2 hybrid orbitals. The large lobes of the hybrid orbitals point toward the corners of an equilateral triangle.

An s orbital can also mix with all three p orbitals in the same subshell. For example, the carbon atom in CH_4 forms four equivalent bonds with the four hydrogen atoms. We envision this process as resulting from the mixing of the $2s$ and all three $2p$ atomic orbitals of carbon to create four equivalent sp^3 (pronounced "s-p-three") hybrid orbitals. Each sp^3 hybrid orbital has a large lobe that points toward a vertex of a tetrahedron, as shown in Figure 9.19 ▼.

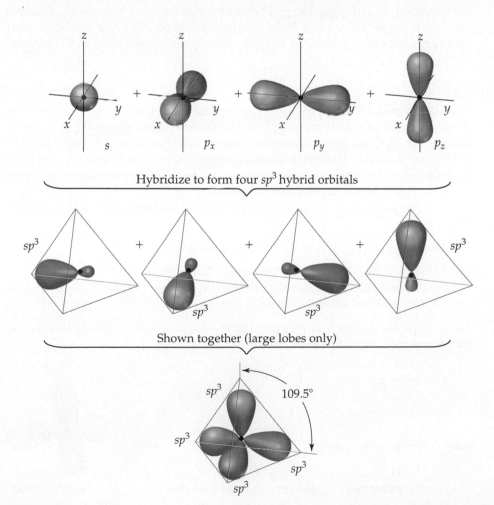

◀ Figure 9.19 **Formation of sp^3 hybrid orbitals.** One s orbital and three p orbitals can hybridize to form four equivalent sp^3 hybrid orbitals. The large lobes of the hybrid orbitals point toward the corners of a tetrahedron.

▲ **Figure 9.20 Valence-bond description of H₂O.** The bonding in a water molecule can be envisioned as sp^3 hybridization of the orbitals on O. Two of the four hybrid orbitals overlap with 1s orbitals of H to form covalent bonds. The other two hybrid orbitals are occupied by nonbonding pairs of electrons.

These hybrid orbitals can be used to form two-electron bonds by overlap with the atomic orbitals of another atom, such as H. Using valence-bond theory, we can describe the bonding in CH_4 as the overlap of four equivalent sp^3 hybrid orbitals on C with the 1s orbitals of the four H atoms to form four equivalent bonds.

The idea of hybridization is used in a similar way to describe the bonding in molecules containing nonbonding pairs of electrons. In H_2O, for example, the electron-domain geometry around the central O atom is approximately tetrahedral. Thus, the four electron pairs can be envisioned as occupying sp^3 hybrid orbitals. Two of the hybrid orbitals contain nonbonding pairs of electrons, while the other two are used to form bonds with hydrogen atoms, as shown in Figure 9.20 ◄.

Hybridization Involving d Orbitals

With the exception of H and He, all atoms have one s and three p orbitals in their valence shell. Because the number of hybrid orbitals must be equal to the number of atomic orbitals that mix to form the hybrids, this fact would seem to limit the maximum number of hybrid orbitals to four. How, then, can we apply the concept of hybridization to molecules where the central atom has more than an octet of electrons around it, such as PF_5 and SF_6? To do so we turn to the unfilled d orbitals with the same value of the principal quantum number, n. Mixing one s orbital, three p orbitals, and one d orbital leads to five sp^3d hybrid orbitals. These hybrid orbitals are directed toward the vertices of a trigonal bipyramid.

Similarly, mixing one s orbital, three p orbitals, and two d orbitals gives six sp^3d^2 hybrid orbitals that are directed toward the vertices of an octahedron. The use of d orbitals in constructing hybrid orbitals nicely corresponds to the notion of an expanded valence shell. ∞ (Section 8.7) Keep in mind, however, that only atoms in the third period and beyond (atoms beyond Ne) possess vacant d orbitals that can be used to form hybrid orbitals of this type. The geometric arrangements characteristic of hybrid orbitals are summarized in Table 9.4▶.

Hybrid Orbital Summary

Overall, hybrid orbitals provide a convenient model for using valence-bond theory to describe covalent bonds in molecules with geometries that conform to the electron-domain geometries predicted by the VSEPR model. The picture of hybrid orbitals has limited predictive value. When we know the electron-domain geometry, however, we can employ hybridization to describe the atomic orbitals used by the central atom in bonding.

The following steps allow us to predict the hybrid orbitals used by an atom in bonding:

1. Draw the *Lewis structure* for the molecule or ion.
2. Determine the electron-domain geometry using the *VSEPR model*.
3. Specify the *hybrid orbitals* needed to accommodate the electron pairs based on their geometric arrangement (Table 9.4).

These steps are illustrated in Figure 9.21 ▼, which shows how the hybridization employed by N in NH_3 is determined.

▼ **Figure 9.21 Bonding in NH₃.** The hybrid orbitals used by N in the NH_3 molecule are predicted by first drawing the Lewis structure, then using the VSEPR model to determine the electron-domain geometry, and then specifying the hybrid orbitals that correspond to that geometry. This is essentially the same procedure as that used to determine molecular structure (Figure 9.6), except we focus on the orbitals used to make bonds and to hold nonbonding pairs.

NH_3 Lewis structure   Electron-domain geometry sp^3 hybridization

TABLE 9.4 ■ Geometric Arrangements Characteristic of Hybrid Orbital Sets

Atomic Orbital Set	Hybrid Orbital Set	Geometry	Examples
s,p	Two sp	 180° Linear	BeF_2, $HgCl_2$
s,p,p	Three sp^2	 120° Trigonal planar	BF_3, SO_3
s,p,p,p	Four sp^3	 109.5° Tetrahedral	CH_4, NH_3, H_2O, NH_4^+
s,p,p,p,d	Five sp^3d	 90° 120° Trigonal bipyramidal	PF_5, SF_4, BrF_3
s,p,p,p,d,d	Six sp^3d^2	 90° 90° Octahedral	SF_6, ClF_5, XeF_4, PF_6^-

■ **SAMPLE EXERCISE 9.5** | **Hybridization**

Indicate the hybridization of orbitals employed by the central atom in **(a)** NH_2^-, **(b)** SF_4 (see Sample Exercise 9.2).

SOLUTION

Analyze: We are given two chemical formulas—one for a polyatomic anion and one for a molecular compound—and asked to describe the type of hybrid orbitals surrounding the central atom in each case.

Plan: To determine the hybrid orbitals used by an atom in bonding, we must know the electron-domain geometry around the atom. Thus, we first draw the Lewis structure to determine the number of electron domains around the central atom. The hybridization conforms to the number and geometry of electron domains around the central atom as predicted by the VSEPR model.

Solve:

(a) The Lewis structure of NH_2^- is

$$\left[H : \ddot{N} : H\right]^-$$

Because there are four electron domains around N, the electron-domain geometry is tetrahedral. The hybridization that gives a tetrahedral electron-domain geometry is sp^3 (Table 9.4). Two of the sp^3 hybrid orbitals contain nonbonding pairs of electrons, and the other two are used to make bonds with the hydrogen atoms.

(b) The Lewis structure and electron-domain geometry of SF_4 are shown in Sample Exercise 9.2. The S atom has five electron domains around it, giving rise to a trigonal-bipyramidal electron-domain geometry. With an expanded octet of ten electrons, a d orbital on the sulfur must be used. The trigonal-bipyramidal electron-domain geometry corresponds to sp^3d hybridization (Table 9.4). One of the hybrid orbitals that points in an equatorial direction contains a nonbonding pair of electrons; the other four are used to form the S—F bonds.

■ **PRACTICE EXERCISE**

Predict the electron-domain geometry and the hybridization of the central atom in **(a)** SO_3^{2-} **(b)** SF_6.
Answers: **(a)** tetrahedral, sp^3; **(b)** octahedral, sp^3d^2

9.6 MULTIPLE BONDS

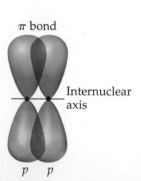

π bond

Internuclear axis

p p

▲ **Figure 9.22 The π bond.** When two p orbitals overlap in a sideways fashion, the result is a π bond. Note that the two regions of overlap constitute a *single π bond*.

In the covalent bonds we have considered thus far, the electron density is concentrated along the line connecting the nuclei (the *internuclear axis*). In other words, the line joining the two nuclei passes through the middle of the overlap region. These bonds are called **sigma (σ) bonds**. The overlap of two s orbitals as in H_2 [Figure 9.14(a)], the overlap of an s and a p orbital as in HCl [Figure 9.14(b)], the overlap between two p orbitals as in Cl_2 [Figure 9.14(c)], and the overlap of a p orbital with an sp hybrid orbital as in BeF_2 (Figure 9.17) are all examples of σ bonds.

To describe multiple bonding, we must consider a second kind of bond that results from the overlap between two p orbitals oriented perpendicularly to the internuclear axis (Figure 9.22 ◄). This sideways overlap of p orbitals produces a **pi (π) bond**. A π bond is a covalent bond in which the overlap regions lie above and below the internuclear axis. Unlike in a σ bond, in a π bond there is no probability of finding the electron on the internuclear axis. Because the p orbitals in a π bond overlap sideways rather than directly facing each other, the total overlap in a π bond tends to be less than that in a σ bond. Consequently, π bonds are generally weaker than σ bonds.

In almost all cases, single bonds are σ bonds. A double bond consists of one σ bond and one π bond, and a triple bond consists of one σ bond and two π bonds:

H—H

One σ bond

One σ bond plus one π bond

One σ bond plus two π bonds

To see how these ideas are used, consider ethylene (C_2H_4), which possesses a C=C double bond. The bond angles in ethylene are all approximately 120° (Figure 9.23▶), suggesting that each carbon atom uses sp^2 hybrid orbitals (Figure 9.18) to form σ bonds with the other carbon and with two hydrogens. Because carbon has four valence electrons, after sp^2 hybridization one electron in each of the carbon atoms remains in the *unhybridized* 2p orbital. The unhybridized 2p orbital is directed perpendicular to the plane that contains the three sp^2 hybrid orbitals.

Each sp^2 hybrid orbital on a carbon atom contains one electron. Figure 9.24▶ shows how the four C—H σ bonds are formed by overlap of sp^2 hybrid orbitals on C with the 1s orbitals on each H atom. We use eight electrons to form these four electron-pair bonds. The C—C σ bond is formed by the overlap of two sp^2 hybrid orbitals, one on each carbon atom, and requires two more electrons. Thus ten of the 12 valence electrons in the C_2H_4 molecule are used to form five σ bonds.

The remaining two valence electrons reside in the unhybridized 2p orbitals, one electron on each carbon atom. These two 2p orbitals can overlap with each other in a sideways fashion, as shown in Figure 9.25▼. The resultant electron density is concentrated above and below the C—C bond axis, which means this is a π bond (Figure 9.22). Thus, the C=C double bond in ethylene consists of one σ bond and one π bond.

Although we cannot experimentally observe a π bond directly (all we can observe are the positions of the atoms), the structure of ethylene provides strong support for its presence. First, the C—C bond length in ethylene (1.34 Å) is much shorter than in compounds with C—C single bonds (1.54 Å), consistent with the presence of a stronger C=C double bond. Second, all six atoms in C_2H_4 lie in the same plane. The 2p orbitals that make up the π bond can achieve a good overlap only when the two CH_2 fragments lie in the same plane. If the π bond were absent, there would be no reason for the two CH_2 fragments of ethylene to lie in the same plane. Because π bonds require that portions of a molecule be planar, they can introduce rigidity into molecules.

▲ **Figure 9.23 The molecular geometry of ethylene.** Ethylene, C_2H_4, has one C—C σ bond and one C—C π bond.

▲ **Figure 9.24 Hybridization in ethylene.** The σ bonding framework is formed from sp^2 hybrid orbitals on the carbon atoms. The unhybridized 2p orbitals on the C atoms are used to make a π bond.

△ **GIVE IT SOME THOUGHT**

The molecule called *diazine* has the formula N_2H_2 and the Lewis structure

$$H—\ddot{N}=\ddot{N}—H$$

Would you expect diazine to be a linear molecule (all four atoms on the same line)? If not, would you expect the molecule to be planar (all four atoms in the same plane)?

Two lobes of one π bond

▲ **Figure 9.25 The π bond in ethylene.** The unhybridized 2p orbitals on each C atom overlap to form a π bond. The electron density in the π bond is above and below the bond axis, whereas in the σ bonds the electron density lies directly along the bond axes. As noted in Figure 9.22, the two lobes constitute one π bond.

▲ **Figure 9.26 Formation of two π bonds.** In acetylene, C_2H_2, the overlap of two sets of unhybridized carbon 2p orbitals leads to the formation of two π bonds.

Triple bonds can also be explained using hybrid orbitals. Acetylene (C_2H_2), for example, is a linear molecule containing a triple bond: $H-C\equiv C-H$. The linear geometry suggests that each carbon atom uses sp hybrid orbitals to form σ bonds with the other carbon and one hydrogen. Each carbon atom thus has two remaining unhybridized 2p orbitals at right angles to each other and to the axis of the sp hybrid set (Figure 9.26◄). These p orbitals overlap to form a pair of π bonds. Thus, the triple bond in acetylene consists of one σ bond and two π bonds.

Although it is possible to make π bonds from d orbitals, the only π bonds we will consider are those formed by the overlap of p orbitals. These π bonds can form only if unhybridized p orbitals are present on the bonded atoms. Therefore, only atoms having sp or sp^2 hybridization can be involved in such π bonding. Further, double and triple bonds (and hence π bonds) are more common in molecules made up of small atoms from the second period, especially C, N, and O. Larger atoms, such as S, P, and Si, form π bonds less readily.

SAMPLE EXERCISE 9.6 | Describing σ and π Bonds in a Molecule

Formaldehyde has the Lewis structure

$$\begin{array}{c} H \\ \diagdown \\ C{=}\ddot{O}: \\ \diagup \\ H \end{array}$$

Describe how the bonds in formaldehyde are formed in terms of overlaps of appropriate hybridized and unhybridized orbitals.

SOLUTION

Analyze: We are asked to describe the bonding in formaldehyde in terms of orbital overlaps.

Plan: Single bonds will be of the σ type, whereas double bonds will consist of one σ bond and one π bond. The ways in which these bonds form can be deduced from the geometry of the molecule, which we predict using the VSEPR model.

Solve: The C atom has three electron domains around it, which suggests a trigonal-planar geometry with bond angles of about 120°. This geometry implies sp^2 hybrid orbitals on C (Table 9.4). These hybrids are used to make the two C—H and one C—O σ bonds to C. There remains an unhybridized 2p orbital on carbon, perpendicular to the plane of the three sp^2 hybrids.

The O atom also has three electron domains around it, and so we will assume that it has sp^2 hybridization as well. One of these hybrids participates in the C—O σ bond, while the other two hybrids hold the two nonbonding electron pairs of the O atom. Like the C atom, therefore, the O atom has an unhybridized 2p orbital that is perpendicular to the plane of the molecule. The unhybridized 2p orbitals on the C and O atoms overlap to form a C—O π bond, as illustrated in Figure 9.27▼.

One π bond

▲ **Figure 9.27 Formation of σ and π bonds in formaldehyde, H_2CO.**

■ **PRACTICE EXERCISE**
Consider the acetonitrile molecule:

(a) Predict the bond angles around each carbon atom; (b) describe the hybridization at each of the carbon atoms; (c) determine the total number of σ and π bonds in the molecule.
Answers: (a) approximately 109° around the left C and 180° on the right C; (b) sp^3, sp; (c) five σ bonds and two π bonds

Resonance Structures, Delocalization, and π Bonding

In the molecules we have discussed thus far in this section, the bonding electrons are *localized*. By this we mean that the σ and π electrons are associated totally with the two atoms that form the bond. In many molecules, however, we cannot adequately describe the bonding as being entirely localized. This situation arises particularly in molecules that have two or more resonance structures involving π bonds.

One molecule that cannot be described with localized π bonds is benzene (C_6H_6), which has two resonance structures: ∞∞ (Section 8.6)

To describe the bonding in benzene using hybrid orbitals, we first choose a hybridization scheme consistent with the geometry of the molecule. Because each carbon is surrounded by three atoms at 120° angles, the appropriate hybrid set is sp^2. Six localized C—C σ bonds and six localized C—H σ bonds are formed from the sp^2 hybrid orbitals, as shown in Figure 9.28(a)▶. This leaves a 2p orbital on each carbon that is oriented perpendicularly to the plane of the molecule. The situation is very much like that in ethylene, except we now have six carbon 2p orbitals arranged in a ring [Figure 9.28(b)]. Each of the unhybridized 2p orbitals is occupied by one electron, leaving a total of six electrons to be accounted for by π bonding.

We could envision using the unhybridized 2p orbitals of benzene to form three localized π bonds. As shown in Figure 9.29(a)▼ and (b), there are two equivalent ways to make these localized bonds, and each corresponds to one of the resonance structures of the molecule. A representation that reflects *both* resonance structures has the

(a) σ bonds

(b) 2p atomic orbitals

▼ **Figure 9.28 The σ and π bond networks in benzene, C_6H_6.** (a) The C—C and C—H σ bonds all lie in the plane of the molecule and are formed by using carbon sp^2 hybrid orbitals. (b) Each carbon atom has an unhybridized 2p orbital that lies perpendicular to the molecular plane. These six 2p orbitals form the π *orbitals* of benzene.

(a) Localized π bonds

(b) Localized π bonds

(c) Delocalized π bonds

◀ **Figure 9.29 Delocalized π bonds.** The six 2p orbitals of benzene, shown in Figure 9.28(b), can be used to make C—C π bonds. (a, b) Two equivalent ways to make localized π bonds. These π bonds correspond to the two resonance structures for benzene. (c) A representation of the smearing out, or delocalization, of the three C—C π bonds among the six C atoms.

six π electrons "smeared out" among all six carbon atoms, as shown in Figure 9.29(c). Notice how this figure corresponds to the "circle-in-a-hexagon" drawing we often use to represent benzene. This model leads to the description of each carbon–carbon bond as having identical bond lengths that are between the length of a C—C single bond (1.54 Å) and the length of a C=C double bond (1.34 Å), consistent with the observed bond lengths in benzene (1.40 Å).

Because we cannot describe the π bonds in benzene as individual electron-pair bonds between neighboring atoms, we say that the π bonds are **delocalized** among the six carbon atoms. Delocalization of the electrons in its π bonds gives benzene a special stability, as we will discuss in Section 25.3. Delocalization of π bonds is also responsible for the color of many organic molecules. A final important point to remember about delocalized π bonds is the constraint they place upon the geometry of a molecule. For optimal overlap of the unhybridized p orbitals, all of the atoms involved in the delocalized π bonding network should lie in the same plane. This location imparts a certain rigidity to the molecule that is absent in molecules containing only σ bonds (see the "Chemistry and Life" box on vision). If you take a course in organic chemistry, you will see many examples of how electron delocalization influences the properties of organic molecules.

(a) N—O π bond in one of the resonance structures of NO$_3^-$.

(b) Delocalization of the π bonds in the NO$_3^-$ ion.

▲ **Figure 9.30 Localized and delocalized π bonds in NO$_3^-$.**

■■ **SAMPLE EXERCISE 9.7** | Delocalized Bonding

Describe the bonding in the nitrate ion, NO$_3^-$. Does this ion have delocalized π bonds?

SOLUTION

Analyze: Given the chemical formula for a polyatomic anion, we are asked to describe the bonding and determine whether the ion has delocalized π bonds.

Plan: Our first step in describing the bonding in NO$_3^-$ is to construct appropriate Lewis structures. If there are multiple resonance structures that involve the placement of the double bonds in different locations, that suggests that the π component of the double bonds is delocalized.

Solve: In Section 8.6 we saw that NO$_3^-$ has three resonance structures:

In each of these structures the electron-domain geometry at nitrogen is trigonal planar, which implies sp^2 hybridization of the N atom. The sp^2 hybrid orbitals are used to construct the three N—O σ bonds that are present in each of the resonance structures.

The unhybridized $2p$ orbital on the N atom can be used to make π bonds. For any one of the three resonance structures shown, we might imagine a single localized N—O π bond formed by the overlap of the unhybridized $2p$ orbital on N and a $2p$ orbital on one of the O atoms, as shown in Figure 9.30(a) ◄. Because each resonance structure contributes equally to the observed structure of NO$_3^-$, however, we represent the π bonding as spread out, or delocalized, over the three N—O bonds, as shown in Figure 9.30(b).

■■ **PRACTICE EXERCISE**

Which of the following molecules or ions will exhibit delocalized bonding: SO$_3$, SO$_3^{2-}$, H$_2$CO, O$_3$, NH$_4^+$?
Answer: SO$_3$ and O$_3$, as indicated by the presence of two or more resonance structures involving π bonding for each of these molecules

Chemistry and Life   THE CHEMISTRY OF VISION

In recent years scientists have begun to understand the complex chemistry of vision. Vision begins when light is focused by the lens of the eye onto the retina, the layer of cells lining the interior of the eyeball. The retina contains *photoreceptor* cells called rods and cones (Figure 9.31 ▼). The human retina contains about 3 million cones and 100 million rods. The rods are sensitive to dim light and are used in night vision. The cones are sensitive to colors. The tops of the rods and cones contain a molecule called *rhodopsin*. Rhodopsin consists of a protein, called *opsin*, bonded to a reddish purple pigment called *retinal*. Structural changes around a double bond in the retinal portion of the molecule trigger a series of chemical reactions that result in vision.

Double bonds between atoms are stronger than single bonds between the same atoms (Table 8.4). For example, a $C=C$ double bond is stronger (614 kJ/mol) than a $C-C$ single bond (348 kJ/mol), though not twice as strong. Our recent discussions now allow us to appreciate another aspect of double bonds: the stiffness or rigidity that they introduce into molecules.

Imagine taking the —CH_2 group of the ethylene molecule and rotating it relative to the other —CH_2 group as shown in Figure 9.32 ▼. This rotation destroys the overlap of *p* orbitals, breaking the π bond, a process that requires considerable energy. Thus, the presence of a double bond restricts the rotation of the bonds in a molecule. In contrast, molecules can rotate almost freely around the bond axis in single (σ) bonds because this motion has no effect on the orbital overlap for a σ bond. This rotation allows molecules with single bonds to twist and fold almost as if their atoms were attached by hinges.

Our vision depends on the rigidity of double bonds in retinal. In its normal form, retinal is held rigid by its double bonds, as shown on the left in Figure 9.33 ▼. Light entering the eye is absorbed by rhodopsin, and the energy is used to break the π-bond portion of the indicated double bond. The molecule then rotates around this bond, changing its geometry. The retinal then separates from the opsin, triggering the reactions that produce a nerve impulse that the brain interprets as the sensation of vision. It takes as few as five closely spaced molecules reacting in this fashion to produce the sensation of vision. Thus, only five photons of light are necessary to stimulate the eye.

The retinal slowly reverts to its original form and reattaches to the opsin. The slowness of this process helps explain why intense bright light causes temporary blindness. The light causes all the retinal to separate from opsin, leaving no further molecules to absorb light.

Related Exercises: 9.99 and 9.100

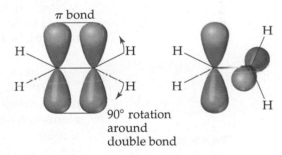

▲ **Figure 9.32 Rotation about the carbon–carbon double bond in ethylene.** The overlap of the *p* orbitals that form the π bond is lost in the rotation. For this reason, rotation about double bonds requires the absorption of energy.

▲ **Figure 9.31 Inside the eye.** A color-enhanced scanning electron micrograph of the rods (yellow) and cones (blue) in the retina of the human eye.

▲ **Figure 9.33 The chemical basis of vision.** When rhodopsin absorbs visible light, the π component of the double bond shown in red breaks, allowing rotation that produces a change in molecular geometry.

General Conclusions

On the basis of the examples we have seen, we can draw a few helpful conclusions for using the concept of hybrid orbitals to describe molecular structures:

1. Every pair of bonded atoms shares one or more pairs of electrons. The lines we draw in Lewis structures represent two electrons each. In every bond at least one pair of electrons is localized in the space between the atoms in a σ bond. The appropriate set of hybrid orbitals used to form the σ bonds between an atom and its neighbors is determined by the observed geometry of the molecule. The correlation between the set of hybrid orbitals and the geometry about an atom is given in Table 9.4.

2. The electrons in σ bonds are localized in the region between two bonded atoms and do not make a significant contribution to the bonding between any other two atoms.

3. When atoms share more than one pair of electrons, one pair is used to form a σ bond; the additional pairs form π bonds. The centers of charge density in a π bond lie above and below the internuclear axis.

4. Molecules with two or more resonance structures can have π bonds that extend over more than two bonded atoms. Electrons in π bonds that extend over more than two atoms are delocalized.

9.7 MOLECULAR ORBITALS

Valence-bond theory and hybrid orbitals allow us to move in a straightforward way from Lewis structures to rationalizing the observed geometries of molecules in terms of atomic orbitals. For example, we can use this theory to understand why methane has the formula CH_4, how the carbon and hydrogen atomic orbitals are used to form electron-pair bonds, and why the arrangement of the C—H bonds about the central carbon is tetrahedral. This model, however, does not explain all aspects of bonding. It is not successful, for example, in describing the excited states of molecules, which we must understand to explain how molecules absorb light, giving them color.

Some aspects of bonding are better explained by a more sophisticated model called **molecular orbital theory**. In Chapter 6 we saw that electrons in atoms can be described by wave functions, which we call atomic orbitals. In a similar way, molecular orbital theory describes the electrons in molecules by using specific wave functions called **molecular orbitals**. Chemists use the abbreviation **MO** for molecular orbital.

Molecular orbitals have many of the same characteristics as atomic orbitals. For example, an MO can hold a maximum of two electrons (with opposite spins), it has a definite energy, and we can visualize its electron-density distribution by using a contour representation, as we did when we discussed atomic orbitals. Unlike atomic orbitals, however, MOs are associated with the entire molecule, not with a single atom.

▼ Figure 9.34 The molecular orbitals of H₂. The combination of two H $1s$ atomic orbitals forms two molecular orbitals (MOs) of H_2. In the bonding MO, σ_{1s}, the atomic orbitals combine constructively, leading to a buildup of electron density between the nuclei. In the antibonding MO, σ_{1s}^*, the orbitals combine destructively in the bonding region. Note that the σ_{1s}^* MO has a node between the two nuclei.

The Hydrogen Molecule

To appreciate the approach taken in MO theory, we will begin with the simplest molecule: the hydrogen molecule, H_2. We will use the two $1s$ atomic orbitals (one on each H atom) to "build" molecular orbitals for the H_2 molecule. *Whenever two atomic orbitals overlap, two molecular orbitals form.* Thus, the overlap of the $1s$ orbitals of two hydrogen atoms to form H_2 produces two MOs (Figure 9.34 ◄). One of the molecular orbitals lies lower in energy from the two atomic orbitals from which it was made; the other molecular orbital lies higher in energy.

H atomic orbitals

H_2 molecular orbitals

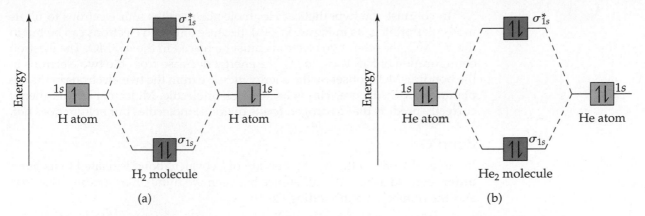

▲ **Figure 9.35 Energy-level diagrams for H_2 and He_2.** (a) The two electrons in the H_2 molecule occupy the σ_{1s} bonding MO. (b) In the (hypothetical) He_2 molecule, both the σ_{1s} bonding MO and the σ_{1s}^* antibonding MO are occupied by two electrons.

The lower-energy MO of H_2 concentrates electron density between the two hydrogen nuclei and is called the **bonding molecular orbital**. This sausage-shaped MO results from summing the two atomic orbitals so that the atomic orbital wave functions combine in the region between the two nuclei. Because an electron in this MO is attracted to both nuclei, the electron is more stable (it has lower energy) than it is in the $1s$ atomic orbital of an isolated hydrogen atom. Further, because this bonding MO concentrates electron density between the nuclei, it holds the atoms together in a covalent bond.

The higher-energy MO in Figure 9.34 has very little electron density between the nuclei and is called the **antibonding molecular orbital**. Instead of combining in the region between the nuclei, the atomic orbital wave functions cancel each other in this region, leaving the greatest electron density on opposite sides of the nuclei. Thus, this MO excludes electrons from the very region in which a bond must be formed. An electron in this MO is repelled from the bonding region and is therefore less stable (it has higher energy) than it is in the $1s$ atomic orbital of a hydrogen atom.

The electron density in both the bonding MO and the antibonding MO of H_2 is centered about the internuclear axis. MOs of this type are called **sigma (σ) molecular orbitals** (by analogy to the σ bonds that were defined in Section 9.6). The bonding sigma MO of H_2 is labeled σ_{1s}; the subscript indicates that the MO is formed from two $1s$ orbitals. The antibonding sigma MO of H_2 is labeled σ_{1s}^* (read "sigma-star-one-s"); the asterisk denotes that the MO is antibonding.

The interaction between two $1s$ atomic orbitals and the molecular orbitals that result can be represented by an **energy-level diagram** (also called a **molecular orbital diagram**), such as those in Figure 9.35▲. Such diagrams show the interacting atomic orbitals in the left and right columns and the MOs in the middle column. Note that the bonding molecular orbital, σ_{1s}, is lower in energy than the atomic $1s$ orbitals, whereas the antibonding orbital, σ_{1s}^*, is higher in energy than the $1s$ orbitals. Like atomic orbitals, each MO can accommodate two electrons with their spins paired (Pauli exclusion principle). ∞ (Section 6.7) The molecular orbital diagram of the H_2 molecule is shown in Figure 9.35(a). Each H atom brings one electron to the molecule, so there are two electrons in H_2. These two electrons occupy the lower-energy bonding (σ_{1s}) MO, and their spins are paired. Electrons occupying a bonding molecular orbital are called *bonding electrons*. Because the σ_{1s} MO is lower in energy than the isolated $1s$ atomic orbitals, the H_2 molecule is more stable than the two separate H atoms. By analogy with atomic electron configurations, the electron configurations for molecules can also be written with superscripts to indicate electron occupancy. The electron configuration for H_2, then, is σ_{1s}^2.

In contrast, the hypothetical He_2 molecule requires four electrons to fill its molecular orbitals, as in Figure 9.35(b). Because only two electrons can be put in the σ_{1s} MO, the other two electrons must be placed in the σ_{1s}^* MO. The electron configuration of He_2 is $\sigma_{1s}^2 \sigma_{1s}^{*2}$. The energy decrease from the two electrons in the bonding MO is offset by the energy increase from the two electrons in the antibonding MO.* Hence, He_2 is an unstable molecule. Molecular orbital theory correctly predicts that hydrogen forms diatomic molecules but helium does not.

Bond Order

In molecular orbital theory the stability of a covalent bond is related to its **bond order**, defined as half the difference between the number of bonding electrons and the number of antibonding electrons:

$$\text{Bond order} = \tfrac{1}{2}(\text{no. of bonding electrons} - \text{no. of antibonding electrons})$$

We take half the difference because we are used to thinking of bonds as pairs of electrons. *A bond order of 1 represents a single bond, a bond order of 2 represents a double bond, and a bond order of 3 represents a triple bond.* Because MO theory also treats molecules containing an odd number of electrons, bond orders of 1/2, 3/2, or 5/2 are possible.

Because H_2 has two bonding electrons and zero antibonding electrons [Figure 9.35(a)], it has a bond order of 1. Because He_2 has two bonding and two antibonding electrons [Figure 9.35(b)], it has a bond order of 0. A bond order of 0 means that no bond exists.

GIVE IT SOME THOUGHT

Suppose light is used to excite one of the electrons in the H_2 molecule from the σ_{1s} MO to the σ_{1s}^* MO. Would you expect the H atoms to remain bonded to each other, or would the molecule fall apart?

■ SAMPLE EXERCISE 9.8 | Bond Order

What is the bond order of the He_2^+ ion? Would you expect this ion to be stable relative to the separated He atom and He^+ ion?

SOLUTION

Analyze: We will determine the bond order for the He_2^+ ion and use it to predict whether the ion is stable.

Plan: To determine the bond order, we must determine the number of electrons in the molecule and how these electrons populate the available MOs. The valence electrons of He are in the $1s$ orbital, and the $1s$ orbitals combine to give an MO diagram like that for H_2 or He_2 (Figure 9.35). If the bond order is greater than 0, we expect a bond to exist, and the ion is stable.

Solve: The energy-level diagram for the He_2^+ ion is shown in Figure 9.36 ◀. This ion has three electrons. Two are placed in the bonding orbital, the third in the antibonding orbital. Thus, the bond order is

$$\text{Bond order} = \tfrac{1}{2}(2 - 1) = \tfrac{1}{2}$$

Because the bond order is greater than 0, we predict the He_2^+ ion to be stable relative to the separated He and He^+. Formation of He_2^+ in the gas phase has been demonstrated in laboratory experiments.

■ PRACTICE EXERCISE

Determine the bond order of the H_2^- ion.
Answer: $\tfrac{1}{2}$

▲ **Figure 9.36 Energy-level diagram for the He_2^+ ion.**

In fact, antibonding MOs are slightly more unfavorable than bonding MOs are favorable. Thus, whenever there is an equal number of electrons in bonding and antibonding orbitals, the energy of the molecule is slightly higher than that for the separated atoms and no bond is formed.

9.8 SECOND-ROW DIATOMIC MOLECULES

Just as we treated the bonding in H_2 by using molecular orbital theory, we can consider the MO description of other diatomic molecules. Initially we will restrict our discussion to *homonuclear* diatomic molecules (those composed of two identical atoms) of elements in the second row of the periodic table. As we will see, the procedure for determining the distribution of electrons in these molecules closely follows the one we used for H_2.

Second-row atoms have valence $2s$ and $2p$ orbitals, and we need to consider how they interact to form MOs. The following rules summarize some of the guiding principles for the formation of MOs and for how they are populated by electrons:

1. The number of MOs formed equals the number of atomic orbitals combined.

2. Atomic orbitals combine most effectively with other atomic orbitals of similar energy.

3. The effectiveness with which two atomic orbitals combine is proportional to their overlap. That is, as the overlap increases, the energy of the bonding MO is lowered and the energy of the antibonding MO is raised.

4. Each MO can accommodate, at most, two electrons, with their spins paired (Pauli exclusion principle).

5. When MOs of the same energy are populated, one electron enters each orbital (with the same spin) before spin pairing occurs (Hund's rule).

Molecular Orbitals for Li_2 and Be_2

Lithium, the first element of the second period, has a $1s^2 2s^1$ electron configuration. When lithium metal is heated above its boiling point (1342 °C), Li_2 molecules are found in the vapor phase. The Lewis structure for Li_2 indicates a Li—Li single bond. We will now use MOs to describe the bonding in Li_2.

Because the $1s$ and $2s$ orbitals of Li are so different in energy, we may assume that the $1s$ orbital on one Li atom interacts only with the $1s$ orbital on the other atom (rule 2). Likewise, the $2s$ orbitals interact only with each other. The resulting energy-level diagram is shown in Figure 9.37▼. Notice that combining four atomic orbitals produces four MOs (rule 1).

The $1s$ orbitals of Li combine to form σ_{1s} and σ_{1s}^* bonding and antibonding MOs, as they did for H_2. The $2s$ orbitals interact with one another in exactly the same way, producing bonding (σ_{2s}) and antibonding (σ_{2s}^*) MOs. Because the $2s$ orbitals of Li extend farther from the nucleus than the $1s$ orbitals do, the $2s$ orbitals overlap more effectively. As a result, the energy separation between the σ_{2s} and σ_{2s}^* orbitals is greater than that for the $1s$-based MOs. The $1s$ orbitals of Li are so much lower in energy than the $2s$ orbitals, however, that the σ_{1s}^* antibonding MO is still well below the σ_{2s} bonding MO.

Each Li atom has three electrons, so six electrons must be placed in the MOs of Li_2. As shown in Figure 9.37, these electrons occupy the σ_{1s}, σ_{1s}^*, and σ_{2s} MOs, each with two electrons. There are four electrons in bonding orbitals and two in antibonding orbitals, so the bond order equals $\frac{1}{2}(4 - 2) = 1$. The molecule has a single bond, in agreement with its Lewis structure.

Because both the σ_{1s} and σ_{1s}^* MOs of Li_2 are completely filled, the $1s$ orbitals contribute almost nothing to the bonding. The single bond in Li_2 is due essentially to the interaction of the valence $2s$ orbitals on the Li atoms. This example

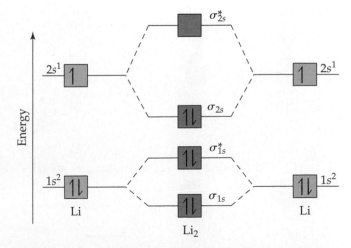

▲ **Figure 9.37 Energy-level diagram for the Li_2 molecule.**

illustrates the general rule that *core electrons usually do not contribute signifi-cantly to bonding in molecule formation.* The rule is equivalent to using only the valence electrons when drawing Lewis structures. Thus, we need not con-sider further the 1s orbitals while discussing the other second-row diatomic molecules.

The MO description of Be_2 follows readily from the energy-level diagram for Li_2. Each Be atom has four electrons ($1s^2 2s^2$), so we must place eight elec-trons in molecular orbitals. Thus, we completely fill the σ_{1s}, σ_{1s}^*, σ_{2s}, and σ_{2s}^* MOs. We have an equal number of bonding and antibonding electrons, so the bond order equals 0. Consistent with this analysis, Be_2 does not exist.

GIVE IT SOME THOUGHT

Would you expect Be_2^+ to be a stable ion?

Molecular Orbitals from 2p Atomic Orbitals

Before we can consider the remaining second-row molecules, we must look at the MOs that result from combining 2p atomic orbitals. The interactions be-tween p orbitals are shown in Figure 9.38▼, where we have arbitrarily chosen the internuclear axis to be the z-axis. The $2p_z$ orbitals face each other in a "head-to-head" fashion. Just as we did for the s orbitals, we can combine the $2p_z$ orbitals in two ways. One combination concentrates electron density between the nuclei and is therefore a bonding molecular orbital. The other combination excludes electron density from the bonding region; it is an antibonding molec-ular orbital. In each of these MOs the electron density lies along the line through the nuclei, so they are σ molecular orbitals: σ_{2p} and σ_{2p}^*.

The other 2p orbitals overlap sideways and thus concentrate electron density above and below a line connecting the nuclei. MOs of this type are called **pi (π) molecular orbitals**, by analogy to π bonds we saw in Section 9.6. We get one π bonding MO by combining the $2p_x$ atomic orbitals and another from the $2p_y$ atomic orbitals. These two π_{2p} molecular orbitals have the same energy; in other words, they are degenerate. Likewise, we get two degenerate π_{2p}^* antibond-ing MOs. The π_{2p}^* orbitals are mutually perpen-dicular to each other, like the 2p orbitals from which they were made. These π_{2p}^* orbitals have four lobes, pointing away from the two nuclei, as shown in the Figure 9.38. The $2p_z$ or-bitals on two atoms point directly at each other. Hence, the overlap of two $2p_z$ orbitals is greater than that for two $2p_x$ or $2p_y$ orbitals. From rule 3 we therefore expect the σ_{2p} MO to be lower in energy (more stable) than the π_{2p} MOs. Simi-larly, the σ_{2p}^* MO should be higher in energy (less stable) than the π_{2p}^* MOs.

(a) "End-on" overlap of p orbitals forms σ and σ^* MOs.

$2p_z$ $2p_z$ σ_{2p}^* σ_{2p}

(b) "Sideways" overlap of p orbitals forms two sets of π and π^* MOs.

$2p_x$ $2p_x$ π_{2p}^* π_{2p}

$2p_y$ $2p_y$ π_{2p}^* π_{2p}

◀ **Figure 9.38 Contour representations of the molecular orbitals formed by 2p orbitals.** Each time we combine two atomic orbitals, we obtain two MOs: one bonding and one antibonding.

A Closer Look | PHASES IN ATOMIC AND MOLECULAR ORBITALS

Our discussion of atomic orbitals in Chapter 6 and molecular orbitals in this chapter highlights some of the most important applications of quantum mechanics in chemistry. In the quantum mechanical treatment of electrons in atoms and molecules, we are mainly interested in obtaining two characteristics of the electrons—namely, their energies and their distribution in space. Recall that solving Schrödinger's wave equation yields the electron's energy, E, and wave function, ψ, but that ψ itself does not have a direct physical meaning. ∞ (Section 6.5) The contour representations of atomic and molecular orbitals that we have presented thus far are based on the square of the wave function, ψ^2 (the *probability density*), which gives the probability of finding the electron at a given point in space.

Because probability densities are squares of functions, their values must be nonnegative (zero or positive) at all points in space. However, the functions themselves can have negative values. Consider, for example, the simple sine function, plotted in Figure 9.39▶. From the drawing on the top, you can see that the sine function has negative values for x between 0 and $-\pi$, and positive values for x between 0 and $+\pi$. We say that the *phase* of the sine function is negative between 0 and $-\pi$ and positive between 0 and $+\pi$. If we square the sine function (bottom graph), we get two peaks that look the same on each side of the origin. Both of the peaks are positive because squaring a negative number produces a positive number—*we lose the phase information of the function upon squaring it.*

The wave functions for the atomic orbitals are much more complicated than the simple sine function, but like the sine function they can have phases. Consider, for example, the representations of the 1s orbital in Figure 9.40a▼. The 1s orbital is plotted a bit differently than we saw in Chapter 6. The origin

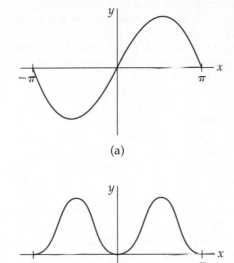

◀ **Figure 9.39 Phases in the sine function.** (a) A full cycle of a sine wave ($y = \sin x$) has two equivalent halves that are opposite in sign (phase). The origin is a node. (b) Squaring the sine function ($y = \sin^2 x$) produces two equivalent positive peaks. The origin is still a node.

▼ **Figure 9.40 Phases in *s* and *p* atomic orbital wavefunctions.** (a) The wave function for a 1s orbital is plotted on top, with its contour representation below. The 1s orbital has no nodes. (b) The wave function for a 2p orbital is plotted on top, with its contour representation below. The 2p orbital has a node, and the phase of the wave function changes on either side of the node. In the contour picture, this change of phase is represented by two colors for the two lobes of the orbital. (c) The square of the wave function gives the probability density for the 2p orbital. The square of the wave function is plotted on top, with the contour representation below. The contour representation has only one color to denote that the phase information has been lost by squaring the wave function.

1s orbital

(a)

2p_z orbital

(b)

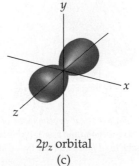

2p_z orbital

(c)

is the point where the nucleus is, and the wave function for the 1s orbital extends from the origin in space. The plot shows the value of the wave function, ψ for a slice taken along the z-axis. Below the plot is a contour representation of the 1s orbital. Notice that the value of the 1s wave function is always a positive number. Thus, it has only one phase. Notice also that the wave function never goes to zero. It therefore has no nodes, as we saw in Chapter 6.

In Figure 9.40b we see a similar set of representations for the $2p_z$ orbital. Now the wave function changes sign upon crossing through $z = 0$. Notice that the two halves of the wave have the same shape except that one has positive values and the other negative values; the wave function changes phase upon crossing through the origin. Mathematically the $2p_z$ wave function is equal to zero whenever $z = 0$. This corresponds to any point on the xy plane, so we say that the xy plane is a *nodal plane* of the $2p_z$ orbital. The wave function for a p orbital is much like a sine function because it has two equal parts that have opposite phases. Figure 9.40b gives a typical representation used by chemists of the wave function for a p_z orbital.* The two different colors for the lobes indicate the different signs or phases of the orbital. Like the sine function, the origin is a node.

What happens if we square the $2p_z$ wave function? Figure 9.40c shows that when we square the wave function of the $2p_z$ orbital, we get two peaks that look the same on each side of the origin. Both of the peaks are positive because squaring a negative number produces a positive number—*we lose the phase information of the function upon squaring it*, just as we did for the sine function. When we square the wave function for the p_x orbital, we get the probability density for the orbital, which is given as a contour representation in Figure 9.40c. For this function, both lobes have the same phase and therefore the same color. We use this representation throughout most of this book.

The lobes of the wave functions for the d orbitals also have different phases. For example, the wave function for a d_{xy} orbital has four lobes that alternate in phase, as shown in Figure 9.41 ▼. The wave functions for the other d orbitals likewise have lobes of alternating phase.

Why do we need to consider the additional complexity that is introduced by considering the sign or phase of the wave function? While it is true that the phase of the wave function is not necessary to visualize the shape of an atomic orbital in an isolated atom, it does become important when we consider overlap of orbitals in molecular orbital theory. Let's use the sine function as an example again. If you add two sine functions together that are of the same phase, they add

constructively, resulting in increased amplitude as shown in the accompanying figure. On the other hand, if you add two sine functions together that are of the opposite phase, they add *destructively* and cancel each other.

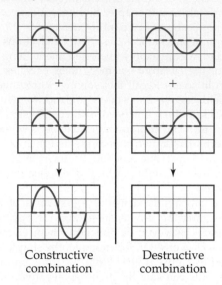

Constructive combination

Destructive combination

The idea of constructive and destructive interactions of wave functions is key to understanding the origin of bonding and antibonding molecular orbitals that arise from combining or overlapping of atomic orbitals. For example, the wave function of the σ_{1s} MO of H_2 is generated by adding the wave function for the 1s orbital on one atom to the wave function for the 1s orbital on the other atom, with both orbitals having the same phase. We can say that the atomic wave functions overlap *constructively* in this case, to increase the electron density between the two atoms (Figure 9.42a ▼). The wave function of the σ_{1s}^* MO of H_2 is generated by adding the wave function for a 1s orbital on one atom to the wave function for a 1s orbital on the other atom, but with the two orbitals having opposite phases. This amounts to subtracting 1s orbital from the other. The atomic orbital wave functions overlap *destructively* in this case, to create a region of zero electron density between the two atoms— a node. Chemists generally sketch the wave functions by simply drawing the contour representations of the 1s orbitals in different colors to denote the two phases (Figure 9.42b).

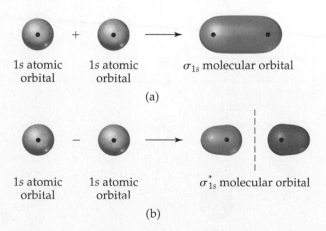

1s atomic orbital   1s atomic orbital σ_{1s} molecular orbital

(a)

1s atomic orbital   1s atomic orbital σ_{1s}^* molecular orbital

(b)

▲ **Figure 9.42 Molecular orbitals from atomic orbital wave functions.** (a) The mixing of two 1s orbitals that are of the same phase produces a σ_{1s} MO. (b) The mixing of two 1s orbitals that are opposite in phase produces a σ_{1s}^* MO.

◄ **Figure 9.41 Phases in d orbitals.** The wave function for the d_{xy} orbital has lobes of alternating sign.

*The mathematical development of this three-dimensional function (and its square) is beyond the scope of this book, and, as is typically done by chemists, we have used lobes that are the same shape as in Figure 6.23.

Notice how this wave function has similarities to both the sine function and to a p orbital—in each of these cases we have two parts of the function of opposite phase separated by a node. When we square the wave function of the σ_{1s}^* MO, we get the probability density representation given in Figure 9.34—notice that we once again lose the phase information when we look at the probability density.

Our brief discussion has given you only an introduction to the mathematical subtleties of atomic and molecular orbitals. Chemists use the wave functions of atomic and molecular orbitals to understand many aspects of chemical bonding and spectroscopy. You will probably see orbitals drawn in color to show phases if you take a course in organic chemistry. **Related Exercises: 9.89, 9.103, 9.105**

Electron Configurations for B_2 Through Ne_2

So far, we have independently considered the MOs that result from s orbitals (Figure 9.34) and from p orbitals (Figure 9.38). We can combine these results to construct an energy-level diagram (Figure 9.43▶) for homonuclear diatomic molecules of the elements boron through neon, all of which have valence $2s$ and $2p$ atomic orbitals. The following features of the diagram are notable:

1. The $2s$ atomic orbitals are lower in energy than the $2p$ atomic orbitals. ∞ (Section 6.7) Consequently, both of the molecular orbitals that result from the $2s$ orbitals, the bonding σ_{2s} and antibonding σ_{2s}^*, are lower in energy than the lowest-energy MO derived from the $2p$ atomic orbitals.

2. The overlap of the two $2p_z$ orbitals is greater than that of the two $2p_x$ or $2p_y$ orbitals. As a result, the bonding σ_{2p} MO is lower in energy than the π_{2p} MOs, and the antibonding σ_{2p}^* MO is higher in energy than the π_{2p}^* MOs.

3. Both the π_{2p} and π_{2p}^* molecular orbitals are *doubly degenerate*; that is, there are two degenerate MOs of each type.

Before we can add electrons to the energy-level diagram in Figure 9.43, we must consider one more effect. We have constructed the diagram assuming there is no interaction between the $2s$ orbital on one atom and the $2p$ orbitals on the other. In fact, such interactions can and do take place. Figure 9.44▶ shows the overlap of a $2s$ orbital on one of the atoms with a $2p_z$ orbital on the other. These interactions affect the energies of the σ_{2s} and σ_{2p} molecular orbitals in such a way that these MOs move further apart in energy, the σ_{2s} falling and the σ_{2p} rising in energy (Figure 9.45▼). These $2s$–$2p$ interactions are strong enough that the energetic ordering of the MOs can be altered: *For B_2, C_2, and N_2, the σ_{2p} MO is above the π_{2p} MOs in energy. For O_2, F_2, and Ne_2, the σ_{2p} MO is below the π_{2p} MOs.*

Given the energy ordering of the molecular orbitals, it is a simple matter to determine the electron configurations for the second-row diatomic molecules B_2 through Ne_2. For example, a boron atom has three valence electrons. (Remember that we are ignoring the inner-shell $1s$ electrons.) Thus, for B_2 we must place six electrons in MOs. Four of these electrons fully occupy the σ_{2s} and σ_{2s}^* MOs, leading to no net bonding. The last two electrons are put in the π_{2p} bonding MOs; one electron is put in one π_{2p} MO and the other electron is put in the other π_{2p} MO, with the two electrons having the same spin. Therefore, B_2 has a bond order of 1. Each time we move one element to the right in the second row, two more electrons must be placed in the diagram. For example, on moving to C_2, we have two more electrons than in B_2, and these electrons are placed in the

▲ **Figure 9.43 Energy-level diagram for MOs of second-row homonuclear diatomic molecules.** The diagram assumes no interaction between the $2s$ atomic orbital on one atom and the $2p$ atomic orbitals on the other atom, and experiment shows that it fits only for O_2, F_2, and Ne_2.

▲ **Figure 9.44 Interaction of 2s and 2p atomic orbitals.** The $2s$ orbital on one atom of a diatomic molecule can overlap with the $2p_z$ orbital on the other atom. These $2s$–$2p$ interactions can alter the energetic ordering of the MOs of the molecule.

▶ **Figure 9.45 The effect of 2s–2p interactions.** When the 2s and 2p orbitals interact, the σ_{2s} MO falls in energy and the σ_{2p} MO rises in energy. For O_2, F_2, and Ne_2, the interaction is small, and the σ_{2p} MO remains below the π_{2p} MOs, as in Figure 9.43. For B_2, C_2, and N_2, the 2s–2p interaction is great enough that the σ_{2p} MO rises above the π_{2p} MOs, as shown on the right.

π_{2p} MOs, completely filling them. The electron configurations and bond orders for the diatomic molecules B_2 through Ne_2 are given in Figure 9.46 ▼.

Electron Configurations and Molecular Properties

The way a substance behaves in a magnetic field provides an important insight into the arrangements of its electrons. Molecules with one or more unpaired electrons are attracted into a magnetic field. The more unpaired electrons in a species, the stronger the force of attraction will be. This type of magnetic behavior is called **paramagnetism**.

Substances with no unpaired electrons are weakly repelled from a magnetic field. This property is called **diamagnetism**. Diamagnetism is a much weaker effect than paramagnetism. A straightforward method for measuring the

	Large 2s–2p interaction			Small 2s–2p interaction		
	B_2	**C_2**	**N_2**	**O_2**	**F_2**	**Ne_2**
σ_{2p}^*	☐	☐	☐	☐	☐	⇅
π_{2p}^*	☐ ☐	☐ ☐	☐ ☐	↑ ↑	⇅ ⇅	⇅ ⇅
σ_{2p}	☐	☐	⇅	⇅ ⇅	⇅ ⇅	⇅ ⇅
π_{2p}	↑ ↑	⇅ ⇅	⇅ ⇅	⇅	⇅	⇅
σ_{2s}^*	⇅	⇅	⇅	⇅	⇅	⇅
σ_{2s}	⇅	⇅	⇅	⇅	⇅	⇅
Bond order	1	2	3	2	1	0
Bond enthalpy (kJ/mol)	290	620	941	495	155	—
Bond length (Å)	1.59	1.31	1.10	1.21	1.43	—
Magnetic behavior	Paramagnetic	Diamagnetic	Diamagnetic	Paramagnetic	Diamagnetic	—

▲ **Figure 9.46 The second-row diatomic molecules.** Molecular orbital electron configurations and some experimental data for several second-row diatomic molecules.

(a) The sample is first weighed in the absence of a magnetic field.

(b) When a field is applied, a diamagnetic sample moves out of the field and thus appears to have a lower mass.

(c) A paramagnetic sample is drawn into the field and thus appears to gain mass.

▲ **Figure 9.47 Determining the magnetic properties of a sample.** The response of a sample to a magnetic field indicates whether it is diamagnetic or paramagnetic. Paramagnetism is a much stronger effect than diamagnetism.

magnetic properties of a substance, illustrated in Figure 9.47 ▲, involves weighing the substance in the presence and absence of a magnetic field. If the substance is paramagnetic, it will appear to weigh more in the magnetic field; if it is diamagnetic, it will appear to weigh less. The magnetic behaviors observed for the diatomic molecules of the second-row elements agree with the electron configurations shown in Figure 9.46.

GIVE IT SOME THOUGHT

Figure 9.46 indicates that the C_2 molecule is diamagnetic. Would that be expected if the σ_{2p} MO were lower in energy than the π_{2p} MOs?

The electron configurations can also be related to the bond distances and bond enthalpies of the molecules. ∞ (Section 8.8) As bond orders increase, bond distances decrease and bond enthalpies increase. N_2, for example, whose bond order is 3, has a short bond distance and a large bond enthalpy. The N_2 molecule does not react readily with other substances to form nitrogen compounds. The high bond order of the molecule helps explain its exceptional stability. We should also note, however, that molecules with the same bond orders do *not* have the same bond distances and bond enthalpies. Bond order is only one factor influencing these properties. Other factors include the nuclear charges and the extent of orbital overlap.

Bonding in the dioxygen molecule, O_2, is especially interesting. Its Lewis structure shows a double bond and complete pairing of electrons:

$$\ddot{O}\!=\!\ddot{O}$$

The short O—O bond distance (1.21 Å) and the relatively high bond enthalpy (495 kJ/mol) are in agreement with the presence of a double bond. However, the molecule contains two unpaired electrons. The paramagnetism of O_2 is demonstrated in Figure 9.48 ▼. Although the Lewis structure fails to account for the paramagnetism of O_2, molecular orbital theory correctly predicts that two unpaired electrons are in the π_{2p}^* orbital of the molecule (Figure 9.46). The MO description also correctly indicates a bond order of 2.

Going from O_2 to F_2, we add two more electrons, completely filling the π_{2p}^* MOs. Thus, F_2 is expected to be diamagnetic and have an F—F single bond, in accord with its Lewis structure. Finally, the addition of two more electrons to make Ne_2 fills all the bonding and antibonding MOs. Therefore, the bond order of Ne_2 is zero, and the molecule is not expected to exist.

▲ **Figure 9.48 Paramagnetism of O₂.** Liquid O_2 being poured between the poles of a magnet. Because each O_2 molecule contains two unpaired electrons, O_2 is paramagnetic. It is therefore attracted into the magnetic field and "sticks" between the magnetic poles.

■ **SAMPLE EXERCISE 9.9** | **Molecular Orbitals of a Second-Row Diatomic Ion**

Predict the following properties of O_2^+: **(a)** number of unpaired electrons, **(b)** bond order, **(c)** bond enthalpy and bond length.

SOLUTION

Analyze: Our task is to predict several properties of the cation O_2^+.

Plan: We will use the MO description of O_2^+ to determine the desired properties. We must first determine the number of electrons in O_2^+ and then draw its MO energy diagram. The unpaired electrons are those without a partner of opposite spin. The bond order is one-half the difference between the number of bonding and antibonding electrons. After calculating the bond order, we can use the data in Figure 9.46 to estimate the bond enthalpy and bond length.

Solve:

(a) The O_2^+ ion has 11 valence electrons, one fewer than O_2. The electron removed from O_2 to form O_2^+ is one of the two unpaired π_{2p}^* electrons (see Figure 9.46). Therefore, O_2^+ has just one unpaired electron.

(b) The molecule has eight bonding electrons (the same as O_2) and three antibonding electrons (one fewer than O_2). Thus, its bond order is

$$\tfrac{1}{2}(8 - 3) = 2\tfrac{1}{2}$$

(c) The bond order of O_2^+ is between that for O_2 (bond order 2) and N_2 (bond order 3). Thus, the bond enthalpy and bond length should be about midway between those for O_2 and N_2, approximately 700 kJ/mol and 1.15 Å, respectively. The experimental bond enthalpy and bond length of the ion are 625 kJ/mol and 1.123 Å, respectively.

■ **PRACTICE EXERCISE**

Predict the magnetic properties and bond orders of **(a)** the peroxide ion, O_2^{2-}; **(b)** the acetylide ion, C_2^{2-}.
Answers: **(a)** diamagnetic, 1; **(b)** diamagnetic, 3

Heteronuclear Diatomic Molecules

The same principles we have used in developing an MO description of homonuclear diatomic molecules can be extended to *heteronuclear* diatomic molecules—those in which the two atoms in the molecule are not the same. We will conclude this section on MO theory with a brief discussion of the MOs of a fascinating heteronuclear diatomic molecule—the nitric oxide, NO, molecule.

The NO molecule controls several important human physiological functions. Our bodies use it, for example, to relax muscles, to kill foreign cells, and to reinforce memory. The 1998 Nobel Prize in Physiology or Medicine was awarded to three scientists for their research that uncovered the importance of NO as a "signalling" molecule in the cardiovascular system. NO also functions as a neurotransmitter and is implicated in many other biological pathways. That NO plays such an important role in human metabolism was unsuspected before 1987 because NO has an odd number of electrons and is highly reactive. The molecule has 11 valence electrons, and two possible Lewis structures can be drawn. The Lewis structure with the lower formal charges places the odd electron on the N atom:

$$\overset{0}{\underset{..}{\text{N}}}{=}\overset{0}{\underset{..}{\ddot{\text{O}}}} \longleftrightarrow \overset{-1}{\underset{..}{\ddot{\text{N}}}}{=}\overset{+1}{\underset{.}{\ddot{\text{O}}}}$$

Both structures indicate the presence of a double bond, but when compared with the molecules in Figure 9.46, the experimental bond length of NO (1.15 Å) suggests a bond order greater than two. How do we treat NO using the MO model?

If the atoms in a heteronuclear diatomic molecule do not differ too greatly in their electronegativities, the description of their MOs will resemble those for homonuclear diatomics, with one important modification: The atomic energies of the more electronegative atom will be lower in energy than those of the less electronegative element. The MO diagram for NO is shown in Figure 9.49▶. You can see that the 2s and 2p atomic orbitals of oxygen are slightly lower than those of nitrogen because oxygen is more electronegative than nitrogen. We see that the MO energy-level diagram is much like that of a homonuclear diatomic molecule—because the 2s and 2p orbitals on the two atoms interact, the same types of MOs are produced.

There is one other important change in the MOs when we consider heteronuclear molecules. The MOs that result are still a mix of the atomic orbitals from both atoms, but in general *an MO will have a greater contribution from the atomic orbital to which it is closer in energy*. In the case of NO, for example, the σ_{2s} bonding MO is closer in energy to the O 2s atomic orbital than to the N 2s atomic orbital. As a result, the σ_{2s} MO has a slightly greater contribution from O than from N—the orbital is no longer an equal mixture of the two atoms, as was the case for the homonuclear diatomic molecules. Similarly, the σ_{2s}^* antibonding MO is weighted more heavily toward the N atom because that MO is closest in energy to the N 2s atomic orbital.

We complete the MO diagram for NO by filling the MOs in Figure 9.49 with the 11 valence electrons. We see that there are eight bonding and three antibonding electrons, giving a bond order of $\frac{1}{2}(8 - 3) = 2\frac{1}{2}$, which agrees better with experiment than the Lewis structures do. The unpaired electron resides in one of the π_{2p}^* MOs, which are more heavily weighted toward the N atom. Thus, the Lewis structure that places the unpaired electron on nitrogen (the one preferred on the basis of formal charge) is a more accurate description of the true electron distribution in the molecule.

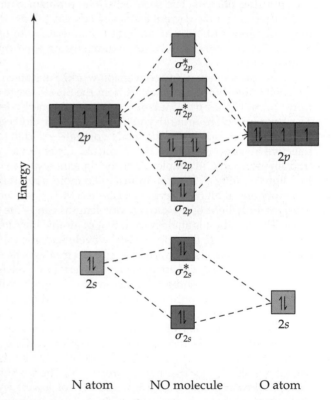

N atom   NO molecule   O atom

▲ Figure 9.49 **The MO energy-level diagram for NO.**

Chemistry Put to Work   ORBITALS AND ENERGY

I f you were asked to identify the major technological challenge for the twenty-first century, you might say "energy." The development of sustainable energy sources is crucial to meet the needs of future generations of people on our planet. Currently, the majority of the world, in one way or another, relies on exothermic combustion reactions of oil, coal, or natural gas to provide heat and ultimately power. Oil, coal, and natural gas are examples of "fossil fuels"—carbon-containing compounds that are the long-term decomposition products of ancient plants and animals. Fossil fuels are not renewable in the several-hundred-year timeframe in which we need them. But every day our planet receives plenty of energy from the Sun to easily power the world for millions of years. Whereas the combustion of fossil fuels results in release of CO_2 into the atmosphere, solar energy represents a renewable energy source that is potentially less harmful to the environment. One way to utilize solar energy is to convert it into electrical energy using photovoltaic solar cells. The problem with this alternative is that the current efficiency of solar cell devices is low; only about 10–15% of sunlight is converted into useful energy. Furthermore, the cost of manufacturing solar cells is relatively high.

How does solar energy conversion work? Fundamentally, we need to be able to use photons from the Sun to excite electrons in molecules and materials to different energy levels. The brilliant colors around you—those of your clothes, the photographs in this book, the foods you eat—are due to the selective absorption of light by chemicals. Light excites electrons in molecules. In a molecular orbital picture, we can envision light exciting an electron from a filled molecular orbital to an empty one at higher energy. Because the MOs have definite energies, only light of the proper wavelengths can excite electrons. The situation is analogous to that of atomic line spectra. ∞ (Section 6.3) If the appropriate wavelength for exciting electrons is in the visible portion of the electromagnetic spectrum, the substance will appear colored: Certain wavelengths of white light are absorbed; others are not. A green leaf appears green because only green light is reflected by the leaf; other wavelengths of visible light are absorbed by the leaf.

In using molecular orbital theory to discuss the absorptions of light by molecules, we can focus on two MOs in particular. The *highest occupied molecular orbital* (HOMO) is the MO of highest energy that has electrons in it. The *lowest unoccupied molecular orbital* (LUMO) is the MO of lowest energy that does not have electrons in it. In N_2, for example, the HOMO is the π_{2p} MO and the LUMO is the π_{2p}^* MO (Figure 9.46). The energy difference between the HOMO and the LUMO—known as the HOMO–LUMO gap—is related to the minimum energy needed to excite an electron in the molecule. Colorless or white substances usually have such a large HOMO–LUMO gap that visible light is not energetic enough to excite an electron to the higher level. The minimum energy

▲ **Figure 9.50 Light into electricity.** The light from a flashlight, to the left of the photograph frame, is sufficient to excite electrons from one energy level to another in this simple solar cell, producing electricity that powers the small fan. To capture more light, a red ruthenium-containing complex is added to the colorless TiO_2 paste between the glass plates.

needed to excite an electron in N_2 corresponds to light with a wavelength of less than 200 nm, which is far into the ultraviolet part of the spectrum (Figure 6.4). As a result, N_2 cannot absorb any visible light and is therefore colorless.

The magnitude of the energy gap between filled and empty electronic states is critical for solar energy conversion. Ideally, we would like to have a substance that absorbs as many solar photons as possible and then convert the energy of those photons into a useful form of energy such as electricity. Titanium dioxide, for example, is a readily available material that can be reasonably efficient at converting light directly into electricity. However, TiO_2 is white and absorbs only a small amount of the Sun's radiant energy output. Scientists are working to make solar cells in which TiO_2 is mixed with highly colored molecules, whose HOMO–LUMO gaps correspond to visible and near-infrared light. That way, the molecules can absorb more of the solar spectrum. The molecule's HOMO must also be higher in energy than the TiO_2's HOMO so that the excited electrons can flow from the molecules into the TiO_2, thereby generating electricity when the device is illuminated with light and connected to an external circuit.

Figure 9.50 ▲ shows a simple solar cell made from ruthenium-containing molecules, which appear red, mixed with TiO_2 in a paste, sandwiched between two glass plates. Light that shines from a flashlight to the left of the device generates enough current to run the small fan that is connected to it with wires.

Related Exercises: 9.93, 9.104

■ SAMPLE INTEGRATIVE EXERCISE | Putting Concepts Together

Elemental sulfur is a yellow solid that consists of S_8 molecules. The structure of the S_8 molecule is a puckered, eight-membered ring (Figure 7.34). Heating elemental sulfur to high temperatures produces gaseous S_2 molecules:

$$S_8\,(s) \longrightarrow 4\,S_2\,(g)$$

(a) With respect to electronic structure, which element in the second row of the periodic table is most similar to sulfur? **(b)** Use the VSEPR model to predict the S—S—S bond angles in S_8 and the hybridization at S in S_8. **(c)** Use MO theory to predict the sulfur–sulfur bond order in S_2. Is the molecule expected to be diamagnetic or paramagnetic? **(d)** Use average bond enthalpies (Table 8.4) to estimate the enthalpy change for the reaction just described. Is the reaction exothermic or endothermic?

SOLUTION

(a) Sulfur is a group 6A element with an $[Ne]3s^23p^4$ electron configuration. It is expected to be most similar electronically to oxygen (electron configuration, $[He]2s^22p^4$), which is immediately above it in the periodic table. **(b)** The Lewis structure of S_8 is

There is a single bond between each pair of S atoms and two nonbonding electron pairs on each S atom. Thus, we see four electron domains around each S atom, and we would expect a tetrahedral electron-domain geometry corresponding to sp^3 hybridization. ∞ (Sections 9.2, 9.5) Because of the nonbonding pairs, we would expect the S—S—S angles to be somewhat less than 109°, the tetrahedral angle. Experimentally, the S—S—S angle in S_8 is 108°, in good agreement with this prediction. Interestingly, if S_8 were a planar ring (like a stop sign), it would have S—S—S angles of 135°. Instead, the S_8 ring puckers to accommodate the smaller angles dictated by sp^3 hybridization. **(c)** The MOs of S_2 are entirely analogous to those of O_2, although the MOs for S_2 are constructed from the $3s$ and $3p$ atomic orbitals of sulfur. Further, S_2 has the same number of valence electrons as O_2. Thus, by analogy to our discussion of O_2, we would expect S_2 to have a bond order of 2 (a double bond) and to be paramagnetic with two unpaired electrons in the π_{3p}^* molecular orbitals of S_2. ∞ (Section 9.8) **(d)** We are considering the reaction in which an S_8 molecule falls apart into four S_2 molecules. From parts (b) and (c), we see that S_8 has S—S single bonds and S_2 has S=S double bonds. During the course of the reaction, therefore, we are breaking eight S—S single bonds and forming four S=S double bonds. We can estimate the enthalpy of the reaction by using Equation 8.12 and the average bond enthalpies in Table 8.4:

$$\Delta H_{rxn} = 8\,D(S—S) - 4\,D(S=S) = 8(266\text{ kJ}) - 4(418\text{ kJ}) = +456\text{ kJ}$$

Because $\Delta H_{rxn} > 0$, the reaction is endothermic. ∞ (Section 5.4) The very positive value of ΔH_{rxn} suggests that high temperatures are required to cause the reaction to occur.

CHAPTER REVIEW

SUMMARY AND KEY TERMS

Introduction and Section 9.1 The three-dimensional shapes and sizes of molecules are determined by their **bond angles** and bond lengths. Molecules with a central atom A surrounded by n atoms B, denoted AB_n, adopt a number of different geometric shapes, depending on the value of n and on the particular atoms involved. In the overwhelming majority of cases, these geometries are related to five basic shapes (linear, trigonal pyramidal, tetrahedral, trigonal bipyramidal, and octahedral).

Section 9.2 The **valence-shell electron-pair repulsion (VSEPR) model** rationalizes molecular geometries based on the repulsions between **electron domains**, which are regions about a central atom in which electrons are likely to be found. **Bonding pairs** of electrons, which are those involved in making bonds, and **nonbonding pairs** of electrons, also called lone pairs, both create electron domains around an atom. According to the VSEPR model, electron domains orient themselves to minimize electrostatic repulsions; that is, they remain as far apart as possible. Electron domains from nonbonding pairs exert slightly greater repulsions than those from bonding pairs, which leads to certain preferred positions for nonbonding pairs and to the departure of bond angles from idealized values. Electron domains from multiple bonds exert slightly greater repulsions than those from single bonds. The arrangement of electron domains around a central atom is called the **electron-domain geometry**; the arrangement of atoms is called the **molecular geometry**.

Section 9.3 The dipole moment of a polyatomic molecule depends on the vector sum of the dipole moments associated with the individual bonds, called the **bond dipoles**. Certain molecular shapes, such as linear AB_2 and trigonal planar AB_3, assure that the bond dipoles cancel, producing a nonpolar molecule, which is one whose dipole moment is zero. In other shapes, such as bent AB_2 and trigonal pyramidal AB_3, the bond dipoles do not cancel and the molecule will be polar (that is, it will have a nonzero dipole moment).

Section 9.4 **Valence-bond theory** is an extension of Lewis's notion of electron-pair bonds. In valence-bond theory, covalent bonds are formed when atomic orbitals on neighboring atoms **overlap** one another. The overlap region is a favorable one for the two electrons because of their attraction to two nuclei. The greater the overlap between two orbitals, the stronger will be the bond that is formed.

Section 9.5 To extend the ideas of valence-bond theory to polyatomic molecules, we must envision mixing s, p, and sometimes d orbitals to form **hybrid orbitals**. The process of **hybridization** leads to hybrid atomic orbitals that have a large lobe directed to overlap with orbitals on another atom to make a bond. Hybrid orbitals can also accommodate nonbonding pairs. A particular mode of hybridization can be associated with each of the five common electron-domain geometries (linear = sp; trigonal planar = sp^2; tetrahedral = sp^3; trigonal bipyramidal = sp^3d; and octahedral = sp^3d^2).

Section 9.6 Covalent bonds in which the electron density lies along the line connecting the atoms (the internuclear axis) are called **sigma (σ) bonds**. Bonds can also be formed from the sideways overlap of p orbitals. Such a bond is called a **pi (π) bond**. A double bond, such as that in C_2H_4, consists of one σ bond and one π bond; a triple bond, such as that in C_2H_2, consists of one σ and two π bonds. The formation of a π bond requires that molecules adopt a specific orientation; the two CH_2 groups in C_2H_4, for example, must lie in the same plane. As a result, the presence of π bonds introduces rigidity into molecules.

In molecules that have multiple bonds and more than one resonance structure, such as C_6H_6, the π bonds are **delocalized**; that is, the π bonds are spread among several atoms.

Section 9.7 **Molecular orbital theory** is another model used to describe the bonding in molecules. In this model the electrons exist in allowed energy states called **molecular orbitals (MOs)**. These orbitals can be spread among all the atoms of a molecule. Like an atomic orbital, a molecular orbital has a definite energy and can hold two electrons of opposite spin. The combination of two atomic orbitals leads to the formation of two MOs, one at lower energy, and one at higher energy relative to the energy of the atomic orbitals. The lower-energy MO concentrates charge density in the region between the nuclei and is called a **bonding molecular orbital**. The higher-energy MO excludes electrons from the region between the nuclei and is called an **antibonding molecular orbital**. Occupation of bonding MOs favors bond formation, whereas occupation of antibonding MOs is unfavorable. The bonding and antibonding MOs formed by the combination of s orbitals are **sigma (σ) molecular orbitals**; like σ bonds, they lie on the internuclear axis.

The combination of atomic orbitals and the relative energies of the molecular orbitals are shown by an **energy-level** (or **molecular orbital**) **diagram**. When the appropriate number of electrons are put into the MOs, we can calculate the **bond order** of a bond, which is half the difference between the number of electrons in bonding MOs and the number of electrons in antibonding MOs. A bond order of 1 corresponds to a single bond, and so forth. Bond orders can be fractional numbers.

Section 9.8 Electrons in core orbitals do not contribute to the bonding between atoms, so a molecular orbital description usually needs to consider only electrons in the outermost electron subshells. In order to describe the MOs of second-row homonuclear diatomic molecules, we need to consider the MOs that can form by the combination of p orbitals. The p orbitals that point directly at one another can form σ bonding and σ^* antibonding MOs. The p orbitals that are oriented perpendicular to the internuclear axis combine to form **pi (π) molecular orbitals**. In diatomic molecules the π molecular orbitals occur as pairs of degenerate (same energy) bonding and antibonding MOs. The σ_{2p} bonding MO is expected to be lower in energy than the π_{2p} bonding MOs because of larger orbital overlap. This ordering is reversed in B_2, C_2, and N_2 because of interaction between the $2s$ and $2p$ atomic orbitals of different atoms.

The molecular orbital description of second-row diatomic molecules leads to bond orders in accord with the Lewis structures of these molecules. Further, the model predicts correctly that O_2 should exhibit **paramagnetism**, an attraction of a molecule by a magnetic field due to unpaired electrons. Those molecules in which all the electrons are paired exhibit **diamagnetism**, a weak repulsion from a magnetic field.

KEY SKILLS

- Be able to describe the three-dimensional shapes of molecules using the VSEPR model.
- Determine whether a molecule is polar or nonpolar based on its geometry and the individual bond dipole moments.
- Be able to identify the hybridization state of atoms in molecules.
- Be able to sketch how orbitals overlap to form sigma (σ) and pi (π) bonds.
- Be able to explain the concept of bonding and antibonding orbitals.
- Be able to draw molecular orbital energy-level diagrams and place electrons into them to obtain the bond orders and electron configurations of diatomic molecules using molecular orbital theory.
- Understand the relationships among bond order, bond strength, and bond length.

KEY EQUATIONS

- Bond order = $\frac{1}{2}$ (no. of bonding electrons − no. of antibonding electrons)

VISUALIZING CONCEPTS

9.1 A certain AB_4 molecule has a "seesaw" shape:

From which of the fundamental geometries shown in Figure 9.3 could you remove one or more atoms to create a molecule having this seesaw shape? [Section 9.1]

9.2 (a) If the three balloons shown on the right are all the same size, what angle is formed between the red one and the green one? **(b)** If additional air is added to the blue balloon so that it gets larger, what happens to the angle between the red and green balloons? **(c)** What aspect of the VSEPR model is illustrated by part (b)? [Section 9.2]

9.3 An AB_5 molecule adopts the geometry shown below. **(a)** What is the name of this geometry? **(b)** Do you think there are any nonbonding electron pairs on atom A? Why or why not? **(c)** Suppose the atoms B are halogen atoms. Can you determine uniquely to which group in the periodic table atom A belongs? [Section 9.2]

9.4 The molecule shown here is *difluoromethane* (CH_2F_2), which is used as a refrigerant called R-32. **(a)** Based on the structure, how many electron domains surround the C atom in this molecule? **(b)** Would the molecule have a nonzero dipole moment? **(c)** If the molecule is polar, in what direction will the overall dipole moment vector point in the molecule? [Sections 9.2 and 9.3]

9.5 The plot below shows the potential energy of two Cl atoms as a function of the distance between them. **(a)** To what does an energy of zero correspond in this diagram? **(b)** According to the valence-bond model, why does the energy decrease as the Cl atoms move from a large separation to a smaller one? **(c)** What is the significance of the Cl–Cl distance at the minimum point in the plot? **(d)** Why does the energy rise at Cl–Cl distances less than that at the minimum point in the plot? **(e)** How can you estimate the bond strength of the Cl–Cl bond from the plot? [Section 9.4]

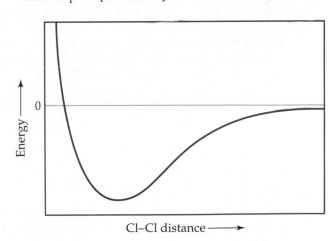

9.6 Shown below are three pairs of hybrid orbitals, with each set at a characteristic angle. For each pair, determine the type or types of hybridization that could lead to hybrid orbitals at the specified angle. [Section 9.5]

9.7 The orbital diagram below presents the final step in the formation of hybrid orbitals by a silicon atom. What type of hybrid orbitals is produced in this hybridization? [Section 9.5]

3s   3p

9.8 Consider the hydrocarbon drawn below. **(a)** What is the hybridization at each carbon atom in the molecule? **(b)** How many σ bonds are there in the molecule? **(c)** How many π bonds? [Section 9.6]

$$H-\underset{\underset{H}{|}}{\overset{\overset{H}{|}}{C}}=\underset{\underset{H}{|}}{\overset{\overset{H}{|}}{C}}-\underset{\underset{H}{|}}{\overset{\overset{H}{|}}{C}}-C\equiv C-\underset{\underset{H}{|}}{\overset{\overset{H}{|}}{C}}-H$$

9.9 For each of the following contour representations of molecular orbitals, identify (i) the atomic orbitals (s or p) used to construct the MO, (ii) the type of MO (σ or π), and (iii) whether the MO is bonding or antibonding. [Sections 9.7 and 9.8]

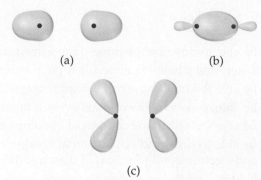

(a)   (b)

(c)

9.10 The diagram below shows the highest occupied MOs of a neutral molecule CX, where element X is in the same row of the periodic table as C. **(a)** Based on the number of electrons, can you determine the identity of X? **(b)** Would the molecule be diamagnetic or paramagnetic? **(c)** Consider the π_{2p} MOs of the molecule. Would you expect them to have a greater atomic orbital contribution from C, have a greater atomic orbital contribution from X, or be an equal mixture of atomic orbitals from the two atoms? [Section 9.8]

σ_{2p}

π_{2p}

EXERCISES

Molecular Shapes; the VSEPR Model

9.11 An AB_2 molecule is described as linear, and the A—B bond length is known. **(a)** Does this information completely describe the geometry of the molecule? **(b)** Can you tell how many nonbonding pairs of electrons are around the A atom from this information?

9.12 **(a)** Methane (CH_4) and the perchlorate ion (ClO_4^-) are both described as tetrahedral. What does this indicate about their bond angles? **(b)** The NH_3 molecule is trigonal pyramidal, while BF_3 is trigonal planar. Which of these molecules is flat?

9.13 **(a)** What is meant by the term *electron domain*? **(b)** Explain in what way electron domains behave like the balloons in Figure 9.5. Why do they do so?

9.14 **(a)** How does one determine the number of electron domains in a molecule or ion? **(b)** What is the difference between a *bonding electron domain* and a *nonbonding electron domain*?

9.15 How many nonbonding electron pairs are there in each of the following molecules: (a) $(CH_3)_2S$; (b) HCN; (c) H_2C_2; (d) CH_3F?

9.16 Describe the characteristic electron-domain geometry of each of the following numbers of electron domains about a central atom: **(a)** 3, **(b)** 4, **(c)** 5, **(d)** 6.

9.17 What is the difference between the electron-domain geometry and the molecular geometry of a molecule? Use the water molecule as an example in your discussion.

9.18 An AB_3 molecule is described as having a trigonal-bipyramidal electron-domain geometry. How many nonbonding domains are on atom A? Explain.

9.19 Give the electron-domain and molecular geometries of a molecule that has the following electron domains on its central atom: **(a)** four bonding domains and no nonbonding domains, **(b)** three bonding domains and two nonbonding domains, **(c)** five bonding domains and one nonbonding domain, **(e)** four bonding domains and two nonbonding domains.

9.20 What are the electron-domain and molecular geometries of a molecule that has the following electron domains on its central atom? **(a)** three bonding domains and no nonbonding domains, **(b)** three bonding domains and one nonbonding domain, **(c)** two bonding domains and two nonbonding domains.

9.21 Give the electron-domain and molecular geometries for the following molecules and ions: **(a)** HCN, **(b)** SO_3^{2-}, **(c)** SF_4, **(d)** PF_6^-, **(e)** NH_3Cl^+, **(f)** N_3^-.

9.22 Draw the Lewis structure for each of the following molecules or ions, and predict their electron-domain and molecular geometries: **(a)** PF_3, **(b)** CH_3^+, **(c)** BrF_3, **(d)** ClO_4^- **(e)** XeF_2, **(f)** BrO_2^-.

9.23 The figure that follows shows ball-and-stick drawings of three possible shapes of an AF_3 molecule. **(a)** For each shape, give the electron-domain geometry on which the molecular geometry is based. **(b)** For each shape, how many nonbonding electron domains are there on atom A? **(c)** Which of the following elements will lead to an AF_3 molecule with the shape in (ii): Li, B, N, Al, P, Cl? **(d)** Name an element A that is expected to lead to the AF_3 structure shown in (iii). Explain your reasoning.

(i) (ii) (iii)

9.24 The figure that follows contains ball-and-stick drawings of three possible shapes of an AF_4 molecule. **(a)** For each shape, give the electron-domain geometry on which the molecular geometry is based. **(b)** For each shape, how many nonbonding electron domains are there on atom A? **(c)** Which of the following elements will lead to an AF_4 molecule with the shape in (iii): Be, C, S, Se, Si, Xe? **(d)** Name an element A that is expected to lead to the AF_4 structure shown in (i).

(i) (ii) (iii)

9.25 Give the approximate values for the indicated bond angles in the following molecules:

(a)

(b)

(c) H—C≡C—H

(d) H—C—Ö—C—H

9.26 Give approximate values for the indicated bond angles in the following molecules:

(a) H—Ö—N=Ö

(b) H—C—C=Ö

(c) H—N—Ö—H

(d) H—C—C≡N:

9.27 Predict the trend in the F(axial)—A—F(equatorial) bond angle in the following AF_n molecules: PF_5, SF_4, and ClF_3.

9.28 The three species NH_2^- NH_3, and NH_4^+ have H—N—H bond angles of 105°, 107°, and 109°, respectively. Explain this variation in bond angles.

9.29 **(a)** Explain why BrF_4^- is square planar, whereas BF_4^- is tetrahedral. **(b)** Water, H_2O, is a bent molecule. Predict the shape of the molecular ion formed from the water molecule if you were able to remove four electrons to make $(H_2O)^{4+}$.

9.30 **(a)** Explain why the following ions have different bond angles: ClO_2^- and NO_2^-. Predict the bond angle in each case. **(b)** Explain why the XeF_2 molecule is linear and not bent.

Polarity of Polyatomic Molecules

9.31 **(a)** Does SCl_2 have a dipole moment? If so, in which direction does the net dipole point? **(b)** Does $BeCl_2$ have a dipole moment? If so, in which direction does the net dipole point?

9.32 **(a)** The PH_3 molecule is polar. How does this offer experimental proof that the molecule cannot be planar? **(b)** It turns out that ozone, O_3, has a small dipole moment. How is this possible, given that all the atoms are the same?

9.33 **(a)** Consider the AF_3 molecules in Exercise 9.23. Which of these will have a nonzero dipole moment? Explain. **(b)** Which of the AF_4 molecules in Exercise 9.24 will have a zero dipole moment?

9.34 **(a)** What conditions must be met if a molecule with polar bonds is nonpolar? **(b)** What geometries will give nonpolar molecules for AB_2, AB_3, and AB_4 geometries?

9.35 Predict whether each of the following molecules is polar or nonpolar: **(a)** IF, **(b)** CS_2, **(c)** SO_3, **(d)** PCl_3, **(e)** SF_6, **(f)** IF_5.

9.36 Predict whether each of the following molecules is polar or nonpolar: **(a)** CCl_4, **(b)** NH_3, **(c)** SF_4, **(d)** XeF_4, **(e)** CH_3Br, **(f)** GaH_3.

9.37 Dichloroethylene ($C_2H_2Cl_2$) has three forms (isomers), each of which is a different substance. **(a)** Draw Lewis structures of the three isomers, all of which have a carbon–carbon double bond. **(b)** Which of these isomers has a zero dipole moment? **(c)** How many isomeric forms can chloroethylene, C_2H_3Cl, have? Would they be expected to have dipole moments?

9.38 Dichlorobenzene, $C_6H_4Cl_2$, exists in three forms (isomers), called *ortho*, *meta*, and *para*:

ortho   *meta*   *para*

Which of these would have a nonzero dipole moment? Explain.

Orbital Overlap; Hybrid Orbitals

9.39 (a) What is meant by the term *orbital overlap*? (b) Describe what a chemical bond is in terms of electron density between two atoms.

9.40 Draw sketches illustrating the overlap between the following orbitals on two atoms: (a) the $2s$ orbital on each atom, (b) the $2p_z$ orbital on each atom (assume both atoms are on the z-axis), (c) the $2s$ orbital on one atom and the $2p_z$ orbital on the other atom.

9.41 Consider the bonding in an MgH_2 molecule. (a) Draw a Lewis structure for the molecule, and predict its molecular geometry. (b) What hybridization scheme is used in MgH_2? (c) Sketch one of the two-electron bonds between an Mg hybrid orbital and an H $1s$ atomic orbital.

9.42 How would you expect the extent of overlap of atomic orbitals to vary in the series IF, ICl, IBr, and I_2?

9.43 Fill in the following chart. If the molecule column is blank, find an example that fulfills the conditions of the rest of the row.

Molecule	Electron-domain Geometry	Hybridization of Central Atom	Dipole moment? Yes or No.
CO_2			
		sp^3	yes
		sp^3	no
	trigonal planar		no
SF_4			
	octahedral		no
		sp^2	yes
	trigonal bipyramidal		no
XeF_2			

9.44 Why are there no sp^4 or sp^5 hybrid orbitals?

9.45 (a) Starting with the orbital diagram of a boron atom, describe the steps needed to construct hybrid orbitals appropriate to describe the bonding in BF_3. (b) What is the name given to the hybrid orbitals constructed in (a)? (c) Sketch the large lobes of the hybrid orbitals constructed in part (a). (d) Are there any valence atomic orbitals of B that are left unhybridized? If so, how are they oriented relative to the hybrid orbitals?

9.46 (a) Starting with the orbital diagram of a sulfur atom, describe the steps needed to construct hybrid orbitals appropriate to describe the bonding in SF_2. (b) What is the name given to the hybrid orbitals constructed in (a)? (c) Sketch the large lobes of the hybrid orbitals constructed in part (a). (d) Would the hybridization scheme in part (a) be appropriate for SF_4? Explain.

9.47 Indicate the hybridization of the central atom in (a) BCl_3, (b) $AlCl_4^-$, (c) CS_2, (d) KrF_2, (e) PF_6^-.

9.48 What is the hybridization of the central atom in (a) $SiCl_4$, (b) HCN, (c) SO_3, (d) ICl_2^-, (e) BrF_4^-?

Multiple Bonds

9.49 (a) Draw a picture showing how two p orbitals on two different atoms can be combined to make a sigma bond. (b) Sketch a π bond that is constructed from p orbitals. (c) Which is generally stronger, a σ bond or a π bond? Explain. (d) Can two s orbitals make a π bond? Explain.

9.50 (a) If the valence atomic orbitals of an atom are sp hybridized, how many unhybridized p orbitals remain in the valence shell? How many π bonds can the atom form? (b) Imagine that you could hold two atoms that are bonded together, twist them, and not change the bond length. Would it be easier to twist (rotate) around a single σ bond or around a double (σ plus π) bond, or would they be the same? Explain.

9.51 (a) Draw Lewis structures for ethane (C_2H_6), ethylene (C_2H_4), and acetylene (C_2H_2). (b) What is the hybridization of the carbon atoms in each molecule? (c) Predict which molecules, if any, are planar. (d) How many σ and π bonds are there in each molecule? (e) Suppose that silicon could form molecules that are precisely the analogs of ethane, ethylene, and acetylene. How would you describe the bonding about Si in terms of hybrid orbitals? Does it make a difference that Si lies in the row below C in the periodic table? Explain.

9.52 The nitrogen atoms in N_2 participate in multiple bonding, whereas those in hydrazine, N_2H_4, do not. (a) Draw Lewis structures for both molecules. (b) What is the hybridization of the nitrogen atoms in each molecule? (c) Which molecule has a stronger N—N bond?

9.53 Propylene, C_3H_6, is a gas that is used to form the important polymer called polypropylene. Its Lewis structure is

$$
\begin{array}{c}
\quad\;\; H \;\;\; H \;\;\; H \\
\quad\;\; | \;\;\;\;\; | \;\;\;\;\; | \\
H-C=C-C-H \\
\quad\quad\quad\quad\quad | \\
\quad\quad\quad\quad\quad H
\end{array}
$$

(a) What is the total number of valence electrons in the propylene molecule? (b) How many valence electrons are used to make σ bonds in the molecule? (c) How many valence electrons are used to make π bonds in the molecule? (d) How many valence electrons remain in nonbonding pairs in the molecule? (e) What is the hybridization at each carbon atom in the molecule?

9.54 Ethyl acetate, $C_4H_8O_2$, is a fragrant substance used both as a solvent and as an aroma enhancer. Its Lewis structure is

$$H-\underset{\underset{H}{|}}{\overset{\overset{H}{|}}{C}}-\underset{}{\overset{\overset{:\ddot{O}:}{\|}}{C}}-\ddot{O}-\underset{\underset{H}{|}}{\overset{\overset{H}{|}}{C}}-\underset{\underset{H}{|}}{\overset{\overset{H}{|}}{C}}-H$$

(a) What is the hybridization at each of the carbon atoms of the molecule? **(b)** What is the total number of valence electrons in ethyl acetate? **(c)** How many of the valence electrons are used to make σ bonds in the molecule? **(d)** How many valence electrons are used to make π bonds? **(e)** How many valence electrons remain in non-bonding pairs in the molecule?

9.55 Consider the Lewis structure for glycine, the simplest amino acid:

$$H-\underset{\underset{H}{|}}{\overset{\overset{H}{|}}{\ddot{N}}}-\underset{\underset{H}{|}}{\overset{\overset{:O:}{\|}}{C}}-\overset{\overset{:O:}{\|}}{C}-\ddot{O}-H$$

(a) What are the approximate bond angles about each of the two carbon atoms, and what are the hybridizations of the orbitals on each of them? **(b)** What are the hybridizations of the orbitals on the two oxygens and the nitrogen atom, and what are the approximate bond angles at the nitrogen? **(c)** What is the total number of σ bonds in the entire molecule, and what is the total number of π bonds?

9.56 The compound with the following Lewis structure is acetylsalicylic acid, better known as aspirin:

(a) What are the approximate values of the bond angles labeled 1, 2, and 3? **(b)** What hybrid orbitals are used about the central atom of each of these angles? **(c)** How many σ bonds are in the molecule?

9.57 **(a)** What is the difference between a localized π bond and a delocalized one? **(b)** How can you determine whether a molecule or ion will exhibit delocalized π bonding? **(c)** Is the π bond in $NO_2{}^-$ localized or delocalized?

9.58 **(a)** Write a single Lewis structure for SO_3, and determine the hybridization at the S atom. **(b)** Are there other equivalent Lewis structures for the molecule? **(c)** Would you expect SO_3 to exhibit delocalized π bonding? Explain.

Molecular Orbitals

9.59 **(a)** What is the difference between hybrid orbitals and molecular orbitals? **(b)** How many electrons can be placed into each MO of a molecule? **(c)** Can antibonding molecular orbitals have electrons in them?

9.60 **(a)** If you combine two atomic orbitals on two different atoms to make a new orbital, is this a hybrid orbital or a molecular orbital? **(b)** If you combine two atomic orbitals on *one* atom to make a new orbital, is this a hybrid orbital or a molecular orbital? **(c)** Does the Pauli exclusion principle (Section 6.7) apply to MOs? Explain.

9.61 Consider the $H_2{}^+$ ion. **(a)** Sketch the molecular orbitals of the ion, and draw its energy-level diagram. **(b)** How many electrons are there in the $H_2{}^+$ ion? **(c)** Draw the electron configuration of the ion in terms of its MOs. **(d)** What is the bond order in $H_2{}^+$? **(e)** Suppose that the ion is excited by light so that an electron moves from a lower-energy to a higher-energy MO. Would you expect the excited-state $H_2{}^+$ ion to be stable or to fall apart? Explain.

9.62 **(a)** Sketch the molecular orbitals of the $H_2{}^-$ ion, and draw its energy-level diagram. **(b)** Write the electron configuration of the ion in terms of its MOs. **(c)** Calculate the bond order in $H_2{}^-$ **(d)** Suppose that the ion is excited by light, so that an electron moves from a lower-energy to a higher-energy molecular orbital. Would you expect the excited-state $H_2{}^-$ ion to be stable? Explain.

9.63 Draw a picture that shows all three $2p$ orbitals on one atom and all three $2p$ orbitals on another atom. **(a)** Imagine the atoms coming close together to bond. How many σ bonds can the two sets of $2p$ orbitals make with each other? **(b)** How many π bonds can the two sets of $2p$ orbitals make with each other? **(c)** How many antibonding orbitals, and of what type, can be made from the two sets of $2p$ orbitals?

9.64 **(a)** What is the probability of finding an electron on the internuclear axis if the electron occupies a π molecular orbital? **(b)** For a homonuclear diatomic molecule, what similarities and differences are there between the π_{2p} MO made from the $2p_x$ atomic orbitals and the π_{2p} MO made from the $2p_y$ atomic orbitals? **(c)** Why are the π_{2p} MOs lower in energy than the π_{2p}^* MOs?

9.65 **(a)** What are the relationships among bond order, bond length, and bond energy? **(b)** According to molecular orbital theory, would either Be_2 or $Be_2{}^+$ be expected to exist? Explain.

9.66 Explain the following: **(a)** The *peroxide* ion, $O_2{}^{2-}$, has a longer bond length than the *superoxide* ion, $O_2{}^-$. **(b)** The magnetic properties of B_2 are consistent with the π_{2p} MOs being lower in energy than the σ_{2p} MO. **(c)** The $O_2{}^{2+}$ ion has a stronger O—O bond than O_2 itself.

9.67 **(a)** What does the term *diamagnetism* mean? **(b)** How does a diamagnetic substance respond to a magnetic field? **(c)** Which of the following ions would you expect to be diamagnetic: $N_2{}^{2-}$, $O_2{}^{2-}$, $Be_2{}^{2+}$, $C_2{}^-$?

9.68 **(a)** What does the term *paramagnetism* mean? **(b)** How can one determine experimentally whether a substance is paramagnetic? **(c)** Which of the following ions would you expect to be paramagnetic: $O_2{}^+$, $N_2{}^{2-}$, $Li_2{}^+$, $O_2{}^{2-}$? For those ions that are paramagnetic, determine the number of unpaired electrons.

9.69 Using Figures 9.37 and 9.45 as guides, draw the molecular orbital electron configuration for **(a)** B_2^+, **(b)** Li_2^+, **(c)** N_2^+, **(d)** Ne_2^{2+}. In each case indicate whether the addition of an electron to the ion would increase or decrease the bond order of the species.

9.70 If we assume that the energy-level diagrams for homonuclear diatomic molecules shown in Figure 9.43 can be applied to heteronuclear diatomic molecules and ions, predict the bond order and magnetic behavior of **(a)** CO^+, **(b)** NO^-, **(c)** OF^+, **(d)** NeF^+.

9.71 Determine the electron configurations for CN^+, CN, and CN^-. **(a)** Which species has the strongest C—N bond? **(b)** Which species, if any, has unpaired electrons?

9.72 **(a)** The nitric oxide molecule, NO, readily loses one electron to form the NO^+ ion. Why is this consistent with the electronic structure of NO? **(b)** Predict the order of the N—O bond strengths in NO, NO^+, and NO^-, and describe the magnetic properties of each. **(c)** With what neutral homonuclear diatomic molecules are the NO^+ and NO^- ions isoelectronic (same number of electrons)?

[9.73] Consider the molecular orbitals of the P_2 molecule. Assume that the MOs of diatomics from the third row of the periodic table are analogous to those from the second row. **(a)** Which valence atomic orbitals of P are used to construct the MOs of P_2? **(b)** The figure below shows a sketch of one of the MOs for P_2. What is the label for this MO? **(c)** For the P_2 molecule, how many electrons occupy the MO in the figure? **(d)** Is P_2 expected to be diamagnetic or paramagnetic? Explain.

[9.74] The iodine bromide molecule, IBr, is an *interhalogen compound*. Assume that the molecular orbitals of IBr are analogous to the homonuclear diatomic molecule F_2. **(a)** Which valence atomic orbitals of I and of Br are used to construct the MOs of IBr? **(b)** What is the bond order of the IBr molecule? **(c)** One of the valence MOs of IBr is sketched below. Why are the atomic orbital contributions to this MO different in size? **(d)** What is the label for the MO? **(e)** For the IBr molecule, how many electrons occupy the MO?

ADDITIONAL EXERCISES

9.75 **(a)** What is the physical basis for the VSEPR model? **(b)** When applying the VSEPR model, we count a double or triple bond as a single electron domain. Why is this justified?

9.76 The molecules SiF_4, SF_4, and XeF_4 have molecular formulas of the type AF_4, but the molecules have different molecular geometries. Predict the shape of each molecule, and explain why the shapes differ.

[9.77] The vertices of a tetrahedron correspond to four alternating corners of a cube. By using analytical geometry, demonstrate that the angle made by connecting two of the vertices to a point at the center of the cube is 109.5°, the characteristic angle for tetrahedral molecules.

9.78 Consider the molecule PF_4Cl. **(a)** Draw a Lewis structure for the molecule, and predict its electron-domain geometry. **(b)** Which would you expect to take up more space, a P—F bond or a P—Cl bond? Explain. **(c)** Predict the molecular geometry of PF_4Cl. How did your answer for part (b) influence your answer here in part (c)? **(d)** Would you expect the molecule to distort from its ideal electron-domain geometry? If so, how would it distort?

9.79 From their Lewis structures, determine the number of σ and π bonds in each of the following molecules or ions: **(a)** CO_2; **(b)** thiocyanate ion, NCS^-; **(c)** formaldehyde, H_2CO; **(d)** formic acid, HCOOH, which has one H and two O atoms attached to C.

9.80 The lactic acid molecule, $CH_3CH(OH)COOH$, gives sour milk its unpleasant, sour taste. **(a)** Draw the Lewis structure for the molecule, assuming that carbon always forms four bonds in its stable compounds. **(b)** How many π and how many σ bonds are in the molecule? **(c)** Which CO bond is shortest in the molecule? **(d)** What

is the hybridization of atomic orbitals around each carbon atom associated with that short bond? **(e)** What are the approximate bond angles around each carbon atom in the molecule?

9.81 The PF_3 molecule has a dipole moment of 1.03 D, but BF_3 has a dipole moment of zero. How can you explain the difference?

9.82 There are two compounds of the formula $Pt(NH_3)_2Cl_2$:

$$Cl-\underset{\underset{NH_3}{|}}{\overset{\overset{NH_3}{|}}{Pt}}-Cl \qquad Cl-\underset{\underset{NH_3}{|}}{\overset{\overset{Cl}{|}}{Pt}}-NH_3$$

The compound on the right, *cisplatin*, is used in cancer therapy. The compound on the left, *transplatin*, is ineffective for cancer therapy. Both compounds have a square-planar geometry. **(a)** Which compound has a nonzero dipole moment? **(b)** The reason cisplatin is a good anticancer drug is that it binds tightly to DNA. Cancer cells are rapidly dividing, producing a lot of DNA. Consequently cisplatin kills cancer cells at a faster rate than normal cells. However, since normal cells also are making DNA, cisplatin also attacks healthy cells, which leads to unpleasant side effects. The way both molecules bind to DNA involves the Cl^- ions leaving the Pt ion, to be replaced by two nitrogens in DNA. Draw a picture in which a long vertical line represents a piece of DNA. Draw the $Pt(NH_3)_2$ fragments of cisplatin and transplatin with the proper shape. Also draw them attaching to your DNA line. Can you explain from your drawing why the shape of the cisplatin causes it to bind to DNA more effectively than transplatin?

[9.83] The O—H bond lengths in the water molecule (H_2O) are 0.96 Å, and the H—O—H angle is 104.5°. The dipole moment of the water molecule is 1.85 D. **(a)** In what directions do the bond dipoles of the O—H bonds point? In what direction does the dipole moment vector of the water molecule point? **(b)** Calculate the magnitude of the bond dipole of the O—H bonds. (*Note:* You will need to use vector addition to do this.) **(c)** Compare your answer from part (b) to the dipole moments of the hydrogen halides (Table 8.3). Is your answer in accord with the relative electronegativity of oxygen?

[9.84] The reaction of three molecules of fluorine gas with a Xe atom produces the substance xenon hexafluoride, XeF_6:

$$Xe(g) + 3\ F_2(g) \longrightarrow XeF_6(s)$$

(a) Draw a Lewis structure for XeF_6. **(b)** If you try to use the VSEPR model to predict the molecular geometry of XeF_6, you run into a problem. What is it? **(c)** What could you do to resolve the difficulty in part (b)? **(d)** Suggest a hybridization scheme for the Xe atom in XeF_6. **(e)** The molecule IF_7 has a pentagonal-bipyramidal structure (five equatorial fluorine atoms at the vertices of a regular pentagon and two axial fluorine atoms). Based on the structure of IF_7, suggest a structure for XeF_6.

[9.85] The Lewis structure for allene is

$$\begin{array}{c} H \\ \diagdown \\ C=C=C \\ \diagup \\ H \end{array} \begin{array}{c} H \\ \diagup \\ \\ \diagdown \\ H \end{array}$$

Make a sketch of the structure of this molecule that is analogous to Figure 9.27. In addition, answer the following three questions: **(a)** Is the molecule planar? **(b)** Does it have a nonzero dipole moment? **(c)** Would the bonding in allene be described as delocalized? Explain.

[9.86] The azide ion, N_3^-, is linear with two N—N bonds of equal length, 1.16 Å. **(a)** Draw a Lewis structure for the azide ion. **(b)** With reference to Table 8.5, is the observed N—N bond length consistent with your Lewis structure? **(c)** What hybridization scheme would you expect at each of the nitrogen atoms in N_3^-? **(d)** Show which hybridized and unhybridized orbitals are involved in the formation of σ and π bonds in N_3^-. **(e)** It is often observed that σ bonds that involve an sp hybrid orbital are shorter than those that involve only sp^2 or sp^3 hybrid orbitals. Can you propose a reason for this? Is this observation applicable to the observed bond lengths in N_3^-?

[9.87] In ozone, O_3, the two oxygen atoms on the ends of the molecule are equivalent to one another. **(a)** What is the best choice of hybridization scheme for the atoms of ozone? **(b)** For one of the resonance forms of ozone, which of the orbitals are used to make bonds and which are used to hold nonbonding pairs of electrons? **(c)** Which of the orbitals can be used to delocalize the π electrons? **(d)** How many electrons are delocalized in the π system of ozone?

9.88 Butadiene, C_4H_6, is a planar molecule that has the following carbon—carbon bond lengths:

$$H_2C \underset{1.34\ \text{Å}}{=\!=\!=} CH \underset{1.48\ \text{Å}}{\rule{1.2cm}{0.4pt}} CH \underset{1.34\ \text{Å}}{=\!=\!=} CH_2$$

(a) Predict the bond angles around each of the carbon atoms, and sketch the molecule. **(b)** Compare the bond lengths to the average bond lengths listed in Table 8.5. Can you explain any differences?

[9.89] The sketches below show the atomic orbital wave functions (with phases) used to construct some of the MOs of a homonuclear diatomic molecule. For each sketch, determine the MO that will result from mixing the atomic orbital wave functions as drawn. Use the same labels for the MOs as in the "Closer Look" box on phases.

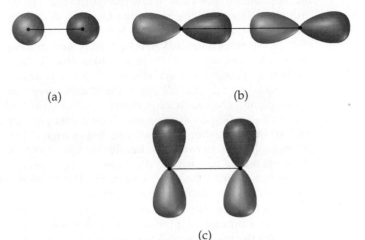

(a)   (b)

(c)

[9.90] The *cyclopentadienide ion* has the formula $C_5H_5^-$. The ion consists of a regular pentagon of C atoms, each bonded to two C neighbors, with a hydrogen atom bonded to each C atom. All the atoms lie in the same plane. **(a)** Draw a Lewis structure for the ion. According to your structure, do all five C atoms have the same hybridization? Explain. **(b)** Chemists generally view this ion as having sp^2 hybridization at each C atom. Is that view consistent with your answer to part (a)? **(c)** Your Lewis structure should show one nonbonding pair of electrons. Under the assumption of part (b), in what type of orbital must this nonbonding pair reside? **(d)** Are there resonance structures equivalent to the Lewis structure you drew in part (a)? If so, how many? **(e)** The ion is often drawn as a pentagon enclosing a circle. Is this representation consistent with your answer to part (d)? Explain. **(f)** Both benzene and the cyclopentadienide ion are often described as systems containing six π electrons. What do you think is meant by this description?

9.91 Write the electron configuration for the first excited state for N_2—that is, the state with the highest-energy electron moved to the next available energy level. **(a)** Is the nitrogen in its first excited state diamagnetic or paramagnetic? **(b)** Is the N—N bond strength in the first excited state stronger or weaker compared to that in the ground state? Explain.

9.92 Figure 9.47 shows how the magnetic properties of a compound can be measured experimentally. When such measurements are made, the sample is generally covered by an atmosphere of pure nitrogen gas rather than air. Why do you suppose this is done?

9.93 *Azo dyes* are organic dyes that are used for many applications, such as the coloring of fabrics. Many azo dyes are derivatives of the organic substance *azobenzene*, $C_{12}H_{10}N_2$.

A closely related substance is *hydrazobenzene*, $C_{12}H_{12}N_2$. The Lewis structures of these two substances are

Azobenzene   Hydrazobenzene

(Recall the shorthand notation used for benzene.) **(a)** What is the hybridization at the N atom in each of the substances? **(b)** How many unhybridized atomic orbitals are there on the N and the C atoms in each of the substances? **(c)** Predict the N—N—C angles in each of the substances. **(d)** Azobenzene is said to have greater delocalization of its π electrons than hydrazobenzene. Discuss this statement in light of your answers to (a) and (b). **(e)** All the atoms of azobenzene lie in one plane, whereas those of hydrazobenzene do not. Is this observation consistent with the statement in part (d)? **(f)** Azobenzene is an intense red-orange color, whereas hydrazobenzene is nearly colorless. Which molecule would be a better one to use in a solar energy conversion device? (See the "Chemistry Put to Work" box for more information about solar cells.)

[9.94] **(a)** Using only the valence atomic orbitals of a hydrogen atom and a fluorine atom, how many MOs would you expect for the HF molecule? **(b)** How many of the MOs from part (a) would be occupied by electrons? **(c)** Do you think the MO diagram shown in Figure 9.49 could be used to describe the MOs of the HF molecule? Why or why not? **(d)** It turns out that the difference in energies between the valence atomic orbitals of H and F are sufficiently different that we can neglect the interaction of the 1s orbital of hydrogen with the 2s orbital of fluorine. The 1s orbital of hydrogen will mix only with one 2p orbital of fluorine. Draw pictures showing the proper orientation of all three 2p orbitals on F interacting with a 1s orbital on H. Which of the 2p orbitals can actually make a bond with a 1s orbital, assuming that the atoms lie on the z-axis? **(e)** In the most accepted picture of HF, all the other atomic orbitals on fluorine move over at the same energy into the molecular orbital energy level diagram for HF. These are called "nonbonding orbitals." Sketch the energy level diagram for HF using this information, and calculate the bond order. (Nonbonding electrons do not contribute to bond order.) **(f)** Look at the Lewis structure for HF. Where are the nonbonding electrons?

[9.95] Carbon monoxide, CO, is isoelectronic to N_2. **(a)** Draw a Lewis structure for CO that satisfies the octet rule. **(b)** Assume that the diagram in Figure 9.49 can be used to describe the MOs of CO. What is the predicted bond order for CO? Is this answer in accord with the Lewis structure you drew in part (a)? **(c)** Experimentally, it is found that the highest-energy electrons in CO reside in a σ-type MO. Is that observation consistent with Figure 9.49? If not, what modification needs to be made to the diagram? How does this modification relate to Figure 9.45? **(d)** Would you expect the π_{2p} MOs of CO to have equal atomic orbital contributions from the C and O atoms? If not, which atom would have the greater contribution?

INTEGRATIVE EXERCISES

9.96 A compound composed of 2.1% H, 29.8% N, and 68.1% O has a molar mass of approximately 50 g/mol. **(a)** What is the molecular formula of the compound? **(b)** What is its Lewis structure if H is bonded to O? **(c)** What is the geometry of the molecule? **(d)** What is the hybridization of the orbitals around the N atom? **(e)** How many σ and how many π bonds are there in the molecule?

9.97 Sulfur tetrafluoride (SF_4) reacts slowly with O_2 to form sulfur tetrafluoride monoxide (OSF_4) according to the following unbalanced reaction:

$$SF_4(g) + O_2(g) \longrightarrow OSF_4(g)$$

The O atom and the four F atoms in OSF_4 are bonded to a central S atom. **(a)** Balance the equation. **(b)** Write a Lewis structure of OSF_4 in which the formal charges of all atoms are zero. **(c)** Use average bond enthalpies (Table 8.4) to estimate the enthalpy of the reaction. Is it endothermic or exothermic? **(d)** Determine the electron-domain geometry of OSF_4, and write two possible molecular geometries for the molecule based on this electron-domain geometry. **(e)** Which of the molecular geometries in part (d) is more likely to be observed for the molecule? Explain.

[9.98] The phosphorus trihalides (PX_3) show the following variation in the bond angle X—P—X: PF_3, 96.3°; PCl_3, 100.3°; PBr_3, 101.0°; PI_3, 102.0°. The trend is generally attributed to the change in the electronegativity of the halogen. **(a)** Assuming that all electron domains are the same size, what value of the X—P—X angle is predicted by the VSEPR model? **(b)** What is the general trend in the X—P—X angle as the electronegativity increases? **(c)** Using the VSEPR model, explain the observed trend in X—P—X angle as the electronegativity of X changes. **(d)** Based on your answer to part (c), predict the structure of $PBrCl_4$.

[9.99] The molecule 2-butene, C_4H_8, can undergo a geometric change called *cis-trans isomerization*:

cis-2-butene   *trans*-2-butene

As discussed in the "Chemistry and Life" box on the chemistry of vision, such transformations can be induced by light and are the key to human vision. **(a)** What is the hybridization at the two central carbon atoms of 2-butene? **(b)** The isomerization occurs by rotation about the central C—C bond. With reference to Figure 9.32, explain why the π bond between the two central carbon atoms is destroyed halfway through the rotation from *cis*- to *trans*-2-butene. **(c)** Based on average bond enthalpies (Table 8.4), how much energy per molecule must be supplied to break the C—C π bond? **(d)** What is the longest wavelength of light that will provide photons of sufficient energy to break the C—C π bond and cause the isomerization? **(e)** Is the wavelength in your answer to part (d) in the visible portion of the electromagnetic spectrum? Comment on the importance of this result for human vision.

9.100 **(a)** Compare the bond enthalpies (Table 8.4) of the carbon–carbon single, double, and triple bonds to deduce an average π-bond contribution to the enthalpy. What fraction of a single bond does this quantity represent? **(b)** Make a similar comparison of nitrogen–nitrogen bonds. What do you observe? **(c)** Write Lewis structures of N_2H_4, N_2H_2, and N_2, and determine the hybridization around nitrogen in each case. **(d)** Propose a reason for the large difference in your observations of parts (a) and (b).

9.101 Use average bond enthalpies (Table 8.4) to estimate ΔH for the atomization of benzene, C_6H_6:

$$C_6H_6(g) \longrightarrow 6\,C(g) + 6\,H(g)$$

Compare the value to that obtained by using ΔH_f° data given in Appendix C and Hess's law. To what do you attribute the large discrepancy in the two values?

9.102 For both atoms and molecules, ionization energies (Section 7.4) are related to the energies of orbitals: The lower the energy of the orbital, the greater the ionization energy. The first ionization energy of a molecule is therefore a measure of the energy of the highest occupied molecular orbital (HOMO). See the "Chemistry Put to Work" box on Orbitals and Energy. The first ionization energies of several diatomic molecules are given in electron-volts in the following table:

Molecule	I_1 (eV)
H_2	15.4
N_2	15.6
O_2	12.1
F_2	15.7

(a) Convert these ionization energies to kJ/mol. **(b)** On the same plot, graph I_1 for the H, N, O, and F atoms (Figure 7.12) and I_1 for the molecules listed. **(c)** Do the ionization energies of the molecules follow the same periodic trends as the ionization energies of the atoms? **(d)** Use molecular orbital energy-level diagrams to explain the trends in the ionization energies of the molecules.

[9.103] Many compounds of the transition-metal elements contain direct bonds between metal atoms. We will assume that the z-axis is defined as the metal–metal bond axis. **(a)** Which of the 3d orbitals (Figure 6.24) can be used to make a σ bond between metal atoms? **(b)** Sketch the σ_{3d} bonding and σ_{3d}^* antibonding MOs. **(c)** With reference to the "Closer Look" box on the phases of orbitals, explain why a node is generated in the σ_{3d}^* MO. **(d)** Sketch the energy-level diagram for the Sc_2 molecule, assuming that only the 3d orbital from part (a) is important. **(e)** What is the bond order in Sc_2?

[9.104] The organic molecules shown below are derivatives of benzene in which additional six-membered rings are "fused" at the edges of the hexagons. The compounds are shown in the usual abbreviated method for organic molecules. **(a)** Determine the empirical formula of benzene and of these three compounds. **(b)** Suppose you are given a sample of one of the compounds. Could combustion analysis be used to determine unambiguously

which of the compounds it is? **(c)** Naphthalene, which is the active ingredient in mothballs, is a white solid. Write a balanced equation for the combustion of naphthalene to $CO_2(g)$ and $H_2O(g)$. **(d)** Using the Lewis structure below for naphthalene and the average bond enthalpies in Table 8.4, estimate the heat of combustion of naphthalene, in kJ/mol. **(e)** Would you expect naphthalene, anthracene, and tetracene to have multiple resonance structures? If so, draw the additional resonance structures for naphthalene. **(f)** Benzene, naphthalene, and anthracene are colorless, but tetracene is orange. What does this imply about the relative HOMO–LUMO energy gaps in these molecules? See the "Chemistry Put to Work" box on Orbitals and Energy.

Naphthalene   Anthracene

Tetracene

[9.105] Antibonding molecular orbitals can be used to make bonds to other atoms in a molecule. For example, metal atoms can use appropriate d orbitals to overlap with the π_{2p}^* orbitals of the carbon monoxide molecule. This is called d–π backbonding. **(a)** Draw a coordinate axis system in which the y-axis is vertical in the plane of the paper and the x-axis horizontal. Write "M" at the origin to denote a metal atom. **(b)** Now, on the x-axis to the right of M, draw the Lewis structure of a CO molecule, with the carbon nearest the M. The CO bond axis should be on the x-axis. **(c)** Draw the CO π_{2p}^* orbital, with phases (see the Closer Look box on phases) in the plane of the paper. Two lobes should be pointing toward M. **(d)** Now draw the d_{xy} orbital of M, with phases. Can you see how they will overlap with the π_{2p}^* orbital of CO? **(e)** What kind of bond is being made with the orbitals between M and C, σ or π? **(f)** Predict what will happen to the strength of the CO bond in a metal–CO complex compared to CO alone.

[9.106] You can think of the bonding in the Cl_2 molecule in several ways. For example, you can picture the Cl—Cl bond containing two electrons that each come from the 3p orbitals of a Cl atom that are pointing in the appropriate direction. However, you can also think about hybrid orbitals. **(a)** Draw the Lewis structure of the Cl_2 molecule. **(b)** What is the hybridization of each Cl atom? **(c)** What kind of orbital overlap, in this view, makes the Cl—Cl bond? **(d)** Imagine if you could measure the positions of the lone pairs of electrons in Cl_2. How would you distinguish between the atomic orbital and hybrid orbital models of bonding using that knowledge? **(e)** You can also treat Cl_2 using molecular orbital theory to obtain an energy level diagram similar to that for F_2. Design an experiment that could tell you if the MO picture of Cl_2 is the best one, assuming you could easily measure bond lengths, bond energies, and the light absorption properties for any ionized species.

CHAPTER 10 GASES

A TORNADO IS A VIOLENT rotating column of air characterized by a funnel-shaped cloud. A tornado can be up to 100 m in diameter with wind speeds approaching 500 km/hr.

WHAT'S AHEAD

10.1 Characteristics of Gases
We begin by comparing the distinguishing characteristics of gases with those of liquids and solids.

10.2 Pressure
We will then study gas *pressure*, how it is measured, and the units used to express it, as well as consider Earth's atmosphere and the pressure it exerts.

10.3 The Gas Laws
We consider that the state of a gas can be expressed in terms of its volume (V), pressure (P), temperature (T), and quantity of gas (n). We will examine several empirical relationships among V, P, T, and n that are called *gas laws*.

10.4 The Ideal-Gas Equation
We find that when considered together, the empirical gas laws yield the *ideal-gas equation*, $PV = nRT$. Although the ideal-gas equation is not obeyed exactly by any real gas, most gases come very close to obeying it at temperature and pressure conditions of greatest interest.

10.5 Further Applications of the Ideal-Gas Equation
We can use the ideal-gas equation to make many useful calculations, such as the calculation of the density or molar mass of a gas.

10.6 Gas Mixtures and Partial Pressures
We recognize that in a mixture of gases, each component exerts a pressure that is part of the total pressure. This *partial pressure* is the pressure that the component would exert if it were by itself.

10.7 Kinetic-Molecular Theory
We explore this theory of gases, which helps us understand the behavior of gases, including their pressure and temperature, on a molecular level. According to the theory, the atoms or molecules that make up the gas move about with an average kinetic energy that is proportional to the gas temperature.

10.8 Molecular Effusion and Diffusion
We next observe that the kinetic-molecular theory helps us account for such gas properties as *effusion* through tiny openings and for *diffusion*, the movement through another substance.

10.9 Real Gases: Deviations from Ideal Behavior
We then consider that real gases deviate from ideal behavior because the gas molecules have finite volume and because attractive forces exist between molecules. The *van der Waals equation* gives a more accurate account of real gas behavior at high pressures and low temperatures.

IN THE PAST SEVERAL CHAPTERS we have learned about electronic structures of atoms and about how atoms combine to form molecules and ionic substances. In everyday life, however, we do not have direct experiences with atoms. Instead, we encounter matter as collections of enormous numbers of atoms or molecules that make up gases, liquids, and solids. Large collections of atoms and molecules in the atmosphere, for example, are responsible for our weather—the gentle breezes and the gales, the humidity and the rain. Tornados, such as the one shown in the chapter-opening photo, form when moist, warm air at lower elevations converges with cooler, dry air above. The resultant air flows produce winds that can approach speeds up to 500 km/hr (300 mph).

It was John Dalton's interest in the weather that motivated him to study gases and eventually to propose the atomic theory of matter. ∞ (Section 2.1) We now know that the properties of gases, liquids, and solids are readily understood in terms of the behavior of their component atoms, ions, and molecules. In this chapter we will examine the physical properties of gases and consider how we can understand these properties in terms of the behavior of gas molecules. In Chapter 11 we will turn our attention to the physical properties of liquids and solids.

10.1 CHARACTERISTICS OF GASES

In many ways gases are the most easily understood form of matter. Even though different gaseous substances may have very different *chemical* properties, they behave quite similarly as far as their *physical* properties are concerned. For example, we live in an atmosphere composed of a mixture of gases that we refer to as air. Air is a complex mixture of several substances, primarily N_2 (78%) and O_2 (21%), with small amounts of several other gases, including Ar (0.9%). We breathe air to absorb oxygen, O_2, which supports human life. Air also contains nitrogen, N_2, which has very different chemical properties from oxygen, yet this mixture behaves physically as one gaseous material.

Only a few elements exist as gases under ordinary conditions of temperature and pressure: The noble gases (He, Ne, Ar, Kr, and Xe) are all monatomic gases, whereas H_2, N_2, O_2, F_2, and Cl_2 are diatomic gases. Many molecular compounds are also gases. Table 10.1▼ lists a few of the more common gaseous compounds. Notice that all of these gases are composed entirely of nonmetallic elements. Furthermore, all have simple molecular formulas and, therefore, low molar masses. Substances that are liquids or solids under ordinary conditions can also exist in the gaseous state, where they are often referred to as **vapors**. The substance H_2O, for example, can exist as liquid water, solid ice, or water vapor.

Gases differ significantly from solids and liquids in several respects. For example, a gas expands spontaneously to fill its container. Consequently, the volume of a gas equals the volume of the container in which it is held. Gases also are highly compressible: When pressure is applied to a gas, its volume readily decreases. Solids and liquids, on the other hand, do not expand to fill their containers, and solids and liquids are not readily compressible.

Gases form homogeneous mixtures with each other regardless of the identities or relative proportions of the component gases. The atmosphere serves as an excellent example. As a further example, when water and gasoline are mixed, the two liquids remain as separate layers. In contrast, the water vapor and gasoline vapors above the liquids form a homogeneous gas mixture.

The characteristic properties of gases arise because the individual molecules are relatively far apart. In the air we breathe, for example, the molecules take up only about 0.1% of the total volume, with the rest being empty space. Thus, each molecule behaves largely as though the others were not present.

TABLE 10.1 ■ Some Common Compounds That Are Gases at Room Temperature

Formula	Name	Characteristics
HCN	Hydrogen cyanide	Very toxic, slight odor of bitter almonds
H_2S	Hydrogen sulfide	Very toxic, odor of rotten eggs
CO	Carbon monoxide	Toxic, colorless, odorless
CO_2	Carbon dioxide	Colorless, odorless
CH_4	Methane	Colorless, odorless, flammable
C_2H_4	Ethylene	Colorless, ripens fruit
C_3H_8	Propane	Colorless, odorless, bottled gas
N_2O	Nitrous oxide	Colorless, sweet odor, laughing gas
NO_2	Nitrogen dioxide	Toxic, red-brown, irritating odor
NH_3	Ammonia	Colorless, pungent odor
SO_2	Sulfur dioxide	Colorless, irritating odor

As a result, different gases behave similarly, even though they are made up of different molecules. In contrast, the individual molecules in a liquid are close together and occupy perhaps 70% of the total space. The attractive forces among the molecules keep the liquid together.

GIVE IT SOME THOUGHT

What is the major reason that physical properties do not differ much from one gaseous substance to another?

10.2 PRESSURE

Among the most readily measured properties of a gas are its temperature, volume, and pressure. Many early studies of gases focused on relationships among these properties. We have already discussed volume and temperature. ∞ (Section 1.4) Let's now consider the concept of pressure.

In general terms, **pressure** conveys the idea of a force, a push that tends to move something in a given direction. Pressure, P, is, in fact, the force, F, that acts on a given area, A.

$$P = \frac{F}{A} \qquad [10.1]$$

Gases exert a pressure on any surface with which they are in contact. The gas in an inflated balloon, for example, exerts a pressure on the inside surface of the balloon.

Atmospheric Pressure and the Barometer

You and I, coconuts and nitrogen molecules, all experience an attractive force that pulls toward the center of Earth. When a coconut comes loose from its place at the top of the tree, for example, the gravitational attractive force causes it to be accelerated toward Earth, increasing in speed as its potential energy is converted into kinetic energy. ∞ (Section 5.1) The atoms and molecules of the atmosphere also experience a gravitational acceleration. Because the gas particles have such tiny masses, however, their thermal energies of motion (their kinetic energies) override the gravitational forces, so all the molecules that make up the atmosphere don't just pile up in a thin layer at Earth's surface. Nevertheless, gravity does operate, and it causes the atmosphere as a whole to press down on Earth's surface, creating an atmospheric pressure.

You can demonstrate the existence of atmospheric pressure to yourself with an empty plastic bottle of the sort used to hold water or soft drinks. If you suck on the mouth of the empty bottle, chances are you can cause it to partially cave in. When you break the partial vacuum you have created, the bottle pops out to its original shape. What causes the bottle to cave in when the pressure inside it is reduced, even by the relatively small amount you can manage with your lungs? The gas molecules in the atmosphere are exerting a force on the outside of the bottle that is greater than the force within the bottle when some of the gas has been sucked out.

We can calculate the magnitude of the atmospheric pressure as follows: The force, F, exerted by any object is the product of its mass, m, times its acceleration, a; that is, $F = ma$. The acceleration given by Earth's gravity to any object located near Earth's surface is 9.8 m/s². Now imagine a column of air 1 m² in cross section extending through the entire atmosphere (Figure 10.1▶). That column has a mass of roughly 10,000 kg. The force exerted by the column on Earth's surface is

$$F = (10,000 \text{ kg})(9.8 \text{ m/s}^2) = 1 \times 10^5 \text{ kg-m/s}^2 = 1 \times 10^5 \text{ N}$$

▼ **Figure 10.1 Atmospheric pressure.** Illustration of the manner in which Earth's atmosphere exerts pressure at the surface of the planet. The mass of a column of atmosphere exactly 1 m² in cross-sectional area and extending to the upper atmosphere exerts a force of 1×10^5 N.

▲ **Figure 10.2 A mercury barometer.** The pressure of the atmosphere on the surface of the mercury (represented by the blue arrow) equals the pressure of the column of mercury (red arrow).

The SI unit for force is kg-m/s² and is called the *newton* (N): $1\text{ N} = 1\text{ kg-m/s}^2$. The pressure exerted by the column is the force divided by the cross-sectional area, A, over which the force is applied. Because our air column has a cross-sectional area of 1 m^2, we have

$$P = \frac{F}{A} = \frac{1 \times 10^5 \text{ N}}{1 \text{ m}^2} = 1 \times 10^5 \text{ N/m}^2 = 1 \times 10^5 \text{ Pa} = 1 \times 10^2 \text{ kPa}$$

The SI unit of pressure is N/m^2. It is given the name **pascal** (Pa) after Blaise Pascal (1623–1662), a French mathematician and scientist who studied pressure: $1\text{ Pa} = 1\text{ N/m}^2$. A related unit sometimes used to report pressures is the **bar**, which equals 10^5 Pa. Atmospheric pressure at sea level is about 100 kPa, or 1 bar. The actual atmospheric pressure at any location depends on weather conditions and altitude.

In the early part of the seventeenth century many scientists and philosophers believed that the atmosphere had no weight. Evangelista Torricelli (1608–1647), who was a student of Galileo's, proved this untrue. He invented the *barometer* (Figure 10.2 ◄), which is made from a glass tube more than 760 mm long that is closed at one end, completely filled with mercury, and inverted into a dish that contains additional mercury. (Care must be taken so that no air gets into the tube.) When the mercury-filled tube is inverted into the dish of mercury, some of the mercury flows out of the tube, but a column of mercury remains inside. Torricelli argued that the mercury surface in the dish experiences the full force, or weight, of Earth's atmosphere, which pushes the mercury up the tube until the pressure exerted by the mercury column equals the atmospheric pressure at the base of the tube. So the height, h, of the mercury column is a measure of the atmosphere's pressure, and it changes as the atmospheric pressure changes.

Torricelli's proposed explanation met with fierce opposition. Some argued that there could not possibly be a vacuum at the top of the tube. They said, "Nature does not permit a vacuum!" But Torricelli also had his supporters. Blaise Pascal, for example, had one of the barometers carried to the top of Puy de Dome, a volcanic mountain in central France, and compared its readings with a duplicate barometer kept at the foot of the mountain. As the barometer ascended, the height of the mercury column diminished, as expected, because the amount of atmosphere pressing down on the surface decreases as one moves higher. These and other experiments by scientists eventually prevailed, and the idea that the atmosphere has weight became accepted over many years.

GIVE IT SOME THOUGHT

What happens to the height of the mercury column in a mercury barometer as you move to higher altitude, and why?

Standard atmospheric pressure, which corresponds to the typical pressure at sea level, is the pressure sufficient to support a column of mercury 760 mm high. In SI units this pressure equals 1.01325×10^5 Pa. Standard atmospheric pressure defines some common non-SI units used to express gas pressures, such as the **atmosphere** (atm) and the *millimeter of mercury* (mm Hg). The latter unit is also called the **torr**, after Torricelli.

$$1\text{ atm} = 760\text{ mm Hg} = 760\text{ torr} = 1.01325 \times 10^5 \text{ Pa} = 101.325\text{ kPa}$$

Note that the units mm Hg and torr are equivalent: 1 torr = 1 mm Hg.

We will usually express gas pressure in units of atm, Pa (or kPa), or torr, so you should be comfortable converting gas pressures from one set of units to another.

■ SAMPLE EXERCISE 10.1 | Converting Units of Pressure

(a) Convert 0.357 atm to torr. **(b)** Convert 6.6×10^{-2} torr to atm. **(c)** Convert 147.2 kPa to torr.

SOLUTION

Analyze: In each case we are given the pressure in one unit and asked to convert it to another unit. Our task, therefore, is to choose the appropriate conversion factors.

Plan: We can use dimensional analysis to perform the desired conversions.

Solve:

(a) To convert atmospheres to torr, we use the relationship 760 torr = 1 atm:

$$(0.357 \; \text{atm})\left(\frac{760 \; \text{torr}}{1 \; \text{atm}}\right) = 271 \; \text{torr}$$

Note that the units cancel in the required manner.

(b) We use the same relationship as in part (a). To get the appropriate units to cancel, we must use the conversion factor as follows:

$$(6.6 \times 10^{-2} \; \text{torr})\left(\frac{1 \; \text{atm}}{760 \; \text{torr}}\right) = 8.7 \times 10^{-5} \; \text{atm}$$

(c) The relationship 760 torr = 101.325 kPa allows us to write an appropriate conversion factor for this problem:

$$(147.2 \; \text{kPa})\left(\frac{760 \; \text{torr}}{101.325 \; \text{kPa}}\right) = 1104 \; \text{torr}$$

Check: In each case look at the magnitude of the answer and compare it with the starting value. The torr is a much smaller unit than the atmosphere, so we expect the *numerical* answer to be larger than the starting quantity in (a) and smaller in (b). In (c) notice that there are nearly 8 torr per kPa, so the numerical answer in torr should be about 8 times larger than its value in kPa, consistent with our calculation.

■ PRACTICE EXERCISE

(a) In countries that use the metric system, such as Canada, atmospheric pressure in weather reports is given in units of kPa. Convert a pressure of 745 torr to kPa. **(b)** An English unit of pressure sometimes used in engineering is pounds per square inch (lb/in.²), or psi: 1 atm = 14.7 lb/in.². If a pressure is reported as 91.5 psi, express the measurement in atmospheres.
Answer: **(a)** 99.3 kPa, **(b)** 6.22 atm

We can use various devices to measure the pressures of enclosed gases. Tire gauges, for example, measure the pressure of air in automobile and bicycle tires. In laboratories we sometimes use a device called a *manometer*. A manometer operates on a principle similar to that of a barometer, as shown in Sample Exercise 10.2.

■ SAMPLE EXERCISE 10.2 | Using a Manometer to Measure Gas Pressure

On a certain day the barometer in a laboratory indicates that the atmospheric pressure is 764.7 torr. A sample of gas is placed in a flask attached to an open-end mercury manometer, shown in Figure 10.3▶. A meter stick is used to measure the height of the mercury above the bottom of the manometer. The level of mercury in the open-end arm of the manometer has a height of 136.4 mm, and the mercury in the arm that is in contact with the gas has a height of 103.8 mm. What is the pressure of the gas **(a)** in atmospheres, **(b)** in kPa?

SOLUTION

Analyze: We are given the atmospheric pressure (764.7 torr) and the heights of the mercury in the two arms of the manometer and asked to determine the gas pressure in the flask. We know that this pressure must be greater than atmospheric because the manometer level on the flask side (103.8 mm) is lower than that on the side open to the atmosphere (136.4 mm), as indicated in Figure 10.3.

Plan: We will use the difference in height between the two arms (h in Figure 10.3) to obtain the amount by which the pressure of the gas exceeds atmospheric pressure. Because an open-end mercury manometer is used, the height difference directly measures the pressure difference in mm Hg or torr between the gas and the atmosphere.

P_{atm}

Open end

h

P_{gas}

$$P_{gas} = P_{atm} + P_h$$

▲ **Figure 10.3 A mercury manometer.** This device is sometimes employed in the laboratory to measure gas pressures near atmospheric pressure.

Solve:
(a) The pressure of the gas equals the atmospheric pressure plus h:

$$P_{gas} = P_{atm} + h$$
$$= 764.7 \text{ torr} + (136.4 \text{ torr} - 103.8 \text{ torr})$$
$$= 797.3 \text{ torr}$$

We convert the pressure of the gas to atmospheres:

$$P_{gas} = (797.3 \text{ torr})\left(\frac{1 \text{ atm}}{760 \text{ torr}}\right) = 1.049 \text{ atm}$$

(b) To calculate the pressure in kPa, we employ the conversion factor between atmospheres and kPa:

$$1.049 \text{ atm}\left(\frac{101.3 \text{ kPa}}{1 \text{ atm}}\right) = 106.3 \text{ kPa}$$

Check: The calculated pressure is a bit more than one atmosphere. This makes sense because we anticipated that the pressure in the flask would be greater than the pressure of the atmosphere acting on the manometer, which is a bit greater than one standard atmosphere.

■ **PRACTICE EXERCISE**
Convert a pressure of 0.975 atm into Pa and kPa.
Answers: 9.88×10^4 Pa and 98.8 kPa

 Chemistry and Life   BLOOD PRESSURE

The human heart pumps blood to the parts of the body through arteries, and the blood returns to the heart through veins. When your blood pressure is measured, two values are reported, such as 120/80 (120 over 80), which is a normal reading. The first measurement is the *systolic pressure*, the maximum pressure when the heart is pumping. The second is the *diastolic pressure*, the pressure when the heart is in the resting part of its pumping cycle. The units associated with these pressure measurements are torr.

Blood pressure is measured using a pressure gauge attached to a closed, air-filled jacket or cuff that is applied like a tourniquet to the arm (Figure 10.4▶). The pressure gauge may be a mercury manometer or some other device. The air pressure in the cuff is increased using a small pump until it is above the systolic pressure and prevents the flow of blood. The air pressure inside the cuff is then slowly reduced until blood just begins to pulse through the artery, as detected by the use of a stethoscope. At this point the pressure in the cuff equals the pressure that the blood exerts inside the arteries. Reading the gauge gives the systolic pressure. The pressure in the cuff is then reduced further until the blood flows freely. The pressure at this point is the diastolic pressure.

▲ **Figure 10.4 Measuring blood pressure.**

Hypertension is the presence of abnormally high blood pressure. The usual criterion for hypertension is a blood pressure greater than 140/90, although recent studies suggest that health risks increase for systolic readings above 120. Hypertension significantly increases the workload on the heart and places a stress on the walls of the blood vessels throughout the body. These effects increase the risk of aneurysms, heart attacks, and strokes.

10.3 THE GAS LAWS

Experiments with a large number of gases reveal that four variables are needed to define the physical condition, or *state*, of a gas: temperature, T, pressure, P, volume, V, and the amount of gas, which is usually expressed as the number of moles, n. The equations that express the relationships among T, P, V, and n are known as the *gas laws*. Because volume is easily measured, the first gas laws to be studied expressed the effect of one of the variables on volume with the remaining two variables held constant.

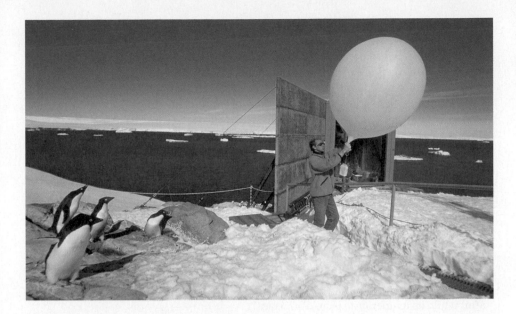

◀ Figure 10.5 **An application of the pressure-volume relationship.** The volume of gas in this weather balloon will increase as it ascends into the high atmosphere, where the atmospheric pressure is lower than on Earth's surface.

The Pressure-Volume Relationship: Boyle's Law

If the pressure on a balloon is decreased, the balloon expands. That is why weather balloons expand as they rise through the atmosphere (Figure 10.5▲). Conversely, when a volume of gas is compressed, the pressure of the gas increases. British chemist Robert Boyle (1627–1691) first investigated the relationship between the pressure of a gas and its volume.

To perform his gas experiments, Boyle used a J-shaped tube like that shown in Figure 10.6▶. In the tube on the left, a quantity of gas is trapped above a column of mercury. Boyle changed the pressure on the gas by adding mercury to the tube. He found that the volume of the gas decreased as the pressure increased. For example, doubling the pressure caused the gas volume to decrease to half its original value.

Boyle's law, which summarizes these observations, states that *the volume of a fixed quantity of gas maintained at constant temperature is inversely proportional to the pressure.* When two measurements are inversely proportional, one gets smaller as the other gets larger. Boyle's law can be expressed mathematically as

$$V = \text{constant} \times \frac{1}{P} \quad \text{or} \quad PV = \text{constant} \qquad [10.2]$$

The value of the constant depends on the temperature and the amount of gas in the sample. The graph of V versus P in Figure 10.7(a)▼ shows the type of curve obtained for a given quantity of gas at a fixed temperature. A linear relationship is obtained when V is plotted versus $1/P$ [Figure 10.7(b)].

(a)   (b)

▲ Figure 10.6 **An illustration of Boyle's experiment relating pressure and volume.** In (a) the volume of the gas trapped in the J-tube is 60 mL when the gas pressure is 760 torr. When additional mercury is added, as shown in (b), the trapped gas is compressed. The volume is 30 mL when its total pressure is 1520 torr, corresponding to atmospheric pressure plus the pressure exerted by the 760-mm column of mercury.

(a)

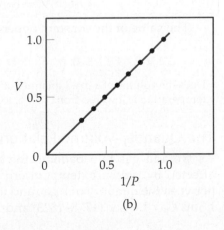

(b)

◀ Figure 10.7 **Graphs based on Boyle's law.** (a) Volume versus pressure. (b) Volume versus $1/P$.

▲ **Figure 10.8 An illustration of the effect of temperature on volume.** As liquid nitrogen (−196 °C) is poured over a balloon, the gas in the balloon is cooled and its volume decreases.

▲ **Figure 10.9 Graph based on Charles's law.** At constant pressure, the volume of an enclosed gas increases as the temperature increases. The dashed line is an extrapolation to temperatures at which the substance is no longer a gas.

Simple though it is, Boyle's law occupies a special place in the history of science. Boyle was the first to carry out a series of experiments in which one variable was systematically changed to determine the effect on another variable. The data from the experiments were then employed to establish an empirical relationship—a "law."

We apply Boyle's law every time we breathe. The rib cage, which can expand and contract, and the diaphragm, a muscle beneath the lungs, govern the volume of the lungs. Inhalation occurs when the rib cage expands and the diaphragm moves downward. Both of these actions increase the volume of the lungs, thus decreasing the gas pressure inside the lungs. The atmospheric pressure then forces air into the lungs until the pressure in the lungs equals atmospheric pressure. Exhalation reverses the process—the rib cage contracts and the diaphragm moves up, both of which decrease the volume of the lungs. Air is forced out of the lungs by the resulting increase in pressure.

▲ GIVE IT SOME THOUGHT

What happens to the volume of a gas if you double its pressure, say from 1 atm to 2 atm, while its temperature is held constant?

The Temperature-Volume Relationship: Charles's Law

Hot-air balloons rise because air expands as it is heated. The warm air in the balloon is less dense than the surrounding cool air at the same pressure. This difference in density causes the balloon to ascend. Conversely, a balloon will shrink when the gas in it is cooled, as seen in Figure 10.8 ◀.

The relationship between gas volume and temperature was discovered in 1787 by the French scientist Jacques Charles (1746–1823). Charles found that the volume of a fixed quantity of gas at constant pressure increases linearly with temperature. Some typical data are shown in Figure 10.9 ◀. Notice that the extrapolated (extended) line (which is dashed) passes through −273 °C. Note also that the gas is predicted to have zero volume at this temperature. This condition is never realized, however, because all gases liquefy or solidify before reaching this temperature.

In 1848 William Thomson (1824–1907), a British physicist whose title was Lord Kelvin, proposed an absolute-temperature scale, now known as the Kelvin scale. On this scale 0 K, which is called *absolute zero*, equals −273.15 °C. ∞ (Section 1.4) In terms of the Kelvin scale, **Charles's law** can be stated as follows: *The volume of a fixed amount of gas maintained at constant pressure is directly proportional to its absolute temperature.* Thus, doubling the absolute temperature, say from 200 K to 400 K, causes the gas volume to double. Mathematically, Charles's law takes the following form:

$$V = \text{constant} \times T \quad \text{or} \quad \frac{V}{T} = \text{constant} \qquad [10.3]$$

The value of the constant depends on the pressure and amount of gas.

▲ GIVE IT SOME THOUGHT

Does the volume of a fixed quantity of gas decrease to half its original value when the temperature is lowered from 100 °C to 50 °C?

The Quantity-Volume Relationship: Avogadro's Law

As we add gas to a balloon, the balloon expands. The volume of a gas is affected by pressure, temperature, and the amount of gas. The relationship between the quantity of a gas and its volume follows from the work of Joseph Louis Gay-Lussac (1778–1823) and Amedeo Avogadro (1776–1856).

Observation	Two volumes hydrogen		One volume oxygen		Two volumes water vapor
Explanation		+		→	
Equation	$2\,H_2(g)$	+	$O_2(g)$	⟶	$2\,H_2O(g)$

◀ Figure 10.10 **The law of combining volumes.** Gay-Lussac's experimental observation of combining volumes is shown together with Avogadro's explanation of this phenomenon.

Gay-Lussac is one of those extraordinary figures in the history of science who could truly be called an adventurer. He was interested in lighter-than-air balloons, and in 1804 he made an ascent to 23,000 ft—an exploit that held the altitude record for several decades. To better control lighter-than-air balloons, Gay-Lussac carried out several experiments on the properties of gases. In 1808 he observed the *law of combining volumes*: At a given pressure and temperature, the volumes of gases that react with one another are in the ratios of small whole numbers. For example, two volumes of hydrogen gas react with one volume of oxygen gas to form two volumes of water vapor, as shown in Figure 10.10▲.

Three years later Amedeo Avogadro interpreted Gay-Lussac's observation by proposing what is now known as **Avogadro's hypothesis:** *Equal volumes of gases at the same temperature and pressure contain equal numbers of molecules.* For example, experiments show that 22.4 L of any gas at 0 °C and 1 atm contain 6.02×10^{23} gas molecules (that is, 1 mol), as depicted in Figure 10.11▶.

Avogadro's law follows from Avogadro's hypothesis: *The volume of a gas maintained at constant temperature and pressure is directly proportional to the number of moles of the gas.* That is,

$$V = \text{constant} \times n \qquad [10.4]$$

Thus, doubling the number of moles of gas will cause the volume to double if T and P remain constant.

	He	N₂	CH₄
Volume	22.4 L	22.4 L	22.4 L
Pressure	1 atm	1 atm	1 atm
Temperature	0 °C	0 °C	0 °C
Mass of gas	4.00 g	28.0 g	16.0 g
Number of gas molecules	6.02×10^{23}	6.02×10^{23}	6.02×10^{23}

▲ Figure 10.11 **A comparison illustrating Avogadro's hypothesis.** Note that helium gas consists of helium atoms. Each gas has the same volume, temperature, and pressure and thus contains the same number of molecules. Because a molecule of one substance differs in mass from a molecule of another, the masses of gas in the three containers differ.

■ SAMPLE EXERCISE 10.3 │ Evaluating the Effects of Changes in P, V, n, and T on a Gas

Suppose we have a gas confined to a cylinder as shown in Figure 10.12▶. Consider the following changes: **(a)** Heat the gas from 298 K to 360 K, while maintaining the piston in the position shown in the drawing. **(b)** Move the piston to reduce the volume of gas from 1 L to 0.5 L. **(c)** Inject additional gas through the gas inlet valve. Indicate how each of these changes will affect the average distance between molecules, the pressure of the gas, and the number of moles of gas present in the cylinder.

SOLUTION

Analyze: We need to think how each of three different changes in the system affects (1) the distance between molecules, (2) the pressure of the gas, and (3) the number of moles of gas in the cylinder.

Plan: We will use our understanding of the gas laws and the general properties of gases to analyze each situation.

Solve:

(a) Heating the gas while maintaining the position of the piston will cause no change in the number of molecules per unit volume. Thus, the distance between molecules and the total moles of gas remain the same. The increase in temperature, however, will cause the pressure to increase (Charles's law).

▲ Figure 10.12 **Cylinder with piston and gas inlet valve.**

(b) Moving the piston compresses the same quantity of gas into a smaller volume. The total number of molecules of gas, and thus the total number of moles, remains the same. The average distance between molecules, however, must decrease because of the smaller volume in which the gas is confined. The reduction in volume causes the pressure to increase (Boyle's law).

(c) Injecting more gas into the cylinder while keeping the volume and temperature the same will result in more molecules and thus a greater number of moles of gas. The average distance between atoms must decrease because their number per unit volume increases. Correspondingly, the pressure increases (Avogadro's law).

■ **PRACTICE EXERCISE**

What happens to the density of a gas as **(a)** the gas is heated in a constant-volume container; **(b)** the gas is compressed at constant temperature; **(c)** additional gas is added to a constant-volume container?
Answer: **(a)** no change, **(b)** increase, **(c)** increase

10.4 THE IDEAL-GAS EQUATION

In Section 10.3 we examined three historically important gas laws that describe the relationships between the four variables, P, V, T, and n, that define the state of a gas. Each law was obtained by holding two variables constant to see how the remaining two variables affect each other. We can express each law as a proportionality relationship. Using the symbol $\propto$, which is read "is proportional to," we have

Boyle's law:	$V \propto \dfrac{1}{P}$	(constant n, T)
Charles's law:	$V \propto T$	(constant n, P)
Avogadro's law:	$V \propto n$	(constant P, T)

We can combine these relationships to make a more general gas law.

$$V \propto \frac{nT}{P}$$

If we call the proportionality constant R, we obtain

$$V = R\left(\frac{nT}{P}\right)$$

Rearranging, we have this relationship in its more familiar form:

$$PV = nRT \qquad [10.5]$$

This equation is known as the **ideal-gas equation**. An **ideal gas** is a hypothetical gas whose pressure, volume, and temperature behavior are described completely by the ideal-gas equation.

The term R in the ideal-gas equation is called the **gas constant**. The value and units of R depend on the units of P, V, n, and T. Temperature must *always* be expressed as an absolute temperature when used in the ideal-gas equation. The quantity of gas, n, is normally expressed in moles. The units chosen for pressure and volume are most often atm and liters, respectively. However, other units can be used. In most countries other than the United States, the SI unit of Pa (or kPa) is most commonly employed. Table 10.2 ◀ shows the numerical value for R in various units. As we saw in the "A Closer Look" box on P-V work in Section 5.3, the product PV has the units of energy. Therefore, the units of R can include joules or calories. In working problems with the ideal-gas equation, the units of P, V, n, and T must agree with the units in the gas constant. In this chapter we will most often use the value $R = 0.08206$ L-atm/mol-K (four significant figures) or 0.0821 L-atm/mol-K (three significant figures) whenever we use the ideal-gas equation, consistent with the units of atm for pressure. Use of the value $R = 8.314$ J/mol-K, consistent with the units of Pa for pressure, is also very common.

TABLE 10.2 ■ Numerical Values of the Gas Constant, *R*, in Various Units

Units	Numerical Value
L-atm/mol-K	0.08206
J/mol-K*	8.314
cal/mol-K	1.987
m³-Pa/mol-K*	8.314
L-torr/mol-K	62.36

*SI unit

Suppose we have 1.000 mol of an ideal gas at 1.000 atm and 0.00 °C (273.15 K). According to the ideal-gas equation, the volume of the gas is

$$V = \frac{nRT}{P} = \frac{(1.000 \text{ mol})(0.08206 \text{ L-atm/mol-K})(273.15 \text{ K})}{1.000 \text{ atm}} = 22.41 \text{ L}$$

The conditions 0 °C and 1 atm are referred to as the **standard temperature and pressure (STP)**. Many properties of gases are tabulated for these conditions. The volume occupied by one mole of ideal gas at STP, 22.41 L, is known as the *molar volume* of an ideal gas at STP.

GIVE IT SOME THOUGHT

How many molecules are in 22.41 L of an ideal gas at STP?

▼ Figure 10.13 **Comparison of molar volumes at STP.** One mole of an ideal gas at STP occupies a volume of 22.41 L. One mole of various real gases at STP occupies close to this ideal volume.

The ideal-gas equation accounts adequately for the properties of most gases under a wide variety of circumstances. The equation is not exactly correct, however, for any real gas. Thus, the measured volume, V, for given conditions of P, n, and T, might differ from the volume calculated from $PV = nRT$. To illustrate, the measured molar volumes of real gases at STP are compared with the calculated volume of an ideal gas in Figure 10.13▶. While these real gases do not match the ideal gas behavior exactly, the differences are so small that we can ignore them for all but the most accurate work. We will have more to say about the differences between ideal and real gases in Section 10.9.

■■ SAMPLE EXERCISE 10.4 │ Using the Ideal-Gas Equation

Calcium carbonate, $CaCO_3(s)$, decomposes upon heating to give $CaO(s)$ and $CO_2(g)$. A sample of $CaCO_3$ is decomposed, and the carbon dioxide is collected in a 250-mL flask. After the decomposition is complete, the gas has a pressure of 1.3 atm at a temperature of 31 °C. How many moles of CO_2 gas were generated?

SOLUTION

Analyze: We are given the volume (250 mL), pressure (1.3 atm), and temperature 31 °C of a sample of CO_2 gas and asked to calculate the number of moles of CO_2 in the sample.

Plan: Because we are given V, P, and T, we can solve the ideal-gas equation for the unknown quantity, n.

Solve: In analyzing and solving gas-law problems, it is helpful to tabulate the information given in the problems and then to convert the values to units that are consistent with those for R (0.0821 L-atm/mol-K). In this case the given values are

$V = 250 \text{ mL} = 0.250 \text{ L}$

$P = 1.3 \text{ atm}$

$T = 31 \text{ °C} = (31 + 273) \text{ K} = 304 \text{ K}$

Remember: *Absolute temperature must always be used when the ideal-gas equation is solved.*

We now rearrange the ideal-gas equation (Equation 10.5) to solve for n

$$n = \frac{PV}{RT}$$

$$n = \frac{(1.3 \text{ atm}) (0.250 \text{ L})}{(0.0821 \text{ L-atm/mol-K}) (304 \text{ K})} = 0.013 \text{ mol } CO_2$$

Check: Appropriate units cancel, thus ensuring that we have properly rearranged the ideal-gas equation and have converted to the correct units.

■■ PRACTICE EXERCISE

Tennis balls are usually filled with air or N_2 gas to a pressure above atmospheric pressure to increase their "bounce." If a particular tennis ball has a volume of 144 cm^3 and contains 0.33 g of N_2 gas, what is the pressure inside the ball at 24 °C?
Answer: 2.0 atm

Strategies in Chemistry   CALCULATIONS INVOLVING MANY VARIABLES

In chemistry and throughout your studies of science and math, you may encounter problems that involve several experimentally measured variables as well as several different physical constants. In this chapter we encounter a variety of problems based on the ideal-gas equation, which consists of four experimental quantities—P, V, n, and T—and one constant, R. Depending on the type of problem, we might need to solve for any of the four quantities.

To avoid any difficulty extracting the necessary information from problems when many variables are involved, we suggest you take the following steps as you analyze, plan, and solve such problems:

1. *Tabulate information.* Read the problems carefully to determine which quantity is the unknown and which quantities are given. Every time you encounter a numerical value, jot it down. In many cases, constructing a table of the given information will be useful.

2. *Convert to consistent units.* As you have already seen, we can use several different units to express the same quantity. Make certain that quantities are converted to the proper units by using the correct conversion factors. In using the ideal-gas equation, for example, we usually use the value of R that has units of L-atm/mol-K. If you are given a pressure in torr, you will need to convert it to atmospheres.

3. *If a single equation relates the variables, rearrange the equation to solve for the unknown.* Make certain that you are comfortable using algebra to solve the equation for the desired variable. In the case of the ideal-gas equation the following algebraic rearrangements will all be used at one time or another:

$$P = \frac{nRT}{V}; \quad V = \frac{nRT}{P}; \quad n = \frac{PV}{RT}; \quad T = \frac{PV}{nR}$$

4. *Use dimensional analysis.* Carry the units through your calculation. Use of dimensional analysis enables you to check that you have solved the equation correctly. If the units of the quantities in the equation cancel properly to give the units of the desired variable, you have probably used the equation correctly.

Sometimes you will not be given values for the necessary variables directly. Rather, you will be given the values of other quantities that can be used to determine the needed variables. For example, suppose you are trying to use the ideal-gas equation to calculate the pressure of a gas. You are given the temperature of the gas, but you are not given explicit values for n and V. However, the problem states that "the sample of gas contains 0.15 mol of gas per liter." We can turn this statement into the expression

$$\frac{n}{V} = 0.15 \text{ mol/L}$$

Solving the ideal-gas equation for pressure yields

$$P = \frac{nRT}{V} = \left(\frac{n}{V}\right)RT$$

Thus, we can solve the equation even though we are not given specific values for n and V. We will examine how to use the density and molar mass of a gas in this fashion in Section 10.5.

As we have continuously stressed, the most important thing you can do to become proficient at solving problems is to practice by using practice exercises and assigned exercises at the end of each chapter. By using systematic procedures, such as those described here, you should be able to minimize difficulties in solving problems involving many variables.

Relating the Ideal-Gas Equation and the Gas Laws

The simple gas laws that we discussed in Section 10.3, such as Boyle's law, are special cases of the ideal-gas equation. For example, when the quantity of gas and the temperature are held constant, n and T have fixed values. Therefore, the product nRT is the product of three constants and must itself be a constant.

$$PV = nRT = \text{constant} \quad \text{or} \quad PV = \text{constant} \qquad [10.6]$$

Thus, we have Boyle's law. We see that if n and T are constant, the individual values of P and V can change, but the product PV must remain constant.

We can use Boyle's law to determine how the volume of a gas changes when its pressure changes. For example, if a metal cylinder holds 50.0 L of O_2 gas at 18.5 atm and 21 °C, what volume will the gas occupy if the temperature is maintained at 21 °C while the pressure is reduced to 1.00 atm? Because the product PV is a constant when a gas is held at constant n and T, we know that

$$P_1V_1 = P_2V_2 \qquad [10.7]$$

where P_1 and V_1 are initial values and P_2 and V_2 are final values. Dividing both sides of this equation by P_2 gives the final volume, V_2.

$$V_2 = V_1 \times \frac{P_1}{P_2}$$

Substituting the given quantities into this equation gives

$$V_2 = (50.0 \text{ L})\left(\frac{18.5 \text{ atm}}{1.00 \text{ atm}}\right) = 925 \text{ L}$$

The answer is reasonable because gases expand as their pressures are decreased.

In a similar way we can start with the ideal-gas equation and derive relationships between any other two variables, V and T (Charles's law), n and V (Avogadro's law), or P and T. Sample Exercise 10.5 illustrates how these relationships can be derived and used.

■■ **SAMPLE EXERCISE 10.5** | **Calculating the Effect of Temperature Changes on Pressure**

The gas pressure in an aerosol can is 1.5 atm at 25 °C. Assuming that the gas inside obeys the ideal-gas equation, what would the pressure be if the can were heated to 450 °C?

SOLUTION

Analyze: We are given the initial pressure (1.5 atm) and temperature (25 °C) of the gas and asked for the pressure at a higher temperature (450 °C).

Plan: The volume and number of moles of gas do not change, so we must use a relationship connecting pressure and temperature. Converting temperature to the Kelvin scale and tabulating the given information, we have

	P	T
INITIAL	1.5 atm	298 K
FINAL	P_2	723 K

Solve: To determine how P and T are related, we start with the ideal-gas equation and isolate the quantities that do not change (n, V, and R) on one side and the variables (P and T) on the other side.

$$\frac{P}{T} = \frac{nR}{V} = \text{constant}$$

Because the quotient P/T is a constant, we can write

$$\frac{P_1}{T_1} = \frac{P_2}{T_2}$$

(where the subscripts 1 and 2 represent the initial and final states, respectively). Rearranging to solve for P_2 and substituting the given data give

$$P_2 = P_1 \times \frac{T_2}{T_1}$$

$$P_2 = (1.5 \text{ atm})\left(\frac{723 \text{ K}}{298 \text{ K}}\right) = 3.6 \text{ atm}$$

Check: This answer is intuitively reasonable—increasing the temperature of a gas increases its pressure.

Comment: It is evident from this example why aerosol cans carry a warning not to incinerate.

■ **PRACTICE EXERCISE**

A large natural-gas storage tank is arranged so that the pressure is maintained at 2.20 atm. On a cold day in December when the temperature is −15 °C (4 °F), the volume of gas in the tank is 3.25×10^3 m^3. What is the volume of the same quantity of gas on a warm July day when the temperature is 31 °C (88 °F)?
Answer: 3.83×10^3 m^3

We are often faced with the situation in which P, V, and T all change for a fixed number of moles of gas. Because n is constant under these circumstances, the ideal-gas equation gives

$$\frac{PV}{T} = nR = \text{constant}$$

If we represent the initial and final conditions of pressure, temperature, and volume by subscripts 1 and 2, respectively, we can write

$$\frac{P_1 V_1}{T_1} = \frac{P_2 V_2}{T_2} \qquad\qquad [10.8]$$

This equation is often called the *combined gas law*.

■ **SAMPLE EXERCISE 10.6** | Calculating the Effect of Changing *P* and *T* on the Volume of a Gas

An inflated balloon has a volume of 6.0 L at sea level (1.0 atm) and is allowed to ascend in altitude until the pressure is 0.45 atm. During ascent the temperature of the gas falls from 22 °C to −21 °C. Calculate the volume of the balloon at its final altitude.

SOLUTION

Analyze: We need to determine a new volume for a gas sample in a situation where both pressure and temperature change.

Plan: Let's again proceed by converting temperature to the Kelvin scale and tabulating the given information.

	P	*V*	*T*
INITIAL	1.0 atm	6.0 L	295 K
FINAL	0.45 atm	V_2	252 K

Because *n* is constant, we can use Equation 10.8.

Solve: Rearranging Equation 10.8 to solve for V_2 gives

$$V_2 = V_1 \times \frac{P_1}{P_2} \times \frac{T_2}{T_1} = (6.0 \text{ L}) \left(\frac{1.0 \text{ atm}}{0.45 \text{ atm}} \right) \left(\frac{252 \text{ K}}{295 \text{ K}} \right) = 11 \text{ L}$$

Check: The result appears reasonable. Notice that the calculation involves multiplying the initial volume by a ratio of pressures and a ratio of temperatures. Intuitively, we expect that decreasing pressure will cause the volume to increase. Similarly, decreasing temperature should cause the volume to decrease. Note that the difference in pressures is more dramatic than the difference in temperatures. Thus, we should expect the effect of the pressure change to predominate in determining the final volume, as it does.

■ **PRACTICE EXERCISE**

A 0.50-mol sample of oxygen gas is confined at 0 °C in a cylinder with a movable piston, such as that shown in Figure 10.12. The gas has an initial pressure of 1.0 atm. The piston then compresses the gas so that its final volume is half the initial volume. The final pressure of the gas is 2.2 atm. What is the final temperature of the gas in degrees Celsius?
Answer: 27 °C

10.5 FURTHER APPLICATIONS OF THE IDEAL-GAS EQUATION

The ideal-gas equation can be used to determine many relationships involving the physical properties of gases. In this section we use it first to define the relationship between the density of a gas and its molar mass, and then to calculate the volumes of gases formed or consumed in chemical reactions.

Gas Densities and Molar Mass

The ideal-gas equation allows us to calculate gas density from the molar mass, pressure, and temperature of the gas. Recall that density has the units of mass per unit volume ($d = m/V$). ∞ (Section 1.4) We can arrange the gas equation to obtain similar units, moles per unit volume, n/V:

$$\frac{n}{V} = \frac{P}{RT}$$

If we multiply both sides of this equation by the molar mass, $\mathcal{M}$, which is the number of grams in one mole of a substance, we obtain the following relationship:

$$\frac{n\mathcal{M}}{V} = \frac{P\mathcal{M}}{RT} \qquad [10.9]$$

The product of the quantities n/V and $\mathcal{M}$ equals the density in g/L, as seen from their units:

$$\frac{\text{moles}}{\text{liter}} \times \frac{\text{grams}}{\text{mole}} = \frac{\text{grams}}{\text{liter}}$$

Thus, the density, d, of the gas is given by the expression on the right in Equation 10.9:

$$d = \frac{P\mathcal{M}}{RT} \qquad\qquad [10.10]$$

From Equation 10.10 we see that the density of a gas depends on its pressure, molar mass, and temperature. The higher the molar mass and pressure, the more dense the gas. The higher the temperature, the less dense the gas. Although gases form homogeneous mixtures regardless of their identities, a less dense gas will lie above a more dense gas in the absence of mixing. For example, CO_2 has a higher molar mass than N_2 or O_2 and is therefore more dense than air. When CO_2 is released from a CO_2 fire extinguisher, as shown in Figure 10.14▶, it blankets a fire, preventing O_2 from reaching the combustible material. The fact that a hotter gas is less dense than a cooler one explains why hot air rises. The difference between the densities of hot and cold air is responsible for the lift of hot-air balloons. It is also responsible for many phenomena in weather, such as the formation of large thunderhead clouds during thunderstorms.

▲ **Figure 10.14 A CO_2 fire extinguisher.** The CO_2 gas from a fire extinguisher is denser than air. The CO_2 cools significantly as it emerges from the tank. Water vapor in the air is condensed by the cool CO_2 gas and forms a white fog accompanying the colorless CO_2.

▲ **GIVE IT SOME THOUGHT**

Is water vapor more or less dense than N_2 under the same conditions of temperature and pressure?

■ **SAMPLE EXERCISE 10.7** | Calculating Gas Density

What is the density of carbon tetrachloride vapor at 714 torr and 125 °C?

SOLUTION

Analyze: We are asked to calculate the density of a gas given its name, its pressure, and its temperature. From the name we can write the chemical formula of the substance and determine its molar mass.

Plan: We can use Equation 10.10 to calculate the density. Before we can use that equation, however, we need to convert the given quantities to the appropriate units. We must convert temperature to the Kelvin scale and pressure to atmospheres. We must also calculate the molar mass of CCl_4.

Solve: The temperature on the Kelvin scale is $125 + 273 = 398$ K.
The pressure in atmospheres is (714 torr)(1 atm/760 torr) = 0.939 atm.
The molar mass of CCl_4 is $12.0 + (4)(35.5) = 154.0$ g/mol.
Using these quantities along with Equation 10.10, we have

$$d = \frac{(0.939 \text{ atm})(154.0 \text{ g/mol})}{(0.0821 \text{ L-atm/mol-K})(398 \text{ K})} = 4.43 \text{ g/L}$$

Check: If we divide the molar mass (g/mol) by the density (g/L), we end up with L/mol. The numerical value is roughly $154/4.4 = 35$. That is in the right ballpark for the molar volume of a gas heated to 125 °C at near atmospheric pressure, so our answer is reasonable.

■ **PRACTICE EXERCISE**

The mean molar mass of the atmosphere at the surface of Titan, Saturn's largest moon, is 28.6 g/mol. The surface temperature is 95 K, and the pressure is 1.6 atm. Assuming ideal behavior, calculate the density of Titan's atmosphere.
Answer: 5.9 g/L

Equation 10.10 can be rearranged to solve for the molar mass of a gas:

$$\mathcal{M} = \frac{dRT}{P} \qquad\qquad [10.11]$$

Thus, we can use the experimentally measured density of a gas to determine the molar mass of the gas molecules, as shown in Sample Exercise 10.8.

■■ **SAMPLE EXERCISE 10.8** | Calculating the Molar Mass of a Gas

A series of measurements are made to determine the molar mass of an unknown gas. First, a large flask is evacuated and found to weigh 134.567 g. It is then filled with the gas to a pressure of 735 torr at 31 °C and reweighed. Its mass is now 137.456 g. Finally, the flask is filled with water at 31 °C and found to weigh 1067.9 g. (The density of the water at this temperature is 0.997 g/mL.) Assume that the ideal-gas equation applies, and calculate the molar mass of the unknown gas.

SOLUTION

Analyze: We are given the temperature (31 °C) and pressure (735 torr) for a gas, together with information to determine its volume and mass, and we are asked to calculate its molar mass.

Plan: We need to use the mass information given to calculate the volume of the container and the mass of the gas within it. From this we calculate the gas density and then apply Equation 10.11 to calculate the molar mass of the gas.

Solve: The mass of the gas is the difference between the mass of the flask filled with gas and that of the empty (evacuated) flask:

137.456 g − 134.567 g = 2.889 g

The volume of the gas equals the volume of water that the flask can hold. The volume of water is calculated from its mass and density. The mass of the water is the difference between the masses of the full and empty flask:

1067.9 g − 134.567 g = 933.3 g

By rearranging the equation for density ($d = m/V$), we have

$$V = \frac{m}{d} = \frac{(933.3 \text{ g})}{(0.997 \text{ g/mL})} = 936 \text{ mL} = 0.936 \text{ L}$$

Knowing the mass of the gas (2.889 g) and its volume (936 mL), we can calculate the density of the gas:

2.889 g/0.936 L = 3.09 g/L

After converting pressure to atmospheres and temperature to kelvins, we can use Equation 10.11 to calculate the molar mass:

$$\mathcal{M} = \frac{dRT}{P}$$

$$= \frac{(3.09 \text{ g/L})(0.0821 \text{ L-atm/mol-K})(304 \text{ K})}{(735/760) \text{ atm}}$$

$$= 79.7 \text{ g/mol}$$

Check: The units work out appropriately, and the value of molar mass obtained is reasonable for a substance that is gaseous near room temperature.

■■ **PRACTICE EXERCISE**

Calculate the average molar mass of dry air if it has a density of 1.17 g/L at 21 °C and 740.0 torr.
Answer: 29.0 g/mol

Volumes of Gases in Chemical Reactions

We are often concerned with knowing the identity of a gas involved as a reactant or product in a chemical reaction, as well as its quantity. Thus, it is useful to be able to calculate the volumes of gases consumed or produced in reactions. Such calculations are based on the use of the mole concept together with balanced chemical equations. ∞ (Section 3.6) We have seen that the coefficients in balanced chemical equations tell us the relative amounts (in moles) of reactants and products in a reaction. The ideal gas equation relates the number of moles of a gas, to P, V, and T.

Most people are quite unaware of the vast network of underground pipelines that undergirds the developed world. Pipelines are used to move massive quantities of liquids and gases over considerable distances. For example, pipelines move natural gas (methane) from huge natural-gas fields in Siberia to Western Europe. Natural gas from Algeria is moved to Italy through a pipeline 120 cm in diameter and 2500 km in length that stretches across the Mediterranean Sea at depths up to 600 m. In the United States the pipeline systems consist of trunk lines, large-diameter pipes for long-distance transport, with branch lines of lower diameter and lower pressure for local transport to and from the trunk lines.

Essentially all substances that are gases at STP are transported commercially by pipeline, including ammonia, carbon dioxide, carbon monoxide, chlorine, ethane, helium, hydrogen, and methane. The largest volume transport by far, though, is natural gas. (Figure 10.15▶) The methane-rich gas from oil and gas wells is processed to remove particulates, water, and various gaseous impurities such as hydrogen sulfide and carbon dioxide. The gas is then compressed to pressures ranging from 3.5 MPa (35 atm) to 10 MPa (100 atm), depending on the age and diameter of the pipe. The long-distance pipelines are about 40 cm in diameter and made of steel. Large compressor stations along the pipeline, spaced at 50- to 100-mile intervals, maintain pressure.

Recall from Figure 5.24 that natural gas is a major source of energy for the United States. To meet this demand, methane must be transported from source wells throughout the United

▲ Figure 10.15 **A natural gas pipeline junction.**

States and Canada to all parts of the nation. The total length of pipeline for natural-gas transport in the United States is about 6×10^5 km, and growing. The United States is divided into seven regions. The total deliverability of natural gas to the seven regions exceeds 2.7×10^{12} L (measured at STP), which is almost 100 billion cubic feet per day! The volume of the pipelines themselves would be entirely inadequate for managing the enormous quantities of natural gas that are placed into and taken out of the system on a continuing basis. For this reason, underground storage facilities, such as salt caverns and other natural formations, are employed to hold large quantities of gas.

Related Exercise: 10.117

■ SAMPLE EXERCISE 10.9 | Relating the Volume of a Gas to the Amount of Another Substance in a Reaction

The safety air bags in automobiles are inflated by nitrogen gas generated by the rapid decomposition of sodium azide, NaN_3:

$$2\,NaN_3(s) \longrightarrow 2\,Na(s) + 3\,N_2(g)$$

If an air bag has a volume of 36 L and is to be filled with nitrogen gas at a pressure of 1.15 atm at a temperature of 26.0 °C, how many grams of NaN_3 must be decomposed?

SOLUTION

Analyze: This is a multistep problem. We are given the volume, pressure, and temperature of the N_2 gas and the chemical equation for the reaction by which the N_2 is generated. We must use this information to calculate the number of grams of NaN_3 needed to obtain the necessary N_2.

Plan: We need to use the gas data (P, V, and T) and the ideal-gas equation to calculate the number of moles of N_2 gas that should be formed for the air bag to operate correctly. We can then use the balanced equation to determine the number of moles of NaN_3. Finally, we can convert the moles of NaN_3 to grams.

Gas data → mol N_2 → mol NaN_3 → g NaN_3

Solve: The number of moles of N_2 is determined using the ideal-gas equation:

$$n = \frac{PV}{RT} = \frac{(1.15\ \text{atm})(36\ \text{L})}{(0.0821\ \text{L-atm/mol-K})(299\ \text{K})} = 1.7\ \text{mol}\ N_2$$

From here we use the coefficients in the balanced equation to calculate the number of moles of NaN_3.

$$(1.7\ \text{mol}\ N_2)\left(\frac{2\ \text{mol}\ NaN_3}{3\ \text{mol}\ N_2}\right) = 1.1\ \text{mol}\ NaN_3$$

Finally, using the molar mass of NaN_3, we convert moles of NaN_3 to grams:

$$(1.1\ \text{mol}\ NaN_3)\left(\frac{65.0\ \text{g}\ NaN_3}{1\ \text{mol}\ NaN_3}\right) = 72\ \text{g}\ NaN_3$$

Check: The best way to check our approach is to make sure the units cancel properly at each step in the calculation, leaving us with the correct units in the answer, g NaN_3.

■ **PRACTICE EXERCISE**

In the first step in the industrial process for making nitric acid, ammonia reacts with oxygen in the presence of a suitable catalyst to form nitric oxide and water vapor:

$$4\,NH_3(g) + 5\,O_2(g) \longrightarrow 4\,NO(g) + 6\,H_2O(g)$$

How many liters of $NH_3(g)$ at 850 °C and 5.00 atm are required to react with 1.00 mol of $O_2(g)$ in this reaction?
Answer: 14.8 L

10.6 GAS MIXTURES AND PARTIAL PRESSURES

Thus far we have considered mainly the behavior of pure gases—those that consist of only one substance in the gaseous state. How do we deal with gases composed of a mixture of two or more different substances? While studying the properties of air, John Dalton (Section 2.1) made an important observation: The *total pressure of a mixture of gases equals the sum of the pressures that each would exert if it were present alone.* The pressure exerted by a particular component of a mixture of gases is called the **partial pressure** of that gas. Dalton's observation is known as **Dalton's law of partial pressures**.

▲ **GIVE IT SOME THOUGHT**

How is the pressure exerted by N_2 gas affected when some O_2 is introduced into a container if the temperature and volume remain constant?

If we let P_t be the total pressure of a mixture of gases and P_1, P_2, P_3, and so forth be the partial pressures of the individual gases, we can write Dalton's law as follows:

$$P_t = P_1 + P_2 + P_3 + \cdots \qquad [10.12]$$

This equation implies that each gas in the mixture behaves independently of the others, as we can see by the following analysis. Let n_1, n_2, n_3, and so forth be the number of moles of each of the gases in the mixture and n_t be the total number of moles of gas ($n_t = n_1 + n_2 + n_3 + \cdots$).

If each of the gases obeys the ideal-gas equation, we can write

$$P_1 = n_1\left(\frac{RT}{V}\right); \quad P_2 = n_2\left(\frac{RT}{V}\right); \quad P_3 = n_3\left(\frac{RT}{V}\right); \quad \text{and so forth}$$

All the gases in the mixture are at the same temperature and occupy the same volume. Therefore, by substituting into Equation 10.12, we obtain

$$P_t = (n_1 + n_2 + n_3 + \cdots)\frac{RT}{V} = n_t\left(\frac{RT}{V}\right) \qquad [10.13]$$

That is, at constant temperature and constant volume the total pressure is determined by the total number of moles of gas present, whether that total represents just one substance or a mixture.

■ **SAMPLE EXERCISE 10.10** | **Applying Dalton's Law of Partial Pressures**

A gaseous mixture made from 6.00 g O_2 and 9.00 g CH_4 is placed in a 15.0-L vessel at 0 °C. What is the partial pressure of each gas, and what is the total pressure in the vessel?

SOLUTION

Analyze: We need to calculate the pressure for two different gases in the same volume and at the same temperature.

Plan: Because each gas behaves independently, we can use the ideal-gas equation to calculate the pressure that each would exert if the other were not present. The total pressure is the sum of these two partial pressures.

Solve: We must first convert the mass of each gas to moles:

$$n_{O_2} = (6.00 \text{ g } O_2)\left(\frac{1 \text{ mol } O_2}{32.0 \text{ g } O_2}\right) = 0.188 \text{ mol } O_2$$

$$n_{CH_4} = (9.00 \text{ g } CH_4)\left(\frac{1 \text{ mol } CH_4}{16.0 \text{ g } CH_4}\right) = 0.563 \text{ mol } CH_4$$

We can now use the ideal-gas equation to calculate the partial pressure of each gas:

$$P_{O_2} = \frac{n_{O_2} RT}{V} = \frac{(0.188 \text{ mol})(0.0821 \text{ L-atm/mol-K})(273 \text{ K})}{15.0 \text{ L}} = 0.281 \text{ atm}$$

$$P_{CH_4} = \frac{n_{CH_4} RT}{V} = \frac{(0.563 \text{ mol})(0.0821 \text{ L-atm/mol-K})(273 \text{ K})}{15.0 \text{ L}} = 0.841 \text{ atm}$$

According to Dalton's law (Equation 10.12), the total pressure in the vessel is the sum of the partial pressures:

$$P_t = P_{O_2} + P_{CH_4} = 0.281 \text{ atm} + 0.841 \text{ atm} = 1.122 \text{ atm}$$

Check: Performing rough estimates is good practice, even when you may not feel that you need to do it to check an answer. In this case a pressure of roughly 1 atm seems right for a mixture of about 0.2 mol O_2 (that is, 6/32) and a bit more than 0.5 mol CH_4 (that is, 9/16), together in a 15-L volume, because one mole of an ideal gas at 1 atm pressure and 0 °C occupies about 22 L.

■ **PRACTICE EXERCISE**

What is the total pressure exerted by a mixture of 2.00 g of H_2 and 8.00 g of N_2 at 273 K in a 10.0-L vessel?
Answer: 2.86 atm

Partial Pressures and Mole Fractions

Because each gas in a mixture behaves independently, we can relate the amount of a given gas in a mixture to its partial pressure. For an ideal gas, $P = nRT/V$ and so we can write

$$\frac{P_1}{P_t} = \frac{n_1 RT/V}{n_t RT/V} = \frac{n_1}{n_t} \qquad [10.14]$$

The ratio n_1/n_t is called the mole fraction of gas 1, which we denote X_1. The **mole fraction**, X, is a dimensionless number that expresses the ratio of the number of moles of one component to the total number of moles in the mixture. We can rearrange Equation 10.14 to give

$$P_1 = \left(\frac{n_1}{n_t}\right) P_t = X_1 P_t \qquad [10.15]$$

Thus, the partial pressure of a gas in a mixture is its mole fraction times the total pressure.

The mole fraction of N_2 in air is 0.78 (that is, 78% of the molecules in air are N_2). If the total barometric pressure is 760 torr, then the partial pressure of N_2 is

$$P_{N_2} = (0.78) (760 \text{ torr}) = 590 \text{ torr}$$

This result makes intuitive sense: Because N_2 makes up 78% of the mixture, it contributes 78% of the total pressure.

■ **SAMPLE EXERCISE 10.11** | Relating Mole Fractions and Partial Pessures

A study of the effects of certain gases on plant growth requires a synthetic atmosphere composed of 1.5 mol percent CO_2, 18.0 mol percent O_2, and 80.5 mol percent Ar. **(a)** Calculate the partial pressure of O_2 in the mixture if the total pressure of the atmosphere is to be 745 torr. **(b)** If this atmosphere is to be held in a 121-L space at 295 K, how many moles of O_2 are needed?

SOLUTION

Analyze: **(a)** We first need to calculate the partial pressure of O_2 given its mole percentage and the total pressure of the mixture. **(b)** We need to calculate the number of moles of O_2 in the mixture given its volume (121 L), temperature (745 torr), and partial pressure (from part (a)).

Plan: **(a)** We will calculate the partial pressures using Equation 10.15. **(b)** We will then use P_{O_2}, V, and T together with the ideal-gas equation to calculate the number of moles of O_2, n_{O_2}.

Solve: **(a)** The mole percent is just the mole fraction times 100. Therefore, the mole fraction of O_2 is 0.180. Using Equation 10.15, we have

$$P_{O_2} = (0.180)(745 \text{ torr}) = 134 \text{ torr}$$

(b) Tabulating the given variables and changing them to appropriate units, we have

$$P_{O_2} = (134 \text{ torr})\left(\frac{1 \text{ atm}}{760 \text{ torr}}\right) = 0.176 \text{ atm}$$

$$V = 121 \text{ L}$$

$$n_{O_2} = ?$$

$$R = 0.0821 \frac{\text{L-atm}}{\text{mol-K}}$$

$$T = 295 \text{ K}$$

Solving the ideal-gas equation for n_{O_2}, we have

$$n_{O_2} = P_{O_2}\left(\frac{V}{RT}\right)$$

$$= (0.176 \text{ atm})\frac{121 \text{ L}}{(0.0821 \text{ L-atm/mol-K})(295 \text{ K})} = 0.879 \text{ mol}$$

Check: The units check out satisfactorily, and the answer seems to be the right order of magnitude.

■ **PRACTICE EXERCISE**

From data gathered by *Voyager 1*, scientists have estimated the composition of the atmosphere of Titan, Saturn's largest moon. The total pressure on the surface of Titan is 1220 torr. The atmosphere consists of 82 mol percent N_2, 12 mol percent Ar, and 6.0 mol percent CH_4. Calculate the partial pressure of each of these gases in Titan's atmosphere.
Answer: 1.0×10^3 torr N_2, 1.5×10^2 torr Ar, and 73 torr CH_4

Collecting Gases over Water

An experiment that is often encountered in general chemistry laboratories involves determining the number of moles of gas collected from a chemical reaction. Sometimes this gas is collected over water. For example, solid potassium chlorate, $KClO_3$, can be decomposed by heating it in a test tube in an arrangement such as that shown in Figure 10.16▶. The balanced equation for the reaction is

$$2 \text{ KClO}_3(s) \longrightarrow 2 \text{ KCl}(s) + 3 \text{ O}_2(g) \qquad [10.16]$$

The oxygen gas is collected in a bottle that is initially filled with water and then inverted in a water pan.

The volume of gas collected is measured by raising or lowering the bottle as necessary until the water levels inside and outside the bottle are the same. When this condition is met, the pressure inside the bottle is equal to the atmospheric pressure outside. The total pressure inside is the sum of the pressure of gas collected and the pressure of water vapor in equilibrium with liquid water.

$$P_{total} = P_{gas} + P_{H_2O} \qquad [10.17]$$

The pressure exerted by water vapor, P_{H_2O}, at various temperatures is listed in Appendix B.

◄ Figure 10.16 **Collecting a water-insoluble gas over water.** (a) A solid is heated, releasing a gas, which is bubbled through water into a collection bottle. (b) When the gas has been collected, the bottle is raised or lowered so that the water levels inside and outside the bottle are equal. The total pressure of the gases inside the bottle is then equal to the atmospheric pressure.

Gas collection

Gas volume measurement

P_{total} P_{atm}

(a)   (b)

■ **SAMPLE EXERCISE 10.12** | Calculating the Amount of Gas Collected over Water

A sample of $KClO_3$ is partially decomposed (Equation 10.16), producing O_2 gas that is collected over water as in Figure 10.16. The volume of gas collected is 0.250 L at 26 °C and 765 torr total pressure. **(a)** How many moles of O_2 are collected? **(b)** How many grams of $KClO_3$ were decomposed?

SOLUTION

(a) Analyze: We need to calculate the number of moles of O_2 gas in a container that also contains water vapor.

Plan: If we tabulate the information presented, we will see that values are given for V and T. To use the ideal-gas equation to calculate the unknown, n_{O_2}, we also must know the partial pressure of O_2 in the system. We can calculate the partial pressure of O_2 from the total pressure (765 torr) and the vapor pressure of water.

Solve: The partial pressure of the O_2 gas is the difference between the total pressure, 765 torr, and the pressure of the water vapor at 26 °C, 25 torr (Appendix B):

$$P_{O_2} = 765 \text{ torr} - 25 \text{ torr} = 740 \text{ torr}$$

We can use the ideal-gas equation to calculate the number of moles of O_2:

$$n_{O_2} = \frac{P_{O_2} V}{RT} = \frac{(740 \text{ torr})(1 \text{ atm}/760 \text{ torr})(0.250 \text{ L})}{(0.0821 \text{ L-atm/mol-K})(299 \text{ K})} = 9.92 \times 10^{-3} \text{ mol } O_2$$

(b) Analyze: We now need to calculate the number of moles of reactant $KClO_3$ decomposed.

Plan: We can use the number of moles of O_2 formed and the balanced chemical equation to determine the number of moles of $KClO_3$ decomposed, which we can then convert to grams of $KClO_3$.

Solve: From Equation 10.16, we have 2 mol $KClO_3 \simeq$ 3 mol O_2. The molar mass of $KClO_3$ is 122.6 g/mol. Thus, we can convert the moles of O_2 that we found in part (a) to moles of $KClO_3$ and then to grams of $KClO_3$:

$$(9.92 \times 10^{-3} \text{ mol } O_2)\left(\frac{2 \text{ mol } KClO_3}{3 \text{ mol } O_2}\right)\left(\frac{122.6 \text{ g } KClO_3}{1 \text{ mol } KClO_3}\right) = 0.811 \text{ g } KClO_3$$

Check: As always, we make sure that the units cancel appropriately in the calculations. In addition, the numbers of moles of O_2 and $KClO_3$ seem reasonable, given the small volume of gas collected.

Comment: Many chemical compounds that react with water and water vapor would be degraded by exposure to wet gas. Thus, in research laboratories gases are often dried by passing wet gas over a substance that absorbs water (a *desiccant*), such as calcium sulfate, $CaSO_4$. Calcium sulfate crystals are sold as a desiccant under the trade name Drierite™.

■ **PRACTICE EXERCISE**

Ammonium nitrite, NH_4NO_2, decomposes upon heating to form N_2 gas:

$$NH_4NO_2(s) \longrightarrow N_2(g) + 2 H_2O(l)$$

When a sample of NH_4NO_2 is decomposed in a test tube, as in Figure 10.16, 511 mL of N_2 gas is collected over water at 26 °C and 745 torr total pressure. How many grams of NH_4NO_2 were decomposed?
Answer: 1.26 g

10.7 KINETIC-MOLECULAR THEORY

The ideal-gas equation describes *how* gases behave, but it does not explain *why* they behave as they do. Why does a gas expand when heated at constant pressure? Or, why does its pressure increase when the gas is compressed at constant temperature? To understand the physical properties of gases, we need a model that helps us picture what happens to gas particles as experimental conditions such as pressure or temperature change. Such a model, known as the **kinetic-molecular theory**, was developed over a period of about 100 years, culminating in 1857 when Rudolf Clausius (1822–1888) published a complete and satisfactory form of the theory.

The kinetic-molecular theory (the theory of moving molecules) is summarized by the following statements:

1. Gases consist of large numbers of molecules that are in continuous, random motion. (The word *molecule* is used here to designate the smallest particle of any gas; some gases, such as the noble gases, consist of individual atoms.)
2. The combined volume of all the molecules of the gas is negligible relative to the total volume in which the gas is contained.
3. Attractive and repulsive forces between gas molecules are negligible.
4. Energy can be transferred between molecules during collisions, but the *average* kinetic energy of the molecules does not change with time, as long as the temperature of the gas remains constant. In other words, the collisions are perfectly elastic.
5. The average kinetic energy of the molecules is proportional to the absolute temperature. At any given temperature the molecules of all gases have the same average kinetic energy.

The kinetic-molecular theory explains both pressure and temperature at the molecular level. The pressure of a gas is caused by collisions of the molecules with the walls of the container, as shown in Figure 10.17 ◄. The magnitude of the pressure is determined by both how often and how forcefully the molecules strike the walls.

The absolute temperature of a gas is a measure of the *average* kinetic energy of its molecules. If two different gases are at the same temperature, their molecules have the same average kinetic energy (statement 5 of the kinetic-molecular theory). If the absolute temperature of a gas is doubled, the average kinetic energy of its molecules doubles. Thus, molecular motion increases with increasing temperature.

▲ Figure 10.17 **The molecular origin of gas pressure.** The pressure exerted by a gas is caused by collisions of the gas molecules with the walls of their container.

GIVE IT SOME THOUGHT

Consider three samples of gas: HCl at 298 K, H_2 at 298 K, and O_2 at 350 K. Compare the average kinetic energies of the molecules in the three samples.

Distributions of Molecular Speed

Although the molecules in a sample of gas have an *average* kinetic energy and hence an average speed, the individual molecules move at varying speeds. The moving molecules collide frequently with other molecules. Momentum is conserved in each collision, but one of the colliding molecules might be deflected off at high speed while the other is nearly stopped. The result is that the molecules at any instant have a wide range of speeds. Figure 10.18 ► illustrates the distribution of molecular speeds for nitrogen gas at 0 °C (blue line) and at 100 °C (red line). The curve shows us the fraction of molecules moving at each speed. At higher temperatures, a larger fraction of molecules moves at greater speeds; the distribution curve has shifted to the right toward higher speeds and hence toward higher average kinetic energy. The peak of each curve represents the most probable speed (the speed of the largest number of molecules). Notice that the blue curve (0 °C) has a peak at about 4×10^2 m/s, whereas the red curve (100 °C) has a peak at a higher speed, about 5×10^2 m/s.

Figure 10.18 also shows the value of the **root-mean-square (rms) speed**, u, of the molecules at each temperature. This quantity is the speed of a molecule possessing average kinetic energy. The rms speed is not quite the same as the average (mean) speed. The difference between the two, however, is small.* Notice that the rms speed is higher at 100 °C than at 0 °C. Notice also that the distribution curve broadens as we go to a higher temperature.

The rms speed is important because the average kinetic energy of the gas molecules in a sample, ε, is related directly to u^2:

$$\varepsilon = \tfrac{1}{2} mu^2 \qquad [10.18]$$

where m is the mass of an individual molecule. Mass does not change with temperature. Thus, the increase in the average kinetic energy as the temperature increases implies that the rms speed (and also the average speed) of molecules likewise increases as temperature increases.

▲ Figure 10.18 **The effect of temperature on molecular speeds.** Distribution of molecular speeds for nitrogen at 0 °C (blue line) and 100 °C (red line). Increasing temperature increases both the most probable speed (curve maximum) and the rms speed, u, which is indicated by the vertical dashed line.

▲ **GIVE IT SOME THOUGHT**

Consider three samples of gas all at 298 K: HCl, H_2, and O_2. List the molecules in order of increasing average speed.

Application to the Gas Laws

The empirical observations of gas properties as expressed by the various gas laws are readily understood in terms of the kinetic-molecular theory. The following examples illustrate this point:

1. *Effect of a volume increase at constant temperature:* A constant temperature means that the average kinetic energy of the gas molecules remains unchanged. This in turn means that the rms speed of the molecules, u, is unchanged. If the volume is increased, however, the molecules must move a longer distance between collisions. Consequently, there are fewer collisions per unit time with the container walls, and pressure decreases. Thus, the model accounts in a simple way for Boyle's law.

2. *Effect of a temperature increase at constant volume:* An increase in temperature means an increase in the average kinetic energy of the molecules and thus an increase in u. If there is no change in volume, there will be more collisions with the walls per unit time. Furthermore, the change in momentum in each collision increases (the molecules strike the walls more forcefully). Hence, the model explains the observed pressure increase.

To illustrate the difference between rms speed and average speed, suppose that we have four objects with speeds of 4.0, 6.0, 10.0, and 12.0 m/s. Their average speed is $\tfrac{1}{4}(4.0 + 6.0 + 10.0 + 12.0) = 8.0$ m/s. The rms speed, u, however, is the square root of the average squared speeds of the molecules:

$$\sqrt{\tfrac{1}{4}(4.0^2 + 6.0^2 + 10.0^2 + 12.0^2)} = \sqrt{74.0} = 8.6 \text{ m/s}$$

For an ideal gas the average speed equals 0.921 × u. Thus, the average speed is directly proportional to the rms speed, and the two are in fact nearly equal.

A Closer Look | THE IDEAL-GAS EQUATION

Beginning with the five statements given in the text for the kinetic-molecular theory, it is possible to derive the ideal-gas equation. Rather than proceed through a derivation, let's consider in somewhat qualitative terms how the ideal-gas equation might follow. As we have seen, pressure is force per unit area. ∞ (Section 10.2) The total force of the molecular collisions on the walls and hence the pressure produced by these collisions depend both on how strongly the molecules strike the walls (impulse imparted per collision) and on the rate at which these collisions occur:

$$P \propto \text{impulse imparted per collision} \times \text{rate of collisions}$$

For a molecule traveling at the rms speed, u, the impulse imparted by a collision with a wall depends on the momentum of the molecule; that is, it depends on the product of its mass and speed, mu. The rate of collisions is proportional to both the number of molecules per unit volume, n/V, and their speed, u. If there are more molecules in a container, there will be more frequent collisions with the container walls. As the molecular speed increases or the volume of the container decreases, the time required for molecules to traverse the distance from one wall to another is reduced, and the molecules collide more frequently with the walls. Thus, we have

$$P \propto mu \times \frac{n}{V} \times u \propto \frac{nmu^2}{V} \qquad [10.19]$$

Because the average kinetic energy, $\frac{1}{2}mu^2$, is proportional to temperature, we have $mu^2 \propto T$. Making this substitution into Equation 10.19 gives

$$P \propto \frac{n(mu^2)}{V} \propto \frac{nT}{V} \qquad [10.20]$$

Let's now convert the proportionality sign to an equal sign by expressing n as the number of moles of gas. We then insert a proportionality constant, R, the gas constant:

$$P = \frac{nRT}{V} \qquad [10.21]$$

This expression is the ideal-gas equation.

An eminent Swiss mathematician, Daniel Bernoulli (1700–1782), conceived of a model for gases that was, for all practical purposes, the same as the kinetic theory model. From this model, Bernoulli derived Boyle's law and the ideal-gas equation. His was one of the first examples in science of developing a mathematical model from a set of assumptions, or hypothetical statements. However, Bernoulli's work on this subject was completely ignored, only to be rediscovered a hundred years later by Clausius and others. It was ignored because it conflicted with popular beliefs and was in conflict with Isaac Newton's incorrect model for gases. Those idols of the times had to fall before the way was clear for the kinetic-molecular theory. As this story illustrates, science is not a straight road running from here to the "truth." The road is built by humans, so it zigs and zags.

Related Exercises: 10.71, 10.72, 10.73, and 10.74

■■■ **SAMPLE EXERCISE 10.13** | Applying the Kinetic-Molecular Theory

A sample of O_2 gas initially at STP is compressed to a smaller volume at constant temperature. What effect does this change have on **(a)** the average kinetic energy of O_2 molecules, **(b)** the average speed of O_2 molecules, **(c)** the total number of collisions of O_2 molecules with the container walls in a unit time, **(d)** the number of collisions of O_2 molecules with a unit area of container wall per unit time?

SOLUTION

Analyze: We need to apply the concepts of the kinetic-molecular theory to a situation in which a gas is compressed at constant temperature.

Plan: We will determine how each of the quantities in (a)–(d) is affected by the change in volume at constant temperature.

Solve: **(a)** The average kinetic energy of the O_2 molecules is determined only by temperature. Thus the average kinetic energy is unchanged by the compression of O_2 at constant temperature. **(b)** If the average kinetic energy of O_2 molecules does not change, the average speed remains constant. **(c)** The total number of collisions with the container walls per unit time must increase because the molecules are moving within a smaller volume but with the same average speed as before. Under these conditions they must encounter a wall more frequently. **(d)** The number of collisions with a unit area of wall per unit time increases because the total number of collisions with the walls per unit time increases and the area of the walls decreases.

Check: In a conceptual exercise of this kind, there is no numerical answer to check. All we can check in such cases is our reasoning in the course of solving the problem.

■■■ **PRACTICE EXERCISE**

How is the rms speed of N_2 molecules in a gas sample changed by **(a)** an increase in temperature, **(b)** an increase in volume, **(c)** mixing with a sample of Ar at the same temperature?

Answers: **(a)** increases, **(b)** no effect, **(c)** no effect

10.8 MOLECULAR EFFUSION AND DIFFUSION

According to the kinetic-molecular theory, the average kinetic energy of *any* collection of gas molecules, $\frac{1}{2}mu^2$, has a specific value at a given temperature. Thus, a gas composed of lightweight particles, such as He, will have the same average kinetic energy as one composed of much heavier particles, such as Xe, provided the two gases are at the same temperature. The mass, m, of the particles in the lighter gas is smaller than that in the heavier gas. Consequently, the particles of the lighter gas must have a higher rms speed, u, than the particles of the heavier one. The following equation, which expresses this fact quantitatively, can be derived from kinetic-molecular theory:

$$u = \sqrt{\frac{3RT}{M}} \qquad [10.22]$$

▲ Figure 10.19 **The effect of molecular mass on molecular speeds.** The distributions of molecular speeds for different gases are compared at 25 °C. The molecules with lower molecular masses have higher rms speeds.

Because the molar mass, M, appears in the denominator, the less massive the gas molecules, the higher the rms speed, u.

Figure 10.19▲ shows the distribution of molecular speeds for several gases at 25 °C. Notice how the distributions are shifted toward higher speeds for gases of lower molar masses.

■■ **SAMPLE EXERCISE 10.14** | Calculating a Root-Mean-Square Speed

Calculate the rms speed, u, of an N_2 molecule at 25 °C.

SOLUTION

Analyze: We are given the identity of the gas and the temperature, the two quantities we need to calculate the rms speed.

Plan: We will calculate the rms speed using Equation 10.22.

Solve: In using Equation 10.22, we should convert each quantity to SI units so that all the units are compatible. We will also use R in units of J/mol-K (Table 10.2) to make the units cancel correctly.

$T = 25 + 273 = 298$ K

$M = 28.0$ g/mol $= 28.0 \times 10^{-3}$ kg/mol

$R = 8.314$ J/mol-K $= 8.314$ kg-m^2/s^2-mol-K   (These units follow from the fact that 1 J $= 1$ kg-m^2/s^2)

$$u = \sqrt{\frac{3(8.314 \text{ kg-m}^2/\text{s}^2\text{-mol-K})(298 \text{ K})}{28.0 \times 10^{-3} \text{ kg/mol}}} = 5.15 \times 10^2 \text{ m/s}$$

Comment: This corresponds to a speed of 1150 mi/hr. Because the average molecular weight of air molecules is slightly greater than that of N_2, the rms speed of air molecules is a little slower than that for N_2. The speed at which sound propagates through air is about 350 m/s, a value about two-thirds the average rms speed for air molecules.

■■ **PRACTICE EXERCISE**

What is the rms speed of an He atom at 25 °C?
Answer: 1.36×10^3 m/s

▲ Figure 10.20 Effusion. The top half of this cylinder is filled with a gas, and the bottom half is an evacuated space. Gas molecules effuse through a pinhole in the partitioning wall only when they happen to hit the hole.

The dependence of molecular speeds on mass has several interesting consequences. The first phenomenon is **effusion**, which is the escape of gas molecules through a tiny hole into an evacuated space, as shown in Figure 10.20 ◄. The second is **diffusion**, which is the spread of one substance throughout a space or throughout a second substance. For example, the molecules of a perfume diffuse throughout a room.

Graham's Law of Effusion

In 1846 Thomas Graham (1805–1869) discovered that the effusion rate of a gas is inversely proportional to the square root of its molar mass. Assume that we have two gases at the same temperature and pressure in containers with identical pinholes. If the rates of effusion of the two substances are r_1 and r_2 and their respective molar masses are $\mathcal{M}_1$ and $\mathcal{M}_2$, **Graham's law** states

$$\frac{r_1}{r_2} = \sqrt{\frac{\mathcal{M}_2}{\mathcal{M}_1}} \qquad\qquad [10.23]$$

Equation 10.23 compares the *rates* of effusion of two different gases under identical conditions, and it indicates that the lighter gas effuses more rapidly.

GRAHAM'S LAW OF EFFUSION

The effusion rate of a gas is inversely proportional to the square root of its molar mass. Gas effuses through pores of a balloon. At identical pressure and temperature, the lighter gas effuses more rapidly.

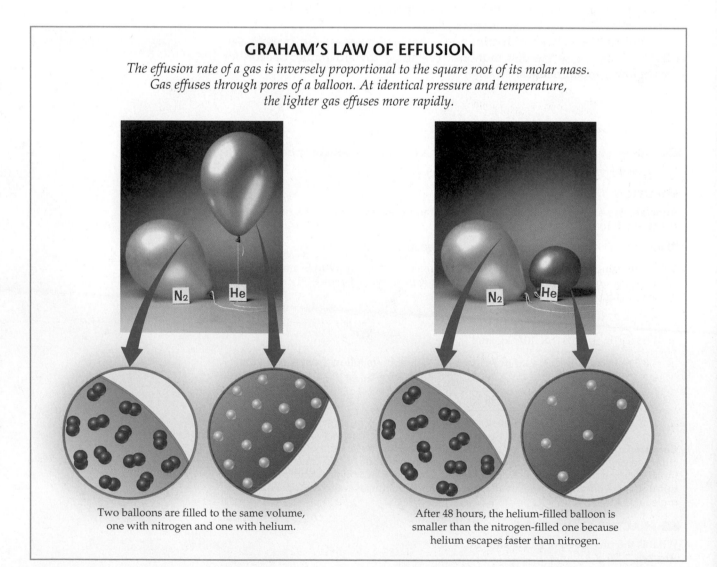

Two balloons are filled to the same volume, one with nitrogen and one with helium.

After 48 hours, the helium-filled balloon is smaller than the nitrogen-filled one because helium escapes faster than nitrogen.

▲ Figure 10.21 An illustration of Graham's law.

Figure 10.20 illustrates the basis of Graham's law. The only way for a molecule to escape from its container is for it to "hit" the hole in the partitioning wall. The faster the molecules are moving, the greater is the likelihood that a molecule will hit the hole and effuse. This implies that the rate of effusion is directly proportional to the rms speed of the molecules. Because R and T are constant, we have, from Equation 10.22

$$\frac{r_1}{r_2} = \frac{u_1}{u_2} = \sqrt{\frac{3RT/\mathcal{M}_1}{3RT/\mathcal{M}_2}} = \sqrt{\frac{\mathcal{M}_2}{\mathcal{M}_1}} \qquad [10.24]$$

As expected from Graham's law, helium escapes from containers through tiny pinhole leaks more rapidly than other gases of higher molecular weight (Figure 10.21 ◄).

■ SAMPLE EXERCISE 10.15 | Applying Graham's Law

An unknown gas composed of homonuclear diatomic molecules effuses at a rate that is only 0.355 times that of O_2 at the same temperature. Calculate the molar mass of the unknown, and identify it.

SOLUTION

Analyze: We are given the rate of effusion of an unknown gas relative to that of O_2, and we are asked to find the molar mass and identity of the unknown. Thus, we need to connect relative rates of effusion to relative molar masses.

Plan: We can use Graham's law of effusion, Equation 10.23, to determine the molar mass of the unknown gas. If we let r_x and $\mathcal{M}_x$ represent the rate of effusion and molar mass of the unknown gas, Equation 10.23 can be written as follows:

$$\frac{r_x}{r_{O_2}} = \sqrt{\frac{\mathcal{M}_{O_2}}{\mathcal{M}_x}}$$

Solve: From the information given,

$$r_x = 0.355 \times r_{O_2}$$

Thus,

$$\frac{r_x}{r_{O_2}} = 0.355 = \sqrt{\frac{32.0 \text{ g/mol}}{\mathcal{M}_x}}$$

We now solve for the unknown molar mass, $\mathcal{M}_x$

$$\frac{32.0 \text{ g/mol}}{\mathcal{M}_x} = (0.355)^2 = 0.126$$

$$\mathcal{M}_x = \frac{32.0 \text{ g/mol}}{0.126} = 254 \text{ g/mol}$$

Because we are told that the unknown gas is composed of homonuclear diatomic molecules, it must be an element. The molar mass must represent twice the atomic weight of the atoms in the unknown gas. We conclude that the unknown gas is I_2.

■ PRACTICE EXERCISE

Calculate the ratio of the effusion rates of N_2 and O_2, r_{N_2}/r_{O_2}.
Answer: $r_{N_2}/r_{O_2} = 1.07$

Diffusion and Mean Free Path

Diffusion, like effusion, is faster for lower mass molecules than for higher mass ones. In fact, Graham's law, Equation 10.23, approximates the ratio of rates of diffusion of two gases under identical experimental conditions. Nevertheless, molecular collisions make diffusion more complicated than effusion.

We can see from the horizontal scale in Figure 10.19 that the speeds of molecules are quite high. For example, the average speed of N_2 at room temperature is 515 m/s (1150 mi/hr). In spite of this high speed, if someone opens a vial of perfume at one end of a room, some time elapses—perhaps a few minutes—before the scent is detected at the other end of the room. The diffusion of gases is much slower than molecular speeds because of molecular collisions.* These

---

*The rate at which the perfume moves across the room also depends on how well stirred the air is from temperature gradients and the movement of people. Nevertheless, even with the aid of these factors, it still takes much longer for the molecules to traverse the room than one would expect from the rms speed alone.

Chemistry Put to Work   GAS SEPARATIONS

The fact that lighter molecules move at higher average speeds than more massive ones has many interesting consequences and applications. For example, the effort to develop the atomic bomb during World War II required scientists to separate the relatively low-abundance uranium isotope ^{235}U (0.7%) from the much more abundant ^{238}U (99.3%). This separation was accomplished by converting the uranium into a volatile compound, UF_6, that was then allowed to pass through porous barriers. Because of the diameters of the pores, this process is not a simple effusion. Nevertheless, the dependence on molar mass is essentially the same. The slight difference in molar mass between the two hexafluorides, $^{235}UF_6$ and $^{238}UF_6$, caused the molecules to move at slightly different rates:

$$\frac{r_{235}}{r_{238}} = \sqrt{\frac{352.04}{349.03}} = 1.0043$$

Thus, the gas initially appearing on the opposite side of the barrier was very slightly enriched in the lighter molecule. The effusion process was repeated thousands of times, leading to a nearly complete separation of the two isotopes of uranium.

Separation of uranium isotopes by effusion has been largely replaced by a technique that uses centrifuges. In this procedure, cylindrical rotors containing UF_6 vapor spin at high speed inside an evacuated casing. Molecules containing the heavier ^{238}U isotope move closer to the spinning walls, whereas the molecules containing the lighter ^{235}U isotope remain in the middle of the cylinders. A stream of gas moves the UF_6 from the center of one centrifuge into another. Plants that use centrifuges consume less energy and can be constructed in a more compact, modular fashion than those that rely on effusion. Such plants are frequently in the news as countries such as Iran and North Korea enrich uranium in the ^{235}U isotope for both nuclear power and nuclear weaponry.

Related Exercises: 10.79 and 10.80

▲ **Figure 10.22 Diffusion of a gas molecule.** For clarity, no other gas molecules in the container are shown. The path of the molecule of interest begins at the dot. Each short segment of line represents travel between collisions. The blue arrow indicates the net distance traveled by the molecule.

collisions occur quite frequently for a gas at atmospheric pressure—about 10^{10} times per second for each molecule. Collisions occur because real gas molecules have finite volumes.

Because of molecular collisions, the direction of motion of a gas molecule is constantly changing. Therefore, the diffusion of a molecule from one point to another consists of many short, straight-line segments as collisions buffet it around in random directions, as depicted in Figure 10.22 ◄. First the molecule moves in one direction, then in another; one instant at high speed, the next at low speed.

The average distance traveled by a molecule between collisions is called the **mean free path** of the molecule. The mean free path varies with pressure as the following analogy illustrates. Imagine walking through a shopping mall. When the mall is very crowded (high pressure), the average distance you can walk before bumping into someone is short (short mean free path). When the mall is empty (low pressure), you can walk a long way (long mean free path) before bumping into someone. The mean free path for air molecules at sea level is about 60 nm (6×10^{-8} m). At about 100 km in altitude, where the air density is much lower, the mean free path is about 10 cm, over 1 million times longer than at Earth's surface.

▲ **GIVE IT SOME THOUGHT**

Will the following changes increase, decrease, or have no effect on the mean free path of the gas molecules in a sample of gas? **(a)** Increasing pressure, **(b)** increasing temperature?

10.9 REAL GASES: DEVIATIONS FROM IDEAL BEHAVIOR

Although the ideal-gas equation is a very useful description of gases, all real gases fail to obey the relationship to some degree. The extent to which a real gas departs from ideal behavior can be seen by rearranging the ideal-gas equation to solve for n:

$$\frac{PV}{RT} = n \qquad [10.25]$$

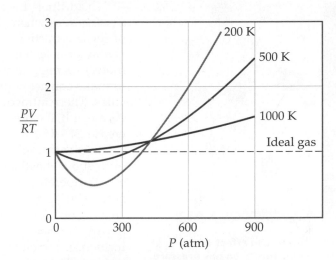

▲ **Figure 10.23 The effect of pressure on the behavior of several gases.** The ratios of PV/RT versus pressure are compared for one mole of several gases at 300 K. The data for CO_2 are at 313 K because under high pressure CO_2 liquefies at 300 K. The dashed horizontal line shows the behavior of an ideal gas.

▲ **Figure 10.24 The effect of temperature and pressure on the behavior of nitrogen gas.** The ratios of PV/RT versus pressure are shown for 1 mol of nitrogen gas at three temperatures. As temperature increases, the gas more closely approaches ideal behavior, which is represented by the dashed horizontal line.

For one mole of ideal gas ($n = 1$) the quantity PV/RT equals 1 at all pressures. In Figure 10.23▲ PV/RT is plotted as a function of P for one mole of several different gases. At high pressures the deviation from ideal behavior ($PV/RT = 1$) is large and is different for each gas. *Real gases, therefore, do not behave ideally at high pressure.* At lower pressures (usually below 10 atm), however, the deviation from ideal behavior is small, and we can use the ideal-gas equation without generating serious error.

The deviation from ideal behavior also depends on temperature. Figure 10.24▲ shows graphs of PV/RT versus P for 1 mol of N_2 at three temperatures. As temperature increases, the behavior of the gas more nearly approaches that of the ideal gas. In general, *the deviations from ideal behavior increase as temperature decreases,* becoming significant near the temperature at which the gas is converted into a liquid.

▼ **Figure 10.25 Comparing the volume of gas molecules to the container volume.** In (a), at low pressure, the combined volume of the gas molecules is small relative to the container volume, and we can approximate the empty space between molecules as being equal to the container volume. In (b), at high pressure, the combined volume of the gas molecules is a larger fraction of the total space available. Now we must account for the volume of the molecules themselves in determining the empty space available for the motion of the gas molecules.

GIVE IT SOME THOUGHT

Would you expect helium gas to deviate from ideal behavior more at **(a)** 100 K and 1 atm, **(b)** 100 K and 5 atm, or **(c)** 300 K and 2 atm?

The basic assumptions of the kinetic-molecular theory give us insight into why real gases deviate from ideal behavior. The molecules of an ideal gas are assumed to occupy no space and have no attractions for one another. *Real molecules, however, do have finite volumes, and they do attract one another.* As shown in Figure 10.25▶, the free, unoccupied space in which molecules can move is somewhat less than the container volume. At relatively low pressures the volume of the gas molecules is negligible relative to the container volume. Thus, the free volume available to the molecules is essentially the entire volume of the container. As the pressure increases, however, the free space in which the molecules can move becomes a smaller fraction of the container volume. Under these conditions, therefore, gas volumes tend to be slightly greater than those predicted by the ideal-gas equation.

(a)   (b)

▲ **Figure 10.26 The effect of intermolecular forces on gas pressure.** The molecule that is about to strike the wall experiences attractive forces from nearby gas molecules, and its impact on the wall is thereby lessened. The lessened impact means the molecule exerts a lower-than-expected pressure on the wall. The attractive forces become significant only under high-pressure conditions, when the average distance between molecules is small.

In addition, the attractive forces between molecules come into play at short distances, as when molecules are crowded together at high pressures. Because of these attractive forces, the impact of a given molecule with the wall of the container is lessened. If we could stop the action in a gas, the positions of the molecules might resemble the illustration in Figure 10.26 ◄. The molecule about to make contact with the wall experiences the attractive forces of nearby molecules. These attractions lessen the force with which the molecule hits the wall. As a result, the pressure is less than that of an ideal gas. This effect serves to decrease PV/RT below its ideal value, as seen in Figure 10.23. When the pressure is sufficiently high, however, the volume effects dominate and PV/RT increases to above the ideal value.

Temperature determines how effective attractive forces between gas molecules are. As a gas is cooled, the average kinetic energy of the molecules decreases, but intermolecular attractions remain constant. In a sense, cooling a gas deprives molecules of the energy they need to overcome their mutual attractive influence. The effects of temperature shown in Figure 10.24 illustrate this point very well. As temperature increases, the negative departure of PV/RT from ideal-gas behavior disappears. The difference that remains at high temperature stems mainly from the effect of the finite volumes of the molecules.

▲ **GIVE IT SOME THOUGHT**

List two reasons why gases deviate from ideal behavior.

The van der Waals Equation

Engineers and scientists who work with gases at high pressures often cannot use the ideal-gas equation to predict the pressure-volume properties of gases because departures from ideal behavior are too large. One useful equation developed to predict the behavior of real gases was proposed by the Dutch scientist Johannes van der Waals (1837–1923).

Van der Waals recognized that the ideal gas equation could be corrected to account for the effects of attractive forces between gas molecules and for molecular volumes. He introduced two constants, a and b, to make these corrections. The constant a is a measure of how strongly the gas molecules attract each other. The constant b is a measure of the small but finite volume occupied by the gas molecules themselves. His description of gas behavior is known as the **van der Waals equation**:

$$\left(P + \frac{n^2a}{V^2}\right)(V - nb) = nRT \qquad [10.26]$$

In this equation the factor n^2a/V^2 accounts for the attractive forces. The van der Waals equation adjusts the pressure, P, upward by adding n^2a/V^2 because attractive forces between molecules tend to reduce the pressure (Figure 10.26). That is, a correction must be added to give the pressure that an ideal gas would have. The form of the correction factor, n^2a/V^2, results because the attractive forces between pairs of molecules increase as the square of the number of molecules per unit volume, $(n/V)^2$.

The factor nb accounts for the small but finite volume occupied by the gas molecules themselves (Figure 10.25). The van der Waals equation adjusts volume, V, downward (subtracting nb), to give the free volume available to the gas molecules. That is, the particles of an ideal gas have the full volume, V, as free space in which to move, whereas in a real gas only the volume $V - nb$ is available as free space.

The constants a and b, which are called *van der Waals constants*, are experimentally determined positive quantities that differ for each gas. Values of these constants for several gases are listed in Table 10.3 ▶. Notice that the values of

TABLE 10.3 ■ van der Waals Constants for Gas Molecules		
Substance	a (L^2-atm/mol^2)	b (L/mol)
He	0.0341	0.02370
Ne	0.211	0.0171
Ar	1.34	0.0322
Kr	2.32	0.0398
Xe	4.19	0.0510
H_2	0.244	0.0266
N_2	1.39	0.0391
O_2	1.36	0.0318
Cl_2	6.49	0.0562
H_2O	5.46	0.0305
CH_4	2.25	0.0428
CO_2	3.59	0.0427
CCl_4	20.4	0.1383

both a and b generally increase with an increase in mass of the molecule and with an increase in the complexity of its structure. Larger, more massive molecules have larger volumes and tend to have greater intermolecular attractive forces. As we will see in the next chapter, substances with large intermolecular attractive forces are relatively easy to liquefy.

■ **SAMPLE EXERCISE 10.16** | Using the van der Waals Equation

If 1.000 mol of an ideal gas were confined to 22.41 L at 0.0 °C, it would exert a pressure of 1.000 atm. Use the van der Waals equation and the constants in Table 10.3 to estimate the pressure exerted by 1.000 mol of $Cl_2(g)$ in 22.41 L at 0.0 °C.

SOLUTION

Analyze: The quantity we need to solve for is pressure. Because we will use the van der Waals equation, we must identify the appropriate values for the constants that appear there.

Plan: Solving Equation 10.26 for P, we have

$$P = \frac{nRT}{V - nb} - \frac{n^2a}{V^2}$$

Solve: Substituting $n = 1.000$ mol, $R = 0.08206$ L-atm/mol-K, $T = 273.2$ K, $V = 22.41$ L, $a = 6.49$ L^2-atm/mol^2, and $b = 0.0562$ L/mol:

$$P = \frac{(1.000 \text{ mol})(0.08206 \text{ L-atm/mol-K})(273.2 \text{ K})}{22.41 \text{ L} - (1.000 \text{ mol})(0.0562 \text{ L/mol})} - \frac{(1.000 \text{ mol})^2(6.49 \text{ L}^2\text{-atm/mol}^2)}{(22.14 \text{ L})^2}$$

$$= 1.003 \text{ atm} - 0.013 \text{ atm} = 0.990 \text{ atm}$$

Check: We expect a pressure not far from 1.000 atm, which would be the value for an ideal gas, so our answer seems very reasonable.

Comment: Notice that the first term, 1.003 atm, is the pressure corrected for molecular volume. This value is higher than the ideal value, 1.000 atm, because the volume in which the molecules are free to move is smaller than the container volume, 22.41 L. Thus, the molecules must collide more frequently with the container walls. The second factor, 0.013 atm, corrects for intermolecular forces. The intermolecular attractions between molecules reduce the pressure to 0.990 atm. We can conclude, therefore, that the intermolecular attractions are the main cause of the slight deviation of $Cl_2(g)$ from ideal behavior under the stated experimental conditions.

■ **PRACTICE EXERCISE**

Consider a sample of 1.000 mol of $CO_2(g)$ confined to a volume of 3.000 L at 0.0 °C. Calculate the pressure of the gas using **(a)** the ideal-gas equation and **(b)** the van der Waals equation.
Answers: **(a)** 7.473 atm, **(b)** 7.182 atm

■■ SAMPLE INTEGRATIVE EXERCISE | Putting Concepts Together

Cyanogen, a highly toxic gas, is composed of 46.2% C and 53.8% N by mass. At 25 °C and 751 torr, 1.05 g of cyanogen occupies 0.500 L. **(a)** What is the molecular formula of cyanogen? **(b)** Predict its molecular structure. **(c)** Predict the polarity of the compound.

SOLUTION

Analyze: First we need to determine the molecular formula of a compound from elemental analysis data and data on the properties of the gaseous substance. Thus, we have two separate calculations to do.

(a) Plan: We can use the percentage composition of the compound to calculate its empirical formula. ∞ (Section 3.5) Then we can determine the molecular formula by comparing the mass of the empirical formula with the molar mass. ∞ (Section 3.5)

Solve: To determine the empirical formula, we assume that we have a 100-g sample of the compound and then calculate the number of moles of each element in the sample:

$$\text{Moles C} = (46.2 \text{ g C})\left(\frac{1 \text{ mol C}}{12.01 \text{ g C}}\right) = 3.85 \text{ mol C}$$

$$\text{Moles N} = (53.8 \text{ g N})\left(\frac{1 \text{ mol N}}{14.01 \text{ g N}}\right) = 3.84 \text{ mol N}$$

Because the ratio of the moles of the two elements is essentially 1:1, the empirical formula is CN.

To determine the molar mass of the compound, we use Equation 10.11.

$$\mathcal{M} = \frac{dRT}{p} = \frac{(1.05 \text{ g}/0.500 \text{ L}) \, (0.0821 \text{ L-atm/mol-K}) \, (298 \text{ K})}{(751/760) \text{ atm}} = 52.0 \text{ g/mol}$$

The molar mass associated with the empirical formula, CN, is 12.0 + 14.0 = 26.0 g/mol. Dividing the molar mass of the compound by that of its empirical formula gives (52.0 g/mol)/(26.0 g/mol) = 2.00. Thus, the molecule has twice as many atoms of each element as the empirical formula, giving the molecular formula C_2N_2.

(b) Plan: To determine the molecular structure of the molecule, we must first determine its Lewis structure. ∞ (Section 8.5) We can then use the VSEPR model to predict the structure. ∞ (Section 9.2)

Solve: The molecule has 2(4) + 2(5) = 18 valence-shell electrons. By trial and error, we seek a Lewis structure with 18 valence electrons in which each atom has an octet and in which the formal charges are as low as possible. The following structure meets these criteria:

$$:N\equiv C-C\equiv N:$$

(This structure has zero formal charge on each atom.)

The Lewis structure shows that each atom has two electron domains. (Each nitrogen has a nonbonding pair of electrons and a triple bond, whereas each carbon has a triple bond and a single bond.) Thus the electron-domain geometry around each atom is linear, causing the overall molecule to be linear.

(c) Plan: To determine the polarity of the molecule, we must examine the polarity of the individual bonds and the overall geometry of the molecule.

Solve: Because the molecule is linear, we expect the two dipoles created by the polarity in the carbon–nitrogen bond to cancel each other, leaving the molecule with no dipole moment.

CHAPTER REVIEW

SUMMARY AND KEY TERMS

Section 10.1 Substances that are gases at room temperatures tend to be molecular substances with low molar masses. Air, a mixture composed mainly of N_2 and O_2, is the most common gas we encounter. Some liquids and solids can also exist in the gaseous state, where they are known as vapors. Gases are compressible; they mix in all proportions because their component molecules are far apart.

Section 10.2 To describe the state, or condition, of a gas, we must specify four variables: pressure (P), volume (V), temperature (T), and quantity (n). Volume is usually measured in liters, temperature in kelvins, and quantity of gas in moles. **Pressure** is the force per unit area. It is expressed in SI units as **pascals**, Pa (1 Pa = 1 N/m^2 = 1 kg/m-s^2). A related unit, the **bar**, equals 10^5 Pa. In chemistry, **standard atmospheric pressure** is used to define the **atmosphere** (atm) and the **torr** (also called the millimeter of mercury). One atmosphere of pressure equals 101.325 kPa, or 760 torr. A barometer is often used to measure the atmospheric pressure. A manometer can be used to measure the pressure of enclosed gases.

Sections 10.3 and 10.4 Studies have revealed several simple gas laws: For a constant quantity of gas at constant temperature, the volume of the gas is inversely proportional to the pressure (**Boyle's law**). For a fixed quantity of gas at constant pressure, the volume is directly proportional to its absolute temperature (**Charles's law**). Equal volumes of gases at the same temperature and pressure contain equal numbers of molecules (**Avogadro's hypothesis**). For a gas at constant temperature and pressure, the volume of the gas is directly proportional to the number of moles of gas (**Avogadro's law**). Each of these gas laws is a special case of the ideal-gas equation.

The **ideal-gas equation**, $PV = nRT$, is the equation of state for an **ideal gas**. The term R in this equation is the **gas constant**. We can use the ideal-gas equation to calculate variations in one variable when one or more of the others are changed. Most gases at pressures of about 1 atm and temperatures near 273 K and above obey the ideal-gas equation reasonably well. The conditions of 273 K (0 °C) and 1 atm are known as the **standard temperature and pressure (STP)**. In all applications of the ideal-gas equation we must remember to convert temperatures to the absolute-temperature scale (the Kelvin scale).

Sections 10.5 and 10.6 Using the ideal-gas equation, we can relate the density of a gas to its molar mass: $\mathcal{M} = dRT/P$. We can also use the ideal-gas equation to solve problems involving gases as reactants or products in chemical reactions.

In gas mixtures the total pressure is the sum of the **partial pressures** that each gas would exert if it were present alone under the same conditions (**Dalton's law of partial pressures**). The partial pressure of a component of a mixture is equal to its mole fraction times the total pressure: $P_1 = X_1 P_t$. The **mole fraction** is the ratio of the moles of one component of a mixture to the total moles of all components. In calculating the quantity of a gas collected over water, correction must be made for the partial pressure of water vapor in the gas mixture.

Section 10.7 The **kinetic-molecular theory** accounts for the properties of an ideal gas in terms of a set of statements about the nature of gases. Briefly, these statements are as follows: Molecules are in continuous chaotic motion. The volume of gas molecules is negligible compared to the volume of their container. The gas molecules have no attractive forces for one another. Their collisions are elastic. The average kinetic energy of the gas molecules is proportional to the absolute temperature.

The molecules of a gas do not all have the same kinetic energy at a given instant. Their speeds are distributed over a wide range; the distribution varies with the molar mass of the gas and with temperature. The **root-mean-square (rms) speed**, u, varies in proportion to the square root of the absolute temperature and inversely with the square root of the molar mass: $u = \sqrt{3RT/\mathcal{M}}$.

Section 10.8 It follows from kinetic-molecular theory that the rate at which a gas undergoes **effusion** (escapes through a tiny hole into a vacuum) is inversely proportional to the square root of its molar mass (**Graham's law**). The **diffusion** of one gas through the space occupied by a second gas is another phenomenon related to the speeds at which molecules move. Because molecules undergo frequent collisions with one another, the **mean free path**—the mean distance traveled between collisions—is short. Collisions between molecules limit the rate at which a gas molecule can diffuse.

Section 10.9 Departures from ideal behavior increase in magnitude as pressure increases and as temperature decreases. The extent of nonideality of a real gas can be seen by examining the quantity $PV = RT$ for one mole of the gas as a function of pressure; for an ideal gas, this quantity is exactly 1 at all pressures. Real gases depart from ideal behavior because the molecules possess finite volume and because the molecules experience attractive forces for one another. The **van der Waals equation** is an equation of state for gases that modifies the ideal-gas equation to account for intrinsic molecular volume and intermolecular forces.

KEY SKILLS

- Convert between pressure units with an emphasis on torr and atmospheres.
- Calculate P, V, n, or T using the ideal gas equation.
- Understand how the gas laws relate to the ideal gas equation and apply the gas laws in calculations.
- Calculate the density or molecular weight of a gas.
- Calculate the volume of gas used or formed in a chemical reaction.
- Calculate the total pressure of a gas mixture given its partial pressures or information for calculating partial pressures.
- Describe the kinetic molecular theory and how it explains the pressure and temperature of a gas, the gas laws, and the rates of effusion and diffusion.
- Explain why intermolecular attractions and molecular volumes cause real gases to deviate from ideal behavior at high pressure or low temperature.

KEY EQUATIONS

- $PV = nRT$ \quad [10.5] \qquad Ideal gas equation

- $\dfrac{P_1 V_1}{T_1} = \dfrac{P_2 V_2}{T_2}$ \quad [10.8] \qquad The combined gas law, showing how P, V, and T are related for a constant n

- $d = \dfrac{P\mathcal{M}}{RT}$ \quad [10.10] \qquad Calculating the density or molar mass of a gas

- $P_t = P_1 + P_2 + P_3 + \cdots$ \quad [10.12] \qquad Relating the total pressure of a gas mixture to the partial pressures of its components (Dalton's law of partial pressures)

- $P_1 = \left(\dfrac{n_1}{n_t}\right) P_t = X_1 P_t$ \quad [10.15] \qquad Relating partial pressure to mole fraction

- $u = \sqrt{\dfrac{3RT}{\mathcal{M}}}$ \quad [10.22] \qquad Relating the root-mean-square (rms) speed of gas molecules to temperature and molar mass

- $\dfrac{r_1}{r_2} = \sqrt{\dfrac{\mathcal{M}_2}{\mathcal{M}_1}}$ \quad [10.23] \qquad Relating the relative rates of effusion of two gases to their molar masses

VISUALIZING CONCEPTS

10.1 Mars has an average atmospheric pressure of 0.007 atm. Would it be easier or harder to drink from a straw on Mars than on Earth? Explain. [Section 10.2]

10.2 Assume that you have a sample of gas in a container with a movable piston, such as the one in the drawing. **(a)** Redraw the container to show what it might look like if the temperature of the gas is increased from 300 K to 500 K while the pressure is kept constant. **(b)** Redraw the container to show what it might look like if the pressure on the piston is increased from 1.0 atm to 2.0 atm while the temperature is kept constant. [Section 10.3]

10.3 Consider the sample of gas depicted below. What would the drawing look like if the volume and temperature remained constant while you removed enough of the gas to decrease the pressure by a factor of 2? [Section 10.3]

10.4 Consider the following reaction:

$$2\,CO(g) + O_2(g) \longrightarrow 2\,CO_2(g)$$

Imagine that this reaction occurs in a container that has a piston that moves to allow a constant pressure to be maintained when the reaction occurs at constant temperature. **(a)** What happens to the volume of the container as a result of the reaction? Explain. **(b)** If the piston is not allowed to move, what happens to the pressure as a result of the reaction? [Sections 10.3 and 10.5]

10.5 Consider the apparatus below, which shows gases in two containers and one empty container. When the stopcocks are opened and the gases are allowed to mix at constant temperature, what is the distribution of atoms in each container? Assume that the containers are of equal volume, and ignore the volume of the tubing connecting them. Which gas has the greater partial pressure after the stopcocks are opened? [Section 10.6]

10.6 The drawing below represents a mixture of three different gases. **(a)** Rank the three components in order of increasing partial pressure. **(b)** If the total pressure of the mixture is 0.90 atm, calculate the partial pressure of each gas. [Section 10.6]

10.7 On a single plot, qualitatively sketch the distribution of molecular speeds for **(a)** $Kr(g)$ at $-50\ ^\circ C$, **(b)** $Kr(g)$ at $0\ ^\circ C$, **(c)** $Ar(g)$ at $0\ ^\circ C$. [Section 10.7]

10.8 Consider the following drawing. **(a)** If the curves A and B refer to two different gases, He and O_2 at the same

temperature, which is which? Explain. **(b)** If A and B refer to the same gas at two different temperatures, which represents the higher temperature? [Section 10.7]

10.9 Consider the following samples of gases:

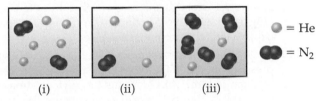

If the three samples are all at the same temperature, rank them with respect to **(a)** total pressure, **(b)** partial pressure of helium, **(c)** density, **(d)** average kinetic energy of particles. [Section 10.6 and 10.7]

10.10 Which of the substances listed in Table 10.3 would you expect to deviate the most from ideal-gas behavior at low temperature and high pressure? Which would deviate the least? Explain. [Section 10.9]

EXERCISES

Gas Characteristics; Pressure

10.11 How does a gas differ from a liquid with respect to each of the following properties: **(a)** density, **(b)** compressibility, **(c)** ability to mix with other substances of the same phase to form homogeneous mixtures?

10.12 **(a)** Both a liquid and a gas are moved to larger containers. How does their behavior differ? Explain the difference in molecular terms. **(b)** Although water and carbon tetrachloride, $CCl_4(l)$, do not mix, their vapors form homogeneous mixtures. Explain. **(c)** The densities of gases are generally reported in units of g/L, whereas those for liquids are reported as g/mL. Explain the molecular basis for this difference.

10.13 Suppose that a woman weighing 130 lb and wearing high-heeled shoes momentarily places all her weight on the heel of one foot. If the area of the heel is 0.50 in.^{2}, calculate the pressure exerted on the underlying surface in kilopascals.

10.14 A set of bookshelves rests on a hard floor surface on four legs, each having a cross-sectional dimension of 3.0×4.1 cm in contact with the floor. The total mass of the shelves plus the books stacked on them is 262 kg. Calculate the pressure in pascals exerted by the shelf footings on the surface.

10.15 **(a)** How high in meters must a column of water be to exert a pressure equal to that of a 760-mm column of mercury? The density of water is 1.0 g/mL, whereas that of mercury is 13.6 g/mL. **(b)** What is the pressure in atmospheres on the body of a diver if he is 39 ft below the surface of the water when atmospheric pressure at the surface is 0.97 atm?

10.16 The compound 1-iodododecane is a nonvolatile liquid with a density of 1.20 g/mL. The density of mercury is 13.6 g/mL. What do you predict for the height of a barometer column based on 1-iodododecane, when the atmospheric pressure is 752 torr?

10.17 Each of the following statements concerns a mercury barometer such as that shown in Figure 10.2. Identify any incorrect statements, and correct them. **(a)** The tube must be 1 cm^2 in cross-sectional area. **(b)** At equilibrium the force of gravity per unit area acting on the mercury column at the level of the outside mercury equals the force of gravity per unit area acting on the atmosphere. **(c)** The column of mercury is held up by the vacuum at the top of the column.

10.18 Suppose you make a mercury barometer using a glass tube about 50 cm in length, closed at one end. What would you expect to see if the tube is filled with mercury and inverted in a mercury dish, as in Figure 10.2? Explain.

10.19 The typical atmospheric pressure on top of Mt. Everest (29,028 ft) is about 265 torr. Convert this pressure to (a) atm, (b) mm Hg, (c) pascals, (d) bars.

10.20 Perform the following conversions: (a) 0.850 atm to torr, (b) 785 torr to kilopascals, (c) 655 mm Hg to atmospheres, (d) 1.323×10^5 Pa to atmospheres, (e) 2.50 atm to bars.

10.21 In the United States, barometric pressures are generally reported in inches of mercury (in. Hg). On a beautiful summer day in Chicago the barometric pressure is 30.45 in. Hg. (a) Convert this pressure to torr. (b) A meteorologist explains the nice weather by referring to a "high-pressure area." In light of your answer to part (a), explain why this term makes sense.

10.22 (a) On Titan, the largest moon of Saturn, the atmospheric pressure is 1.63105 Pa. What is the atmospheric pressure of Titan in atm? (b) On Venus the surface atmospheric pressure is about 90 Earth atmospheres. What is the Venusian atmospheric pressure in kilopascals?

10.23 If the atmospheric pressure is 0.985 atm, what is the pressure of the enclosed gas in each of the three cases depicted in the drawing?

Open end	Open end	Closed end
Gas	Gas	Gas
$h = 52$ cm	$h = 67$ mm	$h = 10.3$ cm
(i)	(ii)	(iii)

10.24 An open-end manometer containing mercury is connected to a container of gas, as depicted in Sample Exercise 10.2. What is the pressure of the enclosed gas in torr in each of the following situations? (a) The mercury in the arm attached to the gas is 15.4 mm higher than in the one open to the atmosphere; atmospheric pressure is 0.966 atm. (b) The mercury in the arm attached to the gas is 8.7 mm lower than in the one open to the atmosphere; atmospheric pressure is 0.99 atm.

The Gas Laws

10.25 Assume that you have a cylinder with a movable piston. What would happen to the gas pressure inside the cylinder if you do the following? (a) Decrease the volume to one-fourth the original volume while holding the temperature constant. (b) Reduce the Kelvin temperature to half its original value while holding the volume constant. (c) Reduce the amount of gas to half while keeping the volume and temperature constant.

10.26 A fixed quantity of gas at 21 °C exhibits a pressure of 752 torr and occupies a volume of 4.38 L. (a) Use Boyle's law to calculate the volume the gas will occupy if the pressure is increased to 1.88 atm while the temperature is held constant. (b) Use Charles's law to calculate the volume the gas will occupy if the temperature is increased to 175 °C while the pressure is held constant.

10.27 (a) How is the law of combining volumes explained by Avogadro's hypothesis? (b) Consider a 1.0-L flask containing neon gas and a 1.5-L flask containing xenon gas. Both gases are at the same pressure and temperature. According to Avogadro's law, what can be said about the ratio of the number of atoms in the two flasks?

10.28 Nitrogen and hydrogen gases react to form ammonia gas as follows:

$$N_2(g) + 3\,H_2(g) \longrightarrow 2\,NH_3(g)$$

At a certain temperature and pressure, 1.2 L of N_2 reacts with 3.6 L of H_2. If all the N_2 and H_2 are consumed, what volume of NH_3, at the same temperature and pressure, will be produced?

The Ideal-Gas Equation

10.29 (a) Write the ideal-gas equation, and give the units used for each term in the equation when $R = 0.0821$ L-atm/mol-K. (b) What is an ideal gas?

10.30 (a) What conditions are represented by the abbreviation STP? (b) What is the molar volume of an ideal gas at STP? (c) Room temperature is often assumed to be 25 °C. Calculate the molar volume of an ideal gas at 25 °C and 1 atm pressure.

10.31 Suppose you are given two 1-L flasks and told that one contains a gas of molar mass 30, the other a gas of molar mass 60, both at the same temperature. The pressure in

flask A is X atm, and the mass of gas in the flask is 1.2 g. The pressure in flask B is 0.5X atm, and the mass of gas in that flask is 1.2 g. Which flask contains gas of molar mass 30, and which contains the gas of molar mass 60?

10.32 Suppose you are given two flasks at the same temperature, one of volume 2 L and the other of volume 3 L. The 2-L flask contains 4.8 g of gas, and the gas pressure is X atm. The 3-L flask contains 0.36 g of gas, and the gas pressure is 0.1X. Do the two gases have the same molar mass? If not, which contains the gas of higher molar mass?

10.33 Complete the following table for an ideal gas:

P	V	n	T
2.00 atm	1.00 L	0.500 mol	? K
0.300 atm	0.250 L	? mol	27 °C
650 torr	? L	0.333 mol	350 K
? atm	585 mL	0.250 mol	295 K

10.34 Calculate each of the following quantities for an ideal gas: **(a)** the volume of the gas, in liters, if 1.50 mol has a pressure of 0.985 atm at a temperature of -6 °C; **(b)** the absolute temperature of the gas at which 3.33×10^{-3} mol occupies 325 mL at 750 torr; **(c)** the pressure, in atmospheres, if 0.0467 mol occupies 413 mL at 138 °C; **(d)** the quantity of gas, in moles, if 55.7 L at 54 °C has a pressure of 11.25 kPa.

10.35 The Goodyear blimps, which frequently fly over sporting events, hold approximately 175,000 ft^3 of helium. If the gas is at 23 °C and 1.0 atm, what mass of helium is in the blimp?

10.36 A neon sign is made of glass tubing whose inside diameter is 2.5 cm and whose length is 5.5 m. If the sign contains neon at a pressure of 1.78 torr at 35 °C, how many grams of neon are in the sign? (The volume of a cylinder is $\pi r^2 h$.)

10.37 Calculate the number of molecules in a deep breath of air whose volume is 2.25 L at body temperature, 37 °C, and a pressure of 735 torr.

10.38 If the pressure exerted by ozone, O_3, in the stratosphere is 3.0×10^{-3} atm and the temperature is 250 K, how many ozone molecules are in a liter?

10.39 A scuba diver's tank contains 0.29 kg of O_2 compressed into a volume of 2.3 L. **(a)** Calculate the gas pressure inside the tank at 9 °C. **(b)** What volume would this oxygen occupy at 26 °C and 0.95 atm?

10.40 An aerosol spray can with a volume of 250 mL contains 2.30 g of propane gas (C_3H_8) as a propellant. **(a)** If the can is at 23 °C, what is the pressure in the can? **(b)** What volume would the propane occupy at STP? **(c)** The can says that exposure to temperatures above 130 °F may cause the can to burst. What is the pressure in the can at this temperature?

10.41 Chlorine is widely used to purify municipal water supplies and to treat swimming pool waters. Suppose that the volume of a particular sample of Cl_2 gas is 8.70 L at 895 torr and 24 °C. **(a)** How many grams of Cl_2 are in the sample? **(b)** What volume will the Cl_2 occupy at STP? **(c)** At what temperature will the volume be 15.00 L if the pressure is 8.76×10^2 torr? **(d)** At what pressure will the volume equal 6.00 L if the temperature is 58 °C?

10.42 Many gases are shipped in high-pressure containers. Consider a steel tank whose volume is 65.0 L and which contains O_2 gas at a pressure of 16,500 kPa at 23 °C. **(a)** What mass of O_2 does the tank contain? **(b)** What volume would the gas occupy at STP? **(c)** At what temperature would the pressure in the tank equal 150.0 atm? **(d)** What would be the pressure of the gas, in kPa, if it were transferred to a container at 24 °C whose volume is 55.0 L?

10.43 In an experiment reported in the scientific literature, male cockroaches were made to run at different speeds on a miniature treadmill while their oxygen consumption was measured. In one hour the average cockroach running at 0.08 km/hr consumed 0.8 mL of O_2 at 1 atm pressure and 24 °C per gram of insect weight. **(a)** How many moles of O_2 would be consumed in 1 hr by a 5.2-g cockroach moving at this speed? **(b)** This same cockroach is caught by a child and placed in a 1-qt fruit jar with a tight lid. Assuming the same level of continuous activity as in the research, will the cockroach consume more than 20% of the available O_2 in a 48-hr period? (Air is 21 mol percent O_2.)

10.44 After the large eruption of Mount St. Helens in 1980, gas samples from the volcano were taken by sampling the downwind gas plume. The unfiltered gas samples were passed over a gold-coated wire coil to absorb mercury (Hg) present in the gas. The mercury was recovered from the coil by heating it, and then analyzed. In one particular set of experiments scientists found a mercury vapor level of 1800 ng of Hg per cubic meter in the plume, at a gas temperature of 10 °C. Calculate **(a)** the partial pressure of Hg vapor in the plume, **(b)** the number of Hg atoms per cubic meter in the gas, **(c)** the total mass of Hg emitted per day by the volcano if the daily plume volume was 1600 km^3.

Further Applications of the Ideal-Gas Equation

10.45 Which gas is most dense at 1.00 atm and 298 K? CO_2, N_2O, or Cl_2. Explain.

10.46 Rank the following gases from least dense at 1.00 atm and 298 K to most dense under these same conditions: SO_2, HBr, CO_2. Explain.

10.47 Which of the following statements best explains why a closed balloon filled with helium gas rises in air?
(a) Helium is a monatomic gas, whereas nearly all the molecules that make up air, such as nitrogen and oxygen, are diatomic.
(b) The average speed of helium atoms is higher than the average speed of air molecules, and the higher speed of collisions with the balloon walls propels the balloon upward.
(c) Because the helium atoms are of lower mass than the average air molecule, the helium gas is less dense than air. The balloon thus weighs less than the air displaced by its volume.
(d) Because helium has a lower molar mass than the average air molecule, the helium atoms are in faster motion. This means that the temperature of the helium is higher than the air temperature. Hot gases tend to rise.

10.48 Which of the following statements best explains why nitrogen gas at STP is less dense than Xe gas at STP?

(a) Because Xe is a noble gas, there is less tendency for the Xe atoms to repel one another, so they pack more densely in the gas state.

(b) Xe atoms have a higher mass than N_2 molecules. Because both gases at STP have the same number of molecules per unit volume, the Xe gas must be denser.

(c) The Xe atoms are larger than N_2 molecules and thus take up a larger fraction of the space occupied by the gas.

(d) Because the Xe atoms are much more massive than the N_2 molecules, they move more slowly and thus exert less upward force on the gas container and make the gas appear denser.

10.49 (a) Calculate the density of NO_2 gas at 0.970 atm and 35 °C. (b) Calculate the molar mass of a gas if 2.50 g occupies 0.875 L at 685 torr and 35 °C.

10.50 (a) Calculate the density of sulfur hexafluoride gas at 707 torr and 21 °C. (b) Calculate the molar mass of a vapor that has a density of 7.135 g/L at 12 °C and 743 torr.

10.51 In the Dumas-bulb technique for determining the molar mass of an unknown liquid, you vaporize the sample of a liquid that boils below 100 °C in a boiling-water bath and determine the mass of vapor required to fill the bulb (see drawing). From the following data, calculate the molar mass of the unknown liquid: mass of unknown vapor, 1.012 g; volume of bulb, 354 cm³; pressure, 742 torr; temperature, 99 °C.

Vapor of unknown

Boiling water

10.52 The molar mass of a volatile substance was determined by the Dumas-bulb method described in Exercise 10.51. The unknown vapor had a mass of 0.846 g; the volume of the bulb was 354 cm³, pressure 752 torr, and temperature 100 °C. Calculate the molar mass of the unknown vapor.

10.53 Magnesium can be used as a "getter" in evacuated enclosures, to react with the last traces of oxygen. (The magnesium is usually heated by passing an electric current through a wire or ribbon of the metal.) If an enclosure of 0.382 L has a partial pressure of O_2 of 3.5×10^{-6} torr at 27 °C, what mass of magnesium will react according to the following equation?

$$2\,Mg(s) + O_2(g) \longrightarrow 2\,MgO(s)$$

10.54 Calcium hydride, CaH_2, reacts with water to form hydrogen gas:

$$CaH_2(s) + 2\,H_2O(l) \longrightarrow Ca(OH)_2\,(aq) + 2\,H_2(g)$$

This reaction is sometimes used to inflate life rafts, weather balloons, and the like, where a simple, compact means of generating H_2 is desired. How many grams of CaH_2 are needed to generate 53.5 L of H_2 gas if the pressure of H_2 is 814 torr at 21 °C?

10.55 The metabolic oxidation of glucose, $C_6H_{12}O_6$, in our bodies produces CO_2, which is expelled from our lungs as a gas:

$$C_6H_{12}O_6(aq) + 6\,O_2(g) \longrightarrow 6\,CO_2(g) + 6\,H_2O(l)$$

Calculate the volume of dry CO_2 produced at body temperature (37 °C) and 0.970 atm when 24.5 g of glucose is consumed in this reaction.

10.56 Both Jacques Charles and Joseph Louis Guy-Lussac were avid balloonists. In his original flight in 1783, Jacques Charles used a balloon that contained approximately 31,150 L of H_2. He generated the H_2 using the reaction between iron and hydrochloric acid:

$$Fe(s) + 2\,HCl(aq) \longrightarrow FeCl_2(aq) + H_2(g)$$

How many kilograms of iron were needed to produce this volume of H_2 if the temperature was 22 °C?

10.57 Hydrogen gas is produced when zinc reacts with sulfuric acid:

$$Zn(s) + H_2SO_4(aq) \longrightarrow ZnSO_4(aq) + H_2(g)$$

If 159 mL of wet H_2 is collected over water at 24 °C and a barometric pressure of 738 torr, how many grams of Zn have been consumed? (The vapor pressure of water is tabulated in Appendix B.)

10.58 Acetylene gas, $C_2H_2(g)$, can be prepared by the reaction of calcium carbide with water:

$$CaC_2(s) + 2\,H_2O(l) \longrightarrow Ca(OH)_2(s) + C_2H_2(g)$$

Calculate the volume of C_2H_2 that is collected over water at 23 °C by reaction of 0.752 g of CaC_2 if the total pressure of the gas is 745 torr. (The vapor pressure of water is tabulated in Appendix B.)

Partial Pressures

10.59 Consider the apparatus shown in the drawing on the next page. (a) When the stopcock between the two containers is opened and the gases allowed to mix, how does the volume occupied by the N_2 gas change? What is the partial pressure of N_2 after mixing? (b) How does the volume of the O_2 gas change when the gases mix? What is the partial pressure of O_2 in the mixture? (c) What is the total pressure in the container after the gases mix?

2.0 L
1.0 atm
25 °C

3.0 L
2.0 atm
25 °C

10.60 Consider a mixture of two gases, A and B, confined in a closed vessel. A quantity of a third gas, C, is added to the same vessel at the same temperature. How does the addition of gas C affect the following: **(a)** the partial pressure of gas A, **(b)** the total pressure in the vessel, **(c)** the mole fraction of gas B?

10.61 A mixture containing 0.477 mol He(g), 0.280 mol Ne(g), and 0.110 mol Ar(g) is confined in a 7.00-L vessel at 25 °C. **(a)** Calculate the partial pressure of each of the gases in the mixture. **(b)** Calculate the total pressure of the mixture.

10.62 A deep-sea diver uses a gas cylinder with a volume of 10.0 L and a content of 51.2 g of O_2 and 32.6 g of He. Calculate the partial pressure of each gas and the total pressure if the temperature of the gas is 19 °C.

10.63 A piece of solid carbon dioxide with a mass of 5.50 g is placed in a 10.0-L vessel that already contains air at 705 torr and 24 °C. After the carbon dioxide has totally vaporized, what is the partial pressure of carbon dioxide and the total pressure in the container at 24 °C?

10.64 A sample of 4.00 mL of diethylether ($C_2H_5OC_2H_5$, density = 0.7134 g/mL) is introduced into a 5.00-L vessel that already contains a mixture of N_2 and O_2, whose partial pressures are $P_{N_2} = 0.751$ atm and $P_{O_2} = 0.208$ atm. The temperature is held at 35.0 °C, and the diethylether totally evaporates. **(a)** Calculate the partial pressure of the diethylether. **(b)** Calculate the total pressure in the container.

10.65 A mixture of gases contains 0.75 mol N_2, 0.30 mol O_2, and 0.15 mol CO_2. If the total pressure of the mixture is 1.56 atm, what is the partial pressure of each component?

10.66 A mixture of gases contains 10.25 g of N_2, 1.83 g of H_2, and 7.95 g of NH_3. If the total pressure of the mixture is 1.85 atm, what is the partial pressure of each component?

10.67 At an underwater depth of 250 ft, the pressure is 8.38 atm. What should the mole percent of oxygen be in the diving gas for the partial pressure of oxygen in the mixture to be 0.21 atm, the same as in air at 1 atm?

10.68 **(a)** What are the mole fractions of each component in a mixture of 5.08 g of O_2, 7.17 g of N_2, and 1.32 g of H_2? **(b)** What is the partial pressure in atm of each component of this mixture if it is held in a 12.40-L vessel at 15 °C?

10.69 A quantity of N_2 gas originally held at 4.75 atm pressure in a 1.00-L container at 26 °C is transferred to a 10.0-L container at 20 °C. A quantity of O_2 gas originally at 5.25 atm and 26 °C in a 5.00-L container is transferred to this same container. What is the total pressure in the new container?

10.70 A sample of 3.00 g of $SO_2(g)$ originally in a 5.00-L vessel at 21 °C is transferred to a 10.0-L vessel at 26 °C. A sample of 2.35 g $N_2(g)$ originally in a 2.50-L vessel at 20 °C is transferred to this same 10.0-L vessel. **(a)** What is the partial pressure of $SO_2(g)$ in the larger container? **(b)** What is the partial pressure of $N_2(g)$ in this vessel? **(c)** What is the total pressure in the vessel?

Kinetic-Molecular Theory; Graham's Law

10.71 What change or changes in the state of a gas bring about each of the following effects? **(a)** The number of impacts per unit time on a given container wall increases. **(b)** The average energy of impact of molecules with the wall of the container decreases. **(c)** The average distance between gas molecules increases. **(d)** The average speed of molecules in the gas mixture is increased.

10.72 Indicate which of the following statements regarding the kinetic-molecular theory of gases are correct. For those that are false, formulate a correct version of the statement. **(a)** The average kinetic energy of a collection of gas molecules at a given temperature is proportional to $m^{1/2}$. **(b)** The gas molecules are assumed to exert no forces on each other. **(c)** All the molecules of a gas at a given temperature have the same kinetic energy. **(d)** The volume of the gas molecules is negligible in comparison to the total volume in which the gas is contained.

10.73 What property or properties of gases can you point to that support the assumption that most of the volume in a gas is empty space?

10.74 Newton had an incorrect theory of gases in which he assumed that all gas molecules repel one another and the walls of their container. Thus, the molecules of a gas are statically and uniformly distributed, trying to get as far apart as possible from one another and the vessel walls. This repulsion gives rise to pressure. Explain why Charles's law argues for the kinetic-molecular theory and against Newton's model.

10.75 The temperature of a 5.00-L container of N_2 gas is increased from 20 °C to 250 °C. If the volume is held constant, predict qualitatively how this change affects the following: **(a)** the average kinetic energy of the molecules; **(b)** the average speed of the molecules; **(c)** the strength of the impact of an average molecule with the container walls; **(d)** the total number of collisions of molecules with walls per second.

10.76 Suppose you have two 1-L flasks, one containing N_2 at STP, the other containing CH_4 at STP. How do these systems compare with respect to **(a)** number of molecules, **(b)** density, **(c)** average kinetic energy of the molecules, **(d)** rate of effusion through a pinhole leak?

10.77 **(a)** Place the following gases in order of increasing average molecular speed at 25 °C: Ne, HBr, SO_2, NF_3, CO. **(b)** Calculate the rms speed of NF_3 molecules at 25 °C.

10.78 **(a)** Place the following gases in order of increasing average molecular speed at 300 K: CO, SF_6, H_2S, Cl_2, HBr. **(b)** Calculate and compare the rms speeds of CO and Cl_2 molecules at 300 K.

10.79 Hydrogen has two naturally occurring isotopes, ^1H and ^2H. Chlorine also has two naturally occurring isotopes, ^{35}Cl and ^{37}Cl. Thus, hydrogen chloride gas consists of four distinct types of molecules: $^1H^{35}Cl$, $^1H^{37}Cl$, $^2H^{35}Cl$, and $^2H^{37}Cl$. Place these four molecules in order of increasing rate of effusion.

10.80 As discussed in the "Chemistry Put to Work" box in Section 10.8, enriched uranium can be produced by gaseous diffusion of UF_6. Suppose a process were developed to allow diffusion of gaseous uranium atoms, $U(g)$.

Calculate the ratio of diffusion rates for ^{235}U and ^{238}U, and compare it to the ratio for UF_6 given in the essay.

10.81 Arsenic(III) sulfide sublimes readily, even below its melting point of 320 °C. The molecules of the vapor phase are found to effuse through a tiny hole at 0.28 times the rate of effusion of Ar atoms under the same conditions of temperature and pressure. What is the molecular formula of arsenic(III) sulfide in the gas phase?

10.82 A gas of unknown molecular mass was allowed to effuse through a small opening under constant-pressure conditions. It required 105 s for 1.0 L of the gas to effuse. Under identical experimental conditions it required 31 s for 1.0 L of O_2 gas to effuse. Calculate the molar mass of the unknown gas. (Remember that the faster the rate of effusion, the shorter the time required for effusion of 1.0 L; that is, rate and time are inversely proportional.)

Nonideal-Gas Behavior

10.83 **(a)** List two experimental conditions under which gases deviate from ideal behavior. **(b)** List two reasons why the gases deviate from ideal behavior. **(c)** Explain how the function PV/RT can be used to show how gases behave nonideally.

10.84 The planet Jupiter has a surface temperature of 140 K and a mass 318 times that of Earth. Mercury has a surface temperature between 600 K and 700 K and a mass 0.05 times that of Earth. On which planet is the atmosphere more likely to obey the ideal-gas law? Explain.

10.85 Based on their respective van der Waals constants (Table 10.3), is Ar or CO_2 expected to behave more nearly like an ideal gas at high pressures? Explain.

10.86 Briefly explain the significance of the constants a and b in the van der Waals equation.

10.87 In Sample Exercise 10.16, we found that one mole of Cl_2 confined to 22.41 L at 0 °C deviated slightly from ideal behavior. Calculate the pressure exerted by 1.00 mol Cl_2 confined to a smaller volume, 5.00 L, at 25 °C. **(a)** First use the ideal gas equation and **(b)** then use van der Waals equation for your calculation. (Values for the van der Waals constants are given in Table 10.3.) **(c)** Why is the difference between the result for an ideal gas and that calculated using van der Waals equation greater when the gas is confined to 5.00 L compared to 22.4L?

10.88 Calculate the pressure that CCl_4 will exert at 40 °C if 1.00 mol occupies 28.0 L, assuming that **(a)** CCl_4 obeys the ideal-gas equation; **(b)** CCl_4 obeys the van der Waals equation. (Values for the van der Waals constants are given in Table 10.3.) **(c)** Which would you expect to deviate more from ideal behavior under these conditions, Cl_2 or CCl_4? Explain.

ADDITIONAL EXERCISES

10.89 Suppose the mercury used to make a barometer has a few small droplets of water trapped in it that rise to the top of the mercury in the tube. Will the barometer show the correct atmospheric pressure? Explain.

10.90 Suppose that when Torricelli had his great idea for constructing a mercury manometer, he had rushed into the laboratory and found the following items of glass:

Which of these would have been satisfactory for his use in forming the first manometer? Explain why the unsatisfactory ones would not have worked.

10.91 A gas bubble with a volume of 1.0 mm^3 originates at the bottom of a lake where the pressure is 3.0 atm. Calculate its volume when the bubble reaches the surface of the lake where the pressure is 695 torr, assuming that the temperature doesn't change.

10.92 A 15.0-L tank is filled with helium gas at a pressure of 1.00×10^2. How many balloons (each 2.00 L) can be inflated to a pressure of 1.00 atm, assuming that the temperature remains constant and that the tank cannot be emptied below 1.00 atm?

10.93 To minimize the rate of evaporation of the tungsten filament, 1.4×10^{-5} mol of argon is placed in a 600-cm^3 lightbulb. What is the pressure of argon in the lightbulb at 23 °C?

(a)   (b)   (c)   (d)   (e)

800 mm

10.94 Carbon dioxide, which is recognized as the major contributor to global warming as a "greenhouse gas," is formed when fossil fuels are combusted, as in electrical power plants fueled by coal, oil, or natural gas. One potential way to reduce the amount of CO_2 added to the atmosphere is to store it as a compressed gas in underground formations. Consider a 1000-megawatt coal-fired power plant that produces about 6×10^6 tons of CO_2 per year. (a) Assuming ideal gas behavior, 1.00 atm, and 27 °C, calculate the volume of CO_2 produced by this power plant. (b) If the CO_2 is stored underground as a liquid at 10 °C and 120 atm and a density of 1.2 g/cm^3, what volume does it possess? (c) If it is stored underground as a gas at 36 °C and 90 atm, what volume does it occupy?

10.95 Propane, C_3H_8, liquefies under modest pressure, allowing a large amount to be stored in a container. (a) Calculate the number of moles of propane gas in a 110-L container at 3.00 atm and 27 °C. (b) Calculate the number of moles of liquid propane that can be stored in the same volume if the density of the liquid is 0.590 g/mL. (c) Calculate the ratio of the number of moles of liquid to moles of gas. Discuss this ratio in light of the kinetic-molecular theory of gases.

10.96 Nickel carbonyl, $Ni(CO)_4$, is one of the most toxic substances known. The present maximum allowable concentration in laboratory air during an 8-hr workday is 1 part in 10^9 parts by volume, which means that there is one mole of $Ni(CO)_4$ for every 10^9 moles of gas. Assume 24 °C and 1.00 atm pressure. What mass of $Ni(CO)_4$ is allowable in a laboratory that is 54 m^2 in area, with a ceiling height of 3.1 m?

10.97 A large flask is evacuated and weighed, filled with argon gas, and then reweighed. When reweighed, the flask is found to have gained 3.224 g. It is again evacuated and then filled with a gas of unknown molar mass. When reweighed, the flask is found to have gained 8.102 g. (a) Based on the molar mass of argon, estimate the molar mass of the unknown gas. (b) What assumptions did you make in arriving at your answer?

10.98 Consider the arrangement of bulbs shown in the drawing. Each of the bulbs contains a gas at the pressure shown. What is the pressure of the system when all the stopcocks are opened, assuming that the temperature remains constant? (We can neglect the volume of the capillary tubing connecting the bulbs.)

Volume	1.0 L	1.0 L	0.5 L
Pressure	265 torr	800 torr	532 torr

10.99 Assume that a single cylinder of an automobile engine has a volume of 524 cm^3. (a) If the cylinder is full of air at 74 °C and 0.980 atm, how many moles of O_2 are present? (The mole fraction of O_2 in dry air is 0.2095.)

(b) How many grams of C_8H_{18} could be combusted by this quantity of O_2, assuming complete combustion with formation of CO_2 and H_2O?

10.100 Assume that an exhaled breath of air consists of 74.8% N_2, 15.3% O_2, 3.7% CO_2, and 6.2% water vapor. (a) If the total pressure of the gases is 0.980 atm, calculate the partial pressure of each component of the mixture. (b) If the volume of the exhaled gas is 455 mL and its temperature is 37 °C, calculate the number of moles of CO_2 exhaled. (c) How many grams of glucose ($C_6H_{12}O_6$) would need to be metabolized to produce this quantity of CO_2? (The chemical reaction is the same as that for combustion of $C_6H_{12}O_6$. See Section 3.2.)

10.101 A sample of 1.42 g of helium and an unweighed quantity of O_2 are mixed in a flask at room temperature. The partial pressure of helium in the flask is 42.5 torr, and the partial pressure of oxygen is 158 torr. What is the mass of the oxygen in the container?

10.102 A gaseous mixture of O_2 and Kr has a density of 1.104 g/L at 435 torr and 300 K. What is the mole percent O_2 in the mixture?

10.103 The density of a gas of unknown molar mass was measured as a function of pressure at 0 °C, as in the table below. (a) Determine a precise molar mass for the gas. *Hint:* Graph d/P versus P. (b) Why is d/P not a constant as a function of pressure?

Pressure (atm)	1.00	0.666	0.500	0.333	0.250
Density (g/L)	2.3074	1.5263	1.1401	0.7571	0.5660

10.104 A glass vessel fitted with a stopcock has a mass of 337.428 g when evacuated. When filled with Ar, it has a mass of 339.854 g. When evacuated and refilled with a mixture of Ne and Ar, under the same conditions of temperature and pressure, it weighs 339.076 g. What is the mole percent of Ne in the gas mixture?

10.105 You have a sample of gas at −33 °C. You wish to increase the rms speed by 10.0%. To what temperature should the gas be heated?

10.106 Consider the following gases, all at STP: Ne, SF_6, N_2, CH_4. (a) Which gas is most likely to depart from assumption 3 of the kinetic molecular theory (Section 10.7)? (b) Which one is closest to an ideal gas in its behavior? (c) Which one has the highest root-mean-square molecular speed? (d) Which one has the highest total molecular volume relative to the space occupied by the gas? (e) Which has the highest average kinetic molecular energy? (f) Which one would effuse more rapidly than N_2?

10.107 Does the effect of intermolecular attraction on the properties of a gas become more significant or less significant if (a) the gas is compressed to a smaller volume at constant temperature; (b) the temperature of the gas is increased at constant volume?

10.108 For nearly all real gases, the quantity PV/RT decreases below the value of 1, which characterizes an ideal gas, as pressure on the gas increases. At much higher pressures, however, PV/RT increases and rises above the value of 1. **(a)** Explain the initial drop in value of PV/RT below 1 and the fact that it rises above 1 for still higher pressures. **(b)** The effects we have just noted are smaller for gases at higher temperature. Why is this so?

10.109 Which of the noble gases other than radon would you expect to depart most readily from ideal behavior? Use the density data in Table 7.8 to show evidence in support of your answer.

10.110 It turns out that the van der Waals constant b equals four times the total volume actually occupied by the molecules of a mole of gas. Using this figure, calculate the fraction of the volume in a container actually occupied by Ar atoms **(a)** at STP, **(b)** at 100 atm pressure and 0 °C. (Assume for simplicity that the ideal-gas equation still holds.)

[10.111] Large amounts of nitrogen gas are used in the manufacture of ammonia, principally for use in fertilizers. Suppose 120.00 kg of $N_2(g)$ is stored in a 1100.0-L metal cylinder at 280 °C. **(a)** Calculate the pressure of the gas, assuming ideal-gas behavior. **(b)** By using data in Table 10.3, calculate the pressure of the gas according to the van der Waals equation. **(c)** Under the conditions of this problem, which correction dominates, the one for finite volume of gas molecules or the one for attractive interactions?

INTEGRATIVE EXERCISES

10.112 Cyclopropane, a gas used with oxygen as a general anesthetic, is composed of 85.7% C and 14.3% H by mass. **(a)** If 1.56 g of cyclopropane has a volume of 1.00 L at 0.984 atm and 50.0 °C, what is the molecular formula of cyclopropane? **(b)** Judging from its molecular formula, would you expect cyclopropane to deviate more or less than Ar from ideal-gas behavior at moderately high pressures and room temperature? Explain.

10.113 Consider the combustion reaction between 25.0 mL of liquid methanol (density = 0.850 g/mL) and 12.5 L of oxygen gas measured at STP. The products of the reaction are $CO_2(g)$ and $H_2O(g)$. Calculate the number of moles of H_2O formed if the reaction goes to completion.

10.114 An herbicide is found to contain only C, H, N, and Cl. The complete combustion of a 100.0-mg sample of the herbicide in excess oxygen produces 83.16 mL of CO_2 and 73.30 mL of H_2O vapor at STP. A separate analysis shows that the sample also contains 16.44 mg of Cl. **(a)** Determine the percent composition of the substance. **(b)** Calculate its empirical formula.

[10.115] A 4.00-g sample of a mixture of CaO and BaO is placed in a 1.00-L vessel containing CO_2 gas at a pressure of 730 torr and a temperature of 25 °C. The CO_2 reacts with the CaO and BaO, forming $CaCO_3$ and $BaCO_3$. When the reaction is complete, the pressure of the remaining CO_2 is 150 torr. **(a)** Calculate the number of moles of CO_2 that have reacted. **(b)** Calculate the mass percentage of CaO in the mixture.

[10.116] Ammonia, $NH_3(g)$, and hydrogen chloride, $HCl(g)$, react to form solid ammonium chloride, $NH_4Cl(s)$:

$$NH_3(g) + HCl(g) \longrightarrow NH_4Cl(s)$$

Two 2.00-L flasks at 25 °C are connected by a stopcock, as shown in the drawing. One flask contains 5.00 g $NH_3(g)$, and the other contains 5.00 g $HCl(g)$. When the stopcock is opened, the gases react until one is completely consumed. **(a)** Which gas will remain in the system after the reaction is complete? **(b)** What will be the final pressure of the system after the reaction is complete? (Neglect the volume of the ammonium chloride formed.)

10.117 The "Chemistry Put to Work" box on pipelines in Section 10.5 mentions that the total deliverability of natural gas (methane, CH_4) to the various regions of the United States is on the order of 2.7×10^{12} L per day, measured at STP. Calculate the total enthalpy change for combustion of this quantity of methane. (*Note*: Less than this amount of methane is actually combusted daily. Some of the delivered gas is passed through to other regions.)

[10.118] A gas forms when elemental sulfur is heated carefully with AgF. The initial product boils at 15 °C. Experiments on several samples yielded a gas density of 0.803 ± 0.010 g/L for the gas at 150 mm pressure and 32 °C. When the gas reacts with water, all the fluorine is converted to aqueous HF. Other products are elemental sulfur, S_8, and other sulfur-containing compounds. A 480-mL sample of the dry gas at 126 mm pressure and 28 °C, when reacted with 80 mL of water, yielded a 0.081 M solution of HF. The initial gaseous product undergoes a transformation over a period of time to a second compound with the same empirical and molecular formula, which boils at −10 °C. **(a)** Determine the empirical and molecular formulas of the first compound formed. **(b)** Draw at least two reasonable Lewis structures that represent the initial compound and the one into which it is transformed over time. **(c)** Describe the likely geometries of these compounds, and estimate the single bond distances, given that the S—S bond distance in S_8 is 2.04 Å and the F—F distance in F_2 is 1.43 Å.

10.119 Chlorine dioxide gas (ClO_2) is used as a commercial bleaching agent. It bleaches materials by oxidizing them. In the course of these reactions, the ClO_2 is itself reduced. **(a)** What is the Lewis structure for ClO_2? **(b)** Why do you think that ClO_2 is reduced so readily? **(c)** When a ClO_2 molecule gains an electron, the chlorite ion, ClO_2^-, forms. Draw the Lewis structure for ClO_2^-. **(d)** Predict the O—Cl—O bond angle in the ClO_2^- ion. **(e)** One method of preparing ClO_2 is by the reaction of chlorine and sodium chlorite:

$$Cl_2(g) + 2\,NaClO_2(s) \longrightarrow 2\,ClO_2(g) + 2\,NaCl(s)$$

If you allow 10.0 g of $NaClO_2$ to react with 2.00 L of chlorine gas at a pressure of 1.50 atm at 21 °C, how many grams of ClO_2 can be prepared?

10.120 Natural gas is very abundant in many Middle Eastern oil fields. However, the costs of shipping the gas to markets in other parts of the world are high because it is necessary to liquefy the gas, which is mainly methane and thus has a boiling point at atmospheric pressure of −164 °C. One possible strategy is to oxidize the methane to methanol, CH_3OH, which has a boiling point of 65 °C and can therefore be shipped more readily. Suppose that 10.7×10^9 ft^3 of methane at atmospheric pressure and 25 °C are oxidized to methanol. **(a)** What volume of methanol is formed if the density of CH_3OH is 0.791 g/mL? **(b)** Write balanced chemical equations for the oxidations of methane and methanol to $CO_2(g)$ and $H_2O(l)$. Calculate the total enthalpy change for complete combustion of the 10.7×10^9 ft^3 of methane described above and for complete combustion of the equivalent amount of methanol, as calculated in part (a). **(c)** Methane, when liquefied, has a density of 0.466 g/mL; the density of methanol at 25 °C is 0.791 g/mL. Compare the enthalpy change upon combustion of a unit volume of liquid methane and liquid methanol. From the standpoint of energy production, which substance has the higher enthalpy of combustion per unit volume?

[10.121] Gaseous iodine pentafluoride, IF_5, can be prepared by the reaction of solid iodine and gaseous fluorine:

$$I_2(s) + 5\,F_2(g) \longrightarrow 2\,IF_5(g)$$

A 5.00-L flask containing 10.0 g I_2 is charged with 10.0 g F_2, and the reaction proceeds until one of the reagents is completely consumed. After the reaction is complete, the temperature in the flask is 125 °C. **(a)** What is the partial pressure of IF_5 in the flask? **(b)** What is the mole fraction of IF_5 in the flask?

[10.122] A 6.53-g sample of a mixture of magnesium carbonate and calcium carbonate is treated with excess hydrochloric acid. The resulting reaction produces 1.72 L of carbon dioxide gas at 28 °C and 743 torr pressure. **(a)** Write balanced chemical equations for the reactions that occur between hydrochloric acid and each component of the mixture. **(b)** Calculate the total number of moles of carbon dioxide that forms from these reactions. **(c)** Assuming that the reactions are complete, calculate the percentage by mass of magnesium carbonate in the mixture.

11

INTERMOLECULAR FORCES, LIQUIDS, AND SOLIDS

THE WATER IN HOT SPRINGS is at least 5–10 °C warmer than the mean annual air temperature where they are located. There are over 1000 hot springs in the United States, the majority in the West. Some hot springs are popular for bathing, while others are far too hot for people to enter them safely.

WHAT'S AHEAD

11.1 A Molecular Comparison of Gases, Liquids, and Solids

We begin with a brief comparison of solids, liquids, and gases from a molecular perspective. This comparison reveals the important roles that temperature and *intermolecular forces* play in determining the physical state of a substance.

11.2 Intermolecular Forces

We then examine the main kinds of intermolecular forces that occur within and between substances: *ion–dipole forces, dipole–dipole forces, London dispersion forces,* and *hydrogen bonds.*

11.3 Some Properties of Liquids

We will learn that the nature and strength of the intermolecular forces between molecules are largely responsible for many properties of liquids, including their *viscosity*, which is a measure of a liquid's resistance to flow, and *surface tension*, which is a measure of a liquid's resistance to increasing its surface area.

11.4 Phase Changes

We will explore *phase changes*—the transitions of matter between the gaseous, liquid, and solid states—and their associated energies.

11.5 Vapor Pressure

We will examine the *dynamic equilibrium* that exists between a liquid and its gaseous state and introduce the idea of *vapor pressure*. A liquid boils when its vapor pressure equals the pressure acting on the surface of the liquid.

11.6 Phase Diagrams

We will recognize that in a *phase diagram* the equilibria among the gaseous, liquid, and solid phases are displayed graphically.

11.7 Structures of Solids

We then observe that the orderly arrangements of molecules or ions in three dimensions characterize *crystalline solids*. We also examine how the structure of a crystalline solid can be conveyed in terms of its *unit cell* and how simple molecules and ions are most efficiently arranged in three dimensions.

11.8 Bonding in Solids

We learn that solids can be characterized and classified according to the attractive forces between their component atoms, molecules, or ions. We examine four such classes: *molecular solids, covalent-network solids, ionic solids,* and *metallic solids.*

SITTING IN A HOT SPRING ON A SNOWY day is not something many of us have experienced. If we were in the hot spring shown in the chapter-opening photo, however, we would be surrounded simultaneously by all three phases of water—gas, liquid, and solid. The water vapor—

or humidity—in the air, the water in the hot spring, and the surrounding snow are all forms of the same substance, H_2O. They all have the same chemical properties. Their physical properties differ greatly, however, because the physical properties of a substance depend on its physical state. In Chapter 10 we discussed the gaseous state in some detail. In this chapter we turn our attention to the physical properties of liquids and solids and to the phase changes that occur between the three states of matter.

Many of the substances that we will consider in this chapter are molecular. In fact, virtually all substances that are liquids at room temperature are molecular substances. The intramolecular forces *within* molecules that give rise to covalent bonding influence molecular shape, bond energies, and many aspects of chemical behavior. The physical properties of molecular liquids and solids, however, are due largely to **intermolecular forces**, the forces that exist *between* molecules. We learned in Section 10.9 that attractions between gas molecules lead to deviations from ideal-gas behavior. But how do these intermolecular attractions arise? By understanding the nature and strength of intermolecular forces, we can begin to relate the composition and structure of molecules to their physical properties.

TABLE 11.1 ■ Some Characteristic Properties of the States of Matter	
Gas	Assumes both the volume and shape of its container
	Is compressible
	Flows readily
	Diffusion within a gas occurs rapidly
Liquid	Assumes the shape of the portion of the container it occupies
	Does not expand to fill container
	Is virtually incompressible
	Flows readily
	Diffusion within a liquid occurs slowly
Solid	Retains its own shape and volume
	Is virtually incompressible
	Does not flow
	Diffusion within a solid occurs extremely slowly

11.1 A MOLECULAR COMPARISON OF GASES, LIQUIDS, AND SOLIDS

Some of the characteristic properties of gases, liquids, and solids are listed in Table 11.1▲. These properties can be understood in terms of the energy of motion (kinetic energy) of the particles (atoms, molecules, or ions) of each state, compared to the energy of the intermolecular interactions between those particles. As we learned from the kinetic-molecular theory of gases in Chapter 10, the average kinetic energy, which is related to the particle's average speed, is proportional to the absolute temperature.

Gases consist of a collection of widely separated particles in constant, chaotic motion. The average energy of the attractions between the particles is much smaller than their average kinetic energy. The lack of strong attractive forces between particles allows a gas to expand to fill its container.

In liquids the intermolecular attractive forces are stronger than in gases and are strong enough to hold particles close together. Thus, liquids are much denser and far less compressible than gases. Unlike gases, liquids have a definite volume, independent of the size and shape of their container. The attractive forces in liquids are not strong enough, however, to keep the particles from moving past one another. Thus, any liquid can be poured, and it assumes the shape of whatever portion of its container it occupies.

In solids the intermolecular attractive forces are strong enough to hold particles close together and to lock them virtually in place. Solids, like liquids, are not very compressible because the particles have little free space between them. Because the particles in a solid or liquid are fairly close together compared with those of a gas, we often refer to solids and liquids as *condensed phases*. Often the particles of a solid take up positions in a highly regular three-dimensional pattern. Solids that possess highly ordered three-dimensional structures are said to be *crystalline*. Because the particles of a solid are not free to undergo long-range movement, solids are rigid. Keep in mind, however, that the units that form the solid, whether ions or molecules, possess thermal energy and vibrate in place. This vibrational motion increases in amplitude as a solid is heated. In fact, the energy may increase to the point that the solid either melts or sublimes.

Figure 11.1 ▶ compares the three states of matter. The particles that compose the substance can be individual atoms, as in Ar; molecules, as in H_2O; or ions, as in NaCl. *The state of a substance depends largely on the balance between the kinetic energies of the particles and the interparticle energies of attraction.* The kinetic energies, which depend on temperature, tend to keep the particles apart and moving. The interparticle attractions tend to draw the particles together.

Gas

Total disorder; much empty space; particles have complete freedom of motion; particles far apart

Liquid

Disorder; particles or clusters of particles are free to move relative to each other; particles close together

Crystalline solid

Ordered arrangement; particles are essentially in fixed positions; particles close together

◀ **Figure 11.1 Molecular-level comparison of gases, liquids, and solids.** The particles can be atoms, ions, or molecules. The density of particles in the gas phase is exaggerated compared with most real situations.

Substances that are gases at room temperature have weaker interparticle attractions than those that are liquids; substances that are liquids have weaker interparticle attractions than those that are solids.

We can change a substance from one state to another by heating or cooling, which changes the average kinetic energy of the particles. NaCl, for example, which is a solid at room temperature, melts at 801 °C and boils at 1413 °C under 1 atm pressure. N_2O, on the other hand, which is a gas at room temperature, liquefies at −88.5 °C and solidifies at −90.8 °C under 1 atm pressure. As the temperature of a gas decreases, the average kinetic energy of its particles decreases, allowing the attractions between the particles to first draw the particles close together, forming a liquid, and then virtually locking them in place, forming a solid.

Increasing the pressure on a gas forces the molecules closer together, which in turn increases the strength of the intermolecular forces of attraction. Propane (C_3H_8) is a gas at room temperature and 1 atm pressure, whereas liquefied propane (LP) gas is a liquid at room temperature because it is stored under much higher pressure.

◤ **GIVE IT SOME THOUGHT**

How does the energy of attraction between particles compare with their kinetic energies in **(a)** a gas, **(b)** a solid?

11.2 INTERMOLECULAR FORCES

The strengths of intermolecular forces of different substances vary over a wide range, but they are generally much weaker than ionic or covalent bonds (Figure 11.2▶). Less energy, therefore, is required to vaporize, or evaporate, a liquid or to melt a solid than to break covalent bonds in molecules. For example, only 16 kJ/mol is required to overcome the intermolecular attractions between HCl molecules in liquid HCl to vaporize it. In contrast, the energy required to break the covalent bond to dissociate HCl into H and Cl atoms is 431 kJ/mol. Thus, when a molecular substance such as HCl changes from solid to liquid to gas, the molecules themselves remain intact.

Many properties of liquids, including their *boiling points*, reflect the strengths of the intermolecular forces. For example, because the forces between HCl molecules are so weak, HCl boils at a very low temperature, −85 °C at atmospheric pressure. A liquid boils when bubbles of its vapor form within the liquid.

▼ **Figure 11.2 Intermolecular attraction.** Comparison of a covalent bond (an intramolecular force) and an intermolecular attraction. Because intermolecular attractions are weaker than covalent bonds, they are usually represented by dashes or dots.

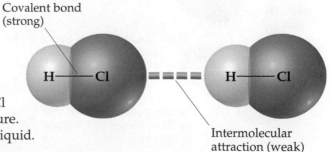

Covalent bond (strong)

H —— Cl

Intermolecular attraction (weak)

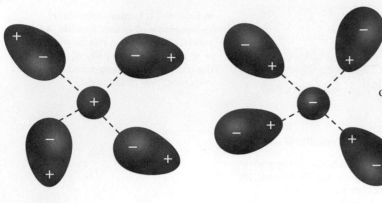

Cation–dipole attractions   Anion–dipole attractions

(a)   (b)

▲ **Figure 11.3 Ion–dipole attractions.** Illustration of the preferred orientations of polar molecules toward ions. The negative end of the dipoles is oriented toward a cation (a), and the positive end of the dipoles is oriented toward an anion (b).

The molecules of a liquid must overcome their attractive forces to separate and form a vapor. The stronger the attractive forces are, the higher the temperature at which the liquid boils. Similarly, the *melting points* of solids increase as the strengths of the intermolecular forces increase.

Three types of intermolecular attractions exist between neutral molecules: dipole–dipole attractions, London dispersion forces, and hydrogen bonding. The first two forms of attraction are collectively called *van der Waals forces* after Johannes van der Waals, who developed the equation for predicting the deviation of gases from ideal behavior. ∞ (Section 10.9) Another kind of attractive force, the ion–dipole force, is important in solutions. All of these intermolecular interactions are electrostatic in nature, involving attractions between positive and negative species. Even at their strongest, these interactions are much weaker than covalent or ionic bonds (<15% as strong).

Ion–Dipole Forces

An **ion–dipole force** exists between an ion and the partial charge on the end of a polar molecule. Polar molecules are dipoles; they have a positive end and a negative end. ∞ (Section 9.3) HCl is a polar molecule, for example, because the electronegativities of the H and Cl atoms differ.

Positive ions are attracted to the negative end of a dipole, whereas negative ions are attracted to the positive end, as shown in Figure 11.3 ▲. The magnitude of the attraction increases as either the charge of the ion or the magnitude of the dipole moment increases. Ion–dipole forces are especially important for solutions of ionic substances in polar liquids, such as a solution of NaCl in water. ∞ (Section 4.1) We will discuss these solutions in more detail in Section 13.1.

> ### GIVE IT SOME THOUGHT
>
> In which of the following mixtures do you encounter ion–dipole forces: CH_3OH in water or $Ca(NO_3)_2$ in water?

▼ **Figure 11.4 Dipole–dipole attractions.** The interaction of many dipoles in a condensed state. There are both repulsive interactions between like charges and attractive interactions between unlike charges, but the attractive interactions predominate.

The interaction between any two opposite charges is attractive (solid red lines).

The interaction between any two like charges is repulsive (dashed blue lines).

Dipole–Dipole Forces

Neutral polar molecules attract each other when the positive end of one molecule is near the negative end of another, as shown in Figure 11.4 ◀. These **dipole–dipole forces** are effective only when polar molecules are very close together. Dipole–dipole forces are generally weaker than ion–dipole forces.

In liquids polar molecules are free to move with respect to one another. As shown in Figure 11.4, the polar molecules will sometimes be in an orientation that is attractive (red solid lines) and sometimes in an orientation that is repulsive (blue dashed lines). Two molecules that are attracting each other spend more time near each other than do two that are repelling each other. Thus, the overall effect is a net attraction. When we examine various liquids, we find that *for molecules of approximately equal mass and size, the strengths of intermolecular attractions increase with increasing polarity*. We can see this trend in Table 11.2 ▶, which lists several substances with similar molecular weights but different dipole moments. Notice that the boiling point increases as the dipole moment increases. For dipole–dipole forces to operate, the molecules must be able to get close together in the correct orientation.

TABLE 11.2 ■ Molecular Weights, Dipole Moments, and Boiling Points of Several Simple Organic Substances

Substance	Molecular Weight (amu)	Dipole Moment μ (D)	Boiling Point (K)
Propane, $CH_3CH_2CH_3$	44	0.1	231
Dimethyl ether, CH_3OCH_3	46	1.3	248
Methyl chloride, CH_3Cl	50	1.9	249
Acetaldehyde, CH_3CHO	44	2.7	294
Acetonitrile, CH_3CN	41	3.9	355

For molecules of comparable polarity, therefore, those with smaller molecular volumes generally experience higher dipole–dipole attractive forces.

GIVE IT SOME THOUGHT

For which of the substances in Table 11.2 are the dipole–dipole attractive forces greatest?

London Dispersion Forces

No dipole–dipole forces exist between nonpolar atoms and molecules, because nonpolar molecules (and atoms) do not have a dipole moment. Some kind of attractive interactions exist, however, because nonpolar gases such as helium and argon can be liquefied. Fritz London, a German-American physicist, first proposed the origin of this attraction in 1930. London recognized that the motion of electrons in an atom or molecule can create an *instantaneous*, or momentary, dipole moment.

In a collection of helium atoms, for example, the *average* distribution of the electrons about each nucleus is spherically symmetrical. The atoms are nonpolar and possess no permanent dipole moment. The *instantaneous* distribution of the electrons, however, can be different from the average distribution. If we could freeze the motion of the electrons in a helium atom at any given instant, both electrons could be on one side of the nucleus. At just that instant, then, the atom would have an instantaneous dipole moment.

The motions of electrons on one atom influence the motions of electrons on its neighbors. The temporary dipole on one atom can induce a similar temporary dipole on an adjacent atom, causing the atoms to be attracted to each other as shown in Figure 11.5▼. This attractive interaction is called the **London dispersion force** (or merely the *dispersion force*). The dispersion force, like dipole–dipole forces, is significant only when molecules are very close together.

The strength of the dispersion force depends on the ease with which the charge distribution in a molecule can be distorted to induce a momentary dipole. The ease with which the electron distribution in a molecule is distorted is called its **polarizability**. We can think of the polarizability of a molecule as a measure of the "squashiness" of its electron cloud; the greater the polarizability of the molecule, the more easily its electron cloud can be distorted to give a momentary dipole. Therefore, more polarizable molecules have larger dispersion forces.

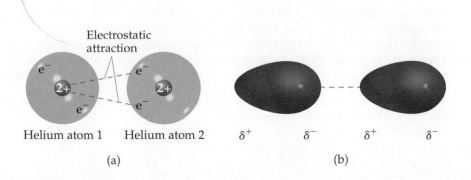

Electrostatic attraction

Helium atom 1   Helium atom 2 δ^+ δ^- δ^+ δ^-

(a)   (b)

◄ **Figure 11.5 Dispersion forces.** On the average, the charge distribution in the helium atoms is spherical, as represented by the spheres in (a). At a particular instant, however, there can be a nonspherical arrangement of the electrons, as indicated by the location of the electrons (e^-) in (a) and by the nonspherical shape of the electron cloud in (b). The nonspherical electron distributions produce momentary dipoles and allow momentary electrostatic attractions between atoms that are called London dispersion forces or merely dispersion forces.

TABLE 11.3 ■ Boiling Points of the Halogens and the Noble Gases					
Halogen	**Molecular Weight (amu)**	**Boiling Point (K)**	**Noble Gas**	**Molecular Weight (amu)**	**Boiling Point (K)**
F_2	38.0	85.1	He	4.0	4.6
Cl_2	71.0	238.6	Ne	20.2	27.3
Br_2	159.8	332.0	Ar	39.9	87.5
I_2	253.8	457.6	Kr	83.8	120.9
			Xe	131.3	166.1

In general, larger molecules tend to have greater polarizabilities because they have a greater number of electrons, and their electrons are farther from the nuclei. The strength of the dispersion forces, therefore, tends to increase with increasing atomic or molecular size. Because molecular size and mass generally parallel each other, *dispersion forces tend to increase in strength with increasing molecular weight*. Thus, the boiling points of the halogens and the noble gases increase with increasing molecular weight (Table 11.3▲).

n-Pentane
(bp = 309.4 K)

Neopentane
(bp = 282.7 K)

▲ **Figure 11.6 Molecular shape affects intermolecular attraction.** The *n*-pentane molecules make more contact with each other than do the neopentane molecules. Thus, *n*-pentane has the greater intermolecular attractive forces and therefore has the higher boiling point (bp).

▲ **GIVE IT SOME THOUGHT**

List the substances CCl_4, CBr_4, and CH_4 in order of increasing **(a)** polarizability, **(b)** strength of dispersion forces, and (c) boiling point.

The shapes of molecules also influence the magnitudes of dispersion forces. For example, *n*-pentane* and neopentane, illustrated in Figure 11.6◄, have the same molecular formula (C_5H_{12}), yet the boiling point of *n*-pentane is 27 K higher than that of neopentane. The difference can be traced to the different shapes of the two molecules. The overall attraction between molecules is greater for *n*-pentane because the molecules can come in contact over the entire length of the long, somewhat cylindrically shaped molecule. Less contact is possible between the more compact and nearly spherical molecules of neopentane.

Dispersion forces operate between all molecules, whether they are polar or nonpolar. Polar molecules experience dipole–dipole interactions, but they also experience dispersion forces at the same time. In fact, dispersion forces between polar molecules commonly contribute more to intermolecular attractions than do dipole–dipole forces. In liquid HCl, for example, dispersion forces are estimated to account for more than 80% of the total attraction between molecules; dipole–dipole attractions account for the rest.

When comparing the relative strengths of intermolecular attractions in two substances, consider the following generalizations:

1. When the molecules of two substances have comparable molecular weights and shapes, dispersion forces are approximately equal in the two substances. In this case, differences in the magnitudes of the attractive forces are due to differences in the strengths of dipole–dipole attractions, with the more polar molecules having the stronger attractions.

2. When the molecules of two substances differ widely in molecular weights, dispersion forces tend to be decisive in determining which substance has the stronger intermolecular attractions. In this case differences in the magnitudes of the attractive forces can usually be associated with differences in molecular weights, with the substance made up of the more massive molecule having the strongest attractions (assuming similar dipole moments).

*The n in n-pentane is an abbreviation for the word normal. A normal hydrocarbon is one in which the carbon atoms are arranged in a straight chain. ∞ (Section 2.9)

The hydrogen bond, which we consider after Sample Exercise 11.1, is a special type of dipole–dipole attraction that is typically stronger than dispersion forces.

■ **SAMPLE EXERCISE 11.1** | **Comparing Intermolecular Forces**

The dipole moments of acetonitrile, CH_3CN, and methyl iodide, CH_3I, are 3.9 D and 1.62 D, respectively. **(a)** Which of these substances has greater dipole–dipole attractions among its molecules? **(b)** Which of these substances has greater London dispersion attractions? **(c)** The boiling points of CH_3CN and CH_3I are 354.8 K and 315.6 K, respectively. Which substance has the greater overall attractive forces?

SOLUTION

(a) Dipole–dipole attractions increase in magnitude as the dipole moment of the molecule increases. Thus, CH_3CN molecules attract each other by stronger dipole–dipole forces than CH_3I molecules do. **(b)** When molecules differ in their molecular weights, the more massive molecule generally has the stronger dispersion attractions. In this case CH_3I (142.0 g/mol) is much more massive than CH_3CN (41.0 g/mol), so the dispersion forces will be stronger for CH_3I. **(c)** Because CH_3CN has the higher boiling point, we can conclude that more energy is required to overcome attractive interactions between CH_3CN molecules. Thus, the total intermolecular attractions are stronger for CH_3CN, suggesting that the energies resulting from dipole–dipole forces are decisive when comparing these two substances. Nevertheless, the attractive interactions due to dispersion forces play an important role in determining the properties of CH_3I.

■ **PRACTICE EXERCISE**

Of Br_2, Ne, HCl, HBr, and N_2, which is likely to have **(a)** the largest intermolecular dispersion forces, **(b)** the largest dipole–dipole attractive forces? *Answers:* **(a)** Br_2 (largest molecular weight), **(b)** HCl (largest polarity)

Hydrogen Bonding

Figure 11.7▶ shows the boiling points of the simple hydrogen compounds of group 4A and 6A elements. In general, the boiling point increases with increasing molecular weight, owing to increased dispersion forces. The notable exception to this trend is H_2O, whose boiling point is much higher than we would expect because of its molecular weight. This observation indicates that there are stronger intermolecular attractions between H_2O molecules than there are between other molecules in the same group. The compounds NH_3 and HF likewise have abnormally high boiling points. In fact, these compounds have many characteristics that distinguish them from other substances of similar molecular weight and polarity. For example, water has a high melting point, a high specific heat, and a high heat of vaporization. Each of these properties indicates that the intermolecular forces between H_2O molecules are abnormally strong.

The strong intermolecular attractions in H_2O result from hydrogen bonding. **Hydrogen bonding** *is a special type of intermolecular attraction between the hydrogen atom in a polar bond (particularly an H—F, H—O, or H—N bond) and nonbonding electron pair on a nearby small electronegative ion or atom (usually an F, O, or N atom in another molecule).* For example, a hydrogen bond exists between the H atom in an HF molecule and the F atom of an adjacent HF molecule, F—H···F—H (where the

▼ **Figure 11.7 Boiling point as a function of molecular weight.** The boiling points of the group 4A (bottom) and 6A (top) hydrides are shown as a function of molecular weight. In general, the boiling points increase with increasing molecular weight, because of the increasing strength of dispersion forces. The very strong hydrogen-bonding forces between H_2O molecules, however, cause water to have an unusually high boiling point.

$$H—\overset{..}{\underset{|}{O}}:\cdots H—\overset{..}{\underset{|}{O}}:$$
$$HH$$

$$\overset{H}{\underset{|}{H}}—\overset{|}{\underset{|}{N}}:\cdots H—\overset{H}{\underset{|}{N}}:$$
$$HH$$

$$\overset{H}{\underset{|}{}}$$
$$H—\overset{|}{\underset{|}{N}}:\cdots H—\overset{..}{\underset{|}{O}}:$$
$$HH$$

$$\overset{H}{\underset{|}{}}$$
$$H—\overset{..}{\underset{|}{O}}:\cdots H—\overset{|}{\underset{|}{N}}:$$
$$HH$$

▲ **Figure 11.8 Examples of hydrogen bonding.** The solid lines represent covalent bonds; the red dotted lines represent hydrogen bonds.

dots represent the hydrogen bond between the molecules). Several additional examples are shown in Figure 11.8 ◄.

Hydrogen bonds can be considered unique dipole–dipole attractions. Because F, N, and O are so electronegative, a bond between hydrogen and any of these three elements is quite polar, with hydrogen at the positive end:

$$\overset{\longleftarrow+}{N—H}\qquad\overset{\longleftarrow+}{O—H}\qquad\overset{\longleftarrow+}{F—H}$$

The hydrogen atom has no inner core of electrons. Thus, the positive side of the bond dipole has the concentrated charge of the partially exposed, nearly bare proton of the hydrogen nucleus. This positive charge is attracted to the negative charge of an electronegative atom in a nearby molecule. Because the electron-poor hydrogen is so small, it can approach an electronegative atom very closely and thus interact strongly with it.

■ SAMPLE EXERCISE 11.2 | Identifying Substances that Can Form Hydrogen Bonds

In which of the following substances is hydrogen bonding likely to play an important role in determining physical properties: methane (CH_4), hydrazine (H_2NNH_2), methyl fluoride (CH_3F), or hydrogen sulfide (H_2S)?

SOLUTION

Analyze: We are given the chemical formulas of four substances and asked to predict whether they can participate in hydrogen bonding. All of these compounds contain H, but hydrogen bonding usually occurs only when the hydrogen is covalently bonded to N, O, or F.

Plan: We can analyze each formula to see if it contains N, O, or F directly bonded to H. There also needs to be a nonbonding pair of electrons on an electronegative atom (usually N, O, or F) in a nearby molecule, which can be revealed by drawing the Lewis structure for the molecule.

Solve: The criteria listed above eliminate CH_4 and H_2S, which do not contain H bonded to N, O, or F. They also eliminate CH_3F, whose Lewis structure shows a central C atom surrounded by three H atoms and an F atom. (Carbon always forms four bonds, whereas hydrogen and fluorine form one each.) Because the molecule contains a C—F bond and not a H—F bond, it does not form hydrogen bonds. In H_2NNH_2, however, we find N—H bonds. If we draw the Lewis structure for the molecule, we see that there is a nonbonding pair of electrons on each N atom. Therefore, hydrogen bonds can exist between the molecules as depicted below.

$$\overset{H}{\underset{|}{:N}}—\overset{H}{\underset{|}{N}}:\cdots H—\overset{H}{\underset{|}{\overset{..}{N}}}—\overset{H}{\underset{|}{N}}:$$

Check: While we can generally identify substances that participate in hydrogen bonding based on their containing N, O, or F covalently bonded to H, drawing the Lewis structure for the interaction, as shown above, provides a way to check the prediction.

■ PRACTICE EXERCISE

In which of the following substances is significant hydrogen bonding possible: methylene chloride (CH_2Cl_2), phosphine (PH_3), hydrogen peroxide (HOOH), or acetone (CH_3COCH_3)?
Answer: HOOH

The energies of hydrogen bonds vary from about 5 kJ/mol to 25 kJ/mol or so, although there are isolated examples of hydrogen bond energies that are close to 100 kJ/mol. Thus, hydrogen bonds are typically much weaker than ordinary chemical bonds, which have bond energies of 200–1100 kJ/mol (see Table 8.4). Nevertheless, because hydrogen bonds are generally stronger than dipole–dipole or dispersion forces, they play important roles in many chemical systems, including those of biological significance. For example, hydrogen bonds help stabilize the structures of proteins, which are key parts of skin, muscles, and other structural components of animal tissues. ∞ (Section 25.9) They are also responsible for the way that DNA is able to carry genetic information. ∞ (Section 25.11)

One of the remarkable consequences of hydrogen bonding is found when the densities of ice and liquid water are compared. In most substances the molecules in the solid are more densely packed than in the liquid. Thus, the solid phase is denser than the liquid phase (Figure 11.9▶). By contrast, the density of ice at 0 °C (0.917 g/mL) is less than that of liquid water at 0 °C (1.00 g/mL), so ice floats on liquid water (Figure 11.9).

The lower density of ice compared to that of water can be understood in terms of hydrogen-bonding interactions between H_2O molecules. In ice, the H_2O molecules assume an ordered, open arrangement as shown in Figure 11.10▼. This arrangement optimizes the hydrogen bonding interactions between molecules, with each H_2O molecule forming hydrogen bonds to four other H_2O molecules. These hydrogen bonds, however, create the open cavities shown in the structure. When the ice melts, the motions of the molecules cause the structure to collapse. The hydrogen bonding in the liquid is more random than in ice, but it is strong enough to hold the molecules close together. Consequently, liquid water has a more dense structure than ice, meaning that a given mass of water occupies a smaller volume than the same mass of ice.

▲ **Figure 11.9 Comparing densities of liquid and solid phases.** As with most other substances, the solid phase of paraffin is denser than the liquid phase, and the solid therefore sinks below the surface of the liquid paraffin in the beaker on the left. In contrast, the solid phase of water, ice, is less dense than its liquid phase (right beaker), causing the ice to float on the water.

(a) (b) (c)

▲ **Figure 11.10 Hydrogen bonding in ice.** (a) The hexagonal shape is characteristic of snowflakes. (b) The arrangement of H_2O molecules in ice. Each hydrogen atom on one H_2O molecule is oriented toward a nonbonding pair of electrons on an adjacent H_2O molecule. As a result, ice has an open, hexagonal arrangement of H_2O molecules. (c) Hydrogen bonding between two water molecules. The distances shown are those found in ice.

▲ **Figure 11.11 Expansion of water upon freezing.** Water is one of the few substances that expand upon freezing. The expansion is due to the open structure of ice relative to that of liquid water.

The lower density of ice compared to liquid water profoundly affects life on Earth. Because ice floats (Figure 11.9), it covers the top of the water when a lake freezes in cold weather, thereby insulating the water below. If ice were more dense than water, ice forming at the top of a lake would sink to the bottom, and the lake could freeze solid. Most aquatic life could not survive under these conditions. The expansion of water upon freezing (Figure 11.11 ◄) is also what causes water pipes to break in freezing weather.

GIVE IT SOME THOUGHT

What is unusual about the relative densities of liquid water and ice?

Comparing Intermolecular Forces

We can identify the intermolecular forces that are operative in a substance by considering its composition and structure. *Dispersion forces are found in all substances.* The strengths of these attractions increase with increasing molecular weight and depend on molecular shapes. Dipole–dipole forces add to the effect of dispersion forces and are found in polar molecules. Hydrogen bonds, which require H atoms bonded to F, O, or N, also add to the effect of dispersion forces. Hydrogen bonds tend to be the strongest type of intermolecular attraction. None of these intermolecular attractions, however, is as strong as ordinary ionic or covalent bonds. In general, the energies associated with the dispersion forces and dipole–dipole forces are in the range of 2–10 kJ/mol, while the energies of hydrogen bonds are in the range 5–25 kJ/mol. Ion–dipole attractions lead to energies of approximately 15 kJ/mol.

It is important to realize that the effects of all these attractions are additive. For example, if a pair of molecules can make one hydrogen bond between them but a second pair can make three equivalent hydrogen bonds between them, the second pair of molecules will be held together with triple the energy of the first pair. Likewise, very large polar molecules such as proteins, which have multiple dipoles over their surfaces, can be held together in solution to a surprisingly high degree by multiple dipole–dipole attractions. Figure 11.12▼ presents a systematic way of identifying the kinds of intermolecular forces in a particular system, including ion–dipole and ion–ion forces.

▼ **Figure 11.12 Flowchart for determining intermolecular forces.** London dispersion forces occur in all instances. The strengths of the other forces generally increase proceeding from left to right across the chart.

■ **SAMPLE EXERCISE 11.3** | Predicting the Types and Relative Strengths of Intermolecular Attractions

List the substances $BaCl_2$, H_2, CO, HF, and Ne in order of increasing boiling points.

SOLUTION

Analyze: We need to relate the properties of the listed substances to boiling point.

Plan: The boiling point depends in part on the attractive forces in the liquid. We need to order these according to the relative strengths of the different kinds of intermolecular attractions.

Solve: The attractive forces are stronger for ionic substances than for molecular ones, so $BaCl_2$ should have the highest boiling point. The intermolecular forces of the remaining substances depend on molecular weight, polarity, and hydrogen bonding. The molecular weights are H_2 (2), CO (28), HF (20), and Ne (20). The boiling point of H_2 should be the lowest because it is nonpolar and has the lowest molecular weight. The molecular weights of CO, HF, and Ne are roughly the same. Because HF can hydrogen bond, however, it should have the highest boiling point of the three. Next is CO, which is slightly polar and has the highest molecular weight. Finally, Ne, which is nonpolar, should have the lowest boiling point of these three. The predicted order of boiling points is therefore

$$H_2 < Ne < CO < HF < BaCl_2$$

Check: The actual normal boiling points are H_2 (20 K), Ne (27 K), CO (83 K), HF (293 K), and $BaCl_2$ (1813 K)—in agreement with our predictions.

■ **PRACTICE EXERCISE**

(a) Identify the intermolecular attractions present in the following substances, and **(b)** select the substance with the highest boiling point: CH_3CH_3, CH_3OH, and CH_3CH_2OH.
Answers: **(a)** CH_3CH_3 has only dispersion forces, whereas the other two substances have both dispersion forces and hydrogen bonds; **(b)** CH_3CH_2OH

11.3 SOME PROPERTIES OF LIQUIDS

The intermolecular attractions we have just discussed can help us understand many familiar properties of liquids and solids. In this section we examine two important properties of liquids: viscosity and surface tension.

Viscosity

Some liquids, such as molasses and motor oil, flow very slowly; others, such as water and gasoline, flow easily. The resistance of a liquid to flow is called its **viscosity**. The greater a liquid's viscosity, the more slowly it flows. Viscosity can be measured by timing how long it takes a certain amount of the liquid to flow through a thin tube under gravitational force. More viscous liquids take longer (Figure 11.13▶). Viscosity can also be determined by measuring the rate at which steel balls fall through the liquid. The balls fall more slowly as the viscosity increases.

Viscosity is related to the ease with which individual molecules of the liquid can move with respect to one another. It thus depends on the attractive forces between molecules, and on whether structural features exist that cause the molecules to become entangled (for example, long molecules might become tangled like spaghetti). For a series of related compounds, therefore, viscosity increases with molecular weight, as illustrated in Table 11.4▼. The SI units for viscosity are kg/m-s. For any given substance, viscosity decreases with increasing temperature. Octane, for example, has a viscosity of 7.06×10^{-4} kg/m-s at 0 °C, and of 4.33×10^{-4} kg/m-s at 40 °C. At higher temperatures the greater average kinetic energy of the molecules more easily overcomes the attractive forces between molecules.

▲ Figure 11.13 **Comparing viscosities.** The Society of Automotive Engineers (SAE) has established numbers to indicate the viscosity of motor oils. The higher the number, the greater the viscosity is at any given temperature. The SAE 40 motor oil on the left is more viscous and flows more slowly than the less viscous SAE 10 oil on the right.

▲ **Figure 11.14 Surface tension.**
Surface tension permits an insect such as the water strider to "walk" on water.

▲ **Figure 11.15 Molecular-level view of surface and interior intermolecular forces in a liquid.** Molecules at the surface are attracted only by other surface molecules and by molecules below the surface. The result is a net downward attraction into the interior of the liquid. Molecules in the interior experience attractions in all directions, resulting in no net attraction in any direction.

▲ **Figure 11.16 Two meniscus shapes.** The water meniscus in a glass tube compared with the mercury meniscus in a similar tube. Water wets the glass, and the bottom of the meniscus is below the level of the water–glass contact line, giving a U-shape to the water surface. Mercury does not wet glass and the meniscus is above the mercury–glass contact line, giving an inverted U-shape to the mercury surface.

TABLE 11.4 ■ Viscosities of a Series of Hydrocarbons at 20 °C

Substance	Formula	Viscosity (kg/m-s)
Hexane	$CH_3CH_2CH_2CH_2CH_2CH_3$	3.26×10^{-4}
Heptane	$CH_3CH_2CH_2CH_2CH_2CH_2CH_3$	4.09×10^{-4}
Octane	$CH_3CH_2CH_2CH_2CH_2CH_2CH_2CH_3$	5.42×10^{-4}
Nonane	$CH_3CH_2CH_2CH_2CH_2CH_2CH_2CH_2CH_3$	7.11×10^{-4}
Decane	$CH_3CH_2CH_2CH_2CH_2CH_2CH_2CH_2CH_2CH_3$	1.42×10^{-3}

Surface Tension

The surface of water behaves almost as if it had an elastic skin, as evidenced by the ability of certain insects to "walk" on water (Figure 11.14 ◄). This behavior is due to an imbalance of intermolecular forces at the surface of the liquid, as shown in Figure 11.15 ◄. Notice that molecules in the interior are attracted equally in all directions, whereas those at the surface experience a net inward force. The resultant inward force pulls molecules from the surface into the interior, thereby reducing the surface area and making the molecules at the surface pack closely together. Because spheres have the smallest surface area for their volume, water droplets assume an almost spherical shape. Similarly, water tends to "bead up" on a newly waxed car because there is little or no attraction between the polar water molecules and the nonpolar wax molecules.

A measure of the inward forces that must be overcome to expand the surface area of a liquid is given by its surface tension. **Surface tension** is the energy required to increase the surface area of a liquid by a unit amount. For example, the surface tension of water at 20 °C is 7.29×10^{-2} J/m^2, which means that an energy of 7.29×10^{-2} J must be supplied to increase the surface area of a given amount of water by 1 m^2. Water has a high surface tension because of its strong hydrogen bonds. The surface tension of mercury is even higher $(4.6 \times 10^{-1}$ J/m$^2)$ because of even stronger metallic bonds between the atoms of mercury. ∞ (Section 11.8)

▲ **GIVE IT SOME THOUGHT**

How do viscosity and surface tension change **(a)** as temperature increases, **(b)** as intermolecular forces of attraction become stronger?

Intermolecular forces that bind similar molecules to one another, such as the hydrogen bonding in water, are also called *cohesive forces*. Intermolecular forces that bind a substance to a surface are called *adhesive forces*. Water placed in a glass tube adheres to the glass because the adhesive forces between the water and glass are even greater than the cohesive forces between water molecules. The curved upper surface, or *meniscus*, of the water is therefore U-shaped (Figure 11.16 ◄). For mercury, however, the meniscus is curved downward where the mercury contacts the glass. In this case the cohesive forces between the mercury atoms are much greater than the adhesive forces between the mercury atoms and the glass.

When a small-diameter glass tube, or capillary, is placed in water, water rises in the tube. The rise of liquids up very narrow tubes is called **capillary action**. The adhesive forces between the liquid and the walls of the tube tend to increase the surface area of the liquid. The surface tension of the liquid tends to reduce the area, thereby pulling the liquid up the tube. The liquid climbs until the force of gravity on the liquid balances the adhesive and cohesive forces. Capillary action helps water and dissolved nutrients move upward through plants.

▲ **GIVE IT SOME THOUGHT**

If a liquid in a thin tube has no meniscus (in other words, the top of the liquid looks completely flat), what does that imply about the relative strengths of cohesive and adhesive forces?

11.4 PHASE CHANGES

Water left uncovered in a glass for several days evaporates. An ice cube left in a warm room quickly melts. Solid CO_2 (sold as dry ice) *sublimes* at room temperature; that is, it changes directly from the solid to the vapor state. In general, each state of matter can change into either of the other two states. Figure 11.17▶ shows the name associated with each of these transformations. These transformations are called either **phase changes** or changes of state.

Energy Changes Accompanying Phase Changes

Every phase change is accompanied by a change in the energy of the system. In a solid, for example, the molecules or ions are in more or less fixed positions with respect to one another and closely arranged to minimize the energy of the system. As the temperature of the solid increases, the units of the solid vibrate about their equilibrium positions with increasingly energetic motion. When the solid melts, the units that made up the solid are freed to move with respect to one another, which ordinarily means that their average separations increase. This melting process is called (somewhat confusingly) *fusion*. The increased freedom of motion of the molecules or ions comes at a price, measured by the **heat of fusion**, or enthalpy of fusion, denoted ΔH_{fus}. The heat of fusion of ice, for example, is 6.01 kJ/mol.

As the temperature of the liquid phase increases, the molecules of the liquid move about with increasing energy. One measure of this increasing energy is that the concentration of gas-phase molecules over the liquid increases with temperature. These molecules exert a pressure called the vapor pressure. We will explore vapor pressure in Section 11.5. For now we just need to understand that the vapor pressure increases with increasing temperature until it equals the external pressure over the liquid, typically atmospheric pressure. At this point the liquid boils—the molecules of the liquid move into the gaseous state, where they are widely separated. The energy required to cause this transition is called the **heat of vaporization**, or enthalpy of vaporization, denoted ΔH_{vap}. For water, the heat of vaporization is 40.7 kJ/mol.

Figure 11.18▼ shows the comparative values of ΔH_{fus} and ΔH_{vap} for four substances. ΔH_{vap} values tend to be larger than ΔH_{fus} because in the transition from the liquid to the vapor state, the molecules must essentially sever all their intermolecular attractive interactions; whereas in melting, many of these attractive interactions remain.

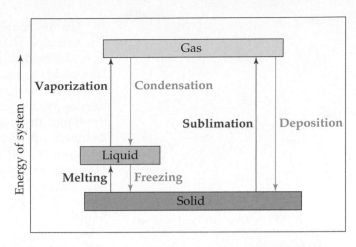

▲ **Figure 11.17 Phase changes and the names associated with them.** Those changes indicated by red arrows and names are endothermic, whereas those in green are exothermic.

▲ **Figure 11.18 Comparing enthalpy changes for fusion and vaporization.** Heats of fusion (violet bars) and heats of vaporization (blue bars) for several substances. Notice that the heat of vaporization for a substance is always larger than its heat of fusion. The heat of sublimation is the sum of the heats of vaporization and fusion.

The molecules of a solid can be transformed directly into the gaseous state. The enthalpy change required for this transition is called the **heat of sublimation**, denoted ΔH_{sub}. For the substances shown in Figure 11.18, ΔH_{sub} is the sum of ΔH_{fus} and ΔH_{vap}. Thus, ΔH_{sub} for water is approximately 47 kJ/mol.

Phase changes of matter show up in important ways in our everyday experiences. We use ice cubes to cool our liquid drinks; the heat of fusion of ice cools the liquid in which the ice is immersed. We feel cool when we step out of a swimming pool or a warm shower because the heat of vaporization is drawn from our bodies as the water evaporates from our skin. Our bodies use the evaporation of water from skin to regulate body temperature, especially when we exercise vigorously in warm weather. A refrigerator also relies on the cooling effects of vaporization. Its mechanism contains an enclosed gas that can be liquefied under pressure. The liquid absorbs heat as it subsequently evaporates, thereby cooling the interior of the refrigerator. The vapor is then recycled through a compressor.

What happens to the heat absorbed when the liquid refrigerant vaporizes? According to the first law of thermodynamics (⚭ Section 5.2), the heat absorbed by the liquid in vaporizing must be released when the reverse process, condensation of the vapor to form the liquid, occurs. As the refrigerator compresses the vapor to form a liquid, the heat released by the condensation process is dissipated through cooling coils in the back of the refrigerator. Just as the heat of condensation is equal in magnitude to the heat of vaporization and has the opposite sign, so also the *heat of deposition* is exothermic to the same degree that the heat of sublimation is endothermic; and the *heat of freezing* is exothermic to the same degree that the heat of fusion is endothermic. These relationships, shown in Figure 11.17, are consequences of the first law of thermodynamics.

GIVE IT SOME THOUGHT

What is the name of the phase change that occurs when ice left at room temperature changes to liquid water? Is this change exothermic or endothermic?

Heating Curves

What happens when we heat an ice cube that is initially at −25 °C and 1 atm pressure? The addition of heat causes the temperature of the ice to increase. As long as the temperature is below 0 °C, the ice cube remains frozen. When the temperature reaches 0 °C, the ice begins to melt. Because melting is an endothermic process, the heat we add at 0 °C is used to convert ice to water; the temperature remains constant until all the ice has melted. Once we reach this point, the further addition of heat causes the temperature of the liquid water to increase.

A graph of the temperature of the system versus the amount of heat added is called a *heating curve*. Figure 11.19 ◄ shows a heating curve for transforming ice at −25 °C to steam at 125 °C under a constant pressure of 1 atm. Heating the ice from −25 °C to 0 °C is represented by the line segment *AB* in Figure 11.19, while converting the ice at 0 °C to water at 0 °C is the horizontal segment *BC*. Additional heat increases the temperature of the water until the temperature reaches 100 °C (segment *CD*). The heat is then used to convert water to steam at a constant temperature of 100 °C (segment *DE*). Once all the water has been converted to steam, the steam is heated to its final temperature of 125 °C (segment *EF*).

▲ **Figure 11.19 Heating curve for water.** This graph indicates the changes that occur when 1.00 mol of water is heated from 25 °C to 125 °C at a constant pressure of 1 atm. Blue lines show the heating of one phase from a lower temperature to a higher one. Red lines show the conversion of one phase to another at constant temperature.

We can calculate the enthalpy change of the system for each of the segments of the heating curve. In segments AB, CD, and EF we are heating a single phase from one temperature to another. As we saw in Section 5.5, the amount of heat needed to raise the temperature of a substance is given by the product of the specific heat, mass, and temperature change (Equation 5.22). The greater the specific heat of a substance, the more heat we must add to accomplish a certain temperature increase. Because the specific heat of water is greater than that of ice, the slope of segment CD is less than that of segment AB; we must add more heat to water to achieve a 1 °C temperature change than is needed to warm the same quantity of ice by 1 °C.

In segments BC and DE we are converting one phase to another at a constant temperature. The temperature remains constant during these phase changes because the added energy is used to overcome the attractive forces between molecules rather than to increase their average kinetic energy. For segment BC, in which ice is converted to water, the enthalpy change can be calculated by using ΔH_{fus}, while for segment DE we can use ΔH_{vap}. In Sample Exercise 11.4 we calculate the total enthalpy change for the heating curve in Figure 11.19.

■ SAMPLE EXERCISE 11.4 | Calculating ΔH for Temperature and Phase Changes

Calculate the enthalpy change upon converting 1.00 mol of ice at −25 °C to water vapor (steam) at 125 °C under a constant pressure of 1 atm. The specific heats of ice, water, and steam are 2.03 J/g-K, 4.18 J/g-K, and 1.84 J/g-K, respectively. For H_2O, ΔH_{fus} = 6.01 kJ/mol and ΔH_{vap} = 40.67 kJ/mol.

SOLUTION

Analyze: Our goal is to calculate the total heat required to convert 1 mol of ice at −25 °C to steam at 125 °C.

Plan: We can calculate the enthalpy change for each segment and then sum them to get the total enthalpy change (Hess's law, Section 5.6).

Solve: For segment AB in Figure 11.19, we are adding enough heat to ice to increase its temperature by 25 °C. A temperature change of 25 °C is the same as a temperature change of 25 K, so we can use the specific heat of ice to calculate the enthalpy change during this process:

AB: ΔH = (1.00 mol)(18.0 g/mol)(2.03 J/g-K)(25 K) = 914 J = 0.91 kJ

For segment BC in Figure 11.19, in which we convert ice to water at 0 °C, we can use the molar enthalpy of fusion directly:

BC: ΔH = (1.00 mol)(6.01 kJ/mol) = 6.01 kJ

The enthalpy changes for segments CD, DE, and EF can be calculated in similar fashion:

CD: ΔH = (1.00 mol)(18.0 g/mol)(4.18 J/g-K)(100 K) = 7520 J = 7.52 kJ

DE: ΔH = (1.00 mol)(40.67 kJ/mol) = 40.7 kJ

EF: ΔH = (1.00 mol)(18.0 g/mol)(1.84 J/g-K)(25 K) = 830 J = 0.83 kJ

The total enthalpy change is the sum of the changes of the individual steps:

ΔH = 0.91 kJ + 6.01 kJ + 7.52 kJ + 40.7 kJ + 0.83 kJ = 56.0 kJ

Check: The components of the total energy change are reasonable in comparison with the lengths of the horizontal segments of the lines in Figure 11.19. Notice that the largest component is the heat of vaporization.

■ PRACTICE EXERCISE

What is the enthalpy change during the process in which 100.0 g of water at 50.0 °C is cooled to ice at −30.0 °C? (Use the specific heats and enthalpies for phase changes given in Sample Exercise 11.4.)
Answer: −20.9 kJ − 33.4 kJ − 6.09 kJ = −60.4 kJ

Cooling a substance has the opposite effect of heating it. Thus, if we start with water vapor and begin to cool it, we would move right to left through the events shown in Figure 11.19. We would first lower the temperature of the vapor ($F \longrightarrow E$), then condense it ($E \longrightarrow D$), and so forth. Sometimes as we remove heat from a liquid, we can temporarily cool it below its freezing point without forming a solid. This phenomenon is called *supercooling*. Supercooling occurs when heat is removed from a liquid so rapidly that the molecules literally have no time to assume the ordered structure of a solid. A supercooled liquid is unstable; particles of dust entering the solution or gentle stirring is often sufficient to cause the substance to solidify quickly.

Critical Temperature and Pressure

A gas normally liquefies at some point when pressure is applied to it. Suppose we have a cylinder with a piston, containing water vapor at 100 °C. If we increase the pressure on the water vapor, liquid water will form when the pressure is 760 torr. On the other hand, if the temperature is 110 °C, the liquid phase does not form until the pressure is 1075 torr. At 374 °C the liquid phase forms only at 1.655×10^5 torr (217.7 atm). Above this temperature no amount of pressure will cause a distinct liquid phase to form. Instead, as pressure increases, the gas merely becomes steadily more compressed. The highest temperature at which a distinct liquid phase can form is referred to as the **critical temperature**. The **critical pressure** is the pressure required to bring about liquefaction at this critical temperature.

The critical temperature is the highest temperature at which a liquid can exist. Above the critical temperature, the motional energies of the molecules are greater than the attractive forces that lead to the liquid state regardless of how much the substance is compressed to bring the molecules closer together. The greater the intermolecular forces, the greater the critical temperature of a substance.

The critical temperatures and pressures are listed for several substances in Table 11.5▼. Notice that nonpolar, low molecular weight substances, which have weak intermolecular attractions, have lower critical temperatures and pressures than those that are polar or of higher molecular weight. Notice also that water and ammonia have exceptionally high critical temperatures and pressures as a consequence of strong intermolecular hydrogen-bonding forces.

The critical temperatures and pressures of substances are often of considerable importance to engineers and other people working with gases, because they provide information about the conditions under which gases liquefy. Sometimes we want to liquefy a gas; other times we want to avoid liquefying it. It is useless to try to liquefy a gas by applying pressure if the gas is above its critical temperature. For example, O_2 has a critical temperature of 154.4 K. It must be cooled below this temperature before it can be liquefied by pressure. In contrast, ammonia has a critical temperature of 405.6 K. Thus, it can be liquefied at room temperature (approximately 295 K) by compressing the gas to a sufficient pressure.

TABLE 11.5 ■ Critical Temperatures and Pressures of Selected Substances		
Substance	**Critical Temperature (K)**	**Critical Pressure (atm)**
Ammonia, NH_3	405.6	111.5
Argon, Ar	150.9	48
Carbon dioxide, CO_2	304.3	73.0
Hydrogen sulfide, H_2S	373.5	88.9
Nitrogen, N_2	126.1	33.5
Oxygen, O_2	154.4	49.7
Phosphine, PH_3	324.4	64.5
Propane, $CH_3CH_2CH_3$	370.0	42.0
Water, H_2O	647.6	217.7

Chemistry Put to Work   SUPERCRITICAL FLUID EXTRACTION

At ordinary pressures, a substance above its critical temperature behaves as an ordinary gas. However, as pressure increases up to several hundred atmospheres, its character changes. Like a gas, the substance still expands to fill the confines of its container, but its density approaches that of a liquid. (For example, the critical temperature of water is 647.6 K, and its critical pressure is 217.7 atm. At this temperature and pressure, the density of water is 0.4 g/mL.) A substance at temperatures and pressures higher than its critical temperature and pressure is better considered a *supercritical fluid* rather than a gas.

Like liquids, supercritical fluids can behave as solvents, dissolving a wide range of substances. Using *supercritical fluid extraction*, the components of mixtures can be separated. The solvent power of a supercritical fluid increases as its density increases. Conversely, lowering its density (by either decreasing pressure or increasing temperature) causes the supercritical fluid and the dissolved material to separate. By appropriate manipulation of pressure, supercritical fluid extraction has been used successfully to separate complex mixtures in the chemical, food, pharmaceutical, and energy industries. Supercritical carbon dioxide, for example, is environmentally benign because there are no problems disposing of solvent and there are no toxic residues resulting from the process. In addition, supercritical CO_2 is inexpensive compared to solvents other than water. Figure 11.20▼ shows what supercritical CO_2 looks like through a window in a high-pressure vessel.

A process for removing caffeine from green coffee beans by extraction with supercritical CO_2, diagrammed in Figure 11.21▼, has been in commercial operation for several years. At the proper temperature and pressure the supercritical CO_2 removes caffeine from the beans by dissolution, but leaves the flavor and aroma components, producing decaffeinated coffee. Supercritical CO_2 is also now used as a solvent in dry cleaning. In 2007, the state of California banned the most common dry-cleaning solvent, perchloroethylene ($Cl_2C{=}CCl_2$), which is an environmental hazard and a suspected human carcinogen. Supercritical CO_2 is a more sustainable alternative.

Related Exercises: 11.41, 11.42, 11.87

▲ **Figure 11.20** Photograph of carbon dioxide under temperature and pressure conditions so that it is a supercritical fluid.

▲ **Figure 11.21 Diagram of a supercritical fluid extraction process.** The material to be processed is placed in the extractor. The desired material dissolves in supercritical CO_2 at high pressure, then is precipitated in the separator when the CO_2 pressure is reduced. The carbon dioxide is then recycled through the compressor with a fresh batch of material in the extractor.

11.5 VAPOR PRESSURE

Molecules can escape from the surface of a liquid into the gas phase by evaporation. Suppose we conduct an experiment in which we place a quantity of ethanol (CH_3CH_2OH) in an evacuated, closed container such as that in Figure 11.22▼. The ethanol will quickly begin to evaporate. As a result, the pressure exerted by the vapor in the space above the liquid will begin to increase. After a short time the pressure of the vapor will attain a constant value, which we call the **vapor pressure** of the substance.

► Figure 11.22 **Illustration of the equilibrium vapor pressure over a liquid.** In (a) we imagine that no molecules of the liquid are in the gas phase initially; there is zero vapor pressure in the vessel. In (b), after equilibrium is reached, the rate at which molecules leave the surface equals the rate at which gas molecules return to the liquid phase. These equal rates produce a stable vapor pressure that does not change as long as the temperature remains constant.

P_{gas} = equilibrium vapor pressure

Liquid ethanol

Liquid before any evaporation

At equilibrium, molecules enter and leave liquid at the same rate.

(a)   (b)

▲ Figure 11.23 **The effect of temperature on the distribution of kinetic energies in a liquid.** The distribution of kinetic energies of surface molecules of a hypothetical liquid are shown at two temperatures. Only the fastest molecules have sufficient kinetic energy to escape the liquid and enter the vapor, as shown by the shaded areas. The higher the temperature, the larger is the fraction of molecules with enough energy to escape from the liquid into the vapor phase.

Explaining Vapor Pressure on the Molecular Level

The molecules of a liquid move at various speeds. Figure 11.23 ◄ shows the distribution of kinetic energies of the particles at the surface of a liquid at two temperatures. The distribution curves are like those shown earlier for gases. ∞ (Section 10.7) At any instant some of the molecules on the surface of the liquid possess sufficient kinetic energy to overcome the attractive forces of their neighbors and escape into the gas phase. The weaker the attractive forces, the larger is the number of molecules that are able to escape and therefore the higher the vapor pressure.

At any particular temperature the movement of molecules from the liquid to the gas phase goes on continuously. As the number of gas-phase molecules increases, however, the probability increases that a molecule in the gas phase will strike the liquid surface and be recaptured by the liquid, as shown in Figure 11.22(b). Eventually, the rate at which molecules return to the liquid exactly equals the rate at which they escape. The number of molecules in the gas phase then reaches a steady value, and the pressure of the vapor at this stage becomes constant.

The condition in which two opposing processes are occurring simultaneously at equal rates is called a **dynamic equilibrium**, but is usually referred to merely as *equilibrium*. A liquid and its vapor are in dynamic equilibrium when evaporation and condensation occur at equal rates. It may appear that nothing is occurring at equilibrium because there is no net change in the system. In fact, a great deal is happening: molecules continuously pass from the liquid state to the gas state and from the gas state to the liquid state. All equilibria between different states of matter possess this dynamic character. *The vapor pressure of a liquid is the pressure exerted by its vapor when the liquid and vapor states are in dynamic equilibrium.*

Volatility, Vapor Pressure, and Temperature

When vaporization occurs in an open container, as when water evaporates from a bowl, the vapor spreads away from the liquid. Little, if any, is recaptured at the surface of the liquid. Equilibrium never occurs, and the vapor continues to form until the liquid evaporates to dryness. Substances with high vapor pressure (such as gasoline) evaporate more quickly than substances with low vapor pressure (such as motor oil). Liquids that evaporate readily are said to be **volatile**.

Hot water evaporates more quickly than cold water because vapor pressure increases with increasing temperature. We see this effect in Figure 11.23: As the temperature of a liquid is increased, the molecules move more energetically and a greater fraction can therefore escape more readily from their neighbors. Figure 11.24▶ depicts the variation in vapor pressure with temperature for four common substances that differ greatly in volatility. Note that the vapor pressure in all cases increases nonlinearly with increasing temperature.

▲ **Figure 11.24 Vapor pressure for four common liquids as a function of temperature.** The temperature at which the vapor pressure is 760 torr is the normal boiling point of each liquid.

▲ **GIVE IT SOME THOUGHT**

Which compound do you think would be more volatile at 25 °C, CCl₄ or CBr₄?

Vapor Pressure and Boiling Point

A liquid boils when its vapor pressure equals the external pressure acting on the surface of the liquid. At this point bubbles of vapor form within the liquid. The temperature at which a given liquid boils increases with increasing external pressure. The boiling point of a liquid at 1 atm (or 760 torr) pressure is called its **normal boiling point**. From Figure 11.24 we see that the normal boiling point of water is 100 °C.

The boiling point is important to many processes that involve heating liquids, including cooking. The time required to cook food depends on the temperature. As long as water is present, the maximum temperature of the food being cooked is the boiling point of water. Pressure cookers work by allowing steam to escape only when it exceeds a predetermined pressure; the pressure above the water can therefore increase above atmospheric pressure. The higher pressure causes the water to boil at a higher temperature, thereby allowing the food to get hotter and to cook more rapidly. The effect of pressure on boiling point also explains why it takes longer to cook food at higher elevations than at sea level. The atmospheric pressure is lower at higher altitudes, so water boils at a lower temperature and foods generally take longer to cook.

■■ **SAMPLE EXERCISE 11.5** | Relating Boiling Point to Vapor Pressure

Use Figure 11.24 to estimate the boiling point of diethyl ether under an external pressure of 0.80 atm.

SOLUTION

Analyze: We are asked to read a graph of vapor pressure versus temperature to determine the boiling point of a substance at a particular pressure. The boiling point is the temperature at which the vapor pressure is equal to the external pressure.

Plan: We need to convert 0.80 atm to torr because that is the pressure scale on the graph. We estimate the location of that pressure on the graph, move horizontally to the vapor pressure curve, and then drop vertically from the curve to estimate the temperature.

Solve: The pressure equals (0.80 atm)(760 torr/atm) = 610 torr. From Figure 11.24 we see that the boiling point at this pressure is about 27 °C, which is close to room temperature.

Comment: We can make a flask of diethyl ether boil at room temperature by using a vacuum pump to lower the pressure above the liquid to about 0.8 atm.

■■ **PRACTICE EXERCISE**

At what external pressure will ethanol have a boiling point of 60 °C?
Answer: about 340 torr (0.45 atm)

A Closer Look THE CLAUSIUS–CLAPEYRON EQUATION

You might have noticed that the plots of the variation of vapor pressure with temperature shown in Figure 11.24 have a distinct shape: Each curves sharply upward to a higher vapor pressure with increasing temperature. The relationship between vapor pressure and temperature is given by an equation called the *Clausius–Clapeyron equation*:

$$\ln P = \frac{-\Delta H_{vap}}{RT} + C \qquad [11.1]$$

In this equation P is the vapor pressure, T is the absolute temperature, R is the gas constant (8.314 J/mol-K), ΔH_{vap} is the molar enthalpy of vaporization, and C is a constant. The Clausius–Clapeyron equation predicts that a graph of $\ln P$ versus $1/T$ should give a straight line with a slope equal to $-\Delta H_{vap}/R$. Thus, we can use such a plot to determine the enthalpy of vaporization of a substance as follows:

$$\Delta H_{vap} = -\text{slope} \times R$$

As an example of the application of the Clausius–Clapeyron equation, the vapor-pressure data for ethanol shown in Figure 11.24 are graphed as $\ln P$ versus $1/T$ in Figure 11.25▶. The data lie on a straight line with a negative slope. We can use the slope of the line to determine ΔH_{vap} for ethanol. We can also extrapolate the line to obtain values for the vapor pressure of ethanol at temperatures above and below the temperature range for which we have data.

Related Exercises: 11.89, 11.90, 11.91

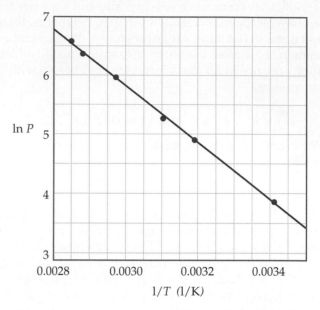

▲ **Figure 11.25 Linear graph of vapor pressure data for ethanol.** The Clausius–Clapeyron equation, Equation 11.1, indicates that a graph of $\ln P$ versus $1/T$ gives a straight line whose slope equals $-\Delta H_{vap}/R$. The slope of this line gives $\Delta H_{vap} = 38.56$ kJ/mol.

11.6 PHASE DIAGRAMS

The equilibrium between a liquid and its vapor is not the only dynamic equilibrium that can exist between states of matter. Under appropriate conditions of temperature and pressure, a solid can be in equilibrium with its liquid state or even with its vapor state. A **phase diagram** is a graphic way to summarize the conditions under which equilibria exist between the different states of matter. Such a diagram also allows us to predict the phase of a substance that is stable at any given temperature and pressure.

The general form of a phase diagram for a substance that can exist in any one of the three phases of matter is shown in Figure 11.26▶. The diagram is a two-dimensional graph, with pressure and temperature as the axes. It contains three important curves, each of which represents the conditions of temperature and pressure at which the various phases can coexist at equilibrium. The only substance present in the system is the one whose phase diagram is under consideration. The pressure shown in the diagram is either the pressure applied to the system or the pressure generated by the substance itself. The curves may be described as follows:

1. The line from T to C is the vapor-pressure curve of the liquid. It represents the equilibrium between the liquid and gas phases. The point on this curve where the vapor pressure is 1 atm is the normal boiling point of the substance. The vapor-pressure curve ends at the *critical point* (C), which is at the critical temperature and critical pressure of the substance. Beyond the critical point the liquid and gas phases become indistinguishable from each other, and the state of the substance is a *supercritical fluid*.

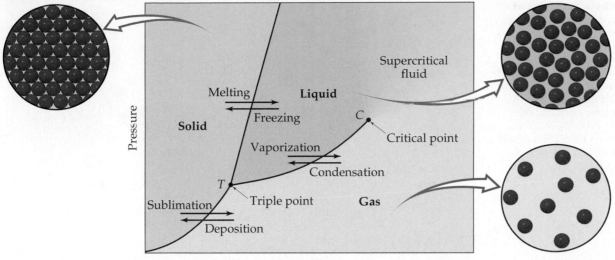

▲ **Figure 11.26 Phase diagram for a three-phase system.** In this generic diagram, the substance being investigated can exist as a solid, a liquid, or a gas, depending on pressure and temperature. Beyond the critical point (C), the distinction between liquid and gas is lost, and the substance is a supercritical fluid.

2. The line that separates the solid phase from the gas phase represents the change in the vapor pressure of the solid as it sublimes at different temperatures.

3. The line that separates the solid phase from the liquid phase corresponds to the change in melting point of the solid with increasing pressure. This line usually slopes slightly to the right as pressure increases, because for most substances the solid form is denser than the liquid form. An increase in pressure usually favors the more compact solid phase; thus, higher temperatures are required to melt the solid at higher pressures. The *melting point* of a substance is identical to its *freezing point*. The two differ only in the direction from which the phase change is approached. The melting point at 1 atm is the **normal melting point**.

Point *T*, where the three curves intersect, is known as the **triple point**. All three phases are in equilibrium at this temperature and pressure. Any other point on the three curves represents equilibrium between two phases. Any point on the diagram that does not fall on a line corresponds to conditions under which only one phase is present. The gas phase, for example, is stable at low pressures and high temperatures, whereas the solid phase is stable at low temperatures and high pressures. Liquids are stable in the region between the other two.

▲ **GIVE IT SOME THOUGHT**

On a phase diagram, which way does the line representing boiling usually slope, and why?

The Phase Diagrams of H₂O and CO₂

Figure 11.27 ▼ shows the phase diagrams of H_2O and CO_2. The solid–liquid equilibrium (melting point) line of CO_2 follows the typical behavior, slanting to the right with increasing pressure, telling you that its melting point increases with increasing pressure. In contrast, the melting point line of H_2O is atypical, slanting to the left with increasing pressure, indicating that for water the melting point *decreases* with increasing pressure. As seen in Figure 11.11, water is among the very few substances whose liquid form is more compact than its solid form. ∞ (Section 11.2)

The triple point of H_2O (0.0098 °C and 0.00603 atm) is at much lower pressure than that of CO_2 (−56.4 °C and 5.11 atm). For CO_2 to exist as a liquid, the pressure must exceed 5.11 atm. Consequently, solid CO_2 does not melt but

▶ **Figure 11.27 Phase diagrams of H₂O and CO₂.** The axes are not drawn to scale in either case. In (a), for water, note the triple point T at 0.0098 °C and 0.00603 atm, the normal melting (or freezing) point of 0 °C at 1 atm, the normal boiling point of 100 °C at 1 atm, and the critical point C (374.4 °C and 217.7 atm). In (b), for carbon dioxide, note the triple point T at −56.4 °C and 5.11 atm, the normal sublimation point of −78.5 °C at 1 atm, and the critical point C (31.1 °C and 73.0 atm).

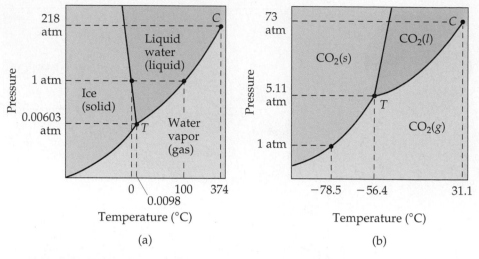

(a)　　　　　(b)

sublimes when heated at 1 atm. Thus, CO_2 does not have a normal melting point; instead, it has a normal sublimation point, −78.5 °C. Because CO_2 sublimes rather than melts as it absorbs energy at ordinary pressures, solid CO_2 (dry ice) is a convenient coolant. For water (ice) to sublime, however, its vapor pressure must be below 0.00603 atm. Food is "freeze-dried" by placing frozen food in a low-pressure chamber (below 0.00603 atm) so that the ice in it sublimes.

■■ SAMPLE EXERCISE 11.6 | Interpreting a Phase Diagram

Referring to Figure 11.28 ◀, describe any changes in the phases present when H_2O is **(a)** kept at 0 °C while the pressure is increased from that at point 1 to that at point 5 (vertical line), **(b)** kept at 1.00 atm while the temperature is increased from that at point 6 to that at point 9 (horizontal line).

SOLUTION

Analyze: We are asked to use the phase diagram provided to deduce what phase changes might occur when specific pressure and temperature changes are brought about.

Plan: Trace the path indicated on the phase diagram, and note what phases and phase changes occur.

Solve:

(a) At point 1, H_2O exists totally as a vapor. At point 2 a solid–vapor equilibrium exists. Above that pressure, at point 3, all the H_2O is converted to a solid. At point 4 some of the solid melts and equilibrium between solid and liquid is achieved. At still higher pressures all the H_2O melts, so only the liquid phase is present at point 5.

(b) At point 6 the H_2O exists entirely as a solid. When the temperature reaches point 4, the solid begins to melt and equilibrium exists between the solid and liquid phases. At an even higher temperature, point 7, the solid has been converted entirely to a liquid. A liquid–vapor equilibrium exists at point 8. Upon further heating to point 9, the H_2O is converted entirely to the vapor phase.

Check: The indicated phases and phase changes are consistent with our knowledge of the properties of water.

■■ PRACTICE EXERCISE

Using Figure 11.27(b), describe what happens when the following changes are made to a CO_2 sample: **(a)** Pressure increases from 1 atm to 60 atm at a constant temperature of −60 °C. **(b)** Temperature increases from −60 °C to −20 °C at a constant pressure of 60 atm.

Answers: **(a)** $CO_2(g) \longrightarrow CO_2(s)$; **(b)** $CO_2(s) \longrightarrow CO_2(l)$

▲ **Figure 11.28 Phase diagram of H₂O.**

11.7 STRUCTURES OF SOLIDS

Throughout the remainder of this chapter we will focus on how the properties of solids relate to their structures and bonding. Solids can be either crystalline or amorphous (noncrystalline). In a **crystalline solid** the atoms, ions, or molecules are ordered in well-defined three-dimensional arrangements.

(a) Pyrite (fool's gold)

(b) Fluorite

(c) Amethyst

▲ **Figure 11.29 Crystalline solids.** Crystalline solids come in a variety of forms and colors: (a) pyrite (fool's gold), (b) fluorite, (c) amethyst.

These solids usually have flat surfaces, or *faces*, that make definite angles with one another. The orderly stacks of particles that produce these faces also cause the solids to have highly regular shapes (Figure 11.29▲). Quartz and diamond are crystalline solids.

An **amorphous solid** (from the Greek words for "without form") is a solid in which particles have no orderly structure. These solids lack well-defined faces and shapes. Many amorphous solids are mixtures of particles that do not stack together well. Most others are composed of large, complicated molecules. Familiar amorphous solids include rubber and glass.

Quartz (SiO_2) is a crystalline solid with a three-dimensional structure like that shown in Figure 11.30(a)▼. When quartz melts (at about 1600 °C), it becomes a viscous, tacky liquid. Although the silicon–oxygen network remains largely intact, many Si—O bonds are broken, and the rigid order of the quartz is lost. If the melt is rapidly cooled, the atoms are unable to return to an orderly arrangement. As a result, an amorphous solid known either as quartz glass or as silica glass results [Figure 11.30(b)].

▼ **Figure 11.30 Schematic comparisons of (a) crystalline SiO_2 (quartz) and (b) amorphous SiO_2 (quartz glass).** The structures are actually three-dimensional and not planar as drawn. The two-dimensional unit shown as the basic building block of the structure (silicon and three oxygens) actually has four oxygens, the fourth coming out of the plane of the paper and capable of bonding to other silicon atoms. The actual three-dimensional building block is shown.

(a) Crystalline SiO_2

(b) Amorphous SiO_2

O

Si

Two-dimensional unit

Actual three-dimensional unit

▲ **Figure 11.31 A two-dimensional analog of a lattice and its unit cell.** The wallpaper design shows a characteristic repeat pattern. Each dashed blue square denotes a unit cell of the pattern. The unit cell could equally well be selected with red figures at the corners.

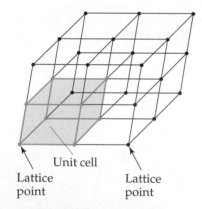

▲ **Figure 11.32 Part of a simple crystal lattice and its associated unit cell.** A lattice is an array of points that define the positions of particles in a crystalline solid. Each lattice point represents an identical environment in the solid. The points here are shown connected by lines to help convey the three-dimensional character of the lattice and to help us see the unit cell.

Because the particles of an amorphous solid lack any long-range order, intermolecular forces vary in strength throughout a sample. Thus, amorphous solids do not melt at specific temperatures. Instead, they soften over a temperature range as intermolecular forces of various strengths are overcome. A crystalline solid, in contrast, melts at a specific temperature.

△ **GIVE IT SOME THOUGHT**

What is the general difference in the melting behaviors of crystalline and amorphous solids?

Unit Cells

The characteristic order of crystalline solids allows us to convey a picture of an entire crystal by looking at only a small part of it. We can think of the solid as being built by stacking together identical building blocks, much as stacking rows of individual "identical" bricks forms a brick wall. The repeating unit of a solid, the crystalline "brick," is known as the **unit cell**. A simple two-dimensional example appears in the sheet of wallpaper shown in Figure 11.31 ◄. There are several ways of choosing a unit cell, but the choice is usually the smallest unit cell that shows clearly the symmetry characteristic of the entire pattern.

A crystalline solid can be represented by a three-dimensional array of points called a **crystal lattice**. Each point in the lattice is called a *lattice point*, and it represents an identical environment within the solid. The crystal lattice is, in effect, an abstract scaffolding for the crystal structure. We can imagine forming the entire crystal structure by arranging the contents of the unit cell repeatedly on the crystal lattice. In the simplest case the crystal structure would consist of identical atoms, and each atom would be centered on a lattice point. This is the case for most metals.

Figure 11.32 ◄ shows a crystal lattice and its associated unit cell. In general, unit cells are parallelepipeds (six-sided figures whose faces are parallelograms). Each unit cell can be described by the lengths of the edges of the cell and by the angles between these edges. Seven basic types of unit cells can describe the lattices of all crystalline compounds. The simplest of these is the cubic unit cell, in which all the sides are equal in length and all the angles are 90°.

Three kinds of cubic unit cells are illustrated in Figure 11.33 ▼. When lattice points are at the corners only, the unit cell is called **primitive cubic** (or *simple cubic*). When a lattice point also occurs at the center of the unit cell, the cell is **body-centered cubic**. When the cell has lattice points at the center of each face, as well as at each corner, it is **face-centered cubic**.

The simplest crystal structures are cubic unit cells with only one atom centered at each lattice point. Most metals have such structures. Nickel, for example, has a face-centered cubic unit cell, whereas sodium has a body-centered cubic one. Figure 11.34 ▶ shows how atoms fill the cubic unit cells. Notice that

Primitive cubic

Body-centered cubic

Face-centered cubic

▲ **Figure 11.33 The three types of unit cells found in cubic lattices.** For clarity, the corner spheres are red and the body-centered and face-centered ones are yellow. Each sphere represents a lattice point (an identical environment in the solid).

$\frac{1}{8}$ atom at 8 corners

Primitive cubic

$\frac{1}{8}$ atom at 8 corners

1 atom at center

Body-centered cubic

$\frac{1}{2}$ atom at 6 faces

$\frac{1}{8}$ atom at 8 corners

Face-centered cubic

▲ Figure 11.34 **Space-filling view of cubic unit cells.** Only the portion of each atom that belongs to the unit cell is shown.

the atoms on the corners and faces do not lie wholly within the unit cell. Instead, these atoms are shared between unit cells. As an example, let's look at the primitive cubic structure in Figure 11.34. In an actual solid, this primitive cubic structure has other primitive cubic unit cells next to it in all directions, on top of it, and underneath it. If you look at any one corner of the primitive cubic unit cell, you will see that is shared by 8 unit cells. Therefore, in an individual primitive cubic unit cell, each corner contains only one-eighth of an atom. Because a cube has eight corners, each primitive cubic unit cell has a total of $1/8 \times 8 = 1$ atom. Similarly, each body-centered cubic unit cell shown in Figure 11.34 contains two atoms ($1/8 \times 8 = 1$ from the corners, and 1 totally inside the cube). Two unit cells share equally atoms that are on the faces of a face-centered cubic unit cell so that only one-half of the atom belongs to each unit cell. Therefore, the total number of atoms in the face-centered cubic unit cell shown in Figure 11.34 is four (that is, $1/8 \times 8 = 1$ from the corners and $1/2 \times 6 = 3$ from the faces). Table 11.6▶ summarizes the fraction of an atom that occupies a unit cell when atoms are shared between unit cells.

TABLE 11.6 ■ Fraction of an Atom That Occupies a Unit Cell for Various Positions in the Unit Cell

Position in Unit Cell	Fraction in Unit Cell
Center	1
Face	$\frac{1}{2}$
Edge	$\frac{1}{4}$
Corner	$\frac{1}{8}$

GIVE IT SOME THOUGHT

If you know the unit cell dimensions for a solid, the number of atoms per unit cell, and the mass of the atoms, show how you can calculate the density of the solid.

The Crystal Structure of Sodium Chloride

In the crystal structure of NaCl (Figure 11.35▼) we can center either the Na^+ ions or the Cl^- ions on the lattice points of a face-centered cubic unit cell. Thus, we can describe the structure as being face-centered cubic.

▼ Figure 11.35 **Two ways of defining the unit cell of NaCl.** A representation of an NaCl crystal lattice can show either (a) Cl^- ions (green spheres) or (b) Na^+ ions (purple spheres) at the lattice points of the unit cell. In both cases, the red lines define the unit cell. Both of these choices for the unit cell are acceptable; both have the same volume, and in both cases identical points are arranged in a face-centered cubic fashion.

(a)

(b)

▶ **Figure 11.36**
Relative size of ions in an NaCl unit cell.
As in Figure 11.35, purple represents Na$^+$ ions and green represents Cl$^-$ ions. Only portions of most of the ions lie within the boundaries of the single unit cell.

In Figure 11.35 the Na$^+$ and Cl$^-$ ions have been moved apart so the symmetry of the structure can be seen more clearly. In this representation no attention is paid to the relative sizes of the ions. The representation in Figure 11.36 ◀, on the other hand, shows the relative sizes of the ions and how they fill the unit cell. Notice that other unit cells share the particles at corners, edges, and faces.

The total cation-to-anion ratio of a unit cell must be the same as that for the entire crystal. Therefore, within the unit cell of NaCl there must be an equal number of Na$^+$ and Cl$^-$ ions. Similarly, the unit cell for CaCl$_2$ would have one Ca^{2+} for every two Cl$^-$, and so forth.

■ **SAMPLE EXERCISE 11.7** | **Determining the Contents of a Unit Cell**

Determine the net number of Na$^+$ and Cl$^-$ ions in the NaCl unit cell (Figure 11.36).

SOLUTION

Analyze: We must sum the various contributing elements to determine the number of Na$^+$ and Cl$^-$ ions within the unit cell.

Plan: To find the total number of ions of each type, we must identify the different locations within the unit cell and determine the fraction of the ion that lies within the unit cell boundaries.

Solve: There is one-fourth of an Na$^+$ on each edge, a whole Na$^+$ in the center of the cube (refer also to Figure 11.35), one-eighth of a Cl$^-$ on each corner, and one-half of a Cl$^-$ on each face. Thus, we have the following:

$$Na^+: \left(\tfrac{1}{4} Na^+ \text{ per edge}\right)(12 \text{ edges}) = 3 \, Na^+$$
$$(1 \, Na^+ \text{ per center})(1 \text{ center}) = 1 \, Na^+$$
$$Cl^-: \left(\tfrac{1}{8} Cl^- \text{ per corner}\right)(8 \text{ corners}) = 1 \, Cl^-$$
$$\left(\tfrac{1}{2} Cl^- \text{ per face}\right)(6 \text{ faces}) = 3 \, Cl^-$$

Thus, the unit cell contains

4 Na$^+$ and 4 Cl$^-$

Check: Since individually the Cl$^-$ ions form a face-centered cubic lattice [see Figure 11.35(a)], as do the Na$^+$ ions [see Figure 11.35(b)], we would expect there to be four ions of each type in the unit cell. More important, the presence of equal amounts of the two ions agree with the compound's stoichiometry:

1 Na$^+$ for each Cl$^-$

■ **PRACTICE EXERCISE**

The element iron crystallizes in a form called α-iron, which has a body-centered cubic unit cell. How many iron atoms are in the unit cell?
Answer: two

■ **SAMPLE EXERCISE 11.8** | **Using the Contents and Dimensions of a Unit Cell to Calculate Density**

The geometric arrangement of ions in crystals of LiF is the same as that in NaCl. The unit cell of LiF is 4.02 Å on an edge. Calculate the density of LiF.

SOLUTION

Analyze: We are asked to calculate the density of LiF from the size of the unit cell.

Plan: Density is mass per volume, and this is true at the unit cell level as well as the bulk level. We need to determine the number of formula units of LiF within the unit cell. From that, we can calculate the total mass within the unit cell. Because we know the mass and can calculate the volume of the unit cell, we can then calculate density.

Solve: The arrangement of ions in LiF is the same as that in NaCl (Sample Exercise 11.7), so a unit cell of LiF contains

4 Li$^+$ ions and 4 F$^-$ ions

Density is mass per unit volume. Thus, we can calculate the density of LiF from the mass contained in a unit cell and the volume of the unit cell. The mass contained in one unit cell is

4(6.94 amu) + 4(19.0 amu) = 103.8 amu

The volume of a cube of length a on an edge is a^3, so the volume of the unit cell is $(4.02 \text{ Å})^3$. We can now calculate the density, converting to the common units of g/cm^3:

$$\text{Density} = \frac{(103.8 \text{ amu})}{(4.02 \text{ Å})^3} \left(\frac{1 \text{ g}}{6.02 \times 10^{23} \text{ amu}}\right) \left(\frac{1 \text{ Å}}{10^{-8} \text{ cm}}\right)^3 = 2.65 \text{ g/cm}^3$$

Check: This value agrees with that found by simple density measurements, 2.640 g/cm^3 at 20 °C. The size and contents of the unit cell are therefore consistent with the macroscopic density of the substance.

■ **PRACTICE EXERCISE**
The body-centered cubic unit cell of a particular crystalline form of iron is 2.8664 Å on each side. Calculate the density of this form of iron.
Answer: 7.8753 g/cm^3

Close Packing of Spheres

The structures adopted by crystalline solids are those that bring particles in closest contact to maximize the attractive forces between them. In many cases the particles that make up the solids are spherical or approximately so. Such is the case for atoms in metallic solids. It is therefore instructive to consider how equal-sized spheres can pack most efficiently (that is, with the minimum amount of empty space).

The most efficient arrangement of a layer of equal-sized spheres is shown in Figure 11.37(a)▼. Each sphere is surrounded by six others in the layer. A second layer of spheres can be placed in the depressions on top of the first layer. A third layer can then be added above the second with the spheres sitting in the depressions of the second layer. However, there are two types of depressions for this third layer, and they result in different structures, as shown in Figure 11.37(b) and (c).

If the spheres of the third layer are placed in line with those of the first layer, as shown in Figure 11.37(b), the structure is known as **hexagonal close packing**. The third layer repeats the first layer, the fourth layer repeats the second layer, and so forth, giving a layer sequence that we denote ABAB.

The spheres of the third layer, however, can be placed so they do not sit above the spheres in the first layer. The resulting structure, shown in Figure 11.37(c), is known as **cubic close packing**. In this case it is the fourth layer that repeats the first layer, and the layer sequence is ABCA. Although it cannot be seen in Figure 11.37(c), the unit cell of the cubic close-packed structure is face-centered cubic.

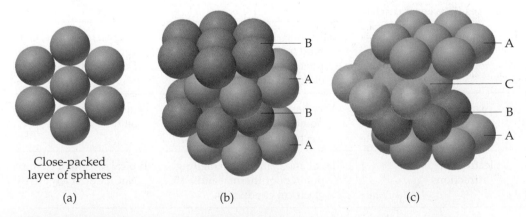

Close-packed layer of spheres

(a)   (b)   (c)

▲ **Figure 11.37 Close packing of equal-sized spheres.** (a) Close packing of a single layer of equal-sized spheres. (b) In the hexagonal close-packed structure the atoms in the third layer lie directly over those in the first layer. The order of layers is ABAB. (c) In the cubic close-packed structure the atoms in the third layer are not over those in the first layer. Instead, they are offset a bit, and it is the fourth layer that lies directly over the first. Thus, the order of layers is ABCA.

In both of the close-packed structures, each sphere has 12 equidistant nearest neighbors: six in one plane, three above that plane, and three below. We say that each sphere has a **coordination number** of 12. The coordination number is the number of particles immediately surrounding a particle in the crystal structure. In both types of close packing, spheres occupy 74% of the total volume of the structure; 26% is empty space between the spheres. By comparison, each sphere in the body-centered cubic structure has a coordination number of 8, and only 68% of the space is occupied. In the primitive cubic structure the coordination number is 6, and only 52% of the space is occupied.

When unequal-sized spheres are packed in a lattice, the larger particles sometimes assume one of the close-packed arrangements, with smaller particles occupying the holes between the large spheres. In Li_2O, for example, the larger oxide ions assume a cubic close-packed structure, and the smaller Li^+ ions occupy small cavities that exist between oxide ions.

> ▲ **GIVE IT SOME THOUGHT**
>
> Based on the information given above for close-packed structures and structures with cubic unit cells, what qualitative relationship exists between coordination numbers and packing efficiencies?

11.8 BONDING IN SOLIDS

The physical properties of crystalline solids, such as melting point and hardness, depend both on the arrangements of particles (atoms, ions, or molecules) and on the attractive forces between them. Table 11.7▼ classifies solids according to the types of forces between particles in solids.

Molecular Solids

Molecular solids consist of atoms or molecules held together by intermolecular forces (dipole–dipole forces, London dispersion forces, and hydrogen bonds). Because these forces are weak, molecular solids are soft. Furthermore, they normally have relatively low melting points (usually below 200 °C). Most substances that are gases or liquids at room temperature form molecular solids at low temperature. Examples include Ar, H_2O, and CO_2.

TABLE 11.7 ■ **Types of Crystalline Solids**

Type of Solid	Form of Unit Particles	Forces Between Particles	Properties	Examples
Molecular	Atoms or molecules	London dispersion forces, dipole–dipole forces, hydrogen bonds	Fairly soft, low to moderately high melting point, poor thermal and electrical conduction	Argon, Ar; methane, CH_4; sucrose, $C_{12}H_{22}O_{11}$; Dry Ice, CO_2
Covalent-network	Atoms connected in a network of covalent bonds	Covalent bonds	Very hard, very high melting point, variable thermal and electrical conduction	Diamond, C; quartz, SiO_2
Ionic	Positive and negative ions	Electrostatic attractions	Hard and brittle, high melting point, poor thermal and electrical conduction	Typical salts—for example, NaCl, $Ca(NO_3)_2$
Metallic	Atoms	Metallic bonds	Soft to very hard, low to very high melting point, excellent thermal and electrical conduction, malleable and ductile	All metallic elements—for example, Cu, Fe, Al, Pt

A Closer Look   X-RAY DIFFRACTION BY CRYSTALS

When light waves pass through a narrow slit, they are scattered in such a way that the wave seems to spread out. This physical phenomenon is called *diffraction*. When light passes through many evenly spaced narrow slits (a *diffraction grating*), the scattered waves interact to form a series of bright and dark bands, known as a diffraction pattern. The bright bands correspond to constructive overlapping of the light waves, and the dark bands correspond to destructive overlapping of the light waves (∞ Section 9.8, "A Closer Look: Phases in Atomic and Molecular Orbitals"). The most effective diffraction of light occurs when the wavelength of the light and the width of the slits are similar in magnitude.

The spacing of the layers of atoms in solid crystals is usually about 2–20 Å. The wavelengths of X-rays are also in this range. Thus, a crystal can serve as an effective diffraction grating for X-rays. X-ray diffraction results from the scattering of X-rays by a regular arrangement of atoms, molecules, or ions. Much of what we know about crystal structures has been obtained from studies of X-ray diffraction by crystals, a technique known as *X-ray crystallography*. Figure 11.38▼ depicts the diffraction of a beam of X-rays as it passes through a crystal. The diffracted X-rays were formerly detected by photographic film. Today, crystallographers use an *array detector*, a device analogous to that used in digital cameras, to capture and measure the intensities of the diffracted rays. The diffraction pattern of spots on the detector in Figure 11.38 depends on the particular arrangement of atoms in the crystal. Thus, different types of crystals give rise to different diffraction

patterns. In 1913 the English scientists William and Lawrence Bragg (father and son) determined for the first time how the spacing of layers in crystals leads to different X-ray diffraction patterns. By measuring the intensities of the diffracted beams and the angles at which they are diffracted, it is possible to reason backward to the structure that must have given rise to the pattern. One of the most famous X-ray diffraction patterns is the one for crystals of the genetic material DNA (Figure 11.39▼), first obtained in the early 1950s. Working from X-ray diffraction data obtained by Rosalind Franklin, Franklin, Maurice Wilkins, James Watson, and Francis Crick determined the double-helix structure of DNA, one of the most important discoveries in molecular biology. For this achievement, Watson, Crick, and Wilkins were awarded the Nobel Prize in Physiology or Medicine in 1962. Franklin died in 1958, at age 37, from cancer; Nobel Prizes are awarded only to the living (and can be shared by three people at the most).

Today X-ray crystallography is used extensively to determine the structures of molecules in crystals. The instruments used to measure X-ray diffraction, known as *X-ray diffractometers*, are now computer-controlled, making the collection of diffraction data highly automated. The diffraction pattern of a crystal can be determined very accurately and quickly (sometimes in a matter of hours), even though thousands of diffraction points are measured. Computer programs are then used to analyze the diffraction data and determine the arrangement and structure of the molecules in the crystal.
Related Exercises: 11.95, 11.96

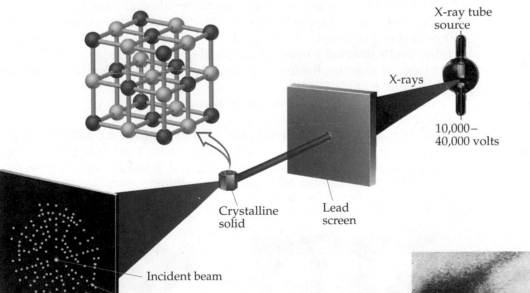

◀ **Figure 11.38 Diffraction of X-rays by a crystal.** In X-ray crystallography, an X-ray beam is diffracted by a crystal. The diffraction pattern can be recorded as spots where the diffracted X-rays strike a detector, which records the positions and intensities of the spots.

X-ray tube source

X-rays

10,000– 40,000 volts

Crystalline solid

Lead screen

Incident beam

Diffracted X-rays

X-ray detector

▶ **Figure 11.39 The X-ray diffraction photograph of one form of crystalline DNA.** This photograph was taken in the early 1950s. From the pattern of dark spots, the double-helical shape of the DNA molecule was deduced.

	Benzene	Toluene	Phenol
Melting point (°C)	5	−95	43
Boiling point (°C)	80	111	182

▲ **Figure 11.40 Comparative melting and boiling points for benzene, toluene, and phenol.**

The properties of molecular solids depend on the strengths of the forces that exist between molecules and on the abilities of the molecules to pack efficiently in three dimensions. Benzene (C_6H_6), for example, is a highly symmetrical planar molecule. ∞ (Section 8.6) It has a higher melting point than toluene, a compound in which one of the hydrogen atoms of benzene has been replaced by a CH_3 group (Figure 11.40 ◀). The lower symmetry of toluene molecules prevents them from packing as efficiently as benzene molecules. As a result, the intermolecular forces that depend on close contact are not as effective and the melting point is lower. In contrast, the boiling point of toluene is higher than that of benzene, indicating that the intermolecular attractive forces are larger in liquid toluene than in liquid benzene. Both the melting and boiling points of phenol, another substituted benzene shown in Figure 11.40, are higher than those of benzene because the OH group of phenol can form hydrogen bonds.

GIVE IT SOME THOUGHT

Which of the following substances would you expect to form molecular solids: Co, C_6H_6, or K_2O?

Covalent-Network Solids

Covalent-network solids consist of atoms held together, throughout the entire sample of material, in large networks or chains by covalent bonds. Because covalent bonds are much stronger than intermolecular forces, these solids are much harder and have higher melting points than molecular solids. Diamond and graphite, two allotropes of carbon, are covalent-network solids. Other examples include silicon and germanium; quartz, SiO_2; silicon carbide, SiC; and boron nitride, BN.

In diamond, each carbon atom is bonded tetrahedrally to four other carbon atoms, as shown in Figure 11.41(a)▼. This interconnected three-dimensional array of strong carbon–carbon single bonds that are sp^3 hybridized contributes to diamond's unusual hardness. Industrial-grade diamonds are employed in the blades of saws for the most demanding cutting jobs. Diamond also has a high melting point, 3550 °C.

In graphite, the carbon atoms are bonded in trigonal planar geometries to three other carbons to form interconnected hexagonal rings, as shown in Figure 11.41(b). The distance between adjacent carbon atoms in the plane, 1.42 Å, is very close to the C—C distance in benzene, 1.395 Å. In fact, the bonding resembles that of benzene, with delocalized π bonds extending over the layers. ∞ (Section 9.6) Electrons move freely through the delocalized orbitals,

▶ **Figure 11.41 Structures of (a) diamond and (b) graphite.** The blue color in (b) is added to emphasize the planarity of the carbon layers.

(a) Diamond   (b) Graphite

(a) CsCl   (b) ZnS   (c) CaF$_2$

▲ Figure 11.42 **Unit cells of some common ionic structures.** The ZnS shown in (b) is called zinc blende, and the CaF$_2$ of (c) is known as fluorite.

making graphite a good conductor of electricity along the layers. (In fact, graphite is used as a conducting electrode in batteries.) These sp^2-hybridized sheets of carbon atoms are separated by 3.35 Å from one another, and the sheets are held together only by weak dispersion forces. Thus, the layers readily slide past one another when rubbed, giving graphite a greasy feel. Graphite is used as a lubricant and in the "lead" in pencils. The enormous differences in physical properties of graphite and diamond—both of which are pure carbon—arise from their differences in their three-dimensional structure and bonding.

Ionic Solids

Ionic solids consist of ions held together by ionic bonds. ∞ (Section 8.2) The strength of an ionic bond depends greatly on the charges of the ions. Thus, NaCl, in which the ions have charges of 1+ and 1−, has a melting point of 801 °C, whereas MgO, in which the charges are 2+ and 2−, melts at 2852 °C.

The structures of simple ionic solids can be classified as a few basic types. The NaCl structure is a representative example of one type. Other compounds that possess this same structure include LiF, KCl, AgCl, and CaO. Three other common types of crystal structures are shown in Figure 11.42 ▲.

The structure adopted by an ionic solid depends largely on the charges and relative sizes of the ions. In the NaCl structure, for example, the Na$^+$ ions have a coordination number of 6 because each Na$^+$ ion is surrounded by six nearest neighbor Cl$^-$ ions. In the CsCl structure [Figure 11.42(a)], by comparison, the Cl$^-$ ions adopt a primitive cubic arrangement with each Cs$^+$ ion surrounded by eight Cl$^-$ ions. The increase in the coordination number as the alkali metal ion changes from Na$^+$ to Cs$^+$ is a consequence of the larger size of Cs$^+$ compared to Na$^+$.

In the zinc blende (ZnS) structure [Figure 11.42(b)], the S^{2-} ions adopt a face-centered cubic arrangement, with the smaller Zn^{2+} ions arranged so they are each surrounded tetrahedrally by four S^{2-} ions (compare with Figure 11.33). CuCl also adopts this structure.

In the fluorite (CaF$_2$) structure [Figure 11.42(c)], the Ca^{2+} ions are shown in a face-centered cubic arrangement. As required by the chemical formula of the substance, there are twice as many F$^-$ ions (blue) in the unit cell as there are Ca^{2+} ions. Other compounds that have the fluorite structure include BaCl$_2$ and PbF$_2$.

Metallic Solids

Metallic solids, also simply called metals, consist entirely of metal atoms. Metallic solids usually have hexagonal close-packed, cubic close-packed (face-centered cubic), or body-centered cubic structures. Thus, each atom typically is surrounded by eight or 12 adjacent atoms.

A Closer Look THE THIRD FORM OF CARBON

Until the mid-1980s pure solid carbon was thought to exist in only two forms: diamond and graphite, both of which are covalent-network solids. In 1985 a group of researchers led by Richard Smalley and Robert Curl of Rice University in Houston and Harry Kroto of the University of Sussex in England made a startling discovery. They vaporized a sample of graphite with an intense pulse of laser light and used a stream of helium gas to carry the vaporized carbon into a mass spectrometer (∞ Section 2.4, "A Closer Look: The Mass Spectrometer"). The mass spectrum showed peaks corresponding to clusters of carbon atoms, with a particularly strong peak corresponding to molecules composed of 60 carbon atoms, C_{60}.

Because C_{60} clusters were so preferentially formed, the group proposed a radically different form of carbon, namely, C_{60} *molecules* that were nearly spherical in shape. They proposed that the carbon atoms of C_{60} form a "ball" with 32 faces, 12 of which are pentagons and 20 are hexagons (Figure 11.43 ▼), exactly like a soccer ball. The shape of this molecule is reminiscent of the geodesic dome invented by the U.S. engineer and philosopher R. Buckminster Fuller, so C_{60} was whimsically named "buckminsterfullerene," or "buckyball" for short. Since the discovery of C_{60}, other related molecules of carbon atoms have been discovered. These molecules are now known as fullerenes.

Appreciable amounts of buckyball can be prepared by electrically evaporating graphite in an atmosphere of helium gas. About 14% of the resulting soot consists of C_{60} and a related molecule, C_{70}, which has a more elongated structure.

The carbon-rich gases from which C_{60} and C_{70} condense also contain other fullerenes, mostly with more carbon atoms such as C_{76} and C_{84}. The smallest possible fullerene, C_{20}, was first detected in 2000. This small, ball-shaped molecule is much more reactive than the larger fullerenes.

Because the fullerenes are composed of individual molecules, they dissolve in various organic solvents, whereas diamond and graphite do not (Figure 11.44 ▼). This solubility permits the fullerenes to be separated from the other components of soot and even from each other. It also allows the study of their reactions in solution. Study of these substances has led to the discovery of some very interesting chemistry. For example, it is possible to place a metal atom inside a buckyball, generating a molecule in which a metal atom is completely enclosed by the carbon sphere. The C_{60} molecules also react with potassium to give K_3C_{60}, which contains a face-centered cubic array of buckyballs with K^+ ions in the cavities between them. This compound is a superconductor at 18 K (to ∞ Section 12.1), suggesting the possibility that other fullerenes may also have interesting electrical, magnetic, or optical properties. For their discovery and pioneering work with fullerenes, Professors Smalley, Curl, and Kroto were awarded the 1996 Nobel Prize in Chemistry.

Related Exercise: 11.97

▲ **Figure 11.43 The buckminsterfullerene molecule, C_{60}.** The molecule has a highly symmetric structure in which the 60 carbon atoms sit at the vertices of a truncated icosahedron—the same geometry as a soccer ball.

▲ **Figure 11.44 Solutions of fullerenes.** Unlike diamond and graphite, the new molecular forms of carbon can be dissolved in organic solvents. The orange solution on the left is a solution of C_{70} in *n*-hexane, which is a colorless liquid. The magenta solution on the right is a solution of buckyball, C_{60}, in *n*-hexane.

The bonding in metals is too strong to be due to London dispersion forces, and yet there are not enough valence electrons for ordinary covalent bonds between atoms. The bonding is due to valence electrons that are delocalized throughout the entire solid. In fact, we can visualize the metal as an array of positive ions immersed in a sea of delocalized valence electrons, as shown in Figure 11.45 ▶. This type of bonding is called *metallic bonding* and will be discussed in more detail in Section 23.5.

Metals vary greatly in the strength of their bonding, as shown by their wide range of physical properties such as hardness and melting point. In general, however, the strength of the bonding increases as the number of electrons available for bonding increases. Thus, sodium, which has only one valence electron per atom, melts at 97.5 °C, whereas chromium, with six electrons beyond the noble-gas core, melts at 1890 °C. The mobility of the electrons explains why metals are good conductors of heat and electricity. The bonding and properties of metals will be examined more closely in Chapter 23.

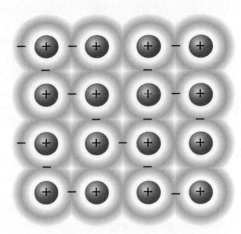

▲ **Figure 11.45 Representation of a cross section of a metal.** Each sphere represents the nucleus and inner-core electrons of a metal atom. The surrounding blue "fog" represents the mobile sea of electrons that binds the atoms together.

■ **SAMPLE INTEGRATIVE EXERCISE** | **Putting Concepts Together**

The substance CS_2 has a melting point of -110.8 °C and a boiling point of 46.3 °C. Its density at 20 °C is 1.26 g/cm^3. It is highly flammable. **(a)** What is the name of this compound? **(b)** List the intermolecular forces that CS_2 molecules would have with each other. **(c)** Predict what type of crystalline solid $CS_2(s)$ would form. **(d)** Write a balanced equation for the combustion of this compound in air. (You will have to decide on the most likely oxidation products.) **(e)** The critical temperature and pressure for CS_2 are 552 K and 78 atm, respectively. Compare these values with those for CO_2 (Table 11.5), and discuss the possible origins of the differences. **(f)** Would you expect the density of CS_2 at 40 °C to be greater or less than at 20 °C? What accounts for the difference?

SOLUTION

(a) The compound is named carbon disulfide, in analogy with the naming of other binary molecular compounds such as carbon dioxide. ∞ (Section 2.8)
(b) Only London dispersion forces affect CS_2; it does not have a dipole moment, based upon its molecular shape, and obviously cannot undergo hydrogen bonding.
(c) Because $CS_2(s)$ consists of individual CS_2 molecules, it will be a molecular solid.
(d) The most likely products of the combustion will be CO_2 and SO_2. ∞ (Sections 3.2 and 7.8) Under some conditions SO_3 might be formed, but this would be the less likely outcome. Thus, we have the following equation for combustion:

$$CS_2(l) + 3\,O_2(g) \longrightarrow CO_2(g) + 2\,SO_2(g)$$

(e) The critical temperature and pressure of CS_2 (552 K and 78 atm) are both higher than those given for CO_2 in Table 11.5 (304 K and 73 atm). The difference in critical temperatures is especially notable. The higher values for CS_2 arise from the greater London dispersion attractions between the CS_2 molecules compared with CO_2. These greater attractions are due to the larger size of the sulfur compared to oxygen and therefore its greater polarizability.
(f) The density would be lower at the higher temperature. Density decreases with increasing temperature because the molecules possess higher kinetic energies. Their more energetic movements result in larger average distances between molecules, which translate into lower densities.

CHAPTER REVIEW

SUMMARY AND KEY TERMS

Introduction and Section 11.1 Substances that are gases or liquids at room temperature are usually composed of molecules. In gases the intermolecular attractive forces are negligible compared to the kinetic energies of the molecules; thus, the molecules are widely separated and undergo constant, chaotic motion. In liquids the **intermolecular forces** are strong enough to keep the molecules in close proximity; nevertheless, the molecules are free to move with respect to one another. In solids the interparticle attractive forces are strong enough to restrain

molecular motion and to force the particles to occupy specific locations in a three-dimensional arrangement.

Section 11.2 Three types of intermolecular forces exist between neutral molecules: **dipole–dipole forces, London dispersion forces,** and **hydrogen bonding. Ion–dipole forces** are important in solutions in which ionic compounds are dissolved in polar solvents. London dispersion forces operate between all molecules (and atoms, for atomic substances such as He, Ne, Ar, and so forth). The relative strengths of the dipole–dipole and dispersion forces depend on the polarity, **polarizability**, size, and shape of the molecule. Dipole–dipole forces increase in strength with increasing polarity. Dispersion forces increase in strength with increasing molecular weight, although molecular shape is also an important factor. Hydrogen bonding occurs in compounds containing O—H, N—H, and F—H bonds. Hydrogen bonds are generally stronger than dipole–dipole or dispersion forces.

Section 11.3 The stronger the intermolecular forces, the greater is the **viscosity**, or resistance to flow, of a liquid. The surface tension of a liquid also increases as intermolecular forces increase in strength. **Surface tension** is a measure of the tendency of a liquid to maintain a minimum surface area. The adhesion of a liquid to the walls of a narrow tube and the cohesion of the liquid account for **capillary action** and the formation of a meniscus at the surface of a liquid.

Section 11.4 A substance may exist in more than one state of matter, or phase. **Phase changes** are transformations from one phase to another. Changes of a solid to liquid (melting), solid to gas (sublimation), and liquid to gas (vaporization) are all endothermic processes. Thus, the **heat of fusion** (melting), the **heat of sublimation**, and the **heat of vaporization** are all positive quantities. The reverse processes are exothermic. A gas cannot be liquefied by application of pressure if the temperature is above its **critical temperature**. The pressure required to liquefy a gas at its critical temperature is called the **critical pressure**.

Section 11.5 The **vapor pressure** of a liquid indicates the tendency of the liquid to evaporate. The vapor pressure is the partial pressure of the vapor when it is in **dynamic equilibrium** with the liquid. At equilibrium the rate of transfer of molecules from the liquid to the vapor equals the rate of transfer from the vapor to the liquid. The higher the vapor pressure of a liquid, the more readily it evaporates and the more **volatile** it is. Vapor pressure increases nonlinearly with temperature. Boiling occurs when the vapor pressure equals the external pressure.

The **normal boiling point** is the temperature at which the vapor pressure equals 1 atm.

Section 11.6 The equilibria between the solid, liquid, and gas phases of a substance as a function of temperature and pressure are displayed on a **phase diagram**. A line indicates equilibria between any two phases. The line through the melting point usually slopes slightly to the right as pressure increases, because the solid is usually more dense than the liquid. The melting point at 1 atm is the **normal melting point**. The point on the diagram at which all three phases coexist in equilibrium is called the **triple point**.

Section 11.7 In a **crystalline solid**, particles are arranged in a regularly repeating pattern. An **amorphous solid** is one whose particles show no such order. The essential structural features of a crystalline solid can be represented by its **unit cell**, the smallest part of the crystal that can, by simple displacement, reproduce the three-dimensional structure. The three-dimensional structures of a crystal can also be represented by its **crystal lattice**. The points in a crystal lattice represent positions in the structure where there are identical environments. The simplest unit cells are cubic. There are three kinds of cubic unit cells: **primitive cubic, body-centered cubic**, and **face-centered cubic**. Many solids have a close-packed structure in which spherical particles are arranged so as to leave the minimal amount of empty space. Two closely related forms of close packing, **cubic close packing** and **hexagonal close packing**, are possible. In both, each sphere has a **coordination number** of 12.

Section 11.8 The properties of solids depend both on the arrangements of particles and on the attractive forces between them. **Molecular solids**, which consist of atoms or molecules held together by intermolecular forces, are soft and low melting. **Covalent-network solids**, which consist of atoms held together by covalent bonds that extend throughout the solid, are hard and high melting. **Ionic solids** are hard and brittle and have high melting points. **Metallic solids**, which consist of metal cations held together by a sea of electrons, exhibit a wide range of properties.

KEY SKILLS

- Understand and be able to describe the intermolecular attractive interactions (ion–dipole, dipole–dipole, London dispersion, hydrogen bonding) that exist between molecules or ions, and be able to compare the relative strengths of intermolecular attractions in substances based on their molecular structure, or physical properties.

- Understand the concept of polarizability.

- Understand the concepts of viscosity and surface tension in liquids.

- Know the names of the various phase changes for a pure substance.

- Interpret heating curves and be able to calculate quantities related to temperature and enthalpies of phase changes.

- Define critical pressure, critical temperature, vapor pressure, normal boiling point, normal melting point, critical point, triple point.

- Be able to interpret and sketch phase diagrams; know how water's phase diagram differs from most other substances, and why.

- Know the difference between crystalline and amorphous solids, and be able to explain the differences between primitive cubic, body-centered cubic, and face-centered cubic unit cells.

- Classify solids based on their bonding/intermolecular forces and understand how difference in bonding relates to physical properties.

VISUALIZING CONCEPTS

11.1 Does the following diagram best describe a crystalline solid, liquid, or gas? Explain. [Section 11.1]

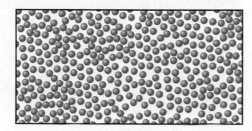

11.2 (a) What kind of intermolecular attractive force is shown in each of the following cases? **(b)** Predict which two interactions are stronger than the other two. [Section 11.2]

(a) H F ····· H F (c) Na^+ ····· O H H

(b) F F ····· F F (d) O S O ····· S O O

11.3 The molecular models of glycerol and 1-propanol are given here.

(a) Glycerol   (b) 1-propanol

Do you expect the viscosity of glycerol to be larger or smaller than that of 1-propanol? Explain. [Section 11.3]

11.4 Using the following graph of CS_2 data, determine **(a)** the approximate vapor pressure of CS_2 at 30 °C, **(b)** the temperature at which the vapor pressure equals 300 torr, **(c)** the normal boiling point of CS_2. [Section 11.5]

11.5 The following molecules have the same molecular formula (C_3H_8O), yet they have different normal boiling points, as shown. Rationalize the difference in boiling points. [Sections 11.2 and 11.5]

(a) Propanol   (b) Ethyl methyl ether
97.2 °C   10.8 °C

11.6 The phase diagram of a hypothetical substance is shown below.

(a) Estimate the normal boiling point and freezing point of the substance.
(b) What is the physical state of the substance under the following conditions?
   (i) $T = 150$ K, $P = 0.2$ atm
   (ii) $T = 100$ K, $P = 0.8$ atm
   (iii) $T = 300$ K, $P = 1.0$ atm
(c) What is the triple point of the substance? [Section 11.6]

11.7 Niobium(II) oxide crystallizes in the following cubic unit cell.

● Oxygen
○ Niobium

(a) How many niobium atoms and how many oxygen atoms are within the unit cell?
(b) What is the empirical formula of niobium oxide?
(c) Is this a molecular, covalent-network, or ionic solid? [Sections 11.7 and 11.8]

11.8 (a) What kind of packing arrangement is seen in the accompanying photo? **(b)** What is the coordination number of each orange in the interior of the stack? **(c)** If each orange represents an argon atom, what category of solid is represented? [Sections 11.7 and 11.8]

EXERCISES

Molecular Comparisons of Gases, Liquids, and Solids

11.9 List the three states of matter in order of **(a)** increasing molecular disorder and **(b)** increasing intermolecular attractions. **(c)** Which state of matter is most easily compressed?

11.10 (a) How does the average kinetic energy of molecules compare with the average energy of attraction between molecules in solids, liquids, and gases? **(b)** Why does increasing the temperature cause a solid substance to change in succession from a solid to a liquid to a gas? **(c)** What happens to a gas if you put it under extremely high pressure?

11.11 If you mix olive oil with water, the olive oil will float on top of the water. The density of water is 1.00 g/cm^3 at room temperature. **(a)** Is the density of olive oil more or less than 1.00 g/cm^3? **(b)** The density of olive oil in its liquid phase does vary with temperature. Do you think olive oil would be more dense or less dense at higher temperatures? Explain.

11.12 Benzoic acid, C_6H_5COOH, melts at 122 °C. The density in the liquid state at 130 °C is 1.08 g/cm^3. The density of solid benzoic acid at 15 °C is 1.266 g/cm^3. **(a)** In which of these two states is the average distance between molecules greater? **(b)** Explain the difference in densities at the two temperatures in terms of the relative kinetic energies of the molecules.

Intermolecular Forces

11.13 Which type of intermolecular attractive force operates between **(a)** all molecules, **(b)** polar molecules, **(c)** the hydrogen atom of a polar bond and a nearby small electronegative atom?

11.14 Based on what you have learned about intermolecular forces, would you say that matter is fundamentally attracted or repulsed by other matter?

11.15 Describe the intermolecular forces that must be overcome to convert each of the following from a liquid or solid to a gas: (a) I_2, (b) CH_3CH_2OH, (c) H_2Se.

11.16 What type of intermolecular force accounts for the following differences in each case? **(a)** CH_3OH boils at 65 °C, CH_3SH boils at 6 °C. **(b)** Xe is liquid at atmospheric pressure and 120 K, whereas Ar is a gas. **(c)** Kr,

atomic weight 84, boils at 120.9 K, whereas Cl_2, molecular weight about 71, boils at 238 K. **(d)** Acetone boils at 56 °C, whereas 2-methylpropane boils at −12 °C.

$$\begin{array}{cc}
\overset{\displaystyle O}{\underset{\displaystyle \|}{}} & CH_3 \\
CH_3-C-CH_3 & CH_3-\overset{|}{C}H-CH_3 \\
\text{Acetone} & \text{2-Methylpropane}
\end{array}$$

11.17 (a) What is meant by the term *polarizability*? **(b)** Which of the following atoms would you expect to be most polarizable: N, P, As, Sb? Explain. **(c)** Put the following molecules in order of increasing polarizability: $GeCl_4$, CH_4, $SiCl_4$, SiH_4, and $GeBr_4$. **(d)** Predict the order of boiling points of the substances in part (c).

11.18 True or false:
- **(a)** The more polarizable the molecules, the stronger the dispersion forces between them.
- **(b)** The boiling points of the noble gases decrease as you go down the column in the periodic table.
- **(c)** In general, the smaller the molecule, the stronger the dispersion forces.
- **(d)** All other factors being the same, dispersion forces between molecules increase with the number of electrons in the molecules.

11.19 Which member of the following pairs has the larger London dispersion forces: **(a)** H_2O or H_2S, **(b)** CO_2 or CO, **(c)** SiH_4 or GeH_4?

11.20 Which member of the following pairs has the stronger intermolecular dispersion forces: **(a)** Br_2 or O_2, **(b)** $CH_3CH_2CH_2CH_2SH$ or $CH_3CH_2CH_2CH_2CH_2SH$, **(c)** $CH_3CH_2CH_2Cl$ or $(CH_3)_2CHCl$?

11.21 Butane and 2-methylpropane, whose space-filling models are shown, are both nonpolar and have the same molecular formula, yet butane has the higher boiling point (-0.5 °C compared to -11.7 °C). Explain.

(a) Butane   (b) 2-Methylpropane

11.22 Propyl alcohol ($CH_3CH_2CH_2OH$) and isopropyl alcohol [$(CH_3)_2CHOH$], whose space-filling models are shown, have boiling points of 97.2 °C and 82.5 °C, respectively. Explain why the boiling point of propyl alcohol is higher, even though both have the molecular formula of C_3H_8O.

(a) Propyl alcohol   (b) Isopropyl alcohol

11.23 **(a)** What atoms must a molecule contain to participate in hydrogen bonding with other molecules of the same kind? **(b)** Which of the following molecules can form hydrogen bonds with other molecules of the same kind: CH_3F, CH_3NH_2, CH_3OH, CH_3Br?

11.24 Rationalize the difference in boiling points between the members of the following pairs of substances: **(a)** HF (20 °C) and HCl (-85 °C), **(b)** $CHCl_3$ (61 °C) and $CHBr_3$ (150 °C), **(c)** Br_2 (59 °C) and ICl (97 °C).

11.25 Ethylene glycol ($HOCH_2CH_2OH$), the major substance in antifreeze, has a normal boiling point of 198 °C. By comparison, ethyl alcohol (CH_3CH_2OH) boils at 78 °C at atmospheric pressure. Ethylene glycol dimethyl ether ($CH_3OCH_2CH_2OCH_3$) has a normal boiling point of 83 °C, and ethyl methyl ether ($CH_3CH_2OCH_3$) has a normal boiling point of 11 °C. **(a)** Explain why replacement of a hydrogen on the oxygen by CH_3 generally results in a lower boiling point. **(b)** What are the major factors responsible for the difference in boiling points of the two ethers?

11.26 Identify the types of intermolecular forces present in each of the following substances, and select the substance in each pair that has the higher boiling point: **(a)** C_6H_{14} or C_8H_{18}, **(b)** C_3H_8 or CH_3OCH_3, **(c)** HOOH or HSSH, **(d)** NH_2NH_2 or CH_3CH_3.

11.27 Look up and compare the normal boiling points and normal melting points of H_2O and H_2S. **(a)** Based on these physical properties, which substance has stronger intermolecular forces? What kind of intermolecular forces exist for each molecule? **(b)** Predict whether solid H_2S is more or less dense than liquid H_2S. How does this compare to H_2O? Explain. **(c)** Water has an unusually high specific heat. Is this related to its intermolecular forces? Explain.

11.28 The following quote about ammonia (NH_3) is from a textbook of inorganic chemistry: "It is estimated that 26% of the hydrogen bonding in NH_3 breaks down on melting, 7% on warming from the melting to the boiling point, and the final 67% on transfer to the gas phase at the boiling point." From the standpoint of the kinetic energy of the molecules, explain **(a)** why there is a decrease of hydrogen-bonding energy on melting and **(b)** why most of the loss in hydrogen bonding occurs in the transition from the liquid to the vapor state.

Viscosity and Surface Tension

11.29 **(a)** Explain why surface tension and viscosity decrease with increasing temperature. **(b)** Why do substances with high surface tensions also tend to have high viscosities?

11.30 **(a)** Distinguish between adhesive forces and cohesive forces. **(b)** What adhesive and cohesive forces are involved when a paper towel absorbs water? **(c)** Explain the cause for the U-shaped meniscus formed when water is in a glass tube.

11.31 Explain the following observations: **(a)** The surface tension of $CHBr_3$ is greater than that of $CHCl_3$. **(b)** As tem-

perature increases, oil flows faster through a narrow tube. **(c)** Raindrops that collect on a waxed automobile hood take on a nearly spherical shape. **(d)** Oil droplets that collect on a waxed automobile hood take on a flat shape.

11.32 Hydrazine (H_2NNH_2), hydrogen peroxide (HOOH), and water (H_2O) all have exceptionally high surface tensions compared with other substances of comparable molecular weights. **(a)** Draw the Lewis structures for these three compounds. **(b)** What structural property do these substances have in common, and how might that account for the high surface tensions?

Phase Changes

11.33 Name the phase transition in each of the following situations, and indicate whether it is exothermic or endothermic: **(a)** When ice is heated, it turns to water. **(b)** Wet clothes dry on a warm summer day. **(c)** Frost appears on a window on a cold winter day. **(d)** Droplets of water appear on a cold glass of beer.

11.34 Name the phase transition in each of the following situations, and indicate whether it is exothermic or endothermic: **(a)** Bromine vapor turns to bromine liquid as it is cooled. **(b)** Crystals of iodine disappear from an evaporating dish as they stand in a fume hood. **(c)** Rubbing alcohol in an open container slowly disappears. **(d)** Molten lava from a volcano turns into solid rock.

11.35 Explain why the heat of fusion of any substance is generally lower than its heat of vaporization.

11.36 Ethyl chloride (C_2H_5Cl) boils at 12 °C. When liquid C_2H_5Cl under pressure is sprayed on a room-temperature (25 °C) surface in air, the surface is cooled considerably. **(a)** What does this observation tell us about the specific heat of $C_2H_5Cl(g)$ as compared with $C_2H_5Cl(l)$? **(b)** Assume that the heat lost by the surface is gained by ethyl chloride. What enthalpies must you consider if you were to calculate the final temperature of the surface?

11.37 For many years drinking water has been cooled in hot climates by evaporating it from the surfaces of canvas bags or porous clay pots. How many grams of water can be cooled from 35 °C to 20 °C by the evaporation of 60 g of water? (The heat of vaporization of water in this temperature range is 2.4 kJ/g. The specific heat of water is 4.18 J/g-K.)

11.38 Compounds like CCl_2F_2 are known as chlorofluorocarbons, or CFCs. These compounds were once widely used as refrigerants but are now being replaced by compounds that are believed to be less harmful to the environment. The heat of vaporization of CCl_2F_2 is 289 J/g. What mass of this substance must evaporate to freeze 200 g of water initially at 15 °C? (The heat of fusion of water is 334 J/g; the specific heat of water is 4.18 J/g-K.)

11.39 Ethanol (C_2H_5OH) melts at −114 °C and boils at 78 °C. Its density is 0.789 g/mL. The enthalpy of fusion of ethanol is 5.02 kJ/mol, and its enthalpy of vaporization is 38.56 kJ/mol. The specific heats of solid and liquid ethanol are 0.97 J/g-K and 2.3 J/g-K, respectively. **(a)** How much heat is required to convert 25.0 g of ethanol at 25 °C to the vapor phase at 78 °C? **(b)** How much heat is required to convert 5.00 L of ethanol at −140 °C to the vapor phase at 78 °C?

11.40 The fluorocarbon compound $C_2Cl_3F_3$ has a normal boiling point of 47.6 °C. The specific heats of $C_2Cl_3F_3(l)$ and $C_2Cl_3F_3(g)$ are 0.91 J/g-K and 0.67 J/g-K, respectively. The heat of vaporization for the compound is 27.49 kJ/mol. Calculate the heat required to convert 50.0 g of $C_2Cl_3F_3$ from a liquid at 10.00 °C to a gas at 85.00 °C.

11.41 **(a)** What is the significance of the critical pressure of a substance? **(b)** What happens to the critical temperature of a series of compounds as the force of attraction between molecules increases? **(c)** Which of the substances listed in Table 11.5 can be liquefied at the temperature of liquid nitrogen (−196 °C)?

11.42 The critical temperatures (K) and pressures (atm) of a series of halogenated methanes are as follows:

Compound	CCl_3F	CCl_2F_2	$CClF_3$	CF_4
Critical Temperature	471	385	302	227
Critical Pressure	43.5	40.6	38.2	37.0

(a) List the intermolecular forces that occur for each compound. **(b)** Predict the order of increasing intermolecular attraction, from least to most, for this series of compounds. **(c)** Predict the critical temperature and pressure for CCl_4 based on the trends in this table. Look up the experimentally determined critical temperatures and pressures for CCl_4, using a source such as the *CRC Handbook of Chemistry and Physics*, and suggest a reason for any discrepancies.

Vapor Pressure and Boiling Point

11.43 Explain how each of the following affects the vapor pressure of a liquid: **(a)** volume of the liquid, **(b)** surface area, **(c)** intermolecular attractive forces, **(d)** temperature, **(e)** density of the liquid.

11.44 A liquid that has an equilibrium vapor pressure of 130 torr at 25 °C is placed into a 1-L vessel like that shown in Figure 11.22. What is the pressure difference shown on the manometer, and what is the composition of the gas in the vessel, under each of the following conditions: **(a)** Two hundred mL of the liquid is introduced into the vessel and frozen at the bottom. The vessel is evacuated and sealed, and the liquid is allowed to warm to 25 °C. **(b)** Two hundred milliliters of the liquid is added to the vessel at 25 °C under atmospheric pressure, and after a

few minutes the vessel is closed off. **(c)** A few mL of the liquid is introduced into the vessel at 25 °C while it has a pressure of 1 atm of air in it, without allowing any of the air to escape. After a few minutes a few drops of liquid remain in the vessel.

11.45 **(a)** Place the following substances in order of increasing volatility: CH_4, CBr_4, CH_2Cl_2, CH_3Cl, $CHBr_3$, and CH_2Br_2. Explain. **(b)** How do the boiling points vary through this series?

11.46 True or false:
(a) CBr_4 is more volatile than CCl_4.
(b) CBr_4 has a higher boiling point than CCl_4.
(c) CBr_4 has weaker intermolecular forces than CCl_4.
(d) CBr_4 has a higher vapor pressure at the same temperature than CCl_4.

11.47 **(a)** Two pans of water are on different burners of a stove. One pan of water is boiling vigorously, while the other is boiling gently. What can be said about the temperature of the water in the two pans? **(b)** A large container of water and a small one are at the same temperature. What can be said about the relative vapor pressures of the water in the two containers?

11.48 Explain the following observations: **(a)** Water evaporates more quickly on a hot, dry day than on a hot, humid day. **(b)** It takes longer to boil water for tea at high altitudes than at lower altitudes.

11.49 Using the vapor-pressure curves in Figure 11.24, **(a)** estimate the boiling point of ethanol at an external pressure of 200 torr; **(b)** estimate the external pressure at which ethanol will boil at 60 °C; **(c)** estimate the boiling point of diethyl ether at 400 torr; **(d)** estimate the external pressure at which diethyl ether will boil at 40 °C.

11.50 Appendix B lists the vapor pressure of water at various external pressures.
 (a) Plot the data in Appendix B, vapor pressure (torr) vs. temperature (°C). From your plot, estimate the vapor pressure of water at body temperature, 37 °C.
 (b) Explain the significance of the data point at 760.0 torr, 100 °C.
 (c) A city at an altitude of 5000 ft above sea level has a barometric pressure of 633 torr. To what temperature would you have to heat water to boil it in this city?
 (d) A city at an altitude of 500 ft below sea level would have a barometric pressure of 774 torr. To what temperature would you have to heat water to boil it in this city?
 (e) For the two cities in parts (c) and (d), compare the average kinetic energies of the water molecules at their boiling points. Are the kinetic energies the same or different? Explain.

Phase Diagrams

11.51 **(a)** What is the significance of the critical point in a phase diagram? **(b)** Why does the line that separates the gas and liquid phases end at the critical point?

11.52 **(a)** What is the significance of the triple point in a phase diagram? **(b)** Could you measure the triple point of water by measuring the temperature in a vessel in which water vapor, liquid water, and ice are in equilibrium under one atmosphere of air? Explain.

11.53 Refer to Figure 11.27(a), and describe all the phase changes that would occur in each of the following cases: **(a)** Water vapor originally at 0.005 atm and −0.5 °C is slowly compressed at constant temperature until the final pressure is 20 atm. **(b)** Water originally at 100.0 °C and 0.50 atm is cooled at constant pressure until the temperature is −10 °C.

11.54 Refer to Figure 11.27(b), and describe the phase changes (and the temperatures at which they occur) when CO_2 is heated from −80 °C to −20 °C at **(a)** a constant pressure of 3 atm, **(b)** a constant pressure of 6 atm.

11.55 Sketch a generic phase diagram for a substance that has a more dense solid phase than a liquid phase. Label all regions, lines, and points.

11.56 The normal melting and boiling points of O_2 are −218 °C and −183 °C respectively. Its triple point is at −219 °C and 1.14 torr, and its critical point is at −119 °C and 49.8 atm. **(a)** Sketch the phase diagram for O_2, showing the four points given and indicating the area in which each phase is stable. **(b)** Will $O_2(s)$ float on $O_2(l)$? Explain. **(c)** As it is heated, will solid O_2 sublime or melt under a pressure of 1 atm?

Structures of Solids

11.57 **(a)** Draw a picture that represents a crystalline solid at the atomic level. **(b)** Now draw a picture that represents an amorphous solid at the atomic level.

11.58 Amorphous silica has a density of about 2.2 g/cm³, whereas the density of crystalline quartz is 2.65 g/cm³. Account for this difference in densities.

11.59 Tausonite, a mineral composed of Sr, O, and Ti, has the cubic unit cell shown in the drawing. **(a)** What is the chemical formula of this mineral? **(b)** It is easy to see that Ti is coordinated by six oxygen atoms, because the Ti atom is located at the center of the unit cell. To see the full coordination environment of the other ions, we have to consider neighboring unit cells. How many oxygens are coordinated to strontium?

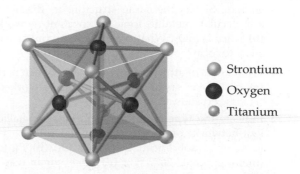

○ Strontium

● Oxygen

○ Titanium

11.60 Rutile is a mineral composed of Ti and O. Its unit cell, shown in the drawing, contains Ti atoms at each corner and a Ti atom at the center of the cell. Four O atoms are on the opposite faces of the cell, and two are entirely within the cell. **(a)** What is the chemical formula of this mineral? **(b)** What is the nature of the bonding that holds the solid together? **(c)** What is the coordination number of each atom?

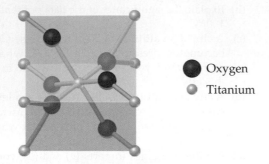

Oxygen

Titanium

11.61 Iridium crystallizes in a face-centered cubic unit cell that has an edge length of 3.833 Å. The atom in the center of the face is in contact with the corner atoms, as shown in the drawing. **(a)** Calculate the atomic radius of an iridium atom. **(b)** Calculate the density of iridium metal.

11.62 Aluminum metal crystallizes in a cubic close-packed structure (face-centered cubic cell, Figure 11.34). **(a)** How many aluminum atoms are in a unit cell? **(b)** What is the coordination number of each aluminum atom? **(c)** Assume that the aluminum atoms can be represented as spheres, as shown in the drawing for

Exercise 11.61. If each Al atom has a radius of 1.43 Å, what is the length of a side of the unit cell? **(d)** Calculate the density of aluminum metal.

11.63 An element crystallizes in a body-centered cubic lattice. The edge of the unit cell is 2.86 Å, and the density of the crystal is 7.92 g/cm^3. Calculate the atomic weight of the element.

11.64 KCl has the same structure as NaCl. The length of the unit cell is 628 pm. The density of KCl is 1.984 g/cm^3, and its formula mass is 74.55 amu. Using this information, calculate Avogadro's number.

11.65 NaF has the same structure as NaCl. **(a)** Use ionic radii from Chapter 7 to estimate the length of the unit cell edge for NaF. **(b)** Use the unit cell size calculated in part (a) to estimate the density of NaF.

11.66 What is the coordination number of **(a)** Na^+ in the NaCl structure, Figure 11.35; **(b)** Zn^{2+} in the ZnS unit cell, Figure 11.42(b); **(c)** Ca^{2+} in the CaF_2 unit cell, Figure 11.42(c)?

11.67 Clausthalite is a mineral composed of lead selenide (PbSe). The mineral adopts a NaCl-type structure. The density of PbSe at 25 °C is 8.27 g/cm^3. Calculate the length of an edge of the PbSe unit cell.

11.68 A particular form of cinnabar (HgS) adopts the zinc blende structure, Figure 11.42(b). The length of the unit cell side is 5.852 Å. **(a)** Calculate the density of HgS in this form. **(b)** The mineral tiemmanite (HgSe) also forms a solid phase with the zinc blende structure. The length of the unit cell side in this mineral is 6.085 Å. What accounts for the larger unit cell length in tiemmanite? **(c)** Which of the two substances has the higher density? How do you account for the difference in densities?

Bonding in Solids

11.69 It is possible to change the temperature and pressure of a vessel containing argon gas so that the gas solidifies. **(a)** What intermolecular forces exist between argon atoms? **(b)** Is the solid argon a "covalent network solid"? Why or why not?

11.70 **(a)** Silicon is the fundamental component of integrated circuits. Si has the same structure as diamond. Is Si a molecular, metallic, ionic, or covalent-network solid? **(b)** Silica is SiO_2. What type of solid would you expect silica to form?

11.71 What kinds of attractive forces exist between particles in **(a)** molecular crystals, **(b)** covalent-network crystals, **(c)** ionic crystals, **(d)** metallic crystals?

11.72 Indicate the type of crystal (molecular, metallic, covalent-network, or ionic) each of the following would form upon solidification: **(a)** $CaCO_3$, **(b)** Pt, **(c)** ZrO_2 (melting point, 2677 °C), **(d)** table sugar ($C_{12}H_{22}O_{11}$), **(e)** benzene, **(f)** I_2.

11.73 Covalent bonding occurs in both molecular and covalent-network solids. Why do these two kinds of solids differ so greatly in their hardness and melting points?

11.74 Which type (or types) of crystalline solid is characterized by each of the following: **(a)** high mobility of electrons throughout the solid; **(b)** softness, relatively low melting point; **(c)** high melting point and poor electrical conductivity; **(d)** network of covalent bonds; **(e)** charged particles throughout the solid.

11.75 A white substance melts with some decomposition at 730 °C. As a solid, it does not conduct electricity, but it dissolves in water to form a conducting solution. Which type of solid (Table 11.7) might the substance be?

11.76 You are given a white substance that sublimes at 3000 °C; the solid is a nonconductor of electricity and is insoluble in water. Which type of solid (Table 11.7) might this substance be?

11.77 For each of the following pairs of substances, predict which will have the higher melting point and indicate why: **(a)** Ar, Xe; **(b)** SiO_2, CO_2; **(c)** KBr, Br_2; **(d)** C_6Cl_6, C_6H_6.

11.78 For each of the following pairs of substances, predict which will have the higher melting point, and indicate why: **(a)** HF, HCl; **(b)** C (graphite), CH_4; **(c)** KCl, Cl_2; **(d)** LiF, MgF_2.

ADDITIONAL EXERCISES

11.79 As the intermolecular attractive forces between molecules increase in magnitude, do you expect each of the following to increase or decrease in magnitude? **(a)** vapor pressure, **(b)** heat of vaporization, **(c)** boiling point, **(d)** freezing point, **(e)** viscosity, **(f)** surface tension, **(g)** critical temperature.

11.80 Suppose you have two colorless molecular liquids, one boiling at -84 °C, the other at 34 °C, and both at atmospheric pressure. Which of the following statements is correct? For those that are not correct, modify the statement so that it is correct. **(a)** The higher-boiling liquid has greater total intermolecular forces than the other. **(b)** The lower boiling liquid must consist of nonpolar molecules. **(c)** The lower-boiling liquid has a lower molecular weight than the higher-boiling liquid. **(d)** The two liquids have identical vapor pressures at their normal boiling points. **(e)** At 34 °C both liquids have vapor pressures of 760 mm Hg.

11.81 Two isomers of the planar compound 1,2-dichloroethylene are shown here, along with their melting and boiling points.

	cis isomer	trans isomer
Melting point (°C)	-80.5	-49.8
Boiling point (°C)	60.3	47.5

(a) Which of the two isomers will have the stronger dipole–dipole forces? Is this prediction borne out by the data presented here? **(b)** Based on the data presented here, which isomer packs more efficiently in the solid phase?

11.82 In dichloromethane, CH_2Cl_2 ($\mu = 1.60$ D), the dispersion force contribution to the intermolecular attractive forces is about five times larger than the dipole–dipole contribution. Would you expect the relative importance of the two kinds of intermolecular attractive forces to differ **(a)** in dibromomethane ($\mu = 1.43$ D), **(b)** in difluoromethane ($\mu = 1.93$ D)? Explain.

11.83 When an atom or group of atoms is substituted for an H atom in benzene (C_6H_6), the boiling point changes. Explain the order of the following boiling points: C_6H_6 (80 °C), C_6H_5Cl (132 °C), C_6H_5Br (156 °C), C_6H_5OH (182 °C).

11.84 The DNA double helix (Figure 25.40) at the atomic level looks like a twisted ladder, where the "rungs" of the ladder consist of molecules that are hydrogen-bonded together. Sugar and phosphate groups make up the sides of the ladder. Shown are the structures of the adenine-thymine (AT) "base pair" and the guanine-cytosine (GC) base pair:

Thymine   Adenine

You can see that AT base pairs are held together by two hydrogen bonds, and the GC base pairs are held together by three hydrogen bonds. Which base pair is more stable to heating?

Cytosine   Guanine

11.85 Ethylene glycol [$CH_2(OH)CH_2(OH)$] is the major component of antifreeze. It is a slightly viscous liquid, not very volatile at room temperature, with a boiling point of 198 °C. Pentane (C_5H_{12}), which has about the same molecular weight, is a nonviscous liquid that is highly volatile at room temperature and whose boiling point is 36.1 °C. Explain the differences in the physical properties of the two substances.

11.86 Using the following list of normal boiling points for a series of hydrocarbons, estimate the normal boiling point for octane, C_8H_{18}: propane (C_3H_8, -42.1 °C), butane (C_4H_{10}, -0.5 °C), pentane (C_5H_{12}, 36.1 °C), hexane (C_6H_{14}, 68.7 °C), heptane (C_7H_{16}, 98.4 °C). Explain the trend in the boiling points.

[11.87] Notice in Figure 11.21 that there is a pressure-reduction valve in the line just before the supercritical CO_2 and dissolved caffeine enter the separator. Suggest an explanation for the function of this valve in the overall process.

11.88 **(a)** When you exercise vigorously, you sweat. How does this help your body cool? **(b)** A flask of water is connected to a vacuum pump. A few moments after the pump is turned on, the water begins to boil. After a few minutes, the water begins to freeze. Explain why these processes occur.

[11.89] The following table gives the vapor pressure of hexafluorobenzene (C_6F_6) as a function of temperature:

Temperature (K)	Vapor Pressure (torr)
280.0	32.42
300.0	92.47
320.0	225.1
330.0	334.4
340.0	482.9

(a) By plotting these data in a suitable fashion, determine whether the Clausius–Clapeyron equation is obeyed. If it is obeyed, use your plot to determine ΔH_{vap} for C_6F_6. **(b)** Use these data to determine the boiling point of the compound.

[11.90] Suppose the vapor pressure of a substance is measured at two different temperatures. **(a)** By using the Clausius–Clapeyron equation, Equation 11.1, derive the following relationship between the vapor pressures, P_1 and P_2, and the absolute temperatures at which they were measured, T_1 and T_2:

$$\ln \frac{P_1}{P_2} = -\frac{\Delta H_{vap}}{R}\left(\frac{1}{T_1} - \frac{1}{T_2}\right)$$

(b) Gasoline is a mixture of hydrocarbons, a major component of which is octane, $CH_3CH_2CH_2CH_2CH_2CH_2CH_2CH_3$. Octane has a vapor pressure of 13.95 torr at 25 °C and a vapor pressure of 144.78 torr at 75 °C. Use these data and the equation in part (a) to calculate the heat of vaporization of octane. (c) By using the equation in part (a) and the data given in part (b), calculate the normal boiling point of octane. Compare your answer to the one you obtained from Exercise 11.86. (d) Calculate the vapor pressure of octane at −30 °C.

[11.91] The following data present the temperatures at which certain vapor pressures are achieved for dichloromethane (CH_2Cl_2) and methyl iodide (CH_3I):

Vapor Pressure (torr):	10.0	40.0	100.0	400.0
T for CH_2Cl_2 (°C):	−43.3	−22.3	−6.3	24.1
T for CH_3I (°C):	−45.8	−24.2	−7.0	25.3

(a) Which of the two substances is expected to have the greater dipole–dipole forces? Which is expected to have the greater London dispersion forces? Based on your answers, explain why it is difficult to predict which compound would be more volatile. (b) Which compound would you expect to have the higher boiling point? Check your answer in a reference book such as the *CRC Handbook of Chemistry and Physics*. (c) The order of volatility of these two substances changes as the temperature is increased. What quantity must be different for the two substances in order for this phenomenon to occur? (d) Substantiate your answer for part (c) by drawing an appropriate graph.

11.92 The elements xenon and gold both have solid-state structures with face-centered cubic unit cells, yet Xe melts at −112 °C and gold melts at 1064 °C. Account for these greatly different melting points.

11.93 In Chapter 7 we saw that Li reacts with oxygen to form lithium oxide, Li_2O, while the larger alkali metals react with oxygen to form peroxides (such as Na_2O_2, K_2O_2, etc.). The origin of this curious pattern of reactivity is related to the relative sizes of the cation and anion and the overall stoichiometry. Li_2O crystallizes with a structure called the antifluorite structure. This structure is identical to the fluorite structure shown in Figure 11.42(c), but the cations and anions have been switched. (a) In Li_2O, what are the coordination numbers for each ion? (b) As the cation radius increases, would you expect the coordination number to increase or decrease (assuming the anion size does not change)? (c) Why do you think the antifluorite structure becomes unstable for A_2O (A = Li, Na, K, Rb, Cs) compounds of the heavier alkali metal ions?

11.94 Gold crystallizes in a face-centered cubic unit cell that has an edge length of 4.078 Å. The atom in the center of the face is in contact with the corner atoms, as shown in the drawing for Exercise 11.61. (a) Calculate the apparent radius of a gold atom in this structure. (b) Calculate the density of gold metal.

11.95 (a) Explain why X-rays can be used to measure atomic distances in crystals but visible light cannot. (b) Why can't $CaCl_2$ have the same crystal structure as NaCl?

11.96 In their study of X-ray diffraction, William and Lawrence Bragg determined that the relationship among the wavelength of the radiation (λ), the angle at which the radiation is diffracted (θ), and the distance between the layers of atoms in the crystal that cause the diffraction (d) is given by $n\lambda = 2d \sin \theta$. (a) X-rays from a copper X-ray tube that have a wavelength of 1.54 Å are diffracted at an angle of 14.22 degrees by crystalline silicon. Using the Bragg equation, calculate the interplanar spacing in the crystal, assuming $n = 1$ (first-order diffraction). (b) Repeat the calculation of part (a) but for the $n = 2$ case (second-order diffraction).

[11.97] (a) The density of diamond [Figure 11.41(a)] is 3.5 g/cm³, and that of graphite [Figure 11.41(b)] is 2.3 g/cm³. Based on the structure of buckminsterfullerene (Figure 11.43), what would you expect its density to be relative to these other forms of carbon? (b) X-ray diffraction studies of buckminsterfullerene show that it has a face-centered cubic lattice of C_{60} molecules. The length of a side of the unit cell is 14.2 Å. Calculate the density of buckminsterfullerene.

INTEGRATIVE EXERCISES

11.98 Spinel is a mineral that contains 37.9% Al, 17.1% Mg, and 45.0% O, by mass, and has a density of 3.57 g/cm³. The unit cell is cubic, with an edge length of 809 pm. How many atoms of each type are in the unit cell?

11.99 (a) At the molecular level, what factor is responsible for the steady increase in viscosity with increasing molecular weight in the hydrocarbon series shown in Table 11.4? (b) Although the viscosity varies over a factor of more than two in the series from hexane to nonane, the surface tension at 25 °C increases by only about 20% in the same series. How do you account for this? (c) *n*-Octyl alcohol, $CH_3(CH_2)_7OH$, has a viscosity of 1.01×10^{-2} kg m^{-1}s^{-1}, much higher than nonane, which has about the same molecular weight. What accounts for this difference? How does your answer relate to the difference in normal boiling points for these two substances?

11.100 Acetone, $(CH_3)_2CO$, is widely used as an industrial solvent. (a) Draw the Lewis structure for the acetone molecule, and predict the geometry around each carbon atom. (b) Is the acetone molecule polar or nonpolar? (c) What kinds of intermolecular attractive forces exist between acetone molecules? (d) 1-Propanol, $CH_3CH_2CH_2OH$, has a molecular weight that is very similar to that of acetone, yet acetone boils at 56.5 °C and 1-propanol boils at 97.2 °C. Explain the difference.

11.101 The table shown here lists the molar heats of vaporization for several organic compounds. Use specific examples from this list to illustrate how the heat of vaporization varies with **(a)** molar mass, **(b)** molecular shape, **(c)** molecular polarity, **(d)** hydrogen-bonding interactions. Explain these comparisons in terms of the nature of the intermolecular forces at work. (You may find it helpful to draw out the structural formula for each compound.)

Compound	Heat of Vaporization (kJ/mol)
$CH_3CH_2CH_3$	19.0
$CH_3CH_2CH_2CH_2CH_3$	27.6
$CH_3CHBrCH_3$	31.8
CH_3COCH_3	32.0
$CH_3CH_2CH_2Br$	33.6
$CH_3CH_2CH_2OH$	47.3

11.102 Liquid butane, C_4H_{10}, is stored in cylinders, to be used as a fuel. The normal boiling point of butane is listed as $-0.5\ °C$. **(a)** Suppose the tank is standing in the sun and reaches a temperature of $35\ °C$. Would you expect the pressure in the tank to be greater or less than atmospheric pressure? How does the pressure within the tank depend on how much liquid butane is in it? **(b)** Suppose the valve to the tank is opened and a few liters of butane are allowed to escape rapidly. What do you expect would happen to the temperature of the remaining liquid butane in the tank? Explain. **(c)** How much heat must be added to vaporize 250 g of butane if its heat of vaporization is 21.3 kJ/mol? What volume does this much butane occupy at 755 torr and $35\ °C$?

[11.103] Using information in Appendices B and C, calculate the minimum number of grams of propane, $C_3H_8(g)$, that must be combusted to provide the energy necessary to convert 5.50 kg of ice at $-20\ °C$ to liquid water at $75\ °C$.

11.104 In a certain type of nuclear reactor, liquid sodium metal is employed as a circulating coolant in a closed system, protected from contact with air or water. Much like the coolant that circulates in an automobile engine, the liquid sodium carries heat from the hot reactor core to heat exchangers. **(a)** What properties of the liquid sodium are of special importance in this application? **(b)** The viscosity of liquid sodium varies with temperature as follows:

Temperature (°C)	Viscosity (kg $m^{-1}s^{-1}$)
100	7.05×10^{-4}
200	4.50×10^{-4}
300	3.45×10^{-4}
600	2.10×10^{-4}

What forces within the liquid sodium are likely to be the major contributors to the viscosity? Why does viscosity decrease with increasing temperature?

11.105 The vapor pressure of a volatile liquid can be determined by slowly bubbling a known volume of gas through it at a known temperature and pressure. In an experiment, 5.00 L of N_2 gas is passed through 7.2146 g of liquid benzene, C_6H_6, at $26.0\ °C$. The liquid remaining after the experiment weighs 5.1493 g. Assuming that the gas becomes saturated with benzene vapor and that the total gas volume and temperature remain constant, what is the vapor pressure of the benzene in torr?

11.106 The relative humidity of air equals the ratio of the partial pressure of water in the air to the equilibrium vapor pressure of water at the same temperature, times 100%. If the relative humidity of the air is 58% and its temperature is $68\ °F$, how many molecules of water are present in a room measuring 12 ft $\times$ 10 ft $\times$ 8 ft?

11.107 Use a reference source such as the *CRC Handbook of Chemistry and Physics* to compare the melting and boiling points of the following pairs of inorganic substances: **(a)** W and WF_6, **(b)** SO_2 and SF_4, **(c)** SiO_2 and $SiCl_4$. Account for the major differences observed in terms of likely structures and bonding.

12 MODERN MATERIALS

HERE, THERE, AND EVERYWHERE. The element silicon is the primary component of computer processor chips and commercial solar panels.

WHAT'S AHEAD

12.1 Classes of Materials
We will characterize materials according to their predominant bonding interactions and their ability to conduct electricity. Ionic solids have localized electrons and are *insulators*. Covalent-network solids also have localized electrons, but in some cases conductivity can result when electrons are excited or impurities are added. Hence they are called *semiconductors*. The valence electrons in metallic solids are collectively shared, resulting in high electrical conductivity.

12.2 Electronic Structure of Materials
We see that the behavior of electrons in bulk solids can be explained by the rules of molecular orbital theory. The large number of atoms in a solid leads to *bands* of allowed energy instead of discrete molecular orbitals. The electrical and optical properties of materials depend upon the filling of these bands.

12.3 Semiconductors
We next study semiconductors, where there is an energy gap between the filled and empty bands called the *band gap*. The size of the band gap can be controlled by manipulating the orbital overlap and bond polarity. The conductivity of semiconductors can be controlled by intentionally adding controlled amounts of impurities called *dopants*. A variety of important technological devices can be built from semiconductors, including computer chips, solar cells, and light-emitting diodes (LEDs).

12.4 Ceramics
We investigate a class of materials called *ceramics*, which are inorganic solids, such as oxides, nitrides, carbides, and silicates, that are highly resistant to heat, corrosion, and wear. Many ceramics are stronger and lighter than metals, but their applications are somewhat limited by their brittle nature.

12.5 Superconductors
We examine materials with the remarkable ability to conduct electricity with no resistance when they are cooled to low temperature. Another unusual property of *superconductors* is that they expel magnetic fields.

12.6 Polymers and Plastics
Next we investigate *polymers*—long chainlike molecules where the motif of a small molecule is repeated many times over. We will see how the shapes, interactions, and bonding between polymer chains affect the physical properties.

12.7 Biomaterials
We recognize that materials used in biomedical applications must meet special requirements to be compatible with living organisms. We will see several examples of *biomaterials* and how they work in such applications as heart valves, artificial tissues, and vascular grafts.

12.8 Liquid Crystals
We will learn that instead of passing directly from the solid to the liquid phase when heated, some substances pass through an intermediate liquid crystalline phase that has some of the structural order of a solid and some of the freedom of motion possessed by liquids. *Liquid crystals* are widely used in electronic displays.

12.9 Nanomaterials
We will discuss how physical and chemical properties of bulk materials change when their crystals become very small. These effects begin to occur when materials have sizes on the order of 1–100 nm. Materials that take advantage of these effects are collectively called *nanomaterials*.

THE MODERN WORLD IS FULL OF "STUFF"—materials that are used to make computers, compact discs, cell phones, contact lenses, skis, furniture, and a host of other things. Chemists have contributed to the discovery and development of new materials by inventing entirely new substances and by developing the means for processing naturally occurring materials to form fibers, films, coatings, adhesives, and substances with special electrical, magnetic, or optical properties. In this chapter we will discuss classes of materials, their properties, and their applications in modern society. Our aim is to show how we can understand the physical or chemical properties of materials by applying the principles discussed in earlier chapters.

This chapter demonstrates the important point that *the observable, macroscopic properties of materials are the result of atomic- and molecular-level structures and processes.*

12.1 CLASSES OF MATERIALS

When we say "material," we generally mean a substance or mixture of substances that is held together by strong chemical bonds throughout the whole sample—in other words, covalent-network solids, ionic solids, and metallic solids. ∞(Section 11.8) The electrical, mechanical, optical, and magnetic properties of these substances find numerous technological uses. Because the types of materials and their applications are so diverse, we need a system for classifying them. One possibility is to classify materials according to the type of bonding that holds the atoms together: ionic, covalent, or metallic. Another possible organization would be to classify materials according to their electrical conductivity: **insulators**, **semiconductors**, or **conductors**. Because bonding and conductivity both depend upon the behavior of the electrons, these two classification schemes are related.

In ionic solids the valence electrons are localized predominantly on the anions. The localized nature of the electrons leads to insulating behavior. In metallic solids the valence electrons are delocalized and shared collectively. The delocalized nature of the valence electrons makes it possible for the electrons to move freely throughout the sample and is responsible for the high electrical and thermal conductivity of metals. In covalent network solids the electrons are localized in covalent bonds. This localization limits the conductivity, but in many such materials conductivity can result if light, heat, or an electrical field excites some of the electrons. Therefore, many covalent-network solids are classified as semiconductors because they have conductivities that are intermediate between metals and insulators. In this chapter we will discuss semiconductors and their applications in some detail. We will also consider ceramics, which are an important class of materials that are normally insulators. We will defer a detailed discussion of the largest class of conductors, metals, to Chapter 23.

Examples of functional and structural materials also exist among molecular solids. In these materials the intermolecular forces we learned about in Chapter 11—London dispersion forces, dipole–dipole interactions, and hydrogen bonds—play an important role. Because intermolecular forces are relatively weak, molecular materials are often softer than materials with extended networks of chemical bonds. Hence, molecular materials are often referred to as "soft" materials, while extended solids are referred to as "hard" materials. Two important types of soft materials are investigated in this chapter: polymers and liquid crystals. Both are examples of molecular solids in which the intermolecular forces give rise to unusual properties that are the basis of important technological applications.

12.2 ELECTRONIC STRUCTURE OF MATERIALS

To understand the relationship between conductivity and bonding, let's consider how the atomic orbitals interact in a crystalline solid. In a crystal there are numerous atoms, each with its own atomic orbitals. These atomic orbitals overlap and combine to give numerous molecular orbitals. Nevertheless, the bonding concepts that apply to a simple diatomic molecule also apply to a crystal with a gigantic number of atoms. Let's briefly review some of the basic rules of molecular orbital theory. ∞(Section 9.7) First, the atomic orbitals combine in a molecule to make molecular orbitals that can extend over the entire molecule. Second, a given molecular orbital can contain up to two electrons, depending on the energy level of the molecular orbital and how many electrons the molecule possesses.

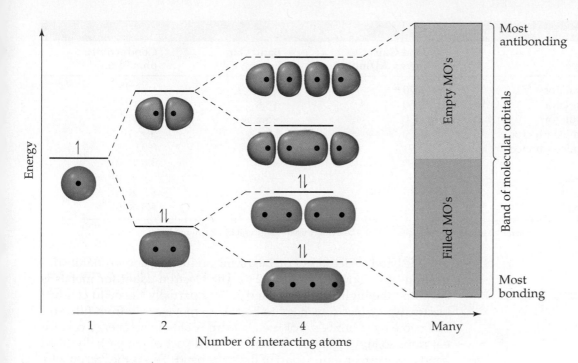

◀ **Figure 12.1 Discrete energy levels in molecules become continuous bands in solids.** A schematic illustration of the energies and filling of molecular orbitals for a linear chain of hydrogen atoms as the number of atoms increases. The molecular orbitals holding electrons are shaded in blue, while the empty ones are shaded in pink. When the chain becomes very long, essentially no separation exists between energy levels, and the chain possesses a continuous band of energy levels.

Finally, the number of molecular orbitals in a molecule is equal to the number of atomic orbitals that combine to make the molecular orbitals.

The electronic structures of bulk solids and small molecules have many similarities as well as some important differences. To illustrate this point, consider the following thought experiment. How does the molecular orbital diagram for a chain of hydrogen atoms change as we increase the length of the chain (Figure 12.1▲)? For one hydrogen atom there is only a single half-filled 1s orbital. For two atoms the molecular orbital diagram corresponds to the molecular orbital diagram of an H_2 molecule that we considered in detail in Chapter 9. ∞ (Section 9.7) For a four-atom chain there are now four molecular orbitals. These range from the lowest energy orbital, where the orbital interactions are completely bonding (0 nodal planes), to the highest energy orbital, where all of the orbital interactions are antibonding (a nodal plane between each pair of atoms).

As the length of the chain increases, the number of molecular orbitals also increases. Fortunately, we can apply a few general concepts regardless of the length of the chain. First, the lowest energy molecular orbitals are always the most bonding, while the highest energy molecular orbitals are the most antibonding. In the middle, the bonding and antibonding interactions tend to balance out. Second, because each hydrogen atom has only one atomic orbital, the number of molecular orbitals is equal to the number of hydrogen atoms in the chain. Given the fact that each hydrogen atom has one electron, this means that half of the molecular orbitals will be fully occupied while the other half will be empty, regardless of chain length.*

If the length of the chain becomes very large, there are so many molecular orbitals that the energy separation between them becomes vanishingly small, which leads to the formation of a continuous **band** of energy states (Figure 12.1). For the bulk solids we will discuss later in this chapter, the number of atoms is very large. Consequently, their electronic structures will consist of bands, as shown for the very long chain of hydrogen atoms, rather than discrete molecular orbitals. The electronic structures of most materials are more complicated because we have to consider more than one type of atomic orbital on each atom. As each type of orbital can give rise to its own band, the electronic structure of a solid will consist of a series of bands. Consequently, the electronic structure of a bulk solid is referred to as a **band structure**.

This is only strictly true for chains with an even number of atoms.

TABLE 12.1 ■ Electronic Properties of Common Materials*

Material	Type	Band Gap Energy, kJ/mol	Band Gap Energy, eV	Conductivity, $ohm^{-1} cm^{-1}$
SiO_2	Insulator	~870	~9	$<10^{-18}$
Al_2O_3	Insulator	~850	~8.8	$<10^{-14}$
C (diamond)	Insulator	~530	~5.5	$<10^{-18}$
Si	Semiconductor	110	1.1	5×10^{-6}
Ge	Semiconductor	65	0.67	0.02
Al	Metal	—	—	3.8×10^5
Cu	Metal	—	—	5.9×10^5
Ag	Metal	—	—	6.3×10^5
Au	Metal	—	—	4.3×10^5

*Band gap energies and conductivities are room temperature values. Electron volts (eV) are a commonly used energy unit in the semiconductor industry; $1 \text{ eV} = 1.602 \times 10^{-19}$ J.

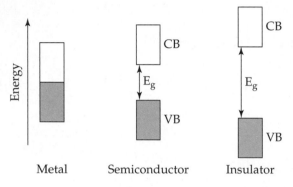

▲ **Figure 12.2 Energy bands in metals, semiconductors, and insulators.** Metals are characterized by the highest-energy electrons occupying a partially filled band. Semiconductors and insulators have an energy gap that separates the completely filled band (shaded in blue) and the empty band (unshaded), known as the band gap and represented by the symbol E_g. The filled band is called the valence band (VB), and the empty band is called the conduction band (CB). Semiconductors have a smaller band gap than insulators.

The band structures of metals, semiconductors, and insulators differ from each other (Figure 12.2 ◄). The electron count for metals is such that the highest energy band is only partially filled. To conduct electricity, the electrons need to be able to move easily from filled orbitals to empty orbitals. For metals there is almost no energy cost for electrons to jump from the lower, occupied part of the partially filled band to the upper, empty part of the same band. This is the reason why metals conduct electricity so easily, as illustrated by the dramatic differences in conductivity between metals, semiconductors, and insulators (Table 12.1 ▲). Based on this classification the semi-infinite chain of hydrogen atoms discussed above would behave as a metal. Under normal conditions this does not happen because hydrogen forms diatomic molecules instead of long chains (or three-dimensional crystals). However, theoretical studies suggest that hydrogen becomes metallic under very high pressures. The hydrogen that surrounds the rocky planetary cores of the gas giant planets, such as Jupiter and Saturn, is thought to be metallic.

In semiconductors and insulators an energy gap, known as the **band gap**, separates a completely filled band from an empty band. The electrons in these materials do not move so easily because they must be excited across the gap before they can become delocalized and move through the solid. The situation is analogous to the energy gap that exists between the highest-occupied molecular orbital and the lowest-unoccupied molecular orbital in a molecule. ∞ (Section 9.8: "Chemistry Put to Work: Orbitals and Energy") Semiconductors are measurably conductive, but far less so than metals because the presence of a band gap lowers the probability that an electron will have enough energy at any given temperature to jump the gap. The distinction between semiconductors and insulators lies in the size of the band gap. If the band gap is much larger than approximately 3 eV, a material is generally classified as an insulator.

12.3 SEMICONDUCTORS

Semiconductors can be divided into two classes, **elemental semiconductors**, which contain only one type of atom, and **compound semiconductors**, which are made up of two or more elements. At room temperature the only elemental semiconductors are silicon, germanium, gray tin, and carbon in its graphite form.* What special feature makes these elements semiconductors, whereas their neighbors are either metals or insulators? First, notice that each of these elements belongs to group 4A. Second, with the exception of graphite, all of these elements adopt the same crystal structure as diamond. ∞ (Section 11.8) In this structure four atoms in a tetrahedral coordination geometry surround each atom (Figure 12.3 ▶).

*Graphite is unusual. Within the layers graphite is nearly metallic, but it is a semiconductor between the layers. ∞ (Section 11.8)

When the atomic s and p orbitals overlap, they form bonding molecular orbitals and antibonding molecular orbitals. Each pair of s orbitals overlaps to give one bonding and one antibonding molecular orbital, while at the same time the p orbitals overlap to give three bonding and three antibonding molecular orbitals. The large numbers of atoms in a bulk crystal lead to the formation of bands (Figure 12.4▼). The band that forms from the bonding molecular orbitals is called the **valence band**, while the band that forms from the antibonding orbitals is called the **conduction band**. In a semiconductor the valence band is filled with electrons and the conduction band is empty. These two bands are separated by the band gap, E_g.

As described in the preceding paragraph every pair of atoms in the solid generates four bonding molecular orbitals and four antibonding molecular orbitals. As each molecular orbital can accommodate two electrons this means that the bonding molecular orbitals can hold eight electrons for every pair of atoms, or four electrons per atom. For the group 4A elements this is equal to the number of valence electrons, which explains why the elemental semiconductors belong to group 4A.

Compound semiconductors maintain the same average valence electron count as elemental semiconductors—four per atom. For example, gallium, Ga, is in column 3A of the periodic table, and arsenic, As, is in column 5A. In gallium arsenide, GaAs, each Ga atom contributes three electrons and each As atom contributes five, which averages out to four per atom—the same number as in silicon or germanium. GaAs is indeed a semiconductor, as are silicon and germanium. Similarly, InP, where indium provides three valence electrons and phosphorus contributes five, and CdTe, where cadmium provides two valence electrons and tellurium contributes six, are also semiconductors because they maintain an average of four valence electrons per atom.

GaAs, InP, and CdTe all crystallize with a zinc blende structure. ∞∞(Section 11.8) While the zinc blende structure can be described as a face-centered cubic lattice of anions, with cations in the tetrahedral voids, as was done in Chapter 11, it also can be described (perhaps more simply) starting from the diamond structure. The zinc blende structure can be generated from the diamond structure if we change half of the carbon atoms into zinc (or gallium) and the other half into sulfur (or arsenic) (Figure 12.3). In this structure each atom is surrounded by a tetrahedron of atoms of the opposite type. That is, in gallium arsenide each gallium atom is surrounded by four arsenic atoms, and vice versa. This structure is commonly observed among compound semiconductors.

Si

GaAs

▲ **Figure 12.3 The structures of silicon and gallium arsenide.** The crystal structure of silicon (above) is the same as the structure of diamond. In this structure, each atom is tetrahedrally coordinated. The crystal structure of gallium arsenide (below) is similar to diamond and silicon, but now gallium (shown in blue) replaces half of the silicon atoms and arsenic (shown in red) replaces the other half.

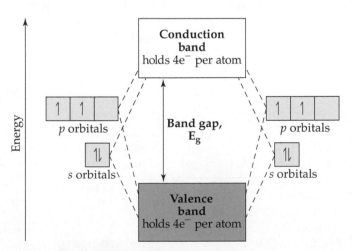

▲ **Figure 12.4 The electronic band structure of semiconductors with the diamond structure.** In semiconductors with the diamond structure each atom is surrounded by a tetrahedron of like atoms. The atomic s and p orbitals form bonding molecular orbitals that make up the valence band and antibonding molecular orbitals that make up the conduction band. The completely filled valence band is separated from the empty conduction band by the band gap, E_g.

TABLE 12.2 ■ Band Gaps of Elemental and Compound Semiconductors*

Material	Structure Type	Band Gap Energy, kJ/mol	Band Gap Energy, eV
Si	Diamond	107	1.11
AlP	Zinc Blende	234	2.43
GaP	Zinc Blende	218	2.26
Ge	Diamond	65	0.67
GaAs	Zinc Blende	138	1.43
ZnSe	Zinc Blende	249	2.58
CuBr	Zinc Blende	294	3.05
Sn**	Diamond	8	0.08
InSb	Zinc Blende	17	0.18
CdTe	Zinc Blende	145	1.50

*Band gap energies and conductivities are room temperature values. Electron volts (eV) are a commonly used energy unit in the semiconductor industry; 1 eV = 1.602 × 10⁻¹⁹ J.
**These data are for gray tin, the semiconducting allotrope of tin. The other allotrope, white tin, is a metal.

The band gaps of semiconductors range from a few tenths of an electron volt to approximately 3 eV (Table 12.2◄). A close look at the values of the band gaps reveals some interesting periodic trends. Upon moving down a group (for example, C ⟶ Si ⟶ Ge ⟶ Sn), the band gap decreases. This trend stems from the fact that as the interatomic distance increases the orbital overlap decreases. The decrease in orbital overlap reduces both the energetic stabilization of the bonding molecular orbitals that make up the valence band and the energetic destabilization of the antibonding molecular orbitals that make up the conduction band. The net result is a decrease in the band gap (Figure 12.5▼). By the time we reach tin the band gap is almost gone, which explains why tin has two allotropes: gray tin, which is a semiconductor with a very small band gap, and white tin, which is a metal. Going one element further down the group we come to lead, where the orbital overlap is not sufficient to separate the valence and conduction bands. Consequently, lead has a crystal structure and physical properties that are characteristic of a metal rather than a semiconductor.

The second periodic trend that can be deduced from Table 12.2 is a tendency for the band gap to increase as the difference in group numbers of the elements increases. For example, upon moving from the elemental semiconductor germanium with a band gap of 0.67 eV to the compound semiconductor GaAs, where gallium is from group 3A and arsenic from group 5A, the band gap effectively doubles to 1.43 eV. If we further increase the difference in group number to four, as is the case for ZnSe, the band gap increases to 2.70 eV. Increasing the horizontal separation of the elements even more leads to a band gap of 3.05 eV in CuBr. This progression is a result of the transition from pure covalent bonding in the elemental semiconductors to polar covalent bonding in compound semiconductors. As the difference in electronegativity of the elements increases, the bonding becomes more polar and the band gap increases (Figure 12.6►). Electrical engineers manipulate both the orbital overlap and the bond polarity to control the band gaps of compound semiconductors for use in a wide range of electrical and optical devices.

▼ **Figure 12.5 The relationship between orbital overlap and band gap.** (a) In diamond the C–C distance is relatively short (1.55 Å). This distance leads to effective overlap of orbitals on neighboring atoms, which in turn leads to a large splitting between the valence and conduction bands (E_g = 5.5 eV). (b) In silicon the Si–Si distance is much longer (2.35 Å), which diminishes the orbital overlap leading to small splitting between the valence and conduction bands (E_g = 1.11 eV).

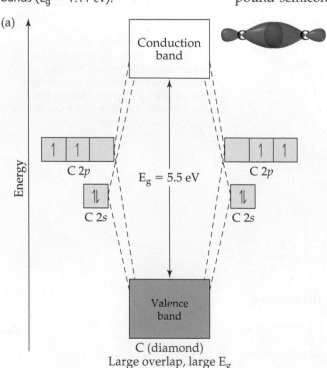

(a) C (diamond)
Large overlap, large E_g

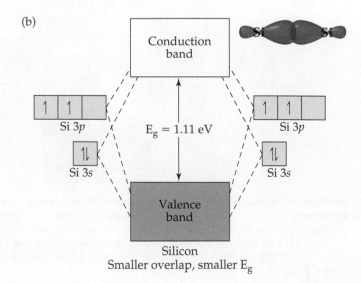

(b) Silicon
Smaller overlap, smaller E_g

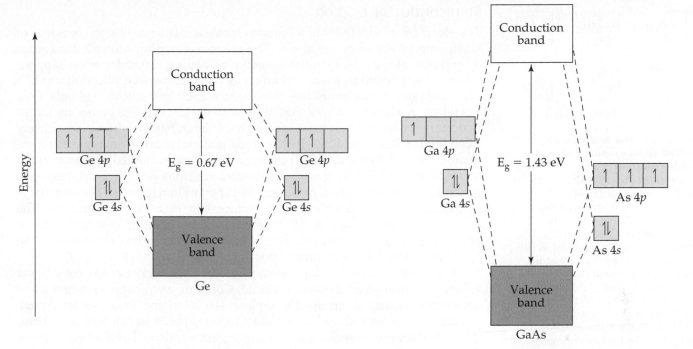

▲ **Figure 12.6 The relationship between bond polarity and band gap.** In germanium the bonding is purely covalent. In gallium arsenide the difference in electronegativity introduces polarity into the bonds. The gallium atoms are less electronegative than germanium, which is reflected in an upward shift of the energies of the gallium atomic orbitals. The arsenic atoms are more electronegative than germanium, which is reflected in a downward shift of the energies of the arsenic atomic orbitals. The introduction of bond polarity increases the band gap from 0.67 eV for Ge to 1.43 eV for GaAs.

■ **SAMPLE EXERCISE 12.1** | Qualitative Comparison of Semiconductor Band Gaps

Will GaP have a larger or smaller band gap than ZnS? Will it have a larger or smaller band gap than GaN?

SOLUTION

Analyze: The size of the band gap depends upon the vertical and horizontal positions of the elements in the periodic table. The band gap will increase when either of the following conditions is met: (1) The elements are located higher up in the periodic table, where enhanced orbital overlap leads to a larger splitting between bonding and antibonding orbitals: or (2) The horizontal separation between the elements increases, which leads to an increase in the electronegativity difference and the bond polarity.

Plan: We must look at the periodic table and compare the relative positions of the elements in each case.

Solve: Gallium is in the fourth period and group 3A. Its electron configuration is $[Ar]3d^{10}4s^24p^1$. Phosphorus is in the third period and group 5A. Its electron configuration is $[Ne]3s^23p^3$. Zinc and sulfur are in the same periods as gallium and phosphorus, respectively. However, zinc, in group 2B, is one element to the left of gallium and sulfur in group 5A, is one element to the right of phosphorus. Thus we would expect the electronegativity difference to be larger for ZnS, which should result in ZnS having a larger band gap than GaP.

For both GaP and GaN the more electropositive element is gallium. So we need only compare the positions of the more electronegative elements, P and N. Nitrogen is located above phosphorus in group 5A. Therefore, based on increased orbital overlap, we would expect GaN to have a larger band gap than GaP. Additionally, nitrogen is more electronegative than phosphorus, which also should result in a larger band gap for GaP.

Check: Looking in Table 12.2 we see that the band gap of GaP is 2.26 eV. ZnS and GaN are not listed in the table, but external references show the band gaps for these compounds to be 3.6 eV for ZnS and 3.4 eV for GaN.

■ **PRACTICE EXERCISE**

Will ZnSe have a larger or smaller band gap than ZnS?
Answer: Because zinc is common to both compounds and selenium is below sulfur in the periodic table, the band gap of ZnSe will be smaller than ZnS.

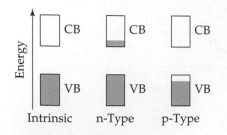

▲ **Figure 12.7 The addition of controlled small amounts of impurities (doping) to a semiconductor changes the electronic properties of the material.** Left: A pure, intrinsic semiconductor has a filled valence band and an empty conduction band (ideally). Middle: The addition of a dopant atom that has more valence electrons than the host atom adds electrons to the conduction band (i.e., phosphorus doped into silicon). The resulting material is an *n-type semiconductor*. Right: The addition of a dopant atom that has fewer valence electrons than the host atom leads to fewer electrons in the valence band or more holes in the valence band (i.e., aluminum doped into silicon). The resulting material is a *p-type semiconductor*.

Semiconductor Doping

The electrical conductivity of a semiconductor is influenced by the presence of small amounts of impurity atoms. The process of adding controlled amounts of impurity atoms to a material is known as **doping**. Consider what happens when a few phosphorus atoms (known as dopants) replace silicon atoms in a silicon crystal. Phosphorus has five valence electrons; silicon has only four. Therefore, the extra electrons that come with the phosphorus atoms are forced to occupy the conduction band because the valence band is already completely filled (Figure 12.7◄, middle). The resulting material is called an *n-type* semiconductor, "n" signifying that the number of negatively charged electrons in the conduction band has increased. These extra electrons can move very easily in the conduction band. Thus, just a few parts per million (ppm) of phosphorus in silicon can increase silicon's intrinsic conductivity by a factor of a million! The dramatic change in conductivity in response to the addition of a trace amount of a dopant means that extreme care must be taken to control the impurities in semiconductors. We will return to this point later.

It is also possible to dope semiconductors with atoms that have fewer valence electrons than the host material. Consider what happens when a few aluminum atoms, in group 3A, replace silicon atoms in a silicon crystal. Aluminum has only three valence electrons compared to silicon's four. Thus, there are electron vacancies, known as *holes*, in the valence band when silicon is doped with aluminum. Since the negatively charged electron is not there, the hole can be thought of as having a positive charge. When an adjacent electron jumps into the hole, the electron leaves behind a new hole. Thus, the positive hole moves about in the lattice like a particle itself. This movement is analogous to watching people changing seats in a classroom; you can watch the people (electrons) move about the seats (atoms), or you can watch the empty seats (holes) "move." Thus, holes can conduct, too, and a material like this is called a *p-type* semiconductor, "p" signifying that the number of positive holes in the material has increased. Like n-type conductivity, only parts-per-million levels of a p-type dopant can lead to a millionfold increase in the conductivity of the semiconductor—but in this case, the holes in the valence band are doing the conduction (Figure 12.7, right). The junction of an n-type semiconductor with a p-type semiconductor forms the basis for diodes, transistors, solar cells, and other devices.

■ **SAMPLE EXERCISE 12.2** | **Identifying Types of Semiconductors**

Which of the following elements, if doped into silicon, would yield an n-type semiconductor? Ga; As; C.

SOLUTION

Analyze: An n-type semiconductor means that the dopant atoms must have more valence electrons than the host material. Silicon is the host material in this case.

Plan: We must look at the periodic table and determine the number of valence electrons associated with Si, Ga, As, and C. The elements with more valence electrons than silicon are the ones that will produce an n-type material upon doping.

Solve: Si is in column 4A, and so has four valence electrons. Ga is in column 3A, and so has three valence electrons. As is in column 5A, and so has five valence electrons; C is in column 4A, and so has four valence electrons. Therefore, As, if doped into silicon, would yield an n-type semiconductor.

■ **PRACTICE EXERCISE**

Suggest an element that could be used to dope silicon to yield a p-type material.
Answer: Because Si is in group 4A, we need to pick an element in group 3A. Boron and aluminum are both good choices—both are in group 3A. In the semiconductor industry boron and aluminum are commonly used dopants for silicon.

▲ Figure 12.8 **Photograph of a Pentium 4 processor computer chip.** On this scale, mostly what you can see is the connections between transistors.

▲ Figure 12.9 **Making silicon chips.** IBM's clean room for the fabrication of 300 mm silicon wafers, in East Fishkill, New York. *Tom Way/Courtesy of International Business Machines Corporation. Unauthorized use not permitted.*

The Silicon Chip

Silicon is the most important commercial semiconductor. It is also one of the most abundant elements on Earth. ∞ (Section 1.2) The semiconductor industry that is responsible for the circuitry in computers, cell phones, and a host of other devices relies on silicon wafers called "chips," on which complex patterns of semiconductors, insulators, and metal wires are assembled (Figure 12.8▲). Although the details of integrated circuit design are beyond the scope of this book, we can say that the process of making chips relies mostly on chemistry.

Silicon is the semiconductor of choice because it is abundant and cheap in its raw form (obtained from sand), it can be made 99.999999999% pure in highly specialized facilities called "clean rooms" (Figure 12.9▲), and it can be grown into enormous crystals that are nearly atomically perfect. If you recall that as little as parts-per-million level of impurities can change the conductivity of silicon by a millionfold, it becomes clear why clean rooms are necessary. In addition, the individual units that make up integrated circuits are about 50 nm wide, which makes them similar in size to an individual virus particle and quite a bit smaller than ordinary dust particles. Thus, to an integrated circuit, a dust particle looks like a boulder.

Silicon has some other important advantages over competing semiconductors. It is nontoxic (compared to GaAs, its closest competitor), and its surface can be chemically protected by SiO_2, the natural product of its reaction with air. SiO_2 is an excellent insulator, and its atoms are in excellent registry with the underlying silicon substrate, which means it is easy to grow crystalline layers of SiO_2 on top of Si. The basic unit of the integrated circuit, the **transistor**, requires a metal/insulator "gate" between a semiconductor "source" and a semiconductor "drain." Electrons move from the source to the drain when voltage is applied to the gate. That silicon's naturally occurring oxide is a very chemically stable insulator provides a huge benefit to silicon technology.

Solar Energy Conversion

The magnitude of the band gap for semiconductors is ~50 kJ/mol to ~300 kJ/mol (Table 12.2)—just into the range of the bond-dissociation energies for single chemical bonds—and corresponds to photon energies of infrared, visible, and low-energy ultraviolet light. ∞ (Section 6.2) Thus, if you shine light of the appropriate wavelength on a semiconductor, you will promote electrons from bonding molecular orbitals in the valence band to antibonding molecular orbitals in the conduction band. The promotion of an electron into the empty conduction band allows it to move freely through the crystal.

A Closer Look | THE TRANSISTOR

The transistor is the heart of integrated circuits. By applying a small electric charge, a transistor controls the flow of a substantially larger current. Therefore, the transistor controls information flow in the form of electrons; it also amplifies the signal.

One common type of transistor is the MOSFET, which stands for *metal-oxide-semiconductor field-effect transistor*. The design of a MOSFET is shown in Figure 12.10▶. A piece of silicon is p-doped to form the substrate, and then patches of n-type doping are created and connected to metal wires. The back of the substrate is also connected to a metal wire. The n-type regions are the *source* and *drain*. They are separated by a p-type *channel*, which is about 65 nm long in a Pentium 4 processor. Deposited on top of the channel is the *gate*, an insulating oxide—traditionally SiO_2, but it can also be silicon nitride, Si_3N_4, or a mixture of these two. In the Pentium 4 chip, the oxide layer is only about 2 nm, or 20 atoms, thick. A metal contact is made to this oxide gate.

When a tiny positive charge is applied to the gate, electrons from the n-type source are drawn into the channel and flow to the drain. The larger the charge applied, the more the channel is "opened" and the greater the number of electrons that flow. The reverse situation, in which the source and drain are p-type and the channel is n-type, can also be made. There are many other transistor designs.

The size of a MOSFET is defined by the gate channel length, which defines the distance electrons must travel to get from the source to the drain. Using SiO_2 as the gate oxide, the smallest circuit elements in commercial use are 65 nm wide. This means that a typical integrated circuit board holds in excess of 65 million transistors over an area the size of your fingernail. The electrical properties of SiO_2 prevent further reduction in the size of the gate channel. To continue shrinking the size of transistors scientists have turned to hafnium oxide, HfO_2, as the gate oxide of choice. This materials advance has enabled chip designers to reduce the gate channel length to 45 nm, resulting

▲ **Figure 12.10 Design of a basic MOSFET.** The substrate is greenish blue, the source and drain are pale gray, the channel is blue, and the gate oxide is pale brown. The blue lines and gray cylinders represent metallic contacts. A small positive voltage applied to the gate creates a positive channel through which electrons (red arrows) from the source are drawn, and go into the drain.

in a twofold increase in the number of transistors that can be manufactured per unit area.

The patterning of 300-mm commercial silicon wafers, which are about the size of a medium pizza, results from a series of chemical reactions that put down layers of thin films of semiconductor, insulator, and metal in patterns that are defined by a series of masks. Figure 12.11▼ shows that the thickness of the layers varies from about 100 nm to 3000 nm. In the research laboratory, film thicknesses have been made as thin as several atoms.

Related Exercises: 12.19, 12.20, 12.83

The First Steps in Making a Transistor:

1. Wafer is oxidized

Silicon dioxide

Silicon

2. Oxidized wafer is covered with photoresist

Photoresist

3. Wafer is exposed to UV light through a photomask

Ultraviolet radiation

Mask

4. Exposed photoresist is dissolved in developer solution

5. Oxide now unprotected by photoresist is etched away in hydrofluoric acid

6. The rest of the photoresist is removed; wafer is now ready for doping

◀ **Figure 12.11 Schematic illustration of the process of photolithography.** Typical film thicknesses are 200–600 μm for the silicon substrate; 80 nm for the silicon dioxide layer; and 0.3–2.5 μm for the photoresist, which is a polymer that degrades when irradiated with ultraviolet light so it will be chemically dissolved by organic solvents in the "developer" solution. The mask protects parts of the photoresist from light and provides patterning to the surface on the ~100-nm scale. Commercial silicon chips can have as many as 30 layers. Other forms of lithography utilize means other than light to pattern the surface (for example, ion-beam lithography and electron-beam lithography use beams of ions or electrons to "write" on the surface).

The same can be said for the hole that is left behind in the valence band. Hence, the material becomes more conductive when irradiated by photons whose energy is greater than the band gap. This property, known as *photoconductivity*, is essential for many solar energy conversion applications.

A **solar cell** is a semiconductor device that converts photons from the sun into electricity. The heart of a solar cell is a junction between an n-type semiconductor and a p-type semiconductor (Figure 12.12▶). When a photon is absorbed in the region near the junction, an electron is excited from the valence band to the conduction band, creating a hole at the same time. The photoexcited electron moves toward the n-type semiconductor and the hole moves toward the p-type semiconductor. In this way the energy of the photon is partially converted into an electrical current (Figure 12.13▶).

The efficiency of a solar cell depends upon a number of factors including the band gap. If the energy of the photon is smaller than the band gap, the photon will not be absorbed and its energy cannot be harvested. On the other hand, if the energy of the photon is larger than the band gap energy, the excess energy of the photon is converted into thermal energy rather than electrical energy. Therefore, a tradeoff in the size of the band gap is necessary. For a solar cell made out of a single material, the maximum efficiency occurs for a material with a band gap of approximately 1.3 eV, where the conversion of optical energy into electrical energy is theoretically predicted to be as high as 31%. In the laboratory single crystal silicon solar cells have been produced with efficiencies of approximately 24%, while the efficiencies of commercial silicon solar cells tend to fall closer to 15%.

Power generation using solar cells has many attractive aspects. Sunlight is renewable, abundant, free, and widely distributed. There are no greenhouse gas emissions. The fundamental limitation is the cost of producing solar cells. Assuming a 20-year system lifetime, the present cost of electricity generated from solar cells is in the range of $0.25–$0.65 per kilowatt hour (kWh). In comparison, electricity generated from a coal-based power plant is closer to $0.04–$0.06 per kWh. This cost comparison does not take into account the long-term environmental costs of burning coal, which are significant. Burning coal releases a number of undesirable pollutants into the atmosphere including sulfur dioxide and nitrogen oxides, which are responsible for acid rain ∞∞ (Section 18.4), as well as fly ash and other particulates. The sheer volume of coal that is burned means that even elements that are present in only trace amounts, such as mercury, arsenic, and uranium, are released in fairly large quantities over time.

In the most advanced coal-fired power plants, modern technologies are employed that significantly reduce the emission of many pollutants. However, it is not possible to remove carbon dioxide, the main product of coal burning. It is estimated that a 500 MW coal-fired power plant, which is large enough to power a city of 140,000, releases 3.7 million tons of CO_2 every year. Given the concerns about global warming and climate change ∞∞ (Section 18.4), this release of CO_2 will continue to be a major concern associated with generating electricity from coal. Driven in large part by environmental concerns, many governments have initiated programs to encourage development of solar power, which has led to an almost exponential growth in the market for solar cells. In 2005 the market exceeded $10 billion.

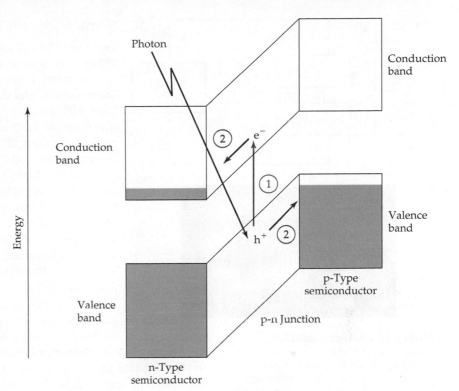

▲ **Figure 12.12 Absorption of light at a p-n junction.** This figure illustrates the process by which light is absorbed at the junction between a p-type and an n-type semiconductor. First, a photon is absorbed in the junction exciting an electron from the valence band into the conduction band creating an electron-hole pair (this process is marked with a 1). Next the electron (e^-) is attracted to the n-type semiconductor and the hole (h^+) toward the p-type semiconductor (this process is marked with a 2). In this way the energy of the photon can be converted into electrical energy.

▲ **Figure 12.13 Electricity from sunlight.** Solar panels, made of silicon, are used both as an energy source and as architectural elements in this apartment building in southern California.

▲ Figure 12.14 Light emitting diodes. Top: The heart of a light emitting diode is a p-n junction where an applied voltage drives electrons and holes to meet. Bottom: The color of light emitted depends upon the band gap of the semiconductor used to form the p-n junction. For display technology red, green, and blue are the most important colors because all other colors can be made by mixing these colors.

Semiconductor Light-Emitting Diodes

Light-emitting diodes (LEDs) are used as displays in clocks; brake lights, turn signals, and dashboards of some cars; as traffic lights; and in many other places. There is a growing interest to replace incandescent and fluorescent lights with white LEDs because of their efficiency and long lifetimes, which are approximately 25 times longer than incandescent lights and roughly twice as long as fluorescent lights. Their efficiency is currently 2–3 times greater than incandescent lights and competitive with fluorescent lights.

The mechanism by which LEDs work is the opposite of the mechanism operating in solar energy conversion devices. In an LED, a small voltage is applied to a semiconductor device that, like a solar cell, has a junction between an n-type semiconductor and a p-type semiconductor. The junction forms a p-n *diode*, in which electrons can flow only one way. When a voltage is applied, electrons in the conduction band from the n-side are forced to the junction where they meet holes from the p-side. The electron falls into the empty hole, and its energy is converted into light whose photons have energy equal to the band gap (Figure 12.14 ◀). In this way electrical energy is converted into optical energy.

Because the wavelength of light that is emitted is inversely proportional to the band gap of the semiconductor, the color of light produced by the LED can be controlled by appropriate choice of semiconductor. Most red LEDs are made of a mixture of GaP and GaAs. The band gap of GaP is 2.26 eV (3.62×10^{-19} J), which corresponds to a green photon with a wavelength of 549 nm. ∞ (Section 6.1) On the other hand, GaAs has a band gap of 1.43 eV (2.29×10^{-19} J), which corresponds to an infrared photon with a wavelength of 867 nm. By forming solid solutions of these two compounds, with stoichiometries of $GaP_{1-x}As_x$, the band gap can be adjusted to any intermediate value. Thus, $GaP_{1-x}As_x$ is the solid solution of choice for red, orange, and yellow LEDs. Green LEDs are made from mixtures of GaP and AlP ($E_g = 2.43$ eV). Blue LEDs, made from GaN ($E_g = 3.4$ eV) and InN ($E_g = 2.4$ eV), are beginning to appear in consumer products as well. White light can be produced from LEDs using different approaches. In some cases, light is combined from blue, green, and red LEDs. More commonly a blue LED is coated with a material that converts some of the blue light into yellow light. In either case the combined colors appear white to the eye.

■ SAMPLE EXERCISE 12.3 | Designing an LED

Green light-emitting diodes can be made from a solid solution of GaP and AlP. These two compounds have band gaps that are 2.26 eV and 2.43 eV, respectively. If we assume that the band gap of a $Ga_{1-x}Al_xP$ solid solution varies linearly from GaP to AlP, what composition would be needed for the emitted light to have a wavelength of 520 nm?

SOLUTION

Analyze: The wavelength of the emitted light has an energy that is nearly equal to the band gap of the semiconductor. The band gap depends upon the composition.

Plan: First we must convert the desired wavelength of emitted light, 520 nm, to an energy in eV. Next we must estimate the value of x that gives a band gap of this size.

Solve: The wavelength of light can be determined from Equation 6.3:

$$E = \frac{hc}{\lambda} = \frac{(6.626 \times 10^{-34} \text{ J-s})(3.00 \times 10^8 \text{ m/s})}{520 \times 10^{-9} \text{ m}}$$

$$E = (3.82 \times 10^{-19} \text{ J}) \times \frac{1 \text{ eV}}{1.602 \times 10^{-19} \text{ J}} = 2.38 \text{ eV}$$

The band gap changes linearly from 2.26 eV for GaP to 2.43 eV for AlP. Therefore, we can estimate the band gap of any composition in the $Ga_{1-x}Al_xP$ solid solution from the following expression:

$$E = 2.26 \text{ eV} + x(2.43 \text{ eV} - 2.26 \text{ eV}) = (2.26 + 0.17x) \text{ eV}$$

We can solve for x by rearranging the equation and inserting the desired value of the band gap:

$$x = \frac{E - 2.26}{0.17} = \frac{2.38 - 2.26}{0.17} = 0.71$$

Thus, the desired composition would be $Ga_{0.29}Al_{0.71}P$.

■ PRACTICE EXERCISE

If we assume that the band gap of an $In_{1-x}Ga_xN$ solid solution varies linearly from InN (2.4 eV) to GaN (3.4 eV), what composition would be needed for the emitted light to have a wavelength of 410 nm?
Answer: The emitted light has an energy of 3.02 eV (485×10^{-19} J). Therefore, the composition of the semiconductor should be $In_{0.37}Ga_{0.63}N$.

12.4 CERAMICS

Ceramics are inorganic solids that are normally hard, brittle, and stable at high temperatures. They are generally electrical insulators. Ceramic materials include such familiar objects as pottery, china, cement, roof tiles, and spark-plug insulators. Ceramic materials (Table 12.3 ▼) come in a variety of chemical forms, including *oxides* (oxygen and metals), *carbides* (carbon and metals), *nitrides* (nitrogen and metals), *silicates* (silica, SiO_2, mixed with metal oxides), and *aluminates* (alumina, Al_2O_3, mixed with metal oxides).

Ceramics are highly resistant to heat, corrosion, and wear; do not readily deform under stress; and are less dense than the metals used for high-temperature applications. Some ceramics used in aircraft, missiles, and spacecraft weigh only 40% as much as the metal components they replace. In spite of these advantages, the use of ceramics as engineering materials has been limited because ceramics are extremely brittle. Whereas a metal component might suffer a dent when struck, a ceramic part typically shatters.

The important differences in the mechanical properties of metals and ceramics arise from bonding interactions at the atomic scale. In a metal, a sea of delocalized electrons holds the atoms together. Metallic bonding tends to be very similar in strength in all directions. This characteristic allows the metal atoms to slip past one another under force, giving rise to the important properties of malleability and ductility. In a ceramic the atoms are held together by ionic or polar covalent bonds. These bonds are typically strong, but they are also highly directional, which prevents the atoms from sliding over one another and is responsible for the brittle nature of ceramics.

TABLE 12.3 ■ Properties of Some Ceramic and Metallic Materials

Material	Melting Point (°C)	Density (g/cm^3)	Hardness (Mohs)[*]	Modulus of Elasticity[**]	Coefficient of Thermal Expansion[***]
		Ceramic materials			
Alumina, Al_2O_3	2050	3.8	9	34	8.1
Silicon carbide, SiC	2800	3.2	9	65	4.3
Silicon nitride, Si_3N_4	1900	3.2	9	31	3.3
Zirconia, ZrO_2	2660	5.6	8	24	6.6
Beryllia, BeO	2550	3.0	9	40	10.4
		Metallic materials			
Mild steel	1370	7.9	5	17	15
Aluminum	660	2.7	3	7	24

[*]The Mohs scale is a logarithmic scale based on the relative ability of a material to scratch another softer material. Diamond, the hardest material, is assigned a value of 10.
[**]A measure of the stiffness of a material when subjected to a load (MPa $\times 10^4$). The larger the number, the stiffer the material.
[***]In units of ($K^{-1} \times 10^{-6}$). The larger the number, the greater is the size change upon heating or cooling.

Applications of Ceramics

Ceramics are used in structural applications such as bricks and concrete, where the major components are SiO_2, Al_2O_3, Fe_2O_3, and CaO. For more specialized applications synthetic ceramics such as silicon carbide and silicon nitride are used. SiC and Si_3N_4, expand more slowly with increasing temperature (Table 12.3) because the bonding is three dimensional and more covalent than most ceramics. Thermal expansion can be an important criterion for high temperature applications.

At high temperatures the oxide ions in zirconia can move through the crystal, a property known as **ionic conductivity**. The ionic conductivity increases when some of the oxygen atoms are missing because the empty sites are used by the oxygen ions as they hop through the crystal. To create oxygen vacancies Zr^{4+} ions are substituted by Y^{3+} ions. Because the charge of the yttrium cation is one less than zirconium, an oxygen vacancy is created for every two Y^{3+} ions that substitute for Zr^{4+}, as reflected in the empirical formula, $Zr_{1-2x}Y_{2x}O_{2-x}$. This compound is used in many important applications such as gas sensors and fuel cells because of its ability to conduct oxide ions.

Ceramic components are difficult to manufacture free of defects. Ceramic parts often develop random, undetectable microcracks and voids (hollow spaces) during processing. These defects are more susceptible to stress than the rest of the ceramic; thus, they are generally the origin of cracking and fractures. To "toughen" a ceramic—to increase its resistance to fracture—scientists produce very pure, uniform particles of the ceramic material that are less than 1 μm (10^{-6} m) in diameter. These are then *sintered* (heated at high temperature under pressure so that the individual particles bond together) to form the desired object.

Making Ceramics

Many ceramics are naturally occurring and can be mined. However, there are inorganic counterparts of the polymerization reactions that we will discuss later in this chapter that are useful for making ceramic materials with increased purity and controlled particle size. The **sol-gel process** is an important method of forming extremely fine particles of uniform size. A typical sol-gel procedure begins with a *metal alkoxide*, which contains organic groups (for example, $-CH_3$, $-C_2H_5$, $-C_3H_7$, etc.) bonded to a metal atom through oxygen atoms. Metal alkoxides are produced when the metal reacts with an alcohol, which is an organic compound containing an OH group bonded to carbon. ∞ (Section 2.9) To illustrate this process, we will use titanium as the metal and ethanol, CH_3CH_2OH, as the alcohol.

$$Ti(s) + 4\,CH_3CH_2OH(l) \longrightarrow Ti(OCH_2CH_3)_4(s) + 2\,H_2(g) \qquad [12.1]$$

The alkoxide product, $Ti(OCH_2CH_3)_4$, is dissolved in an appropriate alcohol solvent. Water is then added, and it reacts with the alkoxide to form Ti—OH groups and regenerate ethanol.

$$Ti(OCH_2CH_3)_4(soln) + 4\,H_2O(l) \longrightarrow Ti(OH)_4(s) + 4\,CH_3CH_2OH(l) \quad [12.2]$$

Even though the ethanol is simply regenerated, the initial reaction with ethanol is important because the direct reaction of $Ti(s)$ with $H_2O(l)$ leads to a complex mixture of titanium oxides and titanium hydroxides. The intermediate formation of $Ti(OC_2H_5)_4(s)$ ensures that a uniform suspension of $Ti(OH)_4$ will be formed. The $Ti(OH)_4$ is present at this stage as a *sol*, a suspension of extremely small particles. The acidity or basicity of the sol is adjusted to split out water from between two of the Ti—OH bonds.

$$(HO)_3Ti-O-H(s) + H-O-Ti(OH)_3(s) \longrightarrow$$
$$(HO)_3Ti-O-Ti(OH)_3(s) + H_2O(l) \qquad\qquad [12.3]$$

This is an example of a *condensation reaction*. We will learn more about condensation reactions when discussing polymers. Condensation also occurs at some of the other OH groups bonded to the central titanium atom, producing a three-dimensional network. The resultant material, called a *gel*, is a porous, interconnected network of extremely small particles with the consistency of gelatin. When this material is heated carefully at 200 °C to 500 °C, all the liquid is removed, and the gel is converted to a finely divided metal oxide powder with particles in the range of 3 to 100 nm in diameter. These particles might be used in an application where small, uniform particle size is required, such as the TiO_2 solar cells discussed in Chapter 9 (Section 9.8: "Chemistry Put to Work: Orbitals and Energy"), or they might be used as the starting point in the sintering process. Figure 12.15▶ shows SiO_2 particles, formed into remarkably uniform spheres by a precipitation process similar to the sol-gel process.

▲ Figure 12.15 **Uniformly sized spheres of amorphous silica, SiO₂.** These are formed by precipitation from a methanol solution of $Si(OCH_3)_4$ upon addition of water and ammonia. The average diameter is 550 nm.

12.5 SUPERCONDUCTORS

Even metals are not infinitely conductive; there is some resistance to electron flow due to the vibrations of atoms and the presence of impurities and defects. However, in 1911 the Dutch physicist H. Kamerlingh Onnes discovered that when mercury is cooled below 4.2 K, it loses all resistance to the flow of an electrical current. Since that discovery, scientists have found that many substances exhibit this "frictionless" flow of electrons, known as **superconductivity**. Substances that exhibit superconductivity do so only when cooled below a particular temperature called the **superconducting transition temperature**, T_c. The observed values of T_c are generally very low. Table 12.4▶ lists numerous superconducting materials, the year of their discovery, and their transition temperature. Some are notable for their relatively high T_c, and others for the mere fact that a material of that composition would superconduct at all. It is surprising to see that the materials with the highest transition temperatures are ceramics rather than metals. Until a few decades ago most scientists working in the area of superconductivity concentrated their studies on metallic elements and compounds.

Superconductivity has tremendous economic potential. If electrical power lines or the conductors in a variety of electrical devices were capable of conducting current without resistance, enormous amounts of energy could be saved. Unfortunately, this advantage is offset by the need to cool the superconducting wires to a temperature below T_c (Figure 12.16▼). However, superconducting power lines have the advantage that they can carry 2–5 times more current than a conventional power

TABLE 12.4 ■ Superconducting Materials: Dates of Discovery and Transition Temperatures

Substance	Discovery Date	T_c (K)
Hg	1911	4.0
NbO	1933	1.5
NbN	1941	16.1
Nb_3Sn	1954	18.0
Nb_3Ge	1973	22.3
$BaPb_{1-x}Bi_xO_3$	1975	13
$La_{2-x}Ba_xCuO_4$	1986	35
$YBa_2Cu_3O_7$	1987	95
$Bi_2Sr_{3-x}Ca_xCu_2O_{8+y}$	1988	95
$Tl_2Ba_2Ca_2Cu_3O_{10}$	1988	125
$HgBa_2Ca_2Cu_3O_{8+x}$	1993	133
$Hg_{0.8}Tl_{0.2}Ba_2Ca_2Cu_3O_{8.33}$	1993	138
Cs_3C_{60}	1995	40
MgB_2	2001	39

◀ Figure 12.16 **Superconducting power transmission cable.** The superconducting wires surround a hollow core through which liquid nitrogen flows, cooling the superconductor to the zero resistance state. An electrical insulator surrounds the superconducting wires.

Electrical insulation

Superconductor

Pipe coating

Steel pipe

Liquid nitrogen flow

▲ **Figure 12.17 Magnetic levitation.** A small permanent magnet is levitated by its interaction with a ceramic superconductor that is cooled to liquid nitrogen temperature, 77 K. The magnet floats in space because the superconductor excludes magnetic field lines, a property known as the Meissner effect.

▲ **Figure 12.18 Superconduction in action.** High speed superconductive magnetic levitation (maglev) train. Shown is the ML002 experimental train being tested in Japan.

transmission cable in the same volume of space. This characteristic makes them attractive alternatives in crowded urban areas where power distribution demands are pushing the capacity of the system, but the population density makes it very expensive to dig up the streets to install new lines. Prototype underground superconducting transmission lines installed in the United States and Japan have demonstrated the potential of this technology.

In addition to their remarkable electrical properties, superconducting materials exhibit a property called the *Meissner effect* (Figure 12.17▲), in which they exclude all magnetic fields from their volume. The Meissner effect is being explored for high-speed trains that are magnetically levitated "maglev" (Figure 12.18▲). Because superconductivity appears in most materials only at very low temperatures, applications of this phenomenon to date have been limited. One important use of superconductors is in the windings of the large magnets that form the magnetic fields needed in magnetic resonance imaging (MRI) instruments, used in medical imaging (Figure 12.19▼). The magnet

▶ **Figure 12.19 A magnetic resonance imaging (MRI) machine used in medical diagnosis.** The magnetic field needed for the procedure is generated by current flowing in superconducting wires, which must be kept below their superconducting transition temperature, T_c, of 18 K. This low temperature requires liquid helium as a coolant.

windings, typically formed from Nb_3Sn, must be kept cooled with liquid helium, which boils at about 4 K. The cost of liquid helium is a significant factor in the cost of using an MRI instrument.

Ceramic Superconductors

Before 1986 the highest value observed for T_c was about 22 K for the compound Nb_3Ge (Table 12.4). While most of the known superconductors at that time were elements and compounds containing only metals, superconductivity at a respectable T_c of 13 K had been found a decade earlier in an oxide material, $BaBi_{1-x}Pb_xO_3$ by A. W. Sleight and coworkers at DuPont in Wilmington, Delaware. In 1986, working on a structurally related oxide system, J. G. Bednorz and K. A. Müller, at the IBM research laboratories in Zürich, Switzerland, discovered superconductivity above 30 K in a ceramic oxide containing lanthanum, barium, and copper. That discovery, for which Bednorz and Müller received the Nobel Prize in Physics in 1987, set off a flurry of research activity all over the world. Before the end of 1987 scientists had verified the onset of superconductivity at 95 K in yttrium-barium-copper oxide, $YBa_2Cu_3O_7$. The highest temperature observed to date for onset of zero resistance at 1 atm pressure is 138 K, which was achieved in another complex copper oxide, $Hg_{0.8}Tl_{0.2}Ba_2Ca_2Cu_3O_{8.33}$.

The discovery of so-called **high-temperature** (high-T_c) **superconductivity** is of great significance. Because maintaining extremely low temperatures is costly, many applications of superconductivity will become feasible only with the development of usable high-temperature superconductors. Liquid nitrogen is the favored coolant because of its low cost (it is cheaper than milk), but it can cool objects to only 77 K. The only readily available safe coolant at temperatures below 77 K is liquid helium, which costs as much as fine wine. Thus, many practical applications of superconductors, such as the power transmission cables discussed in the previous section, will become commercially viable only if T_c is well above 77 K.

What is the mechanism of superconductivity in the materials with unexpectedly high values of T_c? The answer to this question is still being vigorously debated. One of the most widely studied ceramic superconductors is $YBa_2Cu_3O_7$, whose structure is shown in Figure 12.20▶. The copper–oxygen planes present in $YBaCu_3O_7$ (marked in transparent blue) are present in all of the high-T_c oxide superconductors containing copper and oxygen in Table 12.4, as well as many others. Extensive work has gone into studying these materials. This work indicates that the copper–oxygen planes are responsible for the superconductivity, as well as the relatively high metallic-type conductivity observed at temperatures above T_c, where the electrical conductivity parallel to the copper–oxygen planes is 10^4 times greater than in the perpendicular direction. The Cu^{2+} ions have an $[Ar]3d^9$ electron configuration with a single electron in the $3d_{x^2-y^2}$ orbital. ⚬⚬ (Section 6.6) The copper $3d_{x^2-y^2}$ orbital forms strong covalent bonds with the $2p_x$ and $2p_y$ orbitals on the neighboring oxygen atoms. In this way the copper and oxygen atoms form a partially filled band that is responsible for the metallic conductivity of $YBaCu_3O_7$ and related cuprate superconductors.

Our understanding of superconductivity is much better in metals and metal alloys, such as Nb_3Sn, where the mechanism of superconductivity is reasonably well accounted for by *BCS theory*, named after its inventors John Bardeen, Leon Cooper, and Robert Schrieffer, who shared the 1972 Nobel Prize in Physics for their work.* However, after years of research there is still no satisfactory theory of superconductivity in the newer ceramic materials. Because it seems that superconductivity might arise in many different kinds of materials, much empirical research is devoted to discovering new classes of superconductors.

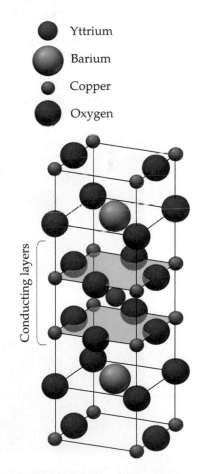

Yttrium

Barium

Copper

Oxygen

Conducting layers

▲ **Figure 12.20 The structure of $YBa_2Cu_3O_7$.** The copper–oxygen planes responsible for the superconductivity are highlighted in blue. The unit cell is defined by the black lines that outline the large rectangular box.

---

*Bardeen had previously received the 1956 Nobel Prize in Physics, together with Walter H. Brattain and William Shockley, for the discovery of the transistor (Section 12.3: "A Closer Look: The Transistor").

As noted in Table 12.4, it was discovered in 1995 that when C_{60} (Section 11.8: "A Closer Look: Buckyball") reacts with an alkali metal, it becomes an electrically conducting material that becomes superconducting at about 40 K. Even more recently, the simple binary compound magnesium diboride, MgB_2, was found to become superconducting at 39 K. This observation is surprising and potentially quite important, since MgB_2, which is a semiconductor somewhat like graphite*, is a relatively cheap material. This material is also easier to machine than the copper containing superconductors. The field of superconductivity has a great deal of promise, and the appearance of superconductors in major commercial products is just beginning (see "Chemistry Put to Work: Cell Phone Tower Range").

Chemistry Put to Work   CELL PHONE TOWER RANGE

Cell phone towers increasingly dot both rural and urban landscapes (Figure 12.21 ▼), yet it can be difficult to maintain contact with a tower in the midst of a telephone conversation. A cell phone communicates with the system by receiving a signal from the tower's transmitter and transmitting signals back to it. Although the tower's transmitter may be quite powerful, the cell phone has very limited power. As the distance from the tower increases or intervening structures interfere, the cell phone signal becomes too weak to be distinguished from general electronic noise.

The amplifiers in the tower's receiver have electronic filters that discriminate between the desired incoming signal and other electronic signals. The sharper the filter, the higher is the discrimination between one channel and another, and therefore, the higher is the capability of detecting the desired signal with clarity. Filters can be made from a high-temperature superconducting oxide that when cooled below its T_c provides much sharper filtering capabilities than conventional filters. By incorporating such filters in the tower receivers, communications companies can extend the range of the tower by up to a factor of two, which saves construction costs and improves the reliability of communication.

Superconducting technology is now being deployed in personal-computer–sized boxes located in cell phone base stations (the little building at the foot of the tower). The filters are formed from a ceramic oxide, typically $YBa_2Cu_3O_7$ or $Tl_2Ba_2CaCu_2O_5$. The requisite cooling is provided by a mechanical cooling device, which is a small refrigeration unit capable of cooling the filter below its T_c (Figure 12.22 ▼).

▲ **Figure 12.22 Cryogenic receiver system employing a cryo-cooled superconducting filter and a low-noise amplifier (LNA).** The cylindrical object to the left is the cryogenic refrigerator used to keep the filter and LNA at a temperature below its T_c value.

▲ **Figure 12.21 Wireless communications tower.** Towers like this (here, disguised as a tree) enable cell phone communicaton.

*The boron atoms in MgB_2 form hexagonal layers identical to the layers of carbon atoms in graphite.

12.6 POLYMERS AND PLASTICS

In nature we find many substances of very high molecular mass, running into millions of amu, that make up much of the structure of living organisms and tissues. Some examples include starch and cellulose, which abound in plants, as well as proteins, which are found in both plants and animals. In 1827 Jons Jakob Berzelius coined the word **polymer** (from the Greek *polys*, "many," and *meros*, "parts") to denote molecular substances of high molecular mass formed by the *polymerization* (joining together) of **monomers**, molecules with low molecular mass.

For a long time humans processed naturally occurring polymers, such as wool, leather, silk, and natural rubber, into usable materials. During the past 70 years or so, chemists have learned to form synthetic polymers by polymerizing monomers through controlled chemical reactions. A great many of these synthetic polymers have a backbone of carbon–carbon bonds because carbon atoms have an exceptional ability to form strong stable bonds with one another.

Plastics are materials that can be formed into various shapes, usually by the application of heat and pressure. **Thermoplastic** materials can be reshaped. For example, plastic milk containers are made from a polymer known as *polyethylene* that has a high molecular mass. These containers can be melted down and the polymer recycled for some other use. In contrast, a **thermosetting plastic** is shaped through irreversible chemical processes and therefore cannot be reshaped readily.

An **elastomer** is a material that exhibits rubbery or elastic behavior. When subjected to stretching or bending, it regains its original shape upon removal of the distorting force, if it has not been distorted beyond some elastic limit. Some polymers, such as nylon and polyesters, can also be formed into fibers that, like hair, are very long relative to their cross-sectional area and are not elastic. These fibers can be woven into fabrics and cords and fashioned into clothing, tire cord, and other useful objects.

Making Polymers

The simplest example of a polymerization reaction is the formation of polyethylene from ethylene molecules. In this reaction the double bond in each ethylene molecule "opens up," and two of the electrons originally in this bond are used to form new C—C single bonds with two other ethylene molecules:

Ethylene   Polyethylene

Polymerization that occurs by the coupling of monomers through their multiple bonds is called **addition polymerization**.

We can write the equation for the polymerization reaction as follows:

$$n\ CH_2{=}CH_2 \longrightarrow \left[\begin{matrix} H & H \\ | & | \\ -C & -C- \\ | & | \\ H & H \end{matrix}\right]_n \qquad [12.4]$$

Here n represents the large number—ranging from hundreds to many thousands—of monomer molecules (ethylene in this case) that react to form one large polymer molecule. Within the polymer a repeat unit (the unit shown in brackets in Equation 12.4) appears along the entire chain. The ends of the chain are capped by carbon–hydrogen bonds or by some other bond, so that the end carbons have four bonds.

Polyethylene is a very important material; the demand in North America is approximately 40 billion pounds each year. Although its composition is simple, the polymer is not easy to make. The right conditions for manufacturing this commercially useful polymer were identified only after many years of research. Today many different forms of polyethylene, varying widely in physical properties, are known. Polymers of other chemical compositions provide still greater variety in physical and chemical properties. Table 12.5▼ lists several other common polymers obtained by addition polymerization.

TABLE 12.5 ■ Polymers of Commercial Importance

Polymer	Structure	Uses
Addition Polymers		
Polyethylene	$-(CH_2{-}CH_2)_n$	Films, packaging, bottles
Polypropylene	$\left[CH_2{-}CH(CH_3)\right]_n$	Kitchenware, fibers, appliances
Polystyrene	$\left[CH_2{-}CH(C_6H_5)\right]_n$	Packaging, disposable food containers, insulation
Polyvinyl chloride (PVC)	$\left[CH_2{-}CHCl\right]_n$	Pipe fittings, clear film for meat packaging
Condensation polymers		
Polyurethane	$\left[\overset{O}{\underset{\parallel}{C}}{-}NH{-}R{-}NH{-}\overset{O}{\underset{\parallel}{C}}{-}O{-}R'{-}O\right]_n$ R, R' = $-CH_2{-}CH_2-$ (for example)	"Foam" furniture stuffing, spray-on insulation, automotive parts, footwear, water-protective coatings
Polyethylene terephthalate (a polyester)	$\left[O{-}CH_2{-}CH_2{-}O{-}\overset{O}{\underset{\parallel}{C}}{-}C_6H_4{-}\overset{O}{\underset{\parallel}{C}}\right]_n$	Tire cord, magnetic tape, apparel, soft-drink bottles
Nylon 6,6	$\left[NH{-}(CH_2)_6{-}NH{-}\overset{O}{\underset{\parallel}{C}}{-}(CH_2)_4{-}\overset{O}{\underset{\parallel}{C}}\right]_n$	Home furnishings, apparel, carpet, fishing line, toothbrush bristles
Polycarbonate	$\left[O{-}C_6H_4{-}\overset{CH_3}{\underset{CH_3}{C}}{-}C_6H_4{-}O{-}\overset{O}{\underset{\parallel}{C}}\right]_n$	Shatterproof eyeglass lenses, CDs, DVDs, bulletproof windows, greenhouses

Chemistry Put to Work   RECYCLING PLASTICS

I f you look at the bottom of a plastic container, you are likely
to see a recycle symbol containing a number, as shown in
Figure 12.23▶. The number and the letter abbreviation below
it indicate the kind of polymer from which the container is
made, as summarized in Table 12.6▼. (The chemical structures
of these polymers are shown in Table 12.5.) These symbols
make it possible to sort containers by composition. In general,
the lower the number, the greater is the ease with which the
material can be recycled.

TABLE 12.6 ■ Categories Used for Recycling Polymeric Materials in the United States

Number	Abbreviation	Polymer
1	PET or PETE	Polyethylene terephthalate
2	HDPE	High-density polyethylene
3	V or PVC	Polyvinyl chloride (PVC)
4	LDPE	Low-density polyethylene
5	PP	Polypropylene
6	PS	Polystyrene

▲ **Figure 12.23 Recycling symbols.**
Most plastic containers manufactured
today have a recycling symbol imprinted
on them, indicating the type of polymer
used to make the container and the
polymer's suitability for recycling.

A second general kind of reaction used to synthesize commercially important polymers is **condensation polymerization**. In a condensation reaction two
molecules are joined to form a larger molecule by elimination of a small molecule such as H_2O. For example, an amine (a compound containing the $-NH_2$
group) will react with a carboxylic acid (a compound containing the $-COOH$
group) to form a bond between N and C along with the formation of H_2O.

$$-\overset{\overset{\displaystyle H}{|}}{N}-\boxed{H + H-O}-\overset{\overset{\displaystyle O}{\|}}{C}- \longrightarrow -\overset{\overset{\displaystyle H}{|}}{N}-\overset{\overset{\displaystyle O}{\|}}{C}- + H_2O$$

[12.5]

Polymers formed from two different monomers are called **copolymers**.
In the formation of many nylons a *diamine*, a compound with an $-NH_2$ group
at each end, is reacted with a *diacid*, a compound with a $-COOH$ group at each
end. For example, nylon 6,6 is formed when a diamine that has six carbon
atoms and an amino group on each end is reacted with adipic acid, which also
has six carbon atoms.

$$n\,H-\overset{\overset{\displaystyle H}{|}}{N}(CH_2)_6\overset{\overset{\displaystyle H}{|}}{N}-H + n\,HO\overset{\overset{\displaystyle O}{\|}}{C}(CH_2)_4\overset{\overset{\displaystyle O}{\|}}{C}OH \longrightarrow \left[\overset{\overset{\displaystyle H}{|}}{N}(CH_2)_6\overset{\overset{\displaystyle H}{|}}{N}-\overset{\overset{\displaystyle O}{\|}}{C}(CH_2)_4\overset{\overset{\displaystyle O}{\|}}{C}\right]_n + 2n\,H_2O$$

[12.6]

$\qquad$ Diamine $\qquad\qquad\qquad$ Adipic acid $\qquad\qquad\qquad$ Nylon 6,6

A condensation reaction occurs on each end of the diamine and the acid. The components of H_2O are split out, and N—C bonds are formed between molecules. Table 12.5 lists nylon 6,6 and some other common polymers obtained by condensation polymerization. Notice that these polymers have backbones containing N or O atoms as well as C atoms. In Chapter 25 we will see that proteins, DNA, and RNA are also condensation polymers.

▲ GIVE IT SOME THOUGHT

Would it be possible to make a condensation polymer out of this molecule alone?

$$H_2N-\bigcirc-\overset{\displaystyle \overset{O}{\|}}{C}-O-H$$

Structure and Physical Properties of Polymers

The simple structural formulas given for polyethylene and other polymers are deceptive. Because four bonds surround each carbon atom in polyethylene, the atoms are arranged in a tetrahedral fashion, so that the chain is not straight as we have depicted it. Furthermore, the atoms are relatively free to rotate around the C—C single bonds. Rather than being straight and rigid, therefore, the chains are very flexible, folding readily (Figure 12.24 ◀). The flexibility in the molecular chains causes the entire polymer material to be very flexible.

Both synthetic and naturally occurring polymers commonly consist of a collection of *macromolecules* (large molecules) of different molecular weights. Depending on the conditions of formation, the molecular weights may be distributed over a wide range or be closely clustered around an average value. In part because of this distribution in molecular weights, polymers are largely amorphous (noncrystalline) materials. Rather than exhibiting a well-defined crystalline phase with a sharp melting point, polymers soften over a range of temperatures. They may, however, possess short-range order in some regions of the solid, with chains lined up in regular arrays as shown in Figure 12.25 ◀. The extent of such ordering is indicated by the degree of **crystallinity** of the polymer. Mechanical stretching or pulling to align the chains as the molten polymer is drawn through small holes can frequently enhance the crystallinity of a polymer. Intermolecular forces between the polymer chains hold the chains together in the ordered crystalline regions, making the polymer denser, harder, less soluble, and more resistant to heat. Table 12.7 ▼ shows how the properties of polyethylene change as the degree of crystallinity increases.

The simple linear structure of polyethylene is conducive to intermolecular interactions that lead to crystallinity. However, the degree of crystallinity in polyethylene strongly depends on the average molecular mass. Polymerization

▲ **Figure 12.24 A segment of a polyethylene chain.** The segment shown here consists of 28 carbon atoms. In commercial polyethylenes the chain lengths range from about 10^3 to 10^5 CH_2 units. As this illustration implies, the chains are flexible and can coil and twist in random fashion, which makes it difficult to see all 28 carbons in this illustration.

Ordered regions

▲ **Figure 12.25 Interactions between polymer chains.** In the circled regions, the forces that operate between adjacent polymer-chain segments lead to ordering analogous to the ordering in crystals, though less regular.

TABLE 12.7 ■ Properties of Polyethylene as a Function of Crystallinity					
	Crystallinity				
	55%	**62%**	**70%**	**77%**	**85%**
Melting point (°C)	109	116	125	130	133
Density (g/cm³)	0.92	0.93	0.94	0.95	0.96
Stiffness*	25	47	75	120	165
Yield stress*	1700	2500	3300	4200	5100

*These test results show that the mechanical strength of the polymer increases with increased crystallinity. The physical units for the stiffness test are psi $\times 10^{-3}$ (psi = pounds per square inch); those for the yield stress test are psi. Discussion of the exact meaning and significance of these tests is beyond the scope of this text.

▲ **Figure 12.26 Two types of polyethylene.** (a) Schematic illustration of the structure of low-density polyethylene (LDPE) and food-storage bags made of LDPE film. (b) Schematic illustration of the structure of high-density polyethylene (HDPE) and containers formed from HDPE.

results in a mixture of macromolecules with varying values of n and hence varying molecular masses. So-called low-density polyethylene (LDPE) used in forming films and sheets has an average molecular mass in the range of 10^4 amu and has substantial chain branching. That is, there are side chains off the main chain of the polymer, much like spur lines that branch from a main railway line. These branches inhibit the formation of crystalline regions, reducing the density of the material. High-density polyethylene (HDPE), used to form bottles, drums, and pipes, has an average molecular mass in the range of 10^6 amu. This form has less branching and thus a higher degree of crystallinity. Low-density and high-density polyethylene are illustrated in Figure 12.26▲, where the higher degree of branching of the low-density form can clearly be seen. Thus, the properties of polyethylene can be "tuned" by varying the average length, crystallinity, and branching of the chains, making it a very versatile material.

GIVE IT SOME THOUGHT

Both the melting point and the degree of crystallinity decrease in copolymers made of ethylene and vinyl acetate monomers as the percentage of vinyl acetate increases. Suggest an explanation.

Ethylene   Vinyl acetate

Various substances may be added to polymers to provide protection against the effects of sunlight or against degradation by oxidation. For example, manganese(II) salts and copper(I) salts, in concentrations as low as 5×10^{-4} mol%, are added to nylons to provide protection from light and oxidation and to help maintain whiteness. In addition, the physical properties of polymeric materials can be modified extensively by adding substances with lower molecular mass, called *plasticizers*, to reduce the extent of interactions between chains and thus to make the polymer more pliable. Polyvinyl chloride (PVC) (Table 12.5), for example, is a hard, rigid material of high molecular mass that is used to manufacture home drainpipes. When blended with a suitable substance of lower molecular mass, however, it forms a flexible polymer that can be used to make rain boots and doll parts. In some applications the plasticizer in a plastic object may be lost over time because of evaporation. As this happens, the plastic loses its flexibility and becomes subject to cracking.

Polymers can be made stiffer by introducing chemical bonds between the polymer chains, as illustrated in Figure 12.27 ◄. Forming bonds between chains is called **cross-linking**. The greater the number of cross-links in a polymer, the more rigid the material. Whereas thermoplastic materials consist of independent polymer chains, thermosetting ones become cross-linked when heated; the cross-links allow them to hold their shapes.

An important example of cross-linking is the **vulcanization** of natural rubber, a process discovered by Charles Goodyear in 1839. Natural rubber is formed from a liquid resin derived from the inner bark of the *Hevea brasiliensis* tree. Chemically, it is a polymer of isoprene, C_5H_8.

▲ **Figure 12.27 Cross-linking of polymer chains.** The cross-linking groups (red) constrain the relative motions of the polymer chains, making the material harder and less flexible than when the cross-links are not present.

$$(n + 2) \quad \begin{matrix} CH_3 & & H \\ & C-C & \\ CH_2 & & CH_2 \end{matrix} \quad \longrightarrow$$

Isoprene

$$\begin{matrix} CH_3 & & H & \left[CH_3 & & H \right] & CH_3 & & H \\ & C=C & & & C=C & & & C=C & \\ \sim\!CH_2 & & CH_2 \!+\! CH_2 & & CH_2 \!+\! CH_2 & & CH_2\!\sim \end{matrix}_n$$

Rubber

[12.7]

Because rotation about the carbon–carbon double bond does not readily occur, the orientation of the groups bound to the carbons is rigid. In naturally occurring rubber, the chain extensions are on the same side of the double bond, as shown in Equation 12.7. This form is called *cis*-polyisoprene; the prefix cis- is derived from a Latin phrase meaning "on this side."

Natural rubber is not a useful polymer because it is too soft and too chemically reactive. Goodyear accidentally discovered that adding sulfur to rubber and then heating the mixture makes the rubber harder and reduces its susceptibility to oxidation or other chemical attack. The sulfur changes rubber into a thermosetting polymer by cross-linking the polymer chains through reactions at some of the double bonds, as shown schematically in Figure 12.28 ►. Cross-linking of about 5% of the double bonds creates a flexible, resilient rubber. When the rubber is stretched, the cross-links help prevent the chains from slipping, so that the rubber retains its elasticity.

(a) (b)

▲ Figure 12.28 **Vulcanization of rubber.** (a) In polymeric natural rubber, there are carbon–carbon double bonds at regular intervals along the chain, as shown in Equation 12.7. (b) Chains of four sulfur atoms have been added across two polymer chains, by the opening of a carbon–carbon double bond on each chain.

■■■ **SAMPLE EXERCISE 12.4** | The Stoichiometry of Cross-Linking

If we assume four sulfur atoms per cross-link connection, what mass of sulfur per gram of isoprene, C_5H_8, is required to establish a cross-link as illustrated in Figure 12.28 with 5% of the isoprene units in rubber?

SOLUTION

Analyze: We are asked to calculate the mass of sulfur required per gram of isoprene.

Plan: We need to evaluate the ratio of sulfur atoms to isoprene units based on Figure 12.15, then scale the required mass of sulfur to take account of the 5% cross-linking.

Solve: We can see from the figure that each cross-link involves eight sulfur atoms for every two isoprene units; this means that the ratio of S to C_5H_8 is four. Thus, with 5% (0.05) of the isoprene units cross-linked, we have

$$(1.0 \text{ g } C_5H_8)\left(\frac{1 \text{ mol } C_5H_8}{68.1 \text{ g } C_5H_8}\right)\left(\frac{4 \text{ mol S}}{1 \text{ mol } C_5H_8}\right)\left(\frac{32.1 \text{ g S}}{1 \text{ mol S}}\right)(0.05) = 0.09 \text{ g S}$$

■■■ **PRACTICE EXERCISE**

How would you expect the properties of vulcanized rubber to vary as the percentage of sulfur increases? Explain.
Answer: The rubber would be harder and less flexible as the percentage of sulfur increases because of an increased degree of cross-linking, which covalently bonds the polymer chains together.

12.7 BIOMATERIALS

Increasingly, modern synthetic materials are being used in medical and biological applications. For the purposes of our discussion here, a **biomaterial** is any material that has a biomedical application. The material might have a therapeutic use, for example, in the treatment of an injury or a disease. Or it might have a diagnostic use, as part of a system for identifying a disease or for monitoring a quantity such as the glucose level in blood. Whether the use is therapeutic or diagnostic, the biomaterial is in contact with biological fluids, and this material must have properties that meet the demands of that application. For example, a polymer employed to form a disposable contact lens must be soft and have an easily wetted surface, whereas the polymer used to fill a dental cavity must be hard and wear-resistant.

Chemistry Put to Work   TOWARD THE PLASTIC CAR

Many polymers can be formulated and processed to have sufficient structural strength, rigidity, and heat stability to replace metals, glass, and other materials in a variety of applications. The housings of electric motors and kitchen appliances such as coffeemakers and can openers, for example, are now commonly formed from specially formulated polymers. *Engineering polymers* are tailored to particular applications through choice of polymers, blending of polymers, and modifications of processing steps. They generally have lower costs or superior performance over the materials they replace. In addition, shaping and coloring of the individual parts and their assembly to form the final product are often much easier.

▲ Figure 12.29 **Plastic engines.** The intake manifold of some Ford V-8 engines is formed from nylon.

The modern automobile provides many examples of the inroads engineering polymers have made in automobile design and construction. Car interiors have long been formed mainly of plastics. Through development of high-performance materials, significant progress has been made in introducing engineering polymers as engine components and car body parts. Figure 12.29 ◄, for example, shows the manifold in a series of Ford V-8 pickup and van engines. The use of an engineering polymer in this application eliminates machining and several assembly steps. The manifold, which is made of nylon, must be stable at high temperatures.

Car body parts can also be formed from engineering polymers. Components formed from engineering polymers usually weigh less than the components they replace, thus improving fuel economy. The fenders of the Volkswagen Beetle (Figure 12.30 ◄), for example, are made of nylon reinforced with a second polymer, polyphenylene ether (PPE), which has the following structure:

$$\left[\text{—}\bigcirc\text{—O—}\bigcirc\text{—O—}\bigcirc\text{—O—} \right]_n$$

Because the polyphenylene ether polymer is linear and rather rigid, the PPE confers rigidity and shape retention.

A big advantage of most engineering polymers over metals is that they eliminate the need for costly corrosion protection steps in manufacture. In addition, some engineering polymer formulations permit manufacturing with the color built in, thus eliminating painting steps (Figure 12.31 ▼).

▲ Figure 12.30 **Plastic fenders.** The fenders of this Volkswagen Beetle are made of General Electric Noryl GTX, a composite of nylon and polyphenylene ether.

▲ Figure 12.31 **A mostly plastic truck.** This truck has door handles and bumper supports made of polycarbonate/polybutylene plastic. Coloring pigments are incorporated into the manufacture of the plastic, eliminating the need for painting.

Characteristics of Biomaterials

The most important characteristics that influence the choice of a biomaterial are **biocompatibility**, **physical requirements**, and **chemical requirements**, as illustrated in Figure 12.32 ▶. Each of these characteristics is examined in turn below.

Living organisms, particularly those in higher animals, have a complex set of protections against invasions by other organisms. Our bodies have an extraordinary ability to determine whether an object is the body's own material or whether it is foreign. Any substance that is foreign to the body has the potential to generate a response from the immune system. Molecular-sized objects are bound by antibodies and rejected, whereas larger objects induce an inflammatory response. Some materials are more **biocompatible** than others; that is, they are more readily integrated into the body without an inflammatory response. The most important determining factors in the biocompatibility of a substance are its chemical nature and the physical texture of its surface.

The physical properties of a biomaterial are often required to meet very exacting demands. Tubing that is to be used to replace a defective blood vessel must be flexible and must not collapse under bending or other distortion. Materials used in joint replacements must be wear-resistant. An artificial heart valve must open and close 70 to 80 times a minute, day after day, for many years. If the valve is to have a life expectancy of 20 years, the valve will have to open and close about 750 million times! Unlike the failure of the valve in a car engine, failure of the heart valve may have fatal consequences for its owner.

The chemical properties of a biomaterial are also critically important. Biomaterials must be *medical grade*, which means that they must be approved for use in any medical application. All ingredients present in a medical-grade biomaterial must remain innocuous over the lifetime of the application. Polymers are important biomaterials, but most commercial polymeric materials contain contaminants such as unreacted monomers, fillers or plasticizers, and antioxidants or other stabilizers. The small amounts of foreign materials present in HDPE used in milk jugs (Figure 12.26) pose no threat in that application, but they might if the same plastic material were implanted in the body.

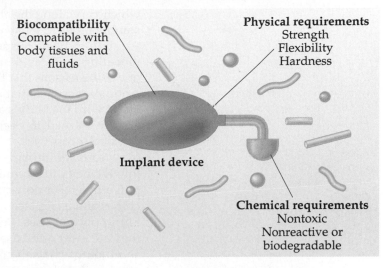

Biocompatibility
Compatible with body tissues and fluids

Physical requirements
Strength
Flexibility
Hardness

Implant device

Chemical requirements
Nontoxic
Nonreactive or biodegradable

▲ **Figure 12.32 Schematic illustration of a human-made device implanted in a biological system.** To function successfully, the device must be biocompatible with its surroundings and meet the necessary physical and chemical requirements, some of which are listed for illustrative purposes.

Polymeric Biomaterials

Specialized polymers have been developed for a variety of biomedical applications. The degree to which the body accepts or rejects the foreign polymer is determined by the nature of the atomic groups along the chain and the possibilities for interactions with the body's own molecules. Our bodies are composed largely of biopolymers such as proteins, polysaccharides (sugars), and polynucleotides (RNA, DNA). We will learn more about these molecules in Chapter 25. For now, we can simply note that the body's own biopolymers have complex structures, with polar groups along the polymer chain. Proteins, for example, are long strings of amino acids (the monomers) that have formed a condensation polymer. The protein chain has the following structure:

$$\left[\begin{array}{c} H\ \ O \\ |\ \ \ \| \\ -C-C-N-C-C-N-C-C-N- \\ |\ \ \ \ \ \ \ \ \ |\ \ \ \ \ \ \ \ \ |\ \ \ \ \ \ \ \ \ | \\ R_1\ \ \ \ H\ \ R_2\ \ \ H\ \ R_3\ \ \ \ H \end{array}\right]_n$$

where R represents a small group of atoms [i.e. $-CH_3$, $-CH(CH_3)_2$, etc.]

which vary along the chain. Twenty different amino acids are used in different combinations to make up proteins in humans. By contrast, synthetic polymers are simpler, being formed from a single repeating unit or perhaps two different repeating units, as described in Section 12.6. This difference in complexity is one of the reasons that the body identifies synthetic polymers as foreign objects. Another reason is that the chain may have few or no polar groups that can interact with the body's aqueous medium. ∞ (Section 11.2)

We learned in Section 12.6 that polymers can be characterized by their physical properties. Elastomers are used as biomaterials in flexible tubing over leads for implanted heart pacemakers and as catheters (tubes implanted into the body to administer a drug or to drain fluids). Thermoplastics, such as polyethylene or polyesters, are employed as membranes in blood dialysis machines and as replacements for blood vessels. Thermoset plastics find limited but important uses. Because they are hard, inflexible, and somewhat brittle, they are most often used in dental devices or in orthopedic applications, such as in joint replacements. To fill a cavity, for example, the dentist may pack some material into the cavity, and then shine an ultraviolet lamp on it. The light initiates a photochemical reaction that forms a hard, highly cross-linked thermoset polymer.

Heart Repair

The term *cardiovascular* pertains to the heart, blood, and blood vessels. The heart is, of course, an absolutely essential organ. A donor organ can replace a heart that fails completely. But, about 60,000 people suffer terminal heart failure each year in the United States while only about 2500 donor hearts become available for transplant. Many attempts have been made—and continue to be made—to produce an artificial heart that can serve over a long time as a replacement for the natural organ. We will not devote attention to these efforts, except to note that recent results are quite promising.

It often happens that only a part of the heart, such as the aortic valve, fails and needs replacement. Repair could be made by using foreign tissue (for example, a pig heart valve) or a mechanical heart valve implant to replace a diseased one. About 250,000 valve replacement procedures are performed annually worldwide. In the United States about 45% of the procedures involve a mechanical valve. The most widely used valve is shown in Figure 12.33 ◀. It has two semicircular discs that move to allow blood to flow in the desired direction as the heart pumps. The discs then fall back together to form a seal against the backflow of blood.

It is vital to minimize fluid disturbance as the blood passes through artificial devices. Surface roughness in a device causes *hemolysis*, the breakdown of red blood cells. Furthermore, surface roughness can serve as a site for invading bacteria to adhere and colonize. Finally, rough surfaces also promote coagulation of the blood, which forms a thrombus, or blood clot. Thus, even though we may have a perfectly fine piece of machinery from a mechanical point of view, the heart valve may not be suitable as a long-term implant. To minimize blood clots, the discs in the heart valve must have a smooth, chemically inert surface.

A second challenge in the use of a heart valve implant is to fix it in place. As shown in Figure 12.33, the retainer ring that forms the body of the valve is covered with a mesh of woven fabric. The material of choice is Dacron®, du Pont's trade name for the fiber formed from polyethylene terephthalate (PET, a polyester; Table 12.5). The mesh acts as a lattice on which the body's tissues can grow, so that the valve becomes incorporated into its surroundings. Tissue grows through the polyester mesh, whereas it does not do so on many other plastics. Apparently the polar, oxygen-containing functional groups along the polyester chain afford attractive interactions to facilitate tissue growth.

▲ **Figure 12.33 A disc heart valve.** Known as the St. Jude valve, this device was named for the medical center at which it was developed. The surfaces of the valve are coated with pyrolytic carbon. The valve is secured to the surrounding tissue via a Dacron® sewing ring.

Vascular Grafts

A vascular graft is a replacement for a segment of diseased blood vessels. Where possible, diseased blood vessels are replaced with vessels taken from the patient's own body. When this procedure is not feasible, artificial materials must be used. Dacron® is used as replacement for large-diameter arteries around the heart. For this purpose, it is fabricated into a crimped, woven, tubular form, as shown in Figure 12.34▶. The tubing is crimped to enable it to bend without serious decrease in its cross-sectional area. The graft must integrate with surrounding tissue after it has been put in place. It must therefore have an open structure, with pores of approximately 10 μm in diameter. During the healing process blood capillaries grow into the graft and new tissues form throughout it. Similarly, polytetrafluoroethylene $[-(CF_2CF_2)_n-]$ is used for the smaller-diameter vascular grafts in the limbs.

Ideally, the inside surface of the graft would become lined with the same sorts of cells that line the native vessel, but this process does not occur with materials currently available. Instead, the inside surface of the tubing is recognized as being foreign to the blood. Platelets, which are circulating components of the blood, normally serve the function of sealing up wounds in the blood vessel walls. Unfortunately, platelets also attach to foreign surfaces and cause blood coagulation. The search for a more biocompatible lining for grafts is an area of active research, but at present there is a continuing risk of blood clots. Excess tissue growth at the intersection of the graft with the native blood vessel is also a frequent problem. Because of the possibility of blood clot formation, patients who have received artificial heart valves or vascular grafts are generally required to take anticoagulation drugs on a continuing basis.

▲ **Figure 12.34 A Dacron® vascular graft.** Grafts like this are used in coronary artery surgery.

Artificial Tissue

The treatment of patients who have lost extensive skin tissue—for example, burn patients or those with skin ulcers—is one of the most difficult problems in therapeutic medicine. Today, laboratory-grown skin can be employed to replace grafts in such patients. Ideally, the "artificial" tissue would be grown from cells taken from the patient. When this procedure is not possible, for example with burn victims, the tissue cells come from another source. When the graft skin is not formed from the patient's own cells, drugs that suppress the body's natural immune defense system must be used, or steps must be taken to modify the new cell line to prevent rejection of the tissue.

The challenge in growing artificial tissue is to get the cells to organize themselves as they would in a living system. The first step in accomplishing this objective is to provide a suitable scaffold on which the cells can grow—one that will keep them in contact with each other and allow them to organize. Such a scaffold must be biocompatible; cells must adhere to the scaffold and differentiate (that is, develop into cells of different types) as the culture grows. The scaffolding must also be mechanically strong and biodegradable over time.

The most successful scaffolds have been lactic acid—glycolic acid copolymers. Formation of the copolymer via a condensation reaction is shown in Equation 12.8.

$$n \text{ HOCH}_2\overset{O}{\underset{\|}{C}}-\text{OH} + n \text{ HOCH}\overset{O}{\underset{\|}{C}}-\text{OH} \longrightarrow -O-CH_2\overset{O}{\underset{\|}{C}}-O\left[-CH\overset{O}{\underset{\|}{C}}-O-CH_2\overset{O}{\underset{\|}{C}}-O\right]_n-CH\overset{O}{\underset{\|}{C}}- \quad [12.8]$$

Glycolic acid   Lactic acid   Copolymer

The copolymer has an abundance of polar carbon–oxygen bonds along the chain, affording many opportunities for hydrogen-bonding interactions. The ester linkages formed in the condensation reaction are susceptible to hydrolysis,

▲ **Figure 12.35 Artificial skin prepared for use as a skin graft.** Burn victims benefit from these biocompatible polymers.

which is just the reverse reaction. When the artificial tissue is deployed in the body, the underlying copolymer scaffold slowly hydrolyzes away as the tissue cells continue to develop and merge with adjacent tissue. An example of a graft skin product is shown in Figure 12.35 ◄.

Until the 1970s, if you had to go to the doctor to get stitches, you later had to go back to get the stitches removed. Stitches that are used to hold living tissue together are called *sutures*. Since the 1970s, biodegradable sutures have been available. After the sutures have been applied to the tissue, they slowly dissolve away and do not release any harmful by-products. Today's biodegradable sutures are made out of lactic acid—glycolic acid copolymers, similar to those just described, which slowly hydrolyze over time.

12.8 LIQUID CRYSTALS

(a)　　　　　　　(b)

▲ **Figure 12.36 Liquid and liquid crystal.** (a) Molten cholesteryl benzoate at a temperature above 179 °C. In this temperature region the substance is a clear liquid. Note that the printing on the surface of the beaker in back of the sample test tube is readable. (b) Cholesteryl benzoate at a temperature between 179 °C and its melting point, 145 °C. In this temperature interval cholesteryl benzoate exists as a milky liquid crystalline phase.

Almost all electronic devices, such as laptops, cellular phones, calculators, and portable media players such as iPods, have to display information. Most electronic displays are based on one of two technologies: light-emitting diodes (LEDs) (Section 12.3) and liquid crystal displays (LCDs). In this section we will explore the molecular principles that give liquid crystals their special optical properties.

When a solid is heated to its melting point, the added thermal energy overcomes the intermolecular attractions that provide molecular order to the solid. ∞ (Section 11.1) The liquid that forms is characterized by random molecular orientations and considerable molecular motion. In a typical liquid the molecules are not ordered at all (zero-dimensional order). Some substances, however, exhibit more complex behavior as their solids are heated. Instead of the three-dimensional order that characterizes a crystalline solid, these substances exhibit phases that have one- or two-dimensional order, meaning that the molecules line up so that they point in one direction (one-dimensional order) or are arranged in layers (two-dimensional order).

In 1888 Frederick Reinitzer, an Austrian botanist, discovered that an organic compound he was studying, cholesteryl benzoate, has interesting and unusual properties. When heated, the substance melts at 145 °C to form a viscous milky liquid; then at 179 °C the milky liquid suddenly becomes clear. When the substance is cooled, the reverse processes occur. The clear liquid turns viscous and milky at 179 °C (Figure 12.36 ◄), and the milky liquid solidifies at 145 °C. Reinitzer's work represents the first systematic report of what we now call a **liquid crystal**.

Instead of passing directly from the solid to the liquid phase when heated, some substances, such as cholesteryl benzoate, pass through an intermediate liquid crystalline phase that has some of the structure of solids and some of the freedom of motion possessed by liquids. Because of the partial ordering, liquid crystals may be very viscous and possess properties intermediate between those of the solid and liquid phases. The region in which they exhibit these properties is marked by sharp transition temperatures, as in Reinitzer's example.

▲ **Figure 12.37 A handheld wireless device with liquid crystal display.** LCDs can be made very thin and light.

From the time of their discovery in 1888 until about 30 years ago, liquid crystals were largely a laboratory curiosity. They are now widely used as pressure and temperature sensors and in the displays of electrical devices such as digital watches, cellular phones, calculators, and laptop computers (Figure 12.37 ◄). Liquid crystals can be used for these applications because the weak intermolecular forces that hold the molecules together in a liquid crystal are easily affected by changes in temperature, pressure, and electric fields.

(a) Normal liquid

(b) Nematic
liquid crystal

(c) Smectic A
liquid crystal

(d) Smectic C
liquid crystal

▲ **Figure 12.38 Ordering in liquid crystalline phases and a normal (non–liquid crystalline) liquid.** The ordering arises from intermolecular forces. Smectic A and smectic C phases are both smectic phases; they differ in the angles that the molecules make with respect to the layers.

Types of Liquid Crystalline Phases

Substances that form liquid crystals are often composed of long, rodlike molecules. In the normal liquid phase these molecules are oriented randomly [Figure 12.38(a)▲]. Liquid crystalline phases, by contrast, exhibit some ordering of the molecules. Depending on the nature of the ordering, liquid crystals can be classified as nematic, smectic, or cholesteric.

In the **nematic liquid crystalline phase** the molecules exhibit one-dimensional ordering. The molecules are aligned along their long axes [Figure 12.38(b)]. The arrangement of the molecules is like that of a handful of pencils whose ends are not aligned. In the **smectic liquid crystalline phases** the molecules exhibit two-dimensional ordering; the molecules are aligned along their long axes and in layers [Figure 12.38(c, d)].

Two molecules that exhibit liquid crystalline phases are shown in Figure 12.39▼. The lengths of these molecules are much greater than their widths. The C=N double bond and the benzene rings add rigidity to the chainlike molecules. The flat benzene rings also help the molecules stack with one another. In addition, many liquid crystalline molecules contain polar groups that give rise to dipole–dipole interactions and promote alignment of the molecules. ∞ (Section 11.2) Thus, the molecules order themselves quite naturally along their long axes. They can, however, rotate around their axes and slide parallel to one another. In smectic phases the intermolecular forces between the molecules (such as London dispersion forces, dipole–dipole attractions, and hydrogen bonding) limit the ability of the molecules to slide past one another.

CH$_3$O—⟨O⟩—CH=N—⟨O⟩—C$_4$H$_9$ 21–47 °C

CH$_3$(CH$_2$)$_7$—O—⟨O⟩—C(=O)—OH   108–147 °C

▲ **Figure 12.39 Structures and liquid crystal temperature intervals of two typical liquid crystalline materials.** The temperature interval indicates the temperature range in which the substance exhibits liquid crystalline behavior.

Cholesteric structure

(a)

▲ Figure 12.40 Ordering in a cholesteric liquid crystal. (a) The molecules in successive layers are oriented at a characteristic angle with respect to those in adjacent layers to avoid repulsive interactions. (b) The result is a screwlike axis.

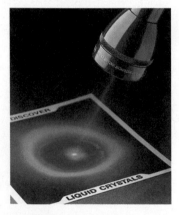

▲ Figure 12.41 Color change in a cholesteric liquid crystalline material as a function of temperature.

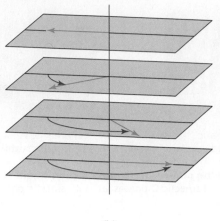

(b)

Figure 12.40 ◄ shows the ordering of the **cholesteric liquid crystalline phase**. The molecules are aligned along their long axes as in nematic liquid crystals, but they are arranged in layers with the molecules in each plane twisted slightly in relation to the molecules in the planes above and below. These liquid crystals are so named because many derivatives of cholesterol adopt this structure. The spiral nature of the molecular ordering produces unusual coloring patterns with visible light. Changes in temperature and pressure change the order and hence the color (Figure 12.41 ◄). Cholesteric liquid crystals have been used to monitor temperature changes in situations where conventional methods are not feasible. For example, they can detect hot spots in microelectronic circuits, which may signal the presence of flaws. They can also be fashioned into thermometers for measuring the skin temperature of infants.

■■■ SAMPLE EXERCISE 12.5 | Properties of Liquid Crystals

Which of the following substances is most likely to exhibit liquid crystalline behavior?

$$CH_3-CH_2-\underset{\underset{CH_3}{|}}{\overset{\overset{CH_3}{|}}{C}}-CH_2-CH_3$$

(i)

$$CH_3CH_2-\bigcirc-N=N-\bigcirc-\overset{\overset{O}{\|}}{C}-OCH_3$$

(ii)

$$\bigcirc-CH_2-\overset{\overset{O}{\|}}{C}-O^-\,Na^+$$

(iii)

SOLUTION

Analyze: We have three molecules of differing molecular structure, and we are asked to determine whether any of them is likely to be a liquid crystalline substance.

Plan: We need to identify the structural features of each case that might induce liquid crystalline behavior.

Solve: Molecule (i) is not likely to be liquid crystalline because it does not have a long axial structure. Molecule (iii) is ionic; the generally high melting points of ionic materials and the absence of a characteristic long axis make it unlikely that this substance will exhibit liquid crystalline behavior. Molecule (ii) possesses the characteristic long axis and the kinds of structural features that are often seen in liquid crystals (Figure 12.39).

■ PRACTICE EXERCISE

Suggest a reason why the following molecule, decane, does not exhibit liquid crystalline behavior:

$$CH_3CH_2CH_2CH_2CH_2CH_2CH_2CH_2CH_2CH_3$$

Answer: Because rotation can occur about carbon–carbon single bonds, molecules whose backbone consists predominantly of C—C single bonds are too flexible; the molecules tend to coil in random ways and thus are not rodlike.

Liquid crystals are widely used in electrically controlled LCD devices in watches, calculators, and computer screens, as illustrated in Figure 12.37. These applications are possible because an applied electrical field changes the orientation of liquid crystal molecules and thus affects the optical properties of the device.

LCDs come in a variety of designs, but the structure illustrated in Figure 12.42▼ is typical. A thin layer (5–20 μm) of liquid crystalline material is placed between electrically conducting, transparent glass electrodes. Ordinary light passes through a vertical polarizer that permits light in only the vertical plane to pass. Through a special process the liquid crystal molecules are oriented so that the molecules at the front plate are oriented vertically and those at the bottom plate horizontally. The molecules in between vary in orientation in a regular way, as shown in Figure 12.42. Displays of this kind are called "twisted nematic." The plane of polarization of the light is turned by 90° as it passes through the device and is thus in the correct orientation to pass

through the horizontal polarizer. In a watch display, a mirror reflects the light back, and the light retraces its path, allowing the device to look bright. When a voltage is applied to the plates, the liquid crystalline molecules align with the voltage [Figure 12.42(b)]. The light rays thus are not properly oriented to pass through the horizontal polarizer, and the device appears dark.

Computer screens employ backlighting rather than reflected light, but the principle is the same. The computer screen is divided into a large number of tiny cells, with the voltages at points on the screen surface controlled by transistors made from thin films of amorphous silicon. Red-green-blue color filters are employed to provide full color. The entire display is refreshed at a frequency of about 60 Hz, so the display can change rapidly with respect to the response time of the human eye. Displays of this kind are remarkable technical achievements, based on a combination of basic scientific discovery and creative engineering.

Related Exercises: 12.64, 12.74, 12.78

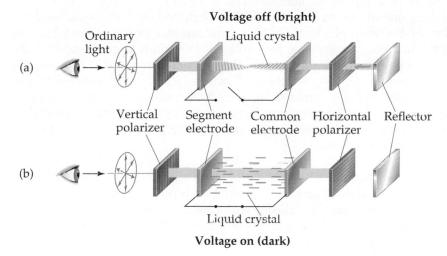

Voltage off (bright)

Ordinary light

(a)

Vertical polarizer · Segment electrode · Common electrode · Horizontal polarizer · Reflector

(b)

Liquid crystal

Voltage on (dark)

◄ **Figure 12.42 Schematic illustration of the operation of a twisted nematic liquid crystal display (LCD).** (a) Ordinary light, which is unpolarized, passes through the vertical polarizer. The vertically polarized light then passes into the liquid crystalline layer, where the plane of polarization is rotated 90°. It passes through the horizontal polarizer, is reflected, and retraces its path to give a bright display. (b) When a voltage is applied to the segment electrode that covers the small area, the liquid crystal molecules align along the direction of the light path. Thus, the vertically polarized light is not turned 90° and is unable to pass through the horizontal polarizer. The area encompassed by the small transparent segment electrode therefore appears dark. Digital watches typically have such displays.

12.9 NANOMATERIALS

The prefix "nano" means 10^{-9}. ∞(Section 1.4) When people speak of "nano-technology," they usually mean making devices that are on the 1–100 nm scale. It turns out that the properties of semiconductors and metals change in this size range. **Nanomaterials**—materials that have dimensions on the 1–100 nm scale—are under intense investigation in the research laboratories of scientists and engineers. Chemistry plays a central role in nanomaterials research.

Semiconductors on the Nanoscale

Look back at Figure 12.1 and consider the pictures of electron energy levels for molecules and bulk solids. Electrons are in discrete molecular orbitals for individual molecules and in delocalized bands for solids. At what point does a molecule get so large that it starts behaving as though it has delocalized bands, rather than localized molecular orbitals? For semiconductors, both theory and experiment now tell us that the answer is roughly at 1 to 10 nm (about 10–100 atoms across). The exact number depends on the specific semiconductor material. The equations of quantum mechanics that were used for electrons in atoms can be applied to electrons (and holes) in semiconductors to estimate the size at

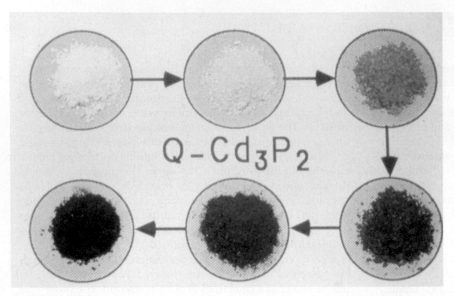

▲ Figure 12.43 Cd₃P₂ powders with different particle sizes. The arrows indicate increasing particle size and a corresponding decrease in the band gap energy, so that the material looks white (smallest nanoparticles), yellow, orange, red, brown, and finally black (the largest nanoparticles).

which unusual effects occur because of the crossover from molecular orbitals to bands. Because these effects become important at 1 to 10 nm, semiconductor particles with diameters in this size range are called *quantum dots*.

One of the most spectacular effects is that the band gap of the semiconductor changes substantially with size in the 1–10-nm range. As the particle gets smaller, the band gap gets bigger. This effect is easily seen by the naked eye, as shown in Figure 12.43◄. On the macro level, the semiconductor cadmium phosphide looks black because its band gap is small ($E_g = 0.5$ eV), and it absorbs all wavelengths of visible light. As the crystals are made smaller, the material progressively changes color until it looks white! It looks white because now no visible light is absorbed. The band gap is so large that only high-energy ultraviolet light can excite electrons to the conduction band ($E_g > 3.0$ eV).

Making quantum dots is most easily accomplished using chemical reactions in solution. For example, to make CdS, you can mix $Cd(NO_3)_2$ and Na_2S in water. If you do not do anything else, you will precipitate large crystals of CdS. However, if you first add a negatively charged polymer to the water (such as polyphosphate, $—(OPO_2)_n—$), the Cd^{2+} will associate with the polymer, like tiny "meatballs" in the polymer "spaghetti." When sulfide is added, CdS particles grow but are kept from forming large crystals by the polymer. A great deal of fine-tuning of reaction conditions is necessary to produce nanocrystals that are of uniform size and shape.

We saw in Section 12.3 that semiconductors can emit light when a voltage is applied. Some semiconductors can also emit light with a wavelength equal to the band gap energy when illuminated by light with photon energies larger than the band gap. This process is called *photoluminescence*. An electron from the valence band first absorbs a photon and is promoted to the conduction band. If the excited electron then falls back down into the hole it can emit a photon having energy equal to the band gap. In the case of quantum dots, the band gap is tunable with the crystal size, and thus all the colors of the rainbow can be obtained from just one material (Figure 12.44◄).

Quantum dots are being explored for applications ranging from electronics to lasers to medical imaging because they are very bright, very stable, and small enough to be taken up by living cells, even after being coated with a biocompatible surface layer. Quantum dots of semiconductors such as InP, CdSe, CdS, and PbS, with sizes in the range of 3–5 nm, have been used in place of conjugated organic dye molecules to absorb light and transfer electrons to somewhat larger TiO_2 nanoparticles (Figure 12.45▶) in a variation on the solar cell described in Chapter 9 (Section 9.8: "Chemistry Put to Work: Orbitals and Energy").

Semiconductors do not have to be shrunk to the nanoscale in all three dimensions to show new properties. They can be laid down in relatively large two-dimensional areas on a substrate, but be only a few nanometers in thickness, or even thinner, to make *quantum wells*. *Quantum wires*, in which the semiconductor wire diameter is only a few nanometers but its length is very long, have also been made by various chemical routes. In both quantum wells and quantum wires, measurements along the nanoscale dimension(s) show quantum behavior, but in the long dimension, the properties seem to be just like those of the bulk material.

▲ Figure 12.44 Semiconductor nanoparticles of CdS of different sizes. When illuminated with ultraviolet light, these solutions of semiconductor nanoparticles emit light that corresponds to their respective band gap energies.

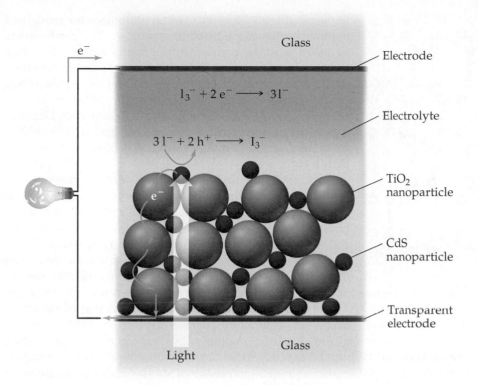

$$I_3^- + 2\,e^- \longrightarrow 3\,I^-$$

$$3\,I^- + 2\,h^+ \longrightarrow I_3^-$$

Glass

Electrode

Electrolyte

TiO$_2$ nanoparticle

CdS nanoparticle

Transparent electrode

Glass

Light

◀ Figure 12.45 **Titanium dioxide solar cell sensitized with semiconductor nanoparticles.** In this solar cell CdS nanoparticles absorb visible photons and transfer the electrons from their conduction band into the conduction band of larger TiO$_2$ nanoparticles. The electrons (e$^-$) then move to through the TiO$_2$ nanoparticles to the transparent electrode and through an external circuit generating electricity. The hole (h$^+$) that is created in the CdS nanoparticle is filled through oxidation of iodide ions, I$^-$, in the electrolyte. The I$^-$ ions are regenerated by reduction reactions at the counter electrode. The wavelengths of light that are absorbed by the CdS nanoparticles can be tuned by controlling their size.

▲ Figure 12.46 **Stained-glass window from Milan Cathedral in Italy.** The red color is due to gold nanoparticles. *Fototeca della Veneranda Fabbrica del Duomo di Milano.*

Metals on the Nanoscale

Metals also have unusual properties on the 1–100-nm length scale. Fundamentally, this is because the *mean free path* of an electron in a metal at room temperature is typically about 1–100 nm. The mean free path of an electron is the average distance it can move before it bumps into something (another electron, an atom, or the surface of the metal) and "scatters." So when the particle size of a metal is 100 nm or less, one might expect unusual effects to occur.

Although it was not fully understood, people have known for hundreds of years that metals are different when they are very finely divided. As early as the 1400s, the makers of stained-glass windows knew that gold dispersed in molten glass made the glass a beautiful deep red (Figure 12.46▶). Much later, in 1857, Michael Faraday ∞ (Chapter 20.5) reported that dispersions of small particles of gold could be made stable and were deeply colored—some of the original colloidal solutions that he made are still in the Royal Institution of Great Britain's Faraday Museum in London (Figure 12.47▶). We will examine colloids more closely in Section 13.6. Other physical and chemical properties of metallic nanoparticles are also different than bulk properties. Gold particles less than 20 nm in diameter melt at a far lower temperature than bulk gold, and between 2 and 3 nm gold is no longer a "noble," unreactive metal; in this size range it becomes chemically reactive.

At nanoscale dimensions, silver has properties analogous to those of gold in its beautiful colors, although it is more reactive than gold. Currently, there is great interest in research laboratories around the world in taking advantage of the unusual optical properties of metal nanoparticles for applications in biomedical imaging and chemical detection.

Carbon Nanotubes

We have already seen that elemental carbon is extraordinarily interesting. In its sp^3-hybridized solid-state form, it is diamond; in its sp^2-hybridized solid-state form, it is graphite; in its molecular form it is C$_{60}$. ∞ (Section 11.8) Soon after the discovery of C$_{60}$, chemists discovered carbon nanotubes. You can think of these as sheets of graphite rolled up and capped at one or both ends by half of a C$_{60}$

▲ Figure 12.47 **The solutions of colloidal gold nanoparticles made by Michael Faraday in the 1850s.** These are on display in the Faraday Museum, London.

▲ Figure 12.48 **Atomic models of carbon nanotubes.** Left: "Armchair" nanotube, which shows metallic behavior. Right: "Zigzag" nanotube, which can be either semiconducting or metallic, depending on tube diameter.

molecule (Figure 12.48 ◄). Carbon nanotubes are made in a manner similar to that used to make C_{60}. They can be made in either *multiwall* or *single-walled* forms. Multiwall carbon nanotubes consist of tubes within tubes, nested together; single-walled carbon nanotubes consist of single tubes. Single-walled carbon nanotubes can be 1000 nm long or even longer, but are only about 1 nm in diameter. Depending on the diameter of the graphite sheet and how it is rolled up, carbon nanotubes can behave as either semiconductors or metals.

The fact that carbon nanotubes can be made either semiconducting or metallic without any doping is unique among all solid-state materials, and a great deal of work is ongoing to make carbon-based electronic devices. In 2003 chemists learned how to separate metallic carbon nanotubes from semiconducting ones. Because the basic elements of the transistor are semiconductors and metals, there is great interest in building nanoscale electronic circuits using only carbon nanotubes. The first report of a field effect transistor (FET, related to the MOS-FET) using a single carbon nanotube as the channel appeared in 1998.

Carbon nanotubes are also being explored for their mechanical properties. The carbon–carbon bonded framework of the nanotubes means that the imperfections that might appear in a metal nanowire of similar dimensions are nearly absent. Experiments on individual carbon nanotubes suggest that they are stronger than steel if steel were the dimensions of a carbon nanotube. Carbon nanotubes have also been spun into fibers with polymers, adding great strength and toughness to the composite material.

■ SAMPLE INTEGRATIVE EXERCISE │ Putting Concepts Together

A *conducting polymer* is a polymer that can conduct electricity. Some polymers can be made semiconducting; others can be nearly metallic. Polyacetylene is an example of a polymer that is a semiconductor. It can also be doped to increase its conductivity.

Polyacetylene is made from acetylene in a reaction that looks simple but is actually tricky to do:

$$H - C \equiv C - H \qquad \left[CH = CH \right]_n$$

Acetylene   Polyacetylene

(a) What is the hybridization of the carbon atoms, and the geometry around those atoms, in acetylene and in polyacetylene?
(b) Write a balanced equation to make polyacetylene from acetylene.
(c) Acetylene is a gas at room temperature and pressure (298 K, 1.00 atm). How many grams of polyacetylene can you make from a 5.00-L vessel of acetylene gas at room temperature and room pressure? Assume acetylene behaves ideally, and that the polymerization reaction occurs with 100% yield.
(d) Using the average bond enthalpies in Table 8.4, predict whether the formation of polyacetylene from acetylene is endothermic or exothermic.

SOLUTION

Analyze: For part (a), we need to recall what we have learned about sp, sp^2, and sp^3 hybridization and geometry. ∞ (Section 9.5) For part (b), we need to write a balanced equation. For part (c), we need to use the ideal-gas equation. ∞ (Section 10.4) For part (d), we need to recall the definitions of endothermic and exothermic, and how bond enthalpies can be used to predict overall reaction enthalpies. ∞ (Section 8.8)

Plan: For part (a), we should draw out the chemical structures of the reactant and product. For part (b), we need to make sure the equation is properly balanced. For part (c), we need to convert from liters of gas to moles of gas, using the ideal-gas equation ($PV = nRT$); then we need to convert from moles of acetylene gas to moles of polyacetylene using the answer from part (b); then, we can convert to grams of polyacetylene. For part (d), we need to recall that $\Delta H_{rxn} = \Sigma$(bond enthalpies of bonds broken) $- \Sigma$(bond enthalpies of bonds formed).

Solve:

(a) Carbon always forms four bonds. Thus, each C atom must have a single bond to H and a triple bond to the other C atom in acetylene. As a result, each C atom has two electron domains and must be sp hybridized. This sp hybridization also means that the H—C—C angles in acetylene are 180°, and the molecule is linear.

We can write out the partial structure of polyacetylene as follows:

$$\begin{array}{ccc} H & & H \\ | & & | \\ \diagdown C & & \diagup C \diagdown \\ \diagup \ \| & & \ \| \\ C & & C \\ | & & | \\ H & & H \end{array}$$

Each carbon is identical but now has three bonding electron domains that surround it. Therefore, the hybridization of each carbon atom is sp^2, and each carbon has local trigonal planar geometry, with 120° angles.

(b) We can write:

$$n\ C_2H_2(g) \longrightarrow -[CH{=}CH]_n-$$

Note that all atoms originally present in acetylene end up in the polyacetylene product.

(c) We can use the ideal-gas equation as follows:

$$PV = nRT$$

$$(1.00\ \text{atm})(5.00\ \text{L}) = n(0.08206\ \text{L atm/K mol})(298\ \text{K})$$

$$n = 0.204\ \text{mol}$$

Acetylene has a molar mass of 26.0 g/mol; therefore, the mass of 0.204 mol is

$$(0.204\ \text{mol})(26.0\ \text{g/mol}) = 5.32\ \text{g acetylene}$$

Note that from the answer to part (b), all the atoms in acetylene go into polyacetylene. Due to conservation of mass, then, the mass of polyacetylene produced must also be 5.32 g, if we assume 100% yield.

(d) Let's consider the case for $n = 1$. We note that the reactant side of the equation in part (b) has one C≡C triple bond and two C—H single bonds. The product side of the equation in part (b) has one C=C double bond, one C—C single bond (to link to the adjacent monomer), and two C—H single bonds. Therefore, we are breaking one C≡C triple bond and are forming one C=C double bond and one C—C single bond. Accordingly, the enthalpy change for polyacetylene formation is:

$$\Delta H_{rxn} = (\text{C}{\equiv}\text{C triple bond enthalpy}) - (\text{C}{=}\text{C double bond enthalpy})$$
$$- (\text{C—C single bond enthalpy})$$

$$\Delta H_{rxn} = (839\ \text{kJ/mol}) - (614\ \text{kJ/mol}) - (348\ \text{kJ/mol})$$
$$= -123\ \text{kJ/mol}$$

Because ΔH is a negative number, the reaction releases heat, and is exothermic.

CHAPTER REVIEW

SUMMARY AND KEY TERMS

Section 12.1 In this chapter we consider the classification of materials according to their bonding (ionic, metallic, or covalent) and electrical conductivity (**insulators**, **metals**, or **semiconductors**). Ionic solids are typically insulators. Metallic solids are typically good conductors of electricity. Covalent-network solids can be either semiconductors or insulators.

Section 12.2 In extended solids, continuous **bands** rather than discrete molecular orbitals are the best way to picture electron energy levels. Metals have a partially filled band, which makes it easy for electrons to move between filled and empty molecular orbitals in the solid. Semiconductors and insulators have an energy gap known as the **band gap** that separates the filled **valence band** and the empty **conduction band**. In such materials electrons have to be excited over the band gap for the material to conduct electricity.

Section 12.3 **Elemental semiconductors** contain a single element while **compound semiconductors** contain two or more elements. In a typical semiconductor the valence band is made up of bonding molecular orbitals and the conduction band of antibonding molecular orbitals. The size of the band gap increases (a) as the bond distance decreases, due to increased orbital overlap, and (b) as the difference in electronegativity between the two elements increases, due to increased bond polarity.

Doping semiconductors changes their ability to conduct electricity by orders of magnitude. An n-type semiconductor is one that is doped so that there are excess electrons in the conduction band; a p-type semiconductor is one that is doped so that there are excess **holes** in the valence band.

The reasons for silicon's preeminence in electronics are its abundance, low cost, and ability to be grown into highly pure crystals, as well as the ability of its native oxide to grow on it in perfect registry with the atoms underneath. The design of the **transistor**, the heart of integrated circuitry, relies on the junction between n-type and p-type silicon. Solar cells are also based on a junction between an n-type and a p-type material. In a **solar cell**, photons with energies larger than its band gap energy can be absorbed near the p-n junction leading to the conversion from optical energy to electrical energy.

Light emitting diodes (LEDs) operate like a solar cell in reverse. Electrons in the conduction band from the n-type semiconductor are forced to the junction where they meet holes from the p-type semiconductor. The electron falls into the empty hole and its energy is converted into light whose photons have energy equal to the band gap.

Section 12.4 Ceramics are inorganic solids with generally high thermal stability, usually formed through three-dimensional network bonding. The bonding in ceramics may be either covalent or ionic, and ceramics may be crystalline or amorphous. The **sol-gel process** plays an important role in the processing of commercial ceramics. The process begins with the formation of a *sol*, a suspension of very small, uniformly sized particles, which is then transformed to a *gel*, where the particles form an interconnected network. Finally, the small particles are compressed and heated at high temperature. They coalesce through a process known as sintering.

Section 12.5 A **superconductor** is a material that is capable of conducting an electrical current without any apparent resistance when cooled below its **superconducting transition temperature**, T_c. Since discovery of the phenomenon in 1911, many elements and compounds have been found to be superconducting when cooled to very low temperatures. **Superconducting ceramics** such as $YBa_2Cu_3O_7$ are capable of superconductivity at temperatures higher than that for any nonceramic superconductor.

Section 12.6 **Polymers** are molecules of high molecular mass formed by joining together large numbers of small molecules, called **monomers**. In an **addition polymerization** reaction, the molecules form new linkages by opening existing π bonds. Polyethylene forms, for example, when the carbon–carbon double bonds of ethylene open up. In a **condensation polymerization** reaction, the monomers are joined by splitting out a small molecule from between them. The various kinds of nylon are formed, for example, by removing a water molecule from between an amine and a carboxylic acid. A polymer formed from two different monomers is called a **copolymer**.

Plastics are materials that can be formed into various shapes, usually by the application of heat and pressure. **Thermoplastic** polymers can be reshaped, perhaps through heating, in contrast to **thermosetting plastics**, which are formed into objects through an irreversible chemical process and cannot readily be reshaped. An **elastomer** is a material that exhibits elastic behavior; that is, it returns to its original shape following stretching or bending.

Polymers are largely amorphous, but some materials possess a degree of **crystallinity**. For a given chemical composition, the crystallinity depends on the molecular mass and the degree of branching along the main polymer chain. High-density polyethylene, for example, with little side-chain branching and a high molecular mass, has a higher degree of crystallinity than low-density polyethylene, which has a lower molecular mass and a relatively high degree of branching. Polymer properties are also strongly affected by **cross-linking**, in which short chains of atoms connect the long polymer chains. Rubber is cross-linked by short chains of sulfur atoms in the process called **vulcanization**.

Section 12.7 A **biomaterial** is any material that has a biomedical application. Biomaterials are typically in contact with body tissues and fluids. They must be **biocompatible**, which means that they are not toxic, nor do they cause an inflammatory response. They must meet physical requirements, such as long-term reliability, strength, and flexibility or hardness, depending on the application. They must also meet chemical requirements of nonreactivity in the biological environment, or of biodegradability. Biomaterials are commonly polymers with special properties matched to the application.

Section 12.8 A **liquid crystal** is a substance that exhibits one or more ordered phases at a temperature above the melting point of the solid. In a **nematic liquid crystalline phase** the molecules are aligned along a common direction, but the ends of the molecules are not lined up. In a **smectic liquid crystalline phase** the ends of the molecules are lined up so that the molecules form sheets. Nematic and smectic liquid crystalline phases are generally composed of molecules with fairly rigid, elongated shapes, with polar groups along the molecules to help retain relative alignment through dipole–dipole interactions. The **cholesteric liquid crystalline phase** is composed of molecules that align as in nematic liquid crystalline phases, but with each molecule twisted with respect to its neighbors, to form a helical structure.

Section 12.9 Nanotechnology is the manipulation of matter, and fabrication of devices, on the 1–100 nm scale. The unusual properties that semiconductors, metals, and carbon have on the size scale of 1–100 nm are explored. Quantum dots are semiconductor particles with diameters of 1–10 nm. In this size range the material's band gap energy becomes size-dependent. Metal nanoparticles have different chemical and physical properties in the 1–100-nm size range. Gold, for example, is more reactive and no longer has a golden color. Carbon nanotubes are sheets of graphite rolled up, and they can behave as either semiconductors or metals depending on how the sheet was rolled. Applications of these **nanomaterials** are being developed now for imaging, electronics, and medicine.

KEY SKILLS

- Understand the relationship between the type of bonding that holds atoms together in a material and the electrical conductivity of the material.
- Understand how molecular orbitals combine to form bands and be able to differentiate the electronic band structures of metals, semiconductors, and insulators.
- Be able to recognize elemental and compound semiconductors from empirical formulas. Be able to use the periodic table to qualitatively compare the band gap energies of semiconductors.
- Understand how n-type and p-type doping can be used to control the conductivity of semiconductors and how the most important semiconductor devices, the silicon chip, LEDs, and solar cells, operate.
- Understand how forming solid solutions can control the band gaps of semiconductors.
- Be familiar with the bonding and properties of ceramics.
- Understand how the electrical and magnetic properties of superconductors differ from ordinary conductors. Understand the differences between conventional superconductors and high T_c superconductors.
- Understand how polymers are formed from monomers. Be able to recognize the features of a molecule that allow it to react to form a polymer. Understand the differences between addition polymerization and condensation polymerization.
- Understand how the interactions between polymer chains impact the physical properties of polymers.
- Be familiar with the characteristics and applications of biomaterials.
- Understand how the molecular arrangements characteristic of nematic, smectic, and cholesteric liquid crystals differ from ordinary liquids and from each other. Be able to recognize the features of a molecule that favor formation of a liquid crystalline phase.
- Understand how the properties of bulk semiconductors and metals change as the size of the crystals decreases into the nanometer length scale.
- Be familiar with the structures and unique properties of carbon nanotubes.

VISUALIZING CONCEPTS

12.1 Based on the following band structures, which material is a metal? [Section 12.2]

Material
A

Material
B

12.2 The band structures below both belong to semiconductors. One is AlAs and the other GaAs. Which is which? [Section 12.3]

Material
X

Material
Y

12.3 Shown below are cartoons of two different polymers. Based on these cartoons, which polymer would you expect to be denser? Which one would have the higher melting point? [Section 12.6]

(a)   (b)

12.4 Which picture best represents molecules that are in a liquid crystalline phase? [Section 12.8]

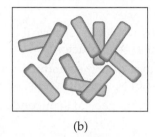

(a)   (b)

12.5 Consider the molecules shown below. (a) Which of these would be most likely to form an addition polymer? (b) Which would be most likely to form a condensation polymer? (c) Which would be most likely to form a liquid crystal? [Sections 12.6 and 12.8]

(i)   (ii)   (iii)

12.6 The three plots to the right show the conductivity as a function of temperature for three different types of materials: a semiconductor with a small band gap, a metallic conductor, and a superconductor. Which plot corresponds to which material? (*Hint*: Remember in a semiconductor thermal energy is needed to excite electrons from the valence band to the conduction band.) [Sections 12.1, 12.3, and 12.5]

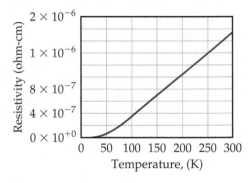

EXERCISES

Classes of Materials and Electronic Structure of Materials

12.7 Classify each of the following materials as metal, semiconductor, or insulator: **(a)** GaN **(b)** B **(c)** ZnO **(d)** Pb

12.8 Classify each of the following materials as metal, semiconductor, or insulator: **(a)** InAs **(b)** MgO **(c)** HgS **(d)** Sn

12.9 The molecular orbital diagrams for linear chains of hydrogen atoms with two and four atoms in the chain are shown in Figure 12.1. Construct a molecular orbital diagram for a chain containing six hydrogen atoms and use it to answer the following questions. **(a)** How many molecular orbitals are there in the diagram? **(b)** How many nodes are there in the lowest energy molecular orbital? **(c)** How many nodes are there in the highest energy molecular orbital? **(d)** How many nodes are there in the highest occupied molecular orbital (HOMO)? **(e)** How many nodes are there in the lowest unoccupied molecular orbital (LUMO)?

12.10 Repeat exercise 12.9 for a linear chain of eight hydrogen atoms.

12.11 State whether each statement is true or false, and why.
  (a) Semiconductors have a larger band gap than insulators.
  (b) Doping a semiconductor makes it more conductive.
  (c) Metals have delocalized electrons.
  (d) Most metal oxides are insulators.

12.12 State whether each statement is true or false, and why.
  (a) A typical band gap energy for an insulator is 400 kJ/mol.
  (b) The conduction band is higher in energy than the valence band.
  (c) Electrons can conduct well if they are in a filled valence band.
  (d) Holes refer to empty atomic sites in a solid crystal.

Semiconductors

12.13 For each of the following pairs of semiconductors, which one will have the larger band gap: **(a)** CdS or CdTe **(b)** GaN or InP **(c)** GaAs or InAs?

12.14 For each of the following pairs of semiconductors, which one will have the larger band gap: **(a)** InP or InAs **(b)** Ge or AlP **(c)** AgI or CdTe?

12.15 If you want to dope GaAs to make an n-type semiconductor with an element to replace Ga, which element(s) would you pick?

12.16 If you want to dope GaAs to make a p-type semiconductor with an element to replace As, which element(s) would you pick?

12.17 What advantages does silicon have over other semiconductors for use in integrated circuits?

12.18 Why is it important for Si crystals to be 99.999999999% pure, as opposed to 99% pure, for silicon chips?

12.19 What material is traditionally used to make the gate in a MOSFET transistor? What material is used in the next generation transistors?

12.20 Look up the diameter of a silicon atom, in Å. The channel length in a Pentium 4 processor chip is 65 nm long. How many silicon atoms does this correspond to?

12.21 Silicon has a band gap of 1.1 eV at room temperature. **(a)** What wavelength of light would a photon of this energy correspond to? **(b)** Draw a vertical line at this wavelength in the figure shown, which shows the light output of the sun as a function of wavelength. Does silicon absorb all, none or a portion of the visible light that comes from the sun?

12.22 Cadmium telluride is an important material for solar cells. **(a)** What is the band gap of CdTe? **(b)** What wavelength of light would a photon of this energy correspond to? **(c)** Draw a vertical line at this wavelength in the figure shown with exercise 12.21, which shows the light output of the Sun as a function of wavelength. **(d)** With respect to silicon, does CdTe absorb a larger or smaller portion of the solar spectrum?

12.23 The semiconductor GaP has a band gap of 2.2 eV. Green LEDs are made from pure GaP. What wavelength of light would be emitted from an LED made from GaP?

12.24 The first LEDs were made from GaAs, which has a band gap of 1.43 eV. What wavelength of light would be emitted from an LED made from GaAs? What region of the electromagnetic spectrum does this light correspond to: UV, Visible, or IR?

12.25 GaAs and GaP make solid solutions that have the same crystal structure as the parent materials, with As and P randomly distributed through the crystal. $GaP_x As_{1-x}$ exists for any value of x. If we assume that the band gap varies linearly with composition between $x = 0$ and $x = 1$, estimate the band gap for $GaP_{0.5}As_{0.5}$. What wavelength of light does this correspond to?

12.26 Red light-emitting diodes are made from GaAs and GaP solid solutions, $GaP_x As_{1-x}$ (see exercise 12.25). The original red LEDs emitted light with a wavelength of 660 nm. If we assume that the band gap varies linearly with composition between $x = 0$ and $x = 1$, estimate the composition (the value of x) that is used in these LEDs.

Ceramics

12.27 Metals, such as Al or Fe, and many plastics are recyclable. With the exception of many glasses, such as bottle glass, ceramic materials in general are not recyclable. What characteristics of ceramics make them less readily recyclable?

12.28 It is desirable to construct automobiles out of lightweight materials to maximize fuel economy. All of the ceramics listed in Table 12.3 are less dense than steel. Why do you think ceramic materials are not more widely used in the construction of automobiles?

12.29 Why is the formation of very small, uniformly sized and shaped particles important for many applications of ceramic materials?

12.30 Describe the general chemical steps in a sol-gel process, beginning with Zr(s) and $CH_3CH_2OH(l)$. Indicate whether each step is an oxidation-reduction reaction (refer to Section 4.4), condensation reaction, or other process.

12.31 The hardnesses of several substances according to a scale known as the Knoop value are as follows:

Substance	Knoop Value
Ag	60
$CaCO_3$	135
MgO	370
Soda-lime glass	530
Cr	935
ZrB_2	1550
Al_2O_3	2100
TaC	2000

Which of the materials in this list would you classify as a ceramic? What were your criteria for making this classification? Does classification as a ceramic correlate with Knoop hardness? If you think it does, is hardness alone a sufficient criterion to determine whether a substance is a ceramic? Explain.

12.32 Silicon carbide, SiC, has the three-dimensional structure shown in the figure.

(a) Name another compound that has the same structure. **(b)** Would you expect the bonding in SiC to be predominantly ionic, metallic, or covalent? **(c)** How do the bonding and structure of SiC lead to its high thermal stability (to 2700 °C) and exceptional hardness?

Superconductors

12.33 To what does the term *superconductivity* refer? Why might superconductive materials be of value?

12.34 What are the differences in the electrical and magnetic properties of an excellent metallic conductor of electricity (such as silver) and a superconducting substance (such as Nb_3Sn) below its superconducting transition temperature?

12.35 The following graph shows the resistivity (a measure of electrical resistance) of MgB_2 as a function of temperature in the region from about 4 K to 100 K. What is the significance of the sharp drop in resistivity below 40 K?

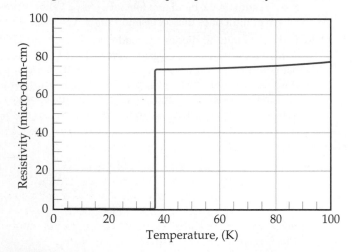

12.36 **(a)** What is the superconducting transition temperature, T_c? **(b)** The discovery by Müller and Bednorz of superconductivity in a copper oxide ceramic at 35 K set off a frantic scramble among physicists and chemists to find materials that exhibit superconductivity at higher temperatures. What is the significance of achieving T_c values above 77 K?

12.37 Explain how the Meissner effect can be used to levitate trains. What can the tracks and train wheels be made of, and which one would be more likely to be cooled?

12.38 The group 4B metal nitrides (TiN, ZrN, and HfN) as well as the group 5B metal nitrides (VN, NbN, and TaN) are all superconductors at low temperature. Niobium(III) nitride, which has the highest T_c, superconducts below 16.1 K. All of these compounds have crystal structures that are analogous to the sodium chloride structure. Scandium nitride also adopts the sodium chloride structure, but it is not a superconductor. **(a)** At room temperature will NbN be a metallic conductor or an insulator? **(b)** At room temperature will ScN be a metallic conductor or an insulator? **(c)** Why do you think the properties of ScN are so different than the group 4B and 5B metal nitrides? (*Hint:* Consider the electron configuration of the metal cation.)

Polymers

12.39 What is a monomer? Give three examples of monomers, taken from the examples given in this chapter.

12.40 The structure of decane is shown in Practice Exercise 12.5. Decane is not considered a polymer, whereas polyethylene is. What is the distinction?

12.41 An ester is a compound formed by a condensation reaction between a carboxylic acid and an alcohol. Use the index to find the discussion of esters in Chapter 25, and give an example of a reaction forming an ester. How might this kind of reaction be extended to form a polymer (a polyester)?

12.42 Write a chemical equation for formation of a polymer via a condensation reaction from the monomers succinic acid ($HOOCCH_2CH_2COOH$) and ethylenediamine ($H_2NCH_2CH_2NH_2$).

12.43 Draw the structure of the monomer(s) employed to form each of the following polymers shown in Table 12.5 **(a)** polyvinyl chloride, **(b)** nylon 6,6, **(c)** polyethylene terephthalate.

12.44 Write the chemical equation that represents the formation of **(a)** polychloroprene from chloroprene

$$CH_2{=}CH{-}\underset{\underset{Cl}{|}}{C}{=}CH_2$$

(Polychloroprene is used in highway-pavement seals, expansion joints, conveyor belts, and wire and cable jackets.); **(b)** polyacrylonitrile from acrylonitrile

$$CH_2{=}\underset{\underset{CN}{|}}{CH}$$

(Polyacrylonitrile is used in home furnishings, craft yarns, clothing, and many other items.)

12.45 The nylon Nomex®, a condensation polymer, has the following structure:

Draw the structures of the two monomers that yield Nomex®.

12.46 Proteins are polymers formed by condensation reactions of amino acids, which have the general structure

In this structure, R represents —H, —CH₃, or another group of atoms. Draw the general structure for a polyamino acid polymer formed by condensation polymerization of the molecule shown here.

12.47 What molecular features make a polymer flexible? Explain how cross-linking affects the chemical and physical properties of the polymer.

12.48 What molecular structural features cause high-density polyethylene to be denser than low-density polyethylene?

12.49 Are high molecular masses and a high degree of crystallinity always desirable properties of a polymer? Explain.

12.50 Briefly describe each of the following: **(a)** elastomer, **(b)** thermoplastic, **(c)** thermosetting plastic, **(d)** plasticizer.

Biomaterials

12.51 Neoprene is a polymer of chlorobutadiene.

$$CH_2{=}C{-}C{=}CH_2$$

$$\underset{H}{\mid} \quad \underset{Cl}{\mid}$$

The polymer can be used to form flexible tubing that is resistant to chemical attack from a variety of chemical reagents. Suppose it is proposed to use neoprene tubing as a coating for the wires running to the heart from an implanted pacemaker. What questions would you ask to determine whether it might be suitable for such an application?

12.52 On the basis of the structure shown in Table 12.5 for polystyrene and polyurethane, which of these two classes of polymer would you expect to form the most effective interface with biological systems? Explain.

12.53 Patients who receive vascular grafts formed from polymer material such as Dacron® are required to take anticoagulation drugs on a continuing basis to prevent blood clots. Why? What advances in such vascular implants are needed to make this precaution unnecessary?

12.54 Several years ago a biomedical company produced and marketed a new, efficient heart valve implant. It was later withdrawn from the market, however, because patients using it suffered from severe loss of red blood cells. Describe what properties of the valve could have been responsible for this result.

12.55 Skin cells from the body do not differentiate when they are simply placed in a tissue culture medium; that is, they do not organize into the structure of skin, with different layers and different cell types. What is needed to cause such differentiation to occur? Indicate the most important requirements on any material used.

12.56 If you were going to attempt to grow skin cells in a medium that affords an appropriate scaffolding for the cells and you had only two fabrics available, one made from polystyrene and the other from polyethyleneterephthalate (Table 12.5), which would you choose for your experiments? Explain.

Liquid Crystals

12.57 In what ways are a nematic liquid crystalline phase and an ordinary liquid the same, and in what ways do they differ?

12.58 In contrast to ordinary liquids, liquid crystals are said to possess "order." What does this mean?

12.59 Describe what is occurring at the molecular level as a substance passes from the solid to the nematic liquid crystalline to the isotropic (normal) liquid phase upon heating.

12.60 What observations made by Reinitzer on cholesteryl benzoate suggested that this substance possesses a liquid crystalline phase?

12.61 The molecules shown in Figure 12.39 possess polar groups (that is, groupings of atoms that give rise to sizable dipole moments within the molecules). How might the presence of polar groups enhance the tendency toward liquid crystal formation?

12.62 Liquid crystalline phases tend to be more viscous than the isotropic, or normal, liquid phase of the same substance. Why?

12.63 The smectic liquid crystalline phase can be said to be more highly ordered than the nematic. In what sense is this true?

12.64 One of the more effective liquid crystalline substances employed in LCDs is this molecule.

By comparing this structure with the structural formulas and models shown in Figure 12.39, describe the features of the molecule that promote its liquid crystalline behavior.

12.65 Describe how a cholesteric liquid crystal phase differs from a nematic phase.

Nanomaterials

12.67 Explain why "bands" may not be the most accurate description of bonding in a solid when the solid has nanoscale dimensions.

12.68 CdS has a band gap of 2.4 eV. If large crystals of CdS are illuminated with ultraviolet light they emit light equal to the band gap energy. **(a)** What color is the emitted light? **(b)** Would appropriately sized CdS quantum dots be able to emit blue light? **(c)** What about red light?

12.69 True or false:
(a) The band gap of a semiconductor decreases as the particle size decreases, in the 1–10-nm range.
(b) The light that is emitted from a semiconductor, upon external stimulation, is longer and longer in wavelength as the particle size of the semiconductor decreases.

12.66 It often happens that a substance possessing a smectic liquid crystalline phase just above the melting point passes into a nematic liquid crystalline phase at a higher temperature. Account for this type of behavior in terms of the ideas developed in Chapter 11 relating molecular energies to temperature.

12.70 True or false:
If you want a semiconductor that emits blue light, you could either use a material that has a band gap corresponding to the energy of a blue photon or you could use a material that has a smaller band gap, but make a nanoparticle out of it of the right size.

12.71 Gold is a face-centered cubic structure that has a unit cell edge length of 4.08 Å (Figure 11.34). How many gold atoms are there in a sphere that is 20 nm in diameter? Recall that the volume of a sphere is $\frac{4}{3}\pi r^3$.

12.72 Cadmium telluride, CdTe, takes the zinc blende structure (Section 11.8) with a unit cell edge length of 6.49 Å. There are four cadmium atoms and four tellurium atoms per unit cell. How many of each type of atom are there in a cubic crystal with an edge length of 5 nm?

ADDITIONAL EXERCISES

12.73 One major difference in the behavior of semiconductors and metals is that semiconductors increase their conductivity as you heat them (up to a point), but the conductivity of a metal decreases as you heat it. Suggest an explanation.

12.74 What properties of the typical nematic liquid crystalline molecule are likely to cause it to reorient when it is placed in an electrical field that is perpendicular to the direction of orientation of the molecules?

12.75 Teflon® is a polymer formed by the polymerization of $F_2C{=}CF_2$. Draw the structure of a section of this polymer. What type of polymerization reaction is required to form it?

12.76 Classify each of the following as a ceramic, polymer, or liquid crystal.

(a) $\left[-CH_2-\underset{\underset{COOCH_3}{|}}{\overset{\overset{CH_3}{|}}{C}}- \right]_n$ (b) $LiNbO_3$

(c) SiC   (d) $\left[-\underset{\underset{CH_3}{|}}{\overset{\overset{CH_3}{|}}{Si}}- \right]_n$

(e) $CH_3O-\bigcirc-N{=}\overset{\overset{O}{|}}{N}-\bigcirc-OCH_3$

12.77 Ceramics are generally brittle, subject to crack failure, and stable to high temperatures. In contrast, plastics are generally deformable under stress and have limited thermal stability. Discuss these differences in terms of the structures and bonding in the two classes of materials.

12.78 A watch with a liquid crystal display (LCD) does not function properly when it is exposed to low temperatures during a trip to Antarctica. Explain why the LCD might not function well at low temperature.

12.79 The temperature range over which a liquid possesses liquid crystalline behavior is rather narrow (for examples, see Figure 12.39). Why?

12.80 Suppose that a liquid crystalline material such as cholesteryl benzoate is warmed to well above its liquid crystalline range and then cooled. On cooling, the sample unexpectedly remains clear until it reaches a temperature just below the melting point, at which time it solidifies. What explanation can you give for this behavior?

12.81 Hydrogen bonding between polyamide chains plays an important role in determining the properties of a nylon such as nylon 6,6 (Table 12.5). Draw the structural formulas for two adjacent chains of nylon 6,6, and show where hydrogen-bonding interactions could occur between them.

[12.82] A particular liquid crystalline substance has the phase diagram shown in the figure. By analogy with the phase diagram for a non–liquid crystalline substance (Section 11.6), identify the phase present in each area.

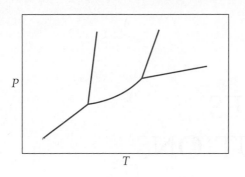

12.83 In fabricating microelectronics circuits, a ceramic conductor such as $TiSi_2$ is employed to connect various regions of a transistor with the outside world, notably aluminum wires. The $TiSi_2$ is deposited as a thin film via chemical vapor deposition, in which $TiCl_4(g)$ and $SiH_4(g)$ are reacted at the Si surface. Write a balanced equation for the reaction, assuming that the other products are H_2 and HCl. Why might $TiSi_2$ behave better as a conducting interconnect on Si than on a metal such as Cu?

INTEGRATIVE EXERCISES

[12.84] Employing the bond enthalpy values listed in Table 8.4, estimate the molar enthalpy change occurring upon **(a)** polymerization of ethylene, **(b)** formation of nylon 6,6, **(c)** formation of polyethylene terephthalate (PET).

[12.85] Although polyethylene can twist and turn in random ways, as illustrated in Figure 12.24, the most stable form is a linear one with the carbon backbone oriented as shown in the figure below:

H H H H H H H H H H H H

H H H H H H H H H H H H

The solid wedges in the figure indicate bonds from carbon that come out of the plane of the page; the dotted wedges indicate bonds that lie behind the plane of the page.
(a) What is the hybridization of orbitals at each carbon atom? What angles do you expect between the bonds?
(b) Now imagine that the polymer is polypropylene rather than polyethylene. Draw structures for polypropylene in which (i) the CH_3 groups all lie on the same side of the plane of the paper (this form is called isotactic polypropylene); (ii) the CH_3 groups lie on alternating sides of the plane (syndiotactic polypropylene); or (iii) the CH_3 groups are randomly distributed on either side (atactic polypropylene). Which of these forms would you expect to have the highest crystallinity and melting point, and which the lowest? Explain in terms of intermolecular interactions and molecular shapes.
(c) Polypropylene fibers have been employed in athletic wear. The product is said to be superior to cotton or polyester clothing in wicking moisture away from the body through the fabric to the outside. Explain the difference between polypropylene and polyester or cotton (which has many —OH groups along the molecular chain), in terms of intermolecular interactions with water.

12.86 In the superconducting ceramic $YBa_2Cu_3O_7$, what is the average oxidation state of copper, assuming that Y and Ba are in their expected oxidation states? Yttrium can be replaced with a rare-earth element such as La, and Ba can be replaced with other similar elements without fundamentally changing the superconducting properties of the material. However, general replacement of copper by any other element leads to a loss of superconductivity.

In what respects is the electron configuration of copper different from that of the other two metallic elements in this compound?

12.87 A sample of the superconducting oxide $HgBa_2Ca_2Cu_3O_{8+x}$ is found to contain 14.99% oxygen by mass. **(a)** Assuming that all other elements are present in the ratios represented by the formula, what is the value of x in the formula? **(b)** Which of the metallic elements in the compound is (are) most likely to have noninteger average charges? Explain your answer. **(c)** Which of the metallic ions in the substance is likely to have the largest ionic radius? Which will have the smallest?

12.88 **(a)** In polyvinyl chloride shown on Table 12.5, which bonds have the lowest average bond enthalpy? **(b)** In the thermal conversion of polyvinyl chloride to diamond, at high pressure, which bonds are most likely to rupture first on heating the material? **(c)** Employing the values of average bond enthalpy in Table 8.4, estimate the overall enthalpy change for converting PVC to diamond.

12.89 Consider *para*-azoxyanisole, which is a nematic liquid crystal in the temperature range of 21 °C to 47 °C:

$$CH_3O-\langle \bigcirc \rangle-N=N-\langle \bigcirc \rangle-OCH_3$$

(a) Write out the Lewis structure for this molecule, showing all lone-pair electrons as well as bonds. **(b)** Describe the hybrid orbitals employed by each of the two nitrogens. What bond angles do you anticipate about the nitrogen atom that is bonded to oxygen? **(c)** Replacing one of the —OCH_3 groups in *para*-azoxyanisole by a —$CH_2CH_2CH_2CH_3$ group causes the melting point of the substance to drop; the liquid crystal range changes to 19 °C to 76 °C. Explain why this substitution produces the observed changes in properties. **(d)** How would you expect the density of *para*-azoxyanisole to change upon melting at 117 °C? Upon passing from the nematic to the isotropic liquid state at 137 °C? Explain.

[12.90] Silicon has diamond structure (Figure 12.3) with unit cell edge length of 5.43 Å and eight atoms per unit cell. **(a)** How many silicon atoms are there in 1 cm^3 of material? **(b)** Suppose you dope that 1-cm^3 sample of silicon with 1 ppm of phosphorus that will increase the conductivity by a factor of a million. How many milligrams of phosphorus are required?

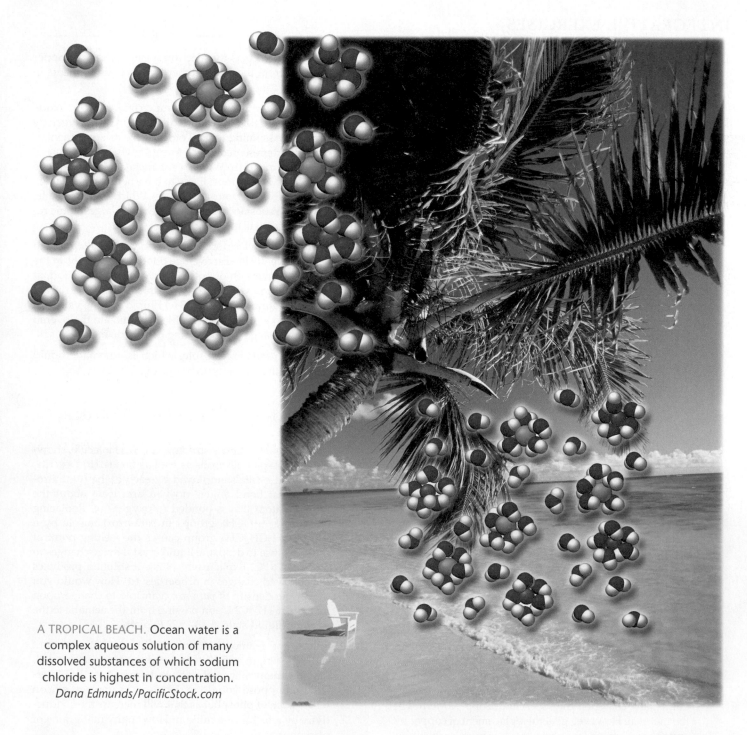

CHAPTER 13

PROPERTIES OF SOLUTIONS

A TROPICAL BEACH. Ocean water is a complex aqueous solution of many dissolved substances of which sodium chloride is highest in concentration.
Dana Edmunds/PacificStock.com

WHAT'S AHEAD

13.1 The Solution Process
We begin by considering what happens at a molecular level when a substance dissolves, paying particular attention to the role of *intermolecular forces* in the process. Two important aspects of the solution process are the changes in *energy* and the changes in how particles are distributed in space as a result of the solution process.

13.2 Saturated Solutions and Solubility
We will see that in *saturated solutions* the dissolved and undissolved solutes are in *equilibrium*. The amount of solute in a saturated solution defines its *solubility*, the extent to which a particular solute dissolves in a particular solvent.

13.3 Factors Affecting Solubility
We next consider the major factors affecting solubility. The nature of the solute and solvent determines the kinds of intermolecular forces between and within them and strongly influences solubility. Temperature also influences solubility: Most solids are more soluble in water at higher temperatures, whereas gases are less soluble in water at higher temperatures. In addition, the solubility of gases increases with increasing pressure.

13.4 Ways of Expressing Concentration
We observe that many physical properties of solutions depend on their concentration, and we examine several common ways of expressing concentration.

13.5 Colligative Properties
We observe that solutes affect the properties of solutions. The physical properties of solutions that depend only on concentration and not on the identity of the solute are called *colligative properties*. These properties include the extent to which the solute lowers the vapor pressure, increases the boiling point, and decreases the freezing point of the solvent. The osmotic pressure of a solution is also a colligative property.

13.6 Colloids
We close the chapter by investigating *colloids*, mixtures in which particles larger than typical molecular sizes are dispersed in another component.

ON A WARM DAY AT THE BEACH, we are unlikely to consider chemistry while we enjoy the water and warmth of the sun. But as we breathe the air, swim in the water, and walk on the sand, we are experiencing the three states of matter. In Chapter 10 and 11, we explored the properties

of gases, liquids, and solids. Most of the discussion focused on pure substances. However, the matter that we encounter in our daily lives, such as air, seawater, and sand, is usually composed of mixtures. In this chapter we examine mixtures, although we limit ourselves to those that are homogeneous. As we have noted in earlier chapters, a homogeneous mixture is called a *solution*. ∞ (Sections 1.2 and 4.1)

When we think of solutions, we usually first think about liquids, such as a solution of salt in water, like the seawater shown in this chapter's opening photo. Sterling silver, which is used in jewelry, is also a solution—a homogeneous distribution of about 7% copper in silver. Sterling silver is an example of a solid solution. Numerous examples of solutions abound in the world around us—some solid, some liquid, and some gas. For example, the air we breathe is a solution of several gases; brass is a solid solution of zinc in copper; and the fluids that run through our bodies are solutions that carry a variety of essential nutrients, salts, and other materials.

Each of the substances in a solution is called a *component* of the solution. As we saw in Chapter 4, the *solvent* is normally the component present in the greatest amount. Other components are called *solutes*. Because liquid solutions are the most common, we will focus our attention on them in this chapter. Our primary goal is to examine the physical properties of solutions, comparing them with the properties of their components. We will be particularly concerned with *aqueous solutions*, which contain water as the solvent and a gas, liquid, or solid as a solute.

13.1 THE SOLUTION PROCESS

A solution is formed when one substance disperses uniformly throughout another. As we noted in the introduction, solutions may be gases, liquids, or solids. Each of these possibilities is listed in Table 13.1 ▼.

The ability of substances to form solutions depends on two general factors: (1) the types of intermolecular interactions involved in the solution process, and (2) the natural tendency of substances to spread into larger volumes when not restrained in some way. We begin our discussion of the solution process by examining the role of intermolecular interactions.

The Effect of Intermolecular Forces

Any of the various kinds of intermolecular forces that we discussed in Chapter 11 can operate between solute and solvent particles in a solution. Ion–dipole forces, for example, dominate in solutions of ionic substances in water. Dispersion forces, on the other hand, dominate when a nonpolar substance such as C_6H_{14} dissolves in another nonpolar one like CCl_4. Indeed, a major factor determining whether a solution forms is the relative strengths of intermolecular forces between and among the solute and solvent particles. That is, the extent to which one substance is able to dissolve in another depends on the relative magnitudes of the solute–solvent, solute–solute, and solvent–solvent interactions involved in the solution process.

Solutions form when the magnitudes of the attractive forces between solute and solvent particles are comparable to or greater than those that exist between the solute particles themselves or between the solvent particles themselves. For example, the ionic substance NaCl dissolves readily in water because the attractive interactions between the ions and the polar H_2O molecules (solute–solvent interactions) overcome the attraction between the ions in the solid NaCl

TABLE 13.1 ■ Examples of Solutions			
State of Solution	**State of Solvent**	**State of Solute**	**Example**
Gas	Gas	Gas	Air
Liquid	Liquid	Gas	Oxygen in water
Liquid	Liquid	Liquid	Alcohol in water
Liquid	Liquid	Solid	Salt in water
Solid	Solid	Gas	Hydrogen in palladium
Solid	Solid	Liquid	Mercury in silver
Solid	Solid	Solid	Silver in gold

(a)   (b)   (c)

▲ **Figure 13.1 Dissolution of an ionic solid in water.** (a) A crystal of the ionic solid is hydrated by water molecules, with the oxygen atoms of the water molecules oriented toward the cations (purple) and the hydrogens oriented toward the anions (green). (b, c) As the solid dissolves, the individual ions are removed from the solid surface and become completely separate hydrated species in solution.

(solute–solute interactions) and between H_2O molecules in the solvent (solvent–solvent interactions). Let's examine this solution process more closely, paying attention to these various attractive forces.

When NaCl is added to water (Figure 13.1▲), the water molecules orient themselves on the surface of the NaCl crystals. The positive end of the water dipole is oriented toward the Cl^- ions, and the negative end of the water dipole is oriented toward the Na^+ ions. The ion–dipole attractions between the ions and water molecules are strong enough to pull the ions from their positions in the crystal.

Once separated from the crystal, the Na^+ and Cl^- ions are surrounded by water molecules, as shown in Figure 13.1(b and c) and Figure 13.2▶. We learned in Section 4.1 that interactions such as this between solute and solvent molecules are known as **solvation**. When the solvent is water, the interactions are also referred to as **hydration**.

In addition to the solvent–solute interactions (the ion–dipole attractions between H_2O molecules and the Na^+ and Cl^- ions) and the solute–solute interactions (between the Na^+ and Cl^- ions in the solid), we must consider one other interaction: the solvent–solvent interaction (in this case the hydrogen-bonding attractions between H_2O molecules). In forming the solution, the water molecules must make room for the hydrated Na^+ and Cl^- ions in their midst, causing some water molecules to move apart.

▲ **GIVE IT SOME THOUGHT**

Why doesn't NaCl dissolve in nonpolar solvents such as hexane, C_6H_{14}?

Energy Changes and Solution Formation

We can analyze the roles played by the solute–solvent, solute–solute, and solvent–solvent interactions by examining the energy changes associated with each. Let's continue to analyze the process of dissolving NaCl in water as an example of how energy considerations provide insight into the solution process.

We have observed that sodium chloride dissolves in water because the water molecules have a strong enough attraction for the Na^+ and Cl^- ions to overcome the attraction of these two ions for one another in the crystal. In addition, water molecules must separate from one another to form spaces in the solvent that the Na^+ and Cl^- ions will occupy. Thus, we can think of the overall energetics of solution formation as having three components—associated

▲ **Figure 13.2 Hydrated Na^+ and Cl^- ions.** The negative ends of the water dipoles point toward the positive ion, and the positive ends point toward the negative ion.

▶ **Figure 13.3 Enthalpy contributions to ΔH_{soln}.** The enthalpy changes ΔH_1 and ΔH_2 represent endothermic processes, requiring an input of energy, whereas ΔH_3 represents an exothermic process.

ΔH_1: Separation of solute molecules

ΔH_2: Separation of solvent molecules

ΔH_3: Formation of solute–solvent interactions

with breaking the solute–solute and the solvent–solvent interactions and forming the solute–solvent interactions—as illustrated schematically in Figure 13.3 ▲. The overall enthalpy change in forming a solution, ΔH_{soln}, is the sum of three terms associated with these three processes:

$$\Delta H_{\text{soln}} = \Delta H_1 + \Delta H_2 + \Delta H_3 \qquad [13.1]$$

Regardless of the particular solute being considered, separation of the solute particles from one another requires an input of energy to overcome their attractive interactions. The process is therefore endothermic ($\Delta H_1 > 0$). Separation of solvent molecules to accommodate the solute also always requires energy ($\Delta H_2 > 0$). The third component, which arises from the attractive interactions between solute and solvent, is always exothermic ($\Delta H_3 < 0$).

As shown in Figure 13.4 ▶, the three enthalpy terms in Equation 13.1 can be added together to give either a negative or a positive sum, depending on the relative magnitudes of the terms. Thus, the formation of a solution can be either exothermic or endothermic. For example, when magnesium sulfate, $MgSO_4$, is added to water, the resultant solution gets quite warm: $\Delta H_{\text{soln}} = -91.2$ kJ/mol. In contrast, the dissolution of ammonium nitrate (NH_4NO_3) is endothermic: $\Delta H_{\text{soln}} = 26.4$ kJ/mol. These particular substances have been used to make the instant heat packs and ice packs that are used to treat athletic injuries (Figure 13.5 ▶). The packs consist of a pouch of water and a dry chemical, $MgSO_4$ for hot packs and NH_4NO_3 for cold packs. When the pack is squeezed, the seal separating the solid from the water is broken and a solution forms, either increasing or decreasing the temperature.

In Chapter 5 we learned that the enthalpy change in a process can provide information about the extent to which a process will occur. ∞ (Section 5.4) Processes that are exothermic tend to proceed spontaneously. A solution will not form if ΔH_{soln} is too endothermic. The solvent–solute interaction must be

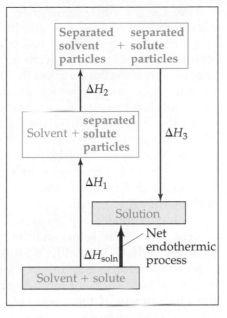

(a) Exothermic   (b) Endothermic

◄ Figure 13.4 Enthalpy changes accompanying the solution process. The three processes are illustrated in Figure 13.3. The diagram on the left (a) illustrates a net exothermic process ($\Delta H_{soln} < 0$); that on the right (b) shows a net endothermic process ($\Delta H_{soln} > 0$).

strong enough to make ΔH_3 comparable in magnitude to $\Delta H_1 + \Delta H_2$. This fact explains why ionic solutes such as NaCl do not dissolve in nonpolar liquids such as gasoline. The nonpolar hydrocarbon molecules of the gasoline would experience only weak attractive interactions with the ions, and these interactions would not compensate for the energies required to separate the ions from one another.

By similar reasoning, a polar liquid such as water does not form solutions with a nonpolar liquid such as octane (C_8H_{18}). The water molecules experience strong hydrogen-bonding interactions with one another. ∞∞ (Section 11.2) These attractive forces must be overcome to disperse the water molecules throughout the nonpolar liquid. The energy required to separate the H_2O molecules is not recovered in the form of attractive interactions between H_2O and C_8H_{18} molecules.

GIVE IT SOME THOUGHT

Label the following processes as exothermic or endothermic: **(a)** forming solvent–solute interactions, **(b)** breaking solvent–solvent interactions.

Solution Formation, Spontaneity, and Entropy

When carbon tetrachloride (CCl_4) and hexane (C_6H_{14}) are mixed, they dissolve in one another in all proportions. Both substances are nonpolar, and they have similar boiling points (77 °C for CCl_4 and 69 °C for C_6H_{14}). It is therefore reasonable to suppose that the magnitudes of the attractive forces (London dispersion forces) among molecules in the two substances and in their solution are comparable. When the two substances are mixed, dissolving occurs spontaneously; that is, it occurs without any extra input of energy from outside the system. Processes that occur spontaneously involve two distinct factors. The most obvious factor is energy, which we have used to analyze the dissolving of NaCl in water. The other factor is the distribution of each component into a larger volume—the tendency in nature for substances to mix and spread out into larger volumes.

▲ Figure 13.5 Endothermic dissolution. Ammonium nitrate instant ice packs are often used to treat athletic injuries. To activate the pack, the container is kneaded. The kneading breaks an interior seal separating solid NH_4NO_3 from water. The heat of solution of NH_4NO_3 is positive, so the temperature of the solution decreases.

▲ **Figure 13.6 Increasing randomness in a solution process.** A homogeneous solution of CCl_4 and C_6H_{14} forms when a barrier separating the two liquids is removed. Each CCl_4 molecule of the solution in (b) is more dispersed in space than it was in the left compartment in (a), and each C_6H_{14} molecule in (b) is more dispersed than it was in the right compartment in (a).

We see the influence of energy all around us. If you let go of a book, it falls to the floor because of gravity. At its initial height, the book has a higher potential energy than it has when it is on the floor. Unless the book is restrained, it falls; and as it does, potential energy is converted into kinetic energy. When the book strikes the floor, the kinetic energy is converted largely into heat energy, which is dispersed throughout the surroundings. The book has lost energy to its surroundings in this process. This fact leads us to the first basic principle identifying spontaneous processes and the direction they take: *Processes in which the energy content of the system decreases tend to occur spontaneously.* Spontaneous processes tend to be exothermic. ∞(Section 5.4, "Strategies in Chemistry: Using Enthalpy as a Guide") Change tends to occur in the direction that leads to a lower energy or lower enthalpy for the system.

Some spontaneous processes, however, do not result in lower energy for a system, and even some endothermic processes occur spontaneously. For example, NH_4NO_3 readily dissolves in water, even though the solution process is endothermic. Processes such as this are characterized by a more dispersed state of one or more components, resulting in an overall increase in the randomness of the system. In this example the densely ordered solid, NH_4NO_3, is dispersed throughout the solution as separated NH_4^+ and NO_3^- ions. The mixing of CCl_4 and C_6H_{14} provides another simple example.

Suppose that we could suddenly remove a barrier that separates 500 mL of CCl_4 from 500 mL of C_6H_{14}, as shown in Figure 13.6(a) ◄. Before the barrier is removed, each liquid occupies a volume of 500 mL. All the CCl_4 molecules are in the 500 mL to the left of the barrier, and all the C_6H_{14} molecules are in the 500 mL to the right. When equilibrium has been established after the barrier has been removed, the two liquids together occupy a volume of about 1000 mL. Formation of a homogeneous solution has increased the degree of dispersal, or randomness, because the molecules of each substance are now mixed and distributed in a volume twice as large as that which they occupied individually before mixing. The degree of randomness in the system, sometimes referred to as disorder, is given by a thermodynamic quantity called **entropy**. This example illustrates our second basic principle: *Processes occurring at a constant temperature in which the randomness or dispersal in space (entropy) of the system increases tend to occur spontaneously.*

When molecules of different types are brought together, mixing—hence an increased dispersal—occurs spontaneously unless the molecules are restrained by sufficiently strong intermolecular forces or by physical barriers. Thus, gases spontaneously mix and expand unless restrained by their containers; in this case intermolecular forces are too weak to restrain the molecules. However, because strong bonds hold sodium and chloride ions together, sodium chloride does not spontaneously dissolve in gasoline.

We will discuss spontaneous processes again in Chapter 19. At that time we will consider the balance between the tendencies toward lower enthalpy and toward increased entropy in greater detail. For the moment, we need to be aware that the solution process involves two factors: a change in enthalpy and a change in entropy. In most cases the *formation of solutions is favored by the increase in entropy that accompanies mixing.* Consequently, a solution will form unless solute–solute or solvent–solvent interactions are too strong relative to the solute–solvent interactions.

GIVE IT SOME THOUGHT

Silver chloride, AgCl, is essentially insoluble in water. Would you expect a significant change in the entropy of the system when 10 g of AgCl is added to 500 mL of water?

(a)

(b)▶

(c)

▲ **Figure 13.7 The nickel-acid reaction is *not* a simple dissolution.** (a) Nickel metal and hydrochloric acid. (b) Nickel reacts slowly with hydrochloric acid, forming $NiCl_2(aq)$ and $H_2(g)$. (c) $NiCl_2 \cdot 6\,H_2O$ is obtained when the solution from (b) is evaporated to dryness. Because the residue left after evaporation is chemically different from either reactant, we know what takes place is a chemical reaction rather than merely a solution process.

Solution Formation and Chemical Reactions

In all our discussions of solutions, we must be careful to distinguish the physical process of solution formation from chemical reactions that lead to a solution. For example, nickel metal is dissolved on contact with hydrochloric acid solution because the following chemical reaction occurs:

$$Ni(s) + 2\,HCl(aq) \longrightarrow NiCl_2(aq) + H_2(g) \qquad [13.2]$$

In this instance the chemical form of the substance being dissolved is changed from Ni to $NiCl_2$. If the solution is evaporated to dryness, $NiCl_2 \cdot 6\,H_2O(s)$, not Ni(s), is recovered (Figure 13.7▶). When NaCl(s) is dissolved in water, on the other hand, no chemical reaction occurs. If the solution is evaporated to dryness, NaCl is recovered. Our focus throughout this chapter is on solutions from which the solute can be recovered unchanged from the solution.

A Closer Look   HYDRATES

Frequently, hydrated ions remain in crystalline salts that are obtained by evaporation of water from aqueous solutions. Common examples include $FeCl_3 \cdot 6\,H_2O$ [iron(III) chloride hexahydrate] and $CuSO_4 \cdot 5\,H_2O$ [copper(II) sulfate pentahydrate]. The $FeCl_3 \cdot 6\,H_2O$ consists of $Fe(H_2O)_6^{3+}$ and Cl^- ions; the $CuSO_4 \cdot 5\,H_2O$ consists of $Cu(H_2O)_4^{2+}$ and $SO_4(H_2O)^{2-}$ ions. Water molecules can also occur in positions in the crystal lattice that are not specifically associated with either a cation or an anion. $BaCl_2 \cdot 2\,H_2O$ (barium chloride dihydrate) is an example. Compounds such as $FeCl_3 \cdot 6\,H_2O$, $CuSO_4 \cdot 5\,H_2O$, and $BaCl_2 \cdot 2\,H_2O$, which contain a salt and water combined in definite proportions, are known as *hydrates*. The water associated with them is called *water of hydration*. Figure 13.8▶ shows an example of a hydrate and the corresponding anhydrous (water-free) substance.

Related Exercises: 13.4, 13.107

▲ **Figure 13.8 Hydrates and anhydrous salts.** Hydrated cobalt(II) chloride, $CoCl_2 \cdot 6\,H_2O$ (left), and anhydrous $CoCl_2$ (right).

▲ **Figure 13.9 Dynamic equilibrium in a saturated solution.** In a solution in which excess ionic solute is present, ions on the surface of the solute are continuously passing into the solution as hydrated species, while hydrated ions from the solution are deposited on the surfaces of the solute. At equilibrium in a saturated solution, the two processes occur at equal rates.

■■ **SAMPLE EXERCISE 13.1** | Assessing Entropy Change

In the process illustrated below, water vapor reacts with excess solid sodium sulfate to form the hydrated form of the salt. The chemical reaction is

$$Na_2SO_4(s) + 10\ H_2O(g) \longrightarrow Na_2SO_4 \cdot 10\ H_2O(s)$$

Essentially all of the water vapor in the closed container is consumed in this reaction. If we consider our system to consist initially of $Na_2SO_4(s)$ and $10\ H_2O(g)$, **(a)** does the system become more or less ordered in this process, and **(b)** does the entropy of the system increase or decrease?

$Na_2SO_4(s)$ $Na_2SO_4(s) + Na_2SO_4{\cdot}10\ H_2O(s)$

SOLUTION

Analyze: We are asked to determine whether the reaction of water vapor with the solid salt causes the system to become more or less dispersed, or random, and to determine whether the process results in a higher or lower entropy for the system.

Plan: We need to examine the initial and final states and judge whether the process has made the system more or less dispersed. Depending on our answer to that question, we can determine whether the entropy has increased or decreased.

Solve: **(a)** In the course of forming the hydrate of $Na_2SO_4(s)$, the water vapor moves from the vapor state, in which it is dispersed throughout the entire volume of the container, to the solid state, where it is confined to the $Na_2SO_4 \cdot 10\ H_2O(s)$ lattice. This means that the water vapor becomes less dispersed (more ordered, or less random). **(b)** When a system becomes less dispersed, or more ordered, its entropy is decreased.

■■ **PRACTICE EXERCISE**

Does the entropy of the system increase or decrease when the stopcock is opened to allow mixing of the two gases in this apparatus?

Answer: The entropy increases because each gas eventually becomes dispersed in twice the volume it originally occupied.

13.2 SATURATED SOLUTIONS AND SOLUBILITY

As a solid solute begins to dissolve in a solvent, the concentration of solute particles in solution increases, thus increasing the chances of the solute particles colliding with the surface of the solid (Figure 13.9 ◄). Because of such a collision, the solute particle may become reattached to the solid. This process, which is the opposite of the solution process, is called **crystallization**. Thus, two opposing processes occur in a solution in contact with undissolved solute. This situation is represented in a chemical equation by use of two half arrows:

$$\text{Solute + solvent} \underset{\text{crystallize}}{\overset{\text{dissolve}}{\rightleftharpoons}} \text{solution} \qquad [13.3]$$

(a) A seed crystal of NaCH₃COO being added to the supersaturated solution.

(b) Excess NaCH₃COO crystallizes from the solution.

(c) The solution arrives at saturation.

When the rates of these opposing processes become equal, there is no further net increase in the amount of solute in solution. A dynamic equilibrium is established similar to the one between evaporation and condensation discussed in Section 11.5.

A solution that is in equilibrium with undissolved solute is **saturated**. Additional solute will not dissolve if added to a saturated solution. The amount of solute needed to form a saturated solution in a given quantity of solvent is known as the **solubility** of that solute. That is, *the solubility is the maximum amount of solute that will dissolve in a given amount of solvent at a specified temperature, given that excess solute is present*. For example, the solubility of NaCl in water at 0 °C is 35.7 g per 100 mL of water. This is the maximum amount of NaCl that can be dissolved in water to give a stable equilibrium solution at that temperature.

If we dissolve less solute than that needed to form a saturated solution, the solution is **unsaturated**. Thus, a solution containing only 10.0 g of NaCl per 100 mL of water at 0 °C is unsaturated because it has the capacity to dissolve more solute.

Under suitable conditions it is sometimes possible to form solutions that contain a greater amount of solute than that needed to form a saturated solution. Such solutions are **supersaturated**. For example, considerably more sodium acetate (NaCH₃COO) can dissolve in water at high temperatures than at low temperatures. When a saturated solution of sodium acetate is made at a high temperature and then slowly cooled, all of the solute may remain dissolved even though the solubility decreases as the temperature decreases. Because the solute in a supersaturated solution is present in a concentration higher than the equilibrium concentration, supersaturated solutions are unstable. Supersaturated solutions result for much the same reason as supercooled liquids (Section 11.4). For crystallization to occur, the molecules or ions of solute must arrange themselves properly to form crystals. The addition of a small crystal of the solute (a seed crystal) provides a template for crystallization of the excess solute, leading to a saturated solution in contact with excess solid (Figure 13.10▲).

▲ **Figure 13.10 Sodium acetate readily forms a supersaturated solution in water.**

GIVE IT SOME THOUGHT

Is a supersaturated solution of sodium acetate a stable equilibrium solution?

13.3 FACTORS AFFECTING SOLUBILITY

The extent to which one substance dissolves in another depends on the nature of both the solute and the solvent. ∞ (Section 13.1) It also depends on temperature and, at least for gases, on pressure. Let's consider these factors more closely.

TABLE 13.2 ■ Solubilities of Gases in Water at 20 °C, with 1 atm Gas Pressure	
Gas	Solubility (M)
N_2	0.69×10^{-3}
CO	1.04×10^{-3}
O_2	1.38×10^{-3}
Ar	1.50×10^{-3}
Kr	2.79×10^{-3}

:O:
‖
CH_3CCH_3

Acetone

Hexane is insoluble in water. Hexane is the top layer because it is less dense than water.

Solute–Solvent Interactions

One factor determining solubility is the natural tendency of substances to mix (the tendency of systems to move toward a more dispersed, or random, state). ∞ (Section 13.2) If this were the only factor involved, however, we would expect all substances to be completely soluble in one another. This is clearly not the case. So what other factors are involved? As we saw in Section 13.1, the relative forces of attraction among the solute and solvent molecules also play very important roles in the solution process.

Although the tendency toward dispersal and the various interactions among solute and solvent particles are all involved in determining the solubilities, considerable insight can often be gained by focusing on the interaction between the solute and solvent. The data in Table 13.2 ◄ show, for example, that the solubilities of various simple gases in water increase with increasing molecular mass or polarity. The attractive forces between the gas and solvent molecules are mainly of the London dispersion type, which increase with increasing size and mass of the gas molecules. ∞ (Section 11.2) Thus, the data indicate that the solubilities of gases in water increase as the attraction between the solute (gas) and solvent (water) increases. In general, when other factors are comparable, *the stronger the attractions are between solute and solvent molecules, the greater the solubility.*

Because of favorable dipole–dipole attractions between solvent molecules and solute molecules, *polar liquids tend to dissolve readily in polar solvents.* Water is both polar and able to form hydrogen bonds. ∞ (Section 11.2) Thus, polar molecules, especially those that can form hydrogen bonds with water molecules, tend to be soluble in water. For example, acetone, a polar molecule with the structural formula shown in the margin, mixes in all proportions with water. Acetone has a strongly polar C=O bond and pairs of nonbonding electrons on the O atom that can form hydrogen bonds with water.

Pairs of liquids such as acetone and water that mix in all proportions are **miscible**, whereas those that do not dissolve in one another are **immiscible**. Gasoline, which is a mixture of hydrocarbons, is immiscible with water. Hydrocarbons are nonpolar substances because of several factors: The C—C bonds are nonpolar, the C—H bonds are nearly nonpolar, and the shapes of the molecules are symmetrical enough to cancel much of the weak C—H bond dipoles. The attraction between the polar water molecules and the nonpolar hydrocarbon molecules is not sufficiently strong to allow the formation of a solution. *Nonpolar liquids tend to be insoluble in polar liquids.* As a result, hexane (C_6H_{14}) does not dissolve in water, as the photo in the margin shows.

The series of compounds in Table 13.3 ▼ demonstrates that polar liquids tend to dissolve in other polar liquids and nonpolar liquids in nonpolar ones. These organic compounds all contain the OH group attached to a C atom. Organic compounds with this molecular feature are called *alcohols.* The O—H bond is polar and is able to form hydrogen bonds. For example, CH_3CH_2OH molecules can form hydrogen bonds with water molecules as well as with each other (Figure 13.11 ▶). As a result, the solute–solute, solvent–solvent, and solute–solvent forces

TABLE 13.3 ■ Solubilities of Some Alcohols in Water and in Hexane*		
Alcohol	Solubility in H_2O	Solubility in C_6H_{14}
CH_3OH (methanol)	∞	0.12
CH_3CH_2OH (ethanol)	∞	∞
$CH_3CH_2CH_2OH$ (propanol)	∞	∞
$CH_3CH_2CH_2CH_2OH$ (butanol)	0.11	∞
$CH_3CH_2CH_2CH_2CH_2OH$ (pentanol)	0.030	∞
$CH_3CH_2CH_2CH_2CH_2CH_2OH$ (hexanol)	0.0058	∞

*Expressed in mol alcohol/100 g solvent at 20 °C. The infinity symbol (∞) indicates that the alcohol is completely miscible with the solvent.

are not greatly different within a mixture of CH_3CH_2OH and H_2O. No major change occurs in the environments of the molecules as they are mixed. Therefore, the increased dispersal (entropy) of the two components into a larger combined volume as they mix plays a significant role in the formation of the solution. Ethanol (CH_3CH_2OH), therefore, is completely miscible with water.

The number of carbon atoms in an alcohol affects its solubility in water. As the length of the carbon chain increases, the polar OH group becomes an ever smaller part of the molecule, and the molecule behaves more like a hydrocarbon. The solubility of the alcohol in water decreases correspondingly. On the other hand, the solubility of the alcohol in a nonpolar solvent like hexane (C_6H_{14}) increases as the nonpolar hydrocarbon chain increases in length.

One way to enhance the solubility of a substance in water is to increase the number of polar groups it contains. For example, increasing the number of OH groups along a carbon chain of a solute increases the extent of hydrogen bonding between that solute and water, thereby increasing solubility. Glucose ($C_6H_{12}O_6$) has five OH groups on a six-carbon framework, which makes the molecule very soluble in water (83 g dissolves in 100 mL of water at 17.5 °C). The glucose molecule is shown in Figure 13.12▼.

▲ Figure 13.11 **Hydrogen-bonding interactions.** (a) Between two ethanol molecules and (b) between an ethanol molecule and a water molecule.

HYDROGEN BONDING AND AQUEOUS SOLUBILITY

The presence of OH groups capable of hydrogen bonding with water enhances the aqueous solubility of organic molecules.

Cyclohexane, C_6H_{12}, which has no polar OH groups, is essentially insoluble in water.

Glucose, $C_6H_{12}O_6$, has five OH groups and is highly soluble in water.

Hydrogen-bonding sites

▲ Figure 13.12 **Structure and solubility.**

Chemistry and Life FAT- AND WATER-SOLUBLE VITAMINS

Vitamins have unique chemical structures that affect their solubilities in different parts of the human body. Vitamins B and C are water soluble, for example, whereas vitamins A, D, E, and K are soluble in nonpolar solvents and in the fatty tissue of the body (which is nonpolar). Because of their water solubility, vitamins B and C are not stored to any appreciable extent in the body, and so foods containing these vitamins should be included in the daily diet. In contrast, the fat-soluble vitamins are stored in sufficient quantities to keep vitamin-deficiency diseases from appearing even after a person has subsisted for a long period on a vitamin-deficient diet.

The different solubility patterns of the water-soluble vitamins and the fat-soluble ones can be rationalized in terms of the structures of the molecules. The chemical structures of vitamin A (retinol) and of vitamin C (ascorbic acid) are shown in Figure 13.13 ▼. Notice that the vitamin A molecule is an alcohol with a very long carbon chain. Because the OH group is such a small part of the molecule, the molecule resembles the long-chain alcohols listed in Table 13.3. This vitamin is nearly nonpolar. In contrast, the vitamin C molecule is smaller and has more OH groups that can form hydrogen bonds with water. It is somewhat like glucose, which was discussed earlier. It is a more polar substance.

Related Exercises: 13.7, 13.44

Vitamin A   Vitamin C
(a)   (b)

▲ **Figure 13.13 Vitamins A and C.** (a) The molecular structure of vitamin A, a fat-soluble vitamin. The molecule is composed largely of carbon–carbon and carbon–hydrogen bonds, so it is nearly nonpolar. (b) The molecular structure of vitamin C, a water-soluble vitamin. Notice the OH groups and the other oxygen atoms in the molecule, which can interact with water molecules by hydrogen bonding.

Examination of different combinations of solvents and solutes such as those considered in the preceding paragraphs has led to an important generalization: *Substances with similar intermolecular attractive forces tend to be soluble in one another.* This generalization is often simply stated as *"like dissolves like."* Nonpolar substances are more likely to be soluble in nonpolar solvents; ionic and polar solutes are more likely to be soluble in polar solvents. Network solids such as diamond and quartz are not soluble in either polar or nonpolar solvents because of the strong bonding forces within the solid.

▲ **GIVE IT SOME THOUGHT**

Suppose the hydrogens on the OH groups in glucose (Figure 13.12) were replaced with methyl groups, CH_3. Would you expect the water solubility of the resulting molecule to be higher than, lower than, or about the same as the solubility of glucose?

■ **SAMPLE EXERCISE 13.2** | Predicting Solubility Patterns

Predict whether each of the following substances is more likely to dissolve in the nonpolar solvent carbon tetrachloride (CCl_4) or in water: C_7H_{16}, Na_2SO_4, HCl, and I_2.

SOLUTION

Analyze: We are given two solvents, one that is nonpolar (CCl_4) and the other that is polar (H_2O), and asked to determine which will be the best solvent for each solute listed.

Plan: By examining the formulas of the solutes, we can predict whether they are ionic or molecular. For those that are molecular, we can predict whether they are polar or nonpolar. We can then apply the idea that the nonpolar solvent will be best for the nonpolar solutes, whereas the polar solvent will be best for the ionic and polar solutes.

Solve: C_7H_{16} is a hydrocarbon, so it is molecular and nonpolar. Na_2SO_4, a compound containing a metal and nonmetals, is ionic. HCl, a diatomic molecule containing two nonmetals that differ in electronegativity, is polar. I_2, a diatomic molecule with atoms of equal electronegativity, is nonpolar. We would therefore predict that C_7H_{16} and I_2 (the nonpolar solutes) would be more soluble in the nonpolar CCl_4 than in polar H_2O, whereas water would be the better solvent for Na_2SO_4 and HCl (the ionic and polar covalent solutes).

■ **PRACTICE EXERCISE**

Arrange the following substances in order of increasing solubility in water:

$$H-\overset{\overset{\displaystyle H}{|}}{\underset{\underset{\displaystyle H}{|}}{C}}-\overset{\overset{\displaystyle H}{|}}{\underset{\underset{\displaystyle H}{|}}{C}}-\overset{\overset{\displaystyle H}{|}}{\underset{\underset{\displaystyle H}{|}}{C}}-\overset{\overset{\displaystyle H}{|}}{\underset{\underset{\displaystyle H}{|}}{C}}-\overset{\overset{\displaystyle H}{|}}{\underset{\underset{\displaystyle H}{|}}{C}}-H \qquad HO-\overset{\overset{\displaystyle H}{|}}{\underset{\underset{\displaystyle H}{|}}{C}}-\overset{\overset{\displaystyle H}{|}}{\underset{\underset{\displaystyle H}{|}}{C}}-\overset{\overset{\displaystyle H}{|}}{\underset{\underset{\displaystyle H}{|}}{C}}-\overset{\overset{\displaystyle H}{|}}{\underset{\underset{\displaystyle H}{|}}{C}}-\overset{\overset{\displaystyle H}{|}}{\underset{\underset{\displaystyle H}{|}}{C}}-OH$$

$$H-\overset{\overset{\displaystyle H}{|}}{\underset{\underset{\displaystyle H}{|}}{C}}-\overset{\overset{\displaystyle H}{|}}{\underset{\underset{\displaystyle H}{|}}{C}}-\overset{\overset{\displaystyle H}{|}}{\underset{\underset{\displaystyle H}{|}}{C}}-\overset{\overset{\displaystyle H}{|}}{\underset{\underset{\displaystyle H}{|}}{C}}-\overset{\overset{\displaystyle H}{|}}{\underset{\underset{\displaystyle H}{|}}{C}}-OH \qquad H-\overset{\overset{\displaystyle H}{|}}{\underset{\underset{\displaystyle H}{|}}{C}}-\overset{\overset{\displaystyle H}{|}}{\underset{\underset{\displaystyle H}{|}}{C}}-\overset{\overset{\displaystyle H}{|}}{\underset{\underset{\displaystyle H}{|}}{C}}-\overset{\overset{\displaystyle H}{|}}{\underset{\underset{\displaystyle H}{|}}{C}}-\overset{\overset{\displaystyle H}{|}}{\underset{\underset{\displaystyle H}{|}}{C}}-Cl$$

Answer: $C_5H_{12} < C_5H_{11}Cl < C_5H_{11}OH < C_5H_{10}(OH)_2$ (in order of increasing polarity and hydrogen-bonding ability)

(a)

(b)

▲ **Figure 13.14 Effect of pressure on gas solubility.** When the pressure is increased, as in (b), the rate at which gas molecules enter the solution increases. The concentration of solute molecules at equilibrium increases in proportion to the pressure.

Pressure Effects

The solubilities of solids and liquids are not appreciably affected by pressure, whereas the solubility of a gas in any solvent is increased as the pressure over the solvent increases. We can understand the effect of pressure on the solubility of a gas by considering the dynamic equilibrium illustrated in Figure 13.14▶. Suppose that we have a gaseous substance distributed between the gas and solution phases. When equilibrium is established, the rate at which gas molecules enter the solution equals the rate at which solute molecules escape from the solution to enter the gas phase. The small arrows in Figure 13.14(a) represent the rates of these opposing processes. Now suppose that we exert added pressure on the piston and compress the gas above the solution, as shown in Figure 13.14(b). If we reduced the volume to half its original value, the pressure of the gas would increase to about twice its original value. The rate at which gas molecules strike the surface to enter the solution phase would therefore increase. As a result, the solubility of the gas in the solution would increase until equilibrium is again established; that is, solubility increases until the rate at which gas molecules enter the solution equals the rate at which solute molecules escape from the solution. Thus, *the solubility of the gas increases in direct proportion to its partial pressure above the solution*.

▲ **Figure 13.15 Solubility decreases as pressure decreases.** CO_2 bubbles out of solution when a carbonated beverage is opened, because the CO_2 partial pressure above the solution is reduced.

The relationship between pressure and the solubility of a gas is expressed by a simple equation known as **Henry's law**:

$$S_g = kP_g \qquad [13.4]$$

Here, S_g is the solubility of the gas in the solution phase (usually expressed as molarity), P_g is the partial pressure of the gas over the solution, and k is a proportionality constant known as the *Henry's law constant*. The Henry's law constant is different for each solute–solvent pair. It also varies with temperature. As an example, the solubility of N_2 gas in water at 25 °C and 0.78 atm pressure is 5.3×10^{-4} M. The Henry's law constant for N_2 in water at 25 °C is thus given by $(5.3 \times 10^{-4}$ mol/L$)/0.78$ atm $= 6.8 \times 10^{-4}$ mol/L-atm. If the partial pressure of N_2 is doubled, Henry's law predicts that the solubility in water at 25 °C will also double, to 1.36×10^{-3} M.

Bottlers use the effect of pressure on solubility in producing carbonated beverages such as beer and many soft drinks. These are bottled under a carbon dioxide pressure greater than 1 atm. When the bottles are opened to the air, the partial pressure of CO_2 above the solution decreases. Hence, the solubility of CO_2 decreases, and $CO_2(g)$ escapes from the solution as bubbles (Figure 13.15◀).

■ **SAMPLE EXERCISE 13.3** │ **A Henry's Law Calculation**

Calculate the concentration of CO_2 in a soft drink that is bottled with a partial pressure of CO_2 of 4.0 atm over the liquid at 25 °C. The Henry's law constant for CO_2 in water at this temperature is 3.1×10^{-2} mol/L-atm.

SOLUTION

Analyze: We are given the partial pressure of CO_2, P_{CO_2}, and the Henry's law constant, k, and asked to calculate the concentration of CO_2 in the solution.

Plan: With the information given, we can use Henry's law, Equation 13.4, to calculate the solubility, S_{CO_2}.

◢ *Chemistry and Life*   BLOOD GASES AND DEEP-SEA DIVING

Because the solubility of gases increases with increasing pressure, divers who breathe compressed air (Figure 13.16▶) must be concerned about the solubility of gases in their blood. Although the gases are not very soluble at sea level, their solubilities can become appreciable at deep levels where their partial pressures are greater. Thus, deep-sea divers must ascend slowly to prevent dissolved gases from being released rapidly from blood and other fluids in the body. These bubbles affect nerve impulses and give rise to the affliction known as decompression sickness, or "the bends," which is painful and potentially fatal. Nitrogen is the main problem because it has the highest partial pressure in air and because it can be removed only through the respiratory system. Oxygen, in contrast, is consumed in metabolism.

Deep-sea divers sometimes substitute helium for nitrogen in the air that they breathe, because helium has a much lower solubility in biological fluids than N_2. For example, divers working at a depth of 100 ft experience a pressure of about 4 atm. At this pressure a mixture of 95% helium and 5% oxygen will give an oxygen partial pressure of about 0.2 atm, which is the partial pressure of oxygen in normal air at 1 atm. If the oxygen partial pressure becomes too great, the urge to breathe is reduced, CO_2 is not removed from the body, and

CO_2 poisoning occurs. At excessive concentrations in the body, carbon dioxide acts as a neurotoxin, interfering with nerve conduction and transmission.
Related Exercises: 13.55, 13.56, 13.105

▲ **Figure 13.16 Solubility increases as pressure increases.** Divers who use compressed gases must be concerned about the solubility of the gases in their blood. *Doug Perrine/PacificStock.com*

Solve: $S_{CO_2} = kP_{CO_2} = (3.1 \times 10^{-2}\,\text{mol/L-atm})(4.0\,\text{atm}) = 0.12\,\text{mol/L} = 0.12\,M$

Check: The units are correct for solubility, and the answer has two significant figures consistent with both the partial pressure of CO_2 and the value of Henry's constant.

■ **PRACTICE EXERCISE**

Calculate the concentration of CO_2 in a soft drink after the bottle is opened and equilibrates at 25 °C under a CO_2 partial pressure of 3.0×10^{-4} atm.
Answer: $9.3 \times 10^{-6}\,M$

Temperature Effects

The solubility of most solid solutes in water increases as the temperature of the solution increases. Figure 13.17▼ shows this effect for several ionic substances in water. There are exceptions to this rule, however, as seen for $Ce_2(SO_4)_3$, whose solubility curve slopes downward with increasing temperature.

In contrast to solid solutes, *the solubility of gases in water decreases with increasing temperature* (Figure 13.18▼). If a glass of cold tap water is warmed, you can see bubbles of air on the inside of the glass. Similarly, carbonated beverages go flat as they are allowed to warm. As the temperature of the solution increases, the solubility of CO_2 decreases, and $CO_2(g)$ escapes from the solution.

The decreased solubility of O_2 in water as temperature increases is one of the effects of *thermal pollution* of lakes and streams. The effect is particularly serious in deep lakes because warm water is less dense than cold water. It therefore tends to remain on top of cold water, at the surface. This situation impedes the dissolving of oxygen into the deeper layers, thus stifling the respiration of all aquatic life needing oxygen. Fish may suffocate and die under these conditions.

◢ **GIVE IT SOME THOUGHT**

Why do bubbles form on the inside wall of a cooking pot when water is heated on the stove, even though the temperature is well below the boiling point of water?

▲ **Figure 13.17 Solubilities of several ionic compounds in water as a function of temperature.**

▲ **Figure 13.18 Variation of gas solubility with temperature.** Note that solubilities are in units of millimoles per liter (mmol/L), for a constant total pressure of 1 atm in the gas phase.

13.4 WAYS OF EXPRESSING CONCENTRATION

The concentration of a solution can be expressed either qualitatively or quantitatively. The terms *dilute* and *concentrated* are used to describe a solution qualitatively. A solution with a relatively small concentration of solute is said to be dilute; one with a large concentration is said to be concentrated. We use several different ways to express concentration in quantitative terms, and we examine four of these in this section: mass percentage, mole fraction, molarity, and molality.

Mass Percentage, ppm, and ppb

One of the simplest quantitative expressions of concentration is the **mass percentage** of a component in a solution, given by

$$\text{Mass \% of component} = \frac{\text{mass of component in soln}}{\text{total mass of soln}} \times 100 \qquad [13.5]$$

where we have abbreviated "solution" as "soln." Percent means per hundred. Thus, a solution of hydrochloric acid that is 36% HCl by mass contains 36 g of HCl for each 100 g of solution.

We often express the concentrations of very dilute solution in **parts per million (ppm)**, or parts per billion (ppb). These quantities are similar to mass percentage but use 10^6 (a million) or 10^9 (a billion), respectively, in place of 100 as a multiplier for the ratio of the mass of solute to the mass of solution. Thus, parts per million is defined as

$$\text{ppm of component} = \frac{\text{mass of component in soln}}{\text{total mass of soln}} \times 10^6 \qquad [13.6]$$

A solution whose solute concentration is 1 ppm contains 1 g of solute for each million (10^6) grams of solution or, equivalently, 1 mg of solute per kilogram of solution. Because the density of water is 1 g/mL, 1 kg of a dilute aqueous solution will have a volume very close to 1 L. Thus, 1 ppm also corresponds to 1 mg of solute per liter of aqueous solution.

The acceptable maximum concentrations of toxic or carcinogenic substances in the environment are often expressed in ppm or ppb. For example, the maximum allowable concentration of arsenic in drinking water in the United States is 0.010 ppm; that is, 0.010 mg of arsenic per liter of water. This concentration corresponds to 10 ppb.

▲ **GIVE IT SOME THOUGHT**

A solution of SO_2 in water contains 0.00023 g of SO_2 per liter of solution. What is the concentration of SO_2 in ppm? In ppb?

■ SAMPLE EXERCISE 13.4 | **Calculation of Mass-Related Concentrations**

(a) A solution is made by dissolving 13.5 g of glucose ($C_6H_{12}O_6$) in 0.100 kg of water. What is the mass percentage of solute in this solution? **(b)** A 2.5-g sample of groundwater was found to contain 5.4 μg of Zn^{2+}. What is the concentration of Zn^{2+} in parts per million?

SOLUTION

(a) Analyze: We are given the number of grams of solute (13.5 g) and the number of grams of solvent (0.100 kg = 100 g). From this we must calculate the mass percentage of solute.

Plan: We can calculate the mass percentage by using Equation 13.5. The mass of the solution is the sum of the mass of solute (glucose) and the mass of solvent (water).

Solve: Mass % of glucose $= \dfrac{\text{mass glucose}}{\text{mass soln}} \times 100 = \dfrac{13.5\ g}{13.5\ g\ +\ 100\ g} \times 100 = 11.9\%$

Comment: The mass percentage of water in this solution is $(100 - 11.9)\% = 88.1\%$.

(b) Analyze: In this case we are given the number of micrograms of solute. Because $1\ \mu g$ is $1 \times 10^{-6}\ g$, $5.4\ \mu g = 5.4 \times 10^{-6}\ g$.

Plan: We calculate the parts per million using Equation 13.6.

Solve: $ppm = \dfrac{\text{mass of solute}}{\text{mass of soln}} \times 10^6 = \dfrac{5.4 \times 10^{-6}\ g}{2.5\ g} \times 10^6 = 2.2\ ppm$

■ PRACTICE EXERCISE

(a) Calculate the mass percentage of NaCl in a solution containing 1.50 g of NaCl in 50.0 g of water. **(b)** A commercial bleaching solution contains 3.62 mass % sodium hypochlorite, NaOCl. What is the mass of NaOCl in a bottle containing 2.50 kg of bleaching solution?
Answers: **(a)** 2.91%, **(b)** 90.5 g of NaOCl

Mole Fraction, Molarity, and Molality

Concentration expressions are often based on the number of moles of one or more components of the solution. The three most commonly used are mole fraction, molarity, and molality.

Recall from Section 10.6 that the *mole fraction* of a component of a solution is given by

$$\text{Mole fraction of component} = \frac{\text{moles of component}}{\text{total moles of all components}} \qquad [13.7]$$

The symbol X is commonly used for mole fraction, with a subscript to indicate the component of interest. For example, the mole fraction of HCl in a hydrochloric acid solution is represented as X_{HCl}. Thus, a solution containing 1.00 mol of HCl (36.5 g) and 8.00 mol of water (144 g) has a mole fraction of HCl of $X_{HCl} = (1.00\ \text{mol})/(1.00\ \text{mol} + 8.00\ \text{mol}) = 0.111$. Mole fractions have no units because the units in the numerator and the denominator cancel. The sum of the mole fractions of all components of a solution must equal 1. Thus, in the aqueous HCl solution, $X_{H_2O} = 1.000 - 0.111 = 0.889$. Mole fractions are very useful when dealing with gases as we saw in Section 10.6 but have limited use when dealing with liquid solutions.

Recall from Section 4.5 that the *molarity* (M) of a solute in a solution is defined as

$$\text{Molarity} = \frac{\text{moles solute}}{\text{liters soln}} \qquad [13.8]$$

For example, if you dissolve 0.500 mol of Na_2CO_3 in enough water to form 0.250 L of solution, then the solution has a concentration of $(0.500\ \text{mol})/(0.250\ \text{L}) = 2.00\ M$ in Na_2CO_3. Molarity is especially useful for relating the volume of a solution to the quantity of solute it contains, as we saw in our discussions of titrations. ∞ (Section 4.6)

The **molality** of a solution, denoted m, is a unit that we have not encountered in previous chapters. This concentration unit equals the number of moles of solute per kilogram of solvent:

$$\text{Molality} = \frac{\text{moles of solute}}{\text{kilograms of solvent}} \qquad [13.9]$$

Thus, if you form a solution by mixing 0.200 mol of NaOH (8.00 g) and 0.500 kg of water (500 g), the concentration of the solution is $(0.200\ \text{mol})/(0.500\ \text{kg}) = 0.400\ m$ (that is, 0.400 molal) in NaOH.

The definitions of molarity and molality are similar enough that they can be easily confused. Molarity depends on the *volume* of *solution*, whereas molality depends on the *mass* of *solvent*. When water is the solvent, the molality and molarity of dilute solutions are numerically about the same because 1 kg of solvent is nearly the same as 1 kg of solution, and 1 kg of the solution has a volume of about 1 L.

The molality of a given solution does not vary with temperature because masses do not vary with temperature. Molarity, however, changes with temperature because the expansion or contraction of the solution changes its volume. Thus molality is often the concentration unit of choice when a solution is to be used over a range of temperatures.

GIVE IT SOME THOUGHT

If an aqueous solution is very dilute, will the molality of the solution be **(a)** greater than its molarity, **(b)** nearly the same as its molarity, or **(c)** smaller than its molarity?

▄▄ SAMPLE EXERCISE 13.5 | Calculation of Molality

A solution is made by dissolving 4.35 g glucose ($C_6H_{12}O_6$) in 25.0 mL of water at 25 °C. Calculate the molality of glucose in the solution. Water has a density of 1.00 g/mL.

SOLUTION

Analyze: We are asked to calculate a molality. To do this, we must determine the number of moles of solute (glucose) and the number of kilograms of solvent (water).

Plan: We use the molar mass of $C_6H_{12}O_6$ to convert grams to moles. We use the density of water to convert milliliters to kilograms. The molality equals the number of moles of solute divided by the number of kilograms of solvent (Equation 13.9).

Solve: Use the molar mass of glucose, 180.2 g/mL, to convert grams to moles:

$$\text{Mol } C_6H_{12}O_6 = (4.35 \text{ g } C_6H_{12}O_6)\left(\frac{1 \text{ mol } C_6H_{12}O_6}{180.2 \text{ g } C_6H_{12}O_6}\right) = 0.0241 \text{ mol } C_6H_{12}O_6$$

Because water has a density of 1.00 g/mL, the mass of the solvent is

$$(25.0 \text{ mL})(1.00 \text{ g/mL}) = 25.0 \text{ g} = 0.0250 \text{ kg}$$

Finally, use Equation 13.9 to obtain the molality:

$$\text{Molality of } C_6H_{12}O_6 = \frac{0.0241 \text{ mol } C_6H_{12}O_6}{0.0250 \text{ kg } H_2O} = 0.964 \text{ } m$$

▄▄ PRACTICE EXERCISE

What is the molality of a solution made by dissolving 36.5 g of naphthalene ($C_{10}H_8$) in 425 g of toluene (C_7H_8)?
Answer: 0.670 *m*

Conversion of Concentration Units

Sometimes the concentration of a given solution needs to be known in several different concentration units. It is possible to interconvert concentration units as shown in Sample Exercises 13.6 and 13.7.

▄▄ SAMPLE EXERCISE 13.6 | Calculation of Mole Fraction and Molality

An aqueous solution of hydrochloric acid contains 36% HCl by mass. **(a)** Calculate the mole fraction of HCl in the solution. **(b)** Calculate the molality of HCl in the solution.

SOLUTION

Analyze: We are asked to calculate the concentration of the solute, HCl, in two related concentration units, given only the percentage by mass of the solute in the solution.

Plan: In converting concentration units based on the mass or moles of solute and solvent (mass percentage, mole fraction, and molality), it is useful to assume a certain total mass of solution. Let's assume that there is exactly 100 g of solution. Because the solution is 36% HCl, it contains 36 g of HCl and $(100 - 36)$ g = 64 g of H_2O. We must convert grams of solute (HCl) to moles to calculate either mole fraction or molality. We must convert grams of solvent (H_2O) to moles to calculate mole fractions, and to kilograms to calculate molality.

Solve: (a) To calculate the mole fraction of HCl, we convert the masses of HCl and H_2O to moles and then use Equation 13.7:

$$\text{Moles HCl} = (36 \text{ g HCl})\left(\frac{1 \text{ mol HCl}}{36.5 \text{ g HCl}}\right) = 0.99 \text{ mol HCl}$$

$$\text{Moles } H_2O = (64 \text{ g } H_2O)\left(\frac{1 \text{ mol } H_2O}{18 \text{ g } H_2O}\right) = 3.6 \text{ mol } H_2O$$

$$X_{HCl} = \frac{\text{moles HCl}}{\text{moles } H_2O + \text{moles HCl}} = \frac{0.99}{3.6 + 0.99} = \frac{0.99}{4.6} = 0.22$$

(b) To calculate the molality of HCl in the solution, we use Equation 13.9. We calculated the number of moles of HCl in part (a), and the mass of solvent is 64 g = 0.064 kg:

$$\text{Molality of HCl} = \frac{0.99 \text{ mol HCl}}{0.064 \text{ kg } H_2O} = 15 \, m$$

■ PRACTICE EXERCISE

A commercial bleach solution contains 3.62 mass % NaOCl in water. Calculate **(a)** the mole fraction and **(b)** the molality of NaOCl in the solution.
Answers: **(a)** 9.00×10^{-3}, **(b)** $0.505 \, m$.

To interconvert molality and molarity, we need to know the density of the solution. Figure 13.19 ▼ outlines the calculation of the molarity and molality of a solution from the mass of solute and the mass of solvent. The mass of the solution is the sum of masses of the solvent and solute. The volume of the solution can be calculated from its mass and density.

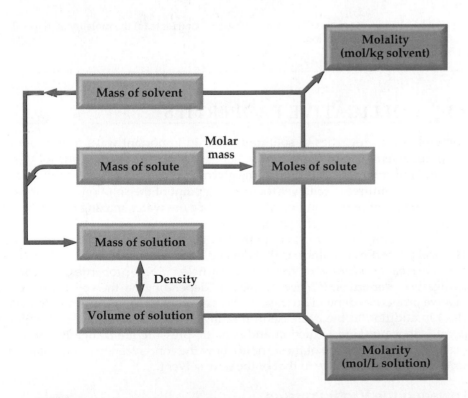

▲ **Figure 13.19 Calculating molality and molarity.** This diagram summarizes the calculation of molality and molarity from the mass of the solute, the mass of the solvent, and the density of the solution.

▇ SAMPLE EXERCISE 13.7 | Calculation of Molarity Using the Density of the Solution

A solution with a density of 0.876 g/mL contains 5.0 g of toluene (C_7H_8) and 225 g of benzene. Calculate the molarity of the solution.

SOLUTION

Analyze: Our goal is to calculate the molarity of a solution, given the masses of solute (5.0 g) and solvent (225 g) and the density of the solution (0.876 g/mL).

Plan: The molarity of a solution is the number of moles of solute divided by the number of liters of solution (Equation 13.8). The number of moles of solute (C_7H_8) is calculated from the number of grams of solute and its molar mass. The volume of the solution is obtained from the mass of the solution (mass of solute + mass of solvent = 5.0 g + 225 g = 230 g) and its density.

Solve: The number of moles of solute is

$$\text{Moles } C_7H_8 = (5.0 \text{ g } C_7H_8)\left(\frac{1 \text{ mol } C_7H_8}{92 \text{ g } C_7H_8}\right) = 0.054 \text{ mol}$$

The density of the solution is used to convert the mass of the solution to its volume:

$$\text{Milliliters soln} = (230 \text{ g})\left(\frac{1 \text{ mL}}{0.876 \text{ g}}\right) = 263 \text{ mL}$$

Molarity is moles of solute per liter of solution:

$$\text{Molarity} = \left(\frac{\text{moles } C_7H_8}{\text{liter soln}}\right) = \left(\frac{0.054 \text{ mol } C_7H_8}{263 \text{ mL soln}}\right)\left(\frac{1000 \text{ mL soln}}{1 \text{ L soln}}\right) = 0.21 \text{ M}$$

Check: The magnitude of our answer is reasonable. Rounding moles to 0.05 and liters to 0.25 gives a molarity of

$$(0.05 \text{ mol})/(0.25 \text{ L}) = 0.2 \text{ M}$$

The units for our answer (mol/L) are correct, and the answer, 0.21 M, has two significant figures, corresponding to the number of significant figures in the mass of solute (2).

Comment: Because the mass of the solvent (0.225 kg) and the volume of the solution (0.263 L) are similar in magnitude, the molarity and molality are also similar in magnitude:

$$(0.054 \text{ mol } C_7H_8)/(0.225 \text{ kg solvent}) = 0.24 \text{ m}$$

▇ PRACTICE EXERCISE

A solution containing equal masses of glycerol ($C_3H_8O_3$) and water has a density of 1.10 g/mL. Calculate **(a)** the molality of glycerol, **(b)** the mole fraction of glycerol, **(c)** the molarity of glycerol in the solution.
Answers: **(a)** 10.9 m, **(b)** $X_{C_3H_8O_3} = 0.163$, **(c)** 5.97 M

13.5 COLLIGATIVE PROPERTIES

Some physical properties of solutions differ in important ways from those of the pure solvent. For example, pure water freezes at 0 °C, but aqueous solutions freeze at lower temperatures. Ethylene glycol is added to the water in radiators of cars as an antifreeze to lower the freezing point of the solution. It also raises the boiling point of the solution above that of pure water, making it possible to operate the engine at a higher temperature.

The lowering of the freezing point and the raising of the boiling point are physical properties of solutions that depend on the *quantity* (concentration) but not the *kind* or *identity* of the solute particles. Such properties are called **colligative properties**. (*Colligative* means "depending on the collection"; colligative properties depend on the collective effect of the number of solute particles.) In addition to the decrease in freezing point and the increase in boiling point, vapor-pressure reduction and osmotic pressure are colligative properties. As we examine each of these, notice how the concentration of the solute affects the property relative to that of the pure solvent.

Lowering the Vapor Pressure

We learned in Section 11.5 that a liquid in a closed container will establish equilibrium with its vapor. When that equilibrium is reached, the pressure exerted by the vapor is called the *vapor pressure*. A substance that has no measurable vapor pressure is *nonvolatile*, whereas one that exhibits a vapor pressure is *volatile*.

When we compare the vapor pressures of various solvents with those of their solutions, we find that adding a nonvolatile solute to a solvent always lowers the vapor pressure. This effect is illustrated in Figure 13.20▶. The extent to which a nonvolatile solute lowers the vapor pressure is proportional to its concentration. This relationship is expressed by **Raoult's law**, which states that the partial pressure exerted by solvent vapor above a solution, P_A, equals the product of the mole fraction of the solvent in the solution, X_A, times the vapor pressure of the pure solvent, P_A°:

$$P_A = X_A P_A^\circ \qquad [13.10]$$

For example, the vapor pressure of water is 17.5 torr at 20 °C. Imagine holding the temperature constant while adding glucose ($C_6H_{12}O_6$) to the water so that the resulting solution has $X_{H_2O} = 0.800$ and $X_{C_6H_{12}O_6} = 0.200$. According to Equation 13.10, the vapor pressure of water over the solution will be 80.0% of that of pure water:

$$P_{H_2O} = (0.800)(17.5 \text{ torr}) = 14.0 \text{ torr}$$

In other words, the presence of the nonvolatile solute lowers the vapor pressure of the volatile solvent by 17.5 torr − 14.0 torr = 3.5 torr.

Raoult's law predicts that when we increase the mole fraction of nonvolatile solute particles in a solution, the vapor pressure over the solution will be reduced. In fact, the reduction in vapor pressure depends on the total concentration of solute particles, regardless of whether they are molecules or ions. Remember that vapor-pressure lowering is a colligative property, so it depends on the concentration of solute particles and not on their kind.

▲ **Figure 13.20 Vapor-pressure lowering.** The vapor pressure over a solution formed by a volatile solvent and a nonvolatile solute (b) is lower than that of the solvent alone (a). The extent of the decrease in the vapor pressure upon addition of the solute depends on the concentration of the solute.

◢ GIVE IT SOME THOUGHT

Adding 1 mol of NaCl to 1 kg of water reduces the vapor pressure of water more than adding 1 mol of $C_6H_{12}O_6$. Explain.

■ SAMPLE EXERCISE 13.8 | Calculation of Vapor-Pressure Lowering

Glycerin ($C_3H_8O_3$) is a nonvolatile nonelectrolyte with a density of 1.26 g/mL at 25 °C. Calculate the vapor pressure at 25 °C of a solution made by adding 50.0 mL of glycerin to 500.0 mL of water. The vapor pressure of pure water at 25 °C is 23.8 torr (Appendix B), and its density is 1.00 g/mL.

SOLUTION

Analyze: Our goal is to calculate the vapor pressure of a solution, given the volumes of solute and solvent and the density of the solute.

Plan: We can use Raoult's law (Equation 13.10) to calculate the vapor pressure of a solution. The mole fraction of the solvent in the solution, X_A, is the ratio of the number of moles of solvent (H_2O) to total moles of solution (moles $C_3H_8O_3$ + moles H_2O).

Solve: To calculate the mole fraction of water in the solution, we must determine the number of moles of $C_3H_8O_3$ and H_2O:

$$\text{Moles } C_3H_8O_3 = (50.0 \text{ mL } C_3H_8O_3)\left(\frac{1.26 \text{ g } C_3H_8O_3}{1 \text{ mL } C_3H_8O_3}\right)\left(\frac{1 \text{ mol } C_3H_8O_3}{92.1 \text{ g } C_3H_8O_3}\right) = 0.684 \text{ mol}$$

$$\text{Moles } H_2O = (500.0 \text{ mL } H_2O)\left(\frac{1.00 \text{ g } H_2O}{1 \text{ mL } H_2O}\right)\left(\frac{1 \text{ mol } H_2O}{18.0 \text{ g } H_2O}\right) = 27.8 \text{ mol}$$

$$X_{H_2O} = \frac{\text{mol } H_2O}{\text{mol } H_2O + \text{mol } C_3H_8O_3} = \frac{27.8}{27.8 + 0.684} = 0.976$$

We now use Raoult's law to calculate the vapor pressure of water for the solution:

$$P_{H_2O} = X_{H_2O}\, P_{H_2O}^\circ = (0.976)(23.8 \text{ torr}) = 23.2 \text{ torr}$$

The vapor pressure of the solution has been lowered by 0.6 torr relative to that of pure water.

■ PRACTICE EXERCISE

The vapor pressure of pure water at 110 °C is 1070 torr. A solution of ethylene glycol and water has a vapor pressure of 1.00 atm at 110 °C. Assuming that Raoult's law is obeyed, what is the mole fraction of ethylene glycol in the solution?
Answer: 0.290

An ideal gas obeys the ideal-gas equation (Section 10.4), and an **ideal solution** obeys Raoult's law. Real solutions best approximate ideal behavior when the solute concentration is low and when the solute and solvent have similar molecular sizes and similar types of intermolecular attractions.

Many solutions do not obey Raoult's law exactly: They are not ideal solutions. If the intermolecular forces between solvent and solute are weaker than those between solvent and solvent and between solute and solute, then the vapor pressure tends to be greater than predicted by Raoult's law. Conversely, when the interactions between solute and solvent are exceptionally strong, as might be the case when hydrogen bonding exists, the vapor pressure is lower than Raoult's law predicts. Although you should be aware that these departures from ideal solution occur, we will ignore them for the remainder of this chapter.

GIVE IT SOME THOUGHT

In Raoult's law, $P_A = X_A P_A^\circ$, what are the meanings of P_A and P_A°?

A Closer Look   IDEAL SOLUTIONS WITH TWO OR MORE VOLATILE COMPONENTS

Solutions sometimes have two or more volatile components. Gasoline, for example, is a complex solution containing several volatile substances. To gain some understanding of such mixtures, consider an ideal solution containing two components, A and B. The partial pressures of A and B vapors above the solution are given by Raoult's law:

$$P_A = X_A P_A^\circ \quad \text{and} \quad P_B = X_B P_B^\circ$$

The total vapor pressure over the solution is the sum of the partial pressures of each volatile component:

$$P_{total} = P_A + P_B = X_A P_A^\circ + X_B P_B^\circ$$

Consider, for example, a mixture of benzene (C_6H_6) and toluene (C_7H_8) containing 1.0 mol of benzene and 2.0 mol of toluene ($X_{ben} = 0.33$ and $X_{tol} = 0.67$). At 20 °C the vapor pressures of the pure substances are

$$\text{Benzene: } P_{ben}^\circ = 75 \text{ torr}$$
$$\text{Toluene: } P_{tol}^\circ = 22 \text{ torr}$$

Thus, the partial pressures of benzene and toluene above the solution are

$$P_{ben} = (0.33)(75 \text{ torr}) = 25 \text{ torr}$$
$$P_{tol} = (0.67)(22 \text{ torr}) = 15 \text{ torr}$$

The total vapor pressure is

$$P_{total} = 25 \text{ torr} + 15 \text{ torr} = 40 \text{ torr}$$

The vapor, therefore, is richer in benzene, the more volatile component. The mole fraction of benzene in the vapor is given by the ratio of its vapor pressure to the total pressure (Equation 10.15):

$$X_{ben} \text{ in vapor} = \frac{P_{ben}}{P_{total}} = \frac{25 \text{ torr}}{40 \text{ torr}} = 0.63$$

Although benzene constitutes only 33% of the molecules in the solution, it makes up 63% of the molecules in the vapor.

When ideal solutions are in equilibrium with their vapor, the more volatile component of the mixture will be relatively richer in the vapor. This fact forms the basis of *distillation*, a technique used to separate (or partially separate) mixtures containing volatile components. Distillation is the procedure by which a moonshiner obtains whiskey using a still and by which petrochemical plants achieve the separation of crude petroleum into gasoline, diesel fuel, lubricating oil, and so forth (Figure 13.21 ▼). Distillation is also used routinely on a small scale in the laboratory. A specially designed *fractional distillation* apparatus can achieve in a single operation a degree of separation that would be equivalent to several successive simple distillations.

Related Exercises: 13.63, 13.64

▲ **Figure 13.21 Separating volatile components.** In an industrial distillation tower, such as the ones shown here, the components of a volatile organic mixture are separated according to boiling-point range.

Boiling-Point Elevation

In Sections 11.5 and 11.6 we examined the vapor pressures of pure substances and how they can be used to construct phase diagrams. How will the phase diagram of a solution, and hence its boiling and freezing points, differ from those of the pure solvent? The addition of a nonvolatile solute lowers the vapor pressure of the solution. Thus, as shown in Figure 13.22▶, the vapor-pressure curve of the solution (blue line) will be shifted downward relative to the vapor-pressure curve of the pure liquid (black line); at any given temperature the vapor pressure of the solution is lower than that of the pure liquid. Recall that the normal boiling point of a liquid is the temperature at which its vapor pressure equals 1 atm. ∞(Section 11.5) At the normal boiling point of the pure liquid, the vapor pressure of the solution will be less than 1 atm (Figure 13.22). Therefore, a higher temperature is required to attain a vapor pressure of 1 atm. Thus, *the boiling point of the solution is higher than that of the pure liquid.*

The increase in boiling point relative to that of the pure solvent, ΔT_b, is a positive quantity obtained by subtracting the boiling point of the pure solvent from the boiling point of the solution. The value of ΔT_b is directly proportional to the concentration of the solution expressed by its molality, m:

$$\Delta T_b = K_b m \qquad [13.11]$$

The magnitude of K_b, which is called the **molal boiling-point-elevation constant**, depends only on the solvent. Some typical values for several common solvents are given in Table 13.4▼. Because solutions generally do not behave ideally, however, the constants listed in Table 13.4 serve well only for solutions that are not too concentrated.

For water, K_b is 0.51 °C/m; therefore, a 1 m aqueous solution of sucrose or any other aqueous solution that is 1 m in nonvolatile solute particles will boil 0.51 °C higher than pure water. The boiling-point elevation is proportional to the concentration of solute particles, regardless of whether the particles are molecules or ions. When NaCl dissolves in water, 2 mol of solute particles (1 mol of Na^+ and 1 mol of Cl^-) are formed for each mole of NaCl that dissolves. Therefore, a 1 m aqueous solution of NaCl is 1 m in Na^+ and 1 m in Cl^-, making it 2 m in total solute particles. As a result, the boiling-point elevation of a 1 m aqueous solution of NaCl is approximately (2 m)(0.51 °C/m) = 1 °C, twice as large as a 1 m solution of a nonelectrolyte such as sucrose. Thus, to properly predict the effect of a particular solute on the boiling point (or any other colligative property), it is important to know whether the solute is an electrolyte or a nonelectrolyte. ∞ (Sections 4.1 and 4.3)

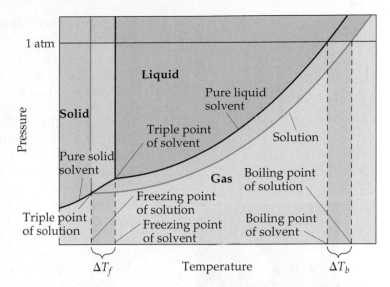

▲ **Figure 13.22 Phase diagrams for a pure solvent and for a solution of a nonvolatile solute.** The vapor pressure of the solid solvent is unaffected by the presence of solute if the solid freezes out without containing a significant concentration of solute, as is usually the case.

GIVE IT SOME THOUGHT

An unknown solute dissolved in water causes the boiling point to increase by 0.51 °C. Does this mean that the concentration of the solute is 1.0 m?

TABLE 13.4 ▪ Molal Boiling-Point-Elevation and Freezing-Point-Depression Constants

Solvent	Normal Boiling Point (°C)	K_b (°C/m)	Normal Freezing Point (°C)	K_f (°C/m)
Water, H_2O	100.0	0.51	0.0	1.86
Benzene, C_6H_6	80.1	2.53	5.5	5.12
Ethanol, C_2H_5OH	78.4	1.22	−114.6	1.99
Carbon tetrachloride, CCl_4	76.8	5.02	−22.3	29.8
Chloroform, $CHCl_3$	61.2	3.63	−63.5	4.68

Freezing-Point Depression

When a solution freezes, crystals of pure solvent usually separate out; the solute molecules are not normally soluble in the solid phase of the solvent. When aqueous solutions are partially frozen, for example, the solid that separates out is almost always pure ice. As a result, the part of the phase diagram in Figure 13.22 that represents the vapor pressure of the solid is the same as that for the pure liquid. The vapor-pressure curves for the liquid and solid phases meet at the triple point. ∞ (Section 11.6) In Figure 13.22 we see that the triple point of the solution must be at a lower temperature than that in the pure liquid because the solution has a lower vapor pressure than the pure liquid.

The freezing point of a solution is the temperature at which the first crystals of pure solvent begin to form in equilibrium with the solution. Recall from Section 11.6 that the line representing the solid-liquid equilibrium rises nearly vertically from the triple point. Because the triple-point temperature of the solution is lower than that of the pure liquid, *the freezing point of the solution is lower than that of the pure liquid*. The decrease in freezing point, ΔT_f, is a positive quantity obtained by subtracting the freezing point of the solution from the freezing point of the pure solvent.

Like the boiling-point elevation, ΔT_f is directly proportional to the molality of the solute:

$$\Delta T_f = K_f m \qquad [13.12]$$

The values of K_f, the **molal freezing-point-depression constant**, for several common solvents are given in Table 13.4. For water, K_f is 1.86 °C/m; therefore, a 1 m aqueous solution of sucrose or any other aqueous solution that is 1 m in nonvolatile solute particles (such as 0.5 m NaCl) will freeze 1.86 °C lower than pure water. The freezing-point lowering caused by solutes explains the use of antifreeze in cars and the use of calcium chloride ($CaCl_2$) to melt ice on roads during winter.

Antifreeze being added to an automobile radiatior. Antifreeze consists of an aqueous solution of ethylene glycol, $CH_2(OH)CH_2(OH)$.

■ **SAMPLE EXERCISE 13.9** | Calculation of Boiling-Point Elevation and Freezing-Point Lowering

Automotive antifreeze consists of ethylene glycol, $CH_2(OH)CH_2(OH)$, a nonvolatile nonelectrolyte. Calculate the boiling point and freezing point of a 25.0 mass % solution of ethylene glycol in water.

SOLUTION

Analyze: We are given that a solution contains 25.0 mass % of a nonvolatile, nonelectrolyte solute and asked to calculate the boiling and freezing points of the solution. To do this, we need to calculate the boiling-point elevation and freezing-point depression.

Plan: To calculate the boiling-point elevation and the freezing-point depression using Equations 13.11 and 12, we must express the concentration of the solution as molality. Let's assume for convenience that we have 1000 g of solution. Because the solution is 25.0 mass % ethylene glycol, the masses of ethylene glycol and water in the solution are 250 and 750 g, respectively. Using these quantities, we can calculate the molality of the solution, which we use with the molal boiling-point-elevation and freezing-point-depression constants (Table 13.4) to calculate ΔT_b and ΔT_f. We add ΔT_b to the boiling point and subtract ΔT_f from the freezing point of the solvent to obtain the boiling point and freezing point of the solution.

Solve: The molality of the solution is calculated as follows:

$$\text{Molality} = \frac{\text{moles } C_2H_6O_2}{\text{kilograms } H_2O} = \left(\frac{250 \text{ g } C_2H_6O_2}{750 \text{ g } H_2O}\right)\left(\frac{1 \text{ mol } C_2H_6O_2}{62.1 \text{ g } C_2H_6O_2}\right)\left(\frac{1000 \text{ g } H_2O}{1 \text{ kg } H_2O}\right)$$

$$= 5.37 \text{ } m$$

We can now use Equations 13.11 and 13.12 to calculate the changes in the boiling and freezing points:

$$\Delta T_b = K_b m = (0.51 \text{ °C}/m)(5.37 \text{ } m) = 2.7 \text{ °C}$$

$$\Delta T_f = K_f m = (1.86 \text{ °C}/m)(5.37 \text{ } m) = 10.0 \text{ °C}$$

Hence, the boiling and freezing points of the solution are

$$\text{Boiling point} = (\text{normal bp of solvent}) + \Delta T_b$$
$$= 100.0\,°C + 2.7\,°C = 102.7\,°C$$

$$\text{Freezing point} = (\text{normal fp of solvent}) - \Delta T_f$$
$$= 0.0\,°C - 10.0\,°C = -10.0\,°C$$

Comment: Notice that the solution is a liquid over a larger temperature range than the pure solvent.

■ PRACTICE EXERCISE
Calculate the freezing point of a solution containing 0.600 kg of $CHCl_3$ and 42.0 g of eucalyptol ($C_{10}H_{18}O$), a fragrant substance found in the leaves of eucalyptus trees. (See Table 13.4.)
Answer: $-65.6\,°C$

■ SAMPLE EXERCISE 13.10 | **Freezing-Point Depression in Aqueous Solutions**

List the following aqueous solutions in order of their expected freezing point: 0.050 *m* $CaCl_2$, 0.15 *m* NaCl, 0.10 *m* HCl, 0.050 *m* CH_3COOH, 0.10 *m* $C_{12}H_{22}O_{11}$.

SOLUTION
Analyze: We must order five aqueous solutions according to expected freezing points, based on molalities and the solute formulas.

Plan: The lowest freezing point will correspond to the solution with the greatest concentration of solute particles. To determine the total concentration of solute particles in each case, we must determine whether the substance is a nonelectrolyte or an electrolyte and consider the number of ions formed when an electrolyte ionizes.

Solve: $CaCl_2$, NaCl, and HCl are strong electrolytes, CH_3COOH (acetic acid) is a weak electrolyte, and $C_{12}H_{22}O_{11}$ is a nonelectrolyte. The molality of each solution in total particles is as follows:

$$0.050\ m\ CaCl_2 \Rightarrow 0.050\ m\ \text{in}\ Ca^{2+}\ \text{and}\ 0.10\ m\ \text{in}\ Cl^- \Rightarrow 0.15\ m\ \text{in particles}$$

$$0.15\ m\ NaCl \Rightarrow 0.15\ m\ \text{in}\ Na^+\ \text{and}\ 0.15\ m\ \text{in}\ Cl^- \Rightarrow 0.30\ m\ \text{in particles}$$

$$0.10\ m\ HCl \Rightarrow 0.10\ m\ \text{in}\ H^+\ \text{and}\ 0.10\ m\ \text{in}\ Cl^- \Rightarrow 0.20\ m\ \text{in particles}$$

$$0.050\ m\ CH_3COOH \Rightarrow \text{weak electrolyte} \Rightarrow \text{between } 0.050\ m\ \text{and } 0.10\ m\ \text{in particles}$$

$$0.10\ m\ C_{12}H_{22}O_{11} \Rightarrow \text{nonelectrolyte} \Rightarrow 0.10\ m\ \text{in particles}$$

Because the freezing points depend on the total molality of particles in solution, the expected ordering is 0.15 *m* NaCl (lowest freezing point), 0.10 *m* HCl, 0.050 *m* $CaCl_2$, 0.10 *m* $C_{12}H_{22}O_{11}$, and 0.050 *m* CH_3COOH (highest freezing point).

■ PRACTICE EXERCISE
Which of the following solutes will produce the largest increase in boiling point upon addition to 1 kg of water: 1 mol of $Co(NO_3)_2$, 2 mol of KCl, 3 mol of ethylene glycol ($C_2H_6O_2$)?
Answer: 2 mol of KCl because it contains the highest concentration of particles, 2 *m* K^+ and 2 *m* Cl^-, giving 4 *m* in all

---

Osmosis

Certain materials, including many membranes in biological systems and synthetic substances such as cellophane, are *semipermeable*. When in contact with a solution, they allow some molecules to pass through their network of tiny pores but not others. Most importantly, semipermeable membranes generally allow small solvent molecules such as water to pass through but block larger solute molecules or ions. This selectivity gives rise to some interesting and important applications.

Consider a situation in which only solvent molecules are able to pass through a membrane. If such a membrane is placed between two solutions of different concentration, solvent molecules move in both directions through the membrane. The concentration of *solvent* is higher in the solution containing less solute, however, so the rate with which solvent passes from the less concentrated (lower solute concentration) to the more concentrated solution (higher solute concentration) is greater than the rate in the opposite direction. Thus, there is a net movement of solvent molecules from the less concentrated solution into the more concentrated one. In this process, called **osmosis**, *the net movement of solvent is always toward the solution with the higher solute concentration.*

Osmosis is illustrated in Figure 13.23▼. Let's begin with two solutions of different concentration separated by a semipermeable membrane. Because the solution on the left is more concentrated than the one on the right, there is a net movement of *solvent* through the membrane from right to left, as if the solutions

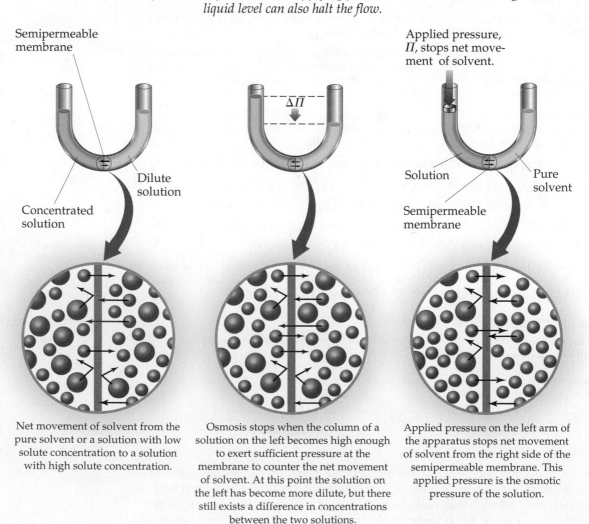

OSMOSIS

In osmosis, the net movement of solvent is always toward the solution with the higher solute concentration. There is a net movement of solvent through the semipermeable membrane, as if the solutions were driven to attain equal concentrations. The difference in liquid level, and thus pressure, eventually causes the flow to cease. Applying pressure to the arm with the higher liquid level can also halt the flow.

Semipermeable membrane

Concentrated solution

Dilute solution

$\Delta\Pi$

Applied pressure, Π, stops net movement of solvent.

Solution

Pure solvent

Semipermeable membrane

Net movement of solvent from the pure solvent or a solution with low solute concentration to a solution with high solute concentration.

Osmosis stops when the column of a solution on the left becomes high enough to exert sufficient pressure at the membrane to counter the net movement of solvent. At this point the solution on the left has become more dilute, but there still exists a difference in concentrations between the two solutions.

Applied pressure on the left arm of the apparatus stops net movement of solvent from the right side of the semipermeable membrane. This applied pressure is the osmotic pressure of the solution.

▲ **Figure 13.23 Osmosis.**

were driven to attain equal concentrations. As a result, the liquid levels in the two arms become unequal. Eventually, the pressure difference resulting from the unequal heights of the liquid in the two arms becomes so large that the net flow of solvent ceases, as shown in the center panel. Alternatively, we may apply pressure to the left arm of the apparatus, as shown in the panel on the right, to halt the net flow of solvent. The pressure required to prevent osmosis by pure solvent is the **osmotic pressure**, Π of the solution. The osmotic pressure obeys a law similar in form to the ideal-gas law, $\Pi V = nRT$ where V is the volume of the solution, n is the number of moles of solute, R is the ideal-gas constant, and T is the temperature on the Kelvin scale. From this equation, we can write

$$\Pi = \left(\frac{n}{V}\right)RT = MRT \qquad [13.13]$$

where M is the molarity of the solution.

If two solutions of identical osmotic pressure are separated by a semipermeable membrane, no osmosis will occur. The two solutions are *isotonic*. If one solution is of lower osmotic pressure, it is *hypotonic* with respect to the more concentrated solution. The more concentrated solution is *hypertonic* with respect to the dilute solution.

GIVE IT SOME THOUGHT

Of two KBr solutions, one 0.5 m and the other 0.20 m, which is hypotonic with respect to the other?

Osmosis plays a very important role in living systems. The membranes of red blood cells, for example, are semipermeable. Placing a red blood cell in a solution that is *hyper*tonic relative to the intracellular solution (the solution within the cells) causes water to move out of the cell, as shown in Figure 13.24(a)▶. This causes the cell to shrivel, a process called *crenation*. Placing the cell in a solution that is *hypotonic* relative to the intracellular fluid causes water to move into the cell, as illustrated in Figure 13.24(b). This may cause the cell to rupture, a process called *hemolysis*.

(a) Crenation

(b) Hemolysis

▲ Figure 13.24 **Osmosis through red blood cell wall.** The blue arrows represent the net movement of water molecules.

People who need body fluids or nutrients replaced but cannot be fed orally are given solutions by intravenous (IV) infusion, which feeds nutrients directly into the veins. To prevent crenation or hemolysis of red blood cells, the IV solutions must be isotonic with the intracellular fluids of the cells.

■ SAMPLE EXERCISE 13.11 | Calculations Involving Osmotic Pressure

The average osmotic pressure of blood is 7.7 atm at 25 °C. What molarity of glucose ($C_6H_{12}O_6$) will be isotonic with blood?

SOLUTION

Analyze: We are asked to calculate the concentration of glucose in water that would be isotonic with blood, given that the osmotic pressure of blood at 25 °C is 7.7 atm.

Plan: Because we are given the osmotic pressure and temperature, we can solve for the concentration, using Equation 13.13.

Solve:

$$\Pi = MRT$$

$$M = \frac{\Pi}{RT} = \frac{7.7 \text{ atm}}{\left(0.0821 \dfrac{\text{L-atm}}{\text{mol-K}}\right)(298 \text{ K})} = 0.31 \ M$$

Comment: In clinical situations the concentrations of solutions are generally expressed as mass percentages. The mass percentage of a 0.31 M solution of glucose is 5.3%. The concentration of NaCl that is isotonic with blood is 0.16 M, because NaCl ionizes to form two particles, Na^+ and Cl^- (a 0.155 M solution of NaCl is 0.310 M in particles). A 0.16 M solution of NaCl is 0.9 mass % in NaCl. This kind of solution is known as a physiological saline solution.

■ PRACTICE EXERCISE

What is the osmotic pressure at 20 °C of a 0.0020 M sucrose ($C_{12}H_{22}O_{11}$) solution?
Answer: 0.048 atm, or 37 torr

A Closer Look   COLLIGATIVE PROPERTIES OF ELECTROLYTE SOLUTIONS

The colligative properties of solutions depend on the total concentration of solute particles, regardless of whether the particles are ions or molecules. Thus, we would expect a 0.100 m solution of NaCl to have a freezing-point depression of (0.200 m)(1.86 °C/m) = 0.372 °C because it is 0.100 m in Na^+ (aq) and 0.100 m in Cl^- (aq). The measured freezing-point depression is only 0.348 °C, however, and the situation is similar for other strong electrolytes. A 0.100 m solution of KCl, for example, freezes at −0.344 °C.

The difference between the expected and observed colligative properties for strong electrolytes is due to electrostatic attractions between ions. As the ions move about in solution, ions of opposite charge collide and "stick together" for brief moments. While they are together, they behave as a single particle called an *ion pair* (Figure 13.25▼). The number of independent particles is thereby reduced, causing a reduction in the freezing-point depression (as well as in the boiling-point elevation, the vapor-pressure reduction, and the osmotic pressure).

One measure of the extent to which electrolytes dissociate is the *van't Hoff factor, i*. This factor is the ratio of the actual value of a colligative property to the value calculated, assuming the substance to be a nonelectrolyte. Using the freezing-point depression, for example, we have

$$i = \frac{\Delta T_f \text{ (measured)}}{\Delta T_f \text{ (calculated for nonelectrolyte)}} \quad [13.14]$$

The ideal value of i can be determined for a salt from the number of ions per formula unit. For NaCl, for example, the ideal van't Hoff factor is 2 because NaCl consists of one Na^+ and one Cl^- per formula unit; for K_2SO_4 it is 3 because K_2SO_4 consists of two K^+ and one SO_4^{2-}. In the absence of any information about the actual value of i for a solution, we will use the ideal value in calculations.

Table 13.5▼ gives the observed van't Hoff factors for several substances at different dilutions. Two trends are evident in these data. First, dilution affects the value of i for electrolytes; the more dilute the solution, the more closely i approaches the ideal or limiting value. Thus, the extent of ion pairing in electrolyte solutions decreases upon dilution. Second, the lower the charges on the ions, the less i departs from the limiting value because the extent of ion pairing decreases as the ionic charges decrease. Both trends are consistent with simple electrostatics: The force of interaction between charged particles decreases as their separation increases and as their charges decrease.
Related Exercises: 13.79, 13.80, 13.99, 13.102

▲ **Figure 13.25 Ion pairing and colligative properties.** A solution of NaCl contains not only separated Na^+(aq) and Cl^-(aq) ions but ion pairs as well. Ion pairing becomes more prevalent as the solution concentration increases and has an effect on all the colligative properties of the solution.

TABLE 13.5 ■ van't Hoff Factors for Several Substances at 25 °C

Compound	Concentration			Limiting Value
	0.100 m	0.0100 m	0.00100 m	
Sucrose	1.00	1.00	1.00	1.00
NaCl	1.87	1.94	1.97	2.00
K_2SO_4	2.32	2.70	2.84	3.00
$MgSO_4$	1.21	1.53	1.82	2.00

There are many interesting biological examples of osmosis. A cucumber placed in concentrated brine loses water via osmosis and shrivels into a pickle. If a carrot that has become limp because of water loss to the atmosphere is placed in water, the water moves into the carrot through osmosis, making it firm once again. People who eat a lot of salty food retain water in tissue cells and intercellular space because of osmosis. The resultant swelling or puffiness is called *edema*. Water moves from soil into plant roots and subsequently into the upper portions of the plant at least in part because of osmosis. Bacteria on salted meat or candied fruit lose water through osmosis, shrivel, and die—thus preserving the food.

The movement of a substance from an area where its concentration is high to an area where it is low is spontaneous. Biological cells transport water and other select materials through their membranes, permitting nutrients to enter and waste materials to exit. In some cases substances must be moved across the cell membrane from an area of low concentration to one of high concentration. This movement—called *active transport*—is not spontaneous, so cells must expend energy to do it.

GIVE IT SOME THOUGHT

Is the osmotic pressure of a 0.10 M solution of NaCl greater than, less than, or equal to that of a 0.10 M solution of KBr?

Determination of Molar Mass

The colligative properties of solutions provide a useful means of experimentally determining molar mass. Any of the four colligative properties can be used, as shown in Sample Exercises 13.12 and 13.13.

■ SAMPLE EXERCISE 13.12 | Molar Mass from Freezing-Point Depression

A solution of an unknown nonvolatile nonelectrolyte was prepared by dissolving 0.250 g of the substance in 40.0 g of CCl_4. The boiling point of the resultant solution was 0.357 °C higher than that of the pure solvent. Calculate the molar mass of the solute.

SOLUTION

Analyze: Our goal is to calculate the molar mass of a solute based on knowledge of the boiling-point elevation of its solution in CCl_4, ΔT_b = 0.357 °C, and the masses of solute and solvent. Table 13.4 gives K_b for the solvent (CCl_4), K_b = 5.02 °C/m.

Plan: We can use Equation 13.11, $\Delta T_b = K_b m$, to calculate the molality of the solution. Then we can use molality and the quantity of solvent (40.0 g CCl_4) to calculate the number of moles of solute. Finally, the molar mass of the solute equals the number of grams per mole, so we divide the number of grams of solute (0.250 g) by the number of moles we have just calculated.

Solve: From Equation 13.11 we have

$$\text{Molality} = \frac{\Delta T_b}{K_b} = \frac{0.357 \text{ °C}}{5.02 \text{ °C}/m} = 0.0711 \ m$$

Thus, the solution contains 0.0711 mol of solute per kilogram of solvent. The solution was prepared using 40.0 g = 0.0400 kg of solvent (CCl_4). The number of moles of solute in the solution is therefore

$$(0.0400 \text{ kg } CCl_4)\left(0.0711 \ \frac{\text{mol solute}}{\text{kg } CCl_4}\right) = 2.84 \times 10^{-3} \text{ mol solute}$$

The molar mass of the solute is the number of grams per mole of the substance:

$$\text{Molar mass} = \frac{0.250 \text{ g}}{2.84 \times 10^{-3} \text{ mol}} = 88.0 \text{ g/mol}$$

■ PRACTICE EXERCISE

Camphor ($C_{10}H_{16}O$) melts at 179.8 °C, and it has a particularly large freezing-point-depression constant, K_f = 40.0 °C/m. When 0.186 g of an organic substance of unknown molar mass is dissolved in 22.01 g of liquid camphor, the freezing point of the mixture is found to be 176.7 °C. What is the molar mass of the solute?
Answer: 110 g/mol

■ SAMPLE EXERCISE 13.13 | Molar Mass from Osmotic Pressure

The osmotic pressure of an aqueous solution of a certain protein was measured to determine the protein's molar mass. The solution contained 3.50 mg of protein dissolved in sufficient water to form 5.00 mL of solution. The osmotic pressure of the solution at 25 °C was found to be 1.54 torr. Treating the protein as a nonelectrolyte, calculate its molar mass.

SOLUTION

Analyze: Our goal is to calculate the molar mass of a high-molecular-mass protein, based on its osmotic pressure and a knowledge of the mass of protein and solution volume.

Plan: The temperature ($T = 25$ °C) and osmotic pressure ($\Pi = 1.54$ torr) are given, and we know the value of R so we can use Equation 13.13 to calculate the molarity of the solution, M. In doing so, we must convert temperature from °C to K and the osmotic pressure from torr to atm. We then use the molarity and the volume of the solution (5.00 mL) to determine the number of moles of solute. Finally, we obtain the molar mass by dividing the mass of the solute (3.50 mg) by the number of moles of solute.

Solve: Solving Equation 13.13 for molarity gives

$$\text{Molarity} = \frac{\Pi}{RT} = \frac{(1.54\ \text{torr})\left(\dfrac{1\ \text{atm}}{760\ \text{torr}}\right)}{\left(0.0821\ \dfrac{\text{L-atm}}{\text{mol-K}}\right)(298\ \text{K})} = 8.28 \times 10^{-5}\ \frac{\text{mol}}{\text{L}}$$

Because the volume of the solution is 5.00 ml = 5.00×10^{-3} L, the number of moles of protein must be

$$\text{Moles} = (8.28 \times 10^{-5}\ \text{mol/L})(5.00 \times 10^{-3}\ \text{L}) = 4.14 \times 10^{-7}\ \text{mol}$$

The molar mass is the number of grams per mole of the substance. The sample has a mass of 3.50 mg = 3.50×10^{-3} g. The molar mass is the number of grams divided by the number of moles:

$$\text{Molar mass} = \frac{\text{grams}}{\text{moles}} = \frac{3.50 \times 10^{-3}\ \text{g}}{4.14 \times 10^{-7}\ \text{mol}} = 8.45 \times 10^{3}\ \text{g/mol}$$

Comment: Because small pressures can be measured easily and accurately, osmotic pressure measurements provide a useful way to determine the molar masses of large molecules.

■ PRACTICE EXERCISE

A sample of 2.05 g of polystyrene of uniform polymer chain length was dissolved in enough toluene to form 0.100 L of solution. The osmotic pressure of this solution was found to be 1.21 kPa at 25 °C. Calculate the molar mass of the polystyrene.
Answer: 4.20×10^{4} g/mol

13.6 COLLOIDS

When finely divided clay particles are dispersed throughout water, they eventually settle out of the water because of gravity. The dispersed clay particles are much larger than most molecules and consist of many thousands or even millions of atoms. In contrast, the dispersed particles of a solution are of molecular size. Between these extremes lie dispersed particles that are larger than typical molecules, but not so large that the components of the mixture separate under the influence of gravity. These intermediate types of dispersions or suspensions are called **colloidal dispersions**, or simply **colloids**. Colloids form the dividing line between solutions and heterogeneous mixtures. Like solutions, colloids can be gases, liquids, or solids. Examples of each are listed in Table 13.6▶.

TABLE 13.6 ■ Types of Colloids

Phase of Colloid	Dispersing (solutelike) Substance	Dispersed (solventlike) Substance	Colloid Type	Example
Gas	Gas	Gas	—	None (all are solutions)
Gas	Gas	Liquid	Aerosol	Fog
Gas	Gas	Solid	Aerosol	Smoke
Liquid	Liquid	Gas	Foam	Whipped cream
Liquid	Liquid	Liquid	Emulsion	Milk
Liquid	Liquid	Solid	Sol	Paint
Solid	Solid	Gas	Solid foam	Marshmallow
Solid	Solid	Liquid	Solid emulsion	Butter
Solid	Solid	Solid	Solid sol	Ruby glass

The size of the dispersed particles is used to classify a mixture as a colloid. Colloid particles range in diameter from approximately 5 to 1000 nm. Solute particles are smaller. The colloid particle may consist of many atoms, ions, or molecules, or it may even be a single giant molecule. The hemoglobin molecule, for example, which carries oxygen in blood, has molecular dimensions of 65 Å × 55 Å × 50 Å and a molar mass of 64,500 g/mol.

Although colloid particles may be so small that the dispersion appears uniform even under a microscope, they are large enough to scatter light very effectively. Consequently, most colloids appear cloudy or opaque unless they are very dilute. (Homogenized milk is a colloid.) Furthermore, because they scatter light, a light beam can be seen as it passes through a colloidal suspension, as shown in Figure 13.26▶. This scattering of light by colloidal particles, known as the **Tyndall effect**, makes it possible to see the light beam of an automobile on a dusty dirt road or the sunlight coming through a forest canopy [Figure 13.27(a)▼]. Not all wavelengths are scattered to the same extent. As a result, brilliant red sunsets are seen when the sun is near the horizon and the air contains dust, smoke, or other particles of colloidal size [Figure 13.27(b)].

▲ **Figure 13.26 Tyndall effect in the laboratory.** The glass on the left contains a colloidal suspension; that on the right contains a solution. The path of the beam through the colloidal suspension is visible because the light is scattered by the colloidal particles. Light is not scattered by the individual solute molecules in the solution.

Hydrophilic and Hydrophobic Colloids

The most important colloids are those in which the dispersing medium is water. These colloids may be **hydrophilic** (water loving) or **hydrophobic** (water fearing). Hydrophilic colloids are most like the solutions that we have previously examined. In the human body the extremely large molecules that make up such important substances as enzymes and antibodies are kept in suspension by interaction with surrounding water molecules. The molecules fold in such a way that the hydrophobic groups are away from the water molecules,

◀ **Figure 13.27 Tyndall effect in nature.** (a) Scattering of sunlight by colloidal particles in the misty air of a forest. (b) The scattering of light by smoke or dust particles produces a rich red sunset.

▲ **Figure 13.28 Hydrophilic colloids.** Examples of hydrophilic groups on the surface of a giant molecule (macromolecule) that help to keep the molecule suspended in water.

on the "inside" of the folded molecule, while the hydrophilic, polar groups are on the surface, interacting with the water molecules. These hydrophilic groups generally contain oxygen or nitrogen and often carry a charge. Some examples are shown in Figure 13.28 ◄.

Hydrophobic colloids can be prepared in water only if they are stabilized in some way. Otherwise, their natural lack of affinity for water causes them to separate from the water. Hydrophobic colloids can be stabilized by adsorption of ions on their surface, as shown in Figure 13.29 ▼. (*Adsorption* means to adhere to a surface. It differs from *absorption*, which means to pass into the interior, as when a sponge absorbs water.) These adsorbed ions can interact with water, thereby stabilizing the colloid. At the same time, the mutual repulsion between colloid particles with adsorbed ions of the same charge keeps the particles from colliding and getting larger.

Hydrophobic colloids can also be stabilized by the presence of hydrophilic groups on their surfaces. Small droplets of oil are hydrophobic, for example, so they do not remain suspended in water. Instead, they aggregate, forming an oil slick on the surface of the water. Sodium stearate (Figure 13.30 ◄), or any similar substance having one end that is hydrophilic (polar, or charged) and one that is hydrophobic (nonpolar), will stabilize a suspension of oil in water. Stabilization results from the interaction of the hydrophobic ends of the stearate ions with the oil droplet and the hydrophilic ends with the water as shown in Figure 13.31 ►.

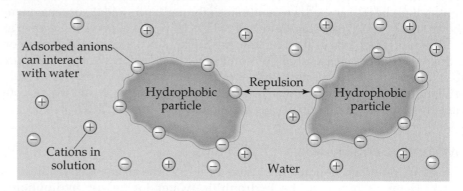

▲ **Figure 13.29 Hydrophobic colloids.** Schematic illustration of how adsorbed ions stabilize a hydrophobic colloid in water.

Sodium stearate

▲ **Figure 13.30 Sodium stearate.**

GIVE IT SOME THOUGHT

Why don't the oil droplets emulsified by sodium stearate coagulate to form larger oil droplets?

The stabilization of colloids has an interesting application in our own digestive system. When fats in our diet reach the small intestine, a hormone causes the gallbladder to excrete a fluid called bile. Among the components of bile are compounds that have chemical structures similar to sodium stearate; that is, they have a hydrophilic (polar) end and a hydrophobic (nonpolar) end. These compounds emulsify the fats present in the intestine and thus permit digestion and absorption of fat-soluble vitamins through the intestinal wall. The term *emulsify* means "to form an emulsion," a suspension of one liquid in another, as in milk, for example (Table 13.6). A substance that aids in the formation of an emulsion is called an emulsifying agent. If you read the labels on foods and other materials, you will find that a variety of chemicals are used as emulsifying agents. These chemicals typically have a hydrophilic end and a hydrophobic end.

◀ **Figure 13.31 Stabilization of an emulsion of oil in water by stearate ions.**

$$CH_3(CH_2)_{16}\overset{\displaystyle O}{\overset{\displaystyle \|}{C}}O^-$$

Hydrophobic   Hydrophilic
end   end

Na⁺ Stearate ion Oil droplet Water

Chemistry and Life   SICKLE-CELL ANEMIA

Our blood contains a complex protein called *hemoglobin* that carries oxygen from our lungs to other parts of our body. In the genetic disease known as *sickle-cell anemia*, hemoglobin molecules are abnormal and have a lower solubility, especially in their unoxygenated form. Consequently, as much as 85% of the hemoglobin in red blood cells crystallizes from solution.

The reason for the insolubility of hemoglobin in sickle-cell anemia can be traced to a structural change in one part of an amino acid side chain. Normal hemoglobin molecules have an amino acid in their makeup that has the following side chain protruding from the main body of the molecule:

$$-CH_2-CH_2-\overset{\displaystyle O}{\overset{\displaystyle \|}{C}}-OH$$

Normal   Normal

This side chain terminates in a polar group, which contributes to the solubility of the hemoglobin molecule in water. In the hemoglobin molecules of persons suffering from sickle-cell anemia, the side chain is of a different type:

$$-\overset{\displaystyle |}{\underset{\displaystyle CH_3}{CH}}-CH_3$$

Abnormal   Abnormal

This abnormal group of atoms is nonpolar (hydrophobic), and its presence leads to the aggregation of this defective form of hemoglobin into particles too large to remain suspended in biological fluids. It also causes the cells to distort into a sickle shape, as shown in Figure 13.32▼. The sickled cells tend to clog the capillaries, causing severe pain, physical weakness, and the gradual deterioration of the vital organs. The disease is hereditary, and if both parents carry the defective genes, it is likely that their children will possess only abnormal hemoglobin.

◀ **Figure 13.32 Normal and sickled red blood cells.** Normal red blood cells are about 1×10^{-3} mm in diameter.

Removal of Colloidal Particles

Colloidal particles frequently must be removed from a dispersing medium, as in the removal of smoke from smokestacks or butterfat from milk. Because colloidal particles are so small, they cannot be separated by simple filtration. Instead, the colloidal particles must be enlarged in a process called *coagulation*. The resultant larger particles can then be separated by filtration or merely by allowing them to settle out of the dispersing medium.

Heating the mixture or adding an electrolyte may bring about coagulation. Heating the colloidal dispersion increases the particle motion and so the number of collisions. The particles increase in size as they stick together after colliding. The addition of electrolytes neutralizes the surface charges of the particles, thereby removing the electrostatic repulsions that prevent them from coming together. Wherever rivers empty into oceans or other salty bodies of water, for example, the suspended clay in the river is deposited as a delta when it mixes with the electrolytes in the salt water.

Semipermeable membranes can also be used to separate ions from colloidal particles because the ions can pass through the membrane but the colloidal particles cannot. This type of separation is known as *dialysis* and is used to purify blood in artificial kidney machines. Our kidneys normally remove the waste products of metabolism from blood. In a kidney machine, blood is circulated through a dialyzing tube immersed in a washing solution. The washing solution contains the same concentrations and kinds of ions as blood but is lacking the molecules and ions that are waste products. Wastes therefore dialyze out of the blood, but the ions and large colloidal particles such as proteins do not.

■ SAMPLE INTEGRATIVE EXERCISE | Putting Concepts Together

A 0.100-L solution is made by dissolving 0.441 g of $CaCl_2(s)$ in water. **(a)** Calculate the osmotic pressure of this solution at 27 °C, assuming that it is completely dissociated into its component ions. **(b)** The measured osmotic pressure of this solution is 2.56 atm at 27 °C. Explain why it is less than the value calculated in (a), and calculate the van't Hoff factor, i, for the solute in this solution. (See the "A Closer Look" box on Colligative Properties of Electrolyte Solutions in Section 13.5.) **(c)** The enthalpy of solution for $CaCl_2$ is $\Delta H = -81.3$ kJ/mol. If the final temperature of the solution is 27.0 °C, what was its initial temperature? (Assume that the density of the solution is 1.00 g/mL, that its specific heat is 4.18 J/g-K, and that the solution loses no heat to its surroundings.)

SOLUTION

(a) The osmotic pressure is given by Equation 13.13, $\Pi = MRT$. We know the temperature, $T = 27\,°C = 300$ K, and the gas constant, $R = 0.0821$ L-atm/mol-K. We can calculate the molarity of the solution from the mass of $CaCl_2$ and the volume of the solution:

$$\text{Molarity} = \left(\frac{0.441 \text{ g } CaCl_2}{0.100 \text{ L}}\right)\left(\frac{1 \text{ mol } CaCl_2}{111.0 \text{ g } CaCl_2}\right) = 0.0397 \text{ mol } CaCl_2/L$$

Soluble ionic compounds are strong electrolytes. ∞ (Sections 4.1 and 4.3) Thus, $CaCl_2$ consists of metal cations (Ca^{2+}) and nonmetal anions (Cl^-). When completely dissociated, each $CaCl_2$ unit forms three ions (one Ca^{2+} and two Cl^-). Hence the total concentration of ions in the solution is $(3)(0.0397\ M) = 0.119\ M$, and the calculated osmotic pressure is

$$\Pi = MRT = (0.119 \text{ mol/L})(0.0821 \text{ L-atm/mol-K})(300 \text{ K}) = 2.93 \text{ atm}$$

(b) The actual values of colligative properties of electrolytes are less than those calculated, because the electrostatic interactions between ions limit their independent movements. In this case the van't Hoff factor, which measures the extent to which electrolytes actually dissociate into ions, is given by

$$i = \frac{\Pi(\text{measured})}{\Pi(\text{calculated for nonelectrolyte})}$$

$$= \frac{2.56 \text{ atm}}{(0.0397 \text{ mol/L})(0.0821 \text{ L-atm/mol-K})(300 \text{ K})} = 2.62$$

Thus, the solution behaves as if the $CaCl_2$ has dissociated into 2.62 particles instead of the ideal 3.

(c) If the solution is 0.0397 M in $CaCl_2$ and has a total volume of 0.100 L, the number of moles of solute is $(0.100 \text{ L})(0.0397 \text{ mol/L}) = 0.00397 \text{ mol}$. Hence the quantity of heat generated in forming the solution is $(0.00397 \text{ mol})(-81.3 \text{ kJ/mol}) = -0.323 \text{ kJ}$. The solution absorbs this heat, causing its temperature to increase. The relationship between temperature change and heat is given by Equation 5.22:

$$q = (\text{specific heat})(\text{grams})(\Delta T)$$

The heat absorbed by the solution is $q = +0.323 \text{ kJ} = 323 \text{ J}$. The mass of the 0.100 L of solution is $(100 \text{ mL})(1.00 \text{ g/mL}) = 100 \text{ g}$ (to 3 significant figures). Thus the temperature change is

$$\Delta T = \frac{q}{(\text{specific heat of solution})(\text{grams of solution})}$$

$$= \frac{323 \text{ J}}{(4.18 \text{ J/g-K})(100 \text{ g})} = 0.773 \text{ K}$$

A kelvin has the same size as a degree Celsius. ∞ (Section 1.4) Because the solution temperature increases by 0.773 °C, the initial temperature was 27.0 °C − 0.773 °C = 26.2 °C.

CHAPTER REVIEW

SUMMARY AND KEY TERMS

Section 13.1 Solutions form when one substance disperses uniformly throughout another. The attractive interaction of solvent molecules with solute is called **solvation**. When the solvent is water, the interaction is called **hydration**. The dissolution of ionic substances in water is promoted by hydration of the separated ions by the polar water molecules. The overall enthalpy change upon solution formation may be either positive or negative. Solution formation is favored both by a negative enthalpy change (exothermic process) and by an increased dispersal in space of the components of the solution, corresponding to a positive **entropy** change.

Section 13.2 The equilibrium between a saturated solution and undissolved solute is dynamic; the process of solution and the reverse process, **crystallization**, occur simultaneously. In a solution in equilibrium with undissolved solute, the two processes occur at equal rates, giving a **saturated** solution. If there is less solute present than is needed to saturate the solution, the solution is **unsaturated**. When solute concentration is greater than the equilibrium concentration value, the solution is **supersaturated**. This is an unstable condition, and separation of some solute from the solution will occur if the process is initiated with a solute seed crystal. The amount of solute needed to form a saturated solution at any particular temperature is the **solubility** of that solute at that temperature.

Section 13.3 The solubility of one substance in another depends on the tendency of systems to become more random, by becoming more dispersed in space, and on the relative intermolecular solute–solute and solvent–solvent energies compared with solute–solvent interactions. Polar and ionic solutes tend to dissolve in polar solvents, and nonpolar solutes tend to dissolve in nonpolar solvents ("like dissolves like"). Liquids that mix in all proportions are **miscible**; those that do not dissolve significantly in one another are **immiscible**. Hydrogen-bonding interactions between solute and solvent often play an important role in determining solubility; for example, ethanol and water, whose molecules form hydrogen bonds with each other, are miscible. The solubilities of gases in a liquid are generally proportional to the pressure of the gas over the solution, as expressed by **Henry's law**: $S_g = kP_g$. The solubilities of most solid solutes in water increase as the temperature of the solution increases. In contrast, the solubilities of gases in water generally decrease with increasing temperature.

Section 13.4 Concentrations of solutions can be expressed quantitatively by several different measures, including **mass percentage** [(mass solute/mass solution) × 10^2], **parts per million (ppm)** [(mass solute/mass solution) × 10^6], **parts per billion (ppb)** [(mass solute/mass solution) × 10^9], and mole fraction [mol solute/(mol solute + mol solvent)]. Molarity, M, is defined as moles of solute per liter of solution; **molality**, m, is defined as moles of solute per kg of solvent. Molarity can be converted to these other concentration units if the density of the solution is known.

Section 13.5 A physical property of a solution that depends on the concentration of solute particles present, regardless of the nature of the solute, is a **colligative property**. Colligative properties include vapor-pressure lowering, freezing-point lowering, boiling-point elevation, and osmotic pressure. **Raoult's law** expresses the lowering of vapor pressure. An **ideal solution** obeys Raoult's law. Differences in solvent–solute as compared with solvent–solvent and solute–solute intermolecular forces cause many solutions to depart from ideal behavior.

A solution containing a nonvolatile solute possesses a higher boiling point than the pure solvent. The **molal boiling-point-elevation constant**, K_b, represents the increase in boiling point for a 1 m solution of solute particles as compared with the pure solvent. Similarly, the **molal freezing-point-depression constant**, K_f, measures the lowering of the freezing point of a solution for a 1 m solution of solute particles. The temperature changes are given by the equations $\Delta T_b = K_b m$ and $\Delta T_f = K_f m$.

When NaCl dissolves in water, two moles of solute particles are formed for each mole of dissolved salt. The boiling point or freezing point is thus elevated or depressed, respectively, approximately twice as much as that of a nonelectrolyte solution of the same concentration. Similar considerations apply to other strong electrolytes.

Osmosis is the movement of solvent molecules through a semipermeable membrane from a less concentrated to a more concentrated solution. This net movement of solvent generates an **osmotic pressure**, Π, which can be measured in units of gas pressure, such as atm. The osmotic pressure of a solution as compared with pure solvent is proportional to the solution molarity: $\Pi = MRT$. Osmosis is a very important process in living systems, in which cell walls act as semipermeable membranes, permitting the passage of water, but restricting the passage of ionic and macromolecular components.

Section 13.6 Particles that are large on the molecular scale but still small enough to remain suspended indefinitely in a solvent system form **colloids**, or **colloidal dispersions**. Colloids, which are intermediate between solutions and heterogeneous mixtures, have many practical applications. One useful physical property of colloids, the scattering of visible light, is referred to as the **Tyndall effect**. Aqueous colloids are classified as **hydrophilic** or **hydrophobic**. Hydrophilic colloids are common in living organisms, in which large molecular aggregates (enzymes, antibodies) remain suspended because they have many polar, or charged, atomic groups on their surfaces that interact with water. Hydrophobic colloids, such as small droplets of oil, may remain in suspension through adsorption of charged particles on their surfaces.

KEY SKILLS

- Understand how enthalpy and entropy changes affect solution formation.

- Understand the relationship between intermolecular forces and solubility, including use of the "like dissolves like" rule.

- Describe the effect of temperature on the solubility of solids and gases.

- Describe the relationship between the partial pressure of a gas and its solubility.

- Be able to calculate the concentration of a solution in terms of molarity, molality, mole fraction, percent composition, and parts per million and be able to interconvert between them.

- Describe what a colligative property is and explain the difference between the effects of nonelectrolytes and electrolyes on colligative properties.

- Be able to calculate the vapor pressure of a solvent over a solution.

- Be able to calculate the boiling point elevation and freezing point depression of a solution.

- Be able to calculate the osmotic pressure of a solution.

KEY EQUATIONS

- $S_g = kP_g$ [13.4]   Henry's law, relating gas solubility to partial pressure

- Mass % of component $= \dfrac{\text{mass of component in soln}}{\text{total mass of soln}} \times 100$ [13.5]   Defining concentration in terms of mass percent

- ppm of component $= \dfrac{\text{mass of component in soln}}{\text{total mass of soln}} \times 10^6$ [13.6]   Defining concentration in terms of parts per million (ppm)

- Mole fraction of component $= \dfrac{\text{moles of component}}{\text{total moles of all components}}$ [13.7]   Defining concentration in terms of mole fraction

- Molarity $= \dfrac{\text{moles solute}}{\text{liters soln}}$ [13.8]   Defining concentration in terms of molarity

- Molality $= \dfrac{\text{moles of solute}}{\text{kilograms of solvent}}$ [13.9]   Defining concentration in terms of molality

- $P_A = X_A P_A^\circ$ [13.10]   Raoult's law, calculating vapor pressure of solvent above a solution

- $\Delta T_b = K_b m$ [13.11]   Calculating the boiling point elevation of a solution

- $\Delta T_f = K_f m$ [13.12]   Calculating the freezing point depression of a solution

- $\Pi = \left(\dfrac{n}{V}\right)RT = MRT$ [13.13]   Calculating the osmotic pressure of a solution

VISUALIZING CONCEPTS

13.1 This figure shows the interaction of a cation with surrounding water molecules.

Would you expect the energy of ion–solvent interaction to be greater for Na^+ or Li^+? Explain. [Section 13.1]

13.2 Why do ionic substances with higher lattice energies tend to be less soluble in water than those with lower lattice energies? [Section 13.1]

13.3 Rank the contents of the following containers in order of increasing entropy: [Section 13.1]

(a)   (b)   (c)

13.4 A quantity of the pink solid on the left in Figure 13.8 is placed in a warming oven and heated for a time. It slowly turns from pink to the deep blue color of the solid on the right. What has occurred? [Section 13.1]

13.5 Which of the following is the best representation of a saturated solution? Explain your reasoning. [Section 13.2]

(a)   (b)   (c)

13.6 The solubility of Xe in water at 20 °C is approximately 5×10^{-3} M. Compare this with the solubilities of Ar and Kr in water (Table 13.2) and explain what properties of the rare gas atoms account for the variation in solubility. [Section 13.3]

13.7 The structures of vitamins E and B$_6$ are shown below. Predict which is largely water soluble and which is largely fat soluble. Explain. [Section 13.3]

Vitamin B$_6$ Vitamin E

13.8 If you wanted to prepare a solution of CO in water at 25 °C in which the CO concentration was 2.5 mM, what pressure of CO would you need to use? (See Figure 13.18.) [Section 13.3]

13.9 The figure shows two identical volumetric flasks containing the same solution at two temperatures.
  (a) Does the molarity of the solution change with the change in temperature? Explain.
  (b) Does the molality of the solution change with the change in temperature? Explain. [Section 13.4]

25 °C   55 °C

13.10 The following diagram shows the vapor pressure curves of a volatile solvent and a solution of that solvent containing a nonvolatile solute. **(a)** Which line represents the solution? **(b)** What are the normal boiling points of the solvent and the solution? [Section 13.5]

13.11 Suppose you had a balloon made of some highly flexible semipermeable membrane. The balloon is filled completely with a 0.2 M solution of some solute and is submerged in a 0.1 M solution of the same solute:

0.1 M

0.2 M

Initially, the volume of solution in the balloon is 0.25 L. Assuming the volume outside the semipermeable membrane is large, as the illustration shows, what would you expect for the solution volume inside the balloon once the system has come to equilibrium through osmosis? [Section 13.5]

13.12 The molecule *n*-octylglucoside, shown here, is widely used in biochemical research as a nonionic detergent for "solubilizing" large hydrophobic protein molecules. What characteristics of this molecule are important for its use in this way? [Section 13.6]

EXERCISES

The Solution Process

13.13 In general, the attractive intermolecular forces between solvent and solute particles must be comparable or greater than solute–solute interactions for significant solubility to occur. Explain this statement in terms of the overall energetics of solution formation.

13.14 **(a)** Considering the energetics of solute–solute, solvent–solvent, and solute–solvent interactions, explain why NaCl dissolves in water but not in benzene (C_6H_6). **(b)** What factors cause a cation to be strongly hydrated?

13.15 Indicate the type of solute–solvent interaction (Section 11.2) that should be most important in each of the following solutions: **(a)** CCl_4 in benzene (C_6H_6), **(b)** methanol (CH_3OH) in water, **(c)** KBr in water, **(d)** HCl in acetonitrile (CH_3CN).

13.16 Indicate the principal type of solute–solvent interaction in each of the following solutions, and rank the solutions from weakest to strongest solute–solvent interaction: **(a)** KCl in water, **(b)** CH_2Cl_2 in benzene (C_6H_6), **(c)** methanol (CH_3OH) in water.

13.17 **(a)** In Equation 13.1 which of the energy terms for dissolving an ionic solid would correspond to the lattice energy? **(b)** Which energy terms in this equation are always exothermic?

13.18 The schematic diagram of the solution process as the net sum of three steps in Figure 13.4 does not show the relative magnitudes of the three components because these will vary from case to case. For the dissolution of NH_4NO_3 in water, which of the three enthalpy changes would you expect to be much smaller than the other two? Explain.

13.19 When two nonpolar organic liquids such as hexane (C_6H_{14}) and heptane (C_7H_{16}) are mixed, the enthalpy change that occurs is generally quite small. **(a)** Use the energy diagram in Figure 13.4 to explain why. **(b)** Given that $\Delta H_{soln} \approx 0$, explain why hexane and heptane spontaneously form a solution.

13.20 The enthalpy of solution of KBr in water is about +198 kJ/mol. Nevertheless, the solubility of KBr in water is relatively high. Why does the solution process occur even though it is endothermic?

Saturated Solutions; Factors Affecting Solubility

13.21 The solubility of $Cr(NO_3)_3 \cdot 9\,H_2O$ in water is 208 g per 100 g of water at 15 °C. A solution of $Cr(NO_3)_3 \cdot 9\,H_2O$ in water at 35 °C is formed by dissolving 324 g in 100 g water. When this solution is slowly cooled to 15 °C, no precipitate forms. **(a)** What term describes this solution? **(b)** What action might you take to initiate crystallization? Use molecular-level processes to explain how your suggested procedure works.

13.22 The solubility of $MnSO_4 \cdot H_2O$ in water at 20 °C is 70 g per 100 mL of water. **(a)** Is a 1.22 M solution of $MnSO_4 \cdot H_2O$ in water at 20 °C saturated, supersaturated, or unsaturated? **(b)** Given a solution of $MnSO_4 \cdot H_2O$ of unknown concentration, what experiment could you perform to determine whether the new solution is saturated, supersaturated, or unsaturated?

13.23 By referring to Figure 13.17, determine whether the addition of 40.0 g of each of the following ionic solids to 100 g of water at 40 °C will lead to a saturated solution: **(a)** $NaNO_3$, **(b)** KCl, **(c)** $K_2Cr_2O_7$, **(d)** $Pb(NO_3)_2$.

13.24 By referring to Figure 13.17, determine the mass of each of the following salts required to form a saturated solution in 250 g of water at 30 °C: **(a)** $KClO_3$, **(b)** $Pb(NO_3)_2$, **(c)** $Ce_2(SO_4)_3$.

13.25 Water and glycerol, $CH_2(OH)CH(OH)CH_2OH$, are miscible in all proportions. What does this mean? How do the OH groups of the alcohol molecule contribute to this miscibility?

13.26 Oil and water are immiscible. What does this mean? Explain in terms of the structural features of their respective molecules and the forces between them.

13.27 **(a)** Would you expect stearic acid, $CH_3(CH_2)_{16}COOH$, to be more soluble in water or in carbon tetrachloride? Explain. **(b)** Which would you expect to be more soluble in water, cyclohexane or dioxane? Explain.

Dioxane   Cyclohexane

13.28 Consider a series of carboxylic acids whose general formula is $CH_3(CH_2)_nCOOH$. How would you expect the solubility of these compounds in water and in hexane to change as n increases? Explain.

13.29 Which of the following in each pair is likely to be more soluble in hexane, C_6H_{14}: **(a)** CCl_4 or $CaCl_2$; **(b)** benzene (C_6H_6) or glycerol, $CH_2(OH)CH(OH)CH_2OH$; **(c)** octanoic acid, $CH_3CH_2CH_2CH_2CH_2CH_2CH_2COOH$, or acetic acid, CH_3COOH. Explain your answer in each case.

13.30 Which of the following in each pair is likely to be more soluble in water: **(a)** cyclohexane (C_6H_{12}) or glucose ($C_6H_{12}O_6$) (Figure 13.12); **(b)** propionic acid (CH_3CH_2COOH) or sodium propionate (CH_3CH_2COONa); **(c)** HCl or ethyl chloride (CH_3CH_2Cl)? Explain in each case.

13.31 **(a)** Explain why carbonated beverages must be stored in sealed containers. **(b)** Once the beverage has been opened, why does it maintain more carbonation when refrigerated than at room temperature?

13.32 Explain why pressure affects the solubility of O_2 in water, but not the solubility of NaCl in water.

13.33 The Henry's law constant for helium gas in water at 30 °C is 3.7×10^{-4} M/atm and the constant for N_2 at 30 °C is 6.0×10^{-4} M/atm. If the two gases are each present at 1.5 atm pressure, calculate the solubility of each gas.

13.34 The partial pressure of O_2 in air at sea level is 0.21 atm. Using the data in Table 13.2, together with Henry's law, calculate the molar concentration of O_2 in the surface water of a mountain lake saturated with air at 20 °C and an atmospheric pressure of 650 torr.

Concentrations of Solutions

13.35 **(a)** Calculate the mass percentage of Na_2SO_4 in a solution containing 10.6 g Na_2SO_4 in 483 g water. **(b)** An ore contains 2.86 g of silver per ton of ore. What is the concentration of silver in ppm?

13.36 **(a)** What is the mass percentage of iodine (I_2) in a solution containing 0.035 mol I_2 in 115 g of CCl_4? **(b)** Seawater contains 0.0079 g Sr^{2+} per kilogram of water. What is the concentration of Sr^{2+} measured in ppm?

13.37 A solution is made containing 14.6 g of CH_3OH in 184 g H_2O. Calculate **(a)** the mole fraction of CH_3OH, **(b)** the mass percent of CH_3OH, **(c)** the molality of CH_3OH.

13.38 A solution is made containing 25.5 g phenol (C_6H_5OH) in 425 g ethanol (C_2H_5OH). Calculate **(a)** the mole fraction of phenol, **(b)** the mass percent of phenol, **(c)** the molality of phenol.

13.39 Calculate the molarity of the following aqueous solutions: **(a)** 0.540 g $Mg(NO_3)_2$ in 250.0 mL of solution, **(b)** 22.4 g $LiClO_4 \cdot 3 H_2O$ in 125 mL of solution, **(c)** 25.0 mL of 3.50 M HNO_3 diluted to 0.250 L.

13.40 What is the molarity of each of the following solutions: **(a)** 15.0 g $Al_2(SO_4)_3$ in 0.350 L solution, **(b)** 5.25 g $Mn(NO_3)_2 \cdot 2 H_2O$ in 175 mL of solution, **(c)** 35.0 mL of 9.00 M H_2SO_4 diluted to 0.500 L?

13.41 Calculate the molality of each of the following solutions: **(a)** 8.66 g benzene (C_6H_6) dissolved in 23.6 g carbon tetrachloride (CCl_4), **(b)** 4.80 g NaCl dissolved in 0.350 L of water.

13.42 **(a)** What is the molality of a solution formed by dissolving 1.25 mol of KCl in 16.0 mol of water? **(b)** How many grams of sulfur (S_8) must be dissolved in 100.0 g naphthalene ($C_{10}H_8$) to make a 0.12 m solution?

13.43 A sulfuric acid solution containing 571.6 g of H_2SO_4 per liter of solution has a density of 1.329 g/cm^3. Calculate **(a)** the mass percentage, **(b)** the mole fraction, **(c)** the molality, **(d)** the molarity of H_2SO_4 in this solution.

13.44 Ascorbic acid (vitamin C, $C_6H_8O_6$) is a water-soluble vitamin. A solution containing 80.5 g of ascorbic acid dissolved in 210 g of water has a density of 1.22 g/mL at 55 °C. Calculate **(a)** the mass percentage, **(b)** the mole fraction, **(c)** the molality, **(d)** the molarity of ascorbic acid in this solution.

13.45 The density of acetonitrile (CH_3CN) is 0.786 g/mL and the density of methanol (CH_3OH) is 0.791 g/mL. A solution is made by dissolving 22.5 mL CH_3OH in 98.7 mL CH_3CN. **(a)** What is the mole fraction of methanol in the solution? **(b)** What is the molality of the solution?

(c) Assuming that the volumes are additive, what is the molarity of CH_3OH in the solution?

13.46 The density of toluene (C_7H_8) is 0.867 g/mL, and the density of thiophene (C_4H_4S) is 1.065 g/mL. A solution is made by dissolving 9.08 g of thiophene in 250.0 mL of toluene. **(a)** Calculate the mole fraction of thiophene in the solution. **(b)** Calculate the molality of thiophene in the solution. **(c)** Assuming that the volumes of the solute and solvent are additive, what is the molarity of thiophene in the solution?

13.47 Calculate the number of moles of solute present in each of the following aqueous solutions: **(a)** 600 mL of 0.250 M $SrBr_2$, **(b)** 86.4 g of 0.180 m KCl, **(c)** 124.0 g of a solution that is 6.45% glucose ($C_6H_{12}O_6$) by mass.

13.48 Calculate the number of moles of solute present in each of the following solutions: **(a)** 185 mL of 1.50 M $HNO_3(aq)$, **(b)** 50.0 mg of an aqueous solution that is 1.25 m NaCl, **(c)** 75.0 g of an aqueous solution that is 1.50% sucrose ($C_{12}H_{22}O_{11}$) by mass.

13.49 Describe how you would prepare each of the following aqueous solutions, starting with solid KBr: **(a)** 0.75 L of 1.5×10^{-2} M KBr, **(b)** 125 g of 0.180 m KBr, **(c)** 1.85 L of a solution that is 12.0% KBr by mass (the density of the solution is 1.10 g/mL), **(d)** a 0.150 M solution of KBr that contains just enough KBr to precipitate 16.0 g of AgBr from a solution containing 0.480 mol of $AgNO_3$.

13.50 Describe how you would prepare each of the following aqueous solutions: **(a)** 1.50 L of 0.110 M $(NH_4)_2SO_4$ solution, starting with solid $(NH_4)_2SO_4$; **(b)** 120 g of a solution that is 0.65 m in Na_2CO_3, starting with the solid solute; **(c)** 1.20 L of a solution that is 15.0% $Pb(NO_3)_2$ by mass (the density of the solution is 1.16 g/mL), starting with solid solute; **(d)** a 0.50 M solution of HCl that would just neutralize 5.5 g of $Ba(OH)_2$ starting with 6.0 M HCl.

13.51 Commercial aqueous nitric acid has a density of 1.42 g/mL and is 16 M. Calculate the percent HNO_3 by mass in the solution.

13.52 Commercial concentrated aqueous ammonia is 28% NH_3 by mass and has a density of 0.90 g/mL. What is the molarity of this solution?

13.53 Brass is a substitutional alloy consisting of a solution of copper and zinc. A particular sample of red brass consisting of 80.0% Cu and 20.0% Zn by mass has a density of 8750 kg/m^3. **(a)** What is the molality of Zn in the solid solution? **(b)** What is the molarity of Zn in the solution?

13.54 Caffeine ($C_8H_{10}N_4O_2$) is a stimulant found in coffee and tea. If a solution of caffeine in chloroform ($CHCl_3$) as a solvent has a concentration of 0.0750 m, calculate **(a)** the percent caffeine by mass, **(b)** the mole fraction of caffeine.

13.55 During a typical breathing cycle the CO_2 concentration in the expired air rises to a peak of 4.6% by volume. Calculate the partial pressure of the CO_2 at this point, assuming 1 atm pressure. What is the molarity of the CO_2 in air at this point, assuming a body temperature of 37 °C?

13.56 Breathing air that contains 4.0% by volume CO_2 over time causes rapid breathing, throbbing headache, and nausea, among other symptoms. What is the concentration of CO_2 in such air in terms of **(a)** mol percentage, **(b)** molarity, assuming 1 atm pressure, and a body temperature of 37 °C?

Colligative Properties

13.57 List four properties of a solution that depend on the total concentration but not the type of particle or particles present as solute. Write the mathematical expression that describes how each of these properties depends on concentration.

13.58 How does increasing the concentration of a nonvolatile solute in water affect the following properties: **(a)** vapor pressure, **(b)** freezing point, **(c)** boiling point; **(d)** osmotic pressure?

13.59 Consider two solutions, one formed by adding 10 g of glucose ($C_6H_{12}O_6$) to 1 L of water and the other formed by adding 10 g of sucrose ($C_{12}H_{22}O_{11}$) to 1 L of water. Are the vapor pressures over the two solutions the same? Why or why not?

13.60 **(a)** What is an *ideal solution*? **(b)** The vapor pressure of pure water at 60 °C is 149 torr. The vapor pressure of water over a solution at 60 °C containing equal numbers of moles of water and ethylene glycol (a nonvolatile solute) is 67 torr. Is the solution ideal according to Raoult's law? Explain.

13.61 **(a)** Calculate the vapor pressure of water above a solution prepared by adding 22.5 g of lactose ($C_{12}H_{22}O_{11}$) to 200.0 g of water at 338 K. (Vapor-pressure data for water are given in Appendix B.) **(b)** Calculate the mass of propylene glycol ($C_3H_8O_2$) that must be added to 0.340 kg of water to reduce the vapor pressure by 2.88 torr at 40 °C.

[13.62] **(a)** Calculate the vapor pressure of water above a solution prepared by dissolving 32.5 g of glycerin ($C_3H_8O_3$) in 125 g of water at 343 K. (The vapor pressure of water is given in Appendix B.) **(b)** Calculate the mass of ethylene glycol ($C_2H_6O_2$) that must be added to 1.00 kg of ethanol (C_2H_5OH) to reduce its vapor pressure by 10.0 torr at 35 °C. The vapor pressure of pure ethanol at 35 °C is 1.00×10^2 torr.

[13.63] At 63.5 °C the vapor pressure of H_2O is 175 torr, and that of ethanol (C_2H_5OH) is 400 torr. A solution is made by mixing equal masses of H_2O and C_2H_5OH. **(a)** What is the mole fraction of ethanol in the solution? **(b)** Assuming ideal-solution behavior, what is the vapor pressure of the solution at 63.5 °C? **(c)** What is the mole fraction of ethanol in the vapor above the solution?

[13.64] At 20 °C the vapor pressure of benzene (C_6H_6) is 75 torr, and that of toluene (C_7H_8) is 22 torr. Assume that benzene and toluene form an ideal solution. **(a)** What is the composition in mole fractions of a solution that has a vapor pressure of 35 torr at 20 °C? **(b)** What is the mole fraction of benzene in the vapor above the solution described in part (a)?

13.65 **(a)** Why does a 0.10 m aqueous solution of NaCl have a higher boiling point than a 0.10 m aqueous solution of $C_6H_{12}O_6$? **(b)** Calculate the boiling point of each solution. **(c)** The experimental boiling point of the NaCl solution is lower than that calculated, assuming that NaCl is completely dissociated in solution. Why is this the case?

13.66 Arrange the following aqueous solutions, each 10% by mass in solute, in order of increasing boiling point: glucose ($C_6H_{12}O_6$), sucrose ($C_{12}H_{22}O_{11}$), sodium nitrate (NaNO$_3$).

13.67 List the following aqueous solutions in order of increasing boiling point: 0.120 m glucose, 0.050 m LiBr, 0.050 m Zn(NO$_3$)$_2$.

13.68 List the following aqueous solutions in order of decreasing freezing point: 0.040 m glycerin ($C_3H_8O_3$), 0.020 m KBr, 0.030 m phenol (C_6H_5OH).

13.69 Using data from Table 13.4, calculate the freezing and boiling points of each of the following solutions: **(a)** 0.22 m glycerol ($C_3H_8O_3$) in ethanol, **(b)** 0.240 mol of naphthalene ($C_{10}H_8$) in 2.45 mol of chloroform, **(c)** 2.04 g KBr and 4.82 g glucose ($C_6H_{12}O_6$) in 188 g of water.

13.70 Using data from Table 13.4, calculate the freezing and boiling points of each of the following solutions: **(a)** 0.30 m glucose in ethanol; **(b)** 20.0 g of decane, $C_{10}H_{22}$, in 45.5 g CHCl$_3$; **(c)** 0.45 mol ethylene glycol and 0.15 mol KBr in 150 g H_2O.

13.71 How many grams of ethylene glycol ($C_2H_6O_2$) must be added to 1.00 kg of water to produce a solution that freezes at $-5.00\,°C$?

13.72 What is the freezing point of an aqueous solution that boils at $105.0\,°C$?

13.73 What is the osmotic pressure formed by dissolving 44.2 mg of aspirin ($C_9H_8O_4$) in 0.358 L of water at 25 °C?

13.74 Seawater contains 3.4 g of salts for every liter of solution. Assuming that the solute consists entirely of NaCl (over 90% is), calculate the osmotic pressure of seawater at 20 °C.

13.75 Adrenaline is the hormone that triggers the release of extra glucose molecules in times of stress or emergency. A solution of 0.64 g of adrenaline in 36.0 g of CCl_4 elevates the boiling point by 0.49 °C. Is the molar mass of adrenaline calculated from the boiling point elevation in agreement with the following structural formula?

13.76 Lauryl alcohol is obtained from coconut oil and is used to make detergents. A solution of 5.00 g of lauryl alcohol in 0.100 kg of benzene freezes at 4.1 °C. What is the approximate molar mass of lauryl alcohol?

13.77 Lysozyme is an enzyme that breaks bacterial cell walls. A solution containing 0.150 g of this enzyme in 210 mL of solution has an osmotic pressure of 0.953 torr at 25 °C. What is the molar mass of lysozyme?

13.78 A dilute aqueous solution of an organic compound soluble in water is formed by dissolving 2.35 g of the compound in water to form 0.250 L solution. The resulting solution has an osmotic pressure of 0.605 atm at 25 °C. Assuming that the organic compound is a nonelectrolyte, what is its molar mass?

[13.79] The osmotic pressure of a 0.010 M aqueous solution of $CaCl_2$ is found to be 0.674 atm at 25 °C. **(a)** Calculate the van't Hoff factor, i, for the solution. **(b)** How would you expect the value of i to change as the solution becomes more concentrated? Explain.

[13.80] Based on the data given in Table 13.5, which solution would give the larger freezing-point lowering, a 0.030 m solution of NaCl or a 0.020 m solution of K_2SO_4? How do you explain the departure from ideal behavior and the differences observed between the two salts?

Colloids

13.81 **(a)** Why is there no colloid in which both the dispersed substance and the dispersing substance are gases? **(b)** Michael Faraday first prepared ruby-red colloids of gold particles in water that were stable for indefinite times. ∞ (Section 12.6) To the unaided eye these brightly colored colloids are not distinguishable from solutions. How could you determine whether a given colored preparation is a solution or colloid?

13.82 **(a)** Many proteins that remain homogeneously distributed in water have molecular masses in the range of 30,000 amu and larger. In what sense is it appropriate to consider such suspensions to be colloids rather than solutions? Explain. **(b)** What general name is given to a colloidal dispersion of one liquid in another? What is an emulsifying agent?

13.83 Indicate whether each of the following is a hydrophilic or a hydrophobic colloid: **(a)** butterfat in homogenized milk, **(b)** hemoglobin in blood, **(c)** vegetable oil in a salad dressing, **(d)** colloidal gold particles in water.

13.84 Explain how each of the following factors helps determine the stability or instability of a colloidal dispersion: **(a)** particulate mass, **(b)** hydrophobic character, **(c)** charges on colloidal particles.

13.85 Colloidal suspensions of proteins, such as a gelatin, can often be caused to separate into two layers by addition of a solution of an electrolyte. Given that protein molecules may carry electrical charges on their outer surface as illustrated in Figure 13.28, what do you believe happens when the electrolyte solution is added?

13.86 Explain how **(a)** a soap such as sodium stearate stabilizes a colloidal dispersion of oil droplets in water; **(b)** milk curdles upon addition of an acid.

ADDITIONAL EXERCISES

13.87 Butylated hydroxytoluene (BHT) has the following molecular structure:

BHT

It is widely used as a preservative in a variety of foods, including dried cereals. Based on its structure, would you expect BHT to be more soluble in water or in hexane (C_6H_{14})? Explain.

13.88 A saturated solution of sucrose ($C_{12}H_{22}O_{11}$) is made by dissolving excess table sugar in a flask of water. There are 50 g of undissolved sucrose crystals at the bottom of the flask in contact with the saturated solution. The flask is stoppered and set aside. A year later a single large crystal of mass 50 g is at the bottom of the flask. Explain how this experiment provides evidence for a dynamic equilibrium between the saturated solution and the undissolved solute.

13.89 Fish need at least 4 ppm dissolved O_2 for survival. **(a)** What is this concentration in mol/L? **(b)** What partial pressure of O_2 above the water is needed to obtain this concentration at 10 °C? (The Henry's law constant for O_2 at this temperature is 1.71×10^{-3} mol/L-atm.)

13.90 The presence of the radioactive gas radon (Rn) in well water obtained from aquifers that lie in rock deposits presents a possible health hazard in parts of the United States. **(a)** Assuming that the solubility of radon in water with 1 atm pressure of the gas over the water at 30 °C is 7.27×10^{-3} M, what is the Henry's law constant for radon in water at this temperature? **(b)** A sample consisting of various gases contains 3.5×10^{-6} mole fraction of radon. This gas at a total pressure of 32 atm is shaken with water at 30 °C. Calculate the molar concentration of radon in the water.

13.91 Glucose makes up about 0.10% by mass of human blood. Calculate the concentration in **(a)** ppm, **(b)** molality. What further information would you need to determine the molarity of the solution?

13.92 The maximum allowable concentration of lead in drinking water is 9.0 ppb. **(a)** Calculate the molarity of lead in a 9.0-ppb solution. What assumption did you have to make in your calculation? **(b)** How many grams of lead are in a swimming pool containing 9.0 ppb lead in 60 m³ of water?

13.93 Acetonitrile (CH_3CN) is a polar organic solvent that dissolves a wide range of solutes, including many salts. The density of a 1.80 M LiBr solution in acetonitrile is 0.826 g/cm³. Calculate the concentration of the solution in **(a)** molality, **(b)** mole fraction of LiBr, **(c)** mass percentage of CH_3CN.

13.94 A "canned heat" product used to warm chafing dishes consists of a homogeneous mixture of ethanol (C_2H_5OH) and paraffin that has an average formula of $C_{24}H_{50}$. What mass of C_2H_5OH should be added to 620 kg of the paraffin in formulating the mixture if the vapor pressure of ethanol at 35 °C over the mixture is to be 8 torr? The vapor pressure of pure ethanol at 35 °C is 100 torr.

[13.95] Two beakers are placed in a sealed box at 25 °C. One beaker contains 30.0 mL of a 0.050 M aqueous solution of a nonvolatile nonelectrolyte. The other beaker contains 30.0 mL of a 0.035 M aqueous solution of NaCl. The water vapor from the two solutions reaches equilibrium. **(a)** In which beaker does the solution level rise, and in which one does it fall? **(b)** What are the volumes in the two beakers when equilibrium is attained, assuming ideal behavior?

13.96 A solution contains 0.115 mol H_2O and an unknown number of moles of sodium chloride. The vapor pressure of the solution at 30 °C is 25.7 torr. The vapor pressure of pure water at this temperature is 31.8 torr. Calculate the number of moles of sodium chloride in the solution. (*Hint:* remember that sodium chloride is a strong electrolyte.)

13.97 Show that the vapor-pressure reduction, $\Delta P_{solvent}$, associated with the addition of a nonvolatile solute to a volatile solvent is given by the equation $\Delta P_{solvent} = X_{solute} \times P^{\circ}_{solvent}$.

13.98 A car owner who knows no chemistry has to put antifreeze in his car's radiator. The instructions recommend a mixture of 30% ethylene glycol and 70% water. Thinking he will improve his protection he uses pure ethylene glycol. He is saddened to find that the solution does not provide as much protection as he hoped. Why not?

13.99 Calculate the freezing point of a 0.100 *m* aqueous solution of K_2SO_4, **(a)** ignoring interionic attractions, and **(b)** taking interionic attractions into consideration by using the van't Hoff factor (Table 13.5).

13.100 When 10.0 g of mercuric nitrate, $Hg(NO_3)_2$, is dissolved in 1.00 kg of water, the freezing point of the solution is −0.162 °C. When 10.0 g of mercuric chloride ($HgCl_2$) is dissolved in 1.00 kg of water, the solution freezes at −0.0685 °C. Use these data to determine which is the stronger electrolyte, $Hg(NO_3)_2$ or $HgCl_2$.

13.101 Carbon disulfide (CS_2) boils at 46.30 °C and has a density of 1.261 g/mL. **(a)** When 0.250 mol of a nondissociating solute is dissolved in 400.0 mL of CS_2, the solution boils at 47.46 °C. What is the molal boiling-point-elevation constant for CS_2? **(b)** When 5.39 g of a nondissociating unknown is dissolved in 50.0 mL of CS_2, the solution boils at 47.08 °C. What is the molecular weight of the unknown?

[13.102] A 40.0% by weight solution of KSCN in water at 20 °C has a density of 1.22 g/mL. **(a)** What is the mole fraction of KSCN in the solution, and what are the molarity and molality? **(b)** Given the calculated mole fraction of salt in the solution, comment on the total number of water molecules available to hydrate each anion and cation.

What ion pairing (if any) would you expect to find in the solution? Would you expect the colligative properties of such a solution to be those predicted by the formulas given in this chapter? Explain.

[13.103] A lithium salt used in lubricating grease has the formula $LiC_nH_{2n+1}O_2$. The salt is soluble in water to the extent of 0.036 g per 100 g of water at 25 °C. The osmotic pressure of this solution is found to be 57.1 torr. Assuming that molality and molarity in such a dilute solution are the same and that the lithium salt is completely dissociated in the solution, determine an appropriate value of n in the formula for the salt.

INTEGRATIVE EXERCISES

13.104 Fluorocarbons (compounds that contain both carbon and fluorine) were, until recently, used as refrigerants. The compounds listed in the following table are all gases at 25 °C, and their solubilities in water at 25 °C and 1 atm fluorocarbon pressure are given as mass percentages. **(a)** For each fluorocarbon, calculate the molality of a saturated solution. **(b)** Explain why the molarity of each of the solutions should be very close numerically to the molality. **(c)** Based on their molecular structures, account for the differences in solubility of the four fluorocarbons. **(d)** Calculate the Henry's law constant at 25 °C for $CHClF_2$, and compare its magnitude to that for N_2 (6.8×10^{-4} mol/L-atm). Can you account for the difference in magnitude?

Fluorocarbon	Solubility (mass %)
CF_4	0.0015
$CClF_3$	0.009
CCl_2F_2	0.028
$CHClF_2$	0.30

[13.105] At ordinary body temperature (37 °C) the solubility of N_2 in water in contact with air at ordinary atmospheric pressure (1.0 atm) is 0.015 g/L. Air is approximately 78 mol % N_2. Calculate the number of moles of N_2 dissolved per liter of blood, which is essentially an aqueous solution. At a depth of 100 ft in water, the pressure is 4.0 atm. What is the solubility of N_2 from air in blood at this pressure? If a scuba diver suddenly surfaces from this depth, how many milliliters of N_2 gas, in the form of tiny bubbles, are released into the bloodstream from each liter of blood?

[13.106] Consider the following values for enthalpy of vaporization (kJ/mol) of several organic substances:

$$CH_3\overset{\displaystyle O}{\overset{\displaystyle \|}{C}}{-}H \quad 30.4 \qquad H_2C\overset{O}{\diagdown\diagup}CH_2 \quad 28.5$$

Acetaldehyde   Ethylene oxide

$$CH_3\overset{\displaystyle O}{\overset{\displaystyle \|}{C}}CH_3 \quad 32.0 \qquad H_2C\overset{CH_2}{\diagup\diagdown}CH_2 \quad 24.7$$

Acetone   Cyclopropane

(a) Use variations in the intermolecular forces operating in these organic liquids to account for their variations in heats of vaporization. **(b)** How would you expect the solubilities of these substances to vary in hexane as solvent? In ethanol? Use intermolecular forces, including hydrogen-bonding interactions where applicable, to explain your responses.

[13.107] The enthalpies of solution of hydrated salts are generally more positive than those of anhydrous materials. For example, ΔH of solution for KOH is -57.3 kJ/mol whereas that for $KOH \cdot H_2O$ is -14.6 kJ/mol. Similarly, ΔH_{soln} for $NaClO_4$ is $+13.8$ kJ/mol, whereas that for $NaClO_4 \cdot H_2O$ is $+22.5$ kJ/mol. Use the enthalpy contributions to the solution process depicted in Figure 13.4 to explain this effect.

[13.108] A textbook on chemical thermodynamics states, "The heat of solution represents the difference between the lattice energy of the crystalline solid and the solvation energy of the gaseous ions." **(a)** Draw a simple energy diagram to illustrate this statement. **(b)** A salt such as NaBr is insoluble in most polar nonaqueous solvents such as acetonitrile (CH_3CN) or nitromethane (CH_3NO_2), but salts of large cations, such as tetramethylammonium bromide [$(CH_3)_4NBr$], are generally more soluble. Use the thermochemical cycle you drew in part (a) and the factors that determine the lattice energy (Section 8.2) to explain this fact.

13.109 **(a)** A sample of hydrogen gas is generated in a closed container by reacting 2.050 g of zinc metal with 15.0 mL of 1.00 M sulfuric acid. Write the balanced equation for the reaction, and calculate the number of moles of hydrogen formed, assuming that the reaction is complete. **(b)** The volume over the solution is 122 mL. Calculate the partial pressure of the hydrogen gas in this volume at 25 °C, ignoring any solubility of the gas in the solution. **(c)** The Henry's law constant for hydrogen in water at 25 °C is 7.8×10^{-4} mol/L-atm. Estimate the number of moles of hydrogen gas that remain dissolved in the solution. What fraction of the gas molecules in the system is dissolved in the solution? Was it reasonable to ignore any dissolved hydrogen in part (b)?

[13.110] The following table presents the solubilities of several gases in water at 25 °C under a total pressure of gas and water vapor of 1 atm. **(a)** What volume of $CH_4(g)$ under standard conditions of temperature and pressure is contained in 4.0 L of a saturated solution at 25 °C? **(b)** Explain the variation in solubility among the hydrocarbons listed (the first three compounds), based on their molecular structures and intermolecular forces. **(c)** Compare the solubilities of O_2, N_2, and NO, and account for the variations based on molecular structures and intermolecular forces. **(d)** Account for the much larger values observed for H_2S and SO_2 as compared with the other gases listed. **(e)** Find several pairs of substances with the same or nearly the same molecular masses (for example, C_2H_4 and N_2), and use intermolecular interactions to explain the differences in their solubilities.

Gas	Solubility (mM)
CH_4(methane)	1.3
C_2H_6 (ethane)	1.8
C_2H_4 (ethylene)	4.7
N_2	0.6
O_2	1.2
NO	1.9
H_2S	99
SO_2	1476

[13.111] Consider the process illustrated in Figure 13.23, the net movement of solvent from the more dilute to the more concentrated solution through the semipermeable membrane. Is there a change in entropy of the system in going from the left to center panels? Explain. (*Hint:* Imagine that the more dilute solution on the right hand side of the apparatus is pure solvent.)

[13.112] When 0.55 g of pure benzoic acid (C_6H_5COOH) is dissolved in 32.0 g of benzene (C_6H_6), the freezing point of the solution is 0.36 °C lower than the freezing point value of 5.5 °C for the pure solvent. **(a)** Calculate the molecular weight of benzoic acid in benzene. **(b)** Use the structure of the solute to account for the observed value:

[13.113] At 35 °C the vapor pressure of acetone, $(CH_3)_2CO$, is 360 torr, and that of chloroform, $CHCl_3$, is 300 torr. Acetone and chloroform can form weak hydrogen bonds between one another as follows:

A solution composed of an equal number of moles of acetone and chloroform has a vapor pressure of 250 torr at 35 °C. **(a)** What would be the vapor pressure of the solution if it exhibited ideal behavior? **(b)** Use the existence of hydrogen bonds between acetone and chloroform molecules to explain the deviation from ideal behavior. **(c)** Based on the behavior of the solution, predict whether the mixing of acetone and chloroform is an exothermic ($\Delta H_{soln} < 0$) or endothermic ($\Delta H_{soln} > 0$) process.

CHEMICAL KINETICS

FIREWORKS RELY ON RAPID CHEMICAL reactions both to propel them skyward and to produce their colorful bursts of light.

WHAT'S AHEAD

14.1 Factors that Affect Reaction Rates
We explore four experimental variables that affect reaction rates: concentration, physical states of reactants, temperature, and catalysts. These factors can be understood in terms of the collisions among reactant molecules that lead to reaction.

14.2 Reaction Rates
We consider how to express *reaction rates* and how the rates of disappearance of reactants and the rates of appearance of products are related to the stoichiometry of the reaction.

14.3 Concentration and Rate
We then examine how the effect of concentration on rate is expressed quantitatively by *rate laws* and how rate laws can be determined experimentally.

14.4 The Change of Concentration with Time
We learn that rate equations can be written to express how concentrations change with time. We will look at two simple kinds of such rate equations.

14.5 Temperature and Rate
We next consider the effect of temperature on rate and the fact that reactions require a minimum input of energy called the *activation energy* to occur.

14.6 Reaction Mechanisms
We then examine the *mechanisms* of reactions, the step-by-step molecular pathways leading from reactants to products.

14.7 Catalysis
We end the chapter with a discussion of how *catalysts* speed reaction rates, including a discussion of biological catalysts, called *enzymes*.

CHEMISTRY IS, BY ITS VERY NATURE, CONCERNED WITH CHANGE.

Chemical reactions convert substances with well-defined properties into other materials with different properties. Much of our study of chemical reactions concerns the formation of new substances from a given set of reactants. It is equally important to understand how rapidly chemical reactions occur.

The rates of reactions span an enormous range, from those that are complete within fractions of seconds, such as explosions, to those that take thousands or even millions of years, such as the formation of diamonds or other minerals in Earth's crust (Figure 14.1 ▼). The fireworks shown in this chapter-opening photograph require very rapid reactions, both to propel them skyward and to produce their colorful bursts of light. The chemicals used in the fireworks are chosen both to give the desired colors and to do so very rapidly. The characteristic red, blue, and green colors are produced by salts of strontium, copper, and barium, respectively.

▲ **Figure 14.1 Reaction rates.** The rates of chemical reactions span a range of time scales. For example, explosions are rapid, occurring in seconds or fractions of seconds; corrosion can take years; and the weathering of rocks takes place over thousands or even millions of years.

The area of chemistry that is concerned with the speeds, or rates, of reactions is called **chemical kinetics**. Chemical kinetics is a subject of broad importance. It relates, for example, to how quickly a medicine is able to work, to whether the formation and depletion of ozone in the upper atmosphere are in balance, and to industrial challenges such as the development of catalysts to synthesize new materials.

Our goal in this chapter is to understand how to determine the rates at which reactions occur and to consider the factors that control these rates. For example, what factors determine how rapidly food spoils? How does one design a fast-setting material for dental fillings? What determines the rate at which steel rusts? What controls the rate at which fuel burns in an automobile engine? Although we will not address these specific questions, we will see that the rates of all chemical reactions are subject to the same basic principles.

14.1 FACTORS THAT AFFECT REACTION RATES

Before we examine the quantitative aspects of chemical kinetics, such as how rates are measured, let's examine the key factors that influence the rates of reactions. Because reactions involve the breaking and forming of bonds, the speeds of reactions depend on the nature of the reactants themselves. Four factors allow us to change the rates at which particular reactions occur:

1. *The physical state of the reactants.* Reactants must come together to react. The more readily molecules collide with each other, the more rapidly they react. Most of the reactions we consider are homogeneous, involving either gases or liquid solutions. When reactants are in different phases, as when one is a gas and another a solid, the reaction is limited to their area of contact. Thus, reactions that involve solids tend to proceed faster if the surface area of the solid is increased. For example, a medicine in the form of a fine powder will dissolve in the stomach and enter the bloodstream more quickly than the same medicine in the form of a tablet.

2. *The concentrations of the reactants.* Most chemical reactions proceed faster if the concentration of one or more of the reactants is increased. For example, steel wool burns with difficulty in air, which contains 20% O_2, but bursts into a brilliant white flame in pure oxygen (Figure 14.2▶). As concentration increases, the frequency with which the reactant molecules collide increases, leading to increased rates.

3. *The temperature at which the reaction occurs.* The rates of chemical reactions increase as temperature is increased. We refrigerate perishable foods such as milk for this reason. The bacterial reactions that lead to the spoiling of milk proceed much more rapidly at room temperature than they do at the lower temperature of a refrigerator. Increasing temperature increases the kinetic energies of molecules. ∞ (Section 10.7) As molecules move more rapidly, they collide more frequently and also with higher energy, leading to increased reaction rates.

4. *The presence of a catalyst.* Catalysts are agents that increase reaction rates without being used up. They affect the kinds of collisions (the mechanism) that lead to reaction. Catalysts play a crucial role in our lives. The physiology of most living species depends on *enzymes*, protein molecules that act as catalysts, increasing the rates of selected biochemical reactions.

On a molecular level, reaction rates depend on the frequency of collisions between molecules. The greater the frequency of collisions, the greater is the rate of reaction. For a collision to lead to a reaction, however, it must occur with sufficient energy to stretch bonds to a critical length and with suitable orientation for new bonds to form in the proper locations. We will consider these factors as we proceed through this chapter.

(a)

(b)

▲ **Figure 14.2 Effect of concentration on rate.** (a) When heated in air, steel wool glows red-hot but oxidizes slowly. (b) When the red-hot steel wool is placed in an atmosphere of pure oxygen, it burns vigorously, forming Fe_2O_3 at a much faster rate. The difference in behavior is due to the different concentrations of O_2 in the two environments.

▲ **GIVE IT SOME THOUGHT**

How does increasing the partial pressures of the reactive components of a gaseous mixture affect the rate at which the components react with one another?

14.2 REACTION RATES

The *speed* of an event is defined as the *change* that occurs in a given interval of *time*: Whenever we talk about speed, we necessarily bring in the notion of time. For example, the speed of a car is expressed as the change in the car's position over a certain period of time. The units of this speed are usually miles per hour (mi/hr)—that is, the quantity that is changing (position, measured in miles) divided by a time interval (hours).

Similarly, the speed of a chemical reaction—its **reaction rate**—is the change in the concentration of reactants or products per unit of time. Thus, the units for reaction rate are usually molarity per second (M/s)—that is, the change in concentration (measured in molarity) divided by a time interval (seconds).

Let's consider a simple hypothetical reaction, A $\longrightarrow$ B, depicted in Figure 14.3▼. Each red sphere represents 0.01 mol of A, each blue sphere represents 0.01 mol of B, and the container has a volume of 1.00 L. At the beginning of the reaction there is 1.00 mol A, so the concentration is 1.00 mol/L = 1.00 M. After 20 s the concentration of A has fallen to 0.54 M, whereas the concentration of B has risen to 0.46 M. The sum of the concentrations is still 1.00 M because 1 mol

0 s

1.00 mol A
0 mol B

(a)

20 s

0.54 mol A
0.46 mol B

(b)

40 s

0.30 mol A
0.70 mol B

(c)

◀ **Figure 14.3 Progress of a hypothetical reaction A $\longrightarrow$ B.** Each red sphere represents 0.01 mol A, each blue sphere represents 0.01 mol B, and the vessel has a volume of 1.00 L. (a) At time zero the vessel contains 1.00 mol A (100 red spheres) and 0 mol B (no blue spheres). (b) After 20 s the vessel contains 0.54 mol A and 0.46 mol B. (c) After 40 s the vessel contains 0.30 mol A and 0.70 mol B.

of B is produced for each mole of A that reacts. After 40 s the concentration of A is 0.30 M and that of B is 0.70 M.

The rate of this reaction can be expressed either as the rate of disappearance of reactant A or as the rate of appearance of product B. The *average* rate of appearance of B over a particular time interval is given by the change in concentration of B divided by the change in time:

$$\text{Average rate of appearance of B} = \frac{\text{change in concentration of B}}{\text{change in time}}$$

$$= \frac{[B]\ at\ t_2 - [B]\ at\ t_1}{t_2 - t_1} = \frac{\Delta[B]}{\Delta t} \qquad [14.1]$$

We use brackets around a chemical formula, as in [B], to indicate the concentration of the substance in molarity. The Greek letter delta, Δ, is read "change in" and is always equal to the final quantity minus the initial quantity. ∞ (Section 5.2) The average rate of appearance of B over the 20-s interval from the beginning of the reaction ($t_1 = 0$ s to $t_2 = 20$ s) is given by

$$\text{Average rate} = \frac{0.46\ M - 0.00\ M}{20\ s - 0\ s} = 2.3 \times 10^{-2}\ M/s$$

We could equally well express the rate of the reaction with respect to the change of concentration of the reactant, A. In this case we would be describing the rate of disappearance of A, which we express as

$$\text{Average rate of disappearance of A} = -\frac{\Delta[A]}{\Delta t} \qquad [14.2]$$

Notice the minus sign in this equation. By convention, *rates are always expressed as positive quantities*. Because [A] is decreasing with time, $\Delta[A]$ is a negative number. We use the negative sign to convert the negative $\Delta[A]$ to a positive rate. Because one molecule of A is consumed for every molecule of B that forms, the average rate of disappearance of A equals the average rate of appearance of B, as calculation shows:

$$\text{Average rate} = -\frac{\Delta[A]}{\Delta t} = -\frac{0.54\ M - 1.00\ M}{20\ s - 0\ s} = 2.3 \times 10^{-2}\ M/s$$

■ SAMPLE EXERCISE 14.1 | Calculating an Average Rate of Reaction

From the data given in the caption of Figure 14.3, calculate the average rate at which A disappears over the time interval from 20 s to 40 s.

SOLUTION

Analyze: We are given the concentration of A at 20 s (0.54 M) and at 40 s (0.30 M) and asked to calculate the average rate of reaction over this time interval.

Plan: The average rate is given by the change in concentration, $\Delta[A]$, divided by the corresponding change in time, Δt. Because A is a reactant, a minus sign is used in the calculation to make the rate a positive quantity.

Solve: Average rate $= -\dfrac{\Delta[A]}{\Delta t} = -\dfrac{0.30\ M - 0.54\ M}{40\ s - 20\ s} = 1.2 \times 10^{-2}\ M/s$

■ PRACTICE EXERCISE

For the reaction pictured in Figure 14.3, calculate the average rate of appearance of B over the time interval from 0 to 40 s.
Answer: $1.8 \times 10^{-2}\ M/s$

TABLE 14.1 ▪ Rate Data for Reaction of C_4H_9Cl with Water		
Time, t(s)	$[C_4H_9Cl](M)$	Average Rate (M/s)
0.0	0.1000	1.9×10^{-4}
50.0	0.0905	1.7×10^{-4}
100.0	0.0820	1.6×10^{-4}
150.0	0.0741	1.4×10^{-4}
200.0	0.0671	1.22×10^{-4}
300.0	0.0549	1.01×10^{-4}
400.0	0.0448	0.80×10^{-4}
500.0	0.0368	0.560×10^{-4}
800.0	0.0200	
10,000	0	

Change of Rate with Time

Now, let's consider an actual chemical reaction, one that occurs when butyl chloride (C_4H_9Cl) is placed in water. The products formed are butyl alcohol (C_4H_9OH) and hydrochloric acid:

$$C_4H_9Cl(aq) + H_2O(l) \longrightarrow C_4H_9OH(aq) + HCl(aq) \qquad [14.3]$$

Suppose that we prepare a 0.1000 M aqueous solution of C_4H_9Cl and then measure the concentration of C_4H_9Cl at various times after time zero (the time the reactants are mixed, thereby initiating the reaction). The resultant data are shown in the first two columns of Table 14.1 ▲. We can use these data to calculate the average rate of disappearance of C_4H_9Cl over the intervals between measurements; these rates are given in the third column. Notice that the average rate decreases over each 50-s interval for the first several measurements and continues to decrease over even larger intervals through the remaining measurements. *It is typical for rates to decrease as a reaction proceeds, because the concentration of reactants decreases.* The change in rate as the reaction proceeds is also seen in a graph of the concentration of C_4H_9Cl versus time (Figure 14.4 ▶). Notice how the steepness of the curve decreases with time, indicating a decreasing rate of reaction.

▲ **GIVE IT SOME THOUGHT**

Why do the rates of reactions decrease as concentrations decrease?

Instantaneous Rate

Graphs showing how the concentration of a reactant or product changes with time, such as the graph in Figure 14.4, allow us to evaluate the **instantaneous rate**, the rate at a particular moment in the reaction. The instantaneous rate is determined from the slope (or tangent) of this curve at the point of interest. We have drawn two tangents in Figure 14.4, one at $t = 0$ and the other at $t = 600$ s. The slopes of these tangents give the instantaneous rates at these times.* For example, to determine the instantaneous rate at 600 s, we draw the tangent to the curve at this time, then construct horizontal and vertical lines to

You may wish to review briefly the idea of graphical determination of slopes by referring to Appendix A. If you are familiar with calculus, you may recognize that the average rate approaches the instantaneous rate as the time interval approaches zero. This limit, in the notation of calculus, is represented as $-d[C_4H_9Cl]/dt$.

▲ **Figure 14.4 Concentration of butyl chloride (C_4H_9Cl) as a function of time.** The dots represent the experimental data from the first two columns of Table 14.1, and the red curve is drawn to connect the data points smoothly. Lines are drawn that are tangent to the curve at $t = 0$ and $t = 600$ s. The slope of each tangent is defined as the vertical change divided by the horizontal change: $\Delta[C_4H_9Cl]/\Delta t$. The reaction rate at any time is related to the slope of the tangent to the curve at that time. Because C_4H_9Cl is disappearing, the rate is equal to the negative of the slope.

form the right triangle shown. The slope is the ratio of the height of the vertical side to the length of the horizontal side:

$$\text{Instantaneous rate} = -\frac{\Delta[C_4H_9Cl]}{\Delta t} = -\frac{(0.017 - 0.042)\ M}{(800 - 400)s}$$

$$= 6.3 \times 10^{-5}\ M/s$$

In discussions that follow, the term "rate" means "instantaneous rate," unless indicated otherwise. The instantaneous rate at $t = 0$ is called the *initial rate* of the reaction.

To understand better the difference between average rate and instantaneous rate, imagine that you have just driven 98 mi in 2.0 hr. Your average speed is 49 mi/hr, whereas your instantaneous speed at any moment is the speedometer reading at that time.

■ SAMPLE EXERCISE 14.2 | **Calculating an Instantaneous Rate of Reaction**

Using Figure 14.4, calculate the instantaneous rate of disappearance of C_4H_9Cl at $t = 0$ (the initial rate).

SOLUTION

Analyze: We are asked to determine an instantaneous rate from a graph of concentration versus time.

Plan: To obtain the instantaneous rate at $t = 0$, we must determine the slope of the curve at $t = 0$. The tangent is drawn on the graph. The slope of this straight line equals the change in the vertical axis divided by the corresponding change in the horizontal axis (that is, change in molarity over change in time).

Solve: The straight line falls from $[C_4H_9Cl] = 0.100\ M$ to $0.060\ M$ in the time change from 0 s to 210 s, as indicated by the tan triangle shown in Figure 14.4. Thus, the initial rate is

$$\text{Rate} = -\frac{\Delta[C_4H_9Cl]}{\Delta t} = -\frac{(0.060 - 0.100)\ M}{(210 - 0)\ s} = 1.9 \times 10^{-4}\ M/s$$

■ PRACTICE EXERCISE

Using Figure 14.4, determine the instantaneous rate of disappearance of C_4H_9Cl at $t = 300$ s.
Answer: $1.1 \times 10^{-4}\ M/s$

▲ **GIVE IT SOME THOUGHT**

Figure 14.4 shows two triangles that are used to determine the slope of the curve at two different times. How do you determine how large a triangle to draw when determining the slope of a curve at a particular point?

Reaction Rates and Stoichiometry

During our earlier discussion of the hypothetical reaction A $\longrightarrow$ B, we saw that the stoichiometry requires that the rate of disappearance of A equals the rate of appearance of B. Likewise, the stoichiometry of Equation 14.3 indicates that 1 mol of C_4H_9OH is produced for each mole of C_4H_9Cl consumed. Therefore, the rate of appearance of C_4H_9OH equals the rate of disappearance of C_4H_9Cl:

$$\text{Rate} = -\frac{\Delta[C_4H_9Cl]}{\Delta t} = \frac{\Delta[C_4H_9OH]}{\Delta t}$$

What happens when the stoichiometric relationships are not one to one? For example, consider this reaction:

$$2 \, HI(g) \longrightarrow H_2(g) + I_2(g)$$

We can measure the rate of disappearance of HI or the rate of appearance of either H_2 or I_2. Because 2 mol of HI disappear for each mole of H_2 or I_2 that forms, the rate of disappearance of HI is twice the rate of appearance of either H_2 or I_2. To equate the rates, we must therefore divide the rate of disappearance of HI by 2 (its coefficient in the balanced chemical equation):

$$Rate = -\frac{1}{2}\frac{\Delta[HI]}{\Delta t} = \frac{\Delta[H_2]}{\Delta t} = \frac{\Delta[I_2]}{\Delta t}$$

In general, for the reaction

$$a \, A + b \, B \longrightarrow c \, C + d \, D$$

the rate is given by

$$Rate = -\frac{1}{a}\frac{\Delta[A]}{\Delta t} = -\frac{1}{b}\frac{\Delta[B]}{\Delta t} = \frac{1}{c}\frac{\Delta[C]}{\Delta t} = \frac{1}{d}\frac{\Delta[D]}{\Delta t} \qquad [14.4]$$

When we speak of the rate of a reaction without specifying a particular reactant or product, we will mean it in this sense.*

■ SAMPLE EXERCISE 14.3 | Relating Rates at Which Products Appear and Reactants Disappear

(a) How is the rate at which ozone disappears related to the rate at which oxygen appears in the reaction $2 \, O_3(g) \longrightarrow 3 \, O_2(g)$?
(b) If the rate at which O_2 appears, $\Delta[O_2]/\Delta t$, is 6.0×10^{-5} M/s at a particular instant, at what rate is O_3 disappearing at this same time, $-\Delta[O_3]/\Delta t$?

SOLUTION

Analyze: We are given a balanced chemical equation and asked to relate the rate of appearance of the product to the rate of disappearance of the reactant.

Plan: We can use the coefficients in the chemical equation as shown in Equation 14.4 to express the relative rates of reactions.

Solve: (a) Using the coefficients in the balanced equation and the relationship given by Equation 14.4, we have:

$$Rate = -\frac{1}{2}\frac{\Delta[O_3]}{\Delta t} = \frac{1}{3}\frac{\Delta[O_2]}{\Delta t}$$

(b) Solving the equation from part (a) for the rate at which O_3 disappears, $-\Delta[O_3]/\Delta t$ we have:

$$-\frac{\Delta[O_3]}{\Delta t} = \frac{2}{3}\frac{\Delta[O_2]}{\Delta t} = \frac{2}{3}(6.0 \times 10^{-5} \, M/s) = 4.0 \times 10^{-5} \, M/s$$

Check: We can directly apply a stoichiometric factor to convert the O_2 formation rate to the rate at which the O_3 disappears:

$$-\frac{\Delta[O_3]}{\Delta t} = \left(6.0 \times 10^{-5} \frac{mol \, O_2/L}{s}\right)\left(\frac{2 \, mol \, O_3}{3 \, mol \, O_2}\right) = 4.0 \times 10^{-5} \frac{mol \, O_3/L}{s}$$

$$= 4.0 \times 10^{-5} \, M/s$$

■ PRACTICE EXERCISE

The decomposition of N_2O_5 proceeds according to the following equation:

$$2 \, N_2O_5(g) \longrightarrow 4 \, NO_2(g) + O_2(g)$$

If the rate of decomposition of N_2O_5 at a particular instant in a reaction vessel is 4.2×10^{-7} M/s, what is the rate of appearance of (a) NO_2, (b) O_2?
Answers: **(a)** 8.4×10^{-7} M/s, **(b)** 2.1×10^{-7} M/s

Equation 14.4 does not hold true if substances other than C and D are formed in significant amounts during the course of the reaction. For example, sometimes intermediate substances build in concentration before forming the final products. In that case, the relationship between the rate of disappearance of reactants and the rate of appearance of products will not be given by Equation 14.4. All reactions whose rates we consider in this chapter obey Equation 14.4.

A variety of techniques can be used to monitor the concentration of a reactant or product during a reaction. Spectroscopic methods, which rely on the ability of substances to absorb (or emit) electromagnetic radiation, are some of the most useful. Spectroscopic kinetic studies are often performed with the reaction mixture in the sample compartment of the spectrometer. The spectrometer is set to measure the light absorbed at a wavelength characteristic of one of the reactants or products. In the decomposition of HI(g) into $H_2(g)$ and $I_2(g)$ for example, both HI and H_2 are colorless, whereas I_2 is violet. During the course of the reaction, the color increases in intensity as I_2 forms. Thus, visible light of appropriate wavelength can be used to monitor the reaction.

Figure 14.5▼ shows the basic components of a spectrometer. The spectrometer measures the amount of light absorbed by the sample by comparing the intensity of the light emitted from the light source with the intensity of the light that emerges from the sample. As the concentration of I_2 increases and its color becomes more intense, the amount of light absorbed by the reaction mixture increases, causing less light to reach the detector.

Beer's law relates the amount of light being absorbed to the concentration of the substance absorbing the light:

$$A = abc \qquad [14.5]$$

In this equation, A is the measured absorbance, a is the molar absorptivity constant (a characteristic of the substance being monitored), b is the path length through which the radiation must pass, and c is the molar concentration of the absorbing substance. Thus, the concentration is directly proportional to absorbance.

Related Exercise: 14.92

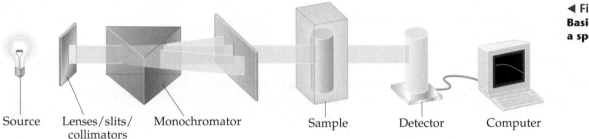

◀ **Figure 14.5**
Basic components of a spectrometer.

Source   Lenses/slits/   Monochromator   Sample   Detector   Computer
          collimators

14.3 THE RATE LAW: THE EFFECT OF CONCENTRATION ON RATE

One way of studying the effect of concentration on reaction rate is to determine the way in which the rate at the beginning of a reaction (the initial rate) depends on the starting concentrations. To illustrate this approach, consider the following reaction:

$$NH_4^+(aq) + NO_2^-(aq) \longrightarrow N_2(g) + 2\,H_2O(l)$$

We might study the rate of this reaction by measuring the concentration of NH_4^+ or NO_2^- as a function of time or by measuring the volume of N_2 collected. Because the stoichiometric coefficients on NH_4^+, NO_2^-, and N_2 are all the same, all of these rates will be equal.

Table 14.2▼ shows the initial reaction rate for various starting concentrations of NH_4^+ and NO_2^-. These data indicate that changing either $[NH_4^+]$ or

TABLE 14.2 ■ Rate Data for the Reaction of Ammonium and Nitrite Ions in Water at 25 °C			
Experiment Number	Initial NH_4^+ Concentration (M)	Initial NO_2^- Concentration (M)	Observed Initial Rate (M/s)
1	0.0100	0.200	5.4×10^{-7}
2	0.0200	0.200	10.8×10^{-7}
3	0.0400	0.200	21.5×10^{-7}
4	0.200	0.0202	10.8×10^{-7}
5	0.200	0.0404	21.6×10^{-7}
6	0.200	0.0808	43.3×10^{-7}

$[NO_2^-]$ changes the reaction rate. Notice that if we double $[NH_4^+]$ while holding $[NO_2^-]$ constant, the rate doubles (compare experiments 1 and 2). If $[NH_4^+]$ is increased by a factor of 4 with $[NO_2^-]$ left unchanged (compare experiments 1 and 3), the rate changes by a factor of 4, and so forth. These results indicate that the rate is proportional to $[NH_4^+]$. When $[NO_2^-]$ is similarly varied while $[NH_4^+]$ is held constant, the rate is affected in the same manner. Thus, the rate is also directly proportional to the concentration of NO_2^-. We can express the way in which the rate depends on the concentrations of the reactants, NH_4^+ and NO_2^-, in terms of the following equation:

$$\text{Rate} = k[NH_4^+][NO_2^-] \qquad [14.6]$$

An equation such as Equation 14.6, which shows how the rate depends on the concentrations of reactants, is called a **rate law**. For a general reaction,

$$a\,A + b\,B \longrightarrow c\,C + d\,D$$

the rate law generally has the form

$$\text{Rate} = k[A]^m[B]^n \qquad [14.7]$$

The constant k in the rate law is called the **rate constant**. The magnitude of k changes with temperature and therefore determines how temperature affects rate, as we will see in Section 14.5. The exponents m and n are typically small whole numbers (usually 0, 1, or 2). We will consider these exponents more closely very shortly.

 If we know the rate law for a reaction and its rate for a set of reactant concentrations, we can calculate the value of the rate constant, k. For example, using the data in Table 14.2 and the results from experiment 1, we can substitute into Equation 14.6

$$5.4 \times 10^{-7}\ M/s = k(0.0100\ M)(0.200\ M)$$

Solving for k gives

$$k = \frac{5.4 \times 10^{-7}\ M/s}{(0.0100\ M)(0.200\ M)} = 2.7 \times 10^{-4}\ M^{-1}\,s^{-1}$$

You may wish to verify that this same value of k is obtained using any of the other experimental results given in Table 14.2.

 Once we have both the rate law and the value of the rate constant for a reaction, we can calculate the rate of reaction for any set of concentrations. For example, using Equation 14.6 and $k = 2.7 \times 10^{-4}\ M^{-1}\,s^{-1}$, we can calculate the rate for $[NH_4^+] = 0.100\ M$ and $[NO_2^-] = 0.100\ M$:

$$\text{Rate} = (2.7 \times 10^{-4}\ M^{-1}\,s^{-1})(0.100\ M)\,(0.100\ M) = 2.7 \times 10^{-6}\ M/s$$

GIVE IT SOME THOUGHT

(a) What is a rate law? **(b)** What is the name of the quantity k in any rate law?

Reaction Orders: The Exponents in the Rate Law

The rate laws for most reactions have the general form

$$\text{Rate} = k[\text{reactant 1}]^m[\text{reactant 2}]^n \ldots \qquad [14.8]$$

The exponents m and n in a rate law are called **reaction orders**. For example, consider again the rate law for the reaction of NH_4^+ with NO_2^-:

$$\text{Rate} = k[NH_4^+][NO_2^-]$$

Because the exponent of $[NH_4^+]$ is 1, the rate is *first order* in NH_4^+. The rate is also first order in NO_2^-. (The exponent "1" is not shown explicitly in rate laws.) The **overall reaction order** is the sum of the orders with respect to each reactant in the rate law. Thus, the rate law has an overall reaction order of $1 + 1 = 2$, and the reaction is *second order overall*.

The exponents in a rate law indicate how the rate is affected by the concentration of each reactant. Because the rate at which NH_4^+ reacts with NO_2^- depends on $[NH_4^+]$ raised to the first power, the rate doubles when $[NH_4^+]$ doubles, triples when $[NH_4^+]$ triples, and so forth. Doubling or tripling $[NO_2^-]$ likewise doubles or triples the rate. If a rate law is second order with respect to a reactant, $[A]^2$, then doubling the concentration of that substance causes the reaction rate to quadruple ($[2]^2 = 4$), whereas tripling the concentration causes the rate to increase ninefold ($[3]^2 = 9$).

The following are some additional examples of rate laws:

$$2\,N_2O_5(g) \longrightarrow 4\,NO_2(g) + O_2(g) \quad \text{Rate} = k[N_2O_5] \qquad [14.9]$$

$$CHCl_3(g) + Cl_2(g) \longrightarrow CCl_4(g) + HCl(g) \quad \text{Rate} = k[CHCl_3][Cl_2]^{1/2} \quad [14.10]$$

$$H_2(g) + I_2(g) \longrightarrow 2\,HI(g) \qquad\qquad \text{Rate} = k[H_2][I_2] \qquad [14.11]$$

Although the exponents in a rate law are sometimes the same as the coefficients in the balanced equation, this is not necessarily the case, as seen in Equations 14.9 and 14.10. *The values of these exponents must be determined experimentally.* In most rate laws, reaction orders are 0, 1, or 2. However, we also occasionally encounter rate laws in which the reaction order is fractional (such as Equation 14.10) or even negative.

▲ **GIVE IT SOME THOUGHT**

The experimentally determined rate law for the reaction $2\,NO(g) + 2\,H_2(g) \longrightarrow N_2(g) + 2\,H_2O(g)$ is rate = $k[NO]^2[H_2]$. **(a)** What are the reaction orders in this rate law? **(b)** Does doubling the concentration of NO have the same effect on rate as doubling the concentration of H_2?

■■■ **SAMPLE EXERCISE 14.4** | Relating a Rate Law to the Effect of Concentration on Rate

Consider a reaction $A + B \longrightarrow C$ for which rate = $k[A][B]^2$. Each of the following boxes represents a reaction mixture in which A is shown as red spheres and B as purple ones. Rank these mixtures in order of increasing rate of reaction.

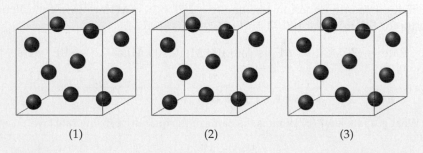

(1)   (2)   (3)

SOLUTION

Analyze: We are given three boxes containing different numbers of spheres representing mixtures containing different reactant concentrations. We are asked to use the given rate law and the compositions of the boxes to rank the mixtures in order of increasing reaction rates.

Plan: Because all three boxes have the same volume, we can put the number of spheres of each kind into the rate law and calculate the rate for each box.

Solve: Box 1 contains 5 red spheres and 5 purple spheres, giving the following rate:

$$\text{Box 1: Rate} = k(5)(5)^2 = 125k$$

Box 2 contains 7 red spheres and 3 purple spheres:

$$\text{Box 2: Rate} = k(7)(3)^2 = 63k$$

Box 3 contains 3 red spheres and 7 purple spheres:

$$\text{Box 3: Rate} = k(3)(7)^2 = 147k$$

The slowest rate is $63k$ (box 2), and the highest is $147k$ (box 3). Thus, the rates vary in the order $2 < 1 < 3$.

Check: Each box contains 10 spheres. The rate law indicates that in this case [B] has a greater influence on rate than [A] because B has a higher reaction order. Hence, the mixture with the highest concentration of B (most purple spheres) should react fastest. This analysis confirms the order $2 < 1 < 3$.

■ PRACTICE EXERCISE
Assuming that rate $= k[A][B]$, rank the mixtures represented in this Sample Exercise in order of increasing rate.
Answer: $2 = 3 < 1$

Units of Rate Constants

The units of the rate constant depend on the overall reaction order of the rate law. In a reaction that is second order overall, for example, the units of the rate constant must satisfy the equation:

$$\text{Units of rate} = (\text{units of rate constant})(\text{units of concentration})^2$$

Hence, in our usual units of concentration and time

$$\text{Units of rate constant} = \frac{\text{units of rate}}{(\text{units of concentration})^2} = \frac{M/s}{M^2} = M^{-1}\,s^{-1}$$

■ SAMPLE EXERCISE 14.5 | **Determining Reaction Orders and Units for Rate Constants**

(a) What are the overall reaction orders for the reactions described in Equations 14.9 and 14.10? (b) What are the units of the rate constant for the rate law in Equation 14.9?

SOLUTION

Analyze: We are given two rate laws and asked to express (a) the overall reaction order for each and (b) the units for the rate constant for the first reaction.

Plan: The overall reaction order is the sum of the exponents in the rate law. The units for the rate constant, k, are found by using the normal units for rate (M/s) and concentration (M) in the rate law and applying algebra to solve for k.

Solve: (a) The rate of the reaction in Equation 14.9 is first order in N_2O_5 and first order overall. The reaction in Equation 14.10 is first order in $CHCl_3$ and one-half order in Cl_2. The overall reaction order is three halves.
(b) For the rate law for Equation 14.9, we have

$$\text{Units of rate} = (\text{units of rate constant})(\text{units of concentration})$$

So

$$\text{Units of rate constant} = \frac{\text{units of rate}}{\text{units of concentration}} = \frac{M/s}{M} = s^{-1}$$

Notice that the units of the rate constant change as the overall order of the reaction changes.

■ PRACTICE EXERCISE
(a) What is the reaction order of the reactant H_2 in Equation 14.11? (b) What are the units of the rate constant for Equation 14.11?
Answers: (a) 1, (b) $M^{-1}s^{-1}$

Using Initial Rates to Determine Rate Laws

The rate law for any chemical reaction must be determined experimentally; it cannot be predicted by merely looking at the chemical equation. We often determine the rate law for a reaction by the same method we applied to the data in Table 14.2: We observe the effect of changing the initial concentrations of the reactants on the initial rate of the reaction.

We have seen that the rate laws for most reactions have the general form

$$\text{Rate} = k[\text{reactant 1}]^m[\text{reactant 2}]^n \dots$$

Thus, the task of determining the rate law becomes one of determining the reaction orders, m and n. In most reactions the reaction orders are 0, 1, or 2. If a reaction is zero order in a particular reactant, changing its concentration will have no effect on rate (as long as some of the reactant is present) because any concentration raised to the zero power equals 1. On the other hand, we have seen that when a reaction is first order in a reactant, changes in the concentration of that reactant will produce proportional changes in the rate. Thus, doubling the concentration will double the rate, and so forth. Finally, when the rate law is second order in a particular reactant, doubling its concentration increases the rate by a factor of $2^2 = 4$, tripling its concentration causes the rate to increase by a factor of $3^2 = 9$, and so forth.

In working with rate laws, it is important to realize that the *rate* of a reaction depends on concentration, but the *rate constant* does not. As we will see later in this chapter, the rate constant (and hence the reaction rate) is affected by temperature and by the presence of a catalyst.

■ SAMPLE EXERCISE 14.6 | Determining a Rate Law from Initial Rate Data

The initial rate of a reaction $A + B \longrightarrow C$ was measured for several different starting concentrations of A and B, and the results are as follows:

Experiment Number	[A] (M)	[B] (M)	Initial Rate (M/s)
1	0.100	0.100	4.0×10^{-5}
2	0.100	0.200	4.0×10^{-5}
3	0.200	0.100	16.0×10^{-5}

Using these data, determine **(a)** the rate law for the reaction, **(b)** the rate constant, **(c)** the rate of the reaction when [A] = 0.050 M and [B] = 0.100 M.

SOLUTION

Analyze: We are given a table of data that relates concentrations of reactants with initial rates of reaction and asked to determine **(a)** the rate law, **(b)** the rate constant, and **(c)** the rate of reaction for a set of concentrations not listed in the table.

Plan: (a) We assume that the rate law has the following form: Rate = $k[\text{A}]^m[\text{B}]^n$ so we must use the given data to deduce the reaction orders m and n. We do so by determining how changes in the concentration change the rate. **(b)** Once we know m and n, we can use the rate law and one of the sets of data to determine the rate constant k. **(c)** Now that we know both the rate constant and the reaction orders, we can use the rate law with the given concentrations to calculate rate.

Solve: (a) As we move from experiment 1 to experiment 2, [A] is held constant and [B] is doubled. Thus, this pair of experiments shows how [B] affects the rate, allowing us to deduce the order of the rate law with respect to B. Because the rate remains the same when [B] is doubled, the concentration of B has no effect on the reaction rate. The rate law is therefore zero order in B (that is, $n = 0$).

In experiments 1 and 3, [B] is held constant so these data show how [A] affects rate. Holding [B] constant while doubling [A] increases the rate fourfold. This result indicates that rate is proportional to $[\text{A}]^2$ (that is, the reaction is second order in A). Hence, the rate law is

$$\text{Rate} = k[\text{A}]^2[\text{B}]^0 = k[\text{A}]^2$$

This rate law could be reached in a more formal way by taking the ratio of the rates from two experiments:

$$\frac{\text{Rate 2}}{\text{Rate 1}} = \frac{4.0 \times 10^{-5}\ M/s}{4.0 \times 10^{-5}\ M/s} = 1$$

Using the rate law, we have

$$1 = \frac{\text{rate 2}}{\text{rate 1}} = \frac{k[0.100\ M]^{m}[0.200\ M]^{n}}{k[0.100\ M]^{m}[0.100\ M]^{n}} = \frac{[0.200]^{n}}{[0.100]^{n}} = 2^{n}$$

2^{n} equals 1 under only one condition:

$$n = 0$$

We can deduce the value of m in a similar fashion:

$$\frac{\text{Rate 3}}{\text{Rate 1}} = \frac{16.0 \times 10^{-5}\ M/s}{4.0 \times 10^{-5}\ M/s} = 4$$

Using the rate law gives

$$4 = \frac{\text{rate 3}}{\text{rate 1}} = \frac{k[0.200\ M]^{m}[0.100\ M]^{n}}{k[0.100\ M]^{m}[0.100\ M]^{n}} = \frac{[0.200]^{m}}{[0.100]^{m}} = 2^{m}$$

Because $2^{m} = 4$, we conclude that

$$m = 2$$

(b) Using the rate law and the data from experiment 1, we have

$$k = \frac{\text{rate}}{[A]^{2}} = \frac{4.0 \times 10^{-5}\ M/s}{(0.100\ M)^{2}} = 4.0 \times 10^{-3}\ M^{-1}\,s^{-1}$$

(c) Using the rate law from part (a) and the rate constant from part (b), we have

$$\text{Rate} = k[A]^{2} = (4.0 \times 10^{-3}\ M^{-1}\,s^{-1})(0.050\ M)^{2} = 1.0 \times 10^{-5}\ M/s$$

Because [B] is not part of the rate law, it is irrelevant to the rate, if there is at least some B present to react with A.

Check: A good way to check our rate law is to use the concentrations in experiment 2 or 3 and see if we can correctly calculate the rate. Using data from experiment 3, we have

$$\text{Rate} = k[A]^{2} = (4.0 \times 10^{-3}\ M^{-1}\,s^{-1})(0.200\ M)^{2} = 1.6 \times 10^{-4}\ M/s$$

Thus, the rate law correctly reproduces the data, giving both the correct number and the correct units for the rate.

■ PRACTICE EXERCISE

The following data were measured for the reaction of nitric oxide with hydrogen:

$$2\ NO(g) + 2\ H_2(g) \longrightarrow N_2(g) + 2\ H_2O(g)$$

Experiment Number	[NO] (M)	[H$_2$] (M)	Initial Rate (M/s)
1	0.10	0.10	1.23×10^{-3}
2	0.10	0.20	2.46×10^{-3}
3	0.20	0.10	4.92×10^{-3}

(a) Determine the rate law for this reaction. **(b)** Calculate the rate constant. **(c)** Calculate the rate when [NO] = 0.050 M and [H$_2$] = 0.150 M

Answers: **(a)** rate = $k[NO]^{2}[H_2]$; **(b)** $k = 1.2\ M^{-2}\,s^{-1}$; **(c)** rate = $4.5 \times 10^{-4}\ M/s$

14.4 THE CHANGE OF CONCENTRATION WITH TIME

The rate laws that we have examined so far enable us to calculate the rate of a reaction from the rate constant and reactant concentrations. These rate laws can also be converted into equations that show the relationship between the concentrations of the reactants or products and time. The mathematics required to accomplish this conversion involves calculus. We do not expect you to be able to perform the calculus operations; however, you should be able to use the resulting equations. We will apply this conversion to two of the simplest rate laws: those that are first order overall and those that are second order overall.

First-Order Reactions

A **first-order reaction** is one whose rate depends on the concentration of a single reactant raised to the first power. For a reaction of the type A ⟶ products the rate law may be first order:

$$\text{Rate} = -\frac{\Delta[A]}{\Delta t} = k[A]$$

This form of a rate law, which expresses how rate depends on concentration, is called the *differential rate law*. Using an operation from calculus called integration, this relationship can be transformed into an equation that relates the concentration of A at the start of the reaction, $[A]_0$, to its concentration at any other time t, $[A]_t$:

$$\ln[A]_t - \ln[A]_0 = -kt \qquad \text{or} \qquad \ln\frac{[A]_t}{[A]_0} = -kt \qquad [14.12]$$

This form of the rate law is called the *integrated rate law*. The function "ln" in Equation 14.12 is the natural logarithm (Appendix A.2). Equation 14.12 can also be rearranged and written as follows:

$$\ln[A]_t = -kt + \ln[A]_0 \qquad [14.13]$$

Equations 14.12 and 14.13 can be used with any concentration units, as long as the units are the same for both $[A]_t$ and $[A]_0$.

For a first-order reaction, Equation 14.12 or 14.13 can be used in several ways. Given any three of the following quantities, we can solve for the fourth: k, t, $[A]_0$, and $[A]_t$. Thus, you can use these equations, for example, to determine (1) the concentration of a reactant remaining at any time after the reaction has started, (2) the time required for a given fraction of a sample to react, or (3) the time required for a reactant concentration to fall to a certain level.

■ SAMPLE EXERCISE 14.7 | Using the Integrated First-Order Rate Law

The decomposition of a certain insecticide in water follows first-order kinetics with a rate constant of 1.45 yr^{-1} at 12 °C. A quantity of this insecticide is washed into a lake on June 1, leading to a concentration of 5.0×10^{-7} g/cm^3. Assume that the average temperature of the lake is 12 °C. **(a)** What is the concentration of the insecticide on June 1 of the following year? **(b)** How long will it take for the concentration of the insecticide to decrease to 3.0×10^{-7} g/cm^3?

SOLUTION

Analyze: We are given the rate constant for a reaction that obeys first-order kinetics, as well as information about concentrations and times, and asked to calculate how much reactant (insecticide) remains after one year. We must also determine the time interval needed to reach a particular insecticide concentration. Because the exercise gives time in (a) and asks for time in (b), we know that the integrated rate law, Equation 14.13, is required.

Plan: **(a)** We are given $k = 1.45$ yr^{-1}, $t = 1.00$ yr and $[\text{insecticide}]_0 = 5.0 \times 10^{-7}$ g/cm^3, and so Equation 14.13 can be solved for $[\text{insecticide}]_t$. **(b)** We have $k = 1.45$ yr^{-1}, $[\text{insecticide}]_0 = 5.0 \times 10^{-7}$ g/cm^3, and $[\text{insecticide}]_t = 3.0 \times 10^{-7}$ g/cm^3, and so we can solve Equation 14.13 for time, t.

Solve: **(a)** Substituting the known quantities into Equation 14.13, we have

$$\ln[\text{insecticide}]_{t=1\,yr} = -(1.45\ yr^{-1})(1.00\ yr) + \ln(5.0 \times 10^{-7})$$

We use the ln function on a calculator to evaluate the second term on the right, giving

$$\ln[\text{insecticide}]_{t=1\,yr} = -1.45 + (-14.51) = -15.96$$

To obtain $[\text{insecticide}]_{t=1\,yr}$, we use the inverse natural logarithm, or e^x, function on the calculator:

$$[\text{insecticide}]_{t=1\,yr} = e^{-15.96} = 1.2 \times 10^{-7}\ g/cm^3$$

Note that the concentration units for $[A]_t$ and $[A]_0$ must be the same.

(b) Again substituting into Equation 14.13, with $[\text{insecticide}]_t = 3.0 \times 10^{-7}\ g/cm^3$, gives

$$\ln(3.0 \times 10^{-7}) = -(1.45\ yr^{-1})(t) + \ln(5.0 \times 10^{-7})$$

Solving for t gives

$$t = -[\ln(3.0 \times 10^{-7}) - \ln(5.0 \times 10^{-7})]/1.45\ yr^{-1}$$
$$= -(-15.02 + 14.51)/1.45\ yr^{-1} = 0.35\ yr$$

Check: In part (a) the concentration remaining after 1.00 yr (that is, 1.2×10^{-7} g/cm^3) is less than the original concentration (5.0×10^{-7} g/cm^3), as it should be. In (b) the given concentration (3.0×10^{-7} g/cm^3) is greater than that remaining after 1.00 yr, indicating that the time must be less than a year. Thus, $t = 0.35$ yr is a reasonable answer.

■ PRACTICE EXERCISE

The decomposition of dimethyl ether, $(CH_3)_2O$, at 510 °C is a first-order process with a rate constant of 6.8×10^{-4} s^{-1}:

$$(CH_3)_2\,O(g) \longrightarrow CH_4(g) + H_2(g) + CO(g)$$

If the initial pressure of $(CH_3)_2O$ is 135 torr, what is its pressure after 1420 s?
Answer: 51 torr

Equation 14.13 can be used to verify whether a reaction is first order and to determine its rate constant. This equation has the form of the general equation for a straight line, $y = mx + b$, in which m is the slope and b is the y-intercept of the line (Appendix A.4):

$$\ln[A]_t = -k\cdot t + \ln[A]_0$$

$$y \;=\; m\cdot x + \; b$$

For a first-order reaction, therefore, a graph of $\ln[A]_t$ versus time gives a straight line with a slope of $-k$ and a y-intercept of $\ln[A]_0$. A reaction that is not first order will not yield a straight line.

As an example, consider the conversion of methyl isonitrile (CH_3NC) to acetonitrile (CH_3CN) (Figure 14.6►). Because experiments show that the reaction is first order, we can write the rate equation:

$$\ln[CH_3NC]_t = -kt + \ln[CH_3NC]_0$$

Figure 14.7(a)▼ shows how the pressure of methyl isonitrile varies with time as it rearranges in the gas phase at 198.9 °C. We can use pressure as a unit of concentration for a gas because from the ideal-gas law the pressure is directly proportional to the number of moles per unit volume. Figure 14.7(b) shows a plot of the natural logarithm of the pressure versus time, a plot that yields a straight line. The slope of this line is -5.1×10^{-5} s^{-1}. (You should verify this for yourself, remembering that your result may vary slightly from ours because of inaccuracies associated with reading the graph.) Because the slope of the line equals $-k$, the rate constant for this reaction equals 5.1×10^{-5} s^{-1}.

Methyl isonitrile

Acetonitrile

▲ **Figure 14.6 A first-order reaction.** The transformation of methyl isonitrile (CH_3NC) to acetonitrile (CH_3CN) is a first-order process. Methyl isonitrile and acetonitrile are isomers, molecules that have the same atoms arranged differently. This reaction is called an isomerization reaction.

⧋ GIVE IT SOME THOUGHT

What do the y-intercepts in Figure 14.7(a) and (b) represent?

(a)

(b)

◄ **Figure 14.7 Kinetic data for conversion of methyl isonitrile.** (a) Variation in the partial pressure of methyl isonitrile (CH_3NC) with time during the reaction $CH_3NC \longrightarrow CH_3CN$ at 198.9 °C. (b) A plot of the natural logarithm of the CH_3NC pressure as a function of time. The fact that a straight line fits the data confirms that the rate law is first order.

Second-Order Reactions

A **second-order reaction** is one whose rate depends on the reactant concentration raised to the second power or on the concentrations of two different reactants, each raised to the first power. For simplicity, let's consider reactions of the type A $\longrightarrow$ products or A + B $\longrightarrow$ products that are second order in just one reactant, A:

$$\text{Rate} = -\frac{\Delta[A]}{\Delta t} = k[A]^2$$

With the use of calculus, this differential rate law can be used to derive the following integrated rate law:

$$\frac{1}{[A]_t} = kt + \frac{1}{[A]_0} \qquad\qquad [14.14]$$

This equation, like Equation 14.13, has four variables, k, t, $[A]_0$, and $[A]_t$, and any one of these can be calculated knowing the other three. Equation 14.14 also has the form of a straight line ($y = mx + b$). If the reaction is second order, a plot of $1/[A]_t$ versus t will yield a straight line with a slope equal to k and a y-intercept equal to $1/[A]_0$. One way to distinguish between first- and second-order rate laws is to graph both $\ln[A]_t$ and $1/[A]_t$ against t. If the $\ln[A]_t$ plot is linear, the reaction is first order; if the $1/[A]_t$ plot is linear, the reaction is second order.

■ SAMPLE EXERCISE 14.8 | Determining Reaction Order from the Integrated Rate Law

The following data were obtained for the gas-phase decomposition of nitrogen dioxide at 300 °C, $NO_2(g) \longrightarrow NO(g) + \frac{1}{2}O_2(g)$:

Time (s)	[NO$_2$] (M)
0.0	0.01000
50.0	0.00787
100.0	0.00649
200.0	0.00481
300.0	0.00380

Is the reaction first or second order in NO_2?

SOLUTION

Analyze: We are given the concentrations of a reactant at various times during a reaction and asked to determine whether the reaction is first or second order.

Plan: We can plot $\ln[NO_2]$ and $1/[NO_2]$ against time. One or the other will be linear, indicating whether the reaction is first or second order.

Solve: To graph $\ln[NO_2]$ and $1/[NO_2]$ against time, we will first prepare the following table from the data given:

Time (s)	[NO$_2$] (M)	ln[NO$_2$]	1/[NO$_2$]
0.0	0.01000	−4.605	100
50.0	0.00787	−4.845	127
100.0	0.00649	−5.037	154
200.0	0.00481	−5.337	208
300.0	0.00380	−5.573	263

As Figure 14.8 ▶ shows, only the plot of $1/[NO_2]$ versus time is linear. Thus, the reaction obeys a second-order rate law: Rate = $k[NO_2]^2$. From the slope of this straight-line graph, we determine that $k = 0.543\ M^{-1}\ s^{-1}$ for the disappearance of NO_2.

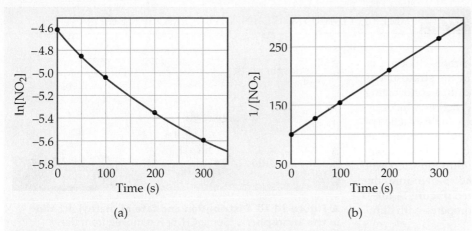

▲ Figure 14.8 Kinetic data for decomposition of NO₂. The reaction is $NO_2(g) \longrightarrow NO(g) + \frac{1}{2} O_2(g)$, and the data were collected at 300 °C. (a) A plot of $\ln[NO_2]$ versus time is not linear, indicating that the reaction is not first order in NO_2. (b) A plot of $1/[NO_2]$ versus time is linear, indicating that the reaction is second order in NO_2.

▨ PRACTICE EXERCISE

Consider again the decomposition of NO_2 discussed in the Sample Exercise. The reaction is second order in NO_2 with $k - 0.543\ M^{-1}\ s^{-1}$. If the initial concentration of NO_2 in a closed vessel is 0.0500 M, what is the remaining concentration after 0.500 h?
Answer: Using Equation 14.14, we find $[NO_2] = 1.00 \times 10^{-3}\ M$

Half-life

The **half-life** of a reaction, $t_{1/2}$, is the time required for the concentration of a reactant to reach one-half of its initial value, $[A]_{t_{1/2}} = \frac{1}{2}[A]_0$. The half-life is a convenient way to describe how fast a reaction occurs, especially if it is a first-order process. A fast reaction will have a short half-life.

We can determine the half-life of a first-order reaction by substituting $[A]_{t_{1/2}}$ into Equation 14.12:

$$\ln \frac{\frac{1}{2}[A]_0}{[A]_0} = -kt_{1/2}$$

$$\ln \frac{1}{2} = -kt_{1/2}$$

$$t_{1/2} = -\frac{\ln \frac{1}{2}}{k} = \frac{0.693}{k} \qquad [14.15]$$

From Equation 14.15, we see that $t_{1/2}$ for a first-order rate law does not depend on the starting concentration. Consequently, the half-life remains constant throughout the reaction. If, for example, the concentration of the reactant is 0.120 M at some moment in the reaction, it will be $\frac{1}{2}(0.120\ M) = 0.060\ M$ after one half-life. After one more half-life passes, the concentration will drop to 0.030 M, and so on. Equation 14.15 also indicates that we can calculate $t_{1/2}$ for a first-order reaction if k is known, or k if $t_{1/2}$ is known.

The change in concentration over time for the first-order rearrangement of methyl isonitrile at 198.9 °C is graphed in Figure 14.9▶. The first half-life is shown at 13,600 s (that is, 3.78 h). At a time 13,600 s later, the isonitrile concentration has decreased to one-half of one-half, or one-fourth the original concentration. *In a first-order reaction, the concentration of the reactant decreases by $\frac{1}{2}$ in each of a series of regularly spaced time intervals, namely, $t_{1/2}$.* The concept of half-life is widely used in describing radioactive decay, a first-order process that we will discuss in detail in Section 21.4.

▼ Figure 14.9 Half-life of a first-order reaction. Pressure of methyl isonitrile as a function of time showing two successive half-lives of the isomerization reaction depicted in Figure 14.6.

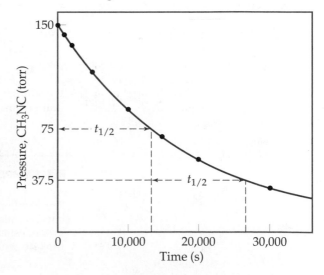

Chemistry Put to Work   METHYL BROMIDE IN THE ATMOSPHERE

Several small molecules containing carbon–chlorine or carbon–bromine bonds, when present in the stratosphere, are capable of reacting with ozone (O_3) and thus contributing to the destruction of Earth's ozone layer. Whether a halogen-containing molecule contributes significantly to destruction of the ozone layer depends in part on the molecule's average lifetime in the atmosphere. It takes quite a long time for molecules formed at Earth's surface to diffuse through the lower atmosphere (called the troposphere) and move into the stratosphere, where the ozone layer is located (Figure 14.10▶). Decomposition in the lower atmosphere competes with diffusion into the stratosphere.

The much-discussed chlorofluorocarbons, or CFCs, contribute to the destruction of the ozone layer because they have long lifetimes in the troposphere. Thus, they persist long enough for a substantial fraction of the molecules to find their way to the stratosphere.

Another simple molecule that has the potential to destroy the stratospheric ozone layer is methyl bromide (CH_3Br). This substance has a wide range of uses, including antifungal treatment of plant seeds, and has therefore been produced in large quantities in the past (about 150 million pounds per year worldwide in 1997). In the stratosphere, the C—Br bond is broken through absorption of short-wavelength radiation. The resultant Br atoms then catalyze decomposition of O_3.

Methyl bromide is removed from the lower atmosphere by a variety of mechanisms, including a slow reaction with ocean water:

$$CH_3Br(g) + H_2O(l) \longrightarrow CH_3OH(aq) + HBr(aq) \qquad [14.16]$$

To determine the potential importance of CH_3Br in destruction of the ozone layer, it is important to know how rapidly the reaction in Equation 14.16 and all other reactions together remove CH_3Br from the atmosphere before it can diffuse into the stratosphere.

Scientists have carried out research to estimate the average lifetime of CH_3Br in Earth's atmosphere. Such an estimate is difficult to make. It cannot be done in laboratory-based experiments because the conditions that exist in the atmosphere above the planet are too complex to be simulated in the laboratory. Instead, scientists gathered nearly 4000 samples of the

▲ **Figure 14.10 Distribution and fate of methyl bromide in the atmosphere.** Some CH_3Br is removed from the atmosphere by decomposition, and some diffuses upward into the stratosphere, where it contributes to destruction of the ozone layer. The relative rates of decomposition and diffusion determine how extensively methyl bromide is involved in destruction of the ozone layer.

atmosphere during aircraft flights all over the Pacific Ocean and analyzed them for the presence of several trace organic substances, including methyl bromide. From a detailed analysis of the concentrations, it was possible to estimate the *atmospheric residence time* for CH_3Br.

The atmospheric residence time is related to the half-life for CH_3Br in the lower atmosphere, assuming that it decomposes by a first-order process. From the experimental data, the half-life for methyl bromide in the lower atmosphere is estimated to be 0.8 ± 0.1 yr. That is, a collection of CH_3Br molecules present at any given time will, on average, be 50% decomposed after 0.8 years, 75% decomposed after 1.6 years, and so on. A half-life of 0.8 years, while comparatively short, is still sufficiently long so that CH_3Br contributes significantly to the destruction of the ozone layer. In 1997 an international agreement was reached to phase out use of methyl bromide in developed countries by 2005. However, in recent years exemptions for critical agricultural use have been requested and granted. Nevertheless, worldwide production was down to 30 million pounds worldwide in 2005, two-thirds of which is used in the United States.

Related Exercise: 14.111

GIVE IT SOME THOUGHT

If a solution containing 10.0 g of a substance reacts by first-order kinetics, how many grams remain after 3 half-lives?

■ **SAMPLE EXERCISE 14.9** | Determining the Half-life of a First-Order Reaction

The reaction of C_4H_9Cl with water is a first-order reaction. Figure 14.4 shows how the concentration of C_4H_9Cl changes with time at a particular temperature. **(a)** From that graph, estimate the half-life for this reaction. **(b)** Use the half-life from (a) to calculate the rate constant.

SOLUTION

Analyze: We are asked to estimate the half-life of a reaction from a graph of concentration versus time and then to use the half-life to calculate the rate constant for the reaction.

Plan: **(a)** To estimate a half-life, we can select a concentration and then determine the time required for the concentration to decrease to half of that value. **(b)** Equation 14.15 is used to calculate the rate constant from the half-life.

Solve: **(a)** From the graph, we see that the initial value of $[C_4H_9Cl]$ is 0.100 M. The half-life for this first-order reaction is the time required for $[C_4H_9Cl]$ to decrease to 0.050 M, which we can read off the graph. This point occurs at approximately 340 s. **(b)** Solving Equation 14.15 for k, we have

$$k = \frac{0.693}{t_{1/2}} = \frac{0.693}{340 \text{ s}} = 2.0 \times 10^{-3} \text{ s}^{-1}$$

Check: At the end of the second half-life, which should occur at 680 s, the concentration should have decreased by yet another factor of 2, to 0.025 M. Inspection of the graph shows that this is indeed the case.

■ PRACTICE EXERCISE

(a) Using Equation 14.15, calculate $t_{1/2}$ for the decomposition of the insecticide described in Sample Exercise 14.7. **(b)** How long does it take for the concentration of the insecticide to reach one-quarter of the initial value?
Answers: **(a)** 0.478 yr = 1.51×10^7 s; **(b)** it takes two half-lives, 2(0.478 yr) = 0.956 yr

---

In contrast to the behavior of first-order reactions, the half-life for second-order and other reactions depends on reactant concentrations and therefore changes as the reaction progresses. Using Equation 14.14, we find that the half-life of a second-order reaction is

$$t_{1/2} = \frac{1}{k[A]_0} \qquad [14.17]$$

In this case the half-life depends on the initial concentration of reactant—the lower the initial concentration, the greater the half-life.

▲ GIVE IT SOME THOUGHT

How does the half-life of a second-order reaction change as the reaction proceeds?

14.5 TEMPERATURE AND RATE

The rates of most chemical reactions increase as the temperature rises. For example, dough rises faster at room temperature than when refrigerated, and plants grow more rapidly in warm weather than in cold. We can literally see the effect of temperature on reaction rate by observing a chemiluminescence reaction (one that produces light). The characteristic glow of fireflies is a familiar example of chemiluminescence. Another is the light produced by Cyalume® light sticks, which contain chemicals that produce chemiluminescence when mixed. As seen in Figure 14.11▶, these light sticks produce a brighter light at higher temperature. The amount of light produced is greater because the rate of the reaction is faster at the higher temperature. Although the light stick glows more brightly initially, its luminescence also dies out more rapidly.

How is this experimentally observed temperature effect reflected in the rate expression? The faster rate at higher temperature is due to an increase in the rate constant with increasing temperature. For example, let's reconsider the first-order reaction $CH_3NC \longrightarrow CH_3CN$

▼ **Figure 14.11 Temperature affects the rate of the chemiluminescence reaction in Cyalume® light sticks.** The light stick in hot water (left) glows more brightly than the one in cold water (right); at the higher temperature, the reaction is initially faster and produces a brighter light.

Higher temperature   Lower temperature

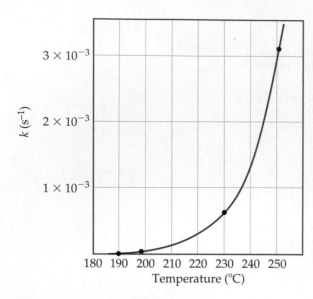

▲ Figure 14.12 Dependence of rate constant on temperature. The data show the variation in the first-order rate constant for the rearrangement of methyl isonitrile as a function of temperature. The four points indicated are used in connection with Sample Exercise 14.11.

(Figure 14.6). Figure 14.12 ◄ shows the rate constant for this reaction as a function of temperature. The rate constant, and hence the rate of the reaction, increases rapidly with temperature, approximately doubling for each 10 °C rise.

The Collision Model

We have seen that reaction rates are affected both by the concentrations of reactants and by temperature. The **collision model**, which is based on the kinetic-molecular theory (Section 10.7), accounts for both of these effects at the molecular level. The central idea of the collision model is that molecules must collide to react. The greater the number of collisions occurring per second, the greater is the reaction rate. As the concentration of reactant molecules increases, therefore, the number of collisions increases, leading to an increase in reaction rate. According to the kinetic-molecular theory of gases, increasing the temperature increases molecular speeds. As molecules move faster, they collide more forcefully (with more energy) and more frequently, increasing reaction rates.

For a reaction to occur, though, more is required than simply a collision. For most reactions, only a tiny fraction of the collisions leads to a reaction. For example, in a mixture of H_2 and I_2 at ordinary temperatures and pressures, each molecule undergoes about 10^{10} collisions per second. If every collision between H_2 and I_2 resulted in the formation of HI, the reaction would be over in much less than a second. Instead, at room temperature the reaction proceeds very slowly. Only about one in every 10^{13} collisions produces a reaction. What keeps the reaction from occurring more rapidly?

▲ GIVE IT SOME THOUGHT

What is the central idea of the collision model?

The Orientation Factor

In most reactions, molecules must be oriented in a certain way during collisions for a reaction to occur. The relative orientations of the molecules during their collisions determine whether the atoms are suitably positioned to form new bonds. For example, consider the reaction of Cl atoms with NOCl:

$$Cl + NOCl \longrightarrow NO + Cl_2$$

The reaction will take place if the collision brings Cl atoms together to form Cl_2 as shown in Figure 14.13(a) ▶. In contrast, the collision shown in Figure 14.13(b) will be ineffective and will not yield products. Indeed, a great many collisions do not lead to reaction, merely because the molecules are not suitably oriented. Another factor, however, is usually even more important in determining whether particular collisions result in reaction.

Activation Energy

In 1888 the Swedish chemist Svante Arrhenius suggested that molecules must possess a certain minimum amount of energy to react. According to the collision model, this energy comes from the kinetic energies of the colliding molecules. Upon collision, the kinetic energy of the molecules can be used to stretch, bend, and ultimately break bonds, leading to chemical reactions. That is, the kinetic energy is used to change the potential energy of the molecule. If molecules are moving too slowly, with too little kinetic energy, they merely bounce off one another without changing. To react, colliding molecules must have a total kinetic energy equal to or greater than some minimum value. The minimum energy required to initiate a chemical reaction is called the **activation energy**, E_a. The value of E_a varies from reaction to reaction.

(a) Effective collision

(b) Ineffective collision

▲ **Figure 14.13 Molecular collisions and chemical reactions.** Two possible ways that Cl atoms and NOCl molecules can collide are shown. (a) If molecules are oriented properly, a sufficiently energetic collision will lead to reaction. (b) If the orientation of the colliding molecules is wrong, no reaction occurs.

The situation during reactions is rather like that shown in Figure 14.14▶. The player on the putting green needs to move her ball over the hill to the vicinity of the cup. To do this, she must impart enough kinetic energy with the putter to move the ball to the top of the hill. If she does not impart enough energy, the ball will roll partway up the hill and then back down. In the same way, molecules may require a certain minimum energy to break existing bonds during a chemical reaction. In the rearrangement of methyl isonitrile to acetonitrile, for example, we might imagine the reaction passing through an intermediate state in which the N≡C portion of the molecule is sitting sideways:

▲ **Figure 14.14 An energy barrier.** To move the golf ball to the vicinity of the cup, the player must impart enough kinetic energy to the ball to enable it to surmount the barrier represented by the hill. This situation is analogous to a chemical reaction, in which molecules must gain enough energy through collisions to enable them to overcome the barrier to chemical reaction.

$$H_3C-N\equiv C: \longrightarrow \left[H_3C\cdots \overset{\ddot{C}}{\underset{\ddot{N}}{\||}} \right] \longrightarrow H_3C-C\equiv N:$$

The change in the potential energy of the molecule during the reaction is shown in Figure 14.15▼. The diagram shows that energy must be supplied to stretch the bond between the H_3C group and the N≡C group to allow the N≡C group to rotate. After the N≡C group has twisted sufficiently, the C—C bond begins to form, and the energy of the molecule drops. Thus, the barrier represents the energy necessary to force the molecule through the relatively unstable intermediate state to the final product. The energy difference between that of the starting molecule and the highest energy along the reaction pathway is the activation energy, E_a. The particular arrangement of atoms at the top of the barrier is called the **activated complex**, or **transition state**.

The conversion of $H_3C-N\equiv C$ to $H_3C-C\equiv N$ is exothermic. Figure 14.15 therefore shows the product as having a lower energy than the reactant. The energy change for the reaction, ΔE, has no effect on the rate of the reaction. *The rate depends on the magnitude of E_a; generally, the lower E_a is, the faster the reaction.* Notice that the reverse reaction is endothermic. The activation barrier for the reverse reaction is equal to the sum of ΔE and E_a for the forward reaction.

▶ **Figure 14.15 Energy profile for methyl isonitrile isomerization.** The methyl isonitrile molecule must surmount the activation-energy barrier before it can form the product, acetonitrile. The horizontal axis is variously labeled "reaction pathway," as here, or "progress of reaction."

▲ **Figure 14.16 The effect of temperature on the distribution of kinetic energies.** At the higher temperature, a larger number of molecules have higher kinetic energies. Thus, a larger fraction at any one instant will have more than the minimum energy required for reaction.

How does any particular methyl isonitrile molecule acquire sufficient energy to overcome the activation barrier? It does so through collisions with other molecules. Recall from the kinetic-molecular theory of gases that, at any given instant, gas molecules are distributed in energy over a wide range. ∞ (Section 10.7) Figure 14.16 ◀ shows the distribution of kinetic energies for two different temperatures, comparing them with the minimum energy needed for reaction, E_a. At the higher temperature a much greater fraction of the molecules has kinetic energy greater than E_a, which leads to a much greater rate of reaction.

The fraction of molecules that has an energy equal to or greater than E_a is given by the expression

$$f = e^{-E_a/RT} \qquad [14.18]$$

In this equation R is the gas constant (8.314 J/mol-K) and T is absolute temperature. To get an idea of the magnitude of f, let's suppose that E_a is 100 kJ/mol, a value typical of many reactions, and that T is 300 K, around room temperature. The calculated value of f is 3.9×10^{-18}, an extremely small number! At 310 K the fraction is $f = 1.4 \times 10^{-17}$. Thus, a 10-degree increase in temperature produces a 3.6-fold increase in the fraction of molecules possessing at least 100 kJ/mol of energy.

▲ **GIVE IT SOME THOUGHT**

Why isn't collision frequency the only factor affecting a reaction rate?

The Arrhenius Equation

Arrhenius noted that for most reactions the increase in rate with increasing temperature is nonlinear, as shown in Figure 14.12. He found that most reaction-rate data obeyed an equation based on three factors: (a) the fraction of molecules possessing an energy of E_a or greater, (b) the number of collisions occurring per second, and (c) the fraction of collisions that have the appropriate orientation. These three factors are incorporated into the **Arrhenius equation**:

$$k = Ae^{-E_a/RT}$$ [14.19]

In this equation k is the rate constant, E_a is the activation energy, R is the gas constant (8.314 J/mol-K), and T is the absolute temperature. The **frequency factor**, A, is constant, or nearly so, as temperature is varied. This factor is related to the frequency of collisions and the probability that the collisions are favorably oriented for reaction.* As the magnitude of E_a increases, k decreases because the fraction of molecules that possess the required energy is smaller. Thus, *reaction rates decrease as E_a increases.*

■■ **SAMPLE EXERCISE 14.10** | Relating Energy Profiles to Activation Energies and Speeds of Reaction

Consider a series of reactions having the following energy profiles:

(1)   (2)   (3)

Rank the reactions from slowest to fastest assuming that they have nearly the same frequency factors.

SOLUTION

The lower the activation energy, the faster the reaction. The value of ΔE does not affect the rate. Hence the order is (2) < (3) < (1).

■■ **PRACTICE EXERCISE**

Imagine that these reactions are reversed. Rank these reverse reactions from slowest to fastest.
Answer: (2) < (1) < (3) because E_a values are 40, 25, and 15 kJ/mol, respectively

---

Determining the Activation Energy

Taking the natural log of both sides of Equation 14.19, we have

$$\ln k = -\frac{E_a}{RT} + \ln A$$ [14.20]

$$\begin{array}{ccc} \updownarrow & \updownarrow & \updownarrow \\ y = & mx + & b \end{array}$$

As shown, Equation 14.20 has the form of the equation for a straight line. A graph of $\ln k$ versus $1/T$ will be a line with a slope equal to $-E_a/R$ and a y-intercept equal to $\ln A$. Thus, the activation energy can be determined by measuring k at a series of temperatures, graphing $\ln k$ versus $1/T$, and then calculating E_a from the slope of the resultant line.

We can also use Equation 14.20 to evaluate E_a in a nongraphical way if we know the rate constant of a reaction at two or more temperatures. For example,

---

*Because the frequency of collisions increases with temperature, A also has some temperature dependence, but it is small compared to the exponential term. Therefore, A is considered approximately constant.

suppose that at two different temperatures, T_1 and T_2, a reaction has rate constants k_1 and k_2. For each condition, we have

$$\ln k_1 = -\frac{E_a}{RT_1} + \ln A \quad \text{and} \quad \ln k_2 = -\frac{E_a}{RT_2} + \ln A$$

Subtracting $\ln k_2$ from $\ln k_1$ gives

$$\ln k_1 - \ln k_2 = \left(-\frac{E_a}{RT_1} + \ln A\right) - \left(-\frac{E_a}{RT_2} + \ln A\right)$$

Simplifying this equation and rearranging it gives

$$\ln \frac{k_1}{k_2} = \frac{E_a}{R}\left(\frac{1}{T_2} - \frac{1}{T_1}\right) \qquad [14.21]$$

Equation 14.21 provides a convenient way to calculate the rate constant, k_1, at some temperature, T_1, when we know the activation energy and the rate constant, k_2, at some other temperature, T_2.

■ **SAMPLE EXERCISE 14.11** | **Determining the Energy of Activation**

The following table shows the rate constants for the rearrangement of methyl isonitrile at various temperatures (these are the data in Figure 14.12):

Temperature (°C)	k (s^{-1})
189.7	2.52×10^{-5}
198.9	5.25×10^{-5}
230.3	6.30×10^{-4}
251.2	3.16×10^{-3}

(a) From these data, calculate the activation energy for the reaction. **(b)** What is the value of the rate constant at 430.0 K?

SOLUTION

Analyze: We are given rate constants, k, measured at several temperatures and asked to determine the activation energy, E_a, and the rate constant, k, at a particular temperature.

Plan: We can obtain E_a from the slope of a graph of $\ln k$ versus $1/T$. Once we know E_a, we can use Equation 14.21 together with the given rate data to calculate the rate constant at 430.0 K.

Solve: (a) We must first convert the temperatures from degrees Celsius to kelvins. We then take the inverse of each temperature, $1/T$, and the natural log of each rate constant, $\ln k$. This gives us the table shown at the right:

T (K)	$1/T$ (K^{-1})	$\ln k$
462.9	2.160×10^{-3}	-10.589
472.1	2.118×10^{-3}	-9.855
503.5	1.986×10^{-3}	-7.370
524.4	1.907×10^{-3}	-5.757

A graph of $\ln k$ versus $1/T$ results in a straight line, as shown in Figure 14.17▶.

◀ **Figure 14.17 Graphical determination of activation energy.** The natural logarithm of the rate constant for the rearrangement of methyl isonitrile is plotted as a function of $1/T$. The linear relationship is predicted by the Arrhenius equation giving a slope equal to $-E_a/R$.

The slope of the line is obtained by choosing two well-separated points, as shown, and using the coordinates of each:

$$\text{Slope} = \frac{\Delta y}{\Delta x} = \frac{-6.6 - (-10.4)}{0.00195 - 0.00215} = -1.9 \times 10^4$$

Because logarithms have no units, the numerator in this equation is dimensionless. The denominator has the units of $1/T$, namely, K^{-1}. Thus, the overall units for the slope are K. The slope equals $-E_a/R$. We use the value for the molar gas constant R in units of J/mol-K (Table 10.2). We thus obtain

$$\text{Slope} = -\frac{E_a}{R}$$

$$E_a = -(\text{slope})(R) = -(-1.9 \times 10^4 \, K)\left(8.314\frac{J}{\text{mol-K}}\right)\left(\frac{1 \, kJ}{1000 \, J}\right)$$

$$= 1.6 \times 10^2 \, kJ/mol = 160 \, kJ/mol$$

We report the activation energy to only two significant figures because we are limited by the precision with which we can read the graph in Figure 14.17.

(b) To determine the rate constant, k_1, at $T_1 = 430.0$ K, we can use Equation 14.21 with $E_a = 160$ kJ/mol, and one of the rate constants and temperatures from the given data, such as $k_2 = 2.52 \times 10^{-5} \, s^{-1}$ and $T_2 = 462.9$ K:

$$\ln\left(\frac{k_1}{2.52 \times 10^{-5} \, s^{-1}}\right) = \left(\frac{160 \, kJ/mol}{8.314 \, J/mol\text{-}K}\right)\left(\frac{1}{462.9 \, K} - \frac{1}{430.0 \, K}\right)\left(\frac{1000 \, J}{1 \, kJ}\right) - -3.18$$

Thus,

$$\frac{k_1}{2.52 \times 10^{-5} \, s^{-1}} = e^{-3.18} = 4.15 \times 10^{-2}$$

$$k_1 = (4.15 \times 10^{-2})(2.52 \times 10^{-5} \, s^{-1}) = 1.0 \times 10^{-6} \, s^{-1}$$

Note that the units of k_1 are the same as those of k_2.

■ PRACTICE EXERCISE

Using the data in Sample Exercise 14.11, calculate the rate constant for the rearrangement of methyl isonitrile at 280 °C.
Answer: $2.2 \times 10^{-2} \, s^{-1}$

14.6 REACTION MECHANISMS

A balanced equation for a chemical reaction indicates the substances present at the start of the reaction and those produced as the reaction proceeds. It provides no information, however, about how the reaction occurs. The process by which a reaction occurs is called the **reaction mechanism**. At the most sophisticated level, a reaction mechanism will describe in great detail the order in which bonds are broken and formed and the changes in relative positions of the atoms in the course of the reaction. We will begin with more rudimentary descriptions of how reactions occur, considering further the nature of the collisions leading to reaction.

Elementary Reactions

We have seen that reactions take place because of collisions between reacting molecules. For example, the collisions between molecules of methyl isonitrile (CH_3NC) can provide the energy to allow the CH_3NC to rearrange:

Similarly, the reaction of NO and O_3 to form NO_2 and O_2 appears to occur as a result of a single collision involving suitably oriented and sufficiently energetic NO and O_3 molecules:

$$NO(g) + O_3(g) \longrightarrow NO_2(g) + O_2(g) \qquad [14.22]$$

Both of these processes occur in a single event or step and are called **elementary reactions** (or elementary processes).

The number of molecules that participate as reactants in an elementary reaction defines its **molecularity**. If a single molecule is involved, the reaction is **unimolecular**. The rearrangement of methyl isonitrile is a unimolecular process. Elementary reactions involving the collision of two reactant molecules are **bimolecular**. The reaction between NO and O_3 (Equation 14.22) is bimolecular. Elementary reactions involving the simultaneous collision of three molecules are **termolecular**. Termolecular reactions are far less probable than unimolecular or bimolecular processes and are rarely encountered. The chance that four or more molecules will collide simultaneously with any regularity is even more remote; consequently, such collisions are never proposed as part of a reaction mechanism.

GIVE IT SOME THOUGHT

What is the molecularity of this elementary reaction?

$$NO(g) + Cl_2(g) \longrightarrow NOCl(g) + Cl(g)$$

Multistep Mechanisms

The net change represented by a balanced chemical equation often occurs by a *multistep mechanism*, which consists of a sequence of elementary reactions. For example, consider the reaction of NO_2 and CO:

$$NO_2(g) + CO(g) \longrightarrow NO(g) + CO_2(g) \qquad [14.23]$$

Below 225 °C, this reaction appears to proceed in two elementary reactions (or two *elementary steps*), each of which is bimolecular. First, two NO_2 molecules collide, and an oxygen atom is transferred from one to the other. The resultant NO_3 then collides with a CO molecule and transfers an oxygen atom to it:

$$NO_2(g) + NO_2(g) \longrightarrow NO_3(g) + NO(g)$$
$$NO_3(g) + CO(g) \longrightarrow NO_2(g) + CO_2(g)$$

Thus, we say that the reaction occurs by a two-step mechanism.

The chemical equations for the elementary reactions in a multistep mechanism must always add to give the chemical equation of the overall process. In the present example the sum of the two elementary reactions is

$$2\,NO_2(g) + NO_3(g) + CO(g) \longrightarrow NO_2(g) + NO_3(g) + NO(g) + CO_2(g)$$

Simplifying this equation by eliminating substances that appear on both sides of the arrow gives Equation 14.23, the net equation for the process. Because NO_3 is neither a reactant nor a product in the overall reaction—it is formed in one elementary reaction and consumed in the next—it is called an **intermediate**. Multistep mechanisms involve one or more intermediates.

■ **SAMPLE EXERCISE 14.12** | Determining Molecularity and Identifying Intermediates

It has been proposed that the conversion of ozone into O_2 proceeds by a two-step mechanism:

$$O_3(g) \longrightarrow O_2(g) + O(g)$$
$$O_3(g) + O(g) \longrightarrow 2\,O_2(g)$$

(a) Describe the molecularity of each elementary reaction in this mechanism.
(b) Write the equation for the overall reaction. (c) Identify the intermediate(s).

SOLUTION

Analyze: We are given a two-step mechanism and asked for (a) the molecularities of each of the two elementary reactions, (b) the equation for the overall process, and (c) the intermediate.

Plan: The molecularity of each elementary reaction depends on the number of reactant molecules in the equation for that reaction. The overall equation is the sum of the equations for the elementary reactions. The intermediate is a substance formed in one step of the mechanism and used in another and therefore not part of the equation for the overall reaction.

Solve: (a) The first elementary reaction involves a single reactant and is consequently unimolecular. The second reaction, which involves two reactant molecules, is bimolecular.
(b) Adding the two elementary reactions gives

$$2\,O_3(g) + O(g) \longrightarrow 3\,O_2(g) + O(g)$$

Because $O(g)$ appears in equal amounts on both sides of the equation, it can be eliminated to give the net equation for the chemical process:

$$2\,O_3(g) \longrightarrow 3\,O_2(g)$$

(c) The intermediate is $O(g)$. It is neither an original reactant nor a final product but is formed in the first step of the mechanism and consumed in the second.

■ PRACTICE EXERCISE

For the reaction

$$Mo(CO)_6 + P(CH_3)_3 \longrightarrow Mo(CO)_5\,P(CH_3)_3 + CO$$

the proposed mechanism is

$$Mo(CO)_6 \longrightarrow Mo(CO)_5 + CO$$
$$Mo(CO)_5 + P(CH_3)_3 \longrightarrow Mo(CO)_5\,P(CH_3)_3$$

(a) Is the proposed mechanism consistent with the equation for the overall reaction?
(b) What is the molecularity of each step of the mechanism? (c) Identify the intermediate(s).
Answers: (a) Yes, the two equations add to yield the equation for the reaction. (b) The first elementary reaction is unimolecular, and the second one is bimolecular. (c) $Mo(CO)_5$

Rate Laws for Elementary Reactions

In Section 14.3 we stressed that rate laws must be determined experimentally; they cannot be predicted from the coefficients of balanced chemical equations. We are now in a position to understand why this is so. Every reaction is made up of a series of one or more elementary steps, and the rate laws and relative speeds of these steps will dictate the overall rate law. Indeed, the rate law for a reaction can be determined from its mechanism, as we will see shortly. Thus, our next challenge in kinetics is to arrive at reaction mechanisms that lead to rate laws that are consistent with those observed experimentally. We will start by examining the rate laws of elementary reactions.

Elementary reactions are significant in a very important way: *If a reaction is an elementary reaction, then its rate law is based directly on its molecularity.* For example, consider a unimolecular process:

$$A \longrightarrow products$$

As the number of A molecules increases, the number that react in a given interval of time will increase proportionally. Thus, the rate of a unimolecular process will be first order:

$$Rate = k[A]$$

TABLE 14.3 ■ Elementary Reactions and Their Rate Laws

Molecularity	Elementary Reaction	Rate Law
*Uni*molecular	A $\longrightarrow$ products	Rate = $k[A]$
*Bi*molecular	A + A $\longrightarrow$ products	Rate = $k[A]^2$
*Bi*molecular	A + B $\longrightarrow$ products	Rate = $k[A][B]$
*Ter*molecular	A + A + A $\longrightarrow$ products	Rate = $k[A]^3$
*Ter*molecular	A + A + B $\longrightarrow$ products	Rate = $k[A]^2[B]$
*Ter*molecular	A + B + C $\longrightarrow$ products	Rate = $k[A][B][C]$

In the case of bimolecular elementary steps, the rate law is second order, as in the following example:

$$A + B \longrightarrow \text{products} \qquad \text{Rate} = k[A][B]$$

The second-order rate law follows directly from the collision theory. If we double the concentration of A, the number of collisions between molecules of A and B will double; likewise, if we double [B], the number of collisions will double. Therefore, the rate law will be first order in both [A] and [B], and second order overall.

The rate laws for all feasible elementary reactions are given in Table 14.3 ▲. Notice how the rate law for each kind of elementary reaction follows directly from the molecularity of that reaction. It is important to remember, however, that we cannot tell by merely looking at a balanced chemical equation whether the reaction involves one or several elementary steps.

■■■ SAMPLE EXERCISE 14.13 | Predicting the Rate Law
for an Elementary Reaction

If the following reaction occurs in a single elementary reaction, predict its rate law:

$$H_2(g) + Br_2(g) \longrightarrow 2\,HBr(g)$$

SOLUTION

Analyze: We are given the equation and asked for its rate law, assuming that it is an elementary process.

Plan: Because we are assuming that the reaction occurs as a single elementary reaction, we are able to write the rate law using the coefficients for the reactants in the equation as the reaction orders.

Solve: The reaction is bimolecular, involving one molecule of H_2 with one molecule of Br_2. Thus, the rate law is first order in each reactant and second order overall:

$$\text{Rate} = k[H_2][Br_2]$$

Comment: Experimental studies of this reaction show that the reaction actually has a very different rate law:

$$\text{Rate} = k[H_2][Br_2]^{1/2}$$

Because the experimental rate law differs from the one obtained by assuming a single elementary reaction, we can conclude that the mechanism cannot occur by a single elementary step. It must, therefore, involve two or more elementary steps.

■■■ PRACTICE EXERCISE

Consider the following reaction: $2\,NO(g) + Br_2(g) \longrightarrow 2\,NOBr(g)$. **(a)** Write the rate law for the reaction, assuming it involves a single elementary reaction. **(b)** Is a single-step mechanism likely for this reaction?
Answers: **(a)** Rate = $k[NO]^2[Br_2]$ **(b)** No, because termolecular reactions are very rare.

The Rate-Determining Step for a Multistep Mechanism

As with the reaction in Sample Exercise 14.13, most chemical reactions occur by mechanisms that involve two or more elementary reactions. Each of these steps of the mechanism has its own rate constant and activation energy. Often one of the steps is much slower than the others. The overall rate of a reaction cannot exceed the rate of the slowest elementary step of its mechanism. Because the slow step limits the overall reaction rate, it is called the **rate-determining step** (or *rate-limiting step*).

To understand the concept of a rate-determining step, consider a toll road with two toll plazas (Figure 14.18 ►). We will measure the rate at which cars pass through the second toll plaza. Cars enter the toll road at point 1 and pass through toll plaza A. They then pass an intermediate point 2 before passing through toll plaza B and arriving at point 3. We can therefore envision this trip along the toll road as occurring in two elementary steps:

Cars slowed at plaza A
(a)

Cars slowed at plaza B
(b)

▲ Figure 14.18 **Rate-determining step.** The flow of traffic on a toll road illustrates how a rate-determining step controls reaction rate. The flow is limited by the flow of traffic through the slower toll plaza. In (a) the rate at which cars can reach point 3 is limited by how quickly they can get through plaza A. In this case, getting from point 1 to point 2 is the rate-determining step. In (b), getting from point 2 to point 3 is the rate-determining step.

Step 1:   Point 1 ⟶ point 2   (through plaza A)

Step 2:   Point 2 ⟶ point 3   (through plaza B)

Overall: Point 1 ⟶ point 3   (through plaza A and B)

Now suppose that several of the gates at toll plaza A are malfunctioning, so that traffic backs up behind it as depicted in Figure 14.18(a). The rate at which cars can get to point 3 is limited by the rate at which they can get through the traffic jam at plaza A. Thus, step 1 is the rate-determining step of the journey along the toll road. If, however, traffic flows quickly through plaza A but gets backed up at plaza B, as depicted in Figure 14.18(b), the number of cars builds up in the intermediate region between the plazas. In this case step 2 is rate determining: The rate at which cars can travel the toll road is limited by the rate at which they can pass through plaza B.

In the same way, *the slowest step in a multistep reaction limits the overall rate.* By analogy to Figure 14.18(a), the rate of a faster step following the rate-determining step does not speed up the overall rate. If the slow step is not the first one, as in Figure 14.18(b), the faster preceding steps produce intermediate products that accumulate before being consumed in the slow step. In either case *the rate-determining step governs the rate law for the overall reaction.*

GIVE IT SOME THOUGHT

Why can't the rate law for a reaction generally be deduced from the balanced equation for the reaction?

Mechanisms with a Slow Initial Step

We can most easily see the relationship between the slow step in a mechanism and the rate law for the overall reaction by considering an example in which the first step in a multistep mechanism is the slow, rate-determining step. As an example, consider the reaction of NO_2 and CO to produce NO and CO_2 (Equation 14.23). Below 225 °C, it is found experimentally that the rate law for this reaction is second order in NO_2 and zero order in CO: Rate $= k[NO_2]^2$. Can we propose a reaction mechanism that is consistent with this rate law? Consider the following two-step mechanism:*

Step 1: $NO_2(g) + NO_2(g) \xrightarrow{k_1} NO_3(g) + NO(g)$ (slow)

Step 2: $NO_3(g) + CO(g) \xrightarrow{k_2} NO_2(g) + CO_2(g)$ (fast)

Overall: $NO_2(g) + CO(g) \longrightarrow NO(g) + CO_2(g)$

Step 2 is much faster than step 1; that is, $k_2 \gg k_1$. The intermediate $NO_3(g)$ is slowly produced in step 1 and is immediately consumed in step 2.

*The subscript on the rate constant identifies the elementary step involved. Thus, k_1 is the rate constant for step 1, k_2 is the rate constant for step 2, and so forth. A negative subscript refers to the rate constant for the reverse of an elementary step. For example, k_{-1} is the rate constant for the reverse of the first step.

Because step 1 is slow and step 2 is fast, step 1 is rate determining. Thus, the rate of the overall reaction depends on the rate of step 1, and the rate law of the overall reaction equals the rate law of step 1. Step 1 is a bimolecular process that has the rate law

$$\text{Rate} = k_1[NO_2]^2$$

Thus, the rate law predicted by this mechanism agrees with the one observed experimentally. CO is absent from the rate law because it reacts in a step that follows the rate-determining step.

■ **SAMPLE EXERCISE 14.14** | **Determining the Rate Law for a Multistep Mechanism**

The decomposition of nitrous oxide, N_2O, is believed to occur by a two-step mechanism:

$$N_2O(g) \longrightarrow N_2(g) + O(g) \quad \text{(slow)}$$
$$N_2O(g) + O(g) \longrightarrow N_2(g) + O_2(g) \quad \text{(fast)}$$

(a) Write the equation for the overall reaction. **(b)** Write the rate law for the overall reaction.

SOLUTION

Analyze: Given a multistep mechanism with the relative speeds of the steps, we are asked to write the overall reaction and the rate law for that overall reaction.

Plan: **(a)** Find the overall reaction by adding the elementary steps and eliminating the intermediates. **(b)** The rate law for the overall reaction will be that of the slow, rate-determining step.

Solve: **(a)** Adding the two elementary reactions gives

$$2 N_2O(g) + O(g) \longrightarrow 2 N_2(g) + O_2(g) + O(g)$$

Omitting the intermediate, $O(g)$, which occurs on both sides of the equation, gives the overall reaction:

$$2 N_2O(g) \longrightarrow 2 N_2(g) + O_2(g)$$

(b) The rate law for the overall reaction is just the rate law for the slow, rate-determining elementary reaction. Because that slow step is a unimolecular elementary reaction, the rate law is first order:

$$\text{Rate} = k[N_2O]$$

■ **PRACTICE EXERCISE**

Ozone reacts with nitrogen dioxide to produce dinitrogen pentoxide and oxygen:

$$O_3(g) + 2 NO_2(g) \longrightarrow N_2O_5(g) + O_2(g)$$

The reaction is believed to occur in two steps:

$$O_3(g) + NO_2(g) \longrightarrow NO_3(g) + O_2(g)$$
$$NO_3(g) + NO_2(g) \longrightarrow N_2O_5(g)$$

The experimental rate law is rate = $k[O_3][NO_2]$. What can you say about the relative rates of the two steps of the mechanism?
Answer: Because the rate law conforms to the molecularity of the first step, that must be the rate-determining step. The second step must be much faster than the first one.

Mechanisms with a Fast Initial Step

It is less straightforward to derive the rate law for a mechanism in which an intermediate is a reactant in the rate-determining step. This situation arises in multistep mechanisms when the first step is fast and therefore *not* the rate-determining one.

Let's consider one example: the gas-phase reaction of nitric oxide (NO) with bromine (Br_2).

$$2\,NO(g) + Br_2(g) \longrightarrow 2\,NOBr(g) \qquad [14.24]$$

The experimentally determined rate law for this reaction is second order in NO and first order in Br_2:

$$Rate = k[NO]^2[Br_2] \qquad [14.25]$$

We seek a reaction mechanism that is consistent with this rate law. One possibility is that the reaction occurs in a single termolecular step:

$$NO(g) + NO(g) + Br_2(g) \longrightarrow 2\,NOBr(g) \qquad Rate = k[NO]^2[Br_2] \qquad [14.26]$$

As noted in Practice Exercise 14.13, this does not seem likely because termolecular processes are so rare.

Let's consider an alternative mechanism that does not invoke a termolecular step:

$$\text{Step 1:} \quad NO(g) + Br_2(g) \underset{k_{-1}}{\overset{k_1}{\rightleftharpoons}} NOBr_2(g) \qquad \text{(fast)}$$

$$\text{Step 2:} \quad NOBr_2(g) + NO(g) \xrightarrow{k_2} 2\,NOBr(g) \quad \text{(slow)}$$

In this mechanism, step 1 actually involves two processes: a forward reaction and its reverse.

Because step 2 is the slow, rate-determining step, the rate law for that step governs the rate of the overall reaction:

$$Rate = k[NOBr_2][NO] \qquad [14.27]$$

However, $NOBr_2$ is an intermediate generated in step 1. Intermediates are usually unstable molecules that have a low, unknown concentration. Thus, our rate law depends on the unknown concentration of an intermediate.

Fortunately, with the aid of some assumptions, we can express the concentration of the intermediate ($NOBr_2$) in terms of the concentrations of the starting reactants (NO and Br_2). We first assume that $NOBr_2$ is intrinsically unstable and that it does not accumulate to a significant extent in the reaction mixture. There are two ways for $NOBr_2$ to be consumed once it is formed: It can either react with NO to form NOBr or reform NO and Br_2. The first of these possibilities is step 2, a slow process. The second is the reverse of step 1, a unimolecular process:

$$NOBr_2(g) \xrightarrow{k_{-1}} NO(g) + Br_2(g) \qquad [14.28]$$

Because step 2 is slow, we assume that most of the $NOBr_2$ falls apart according to Equation 14.28. Thus, we have both the forward and reverse reactions of step 1 occurring much faster than step 2. Because the forward and reverse processes of step 1 occur rapidly with respect to the reaction in step 2, they establish an equilibrium. We have seen examples of dynamic equilibrium before, in the equilibrium between a liquid and its vapor ∞ (Section 11.5) and between a solid solute and its solution. ∞ (Section 13.3) As in any dynamic equilibrium, the rates of the forward and reverse reactions are equal. Thus, we can equate the rate expression for the forward reaction in step 1 with the rate expression for the reverse reaction:

$$\underset{\text{Rate of forward reaction}}{k_1[NO][Br_2]} \quad = \quad \underset{\text{Rate of reverse reaction}}{k_{-1}[NOBr_2]}$$

Solving for $[NOBr_2]$, we have

$$[NOBr_2] = \frac{k_1}{k_{-1}}[NO][Br_2]$$

Substituting this relationship into the rate law for the rate-determining step (Equation 14.27), we have

$$\text{Rate} = k_2 \frac{k_1}{k_{-1}} [NO][Br_2][NO] = k[NO]^2[Br_2]$$

This result is consistent with the experimental rate law (Equation 14.25). The experimental rate constant, k, equals $k_2 k_1 / k_{-1}$. This mechanism, which involves only unimolecular and bimolecular processes, is far more probable than the single termolecular step (Equation 14.26).

In general, *whenever a fast step precedes a slow one, we can solve for the concentration of an intermediate by assuming that an equilibrium is established in the fast step.*

■ **SAMPLE EXERCISE 14.15** | Deriving the Rate Law for a Mechanism with a Fast Initial Step

Show that the following mechanism for Equation 14.24 also produces a rate law consistent with the experimentally observed one:

$$\text{Step 1:} \quad NO(g) + NO(g) \underset{k_{-1}}{\overset{k_1}{\rightleftharpoons}} N_2O_2(g) \quad \text{(fast equilibrium)}$$

$$\text{Step 2:} \quad N_2O_2(g) + Br_2(g) \overset{k_2}{\longrightarrow} 2\,NOBr(g) \quad \text{(slow)}$$

SOLUTION

Analyze: We are given a mechanism with a fast initial step and asked to write the rate law for the overall reaction.

Plan: The rate law of the slow elementary step in a mechanism determines the rate law for the overall reaction. Thus, we first write the rate law based on the molecularity of the slow step. In this case the slow step involves the intermediate N_2O_2 as a reactant. Experimental rate laws, however, do not contain the concentrations of intermediates; instead they are expressed in terms of the concentrations of starting substances. Thus, we must relate the concentration of N_2O_2 to the concentration of NO by assuming that an equilibrium is established in the first step.

Solve: The second step is rate determining, so the overall rate is

$$\text{Rate} = k_2[N_2O_2][Br_2]$$

We solve for the concentration of the intermediate N_2O_2 by assuming that an equilibrium is established in step 1; thus, the rates of the forward and reverse reactions in step 1 are equal:

$$k_1[NO]^2 = k_{-1}[N_2O_2]$$

Solving for the concentration of the intermediate, N_2O_2, gives

$$[N_2O_2] = \frac{k_1}{k_{-1}} [NO]^2$$

Substituting this expression into the rate expression gives

$$\text{Rate} = k_2 \frac{k_1}{k_{-1}} [NO]^2[Br_2] = k[NO]^2[Br_2]$$

Thus, this mechanism also yields a rate law consistent with the experimental one.

■ **PRACTICE EXERCISE**

The first step of a mechanism involving the reaction of bromine is

$$Br_2(g) \underset{k_{-1}}{\overset{k_1}{\rightleftharpoons}} 2\,Br(g) \quad \text{(fast, equilibrium)}$$

What is the expression relating the concentration of $Br(g)$ to that of $Br_2(g)$?

Answer: $[Br] = \left(\dfrac{k_1}{k_{-1}} [Br_2] \right)^{1/2}$

14.7 CATALYSIS

A **catalyst** is a substance that changes the speed of a chemical reaction without undergoing a permanent chemical change itself in the process. Catalysts are very common; most reactions in the body, the atmosphere, and the oceans occur with the help of catalysts. Much industrial chemical research is devoted to the search for new and more effective catalysts for reactions of commercial importance. Extensive research efforts also are devoted to finding means of inhibiting or removing certain catalysts that promote undesirable reactions, such as those that corrode metals, age our bodies, and cause tooth decay.

Homogeneous Catalysis

A catalyst that is present in the same phase as the reacting molecules is called a **homogeneous catalyst**. Examples abound both in solution and in the gas phase. Consider, for example, the decomposition of aqueous hydrogen peroxide, $H_2O_2(aq)$, into water and oxygen:

$$2\,H_2O_2(aq) \longrightarrow 2\,H_2O(l) + O_2(g) \qquad [14.29]$$

In the absence of a catalyst, this reaction occurs extremely slowly.

Many different substances are capable of catalyzing the reaction represented by Equation 14.29, including bromide ion, $Br^-(aq)$, as shown in Figure 14.19▼.

▼ **Figure 14.19 Effect of catalyst.** (H_2O molecules and Na^+ ions are omitted from the molecular art for clarity.)

HOMOGENEOUS CATALYSIS

A catalyst that is present in the same phase as the reacting molecules is a homogeneous catalyst.

In the absence of a catalyst, $H_2O_2(aq)$ decomposes very slowly.

Shortly after the addition of a small amount of NaBr(aq) to $H_2O_2(aq)$, the solution turns brown because Br_2 is generated (Equation 14.30). The buildup of Br_2 leads to rapid evolution of O_2 according to Equation 14.31.

After all of the H_2O_2 has decomposed, a colorless solution of NaBr(aq) remains. Thus, NaBr has catalyzed the reaction even though it is not consumed during the reaction.

The bromide ion reacts with hydrogen peroxide in acidic solution, forming aqueous bromine and water:

$$2\,Br^-(aq) + H_2O_2(aq) + 2\,H^+ \longrightarrow Br_2(aq) + 2\,H_2O(l) \qquad [14.30]$$

The brown color observed in the middle photograph of Figure 14.19 indicates the formation of $Br_2(aq)$. If this were the complete reaction, bromide ion would not be a catalyst, because it undergoes chemical change during the reaction. However, hydrogen peroxide also reacts with the $Br_2(aq)$ generated in Equation 14.30:

$$Br_2(aq) + H_2O_2(aq) \longrightarrow 2\,Br^-(aq) + 2\,H^+(aq) + O_2(g) \qquad [14.31]$$

The bubbling evident in Figure 14.19(b) is due to the formation of $O_2(g)$.

The sum of Equations 14.30 and 14.31 is just Equation 14.29:

$$2\,H_2O_2(aq) \longrightarrow 2\,H_2O(l) + O_2(g)$$

When H_2O_2 has been completely decomposed, we are left with a colorless solution of $Br^-(aq)$, as seen in the photograph on the right in Figure 14.19. Bromide ion, therefore, is indeed a catalyst of the reaction because it speeds the overall reaction without itself undergoing any net change. It is added at the start of the reaction, reacts, and then reforms at the end. In contrast, Br_2 is an intermediate because it is first formed (Equation 14.30) and then consumed (Equation 14.31). Neither the catalyst nor the intermediate appears in the chemical equation for the overall reaction. Notice, however, that *the catalyst is there at the start of the reaction, whereas the intermediate is formed during the course of the reaction.*

On the basis of the Arrhenius equation (Equation 14.19), the rate constant (k) is determined by the activation energy (E_a) and the frequency factor (A). A catalyst may affect the rate of reaction by altering the value of either E_a or A. The most dramatic catalytic effects come from lowering E_a. As a general rule, *a catalyst lowers the overall activation energy for a chemical reaction.*

A catalyst usually lowers the overall activation energy for a reaction by providing a different mechanism for the reaction. In the decomposition of hydrogen peroxide, for example, two successive reactions of H_2O_2, with bromide and then with bromine, take place. Because these two reactions together serve as a catalytic pathway for hydrogen peroxide decomposition, *both* of them must have significantly lower activation energies than the uncatalyzed decomposition, as shown schematically in Figure 14.20 ◄.

▲ **Figure 14.20 Energy profiles for uncatalyzed and catalyzed reactions.** The energy profiles for the uncatalyzed decomposition of hydrogen peroxide and for the reaction as catalyzed by Br^- are compared. The catalyzed reaction involves two successive steps, each of which has a lower activation energy than the uncatalyzed reaction. Notice that the energies of reactants and products are unchanged by the catalyst.

> ### GIVE IT SOME THOUGHT
>
> How does a catalyst increase the rate of a reaction?

Heterogeneous Catalysis

A **heterogeneous catalyst** exists in a different phase from the reactant molecules, usually as a solid in contact with either gaseous reactants or with reactants in a liquid solution. Many industrially important reactions are catalyzed by the surfaces of solids. For example, hydrocarbon molecules are rearranged to form gasoline with the aid of what are called "cracking" catalysts (see the "Chemistry Put to Work" box in Section 25.3). Heterogeneous catalysts are often composed of metals or metal oxides. Because the catalyzed reaction occurs on the surface, special methods are often used to prepare catalysts so that they have very large surface areas.

The initial step in heterogeneous catalysis is usually **adsorption** of reactants. *Adsorption* refers to the binding of molecules to a surface, whereas *absorption* refers to the uptake of molecules into the interior of another substance. ∞ (Section 13.6) Adsorption occurs because the atoms or ions at the surface of a solid are extremely reactive. Unlike their counterparts in the interior of the substance, surface atoms and ions have unused bonding capacity. This unused bonding capability may be used to bond molecules from the gas or solution phase to the surface of the solid.

The reaction of hydrogen gas with ethylene gas to form ethane gas provides an example of heterogeneous catalysis:

$$C_2H_4(g) + H_2(g) \longrightarrow C_2H_6(g) \quad \Delta H° = -137 \text{ kJ/mol} \qquad [14.32]$$

Ethylene   Ethane

Even though this reaction is exothermic, it occurs very slowly in the absence of a catalyst. In the presence of a finely powdered metal, however, such as nickel, palladium, or platinum, the reaction occurs rather easily at room temperature. The mechanism by which the reaction occurs is diagrammed in Figure 14.21 ▼. Both ethylene and hydrogen are adsorbed on the metal surface [Figure 14.21(a)]. Upon adsorption the H—H bond of H_2 breaks, leaving two H atoms that are bonded to the metal surface, as shown in Figure 14.21(b). The hydrogen atoms are relatively free to move about the surface. When a hydrogen encounters an adsorbed ethylene molecule, it can form a σ bond to one of the carbon atoms, effectively destroying the C—C π bond and leaving an *ethyl group* (C_2H_5) bonded to the surface via a metal-to-carbon σ bond [Figure 14.21(c)]. This σ bond is relatively weak, so when the other carbon atom also encounters a hydrogen atom, a sixth C—H σ bond is readily formed and an ethane molecule is released from the metal surface [Figure 14.21(d)]. The site is ready to adsorb another ethylene molecule and thus begin the cycle again.

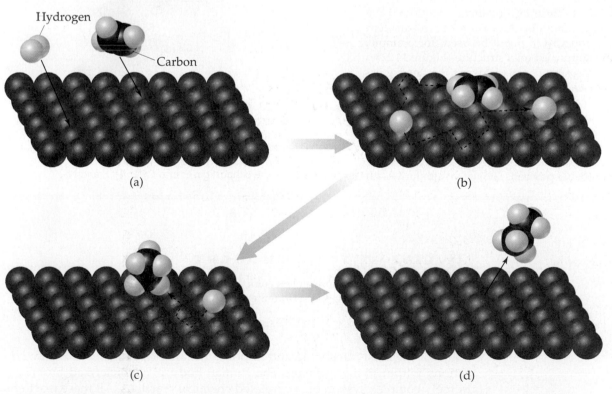

(a)   (b)

(c)   (d)

▲ **Figure 14.21 Mechanism for reaction of ethylene with hydrogen on a catalytic surface.** (a) The hydrogen and ethylene are adsorbed at the metal surface. (b) The H—H bond is broken to give adsorbed hydrogen atoms. (c) These migrate to the adsorbed ethylene and bond to the carbon atoms. (d) As C—H bonds are formed, the adsorption of the molecule to the metal surface is decreased and ethane is released.

Chemistry Put to Work   CATALYTIC CONVERTERS

Heterogeneous catalysis plays a major role in the fight against urban air pollution. Two components of automobile exhausts that help form photochemical smog are nitrogen oxides and unburned hydrocarbons of various types (Section 18.4). In addition, automobile exhausts may contain considerable quantities of carbon monoxide. Even with the most careful attention to engine design, it is impossible under normal driving conditions to reduce the quantity of these pollutants to an acceptable level in the exhaust gases. It is therefore necessary to remove them from the exhaust before they are vented to the air. This removal is accomplished in the *catalytic converter*.

The catalytic converter, which is part of the exhaust system, must perform two distinct functions: (1) oxidation of CO and unburned hydrocarbons (C_xH_y) to carbon dioxide and water, and (2) reduction of nitrogen oxides to nitrogen gas:

$$CO, C_xH_y \xrightarrow{O_2} CO_2 + H_2O$$
$$NO, NO_2 \longrightarrow N_2$$

These two functions require two distinctly different catalysts, so the development of a successful catalyst system is a difficult challenge. The catalysts must be effective over a wide range of operating temperatures. They must continue to be active despite the fact that various components of the exhaust can block the active sites of the catalyst. And the catalysts must be sufficiently rugged to withstand exhaust gas turbulence and the mechanical shocks of driving under various conditions for thousands of miles.

Catalysts that promote the combustion of CO and hydrocarbons are, in general, the transition-metal oxides and the noble metals, such as platinum. A mixture of two different metal oxides, CuO and Cr_2O_3, might be used, for example. These materials are supported on a structure (Figure 14.22▶) that allows the best possible contact between the flowing exhaust gas and the catalyst surface. A honeycomb structure made from alumina (Al_2O_3) and impregnated with the catalyst is employed. Such catalysts operate by first adsorbing oxygen gas, also present in the exhaust gas. This adsorption weakens the O—O bond in O_2, so that oxygen atoms are available for reaction with adsorbed CO to form CO_2. Hydrocarbon oxidation probably proceeds somewhat similarly, with

the hydrocarbons first being adsorbed followed by rupture of a C—H bond.

The most effective catalysts for reduction of NO to yield N_2 and O_2 are transition-metal oxides and noble metals, the same kinds of materials that catalyze the oxidation of CO and hydrocarbons. The catalysts that are most effective in one reaction, however, are usually much less effective in the other. It is therefore necessary to have two different catalytic components.

Catalytic converters are remarkably efficient heterogeneous catalysts. The automotive exhaust gases are in contact with the catalyst for only 100 to 400 ms. In this very short time, 96% of the hydrocarbons and CO is converted to CO_2 and H_2O, and the emission of nitrogen oxides is reduced by 76%.

There are costs as well as benefits associated with the use of catalytic converters. Some of the metals used in the converters are very expensive. Catalytic converters currently account for about 35% of the platinum, 65% of the palladium, and 95% of the rhodium used annually. All of these metals, which come mainly from Russia and South Africa, are far more expensive than gold.

Related Exercises: 14.56, 14.75, and 14.76

▲ **Figure 14.22 Cross section of a catalytic converter.** Automobiles are equipped with catalytic converters, which are part of their exhaust systems. The exhaust gases contain CO, NO, NO_2, and unburned hydrocarbons that pass over surfaces impregnated with catalysts. The catalysts promote the conversion of the exhaust gases into CO_2, H_2O, and N_2.

GIVE IT SOME THOUGHT

How does a homogeneous catalyst compare with a heterogeneous one regarding the ease of recovery of the catalyst from the reaction mixture?

Enzymes

Many of the most interesting and important examples of catalysis involve reactions within living systems. The human body is characterized by an extremely complex system of interrelated chemical reactions. All these reactions must occur at carefully controlled rates to maintain life. A large number of marvelously efficient biological catalysts known as **enzymes** are necessary for many of these reactions to occur at suitable rates. Most enzymes are large protein molecules with molecular weights ranging from about 10,000 to about 1 million amu. They are very selective in the reactions that they catalyze, and some are absolutely specific, operating for only one substance in only one reaction. The decomposition of hydrogen peroxide, for example, is an important biological process. Because hydrogen peroxide is strongly oxidizing, it can be physiologically harmful. For this reason, the blood and livers of mammals contain an enzyme, *catalase*, which catalyzes the decomposition of hydrogen peroxide into water and oxygen (Equation 14.29). Figure 14.23▶ shows the dramatic acceleration of this chemical reaction by the catalase in beef liver.

Although an enzyme is a large molecule, the reaction is catalyzed at a very specific location in the enzyme, called the **active site**. The substances that undergo reaction at this site are called **substrates**. The **lock-and-key model**, illustrated in Figure 14.24▼, provides a simple explanation for the specificity of an enzyme. The substrate is pictured as fitting neatly into a special place on the enzyme (the active site), much like a specific key fits into a lock. The active site is created by coiling and folding of the long protein molecule to form a space, something like a pocket, into which the substrate molecule fits. Figure 14.25▶ shows a model of the enzyme *lysozyme* with and without a bound substrate molecule.

The combination of the enzyme and the substrate is called the *enzyme-substrate complex*. Although Figure 14.24 shows both the active site and its complementary substrate as having rigid shapes, the active site is often fairly flexible. Thus, the active site may change shape as it binds the substrate. The binding between the substrate and the active site involves intermolecular forces such as dipole–dipole attractions, hydrogen bonds, and London dispersion forces. ∞ (Section 11.2)

As the substrate molecules enter the active site, they are somehow activated, so that they are capable of extremely rapid reaction. This activation may result from the withdrawal or donation of electron density at a particular bond by the enzyme. In addition, in the process of fitting into the active site, the substrate molecule may be distorted and thus made more reactive. Once the reaction occurs, the products then depart, allowing another substrate molecule to enter.

▲ **Figure 14.23   Effect of an enzyme.** Ground-up beef liver causes hydrogen peroxide to decompose rapidly into water and oxygen. The decomposition is catalyzed by the enzyme *catalase*. Grinding the liver breaks open the cells, so that the reaction takes place more rapidly. The frothing is due to escape of oxygen gas from the reaction mixture.

(a)

(b)

▲ **Figure 14.25   Molecular model of an enzyme.** (a) A molecular model of the enzyme *lysozyme*. Note the characteristic cleft, which is the location of the active site. (b) Lysozyme with a bound substrate molecule.

▲ **Figure 14.24   The lock-and-key model for enzyme action.** The correct substrate is recognized by its ability to fit the active site of the enzyme, forming the enzyme-substrate complex. After the reaction of the substrate is complete, the products separate from the enzyme.

Chemistry and Life   NITROGEN FIXATION AND NITROGENASE

Nitrogen is one of the most essential elements in living organisms. It is found in many compounds that are vital to life, including proteins, nucleic acids, vitamins, and hormones. Plants use very simple nitrogen-containing compounds, especially NH_3, NH_4^+, and NO_3^-, as starting materials from which such complex, biologically necessary compounds are formed. Animals are unable to synthesize the complex nitrogen compounds they require from the simple substances used by plants. Instead, they rely on more complicated precursors present in vitamin- and protein-rich foods.

Nitrogen is continually cycling through this biological arena in various forms, as shown in the simplified nitrogen cycle in Figure 14.26▼. For example, certain microorganisms convert the nitrogen in animal waste and dead plants and animals into molecular nitrogen, $N_2(g)$, which returns to the atmosphere. For the food chain to be sustained, there must be a means of reincorporating this atmospheric N_2 in a form that plants can utilize. The process of converting N_2 into compounds that plants can use is called *nitrogen fixation*. Fixing nitrogen is difficult; N_2 is an exceptionally unreactive molecule, in large part because of its very strong $N\equiv N$ triple bond. ∞ (Section 8.3) Some fixed nitrogen results from the action of lightning on the atmosphere, and some is produced industrially using a process we will discuss in Chapter 15. About 60% of fixed nitrogen, however, is a consequence of the action of a remarkable and complex enzyme called *nitrogenase*. This enzyme

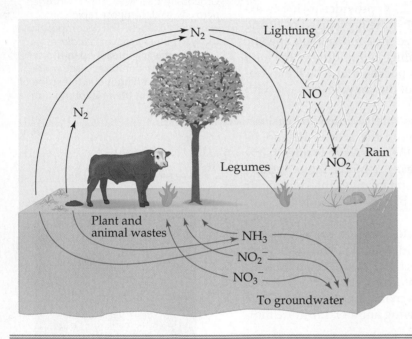

◄ **Figure 14.26 Simplified picture of the nitrogen cycle.** The compounds of nitrogen in the soil are water-soluble species, such as NH_3, NO_2^-, and NO_3^-, which can be washed out of the soil by groundwater. These nitrogen compounds are converted into biomolecules by plants and are incorporated into animals that eat the plants. Certain bacteria that release N_2 to the atmosphere attack animal waste and dead plants and animals. Atmospheric N_2 is fixed in the soil predominantly by the action of certain plants that contain the enzyme nitrogenase, thereby completing the cycle.

The activity of an enzyme is destroyed if some molecule in the solution is able to bind strongly to the active site and block the entry of the substrate. Such substances are called *enzyme inhibitors*. Nerve poisons and certain toxic metal ions such as lead and mercury are believed to act in this way to inhibit enzyme activity. Some other poisons act by attaching elsewhere on the enzyme, thereby distorting the active site so that the substrate no longer fits.

Enzymes are enormously more efficient than ordinary nonbiochemical catalysts. The number of individual catalyzed reaction events occurring at a particular active site, called the *turnover number*, is generally in the range of 10^3 to 10^7 per second. Such large turnover numbers correspond to very low activation energies.

▲ **GIVE IT SOME THOUGHT**

What names are given to the following aspects of enzymes and enzyme catalysis: **(a)** the place on the enzyme where catalysis occurs, **(b)** the substances that undergo catalysis?

▲ **Figure 14.27 The FeMo-cofactor of nitrogenase.** Nitrogenase is found in nodules in the roots of certain plants, such as the white clover roots shown at the left. The cofactor, which is thought to be the active site of the enzyme, contains seven Fe atoms and one Mo atom, linked by sulfur atoms. The molecules on the outside of the cofactor connect it to the rest of the protein.

is *not* present in humans or other animals; rather, it is found in bacteria that live in the root nodules of certain plants such as the legumes clover and alfalfa.

Nitrogenase converts N_2 into NH_3, a process that, in the absence of a catalyst, has a very large activation energy. This process is a *reduction* of nitrogen—during the reaction, its oxidation state is reduced from 0 in N_2 to -3 in NH_3. The mechanism by which nitrogenase reduces N_2 is not fully understood. Like many other enzymes, including catalase, the active site of nitrogenase contains transition-metal atoms; such enzymes are called *metalloenzymes*. Because transition metals can readily change oxidation state, metalloenzymes are especially useful for effecting transformations in which substrates are either oxidized or reduced.

It has been known for nearly 20 years that a portion of nitrogenase contains iron and molybdenum atoms. This portion, called the *FeMo-cofactor*, is thought to serve as the active site of the enzyme. The FeMo-cofactor of nitrogenase is a striking cluster of seven Fe atoms and one Mo atom, all linked by sulfur atoms (Figure 14.27 ▲).

It is one of the wonders of life that simple bacteria can contain beautifully complex and vitally important enzymes such as nitrogenase. Because of this enzyme, nitrogen is continually cycled between its comparatively inert role in the atmosphere and its critical role in living organisms; without it, life as we know it could not exist on Earth.

Related Exercises: 14.79, 14.80, 14.102, 14.109

■ SAMPLE INTEGRATIVE EXERCISE | Putting Concepts Together

Formic acid (HCOOH) decomposes in the gas phase at elevated temperatures as follows:

$$HCOOH(g) \longrightarrow CO_2(g) + H_2(g)$$

The uncatalyzed decomposition reaction is determined to be first order. A graph of the partial pressure of HCOOH versus time for decomposition at 838 K is shown as the red curve in Figure 14.28 ▶. When a small amount of solid ZnO is added to the reaction chamber, the partial pressure of acid versus time varies as shown by the blue curve in Figure 14.28.

(a) Estimate the half-life and first-order rate constant for formic acid decomposition.
(b) What can you conclude from the effect of added ZnO on the decomposition of formic acid?
(c) The progress of the reaction was followed by measuring the partial pressure of formic acid vapor at selected times. Suppose that, instead, we had plotted the concentration of formic acid in units of mol/L. What effect would this have had on the calculated value of k?
(d) The pressure of formic acid vapor at the start of the reaction is 3.00×10^2 torr. Assuming constant temperature and ideal-gas behavior, what is the pressure in the

▲ **Figure 14.28 Variation in pressure of HCOOH(*g*) as a function of time at 838 K.** The red line corresponds to decomposition when only gaseous HCOOH is present. The blue line corresponds to decomposition in the presence of added ZnO(*s*).

system at the end of the reaction? If the volume of the reaction chamber is 436 cm^3, how many moles of gas occupy the reaction chamber at the end of the reaction?

(e) The standard heat of formation of formic acid vapor is $\Delta H_f^\circ = -378.6$ kJ/mol. Calculate ΔH° for the overall reaction. If the activation energy (E_a) for the reaction is 184 kJ/mol, sketch an approximate energy profile for the reaction, and label E_a, ΔH°, and the transition state.

SOLUTION

(a) The initial pressure of HCOOH is 3.00×10^2 torr. On the graph we move to the level at which the partial pressure of HCOOH is 1.50×10^2 torr, half the initial value. This corresponds to a time of about 6.60×10^2 s, which is therefore the half-life. The first-order rate constant is given by Equation 14.15: $k = 0.693/t_{1/2} = 0.693/660$ s $= 1.05 \times 10^{-3}$ s^{-1}.

(b) The reaction proceeds much more rapidly in the presence of solid ZnO, so the surface of the oxide must be acting as a catalyst for the decomposition of the acid. This is an example of heterogeneous catalysis.

(c) If we had graphed the concentration of formic acid in units of moles per liter, we would still have determined that the half-life for decomposition is 660 seconds, and we would have computed the same value for k. Because the units for k are s^{-1}, the value for k is independent of the units used for concentration.

(d) According to the stoichiometry of the reaction, two moles of product are formed for each mole of reactant. When reaction is completed, therefore, the pressure will be 600 torr, just twice the initial pressure, assuming ideal-gas behavior. (Because we are working at quite high temperature and fairly low gas pressure, assuming ideal-gas behavior is reasonable.) The number of moles of gas present can be calculated using the ideal-gas equation (Section 10.4):

$$n = \frac{PV}{RT} = \frac{(600/760 \text{ atm})(0.436 \text{ L})}{(0.0821 \text{ L-atm/mol-K})(838 \text{ K})} = 5.00 \times 10^{-3} \text{ moles}$$

(e) We first calculate the overall change in energy, ΔH° (Section 5.7 and Appendix C), as in

$$\Delta H^\circ = \Delta H_f^\circ(CO_2(g)) + \Delta H_f^\circ(H_2(g)) - \Delta H_f^\circ(HCOOH(g))$$
$$= -393.5 \text{ kJ/mol} + 0 - (-378.6 \text{ kJ/mol})$$
$$= -14.9 \text{ kJ/mol}$$

From this and the given value for E_a, we can draw an approximate energy profile for the reaction, in analogy to Figure 14.15.

CHAPTER REVIEW

SUMMARY AND KEY TERMS

Introduction and Section 14.1 In this chapter we explored **chemical kinetics**, the area of chemistry that studies the rates of chemical reactions and the factors that affect them, namely, concentration, temperature, and catalysts.

Section 14.2 Reaction rates are usually expressed as changes in concentration per unit time: Typically, for reactions in solution, rates are given in units of molarity per second M/s. For most reactions, a plot of molarity versus time shows that the rate slows down as the reaction proceeds. The **instantaneous rate** is the slope of a line drawn tangent to the concentration-versus-time curve at a specific time. Rates can be written in terms of the appearance of products or the disappearance of reactants; the stoichiometry of the reaction dictates the relationship between rates of appearance and disappearance. Spectroscopy is one technique that can be used to monitor the course of a reaction. According to **Beer's law**, the absorption of electromagnetic radiation by a substance at a particular wavelength is directly proportional to its concentration.

Section 14.3 The quantitative relationship between rate and concentration is expressed by a **rate law**, which usually has the following form:

$$\text{Rate} = k[\text{reactant } 1]^m[\text{reactant } 2]^n \ldots$$

The constant k in the rate law is called the **rate constant**; the exponents m, n, and so forth are called **reaction orders** for the reactants. The sum of the reaction orders gives the **overall reaction order**. Reaction orders must be determined experimentally. The units of the rate constant depend on the overall reaction order. For a reaction in which the overall reaction order is 1, k has units of s^{-1}; for one in which the overall reaction order is 2, k has units of $M^{-1}\,s^{-1}$.

Section 14.4 Rate laws can be used to determine the concentrations of reactants or products at any time during a reaction. In a **first-order reaction** the rate is proportional to the concentration of a single reactant raised to the first power: Rate $= k[A]$. In such cases the integrated form of the rate law is $\ln[A]_t = -kt + \ln[A]_0$, where $[A]_t$ is the concentration of reactant A at time t, k is the rate constant, and $[A]_0$ is the initial concentration of A. Thus, for a first-order reaction, a graph of $\ln[A]$ versus time yields a straight line of slope $-k$.

A **second-order reaction** is one for which the overall reaction order is 2. If a second-order rate law depends on the concentration of only one reactant, then rate $= k[A]^2$, and the time dependence of $[A]$ is given by the integrated form of the rate law: $1/[A]_t = 1/[A]_0 + kt$. In this case a graph of $1/[A]_t$ versus time yields a straight line.

The **half-life** of a reaction, $t_{1/2}$, is the time required for the concentration of a reactant to drop to one-half of its original value. For a first-order reaction, the half-life depends only on the rate constant and not on the initial concentration: $t_{1/2} = 0.693/k$. The half-life of a second-order reaction depends on both the rate constant and the initial concentration of A: $t_{1/2} = 1/k[A]_0$.

Section 14.5 The **collision model**, which assumes that reactions occur as a result of collisions between molecules, helps explain why the magnitudes of rate constants increase with increasing temperature. The greater the kinetic energy of the colliding molecules, the greater is the energy of collision. The minimum energy required for a reaction to occur is called the **activation energy**, E_a. A collision with energy E_a or greater can cause the atoms of the colliding molecules to reach the **activated complex** (or **transition state**), which is the highest energy arrangement in the pathway from reactants to products. Even if a collision is energetic enough, it may not lead to reaction; the reactants must also be correctly oriented relative to one another in order for a collision to be effective.

Because the kinetic energy of molecules depends on temperature, the rate constant of a reaction is very dependent on temperature. The relationship between k and temperature is given by the **Arrhenius equation**: $k = Ae^{-E_a/RT}$. The term A is called the **frequency factor**; it relates to the number of collisions that are favorably oriented for reaction. The Arrhenius equation is often used in logarithmic form: $k = \ln A - E_a/RT$. Thus, a graph of $\ln k$ versus $1/T$ yields a straight line with slope $-E_a/R$.

Section 14.6 A **reaction mechanism** details the individual steps that occur in the course of a reaction. Each of these steps, called **elementary reactions**, has a well-defined rate law that depends on the number of molecules (the **molecularity**) of the step. Elementary reactions are defined as either **unimolecular**, **bimolecular**, or **termolecular**, depending on whether one, two, or three reactant molecules are involved, respectively. Termolecular elementary reactions are very rare. Unimolecular, bimolecular, and termolecular reactions follow rate laws that are first order overall, second order overall, and third order overall, respectively. Many reactions occur by a multistep mechanism, involving two or more elementary reactions, or steps. An **intermediate** is produced in one elementary step, is consumed in a later elementary step, and therefore does not appear in the overall equation for the reaction. When a mechanism has several elementary steps, the overall rate is limited by the slowest elementary step, called the **rate-determining step**. A fast elementary step that follows the rate-determining step will have no effect on the rate law of the reaction. A fast step that precedes the rate-determining step often creates an equilibrium that involves an intermediate. For a mechanism to be valid, the rate law predicted by the mechanism must be the same as that observed experimentally.

Section 14.7 A **catalyst** is a substance that increases the rate of a reaction without undergoing a net chemical change itself. It does so by providing a different mechanism for the reaction, one that has a lower activation energy. A **homogeneous catalyst** is one that is in the same phase as the reactants. A **heterogeneous catalyst** has a different phase from the reactants. Finely divided metals are often used as heterogeneous catalysts for solution-

and gas-phase reactions. Reacting molecules can undergo binding, or **adsorption**, at the surface of the catalyst. The adsorption of a reactant at specific sites on the surface makes bond breaking easier, lowering the activation energy. Catalysis in living organisms is achieved by **enzymes**, large protein molecules that usually catalyze a very specific reaction. The specific reactant molecules involved in an enzymatic reaction are called **substrates**. The site of the enzyme where the catalysis occurs is called the **active site**. In the **lock-and-key model** for enzyme catalysis, substrate molecules bind very specifically to the active site of the enzyme, after which they can undergo reaction.

KEY SKILLS

- Understand the factors that affect the rate of chemical reactions.

- Be able to determine the rate of a reaction given time and concentration.

- Be able to relate the rate of formation of products and the rate of disappearance of reactants given the balanced chemical equation for the reaction.

- Understand the form and meaning of a rate law including the ideas of reaction order and rate constant.

- Be able to determine the rate law and rate constant for a reaction from a series of experiments given the measured rates for various concentrations of reactants.

- Be able to use the integrated form of a rate law to determine the concentration of a reactant at a given time.

- Explain how the activation energy affects a rate and be able to use the Arrhenius Equation.

- Be able to predict a rate law for a reaction having a multistep mechanism given the individual steps in the mechanism.

- Explain how a catalyst works.

KEY EQUATIONS

- $\text{Rate} = -\dfrac{1}{a}\dfrac{\Delta[A]}{\Delta t} = -\dfrac{1}{b}\dfrac{\Delta[B]}{\Delta t} = \dfrac{1}{c}\dfrac{\Delta[C]}{\Delta t} = \dfrac{1}{d}\dfrac{\Delta[D]}{\Delta t}$ [14.4]   Relating rates to the components of a balanced chemical equation

- $\text{Rate} = k[A]^m[B]^n$ [14.7]   General form of a rate law

- $\ln[A]_t - \ln[A]_0 = -kt$ or $\ln\dfrac{[A]_t}{[A]_0} = -kt$ [14.12]   The integrated form of a first-order rate law

- $\dfrac{1}{[A]_t} = kt + \dfrac{1}{[A]_0}$ [14.14]   The integrated form of the second-order rate law

- $t_{1/2} = \dfrac{0.693}{k}$ [14.15]   Relating the half-life and rate constant for a first-order reaction

- $k = Ae^{-E_a/RT}$ [14.19]   The Arrhenius equation, which expresses how the rate constant depends on temperature

- $\ln k = -\dfrac{E_a}{RT} + \ln A$ [14.20]   Linear form of the Arrhenius equation

VISUALIZING CONCEPTS

14.1 Consider the following graph of the concentration of a substance over time. **(a)** Is X a reactant or product of the reaction? **(b)** Why is the average rate of the reaction greater between points 1 and 2 than between points 2 and 3? [Section 14.2]

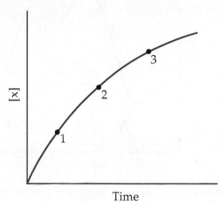

14.2 You study the rate of a reaction, measuring both the concentration of the reactant and the concentration of the product as a function of time, and obtain the following results:

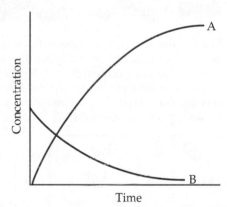

What chemical equation is consistent with this data: **(a)** A $\longrightarrow$ B, **(b)** B $\longrightarrow$ A, **(c)** A $\longrightarrow$ 2 B, **(d)** B $\longrightarrow$ 2 A. Explain your choice. [Section 14.2]

14.3 You perform a series of experiments for the reaction A $\longrightarrow$ B + C and find that the rate law has the form rate = $k[A]^x$. Determine the value of x in each of the following cases: **(a)** There is no rate change when [A] is tripled. **(b)** The rate increases by a factor of 9 when [A] is tripled. **(c)** When [A] is doubled, the rate increases by a factor of 8. [Section 14.3]

14.4 The following diagrams represent mixtures of NO(g) and $O_2(g)$. These two substances react as follows:

$$2\,NO(g) + O_2(g) \longrightarrow 2\,NO_2(g)$$

It has been determined experimentally that the rate is second order in NO and first order in O_2. Based on this fact, which of the following mixtures will have the fastest initial rate? [Section 14.3]

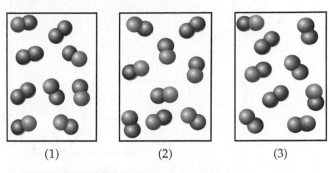

(1) (2) (3)

14.5 A friend studies a first-order reaction and obtains the following three graphs for experiments done at two different temperatures. **(a)** Which two lines represent experiments done at the same temperature? What accounts for the difference in these two lines? In what way are they the same? **(b)** Which two lines represent experiments done with the same starting concentration but at different temperatures? Which line probably represents the lower temperature? How do you know? [Section 14.4]

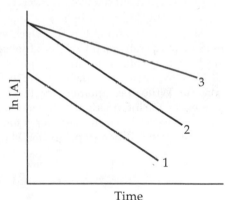

14.6 **(a)** Given the following diagrams at $t = 0$ and $t = 30$, what is the half-life of the reaction if it follows first-order kinetics?

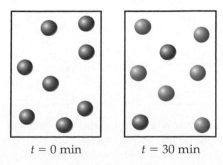

$t = 0$ min $t = 30$ min

(b) After four half-life periods for a first-order reaction, what fraction of reactant remains? [Section 14.4]

14.7 The following diagram shows the reaction profile of a reaction. Label the components indicated by the boxes. [Section 14.5]

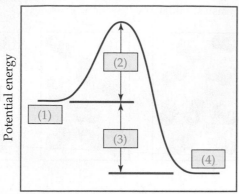

Reaction pathway

14.8 You study the effect of temperature on the rate of two reactions and graph the natural logarithm of the rate constant for each reaction as a function of $1/T$. How do the two graphs compare **(a)** if the activation energy of the second reaction is higher than the activation energy of the first reaction but the two reactions have the same frequency factor, and **(b)** if the frequency factor of the second reaction is higher than the frequency factor of the first reaction but the two reactions have the same activation energy? [Section 14.5]

14.9 Consider the diagram below, which represents two steps in an overall reaction. The red spheres are oxygen, the blue ones nitrogen, and the green ones fluorine. **(a)** Write the chemical equation for each step in the reaction. **(b)** Write the equation for the overall reaction. **(c)** Identify the intermediate in the mechanism. **(d)** Write the rate law for the overall reaction if the first step is the slow, rate-determining step. [Section 14.6]

14.10 Based on the following reaction profile, how many intermediates are formed in the reaction A ⟶ C? How many transition states are there? Which step is the fastest? Is the reaction A ⟶ C exothermic or endothermic? [Section 14.6]

Reaction pathway

14.11 Draw a possible transition state for the bimolecular reaction depicted below. (The blue spheres are nitrogen atoms, and the red ones are oxygen atoms.) Use dashed lines to represent the bonds that are in the process of being broken or made in the transition state. [Section 14.6]

14.12 The following diagram represents an imaginary two-step mechanism. Let the red spheres represent element A, the green ones element B, and the blue ones element C. **(a)** Write the equation for the net reaction that is occurring. **(b)** Identify the intermediate. **(c)** Identify the catalyst. [Sections 14.6 and 14.7]

EXERCISES

Reaction Rates

14.13 (a) What is meant by the term *reaction rate*? **(b)** Name three factors that can affect the rate of a chemical reaction. **(c)** What information is necessary to relate the rate of disappearance of reactants to the rate of appearance of products?

14.14 (a) What are the units usually used to express the rates of reactions occurring in solution? **(b)** From your everyday experience, give two examples of the effects of temperature on the rates of reactions. **(c)** What is the difference between average rate and instantaneous rate?

14.15 Consider the following hypothetical aqueous reaction: A(*aq*) ⟶ B(*aq*). A flask is charged with 0.065 mol of A in a total volume of 100.0 mL. The following data are collected:

Time (min)	0	10	20	30	40
Moles of A	0.065	0.051	0.042	0.036	0.031

(a) Calculate the number of moles of B at each time in the table, assuming that there are no molecules of B at time zero. (b) Calculate the average rate of disappearance of A for each 10-min interval, in units of M/s. (c) Between $t = 10$ min and $t = 30$ min, what is the average rate of appearance of B in units of M/s? Assume that the volume of the solution is constant.

14.16 A flask is charged with 0.100 mol of A and allowed to react to form B according to the hypothetical gas-phase reaction $A(g) \longrightarrow B(g)$. The following data are collected:

Time (s)	0	40	80	120	160
Moles of A	0.100	0.067	0.045	0.030	0.020

(a) Calculate the number of moles of B at each time in the table. (b) Calculate the average rate of disappearance of A for each 40-s interval, in units of mol/s. (c) What additional information would be needed to calculate the rate in units of concentration per time?

14.17 The isomerization of methyl isonitrile (CH_3NC) to acetonitrile (CH_3CN) was studied in the gas phase at 215 °C, and the following data were obtained:

Time (s)	[CH₃NC] (M)
0	0.0165
2,000	0.0110
5,000	0.00591
8,000	0.00314
12,000	0.00137
15,000	0.00074

(a) Calculate the average rate of reaction, in M/s, for the time interval between each measurement. (b) Graph [CH₃NC] versus time, and determine the instantaneous rates in M/s at $t = 5000$ s and $t = 8000$ s.

14.18 The rate of disappearance of HCl was measured for the following reaction:

$$CH_3OH(aq) + HCl(aq) \longrightarrow CH_3Cl(aq) + H_2O(l)$$

The following data were collected:

Time (min)	[HCl] (M)
0.0	1.85
54.0	1.58
107.0	1.36
215.0	1.02
430.0	0.580

(a) Calculate the average rate of reaction, in M/s, for the time interval between each measurement.
(b) Graph [HCl] versus time, and determine the instantaneous rates in M/min and M/s at $t = 75.0$ min and $t = 250$ min.

14.19 For each of the following gas-phase reactions, indicate how the rate of disappearance of each reactant is related to the rate of appearance of each product:
(a) $H_2O_2(g) \longrightarrow H_2(g) + O_2(g)$
(b) $2 N_2O(g) \longrightarrow 2 N_2(g) + O_2(g)$
(c) $N_2(g) + 3 H_2(g) \longrightarrow 2 NH_3(g)$

14.20 For each of the following gas-phase reactions, write the rate expression in terms of the appearance of each product or disappearance of each reactant:
(a) $2 H_2O(g) \longrightarrow 2 H_2(g) + O_2(g)$
(b) $2 SO_2(g) + O_2(g) \longrightarrow 2 SO_3(g)$
(c) $2 NO(g) + 2 H_2(g) \longrightarrow N_2(g) + 2 H_2O(g)$

14.21 (a) Consider the combustion of $H_2(g)$: $2 H_2(g) + O_2(g) \longrightarrow 2 H_2O(g)$. If hydrogen is burning at the rate of 0.85 mol/s, what is the rate of consumption of oxygen? What is the rate of formation of water vapor? (b) The reaction $2 NO(g) + Cl_2(g) \longrightarrow 2 NOCl(g)$ is carried out in a closed vessel. If the partial pressure of NO is decreasing at the rate of 23 torr/min, what is the rate of change of the total pressure of the vessel?

14.22 (a) Consider the combustion of ethylene, $C_2H_4(g) + 3 O_2(g) \longrightarrow 2 CO_2(g) + 2 H_2O(g)$. If the concentration of C_2H_4 is decreasing at the rate of 0.025 M/s, what are the rates of change in the concentrations of CO_2 and H_2O? (b) The rate of decrease in N_2H_4 partial pressure in a closed reaction vessel from the reaction $N_2H_4(g) + H_2(g) \longrightarrow 2 NH_3(g)$ is 63 torr/h. What are the rates of change of NH_3 partial pressure and total pressure in the vessel?

Rate Laws

14.23 A reaction $A + B \longrightarrow C$ obeys the following rate law: Rate $= k[B]^2$. (a) If [A] is doubled, how will the rate change? Will the rate constant change? Explain. (b) What are the reaction orders for A and B? What is the overall reaction order? (c) What are the units of the rate constant?

14.24 Consider a hypothetical reaction between A, B, and C that is first order in A, zero order in B, and second order in C. (a) Write the rate law for the reaction. (b) How does the rate change when [A] is doubled and the other reactant concentrations are held constant? (c) How does the rate change when [B] is tripled and the other reac-

tant concentrations are held constant? (d) How does the rate change when [C] is tripled and the other reactant concentrations are held constant? (e) By what factor does the rate change when the concentrations of all three reactants are tripled?

14.25 The decomposition of N_2O_5 in carbon tetrachloride proceeds as follows: $2 N_2O_5 \longrightarrow 4 NO_2 + O_2$. The rate law is first order in N_2O_5. At 64 °C the rate constant is $4.82 \times 10^{-3} s^{-1}$. (a) Write the rate law for the reaction. (b) What is the rate of reaction when $[N_2O_5] = 0.0240$ M? (c) What happens to the rate when the concentration of N_2O_5 is doubled to 0.0480 M?

14.26 Consider the following reaction:

$$2 NO(g) + 2 H_2(g) \longrightarrow N_2(g) + 2 H_2O(g)$$

(a) The rate law for this reaction is first order in H_2 and second order in NO. Write the rate law. **(b)** If the rate constant for this reaction at 1000 K is $6.0 \times 10^4 \, M^{-2} s^{-1}$, what is the reaction rate when [NO] = 0.035 M and $[H_2]$ = 0.015 M? **(c)** What is the reaction rate at 1000 K when the concentration of NO is increased to 0.10 M, while the concentration of H_2 is 0.010 M?

14.27 Consider the following reaction:

$$CH_3Br(aq) + OH^-(aq) \longrightarrow CH_3OH(aq) + Br^-(aq)$$

The rate law for this reaction is first order in CH_3Br and first order in OH^-. When $[CH_3Br]$ is $5.0 \times 10^{-3} M$ and $[OH^-]$ is 0.050 M, the reaction rate at 298 K is 0.0432 M/s. **(a)** What is the value of the rate constant? **(b)** What are the units of the rate constant? **(c)** What would happen to the rate if the concentration of OH^- were tripled?

14.28 The reaction between ethyl bromide (C_2H_5Br) and hydroxide ion in ethyl alcohol at 330 K, $C_2H_5Br(alc)$ + $OH^-(alc) \longrightarrow C_2H_5OH(l) + Br^-(alc)$, is first order each in ethyl bromide and hydroxide ion. When $[C_2H_5Br]$ is 0.0477 M and $[OH^-]$ is 0.100 M, the rate of disappearance of ethyl bromide is $1.7 \times 10^{-7} M$/s. **(a)** What is the value of the rate constant? **(b)** What are the units of the rate constant? **(c)** How would the rate of disappearance of ethyl bromide change if the solution were diluted by adding an equal volume of pure ethyl alcohol to the solution?

14.29 The iodide ion reacts with hypochlorite ion (the active ingredient in chlorine bleaches) in the following way: $OCl^- + I^- \longrightarrow OI^- + Cl^-$. This rapid reaction gives the following rate data:

$[OCl^-]$ (M)	$[I^-]$ (M)	Rate (M/s)
1.5×10^{-3}	1.5×10^{-3}	1.36×10^{-4}
3.0×10^{-3}	1.5×10^{-3}	2.72×10^{-4}
1.5×10^{-3}	3.0×10^{-3}	2.72×10^{-4}

(a) Write the rate law for this reaction. **(b)** Calculate the rate constant. **(c)** Calculate the rate when $[OCl^-]$ = $2.0 \times 10^{-3} M$ and $[I^-]$ = $5.0 \times 10^{-4} M$.

14.30 The reaction $2 ClO_2(aq) + 2 OH^-(aq) \longrightarrow ClO_3^-(aq) + ClO_2^-(aq) + H_2O(l)$ was studied with the following results:

Experiment	$[ClO_2]$ (M)	$[OH^-]$ (M)	Rate (M/s)
1	0.060	0.030	0.0248
2	0.020	0.030	0.00276
3	0.020	0.090	0.00828

(a) Determine the rate law for the reaction. **(b)** Calculate the rate constant. **(c)** Calculate the rate when $[ClO_2]$ = 0.100 M and $[OH^-]$ = 0.050 M.

14.31 The following data were measured for the reaction $BF_3(g) + NH_3(g) \longrightarrow F_3BNH_3(g)$:

Experiment	$[BF_3]$ (M)	$[NH_3]$ (M)	Initial Rate (M/s)
1	0.250	0.250	0.2130
2	0.250	0.125	0.1065
3	0.200	0.100	0.0682
4	0.350	0.100	0.1193
5	0.175	0.100	0.0596

(a) What is the rate law for the reaction? **(b)** What is the overall order of the reaction? **(c)** What is the value of the rate constant for the reaction? **(d)** What is the rate when $[BF_3]$ = 0.100 M and $[NH_3]$ = 0.500 M?

14.32 The following data were collected for the rate of disappearance of NO in the reaction $2 NO(g) + O_2(g) \longrightarrow 2 NO_2(g)$:

Experiment	[NO] (M)	$[O_2]$ (M)	Initial Rate (M/s)
1	0.0126	0.0125	1.41×10^{-2}
2	0.0252	0.0125	5.64×10^{-2}
3	0.0252	0.0250	1.13×10^{-1}

(a) What is the rate law for the reaction? **(b)** What are the units of the rate constant? **(c)** What is the average value of the rate constant calculated from the three data sets? **(d)** What is the rate of disappearance of NO when [NO] = 0.0750 M and $[O_2]$ = 0.0100 M? **(e)** What is the rate of disappearance of O_2 at the concentrations given in part (d)?

[14.33] Consider the gas-phase reaction between nitric oxide and bromine at 273 °C: $2 NO(g) + Br_2(g) \longrightarrow 2 NOBr(g)$. The following data for the initial rate of appearance of NOBr were obtained:

Experiment	[NO] (M)	$[Br_2]$ (M)	Initial Rate (M/s)
1	0.10	0.20	24
2	0.25	0.20	150
3	0.10	0.50	60
4	0.35	0.50	735

(a) Determine the rate law. **(b)** Calculate the average value of the rate constant for the appearance of NOBr from the four data sets. **(c)** How is the rate of appearance of NOBr related to the rate of disappearance of Br_2? **(d)** What is the rate of disappearance of Br_2 when [NO] = 0.075 M and $[Br_2]$ = 0.25 M?

[14.34] Consider the reaction of peroxydisulfate ion ($S_2O_8^{2-}$) with iodide ion (I^-) in aqueous solution:

$$S_2O_8^{2-}(aq) + 3 I^-(aq) \longrightarrow 2 SO_4^{2-}(aq) + I_3^-(aq)$$

At a particular temperature the rate of disappearance of $S_2O_8^{2-}$ varies with reactant concentrations in the following manner:

Experiment	$[S_2O_8^{2-}]$ (M)	$[I^-]$ (M)	Initial Rate (M/s)
1	0.018	0.036	2.6×10^{-6}
2	0.027	0.036	3.9×10^{-6}
3	0.036	0.054	7.8×10^{-6}
4	0.050	0.072	1.4×10^{-5}

(a) Determine the rate law for the reaction. (b) What is the average value of the rate constant for the disappearance of $S_2O_8^{2-}$ based on the four sets of data? (c) How is the rate of disappearance of $S_2O_8^{2-}$ related to the rate of disappearance of I^-? (d) What is the rate of disappearance of I^- when $[S_2O_8^{2-}] = 0.025$ M and $[I^-] = 0.050$ M?

Change of Concentration with Time

14.35 (a) Define the following symbols that are encountered in rate equations: $[A]_0$, $t_{1/2}$ $[A]_t$, k. (b) What quantity, when graphed versus time, will yield a straight line for a first-order reaction?

14.36 (a) For a second-order reaction, what quantity, when graphed versus time, will yield a straight line? (b) How do the half-lives of first-order and second-order reactions differ?

14.37 (a) The gas-phase decomposition of SO_2Cl_2, $SO_2Cl_2(g)$ $\longrightarrow$ $SO_2(g) + Cl_2(g)$, is first order in SO_2Cl_2. At 600 K the half-life for this process is 2.3×10^5 s. What is the rate constant at this temperature? (b) At 320 °C the rate constant is 2.2×10^{-5} s^{-1}. What is the half-life at this temperature?

14.38 Molecular iodine, $I_2(g)$, dissociates into iodine atoms at 625 K with a first-order rate constant of 0.271 s^{-1}. (a) What is the half-life for this reaction? (b) If you start with 0.050 M I_2 at this temperature, how much will remain after 5.12 s assuming that the iodine atoms do not recombine to form I_2?

14.39 As described in Exercise 14.37, the decomposition of sulfuryl chloride (SO_2Cl_2) is a first-order process. The rate constant for the decomposition at 660 K is 4.5×10^{-2} s^{-1}. (a) If we begin with an initial SO_2Cl_2 pressure of 375 torr, what is the pressure of this substance after 65 s? (b) At what time will the pressure of SO_2Cl_2 decline to one-tenth its initial value?

14.40 The first-order rate constant for the decomposition of N_2O_5, $2 N_2O_5(g)$ $\longrightarrow$ $4 NO_2(g) + O_2(g)$, at 70 °C is 6.82×10^{-3} s^{-1}. Suppose we start with 0.0250 mol of $N_2O_5(g)$ in a volume of 2.0 L. (a) How many moles of N_2O_5 will remain after 5.0 min? (b) How many minutes will it take for the quantity of N_2O_5 to drop to 0.010 mol? (c) What is the half-life of N_2O_5 at 70 °C?

14.41 The reaction

$$SO_2Cl_2(g) \longrightarrow SO_2(g) + Cl_2(g)$$

is first order in SO_2Cl_2. Using the following kinetic data, determine the magnitude of the first-order rate constant:

Time (s)	Pressure SO_2Cl_2 (atm)
0	1.000
2,500	0.947
5,000	0.895
7,500	0.848
10,000	0.803

14.42 From the following data for the first-order gas-phase isomerization of CH_3NC at 215 °C, calculate the first-order rate constant and half-life for the reaction:

Time (s)	Pressure CH_3NC (torr)
0	502
2,000	335
5,000	180
8,000	95.5
12,000	41.7
15,000	22.4

14.43 Consider the data presented in Exercise 14.15. (a) By using appropriate graphs, determine whether the reaction is first order or second order. (b) What is the value of the rate constant for the reaction? (c) What is the half-life for the reaction?

14.44 Consider the data presented in Exercise 14.16. (a) Determine whether the reaction is first order or second order. (b) What is the value of the rate constant? (c) What is the half-life?

14.45 The gas-phase decomposition of NO_2, $2 NO_2(g)$ $\longrightarrow$ $2 NO(g) + O_2(g)$, is studied at 383 °C, giving the following data:

Time (s)	$[NO_2]$ (M)
0.0	0.100
5.0	0.017
10.0	0.0090
15.0	0.0062
20.0	0.0047

(a) Is the reaction first order or second order with respect to the concentration of NO_2? (b) What is the value of the rate constant?

14.46 Sucrose ($C_{12}H_{22}O_{11}$), which is commonly known as table sugar, reacts in dilute acid solutions to form two simpler sugars, glucose and fructose, both of which have the formula $C_6H_{12}O_6$: At 23 °C and in 0.5 M HCl, the following data were obtained for the disappearance of sucrose:

Time (min)	$[C_{12}H_{22}O_{11}]$ (M)
0	0.316
39	0.274
80	0.238
140	0.190
210	0.146

(a) Is the reaction first order or second order with respect to $[C_{12}H_{22}O_{11}]$? (b) What is the value of the rate constant?

Temperature and Rate

14.47 (a) What factors determine whether a collision between two molecules will lead to a chemical reaction? (b) According to the collision model, why does temperature affect the value of the rate constant?

14.48 (a) In which of the following reactions would you expect the orientation factor to be least important in leading to reaction? $NO + O \longrightarrow NO_2$ or $H + Cl \longrightarrow HCl$? (b) How does the kinetic-molecular theory help us understand the temperature dependence of chemical reactions?

14.49 Calculate the fraction of atoms in a sample of argon gas at 400 K that has an energy of 10.0 kJ or greater.

14.50 (a) The activation energy for the isomerization of methyl isonitrile (Figure 14.6) is 160 kJ/mol. Calculate the fraction of methyl isonitrile molecules that has an energy of 160.0 kJ or greater at 500 K. (b) Calculate this fraction for a temperature of 510 K. What is the ratio of the fraction at 510 K to that at 500 K?

14.51 The gas-phase reaction $Cl(g) + HBr(g) \longrightarrow HCl(g) + Br(g)$ has an overall enthalpy change of −66 kJ. The activation energy for the reaction is 7 kJ. (a) Sketch the energy profile for the reaction, and label E_a and ΔE. (b) What is the activation energy for the reverse reaction?

14.52 For the elementary process $N_2O_5(g) \longrightarrow NO_2(g) + NO_3(g)$ the activation energy (E_a) and overall ΔE are 154 kJ/mol, and 136 kJ/mol, respectively. (a) Sketch the energy profile for this reaction, and label E_a and ΔE. (b) What is the activation energy for the reverse reaction?

14.53 Based on their activation energies and energy changes and assuming that all collision factors are the same, which of the following reactions would be fastest and which would be slowest? Explain your answer.
(a) $E_a = 45$ kJ/mol; $\Delta E = -25$ kJ/mol
(b) $E_a = 35$ kJ/mol; $\Delta E = -10$ kJ/mol
(c) $E_a = 55$ kJ/mol; $\Delta E = 10$ kJ/mol

14.54 Which of the reactions in Exercise 14.53 will be fastest in the reverse direction? Which will be slowest? Explain.

14.55 A certain first-order reaction has a rate constant of 2.75×10^{-2} s^{-1} at 20 °C. What is the value of k at 60 °C if (a) $E_a = 75.5$ kJ/mol; (b) $E_a = 125$ kJ/mol?

14.56 Understanding the high-temperature behavior of nitrogen oxides is essential for controlling pollution generated in automobile engines. The decomposition of nitric oxide (NO) to N_2 and O_2 is second order with a rate constant of 0.0796 M^{-1} s^{-1} at 737 °C and 0.0815 M^{-1} s^{-1} at 947 °C. Calculate the activation energy for the reaction.

14.57 The rate of the reaction

$$CH_3COOC_2H_5(aq) + OH^-(aq) \longrightarrow$$
$$CH_3COO^-(aq) + C_2H_5OH(aq)$$

was measured at several temperatures, and the following data were collected:

Temperature (°C)	k (M^{-1} s^{-1})
15	0.0521
25	0.101
35	0.184
45	0.332

Using these data, graph $\ln k$ versus $1/T$. Using your graph, determine the value of E_a.

14.58 The temperature dependence of the rate constant for the reaction is tabulated as follows:

Temperature (K)	k (M^{-1} s^{-1})
600	0.028
650	0.22
700	1.3
750	6.0
800	23

Calculate E_a and A.

[14.59] The activation energy of a certain reaction is 65.7 kJ/mol. How many times faster will the reaction occur at 50 °C than at 0 °C?

[14.60] The following is a quote from an article in the August 18, 1998, issue of The New York Times about the breakdown of cellulose and starch: "A drop of 18 degrees Fahrenheit [from 77 °F to 59 °F] lowers the reaction rate six times; a 36-degree drop [from 77 °F to 41 °F] produces a fortyfold decrease in the rate." (a) Calculate activation energies for the breakdown process based on the two estimates of the effect of temperature on rate. Are the values consistent? (b) Assuming the value of E_a calculated from the 36-degree drop and that the rate of breakdown is first order with a half-life at 25 °C of 2.7 years, calculate the half-life for breakdown at a temperature of −15 °C.

Reaction Mechanisms

14.61 **(a)** What is meant by the term *elementary reaction?* **(b)** What is the difference between a *unimolecular* and a *bimolecular* elementary reaction? **(c)** What is a *reaction mechanism?*

14.62 **(a)** What is meant by the term *molecularity?* **(b)** Why are termolecular elementary reactions so rare? **(c)** What is an *intermediate* in a mechanism?

14.63 What is the molecularity of each of the following elementary reactions? Write the rate law for each.
(a) $Cl_2(g) \longrightarrow 2\, Cl(g)$
(b) $OCl^-(aq) + H_2O(l) \longrightarrow HOCl(aq) + OH^-(aq)$
(c) $NO(g) + Cl_2(g) \longrightarrow NOCl_2(g)$

14.64 What is the molecularity of each of the following elementary reactions? Write the rate law for each.
(a) $2\, NO(g) \longrightarrow N_2O_2(g)$

(b) $H_2C \overset{\displaystyle CH_2}{-\!\!\!-\!\!\!-} CH_2(g) \longrightarrow CH_2{=}CH{-}CH_3(g)$

(c) $SO_3(g) \longrightarrow SO_2(g) + O(g)$

14.65 **(a)** Based on the following reaction profile, how many intermediates are formed in the reaction $A \longrightarrow D$? **(b)** How many transition states are there? **(c)** Which step is the fastest? **(d)** Is the reaction $A \longrightarrow D$ exothermic or endothermic?

Reaction pathway

14.66 Consider the following energy profile.

Reaction pathway

(a) How many elementary reactions are in the reaction mechanism? **(b)** How many intermediates are formed in the reaction? **(c)** Which step is rate limiting? **(d)** Is the overall reaction exothermic or endothermic?

14.67 The following mechanism has been proposed for the gas-phase reaction of H_2 with ICl:
$$H_2(g) + ICl(g) \longrightarrow HI(g) + HCl(g)$$
$$HI(g) + ICl(g) \longrightarrow I_2(g) + HCl(g)$$

(a) Write the balanced equation for the overall reaction. **(b)** Identify any intermediates in the mechanism. **(c)** Write rate laws for each elementary reaction in the mechanism. **(d)** If the first step is slow and the second one is fast, what rate law do you expect to be observed for the overall reaction?

14.68 The decomposition of hydrogen peroxide is catalyzed by iodide ion. The catalyzed reaction is thought to proceed by a two-step mechanism:
$$H_2O_2(aq) + I^-(aq) \longrightarrow H_2O(l) + IO^-(aq) \qquad \text{(slow)}$$
$$IO^-(aq) + H_2O_2(aq) \longrightarrow H_2O(l) + O_2(g) + I^-(aq) \quad \text{(fast)}$$

(a) Write the rate law for each of the elementary reactions of the mechanism. **(b)** Write the chemical equation for the overall process. **(c)** Identify the intermediate, if any, in the mechanism. **(d)** Assuming that the first step of the mechanism is rate determining, predict the rate law for the overall process.

14.69 The reaction $2\, NO(g) + Cl_2(g) \longrightarrow 2\, NOCl(g)$ obeys the rate law, rate $= k[NO]^2[Cl_2]$. The following mechanism has been proposed for this reaction:
$$NO(g) + Cl_2(g) \longrightarrow NOCl_2(g)$$
$$NOCl_2(g) + NO(g) \longrightarrow 2\, NOCl(g)$$

(a) What would the rate law be if the first step were rate determining? **(b)** Based on the observed rate law, what can we conclude about the relative rates of the two steps?

14.70 You have studied the gas-phase oxidation of HBr by O_2:
$$4\, HBr(g) + O_2(g) \longrightarrow 2\, H_2O(g) + 2\, Br_2(g)$$

You find the reaction to be first order with respect to HBr and first order with respect to O_2. You propose the following mechanism:
$$HBr(g) + O_2(g) \longrightarrow HOOBr(g)$$
$$HOOBr(g) + HBr(g) \longrightarrow 2\, HOBr(g)$$
$$HOBr(g) + HBr(g) \longrightarrow H_2O(g) + Br_2(g)$$

(a) Indicate how the elementary reactions add to give the overall reaction. (*Hint:* You will need to multiply the coefficients of one of the equations by 2.) **(b)** Based on the rate law, which step is rate determining? **(c)** What are the intermediates in this mechanism? **(d)** If you are unable to detect HOBr or HOOBr among the products, does this disprove your mechanism?

Catalysis

14.71 (a) What part of the energy profile of a reaction is affected by a catalyst? (b) What is the difference between a homogeneous and a heterogeneous catalyst?

14.72 (a) Most heterogeneous catalysts of importance are extremely finely divided solid materials. Why is particle size important? (b) What role does adsorption play in the action of a heterogeneous catalyst?

14.73 The oxidation of SO_2 to SO_3 is catalyzed by NO_2. The reaction proceeds as follows:

$$NO_2(g) + SO_2(g) \longrightarrow NO(g) + SO_3(g)$$
$$2 NO(g) + O_2(g) \longrightarrow 2 NO_2(g)$$

(a) Show that the two reactions can be summed to give the overall oxidation of SO_2 by O_2 to give SO_3. (*Hint:* The top reaction must be multiplied by a factor so the NO and NO_2 cancel out.) (b) Why do we consider NO_2 a catalyst and not an intermediate in this reaction? (c) Is this an example of homogeneous catalysis or heterogeneous catalysis?

14.74 NO catalyzes the decomposition of N_2O, possibly by the following mechanism:

$$NO(g) + N_2O(g) \longrightarrow N_2(g) + NO_2(g)$$
$$2 NO_2(g) \longrightarrow 2 NO(g) + O_2(g)$$

(a) What is the chemical equation for the overall reaction? Show how the two steps can be added to give the overall equation. (b) Why is NO considered a catalyst and not an intermediate? (c) If experiments show that during the decomposition of N_2O, NO_2 does not accumulate in measurable quantities, does this rule out the proposed mechanism? If you think not, suggest what might be going on.

14.75 Many metallic catalysts, particularly the precious-metal ones, are often deposited as very thin films on a substance of high surface area per unit mass, such as alumina (Al_2O_3) or silica (SiO_2). (a) Why is this an effective way of utilizing the catalyst material? (b) How does the surface area affect the rate of reaction?

14.76 (a) If you were going to build a system to check the effectiveness of automobile catalytic converters on cars, what substances would you want to look for in the car exhaust? (b) Automobile catalytic converters have to work at high temperatures, as hot exhaust gases stream through them. In what ways could this be an advantage? In what ways a disadvantage? (c) Why is the rate of flow of exhaust gases over a catalytic converter important?

14.77 When D_2 reacts with ethylene (C_2H_4) in the presence of a finely divided catalyst, ethane with two deuteriums, $CH_2D—CH_2D$, is formed. (Deuterium, D, is an isotope of hydrogen of mass 2.) Very little ethane forms in which two deuteriums are bound to one carbon (for example, $CH_3—CHD_2$). Use the sequence of steps involved in the reaction to explain why this is so.

14.78 Heterogeneous catalysts that perform hydrogenation reactions, as illustrated in Figure 14.21, are subject to poisoning, which shuts down their catalytic ability. Compounds of sulfur are often poisons. Suggest a mechanism by which such compounds might act as poisons.

14.79 (a) Explain the importance of enzymes in biological systems. (b) What chemical transformations are catalyzed (*i*) by the enzyme catalase, (*ii*) by nitrogenase?

14.80 There are literally thousands of enzymes at work in complex living systems such as human beings. What properties of the enzymes give rise to their ability to distinguish one substrate from another?

[14.81] The activation energy of an uncatalyzed reaction is 95 kJ/mol. The addition of a catalyst lowers the activation energy to 55 kJ/mol. Assuming that the collision factor remains the same, by what factor will the catalyst increase the rate of the reaction at (a) 25 °C, (b) 125 °C?

[14.82] Suppose that a certain biologically important reaction is quite slow at physiological temperature (37 °C) in the absence of a catalyst. Assuming that the collision factor remains the same, by how much must an enzyme lower the activation energy of the reaction in order to achieve a 1×10^5-fold increase in the reaction rate?

ADDITIONAL EXERCISES

14.83 Explain why rate laws generally cannot be written from balanced equations. Under what circumstance is the rate law related directly to the balanced equation for a reaction?

14.84 Hydrogen sulfide (H_2S) is a common and troublesome pollutant in industrial wastewaters. One way to remove H_2S is to treat the water with chlorine, in which case the following reaction occurs:

$$H_2S(aq) + Cl_2(aq) \longrightarrow S(s) + 2 H^+(aq) + 2 Cl^-(aq)$$

The rate of this reaction is first order in each reactant. The rate constant for the disappearance of H_2S at 28 °C is $3.5 \times 10^{-2} M^{-1} s^{-1}$. If at a given time the concentration of H_2S is $2.0 \times 10^{-4} M$ and that of Cl_2 is $0.025 M$, what is the rate of formation of Cl^-?

14.85 The reaction $2 NO(g) + O_2(g) \longrightarrow 2 NO_2(g)$ is second order in NO and first order in O_2. When [NO] = 0.040 M and [O_2] = 0.035 M, the observed rate of disappearance of NO is $9.3 \times 10^{-5} M/s$. (a) What is the rate of disappearance of O_2 at this moment? (b) What is the value of the rate constant? (c) What are the units of the rate constant? (d) What would happen to the rate if the concentration of NO were increased by a factor of 1.8?

14.86 Consider the following reaction between mercury(II) chloride and oxalate ion:

$$2 HgCl_2(aq) + C_2O_4^{2-}(aq) \longrightarrow$$
$$2 Cl^-(aq) + 2 CO_2(g) + Hg_2Cl_2(s)$$

The initial rate of this reaction was determined for several concentrations of $HgCl_2$ and $C_2O_4^{2-}$, and the following rate data were obtained for the rate of disappearance of $C_2O_4^{2-}$:

Experiment	[HgCl$_2$] (M)	[C$_2$O$_4{}^{2-}$] (M)	Rate (M/s)
1	0.164	0.15	3.2×10^{-5}
2	0.164	0.45	2.9×10^{-4}
3	0.082	0.45	1.4×10^{-4}
4	0.246	0.15	4.8×10^{-5}

(a) What is the rate law for this reaction? **(b)** What is the value of the rate constant? **(c)** What is the reaction rate when the concentration of HgCl$_2$ is 0.100 M and that of $(C_2O_4^{2-})$ is 0.25 M, if the temperature is the same as that used to obtain the data shown?

14.87 The reaction $2\,NO_2 \longrightarrow 2\,NO + O_2$ has the rate constant $k = 0.63\ M^{-1}s^{-1}$. Based on the units for k, is the reaction first or second order in NO$_2$? If the initial concentration of NO$_2$ is 0.100 M, how would you determine how long it would take for the concentration to decrease to 0.025 M?

14.88 Consider two reactions. Reaction (1) has a constant half-life, whereas reaction (2) has a half-life that gets longer as the reaction proceeds. What can you conclude about the rate laws of these reactions from these observations?

14.89 **(a)** The reaction $H_2O_2(aq) \longrightarrow H_2O(l) + \frac{1}{2}O_2(g)$, is first order. Near room temperature, the rate constant equals $7.0 \times 10^{-4}s^{-1}$. Calculate the half-life at this temperature. **(b)** At 415 °C, (CH$_2$)$_2$O decomposes in the gas phase, $(CH_2)_2O(g) \longrightarrow CH_4(g) + CO(g)$. If the reaction is first order with a half-life of 56.3 min at this temperature, calculate the rate constant in s^{-1}.

14.90 Americium-241 is used in smoke detectors. It has a rate constant for radioactive decay of $k = 1.6 \times 10^{-3}\ yr^{-1}$. By contrast, iodine-125, which is used to test for thyroid functioning, has a rate constant for radioactive decay of $k = 0.011\ day^{-1}$. **(a)** What are the half-lives of these two isotopes? **(b)** Which one decays at a faster rate? **(c)** How much of a 1.00-mg sample of either isotope remains after three half-lives?

14.91 Urea (NH$_2$CONH$_2$) is the end product in protein metabolism in animals. The decomposition of urea in 0.1 M HCl occurs according to the reaction

$$NH_2CONH_2(aq) + H^+(aq) + 2\,H_2O(l) \longrightarrow 2\,NH_4^+(aq) + HCO_3^-(aq)$$

The reaction is first order in urea and first order overall. When [NH$_2$CONH$_2$] = 0.200 M, the rate at 61.05 °C is $8.56 \times 10^{-5}\ M/s$. **(a)** What is the value for the rate constant, k? **(b)** What is the concentration of urea in this solution after 4.00×10^3 s if the starting concentration is 0.500 M? **(c)** What is the half-life for this reaction at 61.05 °C?

14.92 The rate of a first-order reaction is followed by spectroscopy, monitoring the absorption of a colored reactant. The reaction occurs in a 1.00-cm sample cell, and the only colored species in the reaction has a molar absorptivity constant of $5.60 \times 10^3\ cm^{-1}\ M^{-1}$. **(a)** Calculate the initial concentration of the colored reactant if the absorbance is 0.605 at the beginning of the reaction. **(b)** The absorbance falls to 0.250 within 30.0 min. Calculate the

rate constant in units of s^{-1}. **(c)** Calculate the half-life of the reaction. **(d)** How long does it take for the absorbance to fall to 0.100?

14.93 Cyclopentadiene (C$_5$H$_6$) reacts with itself to form dicyclopentadiene (C$_{10}$H$_{12}$). A 0.0400 M solution of C$_5$H$_6$ was monitored as a function of time as the reaction $2\,C_5H_6 \longrightarrow C_{10}H_{12}$ proceeded. The following data were collected:

Time (s)	[C$_5$H$_6$] (M)
0.0	0.0400
50.0	0.0300
100.0	0.0240
150.0	0.0200
200.0	0.0174

Plot [C$_5$H$_6$] versus time, ln [C$_5$H$_6$] versus time, and 1/[C$_5$H$_6$] versus time. What is the order of the reaction? What is the value of the rate constant?

14.94 **(a)** Two reactions have identical values for E_a. Does this ensure that they will have the same rate constant if run at the same temperature? Explain. **(b)** Two similar reactions have the same rate constant at 25 °C, but at 35 °C one of the reactions has a higher rate constant than the other. Account for these observations.

14.95 The first-order rate constant for reaction of a particular organic compound with water varies with temperature as follows:

Temperature (K)	Rate Constant (s^{-1})
300	3.2×10^{-11}
320	1.0×10^{-9}
340	3.0×10^{-8}
355	2.4×10^{-7}

From these data, calculate the activation energy in units of kJ/mol.

14.96 The following mechanism has been proposed for the reaction of NO with H$_2$ to form N$_2$O and H$_2$O:

$$NO(g) + NO(g) \longrightarrow N_2O_2(g)$$
$$N_2O_2(g) + H_2(g) \longrightarrow N_2O(g) + H_2O(g)$$

(a) Show that the elementary reactions of the proposed mechanism add to provide a balanced equation for the reaction. **(b)** Write a rate law for each elementary reaction in the mechanism. **(c)** Identify any intermediates in the mechanism. **(d)** The observed rate law is rate = $k[NO]^2[H_2]$. If the proposed mechanism is correct, what can we conclude about the relative speeds of the first and second reactions?

14.97 Ozone in the upper atmosphere can be destroyed by the following two-step mechanism:

$$Cl(g) + O_3(g) \longrightarrow ClO(g) + O_2(g)$$
$$ClO(g) + O(g) \longrightarrow Cl(g) + O_2(g)$$

(a) What is the overall equation for this process? **(b)** What is the catalyst in the reaction? How do you know? **(c)** What is the intermediate in the reaction? How do you distinguish it from the catalyst?

14.98 Using Figure 14.20 as your basis, draw the energy profile for the bromide ion–catalyzed decomposition of hydrogen peroxide. **(a)** Label the curve with the activation energies for reactions [14.30] and [14.31]. **(b)** Notice from Figure 14.19(b) that when $Br^-(aq)$ is added initially, Br_2 accumulates to some extent during the reaction. What does this tell us about the relative rates of reactions [14.30] and [14.31]?

[14.99] The following mechanism has been proposed for the gas-phase reaction of chloroform ($CHCl_3$) and chlorine:

Step 1: $Cl_2(g) \underset{k_{-1}}{\overset{k_1}{\rightleftharpoons}} 2\,Cl(g)$ (fast)

Step 2: $Cl(g) + CHCl_3(g) \xrightarrow{k_2} HCl(g) + CCl_3(g)$ (slow)

Step 3: $Cl(g) + CCl_3(g) \xrightarrow{k_3} CCl_4$ (fast)

(a) What is the overall reaction? **(b)** What are the intermediates in the mechanism? **(c)** What is the molecularity of each of the elementary reactions? **(d)** What is the rate-determining step? **(e)** What is the rate law predicted by this mechanism? (*Hint:* The overall reaction order is not an integer.)

[14.100] In a hydrocarbon solution, the gold compound $(CH_3)_3AuPH_3$ decomposes into ethane (C_2H_6) and a different gold compound, $(CH_3)AuPH_3$. The following mechanism has been proposed for the decomposition of $(CH_3)_3AuPH_3$:

Step 1: $(CH_3)_3\,AuPH_3 \underset{k_{-1}}{\overset{k_1}{\rightleftharpoons}} (CH_3)_3\,Au + PH_3$ (fast)

Step 2: $(CH_3)_3\,Au \xrightarrow{k_2} C_2H_6 + (CH_3)Au$ (slow)

Step 3: $(CH_3)Au + PH_3 \xrightarrow{k_3} (CH_3)AuPH_3$ (fast)

(a) What is the overall reaction? **(b)** What are the intermediates in the mechanism? **(c)** What is the molecularity of each of the elementary steps? **(d)** What is the rate-determining step? **(e)** What is the rate law predicted by this mechanism? **(f)** What would be the effect on the reaction rate of adding PH_3 to the solution of $(CH_3)_3AuPH_3$?

14.101 One of the many remarkable enzymes in the human body is carbonic anhydrase, which catalyzes the interconversion of carbonic acid with carbon dioxide and water. If it were not for this enzyme, the body could not rid itself rapidly enough of the CO_2 accumulated by cell metabolism. The enzyme catalyzes the dehydration (release to air) of up to 10^7 CO_2 molecules per second. Which components of this description correspond to the terms *enzyme*, *substrate*, and *turnover number*?

14.102 Enzymes are often described as following the two-step mechanism:

$$E + S \rightleftharpoons ES \quad (fast)$$
$$ES \longrightarrow E + P \quad (slow)$$

Where E = enzyme, S = substrate, and P = product. If an enzyme follows this mechanism, what rate law is expected for the reaction?

14.103 The enzyme *invertase* catalyzes the conversion of sucrose, a disaccharide, to invert sugar, a mixture of glucose and fructose. When the concentration of invertase is 4.2×10^{-7} M and the concentration of sucrose is 0.0077 M, invert sugar is formed at the rate of 1.5×10^{-4} M/s. When the sucrose concentration is doubled, the rate of formation of invert sugar is doubled also. **(a)** Assuming that the enzyme-substrate model is operative, is the fraction of enzyme tied up as a complex large or small? Explain. **(b)** Addition of inositol, another sugar, decreases the rate of formation of invert sugar. Suggest a mechanism by which this occurs.

INTEGRATIVE EXERCISES

14.104 Dinitrogen pentoxide (N_2O_5) decomposes in chloroform as a solvent to yield NO_2 and O_2. The decomposition is first order with a rate constant at 45 °C of 1.0×10^{-5} s^{-1}. Calculate the partial pressure of O_2 produced from 1.00 L of 0.600 M N_2O_5 solution at 45 °C over a period of 20.0 h if the gas is collected in a 10.0-L container. (Assume that the products do not dissolve in chloroform.)

[14.105] The reaction between ethyl iodide and hydroxide ion in ethanol (C_2H_5OH) solution, $C_2H_5I(alc) + OH^-(alc) \longrightarrow C_2H_5OH(l) + I^-(alc)$, has an activation energy of 86.8 kJ/mol and a frequency factor of 2.10×10^{11} $M^{-1}\,s^{-1}$. **(a)** Predict the rate constant for the reaction at 35 °C. **(b)** A solution of KOH in ethanol is made up by dissolving 0.335 g KOH in ethanol to form 250.0 mL of solution. Similarly, 1.453 g of C_2H_5I is dissolved in ethanol to form 250.0 mL of solution. Equal volumes of the two solutions are mixed. Assuming the reaction is first order in each reactant, what is the initial rate at 35 °C? **(c)** Which reagent in the reaction is limiting, assuming the reaction proceeds to completion?

14.106 Zinc metal dissolves in hydrochloric acid according to the reaction

$$Zn(s) + 2\,HCl(aq) \longrightarrow ZnCl_2(aq) + H_2(g)$$

Suppose you are asked to study the kinetics of this reaction by monitoring the rate of production of $H_2(g)$. **(a)** By using a reaction flask, a manometer, and any other common laboratory equipment, design an experimental apparatus that would allow you to monitor the partial pressure of $H_2(g)$ produced as a function of time. **(b)** Explain how you would use the apparatus to determine the rate law of the reaction. **(c)** Explain how you would use the apparatus to determine the reaction order for $[H^+]$ for the reaction. **(d)** How could you use the apparatus to determine the activation energy of the reaction? **(e)** Explain how you would use the apparatus to determine the effects of changing the form of $Zn(s)$ from metal strips to granules.

14.107 The gas-phase reaction of NO with F_2 to form NOF and F has an activation energy of E_a = 6.3 kJ/mol and a frequency factor of $A = 6.0 \times 10^8\ M^{-1}\,s^{-1}$. The reaction is believed to be bimolecular:

$$NO(g) + F_2(g) \longrightarrow NOF(g) + F(g)$$

(a) Calculate the rate constant at 100 °C. **(b)** Draw the Lewis structures for the NO and the NOF molecules, given that the chemical formula for NOF is misleading because the nitrogen atom is actually the central atom in the molecule. **(c)** Predict the structure for the NOF molecule. **(d)** Draw a possible transition state for the formation of NOF, using dashed lines to indicate the weak bonds that are beginning to form. **(e)** Suggest a reason for the low activation energy for the reaction.

14.108 The mechanism for the oxidation of HBr by O_2 to form $2\ H_2O$ and Br_2 is shown in Exercise 14.70. **(a)** Calculate the overall standard enthalpy change for the reaction process. **(b)** HBr does not react with O_2 at a measurable rate at room temperature under ordinary conditions. What can you infer from this about the magnitude of the activation energy for the rate-determining step? **(c)** Draw a plausible Lewis structure for the intermediate HOOBr. To what familiar compound of hydrogen and oxygen does it appear similar?

14.109 Enzymes, the catalysts of biological systems, are high molecular weight protein materials. The active site of the enzyme is formed by the three-dimensional arrangement of the protein in solution. When heated in solution, proteins undergo denaturation, a process in which the three-dimensional structure of the protein unravels or at least partly does so. The accompanying graph shows the variation with temperature of the activity of a typical enzyme. The activity increases with temperature to a point above the usual operating region for the enzyme, then declines rapidly with further temperature increases. What role does denaturation play in determining the shape of this curve? How does your explanation fit in with the lock-and-key model of enzyme action?

[14.110] Metals often form several cations with different charges. Cerium, for example, forms Ce^{3+} and Ce^{4+} ions, and thallium forms Tl^+ and Tl^{3+} ions. Cerium and thallium ions react as follows:

$$2\ Ce^{4+}(aq) + Tl^+(aq) \longrightarrow 2\ Ce^{3+}(aq) + Tl^{3+}(aq)$$

This reaction is very slow and is thought to occur in a single elementary step. The reaction is catalyzed by the

addition of $Mn^{2+}(aq)$, according to the following mechanism:

$$Ce^{4+}(aq) + Mn^{2+}(aq) \longrightarrow Ce^{3+}(aq) + Mn^{3+}(aq)$$
$$Ce^{4+}(aq) + Mn^{3+}(aq) \longrightarrow Ce^{3+}(aq) + Mn^{4+}(aq)$$
$$Mn^{4+}(aq) + Tl^+(aq) \longrightarrow Mn^{2+}(aq) + Tl^{3+}(aq)$$

(a) Write the rate law for the uncatalyzed reaction. **(b)** What is unusual about the uncatalyzed reaction? Why might it be a slow reaction? **(c)** The rate for the catalyzed reaction is first order in $[Ce^{4+}]$ and first order in $[Mn^{2+}]$. Based on this rate law, which of the steps in the catalyzed mechanism is rate determining? **(d)** Use the available oxidation states of Mn to comment on its special suitability to catalyze this reaction.

[14.111] The rates of many atmospheric reactions are accelerated by the absorption of light by one of the reactants. For example, consider the reaction between methane and chlorine to produce methyl chloride and hydrogen chloride:

Reaction 1: $CH_4(g) + Cl_2(g) \longrightarrow CH_3Cl(g) + HCl(g)$

This reaction is very slow in the absence of light. However, $Cl_2(g)$ can absorb light to form Cl atoms:

Reaction 2: $Cl_2(g) + h\nu \longrightarrow 2\ Cl(g)$

Once the Cl atoms are generated, they can catalyze the reaction of CH_4 and Cl_2, according to the following proposed mechanism:

Reaction 3: $CH_4(g) + Cl(g) \longrightarrow CH_3(g) + HCl(g)$
Reaction 4: $CH_3(g) + Cl_2(g) \longrightarrow CH_3Cl(g) + Cl(g)$

The enthalpy changes and activation energies for these two reactions are tabulated as follows:

Reaction	$\Delta H^\circ_{\mathrm{ran}}$ (kJ/mol)	E_a (kJ/mol)
3	+4	17
4	−109	4

(a) By using the bond enthalpy for Cl_2 (Table 8.4), determine the longest wavelength of light that is energetic enough to cause reaction 2 to occur. In which portion of the electromagnetic spectrum is this light found? **(b)** By using the data tabulated here, sketch a quantitative energy profile for the catalyzed reaction represented by reactions 3 and 4. **(c)** By using bond enthalpies, estimate where the reactants, $CH_4(g) + Cl_2(g)$, should be placed on your diagram in part (b). Use this result to estimate the value of E_a for the reaction $CH_4(g) + Cl_2(g) \longrightarrow CH_3(g) + HCl(g) + Cl(g)$. **(d)** The species $Cl(g)$ and $CH_3(g)$ in reactions 3 and 4 are radicals, that is, atoms or molecules with unpaired electrons. Draw a Lewis structure of CH_3, and verify that it is a radical. **(e)** The sequence of reactions 3 and 4 comprise a radical chain mechanism. Why do you think this is called a "chain reaction"? Propose a reaction that will terminate the chain reaction.

CHAPTER 15

CHEMICAL EQUILIBRIUM

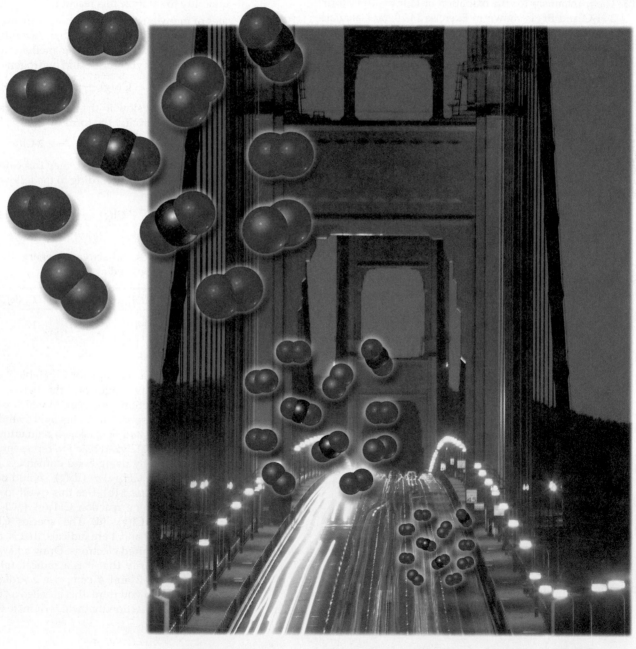

TRAFFIC ENTERING AND LEAVING San Francisco over the Golden Gate Bridge in the early morning.

WHAT'S AHEAD

15.1 The Concept of Equilibrium
We begin by examining the concept of equilibrium.

15.2 The Equilibrium Constant
We then define the *equilibrium constant* and learn how to write *equilibrium-constant expressions* for homogeneous reactions.

15.3 Interpreting and Working with Equilibrium Constants
We also learn how to interpret the magnitude of an equilibrium constant and how to determine the way in which its value is affected when the chemical equation is reversed or changed in some other fashion.

15.4 Heterogeneous Equilibria
We then learn how to write equilibrium-constant expressions for heterogeneous reactions.

15.5 Calculating Equilibrium Constants
We will see that the value of an equilibrium constant can be calculated using equilibrium concentrations of reactants and products.

15.6 Applications of Equilibrium Constants
We will also see that equilibrium constants can be used to predict the equilibrium concentrations of reactants and products and to determine the direction in which a reaction mixture must proceed to achieve equilibrium.

15.7 Le Châtelier's Principle
We conclude the chapter with a discussion of *Le Châtelier's principle*, which predicts how a system at equilibrium responds to changes in concentration, volume, pressure, and temperature.

TO BE IN EQUILIBRIUM IS TO BE IN A STATE OF BALANCE. A tug of war in which the two sides are pulling with equal force so that the rope does not move is an example of a *static* equilibrium, one in which an object is at rest. Equilibria can also be *dynamic*, as illustrated in the

chapter-opening photograph, which shows cars traveling in both directions over a bridge that serves as an entranceway to a city. If the rate at which cars leave the city equals the rate at which they enter, the two opposing processes are in balance, and the net number of cars in the city is constant.

We have already encountered several instances of dynamic equilibrium. For example, the vapor above a liquid is in equilibrium with the liquid phase. ∞ (Section 11.5) The rate at which molecules escape from the liquid into the gas phase equals the rate at which molecules in the gas phase strike the surface and become part of the liquid. Similarly, in a saturated solution of sodium chloride the solid sodium chloride is in equilibrium with the ions dispersed in water. ∞ (Section 13.2) The rate at which ions leave the solid surface equals the rate at which other ions are removed from the liquid to become part of the solid. Both of these examples involve a pair of opposing processes. At equilibrium these opposing processes are occurring at the same rate.

In this chapter we will consider yet another type of dynamic equilibrium, one involving chemical reactions. **Chemical equilibrium** *occurs when opposing reactions are proceeding at equal rates*: The rate at which the products are formed from the reactants equals the rate at which the reactants are formed from the products. As a result, concentrations cease to change, making the reaction appear to be stopped. How fast a reaction reaches equilibrium is a matter of kinetics.

Chemical equilibria are involved in a great many natural phenomena, and they play important roles in many industrial processes. In this and the next two chapters, we will explore chemical equilibrium in some detail. Here we will learn how to express the equilibrium position of a reaction in quantitative terms. We will also study the factors that determine the relative concentrations of reactants and products in equilibrium mixtures.

15.1 THE CONCEPT OF EQUILIBRIUM

Let's examine a simple reaction to see how it reaches an *equilibrium state*—a mixture of reactants and products whose concentrations no longer change with time. We begin with N_2O_4, a colorless substance that dissociates to form NO_2, which is brown. Figure 15.1▶ shows a sample of frozen N_2O_4 inside a sealed tube resting in a beaker. Because the chemical reaction occurs in a closed system, the reaction will eventually reach equilibrium.

The solid N_2O_4 vaporizes as it is warmed above its boiling point (21.2 °C), and the gas turns progressively darker as the colorless N_2O_4 gas dissociates into brown NO_2 gas (Figure 15.1). Eventually, even though there is still N_2O_4 in the tube, the color stops getting darker because the system reaches equilibrium. We are left with an *equilibrium mixture* of N_2O_4 and NO_2 in which the concentrations of the gases no longer change as time passes.

The equilibrium mixture results because the reaction is *reversible*. N_2O_4 can react to form NO_2, and NO_2 can react to form N_2O_4. This situation is represented by writing the equation for the reaction with two half arrows pointing in both directions: ∞ (Section 4.1)

$$N_2O_4(g) \rightleftharpoons 2\,NO_2(g) \tag{15.1}$$
$$\text{Colorless} \qquad\quad \text{Brown}$$

We can analyze this equilibrium using our knowledge of kinetics. Let's call the decomposition of N_2O_4 to form NO_2 the forward reaction and the reaction of NO_2 to re-form N_2O_4 the reverse reaction. In this case both the forward reaction and the reverse reaction are elementary reactions. As we learned in Section 14.6, the rate laws for elementary reactions can be written from their chemical equations:

Forward reaction: $\quad N_2O_4(g) \longrightarrow 2\,NO_2(g) \qquad \text{Rate}_f = k_f[N_2O_4] \quad$ [15.2]

Reverse reaction: $\quad 2\,NO_2(g) \longrightarrow N_2O_4(g) \qquad \text{Rate}_r = k_r[NO_2]^2 \quad$ [15.3]

where k_f and k_r are the rate constants for the forward and reverse reactions, respectively. At equilibrium the rate at which products are produced from reactants equals the rate at which reactants are produced from products:

$$\underbrace{k_f[N_2O_4]}_{\text{Forward reaction}} = \underbrace{k_r[NO_2]^2}_{\text{Reverse reaction}} \tag{15.4}$$

Rearranging this equation gives

$$\frac{[NO_2]^2}{[N_2O_4]} = \frac{k_f}{k_r} = \text{a constant} \tag{15.5}$$

As shown in Equation 15.5, the quotient of two constants, such as k_f and k_r, is itself a constant. Thus, at equilibrium the ratio of the concentration terms involving N_2O_4 and NO_2 equals a constant. (We will consider this constant, called the *equilibrium constant*, in Section 15.2.) It makes no difference whether we start with N_2O_4 or with NO_2, or even with some mixture of the two. At equilibrium the ratio equals a specific value. Thus, there is an important constraint on the proportions of N_2O_4 and NO_2 at equilibrium.

ESTABLISHING EQUILIBRIUM

The condition in which the concentrations of all reactants and products in a closed system cease to change with time is called chemical equilibrium.

$N_2O_4(s)$

Frozen N_2O_4 is nearly colorless.

$N_2O_4(g) \longrightarrow 2 NO_2(g)$

As N_2O_4 is warmed above its boiling point, it starts to dissociate into brown NO_2 gas.

$N_2O_4(g) \rightleftharpoons 2 NO_2(g)$

Eventually the color stops changing as $N_2O_4(g)$ and $NO_2(g)$ reach concentrations at which they are interconverting at the same rate. The two gases are in equilibrium.

▲ **Figure 15.1 The $N_2O_4(g) \rightleftharpoons 2 NO_2(g)$ equilibrium.**

Once equilibrium is established, the concentrations of N_2O_4 and NO_2 no longer change, as shown in Figure 15.2(a) ▼. If the composition of the equilibrium mixture remains constant with time, it does not mean, however, that N_2O_4 and NO_2 stop reacting. On the contrary, the equilibrium is dynamic—some N_2O_4 is still converting to NO_2, and some NO_2 is still converting to N_2O_4. At equilibrium, however, the two processes occur at the same rate, as shown in Figure 15.2(b).

(a)

(b)

◀ **Figure 15.2 Achieving chemical equilibrium for $N_2O_4(g) \rightleftharpoons 2 NO_2(g)$.** (a) The concentration of N_2O_4 decreases while the concentration of NO_2 increases during the course of the reaction. Equilibrium is indicated when the concentrations no longer change with time. (b) The rate of disappearance of N_2O_4 decreases with time as the concentration of N_2O_4 decreases. At the same time, the rate of formation of NO_2 also decreases with time. Equilibrium occurs when these two rates are equal.

We learn several important lessons about equilibrium from this example:

- At equilibrium, the concentrations of reactants and products no longer change with time.
- For equilibrium to occur, neither reactants nor products can escape from the system.
- At equilibrium a particular ratio of concentration terms equals a constant.

We will examine the last of these facts in the next section.

GIVE IT SOME THOUGHT

(a) Which quantities are equal in a dynamic equilibrium? **(b)** If the rate constant for the forward reaction in Equation 15.1 is larger than the rate constant for the reverse reaction, will the constant in Equation 15.5 be greater than 1 or smaller than 1?

15.2 THE EQUILIBRIUM CONSTANT

Opposing reactions naturally lead to an equilibrium, regardless of how complicated the reaction might be and regardless of the nature of the kinetic processes for the forward and reverse reactions. Consider the synthesis of ammonia from nitrogen and hydrogen:

$$N_2(g) + 3\,H_2(g) \rightleftharpoons 2\,NH_3(g) \qquad [15.6]$$

This reaction is the basis for the **Haber process**, which, in the presence of a catalyst, combines N_2 and H_2 at a pressure of several hundred atmospheres and a temperature of several hundred degrees Celsius. The two gases react to form ammonia under these conditions, but in a closed system the reaction does not lead to complete consumption of the N_2 and H_2. Rather, at some point the reaction appears to stop, with all three components of the reaction mixture present at the same time.

The manner in which the concentrations of H_2, N_2, and NH_3 vary with time is shown in Figure 15.3(a) ▼. Notice that an equilibrium mixture is obtained regardless of whether we begin with N_2 and H_2 or only with NH_3. *The equilibrium condition can be reached from either direction.*

GIVE IT SOME THOUGHT

How do we know when equilibrium has been reached in a chemical reaction?

Earlier, we saw that when the reaction $N_2O_4(g) \rightleftharpoons 2\,NO_2(g)$ reaches equilibrium, a ratio based on the equilibrium concentrations of N_2O_4 and NO_2 has a constant value (Equation 15.5). A similar relationship governs the concentrations of N_2, H_2, and NH_3 at equilibrium. If we were to systematically change the relative amounts of the three gases in the starting mixture and then analyze each equilibrium mixture, we could determine the relationship among the equilibrium concentrations.

▶ **Figure 15.3 Concentration changes approaching equilibrium.** (a) Equilibrium for the reaction $N_2 + 3\,H_2 \rightleftharpoons 2\,NH_3$ is approached beginning with H_2 and N_2 present in the ratio 3:1 and no NH_3 present. (b) Equilibrium for the same reaction is approached beginning with only NH_3 in the reaction vessel.

(a)

(b)

Chemistry Put to Work   THE HABER PROCESS

We presented a "Chemistry and Life" box in Section 14.7 that discussed *nitrogen fixation*, the processes that convert N₂ gas into ammonia, which can then be incorporated into living organisms. We learned that the enzyme nitrogenase is responsible for generating most of the fixed nitrogen essential for plant growth. However, the quantity of food required to feed the ever-increasing human population far exceeds that provided by nitrogen-fixing plants, so human agriculture requires substantial amounts of ammonia-based fertilizers that can be applied directly to croplands. Thus, of all the chemical reactions that humans have learned to carry out and control for their own purposes, the synthesis of ammonia from hydrogen and atmospheric nitrogen is one of the most important.

In 1912 the German chemist Fritz Haber (1868–1934) developed a process for synthesizing ammonia directly from nitrogen and hydrogen (Figure 15.4▼). The process is sometimes called the *Haber–Bosch process* to also honor Karl Bosch, the engineer who developed the equipment for the industrial production of ammonia. The engineering needed to implement the Haber process requires the use of temperatures and pressures (approximately 500 °C and 200 atm) that were difficult to achieve at that time.

The Haber process provides a historically interesting example of the complex impact of chemistry on our lives. At the start of World War I, in 1914, Germany depended on nitrate deposits in Chile for the nitrogen-containing compounds needed to manufacture explosives. During the war the Allied naval

▲ **Figure 15.5 Liquid ammonia used as fertilizer.** Ammonia, produced by the Haber process, can be added directly to the soil as a fertilizer. Agricultural use is the largest single application of manufactured NH₃.

blockade of South America cut off this supply. However, by fixing nitrogen from air, Germany was able to continue to produce explosives. Experts have estimated that World War I would have ended before 1918 had it not been for the Haber process.

From these unhappy beginnings as a major factor in international warfare, the Haber process has become the world's principal source of fixed nitrogen. The same process that prolonged World War I has enabled the manufacture of fertilizers that have increased crop yields, thereby saving millions of people from starvation. About 40 billion pounds of ammonia are manufactured annually in the United States, mostly by the Haber process. The ammonia can be applied directly to the soil as fertilizer (Figure 15.5▲). It can also be converted into ammonium salts—for example, ammonium sulfate, (NH₄)₂SO₄, and ammonium hydrogen phosphate, (NH₄)₂HPO₄—which, in turn, are used as fertilizers.

Haber was a patriotic German who gave enthusiastic support to his nation's war effort. He served as chief of Germany's Chemical Warfare Service during World War I and developed the use of chlorine as a poison-gas weapon. Consequently, the decision to award him the Nobel Prize for chemistry in 1918 was the subject of considerable controversy and criticism. The ultimate irony, however, came in 1933 when Haber was expelled from Germany because he was Jewish.

Related Exercises: 15.38 and 15.64

▲ **Figure 15.4 The Haber process.** Used to convert N₂(*g*) and H₂(*g*) to NH₃(*g*), this process, although exothermic, requires breaking the very strong triple bond in N₂.

Chemists carried out studies of this kind on other chemical systems in the nineteenth century, before Haber's work. In 1864, Cato Maximilian Guldberg (1836–1902) and Peter Waage (1833–1900) postulated their **law of mass action**, which expresses, for any reaction, the relationship between the concentrations of the reactants and products present at equilibrium. Suppose we have the following general equilibrium equation:

$$a\,\text{A} + b\,\text{B} \rightleftharpoons d\,\text{D} + e\,\text{E} \qquad [15.7]$$

where A, B, D, and E are the chemical species involved and *a*, *b*, *d*, and *e* are their coefficients in the balanced chemical equation. According to the law of mass action, the equilibrium condition is expressed by the expression

$$K_c = \frac{[\text{D}]^d[\text{E}]^e}{[\text{A}]^a[\text{B}]^b} \quad \begin{array}{l} \longleftarrow \text{ products} \\ \longleftarrow \text{ reactants} \end{array} \qquad [15.8]$$

We call this relationship the **equilibrium-constant expression** (or merely the equilibrium expression) for the reaction. The constant K_c, which we call the **equilibrium constant**, is the numerical value obtained when we substitute equilibrium concentrations into the equilibrium-constant expression. The subscript c on the K indicates that concentrations expressed in molarity are used to evaluate the constant.

In general, the numerator of the equilibrium-constant expression is the product of the concentrations of all substances on the product side of the equilibrium equation, each raised to a power equal to its coefficient in the balanced equation. The denominator is similarly derived from the reactant side of the equilibrium equation. (Remember, the convention is to write the substances on the *product* side in the *numerator* and the substances on the *reactant* side in the *denominator*.) Thus, for the Haber process, $N_2(g) + 3\,H_2(g) \rightleftharpoons 2\,NH_3(g)$, the equilibrium-constant expression is

$$K_c = \frac{[NH_3]^2}{[N_2][H_2]^3} \qquad [15.9]$$

Note that once we know the balanced chemical equation for an equilibrium, we can write the equilibrium-constant expression even if we do not know the reaction mechanism. *The equilibrium-constant expression depends only on the stoichiometry of the reaction, not on its mechanism.*

The value of the equilibrium constant at any given temperature does not depend on the initial amounts of reactants and products. It also does not matter whether other substances are present, as long as they do not react with a reactant or a product. The value of the equilibrium constant depends only on the particular reaction and on the temperature.

■ SAMPLE EXERCISE 15.1 | Writing Equilibrium-Constant Expressions

Write the equilibrium expression for K_c for the following reactions:

(a) $2\,O_3(g) \rightleftharpoons 3\,O_2(g)$

(b) $2\,NO(g) + Cl_2(g) \rightleftharpoons 2\,NOCl(g)$

(c) $Ag^+(aq) + 2\,NH_3(aq) \rightleftharpoons Ag(NH_3)_2{}^+(aq)$

SOLUTION

Analyze: We are given three equations and are asked to write an equilibrium-constant expression for each.

Plan: Using the law of mass action, we write each expression as a quotient having the product concentration terms in the numerator and the reactant concentration terms in the denominator. Each concentration term is raised to the power of its coefficient in the balanced chemical equation.

Solve: (a) $K_c = \dfrac{[O_2]^3}{[O_3]^2}$, (b) $K_c = \dfrac{[NOCl]^2}{[NO]^2[Cl_2]}$, (c) $K_c = \dfrac{[Ag(NH_3)_2{}^+]}{[Ag^+][NH_3]^2}$

■ PRACTICE EXERCISE

Write the equilibrium-constant expression K_c, for (a) $H_2(g) + I_2(g) \rightleftharpoons 2\,HI(g)$, (b) $Cd^{2+}(aq) + 4\,Br^-(aq) \rightleftharpoons CdBr_4{}^{2-}(aq)$

Answers: (a) $K_c = \dfrac{[HI]^2}{[H_2][I_2]}$; (b) $K_c = \dfrac{[CdBr_4{}^{2-}]}{[Cd^{2+}][Br^-]^4}$

Evaluating K_c

We can illustrate how the law of mass action was discovered empirically and demonstrate that the equilibrium constant is independent of starting concentrations by examining a series of experiments involving dinitrogen tetroxide and nitrogen dioxide:

$$N_2O_4(g) \rightleftharpoons 2\,NO_2(g) \qquad K_c = \frac{[NO_2]^2}{[N_2O_4]} \qquad [15.10]$$

TABLE 15.1 ▪ Initial and Equilibrium Concentrations of N_2O_4 and NO_2 in the Gas Phase at 100 °C					
Experiment	Initial $[N_2O_4]$ (M)	Initial $[NO_2]$ (M)	Equilibrium $[N_2O_4]$ (M)	Equilibrium $[NO_2]$ (M)	K_c
1	0.0	0.0200	0.00140	0.0172	0.211
2	0.0	0.0300	0.00280	0.0243	0.211
3	0.0	0.0400	0.00452	0.0310	0.213
4	0.0200	0.0	0.00452	0.0310	0.213

We start with several sealed tubes containing different concentrations of NO_2 and N_2O_4. The tubes are kept at 100 °C until equilibrium is reached. We then analyze the mixtures and determine the equilibrium concentrations of NO_2 and N_2O_4, as shown in Table 15.1 ▲.

To evaluate the equilibrium constant, K_c, we insert the equilibrium concentrations into the equilibrium-constant expression. For example, using the Experiment 1 data, $[NO_2] = 0.0172\ M$ and $[N_2O_4] = 0.00140\ M$, we find

$$K_c = \frac{[NO_2]^2}{[N_2O_4]} = \frac{[0.0172]^2}{0.00140} = 0.211$$

Proceeding in the same way, the values of K_c for the other samples were calculated, as listed in Table 15.1. Note that the value for K_c is constant ($K_c = 0.212$, within the limits of experimental error) even though the initial concentrations vary. Furthermore, the results of Experiment 4 show that equilibrium can be achieved beginning with N_2O_4 rather than with NO_2. That is, equilibrium can be approached from either direction. Figure 15.6 ▶ shows how both Experiments 3 and 4 result in the same equilibrium mixture even though one begins with 0.0400 M NO_2 and the other with 0.0200 M N_2O_4.

Notice that no units are given for the values of K_c either in Table 15.1 or in our calculation using the Experiment 1 data. It is common practice to write equilibrium constants without units for reasons that we address later in this section.

▲ **Figure 15.6 Concentration changes approaching equilibrium.** As seen in Table 15.1, the same equilibrium mixture is produced starting with either 0.0400 M NO_2 (Experiment 3) or 0.0200 M N_2O_4 (Experiment 4).

GIVE IT SOME THOUGHT

How does the value of K_c in Equation 15.10 depend on the starting concentrations of NO_2 and N_2O_4?

Equilibrium Constants in Terms of Pressure, K_p

When the reactants and products in a chemical reaction are gases, we can formulate the equilibrium-constant expression in terms of partial pressures instead of molar concentrations. When partial pressures in atmospheres are used in the equilibrium-constant expression, we can denote the equilibrium constant as K_p (where the subscript p stands for pressure). For the general reaction in Equation 15.7, the expression for K_p is

$$K_p = \frac{(P_D)^d (P_E)^e}{(P_A)^a (P_B)^b} \qquad [15.11]$$

where P_A is the partial pressure of A in atmospheres, and so forth. For example, for $N_2O_4(g) \rightleftharpoons 2\ NO_2(g)$ we have

$$K_p = \frac{(P_{NO_2})^2}{P_{N_2O_4}}$$

GIVE IT SOME THOUGHT

What do the symbols K_c and K_p represent?

For a given reaction, the numerical value of K_c is generally different from the numerical value of K_p. We must therefore take care to indicate, via a subscript c or p, which of these equilibrium constants we are using. It is possible, however, to calculate one from the other using the ideal-gas equation (Section 10.4) to convert between concentration (in molarity, M) and pressure (in atm):

$$PV = nRT, \text{ so } P = \frac{n}{V} RT \qquad [15.12]$$

The usual units for n/V are mol/L, which equals molarity, M. For substance A we therefore see that

$$P_A = \frac{n_A}{V} RT = [A]RT \qquad [15.13]$$

When we substitute Equation 15.13 and like expressions for the other gaseous components of the reaction into the expression for K_p (Equation 15.11), we obtain a general expression relating K_p and K_c.

$$K_p = K_c(RT)^{\Delta n} \qquad [15.14]$$

The quantity Δn is the change in the number of moles of gas in the chemical equation for the reaction. It equals the sum of the coefficients of the gaseous products minus the sum of the coefficients of the gaseous reactants:

$$\Delta n = (\text{moles of gaseous product}) - (\text{moles of gaseous reactant}) \qquad [15.15]$$

For example, in the reaction $N_2O_4(g) \rightleftharpoons 2 NO_2(g)$, there are two moles of the product NO_2 (the coefficient in the balanced equation) and one mole of the reactant N_2O_4. Therefore, $\Delta n = 2 - 1 = 1$, and $K_p = K_c(RT)$ for this reaction. From Equation 15.14, we see that $K_p = K_c$ only when the same number of moles of gas appears on both sides of the balanced chemical equation, which means that $\Delta n = 0$.

■ SAMPLE EXERCISE 15.2 | Converting between K_c and K_p

In the synthesis of ammonia from nitrogen and hydrogen,

$$N_2(g) + 3 H_2(g) \rightleftharpoons 2 NH_3(g)$$

$K_c = 9.60$ at 300 °C. Calculate K_p for this reaction at this temperature.

SOLUTION

Analyze: We are given K_c for a reaction and asked to calculate K_p.

Plan: The relationship between K_c and K_p is given by Equation 15.14. To apply that equation, we must determine Δn by comparing the number of moles of product with the number of moles of reactants (Equation 15.15).

Solve: There are two moles of gaseous products (2 NH_3) and four moles of gaseous reactants (1 N_2 + 3 H_2). Therefore, $\Delta n = 2 - 4 = -2$. (Remember that Δ functions are always based on products minus reactants.) The temperature, T, is $273 + 300 = 573$ K. The value for the ideal-gas constant, R, is 0.0821 L-atm/mol-K. Using $K_c = 9.60$, we therefore have

$$K_p = K_c(RT)^{\Delta n} = (9.60)(0.0821 \times 573)^{-2} = \frac{(9.60)}{(0.0821 \times 573)^2} = 4.34 \times 10^{-3}$$

■ PRACTICE EXERCISE

For the equilibrium $2 SO_3(g) \rightleftharpoons 2 SO_2(g) + O_2(g)$, K_c is 4.08×10^{-3} at 1000 K. Calculate the value for K_p.
Answer: 0.335

Equilibrium Constants and Units

You may wonder why equilibrium constants are reported without units. The equilibrium constant is related to the kinetics of a reaction as well as to the thermodynamics of the process. (We will explore this latter connection in Chapter 19.) Equilibrium constants derived from thermodynamic measurements are defined in terms of *activities* rather than concentrations or partial pressures.

The activity of any substance in an *ideal* mixture is the ratio of the concentration or pressure of the substance to a reference concentration ($1 M$) or a reference pressure (1 atm). For example, if the concentration of a substance in an equilibrium mixture is $0.010 M$, its activity is $0.010 M/1 M = 0.010$. The units of such ratios always cancel, and consequently, activities have no units. Furthermore, the numerical value of the activity equals the concentration. For pure solids and pure liquids, the situation is even simpler because the activities then merely equal 1 (again with no units).

In real systems, activities are also ratios that have no units. Even though these activities may not be exactly numerically equal to concentrations, we will ignore the differences. All we need to know at this point is that activities have no units. As a result, the *thermodynamic equilibrium constant* derived from them also has no units. It is therefore common practice to write all types of equilibrium constants without units as well, a practice that we adhere to in this text.

> ### GIVE IT SOME THOUGHT
>
> If the concentration of N_2O_4 in an equilibrium mixture is $0.00140 M$, what is its activity? (Assume the solution is ideal.)

15.3 INTERPRETING AND WORKING WITH EQUILIBRIUM CONSTANTS

In this section we explore two important ideas. Before doing calculations with equilibrium constants, it is valuable to understand what the magnitude of an equilibrium constant can tell us about the relative concentrations of reactants and products in an equilibrium mixture. It is also useful to consider how the magnitude of any equilibrium constant depends on how the chemical equation is expressed.

The Magnitude of Equilibrium Constants

Equilibrium constants can vary from very large to very small. The magnitude of the constant provides us with important information about the composition of an equilibrium mixture. For example, consider the reaction of carbon monoxide gas and chlorine gas at 100 °C to form phosgene ($COCl_2$), a toxic gas used in the manufacture of certain polymers and insecticides:

$$CO(g) + Cl_2(g) \rightleftharpoons COCl_2(g) \qquad K_c = \;\; = \frac{[COCl_2]}{[CO][Cl_2]} = 4.56 \times 10^9$$

For the equilibrium constant to be so large, the numerator of the equilibrium-constant expression must be much larger than the denominator. Thus, the equilibrium concentration of $COCl_2$ must be much greater than that of CO or Cl_2, and in fact this is just what we find experimentally. We say that this equilibrium *lies to the right* (that is, toward the product side). Likewise, a very small equilibrium constant indicates that the equilibrium mixture contains mostly reactants.

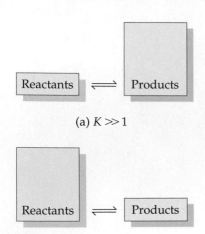

(a) $K \gg 1$

(b) $K \ll 1$

▲ **Figure 15.7 K and the composition of the equilibrium mixture.** The equilibrium expression has products in the numerator and reactants in the denominator. (a) When $K \gg 1$, there are more products than reactants at equilibrium, and the equilibrium is said to lie to the right. (b) When $K \ll 1$, there are more reactants than products at equilibrium, and the equilibrium is said to lie to the left.

We then say that the equilibrium *lies to the left*. In general,

If $K \gg 1$ (that is, large K): Equilibrium lies to the right; products predominate.

If $K \ll 1$ (that is, small K): Equilibrium lies to the left; reactants predominate.

These situations are summarized in Figure 15.7 ◄. Remember, opposing rates, not concentrations, are equal at equilibrium. Thus, equilibrium constants for different reactions can span a very wide range.

■■ **SAMPLE EXERCISE 15.3** | **Interpreting the Magnitude of an Equilibrium Constant**

The following diagrams represent three different systems at equilibrium, all in the same size containers. **(a)** Without doing any calculations, rank the three systems in order of increasing equilibrium constant, K_c. **(b)** If the volume of the containers is 1.0 L and each sphere represents 0.10 mol, calculate K_c for each system.

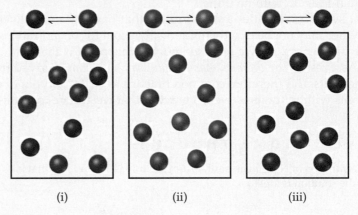

(i)   (ii)   (iii)

SOLUTION

Analyze: We are first asked to judge the relative magnitudes of three equilibrium constants and then to calculate them.

Plan: **(a)** The more product that is present at equilibrium, relative to the reactant, the greater the equilibrium constant. **(b)** The equilibrium constant is given by the concentrations of products over reactants.

Solve:

(a) Each box contains 10 spheres. The amount of product in each varies as follows: (i) 6, (ii) 1, (iii) 8. Thus, the equilibrium constant varies in the order (ii) < (i) < (iii).

(b) In (i) we have 0.60 mol/L product and 0.40 mol/L reactant, giving $K_c = 0.60/0.40 = 1.5$. (You will get the same result by merely dividing the number of spheres of each kind: 6 spheres/4 spheres = 1.5.) In (ii) we have 0.10 mol/L product and 0.90 mol/L reactant, giving $K_c = 0.10/0.90 = 0.11$ (or 1 sphere/9 spheres = 0.11). In (ii) we have 0.80 mol/L product and 0.29 mol/L reactant, giving $K_c = 0.80/0.20 = 4.0$ (or 8 spheres/2 spheres = 4.0). These calculations verify the order in (a).

Comment: Imagine that there was a drawing, like those above, that represents a reaction with a very small or very large value of K_c. For example, what would the drawing look like if $K_c = 1 \times 10^{-5}$? In that case there would need to be 100,000 reactant molecules for only 1 product molecule. But then, that would be impractical to draw.

■■ **PRACTICE EXERCISE**

For the reaction $H_2(g) + I_2(g) \rightleftharpoons 2 HI(g)$, $K_p = 794$ at 298 K and $K_p = 54$ at 700 K. Is the formation of HI favored more at the higher or lower temperature?

Answer: The formation of product, HI, is favored at the lower temperature because K_p is larger at the lower temperature.

The Direction of the Chemical Equation and K

Because an equilibrium can be approached from either direction, the direction in which we write the chemical equation for an equilibrium is arbitrary. For example, we have seen that we can represent the $N_2O_4 - NO_2$ equilibrium as

$$N_2O_4(g) \rightleftharpoons 2 NO_2(g) \quad K_c = \frac{[NO_2]^2}{[N_2O_4]} = 0.212 \quad \text{(at 100 °C)} \quad [15.16]$$

We could equally well consider this same equilibrium in terms of the reverse reaction:

$$2\,NO_2(g) \rightleftharpoons N_2O_4(g)$$

The equilibrium expression is then

$$K_c = \frac{[N_2O_4]}{[NO_2]^2} = \frac{1}{0.212} = 4.72 \quad \text{(at 100 °C)} \qquad [15.17]$$

Equation 15.17 is just the reciprocal of the equilibrium-constant expression in Equation 15.16. *The equilibrium-constant expression for a reaction written in one direction is the reciprocal of the one for the reaction written in the reverse direction.* Consequently, the numerical value of the equilibrium constant for the reaction written in one direction is the reciprocal of that for the reverse reaction. Both expressions are equally valid, but it is meaningless to say that the equilibrium constant for the equilibrium between NO_2 and N_2O_4 is 0.212 or 4.72 unless we indicate how the equilibrium reaction is written and specify the temperature.

■ **SAMPLE EXERCISE 15.4** | **Evaluating an Equilibrium Constant When an Equation Is Reversed**

The equilibrium constant for the reaction of N_2 with O_2 to form NO equals $K_c = 1 \times 10^{-30}$ at 25 °C:

$$N_2(g) + O_2(g) \rightleftharpoons 2\,NO(g) \qquad K_c = 1 \times 10^{-30}$$

Using this information, write the equilibrium constant expression and calculate the equilibrium constant for the following reaction:

$$2\,NO(g) \rightleftharpoons N_2(g) + O_2(g)$$

SOLUTION

Analyze: We are asked to write the equilibrium-constant expression for a reaction and to determine the value of K_c given the chemical equation and equilibrium constant for the reverse reaction.

Plan: The equilibrium-constant expression is a quotient of products over reactants, each raised to a power equal to its coefficient in the balanced equation. The value of the equilibrium constant is the reciprocal of that for the reverse reaction.

Solve:
Writing products over reactants, we have

$$K_c = \frac{[N_2][O_2]}{[NO]^2}$$

Both the equilibrium-constant expression and the numerical value of the equilibrium constant are the reciprocals of those for the formation of NO from N_2 and O_2:

$$K_c = \frac{[N_2][O_2]}{[NO]^2} = \frac{1}{1 \times 10^{-30}} = 1 \times 10^{30}$$

Comment: Regardless of the way we express the equilibrium among NO, N_2, and O_2, at 25 °C it lies on the side that favors N_2 and O_2. Thus, the equilibrium mixture will contain mostly N_2 and O_2, with very little NO present.

■ **PRACTICE EXERCISE**

For the formation of NH_3 from N_2 and H_2, $N_2(g) + 3\,H_2(g) \rightleftharpoons 2\,NH_3(g)$, $K_p = 4.34 \times 10^{-3}$ at 300 °C. What is the value of K_p for the reverse reaction?
Answer: 2.30×10^2

Relating Chemical Equations and Equilibrium Constants

Just as the equilibrium constants of forward and reverse reactions are reciprocals of each other, the equilibrium constants of reactions associated in other ways are also related. For example, if we were to multiply our original $N_2O_4 - NO_2$ equilibrium by 2, we would have

$$2\,N_2O_4(g) \rightleftharpoons 4\,NO_2(g)$$

The equilibrium-constant expression, K_c, for this equation is

$$K_c = \frac{[NO_2]^4}{[N_2O_4]^2}$$

which is simply the square of the equilibrium-constant expression for the original equation, given in Equation 15.10. Because the new equilibrium-constant expression equals the original expression squared, the new equilibrium constant equals the original constant squared: in this case $0.212^2 = 0.0449$ (at 100 °C).

How does the magnitude of the equilibrium constant K_p for the reaction $2\,HI(g) \rightleftharpoons$ $H_2(g) + I_2(g)$ change if the equilibrium is written $6\,HI(g) \rightleftharpoons 3\,H_2(g) + 3\,I_2(g)$?

Sometimes, as in problems that utilize Hess's law (Section 5.6), we must use equations made up of two or more steps in the overall process. We obtain the net equation by adding the individual equations and canceling identical terms. Consider the following two reactions, their equilibrium-constant expressions, and their equilibrium constants at 100 °C:

$$2\,NOBr(g) \rightleftharpoons 2\,NO(g) + Br_2(g) \quad K_c = \frac{[NO]^2[Br_2]}{[NOBr]^2} = 0.014$$

$$Br_2(g) + Cl_2(g) \rightleftharpoons 2\,BrCl(g) \qquad K_c = \frac{[BrCl]^2}{[Br_2][Cl_2]} = 7.2$$

The net sum of these two equations is

$$2\,NOBr(g) + Cl_2(g) \rightleftharpoons 2\,NO(g) + 2\,BrCl(g)$$

and the equilibrium-constant expression for the net equation is the product of the expressions for the individual steps:

$$K_c = \frac{[NO^2][BrCl]^2}{[NOBr]^2[Cl_2]} = \frac{[NO]^2[Br_2]}{[NOBr]^2} \times \frac{[BrCl]^2}{[Br_2][Cl_2]}$$

Because the equilibrium-constant expression for the net equation is the product of two equilibrium-constant expressions, the equilibrium constant for the net equation is the product of the two individual equilibrium constants: $K_c = 0.014 \times 7.2 = 0.10$.

To summarize:

1. The equilibrium constant of a reaction in the *reverse* direction is the *inverse* of the equilibrium constant of the reaction in the forward direction.

2. The equilibrium constant of a reaction that has been *multiplied* by a number is the equilibrium constant raised to a *power* equal to that number.

3. The equilibrium constant for a net reaction made up of *two or more steps* is the *product* of the equilibrium constants for the individual steps.

■ **SAMPLE EXERCISE 15.5** | Combining Equilibrium Expressions

Given the following information,

$$HF(aq) \rightleftharpoons H^+(aq) + F^-(aq) \qquad K_c = 6.8 \times 10^{-4}$$
$$H_2C_2O_4(aq) \rightleftharpoons 2\,H^+(aq) + C_2O_4{}^{2-}(aq) \quad K_c = 3.8 \times 10^{-6}$$

determine the value of K_c for the reaction

$$2\,HF(aq) + C_2O_4{}^{2-}(aq) \rightleftharpoons 2\,F^-(aq) + H_2C_2O_4(aq)$$

SOLUTION

Analyze: We are given two equilibrium equations and the corresponding equilibrium constants and are asked to determine the equilibrium constant for a third equation, which is related to the first two.

Plan: We cannot simply add the first two equations to get the third. Instead, we need to determine how to manipulate the equations to come up with the steps that will add to give us the desired equation.

Solve: If we multiply the first equation by 2 and make the corresponding change to its equilibrium constant (raising to the power 2), we get

$$2\,HF(aq) \rightleftharpoons 2\,H^+(aq) + 2\,F^-(aq) \quad K_c = (6.8 \times 10^{-4})^2 = 4.6 \times 10^{-7}$$

Reversing the second equation and again making the corresponding change to its equilibrium constant (taking the reciprocal) gives

$$2 \, H^+(aq) + C_2O_4{}^{2-}(aq) \rightleftharpoons H_2C_2O_4(aq) \quad K_c = \frac{1}{3.8 \times 10^{-6}} = 2.6 \times 10^5$$

Now we have two equations that sum to give the net equation, and we can multiply the individual K_c values to get the desired equilibrium constant.

$$
\begin{aligned}
2 \, HF(aq) &\rightleftharpoons 2 \, H^+(aq) + 2 \, F^-(aq) & K_c &= 4.6 \times 10^{-7} \\
2 \, H^+(aq) + C_2O_4{}^{2-}(aq) &\rightleftharpoons H_2C_2O_4(aq) & K_c &= 2.5 \times 10^5 \\
\hline
2 \, HF(aq) + C_2O_4{}^{2-}(aq) &\rightleftharpoons 2 \, F^-(aq) + H_2C_2O_4(aq) & K_c &= (4.6 \times 10^{-7})(2.6 \times 10^5) = 0.12
\end{aligned}
$$

◼ PRACTICE EXERCISE

Given that, at 700 K, $K_p = 54.0$ for the reaction $H_2(g) + I_2(g) \rightleftharpoons 2 \, HI(g)$ and $K_p = 1.04 \times 10^{-4}$ for the reaction $N_2(g) + 3 \, H_2(g) \rightleftharpoons 2 \, NH_3(g)$, determine the value of K_p for the reaction $2 \, NH_3(g) + 3 \, I_2(g) \rightleftharpoons 6 \, HI(g) + N_2(g)$ at 700 K.

Answer: $\dfrac{(54.0)^3}{1.04 \times 10^{-4}} = 1.51 \times 10^9$

15.4 HETEROGENEOUS EQUILIBRIA

Many equilibria, such as the hydrogen-nitrogen-ammonia system, involve substances all in the same phase. Such equilibria are called **homogeneous equilibria**. In other cases the substances in equilibrium are in different phases, giving rise to **heterogeneous equilibria**. As an example, consider the equilibrium that occurs when solid lead(II) chloride ($PbCl_2$) dissolves in water to form a saturated solution:

$$PbCl_2(s) \rightleftharpoons Pb^{2+}(aq) + 2 \, Cl^-(aq) \qquad [15.18]$$

This system consists of a solid in equilibrium with two aqueous species. If we write the equilibrium-constant expression for this process, we encounter a problem we have not encountered previously: How do we express the concentration of a solid substance? Although it is possible to express the concentration of a solid in terms of moles per unit volume, it is unnecessary to do so in writing equilibrium-constant expressions. *Whenever a pure solid or a pure liquid is involved in a heterogeneous equilibrium, its concentration is not included in the equilibrium-constant expression for the reaction.* Thus, the equilibrium-constant expression for Equation 15.18 is

$$K_c = [Pb^{2+}][Cl^-]^2 \qquad [15.19]$$

Even though $PbCl_2(s)$ does not appear in the equilibrium-constant expression, it must be present for equilibrium to occur.

The fact that pure solids and pure liquids are excluded from equilibrium-constant expressions can be explained in two ways. First, the concentration of a pure solid or liquid has a constant value. If the mass of a solid is doubled, its volume also doubles. Thus, its concentration, which relates to the ratio of mass to volume, stays the same. Because equilibrium-constant expressions include terms only for reactants and products whose concentrations can change during a chemical reaction, the concentrations of pure solids and pure liquids are omitted.

The omission of pure solids and liquids from equilibrium-constant expressions can also be rationalized in a second way. Recall from the last section that what is substituted into a thermodynamic equilibrium expression is the activity of each substance, which is a ratio of the concentration to a reference value. For a pure substance, the reference value is the concentration of the pure substance itself, so that the activity of any pure solid or liquid is always simply 1.

▲ GIVE IT SOME THOUGHT

Write the equilibrium-constant expression for the evaporation of water, $H_2O(l) \rightleftharpoons H_2O(g)$, in terms of partial pressures, K_p.

▶ **Figure 15.8 A heterogeneous equilibrium.** The equilibrium involving $CaCO_3$, CaO, and CO_2 is a heterogeneous equilibrium. The equilibrium pressure of CO_2 is the same in the two bell jars as long as the two systems are at the same temperature, even though the relative amounts of pure $CaCO_3$ and CaO differ greatly. The equilibrium-constant expression for the reaction is $K_p = P_{CO_2}$.

CaO   CaCO₃   CaO   CaCO₃

(a)   (b)

As a further example of a heterogeneous reaction, consider the decomposition of calcium carbonate:

$$CaCO_3(s) \rightleftharpoons CaO(s) + CO_2(g)$$

Omitting the concentrations of solids from the equilibrium-constant expression gives

$$K_c = [CO_2] \quad \text{and} \quad K_p = P_{CO_2}$$

These equations tell us that at a given temperature, an equilibrium among $CaCO_3$, CaO, and CO_2 will always lead to the same partial pressure of CO_2 as long as all three components are present. As shown in Figure 15.8▲, we would have the same pressure of CO_2 regardless of the relative amounts of CaO and $CaCO_3$.

■■ **SAMPLE EXERCISE 15.6** │ **Writing Equilibrium-Constant Expressions for Heterogeneous Reactions**

Write the equilibrium-constant expression for K_c for each of the following reactions:

(a) $CO_2(g) + H_2(g) \rightleftharpoons CO(g) + H_2O(l)$
(b) $SnO_2(s) + 2\,CO(g) \rightleftharpoons Sn(s) + 2\,CO_2(g)$

SOLUTION

Analyze: We are given two chemical equations, both for heterogeneous equilibria, and asked to write the corresponding equilibrium-constant expressions.

Plan: We use the law of mass action, remembering to omit any pure solids, pure liquids, and solvents from the expressions.

Solve:

(a) The equilibrium-constant expression is

$$K_c = \frac{[CO]}{[CO_2][H_2]}$$

Because H_2O appears in the reaction as a pure liquid, its concentration does not appear in the equilibrium-constant expression.

(b) The equilibrium-constant expression is

$$K_c = \frac{[CO_2]^2}{[CO]^2}$$

Because SnO_2 and Sn are both pure solids, their concentrations do not appear in the equilibrium-constant expression.

■■ **PRACTICE EXERCISE**

Write the following equilibrium-constant expressions:

(a) K_c for $Cr(s) + 3\,Ag^+(aq) \rightleftharpoons Cr^{3+}(aq) + 3\,Ag(s)$
(b) K_p for $3\,Fe(s) + 4\,H_2O(g) \rightleftharpoons Fe_3O_4(s) + 4\,H_2(g)$

Answers: **(a)** $K_c = \dfrac{[Cr^{3+}]}{[Ag^+]^3}$, **(b)** $K_p = \dfrac{(P_{H_2})^4}{(P_{H_2O})^4}$

■ **SAMPLE EXERCISE 15.7** | **Analyzing a Heterogeneous Equilibrium**

Each of the following mixtures was placed in a closed container and allowed to stand. Which is capable of attaining the equilibrium $CaCO_3(s) \rightleftharpoons CaO(s) + CO_2(g)$: **(a)** pure $CaCO_3$, **(b)** CaO and a CO_2 pressure greater than the value of K_p, **(c)** some $CaCO_3$ and a CO_2 pressure greater than the value of K_p, **(d)** $CaCO_3$ and CaO?

SOLUTION

Analyze: We are asked which of several combinations of species can establish an equilibrium between calcium carbonate and its decomposition products, calcium oxide and carbon dioxide.

Plan: For equilibrium to be achieved, it must be possible for both the forward process and the reverse process to occur. For the forward process to occur, there must be some calcium carbonate present. For the reverse process to occur, there must be both calcium oxide and carbon dioxide. In both cases, either the necessary compounds may be present initially, or they may be formed by reaction of the other species.

Solve: Equilibrium can be reached in all cases except (c) as long as sufficient quantities of solids are present. **(a)** $CaCO_3$ simply decomposes, forming $CaO(s)$ and $CO_2(g)$ until the equilibrium pressure of CO_2 is attained. There must be enough $CaCO_3$, however, to allow the CO_2 pressure to reach equilibrium. **(b)** CO_2 continues to combine with CaO until the partial pressure of the CO_2 decreases to the equilibrium value. **(c)** There is no CaO present, so equilibrium cannot be attained because there is no way the CO_2 pressure can decrease to its equilibrium value (which would require some of the CO_2 to react with CaO). **(d)** The situation is essentially the same as in (a): $CaCO_3$ decomposes until equilibrium is attained. The presence of CaO initially makes no difference.

■ **PRACTICE EXERCISE**

When added to $Fe_3O_4(s)$ in a closed container, which one of the following substances— $H_2(g)$, $H_2O(g)$, $O_2(g)$—will allow equilibrium to be established in the reaction $3 Fe(s) + 4 H_2O(g) \rightleftharpoons Fe_3O_4(s) + 4 H_2(g)$?
Answer: only $H_2(g)$

When a solvent is involved as a reactant or product in an equilibrium, its concentration is also excluded from the equilibrium-constant expression, provided the concentrations of reactants and products are low, so that the solvent is essentially a pure substance. Applying this guideline to an equilibrium involving water as a solvent,

$$H_2O(l) + CO_3^{2-}(aq) \rightleftharpoons OH^-(aq) + HCO_3^-(aq) \qquad [15.20]$$

gives an equilibrium-constant expression in which $[H_2O]$ is excluded:

$$K_c = \frac{[OH^-][HCO_3^-]}{[CO_3^{2-}]} \qquad [15.21]$$

◢ **GIVE IT SOME THOUGHT**

Write the equilibrium-constant expression for the following reaction: $NH_3(aq) + H_2O(l) \rightleftharpoons NH_4^+(aq) + OH^-(aq)$

15.5 CALCULATING EQUILIBRIUM CONSTANTS

One of the first tasks confronting Haber when he approached the problem of ammonia synthesis was finding the magnitude of the equilibrium constant for the synthesis of NH_3 at various temperatures. If the value of K for Equation 15.6 was very small, the amount of NH_3 in an equilibrium mixture would be small relative to the amounts of N_2 and H_2. That is, if the equilibrium lies too far to the left, it would be impossible to develop a satisfactory synthesis process for ammonia.

Haber and his coworkers therefore evaluated the equilibrium constants for this reaction at various temperatures. The method they employed is analogous to that described in constructing Table 15.1: They started with various mixtures of N_2, H_2, and NH_3, allowed the mixtures to achieve equilibrium at a specific temperature, and measured the concentrations of all three gases at equilibrium. Because the equilibrium concentrations of all products and reactants were known, the equilibrium constant could be calculated directly from the equilibrium-constant expression.

■■ **SAMPLE EXERCISE 15.8** | Calculating *K* When All Equilibrium
                            Concentrations Are Known

A mixture of hydrogen and nitrogen in a reaction vessel is allowed to attain equilibrium at 472 °C. The equilibrium mixture of gases was analyzed and found to contain 7.38 atm H_2, 2.46 atm N_2, and 0.166 atm NH_3. From these data, calculate the equilibrium constant K_p for the reaction

$$N_2(g) + 3\,H_2(g) \rightleftharpoons 2\,NH_3(g)$$

SOLUTION

Analyze: We are given a balanced equation and equilibrium partial pressures and are asked to calculate the value of the equilibrium constant.

Plan: Using the balanced equation, we write the equilibrium-constant expression. We then substitute the equilibrium partial pressures into the expression and solve for K_p.

Solve:

$$K_p = \frac{(P_{NH_3})^2}{P_{N_2}(P_{H_2})^3} = \frac{(0.166)^2}{(2.46)(7.38)^3} = 2.79 \times 10^{-5}$$

■■ **PRACTICE EXERCISE**

An aqueous solution of acetic acid is found to have the following equilibrium concentrations at 25 °C: $[HC_2H_3O_2] = 1.65 \times 10^{-2}\,M$; $[H^+] = 5.44 \times 10^{-4}\,M$; and $[C_2H_3O_2^-] = 5.44 \times 10^{-4}\,M$. Calculate the equilibrium constant K_c for the ionization of acetic acid at 25 °C. The reaction is

$$HC_2H_3O_2(aq) \rightleftharpoons H^+(aq) + C_2H_3O_2^-(aq)$$

Answer: 1.79×10^{-5}

We often do not know the equilibrium concentrations of all chemical species in an equilibrium mixture. If we know the equilibrium concentration of at least one species, however, we can generally use the stoichiometry of the reaction to deduce the equilibrium concentrations of the others. The following steps outline the procedure we use to do this:

1. Tabulate all the known initial and equilibrium concentrations of the species that appear in the equilibrium-constant expression.

2. For those species for which both the initial and equilibrium concentrations are known, calculate the change in concentration that occurs as the system reaches equilibrium.

3. Use the stoichiometry of the reaction (that is, use the coefficients in the balanced chemical equation) to calculate the changes in concentration for all the other species in the equilibrium.

4. From the initial concentrations and the changes in concentration, calculate the equilibrium concentrations. These are then used to evaluate the equilibrium constant.

◼◼◼ SAMPLE EXERCISE 15.9 | Calculating K from Initial and Equilibrium Concentrations

A closed system initially containing 1.000×10^{-3} M H_2 and 2.000×10^{-3} M I_2 at 448 °C is allowed to reach equilibrium. Analysis of the equilibrium mixture shows that the concentration of HI is 1.87×10^{-3} M. Calculate K_c at 448 °C for the reaction taking place, which is

$$H_2(g) + I_2(g) \rightleftharpoons 2\,HI(g)$$

SOLUTION

Analyze: We are given the initial concentrations of H_2 and I_2 and the equilibrium concentration of HI. We are asked to calculate the equilibrium constant K_c for $H_2(g) + I_2(g) \rightleftharpoons 2\,HI(g)$.

Plan: We construct a table to find equilibrium concentrations of all species and then use the equilibrium concentrations to calculate the equilibrium constant.

Solve: First, we tabulate the initial and equilibrium concentrations of as many species as we can. We also provide space in our table for listing the changes in concentrations. As shown, it is convenient to use the chemical equation as the heading for the table.

	$H_2(g)$	+	$I_2(g)$	$\rightleftharpoons$	$2\,HI(g)$
Initial	1.000×10^{-3} M		2.000×10^{-3} M		0 M
Change					
Equilibrium					1.87×10^{-3} M

Second, we calculate the change in concentration of HI, which is the difference between the equilibrium values and the initial values:

$$\text{Change in [HI]} = 1.87 \times 10^{-3}\ M - 0 - 1.87 \times 10^{-3}\ M$$

Third, we use the coefficients in the balanced equation to relate the change in [HI] to the changes in $[H_2]$ and $[I_2]$:

$$\left(1.87 \times 10^{-3}\ \frac{\text{mol HI}}{\text{L}}\right)\left(\frac{1\ \text{mol } H_2}{2\ \text{mol HI}}\right) = 0.935 \times 10^{-3}\ \frac{\text{mol } H_2}{\text{L}}$$

$$\left(1.87 \times 10^{-3}\ \frac{\text{mol HI}}{\text{L}}\right)\left(\frac{1\ \text{mol } I_2}{2\ \text{mol HI}}\right) = 0.935 \times 10^{-3}\ \frac{\text{mol } I_2}{\text{L}}$$

Fourth, we calculate the equilibrium concentrations of H_2 and I_2, using the initial concentrations and the changes. The equilibrium concentration equals the initial concentration minus that consumed:

$$[H_2] = 1.000 \times 10^{-3}\ M - 0.935 \times 10^{-3}\ M = 0.065 \times 10^{-3}\ M$$

$$[I_2] = 2.000 \times 10^{-3}\ M - 0.935 \times 10^{-3}\ M = 1.065 \times 10^{-3}\ M$$

The completed table now looks like this (with equilibrium concentrations in blue for emphasis):

	$H_2(g)$	+	$I_2(g)$	$\rightleftharpoons$	$2\,HI(g)$
Initial	1.000×10^{-3} M		2.000×10^{-3} M		0 M
Change	-0.935×10^{-3} M		-0.935×10^{-3} M		$+1.87 \times 10^{-3}$ M
Equilibrium	0.065×10^{-3} M		1.065×10^{-3} M		1.87×10^{-3} M

Notice that the entries for the changes are negative when a reactant is consumed and positive when a product is formed.

Finally, now that we know the equilibrium concentration of each reactant and product, we can use the equilibrium-constant expression to calculate the equilibrium constant.

$$K_c = \frac{[HI]^2}{[H_2][I_2]} = \frac{(1.87 \times 10^{-3})^2}{(0.065 \times 10^{-3})(1.065 \times 10^{-3})} = 51$$

Comment: The same method can be applied to gaseous equilibrium problems to calculate K_p, in which case partial pressures are used as table entries in place of molar concentrations.

◼◼◼ PRACTICE EXERCISE

Sulfur trioxide decomposes at high temperature in a sealed container: $2\,SO_3(g) \rightleftharpoons 2\,SO_2(g) + O_2(g)$. Initially, the vessel is charged at 1000 K with $SO_3(g)$ at a partial pressure of 0.500 atm. At equilibrium the SO_3 partial pressure is 0.200 atm. Calculate the value of K_p at 1000 K.
Answer: 0.338

15.6 APPLICATIONS OF EQUILIBRIUM CONSTANTS

We have seen that the magnitude of K indicates the extent to which a reaction will proceed. If K is very large, the equilibrium mixture will contain mostly substances on the product side of the equation. (That is, the reaction will tend to proceed far to the right.) If K is very small (that is, much less than 1), the equilibrium mixture will contain mainly reactants. The equilibrium constant also allows us to (1) predict the direction in which a reaction mixture will proceed to achieve equilibrium and (2) calculate the concentrations of reactants and products when equilibrium has been reached.

Predicting the Direction of Reaction

For the formation of NH_3 from N_2 and H_2 (Equation 15.6), $K_c = 0.105$ at 472 °C. Suppose we place a mixture of 2.00 mol of H_2, 1.00 mol of N_2, and 2.00 mol of NH_3 in a 1.00-L container at 472 °C. How will the mixture react to reach equilibrium? Will N_2 and H_2 react to form more NH_3, or will NH_3 decompose to form N_2 and H_2?

To answer this question, we can substitute the starting concentrations of N_2, H_2, and NH_3 into the equilibrium-constant expression and compare its value to the equilibrium constant:

$$\frac{[NH_3]^2}{[N_2][H_2]^3} = \frac{(2.00)^2}{(1.00)(2.00)^3} = 0.500 \quad \text{whereas} \quad K_c = 0.105$$

To reach equilibrium, the quotient $[NH_3]^2/[N_2][H_2]^3$ will need to decrease from the starting value of 0.500 to the equilibrium value of 0.105. Because the system is closed, this change can happen only if the concentration of NH_3 decreases and the concentrations of N_2 and H_2 increase. Thus, the reaction proceeds toward equilibrium by forming N_2 and H_2 from NH_3; that is, the reaction proceeds from right to left.

The approach we have illustrated can be formalized by defining a quantity called the reaction quotient. The **reaction quotient**, Q, *is a number obtained by substituting reactant and product concentrations or partial pressures at any point during a reaction into an equilibrium-constant expression.* Therefore, for the general reaction

$$a\,A + b\,B \rightleftharpoons d\,D + e\,E$$

the reaction quotient is defined as

$$Q_c = \frac{[D]^d[E]^e}{[A]^a[B]^b} \tag{15.22}$$

(A related quantity Q_p can be written for any reaction that involves gases by using partial pressures instead of concentrations.)

Although we use the equilibrium-constant expression to calculate the reaction quotient, the concentrations that we use are not restricted to the equilibrium state. Thus, in our earlier example, when we substituted the starting concentrations into the equilibrium-constant expression, we obtained $Q_c = 0.500$ whereas $K_c = 0.105$. The equilibrium constant has only one value at each temperature. The reaction quotient, however, varies as the reaction proceeds.

To determine the direction in which a reaction will proceed to achieve equilibrium, we compare the values of Q_c and K_c or Q_p and K_p. Three possible situations arise:

- $Q = K$: The reaction quotient will equal the equilibrium constant only if the system is already at equilibrium.
- $Q > K$: The concentration of products is too large and that of reactants too small. Thus, substances on the right side of the chemical equation will react to form substances on the left; the reaction moves from right to left in approaching equilibrium.

◄ Figure 15.9 Predicting the direction of a reaction by comparing Q and K. The relative magnitudes of the reaction quotient Q and the equilibrium constant K indicate how the reaction mixture changes as it moves toward equilibrium. If Q is smaller than K, the reaction proceeds from left to right until $Q = K$. When $Q = K$, the reaction is at equilibrium and has no tendency to change. If Q is larger than K, the reaction proceeds from right to left until $Q = K$.

- $Q < K$: The concentration of products is too small and that of reactants too large. Thus, the reaction will achieve equilibrium by forming more products; it moves from left to right.

These relationships are summarized in Figure 15.9 ▲.

■■ SAMPLE EXERCISE 15.10 | Predicting the Direction of Approach to Equilibrium

At 448 °C the equilibrium constant K_c for the reaction

$$H_2(g) + I_2(g) \rightleftharpoons 2\,HI(g)$$

is 50.5. Predict in which direction the reaction will proceed to reach equilibrium at 448 °C if we start with 2.0×10^{-2} mol of HI, 1.0×10^{-2} mol of H_2, and 3.0×10^{-2} mol of I_2 in a 2.00-L container.

SOLUTION

Analyze: We are given a volume and initial molar amounts of the species in a reaction and asked to determine in which direction the reaction must proceed to achieve equilibrium.

Plan: We can determine the starting concentration of each species in the reaction mixture. We can then substitute the starting concentrations into the equilibrium-constant expression to calculate the reaction quotient, Q_c. Comparing the magnitudes of the equilibrium constant, which is given, and the reaction quotient will tell us in which direction the reaction will proceed.

Solve: The initial concentrations are

$$[HI] = 2.0 \times 10^{-2}\,\text{mol}/2.00\,\text{L} = 1.0 \times 10^{-2}\,M$$

$$[H_2] = 1.0 \times 10^{-2}\,\text{mol}/2.00\,\text{L} = 5.0 \times 10^{-3}\,M$$

$$[I_2] = 3.0 \times 10^{-2}\,\text{mol}/2.00\,\text{L} = 1.5 \times 10^{-2}\,M$$

The reaction quotient is therefore

$$Q_c = \frac{[HI]^2}{[H_2][I_2]} = \frac{(1.0 \times 10^{-2})^2}{(5.0 \times 10^{-3})(1.5 \times 10^{-2})} = 1.3$$

Because $Q_c < K_c$, the concentration of HI must increase and the concentrations of H_2 and I_2 must decrease to reach equilibrium; the reaction will proceed from left to right as it moves toward equilibrium.

■■ PRACTICE EXERCISE

At 1000 K the value of K_p for the reaction $2\,SO_3(g) \rightleftharpoons 2\,SO_2(g) + O_2(g)$ is 0.338. Calculate the value for Q_p, and predict the direction in which the reaction will proceed toward equilibrium if the initial partial pressures are $P_{SO_3} = 0.16$ atm; $P_{SO_2} = 0.41$ atm; $P_{O_2} = 2.5$ atm.

Answer: $Q_p = 16$; $Q_p > K_p$, and so the reaction will proceed from right to left, forming more SO_3.

Calculating Equilibrium Concentrations

Chemists frequently need to calculate the amounts of reactants and products present at equilibrium. Our approach in solving problems of this type is similar to the one we used for evaluating equilibrium constants: We tabulate the initial concentrations or partial pressures, the changes therein, and the final equilibrium concentrations or partial pressures. Usually we end up using the equilibrium-constant expression to derive an equation that must be solved for an unknown quantity, as demonstrated in Sample Exercise 15.11.

■ SAMPLE EXERCISE 15.11 | **Calculating Equilibrium Concentrations**

For the Haber process, $N_2(g) + 3 H_2(g) \rightleftharpoons 2 NH_3(g)$, $K_p = 1.45 \times 10^{-5}$ at 500 °C. In an equilibrium mixture of the three gases at 500 °C, the partial pressure of H_2 is 0.928 atm and that of N_2 is 0.432 atm. What is the partial pressure of NH_3 in this equilibrium mixture?

SOLUTION

Analyze: We are given an equilibrium constant, K_p, and the equilibrium partial pressures of two of the three substances in the equation (N_2 and H_2), and we are asked to calculate the equilibrium partial pressure for the third substance (NH_3).

Plan: We can set K_p equal to the equilibrium-constant expression and substitute in the partial pressures that we know. Then we can solve for the only unknown in the equation.

Solve: We tabulate the equilibrium pressures as follows:

	$N_2(g)$	$+$ 3 $H_2(g)$	$\rightleftharpoons$ 2 $NH_3(g)$
Equilibrium pressure (atm)	0.432	0.928	x

Because we do not know the equilibrium pressure of NH_3, we represent it with a variable, x. At equilibrium the pressures must satisfy the equilibrium-constant expression:

$$K_p = \frac{(P_{NH_3})^2}{P_{N_2}(P_{H_2})^3} = \frac{x^2}{(0.432)(0.928)^3} = 1.45 \times 10^{-5}$$

We now rearrange the equation to solve for x:

$$x^2 = (1.45 \times 10^{-5})(0.432)(0.928)^3 = 5.01 \times 10^{-6}$$

$$x = \sqrt{5.01 \times 10^{-6}} = 2.24 \times 10^{-3} \text{ atm} = P_{NH_3}$$

Comment: We can always check our answer by using it to recalculate the value of the equilibrium constant:

$$K_p = \frac{(2.24 \times 10^{-3})^2}{(0.432)(0.928)^3} = 1.45 \times 10^{-5}$$

■ PRACTICE EXERCISE

At 500 K the reaction $PCl_5(g) \rightleftharpoons PCl_3(g) + Cl_2(g)$ has $K_p = 0.497$. In an equilibrium mixture at 500 K, the partial pressure of PCl_5 is 0.860 atm and that of PCl_3 is 0.350 atm. What is the partial pressure of Cl_2 in the equilibrium mixture?
Answer: 1.22 atm

In many situations we will know the value of the equilibrium constant and the initial amounts of all species. We must then solve for the equilibrium amounts. Solving this type of problem usually entails treating the change in concentration as a variable. The stoichiometry of the reaction gives us the relationship between the changes in the amounts of all the reactants and products, as illustrated in Sample Exercise 15.12.

■ SAMPLE EXERCISE 15.12 | **Calculating Equilibrium Concentrations from Initial Concentrations**

A 1.000-L flask is filled with 1.000 mol of H_2 and 2.000 mol of I_2 at 448 °C. The value of the equilibrium constant K_c for the reaction

$$H_2(g) + I_2(g) \rightleftharpoons 2 HI(g)$$

at 448 °C is 50.5. What are the equilibrium concentrations of H_2, I_2, and HI in moles per liter?

SOLUTION

Analyze: We are given the volume of a container, an equilibrium constant, and starting amounts of reactants in the container and are asked to calculate the equilibrium concentrations of all species.

Plan: In this case we are not given any of the equilibrium concentrations. We must develop some relationships that relate the initial concentrations to those at equilibrium. The procedure is similar in many regards to that outlined in Sample Exercise 15.9, where we calculated an equilibrium constant using initial concentrations.

Solve: First, we note the initial concentrations of H_2 and I_2 in the 1.000-L flask:

$$[H_2] = 1.000 \, M \quad \text{and} \quad [I_2] = 2.000 \, M$$

Second, we construct a table in which we tabulate the initial concentrations:

	$H_2(g)$	$+$ $I_2(g)$	$\rightleftharpoons$ 2 HI(g)
Initial	1.000 M	2.000 M	0 M
Change			
Equilibrium			

Third, we use the stoichiometry of the reaction to determine the changes in concentration that occur as the reaction proceeds to equilibrium. The concentrations of H_2 and I_2 will decrease as equilibrium is established and that of HI will increase. Let's represent the change in concentration of H_2 by the variable x. The balanced chemical equation tells us the relationship between the changes in the concentrations of the three gases:

For each x mol of H_2 that reacts, x mol of I_2 are consumed and $2x$ mol of HI are produced:

	$H_2(g)$	$+$	$I_2(g)$	$\rightleftharpoons$	$2\ HI(g)$
Initial	1.000 M		2.000 M		0 M
Change	$-x$		$-x$		$+2x$
Equilibrium					

Fourth, we use the initial concentrations and the changes in concentrations, as dictated by stoichiometry, to express the equilibrium concentrations. With all our entries, our table now looks like this:

	$H_2(g)$	$+$	$I_2(g)$	$\rightleftharpoons$	$2\ HI(g)$
Initial	1.000 M		2.000 M		0 M
Change	$-x$		$-x$		$+2x$
Equilibrium	$(1.000 - x)\ M$		$(2.000 - x)\ M$		$2x\ M$

Fifth, we substitute the equilibrium concentrations into the equilibrium constant expression and solve for the unknown, x:

$$K_c = \frac{[HI]^2}{[H_2][I_2]} = \frac{(2x)^2}{(1.000 - x)(2.000 - x)} = 50.5$$

If you have an equation-solving calculator, you can solve this equation directly for x. If not, expand this expression to obtain a quadratic equation in x:

$$4x^2 = 50.5(x^2 - 3.000x + 2.000)$$
$$46.5x^2 - 151.5x + 101.0 = 0$$

Solving the quadratic equation (Appendix A.3) leads to two solutions for x:

$$x = \frac{-(-151.5) \pm \sqrt{1(-151.5)^2 - 4(46.5)(101.0)}}{2(46.5)} = 2.323 \text{ or } 0.935$$

When we substitute $x = 2.323$ into the expressions for the equilibrium concentrations, we find *negative* concentrations of H_2 and I_2. Because a negative concentration is not chemically meaningful, we reject this solution. We then use $x = 0.935$ to find the equilibrium concentrations:

$$[H_2] = 1.000 - x = 0.065\ M$$
$$[I_2] = 2.000 - x = 1.065\ M$$
$$[HI] = 2x = 1.87\ M$$

Check: We can check our solution by putting these numbers into the equilibrium-constant expression to assure that we correctly calculate the equilibrium constant:

$$K_c = \frac{[HI]^2}{[H_2][I_2]} = \frac{(1.87)^2}{(0.065)(1.065)} = 51$$

Comment: Whenever you use a quadratic equation to solve an equilibrium problem, one of the solutions will not be chemically meaningful and should be rejected.

◼ PRACTICE EXERCISE

For the equilibrium $PCl_5(g) \rightleftharpoons PCl_3(g) + Cl_2(g)$, the equilibrium constant K_p has the value 0.497 at 500 K. A gas cylinder at 500 K is charged with $PCl_5(g)$ at an initial pressure of 1.66 atm. What are the equilibrium pressures of PCl_5, PCl_3, and Cl_2 at this temperature?
Answer: $P_{PCl_5} = 0.967$ atm; $P_{PCl_3} = P_{Cl_2} = 0.693$ atm

▲ **Figure 15.10 Effect of temperature and pressure on the percentage of NH₃ in an equilibrium mixture of N₂, H₂, and NH₃.** Each mixture was produced by starting with a 3 : 1 molar mixture of H₂ and N₂. The yield of NH₃ is greatest at the lowest temperature and at the highest pressure.

15.7 LE CHÂTELIER'S PRINCIPLE

When Haber developed his process for making ammonia from N_2 and H_2, he sought the factors that might be varied to increase the yield of NH_3. Using the values of the equilibrium constant at various temperatures, he calculated the equilibrium amounts of NH_3 formed under a variety of conditions. Some of Haber's results are shown in Figure 15.10 ◄. Notice that the percent of NH_3 present at equilibrium decreases with increasing temperature and increases with increasing pressure. We can understand these effects in terms of a principle first put forward by Henri-Louis Le Châtelier* (1850–1936), a French industrial chemist. **Le Châtelier's principle** can be stated as follows: *If a system at equilibrium is disturbed by a change in temperature, pressure, or the concentration of one of the components, the system will shift its equilibrium position so as to counteract the effect of the disturbance.*

In this section we will use Le Châtelier's principle to make qualitative predictions about how a system at equilibrium responds to various changes in external conditions. We will consider three ways that a chemical equilibrium can be disturbed: (1) adding or removing a reactant or product, (2) changing the pressure by changing the volume, and (3) changing the temperature.

Change in Reactant or Product Concentrations

A system at equilibrium is in a dynamic state of balance. When the conditions of the equilibrium are altered, the equilibrium shifts until a new state of balance is attained. Le Châtelier's principle states that the shift will be in the direction that minimizes or reduces the effect of the change. Therefore, *if a chemical system is at equilibrium and we increase the concentration of a substance (either a reactant or a product), the system reacts to consume some of the substance. Conversely, if we decrease the concentration of a substance, the system reacts to produce some of the substance.*

As an example, consider an equilibrium mixture of N_2, H_2, and NH_3:

$$N_2(g) + 3\,H_2(g) \rightleftharpoons 2\,NH_3(g)$$

Adding H_2 would cause the system to shift so as to reduce the newly increased concentration of H_2. This change can occur only by consuming H_2 and simultaneously consuming N_2 to form more NH_3. This situation is illustrated in Figure 15.11 ◄. Adding more N_2 to the equilibrium mixture would likewise cause the direction of the reaction to shift toward forming more NH_3. Removing NH_3 would also cause a shift toward producing more NH_3, whereas *adding* NH_3 to the system at equilibrium would cause the concentrations to shift in the direction that reduces the newly increased NH_3 concentration. Some of the added ammonia would decompose to form N_2 and H_2.

In the Haber reaction, therefore, removing NH_3 from an equilibrium mixture of N_2, H_2, and NH_3 causes the reaction to shift from left to right to form more NH_3. If the NH_3 can be removed continuously, the yield of NH_3 can be increased dramatically. In the industrial production of ammonia, the NH_3 is continuously

▲ **Figure 15.11 Effect of adding H₂ to an equilibrium mixture of N₂, H₂, and NH₃.** When H₂ is added, a portion of the H₂ reacts with N₂ to form NH₃, thereby establishing a new equilibrium position that has the same equilibrium constant. The results shown are in accordance with Le Châtelier's principle.

*Pronounced "le-SHOT-lee-ay."

Pump to circulate
and compress gases

N_2, H_2
inlet

Expanding
gases cool

Recycled
unreacted
N_2 and H_2

Heat
exchanger

Heat
exchanger

Catalyst
(460–550 °C)

Refrigerated
unit

Heating
coil

Liquid NH_3

NH_3 outlet

Preheated
feed gases

◄ **Figure 15.12 Schematic diagram summarizing the industrial production of ammonia.** Incoming N_2 and H_2 gases are heated to approximately 500 °C and passed over a catalyst. The resultant gas mixture is allowed to expand and cool, causing NH_3 to liquefy. Unreacted N_2 and H_2 gases are recycled.

removed by selectively liquefying it; the boiling point of NH_3 (−33 °C) is much higher than that of N_2 (−196 °C) and H_2 (−253 °C). The liquid NH_3 is removed, and the N_2 and H_2 are recycled to form more NH_3, as diagrammed in Figure 15.12▲. By continuously removing the product, the reaction is driven essentially to completion.

GIVE IT SOME THOUGHT

What happens to the equilibrium $2\,NO(g) + O_2(g) \rightleftharpoons 2\,NO_2(g)$ if **(a)** O_2 is added to the system, **(b)** NO is removed?

Effects of Volume and Pressure Changes

If a system is at equilibrium and its volume is decreased, thereby increasing its total pressure, Le Châtelier's principle indicates that the system will respond by shifting its equilibrium position to reduce the pressure. A system can reduce its pressure by reducing the total number of gas molecules (fewer molecules of gas exert a lower pressure). Thus, at constant temperature, *reducing the volume of a gaseous equilibrium mixture causes the system to shift in the direction that reduces the number of moles of gas*. Conversely, increasing the volume causes a shift in the direction that produces more gas molecules.

For example, let's again consider the equilibrium $N_2O_4(g) \rightleftharpoons 2\,NO_2(g)$. What happens if the total pressure of an equilibrium mixture is increased by decreasing the volume as shown in the sequential photos in Figure 15.13▼? According to Le Châtelier's principle, we expect the equilibrium to shift to the side that reduces the total number of moles of gas, which is the reactant side in this case. (Notice the coefficients in the chemical equation; 1 mol of N_2O_4 appears on the reactant side and 2 mol NO_2 appears on the product side.) We therefore expect the equilibrium to shift to the left, so that NO_2 is

LE CHÂTELIER'S PRINCIPLE

If a system at equilibrium is disturbed by a change in temperature, pressure, or the concentration of one of the components, the system will shift its equilibrium position so as to counteract the effect of the disturbance. The equilibrium shown is $N_2O_4(g) \rightleftharpoons 2\,NO_2(g)$.

N₂O₄

NO₂

| An equilibrium mixture of brown $NO_2(g)$ (red) and colorless $N_2O_4(g)$ (gray) held in a gas-tight syringe. | The volume and hence the pressure are changed by moving the plunger. Compression of the mixture temporarily increases the concentration of NO_2. | When the mixture reestablishes equilibrium, the color is lighter than that at the beginning because the formation of $N_2O_4(g)$ is favored by the pressure increase. |

▲ **Figure 15.13 Effect of pressure on an equilibrium.** (The equilibrium shown is $N_2O_4(g) \rightleftharpoons 2\,NO_2(g)$.)

converted into N_2O_4 as equilibrium is reestablished. In Figure 15.13, compressing the gas mixture initially causes the color to darken as the concentration of NO_2 increases. The color then fades as equilibrium is reestablished. The color fades because the pressure increase causes the equilibrium to shift in favor of colorless N_2O_4.

GIVE IT SOME THOUGHT

What happens to the equilibrium $2\,SO_2(g) + O_2(g) \rightleftharpoons 2\,SO_3(g)$ if the volume of the system is increased?

For the reaction $N_2(g) + 3\,H_2(g) \rightleftharpoons 2\,NH_3(g)$, four molecules of reactant are consumed for every two molecules of product produced. Consequently, an increase in pressure (decrease in volume) causes a shift toward the side with fewer gas molecules, which leads to the formation of more NH_3, as indicated in Figure 15.10. In the case of the reaction $H_2(g) + I_2(g) \rightleftharpoons 2\,HI(g)$, the number of molecules of gaseous products (two) equals the number of molecules of gaseous reactants; therefore, changing the pressure will not influence the position of the equilibrium.

Keep in mind that pressure-volume changes do *not* change the value of K as long as the temperature remains constant. Rather, they change the partial pressures of the gaseous substances. In Sample Exercise 15.8 we calculated K_p for an equilibrium mixture at 472 °C that contained 7.38 atm H_2, 2.46 atm N_2, and 0.166 atm NH_3. The value of K_p is 2.79×10^{-5}. Consider what happens when we suddenly reduce the volume of the system by one-half. If there were no shift in equilibrium, this volume change would cause the partial pressures of all substances to double, giving $P_{H_2} = 14.76$ atm, $P_{N_2} = 4.92$ atm, and $P_{NH_3} = 0.332$ atm. The reaction quotient would then no longer equal the equilibrium constant.

$$Q_p = \frac{(P_{NH_3})^2}{P_{N_2}(P_{H_2})^3} = \frac{(0.332)^2}{(4.92)(14.76)^3} = 6.97 \times 10^{-6} \neq K_p$$

Because $Q_p < K_p$, the system is no longer at equilibrium. Equilibrium will be reestablished by increasing P_{NH_3} and decreasing P_{N_2} and P_{H_2} until $Q_p = K_p = 2.79 \times 10^{-5}$. Therefore, the equilibrium shifts to the right as Le Châtelier's principle predicts.

It is possible to change the total pressure of the system without changing its volume. For example, pressure increases if additional amounts of any of the reacting components are added to the system. We have already seen how to deal with a change in concentration of a reactant or product. The total pressure within the reaction vessel might also be increased by adding a gas that is not involved in the equilibrium. For example, argon might be added to the ammonia equilibrium system. The argon would not alter the partial pressures of any of the reacting components and therefore would not cause a shift in equilibrium.

Effect of Temperature Changes

Changes in concentrations or partial pressures cause shifts in equilibrium without changing the value of the equilibrium constant. In contrast, almost every equilibrium constant changes in value as the temperature changes. For example, consider the equilibrium established when cobalt(II) chloride ($CoCl_2$) is dissolved in hydrochloric acid, $HCl(aq)$:

$$\underset{\text{Pale pink}}{Co(H_2O)_6^{2+}(aq)} + 4\,Cl^-(aq) \rightleftharpoons \underset{\text{Deep blue}}{CoCl_4^{2-}(aq)} + 6\,H_2O(l) \qquad \Delta H > 0 \qquad [15.23]$$

The formation of $CoCl_4^{2-}$ from $Co(H_2O)_6^{2+}$ is an endothermic process. We will discuss the significance of this enthalpy change shortly. Because $Co(H_2O)_6^{2+}$ is pink and $CoCl_4^{2-}$ is blue, the position of this equilibrium is readily apparent from the color of the solution. Figure 15.14(left)▼ shows a room-temperature solution of $CoCl_2$ in $HCl(aq)$. Both $Co(H_2O)_6^{2+}$ and $CoCl_4^{2-}$ are present in significant amounts in the solution; the violet color results from the presence of both the pink and blue ions. When the solution is heated [Figure 15.14(middle)], it becomes intensely blue in color, indicating that the equilibrium has shifted to form more $CoCl_4^{2-}$. Cooling the solution, as in Figure 15.14(right), leads to a pink solution, indicating that the equilibrium has shifted to produce more $Co(H_2O)_6^{2+}$. How can we explain the dependence of this equilibrium on temperature?

We can deduce the rules for the temperature dependence of the equilibrium constant by applying Le Châtelier's principle. A simple way to do this is to treat heat as if it were a chemical reagent. In an *endothermic* (heat-absorbing) reaction we can consider heat as a *reactant*, whereas in an *exothermic* (heat-releasing) reaction we can consider heat as a *product*.

Endothermic: Reactants + *heat* $\rightleftharpoons$ products

Exothermic: Reactants $\rightleftharpoons$ products + *heat*

When the temperature of a system at equilibrium is increased, the system reacts as if we added a reactant to an endothermic reaction or a product to an exothermic reaction. The equilibrium shifts in the direction that consumes the excess reactant (or product), namely heat.

EFFECT OF TEMPERATURE CHANGES

Almost every equilibrium constant changes in value as the temperature changes. In an endothermic reaction, such as the one shown, heat is absorbed as reactants are converted to products. Increasing the temperature causes the equilibrium to shift to the right and K to increase. Lowering the temperature shifts the equilibrium in the direction that produces heat, to the left, decreasing K.

$$Co(H_2O)_6{}^{2+}(aq) + 4\,Cl^-(aq) \rightleftharpoons CoCl_4{}^{2-}(aq) + 6\,H_2O(l)$$

$Co(H_2O)_6{}^{2+}$

$CoCl_4{}^{2-}$

At room temperature both the pink $Co(H_2O)_6{}^{2+}$ and blue $CoCl_4{}^{2-}$ ions are present in significant amounts, giving a violet color to the solution.

Heating the solution shifts the equilibrium to the right, forming more blue $CoCl_4{}^{2-}$.

Cooling the solution shifts the equilibrium to the left, toward pink $Co(H_2O)_6{}^{2+}$.

▲ **Figure 15.14 Temperature and equilibrium.** (The reaction shown is $Co(H_2O)_6{}^{2+}(aq) + 4\,Cl^-(aq) \rightleftharpoons CoCl_4{}^{2-}(aq) + 6\,H_2O(l)$.)

▲ **GIVE IT SOME THOUGHT**

Use Le Châtelier's principle to explain why the equilibrium vapor pressure of a liquid increases with increasing temperature.

In an endothermic reaction, such as Equation 15.23, heat is absorbed as reactants are converted to products. Thus, increasing the temperature causes the equilibrium to shift to the right, in the direction of products, and K increases. For Equation 15.23, increasing the temperature leads to the formation of more $CoCl_4{}^{2-}$, as observed in Figure 15.14(b).

In an exothermic reaction the opposite occurs. Heat is absorbed as products are converted to reactants; therefore the equilibrium shifts to the left and K decreases. We can summarize these results as follows:

Endothermic: Increasing T results in an increase in K.

Exothermic: Increasing T results in a decrease in K.

Cooling a reaction has the opposite effect. As we lower the temperature, the equilibrium shifts to the side that produces heat. Thus, cooling an endothermic reaction shifts the equilibrium to the left, decreasing K. We observed this effect in Figure 15.14(c). Cooling an exothermic reaction shifts the equilibrium to the right, increasing K.

■ **SAMPLE EXERCISE 15.13** | **Using Le Châtelier's Principle to Predict Shifts in Equilibrium**

Consider the equilibrium

$$N_2O_4(g) \rightleftharpoons 2\,NO_2(g) \qquad \Delta H^\circ = 58.0 \text{ kJ}$$

In which direction will the equilibrium shift when **(a)** N_2O_4 is added, **(b)** NO_2 is removed, **(c)** the total pressure is increased by addition of $N_2(g)$, **(d)** the volume is increased, **(e)** the temperature is decreased?

SOLUTION

Analyze: We are given a series of changes to be made to a system at equilibrium and are asked to predict what effect each change will have on the position of the equilibrium.

Plan: Le Châtelier's principle can be used to determine the effects of each of these changes.

Solve:
(a) The system will adjust to decrease the concentration of the added N_2O_4, so the equilibrium shifts to the right, in the direction of products.
(b) The system will adjust to the removal of NO_2 by shifting to the side that produces more NO_2; thus, the equilibrium shifts to the right.
(c) Adding N_2 will increase the total pressure of the system, but N_2 is not involved in the reaction. The partial pressures of NO_2 and N_2O_4 are therefore unchanged, and there is no shift in the position of the equilibrium.
(d) If the volume is increased, the system will shift in the direction that occupies a larger volume (more gas molecules); thus, the equilibrium shifts to the right. (This is the opposite of the effect observed in Figure 15.13, where the volume was decreased.)
(e) The reaction is endothermic, so we can imagine heat as a reagent on the reactant side of the equation. Decreasing the temperature will shift the equilibrium in the direction that produces heat, so the equilibrium shifts to the left, toward the formation of more N_2O_4. Note that only this last change also affects the value of the equilibrium constant, K.

■ **PRACTICE EXERCISE**

For the reaction

$$PCl_5(g) \rightleftharpoons PCl_3(g) + Cl_2(g) \qquad \Delta H^\circ = 87.9 \text{ kJ}$$

in which direction will the equilibrium shift when **(a)** $Cl_2(g)$ is removed, **(b)** the temperature is decreased, **(c)** the volume of the reaction system is increased, **(d)** $PCl_3(g)$ is added?
Answers: **(a)** right, **(b)** left, **(c)** right, **(d)** left

---

■ **SAMPLE EXERCISE 15.14** | **Predicting the Effect of Temperature on K**

(a) Using the standard heat of formation data in Appendix C, determine the standard enthalpy change for the reaction

$$N_2(g) + 3\,H_2(g) \rightleftharpoons 2\,NH_3(g)$$

(b) Determine how the equilibrium constant for this reaction should change with temperature.

SOLUTION

Analyze: We are asked to determine the standard enthalpy change of a reaction and how the equilibrium constant for the reaction varies with temperature.

Plan: **(a)** We can use standard enthalpies of formation to calculate ΔH° for the reaction. **(b)** We can then use Le Châtelier's principle to determine what effect temperature will have on the equilibrium constant.

TABLE 15.2 ■ Variation in K_p for the Equilibrium $N_2 + 3 H_2 \rightleftharpoons 2 NH_3$ as a Function of Temperature

Temperature (°C)	K_p
300	4.34×10^{-3}
400	1.64×10^{-4}
450	4.51×10^{-5}
500	1.45×10^{-5}
550	5.38×10^{-6}
600	2.25×10^{-6}

Solve:

(a) Recall that the standard enthalpy change for a reaction is given by the sum of the standard molar enthalpies of formation of the products, each multiplied by its coefficient in the balanced chemical equation, less the same quantities for the reactants. At 25 °C, ΔH_f° for $NH_3(g)$ is -46.19 kJ/mol. The ΔH_f° values for $H_2(g)$ and $N_2(g)$ are zero by definition because the enthalpies of formation of the elements in their normal states at 25 °C are defined as zero (Section 5.7). Because 2 mol of NH_3 is formed, the total enthalpy change is

$$(2 \text{ mol})(-46.19 \text{ kJ/mol}) - 0 = -92.38 \text{ kJ}$$

(b) Because the reaction in the forward direction is exothermic, we can consider heat a product of the reaction. An increase in temperature causes the reaction to shift in the direction of less NH_3 and more N_2 and H_2. This effect is seen in the values for K_p presented in Table 15.2 ◄. Notice that K_p changes markedly with changes in temperature and that it is larger at lower temperatures.

Comment: The fact that K_p for the formation of NH_3 from N_2 and H_2 decreases with increasing temperature is a matter of great practical importance. To form NH_3 at a reasonable rate requires higher temperatures. At higher temperatures, however, the equilibrium constant is smaller, and so the percentage conversion to NH_3 is smaller. To compensate for this, higher pressures are needed because high pressure favors NH_3 formation.

■ PRACTICE EXERCISE

Using the thermodynamic data in Appendix C, determine the enthalpy change for the reaction

$$2 POCl_3(g) \rightleftharpoons 2 PCl_3(g) + O_2(g)$$

Use this result to determine how the equilibrium constant for the reaction should change with temperature.
Answer: $\Delta H° = 508.3$ kJ; the equilibrium constant will increase with increasing temperature

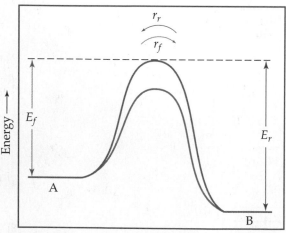

▲ **Figure 15.15 Effect of a catalyst on equilibrium.** At equilibrium for the hypothetical reaction A $\rightleftharpoons$ B, the forward reaction rate, r_f, equals the reverse reaction rate, r_r. The violet curve represents the path over the transition state in the absence of a catalyst. A catalyst lowers the energy of the transition state, as shown by the green curve. Thus, the activation energy is lowered for both the forward and the reverse reactions. As a result, the rates of forward and reverse reactions in the catalyzed reaction are increased.

The Effect of Catalysts

What happens if we add a catalyst to a chemical system that is at equilibrium? As shown in Figure 15.15 ◄, a catalyst lowers the activation barrier between the reactants and products. The activation energy of the forward reaction is lowered to the same extent as that for the reverse reaction. The catalyst thereby increases the rates of both the forward and reverse reactions. As a result, *a catalyst increases the rate at which equilibrium is achieved, but it does not change the composition of the equilibrium mixture.* The value of the equilibrium constant for a reaction is not affected by the presence of a catalyst.

The rate at which a reaction approaches equilibrium is an important practical consideration. As an example, let's again consider the synthesis of ammonia from N_2 and H_2. In designing a process for ammonia synthesis, Haber had to deal with a rapid decrease in the equilibrium constant with increasing temperature, as shown in Table 15.2. At temperatures sufficiently high to give a satisfactory reaction rate, the amount of ammonia formed was too small. The solution to this dilemma was to develop a catalyst that would produce a reasonably rapid approach to equilibrium at a sufficiently low temperature, so that the equilibrium constant was still reasonably large. The development of a suitable catalyst thus became the focus of Haber's research efforts.

After trying different substances to see which would be most effective, Haber finally settled on iron mixed with metal oxides. Variants of the original catalyst formulations are still used. These catalysts make it possible to obtain a reasonably rapid approach to equilibrium at temperatures around 400 °C to 500 °C and with gas pressures of 200 to 600 atm. The high pressures are needed to obtain a satisfactory degree of conversion at equilibrium. You can see from

Figure 15.10 that if an improved catalyst could be found—one that would lead to sufficiently rapid reaction at temperatures lower than 400 °C to 500 °C—it would be possible to obtain the same degree of equilibrium conversion at much lower pressures. This would result in great savings in the cost of equipment for ammonia synthesis. In view of the growing need for nitrogen as fertilizer, the fixation of nitrogen is a process of ever-increasing importance.

GIVE IT SOME THOUGHT

Does the addition of a catalyst have any effect on the position of an equilibrium?

■ SAMPLE INTEGRATIVE EXERCISE | Putting Concepts Together

At temperatures near 800 °C, steam passed over hot coke (a form of carbon obtained from coal) reacts to form CO and H_2:

$$C(s) + H_2O(g) \rightleftharpoons CO(g) + H_2(g)$$

The mixture of gases that results is an important industrial fuel called *water gas*. **(a)** At 800 °C the equilibrium constant for this reaction is $K_p = 14.1$. What are the equilibrium partial pressures of H_2O, CO, and H_2 in the equilibrium mixture at this temperature if we start with solid carbon and 0.100 mol of H_2O in a 1.00-L vessel? **(b)** What is the minimum amount of carbon required to achieve equilibrium under these conditions? **(c)** What is the total pressure in the vessel at equilibrium? **(d)** At 25 °C the value of K_p for this reaction is 1.7×10^{-21}. Is the reaction exothermic or endothermic? **(e)** To produce the maximum amount of CO and H_2 at equilibrium, should the pressure of the system be increased or decreased?

SOLUTION

(a) To determine the equilibrium partial pressures, we use the ideal gas equation, first determining the starting partial pressure of hydrogen.

$$P_{H_2O} = \frac{n_{H_2O}RT}{V} = \frac{(0.100 \text{ mol})(0.0821 \text{ L-atm/mol-K})(1073 \text{ K})}{1.00 \text{ L}} = 8.81 \text{ atm}$$

We then construct a table of starting partial pressures and their changes as equilibrium is achieved:

	C(s)	+	$H_2O(g)$	$\rightleftharpoons$	CO(g)	+	$H_2(g)$
Initial			8.81 atm		0 atm		0 atm
Change			$-x$		$+x$		$+x$
Equilibrium			$8.81 - x$ atm		x atm		x atm

There are no entries in the table under C(s) because the reactant, being a solid, does not appear in the equilibrium-constant expression. Substituting the equilibrium partial pressures of the other species into the equilibrium-constant expression for the reaction gives

$$K_p = \frac{P_{CO}P_{H_2}}{P_{H_2O}} = \frac{(x)(x)}{(8.81 - x)} = 14.1$$

Multiplying through by the denominator gives a quadratic equation in x:

$$x^2 = (14.1)(8.81 - x)$$

$$x^2 + 14.1x - 124.22 = 0$$

Solving this equation for x using the quadratic formula yields $x = 6.14$ atm. Hence, the equilibrium partial pressures are $P_{CO} = x = 6.14$ atm, $P_{H_2} = x = 6.14$ atm, and $P_{H_2O} = (8.81 - x) = 2.67$ atm.
(b) Part (a) shows that $x = 6.14$ atm of H_2O must react for the system to achieve equilibrium. We can use the ideal-gas equation to convert this partial pressure into a mole amount.

$$n = \frac{PV}{RT} = \frac{(6.14 \text{ atm})(1.00 \text{ L})}{(0.0821 \text{ L-atm/mol-K})(1073 \text{ K})} = 0.0697 \text{ mol}$$

Thus, 0.0697 mol of H_2O and the same amount of C must react to achieve equilibrium. As a result, there must be at least 0.0697 mol of C (0.836 g C) present among the reactants at the start of the reaction.

(c) The total pressure in the vessel at equilibrium is simply the sum of the equilibrium partial pressures:

$$P_{total} = P_{H_2O} + P_{CO} + P_{H_2} = 2.67 \text{ atm} + 6.14 \text{ atm} + 6.14 \text{ atm} = 14.95 \text{ atm}$$

(d) In discussing Le Châtelier's principle, we saw that endothermic reactions exhibit an increase in K_p with increasing temperature. Because the equilibrium constant for this reaction increases as temperature increases, the reaction must be endothermic. From the enthalpies of formation given in Appendix C, we can verify our prediction by calculating the enthalpy change for the reaction, $\Delta H° = \Delta H_f°(CO) + \Delta H_f°(H_2) - \Delta H°(C) - \Delta H_f°(H_2O) = +131.3 \text{ kJ}$. The positive sign for $\Delta H°$ indicates that the reaction is endothermic.

(e) According to Le Châtelier's principle, a decrease in the pressure causes a gaseous equilibrium to shift toward the side of the equation with the greater number of moles of gas. In this case there are two moles of gas on the product side and only one on the reactant side. Therefore, the pressure should be reduced to maximize the yield of the CO and H_2.

Chemistry Put to Work CONTROLLING NITRIC OXIDE EMISSIONS

The formation of NO from N_2 and O_2 provides another interesting example of the practical importance of changes in the equilibrium constant and reaction rate with temperature. The equilibrium equation and the standard enthalpy change for the reaction are

$$\tfrac{1}{2} N_2(g) + \tfrac{1}{2} O_2(g) \rightleftharpoons NO(g) \qquad \Delta H° = 90.4 \text{ kJ} \qquad [15.24]$$

The reaction is endothermic; that is, heat is absorbed when NO is formed from the elements. By applying Le Châtelier's principle, we deduce that an increase in temperature will shift the equilibrium in the direction of more NO. The equilibrium constant K_p for formation of 1 mol of NO from the elements at 300 K is only about 10^{-15}. In contrast, at a much higher temperature of about 2400 K the equilibrium constant is 10^{13} times as large, about 0.05. The manner in which K_p for Equation 15.24 varies with temperature is shown in Figure 15.16▶.

This graph helps to explain why NO is a pollution problem. In the cylinder of a modern high-compression auto engine, the temperatures during the fuel-burning part of the cycle may be approximately 2400 K. Also, there is a fairly large excess of air in the cylinder. These conditions favor the formation of some NO. After the combustion, however, the gases are quickly cooled. As the temperature drops, the equilibrium in Equation 15.24 shifts strongly to the left (that is, in the direction of N_2 and O_2). The lower temperatures also mean that the rate of the reaction is decreased, however, so the NO formed at high temperatures is essentially "frozen" in that form as the gas cools.

The gases exhausting from the cylinder are still quite hot, perhaps 1200 K. At this temperature, as shown in Figure 15.16, the equilibrium constant for formation of NO is much smaller. However, the rate of conversion of NO to N_2 and O_2 is too slow to permit much loss of NO before the gases are cooled still further.

As discussed in the "Chemistry Put to Work" box in Section 14.7, one of the goals of automotive catalytic converters is to achieve the rapid conversion of NO to N_2 and O_2 at the temperature of the exhaust gas. Some catalysts for this reaction have been developed that are reasonably effective under the grueling conditions found in automotive exhaust systems. Nevertheless, scientists and engineers are continually searching for new materials that provide even more effective catalysis of the decomposition of nitrogen oxides.

▲ **Figure 15.16 Equilibrium and temperature.** The graph shows how the equilibrium constant for the reaction $\tfrac{1}{2} N_2(g) + \tfrac{1}{2} O_2(g) \rightleftharpoons NO(g)$ varies as a function of temperature. The equilibrium constant increases with increasing temperature because the reaction is endothermic. It is necessary to use a log scale for K_p because the values vary over such a large range.

CHAPTER REVIEW

SUMMARY AND KEY TERMS

Introduction and Section 15.1 A chemical reaction can achieve a state in which the forward and reverse processes are occurring at the same rate. This condition is called **chemical equilibrium**, and it results in the formation of an equilibrium mixture of the reactants and products of the reaction. The composition of an equilibrium mixture does not change with time.

Section 15.2 An equilibrium that is used throughout this chapter is the reaction $N_2(g) + 3 H_2(g) \rightleftharpoons 2 NH_3(g)$. This reaction is the basis of the **Haber process** for the production of ammonia. The relationship between the concentrations of the reactants and products of a system at equilibrium is given by the **law of mass action**. For an equilibrium equation of the form $a A + b B \rightleftharpoons d D + e E$, the **equilibrium-constant expression** is written as

$$K_c = \frac{[D]^d [E]^e}{[A]^a [B]^b}$$

where K_c is a constant called the **equilibrium constant**. When the equilibrium system of interest consists of gases, it is often convenient to express the concentrations of reactants and products in terms of gas pressures:

$$K_p = \frac{(P_D)^d (P_E)^e}{(P_A)^a (P_B)^b}$$

K_c and K_p are related by the expression $K_p = K_c(RT)^{\Delta n}$.

Section 15.3 The value of the equilibrium constant changes with temperature. A large value of K_c indicates that the equilibrium mixture contains more products than reactants and therefore lies toward the product side of the equation. A small value for the equilibrium constant means that the equilibrium mixture contains less products than reactants and therefore lies toward the reactant side. The equilibrium-constant expression and the equilibrium constant of the reverse of a reaction are the reciprocals of those of the forward reaction. If a reaction is the sum of two or more reactions, its equilibrium constant will be the product of the equilibrium constants for the individual reactions.

Section 15.4 Equilibria for which all substances are in the same phase are called **homogeneous equilibria**; in **heterogeneous equilibria** two or more phases are present. The concentrations of pure solids and liquids are left out of the equilibrium-constant expression for a heterogeneous equilibrium.

Section 15.5 If the concentrations of all species in an equilibrium are known, the equilibrium-constant expression can be used to calculate the value of the equilibrium constant. The changes in the concentrations of reactants and products on the way to achieving equilibrium are governed by the stoichiometry of the reaction.

Section 15.6 The **reaction quotient**, Q, is found by substituting reactant and product concentrations or partial pressures at any point during a reaction into the equilibrium-constant expression. If the system is at equilibrium, $Q = K$. If $Q \neq K$, however, the system is not at equilibrium. When $Q < K$, the reaction will move toward equilibrium by forming more products (the reaction moves from left to right); when $Q > K$, the reaction will proceed from right to left. Knowing the value of K makes it possible to calculate the equilibrium amounts of reactants and products, often by the solution of an equation in which the unknown is the change in a partial pressure or concentration.

Section 15.7 **Le Châtelier's principle** states that if a system at equilibrium is disturbed, the equilibrium will shift to minimize the disturbing influence. By this principle, if a reactant or product is added to a system at equilibrium, the equilibrium will shift to consume the added substance. The effects of removing reactants or products and of changing the pressure or volume of a reaction can be similarly deduced. For example, if the volume of the system is reduced, the equilibrium will shift in the direction that decreased the number of gas molecules. The enthalpy change for a reaction indicates how an increase in temperature affects the equilibrium: For an endothermic reaction, an increase in temperature shifts the equilibrium to the right; for an exothermic reaction, a temperature increase shifts the equilibrium to the left. Catalysts affect the speed at which equilibrium is reached but do not affect the magnitude of K.

KEY SKILLS

- Understand what is meant by chemical equilibrium and how it relates to reaction rates.
- Write the equilibrium-constant expression for any reaction.
- Relate K_c and K_p.
- Relate the magnitude of an equilibrium constant to the relative amounts of reactants and products present in an equilibrium mixture.
- Manipulate the equilibrium constant to reflect changes in the chemical equation.

- Write the equilibrium-constant expression for a heterogeneous reaction.
- Calculate an equilibrium constant from concentration measurements.
- Predict the direction of a reaction given the equilibrium constant and the concentrations of reactants and products.
- Calculate equilibrium concentrations given the equilibrium constant and all but one equilibrium concentration.
- Calculate equilibrium concentrations given the equilibrium constant and the starting concentrations.
- Understand how changing the concentrations, volume, or temperature of a system at equilibrium affects the equilibrium position.

KEY EQUATIONS

- $K_c = \dfrac{[D]^d[E]^e}{[A]^a[B]^b}$ [15.8]   The equilibrium-constant expression for a general reaction of the type $a\,A + b\,B \rightleftharpoons d\,D + e\,E$; the concentrations are equilibrium concentrations only

- $K_p = \dfrac{(P_D)^d(P_E)^e}{(P_A)^a(P_B)^b}$ [15.11]   The equilibrium-constant expression in terms of equilibrium partial pressures

- $K_p = K_c(RT)^{\Delta n}$ [15.14]   Relating the equilibrium constant based on pressures to the equilibrium constant based on concentration

- $Q_c = \dfrac{[D]^d[E]^e}{[A]^a[B]^b}$ [15.22]   The reaction quotient. The concentrations are for any time during a reaction. If the concentrations are equilibrium concentrations, then $Q_c = K_c$.

VISUALIZING CONCEPTS

15.1 (a) Based on the following energy profile, predict whether $k_f > k_r$ or $k_f < k_r$. (b) Using Equation 15.5, predict whether the equilibrium constant for the process is greater than 1 or less than 1. [Section 15.1]

Reaction pathway

15.2 The following diagrams represent a hypothetical reaction A $\longrightarrow$ B, with A represented by red spheres and B represented by blue spheres. The sequence from left to right represents the system as time passes. Do the diagrams indicate that the system reaches an equilibrium state? Explain. [Sections 15.1 and 15.2]

15.3 The following diagram represents an equilibrium mixture produced for a reaction of the type A + X $\rightleftharpoons$ AX. If the volume is 1 L, is K greater or smaller than 1? [Section 15.2]

15.4 The following diagram represents a reaction shown going to completion. (a) Letting A = red spheres and B = blue spheres, write a balanced equation for the reaction. (b) Write the equilibrium-constant expression for the reaction. (c) Assuming that all of the molecules are in the gas phase, calculate Δn, the change in the number of gas molecules that accompanies the reaction. (d) How can you calculate K_p if you know K_c at a particular temperature? [Section 15.2]

15.5 Ethene (C_2H_4) reacts with halogens (X_2) by the following reaction:

$$C_2H_4(g) + X_2(g) \rightleftharpoons C_2H_4X_2(g)$$

The following figures represent the concentrations at equilibrium at the same temperature when X_2 is Cl_2 (green), Br_2 (brown), and I_2 (purple). List the equilibria from smallest to largest equilibrium constant. [Section 15.3]

(a)   (b)

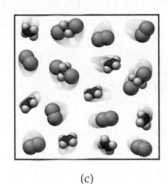

(c)

15.6 The reaction $A_2 + B_2 \rightleftharpoons 2$ AB has an equilibrium constant $K_c = 1.5$. The following diagrams represent reaction mixtures containing A_2 molecules (red), B_2 molecules (blue), and AB molecules. **(a)** Which reaction mixture is at equilibrium? **(b)** For those mixtures that are not at equilibrium, how will the reaction proceed to reach equilibrium? [Sections 15.5 and 15.6]

(i)   (ii)   (iii)

15.7 The reaction $A_2(g) + B(g) \rightleftharpoons A(g) + AB(g)$ has an equilibrium constant of $K_p = 2$. The accompanying diagram shows a mixture containing A atoms (red), A_2 molecules, and AB molecules (red and blue). How many B atoms should be added to the diagram to illustrate an equilibrium mixture? [Section 15.6]

15.8 The following diagram represents the equilibrium state for the reaction $A_2(g) + 2$ B(g) $\rightleftharpoons$ 2 AB(g). **(a)** Assuming the volume is 1 L, calculate the equilibrium constant, K_c, for the reaction. **(b)** If the volume of the equilibrium mixture is decreased, will the number of AB molecules increase or decrease? [Sections 15.5 and 15.7]

15.9 The following diagrams represent equilibrium mixtures for the reaction $A_2 + B \rightleftharpoons A + AB$ at (1) 300 K and (2) 500 K. The A atoms are red, and the B atoms are blue. Is the reaction exothermic or endothermic? [Section 15.7]

(a)   (b)

15.10 The following graph represents the yield of the compound AB at equilibrium in the reaction A(g) + B(g) $\longrightarrow$ AB(g).

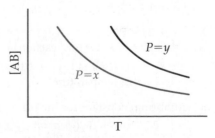

(a) Is this reaction exothermic or endothermic? **(b)** Is $P = x$ greater or smaller than $P = y$? [Section 15.7]

EXERCISES

Equilibrium; The Equilibrium Constant

15.11 Suppose that the gas-phase reactions A $\longrightarrow$ B and B $\longrightarrow$ A are both elementary processes with rate constants of $3.8 \times 10^{-2}\,s^{-1}$ and $3.1 \times 10^{-1}\,s^{-1}$, respectively. **(a)** What is the value of the equilibrium constant for the equilibrium A(g) $\rightleftharpoons$ B(g)? **(b)** Which is greater at equilibrium, the partial pressure of A or the partial pressure of B? Explain.

15.12 Consider the reaction A + B $\rightleftharpoons$ C + D. Assume that both the forward reaction and the reverse reaction are elementary processes and that the value of the equilibrium constant is very large. **(a)** Which species predominate at equilibrium, reactants or products? **(b)** Which reaction has the larger rate constant, the forward or the reverse? Explain.

15.13 Write the expression for K_c for the following reactions. In each case indicate whether the reaction is homogeneous or heterogeneous.
(a) 3 NO(g) $\rightleftharpoons$ N$_2$O(g) + NO$_2$(g)
(b) CH$_4$(g) + 2 H$_2$S(g) $\rightleftharpoons$ CS$_2$(g) + 4 H$_2$(g)
(c) Ni(CO)$_4$(g) $\rightleftharpoons$ Ni(s) + 4 CO(g)
(d) HF(aq) $\rightleftharpoons$ H$^+$(aq) + F$^-$(aq)
(e) 2 Ag(s) + Zn^{2+}(aq) $\rightleftharpoons$ 2 Ag$^+$(aq) + Zn(s)

15.14 Write the expressions for K_c for the following reactions. In each case indicate whether the reaction is homogeneous or heterogeneous.
(a) 2 O$_3$(g) $\rightleftharpoons$ 3 O$_2$(g)
(b) Ti(s) + 2 Cl$_2$(g) $\rightleftharpoons$ TiCl$_4$(l)
(c) 2 C$_2$H$_4$(g) + 2 H$_2$O(g) $\rightleftharpoons$ 2 C$_2$H$_6$(g) + O$_2$(g)
(d) C(s) + 2 H$_2$(g) $\rightleftharpoons$ CH$_4$(g)
(e) 4 HCl(aq) + O$_2$(g) $\rightleftharpoons$ 2 H$_2$O(l) + 2 Cl$_2$(g)

15.15 When the following reactions come to equilibrium, does the equilibrium mixture contain mostly reactants or mostly products?
(a) N$_2$(g) + O$_2$(g) $\rightleftharpoons$ 2 NO(g); $K_c = 1.5 \times 10^{-10}$
(b) 2 SO$_2$(g) + O$_2$(g) $\rightleftharpoons$ 2 SO$_3$(g); $K_p = 2.5 \times 10^9$

15.16 Which of the following reactions lies to the right, favoring the formation of products, and which lies to the left, favoring formation of reactants?
(a) 2 NO(g) + O$_2$(g) $\rightleftharpoons$ 2 NO$_2$(g); $K_p = 5.0 \times 10^{12}$
(b) 2 HBr(g) $\rightleftharpoons$ H$_2$(g) + Br$_2$(g); $K_c = 5.8 \times 10^{-18}$

15.17 If $K_c = 0.042$ for PCl$_3$(g) + Cl$_2$(g) $\rightleftharpoons$ PCl$_5$(g) at 500 K, what is the value of K_p for this reaction at this temperature?

15.18 Calculate K_c at 303 K for SO$_2$(g) + Cl$_2$(g) $\rightleftharpoons$ SO$_2$Cl$_2$(g) if $K_p = 34.5$ at this temperature.

15.19 The equilibrium constant for the reaction
$$2\,NO(g) + Br_2(g) \rightleftharpoons 2\,NOBr(g)$$

is $K_c = 1.3 \times 10^{-2}$ at 1000 K. **(a)** Calculate K_c for 2 NOBr(g) $\rightleftharpoons$ 2 NO(g) + Br$_2$(g). **(b)** At this temperature does the equilibrium favor NO and Br$_2$, or does it favor NOBr?

15.20 Consider the following equilibrium:
$$2\,H_2(g) + S_2(g) \rightleftharpoons 2\,H_2S(g) \quad K_c = 1.08 \times 10^7 \text{ at 700 °C}$$
(a) Calculate K_p. **(b)** Does the equilibrium mixture contain mostly H$_2$ and S$_2$ or mostly H$_2$S?

15.21 At 1000 K, $K_p = 1.85$ for the reaction
$$SO_2(g) + \tfrac{1}{2}O_2(g) \rightleftharpoons SO_3(g)$$
(a) What is the value of K_p for the reaction SO$_3$(g) $\rightleftharpoons$ SO$_2$(g) + $\frac{1}{2}$O$_2$(g)? **(b)** What is the value of K_p for the reaction 2 SO$_2$(g) + O$_2$(g) $\rightleftharpoons$ 2 SO$_3$(g)? **(c)** What is the value of K_c for the reaction in part (b)?

15.22 Consider the following equilibrium, for which $K_p = 0.0752$ at 480 °C:
$$2\,Cl_2(g) + 2\,H_2O(g) \rightleftharpoons 4\,HCl(g) + O_2(g)$$
(a) What is the value of K_p for the reaction 4 HCl(g) + O$_2$(g) $\rightleftharpoons$ 2 Cl$_2$(g) + 2 H$_2$O(g)? **(b)** What is the value of K_p for the reaction Cl$_2$(g) + H$_2$O(g) $\rightleftharpoons$ 2 HCl(g) + $\frac{1}{2}$O$_2$(g)? **(c)** What is the value of K_c for the reaction in part (b)?

15.23 The following equilibria were attained at 823 K:
$$CoO(s) + H_2(g) \rightleftharpoons Co(s) + H_2O(g) \quad K_c = 67$$
$$CoO(s) + CO(g) \rightleftharpoons Co(s) + CO_2(g) \quad K_c = 490$$
Based on these equilibria, calculate the equilibrium constant for H$_2$(g) + CO$_2$(g) $\rightleftharpoons$ CO(g) + H$_2$O(g) at 823 K.

15.24 Consider the equilibrium
$$N_2(g) + O_2(g) + Br_2(g) \rightleftharpoons 2\,NOBr(g)$$
Calculate the equilibrium constant K_p for this reaction, given the following information (at 298 K):
$$2\,NO(g) + Br_2(g) \rightleftharpoons 2\,NOBr(g) \quad K_c = 2.0$$
$$2\,NO(g) \rightleftharpoons N_2(g) + O_2(g) \quad K_c = 2.1 \times 10^{30}$$

15.25 Mercury(I) oxide decomposes into elemental mercury and elemental oxygen: 2 Hg$_2$O(s) $\rightleftharpoons$ 4 Hg(l) + O$_2$(g) **(a)** Write the equilibrium-constant expression for this reaction in terms of partial pressures. **(b)** Explain why we normally exclude pure solids and liquids from equilibrium-constant expressions.

15.26 Consider the equilibrium Na$_2$O(s) + SO$_2$(g) $\rightleftharpoons$ Na$_2$SO$_3$(s). **(a)** Write the equilibrium-constant expression for this reaction in terms of partial pressures. **(b)** Why doesn't the concentration of Na$_2$O appear in the equilibrium-constant expression?

Calculating Equilibrium Constants

15.27 Gaseous hydrogen iodide is placed in a closed container at 425 °C, where it partially decomposes to hydrogen and iodine: 2 HI(g) $\rightleftharpoons$ H$_2$(g) + I$_2$(g). At equilibrium it is found that [HI] $= 3.53 \times 10^{-3}\,M$, [H$_2$] $= 4.79 \times 10^{-4}\,M$, and [I$_2$] $= 4.79 \times 10^{-4}\,M$. What is the value of K_c at this temperature?

15.28 Methanol (CH_3OH) is produced commercially by the catalyzed reaction of carbon monoxide and hydrogen: $CO(g) + 2 H_2(g) \rightleftharpoons CH_3OH(g)$. An equilibrium mixture in a 2.00-L vessel is found to contain 0.0406 mol CH_3OH, 0.170 mol CO, and 0.302 mol H_2 at 500 K. Calculate K_c at this temperature.

15.29 The equilibrium $2 NO(g) + Cl_2(g) \rightleftharpoons 2 NOCl(g)$ is established at 500 K. An equilibrium mixture of the three gases has partial pressures of 0.095 atm, 0.171 atm, and 0.28 atm for NO, Cl_2, and NOCl, respectively. Calculate K_p for this reaction at 500 K.

15.30 Phosphorus trichloride gas and chlorine gas react to form phosphorus pentachloride gas: $PCl_3 + Cl_2(g) \rightleftharpoons PCl_5(g)$. A gas vessel is charged with a mixture of $PCl_3(g)$ and $Cl_2(g)$, which is allowed to equilibrate at 450 K. At equilibrium the partial pressures of the three gases are $P_{PCl_3} = 0.124$ atm, $P_{Cl_2} = 0.157$ atm, and $P_{PCl_5} = 1.30$ atm. **(a)** What is the value of K_p at this temperature? **(b)** Does the equilibrium favor reactants or products?

15.31 A mixture of 0.10 mol of NO, 0.050 mol of H_2, and 0.10 mol of H_2O is placed in a 1.0-L vessel at 300 K. The following equilibrium is established:

$$2 NO(g) + 2 H_2(g) \rightleftharpoons N_2(g) + 2 H_2O(g)$$

At equilibrium [NO] = 0.062 M. **(a)** Calculate the equilibrium concentrations of H_2, N_2, and H_2O. **(b)** Calculate K_c.

15.32 A mixture of 1.374 g of H_2 and 70.31 g of Br_2 is heated in a 2.00-L vessel at 700 K. These substances react as follows:

$$H_2(g) + Br_2(g) \rightleftharpoons 2 HBr(g)$$

At equilibrium the vessel is found to contain 0.566 g of H_2. **(a)** Calculate the equilibrium concentrations of H_2, Br_2, and HBr. **(b)** Calculate K_c.

15.33 A mixture of 0.2000 mol of CO_2, 0.1000 mol of H_2, and 0.1600 mol of H_2O is placed in a 2.000-L vessel. The following equilibrium is established at 500 K:

$$CO_2(g) + H_2(g) \rightleftharpoons CO(g) + H_2O(g)$$

(a) Calculate the initial partial pressures of CO_2, H_2, and H_2O. **(b)** At equilibrium $P_{H_2O} = 3.51$ atm. Calculate the equilibrium partial pressures of CO_2, H_2, and CO. **(c)** Calculate K_p for the reaction.

15.34 A flask is charged with 1.500 atm of $N_2O_4(g)$ and 1.00 atm $NO_2(g)$ at 25 °C, and the following equilibrium is achieved:

$$N_2O_4(g) \rightleftharpoons 2 NO_2(g)$$

After equilibrium is reached, the partial pressure of NO_2 is 0.512 atm. **(a)** What is the equilibrium partial pressure of N_2O_4? **(b)** Calculate the value of K_p for the reaction.

Applications of Equilibrium Constants

15.35 **(a)** How does a reaction quotient differ from an equilibrium constant? **(b)** If $Q_c < K_c$, in which direction will a reaction proceed in order to reach equilibrium? **(c)** What condition must be satisfied so that $Q_c = K_c$?

15.36 **(a)** How is a reaction quotient used to determine whether a system is at equilibrium? **(b)** If $Q_c > K_c$, how must the reaction proceed to reach equilibrium? **(c)** At the start of a certain reaction, only reactants are present; no products have been formed. What is the value of Q_c at this point in the reaction?

15.37 At 100 °C the equilibrium constant for the reaction $COCl_2(g) \rightleftharpoons CO(g) + Cl_2(g)$ has the value $K_c = 2.19 \times 10^{-10}$. Are the following mixtures of $COCl_2$, CO, and Cl_2 at 100 °C at equilibrium? If not, indicate the direction that the reaction must proceed to achieve equilibrium. **(a)** $[COCl_2] = 2.00 \times 10^{-3}$ M, $[CO] = 3.3 \times 10^{-6}$ M, $[Cl_2] = 6.62 \times 10^{-6}$ M; **(b)** $[COCl_2] = 4.50 \times 10^{-2}$ M, $[CO] = 1.1 \times 10^{-7} M$, $[Cl_2] = 2.25 \times 10^{-6} M$; **(c)** $[COCl_2] = 0.0100$ M, $[CO] = [Cl_2] = 1.48 \times 10^{-6}$ M

15.38 As shown in Table 15.2, K_p for the equilibrium

$$N_2(g) + 3 H_2(g) \rightleftharpoons 2 NH_3(g)$$

is 4.51×10^{-5} at 450 °C. For each of the mixtures listed here, indicate whether the mixture is at equilibrium at 450 °C. If it is not at equilibrium, indicate the direction (toward product or toward reactants) in which the mixture must shift to achieve equilibrium.
(a) 98 atm NH_3, 45 atm N_2, 55 atm H_2
(b) 57 atm NH_3, 143 atm N_2, no H_2
(c) 13 atm NH_3, 27 atm N_2, 82 atm H_2.

15.39 At 100 °C, $K_c = 0.078$ for the reaction

$$SO_2Cl_2(g) \rightleftharpoons SO_2(g) + Cl_2(g)$$

In an equilibrium mixture of the three gases, the concentrations of SO_2Cl_2 and SO_2 are 0.108 M and 0.052 M, respectively. What is the partial pressure of Cl_2 in the equilibrium mixture?

15.40 At 900 K the following reaction has $K_p = 0.345$:

$$2 SO_2(g) + O_2(g) \rightleftharpoons 2 SO_3(g)$$

In an equilibrium mixture the partial pressures of SO_2 and O_2 are 0.135 atm and 0.455 atm, respectively. What is the equilibrium partial pressure of SO_3 in the mixture?

15.41 **(a)** At 1285 °C the equilibrium constant for the reaction $Br_2(g) \rightleftharpoons 2 Br(g)$ is $K_c = 1.04 \times 10^{-3}$. A 0.200-L vessel containing an equilibrium mixture of the gases has 0.245 g $Br_2(g)$ in it. What is the mass of Br(g) in the vessel? **(b)** For the reaction $H_2(g) + I_2(g) \rightleftharpoons 2 HI(g)$, $K_c = 55.3$ at 700 K. In a 2.00-L flask containing an equilibrium mixture of the three gases, there are 0.056 g H_2 and 4.36 g I_2. What is the mass of HI in the flask?

15.42 **(a)** At 800 K the equilibrium constant for $I_2(g) \rightleftharpoons 2 I(g)$ is $K_c = 3.1 \times 10^{-5}$. If an equilibrium mixture in a 10.0-L vessel contains 2.67×10^{-2} g of I(g), how many grams of I_2 are in the mixture? **(b)** For $2 SO_2(g) + O_2(g) \rightleftharpoons 2 SO_3(g)$, $K_p = 3.0 \times 10^4$ at 700 K. In a 2.00-L vessel the equilibrium mixture contains 1.17 g of SO_3 and 0.105 g of O_2. How many grams of SO_2 are in the vessel?

15.43 At 2000 °C the equilibrium constant for the reaction

$$2\,NO(g) \rightleftharpoons N_2(g) + O_2(g)$$

is $K_c = 2.4 \times 10^3$. If the initial concentration of NO is 0.200 M, what are the equilibrium concentrations of NO, N_2, and O_2?

15.44 For the equilibrium

$$Br_2(g) + Cl_2(g) \rightleftharpoons 2\,BrCl(g)$$

at 400 K, $K_c = 7.0$. If 0.25 mol of Br_2 and 0.25 mol of Cl_2 are introduced into a 1.0-L container at 400 K, what will be the equilibrium concentrations of Br_2, Cl_2, and BrCl?

15.45 At 373 K, $K_p = 0.416$ for the equilibrium

$$2\,NOBr(g) \rightleftharpoons 2\,NO(g) + Br_2(g)$$

If the pressures of NOBr(g) and NO(g) are equal, what is the equilibrium pressure of $Br_2(g)$?

15.46 At 218 °C, $K_c = 1.2 \times 10^{-4}$ for the equilibrium

$$NH_4HS(s) \rightleftharpoons NH_3(g) + H_2S(g)$$

Calculate the equilibrium concentrations of NH_3 and H_2S if a sample of solid NH_4HS is placed in a closed vessel and decomposes until equilibrium is reached.

15.47 Consider the reaction

$$CaSO_4(s) \rightleftharpoons Ca^{2+}(aq) + SO_4{}^{2-}(aq)$$

At 25 °C the equilibrium constant is $K_c = 2.4 \times 10^{-5}$ for this reaction. **(a)** If excess $CaSO_4(s)$ is mixed with water at 25 °C to produce a saturated solution of $CaSO_4$, what are the equilibrium concentrations of Ca^{2+} and $SO_4{}^{2-}$? **(b)** If the resulting solution has a volume of 3.0 L, what is the minimum mass of $CaSO_4(s)$ needed to achieve equilibrium?

15.48 At 80 °C, $K_c = 1.87 \times 10^{-3}$ for the reaction

$$PH_3BCl_3(s) \rightleftharpoons PH_3(g) + BCl_3(g)$$

(a) Calculate the equilibrium concentrations of PH_3 and BCl_3 if a solid sample of PH_3BCl_3 is placed in a closed vessel and decomposes until equilibrium is reached. **(b)** If the flask has a volume of 0.500 L, what is the minimum mass of $PH_3BCl_3(s)$ that must be added to the flask to achieve equilibrium?

15.49 For the reaction $I_2 + Br_2(g) \rightleftharpoons 2\,IBr(g)$, $K_c = 280$ at 150 °C. Suppose that 0.500 mol IBr in a 1.00-L flask is allowed to reach equilibrium at 150 °C. What are the equilibrium concentrations of IBr, I_2, and Br_2?

15.50 At 25 °C the reaction

$$CaCrO_4(g) \rightleftharpoons Ca^{2+}(aq) + CrO_4{}^{2-}(aq)$$

has an equilibrium constant $K_c = 7.1 \times 10^{-4}$. What are the equilibrium concentrations of Ca^{2+} and $CrO_4{}^{2-}$ in a saturated solution of $CaCrO_4$?

Le Châtelier's Principle

15.51 Consider the following equilibrium, for which $\Delta H < 0$

$$2\,SO_2(g) + O_2(g) \rightleftharpoons 2\,SO_3(g)$$

How will each of the following changes affect an equilibrium mixture of the three gases? **(a)** $O_2(g)$ is added to the system; **(b)** the reaction mixture is heated; **(c)** the volume of the reaction vessel is doubled; **(d)** a catalyst is added to the mixture; **(e)** the total pressure of the system is increased by adding a noble gas; **(f)** $SO_3(g)$ is removed from the system.

15.52 Consider $4\,NH_3(g) + 5\,O_2(g) \rightleftharpoons 4\,NO(g) + 6\,H_2O(g)$, $\Delta H = -904.4$ kJ. How does each of the following changes affect the yield of NO at equilibrium? Answer increase, decrease, or no change: **(a)** increase $[NH_3]$; **(b)** increase $[H_2O]$; **(c)** decrease $[O_2]$; **(d)** decrease the volume of the container in which the reaction occurs; **(e)** add a catalyst; **(f)** increase temperature.

15.53 How do the following changes affect the value of the equilibrium constant for a gas-phase exothermic reaction: **(a)** removal of a reactant or product, **(b)** decrease in the volume, **(c)** decrease in the temperature, **(d)** addition of a catalyst?

15.54 For a certain gas-phase reaction, the fraction of products in an equilibrium mixture is increased by increasing the temperature and increasing the volume of the reaction vessel. **(a)** What can you conclude about the reaction from the influence of temperature on the equilibrium? **(b)** What can you conclude from the influence of increasing the volume?

15.55 Consider the following equilibrium between oxides of nitrogen

$$3\,NO(g) \rightleftharpoons NO_2(g) + N_2O(g)$$

(a) Use data in Appendix C to calculate $\Delta H°$ for this reaction. **(b)** Will the equilibrium constant for the reaction increase or decrease with increasing temperature? Explain. **(c)** At constant temperature would a change in the volume of the container affect the fraction of products in the equilibrium mixture?

15.56 Methanol (CH_3OH) can be made by the reaction of CO with H_2:

$$CO(g) + 2\,H_2(g) \rightleftharpoons CH_3OH(g)$$

(a) Use thermochemical data in Appendix C to calculate $\Delta H°$ for this reaction. **(b)** To maximize the equilibrium yield of methanol, would you use a high or low temperature? **(c)** To maximize the equilibrium yield of methanol, would you use a high or low pressure?

ADDITIONAL EXERCISES

15.57 Both the forward reaction and the reverse reaction in the following equilibrium are believed to be elementary steps:

$$CO(g) + Cl_2(g) \rightleftharpoons COCl(g) + Cl(g)$$

At 25 °C the rate constants for the forward and reverse reactions are 1.4×10^{-28} $M^{-1}\,s^{-1}$ and 9.3×10^{10} $M^{-1}\,s^{-1}$, respectively. **(a)** What is the value for the equilibrium constant at 25 °C? **(b)** Are reactants or products more plentiful at equilibrium?

15.58 If $K_c = 1$ for the equilibrium $2\,A(g) \rightleftharpoons B(g)$, what is the relationship between [A] and [B] at equilibrium?

15.59 A mixture of CH_4 and H_2O is passed over a nickel catalyst at 1000 K. The emerging gas is collected in a 5.00-L flask and is found to contain 8.62 g of CO, 2.60 g of H_2, 43.0 g of CH_4, and 48.4 g of H_2O. Assuming that equilibrium has been reached, calculate K_c and K_p for the reaction.

15.60 When 2.00 mol of SO_2Cl_2 is placed in a 2.00-L flask at 303 K, 56% of the SO_2Cl_2 decomposes to SO_2 and Cl_2:

$$SO_2Cl_2(g) \rightleftharpoons SO_2(g) + Cl_2(g)$$

Calculate K_c for this reaction at this temperature.

15.61 A mixture of H_2, S, and H_2S is held in a 1.0-L vessel at 90 °C until the following equilibrium is achieved:

$$H_2(g) + S(s) \rightleftharpoons H_2S(g)$$

At equilibrium the mixture contains 0.46 g of H_2S and 0.40 g H_2. **(a)** Write the equilibrium-constant expression for this reaction. **(b)** What is the value of K_c for the reaction at this temperature? **(c)** Why can we ignore the amount of S when doing the calculation in part (b)?

15.62 A sample of nitrosyl bromide (NOBr) decomposes according to the equation

$$2\,NOBr(g) \rightleftharpoons 2\,NO(g) + Br_2(g)$$

An equilibrium mixture in a 5.00-L vessel at 100 °C contains 3.22 g of NOBr, 3.08 g of NO, and 4.19 g of Br_2. **(a)** Calculate K_c. **(b)** What is the total pressure exerted by the mixture of gases?

15.63 Consider the hypothetical reaction $A(g) \rightleftharpoons 2\,B(g)$. A flask is charged with 0.75 atm of pure A, after which it is allowed to reach equilibrium at 0 °C. At equilibrium the partial pressure of A is 0.36 atm. **(a)** What is the total pressure in the flask at equilibrium? **(b)** What is the value of K_p?

15.64 As shown in Table 15.2, the equilibrium constant for the reaction $N_2(g) + 3\,H_2(g) \rightleftharpoons 2\,NH_3(g)$ is $K_p = 4.34 \times 10^{-3}$ at 300 °C. Pure NH_3 is placed in a 1.00-L flask and allowed to reach equilibrium at this temperature. There are 1.05 g NH_3 in the equilibrium mixture. **(a)** What are the masses of N_2 and H_2 in the equilibrium mixture? **(b)** What was the initial mass of ammonia placed in the vessel? **(c)** What is the total pressure in the vessel?

15.65 For the equilibrium

$$2\,IBr(g) \rightleftharpoons I_2(g) + Br_2(g)$$

$K_p = 8.5 \times 10^{-3}$ at 150 °C. If 0.025 atm of IBr is placed in a 2.0-L container, what is the partial pressure of this substance after equilibrium is reached?

15.66 For the equilibrium

$$PH_3BCl_3(s) \rightleftharpoons PH_3(g) + BCl_3(g)$$

$K_p = 0.052$ at 60 °C. **(a)** Calculate K_c. **(b)** Some solid PH_3BCl_3 is added to a closed 0.500-L vessel at 60 °C; the vessel is then charged with 0.0128 mol of $BCl_3(g)$. What is the equilibrium concentration of PH_3?

[15.67] Solid NH_4HS is introduced into an evacuated flask at 24 °C. The following reaction takes place:

$$NH_4HS(s) \rightleftharpoons NH_3(g) + H_2S(g)$$

At equilibrium the total pressure (for NH_3 and H_2S taken together) is 0.614 atm. What is K_p for this equilibrium at 24 °C?

[15.68] A 0.831-g sample of SO_3 is placed in a 1.00-L container and heated to 1100 K. The SO_3 decomposes to SO_2 and O_2.

$$2\,SO_3(g) \rightleftharpoons 2\,SO_2(g) + O_2(g)$$

At equilibrium the total pressure in the container is 1.300 atm. Find the values of K_p and K_c for this reaction at 1100 K.

15.69 Nitric oxide (NO) reacts readily with chlorine gas as follows:

$$2\,NO(g) + Cl_2(g) \rightleftharpoons 2\,NOCl(g)$$

At 700 K the equilibrium constant K_p for this reaction is 0.26. Predict the behavior of each of the following mixtures at this temperature: **(a)** $P_{NO} = 0.15$ atm, $P_{Cl_2} = 0.31$ atm, and $P_{NOCl} = 0.11$ atm; **(b)** $P_{NO} = 0.12$ atm, $P_{Cl_2} = 0.10$ atm, and $P_{NOCl} = 0.050$ atm; **(c)** $P_{NO} = 0.15$ atm, $P_{Cl_2} = 0.20$ atm, and $P_{NOCl} = 5.10 \times 10^{-3}$ atm.

15.70 At 900 °C, $K_c = 0.0108$ for the reaction

$$CaCO_3(s) \rightleftharpoons CaO(s) + CO_2(g)$$

A mixture of $CaCO_3$, CaO, and CO_2 is placed in a 10.0-L vessel at 900 °C. For the following mixtures, will the amount of $CaCO_3$ increase, decrease, or remain the same as the system approaches equilibrium? **(a)** 15.0 g $CaCO_3$, 15.0 g CaO, and 4.25 g CO_2 **(b)** 2.50 g $CaCO_3$, 25.0 g CaO, and 5.66 g CO_2 **(c)** 30.5 g $CaCO_3$, 25.5 g CaO, and 6.48 g CO_2.

15.71 When 1.50 mol CO_2 and 1.50 mol H_2 are placed in a 0.750-L container at 395 °C, the following equilibrium is achieved: $CO_2(g) + H_2(g) \rightleftharpoons CO(g) + H_2O(g)$. If $K_c = 0.802$, what are the concentrations of each substance in the equilibrium mixture?

15.72 The equilibrium constant K_c for $C(s) + CO_2(g) \rightleftharpoons 2\,CO(g)$ is 1.9 at 1000 K and 0.133 at 298 K. **(a)** If excess C is allowed to react with 25.0 g of CO_2 in a 3.00-L vessel at 1000 K, how many grams of CO_2 are produced? **(b)** How many grams of C are consumed? **(c)** If a smaller vessel is used for the reaction, will the yield of CO be greater or smaller? **(d)** If the reaction is endothermic, how does increasing the temperature affect the equilibrium constant?

15.73 NiO is to be reduced to nickel metal in an industrial process by use of the reaction

$$NiO(s) + CO(g) \rightleftharpoons Ni(s) + CO_2(g)$$

At 1600 K the equilibrium constant for the reaction is $K_p = 6.0 \times 10^2$. If a CO pressure of 150 torr is to be employed in the furnace and total pressure never exceeds 760 torr, will reduction occur?

[15.74] At 700 K the equilibrium constant for the reaction

$$CCl_4(g) \rightleftharpoons C(s) + 2\,Cl_2(g)$$

is $K_p = 0.76$. A flask is charged with 2.00 atm of CCl_4, which then reaches equilibrium at 700 K. **(a)** What fraction of the CCl_4 is converted into C and Cl_2? **(b)** What are the partial pressures of CCl_4 and Cl_2 at equilibrium?

[15.75] The reaction $PCl_3(g) + Cl_2(g) \rightleftharpoons PCl_5(g)$ has $K_p = 0.0870$ at 300 °C. A flask is charged with 0.50 atm PCl_3, 0.50 atm Cl_2, and 0.20 atm PCl_5 at this temperature. **(a)** Use the reaction quotient to determine the direction the reaction must proceed to reach equilibrium. **(b)** Calculate the equilibrium partial pressures of the gases. **(c)** What effect will increasing the volume of the system have on the mole fraction of Cl_2 in the equilibrium mixture? **(d)** The reaction is exothermic. What effect will increasing the temperature of the system have on the mole fraction of Cl_2 in the equilibrium mixture?

[15.76] An equilibrium mixture of H_2, I_2, and HI at 458 °C contains 0.112 mol H_2, 0.112 mol I_2, and 0.775 mol HI in a 5.00-L vessel. What are the equilibrium partial pressures when equilibrium is reestablished following the addition of 0.100 mol of HI?

[15.77] Consider the hypothetical reaction $A(g) + 2 B(g) \rightleftharpoons 2 C(g)$, for which $K_c = 0.25$ at some temperature. A 1.00-L reaction vessel is loaded with 1.00 mol of compound C, which is allowed to reach equilibrium. Let the variable x represent the number of mol/L of compound A present at equilibrium. (a) In terms of x, what are the equilibrium concentrations of compounds B and C? (b) What limits must be placed on the value of x so that all concentrations are positive? (c) By putting the equilibrium concentrations (in terms of x) into the equilibrium-constant expression, derive an equation that can be solved for x. (d) The equation from part (c) is a cubic equation (one that has the form $ax^3 + bx^2 + cx + d = 0$). In general, cubic equations cannot be solved in closed form. However, you can estimate the solution by plotting the cubic equation in the allowed range of x that you specified in part (b). The point at which the cubic equation crosses the x-axis is the solution. (e) From the plot in part (d), estimate the equilibrium concentrations of A, B, and C. (Hint: You can check the accuracy of your answer by substituting these concentrations into the equilibrium expression.)

15.78 At 1200 K, the approximate temperature of automobile exhaust gases (Figure 15.16), K_p for the reaction

$$2 CO_2(g) \rightleftharpoons 2 CO(g) + O_2(g)$$

is about 1×10^{-13}. Assuming that the exhaust gas (total pressure 1 atm) contains 0.2% CO, 12% CO_2, and 3% O_2 by volume, is the system at equilibrium with respect to the above reaction? Based on your conclusion, would the CO concentration in the exhaust be decreased or increased by a catalyst that speeds up the reaction above?

15.79 Suppose that you worked at the U.S. Patent Office and a patent application came across your desk claiming that a newly developed catalyst was much superior to the Haber catalyst for ammonia synthesis because the catalyst led to much greater equilibrium conversion of N_2 and H_2 into NH_3 than the Haber catalyst under the same conditions. What would be your response?

INTEGRATIVE EXERCISES

15.80 Consider the equilibrium $IO_4^-(aq) + 2 H_2O(l) \rightleftharpoons H_4IO_6^-(aq)$, $K_c = 3.5 \times 10^{-2}$. If you start with 20.0 mL of a 0.905 M solution of $NaIO_4$, and then dilute it with water to 250.0 mL, what is the concentration of $H_4IO_6^-$ at equilibrium?

15.81 Consider the following equilibria in aqueous solution:
  (i) $Na(s) + Ag^+(aq) \rightleftharpoons Na^+(aq) + Ag(s)$
  (ii) $3 Hg(l) + 2 Al^{3+}(aq) \rightleftharpoons 3 Hg^{2+}(aq) + 2 Al(s)$
  (iii) $Zn(s) + 2 H^+(aq) \rightleftharpoons Zn^{2+}(aq) + H_2(g)$

  (a) For each reaction, write the equilibrium-constant expression for K_c. (b) Using information provided in Table 4.5, predict whether K_c is large ($K_c \gg 1$) or small ($K_c \ll 1$). Explain your reasoning. (c) At 25 °C the reaction $Cd(s) + Fe^{2+}(aq) \rightleftharpoons Cd^{2+}(aq) + Fe(s)$ has $K_c = 6 \times 10^{-2}$. If Cd were added to Table 4.5, would you expect it to be above or below iron? Explain.

15.82 Silver chloride, $AgCl(s)$, is an insoluble strong electrolyte. (a) Write the equation for the dissolution of $AgCl(s)$ in $H_2O(l)$ (b) Write the expression for K_c for the reaction in part (a). (c) Based on the thermochemical data in Appendix C and Le Châtelier's principle, predict whether the solubility of AgCl in H_2O increases or decreases with increasing temperature.

15.83 The hypothetical reaction $A + B \rightleftharpoons C$ occurs in the forward direction in a single step. The energy profile of the reaction is shown in the drawing. (a) Is the forward or reverse reaction faster at equilibrium? (b) Would you expect the equilibrium to favor reactants or products? (c) In general, how would a catalyst affect the energy profile shown? (d) How would a catalyst affect the ratio of the rate constants for the forward and reverse reactions? (e) How would you expect the equilibrium constant of the reaction to change with increasing temperature?

Reaction pathway

[15.84] Consider the equilibrium $A \rightleftharpoons B$ in which both the forward and reverse reactions are elementary (single-step) reactions. Assume that the only effect of a catalyst on the reaction is to lower the activation energies of the forward and reverse reactions, as shown in Figure 15.15. Using the Arrhenius equation (Section 14.5), prove that the equilibrium constant is the same for the catalyzed reaction as for the uncatalyzed one.

[15.85] At 25 °C the reaction

$$NH_4HS(s) \rightleftharpoons NH_3(g) + H_2S(g)$$

has $K_p = 0.120$. A 5.00-L flask is charged with 0.300 g of pure $H_2S(g)$ at 25 °C. Solid NH_4HS is then added until there is excess unreacted solid remaining. (a) What is the initial pressure of $H_2S(g)$ in the flask? (b) Why does no reaction occur until NH_4HS is added? (c) What are the partial pressures of NH_3 and H_2S at equilibrium? (d) What is the mole fraction of H_2S in the gas mixture at equilibrium? (e) What is the minimum mass, in grams, of NH_4HS that must be added to the flask to achieve equilibrium?

[15.86] Write the equilibrium-constant expression for the equilibrium

$$C(s) + CO_2(g) \rightleftharpoons 2\,CO(g)$$

The table included below shows the relative mole percentages of $CO_2(g)$ and $CO(g)$ at a total pressure of 1 atm for several temperatures. Calculate the value of K_p at each temperature. Is the reaction exothermic or endothermic? Explain.

Temperature (°C)	CO₂ (mol %)	CO (mol %)
850	6.23	93.77
950	1.32	98.68
1050	0.37	99.63
1200	0.06	99.94

15.87 In Section 11.5 we defined the vapor pressure of a liquid in terms of an equilibrium. (a) Write the equation representing the equilibrium between liquid water and water vapor, and the corresponding expression for K_p. (b) By using data in Appendix B, give the value of K_p for this reaction at 30 °C. (c) What is the value of K_p for any liquid in equilibrium with its vapor at the normal boiling point of the liquid?

[15.88] Polyvinyl chloride (PVC) is one of the most commercially important polymers (Table 12.5). PVC is made by addition polymerization of vinyl chloride (C_2H_3Cl). Vinyl chloride is synthesized from ethylene (C_2H_4) in a two-step process involving the following equilibria:

Equilibrium 1: $C_2H_4(g) + Cl_2(g) \rightleftharpoons C_2H_4Cl_2(g)$

Equilibrium 2: $C_2H_4Cl_2(g) \rightleftharpoons C_2H_3Cl(g) + HCl(g)$

The product of Equilibrium 1 is 1,2-dichloroethane, a compound in which one Cl atom is bonded to each C atom. (a) Draw Lewis structures for $C_2H_4Cl_2$ and C_2H_3Cl. What are the C—C bond orders in these two compounds? (b) Use average bond enthalpies (Table 8.4) to estimate the enthalpy changes in the two equilibria. (c) How would the yield of $C_2H_4Cl_2$ in Equilibrium 1 vary with temperature and volume? (d) How would the yield of C_2H_3Cl in Equilibrium 2 vary with temperature and volume? (e) Look up the normal boiling points of 1,2-dichloroethane and vinyl chloride in a sourcebook, such as the CRC Handbook of Chemistry and Physics. Based on these data, propose a reactor design (analogous to Figure 15.12) that could be used to maximize the amount of C_2H_3Cl produced by using the two equilibria.

16 ACID–BASE EQUILIBRIA

VARIETIES OF CITRUS FRUITS
(oranges, limes, lemons,
grapefruit, and tangerines).

16.1 Acids and Bases: A Brief Review
We start by reviewing the *Arrhenius* definitions of acid and base.

16.2 Brønsted–Lowry Acids and Bases
We then learn the more general *Brønsted–Lowry* definitions for acid and base. A Brønsted–Lowry acid is a *proton donor* and a Brønsted–Lowry base is a *proton acceptor*. Two species that differ by the presence or absence of a proton are known as a *conjugate acid–base pair*.

16.3 The Autoionization of Water
We will see that the *autoionization* of water produces small concentrations of H_3O^+ and OH^- ions. The *equilibrium constant* for autoionization, $K_w = [H_3O^+][OH^-]$, defines the relationship between H_3O^+ and OH^- concentrations in aqueous solutions.

16.4 The pH Scale
We will use the pH scale (pH $= -\log[H^+]$) to describe the acidity or basicity of an aqueous solution. Neutral solutions have a pH $= 7$, acidic solutions have pH below 7, and basic solutions have pH above 7.

16.5 Strong Acids and Bases
We categorize acids and bases as being either strong or weak electrolytes. *Strong* acids and bases are strong electrolytes, ionizing or dissociating completely in aqueous solution. *Weak* acids and bases are weak electrolytes and therefore ionize only partially.

16.6 Weak Acids
We learn that the ionization of a weak acid in water is an equilibrium process with an equilibrium constant K_a that can be used to calculate the pH of a weak acid solution.

16.7 Weak Bases
We learn that the ionization of a weak base in water is an equilibrium process with equilibrium constant K_b that can be used to calculate the pH of a weak base solution.

16.8 Relationship between K_a and K_b
We will see that a relationship exists between the K_a and K_b of any conjugate *acid–base* pair: $K_a \times K_b = K_w$. Thus, the stronger an acid, the weaker its conjugate base.

16.9 Acid–Base Properties of Salt Solutions
We will explore the fact that the ions of a soluble ionic compound (a salt) can serve as Brønsted–Lowry acids or bases.

16.10 Acid–Base Behavior and Chemical Structure
We continue by exploring the relationship between chemical structure and acid–base behavior.

16.11 Lewis Acids and Bases
Finally, we learn the *Lewis* definitions of acid and base. A Lewis acid is an *electron-pair acceptor*, and a Lewis base is an *electron-pair donor*.

WHAT IS THE SOUREST FOOD YOU'VE EVER TASTED? Citrus fruit, such as the lemons shown in the chapter-opening photograph? Sour cherries? Rhubarb? The sour taste of foods is due primarily to the presence of acids. Citric acid ($H_3C_6H_5O_7$), malic acid ($H_2C_4H_4O_5$), oxalic acid ($H_2C_2O_4$), and ascorbic acid, also known as vitamin C ($HC_6H_7O_6$), are present in many fruits as well as in certain vegetables, such as rhubarb and tomatoes.

Acids and bases are important in numerous chemical processes that occur around us—from industrial processes to biological ones, from reactions in the laboratory to those in our environment. The time required for a metal object immersed in water to corrode, the ability of an aquatic environment to support fish and plant life, the fate of pollutants washed out of the air by rain, and even the rates of reactions that maintain our lives all critically depend upon the acidity or basicity of solutions. Indeed, an enormous amount of chemistry can be understood in terms of acid–base reactions.

We have encountered acids and bases many times in earlier discussions. For example, a portion of Chapter 4 focused on their reactions. But what makes a substance behave as an acid or as a base? In this chapter we reexamine acids and bases, taking a closer look at how they are identified and characterized. In doing so, we will consider their behavior both in terms of their structure and bonding and in terms of the chemical equilibria in which they participate.

16.1 ACIDS AND BASES: A BRIEF REVIEW

From the earliest days of experimental chemistry, scientists have recognized acids and bases by their characteristic properties. Acids have a sour taste and cause certain dyes to change color (for example, litmus turns red on contact with acids). Indeed, the word *acid* comes from the Latin word *acidus*, meaning sour or tart. Bases, in contrast, have a bitter taste and feel slippery (soap is a good example). The word *base* comes from an old English meaning of the word, which is "to bring low." (We still use the word *debase* in this sense, meaning to lower the value of something.) When bases are added to acids, they lower the amount of acid. Indeed, when acids and bases are mixed in certain proportions, their characteristic properties disappear altogether. ∞ (Section 4.3)

Historically, chemists have sought to relate the properties of acids and bases to their compositions and molecular structures. By 1830 it was evident that all acids contain hydrogen but not all hydrogen-containing substances are acids. During the 1880s, the Swedish chemist Svante Arrhenius (1859–1927) linked acid behavior with the presence of H^+ ions and base behavior with the presence of OH^- ions in aqueous solution.

Arrhenius defined acids as substances that produce H^+ ions in water and bases as substances that produce OH^- ions in water. Indeed, the properties of aqueous solutions of acids, such as sour taste, are due to $H^+(aq)$, whereas the properties of aqueous solutions of bases are due to $OH^-(aq)$. Over time the Arrhenius concept of acids and bases came to be stated in the following way:

- An *acid* is a substance that, when dissolved in water, increases the concentration of H^+ ions.
- A *base* is a substance that, when dissolved in water, increases the concentration of OH^- ions.

Hydrogen chloride is an Arrhenius acid. Hydrogen chloride gas is highly soluble in water because of its chemical reaction with water, which produces hydrated H^+ and Cl^- ions:

$$HCl(g) \xrightarrow{H_2O} H^+(aq) + Cl^-(aq) \qquad [16.1]$$

The aqueous solution of HCl is known as hydrochloric acid. Concentrated hydrochloric acid is about 37% HCl by mass and is 12 M in HCl. Sodium hydroxide, on the other hand, is an Arrhenius base. Because NaOH is an ionic compound, it dissociates into Na^+ and OH^- ions when it dissolves in water, thereby releasing OH^- ions into the solution.

▲ GIVE IT SOME THOUGHT

What two ions are central to the Arrhenius definitions of acids and bases?

16.2 BRØNSTED–LOWRY ACIDS AND BASES

The Arrhenius concept of acids and bases, while useful, has limitations. For one thing, it is restricted to aqueous solutions. In 1923 the Danish chemist Johannes Brønsted (1879–1947) and the English chemist Thomas Lowry (1874–1936) independently proposed a more general definition of acids and bases. Their concept is based on the fact that *acid–base reactions involve the transfer of H^+ ions from one substance to another.*

The H⁺ Ion in Water

In Equation 16.1 hydrogen chloride is shown ionizing in water to form $H^+(aq)$. *An H^+ ion is simply a proton with no surrounding valence electron.* This small, positively charged particle interacts strongly with the nonbonding electron pairs of water molecules to form hydrated hydrogen ions. For example, the interaction of a proton with one water molecule forms the **hydronium ion**, $H_3O^+(aq)$:

$$H^+ + :\overset{\displaystyle |}{\underset{\displaystyle H}{O}}-H \longrightarrow \left[H-\overset{\displaystyle |}{\underset{\displaystyle H}{O}}-H \right]^+ \qquad [16.2]$$

The formation of hydronium ions is one of the complex features of the interaction of the H^+ ion with liquid water. In fact, the H_3O^+ ion can form hydrogen bonds to additional H_2O molecules to generate larger clusters of hydrated hydrogen ions, such as $H_5O_2^+$ and $H_9O_4^+$ (Figure 16.1▶).

Chemists use $H^+(aq)$ and $H_3O^+(aq)$ interchangeably to represent the same thing—namely the hydrated proton that is responsible for the characteristic properties of aqueous solutions of acids. We often use the $H^+(aq)$ ion for simplicity and convenience, as we did in Equation 16.1. The $H_3O^+(aq)$ ion, however, more closely represents reality.

Proton-Transfer Reactions

When we closely examine the reaction that occurs when HCl dissolves in water, we find that the HCl molecule actually transfers an H^+ ion (a proton) to a water molecule as depicted in Figure 16.2▶. Thus, we can represent the reaction as occurring between an HCl molecule and a water molecule to form hydronium and chloride ions:

$$HCl(g) + H_2O(l) \longrightarrow H_3O^+(aq) + Cl^-(aq) \qquad [16.3]$$

The polar H_2O molecule promotes the ionization of acids in water solution by accepting a proton to form H_3O^+.

Brønsted and Lowry proposed definitions of acids and bases in terms of their ability to transfer protons:

- An *acid* is a substance (molecule or ion) that donates a proton to another substance.

- A *base* is a substance that accepts a proton.

Thus, when HCl dissolves in water (Equation 16.3), HCl acts as a **Brønsted–Lowry acid** (it donates a proton to H_2O), and H_2O acts as a **Brønsted–Lowry base** (it accepts a proton from HCl).

Because the emphasis in the Brønsted–Lowry concept is on proton transfer, the concept also applies to reactions that do not occur in aqueous solution. In the reaction between HCl and NH_3, for example, a proton is transferred from the acid HCl to the base NH_3:

$$:\overset{\cdot\cdot}{\underset{}{Cl}}-H + :\overset{\displaystyle H}{\underset{\displaystyle H}{N}}-H \longrightarrow :\overset{\cdot\cdot}{\underset{\cdot\cdot}{Cl}}:^- + \left[H-\overset{\displaystyle H}{\underset{\displaystyle H}{N}}-H \right]^+ \qquad [16.4]$$

This reaction can occur in the gas phase. The hazy film that forms on the windows of general chemistry laboratories and on glassware in the lab is largely solid NH_4Cl formed by the gas-phase reaction of HCl and NH_3 (Figure 16.3▼).

$H_5O_2^+$

(a)

$H_9O_4^+$

(b)

▲ **Figure 16.1 Hydrated hydronium ions.** Lewis structures and molecular models for $H_5O_2^+$ and $H_9O_4^+$. There is good experimental evidence for the existence of both these species.

▲ **Figure 16.2 A proton-transfer reaction.** When a proton is transferred from HCl to H_2O, HCl acts as the Brønsted–Lowry acid and H_2O acts as the Brønsted–Lowry base.

▲ **Figure 16.3 A gas-phase acid–base reaction.** The HCl(g) escaping from concentrated hydrochloric acid and the $NH_3(g)$ escaping from aqueous ammonia (here labeled ammonium hydroxide) combine to form a white fog of $NH_4Cl(s)$.

Let's consider another example that compares the relationship between the Arrhenius definitions and the Brønsted–Lowry definitions of acids and bases— an aqueous solution of ammonia, in which the following equilibrium occurs:

$$NH_3(aq) + H_2O(l) \rightleftharpoons NH_4^+(aq) + OH^-(aq) \qquad [16.5]$$

Ammonia is an Arrhenius base because adding it to water leads to an increase in the concentration of $OH^-(aq)$. It is a Brønsted–Lowry base because it accepts a proton from H_2O. The H_2O molecule in Equation 16.5 acts as a Brønsted– Lowry acid because it donates a proton to the NH_3 molecule.

An acid and a base always work together to transfer a proton. In other words, a substance can function as an acid only if another substance simultaneously behaves as a base. To be a Brønsted–Lowry acid, a molecule or ion must have a hydrogen atom that it can lose as an H^+ ion. To be a Brønsted–Lowry base, a molecule or ion must have a nonbonding pair of electrons that it can use to bind the H^+ ion.

Some substances can act as an acid in one reaction and as a base in another. For example, H_2O is a Brønsted–Lowry base in its reaction with HCl (Equation 16.3) and a Brønsted–Lowry acid in its reaction with NH_3 (Equation 16.5). A substance that is capable of acting as either an acid or a base is called **amphiprotic**. An amphiprotic substance acts as a base when combined with something more strongly acidic than itself and as an acid when combined with something more strongly basic than itself.

GIVE IT SOME THOUGHT

In the forward reaction, which substance acts as the Brønsted–Lowry base: $HSO_4^-(aq) + NH_3(aq) \rightleftharpoons SO_4^{2-}(aq) + NH_4^+(aq)$?

Conjugate Acid–Base Pairs

In any acid–base equilibrium both the forward reaction (to the right) and the reverse reaction (to the left) involve proton transfers. For example, consider the reaction of an acid, which we will denote HX, with water:

$$HX(aq) + H_2O(l) \rightleftharpoons X^-(aq) + H_3O^+(aq) \qquad [16.6]$$

In the forward reaction HX donates a proton to H_2O. Therefore, HX is the Brønsted–Lowry acid, and H_2O is the Brønsted–Lowry base. In the reverse reaction the H_3O^+ ion donates a proton to the X^- ion, so H_3O^+ is the acid and X^- is the base. When the acid HX donates a proton, it leaves behind a substance, X^-, which can act as a base. Likewise, when H_2O acts as a base, it generates H_3O^+, which can act as an acid.

An acid and a base such as HX and X^- that differ only in the presence or absence of a proton are called a **conjugate acid–base pair**.* Every acid has a **conjugate base**, formed by removing a proton from the acid. For example, OH^- is the conjugate base of H_2O, and X^- is the conjugate base of HX. Similarly, every base has associated with it a **conjugate acid**, formed by adding a proton to the base. Thus, H_3O^+ is the conjugate acid of H_2O, and HX is the conjugate acid of X^-.

In any acid–base (proton-transfer) reaction we can identify two sets of conjugate acid–base pairs. For example, consider the reaction between nitrous acid (HNO_2) and water:

remove H^+

$$HNO_2(aq) + H_2O(l) \rightleftharpoons NO_2^-(aq) + H_3O^+(aq) \qquad [16.7]$$

Acid   Base   Conjugate   Conjugate
                          base   acid

add H^+

*The word conjugate means "joined together as a pair."

Likewise, for the reaction between NH_3 and H_2O (Equation 16.5), we have

$$NH_3(aq) + H_2O(l) \rightleftharpoons NH_4^+(aq) + OH^-(aq) \qquad [16.8]$$

Base   Acid   Conjugate   Conjugate
                       acid   base

SAMPLE EXERCISE 16.1 | Identifying Conjugate Acids and Bases

(a) What is the conjugate base of each of the following acids: $HClO_4$, H_2S, PH_4^+, HCO_3^-? **(b)** What is the conjugate acid of each of the following bases: CN^-, SO_4^{2-}, H_2O, HCO_3^-?

SOLUTION

Analyze: We are asked to give the conjugate base for each of a series of species and to give the conjugate acid for each of another series of species.

Plan: The conjugate base of a substance is simply the parent substance minus one proton, and the conjugate acid of a substance is the parent substance plus one proton.

Solve: (a) $HClO_4$ less one proton (H^+) is ClO_4^-. The other conjugate bases are HS^-, PH_3, and CO_3^{2-}. **(b)** CN^- plus one proton (H^+) is HCN. The other conjugate acids are HSO_4^-, H_3O^+, and H_2CO_3.

Notice that the hydrogen carbonate ion (HCO_3^-) is amphiprotic. It can act as either an acid or a base.

PRACTICE EXERCISE

Write the formula for the conjugate acid of each of the following: HSO_3^-, F^-, PO_4^{3-}, CO.
Answers: H_2SO_3, HF, HPO_4^{2-}, HCO^+

SAMPLE EXERCISE 16.2 | Writing Equations for Proton-Transfer Reactions

The hydrogen sulfite ion (HSO_3^-) is amphiprotic. **(a)** Write an equation for the reaction of HSO_3^- with water, in which the ion acts as an acid. **(b)** Write an equation for the reaction of HSO_3^- with water, in which the ion acts as a base. In both cases identify the conjugate acid–base pairs.

SOLUTION

Analyze and Plan: We are asked to write two equations representing reactions between HSO_3^- and water, one in which HSO_3^- should donate a proton to water, thereby acting as a Brønsted–Lowry acid, and one in which HSO_3^- should accept a proton from water, thereby acting as a base. We are also asked to identify the conjugate pairs in each equation.

Solve:

(a)
$$HSO_3^-(aq) + H_2O(l) \rightleftharpoons SO_3^{2-}(aq) + H_3O^+(aq)$$

The conjugate pairs in this equation are HSO_3^- (acid) and SO_3^{2-} (conjugate base); and H_2O (base) and H_3O^+ (conjugate acid).

(b)
$$HSO_3^-(aq) + H_2O(l) \rightleftharpoons H_2SO_3(aq) + OH^-(aq)$$

The conjugate pairs in this equation are H_2O (acid) and OH^- (conjugate base), and HSO_3^- (base) and H_2SO_3 (conjugate acid).

PRACTICE EXERCISE

When lithium oxide (Li_2O) is dissolved in water, the solution turns basic from the reaction of the oxide ion (O^{2-}) with water. Write the reaction that occurs, and identify the conjugate acid–base pairs.
Answer: $O^{2-}(aq) + H_2O(l) \longrightarrow OH^-(aq) + OH^-(aq)$. OH^- is the conjugate acid of the base O^{2-}. OH^- is also the conjugate base of the acid H_2O.

Relative Strengths of Acids and Bases

Some acids are better proton donors than others; likewise, some bases are better proton acceptors than others. If we arrange acids in order of their ability to donate a proton, we find that the more easily a substance gives up a proton, the less easily its conjugate base accepts a proton. Similarly, the more easily a base accepts a proton, the less easily its conjugate acid gives up a proton. In other words, *the stronger an acid, the weaker is its conjugate base; the stronger a base, the weaker is its conjugate acid.* Thus, if we know something about the strength of an acid (its ability to donate protons), we also know something about the strength of its conjugate base (its ability to accept protons).

The inverse relationship between the strengths of acids and the strengths of their conjugate bases is illustrated in Figure 16.4 ◀. Here we have grouped acids and bases into three broad categories based on their behavior in water.

1. A *strong acid* completely transfers its protons to water, leaving no undissociated molecules in solution. ∞ (Section 4.3) Its conjugate base has a negligible tendency to be protonated (to abstract protons) in aqueous solution.

2. A *weak acid* only partially dissociates in aqueous solution and therefore exists in the solution as a mixture of acid molecules and their constituent ions. The conjugate base of a weak acid shows a slight ability to remove protons from water. (*The conjugate base of a weak acid is a weak base.*)

3. A substance with *negligible acidity*, such as CH_4, contains hydrogen but does not demonstrate any acidic behavior in water. Its conjugate base is a strong base, reacting completely with water, abstracting protons to form OH^- ions.

▲ **Figure 16.4 Relative strengths of some conjugate acid–base pairs.** The two members of each pair are listed opposite each other in the two columns. The acids decrease in strength from top to bottom, whereas their conjugate bases increase in strength from top to bottom.

GIVE IT SOME THOUGHT

Using the three categories above, specify the strength of HNO_3 and the strength of its conjugate base, NO_3^-.

We can think of proton-transfer reactions as being governed by the relative abilities of two bases to abstract protons. For example, consider the proton transfer that occurs when an acid HX dissolves in water:

$$HX(aq) + H_2O(l) \rightleftharpoons H_3O^+(aq) + X^-(aq) \qquad [16.9]$$

If H_2O (the base in the forward reaction) is a stronger base than X^- (the conjugate base of HX), then H_2O will abstract the proton from HX to produce H_3O^+ and X^-. As a result, the equilibrium will lie to the right. This describes the behavior of a strong acid in water. For example, when HCl dissolves in water, the solution consists almost entirely of H_3O^+ and Cl^- ions with a negligible concentration of HCl molecules.

$$HCl(g) + H_2O(l) \longrightarrow H_3O^+(aq) + Cl^-(aq) \qquad [16.10]$$

H_2O is a stronger base than Cl^- (Figure 16.4), so H_2O acquires the proton to become the hydronium ion.

When X^- is a stronger base than H_2O, the equilibrium will lie to the left. This situation occurs when HX is a weak acid. For example, an aqueous solution of acetic acid (CH_3COOH) consists mainly of CH_3COOH molecules with only a relatively few H_3O^+ and CH_3COO^- ions.

$$CH_3COOH(aq) + H_2O(l) \rightleftharpoons H_3O^+(aq) + CH_3COO^-(aq) \quad [16.11]$$

CH_3COO^- is a stronger base than H_2O (Figure 16.4) and therefore abstracts the proton from H_3O^+.

From these examples, we conclude that *in every acid–base reaction the position of the equilibrium favors transfer of the proton from the stronger acid to the stronger base to form the weaker acid and the weaker base.* As a result, the equilibrium mixture contains more of the weaker acid and weaker base and less of the stronger acid and stronger base.

■ **SAMPLE EXERCISE 16.3** | Predicting the Position of a Proton-Transfer Equilibrium

For the following proton-transfer reaction, use Figure 16.4 to predict whether the equilibrium lies predominantly to the left (that is, $K_c < 1$) or to the right ($K_c > 1$):

$$HSO_4^-(aq) + CO_3^{2-}(aq) \rightleftharpoons SO_4^{2-}(aq) + HCO_3^-(aq)$$

SOLUTION

Analyze: We are asked to predict whether the equilibrium shown lies to the right, favoring products, or to the left, favoring reactants.

Plan: This is a proton-transfer reaction, and the position of the equilibrium will favor the proton going to the stronger of two bases. The two bases in the equation are CO_3^{2-}, the base in the forward reaction as written, and SO_4^{2-}, the conjugate base of HSO_4^-. We can find the relative positions of these two bases in Figure 16.4 to determine which is the stronger base.

Solve: CO_3^{2-} appears lower in the right-hand column in Figure 16.4 and is therefore a stronger base than SO_4^{2-}. CO_3^{2-}, therefore, will get the proton preferentially to become HCO_3^-, while SO_4^{2-} will remain mostly unprotonated. The resulting equilibrium will lie to the right, favoring products (that is, $K_c > 1$).

$$\underset{\text{Acid}}{HSO_4^-(aq)} + \underset{\text{Base}}{CO_3^{2-}(aq)} \rightleftharpoons \underset{\text{Conjugate base}}{SO_4^{2-}(aq)} + \underset{\text{Conjugate acid}}{HCO_3^-(aq)} \quad K_c > 1$$

Comment: Of the two acids in the equation, HSO_4^- and HCO_3^-, the stronger one gives up a proton more readily while the weaker one tends to retain its proton. Thus, the equilibrium favors the direction in which the proton moves from the stronger acid and becomes bonded to the stronger base.

■ **PRACTICE EXERCISE**

For each of the following reactions, use Figure 16.4 to predict whether the equilibrium lies predominantly to the left or to the right:
(a) $HPO_4^{2-}(aq) + H_2O(l) \rightleftharpoons H_2PO_4^-(aq) + OH^-(aq)$
(b) $NH_4^+(aq) + OH^-(aq) \rightleftharpoons NH_3(aq) + H_2O(l)$
Answers: **(a)** left, **(b)** right

16.3 THE AUTOIONIZATION OF WATER

One of the most important chemical properties of water is its ability to act as either a Brønsted acid or a Brønsted base, depending on the circumstances. In the presence of an acid, water acts as a proton acceptor; in the presence of a base, water acts as a proton donor. In fact, one water molecule can donate a proton to another water molecule:

$$H-\ddot{O}: + H-\ddot{O}: \rightleftharpoons \left[H-\overset{|}{\underset{|}{\ddot{O}}}-H\right]^+ + :\ddot{O}-H^- \quad [16.12]$$

We call this process the **autoionization** of water. No individual molecule remains ionized for long; the reactions are extremely rapid in both directions.

At room temperature only about two out of every 10^9 molecules are ionized at any given instant. Thus, pure water consists almost entirely of H_2O molecules and is an extremely poor conductor of electricity. Nevertheless, the autoionization of water is very important, as we will soon see.

The Ion Product of Water

Because the autoionization of water (Equation 16.12) is an equilibrium process, we can write the following equilibrium-constant expression for it:

$$K_c = [H_3O^+][OH^-] \qquad [16.13]$$

The term $[H_2O]$ is excluded from the equilibrium-constant expression because we exclude the concentrations of pure solids and liquids. ∞ (Section 15.4) Because this equilibrium-constant expression refers specifically to the autoionization of water, we use the symbol K_w to denote the equilibrium constant, which we call the **ion-product constant** for water. At 25 °C, K_w equals 1.0×10^{-14}. Thus, we have

$$K_w = [H_3O^+][OH^-] = 1.0 \times 10^{-14} \quad \text{(at 25 °C)} \qquad [16.14]$$

Because we use $H^+(aq)$ and $H_3O^+(aq)$ interchangeably to represent the hydrated proton, the autoionization reaction for water can also be written as

$$H_2O(l) \rightleftharpoons H^+(aq) + OH^-(aq) \qquad [16.15]$$

Likewise, the expression for K_w can be written in terms of either H_3O^- or H^+, and K_w has the same value in either case:

$$K_w = [H_3O^+][OH^-] = [H^+][OH^-] = 1.0 \times 10^{-14} \quad \text{(at 25 °C)} \qquad [16.16]$$

This equilibrium-constant expression and the value of K_w at 25 °C are extremely important, and you should commit them to memory.

What makes Equation 16.16 particularly useful is that it is applicable to pure water and to any aqueous solution. Although the equilibrium between $H^+(aq)$ and $OH^-(aq)$ as well as other ionic equilibria are affected somewhat by the presence of additional ions in solution, it is customary to ignore these ionic effects except in work requiring exceptional accuracy. Thus, Equation 16.16 is taken to be valid for any dilute aqueous solution, and it can be used to calculate either $[H^+]$ (if $[OH^-]$ is known) or $[OH^-]$ (if $[H^+]$ is known).

A solution in which $[H^+] = [OH^-]$ is said to be *neutral*. In most solutions H^+ and OH^- concentrations are not equal. As the concentration of one of these ions increases, the concentration of the other must decrease, so that the product of their concentrations equals 1.0×10^{-14}. In acidic solutions $[H^+]$ exceeds $[OH^-]$. In basic solutions $[OH^-]$ exceeds $[H^+]$.

■ **SAMPLE EXERCISE 16.4** | Calculating $[H^+]$ for Pure Water

Calculate the values of $[H^+]$ and $[OH^-]$ in a neutral solution at 25 °C.

SOLUTION

Analyze: We are asked to determine the concentrations of H^+ and OH^- ions in a neutral solution at 25 °C.

Plan: We will use Equation 16.16 and the fact that, by definition, $[H^+] = [OH^-]$ in a neutral solution.

Solve: We will represent the concentration of $[H^+]$ and $[OH^-]$ in neutral solution with x. This gives

$$[H^+][OH^-] = (x)(x) = 1.0 \times 10^{-14}$$
$$x^2 = 1.0 \times 10^{-14}$$
$$x = 1.0 \times 10^{-7} \, M = [H^+] = [OH^-]$$

In an acid solution $[H^+]$ is greater than $1.0 \times 10^{-7} \, M$; in a basic solution $[H^+]$ is less than $1.0 \times 10^{-7} \, M$.

■ **PRACTICE EXERCISE**

Indicate whether solutions with each of the following ion concentrations are neutral, acidic, or basic: **(a)** $[H^+] = 4 \times 10^{-9}\ M$; **(b)** $[OH^-] = 1 \times 10^{-7}\ M$; **(c)** $[OH^-] = 7 \times 10^{-13}\ M$.
Answers: **(a)** basic, **(b)** neutral, **(c)** acidic

■ **SAMPLE EXERCISE 16.5** | **Calculating [H⁺] from [OH⁻]**

Calculate the concentration of $H^+(aq)$ in **(a)** a solution in which $[OH^-]$ is 0.010 M, **(b)** a solution in which $[OH^-]$ is $1.8 \times 10^{-9}\ M$.
Note: In this problem and all that follow, we assume, unless stated otherwise, that the temperature is 25 °C.

SOLUTION

Analyze: We are asked to calculate the hydronium ion concentration in an aqueous solution where the hydroxide concentration is known.

Plan: We can use the equilibrium-constant expression for the autoionization of water and the value of K_w to solve for each unknown concentration.

Solve:

(a) Using Equation 16.16, we have: $[H^+][OH^-] = 1.0 \times 10^{-14}$

$$[H^+] = \frac{(1.0 \times 10^{-14})}{[OH^-]} = \frac{1.0 \times 10^{-14}}{0.010} = 1.0 \times 10^{-12}\ M$$

This solution is basic because $[OH^-] > [H^+]$

(b) In this instance $[H^+] = \frac{(1.0 \times 10^{-14})}{[OH^-]} = \frac{1.0 \times 10^{-14}}{1.8 \times 10^{-9}} = 5.6 \times 10^{-6}\ M$

This solution is acidic because $[H^+] > [OH^-]$

■ **PRACTICE EXERCISE**

Calculate the concentration of $OH^-(aq)$ in a solution in which **(a)** $[H^+] = 2 \times 10^{-6}\ M$; **(b)** $[H^+] = [OH^-]$; **(c)** $[H^+] = 100 \times [OH^-]$.
Answers: **(a)** $5 \times 10^{-9}\ M$, **(b)** $1.0 \times 10^{-7}\ M$, **(c)** $1.0 \times 10^{-8}\ M$

16.4 THE pH SCALE

The molar concentration of $H^+(aq)$ in an aqueous solution is usually very small. For convenience, we therefore usually express $[H^+]$ in terms of **pH**, which is the negative logarithm in base 10 of $[H^+]$.*

$$pH = -\log[H^+] \qquad [16.17]$$

If you need to review the use of logs, see Appendix A.

We can use Equation 16.17 to calculate the pH of a neutral solution at 25 °C (that is, one in which $[H^+] = 1.0 \times 10^{-7}\ M$):

$$pH = -\log(1.0 \times 10^{-7}) = -(-7.00) = 7.00$$

The pH of a neutral solution is 7.00 at 25 °C. Notice that the pH is reported with two decimal places. We do so because only the numbers to the right of the decimal point are the significant figures in a logarithm. Because our original value for the concentration ($1.0 \times 10^{-7}\ M$) has two significant figures, the corresponding pH has two decimal places (7.00).

What happens to the pH of a solution as we make the solution acidic? An acidic solution is one in which $[H^+] > 1.0 \times 10^{-7}\ M$. Because of the negative sign in Equation 16.17, *the pH decreases as $[H^+]$ increases.* For example, the pH of an acidic solution in which $[H^+] = 1.0 \times 10^{-3}\ M$ is

$$pH = -\log(1.0 \times 10^{-3}) = -(-3.00) = 3.00$$

At 25 °C the pH of an acidic solution is less than 7.00.

*Because [H⁺] and [H₃O⁺] are used interchangeably, you might see pH defined as $-\log[H_3O^+]$.

TABLE 16.1 ■ Relationships among [H⁺], [OH⁻], and pH at 25 °C			
Solution Type	$[H^+]$ (M)	$[OH^-]$ (M)	pH Value
Acidic	$>1.0 \times 10^{-7}$	$<1.0 \times 10^{-7}$	<7.00
Neutral	$=1.0 \times 10^{-7}$	$=1.0 \times 10^{-7}$	$=7.00$
Basic	$<1.0 \times 10^{-7}$	$>1.0 \times 10^{-7}$	>7.00

We can also calculate the pH of a basic solution, one in which $[OH^-] > 1.0 \times 10^{-7}$ M. Suppose $[OH^-] = 2.0 \times 10^{-3}$ M. We can use Equation 16.16 to calculate $[H^+]$ for this solution, and Equation 16.17 to calculate the pH:

$$[H^+] = \frac{K_w}{[OH^-]} = \frac{1.0 \times 10^{-14}}{2.0 \times 10^{-3}} = 5.0 \times 10^{-12} \, M$$

$$pH = -\log(5.0 \times 10^{-12}) = 11.30$$

At 25 °C the pH of a basic solution is greater than 7.00. The relationships among $[H^+]$, $[OH^-]$, and pH are summarized in Table 16.1▲ and in Figure 16.5▼.

The pH values characteristic of several familiar solutions are shown in Figure 16.5. Notice that a change in $[H^+]$ by a factor of 10 causes the pH to change by 1. Thus, a solution of pH 6 has 10 times the concentration of $H^+(aq)$ as a solution of pH 7.

GIVE IT SOME THOUGHT

(a) What is the significance of pH = 7? (b) How does the pH change as OH^- is added to the solution?

You might think that when $[H^+]$ is very small, as it is for some of the examples shown in Figure 16.5, it would be unimportant. Nothing is further from the truth. If $[H^+]$ is part of a kinetic rate law, then changing its concentration will

▶ Figure 16.5 **H⁺ concentrations and pH values of some common substances at 25 °C.** The pH of a solution can be estimated using the benchmark concentrations of H⁺ and OH⁻ corresponding to whole-number pH values.

	$[H^+]$ (M)	pH	pOH	$[OH^-]$ (M)
	$1 \,(1 \times 10^{-0})$	0.0	14.0	1×10^{-14}
Gastric juice	1×10^{-1}	1.0	13.0	1×10^{-13}
Lemon juice	1×10^{-2}	2.0	12.0	1×10^{-12}
Cola, vinegar	1×10^{-3}	3.0	11.0	1×10^{-11}
Wine	1×10^{-4}	4.0	10.0	1×10^{-10}
Tomatoes				
Banana				
Black coffee	1×10^{-5}	5.0	9.0	1×10^{-9}
Rain	1×10^{-6}	6.0	8.0	1×10^{-8}
Saliva				
Milk				
Human blood, tears	1×10^{-7}	7.0	7.0	1×10^{-7}
Egg white, seawater	1×10^{-8}	8.0	6.0	1×10^{-6}
Baking soda				
Borax	1×10^{-9}	9.0	5.0	1×10^{-5}
Milk of magnesia	1×10^{-10}	10.0	4.0	1×10^{-4}
Lime water				
	1×10^{-11}	11.0	3.0	1×10^{-3}
Household ammonia	1×10^{-12}	12.0	2.0	1×10^{-2}
Household bleach				
NaOH, 0.1 M	1×10^{-13}	13.0	1.0	1×10^{-1}
	1×10^{-14}	14.0	0.0	$1 \,(1 \times 10^{-0})$

More acidic

More basic

change the rate. ∞∞(Section 14.3) Thus, if the rate law is first order in $[H^+]$ doubling its concentration will double the rate even if the change is merely from 1×10^{-7} M to 2×10^{-7} M. In biological systems many reactions involve proton transfers and have rates that depend on $[H^+]$. Because the speeds of these reactions are crucial, the pH of biological fluids must be maintained within narrow limits. For example, human blood has a normal pH range of 7.35 to 7.45. Illness and even death can result if the pH varies much from this narrow range.

■■ SAMPLE EXERCISE 16.6 | Calculating pH from $[H^+]$

Calculate the pH values for the two solutions described in Sample Exercise 16.5.

SOLUTION

Analyze: We are asked to determine the pH of aqueous solutions for which we have already calculated $[H^+]$.

Plan: We can calculate pH using its defining equation, Equation 16.17.

Solve:

(a) In the first instance we found $[H^+]$ to be 1.0×10^{-12} M.

$$pH = -\log(1.0 \times 10^{-12}) = -(-12.00) = 12.00$$

Because 1.0×10^{-12} has two significant figures, the pH has two decimal places, **12.00**.
(b) For the second solution, $[H^+] = 5.6 \times 10^{-6}$ M. Before performing the calculation, it is helpful to estimate the pH. To do so, we note that $[H^+]$ lies between 1×10^{-6} and 1×10^{-5}.

$$1 \times 10^{-6} < 5.6 \times 10^{-6} < 1 \times 10^{-5}$$

Thus, we expect the pH to lie between 6.0 and 5.0. We use Equation 16.17 to calculate the pH.

$$pH = -\log(5.6 \times 10^{-6}) = 5.25$$

Check: After calculating a pH, it is useful to compare it to your prior estimate. In this case the pH, as we predicted, falls between 6 and 5. Had the calculated pH and the estimate not agreed, we should have reconsidered our calculation or estimate or both.

■■ PRACTICE EXERCISE

(a) In a sample of lemon juice $[H^+]$ is 3.8×10^{-4} M. What is the pH? **(b)** A commonly available window-cleaning solution has $[OH^-] = 1.9 \times 10^{-6}$ M. What is the pH?
Answers: **(a)** 3.42, **(b)** $[H^+] = 5.3 \times 10^{-9}$ M, so pH = 8.28

■■ SAMPLE EXERCISE 16.7 | Calculating $[H^+]$ from pH

A sample of freshly pressed apple juice has a pH of 3.76. Calculate $[H^+]$.

SOLUTION

Analyze: We need to calculate $[H^+]$ from pH.

Plan: We will use Equation 16.17, $pH = -\log[H^+]$, for the calculation.

Solve: From Equation 16.17, we have $\qquad pH = -\log[H^+] = 3.76$

Thus, $\qquad \log[H^+] = -3.76$

To find $[H^+]$, we need to determine the *antilog* of -3.76. Scientific calculators have an antilog function (sometimes labeled INV log or 10^x) that allows us to perform the calculation: $\qquad [H^+] = \text{antilog}(-3.76) = 10^{-3.76} = 1.7 \times 10^{-4}$ M

Comment: Consult the user's manual for your calculator to find out how to perform the antilog operation. The number of significant figures in $[H^+]$ is two because the number of decimal places in the pH is two.

Check: Because the pH is between 3.0 and 4.0, we know that $[H^+]$ will be between 1×10^{-3} and 1×10^{-4} M. Our calculated $[H^+]$ falls within this estimated range.

■■ PRACTICE EXERCISE

A solution formed by dissolving an antacid tablet has a pH of 9.18. Calculate $[H^+]$.
Answer: $[H^+] = 6.6 \times 10^{-10}$ M

pOH and Other "p" Scales

The negative log is also a convenient way of expressing the magnitudes of small quantities. We use the convention that the negative log of a quantity is labeled "p" (quantity). Thus, we can express the concentration of OH^- as pOH:

$$pOH = -\log[OH^-] \qquad [16.18]$$

Likewise, pK_w equals $-\log K_w$.

By taking the negative log of both sides of Equation 16.16,

$$-\log[H^+] + (-\log[OH^-]) = -\log K_w \qquad [16.19]$$

we obtain the following useful expression:

$$pH + pOH = 14.00 \quad (\text{at } 25\,°C) \qquad [16.20]$$

We will see in Section 16.8 that p scales are also useful when working with other equilibrium constants.

> ### GIVE IT SOME THOUGHT
>
> If the pOH for a solution is 3.00, what is the pH of the solution? Is the solution acidic or basic?

Measuring pH

The pH of a solution can be measured quickly and accurately with a *pH meter* (Figure 16.6 ◀). A complete understanding of how this important device works requires a knowledge of electrochemistry, a subject we take up in Chapter 20. In brief, a pH meter consists of a pair of electrodes connected to a meter capable of measuring small voltages, on the order of millivolts. A voltage, which varies with the pH, is generated when the electrodes are placed in a solution. This voltage is read by the meter, which is calibrated to give pH.

The electrodes used with pH meters come in many shapes and sizes, depending on their intended use. Electrodes have even been developed that are so small that they can be inserted into single living cells to monitor the pH of the cell medium. Pocket-size pH meters are also available for use in environmental studies, in monitoring industrial effluents, and in agricultural work.

Although less precise, acid–base indicators can be used to measure pH. An acid–base indicator is a colored substance that itself can exist in either an acid or a base form. The two forms have different colors. Thus, the indicator turns one color in an acid and another color in a base. If you know the pH at which the indicator turns from one form to the other, you can determine whether a solution has a higher or lower pH than this value. Litmus, for example, changes color in the vicinity of pH 7. The color change, however, is not very sharp. Red litmus indicates a pH of about 5 or lower, and blue litmus indicates a pH of about 8 or higher.

Some of the more common indicators are listed in Figure 16.7 ▶. Methyl orange, for example, changes color over the pH interval from 3.1 to 4.4. Below pH 3.1 it is in the acid form, which is red. In the interval between 3.1 and 4.4, it is gradually converted to its basic form, which has a yellow color. By pH 4.4 the conversion is complete, and the solution is yellow. Paper tape that is impregnated with several indicators and comes complete with a comparator color scale is widely used for approximate determinations of pH.

▲ **Figure 16.6 A digital pH meter.** The device is a millivoltmeter, and the electrodes immersed in the solution being tested produce a voltage that depends on the pH of the solution.

> ### GIVE IT SOME THOUGHT
>
> If phenolphthalein turns pink when added to a solution, what can we conclude about the pH of the solution?

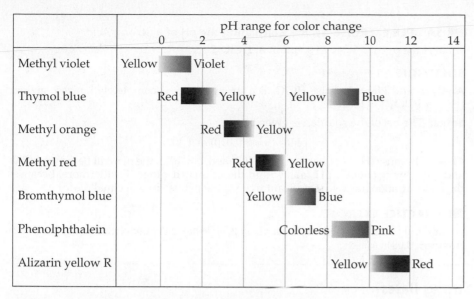

▲ Figure 16.7 Some common acid–base indicators. The pH ranges for the color changes of some common acid–base indicators. Most indicators have a useful range of about 2 pH units.

16.5 STRONG ACIDS AND BASES

The chemistry of an aqueous solution often depends critically on the pH of the solution. It is therefore important to examine how the pH of solutions relates to the concentrations of acids and bases. The simplest cases are those involving strong acids and strong bases. Strong acids and bases are *strong electrolytes*, existing in aqueous solution entirely as ions. There are relatively few common strong acids and bases, and we listed these substances in Table 4.2.

Strong Acids

The seven most common strong acids include six monoprotic acids (HCl, HBr, HI, HNO$_3$, HClO$_3$, and HClO$_4$), and one diprotic acid (H$_2$SO$_4$). Nitric acid (HNO$_3$) exemplifies the behavior of the monoprotic strong acids. For all practical purposes, an aqueous solution of HNO$_3$ consists entirely of H$_3$O$^+$ and NO$_3^-$ ions.

$$\text{HNO}_3(aq) + \text{H}_2\text{O}(l) \longrightarrow \text{H}_3\text{O}^+(aq) + \text{NO}_3^-(aq) \text{ (complete ionization)} \quad [16.21]$$

We have not used equilibrium arrows for Equation 16.21 because the reaction lies entirely to the right, the side with the ions. ∞ (Section 4.1) As noted in Section 16.3, we use H$_3$O$^+$(aq) and H$^+$(aq) interchangeably to represent the hydrated proton in water. Thus, we often simplify the equations for the ionization reactions of acids as follows:

$$\text{HNO}_3(aq) \longrightarrow \text{H}^+(aq) + \text{NO}_3^-(aq)$$

In an aqueous solution of a strong acid, the acid is normally the only significant source of H$^+$ ions.* As a result, calculating the pH of a solution of a strong monoprotic acid is straightforward because [H$^+$] equals the original concentration of acid. In a 0.20 M solution of HNO$_3$(aq), for example, [H$^+$] = [NO$_3^-$] = 0.20 M. The situation with the diprotic acid H$_2$SO$_4$ is more complex, as we will see in Section 16.6.

---

*If the concentration of the acid is 10^{-6} M or less, we also need to consider H$^+$ ions that result from the autoionization of H$_2$O. Normally, the concentration of H$^+$ from H$_2$O is so small that it can be neglected.

■■ **SAMPLE EXERCISE 16.8** | Calculating the pH of a Strong Acid

What is the pH of a 0.040 M solution of $HClO_4$?

SOLUTION

Analyze and Plan: Because $HClO_4$ is a strong acid, it is completely ionized, giving $[H^+] = [ClO_4^-] = 0.040\ M$.

Solve: The pH of the solution is given by

$$pH = -\log(0.040) = 1.40.$$

Check: Because $[H^+]$ lies between 1×10^{-2} and 1×10^{-1}, the pH will be between 2.0 and 1.0. Our calculated pH falls within the estimated range. Furthermore, because the concentration has two significant figures, the pH has two decimal places.

■■ **PRACTICE EXERCISE**

An aqueous solution of HNO_3 has a pH of 2.34. What is the concentration of the acid? *Answer:* 0.0046 M

Strong Bases

There are relatively few common strong bases. The most common soluble strong bases are the ionic hydroxides of the alkali metals (group 1A) and the heavier alkaline earth metals (group 2A), such as NaOH, KOH, and $Ca(OH)_2$. These compounds completely dissociate into ions in aqueous solution. Thus, a solution labeled 0.30 M NaOH consists of 0.30 M $Na^+(aq)$ and 0.30 M $OH^-(aq)$; there is essentially no undissociated NaOH.

■■ **SAMPLE EXERCISE 16.9** | Calculating the pH of a Strong Base

What is the pH of **(a)** a 0.028 M solution of NaOH, **(b)** a 0.0011 M solution of $Ca(OH)_2$?

SOLUTION

Analyze: We are asked to calculate the pH of two solutions of strong bases.

Plan: We can calculate each pH by either of two equivalent methods. First, we could use Equation 16.16 to calculate $[H^+]$ and then use Equation 16.17 to calculate the pH. Alternatively, we could use $[OH^-]$ to calculate pOH and then use Equation 16.20 to calculate the pH.

Solve:

(a) NaOH dissociates in water to give one OH^- ion per formula unit. Therefore, the OH^- concentration for the solution in (a) equals the stated concentration of NaOH, namely 0.028 M.

Method 1:

$$[H^+] = \frac{1.0 \times 10^{-14}}{0.028} = 3.57 \times 10^{-13}\ M \qquad pH = -\log(3.57 \times 10^{-13}) = 12.45$$

Method 2:

$$pOH = -\log(0.028) = 1.55 \qquad pH = 14.00 - pOH = 12.45$$

(b) $Ca(OH)_2$ is a strong base that dissociates in water to give two OH^- ions per formula unit. Thus, the concentration of $OH^-(aq)$ for the solution in part (b) is $2 \times (0.0011\ M) = 0.0022\ M$.

Method 1:

$$[H^+] = \frac{1.0 \times 10^{-14}}{0.0022} = 4.55 \times 10^{-12}\ M \qquad pH = -\log(4.55 \times 10^{-12}) = 11.34$$

Method 2:

$$pOH = -\log(0.0022) = 2.66 \qquad pH = 14.00 - pOH = 11.34$$

■■ **PRACTICE EXERCISE**

What is the concentration of a solution of **(a)** KOH for which the pH is 11.89; **(b)** $Ca(OH)_2$ for which the pH is 11.68?
Answers: **(a)** $7.8 \times 10^{-3}\ M$, **(b)** $2.4 \times 10^{-3}\ M$

Although all the hydroxides of the alkali metals (group 1A) are strong electrolytes, LiOH, RbOH, and CsOH are not commonly encountered in the laboratory. The hydroxides of the heavier alkaline earth metals, $Ca(OH)_2$, $Sr(OH)_2$, and $Ba(OH)_2$, are also strong electrolytes. They have limited solubilities, however, so they are used only when high solubility is not critical.

Strongly basic solutions are also created by certain substances that react with water to form $OH^-(aq)$. The most common of these contain the oxide ion. Ionic metal oxides, especially Na_2O and CaO, are often used in industry when a strong base is needed. The O^{2-} reacts with water to form OH^-, leaving virtually no O^{2-} remaining in the solution:

$$O^{2-}(aq) + H_2O(l) \longrightarrow 2\,OH^-(aq) \qquad [16.22]$$

Thus, a solution formed by dissolving 0.010 mol of $Na_2O(s)$ in enough water to form 1.0 L of solution will have $[OH^-] = 0.020\ M$ and a pH of 12.30.

> ### GIVE IT SOME THOUGHT
>
> The CH_3^- ion is the conjugate base of CH_4, and CH_4 shows no evidence of being an acid in water. What happens when CH_3^- is added to water?

16.6 WEAK ACIDS

Most acidic substances are weak acids and are therefore only partially ionized in aqueous solution. We can use the equilibrium constant for the ionization reaction to express the extent to which a weak acid ionizes. If we represent a general weak acid as HA, we can write the equation for its ionization reaction in either of the following ways, depending on whether the hydrated proton is represented as $H_3O^+(aq)$ or $H^+(aq)$:

$$HA(aq) + H_2O(l) \rightleftharpoons H_3O^+(aq) + A^-(aq) \qquad [16.23]$$

or

$$HA(aq) \rightleftharpoons H^+(aq) + A^-(aq) \qquad [16.24]$$

Because H_2O is the solvent, it is omitted from the equilibrium-constant expression. ∞ (Section 15.4) Thus, we can write the equilibrium-constant expression as either

$$K_c = \frac{[H_3O^+][A^-]}{[HA]} \quad \text{or} \quad K_c = \frac{[H^+][A^-]}{[HA]}$$

As we did for the ion-product constant for the autoionization of water, we change the subscript on this equilibrium constant to indicate the type of equation to which it corresponds.

$$K_a = \frac{[H_3O^+][A^-]}{[HA]} \quad \text{or} \quad K_a = \frac{[H^+][A^-]}{[HA]} \qquad [16.25]$$

The subscript a on K_a denotes that it is an equilibrium constant for the ionization of an acid, so K_a is called the **acid-dissociation constant**.

Table 16.2▼ shows the names, structures, and K_a values for several weak acids. Appendix D provides a more complete list. Many weak acids are organic compounds composed entirely of carbon, hydrogen, and oxygen. These compounds usually contain some hydrogen atoms bonded to carbon atoms and some bonded to oxygen atoms. In almost all cases the hydrogen atoms bonded to carbon do not ionize in water; instead, the acidic behavior of these compounds is due to the hydrogen atoms attached to oxygen atoms.

The magnitude of K_a indicates the tendency of the acid to ionize in water: *The larger the value of* K_a, *the stronger the acid.* Hydrofluoric acid (HF), for example, is the strongest acid listed in Table 16.2, and phenol (HOC_6H_5) is the weakest. Notice that K_a is typically less than 10^{-3}.

TABLE 16.2 ■ Some Weak Acids in Water at 25 °C

Acid	Structural Formula*	Conjugate Base	Equilibrium Reaction	K_a
Hydrofluoric (HF)	H—F	F⁻	$HF(aq) + H_2O(l) \rightleftharpoons H_3O^+(aq) + F^-(aq)$	6.8×10^{-4}
Nitrous (HNO₂)	H—O—N=O	NO₂⁻	$HNO_2(aq) + H_2O(l) \rightleftharpoons H_3O^+(aq) + NO_2^-(aq)$	4.5×10^{-4}
Benzoic (C₆H₅COOH)	H—O—C(=O)—⬡	C₆H₅COO⁻	$C_6H_5COOH(aq) + H_2O(l) \rightleftharpoons$ $H_3O^+(aq) + C_6H_5COO^-(aq)$	6.3×10^{-5}
Acetic (CH₃COOH)	H—O—C(=O)—C(H)(H)—H	CH₃COO⁻	$CH_3COOH(aq) + H_2O(l) \rightleftharpoons$ $H_3O^+(aq) + CH_3COO^-(aq)$	1.8×10^{-5}
Hypochlorous (HClO)	H—O—Cl	ClO⁻	$HClO(aq) + H_2O(l) \rightleftharpoons H_3O^+(aq) + ClO^-(aq)$	3.0×10^{-8}
Hydrocyanic (HCN)	H—C≡N	CN⁻	$HCN(aq) + H_2O(l) \rightleftharpoons H_3O^+(aq) + CN^-(aq)$	4.9×10^{-10}
Phenol (HOC₆H₅)	H—O—⬡	C₆H₅O⁻	$HOC_6H_5(aq) + H_2O(l) \rightleftharpoons$ $H_3O^+(aq) + C_6H_5O^-(aq)$	1.3×10^{-10}

* The proton that ionizes is shown in blue.

Calculating K_a from pH

In order to calculate either the K_a value for a weak acid or the pH of its solutions, we will use many of the skills for solving equilibrium problems that we developed in Section 15.5. In many cases the small magnitude of K_a allows us to use approximations to simplify the problem. In doing these calculations, it is important to realize that proton-transfer reactions are generally very rapid. As a result, the measured or calculated pH for a weak acid always represents an equilibrium condition.

■ SAMPLE EXERCISE 16.10 | Calculating K_a from Measured pH

A student prepared a 0.10 M solution of formic acid (HCOOH) and measured its pH. The pH at 25 °C was found to be 2.38. Calculate K_a for formic acid at this temperature.

SOLUTION

Analyze: We are given the molar concentration of an aqueous solution of weak acid and the pH of the solution, and we are asked to determine the value of K_a for the acid.

Plan: Although we are dealing specifically with the ionization of a weak acid, this problem is very similar to the equilibrium problems we encountered in Chapter 15. We can solve this problem using the method first outlined in Sample Exercise 15.9, starting with the chemical reaction and a tabulation of initial and equilibrium concentrations.

Solve: The first step in solving any equilibrium problem is to write the equation for the equilibrium reaction. The ionization of formic acid can be written as follows:

$$HCOOH(aq) \rightleftharpoons H^+(aq) + HCOO^-(aq)$$

The equilibrium-constant expression is

$$K_a = \frac{[H^+][HCOO^-]}{[HCOOH]}$$

From the measured pH, we can calculate [H⁺]:

$$pH = -\log[H^+] = 2.38$$
$$\log[H^+] = -2.38$$
$$[H^+] = 10^{-2.38} = 4.2 \times 10^{-3}\ M$$

We can do a little accounting to determine the concentrations of the species involved in the equilibrium. We imagine that the solution is initially 0.10 M in HCOOH molecules. We then consider the ionization of the acid into H^+ and $HCOO^-$. For each HCOOH molecule that ionizes, one H^+ ion and one $HCOO^-$ ion are produced in solution. Because the pH measurement indicates that $[H^+] = 4.2 \times 10^{-3}\ M$ at equilibrium, we can construct the following table:

	$HCOOH(aq)$ $\rightleftharpoons$	$H^+(aq)$ +	$HCOO^-(aq)$
Initial	0.10 M	0	0
Change	$-4.2 \times 10^{-3}\ M$	$+4.2 \times 10^{-3}\ M$	$+4.2 \times 10^{-3}\ M$
Equilibrium	$(0.10 - 4.2 \times 10^{-3})\ M$	$4.2 \times 10^{-3}\ M$	$4.2 \times 10^{-3}\ M$

Notice that we have neglected the very small concentration of $H^+(aq)$ that is due to the autoionization of H_2O. Notice also that the amount of HCOOH that ionizes is very small compared with the initial concentration of the acid. To the number of significant figures we are using, the subtraction yields 0.10 M:

$$(0.10 - 4.2 \times 10^{-3})\ M \simeq 0.10\ M$$

We can now insert the equilibrium concentrations into the expression for K_a:

$$K_a = \frac{(4.2 \times 10^{-3})(4.2 \times 10^{-3})}{0.10} = 1.8 \times 10^{-4}$$

Check: The magnitude of our answer is reasonable because K_a for a weak acid is usually between 10^{-3} and 10^{-10}.

■ **PRACTICE EXERCISE**

Niacin, one of the B vitamins, has the following molecular structure:

A 0.020 M solution of niacin has a pH of 3.26. What is the acid-dissociation constant, K_a, for niacin?
Answers: 1.5×10^{-5}

Percent Ionization

We have seen that the magnitude of K_a indicates the strength of a weak acid. Another measure of acid strength is **percent ionization**, which is defined as

$$\text{Percent ionization} = \frac{\text{concentration ionized}}{\text{original concentration}} \times 100\% \qquad [16.26]$$

The stronger the acid, the greater is the percent ionization.

For any acid, the concentration of acid that undergoes ionization equals the concentration of $H^+(aq)$ that forms, assuming that the autoionization of water is negligible. Thus, the percent ionization for an acid HA is also given by

$$\text{Percent ionization} = \frac{[H^+]_{\text{equilibrium}}}{[HA]_{\text{initial}}} \times 100\% \qquad [16.27]$$

For example, a 0.035 M solution of HNO_2 contains $3.7 \times 10^{-3}\ M\ H^+(aq)$. Thus, the percent ionization is

$$\text{Percent ionization} = \frac{[H^+]_{\text{equilibrium}}}{[HNO_2]_{\text{initial}}} \times 100\% = \frac{3.7 \times 10^{-3}\ M}{0.035\ M} \times 100\% = 11\%$$

■ **SAMPLE EXERCISE 16.11** | Calculating Percent Ionization

A 0.10 M solution of formic acid (HCOOH) contains $4.2 \times 10^{-3}\ M$ H$^+$(aq). Calculate the percentage of the acid that is ionized.

SOLUTION

Analyze: We are given the molar concentration of an aqueous solution of weak acid and the equilibrium concentration of H$^+$(aq) and asked to determine the percent ionization of the acid.

Plan: The percent ionization is given by Equation 16.27.

Solve:

$$\text{Percent ionization} = \frac{[\text{H}^+]_{\text{equilibrium}}}{[\text{HCOOH}]_{\text{initial}}} \times 100\% = \frac{4.2 \times 10^{-3}\ M}{0.10\ M} \times 100\% = 4.2\%$$

■ **PRACTICE EXERCISE**

A 0.020 M solution of niacin has a pH of 3.26. Calculate the percent ionization of the niacin.

Answer: 2.7%

Using K_a to Calculate pH

Knowing the value of K_a and the initial concentration of the weak acid, we can calculate the concentration of H$^+$(aq) in a solution of a weak acid. Let's calculate the pH of a 0.30 M solution of acetic acid (CH$_3$COOH), the weak acid responsible for the characteristic odor and acidity of vinegar, at 25 °C.

Our *first* step is to write the ionization equilibrium for acetic acid:

$$\text{CH}_3\text{COOH}(aq) \rightleftharpoons \text{H}^+(aq) + \text{CH}_3\text{COO}^-(aq) \qquad [16.28]$$

Notice that the hydrogen that ionizes is the one attached to an oxygen atom.

The *second* step is to write the equilibrium-constant expression and the value for the equilibrium constant. From Table 16.2, we have $K_a = 1.8 \times 10^{-5}$. Thus, we can write the following:

$$K_a = \frac{[\text{H}^+][\text{CH}_3\text{COO}^-]}{[\text{CH}_3\text{COOH}]} = 1.8 \times 10^{-5} \qquad [16.29]$$

As the *third* step, we need to express the concentrations that are involved in the equilibrium reaction. This can be done with a little accounting, as described in Sample Exercise 16.10. Because we want to find the equilibrium value for [H$^+$], let's call this quantity x. The concentration of acetic acid before any of it ionizes is 0.30 M. The chemical equation tells us that for each molecule of CH$_3$COOH that ionizes, one H$^+$(aq) and one CH$_3$COO$^-$(aq) are formed. Consequently, if x moles per liter of H$^+$(aq) form at equilibrium, x moles per liter of CH$_3$COO$^-$(aq) must also form, and x moles per liter of CH$_3$COOH must be ionized. This gives rise to the following table with the equilibrium concentrations shown on the last line:

	CH$_3$COOH(aq) $\rightleftharpoons$	H$^+$(aq) +	CH$_3$COO$^-$(aq)
Initial	0.30 M	0	0
Change	$-x\ M$	$+x\ M$	$+x\ M$
Equilibrium	$(0.30 - x)\ M$	$x\ M$	$x\ M$

As the *fourth* step of the problem, we need to substitute the equilibrium concentrations into the equilibrium-constant expression. The substitutions give the following equation:

$$K_a = \frac{[\text{H}^+][\text{CH}_3\text{COO}^-]}{[\text{CH}_3\text{COOH}]} = \frac{(x)(x)}{0.30 - x} = 1.8 \times 10^{-5} \qquad [16.30]$$

This expression leads to a quadratic equation in x, which we can solve by using an equation-solving calculator or by using the quadratic formula. We can also simplify the problem, however, by noting that the value of K_a is quite small. As a result, we anticipate that the equilibrium will lie far to the left and that x will be very small compared to the initial concentration of acetic acid. Thus, we will *assume* that x is negligible compared to 0.30, so that $0.30 - x$ is essentially equal to 0.30.

$$0.30 - x \sim 0.30$$

As we will see, we can (and should!) check the validity of this assumption when we finish the problem. By using this assumption, Equation 16.30 now becomes

$$K_a = \frac{x^2}{0.30} = 1.8 \times 10^{-5}$$

Solving for x, we have

$$x^2 = (0.30)(1.8 \times 10^{-5}) = 5.4 \times 10^{-6}$$

$$x = \sqrt{5.4 \times 10^{-6}} = 2.3 \times 10^{-3}$$

$$[H^+] = x = 2.3 \times 10^{-3}\ M$$

$$pH = -\log(2.3 \times 10^{-3}) = 2.64$$

We should now go back and check the validity of our simplifying assumption that $0.30 - x \simeq 0.30$. The value of x we determined is so small that, for this number of significant figures, the assumption is entirely valid. We are thus satisfied that the assumption was a reasonable one to make. Because x represents the moles per liter of acetic acid that ionize, we see that, in this particular case, less than 1% of the acetic acid molecules ionize:

$$\text{Percent ionization of CH}_3\text{COOH} = \frac{0.0023\ M}{0.30\ M} \times 100\% = 0.77\%$$

As a general rule, if the quantity x is more than about 5% of the initial value, it is better to use the quadratic formula. You should always check the validity of any simplifying assumptions after you have finished solving a problem.

GIVE IT SOME THOUGHT

Why can we generally assume that the equilibrium concentration of a weak acid equals its initial concentration?

Finally, we can compare the pH value of this weak acid to a solution of a strong acid of the same concentration. The pH of the 0.30 M solution of acetic acid is 2.64. By comparison, the pH of a 0.30 M solution of a strong acid such as HCl is $-\log(0.30) = 0.52$. As expected, the pH of a solution of a weak acid is higher than that of a solution of a strong acid of the same molarity.

■ SAMPLE EXERCISE 16.12 | Using K_a to Calculate pH

Calculate the pH of a 0.20 M solution of HCN. (Refer to Table 16.2 or Appendix D for the value of K_a.)

SOLUTION

Analyze: We are given the molarity of a weak acid and are asked for the pH. From Table 16.2, K_a for HCN is 4.9×10^{-10}.

Plan: We proceed as in the example just worked in the text, writing the chemical equation and constructing a table of initial and equilibrium concentrations in which the equilibrium concentration of H^+ is our unknown.

Solve: Writing both the chemical equation for the ionization reaction that forms $H^+(aq)$ and the equilibrium-constant (K_a) expression for the reaction:

$$HCN(aq) \rightleftharpoons H^+(aq) + CN^-(aq)$$

$$K_a = \frac{[H^+][CN^-]}{[HCN]} = 4.9 \times 10^{-10}$$

Next, we tabulate the concentration of the species involved in the equilibrium reaction, letting $x = [\mathrm{H}^+]$ at equilibrium:

	HCN(aq) $\rightleftharpoons$	H$^+$(aq) +	CN$^-$(aq)
Initial	0.20 M	0	0
Change	$-x$ M	$+x$ M	$+x$ M
Equilibrium	$(0.20 - x)$ M	x M	x M

Substituting the equilibrium concentrations from the table into the equilibrium-constant expression yields

$$K_a = \frac{(x)(x)}{0.20 - x} = 4.9 \times 10^{-10}$$

We next make the simplifying approximation that x, the amount of acid that dissociates, is small compared with the initial concentration of acid; that is,

$$0.20 - x \simeq 0.20$$

Thus,

$$\frac{x^2}{0.20} = 4.9 \times 10^{-10}$$

Solving for x, we have

$$x^2 = (0.20)(4.9 \times 10^{-10}) = 0.98 \times 10^{-10}$$

$$x = \sqrt{0.98 \times 10^{-10}} = 9.9 \times 10^{-6} \, M = [\mathrm{H}^+]$$

A concentration of $9.9 \times 10^{-6} \, M$ is much smaller than 5% of 0.20, the initial HCN concentration. Our simplifying approximation is therefore appropriate. We now calculate the pH of the solution:

$$\mathrm{pH} = -\log[\mathrm{H}^+] = -\log(9.9 \times 10^{-6}) = 5.00$$

■ PRACTICE EXERCISE

The K_a for niacin (Practice Exercise 16.10) is 1.5×10^{-5}. What is the pH of a 0.010 M solution of niacin?
Answer: 3.41

The properties of the acid solution that relate directly to the concentration of H$^+$(aq), such as electrical conductivity and rate of reaction with an active metal, are much less evident for a solution of a weak acid than for a solution of a strong acid of the same concentration. Figure 16.8▼ presents an experiment that demonstrates this difference by comparing the behavior of 1 M CH$_3$COOH and 1 M HCl. The 1 M CH$_3$COOH contains only 0.004 M H$^+$(aq), whereas the 1 M HCl solution contains 1 M H$^+$(aq). As a result, the rate of reaction with the metal is much faster for the solution of HCl.

As the concentration of a weak acid increases, the equilibrium concentration of H$^+$(aq) increases, as expected. However, as shown in Figure 16.9▶, *the percent ionization decreases as the concentration increases*. Thus, the concentration of H$^+$(aq) is not directly proportional to the concentration of the weak acid. For example, doubling the concentration of a weak acid does not double the concentration of H$^+$(aq). This lack of proportionality between the concentration of a weak acid and the concentration of H$^+$(aq) is demonstrated in Sample Exercise 16.13.

▶ Figure 16.8 **Reaction rates for weak and strong acids.** (a) The flask on the left contains 1 M CH$_3$COOH; the one on the right contains 1 M HCl. Each balloon contains the same amount of magnesium metal. (b) When the Mg metal is dropped into the acid, H$_2$ gas is formed. The rate of H$_2$ formation is higher for the 1 M HCl solution on the right as evidenced by more gas in the balloon. Eventually, the same amount of H$_2$ forms in both cases.

(a)   (b)

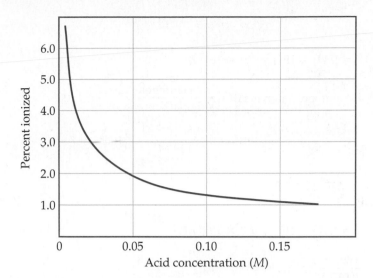

◀ Figure 16.9 **The effect of concentration on ionization of a weak acid.** The percent ionization of a weak acid decreases with increasing concentration. The data shown are for acetic acid.

■ **SAMPLE EXERCISE 16.13** | Using K_a to Calculate Percent Ionization

Calculate the percentage of HF molecules ionized in **(a)** a 0.10 M HF solution, **(b)** a 0.010 M HF solution.

SOLUTION

Analyze: We are asked to calculate the percent ionization of two HF solutions of different concentration. From Appendix D, we find $K_a = 6.8 \times 10^{-4}$.

Plan: We approach this problem as we would previous equilibrium problems. We begin by writing the chemical equation for the equilibrium and tabulating the known and unknown concentrations of all species. We then substitute the equilibrium concentrations into the equilibrium-constant expression and solve for the unknown concentration, that of H^+.

Solve:

(a) The equilibrium reaction and equilibrium concentrations are as follows:

	HF(aq) $\rightleftharpoons$	$H^+(aq)$ +	$F^-(aq)$
Initial	0.10 M	0	0
Change	$-x$ M	$+x$ M	$+x$ M
Equilibrium	$(0.10 - x)$ M	x M	x M

The equilibrium-constant expression is

$$K_a = \frac{[H^+][F^-]}{[HF]} = \frac{(x)(x)}{0.10 - x} = 6.8 \times 10^{-4}$$

When we try solving this equation using the approximation $0.10 - x = 0.10$ (that is, by neglecting the concentration of acid that ionizes in comparison with the initial concentration), we obtain

$$x = 8.2 \times 10^{-3} \, M$$

Because this value is greater than 5% of 0.10 M, we should work the problem without the approximation, using an equation-solving calculator or the quadratic formula. Rearranging our equation and writing it in standard quadratic form, we have

$$x^2 = (0.10 - x)(6.8 \times 10^{-4})$$
$$= 6.8 \times 10^{-5} - (6.8 \times 10^{-4})x$$
$$x^2 + (6.8 \times 10^{-4})x - 6.8 \times 10^{-5} = 0$$

This equation can be solved using the standard quadratic formula.

$$x = \frac{-b \pm \sqrt{b^2 - 4ac}}{2a}$$

Substituting the appropriate numbers gives

$$x = \frac{-6.8 \times 10^{-4} \pm \sqrt{(6.8 \times 10^{-4})^2 + 4(6.8 \times 10^{-5})}}{2}$$
$$= \frac{-6.8 \times 10^{-4} \pm 1.6 \times 10^{-2}}{2}$$

Of the two solutions, only the one that gives a positive value for x is chemically reasonable. Thus,

$$x = [\text{H}^+] = [\text{F}^-] = 7.9 \times 10^{-3}\ M$$

From our result, we can calculate the percent of molecules ionized:

$$\text{Percent ionization of HF} = \frac{\text{concentration ionized}}{\text{original concentration}} \times 100\%$$

$$= \frac{7.9 \times 10^{-3}\ M}{0.10\ M} \times 100\% = 7.9\%$$

(b) Proceeding similarly for the 0.010 M solution, we have

$$\frac{x^2}{0.010 - x} = 6.8 \times 10^{-4}$$

Solving the resultant quadratic expression, we obtain

$$x = [\text{H}^+] = [\text{F}^-] = 2.3 \times 10^{-3}\ M$$

The percentage of molecules ionized is

$$\frac{0.0023\ M}{0.010\ M} \times 100\% = 23\%$$

Comment: Notice that if we do not use the quadratic formula to solve the problem properly, we calculate 8.2% ionization for (a) and 26% ionization for (b). Notice also that in diluting the solution by a factor of 10, the percentage of molecules ionized increases by a factor of 3. This result is in accord with what we see in Figure 16.9. It is also what we would expect from Le Châtelier's principle. ∞ (Section 15.7) There are more "particles" or reaction components on the right side of the equation than on the left. Dilution causes the reaction to shift in the direction of the larger number of particles because this counters the effect of the decreasing concentration of particles.

■■ **PRACTICE EXERCISE**

In Practice Exercise 16.11, we found that the percent ionization of niacin ($K_a = 1.5 \times 10^{-5}$) in a 0.020 M solution is 2.7%. Calculate the percentage of niacin molecules ionized in a solution that is **(a)** 0.010 M, **(b)** $1.0 \times 10^{-3}\ M$.
Answers: **(a)** 3.9%, **(b)** 12%

Polyprotic Acids

Many acids have more than one ionizable H atom. These acids are known as **polyprotic acids**. For example, each of the H atoms in sulfurous acid (H_2SO_3) can ionize in successive steps:

$$\text{H}_2\text{SO}_3(aq) \rightleftharpoons \text{H}^+(aq) + \text{HSO}_3^-(aq) \qquad K_{a1} = 1.7 \times 10^{-2} \qquad [16.31]$$

$$\text{HSO}_3^-(aq) \rightleftharpoons \text{H}^+(aq) + \text{SO}_3^{2-}(aq) \qquad K_{a2} = 6.4 \times 10^{-8} \qquad [16.32]$$

The acid-dissociation constants for these equilibria are labeled K_{a1} and K_{a2}. The numbers on the constants refer to the particular proton of the acid that is ionizing. Thus, K_{a2} always refers to the equilibrium involving removal of the second proton of a polyprotic acid.

In the preceding example K_{a2} is much smaller than K_{a1}. Because of electrostatic attractions, we would expect a positively charged proton to be lost more readily from the neutral H_2SO_3 molecule than from the negatively charged HSO_3^- ion. This observation is general: *It is always easier to remove the first proton from a polyprotic acid than to remove the second.* Similarly, for an acid with three ionizable protons, it is easier to remove the second proton than the third. Thus, the K_a values become successively smaller as successive protons are removed.

▲ **GIVE IT SOME THOUGHT**

What is meant by the symbol K_{a3} for H_3PO_4?

The acid-dissociation constants for a few common polyprotic acids are listed in Table 16.3 ▶. Appendix D provides a more complete list. The structures for ascorbic and citric acids are shown in the margin. Notice that the K_a values for successive losses of protons from these acids usually differ by a factor of at least 10^3. Notice also that the value of K_{a1} for sulfuric acid is listed simply as "large."

TABLE 16.3 ■ Acid-Dissociation Constants of Some Common Polyprotic Acids

Name	Formula	K_{a1}	K_{a2}	K_{a3}
Ascorbic	$H_2C_6H_6O_6$	8.0×10^{-5}	1.6×10^{-12}	
Carbonic	H_2CO_3	4.3×10^{-7}	5.6×10^{-11}	
Citric	$H_3C_6H_5O_7$	7.4×10^{-4}	1.7×10^{-5}	4.0×10^{-7}
Oxalic	$H_2C_2O_4$	5.9×10^{-2}	6.4×10^{-5}	
Phosphoric	H_3PO_4	7.5×10^{-3}	6.2×10^{-8}	4.2×10^{-13}
Sulfurous	H_2SO_3	1.7×10^{-2}	6.4×10^{-8}	
Sulfuric	H_2SO_4	Large	1.2×10^{-2}	
Tartaric	$H_2C_4H_4O_6$	1.0×10^{-3}	4.6×10^{-5}	

Ascorbic acid
(vitamin C)

Citric acid

Sulfuric acid is a strong acid with respect to the removal of the first proton. Thus, the reaction for the first ionization step lies completely to the right:

$$H_2SO_4(aq) \longrightarrow H^+(aq) + HSO_4^-(aq) \quad \text{(complete ionization)}$$

HSO_4^-, on the other hand, is a weak acid for which $K_{a2} = 1.2 \times 10^{-2}$.

Because K_{a1} is so much larger than subsequent dissociation constants for these polyprotic acids, most of the $H^+(aq)$ in the solution comes from the first ionization reaction. As long as successive K_a values differ by a factor of 10^3 or more, it is possible to obtain a satisfactory estimate of the pH of polyprotic acid solutions by treating them as if they were monoprotic acids, considering only K_{a1}.

■ SAMPLE EXERCISE 16.14 | Calculating the pH of a Polyprotic Acid Solution

The solubility of CO_2 in pure water at 25 °C and 0.1 atm pressure is 0.0037 M. The common practice is to assume that all of the dissolved CO_2 is in the form of carbonic acid (H_2CO_3), which is produced by reaction between the CO_2 and H_2O:

$$CO_2(aq) + H_2O(l) \rightleftharpoons H_2CO_3(aq)$$

What is the pH of a 0.0037 M solution of H_2CO_3?

SOLUTION

Analyze: We are asked to determine the pH of a 0.0037 M solution of a polyprotic acid.

Plan: H_2CO_3 is a diprotic acid; the two acid-dissociation constants, K_{a1} and K_{a2} (Table 16.3), differ by more than a factor of 10^3. Consequently, the pH can be determined by considering only K_{a1}, thereby treating the acid as if it were a monoprotic acid.

Solve: Proceeding as in Sample Exercises 16.12 and 16.13, we can write the equilibrium reaction and equilibrium concentrations as follows:

	$H_2CO_3(aq)$ $\rightleftharpoons$	$H^+(aq)$ +	$HCO_3^-(aq)$
Initial	0.0037 M	0	0
Change	$-x$ M	$+x$ M	$+x$ M
Equilibrium	$(0.0037 - x)$ M	x M	x M

The equilibrium-constant expression is as follows:

$$K_{a1} = \frac{[H^+][HCO_3^-]}{[H_2CO_3]} = \frac{(x)(x)}{0.0037 - x} = 4.3 \times 10^{-7}$$

Solving this equation using an equation-solving calculator, we get

$$x = 4.0 \times 10^{-5} M$$

Alternatively, because K_{a1} is small, we can make the simplifying approximation that x is small, so that

$$0.0037 - x \approx 0.0037$$

Thus,

$$\frac{(x)(x)}{0.0037} = 4.3 \times 10^{-7}$$

Solving for x, we have

$$x^2 = (0.0037)(4.3 \times 10^{-7}) = 1.6 \times 10^{-9}$$

$$x = [H^+] = [HCO_3^-] = \sqrt{1.6 \times 10^{-9}} = 4.0 \times 10^{-5} M$$

The small value of x indicates that our simplifying assumption was justified. The pH is therefore

$$pH = -\log[H^+] = -\log(4.0 \times 10^{-5}) = 4.40$$

Comment: If we were asked to solve for $[CO_3^{2-}]$, we would need to use K_{a2}. Let's illustrate that calculation. Using the values of $[HCO_3^-]$ and $[H^+]$ calculated above, and setting $[CO_3^{2-}] = y$, we have the following initial and equilibrium concentration values:

	$HCO_3^-(aq)$	$\rightleftharpoons$	$H^+(aq)$	$+$	$CO_3^{2-}(aq)$
Initial	$4.0 \times 10^{-5}\,M$		$4.0 \times 10^{-5}\,M$		0
Change	$-y\,M$		$+y\,M$		$+y\,M$
Equilibrium	$(4.0 \times 10^{-5} - y)\,M$		$(4.0 \times 10^{-5} + y)\,M$		$y\,M$

Assuming that y is small compared to 4.0×10^{-5}, we have

$$K_{a2} = \frac{[H^+][CO_3^{2-}]}{[HCO_3^-]} = \frac{(4.0 \times 10^{-5})(y)}{4.0 \times 10^{-5}} = 5.6 \times 10^{-11}$$

$$y = 5.6 \times 10^{-11}\,M = [CO_3^{2-}]$$

The value calculated for y is indeed very small compared to 4.0×10^{-5}, showing that our assumption was justified. It also shows that the ionization of HCO_3^- is negligible compared to that of H_2CO_3, as far as production of H^+ is concerned. However, it is the *only* source of CO_3^{2-}, which has a very low concentration in the solution. Our calculations thus tell us that in a solution of carbon dioxide in water, most of the CO_2 is in the form of CO_2 or H_2CO_3, a small fraction ionizes to form H^+ and HCO_3^-, and an even smaller fraction ionizes to give CO_3^{2-}. Notice also that $[CO_3^{2-}]$ is numerically equal to K_{a2}.

■ **PRACTICE EXERCISE**

(a) Calculate the pH of a 0.020 M solution of oxalic acid ($H_2C_2O_4$). (See Table 16.3 for K_{a1} and K_{a2}.) **(b)** Calculate the concentration of oxalate ion, $[C_2O_4^{2-}]$, in this solution.
Answers: **(a)** pH = 1.80, **(b)** $[C_2O_4^{2-}] = 6.4 \times 10^{-5}\,M$

16.7 WEAK BASES

Many substances behave as weak bases in water. Weak bases react with water, abstracting protons from H_2O, thereby forming the conjugate acid of the base and OH^- ions.

$$B(aq) + H_2O(l) \rightleftharpoons HB^+(aq) + OH^-(aq) \qquad [16.33]$$

The equilibrium-constant expression for this reaction can be written as

$$K_b = \frac{[BH^+][OH^-]}{[B]} \qquad [16.34]$$

Water is the solvent, so it is omitted from the equilibrium-constant expression. The most commonly encountered weak base is ammonia.

$$NH_3(aq) + H_2O(l) \rightleftharpoons NH_4^+(aq) + OH^-(aq) \quad K_b = \frac{[NH_4^+][OH^-]}{[NH_3]} \quad [16.35]$$

As with K_w and K_a, the subscript "b" denotes that this equilibrium constant refers to a particular type of reaction, namely the ionization of a weak base in water. The constant K_b is called the **base-dissociation constant**. *The constant K_b always refers to the equilibrium in which a base reacts with H_2O to form the corresponding conjugate acid and OH^-.*

Table 16.4▶ lists the names, formulas, Lewis structures, equilibrium reactions, and values of K_b for several weak bases in water. Appendix D includes a more extensive list. These bases contain one or more lone pairs of electrons because a lone pair is necessary to form the bond with H^+. Notice that in the neutral molecules in Table 16.4, the lone pairs are on nitrogen atoms. The other bases listed are anions derived from weak acids.

TABLE 16.4 ▪ Some Weak Bases and Their Aqueous Solution Equilibria

Base	Lewis Structure	Conjugate Acid	Equilibrium Reaction	K_b
Ammonia (NH_3)	$H-\overset{..}{\underset{H}{N}}-H$	NH_4^+	$NH_3 + H_2O \rightleftharpoons NH_4^+ + OH^-$	1.8×10^{-5}
Pyridine (C_5H_5N)	(ring)$N:$	$C_5H_5NH^+$	$C_5H_5N + H_2O \rightleftharpoons C_5H_5NH^+ + OH^-$	1.7×10^{-9}
Hydroxylamine (H_2NOH)	$H-\overset{..}{\underset{H}{N}}-\overset{..}{\underset{..}{O}}H$	H_3NOH^+	$H_2NOH + H_2O \rightleftharpoons H_3NOH^+ + OH^-$	1.1×10^{-8}
Methylamine (NH_2CH_3)	$H-\overset{..}{\underset{H}{N}}-CH_3$	$NH_3CH_3^+$	$NH_2CH_3 + H_2O \rightleftharpoons NH_3CH_3^+ + OH^-$	4.4×10^{-4}
Hydrosulfide ion (HS^-)	$\left[H-\overset{..}{\underset{..}{S}}:\right]^-$	H_2S	$HS^- + H_2O \rightleftharpoons H_2S + OH^-$	1.8×10^{-7}
Carbonate ion (CO_3^{2-})	$\left[\overset{..}{\underset{..}{O}}=C\overset{:\overset{..}{O}:}{\underset{..}{O}:}\right]^{2-}$	HCO_3^-	$CO_3^{2-} + H_2O \rightleftharpoons HCO_3^- + OH^-$	1.8×10^{-4}
Hypochlorite ion (ClO^-)	$\left[:\overset{..}{\underset{..}{Cl}}-\overset{..}{\underset{..}{O}}:\right]^-$	$HClO$	$ClO^- + H_2O \rightleftharpoons HClO + OH^-$	3.3×10^{-7}

▪▪ **SAMPLE EXERCISE 16.15** | Using K_b to Calculate OH^-

Calculate the concentration of OH^- in a 0.15 M solution of NH_3.

SOLUTION

Analyze: We are given the concentration of a weak base and are asked to determine the concentration of OH^-.

Plan: We will use essentially the same procedure here as used in solving problems involving the ionization of weak acids; that is, we write the chemical equation and tabulate initial and equilibrium concentrations.

Solve: We first write the ionization reaction and the corresponding equilibrium-constant (K_b) expression:

$$NH_3(aq) + H_2O(l) \rightleftharpoons NH_4^+(aq) + OH^-(aq)$$

$$K_b = \frac{[NH_4^+][OH^-]}{[NH_3]} = 1.8 \times 10^{-5}$$

We then tabulate the equilibrium concentrations involved in the equilibrium:

	$NH_3(aq)$	+	$H_2O(l)$	$\rightleftharpoons$	$NH_4^+(aq)$	+	$OH^-(aq)$
Initial	0.15 M		—		0		0
Change	$-x$ M		—		$+x$ M		$+x$ M
Equilibrium	$(0.15 - x)$ M		—		x M		x M

(We ignore the concentration of H_2O because it is not involved in the equilibrium-constant expression.) Inserting these quantities into the equilibrium-constant expression gives the following:

$$K_b = \frac{[NH_4^+][OH^-]}{[NH_3]} = \frac{(x)(x)}{0.15 - x} = 1.8 \times 10^{-5}$$

Because K_b is small, we can neglect the small amount of NH_3 that reacts with water, as compared to the total NH_3 concentration; that is, we can neglect x relative to 0.15 M. Then we have

$$\frac{x^2}{0.15} = 1.8 \times 10^{-5}$$

$$x^2 = (0.15)(1.8 \times 10^{-5}) = 2.7 \times 10^{-6}$$

$$x = [NH_4^+] = [OH^-] = \sqrt{2.7 \times 10^{-6}} = 1.6 \times 10^{-3} \, M$$

Check: The value obtained for x is only about 1% of the NH_3 concentration, 0.15 M. Therefore, neglecting x relative to 0.15 was justified.

Comment: You may be asked to find the pH of a solution of a weak base. Once you have found $[OH^-]$, you can proceed as in Sample Exercise 16.9, where we calculated the pH of a strong base. In the present sample exercise, we have seen that the 0.15 M solution of NH_3 contains $[OH^-] = 1.6 \times 10^{-3}$ M. Thus, $pOH = -\log(1.6 \times 10^{-3}) = 2.80$, and $pH = 14.00 - 2.80 = 11.20$. The pH of the solution is above 7 because we are dealing with a solution of a base.

■ PRACTICE EXERCISE

Which of the following compounds should produce the highest pH as a 0.05 M solution: pyridine, methylamine, or nitrous acid?
Answer: methylamine (because it has the largest K_b value of the two amine bases in the list)

Types of Weak Bases

How can we recognize from a chemical formula whether a molecule or ion is able to behave as a weak base? Weak bases fall into two general categories. The first category contains neutral substances that have an atom with a nonbonding pair of electrons that can serve as a proton acceptor. Most of these bases, including all of the uncharged bases listed in Table 16.4, contain a nitrogen atom. These substances include ammonia and a related class of compounds called **amines**. In organic amines, one or more of the N—H bonds in NH_3 is replaced with a bond between N and C. Thus, the replacement of one N—H bond in NH_3 with a N—CH_3 bond gives methylamine, NH_2CH_3 (usually written CH_3NH_2). Like NH_3, amines can abstract a proton from a water molecule by forming an additional N—H bond, as shown here for methylamine:

$$H\overset{..}{-}N-CH_3(aq) + H_2O(l) \rightleftharpoons \left[H-N-CH_3 \right]^+ (aq) + OH^-(aq) \quad [16.36]$$

The chemical formula for the conjugate acid of methylamine is usually written $CH_3NH_3^+$.

The second general category of weak bases consists of the anions of weak acids. In an aqueous solution of sodium hypochlorite (NaClO), for example, NaClO dissociates to give Na^+ and ClO^- ions. The Na^+ ion is always a spectator ion in acid–base reactions. ∞ (Section 4.3) The ClO^- ion, however, is the conjugate base of a weak acid, hypochlorous acid. Consequently, the ClO^- ion acts as a weak base in water:

$$ClO^-(aq) + H_2O(l) \rightleftharpoons HClO(aq) + OH^-(aq) \quad K_b = 3.3 \times 10^{-7} \quad [16.37]$$

■ SAMPLE EXERCISE 16.16 | Using pH to Determine the Concentration of a Salt

A solution made by adding solid sodium hypochlorite (NaClO) to enough water to make 2.00 L of solution has a pH of 10.50. Using the information in Equation 16.37, calculate the number of moles of NaClO that were added to the water.

SOLUTION

Analyze: We are given the pH of a 2.00-L solution of NaClO and must calculate the number of moles of NaClO needed to raise the pH to 10.50. NaClO is an ionic compound consisting of Na^+ and ClO^- ions. As such, it is a strong electrolyte that completely dissociates in solution into Na^+, which is a spectator ion, and ClO^- ion, which is a weak base with $K_b = 3.3 \times 10^{-7}$ (Equation 16.37).

Plan: From the pH, we can determine the equilibrium concentration of OH^-. We can then construct a table of initial and equilibrium concentrations in which the initial concentration of ClO^- is our unknown. We can calculate $[ClO^-]$ using the equilibrium-constant expression, K_b.

Solve: We can calculate $[OH^-]$ by using either Equation 16.16 or Equation 16.20; we will use the latter method here:

$pOH = 14.00 - pH = 14.00 - 10.50 = 3.50$

$[OH^-] = 10^{-3.50} = 3.2 \times 10^{-4}$ M

This concentration is high enough that we can assume that Equation 16.37 is the only source of OH^-; that is, we can neglect any OH^- produced by the autoionization of H_2O. We now assume a value of x for the initial concentration of ClO^- and solve the equilibrium problem in the usual way.

	$ClO^-(aq)$	$+$ $H_2O(l)$ $\rightleftharpoons$	$HClO(aq)$	$+$	$OH^-(aq)$
Initial	$x\ M$	—	0		0
Change	$-3.2 \times 10^{-4}\ M$	—	$+3.2 \times 10^{-4}\ M$		$+3.2 \times 10^{-4}\ M$
Equilibrium	$(x - 3.2 \times 10^{-4})\ M$	—	$3.2 \times 10^{-4}\ M$		$3.2 \times 10^{-4}\ M$

We now use the expression for the base-dissociation constant to solve for x:

$$K_b = \frac{[HClO][OH^-]}{[ClO^-]} = \frac{(3.2 \times 10^{-4})^2}{x - 3.2 \times 10^{-4}} = 3.3 \times 10^{-7}$$

Thus

$$x = \frac{(3.2 \times 10^{-4})^2}{3.3 \times 10^{-7}} + (3.2 \times 10^{-4}) = 0.31\ M$$

We say that the solution is $0.31\ M$ in NaClO even though some of the ClO^- ions have reacted with water. Because the solution is $0.31\ M$ in NaClO and the total volume of solution is 2.00 L, 0.62 mol of NaClO is the amount of the salt that was added to the water.

■ PRACTICE EXERCISE

A solution of NH_3 in water has a pH of 11.17. What is the molarity of the solution?
Answer: $0.12\ M$

16.8 RELATIONSHIP BETWEEN K_a AND K_b

We have seen in a qualitative way that the stronger acids have the weaker conjugate bases. To see if we can find a corresponding *quantitative* relationship, let's consider the NH_4^+ and NH_3 conjugate acid–base pair. Each of these species reacts with water:

$$NH_4^+(aq) \rightleftharpoons NH_3(aq) + H^+(aq) \qquad [16.38]$$

$$NH_3(aq) + H_2O(l) \rightleftharpoons NH_4^+(aq) + OH^-(aq) \qquad [16.39]$$

Each of these equilibria is expressed by a characteristic dissociation constant:

$$K_a = \frac{[NH_3][H^+]}{[NH_4^+]}$$

$$K_b = \frac{[NH_4^+][OH^-]}{[NH_3]}$$

When Equations 16.38 and 16.39 are added together, the NH_4^+ and NH_3 species cancel and we are left with just the autoionization of water.

$$NH_4^+(aq) \rightleftharpoons NH_3(aq) + H^+(aq)$$

$$NH_3(aq) + H_2O(l) \rightleftharpoons NH_4^+(aq) + OH^-(aq)$$

$$H_2O(l) \rightleftharpoons H^+(aq) + OH^-(aq)$$

Recall that when two equations are added to give a third, the equilibrium constant associated with the third equation equals the product of the equilibrium constants for the two equations added together. ∞ (Section 15.3)

Applying this rule to our present example, when we multiply K_a and K_b, we obtain the following:

$$K_a \times K_b = \left(\frac{[NH_3][H^+]}{[NH_4^+]}\right)\left(\frac{[NH_4^+][OH^-]}{[NH_3]}\right)$$

$$= [H^+][OH^-] = K_w$$

TABLE 16.5 ▪ Some Conjugate Acid–Base Pairs

Acid	K_a	Base	K_b
HNO_3	(Strong acid)	NO_3^-	(Negligible basicity)
HF	6.8×10^{-4}	F^-	1.5×10^{-11}
$HC_2H_3O_2$	1.8×10^{-5}	$C_2H_3O_2^-$	5.6×10^{-10}
H_2CO_3	4.3×10^{-7}	HCO_3^-	2.3×10^{-8}
NH_4^+	5.6×10^{-10}	NH_3	1.8×10^{-5}
HCO_3^-	5.6×10^{-11}	CO_3^{2-}	1.8×10^{-4}
OH^-	(Negligible acidity)	O^{2-}	(Strong base)

Thus, the result of multiplying K_a times K_b is just the ion-product constant for water, K_w (Equation 16.16). We expect this result because adding Equations 16.38 and 16.39 gave us the autoionization equilibrium for water, for which the equilibrium constant is K_w.

This relationship is so important that it should receive special attention: *The product of the acid-dissociation constant for an acid and the base-dissociation constant for its conjugate base equals the ion-product constant for water.*

$$K_a \times K_b = K_w \qquad [16.40]$$

As the strength of an acid increases (larger K_a), the strength of its conjugate base must decrease (smaller K_b) so that the product $K_a \times K_b$ equals 1.0×10^{-14} at 25 °C. The K_a and K_b data in Table 16.5▲ demonstrate this relationship. Remember, this important relationship applies *only* to conjugate acid–base pairs.

Chemistry Put to Work   AMINES AND AMINE HYDROCHLORIDES

Many amines with low molecular weights have unpleasant "fishy" odors. Amines and NH_3 are produced by the anaerobic (absence of O_2) decomposition of dead animal or plant matter. Two such amines with very disagreeable odors are $H_2N(CH_2)_4NH_2$, known as *putrescine*, and $H_2N(CH_2)_5NH_2$, known as *cadaverine*.

Many drugs, including quinine, codeine, caffeine, and amphetamine (Benzedrine®), are amines. Like other amines, these substances are weak bases; the amine nitrogen is readily protonated upon treatment with an acid. The resulting products are called *acid salts*. If we use A as the abbreviation for an amine, the acid salt formed by reaction with hydrochloric acid can be written as AH^+Cl^-. It can also be written as A·HCl and referred to as a hydrochloride. Amphetamine hydrochloride, for example, is the acid salt formed by treating amphetamine with HCl:

Amphetamine

Amphetamine hydrochloride

Such acid salts are much less volatile, more stable, and generally more water soluble than the corresponding neutral amines. Many drugs that are amines are sold and administered as acid salts. Some examples of over-the-counter medications that contain amine hydrochlorides as active ingredients are shown in Figure 16.10▼.

Related Exercises: 16.77, 16.78, 16.108, 16.119, and 16.127

▲ **Figure 16.10** Some over-the-counter medications in which an amine hydrochloride is a major active ingredient.

By using Equation 16.40, we can calculate K_b for any weak base if we know K_a for its conjugate acid. Similarly, we can calculate K_a for a weak acid if we know K_b for its conjugate base. As a practical consequence, ionization constants are often listed for only one member of a conjugate acid–base pair. For example, Appendix D does not contain K_b values for the anions of weak acids because they can be readily calculated from the tabulated K_a values for their conjugate acids.

If you look up the values for acid- or base-dissociation constants in a chemistry handbook, you may find them expressed as pK_a or pK_b (that is, as $-\log K_a$ or $-\log K_b$). ∞ (Section 16.4) Equation 16.40 can be written in terms of pK_a and pK_b by taking the negative log of both sides:

$$pK_a + pK_b = pK_w = 14.00 \quad \text{at } 25\,^\circ\text{C} \qquad [16.41]$$

■ SAMPLE EXERCISE 16.17 | **Calculating K_a or K_b for a Conjugate Acid–Base Pair**

Calculate **(a)** the base-dissociation constant, K_b, for the fluoride ion (F^-); **(b)** the acid-dissociation constant, K_a, for the ammonium ion (NH_4^+).

SOLUTION

Analyze: We are asked to determine dissociation constants for F^-, the conjugate base of HF, and NH_4^+, the conjugate acid of NH_3.

Plan: Although neither F^- nor NH_4^+ appears in the tables, we can find the tabulated values for ionization constants for HF and NH_3, and use the relationship between K_a and K_b to calculate the ionization constants for each of the conjugates.

Solve:

(a) K_a for the weak acid, HF, is given in Table 16.2 and Appendix D as $K_a = 6.8 \times 10^{-4}$. We can use Equation 16.40 to calculate K_b for the conjugate base, F^-:

$$K_b = \frac{K_w}{K_a} = \frac{1.0 \times 10^{-14}}{6.8 \times 10^{-4}} = 1.5 \times 10^{-11}$$

(b) K_b for NH_3 is listed in Table 16.4 and in Appendix D as $K_b = 1.8 \times 10^{-5}$. Using Equation 16.40, we can calculate K_a for the conjugate acid, NH_4^+:

$$K_a = \frac{K_w}{K_b} = \frac{1.0 \times 10^{-14}}{1.8 \times 10^{-5}} = 5.6 \times 10^{-10}$$

■ PRACTICE EXERCISE

(a) Which of the following anions has the largest base-dissociation constant: NO_2^-, PO_4^{3-}, or N_3^-? **(b)** The base quinoline has the following structure:

Its conjugate acid is listed in handbooks as having a pK_a of 4.90. What is the base-dissociation constant for quinoline?

Answers: **(a)** PO_4^{3-} ($K_b = 2.4 \times 10^{-2}$), **(b)** 7.9×10^{-10}

16.9 ACID–BASE PROPERTIES OF SALT SOLUTIONS

Even before you began this chapter, you were undoubtedly aware of many substances that are acidic, such as HNO_3, HCl, and H_2SO_4, and others that are basic, such as NaOH and NH_3. However, our recent discussions have indicated that ions can also exhibit acidic or basic properties. For example, we calculated K_a for NH_4^+ and K_b for F^- in Sample Exercise 16.17. Such behavior implies that salt solutions can be acidic or basic. Before proceeding with further discussions of acids and bases, let's examine the way dissolved salts can affect pH.

Because nearly all salts are strong electrolytes, we can assume that when salts dissolve in water, they are completely dissociated. Consequently, the acid–base properties of salt solutions are due to the behavior of their constituent cations and anions. Many ions are able to react with water to generate $H^+(aq)$ or $OH^-(aq)$. This type of reaction is often called **hydrolysis**. The pH of an aqueous salt solution can be predicted qualitatively by considering the ions of which the salt is composed.

An Anion's Ability to React with Water

In general, an anion, X^-, in solution can be considered the conjugate base of an acid. For example, Cl^- is the conjugate base of HCl, and CH_3COO^- is the conjugate base of CH_3COOH. Whether an anion reacts with water to produce hydroxide depends upon the strength of the acid to which it is conjugate. To identify the acid and assess its strength, we can simply add a proton to the anion's formula:

$$X^- \text{ plus a proton } (H^+) \text{ gives } HX$$

If the acid determined in this way is one of the strong acids listed at the beginning of Section 16.5, then the anion in question will have a negligible tendency to abstract protons from water. ∞ (Section 16.2) Consequently, the anion X^- will not affect the pH of the solution. The presence of Cl^- in an aqueous solution, for example, does not result in the production of any OH^- and does not affect the pH. Thus, Cl^- is always a spectator in acid–base chemistry.

Conversely, if HX is *not* one of the seven strong acids, then it is a weak acid. In this case the conjugate base X^- is a weak base. This anion will therefore react to a small extent with water to produce the weak acid and hydroxide ions:

$$X^-(aq) + H_2O(l) \rightleftharpoons HX(aq) + OH^-(aq) \qquad [16.42]$$

The OH^- ion generated in this way increases the pH of the solution, making it basic. Acetate ion (CH_3COO^-), for example, being the conjugate base of a weak acid, reacts with water to produce acetic acid and hydroxide ions, thereby increasing the pH of the solution.*

$$CH_3COO^-(aq) + H_2O(l) \rightleftharpoons CH_3COOH(aq) + OH^-(aq) \qquad [16.43]$$

▲ **GIVE IT SOME THOUGHT**

What effect will each of the following ions have on the pH of a solution: NO_3^- and CO_3^{2-}?

Anions that still have ionizable protons, such as HSO_3^-, are amphiprotic. ∞ (Section 16.2) They can act as either acids or bases. Their behavior toward water will be determined by the relative magnitudes of K_a and K_b for the ion, as shown in Sample Exercise 16.19. If $K_a > K_b$, the ion will cause the solution to be acidic. If $K_b > K_a$, the solution will be basic.

A Cation's Ability to React with Water

Polyatomic cations whose formulas contain one or more protons can be considered the conjugate acids of weak bases. NH_4^+, for example, is the conjugate acid of the weak base NH_3. Thus, NH_4^+ is a weak acid and will donate a proton to water, producing hydronium ions and thereby lowering the pH:

$$NH_4^+(aq) + H_2O(l) \rightleftharpoons NH_3(aq) + H_3O^+(aq) \qquad [16.44]$$

---

*These rules apply to what are called normal salts. These salts contain no ionizable protons on the anion. The pH of an acid salt (such as $NaHCO_3$ or NaH_2PO_4) is affected by the hydrolysis of the anion and by its acid dissociation, as shown in Sample Exercise 16.19.

Most metal ions can also react with water to decrease the pH of an aqueous solution. The mechanism by which metal ions produce acidic solutions is described in Section 16.11. However, ions of alkali metals and of the heavier alkaline earth metals do not react with water and therefore do not affect pH. Note that these exceptions are the cations found in the strong bases. ∞(Section 16.5)

GIVE IT SOME THOUGHT

Which of the following cations has no effect on the pH of a solution: K^+, Fe^{2+}, or Al^{3+}?

Combined Effect of Cation and Anion in Solution

If an aqueous salt solution contains an anion that does not react with water and a cation that does not react with water, we expect the pH to be neutral. If the solution contains an anion that reacts with water to produce hydroxide and a cation that does not react with water, we expect the pH to be basic. If the solution contains a cation that reacts with water to produce hydronium and an anion that does not react with water, we expect the pH to be acidic. Finally, a solution may contain an anion and a cation *both* capable of reacting with water. In this case both hydroxide and hydronium will be produced. Whether the solution is basic, neutral, or acidic will depend upon the relative abilities of the ions to react with water.

To summarize:

1. An anion that is the conjugate base of a strong acid, for example, Br^-, will not affect the pH of a solution. (It will be a spectator ion in acid–base chemistry.)

2. An anion that is the conjugate base of a weak acid, for example, CN^-, will cause an increase in pH.

3. A cation that is the conjugate acid of a weak base, for example, $CH_3NH_3^+$, will cause a decrease in pH.

4. The cations of group 1A and heavier members of group 2A (Ca^{2+}, Sr^{2+}, and Ba^{2+}) will not affect pH. These are the cations of the strong Arrhenius bases. (They will be spectator ions in acid–base chemistry.)

5. Other metal ions will cause a decrease in pH.

6. When a solution contains both the conjugate base of a weak acid and the conjugate acid of a weak base, the ion with the larger equilibrium constant, K_a or K_b, will have the greater influence on the pH.

Figure 16.11 ▼ demonstrates the influence of several salts on pH.

(a)   (b)   (c)

◀ Figure 16.11 **Salt solutions can be neutral, acidic, or basic.** These three solutions contain the acid–base indicator bromthymol blue. (a) The NaCl solution is neutral (pH = 7.0); (b) the NH₄Cl solution is acidic (pH = 3.5); (c) the NaClO solution is basic (pH = 9.5).

■ **SAMPLE EXERCISE 16.18** | Determining Whether Salt Solutions Are Acidic, Basic, or Neutral

Determine whether aqueous solutions of each of the following salts will be acidic, basic, or neutral: **(a)** $Ba(CH_3COO)_2$, **(b)** NH_4Cl, **(c)** CH_3NH_3Br, **(d)** KNO_3, **(e)** $Al(ClO_4)_3$.

SOLUTION

Analyze: We are given the chemical formulas of five ionic compounds (salts) and asked whether their aqueous solutions will be acidic, basic, or neutral.

Plan: We can determine whether a solution of a salt is acidic, basic, or neutral by identifying the ions in solution and by assessing how each ion will affect the pH.

Solve:

(a) This solution contains barium ions and acetate ions. The cation, Ba^{2+}, is an ion of one of the heavy alkaline earth metals and will therefore not affect the pH (summary point 4). The anion, CH_3COO^-, is the conjugate base of the weak acid CH_3COOH and will hydrolyze to produce OH^- ions, thereby making the solution basic (summary point 2).
(b) This solution contains NH_4^+ and Cl^- ions. NH_4^+ is the conjugate acid of a weak base (NH_3) and is therefore acidic (summary point 3). Cl^- is the conjugate base of a strong acid (HCl) and therefore has no influence on the pH of the solution (summary point 1). Because the solution contains an ion that is acidic (NH_4^+) and one that has no influence on pH (Cl^-), the solution of NH_4Cl will be acidic.
(c) This solution contains $CH_3NH_3^+$ and Br^- ions. $CH_3NH_3^+$ is the conjugate acid of a weak base (CH_3NH_2, an amine) and is therefore acidic (summary point 3). Br^- is the conjugate base of a strong acid (HBr) and is therefore pH-neutral (summary point 1). Because the solution contains one ion that is acidic and one that is neutral, the solution of CH_3NH_3Br will be acidic.
(d) This solution contains the K^+ ion, which is a cation of group 1A, and the NO_3^- ion, which is the conjugate base of the strong acid HNO_3. Neither of the ions will react with water to any appreciable extent (summary points 1 and 4), making the solution neutral.
(e) This solution contains Al^{3+} and ClO_4^- ions. Cations, such as Al^{3+}, that are not in groups 1A or 2A are acidic (summary point 5). The ClO_4^- ion is the conjugate base of a strong acid ($HClO_4$) and therefore does not affect pH (summary point 1). Thus, the solution of $Al(ClO_4)_3$ will be acidic.

■ **PRACTICE EXERCISE**

In each of the following, indicate which salt in each of the following pairs will form the more acidic (or less basic) 0.010 M solution: **(a)** $NaNO_3$, or $Fe(NO_3)_3$; **(b)** KBr, or KBrO; **(c)** CH_3NH_3Cl, or $BaCl_2$, **(d)** NH_4NO_2, or NH_4NO_3.
Answers: **(a)** $Fe(NO_3)_3$, **(b)** KBr, **(c)** CH_3NH_3Cl, **(d)** NH_4NO_3

■ **SAMPLE EXERCISE 16.19** | Predicting Whether the Solution of an Amphiprotic Anion Is Acidic or Basic

Predict whether the salt Na_2HPO_4 will form an acidic solution or a basic solution on dissolving in water.

SOLUTION

Analyze: We are asked to predict whether a solution of Na_2HPO_4 will be acidic or basic. This substance is an ionic compound composed of Na^+ and HPO_4^{2-} ions.

Plan: We need to evaluate each ion, predicting whether each is acidic or basic. Because Na^+ is a cation of group 1A, we know that it has no influence on pH. It is merely a spectator ion in acid–base chemistry. Thus, our analysis of whether the solution is acidic or basic must focus on the behavior of the HPO_4^{2-} ion. We need to consider the fact that HPO_4^{2-} can act as either an acid or a base.

$$HPO_4^{2-}(aq) \rightleftharpoons H^+(aq) + PO_4^{3-}(aq) \qquad [16.45]$$

$$HPO_4^{2-}(aq) + H_2O \rightleftharpoons H_2PO_4^-(aq) + OH^-(aq) \qquad [16.46]$$

The reaction with the larger equilibrium constant will determine whether the solution is acidic or basic.

Solve: The value of K_a for Equation 16.45, as shown in Table 16.3, is 4.2×10^{-13}. We must calculate the value of K_b for Equation 16.46 from the value of K_a for its conjugate acid, $H_2PO_4^-$. We make use of the relationship shown in Equation 16.40.

$$K_a \times K_b = K_w$$

We want to know K_b for the base $HPO_4{}^{2-}$, knowing the value of K_a for the conjugate acid $H_2PO_4{}^-$:

$$K_b(HPO_4{}^{2-}) \times K_a(H_2PO_4{}^-) = K_w = 1.0 \times 10^{-14}$$

Because K_a for $H_2PO_4{}^-$ is 6.2×10^{-8} (Table 16.3), we calculate K_b for $HPO_4{}^{2-}$ to be 1.6×10^{-7}. This is more than 10^5 times larger than K_a for $HPO_4{}^{2-}$; thus, the reaction shown in Equation 16.46 predominates over that in Equation 16.45, and the solution will be basic.

■ PRACTICE EXERCISE

Predict whether the dipotassium salt of citric acid ($K_2HC_6H_5O_7$) will form an acidic or basic solution in water (see Table 16.3 for data).
Answer: acidic

16.10 ACID–BASE BEHAVIOR AND CHEMICAL STRUCTURE

When a substance is dissolved in water, it may behave as an acid, behave as a base, or exhibit no acid–base properties. How does the chemical structure of a substance determine which of these behaviors is exhibited by the substance? For example, why do some substances that contain OH groups behave as bases, releasing OH^- ions into solution, whereas others behave as acids, ionizing to release H^+ ions? Why are some acids stronger than others? In this section we will discuss briefly the effects of chemical structure on acid–base behavior.

Factors That Affect Acid Strength

A molecule containing H will transfer a proton only if the H—X bond is polarized in the following way:

$$\overset{+\longrightarrow}{H-X}$$

In ionic hydrides, such as NaH, the reverse is true; the H atom possesses a negative charge and behaves as a proton acceptor. Essentially nonpolar H—X bonds, such as the H—C bond in CH_4, produce neither acidic nor basic aqueous solutions.

A second factor that helps determine whether a molecule containing an H—X bond will donate a proton is the strength of the bond. Very strong bonds are less easily dissociated than weaker ones. This factor is important, for example, in the case of the hydrogen halides. The H—F bond is the most polar H—X bond. You therefore might expect that HF would be a very strong acid if the first factor were all that mattered. However, HF has the highest bond strength among the hydrogen halides, as seen in Table 8.4. As a result, HF is a weak acid, whereas all the other hydrogen halides are strong acids in water.

A third factor that affects the ease with which a hydrogen atom ionizes from HX is the stability of the conjugate base, X^-. In general, the greater the stability of the conjugate base, the stronger is the acid. The strength of an acid is often a combination of all three factors: (1) the polarity of the H—X bond, (2) the strength of the H—X bond, and (3) the stability of the conjugate base, X^-.

Binary Acids

In general, the H—X bond strength is the most important factor determining acid strength among the binary acids (those containing hydrogen and just one other element) in which X is in the same *group* in the periodic table. The strength of an H—X bond tends to decrease as the element X increases in size. As a result, the bond strength decreases and the acidity increases down a group. Thus, HCl is a stronger acid than HF, and H_2S is a stronger acid than H_2O.

			GROUP		
		4A	5A	6A	7A
Period 2		CH_4 No acid or base properties	NH_3 Weak base	H_2O ---	HF Weak acid
Period 3		SiH_4 No acid or base properties	PH_3 Weak base	H_2S Weak acid	HCl Strong acid

Increasing acid strength →

Increasing base strength ←

▲ **Figure 16.12 Trends in acid–base properties of binary hydrides.** The acidity of the binary compounds of hydrogen and nonmetals increases moving left to right across a period and moving top to bottom down a group.

Bond strengths change less moving across a row in the periodic table than they do down a group. As a result, bond polarity is the major factor determining acidity for binary acids in the same *row*. Thus, acidity increases as the electronegativity of the element X increases, as it generally does moving from left to right in a row. For example, the acidity of the second-row elements varies in the following order: $CH_4 < NH_3 \ll H_2O < HF$. Because the C—H bond is essentially nonpolar, CH_4 shows no tendency to form H^+ and CH_3^- ions. Although the N—H bond is polar, NH_3 has a nonbonding pair of electrons on the nitrogen atom that dominates its chemistry, so NH_3 acts as a base rather than as an acid. The periodic trends in the acid strengths of binary compounds of hydrogen and the nonmetals of periods 2 and 3 are summarized in Figure 16.12 ▲.

GIVE IT SOME THOUGHT

What is the major factor determining the increase in acidity of binary acids going down a column of the periodic table? What is the major factor going across a period?

Oxyacids

Many common acids, such as sulfuric acid, contain one or more O—H bonds:

$$H—\ddot{O}—\underset{\underset{:\ddot{O}:}{|}}{\overset{\overset{:\ddot{O}:}{||}}{S}}—\ddot{O}—H$$

Acids in which OH groups and possibly additional oxygen atoms are bound to a central atom are called **oxyacids**. The OH group is also present in bases. What factors determine whether an OH group will behave as a base or as an acid?

Let's consider an OH group bound to some atom Y, which might in turn have other groups attached to it:

$$\overset{|}{\underset{|}{Y}}—O—H$$

At one extreme, Y might be a metal, such as Na, K, or Mg. Because of their low electronegativities, the pair of electrons shared between Y and O is completely transferred to oxygen, and an ionic compound containing OH^- is formed. Such compounds are therefore sources of OH^- ions and behave as bases.

Shift of electron density

EN = 3.0

$K_a = 3.0 \times 10^{-8}$

EN = 2.5

$K_a = 2.3 \times 10^{-11}$

◀ **Figure 16.13 The acidity of oxyacids increases with increasing electronegativity of the central atom.** As the electronegativity of the atom attached to an OH group increases, the ease with which the hydrogen ion is released increases. The drift of electron density toward the electronegative atom further polarizes the O—H bond, which favors ionization. In addition, the electronegative atom will help stabilize the conjugate base, which also leads to a stronger acid. Because Cl is more electronegative than I, HClO is a stronger acid than HIO.

When Y is a nonmetal, the bond to O is covalent and the substance does not readily lose OH⁻. Instead, these compounds are either acidic or neutral. *Generally, as the electronegativity of Y increases, so will the acidity of the substance.* This happens for two reasons: First, as electron density is drawn toward Y, the O—H bond becomes weaker and more polar, thereby favoring loss of H⁺ (Figure 16.13▲). Second, because the conjugate base is usually an anion, its stability generally increases as the electronegativity of Y increases.

Many oxyacids contain additional oxygen atoms bonded to the central atom Y. The additional electronegative oxygen atoms pull electron density from the O—H bond, further increasing its polarity. Increasing the number of oxygen atoms also helps stabilize the conjugate base by increasing its ability to "spread out" its negative charge. Thus, *the strength of an acid will increase as additional electronegative atoms bond to the central atom Y.*

We can summarize these ideas as two simple rules that relate the acid strength of oxyacids to the electronegativity of Y and to the number of groups attached to Y.

1. For oxyacids that have the same number of OH groups and the same number of O atoms, acid strength increases with increasing electronegativity of the central atom Y. For example, the strength of the hypohalous acids, which have the structure H—O—Y, increases as the electronegativity of Y increases (Table 16.6▶).

2. For oxyacids that have the same central atom Y, acid strength increases as the number of oxygen atoms attached to Y increases. For example, the strength of the oxyacids of chlorine steadily increases from hypochlorous acid (HClO) to perchloric acid (HClO₄):

TABLE 16.6 ▪ Electronegativity Values (EN) of Y and Acid-Dissociation Constants		
Acid	**EN of Y**	**K_a**
HClO	3.0	3.0×10^{-8}
HBrO	2.8	2.5×10^{-9}
HIO	2.5	2.3×10^{-11}

Hypochlorous   Chlorous   Chloric   Perchloric

$$\ddot{\text{O}}:$$ $$\ddot{\text{O}}:$$

$$\text{H}—\ddot{\text{O}}—\ddot{\text{Cl}}:$$ $$\text{H}—\ddot{\text{O}}—\ddot{\text{Cl}}—\ddot{\text{O}}:$$ $$\text{H}—\ddot{\text{O}}—\ddot{\text{Cl}}—\ddot{\text{O}}:$$ $$\text{H}—\ddot{\text{O}}—\text{Cl}—\ddot{\text{O}}:$$

$$:\ddot{\text{O}}:$$

$K_a = 3.0 \times 10^{-8}$ $K_a = 1.1 \times 10^{-2}$ Strong acid   Strong acid

Increasing acid strength

Because the oxidation number of the central atom increases as the number of attached O atoms increases, this correlation can be stated in an equivalent way: In a series of oxyacids, the acidity increases as the oxidation number of the central atom increases.

■ **SAMPLE EXERCISE 16.20** | Predicting Relative Acidities from Composition and Structure

Arrange the compounds in each of the following series in order of increasing acid strength: **(a)** AsH_3, HI, NaH, H_2O; **(b)** H_2SO_4, H_2SeO_3, H_2SeO_4.

SOLUTION

Analyze: We are asked to arrange two sets of compounds in order from weakest acid to strongest acid. In (a), the substances are binary compounds containing H, whereas in (b) the substances are oxyacids.

Plan: For the binary compounds in part (a), we will consider the electronegativities of As, I, Na, and O relative to H. A higher electronegativity will cause the H to have a higher partial positive charge, causing the compound to be more acidic. For the oxyacids in part (b), we will consider both the relative electronegativities of the central atom (S and Se) and the number of oxygen atoms bonded to the central atom.

Solve:

(a) Because Na is on the left side of the periodic table, we know that it has a very low electronegativity. As a result, the hydrogen in NaH carries a negative charge. Thus NaH should be the least acidic (most basic) compound on the list. Because arsenic is less electronegative than oxygen, we might expect that AsH_3 would be a weak base toward water. We would make the same prediction by an extension of the trends shown in Figure 16.12. Further, we expect that the binary hydrogen compounds of the halogens, as the most electronegative element in each period, will be acidic relative to water. In fact, HI is one of the strong acids in water. Thus the order of increasing acidity is $NaH < AsH_3 < H_2O < HI$.

(b) The acids H_2SO_4 and H_2SeO_4 have the same number of O atoms and OH groups. In such cases, the acid strength increases with increasing electronegativity of the central atom. Because S is more electronegative than Se, we predict that H_2SO_4 is more acidic than H_2SeO_4. Next, we can compare H_2SeO_4 and H_2SeO_3. For acids with the same central atom, the acidity increases as the number of oxygen atoms bonded to the central atom increases. Thus, H_2SeO_4 should be a stronger acid than H_2SeO_3. Thus, we predict the order of increasing acidity to be $H_2SeO_3 < H_2SeO_4 < H_2SO_4$.

■ **PRACTICE EXERCISE**

In each of the following pairs choose the compound that leads to the more acidic (or less basic) solution: **(a)** HBr, HF; **(b)** PH_3, H_2S; **(c)** HNO_2, HNO_3; **(d)** H_2SO_3, H_2SeO_3.
Answers: **(a)** HBr, **(b)** H_2S, **(c)** HNO_3, **(d)** H_2SO_3

Carboxylic Acids

Another large group of acids is illustrated by acetic acid:

The portion of the structure shown in blue is called the *carboxyl group*, which is often written as COOH. Thus, the chemical formula of acetic acid is written as CH_3COOH, where only the hydrogen atom in the carboxyl group can be ionized. Acids that contain a carboxyl group are called **carboxylic acids**, and they form the largest category of organic acids. Formic acid and benzoic acid, whose structures are drawn below, are further examples of this large and important category of acids.

Formic acid   Benzoic acid

Acetic acid (CH_3COOH) is a weak acid ($K_a = 1.8 \times 10^{-5}$). Two factors contribute to the acidic behavior of carboxylic acids. First, the additional oxygen atom attached to the carbon of the carboxyl group draws electron density from the O—H bond, increasing its polarity and helping to stabilize the conjugate base.

Chemistry and Life   THE AMPHIPROTIC BEHAVIOR OF AMINO ACIDS

*A*mino acids are the building blocks of proteins. The general structure of amino acids is shown here, where different amino acids have different R groups attached to the central carbon atom:

Amine group (basic)   Carboxyl group (acidic)

For example, in *glycine*, which is the simplest amino acid, R is a hydrogen atom, whereas in *alanine*, R is a CH_3 group.

Glycine   Alanine

Amino acids contain a carboxyl group and can therefore serve as acids. They also contain an NH_2 group, characteristic of amines (Section 16.7), and thus they can also act as bases. Amino acids, therefore, are amphiprotic. For glycine, we might expect that the acid and the base reactions with water would be as follows:

Acid: $H_2N—CH_2—COOH(aq) + H_2O(l) \rightleftharpoons$
$H_2N—CH_2—COO^-(aq) + H_3O^+(aq)$ [16.47]

Base: $H_2N—CH_2—COOH(aq) + H_2O(l) \rightleftharpoons$
$^+H_3N—CH_2—COOH(aq) + OH^-(aq)$ [16.48]

The pH of a solution of glycine in water is about 6.0, indicating that it is a slightly stronger acid than a base.

The acid–base chemistry of amino acids is somewhat more complicated than shown in Equations 16.47 and 16.48, however. Because the COOH can act as an acid and the NH_2 group can act as a base, amino acids undergo a "self-contained" Brønsted–Lowry acid–base reaction in which the proton of the carboxyl group is transferred to the basic nitrogen atom:

proton transfer

Neutral molecule   Zwitterion   [16.49]

Although the form of the amino acid on the right side of Equation 16.49 is electrically neutral overall, it has a positively charged end and a negatively charged end. A molecule of this type is called a *zwitterion* (German for "hybrid ion").

Do amino acids exhibit any properties indicating that they behave as zwitterions? If so, they should behave similar to ionic substances. ∞ (Section 8.2) Crystalline amino acids (Figure 16.14▼) have relatively high melting points, usually above 200 °C, which is characteristic of ionic solids. Amino acids are far more soluble in water than in nonpolar solvents. In addition, the dipole moments of amino acids are large, consistent with a large separation of charge in the molecule. Thus, the ability of amino acids to act simultaneously as acids and bases has important effects on their properties.
Related Exercise: 16.119

▲ **Figure 16.14 Lysine.** One of the amino acids found in proteins, lysine is available as a dietary supplement. The L on the label refers to a specific arrangement of atoms that is found in naturally occurring amino acids. Molecules with the L arrangement are mirror images of molecules with the D arrangement, much like our left hand is a mirror image of our right hand.

Second, the conjugate base of a carboxylic acid (a *carboxylate anion*) can exhibit resonance (Section 8.6), which contributes further to the stability of the anion by spreading the negative charge over several atoms:

resonance

GIVE IT SOME THOUGHT

What group of atoms is present in all carboxylic acids?

16.11 LEWIS ACIDS AND BASES

For a substance to be a proton acceptor (a Brønsted–Lowry base), it must have an unshared pair of electrons for binding the proton. NH_3, for example, acts as a proton acceptor. Using Lewis structures, we can write the reaction between H^+ and NH_3 as follows:

$$H^+ + \overset{\displaystyle H}{\underset{\displaystyle H}{:N-H}} \longrightarrow \left[\overset{\displaystyle H}{\underset{\displaystyle H}{H-N-H}}\right]^+$$

G. N. Lewis was the first to notice this aspect of acid–base reactions. He proposed a definition of acid and base that emphasizes the shared electron pair: A **Lewis acid** is an electron-pair acceptor, and a **Lewis base** is an electron-pair donor.

Every base that we have discussed thus far—whether it be OH^-, H_2O, an amine, or an anion—is an electron-pair donor. Everything that is a base in the Brønsted–Lowry sense (a proton acceptor) is also a base in the Lewis sense (an electron-pair donor). In the Lewis theory, however, a base can donate its electron pair to something other than H^+. The Lewis definition therefore greatly increases the number of species that can be considered acids; H^+ is a Lewis acid, but not the only one. For example, consider the reaction between NH_3 and BF_3. This reaction occurs because BF_3 has a vacant orbital in its valence shell. ∞ (Section 8.7) It therefore acts as an electron-pair acceptor (a Lewis acid) toward NH_3, which donates the electron pair. The curved arrow shows the donation of a pair of electrons from N to B to form a covalent bond:

$$\underset{\text{Lewis base}}{\overset{\displaystyle H}{\underset{\displaystyle H}{H-N:}}} + \underset{\text{Lewis acid}}{\overset{\displaystyle F}{\underset{\displaystyle F}{B-F}}} \longrightarrow \overset{\displaystyle H \quad F}{\underset{\displaystyle H \quad F}{H-N-B-F}}$$

> ▲ **GIVE IT SOME THOUGHT**
>
> What feature must any molecule or ion have to act as a Lewis base?

Our emphasis throughout this chapter has been on water as the solvent and on the proton as the source of acidic properties. In such cases we find the Brønsted–Lowry definition of acids and bases to be the most useful. In fact, when we speak of a substance as being acidic or basic, we are usually thinking of aqueous solutions and using these terms in the Arrhenius or Brønsted–Lowry sense. The advantage of the Lewis theory is that it allows us to treat a wider variety of reactions, including those that do not involve proton transfer, as acid–base reactions. To avoid confusion, a substance such as BF_3 is rarely called an acid unless it is clear from the context that we are using the term in the sense of the Lewis definition. Instead, substances that function as electron-pair acceptors are referred to explicitly as "Lewis acids."

Lewis acids include molecules that, like BF_3, have an incomplete octet of electrons. In addition, many simple cations can function as Lewis acids. For example, Fe^{3+} interacts strongly with cyanide ions to form the ferricyanide ion, $Fe(CN)_6^{3-}$.

$$Fe^{3+} + 6[:C\equiv N:]^- \longrightarrow [Fe(C\equiv N:)_6]^{3-}$$

The Fe^{3+} ion has vacant orbitals that accept the electron pairs donated by the cyanide ions; we will learn more in Chapter 24 about just which orbitals are used by the Fe^{3+} ion. The metal ion is highly charged, too, which contributes to the interaction with CN^- ions.

Some compounds with multiple bonds can behave as Lewis acids. For example, the reaction of carbon dioxide with water to form carbonic acid (H_2CO_3) can be pictured as an attack by a water molecule on CO_2, in which the water acts as an electron-pair donor and the CO_2 as an electron-pair acceptor, as shown in the margin. The electron pair of one of the carbon–oxygen double bonds is moved onto the oxygen, leaving a vacant orbital on the carbon that can act as an electron-pair acceptor. We have shown the shift of these electrons with arrows. After forming the initial acid–base product, a proton moves from one oxygen to another, thereby forming carbonic acid. A similar kind of Lewis acid–base reaction takes place when any oxide of a nonmetal dissolves in water to form an acidic solution.

Hydrolysis of Metal Ions

As we have already seen, most metal ions behave as acids in aqueous solution. ∞ (Section 16.9) For example, an aqueous solution of $Fe(NO_3)_3$ is quite acidic. An aqueous solution of $ZnCl_2$ is also acidic, though to a lesser extent. The Lewis concept helps explain the interactions between metal ions and water molecules that give rise to this acidic behavior.

Because metal ions are positively charged, they attract the unshared electron pairs of water molecules. It is primarily this interaction, referred to as *hydration*, that causes salts to dissolve in water. ∞ (Section 13.1) The process of hydration can be thought of as a Lewis acid–base interaction in which the metal ion acts as a Lewis acid and the water molecules as Lewis bases. When a water molecule interacts with the positively charged metal ion, electron density is drawn from the oxygen, as illustrated in Figure 16.15▶. This flow of electron density causes the O—H bond to become more polarized; as a result, water molecules bound to the metal ion are more acidic than those in the bulk solvent.

The hydrated Fe^{3+} ion, $Fe(H_2O)_6^{3-}$, which we usually represent simply as $Fe^{3+}(aq)$, acts as a source of protons:

$$Fe(H_2O)_6^{3+}(aq) \rightleftharpoons Fe(H_2O)_5(OH)^{2+}(aq) + H^+(aq) \quad [16.50]$$

The acid-dissociation constant for this hydrolysis reaction has the value $K_a = 2 \times 10^{-3}$, so $Fe^{3+}(aq)$ is a fairly strong acid. Acid-dissociation constants for hydrolysis reactions generally increase with increasing charge and decreasing radius of the ion (Figure 16.15). Thus, the Cu^{2+} ion, which has a smaller charge and a larger radius than Fe^{3+}, forms less acidic solutions than Fe^{3+}: The K_a for $Cu^{2+}(aq)$ is 1×10^{-8}. The acid hydrolysis of a number of salts of metal ions is demonstrated in Figure 16.16▼. Note that the Na^+ ion, which is large and has only a 1+ charge (and which we have previously identified as the cation of a strong base), exhibits no acid hydrolysis and yields a neutral solution.

Weak electrostatic interaction

Weak shift of electron density

Strong electron interaction

Strong shift of electron density

▲ Figure 16.15 **The acidity of a hydrated cation depends on cation charge and size.** The interaction between a water molecule and a cation is much stronger when the cation is a smaller ion of higher charge. The pull of electron density toward the cation weakens the polar O—H bond of the water molecule and allows the transfer of a H^+ ion to a nearby water molecule. As a result, hydrated cations tend to be acidic, with their acidities increasing with increasing charge and decreasing size.

Salt:	$NaNO_3$	$Ca(NO_3)_2$	$Zn(NO_3)_2$	$Al(NO_3)_3$
Indicator:	Bromthymol blue	Bromthymol blue	Methyl red	Methyl orange
Estimated pH:	7.0	6.9	5.5	3.5

◀ Figure 16.16 **Effect of cations on the pH of a solution.** The pH values of 1.0 *M* solutions of a series of nitrate salts are estimated using acid–base indicators.

GIVE IT SOME THOUGHT

Which of the following cations will be most acidic and why: Ca^{2+}, Fe^{2+}, Fe^{3+}?

▌ SAMPLE INTEGRATIVE EXERCISE | Putting Concepts Together

Phosphorous acid (H_3PO_3) has the following Lewis structure.

$$
\begin{array}{c}
\overset{\displaystyle H}{\overset{\displaystyle |}{}} \\
:\!\ddot{O}\!-\!P\!-\!\ddot{O}\!-\!H \\
\overset{\displaystyle |}{:\!\ddot{O}\!-\!H}
\end{array}
$$

(a) Explain why H_3PO_3 is diprotic and not triprotic. **(b)** A 25.0-mL sample of a solution of H_3PO_3 is titrated with 0.102 M NaOH. It requires 23.3 mL of NaOH to neutralize both acidic protons. What is the molarity of the H_3PO_3 solution? **(c)** The original solution from part (b) has a pH of 1.59. Calculate the percent ionization and K_{a1} for H_3PO_3, assuming that $K_{a1} \gg K_{a2}$. **(d)** How does the osmotic pressure of a 0.050 M solution of HCl compare qualitatively with that of a 0.050 M solution of H_3PO_3? Explain.

SOLUTION

We will use what we have learned about molecular structure and its impact on acidic behavior to answer part (a). We will then use stoichiometry and the relationship between pH and [H^+] to answer parts (b) and (c). Finally, we will consider acid strength in order to compare the colligative properties of the two solutions in part (d).

(a) Acids have polar H—X bonds. From Figure 8.6 we see that the electronegativity of H is 2.1 and that of P is also 2.1. Because the two elements have the same electronegativity, the H—P bond is nonpolar. ∞ (Section 8.4) Thus, this H cannot be acidic. The other two H atoms, however, are bonded to O, which has an electronegativity of 3.5. The H—O bonds are therefore polar, with H having a partial positive charge. These two H atoms are consequently acidic.

(b) The chemical equation for the neutralization reaction is

$$H_3PO_3(aq) + 2\,NaOH(aq) \longrightarrow Na_2HPO_3(aq) + 2\,H_2O(l)$$

From the definition of molarity, M = mol/L, we see that moles = $M \times$ L. ∞ (Section 4.5) Thus, the number of moles of NaOH added to the solution is (0.0233 L)(0.102 mol/L) = 2.38×10^{-3} mol NaOH. The balanced equation indicates that 2 mol of NaOH is consumed for each mole of H_3PO_3. Thus, the number of moles of H_3PO_3 in the sample is

$$(2.38 \times 10^{-3}\text{ mol NaOH})\left(\frac{1\text{ mol }H_3PO_3}{2\text{ mol NaOH}}\right) = 1.19 \times 10^{-3}\text{ mol }H_3PO_3$$

The concentration of the H_3PO_3 solution, therefore, equals $(1.19 \times 10^{-3}$ mol)/(0.0250 L) = 0.0476 M.

(c) From the pH of the solution, 1.59, we can calculate [H^+] at equilibrium.

$$[H^+] = \text{antilog}(-1.59) = 10^{-1.59} = 0.026\ M \text{ (two significant figures)}$$

Because $K_{a1} \gg K_{a2}$, the vast majority of the ions in solution are from the first ionization step of the acid.

$$H_3PO_3(aq) \rightleftharpoons H^+(aq) + H_2PO_3^-(aq)$$

Because one $H_2PO_3^-$ ion forms for each H^+ ion formed, the equilibrium concentrations of H^+ and $H_2PO_3^-$ are equal: [H^+] = [$H_2PO_3^-$] = 0.026 M. The equilibrium concentration of H_3PO_3 equals the initial concentration minus the amount that ionizes to form H^+ and $H_2PO_3^-$: [H_3PO_3] = 0.0476 M − 0.026 M = 0.022 M (two significant figures). These results can be tabulated as follows:

	$H_3PO_3(aq)$ $\rightleftharpoons$	$H^+(aq)$ +	$H_2PO_3^-(aq)$
Initial	0.0476 M	0	0
Change	−0.026 M	+0.026 M	+0.026 M
Equilibrium	0.022 M	0.026 M	0.026 M

The percent ionization is

$$\text{Percent ionization} = \frac{[\text{H}^+]_{\text{equilibrium}}}{[\text{H}_3\text{PO}_3]_{\text{initial}}} \times 100\% = \frac{0.026\ M}{0.0476\ M} \times 100\% = 55\%$$

The first acid-dissociation constant is

$$K_{a1} = \frac{[\text{H}^+][\text{H}_2\text{PO}_3^-]}{[\text{H}_3\text{PO}_3]} = \frac{(0.026)(0.026)}{0.022} = 0.031$$

(d) Osmotic pressure is a colligative property and depends on the total concentration of particles in solution. ∞ (Section 13.5) Because HCl is a strong acid, a 0.050 M solution will contain 0.050 M $\text{H}^+(aq)$ and 0.050 M $\text{Cl}^-(aq)$, or a total of 0.100 mol/L of particles. Because H_3PO_3 is a weak acid, it ionizes to a lesser extent than HCl, and, hence, there are fewer particles in the H_3PO_3 solution. As a result, the H_3PO_3 solution will have the lower osmotic pressure.

CHAPTER REVIEW

SUMMARY AND KEY TERMS

Section 16.1 Acids and bases were first recognized by the properties of their aqueous solutions. For example, acids turn litmus red, whereas bases turn litmus blue. Arrhenius recognized that the properties of acidic solutions are due to $\text{H}^+(aq)$ ions and those of basic solutions are due to $\text{OH}^-(aq)$ ions.

Section 16.2 The Brønsted–Lowry concept of acids and bases is more general than the Arrhenius concept and emphasizes the transfer of a proton (H^+) from an acid to a base. The H^+ ion, which is merely a proton with no surrounding valence electrons, is strongly bound to water. For this reason, the **hydronium ion**, $\text{H}_3\text{O}^+(aq)$, is often used to represent the predominant form of H^+ in water instead of the simpler $\text{H}^+(aq)$.

A **Brønsted–Lowry acid** is a substance that donates a proton to another substance; a **Brønsted–Lowry base** is a substance that accepts a proton from another substance. Water is an **amphiprotic** substance, one that can function as either a Brønsted–Lowry acid or base, depending on the substance with which it reacts.

The **conjugate base** of a Brønsted–Lowry acid is the species that remains when a proton is removed from the acid. The **conjugate acid** of a Brønsted–Lowry base is the species formed by adding a proton to the base. Together, an acid and its conjugate base (or a base and its conjugate acid) are called a **conjugate acid–base pair**.

The acid–base strengths of conjugate acid–base pairs are related: The stronger an acid, the weaker is its conjugate base; the weaker an acid, the stronger is its conjugate base. In every acid–base reaction, the position of the equilibrium favors the transfer of the proton from the stronger acid to the stronger base.

Section 16.3 Water ionizes to a slight degree, forming $\text{H}^+(aq)$ and $\text{OH}^-(aq)$. The extent of this **autoionization** is expressed by the **ion-product constant** for water:

$K_w = [\text{H}^+][\text{OH}^-] = 1.0 \times 10^{-14}$ (25 °C). This relationship describes both pure water and aqueous solutions. The K_w expression indicates that the product of $[\text{H}^+]$ and $[\text{OH}^-]$ is a constant. Thus, as $[\text{H}^+]$ increases, $[\text{OH}^-]$ decreases. Acidic solutions are those that contain more $\text{H}^+(aq)$ than $\text{OH}^-(aq)$; basic solutions contain more $\text{OH}^-(aq)$ than $\text{H}^+(aq)$.

Section 16.4 The concentration of $\text{H}^+(aq)$ can be expressed in terms of **pH**: $\text{pH} = -\log[\text{H}^+]$. At 25 °C the pH of a neutral solution is 7.00, whereas the pH of an acidic solution is below 7.00, and the pH of a basic solution is above 7.00. The pX notation is also used to represent the negative log of other small quantities, as in pOH and $\text{p}K_w$. The pH of a solution can be measured using a pH meter, or it can be estimated using acid–base indicators.

Section 16.5 Strong acids are strong electrolytes, ionizing completely in aqueous solution. The common strong acids are HCl, HBr, HI, HNO_3, HClO_3, HClO_4, and H_2SO_4. The conjugate bases of strong acids have negligible basicity.

Common strong bases are the ionic hydroxides of alkali metals and the heavy alkaline earth metals. The cations of these metals have negligible acidity.

Section 16.6 Weak acids are weak electrolytes; only some of the molecules exist in solution in ionized form. The extent of ionization is expressed by the **acid-dissociation constant**, K_a, which is the equilibrium constant for the reaction $\text{HA}(aq) \rightleftharpoons \text{H}^+(aq) + \text{A}^-(aq)$, which can also be written $\text{HA}(aq) + \text{H}_2\text{O}(l) \rightleftharpoons \text{H}_3\text{O}^+(aq) + \text{A}^-(aq)$. The larger the value of K_a, the stronger is the acid. For solutions of the same concentration, a stronger acid also has a larger **percent ionization**. The concentration of a weak acid and its K_a value can be used to calculate the pH of a solution.

Polyprotic acids, such as H_2SO_3, have more than one ionizable proton. These acids have acid-dissociation constants that decrease in magnitude in the order

$K_{a1} > K_{a2} > K_{a3}$. Because nearly all the $H^+(aq)$ in a polyprotic acid solution comes from the first dissociation step, the pH can usually be estimated satisfactorily by considering only K_{a1}.

Sections 16.7 and 16.8 Weak bases include NH_3, **amines**, and the anions of weak acids. The extent to which a weak base reacts with water to generate the corresponding conjugate acid and OH^- is measured by the **base-dissociation constant**, K_b. This is the equilibrium constant for the reaction $B(aq) + H_2O(l) \rightleftharpoons HB^+(aq) + OH^-(aq)$, where B is the base.

The relationship between the strength of an acid and the strength of its conjugate base is expressed quantitatively by the equation $K_a \times K_b = K_w$, where K_a and K_b are dissociation constants for conjugate acid–base pairs.

Section 16.9 The acid–base properties of salts can be ascribed to the behavior of their respective cations and anions. The reaction of ions with water, with a resultant change in pH, is called **hydrolysis**. The cations of the alkali metals and the alkaline earth metals and the anions of strong acids do not undergo hydrolysis. They are always spectator ions in acid–base chemistry.

Section 16.10 The tendency of a substance to show acidic or basic characteristics in water can be correlated with its chemical structure. Acid character requires the presence of a highly polar H—X bond. Acidity is also favored when the H—X bond is weak and when the X^- ion is very stable.

For **oxyacids** with the same number of OH groups and the same number of O atoms, acid strength increases with increasing electronegativity of the central atom. For oxyacids with the same central atom, acid strength increases as the number of oxygen atoms attached to the central atom increases. The structures of **carboxylic acids**, which are organic acids containing the COOH group, also help us to understand their acidity.

Section 16.11 The Lewis concept of acids and bases emphasizes the shared electron pair rather than the proton. A **Lewis acid** is an electron-pair acceptor, and a **Lewis base** is an electron-pair donor. The Lewis concept is more general than the Brønsted–Lowry concept because it can apply to cases in which the acid is some substance other than H^+. The Lewis concept helps to explain why many hydrated metal cations form acidic aqueous solutions. The acidity of these cations generally increases as their charge increases and as the size of the metal ion decreases.

KEY SKILLS

- Understand the nature of the hydrated proton, represented as either $H^+(aq)$ or $H_3O^+(aq)$.
- Define and identify Arrhenius acids and bases.
- Define and identify Brønsted–Lowry acids and bases, and identify conjugate acid–base pairs.
- Relate the strength of an acid to the strength of its conjugate base.
- Understand how the equilibrium position of a proton transfer reaction relates the strengths of the acids and bases involved.
- Describe the autoionization of water and understand how $[H_3O^+]$ and $[OH^-]$ are related.
- Calculate the pH of a solution given $[H_3O^+]$ or $[OH^-]$.
- Calculate the pH of a strong acid or strong base given its concentration.
- Calculate K_a or K_b for a weak acid or weak base given its concentration and the pH of the solution.
- Calculate the pH of a weak acid or weak base or its percent ionization given its concentration and K_a or K_b.
- Calculate K_b for a weak base given K_a of its conjugate acid, and similarly calculate K_a from K_b.
- Predict whether an aqueous solution of a salt will be acidic, basic, or neutral.
- Predict the relative strength of a series of acids from their molecular structures.
- Define and identify Lewis acids and bases.

KEY EQUATIONS

- $K_w = [H_3O^+][OH^-] = [H^+][OH^-] = 1.0 \times 10^{-14}$ [16.16]   The ion product of water at 25 °C

- $pH = -\log[H^+]$ [16.17]   Definition of pH

- $pOH = -\log[OH^-]$ [16.18]   Definition of pOH

- pH + pOH = 14.00   [16.20]

  Relationship between pH and pOH

- $K_a = \dfrac{[H_3O^+][A^-]}{[HA]}$ or $K_a = \dfrac{[H^+][A^-]}{[HA]}$ [16.25]

  The acid dissociation constant for a weak acid, HA

- Percent ionization $= \dfrac{[H^+]_{equilibrium}}{[HA]_{initial}} \times 100\%$ [16.27]

  Percent ionization of a weak acid

- $K_b = \dfrac{[BH^+][OH^-]}{[B]}$ [16.34]

  The base dissociation constant for a weak base, B

- $K_a \times K_b = K_w$ [16.40]

  The relationship between the acid and base dissociation constants of a conjugate acid–base pair

VISUALIZING CONCEPTS

16.1 (a) Identify the Brønsted–Lowry acid and the Brønsted–Lowry base in the following reaction:

(b) Identify the Lewis acid and the Lewis base in the reaction. [Sections 16.2 and 16.11]

16.2 The following diagrams represent aqueous solutions of two monoprotic acids, HA (A = X or Y). The water molecules have been omitted for clarity. **(a)** Which is the stronger acid, HX or HY? **(b)** Which is the stronger base, X⁻ or Y⁻? **(c)** If you mix equal concentrations of HX and NaY, will the equilibrium

$$HX(aq) + Y^-(aq) \rightleftharpoons HY(aq) + X^-(aq)$$

lie mostly to the right ($K_c > 1$) or to the left ($K_c < 1$)? [Section 16.2]

16.3 The following diagrams represent aqueous solutions of three acids, HX, HY, and HZ. The water molecules have been omitted for clarity, and the hydrated proton is represented as a simple sphere rather than as a hydronium ion. **(a)** Which of the acids is a strong acid? Explain.

(b) Which acid would have the smallest acid-dissociation constant, K_a? **(c)** Which solution would have the highest pH? [Sections 16.5 and 16.6]

16.4 In which of the following cases is the approximation that the equilibrium concentration of H⁺(aq) is small relative to the initial concentration of HA likely to be most valid: **(a)** initial [HA] = 0.100 M and $K_a = 1.0 \times 10^{-6}$, **(b)** initial [HA] = 0.100 M and $K_a = 1.0 \times 10^{-4}$, **(c)** initial [HA] = 0.100 M and $K_a = 1.0 \times 10^{-3}$? [Section 16.6]

16.5 (a) Which of these three lines represents the effect of concentration on the percent ionization of a weak acid? **(b)** Explain in qualitative terms why the curve you choose has the shape it does. [Section 16.6]

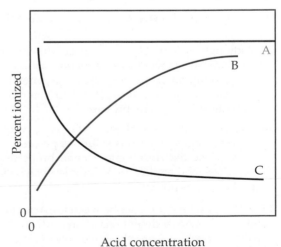

16.6 Refer to the diagrams accompanying Exercise 16.3. **(a)** Rank the anions, X^-, Y^-, and Z^-, in order of increasing basicity. **(b)** Which of the ions would have the largest base-dissociation constant, K_b? [Sections 16.2 and 16.8]

16.7 **(a)** Draw the Lewis structure for the following molecule and explain why it is able to act as a base. **(b)** To what class of organic compounds does this substance belong? (See the color key in Exercise 16.1.) [Section 16.7]

16.8 The following diagram represents an aqueous solution formed by dissolving a sodium salt of a weak acid in water. The diagram shows only the Na^+ ions, the X^- ions, and the HX molecules. What ion is missing from the diagram? If the drawing is completed by drawing all the ions, how many of the missing ions should be shown? [Section 16.9]

16.9 **(a)** What kinds of acids are represented by the following molecular models? **(b)** Indicate how the acidity of each molecule is affected by increasing the electronegativity of the atom X, and explain the origin of the effect. [Section 16.10]

(a)   (b)

16.10 In this model of acetylsalicylic acid (aspirin), identify the carboxyl group in the molecule. [Section 16.10]

16.11 Rank the following acids in order of increasing acidity: CH_3COOH, $CH_2ClCOOH$, $CHCl_2COOH$, CCl_3COOH, CF_3COOH. [Section 16.10]

16.12 **(a)** The following diagram represents the reaction of PCl_4^+ with Cl^-. Draw the Lewis structures for the reactants and products, and identify the Lewis acid and the Lewis base in the reaction.

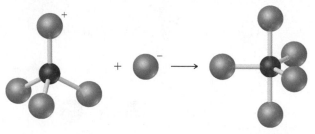

(b) The following reaction represents the acidity of a hydrated cation. How does the equilibrium constant for the reaction change as the charge of the cation increases? [Section 16.11]

EXERCISES

Arrhenius and Brønsted–Lowry Acids and Bases

16.13 Although HCl and H_2SO_4 have very different properties as pure substances, their aqueous solutions possess many common properties. List some general properties of these solutions, and explain their common behavior in terms of the species present.

16.14 Although pure NaOH and NH_3 have very different properties, their aqueous solutions possess many common properties. List some general properties of these solutions, and explain their common behavior in terms of the species present.

16.15 **(a)** What is the difference between the Arrhenius and the Brønsted–Lowry definitions of an acid? **(b)** $NH_3(g)$ and $HCl(g)$ react to form the ionic solid $NH_4Cl(s)$ (Figure 16.3). Which substance is the Brønsted–Lowry acid in this reaction? Which is the Brønsted–Lowry base?

16.16 **(a)** What is the difference between the Arrhenius and the Brønsted–Lowry definitions of a base? **(b)** When ammonia is dissolved in water, it behaves both as an Arrhenius base and as a Brønsted–Lowry base. Explain.

16.17 **(a)** Give the conjugate base of the following Brønsted–Lowry acids: **(i)** HIO_3, **(ii)** NH_4^+. **(b)** Give the conjugate acid of the following Brønsted–Lowry bases: **(i)** O^{2-}, **(ii)** $H_2PO_4^-$.

16.18 **(a)** Give the conjugate base of the following Brønsted–Lowry acids: **(i)** C_6H_5COOH. **(ii)** HPO_4^{2-}. **(b)** Give the conjugate acid of the following Brønsted–Lowry bases: **(i)** CO_3^{2-}, **(ii)** $C_2H_5NH_2$.

16.19 Designate the Brønsted–Lowry acid and the Brønsted–Lowry base on the left side of each of the following equations, and also designate the conjugate acid and conjugate base on the right side:

(a) $NH_4^+(aq) + CN^-(aq) \rightleftharpoons HCN(aq) + NH_3(aq)$

(b) $(CH_3)_3N(aq) + H_2O(l) \rightleftharpoons$
$(CH_3)_3NH^+(aq) + OH^-(aq)$

(c) $HCOOH(aq) + PO_4^{3-}(aq) \rightleftharpoons$
$HCOO^-(aq) + HPO_4^{2-}(aq)$

16.20 Designate the Brønsted–Lowry acid and the Brønsted–Lowry base on the left side of each equation, and also designate the conjugate acid and conjugate base on the right side.

(a) $HBrO(aq) + H_2O(l) \rightleftharpoons H_3O^+(aq) + BrO^-(aq)$

(b) $HSO_4^-(aq) + HCO_3^-(aq) \rightleftharpoons$
$SO_4^{2-}(aq) + H_2CO_3(aq)$

(c) $HSO_3^-(aq) + H_3O^+(aq) \rightleftharpoons H_2SO_3(aq) + H_2O(l)$

16.21 (a) The hydrogen oxalate ion ($HC_2O_4^-$) is amphiprotic. Write a balanced chemical equation showing how it acts as an acid toward water and another equation showing how it acts as a base toward water. (b) What is the conjugate acid of $HC_2O_4^-$? What is its conjugate base?

16.22 (a) Write an equation for the reaction in which $H_2C_6H_7O_5^-(aq)$ acts as a base in $H_2O(l)$. (b) Write an equation for the reaction in which $H_2C_6H_7O_5^-(aq)$ acts as an acid in $H_2O(l)$. (c) What is the conjugate acid of $H_2C_6H_7O_5^-$? What is its conjugate base?

16.23 Label each of the following as being a strong base, a weak base, or a species with negligible basicity. In each case write the formula of its conjugate acid, and indicate whether the conjugate acid is a strong acid, a weak acid, or a species with negligible acidity: (a) CH_3COO^-, (b) HCO_3^-, (c) O^{2-}, (d) Cl^-, (e) NH_3.

16.24 Label each of the following as being a strong acid, a weak acid, or a species with negligible acidity. In each case write the formula of its conjugate base, and indicate whether the conjugate base is a strong base, a weak base, or a species with negligible basicity: (a) HNO_2, (b) H_2SO_4, (c) HPO_4^{2-}, (d) CH_4, (e) $CH_3NH_3^+$ (an ion related to NH_4^+).

16.25 (a) Which of the following is the stronger Brønsted–Lowry acid, HBrO or HBr? (b) Which is the stronger Brønsted–Lowry base, F^- or Cl^-? Briefly explain your choices.

16.26 (a) Which of the following is the stronger Brønsted–Lowry acid, HNO_3 or HNO_2? (b) Which is the stronger Brønsted–Lowry base, NH_3 or H_2O? Briefly explain your choices.

16.27 Predict the products of the following acid–base reactions, and predict whether the equilibrium lies to the left or to the right of the equation:

(a) $O^{2-}(aq) + H_2O(l) \rightleftharpoons$

(b) $CH_3COOH(aq) + HS^-(aq) \rightleftharpoons$

(c) $NO_2^-(aq) + H_2O(l) \rightleftharpoons$

16.28 Predict the products of the following acid–base reactions, and predict whether the equilibrium lies to the left or to the right of the equation:

(a) $NH_4^+(aq) + OH^-(aq) \rightleftharpoons$

(b) $CH_3COO^-(aq) + H_3O^+(aq) \rightleftharpoons$

(c) $HCO_3^-(aq) + F^-(aq) \rightleftharpoons$

Autoionization of Water

16.29 (a) What does the term *autoionization* mean? (b) Explain why pure water is a poor conductor of electricity. (c) You are told that an aqueous solution is acidic. What does this statement mean?

16.30 (a) Write a chemical equation that illustrates the autoionization of water. (b) Write the expression for the ion-product constant for water, K_w. Why is [H_2O] absent from this expression? (c) A solution is described as basic. What does this statement mean?

16.31 Calculate [H^+] for each of the following solutions, and indicate whether the solution is acidic, basic, or neutral:
(a) [OH^-] = 0.00045 M; (b) [OH^-] = 8.8×10^{-9} M; (c) a solution in which [OH^-] is 100 times greater than [H^+].

16.32 Calculate [OH^-] for each of the following solutions, and indicate whether the solution is acidic, basic, or neutral:
(a) [H^+] = 0.0045 M; (b) [H^+] = 1.5×10^{-9} M; (c) a solution in which [H^+] is 10 times greater than [OH^-].

16.33 At the freezing point of water (0 °C), $K_w = 1.2 \times 10^{-15}$. Calculate [$H^+$] and [$OH^-$] for a neutral solution at this temperature.

16.34 Deuterium oxide (D_2O, where D is deuterium, the hydrogen-2 isotope) has an ion-product constant, K_w, of 8.9×10^{-16} at 20 °C. Calculate [D^+] and [OD^-] for pure (neutral) D_2O at this temperature.

The pH Scale

16.35 By what factor does [H^+] change for a pH change of (a) 2.00 units, (b) 0.50 units?

16.36 Consider two solutions, solution A and solution B. [H^+] in solution A is 500 times greater than that in solution B. What is the difference in the pH values of the two solutions?

16.37 (a) If NaOH is added to water, how does [H^+] change? How does pH change? (b) Use the pH values in Figure 16.5 to estimate the pH of a solution with [H^+] = 0.0006 M. Is the solution acidic or basic? (c) If the pH of a solution is 5.2, first estimate and then calculate the molar concentrations of $H^+(aq)$ and $OH^-(aq)$ in the solution.

16.38 **(a)** If HNO_3 is added to water, how does $[OH^-]$ change? How does pH change? **(b)** Use the pH values in Figure 16.5 to estimate the pH of a solution with $[OH^-] = 0.014\ M$. Is the solution acidic or basic? **(c)** If pH = 6.6, first estimate and then calculate the molar concentrations of $H^+(aq)$ and $OH^-(aq)$ in the solution.

16.39 Complete the following table by calculating the missing entries and indicating whether the solution is acidic or basic.

$[H^+]$	$OH^-(aq)$	pH	pOH	Acidic or basic?
$7.5 \times 10^{-3}\ M$				
	$3.6 \times 10^{-10}\ M$			
		8.25		
			5.70	

16.40 Complete the following table by calculating the missing entries. In each case indicate whether the solution is acidic or basic.

pH	pOH	$[H^+]$	$[OH^-]$	Acidic or basic?
11.25				
	6.02			
		$4.4 \times 10^{-4}\ M$		
			$8.5 \times 10^{-3}\ M$	

16.41 The average pH of normal arterial blood is 7.40. At normal body temperature (37 °C), $K_w = 2.4 \times 10^{-14}$. Calculate $[H^+]$, $[OH^-]$, and pOH for blood at this temperature.

16.42 Carbon dioxide in the atmosphere dissolves in raindrops to produce carbonic acid (H_2CO_3), causing the pH of clean, unpolluted rain to range from about 5.2 to 5.6. What are the ranges of $[H^+]$ and $[OH^-]$ in the raindrops?

Strong Acids and Bases

16.43 **(a)** What is a strong acid? **(b)** A solution is labeled $0.500\ M$ HCl. What is $[H^+]$ for the solution? **(c)** Which of the following are strong acids: HF, HCl, HBr, HI?

16.44 **(a)** What is a strong base? **(b)** A solution is labeled $0.035\ M$ $Sr(OH)_2$. What is $[OH^-]$ for the solution? **(c)** Is the following statement true or false? Because $Mg(OH)_2$ is not very soluble, it cannot be a strong base. Explain.

16.45 Calculate the pH of each of the following strong acid solutions: **(a)** $8.5 \times 10^{-3}\ M$ HBr, **(b)** 1.52 g of HNO_3 in 575 mL of solution, **(c)** 5.00 mL of $0.250\ M$ $HClO_4$ diluted to 50.0 mL, **(d)** a solution formed by mixing 10.0 mL of $0.100\ M$ HBr with 20.0 mL of $0.200\ M$ HCl.

16.46 Calculate the pH of each of the following strong acid solutions: **(a)** $0.00135\ M$ HNO_3, **(b)** 0.425 g of $HClO_4$ in 2.00 L of solution, **(c)** 5.00 mL of $1.00\ M$ HCl diluted to 0.500 L, **(d)** a mixture formed by adding 50.0 mL of $0.020\ M$ HCl to 150 mL of $0.010\ M$ HI.

16.47 Calculate $[OH^-]$ and pH for **(a)** $1.5 \times 10^{-3}\ M$ $Sr(OH)_2$, **(b)** 2.250 g of LiOH in 250.0 mL of solution, **(c)** 1.00 mL of $0.175\ M$ NaOH diluted to 2.00 L, **(d)** a solution formed by adding 5.00 mL of $0.105\ M$ KOH to 15.0 mL of $9.5 \times 10^{-2}\ M$ $Ca(OH)_2$.

16.48 Calculate $[OH^-]$ and pH for each of the following strong base solutions: **(a)** $0.082\ M$ KOH, **(b)** 1.065 g of KOH in 500.0 mL of solution, **(c)** 10.0 mL of $0.0105\ M$ $Ca(OH)_2$ diluted to 500.0 mL, **(d)** a solution formed by mixing 10.0 mL of $0.015\ M$ $Ba(OH)_2$ with 40.0 mL of $7.5 \times 10^{-3}\ M$ NaOH.

16.49 Calculate the concentration of an aqueous solution of NaOH that has a pH of 11.50.

16.50 Calculate the concentration of an aqueous solution of $Ca(OH)_2$ that has a pH of 12.05.

Weak Acids

16.51 Write the chemical equation and the K_a expression for the ionization of each of the following acids in aqueous solution. First show the reaction with $H^+(aq)$ as a product and then with the hydronium ion: **(a)** $HBrO_2$, **(b)** C_2H_5COOH.

16.52 Write the chemical equation and the K_a expression for the acid dissociation of each of the following acids in aqueous solution. First show the reaction with $H^+(aq)$ as a product and then with the hydronium ion: **(a)** C_6H_5COOH, **(b)** HCO_3^-.

16.53 Lactic acid ($CH_3CH(OH)COOH$) has one acidic hydrogen. A $0.10\ M$ solution of lactic acid has a pH of 2.44. Calculate K_a.

16.54 Phenylacetic acid ($C_6H_5CH_2COOH$) is one of the substances that accumulates in the blood of people with phenylketonuria, an inherited disorder that can cause mental retardation or even death. A $0.085\ M$ solution of $C_6H_5CH_2COOH$ has a pH of 2.68. Calculate the K_a value for this acid.

16.55 A 0.100 M solution of chloroacetic acid ($ClCH_2COOH$) is 11.0% ionized. Using this information, calculate [$ClCH_2COO^-$], [H^+], [$ClCH_2COOH$)], and K_a for chloroacetic acid.

16.56 A 0.100 M solution of bromoacetic acid ($BrCH_2COOH$) is 13.2% ionized. Calculate [H^+], [$BrCH_2COO^-$], and [$BrCH_2COOH$].

16.57 A particular sample of vinegar has a pH of 2.90. If acetic acid is the only acid that vinegar contains ($K_a = 1.8 \times 10^{-5}$), calculate the concentration of acetic acid in the vinegar.

16.58 How many moles of HF ($K_a = 6.8 \times 10^{-4}$) must be present in 0.200 L to form a solution with a pH of 3.25?

16.59 The acid-dissociation constant for benzoic acid (C_6H_5COOH) is 6.3×10^{-5}. Calculate the equilibrium concentrations of H_3O^+, $C_6H_5COO^-$, and C_6H_5COOH in the solution if the initial concentration of C_6H_5COOH is 0.050 M.

16.60 The acid-dissociation constant for hypochlorous acid (HClO) is 3.0×10^{-8}. Calculate the concentrations of H_3O^+, ClO^-, and HClO at equilibrium if the initial concentration of HClO is 0.0090 M.

16.61 Calculate the pH of each of the following solutions (K_a and K_b values are given in Appendix D): (a) 0.095 M propionic acid (C_2H_5COOH), (b) 0.100 M hydrogen chromate ion ($HCrO_4^-$), (c) 0.120 M pyridine (C_5H_5N).

16.62 Determine the pH of each of the following solutions (K_a and K_b values are given in Appendix D): (a) 0.095 M hypochlorous acid, (b) 0.0085 M phenol, (c) 0.095 M hydroxylamine.

16.63 Saccharin, a sugar substitute, is a weak acid with $pK_a = 2.32$ at 25 °C. It ionizes in aqueous solution as follows:

$$HNC_7H_4SO_3(aq) \rightleftharpoons H^+(aq) + NC_7H_4SO_3^-(aq)$$

What is the pH of a 0.10 M solution of this substance?

16.64 The active ingredient in aspirin is acetylsalicylic acid ($HC_9H_7O_4$), a monoprotic acid with $K_a = 3.3 \times 10^{-4}$ at 25 °C. What is the pH of a solution obtained by dissolving two extra-strength aspirin tablets, containing 500 mg of acetylsalicylic acid each, in 250 mL of water?

16.65 Calculate the percent ionization of hydrazoic acid (HN_3) in solutions of each of the following concentrations (K_a is given in Appendix D): (a) 0.400 M, (b) 0.100 M, (c) 0.0400 M.

16.66 Calculate the percent ionization of propionic acid (C_2H_5COOH) in solutions of each of the following concentrations (K_a is given in Appendix D): (a) 0.250 M, (b) 0.0800 M, (c) 0.0200 M.

[16.67] Show that for a weak acid, the percent ionization should vary as the inverse square root of the acid concentration.

[16.68] For solutions of a weak acid, a graph of pH versus the log of the initial acid concentration should be a straight line. What is the magnitude of the slope of that line?

[16.69] Citric acid, which is present in citrus fruits, is a triprotic acid (Table 16.3). Calculate the pH and the citrate ion ($C_6H_5O_7^{3-}$) concentration for a 0.050 M solution of citric acid. Explain any approximations or assumptions that you make in your calculations.

[16.70] Tartaric acid is found in many fruits, including grapes, and is partially responsible for the dry texture of certain wines. Calculate the pH and the tartarate ion ($C_4H_4O_6^{2-}$) concentration for a 0.250 M solution of tartaric acid, for which the acid-dissociation constants are listed in Table 16.3. Explain any approximations or assumptions that you make in your calculation.

Weak Bases

16.71 What is the essential structural feature of all Brønsted–Lowry bases?

16.72 What are two kinds of molecules or ions that commonly function as weak bases?

16.73 Write the chemical equation and the K_b expression for the ionization of each of the following bases in aqueous solution: (a) dimethylamine, $(CH_3)_2NH$; (b) carbonate ion, CO_3^{2-}; (c) formate ion, CHO_2^-.

16.74 Write the chemical equation and the K_b expression for the reaction of each of the following bases with water: (a) propylamine, $C_3H_7NH_2$; (b) monohydrogen phosphate ion, HPO_4^{2-}; (c) benzoate ion, $C_6H_5CO_2^-$.

16.75 Calculate the molar concentration of OH^- ions in a 0.075 M solution of ethylamine ($C_2H_5NH_2$; $K_b = 6.4 \times 10^{-4}$). Calculate the pH of this solution.

16.76 Calculate the molar concentration of OH^- ions in a 0.550 M solution of hypobromite ion (BrO^-; $K_b = 4.0 \times 10^{-6}$). What is the pH of this solution?

16.77 Ephedrine, a central nervous system stimulant, is used in nasal sprays as a decongestant. This compound is a weak organic base:

$$C_{10}H_{15}ON(aq) + H_2O(l) \rightleftharpoons C_{10}H_{15}ONH^+(aq) + OH^-(aq)$$

A 0.035 M solution of ephedrine has a pH of 11.33. (a) What are the equilibrium concentrations of $C_{10}H_{15}ON$, $C_{10}H_{15}ONH^+$, and OH^-? (b) Calculate K_b for ephedrine.

16.78 Codeine ($C_{18}H_{21}NO_3$) is a weak organic base. A $5.0 \times 10^{-3} M$ solution of codeine has a pH of 9.95. Calculate the value of K_b for this substance. What is the pK_b for this base?

The K_a–K_b Relationship; Acid–Base Properties of Salts

16.79 Although the acid-dissociation constant for phenol (C_6H_5OH) is listed in Appendix D, the base-dissociation constant for the phenolate ion ($C_6H_5O^-$) is not. (a) Explain why it is not necessary to list both K_a for phenol and K_b for the phenolate ion. (b) Calculate K_b for the phenolate ion. (c) Is the phenolate ion a weaker or stronger base than ammonia?

16.80 We can calculate K_b for the carbonate ion if we know the K_a values of carbonic acid (H_2CO_3). (a) Is K_{a1} or K_{a2} of carbonic acid used to calculate K_b for the carbonate ion? Explain. (b) Calculate K_b for the carbonate ion. (c) Is the carbonate ion a weaker or stronger base than ammonia?

16.81 (a) Given that K_a for acetic acid is 1.8×10^{-5} and that for hypochlorous acid is 3.0×10^{-8}, which is the stronger acid? (b) Which is the stronger base, the acetate ion or the hypochlorite ion? (c) Calculate K_b values for CH_3COO^- and ClO^-.

16.82 (a) Given that K_b for ammonia is 1.8×10^{-5} and that for hydroxylamine is 1.1×10^{-8}, which is the stronger base? (b) Which is the stronger acid, the ammonium ion or the hydroxylammonium ion? (c) Calculate K_a values for NH_4^+ and H_3NOH^+.

16.83 Using data from Appendix D, calculate $[OH^-]$ and pH for each of the following solutions: (a) 0.10 M NaCN, (b) 0.080 M Na_2CO_3, (c) a mixture that is 0.10 M in $NaNO_2$ and 0.20 M in $Ca(NO_2)_2$.

16.84 Using data from Appendix D, calculate $[OH^-]$ and pH for each of the following solutions: (a) 0.105 M NaF, (b) 0.035 M Na_2S, (c) a mixture that is 0.045 M in CH_3COONa and 0.055 M in $(CH_3COO)_2Ba$.

16.85 Predict whether aqueous solutions of the following compounds are acidic, basic, or neutral: (a) NH_4Br, (b) $FeCl_3$, (c) Na_2CO_3, (d) $KClO_4$, (e) $NaHC_2O_4$.

16.86 Predict whether aqueous solutions of the following substances are acidic, basic, or neutral: (a) $CrBr_3$, (b) LiI, (c) K_3PO_4, (d) $[CH_3NH_3]Cl$, (e) $KHSO_4$.

16.87 An unknown salt is either NaF, NaCl, or NaOCl. When 0.050 mol of the salt is dissolved in water to form 0.500 L of solution, the pH of the solution is 8.08. What is the identity of the salt?

16.88 An unknown salt is either KBr, NH_4Cl, KCN, or K_2CO_3. If a 0.100 M solution of the salt is neutral, what is the identity of the salt?

16.89 Sorbic acid (C_5H_7COOH) is a weak monoprotic acid with $K_a = 1.7 \times 10^{-5}$. Its salt (potassium sorbate) is added to cheese to inhibit the formation of mold. What is the pH of a solution containing 11.25 g of potassium sorbate in 1.75 L of solution?

16.90 Trisodium phosphate (Na_3PO_4) is available in hardware stores as TSP and is used as a cleaning agent. The label on a box of TSP warns that the substance is very basic (caustic or alkaline). What is the pH of a solution containing 35.0 g of TSP in a liter of solution?

Acid–Base Character and Chemical Structure

16.91 How does the acid strength of an oxyacid depend on (a) the electronegativity of the central atom; (b) the number of nonprotonated oxygen atoms in the molecule?

16.92 (a) How does the strength of an acid vary with the polarity and strength of the H—X bond? (b) How does the acidity of the binary acid of an element vary as a function of the electronegativity of the element? How does this relate to the position of the element in the periodic table?

16.93 Explain the following observations: (a) HNO_3 is a stronger acid than HNO_2; (b) H_2S is a stronger acid than H_2O; (c) H_2SO_4 is a stronger acid than HSO_4^-; (d) H_2SO_4 is a stronger acid than H_2SeO_4; (e) CCl_3COOH is a stronger acid than CH_3COOH.

16.94 Explain the following observations: (a) HCl is a stronger acid than H_2S; (b) H_3PO_4 is a stronger acid than H_3AsO_4; (c) HBrO$_3$ is a stronger acid than $HBrO_2$; (d) $H_2C_2O_4$ is a stronger acid than $HC_2O_4^-$; (e) benzoic acid (C_6H_5COOH) is a stronger acid than phenol (C_6H_5OH).

16.95 Based on their compositions and structures and on conjugate acid–base relationships, select the stronger base in each of the following pairs: (a) BrO^- or ClO^-, (b) BrO^- or BrO_2^-, (c) HPO_4^{2-} or $H_2PO_4^-$.

16.96 Based on their compositions and structures and on conjugate acid–base relationships, select the stronger base in each of the following pairs: (a) NO_3^- or NO_2^-, (b) PO_4^{3-} or AsO_4^{3-}, (c) HCO_3^- or CO_3^{2-}.

16.97 Indicate whether each of the following statements is true or false. For each statement that is false, correct the statement to make it true. (a) In general, the acidity of binary acids increases from left to right in a given row of the periodic table. (b) In a series of acids that have the same central atom, acid strength increases with the number of hydrogen atoms bonded to the central atom. (c) Hydrotelluric acid (H_2Te) is a stronger acid than H_2S because Te is more electronegative than S.

16.98 Indicate whether each of the following statements is true or false. For each statement that is false, correct the statement to make it true. **(a)** Acid strength in a series of H—X molecules increases with increasing size of X. **(b)** For acids of the same general structure but differing electronegativities of the central atoms, acid strength decreases with increasing electronegativity of the central atom. **(c)** The strongest acid known is HF because fluorine is the most electronegative element.

Lewis Acids and Bases

16.99 If a substance is an Arrhenius base, is it necessarily a Brønsted–Lowry base? Is it necessarily a Lewis base? Explain.

16.100 If a substance is a Lewis acid, is it necessarily a Brønsted–Lowry acid? Is it necessarily an Arrhenius acid? Explain.

16.101 Identify the Lewis acid and Lewis base among the reactants in each of the following reactions:
(a) $Fe(ClO_4)_3(s) + 6 H_2O(l) \rightleftharpoons$
$$Fe(H_2O)_6^{3+}(aq) + 3 ClO_4^-(aq)$$
(b) $CN^-(aq) + H_2O(l) \rightleftharpoons HCN(aq) + OH^-(aq)$
(c) $(CH_3)_3N(g) + BF_3(g) \rightleftharpoons (CH_3)_3NBF_3(s)$
(d) $HIO(lq) + NH_2^-(lq) \rightleftharpoons NH_3(lq) + IO^-(lq)$
(*lq* denotes liquid ammonia as solvent)

16.102 Identify the Lewis acid and Lewis base in each of the following reactions:
(a) $HNO_2(aq) + OH^-(aq) \rightleftharpoons NO_2^-(aq) + H_2O(l)$
(b) $FeBr_3(s) + Br^-(aq) \rightleftharpoons FeBr_4^-(aq)$
(c) $Zn^{2+}(aq) + 4 NH_3(aq) \rightleftharpoons Zn(NH_3)_4^{2+}(aq)$
(d) $SO_2(g) + H_2O(l) \rightleftharpoons H_2SO_3(aq)$

16.103 Predict which member of each pair produces the more acidic aqueous solution: **(a)** K^+ or Cu^{2+}, **(b)** Fe^{2+} or Fe^{3+}, **(c)** Al^{3+} or Ga^{3+}. Explain.

16.104 Which member of each pair produces the more acidic aqueous solution: **(a)** $ZnBr_2$ or $CdCl_2$, **(b)** $CuCl$ or $Cu(NO_3)_2$, **(c)** $Ca(NO_3)_2$ or $NiBr_2$? Explain.

ADDITIONAL EXERCISES

16.105 In your own words, define or explain **(a)** K_w, **(b)** K_a, **(c)** pOH, **(d)** pK_b.

16.106 Indicate whether each of the following statements is correct or incorrect. For those that are incorrect, explain why they are wrong.
(a) Every Brønsted–Lowry acid is also a Lewis acid.
(b) Every Lewis acid is also a Brønsted–Lowry acid.
(c) Conjugate acids of weak bases produce more acidic solutions than conjugate acids of strong bases.
(d) K^+ ion is acidic in water because it causes hydrating water molecules to become more acidic.
(e) The percent ionization of a weak acid in water increases as the concentration of acid decreases.

16.107 Predict whether the equilibrium lies to the right or to the left in the following reactions:
(a) $NH_4^+(aq) + PO_4^{3-}(aq) \rightleftharpoons$
$NH_3(aq) + HPO_4^{2-}(aq)$ (The ammonium ion is a stronger acid than the hydrogen phosphate ion.)
(b) $CH_3COOH(aq) + CN^-(aq) \rightleftharpoons$
$CH_3COO^-(aq) + HCN(aq)$ (The cyanide ion is a stronger base than the acetate ion.)

16.108 The odor of fish is due primarily to amines, especially methylamine (CH_3NH_2). Fish is often served with a wedge of lemon, which contains citric acid. The amine and the acid react forming a product with no odor, thereby making the less-than-fresh fish more appetizing.

Using data from Appendix D, calculate the equilibrium constant for the reaction of citric acid with methylamine, if only the first proton of the citric acid (K_{a1}) is important in the neutralization reaction.

16.109 Hemoglobin plays a part in a series of equilibria involving protonation-deprotonation and oxygenation-deoxygenation. The overall reaction is approximately as follows:

$$HbH^+(aq) + O_2(aq) \rightleftharpoons HbO_2(aq) + H^+(aq)$$

where Hb stands for hemoglobin, and HbO_2 for oxyhemoglobin. **(a)** The concentration of O_2 is higher in the lungs and lower in the tissues. What effect does high $[O_2]$ have on the position of this equilibrium? **(b)** The normal pH of blood is 7.4. Is the blood acidic, basic, or neutral? **(c)** If the blood pH is lowered by the presence of large amounts of acidic metabolism products, a condition known as acidosis results. What effect does lowering blood pH have on the ability of hemoglobin to transport O_2?

[16.110] Calculate the pH of a solution made by adding 2.50 g of lithium oxide (Li_2O) to enough water to make 1.500 L of solution.

16.111 Which of the following solutions has the higher pH? **(a)** a 0.1 *M* solution of a strong acid or a 0.1 *M* solution of a weak acid, **(b)** a 0.1 *M* solution of an acid with $K_a = 2 \times 10^{-3}$ or one with $K_a = 8 \times 10^{-6}$, **(c)** a 0.1 *M* solution of a base with $pK_b = 4.5$ or one with $pK_b = 6.5$.

[16.112] What is the pH of a solution that is 2.5×10^{-9} M in NaOH? Does your answer make sense?

16.113 Caproic acid ($C_5H_{11}COOH$) is found in small amounts in coconut and palm oils and is used in making artificial flavors. A saturated solution of the acid contains 11 g/L and has a pH of 2.94. Calculate K_a for the acid.

[16.114] A hypothetical acid H_2X is both a strong acid and a diprotic acid. (a) Calculate the pH of a 0.050 M solution of H_2X, assuming that only one proton ionizes per acid molecule. (b) Calculate the pH of the solution from part (a), now assuming that both protons of each acid molecule completely ionize. (c) In an experiment it is observed that the pH of a 0.050 M solution of H_2X is 1.27. Comment on the relative acid strengths of H_2X and HX^-. (d) Would a solution of the salt NaHX be acidic, basic, or neutral? Explain.

16.115 Butyric acid is responsible for the foul smell of rancid butter. The pK_a of butyric acid is 4.84. (a) Calculate the pK_b for the butyrate ion. (b) Calculate the pH of a 0.050 M solution of butyric acid. (c) Calculate the pH of a 0.050 M solution of sodium butyrate.

16.116 Arrange the following 0.10 M solutions in order of increasing acidity (decreasing pH): (i) NH_4NO_3, (ii) $NaNO_3$, (iii) CH_3COONH_4, (iv) NaF, (v) CH_3COONa.

[16.117] What are the concentrations of H^+, $H_2PO_4^-$, HPO_4^{2-}, and PO_4^{3-} in a 0.0250 M solution of H_3PO_4?

[16.118] Many moderately large organic molecules containing basic nitrogen atoms are not very soluble in water as neutral molecules, but they are frequently much more soluble as their acid salts. Assuming that pH in the stomach is 2.5, indicate whether each of the following compounds would be present in the stomach as the neutral base or in the protonated form: nicotine, $K_b = 7 \times 10^{-7}$; caffeine, $K_b = 4 \times 10^{-14}$; strychnine, $K_b = 1 \times 10^{-6}$; quinine, $K_b = 1.1 \times 10^{-6}$.

[16.119] The amino acid glycine (H_2N-CH_2-COOH) can participate in the following equilibria in water:

$$H_2N-CH_2-COOH + H_2O \rightleftharpoons$$
$$H_2N-CH_2-COO^- + H_3O^+ \quad K_a = 4.3 \times 10^{-3}$$

$$H_2N-CH_2-COOH + H_2O \rightleftharpoons$$
$${}^+H_3N-CH_2-COOH + OH^- \quad K_b = 6.0 \times 10^{-5}$$

(a) Use the values of K_a and K_b to estimate the equilibrium constant for the intramolecular proton transfer to form a zwitterion:

$$H_2N-CH_2-COOH \rightleftharpoons {}^+H_3N-CH_2-COO^-$$

What assumptions did you need to make? (b) What is the pH of a 0.050 M aqueous solution of glycine? (c) What would be the predominant form of glycine in a solution with pH 13? With pH 1?

16.120 The structural formula for acetic acid is shown in Table 16.2. Replacing hydrogen atoms on the carbon with chlorine atoms causes an increase in acidity, as follows:

Acid	Formula	K_a (25 °C)
Acetic	CH_3COOH	1.8×10^{-5}
Chloroacetic	$CH_2ClCOOH$	1.4×10^{-3}
Dichloroacetic	$CHCl_2COOH$	3.3×10^{-2}
Trichloroacetic	CCl_3COOH	2×10^{-1}

Using Lewis structures as the basis of your discussion, explain the observed trend in acidities in the series. Calculate the pH of a 0.010 M solution of each acid.

INTEGRATIVE EXERCISES

16.121 Calculate the number of $H^+(aq)$ ions in 1.0 mL of pure water at 25 °C.

16.122 How many milliliters of concentrated hydrochloric acid solution (36.0% HCl by mass, density = 1.18 g/mL) are required to produce 10.0 L of a solution that has a pH of 2.05?

16.123 The volume of an adult's stomach ranges from about 50 mL when empty to 1 L when full. If the stomach volume is 400 mL and its contents have a pH of 2, how many moles of H^+ does the stomach contain? Assuming that all the H^+ comes from HCl, how many grams of sodium hydrogen carbonate will totally neutralize the stomach acid?

16.124 Atmospheric CO_2 levels have risen by nearly 20% over the past 40 years from 315 ppm to 380 ppm. (a) Given that the average pH of clean, unpolluted rain today is 5.4, determine the pH of unpolluted rain 40 years ago. Assume that carbonic acid (H_2CO_3) formed by the reaction of CO_2 and water is the only factor influencing pH.

$$CO_2(g) + H_2O(l) \rightleftharpoons H_2CO_3(aq)$$

(b) What volume of CO_2 at 25 °C and 1.0 atm is dissolved in a 20.0-L bucket of today's rainwater?

[16.125] In many reactions the addition of $AlCl_3$ produces the same effect as the addition of H^+. (a) Draw a Lewis structure for $AlCl_3$ in which no atoms carry formal charges, and determine its structure using the VSEPR method. (b) What characteristic is notable about the structure in part (a) that helps us understand the acidic character of $AlCl_3$? (c) Predict the result of the reaction between $AlCl_3$ and NH_3 in a solvent that does not participate as a reactant. (d) Which acid–base theory is most suitable for discussing the similarities between $AlCl_3$ and H^+?

[16.126] What is the boiling point of a 0.10 M solution of $NaHSO_4$ if the solution has a density of 1.002 g/mL?

[16.127] Cocaine is a weak organic base whose molecular formula is $C_{17}H_{21}NO_4$. An aqueous solution of cocaine was found to have a pH of 8.53 and an osmotic pressure of 52.7 torr at 15 °C. Calculate K_b for cocaine.

[16.128] The iodate ion is reduced by sulfite according to the following reaction:

$$IO_3^-(aq) + 3\,SO_3^{2-}(aq) \longrightarrow I^-(aq) + 3\,SO_4^{2-}(aq)$$

The rate of this reaction is found to be first order in IO_3^-, first order in SO_3^{2-}, and first order in H^+. **(a)** Write the rate law for the reaction. **(b)** By what factor will the rate of the reaction change if the pH is lowered from 5.00 to 3.50? Does the reaction proceed faster or slower at the lower pH? **(c)** By using the concepts discussed in Section 14.6, explain how the reaction can be pH-dependent even though H^+ does not appear in the overall reaction.

[16.129] **(a)** Using dissociation constants from Appendix D, determine the value for the equilibrium constant for each of the following reactions. (Remember that when reactions are added, the corresponding equilibrium constants are multiplied.)

(i) $HCO_3^-(aq) + OH^-(aq) \rightleftharpoons CO_3^{2-}(aq) + H_2O(l)$

(ii) $NH_4^+(aq) + CO_3^{2-}(aq) \rightleftharpoons NH_3(aq) + HCO_3^-(aq)$

(b) We usually use single arrows for reactions when the forward reaction is appreciable (K much greater than 1) or when products escape from the system, so that equilibrium is never established. If we follow this convention, which of these equilibria might be written with a single arrow?

[16.130] Lactic acid, $CH_3CH(OH)COOH$, received its name because it is present in sour milk as a product of bacterial action. It is also responsible for the soreness in muscles after vigorous exercise. **(a)** The pK_a of lactic acid is 3.85. Compare this with the value for propionic acid (CH_3CH_2COOH, $pK_a = 4.89$), and explain the difference. **(b)** Calculate the lactate ion concentration in a 0.050 M solution of lactic acid. **(c)** When a solution of sodium lactate, $CH_3CH(OH)COONa$, is mixed with an aqueous copper(II) solution, it is possible to obtain a solid salt of copper(II) lactate as a blue-green hydrate, $(CH_3CH(OH)COO)_2Cu \cdot xH_2O$. Elemental analysis of the solid tells us that the solid is 22.9% Cu and 26.0% C by mass. What is the value for x in the formula for the hydrate? **(d)** The acid-dissociation constant for the $Cu^{2+}(aq)$ ion is 1.0×10^{-8}. Based on this value and the acid-dissociation constant of lactic acid, predict whether a solution of copper(II) lactate will be acidic, basic, or neutral. Explain your answer.

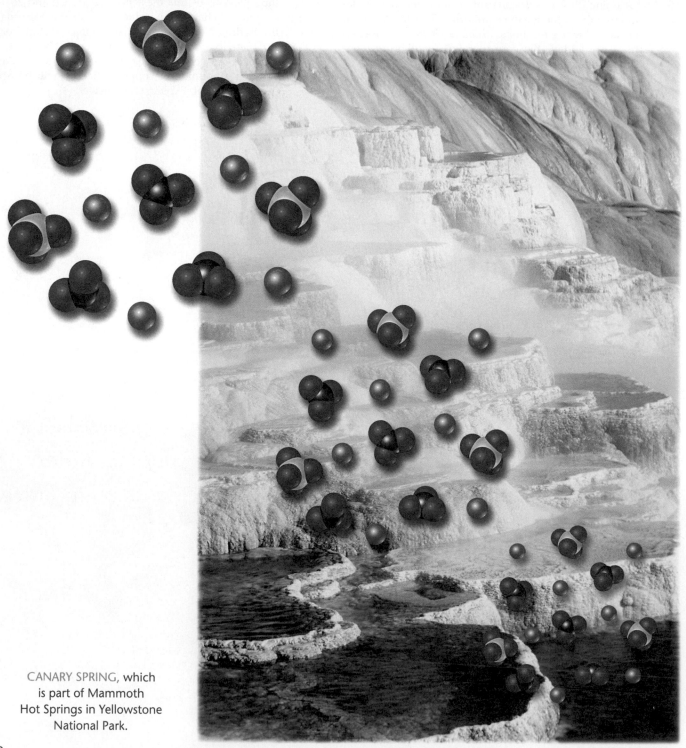

17 ADDITIONAL ASPECTS OF AQUEOUS EQUILIBRIA

CANARY SPRING, which
is part of Mammoth
Hot Springs in Yellowstone
National Park.

WHAT'S AHEAD

17.1 The Common-Ion Effect
We begin by considering a specific example of Le Châtelier's principle known as the *common-ion effect*.

17.2 Buffered Solutions
We then consider the composition of *buffered solutions*, or *buffers*, and learn how they resist pH change upon the addition of small amounts of a strong acid or a strong base.

17.3 Acid–Base Titrations
We continue by examining acid–base titration in detail, and we explore how to determine pH at any point in an acid–base titration.

17.4 Solubility Equilibria
Next, we learn how to use equilibrium constants known as *solubility-product constants* to determine to what extent a sparingly soluble salt will dissolve in water.

17.5 Factors That Affect Solubility
We investigate some of the factors that affect solubility, including the common-ion effect and the effect of acids.

17.6 Precipitation and Separation of Ions
Continuing the discussion of solubility equilibria, we learn how to precipitate ions selectively.

17.7 Qualitative Analysis for Metallic Elements
We conclude the chapter with an explanation of how the principles of solubility and complexation equilibria can be used to identify ions qualitatively in solution.

WATER IS THE MOST COMMON AND MOST IMPORTANT SOLVENT ON EARTH. In a sense, it is the solvent of life. It is difficult to imagine how living matter in all its complexity could exist with any liquid other than water as the solvent. Water occupies its position of importance because of its abundance and its exceptional ability to dissolve a wide variety of substances. For example, the chapter-opening photograph shows a hot spring; this water contains a high concentration of ions (especially Mg^{2+}, Ca^{2+}, Fe^{2+}, CO_3^{2-}, and SO_4^{2-}). The ions are dissolved as the hot water, initially underground, passes through various rocks on its way to the surface and dissolves minerals in the rocks. When the solution reaches the surface and cools, the minerals deposit and make the terracelike formations seen in the photograph.

The various aqueous solutions encountered in nature typically contain many solutes. For example, the aqueous solutions in hot springs and oceans, as well as those in biological fluids, contain a variety of dissolved ions and molecules. Consequently, many equilibria can occur simultaneously in these solutions.

In this chapter we take a step toward understanding such complex solutions by looking first at further applications of acid–base equilibria. The idea is to consider not only solutions in which there is a single solute but also those containing a mixture of solutes. We then broaden our discussion to include two additional types of aqueous equilibria: those involving slightly soluble salts and those involving the formation of metal complexes in solution. For the most part, the discussions and calculations in this chapter are an extension of those in Chapters 15 and 16.

17.1 THE COMMON-ION EFFECT

In Chapter 16 we examined the equilibrium concentrations of ions in solutions containing a weak acid or a weak base. We now consider solutions that contain a weak acid, such as acetic acid (CH_3COOH), and a soluble salt of that acid, such as sodium acetate (CH_3COONa). Notice that these solutions contain two substances that share a *common ion* CH_3COO^-. It is instructive to view these solutions from the perspective of Le Châtelier's principle. ∞ (Section 15.7) Sodium acetate is a soluble ionic compound and is therefore a strong electrolyte. ∞ (Section 4.1) Consequently, it dissociates completely in aqueous solution to form Na^+ and CH_3COO^- ions:

$$CH_3COONa(aq) \longrightarrow Na^+(aq) + CH_3COO^-(aq)$$

In contrast, CH_3COOH is a weak electrolyte that ionizes as follows:

$$CH_3COOH(aq) \rightleftharpoons H^+(aq) + CH_3COO^-(aq) \qquad [17.1]$$

The CH_3COO^- from CH_3COONa causes this equilibrium to shift to the left, thereby decreasing the equilibrium concentration of $H^+(aq)$.

$$CH_3COOH(aq) \rightleftharpoons H^+(aq) + CH_3COO^-(aq)$$

Addition of CH_3COO^- shifts equilibrium, reducing $[H^+]$

In other words, the presence of the added acetate ion causes the acetic acid to ionize less than it normally would.

Whenever a weak electrolyte and a strong electrolyte contain a common ion, the weak electrolyte ionizes less than it would if it were alone in solution. We call this observation the **common-ion effect**. Sample Exercises 17.1 and 17.2 illustrate how equilibrium concentrations may be calculated when a solution contains a mixture of a weak electrolyte and a strong electrolyte that have a common ion. The procedures are similar to those encountered for weak acids and weak bases in Chapter 16.

■ **SAMPLE EXERCISE 17.1** | Calculating the pH When a Common Ion Is Involved

What is the pH of a solution made by adding 0.30 mol of acetic acid and 0.30 mol of sodium acetate to enough water to make 1.0 L of solution?

SOLUTION

Analyze: We are asked to determine the pH of a solution of a weak electrolyte (CH_3COOH) and a strong electrolyte (CH_3COONa) that share a common ion, CH_3COO^-.

Plan: In any problem in which we must determine the pH of a solution containing a mixture of solutes, it is helpful to proceed by a series of logical steps:

1. Consider which solutes are strong electrolytes and which are weak electrolytes, and identify the major species in solution.
2. Identify the important equilibrium that is the source of H^+ and therefore determines pH.
3. Tabulate the concentrations of ions involved in the equilibrium.
4. Use the equilibrium-constant expression to calculate $[H^+]$ and then pH.

Solve: First, because CH_3COOH is a weak electrolyte and CH_3COONa is a strong electrolyte, the major species in the solution are CH_3COOH (a weak acid), Na^+ (which is neither acidic nor basic and is therefore a spectator in the acid–base chemistry), and CH_3COO^- (which is the conjugate base of CH_3COOH).

Second, $[H^+]$ and, therefore, the pH are controlled by the dissociation equilibrium of CH_3COOH:

$$CH_3COOH(aq) \rightleftharpoons H^+(aq) + CH_3COO^-(aq)$$

(We have written the equilibrium using $H^+(aq)$ rather than $H_3O^+(aq)$ but both representations of the hydrated hydrogen ion are equally valid.)

Third, we tabulate the initial and equilibrium concentrations as we did in solving other equilibrium problems in Chapters 15 and 16:

	$CH_3COOH(aq)$	$\rightleftharpoons$	$H^+(aq)$	$+$	$CH_3COO^-(aq)$
Initial	0.30 M		0		0.30 M
Change	$-x$ M		$+x$ M		$+x$ M
Equilibrium	$(0.30 - x)$ M		x M		$(0.30 + x)$ M

The equilibrium concentration of CH_3COO^- (the common ion) is the initial concentration that is due to CH_3COONa (0.30 M) plus the change in concentration (x) that is due to the ionization of CH_3COOH.

Now we can use the equilibrium-constant expression:

$$K_a = 1.8 \times 10^{-5} = \frac{[H^+][CH_3COO^-]}{[CH_3COOH]}$$

(The dissociation constant for CH_3COOH at 25 °C is from Appendix D; addition of CH_3COONa does not change the value of this constant.)

Substituting the equilibrium-constant concentrations from our table into the equilibrium expression gives

$$K_a = 1.8 \times 10^{-5} = \frac{x(0.30 + x)}{0.30 - x}$$

Because K_a is small, we assume that x is small compared to the original concentrations of CH_3COOH and CH_3COO^- (0.30 M each). Thus, we can ignore the very small x relative to 0.30 M, giving

$$K_a = 1.8 \times 10^{-5} = \frac{x(0.30)}{0.30}$$

$$x = 1.8 \times 10^{-5} M = [H^+]$$

The resulting value of x is indeed small relative to 0.30, justifying the approximation made in simplifying the problem.

Finally, we calculate the pH from the equilibrium concentration of $H^+(aq)$:

$$pH = -\log(1.8 \times 10^{-5}) = 4.74$$

Comment: In Section 16.6 we calculated that a 0.30 M solution of CH_3COOH has a pH of 2.64, corresponding to $[H^+] = 2.3 \times 10^{-3}$ M. Thus, the addition of CH_3COONa has substantially decreased $[H^+]$, as we would expect from Le Châtelier's principle.

■ PRACTICE EXERCISE

Calculate the pH of a solution containing 0.085 M nitrous acid (HNO_2; $K_a = 4.5 \times 10^{-4}$) and 0.10 M potassium nitrite (KNO_2).
Answer: 3.42

■ **SAMPLE EXERCISE 17.2** | **Calculating Ion Concentrations When a Common Ion Is Involved**

Calculate the fluoride ion concentration and pH of a solution that is 0.20 M in HF and 0.10 M in HCl.

SOLUTION

Analyze: We are asked to determine the concentration of F^- and the pH in a solution containing the weak acid HF and the strong acid HCl. In this case the common ion is H^+.

Plan: We can again use the four steps outlined in Sample Exercise 17.1.

Solve: Because HF is a weak acid and HCl is a strong acid, the major species in solution are HF, H^+, and Cl^-. The Cl^-, which is the conjugate base of a strong acid, is merely a spectator ion in any acid–base chemistry. The problem asks for $[F^-]$, which is formed by ionization of HF. Thus, the important equilibrium is

$$HF(aq) \rightleftharpoons H^+(aq) + F^-(aq)$$

The common ion in this problem is the hydrogen (or hydronium) ion. Now we can tabulate the initial and equilibrium concentrations of each species involved in this equilibrium:

	$HF(aq)$	$\rightleftharpoons$	$H^+(aq)$	+	$F^-(aq)$
Initial	0.20 M		0.10 M		0
Change	$-x$ M		$+x$ M		$+x$ M
Equilibrium	$(0.20 - x)$ M		$(0.10 + x)$ M		x M

The equilibrium constant for the ionization of HF, from Appendix D, is 6.8×10^{-4}. Substituting the equilibrium-constant concentrations into the equilibrium expression gives

$$K_a = 6.8 \times 10^{-4} = \frac{[H^+][F^-]}{[HF]} = \frac{(0.10 + x)(x)}{0.20 - x}$$

If we assume that x is small relative to 0.10 or 0.20 M, this expression simplifies to

$$\frac{(0.10)(x)}{0.20} = 6.8 \times 10^{-4}$$

$$x = \frac{0.20}{0.10}(6.8 \times 10^{-4}) = 1.4 \times 10^{-3} M = [F^-]$$

This F^- concentration is substantially smaller than it would be in a 0.20 M solution of HF with no added HCl. The common ion, H^+, suppresses the ionization of HF. The concentration of $H^+(aq)$ is

$$[H^+] = (0.10 + x) M \simeq 0.10 M$$

Thus,

$$pH = 1.00$$

Comment: Notice that for all practical purposes, $[H^+]$ is due entirely to the HCl; the HF makes a negligible contribution by comparison.

■ **PRACTICE EXERCISE**

Calculate the formate ion concentration and pH of a solution that is 0.050 M in formic acid (HCOOH; $K_a = 1.8 \times 10^{-4}$) and 0.10 M in HNO$_3$.
Answer: $[HCOO^-] = 9.0 \times 10^{-5}$; pH = 1.00

Sample Exercises 17.1 and 17.2 both involve weak acids. The ionization of a weak base is also decreased by the addition of a common ion. For example, the addition of NH_4^+ (as from the strong electrolyte NH_4Cl) causes the base-dissociation equilibrium of NH_3 to shift to the left, decreasing the equilibrium concentration of OH^- and lowering the pH:

$$NH_3(aq) + H_2O(l) \rightleftharpoons NH_4^+(aq) + OH^-(aq) \qquad [17.2]$$

Addition of NH_4^+ shifts equilibrium, reducing $[OH^-]$

A mixture of 0.10 mol of NH_4Cl and 0.12 mol of NH_3 is added to enough water to make 1.0 L of solution. **(a)** What are the initial concentrations of the major species in the solution? **(b)** Which of the ions in this solution is a spectator ion in any acid–base chemistry occurring in the solution? **(c)** What equilibrium reaction determines $[OH^-]$ and therefore the pH of the solution?

17.2 BUFFERED SOLUTIONS

Solutions such as those discussed in Section 17.1, which contain a weak conjugate acid–base pair, can resist drastic changes in pH upon the addition of small amounts of strong acid or strong base. These solutions are called **buffered solutions** (or merely **buffers**). Human blood, for example, is a complex aqueous mixture with a pH buffered at about 7.4 (see the "Chemistry and Life" box near the end of this section). Much of the chemical behavior of seawater is determined by its pH, buffered at about 8.1 to 8.3 near the surface. Buffered solutions find many important applications in the laboratory and in medicine (Figure 17.1▶).

Composition and Action of Buffered Solutions

A buffer resists changes in pH because it contains both an acid to neutralize OH^- ions and a base to neutralize H^+ ions. The acid and base that make up the buffer, however, must not consume each other through a neutralization reaction. These requirements are fulfilled by a weak acid–base conjugate pair such as CH_3COOH–CH_3COO^- or NH_4^+–NH_3. Thus, buffers are often prepared by mixing a weak acid or a weak base with a salt of that acid or base. The CH_3COOH–CH_3COO^- buffer can be prepared, for example, by adding CH_3COONa to a solution of CH_3COOH. The NH_4^+–NH_3 buffer can be prepared by adding NH_4Cl to a solution of NH_3. By choosing appropriate components and adjusting their relative concentrations, we can buffer a solution at virtually any pH.

▲ **Figure 17.1 Buffer solutions.** Prepackaged buffer solutions and ingredients for making up buffer solutions of predetermined pH can be purchased.

Which of the following conjugate acid–base pairs will *not* function as a buffer: C_2H_5COOH and $C_2H_5COO^-$; HCO_3^- and CO_3^{2-}; HNO_3 and NO_3^-? Explain.

To understand better how a buffer works, let's consider a buffer composed of a weak acid (HX) and one of its salts (MX, where M^+ could be Na^+, K^+, or another cation). The acid-dissociation equilibrium in this buffered solution involves both the acid and its conjugate base:

$$HX(aq) \rightleftharpoons H^+(aq) + X^-(aq) \qquad [17.3]$$

The corresponding acid-dissociation-constant expression is

$$K_a = \frac{[H^+][X^-]}{[HX]} \qquad [17.4]$$

Solving this expression for $[H^+]$, we have

$$[H^+] = K_a \frac{[HX]}{[X^-]} \qquad [17.5]$$

We see from this expression that $[H^+]$, and thus the pH, is determined by two factors: the value of K_a for the weak-acid component of the buffer and the ratio of the concentrations of the conjugate acid–base pair, $[HX]/[X^-]$.

▶ **Figure 17.2 Buffer action.** When a small portion of OH⁻ is added to a buffer consisting of a mixture of the weak acid HF and its conjugate base (left), the OH⁻ reacts with the HF, decreasing [HF] and increasing [F⁻] in the buffer. Conversely, when a small portion of H⁺ is added to the buffer (right), the H⁺ reacts with the F⁻, decreasing [F⁻] and increasing [HF] in the buffer. Because pH depends on the ratio of F⁻ to HF, the resulting pH change is small.

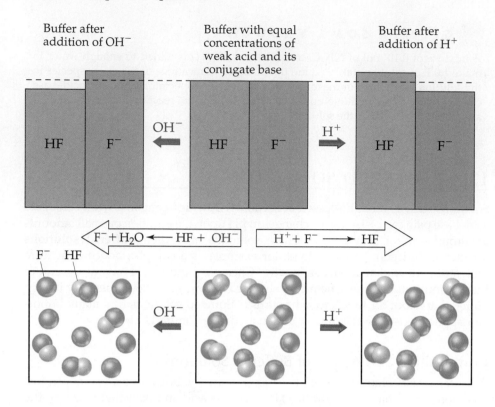

If OH⁻ ions are added to the buffered solution, they react with the acid component of the buffer to produce water and X⁻:

$$\underset{\text{added base}}{\text{OH}^-(aq)} + \underset{\text{weak acid in buffer}}{\text{HX}(aq)} \longrightarrow \text{H}_2\text{O}(l) + \text{X}^-(aq) \qquad [17.6]$$

This reaction causes [HX] to decrease and [X⁻] to increase. As long as the amounts of HX and X⁻ in the buffer are large compared to the amount of OH⁻ added, however, the ratio [HX]/[X⁻] does not change much, and thus the change in pH is small. A specific example of such a buffer, the HF/F⁻ buffer, is shown in Figure 17.2 ▲.

If H⁺ ions are added, they react with the base component of the buffer:

$$\underset{\text{added base}}{\text{H}^+(aq)} + \underset{\text{weak acid in buffer}}{\text{X}^-(aq)} \longrightarrow \text{HX}(aq) \qquad [17.7]$$

This reaction can also be represented using H₃O⁺:

$$\text{H}_3\text{O}^+(aq) + \text{X}^-(aq) \longrightarrow \text{HX}(aq) + \text{H}_2\text{O}(l)$$

Using either equation, we see that the reaction causes [X⁻] to decrease and [HX] to increase. As long as the change in the ratio [HX]/[X⁻] is small, the change in pH will be small.

Figure 17.2 shows a buffer consisting of equal concentrations of hydrofluoric acid and fluoride ion (center). The addition of OH⁻ (left) reduces [HF] and increases [F⁻]. The addition of H⁺ (right) reduces [F⁻] and increases [HF].

◣ **GIVE IT SOME THOUGHT**

(a) What happens when NaOH is added to a buffer composed of CH₃COOH and CH₃COO⁻? **(b)** What happens when HCl is added to this buffer?

Calculating the pH of a Buffer

Because conjugate acid–base pairs share a common ion, we can use the same procedures to calculate the pH of a buffer that we used to treat the common-ion effect (see Sample Exercise 17.1). However, we can sometimes take an alternate approach that is based on an equation derived from Equation 17.5.

Taking the negative log of both sides of Equation 17.5, we have

$$-\log[H^+] = -\log\left(K_a \frac{[HX]}{[X^-]}\right) = -\log K_a - \log \frac{[HX]}{[X^-]}$$

Because $-\log[H^+] = pH$ and $-\log K_a = pK_a$, we have

$$pH = pK_a - \log \frac{[HX]}{[X^-]} = pK_a + \log \frac{[X^-]}{[HX]} \qquad [17.8]$$

In general,

$$pH = pK_a + \log \frac{[base]}{[acid]} \qquad [17.9]$$

where [acid] and [base] refer to the equilibrium concentrations of *the conjugate acid–base pair*. Note that when [base] = [acid], $pH = pK_a$.

Equation 17.9 is known as the **Henderson–Hasselbalch equation**. Biologists, biochemists, and others who work frequently with buffers often use this equation to calculate the pH of buffers. In doing equilibrium calculations, we have seen that we can normally neglect the amounts of the acid and base of the buffer that ionize. Therefore, we can usually use the starting concentrations of the acid and base components of the buffer directly in Equation 17.9.

■ SAMPLE EXERCISE 17.3 | Calculating the pH of a Buffer

What is the pH of a buffer that is 0.12 M in lactic acid [$CH_3CH(OH)COOH$, or $HC_3H_5O_3$] and 0.10 M in sodium lactate [$CH_3CH(OH)COONa$ or $NaC_3H_5O_3$]? For lactic acid, $K_a = 1.4 \times 10^{-4}$.

SOLUTION

Analyze: We are asked to calculate the pH of a buffer containing lactic acid $HC_3H_5O_3$ and its conjugate base, the lactate ion ($C_3H_5O_3^-$).

Plan: We will first determine the pH using the method described in Section 17.1. Because $HC_3H_5O_3$ is a weak electrolyte and $NaC_3H_5O_3$ is a strong electrolyte, the major species in solution are $HC_3H_5O_3$, Na^+, and $C_3H_5O_3^-$. The Na^+ ion is a spectator ion. The $HC_3H_5O_3$–$C_3H_5O_3^-$ conjugate acid–base pair determines [H^+] and thus pH; [H^+] can be determined using the acid-dissociation equilibrium of lactic acid.

Solve: The initial and equilibrium concentrations of the species involved in this equilibrium are

	$HC_3H_5O_3(aq)$	$\rightleftharpoons$ $H^+(aq)$	+ $C_3H_5O_3^-(aq)$
Initial	0.12 M	0	0.10 M
Change	$-x\ M$	$+x\ M$	$+x\ M$
Equilibrium	$(0.12 - x)\ M$	$x\ M$	$(0.10 + x)\ M$

The equilibrium concentrations are governed by the equilibrium expression:

$$K_a = 1.4 \times 10^{-4} = \frac{[H^+][C_3H_5O_3]}{[HC_3H_5O_3]} = \frac{x(0.10 + x)}{(0.12 - x)}$$

Because K_a is small and a common ion is present, we expect x to be small relative to either 0.12 or 0.10 M. Thus, our equation can be simplified to give

$$K_a = 1.4 \times 10^{-4} = \frac{x(0.10)}{0.12}$$

Solving for x gives a value that justifies our approximation:

$$[H^+] = x = \left(\frac{0.12}{0.10}\right)(1.4 \times 10^{-4}) = 1.7 \times 10^{-4}\ M$$

$$pH = -\log(1.7 \times 10^{-4}) = 3.77$$

Alternatively, we could have used the Henderson–Hasselbalch equation to calculate pH directly:

$$pH = pK_a + \log\left(\frac{[base]}{[acid]}\right) = 3.85 + \log\left(\frac{0.10}{0.12}\right)$$

$$= 3.85 + (-0.08) = 3.77$$

■ PRACTICE EXERCISE

Calculate the pH of a buffer composed of 0.12 M benzoic acid and 0.20 M sodium benzoate. (Refer to Appendix D.)
Answer: 4.42

■ SAMPLE EXERCISE 17.4 | Preparing a Buffer

How many moles of NH_4Cl must be added to 2.0 L of 0.10 M NH_3 to form a buffer whose pH is 9.00? (Assume that the addition of NH_4Cl does not change the volume of the solution.)

SOLUTION

Analyze: Here we are asked to determine the amount of NH_4^+ ion required to prepare a buffer of a specific pH.

Plan: The major species in the solution will be NH_4^+, Cl^-, and NH_3. Of these, the Cl^- ion is a spectator (it is the conjugate base of a strong acid). Thus, the NH_4^+–NH_3 conjugate acid–base pair will determine the pH of the buffer solution. The equilibrium relationship between NH_4^+ and NH_3 is given by the base-dissociation constant for NH_3:

$$NH_3(aq) + H_2O(l) \rightleftharpoons NH_4^+(aq) + OH^-(aq) \qquad K_b = \frac{[NH_4^+][OH^-]}{[NH_3]} = 1.8 \times 10^{-5}$$

The key to this exercise is to use this K_b expression to calculate $[NH_4^+]$.

Solve: We obtain $[OH^-]$ from the given pH:

$$pOH = 14.00 - pH = 14.00 - 9.00 = 5.00$$

and so

$$[OH^-] = 1.0 \times 10^{-5} M$$

Because K_b is small and the common ion NH_4^+ is present, the equilibrium concentration of NH_3 will essentially equal its initial concentration:

$$[NH_3] = 0.10 M$$

We now use the expression for K_b to calculate $[NH_4^+]$:

$$[NH_4^+] = K_b \frac{[NH_3]}{[OH^-]} = (1.8 \times 10^{-5}) \frac{(0.10 \ M)}{(1.0 \times 10^{-5} \ M)} = 0.18 M$$

Thus, for the solution to have pH = 9.00, $[NH_4^+]$ must equal 0.18 M. The number of moles of NH_4Cl needed to produce this concentration is given by the product of the volume of the solution and its molarity:

$$(2.0 \ L)(0.18 \ mol \ NH_4Cl/L) = 0.36 \ mol \ NH_4Cl$$

Comment: Because NH_4^+ and NH_3 are a conjugate acid–base pair, we could use the Henderson–Hasselbalch equation (Equation 17.9) to solve this problem. To do so requires first using Equation 16.41 to calculate pK_a for NH_4^+ from the value of pK_b for NH_3. We suggest you try this approach to convince yourself that you can use the Henderson–Hasselbalch equation for buffers for which you are given K_b for the conjugate base rather than K_a for the conjugate acid.

■ PRACTICE EXERCISE

Calculate the concentration of sodium benzoate that must be present in a 0.20 M solution of benzoic acid (C_6H_5COOH) to produce a pH of 4.00.
Answer: 0.13 M

Buffer Capacity and pH Range

Two important characteristics of a buffer are its capacity and its effective pH range. **Buffer capacity** is the amount of acid or base the buffer can neutralize before the pH begins to change to an appreciable degree. The buffer capacity depends on the amount of acid and base from which the buffer is made. The pH of the buffer depends on the K_a for the acid and on the relative concentrations of the acid and base that comprise the buffer. According to Equation 17.5, for example, $[H^+]$ for a 1-L solution that is 1 M in CH_3COOH and 1 M in CH_3COONa will be the same as for a 1-L solution that is 0.1 M in CH_3COOH and 0.1 M in CH_3COONa. The first solution has a greater buffering capacity, however, because it contains more CH_3COOH and CH_3COO^-. The greater the amounts of the conjugate acid–base pair, the more resistant is the ratio of their concentrations, and hence the pH, is to change.

The **pH range** of any buffer is the pH range over which the buffer acts effectively. Buffers most effectively resist a change in pH in *either* direction when the concentrations of weak acid and conjugate base are about the same. From Equation 17.9 we see that when the concentrations of weak acid and conjugate base are equal, pH = pK_a. This relationship gives the optimal pH of any buffer. Thus, we usually try to select a buffer whose acid form has a pK_a close to the desired pH. In practice, we find that if the concentration of one component of the buffer is more than 10 times the concentration of the other component, the buffering action is poor. Because log 10 = 1, *buffers usually have a usable range within ±1 pH unit of* pK_a (that is, a range of pH = pK_a ± 1).

◣ GIVE IT SOME THOUGHT

What is the optimal pH buffered by a solution containing CH_3COOH and CH_3COONa? (K_a for CH_3COOH is 1.8×10^{-5}.)

Addition of Strong Acids or Bases to Buffers

Let's now consider in a more quantitative way the response of a buffered solution to the addition of a strong acid or base. In solving these problems, it is important to understand that reactions between strong acids and weak bases proceed essentially to completion, as do those between strong bases and weak acids. Thus, as long as we do not exceed the buffering capacity of the buffer, we can assume that the strong acid or strong base is completely consumed by reaction with the buffer.

Consider a buffer that contains a weak acid HX and its conjugate base X^-. When a strong acid is added to this buffer, the added H^+ is consumed by X^- to produce HX; thus, [HX] increases and [X^-] decreases. (See Equation 17.7.) When a strong base is added to the buffer, the added OH^- is consumed by HX to produce X^-; in this case [HX] decreases and [X^-] increases. (See Equation 17.6.) These two situations are summarized in Figure 17.2.

To calculate how the pH of the buffer responds to the addition of a strong acid or a strong base, we follow the strategy outlined in Figure 17.3▼:

1. Consider the acid–base neutralization reaction, and determine its effect on [HX] and [X^-]. This step of the procedure is a *stoichiometry calculation*.

2. Use K_a and the new concentrations of [HX] and [X^-] from step 1 to calculate [H^+]. This second step of the procedure is a standard *equilibrium calculation* and is most easily done using the Henderson–Hasselbalch equation.

The complete procedure is illustrated in Sample Exercise 17.5.

▲ **Figure 17.3 Calculation of the pH of a buffer after the addition of acid or base.** First consider how the neutralization reaction between the added strong acid or strong base and the buffer affects the composition of the buffer (stoichiometry calculation). Then calculate the pH of the remaining buffer (equilibrium calculation). As long as the amount of added acid or base does not exceed the buffer capacity, the Henderson–Hasselbalch equation, Equation 17.9, can be used for the equilibrium calculation.

■■■ **SAMPLE EXERCISE 17.5** | Calculating pH Changes in Buffers

A buffer is made by adding 0.300 mol CH_3COOH and 0.300 mol CH_3COONa to enough water to make 1.00 L of solution. The pH of the buffer is 4.74 (Sample Exercise 17.1). **(a)** Calculate the pH of this solution after 0.020 mol of NaOH is added. **(b)** For comparison, calculate the pH that would result if 0.020 mol of NaOH were added to 1.00 L of pure water (neglect any volume changes).

SOLUTION

Analyze: We are asked to determine the pH of a buffer after addition of a small amount of strong base and to compare the pH change to the pH that would result if we were to add the same amount of strong base to pure water.

Plan: (a) Solving this problem involves the two steps outlined in Figure 17.3. Thus, we must first do a stoichiometry calculation to determine how the added OH^- reacts with the buffer and affects its composition. Then we can use the resultant composition of the buffer and either the Henderson–Hasselbalch equation or the equilibrium-constant expression for the buffer to determine the pH.

Solve: *Stoichiometry Calculation:* The OH^- provided by NaOH reacts with CH_3COOH, the weak acid component of the buffer. Prior to this neutralization reaction, there are 0.300 mol each of CH_3COOH and CH_3COO^-. Neutralizing the 0.020 mol OH^- requires 0.020 mol of CH_3COOH. Consequently, the amount of CH_3COOH *decreases* by 0.020 mol, and the amount of the product of the neutralization, CH_3COO^-, *increases* by 0.020 mol. We can create a table to see how the composition of the buffer changes as a result of its reaction with OH^-:

$$CH_3COOH(aq) + OH^-(aq) \longrightarrow H_2O(l) + CH_3COO^-(aq)$$

	CH_3COOH	OH^-		CH_3COO^-
Buffer before addition	0.300 mol	0	—	0.300 mol
Addition	—	0.020 mol		—
Buffer after addition	0.280 mol	0	—	0.320 mol

0.280 *M*
CH_3COOH
0.320 *M*
CH_3COONa

pH = 4.80

add 0.020 mol OH^-

Buffer

0.300 *M*
CH_3COOH
0.300 *M*
CH_3COONa

pH = 4.74

add 0.020 mol H^+

0.320 *M*
CH_3COOH
0.280 *M*
CH_3COONa

pH = 4.68

Equilibrium Calculation: We now turn our attention to the equilibrium that will determine the pH of the buffer, namely the ionization of acetic acid.

$$CH_3COOH(aq) \rightleftharpoons H^+(aq) + CH_3COO^-(aq)$$

Using the quantities of CH_3COOH and CH_3COO^- remaining in the buffer, we can determine the pH using the Henderson–Hasselbalch equation.

$$pH = 4.74 + \log \frac{0.320 \text{ mol}/1.00 \text{ L}}{0.280 \text{ mol}/1.00 \text{ L}} = 4.80$$

Comment Notice that we could have used mole amounts in place of concentrations in the Henderson–Hasselbalch equation and gotten the same result. The volumes of the acid and base are equal and cancel.

If 0.020 mol of H^+ was added to the buffer, we would proceed in a similar way to calculate the resulting pH of the buffer. In this case the pH decreases by 0.06 units, giving pH = 4.68, as shown in the figure in the margin.

(b) To determine the pH of a solution made by adding 0.020 mol of NaOH to 1.00 L of pure water, we can first determine pOH using Equation 16.18 and subtracting from 14.

$$pH = 14 - (-\log 0.020) = 12.30$$

Note that although the small amount of NaOH changes the pH of water significantly, the pH of the buffer changes very little.

■■■ **PRACTICE EXERCISE**

Determine **(a)** the pH of the original buffer described in Sample Exercise 17.5 after the addition of 0.020 mol HCl and **(b)** the pH of the solution that would result from the addition of 0.020 mol HCl to 1.00 L of pure water.
Answers: **(a)** 4.68, **(b)** 1.70

Chemistry and Life BLOOD AS A BUFFERED SOLUTION

Many of the chemical reactions that occur in living systems are extremely sensitive to pH. Many of the enzymes that catalyze important biochemical reactions, for example, are effective only within a narrow pH range. For this reason the human body maintains a remarkably intricate system of buffers, both within tissue cells and in the fluids that transport cells. Blood, the fluid that transports oxygen to all parts of the body (Figure 17.4▶), is one of the most prominent examples of the importance of buffers in living beings.

Human blood is slightly basic with a normal pH of 7.35 to 7.45. Any deviation from this normal pH range can have extremely disruptive effects on the stability of cell membranes, the structures of proteins, and the activities of enzymes. Death may result if the blood pH falls below 6.8 or rises above 7.8. When the pH falls below 7.35, the condition is called *acidosis*; when it rises above 7.45, the condition is called *alkalosis*. Acidosis is the more common tendency because ordinary metabolism generates several acids within the body.

The major buffer system that is used to control the pH of blood is the *carbonic acid–bicarbonate buffer system*. Carbonic acid (H_2CO_3) and bicarbonate ion (HCO_3^-) are a conjugate acid–base pair. In addition, carbonic acid can decompose into carbon dioxide gas and water. The important equilibria in this buffer system are

$$H^+(aq) + HCO_3^-(aq) \rightleftharpoons H_2CO_3(aq) \rightleftharpoons H_2O(l) + CO_2(g)$$
[17.10]

Several aspects of these equilibria are notable. First, although carbonic acid is a diprotic acid, the carbonate ion (CO_3^{2-}) is unimportant in this system. Second, one of the components of this equilibrium, CO_2, is a gas, which provides a mechanism for the body to adjust the equilibria. Removal of CO_2 via exhalation shifts the equilibria to the right, consuming H^+ ions. Third, the buffer system in blood operates at a pH of 7.4, which is fairly far removed from the pK_{a1} value of H_2CO_3 (6.1 at physiological temperatures). For the buffer to have a pH of 7.4, the ratio [base]/[acid] must have a value of about 20. In normal blood plasma the concentrations of HCO_3^- and H_2CO_3 are about 0.024 M and 0.0012 M, respectively. Consequently, the buffer has a high capacity to neutralize additional acid, but only a low capacity to neutralize additional base.

The principal organs that regulate the pH of the carbonic acid–bicarbonate buffer system are the lungs and kidneys. Some of the receptors in the brain are sensitive to the concentrations of H^+ and CO_2 in bodily fluids. When the concentration of CO_2 rises, the equilibria in Equation 17.10 shift to the left, which leads to the formation of more H^+. The receptors trigger a reflex to breathe faster and deeper, increasing the rate of elimination of CO_2 from the lungs and shifting the equilibria back to the right. The kidneys absorb or release H^+ and HCO_3^-; much of the excess acid leaves the body in urine, which normally has a pH of 5.0 to 7.0.

▲ Figure 17.4 **Red blood cells.** A scanning electromicrograph of a group of red blood cells traveling through a small branch of an artery. Blood is a buffered solution whose pH is maintained between 7.35 and 7.45.

The regulation of the pH of blood plasma relates directly to the effective transport of O_2 to bodily tissues. The protein hemoglobin, which is found in red blood cells, carries oxygen. Hemoglobin (Hb) reversibly binds both H^+ and O_2. These two substances compete for the Hb, which can be represented approximately by the following equilibrium:

$$HbH^+ + O_2 \rightleftharpoons HbO_2 + H^+$$
[17.11]

Oxygen enters the blood through the lungs, where it passes into the red blood cells and binds to Hb. When the blood reaches tissue in which the concentration of O_2 is low, the equilibrium in Equation 17.11 shifts to the left and O_2 is released. An increase in H^+ ion concentration (decrease in blood pH) also shifts this equilibrium to the left, as does increasing temperature.

During periods of strenuous exertion, three factors work together to ensure the delivery of O_2 to active tissues: (1) As O_2 is consumed, the equilibrium in Equation 17.11 shifts to the left according to Le Châtelier's principle. (2) Exertion raises the temperature of the body, also shifting the equilibrium to the left. (3) Large amounts of CO_2 are produced by metabolism, which shifts the equilibrium in Equation 17.10 to the left, thus decreasing the pH. Other acids, such as lactic acid, are also produced during strenuous exertion as tissues become starved for oxygen. The decrease in pH shifts the hemoglobin equilibrium to the left, delivering more O_2. In addition, the decrease in pH stimulates an increase in the rate of breathing, which furnishes more O_2 and eliminates CO_2. Without this elaborate arrangement, the O_2 in tissues would be rapidly depleted, making further activity impossible.

Related Exercises: 17.29 and 17.90

▲ **Figure 17.5 Measuring pH during a titration.** A typical setup for using a pH meter to measure data for a titration curve. In this case a standard solution of NaOH (the titrant) is added by buret to a solution of HCl. The HCl solution is stirred during the titration to ensure uniform composition.

17.3 ACID–BASE TITRATIONS

In Section 4.6 we briefly described *titrations*. In an acid–base titration, a solution containing a known concentration of base is slowly added to an acid (or the acid is added to the base). Acid–base indicators can be used to signal the *equivalence point* of a titration (the point at which stoichiometrically equivalent quantities of acid and base have been brought together). Alternatively, a pH meter can be used to monitor the progress of the reaction producing a **pH titration curve**, a graph of the pH as a function of the volume of the added titrant. The shape of the titration curve makes it possible to determine the equivalence point in the titration. The titration curve can also be used to select suitable indicators and to determine the K_a of the weak acid or the K_b of the weak base being titrated.

A typical apparatus for measuring pH during a titration is illustrated in Figure 17.5 ◄. The titrant is added to the solution from a buret, and the pH is continually monitored using a pH meter. To understand why titration curves have certain characteristic shapes, we will examine the curves for three kinds of titrations: (1) strong acid–strong base, (2) weak acid–strong base, and (3) polyprotic acid–strong base. We will also briefly consider how these curves relate to those involving weak bases.

▲ **GIVE IT SOME THOUGHT**

For the setup shown in Figure 17.5, will pH increase or decrease as titrant is added?

Strong Acid–Strong Base Titrations

The titration curve produced when a strong base is added to a strong acid has the general shape shown in Figure 17.6 ▶. This curve depicts the pH change that occurs as 0.100 *M* NaOH is added to 50.0 mL of 0.100 *M* HCl. The pH can be calculated at various stages of the titration. To help understand these calculations, we can divide the curve into four regions:

1. *The initial pH (initial acid):* The pH of the solution before the addition of any base is determined by the initial concentration of the strong acid. For a solution of 0.100 *M* HCl, $[H^+] = 0.100$ *M*, and hence pH $= -\log(0.100) = 1.000$. Thus, the initial pH is low.

2. *Between the initial pH and the equivalence point (remaining acid):* As NaOH is added, the pH increases slowly at first and then rapidly in the vicinity of the equivalence point. The pH of the solution before the equivalence point is determined by the concentration of acid that has not yet been neutralized. This calculation is illustrated in Sample Exercise 17.6(a).

3. *The equivalence point:* At the equivalence point an equal number of moles of the NaOH and HCl have reacted, leaving only a solution of their salt, NaCl. The pH of the solution is 7.00 because the cation of a strong base (in this case Na^+) and the anion of a strong acid (in this case Cl^-) do not hydrolyze and therefore have no appreciable effect on pH. ∞ (Section 16.9)

4. *After the equivalence point (excess base):* The pH of the solution after the equivalence point is determined by the concentration of the excess NaOH in the solution. This calculation is illustrated in Sample Exercise 17.6(b).

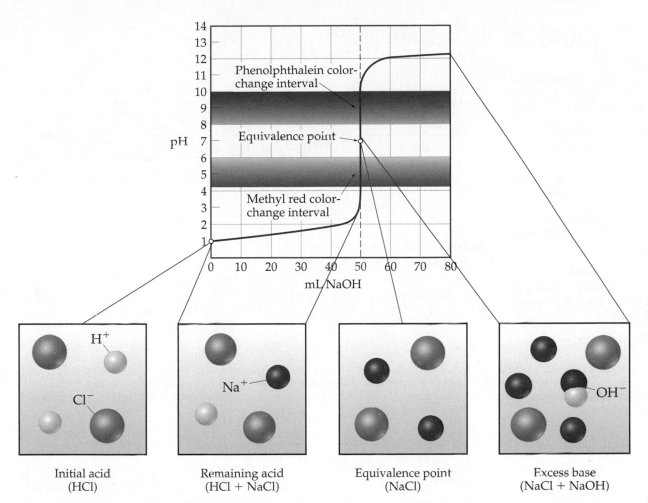

▲ **Figure 17.6 Adding a strong base to a strong acid.** The pH curve for titration of 50.0 mL of a 0.100 M solution of a strong acid with a 0.100 M solution of a strong base. In this case the acid is HCl and the base is NaOH. The pH starts out at a low value characteristic of the acid and then increases as base is added, rising rapidly at the equivalence point. Both phenolphthalein and methyl red change color at the equivalence point. (For clarity, water molecules have been omitted from the molecular art.)

■■■ **SAMPLE EXERCISE 17.6** | Calculating pH for a Strong Acid–Strong Base Titration

Calculate the pH when the following quantities of 0.100 M NaOH solution have been added to 50.0 mL of 0.100 M HCl solution: **(a)** 49.0 mL, **(b)** 51.0 mL.

SOLUTION

Analyze: We are asked to calculate the pH at two points in the titration of a strong acid with a strong base. The first point is just before the equivalence point, so we expect the pH to be determined by the small amount of strong acid that has not yet been neutralized. The second point is just after the equivalence point, so we expect this pH to be determined by the small amount of excess strong base.

Plan: **(a)** As the NaOH solution is added to the HCl solution, $H^+(aq)$ reacts with $OH^-(aq)$ to form H_2O. Both Na^+ and Cl^- are spectator ions, having negligible effect on the pH. To determine the pH of the solution, we must first determine how many moles of H^+ were originally present and how many moles of OH^- were added. We can then calculate how many moles of each ion remain after the neutralization reaction. To calculate $[H^+]$, and hence pH, we must also remember that the volume of the solution increases as we add titrant, thus diluting the concentration of all solutes present.

Solve: The number of moles of H^+ in the original HCl solution is given by the product of the volume of the solution (50.0 mL = 0.0500 L) and its molarity (0.100 M):

$$(0.0500 \text{ L soln})\left(\frac{0.100 \text{ mol } H^+}{1 \text{ L soln}}\right) = 5.00 \times 10^{-3} \text{ mol } H^+$$

Likewise, the number of moles of OH^- in 49.0 mL of 0.100 M NaOH is

$$(0.0490 \text{ L soln})\left(\frac{0.100 \text{ mol } OH^-}{1 \text{ L soln}}\right) = 4.90 \times 10^{-3} \text{ mol } OH^-$$

Because we have not yet reached the equivalence point, there are more moles of H^+ present than OH^-. Each mole of OH^- will react with one mole of H^+. Using the convention introduced in Sample Exercise 17.5,

	$H^+(aq)$	+	$OH^-(aq)$	$\longrightarrow$	$H_2O(l)$
Before addition	5.00×10^{-3} mol		0		—
Addition			4.90×10^{-3} mol		
After addition	0.10×10^{-3} mol		0		—

During the course of the titration, the volume of the reaction mixture increases as the NaOH solution is added to the HCl solution. Thus, at this point in the titration, the total volume of the solutions is

$$50.0 \text{ mL} + 49.0 \text{ mL} = 99.0 \text{ mL} = 0.0990 \text{ L}$$

(We assume that the total volume is the sum of the volumes of the acid and base solutions.) Thus, the concentration of $H^+(aq)$ is

$$[H^+] = \frac{\text{moles } H^+(aq)}{\text{liters soln}} = \frac{0.10 \times 10^{-3} \text{ mol}}{0.09900 \text{ L}} = 1.0 \times 10^{-3} \text{ M}$$

The corresponding pH equals

$$-\log(1.0 \times 10^{-3}) = 3.00$$

Plan: (b) We proceed in the same way as we did in part (a), except we are now past the equivalence point and have more OH^- in the solution than H^+. As before, the initial number of moles of each reactant is determined from their volumes and concentrations. The reactant present in smaller stoichiometric amount (the limiting reactant) is consumed completely, leaving an excess of hydroxide ion.

Solve:

	$H^+(aq)$	+	$OH^-(aq)$	$\longrightarrow$	$H_2O(l)$
Before addition	5.00×10^{-3} mol		0		—
Addition			5.10×10^{-3} mol		
After addition	0		0.10×10^{-3} mol		—

In this case the total volume of the solution is

$$50.0 \text{ mL} + 51.0 \text{ mL} = 101.0 \text{ mL} = 0.1010 \text{ L}$$

Hence, the concentration of $OH^-(aq)$ in the solution is

$$[OH^-] = \frac{\text{moles } OH^-(aq)}{\text{liters soln}} = \frac{0.10 \times 10^{-3} \text{ mol}}{0.1010 \text{ L}} = 1.0 \times 10^{-3} \text{ M}$$

Thus, the pOH of the solution equals

$$pOH = -\log(1.0 \times 10^{-3}) = 3.00$$

and the pH equals

$$pH = 14.00 - pOH = 14.00 - 3.00 = 11.00$$

■ **PRACTICE EXERCISE**

Calculate the pH when the following quantities of 0.100 M HNO_3 have been added to 25.0 mL of 0.100 M KOH solution: **(a)** 24.9 mL, **(b)** 25.1 mL.
Answers: **(a)** 10.30, **(b)** 3.70

Optimally, an indicator would change color at the equivalence point in a titration. In practice, however, that is unnecessary. The pH changes very rapidly near the equivalence point, and in this region merely a drop of titrant can change the pH by several units. Thus, an indicator beginning and ending its color change anywhere on this rapid-rise portion of the titration curve will give a sufficiently accurate measure of the volume of titrant needed to reach the equivalence point. The point in a titration where the indicator changes color is called the *end point* to distinguish it from the actual equivalence point that it closely approximates.

(a)   (b)

◀ **Figure 17.7 Methyl red indicator.** Change in appearance of a solution containing methyl red indicator in the pH range 4.2 to 6.3. The characteristic acidic color is shown in (a), and the characteristic basic color in (b).

In Figure 17.6 we see that the pH changes very rapidly from about 4 to about 10 near the equivalence point. Consequently, an indicator for this strong acid–strong base titration can change color anywhere in this range. Most strong acid–strong base titrations are carried out using phenolphthalein as an indicator (Figure 4.20) because it dramatically changes color in this range. From Figure 16.7, we see that phenolphthalein changes color from pH 8.3 to 10.0. Several other indicators would also be satisfactory, including methyl red, which changes color from pH 4.2 to 6.0 (Figure 17.7▲).

Titration of a solution of a strong base with a solution of a strong acid would yield an analogous curve of pH versus added acid. In this case, however, the pH would be high at the outset of the titration and low at its completion, as shown in Figure 17.8▶.

▲ **Figure 17.8 Adding a strong acid to a strong base.** The shape of a pH curve for titration of a strong base with a strong acid. The pH starts out at a high value characteristic of the base and then decreases as acid is added, dropping rapidly at the equivalence point.

◢ **GIVE IT SOME THOUGHT**

What is the pH at the equivalence point when 0.10 M HNO_3 is added to a solution containing 0.30 g of KOH?

Weak Acid–Strong Base Titrations

The curve for the titration of a weak acid by a strong base is very similar in shape to that for the titration of a strong acid by a strong base. Consider, for example, the titration curve for the titration of 50.0 mL of 0.100 M acetic acid (CH_3COOH) with 0.100 M NaOH shown in Figure 17.9▼. We can calculate the pH at points along this curve, using principles we have discussed earlier. As in the case of the titration of a strong acid by a strong base, we can divide the curve into four regions:

1. *The initial pH (initial acid):* We use K_a of the acid to calculate this pH, as shown in Section 16.6. The calculated pH of 0.100 M CH_3COOH is 2.89.

2. *Between the initial pH and the equivalence point (buffer mixture):* Prior to reaching the equivalence point, the acid is being neutralized, and its conjugate base is being formed:

$$CH_3COOH(aq) + OH^-(aq) \longrightarrow CH_3COO^-(aq) + H_2O(l) \quad [17.12]$$

Thus, the solution contains a mixture of CH_3COOH and CH_3COO^-.

The approach we take in calculating the pH in this region of the titration curve involves two main steps. First, we consider the neutralization reaction between CH_3COOH and OH^- to determine the concentrations of

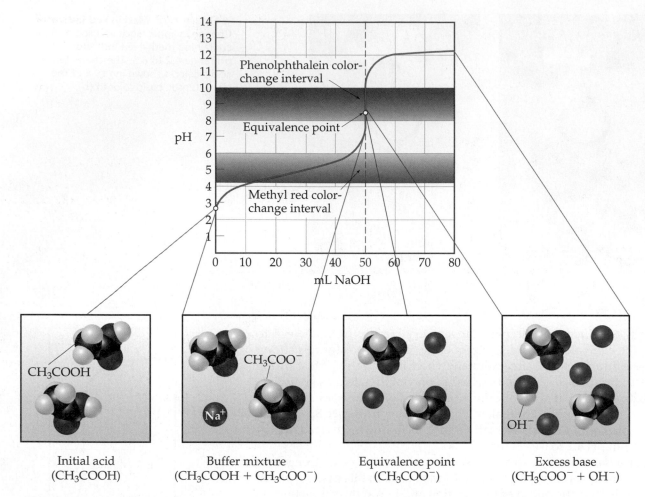

▲ **Figure 17.9 Adding a strong base to a weak acid.** The curve shows the variation in pH as 0.100 *M* NaOH solution is added to 50.0 mL of 0.100 *M* acetic acid solution. Phenolphthalein changes color at the equivalence point, but methyl red does not. (For clarity, water molecules have been omitted from the molecular art.)

▼ **Figure 17.10 Procedure for calculating the pH when a weak acid is partially neutralized by a strong base.** First consider the effect of the neutralization reaction (stoichiometry calculation). Then determine the pH of the resultant buffer mixture (equilibrium calculation). An analogous procedure can be used for the addition of strong acid to a weak base.

CH_3COOH and CH_3COO^- in the solution. Next, we calculate the pH of this buffer pair using procedures developed in Sections 17.1 and 17.2. The general procedure is diagrammed in Figure 17.10▼ and illustrated in Sample Exercise 17.7.

3. *The equivalence point:* The equivalence point is reached after adding 50.0 mL of 0.100 *M* NaOH to the 50.0 mL of 0.100 *M* CH_3COOH. At this point the 5.00×10^{-3} mol of NaOH completely reacts with the 5.00×10^{-3} mol of CH_3COOH to form 5.00×10^{-3} mol of their salt, CH_3COONa. The Na^+ ion of this salt has no significant effect on the pH. The CH_3COO^- ion, however, is a weak base, and the pH at the equivalence point is therefore greater than 7. Indeed, the pH at the equivalence point is always above 7 in a weak acid–strong base titration because the anion of the salt formed is a weak base. The procedure for calculating the pH of the solution of a weak base is described in Section 16.7 and is shown in Sample Exercise 17.8.

4. *After the equivalence point (excess base):* In this region of the titration curve, $[OH^-]$ from the reaction of CH_3COO^- with water is negligible compared to $[OH^-]$ from the excess NaOH. Thus, the pH is determined by the concentration of OH^- from the excess NaOH. The method for calculating pH

in this region is therefore like that for the strong acid–strong base titration illustrated in Sample Exercise 17.6(b). Thus, the addition of 51.0 mL of 0.100 M NaOH to 50.0 mL of either 0.100 M HCl or 0.100 M CH$_3$COOH yields the same pH, 11.00. Notice in Figures 17.6 and 17.9 that the titration curves for the titrations of both the strong acid and the weak acid are the same after the equivalence point.

■ SAMPLE EXERCISE 17.7 | Calculating pH for a Weak Acid–Strong Base Titration

Calculate the pH of the solution formed when 45.0 mL of 0.100 M NaOH is added to 50.0 mL of 0.100 M CH$_3$COOH ($K_a = 1.8 \times 10^{-5}$).

SOLUTION

Analyze: We are asked to calculate the pH before the equivalence point of the titration of a weak acid with a strong base.

Plan: We first must determine the number of moles of CH$_3$COOH and CH$_3$COO$^-$ that are present after the neutralization reaction. We then calculate pH using K_a together with [CH$_3$COOH] and [CH$_3$COO$^-$].

Solve: *Stoichiometry Calculation:* The product of the volume and concentration of each solution gives the number of moles of each reactant present before the neutralization:

$$(0.0500 \text{ L soln})\left(\frac{0.100 \text{ mol CH}_3\text{COOH}}{1 \text{ L soln}}\right) = 5.00 \times 10^{-3} \text{ mol CH}_3\text{COOH}$$

$$(0.0450 \text{ L soln})\left(\frac{0.100 \text{ mol NaOH}}{1 \text{ L soln}}\right) = 4.50 \times 10^{-3} \text{ mol NaOH}$$

The 4.50×10^{-3} mol of NaOH consumes 4.50×10^{-3} mol of CH$_3$COOH:

	CH$_3$COOH(aq) +	OH$^-$(aq)	⟶ CH$_3$COO$^-$(aq) +	H$_2$O(l)
Before addition	5.00×10^{-3} mol	0	0	—
Addition		4.50×10^{-3} mol		
After addition	0.50×10^{-3} mol	0	4.50×10^{-3} mol	—

The total volume of the solution is

$$45.0 \text{ mL} + 50.0 \text{ mL} = 95.0 \text{ mL} = 0.0950 \text{ L}$$

The resulting molarities of CH$_3$COOH and CH$_3$COO$^-$ after the reaction are therefore

$$[\text{CH}_3\text{COOH}] = \frac{0.50 \times 10^{-3} \text{ mol}}{0.0950 \text{ L}} = 0.0053 \ M$$

$$[\text{CH}_3\text{COO}^-] = \frac{4.50 \times 10^{-3} \text{ mol}}{0.0950 \text{ L}} = 0.0474 \ M$$

Equilibrium Calculation: The equilibrium between CH$_3$COOH and CH$_3$COO$^-$ must obey the equilibrium-constant expression for CH$_3$COOH

$$K_a = \frac{[\text{H}^+][\text{CH}_3\text{COO}^-]}{[\text{CH}_3\text{COOH}]} = 1.8 \times 10^{-5}$$

Solving for [H$^+$] gives

$$[\text{H}^+] = K_a \times \frac{[\text{CH}_3\text{COOH}]}{[\text{CH}_3\text{COO}^-]} = (1.8 \times 10^{-5}) \times \left(\frac{0.0053}{0.0474}\right) = 2.0 \times 10^{-6} \ M$$

$$\text{pH} = -\log(2.0 \times 10^{-6}) = 5.70$$

Comment: We could have solved for pH equally well using the Henderson–Hasselbalch equation.

■ PRACTICE EXERCISE

(a) Calculate the pH in the solution formed by adding 10.0 mL of 0.050 M NaOH to 40.0 mL of 0.0250 M benzoic acid (C$_6$H$_5$COOH, $K_a = 6.3 \times 10^{-5}$). **(b)** Calculate the pH in the solution formed by adding 10.0 mL of 0.100 M HCl to 20.0 mL of 0.100 M NH$_3$.
Answers: **(a)** 4.20, **(b)** 9.26

■ SAMPLE EXERCISE 17.8 | Calculating the pH at the Equivalence Point

Calculate the pH at the equivalence point in the titration of 50.0 mL of 0.100 M CH$_3$COOH with 0.100 M NaOH.

SOLUTION

Analyze: We are asked to determine the pH at the equivalence point of the titration of a weak acid with a strong base. Because the neutralization of a weak acid produces its anion, which is a weak base, we expect the pH at the equivalence point to be greater than 7.

Plan: The initial number of moles of acetic acid will equal the number of moles of acetate ion at the equivalence point. We use the volume of the solution at the equivalence point to calculate the concentration of acetate ion. Because the acetate ion is a weak base, we can calculate the pH using K_b and $[CH_3COO^-]$.

Solve: The number of moles of acetic acid in the initial solution is obtained from the volume and molarity of the solution:

$$Moles = M \times L = (0.100 \text{ mol/L})(0.0500 \text{ L}) = 5.00 \times 10^{-3} \text{ mol } CH_3COOH$$

Hence 5.00×10^{-3} mol of CH_3COO^- is formed. It will take 50.0 mL of NaOH to reach the equivalence point (Figure 17.9). The volume of this salt solution at the equivalence point is the sum of the volumes of the acid and base, 50.0 mL + 50.0 mL = 100.0 mL = 0.1000 L. Thus, the concentration of CH_3COO^- is

$$[CH_3COO^-] = \frac{5.00 \times 10^{-3} \text{ mol}}{0.1000 \text{ L}} = 0.0500 \text{ } M$$

The CH_3COO^- ion is a weak base.

$$CH_3COO^-(aq) + H_2O(l) \rightleftharpoons CH_3COOH(aq) + OH^-(aq)$$

The K_b for CH_3COO^- can be calculated from the K_a value of its conjugate acid, $K_b = K_w/K_a = (1.0 \times 10^{-14})/(1.8 \times 10^{-5}) = 5.6 \times 10^{-10}$. Using the K_b expression, we have

$$K_b = \frac{[CH_3COOH][OH^-]}{[CH_3COO^-]} = \frac{(x)(x)}{0.0500 - x} = 5.6 \times 10^{-10}$$

Making the approximation that $0.0500 - x \simeq 0.0500$, and then solving for x, we have $x = [OH^-] = 5.3 \times 10^{-6} M$, which gives pOH = 5.28 and pH = 8.72.

Check: The pH is above 7, as expected for the salt of a weak acid and strong base.

■ PRACTICE EXERCISE

Calculate the pH at the equivalence point when **(a)** 40.0 mL of 0.025 M benzoic acid (C_6H_5COOH, $K_a = 6.3 \times 10^{-5}$) is titrated with 0.050 M NaOH; **(b)** 40.0 mL of 0.100 M NH_3 is titrated with 0.100 M HCl.
Answers: **(a)** 8.21, **(b)** 5.28

▲ **Figure 17.11 Effect of K_a on titration curves.** This set of curves shows the influence of acid strength (K_a) on the shape of the curve for titration with NaOH. Each curve represents titration of 50.0 mL of 0.10 M acid with 0.10 M NaOH. The weaker the acid, the higher the initial pH and the smaller the pH change at the equivalence point.

The pH titration curves for weak acid–strong base titrations differ from those for strong acid–strong base titrations in three noteworthy ways:

1. The solution of the weak acid has a higher initial pH than a solution of a strong acid of the same concentration.

2. The pH change at the rapid-rise portion of the curve near the equivalence point is smaller for the weak acid than it is for the strong acid.

3. The pH at the equivalence point is above 7.00 for the weak acid–strong base titration.

To illustrate these differences further, consider the family of titration curves shown in Figure 17.11 ◄. Notice that the initial pH increases as the acid becomes weaker (that is, as K_a becomes smaller), and that the pH change near the equivalence point becomes less marked. Notice also that the pH at the equivalence point steadily increases as K_a decreases. It is virtually impossible to determine the equivalence point when pK_a is 10 or higher because the pH change is too small and gradual.

Because the pH change near the equivalence point becomes smaller as K_a decreases, the choice of indicator for a weak acid–strong base titration is more critical than it is for a strong acid–strong base titration. When 0.100 M CH_3COOH ($K_a = 1.8 \times 10^{-5}$) is titrated with 0.100 M NaOH, for example, as shown in Figure 17.9, the pH increases rapidly only over the pH range of about 7 to 10. Phenolphthalein is therefore an ideal indicator because it changes color from pH 8.3 to 10.0, close to the pH at the equivalence point. Methyl red is a poor choice, however, because its color change occurs from 4.2 to 6.0, which begins well before the equivalence point is reached.

Titration of a weak base (such as 0.100 M NH$_3$) with a strong acid solution (such as 0.100 M HCl) leads to the titration curve shown in Figure 17.12▶. In this particular example the equivalence point occurs at pH 5.28. Thus, methyl red would be an ideal indicator, but phenolphthalein would be a poor choice.

GIVE IT SOME THOUGHT

Why is the choice of indicator more crucial for a weak acid–strong base titration than for a strong acid–strong base titration?

Titrations of Polyprotic Acids

When weak acids contain more than one ionizable H atom, as in phosphorous acid (H$_3$PO$_3$), reaction with OH$^-$ occurs in a series of steps. Neutralization of H$_3$PO$_3$ proceeds in two stages.
∞ (Chapter 16 Sample Integrative Exercise)

$$H_3PO_3(aq) + OH^-(aq) \longrightarrow H_2PO_3^-(aq) + H_2O(l) \quad [17.13]$$

$$H_2PO_3^-(aq) + OH^-(aq) \longrightarrow HPO_3^{2-}(aq) + H_2O(l) \quad [17.14]$$

When the neutralization steps of a polyprotic acid or polybasic base are sufficiently separated, the substance exhibits a titration curve with multiple equivalence points. Figure 17.13▶ shows the two distinct equivalence points in the titration curve for the H$_3$PO$_3$–H$_2$PO$_3^-$–HPO$_3^{2-}$ system.

GIVE IT SOME THOUGHT

Sketch the titration curve for the titration of Na$_2$CO$_3$ with HCl.

17.4 SOLUBILITY EQUILIBRIA

The equilibria that we have considered thus far in this chapter have involved acids and bases. Furthermore, they have been homogeneous; that is, all the species have been in the same phase. Through the rest of this chapter we will consider the equilibria involved in the dissolution or precipitation of ionic compounds. These reactions are heterogeneous.

The dissolving and precipitating of compounds are phenomena that occur both within us and around us. Tooth enamel dissolves in acidic solutions, for example, causing tooth decay. The precipitation of certain salts in our kidneys produces kidney stones. The waters of Earth contain salts dissolved as water passes over and through the ground. Precipitation of CaCO$_3$ from groundwater is responsible for the formation of stalactites and stalagmites within limestone caves (Figure 4.1).

In our earlier discussion of precipitation reactions, we considered some general rules for predicting the solubility of common salts in water. ∞ (Section 4.2) These rules give us a qualitative sense of whether a compound will have a low or high solubility in water. By considering solubility equilibria, in contrast, we can make quantitative predictions about the amount of a given compound that will dissolve. We can also use these equilibria to analyze the factors that affect solubility.

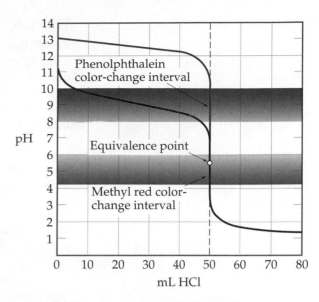

▲ **Figure 17.12 Adding a strong acid to a base.** The blue curve shows pH versus volume of added HCl in the titration of 50.0 mL of 0.10 M ammonia (weak base) with 0.10 M HCl. The red curve shows pH versus added acid for the titration of 0.10 M NaOH (strong base). Both phenolphthalein and methyl red change color at the equivalence point in the titration of the strong base. Phenolphthalein changes color before the equivalence point in the titration of the weak base.

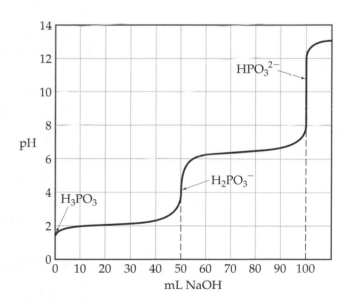

▲ **Figure 17.13 Diprotic acid.** Titration curve for the reaction of 50.0 mL of 0.10 M H$_3$PO$_3$ with 0.10 M NaOH.

The Solubility-Product Constant, K_{sp}

Recall that a *saturated solution* is one in which the solution is in contact with undissolved solute. ∞ (Section 13.2) Consider, for example, a saturated aqueous solution of $BaSO_4$ that is in contact with solid $BaSO_4$. Because the solid is an ionic compound, it is a strong electrolyte and yields $Ba^{2+}(aq)$ and $SO_4^{2-}(aq)$ ions upon dissolving. The following equilibrium is readily established between the undissolved solid and hydrated ions in solution:

$$BaSO_4(s) \rightleftharpoons Ba^{2+}(aq) + SO_4^{2-}(aq) \qquad [17.15]$$

As with any other equilibrium, the extent to which this dissolution reaction occurs is expressed by the magnitude of its equilibrium constant. Because this equilibrium equation describes the dissolution of a solid, the equilibrium constant indicates how soluble the solid is in water and is referred to as the **solubility-product constant** (or simply the **solubility product**). It is denoted K_{sp}, where sp stands for solubility product.

The equilibrium-constant expression for the equilibrium between a solid and an aqueous solution of its component ions is written according to the rules that apply to any equilibrium-constant expression. Remember, however, that solids do not appear in the equilibrium-constant expressions for heterogeneous equilibria. ∞ (Section 15.4) Thus, the solubility-product expression for $BaSO_4$, which is based on Equation 17.15, is

$$K_{sp} = [Ba^{2+}][SO_4^{2-}] \qquad [17.16]$$

In general, *the solubility product of a compound equals the product of the concentration of the ions involved in the equilibrium, each raised to the power of its coefficient in the equilibrium equation.* The coefficient for each ion in the equilibrium equation also equals its subscript in the compound's chemical formula.

The values of K_{sp} at 25 °C for many ionic solids are tabulated in Appendix D. The value of K_{sp} for $BaSO_4$ is 1.1×10^{-10}, a very small number, indicating that only a very small amount of the solid will dissolve in water.

■ SAMPLE EXERCISE 17.9 | Writing Solubility-Product (K_{sp}) Expressions

Write the expression for the solubility-product constant for CaF_2, and look up the corresponding K_{sp} value in Appendix D.

SOLUTION

Analyze: We are asked to write an equilibrium-constant expression for the process by which CaF_2 dissolves in water.

Plan: We apply the same rules for writing any equilibrium-constant expression, excluding the solid reactant from the expression. We assume that the compound dissociates completely into its component ions.

$$CaF_2(s) \rightleftharpoons Ca^{2+}(aq) + 2 F^-(aq)$$

Solve: Following the italicized rule stated previously, the expression for K_{sp} is

$$K_{sp} = [Ca^{2+}][F^-]^2$$

In Appendix D we see that this K_{sp} has a value of 3.9×10^{-11}.

■ PRACTICE EXERCISE

Give the solubility-product-constant expressions and the values of the solubility-product constants (from Appendix D) for the following compounds: **(a)** barium carbonate, **(b)** silver sulfate.

Answers: **(a)** $K_{sp} = [Ba^{2+}][CO_3^{2-}] = 5.0 \times 10^{-9}$; **(b)** $K_{sp} = [Ag^+]^2[SO_4^{2-}] = 1.5 \times 10^{-5}$

Solubility and K_{sp}

It is important to distinguish carefully between solubility and the solubility-product constant. The solubility of a substance is the quantity that dissolves to form a saturated solution. ∞ (Section 13.2) Solubility is often expressed as grams of solute per liter of solution (g/L). *Molar solubility* is the number of moles of the solute that dissolve in forming a liter of saturated solution of the solute (mol/L). The solubility-product constant (K_{sp}) is the equilibrium constant for the equilibrium between an ionic solid and its saturated solution and is a unitless number. Thus, the magnitude of K_{sp} is a measure of how much of the solid dissolves to form a saturated solution.

▲ **Figure 17.14 Relationships between solubility and K_{sp}.** The solubility of any compound in grams per liter can be converted to molar solubility. The molar solubility can be used to determine the concentrations of ions in solution. The concentration of ions can be used to calculate K_{sp}. The steps can be reversed, and solubility calculated from K_{sp}.

GIVE IT SOME THOUGHT

Without doing a calculation, predict which of the following compounds will have the greatest molar solubility in water: AgCl ($K_{sp} = 1.8 \times 10^{-10}$), AgBr ($K_{sp} = 5.0 \times 10^{-13}$), or AgI ($K_{sp} = 8.3 \times 10^{-17}$).

The solubility of a substance can change considerably as the concentrations of other solutes change. The solubility of $Mg(OH)_2$, for example, depends highly on pH. The solubility is also affected by the concentrations of other ions in solution, especially Mg^{2+}. In contrast, the solubility-product constant, K_{sp}, has only one value for a given solute at any specific temperature.*

In principle, it is possible to use the K_{sp} value of a salt to calculate solubility under a variety of conditions. In practice, great care must be taken in doing so for the reasons indicated in "A Closer Look: Limitations of Solubility Products" at the end of this section. Agreement between measured solubility and that calculated from K_{sp} is usually best for salts whose ions have low charges (1+ and 1−) and do not hydrolyze. Figure 17.14▲ summarizes the relationships among various expressions of solubility and K_{sp}.

■ SAMPLE EXERCISE 17.10 | Calculating K_{sp} from Solubility

Solid silver chromate is added to pure water at 25 °C. Some of the solid remains undissolved at the bottom of the flask. The mixture is stirred for several days to ensure that equilibrium is achieved between the undissolved $Ag_2CrO_4(s)$ and the solution. Analysis of the equilibrated solution shows that its silver ion concentration is 1.3×10^{-4} M. Assuming that Ag_2CrO_4 dissociates completely in water and that there are no other important equilibria involving the Ag^+ or CrO_4^{2-} ions in the solution, calculate K_{sp} for this compound.

SOLUTION

Analyze: We are given the equilibrium concentration of Ag^+ in a saturated solution of Ag_2CrO_4. From this information, we are asked to determine the value of the solubility-product constant, K_{sp}, for Ag_2CrO_4.

Plan: The equilibrium equation and the expression for K_{sp} are

$$Ag_2CrO_4(s) \rightleftharpoons 2\,Ag^+(aq) + CrO_4^{2-}(aq) \qquad K_{sp} = [Ag^+]^2[CrO_4^{2-}]$$

To calculate K_{sp}, we need the equilibrium concentrations of Ag^+ and CrO_4^{2-}. We know that at equilibrium $[Ag^+] = 1.3 \times 10^{-4}$ M. All the Ag^+ and CrO_4^{2-} ions in the solution come from the Ag_2CrO_4 that dissolves. Thus, we can use $[Ag^+]$ to calculate $[CrO_4^{2-}]$.

Solve: From the chemical formula of silver chromate, we know that there must be 2 Ag^+ ions in solution for each CrO_4^{2-} ion in solution. Consequently, the concentration of CrO_4^{2-} is half the concentration of Ag^+:

$$[CrO_4^{2-}] = \left(\frac{1.3 \times 10^{-4}\ \text{mol Ag}^+}{L}\right)\left(\frac{1\ \text{mol CrO}_4^{2-}}{2\ \text{mol Ag}^+}\right) = 6.5 \times 10^{-5}\ M$$

We can now calculate the value of K_{sp}.

$$K_{sp} = [Ag^+]^2[CrO_4^{2-}] = (1.3 \times 10^{-4})^2(6.5 \times 10^{-5}) = 1.1 \times 10^{-12}$$

Check: We obtain a small value, as expected for a slightly soluble salt. Furthermore, the calculated value agrees well with the one given in Appendix D, 1.2×10^{-12}.

---

*This is strictly true only for very dilute solutions. The values of equilibrium constants are somewhat altered when the total concentration of ionic substances in water is increased. However, we will ignore these effects, which are taken into consideration only for work that requires exceptional accuracy.

■ **PRACTICE EXERCISE**

A saturated solution of $Mg(OH)_2$ in contact with undissolved solid is prepared at 25 °C. The pH of the solution is found to be 10.17. Assuming that $Mg(OH)_2$ dissociates completely in water and that there are no other simultaneous equilibria involving the Mg^{2+} or OH^- ions in the solution, calculate K_{sp} for this compound.
Answer: 1.6×10^{-12}

■ **SAMPLE EXERCISE 17.11** | Calculating Solubility from K_{sp}

The K_{sp} for CaF_2 is 3.9×10^{-11} at 25 °C. Assuming that CaF_2 dissociates completely upon dissolving and that there are no other important equilibria affecting its solubility, calculate the solubility of CaF_2 in grams per liter.

SOLUTION

Analyze: We are given K_{sp} for CaF_2 and are asked to determine solubility. Recall that the *solubility* of a substance is the quantity that can dissolve in solvent, whereas the *solubility-product constant*, K_{sp}, is an equilibrium constant.

Plan: We can approach this problem by using our standard techniques for solving equilibrium problems. We write the chemical equation for the dissolution process and set up a table of the initial and equilibrium concentrations. We then use the equilibrium-constant expression. In this case we know K_{sp}, and so we solve for the concentrations of the ions in solution.

Solve: Assume initially that none of the salt has dissolved, and then allow x moles/liter of CaF_2 to dissociate completely when equilibrium is achieved.

	$CaF_2(s)$	$\rightleftharpoons$	Ca^{2+}	+	$2\,F^-(aq)$
Initial	—		0		0
Change	—		$+x\;M$		$+2x\;M$
Equilibrium	—		$x\;M$		$2x\;M$

The stoichiometry of the equilibrium dictates that $2x$ moles/liter of F^- are produced for each x moles/liter of CaF_2 that dissolve. We now use the expression for K_{sp} and substitute the equilibrium concentrations to solve for the value of x:

$$K_{sp} = [Ca^{2+}][F^-]^2 = (x)(2x)^2 = 4x^3 = 3.9 \times 10^{-11}$$

$$x = \sqrt[3]{\frac{3.9 \times 10^{-11}}{4}} = 2.1 \times 10^{-4}\;M$$

(Remember that $\sqrt[3]{y} = y^{1/3}$; to calculate the cube root of a number, you can use the y^x function on your calculator, with $x = \frac{1}{3}$.) Thus, the molar solubility of CaF_2 is 2.1×10^{-4} mol/L. The mass of CaF_2 that dissolves in water to form a liter of solution is

$$\left(\frac{2.1 \times 10^{-4}\;mol\;CaF_2}{1\;L\;soln}\right)\left(\frac{78.1\;g\;CaF_2}{1\;mol\;CaF_2}\right) = 1.6 \times 10^{-2}\;g\;CaF_2/L\;soln$$

Check: We expect a small number for the solubility of a slightly soluble salt. If we reverse the calculation, we should be able to recalculate the solubility product: $K_{sp} = (2.1 \times 10^{-4})(4.2 \times 10^{-4})^2 = 3.7 \times 10^{-11}$, close to the starting value for K_{sp}, 3.9×10^{-11}.

Comment: Because F^- is the anion of a weak acid, you might expect that the hydrolysis of the ion would affect the solubility of CaF_2. The basicity of F^- is so small ($K_b = 1.5 \times 10^{-11}$), however, that the hydrolysis occurs to only a slight extent and does not significantly influence the solubility. The reported solubility is 0.017 g/L at 25 °C, in good agreement with our calculation.

■ **PRACTICE EXERCISE**

The K_{sp} for LaF_3 is 2×10^{-19}. What is the solubility of LaF_3 in water in moles per liter?
Answer: 9×10^{-6} mol/L

The concentrations of ions calculated from K_{sp} sometimes deviate appreciably from those found experimentally. In part, these deviations are due to electrostatic interactions between ions in solution, which can lead to ion pairs. (See Section 13.5, "A Closer Look: Colligative Properties of Electrolyte Solutions.") These interactions increase in magnitude both as the concentrations of the ions increase and as their charges increase. The solubility calculated from K_{sp} tends to be low unless it is corrected to account for these interactions between ions. Chemists have developed procedures for correcting for these "ionic-strength" or "ionic-activity" effects, and these procedures are examined in more advanced chemistry courses. As an example of the effect of these interionic interactions, consider $CaCO_3$ (calcite), whose solubility product, $K_{sp} = 4.5 \times 10^{-9}$, gives a calculated solubility of 6.7×10^{-5} mol/L. Making corrections for the interionic interactions in the solution yields a higher solubility, 7.3×10^{-5} mol/L. The reported solubility, however, is twice as high (1.4×10^{-4} mol/L), so there must be one or more additional factors involved.

Another common source of error in calculating ion concentrations from K_{sp} is ignoring other equilibria that occur simultaneously in the solution. It is possible, for example, that acid–base equilibria take place simultaneously with solubility equilibria. In particular, both basic anions and cations with high charge-to-size ratios undergo hydrolysis reactions that can measurably increase the solubilities of their salts. For example, $CaCO_3$ contains the basic carbonate ion ($K_b = 1.8 \times 10^{-4}$), which hydrolyzes in water: $CO_3{}^{2-}(aq) + H_2O(l) \rightleftharpoons HCO_3{}^-(aq) + OH^-(aq)$. If we consider both the effect of the interionic interactions in the solution and the effect of the simultaneous solubility and hydrolysis equilibria, we calculate a solubility of 1.4×10^{-4} mol/L, in agreement with the measured value.

Finally, we generally assume that ionic compounds dissociate completely into their component ions when they dissolve. This assumption is not always valid. When MgF_2 dissolves, for example, it yields not only Mg^{2+} and F^- ions but also MgF^+ ions in solution. Thus, we see that calculating solubility using K_{sp} can be more complicated than it first appears and it requires considerable knowledge of the equilibria occurring in solution.

17.5 FACTORS THAT AFFECT SOLUBILITY

The solubility of a substance is affected by temperature as well as by the presence of other solutes. The presence of an acid, for example, can have a major influence on the solubility of a substance. In Section 17.4 we considered the dissolving of ionic compounds in pure water. In this section we examine three factors that affect the solubility of ionic compounds: (1) the presence of common ions, (2) the pH of the solution, and (3) the presence of complexing agents. We will also examine the phenomenon of *amphoterism*, which is related to the effects of both pH and complexing agents.

Common-Ion Effect

The presence of either $Ca^{2+}(aq)$ or $F^-(aq)$ in a solution reduces the solubility of CaF_2, shifting the solubility equilibrium of CaF_2 to the left.

$$CaF_2(s) \rightleftharpoons Ca^{2+}(aq) + 2\,F^-(aq)$$

Addition of Ca^{2+} or F^- shifts equilibrium, reducing solubility

This reduction in solubility is another application of the common-ion effect. ∞ (Section 17.1) In general, *the solubility of a slightly soluble salt is decreased by the presence of a second solute that furnishes a common ion.* Figure 17.15▶ shows how the solubility of CaF_2 decreases as NaF is added to the solution. Sample Exercise 17.12 shows how the K_{sp} can be used to calculate the solubility of a slightly soluble salt in the presence of a common ion.

▲ **Figure 17.15 Common-ion effect.** The way in which NaF concentration affects the solubility of CaF_2 demonstrates the common-ion effect. Notice that the CaF_2 solubility is on a logarithmic scale.

■ **SAMPLE EXERCISE 17.12** | Calculating the Effect of a Common Ion on Solubility

Calculate the molar solubility of CaF_2 at 25 °C in a solution that is **(a)** 0.010 M in $Ca(NO_3)_2$, **(b)** 0.010 M in NaF.

SOLUTION

Analyze: We are asked to determine the solubility of CaF_2 in the presence of two strong electrolytes, each of which contains an ion common to CaF_2. In (a) the common ion is Ca^{2+}, and NO_3^- is a spectator ion. In (b) the common ion is F^-, and Na^+ is a spectator ion.

Plan: Because the slightly soluble compound is CaF_2, we need to use the K_{sp} for this compound, which is available in Appendix D:

$$K_{sp} = [Ca^{2+}][F^-]^2 = 3.9 \times 10^{-11}$$

The value of K_{sp} is unchanged by the presence of additional solutes. Because of the common-ion effect, however, the solubility of the salt will decrease in the presence of common ions. We can again use our standard equilibrium techniques of starting with the equation for CaF_2 dissolution, setting up a table of initial and equilibrium concentrations, and using the K_{sp} expression to determine the concentration of the ion that comes only from CaF_2.

Solve: (a) In this instance the initial concentration of Ca^{2+} is 0.010 M because of the dissolved $Ca(NO_3)_2$:

	$CaF_2(s)$	$\rightleftharpoons$	$Ca^{2+}(aq)$	+	$2 F^-(aq)$
Initial	—		0.010 M		0
Change	—		$+x\ M$		$+2x\ M$
Equilibrium	—		$(0.010 + x)\ M$		$2x\ M$

Substituting into the solubility-product expression gives

$$K_{sp} = 3.9 \times 10^{-11} = [Ca^{2+}][F^-]^2 = (0.010 + x)(2x)^2$$

This would be a messy problem to solve exactly, but fortunately it is possible to simplify matters greatly. Even without the common-ion effect, the solubility of CaF_2 is very small ($2.1 \times 10^{-4}\ M$). Thus, we assume that the 0.010 M concentration of Ca^{2+} from $Ca(NO_3)_2$ is very much greater than the small additional concentration resulting from the solubility of CaF_2; that is, x is small compared to 0.010 M, and $0.010 + x \simeq 0.010$. We then have

$$3.9 \times 10^{-11} = (0.010)(2x)^2$$

$$x^2 = \frac{3.9 \times 10^{-11}}{4(0.010)} = 9.8 \times 10^{-10}$$

$$x = \sqrt{9.8 \times 10^{-10}} = 3.1 \times 10^{-5}\ M$$

The very small value for x validates the simplifying assumption we have made. Our calculation indicates that 3.1×10^{-5} mol of solid CaF_2 dissolves per liter of the 0.010 M $Ca(NO_3)_2$ solution.

(b) In this case the common ion is F^-, and at equilibrium we have

$$[Ca^{2+}] = x \quad \text{and} \quad [F^-] = 0.010 + 2x$$

Assuming that $2x$ is small compared to 0.010 M (that is, $0.010 + 2x \simeq 0.010$), we have

$$3.9 \times 10^{-11} = x(0.010)^2$$

$$x = \frac{3.9 \times 10^{-11}}{(0.010)^2} = 3.9 \times 10^{-7}\ M$$

Thus, 3.9×10^{-7} mol of solid CaF_2 should dissolve per liter of 0.010 M NaF solution.

Comment: The molar solubility of CaF_2 in pure water is $2.1 \times 10^{-4}\ M$ (Sample Exercise 17.11). By comparison, our calculations above show that the solubility of CaF_2 in the presence of 0.010 M Ca^{2+} is $3.1 \times 10^{-5}\ M$, and in the presence of 0.010 M F^- ion it is $3.9 \times 10^{-7}\ M$. Thus, the addition of either Ca^{2+} or F^- to a solution of CaF_2 decreases the solubility. However, the effect of F^- on the solubility is more pronounced than that of Ca^{2+} because $[F^-]$ appears to the second power in the K_{sp} expression for CaF_2, whereas Ca^{2+} appears to the first power.

■ **PRACTICE EXERCISE**

The value for K_{sp} for manganese(II) hydroxide, $Mn(OH)_2$, is 1.6×10^{-13}. Calculate the molar solubility of $Mn(OH)_2$ in a solution that contains 0.020 M NaOH.
Answer: $4.0 \times 10^{-10}\ M$

Solubility and pH

The pH of a solution will affect the solubility of any substance whose anion is basic. Consider $Mg(OH)_2$, for example, for which the solubility equilibrium is

$$Mg(OH)_2(s) \rightleftharpoons Mg^{2+}(aq) + 2 OH^-(aq) \quad K_{sp} = 1.8 \times 10^{-11} \quad [17.17]$$

A saturated solution of $Mg(OH)_2$ has a calculated pH of 10.52 and contains $[Mg^{2+}] = 1.7 \times 10^{-4}$ M. Now suppose that solid $Mg(OH)_2$ is equilibrated with a solution buffered at a more acidic pH of 9.0. The pOH, therefore, is 5.0, so $[OH^-] = 1.0 \times 10^{-5}$. Inserting this value for $[OH^-]$ into the solubility-product expression, we have

$$K_{sp} = [Mg^{2+}][OH^-]^2 = 1.8 \times 10^{-11}$$

$$[Mg^{2+}](1.0 \times 10^{-5})^2 = 1.8 \times 10^{-11}$$

$$[Mg^{2+}] = \frac{1.8 \times 10^{-11}}{(1.0 \times 10^{-5})^2} = 0.18 \ M$$

Thus, $Mg(OH)_2$ dissolves in the solution until $[Mg^{2+}] = 0.18$ M. It is apparent that $Mg(OH)_2$ is quite soluble in this solution. If the concentration of OH^- were reduced even further by making the solution more acidic, the Mg^{2+} concentration would have to increase to maintain the equilibrium condition. Thus, a sample of $Mg(OH)_2$ will dissolve completely if sufficient acid is added (Figure 17.16▼).

▼ Figure 17.16 **Dissolution of a precipitate in acid.** A white precipitate of $Mg(OH)_2(s)$ in contact with its saturated solution is in the test tube on the left. The dropper poised above the solution surface contains hydrochloric acid. (The anions accompanying the acid have been omitted to simplify the art.)

SOLUBILITY AND pH

The solubility of a substance whose anion is basic will be affected to some extent by the pH of the solution. The solubility of $Mg(OH)_2$ greatly increases as the acidity of the solution increases.

H^+

OH^- H^+

Mg^{2+} H_2O

$2 H^+(aq) + Mg(OH)_2(s)$ $\longrightarrow$ $Mg^{2+}(aq) + 2 H_2O(l)$

A precipitate of $Mg(OH)_2(s)$

The precipitate dissolves upon addition of acid.

A sample of $Mg(OH)_2$ will dissolve completely if sufficient acid is added.

Chemistry and Life   SINKHOLES

A principal cause of sinkholes is the dissolution of limestone, which is calcium carbonate, by groundwater. Although $CaCO_3$ has a relatively small solubility-product constant, it is quite soluble in the presence of acid.

$$CaCO_3(s) \rightleftharpoons Ca^{2+}(aq) + CO_3^{2-}(aq)$$
$$K_{sp} = 4.5 \times 10^{-9}$$

Rainwater is naturally acidic, with a pH range of 5 to 6, and can become more acidic when it comes into contact with decaying plant matter. Because carbonate ion is the conjugate base of the weak acid, hydrogen carbonate ion (HCO_3^-), it readily combines with hydrogen ion.

$$CO_3^{2-}(aq) + H^+(aq) \longrightarrow HCO_3^-(aq)$$

The consumption of carbonate ion shifts the dissolution equilibrium to the right, thus increasing the solubility of $CaCO_3$. This can have profound consequences in areas where the terrain consists of porous calcium carbonate bedrock covered by a relatively thin layer of clay and/or topsoil. As acidic water percolates through and gradually dissolves the limestone, it creates underground voids. A sinkhole results when the overlying ground can no longer be supported by the remaining bedrock and collapses into the underground cavity [Figure 17.17▶].

The sudden formation of large sinkholes can pose a serious threat to life and property. The existence of deep sinkholes also increases the risk of contamination of the aquifer.

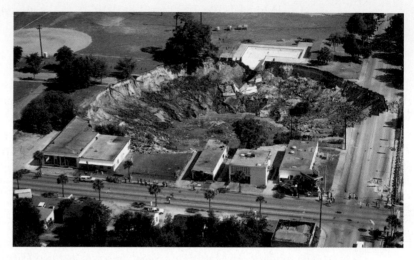

▲ **Figure 17.17 Sinkhole formation.** An underground void develops as limestone, $CaCO_3(s)$, dissolves. Collapse of the overlying ground into an underground cavity causes sinkhole formation. The large sinkhole shown here occured in Orlando, Florida and destroyed several buildings and part of a highway.

▲ **Figure 17.18 The effect of pH on the solubility of CaF_2.** The solubility increases as the solution becomes more acidic (lower pH). Notice that the vertical scale has been multiplied by 10^3.

The solubility of almost any ionic compound is affected if the solution is made sufficiently acidic or basic. The effects are very noticeable, however, only when one or both ions involved are at least moderately acidic or basic. The metal hydroxides, such as $Mg(OH)_2$, are examples of compounds containing a strongly basic ion, the hydroxide ion.

In general, *if a compound contains a basic anion (that is, the anion of a weak acid), its solubility will increase as the solution becomes more acidic.* As we have seen, the solubility of $Mg(OH)_2$ greatly increases as the acidity of the solution increases. The solubility of CaF_2 increases as the solution becomes more acidic, too, because the F^- ion is a weak base; it is the conjugate base of the weak acid HF. As a result, the solubility equilibrium of CaF_2 is shifted to the right as the concentration of F^- ions is reduced by protonation to form HF. Thus, the solution process can be understood in terms of two consecutive reactions:

$$CaF_2(s) \rightleftharpoons Ca^{2+}(aq) + 2\,F^-(aq) \qquad [17.18]$$

$$F^-(aq) + H^+(aq) \rightleftharpoons HF(aq) \qquad [17.19]$$

The equation for the overall process is

$$CaF_2(s) + 2\,H^+(aq) \rightleftharpoons Ca^{2+}(aq) + 2\,HF(aq) \qquad [17.20]$$

Figure 17.18◄ shows how the solubility of CaF_2 changes with pH.

Other salts that contain basic anions, such as CO_3^{2-}, PO_4^{3-}, CN^-, or S^{2-}, behave similarly. These examples illustrate a general rule: *The solubility of slightly soluble salts containing basic anions increases as $[H^+]$ increases (as pH is lowered).* The more basic the anion, the more the solubility is influenced by pH. Salts with anions of negligible basicity (the anions of strong acids) are unaffected by pH changes.

■ **SAMPLE EXERCISE 17.13** | **Predicting the Effect of Acid on Solubility**

Which of the following substances will be more soluble in acidic solution than in basic solution: **(a)** $Ni(OH)_2(s)$, **(b)** $CaCO_3(s)$, **(c)** $BaF_2(s)$, **(d)** $AgCl(s)$?

SOLUTION

Analyze: The problem lists four sparingly soluble salts, and we are asked to determine which will be more soluble at low pH than at high pH.

Plan: Ionic compounds that dissociate to produce a basic anion will be more soluble in acid solution.

Solve:

(a) $Ni(OH)_2(s)$ will be more soluble in acidic solution because of the basicity of OH^-; the H^+ ion reacts with the OH^- ion, forming water.

$$Ni(OH)_2(s) \rightleftharpoons Ni^{2+}(aq) + 2\,OH^-(aq)$$

$$\underline{2\,OH^-(aq) + 2\,H^+(aq) \rightleftharpoons 2\,H_2O(l)}$$

Overall: $Ni(OH)_2(s) + 2\,H^+(aq) \rightleftharpoons Ni^{2+}(aq) + 2\,H_2O(l)$

(b) Similarly, $CaCO_3(s)$ dissolves in acid solutions because CO_3^{2-} is a basic anion.

$$CaCO_3(s) \rightleftharpoons Ca^{2+}(aq) + CO_3^{2-}(aq)$$

$$CO_3^{2-}(aq) + 2\,H^+(aq) \rightleftharpoons H_2CO_3(aq)$$

$$\underline{H_2CO_3(aq) \rightleftharpoons CO_2(g) + H_2O(l)}$$

Overall: $CaCO_3(s) + 2\,H^+(aq) \rightleftharpoons Ca^{2+}(aq) + CO_2(g) + H_2O(l)$

The reaction between CO_3^{2-} and H^+ occurs in a stepwise fashion, first forming HCO_3^-. H_2CO_3 forms in appreciable amounts only when the concentration of H^+ is sufficiently high.

(c) The solubility of BaF_2 is also enhanced by lowering the pH, because F^- is a basic anion.

$$BaF_2(s) \rightleftharpoons Ba^{2+}(aq) + 2\,F^-(aq)$$

$$\underline{2\,F^-(aq) + 2\,H^+(aq) \rightleftharpoons 2\,HF(aq)}$$

Overall: $BaF_2(s) + 2\,H^+(aq) \rightleftharpoons Ba^{2+}(aq) + 2\,HF(aq)$

(d) The solubility of $AgCl$ is unaffected by changes in pH because Cl^- is the anion of a strong acid and therefore has negligible basicity.

■ **PRACTICE EXERCISE**

Write the net ionic equation for the reaction of the following copper(II) compounds with acid: **(a)** CuS, **(b)** $Cu(N_3)_2$.
Answers: **(a)** $CuS(s) + H^+(aq) \rightleftharpoons Cu^{2+}(aq) + HS^-(aq)$
(b) $Cu(N_3)_2(s) + 2\,H^+(aq) \rightleftharpoons Cu^{2+}(aq) + 2\,HN_3(aq)$

Formation of Complex Ions

A characteristic property of metal ions is their ability to act as Lewis acids, or electron-pair acceptors, toward water molecules, which act as Lewis bases, or electron-pair donors. ∞(Section 16.11) Lewis bases other than water can also interact with metal ions, particularly with transition-metal ions. Such interactions can dramatically affect the solubility of a metal salt. AgCl, for example, which has $K_{sp} = 1.8 \times 10^{-10}$, will dissolve in the presence of aqueous ammonia because Ag^+ interacts with the Lewis base NH_3, as shown in Figure 17.19▼. This process can be viewed as the sum of two reactions, the dissolution of AgCl and the Lewis acid–base interaction between Ag^+ and NH_3.

$$AgCl(s) \rightleftharpoons Ag^+(aq) + Cl^-(aq) \qquad [17.21]$$

$$\underline{Ag^+(aq) + 2\,NH_3(aq) \rightleftharpoons Ag(NH_3)_2^+(aq)} \qquad [17.22]$$

Overall: $AgCl(s) + 2\,NH_3(aq) \rightleftharpoons Ag(NH_3)_2^+(aq) + Cl^-(aq) \qquad [17.23]$

FORMATION OF COMPLEX IONS

Lewis bases can interact with metal ions, particularly with transition-metal ions, which can dramatically affect the solubility of a metal salt. AgCl, for example, will dissolve in the presence of aqueous ammonia because Ag$^+$ interacts with the Lewis base NH$_3$.

AgCl(s)   +   2 NH$_3$(aq) $\longrightarrow$ Ag(NH$_3$)$_2$$^+$(aq) + Cl$^-$(aq)

A saturated solution of AgCl in contact with solid AgCl.

When concentrated ammonia is added, Ag$^+$ ions are consumed in the formation of the complex ion Ag(NH$_3$)$_2$$^+$. The AgCl solid is being dissolved by the addition of NH$_3$.

Removal of Ag$^+$ ions from the solution shifts the dissolution equilibrium to the right, causing AgCl to dissolve. Addition of sufficient ammonia results in complete dissolution of the AgCl solid.

▲ **Figure 17.19 Using NH$_3$(aq) to dissolve AgCl(s).**

The presence of NH$_3$ drives the reaction, the dissolution of AgCl, to the right as Ag$^+$(aq) is consumed to form Ag(NH$_3$)$_2$$^+$.

For a Lewis base such as NH$_3$ to increase the solubility of a metal salt, it must be able to interact more strongly with the metal ion than water does. The NH$_3$ must displace solvating H$_2$O molecules (Sections 13.1 and 16.11) in order to form Ag(NH$_3$)$_2$$^+$:

$$Ag^+(aq) + 2\,NH_3(aq) \rightleftharpoons Ag(NH_3)_2^+(aq)$$ [17.24]

Chemistry and Life   TOOTH DECAY AND FLUORIDATION

Tooth enamel consists mainly of a mineral called hydroxyapatite, $Ca_{10}(PO_4)_6(OH)_2$. It is the hardest substance in the body. Tooth cavities are caused when acids dissolve tooth enamel.

$$Ca_{10}(PO_4)_6(OH)_2(s) + 8\,H^+(aq) \longrightarrow$$
$$10\,Ca^{2+}(aq) + 6\,HPO_4^{2-}(aq) + 2\,H_2O(l)$$

The resultant Ca^{2+} and HPO_4^{2-} ions diffuse out of the tooth enamel and are washed away by saliva. The acids that attack the hydroxyapatite are formed by the action of specific bacteria on sugars and other carbohydrates present in the plaque adhering to the teeth.

Fluoride ion, present in drinking water, toothpaste, and other sources, can react with hydroxyapatite to form fluorapatite, $Ca_{10}(PO_4)_6F_2$. This mineral, in which F^- has replaced OH^-, is much more resistant to attack by acids because the fluoride ion is a much weaker Brønsted–Lowry base than the hydroxide ion.

Because the fluoride ion is so effective in preventing cavities, it is added to the public water supply in many places to give a concentration of 1 mg/L (1 ppm). The compound added may be NaF or Na_2SiF_6. Na_2SiF_6 reacts with water to release fluoride ions by the following reaction:

$$SiF_6^{2-}(aq) + 2\,H_2O(l) \longrightarrow 6\,F^-(aq) + 4\,H^+(aq) + SiO_2(s)$$

About 80% of all toothpastes now sold in the United States contain fluoride compounds, usually at the level of 0.1% fluoride by mass. The most common compounds in toothpastes are sodium fluoride (NaF), sodium monofluorophosphate (Na_2PO_3F), and stannous fluoride (SnF_2).

Related Exercise: 17.110

An assembly of a metal ion and the Lewis bases bonded to it, such as $Ag(NH_3)_2^+$, is called a **complex ion**. The stability of a complex ion in aqueous solution can be judged by the size of the equilibrium constant for its formation from the hydrated metal ion. For example, the equilibrium constant for formation of $Ag(NH_3)_2^+$ (Equation 17.24) is 1.7×10^7:

$$K_f = \frac{[Ag(NH_3)_2^+]}{[Ag^+][NH_3]^2} = 1.7 \times 10^7 \qquad [17.25]$$

The equilibrium constant for this kind of reaction is called a **formation constant**, K_f. The formation constants for several complex ions are listed in Table 17.1 ▼.

TABLE 17.1 ■ Formation Constants for Some Metal Complex Ions in Water at 25 °C

Complex Ion	K_f	Equilibrium Equation
$Ag(NH_3)_2^+$	1.7×10^7	$Ag^+(aq) + 2\,NH_3(aq) \rightleftharpoons Ag(NH_3)_2^+(aq)$
$Ag(CN)_2^-$	1×10^{21}	$Ag^+(aq) + 2\,CN^-(aq) \rightleftharpoons Ag(CN)_2^-(aq)$
$Ag(S_2O_3)_2^{3-}$	2.9×10^{13}	$Ag^+(aq) + 2\,S_2O_3^{2-}(aq) \rightleftharpoons Ag(S_2O_3)_2^{3-}(aq)$
$CdBr_4^{2-}$	5×10^3	$Cd^{2+}(aq) + 4\,Br^-(aq) \rightleftharpoons CdBr_4^{2-}(aq)$
$Cr(OH)_4^-$	8×10^{29}	$Cr^{3+}(aq) + 4\,OH^-(aq) \rightleftharpoons Cr(OH)_4^-(aq)$
$Co(SCN)_4^{2-}$	1×10^3	$Co^{2+}(aq) + 4\,SCN^-(aq) \rightleftharpoons Co(SCN)_4^{2-}(aq)$
$Cu(NH_3)_4^{2+}$	5×10^{12}	$Cu^{2+}(aq) + 4\,NH_3(aq) \rightleftharpoons Cu(NH_3)_4^{2+}(aq)$
$Cu(CN)_4^{2-}$	1×10^{25}	$Cu^{2+}(aq) + 4\,CN^-(aq) \rightleftharpoons Cu(CN)_4^{2-}(aq)$
$Ni(NH_3)_6^{2+}$	1.2×10^9	$Ni^{2+}(aq) + 6\,NH_3(aq) \rightleftharpoons Ni(NH_3)_6^{2+}(aq)$
$Fe(CN)_6^{4-}$	1×10^{35}	$Fe^{2+}(aq) + 6\,CN^-(aq) \rightleftharpoons Fe(CN)_6^{4-}(aq)$
$Fe(CN)_6^{3-}$	1×10^{42}	$Fe^{3+}(aq) + 6\,CN^-(aq) \rightleftharpoons Fe(CN)_6^{3-}(aq)$

■ **SAMPLE EXERCISE 17.14** | Evaluating an Equilibrium Involving a Complex Ion

Calculate the concentration of Ag^+ present in solution at equilibrium when concentrated ammonia is added to a 0.010 M solution of $AgNO_3$ to give an equilibrium concentration of $[NH_3] = 0.20 \, M$. Neglect the small volume change that occurs when NH_3 is added.

SOLUTION

Analyze: When $NH_3(aq)$ is added to $Ag^+(aq)$, a reaction occurs forming $Ag(NH_3)_2^+$ as shown in Equation 17.22. We are asked to determine what concentration of $Ag^+(aq)$ will remain uncombined when the NH_3 concentration is brought to 0.20 M in a solution originally 0.010 M in $AgNO_3$.

Plan: We first assume that the $AgNO_3$ is completely dissociated, giving 0.10 M Ag^+. Because K_f for the formation of $Ag(NH_3)_2^+$ is quite large, we assume that essentially all the Ag^+ is then converted to $Ag(NH_3)_2^+$ and approach the problem as though we are concerned with the *dissociation* of $Ag(NH_3)_2^+$ rather than its *formation*. To facilitate this approach, we will need to reverse the equation to represent the formation of Ag^+ and NH_3 from $Ag(NH_3)_2^+$ and also make the corresponding change to the equilibrium constant.

$$Ag(NH_3)_2^+(aq) \rightleftharpoons Ag^+(aq) + 2\,NH_3(aq)$$

$$\frac{1}{K_f} = \frac{1}{1.7 \times 10^7} = 5.9 \times 10^{-8}$$

Solve: If $[Ag^+]$ is 0.010 M initially, then $[Ag(NH_3)_2^+]$ will be 0.010 M following addition of the NH_3. We now construct a table to solve this equilibrium problem. Note that the NH_3 concentration given in the problem is an *equilibrium* concentration rather than an initial concentration.

	$Ag(NH_3)_2^+(aq)$	$\rightleftharpoons$	$Ag^+(aq)$	+	$2\,NH_3(aq)$
Initial	0.010 M		0 M		
Change	$-x \, M$		$+x \, M$		
Equilibrium	$0.010 - x \, M$		$x \, M$		0.20 M

Because the concentration of Ag^+ is very small, we can ignore x in comparison with 0.010. Thus, $0.010 - x \approx 0.010 \, M$. Substituting these values into the equilibrium-constant expression for the dissociation of $Ag(NH_3)_2^+$, we obtain

$$\frac{[Ag^+][NH_3]^2}{[Ag(NH_3)_2^+]} = \frac{(x)(0.20)^2}{0.010} = 5.9 \times 10^{-8}$$

Solving for x, we obtain $x = 1.5 \times 10^{-8} \, M = [Ag^+]$. Thus, formation of the $Ag(NH_3)_2^+$ complex drastically reduces the concentration of free Ag^+ ion in solution.

■ **PRACTICE EXERCISE**

Calculate $[Cr^{3+}]$ in equilibrium with $Cr(OH)_4^-$ when 0.010 mol of $Cr(NO_3)_3$ is dissolved in a liter of solution buffered at pH 10.0.
Answer: $1 \times 10^{-16} \, M$

The general rule is that the solubility of metal salts increases in the presence of suitable Lewis bases, such as NH_3, CN^-, or OH^-, if the metal forms a complex with the base. The ability of metal ions to form complexes is an extremely important aspect of their chemistry. In Chapter 24 we will take a much closer look at complex ions. In that chapter and others we will see applications of complex ions to areas such as biochemistry, metallurgy, and photography.

Amphoterism

Some metal oxides and hydroxides that are relatively insoluble in neutral water dissolve in strongly acidic and strongly basic solutions. These substances are soluble in strong acids and bases because they themselves are capable of behaving as either an acid or base; they are **amphoteric oxides and hydroxides**.

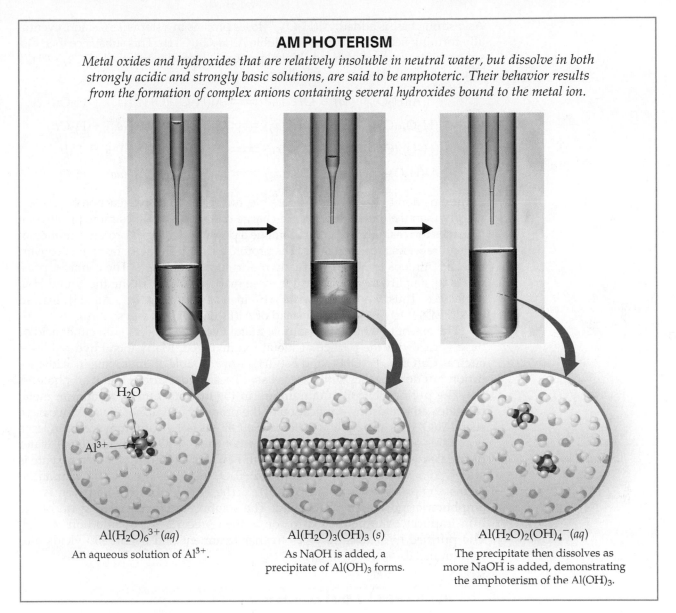

AMPHOTERISM

Metal oxides and hydroxides that are relatively insoluble in neutral water, but dissolve in both strongly acidic and strongly basic solutions, are said to be amphoteric. Their behavior results from the formation of complex anions containing several hydroxides bound to the metal ion.

$Al(H_2O)_6{}^{3+}(aq)$

An aqueous solution of Al^{3+}.

$Al(H_2O)_3(OH)_3$ (s)

As NaOH is added, a precipitate of $Al(OH)_3$ forms.

$Al(H_2O)_2(OH)_4{}^-(aq)$

The precipitate then dissolves as more NaOH is added, demonstrating the amphoterism of the $Al(OH)_3$.

▲ Figure 17.20 **Amphoterism.**

Amphoteric oxides and hydroxides include those of Al^{3+}, Cr^{3+}, Zn^{2+}, and Sn^{2+}. Notice that the term *amphoteric* is applied to the behavior of insoluble oxides and hydroxides that can be made to dissolve in either acidic or basic solutions. The similar term *amphiprotic*, which we encountered in Section 16.2, relates more generally to any molecule or ion that can either gain or lose a proton.

Amphoteric species dissolve in acidic solutions because they contain basic anions. What makes amphoteric oxides and hydroxides special, though, is that they also dissolve in strongly basic solutions (Figure 17.20 ▲). This behavior results from the formation of complex anions containing several (typically four) hydroxides bound to the metal ion.

$$Al(OH)_3(s) + OH^-(aq) \rightleftharpoons Al(OH)_4{}^-(aq) \qquad [17.26]$$

Amphoterism is often explained by the behavior of the water molecules that surround the metal ion and that are bonded to it by Lewis acid–base interactions. ∞ (Section 16.11) For example, $Al^{3+}(aq)$ is more accurately represented as $Al(H_2O)_6{}^{3+}(aq)$ because six water molecules are bonded to the Al^{3+} in aqueous solution. Recall from Section 16.11 that this hydrated ion is a weak acid.

As a strong base is added, $Al(H_2O)_6^{3+}$ loses protons in a stepwise fashion, eventually forming neutral and water-insoluble $Al(H_2O)_3(OH)_3$. This substance then dissolves upon removal of an additional proton to form the anion $Al(H_2O)_2(OH)_4^-$. The reactions that occur are as follows:

$$Al(H_2O)_6^{3+}(aq) + OH^-(aq) \rightleftharpoons Al(H_2O)_5(OH)^{2+}(aq) + H_2O(l)$$

$$Al(H_2O)_5(OH)^{2+}(aq) + OH^-(aq) \rightleftharpoons Al(H_2O)_4(OH)_2^+(aq) + H_2O(l)$$

$$Al(H_2O)_4(OH)_2^+(aq) + OH^-(aq) \rightleftharpoons Al(H_2O)_3(OH)_3(s) + H_2O(l)$$

$$Al(H_2O)_3(OH)_3(s) + OH^-(aq) \rightleftharpoons Al(H_2O)_2(OH)_4^-(aq) + H_2O(l)$$

Removing additional protons is possible, but each successive reaction occurs less readily than the one before. As the charge on the ion becomes more negative, it becomes increasingly difficult to remove a positively charged proton. Addition of an acid reverses these reactions. The proton adds in a stepwise fashion to convert the OH^- groups to H_2O, eventually re-forming $Al(H_2O)_6^{3+}$. The common practice is to simplify the equations for these reactions by excluding the bound H_2O molecules. Thus, we usually write Al^{3+} instead of $Al(H_2O)_6^{3+}$, $Al(OH)_3$ instead of $Al(H_2O)_3(OH)_3$, $Al(OH)_4^-$ instead of $Al(H_2O)_2(OH)_4^-$, and so forth.

The extent to which an insoluble metal hydroxide reacts with either acid or base varies with the particular metal ion involved. Many metal hydroxides—such as $Ca(OH)_2$, $Fe(OH)_2$, and $Fe(OH)_3$—are capable of dissolving in acidic solution but do not react with excess base. These hydroxides are not amphoteric.

The purification of aluminum ore in the manufacture of aluminum metal provides an interesting application of the property of amphoterism. As we have seen, $Al(OH)_3$ is amphoteric, whereas $Fe(OH)_3$ is not. Aluminum occurs in large quantities as the ore *bauxite*, which is essentially Al_2O_3 with additional water molecules. The ore is contaminated with Fe_2O_3 as an impurity. When bauxite is added to a strongly basic solution, the Al_2O_3 dissolves because the aluminum forms complex ions, such as $Al(OH)_4^-$. The Fe_2O_3 impurity, however, is not amphoteric and remains as a solid. The solution is filtered, getting rid of the iron impurity. Aluminum hydroxide is then precipitated by addition of acid. The purified hydroxide receives further treatment and eventually yields aluminum metal. ∞ (Section 23.3)

GIVE IT SOME THOUGHT

What kind of behavior characterizes an amphoteric oxide or an amphoteric hydroxide?

17.6 PRECIPITATION AND SEPARATION OF IONS

Equilibrium can be achieved starting with the substances on either side of a chemical equation. The equilibrium among $BaSO_4(s)$, $Ba^{2+}(aq)$, and $SO_4^{2-}(aq)$ (Equation 17.15) can be achieved starting with solid $BaSO_4$. It can also be reached starting with solutions of salts containing Ba^{2+} and SO_4^{2-}, say $BaCl_2$ and Na_2SO_4. When these two solutions are mixed, $BaSO_4$ will precipitate if the product of the initial ion concentrations, $Q = [Ba^{2+}][SO_4^{2-}]$, is greater than K_{sp}.

The use of the reaction quotient, Q, to determine the direction in which a reaction must proceed to reach equilibrium was discussed earlier. ∞ (Section 15.6) The possible relationships between Q and K_{sp} are summarized as follows:

- If $Q > K_{sp}$, precipitation occurs until $Q = K_{sp}$.

- If $Q = K_{sp}$, equilibrium exists (saturated solution).

- If $Q < K_{sp}$, solid dissolves until $Q = K_{sp}$.

■ **SAMPLE EXERCISE 17.15** | Predicting Whether a Precipitate Will Form

Will a precipitate form when 0.10 L of $8.0 \times 10^{-3}\,M\,Pb(NO_3)_2$ is added to 0.40 L of $5.0 \times 10^{-3}\,M\,Na_2SO_4$?

SOLUTION

Analyze: The problem asks us to determine whether a precipitate will form when two salt solutions are combined.

Plan: We should determine the concentrations of all ions immediately upon mixing of the solutions and compare the value of the reaction quotient, Q, to the solubility-product constant, K_{sp}, for any potentially insoluble product. The possible metathesis products are $PbSO_4$ and $NaNO_3$. Sodium salts are quite soluble; $PbSO_4$ has a K_{sp} of 6.3×10^{-7} (Appendix D), however, and will precipitate if the Pb^{2+} and $SO_4{}^{2-}$ ion concentrations are high enough for Q to exceed K_{sp} for the salt.

Solve: When the two solutions are mixed, the total volume becomes $0.10\,L + 0.40\,L = 0.50\,L$. The number of moles of Pb^{2+} in 0.10 L of $8.0 \times 10^{-3}\,M\,Pb(NO_3)_2$ is

$$(0.10\,L)\left(8.0 \times 10^{-3}\,\frac{mol}{L}\right) = 8.0 \times 10^{-4}\,mol$$

The concentration of Pb^{2+} in the 0.50-L mixture is therefore

$$[Pb^{2+}] = \frac{8.0 \times 10^{-4}\,mol}{0.50\,L} = 1.6 \times 10^{-3}\,M$$

The number of moles of $SO_4{}^{2-}$ in 0.40 L of $5.0 \times 10^{-3}\,M\,Na_2SO_4$ is

$$(0.40\,L)\left(5.0 \times 10^{-3}\,\frac{mol}{L}\right) = 2.0 \times 10^{-3}\,mol$$

Therefore, $[SO_4{}^{2-}]$ in the 0.50-L mixture is

$$[SO_4{}^{2-}] = \frac{2.0 \times 10^{-3}\,mol}{0.50\,L} = 4.0 \times 10^{-3}\,M$$

We then have

$$Q = [Pb^{2+}][SO_4{}^{2-}] = (1.6 \times 10^{-3})(4.0 \times 10^{-3}) = 6.4 \times 10^{-6}$$

Because $Q > K_{sp}$, $PbSO_4$ will precipitate.

■ **PRACTICE EXERCISE**

Will a precipitate form when 0.050 L of $2.0 \times 10^{-2}\,M\,NaF$ is mixed with 0.010 L of $1.0 \times 10^{-2}\,M\,Ca(NO_3)_2$?
Answer: Yes, CaF_2 precipitates because $Q = 4.6 \times 10^{-8}$ is larger than $K_{sp} = 3.9 \times 10^{-11}$

Selective Precipitation of Ions

Ions can be separated from each other based on the solubilities of their salts. Consider a solution containing both Ag^+ and Cu^{2+}. If HCl is added to the solution, $AgCl$ ($K_{sp} = 1.8 \times 10^{-10}$) precipitates, while Cu^{2+} remains in solution because $CuCl_2$ is soluble. Separation of ions in an aqueous solution by using a reagent that forms a precipitate with one or a few of the ions is called *selective precipitation*.

■ **SAMPLE EXERCISE 17.16** | Calculating Ion Concentrations for Precipitation

A solution contains $1.0 \times 10^{-2}\,M\,Ag^+$ and $2.0 \times 10^{-2}\,M\,Pb^{2+}$. When Cl^- is added to the solution, both $AgCl$ ($K_{sp} = 1.8 \times 10^{-10}$) and $PbCl_2$ ($K_{sp} = 1.7 \times 10^{-5}$) precipitate from the solution. What concentration of Cl^- is necessary to begin the precipitation of each salt? Which salt precipitates first?

SOLUTION

Analyze: We are asked to determine the concentration of Cl^- necessary to begin the precipitation from a solution containing Ag^+ and Pb^{2+} ions, and to predict which metal chloride will begin to precipitate first.

Plan: We are given K_{sp} values for the two possible precipitates. Using these and the metal ion concentrations, we can calculate what concentration of Cl^- ion would be necessary to begin precipitation of each. The salt requiring the lower Cl^- ion concentration will precipitate first.

Solve: For AgCl we have

$$K_{sp} = [Ag^+][Cl^-] = 1.8 \times 10^{-10}$$

Because $[Ag^+] = 1.0 \times 10^{-2}\,M$, the greatest concentration of Cl^- that can be present without causing precipitation of AgCl can be calculated from the K_{sp} expression:

$$K_{sp} = [1.0 \times 10^{-2}][Cl^-] = 1.8 \times 10^{-10}$$

$$[Cl^-] = \frac{1.8 \times 10^{-10}}{1.0 \times 10^{-2}} = 1.8 \times 10^{-8}\,M$$

Any Cl^- in excess of this very small concentration will cause AgCl to precipitate from solution. Proceeding similarly for $PbCl_2$, we have

$$K_{sp} = [Pb^{2+}][Cl^-]^2 = 1.7 \times 10^{-5}$$
$$[2.0 \times 10^{-2}][Cl^-]^2 = 1.7 \times 10^{-5}$$
$$[Cl^-]^2 = \frac{1.7 \times 10^{-5}}{2.0 \times 10^{-2}} = 8.5 \times 10^{-4}$$
$$[Cl^-] = \sqrt{8.5 \times 10^{-4}} = 2.9 \times 10^{-2} \, M$$

Thus, a concentration of Cl^- in excess of $2.9 \times 10^{-2} \, M$ will cause $PbCl_2$ to precipitate.

Comparing the concentrations of Cl^- required to precipitate each salt, we see that as Cl^- is added to the solution, AgCl will precipitate first because it requires a much smaller concentration of Cl^-. Thus, Ag^+ can be separated from Pb^{2+} by slowly adding Cl^- so $[Cl^-]$ is between $1.8 \times 10^{-8} \, M$ and $2.9 \times 10^{-2} \, M$.

■ PRACTICE EXERCISE

A solution consists of 0.050 M Mg^{2+} and 0.020 M Cu^{2+}. Which ion will precipitate first as OH^- is added to the solution? What concentration of OH^- is necessary to begin the precipitation of each cation? [$K_{sp} = 1.8 \times 10^{-11}$ for $Mg(OH)_2$, and $K_{sp} = 4.8 \times 10^{-20}$ for $Cu(OH)_2$.]
Answer: $Cu(OH)_2$ precipitates first. $Cu(OH)_2$ begins to precipitate when $[OH^-]$ exceeds $1.5 \times 10^{-9} \, M$; $Mg(OH)_2$ begins to precipitate when $[OH^-]$ exceeds $1.9 \times 10^{-5} \, M$.

Sulfide ion is often used to separate metal ions because the solubilities of sulfide salts span a wide range and depend greatly on the pH of the solution. Cu^{2+} and Zn^{2+}, for example, can be separated by bubbling H_2S gas through an acidified solution. Because CuS ($K_{sp} = 6 \times 10^{-37}$) is less soluble than ZnS ($K_{sp} = 2 \times 10^{-25}$), CuS precipitates from an acidified solution (pH = 1) while ZnS does not (Figure 17.21 ▼):

$$Cu^{2+}(aq) + H_2S(aq) \rightleftharpoons CuS(s) + 2\,H^+(aq) \qquad [17.27]$$

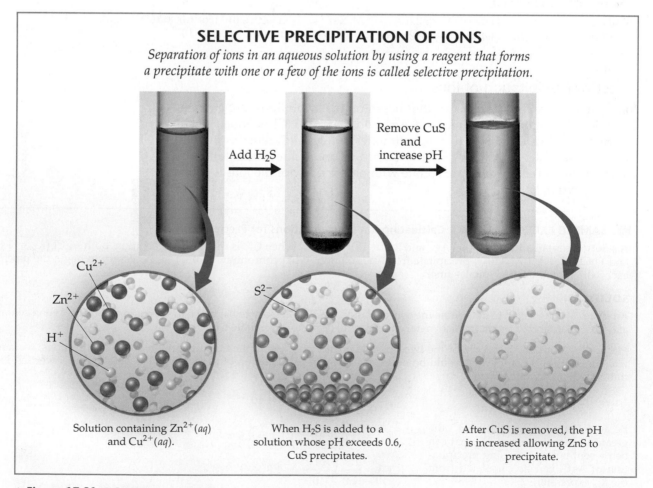

SELECTIVE PRECIPITATION OF IONS

Separation of ions in an aqueous solution by using a reagent that forms a precipitate with one or a few of the ions is called selective precipitation.

Add H_2S → Remove CuS and increase pH →

Cu^{2+}
Zn^{2+}
H^+
S^{2-}

| Solution containing $Zn^{2+}(aq)$ and $Cu^{2+}(aq)$. | When H_2S is added to a solution whose pH exceeds 0.6, CuS precipitates. | After CuS is removed, the pH is increased allowing ZnS to precipitate. |

▲ Figure 17.21 **Selective precipitation.**

The CuS can be separated from the Zn^{2+} solution by filtration. The CuS can then be dissolved by using a high concentration of H^+, shifting the equilibrium shown in Equation 17.27 to the left.

GIVE IT SOME THOUGHT

What experimental conditions will leave the smallest concentration of Cu^{2+} ions in solution according to Equation 17.27?

17.7 QUALITATIVE ANALYSIS FOR METALLIC ELEMENTS

In this chapter we have seen several examples of equilibria involving metal ions in aqueous solution. In this final section we look briefly at how solubility equilibria and complex-ion formation can be used to detect the presence of particular metal ions in solution. Before the development of modern analytical instrumentation, it was necessary to analyze mixtures of metals in a sample by so-called wet chemical methods. For example, a metallic sample that might contain several metallic elements was dissolved in a concentrated acid solution. This solution was then tested in a systematic way for the presence of various metal ions.

Qualitative analysis determines only the presence or absence of a particular metal ion, whereas **quantitative analysis** determines how much of a given substance is present. Wet methods of qualitative analysis have become less important as a means of analysis. They are frequently used in general chemistry laboratory programs, however, to illustrate equilibria, to teach the properties of common metal ions in solution, and to develop laboratory skills. Typically, such analyses proceed in three stages: (1) The ions are separated into broad groups on the basis of solubility properties. (2) The individual ions within each group are then separated by selectively dissolving members in the group. (3) The ions are then identified by means of specific tests.

A scheme in general use divides the common cations into five groups, as shown in Figure 17.22▶. The order of addition of reagents is important. The most selective separations—those that involve the smallest number of ions—are carried out first. The reactions that are used must proceed so far toward completion that any concentration of cations remaining in the solution is too small to interfere with subsequent tests. Let's take a closer look at each of these five groups of cations, briefly examining the logic used in this qualitative analysis scheme.

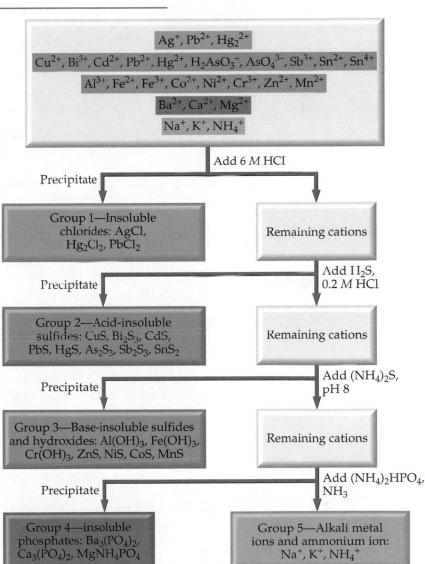

▲ **Figure 17.22 Qualitative analysis.** A flowchart showing the separation of cations into groups as a part of a common scheme for identifying cations.

1. *Insoluble chlorides:* Of the common metal ions, only Ag^+, Hg_2^{2+}, and Pb^{2+} form insoluble chlorides. When dilute HCl is added to a mixture of cations, therefore, only AgCl, Hg_2Cl_2, and $PbCl_2$ will precipitate, leaving the other cations in solution. The absence of a precipitate indicates that the starting solution contains no Ag^+, Hg_2^{2+}, or Pb^{2+}.

2. *Acid-insoluble sulfides:* After any insoluble chlorides have been removed, the remaining solution, now acidic, is treated with H_2S. Only the most insoluble metal sulfides—CuS, Bi_2S_3, CdS, PbS, HgS, As_2S_3, Sb_2S_3, and SnS_2—can precipitate. (Note the very small values of K_{sp} for some of these sulfides in Appendix D.) Those metal ions whose sulfides are somewhat more soluble—for example, ZnS or NiS—remain in solution.

3. *Base-insoluble sulfides and hydroxides:* After the solution is filtered to remove any acid-insoluble sulfides, the remaining solution is made slightly basic, and $(NH_4)_2S$ is added. In basic solutions the concentration of S^{2-} is higher than in acidic solutions. Thus, the ion products for many of the more soluble sulfides are made to exceed their K_{sp} values and precipitation occurs. The metal ions precipitated at this stage are Al^{3+}, Cr^{3+}, Fe^{3+}, Zn^{2+}, Ni^{2+}, Co^{2+}, and Mn^{2+}. (Actually, the Al^{3+}, Fe^{3+}, and Cr^{3+} ions do not form insoluble sulfides; instead they are precipitated as insoluble hydroxides at the same time.)

4. *Insoluble phosphates:* At this point the solution contains only metal ions from periodic table groups 1A and 2A. Adding $(NH_4)_2HPO_4$ to a basic solution precipitates the group 2A elements Mg^{2+}, Ca^{2+}, Sr^{2+}, and Ba^{2+} because these metals form insoluble phosphates.

5. *The alkali metal ions and NH_4^+:* The ions that remain after removing the insoluble phosphates form a small group. We can test the original solution for each ion individually. A flame test can be used to determine the presence of K^+, for example, because the flame turns a characteristic violet color if K^+ is present.

GIVE IT SOME THOUGHT

If a precipitate forms upon addition of HCl to an aqueous solution, what conclusions can you draw about the contents of the solution?

Additional separation and testing is necessary to determine which ions are present within each of the groups. Consider, for example, the ions of the insoluble chloride group. The precipitate containing the metal chlorides is boiled in water. The $PbCl_2$ is relatively soluble in hot water, whereas $AgCl$ and Hg_2Cl_2 are not. The hot solution is filtered, and a solution of Na_2CrO_4 is added to the filtrate. If Pb^{2+} is present, a yellow precipitate of $PbCrO_4$ forms. The test for Ag^+ consists of treating the metal chloride precipitate with dilute ammonia. Only Ag^+ forms an ammonia complex. If $AgCl$ is present in the precipitate, it will dissolve in the ammonia solution.

$$AgCl(s) + 2\,NH_3(aq) \rightleftharpoons Ag(NH_3)_2^+(aq) + Cl^-(aq) \qquad [17.28]$$

After treatment with ammonia, the solution is filtered and the filtrate made acidic by adding nitric acid. The nitric acid removes ammonia from solution by forming NH_4^+, thus releasing Ag^+, which re-forms the $AgCl$ precipitate.

$$Ag(NH_3)_2^+(aq) + Cl^-(aq) + 2\,H^+(aq) \rightleftharpoons AgCl(s) + 2\,NH_4^+(aq) \qquad [17.29]$$

The analyses for individual ions in the acid-insoluble and base-insoluble sulfides are a bit more complex, but the same general principles are involved. The detailed procedures for carrying out such analyses are given in many laboratory manuals.

■■■ **SAMPLE INTEGRATIVE EXERCISE** | **Putting Concepts Together**

A sample of 1.25 L of HCl gas at 21 °C and 0.950 atm is bubbled through 0.500 L of 0.150 M NH$_3$ solution. Calculate the pH of the resulting solution assuming that all the HCl dissolves and that the volume of the solution remains 0.500 L.

SOLUTION

The number of moles of HCl gas is calculated from the ideal-gas law.

$$n = \frac{PV}{RT} = \frac{(0.950 \text{ atm})(1.25 \text{ L})}{(0.0821 \text{ L-atm/mol-K})(294 \text{ K})} = 0.0492 \text{ mol HCl}$$

The number of moles of NH$_3$ in the solution is given by the product of the volume of the solution and its concentration.

$$\text{Moles NH}_3 = (0.500 \text{ L})(0.150 \text{ mol NH}_3/\text{L}) = 0.0750 \text{ mol NH}_3$$

The acid HCl and base NH$_3$ react, transferring a proton from HCl to NH$_3$, producing NH$_4^+$ and Cl$^-$ ions.

$$HCl(g) + NH_3(aq) \longrightarrow NH_4^+(aq) + Cl^-(aq)$$

To determine the pH of the solution, we first calculate the amount of each reactant and each product present at the completion of the reaction.

	HCl(g) +	NH$_3$(aq) $\longrightarrow$	NH$_4^+$(aq) +	Cl$^-$(aq)
Before addition	0.0492 mol	0 mol	0 mol	0 mol
Addition		0.0750 mol		
After addition	0 mol	0.0258 mol	0.0492 mol	0.0492 mol

Thus, the reaction produces a solution containing a mixture of NH$_3$, NH$_4^+$, and Cl$^-$. The NH$_3$ is a weak base ($K_b = 1.8 \times 10^{-5}$), NH$_4^+$ is its conjugate acid, and Cl$^-$ is neither acidic nor basic. Consequently, the pH depends on [NH$_3$] and [NH$_4^+$].

$$[NH_3] = \frac{0.0258 \text{ mol NH}_3}{0.500 \text{ L soln}} = 0.0516 \ M$$

$$[NH_4^+] = \frac{0.0492 \text{ mol NH}_4^+}{0.500 \text{ L soln}} = 0.0984 \ M$$

We can calculate the pH using either K_b for NH$_3$ or K_a for NH$_4^+$. Using the K_b expression, we have

	NH$_3$(aq) +	H$_2$O(l) $\rightleftharpoons$	NH$_4^+$(aq) +	OH$^-$(aq)
Initial	0.0516 M	—	0.0984 M	0
Change	$-x$ M	—	$+x$ M	$+x$ M
Equilibrium	$(0.0516 - x)$ M	—	$(0.0984 + x)$ M	x M

$$K_b = \frac{[NH_4^+][OH^-]}{[NH_3]} = \frac{(0.0984 + x)(x)}{(0.0516 - x)} \simeq \frac{(0.0984)x}{0.0516} = 1.8 \times 10^{-5}$$

$$x = [OH^-] = \frac{(0.0516)(1.8 \times 10^{-5})}{0.0984} = 9.4 \times 10^{-6} \ M$$

Hence, pOH $= -\log(9.4 \times 10^{-6}) = 5.03$ and pH $= 14.00 - \text{pOH} = 14.00 - 5.03 = 8.97$.

CHAPTER REVIEW

SUMMARY AND KEY TERMS

Section 17.1 In this chapter we have considered several types of important equilibria that occur in aqueous solution. Our primary emphasis has been on acid–base equilibria in solutions containing two or more solutes and on solubility equilibria. The dissociation of a weak acid or weak base is repressed by the presence of a strong electrolyte that provides an ion common to the equilibrium. This phenomenon is called the **common-ion effect**.

Section 17.2 A particularly important type of acid–base mixture is that of a weak conjugate acid–base pair. Such mixtures function as **buffered solutions (buffers)**. Addition of small amounts of a strong acid or a strong base to a buffered solution causes only small changes in pH because the buffer reacts with the added acid or base. (Strong acid–strong base, strong acid–weak base, and weak acid–strong base reactions proceed essentially to completion.) Buffered solutions are usually prepared from a weak acid and a salt of that acid or from a weak base and a salt of that base. Two important characteristics of a buffered solution are its **buffer capacity** and its pH. The pH can be calculated using K_a or K_b. The relationship between pH, pK_a, and the concentrations of an acid and its conjugate base can be expressed by the **Henderson–Hasselbalch equation**: $pH = pK_a + \log \dfrac{[\text{base}]}{[\text{acid}]}$.

Section 17.3 The plot of the pH of an acid (or base) as a function of the volume of added base (or acid) is called a **pH titration curve**. Titration curves aid in selecting a proper pH indicator for an acid–base titration. The titration curve of a strong acid–strong base titration exhibits a large change in pH in the immediate vicinity of the equivalence point; at the equivalence point for this titration, pH = 7. For strong acid–weak base or weak acid–strong base titrations, the pH change in the vicinity of the equivalence point is not as large. Furthermore, the pH at the equivalence point is not 7 in either of these cases. Rather, it is the pH of the salt solution that results from the neutralization reaction. It is possible to calculate the pH at any point of the titration curve by first considering the effects of the reaction between the acid and base on solution concentrations and then examining equilibria involving remaining solute species.

Section 17.4 The equilibrium between a solid compound and its ions in solution provides an example of heterogeneous equilibrium. The **solubility-product constant** (or simply the **solubility product**), K_{sp}, is an equilibrium constant that expresses quantitatively the extent to which the compound dissolves. The K_{sp} can be used to calculate the solubility of an ionic compound, and the solubility can be used to calculate K_{sp}.

Section 17.5 Several experimental factors, including temperature, affect the solubilities of ionic compounds in water. The solubility of a slightly soluble ionic compound is decreased by the presence of a second solute that furnishes a common ion (the common-ion effect). The solubility of compounds containing basic anions increases as the solution is made more acidic (as pH decreases). Salts with anions of negligible basicity (the anions of strong acids) are unaffected by pH changes.

The solubility of metal salts is also affected by the presence of certain Lewis bases that react with metal ions to form stable **complex ions**. Complex-ion formation in aqueous solution involves the displacement by Lewis bases (such as NH_3 and CN^-) of water molecules attached to the metal ion. The extent to which such complex formation occurs is expressed quantitatively by the **formation constant** for the complex ion. **Amphoteric oxides and hydroxides** are those that are only slightly soluble in water but dissolve on addition of either acid or base. Acid–base reactions involving the OH^- or H_2O groups bound to the metal ions give rise to the amphoterism.

Section 17.6 Comparison of the ion product, Q, with the value of K_{sp} can be used to judge whether a precipitate will form when solutions are mixed or whether a slightly soluble salt will dissolve under various conditions. Precipitates form when $Q > K_{sp}$. Ions can be separated from each other based on the solubilities of their salts.

Section 17.7 Metallic elements vary a great deal in the solubilities of their salts, in their acid–base behavior, and in their tendencies to form complex ions. These differences can be used to separate and detect the presence of metal ions in mixtures. **Qualitative analysis** determines the presence or absence of species in a sample, whereas **quantitative analysis** determines how much of each species is present. The qualitative analysis of metal ions in solution can be carried out by separating the ions into groups on the basis of precipitation reactions and then analyzing each group for individual metal ions.

KEY SKILLS

- Describe the common-ion effect.
- Explain how a buffer functions.
- Calculate the pH of a buffer solution.
- Calculate the pH of a buffer after the addition of small amounts of a strong acid or a strong base.
- Calculate the pH at any point in an acid–base titration of a strong acid and strong base.
- Calculate the pH at any point in a titration of a weak acid with a strong base or a weak base with a strong acid.
- Understand the differences between the titration curves for a strong acid–strong base titration and those when either the acid or base is weak.
- Calculate K_{sp} from molar solubility and molar solubility from K_{sp}.
- Calculate molar solubility in the presence of a common ion.
- Predict the effect of pH on solubility.
- Predict whether a precipitate will form when solutions are mixed by comparing Q and K_{sp}.
- Calculate the ion concentrations required to begin precipitation.
- Explain the effect of complex-ion formation on solubility.

KEY EQUATIONS

- $$pH = pK_a + \log \frac{[\text{base}]}{[\text{acid}]} \qquad [17.9]$$

  The Henderson–Hasselbalch equation, used to calculate the pH of a buffer from the concentrations of a conjugate acid–base pair

VISUALIZING CONCEPTS

17.1 The following boxes represent aqueous solutions containing a weak acid, HX, and its conjugate base, X⁻. Water molecules and cations are not shown. Which solution has the highest pH? Explain. [Section 17.1]

= HX   = X⁻

17.2 The beaker on the right contains 0.1 M acetic acid solution with methyl orange as an indicator. The beaker on the left contains a mixture of 0.1 M acetic acid and 0.1 M sodium acetate with methyl orange. **(a)** Using Figure 16.7, estimate the pH of each solution, and explain the difference. **(b)** Which solution is better able to maintain its pH when small amounts of NaOH are added? Explain. [Sections 17.1 and 17.2]

17.3 A buffer contains a weak acid, HX, and its conjugate base. The weak acid has a pK_a of 4.5, and the buffer solution has a pH of 4.3. Without doing a calculation, predict whether [HX] = [X⁻], [HX] > [X⁻], or [HX] < [X⁻]. Explain. [Section 17.2]

17.4 The drawing on the left represents a buffer composed of equal concentrations of a weak acid, HX, and its conjugate base, X⁻. The heights of the columns are proportional to the concentrations of the components of the buffer. **(a)** Which of the three drawings, (1), (2), or (3), represents the buffer after the addition of a strong acid? **(b)** Which of the three represents the buffer after the addition of a strong base? **(c)** Which of the three represents a situation that cannot arise from the addition of either an acid or a base? [Section 17.2]

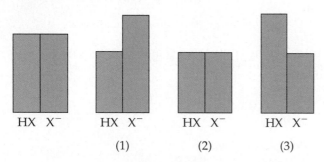

17.5 The following drawings represent solutions at various stages of the titration of a weak acid, HA, with NaOH. (The Na⁺ ions and water molecules have been omitted for clarity.) To which of the following regions of the titration curve does each drawing correspond: **(a)** before addition of NaOH, **(b)** after addition of NaOH but before equivalence point, **(c)** at equivalence point, **(d)** after equivalence point? [Section 17.3]

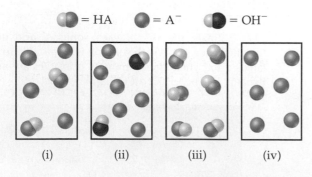

17.6 Match the following descriptions of titration curves with the diagrams: **(a)** strong acid added to strong base, **(b)** strong base added to weak acid, **(c)** strong base added to strong acid, **(d)** strong base added to polyprotic acid. [Section 17.3]

17.7 Equal volumes of two acids are titrated with 0.10 M NaOH resulting in the two titration curves shown in the following figure. **(a)** Which curve corresponds to the more concentrated acid solution? **(b)** Which corresponds to the acid with the largest K_a? Explain. [Section 17.3]

17.8 The following drawings represent saturated solutions of three ionic compounds of silver—AgX, AgY, and AgZ. (Na⁺ cations, which might also be present for charge balance, are not shown.) Which compound has the smallest K_{sp}? [Section 17.4]

17.9 The figures below represent the ions in a saturated aqueous solution of the slightly soluble ionic compound MX: MX(s) $\rightleftharpoons$ M²⁺(aq) + X²⁻(aq). (Only the M²⁺ and X²⁻ ions are shown.) **(a)** Which figure represents a solution prepared by dissolving MX in water? **(b)** Which figure represents a solution prepared by dissolving MX in a solution containing Na₂X? **(c)** If X²⁻ is a basic anion, which figure represents a saturated solution with the lowest pH? **(d)** If you were to calculate the K_{sp} for MX, would you get the same value in each of the three scenarios? Why or why not? [Sections 17.4 and 17.5]

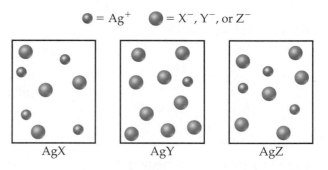

17.10 The following graphs represent the behavior of BaCO₃ under different circumstances. In each case the vertical axis indicates the solubility of the BaCO₃ and the horizontal axis represents the concentration of some other reagent. **(a)** Which graph represents what happens to the solubility of $BaCO_3$ as HNO_3 is added? **(b)** Which graph represents what happens to the $BaCO_3$ solubility as Na_2CO_3 is added? **(c)** Which represents what happens to the $BaCO_3$ solubility as $NaNO_3$ is added? [Section 17.5]

17.11 What is the name given to the kind of behavior demonstrated by a metal hydroxide in this graph? [Section 17.5]

17.12 Three cations, Ni^{2+}, Cu^{2+}, and Ag^+, are separated using two different precipitating agents. Based on Figure 17.22, what two precipitating agents could be used? Using these agents, indicate which of the cations is A, which is B, and which is C. [Section 17.7]

EXERCISES

Common-Ion Effect

17.13 **(a)** What is the common-ion effect? **(b)** Give an example of a salt that can decrease the ionization of HNO_2 in solution.

17.14 **(a)** Consider the equilibrium $B(aq) + H_2O(l) \rightleftharpoons HB^+(aq) + OH^-(aq)$. Using Le Châtelier's principle, explain the effect of the presence of a salt of HB^+ on the ionization of B. **(b)** Give an example of a salt that can decrease the ionization of NH_3 in solution.

17.15 Use information from Appendix D to calculate the pH of **(a)** a solution that is 0.060 M in potassium propionate (C_2H_5COOK or $KC_3H_5O_2$) and 0.085 M in propionic acid (C_2H_5COOH or $HC_3H_5O_2$); **(b)** a solution that is 0.075 M in trimethylamine, $(CH_3)_3N$, and 0.10 M in trimethylammonium chloride, $(CH_3)_3NHCl$; **(c)** a solution that is made by mixing 50.0 mL of 0.15 M acetic acid and 50.0 mL of 0.20 M sodium acetate.

17.16 Use information from Appendix D to calculate the pH of **(a)** a solution that is 0.150 M in sodium formate (HCOONa) and 0.200 M in formic acid (HCOOH); **(b)** a solution that is 0.210 M in pyridine (C_5H_5N) and 0.350 M in pyridinium chloride (C_5H_5NHCl); **(c)** a solution that is made by combining 125 mL of 0.050 M hydrofluoric acid with 50.0 mL of 0.10 M sodium fluoride.

17.17 **(a)** Calculate the percent ionization of 0.0075 M butanoic acid ($K_a = 1.5 \times 10^{-5}$). **(b)** Calculate the percent ionization of 0.0075 M butanoic acid in a solution containing 0.085 M sodium butanoate.

17.18 **(a)** Calculate the percent ionization of 0.085 M lactic acid ($K_a = 1.4 \times 10^{-4}$). **(b)** Calculate the percent ionization of 0.095 M lactic acid in a solution containing 0.0075 M sodium lactate.

Buffers

17.19 Explain why a mixture of CH_3COOH and CH_3COONa can act as a buffer while a mixture of HCl and NaCl cannot.

17.20 Explain why a mixture formed by mixing 100 mL of 0.100 M CH_3COOH and 50 mL of 0.100 M NaOH will act as a buffer.

17.21 **(a)** Calculate the pH of a buffer that is 0.12 M in lactic acid and 0.11 M in sodium lactate. **(b)** Calculate the pH of a buffer formed by mixing 85 mL of 0.13 M lactic acid with 95 mL of 0.15 M sodium lactate.

17.22 **(a)** Calculate the pH of a buffer that is 0.105 M in $NaHCO_3$ and 0.125 M in Na_2CO_3. **(b)** Calculate the pH of a solution formed by mixing 65 mL of 0.20 M $NaHCO_3$ with 75 mL of 0.15 M Na_2CO_3.

17.23 A buffer is prepared by adding 20.0 g of acetic acid (CH_3COOH) and 20.0 g of sodium acetate (CH_3COONa) to enough water to form 2.00 L of solution. **(a)** Determine the pH of the buffer. **(b)** Write the complete ionic equation for the reaction that occurs when a few drops of hydrochloric acid are added to the buffer. **(c)** Write the complete ionic equation for the reaction that occurs when a few drops of sodium hydroxide solution are added to the buffer.

17.24 A buffer is prepared by adding 7.00 g of ammonia (NH_3) and 20.0 g of ammonium chloride (NH_4Cl) to enough water to form 2.50 L of solution. **(a)** What is the pH of this buffer? **(b)** Write the complete ionic equation for the reaction that occurs when a few drops of nitric acid are added to the buffer. **(c)** Write the complete ionic equation for the reaction that occurs when a few drops of potassium hydroxide solution are added to the buffer.

17.25 How many moles of sodium hypobromite (NaBrO) should be added to 1.00 L of 0.050 M hypobromous acid (HBrO) to form a buffer solution of pH 9.15? Assume that no volume change occurs when the NaBrO is added.

17.26 How many grams of sodium lactate [$CH_3CH(OH)COONa$ or $NaC_3H_5O_3$] should be added to 1.00 L of 0.150 M lactic acid [$CH_3CH(OH)COOH$ or $HC_3H_5O_3$] to form a buffer solution with pH 4.00? Assume that no volume change occurs when the sodium lactate is added.

17.27 A buffer solution contains 0.10 mol of acetic acid and 0.13 mol of sodium acetate in 1.00 L. (a) What is the pH of this buffer? (b) What is the pH of the buffer after the

addition of 0.02 mol of KOH? **(c)** What is the pH of the buffer after the addition of 0.02 mol of HNO_3?

17.28 A buffer solution contains 0.10 mol of propionic acid (C_2H_5COOH) and 0.13 mol of sodium propionate (C_2H_5COONa) in 1.50 L. **(a)** What is the pH of this buffer? **(b)** What is the pH of the buffer after the addition of 0.01 mol of NaOH? **(c)** What is the pH of the buffer after the addition of 0.01 mol of HI?

17.29 **(a)** What is the ratio of HCO_3^- to H_2CO_3 in blood of pH 7.4? **(b)** What is the ratio of HCO_3^- to H_2CO_3 in an exhausted marathon runner whose blood pH is 7.1?

17.30 A buffer, consisting of $H_2PO_4^-$ and HPO_4^{2-}, helps control the pH of physiological fluids. Many carbonated soft drinks also use this buffer system. What is the pH of a soft drink in which the major buffer ingredients are 6.5 g of NaH_2PO_4 and 8.0 g of Na_2HPO_4 per 355 mL of solution?

17.31 You have to prepare a pH 3.50 buffer, and you have the following 0.10 M solutions available: HCOOH, CH_3COOH, H_3PO_4, HCOONa, CH_3COONa, and NaH_2PO_4. Which solutions would you use? How many milliliters of each solution would you use to make approximately a liter of the buffer?

17.32 You have to prepare a pH 4.80 buffer, and you have the following 0.10 M solutions available: formic acid, sodium formate, propionic acid, sodium propionate, phosphoric acid, and sodium dihydrogen phosphate. Which solutions would you use? How many milliliters of each solution would you use to make approximately a liter of the buffer?

Acid–Base Titrations

17.33 The accompanying graph shows the titration curves for two monoprotic acids. **(a)** Which curve is that of a strong acid? **(b)** What is the approximate pH at the equivalence point of each titration? **(c)** How do the original concentrations of the two acids compare if 40.0 mL of each is titrated to the equivalence point with 0.100 M base?

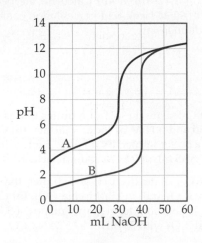

17.34 How does titration of a strong, monoprotic acid with a strong base differ from titration of a weak, monoprotic acid with a strong base with respect to the following: **(a)** quantity of base required to reach the equivalence point, **(b)** pH at the beginning of the titration, **(c)** pH at the equivalence point, **(d)** pH after addition of a slight excess of base, **(e)** choice of indicator for determining the equivalence point?

17.35 Predict whether the equivalence point of each of the following titrations is below, above, or at pH 7: **(a)** $NaHCO_3$ titrated with NaOH, **(b)** NH_3 titrated with HCl, **(c)** KOH titrated with HBr.

17.36 Predict whether the equivalence point of each of the following titrations is below, above, or at pH 7: **(a)** formic acid titrated with NaOH, **(b)** calcium hydroxide titrated with perchloric acid, **(c)** pyridine titrated with nitric acid.

17.37 Two monoprotic acids, both 0.100 M in concentration, are titrated with 0.100 M NaOH. The pH at the equivalence point for HX is 8.8, and that for HY is 7.9. **(a)** Which is the weaker acid? **(b)** Which indicators in Figure 16.7 could be used to titrate each of these acids?

17.38 Assume that 30.0 mL of a 0.10 M solution of a weak base B that accepts one proton is titrated with a 0.10 M solution of the monoprotic strong acid HX. **(a)** How many moles of HX have been added at the equivalence point? **(b)** What is the predominant form of B at the equivalence point? **(c)** What factor determines the pH at the equivalence point? **(d)** Which indicator, phenolphthalein or methyl red, is likely to be the better choice for this titration?

17.39 How many milliliters of 0.0850 M NaOH are required to titrate each of the following solutions to the equivalence point: **(a)** 40.0 mL of 0.0900 M HNO$_3$, **(b)** 35.0 mL of 0.0850 M CH$_3$COOH, **(c)** 50.0 mL of a solution that contains 1.85 g of HCl per liter?

17.40 How many milliliters of 0.105 M HCl are needed to titrate each of the following solutions to the equivalence point: **(a)** 45.0 mL of 0.0950 M NaOH, **(b)** 22.5 mL of 0.118 M NH$_3$, **(c)** 125.0 mL of a solution that contains 1.35 g of NaOH per liter?

17.41 A 20.0-mL sample of 0.200 M HBr solution is titrated with 0.200 M NaOH solution. Calculate the pH of the solution after the following volumes of base have been added: **(a)** 15.0 mL, **(b)** 19.9 mL, **(c)** 20.0 mL, **(d)** 20.1 mL, **(e)** 35.0 mL.

17.42 A 30.0-mL sample of 0.150 M KOH is titrated with 0.125 M HClO$_4$ solution. Calculate the pH after the following volumes of acid have been added: **(a)** 30.0 mL, **(b)** 35.0 mL, **(c)** 36.0 mL, **(d)** 37.0 mL, **(e)** 40.0 mL.

17.43 A 35.0-mL sample of 0.150 M acetic acid (CH$_3$COOH) is titrated with 0.150 M NaOH solution. Calculate the pH after the following volumes of base have been added: **(a)** 0 mL, **(b)** 17.5 mL, **(c)** 34.5 mL, **(d)** 35.0 mL, **(e)** 35.5 mL, **(f)** 50.0 mL.

17.44 Consider the titration of 30.0 mL of 0.030 M NH$_3$ with 0.025 M HCl. Calculate the pH after the following volumes of titrant have been added: **(a)** 0 mL, **(b)** 10.0 mL, **(c)** 20.0 mL, **(d)** 35.0 mL, **(e)** 36.0 mL, **(f)** 37.0 mL.

17.45 Calculate the pH at the equivalence point for titrating 0.200 M solutions of each of the following bases with 0.200 M HBr: **(a)** sodium hydroxide (NaOH), **(b)** hydroxylamine (NH$_2$OH), **(c)** aniline (C$_6$H$_5$NH$_2$).

17.46 Calculate the pH at the equivalence point in titrating 0.100 M solutions of each of the following with 0.080 M NaOH: **(a)** hydrobromic acid (HBr), **(b)** lactic acid [CH$_3$CH(OH)COOH], **(c)** sodium hydrogen chromate (NaHCrO$_4$).

Solubility Equilibria and Factors Affecting Solubility

17.47 **(a)** Why is the concentration of undissolved solid not explicitly included in the expression for the solubility-product constant? **(b)** Write the expression for the solubility-product constant for each of the following strong electrolytes: AgI, SrSO$_4$, Fe(OH)$_2$, and Hg$_2$Br$_2$.

17.48 **(a)** Explain the difference between solubility and solubility-product constant. **(b)** Write the expression for the solubility-product constant for each of the following ionic compounds: MnCO$_3$, Hg(OH)$_2$, and Cu$_3$(PO$_4$)$_2$.

17.49 **(a)** If the molar solubility of CaF$_2$ at 35 °C is 1.24×10^{-3} mol/L, what is K_{sp} at this temperature? **(b)** It is found that 1.1×10^{-2} g of SrF$_2$ dissolves per 100 mL of aqueous solution at 25 °C. Calculate the solubility product for SrF$_2$. **(c)** The K_{sp} of Ba(IO$_3$)$_2$ at 25 °C is 6.0×10^{-10}. What is the molar solubility of Ba(IO$_3$)$_2$?

17.50 **(a)** The molar solubility of PbBr$_2$ at 25 °C is 1.0×10^{-2} mol/L. Calculate K_{sp}. **(b)** If 0.0490 g of AgIO$_3$ dissolves per liter of solution, calculate the solubility-product constant. **(c)** Using the appropriate K_{sp} value from Appendix D, calculate the solubility of Cu(OH)$_2$ in grams per liter of solution.

17.51 A 1.00-L solution saturated at 25 °C with calcium oxalate (CaC$_2$O$_4$) contains 0.0061 g of CaC$_2$O$_4$. Calculate the solubility-product constant for this salt at 25 °C.

17.52 A 1.00-L solution saturated at 25 °C with lead(II) iodide contains 0.54 g of PbI$_2$. Calculate the solubility-product constant for this salt at 25 °C.

17.53 Using Appendix D, calculate the molar solubility of AgBr in **(a)** pure water, **(b)** 3.0×10^{-2} M AgNO$_3$ solution, **(c)** 0.10 M NaBr solution.

17.54 Calculate the solubility of LaF$_3$ in grams per liter in **(a)** pure water, **(b)** 0.010 M KF solution, **(c)** 0.050 M LaCl$_3$ solution.

17.55 Calculate the solubility of Mn(OH)$_2$ in grams per liter when buffered at pH **(a)** 7.0, **(b)** 9.5, **(c)** 11.8.

17.56 Calculate the molar solubility of Fe(OH)$_2$ when buffered at pH **(a)** 8.0, **(b)** 10.0, **(c)** 12.0.

17.57 Which of the following salts will be substantially more soluble in acidic solution than in pure water: **(a)** ZnCO$_3$, **(b)** ZnS, **(c)** BiI$_3$, **(d)** AgCN, **(e)** Ba$_3$(PO$_4$)$_2$?

17.58 For each of the following slightly soluble salts, write the net ionic equation, if any, for reaction with acid: **(a)** MnS, **(b)** PbF$_2$, **(c)** AuCl$_3$, **(d)** Hg$_2$C$_2$O$_4$, **(e)** CuBr.

17.59 From the value of K_f listed in Table 17.1, calculate the concentration of Cu^{2+} in 1.0 L of a solution that contains a total of 1×10^{-3} mol of copper(II) ion and that is 0.10 M in NH$_3$.

17.60 To what final concentration of NH_3 must a solution be adjusted to just dissolve 0.020 mol of NiC_2O_4 ($K_{sp} = 4 \times 10^{-10}$) in 1.0 L of solution? (*Hint:* You can neglect the hydrolysis of $C_2O_4^{2-}$ because the solution will be quite basic.)

17.61 By using the values of K_{sp} for AgI and K_f for $Ag(CN)_2^-$, calculate the equilibrium constant for the reaction

$$AgI(s) + 2\,CN^-(aq) \rightleftharpoons Ag(CN)_2^-(aq) + I^-(aq)$$

17.62 Using the value of K_{sp} for Ag_2S, K_{a1} and K_{a2} for H_2S, and $K_f = 1.1 \times 10^5$ for $AgCl_2^-$, calculate the equilibrium constant for the following reaction:

$$Ag_2S(s) + 4\,Cl^-(aq) + 2\,H^+(aq) \rightleftharpoons 2\,AgCl_2^-(aq) + H_2S(aq)$$

Precipitation; Qualitative Analysis

17.63 (a) Will $Ca(OH)_2$ precipitate from solution if the pH of a 0.050 M solution of $CaCl_2$ is adjusted to 8.0? (b) Will Ag_2SO_4 precipitate when 100 mL of 0.050 M $AgNO_3$ is mixed with 10 mL of 5.0×10^{-2} M Na_2SO_4 solution?

17.64 (a) Will $Co(OH)_2$ precipitate from solution if the pH of a 0.020 M solution of $Co(NO_3)_2$ is adjusted to 8.5? (b) Will $AgIO_3$ precipitate when 20 mL of 0.010 M $AgNO_3$ is mixed with 10 mL of 0.015 M $NaIO_3$? (K_{sp} of $AgIO_3$ is 3.1×10^{-8}.)

17.65 Calculate the minimum pH needed to precipitate $Mn(OH)_2$ so completely that the concentration of Mn^{2+} is less than 1 μg per liter [1 part per billion (ppb)].

17.66 Suppose that a 10-mL sample of a solution is to be tested for Cl^- ion by addition of 1 drop (0.2 mL) of 0.10 M $AgNO_3$. What is the minimum number of grams of Cl^- that must be present for $AgCl(s)$ to form?

17.67 A solution contains 2.0×10^{-4} M Ag^+ and 1.5×10^{-3} M Pb^{2+}. If NaI is added, will AgI ($K_{sp} = 8.3 \times 10^{-17}$) or PbI_2 ($K_{sp} = 7.9 \times 10^{-9}$) precipitate first? Specify the concentration of I^- needed to begin precipitation.

17.68 A solution of Na_2SO_4 is added dropwise to a solution that is 0.010 M in Ba^{2+} and 0.010 M in Sr^{2+}. (a) What concentration of SO_4^{2-} is necessary to begin precipitation? (Neglect volume changes. $BaSO_4$: $K_{sp} = 1.1 \times 10^{-10}$; $SrSO_4$: $K_{sp} = 3.2 \times 10^{-7}$.) (b) Which cation precipitates first? (c) What is the concentration of SO_4^{2-} when the second cation begins to precipitate?

17.69 A solution containing an unknown number of metal ions is treated with dilute HCl; no precipitate forms. The pH is adjusted to about 1, and H_2S is bubbled through. Again, no precipitate forms. The pH of the solution is then adjusted to about 8. Again, H_2S is bubbled through. This time a precipitate forms. The filtrate from this solution is treated with $(NH_4)_2HPO_4$. No precipitate forms.

Which metal ions discussed in Section 17.7 are possibly present? Which are definitely absent within the limits of these tests?

17.70 An unknown solid is entirely soluble in water. On addition of dilute HCl, a precipitate forms. After the precipitate is filtered off, the pH is adjusted to about 1 and H_2S is bubbled in; a precipitate again forms. After filtering off this precipitate, the pH is adjusted to 8 and H_2S is again added; no precipitate forms. No precipitate forms upon addition of $(NH_4)_2HPO_4$. The remaining solution shows a yellow color in a flame test. Based on these observations, which of the following compounds might be present, which are definitely present, and which are definitely absent: CdS, $Pb(NO_3)_2$, HgO, $ZnSO_4$, $Cd(NO_3)_2$, and Na_2SO_4?

17.71 In the course of various qualitative analysis procedures, the following mixtures are encountered: (a) Zn^{2+} and Cd^{2+}, (b) $Cr(OH)_3$ and $Fe(OH)_3$, (c) Mg^{2+} and K^+, (d) Ag^+ and Mn^{2+}. Suggest how each mixture might be separated.

17.72 Suggest how the cations in each of the following solution mixtures can be separated: (a) Na^+ and Cd^{2+}, (b) Cu^{2+} and Mg^{2+}, (c) Pb^{2+} and Al^{3+}, (d) Ag^+ and Hg^{2+}.

17.73 (a) Precipitation of the group 4 cations (Figure 17.22) requires a basic medium. Why is this so? (b) What is the most significant difference between the sulfides precipitated in group 2 and those precipitated in group 3? (c) Suggest a procedure that would serve to redissolve the group 3 cations following their precipitation.

17.74 A student who is in a great hurry to finish his laboratory work decides that his qualitative analysis unknown contains a metal ion from the insoluble phosphate group, group 4 (Figure 17.22). He therefore tests his sample directly with $(NH_4)_2HPO_4$, skipping earlier tests for the metal ions in groups 1, 2, and 3. He observes a precipitate and concludes that a metal ion from group 4 is indeed present. Why is this possibly an erroneous conclusion?

ADDITIONAL EXERCISES

17.75 Derive an equation similar to the Henderson–Hasselbalch equation relating the pOH of a buffer to the pK_b of its base component.

17.76 Benzenesulfonic acid is a monoprotic acid with $pK_a = 2.25$. Calculate the pH of a buffer composed of 0.150 M benzenesulfonic acid and 0.125 M sodium benzensulfonate.

17.77 Furoic acid ($HC_5H_3O_3$) has a K_a value of 6.76×10^{-4} at 25 °C. Calculate the pH at 25 °C of (a) a solution formed by adding 25.0 g of furoic acid and 30.0 g of sodium furoate ($NaC_5H_3O_3$) to enough water to form 0.250 L of solution; (b) a solution formed by mixing 30.0 mL of 0.250 M $HC_5H_3O_3$ and 20.0 mL of 0.22 M $NaC_5H_3O_3$ and diluting the total volume to 125 mL; (c) a solution prepared by adding 50.0 mL of 1.65 M NaOH solution to 0.500 L of 0.0850 M $HC_5H_3O_3$.

17.78 The acid–base indicator bromcresol green is a weak acid. The yellow acid and blue base forms of the indicator are present in equal concentrations in a solution when the pH is 4.68. What is the pK_a for bromcresol green?

17.79 Equal quantities of 0.010 M solutions of an acid HA and a base B are mixed. The pH of the resulting solution is 9.2. (a) Write the equilibrium equation and equilibrium-constant expression for the reaction between HA and B. (b) If K_a for HA is 8.0×10^{-5}, what is the value of the equilibrium constant for the reaction between HA and B? (c) What is the value of K_b for B?

17.80 Two buffers are prepared by adding an equal number of moles of formic acid (HCOOH) and sodium formate (HCOONa) to enough water to make 1.00 L of solution. Buffer A is prepared using 1.00 mol each of formic acid and sodium formate. Buffer B is prepared by using 0.010 mol of each. (a) Calculate the pH of each buffer, and explain why they are equal. (b) Which buffer will have the greater buffer capacity? Explain. (c) Calculate the change in pH for each buffer upon the addition of 1.0 mL of 1.00 M HCl. (d) Calculate the change in pH for each buffer upon the addition of 10 mL of 1.00 M HCl. (e) Discuss your answers for parts (c) and (d) in light of your response to part (b).

17.81 A biochemist needs 750 mL of an acetic acid–sodium acetate buffer with pH 4.50. Solid sodium acetate (CH_3COONa) and glacial acetic acid (CH_3COOH) are available. Glacial acetic acid is 99% CH_3COOH by mass and has a density of 1.05 g/mL. If the buffer is to be 0.15 M in CH_3COOH, how many grams of CH_3COONa and how many milliliters of glacial acetic acid must be used?

17.82 A sample of 0.2140 g of an unknown monoprotic acid was dissolved in 25.0 mL of water and titrated with 0.0950 M NaOH. The acid required 27.4 mL of base to reach the equivalence point. (a) What is the molar mass of the acid? (b) After 15.0 mL of base had been added in the titration, the pH was found to be 6.50. What is the K_a for the unknown acid?

17.83 Show that the pH at the halfway point of a titration of a weak acid with a strong base (where the volume of added base is half of that needed to reach the equivalence point) is equal to pK_a for the acid.

17.84 Potassium hydrogen phthalate, often abbreviated KHP, can be obtained in high purity and is used to determine the concentrations of solutions of strong bases. Strong bases react with the hydrogen phthalate ion as follows:

$$HP^-(aq) + OH^-(aq) \longrightarrow H_2O(l) + P^{2-}(aq)$$

The molar mass of KHP is 204.2 g/mol and K_a for the HP^- ion is 3.1×10^{-6}. (a) If a titration experiment begins with 0.4885 g of KHP and has a final volume of about 100 mL, which indicator from Figure 16.7 would be most appropriate? (b) If the titration required 38.55 mL of NaOH solution to reach the end point, what is the concentration of the NaOH solution?

17.85 If 40.00 mL of 0.100 M Na_2CO_3 is titrated with 0.100 M HCl, calculate (a) the pH at the start of the titration; (b) the volume of HCl required to reach the first equivalence point and the predominant species present at this point; (c) the volume of HCl required to reach the second equivalence point and the predominant species present at this point; (d) the pH at the second equivalence point.

17.86 A hypothetical weak acid, HA, was combined with NaOH in the following proportions: 0.20 mol of HA, 0.080 mol of NaOH. The mixture was diluted to a total volume of 1.0 L, and the pH measured. (a) If pH = 4.80, what is the pK_a of the acid? (b) How many additional moles of NaOH should be added to the solution to increase the pH to 5.00?

[17.87] What is the pH of a solution made by mixing 0.30 mol NaOH, 0.25 mol Na_2HPO_4, and 0.20 mol H_3PO_4 with water and diluting to 1.00 L?

[17.88] Suppose you want to do a physiological experiment that calls for a pH 6.5 buffer. You find that the organism with which you are working is not sensitive to the weak acid H_2X ($K_{a1} = 2 \times 10^{-2}$; $K_{a2} = 5.0 \times 10^{-7}$) or its sodium salts. You have available a 1.0 M solution of this acid and a 1.0 M solution of NaOH. How much of the NaOH solution should be added to 1.0 L of the acid to give a buffer at pH 6.50? (Ignore any volume change.)

[17.89] How many microliters of 1.000 M NaOH solution must be added to 25.00 mL of a 0.1000 M solution of lactic acid [$CH_3CH(OH)COOH$ or $HC_3H_5O_3$] to produce a buffer with pH = 3.75?

17.90 A person suffering from anxiety begins breathing rapidly and as a result suffers alkalosis, an increase in blood pH. **(a)** Using Equation 17.10, explain how rapid breathing can cause the pH of blood to increase. **(b)** One cure for this problem is breathing in a paper bag. Why does this procedure lower blood pH?

17.91 For each pair of compounds, use K_{sp} values to determine which has the greater molar solubility: **(a)** CdS or CuS, **(b)** $PbCO_3$ or $BaCrO_4$, **(c)** $Ni(OH)_2$ or $NiCO_3$, **(d)** AgI or Ag_2SO_4.

17.92 Describe the solubility of $CaCO_3$ in each of the following solutions compared to its solubility in water: **(a)** in 0.10 M NaCl solution; **(b)** in 0.10 M $Ca(NO_3)_2$ solution; **(c)** 0.10 M Na_2CO_3; **(d)** 0.10 M HCl solution. (Answer same, less soluble, or more soluble.)

17.93 Tooth enamel is composed of hydroxyapatite, whose simplest formula is $Ca_5(PO_4)_3OH$, and whose corresponding $K_{sp} = 6.8 \times 10^{-27}$. As discussed in the "Chemistry and Life" box in Section 17.5, fluoride in fluorinated water or in toothpaste reacts with hydroxyapatite to form fluoroapatite, $Ca_5(PO_4)_3F$, whose $K_{sp} = 1.0 \times 10^{-60}$. **(a)** Write the expression for the solubility-constant for hydroxyapatite and for fluoroapatite. **(b)** Calculate the molar solubility of each of these compounds.

17.94 Calculate the solubility of $Mg(OH)_2$ in 0.50 M NH_4Cl.

[17.95] Seawater contains 0.13% magnesium by mass, and has a density of 1.025 g/mL. What fraction of the magnesium can be removed by adding a stoichiometric quantity of CaO (that is, one mole of CaO for each mole of Mg^{2+})?

17.96 The solubility-product constant for barium permanganate, $Ba(MnO_4)_2$, is 2.5×10^{-10}. Assume that solid $Ba(MnO_4)_2$ is in equilibrium with a solution of $KMnO_4$. What concentration of $KMnO_4$ is required to establish a concentration of $2.0 \times 10^{-8} M$ for the Ba^{2+} ion in solution?

17.97 Calculate the ratio of $[Ca^{2+}]$ to $[Fe^{2+}]$ in a lake in which the water is in equilibrium with deposits of both $CaCO_3$ and $FeCO_3$. Assume that the water is slightly basic and that the hydrolysis of the carbonate ion can therefore be ignored.

[17.98] The solubility products of $PbSO_4$ and $SrSO_4$ are 6.3×10^{-7} and 3.2×10^{-7}, respectively. What are the values of $[SO_4^{2-}]$, $[Pb^{2+}]$, and $[Sr^{2+}]$ in a solution at equilibrium with both substances?

[17.99] What pH buffer solution is needed to give a Mg^{2+} concentration of $3.0 \times 10^{-2} M$ in equilibrium with solid magnesium oxalate?

[17.100] The value of K_{sp} for $Mg_3(AsO_4)_2$ is 2.1×10^{-20}. The AsO_4^{3-} ion is derived from the weak acid H_3AsO_4 ($pK_{a1} = 2.22$; $pK_{a2} = 6.98$; $pK_{a3} = 11.50$). When asked to calculate the molar solubility of $Mg_3(AsO_4)_2$ in water, a student used the K_{sp} expression and assumed that $[Mg^{2+}] = 1.5[AsO_4^{3-}]$. Why was this a mistake?

[17.101] The solubility product for $Zn(OH)_2$ is 3.0×10^{-16}. The formation constant for the hydroxo complex, $Zn(OH)_4^{2-}$, is 4.6×10^{17}. What concentration of OH^- is required to dissolve 0.015 mol of $Zn(OH)_2$ in a liter of solution?

INTEGRATIVE EXERCISES

17.102 **(a)** Write the net ionic equation for the reaction that occurs when a solution of hydrochloric acid (HCl) is mixed with a solution of sodium formate ($NaCHO_2$). **(b)** Calculate the equilibrium constant for this reaction. **(c)** Calculate the equilibrium concentrations of Na^+, Cl^-, H^+, CHO_2^-, and $HCHO_2$ when 50.0 mL of 0.15 M HCl is mixed with 50.0 mL of 0.15 M $NaCHO_2$.

17.103 **(a)** A 0.1044-g sample of an unknown monoprotic acid requires 22.10 mL of 0.0500 M NaOH to reach the end point. What is the molecular weight of the unknown? **(b)** As the acid is titrated, the pH of the solution after the addition of 11.05 mL of the base is 4.89. What is the K_a for the acid? **(c)** Using Appendix D, suggest the identity of the acid. Do both the molecular weight and K_a value agree with your choice?

17.104 A sample of 7.5 L of NH_3 gas at 22 °C and 735 torr is bubbled into a 0.50-L solution of 0.40 M HCl. Assuming that all the NH_3 dissolves and that the volume of the solution remains 0.50 L, calculate the pH of the resulting solution.

17.105 Aspirin has the structural formula

At body temperature (37 °C), K_a for aspirin equals 3×10^{-5}. If two aspirin tablets, each having a mass of 325 mg, are dissolved in a full stomach whose volume is 1 L and whose pH is 2, what percent of the aspirin is in the form of neutral molecules?

17.106 What is the pH at 25 °C of water saturated with CO_2 at a partial pressure of 1.10 atm? The Henry's law constant for CO_2 at 25 °C is 3.1×10^{-2} mol/L-atm. The CO_2 is an acidic oxide, reacting with H_2O to form H_2CO_3.

17.107 Excess Ca(OH)$_2$ is shaken with water to produce a saturated solution. The solution is filtered, and a 50.00-mL sample titrated with HCl requires 11.23 mL of 0.0983 M HCl to reach the end point. Calculate K_{sp} for Ca(OH)$_2$. Compare your result with that in Appendix D. Do you think the solution was kept at 25 °C?

17.108 The osmotic pressure of a saturated solution of strontium sulfate at 25 °C is 21 torr. What is the solubility product of this salt at 25 °C?

17.109 A concentration of 10–100 parts per billion (by mass) of Ag$^+$ is an effective disinfectant in swimming pools. However, if the concentration exceeds this range, the Ag$^+$ can cause adverse health effects. One way to main-tain an appropriate concentration of Ag$^+$ is to add a slightly soluble salt to the pool. Using K_{sp} values from Appendix D, calculate the equilibrium concentration of Ag$^+$ in parts per billion that would exist in equilibrium with **(a)** AgCl, **(b)** AgBr, **(c)** AgI.

[17.110] Fluoridation of drinking water is employed in many places to aid in the prevention of tooth decay. Typically the F$^-$ ion concentration is adjusted to about 1 ppb. Some water supplies are also "hard"; that is, they contain certain cations such as Ca^{2+} that interfere with the action of soap. Consider a case where the concentration of Ca^{2+} is 8 ppb. Could a precipitate of CaF$_2$ form under these conditions? (Make any necessary approximations.)

CHEMISTRY OF THE ENVIRONMENT

PHOTOGRAPH OF EARTH taken from Space looking south along the Red Sea.

WHAT'S AHEAD

18.1 Earth's Atmosphere
We begin this chapter with a general look at Earth's atmosphere. We will consider the temperature profile of the atmosphere, its pressure profile, and its chemical composition.

18.2 Outer Regions of the Atmosphere
We next examine the outer (or upper) regions of the atmosphere, where the pressure is very low. These regions absorb a great deal of high-energy radiation from the Sun through *photoionization* and *photodissociation* reactions. By filtering out high-energy radiation, these processes make it possible for life as we know it to exist on Earth.

18.3 Ozone in the Upper Atmosphere
We will see that ozone in the *stratosphere* acts as a filter of high-energy ultraviolet light. Human activities have contributed to the depletion of the ozone layer by introducing into the stratosphere chemicals that perturb the natural cycle of ozone formation and decomposition. Notable among these chemicals are *chlorofluorocarbons*.

18.4 Chemistry of the Troposphere
We complete our discussion of the atmosphere by examining its innermost region, the *troposphere*, which is the region in which we live. Many minor constituents of the troposphere affect air quality and the acidity of rainwater. The concentrations of many of these minor constituents, including those that cause *acid rain* and *photochemical smog*, have been increased by human activities. Carbon dioxide is an important minor constituent of the atmosphere because it acts as a *"greenhouse" gas* that warms Earth's atmosphere.

18.5 The World Ocean
We next turn our attention to our water environment. Almost all the water on Earth is in the world ocean. Seawater contains many salts that participate in the global cycling of the elements and nutrients.

18.6 Freshwater
Our discussion of water concludes with a look at freshwater. Processing seawater to obtain freshwater is energy intensive. We count on freshwater sources to supply most of our needs, but these sources often require treatment to render them usable.

18.7 Green Chemistry
We conclude our discussion of the environment by examining *green chemistry*, which seeks to eliminate pollution at its source. Green chemistry is an international initiative to make industrial products, processes, and chemical reactions compatible with a sustainable society and environment.

IN 1997, REPRESENTATIVES OF 130 NATIONS MET IN KYOTO, JAPAN, to discuss the impact of human activities on global warming. Out of that meeting came an initiative to work toward a global treaty that would, among other things, spell out actions to be taken to reduce emissions of gases that cause

global warming. In 2001 in Bonn, Germany, 178 nations signed a treaty based on the so-called Kyoto Protocols. In 2007, the United Nations Intergovernmental Panel on Climate Change issued a report declaring that human activity has "very likely" been the driving force for the warming trend observed worldwide over the past 50 years. These efforts to address environmental concerns at the international level indicate that many of the most urgent environmental problems are global in nature.

As technology has advanced and the world human population has increased, we have put new and greater stresses on our environment. Paradoxically, the very technology that can cause pollution also provides the tools to help us understand and manage our environment in a beneficial way.

Chemistry is often at the heart of these issues. The economic growth of both developed and developing nations depends critically on chemical processes that range from treatment of water supplies to industrial processes. Some of these processes produce products or by-products that are harmful to the environment.

We are now in a position to apply the principles we have learned in earlier chapters to an understanding of how our environment operates and how human activities affect it. In this chapter we consider some aspects of the chemistry of our environment, focusing on Earth's atmosphere and water. Both the air and water of our planet make life as we know it possible. To understand and maintain the environment in which we live, we must understand how human-made and natural chemical compounds interact on land and in the sea and sky. Our daily decisions as consumers mirror those of the leaders meeting in Bonn and similar international meetings: We must weigh the costs versus the benefits of our actions. Unfortunately, the environmental impacts of our decisions are often very subtle and not immediately evident.

18.1 EARTH'S ATMOSPHERE

Because most of us have never been very far from Earth's surface, we often take for granted the many ways in which the atmosphere determines the environment in which we live. In this section we will examine some of the important characteristics of our planet's atmosphere.

The temperature of the atmosphere varies in a complex manner as a function of altitude, as shown in Figure 18.1(a) ◄. The atmosphere is divided into four regions based on this temperature profile. Just above the surface, in the **troposphere**, the temperature normally decreases with increasing altitude, reaching a minimum of about 215 K at about 12 km. Nearly all of us live our entire lives in the troposphere. Howling winds and soft breezes, rain, and sunny skies—all that we normally think of as "weather"—occur in this region. Commercial jet aircraft typically fly about 10 km (33,000 ft) above Earth, an altitude that approaches the upper limit of the troposphere, which we call the *tropopause*.

Above the tropopause the temperature increases with altitude, reaching a maximum of about 275 K at about 50 km. The region from 10 km to 50 km is called the **stratosphere**. Beyond the stratosphere are the *mesosphere* and the *thermosphere*. Notice in Figure 18.1(a) that the temperature extremes that form the boundaries between adjacent regions are denoted by the suffix *-pause*. The boundaries are important because gases mix across them relatively slowly. For example, pollutant gases generated in the troposphere pass through the tropopause and find their way into the stratosphere only very slowly.

Unlike temperature, which varies in a complex way with increased altitude, the pressure of the atmosphere decreases in a regular way with increasing elevation, as shown in Figure 18.1(b). Atmospheric pressure declines much more rapidly at lower elevations than at higher ones because of the atmosphere's compressibility. Thus, the pressure decreases from an average value of 760 torr (101 kPa) at sea level to 2.3×10^{-3} torr (3.1×10^{-4} kPa) at 100 km, to only 1.0×10^{-6} torr (1.3×10^{-7} kPa) at 200 km. The troposphere and stratosphere

▼ **Figure 18.1 Temperature and pressure in the atmosphere.**
(a) Temperature variations in the atmosphere at altitudes below 110 km.
(b) Variations in atmospheric pressure with altitude. At 80 km the pressure is approximately 0.01 torr.

(a)

(b)

together account for 99.9% of the mass of the atmosphere, 75% of which is the mass in the troposphere. Consequently, most of the chemistry that follows focuses on the troposphere and stratosphere.

Composition of the Atmosphere

Earth's atmosphere is an extremely complex system. Its temperature and pressure change over a wide range with altitude, as we have just seen. The atmosphere is constantly bombarded by radiation and energetic particles from the Sun. This barrage of energy has profound chemical effects, especially on the outer reaches of the atmosphere (Figure 18.2▶). In addition, because of Earth's gravitational field, lighter atoms and molecules tend to rise to the top of the atmosphere, especially in the stratosphere and upper layers that are not as turbulent as the troposphere. Because of all these factors, the composition of the atmosphere is not uniform.

Table 18.1▼ shows the composition by mole fraction of dry air near sea level. Although traces of many substances are present, N_2 and O_2 make up about 99% of the atmosphere. The noble gases and CO_2 make up most of the remainder.

When speaking of trace constituents, we commonly use *parts per million* (ppm) as the unit of concentration. When applied to substances in aqueous solution, parts per million refers to grams of substance per million grams of solution. ∞ (Section 13.4) When dealing with gases, however, one part per million refers to one part by *volume* in 1 million volume units of the whole. Because volume (V) is proportional to the number of moles n of gas via the ideal-gas equation ($PV = nRT$), volume fraction and mole fraction are the same. Thus, 1 ppm of a trace constituent of the atmosphere amounts to one mole of that constituent in 1 million moles of total gas; that is, the concentration in ppm is equal to the mole fraction times 10^6. Table 18.1 lists the mole fraction of CO_2 in the atmosphere as 0.000375. Its concentration in ppm is therefore $0.000375 \times 10^6 = 375$ ppm.

Before we consider the chemical processes that occur in the atmosphere, let's review some of the important chemical properties of the two major components, N_2 and O_2. Recall that the N_2 molecule possesses a triple bond between the nitrogen atoms. ∞ (Section 8.3) This very strong bond is largely responsible for the very low reactivity of N_2, which undergoes reaction only under extreme conditions. The bond energy in O_2, 495 kJ/mol, is much lower than that in N_2, which is 941 kJ/mol. ∞ (Table 8.4) Therefore, O_2 is much more reactive than N_2. For example, oxygen reacts with many substances to form oxides. The oxides of nonmetals, for example SO_2, usually form acidic solutions when dissolved in water. The oxides of active metals, for example CaO, form basic solutions when dissolved in water. ∞ (Section 7.7)

▲ **Figure 18.2 The aurora borealis.** This luminous display in the northern sky, also called the northern lights, is produced by collisions of high-speed electrons and protons from the Sun with air molecules. The charged particles are channeled toward the polar regions by Earth's magnetic field.

TABLE 18.1 ■ Composition of Dry Air Near Sea Level		
Component*	**Content (mole fraction)**	**Molar Mass**
Nitrogen	0.78084	28.013
Oxygen	0.20948	31.998
Argon	0.00934	39.948
Carbon dioxide	0.000382	44.0099
Neon	0.00001818	20.183
Helium	0.00000524	4.003
Methane	0.000002	16.043
Krypton	0.00000114	83.80
Hydrogen	0.0000005	2.0159
Nitrous oxide	0.0000005	44.0128
Xenon	0.000000087	131.30

*Ozone, sulfur dioxide, nitrogen dioxide, ammonia, and carbon monoxide are present as trace gases in variable amounts.

■ SAMPLE EXERCISE 18.1 | Calculating the Concentration of Water in Air

What is the concentration, in parts per million, of water vapor in a sample of air if the partial pressure of the water is 0.80 torr and the total pressure of the air is 735 torr?

SOLUTION

Analyze: We are given the partial pressure of water vapor and the total pressure of an air sample and asked to determine the water vapor concentration.

Plan: Recall that the partial pressure of a component in a mixture of gases is given by the product of its mole fraction and the total pressure of the mixture ∞ (Section 10.6):

$$P_{H_2O} = X_{H_2O}P_t$$

Solve: Solving for the mole fraction of water vapor in the mixture, X_{H_2O}, gives:

$$X_{H_2O} = \frac{P_{H_2O}}{P_t} = \frac{0.80 \text{ torr}}{735 \text{ torr}} = 0.0011$$

The concentration in ppm is the mole fraction times 10^6

$$0.0011 \times 10^6 = 1100 \text{ ppm}$$

■ PRACTICE EXERCISE

The concentration of CO in a sample of air is found to be 4.3 ppm. What is the partial pressure of the CO if the total air pressure is 695 torr?
Answer: 3.0×10^{-3} torr

18.2 OUTER REGIONS OF THE ATMOSPHERE

Although the outer portion of the atmosphere, beyond the stratosphere, contains only a small fraction of the atmospheric mass, it forms the outer defense against the hail of radiation and high-energy particles that continuously bombard Earth. As the bombarding radiation passes through the upper atmosphere, it causes two basic kinds of chemical changes: photodissociation and photoionization. These processes protect us from high-energy radiation by absorbing most of the radiation before it reaches the troposphere where we live. If it were not for these photochemical processes, plant and animal life as we know it could not exist on Earth.

Photodissociation

The Sun emits radiant energy over a wide range of wavelengths (Figure 18.3◀). To understand the connection between the wavelength of radiation and its effect on atoms and molecules, recall that electromagnetic radiation can be pictured as a stream of photons. ∞ (Section 6.2) The energy of each photon is given by the relationship $E = h\nu$, where h is Planck's constant and ν is the frequency of the radiation. For a chemical change to occur when radiation strikes atoms or molecules, two conditions must be met. First, there must be photons with energy sufficient to accomplish some chemical process such as breaking a chemical bond or removing an electron. Second, molecules must absorb these photons. When these requirements are met, the energy of the photons is used to do the work associated with some chemical change.

The rupture of a chemical bond resulting from absorption of a photon by a molecule is called **photodissociation**.

▼ **Figure 18.3 The solar spectrum.** Shown in this graph is the amount of sunlight (in light energy per area per time, known as flux) at different wavelengths that reaches the top of Earth's atmosphere. For comparison, the corresponding data for the amount of sunlight that reaches sea level is shown. The atmosphere absorbs much of the ultraviolet and visible light emitted by the Sun.

No ions are formed when the bond between two atoms is cleaved by photodisso-ciation. Instead, half the bonding electrons stay with one of the atoms and half stay with the other atom. The result is two neutral particles.

One of the most important processes occurring in the upper atmosphere above about 120-km elevation is the photodissociation of the oxygen molecule:

$$:\overset{..}{O}=\overset{..}{O}: + h\nu \longrightarrow :\overset{..}{\underset{.}{O}} + \overset{..}{\underset{.}{O}}: \qquad\qquad [18.1]$$

The minimum energy required to cause this change is determined by the bond energy (or *dissociation energy*) of O_2, 495 kJ/mol. In Sample Exercise 18.2 we calcu-late the longest wavelength photon having sufficient energy to photodissociate the O_2 molecule.

▮ SAMPLE EXERCISE 18.2 | Calculating the Wavelength Required to Break a Bond

What is the maximum wavelength of light, in nanometers, that has enough energy per photon to dissociate the O_2 molecule?

SOLUTION

Analyze: We are asked to determine the wavelength of a photon that has just enough energy to break the $O=O$ double bond in O_2.

Plan: We first need to calculate the energy required to break the $O=O$ double bond in one molecule, and then find the wave-length of a photon of this energy.

Solve: The dissociation energy of O_2 is 495 kJ/mol. Using this value and Avogadro's number, we can calculate the amount of energy needed to break the bond in a single O_2 molecule:

$$\left(495 \times 10^3 \frac{J}{mol}\right)\left(\frac{1\ mol}{6.022 \times 10^{23}\ molecules}\right) = 8.22 \times 10^{-19} \frac{J}{molecule}$$

We next use the Planck relationship, $E = h\nu$, ∞ (Equation 6.2) to calculate the frequency, ν of a photon that has this amount of energy:

$$\nu = \frac{E}{h} = \frac{8.22 \times 10^{-19}\ J}{6.626 \times 10^{-34}\ J\text{-}s} = 1.24 \times 10^{15}\ s^{-1}$$

Finally, we use the relationship be-tween the frequency and wavelength of light ∞ (Section 6.1) to calculate the wavelength of the light:

$$\lambda = \frac{c}{\nu} = \left(\frac{3.00 \times 10^8\ m/s}{1.24 \times 10^{15}/s}\right)\left(\frac{10^9\ nm}{1\ m}\right) = 242\ nm$$

Thus, light of wavelength 242 nm, which is in the ultraviolet region of the electromagnetic spectrum, has sufficient energy per photon to photodissociate an O_2 molecule. Because photon energy increases as wavelength *decreases*, any photon of wavelength *shorter* than 242 nm will have sufficient energy to dissociate O_2.

▮ PRACTICE EXERCISE

The bond energy in N_2 is 941 kJ/mol (Table 8.4). What is the longest wavelength a photon can have and still have sufficient energy to dissociate N_2?
Answer: 127 nm

Fortunately for us, O_2 absorbs much of the high-energy, short-wavelength radiation from the solar spectrum before that radiation reaches the lower atmosphere. As it does, atomic oxygen, $:\overset{..}{O}$, is formed. At higher elevations the dissociation of O_2 is very extensive. At 400 km, for example, only 1% of the oxy-gen is in the form of O_2; the other 99% is atomic oxygen. At 130 km, O_2 and atomic oxygen are just about equally abundant. Below 130 km, O_2 is more abundant than atomic oxygen because most of the solar energy has been ab-sorbed in the upper atmosphere.

The dissociation energy of N_2 is very high (Table 8.4). As shown in the Practice Exercise in Sample Exercise 18.2, only photons of very short wave-length possess sufficient energy to dissociate N_2. Furthermore, N_2 does not readily absorb photons, even when they possess sufficient energy. As a result, very little atomic nitrogen is formed in the upper atmosphere by photodisso-ciation of N_2.

Photoionization

In 1901 Guglielmo Marconi received a radio signal in St. John's, Newfoundland, that had been transmitted from Land's End, England, some 2900 km away. Because people at the time thought radio waves traveled in straight lines, they assumed that the curvature of Earth's surface would make radio communication over large distances impossible. Marconi's successful experiment suggested that Earth's atmosphere in some way substantially affects radio-wave propagation. His discovery led to intensive study of the upper atmosphere. In about 1924, the existence of electrons in the upper atmosphere was established by experimental studies.

For each electron present in the upper atmosphere, there must be a corresponding positively charged particle. The electrons in the upper atmosphere result mainly from the **photoionization** of molecules, caused by solar radiation. Photoionization occurs when a molecule absorbs radiation and the absorbed energy causes an electron to be ejected from the molecule. The molecule then becomes a positively charged ion. For photoionization to occur, therefore, a molecule must absorb a photon, and the photon must have enough energy to remove an electron. ∞ (Section 7.4)

Some of the more important ionization processes occurring in the atmosphere above about 90 km are shown in Table 18.2 ◄, together with the ionization energies and λ_{max}, the maximum wavelength of a photon capable of causing ionization. Photons with energies sufficient to cause ionization have wavelengths in the high-energy end of the ultraviolet region of the electromagnetic spectrum. These wavelengths are completely filtered out of the radiation reaching Earth, because they are absorbed by the upper atmosphere.

TABLE 18.2 ■ Ionization Processes, Ionization Energies, and Maximum Wavelengths Capable of Causing Ionization

Process	Ionization Energy (kJ/mol)	λ_{max} (nm)
$N_2 + h\nu \longrightarrow N_2^+ + e^-$	1495	80.1
$O_2 + h\nu \longrightarrow O_2^+ + e^-$	1205	99.3
$O + h\nu \longrightarrow O^+ + e^-$	1313	91.2
$NO + h\nu \longrightarrow NO^+ + e^-$	890	134.5

18.3 OZONE IN THE UPPER ATMOSPHERE

While N_2, O_2, and atomic oxygen absorb photons having wavelengths shorter than 240 nm, ozone, O_3, is the key absorber of photons having wavelengths ranging from 240 to 310 nm, in the ultraviolet region of the electromagnetic spectrum. Ozone protects us from these harmful high-energy photons, which would otherwise penetrate to Earth's surface. Let's consider how ozone forms in the upper atmosphere and how it absorbs photons.

Below an altitude of 90 km, most of the short-wavelength radiation capable of photoionization has been absorbed. Radiation capable of dissociating the O_2 molecule is sufficiently intense, however, for photodissociation of O_2 (Equation 18.1) to remain important down to an altitude of 30 km. In the region between 30 and 90 km, the concentration of O_2 is much greater than that of atomic oxygen. Therefore, the oxygen atoms that form in this region undergo frequent collisions with O_2 molecules, resulting in the formation of ozone, O_3:

$$:\ddot{O} + O_2 \longrightarrow O_3^* \qquad [18.2]$$

The asterisk over the O_3 denotes that the ozone molecule contains an excess of energy. The reaction in Equation 18.2 releases 105 kJ/mol. This energy must be transferred away from the O_3^* molecule in a very short time or else the molecule will fly apart again into O_2 and $:\ddot{O}$ —a decomposition that is the reverse of the process by which O_3^* is formed.

An energy-rich O_3^* molecule can release its excess energy by colliding with another atom or molecule and transferring some of the excess energy to it. Let's represent the atom or molecule with which O_3^* collides as M. (Usually M is N_2 or O_2 because these are the most abundant molecules in the atmosphere.)

The formation of O_3^* and the transfer of excess energy to M are summarized by the following equations (where O atoms are shown without valence electrons):

$$O(g) + O_2(g) \rightleftharpoons O_3^*(g) \qquad [18.3]$$

$$O_3^*(g) + M(g) \longrightarrow O_3(g) + M^*(g) \qquad [18.4]$$

$$O(g) + O_2(g) + M(g) \longrightarrow O_3(g) + M^*(g) \qquad [18.5]$$

The rate at which O_3 forms according to Equations 18.3 and 18.4 depends on two factors that vary in opposite directions with increasing altitude. First, the formation of O_3^*, according to Equation 18.3, depends on the presence of O atoms. At low altitudes most of the radiation energetic enough to dissociate O_2 has been absorbed; thus, the formation of O is favored at higher altitudes. Second, both Equations 18.3 and 18.4 depend on molecular collisions. ∞ (Section 14.5) The concentration of molecules is greater at low altitudes, and so the frequency of collisions between O and O_2 (Equation 18.3) and between O_3^* and M (Equation 18.4) are both greater at lower altitudes. Because these processes vary with altitude in opposite directions, the highest rate of O_3 formation occurs in a band at an altitude of about 50 km, near the stratopause [Figure 18.1(a)]. Overall, roughly 90% of Earth's ozone is found in the stratosphere, between the altitudes of 10 and 50 km.

Once formed, an ozone molecule does not last long. Ozone is capable of absorbing solar radiation, which decomposes the molecule back into O_2 and O. Because only 105 kJ/mol is required for this process, photons of wavelength shorter than 1140 nm are sufficiently energetic to photodissociate O_3. However, most of the output from the Sun is concentrated in the visible and ultraviolet portions of the electromagnetic spectrum (Figure 18.3). Photons with wavelengths shorter than about 300 nm are energetic enough to break many kinds of single chemical bonds. Thus, the "ozone shield" is essential for our continued well-being. The ozone molecules that form this essential shield against high-energy radiation represent only a tiny fraction of the oxygen atoms present in the stratosphere, however, because they are continually destroyed even as they are formed.

GIVE IT SOME THOUGHT

Look at Figure 18.3. Based on the relative areas under the curves, what fraction of the radiation with wavelengths in the ultraviolet portion of the spectrum is absorbed by the upper atmosphere?

The photodecomposition of ozone reverses the reaction that forms it. We thus have a cyclic process of ozone formation and decomposition, summarized as follows:

$$O_2(g) + h\nu \longrightarrow O(g) + O(g)$$

$$O(g) + O_2(g) + M(g) \longrightarrow O_3(g) + M^*(g) \qquad \text{(heat released)}$$

$$O_3(g) + h\nu \longrightarrow O_2(g) + O(g)$$

$$O(g) + O(g) + M(g) \longrightarrow O_2(g) + M^*(g) \qquad \text{(heat released)}$$

The first and third processes are photochemical; they use a solar photon to initiate a chemical reaction. The second and fourth processes are exothermic chemical reactions. The net result of all four processes is a cycle in which solar radiant energy is converted into thermal energy. The ozone cycle in the stratosphere is responsible for the rise in temperature that reaches its maximum at the stratopause, as illustrated in Figure 18.1(a).

▲ Figure 18.4 Ozone depletion.
Variation in ozone concentration in the atmosphere, as a function of altitude.

▲ Figure 18.5 Map of the total ozone present in the Southern Hemisphere, taken on September 24, 2006 from an orbiting satellite. The different colors represent different ozone concentrations. The center area, which is over Antarctica, is the ozone hole—the area of lowest ozone.

The scheme described for the formation and decomposition of ozone molecules accounts for some, but not all, of the facts about the ozone layer. Many chemical reactions occur that involve substances other than just oxygen. We must also consider the effects of turbulence and winds that mix up the stratosphere. A very complicated picture results. The overall result of ozone formation and removal reactions, coupled with atmospheric turbulence and other factors, is to produce an ozone profile in the upper atmosphere as shown on the left in Figure 18.4 ◄, with a maximum ozone concentration occurring at an altitude of about 30 km.

Depletion of the Ozone Layer

The ozone layer protects Earth's surface from damaging ultraviolet radiation. Satellite monitoring of ozone, which began in 1978, has revealed a depletion of ozone in the stratosphere that is particularly severe over Antarctica, a phenomenon known as the *ozone hole* (Figure 18.5 ◄). The first scientific paper on this phenomenon appeared in 1985.

In 1995 the Nobel Prize in Chemistry was awarded to F. Sherwood Rowland, Mario Molina, and Paul Crutzen for their studies of ozone depletion. In 1970 Crutzen showed that naturally occurring nitrogen oxides catalytically destroy ozone. Rowland and Molina recognized in 1974 that chlorine from **chlorofluorocarbons** (CFCs) may deplete the ozone layer. These substances, principally $CFCl_3$ and CF_2Cl_2, have been widely used as propellants in spray cans, as refrigerant and air-conditioner gases, and as foaming agents for plastics. They are virtually unreactive in the lower atmosphere. Furthermore, they are relatively insoluble in water and are therefore not removed from the atmosphere by rainfall or by dissolution in the oceans. Unfortunately, the lack of reactivity that makes them commercially useful also allows them to survive in the atmosphere and to diffuse eventually into the stratosphere. It is estimated that several million tons of chlorofluorocarbons are now present in the atmosphere.

As CFCs diffuse into the stratosphere, they are exposed to high-energy radiation, which can cause photodissociation. The C—Cl bonds are considerably weaker than the C—F bonds (Table 8.4). As a result, free chlorine atoms are formed readily in the presence of light with wavelengths in the range of 190 to 225 nm, as shown in this equation for CF_2Cl_2:

$$CF_2Cl_2(g) + h\nu \longrightarrow CF_2Cl(g) + Cl(g) \qquad [18.6]$$

Calculations suggest that chlorine atom formation occurs at the greatest rate at an altitude of about 30 km, the altitude at which ozone is at its highest concentration.

Atomic chlorine reacts rapidly with ozone to form chlorine monoxide (ClO) and molecular oxygen (O_2):

$$Cl(g) + O_3(g) \longrightarrow ClO(g) + O_2(g) \qquad [18.7]$$

Equation 18.7 follows a second-order rate law with a very large rate constant:

$$\text{Rate} = k[Cl][O_3] \qquad k = 7.2 \times 10^9 \ M^{-1}\,s^{-1} \text{ at 298 K} \qquad [18.8]$$

Under certain conditions the ClO generated in Equation 18.7 can react to regenerate free Cl atoms. One way that this can happen is by photodissociation of the ClO:

$$ClO(g) + h\nu \longrightarrow Cl(g) + O(g) \qquad [18.9]$$

The Cl atoms generated in Equations 18.6 and 18.9 can react with more O_3, according to Equation 18.7. The result is a sequence of reactions that accomplishes the Cl atom–catalyzed decomposition of O_3 to O_2, as shown in the following multistep mechanism:

$$2\,Cl(g) + 2\,O_3(g) \longrightarrow 2\,ClO(g) + 2\,O_2(g)$$

$$2\,ClO(g) + h\nu \longrightarrow 2\,Cl(g) + 2\,O(g)$$

$$O(g) + O(g) \longrightarrow O_2(g)$$

$$2\,Cl(g) + 2\,O_3(g) + 2\,ClO(g) + 2\,O(g) \longrightarrow 2\,Cl(g) + 2\,ClO(g) + 3\,O_2(g) + 2\,O(g)$$

The equation can be simplified by eliminating like species from each side to give

$$2\,O_3(g) \xrightarrow{\text{Cl}} 3\,O_2(g) \qquad\qquad [18.10]$$

Because the rate of Equation 18.7 increases linearly with [Cl], the rate at which ozone is destroyed increases as the quantity of Cl atoms increases. Thus, the greater the amount of CFCs that diffuse into the stratosphere, the faster the destruction of the ozone layer. Rates of diffusion of molecules from the troposphere into the stratosphere are slow. Nevertheless, a thinning of the ozone layer over the South Pole has been observed, particularly during the months of September and October (Figure 18.5).

Because of the environmental problems associated with CFCs, steps have been taken to limit their manufacture and use. A major step was the signing in 1987 of the Montreal Protocol on Substances That Deplete the Ozone Layer, in which participating nations agreed to reduce CFC production. More stringent limits were set in 1992, when representatives of approximately 100 nations agreed to ban the production and use of CFCs by 1996. Since then, the size of the ozone hole has leveled off. Nevertheless, because CFCs are unreactive and because they diffuse so slowly into the stratosphere, scientists estimate that ozone depletion will continue for many years to come. ∞(Section 1.4, "Chemistry Put to Work: The Hole Story")

What substances have replaced CFCs? At this time the main alternatives are hydrofluorocarbons, compounds in which C—H bonds replace the C—Cl bonds of CFCs. One such compound in current use is CH_2FCF_3, known as HFC-134a.

There are no naturally occurring CFCs, but some natural sources contribute chlorine and bromine to the atmosphere, and, just like halogens from CFC, these naturally occurring Cl and Br atoms can participate in ozone-depleting reactions. The principal naturally occurring sources are methyl bromide and methyl chloride, CH_3Br and CH_3Cl. It is estimated that these molecules contribute less than a third to the total Cl and Br in the atmosphere; the remaining two-thirds is a result of human activities. Volcanoes are a source of HCl, but, generally, the HCl they release reacts with water in the troposphere and does not make it to the upper atmosphere.

▲ GIVE IT SOME THOUGHT

What process involving CFCs generates a substance that initiates the catalytic decomposition of ozone? Identify the catalyst.

18.4 CHEMISTRY OF THE TROPOSPHERE

The troposphere consists primarily of N_2 and O_2, which together make up 99% of Earth's atmosphere at sea level (Table 18.1). Other gases, although present only at very low concentrations, can have major effects on our environment. Table 18.3 ▼ lists the major sources and typical concentrations of some of the important minor constituents of the troposphere. Many of these substances occur to only a slight extent in the natural environment but exhibit much higher concentrations in certain areas because of human activities. In this section we will discuss the most important characteristics of a few of these substances and their chemical roles as air pollutants. As we will see, most form as either a direct or an indirect result of our widespread use of combustion reactions.

TABLE 18.3 ■ Sources and Typical Concentrations of Some Minor Atmospheric Constituents

Minor Constituent	Sources	Typical Concentrations
Carbon dioxide, CO_2	Decomposition of organic matter; release from the oceans; fossil-fuel combustion	375 ppm throughout the troposphere
Carbon monoxide, CO	Decomposition of organic matter; industrial processes; fossil-fuel combustion	0.05 ppm in unpolluted air; 1–50 ppm in urban traffic areas
Methane, CH_4	Decomposition of organic matter; natural-gas seepage	1.77 ppm throughout the troposphere
Nitric oxide, NO	Electrical discharges; internal combustion engines; combustion of organic matter	0.01 ppm in unpolluted air; 0.2 ppm in smog
Ozone, O_3	Electrical discharges; diffusion from the stratosphere; photochemical smog	0 to 0.01 ppm in unpolluted air; 0.5 ppm in photochemical smog
Sulfur dioxide, SO_2	Volcanic gases; forest fires; bacterial action; fossil-fuel combustion; industrial processes	0 to 0.01 ppm in unpolluted air; 0.1–2 ppm in polluted urban environment

TABLE 18.4 ■ Median Concentrations of Atmospheric Pollutants in a Typical Urban Atmosphere

Pollutant	Concentration (ppm)
Carbon monoxide	10
Hydrocarbons	3
Sulfur dioxide	0.08
Nitrogen oxides	0.05
Total oxidants (ozone and others)	0.02

Sulfur Compounds and Acid Rain

Sulfur-containing compounds are present to some extent in the natural, unpolluted atmosphere. They originate in the bacterial decay of organic matter, in volcanic gases, and from other sources listed in Table 18.3. The amount of sulfur-containing compounds released into the atmosphere worldwide from natural sources is about 24×10^{12} g per year, which is less than the amount from human activities (about 79×10^{12} g per year). Sulfur compounds, chiefly sulfur dioxide, SO_2, are among the most unpleasant and harmful of the common pollutant gases. Table 18.4 ◀ shows the concentrations of several pollutant gases in a *typical* urban environment (not one that is particularly affected by smog). According to these data, the level of sulfur dioxide is 0.08 ppm or higher about half the time. This concentration is considerably lower than that of other pollutants, notably carbon monoxide. Nevertheless, SO_2 is regarded as the most serious health hazard among the pollutants shown, especially for people with respiratory difficulties.

Combustion of coal accounts for about 65% of the SO_2 released annually in the United States. The majority of this amount is from coal-burning electrical power plants, which generate about 50% of our electricity. The extent to which SO_2 emissions are a problem in the burning of coal depends on the level of its sulfur concentration. Because of concern about SO_2 pollution, low-sulfur coal is in greater demand and is thus more expensive. Much of the coal from east of the Mississippi is relatively high in sulfur content, up to 6% by mass. Much of the coal from the western states has a lower sulfur content. This coal, however, also has a lower heat content per unit mass of coal, so the difference in sulfur content per unit of heat produced is not as large as is often assumed.

China, which gets 70 percent of its total energy from coal, is the world's largest generator of SO_2, producing about 24×10^{12} g of SO_2 annually by official Chinese counts and as much as 34×10^{12} g by other counts. As a result, that nation has a major problem with SO_2 pollution.

Sulfur dioxide is harmful to both human health and property; furthermore, atmospheric SO_2 can be oxidized to SO_3 by several pathways (such as reaction with O_2 or O_3). When SO_3 dissolves in water, it produces sulfuric acid, H_2SO_4:

$$SO_3(g) + H_2O(l) \longrightarrow H_2SO_4(aq)$$

Many of the environmental effects ascribed to SO_2 are actually due to H_2SO_4.

The presence of SO_2 in the atmosphere and the sulfuric acid that it produces result in the phenomenon of **acid rain**. (Nitrogen oxides, which form nitric acid, are also major contributors to acid rain.) Uncontaminated rainwater is

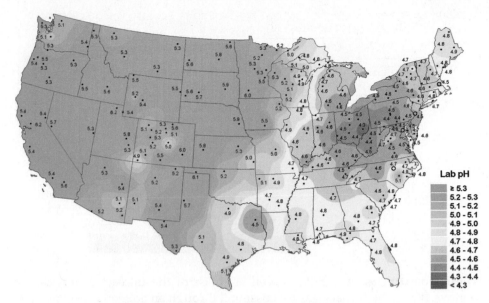

▲ **Figure 18.6 pH values from freshwater sites across the United States, 2005. The numbered dots indicate the locations of monitoring stations.**

Lab pH

	≥ 5.3
	5.2 - 5.3
	5.1 - 5.2
	5.0 - 5.1
	4.9 - 5.0
	4.8 - 4.9
	4.7 - 4.8
	4.6 - 4.7
	4.5 - 4.6
	4.4 - 4.5
	4.3 - 4.4
	< 4.3

naturally acidic and generally has a pH value of about 5.6. The primary source of this natural acidity is CO_2, which reacts with water to form carbonic acid, H_2CO_3. Acid rain typically has a pH value of about 4. This acidity has affected many lakes in northern Europe, the northern United States, and Canada, reducing fish populations and affecting other parts of the ecological network within the lakes and surrounding forests.

The pH of most natural waters containing living organisms is between 6.5 and 8.5, but as Figure 18.6▲ shows, freshwater pH values are far below 6.5 in many parts of the continental United States. At pH levels below 4.0, all vertebrates, most invertebrates, and many microorganisms are destroyed. The lakes that are most susceptible to damage are those with low concentrations of basic ions, such as HCO_3^-, that buffer them against changes in pH. More than 300 lakes in New York State contain no fish, and 140 lakes in Ontario, Canada, are devoid of life. The acid rain that appears to have killed the organisms in these lakes originates hundreds of kilometers upwind in the Ohio Valley and Great Lakes regions. Some of these regions are recovering as sulfur emissions from fossil fuel combustion decrease, in part because of the Clean Air Act, which has resulted in a reduction of more than 40% in SO_2 emissions from power plants since 1980.

Because acids react with metals and with carbonates, acid rain is corrosive both to metals and to stone building materials. Marble and limestone, for example, whose major constituent is $CaCO_3$, are readily attacked by acid rain (Figure 18.7▶). Billions of dollars each year are lost because of corrosion due to SO_2 pollution.

One way to reduce the quantity of SO_2 released into the environment is to remove sulfur from coal and oil before it is burned. Although difficult and expensive, several methods have been developed for removing SO_2 from the gases formed when coal and oil are combusted. Powdered limestone ($CaCO_3$), for example, can be injected into the furnace of a power plant, where it decomposes into lime (CaO) and carbon dioxide:

$$CaCO_3(s) \longrightarrow CaO(s) + CO_2(g)$$

The CaO then reacts with SO_2 to form calcium sulfite:

$$CaO(s) + SO_2(g) \longrightarrow CaSO_3(s)$$

(a)

(b)

▲ **Figure 18.7 Damage from acid rain.** (a) This statue at the Field Museum in Chicago shows the effects of corrosion from acid rain and atmospheric pollutants. (b) The same statue after restoration.

▶ **Figure 18.8 Common method for removing SO₂ from combusted fuel.** Powdered limestone decomposes into CaO, which reacts with SO₂ to form CaSO₃. The CaSO₃ and any unreacted SO₂ enter a purification chamber called a scrubber, where a shower of CaO and water converts the remaining SO₂ into CaSO₃ and precipitates the CaSO₃ into a watery residue called a slurry.

The solid particles of CaSO₃ as well as much of the unreacted SO₂ can be removed from the furnace gas by passing it through an aqueous suspension of lime (Figure 18.8▲). Not all the SO₂ is removed, however, and given the enormous quantities of coal and oil burned worldwide, pollution by SO₂ will probably remain a problem for some time.

GIVE IT SOME THOUGHT

What chemical behavior associated with sulfur oxides gives rise to acid rain?

Carbon Monoxide

Carbon monoxide is formed by the incomplete combustion of carbon-containing material, such as fossil fuels. In terms of total mass, CO is the most abundant of all the pollutant gases. The level of CO present in unpolluted air is very low, approximately 0.05 ppm. The estimated total amount of CO in the atmosphere is about 5.2×10^{14} g. In the United States alone, roughly 1×10^{14} g of CO is produced each year, about two-thirds of which comes from automobiles.

Carbon monoxide is a relatively unreactive molecule and consequently poses no direct threat to vegetation or materials. It does affect humans, however. It has the unusual ability to bind very strongly to **hemoglobin**, the iron-containing protein in red blood cells [Figure 18.9(a) ◀] that transports oxygen in blood. Hemoglobin consists of four protein chains held together by weak intermolecular forces in a cluster [Figure 18.9(b)]. Each protein chain has a heme molecule within its folds. A schematic structure of heme is shown in Figure 18.9(c). Note that iron is situated in the center of a plane of four nitrogen atoms. A hemoglobin molecule in the lungs picks up an O₂ molecule, which reacts with the iron atom to form a species called *oxyhemoglobin*, abbreviated as HbO₂.

▲ **Figure 18.9 Hemoglobin binds O₂ and CO.** Red blood cells (a) contain hemoglobin (b). The hemoglobin contains four heme units, each of which can bind an O₂ molecule (c). When exposed to CO, the heme binds CO in preference to O₂.

As the blood circulates, the oxygen molecule is released in tissues as needed for cell metabolism, that is, for the chemical processes occurring in the cell. ∞ (Section 17.2, "Chemistry and Life: Blood as a Buffered Solution")

Like O_2, CO also binds very strongly to the iron in hemoglobin. The complex is called *carboxyhemoglobin* and is represented as COHb. The equilibrium binding constant of human hemoglobin for CO is about 210 times greater than that for O_2. As a result, a relatively small quantity of CO can inactivate a substantial fraction of the hemoglobin in the blood for oxygen transport. For example, a person breathing air that contains only 0.1% of CO takes in enough CO after a few hours of breathing to convert up to 60% of the hemoglobin into COHb, thereby reducing the blood's normal oxygen-carrying capacity by 60%.

Under normal conditions a nonsmoker breathing unpolluted air has about 0.3 to 0.5% COHb in the bloodstream. This amount arises mainly from the production of small quantities of CO in the course of normal body chemistry and from the small amount of CO present in clean air. Exposure to higher concentrations of CO causes the COHb level to increase, which in turn leaves fewer Hb sites to which O_2 can bind. If the level of COHb becomes too high, oxygen transport is effectively shut down and death occurs. Because CO is colorless and odorless, CO poisoning occurs with very little warning. Improperly ventilated combustion devices, such as kerosene lanterns and stoves, thus pose a potential health hazard (Figure 18.10▶).

▲ Figure 18.10 **Carbon monoxide warnings.** Kerosene lamps and stoves have warning labels concerning use in enclosed spaces, such as an indoor room. Incomplete combustion can produce colorless, odorless carbon monoxide, CO, which is toxic.

Nitrogen Oxides and Photochemical Smog

Nitrogen oxides are primary components of smog, a phenomenon with which city dwellers are all too familiar. The term *smog* refers to a particularly unpleasant condition of pollution in certain urban environments that occurs when weather conditions produce a relatively stagnant air mass. The smog made famous by Los Angeles, but now common in many other urban areas as well, is more accurately described as **photochemical smog** because photochemical processes play a major role in its formation (Figure 18.11▶).

The majority of nitrogen oxide emissions (about 50%) comes from cars, buses, and other forms of transportation. Nitric oxide, NO, forms in small quantities in the cylinders of internal combustion engines by the direct combination of nitrogen and oxygen:

$$N_2(g) + O_2(g) \rightleftharpoons 2\,NO(g) \quad \Delta H = 180.8 \text{ kJ} \qquad [18.11]$$

As noted in the "Chemistry Put to Work" box in Section 15.7, the equilibrium constant K for this reaction increases from about 10^{-15} at 300 K (near room temperature) to about 0.05 at 2400 K (approximately the temperature in the cylinder of an engine during combustion). Thus, the reaction is more favorable at higher temperatures. In fact, some NO is formed in any high-temperature combustion. As a result, electrical power plants are also major contributors to nitrogen oxide pollution.

▲ Figure 18.11 **Photochemical smog.** Smog is produced largely by the action of sunlight on automobile exhaust gases.

Before the installation of pollution-control devices on automobiles, typical emission levels of NO_x were 4 grams per mile (g/mi). (The x is either 1 or 2 because both NO and NO_2 are formed, although NO predominates.) Starting in 2004, the auto emission standards for NO_x call for a phased-in reduction to 0.07 g/mi by 2009. Table 18.5▶ summarizes the federal standards for hydrocarbons and NO_x emissions since 1975 as well as the more restrictive standards enforced in California.

In air, nitric oxide (NO) is rapidly oxidized to nitrogen dioxide (NO_2):

$$2\,NO(g) + O_2(g) \rightleftharpoons 2\,NO_2(g) \quad \Delta H = -113.1 \text{ kJ}$$
$$[18.12]$$

TABLE 18.5 ■ National Tailpipe Emission Standards*		
Year	Hydrocarbons (g/mi)	Nitrogen Oxides (g/mi)
1975	1.5 (0.9)	3.1 (2.0)
1980	0.41 (0.41)	2.0 (1.0)
1985	0.41 (0.41)	1.0 (0.4)
1990	0.41 (0.41)	1.0 (0.4)
1995	0.41 (0.25)	0.4 (0.4)
2004	0.075 (0.05)	0.07 (0.05)

*California standards in parentheses

The equilibrium constant for this reaction decreases from about 10^{12} at 300 K to about 10^{-5} at 2400 K. The photodissociation of NO_2 initiates the reactions associated with photochemical smog. The dissociation of NO_2 into NO and O requires 304 kJ/mol, which corresponds to a photon wavelength of 393 nm. In sunlight, therefore, NO_2 undergoes dissociation to NO and O:

$$NO_2(g) + h\nu \longrightarrow NO(g) + O(g) \qquad [18.13]$$

The atomic oxygen formed undergoes several possible reactions, one of which gives ozone, as described earlier:

$$O(g) + O_2 + M(g) \longrightarrow O_3(g) + M^*(g) \qquad [18.14]$$

Ozone is a key component of photochemical smog. Although it is an essential UV screen in the upper atmosphere, ozone is an undesirable pollutant in the troposphere. It is extremely reactive and toxic, and breathing air that contains appreciable amounts of ozone can be especially dangerous for asthma sufferers, exercisers, and the elderly. We therefore have two ozone problems: excessive amounts in many urban environments, where it is harmful, and depletion in the stratosphere, where it is vital.

In addition to nitrogen oxides and carbon monoxide, an automobile engine also emits unburned *hydrocarbons* as pollutants. These organic compounds, which are composed entirely of carbon and hydrogen, are the principal components of gasoline (Section 25.3) and are major ingredients of smog. A typical engine without effective emission controls emits about 10 to 15 grams of these compounds per mile. Current standards require that hydrocarbon emissions be less than 0.075 grams per mile. Hydrocarbons are also emitted naturally from living organisms (see "A Closer Look" box later in this section).

Reduction or elimination of smog requires that the essential ingredients for its formation be removed from automobile exhaust. Catalytic converters are designed to drastically reduce the levels of NO_x and hydrocarbons, two of the major ingredients of smog. ∞ (Section 14.7, "Chemistry Put to Work: Catalytic Converters")

▲ GIVE IT SOME THOUGHT

What photochemical reaction involving nitrogen oxides initiates the formation of photochemical smog?

Water Vapor, Carbon Dioxide, and Climate

We have seen how the atmosphere makes life as we know it possible on Earth by screening out harmful short-wavelength radiation. In addition, the atmosphere is essential in maintaining a reasonably uniform and moderate temperature on the surface of the planet. The two atmospheric components of greatest importance in maintaining Earth's surface temperature are carbon dioxide and water. Without them, the average surface temperature of Earth would be 254 K, a temperature too low to sustain life.

Earth is in overall thermal balance with its surroundings. This means that Earth radiates energy into space at a rate equal to the rate at which it absorbs energy from the Sun. The Sun has a surface temperature of about 6000 K. As seen from outer space, Earth is relatively cold (254 K). The temperature of an object determines the distribution of wavelengths in the radiation it emits. ∞ (Section 6.2) Why does Earth, viewed from outside its atmosphere, appear so much colder than the temperature we usually experience at its surface? The troposphere is transparent to visible light but not to infrared radiation. Figure 18.12▶ shows the distribution of radiation from Earth's surface and the wavelengths absorbed by atmospheric water vapor and carbon dioxide. According to the graph, these atmospheric gases absorb much of the outgoing radiation from Earth's surface. In doing so, they help to maintain a livable uniform temperature at the surface by holding in, as it were, the infrared radiation, that we feel as heat.

◀ **Figure 18.12 Why Earth appears so cold from outer space.** (a) Carbon dioxide and water in the atmosphere absorb certain wavelengths of infrared radiation, which helps keep energy from escaping from Earth's surface. (b) The distribution of the wavelengths absorbed by CO_2 and H_2O compared to the wavelengths emitted by Earth's surface.

(a)

(b)

The influence of H_2O, CO_2, and certain other atmospheric gases on Earth's temperature is called the *greenhouse effect*, because these heat-trapping gases act much like the glass of a greenhouse. Correspondingly, these gases are called *greenhouse gases*.

Water vapor makes the largest contribution to the greenhouse effect. The partial pressure of water vapor in the atmosphere varies greatly from place to place and time to time, but it is generally highest near Earth's surface and drops off very sharply with increased elevation. Because water vapor absorbs infrared radiation so strongly, it plays the major role in maintaining the atmospheric temperature at night, when the surface is emitting radiation into space and not receiving energy from the Sun. In very dry desert climates, where the water-vapor concentration is unusually low, it may be extremely hot during the day but very cold at night. In the absence of an extensive layer of water vapor to absorb and then radiate part of the infrared radiation back to Earth, the surface loses this radiation into space and cools off very rapidly.

Carbon dioxide plays a secondary, but very important, role in maintaining the surface temperature. The worldwide combustion of fossil fuels, principally coal and oil, on a prodigious scale in the modern era has sharply increased the carbon dioxide level of the atmosphere. To get a sense of the amount of CO_2 produced, for example, by the combustion of hydrocarbons and other carbon-containing substances, which are the components of fossil fuels, consider the combustion of butane, C_4H_{10}. Combustion of 1.00 g of C_4H_{10} produces 3.03 g of CO_2. ∞ (Section 3.6) Similarly, a gallon (3.78 L) of gasoline (density = 0.70 g/mL and approximate composition C_8H_{18}) produces about 8 kg (18 lb) of CO_2. Combustion of fossil fuels releases about 2.2×10^{16} g (24 billion tons) of CO_2 into the atmosphere annually with the largest quantity coming from transportation vehicles.

Much CO_2 is absorbed into oceans or used by plants in photosynthesis. Nevertheless, we are now generating CO_2 much faster than it is being absorbed or used. Chemists have monitored atmospheric CO_2 concentrations since 1958. Analysis of air trapped in ice cores taken from Antarctica and Greenland makes it possible to determine the atmospheric levels of CO_2 during the past 160,000 years. These measurements reveal that the level of CO_2 remained fairly constant from the last Ice Age, some 10,000 years ago, until roughly the beginning of the Industrial Revolution, about 300 years ago. Since that time, the concentration of CO_2 has increased by about 30% to a current high of about 386 ppm (Figure 18.13▶).

▼ **Figure 18.13 Rising CO_2 levels.** The concentration of atmospheric CO_2 has risen more than 15% since the late 1950s. These data were recorded at the Mauna Loa Observatory in Hawaii by monitoring the absorption of infrared radiation. The sawtooth shape of the graph is due to regular seasonal variations in CO_2 concentration for each year.

A consensus is emerging among scientists that this increase in the concentration of CO_2 in the atmosphere is already perturbing Earth's climate and that it may be responsible for the observed increase in the average global air temperature of 0.3 °C to 0.6 °C over the past century. Scientists often use the term climate change instead of global warming to refer to this effect, because as the Earth's temperature increases, it affects winds and ocean currents in ways that can cool some areas and warm others.

On the basis of present and expected future rates of fossil-fuel use, the atmospheric CO_2 level is expected to double from its present level sometime between 2050 and 2100. Computer models predict that this increase will result in an average global temperature increase of 1 °C to 3 °C. Major changes in global climate could result from a temperature change of this magnitude. Because so many factors go into determining climate, we cannot predict with certainty what changes will occur. Clearly, however, humanity has acquired the potential, by changing the concentration of CO_2 and other heat-trapping gases in the atmosphere, to substantially alter the climate of the planet.

The global warming threat posed by atmospheric CO_2 has sparked considerable research into ways of capturing the gas at its largest combustion sources and storing it deep underground or under the seafloor. There is also much interest in developing new ways to use CO_2 as a chemical feedstock.

A Closer Look   OTHER GREENHOUSE GASES

Although CO_2 receives most of the attention, other gases in total make an approximately equal contribution to the greenhouse effect. These gases include methane, CH_4, the hydrofluorocarbons (HFCs) such as CH_2FCF_3, and the chlorofluorocarbons (CFCs) such as CF_2Cl_2. The HFCs have replaced CFCs in a host of applications including refrigerants and air-conditioner gases. ∞ (Section 18.3) Although they do not contribute to the depletion of the ozone layer, HFCs are nevertheless strong greenhouse gases. Their total concentration in the atmosphere is still small (40 parts per trillion), but this amount is increasing about 10 % per year. Thus, these substances are becoming increasingly important contributors to the greenhouse effect.

Methane already makes a significant contribution to the greenhouse effect. Each methane molecule has about 25 times the greenhouse effect of a CO_2 molecule. Studies of atmospheric gas trapped long ago in the Greenland and Antarctic ice sheets show that the concentration of methane in the atmosphere has increased during the industrial age, from preindustrial values in the range of 0.3 to 0.7 ppm to the present value of about 1.8 ppm. The major sources of methane are associated with agriculture and fossil fuel use.

Methane is formed in biological processes that occur in low-oxygen environments. Anaerobic bacteria, which flourish in swamps and landfills, near the roots of rice plants, and in the digestive systems of cows and other ruminant animals, produce methane (Figure 18.14 ▶). It also leaks into the atmosphere during natural-gas extraction and transport (see "Chemistry Put to Work" box, Section 10.5). It is estimated that about two-thirds of present-day methane emissions, which are increasing by about 1% per year, are related to human activities.

Methane has a half-life in the atmosphere of about 10 years, whereas CO_2 is much longer-lived. This might at

first seem a good thing, but there are indirect effects to consider. Some methane is oxidized in the stratosphere, producing water vapor, a powerful greenhouse gas that is otherwise virtually absent from the stratosphere. In the troposphere methane is attacked by reactive species such as OH radicals or nitrogen oxides, eventually producing other greenhouse gases such as O_3. It has been estimated that the climate-changing effects of CH_4 are more than half those of CO_2. Given this large contribution, important reductions of the greenhouse effect could be achieved by reducing methane emissions or capturing the emissions for use as a fuel.

▲ **Figure 18.14 Methane production.** Ruminant animals, such as cows and sheep, produce methane in their digestive systems. In Australia sheep and cattle produce about 14% of the country's total greenhouse emissions.

The approximately 115 million tons of CO_2 used annually by the global chemical industry is but a small fraction of the approximately 24 billion tons of annual CO_2 emissions. However, the use of CO_2 as a raw material will probably never be great enough to reduce its atmospheric concentration.

▲ **GIVE IT SOME THOUGHT**

Explain why nighttime temperatures remain higher in locations where there is higher humidity.

18.5 THE WORLD OCEAN

Water is the most common liquid on Earth. It covers 72% of Earth's surface and is essential to life. Our bodies are about 65% water by mass. Because of extensive hydrogen bonding, water has unusually high melting and boiling points and a high heat capacity. ∞(Section 11.2) Water's highly polar character is responsible for its exceptional ability to dissolve a wide range of ionic and polar-covalent substances. Many reactions occur in water, including reactions in which H_2O itself is a reactant. Recall, for example, that H_2O can participate in acid–base reactions as either a proton donor or a proton acceptor. ∞(Section 16.3) In Chapter 20 we will see that H_2O can also participate in oxidation-reduction reactions as either a donor or an acceptor of electrons. All these properties play a role in our environment.

Seawater

The vast layer of salty water that covers so much of the planet is connected and is generally constant in composition. For this reason, oceanographers speak of a world ocean rather than of the separate oceans we learn about in geography books. The world ocean is huge. Its volume is $1.35 \times 10^9 \, \text{km}^3$. Almost all the water on Earth, 97.2%, is in the world ocean. Of the remaining 2.8%, 2.1% is in the form of ice caps and glaciers. All the freshwater—in lakes, rivers, and groundwater—amounts to only 0.6%. Most of the remaining 0.1% is in brackish (salty) water, such as that in the Great Salt Lake in Utah.

Seawater is often referred to as saline water. The **salinity** of seawater is the mass in grams of dry salts present in 1 kg of seawater. In the world ocean the salinity averages about 35. To put it another way, seawater contains about 3.5% dissolved salts by mass. The list of elements present in seawater is very long. Most, however, are present only in very low concentrations. Table 18.6▼ lists the 11 ionic species that are most abundant in seawater.

The properties of seawater—its salinity, density, and temperature—vary as a function of depth (Figure 18.15▼). Sunlight penetrates well only 200 m into the sea; the region between 200 m and 1000 m deep is the "twilight zone," where visible light is faint. Below 1000 m in depth, the ocean is pitch-black and cold, about 4 °C. The transport of heat, salt, and other chemicals throughout the ocean is influenced by these changes in the physical properties of seawater, and in turn the changes in the way heat and substances are transported affects ocean currents and the global climate.

The sea is so vast that if a substance is present in seawater to the extent of only 1 part per billion (ppb, that is, 1×10^{-6} g per kilogram of water), there is still 1×10^{12} kg of it in the world ocean. Nevertheless, the ocean is rarely used as a source of raw

TABLE 18.6 ■ Ionic Constituents of Seawater Present in Concentrations Greater than 0.001 g/kg (1 ppm)

Ionic Constituent	g/kg Seawater	Concentration (*M*)
Chloride, Cl^-	19.35	0.55
Sodium, Na^+	10.76	0.47
Sulfate, SO_4^{2-}	2.71	0.028
Magnesium, Mg^{2+}	1.29	0.054
Calcium, Ca^{2+}	0.412	0.010
Potassium, K^+	0.40	0.010
Carbon dioxide*	0.106	2.3×10^{-3}
Bromide, Br^-	0.067	8.3×10^{-4}
Boric acid, H_3BO_3	0.027	4.3×10^{-4}
Strontium, Sr^{2+}	0.0079	9.1×10^{-5}
Fluoride, F^-	0.0013	7.0×10^{-5}

*CO_2 is present in seawater as HCO_3^- and CO_3^{2-}.

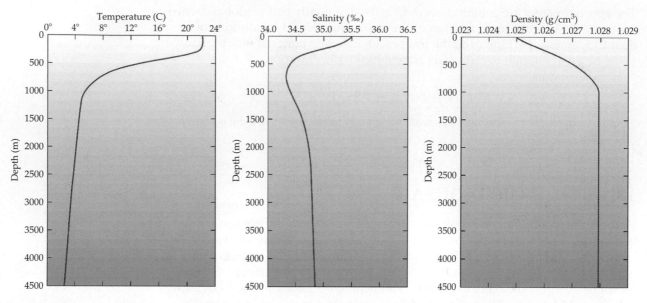

materials because the cost of extracting the desired substances is too high. Only three substances are obtained from seawater in commercially important amounts: sodium chloride, bromine (from bromide salts), and magnesium (from its salts).

Absorption of CO_2 by the ocean plays a large role in global climate. Carbon dioxide reacts with water to form carbonic acid, H_2CO_3 ∞ (Section 16.6), and so as CO_2 from the atmosphere is absorbed by the world ocean, the concentration of H_2CO_3 in the ocean increases. Most of the carbon in the ocean, however, is in the form of HCO_3^- and CO_3^{2-} ions. These ions form a buffer system that maintains the ocean's average pH between 8.0 and 8.3. The buffering capacity of the world ocean is predicted to decrease as the concentration of CO_2 in the atmosphere increases, because of the increase in H_2CO_3 concentration. With less buffering capacity, the carbonate ion can precipitate out as $CaCO_3$, one of the main ingredients in seashells. Thus, both acid–base equilibrium reactions and solubility equilibrium reactions ∞ (Section 17.4) form a complicated web of interactions that tie the ocean to the atmosphere and to the global climate.

Desalination

Because of its high salt content, seawater is unfit for human consumption and for most of the uses to which we put water. In the United States the salt content of municipal water supplies is restricted by health codes to no more than about 500 ppm (0.05% by mass). This amount is much lower than the 3.5% dissolved salts present in seawater and the 0.5% or so present in brackish water found underground in some regions. The removal of salts from seawater or brackish water to make the water usable is called **desalination**.

Water can be separated from dissolved salts by *distillation* because water is a volatile substance and the salts are nonvolatile. ∞ (Section 13.5: "A Closer Look: Ideal Solutions with Two or More Volatile Components") The principle of distillation is simple enough, but carrying out the process on a large scale presents many problems. As water is distilled from seawater, for example, the salts become more and more concentrated and eventually precipitate out.

Seawater can also be desalinated using **reverse osmosis**. Recall that osmosis is the net movement of solvent molecules, but not solute molecules, through a semipermeable membrane. ∞ (Section 13.5) In osmosis the solvent passes from the more dilute solution into the more concentrated one. However, if sufficient external pressure is applied, osmosis can be stopped and, at still higher pressures, reversed. When this occurs, solvent passes from the more concentrated into the more dilute solution. In a modern reverse-osmosis facility tiny

◀ **Figure 18.16 Reverse osmosis.** (a) A room inside a reverse-osmosis desalination plant. (b) Each cylinder shown in (a) is called a permeator and contains several million tiny hollow fibers. (c) When seawater is introduced under pressure into a permeator, water passes through the fiber wall into the fibers and is thereby separated from all the ions initially present in the seawater.

hollow fibers are used as the semipermeable membrane. Water is introduced under pressure into the fibers, and desalinated water is recovered, as illustrated in Figure 18.16▲.

Research and development into improving desalination technologies is ongoing, and already an estimated 11,000 desalination plants exist in some 120 countries around the world. The world's largest desalination plant is located in Jubail, Saudi Arabia. This plant provides 50% of that country's drinking water by using reverse osmosis to desalinate seawater from the Persian Gulf. Such plants are becoming increasingly common in the United States. The largest, near Tampa Bay, Florida, is scheduled to be functional in 2008 and will produce 35 million gallons of drinking water a day by reverse osmosis when fully constructed. Small-scale, manually operated reverse-osmosis desalinators are now also available for use in camping and traveling, and at sea (Figure 18.17▶).

18.6 FRESHWATER

The United States is fortunate in its abundance of freshwater—1.7×10^{15} liters (660 trillion gallons) is the estimated reserve, which is renewed by rainfall. An estimated 9×10^{11} liters of freshwater is used every day in the United States. Most of this is used for agriculture (41%) and hydroelectric power (39%), with small amounts for industry (6%), household needs (6%), and drinking water (1%). An adult needs about 2 liters of water per day for drinking. In the United States our daily use of water per person far exceeds this subsistence level, amounting to an average of about 300 L/day for personal consumption and hygiene. We use about 8 L/person for cooking and drinking, about 120 L/person for cleaning (bathing, laundering, and housecleaning), 80 L/person for flushing toilets, and 80 L/person for watering lawns.

The total amount of freshwater on Earth is not a very large fraction of the total water present. Indeed, freshwater is one of our most precious resources. It forms by evaporation from the oceans and the land. The water vapor that accumulates in the atmosphere is transported by global atmospheric circulation, eventually returning to Earth as rain, snow, and other forms of precipitation.

▲ **Figure 18.17 A portable desalination device.** This hand-operated water desalinator works by reverse osmosis. It can produce 4.5 L (1.2 gal) of pure water from seawater in an hour.

As rain falls and as water runs off the land on its way to the oceans, it dissolves a variety of cations (mainly Na^+, K^+, Mg^{2+}, Ca^{2+}, and Fe^{2+}), anions (mainly Cl^-, SO_4^{2-}, and HCO_3^-), and gases (principally O_2, N_2, and CO_2). As we use water, it becomes laden with additional dissolved material, including the wastes of human society. As our population and output of environmental pollutants increase, we find that we must spend ever-increasing amounts of money and resources to guarantee a supply of freshwater.

Dissolved Oxygen and Water Quality

The amount of dissolved O_2 in water is an important indicator of water quality. Water fully saturated with air at 1 atm and 20 °C contains about 9 ppm of O_2. Oxygen is necessary for fish and most other aquatic life. Cold-water fish require that the water contain at least 5 ppm of dissolved oxygen for survival. Aerobic bacteria consume dissolved oxygen to oxidize organic materials and so meet their energy requirements. The organic material that the bacteria are able to oxidize is said to be **biodegradable**. This oxidation occurs by a complex set of chemical reactions, and the organic material disappears gradually.

Excessive quantities of biodegradable organic materials in water are detrimental because they deplete the water of the oxygen necessary to sustain normal animal life. Typical sources of these biodegradable materials, which are called *oxygen-demanding wastes*, include sewage, industrial wastes from food-processing plants and paper mills, and effluent (liquid waste) from meatpacking plants.

In the presence of oxygen, the carbon, hydrogen, nitrogen, sulfur, and phosphorus in biodegradable material end up mainly as CO_2, HCO_3^-, H_2O, NO_3^-, SO_4^{2-}, and phosphates. The formation of these oxidation products sometimes reduces the amount of dissolved oxygen to the point where aerobic bacteria can no longer survive. Anaerobic bacteria then take over the decomposition process, forming CH_4, NH_3, H_2S, PH_3, and other products, several of which contribute to the offensive odors of some polluted waters.

Plant nutrients, particularly nitrogen and phosphorus, contribute to water pollution by stimulating excessive growth of aquatic plants. The most visible results of excessive plant growth are floating algae and murky water. More significantly, however, as plant growth becomes excessive, the amount of dead and decaying plant matter increases rapidly, a process called *eutrophication* (Figure 18.18 ◄). The decay of plants consumes O_2 as the plants are biodegraded, leading to the depletion of oxygen in the water. Without sufficient supplies of oxygen, the water, in turn, cannot sustain any form of animal life. The most significant sources of nitrogen and phosphorus compounds in water are domestic sewage (phosphate-containing detergents and nitrogen-containing body wastes), runoff from agricultural land (fertilizers containing both nitrogen and phosphorus), and runoff from livestock areas (animal wastes containing nitrogen).

▲ Figure 18.18 Eutrophication. The growth of algae and duckweed in this pond is due to agricultural wastes. The wastes feed the growth of the algae and weeds, which deplete the oxygen in the water, a process called eutrophication. A eutrophic lake cannot support fish.

▲ GIVE IT SOME THOUGHT

One common test for water quality involves measuring dissolved oxygen, storing the water in a closed container at a constant temperature for five days, and then remeasuring the dissolved oxygen. If such a test shows a considerable decrease in dissolved oxygen over the five-day period, what can we conclude about the nature of the pollutants present?

Treatment of Municipal Water Supplies

The water needed for domestic uses, agriculture, and industrial processes is taken either from naturally occurring lakes, rivers, and underground sources or from reservoirs. Much of the water that finds its way into municipal water

▲ Figure 18.19 **Common steps in treating water for a public water system.**

systems is "used" water: It has already passed through one or more sewage systems or industrial plants. Consequently, this water must be treated before it is distributed to our faucets. Municipal water treatment usually involves five steps: coarse filtration, sedimentation, sand filtration, aeration, and sterilization. Figure 18.19▲ shows a typical treatment process.

After coarse filtration through a screen, the water is allowed to stand in large settling tanks in which finely divided sand and other minute particles can settle out. To aid in removing very small particles, the water may first be made slightly basic by adding CaO. Then $Al_2(SO_4)_3$ is added. The aluminum sulfate reacts with OH^- ions to form a spongy, gelatinous precipitate of $Al(OH)_3$ ($K_{sp} = 1.3 \times 10^{-33}$). This precipitate settles slowly, carrying suspended particles down with it, thereby removing nearly all finely divided matter and most bacteria. The water is then filtered through a sand bed. Following filtration, the water may be sprayed into the air to hasten the oxidation of dissolved organic substances.

The final stage of the operation normally involves treating the water with a chemical agent to ensure the destruction of bacteria. Ozone is most effective, but it must be generated at the place where it is used. Chlorine, Cl_2, is therefore more convenient. Chlorine can be shipped in tanks as a liquefied gas and dispensed from the tanks through a metering device directly into the water supply. The amount used depends on the presence of other substances with which the chlorine might react and on the concentrations of bacteria and viruses to be removed. The sterilizing action of chlorine is probably due not to Cl_2 itself, but to hypochlorous acid, which forms when chlorine reacts with water:

$$Cl_2(aq) + H_2O(l) \longrightarrow HClO(aq) + H^+(aq) + Cl^-(aq) \qquad [18.15]$$

As many as a billion people worldwide lack access to clean water. According to the United Nations, 95% of the world's cities still dump raw sewage into their water supplies. Thus, it should come as no surprise to know that 80% of all the health maladies in developing countries can be traced back to waterborne diseases associated with unsanitary water.

One promising development is a device called the LifeStraw (Figure 18.20▶). When a person sucks water through the straw, the water first encounters a textile prefilter with a mesh opening of 100 μm followed by a second textile filter with a mesh opening of 15 μm. These filters remove debris and even clusters of bacteria. The water next encounters a chamber of iodine-impregnated beads, where bacteria, viruses, and parasites are killed. Finally, the water passes through granulated active carbon, which removes the smell of iodine as well as the parasites that have not been taken by the prefilter or killed by the iodine.

▲ Figure 18.20 **LifeStraw.** A LifeStraw is a drinking straw that purifies water as it is drunk.

ater containing a relatively high concentration of Ca^{2+}, Mg^{2+}, and other divalent cations is called **hard water**. Although the presence of these ions is generally not a health threat, they can make water unsuitable for some household and industrial uses. For example, these ions react with soaps to form an insoluble soap scum, the stuff of bathtub rings. In addition, mineral deposits may form when water containing these ions is heated. When water containing calcium ions and bicarbonate ions is heated, some carbon dioxide is driven off. As a result, the solution becomes less acidic and insoluble calcium carbonate forms:

$$Ca^{2+}(aq) + 2\,HCO_3^-(aq) \longrightarrow CaCO_3(s) + CO_2(g) + H_2O(l)$$

The solid $CaCO_3$ coats the surface of hot-water systems and teakettles, thereby reducing heating efficiency. These deposits, called *scale*, can be especially serious in boilers where water is heated under pressure in pipes running through a furnace. Formation of scale reduces the efficiency of heat transfer and reduces the flow of water through pipes (Figure 18.21 ▼).

The removal of the ions that cause hard water is called *water softening*. Not all municipal water supplies require water softening. In those that do, the water is generally taken from underground sources in which it has had considerable contact with limestone, $CaCO_3$, and other minerals containing Ca^{2+}, Mg^{2+}, and Fe^{2+}. The **lime-soda process** is used for

◀ **Figure 18.21 Scale formation.** A section of water pipe that has been coated on the inside with $CaCO_3$ and other insoluble salts deposited from hard water.

large-scale municipal water-softening operations. The water is treated with lime, CaO [or slaked lime, $Ca(OH)_2$], and soda ash, Na_2CO_3. These chemicals precipitate Ca^{2+} as $CaCO_3$ ($K_{sp} = 4.5 \times 10^{-9}$) and Mg^{2+} as $Mg(OH)_2$ ($K_{sp} = 1.6 \times 10^{-12}$):

$$Ca^{2+}(aq) + CO_3^{2-}(aq) \longrightarrow CaCO_3(s)$$

$$Mg^{2+}(aq) + 2\,OH^-(aq) \longrightarrow Mg(OH)_2(s)$$

Ion exchange is a typical household method for water softening. In this procedure the hard water is passed through a bed of an ion-exchange resin: plastic beads with covalently bound anion groups such as $-COO^-$ or $-SO_3^-$. These negatively charged groups have Na^+ ions attached to balance their charges. The Ca^{2+} and other cations in the hard water are attracted to the anionic groups and displace the lower-charged Na^+ into the water. Thus, one type of ion is exchanged for another. To maintain charge balance, 2 Na^+ enter the water for each Ca^{2+} removed. If we represent the resin with its anionic site as $R-COO^-$, we can write the equation for the process as follows:

$$2\,Na(R\text{--}COO)(s) + Ca^{2+}(aq) \rightleftharpoons$$
$$Ca(R\text{--}COO)_2(s) + 2\,Na^+(aq)$$

Water softened in this way contains an increased concentration of Na^+. Although Na^+ does not form precipitates or cause other problems associated with hard-water cations, individuals concerned about their sodium intake, such as those who have high blood pressure (hypertension), should avoid drinking water softened in this way.

When all the available Na^+ have been displaced from the ion-exchange resin, the resin is regenerated by flushing it with a concentrated solution of NaCl. Homeowners can do this by charging their units with large amounts of NaCl(s), which can be purchased at most grocery stores. The high concentration of Na^+ forces the equilibrium shown in the earlier equation to shift to the left, causing the Na^+ to displace the hard-water cations, which are flushed down the drain.

18.7 GREEN CHEMISTRY

The planet on which we live is, to a large extent, a closed system, one that exchanges energy but not matter with its environment. If humankind is to thrive in the future, all the processes we carry out should be in balance with Earth's natural processes and physical resources. This goal requires that no toxic materials are released to the environment, that our needs are met with renewable resources, and that we consume the least possible amount of energy. Although the chemical industry is but a small part of human activity, chemical processes are involved in nearly all aspects of modern life. Chemistry is therefore at the heart of efforts to accomplish these goals.

Green chemistry is an initiative that promotes the design and application of chemical products and processes that are compatible with human health and that preserve the environment. Some of the major principles that govern green chemistry are the following:

- The potential hazards of substances should be a major consideration in the design of chemical processes. It is better to prevent waste than to treat or clean it up after it has been created.

- The methods employed in synthesizing substances should generate as little waste product as possible.
- Chemical processes should be designed to be as energy efficient as possible, avoiding high temperatures and pressures. Thus, catalysts should be sought that facilitate reactions under mild reaction conditions.
- The raw materials for chemical processes should be renewable whenever technically and economically feasible.

To illustrate how green chemistry works in practice, consider the manufacture of styrene, a very important building block in the formation of many polymers, including the expanded polystyrene packages used to pack eggs and restaurant takeout food. The global demand for styrene is more than 25 million metric tons per year. For many years styrene has been produced in a two-step process. In the first step, benzene and ethylene react in the presence of a solid catalyst, forming ethyl benzene. In the second step, the ethyl benzene is mixed with high temperature steam and passed over an iron oxide catalyst at 625 °C yielding styrene:

Benzene   Ethyl benzene   Styrene

This process has several shortcomings. One is that both benzene, which is formed from crude oil, and ethylene, which is formed from natural gas, are high-priced starting materials for a product that should be a low-priced commodity. From an environmental point of view, the high toxicity of benzene and the high energy cost overall for the process are both undesirable.

Researchers at Exelus Inc. have demonstrated a process that bypasses some of these shortcomings. The two-step process is replaced by a one-step process in which toluene is reacted with methanol at 425 °C over a special catalyst:

Toluene   Styrene

The new process saves money because toluene and methanol are less expensive than benzene and ethylene, and the reaction conditions require much lower energy input. Additional benefits are that the methanol could be produced from biomass, and that benzene is replaced by less toxic toluene. The hydrogen formed in the reaction can be recycled as a source of energy. This example demonstrates how finding the right catalyst was key in discovering the new process.

Let's consider some other examples in which green chemistry can operate to improve environmental quality.

Solvents and Reagents

A major area of concern in chemical processes is the use of volatile organic compounds as solvents for reactions. Generally, the solvent in which a reaction is run is not consumed in the reaction, and there are unavoidable releases of solvent into the atmosphere even in the most carefully controlled processes. Further, the solvent may be toxic or may decompose to some extent during the reaction, thus creating waste products. The use of supercritical fluids (Section 11.4: "Chemistry Put to Work: Supercritical Fluid Extraction") represents a way to replace conventional solvents with other reagents. Recall that a supercritical fluid is an unusual state of matter that has properties of both a gas and a liquid. ∞ (Section 11.4) Water and carbon dioxide are the two most popular choices as supercritical fluid solvents. One recently developed industrial process, for example, replaces chlorofluorocarbon solvents with liquid or supercritical CO_2 in the production of polytetrafluoroethylene ($[CF_2CF_2]_n$, sold as Teflon®). The chlorofluorocarbon solvents have harmful effects on Earth's ozone layer (Section 18.3). Though CO_2 is a greenhouse gas, no new CO_2 need be manufactured for use as a supercritical fluid solvent.

As a further example, *para*-xylene is oxidized to form terephthalic acid, which in turn is used to make polyethylene terephthalate (PET) plastic and polyester fiber (Section 12.6, Table 12.5):

$$CH_3-\langle\bigcirc\rangle-CH_3 + 3\,O_2 \xrightarrow[\text{Catalyst}]{190°C,\ 20\ atm} HO-\overset{\displaystyle O}{\overset{\|}{C}}-\langle\bigcirc\rangle-\overset{\displaystyle O}{\overset{\|}{C}}-OH + 2\,H_2O$$

para-Xylene   Terephthalic acid

This commercial process requires pressurization and a relatively high temperature. Oxygen is the oxidizing agent, and acetic acid (CH_3COOH) is the solvent. An alternative route employs supercritical water as the solvent (Table 11.5) and hydrogen peroxide as the oxidant. This alternative process has several potential advantages, most particularly the elimination of acetic acid as solvent and the use of a different oxidizing agent. Whether it can successfully replace the existing commercial process, however, depends on many factors, which will require further study.

▲ GIVE IT SOME THOUGHT

We noted earlier that increasing carbon dioxide levels contribute to global warming, which seems like a bad thing, but now we are saying that using carbon dioxide in industrial processes is a good thing for the environment. Explain this seeming contradiction.

Other Processes

Many processes that are important in modern society use chemicals not found in nature. Let's briefly examine two of these processes, dry cleaning and the coating of automotive bodies to prevent corrosion, and consider alternatives being developed to reduce harmful environmental impacts.

Dry cleaning of clothing typically uses a chlorinated organic solvent such as tetrachloroethylene ($Cl_2C{=}CCl_2$), which may cause cancer in humans. The widespread use of this and related solvents in dry cleaning, metal cleaning, and other industrial processes has contaminated groundwater in some areas. Alternative dry-cleaning methods that employ liquid CO_2, along with special cleaning agents, are being successfully commercialized (Figure 18.22 ◄).

The metal bodies of cars are coated extensively during manufacture to prevent corrosion. One of the key steps is the electrodeposition of a layer of metal ions that creates an interface between the vehicle body and the polymeric coatings that serve as the undercoat for painting. In the past, lead was the metal of choice for inclusion in the electrodeposition mixture. Because lead is highly toxic, however, its use in other paints and coatings has been virtually eliminated, and a relatively nontoxic yttrium hydroxide alternative to lead has been developed as an automotive coating (Figure 18.23 ▶). When this coating is subsequently heated, the hydroxide is converted to the oxide, producing an insoluble ceramiclike coating. ∞ (Section 12.4)

▲ Figure 18.22 Green chemistry for your clothes. This dry-cleaning apparatus employs liquid CO_2 as the solvent.

The Challenges of Water Purification

Access to clean water is essential to the workings of a stable, thriving society. We have seen in the previous section that disinfection of water is an important step in water treatment for human consumption. Water disinfection is one of the greatest public health innovations in human history. It has dramatically decreased the incidences of waterborne bacterial diseases such as cholera and typhus. But this great benefit comes at a price.

In 1974 scientists in both Europe and the United States discovered that chlorination of water produces a group of byproducts that had previously gone undetected. These by-products are called *trihalomethanes* (THMs) because all have a single carbon atom and three halogen atoms: $CHCl_3$, $CHCl_2Br$, $CHClBr_2$, and $CHBr_3$. These and many other chlorine- and bromine-containing organic substances are produced by the reaction of aqueous chlorine with organic materials present in nearly all natural waters, as well as with substances that are by-products of human activity. Recall that chlorine dissolves in water to form HClO, which is the active oxidizing agent ∞ (Section 7.8):

▲ Figure 18.23 **Green chemistry for your car.** An automobile body receives a corrosion-protection coating containing yttrium in place of lead.

$$Cl_2(g) + H_2O(l) \longrightarrow HClO(aq) + H^+(aq) + Cl^-(aq) \qquad [18.16]$$

HClO in turn reacts with organic substances to form the THMs. Bromine enters through the reaction of HClO with dissolved bromide ion:

$$HOCl(aq) + Br^-(aq) \longrightarrow HBrO(aq) + Cl^-(aq) \qquad [18.17]$$

$HBrO(aq)$ halogenates organic substances analogously to $HClO(aq)$.

Some THMs and other halogenated organic substances are suspected carcinogens; others interfere with the body's endocrine system. As a result, the World Health Organization and the Environmental Protection Agency (EPA) have placed concentration limits of 80 $\mu g/L$ (80 ppb) on the total quantity of such substances in drinking water. The goal is to reduce the levels of THMs and related substances in the drinking water supply while preserving the antibacterial effectiveness of the water treatment. In some cases simply lowering the concentration of chlorine may provide adequate disinfection while reducing the concentrations of THMs formed. Alternative oxidizing agents, such as ozone (O_3) or chlorine dioxide (ClO_2), produce less of the halogenated substances, but they have their own disadvantages. For example, each is capable of oxidizing aqueous bromide, as shown here for ozone:

$$O_3(aq) + Br^-(aq) + H_2O(l) \longrightarrow HBrO(aq) + O_2(aq) + OH^-(aq) \qquad [18.18]$$

$$HBrO(aq) + 2\,O_3(aq) \longrightarrow BrO_3^-(aq) + 2\,O_2(aq) + H^+(aq) \qquad [18.19]$$

As we have seen, $HBrO(aq)$ is capable of reacting with dissolved organic substances to form halogenated organic compounds. Furthermore, bromate ion has been shown to cause cancer in animal tests.

There seem to be no completely satisfactory alternatives to chlorination or ozonation at present. The risks of cancer from THMs and related substances in municipal water are very low, however, compared to the risks of cholera, typhus, and gastrointestinal disorders from untreated water. When the water supply is cleaner to begin with, less disinfectant is needed; thus, the danger of contamination through disinfection is reduced. Once the THMs are formed, their concentrations in the water supply can be reduced by aeration because the THMs are more volatile than water. Alternatively, they can be removed by adsorption onto activated charcoal or other adsorbents.

■ **SAMPLE INTEGRATIVE EXERCISE** | Putting Concepts Together

(a) Acids from acid rain or other sources are no threat to lakes in areas where the rock is limestone (calcium carbonate), which can neutralize the excess acid. Where the rock is granite, however, no such neutralization occurs. How does the limestone neutralize the acid? (b) Acidic water can be treated with basic substances to increase the pH, although such a procedure is usually only a temporary cure. Calculate the minimum mass of lime, CaO, needed to adjust the pH of a small lake ($V = 4 \times 10^9$ L) from 5.0 to 6.5. Why might more lime be needed?

SOLUTION

Analyze: We need to remember what a neutralization reaction is, and calculate the amount of a substance needed to effect a certain change in pH.

Plan: For (a), we need to think about how acid can react with calcium carbonate, which evidently does not happen with acid and granite. For (b), we need to think about what reaction would happen with acid and CaO and do stoichiometric calculations. From the proposed change in pH, we can calculate the change in proton concentration needed and then figure out how much CaO it would take to do the reaction.

Solve:

(a) The carbonate ion, which is the anion of a weak acid, is basic. ∞ (Sections 16.2 and 16.7) Thus, the carbonate ion, CO_3^{2-}, reacts with $H^+(aq)$. If the concentration of $H^+(aq)$ is small, the major product is the bicarbonate ion, HCO_3^-. If the concentration of $H^+(aq)$ is higher, however, H_2CO_3 forms and decomposes to CO_2 and H_2O. ∞ (Section 4.3)

(b) The initial and final concentrations of $H^+(aq)$ in the lake are obtained from their pH values:

$$[H^+]_{initial} = 10^{-5.0} = 1 \times 10^{-5}\ M \quad \text{and} \quad [H^+]_{final} = 10^{-6.5} = 3 \times 10^{-7}\ M$$

Using the volume of the lake, we can calculate the number of moles of $H^+(aq)$ at both pH values:

$$(1 \times 10^{-5}\ \text{mol/L})(4.0 \times 10^9\ \text{L}) = 4 \times 10^4\ \text{mol}$$

$$(3 \times 10^{-7}\ \text{mol/L})(4.0 \times 10^9\ \text{L}) = 1 \times 10^3\ \text{mol}$$

Hence, the change in the amount of $H^+(aq)$ is 4×10^4 mol $- 1 \times 10^3$ mol $\approx 4 \times 10^4$ mol.

Let's assume that all the acid in the lake is completely ionized, so that only the free $H^+(aq)$ measured by the pH needs to be neutralized. We will need to neutralize at least that much acid, although there may be a great deal more acid in the lake than that.

The oxide ion of CaO is very basic. ∞ (Section 16.5) In the neutralization reaction 1 mol of O^{2-} reacts with 2 mol of H^+ to form H_2O. Thus, 4×10^4 mol of H^+ requires the following number of grams of CaO:

$$(4 \times 10^4\ \text{mol}\ H^+)\left(\frac{1\ \text{mol CaO}}{2\ \text{mol}\ H^+}\right)\left(\frac{56.1\ \text{g CaO}}{1\ \text{mol CaO}}\right) = 1 \times 10^6\ \text{g CaO}$$

This amounts to slightly more than a ton of CaO. That would not be very costly because CaO is an inexpensive base, selling for less than $100 per ton when purchased in large quantities. The amount of CaO calculated above, however, is the very minimum amount needed because there are likely to be weak acids in the water that must also be neutralized. This liming procedure has been used to adjust the pH of some small lakes to bring their pH into the range necessary for fish to live. The lake in our example would be about a half mile long, a half mile wide, and have an average depth of 20 ft.

CHAPTER REVIEW

SUMMARY AND KEY TERMS

Sections 18.1 and 18.2 In these sections we examined the physical and chemical properties of Earth's atmosphere. The complex temperature variations in the atmosphere give rise to four regions, each with characteristic properties. The lowest of these regions, the **troposphere**, extends from Earth's surface up to an altitude of about 12 km. Above the troposphere, in order of increasing altitude, are the **stratosphere**, mesosphere, and thermosphere. In the upper reaches of the atmosphere, only the simplest chemical species can survive the bombardment of highly energetic particles and radiation from the Sun. The average molecular weight of the atmosphere at high elevations is lower than that at Earth's surface because the lightest atoms and molecules diffuse upward and also because of **photodissociation**, which is the breaking of bonds in molecules because of the absorption of light. Absorption of radiation may also lead to the formation of ions via **photoionization**.

Section 18.3 Ozone is produced in the upper atmosphere from the reaction of atomic oxygen with O_2. Ozone is itself decomposed by absorption of a photon or by reaction with an active species such as Cl. **Chlorofluorocarbons** can undergo photodissociation in the stratosphere, introducing atomic chlorine, which is capable of catalytically destroying ozone. A marked reduction in the ozone level in the upper atmosphere would have serious adverse consequences because the ozone layer filters out certain wavelengths of ultraviolet light that are not removed by any other atmospheric component.

Section 18.4 In the troposphere the chemistry of trace atmospheric components is of major importance. Many of these minor components are pollutants. Sulfur dioxide is one of the more noxious and prevalent examples. It is oxidized in air to form sulfur trioxide, which, upon dissolving in water, forms sulfuric acid. The oxides of sulfur are major contributors to **acid rain**. One method of preventing the escape of SO_2 from industrial operations is to react the SO_2 with CaO to form calcium sulfite ($CaSO_3$).

Carbon monoxide (CO) is found in high concentrations in automobile engine exhaust and in cigarette smoke. CO is a health hazard because it can form a strong bond with **hemoglobin** and thus reduce the capacity for blood to transfer oxygen from the lungs.

Photochemical smog is a complex mixture of components in which both nitrogen oxides and ozone play important roles. Smog components are generated mainly in automobile engines, and smog control consists largely of controlling auto emissions.

Carbon dioxide and water vapor are the major components of the atmosphere that strongly absorb infrared radiation. CO_2 and H_2O are therefore critical in maintaining Earth's temperature. The concentrations of CO_2 and other so-called greenhouse gases in the atmosphere are thus important in determining worldwide climate. Because of the extensive combustion of fossil fuels (coal, oil, and natural gas), the concentration of carbon dioxide in the atmosphere is steadily increasing.

Section 18.5 Seawater contains about 3.5% by mass of dissolved salts and is described as having a **salinity** of 35. Seawater's density and salinity vary with depth. Because most of the world's water is in the oceans, humans may eventually look to the seas for freshwater. **Desalination** is the removal of dissolved salts from seawater or brackish water to make it fit for human consumption. Desalination may be accomplished by distillation or by **reverse osmosis**.

Section 18.6 Freshwater contains many dissolved substances, including dissolved oxygen, which is necessary for fish and other aquatic life. Substances that are decomposed by bacteria are said to be **biodegradable**. Because the oxidation of biodegradable substances by aerobic bacteria consumes dissolved oxygen, these substances are called oxygen-demanding wastes. The presence of an excess amount of oxygen-demanding wastes in water can sufficiently deplete the dissolved oxygen to kill fish and produce offensive odors. Plant nutrients can contribute to the problem by stimulating the growth of plants that become oxygen-demanding wastes when they die.

The water available from freshwater sources may require treatment before it can be used domestically. The several steps generally used in municipal water treatment include coarse filtration, sedimentation, sand filtration, aeration, sterilization, and sometimes water softening. Water softening is required when the water contains ions such as Mg^{2+} and Ca^{2+}, which react with soap to form soap scum. Water containing such ions is called **hard water**. The **lime-soda process**, which involves adding CaO and Na_2CO_3 to hard water, is sometimes used for large-scale municipal water softening. Individual homes usually rely on **ion exchange**, a process by which hard-water ions are exchanged for Na^+ ions.

Section 18.7 The **green chemistry** initiative promotes the design and application of chemical products and processes that are compatible with human health and that preserve the environment. The areas in which the principles of green chemistry can operate to improve environmental quality include choices of solvents and reagents for chemical reactions, development of alternative processes, and improvements in existing systems and practices.

KEY SKILLS

- Describe the regions of Earth's atmosphere in terms of how temperature varies with altitude.
- Describe the composition of the atmosphere in terms of the major components in dry air at sea level.
- Calculate concentrations of gases in parts per million (ppm).
- Describe the processes of photodissociation and photoionization and their role in the upper atmosphere.
- Use bond energies and ionization energies to calculate the minimum frequency or maximum wavelength needed to cause photodissociation or photoionization.
- Explain the role of ozone in the upper atmosphere.
- Explain how chlorofluorocarbons (CFCs) are involved in depleting the ozone layer.

- Describe the origins and behavior of sulfur oxides, carbon monoxide, and nitrogen oxides as air pollutants, including the generation of acid rain and photochemical smog.
- Describe how water and carbon dioxide in the atmosphere affect atmospheric temperature via the greenhouse effect.
- Explain what is meant by the salinity of water and describe the process of reverse osmosis as a means of desalination.
- List the major cations, anions, and gases present in natural waters and describe the relationship between dissolved oxygen and water quality.
- List the main steps involved in treating water for domestic uses.
- Describe the main goals of green chemistry.

VISUALIZING CONCEPTS

18.1 At room temperature (298 K) and 1 atm pressure, one mole of an ideal gas occupies 22.4 L. ∞ (Section 10.4) **(a)** Looking back at Figure 18.1, do you predict that 1 mole of an ideal gas in the middle of the stratosphere would occupy a greater or smaller volume than 22.4 L? **(b)** Looking at Figure 18.1(a), we see that the temperature is lower at 85 km altitude than at 50 km. Does this mean that one mole of an ideal gas would occupy less volume at 85 km than at 50 km? Explain. [Section 18.1]

18.2 Molecules in the upper atmosphere tend to contain double and triple bonds rather than single bonds. Suggest an explanation. [Section 18.2]

18.3 Why does ozone concentration in the atmosphere vary as a function of altitude (see Figure 18.4)? [Section 18.3]

18.4 You are working with an artist who has been commissioned to make a sculpture for a big city in the eastern United States. The artist is wondering what material to use to make her sculpture, because she has heard that acid rain in the eastern U.S. might destroy it over time. You take samples of granite, marble, bronze, and other materials, and place them outdoors for a long time in the big city. You periodically examine the appearance and measure the mass of the samples. **(a)** What observations would lead you to conclude that one, or more, of the materials were well-suited for the sculpture? **(b)** What chemical process (or processes) is (are) the most likely responsible for any observed changes in the materials? [Section 18.4]

18.5 How does carbon dioxide interact with the world ocean? [Section 18.5]

18.6 The following picture represents an ion-exchange column, in which water containing "hard" ions, such as Ca^{2+}, is added to the top of the column, and water containing "soft" ions such as Na^+ come out the bottom. Explain what is happening in the column. [Section 18.6]

Add hard water to top of column

Special plastic bead

Soft water comes out the bottom

18.7 Describe the basic goals of green chemistry. [Section 18.7]

18.8 One mystery in environmental science is the imbalance in the "carbon dioxide budget." Considering only human activities, scientists have estimated that 1.6 billion metric tons of CO_2 is added to the atmosphere every year because of deforestation (plants use CO_2, and fewer plants will leave more CO_2 in the atmosphere). Another 5.5 billion tons per year is put into the atmosphere because of burning fossil fuels. It is further estimated (again, considering only human activities) that the atmosphere actually takes up about 3.3 billion tons of CO_2 per year, while the oceans take up 2 billion tons per year, leaving about 1.8 billion tons of CO_2 per year unaccounted for. This "missing" CO_2 is assumed to be taken up by the "land." What do you think might be happening?

EXERCISES

Earth's Atmosphere

18.9 **(a)** What is the primary basis for the division of the atmosphere into different regions? **(b)** Name the regions of the atmosphere, indicating the altitude interval for each one.

18.10 **(a)** How are the boundaries between the regions of the atmosphere determined? **(b)** Explain why the stratosphere, which is more than 20 miles thick, has a smaller total mass than the troposphere, which is less than 10 miles thick.

18.11 Air pollution in the Mexico City metropolitan area is among the worst in the world. The concentration of ozone in Mexico City has been measured at 441 ppb (0.441 ppm). Mexico City sits at an altitude of 7400 feet, which means its atmospheric pressure is only 0.67 atm. Calculate the partial pressure of ozone at 441 ppb if the atmospheric pressure is 0.67 atm.

18.12 From the data in Table 18.1, calculate the partial pressures of carbon dioxide and argon when the total atmospheric pressure is 96.5 kPa.

18.13 The average concentration of carbon monoxide in air in an Ohio city in 2006 was 3.5 ppm. Calculate the number of CO molecules in 1.0 L of this air at a pressure of 755 torr and a temperature of 18 °C.

18.14 (a) From the data in Table 18.1, what is the concentration of neon in the atmosphere in ppm? (b) What is the concentration of neon in the atmosphere in molecules per L, assuming an atmospheric pressure of 733 torr and a temperature of 292 K?

The Upper Atmosphere; Ozone

18.15 The dissociation energy of a carbon–bromine bond is typically about 210 kJ/mol. What is the maximum wavelength of photons that can cause C—Br bond dissociation?

18.16 In CF_3Cl the C—Cl bond-dissociation energy is 339 kJ/mol. In CCl_4 the C—Cl bond-dissociation energy is 293 kJ/mol. What is the range of wavelengths of photons that can cause C—Cl bond rupture in one molecule but not in the other?

18.17 (a) Distinguish between *photodissociation* and *photoionization*. (b) Use the energy requirements of these two processes to explain why photodissociation of oxygen is more important than photoionization of oxygen at altitudes below about 90 km.

18.18 Why is the photodissociation of N_2 in the atmosphere relatively unimportant compared with the photodissociation of O_2?

18.19 What is a hydrofluorocarbon? Why are these compounds potentially less harmful to the ozone layer than CFCs?

18.20 Draw the Lewis structure for the chlorofluorocarbon CFC-11, $CFCl_3$. What chemical characteristics of this substance allow it to effectively deplete stratospheric ozone?

18.21 (a) Why is the fluorine present in chlorofluorocarbons not a major contributor to depletion of the ozone layer? (b) What are the chemical forms in which chlorine exists in the stratosphere following cleavage of the carbon–chlorine bond?

18.22 Would you expect the substance $CFBr_3$ to be effective in depleting the ozone layer, assuming that it is present in the stratosphere? Explain.

Chemistry of the Troposphere

18.23 For each of the following gases, make a list of known or possible naturally occurring sources: (a) CH_4, (b) SO_2, (c) NO, (d) CO.

18.24 Why is rainwater naturally acidic, even in the absence of polluting gases such as SO_2?

18.25 (a) Write a chemical equation that describes the attack of acid rain on limestone, $CaCO_3$. (b) If a limestone sculpture were treated to form a surface layer of calcium sulfate, would this help to slow down the effects of acid rain? Explain.

18.26 The first stage in corrosion of iron upon exposure to air is oxidation to Fe^{2+}. (a) Write a balanced chemical equation to show the reaction of iron with oxygen and protons from acid rain. (b) Would you expect the same sort of reaction to occur with a silver surface? Explain.

18.27 Alcohol-based fuels for automobiles lead to the production of formaldehyde (CH_2O) in exhaust gases. Formaldehyde undergoes photodissociation, which contributes to photochemical smog:

$$CH_2O + h\nu \longrightarrow CHO + H$$

The maximum wavelength of light that can cause this reaction is 335 nm. (a) In what part of the electromagnetic spectrum is light with this wavelength found?

(b) What is the maximum strength of a bond, in kJ/mol, that can be broken by absorption of a photon of 335-nm light? (c) Compare your answer from part (b) to the appropriate value from Table 8.4. What do you conclude about the C—H bond energy in formaldehyde? (d) Write out the formaldehyde photodissociation reaction, showing Lewis-dot structures.

18.28 An important reaction in the formation of photochemical smog is the photodissociation of NO_2:

$$NO_2 + h\nu \longrightarrow NO(g) + O(g)$$

The maximum wavelength of light that can cause this reaction is 420 nm. (a) In what part of the electromagnetic spectrum is light with this wavelength found? (b) What is the maximum strength of a bond, in kJ/mol, that can be broken by absorption of a photon of 420-nm light? (c) Write out the photodissociation reaction showing Lewis-dot structures.

18.29 Explain why increasing concentrations of CO_2 in the atmosphere affect the quantity of energy leaving Earth but do not affect the quantity entering from the Sun.

18.30 (a) With respect to absorption of radiant energy, what distinguishes a greenhouse gas from a nongreenhouse gas? (b) CH_4 is a greenhouse gas, but Ar is not. How might the molecular structure of CH_4 explain why it is a greenhouse gas?

The World Ocean

18.31 What is the molarity of Na^+ in a solution of NaCl whose salinity is 5.6 if the solution has a density of 1.03 g/mL?

18.32 Phosphorus is present in seawater to the extent of 0.07 ppm by mass. If the phosphorus is present as phosphate, PO_4^{3-}, calculate the corresponding molar concentration of phosphate in seawater.

18.33 A first-stage recovery of magnesium from seawater is precipitation of $Mg(OH)_2$ with CaO:

$$Mg^{2+}(aq) + CaO(s) + H_2O(l) \longrightarrow Mg(OH)_2(s) + Ca^{2+}(aq)$$

What mass of CaO, in grams, is needed to precipitate 1000 lb of $Mg(OH)_2$?

18.34 Gold is found in seawater at very low levels, about 0.05 ppb by mass. Assuming that gold is worth about $800 per troy ounce, how many liters of seawater would you have to process to obtain $1,000,000 worth of gold?

Assume the density of seawater is 1.03 g/mL and that your gold recovery process is 50% efficient.

18.35 Suppose that one wishes to use reverse osmosis to reduce the salt content of brackish water containing 0.22 M total salt concentration to a value of 0.01 M, thus rendering it usable for human consumption. What is the minimum pressure that needs to be applied in the permeators (Figure 18.16) to achieve this goal, assuming that the operation occurs at 298 K? (*Hint:* Refer to Section 13.5.)

18.36 Assume that a portable reverse-osmosis apparatus such as that shown in Figure 18.17 operates on seawater, whose concentrations of constituent ions are listed in Table 18.6, and that the desalinated water output has an effective molarity of about 0.02 M. What minimum pressure must be applied by hand pumping at 297 K to cause reverse osmosis to occur? (*Hint:* Refer to Section 13.5.)

Freshwater

18.37 List the common products formed when an organic material containing the elements carbon, hydrogen, oxygen, sulfur, and nitrogen decomposes **(a)** under aerobic conditions, **(b)** under anaerobic conditions.

18.38 **(a)** Explain why the concentration of dissolved oxygen in freshwater is an important indicator of the quality of the water. **(b)** How is the solubility of oxygen in water affected by increasing temperature?

18.39 The organic anion

$$H_3C-(CH_2)_9-\overset{\overset{\displaystyle H}{|}}{\underset{\underset{\displaystyle CH_3}{|}}{C}}-\!\!\left\langle\bigcirc\right\rangle\!\!-SO_3^-$$

is found in most detergents. Assume that the anion undergoes aerobic decomposition in the following manner:

$$2 \, C_{18}H_{29}SO_3^-(aq) + 51 \, O_2(aq) \longrightarrow$$
$$36 \, CO_2(aq) + 28 \, H_2O(l) + 2 \, H^+(aq) + 2 \, SO_4^{2-}(aq)$$

What is the total mass of O_2 required to biodegrade 1.0 g of this substance?

18.40 The average daily mass of O_2 taken up by sewage discharged in the United States is 59 g per person. How many liters of water at 9 ppm O_2 are totally depleted of oxygen in 1 day by a population of 120,000 people?

18.41 Write a balanced chemical equation to describe how magnesium ions are removed in water treatment by the addition of slaked lime, $Ca(OH)_2$.

18.42 **(a)** Which of the following ionic species could be, responsible for hardness in a water supply: Ca^{2+}, K^+, Mg^{2+}, Fe^{2+}, Na^+? **(b)** What properties of an ion determine whether it will contribute to water hardness?

18.43 How many moles of $Ca(OH)_2$ and Na_2CO_3 should be added to soften 1200 L of water in which $[Ca^{2+}] = 5.0 \times 10^{-4} \, M$ and $[HCO_3^-] = 7.0 \times 10^{-4} \, M$?

18.44 The concentration of Ca^{2+} in a particular water supply is $5.7 \times 10^{-3} \, M$. The concentration of bicarbonate ion, HCO_3^-, in the same water is $1.7 \times 10^{-3} \, M$. What masses of $Ca(OH)_2$ and Na_2CO_3 must be added to $5.0 \times 10^7 \, L$ of this water to reduce the level of Ca^{2+} to 20% of its original level?

18.45 Ferrous sulfate ($FeSO_4$) is often used as a coagulant in water purification. The iron(II) salt is dissolved in the water to be purified, then oxidized to the iron(III) state by dissolved oxygen, at which time gelatinous $Fe(OH)_3$ forms, assuming the pH is above approximately 6. Write balanced chemical equations for the oxidation of Fe^{2+} to Fe^{3+} by dissolved oxygen, and for the formation of $Fe(OH)_3(s)$ by reaction of $Fe^{3+}(aq)$ with $HCO_3^-(aq)$.

18.46 What properties make a substance a good coagulant for water purification?

Green Chemistry

18.47 One of the principles of green chemistry is that it is better to use as few steps as possible in making new chemicals. How does this principle relate to energy efficiency?

18.48 Discuss how catalysts can make processes more energy efficient.

18.49 The Baeyer–Villiger reaction is a classic organic oxidation reaction for converting ketones to lactones, as in this reaction:

Ketone   3-Chloroperbenzoic acid

Lactone   3-Chlorobenzoic acid

The reaction is used in the manufacture of plastics and pharmaceuticals. The reactant 3-chloroperbenzoic acid is somewhat shock sensitive, however, and prone to explode. Also, 3-chlorobenzoic acid is a waste product. An alternative process being developed uses hydrogen peroxide and a catalyst consisting of tin deposited within a solid support. The catalyst is readily recovered from the reaction mixture. **(a)** What would you expect to be the other product of oxidation of the ketone to lactone by hydrogen peroxide? **(b)** What principles of green chemistry are addressed by use of the proposed process?

18.50 The reaction shown here was performed with an iridium catalyst, both in supercritical CO_2 (scCO_2) and in the chlorinated solvent CH_2Cl_2. The kinetic data for the reaction in both solvents are plotted in the graph. Why is this a good example of a green chemical reaction?

ADDITIONAL EXERCISES

18.51 A friend of yours has seen each of the following items in newspaper articles and would like an explanation: **(a)** acid rain, **(b)** greenhouse gas, **(c)** photochemical smog, **(d)** ozone depletion. Give a brief explanation of each term, and identify one or two of the chemicals associated with each.

18.52 Suppose that on another planet the atmosphere consists of 17% Kr, 38% CH_4, and 45% O_2. What is the average molar mass at the surface? What is the average molar mass at an altitude at which all the O_2 is photodissociated?

18.53 If an average O_3 molecule "lives" only 100–200 seconds in the stratosphere before undergoing dissociation, how can O_3 offer any protection from ultraviolet radiation?

18.54 Show how Equations 18.7 and 18.9 can be added to give Equation 18.10. (You may need to multiply one of the reactions by a factor to have them add properly.)

18.55 What properties of CFCs make them ideal for various commercial applications but also make them a long-term problem in the stratosphere?

18.56 *Halons* are fluorocarbons that contain bromine, such as $CBrF_3$. They are used extensively as foaming agents for fighting fires. Like CFCs, halons are very unreactive and ultimately can diffuse into the stratosphere. **(a)** Based on the data in Table 8.4, would you expect photodissociation of Br atoms to occur in the stratosphere?

(b) Propose a mechanism by which the presence of halons in the stratosphere could lead to the depletion of stratospheric ozone.

18.57 It is estimated that the lifetime for HFCs in the stratosphere is 2–7 years. If HFCs have such long lifetimes, why are they being used to replace CFCs?

[18.58] The *hydroxyl radical*, OH, is formed at low altitudes via the reaction of excited oxygen atoms with water:

$$O^*(g) + H_2O(g) \longrightarrow 2\,OH(g)$$

(a) Write the Lewis structure for the hydroxyl radical. (*Hint:* It has one unpaired electron.)

Once produced, the hydroxyl radical is very reactive. Explain why each of the following series of reactions affects the pollution in the troposphere:

(b) OH + $NO_2 \longrightarrow HNO_3$

(c) OH + CO + $O_2 \longrightarrow CO_2$ + OOH
OOH + NO $\longrightarrow$ OH + NO_2

(d) OH + $CH_4 \longrightarrow H_2O$ + CH_3
$CH_3 + O_2 \longrightarrow OOCH_3$
$OOCH_3$ + NO $\longrightarrow OCH_3 + NO_2$

18.59 Explain, using Le Châtelier's principle, why the equilibrium constant for the formation of NO from N_2 and O_2 increases with increasing temperature, whereas the equilibrium constant for the formation of NO_2 from NO and O_2 decreases with increasing temperature.

18.60 The affinity of carbon monoxide for hemoglobin is about 210 times that of O_2. Assume a person is inhaling air that contains 125 ppm of CO. If all the hemoglobin leaving the lungs carries either oxygen or CO, calculate the fraction in the form of carboxyhemoglobin.

18.61 Natural gas consists primarily of methane, $CH_4(g)$. **(a)** Write a balanced chemical equation for the complete combustion of methane to produce $CO_2(g)$ as the only carbon-containing product. **(b)** Write a balanced chemical equation for the incomplete combustion of methane to produce $CO(g)$ as the only carbon-containing product. **(c)** At 25 °C and 1.0 atm pressure, what is the minimum quantity of dry air needed to combust 1.0 L of $CH_4(g)$ completely to $CO_2(g)$?

18.62 One of the possible consequences of global warming is an increase in the temperature of ocean water. The oceans serve as a "sink" for CO_2 by dissolving large amounts of it. **(a)** How would the solubility of CO_2 in the oceans be affected by an increase in the temperature of the water? **(b)** Discuss the implications of your answer to part (a) for the problem of global warming.

18.63 The rate of solar energy striking Earth averages 169 watts per square meter. The rate of energy radiated from Earth's surface averages 390 watts per square meter. Comparing these numbers, one might expect that the planet would cool quickly, yet it does not. Why not?

18.64 The solar power striking Earth every day averages 169 watts per square meter. The peak electrical power usage in New York City is 12,000 megawatts. Considering that present technology for solar energy conversion is only about 10% efficient, from how many square meters of land must sunlight be collected in order to provide this peak power? (For comparison, the total area of the city is 830 km².)

18.65 Write balanced chemical equations for each of the following reactions: **(a)** The nitric oxide molecule undergoes photodissociation in the upper atmosphere. **(b)** The nitric oxide molecule undergoes photoionization in the upper atmosphere. **(c)** Nitric oxide undergoes oxidation by ozone in the stratosphere. **(d)** Nitrogen dioxide dissolves in water to form nitric acid and nitric oxide.

18.66 **(a)** Explain why $Mg(OH)_2$ precipitates when CO_3^{2-} ion is added to a solution containing Mg^{2+}. **(b)** Will $Mg(OH)_2$ precipitate when 4.0 g of Na_2CO_3 is added to 1.00 L of a solution containing 125 ppm of Mg^{2+}?

[18.67] It has been pointed out that there may be increased amounts of NO in the troposphere as compared with the past because of massive use of nitrogen-containing compounds in fertilizers. Assuming that NO can eventually diffuse into the stratosphere, how might it affect the conditions of life on Earth? Using the index to this text, look up the chemistry of nitrogen oxides. What chemical pathways might NO in the troposphere follow?

[18.68] As of the writing of this text, EPA standards limit atmospheric ozone levels in urban environments to 84 ppb. How many moles of ozone would there be in the air above Los Angeles County (area about 4000 square miles; consider a height of 10 m above the ground) if ozone was at this concentration?

INTEGRATIVE EXERCISES

18.69 The estimated average concentration of NO_2 in air in the United States in 2006 was 0.016 ppm. **(a)** Calculate the partial pressure of the NO_2 in a sample of this air when the atmospheric pressure is 755 torr (99.1 kPa). **(b)** How many molecules of NO_2 are present under these conditions at 20 °C in a room that measures 15 × 14 × 8 ft?

[18.70] In 1986 an electrical power plant in Taylorsville, Georgia, burned 8,376,726 tons of coal, a national record at that time. **(a)** Assuming that the coal was 83% carbon and 2.5% sulfur and that combustion was complete, calculate the number of tons of carbon dioxide and sulfur dioxide produced by the plant during the year. **(b)** If 55% of the SO_2 could be removed by reaction with powdered CaO to form $CaSO_3$, how many tons of $CaSO_3$ would be produced?

18.71 The water supply for a midwestern city contains the following impurities: coarse sand, finely divided particulates, nitrate ion, trihalomethanes, dissolved phosphorus in the form of phosphates, potentially harmful bacterial strains, dissolved organic substances. Which of the following processes or agents, if any, is effective in removing each of these impurities: coarse sand filtration, activated carbon filtration, aeration, ozonization, precipitation with aluminum hydroxide?

18.72 The concentration of H_2O in the stratosphere is about 5 ppm. It undergoes photodissociation as follows:

$$H_2O(g) \longrightarrow H(g) + OH(g)$$

(a) Write out the Lewis-dot structures for both products and reactant.

(b) Using Table 8.4, calculate the wavelength required to cause this dissociation.

(c) The hydroxyl radicals, OH, can react with ozone, giving the following reactions:

$$OH(g) + O_3(g) \longrightarrow HO_2(g) + O_2(g)$$
$$HO_2(g) + O(g) \longrightarrow OH(g) + O_2(g)$$

What overall reaction results from these two elementary reactions? What is the catalyst in the overall reaction? Explain.

18.73 Bioremediation is the process by which bacteria repair their environment in response, for example, to an oil spill. The efficiency of bacteria for "eating" hydrocarbons depends on the amount of oxygen in the system, pH, temperature, and many other factors. In a certain oil spill, hydrocarbons from the oil disappeared with a first-order rate constant of $2 \times 10^{-6}\,s^{-1}$. How many days did it take for the hydrocarbons to decrease to 10% of their initial value?

18.74 The standard enthalpies of formation of ClO and ClO_2 are 101 and 102 kJ/mol, respectively. Using these data and the thermodynamic data in Appendix C, calculate the overall enthalpy change for each step in the following catalytic cycle:

$$ClO(g) + O_3(g) \longrightarrow ClO_2(g) + O_2(g)$$
$$ClO_2(g) + O(g) \longrightarrow ClO(g) + O_2(g)$$

What is the enthalpy change for the overall reaction that results from these two steps?

18.75 The main reason that distillation is a costly method for purifying water is the high energy required to heat and vaporize water. (a) Using the density, specific heat, and heat of vaporization of water from Appendix B, calculate the amount of energy required to vaporize 1.00 gal of water beginning with water at 20 °C. (b) If the energy is provided by electricity costing $0.085/kWh, calculate its cost. (c) If distilled water sells in a grocery store for $1.26 per gal, what percentage of the sales price is represented by the cost of the energy?

[18.76] A reaction that contributes to the depletion of ozone in the stratosphere is the direct reaction of oxygen atoms with ozone:

$$O(g) + O_3(g) \longrightarrow 2\,O_2(g)$$

At 298 K the rate constant for this reaction is $4.8 \times 10^5\,M^{-1}s^{-1}$. (a) Based on the units of the rate constant, write the likely rate law for this reaction. (b) Would you expect this reaction to occur via a single elementary process? Explain why or why not. (c) From the magnitude of the rate constant, would you expect the activation energy of this reaction to be large or small? Explain. (d) Use ΔH_f° values from Appendix C to estimate the enthalpy change for this reaction. Would this reaction raise or lower the temperature of the stratosphere?

18.77 Nitrogen dioxide (NO_2) is the only important gaseous species in the lower atmosphere that absorbs visible light. (a) Write the Lewis structure(s) for NO_2. (b) How does this structure account for the fact that NO_2 dimerizes to form N_2O_4? Based on what you can find about this dimerization reaction in the text, would you expect to find the NO_2 that forms in an urban environment to be in the form of dimer? Explain. (c) What would you expect as products, if any, for the reaction of NO_2 with CO? (d) Would you expect NO_2 generated in an urban environment to migrate to the stratosphere? Explain.

18.78 The following data was collected for the destruction of O_3 by H ($O_3 + H \rightarrow O_2 + OH$) at very low concentrations:

Experiment	$[O_3]$, M	$[H]$, M	Initial Rate, M/s
1	5.17×10^{-33}	3.22×10^{-26}	1.88×10^{-14}
2	2.59×10^{-33}	3.25×10^{-26}	9.44×10^{-15}
3	5.19×10^{-33}	6.46×10^{-26}	3.77×10^{-14}

(a) Write the rate law for the reaction.
(b) Calculate the rate constant.

18.79 The degradation of CF_3CH_2F (an HFC) by OH radicals in the troposphere is first order in each reactant and has a rate constant of $k = 1.6 \times 10^8\,M^{-1}s^{-1}$ at 4 °C. If the tropospheric concentrations of OH and CF_3CH_2F are 8.1×10^5 and 6.3×10^8 molecules cm^{-3}, respectively, what is the rate of reaction at this temperature in M/s?

[18.80] The Henry's law constant for CO_2 in water at 25 °C is $3.1 \times 10^{-2}\,M$ atm^{-1}. (a) What is the solubility of CO_2 in water at this temperature if the solution is in contact with air at normal atmospheric pressure? (b) Assume that all of this CO_2 is in the form of H_2CO_3 produced by the reaction between CO_2 and H_2O:

$$CO_2(aq) + H_2O(l) \longrightarrow H_2CO_3(aq)$$

What is the pH of this solution?

[18.81] If the pH of a 1.0-in. rainfall over 1500 mi^2 is 3.5, how many kilograms of H_2SO_4 are present, assuming that it is the only acid contributing to the pH?

[18.82] The precipitation of $Al(OH)_3$ ($K_{sp} = 1.3 \times 10^{-33}$) is sometimes used to purify water. (a) Estimate the pH at which precipitation of $Al(OH)_3$ will begin if 5.0 lb of $Al_2(SO_4)_3$ is added to 2000 gal of water. (b) Approximately how many pounds of CaO must be added to the water to achieve this pH?

CHEMICAL THERMODYNAMICS

THE ORANGE EXTERNAL TANK under the Space Shuttle holds tanks of hydrogen and oxygen. The reaction of hydrogen and oxygen producing water propels the Space Shuttle into orbit. Once the mixture is ignited, the reaction proceeds spontaneously to completion, generating heat and light. The reverse process, formation of hydrogen and oxygen from water, does not occur of its own accord.

WHAT'S AHEAD

19.1 Spontaneous Processes
In this chapter we will see why changes that occur in nature have a directional character. They move *spontaneously* in one direction but not in the reverse direction.

19.2 Entropy and the Second Law of Thermodynamics
We will discuss *entropy*, which is a thermodynamic state function that is important in determining whether a process is spontaneous. The *second law of thermodynamics* tells us that in any spontaneous process the entropy of the universe (system plus surroundings) increases.

19.3 The Molecular Interpretation of Entropy
On a molecular level, we will see that an entropy increase is associated with an increase in the number of accessible *microstates*. Entropy change can also be thought of as a measure of a system's randomness at a stated temperature. The *third law of thermodynamics* states that the entropy of a perfect crystalline solid at 0 K is zero.

19.4 Entropy Changes in Chemical Reactions
Using tabulated *standard molar entropies*, we can calculate the standard entropy changes for systems undergoing reaction.

19.5 Gibbs Free Energy
We next encounter another thermodynamic state function, *free energy* (or *Gibbs free energy*), which is a measure of how far removed the system is from equilibrium. Free energy change measures the maximum amount of useful work obtainable from a process and yields information on the direction in which a chemical reaction will proceed spontaneously.

19.6 Free Energy and Temperature
We will consider how the relationship among free-energy change, enthalpy change, and entropy change provides insight into how temperature affects the spontaneity of a process.

19.7 Free Energy and the Equilibrium Constant
Finally, we will consider how the standard free-energy change for a chemical reaction can be used to calculate the equilibrium constant for the process.

THE ENERGY REQUIRED TO PROPEL THE SPACE SHUTTLE INTO SPACE is obtained from two solid-fuel booster rockets and a rocket engine that relies on the combustion of hydrogen and oxygen to form water. The hydrogen and oxygen are stored as liquids at very low temperatures in tanks mounted below the

Space Shuttle. As the hydrogen and oxygen vapors are ingnited, they react very rapidly and virtually completely, producing enormous quantities of water vapor and heat. Two of the most important questions chemists ask when designing and using chemical reactions are "How fast is the reaction?" and "How far does it proceed?" The first question is addressed by the study of chemical kinetics, which we discussed in Chapter 14. The second question involves the equilibrium constant, which was the focus of Chapter 15.

In Chapter 14 we learned that the rates of chemical reactions are controlled largely by a factor related to energy, namely the activation energy of the reaction. ∞ (Section 14.5) In general, the lower the activation energy, the faster a reaction proceeds. In Chapter 15 we saw that equilibrium depends on the rates of the forward and reverse reactions: Equilibrium is reached when the opposing reactions occur at equal rates. ∞ (Section 15.1) Because reaction rates are closely tied to energy, it is logical that equilibrium also depends in some way on energy.

In this chapter we will explore the connection between energy and the extent of a reaction. Doing so requires us to take a deeper look at *chemical thermodynamics*, the area of chemistry that deals with energy relationships. We first encountered thermodynamics in Chapter 5, where we discussed the nature of energy, the first law of thermodynamics, and the concept of enthalpy. Recall that the enthalpy change is the heat transferred between the system and its surroundings during a constant-pressure process. ∞ (Section 5.3)

Now we will see that reactions involve not only changes in enthalpy but also changes in *entropy*—another important thermodynamic quantity. Our discussion of entropy will lead us to the second law of thermodynamics, which provides insight into why physical and chemical changes tend to favor one direction over another. We drop a brick, for example, and it falls to the ground. We do not expect bricks to spontaneously rise from the ground to our outstretched hand. We light a candle, and it burns down. We do not expect a half-consumed candle to regenerate itself spontaneously, even if we have kept all the gases produced when the candle burned. Thermodynamics helps us understand the significance of this directional character of processes, whether they are exothermic or endothermic.

19.1 SPONTANEOUS PROCESSES

The first law of thermodynamics states that *energy is conserved*. ∞ (Section 5.2) In other words, energy is neither created nor destroyed in any process, whether it is the falling of a brick, the burning of a candle, or the melting of an ice cube. Energy can be transferred between a system and the surroundings or can be converted from one form to another, but the total energy remains constant. We expressed the first law of thermodynamics mathematically as $\Delta E = q + w$, where ΔE is the change in the internal energy of a system, q is the heat absorbed by the system from the surroundings, and w is the work done on the system by the surroundings.

The first law helps us balance the books, so to speak, on the heat transferred between a system and its surroundings and the work done by a particular process or reaction. But the first law does not address another important feature of reactions—the extent to which they occur. As we noted in the introduction, our experience tells us that physical and chemical processes have a directional character. For instance, sodium metal and chlorine gas combine readily to form sodium chloride, which we also know as table salt. We never find table salt decomposing of its own accord to form sodium and chlorine. (Have you ever smelled chlorine gas in the kitchen or seen sodium metal on your table salt?) In both processes—the formation of sodium chloride from sodium and chlorine and the decomposition of sodium chloride into sodium and chlorine—energy is conserved, as it must be according to the first law of thermodynamics. Yet one process occurs, and the other does not. A process that occurs of its own accord without any ongoing outside intervention is said to be spontaneous. A **spontaneous process** is one that proceeds on its own without any outside assistance.

A spontaneous process occurs in a definite direction. Imagine you were to see a video clip in which a brick rises from the ground. You would conclude that the video is running in reverse—bricks do not magically rise from the ground! A brick falling is a spontaneous process, whereas the reverse process is *nonspontaneous*.

A gas will expand into a vacuum as shown in Figure 19.1 ◄, but the process will never reverse itself. The expansion of the gas is spontaneous. Likewise, a nail left out in the weather will rust (Figure 19.2 ►). In this process the iron in the nail reacts with oxygen from the air to form an iron oxide. We would never expect the rusty nail to reverse this process and become shiny. The rusting process

▲ Figure 19.1 Spontaneous expansion of an ideal gas into an evacuated space. In (a) flask *B* holds an ideal gas at 1 atm pressure and flask *A* is evacuated. In (b) the stopcock connecting the flasks has been opened. The ideal gas expands to occupy both flasks *A* and *B* at a pressure of 0.5 atm. The reverse process—all the gas molecules moving back into flask *B*— is not spontaneous.

is spontaneous, whereas the reverse process is nonspontaneous. There are countless other examples we could cite that illustrate the same idea: *Processes that are spontaneous in one direction are nonspontaneous in the opposite direction.*

Experimental conditions, such as temperature and pressure, are often important in determining whether a process is spontaneous. Consider, for example, the melting of ice. When the temperature of the surroundings is above 0 °C at ordinary atmospheric pressures, ice melts spontaneously and the reverse process—liquid water turning into ice—is not spontaneous. However, when the surroundings are below 0 °C, the opposite is true. Liquid water converts into ice spontaneously, and the conversion of ice into water is *not* spontaneous (Figure 19.3▼).

What happens at $T = 0$ °C, the normal melting point of water, when the flask of Figure 19.3 contains both water and ice? At the normal melting point of a substance, the solid and liquid phases are in equilibrium. ∞(Section 11.5) At this temperature the two phases are interconverting at the same rate and there is no preferred direction for the process.

It is important to realize that the fact that a process is spontaneous does not necessarily mean that it will occur at an observable rate. A chemical reaction is spontaneous if it occurs on its own accord, regardless of its speed. A spontaneous reaction can be very fast, as in the case of acid–base neutralization, or very slow, as in the rusting of iron. Thermodynamics can tell us the *direction* and *extent* of a reaction but tells us nothing about the *speed* of the reaction.

Spontaneous   Non-spontaneous

▲ **Figure 19.2 A spontaneous process.** Elemental iron in the shiny nail in the top photograph spontaneously combines with H_2O and O_2 in the surrounding air to form a layer of rust—Fe_2O_3—on the nail surface.

Spontaneous for $T > 0$°C

Spontaneous for $T < 0$°C

▲ **Figure 19.3 Spontaneity can depend on the temperature.** At $T > 0$ °C ice melts spontaneously to liquid water. At $T < 0$ °C the reverse process, water freezing to ice, is spontaneous. At $T = 0$ °C the two states are in equilibrium.

■■ **SAMPLE EXERCISE 19.1** | Identifying Spontaneous Processes

Predict whether the following processes are spontaneous as described, spontaneous in the reverse direction, or in equilibrium: **(a)** When a piece of metal heated to 150 °C is added to water at 40 °C, the water gets hotter. **(b)** Water at room temperature decomposes into $H_2(g)$ and $O_2(g)$. **(c)** Benzene vapor, $C_6H_6(g)$, at a pressure of 1 atm condenses to liquid benzene at the normal boiling point of benzene, 80.1 °C.

SOLUTION

Analyze: We are asked to judge whether each process will proceed spontaneously in the direction indicated, in the reverse direction, or in neither direction.

Plan: We need to think about whether each process is consistent with our experience about the natural direction of events or whether we expect the reverse process to occur.

Solve: (a) This process is spontaneous. Whenever two objects at different temperatures are brought into contact, heat is transferred from the hotter object to the colder one. Thus, heat is transferred from the hot metal to the cooler water. The final temperature, after the metal and water achieve the same temperature (thermal equilibrium), will be somewhere between the initial temperatures of the metal and the water. **(b)** Experience tells us that this process is not spontaneous—we certainly have never seen hydrogen and oxygen gases spontaneously bubbling up out of water! Rather, the *reverse* process—the reaction of H_2 and O_2 to form H_2O—is spontaneous. **(c)** By definition, the normal boiling point is the temperature at which a vapor at 1 atm is in equilibrium with its liquid. Thus, this is an equilibrium situation. If the temperature were below 80.1 °C, condensation would be spontaneous.

■ **PRACTICE EXERCISE**

Under 1 atm pressure $CO_2(s)$ sublimes at -78 °C. Is the transformation of $CO_2(s)$ to $CO_2(g)$ a spontaneous process at -100 °C and 1 atm pressure?
Answer: No, the reverse process is spontaneous at this temperature.

Seeking a Criterion for Spontaneity

A marble rolling down an incline or a brick falling from your hand loses energy. The loss of energy is a common feature of spontaneous change in mechanical systems. During the 1870s Marcellin Bertholet (1827–1907), a famous chemist of that era, suggested that the direction of spontaneous changes in chemical systems was also determined by the loss of energy. He proposed that all spontaneous chemical and physical changes were exothermic. It takes only a few moments, however, to find exceptions to this generalization. For example, the melting of ice at room temperature is spontaneous even though melting is an endothermic process. Similarly, many spontaneous solution processes, such as the dissolving of NH_4NO_3, are endothermic, as we discovered in Section 13.1. We conclude therefore that, although the majority of spontaneous reactions are exothermic, there are spontaneous endothermic ones as well. Clearly, some other factor must be at work in determining the natural direction of processes. What is this factor?

To understand why certain processes are spontaneous, we need to consider more closely the ways in which the state of a system can change. Recall that quantities such as temperature, internal energy, and enthalpy are *state functions*, properties that define a state and do not depend on how we reach that state. ∞ (Section 5.2) The heat transferred between a system and its surroundings, q, and the work done by or on the system, w, are *not* state functions. The values of q and w depend on the specific path taken from one state to another. One of the keys to understanding spontaneity is distinguishing between reversible and irreversible paths between states.

▲ **GIVE IT SOME THOUGHT**

If a process is nonspontaneous, does that mean the process cannot occur under any circumstances?

Reversible and Irreversible Processes

In 1824 a 28-year-old French engineer named Sadi Carnot (1796–1832) published an analysis of the factors that determine how efficiently a steam engine can convert heat to work. Carnot considered what an ideal engine, one with the highest possible efficiency, would be like. He observed that it is impossible to convert the energy content of a fuel completely to work because a significant amount of heat is always lost to the surroundings. Carnot's analysis gave insight into how to build better, more efficient engines, and it was one of the earliest studies in what has developed into the discipline of thermodynamics.

About 40 years later, Rudolph Clausius (1822–1888), a German physicist, extended Carnot's work in an important way. Clausius concluded that a special significance could be ascribed to the ratio of the heat delivered to an ideal engine and the temperature at which it is delivered, q/T. He was so convinced of the importance of this ratio that he gave it a special name, *entropy*. He deliberately selected the name to sound like energy to emphasize his belief that the importance of entropy was comparable to that of energy.

An ideal engine, one with the maximum efficiency, operates under an ideal set of conditions in which all the processes are reversible. In a **reversible process**, a system is changed in such a way that the system and surroundings can be restored to their original state by *exactly* reversing the change. In other words, we can completely restore the system to its original condition with no net change to either the system or its surroundings. An **irreversible process** is one that cannot simply be reversed to restore the system and its surroundings to their original states. What Carnot discovered is that the amount of work we can extract from any spontaneous process depends on the manner in which the process is carried out. *A reversible change produces the maximum amount of work that can be achieved by the system on the surroundings* ($w_{rev} = w_{max}$).

▲ **GIVE IT SOME THOUGHT**

If you evaporate water and then condense it, have you necessarily performed a reversible process?

Let's examine some examples of reversible and irreversible processes. When two objects at different temperatures are in contact, heat will flow spontaneously from the hotter object to the colder one. Because it is impossible to make heat flow in the opposite direction, the flow of heat is irreversible. Given these facts, can we imagine any conditions under which heat transfer can be made reversible? Consider two objects or a system and its surroundings that are at essentially the same temperature, with just an infinitesimal difference (an extremely small temperature difference, ΔT) to make heat flow in the desired direction (Figure 19.4▶). We can then reverse the direction of heat flow by making an infinitesimal change of temperature in the opposite direction. *Reversible processes are those that reverse direction whenever an infinitesimal change is made in some property of the system.*

Now consider another example, the expansion of an ideal gas at constant temperature. A constant-temperature process such as this is said to be **isothermal**. To keep the example simple, consider the gas in the cylinder-and-piston arrangement shown in Figure 19.5▼. When the partition is removed, the gas expands spontaneously to fill the evacuated space. Because the gas is expanding into a vacuum with no external pressure, it does no *P-V* work on the surroundings ($w = 0$). ∞ (Section 5.3) We can use the piston to compress the gas back to its original state, but doing so requires that the surroundings do work on the system ($w > 0$). That is, reversing the process has produced a change in the surroundings as energy is used to do work on the system. The fact that the system and the surroundings are not both returned to their original conditions indicates that the process is irreversible.

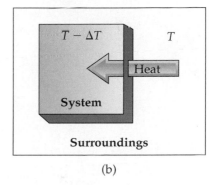

(b)

▲ **Figure 19.4 Reversible flow of heat.** Heat can flow reversibly between a system and its surroundings if the two have only an infinitesimally small difference in temperature, ΔT. The direction of heat flow can be changed by increasing or decreasing the temperature of the system by ΔT. (a) Increasing the temperature of the system by ΔT causes heat to flow from the system to the surroundings. (b) Decreasing the temperature of the system by ΔT causes heat to flow from the surroundings into the system.

▲ **Figure 19.5 An irreversible process.** Restoring the system to its original state after an irreversible process changes the surroundings. In (a) the gas is confined to the right half of the cylinder by a partition. When the partition is removed (b), the gas spontaneously (irreversibly) expands to fill the whole cylinder. No work is done by the system during this expansion. In (c) we can use the piston to compress the gas back to its original state. Doing so requires that the surroundings do work on the system, which changes the surroundings forever.

What might a reversible, isothermal expansion of an ideal gas be like? It will occur only if the external pressure acting on the piston exactly balances the pressure exerted by the gas. Under these conditions, the piston will not move unless the external pressure is reduced infinitely slowly, allowing the pressure of the confined gas to readjust to maintain a balance in the two pressures. This gradual, infinitely slow process in which the external pressure and internal pressure are always in equilibrium is reversible. If we reverse the process and compress the gas in the same infinitely slow manner, we can return the gas to its original volume. The complete cycle of expansion and compression in this hypothetical process, moreover, is accomplished without any net change to the surroundings.

Because real processes can at best only approximate the slow, ever-in-equilibrium change associated with reversible processes, all real processes are irreversible. Further, the reverse of any spontaneous process is a nonspontaneous process. A nonspontaneous process can occur only if the surroundings do work on the system. Thus, *any spontaneous process is irreversible*. Even if we return the system to the original condition, the surroundings will have changed.

19.2 ENTROPY AND THE SECOND LAW OF THERMODYNAMICS

We are now closer to understanding spontaneity because we know that any spontaneous process is irreversible. But how can we use this idea to make predictions about the spontaneity of an unfamiliar process? Understanding spontaneity requires us to examine the thermodynamic quantity called **entropy**. Entropy has been variously associated with the extent of *randomness* in a system or with the extent to which energy is distributed or dispersed among the various motions of the molecules of the system. In fact, entropy is a multifaceted concept whose interpretations are not so quickly summarized by a simple definition. In this section we consider how we can relate entropy changes to heat transfer and temperature. Our analysis will bring us to a profound statement about spontaneity that we call the second law of thermodynamics. In Section 19.3 we examine the molecular significance of entropy.

Entropy Change

The entropy, S, of a system is a state function just like the internal energy, E, and enthalpy, H. As with these other quantities, the value of S is a characteristic of the state of a system. ∞ (Section 5.2) Thus, the change in entropy, ΔS, in a system depends only on the initial and final states of the system and not on the path taken from one state to the other:

$$\Delta S = S_{\text{final}} - S_{\text{initial}} \qquad [19.1]$$

For the special case of an isothermal process, ΔS is equal to the heat that would be transferred if the process were reversible, q_{rev}, divided by the temperature at which the process occurs:

$$\Delta S = \frac{q_{\text{rev}}}{T} \qquad \text{(constant } T) \qquad [19.2]$$

Because S is a state function, we can use Equation 19.2 to calculate ΔS for *any* isothermal process, not just those that are reversible. If a change between two states is irreversible, we calculate ΔS by using a reversible path between the states.

▲ **GIVE IT SOME THOUGHT**

How do we reconcile the fact that S is a state function but that ΔS depends on q, which is not a state function?

ΔS for Phase Changes

The melting of a substance at its melting point and the vaporization of a substance at its boiling point are isothermal processes. Consider the melting of ice. At 1 atm pressure, ice and liquid water are in equilibrium with each other at 0 °C. Imagine that we melt one mole of ice at 0 °C, 1 atm to form one mole of liquid water at 0 °C, 1 atm. We can achieve this change by adding a certain amount of heat to the system from the surroundings: $q = \Delta H_{fusion}$. Now imagine that we carry out the change by adding the heat infinitely slowly, raising the temperature of the surroundings only infinitesimally above 0 °C. When we make the change in this fashion, the process is reversible. We can reverse the process simply by infinitely slowly removing the same amount of heat, ΔH_{fusion}, from the system, using immediate surroundings that are infinitesimally below 0 °C. Thus, $q_{rev} = \Delta H_{fusion}$ and $T = 0\ °C = 273\ K$.

The enthalpy of fusion for H_2O is $\Delta H_{fusion} = 6.01$ kJ/mol. (The melting is an endothermic process, and so the sign of ΔH is positive.) Thus, we can use Equation 19.2 to calculate ΔS_{fusion} for melting one mole of ice at 273 K:

$$\Delta S_{fusion} = \frac{q_{rev}}{T} = \frac{\Delta H_{fusion}}{T} = \frac{(1\ mol)(6.01 \times 10^3\ J/mol)}{273\ K} = 22.0\ \frac{J}{K}$$

Notice that the units for ΔS, J/K, are energy divided by absolute temperature, as we expect from Equation 19.2.

■■ **SAMPLE EXERCISE 19.2** | Calculating ΔS for a Phase Change

The element mercury, Hg, is a silvery liquid at room temperature. The normal freezing point of mercury is −38.9 °C, and its molar enthalpy of fusion is $\Delta H_{fusion} = 2.29$ kJ/mol. What is the entropy change of the system when 50.0 g of Hg(*l*) freezes at the normal freezing point?

SOLUTION

Analyze: We first recognize that freezing is an *exothermic* process; heat is transferred from the system to the surroundings when a liquid freezes ($q < 0$). The enthalpy of fusion is ΔH for the melting process. Because freezing is the reverse of melting, the enthalpy change that accompanies the freezing of 1 mol of Hg is $-\Delta H_{fusion} = -2.29$ kJ/mol.

Plan: We can use $-\Delta H_{fusion}$ and the atomic weight of Hg to calculate q for freezing 50.0 g of Hg:

$$q = (50.0\ g\ Hg)\left(\frac{1\ mol\ Hg}{200.59\ g\ Hg}\right)\left(\frac{-2.29\ kJ}{1\ mol\ Hg}\right)\left(\frac{1000\ J}{1\ kJ}\right) = -571\ J$$

We can use this value of q as q_{rev} in Equation 19.2. We must first, however, convert the temperature to K:

$$-38.9\ °C = (-38.9 + 273.15)\ K = 234.3\ K$$

Solve: We can now calculate the value of ΔS_{sys}

$$\Delta S_{sys} = \frac{q_{rev}}{T} = \frac{-571\ J}{234.3\ K} = -2.44\ J/K$$

Check: The entropy change is negative because heat flows from the system, making q_{rev} negative.

Comment: The procedure we have used here can be used to calculate ΔS for other isothermal phase changes, such as the vaporization of a liquid at its boiling point.

■ **PRACTICE EXERCISE**

The normal boiling point of ethanol, C_2H_5OH, is 78.3 °C, and its molar enthalpy of vaporization is 38.56 kJ/mol. What is the change in entropy in the system when 68.3 g of $C_2H_5OH(g)$ at 1 atm condenses to liquid at the normal boiling point?
Answer: −163 J/K

In general, we will see that if a system becomes more spread out, or more random, the system's entropy increases. Thus, we expect the spontaneous expansion of a gas to result in an increase in entropy. To illustrate how the entropy change associated with an expanding gas can be calculated, consider the expansion of an ideal gas that is initially constrained by a piston, as in Figure 19.5(c). If the gas undergoes a reversible isothermal expansion, the work done on the surroundings by the moving piston can be calculated with the aid of calculus:

$$w_{rev} = -nRT \ln \frac{V_2}{V_1}$$

In this equation, n is the number of moles of gas, R is the gas constant, T is the absolute temperature, V_1 is the initial volume, and V_2 is the final volume. Notice that if $V_2 > V_1$, as it must be in our expansion, then $w_{rev} < 0$, meaning that the expanding gas does work on the surroundings.

One of the characteristics of an ideal gas is that its internal energy depends only on temperature, not on pressure. Thus, when an ideal gas expands at a constant temperature, $\Delta E = 0$. Because $\Delta E = q_{rev} + w_{rev} = 0$, we see that $q_{rev} = -w_{rev} = nRT \ln(V_2/V_1)$. Then, using Equation 19.2, we can calculate the entropy change in the system:

$$\Delta S_{sys} = \frac{q_{rev}}{T} = \frac{nRT \ln \dfrac{V_2}{V_1}}{T} = nR \ln \frac{V_2}{V_1} \quad [19.3]$$

For 1.00 L of an ideal gas at 1.00 atm and 0 °C, we can calculate the number of moles, $n = 4.46 \times 10^{-2}$ mol. The gas constant, R, can be expressed in units of J/mol-K, 8.314 J/mol-K (Table 10.2). Thus, for the expansion of the gas from 1.00 L to 2.00 L, we have

$$\Delta S_{sys} = (4.46 \times 10^{-2}\,\text{mol})\left(8.314 \frac{\text{J}}{\text{mol-K}}\right)\left(\ln \frac{2.00\,\text{L}}{1.00\,\text{L}}\right)$$

$$= 0.26\,\text{J/K}$$

In Section 19.3 we will see that this increase in entropy is a measure of the increased randomness of the molecules because of the expansion.

Related Exercises: 19.27 and 19.28

The Second Law of Thermodynamics

The key idea of the first law of thermodynamics is that energy is conserved in any process. Thus, the quantity of energy lost by a system equals the quantity gained by its surroundings. ∞(Section 5.1) We will see, however, that entropy is different because it actually increases in any spontaneous process. Thus, the sum of the entropy change of the system and surroundings for any spontaneous process is always greater than zero. Entropy change is like a signpost indicating whether a process is spontaneous. Let's illustrate this generalization by again considering the melting of ice, designating the ice and water as our system.

Let's calculate the entropy change of the system and the entropy change of the surroundings when a mole of ice (a piece roughly the size of an ordinary ice cube) melts in the palm of your hand. The process is not reversible because the system and surroundings are at different temperatures. Nevertheless, because ΔS is a state function, the entropy change of the system is the same regardless of whether the process is reversible or irreversible. We calculated the entropy change of the system just before Sample Exercise 19.2:

$$\Delta S_{sys} = \frac{q_{rev}}{T} = \frac{(1\,\text{mol})(6.01 \times 10^3\,\text{J/mol})}{273\,\text{K}} = 22.0\,\frac{\text{J}}{\text{K}}$$

The surroundings immediately in contact with the ice are your hand, which we will assume is at body temperature, 37 °C = 310 K. The heat lost by your hand is equal in magnitude to the heat gained by the ice but has the opposite sign, -6.01×10^3 J/mol. Hence the entropy change of the surroundings is

$$\Delta S_{surr} = \frac{q_{rev}}{T} = \frac{(1\,\text{mol})(-6.01 \times 10^3\,\text{J/mol})}{310\,\text{K}} = -19.4\,\frac{\text{J}}{\text{K}}$$

Thus, the total entropy change is positive:

$$\Delta S_{total} = \Delta S_{sys} + \Delta S_{surr} = \left(22.0\,\frac{\text{J}}{\text{K}}\right) + \left(-19.4\,\frac{\text{J}}{\text{K}}\right) = 2.6\,\frac{\text{J}}{\text{K}}$$

If the temperature of the surroundings were not 310 K but rather some temperature infinitesimally above 273 K, the melting would be reversible instead of irreversible. In that case the entropy change of the surroundings would equal -22.0 J/K and ΔS_{total} would be zero.

In general, any irreversible process results in an overall increase in entropy, whereas a reversible process results in no overall change in entropy. This general statement is known as the **second law of thermodynamics**. The sum of the entropy of a system plus the entropy of the surroundings is everything there is, and so we refer to the total entropy change as the entropy change of the universe, ΔS_{univ}. We can therefore state the second law of thermodynamics in terms of the following equations:

Reversible process: $\Delta S_{univ} = \Delta S_{sys} + \Delta S_{surr} = 0$

Irreversible process: $\Delta S_{univ} = \Delta S_{sys} + \Delta S_{surr} > 0$ [19.4]

All real processes that occur of their own accord are irreversible (with reversible processes being a useful idealization). These processes are also spontaneous. Thus, *the total entropy of the universe increases in any spontaneous process.* This profound generalization is yet another way of expressing the second law of thermodynamics.

GIVE IT SOME THOUGHT

The rusting of iron is accompanied by a decrease in the entropy of the system (the iron and oxygen). What can we conclude about the entropy change of the surroundings?

The second law of thermodynamics tells us the essential character of any spontaneous change—it is always accompanied by an overall increase in entropy. We can, in fact, use this criterion to predict whether processes will be spontaneous. Before beginning to use the second law to predict spontaneity, however, we will find it useful to explore further the meaning of entropy from a molecular perspective.

Throughout most of the remainder of this chapter, we will focus mainly on the systems we encounter rather than on their surroundings. To simplify the notation, we will usually refer to the entropy change of the system merely as ΔS rather than explicitly indicating ΔS_{sys}.

19.3 THE MOLECULAR INTERPRETATION OF ENTROPY

As chemists, we are interested in molecules. What does entropy have to do with them and their transformations? What molecular property does entropy reflect? Ludwig Boltzmann (1844–1906) gave conceptual meaning to entropy. To understand Boltzmann's contribution, we need to examine the ways in which molecules can store energy.

Molecular Motions and Energy

When a substance is heated, the motion of its molecules increases. When we studied the kinetic-molecular theory of gases, we found that the average kinetic energy of the molecules of an ideal gas is directly proportional to the absolute temperature of the gas. ∞ (Section 10.7) That means the higher the temperature, the faster the molecules are moving and the more kinetic energy they possess. Moreover, hotter systems have a *broader distribution* of molecular speeds, as you can see by referring back to Figure 10.18. The particles of an ideal gas, however, are merely idealized points with no volume and no bonds, points that we visualize as flitting around through space. Real molecules can undergo more complex kinds of motions.

— Vibrations — | — Rotation —

▲ **Figure 19.6 Vibrational and rotational motions in a water molecule.** Vibrational motions in the molecule involve periodic displacements of the atoms with respect to one another. Rotational motions involve the spinning of a molecule about an axis.

Molecules can undergo three kinds of motion. The entire molecule can move in one direction, as in the motions of the particles of an ideal gas or the motions of larger objects, such as a baseball being thrown around a baseball field. We call such movement **translational motion**. The molecules in a gas have more freedom of translational motion than those in a liquid, which in turn have more translational freedom than the molecules of a solid.

A molecule may also undergo **vibrational motion**, in which the atoms in the molecule move periodically toward and away from one another, much as a tuning fork vibrates about its equilibrium shape. In addition, molecules may possess **rotational motion**, as though they were spinning like tops. Figure 19.6▲ shows the vibrational motions and one of the rotational motions possible for the water molecule. These different forms of motion are ways in which a molecule can store energy, and we refer to the various forms collectively as the "motional energy" of the molecule.

▲ **GIVE IT SOME THOUGHT**

What kinds of motion can a molecule undergo that a single atom cannot?

Boltzmann's Equation and Microstates

The science of thermodynamics developed as a means of describing the properties of matter in our macroscopic world without regard to the microscopic structure of the matter. In fact, thermodynamics was a well-developed field before the modern views of atomic and molecular structure were even known. The thermodynamic properties of water, for example, addressed the behavior of bulk water (or ice or water vapor) as a substance, without considering any specific properties of individual H_2O molecules.

To connect the microscopic and macroscopic descriptions of matter, scientists have developed the field of *statistical thermodynamics*, which uses the tools of statistics and probability to provide the link between the microscopic and macroscopic worlds. Here we will show how entropy, which is a property of bulk matter, can be connected to the behavior of atoms and molecules. Because the mathematics of statistical thermodynamics is quite complex, our discussion will be largely conceptual.

Let's begin by considering one mole of an ideal gas in a particular thermodynamic state, which we can define by specifying the temperature, T, and volume, V, of the gas. (Remember that the energy, E, of an ideal gas depends only on its temperature and that by fixing the values of n, T, and V, we also fix the value of the pressure, P.) What is happening to our ideal gas sample at the microscopic level, and how does what is going on at the microscopic level relate to the entropy of the sample? To address these questions, we need to consider both the positions of the gas molecules and their individual kinetic energies, which depend on the speeds of the molecules. In our discussion of the kinetic-molecular theory, we considered the gas molecules to be in constant motion within the entire volume of the container. We also saw that the speeds of the gas molecules follow a well-defined distribution at a given temperature, such as that shown in Figure 10.18. ∞ (Section 10.7)

Now imagine that we could take a "snapshot" of the positions and speeds of all of the molecules at a given instant. That particular set of 6×10^{23} positions and energies of the individual gas molecules is what we call a **microstate** of the thermodynamic system. A microstate is a single possible arrangement of the positions and kinetic energies of the gas molecules when the gas is in a specific thermodynamic state. We could envision continuing to take snapshots of our system to see other possible microstates. In fact, as you no doubt see, there would be such a staggeringly large number of microstates that taking individual snapshots of all of them is not feasible. Because we are examining such a large number of particles, however, we can use the tools of statistics and probability to determine the total number of microstates for the thermodynamic state. (That is where the *statistical* part of statistical thermodynamics comes in.) Each thermodynamic state has a characteristic number of microstates associated with it, and we will use the symbol W for that number.

The connection between the number of microstates of a system, W, and its entropy, S, is expressed in a beautifully simple equation developed by Boltzmann:

$$S = k \ln W \qquad [19.5]$$

In this equation, k is Boltzmann's constant, 1.38×10^{-23} J/K. Thus, *entropy is a measure of how many microstates are associated with a particular macroscopic state.* Equation 19.5 appears on Boltzmann's gravestone (Figure 19.7▶).

▲ Figure 19.7 **Ludwig Boltzmann's gravestone.** Boltzmann's gravestone in Vienna is inscribed with his famous relationship between the entropy of a state and the number of available microstates. (In Boltzmann's time, "log" was used to represent the natural logarithm.)

GIVE IT SOME THOUGHT

What is the entropy of a system that has only a single microstate?

The entropy change accompanying any process is

$$\Delta S = k \ln W_{\text{final}} - k \ln W_{\text{initial}} = k \ln \frac{W_{\text{final}}}{W_{\text{initial}}} \qquad [19.6]$$

Thus, any change in the system that leads to an increase in the number of microstates leads to a positive value of ΔS: *Entropy increases with the number of microstates of the system.*

Let's briefly consider two simple changes to our ideal-gas sample and see how the entropy changes in each case. First, suppose we increase the volume of the system, which is analogous to allowing the gas to expand isothermally. A greater volume means that there are a greater number of positions available to the gas atoms. Thus, there will be a greater number of microstates for the system after the increase in volume. The entropy therefore increases as the volume increases, as we saw in the "A Closer Look" box in Section 19.2. Second, suppose we keep the volume fixed but increase the temperature. How will this change affect the entropy of the system? Recall the distribution of molecular speeds presented in Figure 10.18. An increase in temperature increases the average (rms) speed of the molecules and broadens the distribution of speeds. Hence, the molecules have a greater number of possible kinetic energies, and the number of microstates will once again increase. The entropy of the system will therefore increase with increasing temperature.

If we consider real molecules instead of ideal-gas particles, we must also consider the different amounts of vibrational and rotational energies the molecules have in addition to their kinetic energies. A collection of real molecules therefore has a greater number of microstates available than does the same number of ideal-gas particles. In general, *the number of microstates available to a system increases with an increase in volume, an increase in temperature, or an increase in the number of molecules because any of these changes increases the possible positions and energies of the molecules of the system.*

A Closer Look ENTROPY AND PROBABILITY

The game of poker is sometimes used as an analogy to explore the idea of the microstates associated with a particular state. There are about 2.6 million different five-card poker hands that can be dealt, and each of these hands can be viewed as a possible "microstate" for the hand dealt to any one player in a game. Table 19.1▼ shows two poker hands. The probability that a particular hand will contain five *specific* cards is the same regardless of which five cards are specified. Thus, there is an equal probability of dealing either of the hands shown in Table 19.1. However, the first hand, a royal flush (the ten through ace of a single suit), strikes us as much more highly ordered than the second hand, a "nothing." The reason for this is clear if we compare the number of five-card arrangements that correspond to a royal flush to the number corresponding to a nothing: only four hands (microstates) for a royal flush but more than 1.3 million for a nothing hand. The nothing state has a higher probability of being dealt from a shuffled deck than the royal-flush state because there are so many more arrangements of cards that correspond to the nothing state. In other words, the value of W in Boltzmann's equation (Equation 19.5) is much greater for a nothing than for a royal flush. This example teaches us that there is a connection between probability and entropy.

The entropy of any system has a natural tendency to increase, because increased entropy represents a movement toward a state of higher probability. Let's use this reasoning to explain the isothermal expansion of a gas, such as that depicted in Figure 19.1. When the stopcock is opened, the gas molecules are less constrained, and there are more possible arrangements for them (more microstates) in the larger volume. The various microstates are depicted in a schematic way in Figure 19.8▼. In this figure, we make no attempt to describe the motion of the particles, focusing instead only on their locations. The spreading of the molecules over the larger volume represents movement to the more probable state.

When we use the terms randomness and disorder to describe entropy, we have to be careful not to carry an aesthetic sense of what we mean. What we must remember is that the fundamental connection to entropy is not tied directly with randomness, disorder, or energy dispersal, but with the number of available microstates.

(a)   (b)

▲ **Figure 19.8 Probability and the locations of gas molecules.** The two molecules are colored red and blue to keep track of them. (a) Before the stopcock is opened, both molecules are in the right-hand flask. (b) After the stopcock is opened, there are four possible arrangements of the two molecules. Only one of the four arrangements corresponds to both molecules being in the right-hand flask. The greater number of possible arrangements corresponds to greater disorder in the system. In general, the probability that the molecules will stay in the original flask is $\left(\frac{1}{2}\right)^n$, where n is the number of molecules.

TABLE 19.1 ■ A Comparison of the Number of Combinations that Can Lead to a Royal Flush and to a "Nothing" Hand in Poker

Hand					State	Number of Hands that Lead to This State
10♠	J♠	Q♠	K♠	A♠	Royal flush	4
2♣	5♦	6♥	10♦	J♠	"Nothing"	1,302,540

Chemists use several different ways to describe an increase in the number of microstates and therefore an increase in the entropy for a system. Each of these ways seeks to capture a sense of the increased freedom of motion that causes molecules to spread out if not restrained by physical barriers or chemical bonds.

Some say the increase in entropy represents an increase in the *randomness* or *disorder* of the system. Others liken an increase in entropy to an increased *dispersion (spreading out) of energy* because there is an increase in the number of ways the positions and energies of the molecules can be distributed throughout the system. Each of these descriptions (randomness, disorder, and energy dispersal) is conceptually helpful if applied correctly. Indeed, you will find it useful to keep these descriptions in mind as you evaluate entropy changes.

▲ **Figure 19.9 Structure of Ice.** The intermolecular attractions in the three-dimensional lattice restrict the molecules to vibrational motion only.

Making Qualitative Predictions About ΔS

It is usually not difficult to construct a mental picture to estimate qualitatively how the entropy of a system changes during a simple process. In most instances, an increase in the number of microstates, and hence an increase in entropy, parallels an increase in

1. temperature
2. volume
3. number of independently moving particles

Thus, we can usually make qualitative predictions about entropy changes by focusing on these factors. For example, when water vaporizes, the molecules spread out into a larger volume. Because they occupy a larger space, there is an increase in their freedom of motion, giving rise to more accessible microstates and hence to an increase in entropy.

Consider the melting of ice. The rigid structure of the water molecules, shown in Figure 19.9▲, restricts motion to only tiny vibrations throughout the crystal. In contrast, the molecules in liquid water are free to move about with respect to one another (translation) and to tumble around (rotation) as well as vibrate. During melting, therefore, the number of accessible microstates increases and so does the entropy.

When an ionic solid, such as KCl, dissolves in water, a mixture of water and ions replaces the pure solid and pure water (Figure 19.10▼). The ions now move in a larger volume and possess more motional energy than in the rigid solid. We have to be careful, however, because water molecules are held around the ions as water of hydration. ∞ (Section 13.1) These water molecules have less motional energy than before because they are now confined to the immediate environment of the ions. The greater the charge of an ion, the greater are the ion–dipole attractions that hold the ion and the water together, thereby restricting motions. Thus, even though the solution process is normally accompanied by a net increase in entropy, the dissolving of salts with highly charged ions can result in a net *decrease* in entropy.

▶ **Figure 19.10 Dissolving an ionic solid in water.** The ions become more spread out and random in their motions, but the water molecules that hydrate the ions become less random.

2 NO(g) + O₂(g)   2 NO₂(g)

▲ **Figure 19.11 Entropy change for a reaction.** A decrease in the number of gaseous molecules leads to a decrease in the entropy of the system. When the NO(g) and O₂(g) (left) react to form the NO₂(g) (right), the number of gaseous molecules decreases. The atoms have fewer degrees of freedom because new N—O bonds form and the entropy decreases.

The same ideas apply to systems involving chemical reactions. Consider the reaction between nitric oxide gas and oxygen gas to form nitrogen dioxide gas:

$$2\,NO(g) + O_2(g) \longrightarrow 2\,NO_2(g) \qquad [19.7]$$

In this case the reaction results in a decrease in the number of molecules— three molecules of gaseous reactants form two molecules of gaseous products (Figure 19.11 ◄). The formation of new N—O bonds reduces the motions of the atoms in the system. The formation of new bonds decreases the *number of degrees of freedom*, or forms of motion, available to the atoms. That is, the atoms are less free to move in random fashion because of the formation of new bonds. The decrease in the number of molecules and the resultant decrease in motion result in fewer accessible microstates and therefore a decrease in the entropy of the system.

In summary, we generally expect the entropy of the system to increase for processes in which

1. Gases are formed from either solids or liquids.
2. Liquids or solutions are formed from solids.
3. The number of gas molecules increases during a chemical reaction.

■■ **SAMPLE EXERCISE 19.3** | Predicting the Sign of ΔS

Predict whether ΔS is positive or negative for each of the following processes, assuming each occurs at constant temperature:

(a) $H_2O(l) \longrightarrow H_2O(g)$
(b) $Ag^+(aq) + Cl^-(aq) \longrightarrow AgCl(s)$
(c) $4\,Fe(s) + 3\,O_2(g) \longrightarrow 2\,Fe_2O_3(s)$
(d) $N_2(g) + O_2(g) \longrightarrow 2\,NO(g)$

SOLUTION

Analyze: We are given four equations and asked to predict the sign of ΔS for each chemical reaction.

Plan: The sign of ΔS will be positive if there is an increase in temperature, an increase in the volume in which the molecules move, or an increase in the number of gas particles in the reaction. The question states that the temperature is constant. Thus, we need to evaluate each equation with the other two factors in mind.

Solve:

(a) The evaporation of a liquid is accompanied by a large increase in volume. One mole of water (18 g) occupies about 18 mL as a liquid and if it could exist as a gas at STP it would occupy 22.4 L. Because the molecules are distributed throughout a much larger volume in the gaseous state than in the liquid state, an increase in motional freedom accompanies vaporization. Therefore, ΔS is positive.

(b) In this process the ions, which are free to move throughout the volume of the solution, form a solid in which they are confined to a smaller volume and restricted to more highly constrained positions. Thus, ΔS is negative.

(c) The particles of a solid are confined to specific locations and have fewer ways to move (fewer microstates) than do the molecules of a gas. Because O_2 gas is converted into part of the solid product Fe_2O_3, ΔS is negative.

(d) The number of moles of gases is the same on both sides of the equation, and so the entropy change will be small. The sign of ΔS is impossible to predict based on our discussions thus far, but we can predict that ΔS will be close to zero.

■■ **PRACTICE EXERCISE**

Indicate whether each of the following processes produces an increase or decrease in the entropy of the system:

(a) $CO_2(s) \longrightarrow CO_2(g)$
(b) $CaO(s) + CO_2(g) \longrightarrow CaCO_3(s)$
(c) $HCl(g) + NH_3(g) \longrightarrow NH_4Cl(s)$
(d) $2\,SO_2(g) + O_2(g) \longrightarrow 2\,SO_3(g)$

Answers: (a) increase, (b) decrease, (c) decrease, (d) decrease

■ **SAMPLE EXERCISE 19.4** | **Predicting Which Sample of Matter Has the Higher Entropy**

Choose the sample of matter that has greater entropy in each pair, and explain your choice: **(a)** 1 mol of NaCl(s) or 1 mol of HCl(g) at 25 °C, **(b)** 2 mol of HCl(g) or 1 mol of HCl(g) at 25 °C, **(c)** 1 mol of HCl(g) or 1 mol of Ar(g) at 298 K.

SOLUTION

Analyze: We need to select the system in each pair that has the greater entropy.

Plan: To do this, we examine the state of the system and the complexity of the molecules it contains.

Solve: (a) Gaseous HCl has the higher entropy because gases have more available motions than solids. **(b)** The sample containing 2 mol of HCl has twice the number of molecules as the sample containing 1 mol. Thus, the 2-mol sample has twice the number of microstates and twice the entropy when they are at the same pressure. **(c)** The HCl sample has the higher entropy because the HCl molecule is capable of storing energy in more ways than is Ar. HCl molecules can rotate and vibrate; Ar atoms cannot.

■ **PRACTICE EXERCISE**

Choose the substance with the greater entropy in each case: **(a)** 1 mol of $H_2(g)$ at STP or 1 mol of $H_2(g)$ at 100 °C and 0.5 atm, **(b)** 1 mol of $H_2O(s)$ at 0 °C or 1 mol of $H_2O(l)$ at 25 °C, **(c)** 1 mol of $H_2(g)$ at STP or 1 mol of $SO_2(g)$ at STP, **(d)** 1 mol of $N_2O_4(g)$ at STP or 2 mol of $NO_2(g)$ at STP.
Answers: **(a)** 1 mol of $H_2(g)$ at 100 °C and 0.5 atm, **(b)** 1 mol of $H_2O(l)$ at 25 °C, **(c)** 1 mol of $SO_2(g)$ at STP, **(d)** 2 mol of $NO_2(g)$ at STP

◄ *Chemistry and Life*   ENTROPY AND LIFE

The ginkgo leaf shown in Figure 19.12(a) ▶ reveals beautiful patterns of form and color. Both plant systems and animal systems, including those of humans, are incredibly complex structures in which a host of substances come together in organized ways to form cells, tissue, organ systems, and so on. These various components must operate in synchrony for the organism as a whole to be viable. If even one key system strays far from its optimal state, the organism as a whole may die.

To make a living system from its component molecules—such as a ginkgo leaf from sugar molecules, cellulose molecules, and the other substances present in the leaf—requires a very large reduction in entropy. It would seem, then, that living systems might violate the second law of thermodynamics. They seem spontaneously to become more, not less, organized as they develop. To get the full picture, however, we must take into account the surroundings.

We know that a system can move toward lower entropy if we do work on it. (That is, if we supply energy to the system in a very specific way.) When we do work on a gas, for example, by compressing it isothermally, the entropy of the gas is lowered. The energy for the work done is provided by the surroundings, and in the process the net entropy change in the universe is positive.

The striking thing about living systems is that they are organized to recruit energy from their surroundings spontaneously. Some single-celled organisms, called *autotrophs*, capture energy from sunlight and store it in molecules such as sugars and fats [Figure 19.12(b)]. Others, called *heterotrophs*, absorb food molecules from their surroundings and then break down the molecules to provide needed energy. Whatever their mode of existence, however, living systems gain their order at the expense of the surroundings. Each cell exists at the expense of an increase in the entropy of the universe.

▲ **Figure 19.12 Entropy and life.** (a) This ginkgo leaf represents a highly organized living system. (b) Cyanobacteria absorb light energy and utilize it to synthesize the substances needed for growth.

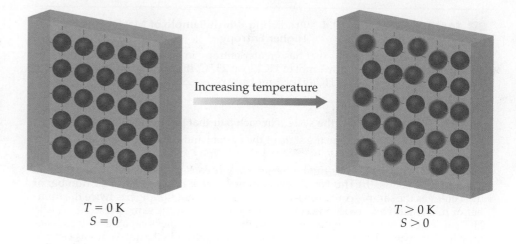

▶ **Figure 19.13 A perfectly ordered crystalline solid at and above 0 K.** At absolute zero (left), all lattice units are in their lattice sites, devoid of thermal motion. As the temperature rises above 0 K (right), the atoms or molecules gain energy and their vibrational motion increases.

$T = 0\,K$
$S = 0$

Increasing temperature

$T > 0\,K$
$S > 0$

The Third Law of Thermodynamics

If we decrease the thermal energy of a system by lowering the temperature, the energy stored in translational, vibrational, and rotational forms of motion decreases. As less energy is stored, the entropy of the system decreases. If we keep lowering the temperature, do we reach a state in which these motions are essentially shut down, a point described by a single microstate? This question is addressed by the **third law of thermodynamics**, which states that *the entropy of a pure crystalline substance at absolute zero is zero*: $S(0\,K) = 0$.

Figure 19.13▲ shows schematically a pure crystalline solid. At absolute zero all the units of the lattice have no thermal motion. There is, therefore, only one microstate. As a result $S = k \ln W = k \ln 1 = 0$. As the temperature is increased from absolute zero, the atoms or molecules in the crystal gain energy in the form of vibrational motion about their lattice positions. Thus, the degrees of freedom of the crystal increase. The entropy of the lattice therefore increases with temperature because vibrational motion causes the atoms or molecules to have a greater number of accessible microstates.

What happens to the entropy of the substance as we continue to heat it? Figure 19.14◀ is a plot of how the entropy of a typical substance varies with temperature. We see that the entropy of the solid continues to increase steadily with increasing temperature up to the melting point of the solid. When the solid melts, the bonds holding the atoms or molecules are broken and the particles are free to move about the entire volume of the substance. The added degrees of freedom for the individual molecules allow greater dispersal of the substance's energy thereby increasing its entropy. We therefore see a sharp increase in the entropy at the melting point. After all the solid has melted to liquid, the temperature again increases and with it, the entropy.

At the boiling point of the liquid, another abrupt increase in entropy occurs. We can understand this increase as resulting from the increased volume in which the molecules may be found. When the gas is heated further, the entropy increases steadily as more energy is stored in the translational motion of the gas molecules. At higher temperatures, the distribution of molecular speeds is spread out toward higher values. ∞(Figure 10.18) More of the molecules have speeds that differ greatly from the most probable value. The expansion of the range of speeds of the gas molecules leads to an increased entropy.

The general conclusions we reach in examining Figure 19.14 are consistent with what we noted earlier: Entropy generally increases with increasing temperature because the increased motional energy can be dispersed in more ways. Further, the entropies of the phases of a given substance follow the order $S_{solid} < S_{liquid} < S_{gas}$. This ordering fits in nicely with our picture of the number of microstates available to solids, liquids, and gases.

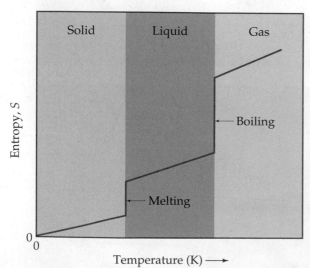

▲ **Figure 19.14 Entropy as a function of temperature.** Entropy increases as the temperature of a crystalline solid is increased from absolute zero. The vertical jumps in entropy correspond to phase changes.

GIVE IT SOME THOUGHT

If you are told that the entropy of a certain system is zero, what do you know about the system?

19.4 ENTROPY CHANGES IN CHEMICAL REACTIONS

In Section 5.5 we discussed how calorimetry can be used to measure ΔH for chemical reactions. No comparable, easy method exists for measuring ΔS for a reaction. By using experimental measurements of the variation of heat capacity with temperature, however, we can determine the absolute value of the entropy, S, for many substances at any temperature. (The theory and the methods used for these measurements and calculations are beyond the scope of this text.) Absolute entropies are based on the reference point of zero entropy for perfect crystalline solids at 0 K (the third law). Entropies are usually tabulated as molar quantities, in units of joules per mole-kelvin (J/mol-K).

The molar entropy values of substances in their standard states are known as **standard molar entropies** and are denoted $S°$. The standard state for any substance is defined as the pure substance at 1 atm pressure.* Table 19.2 ▶ lists the values of $S°$ for several substances at 298 K; Appendix C gives a more extensive list.

We can make several observations about the $S°$ values in Table 19.2:

1. Unlike enthalpies of formation, the standard molar entropies of elements at the reference temperature of 298 K are *not* zero.

2. The standard molar entropies of gases are greater than those of liquids and solids, consistent with our interpretation of experimental observations, as represented in Figure 19.14

3. Standard molar entropies generally increase with increasing molar mass. [Compare Li(s), Na(s), and K(s).]

4. Standard molar entropies generally increase with an increasing number of atoms in the formula of a substance.

Point 4 is related to molecular motion (Section 19.3). In general, the number of degrees of freedom for a molecule increases with increasing number of atoms, and thus the number of accessible microstates also increases. Figure 19.15 ▼ compares the standard molar entropies of three hydrocarbons. Notice how the entropy increases as the number of atoms in the molecule increases.

The entropy change in a chemical reaction equals the sum of the entropies of the products less the sum of the entropies of the reactants:

$$\Delta S° = \sum nS°(\text{products}) - \sum mS°(\text{reactants}) \qquad [19.8]$$

As in Equation 5.31, the coefficients n and m are the coefficients in the chemical equation, as illustrated in Sample Exercise 19.5.

TABLE 19.2 ■ Standard Molar Entropies of Selected Substances at 298 K	
Substance	$S°$, J/mol-K
Gases	
$H_2(g)$	130.6
$N_2(g)$	191.5
$O_2(g)$	205.0
$H_2O(g)$	188.8
$NH_3(g)$	192.5
$CH_3OH(g)$	237.6
$C_6H_6(g)$	269.2
Liquids	
$H_2O(l)$	69.9
$CH_3OH(l)$	126.8
$C_6H_6(l)$	172.8
Solids	
$Li(s)$	29.1
$Na(s)$	51.4
$K(s)$	64.7
$Fe(s)$	27.23
$FeCl_3(s)$	142.3
$NaCl(s)$	72.3

▼ **Figure 19.15 Molar entropies.** In general, the more complex a molecule (that is, the greater the number of atoms present), the greater the molar entropy of the substance, as illustrated here by the molar entropies of three simple hydrocarbons.

Methane, CH_4
$S° = 186.3$ J mol^{-1} K^{-1}

Ethane, C_2H_6
$S° = 229.6$ J mol^{-1} K^{-1}

Propane, C_3H_8
$S° = 270.3$ J mol^{-1} K^{-1}

*The standard pressure used in thermodynamics is no longer 1 atm but is based on the SI unit for pressure, the pascal (Pa). The standard pressure is 10^5 Pa, a quantity known as a bar: 1 bar = 10^5 Pa = 0.987 atm. Because 1 bar differs from 1 atm by only 1.3%, we will continue to refer to the standard pressure as 1 atm.

■ **SAMPLE EXERCISE 19.5** | Calculating ΔS from Tabulated Entropies

Calculate $\Delta S°$ for the synthesis of ammonia from $N_2(g)$ and $H_2(g)$ at 298 K:

$$N_2(g) + 3\,H_2(g) \longrightarrow 2\,NH_3(g)$$

SOLUTION

Analyze: We are asked to calculate the entropy change for the synthesis of $NH_3(g)$ from its constituent elements.

Plan: We can make this calculation using Equation 19.8 and the standard molar entropy values for the reactants and the products that are given in Table 19.2 and in Appendix C.

Solve: Using Equation 19.8, we have

$$\Delta S° = 2S°(NH_3) - [S°(N_2) + 3S°(H_2)]$$

Substituting the appropriate $S°$ values from Table 19.2 yields

$$\Delta S° = (2\text{ mol})(192.5\text{ J/mol-K}) - [(1\text{ mol})(191.5\text{ J/mol-K}) + (3\text{ mol})(130.6\text{ J/mol-K})]$$
$$= -198.3\text{ J/K}$$

Check: The value for $\Delta S°$ is negative, in agreement with our qualitative prediction based on the decrease in the number of molecules of gas during the reaction.

■ **PRACTICE EXERCISE**

Using the standard entropies in Appendix C, calculate the standard entropy change, $\Delta S°$, for the following reaction at 298 K:

$$Al_2O_3(s) + 3\,H_2(g) \longrightarrow 2\,Al(s) + 3\,H_2O(g)$$

Answer: 180.39 J/K

Entropy Changes in the Surroundings

We can use tabulated absolute entropy values to calculate the standard entropy change in a system, such as a chemical reaction, as just described. But what about the entropy change in the surroundings? We encountered this situation in Section 19.2, but it is good to revisit it now that we are examining chemical reactions.

We should recognize that the surroundings serve essentially as a large, constant-temperature heat source (or heat sink if the heat flows from the system to the surroundings). The change in entropy of the surroundings will depend on how much heat is absorbed or given off by the system. For an isothermal process, the entropy change of the surroundings is given by

$$\Delta S_{surr} = \frac{-q_{sys}}{T}$$

For a reaction occurring at constant pressure, q_{sys} is simply the enthalpy change for the reaction, ΔH. Thus, we can write

$$\Delta S_{surr} = \frac{-\Delta H_{sys}}{T} \qquad [19.9]$$

For the reaction in Sample Exercise 19.5, the formation of ammonia from $H_2(g)$ and $N_2(g)$ at 298 K, q_{sys} is the enthalpy change for reaction under standard conditions, $\Delta H°$. ∞ (Section 5.7) Using the procedures described in Section 5.7, we have

$$\Delta H°_{rxn} = 2\Delta H°_f[NH_3(g)] - 3\Delta H°_f[H_2(g)] - \Delta H°_f[N_2(g)]$$
$$= 2(-46.19\text{ kJ}) - 3(0\text{ kJ}) - (0\text{ kJ}) = -92.38\text{ kJ}$$

Thus at 298 K the formation of ammonia from $H_2(g)$ and $N_2(g)$ is exothermic. Absorption of the heat given off by the system results in an increase in the entropy of the surroundings:

$$\Delta S°_{surr} = \frac{92.38\text{ kJ}}{298\text{ K}} = 0.310\text{ kJ/K} = 310\text{ J/K}$$

Notice that the magnitude of the entropy gained by the surroundings is greater than that lost by the system (as calculated in Sample Exercise 19.5):

$$\Delta S^\circ_{univ} = \Delta S^\circ_{sys} + \Delta S^\circ_{surr} = -198.3 \text{ J/K} + 310 \text{ J/K} = 112 \text{ J/K}$$

Because ΔS°_{univ} is positive for any spontaneous reaction, this calculation indicates that when $NH_3(g)$, $H_2(g)$, and $N_2(g)$ are together at 298 K in their standard states (each at 1 atm pressure), the reaction system will move spontaneously toward formation of $NH_3(g)$. Keep in mind that while the thermodynamic calculations indicate that formation of ammonia is spontaneous, they do not tell us anything about the rate at which ammonia is formed. Establishing equilibrium in this system within a reasonable period requires a catalyst, as discussed in Section 15.7.

GIVE IT SOME THOUGHT

If a process is exothermic, does the entropy of the surroundings (1) always increase, (2) always decrease, or (3) sometimes increase and sometimes decrease, depending on the process?

19.5 GIBBS FREE ENERGY

We have seen examples of endothermic processes that are spontaneous, such as the dissolution of ammonium nitrate in water. ⚮ (Section 13.1) We learned in our discussion of the solution process that a spontaneous, endothermic process must be accompanied by an increase in the entropy of the system. However, we have also encountered processes that are spontaneous and yet proceed with a *decrease* in the entropy of the system, such as the highly exothermic formation of sodium chloride from its constituent elements. ⚮ (Section 8.2) Spontaneous processes that result in a decrease in the system's entropy are always exothermic. Thus, the spontaneity of a reaction seems to involve two thermodynamic concepts, enthalpy and entropy.

There should be a way to use ΔH and ΔS to predict whether a given reaction occurring at constant temperature and pressure will be spontaneous. The means for doing so was first developed by the American mathematician J. Willard Gibbs (1839–1903). Gibbs (Figure 19.16▶) proposed a new state function, now called the **Gibbs free energy** (or just **free energy**). The Gibbs free energy, G, of a state is defined as

$$G = H - TS \qquad [19.10]$$

where T is the absolute temperature. For a process occurring at constant temperature, the change in free energy of the system, ΔG, is given by the expression

$$\Delta G = \Delta H - T\Delta S \qquad [19.11]$$

Under standard conditions, this equation becomes

$$\Delta G^\circ = \Delta H^\circ - T\Delta S^\circ \qquad [19.12]$$

To see how the state function G relates to reaction spontaneity, recall that for a reaction occurring at constant temperature and pressure

$$\Delta S_{univ} = \Delta S_{sys} + \Delta S_{surr} = \Delta S_{sys} + \left(\frac{-\Delta H_{sys}}{T}\right)$$

Multiplying both sides by $(-T)$ gives us

$$-T\Delta S_{univ} = \Delta H_{sys} - T\Delta S_{sys} \qquad [19.13]$$

▲ **Figure 19.16 Josiah Willard Gibbs (1839–1903).** Gibbs was the first person to be awarded a Ph.D. in science from an American university (Yale, 1863). From 1871 until his death, he held the chair of mathematical physics at Yale. He developed much of the theoretical foundation that led to the development of chemical thermodynamics.

Position

(a)

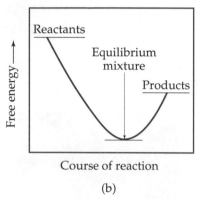

Course of reaction

(b)

▲ **Figure 19.17 Potential energy and free energy.** An analogy is shown between the gravitational potential-energy change of a boulder rolling down a hill (a) and the free-energy change in a spontaneous reaction (b). The equilibrium position in (a) is given by the minimum gravitational potential energy available to the system. The equilibrium position in (b) is given by the minimum free energy available to the system.

Comparing Equation 19.13 with Equation 19.11, we see that the free-energy change in a process occurring at constant temperature and pressure, ΔG, is equal to $-T\Delta S_{univ}$. We know that for spontaneous processes, ΔS_{univ} is always positive, and therefore $-T\Delta S_{univ}$ will be negative. Thus, the sign of ΔG provides us with extremely valuable information about the spontaneity of processes that occur at constant temperature and pressure. If both T and P are constant, the relationship between the sign of ΔG and the spontaneity of a reaction is as follows:

1. If ΔG is negative, the reaction is spontaneous in the forward direction.

2. If ΔG is zero, the reaction is at equilibrium.

3. If ΔG is positive, the reaction in the forward direction is nonspontaneous; work must be supplied from the surroundings to make it occur. However, the reverse reaction will be spontaneous.

It is more convenient to use ΔG as a criterion for spontaneity than to use ΔS_{univ}, because ΔG relates to the system alone and avoids the complication of having to examine the surroundings.

An analogy is often drawn between the free-energy change during a spontaneous reaction and the potential-energy change when a boulder rolls down a hill. Potential energy in a gravitational field "drives" the boulder until it reaches a state of minimum potential energy in the valley [Figure 19.17(a) ◄]. Similarly, the free energy of a chemical system decreases until it reaches a minimum value [Figure 19.17(b)]. When this minimum is reached, a state of equilibrium exists. *In any spontaneous process at constant temperature and pressure, the free energy always decreases.*

As a specific illustration of these ideas, let's return to the Haber process for the synthesis of ammonia from nitrogen and hydrogen, which we discussed extensively in Chapter 15:

$$N_2(g) + 3\,H_2(g) \rightleftharpoons 2\,NH_3(g)$$

Imagine that we have a reaction vessel that allows us to maintain a constant temperature and pressure and that we have a catalyst that allows the reaction to proceed at a reasonable rate. What will happen if we charge the vessel with a certain number of moles of N_2 and three times that number of moles of H_2? As we saw in Figure 15.3(a), the N_2 and H_2 will react spontaneously to form NH_3 until equilibrium is achieved. Similarly, Figure 15.3(b) demonstrates that if we charge the vessel with pure NH_3, it will decompose spontaneously to form N_2 and H_2 until equilibrium is reached. In each case the free energy of the system is lowered on the way to equilibrium, which represents a minimum in the free energy. We illustrate these cases in Figure 19.18 ▶.

◢ **GIVE IT SOME THOUGHT**

Give the criterion for spontaneity first in terms of entropy and then in terms of free energy.

This is a good time to remind ourselves of the significance of the reaction quotient, Q, for a system that is not at equilibrium. ∞ (Section 15.6) Recall that when $Q < K$, there is an excess of reactants relative to products. The reaction will proceed spontaneously in the forward direction to reach equilibrium. When $Q > K$, the reaction will proceed spontaneously in the reverse direction. At equilibrium $Q = K$. We have illustrated these points in Figure 19.18. In Section 19.7 we will see how to use the value of Q to calculate the value of ΔG for systems that are not at equilibrium.

$$N_2(g) + 3H_2(g) \rightleftharpoons 2NH_3(g)$$

Free energy →

Spontaneous

Spontaneous

$Q < K$ $Q > K$

Pure
$N_2 + H_2$

Equilibrium
mixture
$(Q = K, \Delta G = 0)$

Pure
NH_3

◀ Figure 19.18 **Free energy and equilibrium.** In the reaction $N_2(g) + 3\ H_2(g) \rightleftharpoons 2\ NH_3(g)$, if the reaction mixture has too much N_2 and H_2 relative to NH_3 (left), the equilibrium lies too far to the left ($Q < K$) and NH_3 forms spontaneously. If there is too much NH_3 in the mixture (right), the equilibrium lies too far to the right ($Q > K$) and the NH_3 decomposes spontaneously into N_2 and H_2. Both of these spontaneous processes are "downhill" in free energy. At equilibrium (center), $Q = K$ and the free energy is at a minimum ($\Delta G = 0$).

███ **SAMPLE EXERCISE 19.6** | Calculating Free-Energy Change from $\Delta H°$, T, and $\Delta S°$

Calculate the standard free energy change for the formation of $NO(g)$ from $N_2(g)$ and $O_2(g)$ at 298 K:

$$N_2(g) + O_2(g) \longrightarrow 2\ NO(g)$$

given that $\Delta H° = 180.7$ kJ and $\Delta S° = 24.7$ J/K. Is the reaction spontaneous under these circumstances?

SOLUTION

Analyze: We are asked to calculate $\Delta G°$ for the indicated reaction (given $\Delta H°$, $\Delta S°$, and T) and to predict whether the reaction is spontaneous under standard conditions at 298 K.

Plan: To calculate $\Delta G°$, we use Equation 19.12, $\Delta G° = \Delta H° - T\Delta S°$. To determine whether the reaction is spontaneous under standard conditions, we look at the sign of $\Delta G°$.

Solve:

$$\Delta G° = \Delta H° - T\Delta S°$$
$$= 180.7 \text{ kJ} - (298 \text{ K})(24.7 \text{ J/K})\left(\frac{1 \text{ kJ}}{10^3 \text{ J}}\right)$$
$$= 180.7 \text{ kJ} - 7.4 \text{ kJ}$$
$$= 173.3 \text{ kJ}$$

Because $\Delta G°$ is positive, the reaction is not spontaneous under standard conditions at 298 K.

Comment: Notice that we had to convert the units of the $T\Delta S°$ term to kJ so that they could be added to the $\Delta H°$ term, whose units are kJ.

███ **PRACTICE EXERCISE**

A particular reaction has $\Delta H° = 24.6$ kJ and $\Delta S° = 132$ J/K at 298 K. Calculate $\Delta G°$. Is the reaction spontaneous under these conditions?
Answer: $\Delta G° = -14.7$ kJ; the reaction is spontaneous.

TABLE 19.3 ■ Conventions Used in Establishing Standard Free Energies	
State of Matter	**Standard State**
Solid	Pure solid
Liquid	Pure liquid
Gas	1 atm pressure
Solution	1 M concentration
Elements	Standard free energy of formation of an element in its standard state is defined as zero

Standard Free Energy of Formation

Free energy is a state function, like enthalpy. We can tabulate **standard free energies of formation** for substances, just as we can tabulate standard enthalpies of formation. ∞ (Section 5.7) It is important to remember that standard values for these functions imply a particular set of conditions, or standard states. The standard state for gaseous substances is 1 atm pressure. For solid substances, the standard state is the pure solid; for liquids, the pure liquid. For substances in solution, the standard state is normally a concentration of 1 M. (In very accurate work it may be necessary to make certain corrections, but we need not worry about these.) The temperature usually chosen for purposes of tabulating data is 25 °C, but we will calculate $\Delta G°$ at other temperatures as well. Just as for the standard heats of formation, the free energies of elements in their standard states are set to zero. This arbitrary choice of a reference point has no effect on the quantity in which we are interested, namely, the *difference* in free energy between reactants and products. The rules about standard states are summarized in Table 19.3 ◄. A listing of standard free energies of formation, denoted $\Delta G_f°$, appears in Appendix C.

GIVE IT SOME THOUGHT

What does the superscript ° indicate when associated with a thermodynamic quantity, as in $\Delta H°$, $\Delta S°$, or $\Delta G°$?

The standard free energies of formation are useful in calculating the *standard free-energy change* for chemical processes. The procedure is analogous to the calculation of $\Delta H°$ (Equation 5.31) and $\Delta S°$ (Equation 19.8):

$$\Delta G° = \sum n\Delta G_f°(\text{products}) - \sum m\Delta G_f°(\text{reactants}) \qquad [19.14]$$

A Closer Look WHAT'S "FREE" ABOUT FREE ENERGY?

The Gibbs free energy is a remarkable thermodynamic quantity. Because so many chemical reactions are carried out under conditions of near-constant pressure and temperature, chemists, biochemists, and engineers use the sign and magnitude of ΔG as exceptionally useful tools in the design and implementation of chemical and biochemical reactions. We will see examples of the usefulness of ΔG throughout the remainder of this chapter and this text. But what is "free" about free energy?

We have seen that we can use the sign of ΔG to conclude whether a reaction is spontaneous, nonspontaneous, or at equilibrium. The magnitude of ΔG is also significant. A reaction for which ΔG is large and negative, such as the burning of gasoline, is much more capable of doing work on the surroundings than is a reaction for which ΔG is small and negative, such as ice melting at room temperature. In fact, thermodynamics tells us that *the change in free energy for a process, ΔG, equals the maximum useful work that can be done by the system on its surroundings in a spontaneous process occurring at constant temperature and pressure:*

$$\Delta G = -w_{max} \qquad [19.15]$$

In other words, ΔG gives the theoretical limit to how much work can be done by a process. This relationship explains why ΔG is called the *free* energy. It is the portion of the energy change of a spontaneous reaction that is free to do useful work. The remainder of the energy enters the environment as heat.

For example, when we burn gasoline to move a car, only part of the energy of the gasoline is used to drive the car forward, performing useful work. The rest of the energy is dissipated to the surroundings as heat, accomplishing no useful work, as illustrated in Figure 19.19▼. The efficiency of energy conversion is given by the ratio of the work accomplished compared to the total energy used.

For processes that are not spontaneous ($\Delta G > 0$), the free-energy change is a measure of the *minimum* amount of work that must be done to cause the process to occur. In actual cases we always need to do more than this theoretical minimum amount because of the inefficiencies in the way the changes occur.

▲ **Figure 19.19 Energy conversion.** All conversions of energy are accompanied by production of heat, which enters the surroundings without accomplishing useful work.

SAMPLE EXERCISE 19.7 | **Calculating Standard Free-Energy Change from Free Energies of Formation**

(a) Use data from Appendix C to calculate the standard free-energy change for the following reaction at 298 K:

$$P_4(g) + 6\,Cl_2(g) \longrightarrow 4\,PCl_3(g)$$

(b) What is $\Delta G°$ for the reverse of the above reaction?

SOLUTION

Analyze: We are asked to calculate the free-energy change for the indicated reaction and then to determine the free-energy change of its reverse.

Plan: To accomplish our task, we look up the free-energy values for the products and reactants and use Equation 19.14: We multiply the molar quantities by the coefficients in the balanced equation, and subtract the total for the reactants from that for the products.

Solve:

(a) $Cl_2(g)$ is in its standard state, so $\Delta G_f°$ is zero for this reactant. $P_4(g)$, however, is not in its standard state, so $\Delta G_f°$ is not zero for this reactant. From the balanced equation and using Appendix C, we have:

$$\Delta G_{rxn}° = 4\,\Delta G_f°[PCl_3(g)] - \Delta G_f°[P_4(g)] - 6\,\Delta G_f°[Cl_2(g)]$$
$$= (4\;mol)(-269.6\;kJ/mol) - (1\;mol)(24.4\;kJ/mol) - 0$$
$$= -1102.8\;kJ$$

The fact that $\Delta G°$ is negative tells us that a mixture of $P_4(g)$, $Cl_2(g)$, and $PCl_3(g)$ at 25 °C, each present at a partial pressure of 1 atm, would react spontaneously in the forward direction to form more PCl_3. Remember, however, that the value of $\Delta G°$ tells us nothing about the rate at which the reaction occurs.

(b) Remember that $\Delta G = G$ (products) $- G$ (reactants). If we reverse the reaction, we reverse the roles of the reactants and products. Thus, reversing the reaction changes the sign of ΔG, just as reversing the reaction changes the sign of ΔH ∞ (Section 5.4) Hence, using the result from part (a):

$$4\,PCl_3(g) \longrightarrow P_4(g) + 6\,Cl_2(g) \qquad \Delta G° = +1102.8\;kJ$$

PRACTICE EXERCISE

By using data from Appendix C, calculate $\Delta G°$ at 298 K for the combustion of methane: $CH_4(g) + 2\,O_2(g) \longrightarrow CO_2(g) + 2\,H_2O(g)$.
Answer: -800.7 kJ

SAMPLE EXERCISE 19.8 | **Estimating and Calculating $\Delta G°$**

In Section 5.7 we used Hess's law to calculate $\Delta H°$ for the combustion of propane gas at 298 K:

$$C_3H_8(g) + 5\,O_2(g) \longrightarrow 3\,CO_2(g) + 4\,H_2O(l) \qquad \Delta H° = -2220\;kJ$$

(a) *Without using data from Appendix C*, predict whether $\Delta G°$ for this reaction is more negative or less negative than $\Delta H°$. (b) Use data from Appendix C to calculate the standard free-energy change for the reaction at 298 K. Is your prediction from part (a) correct?

SOLUTION

Analyze: In part (a) we must predict the value for $\Delta G°$ relative to that for $\Delta H°$ on the basis of the balanced equation for the reaction. In part (b) we must calculate the value for $\Delta G°$ and compare with our qualitative prediction.

Plan: The free-energy change incorporates both the change in enthalpy and the change in entropy for the reaction (Equation 19.11), so under standard conditions:

$$\Delta G° = \Delta H° - T\Delta S°$$

To determine whether $\Delta G°$ is more negative or less negative than $\Delta H°$, we need to determine the sign of the term $T\Delta S°$. T is the absolute temperature, 298 K, so it is a positive number. We can predict the sign of $\Delta S°$ by looking at the reaction.

Solve:

(a) We see that the reactants consist of six molecules of gas, and the products consist of three molecules of gas and four molecules of liquid. Thus, the number of molecules of gas has decreased significantly during the reaction. By using the general rules we discussed in Section 19.3, we would expect a decrease in the number of gas molecules to lead to a decrease in the entropy of the system—the products have fewer accessible microstates than the reactants. We therefore expect $\Delta S°$ and therefore $T\Delta S°$ to be negative numbers. Because we are subtracting $T\Delta S°$, which is a negative number, we would predict that $\Delta G°$ is *less negative* than $\Delta H°$.

(b) Using Equation 19.14 and values from Appendix C, we can calculate the value of $\Delta G°$

$$\Delta G° = 3\Delta G_f°[CO_2(g)] + 4\Delta G_f°[H_2O(l)] - \Delta G_f°[C_3H_8(g)] - 5\Delta G_f°[O_2(g)]$$

$$= 3 \text{ mol}(-394.4 \text{ kJ/mol}) + 4 \text{ mol}(-237.13 \text{ kJ/mol}) -$$

$$1 \text{ mol}(-23.47 \text{ kJ/mol}) - 5 \text{ mol}(0 \text{ kJ/mol}) = -2108 \text{ kJ}$$

Notice that we have been careful to use the value of $\Delta G_f°$ for $H_2O(l)$, as in the calculation of ΔH values, the phases of the reactants and products are important. As we predicted, $\Delta G°$ is less negative than $\Delta H°$ because of the decrease in entropy during the reaction.

■ PRACTICE EXERCISE

Consider the combustion of propane to form $CO_2(g)$ and $H_2O(g)$ at 298 K: $C_3H_8(g) + 5 O_2(g) \longrightarrow 3 CO_2(g) + 4 H_2O(g)$. Would you expect $\Delta G°$ to be more negative or less negative than $\Delta H°$?
Answer: more negative

19.6 FREE ENERGY AND TEMPERATURE

We have seen that tabulations of $\Delta G_f°$, such as those in Appendix C, make it possible to calculate $\Delta G°$ for reactions at the standard temperature of 25° C. However, we are often interested in examining reactions at other temperatures. How is the change in free energy affected by the change in temperature? Let's look again at Equation 19.11:

$$\Delta G = \Delta H - T\Delta S = \underset{\substack{\text{Enthalpy} \\ \text{term}}}{\Delta H} + \underset{\substack{\text{Entropy} \\ \text{term}}}{(-T\Delta S)}$$

Notice that we have written the expression for ΔG as a sum of two contributions, an enthalpy term, ΔH, and an entropy term, $-T\Delta S$. Because the value of $-T\Delta S$ depends directly on the absolute temperature T, ΔG will vary with temperature. T is a positive number at all temperatures other than absolute zero. We know that the enthalpy term, ΔH, can be positive or negative. The entropy term, $-T\Delta S$, can also be positive or negative. When ΔS is positive, which means that the final state has greater randomness (a greater number of microstates) than the initial state, the term $-T\Delta S$ is negative. When ΔS is negative, the term $-T\Delta S$ is positive.

The sign of ΔG, which tells us whether a process is spontaneous, will depend on the signs and magnitudes of ΔH and $-T\Delta S$. When both ΔH and $-T\Delta S$ are negative, ΔG will always be negative and the process will be spontaneous at all temperatures. Likewise, when both ΔH and $-T\Delta S$ are positive, ΔG will always be positive and the process will be nonspontaneous at all temperatures (the reverse process will be spontaneous at all temperatures). When ΔH and $-T\Delta S$ have opposite signs, however, the sign of ΔG will depend on the magnitudes of these two terms. In these instances temperature is an important consideration. Generally, ΔH and ΔS change very little with temperature.

However, the value of T directly affects the magnitude of $-T\Delta S$. As the temperature increases, the magnitude of the term $-T\Delta S$ increases and it will become relatively more important in determining the sign and magnitude of ΔG.

For example, let's consider once more the melting of ice to liquid water at 1 atm pressure:

$$H_2O(s) \longrightarrow H_2O(l) \quad \Delta H > 0, \Delta S > 0$$

This process is endothermic, which means that ΔH is positive. We also know that the entropy increases during this process, so ΔS is positive and $-T\Delta S$ is negative. At temperatures below 0 °C (273 K) the magnitude of ΔH is greater than that of $-T\Delta S$. Hence, the positive enthalpy term dominates, leading to a positive value for ΔG. The positive value of ΔG means that the melting of ice is not spontaneous at $T < 0$ °C; rather, the reverse process, the freezing of liquid water into ice, is spontaneous at these temperatures.

What happens at temperatures greater than 0 °C? As the temperature increases, so does the magnitude of the entropy term $-T\Delta S$. When $T > 0$ °C, the magnitude of $-T\Delta S$ is greater than the magnitude of ΔH. At these temperatures the negative entropy term dominates, which leads to a negative value for ΔG. The negative value of ΔG tells us that the melting of ice is spontaneous at $T > 0$ °C. At the normal melting point of water, $T = 0$ °C, the two phases are in equilibrium. Recall that $\Delta G = 0$ at equilibrium; at $T = 0$ °C, ΔH and $-T\Delta S$, are equal in magnitude and opposite in sign, so they cancel one another and give $\Delta G = 0$.

GIVE IT SOME THOUGHT

The normal boiling point of benzene is 80 °C. At 100 °C and 1 atm, which term is greater for the vaporization of benzene, ΔH or $T\Delta S$?

The possible situations for the relative signs of ΔH and ΔS are given in Table 19.4▼, along with examples of each. By applying the concepts we have developed for predicting entropy changes, we often can predict how ΔG will change with temperature.

Our discussion of the temperature dependence of ΔG is also relevant to standard free-energy changes. As we saw earlier in Equation 19.12, we can calculate $\Delta G°$ from $\Delta H°$ and $T\Delta S°$: $\Delta G° = \Delta H° - T\Delta S°$. We can readily calculate the values of $\Delta H°$ and $\Delta S°$ at 298 K from the data tabulated in Appendix C. If we assume that the values of $\Delta H°$ and $\Delta S°$ do not change with temperature, we can then use Equation 19.12 to estimate the value of $\Delta G°$ at temperatures other than 298 K.

TABLE 19.4 ■ Effect of Temperature on the Spontaneity of Reactions

ΔH	ΔS	$-T\Delta S$	$\Delta G = \Delta H - T\Delta S$	Reaction Characteristics	Example
−	+	−	−	Spontaneous at all temperatures	$2\,O_3(g) \longrightarrow 3\,O_2(g)$
+	−	+	+	Nonspontaneous at all temperatures	$3\,O_2(g) \longrightarrow 2\,O_3(g)$
−	−	+	+ or −	Spontaneous at low T; nonspontaneous at high T	$H_2O(l) \longrightarrow H_2O(s)$
+	+	−	+ or −	Spontaneous at high T; nonspontaneous at low T	$H_2O(s) \longrightarrow H_2O(l)$

■ SAMPLE EXERCISE 19.9 | Determining the Effect of Temperature on Spontaneity

The Haber process for the production of ammonia involves the equilibrium

$$N_2(g) + 3 H_2(g) \rightleftharpoons 2 NH_3(g)$$

Assume that $\Delta H°$ and $\Delta S°$ for this reaction do not change with temperature. **(a)** Predict the direction in which $\Delta G°$ for this reaction changes with increasing temperature. **(b)** Calculate the values of $\Delta G°$ for the reaction at 25 °C and 500 °C.

SOLUTION

Analyze: In part (a) we are asked to predict the direction in which $\Delta G°$ for the ammonia synthesis reaction changes as temperature increases. In part (b) we need to determine $\Delta G°$ for the reaction at two different temperatures.

Plan: In part (a) we can make this prediction by determining the sign of ΔS for the reaction and then using that information to analyze Equation 19.12. In part (b) we need to calculate $\Delta H°$ and $\Delta S°$ for the reaction by using the data in Appendix C. We can then use Equation 19.12 to calculate $\Delta G°$.

Solve:

(a) Equation 19.12 tells us that $\Delta G°$ is the sum of the enthalpy term $\Delta H°$ and the entropy term $-T\Delta S°$. The temperature dependence of $\Delta G°$ comes from the entropy term. We expect $\Delta S°$ for this reaction to be negative because the number of molecules of gas is smaller in the products. Because $\Delta S°$ is negative, the term $-T\Delta S°$ is positive and grows larger with increasing temperature. As a result, $\Delta G°$ becomes less negative (or more positive) with increasing temperature. Thus, the driving force for the production of NH_3 becomes smaller with increasing temperature.

(b) We calculated the value of $\Delta H°$ in Sample Exercise 15.14, and the value of $\Delta S°$ was determined in Sample Exercise 19.5: $\Delta H° = -92.38$ kJ and $\Delta S° = -198.3$ J/K. If we assume that these values do not change with temperature, we can calculate $\Delta G°$ at any temperature by using Equation 19.12. At $T = 298$ K we have:

$$\Delta G° = -92.38 \text{ kJ} - (298 \text{ K})(-198.3 \text{ J/K})\left(\frac{1 \text{ kJ}}{1000 \text{ J}}\right)$$

$$= -92.38 \text{ kJ} + 59.1 \text{ kJ} = -33.3 \text{ kJ}$$

At $T = 500 + 273 = 773$ K we have

$$\Delta G° = -92.38 \text{ kJ} - (773 \text{ K})\left(-198.3 \frac{\text{J}}{\text{K}}\right)\left(\frac{1 \text{ kJ}}{1000 \text{ J}}\right)$$

$$= -92.38 \text{ kJ} + 153 \text{ kJ} = 61 \text{ kJ}$$

Notice that we have been careful to convert $-T\Delta S°$ into units of kJ so that it can be added to $\Delta H°$, which has units of kJ.

Comment: Increasing the temperature from 298 K to 773 K changes $\Delta G°$ from -33.3 kJ to $+61$ kJ. Of course, the result at 773 K depends on the assumption that $\Delta H°$ and $\Delta S°$ do not change with temperature. In fact, these values do change slightly with temperature. Nevertheless, the result at 773 K should be a reasonable approximation. The positive increase in $\Delta G°$ with increasing T agrees with our prediction in part (a) of this exercise. Our result indicates that a mixture of $N_2(g)$, $H_2(g)$, and $NH_3(g)$, each present at a partial pressure of 1 atm, will react spontaneously at 298 K to form more $NH_3(g)$. In contrast, at 773 K the positive value of $\Delta G°$ tells us that the reverse reaction is spontaneous. Thus, when the mixture of three gases, each at a partial pressure of 1 atm, is heated to 773 K, some of the $NH_3(g)$ spontaneously decomposes into $N_2(g)$ and $H_2(g)$.

■ PRACTICE EXERCISE

(a) Using standard enthalpies of formation and standard entropies in Appendix C, calculate $\Delta H°$ and $\Delta S°$ at 298 K for the following reaction: $2 SO_2(g) + O_2(g) \longrightarrow 2 SO_3(g)$. **(b)** Using the values obtained in part (a), estimate $\Delta G°$ at 400 K.
Answers: **(a)** $\Delta H° = -196.6$ kJ, $\Delta S° = -189.6$ J/K; **(b)** $\Delta G° = -120.8$ kJ

19.7 FREE ENERGY AND THE EQUILIBRIUM CONSTANT

In Section 19.5 we saw a special relationship between ΔG and equilibrium: For a system at equilibrium, $\Delta G = 0$. We have also seen how we can use tabulated thermodynamic data, such as those in Appendix C, to calculate values of the standard free-energy change, $\Delta G°$. In this final section of this chapter, we will learn two more ways in which we can use free energy as a powerful tool in our analysis of chemical reactions. First, we will learn how to use the value of $\Delta G°$ to calculate the value of ΔG under *nonstandard* conditions. Second, we will see how we can directly relate the value of $\Delta G°$ for a reaction to the value of the equilibrium constant for the reaction.

The set of standard conditions for which $\Delta G°$ values pertain are given in Table 19.3. Most chemical reactions occur under nonstandard conditions. For any chemical process the general relationship between the standard free-energy change, $\Delta G°$, and the free-energy change under any other conditions, ΔG, is given by the expression

$$\Delta G = \Delta G° + RT \ln Q \qquad [19.16]$$

In this equation R is the ideal-gas constant, 8.314 J/mol-K, T is the absolute temperature; and Q is the reaction quotient that corresponds to the particular reaction mixture of interest. ∞ (Section 15.6) Recall that the expression for Q is identical to the equilibrium-constant expression except that the reactants and products need not necessarily be at equilibrium.

Under standard conditions the concentrations of all the reactants and products are equal to 1. Thus, under standard conditions $Q = 1$ and therefore $\ln Q = 0$. We see that Equation 19.16 therefore reduces to $\Delta G = \Delta G°$ under standard conditions, as it should.

■■■ **SAMPLE EXERCISE 19.10** | Relating ΔG to a Phase Change at Equilibrium

As we saw in Section 11.5, the *normal boiling point* is the temperature at which a pure liquid is in equilibrium with its vapor at a pressure of 1 atm. **(a)** Write the chemical equation that defines the normal boiling point of liquid carbon tetrachloride, $CCl_4(l)$. **(b)** What is the value of $\Delta G°$ for the equilibrium in part (a)? **(c)** Use thermodynamic data in Appendix C and Equation 19.12 to estimate the normal boiling point of CCl_4.

SOLUTION

Analyze: (a) We must write a chemical equation that describes the physical equilibrium between liquid and gaseous CCl_4 at the normal boiling point. **(b)** We must determine the value of $\Delta G°$ for CCl_4, in equilibrium with its vapor at the normal boiling point. **(c)** We must estimate the normal boiling point of CCl_4, based on available thermodynamic data.

Plan: (a) The chemical equation will merely show the change of state of CCl_4 from liquid to solid. **(b)** We need to analyze Equation 19.16 at equilibrium ($\Delta G = 0$). **(c)** We can use Equation 19.12 to calculate T when $\Delta G = 0$.

Solve: (a) The normal boiling point of CCl_4 is the temperature at which pure liquid CCl_4 is in equilibrium with its vapor at a pressure of 1 atm:

$$CCl_4(l) \rightleftharpoons CCl_4\,(g, 1\ atm)$$

(b) At equilibrium $\Delta G = 0$. In any normal boiling-point equilibrium both the liquid and the vapor are in their standard states (Table 19.2). Consequently, $Q = 1$, $\ln Q = 0$, and $\Delta G = \Delta G°$ for this process. Thus, we conclude that $\Delta G° = 0$ for the equilibrium involved in the normal boiling point of any liquid. We would also find that $\Delta G° = 0$ for the equilibria relevant to normal melting points and normal sublimation points of solids.

(c) Combining Equation 19.12 with the result from part (b), we see that the equality at the normal boiling point, T_b, of $CCl_4(l)$ or any other pure liquid is

$$\Delta G° = \Delta H° - T_b \Delta S° = 0$$

Solving the equation for T_b, we obtain

$$T_b = \Delta H°/\Delta S°$$

Strictly speaking, we would need the values of $\Delta H°$ and $\Delta S°$ for the equilibrium between $CCl_4(l)$ and $CCl_4(g)$ at the normal boiling point to do this calculation. However, we can *estimate* the boiling point by using the values of $\Delta H°$ and $\Delta S°$ for CCl_4 at 298 K, which we can obtain from the data in Appendix C and Equations 5.31 and 19.8:

$$\Delta H° = (1\ mol)(-106.7\ kJ/mol) - (1\ mol)(-139.3\ kJ/mol) = +32.6\ kJ$$

$$\Delta S° = (1\ mol)(309.4\ J/mol\text{-}K) - (1\ mol)(214.4\ J/mol\text{-}K) = +95.0\ J/K$$

Notice that, as expected, the process is endothermic ($\Delta H > 0$) and produces a gas in which energy can be more spread out ($\Delta S > 0$). We can now use these values to estimate T_b for $CCl_4(l)$:

$$T_b = \frac{\Delta H°}{\Delta S°} = \left(\frac{32.6\ kJ}{95.0\ J/K}\right)\left(\frac{1000\ J}{1\ kJ}\right) = 343\ K = 70°\ C$$

Note also that we have used the conversion factor between J and kJ to make sure that the units of $\Delta H°$ and $\Delta S°$ match.

Check: The experimental normal boiling point of $CCl_4(l)$ is 76.5 °C. The small deviation of our estimate from the experimental value is due to the assumption that $\Delta H°$ and $\Delta S°$ do not change with temperature.

■ **PRACTICE EXERCISE**

Use data in Appendix C to estimate the normal boiling point, in K, for elemental bromine, $Br_2(l)$. (The experimental value is given in Table 11.3.)
Answer: 330 K

When the concentrations of reactants and products are nonstandard, we must calculate the value of Q to determine the value of ΔG. We illustrate how this is done in Sample Exercise 19.11. At this stage in our discussion, it becomes important to note the units associated with Q in Equation 19.16. The convention used for standard states imposes itself into the way Q is expressed: In Equation 19.16 the concentrations of gases are always expressed by their partial pressures in atmospheres, and solutes are expressed by their concentrations in molarity.

■ **SAMPLE EXERCISE 19.11** | **Calculating the Free-Energy Change under Nonstandard Conditions**

We will continue to explore the Haber process for the synthesis of ammonia:

$$N_2(g) + 3 H_2(g) \rightleftharpoons 2 NH_3(g)$$

Calculate ΔG at 298 K for a reaction mixture that consists of 1.0 atm N_2, 3.0 atm H_2, and 0.50 atm NH_3.

SOLUTION

Analyze: We are asked to calculate ΔG under nonstandard conditions.

Plan: We can use Equation 19.16 to calculate ΔG. Doing so requires that we calculate the value of the reaction quotient Q for the specified partial pressures of the gases and evaluate $\Delta G°$, using a table of standard free energies of formation.

Solve: Solving for the reaction quotient gives:

$$Q = \frac{P_{NH_3}^2}{P_{N_2} P_{H_2}^3} = \frac{(0.50)^2}{(1.0)(3.0)^3} = 9.3 \times 10^{-3}$$

In Sample Exercise 19.9 we calculated $\Delta G° = -33.3$ kJ for this reaction. We will have to change the units of this quantity in applying Equation 19.16, however. For the units in Equation 19.16 to work out, we will use kJ/mol as our units for $\Delta G°$, where "per mole" means "per mole of the reaction as written." Thus, $\Delta G° = -33.3$ kJ/mol implies per 1 mol of N_2, per 3 mol of H_2, and per 2 mol of NH_3.

We can now use Equation 19.16 to calculate ΔG for these nonstandard conditions:

$$\Delta G = \Delta G° + RT \ln Q$$

$$= (-33.3 \text{ kJ/mol}) + (8.314 \text{ J/mol-K})(298 \text{ K})(1 \text{ kJ}/1000 \text{ J}) \ln(9.3 \times 10^{-3})$$

$$= (-33.3 \text{ kJ/mol}) + (-11.6 \text{ kJ/mol}) = -44.9 \text{ kJ/mol}$$

Comment: We see that ΔG becomes more negative, changing from -33.3 kJ/mol to -44.9 kJ/mol, as the pressures of N_2, H_2, and NH_3 are changed from 1.0 atm each (standard conditions, $\Delta G°$) to 1.0 atm, 3.0 atm, and 0.50 atm, respectively. The larger negative value for ΔG indicates a larger "driving force" to produce NH_3.

We would have made the same prediction based on Le Châtelier's principle. ∞ (Section 15.7) Relative to standard conditions, we have increased the pressure of a reactant (H_2) and decreased the pressure of the product (NH_3). Le Châtelier's principle predicts that both of these changes should shift the reaction more to the product side, thereby forming more NH_3.

■ **PRACTICE EXERCISE**

Calculate ΔG at 298 K for the reaction of nitrogen and hydrogen to form ammonia if the reaction mixture consists of 0.50 atm N_2, 0.75 atm H_2, and 2.0 atm NH_3.
Answer: -26.0 kJ/mol

We can now use Equation 19.16 to derive the relationship between $\Delta G°$ and the equilibrium constant, K. At equilibrium $\Delta G = 0$. Further, recall that the reaction quotient, Q, equals the equilibrium constant, K, when the system is at equilibrium. Thus, at equilibrium, Equation 19.16 transforms as follows:

$$\Delta G = \Delta G° + RT \ln Q$$

$$0 = \Delta G° + RT \ln K$$

$$\Delta G° = -RT \ln K \qquad [19.17]$$

Equation 19.17 also allows us to calculate the value of K if we know the value of $\Delta G°$. If we solve the equation for K, we obtain

$$K = e^{-\Delta G°/RT} \qquad [19.18]$$

As we pointed out in discussing Equation 19.16, some care is necessary in the choice of units. Thus, in Equations 19.17 and 19.18 we again express $\Delta G°$ in kJ/mol. For the reactants and products in the equilibrium-constant expression, we use the following conventions: Gas pressures are given in atm; solution concentrations are given in moles per liter (molarity); and solids, liquids, and solvents do not appear in the expression. ⚭(Section 15.4) Thus, for gas-phase reactions the equilibrium constant is K_p, whereas for reactions in solution it is K_c. ⚭(Section 15.2)

From Equation 19.17 we can see that if $\Delta G°$ is negative, then $\ln K$ must be positive. A positive value for $\ln K$ means $K > 1$. Therefore, the more negative $\Delta G°$ is, the larger the equilibrium constant, K. Conversely, if $\Delta G°$ is positive, then $\ln K$ is negative, which means that $K < 1$. Table 19.5▶ summarizes these conclusions by comparing $\Delta G°$ and K for both positive and negative values of $\Delta G°$.

TABLE 19.5 ■ Relationship between $\Delta G°$ and K at 298 K

$\Delta G°$, kJ/mol	K
+200	8.7×10^{-36}
+100	3.0×10^{-18}
+50	1.7×10^{-9}
+10	1.8×10^{-2}
+1.0	6.7×10^{-1}
0	1.0
−1.0	1.5
−10	5.7×10^{1}
−50	5.8×10^{8}
−100	3.4×10^{17}
−200	1.1×10^{35}

■ **SAMPLE EXERCISE 19.12** | Calculating an Equilibrium Constant from $\Delta G°$

Use standard free energies of formation to calculate the equilibrium constant, K, at 25 °C for the reaction involved in the Haber process:

$$N_2(g) + 3 H_2(g) \rightleftharpoons 2 NH_3(g)$$

The standard free-energy change for this reaction was calculated in Sample Exercise 19.9: $\Delta G° = -33.3$ kJ/mol $= -33,300$ J/mol

SOLUTION

Analyze: We are asked to calculate K for a reaction, given $\Delta G°$.

Plan: We can use Equation 19.18 to evaluate the equilibrium constant, which in this case takes the form

$$K = \frac{P_{NH_3}^2}{P_{N_2} P_{H_2}^3}$$

In this expression the gas pressures are expressed in atmospheres. (Remember that we use kJ/mol as the units of $\Delta G°$ when using Equations 19.16, 19.17, or 19.18.)

Solve: Solving Equation 19.17 for the exponent $-\Delta G°/RT$, we have

$$\frac{-\Delta G°}{RT} = \frac{-(-33,300 \text{ J/mol})}{(8.314 \text{ J/mol-K})(298\text{K})} = 13.4$$

We insert this value into Equation 19.18 to obtain K:

$$K = e^{-\Delta G°/RT} = e^{13.4} = 7 \times 10^5$$

Comment: This is a large equilibrium constant, which indicates that the product, NH_3, is greatly favored in the equilibrium mixture at 25 °C. The equilibrium constants for temperatures in the range of 300 °C to 600 °C, given in Table 15.2, are much smaller than the value at 25 °C. Clearly, a low-temperature equilibrium favors the production of ammonia more than a high-temperature one. Nevertheless, the Haber process is carried out at high temperatures because the reaction is extremely slow at room temperature.

Remember: Thermodynamics can tell us the direction and extent of a reaction, but tells us nothing about the rate at which it will occur. If a catalyst were found that would permit the reaction to proceed at a rapid rate at room temperature, high pressures would not be needed to force the equilibrium toward NH_3.

■ **PRACTICE EXERCISE**

Use data from Appendix C to calculate the standard free-energy change, $\Delta G°$, and the equilibrium constant, K, at 298 K for the reaction $H_2(g) + Br_2(l) \rightleftharpoons 2 HBr(g)$.
Answer: $\Delta G° = -106.4$ kJ/mol, $K = 4 \times 10^{18}$

Chemistry and Life   DRIVING NONSPONTANEOUS REACTIONS

Many desirable chemical reactions, including a large number that are central to living systems, are nonspontaneous as written. For example, consider the extraction of copper metal from the mineral *chalcocite*, which contains Cu_2S. The decomposition of Cu_2S to its elements is nonspontaneous:

$$Cu_2S(s) \longrightarrow 2\,Cu(s) + S(s) \qquad \Delta G^\circ = +86.2 \text{ kJ}$$

Because ΔG° is very positive, we cannot obtain $Cu(s)$ directly via this reaction. Instead, we must find some way to "do work" on the reaction to force it to occur as we wish. We can do this by coupling the reaction to another one so that the overall reaction *is* spontaneous. For example, we can envision the $S(s)$ reacting with $O_2(g)$ to form $SO_2(g)$:

$$S(s) + O_2(g) \longrightarrow SO_2(g) \qquad \Delta G^\circ = -300.4 \text{ kJ}$$

By coupling these reactions, we can extract much of the copper metal via a spontaneous reaction:

$$Cu_2S(s) + O_2(g) \longrightarrow 2\,Cu(s) + SO_2(g)$$
$$\Delta G^\circ = (+86.2 \text{ kJ}) + (-300.4 \text{ kJ}) = -214.2 \text{ kJ}$$

In essence, we have used the spontaneous reaction of $S(s)$ with $O_2(g)$ to provide the free energy needed to extract the copper metal from the mineral.

· Biological systems employ the same principle of using spontaneous reactions to drive nonspontaneous ones. Many of the biochemical reactions that are essential for the formation and maintenance of highly ordered biological structures are not spontaneous. These necessary reactions are made to

occur by coupling them with spontaneous reactions that release energy. The metabolism of food is the usual source of the free energy needed to do the work of maintaining biological systems. For example, complete oxidation of the sugar *glucose*, $C_6H_{12}O_6$, to CO_2 and H_2O yields substantial free energy:

$$C_6H_{12}O_6(s) + 6\,O_2(g) \longrightarrow 6\,CO_2(g) + 6\,H_2O(l)$$
$$\Delta G^\circ = -2880 \text{ kJ}$$

This energy can be used to drive nonspontaneous reactions in the body. However, a means is necessary to transport the energy released by glucose metabolism to the reactions that require energy. One way, shown in Figure 19.20▼, involves the interconversion of adenosine triphosphate (ATP) and adenosine diphosphate (ADP), molecules that are related to the building blocks of nucleic acids. The conversion of ATP to ADP releases free energy ($\Delta G^\circ = -30.5$ kJ) that can be used to drive other reactions.

In the human body the metabolism of glucose occurs via a complex series of reactions, most of which release free energy. The free energy released during these steps is used in part to reconvert lower-energy ADP back to higher-energy ATP. Thus, the ATP–ADP interconversions are used to store energy during metabolism and to release it as needed to drive nonspontaneous reactions in the body. If you take a course in biochemistry, you will have the opportunity to learn more about the remarkable sequence of reactions used to transport free energy throughout the human body.

Related Exercises: 19.94 and 19.95

▲ **Figure 19.20 Free energy and cell metabolism.** This schematic representation shows part of the free-energy changes that occur in cell metabolism. The oxidation of glucose to CO_2 and H_2O produces free energy that is then used to convert ADP into the more energetic ATP. The ATP is then used, as needed, as an energy source to convert simple molecules into more complex cell constituents. When it releases its stored free energy, ATP is converted back into ADP.

■■ **SAMPLE INTEGRATIVE EXERCISE** | **Putting Concepts Together**

Consider the simple salts NaCl(s) and AgCl(s). We will examine the equilibria in which these salts dissolve in water to form aqueous solutions of ions:

$$NaCl(s) \rightleftharpoons Na^+(aq) + Cl^-(aq)$$
$$AgCl(s) \rightleftharpoons Ag^+(aq) + Cl^-(aq)$$

(a) Calculate the value of $\Delta G°$ at 298 K for each of the preceding reactions. **(b)** The two values from part (a) are very different. Is this difference primarily due to the enthalpy term or the entropy term of the standard free-energy change? **(c)** Use the values of $\Delta G°$ to calculate the K_{sp} values for the two salts at 298 K. **(d)** Sodium chloride is considered a soluble salt, whereas silver chloride is considered insoluble. Are these descriptions consistent with the answers to part (c)? **(e)** How will $\Delta G°$ for the solution process of these salts change with increasing T? What effect should this change have on the solubility of the salts?

SOLUTION

(a) We will use Equation 19.14 along with $\Delta G_f°$ values from Appendix C to calculate the $\Delta G°_{soln}$ values for each equilibrium. (As we did in Section 13.1, we use the subscript "soln" to indicate that these are thermodynamic quantities for the formation of a solution.) We find

$$\Delta G°_{soln}(NaCl) = (-261.9 \text{ kJ/mol}) + (-131.2 \text{ kJ/mol}) - (-384.0 \text{ kJ/mol})$$
$$= -9.1 \text{ kJ/mol}$$

$$\Delta G°_{soln}(AgCl) = (+77.11 \text{ kJ/mol}) + (-131.2 \text{ kJ/mol}) - (-109.70 \text{ kJ/mol})$$
$$= +55.6 \text{ kJ/mol}$$

(b) We can write $\Delta G°_{soln}$ as the sum of an enthalpy term, $\Delta H°_{soln}$, and an entropy term, $-T\Delta S°_{soln}$: $\Delta G°_{soln} = \Delta H°_{soln} + (-T\Delta S°_{soln})$. We can calculate the values of $\Delta H°_{soln}$ and $\Delta S°_{soln}$ by using Equations 5.31 and 19.8. We can then calculate $-T\Delta S°_{soln}$ at $T = 298$K. All these calculations are now familiar to us. The results are summarized in the following table:

Salt	$\Delta H°_{soln}$	$\Delta S°_{soln}$	$-T\Delta S°_{soln}$
NaCl	+3.6 kJ/mol	+43.2 J/mol-K	−12.9 kJ/mol
AgCl	+65.7 kJ/mol	+34.3 J/mol-K	−10.2 kJ/mol

The entropy terms for the solution of the two salts are very similar. That seems sensible because each solution process should lead to a similar increase in randomness as the salt dissolves, forming hydrated ions. ∞ (Section 13.1) In contrast, we see a very large difference in the enthalpy term for the solution of the two salts. The difference in the values of $\Delta G°_{soln}$ is dominated by the difference in the values of $\Delta H°_{soln}$.

(c) The solubility product, K_{sp}, is the equilibrium constant for the solution process. ∞ (Section 17.4) As such, we can relate K_{sp} directly to $\Delta G°_{soln}$ by using Equation 19.18:

$$K_{sp} = e^{-\Delta G°_{soln}/RT}$$

We can calculate the K_{sp} values in the same way we applied Equation 19.18 in Sample Exercise 19.12. We use the $\Delta G°_{soln}$ values we obtained in part (a), remembering to convert them from kJ/mol to J/mol:

NaCl: $K_{sp} = [Na^+(aq)][Cl^-(aq)] = e^{-(-9100)/[(8.314)(298)]} = e^{+3.7} = 40$

AgCl: $K_{sp} = [Ag^+(aq)][Cl^-(aq)] = e^{-(+55,600)/[(8.314)(298)]} = e^{-22.4} = 1.9 \times 10^{-10}$

The value calculated for the K_{sp} of AgCl is very close to that listed in Appendix D.

(d) A soluble salt is one that dissolves appreciably in water. ∞ (Section 4.2) The K_{sp} value for NaCl is greater than 1, indicating that NaCl dissolves to a great extent. The K_{sp} value for AgCl is very small, indicating that very little dissolves in water. Silver chloride should indeed be considered an insoluble salt.

(e) As we expect, the solution process has a positive value of ΔS for both salts (see the table in part b). As such, the entropy term of the free-energy change, $-T\Delta S°_{soln}$, is negative. If we assume that $\Delta H°_{soln}$ and $\Delta S°_{soln}$ do not change much with temperature, then an increase in T will serve to make $\Delta G°_{soln}$ more negative. Thus, the driving force for dissolution of the salts will increase with increasing T, and we therefore expect the solubility of the salts to increase with increasing T. In Figure 13.17 we see that the solubility of NaCl (and that of nearly any salt) increases with increasing temperature. ∞ (Section 13.3)

CHAPTER REVIEW

SUMMARY AND KEY TERMS

Section 19.1 Most reactions and chemical processes have an inherent directionality: They are **spontaneous** in one direction and nonspontaneous in the reverse direction. The spontaneity of a process is related to the thermodynamic path the system takes from the initial state to the final state. In a **reversible process**, both the system and its surroundings can be restored to their original state by exactly reversing the change. In an **irreversible process** the system cannnot return to its original state without there being a permanent change in the surroundings. Any spontaneous process is irreversible. A process that occurs at a constant temperature is said to be **isothermal**.

Section 19.2 The spontaneous nature of processes is related to a thermodynamic state function called **entropy**, denoted S. For a process that occurs at constant temperature, the entropy change of the system is given by the heat absorbed by the system along a reversible path, divided by the temperature: $\Delta S = q_{rev}/T$. The way entropy controls the spontaneity of processes is given by the **second law of thermodynamics**, which governs the change in the entropy of the universe, $\Delta S_{univ} = \Delta S_{sys} + \Delta S_{surr}$. The second law states that in a reversible process $\Delta S_{univ} = 0$; in an irreversible (spontaneous) process $\Delta S_{univ} > 0$. Entropy values are usually expressed in units of joules per kelvin, J/K.

Section 19.3 Molecules can undergo three kinds of motion: In **translational motion** the entire molecule moves in space. Molecules can also undergo **vibrational motion**, in which the atoms of the molecule move toward and away from one another in periodic fashion, and **rotational motion**, in which the entire molecule spins like a top. A particular combination of motions and locations of the atoms and molecules of a system at a particular instant is called a **microstate**. Entropy is a measure of the number of microstates, W, over which the energy of the system is distributed: $S = k \ln W$. The number of available microstates, and therefore the entropy, increases with an increase in volume, temperature, or motion of molecules because any of these changes increases the possible motions and locations of the molecules. As a result, entropy generally increases when liquids or solutions are formed from solids, gases are formed from either solids or liquids, or the number of molecules of gas increases during a chemical reaction. The **third law of**

thermodynamics states that the entropy of a pure crystalline solid at 0 K is zero.

Section 19.4 The third law allows us to assign entropy values for substances at different temperatures. Under standard conditions the entropy of a mole of a substance is called its **standard molar entropy**, denoted $S°$. From tabulated values of $S°$, we can calculate the entropy change for any process under standard conditions. For an isothermal process, the entropy change in the surroundings is equal to $-\Delta H/T$.

Section 19.5 The **Gibbs free energy** (or just **free energy**), G, is a thermodynamic state function that combines the two state functions enthalpy and entropy: $G = H - TS$. For processes that occur at constant temperature, $\Delta G = \Delta H - T\Delta S$. For a process occurring at constant temperature and pressure, the sign of ΔG relates to the spontaneity of the process. When ΔG is negative, the process is spontaneous. When ΔG is positive, the process is nonspontaneous but the reverse process is spontaneous. At equilibrium the process is reversible and ΔG is zero. The free energy is also a measure of the maximum useful work that can be performed by a system in a spontaneous process. The standard free-energy change, $\Delta G°$, for any process can be calculated from tabulations of **standard free energies of formation**, $\Delta G_f°$, which are defined in a fashion analogous to standard enthalpies of formation, $\Delta H_f°$. The value of $\Delta G_f°$ for a pure element in its standard state is defined to be zero.

Sections 19.6 and 19.7 The values of ΔH and ΔS generally do not vary much with temperature. Therefore, the dependence of ΔG with temperature is governed mainly by the value of T in the expression $\Delta G = \Delta H - T\Delta S$. The entropy term $-T\Delta S$ has the greater effect on the temperature dependence of ΔG and, hence, on the spontaneity of the process. For example, a process for which $\Delta H > 0$ and $\Delta S > 0$, such as the melting of ice, can be nonspontaneous ($\Delta G > 0$) at low temperatures and spontaneous ($\Delta G < 0$) at higher temperatures. Under nonstandard conditions ΔG is related to $\Delta G°$ and the value of the reaction quotient, Q: $\Delta G = \Delta G° + RT \ln Q$. At equilibrium ($\Delta G = 0, Q = K$), $\Delta G° = -RT \ln K$. Thus, the standard free-energy change is directly related to the equilibrium constant for the reaction. This relationship expresses the temperature dependence of equilibrium constants.

KEY SKILLS

- Understand the meaning of *spontaneous process, reversible process, irreversible process,* and *isothermal process.*
- State the second law of thermodynamics.
- Describe the kinds of molecular motion that a molecule can possess.

- Explain how the entropy of a system is related to the number of accessible microstates.
- Predict the sign of ΔS for physical and chemical processes.
- State the third law of thermodynamics.
- Calculate standard entropy changes for a system from standard molar entropies.
- Calculate entropy changes in the surroundings for isothermal processes.
- Calculate the Gibbs free energy from the enthalpy change and entropy change at a given temperature.
- Use free energy changes to predict whether reactions are spontaneous.
- Calculate standard free energy changes using standard free energies of formation.
- Predict the effect of temperature on spontaneity given ΔH and ΔS.
- Calculate ΔG under nonstandard conditions.
- Relate $\Delta G°$ and equilibrium constant.

KEY EQUATIONS

- $\Delta S = \dfrac{q_{rev}}{T}$ (constant T)   [19.2]   Relating entropy change to the heat absorbed or released in a reversible process

- $\Delta S_{univ} = \Delta S_{sys} + \Delta S_{surr} > 0$ [19.4]   The second law of thermodynamics (spontaneous process)

- $S = k \ln W$ [19.5]   Relating entropy to the number of microstates

- $\Delta S° = \sum n S°(\text{products}) - \sum m S°(\text{reactants})$ [19.8]   Calculating the standard entropy change from standard molar entropies

- $\Delta S_{surr} = \dfrac{-\Delta H_{sys}}{T}$ [19.9]   The entropy change of the surroundings for a process at constant temperature and pressure

- $\Delta G = \Delta H - T\Delta S$ [19.11]   Calculating the Gibbs free-energy change from enthalpy and entropy changes at constant temperature

- $\Delta G° = \sum n\Delta G_f°(\text{products}) - \sum m\Delta G_f°(\text{reactants})$ [19.14]   Calculating the standard free-energy change from standard free energies of formation

- $\Delta G = -w_{max}$ [19.15]   Relating the free-energy change to the maximum work a process can perform.

- $\Delta G = \Delta G° + RT \ln Q$ [19.16]   Calculating free-energy change under nonstandard conditions.

- $\Delta G° = -RT \ln K$ [19.17]   Relating the standard free-energy change and the equilibrium constant.

VISUALIZING CONCEPTS

19.1 Two different gases occupy two separate bulbs. Consider the process that occurs when the stopcock separating the gases is opened, assuming the gases behave ideally. **(a)** Draw the final (equilibrium) state. **(b)** Predict the signs of ΔH and ΔS for the process. **(c)** Is the process that occurs when the stopcock is opened a reversible one? **(d)** How does the process affect the entropy of the surroundings? [Sections 19.1 and 19.2]

19.2 (a) What are the signs of ΔS and ΔH for the process depicted to the right? **(b)** How might temperature affect the sign of ΔG? **(c)** If energy can flow in and out of the system to maintain a constant temperature during the process, what can you say about the entropy change of the surroundings as a result of this process? [Sections 19.2 and 19.5]

19.3 Predict the sign of ΔS accompanying this reaction. Explain your choice. [Section 19.3]

19.4 The diagram below shows the variation in entropy with temperature for a substance that is a gas at the highest temperature shown. **(a)** What processes correspond to the entropy increases along the vertical lines labeled 1 and 2 in this diagram? **(b)** Why is the entropy change for 2 larger than that for 1? [Section 19.3]

Temperature (K)

19.5 The diagram below shows how ΔH (red line) and $T\Delta S$ (blue line) change with temperature for a hypothetical reaction. **(a)** What is the significance of the point at 300 K, where ΔH and $T\Delta S$ are equal? **(b)** In what temperature range is this reaction spontaneous? [Section 19.6]

19.6 Consider the accompanying diagram, which represents how ΔG for a hypothetical reaction responds to temperature. **(a)** At what temperature is the system at equilibrium? **(b)** In what temperature range is the reaction spontaneous? **(c)** Is ΔH positive or negative? **(d)** Is ΔS positive or negative? [Sections 19.5 and 19.6]

19.7 Consider a reaction $A_2(g) + B_2(g) \rightleftharpoons 2\,AB(g)$, with atoms of A shown in red and atoms of B shown in blue. **(a)** If $K_c = 1$, which system is at equilibrium? **(b)** What is the sign of ΔG for any process in which the contents of a reaction vessel move to equilibrium? **(c)** Rank the boxes in order of increasing magnitude of ΔG for the reaction. [Sections 19.5 and 19.7]

(1)   (2)   (3)

19.8 The diagram below shows how the free energy, G, changes during a hypothetical reaction $A(g) + B(g) \longrightarrow AB(g)$. On the left are pure reactants, each at 1 atm, and on the right is the pure product, also at 1 atm. **(a)** What is the significance of the minimum in the plot? **(b)** What does the quantity x, shown on the right side of the diagram, represent? [Section 19.7]

Progress of reaction

EXERCISES

Spontaneous Processes

19.9 Which of the following processes are spontaneous, and which are nonspontaneous: **(a)** the ripening of a banana, **(b)** dissolution of sugar in a cup of hot coffee, **(c)** the reaction of nitrogen atoms to form N_2 molecules at 25 °C and 1 atm, **(d)** lightning, **(e)** formation of CH_4 and O_2 molecules from CO_2 and H_2O at room temperature and 1 atm of pressure?

19.10 Which of the following processes are spontaneous: **(a)** the melting of ice cubes at 10 °C and 1 atm pressure; **(b)** separating a mixture of N_2 and O_2 into two separate samples, one that is pure N_2 and one that is pure O_2; **(c)** alignment of iron filings in a magnetic field; **(d)** the reaction of sodium metal with chlorine gas to form sodium chloride; **(e)** the dissolution of $HCl(g)$ in water to form concentrated hydrochloric acid?

19.11 **(a)** Give two examples of endothermic processes that are spontaneous. **(b)** Give an example of a process that is spontaneous at one temperature but nonspontaneous at a different temperature.

19.12 The crystalline hydrate $Cd(NO_3)_2 \cdot 4H_2O(s)$ loses water when placed in a large, closed, dry vessel:

$$Cd(NO_3)_2 \cdot 4H_2O(s) \longrightarrow Cd(NO_3)_2(s) + 4H_2O(g)$$

This process occurs even though it is endothermic; that is, ΔH is positive. Is this process an exception to Bertholet's generalization? Explain.

19.13 Consider the vaporization of liquid water to steam at a pressure of 1 atm. **(a)** Is this process endothermic or exothermic? **(b)** In what temperature range is it a spontaneous process? **(c)** In what temperature range is it a nonspontaneous process? **(d)** At what temperature are the two phases in equilibrium?

19.14 The normal freezing point of 1-propanol (C_3H_8O) is −127 °C. **(a)** Is the freezing of 1-propanol an endothermic or exothermic process? **(b)** In what temperature range is the freezing of 1-propanol a spontaneous process? **(c)** In what temperature range is it a nonspontaneous process? **(d)** Is there any temperature at which liquid and solid 1-propanol are in equilibrium? Explain.

19.15 **(a)** What is special about a *reversible* process? **(b)** Suppose a reversible process is reversed, restoring the system to its original state. What can be said about the surroundings after the process is reversed? **(c)** Under what circumstances will the vaporization of water to steam be a reversible process? **(d)** Are any of the

processes that occur in the world around us reversible in nature? Explain.

19.16 **(a)** What is meant by calling a process *irreversible*? **(b)** After an irreversible process the system is restored to its original state. What can be said about the condition of the surroundings after the system is restored to its original state? **(c)** Under what conditions will the condensation of a liquid be an irreversible process?

19.17 Consider a process in which an ideal gas changes from state 1 to state 2 in such a way that its temperature changes from 300 K to 200 K. **(a)** Describe how this change might be carried out while keeping the volume of the gas constant. **(b)** Describe how it might be carried out while keeping the pressure of the gas constant. **(c)** Does the change in ΔE depend on the particular pathway taken to carry out this change of state? Explain.

19.18 A system goes from state 1 to state 2 and back to state 1. **(a)** What is the relationship between the value of ΔE for going from state 1 to state 2 to that for going from state 2 back to state 1? **(b)** Without further information, can you conclude anything about the amount of heat transferred to the system as it goes from state 1 to state 2 as compared to that upon going from state 2 back to state 1? **(c)** Suppose the changes in state are reversible processes. Can you conclude anything about the work done by the system upon going from state 1 to state 2 as compared to that upon going from state 2 back to state 1?

19.19 Consider a system consisting of an ice cube. **(a)** Under what conditions can the ice cube melt reversibly? **(b)** If the ice cube melts reversibly, is ΔE zero for the process? Explain.

19.20 Consider what happens when a sample of the explosive TNT (Section 8.8: "Chemistry Put to Work: Explosives and Alfred Nobel") is detonated. **(a)** Is the detonation a spontaneous process? **(b)** What is the sign of q for this process? **(c)** Can you determine whether w is positive, negative, or zero for the process? Explain. **(d)** Can you determine the sign of ΔE for the process? Explain.

Entropy and the Second Law of Thermodynamics

19.21 **(a)** How can we calculate ΔS for an isothermal process? **(b)** Does ΔS for a process depend on the path taken from the initial to the final state of the system? Explain.

19.22 Suppose we vaporize a mole of liquid water at 25 °C and another mole of water at 100 °C. **(a)** Assuming that the enthalpy of vaporization of water does not change much between 25 °C and 100 °C, which process involves the larger change in entropy? **(b)** Does the entropy change in either process depend on whether we carry out the process reversibly or not? Explain.

19.23 The normal boiling point of methanol (CH_3OH) is 64.7 °C, and its molar enthalpy of vaporization is $\Delta H_{vap} = 71.8$ kJ/mol. **(a)** When $CH_3OH(l)$ boils at its normal boiling point, does its entropy increase or decrease?

(b) Calculate the value of ΔS when 1.00 mol of $CH_3OH(l)$ is vaporized at 64.7 °C.

19.24 The element cesium (Cs) freezes at 28.4 °C, and its molar enthalpy of fusion is $\Delta H_{fus} = 2.09$ kJ/mol. **(a)** When molten cesium solidifies to $Cs(s)$ at its normal melting point, is ΔS positive or negative? **(b)** Calculate the value of ΔS when 15.0 g of $Cs(l)$ solidifies at 28.4 °C.

19.25 **(a)** Express the second law of thermodynamics in words. **(b)** If the entropy of the system increases during a reversible process, what can you say about the entropy change of the surroundings? **(c)** In a certain spontaneous process the system undergoes an entropy change, $\Delta S = 42$ J/K. What can you conclude about ΔS_{surr}?

19.26 **(a)** Express the second law of thermodynamics as a mathematical equation. **(b)** In a particular spontaneous process the entropy of the system decreases. What can you conclude about the sign and magnitude of ΔS_{surr}? **(c)** During a certain reversible process, the surroundings undergo an entropy change, $\Delta S_{surr} = -78$ J/K. What is the entropy change of the system for this process?

19.27 The volume of 0.100 mol of helium gas at 27 °C is increased isothermally from 2.00 L to 5.00 L. Assuming the gas to be ideal, calculate the entropy change for the process.

19.28 The pressure on 0.850 mol of neon gas is increased from 1.25 atm to 2.75 atm at 100 °C. Assuming the gas to be ideal, calculate ΔS for this process.

The Molecular Interpretation of Entropy

19.29 How would each of the following changes affect the number of microstates available to a system: **(a)** increase in temperature, **(b)** decrease in volume, **(c)** change of state from liquid to gas?

19.30 **(a)** Using the heat of vaporization in Appendix B, calculate the entropy change for the vaporization of water at 25 °C and at 100 °C. **(b)** From your knowledge of microstates and the structure of liquid water, explain the difference in these two values.

19.31 **(a)** What do you expect for the sign of ΔS in a chemical reaction in which two moles of gaseous reactants are converted to three moles of gaseous products? **(b)** For which of the processes in Exercise 19.9 does the entropy of the system increase?

19.32 **(a)** In a chemical reaction two gases combine to form a solid. What do you expect for the sign of ΔS? **(b)** How does the entropy of the system change in the processes described in Exercise 19.10?

19.33 How does the entropy of the system change when **(a)** a solid melts, **(b)** a gas liquefies, **(c)** a solid sublimes?

19.34 How does the entropy of the system change when **(a)** the temperature of the system increases, **(b)** the volume of a gas increases, **(c)** equal volumes of ethanol and water are mixed to form a solution.

19.35 **(a)** State the third law of thermodynamics. **(b)** Distinguish between translational motion, vibrational motion, and rotational motion of a molecule. **(c)** Illustrate these three kinds of motion with sketches for the HCl molecule.

19.36 **(a)** The energy of a gas is increased by heating it. Using CO_2 as an example, illustrate the different ways in which additional energy can be distributed among the molecules of the gas. **(b)** You are told that the number of microstates for a system increases. What does this tell you about the entropy of the system?

19.37 **(a)** Using Figure 19.14 as a model, sketch how the entropy of water changes as it is heated from −50 °C to 110 °C at sea level. Show the temperatures at which there are vertical increases in entropy. **(b)** Which process has the larger entropy change: melting ice or boiling water? Explain.

19.38 Propanol (C_3H_7OH) melts at −126.5 °C and boils at 97.4 °C. Draw a qualitative sketch of how the entropy changes as propanol vapor at 150 °C and 1 atm is cooled to solid propanol at −150 °C and 1 atm.

19.39 For each of the following pairs, choose the substance with the higher entropy per mole at a given temperature: **(a)** Ar(l) or Ar(g), **(b)** He(g) at 3 atm pressure or He(g) at 1.5 atm pressure, **(c)** 1 mol of Ne(g) in 15.0 L or 1 mol of Ne(g) in 1.50 L, **(d)** $CO_2(g)$ or $CO_2(s)$.

19.40 For each of the following pairs, indicate which substance possesses the larger standard entropy: **(a)** 1 mol of $P_4(g)$ at 300 °C, 0.01 atm, or 1 mol of $As_4(g)$ at 300 °C, 0.01 atm; **(b)** 1 mol of $H_2O(g)$ at 100 °C, 1 atm, or 1 mol of $H_2O(l)$ at 100 °C, 1 atm; **(c)** 0.5 mol of $N_2(g)$ at 298 K, 20-L volume, or 0.5 mol $CH_4(g)$ at 298 K, 20-L volume; **(d)** 100 g $Na_2SO_4(s)$ at 30 °C or 100 g $Na_2SO_4(aq)$ at 30 °C.

19.41 Predict the sign of the entropy change of the system for each of the following reactions:
(a) $2\,SO_2(g) + O_2(g) \longrightarrow 2\,SO_3(g)$
(b) $Ba(OH)_2(s) \longrightarrow BaO(s) + H_2O(g)$
(c) $CO(g) + 2\,H_2(g) \longrightarrow CH_3OH(l)$
(d) $FeCl_2(s) + H_2(g) \longrightarrow Fe(s) + 2\,HCl(g)$

19.42 Predict the sign of ΔS_{sys} for each of the following processes: **(a)** Gaseous Ar is liquefied at 80 K. **(b)** Gaseous N_2O_4 dissociates to form gaseous NO_2. **(c)** Solid potassium reacts with gaseous O_2 to form solid potassium superoxide, KO_2. **(d)** Lead bromide precipitates upon mixing $Pb(NO_3)_2(aq)$ and $KBr(aq)$.

Entropy Changes in Chemical Reactions

19.43 In each of the following pairs, which compound would you expect to have the higher standard molar entropy: **(a)** $C_2H_2(g)$ or $C_2H_6(g)$; **(b)** $CO_2(g)$ or $CO(g)$? Explain.

19.44 Cyclopropane and propylene isomers both have the formula C_3H_6. Based on the molecular structures shown, which of these isomers would you expect to have the higher standard molar entropy at 25 °C?

Cyclopropane   Propylene

19.45 Use Appendix C to compare the standard entropies at 25 °C for the following pairs of substances: **(a)** Sc(s) and Sc(g); **(b)** $NH_3(g)$ and $NH_3(aq)$; **(c)** 1 mol $P_4(g)$ and 2 mol $P_2(g)$; **(d)** C(graphite) and C(diamond). In each case explain the difference in the entropy values.

19.46 Using Appendix C, compare the standard entropies at 25 °C for the following pairs of substances: **(a)** CuO(s) and $Cu_2O(s)$; **(b)** 1 mol $N_2O_4(g)$ and 2 mol $NO_2(g)$; **(c)** $SiO_2(s)$ and $CO_2(g)$; **(d)** CO(g) and $CO_2(g)$. For each pair, explain the difference in the entropy values.

[19.47] The standard entropies at 298 K for certain of the group 4A elements are as follows: C(s, diamond) = 2.43 J/mol-K; Si(s) = 18.81 J/mol-K; Ge(s) = 31.09 J/mol-K; and Sn(s) = 51.18 J/mol-K. All but Sn have the diamond structure. How do you account for the trend in the $S°$ values?

[19.48] Three of the forms of elemental carbon are graphite, diamond, and buckminsterfullerene. The entropies at 298 K for graphite and diamond are listed in Appendix C.

(a) Account for the difference in the $S°$ values of graphite and diamond in light of their structures (Figure 11.41). **(b)** What would you expect for the $S°$ value of buckminsterfullerene (Figure 11.43) relative to the values for graphite and diamond? Explain.

19.49 Using $S°$ values from Appendix C, calculate $\Delta S°$ values for the following reactions. In each case account for the sign of $\Delta S°$.
(a) $C_2H_4(g) + H_2(g) \longrightarrow C_2H_6(g)$
(b) $N_2O_4(g) \longrightarrow 2\ NO_2(g)$
(c) $Be(OH)_2(s) \longrightarrow BeO(s) + H_2O(g)$
(d) $2\ CH_3OH(g) + 3\ O_2(g) \longrightarrow 2\ CO_2(g) + 4\ H_2O(g)$

19.50 Calculate $\Delta S°$ values for the following reactions by using tabulated $S°$ values from Appendix C. In each case explain the sign of $\Delta S°$.
(a) $N_2H_4(g) + H_2(g) \longrightarrow 2\ NH_3(g)$
(b) $K(s) + O_2(g) \longrightarrow KO_2(s)$
(c) $Mg(OH)_2(s) + 2\ HCl(g) \longrightarrow MgCl_2(s) + 2\ H_2O(l)$
(d) $CO(g) + 2\ H_2(g) \longrightarrow CH_3OH(g)$

Gibbs Free Energy

19.51 **(a)** For a process that occurs at constant temperature, express the change in Gibbs free energy in terms of changes in the enthalpy and entropy of the system. **(b)** For a certain process that occurs at constant T and P, the value of ΔG is positive. What can you conclude? **(c)** What is the relationship between ΔG for a process and the rate at which it occurs?

19.52 **(a)** What is the meaning of the standard free-energy change, $\Delta G°$, as compared with ΔG? **(b)** For any process that occurs at constant temperature and pressure, what is the significance of $\Delta G = 0$? **(c)** For a certain process, ΔG is large and negative. Does this mean that the process necessarily occurs rapidly?

19.53 For a certain chemical reaction, $\Delta H° = -35.4$ kJ and $\Delta S° = -85.5$ J/K. **(a)** Is the reaction exothermic or endothermic? **(b)** Does the reaction lead to an increase or decrease in the randomness or disorder of the system? **(c)** Calculate $\Delta G°$ for the reaction at 298 K. **(d)** Is the reaction spontaneous at 298 K under standard conditions?

19.54 A certain reaction has $\Delta H° = -19.5$ kJ and $\Delta S° = +42.7$ J/K. **(a)** Is the reaction exothermic or endothermic? **(b)** Does the reaction lead to an increase or decrease in the randomness or disorder of the system? **(c)** Calculate $\Delta G°$ for the reaction at 298 K. **(d)** Is the reaction spontaneous at 298 K under standard conditions?

19.55 Using data in Appendix C, calculate $\Delta H°$, $\Delta S°$, and $\Delta G°$ at 298 K for each of the following reactions. In each case show that $\Delta G° = \Delta H° - T\Delta S°$.
(a) $H_2(g) + F_2(g) \longrightarrow 2\ HF(g)$
(b) $C(s, graphite) + 2\ Cl_2(g) \longrightarrow CCl_4(g)$
(c) $2\ PCl_3(g) + O_2(g) \longrightarrow 2\ POCl_3(g)$
(d) $2\ CH_3OH(g) + H_2(g) \longrightarrow C_2H_6(g) + 2\ H_2O(g)$

19.56 Use data in Appendix C to calculate $\Delta H°$, $\Delta S°$, and $\Delta G°$ at 25 °C for each of the following reactions. In each case show that $\Delta G° = \Delta H° - T\Delta S°$.
(a) $2\ Cr(s) + 3\ O_2(g) \longrightarrow 2\ Cr_2O_3(s)$
(b) $BaCO_3(s) \longrightarrow BaO(s) + CO_2(g)$
(c) $2\ P(s) + 10\ HF(g) \longrightarrow 2\ PF_5(g) + 5\ H_2(g)$
(d) $K(s) + O_2(g) \longrightarrow KO_2(s)$

19.57 Using data from Appendix C, calculate $\Delta G°$ for the following reactions. Indicate whether each reaction is spontaneous under standard conditions.
(a) $2\ SO_2(g) + O_2(g) \longrightarrow 2\ SO_3(g)$
(b) $NO_2(g) + N_2O(g) \longrightarrow 3\ NO(g)$
(c) $6\ Cl_2(g) + 2\ Fe_2O_3(s) \longrightarrow 4\ FeCl_3(s) + 3\ O_2(g)$
(d) $SO_2(g) + 2\ H_2(g) \longrightarrow S(s) + 2\ H_2O(g)$

19.58 Using data from Appendix C, calculate the change in Gibbs free energy for each of the following reactions. In each case indicate whether the reaction is spontaneous under standard conditions.
(a) $H_2(g) + Cl_2(g) \longrightarrow 2\ HCl(g)$
(b) $MgCl_2(s) + H_2O(l) \longrightarrow MgO(s) + 2\ HCl(g)$
(c) $2\ NH_3(g) \longrightarrow N_2H_4(g) + H_2(g)$
(d) $2\ NOCl(g) \longrightarrow 2\ NO(g) + Cl_2(g)$

19.59 Cyclohexane (C_6H_{12}) is a liquid hydrocarbon at room temperature. **(a)** Write a balanced equation for the combustion of $C_6H_{12}(l)$ to form $CO_2(g)$ and $H_2O(l)$. **(b)** Without using thermochemical data, predict whether $\Delta G°$ for this reaction is more negative or less negative than $\Delta H°$.

19.60 Sulfur dioxide reacts with strontium oxide as follows:

$$SO_2(g) + SrO(s) \longrightarrow SrSO_3(s)$$

(a) Without using thermochemical data, predict whether $\Delta G°$ for this reaction is more negative or less negative than $\Delta H°$. **(b)** If you had only standard enthalpy data for this reaction, how would you go about making a rough estimate of the value of $\Delta G°$ at 298 K, using data from Appendix C on other substances?

19.61 Classify each of the following reactions as one of the four possible types summarized in Table 19.4:
(a) $N_2(g) + 3 F_2(g) \longrightarrow 2 NF_3(g)$
$$\Delta H° = -249 \text{ kJ}; \Delta S° = -278 \text{ J/K}$$
(b) $N_2(g) + 3 Cl_2(g) \longrightarrow 2 NCl_3(g)$
$$\Delta H° = 460 \text{ kJ}; \Delta S° = -275 \text{ J/K}$$
(c) $N_2F_4(g) \longrightarrow 2 NF_2(g)$
$$\Delta H° = 85 \text{ kJ}; \Delta S° = 198 \text{ J/K}$$

19.62 From the values given for $\Delta H°$ and $\Delta S°$, calculate $\Delta G°$ for each of the following reactions at 298 K. If the reaction is not spontaneous under standard conditions at 298 K, at what temperature (if any) would the reaction become spontaneous?
(a) $2 PbS(s) + 3 O_2(g) \longrightarrow 2 PbO(s) + 2 SO_2(g)$
$$\Delta H° = -844 \text{ kJ}; \Delta S° = -165 \text{ J/K}$$
(b) $2 POCl_3(g) \longrightarrow 2 PCl_3(g) + O_2(g)$
$$\Delta H° = 572 \text{ kJ}; \Delta S° = 179 \text{ J/K}$$

19.63 A particular reaction is spontaneous at 450 K. The enthalpy change for the reaction is +34.5 kJ. What can you conclude about the sign and magnitude of ΔS for the reaction?

19.64 A certain reaction is nonspontaneous at -25 °C. The entropy change for the reaction is 95 J/K. What can you conclude about the sign and magnitude of ΔH?

19.65 For a particular reaction, $\Delta H = -32$ kJ and $\Delta S = -98$ J/K. Assume that ΔH and ΔS do not vary with temperature. **(a)** At what temperature will the reaction have $\Delta G = 0$? **(b)** If T is increased from that in part (a), will the reaction be spontaneous or nonspontaneous?

19.66 Reactions in which a substance decomposes by losing CO are called *decarbonylation* reactions. The decarbonylation of acetic acid proceeds as follows:

$$CH_3COOH(l) \longrightarrow CH_3OH(g) + CO(g)$$

By using data from Appendix C, calculate the minimum temperature at which this process will be spontaneous under standard conditions. Assume that $\Delta H°$ and $\Delta S°$ do not vary with temperature.

19.67 Consider the following reaction between oxides of nitrogen:

$$NO_2(g) + N_2O(g) \longrightarrow 3 NO(g)$$

(a) Use data in Appendix C to predict how $\Delta G°$ for the reaction varies with increasing temperature. **(b)** Calculate $\Delta G°$ at 800 K, assuming that $\Delta H°$ and $\Delta S°$ do not change with temperature. Under standard conditions is the reaction spontaneous at 800 K? **(c)** Calculate $\Delta G°$ at 1000 K. Is the reaction spontaneous under standard conditions at this temperature?

19.68 Methanol (CH_3OH) can be made by the controlled oxidation of methane:

$$CH_4(g) + \tfrac{1}{2} O_2(g) \longrightarrow CH_3OH(g)$$

(a) Use data in Appendix C to calculate $\Delta H°$ and $\Delta S°$ for this reaction. **(b)** How is $\Delta G°$ for the reaction expected to vary with increasing temperature? **(c)** Calculate $\Delta G°$ at 298 K. Under standard conditions, is the reaction spontaneous at this temperature? **(d)** Is there a temperature at which the reaction would be at equilibrium under standard conditions and that is low enough so that the compounds involved are likely to be stable?

[19.69] **(a)** Use data in Appendix C to estimate the boiling point of benzene, $C_6H_6(l)$. **(b)** Use a reference source, such as the *CRC Handbook of Chemistry and Physics*, to find the experimental boiling point of benzene. How do you explain any deviation between your answer in part (a) and the experimental value?

[19.70] **(a)** Using data in Appendix C, estimate the temperature at which the free-energy change for the transformation from $I_2(s)$ to $I_2(g)$ is zero. What assumptions must you make in arriving at this estimate? **(b)** Use a reference source, such as WebElements (www.webelements.com), to find the experimental melting and boiling points of I_2. **(c)** Which of the values in part (b) is closer to the value you obtained in part (a)? Can you explain why this is so?

19.71 Acetylene gas, $C_2H_2(g)$, is used in welding. **(a)** Write a balanced equation for the combustion of acetylene gas to $CO_2(g)$ and $H_2O(l)$. **(b)** How much heat is produced in burning 1 mol of C_2H_2 under standard conditions if both reactants and products are brought to 298 K? **(c)** What is the maximum amount of useful work that can be accomplished under standard conditions by this reaction?

19.72 **(a)** How much heat is produced in burning 1 mol of ethane (C_2H_6) under standard conditions if reactants and products are brought to 298 K and $H_2O(l)$ is formed? **(b)** What is the maximum amount of useful work that can be accomplished under standard conditions by this system?

Free Energy and Equilibrium

19.73 Explain qualitatively how ΔG changes for each of the following reactions as the partial pressure of O_2 is increased:
(a) $2 CO(g) + O_2(g) \longrightarrow 2 CO_2(g)$
(b) $2 H_2O_2(l) \longrightarrow 2 H_2O(l) + O_2(g)$
(c) $2 KClO_3(s) \longrightarrow 2 KCl(s) + 3 O_2(g)$

19.74 Indicate whether ΔG increases, decreases, or does not change when the partial pressure of H_2 is increased in each of the following reactions:

(a) $N_2(g) + 3 H_2(g) \longrightarrow 2 NH_3(g)$
(b) $2 HBr(g) \longrightarrow H_2(g) + Br_2(g)$
(c) $2 H_2(g) + C_2H_2(g) \longrightarrow C_2H_6(g)$

19.75 Consider the reaction $2 NO_2(g) \longrightarrow N_2O_4(g)$. **(a)** Using data from Appendix C, calculate $\Delta G°$ at 298 K. **(b)** Calculate ΔG at 298 K if the partial pressures of NO_2 and N_2O_4 are 0.40 atm and 1.60 atm, respectively.

19.76 Consider the reaction $6 H_2(g) + P_4(g) \longrightarrow 4 PH_3(g)$. **(a)** Using data from Appendix C, calculate $\Delta G°$ at 298 K. **(b)** Calculate ΔG at 298 K if the reaction mixture consists of 8.0 atm of H_2, 0.050 atm of P_4, and 0.22 atm of PH_3.

19.77 Use data from Appendix C to calculate the equilibrium constant, K, at 298 K for each of the following reactions:
(a) $H_2(g) + I_2(g) \rightleftharpoons 2 HI(g)$
(b) $C_2H_5OH(g) \rightleftharpoons C_2H_4(g) + H_2O(g)$
(c) $3 C_2H_2(g) \rightleftharpoons C_6H_6(g)$

19.78 Write the equilibrium-constant expression and calculate the value of the equilibrium constant for each of the following reactions at 298 K, using data from Appendix C:
(a) $NaHCO_3(s) \rightleftharpoons NaOH(s) + CO_2(g)$
(b) $2 HBr(g) + Cl_2(g) \rightleftharpoons 2 HCl(g) + Br_2(g)$
(c) $2 SO_2(g) + O_2(g) \rightleftharpoons 2 SO_3(g)$

19.79 Consider the decomposition of barium carbonate:

$$BaCO_3(s) \rightleftharpoons BaO(s) + CO_2(g)$$

Using data from Appendix C, calculate the equilibrium pressure of CO_2 at **(a)** 298 K and **(b)** 1100 K.

19.80 Consider the following reaction:

$$PbCO_3(s) \rightleftharpoons PbO(s) + CO_2(g)$$

Using data in Appendix C, calculate the equilibrium pressure of CO_2 in the system at **(a)** 400 °C and **(b)** 180 °C.

19.81 The value of K_a for nitrous acid (HNO_2) at 25 °C is given in Appendix D. **(a)** Write the chemical equation for the equilibrium that corresponds to K_a. **(b)** By using the value of K_a, calculate $\Delta G°$ for the dissociation of nitrous acid in aqueous solution. **(c)** What is the value of ΔG at equilibrium? **(d)** What is the value of ΔG when $[H^+] = 5.0 \times 10^{-2} M$, $[NO_2^-] = 6.0 \times 10^{-4} M$, and $[HNO_2] = 0.20 M$?

19.82 The K_b for methylamine (CH_3NH_2) at 25 °C is given in Appendix D. **(a)** Write the chemical equation for the equilibrium that corresponds to K_b. **(b)** By using the value of K_b, calculate $\Delta G°$ for the equilibrium in part (a). **(c)** What is the value of ΔG at equilibrium? **(d)** What is the value of ΔG when $[CH_3NH_3^+] = [H^+] = 1.5 \times 10^{-8} M$, $[CH_3NH_3^+] = 5.5 \times 10^{-4} M$, and $[CH_3NH_2] = 0.120 M$?

ADDITIONAL EXERCISES

19.83 Indicate whether each of the following statements is true or false. If it is false, correct it. **(a)** The feasibility of manufacturing NH_3 from N_2 and H_2 depends entirely on the value of ΔH for the process $N_2(g) + 3 H_2(g) \longrightarrow 2 NH_3(g)$. **(b)** The reaction of $Na(s)$ with $Cl_2(g)$ to form $NaCl(s)$ is a spontaneous process. **(c)** A spontaneous process can in principle be conducted reversibly. **(d)** Spontaneous processes in general require that work be done to force them to proceed. **(e)** Spontaneous processes are those that are exothermic and that lead to a higher degree of order in the system.

19.84 For each of the following processes, indicate whether the signs of ΔS and ΔH are expected to be positive, negative, or about zero. **(a)** A solid sublimes. **(b)** The temperature of a sample of $Co(s)$ is lowered from 60 °C to 25 °C. **(c)** Ethyl alcohol evaporates from a beaker. **(d)** A diatomic molecule dissociates into atoms. **(e)** A piece of charcoal is combusted to form $CO_2(g)$ and $H_2O(g)$.

19.85 The reaction $2 Mg(s) + O_2(g) \longrightarrow 2 MgO(s)$ is highly spontaneous and has a negative value for $\Delta S°$. The second law of thermodynamics states that in any spontaneous process there is always an increase in the entropy of the universe. Is there an inconsistency between the above reaction and the second law?

19.86 Ammonium nitrate dissolves spontaneously and endothermally in water at room temperature. What can you deduce about the sign of ΔS for this solution process?

[19.87] *Trouton's rule* states that for many liquids at their normal boiling points, the standard molar entropy of vaporization is about 88 J/mol-K. **(a)** Estimate the normal boiling point of bromine, Br_2, by determining $\Delta H°_{vap}$ for Br_2 using data from Appendix C. Assume that $\Delta H°_{vap}$ remains constant with temperature and that Trouton's rule holds. **(b)** Look up the normal boiling point of Br_2 in a chemistry handbook or at the WebElements web site (www.webelements.com).

[19.88] For the majority of the compounds listed in Appendix C, the value of $\Delta G_f°$ is more positive (or less negative) than the value of $\Delta H_f°$. **(a)** Explain this observation, using $NH_3(g)$, $CCl_4(l)$, and $KNO_3(s)$ as examples. **(b)** An exception to this observation is $CO(g)$. Explain the trend in the $\Delta H_f°$ and $\Delta G_f°$ values for this molecule.

19.89 Consider the following three reactions:
(i) $Ti(s) + 2 Cl_2(g) \longrightarrow TiCl_4(g)$
(ii) $C_2H_6(g) + 7 Cl_2(g) \longrightarrow 2 CCl_4(g) + 6 HCl(g)$
(iii) $BaO(s) + CO_2(g) \longrightarrow BaCO_3(s)$
(a) For each of the reactions, use data in Appendix C to calculate $\Delta H°$, $\Delta G°$, and $\Delta S°$ at 25 °C. **(b)** Which of these reactions are spontaneous under standard conditions at 25 °C? **(c)** For each of the reactions, predict the manner in which the change in free energy varies with an increase in temperature.

19.90 Using the data in Appendix C and given the pressures listed, calculate ΔG for each of the following reactions:
(a) $N_2(g) + 3 H_2(g) \longrightarrow 2 NH_3(g)$
$P_{N_2} = 2.6$ atm, $P_{H_2} = 5.9$ atm, $P_{NH_3} = 1.2$ atm
(b) $2 N_2H_4(g) + 2 NO_2(g) \longrightarrow 3 N_2(g) + 4 H_2O(g)$
$P_{N_2H_4} = P_{NO_2} = 5.0 \times 10^{-2}$ atm, $P_{N_2} = 0.5$ atm, $P_{H_2O} = 0.3$ atm
(c) $N_2H_4(g) \longrightarrow N_2(g) + 2 H_2(g)$
$P_{N_2H_4} = 0.5$ atm, $P_{N_2} = 1.5$ atm, $P_{H_2} = 2.5$ atm

19.91 **(a)** For each of the following reactions, predict the sign of $\Delta H°$ and $\Delta S°$ and discuss briefly how these factors determine the magnitude of K. **(b)** Based on your general chemical knowledge, predict which of these reactions will have $K > 0$. **(c)** In each case indicate whether K should increase or decrease with increasing temperature.
(i) $2 Mg(s) + O_2(g) \rightleftharpoons 2 MgO(s)$
(ii) $2 KI(s) \rightleftharpoons 2 K(g) + I_2(g)$
(iii) $Na_2(g) \rightleftharpoons 2 Na(g)$
(iv) $2 V_2O_5(s) \rightleftharpoons 4 V(s) + 5 O_2(g)$

19.92 Acetic acid can be manufactured by combining methanol with carbon monoxide, an example of a *carbonylation* reaction:

$$CH_3OH(l) + CO(g) \longrightarrow CH_3COOH(l)$$

(a) Calculate the equilibrium constant for the reaction at 25 °C. **(b)** Industrially, this reaction is run at temperatures above 25 °C. Will an increase in temperature produce an increase or decrease in the mole fraction of acetic acid at equilibrium? Why are elevated temperatures used? **(c)** At what temperature will this reaction have an equilibrium constant equal to 1? (You may assume that $\Delta H°$ and $\Delta S°$ are temperature independent, and you may ignore any phase changes that might occur.)

19.93 The oxidation of glucose ($C_6H_{12}O_6$) in body tissue produces CO_2 and H_2O. In contrast, anaerobic decomposition, which occurs during fermentation, produces ethanol (C_2H_5OH) and CO_2. **(a)** Using data given in Appendix C, compare the equilibrium constants for the following reactions:

$$C_6H_{12}O_6(s) + 6\,O_2(g) \rightleftharpoons 6\,CO_2(g) + 6\,H_2O(l)$$
$$C_6H_{12}O_6(s) \rightleftharpoons 2\,C_2H_5OH(l) + 2\,CO_2(g)$$

(b) Compare the maximum work that can be obtained from these processes under standard conditions.

[19.94] The conversion of natural gas, which is mostly methane, into products that contain two or more carbon atoms, such as ethane (C_2H_6), is a very important industrial chemical process. In principle, methane can be converted into ethane and hydrogen:

$$2\,CH_4(g) \longrightarrow C_2H_6(g) + H_2(g)$$

In practice, this reaction is carried out in the presence of oxygen:

$$2\,CH_4(g) + \tfrac{1}{2}O_2(g) \longrightarrow C_2H_6(g) + H_2O(g)$$

(a) Using the data in Appendix C, calculate K for these reactions at 25 °C and 500 °C. **(b)** Is the difference in $\Delta G°$ for the two reactions due primarily to the enthalpy term (ΔH) or the entropy term ($-T\Delta S$)? **(c)** Explain how the preceding reactions are an example of driving a nonspontaneous reaction, as discussed in the "Chemistry and Life" box in Section 19.7. **(d)** The reaction of CH_4 and O_2 to form C_2H_6 and H_2O must be carried out carefully to avoid a competing reaction. What is the most likely competing reaction?

[19.95] Cells use the hydrolysis of adenosine triphosphate (ATP) as a source of energy (Figure 19.20). The conversion of ATP to ADP has a standard free-energy change of −30.5 kJ/mol. If all the free energy from the metabolism of glucose,

$$C_6H_{12}O_6(s) + 6\,O_2(g) \longrightarrow 6\,CO_2(g) + 6\,H_2O(l)$$

goes into the conversion of ADP to ATP, how many moles of ATP can be produced for each mole of glucose?

[19.96] The potassium-ion concentration in blood plasma is about 5.0×10^{-3} M, whereas the concentration in muscle-cell fluid is much greater (0.15 M). The plasma and intracellular fluid are separated by the cell membrane, which we assume is permeable only to K^+. **(a)** What is ΔG for the transfer of 1 mol of K^+ from blood plasma to the cellular fluid at body temperature 37 °C? **(b)** What is the minimum amount of work that must be used to transfer this K^+?

[19.97] The relationship between the temperature of a reaction, its standard enthalpy change, and the equilibrium constant at that temperature can be expressed as the following linear equation:

$$\ln K = \frac{-\Delta H°}{RT} + \text{constant}$$

(a) Explain how this equation can be used to determine $\Delta H°$ experimentally from the equilibrium constants at several different temperatures. **(b)** Derive the preceding equation using relationships given in this chapter. To what is the constant equal?

[19.98] One way to derive Equation 19.3 depends on the observation that at constant T the number of ways, W, of arranging m ideal-gas particles in a volume V is proportional to the volume raised to the m power:

$$W \propto V^m$$

Use this relationship and Boltzmann's relationship between entropy and number of arrangements (Equation 19.5) to derive the equation for the entropy change for the isothermal expansion or compression of n moles of an ideal gas.

[19.99] About 86% of the world's electrical energy is produced by using steam turbines, a form of heat engine. In his analysis of an ideal heat engine, Sadi Carnot concluded that the maximum possible efficiency is defined by the total work that could be done by the engine, divided by the quantity of heat available to do the work (for example from hot steam produced by combustion of a fuel such as coal or methane). This efficiency is given by the ratio $(T_{high} - T_{low})/T_{high}$, where T_{high} is the temperature of the heat going into the engine and T_{low} is that of the heat leaving the engine. **(a)** What is the maximum possible efficiency of a heat engine operating between an input temperature of 700 K and an exit temperature of 288 K? **(b)** Why is it important that electrical power plants be located near bodies of relatively cool water? **(c)** Under what conditions could a heat engine operate at or near 100% efficiency? **(d)** It is often said that if the energy of combustion of a fuel such as methane were captured in an electrical fuel cell instead of by burning the fuel in a heat engine, a greater fraction of the energy could be put to useful work. Make a qualitative drawing like that in Figure 5.10 that illustrates the fact that in principle the fuel cell route will produce more useful work than the heat engine route from combustion of methane.

INTEGRATIVE EXERCISES

19.100 Most liquids follow Trouton's rule, which states that the molar entropy of vaporization lies in the range of 88 ± 5 J/mol-K. The normal boiling points and enthalpies of vaporization of several organic liquids are as follows:

Substance	Normal Boiling Point (°C)	ΔH_{vap} (kJ/mol)
Acetone, $(CH_3)_2CO$	56.1	29.1
Dimethyl ether, $(CH_3)_2O$	−24.8	21.5
Ethanol, C_2H_5OH	78.4	38.6
Octane, C_8H_{18}	125.6	34.4
Pyridine, C_5H_5N	115.3	35.1

(a) Calculate ΔS_{vap} for each of the liquids. Do all of the liquids obey Trouton's rule? (b) With reference to intermolecular forces (Section 11.2), can you explain any exceptions to the rule? (c) Would you expect water to obey Trouton's rule? By using data in Appendix B, check the accuracy of your conclusion. (d) Chlorobenzene (C_6H_5Cl) boils at 131.8 °C. Use Trouton's rule to estimate ΔH_{vap} for this substance.

19.101 Consider the polymerization of ethylene to polyethylene. ∞ (Section 12.6) (a) What would you predict for the sign of the entropy change during polymerization (ΔS_{poly})? Explain your reasoning. (b) The polymerization of ethylene is a spontaneous process at room temperature. What can you conclude about the enthalpy change during polymerization (ΔH_{poly})? (c) Use average bond enthalpies (Table 8.4) to estimate the value of ΔH_{poly} per ethylene monomer added. (d) Polyethylene is an *addition polymer*. By comparison, Nylon 66 is a *condensation polymer*. How would you expect ΔS_{poly} for a condensation polymer to compare to that for an addition polymer? Explain.

19.102 In chemical kinetics the *entropy of activation* is the entropy change for the process in which the reactants reach the activated complex. The entropy of activation for bimolecular processes is usually negative. Explain this observation with reference to Figure 14.15.

19.103 The following processes were all discussed in Chapter 18, "Chemistry of the Environment." Estimate whether the entropy of the system increases or decreases during each process: (a) photodissociation of $O_2(g)$, (b) formation of ozone from oxygen molecules and oxygen atoms, (c) diffusion of CFCs into the stratosphere, (d) desalination of water by reverse osmosis.

19.104 Carbon disulfide (CS_2) is a toxic, highly flammable substance. The following thermodynamic data are available for $CS_2(l)$ and $CS_2(g)$ at 298 K:

	ΔH_f° (kJ/mol)	ΔG_f° (kJ/mol)
$CS_2(l)$	89.7	65.3
$CS_2(g)$	117.4	67.2

(a) Draw the Lewis structure of the molecule. What do you predict for the bond order of the C—S bonds? (b) Use the VSEPR method to predict the structure of the CS_2 molecule. (c) Liquid CS_2 burns in O_2 with a blue flame, forming $CO_2(g)$ and $SO_2(g)$. Write a balanced equation for this reaction. (d) Using the data in the preceding table and in Appendix C, calculate ΔH° and ΔG° for the reaction in part (c). Is the reaction exothermic? Is it spontaneous at 298 K? (e) Use the data in the preceding table to calculate ΔS° at 298 K for the vaporization of $CS_2(l)$. Is the sign of ΔS° as you would expect for a vaporization? (f) Using data in the preceding table and your answer to part (e), estimate the boiling point of $CS_2(l)$. Do you predict that the substance will be a liquid or a gas at 298 K and 1 atm?

[19.105] The following data compare the standard enthalpies and free energies of formation of some crystalline ionic substances and aqueous solutions of the substances:

Substance	ΔH_f° (kJ/mol)	ΔG_f° (kJ/mol)
$AgNO_3(s)$	−124.4	−33.4
$AgNO_3(aq)$	−101.7	−34.2
$MgSO_4(s)$	−1283.7	−1169.6
$MgSO_4(aq)$	−1374.8	−1198.4

(a) Write the formation reaction for $AgNO_3(s)$. Based on this reaction, do you expect the entropy of the system to increase or decrease upon the formation of $AgNO_3(s)$? (b) Use ΔH_f° and ΔG_f° of $AgNO_3(s)$ to determine the entropy change upon formation of the substance. Is your answer consistent with your reasoning in part (a)? (c) Is dissolving $AgNO_3$ in water an exothermic or endothermic process? What about dissolving $MgSO_4$ in water? (d) For both $AgNO_3$ and $MgSO_4$, use the data to calculate the entropy change when the solid is dissolved in water. (e) Discuss the results from part (d) with reference to material presented in this chapter and in the second "Closer Look" box in Section 13.5.

[19.106] Consider the following equilibrium:

$$N_2O_4(g) \rightleftharpoons 2\,NO_2(g)$$

Thermodynamic data on these gases are given in Appendix C. You may assume that ΔH° and ΔS° do not vary with temperature. (a) At what temperature will an equilibrium mixture contain equal amounts of the two gases? (b) At what temperature will an equilibrium mixture of 1 atm total pressure contain twice as much NO_2 as N_2O_4? (c) At what temperature will an equilibrium mixture of 10 atm total pressure contain twice as much NO_2 as N_2O_4? (d) Rationalize the results from parts (b) and (c) by using Le Châtelier's principle. ∞ (Section 15.7)

[19.107] The reaction

$$SO_2(g) + 2\,H_2S(g) \rightleftharpoons 3\,S(s) + 2\,H_2O(g)$$

is the basis of a suggested method for removal of SO_2 from power-plant stack gases. The standard free energy of each substance is given in Appendix C. (a) What is the equilibrium constant for the reaction at 298 K? (b) In principle, is this reaction a feasible method of removing SO_2? (c) If $P_{SO_2} = P_{H_2S}$ and the vapor pressure of water is 25 torr, calculate the equilibrium SO_2 pressure in the system at 298 K. (d) Would you expect the process to be more or less effective at higher temperatures?

19.108 When most elastomeric polymers (e.g., a rubber band) are stretched, the molecules become more ordered, as illustrated here:

Suppose you stretch a rubber band. (a) Do you expect the entropy of the system to increase or decrease? (b) If the rubber band were stretched isothermally, would heat need to be absorbed or emitted to maintain constant temperature?

A VARIETY OF BATTERIES of different sizes, composition, and voltages.

WHAT'S AHEAD

20.1 Oxidation States and Oxidation-Reduction Reactions
We briefly review oxidation states and *oxidation-reduction (redox) reactions*.

20.2 Balancing Oxidation-Reduction Equations
We learn how to balance redox equations using the method of *half-reactions*.

20.3 Voltaic Cells
We next consider *voltaic cells*, which produce electricity from spontaneous redox reactions. In these cells, solid electrodes serve as the surfaces at which oxidation and reduction reactions take place. The electrode where oxidation occurs is the *anode*, and the electrode where reduction occurs is the *cathode*.

20.4 Cell EMF under Standard Conditions
We see that one of the important characteristics of a voltaic cell is its *emf*, or voltage, which is the difference in the electrical potentials at the two electrodes. Electrode potentials are tabulated for reduction half-reactions under standard conditions (*standard reduction potentials*). They are used to calculate cell voltages, to determine the relative strengths of oxidizing agents and reducing agents, and to predict whether particular redox reactions are spontaneous.

20.5 Free Energy and Redox Reactions
We relate the Gibbs free energy, $\Delta G°$, to the emf of an electrochemical cell.

20.6 Cell EMF under Nonstandard Conditions
We can calculate the voltage of cells that run under nonstandard conditions by using standard voltages and the Nernst equation.

20.7 Batteries and Fuel Cells
We describe batteries and fuel cells, which are commercially important energy sources based on electrochemical reactions.

20.8 Corrosion
Next, we discuss *corrosion*, a spontaneous electrochemical process involving metals.

20.9 Electrolysis
Finally, we focus on nonspontaneous redox reactions, examining *electrolytic cells*, which use electricity to perform chemical reactions.

WE ARE SURROUNDED BY AN AMAZING ARRAY of portable electronic gadgets including cell phones, portable music players, laptop computers, and gaming devices. In the absence of batteries, however, all of today's wirelessly connected electronic gadgetry would be nothing more than extra weight.

Thus, a variety of batteries of different sizes, compositions, and voltages have been developed, as shown in the chapter-opening photograph. Considerable research is in progress to develop new batteries with more power, faster recharging ability, lighter weight, or cheaper price. For example, in an effort to develop a battery that lasts long enough for all-day computing, battery makers are trying to increase the capacity of batteries without increasing their mass. At the heart of such development is the utilization of oxidation-reduction (redox) reactions, which are the chemical reactions that power batteries.

Redox reactions are among the most common and important chemical reactions. They are involved in the operation of batteries and in a wide variety of important natural processes, including the rusting of iron, the browning of foods, and the respiration of animals. As we discussed in Chapter 4, *oxidation* refers to the loss of electrons, and *reduction* refers to the gain of electrons. ∞ (Section 4.4) Thus, oxidation-reduction reactions occur when electrons are transferred from an atom that is oxidized to an atom that is reduced.

The examples of redox reactions that we have cited in this introduction are all spontaneous processes. We can also use electrical energy to make certain nonspontaneous redox reactions occur. A common example is electroplating, in which layers of one metal are coated onto another by the application of a voltage. **Electrochemistry** is the study of the relationships between electricity and chemical reactions. It includes the study of both spontaneous and nonspontaneous processes.

20.1 OXIDATION STATES AND OXIDATION-REDUCTION REACTIONS

How do we determine whether a given chemical reaction is an oxidation-reduction reaction? We can do so by keeping track of the *oxidation numbers* (*oxidation states*) of all the elements involved in the reaction. ∞ (Section 4.4) This procedure identifies whether any elements are changing oxidation state. For example, consider the reaction that occurs when zinc metal is added to a strong acid (Figure 20.1 ▼):

$$\text{Zn}(s) + 2\,\text{H}^+(aq) \longrightarrow \text{Zn}^{2+}(aq) + \text{H}_2(g) \qquad [20.1]$$

The reaction proceeds spontaneously and produces energy in the form of heat. The chemical equation for this reaction can be written as

$$\text{Zn}(s) + 2\,\text{H}^+(aq) \longrightarrow \text{Zn}^{2+}(aq) + \text{H}_2(g) \qquad [20.2]$$
$$\quad\;\,\bigcirc{0} \quad\;\; \bigcirc{+1} \qquad\qquad \bigcirc{+2} \quad\;\; \bigcirc{0}$$

▶ **Figure 20.1 Redox reaction in acidic solution.** The addition of zinc metal to hydrochloric acid leads to a spontaneous oxidation-reduction reaction: Zinc metal is oxidized to $\text{Zn}^{2+}(aq)$, and $\text{H}^+(aq)$ is reduced to $\text{H}_2(g)$, which produces the vigorous bubbling.

Zn(s)   2 HCl(aq)   ZnCl₂(aq)   H₂(g)

By writing the oxidation number of each element above or below the equation, we can see how the oxidation numbers change. The oxidation number of Zn changes from 0 to +2, and that of H changes from +1 to 0. Thus, this is an oxidation-reduction reaction. Electrons are transferred from zinc atoms to hydrogen ions, and therefore Zn is oxidized and H^+ is reduced.

In a reaction such as Equation 20.2, a clear transfer of electrons occurs. Zinc loses electrons as $Zn(s)$ is converted to $Zn^{2+}(aq)$, and hydrogen gains electrons as $H^+(aq)$ is turned into $H_2(g)$. In some reactions, however, the oxidation numbers change, but we cannot say that any substance literally gains or loses electrons. For example, consider the combustion of hydrogen gas:

$$2 H_2(g) + O_2(g) \longrightarrow 2 H_2O(g) \qquad [20.3]$$
$$\underset{0}{} \quad \underset{0}{} \qquad \underset{+1}{} \underset{-2}{}$$

In this reaction, hydrogen is oxidized from the 0 to the +1 oxidation state and oxygen is reduced from the 0 to the −2 oxidation state. Therefore, Equation 20.3 is an oxidation-reduction reaction. Water is not an ionic substance, however, and so there is not a complete transfer of electrons from hydrogen to oxygen as water is formed. Using oxidation states, therefore, is a convenient form of "bookkeeping," but you should not generally equate the oxidation state of an atom with its actual charge in a chemical compound. ∞ (Section 8.5 "A Closer Look: Oxidation Numbers, Formal Charges, and Actual Partial Charges")

▲ GIVE IT SOME THOUGHT

What are the oxidation numbers, or states, of the elements in the nitrite ion, NO_2^-?

In any redox reaction, both oxidation and reduction must occur. In other words, if one substance is oxidized, then another must be reduced: The electrons formally have to go from one place to another. The substance that makes it possible for another substance to be oxidized is called either the **oxidizing agent** or the **oxidant**. The oxidizing agent removes electrons from another substance by acquiring them itself; thus, the oxidizing agent is itself reduced. Similarly, a **reducing agent**, or **reductant**, is a substance that gives up electrons, thereby causing another substance to be reduced. The reducing agent is therefore oxidized in the process. In Equation 20.2, $H^+(aq)$ is the oxidizing agent and $Zn(s)$ is the reducing agent.

■ SAMPLE EXERCISE 20.1 | Identifying Oxidizing and Reducing Agents

The nickel-cadmium (nicad) battery, a rechargeable "dry cell" used in battery-operated devices, uses the following redox reaction to generate electricity:

$$Cd(s) + NiO_2(s) + 2 H_2O(l) \longrightarrow Cd(OH)_2(s) + Ni(OH)_2(s)$$

Identify the substances that are oxidized and reduced, and indicate which is the oxidizing agent and which is the reducing agent.

SOLUTION

Analyze: We are given a redox equation and asked to identify the substance oxidized and the substance reduced and to label one as the oxidizing agent and the other as the reducing agent.

Plan: First, we assign oxidation states, or numbers, to all the atoms in the reaction and determine the elements that are changing oxidation state. Second, we apply the definitions of oxidation and reduction.

Solve: $Cd(s) + NiO_2(s) + 2 H_2O(l) \longrightarrow Cd(OH)_2(s) + Ni(OH)_2(s)$
$$\underset{0}{} \quad \underset{+4}{}\underset{-2}{} \quad \underset{+1}{}\underset{-2}{} \qquad \underset{+2}{}\underset{-2}{}\underset{+1}{} \quad \underset{+2}{}\underset{-2}{}\underset{+1}{}$$

Cd increases in oxidation state from 0 to +2, and Ni decreases from +4 to +2.

Because the Cd atom increases in oxidation state, it is oxidized (loses electrons) and therefore serves as the reducing agent. The Ni atom decreases in oxidation state as NiO_2 is converted into $Ni(OH)_2$. Thus, NiO_2 is reduced (gains electrons) and therefore serves as the oxidizing agent.

Comment: A common mnemonic for remembering oxidation and reduction is "LEO the lion says GER": *l*osing *e*lectrons is *o*xidation; *g*aining *e*lectrons is *r*eduction.

■ **PRACTICE EXERCISE**

Identify the oxidizing and reducing agents in the oxidation-reduction reaction

$$2\,H_2O(l) + Al(s) + MnO_4^-(aq) \longrightarrow Al(OH)_4^-(aq) + MnO_2(s)$$

Answer: $Al(s)$ is the reducing agent; $MnO_4^-(aq)$ is the oxidizing agent.

20.2 BALANCING OXIDATION-REDUCTION EQUATIONS

Whenever we balance a chemical equation, we must obey the law of conservation of mass: The amount of each element must be the same on both sides of the equation. (Atoms are neither created nor destroyed in any chemical reaction.) As we balance oxidation-reduction reactions, there is an additional requirement: The gains and losses of electrons must be balanced. In other words, if a substance loses a certain number of electrons during a reaction, then another substance must gain that same number of electrons. (Electrons are neither created nor destroyed in any chemical reaction.)

In many simple chemical reactions, such as Equation 20.2, balancing the electrons is handled "automatically"; we can balance the equation without explicitly considering the transfer of electrons. Many redox reactions are more complex than Equation 20.2, however, and cannot be balanced easily without taking into account the number of electrons lost and gained in the course of the reaction. In this section we examine the method of half reactions, which is a systematic procedure for balancing redox equations.

Half-Reactions

Although oxidation and reduction must take place simultaneously, it is often convenient to consider them as separate processes. For example, the oxidation of Sn^{2+} by Fe^{3+}:

$$Sn^{2+}(aq) + 2\,Fe^{3+}(aq) \longrightarrow Sn^{4+}(aq) + 2\,Fe^{2+}(aq)$$

can be considered to consist of two processes: (1) the oxidation of Sn^{2+} (Equation 20.4) and (2) the reduction of Fe^{3+} (Equation 20.5):

Oxidation:	$Sn^{2+}(aq) \longrightarrow Sn^{4+}(aq) + 2\,e^-$	[20.4]
Reduction:	$2\,Fe^{3+}(aq) + 2\,e^- \longrightarrow 2\,Fe^{2+}(aq)$	[20.5]

Notice that electrons are shown as products in the oxidation process, whereas electrons are shown as reactants in the reduction process.

Equations that show either oxidation or reduction alone, such as Equations 20.4 and 20.5, are called **half-reactions**. In the overall redox reaction, the number of electrons lost in the oxidation half-reaction must equal the number of electrons gained in the reduction half-reaction. When this condition is met and each half-reaction is balanced, the electrons on the two sides cancel when the two half-reactions are added to give the overall balanced oxidation-reduction equation.

Balancing Equations by the Method of Half-Reactions

The use of half-reactions provides a general method for balancing oxidation–reduction equations. We usually begin with a "skeleton" ionic equation that shows only the substances undergoing oxidation and reduction. In such cases, we usually do not need to assign oxidation numbers unless we are unsure whether the reaction actually involves oxidation-reduction. We will find that H^+ (for acidic solutions), OH^- (for basic solutions), and H_2O are often involved as reactants or products in redox reactions. Unless H^+, OH^-, or H_2O are being oxidized or reduced, they do not appear in the skeleton equation. Their presence, however, can be deduced during the course of balancing the equation.

For balancing a redox reaction that occurs in acidic aqueous solution, the procedure is as follows:

1. Divide the equation into two half-reactions, one for oxidation and the other for reduction.
2. Balance each half-reaction.
   (a) First, balance the elements other than H and O.
   (b) Next, balance the O atoms by adding H_2O as needed.
   (c) Then, balance the H atoms by adding H^+ as needed.
   (d) Finally, balance the charge by adding e^- as needed.
   This specific sequence is important, and it is summarized in the diagram in the margin. At this point, you can check whether the number of electrons in each half-reaction corresponds to the changes in oxidation state.
3. Multiply the half-reactions by integers, if necessary, so that the number of electrons lost in one half-reaction equals the number of electrons gained in the other.
4. Add the two half-reactions and, if possible, simplify by canceling species appearing on both sides of the combined equation.
5. Check to make sure that atoms and charges are balanced.

As an example, let's consider the reaction between permanganate ion (MnO_4^-) and oxalate ion ($C_2O_4^{2-}$) in acidic aqueous solution. When MnO_4^- is added to an acidified solution of $C_2O_4^{2-}$, the deep purple color of the MnO_4^- ion fades, as illustrated in Figure 20.2▼. Bubbles of CO_2 form, and the solution takes on the pale pink color of Mn^{2+}. We can therefore write the skeleton equation as

$$MnO_4^-(aq) + C_2O_4^{2-}(aq) \longrightarrow Mn^{2+}(aq) + CO_2(aq) \qquad [20.6]$$

Experiments show that H^+ is consumed and H_2O is produced in the reaction. We will see that their involvement in the reaction is deduced in the course of balancing the equation.

To complete and balance this equation, we first write the two half-reactions (step 1). One half-reaction must have Mn on both sides of the arrow, and the other must have C on both sides of the arrow:

$$MnO_4^-(aq) \longrightarrow Mn^{2+}(aq)$$
$$C_2O_4^{2-}(aq) \longrightarrow CO_2(g)$$

(a) (b) (c)

◄ Figure 20.2 **Titration of an acidic solution of $Na_2C_2O_4$ with $KMnO_4(aq)$.** (a) As it moves from the buret to the reaction flask, the deep purple MnO_4^- is rapidly reduced to extremely pale pink Mn^{2+} by $C_2O_4^{2-}$. (b) Once all the $C_2O_4^{2-}$ in the flask has been consumed, any MnO_4^- added to the flask retains its purple color, and the end point corresponds to the faintest discernible purple color in the solution. (c) Beyond the end point the solution in the flask becomes deep purple because of excess MnO_4^-.

We next complete and balance each half-reaction. First, we balance all the atoms except for H and O (step 2a). In the permanganate half-reaction we already have one manganese atom on each side of the equation and so need to do nothing. In the oxalate half-reaction we need to add a coefficient 2 to the right to balance the two carbons on the left:

$$MnO_4^-(aq) \longrightarrow Mn^{2+}(aq)$$

$$C_2O_4^{2-}(aq) \longrightarrow 2\,CO_2(g)$$

Next we balance O (step 2b). The permanganate half-reaction has four oxygens on the left and none on the right; therefore four H_2O molecules are needed as product to balance the four oxygen atoms in MnO_4^-:

$$MnO_4^-(aq) \longrightarrow Mn^{2+}(aq) + 4\,H_2O(l)$$

The eight hydrogen atoms now in the products must be balanced by adding $8\,H^+$ to the reactants (step 2c):

$$8\,H^+(aq) + MnO_4^-(aq) \longrightarrow Mn^{2+}(aq) + 4\,H_2O(l)$$

There are now equal numbers of each type of atom on the two sides of the equation, but the charge still needs to be balanced. The total charge of the reactants is $8(1+) + (1-) = 7+$, and that of the products is $(2+) + 4(0) = 2+$. To balance the charge, we must add five electrons to the reactant side (step 3d):

$$5\,e^- + 8\,H^+(aq) + MnO_4^-(aq) \longrightarrow Mn^{2+}(aq) + 4\,H_2O(l)$$

We can check our result using oxidation states. In this half-reaction Mn goes from the +7 oxidation state in MnO_4^- to the +2 oxidation state in Mn^{2+}. Therefore, each Mn atom gains five electrons, in agreement with our balanced half-reaction.

Now that the permanganate half-reaction is balanced, let's return to the oxalate half-reaction. We already balanced the C and O atoms in step 2a. We can balance the charge (step 2d) by adding two electrons to the products:

$$C_2O_4^{2-}(aq) \longrightarrow 2\,CO_2(g) + 2\,e^-$$

We can check this result using oxidation states. Carbon goes from the +3 oxidation state in $C_2O_4^{2-}$ to the +4 oxidation state in CO_2. Thus, each C atom loses one electron, and therefore the two C atoms in $C_2O_4^{2-}$ lose two electrons, in agreement with our balanced half-reaction.

Now we need to multiply each half-reaction by an appropriate integer so that the number of electrons gained in one half-reaction equals the number of electrons lost in the other (step 3). We must multiply the MnO_4^- half-reaction by 2 and the $C_2O_4^{2-}$ half-reaction by 5 so that the same number of electrons (10) appears in both equations:

$$10\,e^- + 16\,H^+(aq) + 2\,MnO_4^-(aq) \longrightarrow 2\,Mn^{2+}(aq) + 8\,H_2O(l)$$

$$5\,C_2O_4^{2-}(aq) \longrightarrow 10\,CO_2(g) + 10\,e^-$$

$$\overline{16\,H^+(aq) + 2\,MnO_4^-(aq) + 5\,C_2O_4^{2-}(aq) \longrightarrow}$$
$$2\,Mn^{2+}(aq) + 8\,H_2O(l) + 10\,CO_2(g)$$

The balanced equation is the sum of the balanced half-reactions (step 4). Note that the electrons on the reactant and product sides of the equation cancel each other.

We can check the balanced equation by counting atoms and charges (step 5): There are 16 H, 2 Mn, 28 O, 10 C, and a net charge of 4+ on each side of the equation, confirming that the equation is correctly balanced.

GIVE IT SOME THOUGHT

If the equation for a redox reaction is balanced, will free electrons appear in the equation as either reactants or products?

■ SAMPLE EXERCISE 20.2 | Balancing Redox Equations in Acidic Solution

Complete and balance this equation by the method of half-reactions:

$$Cr_2O_7^{2-}(aq) + Cl^-(aq) \longrightarrow Cr^{3+}(aq) + Cl_2(g) \qquad \text{(acidic solution)}$$

SOLUTION

Analyze: We are given an incomplete, unbalanced (skeleton) equation for a redox reaction occurring in acidic solution and asked to complete and balance it.

Plan: We use the half-reaction procedure we just learned.

Solve: Step 1: We divide the equation into two half-reactions:

$$Cr_2O_7^{2-}(aq) \longrightarrow Cr^{3+}(aq)$$
$$Cl^-(aq) \longrightarrow Cl_2(g)$$

Step 2: We balance each half-reaction. In the first half-reaction the presence of one $Cr_2O_7^{2-}$ among the reactants requires two Cr^{3+} among the products. The seven oxygen atoms in $Cr_2O_7^{2-}$ are balanced by adding seven H_2O to the products. The 14 hydrogen atoms in 7 H_2O are then balanced by adding 14 H^+ to the reactants:

$$14\,H^+(aq) + Cr_2O_7^{2-}(aq) \longrightarrow 2\,Cr^{3+}(aq) + 7\,H_2O(l)$$

We then balance the charge by adding electrons to the left side of the equation so that the total charge is the same on the two sides:

$$6\,e^- + 14\,H^+(aq) + Cr_2O_7^{2-}(aq) \longrightarrow 2\,Cr^{3+}(aq) + 7\,H_2O(l)$$

We can check this result by looking at the oxidation state changes. Each chromium atom goes from +6 to +3, gaining three electrons, and therefore the two Cr atoms in $Cr_2O_7^{2-}$ gains six electrons, in agreement with our half-reaction.

In the second half-reaction, two Cl^- are required to balance one Cl_2:

$$2\,Cl^-(aq) \longrightarrow Cl_2(g)$$

We add two electrons to the right side to attain charge balance:

$$2\,Cl^-(aq) \longrightarrow Cl_2(g) + 2\,e^-$$

This result agrees with the oxidation state changes. Each chlorine atom goes from −1 to 0, losing one electron, and therefore the two chlorine atoms lose two electrons.

Step 3: We equalize the number of electrons transferred in the two half-reactions. To do so, we multiply the Cl half-reaction by 3 so that the number of electrons gained in the Cr half-reaction (6) equals the number lost in the Cl half-reaction, allowing the electrons to cancel when the half-reactions are added:

$$6\,Cl^-(aq) \longrightarrow 3\,Cl_2(g) + 6\,e^-$$

Step 4: The equations are added to give the balanced equation:

$$14\,H^+(aq) + Cr_2O_7^{2-}(aq) + 6\,Cl^-(aq) \longrightarrow 2\,Cr^{3+}(aq) + 7\,H_2O(l) + 3\,Cl_2(g)$$

Step 5: There are equal numbers of atoms of each kind on the two sides of the equation (14 H, 2 Cr, 7 O, 6 Cl). In addition, the charge is the same on the two sides (6+). Thus, the equation is balanced.

■ PRACTICE EXERCISE

Complete and balance the following equations using the method of half-reactions. Both reactions occur in acidic solution.

(a) $Cu(s) + NO_3^-(aq) \longrightarrow Cu^{2+}(aq) + NO_2(g)$

(b) $Mn^{2+}(aq) + NaBiO_3(s) \longrightarrow Bi^{3+}(aq) + MnO_4^-(aq)$

Answers: **(a)** $Cu(s) + 4\,H^+(aq) + 2\,NO_3^-(aq) \longrightarrow Cu^{2+}(aq) + 2\,NO_2(g) + 2\,H_2O(l)$

(b) $2\,Mn^{2+}(aq) + 5\,NaBiO_3(s) + 14\,H^+(aq) \longrightarrow 2\,MnO_4^-(aq) + 5\,Bi^{3+}(aq) + 5\,Na^+(aq) + 7\,H_2O(l)$

Balancing Equations for Reactions Occurring in Basic Solution

If a redox reaction occurs in basic solution, the equation must be completed by using OH^- and H_2O rather than H^+ and H_2O. One way to balance these reactions is to balance the half-reactions initially as if they occurred in acidic solution. Then, count the H^+ in each half-reaction, and add the same number of OH^- *to each side* of the half-reaction. This way the reaction is still mass-balanced because you are adding the same thing to both sides. In essence, what you are doing is "neutralizing" the protons to form water ($H^+ + OH^- \longrightarrow H_2O$) on the side containing H^+, and the other side ends up with the OH^-. The resulting water molecules can be canceled as needed. This procedure is shown in Sample Exercise 20.3.

■ SAMPLE EXERCISE 20.3 | Balancing Redox Equations in Basic Solution

Complete and balance this equation for a redox reaction that takes place in basic solution:

$$CN^-(aq) + MnO_4^-(aq) \longrightarrow CNO^-(aq) + MnO_2(s) \qquad \text{(basic solution)}$$

SOLUTION

Analyze: We are given an incomplete equation for a basic redox reaction and asked to balance it.

Plan: We go through the first steps of our procedure as if the reaction were occurring in acidic solution. We then add the appropriate number of OH^- ions to each side of the equation, combining H^+ and OH^- to form H_2O. We complete the process by simplifying the equation.

Solve: Step 1: We write the incomplete, unbalanced half-reactions:

$$CN^-(aq) \longrightarrow CNO^-(aq)$$
$$MnO_4^-(aq) \longrightarrow MnO_2(s)$$

Step 2: We balance each half-reaction as if it took place in acidic solution. The H^+ ions are set in red for emphasis:

$$CN^-(aq) + H_2O(l) \longrightarrow CNO^-(aq) + 2\,H^+(aq) + 2\,e^-$$
$$3\,e^- + 4\,H^+(aq) + MnO_4^-(aq) \longrightarrow MnO_2(s) + 2\,H_2O(l)$$

Now we need to take into account that the reaction occurs in basic solution, adding OH^- to both sides of both half-reactions to neutralize H^+. The OH^- ions are set in blue for emphasis.

$$CN^-(aq) + H_2O(l) + 2\,OH^-(aq) \longrightarrow CNO^-(aq) + 2\,H^+(aq) + 2\,e^- + 2\,OH^-(aq)$$
$$3\,e^- + 4\,H^+(aq) + MnO_4^-(aq) + 4\,OH^-(aq) \longrightarrow MnO_2(s) + 2\,H_2O(l) + 4\,OH^-(aq)$$

We now "neutralize" H^+ and OH^- by forming H_2O when they are on the same side of either half-reaction:

$$CN^-(aq) + H_2O(l) + 2\,OH^-(aq) \longrightarrow CNO^-(aq) + 2\,H_2O(l) + 2\,e^-$$
$$3\,e^- + 4\,H_2O(l) + MnO_4^-(aq) \longrightarrow MnO_2(s) + 2\,H_2O(l) + 4\,OH^-(aq)$$

Next, we cancel water molecules that appear as both reactants and products:

$$CN^-(aq) + 2\,OH^-(aq) \longrightarrow CNO^-(aq) + H_2O(l) + 2\,e^-$$
$$3\,e^- + 2\,H_2O(l) + MnO_4^-(aq) \longrightarrow MnO_2(s) + 4\,OH^-(aq)$$

Both half-reactions are now balanced. You can check the atoms and the overall charge.

Step 3: Now we multiply the cyanide half-reaction through by 3, which will give 6 electrons on the product side; and multiply the permanganate half-reaction through by 2, which will give 6 electrons on the reactant side:

$$3\,CN^-(aq) + 6\,OH^-(aq) \longrightarrow 3\,CNO^-(aq) + 3\,H_2O(l) + 6\,e^-$$
$$6\,e^- + 4\,H_2O(l) + 2\,MnO_4^-(aq) \longrightarrow 2\,MnO_2(s) + 8\,OH^-(aq)$$

Step 4: Now we can add the two half-reactions together and simplify by canceling species that appear as both reactants and products:

$$3\,CN^-(aq) + H_2O(l) + 2\,MnO_4^-(aq) \longrightarrow 3\,CNO^-(aq) + 2\,MnO_2(s) + 2\,OH^-(aq)$$

Step 5: Check that the atoms and charges are balanced.

There are 3 C, 3 N, 2 H, 9 O, 2 Mn, and a charge of 5− on both sides of the equation.

■ PRACTICE EXERCISE

Complete and balance the following equations for oxidation-reduction reactions that occur in basic solution:

(a) $NO_2^-(aq) + Al(s) \longrightarrow NH_3(aq) + Al(OH)_4^-(aq)$

(b) $Cr(OH)_3(s) + ClO^-(aq) \longrightarrow CrO_4^{2-}(aq) + Cl_2(g)$

Answers: **(a)** $NO_2^-(aq) + 2\,Al(s) + 5\,H_2O(l) + OH^-(aq) \longrightarrow NH_3(aq) + 2\,Al(OH)_4^-(aq)$

(b) $2\,Cr(OH)_3(s) + 6\,ClO^-(aq) \longrightarrow 2\,CrO_4^{2-}(aq) + 3\,Cl_2(g) + 2\,OH^-(aq) + 2\,H_2O(l)$

20.3 VOLTAIC CELLS

The energy released in a spontaneous redox reaction can be used to perform electrical work. This task is accomplished through a **voltaic** (or **galvanic**) **cell**, a device in which the transfer of electrons takes place through an external pathway rather than directly between reactants.

One such spontaneous reaction occurs when a strip of zinc is placed in contact with a solution containing Cu^{2+}. As the reaction proceeds, the blue color of $Cu^{2+}(aq)$ ions fades, and copper metal deposits on the zinc. At the same time, the zinc begins to dissolve. These transformations are shown in Figure 20.3▼ and are summarized by the equation

$$Zn(s) + Cu^{2+}(aq) \longrightarrow Zn^{2+}(aq) + Cu(s) \qquad [20.7]$$

A SPONTANEOUS OXIDATION-REDUCTION REACTION
An atomic-level view of how electrons are transferred in a spontaneous redox reaction.

Atoms in Zn strip
Cu^{2+} ions in solution
$2\,e^-$

Zn^{2+} ion
Cu atom

$Zn(s) + Cu^{2+}(aq)$

$Zn^{2+}(aq) + Cu(s)$

A strip of zinc is placed in a solution of copper (II) sulfate. A Cu^{2+} ion comes in contact with the surface of the Zn strip and gains two elecrons from a Zn atom; the Cu^{2+} is reduced, and the Zn atom is oxidized.

As the reaction proceeds, the zinc dissolves, the blue color due to $Cu^{2+}(aq)$ fades, and copper metal (the dark material on the zinc strip and on the bottom of the beaker) is deposited.

Electrons are transferred from the zinc to the Cu^{2+} ion, forming Zn^{2+} ions and Cu(s). The resulting colorless Zn^{2+} ion enters the solution, and the Cu atom remains deposited on the zinc strip.

▲ Figure 20.3 **A spontaneous oxidation-reduction reaction.**

▲ **Figure 20.4 A voltaic cell based on the reaction in Equation 20.7.** The left compartment contains 1 *M* CuSO₄ and a copper electrode. The one on the right contains 1 *M* ZnSO₄ and a zinc electrode. The solutions are connected by a porous glass disc, which permits contact of the two solutions. The metal electrodes are connected through a voltmeter, which reads the potential of the cell, 1.10 V.

Figure 20.4 ◄ shows a voltaic cell that uses the redox reaction between Zn and Cu^{2+} given in Equation 20.7. Although the setup shown in Figure 20.4 is more complex than that in Figure 20.3, the reaction is the same in both cases. The significant difference is that the Zn metal and $Cu^{2+}(aq)$ are not in direct contact in the voltaic cell. Instead, the Zn metal is in contact with $Zn^{2+}(aq)$ in one compartment of the cell, and Cu metal is in contact with $Cu^{2+}(aq)$ in another compartment. Consequently, the reduction of the Cu^{2+} can occur only by a flow of electrons through an external circuit, namely, the wire that connects the Zn and Cu strips. In other words, by physically separating the reduction half of a redox reaction from the oxidation half, we create a flow of electrons through an external circuit. The electron flow can be used to accomplish electrical work.

The two solid metals that are connected by the external circuit are called *electrodes*. By definition, the electrode at which oxidation occurs is called the **anode**; the electrode at which reduction occurs is called the **cathode**.* The electrodes can be made of materials that participate in the reaction, as in the present example. Over the course of the reaction, the Zn electrode will gradually disappear and the copper electrode will gain mass. More typically, the electrodes are made of a conducting material, such as platinum or graphite, that does not gain or lose mass during the reaction but serves as a surface at which electrons are transferred.

Each of the two compartments of a voltaic cell is called a *half-cell*. One half-cell is the site of the oxidation half-reaction, and the other is the site of the reduction half-reaction. In our present example Zn is oxidized and Cu^{2+} is reduced:

Anode (oxidation half-reaction) $Zn(s) \longrightarrow Zn^{2+}(aq) + 2\,e^{-}$

Cathode (reduction half-reaction) $Cu^{2+}(aq) + 2\,e^{-} \longrightarrow Cu(s)$

Electrons become available as zinc metal is oxidized at the anode. They flow through the external circuit to the cathode, where they are consumed as $Cu^{2+}(aq)$ is reduced. Because $Zn(s)$ is oxidized in the cell, the zinc electrode loses mass, and the concentration of the Zn^{2+} solution increases as the cell operates. Similarly, the Cu electrode gains mass, and the Cu^{2+} solution becomes less concentrated as Cu^{2+} is reduced to $Cu(s)$.

For a voltaic cell to work, the solutions in the two half-cells must remain electrically neutral. As Zn is oxidized in the anode compartment, Zn^{2+} ions enter the solution. Thus, there must be some means for positive ions to migrate out of the anode compartment and for negative ions to migrate in to keep the solution electrically neutral. Similarly, the reduction of Cu^{2+} at the cathode removes positive charge from the solution, leaving an excess of negative charge in that half-cell. Thus, positive ions must migrate into the compartment and negative ions must migrate out. In fact, no measurable electron flow will occur between electrodes unless a means is provided for ions to migrate through the solution from one electrode compartment to the other, thereby completing the circuit.

In Figure 20.4, a porous glass disc separating the two compartments allows a migration of ions that maintains the electrical neutrality of the solutions. In Figure 20.5 ◄, a *salt bridge* serves this purpose. A salt bridge consists of a U-shaped tube that contains an electrolyte solution, such as $NaNO_3(aq)$, whose ions will not react with other ions in the cell or with the electrode materials. The electrolyte is often incorporated into a paste or gel so that the electrolyte solution does

▲ **Figure 20.5 A voltaic cell that uses a salt bridge to complete the electrical circuit.**

To help remember these definitions, note that anode and oxidation both begin with a vowel, and cathode and reduction both begin with a consonant.

not pour out when the U-tube is inverted. As oxidation and reduction proceed at the electrodes, ions from the salt bridge migrate to neutralize charge in the cell compartments. Whatever means is used to allow ions to migrate between half-cells, *anions always migrate toward the anode and cations toward the cathode.*

> ### GIVE IT SOME THOUGHT
>
> Why do anions in a salt bridge migrate toward the anode?

Figure 20.6▶ summarizes the relationships among the anode, the cathode, the chemical process occurring in a voltaic cell, the direction of migration of ions in solution, and the motion of electrons between electrodes in the external circuit. Notice in particular that *in any voltaic cell the electrons flow from the anode through the external circuit to the cathode.* Because the negatively charged electrons flow from the anode to the cathode, the anode in a voltaic cell is labeled with a negative sign and the cathode with a positive sign; we can envision the electrons as being attracted to the positive cathode from the negative anode through the external circuit.

▲ Figure 20.6 A summary of the terminology used to describe voltaic cells. Oxidation occurs at the anode; reduction occurs at the cathode. The electrons flow spontaneously from the negative anode to the positive cathode. The movement of ions in solution completes the electrical circuit. Anions move toward the anode, whereas cations move toward the cathode. The cell compartments can be separated by either a porous glass barrier (as in Figure 20.4) or by a salt bridge (as in Figure 20.5).

■ **SAMPLE EXERCISE 20.4** | **Describing a Voltaic Cell**

The oxidation-reduction reaction

$$Cr_2O_7^{2-}(aq) + 14\,H^+(aq) + 6\,I^-(aq) \longrightarrow 2\,Cr^{3+}(aq) + 3\,I_2(s) + 7\,H_2O(l)$$

is spontaneous. A solution containing $K_2Cr_2O_7$ and H_2SO_4 is poured into one beaker, and a solution of KI is poured into another. A salt bridge is used to join the beakers. A metallic conductor that will not react with either solution (such as platinum foil) is suspended in each solution, and the two conductors are connected with wires through a voltmeter or some other device to detect an electric current. The resultant voltaic cell generates an electric current. Indicate the reaction occurring at the anode, the reaction at the cathode, the direction of electron migration, the direction of ion migration, and the signs of the electrodes.

SOLUTION

Analyze: We are given the equation for a spontaneous reaction that takes place in a voltaic cell and a description of how the cell is constructed. We are asked to write the half-reactions occurring at the anode and at the cathode, as well as the directions of electron and ion movements and the signs assigned to the electrodes.

Plan: Our first step is to divide the chemical equation into half-reactions so that we can identify the oxidation and the reduction processes. We then use the definitions of anode and cathode and the other terminology summarized in Figure 20.6.

Solve: In one half-reaction, $Cr_2O_7^{2-}(aq)$ is converted into $Cr^{3+}(aq)$. Starting with these ions and then completing and balancing the half-reaction, we have

$$Cr_2O_7^{2-}(aq) + 14\,H^+(aq) + 6\,e^- \longrightarrow 2\,Cr^{3+}(aq) + 7\,H_2O(l)$$

In the other half-reaction, $I^-(aq)$ is converted to $I_2(s)$:

$$6\,I^-(aq) \longrightarrow 3\,I_2(s) + 6\,e^-$$

Now we can use the summary in Figure 20.6 to help us describe the voltaic cell. The first half-reaction is the reduction process (electrons on the reactant side of the equation). By definition, the reduction process occurs at the cathode. The second half-reaction is the oxidation process (electrons on the product side of the equation), which occurs at the anode.

The I^- ions are the source of electrons, and the $Cr_2O_7^{2-}$ ions accept the electrons. Hence, the electrons flow through the external circuit from the electrode immersed in the KI solution (the anode) to the electrode immersed in the $K_2Cr_2O_7$–H_2SO_4 solution (the cathode). The electrodes themselves do not react in any way; they merely provide a means of transferring electrons from or to the solutions. The cations move through the solutions toward the cathode, and the anions move toward the anode. The anode (from which the electrons move) is the negative electrode, and the cathode (toward which the electrons move) is the positive electrode.

▼ **Figure 20.7 Atomic-level depiction of the Zn(s)-Cu²⁺(aq) reaction.** The water molecules and anions in the solution are not shown. (a) A Cu^{2+} ion comes in contact with the surface of the Zn strip and gains two electrons from a Zn atom; the Cu^{2+} ion is reduced, and the Zn atom is oxidized. (b) The resulting Zn^{2+} ion enters the solution, and the Cu atom remains deposited on the strip.

Atoms in Zn strip   Cu²⁺ ions in solution

2e⁻

(a)

Zn²⁺ ion

Cu atom

(b)

■ PRACTICE EXERCISE

The two half-reactions in a voltaic cell are

$$Zn(s) \longrightarrow Zn^{2+}(aq) + 2\,e^-$$

$$ClO_3^-(aq) + 6\,H^+(aq) + 6\,e^- \longrightarrow Cl^-(aq) + 3\,H_2O(l)$$

(a) Indicate which reaction occurs at the anode and which at the cathode. **(b)** Which electrode is consumed in the cell reaction? **(c)** Which electrode is positive?
Answers: **(a)** The first reaction occurs at the anode and the second reaction at the cathode. **(b)** The anode (Zn) is consumed in the cell reaction. **(c)** The cathode is positive.

A Molecular View of Electrode Processes

To better understand the relationship between voltaic cells and spontaneous redox reactions, let's look at what happens at the atomic or molecular level. The actual processes involved in the transfer of electrons are quite complex; nevertheless, we can learn much by examining these processes in a simplified way.

Consider the spontaneous reaction between Zn(s) and $Cu^{2+}(aq)$, in which Zn(s) is oxidized to $Zn^{2+}(aq)$ and $Cu^{2+}(aq)$ is reduced to Cu(s), as shown in Figure 20.3. Figure 20.7 ◀ shows a schematic diagram of these processes at the atomic level. We can envision a Cu^{2+} ion coming into contact with the strip of Zn metal, as in Figure 20.7(a). Two electrons are transferred from a Zn atom to the Cu^{2+} ion, leading to a Zn^{2+} ion and a Cu atom. The Zn^{2+} ion migrates away into the aqueous solution while the Cu atom remains deposited on the metal strip [Figure 20.7(b)]. As the reaction proceeds, we produce more and more Cu(s) and deplete the $Cu^{2+}(aq)$.

▼ **Figure 20.8 Atomic-level depiction of the voltaic cell in Figure 20.5.** At the anode a Zn atom loses two electrons and becomes a Zn^{2+} ion; the Zn atom is oxidized. The electrons travel through the external circuit to the cathode. At the cathode a Cu^{2+} ion gains the two electrons, forming a Cu atom; the Cu^{2+} ion is reduced. Ions migrate through the porous barrier to maintain charge balance between the compartments.

▲ GIVE IT SOME THOUGHT

What happens to the surface Zn atoms as they lose electrons?

The voltaic cell in Figure 20.5 is also based on the oxidation of Zn(s) and the reduction of $Cu^{2+}(aq)$. In this case, however, the electrons are not transferred directly from one reacting species to the other. As shown in Figure 20.8 ◀, a Zn atom at the surface of the anode "loses" two electrons and becomes a $Zn^{2+}(aq)$ ion in the anode compartment. We envision the two electrons traveling from the anode through the wire to the cathode. At the surface of the cathode, a Cu^{2+} ion from the solution gains the two electrons to form a Cu atom, which is deposited on the cathode.

The redox reaction between Zn and Cu^{2+} is spontaneous regardless of whether they react directly or in the separate compartments of a voltaic cell. In each case the overall reaction is the same—only the path by which the electrons are transferred from the Zn atom to a Cu^{2+} ion is different.

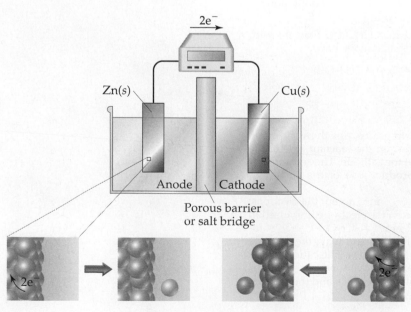

2e⁻

Zn(s)   Cu(s)

Anode | Cathode

Porous barrier or salt bridge

2e⁻   2e⁻

20.4 CELL EMF UNDER STANDARD CONDITIONS

Why do electrons transfer spontaneously from a Zn atom to a Cu^{2+} ion, either directly as in the reaction of Figure 20.3 or through an external circuit as in the voltaic cell of Figure 20.4? In a simple sense, we can compare the electron flow to the flow of water in a waterfall (Figure 20.9▶). Water flows spontaneously over a waterfall because of a difference in potential energy between the top of the falls and the stream below. ∞ (Section 5.1) In a similar fashion, electrons flow from the anode of a voltaic cell to the cathode because of a difference in potential energy. The potential energy of electrons is higher in the anode than in the cathode, and they spontaneously flow through an external circuit from the anode to the cathode. That is, electrons flow spontaneously toward the electrode with the more positive electrical potential.

▲ **Figure 20.9 Water analogy for electron flow.** The flow of electrons from the anode to the cathode of a voltaic cell can be likened to the flow of water over a waterfall. Water flows over the waterfall because its potential energy is lower at the bottom of the falls than at the top. Likewise, if there is an electrical connection between the anode and cathode of a voltaic cell, electrons flow from the anode to the cathode to lower their potential energy.

The difference in potential energy per electrical charge (the *potential difference*) between two electrodes is measured in units of volts. One volt (V) is the potential difference required to impart 1 joule (J) of energy to a charge of 1 coulomb (C).

$$1 \text{ V} = 1 \frac{\text{J}}{\text{C}}$$

Recall that one electron has a charge of 1.60×10^{-19} C. ∞ (Section 2.2)

The potential difference between the two electrodes of a voltaic cell provides the driving force that pushes electrons through the external circuit. Therefore, we call this potential difference the **electromotive** ("causing electron motion") **force**, or **emf**. The emf of a cell, denoted E_{cell}, is also called the **cell potential**. Because E_{cell} is measured in volts, we often refer to it as the *cell voltage*. For any cell reaction that proceeds spontaneously, such as that in a voltaic cell, the cell potential will be *positive*.

The emf of a particular voltaic cell depends on the specific reactions that occur at the cathode and anode, the concentrations of reactants and products, and the temperature, which we will assume to be 25 °C unless otherwise noted. In this section we will focus on cells that are operated at 25 °C under *standard conditions*. Recall from Section 19.5 that standard conditions include 1 M concentrations for reactants and products in solution and 1 atm pressure for those that are gases (Table 19.3). Under standard conditions the emf is called the **standard emf**, or the **standard cell potential**, and is denoted E°_{cell}. For the Zn-Cu voltaic cell in Figure 20.5, for example, the standard cell potential at 25 °C is +1.10 V.

$$Zn(s) + Cu^{2+}(aq, 1 \, M) \longrightarrow Zn^{2+}(aq, 1 \, M) + Cu(s) \qquad E^{\circ}_{cell} = +1.10 \text{ V}$$

Recall that the superscript ° indicates standard-state conditions. ∞ (Section 5.7)

GIVE IT SOME THOUGHT

If a standard cell potential is +0.85 V at 25 °C, is the redox reaction of the cell spontaneous?

Standard Reduction (Half-Cell) Potentials

The emf, or cell potential, of a voltaic cell, E°_{cell}, depends on the particular cathode and anode half-cells involved. We could, in principle, tabulate the standard cell potentials for all possible cathode/anode combinations. However, it is not necessary to undertake this arduous task. Rather, we can assign a standard potential to each individual half-cell and then use these half-cell potentials to determine E°_{cell}.

The cell potential is the difference between two electrode potentials, one associated with the cathode and the other associated with the anode. By convention, the potential associated with each electrode is chosen to be the potential for reduction to occur at that electrode. Thus, standard electrode potentials are tabulated for reduction reactions; they are **standard reduction potentials**, denoted E°_{red}. The cell potential, E°_{cell}, is given by the standard reduction potential of the cathode reaction, E°_{red} (cathode), *minus* the standard reduction potential of the anode reaction, E°_{red} (anode):

$$E^\circ_{cell} = E^\circ_{red} \text{ (cathode)} - E^\circ_{red} \text{ (anode)} \qquad [20.8]$$

For all spontaneous reactions at standard conditions, $E^\circ_{cell} > 0$.

Because every voltaic cell involves two half-cells, it is not possible to measure the standard reduction potential of a half-reaction directly. If we assign a standard reduction potential to a certain reference half-reaction, however, we can then determine the standard reduction potentials of other half-reactions relative to that reference. The reference half-reaction is the reduction of $H^+(aq)$ to $H_2(g)$ under standard conditions, which is assigned a standard reduction potential of exactly 0 V.

$$2\,H^+(aq, 1\,M) + 2\,e^- \longrightarrow H_2(g, 1\,atm) \qquad E^\circ_{red} = 0\,V \qquad [20.9]$$

An electrode designed to produce this half-reaction is called a **standard hydrogen electrode** (SHE), or the normal hydrogen electrode (NHE). An SHE consists of a platinum wire connected to a piece of platinum foil covered with finely divided platinum that serves as an inert surface for the reaction. The electrode is encased in a glass tube so that hydrogen gas under standard conditions (1 atm) can bubble over the platinum, and the solution contains $H^+(aq)$ under standard (1 M) conditions (Figure 20.10 ▼).

(a)   (b)

▲ Figure 20.10 **The standard hydrogen electrode (SHE) used as a reference electrode.** (a) An SHE consists of an electrode with finely divided Pt in contact with $H_2(g)$ at 1 atm pressure and an acidic solution with $[H^+] = 1\,M$. (b) Molecular depiction of the processes that occur at the SHE. When the SHE is the cathode of a cell, two H^+ ions each accept an electron from the Pt electrode and are reduced to H atoms. The H atoms bond together to form H_2. When the SHE is the anode of a cell, the reverse process occurs: An H_2 molecule at the electrode surface loses two electrons and is oxidized to H^+. The H^+ ions in solution are hydrated.

Figure 20.11 ▶ shows a voltaic cell using an SHE and a standard electrode. The spontaneous reaction is the one shown in Figure 20.1, namely, the oxidation of Zn and the reduction of H^+:

$$Zn(s) + 2\,H^+(aq) \longrightarrow Zn^{2+}(aq) + H_2(g)$$

Notice that the Zn^{2+}/Zn electrode is the anode and the SHE is the cathode and that the cell voltage is +0.76 V. By using the defined standard reduction potential of H^+ ($E^\circ_{red} = 0$) and Equation 20.8, we can determine the standard reduction potential for the Zn^{2+}/Zn half-reaction:

$$E^\circ_{cell} = E^\circ_{red}\,(\text{cathode}) - E^\circ_{red}\,(\text{anode})$$

$$+0.76\,\text{V} = 0\,\text{V} - E^\circ_{red}\,(\text{anode})$$

$$E^\circ_{red}\,(\text{anode}) = -0.76\,\text{V}$$

Zn(s) ⟶ Zn²⁺(aq) + 2 e⁻   2 H⁺(aq) + 2 e⁻ ⟶ H₂(g)

▲ Figure 20.11 A voltaic cell using a standard hydrogen electrode.

Thus, a standard reduction potential of -0.76 V can be assigned to the reduction of Zn^{2+} to Zn.

$$Zn^{2+}(aq, 1\,M) + 2\,e^- \longrightarrow Zn(s) \qquad E^\circ_{red} = -0.76\,\text{V}$$

We write the reaction as a reduction even though it is "running in reverse" as an oxidation in the cell in Figure 20.11. *Whenever we assign an electrical potential to a half-reaction, we write the reaction as a reduction.*

The standard reduction potentials for other half-reactions can be established from other cell potentials in a fashion analogous to that used for the Zn^{2+}/Zn half-reaction. Table 20.1 ▼ lists some standard reduction potentials; a more complete list is found in Appendix E. These standard reduction potentials, often called *half-cell potentials*, can be combined to calculate the emfs of a large variety of voltaic cells.

TABLE 20.1 ■ Standard Reduction Potentials in Water at 25 °C	
Potential (V)	**Reduction Half-Reaction**
+2.87	$F_2(g) + 2\,e^- \longrightarrow 2\,F^-(aq)$
+1.51	$MnO_4^-(aq) + 8\,H^+(aq) + 5\,e^- \longrightarrow Mn^{2+}(aq) + 4\,H_2O(l)$
+1.36	$Cl_2(g) + 2\,e^- \longrightarrow 2\,Cl^-(aq)$
+1.33	$Cr_2O_7^{2-}(aq) + 14\,H^+(aq) + 6\,e^- \longrightarrow 2\,Cr^{3+}(aq) + 7\,H_2O(l)$
+1.23	$O_2(g) + 4\,H^+(aq) + 4\,e^- \longrightarrow 2\,H_2O(l)$
+1.06	$Br_2(l) + 2\,e^- \longrightarrow 2\,Br^-(aq)$
+0.96	$NO_3^-(aq) + 4\,H^+(aq) + 3\,e^- \longrightarrow NO(g) + 2\,H_2O(l)$
+0.80	$Ag^+(aq) + e^- \longrightarrow Ag(s)$
+0.77	$Fe^{3+}(aq) + e^- \longrightarrow Fe^{2+}(aq)$
+0.68	$O_2(g) + 2\,H^+(aq) + 2\,e^- \longrightarrow H_2O_2(aq)$
+0.59	$MnO_4^-(aq) + 2\,H_2O(l) + 3\,e^- \longrightarrow MnO_2(s) + 4\,OH^-(aq)$
+0.54	$I_2(s) + 2\,e^- \longrightarrow 2\,I^-(aq)$
+0.40	$O_2(g) + 2\,H_2O(l) + 4\,e^- \longrightarrow 4\,OH^-(aq)$
+0.34	$Cu^{2+}(aq) + 2\,e^- \longrightarrow Cu(s)$
0 [defined]	$2\,H^+(aq) + 2\,e^- \longrightarrow H_2(g)$
−0.28	$Ni^{2+}(aq) + 2\,e^- \longrightarrow Ni(s)$
−0.44	$Fe^{2+}(aq) + 2\,e^- \longrightarrow Fe(s)$
−0.76	$Zn^{2+}(aq) + 2\,e^- \longrightarrow Zn(s)$
−0.83	$2\,H_2O(l) + 2\,e^- \longrightarrow H_2(g) + 2\,OH^-(aq)$
−1.66	$Al^{3+}(aq) + 3\,e^- \longrightarrow Al(s)$
−2.71	$Na^+(aq) + e^- \longrightarrow Na(s)$
−3.05	$Li^+(aq) + e^- \longrightarrow Li(s)$

▲ GIVE IT SOME THOUGHT

For the half-reaction $Cl_2(g) + 2\,e^- \longrightarrow 2\,Cl^-(aq)$, what are the standard conditions for the reactant and product?

Because electrical potential measures potential energy per electrical charge, standard reduction potentials are intensive properties. ∞(Section 1.3). In other words, if we increased the amount of substances in a redox reaction, we would increase both the energy and charges involved, but the ratio of energy (joules) to electrical charge (coulombs) would remain constant ($V = J/C$). Thus, *changing the stoichiometric coefficient in a half-reaction does not affect the value of the standard reduction potential.* For example, $E°_{red}$ for the reduction of 10 mol Zn^{2+} is the same as that for the reduction of 1 mol Zn^{2+}:

$$10\,Zn^{2+}(aq, 1\,M) + 20\,e^- \longrightarrow 10\,Zn(s) \qquad E°_{red} = -0.76\,V$$

■ SAMPLE EXERCISE 20.5 | Calculating $E°_{red}$ from $E°_{cell}$

For the Zn-Cu^{2+} voltaic cell shown in Figure 20.5, we have

$$Zn(s) + Cu^{2+}(aq, 1\,M) \longrightarrow Zn^{2+}(aq, 1\,M) + Cu(s) \qquad E°_{cell} = 1.10\,V$$

Given that the standard reduction potential of Zn^{2+} to Zn(s) is -0.76 V, calculate the $E°_{red}$ for the reduction of Cu^{2+} to Cu:

$$Cu^{2+}(aq, 1\,M) + 2\,e^- \longrightarrow Cu(s)$$

SOLUTION

Analyze: We are given $E°_{cell}$ and $E°_{red}$ for Zn^{2+} and asked to calculate $E°_{red}$ for Cu^{2+}.

Plan: In the voltaic cell, Zn is oxidized and is therefore the anode. Thus, the given $E°_{red}$ for Zn^{2+} is $E°_{red}$ (anode). Because Cu^{2+} is reduced, it is in the cathode half-cell. Thus, the unknown reduction potential for Cu^{2+} is $E°_{red}$ (cathode). Knowing $E°_{cell}$ and $E°_{red}$ (anode), we can use Equation 20.8 to solve for $E°_{red}$ (cathode).

Solve:

$$E°_{cell} = E°_{red}\,(\text{cathode}) - E°_{red}\,(\text{anode})$$
$$1.10\,V = E°_{red}\,(\text{cathode}) - (-0.76\,V)$$
$$E°_{red}\,(\text{cathode}) = 1.10\,V - 0.76\,V = 0.34\,V$$

Check: This standard reduction potential agrees with the one listed in Table 20.1.

Comment: The standard reduction potential for Cu^{2+} can be represented as $E°_{Cu^{2+}} = 0.34$ V, and that for Zn^{2+} as $E°_{Zn^{2+}} = -0.76$ V. The subscript identifies the ion that is reduced in the reduction half-reaction.

■ PRACTICE EXERCISE

A voltaic cell is based on the half-reactions

$$In^+(aq) \longrightarrow In^{3+}(aq) + 2\,e^-$$
$$Br_2(l) + 2\,e^- \longrightarrow 2\,Br^-(aq)$$

The standard emf for this cell is 1.46 V. Using the data in Table 20.1, calculate $E°_{red}$ for the reduction of In^{3+} to In^+.
Answer: -0.40 V

■ SAMPLE EXERCISE 20.6 | Calculating $E°_{cell}$ from $E°_{red}$

Using the standard reduction potentials listed in Table 20.1, calculate the standard emf for the voltaic cell described in Sample Exercise 20.4, which is based on the reaction

$$Cr_2O_7{}^{2-}(aq) + 14\,H^+(aq) + 6\,I^-(aq) \longrightarrow 2\,Cr^{3+}(aq) + 3\,I_2(s) + 7\,H_2O(l)$$

SOLUTION

Analyze: We are given the equation for a redox reaction and asked to use data in Table 20.1 to calculate the standard emf (standard potential) for the associated voltaic cell.

Plan: Our first step is to identify the half-reactions that occur at the cathode and the anode, which we did in Sample Exercise 20.4. Then we can use data from Table 20.1 and Equation 20.8 to calculate the standard emf.

Solve: The half-reactions are

$Cathode:$ $Cr_2O_7^{2-}(aq) + 14\,H^+(aq) + 6\,e^- \longrightarrow 2\,Cr^{3+}(aq) + 7\,H_2O(l)$

$Anode:$ $6\,I^-(aq) \longrightarrow 3\,I_2(s) + 6\,e^-$

According to Table 20.1, the standard reduction potential for the reduction of $Cr_2O_7^{2-}$ to Cr^{3+} is +1.33 V, and the standard reduction potential for the reduction of I_2 to I^- (the reverse of the oxidation half-reaction) is +0.54 V. We then use these values in Equation 20.8.

$$E^\circ_{cell} = E^\circ_{red}\,(cathode) - E^\circ_{red}\,(anode) = 1.33\,V - 0.54\,V = 0.79\,V$$

Although we must multiply the iodide half-reaction at the anode by 3 to obtain a balanced equation for the reaction, the value of E°_{red} is *not* multiplied by 3. As we have noted, the standard reduction potential is an intensive property, so it is independent of the specific stoichiometric coefficients.

Check: The cell potential, 0.79 V, is a positive number. As noted earlier, a voltaic cell must have a positive emf in order to operate.

■ PRACTICE EXERCISE

Using data in Table 20.1, calculate the standard emf for a cell that employs the following overall cell reaction:

$$2\,Al(s) + 3\,I_2(s) \longrightarrow 2\,Al^{3+}(aq) + 6\,I^-(aq)$$

Answer: 0.54 V − (−1.66 V) = 2.20 V

For each of the half-cells in a voltaic cell, the standard reduction potential provides a measure of the driving force for reduction to occur: *The more positive the value of* E°_{red}*, the greater the driving force for reduction under standard conditions.* In any voltaic cell operating under standard conditions, the reaction at the cathode has a more positive value of E°_{red} than does the reaction at the anode.

Equation 20.8, which indicates that the standard cell potential is the difference between the standard reduction potential of the cathode and the standard reduction potential of the anode, is illustrated graphically in Figure 20.12▶. The standard reduction potentials of the cathode and anode are shown on a scale. The more positive E°_{red} value identifies the cathode, and the difference between the two standard reduction potentials is the standard cell potential. Figure 20.13▶ shows the specific values of E°_{red} for the two half-reactions in the Zn-Cu voltaic cell illustrated in Figure 20.5.

▲ GIVE IT SOME THOUGHT

Is the following statement true or false? The smaller the difference is between the standard reduction potentials of the cathode and anode, the smaller the driving force for the overall redox reaction.

■ SAMPLE EXERCISE 20.7 | Determining Half-Reactions at Electrodes and Calculating Cell EMF

A voltaic cell is based on the following two standard half-reactions:

$$Cd^{2+}(aq) + 2\,e^- \longrightarrow Cd(s)$$
$$Sn^{2+}(aq) + 2\,e^- \longrightarrow Sn(s)$$

By using the data in Appendix E, determine **(a)** the half-reactions that occur at the cathode and the anode, and **(b)** the standard cell potential.

SOLUTION

Analyze: We have to look up E°_{red} for two half-reactions and use these values to predict the cathode and anode of the cell and to calculate its standard cell potential, E°_{cell}.

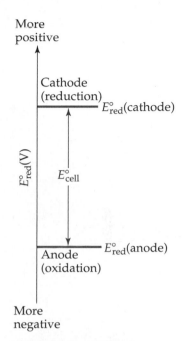

▲ **Figure 20.12 Standard cell potential of a voltaic cell.** The cell potential measures the difference in the standard reduction potentials of the cathode and the anode reactions: $E^\circ_{cell} = E^\circ_{red}\,(cathode) - E^\circ_{red}\,(anode)$. In a voltaic cell the cathode reaction is always the one that has the more positive (or less negative) value for E°_{red}.

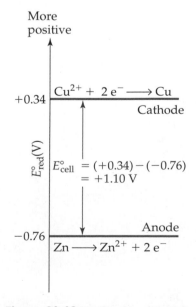

▲ **Figure 20.13 Half-cell potentials.** The half-cell potentials for the voltaic cell in Figure 20.5, diagrammed in the style of Figure 20.12.

Plan: The cathode will have the reduction with the more positive E_{red}° value. The anode will have the less positive E_{red}°. To write the half-reaction at the anode, we reverse the half-reaction written for the reduction, so that the half-reaction is written as an oxidation.

Solve:

(a) According to Appendix E, $E_{red}^\circ(Cd^{2+}/Cd) = -0.403$ V and $E_{red}^\circ(Sn^{2+}/Sn) = -0.136$ V. The standard reduction potential for Sn^{2+} is more positive (less negative) than that for Cd^{2+}; hence, the reduction of Sn^{2+} is the reaction that occurs at the cathode.

Cathode: $\qquad\qquad\qquad\qquad Sn^{2+}(aq) + 2\,e^- \longrightarrow Sn(s)$

The anode reaction therefore is the loss of electrons by Cd.

Anode: $\qquad\qquad\qquad\qquad Cd(s) \longrightarrow Cd^{2+}(aq) + 2\,e^-$

(b) The cell potential is given by Equation 20.8.

$$E_{cell}^\circ = E_{red}^\circ \text{ (cathode)} - E_{red}^\circ \text{ (anode)} = (-0.136 \text{ V}) - (-0.403 \text{ V}) = 0.267 \text{ V}$$

Notice that it is unimportant that the E_{red}° values of both half-reactions are negative; the negative values merely indicate how these reductions compare to the reference reaction, the reduction of $H^+(aq)$.

Check: The cell potential is positive, as it must be for a voltaic cell.

■ **PRACTICE EXERCISE**

A voltaic cell is based on a Co^{2+}/Co half-cell and an $AgCl/Ag$ half-cell.
(a) What half-reaction occurs at the anode? **(b)** What is the standard cell potential?
Answers: **(a)** $Co \longrightarrow Co^{2+} + 2\,e^-$; **(b)** $+0.499$ V

Strengths of Oxidizing and Reducing Agents

We have thus far used the tabulation of standard reduction potentials to examine voltaic cells. We can also use E_{red}° values to understand aqueous reaction chemistry. Recall, for example, the reaction between $Zn(s)$ and $Cu^{2+}(aq)$ shown in Figure 20.3.

$$Zn(s) + Cu^{2+}(aq) \longrightarrow Zn^{2+}(aq) + Cu(s)$$

Zinc metal is oxidized, and $Cu^{2+}(aq)$ is reduced in this reaction. These substances are in direct contact, however, so we are not producing usable electrical work; the direct contact essentially "short-circuits" the cell. Nevertheless, the driving force for the reaction is the same as that in a voltaic cell, as in Figure 20.5. Because the E_{red}° value for the reduction of $Cu^{2+}(0.34$ V$)$ is more positive than the E_{red}° value for the reduction of $Zn^{2+}(-0.76$ V$)$, the reduction of $Cu^{2+}(aq)$ by $Zn(s)$ is a spontaneous process.

We can generalize the relationship between the value of E_{red}° and the spontaneity of redox reactions: *The more positive the E_{red}° value for a half-reaction, the greater the tendency for the reactant of the half-reaction to be reduced and, therefore, to oxidize another species.* In Table 20.1, for example, F_2 is the most easily reduced species, so it is the strongest oxidizing agent listed.

$$F_2(g) + 2\,e^- \longrightarrow 2\,F^-(aq) \qquad E_{red}^\circ = 2.87 \text{ V}$$

Among the most frequently used oxidizing agents are the halogens, O_2, and oxyanions such as MnO_4^-, $Cr_2O_7^{2-}$, and NO_3^-, whose central atoms have high positive oxidation states. According to Table 20.1, all these species undergo reduction with large positive values of E_{red}°.

Lithium ion (Li^+) is the most difficult species to reduce and is therefore the poorest oxidizing agent:

$$Li^+(aq) + e^- \longrightarrow Li(s) \qquad E_{red}^\circ = -3.05 \text{ V}$$

Because Li^+ is so difficult to reduce, the reverse reaction, the oxidation of $Li(s)$ to $Li^+(aq)$, is a highly favorable reaction. *The half-reaction with the smallest reduction potential is most easily reversed as an oxidation.* Thus, lithium metal has a great tendency to transfer electrons to other species. In water, Li is the strongest reducing agent among the substances listed in Table 20.1. Because Table 20.1 lists half-reactions as reductions, only the substances on the reactant side of these equations can serve as oxidizing agents; only those on the product side can serve as reducing agents.

Commonly used reducing agents include H_2 and the active metals such as the alkali metals and the alkaline earth metals. Other metals whose cations have negative $E°_{red}$ values—Zn and Fe, for example—are also used as reducing agents. Solutions of reducing agents are difficult to store for extended periods because of the ubiquitous presence of O_2, a good oxidizing agent. For example, developer solutions used in film photography are mild reducing agents; they have only a limited shelf life because they are readily oxidized by O_2 from the air.

The list of $E°_{red}$ values in Table 20.1 orders the ability of substances to act as oxidizing or reducing agents and is summarized in Figure 20.14▶. The substances that are most readily reduced (strong oxidizing agents) are the reactants on the top left of the table. Their products, on the top right of the table, are oxidized with difficulty (weak reducing agents). The substances on the bottom left of the table are reduced with difficulty, but their products are readily oxidized. This inverse relationship between oxidizing and reducing strength is similar to the inverse relationship between the strengths of conjugate acids and bases. ∞ (Section 16.2 and Figure 16.4)

To help you remember the relationships between the strengths of oxidizing and reducing agents, recall the very exothermic reaction between sodium metal and chlorine gas to form sodium chloride (Figure 8.2). In this reaction $Cl_2(g)$ is reduced (it serves as a strong oxidizing agent) and Na(s) is oxidized (it serves as a strong reducing agent). The products of this reaction—Na^+ and Cl^- ions—are very weak oxidizing and reducing agents, respectively.

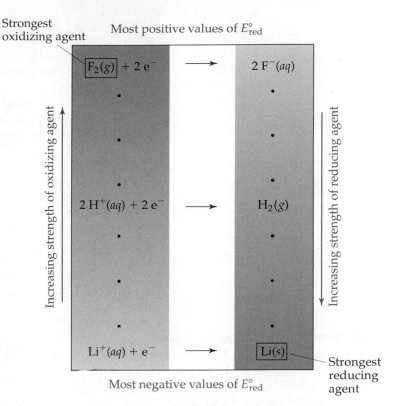

▲ Figure 20.14 **Relative strengths of oxidizing agents.** The standard reduction potentials, $E°_{red}$, listed in Table 20.1 are related to the ability of substances to serve as oxidizing or reducing agents. Species on the left side of the half-reactions can act as oxidizing agents, and those on the right side can act as reducing agents. As $E°_{red}$ becomes more positive, the species on the left become stronger and stronger oxidizing agents. As $E°_{red}$ becomes more negative, the species on the right become stronger and stronger reducing agents.

■ SAMPLE EXERCISE 20.8 | Determining the Relative Strengths of Oxidizing Agents

Using Table 20.1, rank the following ions in order of increasing strength as oxidizing agents: $NO_3^-(aq)$, $Ag^+(aq)$, $Cr_2O_7^{2-}(aq)$.

SOLUTION

Analyze: We are given several ions and asked to rank their abilities to act as oxidizing agents.

Plan: The more readily an ion is reduced (the more positive its $E°_{red}$ value), the stronger it is as an oxidizing agent.

Solve: From Table 20.1, we have

$$NO_3^-(aq) + 4 H^+(aq) + 3 e^- \longrightarrow NO(g) + 2 H_2O(l) \qquad E°_{red} = +0.96 \text{ V}$$

$$Ag^+(aq) + e^- \longrightarrow Ag(s) \qquad E°_{red} = +0.80 \text{ V}$$

$$Cr_2O_7^{2-}(aq) + 14 H^+(aq) + 6 e^- \longrightarrow 2 Cr^{3+}(aq) + 7 H_2O(l) \qquad E°_{red} = +1.33 \text{ V}$$

Because the standard reduction potential of $Cr_2O_7^{2-}$ is the most positive, $Cr_2O_7^{2-}$ is the strongest oxidizing agent of the three. The rank order is $Ag^+ < NO_3^- < Cr_2O_7^{2-}$.

■ PRACTICE EXERCISE

Using Table 20.1, rank the following species from the strongest to the weakest reducing agent: $I^-(aq)$, Fe(s), Al(s).
Answer: Al(s) > Fe(s) > $I^-(aq)$

20.5 FREE ENERGY AND REDOX REACTIONS

We have observed that voltaic cells use redox reactions that proceed spontaneously. Any reaction that can occur in a voltaic cell to produce a positive emf must be spontaneous. Consequently, it is possible to decide whether a redox reaction will be spontaneous by using half-cell potentials to calculate the emf associated with it.

The following discussion will pertain to general redox reactions, not just reactions in voltaic cells. Thus, we will make Equation 20.8 more general by writing it as

$$E° = E°_{red} \text{ (reduction process)} - E°_{red} \text{ (oxidation process)} \qquad [20.10]$$

In modifying Equation 20.8, we have dropped the subscript "cell" to indicate that the calculated emf does not necessarily refer to a voltaic cell. Similarly, we have generalized the standard reduction potentials on the right side of the equation by referring to the reduction and oxidation processes, rather than the cathode and the anode. We can now make a general statement about the spontaneity of a reaction and its associated emf, E: *A positive value of* E *indicates a spontaneous process; a negative value of* E *indicates a nonspontaneous one.* We will use E to represent the emf under nonstandard conditions and $E°$ to indicate the standard emf.

■ SAMPLE EXERCISE 20.9 | Spontaneous or Not?

Using standard reduction potentials in Table 20.1, determine whether the following reactions are spontaneous under standard conditions.

(a) $Cu(s) + 2 H^+(aq) \longrightarrow Cu^{2+}(aq) + H_2(g)$
(b) $Cl_2(g) + 2 I^-(aq) \longrightarrow 2 Cl^-(aq) + I_2(s)$

SOLUTION

Analyze: We are given two equations and must determine whether each is spontaneous.

Plan: To determine whether a redox reaction is spontaneous under standard conditions, we first need to write its reduction and oxidation half-reactions. We can then use the standard reduction potentials and Equation 20.10 to calculate the standard emf, $E°$, for the reaction. If a reaction is spontaneous, its standard emf must be a positive number.

Solve:

(a) In this reaction Cu is oxidized to Cu^{2+} and H^+ is reduced to H_2. The corresponding half-reactions and associated standard reduction potentials are

Reduction:	$2 H^+(aq) + 2 e^- \longrightarrow H_2(g)$	$E°_{red} = 0 V$
Oxidation:	$Cu(s) \longrightarrow Cu^{2+}(aq) + 2 e^-$	$E°_{red} = +0.34 V$

Notice that for the oxidation, we use the standard reduction potential from Table 20.1 for the reduction of Cu^{2+} to Cu. We now calculate $E°$ by using Equation 20.10:

$$E° = E°_{red} \text{ (reduction process)} - E°_{red} \text{ (oxidation process)}$$
$$= (0 V) - (0.34 V) = -0.34 V$$

Because $E°$ is negative, the reaction is not spontaneous in the direction written. Copper metal does not react with acids in this fashion. The reverse reaction, however, *is* spontaneous and would have an $E°$ of +0.34V.

$$Cu^{2+}(aq) + H_2(g) \longrightarrow Cu(s) + 2 H^+(aq) \quad E° = +0.34 V$$

Cu^{2+} can be reduced by H_2.

(b) We follow a procedure analogous to that in (a):

Reduction:	$Cl_2(g) + 2 e^- \longrightarrow 2 Cl^-(aq)$	$E°_{red} = +1.36 V$
Oxidation:	$2 I^-(aq) \longrightarrow I_2(s) + 2 e^-$	$E°_{red} = +0.54 V$

In this case

$$E° = (1.36 V) - (0.54 V) = +0.82 V$$

Because the value of $E°$ is positive, this reaction is spontaneous and could be used to build a voltaic cell.

■ PRACTICE EXERCISE

Using the standard reduction potentials listed in Appendix E, determine which of the following reactions are spontaneous under standard conditions:

(a) $I_2(s) + 5 Cu^{2+}(aq) + 6 H_2O(l) \longrightarrow 2 IO_3^-(aq) + 5 Cu(s) + 12 H^+(aq)$
(b) $Hg^{2+}(aq) + 2 I^-(aq) \longrightarrow Hg(l) + I_2(s)$
(c) $H_2SO_3(aq) + 2 Mn(s) + 4 H^+(aq) \longrightarrow S(s) + 2 Mn^{2+}(aq) + 3 H_2O(l)$

Answer: Reactions (b) and (c) are spontaneous.

We can use standard reduction potentials to understand the activity series of metals. ∞ (Section 4.4) Recall that any metal in the activity series will be oxidized by the ions of any metal below it. We can now recognize the origin of this rule based on standard reduction potentials. The activity series, tabulated in Table 4.5, consists of the oxidation reactions of the metals, ordered from the strongest reducing agent at the top to the weakest reducing agent at the bottom. (Thus, the ordering is "flipped over" relative to that in Table 20.1.) For example, nickel lies above silver in the activity series. We therefore expect nickel to displace silver, according to the net reaction

$$\text{Ni}(s) + 2\,\text{Ag}^+(aq) \longrightarrow \text{Ni}^{2+}(aq) + 2\,\text{Ag}(s)$$

In this reaction Ni is oxidized and Ag^+ is reduced. Therefore, using data from Table 20.1, the standard emf for the reaction is

$$E^\circ = E^\circ_{\text{red}}\,(\text{Ag}^+/\text{Ag}) - E^\circ_{\text{red}}\,(\text{Ni}^{2+}/\text{Ni})$$
$$= (+0.80\ \text{V}) - (-0.28\ \text{V}) = +1.08\ \text{V}$$

The positive value of E° indicates that the displacement of silver by nickel is a spontaneous process. Remember that although we multiply the silver half-reaction by 2, the reduction potential is not multiplied.

▲ GIVE IT SOME THOUGHT

Based on Table 4.5, which is the stronger reducing agent, Hg(*l*) or Pb(*s*)?

EMF and ΔG

The change in the Gibbs free energy, ΔG, is a measure of the spontaneity of a process that occurs at constant temperature and pressure. ∞ (Section 19.5) The emf, E, of a redox reaction also indicates whether the reaction is spontaneous. The relationship between emf and the free-energy change is

$$\Delta G = -nFE \qquad [20.11]$$

In this equation n is a positive number without units that represents the number of electrons transferred in the reaction. The constant F is called *Faraday's constant*, named after Michael Faraday (Figure 20.15 ▶). Faraday's constant is the quantity of electrical charge on one mole of electrons. This quantity of charge is called a **faraday** (*F*).

$$1\ F = 96{,}485\ \text{C/mol} = 96{,}485\ \text{J/V-mol}$$

The units of ΔG calculated by using Equation 20.11 are J/mol; as in Equation 19.16, we use "per mole" to mean "per mole of reaction as written." ∞ (Section 19.7)

Both n and F are positive numbers. Thus, a positive value of E in Equation 20.11 leads to a negative value of ΔG. Remember: *A positive value of* E *and a negative value of* ΔG *both indicate that a reaction is spontaneous.* When the reactants and products are all in their standard states, Equation 20.11 can be modified to relate ΔG° and E°.

$$\Delta G^\circ = -nFE^\circ \qquad [20.12]$$

Equation 20.12 is very important. It relates the standard emf, E°, of an electrochemical reaction to the standard free-energy change, ΔG°, for the reaction. Because ΔG° is related to the equilibrium constant, K, for a reaction by the expression $\Delta G^\circ = -RT \ln K$ ∞ (Section 19.7), we can relate the standard emf to the equilibrium constant for the reaction.

▲ **Figure 20.15 Michael Faraday.** Faraday (1791–1867) was born in England, a child of a poor blacksmith. At the age of 14 he was apprenticed to a bookbinder who gave Faraday time to read and to attend lectures. In 1812 he became an assistant in Humphry Davy's laboratory at the Royal Institution. He succeeded Davy as the most famous and influential scientist in England, making an amazing number of important discoveries, including his formation of the quantitative relationships between electrical current and the extent of chemical reaction in electrochemical cells.

■ SAMPLE EXERCISE 20.10 | **Determining $\Delta G°$ and K**

(a) Use the standard reduction potentials listed in Table 20.1 to calculate the standard free-energy change, $\Delta G°$, and the equilibrium constant, K, at 298 K for the reaction

$$4\,Ag(s) + O_2(g) + 4\,H^+(aq) \longrightarrow 4\,Ag^+(aq) + 2\,H_2O(l)$$

(b) Suppose the reaction in part (a) was written

$$2\,Ag(s) + \tfrac{1}{2}O_2(g) + 2\,H^+(aq) \longrightarrow 2\,Ag^+(aq) + H_2O(l)$$

What are the values of $E°$, $\Delta G°$, and K when the reaction is written in this way?

SOLUTION

Analyze: We are asked to determine $\Delta G°$ and K for a redox reaction, using standard reduction potentials.

Plan: We use the data in Table 20.1 and Equation 20.10 to determine $E°$ for the reaction and then use $E°$ in Equation 20.12 to calculate $\Delta G°$. We will then use Equation 19.17, $\Delta G° = -RT \ln K$, to calculate K.

Solve:

(a) We first calculate $E°$ by breaking the equation into two half-reactions, as we did in Sample Exercise 20.9, and then obtain $E°_{red}$ values from Table 20.1 (or Appendix E):

Reduction:	$O_2(g) + 4\,H^+(aq) + 4\,e^- \longrightarrow 2\,H_2O(l)$	$E°_{red} = +1.23\,V$
Oxidation:	$4\,Ag(s) \longrightarrow 4\,Ag^+(aq) + 4\,e^-$	$E°_{red} = +0.80\,V$

Even though the second half-reaction has 4 Ag, we use the $E°_{red}$ value directly from Table 20.1 because emf is an intensive property.

Using Equation 20.10, we have

$$E° = (1.23\,V) - (0.80\,V) = 0.43\,V$$

The half-reactions show the transfer of four electrons. Thus, for this reaction $n = 4$. We now use Equation 20.12 to calculate $\Delta G°$:

$$\Delta G° = -nFE°$$
$$= -(4)(96{,}485\,J/V\text{-}mol)(+0.43)\,V$$
$$= -1.7 \times 10^5\,J/mol = -170\,kJ/mol$$

The positive value of $E°$ leads to a negative value of $\Delta G°$.

Now we need to calculate the equilibrium constant, K, using $\Delta G° = -RT \ln K$. Because $\Delta G°$ is a large negative number, which means the reaction is thermodynamically very favorable, we expect K to be large.

$$\Delta G° = -RT \ln K$$
$$-1.7 \times 10^5\,J/mol = -(8.314\,J/K\,mol)\,(298\,K)\ln K$$
$$\ln K = \frac{-1.7 \times 10^5\,J/mol}{-(8.314\,J/K\,mol)(298\,K)}$$
$$\ln K = 69$$
$$K = 9 \times 10^{29}$$

K is indeed very large! This means that we expect silver metal to oxidize in acidic aqueous environments, in air, to Ag^+.

Notice that the voltage calculated for the reaction was 0.43 V, which is easy to measure. Directly measuring such a large equilibrium constant by measuring reactant and product concentrations at equilibrium, on the other hand, would be very difficult.

(b) The overall equation is the same as that in part (a), multiplied by $\tfrac{1}{2}$. The half-reactions are

Reduction:	$\tfrac{1}{2}O_2(g) + 2\,H^+(aq) + 2\,e^- \longrightarrow H_2O(l)$	
Oxidation:	$2\,Ag(s) \longrightarrow 2\,Ag^+(aq) + 2\,e^-$	$E°_{red} = +0.80\,V$

The values of $E°_{red}$ are the same as they were in part (a); they are not changed by multiplying the half-reactions by $\tfrac{1}{2}$. Thus, $E°$ has the same value as in part (a):

$$E° = +0.43\,V$$

Notice, though, that the value of n has changed to $n = 2$, which is $\tfrac{1}{2}$ the value in part (a). Thus, $\Delta G°$ is half as large as in part (a).

$$\Delta G° = -(2)(96{,}485\,J/V\text{-}mol)(+0.43\,V) = -83\,kJ/mol$$

Now we can calculate K as before:

$$-8.3 \times 10^4\,J/mol = -(8.314\,J/K\,mol)(298\,K)\ln K$$
$$K = 4 \times 10^{14}$$

Comment: $E°$ is an *intensive* quantity, so multiplying a chemical equation by a certain factor will not affect the value of $E°$. Multiplying an equation will change the value of n, however, and hence the value of $\Delta G°$. The change in free energy, in units of J/mol of reaction as written, is an *extensive* quantity. The equilibrium constant is also an extensive quantity.

■ PRACTICE EXERCISE

For the reaction

$$3\,Ni^{2+}(aq) + 2\,Cr(OH)_3(s) + 10\,OH^-(aq) \longrightarrow 3\,Ni(s) + 2\,CrO_4^{2-}(aq) + 8\,H_2O(l)$$

(a) What is the value of n? (b) Use the data in Appendix E to calculate $\Delta G°$. (c) Calculate K at $T = 298$ K.

Answers: (a) 6, (b) +87 kJ/mol, (c) $K = 6 \times 10^{-16}$

20.6 CELL EMF UNDER NONSTANDARD CONDITIONS

We have seen how to calculate the emf of a cell when the reactants and products are under standard conditions. As a voltaic cell is discharged, however, the reactants of the reaction are consumed, and the products are generated, so the concentrations of these substances change. The emf progressively drops until $E = 0$, at which point we say the cell is "dead." At that point the concentrations of the reactants and products cease to change; they are at equilibrium. In this section we will examine how the cell emf depends on the concentrations of the reactants and products of the cell reaction. The emf generated under nonstandard conditions can be calculated by using an equation first derived by Walther Nernst (1864–1941), a German chemist who established many of the theoretical foundations of electrochemistry.

The Nernst Equation

The dependence of the cell emf on concentration can be obtained from the dependence of the free-energy change on concentration. ∞ (Section 19.7) Recall that the free-energy change, ΔG, is related to the standard free-energy change, $\Delta G°$:

$$\Delta G = \Delta G° + RT \ln Q \qquad [20.13]$$

The quantity Q is the reaction quotient, which has the form of the equilibrium-constant expression except that the concentrations are those that exist in the reaction mixture at a given moment. ∞ (Section 15.6)

Substituting $\Delta G = -nFE$ (Equation 20.11) into Equation 20.13 gives

$$-nFE = -nFE° + RT \ln Q$$

Solving this equation for E gives the **Nernst equation**:

$$E = E° - \frac{RT}{nF} \ln Q \qquad [20.14]$$

This equation is customarily expressed in terms of common (base 10) logarithms, which is related to natural logarithms by a factor of 2.303:

$$E = E° - \frac{2.303\, RT}{nF} \log Q \qquad [20.15]$$

At $T = 298$ K the quantity $2.303\, RT/F$ equals 0.0592, with units of volts (V), and so the equation simplifies to

$$E = E° - \frac{0.0592\ \text{V}}{n} \log Q \qquad (T = 298\ \text{K}) \qquad [20.16]$$

We can use this equation to find the emf produced by a cell under nonstandard conditions or to determine the concentration of a reactant or product by measuring the emf of the cell.

To show how Equation 20.16 might be used, consider the following reaction, which we discussed earlier:

$$\text{Zn}(s) + \text{Cu}^{2+}(aq) \longrightarrow \text{Zn}^{2+}(aq) + \text{Cu}(s)$$

In this case $n = 2$ (two electrons are transferred from Zn to Cu^{2+}), and the standard emf is +1.10 V. Thus, at 298 K the Nernst equation gives

$$E = 1.10\ \text{V} - \frac{0.0592\ \text{V}}{2} \log \frac{[\text{Zn}^{2+}]}{[\text{Cu}^{2+}]} \qquad [20.17]$$

Recall that pure solids are excluded from the expression for Q. ∞ (Section 15.6) According to Equation 20.17, the emf increases as $[\text{Cu}^{2+}]$ increases and as $[\text{Zn}^{2+}]$ decreases. For example, when $[\text{Cu}^{2+}]$ is 5.0 M and $[\text{Zn}^{2+}]$ is 0.050 M, we have

$$E = 1.10\ \text{V} - \frac{0.0592\ \text{V}}{2} \log \left(\frac{0.050}{5.0} \right)$$

$$= 1.10\ \text{V} - \frac{0.0592\ \text{V}}{2} (-2.00) = 1.16\ \text{V}$$

Thus, increasing the concentration of the reactant (Cu^{2+}) and decreasing the concentration of the product (Zn^{2+}) relative to standard conditions increases the emf of the cell ($E = +1.16$ V) relative to standard conditions ($E° = +1.10$ V).

The Nernst equation helps us understand why the emf of a voltaic cell drops as the cell discharges. As reactants are converted to products, the value of Q increases, so the value of E decreases, eventually reaching $E = 0$. Because $\Delta G = -nFE$ (Equation 20.11), it follows that $\Delta G = 0$ when $E = 0$. Recall that a system is at equilibrium when $\Delta G = 0$. ⚏ (Section 19.7) Thus, when $E = 0$, the cell reaction has reached equilibrium, and no net reaction is occurring.

In general, increasing the concentration of reactants or decreasing the concentration of products increases the driving force for the reaction, resulting in a higher emf. Conversely, decreasing the concentration of reactants or increasing the concentration of products causes the emf to decrease.

◼◼◼ SAMPLE EXERCISE 20.11 | Voltaic Cell EMF under Nonstandard Conditions

Calculate the emf at 298 K generated by the cell described in Sample Exercise 20.4 when $[Cr_2O_7^{2-}] = 2.0$ M, $[H^+] = 1.0$ M, $[I^-] = 1.0$ M, and $[Cr^{3+}] = 1.0 \times 10^{-5}$ M.

$$Cr_2O_7^{2-}(aq) + 14\,H^+(aq) + 6\,I^-(aq) \longrightarrow 2\,Cr^{3+}(aq) + 3\,I_2(s) + 7\,H_2O(l)$$

SOLUTION

Analyze: We are given a chemical equation for a voltaic cell and the concentrations of reactants and products under which it operates. We are asked to calculate the emf of the cell under these nonstandard conditions.

Plan: To calculate the emf of a cell under nonstandard conditions, we use the Nernst equation in the form of Equation 20.16.

Solve: We first calculate $E°$ for the cell from standard reduction potentials (Table 20.1 or Appendix E). The standard emf for this reaction was calculated in Sample Exercise 20.6: $E° = 0.79$ V. As you will see if you refer back to that exercise, the balanced equation shows six electrons transferred from reducing agent to oxidizing agent, so $n = 6$. The reaction quotient, Q, is

$$Q = \frac{[Cr^{3+}]^2}{[Cr_2O_7^{2-}][H^+]^{14}\,[I^-]^6} = \frac{(1.0 \times 10^{-5})^2}{(2.0)(1.0)^{14}\,(1.0)^6} = 5.0 \times 10^{-11}$$

Using Equation 20.16, we have

$$E = 0.79\text{ V} - \frac{0.0592\text{ V}}{6}\log(5.0 \times 10^{-11})$$

$$= 0.79\text{ V} - \frac{0.0592\text{ V}}{6}(-10.30)$$

$$= 0.79\text{ V} + 0.10\text{ V} = 0.89\text{ V}$$

Check: This result is qualitatively what we expect: Because the concentration of $Cr_2O_7^{2-}$ (a reactant) is greater than 1 M and the concentration of Cr^{3+} (a product) is less than 1 M, the emf is greater than $E°$. Q is about 10^{-10}, so log Q is about -10. Thus, the correction to $E°$ is about $0.06 \times (10)/6$, which is 0.1, in agreement with the more detailed calculation.

◼◼◼ PRACTICE EXERCISE

Calculate the emf generated by the cell described in the practice exercise accompanying Sample Exercise 20.6 when $[Al^{3+}] = 4.0 \times 10^{-3}$ M and $[I^-] = 0.010$ M.
Answer: $E = +2.36$ V

◼◼◼ SAMPLE EXERCISE 20.12 | Calculating Concentrations in a Voltaic Cell

If the voltage of a $Zn–H^+$ cell (like that in Figure 20.11) is 0.45 V at 25 °C when $[Zn^{2+}] = 1.0$ M and $P_{H_2} = 1.0$ atm, what is the concentration of H^+?

SOLUTION

Analyze: We are given a description of a voltaic cell, its emf, and the concentrations of all reactants and products except H^+, which we are asked to calculate.

Plan: First, we write the equation for the cell reaction and use standard reduction potentials from Table 20.1 to calculate $E°$ for the reaction. After determining the value of n from our reaction equation, we solve the Nernst equation for Q. Finally, we use the equation for the cell reaction to write an expression for Q that contains $[H^+]$ to determine $[H^+]$.

Solve: The cell reaction is

$$Zn(s) + 2\,H^+(aq) \longrightarrow Zn^{2+}(aq) + H_2(g)$$

The standard emf is

$$E° = E°_{red}\,(reduction) - E°_{red}\,(oxidation)$$

$$= 0\,V - (-0.76\,V) = +0.76\,V$$

Because each Zn atom loses two electrons, $n = 2$

Using Equation 20.16, we can solve for Q:

$$0.45\,V = 0.76\,V - \frac{0.0592\,V}{2}\log Q$$

$$\log Q = (0.76\,V - 0.45\,V)\left(\frac{2}{0.0592\,V}\right) = 10.5$$

$$Q = 10^{10.5} = 3 \times 10^{10}$$

Q has the form of the equilibrium constant for the reaction

$$Q = \frac{[Zn^{2+}]P_{H_2}}{[H^+]^2} = \frac{(1.0)(1.0)}{[H^+]^2} = 3 \times 10^{10}$$

Solving for $[H^+]$, we have

$$[H^+]^2 = \frac{1.0}{3 \times 10^{10}} = 3 \times 10^{-11}$$

$$[H^+] = \sqrt{3 \times 10^{-11}} = 6 \times 10^{-6}\,M$$

Comment: A voltaic cell whose cell reaction involves H^+ can be used to measure $[H^+]$ or pH. A pH meter is a specially designed voltaic cell with a voltmeter calibrated to read pH directly. ∞ (Section 16.4)

▬ PRACTICE EXERCISE

What is the pH of the solution in the cathode compartment of the cell pictured in Figure 20.11 when $P_{H_2} = 1.0$ atm, $[Zn^{2+}]$ in the anode compartment is 0.10 M, and cell emf is 0.542 V?
Answer: pH = 4.23 (using data from Appendix E to obtain $E°$ to three significant figures)

Concentration Cells

In each of the voltaic cells that we have looked at thus far, the reactive species at the anode has been different from the one at the cathode. Cell emf depends on concentration, however, so a voltaic cell can be constructed using the *same* species in both the anode and cathode compartments as long as the concentrations are different. A cell based solely on the emf generated because of a difference in a concentration is called a **concentration cell**.

A diagram of a concentration cell is shown in Figure 20.16(a)▼. One compartment consists of a strip of nickel metal immersed in a 1.00 M solution of $Ni^{2+}(aq)$. The other compartment also has an Ni(s) electrode, but it is

▲ **Figure 20.16 Concentration cell based on the Ni^{2+}-Ni cell reaction.** In (a) the concentrations of $Ni^{2+}(aq)$ in the two compartments are unequal, and the cell generates an electrical current. The cell operates until the concentrations of $Ni^{2+}(aq)$ in the two compartments become equal, (b) at which point the cell has reached equilibrium and is "dead."

immersed in a 1.00×10^{-3} M solution of $Ni^{2+}(aq)$. The two compartments are connected by a salt bridge and by an external wire running through a voltmeter. The half-cell reactions are the reverse of each other:

$$\text{Anode:} \qquad Ni(s) \longrightarrow Ni^{2+}(aq) + 2\,e^{-} \qquad E^{\circ}_{red} = -0.28 \text{ V}$$

$$\text{Cathode:} \quad Ni^{2+}(aq) + 2\,e^{-} \longrightarrow Ni(s) \qquad E^{\circ}_{red} = -0.28 \text{ V}$$

Although the *standard* emf for this cell is zero, $E^{\circ}_{cell} = E^{\circ}_{red}$ (cathode) $- E^{\circ}_{red}$ (anode) $= (-0.28 \text{ V}) - (-0.28 \text{ V}) = 0 \text{ V}$, the cell operates under *nonstandard* conditions because the concentration of $Ni^{2+}(aq)$ is different in the two compartments. In fact, the cell will operate until the concentrations of Ni^{2+} in both compartments are equal. Oxidation of $Ni(s)$ occurs in the half-cell containing the more dilute solution, thereby increasing the concentration of $Ni^{2+}(aq)$. It is therefore the anode compartment of the cell. Reduction of $Ni^{2+}(aq)$ occurs in the half-cell containing the more concentrated solution, thereby decreasing the concentration of $Ni^{2+}(aq)$, making it the cathode compartment. The *overall* cell reaction is therefore

$$\text{Anode:} \qquad Ni(s) \longrightarrow Ni^{2+}(aq, \text{dilute}) + 2\,e^{-}$$

$$\text{Cathode:} \quad Ni^{2+}(aq, \text{concentrated}) + 2\,e^{-} \longrightarrow Ni(s)$$

$$\text{Overall:} \qquad Ni^{2+}(aq, \text{concentrated}) \longrightarrow Ni^{2+}(aq, \text{dilute})$$

Chemistry and Life   HEARTBEATS AND ELECTROCARDIOGRAPHY

The human heart is a marvel of efficiency and dependability. In a typical day an adult's heart pumps more than 7000 L of blood through the circulatory system, usually with no maintenance required beyond a sensible diet and lifestyle. We generally think of the heart as a mechanical device, a muscle that circulates blood via regularly spaced muscular contractions. However, more than two centuries ago, two pioneers in electricity, Luigi Galvani (1729–1787) and Alessandro Volta (1745–1827), discovered that the contractions of the heart are controlled by electrical phenomena, as are nerve impulses throughout the body. The pulses of electricity that cause the heart to beat result from a remarkable combination of electrochemistry and the properties of semipermeable membranes. ∞ (Section 13.5)

Cell walls are membranes with variable permeability with respect to a number of physiologically important ions (especially Na^+, K^+, and Ca^{2+}). The concentrations of these ions are different for the fluids inside the cells (the *intracellular fluid*, or ICF) and outside the cells (the *extracellular fluid*, or ECF). In cardiac muscle cells, for example, the concentrations of K^+ in the ICF and ECF are typically about 135 millimolar (mM) and 4 mM, respectively. For Na^+, however, the concentration difference between the ICF and ECF is opposite that for K^+; typically, $[Na^+]_{ICF} = 10$ mM and $[Na^+]_{ECF} = 145$ mM.

The cell membrane is initially permeable to K^+ ions, but is much less so to Na^+ and Ca^{2+}. The difference in concentration of K^+ ions between the ICF and ECF generates a concentration cell: Even though the same ions are present on both sides of the membrane, there is a potential difference between the two fluids that we can calculate using the Nernst equation with $E^{\circ} = 0$. At physiological temperature (37 °C) the potential in millivolts for moving K^+ from the ECF to the ICF is

$$E = E^{\circ} - \frac{2.30\,RT}{nF} \log \frac{[K^+]_{ICF}}{[K^+]_{ECF}}$$

$$= 0 - (61.5 \text{ mV}) \log\left(\frac{135 \text{ m}M}{4 \text{ m}M}\right) = -94 \text{ mV}$$

In essence, the interior of the cell and the ECF together serve as a voltaic cell. The negative sign for the potential indicates that work is required to move K^+ into the intracellular fluid.

Changes in the relative concentrations of the ions in the ECF and ICF lead to changes in the emf of the voltaic cell. The cells of the heart that govern the rate of heart contraction are called the *pacemaker cells*. The membranes of the cells regulate the concentrations of ions in the ICF, allowing them to change in a systematic way. The concentration changes cause the emf to change in a cyclic fashion, as shown in Figure 20.17▶.

We can calculate the emf of a concentration cell by using the Nernst equation. For this particular cell, we see that $n = 2$. The expression for the reaction quotient for the overall reaction is $Q = [Ni^{2+}]_{dilute}/[Ni^{2+}]_{concentrated}$. Thus, the emf at 298 K is

$$E = E° - \frac{0.0592 \text{ V}}{n} \log Q$$

$$= 0 - \frac{0.0592 \text{ V}}{2} \log \frac{[Ni^{2+}]_{dilute}}{[Ni^{2+}]_{concentrated}} = -\frac{0.0592 \text{ V}}{2} \log \frac{1.00 \times 10^{-3} \text{ M}}{1.00 \text{ M}}$$

$$= +0.0888 \text{ V}$$

This concentration cell generates an emf of nearly 0.09 V even though $E° = 0$. The difference in concentration provides the driving force for the cell. When the concentrations in the two compartments become the same, $Q = 1$ and $E = 0$.

The idea of generating a potential by a difference in concentration is the basis for the operation of pH meters (Figure 16.6). It is also a critical aspect in biology. For example, nerve cells in the brain generate a voltage across the cell membrane by having different concentrations of ions on either side of the membrane. The regulation of the heartbeat in mammals, as discussed in the "Chemistry and Life" box in this section, is another example of the importance of electrochemistry to living organisms.

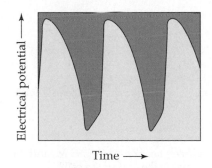

▲ Figure 20.17 **Ion concentration and emf in the human heart.** Variation of the electrical potential caused by changes of ion concentrations in the pacemaker cells of the heart.

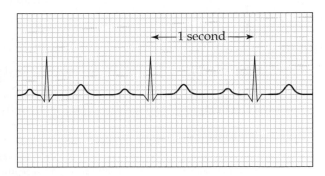

▲ Figure 20.18 **A typical electrocardiogram.** An electrocardiogram (ECG) printout records the electrical events monitored by electrodes attached to the body surface. The horizontal axis is time, and the vertical displacement is the emf.

The emf cycle determines the rate at which the heart beats. If the pacemaker cells malfunction because of disease or injury, an artificial pacemaker can be surgically implanted. The artificial pacemaker is a small battery that generates the electrical pulses needed to trigger the contractions of the heart.

During the late 1800s scientists discovered that the electrical impulses that cause the contraction of the heart muscle are strong enough to be detected at the surface of the body.

This observation formed the basis for *electrocardiography*, noninvasive monitoring of the heart by using a complex array of electrodes on the skin to measure voltage changes during heartbeats. A typical electrocardiogram is shown in Figure 20.18▲. It is quite striking that, although the heart's major function is the *mechanical* pumping of blood, it is most easily monitored by using the *electrical* impulses generated by tiny voltaic cells.

■ **SAMPLE EXERCISE 20.13** | Determining pH Using a Concentration Cell

A voltaic cell is constructed with two hydrogen electrodes. Electrode 1 has $P_{H_2} = 1.00$ atm and an unknown concentration of $H^+(aq)$. Electrode 2 is a standard hydrogen electrode ($[H^+] = 1.00$ M, $P_{H_2} = 1.00$ atm). At 298 K the measured cell voltage is 0.211 V, and the electrical current is observed to flow from electrode 1 through the external circuit to electrode 2. Calculate $[H^+]$ for the solution at electrode 1. What is its pH?

SOLUTION

Analyze: We are given the voltage of a concentration cell and the direction in which the current flows. We also have the concentrations of all reactants and products except for $[H^+]$ in half-cell 1, which is our unknown.

Plan: We can use the Nernst equation to determine Q and then use Q to calculate the unknown concentration. Because this is a concentration cell, $E^\circ_{cell} = 0$ V.

Solve: Using the Nernst equation, we have

$$0.211 \text{ V} = 0 - \frac{0.0592 \text{ V}}{2} \log Q$$

$$\log Q = -(0.211 \text{ V})\left(\frac{2}{0.0592 \text{ V}}\right) = -7.13$$

$$Q = 10^{-7.13} = 7.4 \times 10^{-8}$$

Because electrons flow from electrode 1 to electrode 2, electrode 1 is the anode of the cell and electrode 2 is the cathode. The electrode reactions are therefore as follows, with the concentration of $H^+(aq)$ in electrode 1 represented with the unknown x:

Electrode 1: $\quad H_2(g, 1.00 \text{ atm}) \longrightarrow 2 H^+(aq, x \text{ M}) + 2 e^- \quad E^\circ_{red} = 0$ V

Electrode 2: $\quad 2 H^+(aq; 1.00 \text{ M}) + 2 e^- \longrightarrow H_2(g, 1.00 \text{ atm}) \quad E^\circ_{red} = 0$ V

Thus,

$$Q = \frac{[H^+ \text{ (electrode 1)}]^2 \, P_{H_2} \text{ (electrode 2)}}{[H^+ \text{ (electrode 2)}]^2 \, P_{H_2} \text{ (electrode 1)}}$$

$$= \frac{x^2(1.00)}{(1.00)^2(1.00)} = x^2 = 7.4 \times 10^{-8}$$

$$x = \sqrt{7.4 \times 10^{-8}} = 2.7 \times 10^{-4}$$

At electrode 1, therefore,

$$[H^+] = 2.7 \times 10^{-4} \, M$$

and the pH of the solution is

$$pH = -\log[H^+] = -\log(2.7 \times 10^{-4}) = 3.57$$

Comment: The concentration of H^+ at electrode 1 is lower than that in electrode 2, which is why electrode 1 is the anode of the cell: The oxidation of H_2 to $H^+(aq)$ increases $[H^+]$ at electrode 1.

■ **PRACTICE EXERCISE**

A concentration cell is constructed with two $Zn(s)$-$Zn^{2+}(aq)$ half-cells. The first half-cell has $[Zn^{2+}] = 1.35$ M, and the second half-cell has $[Zn^{2+}] = 3.75 \times 10^{-4}$ M. **(a)** Which half-cell is the anode of the cell? **(b)** What is the emf of the cell?
Answers: **(a)** the second half-cell, **(b)** 0.105 V

20.7 BATTERIES AND FUEL CELLS

A **battery** is a portable, self-contained electrochemical power source that consists of one or more voltaic cells. For example, the common 1.5-V batteries used to power flashlights and many consumer electronic devices are single voltaic cells. Greater voltages can be achieved by using multiple voltaic cells in a single battery, as is the case in 12-V automotive batteries. When cells are connected in series (with the cathode of one attached to the anode of another), the battery produces a voltage that is the sum of the emfs of the individual cells. Higher emfs can also be achieved by using multiple batteries in series (Figure 20.19▶). The electrodes of batteries are marked following the convention of Figure 20.6—the cathode is labeled with a plus sign and the anode with a minus sign.

Although any spontaneous redox reaction can serve as the basis for a voltaic cell, making a commercial battery that has specific performance characteristics can require considerable ingenuity. The substances that are oxidized at the anode and reduced by the cathode determine the emf of a battery, and the usable life of the battery depends on the quantities of these substances packaged in the battery. Usually a barrier analogous to the porous barrier of Figure 20.6 separates the anode and the cathode compartments.

Different applications require batteries with different properties. The battery required to start a car, for example, must be capable of delivering a large electrical current for a short time period. The battery that powers a heart pacemaker, on the other hand, must be very small and capable of delivering a small but steady current over an extended time period. Some batteries are *primary* cells, meaning that they cannot be recharged. A primary cell must be discarded or recycled after its emf drops to zero. A *secondary* cell can be recharged from an external power source after its emf has dropped.

In this section we will briefly discuss some common batteries. As we do so, notice how the principles we have discussed so far in this chapter help us understand these important sources of portable electrical energy.

▲ Figure 20.19 **Combining batteries.** When batteries are connected in series, as in most flashlights, the total emf is the sum of the individual emfs.

Lead-Acid Battery

A 12-V lead-acid automotive battery consists of six voltaic cells in series, each producing 2 V. The cathode of each cell consists of lead dioxide (PbO_2) packed on a metal grid. The anode of each cell is composed of lead. Both electrodes are immersed in sulfuric acid. The electrode reactions that occur during discharge are

Cathode: $PbO_2(s) + HSO_4^-(aq) + 3\,H^+(aq) + 2\,e^- \longrightarrow PbSO_4(s) + 2\,H_2O(l)$

Anode: $\underline{\hspace{5cm} Pb(s) + HSO_4^-(aq) \longrightarrow PbSO_4(s) + H^+(aq) + 2\,e^-}$

$PbO_2(s) + Pb(s) + 2\,HSO_4^-(aq) + 2\,H^+(aq) \longrightarrow 2\,PbSO_4(s) + 2\,H_2O(l)$

The standard cell potential can be obtained from the standard reduction potentials in Appendix E:

$$E^\circ_{cell} = E^\circ_{red}\,(\text{cathode}) - E^\circ_{red}\,(\text{anode}) = (+1.685\ \text{V}) - (-0.356\ \text{V}) = +2.041\ \text{V}$$

The reactants Pb and PbO_2 serve as the electrodes. Because the reactants are solids, there is no need to separate the cell into anode and cathode compartments; the Pb and PbO_2 cannot come into direct physical contact unless one electrode plate touches another. To keep the electrodes from touching, wood or glass-fiber spacers are placed between them (Figure 20.20 ▶).

Using a reaction whose reactants and products are solids has another benefit. Because solids are excluded from the reaction quotient Q, the relative amounts of $Pb(s)$, $PbO_2(s)$, and $PbSO_4(s)$ have no effect on the emf of the lead storage battery, helping the battery maintain a relatively constant emf during discharge. The emf does vary somewhat with use because the concentration of H_2SO_4 varies with the extent of cell discharge. As the equation for the overall cell reaction indicates, H_2SO_4 is consumed during the discharge.

One advantage of a lead-acid battery is that it can be recharged. During recharging, an external source of energy is used to reverse the direction of the overall cell reaction, regenerating $Pb(s)$ and $PbO_2(s)$.

$$2\,PbSO_4(s) + 2\,H_2O(l) \longrightarrow PbO_2(s) + Pb(s) + 2\,HSO_4^-(aq) + 2\,H^+(aq)$$

In an automobile the alternator, driven by the engine, provides the energy necessary for recharging the battery. Recharging is possible because $PbSO_4$ formed during discharge adheres to the electrodes. As the external source forces electrons from one electrode to another, the $PbSO_4$ is converted to Pb at one electrode and to PbO_2 at the other.

H_2SO_4 electrolyte

Lead grid filled with spongy lead (anode)

Lead grid filled with PbO_2 (cathode)

▲ Figure 20.20 **A 12-V automotive lead-acid battery.** Each anode/cathode pair of electrodes in this schematic cutaway produces a potential of about 2 V. Six pairs of electrodes are connected in series, producing the desired battery voltage.

▲ Figure 20.21 Cutaway view of a miniature alkaline battery.

Alkaline Battery

The most common primary (nonrechargeable) battery is the alkaline battery. More than 10^{10} alkaline batteries are produced annually. The anode of this battery consists of powdered zinc metal immobilized in a gel in contact with a concentrated solution of KOH (hence the name *alkaline* battery). The cathode is a mixture of $MnO_2(s)$ and graphite, separated from the anode by a porous fabric. The battery is sealed in a steel can to reduce the risk of leakage of the concentrated KOH. A schematic view of an alkaline battery is shown in Figure 20.21 ◄. The cell reactions are complex, but can be approximately represented as follows:

Cathode: $2\,MnO_2(s) + 2\,H_2O(l) + 2\,e^- \longrightarrow 2\,MnO(OH)(s) + 2\,OH^-(aq)$

Anode: $Zn(s) + 2\,OH^-(aq) \longrightarrow Zn(OH)_2(s) + 2\,e^-$

The emf of an alkaline battery is 1.55 V at room temperature. The alkaline battery provides far superior performance over the older "dry cells" that were also based on MnO_2 and Zn as the electrochemically active substances.

Nickel-Cadmium, Nickel-Metal-Hydride, and Lithium-Ion Batteries

The tremendous growth in high-power-demand portable electronic devices, such as cell phones, notebook computers, and video recorders, has increased the demand for lightweight, readily recharged batteries. One of the most common rechargeable batteries is the nickel-cadmium (nicad) battery. During discharge, cadmium metal is oxidized at the anode of the battery while nickel oxyhydroxide $[NiO(OH)(s)]$ is reduced at the cathode.

Cathode: $2\,NiO(OH)(s) + 2\,H_2O(l) + 2\,e^- \longrightarrow 2\,Ni(OH)_2(s) + 2\,OH^-(aq)$

Anode: $Cd(s) + 2\,OH^-(aq) \longrightarrow Cd(OH)_2(s) + 2\,e^-$

As in the lead-acid battery, the solid reaction products adhere to the electrodes, which permits the electrode reactions to be reversed during charging. A single nicad voltaic cell has an emf of 1.30 V. Nicad battery packs typically contain three or more cells in series to produce the higher emfs needed by most electronic devices.

There are drawbacks to nickel-cadmium batteries. Cadmium is a toxic heavy metal. Its use increases the weight of batteries and provides an environmental hazard—roughly 1.5 billion nickel-cadmium batteries are produced annually, and these must eventually be recycled as they lose their ability to be recharged. Some of these problems have been alleviated by the development of nickel-metal-hydride (NiMH) batteries. The cathode reaction of NiMH batteries is the same as that for nickel-cadmium batteries, but the anode reaction is very different. The anode consists of a metal *alloy*, such as $ZrNi_2$, that has the ability to absorb hydrogen atoms (we will discuss alloys in Section 23.6). During the oxidation at the anode, the hydrogen atoms lose electrons, and the resultant H^+ ions react with OH^- ions to form H_2O, a process that is reversed during charging. The current generation of hybrid gas-electric automobiles, which are powered by both a gasoline engine and an electric motor, use NiMH batteries to store the electrical power. The batteries are recharged by the electric motor while braking. Due to the robustness of the batteries toward discharge and recharge, the batteries can last up to 8 years.

The newest rechargeable battery to receive large use in consumer electronic devices is the lithium-ion (Li-ion) battery. You will find this battery in cell phones and laptop computers. Because lithium is a very light element, Li-ion batteries achieve a greater *energy density*—the amount of energy stored per unit mass—than nickel-based batteries. The technology of Li-ion batteries is very different from that of the other batteries we have described here; it is based on the ability of Li^+ ions to be inserted into and removed from certain layered solids.

For example, Li^+ ions can be inserted reversibly into layers of graphite (Figure 11.41). In most commercial cells, one electrode is graphite or some other carbon-based material, and the other is usually made of lithium cobalt oxide ($LiCoO_2$). When the cell is charged, cobalt ions are oxidized, and Li^+ ions migrate into the graphite. During discharge, when the battery is producing electricity for use, the Li^+ ions spontaneously migrate from the graphite anode to the cathode, enabling electrons to flow through the external circuit. An Li-ion battery produces a maximum voltage of 3.7 V, considerably higher than typical 1.5-V alkaline batteries.

Hydrogen Fuel Cells

The thermal energy released by the combustion of fuels can be converted to electrical energy. The heat may convert water to steam, which drives a turbine that in turn drives a generator. Typically, a maximum of only 40% of the energy from combustion is converted to electricity; the remainder is lost as heat. The direct production of electricity from fuels by a voltaic cell could, in principle, yield a higher rate of conversion of the chemical energy of the reaction. Voltaic cells that perform this conversion using conventional fuels, such as H_2 and CH_4, are called **fuel cells**. Strictly speaking, fuel cells are *not* batteries because they are not self-contained systems—the fuel must be continuously supplied to generate electricity.

The most promising fuel-cell system involves the reaction of $H_2(g)$ and $O_2(g)$ to form $H_2O(l)$ as the only product. These cells can generate electricity twice as efficiently as the best internal combustion engine. Under acidic conditions, the electrode reactions are

Cathode: $\qquad O_2(g) + 4\,H^+ + 4\,e^- \longrightarrow 2\,H_2O(l)$

Anode: $\qquad\qquad\quad 2\,H_2(g) \longrightarrow 4\,H^+ + 4\,e^-$

Overall: $\qquad 2\,H_2(g) + O_2(g) \longrightarrow 2\,H_2O(l)$

The standard emf of an H_2-O_2 fuel cell is +1.23 V, reflecting the large driving force for the reaction of H_2 and O_2 to form H_2O.

In this fuel cell (known as the PEM fuel cell, for "proton-exchange membrane"), the anode and cathode are separated by a thin polymer membrane that is permeable to protons but not to electrons. The polymer membrane, therefore, acts as a salt bridge. The electrodes are typically made from graphite. A PEM cell operates at a temperature of around 80 °C. At this low temperature the electrochemical reactions would normally occur very slowly, and so a thin layer of platinum on each electrode catalyzes the reactions.

Under basic conditions the electrode reactions in the hydrogen fuel cell are

Cathode: $\quad 4\,e^- + O_2(g) + 2\,H_2O(l) \longrightarrow 4\,OH^-(aq)$

Anode: $\qquad 2\,H_2(g) + 4\,OH^-(aq) \longrightarrow 4\,H_2O(l) + 4\,e^-$

$\qquad\qquad\quad 2\,H_2(g) + O_2(g) \longrightarrow 2\,H_2O(l)$

NASA has used this basic hydrogen fuel cell as the energy source for its spacecraft. Liquid hydrogen and oxygen are stored as fuel, and water, the product of the reaction, is drunk by the spacecraft crew.

A schematic drawing of a low-temperature H_2-O_2 fuel cell is shown in Figure 20.22▶. This technology is the basis for fuel cell–powered vehicles that are under study by major automobile manufacturers. Currently a great deal of research is going into improving fuel cells. Much effort is being directed toward developing fuel cells that use conventional fuels such as hydrocarbons and alcohols, which are not as difficult to handle and distribute as hydrogen gas.

▼ **Figure 20.22   A low-temperature H_2-O_2 fuel cell.** The porous membrane allows the H^+ ions generated by the oxidation of H_2 at the anode to migrate to the cathode, where H_2O is formed.

Chemistry Put to Work   DIRECT METHANOL FUEL CELLS

While the hydrogen fuel cell has been widely proposed as a clean and efficient alternative to gasoline-powered internal combustion engines, liquid methanol, CH_3OH, is a far more attractive fuel to store and transport than hydrogen gas. Furthermore, methanol is a clean-burning liquid, and its use would require only minor modifications to existing engines and to fuel-delivery infrastructure.

One of the intriguing aspects of methanol as a fuel is that manufacturing it could make use of carbon dioxide, a source of global warming. ∞ (Section 18.4) Methanol can be made by combining CO_2 and H_2, although the process is presently costly. Imagine, though, that the synthesis can be improved and that the CO_2 used in the synthesis is captured from exhaust gases from power plants or even directly from the atmosphere. In such cases, the CO_2 released by subsequently burning the methanol would be cancelled by the carbon dioxide captured to make it. Thus, the process would be carbon neutral, meaning that it would not increase the concentration of CO_2 in the atmosphere. The prospect of a liquid fuel that could replace conventional fuels without contributing to the greenhouse effect has spurred considerable research to reduce the cost of methanol synthesis and to develop and improve methanol fuel cell technology.

A direct methanol fuel cell has been developed that is similar to the PEM hydrogen fuel cell. The reactions in the cell are

Cathode: $\frac{3}{2}O_2(g) + 6\,H^+ + 6\,e^- \longrightarrow 3\,H_2O(g)$

Anode: $\underline{\quad CH_3OH(l) + H_2O(g) \longrightarrow CO_2(g) + 6\,H^+ + 6\,e^-}$

Overall: $\quad CH_3OH(g) + \frac{3}{2}O_2(g) \longrightarrow CO_2(g) + 2\,H_2O(g)$

The direct methanol fuel cell is currently too expensive to be used in passenger cars because of the quantity of platinum catalyst it requires to operate. Nevertheless, small methanol fuel cells could appear in mobile devices such as computers or cell phones in the near future.

20.8 CORROSION

Batteries are examples of how spontaneous redox reactions can be used productively. In this section we will examine the undesirable redox reactions that lead to the **corrosion** of metals. Corrosion reactions are spontaneous redox reactions in which a metal is attacked by some substance in its environment and converted to an unwanted compound.

For nearly all metals, oxidation is a thermodynamically favorable process in air at room temperature. When the oxidation process is not inhibited in some way, it can be very destructive to whatever object is made from the metal. Oxidation can form an insulating protective oxide layer, however, that prevents further reaction of the underlying metal. Based on the standard reduction potential for Al^{3+}, for example, we would expect aluminum metal to be very readily oxidized. The many aluminum soft-drink and beer cans that litter the environment are ample evidence, however, that aluminum undergoes only very slow chemical corrosion. The exceptional stability of this active metal in air is due to the formation of a thin protective coat of oxide—a hydrated form of Al_2O_3—on the surface of the metal. The oxide coat is impermeable to O_2 or H_2O and so protects the underlying metal from further corrosion. Magnesium metal is similarly protected. Some metal alloys, such as stainless steel, likewise form protective impervious oxide coats. The semiconductor silicon, as we saw in Chapter 12, also readily forms a protective SiO_2 coating that is important to its use in electronic circuits.

Corrosion of Iron

The rusting of iron (Figure 20.23 ◄) is a familiar corrosion process that carries a significant economic impact. Up to 20% of the iron produced annually in the United States is used to replace iron objects that have been discarded because of rust damage.

The rusting of iron requires both oxygen and water. Other factors—such as the pH of the solution, the presence of salts, contact with metals more difficult to oxidize than iron, and stress on the iron—can accelerate rusting.

The corrosion of iron is electrochemical in nature. The corrosion process involves oxidation and reduction, and the metal itself conducts electricity. Thus, electrons can move through the metal from a region where oxidation occurs to another region where reduction occurs, as in voltaic cells.

▲ **Figure 20.23 Corrosion.** The corrosion of iron is an electrochemical process of great economic importance. The annual cost of metallic corrosion in the United States is estimated to be $70 billion.

Because the standard reduction potential for the reduction of $Fe^{2+}(aq)$ is less positive than that for the reduction of O_2, $Fe(s)$ can be oxidized by $O_2(g)$.

Cathode: $O_2(g) + 4 H^+(aq) + 4 e^- \longrightarrow 2 H_2O(l)$ $E^\circ_{red} = 1.23$ V

Anode: $Fe(s) \longrightarrow Fe^{2+}(aq) + 2 e^-$ $E^\circ_{red} = -0.44$ V

A portion of the iron can serve as an anode at which the oxidation of Fe to Fe^{2+} occurs. The electrons produced migrate through the metal to another portion of the surface that serves as the cathode, at which O_2 is reduced. The reduction of O_2 requires H^+, so lowering the concentration of H^+ (increasing the pH) makes the reduction of O_2 less favorable. Iron in contact with a solution whose pH is greater than 9 does not corrode.

The Fe^{2+} formed at the anode is eventually oxidized further to Fe^{3+}, which forms the hydrated iron(III) oxide known as rust:*

$$4 Fe^{2+}(aq) + O_2(g) + 4 H_2O(l) + 2\, xH_2O(l) \longrightarrow 2 Fe_2O_3 \cdot xH_2O(s) + 8 H^+(aq)$$

Because the cathode is generally the area having the largest supply of O_2, rust often deposits there. If you look closely at a shovel after it has stood outside in the moist air with wet dirt adhered to its blade, you may notice that pitting has occurred under the dirt but that rust has formed elsewhere, where O_2 is more readily available. The corrosion process is summarized in Figure 20.24▶.

The enhanced corrosion caused by the presence of salts is usually evident on autos in areas where roads are heavily salted during winter. Like a salt bridge in a voltaic cell, the ions of the salt provide the electrolyte necessary to complete the electrical circuit.

▲ **Figure 20.24 Corrosion of iron in contact with water.**

Preventing the Corrosion of Iron

Iron is often covered with a coat of paint or another metal such as tin or zinc to protect its surface against corrosion. Covering the surface with paint or tin is simply a means of preventing oxygen and water from reaching the iron surface. If the coating is broken and the iron is exposed to oxygen and water, corrosion will begin.

Galvanized iron, which is iron coated with a thin layer of zinc, uses the principles of electrochemistry to protect the iron from corrosion even after the surface coat is broken. The standard reduction potentials for iron and zinc are

$$Fe^{2+}(aq) + 2 e^- \longrightarrow Fe(s)   E^\circ_{red} = -0.44$$ V
$$Zn^{2+}(aq) + 2 e^- \longrightarrow Zn(s)   E^\circ_{red} = -0.76$$ V

Because the E°_{red} value for the reduction of Fe^{2+} is less negative (more positive) than that for the reduction of Zn^{2+}, Fe^{2+} is easier to reduce than Zn^{2+}. Conversely, $Zn(s)$ is easier to oxidize than $Fe(s)$. Thus, even if the zinc coating is broken and the galvanized iron is exposed to oxygen and water, the zinc, which is most easily oxidized, serves as the anode and is corroded instead of the iron. The iron serves as the cathode at which O_2 is reduced, as shown in Figure 20.25▶.

Protecting a metal from corrosion by making it the cathode in an electrochemical cell is known as **cathodic protection**. The metal that is oxidized while protecting the cathode is called the *sacrificial anode*. Underground pipelines are often protected against corrosion by making the pipeline the

▲ **Figure 20.25 Cathodic protection of iron in contact with zinc.**

---

*Frequently, metal compounds obtained from aqueous solution have water associated with them. For example, copper(II) sulfate crystallizes from water with five moles of water per mole of $CuSO_4$. We represent this formula as $CuSO_4 \cdot 5H_2O$. Such compounds are called hydrates. ∞ (Section 13.1) Rust is a hydrate of iron(III) oxide with a variable amount of water of hydration. We represent this variable water content by writing the formula as $Fe_2O_3 \cdot xH_2O$.

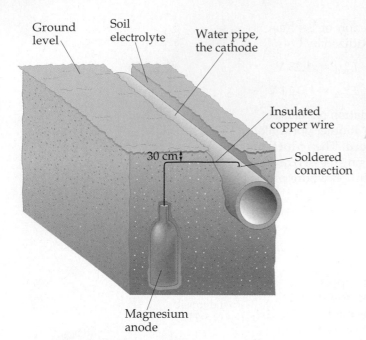

Ground level

Soil electrolyte

Water pipe, the cathode

Insulated copper wire

30 cm

Soldered connection

Magnesium anode

▲ **Figure 20.26 Cathodic protection of an iron water pipe.** A mixture of gypsum, sodium sulfate, and clay surrounds the magnesium anode to promote conductivity of ions. The pipe, in effect, is the cathode of a voltaic cell.

cathode of a voltaic cell. Pieces of an active metal such as magnesium are buried along the pipeline and connected to it by wire, as shown in Figure 20.26 ◄. In moist soil, where corrosion can occur, the active metal serves as the anode, and the pipe experiences cathodic protection.

▲ **GIVE IT SOME THOUGHT**

Based on the standard reduction potentials in Table 20.1, which of the following metals could provide cathodic protection to iron: Al, Cu, Ni, Zn?

20.9 ELECTROLYSIS

Voltaic cells are based on spontaneous oxidation-reduction reactions. Conversely, it is possible to use electrical energy to cause nonspontaneous redox reactions to occur. For example, electricity can be used to decompose molten sodium chloride into its component elements:

$$2\,NaCl(l) \longrightarrow 2\,Na(l) + Cl_2(g)$$

Such processes, which are driven by an outside source of electrical energy, are called **electrolysis reactions** and take place in **electrolytic cells**.

An electrolytic cell consists of two electrodes in a molten salt or a solution. A battery or some other source of direct electrical current acts as an electron pump, pushing electrons into one electrode and pulling them from the other. Just as in voltaic cells, the electrode at which the reduction occurs is called the cathode, and the electrode at which oxidation occurs is called the anode.

In the electrolysis of molten NaCl, shown in Figure 20.27 ▼, Na^+ ions pick up electrons and are reduced to Na at the cathode. As the Na^+ ions near the cathode are depleted, additional Na^+ ions migrate in. Similarly, there is net movement of Cl^- ions to the anode, where they are oxidized. The electrode reactions for the electrolysis of molten NaCl are summarized as follows:

Cathode: $2\,Na^+(l) + 2\,e^- \longrightarrow 2\,Na(l)$

Anode: $2\,Cl^-(l) \longrightarrow Cl_2(g) + 2\,e^-$

$$2\,Na^+(l) + 2\,Cl^-(l) \longrightarrow 2\,Na(l) + Cl_2(g)$$

▶ **Figure 20.27 Electrolysis of molten sodium chloride.** Cl^- ions are oxidized to $Cl_2(g)$ at the anode, and Na^+ ions are reduced to $Na(l)$ at the cathode. Pure NaCl melts at 801 °C.

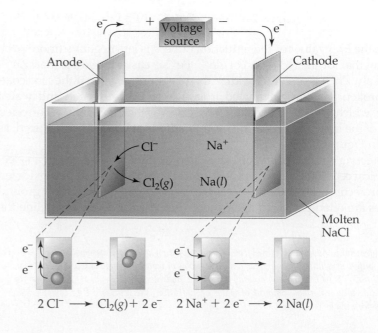

$2\,Cl^- \longrightarrow Cl_2(g) + 2\,e^-$ $2\,Na^+ + 2\,e^- \longrightarrow 2\,Na(l)$

Notice the manner in which the voltage source is connected to the electrodes in Figure 20.27. In a voltaic cell (or any other source of direct current) the electrons move from the negative terminal (Figure 20.6). Thus, the electrode of the electrolytic cell that is connected to the negative terminal of the voltage source is the cathode of the cell; it receives electrons that are used to reduce a substance. The electrons that are removed during the oxidation process at the anode travel to the positive terminal of the voltage source, thus completing the circuit of the cell.

(a)

(b)

▲ **Figure 20.28 Electroplating of silverware.** (a) The silverware is being withdrawn from the electroplating bath. (b) The polished final product.

The electrolysis of molten salts is an important industrial process for the production of active metals such as magnesium, sodium, and aluminum. We will have more to say about it in Chapter 23, when we discuss how ores are refined into metals.

Because of the high melting points of ionic substances, the electrolysis of molten salts requires very high temperatures. ∞ (Section 11.8) Do we obtain the same products if we electrolyze the aqueous solution of a salt instead of the molten salt? Frequently the answer is no; the electrolysis of an aqueous solution is complicated by the presence of water, because we must consider whether the water is oxidized to form O_2 or reduced to form H_2 rather than the ions of the salt. These reactions also depend on pH.

So far in our discussion of electrolysis, we have encountered only electrodes that were *inert*; they did not undergo reaction but merely served as the surface where oxidation and reduction occurred. Several practical applications of electrochemistry, however, are based on *active* electrodes—electrodes that participate in the electrolysis process. *Electroplating*, for example, uses electrolysis to deposit a thin layer of one metal on another metal to improve beauty or resistance to corrosion (Figure 20.28▲). We can illustrate the principles of electrolysis with active electrodes by describing how to electroplate nickel on a piece of steel.

Figure 20.29▶ illustrates the electrolytic cell for our electroplating experiment. The anode of the cell is a strip of nickel metal, and the cathode is the piece of steel that will be electroplated. The electrodes are immersed in a solution of $NiSO_4(aq)$. What happens at the electrodes when the external voltage source is turned on? Reduction will occur at the cathode. The standard reduction potential of Ni^{2+} ($E°_{red} = -0.28$ V) is less negative than that of H_2O ($E°_{red} = -0.83$ V), so Ni^{2+} will be preferentially reduced at the cathode.

At the anode we need to consider which substances can be oxidized. For the $NiSO_4(aq)$ solution, only the H_2O solvent is readily oxidized because neither Ni^{2+} nor the $SO_4{}^{2-}$ can be oxidized (both already have their elements in their highest possible oxidation state). The Ni atoms in the anode, however, can undergo oxidation. Thus, the two possible oxidation processes are

$$2\,H_2O(l) \longrightarrow O_2(g) + 4\,H^+(aq) + 4\,e^- \qquad E°_{red} = +1.23\ \text{V}$$

$$Ni(s) \longrightarrow Ni^{2+}(aq) + 2\,e^- \qquad E°_{red} = -0.28\ \text{V}$$

where the potentials are the standard *reduction* potentials of these reactions. Because this is an oxidation, the half-reaction with the more negative value of $E°_{red}$ is favored. (Remember the behavior summarized in Figure 20.14: The strongest reducing agents, which are the substances oxidized most readily, have the most negative values of $E°_{red}$.) We can summarize the electrode reactions as

Cathode (steel strip): $Ni^{2+}(aq) + 2\,e^- \longrightarrow Ni(s)$ $\qquad E°_{red} = -0.28$ V

Anode (nickel strip): $\qquad\qquad Ni(s) \longrightarrow Ni^{2+}(aq) + 2\,e^-$ $\quad E°_{red} = -0.28$ V

If we look at the overall reaction, it appears as if nothing has been accomplished. During the electrolysis, however, we are transferring Ni atoms from the Ni anode to the steel cathode, plating the steel electrode with a thin layer of nickel atoms. The standard emf for the overall reaction is $E°_{cell} = E°_{red}$ (cathode) $- E°_{red}$ (anode) $= 0$.

▲ **Figure 20.29 Electrolytic cell with an active metal electrode.** Nickel dissolves from the anode to form $Ni^{2+}(aq)$. At the cathode $Ni^{2+}(aq)$ is reduced and forms a nickel "plate" on the cathode.

▲ Figure 20.30 **Relationship between charge and amount of reactant and product in electrolysis reactions.** This flowchart shows the steps relating the quantity of electrical charge used in electrolysis to the amounts of substances oxidized or reduced.

Only a small emf is needed to provide the "push" to transfer the nickel atoms from one electrode to the other. In Chapter 23 we will explore further the utility of electrolysis with active electrodes as a means of purifying crude metals.

Quantitative Aspects of Electrolysis

The stoichiometry of a half-reaction shows how many electrons are needed to achieve an electrolytic process. For example, the reduction of Na^+ to Na is a one-electron process:

$$Na^+ + e^- \longrightarrow Na$$

Thus, one mole of electrons will plate out 1 mol of Na metal, two moles of electrons will plate out 2 mol of Na metal, and so forth. Similarly, two moles of electrons are required to produce 1 mol of copper from Cu^{2+}, and three moles of electrons are required to produce 1 mol of aluminum from Al^{3+}.

$$Cu^{2+} + 2\,e^- \longrightarrow Cu$$

$$Al^{3+} + 3\,e^- \longrightarrow Al$$

For any half-reaction, the amount of a substance that is reduced or oxidized in an electrolytic cell is directly proportional to the number of electrons passed into the cell.

The quantity of charge passing through an electrical circuit, such as that in an electrolytic cell, is generally measured in *coulombs*. As noted in Section 20.5, the charge on one mole of electrons is 96,485 C (1 faraday). A coulomb is the quantity of charge passing a point in a circuit in 1 s when the current is 1 ampere (A). Therefore, the number of coulombs passing through a cell can be obtained by multiplying the amperage and the elapsed time in seconds.

$$\text{Coulombs} = \text{amperes} \times \text{seconds} \qquad [20.18]$$

Figure 20.30▲ shows how the quantities of substances produced or consumed in electrolysis are related to the quantity of electrical charge that is used. The same relationships can also be applied to voltaic cells. In other words, electrons can be thought of as reagents in electrolysis reactions.

■ **SAMPLE EXERCISE 20.14** | Relating Electrical Charge and Quantity of Electrolysis

Calculate the number of grams of aluminum produced in 1.00 h by the electrolysis of molten $AlCl_3$ if the electrical current is 10.0 A.

SOLUTION

Analyze: We are told that $AlCl_3$ is electrolyzed to form Al and asked to calculate the number of grams of Al produced in 1.00 h with 10.0 A.

Plan: Figure 20.30 provides a road map of the problem. First, the product of the amperage and the time in seconds gives the number of coulombs of electrical charge being used (Equation 20.18). Second, the coulombs can be converted with the Faraday constant ($F = 96,485$ C/mole electrons) to tell us the number of moles of electrons being supplied. Third, reduction of 1 mol of Al^{3+} to Al requires three moles of electrons. Hence we can use the number of moles of electrons to calculate the number of moles of Al metal it produces. Finally, we convert moles of Al into grams.

Solve: First, we calculate the coulombs of electrical charge that are passed into the electrolytic cell:

$$\text{Coulombs} = \text{amperes} \times \text{seconds} = (10.0\ \text{A})(1.00\ \text{h})\frac{(3600\ \text{s})}{\text{h}} = 3.60 \times 10^4\,\text{C}$$

Second, we calculate the number of moles of electrons that pass into the cell:

$$\text{Moles}\ e^- = (3.60 \times 10^4\,\text{C})\left(\frac{1\ \text{mol}\ e^-}{96,485\ \text{C}}\right) = 0.373\ \text{mol}\ e^-$$

Third, we relate the number of moles of electrons to the number of moles of aluminum being formed, using the half-reaction for the reduction of Al^{3+}:

$$Al^{3+} + 3\,e^- \longrightarrow Al$$

Thus, three moles of electrons (3 F of electrical charge) are required to form 1 mol of Al:

$$\text{Moles Al} = (0.373\text{ mol e}^-)\left(\frac{1\text{ mol Al}}{3\text{ mol e}^-}\right) = 0.124\text{ mol Al}$$

Finally, we convert moles to grams:

$$\text{Grams Al} = (0.124\text{ mol Al})\left(\frac{27.0\text{ g Al}}{1\text{ mol Al}}\right) = 3.36\text{ g Al}$$

Because each step involves a multiplication by a new factor, the steps can be combined into a single sequence of factors:

$$\text{Grams Al} = (3.60\times10^4\,C)\left(\frac{1\text{ mole e}^-}{96,485\text{ C}}\right)\left(\frac{1\text{ mol Al}}{3\text{ mol e}^-}\right)\left(\frac{27.0\text{ g Al}}{1\text{ mol Al}}\right) = 3.36\text{ g Al}$$

■ PRACTICE EXERCISE

(a) The half-reaction for formation of magnesium metal upon electrolysis of molten $MgCl_2$ is $Mg^{2+} + 2\,e^- \longrightarrow Mg$. Calculate the mass of magnesium formed upon passage of a current of 60.0 A for a period of 4.00×10^3 s.
(b) How many seconds would be required to produce 50.0 g of Mg from $MgCl_2$ if the current is 100.0 A?
Answers: (a) 30.2 g of Mg, (b) 3.97×10^3 s

Electrical Work

We have already seen that a positive value of E is associated with a negative value for the free-energy change and, thus, with a spontaneous process. We also know that for any spontaneous process, ΔG is a measure of the maximum useful work, w_{max}, that can be extracted from the process: $\Delta G = w_{max}$. ∞(Section 5.2) Because $\Delta G = -nFE$, the maximum useful electrical work obtainable from a voltaic cell is

$$w_{max} = -nFE \qquad [20.19]$$

The cell emf, E, is a positive number for a voltaic cell, so w_{max} will be a negative number for a voltaic cell. Work done *by* a system *on* its surroundings is indicated by a negative sign for w. ∞ (Section 5.2) Thus, the negative value for w_{max} means that a voltaic cell does work on its surroundings.

In an electrolytic cell we use an external source of energy to bring about a nonspontaneous electrochemical process. In this case ΔG is positive and E_{cell} is negative. To force the process to occur, we need to apply an external potential, E_{ext}, which must be larger in magnitude than E_{cell}: $E_{ext} > -E_{cell}$. For example, if a nonspontaneous process has $E_{cell} = -0.9$ V, then the external potential E_{ext} must be greater than 0.9 V in order for the process to occur.

When an external potential E_{ext} is applied to a cell, the surroundings are doing work on the system. The amount of work performed is given by

$$w = nFE_{ext} \qquad [20.20]$$

Unlike Equation 20.19, there is no minus sign in Equation 20.20. The work calculated in Equation 20.20 will be a positive number because the surroundings are doing work on the system. The quantity n in Equation 20.20 is the number of moles of electrons forced into the system by the external potential. The product $n\times F$ is the total electrical charge supplied to the system by the external source of electricity.

Electrical work can be expressed in energy units of watts times time. The **watt** (W) is a unit of electrical power (that is, the rate of energy expenditure).

$$1\text{ W} = 1\text{ J/s}$$

Thus, a watt-second is a joule. The unit employed by electric utilities is the kilowatt-hour (kWh), which equals 3.6×10^6 J.

$$1\text{ kWh} = (1000\text{ W})(1\text{ hr})\left(\frac{3600\text{ s}}{1\text{ hr}}\right)\left(\frac{1\text{ J/s}}{1\text{ W}}\right) = 3.6\times10^6\text{ J} \qquad [20.21]$$

Using these considerations, we can calculate the maximum work obtainable from the voltaic cells and the minimum work required to bring about desired electrolysis reactions.

SAMPLE EXERCISE 20.15 | Calculating Energy in Kilowatt-hours

Calculate the number of kilowatt-hours of electricity required to produce 1.0×10^3 kg of aluminum by electrolysis of Al^{3+} if the applied voltage is 4.50 V.

SOLUTION

Analyze: We are given the mass of Al produced from Al^{3+} and the applied voltage and asked to calculate the energy, in kilowatt-hours, required for the reduction.

Plan: From the mass of Al, we can calculate first the number of moles of Al and then the number of coulombs required to obtain that mass. We can then use Equation 20.20, $w = nFE_{ext}$, where nF is the total charge in coulombs and E_{ext} is the applied potential, 4.50 V.

Solve: First, we need to calculate nF, the number of coulombs required:

$$\text{Coulombs} = (1.00 \times 10^3 \text{ kg Al})\left(\frac{1000 \text{ g Al}}{1 \text{ kg Al}}\right)\left(\frac{1 \text{ mol Al}}{27.0 \text{ g Al}}\right)\left(\frac{3 \text{ mol e}^-}{1 \text{ mol Al}}\right)\left(\frac{96{,}485 \text{ C}}{1 \text{ mole e}^-}\right)$$

$$= 1.07 \times 10^{10} \text{ C}$$

We can now calculate w. In doing so, we must apply several conversion factors, including Equation 20.21, which gives the conversion between kilowatt-hours and joules:

$$\text{Kilowatt-hours} = (1.07 \times 10^{10} \text{ C})(4.50 \text{ V})\left(\frac{1 \text{ J}}{1 \text{ C-V}}\right)\left(\frac{1 \text{ kWh}}{3.6 \times 10^6 \text{ J}}\right)$$

$$= 1.34 \times 10^4 \text{ kWh}$$

Comment: This quantity of energy does not include the energy used to mine, transport, and process the aluminum ore, and to keep the electrolysis bath molten during electrolysis. A typical electrolytic cell used to reduce aluminum ore to aluminum metal is only 40% efficient, with 60% of the electrical energy being dissipated as heat. It therefore requires approximately 33 kWh of electricity to produce 1 kg of aluminum. The aluminum industry consumes about 2% of the electrical energy generated in the United States. Because this energy is used mainly to reduce aluminum, recycling this metal saves large quantities of energy.

PRACTICE EXERCISE

Calculate the number of kilowatt-hours of electricity required to produce 1.00 kg of Mg from electrolysis of molten $MgCl_2$ if the applied emf is 5.00 V. Assume that the process is 100% efficient.
Answer: 11.0 kWh

SAMPLE INTEGRATIVE EXERCISE | Putting Concepts Together

The K_{sp} at 298 K for iron(II) fluoride is 2.4×10^{-6}. **(a)** Write a half-reaction that gives the likely products of the two-electron reduction of $FeF_2(s)$ in water. **(b)** Use the K_{sp} value and the standard reduction potential of Fe^{2+} (aq) to calculate the standard reduction potential for the half-reaction in part (a). **(c)** Rationalize the difference in the reduction potential for the half-reaction in part (a) with that for Fe^{2+} (aq).

SOLUTION

Analyze: We are going to have to combine what we know about equilibrium constants and electrochemistry to obtain reduction potentials.

Plan: For (a) we need to determine which ion, Fe^{2+} or F^-, is more likely to be reduced by 2 electrons and write the overall reaction for $FeF_2 + 2 \text{ e}^- \longrightarrow$? For (b) we need to write the K_{sp} reaction and manipulate it to get $E°$ for the reaction in (a). For (c) we need to see what we get for (a) and (b).

Solve:

(a) Iron(II) fluoride is an ionic substance that consists of Fe^{2+} and F^- ions. We are asked to predict where two electrons could be added to FeF_2. We can't envision adding the electrons to the F^- ions to form F^{2-}, so it seems likely that we could reduce the Fe^{2+} ions to Fe(s). We therefore predict the half-reaction

$$FeF_2(s) + 2 \text{ e}^- \longrightarrow Fe(s) + 2 \text{ F}^-(aq)$$

(b) The K_{sp} value refers to the following equilibrium ∞ (Section 17.4):

$$FeF_2(s) \rightleftharpoons Fe^{2+}(aq) + 2 \text{ F}^-(aq) \quad K_{sp} = [Fe^{2+}][F^-]^2 = 2.4 \times 10^{-6}$$

We were also asked to use the standard reduction potential of Fe^{2+}, whose half-reaction and standard voltage are listed in Appendix E:

$$Fe^{2+}(aq) + 2 \text{ e}^- \longrightarrow Fe(s) \quad E° = -0.440 \text{ V}$$

Recall that according to Hess's law, we can add reactions to get the one we want and we can add thermodynamic quantities like ΔH and ΔG to solve for the enthalpy or free energy of the reaction we want. ∞ (Section 5.6) In this case notice that if we add the K_{sp} reaction to the standard reduction half-reaction for Fe^{2+}, we get the half-reaction we want:

$$1. \quad FeF_2(s) \longrightarrow Fe^{2+}(aq) + 2 F^-(aq)$$

$$2. \quad Fe^{2+}(aq) + 2 e^- \longrightarrow Fe(s)$$

Overall:

$$3. \quad FeF_2(s) + 2 e^- \longrightarrow Fe(s) + 2 F^-(aq)$$

Reaction 3 is still a half-reaction, so we do see the free electrons.

If we knew $\Delta G°$ for reactions 1 and 2, we could add them to get $\Delta G°$ for reaction 3. Recall that we can relate $\Delta G°$ to $E°$ by $\Delta G° = -nFE°$ and to K by $\Delta G° = -RT \ln K$. We know K for reaction 1; it is K_{sp}. We know $E°$ for reaction 2. Therefore we can calculate $\Delta G°$ for reactions 1 and 2:

Reaction 1:

$$\Delta G° = -RT \ln K = -(8.314 \text{ J/K mol})(298 \text{ K}) \ln(2.4 \times 10^{-6}) = 3.21 \times 10^4 \text{ J/mol}$$

Reaction 2:

$$\Delta G° = -nFE° = -(2 \text{ mol})(96,485 \text{ C/mol})(-0.440 \text{ J/C}) = 8.49 \times 10^4 \text{ J}$$

(Recall that 1 volt is 1 joule per coulomb.)

Then, $\Delta G°$ for reaction 3, the one we want, is 3.21×10^4 J (for one mole of FeF_2) + 8.49×10^4 J = 1.17×10^5 J. We can convert this to $E°$ easily from the relationship $\Delta G° = -nFE°$:

$$1.17 \times 10^5 \text{ J} = -(2 \text{ mol})(96,485 \text{ C/mol}) E°$$

$$E° = -0.606 \text{ J/C} = -0.606 \text{ V}.$$

(c) The standard reduction potential for FeF_2 (−0.606 V) is more negative than that for Fe^{2+} (−0.440 V), telling us that the reduction of FeF_2 is the less favorable process. When FeF_2 is reduced, we not only reduce the Fe^{2+} ions but also break up the ionic solid. Because this additional energy must be overcome, the reduction of FeF_2 is less favorable than the reduction of Fe^{2+}.

CHAPTER REVIEW

SUMMARY AND KEY TERMS

Introduction and Section 20.1 In this chapter we have focused on **electrochemistry**, the branch of chemistry that relates electricity and chemical reactions. Electrochemistry involves oxidation-reduction reactions, also called redox reactions. These reactions involve a change in the oxidation state of one or more elements. In every oxidation-reduction reaction one substance is oxidized (its oxidation state, or number, increases) and one substance is reduced (its oxidation state, or number, decreases). The substance that is oxidized is referred to as a **reducing agent**, or **reductant**, because it causes the reduction of some other substance. Similarly, the substance that is reduced is referred to as an **oxidizing agent**, or **oxidant**, because it causes the oxidation of some other substance.

Section 20.2 An oxidization-reduction reaction can be balanced by dividing the reaction into two **half-reactions**, one for oxidation and one for reduction. A half-reaction is a balanced chemical equation that includes electrons. In oxidation half-reactions the electrons are on the product

(right) side of the reaction; we can envision that these electrons are transferred from a substance when it is oxidized. In reduction half-reactions the electrons are on the reactant (left) side of the reaction. Each half-reaction is balanced separately, and the two are brought together with proper coefficients to balance the electrons on each side of the equation.

Section 20.3 A **voltaic** (or **galvanic**) **cell** uses a spontaneous oxidation-reduction reaction to generate electricity. In a voltaic cell the oxidation and reduction half-reactions often occur in separate compartments. Each compartment has a solid surface called an electrode, where the half-reaction occurs. The electrode where oxidation occurs is called the **anode**; reduction occurs at the **cathode**. The electrons released at the anode flow through an external circuit (where they do electrical work) to the cathode. Electrical neutrality in the solution is maintained by the migration of ions between the two compartments through a device such as a salt bridge.

Section 20.4 A voltaic cell generates an **electromotive force (emf)** that moves the electrons from the anode to the cathode through the external circuit. The origin of emf is a difference in the electrical potential energy of the two electrodes in the cell. The emf of a cell is called its **cell potential**, E_{cell}, and is measured in volts. The cell potential under standard conditions is called the **standard emf**, or the **standard cell potential**, and is denoted $E°_{cell}$.

A **standard reduction potential**, $E°_{red}$, can be assigned for an individual half-reaction. This is achieved by comparing the potential of the half-reaction to that of the **standard hydrogen electrode** (SHE), which is defined to have $E°_{red} = 0$ V and is based on the following half-reaction:

$$2\,H^+(aq, 1\,M) + 2\,e^- \longrightarrow H_2(g, 1\,atm) \qquad E°_{red} = 0\,V$$

The standard cell potential of a voltaic cell is the difference between the standard reduction potentials of the half-reactions that occur at the cathode and the anode: $E°_{cell} = E°_{red}$ (cathode) $- E°_{red}$ (anode). The value of $E°_{cell}$ is positive for a voltaic cell.

For a reduction half-reaction, $E°_{red}$ is a measure of the tendency of the reduction to occur; the more positive the value for $E°_{red}$, the greater the tendency of the substance to be reduced. Thus, $E°_{red}$ provides a measure of the oxidizing strength of a substance. Fluorine (F_2) has the most positive value for $E°_{red}$ and is the strongest oxidizing agent. Substances that are strong oxidizing agents produce products that are weak reducing agents, and vice versa.

Section 20.5 The emf, E, is related to the change in the Gibbs free energy, ΔG: $\Delta G = -nFE$, where n is the number of electrons transferred during the redox process and F is *Faraday's constant*, defined as the quantity of electrical charge on one mole of electrons. This amount of charge is 1 **faraday** (F): $1\,F = 96{,}485$ C/mol. Because E is related to ΔG, the sign of E indicates whether a redox process is spontaneous: $E > 0$ indicates a spontaneous process, and $E < 0$ indicates a nonspontaneous one. Because ΔG is also related to the equilibrium constant for a reaction ($\Delta G = -RT \ln K$), we can relate E to K as well.

Section 20.6 The emf of a redox reaction varies with temperature and with the concentrations of reactants and products. The **Nernst equation** relates the emf under nonstandard conditions to the standard emf and the reaction quotient Q:

$$E = E° - (RT/nF) \ln Q = E° - (0.0592/n) \log Q$$

The factor 0.0592 is valid when $T = 298$ K. A **concentration cell** is a voltaic cell in which the same half-reaction occurs at both the anode and cathode but with different concentrations of reactants in each compartment.

At equilibrium, $Q = K$ and $E = 0$. The standard emf is therefore related to the equilibrium constant.

Section 20.7 A **battery** is a self-contained electrochemical power source that contains one or more voltaic cells. Batteries are based on a variety of different redox reactions. Several common batteries were discussed. The lead-acid battery, the nickel-cadmium battery, the nickel-metal-hydride battery, and the lithium-ion battery are examples of rechargeable batteries. The common alkaline dry cell is not rechargeable. **Fuel cells** are voltaic cells that utilize redox reactions in which reactants such as H_2 have to be continuously supplied to the cell to generate voltage.

Section 20.8 Electrochemical principles help us understand **corrosion**, undesirable redox reactions in which a metal is attacked by some substance in its environment. The corrosion of iron into rust is caused by the presence of water and oxygen, and it is accelerated by the presence of electrolytes, such as road salt. The protection of a metal by putting it in contact with another metal that more readily undergoes oxidation is called **cathodic protection**. Galvanized iron, for example, is coated with a thin layer of zinc; because zinc is oxidized more readily than iron, the zinc serves as a sacrificial anode in the redox reaction.

Section 20.9 An **electrolysis reaction**, which is carried out in an **electrolytic cell**, employs an external source of electricity to drive a nonspontaneous electrochemical reaction. The negative terminal of the external source is connected to the cathode of the cell, and the positive terminal to the anode. The current-carrying medium within an electrolytic cell may be either a molten salt or an electrolyte solution. The products of electrolysis can generally be predicted by comparing the reduction potentials associated with possible oxidation and reduction processes. The electrodes in an electrolytic cell can be active, meaning that the electrode can be involved in the electrolysis reaction. Active electrodes are important in electroplating and in metallurgical processes.

The quantity of substances formed during electrolysis can be calculated by considering the number of electrons involved in the redox reaction and the amount of electrical charge that passes into the cell. The maximum amount of electrical work produced by a voltaic cell is given by the product of the total charge delivered, nF, and the emf, E: $w_{max} = -nFE$. The work performed in an electrolysis is given by $w = nFE_{ext}$, where E_{ext} is the applied external potential. The **watt** is a unit of power: $1\,W = 1$ J/s. Electrical work is often measured in kilowatt-hours.

KEY SKILLS

- Identify oxidation, reduction, oxidizing agent, and reducing agent in a chemical equation.
- Complete and balance redox equations using the method of half-reactions.
- Sketch a voltaic cell and identify its cathode, anode, and the directions that electrons and ions move.
- Calculate standard emfs (cell potentials), $E°_{cell}$, from standard reduction potentials.

- Use reduction potentials to predict whether a redox reaction is spontaneous.
- Relate $E°_{cell}$ to $\Delta G°$ and equilibrium constants.
- Calculate emf under nonstandard conditions.
- Describe the reactions in electrolytic cells.
- Relate amounts of products and reactants in redox reactions to electrical charge.

KEY EQUATIONS

- $E°_{cell} = E°_{red} \text{ (cathode)} - E°_{red} \text{ (anode)}$ [20.8]   Relating standard emf to standard reduction potentials of the reduction (cathode) and oxidation (anode) half reactions

- $\Delta G = -nFE$ [20.11]   Relating free energy change and emf

- $E = E° - \dfrac{0.0592 \text{ V}}{n} \log Q$ (at 298 K)   [20.16]   The Nernst equation, expressing the effect of concentration on cell potential

VISUALIZING CONCEPTS

20.1 In the Brønsted-Lowry concept of acids and bases, acid–base reactions are viewed as proton-transfer reactions. The stronger the acid, the weaker is its conjugate base. In what ways are redox reactions analogous? [Sections 20.1 and 20.2]

20.2 Consider the reaction in Figure 20.3. Describe what would happen if **(a)** the solution contained cadmium(II) sulfate and the metal was zinc, **(b)** the solution contained silver nitrate and the metal was copper. [Section 20.3]

20.3 The diagram below represents a molecular view of a process occurring at an electrode in a voltaic cell.

(a) Does the process represent oxidation or reduction? **(b)** Is the electrode the anode or cathode? **(c)** Why are the atoms in the electrode represented by larger spheres than the ions in the solution? [Section 20.3]

20.4 Assume that you want to construct a voltaic cell that uses the following half reactions:

$$A^{2+}(aq) + 2\,e^- \longrightarrow A(s) \quad E°_{red} = -0.10 \text{ V}$$
$$B^{2+}(aq) + 2\,e^- \longrightarrow B(s) \quad E°_{red} = -1.10 \text{ V}$$

You begin with the incomplete cell pictured below, in which the electrodes are immersed in water.

(a) What additions must you make to the cell for it to generate a standard emf? **(b)** Which electrode functions as the cathode? **(c)** Which direction do electrons move through the external circuit? **(d)** What voltage will the cell generate under standard conditions? [Sections 20.3 and 20.4]

20.5 Where on Figure 20.14 would you find **(a)** the chemical species that is the easiest to oxidize, and **(b)** the chemical species that is the easiest to reduce? [Section 20.4]

20.6 For the generic reaction $A(aq) + B(aq) \longrightarrow A^-(aq) + B^+(aq)$ for which $E°$ is a positive number, answer the following questions:
  (a) What is being oxidized, and what is being reduced?
  (b) If you made a voltaic cell out of this reaction, what half-reaction would be occurring at the cathode, and what half-reaction would be occurring at the anode?
  (c) Which half-reaction from (b) is higher in potential energy?
  (d) What is the sign of the free energy change for the reaction? [Sections 20.4 and 20.5]

20.7 Consider the half reaction $Ag^+(aq) + e^- \longrightarrow Ag(s)$.
(a) Which of the lines in the following diagram indicates how the reduction potential varies with the concentration of Ag^+? **(b)** What is the value of E_{red} when $\log[Ag^+] = 0$? [Section 20.6]

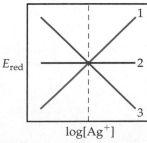

20.8 Draw a generic picture of a fuel cell. What is the main difference between it and a battery, regardless of the redox reactions that occur inside? [Section 20.7]

20.9 How does a zinc coating on iron protect the iron from unwanted oxidation? [Section 20.8]

20.10 You may have heard that "antioxidants" are good for your health. Based on what you have learned in this chapter, what do you deduce an "antioxidant" is? [Sections 20.1 and 20.2]

EXERCISES

Oxidation-Reduction Reactions

20.11 **(a)** What is meant by the term *oxidation*? **(b)** On which side of an oxidation half-reaction do the electrons appear? **(c)** What is meant by the term *oxidant*? **(d)** What is meant by the term *oxidizing agent*?

20.12 **(a)** What is meant by the term *reduction*? **(b)** On which side of a reduction half-reaction do the electrons appear? **(c)** What is meant by the term *reductant*? **(d)** What is meant by the term *reducing agent*?

20.13 Indicate whether each of the following statements is true or false:
(a) If something is oxidized, it is formally losing electrons.
(b) For the reaction $Fe^{3+}(aq) + Co^{2+}(aq) \longrightarrow$ $Fe^{2+}(aq) + Co^{3+}(aq)$, $Fe^{3+}(aq)$ is the reducing agent and $Co^{2+}(aq)$ is the oxidizing agent.
(c) If there are no changes in the oxidation state of the reactants or products of a particular reaction, that reaction is not a redox reaction.

20.14 Indicate whether each of the following statements is true or false:
(a) If something is reduced, it is formally losing electrons.
(b) A reducing agent gets oxidized as it reacts.
(c) Oxidizing agents can convert CO into CO_2.

20.15 In each of the following balanced oxidation-reduction equations, identify those elements that undergo changes in oxidation number and indicate the magnitude of the change in each case.
(a) $I_2O_5(s) + 5\,CO(g) \longrightarrow I_2(s) + 5\,CO_2(g)$
(b) $2\,Hg^{2+}(aq) + N_2H_4(aq) \longrightarrow$
$2\,Hg(l) + N_2(g) + 4\,H^+(aq)$
(c) $3\,H_2S(aq) + 2\,H^+(aq) + 2\,NO_3^-(aq) \longrightarrow$
$3\,S(s) + 2\,NO(g) + 4\,H_2O(l)$
(d) $Ba^{2+}(aq) + 2\,OH^-(aq) +$
$H_2O_2(aq) + 2\,ClO_2(aq) \longrightarrow$
$Ba(ClO_2)_2(s) + 2\,H_2O(l) + O_2(g)$

20.16 Indicate whether the following balanced equations involve oxidation-reduction. If they do, identify the elements that undergo changes in oxidation number.
(a) $PBr_3(l) + 3\,H_2O(l) \longrightarrow H_3PO_3(aq) + 3\,HBr(aq)$
(b) $NaI(aq) + 3\,HOCl(aq) \longrightarrow NaIO_3(aq) + 3\,HCl(aq)$
(c) $3\,SO_2(g) + 2\,HNO_3(aq) + 2\,H_2O(l) \longrightarrow$
$3\,H_2SO_4(aq) + 2\,NO(g)$
(d) $2\,H_2SO_4(aq) + 2\,NaBr(s) \longrightarrow$
$Br_2(l) + SO_2(g) + Na_2SO_4(aq) + 2\,H_2O(l)$

Balancing Oxidation-Reduction Reactions

20.17 At 900 °C titanium tetrachloride vapor reacts with molten magnesium metal to form solid titanium metal and molten magnesium chloride. **(a)** Write a balanced equation for this reaction. **(b)** What is being oxidized, and what is being reduced? **(c)** Which substance is the reductant, and which is the oxidant?

20.18 Hydrazine (N_2H_4) and dinitrogen tetroxide (N_2O_4) form a self-igniting mixture that has been used as a rocket propellant. The reaction products are N_2 and H_2O.
(a) Write a balanced chemical equation for this reaction.
(b) What is being oxidized, and what is being reduced?
(c) Which substance serves as the reducing agent, and which as the oxidizing agent?

20.19 Complete and balance the following half-reactions. In each case indicate whether the half-reaction is an oxidation or a reduction.
(a) $Sn^{2+}(aq) \longrightarrow Sn^{4+}(aq)$ (acidic or basic solution)
(b) $TiO_2(s) \longrightarrow Ti^{2+}(aq)$ (acidic solution)
(c) $ClO_3^-(aq) \longrightarrow Cl^-(aq)$ (acidic solution)
(d) $N_2(g) \longrightarrow NH_4^+(aq)$ (acidic solution)
(e) $OH^-(aq) \longrightarrow O_2(g)$ (basic solution)
(f) $SO_3^{2-}(aq) \longrightarrow SO_4^{2-}(aq)$ (basic solution)
(g) $N_2(g) \longrightarrow NH_3(g)$ (basic solution)

20.20 Complete and balance the following half-reactions. In each case indicate whether the half-reaction is an oxidation or a reduction.
(a) $Mo^{3+}(aq) \longrightarrow Mo(s)$ (acidic or basic solution)
(b) $H_2SO_3(aq) \longrightarrow SO_4^{2-}(aq)$ (acidic solution)

(c) $NO_3^-(aq) \longrightarrow NO(g)$ (acidic solution)
(d) $O_2(g) \longrightarrow H_2O(l)$ (acidic solution)
(e) $Mn^{2+}(aq) \longrightarrow MnO_2(s)$ (basic solution)
(f) $Cr(OH)_3(s) \longrightarrow CrO_4^{2-}(aq)$ (basic solution)
(g) $O_2(g) \longrightarrow H_2O(l)$ (basic solution)

20.21 Complete and balance the following equations, and identify the oxidizing and reducing agents:
(a) $Cr_2O_7^{2-}(aq) + I^-(aq) \longrightarrow Cr^{3+}(aq) + IO_3^-(aq)$
(acidic solution)
(b) $MnO_4^-(aq) + CH_3OH(aq) \longrightarrow$
$Mn^{2+}(aq) + HCO_2H(aq)$ (acidic solution)
(c) $I_2(s) + OCl^-(aq) \longrightarrow IO_3^-(aq) + Cl^-(aq)$
(acidic solution)
(d) $As_2O_3(s) + NO_3^-(aq) \longrightarrow$
$H_3AsO_4(aq) + N_2O_3(aq)$ (acidic solution)
(e) $MnO_4^-(aq) + Br^-(aq) \longrightarrow MnO_2(s) + BrO_3^-(aq)$
(basic solution)
(f) $Pb(OH)_4^{2-}(aq) + ClO^-(aq) \longrightarrow PbO_2(s) + Cl^-(aq)$
(basic solution)

20.22 Complete and balance the following equations, and identify the oxidizing and reducing agents. Recall that the O atoms in hydrogen peroxide, H_2O_2, have an atypical oxidation state. ∞ (Table 2.5)
(a) $NO_2^-(aq) + Cr_2O_7^{2-}(aq) \longrightarrow$
$Cr^{3+}(aq) + NO_3^-(aq)$ (acidic solution)
(b) $S(s) + HNO_3(aq) \longrightarrow H_2SO_3(aq) + N_2O(g)$
(acidic solution)

(c) $Cr_2O_7{}^{2-}(aq) + CH_3OH(aq) \longrightarrow$
$\qquad HCO_2H(aq) + Cr^{3+}(aq)$ (acidic solution)

(d) $MnO_4{}^-(aq) + Cl^-(aq) \longrightarrow Mn^{2+}(aq) + Cl_2(aq)$
$\qquad$ (acidic solution)

(e) $NO_2{}^-(aq) + Al(s) \longrightarrow NH_4{}^+(aq) + AlO_2{}^-(aq)$
$\qquad$ (basic solution)

(f) $H_2O_2(aq) + ClO_2(aq) \longrightarrow ClO_2{}^-(aq) + O_2(g)$
$\qquad$ (basic solution)

Voltaic Cells

20.23 **(a)** What are the similarities and differences between Figure 20.3 and Figure 20.4? **(b)** Why are Na^+ ions drawn into the cathode compartment as the voltaic cell shown in Figure 20.5 operates?

20.24 **(a)** What is the role of the porous glass disc shown in Figure 20.4? **(b)** Why do $NO_3{}^-$ ions migrate into the anode compartment as the voltaic cell shown in Figure 20.5 operates?

20.25 A voltaic cell similar to that shown in Figure 20.5 is constructed. One electrode compartment consists of a silver strip placed in a solution of $AgNO_3$, and the other has an iron strip placed in a solution of $FeCl_2$. The overall cell reaction is

$$Fe(s) + 2 Ag^+(aq) \longrightarrow Fe^{2+}(aq) + 2 Ag(s)$$

(a) What is being oxidized, and what is being reduced? **(b)** Write the half-reactions that occur in the two electrode compartments. **(c)** Which electrode is the anode, and which is the cathode? **(d)** Indicate the signs of the electrodes. **(e)** Do electrons flow from the silver electrode to the iron electrode, or from the iron to the silver? **(f)** In which directions do the cations and anions migrate through the solution?

20.26 A voltaic cell similar to that shown in Figure 20.5 is constructed. One electrode compartment consists of an aluminum strip placed in a solution of $Al(NO_3)_3$, and the other has a nickel strip placed in a solution of $NiSO_4$. The overall cell reaction is

$$2 Al(s) + 3 Ni^{2+}(aq) \longrightarrow 2 Al^{3+}(aq) + 3 Ni(s)$$

(a) What is being oxidized, and what is being reduced? **(b)** Write the half-reactions that occur in the two electrode compartments. **(c)** Which electrode is the anode, and which is the cathode? **(d)** Indicate the signs of the electrodes. **(e)** Do electrons flow from the aluminum electrode to the nickel electrode, or from the nickel to the aluminum? **(f)** In which directions do the cations and anions migrate through the solution? Assume the Al is not coated with its oxide.

Cell EMF under Standard Conditions

20.27 **(a)** What does the term *electromotive force* mean? **(b)** What is the definition of the *volt*? **(c)** What does the term *cell potential* mean?

20.28 **(a)** Which electrode of a voltaic cell, the cathode or the anode, corresponds to the higher potential energy for the electrons? **(b)** What are the units for electrical potential? How does this unit relate to energy expressed in joules? **(c)** What is special about a *standard* cell potential?

20.29 **(a)** Write the half-reaction that occurs at a hydrogen electrode in acidic aqueous solution when it serves as the cathode of a voltaic cell. **(b)** What is *standard* about the standard hydrogen electrode? **(c)** What is the role of the platinum foil in a standard hydrogen electrode?

20.30 **(a)** Write the half-reaction that occurs at a hydrogen electrode in acidic aqueous solution when it serves as the anode of a voltaic cell. **(b)** The platinum electrode in a standard hydrogen electrode is specially prepared to have a large surface area. Why is this important? **(c)** Sketch a standard hydrogen electrode.

20.31 **(a)** What is a *standard reduction potential*? **(b)** What is the standard reduction potential of a standard hydrogen electrode?

20.32 **(a)** Why is it impossible to measure the standard reduction potential of a single half-reaction? **(b)** Describe how the standard reduction potential of a half-reaction can be determined.

20.33 A voltaic cell that uses the reaction

$$Tl^{3+}(aq) + 2 Cr^{2+}(aq) \longrightarrow Tl^+(aq) + 2 Cr^{3+}(aq)$$

has a measured standard cell potential of +1.19 V.

(a) Write the two half-cell reactions. **(b)** By using data from Appendix E, determine E°_{red} for the reduction of $Tl^{3+}(aq)$ to $Tl^+(aq)$. **(c)** Sketch the voltaic cell, label the anode and cathode, and indicate the direction of electron flow.

20.34 A voltaic cell that uses the reaction

$$PdCl_4{}^{2-}(aq) + Cd(s) \longrightarrow Pd(s) + 4 Cl^-(aq) + Cd^{2+}(aq)$$

has a measured standard cell potential of +1.03 V. **(a)** Write the two half-cell reactions. **(b)** By using data from Appendix E, determine E°_{red} for the reaction involving Pd. **(c)** Sketch the voltaic cell, label the anode and cathode, and indicate the direction of electron flow.

20.35 Using standard reduction potentials (Appendix E), calculate the standard emf for each of the following reactions:
(a) $Cl_2(g) + 2 I^-(aq) \longrightarrow 2 Cl^-(aq) + I_2(s)$
(b) $Ni(s) + 2 Ce^{4+}(aq) \longrightarrow Ni^{2+}(aq) + 2 Ce^{3+}(aq)$
(c) $Fe(s) + 2 Fe^{3+}(aq) \longrightarrow 3 Fe^{2+}(aq)$
(d) $2 Al^{3+}(aq) + 3 Ca(s) \longrightarrow 2 Al(s) + 3 Ca^{2+}(aq)$

20.36 Using data in Appendix E, calculate the standard emf for each of the following reactions:
(a) $H_2(g) + F_2(g) \longrightarrow 2 H^+(aq) + 2 F^-(aq)$
(b) $Cu^{2+}(aq) + Ca(s) \longrightarrow Cu(s) + Ca^{2+}(aq)$
(c) $3 Fe^{2+}(aq) \longrightarrow Fe(s) + 2 Fe^{3+}(aq)$
(d) $Hg_2{}^{2+}(aq) + 2 Cu^+(aq) \longrightarrow 2 Hg(l) + 2 Cu^{2+}(aq)$

20.37 The standard reduction potentials of the following half-reactions are given in Appendix E:

$$Ag^+(aq) + e^- \longrightarrow Ag(s)$$
$$Cu^{2+}(aq) + 2 e^- \longrightarrow Cu(s)$$
$$Ni^{2+}(aq) + 2 e^- \longrightarrow Ni(s)$$
$$Cr^{3+}(aq) + 3 e^- \longrightarrow Cr(s)$$

(a) Determine which combination of these half-cell reactions leads to the cell reaction with the largest positive cell emf, and calculate the value. (b) Determine which combination of these half-cell reactions leads to the cell reaction with the smallest positive cell emf, and calculate the value.

20.38 Given the following half-reactions and associated standard reduction potentials:

$$AuBr_4^-(aq) + 3\,e^- \longrightarrow Au(s) + 4\,Br^-(aq)$$
$$E°_{red} = -0.858\ V$$

$$Eu^{3+}(aq) + e^- \longrightarrow Eu^{2+}(aq)$$
$$E°_{red} = -0.43\ V$$

$$IO^-(aq) + H_2O(l) + 2\,e^- \longrightarrow I^-(aq) + 2\,OH^-(aq)$$
$$E°_{red} = +0.49\ V$$

$$Sn^{2+}(aq) + 2\,e^- \longrightarrow Sn(s)$$
$$E°_{red} = -0.14\ V$$

(a) Write the cell reaction for the combination of these half-cell reactions that leads to the largest positive cell emf, and calculate the value. (b) Write the cell reaction for the combination of half-cell reactions that leads to the smallest positive cell emf, and calculate that value.

20.39 A 1 M solution of $Cu(NO_3)_2$ is placed in a beaker with a strip of Cu metal. A 1 M solution of $SnSO_4$ is placed in a second beaker with a strip of Sn metal. A salt bridge connects the two beakers, and wires to a voltmeter link the two metal electrodes. (a) Which electrode serves as the anode, and which as the cathode? (b) Which electrode gains mass and which loses mass as the cell reaction proceeds? (c) Write the equation for the overall cell reaction. (d) What is the emf generated by the cell under standard conditions?

20.40 A voltaic cell consists of a strip of cadmium metal in a solution of $Cd(NO_3)_2$ in one beaker, and in the other beaker a platinum electrode is immersed in a NaCl solution, with Cl_2 gas bubbled around the electrode. A salt bridge connects the two beakers. (a) Which electrode serves as the anode, and which as the cathode? (b) Does the Cd electrode gain or lose mass as the cell reaction proceeds? (c) Write the equation for the overall cell reaction. (d) What is the emf generated by the cell under standard conditions?

Strengths of Oxidizing and Reducing Agents

20.41 From each of the following pairs of substances, use data in Appendix E to choose the one that is the stronger reducing agent:
(a) Fe(s) or Mg(s)
(b) Ca(s) or Al(s)
(c) $H_2(g$, acidic solution) or $H_2S(g)$
(d) $H_2SO_3(aq)$ or $H_2C_2O_4(aq)$

20.42 From each of the following pairs of substances, use data in Appendix E to choose the one that is the stronger oxidizing agent:
(a) $Cl_2(g)$ or $Br_2(l)$
(b) $Zn^{2+}(aq)$ or $Cd^{2+}(aq)$
(c) $BrO_3^-(aq)$ or $IO_3^-(aq)$
(d) $H_2O_2(aq)$ or $O_3(g)$

20.43 By using the data in Appendix E, determine whether each of the following substances is likely to serve as an oxidant or a reductant: (a) $Cl_2(g)$, (b) $MnO_4^-(aq$, acidic solution), (c) Ba(s), (d) Zn(s).

20.44 Is each of the following substances likely to serve as an oxidant or a reductant: (a) $Ce^{3+}(aq)$, (b) Ca(s), (c) $ClO_3^-(aq)$, (d) $N_2O_5(g)$?

20.45 (a) Assuming standard conditions, arrange the following in order of increasing strength as oxidizing agents in acidic solution: $Cr_2O_7^{2-}$, H_2O_2, Cu^{2+}, Cl_2, O_2. (b) Arrange the following in order of increasing strength as reducing agents in acidic solution: Zn, I^-, Sn^{2+}, H_2O_2, Al.

20.46 Based on the data in Appendix E, (a) which of the following is the strongest oxidizing agent and which is the weakest in acidic solution: Ce^{4+}, Br_2, H_2O_2, Zn? (b) Which of the following is the strongest reducing agent, and which is the weakest in acidic solution: F^-, Zn, $N_2H_5^+$, I_2, NO?

20.47 The standard reduction potential for the reduction of $Eu^{3+}(aq)$ to $Eu^{2+}(aq)$ is -0.43 V. Using Appendix E, which of the following substances is capable of reducing $Eu^{3+}(aq)$ to $Eu^{2+}(aq)$ under standard conditions: Al, Co, H_2O_2, $N_2H_5^+$, $H_2C_2O_4$?

20.48 The standard reduction potential for the reduction of $RuO_4^-(aq)$ to $RuO_4^{2-}(aq)$ is $+0.59$ V. By using Appendix E, which of the following substances can oxidize $RuO_4^{2-}(aq)$ to $RuO_4^-(aq)$ under standard conditions: $Br_2(l)$, $BrO_3^-(aq)$, $Mn^{2+}(aq)$, $O_2(g)$, $Sn^{2+}(aq)$?

Free Energy and Redox Reactions

20.49 Given the following reduction half-reactions:

$$Fe^{3+}(aq) + e^- \longrightarrow Fe^{2+}(aq)$$
$$E°_{red} = +0.77\ V$$

$$S_2O_6^{2-}(aq) + 4\,H^+(aq) + 2\,e^- \longrightarrow 2\,H_2SO_3(aq)$$
$$E°_{red} = +0.60\ V$$

$$N_2O(g) + 2\,H^+(aq) + 2\,e^- \longrightarrow N_2(g) + H_2O(l)$$
$$E°_{red} = -1.77\ V$$

$$VO_2^+(aq) + 2\,H^+(aq) + e^- \longrightarrow VO^{2+}(aq) + H_2O(l)$$
$$E°_{red} = +1.00\ V$$

(a) Write balanced chemical equations for the oxidation of $Fe^{2+}(aq)$ by $S_2O_6^{2-}(aq)$, by $N_2O(aq)$, and by $VO_2^+(aq)$. (b) Calculate $\Delta G°$ for each reaction at 298 K. (c) Calculate the equilibrium constant K for each reaction at 298 K.

20.50 For each of the following reactions, write a balanced equation, calculate the standard emf, calculate $\Delta G°$ at 298 K, and calculate the equilibrium constant K at 298 K. (a) Aqueous iodide ion is oxidized to $I_2(s)$ by $Hg_2^{2+}(aq)$. (b) In acidic solution, copper(I) ion is oxidized to copper(II) ion by nitrate ion. (c) In basic solution, $Cr(OH)_3(s)$ is oxidized to $CrO_4^{2-}(aq)$ by $ClO^-(aq)$.

20.51 If the equilibrium constant for a two-electron redox reaction at 298 K is 1.5×10^{-4}, calculate the corresponding $\Delta G°$ and $E°_{cell}$.

20.52 If the equilibrium constant for a one-electron redox reaction at 298 K is 8.7×10^4, calculate the corresponding $\Delta G°$ and $E°_{cell}$.

20.53 Using the standard reduction potentials listed in Appendix E, calculate the equilibrium constant for each of the following reactions at 298 K:
(a) $Fe(s) + Ni^{2+}(aq) \longrightarrow Fe^{2+}(aq) + Ni(s)$
(b) $Co(s) + 2H^+(aq) \longrightarrow Co^{2+}(aq) + H_2(g)$
(c) $10 Br^-(aq) + 2 MnO_4^-(aq) + 16 H^+(aq) \longrightarrow$
$2 Mn^{2+}(aq) + 8 H_2O(l) + 5 Br_2(l)$

20.54 Using the standard reduction potentials listed in Appendix E, calculate the equilibrium constant for each of the following reactions at 298 K:
(a) $Cu(s) + 2 Ag^+(aq) \longrightarrow Cu^{2+}(aq) + 2 Ag(s)$
(b) $3 Ce^{4+}(aq) + Bi(s) + H_2O(l) \longrightarrow$
$3 Ce^{3+}(aq) + BiO^+(aq) + 2 H^+(aq)$
(c) $N_2H_5^+(aq) + 4 Fe(CN)_6^{3-}(aq) \longrightarrow$
$N_2(g) + 5 H^+(aq) + 4 Fe(CN)_6^{4-}(aq)$

20.55 A cell has a standard emf of +0.177 V at 298 K. What is the value of the equilibrium constant for the cell reaction (a) if $n = 1$? (b) if $n = 2$? (c) if $n = 3$?

20.56 At 298 K a cell reaction has a standard emf of +0.17 V. The equilibrium constant for the cell reaction is 5.5×10^5. What is the value of n for the cell reaction?

Cell EMF under Nonstandard Conditions

20.57 (a) Under what circumstances is the Nernst equation applicable? (b) What is the numerical value of the reaction quotient, Q, under standard conditions? (c) What happens to the emf of a cell if the concentrations of the reactants are increased?

20.58 (a) A voltaic cell is constructed with all reactants and products in their standard states. Will this condition hold as the cell operates? Explain. (b) Can the Nernst equation be used at temperatures other than room temperature? Explain. (c) What happens to the emf of a cell if the concentrations of the products are increased?

20.59 What is the effect on the emf of the cell shown in Figure 20.11, which has the overall reaction $Zn(s) + 2 H^+(aq) \longrightarrow Zn^{2+}(aq) + H_2(g)$, for each of the following changes? (a) The pressure of the H_2 gas is increased in the cathode compartment. (b) Zinc nitrate is added to the anode compartment. (c) Sodium hydroxide is added to the cathode compartment, decreasing $[H^+]$. (d) The surface area of the anode is doubled.

20.60 A voltaic cell utilizes the following reaction:
$$Al(s) + 3 Ag^+(aq) \longrightarrow Al^{3+}(aq) + 3 Ag(s)$$
What is the effect on the cell emf of each of the following changes? (a) Water is added to the anode compartment, diluting the solution. (b) The size of the aluminum electrode is increased. (c) A solution of $AgNO_3$ is added to the cathode compartment, increasing the quantity of Ag^+ but not changing its concentration. (d) HCl is added to the $AgNO_3$ solution, precipitating some of the Ag^+ as AgCl.

20.61 A voltaic cell is constructed that uses the following reaction and operates at 298 K:
$$Zn(s) + Ni^{2+}(aq) \longrightarrow Zn^{2+}(aq) + Ni(s)$$
(a) What is the emf of this cell under standard conditions? (b) What is the emf of this cell when $[Ni^{2+}] = 3.00 M$ and $[Zn^{2+}] = 0.100 M$? (c) What is the emf of the cell when $[Ni^{2+}] = 0.200 M$, and $[Zn^{2+}] = 0.900 M$?

20.62 A voltaic cell utilizes the following reaction and operates at 298 K:
$$3 Ce^{4+}(aq) + Cr(s) \longrightarrow 3 Ce^{3+}(aq) + Cr^{3+}(aq)$$

(a) What is the emf of this cell under standard conditions?
(b) What is the emf of this cell when $[Ce^{4+}] = 3.0 M$, $[Ce^{3+}] = 0.10 M$, and $[Cr^{3+}] = 0.010 M$? (c) What is the emf of the cell when $[Ce^{4+}] = 0.10 M$, $[Ce^{3+}] = 1.75 M$, and $[Cr^{3+}] = 2.5 M$?

20.63 A voltaic cell utilizes the following reaction:
$$4 Fe^{2+}(aq) + O_2(g) + 4 H^+(aq) \longrightarrow 4 Fe^{3+}(aq) + 2 H_2O(l)$$
(a) What is the emf of this cell under standard conditions?
(b) What is the emf of this cell when $[Fe^{2+}] = 1.3 M$, $[Fe^{3+}] = 0.010 M$, $P_{O_2} = 0.50$ atm, and the pH of the solution in the cathode compartment is 3.50?

20.64 A voltaic cell utilizes the following reaction:
$$2 Fe^{3+}(aq) + H_2(g) \longrightarrow 2 Fe^{2+}(aq) + 2 H^+(aq)$$
(a) What is the emf of this cell under standard conditions?
(b) What is the emf for this cell when $[Fe^{3+}] = 2.50 M$, $P_{H_2} = 0.85$ atm, $[Fe^{2+}] = 0.0010 M$, and the pH in both compartments is 5.00?

20.65 A voltaic cell is constructed with two Zn^{2+}-Zn electrodes. The two cell compartments have $[Zn^{2+}] = 1.8 M$ and $[Zn^{2+}] = 1.00 \times 10^{-2} M$, respectively. (a) Which electrode is the anode of the cell? (b) What is the standard emf of the cell? (c) What is the cell emf for the concentrations given? (d) For each electrode, predict whether $[Zn^{2+}]$ will increase, decrease, or stay the same as the cell operates.

20.66 A voltaic cell is constructed with two silver–silver chloride electrodes, each of which is based on the following half-reaction:
$$AgCl(s) + e^- \longrightarrow Ag(s) + Cl^-(aq)$$
The two cell compartments have $[Cl^-] = 0.0150 M$ and $[Cl^-] = 2.55 M$, respectively. (a) Which electrode is the cathode of the cell? (b) What is the standard emf of the cell? (c) What is the cell emf for the concentrations given? (d) For each electrode, predict whether $[Cl^-]$ will increase, decrease, or stay the same as the cell operates.

20.67 The cell in Figure 20.11 could be used to provide a measure of the pH in the cathode compartment. Calculate the pH of the cathode compartment solution if the cell emf at 298 K is measured to be +0.684 V when $[Zn^{2+}] = 0.30 M$ and $P_{H_2} = 0.90$ atm.

20.68 A voltaic cell is constructed that is based on the following reaction:

$$Sn^{2+}(aq) + Pb(s) \longrightarrow Sn(s) + Pb^{2+}(aq)$$

(a) If the concentration of Sn^{2+} in the cathode compartment is 1.00 M and the cell generates an emf of +0.22 V, what is the concentration of Pb^{2+} in the anode compartment? **(b)** If the anode compartment contains $[SO_4^{2-}] = 1.00\ M$ in equilibrium with $PbSO_4(s)$, what is the K_{sp} of $PbSO_4$?

Batteries and Fuel Cells

20.69 **(a)** What happens to the emf of a battery as it is used? Why does this happen? **(b)** The AA-size and D-size alkaline batteries are both 1.5-V batteries that are based on the same electrode reactions. What is the major difference between the two batteries? What performance feature is most affected by this difference?

20.70 **(a)** Suggest an explanation for why liquid water is needed in an alkaline battery. **(b)** What is the advantage of using highly concentrated or solid reactants in a voltaic cell?

20.71 During a period of discharge of a lead-acid battery, 402 g of Pb from the anode is converted into $PbSO_4(s)$. What mass of $PbO_2(s)$ is reduced at the cathode during this same period?

20.72 During the discharge of an alkaline battery, 4.50 g of Zn are consumed at the anode of the battery. What mass of MnO_2 is reduced at the cathode during this discharge?

20.73 Heart pacemakers are often powered by lithium–silver chromate "button" batteries. The overall cell reaction is:

$$2\ Li(s) + Ag_2CrO_4(s) \longrightarrow Li_2CrO_4(s) + 2\ Ag(s)$$

(a) Lithium metal is the reactant at one of the electrodes of the battery. Is it the anode or the cathode? **(b)** Choose the two half-reactions from Appendix E that *most closely approximate* the reactions that occur in the battery. What standard emf would be generated by a voltaic cell based on these half-reactions? **(c)** The battery generates an emf of +3.5 V. How close is this value to the one calculated in part (b)? **(d)** Calculate the emf that would be generated at body temperature, 37 °C. How does this compare to the emf you calculated in part (b)?

20.74 Mercuric oxide dry-cell batteries are often used where a high energy density is required, such as in watches and cameras. The two half-cell reactions that occur in the battery are

$$HgO(s) + H_2O(l) + 2\ e^- \longrightarrow Hg(l) + 2\ OH^-(aq)$$
$$Zn(s) + 2\ OH^-(aq) \longrightarrow ZnO(s) + H_2O(l) + 2\ e^-$$

(a) Write the overall cell reaction. **(b)** The value of E°_{red} for the cathode reaction is +0.098 V. The overall cell potential is +1.35 V. Assuming that both half-cells operate under standard conditions, what is the standard reduction potential for the anode reaction? **(c)** Why is the potential of the anode reaction different than would be expected if the reaction occurred in an acidic medium?

20.75 **(a)** Suppose that an alkaline battery was manufactured using cadmium metal rather than zinc. What effect would this have on the cell emf? **(b)** What environmental advantage is provided by the use of nickel-metal-hydride batteries over the nickel-cadmium batteries?

20.76 **(a)** The nonrechargeable lithium batteries used for photography use lithium metal as the anode. What advantages might be realized by using lithium rather than zinc, cadmium, lead, or nickel? **(b)** The rechargeable lithium-ion battery does not use lithium metal as an electrode material. Nevertheless, it still has a substantial advantage over nickel-based batteries. Suggest an explanation.

20.77 The hydrogen-oxygen fuel cell has a standard emf of 1.23 V. What advantages and disadvantages are there to using this device as a source of power, compared to a 1.55-V alkaline battery?

20.78 **(a)** What is the difference between a battery and a fuel cell? **(b)** Can the "fuel" of a fuel cell be a solid? Explain.

Corrosion

20.79 **(a)** Write the anode and cathode reactions that cause the corrosion of iron metal to aqueous iron(II). **(b)** Write the balanced half-reactions involved in the air oxidation of $Fe^{2+}(aq)$ to $Fe_2O_3 \cdot 3\ H_2O$.

20.80 **(a)** Based on standard reduction potentials, would you expect copper metal to oxidize under standard conditions in the presence of oxygen and hydrogen ions? **(b)** When the Statue of Liberty was refurbished, Teflon spacers were placed between the iron skeleton and the copper metal on the surface of the statue. What role do these spacers play?

20.81 **(a)** Magnesium metal is used as a sacrificial anode to protect underground pipes from corrosion. Why is the magnesium referred to as a "sacrificial anode"? **(b)** Looking in Appendix E, suggest what metal the underground pipes could be made from in order for magnesium to be successful as a sacrificial anode.

20.82 An iron object is plated with a coating of cobalt to protect against corrosion. Does the cobalt protect iron by cathodic protection? Explain.

20.83 A plumber's handbook states that you should not connect a brass pipe directly to a galvanized steel pipe because electrochemical reactions between the two metals will cause corrosion. The handbook recommends you use, instead, an insulating fitting to connect them. Brass is a mixture of copper and zinc. What spontaneous redox reaction(s) might cause the corrosion? Justify your answer with standard emf calculations.

20.84 A plumber's handbook states that you should not connect a copper pipe directly to a steel pipe because electrochemical reactions between the two metals will cause corrosion. The handbook recommends you use, instead, an insulating fitting to connect them. What spontaneous redox reaction(s) might cause the corrosion? Justify your answer with standard emf calculations.

Electrolysis; Electrical Work

20.85 **(a)** What is *electrolysis*? **(b)** Are electrolysis reactions thermodynamically spontaneous? Explain. **(c)** What process occurs at the anode in the electrolysis of molten NaCl?

20.86 **(a)** What is an *electrolytic cell*? **(b)** The negative terminal of a voltage source is connected to an electrode of an electrolytic cell. Is the electrode the anode or the cathode of the cell? Explain. **(c)** The electrolysis of water is often done with a small amount of sulfuric acid added to the water. What is the role of the sulfuric acid?

20.87 **(a)** A $Cr^{3+}(aq)$ solution is electrolyzed, using a current of 7.60 A. What mass of $Cr(s)$ is plated out after 2.00 days? **(b)** What amperage is required to plate out 0.250 mol Cr from a Cr^{3+} solution in a period of 8.00 h?

20.88 Metallic magnesium can be made by the electrolysis of molten $MgCl_2$. **(a)** What mass of Mg is formed by passing a current of 4.55 A through molten $MgCl_2$, for 3.50 days? **(b)** How many minutes are needed to plate out 10.00 g Mg from molten $MgCl_2$, using 3.50 A of current?

20.89 A voltaic cell is based on the reaction

$$Sn(s) + I_2(s) \longrightarrow Sn^{2+}(aq) + 2 I^-(aq)$$

Under standard conditions, what is the maximum electrical work, in joules, that the cell can accomplish if 75.0 g of Sn is consumed?

20.90 Consider the voltaic cell illustrated in Figure 20.5, which is based on the cell reaction

$$Zn(s) + Cu^{2+}(aq) \longrightarrow Zn^{2+}(aq) + Cu(s)$$

Under standard conditions, what is the maximum electrical work, in joules, that the cell can accomplish if 50.0 g of copper is plated out?

20.91 **(a)** Calculate the mass of Li formed by electrolysis of molten LiCl by a current of 7.5×10^4 A flowing for a period of 24 h. Assume the electrolytic cell is 85% efficient. **(b)** What is the energy requirement for this electrolysis per mole of Li formed if the applied emf is +7.5 V?

20.92 Elemental calcium is produced by the electrolysis of molten $CaCl_2$. **(a)** What mass of calcium can be produced by this process if a current of 7.5×10^3 A is applied for 48 h? Assume that the electrolytic cell is 68% efficient. **(b)** What is the total energy requirement for this electrolysis if the applied emf is +5.00 V?

ADDITIONAL EXERCISES

20.93 A *disproportionation* reaction is an oxidation-reduction reaction in which the same substance is oxidized and reduced. Complete and balance the following disproportionation reactions:
(a) $Ni^+(aq) \longrightarrow Ni^{2+}(aq) + Ni(s)$ (acidic solution)
(b) $MnO_4^{2-}(aq) \longrightarrow MnO_4^-(aq) + MnO_2(s)$
(acidic solution)
(c) $H_2SO_3(aq) \longrightarrow S(s) + HSO_4^-(aq)$
(acidic solution)
(d) $Cl_2(aq) \longrightarrow Cl^-(aq) + ClO^-(aq)$ (basic solution)

20.94 This oxidation-reduction reaction in acidic solution is spontaneous:

$$5 Fe^{2+}(aq) + MnO_4^-(aq) + 8 H^+(aq) \longrightarrow$$
$$5 Fe^{3+}(aq) + Mn^{2+}(aq) + 4 H_2O(l)$$

A solution containing $KMnO_4$ and H_2SO_4 is poured into one beaker, and a solution of $FeSO_4$ is poured into another. A salt bridge is used to join the beakers. A platinum foil is placed in each solution, and a wire that passes through a voltmeter connects the two solutions. **(a)** Sketch the cell, indicating the anode and the cathode, the direction of electron movement through the external circuit, and the direction of ion migrations through the solutions. **(b)** Sketch the process that occurs at the atomic level at the surface of the anode. **(c)** Calculate the emf of the cell under standard conditions. **(d)** Calculate the emf of the cell at 298 K when the concentrations are the following: pH = 0.0, $[Fe^{2+}] = 0.10\ M$, $[MnO_4^-] = 1.50\ M$, $[Fe^{3+}] = 2.5 \times 10^{-4}\ M$, $[Mn^{2+}] = 0.001\ M$.

20.95 A common shorthand way to represent a voltaic cell is to list its components as follows:

anode|anode solution||cathode solution|cathode

A double vertical line represents a salt bridge or a porous barrier. A single vertical line represents a change in phase, such as from solid to solution. **(a)** Write the half-reactions and overall cell reaction represented by $Fe|Fe^{2+}||Ag^+|Ag$; sketch the cell. **(b)** Write the half-reactions and overall cell reaction represented by $Zn|Zn^{2+}||H^+|H_2$; sketch the cell. **(c)** Using the notation just described, represent a cell based on the following reaction:

$$ClO_3^-(aq) + 3 Cu(s) + 6 H^+(aq) \longrightarrow$$
$$Cl^-(aq) + 3 Cu^{2+}(aq) + 3 H_2O(l)$$

Pt is used as an inert electrode in contact with the ClO_3^- and Cl^-. Sketch the cell.

20.96 Predict whether the following reactions will be spontaneous in acidic solution under standard conditions: **(a)** oxidation of Sn to Sn^{2+} by I_2 (to form I^-), **(b)** reduction of Ni^{2+} to Ni by I^- (to form I_2), **(c)** reduction of Ce^{4+} to Ce^{3+} by H_2O_2, **(d)** reduction of Cu^{2+} to Cu by Sn^{2+} (to form Sn^{4+}).

[20.97] Gold exists in two common positive oxidation states, +1 and +3. The standard reduction potentials for these oxidation states are

$$Au^+(aq) + e^- \longrightarrow Au(s) \quad E^\circ_{red} = +1.69\ V$$
$$Au^{3+}(aq) + 3 e^- \longrightarrow Au(s) \quad E^\circ_{red} = +1.50\ V$$

(a) Can you use these data to explain why gold does not tarnish in the air? **(b)** Suggest several substances that should be strong enough oxidizing agents to oxidize gold metal. **(c)** Miners obtain gold by soaking gold-containing ores in an aqueous solution of sodium cyanide. A very soluble complex ion of gold forms in the aqueous solution because of the redox reaction

$$4 Au(s) + 8 NaCN(aq) + 2 H_2O(l) + O_2(g) \longrightarrow$$
$$4 Na[Au(CN)_2](aq) + 4 NaOH(aq)$$

What is being oxidized, and what is being reduced, in this reaction? **(d)** Gold miners then react the basic aqueous product solution from part (c) with Zn dust to get gold metal. Write a balanced redox reaction for this process. What is being oxidized, and what is being reduced?

20.98 Two important characteristics of voltaic cells are their cell potential and the total charge that they can deliver. Which of these characteristics depends on the amount of reactants in the cell, and which one depends on their concentration?

20.99 A voltaic cell is constructed from an $Ni^{2+}(aq)$-$Ni(s)$ half-cell and an $Ag^+(aq)$-$Ag(s)$ half-cell. The initial concentration of $Ni^{2+}(aq)$ in the Ni^{2+}-Ni half-cell is $[Ni^{2+}]$ = 0.0100 M. The initial cell voltage is +1.12 V. **(a)** By using data in Table 20.1, calculate the standard emf of this voltaic cell. **(b)** Will the concentration of $Ni^{2+}(aq)$ increase or decrease as the cell operates? **(c)** What is the initial concentration of $Ag^+(aq)$ in the Ag^+-Ag half-cell?

[20.100] A voltaic cell is constructed that uses the following half-cell reactions:

$$Cu^+(aq) + e^- \longrightarrow Cu(s)$$
$$I_2(s) + 2\,e^- \longrightarrow 2\,I^-(aq)$$

The cell is operated at 298 K with $[Cu^+]$ = 0.25 M and $[I^-]$ = 3.5 M. **(a)** Determine E for the cell at these concentrations. **(b)** Which electrode is the anode of the cell? **(c)** Is the answer to part (b) the same as it would be if the cell were operated under standard conditions? **(d)** If $[Cu^+]$ was equal to 0.15 M, at what concentration of I^- would the cell have zero potential?

20.101 Derive an equation that directly relates the standard emf of a redox reaction to its equilibrium constant.

20.102 Using data from Appendix E, calculate the equilibrium constant for the disproportionation of the copper(I) ion at room temperature: $2\,Cu^+(aq) \longrightarrow Cu^{2+}(aq) + Cu(s)$

20.103 **(a)** Write the reactions for the discharge and charge of a nickel-cadmium rechargeable battery. **(b)** Given the following reduction potentials, calculate the standard emf of the cell:

$$Cd(OH)_2(s) + 2\,e^- \longrightarrow Cd(s) + 2\,OH^-(aq)$$
$$E^\circ_{red} = -0.76 \text{ V}$$
$$NiO(OH)(s) + H_2O(l) + e^- \longrightarrow Ni(OH)_2(s) + OH^-(aq)$$
$$E^\circ_{red} = +0.49 \text{ V}$$

(c) A typical nicad voltaic cell generates an emf of +1.30 V. Why is there a difference between this value and the one you calculated in part (b)? **(d)** Calculate the equilibrium constant for the overall nicad reaction based on this typical emf value.

20.104 The capacity of batteries such as the typical AA alkaline battery is expressed in units of milliamp-hours (mAh).

An "AA" alkaline battery yields a nominal capacity of 2850 mAh. **(a)** What quantity of interest to the consumer is being expressed by the units of mAh? **(b)** The starting voltage of a fresh alkaline battery is 1.55 V. The voltage decreases during discharge and is 0.80 V when the battery has delivered its rated capacity. If we assume that the voltage declines linearly as current is withdrawn, estimate the total maximum electrical work the battery could perform during discharge.

20.105 If you were going to apply a small potential to a steel ship resting in the water as a means of inhibiting corrosion, would you apply a negative or a positive charge? Explain.

20.106 The following quotation is taken from an article dealing with corrosion of electronic materials: "Sulfur dioxide, its acidic oxidation products, and moisture are well established as the principal causes of outdoor corrosion of many metals." Using Ni as an example, explain why the factors cited affect the rate of corrosion. Write chemical equations to illustrate your points. (*Note:* $NiO(s)$ is soluble in acidic solution.)

[20.107] **(a)** How many coulombs are required to plate a layer of chromium metal 0.25 mm thick on an auto bumper with a total area of 0.32 m^2 from a solution containing CrO_4^{2-}? The density of chromium metal is 7.20 g/cm^3. **(b)** What current flow is required for this electroplating if the bumper is to be plated in 10.0 s? **(c)** If the external source has an emf of +6.0 V and the electrolytic cell is 65% efficient, how much electrical power is expended to electroplate the bumper?

20.108 **(a)** What is the maximum amount of work that a 6-V lead-acid battery of a golf cart can accomplish if it is rated at 300 A-h? **(b)** List some of the reasons why this amount of work is never realized.

[20.109] Some years ago a unique proposal was made to raise the *Titanic*. The plan involved placing pontoons within the ship using a surface-controlled submarine-type vessel. The pontoons would contain cathodes and would be filled with hydrogen gas formed by the electrolysis of water. It has been estimated that it would require about 7×10^8 mol of H_2 to provide the buoyancy to lift the ship (*J. Chem. Educ.*, Vol. 50, 1973, 61). **(a)** How many coulombs of electrical charge would be required? **(b)** What is the minimum voltage required to generate H_2 and O_2 if the pressure on the gases at the depth of the wreckage (2 mi) is 300 atm? **(c)** What is the minimum electrical energy required to raise the *Titanic* by electrolysis? **(d)** What is the minimum cost of the electrical energy required to generate the necessary H_2 if the electricity costs 85 cents per kilowatt-hour to generate at the site?

INTEGRATIVE EXERCISES

20.110 Two wires from a battery are tested with a piece of filter paper moistened with NaCl solution containing phenolphthalein, an acid–base indicator that is colorless in acid and pink in base. When the wires touch the paper about an inch apart, the rightmost wire produces a pink coloration on the filter paper and the leftmost produces none. Which wire is connected to the positive terminal of the battery? Explain.

20.111 The Haber process is the principal industrial route for converting nitrogen into ammonia:

$$N_2(g) + 3\,H_2(g) \longrightarrow 2\,NH_3(g)$$

(a) What is being oxidized, and what is being reduced? **(b)** Using the thermodynamic data in Appendix C, calculate the equilibrium constant for the process at room temperature. **(c)** Calculate the standard emf of the Haber process at room temperature.

[20.112] In a galvanic cell the cathode is an Ag^+ (1.00 M)/Ag(s) half-cell. The anode is a standard hydrogen electrode immersed in a buffer solution containing 0.10 M benzoic acid (C_6H_5COOH) and 0.050 M sodium benzoate ($C_6H_5COO^-Na^+$). The measured cell voltage is 1.030 V. What is the pK_a of benzoic acid?

20.113 Consider the general oxidation of a species A in solution: $A \longrightarrow A^+ + e^-$. The term "oxidation potential" is sometimes used to describe the ease with which species A is oxidized—the easier a species is to oxidize, the greater its oxidation potential. (a) What is the relationship between the standard oxidation potential of A and the standard reduction potential of A^+? (b) Which of the metals listed in Table 4.5 has the highest standard oxidation potential? Which has the lowest? (c) For a series of substances, the trend in oxidation potential is often related to the trend in the first ionization energy. Explain why this relationship makes sense.

[20.114] Gold metal dissolves in aqua regia, a mixture of concentrated hydrochloric acid and concentrated nitric acid. The standard reduction potentials

$$Au^{3+}(aq) + 3\,e^- \longrightarrow Au(s) \qquad E^\circ_{red} = +1.498\ V$$
$$AuCl_4^-(aq) + 3\,e^- \longrightarrow Au(s) + 4\,Cl^-(aq)$$
$$E^\circ_{red} = +1.002\ V$$

are important in gold chemistry. (a) Use half-reactions to write a balanced equation for the reaction of Au and nitric acid to produce Au^{3+} and NO(g), and calculate the standard emf of this reaction. Is this reaction spontaneous? (b) Use half-reactions to write a balanced equation for the reaction of Au and hydrochloric acid to produce $AuCl_4^-(aq)$ and $H_2(g)$, and calculate the standard emf of this reaction. Is this reaction spontaneous? (c) Use half-reactions to write a balanced equation for the reaction of Au and aqua regia to produce $AuCl_4^-(aq)$ and NO(g), and calculate the standard emf of this reaction. Is this reaction spontaneous under standard conditions? (d) Use the Nernst equation to explain why aqua regia made from *concentrated* hydrochloric and nitric acids is able to dissolve gold.

20.115 A voltaic cell is based on $Ag^+(aq)/Ag(s)$ and $Fe^{3+}(aq)/Fe^{2+}(aq)$ half-cells. (a) What is the standard emf of the cell? (b) Which reaction occurs at the cathode, and which at the anode of the cell? (c) Use S° values in Appendix C and the relationship between cell potential and free-energy change to predict whether the standard cell potential increases or decreases when the temperature is raised above 25 °C.

20.116 Hydrogen gas has the potential as a clean fuel in reaction with oxygen. The relevant reaction is

$$2\,H_2(g) + O_2(g) \longrightarrow 2\,H_2O(l)$$

Consider two possible ways of utilizing this reaction as an electrical energy source: (i) Hydrogen and oxygen gases are combusted and used to drive a generator, much as coal is currently used in the electric power industry; (ii) hydrogen and oxygen gases are used to generate electricity directly by using fuel cells that operate at 85 °C.

(a) Use data in Appendix C to calculate ΔH° and ΔS° for the above reaction. We will assume that these values do not change appreciably with temperature. (b) Based on the values from part (a), what trend would you expect for the magnitude of ΔG for the above reaction as the temperature increases? (c) What is the significance of the change in the magnitude of ΔG with temperature with respect to the utility of hydrogen as a fuel (recall Equation 19.15)? (d) Based on the analysis here, would it be more efficient to use the combustion method or the fuel-cell method to generate electrical energy from hydrogen?

20.117 Cytochrome, a complicated molecule that we will represent as $CyFe^{2+}$, reacts with the air we breathe to supply energy required to synthesize adenosine triphosphate (ATP). The body uses ATP as an energy source to drive other reactions. (Section 19.7) At pH 7.0 the following reduction potentials pertain to this oxidation of $CyFe^{2+}$:

$$O_2(g) + 4\,H^+(aq) + 4\,e^- \longrightarrow 2\,H_2O(l) \qquad E^\circ_{red} = +0.82\ V$$
$$CyFe^{3+}(aq) + e^- \longrightarrow CyFe^{2+}(aq) \qquad E^\circ_{red} = +0.22\ V$$

(a) What is ΔG for the oxidation of $CyFe^{2+}$ by air? (b) If the synthesis of 1.00 mol of ATP from adenosine diphosphate (ADP) requires a ΔG of 37.7 kJ, how many moles of ATP are synthesized per mole of O_2?

[20.118] The standard potential for the reduction of AgSCN(s) is +0.0895 V.

$$AgSCN(s) + e^- \longrightarrow Ag(s) + SCN^-(aq)$$

Using this value and the electrode potential for $Ag^+(aq)$, calculate the K_{sp} for AgSCN.

[20.119] The K_{sp} value for PbS(s) is 8.0×10^{-28}. By using this value together with an electrode potential from Appendix E, determine the value of the standard reduction potential for the reaction

$$PbS(s) + 2\,e^- \longrightarrow Pb(s) + S^{2-}(aq)$$

[20.120] A pH meter (Figure 16.6) employs a voltaic cell for which the cell potential is very sensitive to pH. A simple (but impractical) pH meter can be constructed by using two hydrogen electrodes: one standard hydrogen electrode (Figure 20.10) and a hydrogen electrode (with 1 atm pressure of H_2 gas) dipped into the solution of unknown pH. The two half-cells are connected by a salt bridge or porous glass disk. (a) Sketch the cell described above. (b) Write the half-cell reactions for the cell, and calculate the standard emf. (c) What is the pH of the solution in the half-cell that has the standard hydrogen electrode? (d) What is the cell emf when the pH of the unknown solution is 5.0? (e) How precise would a voltmeter have to be in order to detect a change in the pH of 0.01 pH units?

[20.121] A student designs an ammeter (a device that measures electrical current) that is based on the electrolysis of water into hydrogen and oxygen gases. When electrical current of unknown magnitude is run through the device for 2.00 min, 12.3 mL of water-saturated $H_2(g)$ is collected. The temperature of the system is 25.5 °C, and the atmospheric pressure is 768 torr. What is the magnitude of the current in amperes?

NUCLEAR CHEMISTRY

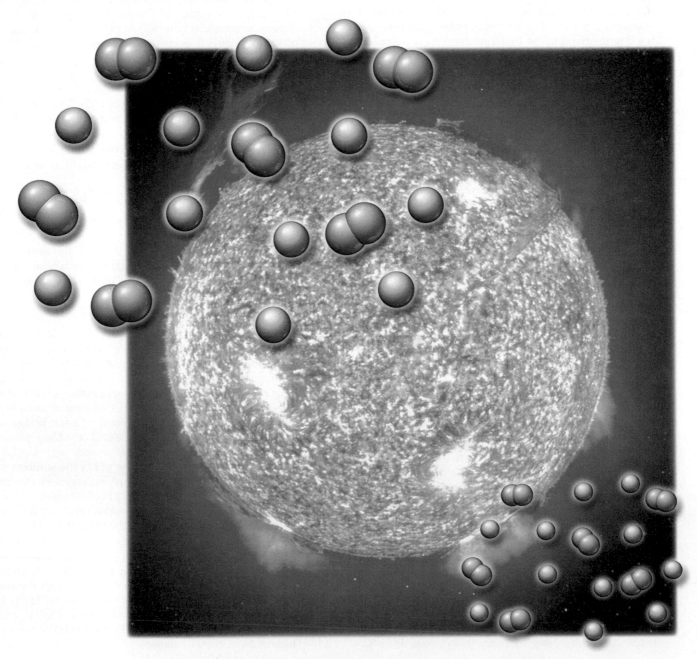

THE SUN'S ENERGY COMES FROM NUCLEAR reactions of hydrogen atoms in its core.

21.1 Radioactivity
In this chapter we will learn how to describe nuclear reactions by equations analogous to chemical equations, in which the nuclear charges and masses of reactants and products are in balance. Radioactive nuclei most commonly decay by emission of *alpha*, *beta*, or *gamma radiations*.

21.2 Patterns of Nuclear Stability
We recognize that nuclear stability is determined largely by the *neutron-to-proton ratio*. For stable nuclei, this ratio increases with increasing atomic number. All nuclei with 84 or more protons are radioactive. Heavy nuclei gain stability by a series of nuclear disintegrations leading to stable nuclei.

21.3 Nuclear Transmutations
We study *nuclear transmutations*, which are nuclear reactions induced by bombardment of a nucleus by a neutron or an accelerated charged particle, such as an alpha particle.

21.4 Rates of Radioactive Decay
We also learn that radioisotope decays are first-order kinetic processes that exhibit characteristic half-lives. The decays of radioisotopes can be used to determine the ages of ancient artifacts and geological formations.

21.5 Detection of Radioactivity
We see that the radiation emitted by a radioactive substance is detected with Geiger counters and scintillation counters.

21.6 Energy Changes in Nuclear Reactions
We recognize that energy changes in nuclear reactions are related to mass changes via Einstein's famous equation, $E = mc^2$. The *binding energies of nuclei* are reflected in the difference between the masses of nuclei and the sum of the masses of the nucleons of which they are composed.

21.7 Nuclear Power: Fission
We learn that in a *nuclear fission* reaction a heavy nucleus splits under nuclear bombardment to form two or more product nuclei, with release of energy. This type of nuclear reaction is the energy source for nuclear power plants.

21.8 Nuclear Power: Fusion
We observe that *nuclear fusion* results from the fusion of two light nuclei to form a more stable, heavier one.

21.9 Radiation in the Environment and Living Systems
We discover that naturally occuring radioisotopes bathe our planet, and us, with low levels of radiation. The radiation emitted in nuclear reactions has the potential to cause cell damage in animals that can, in turn, lead to cancer and other illnesses. Appropriate use of radiation, on the other hand, can be used medically to diagnose and treat cancer.

THE SUN IS THE MAJOR ENERGY SOURCE FOR OUR PLANET AND IS ESSENTIAL TO LIFE ON EARTH. Sunlight provides approximately 1000 watts (1 watt = 1 J/s) of power per square meter at Earth's surface. Sunlight is used by plants for photosynthesis, the process that produces food for plants and oxygen that most life on Earth needs to survive. What makes the Sun shine? Hydrogen fuses in the Sun's core to make helium, releasing tremendous amounts of energy in the process. This hydrogen fusion reaction is an example of a *nuclear reaction.*

A nuclear reaction involves changes in the nucleus of an atom. *Nuclear chemistry* is the study of nuclear reactions, with an emphasis on their uses in chemistry and their effects on biological systems. Nuclear chemistry affects our lives in many ways, particularly in energy and medical applications. For example, radioactive elements can be used as therapeutic and diagnostic tools.

▲ **Figure 21.1 Radioisotope scanning.** A gamma ray detector scan of a normal human heart, obtained following intravenous injection of the radioisotope thallium-201, a gamma emitter. The donut-shaped pink and red area represents uptake of the radioisotope by healthy heart muscles.

Radiation therapy is a common form of cancer therapy in which gamma rays, emitted from a radioactive substance such as cobalt-60, are directed to cancerous tumors to destroy them. In the widely used thallium "stress test," a compound containing thallium-201 (another radioactive substance that emits gamma rays) is injected intravenously into the patient. A scanner detects the gamma emissions from the thallium that has concentrated in the heart muscle and reveals the condition of the heart and surrounding arteries (Figure 21.1 ◄). Radioactivity is also used to help determine the mechanisms of chemical reactions, to trace the movement of atoms in biological systems and the environment, and to date important historical artifacts.

Nuclear reactions are also used to generate electricity. Roughly 10% of the total electricity generated in the world comes from nuclear power plants. The percentage of electricity generated from nuclear power varies widely among countries: 80% in France, 50% in Sweden, 20% in the United States, less than 5% each in India and China, and essentially zero in Italy (Figure 21.2▼). The use of nuclear energy for power generation and the disposal of nuclear wastes from power plants are controversial social and political issues. It is imperative, therefore, that as a citizen with a stake in these matters, you have some understanding of nuclear reactions and the uses of radioactive substances.

21.1 RADIOACTIVITY

To understand nuclear reactions, we must review and develop some ideas introduced in Section 2.3. First, recall that two types of subatomic particle reside in the nucleus: the *proton* and the *neutron*. We will refer to these particles as **nucleons**.

▶ **Figure 21.2 Nuclear energy usage for electricity, 2004.** This bar graph shows, for selected countries, the percentage of electricity supplied by nuclear power.

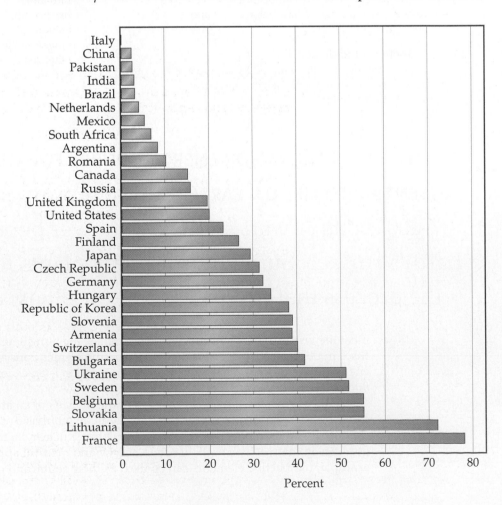

Recall also that all atoms of a given element have the same number of protons; this number is the element's *atomic number*. The atoms of a given element can have different numbers of neutrons, however, so they can have different *mass numbers*; the mass number is the total number of nucleons in the nucleus. Atoms with the same atomic number but different mass numbers are known as *isotopes*.

The different isotopes of an element are distinguished by their mass numbers. For example, the three naturally occurring isotopes of uranium are uranium-234, uranium-235, and uranium-238, where the numerical suffixes represent the mass numbers. These isotopes are also labeled, using chemical symbols, as $^{234}_{92}U$, $^{235}_{92}U$, and $^{238}_{92}U$. The superscript is the mass number; the subscript is the atomic number.

Different isotopes have different natural abundances. For example, 99.3% of naturally occurring uranium is uranium-238, 0.7% is uranium-235, and only a trace is uranium-234. Different nuclei also have different stabilities. Indeed, the nuclear properties of an atom depend on the number of protons and neutrons in its nucleus. A *nuclide* is a nucleus with a specified number of protons and neutrons. Nuclei that are radioactive are called **radionuclides**, and atoms containing these nuclei are called **radioisotopes**.

Nuclear Equations

Most nuclei found in nature are stable and remain intact indefinitely. Radionuclides, however, are unstable and spontaneously emit particles and electromagnetic radiation. Emission of radiation is one of the ways in which an unstable nucleus is transformed into a more stable one with less energy. The emitted radiation is the carrier of the excess energy. Uranium-238, for example, is radioactive, undergoing a nuclear reaction in which helium-4 nuclei are spontaneously emitted. The helium-4 particles are known as **alpha (α) particles**, and a stream of these particles is called *alpha radiation*. When a uranium-238 nucleus loses an alpha particle, the remaining fragment has an atomic number of 90 and a mass number of 234. If you look at the periodic table, you will find that the element with atomic number 90 is Th, thorium. Therefore, the products of uranium-238 decomposition are an alpha particle and a thorium-234 nucleus. We represent this reaction by the following *nuclear equation*:

$$^{238}_{92}U \longrightarrow \ ^{234}_{90}Th + \ ^{4}_{2}He \qquad\qquad [21.1]$$

When a nucleus spontaneously decomposes in this way, it is said to have decayed, or to have undergone *radioactive decay*. Because an alpha particle is involved in this reaction, scientists also describe the process as alpha decay.

GIVE IT SOME THOUGHT

What change in the mass number of a nucleus occurs when the nucleus emits an alpha particle?

In Equation 21.1 the sum of the mass numbers is the same on both sides of the equation (238 = 234 + 4). Likewise, the sum of the atomic numbers on both sides of the equation is equal (92 = 90 + 2). Mass numbers and atomic numbers must be balanced in all nuclear equations.

The radioactive properties of the nucleus are independent of the chemical state of the atom. In writing nuclear equations, therefore, we are not concerned with the chemical form (element or compound) of the atom in which the nucleus resides.

■■■ **SAMPLE EXERCISE 21.1** | Predicting the Product of a Nuclear Reaction

What product is formed when radium-226 undergoes alpha decay?

SOLUTION

Analyze: We are asked to determine the nucleus that results when radium-226 loses an alpha particle.

Plan: We can best do this by writing a balanced nuclear reaction for the process.

Solve: The periodic table shows that radium has an atomic number of 88. The complete chemical symbol for radium-226 is therefore $^{226}_{88}Ra$. An alpha particle is a helium-4 nucleus, and so its symbol is $^{4}_{2}He$ (sometimes written as $^{4}_{2}\alpha$). The alpha particle is a product of the nuclear reaction, and so the equation is of the form

$$^{226}_{88}Ra \longrightarrow ^{A}_{Z}X + ^{4}_{2}He$$

where A is the mass number of the product nucleus and Z is its atomic number. Mass numbers and atomic numbers must balance, so

$$226 = A + 4$$

and

$$88 = Z + 2$$

Hence,

$$A = 222 \quad \text{and} \quad Z = 86$$

Again, from the periodic table, the element with $Z = 86$ is radon (Rn). The product, therefore, is $^{222}_{86}Rn$, and the nuclear equation is

$$^{226}_{88}Ra \longrightarrow ^{222}_{86}Rn + ^{4}_{2}He$$

■■■ **PRACTICE EXERCISE**

Which element undergoes alpha decay to form lead-208?

Answer: $^{212}_{84}Po$

Types of Radioactive Decay

The three most common kinds of radioactive decay are alpha (α), beta (β), and gamma (γ) radiation. ∞∞ (Section 2.2) Table 21.1 ▼ summarizes some of the important properties of these kinds of radiation. As we have just discussed, alpha radiation consists of a stream of helium-4 nuclei known as alpha particles, which we denote as $^{4}_{2}He$ or $^{4}_{2}\alpha$.

Beta radiation consists of streams of **beta (β) particles**, which are high-speed electrons emitted by an unstable nucleus. Beta particles are represented in nuclear equations by the symbol $^{0}_{-1}e$ or sometimes $^{0}_{-1}\beta$. The superscript zero indicates that the mass of the electron is exceedingly small compared to the mass of a nucleon. The subscript -1 represents the negative charge of the particle, which is opposite that of the proton. Iodine-131 is an isotope that undergoes decay by beta emission:

$$^{131}_{53}I \longrightarrow ^{131}_{54}Xe + ^{0}_{-1}e \qquad [21.2]$$

TABLE 21.1 ■ Properties of Alpha, Beta, and Gamma Radiation

Property	Type of Radiation		
	α	β	γ
Charge	2+	1−	0
Mass	6.64×10^{-24} g	9.11×10^{-28} g	0
Relative penetrating power	1	100	10,000
Nature of radiation	$^{4}_{2}He$ nuclei	Electrons	High-energy photons

You can see from Equation 21.2 that beta decay causes the atomic number to increase from 53 to 54. Therefore, beta emission is equivalent to the conversion of a neutron ($_0^1\text{n}$) to a proton ($_1^1\text{p}$ or $_1^1\text{H}$), thereby increasing the atomic number by 1:

$$_0^1\text{n} \longrightarrow {}_1^1\text{p} + {}_{-1}^0\text{e} \qquad [21.3]$$

Just because an electron is ejected from the nucleus, however, we should not think that the nucleus is composed of these particles, any more than we consider a match to be composed of sparks simply because it gives them off when struck. The electron comes into being only when the nucleus undergoes a nuclear reaction.

Gamma radiation (or gamma rays) consists of high-energy photons (that is, electromagnetic radiation of very short wavelength). It changes neither the atomic number nor the mass number of a nucleus and is represented as $_0^0\gamma$, or merely γ. Gamma radiation usually accompanies other radioactive emission because it represents the energy lost when the remaining nucleons reorganize into more stable arrangements. Generally, the gamma rays are not shown when writing nuclear equations.

Two other types of radioactive decay are positron emission and electron capture. A **positron** is a particle that has the same mass as an electron, but an opposite charge.* The positron is represented as $_1^0\text{e}$. The isotope carbon-11 decays by positron emission:

$$_6^{11}\text{C} \longrightarrow {}_5^{11}\text{B} + {}_1^0\text{e} \qquad [21.4]$$

Positron emission causes the atomic number to decrease from 6 to 5. The emission of a positron has the effect of converting a proton to a neutron, thereby decreasing the atomic number of the nucleus by 1:

$$_1^1\text{p} \longrightarrow {}_0^1\text{n} + {}_1^0\text{e} \qquad [21.5]$$

Electron capture is the capture by the nucleus of an electron from the electron cloud surrounding the nucleus. Rubidium-81 undergoes decay in this fashion, as shown in Equation 21.6:

$$_{37}^{81}\text{Rb} + {}_{-1}^0\text{e} \text{ (orbital electron)} \longrightarrow {}_{36}^{81}\text{Kr} \qquad [21.6]$$

Because the electron is consumed rather than formed in the process, it is shown on the reactant side of the equation. Electron capture, like positron emission, has the effect of converting a proton to a neutron:

$$_1^1\text{p} + {}_{-1}^0\text{e} \longrightarrow {}_0^1\text{n} \qquad [21.7]$$

Table 21.2 ▶ summarizes the symbols used to represent the various elementary particles commonly encountered in nuclear reactions.

TABLE 21.2 ■ Common Particles in Radioactive Decay and Nuclear Transformations	
Particle	**Symbol**
Neutron	$_0^1\text{n}$
Proton	$_1^1\text{H}$ or $_1^1\text{p}$
Electron	$_{-1}^0\text{e}$
Alpha particle	$_2^4\text{He}$ or $_2^4\alpha$
Beta particle	$_{-1}^0\text{e}$ or $_{-1}^0\beta$
Positron	$_1^0\text{e}$

GIVE IT SOME THOUGHT

Which of the particles listed in Table 21.2 result in no change in nuclear charge when emitted in nuclear decay?

■ SAMPLE EXERCISE 21.2 | Writing Nuclear Equations

Write nuclear equations for the following processes: **(a)** mercury-201 undergoes electron capture; **(b)** thorium-231 decays to form protactinium-231.

SOLUTION

Analyze: We must write balanced nuclear equations in which the masses and charges of reactants and products are equal.

Plan: We can begin by writing the complete chemical symbols for the nuclei and decay particles that are given in the problem.

*The positron has a very short life because it is annihilated when it collides with an electron, producing gamma rays: $_1^0\text{e} + {}_{-1}^0\text{e} \longrightarrow 2\,_0^0\gamma$.

Solve:

(a) The information given in the question can be summarized as

$$^{201}_{80}Hg + ^{0}_{-1}e \longrightarrow ^{A}_{Z}X$$

The mass numbers must have the same sum on both sides of the equation:

$$201 + 0 = A$$

Thus, the product nucleus must have a mass number of 201. Similarly, balancing the atomic numbers gives

$$80 - 1 = Z$$

Thus, the atomic number of the product nucleus must be 79, which identifies it as gold (Au):

$$^{201}_{80}Hg + ^{0}_{-1}e \longrightarrow ^{201}_{79}Au$$

(b) In this case we must determine what type of particle is emitted in the course of the radioactive decay:

$$^{231}_{90}Th \longrightarrow ^{231}_{91}Pa + ^{A}_{Z}X$$

From $231 = 231 + A$ and $90 = 91 + Z$, we deduce $A = 0$ and $Z = -1$. According to Table 21.2, the particle with these characteristics is the beta particle (electron). We therefore write

$$^{231}_{90}Th \longrightarrow ^{231}_{91}Pa + ^{0}_{-1}e$$

■ **PRACTICE EXERCISE**

Write a balanced nuclear equation for the reaction in which oxygen-15 undergoes positron emission.

Answer: $^{15}_{8}O \longrightarrow ^{15}_{7}N + ^{0}_{1}e$

21.2 PATTERNS OF NUCLEAR STABILITY

The stability of a particular nucleus depends on a variety of factors. No single rule allows us to predict whether a particular nucleus is radioactive and how it might decay. However, several empirical observations will help you predict the stability of a nucleus.

Neutron-to-Proton Ratio

Because like charges repel each other, it may seem surprising that a large number of protons can reside within the small volume of the nucleus. At close distances, however, a strong force of attraction, called the *nuclear force*, exists between nucleons. Neutrons are intimately involved in this attractive force. All nuclei other than $^{1}_{1}H$ contain neutrons. As the number of protons in the nucleus increases, there is an ever-greater need for neutrons to counteract the effect of proton–proton repulsions. Stable nuclei with low atomic numbers (up to about 20) have approximately equal numbers of neutrons and protons. For nuclei with higher atomic numbers, the number of neutrons exceeds the number of protons. Indeed, the number of neutrons necessary to create a stable nucleus increases more rapidly than the number of protons, as shown in Figure 21.3▶. Thus, the neutron-to-proton ratios of stable nuclei increase with increasing atomic number.

The colored band in Figure 21.3 is the area within which all stable nuclei are found. This area is known as the *belt of stability*. The belt of stability ends at element 83 (bismuth). *All nuclei with 84 or more protons (atomic number ≥ 84) are radioactive.* For example, all isotopes of uranium, atomic number 92, are radioactive.

▲ **GIVE IT SOME THOUGHT**

Using Figure 21.3, estimate the optimal number of neutrons for a nucleus containing 70 protons.

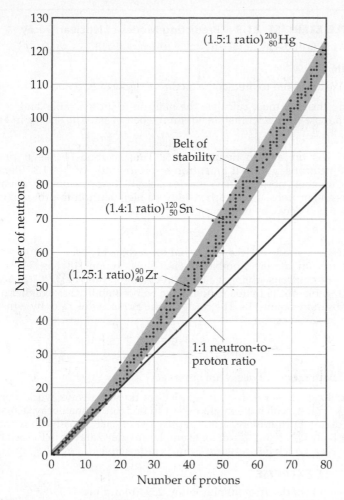

▲ **Figure 21.3 The belt of stability.** The number of neutrons is graphed versus the number of protons for stable nuclei. As the atomic number increases, the neutron-to-proton ratio of the stable nuclei increases. The stable nuclei are located in the shaded area of the graph known as the belt of stability. The majority of radioactive nuclei occur outside this belt.

The type of radioactive decay that a particular radionuclide undergoes depends largely on how its neutron-to-proton ratio compares to those of nearby nuclei within the belt of stability. We can envision three general situations:

1. *Nuclei above the belt of stability (high neutron-to-proton ratios).* These neutron-rich nuclei can lower their ratio and move toward the belt of stability by emitting a beta particle. Beta emission decreases the number of neutrons and increases the number of protons in a nucleus, as shown in Equation 21.3.

2. *Nuclei below the belt of stability (low neutron-to-proton ratios).* These proton-rich nuclei can increase their ratio by either positron emission or electron capture. Both kinds of decay increase the number of neutrons and decrease the number of protons, as shown in Equations 21.5 and 21.7. Positron emission is more common than electron capture among the lighter nuclei. However, electron capture becomes increasingly common as nuclear charge increases.

3. *Nuclei with atomic numbers ≥ 84.* These heavy nuclei, which lie beyond the upper right edge of the band of stability, tend to undergo alpha emission. Emission of an alpha particle decreases both the number of neutrons and the number of protons by 2, moving the nucleus diagonally toward the belt of stability.

These three situations are summarized in Figure 21.4▶.

▲ **Figure 21.4 Proton and neutron changes in nuclear processes.** The graph shows the results of alpha emission ($^4_2$He), beta emission ($^0_{-1}$e), positron emission ($^0_1$e), and electron capture on the number of protons and neutrons in a nucleus. Moving from left to right or from bottom to top, each square represents an additional proton or neutron, respectively. Moving in the reverse direction indicates the loss of a proton or neutron.

■ SAMPLE EXERCISE 21.3 | Predicting Modes of Nuclear Decay

Predict the mode of decay of **(a)** carbon-14, **(b)** xenon-118.

SOLUTION

Analyze: We are asked to predict the modes of decay of two nuclei.

Plan: To do this, we must calculate the neutron-to-proton ratios and compare the values with those for nuclei that lie within the belt of stability shown in Figure 21.3.

Solve:

(a) Carbon has an atomic number of 6. Thus, carbon-14 has 6 protons and $14 - 6 = 8$ neutrons, giving it a neutron-to-proton ratio of $\frac{8}{6} = 1.3$. Elements with low atomic numbers normally have stable nuclei with approximately equal numbers of neutrons and protons. Thus, carbon-14 has a high neutron-to-proton ratio, and we expect that it will decay by emitting a beta particle:

$$^{14}_{6}\text{C} \longrightarrow \, ^{0}_{-1}\text{e} + \, ^{14}_{7}\text{N}$$

This is indeed the mode of decay observed for carbon-14.

(b) Xenon has an atomic number of 54. Thus, xenon-118 has 54 protons and $118 - 54 = 64$ neutrons, giving it a neutron-to-proton ratio of $\frac{64}{54} = 1.2$. According to Figure 21.3, stable nuclei in this region of the belt of stability have higher neutron-to-proton ratios than xenon-118. The nucleus can increase this ratio by either positron emission or electron capture:

$$^{118}_{54}\text{Xe} \longrightarrow \, ^{0}_{1}\text{e} + \, ^{118}_{53}\text{I}$$

$$^{118}_{54}\text{Xe} + \, ^{0}_{-1}\text{e} \longrightarrow \, ^{118}_{53}\text{I}$$

In this case both modes of decay are observed.

Comment: Keep in mind that our guidelines do not always work. For example, thorium-233, $^{233}_{90}\text{Th}$, which we might expect to undergo alpha decay, actually undergoes beta decay. Furthermore, a few radioactive nuclei actually lie within the belt of stability. Both $^{146}_{60}\text{Nd}$ and $^{148}_{60}\text{Nd}$, for example, are stable and lie in the belt of stability. $^{147}_{60}\text{Nd}$, however, which lies between them, is radioactive.

■ PRACTICE EXERCISE

Predict the mode of decay of **(a)** plutonium-239, **(b)** indium-120.
Answers: **(a)** α decay, **(b)** β decay

81 82 83 84 85 86 87 88 89 90 91 92 93
Tl Pb Bi Po At Rn Fr Ra Ac Th Pa U Np

Atomic number ⟶

Radioactive Series

Some nuclei, such as uranium-238, cannot gain stability by a single emission. Consequently, a series of successive emissions occurs. As shown in Figure 21.5 ◄, uranium-238 decays to thorium-234, which is radioactive and decays to protactinium-234. This nucleus is also unstable and subsequently decays. Such successive reactions continue until a stable nucleus, lead-206, is formed. A series of nuclear reactions that begins with an unstable nucleus and terminates with a stable one is known as a **radioactive series**, or a **nuclear disintegration series**. Three such series occur in nature. In addition to the series that begins with uranium-238 and terminates with lead-206, one begins with uranium-235 and ends with lead-207, and another begins with thorium-232 and ends with lead-208.

◄ **Figure 21.5 Nuclear disintegration series for uranium-238.** The $^{238}_{92}\text{U}$ nucleus decays to $^{234}_{90}\text{Th}$. Subsequent decay processes eventually form the stable $^{206}_{82}\text{Pb}$ nucleus. Each blue arrow corresponds to the loss of an alpha particle; each red arrow corresponds to the loss of a beta particle.

Further Observations

Two further observations can help you predict nuclear stability:

- Nuclei with 2, 8, 20, 28, 50, or 82 protons or 2, 8, 20, 28, 50, 82, or 126 neutrons are generally more stable than nuclei that do not contain these numbers of nucleons. These numbers of protons and neutrons are called **magic numbers**.

- Nuclei with even numbers of both protons and neutrons are generally more stable than those with odd numbers of nucleons, as shown in Table 21.3 ▶.

These observations can be understood in terms of the *shell model of the nucleus*, in which nucleons are described as residing in shells analogous to the shell structure for electrons in atoms. Just as certain numbers of electrons (2, 8, 18, 36, 54, and 86) correspond to stable closed-shell electron configurations, so also the magic numbers of nucleons represent closed shells in nuclei. As an example of the stability of nuclei with magic numbers of nucleons, note that the radioactive series depicted in Figure 21.5 ends with formation of the stable $^{206}_{82}Pb$ nucleus, which has a magic number of protons (82).

Evidence also suggests that pairs of protons and pairs of neutrons have a special stability, analogous to the pairs of electrons in molecules. Thus, stable nuclei with an even number of protons and an even number of neutrons are far more numerous than those with odd numbers (Table 21.3).

TABLE 21.3 ■ The Number of Stable Isotopes with Even and Odd Numbers of Protons and Neutrons

Number of Stable Isotopes	Protons	Neutrons
157	Even	Even
53	Even	Odd
50	Odd	Even
5	Odd	Odd

■ SAMPLE EXERCISE 21.4 | Predicting Nuclear Stability

Predict which of these nuclei are especially stable: ^4_2He, $^{40}_{20}Ca$, $^{98}_{43}Tc$?

SOLUTION

Analyze: We are asked to identify especially stable nuclei, given their mass numbers and atomic numbers.

Plan: We look to see whether the numbers of protons and neutrons correspond to magic numbers.

Solve: The ^4_2He nucleus (the alpha particle) has a magic number of both protons (2) and neutrons (2) and is very stable. The $^{40}_{20}Ca$ nucleus also has a magic number of both protons (20) and neutrons (20) and is especially stable.

The $^{98}_{43}Tc$ nucleus does not have a magic number of either protons or neutrons. In fact, it has an odd number of both protons (43) and neutrons (55). There are very few stable nuclei with odd numbers of both protons and neutrons: Indeed, technetium-98 is radioactive.

■ PRACTICE EXERCISE

Which of the following nuclei would you expect to exhibit a special stability: $^{118}_{50}Sn$, $^{210}_{85}At$, $^{208}_{82}Pb$?

Answer: $^{118}_{50}Sn$, $^{208}_{82}Pb$

21.3 NUCLEAR TRANSMUTATIONS

Thus far we have examined nuclear reactions in which a nucleus spontaneously decays. A nucleus can also change identity if it is struck by a neutron or by another nucleus. Nuclear reactions that are induced in this way are known as **nuclear transmutations**.

In 1919, Ernest Rutherford performed the first conversion of one nucleus into another. He succeeded in converting nitrogen-14 into oxygen-17, plus a proton, using the high-velocity alpha particles emitted by radium. The reaction is

$$^{14}_{7}N + ^4_2He \longrightarrow ^{17}_{8}O + ^1_1H \qquad [21.8]$$

This reaction demonstrated that striking nuclei with particles such as alpha particles could induce nuclear reactions. Such reactions made it possible to synthesize hundreds of radioisotopes in the laboratory.

Nuclear transmutations are sometimes represented by listing, in order, the target nucleus, the bombarding particle, the ejected particle, and the product nucleus. Written in this fashion, Equation 21.8 is $^{14}_{7}N(\alpha, p)^{17}_{8}O$. The alpha particle, proton, and neutron are abbreviated as α, p, and n, respectively.

▲ **Figure 21.6 Schematic drawing of a cyclotron.** Charged particles are accelerated around the ring by applying alternating voltage to the dees.

■ SAMPLE EXERCISE 21.5 | Writing a Balanced Nuclear Equation

Write the balanced nuclear equation for the process summarized as $^{27}_{13}Al(n, \alpha)^{24}_{11}Na$.

SOLUTION

Analyze: We must go from the condensed descriptive form of the nuclear reaction to the balanced nuclear equation.

Plan: We arrive at the full nuclear equation by writing n and α, each with its associated subscripts and superscripts.

Solve: The n is the abbreviation for a neutron ($^{1}_{0}n$), and α represents an alpha particle ($^{4}_{2}He$). The neutron is the bombarding particle, and the alpha particle is a product. Therefore, the nuclear equation is

$$^{27}_{13}Al + ^{1}_{0}n \longrightarrow ^{24}_{11}Na + ^{4}_{2}He$$

■ PRACTICE EXERCISE

Using a shorthand notation, write the nuclear reaction

$$^{16}_{8}O + ^{1}_{1}H \longrightarrow ^{13}_{7}N + ^{4}_{2}He$$

Answer: $^{16}_{8}O(p, \alpha)^{13}_{7}N$

▲ **Figure 21.7 CERN (Conseil Européen pour la Recherche Nucléaire), Geneva, Switzerland.** Particles are accelerated to very high energies by circulating them through magnets in the ring, which has a circumference of 27 km.

Accelerating Charged Particles

Charged particles, such as alpha particles, must be moving very fast to overcome the electrostatic repulsion between them and the target nucleus. The higher the nuclear charge on either the projectile or the target, the faster the projectile must be moving to bring about a nuclear reaction. Many methods have been devised to accelerate charged particles, using strong magnetic and electrostatic fields. These **particle accelerators**, popularly called "atom smashers," bear such names as *cyclotron* and *synchrotron*. The cyclotron is illustrated in Figure 21.6 ◄. The hollow D-shaped electrodes are called "dees." The projectile particles are introduced into a vacuum chamber within the cyclotron. The particles are then accelerated by making the dees alternately positively and negatively charged. Magnets placed above and below the dees keep the particles moving in a spiral path until they are finally deflected out of the cyclotron and emerge to strike a target substance. Particle accelerators are used to synthesize heavy elements, to investigate the fundamental structure of matter, and ultimately answer questions about the beginning of the universe. Figure 21.7 ◄ shows an aerial view of CERN (Conseil Européen pour la Recherche Nucléaire), the European Organization for Nuclear Research, near Geneva, Switzerland.

Using Neutrons

Most synthetic isotopes used in medicine and scientific research are made using neutrons as projectiles. Because neutrons are neutral, they are not repelled by the nucleus. Consequently, they do not need to be accelerated, as do charged particles, to cause nuclear reactions. The necessary neutrons are produced by the reactions that occur in nuclear reactors. For example, cobalt-60, which is used in radiation therapy for cancer, is produced by neutron capture. Iron-58 is placed in a nuclear reactor, where neutrons bombard it. The following sequence of reactions takes place:

$$_{26}^{58}\text{Fe} + _{0}^{1}\text{n} \longrightarrow _{26}^{59}\text{Fe} \quad\quad [21.9]$$

$$_{26}^{59}\text{Fe} \longrightarrow _{27}^{59}\text{Co} + _{-1}^{0}\text{e} \quad\quad [21.10]$$

$$_{27}^{59}\text{Co} + _{0}^{1}\text{n} \longrightarrow _{27}^{60}\text{Co} \quad\quad [21.11]$$

▲ **GIVE IT SOME THOUGHT**

Can neutrons be accelerated in a particle accelerator, using electrostatic or magnetic fields? Why or why not?

Transuranium Elements

Artificial transmutations have been used to produce the elements with atomic number above 92. These elements are known as the **transuranium elements** because they occur immediately following uranium in the periodic table. Elements 93 (neptunium, Np) and 94 (plutonium, Pu) were discovered in 1940. They were produced by bombarding uranium-238 with neutrons:

$$_{92}^{238}\text{U} + _{0}^{1}\text{n} \longrightarrow _{92}^{239}\text{U} \longrightarrow _{93}^{239}\text{Np} + _{-1}^{0}\text{e} \quad\quad [21.12]$$

$$_{93}^{239}\text{Np} \longrightarrow _{94}^{239}\text{Pu} + _{-1}^{0}\text{e} \quad\quad [21.13]$$

Elements with still larger atomic numbers are normally formed in small quantities in particle accelerators. Curium-242, for example, is formed when a plutonium-239 target is struck with accelerated alpha particles:

$$_{94}^{239}\text{Pu} + _{2}^{4}\text{He} \longrightarrow _{96}^{242}\text{Cm} + _{0}^{1}\text{n} \quad\quad [21.14]$$

In 1994 a team of European scientists synthesized element 111 by bombarding a bismuth target for several days with a beam of nickel atoms:

$$_{83}^{209}\text{Bi} + _{28}^{64}\text{Ni} \longrightarrow _{111}^{272}\text{X} + _{0}^{1}\text{n}$$

Amazingly, their discovery was based on the detection of only three atoms of the new element. The nuclei are very short lived, and they undergo alpha decay within milliseconds of their synthesis. In 2004 scientists reported the synthesis of elements 113 and 115. As of this writing, these results have yet to be confirmed, although the results look promising. Evidence for element 118 was reported in 2006. Names and symbols have not yet been chosen for these new elements.

21.4 RATES OF RADIOACTIVE DECAY

Some radioisotopes, such as uranium-238, are found in nature, although they are not stable. Other radioisotopes do not exist in nature, even though they can be synthesized in nuclear reactions. To understand this distinction, we must realize that different nuclei undergo radioactive decay at different rates. Many radioisotopes decay essentially completely in a matter of seconds or less, so we do not find them in nature. Uranium-238, on the other hand, decays very slowly. Therefore, despite its instability, we can still observe what remains from its formation in the early history of the universe.

TABLE 21.4 ■ The Half-lives and Type of Decay for Several Radioisotopes

	Isotope	Half-life (yr)	Type of Decay
Natural radioisotopes	$^{238}_{92}U$	4.5×10^9	Alpha
	$^{235}_{92}U$	7.0×10^8	Alpha
	$^{232}_{90}Th$	1.4×10^{10}	Alpha
	$^{40}_{19}K$	1.3×10^9	Beta
	$^{14}_{6}C$	5715	Beta
Synthetic radioisotopes	$^{239}_{94}Pu$	24,000	Alpha
	$^{137}_{55}Cs$	30	Beta
	$^{90}_{38}Sr$	28.8	Beta
	$^{131}_{53}I$	0.022	Beta

▲ **Figure 21.8 Decay of a 10.0-g sample of $^{90}_{38}Sr$ ($t_{1/2}$ = 28.8 yr).**

Radioactive decay is a first-order kinetic process. Recall that a first-order process has a characteristic **half-life**, which is the time required for half of any given quantity of a substance to react. ∞ (Section 14.4) The rates of decay of nuclei are commonly expressed in terms of their half-lives. Each isotope has its own characteristic half-life. For example, the half-life of strontium-90 is 28.8 yr. If we started with 10.0 g of strontium-90, only 5.0 g of that isotope would remain after 28.8 yr, 2.5 g would remain after another 28.8 yr, and so on. Strontium-90 decays to yttrium-90:

$$^{90}_{38}Sr \longrightarrow {}^{90}_{39}Y + {}^{0}_{-1}e \qquad [21.15]$$

The loss of strontium-90 as a function of time is shown in Figure 21.8 ◄.

Half-lives as short as millionths of a second and as long as billions of years are known. The half-lives of some radioisotopes are listed in Table 21.4▲. One important feature of half-lives for nuclear decay is that they are unaffected by external conditions such as temperature, pressure, or state of chemical combination. Unlike toxic chemicals, therefore, radioactive atoms cannot be rendered harmless by chemical reaction or by any other practical treatment. At this point we can do nothing but allow these nuclei to lose radioactivity at their characteristic rates. In the meantime, we must take precautions to prevent radioisotopes, such as those produced in nuclear power plants ∞ (Section 21.7), from entering the environment because of the damage radiation can cause.

■ SAMPLE EXERCISE 21.6 | Calculation Involving Half-Lives

The half-life of cobalt-60 is 5.3 yr. How much of a 1.000-mg sample of cobalt-60 is left after a 15.9-yr period?

SOLUTION

Analyze: We are given the half-life for cobalt-60 and asked to calculate the amount of cobalt-60 remaining from an initial 1.000-mg sample after a 15.9-yr period.

Plan: We will use the fact that the half-life of a first-order decay process is a constant.

Solve: We notice that $5.3 \times 3 = 15.9$. Therefore, a period of 15.9 yr is three half-lives for cobalt-60. At the end of one half-life, 0.500 mg of cobalt-60 remains, 0.250 mg at the end of two half-lives, and 0.125 mg at the end of three half-lives.

■ PRACTICE EXERCISE

Carbon-11, used in medical imaging, has a half-life of 20.4 min. The carbon-11 nuclides are formed, and the carbon atoms are then incorporated into an appropriate compound. The resulting sample is injected into a patient, and the medical image is obtained. If the entire process takes five half-lives, what percentage of the original carbon-11 remains at this time?
Answer: 3.12%

Radiometric Dating

Because the half-life of any particular nuclide is constant, the half-life can serve as a nuclear clock to determine the ages of different objects. The method of dating objects based on their isotopes and isotope abundances is called radiometric dating.

Carbon-14, for example, has been used to determine the age of organic materials (Figure 21.9▶). This technique is known as radiocarbon dating. The procedure is based on the formation of carbon-14 by capture of solar neutrons in the upper atmosphere:

$$^{14}_{7}\text{N} + ^{1}_{0}\text{n} \longrightarrow ^{14}_{6}\text{C} + ^{1}_{1}\text{p} \qquad [21.16]$$

This reaction provides a small but reasonably constant source of carbon-14. The carbon-14 is radioactive, undergoing beta decay with a half-life of 5715 yr:

$$^{14}_{6}\text{C} \longrightarrow ^{14}_{7}\text{N} + ^{0}_{-1}\text{e} \qquad [21.17]$$

The carbon-14 is incorporated into carbon dioxide, which is in turn incorporated, through photosynthesis, into more complex carbon-containing molecules within plants. When animals eat the plants, the carbon-14 becomes incorporated within them. Because a living plant or animal has a constant intake of carbon compounds, it is able to maintain a ratio of carbon-14 to carbon-12 that is nearly identical with that of the atmosphere. Once the organism dies, however, it no longer ingests carbon compounds to replenish the carbon-14 that is lost through radioactive decay. The ratio of carbon-14 to carbon-12 therefore decreases. By measuring this ratio and comparing it to that of the atmosphere, we can estimate the age of an object. For example, if the ratio diminishes to half that of the atmosphere, we can conclude that the object is one half-life, or 5715 yr, old. This method cannot be used to date objects older than about 50,000 yr. After this length of time the radioactivity is too low to be measured accurately.

In radiocarbon dating, a reasonable assumption is that the ratio of carbon-14 to carbon-12 in the atmosphere has been relatively constant for the past 50,000 yr. However, because variations in solar activity control the amount of carbon-14 produced in the atmosphere, the carbon-14/carbon-12 ratio can fluctuate. We can correct for this effect by using other kinds of data. Recently scientists have compared carbon-14 data with data from tree rings, corals, lake sediments, ice cores, and other natural sources to correct variations in the carbon-14 "clock" back to 26,000 yr.

Other isotopes can be similarly used to date other types of objects. For example, it takes 4.5×10^9 yr for half of a sample of uranium-238 to decay to lead-206. The age of rocks containing uranium can therefore be determined by measuring the ratio of lead-206 to uranium-238. If the lead-206 had somehow become incorporated into the rock by normal chemical processes instead of by radioactive decay, the rock would also contain large amounts of the more abundant isotope lead-208. In the absence of large amounts of this "geonormal" isotope of lead, it is assumed that all of the lead-206 was at one time uranium-238.

The oldest rocks found on Earth are approximately 3×10^9 yr old. This age indicates that Earth's crust has been solid for at least this length of time. Scientists estimate that it required 1×10^9 to 1.5×10^9 yr for Earth to cool and its surface to become solid, making the age of Earth 4.0 to 4.5×10^9 (between four to four-and-a-half billion) yr.

▲ Figure 21.9 **Radiocarbon dating.** This machine, a linear accelerator, analyzes the amount of carbon-14 in carbon dioxide, produced by combustion of a sample in oxygen. The ratio of carbon-14 to carbon-12 is related to the time since the death of the animal or plant being investigated.

Calculations Based on Half-life

So far, our discussion has been mainly qualitative. We now consider the topic of half-lives from a more quantitative point of view. This approach enables us to answer questions of the following types: How do we determine the half-life of uranium-238? Similarly, how do we quantitatively determine the age of an object?

Radioactive decay is a first-order kinetic process. Its rate, therefore, is proportional to the number of radioactive nuclei N in the sample:

$$\text{Rate} = kN \qquad \text{[21.18]}$$

The first-order rate constant, k, is called the *decay constant*. The rate at which a sample decays is called its **activity**, and it is often expressed as the number of disintegrations observed per unit time. The **becquerel** (Bq) is the SI unit for expressing the activity of a particular radiation source (that is, the rate at which nuclear disintegrations are occurring). A becquerel is defined as one nuclear disintegration per second. An older, but still widely used, unit of activity is the **curie** (Ci), defined as 3.7×10^{10} disintegrations per second, which is the rate of decay of 1 g of radium. Thus, a 4.0-mCi sample of cobalt-60 undergoes $(4.0 \times 10^{-3}) \times (3.7 \times 10^{10}$ disintegrations per second$) = 1.5 \times 10^8$ disintegrations per second and has an activity of 1.5×10^8 Bq. As a radioactive sample decays, the amount of radiation emanating from the sample decays as well. For example, the half-life of cobalt-60 is 5.26 yr. The 4.0-mCi sample of cobalt-60 would, after 5.26 yr, have a radiation activity of 2.0 mCi, or 7.5×10^7 Bq.

◢ GIVE IT SOME THOUGHT

Why can't spontaneous radioactive decay be a zero-order or second-order kinetic process?

As we saw in Section 14.4, a first-order rate law can be transformed into the following equation:

$$\ln \frac{N_t}{N_0} = -kt \qquad \text{[21.19]}$$

In this equation t is the time interval of decay, k is the decay constant, N_0 is the initial number of nuclei (at time zero), and N_t is the number remaining after the time interval. Both the mass of a particular radioisotope and its activity are proportional to the number of radioactive nuclei. Thus, either the ratio of the mass at any time t to the mass at time $t = 0$ or the ratio of the activities at time t and $t = 0$ can be substituted for N_t/N_0 in Equation 21.19.

From Equation 21.19 we can obtain the relationship between the decay constant, k, and half-life, $t_{1/2}$. ∞ (Section 14.4)

$$k = \frac{0.693}{t_{1/2}} \qquad \text{[21.20]}$$

Thus, if we know the value of either the decay constant or the half-life, we can calculate the value of the other.

◢ GIVE IT SOME THOUGHT

(a) Would doubling the mass of a radioactive sample change the amount of radioactivity the sample shows? (b) Would doubling the mass change the half-life for the radioactive decay?

■ SAMPLE EXERCISE 21.7 | Calculating the Age of a Mineral

A rock contains 0.257 mg of lead-206 for every milligram of uranium-238. The half-life for the decay of uranium-238 to lead-206 is 4.5×10^9 yr. How old is the rock?

SOLUTION

Analyze: We are told that a rock sample has a certain amount of lead-206 for every unit weight of uranium-238 and asked to estimate the age of the rock.

Plan: Lead-206 is the product of the radioactive decay of uranium-228. We will assume that the only source of lead-206 in the rock is from the decay of uranium-238, with a known half-life. To apply first-order kinetics expressions (Equations 21.19 and 21.20) to calculate the time elapsed since the rock was formed, we first need to calculate how much initial uranium-238 there was for every 1 milligram that remains today.

Solve: Let's assume that the rock currently contains 1.000 mg of uranium-238 and therefore 0.257 mg of lead-206. The amount of uranium-238 in the rock when it was first formed therefore equals 1.000 mg plus the quantity that has decayed to lead-206. Because the mass of lead atoms is not the same as uranium atoms, we cannot just add 1.000 mg and 0.257 mg. We have to multiply the present mass of lead-206 (0.257 mg) by the ratio of the mass number of uranium to that of lead, into which it has decayed. The total original $^{238}_{92}U$ was thus

$$\text{Original } ^{238}_{92}U = 1.000 \text{ mg} + \frac{238}{206} (0.257 \text{ mg})$$

$$= 1.297 \text{ mg}$$

Using Equation 21.20, we can calculate the decay constant for the process from its half-life:

$$k = \frac{0.693}{4.5 \times 10^9 \text{ yr}} = 1.5 \times 10^{-10} \text{ yr}^{-1}$$

Rearranging Equation 21.19 to solve for time, t, and substituting known quantities gives

$$t = -\frac{1}{k} \ln \frac{N_t}{N_0} = -\frac{1}{1.5 \times 10^{-10} \text{ yr}^{-1}} \ln \frac{1.000}{1.297} = 1.7 \times 10^9 \text{ yr}$$

Comment: If you want to cross-check this result, you could also use the fact that uranium-237 decays to lead-207 with a half-life of 7×10^8 yr, and measure the relative amounts of uranium-237 and lead-207 in the rock.

■ PRACTICE EXERCISE

A wooden object from an archeological site is subjected to radiocarbon dating. The activity of the sample that is due to ^{14}C is measured to be 11.6 disintegrations per second. The activity of a carbon sample of equal mass from fresh wood is 15.2 disintegrations per second. The half-life of ^{14}C is 5715 yr. What is the age of the archeological sample?
Answer: 2230 yr

■ SAMPLE EXERCISE 21.8 | Calculations Involving Radioactive Decay

If we start with 1.000 g of strontium-90, 0.953 g will remain after 2.00 yr. **(a)** What is the half-life of strontium-90? **(b)** How much strontium-90 will remain after 5.00 yr? **(c)** What is the initial activity of the sample in Bq and in Ci?

SOLUTION

(a) Analyze: We are asked to calculate a half-life, $t_{1/2}$, based on data that tell us how much of a radioactive nucleus has decayed in a given period of time and ($N_0 = 1.000$ g, $N_t = 0.953$ g, and $t = 2.00$ yr).

Plan: We first calculate the rate constant for the decay, k, then use that to compute $t_{1/2}$.

Solve: Equation 21.19 is solved for the decay constant, k, and then Equation 21.20 is used to calculate half-life, $t_{1/2}$:

$$k = -\frac{1}{t} \ln \frac{N_t}{N_0} = -\frac{1}{2.00 \text{ yr}} \ln \frac{0.953 \text{ g}}{1.000 \text{ g}}$$

$$= -\frac{1}{2.00 \text{ yr}} (-0.0481) = 0.0241 \text{ yr}^{-1}$$

$$t_{1/2} = \frac{0.693}{k} = \frac{0.693}{0.0241 \text{ yr}^{-1}} = 28.8 \text{ yr}$$

(b) Analyze: We are asked to calculate the amount of a radionuclide remaining after a given period of time.

Plan: We need to calculate the amount of strontium at time t, N_t, using the initial quantity, N_0, and the rate constant for decay, k, calculated in part (a).

Solve: Again using Equation 21.19, with $k = 0.0241$ yr^{-1}, we have

$$\ln \frac{N_t}{N_0} = -kt = -(0.0241 \text{ yr}^{-1})(5.00 \text{ yr}) = -0.120$$

N_t/N_0 is calculated from $\ln(N_t/N_0) = -0.120$ using the e^x or INV LN function of a calculator:

$$\frac{N_t}{N_0} = e^{-0.120} = 0.887$$

Because $N_0 = 1.000$ g, we have

$$N_t = (0.887)N_0 = (0.887)(1.000 \text{ g}) = 0.887 \text{ g}$$

(c) Analyze: We are asked to calculate the activity of the sample in becquerels and curies.

Plan: We must calculate the number of disintegrations per second per atom, and then multiply by the number of atoms in the sample.

Solve: The number of disintegrations per atom per second is given by the rate constant, k.

$$k = \left(\frac{0.0241}{\text{yr}}\right)\left(\frac{1 \text{ yr}}{365 \text{ days}}\right)\left(\frac{1 \text{ day}}{24 \text{ h}}\right)\left(\frac{1 \text{ h}}{3600 \text{ s}}\right) = 7.64 \times 10^{-10} \text{ s}^{-1}$$

To obtain the total number of disintegrations per second, we calculate the number of atoms in the sample. We multiply this quantity by k, where we express k as the number of disintegrations per atom per second, to obtain the number of disintegrations per second:

$$(1.000 \text{ g } ^{90}\text{Sr})\left(\frac{1 \text{ mol } ^{90}\text{Sr}}{90 \text{ g } ^{90}\text{Sr}}\right)\left(\frac{6.022 \times 10^{23} \text{ atoms Sr}}{1 \text{ mol } ^{90}\text{Sr}}\right) = 6.7 \times 10^{21} \text{ atoms } ^{90}\text{Sr}$$

$$\text{Total disintegrations/s} = \left(\frac{7.64 \times 10^{-10} \text{ disintegrations}}{\text{atom} \cdot \text{s}}\right)(6.7 \times 10^{21} \text{ atoms})$$

$$= 5.1 \times 10^{12} \text{ disintegrations/s}$$

Because a Bq is one disintegration per second, the activity is just 5.1×10^{12} Bq. The activity in Ci is given by

$$(5.1 \times 10^{12} \text{ disintegrations/s})\left(\frac{1 \text{ Ci}}{3.7 \times 10^{10} \text{ disintegrations/s}}\right) = 1.4 \times 10^2 \text{ Ci}$$

We have used only two significant figures in products of these calculations because we do not know the atomic weight of ^{90}Sr to more than two significant figures without looking it up in a special source.

■ PRACTICE EXERCISE

A sample to be used for medical imaging is labeled with ^{18}F, which has a half-life of 110 min. What percentage of the original activity in the sample remains after 300 min?
Answer: 15.1%

21.5 DETECTION OF RADIOACTIVITY

A variety of methods have been devised to detect emissions from radioactive substances. Henri Becquerel discovered radioactivity because radiation caused fogging of photographic plates. Photographic plates and film have long been used to detect radioactivity. The radiation affects photographic film in much the same way as X-rays do. With care, film can be used to give a quantitative

▲ Figure 21.10 **Badge dosimeter.**
Badges such as the one on this worker's
lapel are used to monitor the extent to
which the individual has been exposed to
high-energy radiation. The radiation dose is
determined from the extent of fogging of
the film in the dosimeter. Monitoring the
radiation in this way helps prevent
overexposure for people whose jobs require
them to use radioactive materials or X-rays.

▲ Figure 21.11 **Schematic drawing of a Geiger counter.**

measure of activity. The greater the extent of exposure to radiation, the darker
is the area of the developed negative. People who work with radioactive substances carry film badges to record the extent of their exposure to radiation
(Figure 21.10▲).

Radioactivity can also be detected and measured using a device known as a
Geiger counter. The operation of a Geiger counter is based on the ionization of
matter caused by radiation. The ions and electrons produced by the ionizing radiation permit conduction of an electrical current. The basic design of a Geiger
counter is shown in Figure 21.11▲. It consists of a metal tube filled with gas.
The cylinder has a "window" made of material that can be penetrated by alpha,
beta, or gamma rays. In the center of the tube is a wire. The wire is connected to
one terminal of a source of direct current, and the metal cylinder is attached to
the other terminal. Current flows between the wire and metal cylinder whenever entering radiation produces ions. The current pulse created when radiation
enters the tube is amplified; each pulse is counted as a measure of the amount
of radiation.

▲ **GIVE IT SOME THOUGHT**

Will alpha, beta, and gamma rays pass through the window of a Geiger counter detection tube with equal efficiency?

Certain substances that are electronically excited by radiation can also be
used to detect and measure radiation. For example, some substances excited by
radiation give off light as electrons return to their lower-energy states. These
substances are called *phosphors*. Different substances respond to different particles. Zinc sulfide, for example, responds to alpha particles. An instrument
called a scintillation counter (Figure 21.12▶) is used to detect and measure radiation, based on the tiny flashes of light produced when radiation strikes a suitable phosphor. The flashes are magnified electronically and counted to measure
the amount of radiation.

▲ Figure 21.12 **A scintillation
counter.** Such counters are used to
quantitatively measure the level of
radiation.

Radiotracers

Because radioisotopes can be detected so readily, they can be used to follow an element through its chemical reactions. The incorporation of carbon atoms from CO_2 into glucose during photosynthesis, for example, has been studied using CO_2 enriched in carbon-14:

$$6\ ^{14}CO_2 + 6\ H_2O \xrightarrow[\text{Chlorophyll}]{\text{Sunlight}} {}^{14}C_6H_{12}O_6 + 6\ O_2 \qquad [21.21]$$

The CO_2 is said to be "labeled" with the carbon-14. The use of the carbon-14 label provides direct experimental evidence that carbon dioxide in the environment is chemically converted to glucose in plants. Analogous labeling experiments for oxygen, using oxygen-18, show that the O_2 produced during photosynthesis comes from water, not carbon dioxide. When it is possible to isolate and purify intermediates and products from reactions, detection devices such as scintillation counters can be used to "follow" the radioisotope as it moves from the starting material through the various intermediate compounds to the final product. These types of experiments are useful for identifying elementary steps in a reaction mechanism. ∞ (Section 14.6)

The use of radioisotopes is possible because all isotopes of an element have essentially identical chemical properties. When a small quantity of a radioisotope is mixed with the naturally occurring stable isotopes of the same element, all the isotopes go through the same reactions together. The element's path is revealed by the radioactivity of the radioisotope. Because the radioisotope can be used to trace the path of the element, it is called a **radiotracer**.

Chemistry and Life   MEDICAL APPLICATIONS OF RADIOTRACERS

Radiotracers have found wide use as diagnostic tools in medicine. Table 21.5▼ lists some radiotracers and their uses. These radioisotopes are incorporated into a compound that is administered to the patient, usually intravenously. The diagnostic use of these isotopes is based upon the ability of the radioactive compound to localize and concentrate in the organ or tissue under investigation. Iodine-131, for example, has been used to test the activity of the thyroid gland. This gland is the only important user of iodine in the body. The patient drinks a solution of NaI containing iodine-131. Only a very small amount is used so that the patient does not receive a harmful dose of radioactivity. A Geiger counter placed close to the thyroid, in the neck region, determines the ability of the thyroid to take up the iodine. A normal thyroid will absorb about 12% of the iodine within a few hours.

The medical applications of radiotracers are further illustrated by positron emission tomography (PET). PET is used for clinical diagnosis of many diseases. In this method compounds containing radionuclides that decay by positron emission are injected into a patient. These compounds are chosen to enable researchers to monitor blood flow, oxygen and glucose metabolic rates, and other biological functions. Some of the most interesting work involves the study of the brain, which depends on glucose for most of its energy. Changes in how this sugar is metabolized or used by the brain may signal a disease such as cancer, epilepsy, Parkinson's disease, or schizophrenia.

The compound to be detected in the patient must be labeled with a radionuclide that is a positron emitter. The most widely used nuclides are carbon-11 (half-life 20.4 min), fluorine-18 (half-life 110 min), oxygen-15 (half-life 2 min), and nitrogen-13 (half-life 10 min). Glucose, for example, can be labeled with ^{11}C. Because the half-lives of positron emitters are so short, the chemist must quickly incorporate the radionuclide into the sugar (or other appropriate) molecule and inject the compound immediately. The patient is placed in an elaborate instrument that measures the positron emission and constructs a computer-based image of the organ in which the emitting compound is localized. The nature of this image provides clues to the presence of disease or other abnormality and helps medical researchers understand how a particular disease affects the functioning of the brain.
Related Exercise: 21.53

TABLE 21.5 ▪ Some Radionuclides Used as Radiotracers

Nuclide	Half-life	Area of the Body Studied
Iodine-131	8.04 days	Thyroid
Iron-59	44.5 days	Red blood cells
Phosphorus-32	14.3 days	Eyes, liver, tumors
Technetium-99	6.0 hours	Heart, bones, liver, and lungs
Thallium-201	73 hours	Heart, arteries
Sodium-24	14.8 hours	Circulatory system

21.6 ENERGY CHANGES IN NUCLEAR REACTIONS

The energies associated with nuclear reactions can be considered with the aid of Einstein's famous equation relating mass and energy:

$$E = mc^2 \qquad [21.22]$$

In this equation E stands for energy, m for mass, and c for the speed of light, 2.9979×10^8 m/s. This equation states that the mass and energy of an object are proportional. If a system loses mass, it loses energy (exothermic); if it gains mass, it gains energy (endothermic). Because the proportionality constant in the equation, c^2, is such a large number, even small changes in mass are accompanied by large changes in energy.

The mass changes in chemical reactions are too small to detect. For example, the mass change associated with the combustion of 1 mol of CH_4 (an exothermic process) is -9.9×10^{-9} g. Because the mass change is so small, it is possible to treat chemical reactions as though mass is conserved.

The mass changes and the associated energy changes in nuclear reactions are much greater than those in chemical reactions. The mass change accompanying the radioactive decay of a mole of uranium-238, for example, is 50,000 times greater than that for the combustion of 1 mol of CH_4. Let's examine the energy change for the nuclear reaction

$$^{238}_{92}\text{U} \longrightarrow {}^{234}_{90}\text{Th} + {}^{4}_{2}\text{He}$$

The nuclei in this reaction have the following masses: $^{238}_{92}\text{U}$, 238.0003 amu; $^{234}_{90}\text{Th}$, 233.9942 amu; and $^{4}_{2}\text{He}$, 4.0015 amu. The mass change, Δm, is the total mass of the products minus the total mass of the reactants. The mass change for the decay of a *mole* of uranium-238 can then be expressed in grams:

$$233.9942 \text{ g} + 4.0015 \text{ g} - 238.0003 \text{ g} = -0.0046 \text{ g}$$

The fact that the system has lost mass indicates that the process is exothermic. All spontaneous nuclear reactions are exothermic.

The energy change per mole associated with this reaction can be calculated using Einstein's equations:

$$\Delta E = \Delta(mc^2) = c^2 \, \Delta m$$

$$= (2.9979 \times 10^8 \text{ m/s})^2 (-0.0046 \text{ g}) \left(\frac{1 \text{ kg}}{1000 \text{ g}} \right)$$

$$= -4.1 \times 10^{11} \frac{\text{kg-m}^2}{\text{s}^2} = -4.1 \times 10^{11} \text{ J}$$

Notice that Δm is converted to kilograms, the SI unit of mass, to obtain ΔE in joules, the SI unit for energy. The negative sign for the energy change indicates that energy is released in the reaction—in this case, over 400 billion joules per mole of uranium!

■■ **SAMPLE EXERCISE 21.9** | Calculating Mass Change in a Nuclear Reaction

How much energy is lost or gained when a mole of cobalt-60 undergoes beta decay: $^{60}_{27}\text{Co} \longrightarrow {}^{0}_{-1}\text{e} + {}^{60}_{28}\text{Ni}$? The mass of the $^{60}_{27}\text{Co}$ atom is 59.933819 amu, and that of a $^{60}_{28}\text{Ni}$ atom is 59.930788 amu.

SOLUTION

Analyze: We are asked to calculate the energy change in a nuclear reaction.

Plan: We must first calculate the mass change in the process. We are given atomic masses, but we need the masses of the nuclei in the reaction. We calculate these by taking account of the masses of the electrons that contribute to the atomic masses.

Solve: A $^{60}_{27}$Co atom has 27 electrons. The mass of an electron is 5.4858×10^{-4} amu. (See the list of fundamental constants in the back inside cover.) We subtract the mass of the 27 electrons from the mass of the $^{60}_{27}$Co *atom* to find the mass of the $^{60}_{27}$Co *nucleus*:

$$59.933819 \text{ amu} - (27)(5.4858 \times 10^{-4} \text{ amu}) = 59.919007 \text{ amu (or } 59.919007 \text{ g/mol)}$$

Likewise, for $^{60}_{28}$Ni, the mass of the nucleus is

$$59.930788 \text{ amu} - (28)(5.4858 \times 10^{-4} \text{ amu}) = 59.915428 \text{ amu (or } 59.915428 \text{ g/mol)}$$

The mass change in the nuclear reaction is the total mass of the products minus the mass of the reactant:

$$\Delta m = \text{mass of electron} + \text{mass } {}^{60}_{28}\text{Ni nucleus} - \text{mass of } {}^{60}_{27}\text{Co nucleus}$$

$$= 0.00054858 \text{ amu} + 59.915428 \text{ amu} - 59.919007 \text{ amu}$$

$$= -0.003030 \text{ amu}$$

Thus, when a mole of cobalt-60 decays,

$$\Delta m = -0.003030 \text{ g}$$

Because the mass decreases ($\Delta m < 0$), energy is released ($\Delta E < 0$). The quantity of energy released *per mole* of cobalt-60 is calculated using Equation 21.22:

$$\Delta E = c^2 \, \Delta m$$

$$= (2.9979 \times 10^8 \text{ m/s})^2 (-0.003030 \text{ g}) \left(\frac{1 \text{ kg}}{1000 \text{ g}} \right)$$

$$= -2.723 \times 10^{11} \frac{\text{kg-m}^2}{\text{s}^2} = -2.723 \times 10^{11} \text{ J}$$

■■ **PRACTICE EXERCISE**

Positron emission from ^{11}C, $^{11}_{6}$C $\longrightarrow$ $^{11}_{5}$B $+$ $^{0}_{1}$e, occurs with release of 2.87×10^{11} J per mole of ^{11}C. What is the mass change per mole of ^{11}C in this nuclear reaction?
Answer: -3.19×10^{-3} g

Nuclear Binding Energies

Scientists discovered in the 1930s that the masses of nuclei are always less than the masses of the individual nucleons of which they are composed. For example, the helium-4 nucleus has a mass of 4.00150 amu. The mass of a proton is 1.00728 amu, and that of a neutron is 1.00866 amu. Consequently, two protons and two neutrons have a total mass of 4.03188 amu:

$$\text{Mass of two protons} = 2(1.00728 \text{ amu}) = 2.01456 \text{ amu}$$

$$\text{Mass of two neutrons} = 2(1.00866 \text{ amu}) = \underline{2.01732 \text{ amu}}$$

$$\text{Total mass} = 4.03188 \text{ amu}$$

The mass of the individual nucleons is 0.03038 amu greater than that of the helium-4 nucleus:

$$\text{Mass of two protons and two neutrons} = 4.03188 \text{ amu}$$

$$\text{Mass of } {}^{4}_{2}\text{He nucleus} = \underline{4.00150 \text{ amu}}$$

$$\text{Mass difference} = 0.03038 \text{ amu}$$

The mass difference between a nucleus and its constituent nucleons is called the **mass defect**. The origin of the mass defect is readily understood if we consider that energy must be added to a nucleus to break it into separated protons and neutrons:

$$\text{Energy} + {}^{4}_{2}\text{He} \longrightarrow 2 \, {}^{1}_{1}\text{p} + 2 \, {}^{1}_{0}\text{n} \qquad [21.23]$$

The addition of energy to a system must be accompanied by a proportional increase in mass. The mass change for the conversion of helium-4 into separated

TABLE 21.6 ■ Mass Defects and Binding Energies for Three Nuclei

Nucleus	Mass of Nucleus (amu)	Mass of Individual Nucleons (amu)	Mass Defect (amu)	Binding Energy (J)	Binding Energy per Nucleon (J)
$^{4}_{2}$He	4.00150	4.03188	0.03038	4.53×10^{-12}	1.13×10^{-12}
$^{56}_{26}$Fe	55.92068	56.44914	0.52846	7.90×10^{-11}	1.41×10^{-12}
$^{238}_{92}$U	238.00031	239.93451	1.93420	2.89×10^{-10}	1.21×10^{-12}

nucleons is $\Delta m = 0.03038$ amu, as shown in these calculations. The energy required for this process is

$$\Delta E = c^2 \, \Delta m$$

$$= (2.9979 \times 10^8 \text{ m/s})^2 (0.03038 \text{ amu}) \left(\frac{1 \text{ g}}{6.022 \times 10^{23} \text{ amu}} \right) \left(\frac{1 \text{ kg}}{1000 \text{ g}} \right)$$

$$= 4.534 \times 10^{-12} \text{ J}$$

The energy required to separate a nucleus into its individual nucleons is called the **nuclear binding energy**. The larger the binding energy, the more stable is the nucleus toward decomposition. The nuclear binding energies of helium-4 and two other nuclei (iron-56 and uranium-238) are compared in Table 21.6▲. The binding energies per nucleon (that is, the binding energy of each nucleus divided by the total number of nucleons in that nucleus) are also compared in the table.

The binding energies per nucleon can be used to compare the stabilities of different combinations of nucleons (such as 2 protons and 2 neutrons arranged either as $^{4}_{2}$He or 2 $^{2}_{1}$H). Figure 21.13▶ shows the average binding energy per nucleon plotted against mass number. The binding energy per nucleon at first increases in magnitude as the mass number increases, reaching about 1.4×10^{-12} J for nuclei whose mass numbers are in the vicinity of iron-56. It then decreases slowly to about 1.2×10^{-12} J for very heavy nuclei. This trend indicates that nuclei of intermediate mass numbers are more tightly bound (and therefore more stable) than those with either smaller or larger mass numbers. This trend has two significant consequences: First, heavy nuclei gain stability and therefore give off energy if they are fragmented into two mid-sized nuclei. This process, known as **fission**, is used to generate energy in nuclear power plants. Second, even greater amounts of energy are released if very light nuclei are combined or fused together to give more massive nuclei. This **fusion** process is the essential energy-producing process in the Sun.

▲ **Figure 21.13 Nuclear binding energies.** The average binding energy per nucleon increases to a maximum at a mass number of 50 to 60 and decreases slowly thereafter. Because of these trends, fusion of light nuclei and fission of heavy nuclei are exothermic processes.

▲ GIVE IT SOME THOUGHT

Could fusing two stable nuclei that have mass numbers in the vicinity of 100 cause an energy-releasing process?

21.7 NUCLEAR POWER: FISSION

According to our discussion of the energy changes in nuclear reactions, both the splitting of heavy nuclei (fission) and the union of light nuclei (fusion) are exothermic processes. Commercial nuclear power plants and the most common forms of nuclear weaponry depend on the process of nuclear fission for their operation. The first nuclear fission to be discovered was that of uranium-235.

▶ Figure 21.14 **Diagram of uranium-235 fission.** The diagram shows just one of many fission patterns. In the process shown, 3.5×10^{-11} J of energy is produced per ^{235}U nucleus.

This nucleus, as well as those of uranium-233 and plutonium-239, undergoes fission when struck by a slow-moving neutron.* The induced fission process is illustrated in Figure 21.14▲. A heavy nucleus can split in many different ways. Two ways that the uranium-235 nucleus splits are shown in Equations 21.24 and 21.25:

$$^1_0\text{n} + {}^{235}_{92}\text{U} \longrightarrow {}^{137}_{52}\text{Te} + {}^{97}_{40}\text{Zr} + 2\,{}^1_0\text{n} \qquad [21.24]$$

$$\longrightarrow {}^{142}_{56}\text{Ba} + {}^{91}_{36}\text{Kr} + 3\,{}^1_0\text{n} \qquad [21.25]$$

More than 200 isotopes of 35 elements have been found among the fission products of uranium-235. Most of them are radioactive.

On average, 2.4 neutrons are produced by every fission of a uranium-235 nucleus. If one fission produces two neutrons, these two neutrons can cause two additional fissions. The four neutrons thereby released can produce four fissions, and so forth, as shown in Figure 21.15◀. The number of fissions and the energy released quickly escalate, and if the process is unchecked, the result is a violent explosion. Reactions that multiply in this fashion are called **chain reactions**.

For a fission chain reaction to occur, the sample of fissionable material must have a certain minimum mass. Otherwise, neutrons escape from the sample before they have the opportunity to strike other nuclei and cause additional fission. The chain stops if too many neutrons are lost. The amount of fissionable material large enough to maintain the chain reaction with a constant rate of fission is called the **critical mass**. When a critical mass of material is present, one neutron on average from each fission is subsequently effective in producing another fission. The critical mass of uranium-235 is about 1 kg. If more than a critical mass of fissionable material is present, very few neutrons escape. The chain reaction thus multiplies the number of fissions, which can lead to a nuclear explosion. A mass in excess of a critical mass is referred to as a **supercritical mass**. The effect of mass on a fission reaction is illustrated in Figure 21.16▼.

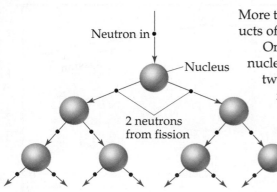

▲ Figure 21.15 **Chain fission reaction.** Assuming that each fission produces two neutrons, the process leads to an accelerating rate of fission, with the number of fissions potentially doubling at each stage.

▶ Figure 21.16 **Subcritical and supercritical fission.** The chain reaction in a subcritical mass soon stops because neutrons are lost from the mass without causing fission. As the size of the mass increases, fewer neutrons are able to escape. In a supercritical mass, the chain reaction is able to accelerate.

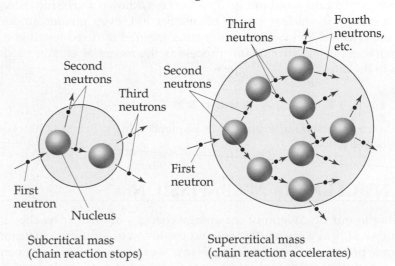

Other heavy nuclei can be induced to undergo fission. However, these three are the only ones of practical importance.

Figure 21.17▶ shows a schematic diagram of the first atomic bomb used in warfare, the bomb that was dropped on Hiroshima, Japan, on August 6, 1945. To trigger a fission reaction, two subcritical masses of uranium-235 are slammed together using chemical explosives. The combined masses of the uranium form a supercritical mass, which leads to a rapid, uncontrolled chain reaction and, ultimately, a nuclear explosion. The energy released by the bomb dropped on Hiroshima was equivalent to that of 20,000 tons of TNT (it therefore is called a *20-kiloton* bomb). Unfortunately, the basic design of a fission-based atomic bomb is quite simple. The fissionable materials are potentially available to any nation with a nuclear reactor. This simplicity has resulted in the proliferation of atomic weapons.

Nuclear Reactors

Nuclear fission produces the energy generated by nuclear power plants. The "fuel" of the nuclear reactor is a fissionable substance, such as uranium-235. Typically, uranium is enriched to about 3% uranium-235 and then used in the form of UO_2 pellets. These enriched uranium pellets are encased in zirconium or stainless steel tubes. Rods composed of materials such as cadmium or boron control the fission process by absorbing neutrons. These *control rods* regulate

▲ Figure 21.17 An atomic bomb design. A conventional explosive is used to bring two subcritical masses together to form a supercritical mass.

Subcritical U-235 target / Bomb casing / Gun barrel / Subcritical U-235 wedge / Chemical explosive

A Closer Look THE DAWNING OF THE NUCLEAR AGE

The fission of uranium-235 was first achieved during the late 1930s by Enrico Fermi and his colleagues in Rome, and shortly thereafter by Otto Hahn and his coworkers in Berlin. Both groups were trying to produce transuranium elements. In 1938 Hahn identified barium among his reaction products. He was puzzled by this observation and questioned the identification because the presence of barium was so unexpected. He sent a detailed letter describing his experiments to Lise Meitner, a former coworker. Meitner had been forced to leave Germany because of the anti-Semitism of the Third Reich and had settled in Sweden. She surmised that Hahn's experiment indicated a new nuclear process was occurring in which the uranium-235 split. She called this process *nuclear fission*.

Meitner passed word of this discovery to her nephew, Otto Frisch, a physicist working at Niels Bohr's institute in Copenhagen. He repeated the experiment, verifying Hahn's observations and finding that tremendous energies were involved. In January 1939, Meitner and Frisch published a short article describing this new reaction. In March 1939, Leo Szilard and Walter Zinn at Columbia University discovered that more neutrons are produced than are used in each fission. As we have seen, this result allows a chain reaction process to occur.

News of these discoveries and an awareness of their potential use in explosive devices spread rapidly within the scientific community. Several scientists finally persuaded Albert Einstein, the most famous physicist of the time, to write a letter to President Franklin D. Roosevelt explaining the implications of these discoveries. Einstein's letter, written in August 1939, outlined the possible military applications of nuclear fission and emphasized the danger that weapons based on fission would pose if they were to be developed by the Nazis.

Roosevelt judged it imperative that the United States investigate the possibility of such weapons. Late in 1941 the decision was made to build a bomb based on the fission reaction. An enormous research project, known as the "Manhattan Project," began.

On December 2, 1942, the first artificial self-sustaining nuclear fission chain reaction was achieved in an abandoned squash court at the University of Chicago (Figure 21.18▼). This accomplishment led to the development of the first atomic bomb, at Los Alamos National Laboratory in New Mexico in July 1945. In August 1945 the United States dropped atomic bombs on two Japanese cities, Hiroshima and Nagasaki. The nuclear age had arrived.

◀ Figure 21.18 The first self-sustaining nuclear fission reactor. This photograph shows part of the original nuclear reactor at the University of Chicago.

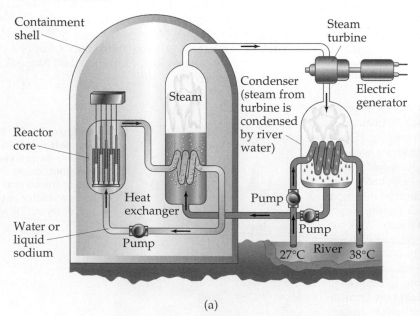

the flux of neutrons to keep the reaction chain self-sustaining, while preventing the reactor core from overheating.*

The reactor is started by a neutron-emitting source; it is stopped by inserting the control rods more deeply into the reactor core at the site of the fission (Figure 21.19 ◀). The reactor core also contains a *moderator*, which acts to slow down neutrons so that they can be captured more readily by the fuel. Graphite is commonly used as a moderator. A *cooling liquid* circulates through the reactor core to carry away the heat generated by the nuclear fission. In some reactor designs, the cooling liquid also serves as the neutron moderator.

The design of a nuclear power plant is basically the same as that of a power plant that burns fossil fuel (except that the burner is replaced by a reactor core). In both instances steam is used to drive a turbine connected to an electrical generator. The steam must be condensed; therefore, additional cooling water, generally obtained from a large source such as a river or lake, is needed. The nuclear power plant design shown in Figure 21.20 ▼ is currently the most popular. The primary coolant, which passes through the core, is in a closed system, which lessens the chance that radioactive products could escape the core. Additionally, the reactor is surrounded by a reinforced concrete shell to shield personnel and nearby residents from radiation and to protect the reactor from external forces. After passing through the reactor core, the primary coolant is very hot. The heat content of the coolant is used to generate high-pressure steam in a heat exchanger. In the United States, more recent nuclear reactor designs use gas cooling instead of liquid cooling, which means fewer pipes and valves; but the basic operating principle is unchanged.

Newer designs for nuclear reactors are being developed. South Africa is building a nuclear power plant near Capetown, expected to be operational in 2010, that is based on the "pebble-bed" reactor design that was originally conceived in the United States just after World War II. In the pebble-bed design, small specks of low-enriched uranium are coated with ceramics and graphite

▲ Figure 21.19 Diagram of a reactor core. The diagram shows fuel elements, control rods, and cooling fluid. A moderator, used to slow the neutrons, is also present.

▼ Figure 21.20 The basic design of a nuclear power plant. (a) A fluid such as water or liquid sodium carries heat produced by the reactor core to a heat exchanger in which steam is generated. The heat exchange liquid moves in a closed loop. The steam is used to drive an electrical generator. (b) A nuclear power plant in Salem, New Jersey. Notice the dome-shaped concrete containment shell.

(a)   (b)

The reactor core cannot reach supercritical levels and explode with the violence of an atomic bomb because the concentration of uranium-235 is too low. However, if the core overheats, sufficient damage can lead to release of radioactive materials into the environment.

and pressed into balls that are about the size of an orange. These balls are the "pebbles," and contain only a few percent uranium. When ready to run, about half a million pebbles are placed in the reactor core along with the control rods. Helium gas is the coolant. As the nuclear reactions proceed, the pebbles heat the helium, and the helium drives a turbine connected to a generator. One advantage of the pebble-bed reactor is that helium is not subject to steam explosions, like water, and does not readily absorb neutrons to produce radioactive products. Another advantage is that the pebble-bed reactor does not need to be shut down to add more fuel; engineers can take some pebbles out of the bottom and weigh them, to see if there is still usable fuel inside, and add usable pebbles to the top of the pile.

Fission products accumulate as a reactor operates. These products decrease the efficiency of the reactor by capturing neutrons. Current commercial reactors must be stopped periodically to replace or reprocess the nuclear fuel. When the fuel rods are removed from the reactor, they are initially very radioactive. It was originally intended that they be stored for several months in pools at the reactor site to allow decay of short-lived radioactive nuclei. They were then to be transported in shielded containers to reprocessing plants where the fuel would be separated from the fission products. Reprocessing plants have been plagued with operational difficulties, however, and there is intense opposition in the United States to the transport of nuclear wastes on the nation's roads. Even if the transportation difficulties could be overcome, the high level of radioactivity of the spent fuel makes reprocessing a hazardous operation. At present in the United States, the spent fuel rods are simply being kept in storage at reactor sites. However, in 2007, the U.S. federal government proposed initiatives that would allow for reprocessing of spent nuclear fuel, overturning a 30-year ban. Russia, the United Kingdom, France, China, and Japan have already developed reprocessing methods and sites.

Storage poses a major problem because the fission products are extremely radioactive. It is estimated that 20 half-lives are required for their radioactivity to reach levels acceptable for biological exposure. Based on the 28.8-yr half-life of strontium-90, one of the longer-lived and most dangerous of the products, the wastes must be stored for 600 years. Plutonium-239 is one of the by-products present in the expended fuel rods. It is formed by absorption of a neutron by uranium-238, followed by two successive beta emissions. (Remember that most of the uranium in the fuel rods is uranium-238.) If the fuel rods are processed, the plutonium-239 is largely recovered because it can be used as a nuclear fuel. However, if the plutonium is not removed, fuel rod storage must be for a very long time because plutonium-239 has a half-life of 24,000 yr.

A considerable amount of research is being devoted to disposal of radioactive wastes. At present, the most attractive possibilities appear to be formation of glass, ceramic, or synthetic rock from the wastes, as a means of immobilizing them. These solid materials would then be placed in containers of high corrosion resistance and durability and buried deep underground. The U.S. Department of Energy has designated Yucca Mountain in Nevada as a disposal site, and extensive construction has been done there. However, because the radioactivity will persist for a long time, there must be assurances that the solids and their containers will not crack from the heat generated by nuclear decay, allowing radioactivity to find its way into underground water supplies.

In spite of all these difficulties, nuclear power is making a modest comeback as an energy source. In 2005, there were 441 working nuclear power plants in the world, most of which are at least 20 years old. Another 26 nuclear power plants are under construction, mostly in Asia, and plans for many more are in progress. The threat of global warming has moved some organizations to propose nuclear power as a major energy source for the planet in the future. However, this goal would require the construction of at least a thousand new nuclear power plants over the next 50 years.

21.8 NUCLEAR POWER: FUSION

Energy is produced when light nuclei are fused into heavier ones. Reactions of this type are responsible for the energy produced by the Sun. Spectroscopic studies indicate that the Sun is composed of 73% H, 26% He, and only 1% of all other elements, by mass. Among the several fusion processes that are believed to occur are:

$$^{1}_{1}\text{H} + ^{1}_{1}\text{H} \longrightarrow ^{2}_{1}\text{H} + ^{0}_{1}\text{e} \qquad [21.26]$$

$$^{1}_{1}\text{H} + ^{2}_{1}\text{H} \longrightarrow ^{3}_{2}\text{He} \qquad [21.27]$$

$$^{3}_{2}\text{He} + ^{3}_{2}\text{He} \longrightarrow ^{4}_{2}\text{He} + 2\,^{1}_{1}\text{H} \qquad [21.28]$$

$$^{3}_{2}\text{He} + ^{1}_{1}\text{H} \longrightarrow ^{4}_{2}\text{He} + ^{0}_{1}\text{e} \qquad [21.29]$$

Fusion is appealing as an energy source because of the availability of light isotopes and because fusion products are generally not radioactive. Despite this fact, fusion is not presently used to generate energy. The problem is that high energies, achieved by high temperatures, are needed to overcome the repulsion between nuclei. Fusion reactions are therefore also known as **thermonuclear reactions**. The lowest temperature required for any fusion is that needed to fuse deuterium ($^{2}_{1}\text{H}$) and tritium ($^{3}_{1}\text{H}$), shown in Equation 21.30. This reaction requires a temperature of about 40,000,000 K:

$$^{2}_{1}\text{H} + ^{3}_{1}\text{H} \longrightarrow ^{4}_{2}\text{He} + ^{1}_{0}\text{n} \qquad [21.30]$$

Such high temperatures have been achieved by using an atomic bomb to initiate the fusion process. This is done in the thermonuclear, or hydrogen, bomb. This approach is unacceptable, however, for controlled power generation.

Numerous problems must be overcome before fusion becomes a practical energy source. In addition to the high temperatures necessary to initiate the reaction, there is the problem of confining the reaction. No known structural material is able to withstand the enormous temperatures necessary for fusion. Research has centered on the use of an apparatus called a *tokamak*, which uses strong magnetic fields to contain and to heat the reaction (Figure 21.21 ▼). Temperatures of over 100,000,000 K have been achieved in a tokamak. In the interior of the Sun, where hydrogen nuclei fuse to form helium nuclei, the temperature is estimated to be 13,600,000 K. This temperature is "low" enough that the reaction is slow—otherwise, the Sun would have consumed its hydrogen long ago.

▶ **Figure 21.21 A tokamak fusion test reactor.** A tokamak is essentially a magnetic "bottle" for confining and heating nuclei in an effort to cause them to fuse.

Vacuum vessel

Toroidal field coil

Poloidal field coils

Plasma

21.9 RADIATION IN THE ENVIRONMENT AND LIVING SYSTEMS

We are continually bombarded by radiation from both natural and artificial sources. We are exposed to infrared, ultraviolet, and visible radiation from the Sun, in addition to radio waves from radio and television stations, microwaves from microwave ovens, and X-rays from various medical procedures. We are also exposed to radioactivity from the soil and other natural materials (Table 21.7▼). Understanding the different energies of these various kinds of radiation is necessary to understand their different effects on matter.

When matter absorbs radiation, the energy of the radiation can cause either excitation or ionization of the matter. Excitation occurs when the absorbed radiation excites electrons to higher energy states or increases the motion of molecules, causing them to move, vibrate, or rotate. Ionization occurs when the radiation removes an electron from an atom or molecule. In general, radiation that causes ionization, called **ionizing radiation**, is far more harmful to biological systems than radiation that does not cause ionization. The latter, called **nonionizing radiation**, is generally of lower energy, such as radiofrequency electromagnetic radiation ∞(Section 6.2), or slow-moving neutrons. Most living tissue contains at least 70% water by mass. When living tissue is irradiated, water molecules absorb most of the energy of the radiation. Thus, it is common to define ionizing radiation as radiation that can ionize water, a process requiring a minimum energy of 1216 kJ/mol. Alpha, beta, and gamma rays (as well as X-rays and higher-energy ultraviolet radiation) possess energies in excess of this quantity and are therefore forms of ionizing radiation.

When ionizing radiation passes through living tissue, electrons are removed from water molecules, forming highly reactive H_2O^+ ions. An H_2O^+ ion can react with another water molecule to form an H_3O^+ ion and a neutral OH molecule:

$$H_2O^+ + H_2O \longrightarrow H_3O^+ + OH \qquad [21.31]$$

The unstable and highly reactive OH molecule is a **free radical**, a substance with one or more unpaired electrons, as seen in the Lewis structure: $\cdot \ddot{O}\!-\!H$. The OH molecule is also called the hydroxyl radical. The presence of the unpaired electron is often emphasized by writing the species with a single dot, $\cdot OH$. In cells and tissues such particles can attack a host of surrounding biomolecules to produce new free radicals, which, in turn, attack yet other compounds. Thus, the formation of a single free radical can initiate a large number of chemical reactions that are ultimately able to disrupt the normal operations of cells.

The damage produced by radiation depends on the activity and energy of the radiation, the length of exposure, and whether the source is inside or outside the body. Gamma rays are particularly harmful outside the body because they penetrate human tissue very effectively, just as X-rays do. Consequently, their damage is not limited to the skin. In contrast, most alpha rays are stopped by skin, and beta rays are able to penetrate only about 1 cm beyond the surface of the skin (Figure 21.22▶). Neither is as dangerous as gamma rays, therefore, *unless*

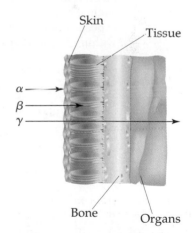

▲ **Figure 21.22 The relative penetrating abilities of alpha, beta, and gamma radiation.**

TABLE 21.7 ■ Average Abundances and Activities of Natural Radionuclides*				
	Potassium-40	**Rubidium-87**	**Thorium-232**	**Uranium-238**
Land elemental abundance (ppm)	28000	112	10.7	2.8
Land activity (Bq/kg)	870	102	43	35
Ocean elemental concentration (mg/L)	339	0.12	1×10^{-7}	0.0032
Ocean activity (Bq/L)	12	0.11	4×10^{-7}	0.040
Ocean sediments elemental abundance (ppm)	17000	—	5.0	1.0
Ocean sediments activity (Bq/kg)	500	—	20	12
Human body activity (Bq)	4000	600	0.08	0.4**

*Data from "Ionizing Radiation Exposure of the Population of the United States," Report 93, 1987, National Council on Radiation Protection.
**Includes lead-210 and polonium-210, daughter nuclei of uranium-238.

the radiation source somehow enters the body. Within the body, alpha rays are particularly dangerous because they transfer their energy efficiently to the surrounding tissue, initiating considerable damage.

In general, the tissues that show the greatest damage from radiation are those that reproduce rapidly, such as bone marrow, blood-forming tissues, and lymph nodes. The principal effect of extended exposure to low doses of radiation is to cause cancer. Cancer is caused by damage to the growth-regulation mechanism of cells, inducing cells to reproduce in an uncontrolled manner. Leukemia, which is characterized by excessive growth of white blood cells, is probably the major type of cancer caused by radiation.

In light of the biological effects of radiation, it is important to determine whether any levels of exposure are safe. Unfortunately, we are hampered in our attempts to set realistic standards because we do not fully understand the effects of long-term exposure to radiation. Scientists concerned with setting health standards have used the hypothesis that the effects of radiation are proportional to exposure, even down to low doses. Any amount of radiation is assumed to cause some finite risk of injury, and the effects of high dosage rates are extrapolated to those of lower ones. Other scientists believe, however, that there is a threshold below which there are no radiation risks. Until scientific evidence enables us to settle the matter with some confidence, it is safer to assume that even low levels of radiation present some danger.

Radiation Doses

Two units commonly used to measure the amount of exposure to radiation are the *gray* and the *rad*. The **gray** (Gy), which is the SI unit of absorbed dose, corresponds to the absorption of 1 J of energy per kilogram of tissue. The **rad** (*r*adiation *a*bsorbed *d*ose) corresponds to the absorption of 1×10^{-2} J of energy per kilogram of tissue. Thus, 1 Gy = 100 rads. The rad is the unit most often used in medicine.

Not all forms of radiation harm biological materials with the same efficiency. A rad of alpha radiation, for example, can produce more damage than a rad of beta radiation. To correct for these differences, the radiation dose is multiplied by a factor that measures the relative biological damage caused by the radiation. This multiplication factor is known as the *relative biological effectiveness* of the radiation, abbreviated *RBE*. The RBE is approximately 1 for gamma and beta radiation, and 10 for alpha radiation. The exact value of the RBE varies with dose rate, total dose, and the type of tissue affected. The product of the radiation dose in rads and the RBE of the radiation gives the effective dosage in units of **rem** (*r*oentgen *e*quivalent for *man*):

Number of rems = (number of rads)(RBE)

[21.32]

The SI unit for effective dosage is the sievert (Sv), obtained by multiplying the RBE times the SI unit for radiation dose, the gray; hence, 1 Sv = 100 rem. The rem is the unit of radiation damage that is usually used in medicine.

The effects of short-term exposures to radiation appear in Table 21.8◀. An exposure of 600 rem is fatal to most humans. To put this number in perspective, a typical dental X-ray entails an exposure of about 0.5 mrem. The average exposure for a person in 1 year due to all natural sources of ionizing radiation (called *background radiation*) is about 360 mrem (Figure 21.23◀).

TABLE 21.8 ■ Effects of Short-Term Exposures to Radiation

Dose (rem)	Effect
0–25	No detectable clinical effects
25–50	Slight, temporary decrease in white blood cell counts
100–200	Nausea; marked decrease in white blood cells
500	Death of half the exposed population within 30 days after exposure

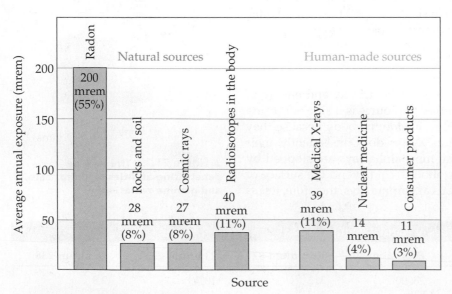

▲ **Figure 21.23 Sources of the average annual exposure of the U.S. population to high-energy radiation.** The total average annual exposure is 360 mrem. *Data from "Ionizing Radiation Exposure of the Population of the United States," Report 93, 1987, National Council on Radiation Protection.*

Naturally occuring radioisotopes are of great value to environmental chemists, geologists, oceanographers, and ecologists. We have already seen how the relative amounts of carbon-14 in organic materials leads to estimates of their age, up to tens of thousands of years old. Different radioisotopes in other systems—layers of ocean sediment, gas bubbles trapped in the deep ice in the Arctic—can be used in a similar manner to track changes in the environment over shorter or longer time scales, depending on the half-life of the radionuclide.

For example, phosphorus-32 and phosphorus-33 are radioisotopes formed in the atmosphere from cosmic ray bombardment of argon gas. These phosphorus isotopes have half-lives of 14 days and 25 days, respectively. They enter the ocean in the form of rain. Chemical oceanographers can learn how phosphorus is distributed in the marine environment (as dissolved inorganic phosphates, as taken up by sea creatures, and so forth) and how rapidly it is distributed over the course of two months by measuring the relative amounts of these two isotopes.

Radon-222 is a naturally occuring gas that is a disintegration product of uranium-238, and therefore is far more abundant on land than at sea (Table 21.7). Figure 21.24▶ shows a map of Biscayne Bay, Florida, and the relative concentrations of radon-222 in units of disintegrations per minute per liter measured along the coast. The "hot spot" in red is evidence of an underground river that is discharging freshwater into the ocean. Results like this help scientists take into account more

and more features of the natural world to predict rates of change of salinity, temperature, nutrients, and many other interrelated parameters across our planet.

▲ **Figure 21.24 Map of radon-222 concentrations measured in Biscayne Bay, Florida.** The concentrations are measured in units of disintegrations per minute per liter. Higher levels of activity are associated with water that has recently flowed from geological formations.

Radon

The radioactive noble gas radon has been much publicized in recent years as a potential risk to health. Radon-222 is a product of the nuclear disintegration series of uranium-238 (Figure 21.5) and is continually generated as uranium in rocks and soil decays. As Figure 21.23 indicates, radon exposure is estimated to account for more than half the 360-mrem average annual exposure to ionizing radiation.

The interplay between the chemical and nuclear properties of radon makes it a health hazard. Because radon is a noble gas, it is extremely unreactive and is therefore free to escape from the ground without chemically reacting along the way. It is readily inhaled and exhaled with no direct chemical effects. The half-life of ^{222}Rn, however, is only 3.82 days. It decays, by losing an alpha particle, into a radioisotope of polonium:

$$^{222}_{86}\text{Rn} \longrightarrow {}^{218}_{84}\text{Po} + {}^{4}_{2}\text{He} \qquad [21.33]$$

Because radon has such a short half-life and alpha particles have a high RBE, inhaled radon is considered a probable cause of lung cancer. Even worse, however, is the decay product, polonium-218, which is an alpha-emitting chemically active element that has an even shorter half-life (3.11 min) than radon-222:

$$^{218}_{84}\text{Po} \longrightarrow {}^{214}_{82}\text{Pb} + {}^{4}_{2}\text{He} \qquad [21.34]$$

The atoms of polonium-218 can become trapped in the lungs, where they bathe the delicate tissue with harmful alpha radiation. The resulting damage is estimated to contribute to 10% of all lung cancer deaths in the United States.

The U.S. Environmental Protection Agency (EPA) has recommended that radon-222 levels not exceed 4 pCi per liter of air in homes. Homes located in areas where the natural uranium content of the soil is high often have levels much greater than that. Because of public awareness, radon-testing kits are readily available in many parts of the country (Figure 21.25▶).

▲ **Figure 21.25 Radon home-testing kit.** Kits such as this are available for measuring radon levels in the home.

Chemistry and Life   RADIATION THERAPY

High-energy radiation poses a health hazard because of the damage it does to cells. Healthy cells are either destroyed or damaged by radiation, leading to physiological disorders. Radiation can also destroy *unhealthy* cells, however, including cancerous cells. All cancers are characterized by the runaway growth of abnormal cells. This growth can produce masses of abnormal tissue, called *malignant tumors*. Malignant tumors can be caused by the exposure of healthy cells to high-energy radiation. Somewhat paradoxically, however, malignant tumors can be destroyed by exposing them to the same radiation that caused them, because rapidly reproducing cells are very susceptible to radiation damage. Thus, cancerous cells are more susceptible to destruction by radiation than healthy ones, allowing radiation to be used effectively in the treatment of cancer. As early as 1904, physicians attempted to use the radiation emitted by radioactive substances to treat tumors by destroying the mass of unhealthy tissue. The treatment of disease by high-energy radiation is called *radiation therapy*.

Many different radionuclides are currently used in radiation therapy. Some of the more commonly used ones are listed in Table 21.9 ▶, along with their half-lives. Most of the half-lives are quite short, meaning that these radioisotopes emit a great deal of radiation in a short period of time.

The radiation source used in radiation therapy may be inside or outside the body. In almost all cases radiation therapy is designed to use the high-energy gamma radiation emitted by radioisotopes. Alpha and beta radiation, which are not as penetrating as gamma radiation, can be blocked by appropriate packaging. For example, ^{192}Ir is often administered as "seeds" consisting of a core of radioactive isotope

coated with 0.1 mm of platinum metal. The platinum coating stops the alpha and beta rays, but the gamma rays penetrate it readily. The radioactive seeds can be surgically implanted in a tumor. In other cases human physiology allows the radioisotope to be ingested. For example, most of the iodine in the human body ends up in the thyroid gland, so thyroid cancer can be treated by using large doses of ^{131}I. Radiation therapy on deep organs, where a surgical implant is impractical, often uses a ^{60}Co "gun" outside the body to shoot a beam of gamma rays at the tumor. Particle accelerators are also used as an external source of high-energy radiation for radiation therapy.

Because gamma radiation is so strongly penetrating, it is nearly impossible to avoid damaging healthy cells during radiation therapy. Many cancer patients undergoing radiation treatment experience unpleasant and dangerous side effects such as fatigue, nausea, hair loss, a weakened immune system, and occasionally even death; but if other treatments such as *chemotherapy* (the use of drugs to combat cancer) fail, radiation therapy is a good option.

TABLE 21.9 ■ Some Radioisotopes Used in Radiation Therapy

Isotope	Half-life	Isotope	Half-life
^{32}P	14.3 days	^{137}Cs	30 yr
^{60}Co	5.27 yr	^{192}Ir	74.2 days
^{90}Sr	28.8 yr	^{198}Au	2.7 days
^{125}I	60.25 days	^{222}Rn	3.82 days
^{131}I	8.04 days	^{226}Ra	1600 yr

■ SAMPLE INTEGRATIVE EXERCISE | Putting Concepts Together

Potassium ion is present in foods and is an essential nutrient in the human body. One of the naturally occurring isotopes of potassium, potassium-40, is radioactive. Potassium-40 has a natural abundance of 0.0117% and a half-life of $t_{1/2} = 1.28 \times 10^9$ yr. It undergoes radioactive decay in three ways: 98.2% is by electron capture, 1.35% is by beta emission, and 0.49% is by positron emission. **(a)** Why should we expect ^{40}K to be radioactive? **(b)** Write the nuclear equations for the three modes by which ^{40}K decays. **(c)** How many ^{40}K$^+$ ions are present in 1.00 g of KCl? **(d)** How long does it take for 1.00% of the ^{40}K in a sample to undergo radioactive decay?

SOLUTION

(a) The ^{40}K nucleus contains 19 protons and 21 neutrons. There are very few stable nuclei with odd numbers of both protons and neutrons. ∞ (Section 21.2).

(b) Electron capture is capture of an inner-shell electron ($_{-1}^{0}$e) by the nucleus:

$$^{40}_{19}\text{K} + {}^{0}_{-1}\text{e} \longrightarrow {}^{40}_{18}\text{Ar}$$

Beta emission is loss of a beta particle ($_{-1}^{0}$e) by the nucleus:

$$^{40}_{19}\text{K} \longrightarrow {}^{40}_{20}\text{Ca} + {}^{0}_{-1}\text{e}$$

Positron emission is loss of $_{1}^{0}$e by the nucleus:

$$^{40}_{19}\text{K} \longrightarrow {}^{40}_{18}\text{Ar} + {}^{0}_{1}\text{e}$$

(c) The total number of K^+ ions in the sample is

$$(1.00 \text{ g KCl})\left(\frac{1 \text{ mol KCl}}{74.55 \text{ g KCl}}\right)\left(\frac{1 \text{ mol } K^+}{1 \text{ mol KCl}}\right)\left(\frac{6.022 \times 10^{23} \text{ } K^+}{1 \text{ mol } K^+}\right) = 8.08 \times 10^{21} \text{ } K^+ \text{ ions}$$

Of these, 0.0117% are $^{40}K^+$ ions:

$$(8.08 \times 10^{21} \text{ } K^+ \text{ ions})\left(\frac{0.0117 \text{ } ^{40}K^+ \text{ ions}}{100 \text{ } K^+ \text{ ions}}\right) - 9.45 \times 10^{17} \text{ } ^{40}K^+ \text{ ions}$$

(d) The decay constant (the rate constant) for the radioactive decay can be calculated from the half-life, using Equation 21.20:

$$k = \frac{0.693}{t_{1/2}} = \frac{0.693}{1.28 \times 10^9 \text{ yr}} = (5.41 \times 10^{-10})/\text{yr}$$

The rate equation, Equation 21.19, then allows us to calculate the time required:

$$\ln \frac{N_t}{N_0} = -kt$$

$$\ln \frac{99}{100} = -((5.41 \times 10^{-10})/\text{yr})t$$

$$-0.01005 = -((5.41 \times 10^{-10})/\text{yr})t$$

$$t = \frac{-0.01005}{(-5.41 \times 10^{-10})/\text{yr}} = 1.86 \times 10^7 \text{ yr}$$

That is, it would take 18.6 million years for just 1.00% of the ^{40}K in a sample to decay.

CHAPTER REVIEW

SUMMARY AND KEY TERMS

Introduction and Section 21.1 The nucleus contains protons and neutrons, both of which are called **nucleons**. Nuclei that are **radioactive** spontaneously emit radiation. These radioactive nuclei are called **radionuclides**, and the atoms containing them are called **radioisotopes**. When a radionuclide decomposes, it is said to undergo radioactive decay. In nuclear equations, reactant and product nuclei are represented by giving their mass numbers and atomic numbers, as well as their chemical symbol. The totals of the mass numbers on both sides of the equation are equal; the totals of the atomic numbers on both sides are also equal. There are five common kinds of radioactive decay: emission of **alpha (α) particles** (^4_2He), emission of **beta (β) particles** ($^0_{-1}e$), **positron** emission (^0_1e), **electron capture**, and emission of **gamma (γ) radiation** ($^0_0\gamma$).

Section 21.2 The neutron-to-proton ratio is an important factor determining nuclear stability. By comparing a nuclide's neutron-to-proton ratio with those in the band of stability, we can predict the mode of radioactive decay. In general, neutron-rich nuclei tend to emit beta particles; proton-rich nuclei tend to either emit positrons or undergo

electron capture; and heavy nuclei tend to emit alpha particles. The presence of **magic numbers** of nucleons and an even number of protons and neutrons also help determine the stability of a nucleus. A nuclide may undergo a series of decay steps before a stable nuclide forms. This series of steps is called a **radioactive series**, or a **nuclear disintegration series**.

Section 21.3 **Nuclear transmutations**, induced conversions of one nucleus into another, can be brought about by bombarding nuclei with either charged particles or neutrons. **Particle accelerators** increase the kinetic energies of positively charged particles, allowing these particles to overcome their electrostatic repulsion by the nucleus. Nuclear transmutations are used to produce the **transuranium elements**, those elements with atomic numbers greater than that of uranium.

Sections 21.4 and 21.5 The SI unit for the activity of a radioactive source is the **becquerel** (Bq), defined as one nuclear disintegration per second. A related unit, the **curie** (Ci), corresponds to 3.7×10^{10} disintegrations per second.

Nuclear decay is a first-order process. The decay rate (**activity**) is therefore proportional to the number of radioactive nuclei, and radionuclides have constant **half-lives**. Some can therefore be used to date objects; ^{14}C, for example, is used to date organic objects. Geiger counters and scintillation counters count the emissions from radioactive samples. The ease of detection of radioisotopes also permits their use as **radiotracers**, to follow elements through their reactions.

Section 21.6 The energy produced in nuclear reactions is accompanied by measurable losses of mass in accordance with Einstein's relationship, $\Delta E = c^2 \Delta m$. The difference in mass between nuclei and the nucleons of which they are composed is known as the **mass defect**. The mass defect of a nuclide makes it possible to calculate its **nuclear binding energy**, the energy required to separate the nucleus into individual nucleons. Energy is produced when heavy nuclei split (**fission**) and when light nuclei fuse (**fusion**).

Sections 21.7 and 21.8 Uranium-235, uranium-233, and plutonium-239 undergo fission when they capture a neutron. The resulting nuclear reaction is a nuclear **chain reaction**, a reaction in which the neutrons produced in one fission cause further fission reactions. A reaction that maintains a constant rate is said to be critical, and the mass necessary to maintain this constant rate is called a **critical mass**. A mass in excess of the critical mass is termed a **supercritical mass**. In nuclear reactors the fission rate is controlled to generate a constant power.

The reactor core consists of fissionable fuel, control rods, a moderator, and cooling fluid. The nuclear power plant resembles a conventional power plant except that the reactor core replaces the fuel burner. There is concern about the disposal of highly radioactive nuclear wastes that are generated in nuclear power plants.

Nuclear fusion requires high temperatures because nuclei must have large kinetic energies to overcome their mutual repulsions. Fusion reactions are therefore called **thermonuclear reactions**. It is not yet possible to generate a controlled fusion process.

Section 21.9 **Ionizing radiation** is energetic enough to remove an electron from a water molecule; radiation with less energy is called **nonionizing radiation**. Ionizing radiation generates **free radicals**, reactive substances with one or more unpaired electrons. The effects of long-term exposure to low levels of radiation are not completely understood, but it is usually assumed that the extent of biological damage varies in direct proportion to the level of exposure.

The amount of energy deposited in biological tissue by radiation is called the radiation dose and is measured in units of grays or rads. One **gray** (Gy) corresponds to a dose of 1 J/kg of tissue. The **rad** is a smaller unit; 100 rads = 1 Gy. The effective dose, which measures the biological damage created by the deposited energy, is measured in units of rems or sieverts (Sv). The **rem** is obtained by multiplying the number of rads by the relative biological effectiveness (RBE); 100 rem = 1 Sv.

KEY SKILLS

- Write balanced nuclear equations.
- Know the difference between fission and fusion.
- Predict nuclear stability in terms of neutron-to-proton ratio.
- Calculate ages of objects or amounts of material from data on nucleon abundances using the half-life of a radioactive material.
- Convert between nuclear activity units.
- Calculate mass and energy changes for nuclear reactions.
- Understand the meaning of radiation dosage terms.
- Understand the biological effects of different kinds of radiation.

KEY EQUATIONS

- $\ln \dfrac{N_t}{N_0} = -kt$ [21.19]   First-order rate law for nuclear decay

- $k = \dfrac{0.693}{t_{1/2}}$ [21.20]   Relationship between nuclear decay constant and half-life; this is derived from the previous equation at $N_t = (1/2)N_0$

- $E = mc^2$ [21.22]   Einstein's equation that relates mass and energy

VISUALIZING CONCEPTS

21.1 Indicate whether each of the following nuclides lies within the belt of stability in Figure 21.3: **(a)** neon-24, **(b)** chlorine-32, **(c)** tin-108, **(d)** polonium-216. For any that do not, describe a nuclear decay process that would alter the neutron-to-proton ratio in the direction of increased stability. [Section 21.2]

21.2 Write the balanced nuclear equation for the reaction represented by the diagram shown here. [Section 21.2]

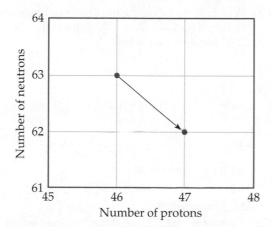

21.3 Draw a diagram similar to that shown in Exercise 21.2 that illustrates the nuclear reaction $^{211}_{83}Bi \longrightarrow ^{4}_{2}He + ^{207}_{81}Tl$. [Section 21.2]

21.4 The accompanying graph illustrates the decay of $^{88}_{42}Mo$, which decays via positron emission. **(a)** What is the half-life of the decay? **(b)** What is the rate constant for the decay? **(c)** What fraction of the original sample of $^{88}_{42}Mo$ remains after 12 minutes? **(d)** What is the product of the decay process? [Section 21.4]

21.5 A $^{100}_{44}Ru$ atom has a mass of 99.90422 amu. **(a)** Why is the mass of this nuclide not much closer to 100? **(b)** Explain the observed mass using Figure 21.13. **(c)** Calculate the binding energy per nucleon for $^{100}_{44}Ru$. [Section 21.6]

21.6 The diagram shown below illustrates a fission process. **(a)** What is the second nuclear product of the fission? **(b)** Use Figure 21.3 to predict whether the nuclear products of this fission reaction are stable. [Section 21.7]

EXERCISES

Radioactivity

21.7 Indicate the number of protons and neutrons in the following nuclei: **(a)** $^{55}_{25}Mn$, **(b)** ^{201}Hg, **(c)** potassium-39.

21.8 Indicate the number of protons and neutrons in the following nuclei: **(a)** $^{126}_{55}Cs$, **(b)** ^{119}Sn, **(c)** barium-141.

21.9 Give the symbol for **(a)** a neutron, **(b)** an alpha particle, **(c)** gamma radiation.

21.10 Give the symbol for **(a)** a proton, **(b)** a beta particle, **(c)** a positron.

21.11 Write balanced nuclear equations for the following processes: **(a)** rubidium-90 undergoes beta decay; **(b)** selenium-72 undergoes electron capture; **(c)** krypton-76 undergoes positron emission; **(d)** radium-226 emits alpha radiation.

21.12 Write balanced nuclear equations for the following transformations: **(a)** gold-191 undergoes electron capture; **(b)** gold-201 decays to a mercury isotope; **(c)** gold-198 undergoes beta decay; **(d)** gold-188 decays by positron emission.

21.13 Decay of which nucleus will lead to the following products: **(a)** bismuth-211 by beta decay; **(b)** chromium-50 by positron emission; **(c)** tantalum-179 by electron capture; **(d)** radium-226 by alpha decay?

21.14 What particle is produced during the following decay processes: **(a)** sodium-24 decays to magnesium-24; **(b)** mercury-188 decays to gold-188; **(c)** iodine-122 decays to xenon-122; **(d)** plutonium-242 decays to uranium-238?

21.15 The naturally occurring radioactive decay series that begins with $^{235}_{92}U$ stops with formation of the stable $^{207}_{82}Pb$ nucleus. The decays proceed through a series of alpha-particle and beta-particle emissions. How many of each type of emission are involved in this series?

21.16 A radioactive decay series that begins with $^{232}_{90}Th$ ends with formation of the stable nuclide $^{208}_{82}Pb$. How many alpha-particle emissions and how many beta-particle emissions are involved in the sequence of radioactive decays?

Nuclear Stability

21.17 Predict the type of radioactive decay process for the following radionuclides: **(a)** $^{8}_{5}B$, **(b)** $^{68}_{29}Cu$, **(c)** phosphorus-32, **(d)** chlorine-39.

21.18 Each of the following nuclei undergoes either beta or positron emission. Predict the type of emission for each: **(a)** $^{66}_{32}Ge$, **(b)** $^{105}_{45}Rh$, **(c)** iodine-137, **(d)** cerium-133.

21.19 One of the nuclides in each of the following pairs is radioactive. Predict which is radioactive and which is stable: **(a)** $^{39}_{19}K$ and $^{40}_{19}K$, **(b)** ^{209}Bi and ^{208}Bi, **(c)** nickel-58 and nickel-65. Explain.

21.20 In each of the following pairs, which nuclide would you expect to be the more abundant in nature: **(a)** $^{115}_{48}Cd$ or $^{112}_{48}Cd$, **(b)** $^{30}_{13}Al$ or $^{27}_{13}Al$, **(c)** palladium-106 or palladium-113, **(d)** xenon-128 or cesium-128? Justify your choices.

21.21 Which of the following nuclides have magic numbers of both protons and neutrons: **(a)** helium-4, **(b)** oxygen-18, **(c)** calcium-40, **(d)** zinc-66, **(e)** lead-208?

21.22 Tin has 10 stable isotopes, but antimony only has two. How can we explain this difference?

21.23 Using the concept of magic numbers, explain why alpha emission is relatively common, but proton emission is nonexistent.

21.24 Which of the following nuclides would you expect to be radioactive: $^{62}_{28}Ni$, $^{58}_{29}Cu$, $^{108}_{47}Ag$, tungsten-184, polonium-206? Justify your choices.

Nuclear Transmutations

21.25 Why are nuclear transmutations involving neutrons generally easier to accomplish than those involving protons or alpha particles?

21.26 Rutherford was able to carry out the first nuclear transmutation reactions by bombarding nitrogen-14 nuclei with alpha particles. In the famous experiment on scattering of alpha particles by gold foil (Section 2.2), however, a nuclear transmutation reaction did not occur. What is the difference between the two experiments? What would one need to do to carry out a successful nuclear transmutation reaction involving gold nuclei and alpha particles?

21.27 Complete and balance the following nuclear equations by supplying the missing particle:

(a) $^{252}_{98}Cf + ^{10}_{5}B \longrightarrow 3\,^{1}_{0}n + ?$

(b) $^{2}_{1}H + ^{3}_{2}He \longrightarrow ^{4}_{2}He + ?$

(c) $^{1}_{1}H + ^{11}_{5}B \longrightarrow 3?$

(d) $^{122}_{53}I \longrightarrow ^{122}_{54}Xe + ?$

(e) $^{59}_{26}Fe \longrightarrow ^{0}_{-1}e + ?$

21.28 Complete and balance the following nuclear equations by supplying the missing particle:

(a) $^{32}_{16}S + ^{1}_{0}n \longrightarrow ^{1}_{1}p + ?$

(b) $^{7}_{4}Be + ^{0}_{-1}e$ (orbital electron) $\longrightarrow ?$

(c) $? \longrightarrow ^{187}_{76}Os + ^{0}_{-1}e$

(d) $^{98}_{42}Mo + ^{2}_{1}H \longrightarrow ^{1}_{0}n + ?$

(e) $^{235}_{92}U + ^{1}_{0}n \longrightarrow ^{135}_{54}Xe + 2\,^{1}_{0}n + ?$

21.29 Write balanced equations for **(a)** $^{238}_{92}U(\alpha, n)^{241}_{94}Pu$, **(b)** $^{14}_{7}N(\alpha, p)^{17}_{8}O$, **(c)** $^{56}_{26}Fe(\alpha, \beta)^{60}_{29}Cu$.

21.30 Write balanced equations for each of the following nuclear reactions: **(a)** $^{238}_{92}U(n, \gamma)^{239}_{92}U$, **(b)** $^{14}_{7}N(p, \alpha)^{11}_{6}C$, **(c)** $^{18}_{8}O(n, \beta)^{19}_{9}F$.

Rates of Radioactive Decay

21.31 Each statement below refers to a comparison between two radioisotopes, A and X. Indicate whether each of the following statements is true or false, and why.

(a) If the half-life for A is shorter than the half-life for X, A has a larger decay rate constant.

(b) If X is "not radioactive," its half-life is essentially zero.

(c) If A has a half-life of 10 years, and X has a half-life of 10,000 years, A would be a more suitable radioisotope to measure processes occurring on the 40-year time scale.

21.32 It has been suggested that strontium-90 (generated by nuclear testing) deposited in the hot desert will undergo radioactive decay more rapidly because it will be exposed to much higher average temperatures. **(a)** Is this a reasonable suggestion? **(b)** Does the process of radioactive decay have an activation energy, like the Arrhenius behavior of many chemical reactions ∞ (Section 14.5)? Discuss.

21.33 The half-life of tritium (hydrogen-3) is 12.3 yr. If 56.2 mg of tritium is released from a nuclear power plant during the course of an accident, what mass of this nuclide will remain after 12.3 yr? After 100 yr?

21.34 It takes 5.2 minutes for a 1.000-g sample of ^{210}Fr to decay to 0.250 g. What is the half-life of ^{210}Fr?

21.35 Cobalt-60 is a strong gamma emitter that has a half-life of 5.26 yr. The cobalt-60 in a radiotherapy unit must be replaced when its radioactivity falls to 75% of the original sample. **(a)** If an original sample was purchased in June 2006, when will it be necessary to replace the cobalt-60? **(b)** How can you store cobalt-60 so that it is safe to handle?

21.36 How much time is required for a 6.25-mg sample of ^{51}Cr to decay to 0.75 mg if it has a half-life of 27.8 days?

[21.37] Radium-226, which undergoes alpha decay, has a half-life of 1600 yr. **(a)** How many alpha particles are emitted in 5.0 min by a 10.0-mg sample of ^{226}Ra? **(b)** What is the activity of the sample in mCi?

[21.38] Cobalt-60, which undergoes beta decay, has a half-life of 5.26 yr. **(a)** How many beta particles are emitted in 180 s by a 3.75-mg sample of ^{60}Co? **(b)** What is the activity of the sample in Bq?

21.39 The cloth shroud from around a mummy is found to have a ^{14}C activity of 9.7 disintegrations per minute per gram of carbon as compared with living organisms that undergo 16.3 disintegrations per minute per gram of carbon. From the half-life for ^{14}C decay, 5715 yr, calculate the age of the shroud.

21.40 A wooden artifact from a Chinese temple has a ^{14}C activity of 38.0 counts per minute as compared with an activity of 58.2 counts per minute for a standard of zero age. From the half-life for ^{14}C decay, 5715 yr, determine the age of the artifact.

21.41 Potassium-40 decays to argon-40 with a half-life of 1.27×10^9 yr. What is the age of a rock in which the mass ratio of ^{40}Ar to ^{40}K is 4.2?

21.42 The half-life for the process $^{238}U \longrightarrow {}^{206}Pb$ is 4.5×10^9 yr. A mineral sample contains 75.0 mg of ^{238}U and 18.0 mg of ^{206}Pb. What is the age of the mineral?

Energy Changes

21.43 The thermite reaction, $Fe_2O_3(s) + 2\ Al(s) \longrightarrow 2\ Fe(s) + Al_2O_3(s)$, is one of the most exothermic reactions known. The heat released is sufficient to melt the iron product; consequently, the thermite reaction is used to weld metal under the ocean. $\Delta H°$ for the thermite reaction is -851.5 kJ/mole. What is the mass change per mole of Fe_2O_3 that accompanies this reaction?

21.44 An analytical laboratory balance typically measures mass to the nearest 0.1 mg. What energy change would accompany the loss of 0.1 mg in mass?

21.45 How much energy must be supplied to break a single aluminum-27 nucleus into separated protons and neutrons if an aluminum-27 atom has a mass of 26.9815386 amu? How much energy is required for 100.0 grams of aluminum-27? (The mass of an electron is given on the inside back cover.)

21.46 How much energy must be supplied to break a single ^{21}Ne nucleus into separated protons and neutrons if the nucleus has a mass of 20.98846 amu? What is the nuclear binding energy for 1 mol of ^{21}Ne?

21.47 Calculate the binding energy per nucleon for the following nuclei. **(a)** $^{12}_{6}C$ (nuclear mass, 11.996708 amu); **(b)** ^{37}Cl (nuclear mass, 36.956576 amu); **(c)** rhodium-103 (atomic mass, 102.905504 amu).

21.48 Calculate the binding energy per nucleon for the following nuclei: **(a)** $^{14}_{7}N$ (nuclear mass, 13.999234 amu); **(b)** ^{48}Ti (nuclear mass, 47.935878 amu); **(c)** xenon-129 (atomic mass, 128.904779 amu).

21.49 The energy from solar radiation falling on Earth is 1.07×10^{16} kJ/min. **(a)** How much loss of mass from the Sun occurs in one day from just the energy falling on Earth? **(b)** If the energy released in the reaction

$$^{235}U + {}^{1}_{0}n \longrightarrow {}^{141}_{56}Ba + {}^{92}_{36}Kr + 3\ {}^{1}_{0}n$$

(^{235}U nuclear mass, 234.9935 amu; ^{141}Ba nuclear mass, 140.8833 amu; ^{92}Kr nuclear mass, 91.9021 amu) is taken as typical of that occurring in a nuclear reactor, what mass of uranium-235 is required to equal 0.10% of the solar energy that falls on Earth in 1.0 day?

21.50 Based on the following atomic mass values—^{1}H, 1.00782 amu; ^{2}H, 2.01410 amu; ^{3}H, 3.01605 amu; ^{3}He, 3.01603 amu; ^{4}He, 4.00260 amu—and the mass of the neutron given in the text, calculate the energy released per mole in each of the following nuclear reactions, all of which are possibilities for a controlled fusion process:

(a) $^2_1H + {}^3_1H \longrightarrow {}^4_2He + {}^1_0n$

(b) $^2_1H + {}^2_1H \longrightarrow {}^3_2He + {}^1_0n$

(c) $^2_1H + {}^3_2He \longrightarrow {}^4_2He + {}^1_1H$

21.51 Which of the following nuclei is likely to have the largest mass defect per nucleon: **(a)** ^{59}Co, **(b)** ^{11}B, **(c)** ^{118}Sn, **(d)** ^{243}Cm? Explain your answer.

21.52 Based on Figure 21.13, what is the most stable nucleus in the periodic table?

Effects and Uses of Radioisotopes

21.53 Iodine-131 is a convenient radioisotope to monitor thyroid activity in humans. It is a beta emitter with a half-life of 8.02 days. The thyroid is the only gland in the body that uses iodine. A person undergoing a test of thyroid activity drinks a solution of NaI, in which only a small fraction of the iodide is radioactive. **(a)** Why is NaI a good choice for the source of iodine? **(b)** If a Geiger counter is placed near the person's thyroid (which is near the neck) right after the sodium iodide solution is taken, what will the data look like as a function of time? **(c)** A normal thyroid will take up about 12% of the ingested iodide in a few hours. How long will it take for the radioactive iodide taken up and held by the thyroid to decay to 0.01% of the original amount?

21.54 Chlorine-36 is a convenient radiotracer. It is a weak beta emitter, with $t_{1/2} = 3 \times 10^5$ yr. Describe how you would use this radiotracer to carry out each of the following experiments. **(a)** Determine whether trichloroacetic acid, CCl_3COOH, undergoes any ionization of its chlorines as chloride ion in aqueous solution. **(b)** Demonstrate that the equilibrium between dissolved $BaCl_2$ and solid $BaCl_2$ in a saturated solution is a dynamic process. **(c)** Determine the effects of soil pH on the uptake of chloride ion from the soil by soybeans.

21.55 Explain the following terms that apply to fission reactions: **(a)** chain reaction, **(b)** critical mass.

21.56 Explain the function of the following components of a nuclear reactor: **(a)** control rods, **(b)** moderator.

21.57 Complete and balance the nuclear equations for the following fission or fusion reactions:

(a) $^{2}_{1}H + ^{2}_{1}H \longrightarrow ^{3}_{2}He + \_\_$

(b) $^{233}_{92}U + ^{1}_{0}n \longrightarrow ^{133}_{51}Sb + ^{98}_{41}Nb + \_\_ ^{1}_{0}n$

21.58 Complete and balance the nuclear equations for the following fission reactions:

(a) $^{235}_{92}U + ^{1}_{0}n \longrightarrow ^{160}_{62}Sm + ^{72}_{30}Zn + \_\_ ^{1}_{0}n$

(b) $^{239}_{94}Pu + ^{1}_{0}n \longrightarrow ^{144}_{58}Ce + \_\_ + 2 ^{1}_{0}n$

21.59 A portion of the Sun's energy comes from the reaction

$$4 ^{1}_{1}H \longrightarrow ^{4}_{2}He + 2 ^{0}_{1}e$$

This reaction requires a temperature of about 10^6 to 10^7 K. **(a)** Why is such a high temperature required? **(b)** Is the Sun solid?

[21.60] The spent fuel rods from a fission reactor are much more intensely radioactive than the original fuel rods. **(a)** What does this tell you about the products of the fission process in relationship to the belt of stability, Figure 21.3? **(b)** Given that only two or three neutrons are released per fission event and knowing that the nucleus undergoing fission has a neutron-to-proton ratio characteristic of a heavy nucleus, what sorts of decay would you expect to be dominant among the fission products?

21.61 Hydroxyl radicals can pluck hydrogen atoms from molecules ("hydrogen abstraction"), and hydroxide ions can pluck protons from molecules ("deprotonation"). Write the reaction equations and Lewis dot structures for the hydrogen abstraction and deprotonation reactions for the generic carboxylic acid R–COOH with hydroxyl radical and hydroxide ion, respectively. Why is hydroxyl radical more toxic to living systems than hydroxide ion?

21.62 Use Lewis structures to represent the reactants and products in Equation 21.31. Why is the H_2O^+ ion a free radical species?

21.63 A laboratory rat is exposed to an alpha-radiation source whose activity is 14.3 mCi. **(a)** What is the activity of the radiation in disintegrations per second? In becquerels? **(b)** The rat has a mass of 385 g and is exposed to the radiation for 14.0 s, absorbing 35% of the emitted alpha particles, each having an energy of 9.12×10^{-13} J. Calculate the absorbed dose in millirads and grays. **(c)** If the RBE of the radiation is 9.5, calculate the effective absorbed dose in mrem and Sv.

21.64 A 65-kg person is accidentally exposed for 240 s to a 15-mCi source of beta radiation coming from a sample of ^{90}Sr. **(a)** What is the activity of the radiation source in disintegrations per second? In becquerels? **(b)** Each beta particle has an energy of 8.75×10^{-14} J, and 7.5% of the radiation is absorbed by the person. Assuming that the absorbed radiation is spread over the person's entire body, calculate the absorbed dose in rads and in grays. **(c)** If the RBE of the beta particles is 1.0, what is the effective dose in mrem and in sieverts? **(d)** How does the magnitude of this dose of radiation compare with that of a mammogram (300 mrem)?

ADDITIONAL EXERCISES

21.65 Radon-222 decays to a stable nucleus by a series of three alpha emissions and two beta emissions. What is the stable nucleus that is formed?

21.66 A free neutron is unstable and decays into a proton with a half-life of 10.4 min. **(a)** What other particle forms? **(b)** Why don't neutrons in atomic nuclei decay at the same rate?

21.67 Americium-241 is an alpha emitter used in smoke detectors. The alpha radiation ionizes molecules in an air-filled gap between two electrodes in the smoke detector, leading to current. When smoke is present, the ionized molecules bind to smoke particles and the current decreases; when the current is reduced sufficiently, an alarm sounds. **(a)** Write the nuclear equation corresponding to the alpha decay of americium-241. **(b)** Why is an alpha emitter a better choice than a gamma emitter for a smoke detector? **(c)** In a commercial smoke detector, only 0.2 micrograms of americium are present. Calculate the energy that is equivalent to the mass loss of this amount of americium due to alpha radiation. The atomic mass of americium-241 is 241.056829 amu. **(d)** The half-life of americium-241 is 432 years; the half life of americium-240 is 2.12 days. Why is the 241 isotope a better choice for a smoke detector?

21.68 Chlorine has two stable nuclides, ^{35}Cl and ^{37}Cl. In contrast, ^{36}Cl is a radioactive nuclide that decays by beta emission. **(a)** What is the product of decay of ^{36}Cl? **(b)** Based on the empirical rules about nuclear stability, explain why the nucleus of ^{36}Cl is less stable than either ^{35}Cl or ^{37}Cl.

21.69 Nuclear scientists have synthesized approximately 1600 nuclei not known in nature. More might be discovered with heavy-ion bombardment using high-energy particle accelerators. Complete and balance the following reactions, which involve heavy-ion bombardments:

(a) $^{6}_{3}Li + ^{56}_{28}Ni \longrightarrow ?$

(b) $^{40}_{20}Ca + ^{248}_{96}Cm \longrightarrow ^{147}_{62}Sm + ?$

(c) $^{88}_{38}Sr + ^{84}_{36}Kr \longrightarrow ^{116}_{46}Pd + ?$

(d) $^{40}_{20}Ca + ^{238}_{92}U \longrightarrow ^{70}_{30}Zn + 4 ^{1}_{0}n + 2?$

[21.70] The synthetic radioisotope technetium-99, which decays by beta emission, is the most widely used isotope in nuclear medicine. The following data were collected on a sample of ^{99}Tc:

Disintegrations per Minute	Time (h)
180	0
130	2.5
104	5.0
77	7.5
59	10.0
46	12.5
24	17.5

Using these data, make an appropriate graph and curve fit to determine the half-life.

[21.71] According to current regulations, the maximum permissible dose of strontium-90 in the body of an adult is $1 \mu Ci$ (1×10^{-6} Ci). Using the relationship rate = kN, calculate the number of atoms of strontium-90 to which this dose corresponds. To what mass of strontium-90 does this correspond? The half-life for strontium-90 is 28.8 yr.

[21.72] Suppose you had a detection device that could count every decay event from a radioactive sample of plutonium-239 ($t_{1/2}$ is 24,000 yr). How many counts per second would you obtain from a sample containing 0.385 g of plutonium-239?

21.73 Methyl acetate (CH_3COOCH_3) is formed by the reaction of acetic acid with methyl alcohol. If the methyl alcohol is labeled with oxygen-18, the oxygen-18 ends up in the methyl acetate:

$$CH_3\overset{O}{\overset{\|}{C}}OH + H^{18}OCH_3 \longrightarrow CH_3\overset{O}{\overset{\|}{C}}{}^{18}OCH_3 + H_2O$$

Do the C—OH bond of the acid and the O—H bond of the alcohol break in the reaction, or do the O—H bond of the acid and the C—OH bond of the alcohol break? Explain.

21.74 An experiment was designed to determine whether an aquatic plant absorbed iodide ion from water. Iodine-131 ($t_{1/2} = 8.02$ days) was added as a tracer, in the form of iodide ion, to a tank containing the plants. The initial activity of a 1.00-μL sample of the water was 214 counts per minute. After 30 days the level of activity in a 1.00-μL sample was 15.7 counts per minute. Did the plants absorb iodide from the water? Explain.

21.75 The nuclear masses of ^{7}Be, ^{9}Be, and ^{10}Be are 7.0147, 9.0100, and 10.0113 amu, respectively. Which of these nuclei has the largest binding energy per nucleon?

[21.76] A 26.00-g sample of water containing tritium, $^{3}_{1}H$, emits 1.50×10^3 beta particles per second. Tritium is a weak beta emitter, with a half-life of 12.3 yr. What fraction of all the hydrogen in the water sample is tritium?

21.77 The Sun radiates energy into space at the rate of 3.9×10^{26} J/s. **(a)** Calculate the rate of mass loss from the Sun in kg/s. **(b)** How does this mass loss arise? **(c)** It is estimated that the Sun contains 9×10^{56} free protons. How many protons per second are consumed in nuclear reactions in the Sun?

[21.78] The average energy released in the fission of a single uranium-235 nucleus is about 3×10^{-11} J. If the conversion of this energy to electricity in a nuclear power plant is 40% efficient, what mass of uranium-235 undergoes fission in a year in a plant that produces 1000 MW (megawatts)? Recall that a watt is 1 J/s.

[21.79] Tests on human subjects in Boston in 1965 and 1966, following the era of atomic bomb testing, revealed average quantities of about 2 pCi of plutonium radioactivity in the average person. How many disintegrations per second does this level of activity imply? If each alpha particle deposits 8×10^{-13} J of energy and if the average person weighs 75 kg, calculate the number of rads and rems of radiation in 1 yr from such a level of plutonium.

INTEGRATIVE EXERCISES

21.80 A 53.8-mg sample of sodium perchlorate contains radioactive chlorine-36 (whose atomic mass is 36.0 amu). If 29.6% of the chlorine atoms in the sample are chlorine-36 and the remainder is naturally occurring nonradioactive chlorine atoms, how many disintegrations per second are produced by this sample? The half-life of chlorine-36 is 3.0×10^5 yr.

21.81 Calculate the mass of octane, $C_8H_{18}(l)$, that must be burned in air to evolve the same quantity of energy as produced by the fusion of 1.0 g of hydrogen in the following fusion reaction:

$$4\,^{1}_{1}H \longrightarrow\, ^{4}_{2}He + 2\,^{0}_{1}e$$

Assume that all the products of the combustion of octane are in their gas phases. Use data from Exercise 21.50, Appendix C, and the inside covers of the text. The standard enthalpy of formation of octane is -250.1 kJ/mol.

21.82 A sample of an alpha emitter having an activity of 0.18 Ci is stored in a 25.0-mL sealed container at 22 °C for 245 days. **(a)** How many alpha particles are formed during this time? **(b)** Assuming that each alpha particle is converted to a helium atom, what is the partial pressure of helium gas in the container after this 245-day period?

[21.83] Charcoal samples from Stonehenge in England were burned in O_2, and the resultant CO_2 gas bubbled into a solution of $Ca(OH)_2$ (limewater), resulting in the precipitation of $CaCO_3$. The $CaCO_3$ was removed by filtration and dried. A 788-mg sample of the $CaCO_3$ had a radioactivity of 1.5×10^{-2} Bq due to carbon-14. By comparison, living organisms undergo 15.3 disintegrations per minute per gram of carbon. Using the half-life of carbon-14, 5715 yr, calculate the age of the charcoal sample.

21.84 When a positron is annihilated by combination with an electron, two photons of equal energy result. What is the wavelength of these photons? Are they gamma ray photons?

21.85 A 25.0-mL sample of 0.050 M barium nitrate solution was mixed with 25.0 mL of 0.050 M sodium sulfate solution labeled with radioactive sulfur-35. The activity of the initial sodium sulfate solution was 1.22×10^6 Bq/mL. After the resultant precipitate was removed by filtration, the remaining filtrate was found to have an activity of 250 Bq/mL. **(a)** Write a balanced chemical equation for the reaction that occurred. **(b)** Calculate the K_{sp} for the precipitate under the conditions of the experiment.

CHEMISTRY
OF THE NONMETALS

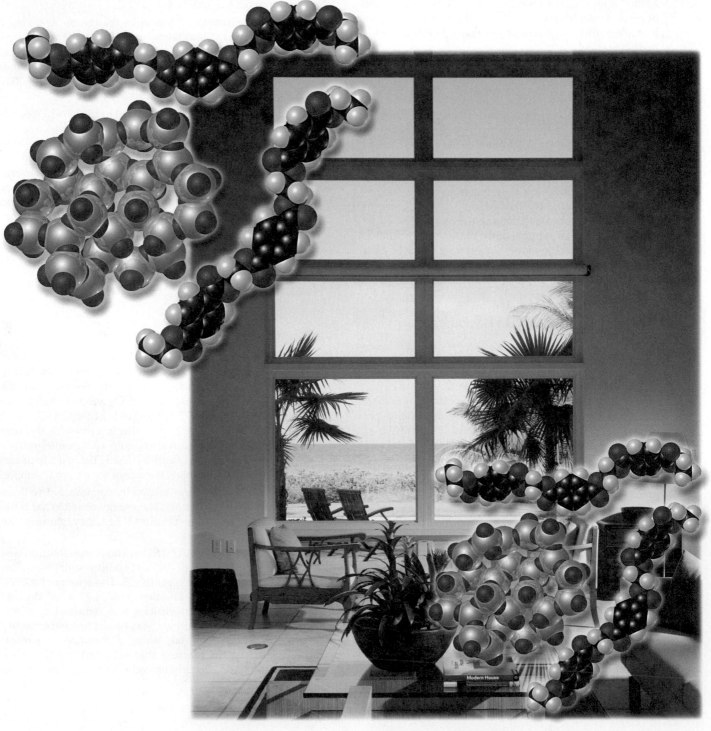

A MODERN LIVING SPACE

WHAT'S AHEAD

22.1 General Concepts: Periodic Trends and Chemical Reactions
We begin with a review of general periodic trends and types of chemical reactions, which will help us to focus on general patterns of behavior as we examine each family in the periodic table.

22.2 Hydrogen
The first nonmetal we consider is hydrogen, an element that forms compounds with most other nonmetals and many metals.

22.3 Group 8A: The Noble Gases
Next, we consider the noble gases, the elements of group 8A, which exhibit very limited chemical reactivity (the fluorides and oxides of Xe are the most numerous).

22.4 Group 7A: The Halogens
We then explore the most electronegative group of elements: the halogens, group 7A. They have a rich and important chemistry, with F and Cl being commercially very important.

22.5 Oxygen
We next consider oxygen, the most abundant element by mass in both Earth's crust and the human body. The chemistry of oxygen includes oxide and peroxide compounds.

22.6 The Other Group 6A Elements: S, Se, Te, and Po
We study the other members of group 6A (S, Se, Te, and Po), of which sulfur is the most important.

22.7 Nitrogen
We next consider nitrogen, a key component of our atmosphere. It forms compounds in which its oxidation number ranges from -3 to $+5$ including such important compounds as NH_3 and HNO_3.

22.8 The Other Group 5A Elements: P, As, Sb, and Bi
Of the other members of group 5A (P, As, Sb, and Bi), we take a closer look at phosphorus—the only one that plays an important and beneficial role in biological systems. Phosphorus is also the most important commercially.

22.9 Carbon
We next focus on the inorganic compounds of carbon.

22.10 The Other Group 4A Elements: Si, Ge, Sn, and Pb
We then consider silicon, the most abundant and significant element of the other members of group 4A. Silicon oxides are the framework substances in Earth's crust.

22.11 Boron
Finally, we examine boron—the sole nonmetallic element of group 3A.

THE CHAPTER-OPENING PHOTOGRAPH shows an attractive modern living space. A great number of chemical elements have gone into making up the space and all its decorations. Chemical innovation has resulted in a host of materials that form the stuff of modern life.

In earlier chapters we have learned about a variety of chemical reaction types that allow chemists to prepare and purify materials. We have also learned how the properties of pure substances and mixtures depend on underlying molecular properties. ∞ (Chapter 11) But what are the chemical and physical properties of the elements that make up all these materials? In this and the next chapter, we will be looking at the properties of many elements and at how these properties determine the possibilities for useful applications.

As we look at the chapter-opening photograph, it is striking that so much of what we see is formed from nonmetallic elements. The pottery objects on the table are formed from clay, which is largely silicon dioxide. The seat cushions are almost certainly woven from a polymeric fiber that is essentially all non-metallic. ∞∞(Section 12.6) The paint on the walls, the chair coverings, the glass of the windows, the lamp shades, and most of the furniture are mainly nonmetallic. Metals are certainly important, as they are necessary to manufacture the electrical wiring in the walls, perhaps steel girders in the building's frame, perhaps plumbing fixtures. Overall, however, the stuff of which this scene is constructed is largely nonmetallic in origin.

In this chapter we will take a panoramic view of the descriptive chemistry of the nonmetallic elements, starting with hydrogen and then progressing, group by group, from right to left in the periodic table. We will consider how the elements occur in nature, how they are isolated from their sources, and how they are used. We will emphasize hydrogen, oxygen, nitrogen, and carbon. These four nonmetals form many commercially important compounds and account for 99% of the atoms required by living cells.

As you study this so-called *descriptive chemistry*, it is important to look for trends rather than trying to memorize all the facts presented. The periodic table is your most valuable tool in this task.

▼ **Figure 22.1 Trends in elemental properties.** Key properties of the elements are a function of position in the periodic table.

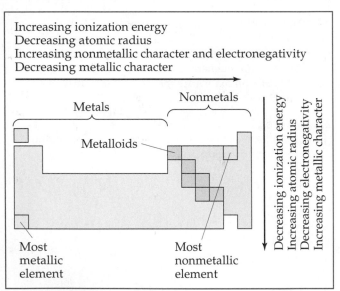

Increasing ionization energy
Decreasing atomic radius
Increasing nonmetallic character and electronegativity
Decreasing metallic character

Metals   Nonmetals

Metalloids

Most metallic element

Most nonmetallic element

Decreasing ionization energy
Increasing atomic radius
Decreasing electronegativity
Increasing metallic character

C—C   Si—Si

Strong overlap   Weak overlap

▲ **Figure 22.2 π Bonds in second- and third-row elements.** Comparison of π-bond formation by sideways overlap of *p* orbitals between two carbon atoms and between two silicon atoms. The distance between nuclei increases as we move from carbon to silicon. The *p* orbitals do not overlap as effectively between two silicon atoms because of this greater separation.

22.1 GENERAL CONCEPTS: PERIODIC TRENDS AND CHEMICAL REACTIONS

Recall that we can classify elements as metals, metalloids, and nonmetals. ∞∞(Section 7.6) Except for hydrogen, which is a special case, the nonmetals occupy the upper right portion of the periodic table. This division of elements relates nicely to trends in the properties of the elements as summarized in Figure 22.1 ◄. Electronegativity, for example, increases as we move left to right across a row of the table, and it decreases as we move down a particular group. The nonmetals thus have higher electronegativities than the metals. This difference leads to the formation of ionic solids in reactions between metals and nonmetals. ∞∞(Sections 7.6, 8.2, and 8.4) In contrast, compounds formed between nonmetals are molecular substances and are often gases, liquids, or volatile solids at room temperature. ∞∞(Sections 7.8 and 8.4)

Among the nonmetals, the chemistry exhibited by the first member of a group can differ in several important ways from that of subsequent members. For example, nonmetals in the third period and below can accommodate a larger number of bonded neighbors. ∞∞(Section 8.7) Another important difference is that the first element in any group can more readily form π bonds than members lower in the group. This trend is due, in part, to atomic size. Small atoms are able to approach each other more closely. As a result, the sideways overlap of *p* orbitals, which results in the formation of π bonds, is more effective for the first element in each group (Figure 22.2 ◄). More effective overlap means stronger π bonds, as is reflected in the bond enthalpies of their multiple bonds. ∞∞(Section 8.8) For example, the difference in the bond enthalpies of C—C and C=C bonds is about 270 kJ/mol (Table 8.4); this value reflects the "strength" of a carbon–carbon π bond. By comparison, the strength of a silicon–silicon π bond is only about 100 kJ/mol, significantly lower than that for carbon. As we shall see, π bonds are particularly important in the chemistry of carbon, nitrogen, and oxygen, which frequently form multiple bonds. The elements in periods 3, 4, 5, and 6 of the periodic table, on the other hand, have a tendency to form only single bonds.

The ready ability of second-period elements to form π bonds is an important factor in determining the structures of these elements and their compounds.

Compare, for example, the elemental forms of carbon and silicon. Carbon has four major crystalline allotropes: diamond, graphite, buckminsterfullerene, and carbon nanotubes. ∞∞(Section 11.8) Diamond is a covalent-network solid that has C—C σ bonds but no π bonds. Graphite, buckminsterfullerene, and carbon nanotubes have π bonds that result from the sideways overlap of p orbitals. Elemental silicon exists only as a diamondlike covalent-network solid with σ bonds; silicon exhibits no form that is analogous to graphite, buckminster-fullerene or carbon nanotubes, apparently because Si—Si π bonds are weak.

We likewise see significant differences in the dioxides of carbon and silicon (Figure 22.3▶). CO_2 is a molecular substance with C—O double bonds, where-as SiO_2 contains no double bonds. SiO_2 is a covalent-network solid in which four oxygen atoms are bonded to each silicon atom by single bonds, forming an extended structure that has the empirical formula SiO_2.

SiO_2

CO_2

▲ **Figure 22.3 Comparison of SiO₂ and CO₂.** SiO₂ has only single bonds, whereas CO₂ has double bonds.

▲ **GIVE IT SOME THOUGHT**

The element nitrogen is found in nature as $N_2(g)$. Would you expect phosphorus to be found in nature as $P_2(g)$? Explain.

■ **SAMPLE EXERCISE 22.1** | Identifying Elemental Properties

Consider the elements Li, K, N, P, and Ne. From this list select the element that **(a)** is the most electronegative, **(b)** has the greatest metallic character, **(c)** can bond to more than four surrounding atoms in a molecule, **(d)** forms π bonds most readily.

SOLUTION

Analyze: We are given a list of elements and asked to predict several properties that can be related to periodic trends.

Plan: We can use the preceding discussion, particularly the summary in Figure 22.1, to guide us to the answers. Thus, we need first to locate each element in the periodic table.

Solve:

(a) Electronegativity increases as we proceed toward the upper right portion of the periodic table, excluding the noble gases. Thus, nitrogen (N) is the most electronega-tive element among those listed.
(b) Metallic character correlates inversely with electronegativity—the less elec-tronegative an element, the greater its metallic character. The element with the great-est metallic character is therefore potassium (K), which is closest to the lower left corner of the periodic table.
(c) Nonmetals tend to form molecular compounds, so we can narrow our choice to the three nonmetals on the list: N, P, and Ne. To form more than four bonds, an ele-ment must be able to expand its valence shell to allow more than an octet of electrons around it. Valence-shell expansion occurs for elements in the third period of the peri-odic table and below; nitrogen and neon are both in the second period and will not undergo valence-shell expansion. Thus, the answer is phosphorus (P).
(d) Nonmetals in the second period form π bonds more readily than elements in the third period and below. There are no compounds known that contain covalent bonds to the noble gas Ne. Thus, the other second-row element, N, is the element from the list that forms π bonds most readily.

■ **PRACTICE EXERCISE**

Consider the elements Be, C, Cl, Sb, and Cs. Select the element that **(a)** has the lowest electronegativity, **(b)** has the greatest nonmetallic character, **(c)** is most likely to par-ticipate in extensive π bonding, **(d)** is most likely to be a metalloid.
Answers: **(a)** Cs, **(b)** Cl, **(c)** C, **(d)** Sb

Chemical Reactions

We will present a large number of chemical reactions in this and later chapters. You will find it helpful to observe general trends in the patterns of reactivity. We have already encountered several general categories of reactions: combus-tion reactions (Section 3.2), metathesis reactions (Section 4.2), Brønsted–Lowry acid–base (proton-transfer) reactions (Section 16.2), Lewis acid–base reactions

(Section 16.11), and redox reactions (Section 20.1). Because O_2 and H_2O are abundant in our environment, it is particularly important to consider the possible reactions of these substances with other compounds. About one-third of the reactions discussed in this chapter involve either O_2 (oxidation or combustion reactions) or H_2O (especially proton-transfer reactions).

In combustion reactions with O_2, hydrogen-containing compounds produce H_2O. Carbon-containing ones produce CO_2 (unless the amount of O_2 is insufficient, in which case CO or even C can form). Nitrogen-containing compounds tend to form N_2, although NO can form in special cases or in small amounts. The following reaction illustrates these generalizations:

$$4\,CH_3NH_2(g) + 9\,O_2(g) \longrightarrow 4\,CO_2(g) + 10\,H_2O(g) + 2\,N_2(g) \qquad [22.1]$$

The formation of H_2O, CO_2, and N_2 reflects the high thermodynamic stabilities of these substances, which are indicated by the large bond energies for the $O-H$, $C=O$, and $N\equiv N$ bonds that they contain (463, 799, and 941 kJ/mol, respectively). ∞ (Section 8.8)

When dealing with proton-transfer reactions, remember that the weaker a Brønsted–Lowry acid, the stronger its conjugate base. ∞ (Section 16.2) For example, H_2, OH^-, NH_3, and CH_4 are exceedingly weak proton donors that have *no* tendency to act as acids in water. Thus, species formed from them by removing one or more protons (such as H^-, O^{2-}, and NH_2^-) are extremely strong bases. All react readily with water, removing protons from H_2O to form OH^-. The following reactions are illustrative:

$$CH_3^-(aq) + H_2O(l) \longrightarrow CH_4(g) + OH^-(aq) \qquad [22.2]$$

$$N^{3-}(aq) + 3\,H_2O(l) \longrightarrow NH_3(aq) + 3\,OH^-(aq) \qquad [22.3]$$

■ SAMPLE EXERCISE 22.2 | Predicting the Products of Chemical Reactions

Predict the products formed in each of the following reactions, and write a balanced equation:

(a) $CH_3NHNH_2(g) + O_2(g) \longrightarrow$

(b) $Mg_3P_2(s) + H_2O(l) \longrightarrow$

SOLUTION

Analyze: We are given the reactants for two chemical equations and asked to predict the products and then balance the equations.

Plan: We need to examine the reactants to see if we might recognize a reaction type. In (a) the carbon compound is reacting with O_2, which suggests a combustion reaction. In (b) water reacts with an ionic compound. The anion, P^{3-}, is a strong base and H_2O is able to act as an acid, so the reactants suggest an acid–base (proton-transfer) reaction.

Solve: (a) Based on the elemental composition of the carbon compound, this combustion reaction should produce CO_2, H_2O, and N_2:

$$2\,CH_3NHNH_2(g) + 5\,O_2(g) \longrightarrow 2\,CO_2(g) + 6\,H_2O(g) + 2\,N_2(g)$$

(b) Mg_3P_2 is ionic, consisting of Mg^{2+} and P^{3-} ions. The P^{3-} ion, like N^{3-}, has a strong affinity for protons and reacts with H_2O to form OH^- and PH_3 (PH^{2-}, PH^{2-}, and PH_3 are all exceedingly weak proton donors).

$$Mg_3P_2(s) + 6\,H_2O(l) \longrightarrow 2\,PH_3(g) + 3\,Mg(OH)_2(s)$$

$Mg(OH)_2$ has low solubility in water and will precipitate.

■ PRACTICE EXERCISE

Write a balanced equation for the reaction of solid sodium hydride with water.

Answer: $NaH(s) + H_2O(l) \longrightarrow NaOH(aq) + H_2(g)$

22.2   HYDROGEN

The English chemist Henry Cavendish (1731–1810) was the first to isolate pure hydrogen. Because the element produces water when burned in air, the French chemist Lavoisier gave it the name *hydrogen*, which means "water producer" (Greek: *hydro*, water; *gennao*, to produce).

Hydrogen is the most abundant element in the universe. It is the nuclear fuel consumed by our Sun and other stars to produce energy. ∞ (Section 21.8) Although about 75% of the known mass of the universe is composed of hydrogen, it constitutes only 0.87% of Earth's mass. Most of the hydrogen on our planet is found associated with oxygen. Water, which is 11% hydrogen by mass, is the most abundant hydrogen compound.

Isotopes of Hydrogen

The most common isotope of hydrogen, $_1^1H$, has a nucleus consisting of a single proton. This isotope, sometimes referred to as **protium**,* comprises 99.9844% of naturally occurring hydrogen.

Two other isotopes are known: $_1^2H$, whose nucleus contains a proton and a neutron, and $_1^3H$, whose nucleus contains a proton and two neutrons (Figure 22.4 ▶). The $_1^2H$ isotope, called **deuterium**, comprises 0.0156% of naturally occurring hydrogen. It is not radioactive. Deuterium is often given the symbol D in chemical formulas, as in D_2O (deuterium oxide), which is known as *heavy water*.

Because an atom of deuterium is about twice as massive as an atom of protium, the properties of deuterium-containing substances vary somewhat from those of the "normal" protium-containing analogs. For example, the normal melting and boiling points of D_2O are 3.81 °C and 101.42 °C, respectively, whereas they are 0.00 °C and 100.00 °C for H_2O. Not surprisingly, the density of D_2O at 25 °C (1.104 g/mL) is greater than that of H_2O (0.997 g/mL). Replacing protium with deuterium (a process called *deuteration*) can also have a profound effect on the rates of reactions, a phenomenon called a *kinetic-isotope effect*. In fact, heavy water can be obtained by the electrolysis of ordinary water, because D_2O undergoes electrolysis at a slower rate and therefore becomes concentrated during the electrolysis.

The third isotope, $_1^3H$, is known as **tritium**. It is radioactive, with a half-life of 12.3 yr.

$$_1^3H \longrightarrow {}_2^3He + {}_{-1}^0e \quad t_{1/2} = 12.3 \text{ yr} \qquad [22.4]$$

Because of its short half-life, only trace quantities of tritium exist naturally. The isotope can be synthesized in nuclear reactors by neutron bombardment of lithium-6.

$$_3^6Li + {}_0^1n \longrightarrow {}_1^3H + {}_2^4He \qquad [22.5]$$

Deuterium and tritium have proved valuable in studying the reactions of compounds containing hydrogen. A compound can be "labeled" by replacing one or more ordinary hydrogen atoms at specific locations within a molecule with deuterium or tritium. By comparing the location of the label in the reactants with that in the products, the mechanism of the reaction can often be inferred. When methyl alcohol (CH_3OH) is placed in D_2O, for example, the H atom of the O—H bond exchanges rapidly with the D atoms in D_2O, forming CH_3OD. The H atoms of the CH_3 group do not exchange. This experiment demonstrates the kinetic stability of C—H bonds and reveals the speed at which the O—H bond in the molecule breaks and re-forms.

(a) Protium

(b) Deuterium

(c) Tritium

▲ **Figure 22.4 Nuclei of the three isotopes of hydrogen.** (a) Protium, $_1^1H$, has a single proton (depicted as a red sphere) in its nucleus. (b) Deuterium, $_1^2H$, has a proton and a neutron (depicted as a gray sphere). (c) Tritium, $_1^3H$, has a proton and two neutrons. Remember that the nuclei are very small! ∞ (Section 2.3)

---

*Giving unique names to isotopes is limited to hydrogen. Because of the proportionally large differences in their masses, the isotopes of H show appreciably more differences in their properties than isotopes of heavier elements.

Properties of Hydrogen

Hydrogen is the only element that is not a member of any family in the periodic table. Because of its $1s^1$ electron configuration, it is generally placed above lithium in the periodic table. However, it is definitely *not* an alkali metal. It forms a positive ion much less readily than any alkali metal. The ionization energy of the hydrogen atom is 1312 kJ/mol, whereas that of lithium is 520 kJ/mol.

Hydrogen is sometimes placed above the halogens in the periodic table because the hydrogen atom can pick up one electron to form the *hydride ion*, H^-, which has the same electron configuration as helium. The electron affinity of hydrogen ($E = -73$ kJ/mol), however, is not as large as that of any halogen; the electron affinity of fluorine is -328 kJ/mol, and that of iodine is -295 kJ/mol. ∞ (Section 7.5) In general, hydrogen shows no closer resemblance to the halogens than it does to the alkali metals.

Elemental hydrogen exists at room temperature as a colorless, odorless, tasteless gas composed of diatomic molecules. We can call H_2 dihydrogen, but it is more commonly referred to as molecular hydrogen or simply hydrogen. Because H_2 is nonpolar and has only two electrons, attractive forces between molecules are extremely weak. As a result, the melting point ($-259\,°C$) and boiling point ($-253\,°C$) of H_2 are very low.

The H—H bond enthalpy (436 kJ/mol) is high for a single bond (Table 8.4). By comparison, the Cl—Cl bond enthalpy is only 242 kJ/mol. Because H_2 has a strong bond, most reactions of H_2 are slow at room temperature. However, the molecule is readily activated by heat, irradiation, or catalysis. The activation process generally produces hydrogen atoms, which are very reactive. Once H_2 is activated, it reacts rapidly and exothermically with a wide variety of substances.

Hydrogen forms strong covalent bonds with many elements, including oxygen; the O—H bond enthalpy is 463 kJ/mol. The formation of the strong O—H bond makes hydrogen an effective reducing agent for many metal oxides. When H_2 is passed over heated CuO, for example, copper is produced.

$$CuO(s) + H_2(g) \longrightarrow Cu(s) + H_2O(g) \qquad [22.6]$$

When H_2 is ignited in air, a vigorous reaction occurs, forming H_2O. Air containing as little as 4% H_2 (by volume) is potentially explosive. Combustion of hydrogen-oxygen mixtures is commonly used in liquid-fuel rocket engines such as those of the Space Shuttles. The hydrogen and oxygen are stored at low temperatures in liquid form.

Preparation of Hydrogen

When a small quantity of H_2 is needed in the laboratory, it is usually obtained by the reaction between an active metal such as zinc and a dilute strong acid such as HCl or H_2SO_4.

$$Zn(s) + 2\,H^+(aq) \longrightarrow Zn^{2+}(aq) + H_2(g) \qquad [22.7]$$

Large quantities of H_2 are produced by reacting methane (CH_4, the principal component of natural gas) with steam at 1100 °C. We can view this process as involving the following reactions:

$$CH_4(g) + H_2O(g) \longrightarrow CO(g) + 3\,H_2(g) \qquad [22.8]$$
$$CO(g) + H_2O(g) \longrightarrow CO_2(g) + H_2(g) \qquad [22.9]$$

When heated to about 1000 °C, carbon also reacts with steam to produce a mixture of H_2 and CO gases.

$$C(s) + H_2O(g) \longrightarrow H_2(g) + CO(g) \qquad [22.10]$$

This mixture, known as *water gas*, is used as an industrial fuel.

A Closer Look   THE HYDROGEN ECONOMY

The reaction of hydrogen with oxygen is highly exothermic:

$$2\,H_2(g) + O_2(g) \longrightarrow 2\,H_2O(g) \quad \Delta H = -483.6\ kJ \quad [22.11]$$

Because the only product of the reaction is water vapor, the prospect of using hydrogen as a fuel in an automotive fuel cell is attractive. ∞ (Section 20.7) Alternatively, hydrogen could be combusted directly with oxygen from the atmosphere in an internal combustion engine. In either case, it would be necessary to generate elemental hydrogen on a large scale and arrange for its transport and storage. The advantages of the so-called hydrogen economy must be viewed, therefore, in light of the potential difficulties and costs. The generation of hydrogen from fossil fuels is feasible, but that strategy may create difficulties in transport and handling of hydrogen gas, a dangerously combustible substance.

The generation of H_2 through electrolysis of water is in principle the cleanest route, because this process produces only hydrogen and oxygen. ∞ (Figure 1.7 and Section 20.9) However, the energy required to electrolyze water must come from somewhere. If we burn fossil fuels to generate the electricity used for the production of hydrogen, we have not advanced very far toward a true hydrogen economy. If the energy for electrolysis came instead from a hydroelectric or nuclear power plant, from solar cells or via wind generators (Figure 22.5▼), consumption of nonrenewable energy sources and undesired production of CO_2 could be avoided.

Related Exercises: 22.29, 22.30, 22.99

▲ **Figure 22.5 Wind power.** Windmills represent one pathway toward electrical energy that can drive a hydrogen economy.

Simple electrolysis of water consumes too much energy and is consequently too costly to be used commercially to produce H_2. However, H_2 is produced as a by-product in the electrolysis of brine (NaCl) solutions in the course of Cl_2 and NaOH manufacture:

$$2\,NaCl(aq) + 2\,H_2O(l) \xrightarrow{\text{electrolysis}} H_2(g) + Cl_2(g) + 2\,NaOH(aq) \quad [22.12]$$

Uses of Hydrogen

Hydrogen is a commercially important substance: About $2 \times 10^8\ kg$ (200,000 tons) is produced annually in the United States. Over two-thirds of the H_2 produced is used to synthesize ammonia by the Haber process. ∞ (Section 15.2) Hydrogen is also used to manufacture methanol (CH_3OH) via the catalytic reaction of CO and H_2 at high pressure and temperature.

$$CO(g) + 2\,H_2(g) \longrightarrow CH_3OH(g) \quad [22.13]$$

Binary Hydrogen Compounds

Hydrogen reacts with other elements to form compounds of three general types: (1) ionic hydrides, (2) metallic hydrides, and (3) molecular hydrides.

The **ionic hydrides** are formed by the alkali metals and by the heavier alkaline earths (Ca, Sr, and Ba). These active metals are much less electronegative than hydrogen. Consequently, hydrogen acquires electrons from them to form hydride ions (H^-):

$$Ca(s) + H_2(g) \longrightarrow CaH_2(s) \quad [22.14]$$

The hydride ion is very basic and reacts readily with compounds having even weakly acidic protons to form H_2. For example, H^- reacts readily with H_2O.

$$H^-(aq) + H_2O(l) \longrightarrow H_2(g) + OH^-(aq) \quad [22.15]$$

Ionic hydrides can therefore be used as convenient (although expensive) sources of H_2. Calcium hydride (CaH_2) is sold commercially and is used to inflate life

(a) (b)

▲ **Figure 22.6 The reaction of CaH₂ with water.** (a) Powdered CaH_2 about to be added to water. (b) The reaction is vigorous and exothermic. The reddish purple color is due to phenolphthalein added to the water, which indicates the formation of OH^- ions. The bubbling is due to the formation of H_2 gas.

4A	5A	6A	7A
$CH_4(g)$	$NH_3(g)$	$H_2O(l)$	$HF(g)$
−50.8	−16.7	−237	−271
$SiH_4(g)$	$PH_3(g)$	$H_2S(g)$	$HCl(g)$
+56.9	+18.2	−33.0	−95.3
$GeH_4(g)$	$AsH_3(g)$	$H_2Se(g)$	$HBr(g)$
+117	+111	+71	−53.2
	$SbH_3(g)$	$H_2Te(g)$	$HI(g)$
	+187	+138	+1.30

▲ **Figure 22.7 Standard free energies of formation of molecular hydrides.** All values are kilojoules per mole of hydride.

8A

| 2 He |
| 10 Ne |
| 18 Ar |
| 36 Kr |
| 54 Xe |
| 86 Rn |

rafts, weather balloons, and the like, where a simple, compact means of generating H_2 is desired. The reaction of CaH_2 with H_2O is shown in Figure 22.6 ◄.

The reaction between H^- and H_2O (Equation 22.15) is an acid–base reaction and a redox reaction. The H^- ion, therefore, is a good base *and* a good reducing agent. In fact, hydrides are able to reduce O_2 to OH^-.

$$2\,NaH(s) + O_2(g) \longrightarrow 2\,NaOH(s) \qquad [22.16]$$

Thus, hydrides are normally stored in an environment that is free of both moisture and air.

Metallic hydrides are formed when hydrogen reacts with transition metals. These compounds are so named because they retain their metallic conductivity and other metallic properties. In many metallic hydrides, the ratio of metal atoms to hydrogen atoms is not fixed or in small whole numbers. The composition can vary within a range, depending on the conditions of synthesis. TiH_2 can be produced, for example, but preparations usually yield $TiH_{1.8}$, which has about 10% less hydrogen than TiH_2. These nonstoichiometric metallic hydrides are sometimes called *interstitial hydrides*.

The **molecular hydrides,** formed by nonmetals and semimetals, are either gases or liquids under standard conditions. The simple molecular hydrides are listed in Figure 22.7 ◄, together with their standard free energies of formation, ΔG_f°. In each family the thermal stability (measured by ΔG_f°) decreases as we move down the family. (Recall that the more stable a compound is with respect to its elements under standard conditions, the more negative ΔG_f° is.) We will discuss the molecular hydrides further in the course of examining the other nonmetallic elements.

▲ **GIVE IT SOME THOUGHT**

If the reaction $H_2(g) + Se(s) \rightleftharpoons H_2Se(g)$ were at equilibrium with an $H_2(g)$ pressure of 1 atm, would there be much $H_2Se(g)$ present?

22.3 GROUP 8A: THE NOBLE GASES

The elements of group 8A are chemically unreactive. Indeed, most of our references to these elements have been in relation to their physical properties, as when we discussed intermolecular forces. ∞(Section 11.2) The relative inertness of these elements is due to the presence of a completed octet of valence-shell electrons (except He, which has a filled 1s shell). The stability of such an arrangement is reflected in the high ionization energies of the group 8A elements. ∞(Section 7.4)

The group 8A elements are all gases at room temperature. They are components of Earth's atmosphere, except for radon, which exists only as a short-lived radioisotope. ∞(Section 21.9) Only argon is relatively abundant (Table 18.1). Neon, argon, krypton, and xenon are recovered from liquid air by distillation. Argon is used as a blanketing atmosphere in electric lightbulbs. The gas conducts heat away from the filament but does not react with it. It is also used as a protective atmosphere to prevent oxidation in welding and certain high-temperature metallurgical processes.

Helium is, in many ways, the most important of the noble gases. Liquid helium is used as a coolant to conduct experiments at very low temperatures. Helium boils at 4.2 K under 1 atm pressure, the lowest boiling point of any substance. Fortunately, helium is found in relatively high concentrations in many natural-gas wells. Some of this helium is separated to meet current demands, and some is kept for later use.

TABLE 22.1 ■ Properties of Xenon Compounds			
Compound	Oxidation State of Xe	Melting Point (°C)	ΔH_f° (kJ/mol)[a]
XeF_2	+2	129	−109(g)
XeF_4	+4	117	−218(g)
XeF_6	+6	49	−298(g)
$XeOF_4$	+6	−41 to −28	+146(l)
XeO_3	+6	—[b]	+402(s)
XeO_2F_2	+6	31	+145(s)
XeO_4	+8	—[c]	—

[a]At 25 °C, for the compound in the state indicated.
[b]A solid; decomposes at 40 °C.
[c]A solid; decomposes at −40 °C.

Noble-Gas Compounds

Because the noble gases are exceedingly stable, they will undergo reaction only under rigorous conditions. Furthermore, we might expect that the heavier noble gases would be most likely to form compounds because their ionization energies are lower (Figure 7.11). A lower ionization energy suggests the possibility of sharing an electron with another atom, leading to a chemical bond. In addition, because the group 8A elements (except helium) already contain eight electrons in their valence shell, formation of covalent bonds will require an expanded valence shell. Valence-shell expansion occurs most readily with larger atoms. ∞ (Section 8.7)

The first noble-gas compound was reported in 1962. This discovery caused a sensation because it undercut the belief that the noble-gas elements were truly chemically inert. The initial study involved xenon in combination with fluorine, the element we would expect to be most reactive. Since that time chemists have prepared several xenon compounds of fluorine and oxygen. Some properties of these substances are listed in Table 22.1▲. The three fluorides (XeF_2, XeF_4, and XeF_6) are made by direct reaction of the elements. By varying the ratio of reactants and altering reaction conditions, one of the three compounds can be obtained. The oxygen-containing compounds are formed when the fluorides are reacted with water, as in Equation 22.17:

$$XeF_6(s) + 3\ H_2O(l) \longrightarrow XeO_3(aq) + 6\ HF(aq) \qquad [22.17]$$

■ SAMPLE EXERCISE 22.3 | Predicting a Molecular Structure
Use the VSEPR model to predict the structure of XeF_4.

SOLUTION

Analyze: We must predict the geometrical structure, given only the molecular formula.

Plan: To predict the structure, we must first write the Lewis structure for the molecule. We then count the number of electron pairs (domains) around the central Xe atom and use that number and the number of bonds to predict the geometry, as discussed in Section 9.2.

Solve: The total number of valence-shell electrons involved is 36 (8 from xenon and 7 from each of the four fluorines). This leads to the Lewis structure shown in Figure 22.8(a)▶. Xe has 12 electrons in its valence shell, so we expect an octahedral disposition of 6 electron pairs. Two of these are nonbonded pairs. Because nonbonded pairs have a larger volume requirement than bonded pairs (Section 9.2), it is reasonable to expect these nonbonded pairs to be opposite one another. The expected structure is square planar, as shown in Figure 22.8(b).

Comment: The experimentally determined structure agrees with this prediction.

■ PRACTICE EXERCISE
Describe the electron-domain geometry and the molecular geometry of XeF_2.
Answer: trigonal bipyramidal, linear

(a)

(b)

▲ **Figure 22.8 Xenon tetrafluoride.**
(a) Lewis structure. (b) Molecular geometry.

The other noble-gas elements form compounds much less readily than xenon. For many years, only one binary krypton compound, KrF_2, was known with certainty, and it decomposes to its elements at $-10\ °C$. Other compounds of krypton have been isolated at very low temperatures (40 K).

22.4 GROUP 7A: THE HALOGENS

The elements of group 7A, the halogens, have outer electron configurations of ns^2np^5, where n ranges from 2 through 6. The halogens have large negative electron affinities (Section 7.5), and they most often achieve a noble-gas configuration by gaining an electron, which results in a -1 oxidation state. Fluorine, being the most electronegative element, exists in compounds only in the -1 state. The other halogens also exhibit positive oxidation states up to $+7$ in combination with more electronegative atoms such as O. In the positive oxidation states the halogens tend to be good oxidizing agents, readily accepting electrons.

Chlorine, bromine, and iodine are found as the halides in seawater and in salt deposits. Fluorine occurs in the minerals fluorspar (CaF_2), cryolite (Na_3AlF_6), and fluorapatite $[Ca_5(PO_4)_3F]$.* Only fluorspar is an important commercial source of fluorine.

All isotopes of astatine are radioactive. The longest lived isotope is astatine-210, which has a half-life of 8.1 hr and decays mainly by electron capture. Because astatine is so unstable to nuclear decay, very little is known about its chemistry.

Properties and Preparation of the Halogens

Some of the properties of the halogens are summarized in Table 22.2▼. Most of the properties vary in a regular fashion as we go from fluorine to iodine. The electronegativity steadily decreases, for example, from 4.0 for fluorine to 2.5 for iodine.

Under ordinary conditions the halogens exist as diatomic molecules. The molecules are held together in the solid and liquid states by London dispersion forces. ∞ (Section 11.2) Because I_2 is the largest and most polarizable of the halogen molecules, the intermolecular forces between I_2 molecules are the strongest. Thus, I_2 has the highest melting point and boiling point. At room temperature and 1 atm pressure, I_2 is a solid, Br_2 is a liquid, and Cl_2 and F_2 are gases. Chlorine readily liquefies upon compression at room temperature and is normally stored and handled in liquid form under pressure in steel containers.

The comparatively low bond enthalpy in F_2 (155 kJ/mol) accounts in part for the extreme reactivity of elemental fluorine. Because of its high reactivity, F_2 is very difficult to work with. Certain metals, such as copper and nickel, can be used to contain F_2 because their surfaces form a protective coating of metal fluoride. Chlorine and the heavier halogens are also reactive, although less so than fluorine.

TABLE 22.2 ■ Some Properties of the Halogens				
Property	F	Cl	Br	I
Atomic radius (Å)	0.71	0.99	1.14	1.33
Ionic radius, X^- (Å)	1.33	1.81	1.96	2.20
First ionization energy (kJ/mol)	1681	1251	1140	1008
Electron affinity (kJ/mol)	-328	-349	-325	-295
Electronegativity	4.0	3.0	2.8	2.5
X—X single-bond enthalpy (kJ/mol)	155	242	193	151
Reduction potential (V):				
$\frac{1}{2}X_2(aq) + e^- \longrightarrow X^-(aq)$	2.87	1.36	1.07	0.54

Minerals are solid substances that occur in nature. They are usually known by their common names rather than by their chemical names. What we know as rock is merely an aggregate of different kinds of minerals.

Because of their high electronegativities, the halogens tend to gain electrons from other substances and thereby serve as oxidizing agents. The oxidizing ability of the halogens, which is indicated by their standard reduction potentials, decreases going down the group. As a result, a given halogen is able to oxidize the anions of the halogens below it in the group. For example, Cl_2 will oxidize Br^- and I^-, but not F^-, as seen in Figure 22.9▶.

▲ **Figure 22.9 Reaction of Cl_2 with aqueous solutions of NaF, NaBr, and NaI.** Each solution is in contact with carbon tetrachloride (CCl_4), which forms the lower layer in each container. The halogens are more soluble in CCl_4 than in H_2O. Because the F^- ion in the NaF solution (left) does not react with Cl_2, both the aqueous layer and the CCl_4 layer remain colorless. The Br^- ion in the NaBr solution (center) is oxidized by Cl_2 to form Br_2, producing a yellow aqueous layer and an orange CCl_4 layer. The I^- ion in the NaI solution (right) is oxidized to I_2, producing an amber aqueous layer and a violet CCl_4 layer.

■ **SAMPLE EXERCISE 22.4** | **Predicting Chemical Reactions among the Halogens**

Write the balanced equation for the reaction, if any, that occurs between **(a)** $I^-(aq)$ and $Br_2(l)$, **(b)** $Cl^-(aq)$ and $I_2(s)$.

SOLUTION

Analyze: We are asked to determine whether a reaction occurs when a particular halide and halogen are combined.

Plan: A given halogen is able to reduce anions of the halogens below it in the periodic table. Thus, the smaller (lower atomic number) halogen will end up as the halide ion. If the halogen with smaller atomic number is already the halide, there will be no reaction. Thus, the key to determining whether a reaction will occur is locating the elements in the periodic table.

Solve:

(a) Br_2 is able to oxidize (remove electrons from) the anions of the halogens below it in the periodic table. Thus, it will oxidize I^-

$$2\,I^-(aq) + Br_2(aq) \longrightarrow I_2(s) + 2\,Br^-(aq)$$

(b) Cl^- is the anion of a halogen above iodine in the periodic table. Thus, I_2 cannot oxidize Cl^-; there is no reaction.

■ **PRACTICE EXERCISE**

Write the balanced chemical equation for the reaction that occurs between $Br^-(aq)$ and $Cl_2(aq)$.

Answer: $2\,Br^-(aq) + Cl_2(aq) \longrightarrow Br_2(aq) + 2\,Cl^-(aq)$

Notice in Table 22.2 that the standard reduction potential of F_2 is exceptionally high. Fluorine gas readily oxidizes water:

$$F_2(aq) + H_2O(l) \longrightarrow 2\,HF(aq) + \tfrac{1}{2}O_2(g) \qquad E^\circ = 1.80\ \text{V} \qquad [22.18]$$

Fluorine cannot be prepared by electrolytic oxidation of aqueous solutions of fluoride salts because water itself is oxidized more readily than F^-. ∞ (Section 20.9) In practice, the element is formed by electrolytic oxidation of a solution of KF in anhydrous HF.

Chlorine is produced mainly by electrolysis of either molten or aqueous sodium chloride, as described in Sections 20.9 and 23.4. Both bromine and iodine are obtained commercially from brines containing the halide ions by oxidation with Cl_2.

Uses of the Halogens

Fluorine is an important industrial chemical. It is used, for example, to prepare fluorocarbons—very stable carbon–fluorine compounds used as refrigerants, lubricants, and plastics. Teflon® (Figure 22.10▶) is a polymeric fluorocarbon noted for its high thermal stability and lack of chemical reactivity.

Chlorine is by far the most commercially important halogen. About 1×10^{10} kg (10 million tons) of Cl_2 is produced annually in the United States. In addition, hydrogen chloride production is about 4.0×10^9 kg (4.4 million tons) annually. About half of this chlorine finds its way eventually into the manufacture of chlorine-containing organic compounds such as vinyl chloride (C_2H_3Cl), used in making polyvinyl chloride (PVC) plastics. ∞ (Section 12.6)

▲ **Figure 22.10 Structure of Teflon®, a fluorocarbon polymer.** This polymer is an analog of polyethylene (Section 12.6) in which the H atoms of polyethylene have been replaced with F atoms.

▲ Figure 22.11 **Iodized salt.**
Common table salt that has been iodized contains 0.02% KI by mass.

(a)

(b)

▲ Figure 22.12 **Reaction of H₂SO₄ with NaI and NaBr.** (a) Sodium iodide is in the left test tube, and sodium bromide is in the tube on the right. Sulfuric acid is in the pipet. (b) Addition of sulfuric acid to the test tubes oxidizes sodium iodide to the dark-colored iodine seen in the tube on the left. Sodium bromide is oxidized to the yellow-brown bromine in the right tube. When more concentrated, bromine has a reddish brown color.

Much of the remainder is used as a bleaching agent in the paper and textile industries. When Cl_2 dissolves in cold dilute base, it disproportionates into Cl^- and hypochlorite, ClO^-.

$$Cl_2(aq) + 2\,OH^-(aq) \rightleftharpoons Cl^-(aq) + ClO^-(aq) + H_2O(l) \qquad [22.19]$$

Sodium hypochlorite (NaClO) is the active ingredient in many liquid bleaches. Chlorine is also used in water treatment to oxidize and thereby destroy bacteria. ∞ (Section 18.6)

A common use of iodine is as KI in table salt. Iodized salt (Figure 22.11 ◄) provides the small amount of iodine necessary in our diets; it is essential for the formation of thyroxin, a hormone secreted by the thyroid gland. Lack of iodine in the diet results in an enlarged thyroid gland, a condition called *goiter*.

The Hydrogen Halides

All of the halogens form stable diatomic molecules with hydrogen. Aqueous solutions of HCl, HBr, and HI are strong acids. The hydrogen halides can be formed by direct reaction of the elements. The most important means of preparing them, however, is by reacting a salt of the halide with a strong nonvolatile acid. Both HF and HCl are prepared in this manner by reaction of an inexpensive, readily available salt with concentrated sulfuric acid as illustrated in the following equation:

$$CaF_2(s) + H_2SO_4(l) \xrightarrow{\Delta} 2\,HF(g) + CaSO_4(s) \qquad [22.20]$$

Neither HBr nor HI can be prepared by analogous reactions of salts with H_2SO_4 because H_2SO_4 oxidizes Br^- and I^- (Figure 22.12 ◄). This difference in reactivity reflects the greater ease of oxidation of Br^- and I^- relative to F^- and Cl^-. These undesirable oxidations are avoided by using a nonvolatile acid, such as H_3PO_4, that is a weaker oxidizing agent than H_2SO_4.

■ **SAMPLE EXERCISE 22.5** | Writing a Balanced Chemical Equation

Write a balanced equation for the formation of hydrogen bromide gas from the reaction of solid sodium bromide with phosphoric acid.

SOLUTION

Analyze: We are asked to write a balanced equation for the reaction between NaBr and H_3PO_4 to form HBr and another product.

Plan: As in Equation 22.20, a metathesis reaction takes place. ∞ (Section 4.2) Let's assume that only one of the hydrogens of H_3PO_4 undergoes reaction. (The actual number depends on the reaction conditions.) Thus, the remaining $H_2PO_4^-$ ion will be associated with the Na^+ ion as NaH_2PO_4 among the products of the equation.

Solve: The balanced equation is

$$NaBr(s) + H_3PO_4(aq) \longrightarrow NaH_2PO_4(aq) + HBr(aq)$$

■ **PRACTICE EXERCISE**

Write the balanced equation for the preparation of HI from NaI and H_3PO_4.
Answer: $NaI(s) + H_3PO_4(l) \longrightarrow NaH_2PO_4(s) + HI(g)$

The hydrogen halides form hydrohalic acid solutions when dissolved in water. These solutions exhibit the characteristic properties of acids, such as reactions with active metals to produce hydrogen gas. ∞ (Section 4.4) Hydrofluoric acid also reacts readily with silica (SiO_2) and with silicates to form hexafluorosilicic acid (H_2SiF_6):

$$SiO_2(s) + 6\,HF(aq) \longrightarrow H_2SiF_6(aq) + 2\,H_2O(l) \qquad [22.21]$$

Glass consists mostly of silicate structures (Section 22.10), and these reactions allow HF to etch, or frost, glass (Figure 22.13 ▶).

Interhalogen Compounds

Because the halogens exist as diatomic molecules, diatomic molecules of two different halogen atoms exist. These compounds are the simplest examples of **interhalogens**, compounds, such as ClF and IF_5, formed between two different halogen elements.

With one exception, the higher interhalogen compounds have a central Cl, Br, or I atom surrounded by 3, 5, or 7 fluorine atoms. The large size of the iodine atom allows the formation of IF_3, IF_5, I_7, in which the oxidation state of I is +3, +5, and +7, respectively. With the smaller bromine atom and chorine atom, only compounds with 3 or 5 fluorines form. The only higher interhalogen compounds that do not have outer F atoms are ICl_3 and ICl_5; the large size of the I atom can accommodate five Cl atoms, whereas Br is not large enough to allow even $BrCl_3$ to form. All of the interhalogen compounds are powerful oxidizing agents.

▲ **Figure 22.13 Etched, or frosted, glass.** Designs such as this are produced by first coating the glass with wax and then removing the wax only in the areas to be etched. When treated with hydrofluoric acid, the exposed areas of the glass are attacked, producing the etching effect.

Oxyacids and Oxyanions

Table 22.3▼ summarizes the formulas of the known oxyacids of the halogens and the way they are named.* ∞ (Section 2.8) The acid strengths of the oxyacids increase with increasing oxidation state of the central halogen atom. ∞ (Section 16.10) All the oxyacids are strong oxidizing agents. The oxyanions, formed on removal of H^+ from the oxyacids, are generally more stable than the oxyacids. Hypochlorite salts are used as bleaches and disinfectants because of the powerful oxidizing capabilities of the ClO^- ion. Sodium chlorite is used as a bleaching agent. Chlorate salts are similarly very reactive. For example, potassium chlorate is used to make matches and fireworks.

GIVE IT SOME THOUGHT

Which would you expect to be the stronger oxidizing agent, $NaBrO_3$ or $NaClO_3$?

Perchloric acid and its salts are the most stable of the oxyacids and oxyanions. Dilute solutions of perchloric acid are quite safe, and many perchlorate salts are stable except when heated with organic materials. When heated, perchlorates can become vigorous, even violent, oxidizers. Considerable caution should be exercised, therefore, when handling these substances, and it is crucial to avoid contact between perchlorates and readily oxidized material such as active metals and combustible organic compounds. The use of ammonium perchlorate (NH_4ClO_4) as the oxidizer in the solid booster rockets for the Space Shuttle demonstrates the oxidizing power of perchlorates. The solid propellant contains a mixture of NH_4ClO_4 and powdered aluminum, the reducing agent. Each shuttle launch requires about 6×10^5 kg (700 tons) of NH_4ClO_4 (Figure 22.14▶).

▲ **Figure 22.14 Launch of the Space Shuttle *Columbia* from the Kennedy Space Center.**

TABLE 22.3 ■ The Stable Oxyacids of the Halogens

Oxidation State of Halogen	Formula of Acid			Acid Name
	Cl	Br	I	
+1	HClO	HBrO	HIO	*Hypo*hal*ous* acid
+3	$HClO_2$	—	—	Hal*ous* acid
+5	$HClO_3$	$HBrO_3$	HIO_3	Hal*ic* acid
+7	$HClO_4$	$HBrO_4$	HIO_4, H_5IO_6	*Per*hal*ic* acid

*Fluorine forms one oxyacid, HOF. Because the electronegativity of fluorine is greater than that of oxygen, we must consider fluorine to be in a −1 oxidation state and oxygen to be in the 0 oxidation state in this compound.

Since the 1950s both NASA and the Pentagon have used ammonium perchlorate, NH_4ClO_4, as a rocket fuel. The result is that traces of perchlorate ion are found in groundwater in many regions of the United States, with levels ranging from about 4 to 100 ppb.

Perchlorate is known to suppress hormone levels in humans by acting in the thyroid gland. However, it is disputed whether the amounts found in drinking water are sufficiently high to cause health problems. The Environmental Protection Agency currently states that a dose of 0.007 mg per kg of body weight per day is not expected to cause adverse health effects in humans. For a 70 kg (154 lb.) person drinking 2 L of water per day, that amounts to a concentration of 25 ppb. California has proposed setting a standard of 6 ppb.

Removal of perchlorate ion from water supplies is not an easy matter. Although perchlorate is an oxidizing agent, the ClO_4^- ion is quite stable in aqueous solution. One promising avenue is to employ biological reduction, using microorganisms. While research continues on the best means of reducing perchlorate levels in drinking water, agencies of the federal government continue to explore the level that is safe.

Related Exercises: 22.39, 22.40, 22.101

▲ **Figure 22.15 Joseph Priestley (1733–1804).** Priestley became interested in chemistry at the age of 39. Priestley lived next door to a brewery from which he could obtain carbon dioxide, so his studies focused on this gas first and were later extended to other gases. Because he was suspected of sympathizing with the American and French revolutions, his church, home, and laboratory in Birmingham, England, were burned by a mob in 1791. Priestley had to flee in disguise. He eventually emigrated to the United States in 1794, where he lived his remaining years in relative seclusion in Pennsylvania.

22.5 OXYGEN

By the middle of the seventeenth century, scientists recognized that air contained a component associated with burning and breathing. That component was not isolated until 1774, however, when Joseph Priestley (Figure 22.15 ◄) discovered oxygen. Lavoisier subsequently named the element *oxygen*, meaning "acid former."

Oxygen is found in combination with other elements in a great variety of compounds. Indeed, oxygen is the most abundant element by mass both in Earth's crust and in the human body. It is the oxidizing agent for the metabolism of our foods and is crucial to human life.

Properties of Oxygen

Oxygen has two allotropes, O_2 and O_3. When we speak of molecular oxygen or simply oxygen, it is usually understood that we are speaking of *dioxygen* (O_2), the normal form of the element; O_3 is called *ozone*.

At room temperature dioxygen is a colorless and odorless gas. It condenses to a liquid at $-183\,°C$ and freezes at $-218\,°C$. It is only slightly soluble in water (0.04 g/L, or 0.001 M at 25 °C), but its presence in water is essential to marine life.

The electron configuration of the oxygen atom is $[He]2s^22p^4$. Thus, oxygen can complete its octet of electrons either by picking up two electrons to form the oxide ion (O^{2-}) or by sharing two electrons. In its covalent compounds it tends to form two bonds: either as two single bonds, as in H_2O, or as a double bond, as in formaldehyde ($H_2C{=}O$). The O_2 molecule itself contains a double bond. ∞ (Section 9.8)

The bond in O_2 is very strong (the bond enthalpy is 495 kJ/mol). Oxygen also forms strong bonds with many other elements. Consequently, many oxygen-containing compounds are thermodynamically more stable than O_2. In the absence of a catalyst, however, most reactions of O_2 have high activation energies and thus require high temperatures to proceed at a suitable rate. Once a sufficiently exothermic reaction begins, however, it may accelerate rapidly, producing a reaction of explosive violence.

Preparation of Oxygen

Nearly all commercial oxygen is obtained from air. The normal boiling point of O_2 is $-183\,°C$, whereas that of N_2, the other principal component of air, is $-196\,°C$. Thus, when air is liquefied and then allowed to warm, the N_2 boils off, leaving liquid O_2 contaminated mainly by small amounts of N_2 and Ar.

In the laboratory, O_2 can be obtained by heating either aqueous hydrogen peroxide or solid potassium chlorate ($KClO_3$):

$$2 \, KClO_3(s) \longrightarrow 2 \, KCl(s) + 3 \, O_2(g) \qquad [22.22]$$

Manganese dioxide (MnO_2) catalyzes both reactions.

Much of the O_2 in the atmosphere is replenished through the process of photosynthesis, in which green plants use the energy of sunlight to generate O_2 from atmospheric CO_2. Photosynthesis, therefore, regenerates O_2 and uses up CO_2.

Uses of Oxygen

Oxygen is one of the most widely used industrial chemicals, ranking behind only sulfuric acid (H_2SO_4) and nitrogen (N_2). About 3×10^{10} kg (30 million tons) of O_2 is used annually in the United States. Oxygen can be shipped and stored either as a liquid or in steel containers as a compressed gas. About 70% of the O_2 output, however, is generated where it is needed.

Oxygen is by far the most widely used oxidizing agent. Over half of the O_2 produced is used in the steel industry, mainly to remove impurities from steel. It is also used to bleach pulp and paper. (Oxidation of colored compounds often gives colorless products.) In medicine, oxygen eases breathing difficulties. It is also used together with acetylene (C_2H_2) in oxyacetylene welding (Figure 22.16▶). The reaction between C_2H_2 and O_2 is highly exothermic, producing temperatures in excess of 3000 °C:

$$2 \, C_2H_2(g) + 5 \, O_2(g) \longrightarrow 4 \, CO_2(g) + 2 \, H_2O(g) \qquad \Delta H° = -2510 \text{ kJ} \qquad [22.23]$$

Ozone

Ozone is a pale blue poisonous gas with a sharp, irritating odor. Most people can detect about 0.01 ppm in air. Exposure to 0.1 to 1 ppm produces headaches, burning eyes, and irritation to the respiratory passages.

The structure of the O_3 molecule is shown in Figure 22.17▶. The molecule possesses a π bond that is delocalized over the three oxygen atoms. ∞ (Section 8.6) The molecule dissociates readily, forming reactive oxygen atoms:

$$O_3(g) \longrightarrow O_2(g) + O(g) \qquad \Delta H° = 105 \text{ kJ} \qquad [22.24]$$

Ozone is a stronger oxidizing agent than dioxygen. Ozone forms oxides with many elements under conditions where O_2 will not react; indeed, it oxidizes all the common metals except gold and platinum.

Ozone can be prepared by passing electricity through dry O_2 in a flow-through apparatus. The electrical discharge causes rupture of the O_2 bond, resulting in reactions like those described in Section 18.3.

$$3 \, O_2(g) \xrightarrow{\text{electricity}} 2 \, O_3(g) \qquad \Delta H° = 285 \text{ kJ} \qquad [22.25]$$

Ozone cannot be stored for long, except at low temperature, because it readily decomposes to O_2. The decomposition is catalyzed by certain metals, such as Ag, Pt, and Pd, and by many transition-metal oxides.

▲ Figure 22.16 **Welding with an oxyacetylene torch.** The heat of combustion of acetylene is exceptionally high, thus giving rise to a very high flame temperature.

1.278 Å   1.278 Å

117°

▲ Figure 22.17 **Structure of the ozone molecule.**

■■ **SAMPLE EXERCISE 22.6** | Calculating an Equilibrium Constant

Using $\Delta G_f°$ for ozone from Appendix C, calculate the equilibrium constant, K, for Equation 22.25 at 298.0 K, assuming no electrical input.

SOLUTION

Analyze: We are asked to calculate the equilibrium constant for the formation of O_3 from O_2 (Equation 22.25), given the temperature and $\Delta G_f°$.

Plan: The relationship between the standard free-energy change, $\Delta G_f°$, for a reaction and the equilibrium constant for the reaction was given in Section 19.7, Equation 19.17.

Solve: From Appendix C we have

$$\Delta G_f^\circ(O_3) = 163.4 \text{ kJ/mol}$$

Thus, for Equation 22.25,

$$\Delta G^\circ = (2 \text{ mol } O_3)(163.4 \text{ kJ/mol } O_3) = 326.8 \text{ kJ}$$

From Equation 19.17 we have

$$\Delta G^\circ = -RT \ln K$$

Thus,

$$\ln K = \frac{-\Delta G^\circ}{RT} = \frac{-326.8 \times 10^3 \text{ J}}{(8.314 \text{ J/K-mol})(298.0 \text{ K})} = -131.9$$

$$K = e^{-131.9} = 5 \times 10^{-58}$$

Comment: In spite of the unfavorable equilibrium constant, ozone can be prepared from O_2 as described in the preceding text. The unfavorable free energy of formation is overcome by energy from the electrical discharge, and O_3 is removed before the reverse reaction can occur, so a nonequilibrium mixture results.

■ **PRACTICE EXERCISE**

Using the data in Appendix C, calculate ΔG° and the equilibrium constant (K) for Equation 22.24 at 298.0 K.
Answer: $\Delta G^\circ = 66.7 \text{ kJ}$, $K = 2 \times 10^{-12}$

Ozone is sometimes used to treat domestic water in place of chlorine. Like Cl_2, it kills bacteria and oxidizes organic compounds. The largest use of ozone, however, is in the preparation of pharmaceuticals, synthetic lubricants, and other commercially useful organic compounds, where O_3 is used to sever carbon–carbon double bonds.

Ozone is an important component of the upper atmosphere, where it screens out ultraviolet radiation. In this way ozone protects Earth from the effects of these high-energy rays. For this reason, depletion of stratospheric ozone is a major scientific concern. ∞ (Section 18.3) In the lower atmosphere, however, ozone is considered an air pollutant. It is a major constituent of smog. ∞ (Section 18.4) Because of its oxidizing power, ozone damages living systems and structural materials, especially rubber.

Oxides

The electronegativity of oxygen is second only to that of fluorine. As a result, oxygen exhibits negative oxidation states in all compounds except those with fluorine, OF_2 and O_2F_2. The -2 oxidation state is by far the most common. Compounds in this oxidation state are called *oxides*.

Nonmetals form covalent oxides. Most of these oxides are simple molecules with low melting and boiling points. SiO_2 and B_2O_3, however, have polymeric structures. ∞ (Sections 22.10 and 22.11) Most nonmetal oxides combine with water to give oxyacids. Sulfur dioxide (SO_2), for example, dissolves in water to give sulfurous acid (H_2SO_3):

$$SO_2(g) + H_2O(l) \longrightarrow H_2SO_3(aq) \qquad [22.26]$$

This reaction and that of SO_3 with H_2O to form H_2SO_4 are largely responsible for acid rain. ∞ (Section 18.4) The analogous reaction of CO_2 with H_2O to form carbonic acid (H_2CO_3) causes the acidity of carbonated water.

Oxides that react with water to form acids are called **acidic anhydrides** (anhydride means "without water") or **acidic oxides**. A few nonmetal oxides, especially ones with the nonmetal in a low oxidation state—such as N_2O, NO, and CO—do not react with water and are not acidic anhydrides.

▲ **GIVE IT SOME THOUGHT**

What acid is produced by the reaction of I_2O_5 with water?

BASIC ANHYDRIDES (BASIC OXIDES)

Most metal oxides are ionic compounds, which behave as a base in water. Ionic oxides that dissolve in water react to form hydroxides. Here, barium oxide (BaO), reacts with water to form barium hydroxide [Ba(OH)₂].

| BaO(s) | + | H₂O(l) | $\longrightarrow$ | Ba(OH)₂(aq) |
| Barium oxide | | reacts with water to produce | | Barium |

▲ **Figure 22.18 Reaction of a basic oxide with water.** (The reddish purple color of the solution is caused by phenolphthalein and indicates the presence of OH⁻ ions in solution.)

Most metal oxides are ionic compounds. Those ionic oxides that dissolve in water react to form hydroxides are consequently called **basic anhydrides** or **basic oxides**. Barium oxide (BaO), for example, reacts with water to form barium hydroxide [Ba(OH)₂], as shown in Figure 22.18▲. These kinds of reactions are due to the high basicity of the O^{2-} ion and its virtually complete hydrolysis in water:

$$O^{2-}(aq) + H_2O(l) \longrightarrow 2\,OH^-(aq) \qquad [22.27]$$

Even those ionic oxides that are water insoluble tend to dissolve in strong acids. Iron(III) oxide, for example, dissolves in acids:

$$Fe_2O_3(s) + 6\,H^+(aq) \longrightarrow 2\,Fe^{3+}(aq) + 3\,H_2O(l) \qquad [22.28]$$

This reaction is used to remove rust (Fe₂O₃·nH₂O) from iron or steel before a protective coat of zinc or tin is applied.

▲ **Figure 22.19 A self-contained breathing apparatus.** The source of oxygen in this apparatus, used by firefighters and rescue workers, is the reaction between potassium superoxide (KO_2) and water in the breath.

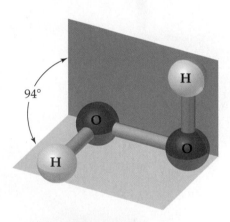

▲ **Figure 22.20 The molecular structure of hydrogen peroxide.** Note that the atoms of H_2O_2 do not lie in a single plane.

TABLE 22.4 ■ Acid–Base Character of Chromium Oxides		
Oxide	Oxidation State of Cr	Nature of Oxide
CrO	+2	Basic
Cr_2O_3	+3	Amphoteric
CrO_3	+6	Acidic

Oxides that can exhibit both acidic and basic characters are said to be *amphoteric*. ∞ (Section 17.5) If a metal forms more than one oxide, the basic character of the oxide decreases as the oxidation state of the metal increases, as illustrated in Table 22.4▲.

Peroxides and Superoxides

Compounds containing O—O bonds and oxygen in an oxidation state of −1 are called *peroxides*. Oxygen has an oxidation state of $-\frac{1}{2}$ in O_2^-, which is called the *superoxide* ion. The most active metals (K, Rb, and Cs) react with O_2 to give superoxides (KO_2, RbO_2, and CsO_2). Their active neighbors in the periodic table (Na, Ca, Sr, and Ba) react with O_2, producing peroxides (Na_2O_2, CaO_2, SrO_2, and BaO_2). Less active metals and nonmetals produce normal oxides. ∞ (Section 7.6)

When superoxides dissolve in water, O_2 is produced:

$$4\,KO_2(s) + 2\,H_2O(l) \longrightarrow 4\,K^+(aq) + 4\,OH^-(aq) + 3\,O_2(g) \qquad [22.29]$$

Because of this reaction, potassium superoxide (KO_2) is used as an oxygen source in masks worn by rescue workers (Figure 22.19◄). Moisture in the breath causes the compound to decompose to form O_2 and KOH. The KOH so formed removes CO_2 from the exhaled breath:

$$2\,OH^-(aq) + CO_2(g) \longrightarrow H_2O(l) + CO_3^{2-}(aq) \qquad [22.30]$$

Hydrogen peroxide (H_2O_2) is the most familiar and commercially important peroxide. The structure of H_2O_2 is shown in Figure 22.20◄. Pure hydrogen peroxide is a clear, syrupy liquid that has a melting point of −0.4 °C. Concentrated hydrogen peroxide is a dangerously reactive substance because the decomposition to form water and oxygen gas is very exothermic.

$$2\,H_2O_2(l) \longrightarrow 2\,H_2O(l) + O_2(g) \qquad \Delta H° = -196.1 \text{ kJ} \qquad [22.31]$$

This is a **disproportionation reaction**, one in which an element is simultaneously oxidized and reduced. The oxidation number of oxygen goes from −1 to −2 and 0.

Hydrogen peroxide is marketed as a chemical reagent in aqueous solutions of up to about 30% by mass. A solution containing about 3% H_2O_2 by mass is sold in drugstores and used as a mild antiseptic. Somewhat more concentrated solutions are employed to bleach fabrics.

The peroxide ion is also a by-product of metabolism that results from the reduction of molecular oxygen (O_2). The body disposes of this reactive species with enzymes such as peroxidase and catalase.

22.6 THE OTHER GROUP 6A ELEMENTS: S, Se, Te, AND Po

In addition to oxygen, the other group 6A elements are sulfur, selenium, tellurium, and polonium. In this section we will survey the properties of the group as a whole and then examine the chemistry of sulfur, selenium, and tellurium. We will not say much about polonium, which has no stable isotopes and is found only in minute quantities in radium-containing minerals.

TABLE 22.5 ■ Some Properties of the Group 6A Elements				
Property	O	S	Se	Te
Atomic radius (Å)	0.73	1.04	1.17	1.43
X^{2-} ionic radius (Å)	1.40	1.84	1.98	2.21
First ionization energy (kJ/mol)	1314	1000	941	869
Electron affinity (kJ/mol)	−141	−200	−195	−190
Electronegativity	3.5	2.5	2.4	2.1
X—X single-bond enthalpy (kJ/mol)	146*	266	172	126
Reduction potential to H_2X in acidic solution (V)	1.23	0.14	−0.40	−0.72

*Based on O—O bond energy in H_2O_2.

General Characteristics of the Group 6A Elements

The group 6A elements possess the general outer-electron configuration ns^2np^4. where n has values ranging from 2 through 6. Thus, these elements may attain a noble-gas electron configuration by the addition of two electrons, which results in a −2 oxidation state. Except for oxygen, however, the group 6A elements are also commonly found in positive oxidation states up to +6, and they can have expanded valence shells. Thus, compounds such as SF_6, SeF_6, and TeF_6 occur in which the central atom is in the +6 oxidation state with more than an octet of valence electrons.

Table 22.5▲ summarizes some of the more important properties of the atoms of the group 6A elements. In most of the properties listed in Table 22.5, we see a regular variation as a function of increasing atomic number. For example, atomic and ionic radii increase and ionization energies decrease, as expected, as we move down the family.

Occurrence and Preparation of S, Se, and Te

Large underground deposits are the principal source of elemental sulfur. Sulfur also occurs widely as sulfide and sulfate minerals. Its presence as a minor component of coal and petroleum poses a major problem. Combustion of these "unclean" fuels leads to serious pollution by sulfur oxides. ∞ (Section 18.4) Much effort has been directed at removing this sulfur, and these efforts have increased the availability of sulfur. The sale of this sulfur helps to partially offset the costs of the desulfurizing processes and equipment.

Selenium and tellurium occur in rare minerals such as Cu_2Se, PbSe, Cu_2Te, and PbTe. They also occur as minor constituents in sulfide ores of copper, iron, nickel, and lead.

Properties and Uses of Sulfur, Selenium, and Tellurium

As we normally encounter it, sulfur is yellow, tasteless, and nearly odorless. It is insoluble in water and exists in several allotropic forms. The thermodynamically stable form at room temperature is rhombic sulfur, which consists of puckered S_8 rings, as shown in Figure 22.21▲. When heated above its melting point (113 °C), sulfur undergoes a variety of changes.

▼ **Figure 22.21 Elemental sulfur.** The common yellow crystalline form of rhombic sulfur, S_8, consists of eight-membered puckered rings of S atoms.

(a)   (b)

▲ **Figure 22.22 Heating sulfur.** When sulfur is heated above its melting point (113 °C), it becomes dark and viscous. Here the liquid is shown falling into cold water, where it again solidifies.

▲ **Figure 22.23 Portion of the structure of crystalline selenium.** The dashed lines represent weak bonding interactions between atoms in adjacent chains. Tellurium has the same structure.

▲ **Figure 22.24 Iron pyrite (FeS₂).** This substance is also known as fool's gold because its color has fooled people into thinking it was gold. Gold is much denser and softer than iron pyrite.

The molten sulfur first contains S_8 molecules and is fluid because the rings readily slip over one another. Further heating of this straw-colored liquid causes the rings to break; the fragments then join to form very long molecules that can become entangled. The sulfur consequently becomes highly viscous. This change is marked by a color change to dark reddish brown (Figure 22.22 ◄). Further heating breaks the chains, and the viscosity again decreases.

Most of the approximately 1×10^{10} kg (10 million tons) of sulfur produced in the United States each year is used to manufacture sulfuric acid. Sulfur is also used to vulcanize rubber, a process that toughens rubber by introducing cross-linking between polymer chains. ∞ (Section 12.6)

The most stable allotropes of both selenium and tellurium are crystalline substances containing helical chains of atoms, as illustrated in Figure 22.23 ◄. Each atom of the chain is close to atoms in adjacent chains, and it appears that some sharing of electron pairs between these atoms occurs.

The electrical conductivity of selenium is very low in the dark, but increases greatly upon exposure to light. This property of the element is utilized in photoelectric cells and light meters. Photocopiers also depend on the photoconductivity of selenium. Photocopy machines contain a belt or drum coated with a film of selenium. This drum is electrostatically charged and then exposed to light reflected from the image being photocopied. The electric charge drains from the selenium where it has been made conductive by exposure to light. A black powder (the toner) sticks only to the areas that remain charged. The photocopy is made when the toner is transferred to a sheet of plain paper, which is heated to fuse the toner to the paper.

Sulfides

Sulfur forms compounds by direct combination with many elements. When the element is less electronegative than sulfur, *sulfides*, which contain S^{2-}, form. Iron(II) sulfide (FeS) forms, for example, by direct combination of iron and sulfur. Many metallic elements are found in the form of sulfide ores, such as PbS (galena) and HgS (cinnabar). A series of related ores containing the disulfide ion, S_2^{2-} (analogous to the peroxide ion), are known as *pyrites*. Iron pyrite, FeS_2, occurs as golden yellow cubic crystals (Figure 22.24 ◄). Because it has been occasionally mistaken for gold by miners, iron pyrite is often called fool's gold.

One of the most important sulfides is hydrogen sulfide (H_2S). This substance is not normally produced by direct union of the elements because it is unstable at elevated temperatures and decomposes into the elements. It is normally prepared by action of dilute acid on iron(II) sulfide.

$$FeS(s) + 2\,H^+(aq) \longrightarrow H_2S(aq) + Fe^{2+}(aq) \qquad [22.32]$$

One of hydrogen sulfide's most readily recognized properties is its odor; H_2S is largely responsible for the offensive odor of rotten eggs. Hydrogen sulfide is actually quite toxic. Fortunately, our noses are able to detect H_2S in extremely low, nontoxic concentrations. A sulfur-containing organic molecule, such as dimethyl sulfide, $(CH_3)_2S$, which is similarly odoriferous and can be detected by smell at a level of one part per trillion, is added to natural gas as a safety factor to give it a detectable odor.

Oxides, Oxyacids, and Oxyanions of Sulfur

Sulfur dioxide is formed when sulfur is combusted in air; it has a choking odor and is poisonous. The gas is particularly toxic to lower organisms, such as fungi, so it is used to sterilize dried fruit and wine. At 1 atm pressure and room temperature, SO_2 dissolves in water to produce a solution of about 1.6 M concentration. The SO_2 solution is acidic, and we describe it as sulfurous acid (H_2SO_3).

Salts of SO_3^{2-} (sulfites) and HSO_3^- (hydrogen sulfites or bisulfites) are well known. Small quantities of Na_2SO_3 or $NaHSO_3$ are used as food additives to prevent bacterial spoilage. Because some people are extremely allergic to sulfites, all food products with sulfites must now carry a warning label disclosing their presence.

Although combustion of sulfur in air produces mainly SO_2, small amounts of SO_3 are also formed. The reaction produces chiefly SO_2 because the activation-energy barrier for further oxidation to SO_3 is very high unless the reaction is catalyzed. Sulfur trioxide is of great commercial importance because it is the anhydride of sulfuric acid. In the manufacture of sulfuric acid, SO_2 is first obtained by burning sulfur. The SO_2 is then oxidized to SO_3, using a catalyst such as V_2O_5 or platinum. The SO_3 is dissolved in H_2SO_4 because it does not dissolve quickly in water, and then the $H_2S_2O_7$ formed in this reaction, called pyrosulfuric acid, is added to water to form H_2SO_4:

$$SO_3(g) + H_2SO_4(l) \longrightarrow H_2S_2O_7(l) \qquad [22.33]$$

$$H_2S_2O_7(l) + H_2O(l) \longrightarrow 2\,H_2SO_4(l) \qquad [22.34]$$

GIVE IT SOME THOUGHT

Is Equation 22.34 an example of an oxidation-reduction reaction?

Commercial sulfuric acid is 98% H_2SO_4. It is a dense, colorless, oily liquid that boils at 340 °C. Sulfuric acid has many useful properties: It is a strong acid, a good dehydrating agent, and a moderately good oxidizing agent. Its dehydrating ability is demonstrated in Figure 22.25▼.

Year after year, the production of sulfuric acid is the largest of any chemical produced in the United States. About 4×10^{10} kg (40 million tons) is produced annually in this country. Sulfuric acid is employed in some way in almost all manufacturing. Consequently, its consumption is considered a measure of industrial activity.

Sulfuric acid is a strong acid, but only the first hydrogen is completely ionized in aqueous solution:

$$H_2SO_4(aq) \longrightarrow H^+(aq) + HSO_4^-(aq) \qquad [22.35]$$

$$HSO_4^-(aq) \rightleftharpoons H^+(aq) + SO_4^{2-}(aq) \qquad K_a = 1.1 \times 10^{-2} \qquad [22.36]$$

(a)　　　　　　　　　(b)　　　　　　　　　(c)

▲ **Figure 22.25 A dehydration reaction.** Sucrose ($C_{12}H_{22}O_{11}$) is a carbohydrate, containing two H atoms for each O atom. (a) The beaker contains solid sucrose (table sugar), initially white. Adding concentrated sulfuric acid, which is an excellent dehydrating agent, removes H_2O from the sucrose. (b, c) The product of the dehydration is carbon, the black mass remaining at the end of the reaction.

▲ **Figure 22.26 Structures of sulfate ions.** (a) The sulfate ion (SO_4^{2-}), and (b) the thiosulfate ion ($S_2O_3^{2-}$).

TABLE 22.6 ■ Oxidation States of Nitrogen	
Oxidation State	**Examples**
+5	N_2O_5, HNO_3, NO_3^-
+4	NO_2, N_2O_4
+3	HNO_2, NO_2^-, NF_3
+2	NO
+1	N_2O, $H_2N_2O_2$, $N_2O_2^{2-}$, HNF_2
0	N_2
−1	NH_2OH, NH_2F
−2	N_2H_4
−3	NH_3, NH_4^+, NH_2^-

Consequently, sulfuric acid forms two series of compounds: sulfates and bisulfates (or hydrogen sulfates). Bisulfate salts are common components of the "dry acids" used for adjusting the pH of swimming pools and hot tubs; they are also components of many toilet bowl cleaners.

The thiosulfate ion ($S_2O_3^{2-}$) is related to the sulfate ion and is formed by boiling an alkaline solution of SO_3^{2-} with elemental sulfur.

$$8\, SO_3^{2-}(aq) + S_8(s) \longrightarrow 8\, S_2O_3^{2-}(aq) \qquad [22.37]$$

The term *thio* indicates substitution of sulfur for oxygen. The structures of the sulfate and thiosulfate ions are compared in Figure 22.26 ◄.

22.7 NITROGEN

Nitrogen was discovered in 1772 by the Scottish botanist Daniel Rutherford. He found that when a mouse was enclosed in a sealed jar, the animal quickly consumed the life-sustaining component of air (oxygen) and died. When the "fixed air" (CO_2) in the container was removed, a "noxious air" remained that would not sustain combustion or life. We now know that gas as nitrogen.

Nitrogen constitutes 78% by volume of Earth's atmosphere, where it occurs as N_2 molecules. Although nitrogen is a key element in living organisms, compounds of nitrogen are not abundant in Earth's crust. The major natural deposits of nitrogen compounds are those of KNO_3 (saltpeter) in India and $NaNO_3$ (Chile saltpeter) in Chile and other desert regions of South America.

Properties of Nitrogen

Nitrogen is a colorless, odorless, and tasteless gas composed of N_2 molecules. Its melting point is −210 °C, and its normal boiling point is −196 °C.

The N_2 molecule is very unreactive because of the strong triple bond between nitrogen atoms (the $N\equiv N$ bond enthalpy is 941 kJ/mol, nearly twice that for the bond in O_2; see Table 8.4). When substances burn in air, they normally react with O_2 but not with N_2. When magnesium burns in air, however, it also reacts with N_2 to form magnesium nitride (Mg_3N_2). A similar reaction occurs with lithium, forming Li_3N.

$$3\, Mg(s) + N_2(g) \longrightarrow Mg_3N_2(s) \qquad [22.38]$$

The nitride ion is a strong Brønsted–Lowry base. It reacts with water to form ammonia (NH_3), as in the following reaction:

$$Mg_3N_2(s) + 6\, H_2O(l) \longrightarrow 2\, NH_3(aq) + 3\, Mg(OH)_2(s) \qquad [22.39]$$

The electron configuration of the nitrogen atom is $[He]2s^2 2p^3$. The element exhibits all formal oxidation states from +5 to −3, as shown in Table 22.6 ◄. The +5, 0, and −3 oxidation states are the most common and generally the most stable of these. Because nitrogen is more electronegative than all elements except fluorine, oxygen, and chlorine, it exhibits positive oxidation states only in combination with these three elements.

Preparation and Uses of Nitrogen

Elemental nitrogen is obtained in commercial quantities by fractional distillation of liquid air. About 4×10^{10} kg (40 million tons) of N_2 is produced annually in the United States.

Because of its low reactivity, large quantities of N_2 are used as an inert gaseous blanket to exclude O_2 during the processing and packaging of foods, the manufacture of chemicals, the fabrication of metals, and the production of electronic devices. Liquid N_2 is employed as a coolant to freeze foods rapidly.

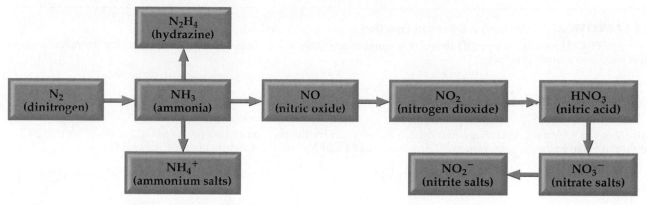

▲ **Figure 22.27 Nitrogen conversions.** Sequence of conversion of N_2 into common nitrogen compounds.

The largest use of N_2 is in the manufacture of nitrogen-containing fertilizers, which provide a source of *fixed* nitrogen. We have previously discussed nitrogen fixation in the "Chemistry and Life" box in Section 14.7 and in the "Chemistry Put to Work" box in Section 15.2. Our starting point in fixing nitrogen is the manufacture of ammonia via the Haber process. ∞(Section 15.2) The ammonia can then be converted into a variety of useful, simple nitrogen-containing species, as shown in Figure 22.27 ▲. Many of the reactions along this chain of conversion are discussed in more detail later in this section.

Hydrogen Compounds of Nitrogen

Ammonia is one of the most important compounds of nitrogen. It is a colorless toxic gas that has a characteristic irritating odor. As we have noted in previous discussions, the NH_3 molecule is basic ($K_b = 1.8 \times 10^{-5}$). ∞ (Section 16.7)

In the laboratory NH_3 can be prepared by the action of NaOH on an ammonium salt. The NH_4^+ ion, which is the conjugate acid of NH_3, transfers a proton to OH^-. The resultant NH_3 is volatile and is driven from the solution by mild heating.

$$NH_4Cl(aq) + NaOH(aq) \longrightarrow NH_3(g) + H_2O(l) + NaCl(aq) \quad [22.40]$$

Commercial production of NH_3 is achieved by the Haber process.

$$N_2(g) + 3\,H_2(g) \longrightarrow 2\,NH_3(g) \quad [22.41]$$

About 1×10^{10} kg (10 million tons) of ammonia is produced annually in the United States. About 75% is used for fertilizer.

Hydrazine (N_2H_4) is another important hydride of nitrogen. As shown in Figure 22.28 ▶, the hydrazine molecule contains an N—N single bond. Hydrazine is quite poisonous. It can be prepared by the reaction of ammonia with hypochlorite ion (OCl^-) in aqueous solution.

$$2\,NH_3(aq) + OCl^-(aq) \longrightarrow N_2H_4(aq) + Cl^-(aq) + H_2O(l) \quad [22.42]$$

The reaction is complex, involving several intermediates, including chloramine (NH_2Cl). The poisonous NH_2Cl bubbles out of solution when household ammonia and chlorine bleach (which contains OCl^-) are mixed. This reaction is one reason for the frequently cited warning not to mix bleach and household ammonia.

Pure hydrazine is a strong and versatile reducing agent. The major use of hydrazine and compounds related to it, such as methylhydrazine (Figure 22.28), is as rocket fuels.

▲ **Figure 22.28 Structures of hydrazine (N_2H_4) and methylhydrazine (CH_3NHNH_2).**

◼ SAMPLE EXERCISE 22.7 | Writing a Balanced Equation

Hydroxylamine (NH_2OH) reduces copper(II) to the free metal in acid solutions. Write a balanced equation for the reaction, assuming that N_2 is the oxidation product.

SOLUTION

Analyze: We are asked to write a balanced oxidation-reduction equation in which NH_2OH is converted to N_2 and Cu^{2+} is converted to Cu.

Plan: Because this is a redox reaction, the equation can be balanced by the method of half-reactions discussed in Section 20.2. Thus, we begin with two half-reactions, one involving the NH_2OH and N_2 and the other involving Cu^{2+} and Cu.

Solve: The unbalanced and incomplete half-reactions are

$$Cu^{2+}(aq) \longrightarrow Cu(s)$$

$$NH_2OH(aq) \longrightarrow N_2(g)$$

Balancing these equations as described in Section 20.2 gives

$$Cu^{2+}(aq) + 2\,e^- \longrightarrow Cu(s)$$

$$2\,NH_2OH(aq) \longrightarrow N_2(g) + 2\,H_2O(l) + 2\,H^+(aq) + 2\,e^-$$

Adding these half-reactions gives the balanced equation:

$$Cu^{2+}(aq) + 2\,NH_2OH(aq) \longrightarrow Cu(s) + N_2(g) + 2\,H_2O(l) + 2\,H^+(aq)$$

◼ PRACTICE EXERCISE

(a) In power plants, hydrazine is used to prevent corrosion of the metal parts of steam boilers by the O_2 dissolved in the water. The hydrazine reacts with O_2 in water to give N_2 and H_2O. Write a balanced equation for this reaction. **(b)** Methylhydrazine, $N_2H_3CH_3(l)$, is used with the oxidizer dinitrogen tetroxide, $N_2O_4(l)$, to power the steering rockets of the Space Shuttle orbiter. The reaction of these two substances produces N_2, CO_2, and H_2O. Write a balanced equation for this reaction.
Answers: **(a)** $N_2H_4(aq) + O_2(aq) \longrightarrow N_2(g) + 2\,H_2O(l)$; **(b)** $5\,N_2O_4(l) + 4\,N_2H_3CH_3(l) \longrightarrow 9\,N_2(g) + 4\,CO_2(g) + 12\,H_2O(g)$

Oxides and Oxyacids of Nitrogen

Nitrogen forms three common oxides: N_2O (nitrous oxide), NO (nitric oxide), and NO_2 (nitrogen dioxide). It also forms two unstable oxides that we will not discuss, N_2O_3 (dinitrogen trioxide) and N_2O_5 (dinitrogen pentoxide).

Nitrous oxide (N_2O) is also known as laughing gas because a person becomes somewhat giddy after inhaling only a small amount of it. This colorless gas was the first substance used as a general anesthetic. It is used as the compressed gas propellant in several aerosols and foams, such as in whipped cream. It can be prepared in the laboratory by carefully heating ammonium nitrate to about 200 °C.

$$NH_4NO_3(s) \xrightarrow{\Delta} N_2O(g) + 2\,H_2O(g) \qquad [22.43]$$

Nitric oxide (NO) is also a colorless gas, but, unlike N_2O, it is slightly toxic. It can be prepared in the laboratory by reduction of dilute nitric acid, using copper or iron as a reducing agent, as shown in Figure 22.29▼.

$$3\,Cu(s) + 2\,NO_3^-(aq) + 8\,H^+(aq) \longrightarrow 3\,Cu^{2+}(aq) + 2\,NO(g) + 4\,H_2O(l)$$
$$[22.44]$$

▶ **Figure 22.29 Nitric oxide formation.** (a) Nitric oxide (NO) can be prepared by the reaction of copper with 6 *M* nitric acid. In this photograph a jar containing 6 *M* HNO_3 has been inverted over some pieces of copper. Colorless NO, which is only slightly soluble in water, is collected in the jar. The blue color of the solution is due to the presence of Cu^{2+} ions. (b) Colorless NO gas, collected as shown in (a). (c) When the stopper is removed from the jar of NO, the NO reacts with oxygen in the air to form yellow-brown NO_2.

(a)

(b)

(c)

Nitric oxide is also produced by direct reaction of N_2 and O_2 at high temperatures. This reaction is a significant source of nitrogen oxide air pollutants. ∞(Section 18.4) The direct combination of N_2 and O_2 is not used for commercial production of NO, however, because the yield is low; the equilibrium constant K_p at 2400 K is only 0.05. ∞(Section 15.7, "Chemistry Put to Work: Controlling Nitric Oxide Emissions")

The commercial route to NO (and hence to other oxygen-containing compounds of nitrogen) is via the catalytic oxidation of NH_3.

$$4\,NH_3(g) + 5\,O_2(g) \xrightarrow[850\,°C]{Pt\,catalyst} 4\,NO(g) + 6\,H_2O(g) \qquad [22.45]$$

The catalytic conversion of NH_3 to NO is the first step in a three-step process known as the **Ostwald process**, by which NH_3 is converted commercially into nitric acid (HNO_3). Nitric oxide reacts readily with O_2, forming NO_2 when exposed to air (Figure 22.29).

$$2\,NO(g) + O_2(g) \longrightarrow 2\,NO_2(g) \qquad [22.46]$$

When dissolved in water, NO_2 forms nitric acid.

$$3\,NO_2(g) + H_2O(l) \longrightarrow 2\,H^+(aq) + 2\,NO_3^-(aq) + NO(g) \qquad [22.47]$$

Nitrogen is both oxidized and reduced in this reaction, so it disproportionates. The reduction product NO can be converted back into NO_2 by exposure to air and thereafter dissolved in water to prepare more HNO_3.

Recently NO has been found to be an important neurotransmitter in the human body. It causes the muscles that line blood vessels to relax, thus allowing an increased passage of blood (see the "Chemistry and Life" box on page 956).

Nitrogen dioxide (NO_2) is a yellow-brown gas (Figure 22.29). Like NO, it is a major constituent of smog. ∞(Section 18.4) It is poisonous and has a choking odor. As discussed in the introduction of Chapter 15, NO_2 and N_2O_4 exist in equilibrium (Figures 15.1 and 15.2):

$$2\,NO_2(g) \rightleftharpoons N_2O_4(g) \qquad \Delta H° = -58\,kJ \qquad [22.48]$$

The two common oxyacids of nitrogen are nitric acid (HNO_3) and nitrous acid (HNO_2) (Figure 22.30▶). *Nitric acid* is a strong acid. It is also a powerful oxidizing agent, as the following standard reduction potential indicates:

$$NO_3^-(aq) + 4\,H^+(aq) + 3\,e^- \longrightarrow NO(g) + 2\,H_2O(l) \qquad E° = +0.96\,V$$
$$[22.49]$$

Concentrated nitric acid will attack and oxidize most metals except Au, Pt, Rh, and Ir.

About 7×10^9 kg (8 million tons) of nitric acid is produced annually in the United States. Its largest use is in the manufacture of NH_4NO_3 for fertilizers, which accounts for about 80% of that produced. HNO_3 is also used in the production of plastics, drugs, and explosives. Among the explosives made from nitric acid are nitroglycerin, trinitrotoluene (TNT), and nitrocellulose. The following reaction occurs when nitroglycerin explodes:

$$4\,C_3H_5N_3O_9(l) \longrightarrow 6\,N_2(g) + 12\,CO_2(g) + 10\,H_2O(g) + O_2(g) \qquad [22.50]$$

All the products of this reaction contain very strong bonds. As a result, the reaction is very exothermic. Furthermore, a tremendous amount of gaseous products forms from the liquid. The sudden formation of these gases, together with their expansion resulting from the heat generated by the reaction, produces the explosion. ∞(Section 8.8: "Chemistry Put to Work: Explosives and Alfred Nobel")

Nitrous acid (HNO_2) (Figure 22.30) is considerably less stable than HNO_3 and tends to disproportionate into NO and HNO_3. It is normally made by action of a strong acid, such as H_2SO_4, on a cold solution of a nitrite salt, such as $NaNO_2$. Nitrous acid is a weak acid ($K_a = 4.5 \times 10^{-4}$).

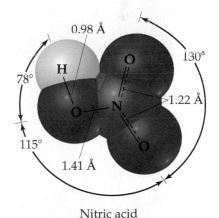

Nitric acid

Nitrous acid

▲ **Figure 22.30 Structures of nitric acid and nitrous acid.**

Chemistry and Life   NITROGLYCERIN AND HEART DISEASE

During the 1870s an interesting observation was made in Alfred Nobel's dynamite factories. Workers who suffered from heart disease that caused chest pains when they exerted themselves found relief from the pains during the workweek. It quickly became apparent that nitroglycerin, present in the air of the factory, acted to enlarge blood vessels. Thus, this powerfully explosive chemical became a standard treatment for angina pectoris, the chest pains accompanying heart failure (Figure 22.31▶). It took more than 100 years to discover that nitroglycerin was converted in the vascular smooth muscle into NO, which is the chemical agent actually causing dilation of the blood vessels. In 1998 the Nobel Prize in medicine and physiology was awarded to Robert F. Furchgott, Louis J. Ignarro, and Ferid Murad for their discoveries of the detailed pathways by which NO acts in the cardiovascular system. It was a sensation that this simple, common air pollutant could exert important functions in the organism.

As useful as nitroglycerin is to this day in treating angina pectoris, it has a limitation in that prolonged administration results in development of tolerance, or desensitization, of the vascular muscle to further vasorelaxation by nitroglycerin. The bioactivation of nitroglycerin is the subject of active research in the hope that a means of circumventing desensitization can be found.

▲ **Figure 22.31**
Nitroglycerin tablets.

GIVE IT SOME THOUGHT

What are the oxidation numbers of the nitrogen atoms in **(a)** nitric acid; **(b)** nitrous acid?

22.8 THE OTHER GROUP 5A ELEMENTS: P, As, Sb, AND Bi

5A
7 N
15 P
33 As
51 Sb
83 Bi

Of the other group 5A elements—phosphorus, arsenic, antimony, and bismuth—phosphorus has a central role in several aspects of biochemistry and environmental chemistry. In this section we will explore the chemistry of these other group 5A elements, with an emphasis on the chemistry of phosphorus.

General Characteristics of the Group 5A Elements

The group 5A elements possess the outer-shell electron configuration ns^2np^3, where n has values ranging from 2 to 6. A noble-gas configuration results from the addition of three electrons to form the -3 oxidation state. Ionic compounds containing X^{3-} ions are not common, however. More commonly, the group 5A element acquires an octet of electrons via covalent bonding. The oxidation number may range from -3 to $+5$, depending on the nature and number of the atoms to which the group 5A element is bonded.

Because of its lower electronegativity, phosphorus is found more frequently in positive oxidation states than is nitrogen. Furthermore, compounds in which phosphorus has the $+5$ oxidation state are not as strongly oxidizing as the corresponding compounds of nitrogen. Conversely, compounds in which phosphorus has a -3 oxidation state are much stronger reducing agents than are the corresponding compounds of nitrogen.

Some of the important properties of the group 5A elements are listed in Table 22.7▶. The general pattern that emerges from these data is similar to what we have seen before with other groups: Size and metallic character increase as atomic number increases within the group.

The variation in properties among the elements of group 5A is more striking than that seen in groups 6A and 7A. Nitrogen at the one extreme exists as a gaseous diatomic molecule; it is clearly nonmetallic in character. At the other extreme, bismuth is a reddish white, metallic-looking substance that has most of the characteristics of a metal.

Property	N	P	As	Sb	Bi
Atomic radius (Å)	0.75	1.10	1.21	1.41	1.55
First ionization energy (kJ/mol)	1402	1012	947	834	703
Electron affinity (kJ/mol)	>0	−72	−78	−103	−91
Electronegativity	3.0	2.1	2.0	1.9	1.9
X—X single-bond enthalpy (kJ/mol)*	163	200	150	120	—
X≡X triple-bond enthalpy (kJ/mol)	941	490	380	295	192

TABLE 22.7 ■ Properties of the Group 5A Elements

*Approximate values only.

The values listed for X—X single-bond enthalpies are not very reliable because it is difficult to obtain such data from thermochemical experiments. However, there is no doubt about the general trend: a low value for the N—N single bond, an increase at phosphorus, and then a gradual decline to arsenic and antimony. From observations of the elements in the gas phase, it is possible to estimate the X≡X triple-bond enthalpies, as listed in Table 22.7. Here we see a trend that is different from that for the X—X single bond. Nitrogen forms a much stronger triple bond than the other elements, and there is a steady decline in the triple-bond enthalpy down through the group. These data help us to appreciate why nitrogen alone of the group 5A elements exists as a diatomic molecule in its stable state at 25 °C. All the other elements exist in structural forms with single bonds between the atoms.

Occurrence, Isolation, and Properties of Phosphorus

Phosphorus occurs mainly in the form of phosphate minerals. The principal source of phosphorus is phosphate rock, which contains phosphate principally as $Ca_3(PO_4)_2$. The element is produced commercially by reduction of calcium phosphate with carbon in the presence of SiO_2:

$$2\,Ca_3(PO_4)_2(s) + 6\,SiO_2(s) + 10\,C(s) \xrightarrow{1500\,°C} P_4(g) + 6\,CaSiO_3(l) + 10\,CO(g) \tag{22.51}$$

The phosphorus produced in this fashion is the allotrope known as white phosphorus. This form distills from the reaction mixture as the reaction proceeds.

White phosphorus consists of P_4 tetrahedra (Figure 22.32▶). The 60° bond angles in P_4 are unusually small for molecules, so there is much strain in the bonding, which is consistent with the high reactivity of white phosphorus. This allotrope bursts spontaneously into flames if exposed to air. When heated in the absence of air to about 400 °C, white phosphorus is converted to a more stable allotrope known as red phosphorus, which does not ignite on contact with air. Red phosphorus is also considerably less poisonous than the white form. Both allotropes are shown in Figure 22.33▶. We will denote elemental phosphorus as simply P(s).

Phosphorus Halides

Phosphorus forms a wide range of compounds with the halogens, the most important of which are the trihalides and pentahalides. Phosphorus trichloride (PCl_3) is commercially the most significant of these compounds and is used to prepare a wide variety of products, including soaps, detergents, plastics, and insecticides.

Phosphorus chlorides, bromides, and iodides can be made by the direct oxidation of elemental phosphorus with the elemental halogen. PCl_3, for example, which is a liquid at room temperature, is made by passing a stream of dry chlorine gas over white or red phosphorus.

$$2\,P(s) + 3\,Cl_2(g) \longrightarrow 2\,PCl_3(l) \tag{22.52}$$

If excess chlorine gas is present, an equilibrium is established between PCl_3 and PCl_5.

$$PCl_3(l) + Cl_2(g) \rightleftharpoons PCl_5(s) \tag{22.53}$$

▲ Figure 22.32 **Structure of white phosphorus.** Tetrahedral structure of the P_4 molecule.

▲ Figure 22.33 **Phosphorus allotropes.** White phosphorus is very reactive and is normally stored under water to protect it from oxygen. Red phosphorus is much less reactive than white phosphorus, and it is not necessary to store it under water.

P

O

P_4O_6

P_4O_{10}

▲ Figure 22.34 Structures of P_4O_6 and P_4O_{10}.

H
O
P
O
H
O
O
H

H_3PO_4 H

H
H
P
O
O
H
O

H_3PO_3

▲ Figure 22.35 Structures of H_3PO_4 and H_3PO_3.

The phosphorus halides hydrolyze on contact with water. The reactions occur readily, and most of the phosphorus halides fume in air because of reaction with water vapor. In the presence of excess water the products are the corresponding phosphorus oxyacid and hydrogen halide.

$$PBr_3(l) + 3\ H_2O(l) \longrightarrow H_3PO_3(aq) + 3\ HBr(aq) \qquad [22.54]$$

$$PCl_5(l) + 4\ H_2O(l) \longrightarrow H_3PO_4(aq) + 5\ HCl(aq) \qquad [22.55]$$

Oxy Compounds of Phosphorus

Probably the most significant compounds of phosphorus are those in which the element is combined in some way with oxygen. Phosphorus(III) oxide (P_4O_6) is obtained by allowing white phosphorus to oxidize in a limited supply of oxygen. When oxidation takes place in the presence of excess oxygen, phosphorus(V) oxide (P_4O_{10}) forms. This compound is also readily formed by oxidation of P_4O_6. These two oxides represent the two most common oxidation states for phosphorus, +3 and +5. The structural relationship between P_4O_6 and P_4O_{10} is shown in Figure 22.34 ◄. Notice the resemblance these molecules have to the P_4 molecule, shown in Figure 22.32; all three substances have a P_4 core.

■ SAMPLE EXERCISE 22.8 | Calculating a Standard Enthalpy Change

The reactive chemicals on the tip of a "strike anywhere" match are usually P_4S_3 and an oxidizing agent such as $KClO_3$. When the match is struck on a rough surface, the heat generated by the friction ignites the P_4S_3, and the oxidizing agent brings about rapid combustion. The products of the combustion of P_4S_3 are P_4O_{10} and SO_2. Calculate the standard enthalpy change for the combustion of P_4S_3 in air, given the following standard enthalpies of formation: P_4S_3 (-154.4 kJ/mol), P_4O_{10} (-2940.1 kJ/mol), SO_2 (-296.9 kJ/mol).

SOLUTION

Analyze: We are given the reactants (P_4S_3 and O_2 from air) and the products (P_4O_{10} and SO_2) for a reaction, together with their standard enthalpies of formation, and asked to calculate the standard enthalpy change for the reaction.

Plan: We first need a balanced chemical equation for the reaction. The enthalpy change for the reaction is then equal to the enthalpies of formation of products minus those of reactants (Equation 5.31, Section 5.7). We also need to recall that the standard enthalpy of formation of any element in its standard state is zero. Thus, $\Delta H_f^\circ(O_2) = 0$.

Solve: The balanced chemical equation for the combustion is

$$P_4S_3(s) + 8\ O_2(g) \longrightarrow P_4O_{10}(s) + 3\ SO_2(g)$$

Thus, we can write

$$\begin{aligned}
\Delta H^\circ &= \Delta H_f^\circ(P_4O_{10}) + 3\ \Delta H_f^\circ(SO_2) - \Delta H_f^\circ(P_4S_3) - 8\ \Delta H_f^\circ(O_2) \\
&= -2940.1\ \text{kJ} + 3(-296.9)\ \text{kJ} - (-154.4\ \text{kJ}) - 8(0) \\
&= -3676.4\ \text{kJ}
\end{aligned}$$

Comment: The reaction is strongly exothermic, making it evident why P_4S_3 is used on match tips.

■ PRACTICE EXERCISE

Write the balanced equation for the reaction of P_4O_{10} with water, and calculate ΔH° for this reaction using data from Appendix C.
Answer: $P_4O_{10}(s) + 6\ H_2O(l) \longrightarrow 4\ H_3PO_4(aq)$; -498.1 kJ

Phosphorus(V) oxide is the anhydride of phosphoric acid (H_3PO_4), a weak triprotic acid. In fact, P_4O_{10} has a very high affinity for water and is consequently used as a drying agent. Phosphorus(III) oxide is the anhydride of phosphorous acid (H_3PO_3), a weak diprotic acid.* The structures of H_3PO_4 and H_3PO_3 are shown in Figure 22.35 ◄. The hydrogen atom that is attached directly to phosphorus in H_3PO_3 is not acidic because the P—H bond is essentially nonpolar.

*Note that the element phosphorus (FOS·for·us) has a -us suffix, whereas phosphorous (fos·FOR·us) acid has an -ous suffix.

One characteristic of phosphoric and phosphorous acids is their tendency to undergo condensation reactions when heated. A **condensation reaction** is one in which two or more molecules combine to form a larger molecule by eliminating a small molecule, such as H_2O. ∞ (Section 12.6) For example, two H_3PO_4 molecules are joined by the elimination of one H_2O molecule to form $H_4P_2O_7$:

These atoms are eliminated as H_2O

$$\text{[22.56]}$$

Phosphoric acid and its salts find their most important uses in detergents and fertilizers. The phosphates in detergents are often in the form of sodium tripolyphosphate ($Na_5P_3O_{10}$). A typical detergent formulation contains 47% phosphate; 16% bleaches, perfumes, and abrasives; and 37% linear alkylsulfonate (LAS) surfactant, whose structural formula is

(We have used the notation for the benzene ring described in Section 8.6.) The phosphate ions form bonds with metal ions that contribute to the hardness of water. This keeps the metal ions from interfering with the action of the surfactant. The phosphates also keep the pH above 7 and thus prevent the surfactant molecules from becoming protonated (gaining an H^+ ion).

Most mined phosphate rock is converted to fertilizers. The $Ca_3(PO_4)_2$ in phosphate rock is insoluble ($K_{sp} = 2.0 \times 10^{-29}$). It is converted to a soluble form for use in fertilizers by treating the phosphate rock with sulfuric or phosphoric acid. The reaction with phosphoric acid yields $Ca(H_2PO_4)_2$:

$$Ca_3(PO_4)_2(s) + 4\,H_3PO_4(aq) \longrightarrow 3\,Ca^{2+}(aq) + 6\,H_2PO_4^-(aq) \quad \text{[22.57]}$$

Although the solubility of $Ca(H_2PO_4)_2$ allows it to be assimilated by plants, it also allows it to be washed from the soil and into bodies of water, thereby contributing to water pollution. ∞ (Section 18.6)

Phosphorus compounds are important in biological systems. The element occurs in phosphate groups in RNA and DNA, the molecules responsible for the control of protein biosynthesis and transmission of genetic information. ∞ (Section 25.10) It also occurs in adenosine triphosphate (ATP), which stores energy in biological cells and has the structure:

Adenosine

Chemistry and Life   ARSENIC IN DRINKING WATER

The present Environmental Protection Agency (EPA) standard for arsenic in public water supplies is 10 ppb (equivalent to 10 μg/L). Most regions of the United States tend to have low to moderate (2–10 ppb) groundwater arsenic levels (Figure 22.36▶). The western region tends to have higher levels, coming mainly from natural geological sources in the area. Estimates, for example, indicate that 35% of water-supply wells in Arizona have arsenic concentrations above 10 ppb.

The problem of arsenic in drinking water in the United States is dwarfed by the problem in other parts of the world—especially in Bangladesh, where the problem is tragic. Historically, surface water sources in that country have been contaminated with microorganisms, causing significant health problems, including one of the highest infant mortality rates in the world. During the 1970s, international agencies, headed by the United Nations Children's Fund (UNICEF), began investing millions of dollars of aid money in Bangladesh for wells to provide "clean" drinking water. Unfortunately, no one tested the well water for the presence of arsenic; the problem was not discovered until the 1980s. The result has been the biggest outbreak of mass poisoning in history. Up to half of the country's estimated 10 million wells have arsenic concentrations above 50 ppb.

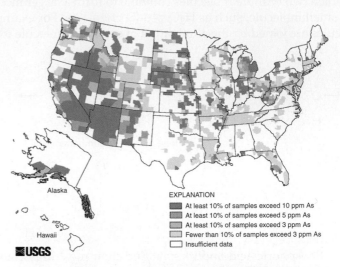

▲ **Figure 22.36 Geographic distribution of arsenic.** Counties in which at least 10% of the groundwater samples exceeded 10 ppm As are indicated by the darkest color. As the color becomes lighter, the scale moves from 10 ppm to 5 ppm to 3 ppm, and then to cases where fewer than 10% of samples exceed 3 ppm. The white areas are those for which there were insufficient data.

The P—O—P bond of the end phosphate group is broken by hydrolysis with water, forming adenosine diphosphate (ADP). This reaction releases 33 kJ of energy.

$$H_2O + O^- \overset{\displaystyle O}{\underset{\displaystyle O^-}{\overset{\|}{P}}} - O - \overset{\displaystyle O}{\underset{\displaystyle O^-}{\overset{\|}{P}}} - O - \overset{\displaystyle O}{\underset{\displaystyle O^-}{\overset{\|}{P}}} - O - \boxed{\text{Adenosine}} \longrightarrow$$

ATP

$$O^- \overset{\displaystyle O}{\underset{\displaystyle O^-}{\overset{\|}{P}}} - O - H \ + \ H - O - \overset{\displaystyle O}{\underset{\displaystyle O^-}{\overset{\|}{P}}} - O - \overset{\displaystyle O}{\underset{\displaystyle O^-}{\overset{\|}{P}}} - O - \boxed{\text{Adenosine}} \qquad [22.58]$$

ADP

This energy is used to perform the mechanical work in muscle contraction and in many other biochemical reactions (Figure 19.20).

22.9 CARBON

Carbon constitutes only 0.027% of Earth's crust, so it is not an abundant element. Although some carbon occurs in elemental form as graphite and diamond, most is found in combined form. Over half occurs in carbonate compounds, such as $CaCO_3$. Carbon is also found in coal, petroleum, and natural gas. The importance of the element stems in large part from its occurrence in all living organisms: Life as we know it is based on carbon compounds. In this section we will take a brief look at carbon and its most common inorganic compounds. We will discuss organic chemistry in Chapter 25.

In water the most common forms of arsenic are the arsenate ion and its protonated hydrogen anions (AsO_4^{3-}, $HAsO_4^{2-}$, and $H_2AsO_4^-$) and the arsenite ion and its protonated forms (AsO_3^{3-}, $HAsO_3^{2-}$, $H_2AsO_3^-$, and H_3AsO_3). These species are collectively referred to by the oxidation number of the arsenic as arsenic(V) and arsenic(III), respectively. Arsenic(V) is more prevalent in oxygen-rich (aerobic) surface waters, whereas arsenic(III) is more likely to occur in oxygen-poor (anaerobic) groundwaters. In the pH range from 4 to 10, the arsenic(V) is present primarily as $HAsO_4^{2-}$ and $H_2AsO_4^-$, and the arsenic(III) is present primarily as the neutral acid H_3AsO_3.

One of the challenges in determining the health effects of arsenic in drinking waters is the different chemistry of arsenic(V) and arsenic(III), as well as the different concentrations required for physiological responses in different individuals. In Bangladesh, skin lesions (Figure 22.37▶) were the first sign of the arsenic problem. Statistical studies correlating arsenic levels with the occurrence of disease indicate a lung and bladder cancer risk arising from even low levels of arsenic.

The current technologies for removing arsenic perform most effectively when treating arsenic in the form of arsenic(V),

so water treatment strategies require preoxidation of the drinking water. Once in the form of arsenic(V), there are a number of possible removal strategies. For example, $Fe_2(SO_4)_3$ could be added to precipitate $FeAsO_4$, which is then removed by filtration.

▲ **Figure 22.37 Lesions and pigmentation changes resulting from arsenic poisoning.**

Elemental Forms of Carbon

Carbon exists in four allotropic crystalline forms: graphite, diamond, fullerenes, and carbon nanotubes. ∞(Section 11.8) *Graphite* is a soft, black, slippery solid that has a metallic luster and conducts electricity. It consists of parallel sheets of carbon atoms that are held together by London forces [Figure 11.41(b)].

Diamond is a clear, hard solid in which the carbon atoms form a covalent network [Figure 11.41(a)]. Diamond is denser than graphite ($d = 2.25$ g/cm^3 for graphite; $d = 3.51$ g/cm^3 for diamond). At very high pressures and temperatures (approximately 100,000 atm at 3000 °C) graphite converts to diamond (Figure 22.38▶). About 3×10^4 kg of industrial-grade diamonds are synthesized each year, mainly for use in cutting, grinding, and polishing tools.

Fullerenes are molecular forms of carbon that were discovered in the mid-1980s. ∞(Section 11.8, "A Closer Look: The Third Form of Carbon") Fullerenes consist of individual molecules like C_{60} and C_{70}. The C_{60} molecules resemble soccer balls (Figure 11.43). Numerous research groups are currently exploring the chemical properties of these substances. Closely related to these molecular forms of carbon are carbon nanotubes, which consist of single or multiple layers of carbon rolled into cylindrical form, as shown in Figure 22.39▶. ∞(Section 12.9)

Graphite also exists in two common amorphous forms. **Carbon black** is formed when hydrocarbons such as methane are heated in a very limited supply of oxygen.

$$CH_4(g) + O_2(g) \longrightarrow C(s) + 2 H_2O(g) \qquad [22.59]$$

It is used as a pigment in black inks; large amounts are also used in making automobile tires. **Charcoal** is formed when wood is heated strongly in the absence of air. Charcoal has a very open structure, giving it an enormous surface area per unit mass. Activated charcoal, a pulverized form whose surface is cleaned by heating with steam, is widely used to adsorb molecules. It is used in filters to remove offensive odors from air and colored or bad-tasting impurities from water.

▲ **Figure 22.38 Synthetic diamonds.** Graphite and synthetic diamonds prepared from graphite. Most synthetic diamonds lack the size, color, and clarity of natural diamonds and are therefore not used in jewelry.

▲ **Figure 22.39 A section of a carbon nanotube.** Each intersection in the network represents a carbon atom bonded to three others.

Oxides of Carbon

Carbon forms two principal oxides: carbon monoxide (CO) and carbon dioxide (CO_2). *Carbon monoxide* is formed when carbon or hydrocarbons are burned in a limited supply of oxygen.

$$2\,C(s) + O_2(g) \longrightarrow 2\,CO(g) \qquad [22.60]$$

It is a colorless, odorless, and tasteless gas. It is toxic because it can bind to hemoglobin in the blood and thus interfere with oxygen transport. ∞(Section 18.4) Low-level poisoning results in headaches and drowsiness; high-level poisoning can cause death. Automobile engines produce carbon monoxide, which is a major air pollutant.

Carbon monoxide is unusual in that it has a lone pair of electrons on carbon: $:C{\equiv}O:$. It is also isoelectronic with N_2, so you might imagine that CO would be equally unreactive. Moreover, both substances have high bond energies (1072 kJ/mol for $C{\equiv}O$ and 941 kJ/mol for $N{\equiv}N$). Because of the lower nuclear charge on carbon (compared with either N or O), however, the lone pair on carbon is not held as strongly as that on N or O. Consequently, CO is better able to function as an electron-pair donor (Lewis base) than is N_2. It forms a wide variety of covalent compounds, known as metal carbonyls, with transition metals. $Ni(CO)_4$, for example, is a volatile, toxic compound that is formed by simply warming metallic nickel in the presence of CO. The formation of metal carbonyls is the first step in the transition-metal catalysis of a variety of reactions of CO.

Carbon monoxide has several commercial uses. Because it burns readily, forming CO_2, it is employed as a fuel.

$$2\,CO(g) + O_2(g) \longrightarrow 2\,CO_2(g) \qquad \Delta H^\circ = -566\ \text{kJ} \qquad [22.61]$$

Chemistry Put to Work CARBON FIBERS AND COMPOSITES

The properties of graphite are anisotropic; that is, they differ in different directions through the solid. Along the carbon planes, graphite possesses great strength because of the number and strength of the carbon–carbon bonds in this direction. The bonds between planes are relatively weak, however, making graphite weak in that direction.

Fibers of graphite can be prepared in which the carbon planes are aligned to varying extents parallel to the fiber axis. These fibers are lightweight (density of about 2 g/cm^3) and chemically quite unreactive. The oriented fibers are made by first slowly pyrolyzing (decomposing by action of heat) organic fibers at about 150 °C to 300 °C. These fibers are then heated to about 2500 °C to graphitize them (convert amorphous carbon to graphite). Stretching the fiber during pyrolysis helps orient the graphite planes parallel to the fiber axis. More amorphous carbon fibers are formed by pyrolysis of organic fibers at lower temperatures (1200 °C to 400 °C). These amorphous materials, commonly called *carbon fibers*, are the type most commonly used in commercial materials.

Composite materials that take advantage of the strength, stability, and low density of carbon fibers are widely used. Composites are combinations of two or more materials. These materials are present as separate phases and are combined to form structures that take advantage of certain desirable properties of each component. In carbon composites the graphite fibers are often woven into a fabric that is embedded in a matrix that binds them into a solid structure. The fibers transmit loads evenly throughout the matrix. The finished composite is thus stronger than any one of its components.

Epoxy systems are useful matrices because of their excellent adherence. They are used widely in a number of applications, including high-performance graphite sports equipment such as tennis racquets, golf clubs, and, more recently, bicycle wheels and frames (Figure 22.40▼). Epoxy systems can be used only when the temperature remains below 150 °C. More heat-resistant resins are required for many aerospace applications, where carbon composites now find wide use.

▲ **Figure 22.40 Carbon composites.** Carbon composites are used extensively in aerospace and automotive applications and in sporting goods. These high-performance (and high-priced!) bike wheels have a carbon-fiber-composite frame, which makes them very lightweight and helps them absorb road shocks.

It is also an important reducing agent, widely used in metallurgical operations to reduce metal oxides, such as the iron oxides.

$$Fe_3O_4(s) + 4\,CO(g) \longrightarrow 3\,Fe(s) + 4\,CO_2(g) \qquad [22.62]$$

This reaction is discussed in greater detail in Section 23.2. Carbon monoxide is also used in the preparation of several organic compounds such as methanol (CH_3OH) (Equation 22.13).

Carbon dioxide is produced when carbon-containing substances are burned in excess oxygen.

$$C_2H_5OH(l) + 3\,O_2(g) \longrightarrow 2\,CO_2(g) + 3\,H_2O(g) \qquad [22.63]$$

It is also produced when many carbonates are heated.

$$CaCO_3(s) \xrightarrow{\Delta} CaCO(s) + CO_2(g) \qquad [22.64]$$

Large quantities are also obtained as a by-product of the fermentation of sugar during the production of ethanol.

$$\underset{\text{Glucose}}{C_6H_{12}O_6(aq)} \xrightarrow{\text{yeast}} 2\,\underset{\text{Ethanol}}{C_2H_5OH(aq)} + 2\,CO_2(g) \qquad [22.65]$$

In the laboratory, CO_2 can be produced by the action of acids on carbonates, as shown in Figure 22.41 ▶:

$$CO_3{}^{2-}(aq) + 2\,H^+(aq) \longrightarrow CO_2(g) + H_2O(l) \qquad [22.66]$$

Carbon dioxide is a colorless and odorless gas. It is a minor component of Earth's atmosphere but a major contributor to the so-called greenhouse effect. ∞ (Section 18.4) Although it is not toxic, high concentrations of CO_2 increase respiration rate and can cause suffocation. It is readily liquefied by compression. When cooled at atmospheric pressure, however, CO_2 condenses as a solid rather than as a liquid. The solid sublimes at atmospheric pressure at $-78\,°C$. This property makes solid CO_2 valuable as a refrigerant that is always free of the liquid form. Solid CO_2 is known as dry ice. About half of the CO_2 consumed annually is used for refrigeration. The other major use of CO_2 is in the production of carbonated beverages. Large quantities are also used to manufacture *washing soda* ($Na_2CO_3 \cdot 10\,H_2O$) and *baking soda* ($NaHCO_3$). Baking soda is so named because the following reaction occurs during baking:

$$NaHCO_3(s) + H^+(aq) \longrightarrow Na^+(aq) + CO_2(g) + H_2O(l) \qquad [22.67]$$

The $H^+(aq)$ is provided by vinegar, sour milk, or the hydrolysis of certain salts. The bubbles of CO_2 that form are trapped in the dough, causing it to rise. Washing soda is used to precipitate metal ions that interfere with the cleansing action of soap.

▲ **Figure 22.41 CO_2 formation.** Solid $CaCO_3$ reacts with a solution of hydrochloric acid to produce CO_2 gas, seen here as the bubbles.

▲ GIVE IT SOME THOUGHT

What solid would precipitate if you treated a solution containing Mg^{2+} ions with washing soda?

Carbonic Acid and Carbonates

Carbon dioxide is moderately soluble in H_2O at atmospheric pressure. The resultant solutions are moderately acidic because of the formation of carbonic acid (H_2CO_3).

$$CO_2(aq) + H_2O(l) \rightleftharpoons H_2CO_3(aq) \qquad [22.68]$$

Carbonic acid is a weak diprotic acid. Its acidic character causes carbonated beverages to have a sharp, slightly acidic taste.

Although carbonic acid cannot be isolated as a pure compound, hydrogen carbonates (bicarbonates) and carbonates can be obtained by neutralizing carbonic acid solutions. Partial neutralization produces $HCO_3{}^-$, and complete

▲ **Figure 22.42 Carlsbad Caverns, New Mexico.**

neutralization gives CO_3^{2-}. The HCO_3^- ion is a stronger base than acid ($K_b = 2.3 \times 10^{-8}$; $K_a = 5.6 \times 10^{-11}$). The carbonate ion is much more strongly basic ($K_b = 1.8 \times 10^{-4}$).

Minerals containing the carbonate ion are plentiful. The principal carbonate minerals are calcite ($CaCO_3$), magnesite ($MgCO_3$), dolomite [$MgCa(CO_3)_2$], and siderite ($FeCO_3$). Calcite is the principal mineral in limestone rock, large deposits of which occur in many parts of the world. It is also the main constituent of marble, chalk, pearls, coral reefs, and the shells of marine animals such as clams and oysters. Although $CaCO_3$ has low solubility in pure water, it dissolves readily in acidic solutions with evolution of CO_2.

$$CaCO_3(s) + 2\,H^+(aq) \rightleftharpoons Ca^{2+}(aq) + H_2O(l) + CO_2(g) \qquad [22.69]$$

Because water containing CO_2 is slightly acidic (Equation 22.68), $CaCO_3$ dissolves slowly in this medium:

$$CaCO_3(s) + H_2O(l) + CO_2(g) \longrightarrow Ca^{2+}(aq) + 2\,HCO_3^-(aq) \qquad [22.70]$$

This reaction occurs when surface waters move underground through limestone deposits. It is the principal way that Ca^{2+} enters groundwater, producing "hard water." ∞ (Section 18.6) If the limestone deposit is deep enough underground, the dissolution of the limestone produces a cave. Two well-known limestone caves are Mammoth Cave in Kentucky and Carlsbad Caverns in New Mexico (Figure 22.42 ◄).

One of the most important reactions of $CaCO_3$ is its decomposition into CaO and CO_2 at elevated temperatures, given earlier in Equation 22.64. About 2×10^{10} kg (20 million tons) of calcium oxide, known as lime or quicklime, is produced in the United States annually. Because calcium oxide reacts with water to form $Ca(OH)_2$, it is an important commercial base. It is also important in making mortar, which is a mixture of sand, water, and CaO used in construction to bind bricks, blocks, and rocks together. Calcium oxide reacts with water and CO_2 to form $CaCO_3$, which binds the sand in the mortar.

$$CaO(s) + H_2O(l) \rightleftharpoons Ca^{2+}(aq) + 2\,OH^-(aq) \qquad [22.71]$$

$$Ca^{2+}(aq) + 2\,OH^-(aq) + CO_2(aq) \longrightarrow CaCO_3(s) + H_2O(l) \qquad [22.72]$$

Carbides

The binary compounds of carbon with metals, metalloids, and certain nonmetals are called **carbides**. There are three types: ionic, interstitial, and covalent. The more active metals form the ionic carbides. The most common ionic carbides contain the *acetylide* ion (C_2^{2-}). This ion is isoelectronic with N_2, and its Lewis structure, [:C≡C:]$^{2-}$, has a carbon–carbon triple bond. The most important ionic carbide is calcium carbide (CaC_2), which is produced by the reduction of CaO with carbon at high temperature:

$$2\,CaO(s) + 5\,C(s) \longrightarrow 2\,CaC_2(s) + CO_2(g) \qquad [22.73]$$

The carbide ion is a very strong base that reacts with water to form acetylene (H—C≡C—H), as in the following reaction:

$$CaC_2(s) + 2\,H_2O(l) \longrightarrow Ca(OH)_2(aq) + C_2H_2(g) \qquad [22.74]$$

Calcium carbide is therefore a convenient solid source of acetylene, which is used in welding (Figure 22.16).

Interstitial carbides are formed by many transition metals. The carbon atoms occupy open spaces (interstices) between metal atoms in a manner analogous to the interstitial hydrides. ∞ (Section 22.2) Tungsten carbide, for example, is very hard and very heat-resistant, and is thus used to make cutting tools.

Covalent carbides are formed by boron and silicon. Silicon carbide (SiC), known as Carborundum®, is used as an abrasive and in cutting tools. Almost as hard as diamond, SiC has a diamondlike structure with alternating Si and C atoms.

Other Inorganic Compounds of Carbon

Hydrogen cyanide, HCN, is an extremely toxic gas that has the odor of bitter almonds. It is produced by the reaction of a cyanide salt, such as NaCN, with an acid. Aqueous solutions of HCN are known as hydrocyanic acid. Neutralization with a base, such as NaOH, produces cyanide salts, such as NaCN. Cyanides find use in the manufacture of several well-known plastics, including nylon and Orlon®. The CN^- ion forms very stable complexes with most transition metals. ∞ (Section 17.5)

Carbon disulfide, CS_2, is an important industrial solvent for waxes, greases, celluloses, and other nonpolar substances. It is a colorless volatile liquid (bp 46.3 °C). The vapor is very poisonous and highly flammable.

22.10 THE OTHER GROUP 4A ELEMENTS: Si, Ge, Sn, AND Pb

The other elements of group 4A, in addition to carbon, are silicon, germanium, tin, and lead. The general trend from nonmetallic to metallic character as we go down a family is strikingly evident in group 4A. Carbon is a nonmetal; silicon and germanium are metalloids; tin and lead are metals. In this section we will consider a few general characteristics of group 4A and then look more thoroughly at silicon.

General Characteristics of the Group 4A Elements

Some properties of the group 4A elements are given in Table 22.8▼. The elements possess the outer-shell electron configuration ns^2np^2. The electronegativities of the elements are generally low; carbides that formally contain C^{4-} ions are observed only in the case of a few compounds of carbon with very active metals. Formation of 4+ ions by electron loss is not observed for any of these elements; the ionization energies are too high. The +2 oxidation state is found in the chemistry of germanium, tin, and lead, however, and it is the principal oxidation state for lead. The vast majority of the compounds of the group 4A elements are covalently bonded. Carbon forms a maximum of four bonds. The other members of the family are able to form higher coordination numbers through valence-shell expansion.

Carbon differs from the other group 4A elements in its pronounced ability to form multiple bonds both with itself and with other nonmetals, especially N, O, and S. The origin of this behavior was considered earlier. ∞ (Section 22.1)

TABLE 22.8 ■ **Some Properties of the Group 4A Elements**

Property	C	Si	Ge	Sn	Pb
Atomic radius (Å)	0.77	1.17	1.22	1.40	1.46
First ionization energy (kJ/mol)	1086	786	762	709	716
Electronegativity	2.5	1.8	1.8	1.8	1.9
X—X single-bond enthalpy (kJ/mol)	348	226	188	151	—

Table 22.8 shows that the strength of a bond between two atoms of a given element decreases as we go down group 4A. Carbon–carbon bonds are quite strong. Carbon, therefore, has a striking ability to form compounds in which carbon atoms are bonded to one another in extended chains and rings, which accounts for the large number of organic compounds that exist. Other elements, especially those near carbon in the periodic table, can also form chains and rings, but these bonds are far less important in the chemistries of these other elements. The Si—Si bond strength (226 kJ/mol), for example, is much smaller than the Si—O bond strength (386 kJ/mol). As a result, the chemistry of silicon is dominated by the formation of Si—O bonds, and Si—Si bonds play a rather minor role.

Occurrence and Preparation of Silicon

Silicon is the second most abundant element, after oxygen, in Earth's crust. It occurs in SiO_2 and in an enormous variety of silicate minerals. The element is obtained by the reduction of molten silicon dioxide with carbon at high temperature.

$$SiO_2(l) + 2\ C(s) \longrightarrow Si(l) + 2\ CO(g) \qquad [22.75]$$

Elemental silicon has a diamond-type structure [see Figure 11.41(a)]. Crystalline silicon is a gray metallic-looking solid that melts at 1410 °C (Figure 22.43 ◀). The element is a semiconductor and is thus used in making transistors and solar cells. ∞ (Section 12.3) To be used as a semiconductor, it must be extremely pure, possessing less than $10^{-7}\%$ (1 ppb) impurities. One method of purification is to treat the element with Cl_2 to form $SiCl_4$. The $SiCl_4$ is a volatile liquid that is purified by fractional distillation and then converted back to elemental silicon by reduction with H_2:

$$SiCl_4(g) + 2\ H_2(g) \longrightarrow Si(s) + 4\ HCl(g) \qquad [22.76]$$

The process of zone refining can further purify the element. In the zone-refining process, a heated coil is passed slowly along a silicon rod, as shown in Figure 22.44 ◀. A narrow band of the element is thereby melted. As the molten area is swept slowly along the length of the rod, the impurities concentrate in the molten region, following it to the end of the rod. The purified top portion of the rod is retained for manufacture of electronic devices.

Silicates

Silicon dioxide and other compounds that contain silicon and oxygen comprise over 90% of Earth's crust. **Silicates** are compounds in which a silicon atom is surrounded in a tetrahedral fashion by four oxygens, as shown in Figure 22.45(a) ▶. In silicates, silicon is found in its most common oxidation state, +4. The simple $SiO_4{}^{4-}$ ion, which is known as the orthosilicate ion, is found in very few silicate minerals. We can view the silicate tetrahedra, however, as "building blocks" that are used to build mineral structures. The individual tetrahedra are linked together by a common oxygen atom that serves as a vertex of both tetrahedra.

We can link two silicate tetrahedra together, for example, by sharing one oxygen atom, as shown in Figure 22.45(b). The resultant structure, called the *disilicate* ion, has two Si atoms and seven O atoms. Si and O are in the +4 and −2 oxidation states, respectively, in all silicates, so the overall charge of the ion must be consistent with these oxidation states. Thus, the charge on Si_2O_7 is $(2)(+4) + (7)(-2) = -6$; it is the $Si_2O_7{}^{6-}$ ion. The mineral *thortveitite* $(Sc_2Si_2O_7)$ contains $Si_2O_7{}^{6-}$ ions.

In most silicate minerals a large number of silicate tetrahedra are linked together to form chains, sheets, or three-dimensional structures. We can connect two vertices of each tetrahedron to two other tetrahedra, for example, leading to an infinite chain with an $\cdots O$—Si—O—Si$\cdots$ backbone. This structure,

▲ **Figure 22.43 Elemental silicon.** To prepare electronic devices, silicon powder is melted and drawn into a single crystal by zone refining (top). Wafers of silicon (bottom) cut from the crystal are subsequently treated by a series of elegant techniques to produce various electronic devices.

Molten section

Heating coil moved slowly along

Silicon rod

Inert atmosphere

▲ **Figure 22.44 Zone-refining apparatus.**

▶ **Figure 22.45 Silicate structures.** (a) Structure of the SiO_4 tetrahedron of the SiO_4^{4-} ion. This ion is found in several minerals, including zircon ($ZrSiO_4$). (b) Geometrical structure of the $Si_2O_7^{6-}$ ion, which is formed when two SiO_4 tetrahedra share a corner oxygen atom. This ion occurs in several minerals, including hardystonite, $Ca_2Zn(Si_2O_7)$.

(a)

(b)

called a single-strand silicate chain, is represented in Figure 22.46(a)▼. As shown, this chain can be viewed as repeating units of the $Si_2O_6^{4-}$ ion or, in terms of its simplest formula, SiO_3^{2-}. The mineral *enstatite* ($MgSiO_3$) consists of rows of single-strand silicate chains with Mg^{2+} ions between the strands to balance charge.

In Figure 22.46(b) each silicate tetrahedron is linked to three others, forming an infinite two-dimensional sheet structure. The simplest formula of this infinite sheet is $Si_2O_5^{2-}$. The mineral *talc*, also known as talcum powder, has the formula $Mg_3(Si_2O_5)_2(OH)_2$ and is based on this sheet structure. The Mg^{2+} and OH^- ions lie between the silicate sheets. The slippery feel of talcum powder is due to the silicate sheets sliding relative to one another, much like the sheets of carbon atoms slide in graphite, giving graphite its lubricating properties. ∞ (Section 11.8)

Asbestos is a general term applied to a group of fibrous silicate minerals. These minerals possess chainlike arrangements of the silicate tetrahedra or sheet structures in which the sheets are formed into rolls. The result is that the minerals have a fibrous character, as shown in Figure 22.47▼. Asbestos minerals have been widely used as thermal insulation, especially in high-temperature applications, because of the great chemical stability of the silicate structure. In addition, the fibers can be woven into asbestos cloth, which can be used for fireproof curtains and other applications. However, the fibrous structure of asbestos minerals poses a health risk. Tiny asbestos fibers readily penetrate soft tissues, such as the lungs, where they can cause diseases, including cancer. The use of asbestos as a common building material has therefore been discontinued.

Single-strand silicate chain, $Si_2O_6^{4-}$

(a)

Two-dimensional silicate sheet, $Si_2O_5^{2-}$

(b)

▲ **Figure 22.46 Silicate chains and sheets.** Silicate structures consist of tetrahedra linked together through their vertices by a shared oxygen atom. (a) Representation of an infinite single-strand silicate chain. Each tetrahedron is linked to two others. The box shows the repeating unit of the chain, which is similar to the unit cell of solids (Section 11.7); the chain can be viewed as an infinite number of repeating units, laid side by side. The repeating unit has a formula of $Si_2O_6^{4-}$ or has the simplest formula, SiO_3^{2-}. (b) Representation of a two-dimensional sheet structure. Each tetrahedron is linked to three others. The repeating unit of the sheet has the formula $Si_2O_5^{2-}$.

▲ **Figure 22.47 Serpentine asbestos.**
Note the fibrous character of this silicate mineral.

When all four vertices of each SiO_4 tetrahedron are linked to other tetrahedra, the structure extends in three dimensions. This linking of the tetrahedra forms quartz (SiO_2), which was depicted two-dimensionally in Figure 11.30(a). Because the structure is locked together in a three-dimensional array much like diamond [Figure 11.41(a)], quartz is harder than strand- or sheet-type silicates.

■ SAMPLE EXERCISE 22.10 | Determining an Empirical Formula

The mineral *chrysotile* is a noncarcinogenic asbestos mineral that is based on the sheet structure shown in Figure 22.46(b). In addition to silicate tetrahedra, the mineral contains Mg^{2+} and OH^- ions. Analysis of the mineral shows that there are 1.5 Mg atoms per Si atom. What is the empirical formula for chrysotile?

SOLUTION

Analyze: A mineral is described that has a sheet silicate structure with Mg^{2+} and OH^- ions to balance charge and 1.5 Mg for each 1 Si. We are asked to write the empirical formula for the mineral.

Plan: As shown in Figure 22.46(b), the silicate sheet structure is based on the $Si_2O_5^{2-}$ ion. We first add Mg^{2+} to give the proper Mg/Si ratio. We then add OH^- ions to obtain a neutral compound.

Solve: The observation that the Mg : Si ratio equals 1.5 is consistent with three Mg^{2+} ions per $Si_2O_5^{2-}$ ion. The addition of three Mg^{2+} ions would make $Mg_3(Si_2O_5)^{4+}$. In order to achieve charge balance in the mineral, there must be four OH^- ions per $Si_2O_5^{2-}$ ion. Thus, the formula of chrysotile is $Mg_3(Si_2O_5)(OH)_4$. Since this is not reducible to a simpler formula, this is the empirical formula.

■ PRACTICE EXERCISE

The cyclosilicate ion consists of three silicate tetrahedra linked together in a ring. The ion contains three Si atoms and nine O atoms. What is the overall charge on the ion?
Answer: 6−

Glass

Quartz melts at approximately 1600 °C, forming a tacky liquid. In the course of melting, many silicon–oxygen bonds are broken. When the liquid is rapidly cooled, silicon–oxygen bonds are re-formed before the atoms are able to arrange themselves in a regular fashion. An amorphous solid, known as quartz glass or silica glass, results (see Figure 11.30). Many different substances can be added to SiO_2 to cause it to melt at a lower temperature. The common **glass** used in windows and bottles is known as soda-lime glass. It contains CaO and Na_2O in addition to SiO_2 from sand. The CaO and Na_2O are produced by heating two inexpensive chemicals, limestone ($CaCO_3$) and soda ash (Na_2CO_3). These carbonates decompose at elevated temperatures:

$$CaCO_3(s) \longrightarrow CaO(s) + CO_2(g) \qquad [22.77]$$

$$Na_2CO_3(s) \longrightarrow Na_2O(s) + CO_2(g) \qquad [22.78]$$

Other substances can be added to soda-lime glass to produce color or to change the properties of the glass in various ways. The addition of CoO, for example, produces the deep blue color of "cobalt glass." Replacing Na_2O with K_2O results in a harder glass that has a higher melting point. Replacing CaO with PbO results in a denser "lead crystal" glass with a higher refractive index. Lead crystal is used for decorative glassware; the higher refractive index gives this glass a particularly sparkling appearance. Addition of nonmetal oxides, such as B_2O_3 and P_4O_{10}, which form network structures related to the silicates, also changes the properties of the glass. Adding B_2O_3 creates a glass with a higher melting point and a greater ability to withstand temperature changes. Such glasses, sold commercially under trade names such as Pyrex® and Kimax®, are used where resistance to thermal shock is important, such as in laboratory glassware or coffeemakers.

Silicones

Silicones consist of O—Si—O chains in which the remaining bonding positions on each silicon are occupied by organic groups such as CH_3. The chains are terminated by —$Si(CH_3)_3$ groups:

Depending on the length of the chain and the degree of cross-linking between chains, silicones can be either oils or rubber-like materials. Silicones are nontoxic and have good stability toward heat, light, oxygen, and water. They are used commercially in a wide variety of products, including lubricants, car polishes, sealants, and gaskets. They are also used for waterproofing fabrics. When applied to a fabric, the oxygen atoms form hydrogen bonds with the molecules on the surface of the fabric. The hydrophobic (water-repelling) organic groups of the silicone are then left pointing away from the surface as a barrier.

22.11 BORON

Boron is the only element of group 3A that can be considered nonmetallic. The element has an extended network structure. Its melting point (2300 °C) is intermediate between that of carbon (3550 °C) and that of silicon (1410 °C). The electron configuration of boron is $[He]2s^2 2p^1$.

In a family of compounds called **boranes,** the molecules contain only boron and hydrogen. The simplest borane is BH_3. This molecule contains only six valence electrons and is therefore an exception to the octet rule. ∞ (Section 8.7) As a result, BH_3 reacts with itself to form *diborane* (B_2H_6). This reaction can be viewed as a Lewis acid–base reaction (Section 16.11), in which one B—H bonding pair of electrons in each BH_3 molecule is donated to the other. As a result, diborane is an unusual molecule in which hydrogen atoms appear to form two bonds (Figure 22.48▶).

Sharing hydrogen atoms between the two boron atoms compensates somewhat for the deficiency in valence electrons around each boron. Nevertheless, diborane is an extremely reactive molecule, spontaneously flammable in air. The reaction of B_2H_6 with O_2 is highly exothermic.

$$B_2H_6(g) + 3\,O_2(g) \longrightarrow B_2O_3(s) + 3\,H_2O(g) \qquad \Delta H° = -2030 \text{ kJ} \qquad [22.79]$$

Other boranes, such as pentaborane (B_5H_9), are also very reactive. Boranes were at one time explored as solid fuels for rockets.

Boron and hydrogen also form a series of anions, called *borane anions*. Salts of the borohydride ion (BH_4^-) are widely used as reducing agents. For example, sodium borohydride ($NaBH_4$) is a commonly used reducing agent for certain organic compounds, and you will encounter it again should you take a course in organic chemistry.

The only important oxide of boron is boric oxide (B_2O_3). This substance is the anhydride of boric acid, which we may write as H_3BO_3 or $B(OH)_3$. Boric acid is so weak an acid ($K_a = 5.8 \times 10^{-10}$) that solutions of H_3BO_3 are used as an eyewash. Upon heating, boric acid loses water by a condensation reaction similar to that described for phosphorus in Section 22.8:

$$4\,H_3BO_3(s) \longrightarrow H_2B_4O_7(s) + 5\,H_2O(g) \qquad [22.80]$$

Diborane B_2H_6

▲ **Figure 22.48 The structure of diborane (B_2H_6).** Two of the H atoms bridge the two B atoms, giving a planar B_2H_2 core to the molecule. Two of the remaining H atoms lie on either side of the B_2H_2 core, giving a nearly tetrahedral bonding environment about the B atoms.

The diprotic acid $H_2B_4O_7$ is called tetraboric acid. The hydrated sodium salt $Na_2B_4O_7 \cdot 10\,H_2O$, called borax, occurs in dry lake deposits in California and can also be readily prepared from other borate minerals. Solutions of borax are alkaline, and the substance is used in various laundry and cleaning products.

■ SAMPLE INTEGRATIVE EXERCISE | Putting Concepts Together

The interhalogen compound BrF_3 is a volatile, straw-colored liquid. The compound exhibits appreciable electrical conductivity because of autoionization.

$$2\,BrF_3(l) \rightleftharpoons BrF_2{}^+(solv) + BrF_4{}^-(solv)$$

(a) What are the molecular structures of the $BrF_2{}^+$ and $BrF_4{}^-$ ions? **(b)** The electrical conductivity of BrF_3 decreases with increasing temperature. Is the autoionization process exothermic or endothermic? **(c)** One chemical characteristic of BrF_3 is that it acts as a Lewis acid toward fluoride ions. What do we expect will happen when KBr is dissolved in BrF_3?

SOLUTION

(a) The $BrF_2{}^+$ ion has a total of $7 + 2(7) - 1 = 20$ valence-shell electrons. The Lewis structure for the ion is

$$\left[:\ddot{F}-\ddot{B}r-\ddot{F}:\right]^+$$

Because there are four electron-pair domains around the central Br atom, the resulting electron-pair geometry is tetrahedral. ∞ (Section 9.2) Because bonding pairs of electrons occupy two of these domains, the molecular geometry is nonlinear.

The $BrF_4{}^-$ ion has a total of $7 + 4(7) + 1 = 36$ electrons, leading to the following Lewis structure.

$$\left[\begin{array}{c}:\ddot{F}:\\ :\ddot{F}-\ddot{B}r-\ddot{F}:\\ :\ddot{F}:\end{array}\right]^-$$

Because there are six electron-pair domains around the central Br atom in this ion, the electron-pair geometry is octahedral. The two nonbonding pairs of electrons are located opposite each other on the octahedron, leading to a square-planar molecular geometry.

(b) The observation that conductivity decreases as temperature increases indicates that there are fewer ions present in the solution at the higher temperature. Thus, increasing the temperature causes the equilibrium to shift to the left. According to Le Châtelier's principle, this shift indicates that the reaction is exothermic as it proceeds from left to right. ∞ (Section 15.7)

(c) A Lewis acid is an electron-pair acceptor. ∞ (Section 16.11) The fluoride ion has four valence-shell electron pairs and can act as a Lewis base (an electron-pair donor). Thus, we can envision the following reaction occurring:

$$F^- + BrF_3 \longrightarrow BrF_4^-$$

CHAPTER REVIEW

SUMMARY AND KEY TERMS

Introduction and Section 22.1 The periodic table is useful for organizing and remembering the descriptive chemistry of the elements. Among elements of a given group, size increases with increasing atomic number, and electronegativity and ionization energy decrease. Nonmetallic character parallels electronegativity, so the most nonmetallic elements are found in the upper right portion of the periodic table.

Among the nonmetallic elements, the first member of each group differs dramatically from the other members; it forms a maximum of four bonds to other atoms and exhibits a much greater tendency to form π bonds than the heavier elements in its group.

Because O_2 and H_2O are abundant in our world, we focus on two important and general reaction types as we discuss the nonmetals: oxidation by O_2 and proton-transfer reactions involving H_2O or aqueous solutions.

Section 22.2 Hydrogen has three isotopes: **protium** (^1_1H), **deuterium** (^2_1H), and **tritium** (^3_1H). Hydrogen is not a member of any particular periodic group, although it is usually placed above lithium. The hydrogen atom can either lose an electron, forming H^+, or gain one, forming H^- (the hydride ion). Because the H—H bond is relatively strong, H_2 is fairly unreactive unless activated by heat or a catalyst. Hydrogen forms a very strong bond to oxygen, so the reactions of H_2 with oxygen-containing compounds usually lead to the formation of H_2O. Because the bonds in CO and CO_2 are even stronger than the O—H bond, the reaction of H_2O with carbon or certain organic compounds leads to the formation of H_2. The $H^+(aq)$ ion is able to oxidize many metals, forming $H_2(g)$. The electrolysis of water also forms $H_2(g)$.

The binary compounds of hydrogen are of three general types: **ionic hydrides** (formed by active metals), **metallic hydrides** (formed by transition metals), and **molecular hydrides** (formed by nonmetals). The ionic hydrides contain the H^- ion; because this ion is extremely basic, ionic hydrides react with H_2O to form H_2 and OH^-.

Sections 22.3 and 22.4 The noble gases (group 8A) exhibit a very limited chemical reactivity because of the exceptional stability of their electron configurations. The xenon fluorides and oxides and KrF_2 are the best-established compounds of the noble gases.

The halogens (group 7A) occur as diatomic molecules. All except fluorine exhibit oxidation states varying from -1 to +7. Fluorine is the most electronegative element, so it is restricted to the oxidation states 0 and -1. The oxidizing power of the element (the tendency to form the -1 oxidation state) decreases as we proceed down the group. The hydrogen halides are among the most useful compounds of these elements; these gases dissolve in water to form the hydrohalic acids, such as HCl(aq). Hydrofluoric acid reacts with silica. The **interhalogens** are compounds formed between two different halogen elements. Chlorine, bromine, and iodine form a series of oxyacids, in which the halogen atom is in a positive oxidation state. These compounds and their associated oxyanions are strong oxidizing agents.

Sections 22.5 and 22.6 Oxygen has two allotropes, O_2 and O_3 (ozone). Ozone is unstable compared to O_2, and it is a stronger oxidizing agent than O_2. Most reactions of O_2 lead to oxides, compounds in which oxygen is in the

−2 oxidation state. The soluble oxides of nonmetals generally produce acidic aqueous solutions; they are called **acidic anhydrides** or **acidic oxides**. In contrast, soluble metal oxides produce basic solutions and are called **basic anhydrides** or **basic oxides**. Many metal oxides that are insoluble in water dissolve in acid, accompanied by the formation of H_2O. Peroxides contain O—O bonds and oxygen in the −1 oxidation state. Peroxides are unstable, decomposing to O_2 and oxides. In such reactions peroxides are simultaneously oxidized and reduced, a process called **disproportionation**. Superoxides contain the O_2^- ion in which oxygen is in the $-\frac{1}{2}$ oxidation state.

Sulfur is the most important of the other group 6A elements. It has several allotropic forms; the most stable one at room temperature consists of S_8 rings. Sulfur forms two oxides, SO_2 and SO_3, and both are important atmospheric pollutants. Sulfur trioxide is the anhydride of sulfuric acid, the most important sulfur compound and the most-produced industrial chemical. Sulfuric acid is a strong acid and a good dehydrating agent. Sulfur forms several oxyanions as well, including the SO_3^{2-} (sulfite), SO_4^{2-} (sulfate), and $S_2O_3^{2-}$ (thiosulfate) ions. Sulfur is found combined with many metals as a sulfide, in which sulfur is in the −2 oxidation state. These compounds often react with acids to form hydrogen sulfide (H_2S), which smells like rotten eggs.

Sections 22.7 and 22.8 Nitrogen is found in the atmosphere as N_2 molecules. Molecular nitrogen is chemically very stable because of the strong $N\equiv N$ bond. Molecular nitrogen can be converted into ammonia via the Haber process. Once the ammonia is made, it can be converted into a variety of different compounds that exhibit nitrogen oxidation states ranging from −3 to +5. The most important industrial conversion of ammonia is the **Ostwald process**, in which ammonia is oxidized to nitric acid (HNO_3). Nitrogen has three important oxides: nitrous oxide (N_2O), nitric oxide (NO), and nitrogen dioxide (NO_2). Nitrous acid (HNO_2) is a weak acid; its conjugate base is the nitrite ion (NO_2^-). Another important nitrogen compound is hydrazine (N_2H_4).

Phosphorus is the most important of the remaining group 5A elements. It occurs in nature in phosphate minerals. Phosphorus has several allotropes, including white phosphorus, which consists of P_4 tetrahedra. In reaction with the halogens, phosphorus forms trihalides (PX_3)

and pentahalides (PX_5). These compounds undergo hydrolysis to produce an oxyacid of phosphorus and HX. Phosphorus forms two oxides, P_4O_6 and P_4O_{10}. Their corresponding acids, phosphorous acid and phosphoric acid, undergo **condensation reactions** when heated. Phosphorus compounds are important in biochemistry and as fertilizers.

Sections 22.9 and 22.10 The allotropes of carbon include diamond, graphite, and the fullerene family of compounds, notably buckminsterfullerene. Carbon nanotubes are a less well-defined form of carbon related to the fullerenes. Amorphous forms of carbon include **charcoal** and **carbon black**. Carbon forms two common oxides, CO and CO_2. Aqueous solutions of CO_2 produce the weak diprotic acid carbonic acid (H_2CO_3), which is the parent acid of hydrogen carbonate and carbonate salts. Binary compounds of carbon are called **carbides**. Carbides may be ionic, interstitial, or covalent. Calcium carbide (CaC_2) contains the strongly basic acetylide ion (C_2^{2-}), which reacts with water to form acetylene. Other important inorganic carbon compounds include hydrogen cyanide (HCN) and carbon disulfide (CS_2).

The other group 4A elements show great diversity in physical and chemical properties. Silicon, the second most abundant element, is a semiconductor. It reacts with Cl_2 to form $SiCl_4$, a liquid at room temperature. Silicon forms strong Si—O bonds and therefore occurs in a variety of silicate minerals. **Silicates** consist of SiO_4 tetrahedra, linked together at their vertices to form chains, sheets, or three-dimensional structures. The most common three-dimensional silicate is quartz (SiO_2). **Glass** is an amorphous (noncrystalline) form of SiO_2. Silicones contain O—Si—O chains with organic groups bonded to the Si atoms. Like silicon, germanium is a metalloid; tin and lead are metallic.

Section 22.11 Boron is the only group 3A element that is a nonmetal. It forms a variety of compounds with hydrogen called boron hydrides, or **boranes**. Diborane (B_2H_6) has an unusual structure with two hydrogen atoms that bridge between the two boron atoms. Boranes react with oxygen to form boric oxide (B_2O_3), in which boron is in the +3 oxidation state. Boric oxide is the anhydride of boric acid (H_3BO_3). Boric acid readily undergoes condensation reactions.

KEY SKILLS

- Be able to use periodic trends to explain the basic differences between the elements of a group or period.
- Explain the ways in which the first element in a group differs from subsequent elements in the group.
- Be able to determine electron configurations, oxidation numbers, and molecular shapes.
- Know the sources of the common nonmetals, how they are obtained, and how they are used.
- Understand how phosphoric and phosphorous acids undergo condensation reactions.
- Explain how the bonding and structures of silicates relate to their chemical formulas and properties.

VISUALIZING CONCEPTS

22.1 One of these structures is a stable compound; the other is not. Identify the stable compound, and explain why it is stable. Explain why the other compound is not stable. [Section 22.1]

22.2 **(a)** Identify the *type* of chemical reaction represented by the diagram below. **(b)** Place appropriate charges on the species on both sides of the equation. **(c)** Write the chemical equation for the reaction. [Section 22.1]

22.3 Which of the following species (there may be more than one) is/are likely to have the structure shown below: **(a)** XeF_4, **(b)** BrF_4^+, **(c)** SiF_4, **(d)** $TeCl_4$, **(e)** $HClO_4$? (The colors shown do not reflect the identity of any element.) [Sections 22.3, 22.4, 22.6, and 22.10]

22.4 Draw an energy profile for the reaction shown in Equation 22.24, assuming that the barrier to dissociation of $O_3(g)$ is 115 kJ. [Section 22.5]

22.5 Write the molecular formula and Lewis structure for each of the following oxides of nitrogen: [Section 22.7]

22.6 Which property of the group 6A elements might be the one depicted in the following graph: **(a)** electronegativity, **(b)** first ionization energy, **(c)** density, **(d)** X—X single-bond enthalpy, **(e)** electron affinity? Explain your answer. [Sections 22.5 and 22.6]

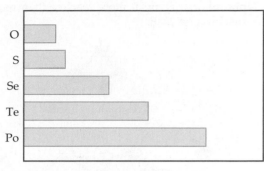

22.7 The atomic and ionic radii of the first three group 6A elements are

(a) Explain why the atomic radius increases in moving downward in the group. **(b)** Explain why the ionic radii are larger than the atomic radii. **(c)** Which of the three anions would you expect to be the strongest base in water? Explain. [Sections 22.5 and 22.6]

22.8 Which property of the third-row nonmetallic elements might be the one depicted in the graph below: **(a)** first ionization energy, **(b)** atomic radius, **(c)** electronegativity, **(d)** melting point, **(e)** X—X single-bond enthalpy? Explain both your choice and why the other choices would not be correct. [Sections 22.3, 22.4, 22.6, 22.8, and 22.10]

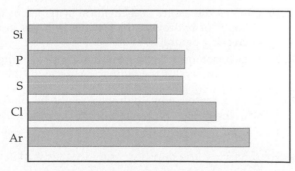

22.9 The structures of white and red phosphorus are shown here. Explain, using these structural models, why white phosphorus is much more reactive than red phosphorus. [Section 22.8]

White phosphorus

Red phosphorus

22.10 (a) Draw the Lewis structures for at least four species that have the general formula

$$\left[:X\equiv Y:\right]^{n}$$

where X and Y may be the same or different, and n may have a value from +1 to −2. **(b)** Which of the compounds is likely to be the strongest Brønsted base? Explain. [Sections 22.1, 22.7, and 22.9]

EXERCISES

Periodic Trends and Chemical Reactions

22.11 Identify each of the following elements as a metal, nonmetal, or metalloid: **(a)** phosphorus, **(b)** strontium, **(c)** manganese, **(d)** selenium, **(e)** rhodium, **(f)** krypton.

22.12 Identify each of the following elements as a metal, nonmetal, or metalloid: **(a)** gallium, **(b)** molybdenum, **(c)** tellurium, **(d)** arsenic, **(e)** xenon, **(f)** cadmium.

22.13 Consider the elements O, Ba, Co, Be, Br, and Se. From this list select the element that **(a)** is most electronegative, **(b)** exhibits a maximum oxidation state of +7, **(c)** loses an electron most readily, **(d)** forms π bonds most readily, **(e)** is a transition metal.

22.14 Consider the elements Li, K, Cl, C, Ne, and Ar. From this list select the element that **(a)** is most electronegative, **(b)** has the greatest metallic character, **(c)** most readily forms a positive ion, **(d)** has the smallest atomic radius, **(e)** forms π bonds most readily.

22.15 Explain the following observations: **(a)** The highest fluoride compound formed by nitrogen is NF_3, whereas phosphorus readily forms PF_5. **(b)** Although CO is a well-known compound, SiO does not exist under ordinary conditions. **(c)** AsH_3 is a stronger reducing agent than NH_3.

22.16 Explain the following observations: **(a)** HNO_3 is a stronger oxidizing agent than H_3PO_4. **(b)** Silicon can form an ion with six fluorine atoms, SiF_6^{2-}, whereas carbon is able to bond to a maximum of four, CF_4. **(c)** There are three compounds formed by carbon and hydrogen that contain two carbon atoms each (C_2H_2, C_2H_4, and C_2H_6), whereas silicon forms only one analogous compound (Si_2H_6).

22.17 Complete and balance the following equations:

(a) $Mg_3N_2(s) + H_2O(l) \longrightarrow$

(b) $C_3H_7OH(l) + O_2(g) \longrightarrow$

(c) $MnO_2(s) + C(s) \xrightarrow{\Delta}$

(d) $AlP(s) + H_2O(l) \longrightarrow$

(e) $Na_2S(s) + HCl(aq) \longrightarrow$

22.18 Complete and balance the following equations:

(a) $NaOCH_3(s) + H_2O(l) \longrightarrow$

(b) $CuO(s) + HNO_3(aq) \longrightarrow$

(c) $WO_3(s) + H_2(g) \xrightarrow{\Delta}$

(d) $NH_2OH(l) + O_2(g) \longrightarrow$

(e) $Al_4C_3(s) + H_2O(l) \longrightarrow$

Hydrogen, the Noble Gases, and the Halogens

22.19 (a) Give the names and chemical symbols for the three isotopes of hydrogen. **(b)** List the isotopes in order of decreasing natural abundance.

22.20 Which isotope of hydrogen is radioactive? Write the nuclear equation for the radioactive decay of this isotope.

22.21 Give a reason why hydrogen might be placed along with the group 1A elements of the periodic table.

22.22 Why are the properties of hydrogen different from those of either the group 1A or 7A elements?

22.23 Write a balanced equation for the preparation of H_2 using **(a)** Mg and an acid, **(b)** carbon and steam, **(c)** methane and steam.

22.24 List **(a)** three commercial means of producing H_2, **(b)** three industrial uses of H_2.

22.25 Complete and balance the following equations:
- **(a)** $NaH(s) + H_2O(l) \longrightarrow$
- **(b)** $Fe(s) + H_2SO_4(aq) \longrightarrow$
- **(c)** $H_2(g) + Br_2(g) \longrightarrow$
- **(d)** $Na(l) + H_2(g) \longrightarrow$
- **(e)** $PbO(s) + H_2(g) \longrightarrow$

22.26 Write balanced equations for each of the following reactions (some of these are analogous to reactions shown in the chapter). **(a)** Aluminum metal reacts with acids to form hydrogen gas. **(b)** Steam reacts with magnesium metal to give magnesium oxide and hydrogen. **(c)** Manganese(IV) oxide is reduced to manganese(II) oxide by hydrogen gas. **(d)** Calcium hydride reacts with water to generate hydrogen gas.

22.27 Identify the following hydrides as ionic, metallic, or molecular: **(a)** BaH_2, **(b)** H_2Te, **(c)** $TiH_{1.7}$.

22.28 Identify the following hydrides as ionic, metallic, or molecular: **(a)** B_2H_6, **(b)** RbH, **(c)** Th_4H_{15}.

22.29 Describe two characteristics of hydrogen that are favorable for its use as a general energy source in vehicles.

22.30 Would hydrogen be a satisfactory basis for a fuel economy if the only available sources were cracking of natural gas and conversion of grain-based ethanol? Explain.

22.31 Why does xenon form stable compounds with fluorine, whereas argon does not?

22.32 Why are there so few compounds of the noble gases?

22.33 Write the chemical formula for each of the following compounds, and indicate the oxidation state of the halogen or noble-gas atom in each: **(a)** chlorate ion, **(b)** hydroiodic acid, **(c)** iodine trichloride, **(d)** sodium hypochlorite, **(e)** perchloric acid, **(f)** xenon tetrafluoride.

22.34 Write the chemical formula for each of the following, and indicate the oxidation state of the halogen or noble-gas atom in each, **(a)** calcium hypobromite, **(b)** bromic acid, **(c)** xenon trioxide, **(d)** perchlorate ion, **(e)** iodous acid, **(f)** iodine pentafluoride.

22.35 Name the following compounds: **(a)** $Fe(ClO_3)_3$, **(b)** $HClO_2$, **(c)** XeF_6, **(d)** BrF_5, **(e)** $XeOF_4$, **(f)** HIO_3 (named as an acid).

22.36 Name the following compounds: **(a)** $KClO_3$, **(b)** $Ca(IO_3)_2$, **(c)** $AlCl_3$, **(d)** $HBrO_3$, **(e)** H_5IO_6, **(f)** XeF_4.

22.37 Explain each of the following observations: **(a)** At room temperature I_2 is a solid, Br_2 is a liquid, and Cl_2 and F_2 are both gases. **(b)** F_2 cannot be prepared by electrolytic oxidation of aqueous F^- solutions. **(c)** The boiling point of HF is much higher than those of the other hydrogen halides. **(d)** The halogens decrease in oxidizing power in the order $F_2 > Cl_2 > Br_2 > I_2$.

22.38 Explain the following observations: **(a)** for a given oxidation state, the acid strength of the oxyacid in aqueous solution decreases in the order chlorine > bromine > iodine. **(b)** Hydrofluoric acid cannot be stored in glass bottles. **(c)** HI cannot be prepared by treating NaI with sulfuric acid. **(d)** The interhalogen ICl_3 is known, but $BrCl_3$ is not.

[22.39] Ammonium perchlorate remains a significant pollutant in soils and water supplies even decades after it is placed into the environment. What can you conclude from this about the stability of the perchlorate anion? What feature of the anion's structure might promote this high stability, even though perchlorate is a strong oxidizing agent?

[22.40] Removal of perchlorate from water supplies is difficult. Naturally occurring microorganisms are, however, capable of destroying perchlorate in solution in minutes. What do you think might be the type of reaction occurring in the microorganisms, and what do you predict might be the fate of the perchlorate ion in the reaction?

Oxygen and the Group 6A Elements

22.41 **(a)** List three industrial uses of O_2. **(b)** List two industrial uses of O_3.

22.42 Draw the Lewis structure of ozone. Explain why the O—O bond (1.28 Å) is longer in ozone than in O_2 (1.21 Å).

22.43 Write balanced equations for each of the following reactions. **(a)** When mercury(II) oxide is heated, it decomposes to form O_2 and mercury metal. **(b)** When copper(II) nitrate is heated strongly, it decomposes to form copper(II) oxide, nitrogen dioxide, and oxygen. **(c)** Lead(II) sulfide, $PbS(s)$, reacts with ozone to form $PbSO_4(s)$ and $O_2(g)$. **(d)** When heated in air, $ZnS(s)$ is converted to ZnO. **(e)** Potassium peroxide reacts with $CO_2(g)$ to give potassium carbonate and O_2.

22.44 Complete and balance the following equations:
- **(a)** $CaO(s) + H_2O(l) \longrightarrow$
- **(b)** $Al_2O_3(s) + H^+(aq) \longrightarrow$
- **(c)** $Na_2O_2(s) + H_2O(l) \longrightarrow$
- **(d)** $N_2O_3(g) + H_2O(l) \longrightarrow$
- **(e)** $KO_2(s) + H_2O(l) \longrightarrow$
- **(f)** $NO(g) + O_3(g) \longrightarrow$

22.45 Predict whether each of the following oxides is acidic, basic, amphoteric, or neutral: **(a)** NO_2, **(b)** CO_2, **(c)** Al_2O_3, **(d)** CaO.

22.46 Select the more acidic member of each of the following pairs: **(a)** Mn_2O_7 and MnO_2, **(b)** SnO and SnO_2, **(c)** SO_2 and SO_3, **(d)** SiO_2 and SO_2, **(e)** Ga_2O_3 and In_2O_3, **(f)** SO_2 and SeO_2.

22.47 Write the chemical formula for each of the following compounds, and indicate the oxidation state of the group 6A element in each: **(a)** selenous acid, **(b)** potassium hydrogen sulfite, **(c)** hydrogen telluride, **(d)** carbon disulfide, **(e)** calcium sulfate.

22.48 Write the chemical formula for each of the following compounds, and indicate the oxidation state of the group 6A element in each: **(a)** sulfur tetrachloride, **(b)** selenium trioxide, **(c)** sodium thiosulfate, **(d)** hydrogen sulfide, **(e)** sulfuric acid.

22.49 In aqueous solution, hydrogen sulfide reduces **(a)** Fe^{3+} to Fe^{2+}, **(b)** Br_2 to Br^-, **(c)** MnO_4^- to Mn^{2+}, **(d)** HNO_3 to NO_2. In all cases, under appropriate conditions, the product is elemental sulfur. Write a balanced net ionic equation for each reaction.

22.50 An aqueous solution of SO_2 reduces **(a)** aqueous $KMnO_4$ to $MnSO_4(aq)$, **(b)** acidic aqueous $K_2Cr_2O_7$ to aqueous Cr^{3+}, **(c)** aqueous $Hg_2(NO_3)_2$ to mercury metal. Write balanced equations for these reactions.

22.51 Write the Lewis structure for each of the following species, and indicate the structure of each: **(a)** SeO_3^{2-}; **(b)** S_2Cl_2; **(c)** chlorosulfonic acid, HSO_3Cl (chlorine is bonded to sulfur).

22.52 The SF_5^- ion is formed when $SF_4(g)$ reacts with fluoride salts containing large cations, such as $CsF(s)$. Draw the Lewis structures for SF_4 and SF_5^-, and predict the molecular structure of each.

22.53 Write a balanced equation for each of the following reactions: **(a)** Sulfur dioxide reacts with water. **(b)** Solid zinc sulfide reacts with hydrochloric acid. **(c)** Elemental sulfur reacts with sulfite ion to form thiosulfate. **(d)** Sulfur trioxide is dissolved in sulfuric acid.

22.54 Write a balanced equation for each of the following reactions. (You may have to guess at one or more of the reaction products, but you should be able to make a reasonable guess, based on your study of this chapter.) **(a)** Hydrogen selenide can be prepared by reaction of an aqueous acid solution on aluminum selenide. **(b)** Sodium thiosulfate is used to remove excess Cl_2 from chlorine-bleached fabrics. The thiosulfate ion forms SO_4^{2-} and elemental sulfur, while Cl_2 is reduced to Cl^-.

Nitrogen and the Group 5A Elements

22.55 Write the chemical formula for each of the following compounds, and indicate the oxidation state of nitrogen in each: **(a)** sodium nitrite, **(b)** ammonia, **(c)** nitrous oxide, **(d)** sodium cyanide, **(e)** nitric acid, **(f)** nitrogen dioxide.

22.56 Write the chemical formula for each of the following compounds, and indicate the oxidation state of nitrogen in each: **(a)** nitric oxide, **(b)** hydrazine, **(c)** potassium cyanide, **(d)** sodium nitrite, **(e)** ammonium chloride, **(f)** lithium nitride.

22.57 Write the Lewis structure for each of the following species, and describe its geometry: **(a)** HNO_2, **(b)** N_3^-, **(c)** $N_2H_5^+$, **(d)** NO_3^-.

22.58 Write the Lewis structure for each of the following species, and describe its geometry: **(a)** NH_4^+, **(b)** NO_2^-, **(c)** N_2O, **(d)** NO_2.

22.59 Complete and balance the following equations:
(a) $Mg_3N_2(s) + H_2O(l) \longrightarrow$
(b) $NO(g) + O_2(g) \longrightarrow$
(c) $N_2O_5(g) + H_2O(l) \longrightarrow$
(d) $NH_3(aq) + H^+(aq) \longrightarrow$
(e) $N_2H_4(l) + O_2(g) \longrightarrow$

22.60 Write a balanced net ionic equation for each of the following reactions: **(a)** Dilute nitric acid reacts with zinc metal with formation of nitrous oxide. **(b)** Concentrated nitric acid reacts with sulfur with formation of nitrogen dioxide. **(c)** Concentrated nitric acid oxidizes sulfur dioxide with formation of nitric oxide. **(d)** Hydrazine is burned in excess fluorine gas, forming NF_3. **(e)** Hydrazine reduces CrO_4^{2-} to $Cr(OH)_4^-$ in base (hydrazine is oxidized to N_2).

22.61 Write complete balanced half-reactions for **(a)** oxidation of nitrous acid to nitrate ion in acidic solution, **(b)** oxidation of N_2 to N_2O in acidic solution.

22.62 Write complete balanced half-reactions for **(a)** reduction of nitrate ion to NO in acidic solution, **(b)** oxidation of HNO_2 to NO_2 in acidic solution.

22.63 Write a molecular formula for each compound, and indicate the oxidation state of the group 5A element in each formula: **(a)** phosphorous acid, **(b)** pyrophosphoric acid, **(c)** antimony trichloride, **(d)** magnesium arsenide, **(e)** diphosphorus pentoxide.

22.64 Write a chemical formula for each compound or ion, and indicate the oxidation state of the group 5A element in each formula: **(a)** phosphate ion, **(b)** arsenous acid, **(c)** antimony(III) sulfide, **(d)** calcium dihydrogen phosphate, **(e)** potassium phosphide.

22.65 Account for the following observations: **(a)** Phosphorus forms a pentachloride, but nitrogen does not. **(b)** H_3PO_2 is a monoprotic acid. **(c)** Phosphonium salts, such as PH_4Cl, can be formed under anhydrous conditions, but they can't be made in aqueous solution. **(d)** White phosphorus is extremely reactive.

22.66 Account for the following observations: **(a)** H_3PO_3 is a diprotic acid. **(b)** Nitric acid is a strong acid, whereas phosphoric acid is weak. **(c)** Phosphate rock is ineffective as a phosphate fertilizer. **(d)** Phosphorus does not exist at room temperature as diatomic molecules, but nitrogen does. **(e)** Solutions of Na_3PO_4 are quite basic.

22.67 Write a balanced equation for each of the following reactions: **(a)** preparation of white phosphorus from calcium phosphate, **(b)** hydrolysis of PBr_3, **(c)** reduction of PBr_3 to P_4 in the gas phase, using H_2.

22.68 Write a balanced equation for each of the following reactions: **(a)** hydrolysis of PCl_5, **(b)** dehydration of phosphoric acid (also called orthophosphoric acid) to form pyrophosphoric acid, **(c)** reaction of P_4O_{10} with water.

Carbon, the Other Group 4A Elements, and Boron

22.69 Give the chemical formula for **(a)** hydrocyanic acid, **(b)** nickel tetracarbonyl, **(c)** barium bicarbonate, **(d)** calcium acetylide.

22.70 Give the chemical formula for **(a)** carbonic acid, **(b)** sodium cyanide, **(c)** potassium hydrogen carbonate, **(d)** acetylene.

22.71 Complete and balance the following equations:

(a) $ZnCO_3(s) \xrightarrow{\Delta}$

(b) $BaC_2(s) + H_2O(l) \longrightarrow$

(c) $C_2H_2(g) + O_2(g) \longrightarrow$

(d) $CS_2(g) + O_2(g) \longrightarrow$

(e) $Ca(CN)_2(s) + HBr(aq) \longrightarrow$

22.72 Complete and balance the following equations:

(a) $CO_2(g) + OH^-(aq) \longrightarrow$

(b) $NaHCO_3(s) + H^+(aq) \longrightarrow$

(c) $CaO(s) + C(s) \xrightarrow{\Delta}$

(d) $C(s) + H_2O(g) \xrightarrow{\Delta}$

(e) $CuO(s) + CO(g) \longrightarrow$

22.73 Write a balanced equation for each of the following reactions: **(a)** Hydrogen cyanide is formed commercially by passing a mixture of methane, ammonia, and air over a catalyst at 800 °C. Water is a by-product of the reaction. **(b)** Baking soda reacts with acids to produce carbon dioxide gas. **(c)** When barium carbonate reacts in air with sulfur dioxide, barium sulfate and carbon dioxide form.

22.74 Write a balanced equation for each of the following reactions: **(a)** Burning magnesium metal in a carbon dioxide atmosphere reduces the CO_2 to carbon. **(b)** In photosynthesis, solar energy is used to produce glucose ($C_6H_{12}O_6$) and O_2 out of carbon dioxide and water. **(c)** When carbonate salts dissolve in water, they produce basic solutions.

22.75 Write the formulas for the following compounds, and indicate the oxidation state of the group 4A element or of boron in each: **(a)** boric acid, **(b)** silicon tetrabromide, **(c)** lead(II) chloride, **(d)** sodium tetraborate decahydrate (borax), **(e)** boric oxide.

22.76 Write the formulas for the following compounds, and indicate the oxidation state of the group 4A element or of boron in each: **(a)** silicon dioxide, **(b)** germanium tetrachloride, **(c)** sodium borohydride, **(d)** stannous chloride, **(e)** diborane.

22.77 Select the member of group 4A that best fits each description: **(a)** has the lowest first ionization energy, **(b)** is found in oxidation states ranging from −4 to +4, **(c)** is most abundant in Earth's crust.

22.78 Select the member of group 4A that best fits each description: **(a)** forms chains to the greatest extent, **(b)** forms the most basic oxide, **(c)** is a metalloid that can form 2+ ions.

22.79 **(a)** What is the characteristic geometry about silicon in all silicate minerals? **(b)** Metasilicic acid has the empirical formula H_2SiO_3. Which of the structures shown in Figure 22.46 would you expect metasilicic acid to have?

22.80 Two silicate anions are known in which the linking of the tetrahedra forms a closed ring. One of these cyclic silicate anions contains three silicate tetrahedra, linked into a ring. The other contains six silicate tetrahedra. **(a)** Sketch these cyclic silicate anions. **(b)** Determine the formula and charge of each of the anions.

22.81 **(a)** How does the structure of diborane (B_2H_6) differ from that of ethane (C_2H_6)? **(b)** By using concepts discussed in Chapter 8, explain why diborane adopts the geometry that it does. **(c)** What is the significance of the statement that the hydrogen atoms in diborane are described as hydridic?

22.82 Write a balanced equation for each of the following reactions: **(a)** Diborane reacts with water to form boric acid and molecular hydrogen. **(b)** Upon heating, boric acid undergoes a condensation reaction to form tetraboric acid. **(c)** Boron oxide dissolves in water to give a solution of boric acid.

ADDITIONAL EXERCISES

22.83 In your own words, define the following terms: **(a)** allotrope, **(b)** disproportionation, **(c)** interhalogen, **(d)** acidic anhydride, **(e)** condensation reaction.

22.84 Starting with D_2O, suggest preparations of **(a)** ND_3, **(b)** D_2SO_4, **(c)** $NaOD$, **(d)** DNO_3, **(e)** C_2D_2, **(f)** DCN.

22.85 Although the ClO_4^- and IO_4^- ions have been known for a long time, BrO_4^- was not synthesized until 1965. The ion was synthesized by oxidizing the bromate ion with xenon difluoride, producing xenon, hydrofluoric acid, and the perbromate ion. Write the balanced equation for this reaction.

22.86 Write a balanced equation for the reaction of each of the following compounds with water: **(a)** $SO_2(g)$, **(b)** $Cl_2O_7(g)$, **(c)** $Na_2O_2(s)$, **(d)** $BaC_2(s)$, **(e)** $RbO_2(s)$, **(f)** $Mg_3N_2(s)$, **(g)** $NaH(s)$.

22.87 What is the anhydride for each of the following acids: **(a)** H_2SO_4, **(b)** $HClO_3$, **(c)** HNO_2, **(d)** H_2CO_3, **(e)** H_3PO_4?

22.88 Explain why SO_2 can be used as a reducing agent but SO_3 cannot.

22.89 A sulfuric acid plant produces a considerable amount of heat. This heat is used to generate electricity, which helps reduce operating costs. The synthesis of H_2SO_4

consists of three main chemical processes: (1) oxidation of S to SO_2, (2) oxidation of SO_2 to SO_3, (3) the dissolving of SO_3 in H_2SO_4 and its reaction with water to form H_2SO_4. If the third process produces 130 kJ/mol, how much heat is produced in preparing a mole of H_2SO_4 from a mole of S? How much heat is produced in preparing a ton of H_2SO_4?

22.90 **(a)** What is the oxidation state of P in PO_4^{3-} and of N in NO_3^-? **(b)** Why doesn't N form a stable NO_4^{3-} ion analogous to P?

22.91 **(a)** The P_4, P_4O_6, and P_4O_{10} molecules have a common structural feature of four P atoms arranged in a tetrahedron (Figures 22.32 and 22.34). Does this mean that the bonding between the P atoms is the same in all these cases? Explain. **(b)** Sodium trimetaphosphate ($Na_3P_3O_9$) and sodium tetrametaphosphate ($Na_4P_4O_{12}$) are used as water-softening agents. They contain cyclic $P_3O_9^{3-}$ and $P_4O_{12}^{4-}$ ions, respectively. Propose reasonable structures for these ions.

22.92 Silicon has a limited capacity to form linear, $Si-Si$ bonded structures similar to those formed by carbon. **(a)** Predict the molecular formula of a hydride of silicon that contains a chain of three silicon atoms. **(b)** Write a balanced equation for the reaction between oxygen and the compound you predicted in part (a).

22.93 Ultrapure germanium, like silicon, is used in semiconductors. Germanium of "ordinary" purity is prepared by the high-temperature reduction of GeO_2 with carbon. The Ge is converted to $GeCl_4$ by treatment with Cl_2 and then purified by distillation; $GeCl_4$ is then hydrolyzed in

water to GeO_2 and reduced to the elemental form with H_2. The element is then zone refined. Write a balanced chemical equation for each of the chemical transformations in the course of forming ultrapure Ge from GeO_2.

22.94 Complete and balance the following equations:
(a) $MnO_4^-(aq) + H_2O_2(aq) + H^+(aq) \longrightarrow$
(b) $Fe^{2+}(aq) + H_2O_2(aq) \longrightarrow$
(c) $I^-(aq) + H_2O_2(aq) + H^+(aq) \longrightarrow$
(d) $Cu(s) + H_2O_2(aq) + H^+(aq) \longrightarrow$
(e) $I^-(aq) + O_3(g) \longrightarrow I_2(s) + O_2(g) + OH^-(aq)$

22.95 Hydrogen peroxide is capable of oxidizing **(a)** hydrazine to N_2 and H_2O, **(b)** SO_2 to SO_4^{2-}, **(c)** NO_2^- to NO_3^-, **(d)** $H_2S(g)$ to $S(s)$, **(e)** Fe^{2+} to Fe^{3+}. Write a balanced net ionic equation for each of these redox reactions.

22.96 Complete and balance the following equations:
(a) $Li_3N(s) + H_2O(l) \longrightarrow$
(b) $NH_3(aq) + H_2O(l) \longrightarrow$
(c) $NO_2(g) + H_2O(l) \longrightarrow$
(d) $NH_3(g) + O_2(g) \xrightarrow{\text{catalyst}}$
(e) $H_2CO_3(aq) \xrightarrow{\Delta}$
(f) $Ni(s) + CO(g) \longrightarrow$
(h) $CS_2(g) + O_2(g) \longrightarrow$
(i) $CaO(s) + SO_2(g) \longrightarrow$
(j) $CH_4(g) + H_2O(g) \xrightarrow{\Delta}$
(k) $LiH(s) + H_2O(l) \longrightarrow$
(l) $Fe_2O_3(s) + 3 H_2(g) \longrightarrow$

INTEGRATIVE EXERCISES

22.97 **(a)** How many grams of H_2 can be stored in 10.0 lb of the alloy FeTi if the hydride $FeTiH_2$ is formed? **(b)** What volume does this quantity of H_2 occupy at STP?

[22.98] Using the thermochemical data in Table 22.1 and Appendix C, calculate the average $Xe-F$ bond enthalpies in XeF_2, XeF_4, and XeF_6, respectively. What is the significance of the trend in these quantities?

22.99 Hydrogen gas has a higher fuel value than natural gas on a mass basis but not on a volume basis. Thus, hydrogen is not competitive with natural gas as a fuel transported long distances through pipelines. Calculate the heats of combustion of H_2 and CH_4 (the principal component of natural gas) **(a)** per mole of each, **(b)** per gram of each, **(c)** per cubic meter of each at STP. Assume $H_2O(l)$ as a product.

22.100 The solubility of Cl_2 in 100 g of water at STP is 310 cm^3. Assume that this quantity of Cl_2 is dissolved and equilibrated as follows:

$$Cl_2(aq) + H_2O \rightleftharpoons Cl^-(aq) + HClO(aq) + H^+(aq)$$

If the equilibrium constant for this reaction is 4.7×10^{-4}, calculate the equilibrium concentration of HClO formed.

[22.101] When ammonium perchlorate decomposes thermally, the products of the reaction are $N_2(g)$, $O_2(g)$, $H_2O(g)$, and $HCl(g)$. **(a)** Write a balanced equation for the reaction. [*Hint:* You might find it easier to use fractional coefficients for the products.] **(b)** Calculate the enthalpy change in the reaction per mole of NH_4ClO_4. The standard enthalpy of formation of $NH_4ClO_4(s)$ is -295.8 kJ. **(c)** When $NH_4ClO_4(s)$ is employed in solid-fuel booster rockets, it is packed with powdered aluminum. Given the high temperature needed for $NH_4ClO_4(s)$ decomposition and what the products of the reaction are, what role does the aluminum play?

22.102 The dissolved oxygen present in any highly pressurized, high-temperature steam boiler can be extremely corrosive to its metal parts. Hydrazine, which is completely miscible with water, can be added to remove oxygen by reacting with it to form nitrogen and water. **(a)** Write the balanced equation for the reaction between gaseous hydrazine and oxygen. **(b)** Calculate the enthalpy change accompanying this reaction. **(c)** Oxygen in air dissolves in water to the extent of 9.1 ppm at 20 °C at sea level. How many grams of hydrazine are required to react with all the oxygen in 3.0×10^4 L (the volume of a small swimming pool) under these conditions?

22.103 One method proposed for removing SO_2 from the flue gases of power plants involves reaction with aqueous H_2S. Elemental sulfur is the product. **(a)** Write a balanced chemical equation for the reaction. **(b)** What volume of H_2S at 27 °C and 740 torr would be required to remove the SO_2 formed by burning 1.0 ton of coal containing 3.5% S by mass? **(c)** What mass of elemental sulfur is produced? Assume that all reactions are 100% efficient.

22.104 The maximum allowable concentration of $H_2S(g)$ in air is 20 mg per kilogram of air (20 ppm by mass). How many grams of FeS would be required to react with hydrochloric acid to produce this concentration at 1.00 atm and 25 °C in an average room measuring $2.7 \times 4.3 \times 4.3$ m? (Under these conditions, the average molar mass of air is 29.0 g/mol.)

22.105 The standard heats of formation of $H_2O(g)$, $H_2S(g)$, $H_2Se(g)$, and $H_2Te(g)$ are −241.8, −20.17, +29.7, and +99.6 kJ/mol, respectively. The enthalpies necessary to convert the elements in their standard states to one mole of gaseous atoms are 248, 277, 227, and 197 kJ/mol of atoms for O, S, Se, and Te, respectively. The enthalpy for dissociation of H_2 is 436 kJ/mol. Calculate the average H—O, H—S, H—Se, and H—Te bond enthalpies, and comment on their trend.

22.106 Manganese silicide has the empirical formula MnSi and melts at 1280 °C. It is insoluble in water but does dissolve in aqueous HF. **(a)** What type of compound do you expect MnSi to be, in terms of Table 11.7? **(b)** Write a likely balanced chemical equation for the reaction of MnSi with concentrated aqueous HF.

22.107 Hydrazine has been employed as a reducing agent for metals. Using standard reduction potentials, predict whether the following metals can be reduced to the metallic state by hydrazine under standard conditions in acidic solution: **(a)** Fe^{2+}, **(b)** Sn^{2+}, **(c)** Cu^{2+}, **(d)** Ag^+, **(e)** Cr^{3+}.

22.108 Both dimethylhydrazine, $(CH_3)_2NNH_2$, and methylhydrazine, CH_3NHNH_2, have been used as rocket fuels. When dinitrogen tetroxide (N_2O_4) is used as the oxidizer, the products are H_2O, CO_2, and N_2. If the thrust of the rocket depends on the volume of the products produced, which of the substituted hydrazines produces a greater thrust per gram total mass of oxidizer plus fuel? (Assume that both fuels generate the same temperature and that $H_2O(g)$ is formed.)

22.109 Carbon forms an unusual, unstable oxide of formula C_3O_2, called carbon suboxide. Carbon suboxide is made by using P_2O_5 to dehydrate the dicarboxylic acid called malonic acid, which has the formula HOOC—CH_2—COOH. **(a)** Write a balanced reaction for the production of carbon suboxide from malonic acid. **(b)** Suggest a Lewis structure for C_3O_2. (*Hint:* The Lewis structure of malonic acid suggests which atoms are connected to which.) **(c)** By using the information in Table 8.5, predict the C—C and C—O bond lengths in C_3O_2. **(d)** Sketch the Lewis structure of a product that could result by the addition of 2 mol of H_2 to 1 mol of C_3O_2.

22.110 Boron nitride has a graphite-like structure with B—N bond distances of 1.45 Å within sheets and a separation of 3.30 Å between sheets. At high temperatures the BN assumes a diamondlike form that is almost as hard as diamond. Rationalize the similarity between BN and elemental carbon.

23 METALS AND METALLURGY

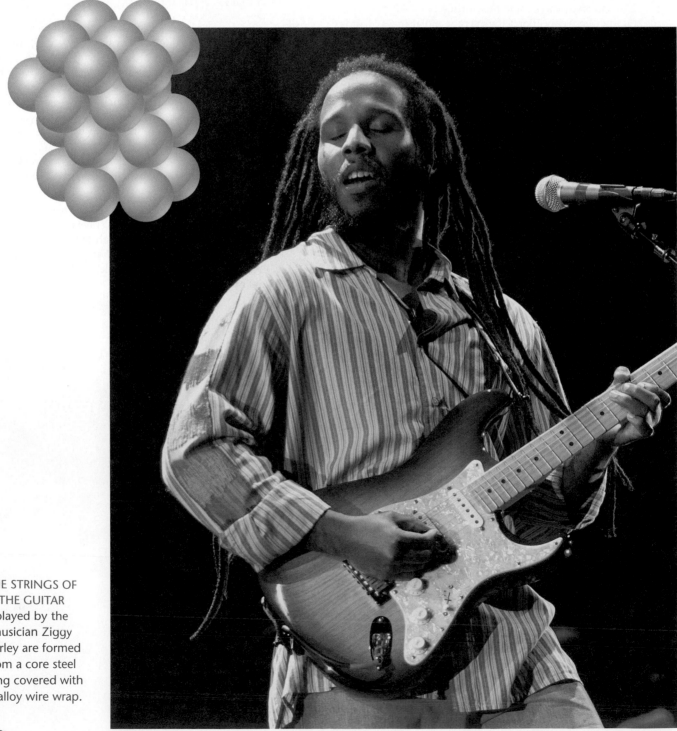

THE STRINGS OF THE GUITAR played by the musician Ziggy Marley are formed from a core steel string covered with an alloy wire wrap.

WHAT'S AHEAD

23.1 Occurrence and Distribution of Metals
We begin by examining the occurrence of metals in the *lithosphere* along with an introduction to *metallurgy*, which is the technology associated with extracting metals from their natural mineral sources and preparing them for use.

23.2 Pyrometallurgy
We see how high temperatures can be used to extract metals from minerals. Such processes play a particularly important role in the production of iron and steel.

23.3 Hydrometallurgy
We see how chemical processes that occur in aqueous solutions can be used to separate metals, with an emphasis on gold and aluminum.

23.4 Electrometallurgy
We investigate the use of electricity to reduce and refine metals via electrolysis. We also learn the importance of this approach for the production of sodium, aluminum, and copper.

23.5 Metallic Bonding
We review the physical properties of metals and then account for them in terms of two models for *metallic bonding*—the electron-sea model and the molecular-orbital model.

23.6 Alloys
We examine three different types of alloys, or metals that contain more than one element: solution alloys, heterogeneous alloys, and intermetallic compounds.

23.7 Transition Metals
We examine the physical properties, electron configurations, oxidation states, and magnetic properties of transition metals and their compounds.

23.8 Chemistry of Selected Transition Metals
To finish, we take a closer look at the chemistry of three key transition metals: chromium, iron, and copper.

METALS HAVE PLAYED A MAJOR ROLE IN THE DEVELOPMENT OF CIVILIZATION, and they continue to do so. For example, without the contributions of metallurgy, rock music as we know it would not exist. The strings of the guitar being played in the chapter-opening photograph are the product of complex technology. The core of the strings is made of mild steel, varying in hardness and ductility, depending on the type of guitar on which it will be used. The core string is wound with fine *wrap wire*, composed typically of an alloy of copper and zinc for acoustic guitar use, or stainless steel for electric guitars. Artists can be very passionate about the particular makes and models of strings they use.

When we think of metals in everyday applications, we tend to think of iron and aluminum, perhaps also chromium or nickel. But even metals of very low natural abundance play vital roles in modern technology. To illustrate this point, Figure 23.1 ▼ shows the approximate composition of a high-performance jet engine. Notice that iron, long the dominant metal of technology, is not present to any significant extent.

In this chapter we will consider the chemical forms in which metallic elements occur in nature and the means by which we obtain metals from these sources. We will also examine how metals bond in solids and see how metals and mixtures of metals, called *alloys*, are employed in modern technology. Finally, we will look specifically at the properties of transition metals. As we will see, metals have a varied and interesting chemistry.

38% Titanium
37% Nickel
12% Chromium
6% Cobalt
5% Aluminum
1% Niobium
0.02% Tantalum

▲ **Figure 23.1 Composition of a jet engine.** The chart on the left shows the metallic elements employed in manufacturing a modern jet engine.

23.1 OCCURRENCE AND DISTRIBUTION OF METALS

The portion of our environment that constitutes the solid earth beneath our feet is called the **lithosphere**. The lithosphere provides most of the materials we use to feed, clothe, shelter, and entertain ourselves. Although the bulk of Earth is solid, we have access to only a small region near the surface. Whereas Earth's radius is 6370 km, the deepest mine extends only about 4 km into Earth.

Many of the metals that are most useful to us are not especially abundant in that portion of the lithosphere to which we have ready access. Deposits that contain metals in economically exploitable quantities are known as **ores**. Usually, the desired compounds or elements must be separated from a large quantity of unwanted material and then chemically processed to make them useful. The costs, in terms of money, energy, and environmental impact, associated with extraction of metals from their ores is an important issue for society. The occurrence and distribution of *concentrated* deposits of metals often play a role in international politics as nations compete for access to these materials. Advances in technology also drive demand for metals that are either rare or difficult to extract. As an illustration, consider the fact that over one-half of the world's supply of cobalt comes from a politically unstable region of Africa (Congo and Zambia). The supply and price of cobalt are increasingly important because cobalt is an important component of superalloys used in aircraft engines (Figure 23.1), as well as the cathodes of Li-ion batteries that power most portable electronic devices. Hence the price of your cell phone may depend in part on the policies of the government in Kinshasa.

Minerals

With the exception of gold and the platinum-group metals (Ru, Rh, Pd, Os, Ir, and Pt), most metallic elements are found in nature in solid inorganic compounds called **minerals**. Table 23.1▶ lists the principal mineral sources of several common metals, three of which are shown in Figure 23.2▶. Notice that minerals are identified by common names rather than by chemical names. Names of minerals are usually based on the locations where they were discovered, the person who discovered them, or some characteristic such as color.

Commercially, the most important sources of metals are oxide, sulfide, and carbonate minerals. Silicate minerals are very abundant, but they are generally difficult to concentrate and reduce. ∞ (Section 22.10) Therefore, most silicates are not economical sources of metals.

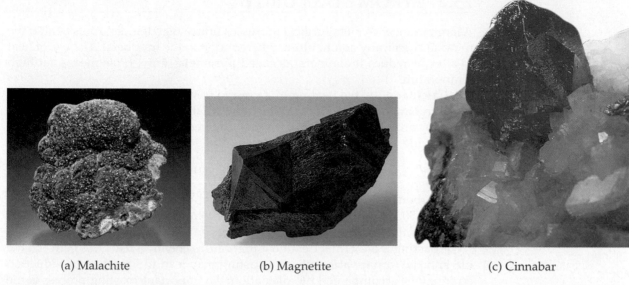

(a) Malachite   (b) Magnetite   (c) Cinnabar

▲ Figure 23.2 **Three common minerals.**

TABLE 23.1 ■ Principal Mineral Sources of Some Common Metals

Metal	Mineral	Composition
Aluminum	Corundum	Al_2O_3
	Gibbsite	$Al(OH)_3$
Chromium	Chromite	$FeCr_2O_4$
Copper	Chalcocite	Cu_2S
	Chalcopyrite	$CuFeS_2$
	Malachite	$Cu_2CO_3(OH)_2$
Iron	Hematite	Fe_2O_3
	Magnetite	Fe_3O_4
Lead	Galena	PbS
Manganese	Pyrolusite	MnO_2
Mercury	Cinnabar	HgS
Molybdenum	Molybdenite	MoS_2
Tin	Cassiterite	SnO_2
Titanium	Rutile	TiO_2
	Ilmenite	$FeTiO_3$
Zinc	Sphalerite	ZnS

Metallurgy

Metallurgy is the science and technology of extracting metals from their natural sources and preparing them for practical use. It usually involves several steps: (1) mining, (2) concentrating the ore or otherwise preparing it for further treatment, (3) reducing the ore to obtain the free metal, (4) refining or purifying the metal, and (5) mixing the metal with other elements to modify its properties. This last process produces an *alloy*, a metallic material that is composed of two or more elements (Section 23.6).

After being mined, an ore is usually crushed and ground and then treated to concentrate the desired metal. The concentration stage relies on differences in the properties of the mineral and the undesired material that accompanies it, which is called *gangue* (pronounced "gang"). After an ore is concentrated, a variety of chemical processes are used to obtain the metal in suitable purity. In Sections 23.2–23.4 we will examine some of the most common metallurgical processes. You will see that these techniques depend in large part on basic chemical concepts discussed earlier in the text.

23.2 PYROMETALLURGY

A large number of metallurgical processes utilize high temperatures to alter the mineral chemically and to ultimately reduce it to the free metal. The use of heat to alter or reduce the mineral is called **pyrometallurgy**. (*Pyro* means "at high temperature.")

Calcination is the heating of an ore to bring about its decomposition and the elimination of a volatile product. The volatile product could be, for example, CO_2 or H_2O. Carbonates are often calcined to drive off CO_2, forming the metal oxide. For example,

$$PbCO_3(s) \xrightarrow{\Delta} PbO(s) + CO_2(g) \qquad [23.1]$$

Most carbonates decompose reasonably rapidly at temperatures in the range of 400 °C to 500 °C, although $CaCO_3$ requires a temperature of about 1000 °C. Most hydrated minerals lose H_2O at temperatures on the order of 100 °C to 300 °C.

Roasting is a thermal treatment that causes chemical reactions between the ore and the furnace atmosphere. Roasting may lead to oxidation or reduction and may be accompanied by calcination. An important roasting process is the oxidation of sulfide ores, in which the metal sulfide is converted to the metal oxide, as in the following examples:

$$2\,ZnS(s) + 3\,O_2(g) \longrightarrow 2\,ZnO(s) + 2\,SO_2(g) \qquad [23.2]$$

$$2\,MoS_2(s) + 7\,O_2(g) \longrightarrow 2\,MoO_3(s) + 4\,SO_2(g) \qquad [23.3]$$

The sulfide ore of a less active metal, such as mercury, can be roasted to the free metal:

$$HgS(s) + O_2(g) \longrightarrow Hg(g) + SO_2(g) \qquad [23.4]$$

In many instances the free metal can be obtained by using a reducing atmosphere during roasting. Carbon monoxide provides such an atmosphere and is frequently used to reduce metal oxides:

$$PbO(s) + CO(g) \longrightarrow Pb(l) + CO_2(g) \qquad [23.5]$$

This method of reduction is not always feasible, however, especially with active metals, which are difficult to reduce.

GIVE IT SOME THOUGHT

Would you expect CaO to reduce to the free metal if heated in the presence of CO? What about Ag_2O?

Smelting is a melting process in which the materials formed in the course of chemical reactions separate into two or more layers. Smelting often involves a roasting stage in the same furnace. Two of the important types of layers formed in smelters are molten metal and slag. The molten metal may consist almost entirely of a single metal, or it may be a solution of two or more metals.

Slag consists mainly of molten silicate minerals, with aluminates, phosphates, and other ionic compounds as constituents. A slag is formed when a basic metal oxide such as CaO reacts at high temperatures with molten silica (SiO_2):

$$CaO(l) + SiO_2(l) \longrightarrow CaSiO_3(l) \qquad [23.6]$$

Pyrometallurgical operations may involve the concentration and reduction of a mineral and the refining of the metal. **Refining** is the treatment of a crude, relatively impure metal product from a metallurgical process to improve its purity and to define its composition better. Sometimes the goal of the refining process is to obtain the metal itself in pure form. The goal may also be to produce a mixture with a well-defined composition, as in the production of steels from crude iron.

The Pyrometallurgy of Iron

The most important pyrometallurgical operation is the reduction of iron. Iron occurs in many different minerals, but the most important sources are two iron oxide minerals—hematite (Fe_2O_3) and magnetite (Fe_3O_4). Figure 23.3▶ shows the world production of iron ore as a function of country. China, Brazil, and Australia account for nearly two-thirds of the world's iron ore. As the higher-grade deposits of iron ore have become depleted, lower-grade ores have been tapped. *Taconite*, which consists of fine-grained silica with variable ratios of iron, present mainly as magnetite, has increased in importance as a source of iron from the great Mesabi Range west of Lake Superior. These deposits account for almost all of the iron ore production in the United States.

The reduction of iron oxides can be accomplished in a *blast furnace*, such as the one illustrated in Figure 23.4▶. A blast furnace is essentially a huge chemical reactor capable of continuous operation. The furnace is charged at the top with a mixture of iron ore, coke, and limestone. Coke is coal that has been heated in the absence of air to drive off volatile components. It is about 85% to 90% carbon. Coke serves as the fuel, producing heat as it is burned in the lower part of the furnace. Through reactions with oxygen and water, coke also serves as the source of the reducing gases CO and H_2. Limestone ($CaCO_3$) serves as the source of the basic oxide CaO, which reacts with silicates and other components of the ore to form slag. Air, which enters the blast furnace at the bottom after preheating, is also an important raw material; it is required for combustion of the coke. Production of 1 kg of crude iron, called *pig iron*, requires about 2 kg of ore, 1 kg of coke, 0.3 kg of limestone, and 1.5 kg of air.

In the furnace, oxygen reacts with the carbon in the coke to form carbon monoxide:*

$$2\ C(s) + O_2(g) \longrightarrow 2\ CO(g) \qquad \Delta H° = -221\ \text{kJ} \qquad [23.7]$$

Water vapor present in the air also reacts with carbon, forming both carbon monoxide and hydrogen:

$$C(s) + H_2O(g) \longrightarrow CO(g) + H_2(g) \qquad \Delta H° = +131\ \text{kJ} \qquad [23.8]$$

The reaction of coke with oxygen is exothermic and provides heat for furnace operation, whereas its reaction with water vapor is endothermic. Addition of water vapor to the air thus provides a means of controlling furnace temperature.

In the upper part of the furnace, limestone decomposes to form CaO and CO_2. Here, also, the iron oxides are reduced by CO and H_2. For example, the important reactions for Fe_3O_4 are

$$Fe_3O_4(s) + 4\ CO(g) \longrightarrow 3\ Fe(s) + 4\ CO_2(g) \qquad \Delta H° = -15\ \text{kJ} \qquad [23.9]$$

$$Fe_3O_4(s) + 4\ H_2(g) \longrightarrow 3\ Fe(s) + 4\ H_2O(g) \qquad \Delta H° = +150\ \text{kJ} \qquad [23.10]$$

*In the reactions shown the enthalpy changes are given as $\Delta H°$ values. Note, however, that the conditions in the furnace are far from standard.

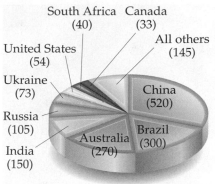

South Africa (40)   Canada (33)
United States (54)
All others (145)
Ukraine (73)
China (520)
Russia (105)
India (150)
Australia (270)
Brazil (300)

Usable ore (in metric tons × 10^6)

▲ **Figure 23.3   Worldwide mine production of iron ores.** The total worldwide production of iron ore was estimated to be $1.69 × 10^{12}$ kg in 2006.

Ore, limestone, and coke

CO, CO_2, NO_2

250 °C

600 °C

Hot-air supply pipe

Hot-air blast nozzle (one of many)

1000 °C

1600 °C

Slag

Molten iron

Molten-iron outlet

▲ **Figure 23.4   A blast furnace.** The furnace is used for reducing iron ore. Notice the increasing temperatures as the materials pass downward through the furnace.

▲ **Figure 23.5 Molten iron from a blast furnace.** The iron is poured for transport to a basic oxygen converter. Steelmakers convert iron to steel by adding scrap steel and other metals as alloying agents.

▲ **Figure 23.6 Refining iron.** In this converter for refining iron, a mixture of oxygen and argon is blown through the molten iron and slag. The heat generated by the oxidation of impurities maintains the mixture in a molten state. When the desired composition is attained, the converter is tilted to pour out its contents.

Reduction of other elements present in the ore also occurs in the hottest parts of the furnace, where carbon is the major reducing agent.

Molten iron collects at the base of the furnace, as shown in Figure 23.4. It is overlaid with a layer of molten slag formed by the reaction of CaO with the silica present in the ore (Equation 23.6). The layer of slag over the molten iron helps to protect it from reacting with the incoming air. Periodically the furnace is tapped to drain off slag and molten iron. The iron produced in the furnace may be cast into solid ingots. Most of the iron, however, is used directly in the manufacture of steel. For this purpose, it is transported, while still liquid, to the steelmaking shop (Figure 23.5 ◀). The production of pig iron using blast furnaces has decreased in recent years because of alternative reduction processes and the increased use of iron scrap in steelmaking. Blast furnaces nevertheless remain a significant means of reducing iron oxides.

Formation of Steel

As we will see in Section 23.6, steel is an alloy of iron. To produce steel with favorable properties, it is necessary to remove undesired impurities from the crude iron. Iron from a blast furnace typically contains 0.6–1.2% silicon, 0.4–2.0% manganese, and lesser amounts of phosphorus and sulfur. In addition, there is considerable dissolved carbon. In the production of steel these impurities are removed by oxidation in a vessel called a *converter*. In modern steelmaking the oxidizing agent is either pure O_2 or O_2 diluted with argon. Air cannot be used directly as the source of O_2 because N_2 reacts with the molten iron to form iron nitride, which causes the steel to become brittle.

A cross-sectional view of one converter design appears in Figure 23.6 ◀. In this converter O_2, diluted with argon, is blown directly into the molten metal. The oxygen reacts exothermically with carbon, silicon, and many metal impurities, reducing the concentrations of these elements in the iron. Carbon and sulfur are expelled as CO and SO_2 gases, respectively. Silicon is oxidized to SiO_2 and adds to whatever slag may have been present initially in the melt. Metal oxides react with the SiO_2 to form silicates. The presence of a basic slag is also important for removal of phosphorus:

$$3\,CaO(l) + P_2O_5(l) \longrightarrow Ca_3(PO_4)_2(l) \qquad [23.11]$$

Nearly all of the O_2 blown into the converter is consumed in the oxidation reactions. By monitoring the O_2 concentration in the gas coming from the converter, it is possible to tell when the oxidation is essentially complete. Oxidation of the impurities present in the iron normally requires about 20 min. When the desired composition is attained, the contents of the converter are dumped into a large ladle. To produce steels with various kinds of properties, alloying elements are added as the ladle is being filled. The still-molten mixture is then poured into molds, where it solidifies.

■ **SAMPLE EXERCISE 23.1** | Enthalpy Changes in a Converter

Which produces more heat in a converter, oxidation of 1 mol of C or oxidation of 1 mol of Si?

SOLUTION

Analyze: We must compare the enthalpy change for 1 mol of C oxidized to form CO_2 with the enthalpy change for 1 mol of Si oxidized to form SiO_2.

Plan: The reactions occur under conditions far different from the standard-state conditions for the substances. Nevertheless, we can estimate the enthalpy changes by using the thermodynamic values of Appendix C.

Solve: The oxidation reactions and enthalpy changes under standard conditions are

$$C(s) + O_2(g) \longrightarrow CO_2(g) \qquad \Delta H^\circ_{rxn} = -393.5 \text{ kJ} - (0 + 0) = -393.5 \text{ kJ}$$
$$Si(s) + O_2(g) \longrightarrow SiO_2(s) \qquad \Delta H^\circ_{rxn} = -910.9 \text{ kJ} - (0 + 0) = -910.9 \text{ kJ}$$

The actual numbers will differ substantially from these because the temperature is far from 298 K, because both C and Si are dissolved in the molten iron, and because SiO_2 will be incorporated in the slag. Nevertheless, the differences in the enthalpies of reaction are so great that it seems certain that the enthalpy change will be larger for Si.

■ PRACTICE EXERCISE
(a) What would you expect as the product when any dissolved manganese in the molten mixture is oxidized? **(b)** Would the manganese oxidation be exothermic or endothermic?
Answers: **(a)** MnO_2, **(b)** exothermic

23.3 HYDROMETALLURGY

Pyrometallurgical operations require large quantities of energy and are often a source of atmospheric pollution, especially by sulfur dioxide. For some metals, other techniques have been developed in which the metal is extracted from its ore by use of aqueous reactions. These processes are called **hydrometallurgy** (*hydro* means "water").

The most important hydrometallurgical process is **leaching**, in which the desired metal-containing compound is selectively dissolved. If the compound is water soluble, water by itself is a suitable leaching agent. More commonly, the agent is an aqueous solution of an acid, a base, or a salt. Often the dissolving process involves formation of a complex ion. ∞∞ (Section 17.5) As an example, we can consider the leaching of gold.

Hydrometallurgy of Gold

As noted in the "A Closer Look" box in Section 4.5, gold metal is often found relatively pure in nature. As concentrated deposits of elemental gold have been depleted, lower-grade sources have become more important. Gold from low-grade ores can be concentrated by placing the crushed ore on large concrete slabs and spraying a solution of NaCN over it. In the presence of CN^- and air, the gold is oxidized, forming the stable $Au(CN)_2^-$ ion, which is soluble in water:

$$4 Au(s) + 8 CN^-(aq) + O_2(g) + 2 H_2O(l) \longrightarrow 4 Au(CN)_2^-(aq) + 4 OH^-(aq)$$
[23.12]

After a metal ion is selectively leached from an ore, it is precipitated from solution as the free metal or as an insoluble ionic compound. Gold, for example, is obtained from its cyanide complex by reduction with zinc powder:

$$2 Au(CN)_2^-(aq) + Zn(s) \longrightarrow Zn(CN)_4^{2-}(aq) + 2 Au(s)$$
[23.13]

The cyanide process for gold recovery has been widely criticized because of the potential for pollution of groundwater, rivers, and streams. For example, leakage from containment ponds into nearby streams in Romania several years ago resulted in poisoning of all aquatic life in small rivers that flow into the Danube. Alternatives to cyanide are being investigated. Thiosulfate ion, $S_2O_3^{2-}$, ∞∞ (Section 22.6) has been studied as an alternative, but the costs are higher than for the cyanide process.

▲ GIVE IT SOME THOUGHT

Gold dissolves in the presence of thiosulfate as follows:

$$4 Au + 8 S_2O_3^{2-}(aq) + 2 H_2O(l) + O_2(aq) \longrightarrow 4 Au(S_2O_3)_2^{3-}(aq) + 4 OH^-(aq)$$

What is the oxidizing agent in this reaction, what is the reducing agent, and what oxidation states of the reducing and oxidizing agents are involved?

Hydrometallurgy of Aluminum

Among metals, aluminum is second only to iron in commercial use. Worldwide production of aluminum is about 2.4×10^{10} kg per year. The most useful ore of aluminum is *bauxite*, in which Al is present as hydrated oxides, $Al_2O_3 \cdot xH_2O$. The value of x varies, depending on the particular mineral present. Because bauxite deposits in the United States are limited, most of the ore used in the production of aluminum must be imported.

The major impurities found in bauxite are SiO_2 and Fe_2O_3. It is essential to separate Al_2O_3 from these impurities before the metal is recovered by electrochemical reduction, as described in Section 23.4. The process used to purify bauxite, called the **Bayer process**, is a hydrometallurgical procedure. The ore is first crushed and ground, then digested in a concentrated aqueous NaOH solution, about 30% NaOH by mass, at a temperature in the range of 150 °C to 230 °C. Sufficient pressure, up to 30 atm, is maintained to prevent boiling. The Al_2O_3 dissolves in this solution, forming the complex aluminate ion, $Al(OH)_4^-$:

$$Al_2O_3 \cdot H_2O(s) + 2\,H_2O(l) + 2\,OH^-(aq) \longrightarrow 2\,Al(OH)_4^-(aq) \quad [23.14]$$

The iron(III) oxides do not dissolve in the strongly basic solution. This difference in the behavior of the aluminum and iron compounds arises because Al^{3+} is amphoteric, whereas Fe^{3+} is not. ∞ (Section 17.5) Thus, the aluminate solution can be separated from the iron-containing solids by filtration. The pH of the solution is then lowered, causing the aluminum hydroxide to precipitate.

After the aluminum hydroxide precipitate has been filtered, it is calcined in preparation for electroreduction to the metal. The solution recovered from the filtration is reconcentrated so that it can be used again. This task is accomplished by heating to evaporate water from the solution, a procedure that requires much energy and is the most costly part of the Bayer process.

▲ GIVE IT SOME THOUGHT

What do you expect to be the product of the calcination of $Al(OH)_3$?

23.4 ELECTROMETALLURGY

Many processes that are used to reduce metal ores or to refine metals are based on electrolysis. ∞ (Section 20.9) Collectively these processes are referred to as **electrometallurgy**. Electrometallurgical procedures can be broadly differentiated according to whether they involve electrolysis of a molten salt or of an aqueous solution.

Electrolytic methods are important for obtaining the more active metals, such as sodium, magnesium, and aluminum. These metals cannot be obtained from aqueous solution because water is more easily reduced than the metal ions. The standard reduction potentials of water under both acidic and basic conditions are more positive than those of Na^+ ($E_{red}^\circ = -2.71$ V), Mg^{2+} ($E_{red}^\circ = -2.37$ V), and Al^{3+} ($E_{red}^\circ = -1.66$ V):

$$2\,H^+(aq) + 2\,e^- \longrightarrow H_2(g) \qquad\qquad E_{red}^\circ = 0.00\ V \quad [23.15]$$

$$2\,H_2O(l) + 2\,e^- \longrightarrow H_2(g) + 2\,OH^-(aq) \qquad E_{red}^\circ = -0.83\ V \quad [23.16]$$

To form such metals by electrochemical reduction, therefore, we must employ a nonaqueous, molten-salt medium in which the metal ion of interest is the most readily reduced species.

Electrometallurgy of Sodium

In the commercial preparation of sodium, molten NaCl is electrolyzed in a specially designed cell called the **Downs cell**, illustrated in Figure 23.7►. Calcium chloride ($CaCl_2$) is added to lower the melting point of the molten NaCl from the

NaCl inlet

$Cl_2(g)$

Molten NaCl

Iron screen to
prevent Na and
Cl_2 from coming
together

Na(l)

Carbon anode   +   Iron cathode
$2\ Cl^- \longrightarrow Cl_2(g) + 2e^-$ $2\ Na^+ + 2e^- \longrightarrow 2\ Na(l)$

▲ **Figure 23.7 Production of sodium.**
The Downs cell is used in the commercial
production of sodium.

normal melting point of 804 °C to around 600 °C. The Na(l) and $Cl_2(g)$ produced
in the electrolysis are kept from coming in contact and re-forming NaCl. In addition, the Na must be prevented from contact with oxygen because the metal
would quickly oxidize under the high-temperature conditions of the cell reaction.

GIVE IT SOME THOUGHT

Which ions carry the current in a Downs cell?

Electrometallurgy of Aluminum

In Section 23.3 we discussed the Bayer process, in
which bauxite is concentrated to produce aluminum
hydroxide. When this concentrate is calcined at temperatures in excess of 1000 °C, anhydrous aluminum
oxide (Al_2O_3) is formed. Anhydrous aluminum oxide
melts at over 2000 °C. This is too high to permit its use
as a molten medium for electrolytic formation of free
aluminum. The electrolytic process commercially used
to produce aluminum is known as the **Hall-Héroult
process**, named after its inventors, Charles M. Hall and
Paul Héroult (see the "A Closer Look" box in this section). The purified Al_2O_3 is dissolved in molten cryolite
(Na_3AlF_6), which has a melting point of 1012 °C and is
an effective conductor of electric current. A schematic diagram of the electrolysis
cell is shown in Figure 23.8▶. Graphite rods are employed as anodes and are
consumed in the electrolysis process. The electrode reactions are as follows:

Graphite
anodes

Molten
aluminum

Al_2O_3 dissolved
in molten
cryolite

Carbon-lined
iron

▲ **Figure 23.8 The Hall-Héroult
process.** An electrolysis cell used to
form aluminum metal through
reduction. Because molten aluminum is
denser than the mixture of cryolite
(Na_3AlF_6) and Al_2O_3, the metal collects
at the bottom of the cell.

Anode:	$C(s) + 2\ O^{2-}(l) \longrightarrow CO_2(g) + 4\ e^-$	[23.17]
Cathode:	$3\ e^- + Al^{3+}(l) \longrightarrow Al(l)$	[23.18]

The amounts of raw materials and energy required to produce 1000 kg of aluminum metal from bauxite by this procedure are summarized in Figure 23.9▼.
Due to the large amount of electrical energy needed in the Hall-Héroult process,
aluminum smelters are invariably located in areas with access to low cost sources
of electricity. In fact, the aluminum industry consumes about 10% of the world's
hydroelectric power. However, recycled aluminum only requires 5% of the
energy needed to produce "new" aluminum. Therefore, considerable energy

▲ Figure 23.9 What it takes to produce 1000 kg of aluminum.

savings can be realized by increasing the amount of aluminum that is recycled. In the United States approximately 42% of post-consumer aluminum is recycled, while in Europe the rate is estimated to be 52%. Much higher rates of recycling are possible. In Norway, for example, the aluminum recycling rate is 93%.

The electrodes represent another challenge surrounding aluminum manufacture. The carbon anode is consumed in the process, generating a variety of pollutants, including CO_2. Approximately 3.8×10^{10} kg of CO_2 are emitted each year from aluminum manufacture. It is a significant challenge to find electrodes that are conducting and chemically inert at elevated temperatures in molten cryolite.

GIVE IT SOME THOUGHT

Does molten cryolite conduct electricity through the movement of electrons (electronic conductivity) or the movement of ions (ionic conductivity)?

Electrorefining of Copper

Copper is widely used to make electrical wiring and in other applications that utilize its high electrical conductivity. Crude copper, which is usually obtained by pyrometallurgical methods, is not suitable to serve in electrical applications because impurities greatly reduce the metal's conductivity.

Purification of copper is achieved by electrolysis, as illustrated in Figure 23.11 ▶. Large slabs of crude copper serve as the anodes in the cell, and thin sheets of pure copper serve as the cathodes. The electrolyte consists of an acidic solution of $CuSO_4$. Application of a suitable voltage to the electrodes causes oxidation of copper metal at the anode and reduction of Cu^{2+} to form copper metal at the cathode. This strategy can be used because copper is both oxidized and reduced more readily than water. The relative ease of reduction of Cu^{2+} and H_2O is seen by comparing their standard reduction potentials:

$$Cu^{2+}(aq) + 2\,e^- \longrightarrow Cu(s) \qquad E^\circ_{red} = +0.34\text{ V} \qquad [23.19]$$

$$2\,H_2O(l) + 2\,e^- \longrightarrow H_2(g) + 2\,OH^-(aq) \qquad E^\circ_{red} = -0.83\text{ V} \qquad [23.20]$$

A Closer Look   CHARLES M. HALL

Charles M. Hall (Figure 23.10 ▶) began to work on the problem of reducing aluminum in about 1885 after he had learned from a professor of the difficulty of reducing ores of very active metals. Before the development of Hall's electrolytic process, aluminum was obtained by a chemical reduction using sodium or potassium as the reducing agent. Because the procedure was very costly, aluminum metal was very expensive. As late as 1852, the cost of aluminum was $545 per pound, far greater than the cost of gold. During the Paris Exposition in 1855, aluminum was exhibited as a rare metal, even though it is the third most abundant element in Earth's crust.

Hall, who was 21 years old when he began his research, utilized handmade and borrowed equipment in his studies and used a woodshed near his Ohio home as his laboratory. In about a year's time he was able to solve the problem of reducing aluminum. His procedure consisted of finding an ionic compound that could be melted to form a conducting medium that would dissolve Al_2O_3 but would not interfere in the electrolysis reactions. The relatively rare mineral cryolite, (Na_3AlF_6), found in Greenland, met these criteria. Ironically,

◀ Figure 23.10 Charles M. Hall (1863–1914) as a young man.

Paul Héroult, who was the same age as Hall, independently made the same discovery in France at about the same time. Because of the research of these two unknown young scientists, large-scale production of aluminum became commercially feasible, and aluminum became a common and familiar metal.

The impurities in the copper anode include lead, zinc, nickel, arsenic, selenium, tellurium, and several precious metals, including gold and silver. Metallic impurities that are more active than copper are readily oxidized at the anode but do not plate out at the cathode because their reduction potentials are more negative than that for Cu^{2+}. Less active metals, however, are not oxidized at the anode. Instead they collect below the anode as a sludge that is collected and processed to recover the valuable metals. The anode sludges from copper-refining cells provide one-fourth of U.S. silver production and about one-eighth of U.S. gold production.

▲ **Figure 23.11 Electrolysis cell for refining copper.** As the anodes dissolve away, the cathodes on which the pure metal is deposited grow larger.

GIVE IT SOME THOUGHT

Which ions carry the current in the electrorefining of copper?

■ SAMPLE EXERCISE 23.2 | Impurities in Electrorefining

Nickel is one of the chief impurities in the crude copper that is subjected to electrorefining. What happens to this nickel in the course of the electrolytic process?

SOLUTION

Analyze: We are asked to predict whether nickel can be oxidized at the anode and reduced at the cathode during the electrorefining of copper.

Plan: We need to compare the standard reduction potentials of Ni^{2+} and Cu^{2+}. The more negative the reduction potential, the less readily the ion is reduced but the more readily the metal itself is oxidized. ⬡ (Section 20.4)

Solve: The standard reduction potential for Ni^{2+} is more negative than that for Cu^{2+}:

$$Ni^{2+}(aq) + 2\,e^- \longrightarrow Ni(s) \qquad E^\circ_{red} = -0.28\text{ V}$$

$$Cu^{2+}(aq) + 2\,e^- \longrightarrow Cu(s) \qquad E^\circ_{red} = +0.34\text{ V}$$

As a result, nickel is more readily oxidized than copper, assuming standard conditions. Although we do not have standard conditions in the electrolytic cell, we nevertheless expect that nickel is preferentially oxidized at the anode. Because the reduction of Ni^{2+} occurs less readily than the reduction of Cu^{2+}, the Ni^{2+} accumulates in the electrolyte solution, while the Cu^{2+} is reduced at the cathode. After a time it is necessary to recycle the electrolyte solution to remove the accumulated metal ion impurities, such as Ni^{2+}.

■ PRACTICE EXERCISE

Zinc is another common impurity in copper. Using standard reduction potentials, determine whether zinc will accumulate in the anode sludge or in the electrolytic solution during the electrorefining of copper.
Answer: It is found in the electrolytic solution because the standard reduction potential of Zn^{2+} is more negative than that of Cu^{2+}.

23.5 METALLIC BONDING

In our discussion of metallurgy we have confined ourselves to discussing the methods employed for obtaining metals in pure form. Metallurgy is also concerned with understanding the properties of metals and with developing useful new materials. As with any branch of science and engineering, our ability to make advances is coupled with our understanding of the fundamental properties of the systems with which we work. At several places in the text we have referred to the differences in physical and chemical behavior between metals and nonmetals. Let's now consider the distinctive properties of metals and relate these properties to a model for metallic bonding.

▲ Figure 23.12 Malleability and ductility. Gold leaf (left) demonstrates the characteristic malleability of metals, and copper wire (right) demonstrates their ductility.

Physical Properties of Metals

You have probably held a length of copper wire or an iron bolt. Perhaps you have even seen the surface of a freshly cut piece of sodium metal. These substances, although distinct from one another, share certain similarities that enable us to classify them as metallic. A clean metal surface has a characteristic luster. In addition, metals that we can handle with bare hands have a characteristic cold feeling related to their high heat conductivity. Metals also have high electrical conductivities; electrical current flows easily through them. Current flow occurs without any displacement of atoms within the metal structure and is due to the flow of electrons within the metal. The heat conductivity of a metal usually parallels its electrical conductivity. Silver and copper, for example, which possess the highest electrical conductivities among the elements, also possess the highest heat conductivities. This observation suggests that the two types of conductivity have the same origin in metals, which we will soon discuss.

Most metals are *malleable*, which means that they can be hammered into thin sheets, and *ductile*, which means that they can be drawn into wires (Figure 23.12 ◄). These properties indicate that the atoms are capable of slipping with respect to one another. Ionic solids or crystals of most covalent compounds do not exhibit such behavior. These types of solids are typically brittle and fracture easily. Consider, for example, the difference between dropping an ice cube and a block of aluminum metal onto a concrete floor.

Most metals form solid structures in which the atoms are arranged as close-packed spheres. In these structures each atom is in contact with 12 neighboring atoms. ∞∞(Section 11.7) This arrangement of atoms is very different from the structures of the nonmetallic elements. Consider for example the structures of elements of the third period (Na–Ar). Argon with eight valence electrons has a complete octet; as a result it does not form any bonds. Chlorine, sulfur, and phosphorous form molecules (Cl_2, S_8, and P_4) where the atoms make one, two, and three bonds, respectively. ∞∞(Chapter 22) Silicon forms an extended network solid where each atom is bonded to four equidistant neighbors (Figure 12.3). Each of these elements forms 8–N bonds, where N is the number of valence electrons. This behavior can easily be understood through application of the octet rule.

If the trend were to continue as we move left across the periodic table, we would expect aluminum to form five bonds. But, like many other metals including magnesium, titanium and gold, aluminum adopts a close-packed structure with 12 near neighbors. Clearly there is a change in the preferred bonding mechanism as the number of valence electrons decreases. As a general rule, metals do not have enough valence-shell electrons to satisfy their bonding requirements through the formation of localized electron pair bonds. In response to this deficiency, the valence electrons are collectively shared. A close-packed lattice of atoms facilitates a delocalized sharing of electrons among all atoms in the lattice.

Metal ion (+)

▲ Figure 23.13 Electron-sea model. Schematic illustration of the electron-sea model of the electronic structure of metals. Each gray sphere is a positively charged metal ion.

Electron-Sea Model for Metallic Bonding

One very simple model that accounts for some of the most important characteristics of metals is the **electron-sea model**. ∞∞(Section 11.8) In this model the metal is pictured as an array of metal cations in a "sea" of valence electrons, as illustrated in Figure 23.13 ◄. The electrons are confined to the metal by electrostatic attractions to the cations, and they are uniformly distributed throughout the structure. The electrons are mobile, however, and no individual electron is confined to any particular metal ion. When a metal wire is connected to the terminals of a battery, electrons flow through the metal toward the positive terminal and into the metal from the battery at the negative terminal. The high heat conductivity of metals is also accounted for by the mobility of the electrons, which permits ready transfer of kinetic energy throughout the solid. The ability of metals to deform (their malleability and ductility) can be explained by the fact that metal atoms form bonds to many neighbors. Changes in the positions of the atoms brought about in reshaping the metal are partly accommodated by a redistribution of electrons.

▲ Figure 23.14 The (a) melting and (b) boiling points of metallic elements.

The electron-sea model, however, does not adequately explain all properties. According to the model, for example, the strength of bonding between metal atoms should increase as the number of valence electrons increases, resulting in a corresponding increase in the melting and boiling points. However, the elements found near the middle of the transition metal series, rather than those at the end, have the highest melting and boiling points in their respective periods (Figure 23.14▲), which implies that the strength of metallic bonding first increases with increasing number of electrons and then decreases. Similar trends are seen in other physical properties of the metals, such as the heat of fusion and hardness.

Molecular-Orbital Model for Metals

To obtain a more accurate picture of the bonding in metals we turn to molecular-orbital theory. In Sections 9.7 and 9.8 we learned how molecular orbitals are created from the overlap of atomic orbitals. In Section 12.2 we saw that in extended solids (such as covalent network solids and metals) the presence of a large number of atoms results in a large number of closely spaced molecular orbitals. These closely spaced orbitals form continuous *bands* of allowed energy states. ∞ (Section 12.2)

The band structure for a typical metal is shown schematically in Figure 23.15▶. The electron filling depicted corresponds to nickel metal, but the basic features of other metals are similar. The electron configuration of a nickel atom is $[Ar]3d^84s^2$, as shown on the left side of the figure. The energy bands that form from these orbitals are shown on the right side of the figure. The $4s$, $4p$, and $3d$ orbitals are treated independently, each giving rise to a band of molecular orbitals. In practice, these overlapping bands are not independent of each other, but for our purposes this simplification is reasonable.

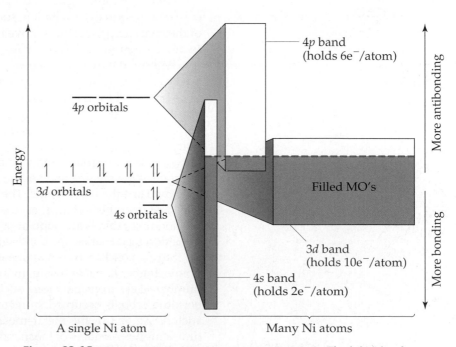

▲ Figure 23.15 **The electronic band structure of nickel.** The left side of the figure shows the electron configuration of a single Ni atom, while the right-hand side of the figure shows how these orbital energy levels broaden into energy bands in bulk nickel. The horizontal dashed gray line denotes the position of the Fermi Level, which separates the occupied molecular orbitals (shaded in blue) from the unoccupied molecular orbitals.

When two atomic orbitals overlap they form two molecular orbitals: (1) a bonding molecular orbital whose energy is lowered with respect to the parent atomic orbitals and (2) an antibonding molecular orbital whose energy is raised with respect to the parent atomic orbitals. ∞∞(Chapters 9 and 12) As the overlap increases, the energy splitting between bonding and antibonding molecular orbitals increases. The 4s and 4p orbitals are considerably larger than the 3d orbitals, which means that they overlap more effectively. This characteristic leads to a large difference in energy between the molecular orbitals at the bottom of the band, where the interactions are strongly bonding, and the molecular orbitals at the top of the band, where the interactions are strongly antibonding. The result is 4s and 4p bands that span a relatively wide range of energy, but due to the Pauli Exclusion Principle ∞∞(Section 6.7) these bands can hold only two and six electrons per atom, respectively.

The band that arises from overlap of the 3d orbitals differs in two important ways. Because the 3d orbitals are smaller, yet the interatomic distance is the same, they overlap less effectively, which reduces the strength of both the stabilizing bonding interactions and the destabilizing antibonding interactions. The result is a band that spans a much narrower range of energy. But because there are five 3d orbitals on each atom, this band can hold up to ten electrons per atom. The filling of bands with electrons is akin to filling a reservoir with water. In a solid the energy of the highest filled molecular orbital is called the *Fermi level*. Only a tiny energy separates it from the energy of the lowest empty orbital. The electrons occupy the lowest energy molecular orbitals available to them, irrespective of the orbitals populated in an individual atom. The occupied molecular orbitals are shaded in blue in Figure 23.15.

Many properties of metals can be understood from the simple band structure diagram shown in Figure 23.15. Because the electrons available for bonding do not completely fill the available molecular orbitals, we can think of the energy band as a partially filled container for electrons. The incomplete filling of the energy band gives rise to characteristic metallic properties. The electrons in orbitals near the top of the occupied levels require very little energy input to be "promoted" to still higher energy orbitals that are unoccupied. Under the influence of any source of excitation, such as an applied electrical potential or an input of thermal energy, electrons move into previously vacant levels and are thus freed to move through the lattice, giving rise to electrical and thermal conductivity.

Without the overlap of energy bands, the periodic properties of metals could not be explained. In the absence of the d- and p-bands we would expect the s-band to be half filled for the alkali metals (group 1A) and completely filled for the alkaline-earth metals (group 2A). If that were true, metals like magnesium, calcium, and strontium would be expected to behave as semiconductors! While the conductivity of metals can be qualitatively understood using either the electron-sea model or the molecular-orbital model, many of the physical properties of transition metals, such as the melting and boiling points plotted in Figure 23.14, can be explained only by using the latter model. The molecular-orbital model predicts that bonding will first become stronger as the number of valence electrons increases and the bonding orbitals are populated. Upon moving past the middle elements of the transition metal series, the bonds grow weaker as we fill the antibonding orbitals.

Strong bonds between atoms lead to metals with higher melting and boiling points, higher heats of fusion, higher hardness, and so forth. Metals with a small number of electrons per atom, such as Rb or Cs, have relatively few metal–metal bonding orbitals occupied. In metals with a large number of electrons per atom, such as Zn or Cd, the metal–metal bonding orbitals are filled, but at the same time a high percentage of the metal–metal antibonding orbitals are also occupied. In either case, the bonding will be much weaker compared to elements in the middle of the transition series, dramatically impacting a number of physical properties. However, we need to remember that factors other than the number of electrons (such as atomic radius, nuclear charge, and the particular packing structure of the metal) also play a role in determining the properties of metals.

23.6 ALLOYS

An **alloy** is a material that contains more than one element and has the characteristic properties of metals. The alloying of metals is of great importance because it is one of the primary ways of modifying the properties of pure metallic elements. Nearly all the common uses of iron, for example, involve alloy compositions. Bronze is formed by alloying copper and tin, while brass is an alloy of copper and zinc. Copper is the majority element in both alloys. Pure gold is too soft to be used in jewelry, but alloys of gold and copper are quite hard. Pure gold is termed 24 karat; the common alloy used in jewelry is 14 karat, meaning that it is 58% gold $\left(\frac{14}{24} \times 100\%\right)$. A gold alloy of this composition has suitable hardness to be used in jewelry. The alloy can be either yellow or white, depending on the elements added. Some further examples of alloys are given in Table 23.2▼.

Alloys can be classified as solution alloys, heterogeneous alloys, and intermetallic compounds. **Solution alloys** are homogeneous mixtures in which components are dispersed randomly and uniformly. Atoms of the solute can take positions normally occupied by a solvent atom, thereby forming a *substitutional alloy*. Or they can occupy interstitial positions in the "holes" between the solvent atoms, thereby forming an *interstitial alloy*. These types are diagrammed in Figure 23.16▼.

Substitutional alloys are formed when the two metallic components have similar atomic radii and chemical-bonding characteristics. For example, silver and gold form such an alloy over the entire range of possible compositions. When two metals differ in radii by more than about 15%, solubility is more limited.

For an interstitial alloy to form, the component present in the interstitial positions between the solvent atoms must have a much smaller bonding atomic radius than the solvent atoms. Typically, an interstitial element is a nonmetal that bonds to neighboring atoms. The presence of the extra bonds provided by the interstitial component causes the metal lattice to become harder, stronger, and less ductile. For example, steel, which is much harder and stronger than pure iron, is an alloy of iron that contains up to 3% carbon. *Mild steels* contain less than 0.2% carbon; they are malleable and ductile and are used to make cables, nails, and chains. *Medium steels* contain 0.2–0.6% carbon; they are tougher than mild steels and are used to make girders and rails. *High-carbon steel*, used in cutlery, tools, and springs, contains 0.6–1.5% carbon. In all three cases other elements may be added to form *alloy steels*. Vanadium and chromium may be added to impart strength and to increase resistance to fatigue and corrosion. For example, a rail steel used in Sweden on lines bearing heavy ore carriers contains 0.7% carbon, 1% chromium, and 0.1% vanadium.

TABLE 23.2 ■ Some Common Alloys

Primary Element	Name of Alloy	Composition by Mass	Properties	Uses
Bismuth	Wood's metal	50% Bi, 25% Pb, 12.5% Sn, 12.5% Cd	Low melting point (70 °C)	Fuse plugs, automatic sprinklers
Copper	Yellow brass	67% Cu, 33% Zn	Ductile, takes polish	Hardware items
Iron	Stainless steel	80.6% Fe, 0.4% C, 18% Cr, 1% Ni	Resists corrosion	Tableware
Lead	Plumber's solder	67% Pb, 33% Sn	Low melting point (275 °C)	Soldering joints
Silver	Sterling silver	92.5% Ag, 7.5% Cu	Bright surface	Tableware
	Dental amalgam	70% Ag, 18% Sn, 10% Cu, 2% Hg	Easily worked	Dental fillings

THE ARRANGEMENT OF ATOMS IN TWO TYPES OF SOLUTION ALLOYS

Solution alloys are homogenous mixtures in which the components are dispersed randomly and uniformly.

In a **Substitutional Alloy**, atoms of the solute take positions normally occupied by a solvent atom.

In an **Interstitial Alloy**, solute atoms occupy interstitial positions in the "holes" between the solvent atoms.

▲ **Figure 23.16 Substitutional and interstitial alloys.** The blue spheres represent host metal; the yellow spheres represent the other components of the alloy.

One of the most important iron alloys is stainless steel, which contains about 0.4% carbon, 18% chromium, and 1% nickel. The chromium is obtained by carbon reduction of chromite ($FeCr_2O_4$) in an electric furnace. The product of the reduction is *ferrochrome* ($FeCr_2$), which is then added in the appropriate amount to molten iron that comes from the converter, to achieve the desired steel composition. The ratio of elements present in the steel may vary over a wide range, imparting a variety of specific physical and chemical properties to the materials.

In a **heterogeneous alloy** the components are not dispersed uniformly. In the form of steel known as pearlite, for example, two distinct phases—essentially pure iron and the compound Fe_3C, known as cementite—are present in alternating layers. In general, the properties of heterogeneous alloys depend on both the composition and the manner in which the solid is formed from the molten mixture. Rapid cooling leads to distinctly different properties than are obtained by slow cooling.

Intermetallic Compounds

Intermetallic compounds are homogeneous alloys that have definite properties and compositions. Unlike substitutional and interstitial alloys, the different types of atoms in an intermetallic compound are ordered rather than randomly distributed. Examples of the atomic ordering seen in some important cubic intermetallic phases are shown in Figure 23.17▶. The ordering of atoms in an intermetallic compound generally leads to better structural stability and higher

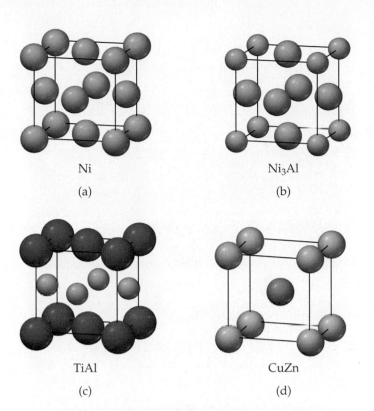

Ni

(a)

Ni$_3$Al

(b)

TiAl

(c)

CuZn

(d)

◀ **Figure 23.17 Ordered intermetallic structures.** (a) The face-centered cubic unit cell of a cubic close-packed metal. (b) The ordered structure of Ni$_3$Al with nickel atoms shown in gray and aluminum atoms in blue. (c) The ordered structure of TiAl with titanium atoms shown in red and aluminum atoms in blue. (d) The ordered structure of ZnCu with copper atoms shown in copper and zinc atoms in gray.

melting points than the constituent metals. These features can be attractive for high temperature applications. On the other hand, intermetallic compounds are often more brittle than substitutional alloys.

Intermetallic compounds play many important roles in modern society. The intermetallic compound Ni$_3$Al is a major component of jet aircraft engines because of its strength at high temperature and its low density. Razor blades are often coated with Cr$_3$Pt, which adds hardness, allowing the blade to stay sharp longer. Both compounds have the ordered structure shown in Figure 23.17(b). The compound Co$_5$Sm is used in the permanent magnets in lightweight headsets and high-fidelity speakers because of its high magnetic strength per unit weight. A related compound with the same structure, LaNi$_5$, is used as the anode in nickel-metal hydride batteries.

A Closer Look   SHAPE-MEMORY ALLOYS

In 1961 a naval engineer, William J. Buechler, made an unexpected and fortunate discovery. In searching for the best metal to use in missile nose cones, he tested many metal alloys. One of them, an intermetallic compound of nickel and titanium, NiTi, behaved very oddly. When he struck the cold metal, the sound was a dull thud. When he struck the metal at a higher temperature, however, it resonated like a bell. Mr. Buechler knew that the way in which sound propagates in a metal is related to its metallic structure. Clearly, the structure of the NiTi alloy had changed as it went from cold to warm. As it turned out, he had discovered an alloy that has shape memory.

Metals and metal alloys consist of many tiny crystalline areas (*crystallites*). When a metal is formed into a certain shape at high temperature, the crystallites are forced into a particular arrangement with respect to one another. Upon cooling a normal metal, the crystallites are "locked" in place by the

bonds between them. When the metal is subsequently bent, the resulting stresses are sometimes elastic, as in a spring. Often, however, the metal merely deforms (for example, when we bend a nail or crumple a sheet of aluminum foil). In these cases bending weakens the bonds that tie the crystallites together, and upon repeated flexing, the metal breaks apart.

In a shape-memory alloy, the atoms can exist in two different bonding arrangements, representing two different solid-state phases. The higher-temperature phase has strong, fixed bonds between the atoms in the crystallites. By contrast, the lower-temperature phase is quite flexible with respect to the arrangements between the atoms. Thus, when the metal is distorted at low temperature, the stresses of the distortion are absorbed *within* the crystallites by changes in the atomic lattice. In the higher-temperature phase, however, the atomic lattice is stiff, and the bonds between crystallites, as in a normal metal, absorb stresses that are due to bending.

continues on next page

continued

To see how a shape-memory metal behaves, suppose we bend a bar of NiTi alloy into a semicircle [Figure 23.18(a)▶] and then heat it to about 500 °C. We then cool the metal to below the transition temperature for the phase change to the low-temperature, flexible form. Although the cold metal remains in the semicircular form, as in Figure 23.18(b), it is now quite flexible and can readily be straightened or bent into another shape. When the metal is subsequently warmed and passes through the phase change to the "stiff" phase, it "remembers" its original shape and immediately returns to it as shown in Figure 23.18(c).

There are many uses for such shape-memory alloys. The curved shape in a dental brace, for example, can be formed at high temperature into the curve that the teeth are desired to follow. Then at low temperature, where the metal is flexible, it can be shaped to fit the mouth of the wearer of the braces. When the brace is inserted in the mouth and warms to body temperature, the metal passes into the stiff phase and exerts a force against the teeth as it attempts to return to its original shape. Other uses for shape-memory metals include heat-actuated shutoff valves in industrial process lines, which need no outside power source. Shape-memory alloys are also being evaluated for use in metallic stents, inserted into narrow veins to hold them open. The chilled NiTi alloy stent is in a collapsed form. When it warms in the body, the stent expands to the original size that it "remembers."

▲ **Figure 23.18 Behavior of a shape-memory alloy.**

23.7 TRANSITION METALS

Many of the most important metals of modern society are transition metals. Transition metals, which occupy the *d* block of the periodic table (Figure 23.19▼), include such familiar elements as chromium, iron, nickel, and copper. They also include less familiar elements that have come to play important roles in modern technology, such as those in the high-performance jet engine pictured in Figure 23.1. In this section we consider some of the physical and chemical properties of transition metals.

Physical Properties

Several physical properties of the elements of the first transition series are listed in Table 23.3▶. Some of these properties, such as ionization energy and atomic radius, are characteristic of isolated atoms. Others, including density and melting point, are characteristic of the bulk solid metal.

▲ **Figure 23.19 Transition elements in the periodic table.** Transition metals are those elements that occupy the *d* block of the periodic table.

TABLE 23.3 ▪ Properties of the First Transition-Series Elements

Group	3B	4B	5B	6B	7B	8B			1B	2B
Element:	Sc	Ti	V	Cr	Mn	Fe	Co	Ni	Cu	Zn
Electron configuration	$3d^14s^2$	$3d^24s^2$	$3d^34s^2$	$3d^54s^1$	$3d^54s^2$	$3d^64s^2$	$3d^74s^2$	$3d^84s^2$	$3d^{10}4s^1$	$3d^{10}4s^2$
First ionization energy (kJ/mol)	631	658	650	653	717	759	758	737	745	906
Radius in metallic substances (Å)	1.64	1.47	1.35	1.29	1.37	1.26	1.25	1.25	1.28	1.37
Density (g/cm³)	3.0	4.5	6.1	7.9	7.2	7.9	8.7	8.9	8.9	7.1
Melting point (°C)	1541	1660	1917	1857	1244	1537	1494	1455	1084	420

The properties of metals vary in similar ways across each series (row). We have already seen evidence for this in the melting and boiling points of metals (Figure 23.14). Another example is shown in Figure 23.20▶ where the atomic radius observed in close packed metallic structures is plotted versus group number*. The variation seen in Figure 23.20 is a result of two competing forces. On the one hand, increasing effective nuclear charge favors a steady decrease in the radius as we proceed from left to right across each transition series. On the other hand, the metallic bonding strength increases until we reach the middle of each transition series and then decreases as we fill antibonding orbitals (Section 23.5). As a general principle a bond shortens as it becomes stronger. ∞(Section 8.8) For groups 3B through 6B these two effects work cooperatively and the result is a rapid decrease in radius. Upon moving past group 6B the two effects counteract each other, slowing the decrease and eventually leading to an increase in radius.

▲ Figure 23.20 **Radii of transition metals.** The variation in the radius of transition metal atoms in close packed metallic substances as a function of position in the periodic table.

GIVE IT SOME THOUGHT

Which of the following elements will have the largest bonding atomic radius: Cr, Mn, Re? Which will have the smallest?

The incomplete screening of the nuclear charge by added electrons produces an interesting and important effect in the third transition-metal series. In general, the radius increases as we move down in a family because of the increasing principal quantum number of the outer-shell electrons. ∞(Section 7.3) Once we move beyond the group 3B elements, however, the second and third transition-series elements have virtually the same radii. In group 5B, for example, tantalum has virtually the same radius as niobium. This effect has its origin in the lanthanide series, the elements with atomic numbers 57 through 70, which occur between Ba and Lu (inside front cover). The filling of $4f$ orbitals through the lanthanide elements causes a steady increase in the effective nuclear charge, producing a contraction in size called the **lanthanide contraction**. This contraction just offsets the increase we would expect as we go from the second to the third series. Thus, the second- and third-series transition metals in each group have about the same radii all the way across the series. Consequently, the second- and third-series metals in a given group have great similarity in their chemical properties. For example, the chemical properties of zirconium and hafnium are remarkably similar. They always occur together in nature, and they are very difficult to separate.

*Note that the radii defined in this way, often referred to as metallic radii, differ somewhat from the covalent radii defined in Section 7.3.

Electron Configurations and Oxidation States

Transition metals owe their location in the periodic table to the filling of the d subshells. Many of the chemical and physical properties of transition metals, result from the unique characteristics of the d orbitals. For a given transition metal atom, the valence $(n–1)d$ orbitals are smaller than the corresponding valence shell ns and np orbitals. In quantum mechanical terms the $(n–1)d$ orbital wave functions drop off more rapidly upon moving away from the nucleus than do the ns and np orbital wave functions. This characteristic feature of the d orbitals limits their interaction with orbitals on neighboring atoms, but not so much that they are insensitive to surrounding atoms. As a result they behave like valence electrons in many instances but more like core electrons in other instances. The details depend upon the location of the element in the periodic table and the environment of the atom.

When transition metals are oxidized, they lose their outer s electrons before they lose electrons from the d subshell. ∞ (Section 7.4) The electron configuration of Fe is $[Ar]3d^64s^2$, for example, whereas that of Fe^{2+} is $[Ar]3d^6$. Formation of Fe^{3+} requires loss of one $3d$ electron, giving $[Ar]3d^5$. Most transition-metal ions contain partially occupied d subshells, which are responsible in large part for several characteristics of transition metals:

1. They often exhibit more than one stable oxidation state.

2. Many of their compounds are colored, as shown in Figure 23.21 ◄. (We will discuss the origin of these colors in Chapter 24.)

3. Transition metals and their compounds exhibit interesting and important magnetic properties.

▲ **Figure 23.21 Salts of transition-metal ions and their solutions.** From left to right: Mn^{2+}, Fe^{2+}, Co^{2+}, Ni^{2+}, Cu^{2+}, and Zn^{2+}.

▲ **Figure 23.22 Nonzero oxidation states of the first transition series.** The larger circles indicate the most common oxidation states.

Figure 23.22 ◄ summarizes the common nonzero oxidation states for the first transition series. The oxidation states shown as large blue circles are those most frequently encountered either in solution or in solid compounds. The ones shown as small green circles are less common. Notice that Sc occurs only in the +3 oxidation state and Zn occurs only in the +2 oxidation state. The other metals, however, exhibit a variety of oxidation states. For example, Mn is commonly found in solution in the +2 (Mn^{2+}) and +7 (MnO_4^-) oxidation states. In the solid state the +4 oxidation state (as in MnO_2) is common. The +3, +5, and +6 oxidation states are less common.

The +2 oxidation state, which commonly occurs for most of these metals, is due to the loss of their two outer $4s$ electrons. This oxidation state is found for all these elements except Sc, where the 3+ ion with an [Ar] configuration is particularly stable.

Oxidation states above +2 are due to successive losses of $3d$ electrons. From Sc through Mn the maximum oxidation state increases from +3 to +7, equaling in each case the total number of $4s$ plus $3d$ electrons in the atom. Thus, manganese has a maximum oxidation state of $2 + 5 = +7$. As we move to the right beyond Mn in the first transition series, the maximum oxidation state decreases. This decrease is due in part to the attraction of d orbital electrons to the nucleus, which increases faster than the attraction of the s orbital electrons to the nucleus as we move left-to-right across the periodic table. Thus, in each period the d electrons become more corelike as the atomic number increases. By the time we get to zinc, it is not possible to remove electrons from the $3d$ orbitals through chemical oxidation. In the second and third transition series the increased size of the 4d and 5d orbitals makes it possible to attain maximum oxidation states as high as +8, which is achieved in RuO_4 and OsO_4. In general, the maximum oxidation states are found only when the metals are combined with the most electronegative elements, especially O, F, and in some cases Cl.

Magnetism

The magnetic properties of transition metals and their compounds are both interesting and important. Measurements of magnetic properties provide information about chemical bonding. In addition, many important uses are made of magnetic properties in modern technology.

An electron possesses a "spin" that gives it a magnetic moment, causing it to behave like a tiny magnet. ∞ (Section 9.8) In a diamagnetic solid, one in which all of the electrons in the solid are paired, the up-spin and down-spin electrons effectively cancel each other. Diamagnetic substances are generally described as nonmagnetic, but when a diamagnetic substance is placed in a magnetic field, the motions of the electrons cause the substance to be very weakly repelled by the magnet.

When an atom or ion possesses one or more unpaired electrons, the substance is said to be *paramagnetic*. ∞ (Section 9.8) In a paramagnetic solid the electrons on adjacent atoms or ions do not influence the unpaired electrons on neighboring atoms or ions of the solid. As a result the magnetic moments on the individual atoms or ions are randomly oriented, as shown in Figure 23.23(a)▶. When placed in a magnetic field, however, the magnetic moments become aligned roughly parallel to one another, producing a net attractive interaction with the magnet. Thus, a paramagnetic substance is drawn into a magnetic field.

When you think of a magnet you probably conjure up an image of a simple iron magnet (Figure 23.24▶). Iron exhibits **ferromagnetism**, a much stronger form of magnetism than paramagnetism. Ferromagnetism arises when the unpaired electrons of the atoms or ions in a solid are influenced by the orientations of the electrons of their neighbors. The most stable (lowest-energy) arrangement results when the spins of electrons on adjacent atoms or ions are aligned in the same direction, as shown in Figure 23.23(b). When a ferromagnetic solid is placed in a magnetic field, the electrons tend to align strongly in a direction parallel to the magnetic field. The attraction to the magnetic field that results may be as much as one million times stronger than that for a simple paramagnetic substance. When the external magnetic field is removed, the interactions between the electrons cause the solid as a whole to maintain a magnetic moment. We then refer to it as a *permanent magnet*. The only transition metal elements that display ferromagnetism are Fe, Co, and Ni. Many alloys also exhibit ferromagnetism, which is in some cases stronger than the ferromagnetism of the pure metals. Particularly powerful magnetism is found in certain intermetallic compounds containing both transition metals and lanthanide metals. Two of the most important examples are $SmCo_5$ and $Nd_2Fe_{14}B$.

Two additional types of magnetism involving ordered arrangements of unpaired electrons are depicted in Figure 23.23. In materials that exhibit **antiferromagnetism** the unpaired electrons on a given atom align so that their spins are oriented in the opposite direction as the spins on neighboring atoms, as shown in Figure 23.23(c). In an antiferromagnetic substance the up-spin and down-spin electrons cancel each other. Examples of antiferromagnetism are found among metals (such as Cr), alloys (such as FeMn), and transition metal oxides (such as Fe_2O_3, $LaFeO_3$ and MnO).

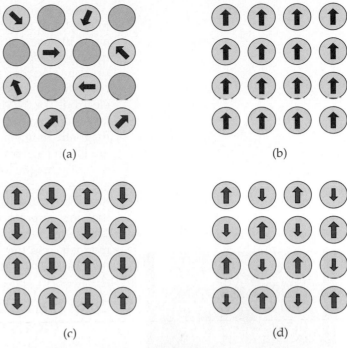

▲ **Figure 23.23 Types of magnetic behavior.** (a) Simple paramagnetic: centers with magnetic moments not aligned unless the substance is in a magnetic field. (b) Ferromagnetic: where the spins of coupled centers align in a common direction. (c) Antiferromagnetic: where the spins of coupled centers align in opposite directions. (d) Ferrimagnetic: where the spins of coupled centers align in opposite directions, but because the two magnetic centers have different numbers of unpaired electrons, the magnetic fields do not fully cancel.

▲ **Figure 23.24 Permanent magnets.** Permanent magnets are made from ferromagnetic and ferrimagnetic materials.

A substance that exhibits **ferrimagnetism** has characteristics of both a ferromagnet and an antiferromagnet. Like an antiferromagnet, the unpaired electrons align themselves so that they are pointing in the opposite direction of their neighbors. However, unlike an antiferromagnet the net magnetic moments of the up-spin electrons are not fully cancelled by the down-spin electrons. This can happen either because the magnetic centers have different numbers of unpaired electrons (i.e. $NiMnO_3$), the number of magnetic sites aligned in one direction is larger than in the other direction (i.e. $Y_3Fe_5O_{12}$), or both (i.e. Fe_3O_4). Because the magnetic moments do not cancel, the bulk properties of ferrimagnetic materials are much like ferromagnetic materials.

All magnetically ordered materials—ferromagnets, ferrimagnets, and antiferromagnets—become paramagnetic when heated above a critical temperature. This happens when the thermal energy is sufficient to overcome the forces that are responsible for orienting the spins with respect to their neighbors. This temperature is called the *Curie temperature*, T_C, for ferromagnets and ferrimagnets, and the *Néel temperature*, T_N, for antiferromagnets. For Fe, Co, and Ni the Curie temperatures are 770 °C, 1115 °C, and 354 °C, respectively.

23.8 CHEMISTRY OF SELECTED TRANSITION METALS

Let's now briefly consider some of the chemistry of three common elements from the first transition series: chromium, iron, and copper. As you read this material, look for the trends that illustrate the generalizations outlined earlier.

Chromium

Chromium dissolves slowly in dilute hydrochloric or sulfuric acid, liberating hydrogen. In the absence of air, the reaction results in the formation of a sky-blue solution of the chromium(II) or chromous ion:

$$Cr(s) + 2\,H^+(aq) \longrightarrow Cr^{2+}(aq) + 2\,H_2(g) \qquad [23.21]$$

In the presence of air, the chromium(II) ion is rapidly oxidized by O_2 to form the chromium(III) ion. The reaction produces the green $[(H_2O)_4Cr(OH)_2Cr(H_2O)_4]^{4+}$ ion (Figure 23.25 ◀). In a strongly acidic solution, this ion reacts slowly with H^+ ions to form the violet $[Cr(H_2O)_6]^{3+}$ ion (Figure 23.25), which is often represented simply as $Cr^{3+}(aq)$. The overall reaction in acid solution is often given simply as shown in Equation 23.22.

$$4\,Cr^{2+}(aq) + O_2(g) + 4\,H^+(aq) \longrightarrow 4\,Cr^{3+}(aq) + 2\,H_2O(l) \qquad [23.22]$$

▲ Figure 23.25 **Two forms of the Cr(III) ion.** The tube on the left contains the violet hydrated chromium(III) ion, $Cr(H_2O)_6{}^{3+}$. The tube on the right contains the green $[(H_2O)_4Cr(OH)_2Cr(H_2O)_4]^{4+}$ ion.

▲ **GIVE IT SOME THOUGHT**

Would you expect the Cr^{3+} ion to contain any unpaired electrons?

Chromium is frequently encountered in aqueous solution in the +6 oxidation state. In basic solution the yellow chromate ion ($CrO_4{}^{2-}$) is the most stable. In acidic solution the dichromate ion ($Cr_2O_7{}^{2-}$) is formed:

$$CrO_4{}^{2-}(aq) + H^+(aq) \rightleftharpoons HCrO_4{}^-(aq) \qquad [23.23]$$

$$2\,HCrO_4{}^-(aq) \rightleftharpoons Cr_2O_7{}^{2-}(aq) + H_2O(l) \qquad [23.24]$$

Equation 23.24 is a condensation reaction in which water is split out from two $HCrO_4{}^-$ ions. Similar reactions occur among the oxyanions of other elements, such as phosphorus. ⟳ (Section 22.8) The equilibrium between the dichromate and chromate ions is readily observable because $CrO_4{}^{2-}$ is bright yellow and $Cr_2O_7{}^{2-}$ is deep orange, as seen in Figure 23.26 ◀. The dichromate ion in acidic solution is a strong oxidizing agent, as evidenced by its large, positive reduction potential. By contrast, the chromate ion in basic solution is not a particularly strong oxidizing agent.

▲ Figure 23.26 **Chromate and dichromate solutions.** Solutions of sodium chromate, Na_2CrO_4 (left), and potassium dichromate, $K_2Cr_2O_7$ (right), illustrate the difference in color of the chromate and dichromate ions.

Iron

We have already discussed the metallurgy of iron in considerable detail in Section 23.2. Here we consider some of its important aqueous solution chemistry. Iron exists in aqueous solution in either the +2 (ferrous) or +3 (ferric) oxidation states. It often appears in natural waters because these waters come into contact with deposits of $FeCO_3$ ($K_{sp} = 3.2 \times 10^{-11}$). Dissolved CO_2 in the water can then help dissolve the mineral:

$$FeCO_3(s) + CO_2(aq) + H_2O(l) \longrightarrow Fe^{2+}(aq) + 2\,HCO_3^-(aq) \quad [23.25]$$

The dissolved Fe^{2+}, together with Ca^{2+} and Mg^{2+}, contributes to water hardness. ∞ (Section 18.6)

The standard reduction potentials in Appendix E reveal much about the kind of chemical behavior we should expect to observe for iron. The potential for reduction from the +2 state to the metal is negative; however, the reduction from the +3 to the +2 state is positive. Iron, therefore, should react with acids that are not highly oxidizing such as dilute sulfuric acid or acetic acid to form $Fe^{2+}(aq)$, as indeed it does. In the presence of air, however, $Fe^{2+}(aq)$ tends to be oxidized to $Fe^{3+}(aq)$, as shown by the positive standard emf for Equation 23.26:

$$4\,Fe^{2+}(aq) + O_2(g) + 4\,H^+(aq) \longrightarrow 4\,Fe^{3+}(aq) + 2\,H_2O(l) \quad E° = +0.46\text{ V}$$
$$[23.26]$$

■ SAMPLE EXERCISE 23.3 | Writing Half-Reactions for an Oxidation-Reduction Reaction

Write the two balanced half-reactions for the reaction in Equation 23.26.

SOLUTION

Analyze: We are asked to write two oxidation-reduction half-reactions that together make up a given oxidation-reduction reaction.

Plan: We need to separate the two half-reactions that make up the given balanced reaction. We can best start by noting that Fe appears in the +2 and +3 states and that oxygen appears in the 0 and −2 oxidation states.

Solve: We note that iron is oxidized from the +2 to +3 oxidation state. The half-reaction for this process is

$$Fe^{2+}(aq) \longrightarrow Fe^{3+}(aq) + e^-$$

Oxygen in acidic medium is reduced:

$$O_2(g) + 4\,H^+(aq) + 4\,e^- \longrightarrow 2\,H_2O(l)$$

To achieve a balanced equation, we need four Fe^{2+} for each O_2. Therefore, all we need to do to obtain Equation 23.26 is to multiply the oxidation half-reaction through by 4.

Comment: As noted in the text, the potential for the oxidation of $Fe^{2+}(aq)$ is negative, but the potential for the reduction of $O_2(g)$ in acidic aqueous solution is sufficiently positive to overcome that, leading to an overall positive potential for the reaction of Equation 23.26.

■ PRACTICE EXERCISE

Calculate the standard potential for the reaction of Equation 23.22.
Answer: 1.64 V

You may have seen instances in which water dripping from a faucet or other outlet has left a brown stain. The brown color is due to insoluble iron(III) oxide, formed by oxidation of iron(II) present in the water:

$$4\,Fe^{2+}(aq) + 8\,HCO_3^-(aq) + O_2(g) \longrightarrow 2\,Fe_2O_3(s) + 8\,CO_2(g) + 4\,H_2O(l)$$
$$[23.27]$$

When iron metal reacts with an oxidizing acid such as warm, dilute nitric acid, $Fe^{3+}(aq)$ is formed directly:

$$Fe(s) + NO_3^-(aq) + 4\,H^+(aq) \longrightarrow Fe^{3+}(aq) + NO(g) + 2\,H_2O(l) \quad [23.28]$$

▲ **Figure 23.27 Precipitation of Fe(OH)₃.** Addition of NaOH solution to an aqueous solution of Fe^{3+} causes $Fe(OH)_3$ to precipitate.

▲ **Figure 23.28 Crystals of copper(II) sulfate pentahydrate, $CuSO_4 \cdot 5\,H_2O$.**

▲ **Figure 23.29 Precipitation of $Cu(OH)_2$.** Addition of NaOH solution to an aqueous solution of Cu^{2+} causes $Cu(OH)_2$ to precipitate.

In the +3 oxidation state iron is soluble in acidic solution as the hydrated ion, $Fe(H_2O)_6^{3+}$. However, this ion hydrolyzes readily ∞ (Section 16.11):

$$Fe(H_2O)_6^{3+}(aq) \rightleftharpoons Fe(H_2O)_5(OH)^{2+}(aq) + H^+(aq) \qquad [23.29]$$

When an acidic solution of iron(III) is made more basic, a gelatinous red-brown precipitate, most accurately described as a hydrous oxide, $Fe_2O_3 \cdot nH_2O$, is formed (Figure 23.27 ◄). In this formulation n represents an indefinite number of water molecules, depending on the precise conditions of the precipitation. Usually, the precipitate that forms is represented merely as $Fe(OH)_3$. The solubility of $Fe(OH)_3$ is very low ($K_{sp} = 4 \times 10^{-38}$). It dissolves in strongly acidic solution but not in basic solution. The fact that it does *not* dissolve in basic solution is the basis of the Bayer process, in which aluminum is separated from impurities, primarily iron(III). ∞ (Section 23.3)

Copper

In its aqueous solution chemistry, copper exhibits two oxidation states: +1 (cuprous) and +2 (cupric). In the +1 oxidation state copper possesses a $3d^{10}$ electron configuration. Salts of Cu^+ are often water insoluble and are mostly white in color. In aqueous solution the Cu^+ ion readily disproportionates:

$$2\,Cu^+(aq) \rightleftharpoons Cu^{2+}(aq) + Cu(s) \qquad K = 1.2 \times 10^6 \qquad [23.30]$$

Because of this reaction and because copper(I) is readily oxidized to copper(II) under most solution conditions, the +2 oxidation state is by far the more common.

Many salts of Cu^{2+}, including $Cu(NO_3)_2$, $CuSO_4$, and $CuCl_2$, are water soluble. Copper sulfate pentahydrate ($CuSO_4 \cdot 5\,H_2O$), a widely used salt, has four water molecules bound to the copper ion and a fifth held to the SO_4^{2-} ion by hydrogen bonding. The salt is blue and is often called *blue vitriol* (Figure 23.28 ◄). Aqueous solutions of Cu^{2+}, in which the copper ion is coordinated by water molecules, are also blue. Among the insoluble compounds of copper(II) is $Cu(OH)_2$, which is formed when NaOH is added to an aqueous Cu^{2+} solution (Figure 23.29 ◄). This blue compound readily loses water on heating to form black copper(II) oxide:

$$Cu(OH)_2(s) \longrightarrow CuO(s) + H_2O(l) \qquad [23.31]$$

CuS is one of the least soluble copper(II) compounds ($K_{sp} = 6.3 \times 10^{-36}$). This black substance does not dissolve in NaOH, NH_3, or nonoxidizing acids such as HCl. It does dissolve in HNO_3, however, which oxidizes the sulfide to sulfur:

$$3\,CuS(s) + 8\,H^+(aq) + 2\,NO_3^-(aq) \longrightarrow$$
$$3\,Cu^{2+}(aq) + 3\,S(s) + 2\,NO(g) + 4\,H_2O(l) \qquad [23.32]$$

$CuSO_4$ is often added to water to stop algae or fungal growth, and other copper preparations are used to spray or dust plants to protect them from lower organisms and insects. Copper compounds are not generally toxic to human beings, except in massive quantities. Our daily diet normally includes from 2 to 5 mg of copper.

■ SAMPLE INTEGRATIVE EXERCISE | Putting Concepts Together

The most important commercial ore of chromium is *chromite* ($FeCr_2O_4$). **(a)** What is the most reasonable assignment of oxidation states to Fe and Cr in this ore? **(b)** Below 74 K the $FeCr_2O_4$ magnetically orders so that the unpaired electrons on chromium point in the opposite direction as those on iron. What type of magnetic state describes $FeCr_2O_4$ below 74 K? Would you expect the magnetism to increase or decrease as $FeCr_2O_4$ is cooled through the 74 K transition? **(c)** Chromite can be reduced in an electric arc furnace (which provides the required heat) using coke (carbon). Write a balanced chemical equation for this reduction, which forms ferrochrome ($FeCr_2$). **(d)** Two of the major forms of chromium in the +6 oxidation state are CrO_4^{2-} and $Cr_2O_7^{2-}$. Draw Lewis structures for these species. (*Hint:* You may find it helpful to consider the Lewis structures of nonmetallic anions of the same formula.) **(e)** Chromium metal is used in alloys (for example, stainless steel) and in electroplating, but chromium is not widely used by itself, in part because it is not ductile at ordinary temperatures. From what we have learned in this chapter about metallic bonding and properties, suggest why chromium is less ductile than most metals.

SOLUTION

(a) Because each oxygen has an oxidation number of -2, the four oxygens represent a total of -8. If the metals have whole-number oxidation numbers, our choices are Fe $= +4$ and Cr $= +2$, or Fe $= +2$ and Cr $= +3$. The latter choice seems the more reasonable because a $+4$ oxidation number for iron is unusual. (Although one alternative would be that Fe is $+3$ and the two Cr have different oxidation states of $+2$ and $+3$, the properties of chromite indicate that the two Cr have the same oxidation number.)

(b) Fe^{2+} and Cr^{3+} have electron configurations of $[Ar]3d^6$ and $[Ar]3d^3$, respectively. Therefore, Fe^{2+} will have four unpaired electrons, and Cr^{3+} will have three unpaired electrons. Because they line up in opposite directions, $FeCr_2O_4$ will either be antiferromagnetic or ferrimagnetic. Because the ions possess different numbers of unpaired electrons as well as the fact that there are twice as many Cr^{3+} ions as there are Fe^{2+} ions, their spins will not cancel each other out, and $FeCr_2O_4$ will be ferrimagnetic below 74 K. The magnetic behavior of a ferrimagnet is similar to that of a ferromagnet, so we would expect to see a large increase in the magnetism below 74 K.

(c) The balanced equation is

$$2\,C(s) + FeCr_2O_4(s) \longrightarrow FeCr_2(s) + 2\,CO_2(g)$$

(d) We expect that in $CrO_4{}^{2-}$, the Cr will be surrounded tetrahedrally by four oxygens. The electron configuration of the Cr atom is $[Ar]3d^54s^1$, giving it six electrons that can be used in bonding, much like the S atom in $SO_4{}^{2-}$. These six electrons must be shared with four O atoms, each of which has six valence-shell electrons. In addition, the ion has a $2-$ charge. Thus, we have a total of $6 + 4(6) + 2 = 32$ valence electrons to place in the Lewis structure. Putting one electron pair in each Cr—O bond and adding unshared electron pairs to the oxygens, we require precisely 32 electrons to achieve an octet around each atom:

In $Cr_2O_7{}^{2-}$ the structure is analogous to that of the diphosphate ion ($P_2O_7{}^{4-}$), which we discussed in Section 22.8. We can think of the $Cr_2O_7{}^{2-}$ ion as formed by a condensation reaction as shown in Equation 23.24.

(e) Recall that chromium, with six electrons available for bonding, has relatively strong metallic bonding among the metals of the transition series, as evidenced by its high melting point (Figure 23.14). This means that distortions of the metallic lattice of the sort that occur when metals are drawn into wires will require more energy than for other metals with weaker metallic bonding.

CHAPTER REVIEW

SUMMARY AND KEY TERMS

Section 23.1 The metallic elements are extracted from the **lithosphere**, the uppermost solid portion of our planet. Metallic elements occur in nature in **minerals**, which are solid inorganic compounds found in various deposits, or **ores**. The desired components of an ore must be separated from the undesired components, called *gangue*. **Metallurgy** is concerned with obtaining metals from these sources and with understanding and modifying the properties of metals.

Section 23.2 **Pyrometallurgy** is the use of heat to bring about chemical reactions that convert an ore from one chemical form to another. In **calcination** an ore is heated to drive off a volatile substance, as in heating a carbonate ore to drive off CO_2. In **roasting**, the ore is heated under conditions that bring about reaction with the furnace atmosphere. For example, sulfide ores might be heated to oxidize sulfur to SO_2. In a **smelting** operation two or more layers of mutually insoluble materials form in the furnace. One layer consists of molten metal, and the other layer (**slag**) is composed of molten silicate minerals and other ionic materials such as phosphates.

Iron, the most extensively used metal in modern society, is obtained from its oxide ores by reduction in a blast furnace. The reducing agent is carbon, in the form of coke. Limestone ($CaCO_3$) is added to react with the silicates present in the crude ore to form slag. The raw iron from the blast furnace, called pig iron, is usually taken directly to a converter, where **refining** takes place to produce various kinds of steel. In the converter the molten iron reacts with pure oxygen to oxidize impurity elements.

Section 23.3 **Hydrometallurgy** is the use of chemical processes occurring in aqueous solution to separate a mineral from its ore or one particular element from others. In **leaching**, an ore is treated with an aqueous reagent to dissolve one component selectively. In the **Bayer process** aluminum is selectively dissolved from bauxite by treatment with concentrated NaOH solution.

Section 23.4 **Electrometallurgy** is the use of electrolytic methods to prepare or purify a metallic element. Sodium is prepared by electrolysis of molten NaCl in a **Downs cell**. Aluminum is obtained in the **Hall-Héroult process** by electrolysis of Al_2O_3 in molten cryolite (Na_3AlF_6). Copper is purified by electrolysis of aqueous copper sulfate solution using anodes composed of impure copper.

Section 23.5 The properties of metals can be accounted for in a qualitative way by the **electron-sea model**, in which the electrons are visualized as being free to move throughout the metal structure. In the molecular-orbital model the valence atomic orbitals of the metal atoms interact to form energy bands that are incompletely filled by valence electrons. The bands that form from the valence shell s, p, and d orbitals overlap so that all metals have partially filled bands. The orbitals that constitute the energy band are delocalized over the atoms of the metal, and their energies are closely spaced. Because the energy differences between orbitals in the band are so small, promoting electrons to higher-energy orbitals requires very little energy. This gives rise to high electrical and thermal conductivity, as well as other characteristic metallic properties.

Section 23.6 **Alloys** are materials that possess characteristic metallic properties and are composed of more than one element. Usually, one or more metallic elements are major components. **Solution alloys** are homogeneous alloys in which the components are distributed uniformly, while in **heterogeneous alloys** the components are not distributed uniformly; instead, two or more distinct phases with characteristic compositions are present. Solution alloys can be either *substitutional alloys*, where atoms of the different metals randomly occupy metal atoms sites in the lattice, or *interstitial alloys*, where smaller often nonmetallic atoms occupy interstitial sites between metal atoms. **Intermetallic compounds** are homogeneous alloys that have definite properties and compositions.

Sections 23.7 and 23.8 Transition metals are characterized by incomplete filling of the d orbitals. The presence of d electrons in transition elements leads to multiple oxidation states. As we proceed through a given series of transition metals, the attraction between the nucleus and the valence electrons increases faster for electrons occupying the d orbitals than for electrons occupying the s orbitals. As a result, the later transition elements in a given row tend to adopt lower oxidation states. Although the atomic and ionic radii increase in the second series as compared to the first, the elements of the second and third series are similar with respect to these and other properties. This similarity is due to the **lanthanide contraction**.

The presence of unpaired electrons in valence orbitals leads to interesting magnetic behavior in transition metals and their compounds. In **ferromagnetic**, **ferrimagnetic**, and **antiferromagnetic** substances the unpaired electron spins on atoms in a solid are affected by those on neighboring atoms. In a ferromagnetic substance the spins all point in the same direction. In an antiferromagnetic substance the spins point in opposite directions and cancel each other. In a ferrimagnetic substance the spins point in opposite directions but do not fully cancel. Ferromagnetic and ferrimagnetic substances are used to make permanent magnets.

The chapter concluded with a concise overview of the chemistry of three of the common transition metals: chromium, iron, and copper.

KEY SKILLS

- Understand the differences between metals, minerals, and ores.
- Be able to write balanced chemical equations to illustrate the kinds of reactions that result when a mineral undergoes calcination or roasting.
- Understand and be able to write balanced chemical equations for the processes that are used to convert iron ore into steel.
- Understand and be able to write balanced chemical equations for the Bayer process whereby aluminum is separated from other metals, most notably iron.
- Describe the electrometallurgical processes used to produce sodium, aluminum, and copper, and write balanced chemical equations for them.
- Explain the relationship between the electronic structure of metals and their physical properties.

- Use the molecular-orbital model to qualitatively predict the trends in melting point, boiling point, and hardness of metals.
- Explain how solution alloys and heterogeneous alloys differ.
- Describe the differences between substitutional alloys, interstitial alloys, and intermetallic compounds.
- Be familiar with the periodic trends in radii and oxidation states of the transition metal ions, including the origin and effect of the lanthanide contraction.
- Understand the differences between paramagnetism, ferromagnetism, antiferromagnetism, and ferrimagnetism.

VISUALIZING CONCEPTS

23.1 This diagram shows the approximate enthalpy change in the roasting of ZnS. Does the roasting reaction cause an increase or a decrease in the temperature of the roasting oven? Explain. In light of your answer, is it necessary to heat the ore to produce roasting? [Section 23.2]

$$2 \, ZnS(s) + 3 \, O_2(g)$$

590 kJ

$$2 \, ZnO(s) + SO_2(g)$$

23.2 Magnesium is produced commercially by electrolysis from a molten salt using a cell similar to the one shown below. **(a)** What salt is used as the electrolyte? **(b)** Which electrode is the anode, and which one is the cathode? **(c)** Write the overall cell reaction and individual half-reactions. **(d)** What precautions would need to be taken with respect to the magnesium formed? [Section 23.4]

23.3 The standard enthalpy of formation of the gaseous element varies among some fourth-row metals as shown in the figure below. Would you expect gaseous molybdenum to have a larger or smaller standard enthalpy of formation than zirconium? What about gaseous cadmium? [Section 23.5]

ΔH_f^o

Rb

Sr

Y

Zr

23.4 Nb_3Sn and $SmCo_5$ are two important intermetallic compounds. Nb_3Sn is a superconductor from which the electromagnets of a magnetic resonance imaging (MRI) instrument are built, while $SmCo_5$ is used to make powerful permanent magnets. The unit cells of Nb_3Sn and $SmCo_5$ are shown below. By counting the atoms in the unit cell, determine the empirical formulas corresponding to each structure. Then use this information to determine which structure corresponds to which intermetallic compound. (*Hint:* In both structures all of the atoms shown are located either at the corners of the unit cell, the body center of the unit cell, or the faces of the unit cell.) [Section 23.6]

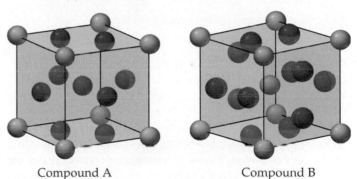

Compound A   Compound B

23.5 This chart shows the variation in an important property of the metals from K through Ge. Is the property atomic radius, electronegativity, or first ionization energy? Explain your choice. [Section 23.7]

K Ca Sc Ti V Cr Mn Fe Co Ni Cu Zn Ga Ge

23.6 **(a)** Except for scandium, chromium is the only element in Figure 23.22 for which the +3 oxidation state is more stable in general than the +2. Explain why the +3 oxidation state is most characteristic of scandium. **(b)** What type of magnetism would you expect from $SrCrO_4$? **(c)** By analogy with inorganic ions of the same formula type, predict the geometrical structure of the permanganate ion, MnO_4^-. [Sections 23.7 and 23.8]

EXERCISES

Metallurgy

23.7 Two of the most heavily utilized metals are aluminum and iron. What are the most important natural sources of these elements? In what oxidation state is each metal found in nature?

23.8 **(a)** Pyrolusite (MnO_2) is a commercially important mineral of manganese. What is the oxidation state of Mn in this metal? **(b)** Name some reagents that might be used to reduce this ore to the metal.

23.9 Explain in your own words what is meant by the statement, "This ore consists of a small concentration of chalcopyrite, together with considerable gangue."

23.10 What is meant by the following terms: **(a)** calcination, **(b)** leaching, **(c)** smelting, **(d)** slag?

23.11 Complete and balance each of the following equations:

(a) $Cr_2O_3(s) + Na(l) \longrightarrow$

(b) $PbCO_3(s) \xrightarrow{\Delta}$

(c) $CdS(s) + O_2(g) \xrightarrow{\Delta}$

(d) $ZnO(s) + CO(g) \xrightarrow{\Delta}$

23.12 Complete and balance each equation:

(a) $PbS(s) + O_2(g) \xrightarrow{\Delta}$

(b) $CoCO_3(s) \xrightarrow{\Delta}$

(c) $WO_3(s) + H_2(g) \xrightarrow{\Delta}$

(d) $VCl_3(g) + K(l) \longrightarrow$

(e) $BaO(s) + P_2O_5(l) \longrightarrow$

23.13 A sample containing $PbSO_4$ is to be refined to Pb metal via calcination, followed by roasting. **(a)** What volatile product would you expect to be produced by calcination? **(b)** Propose an appropriate atmosphere to accompany the roasting. **(c)** Write balanced chemical equations for the two steps.

23.14 Consider the thermodynamics of calcination. **(a)** The following equation is a generic reaction representing calcination of a metal carbonate:

$$MCO_3(s) \longrightarrow MO(s) + CO_2(g)$$

Would you expect this reaction to become more or less spontaneous as the temperature increases? **(b)** What is the standard free energy, $\Delta G°$, for the reaction corresponding to the calcination of $PbCO_3$ (Equation 23.1) at 25 °C under standard conditions? Is this reaction

spontaneous? If not, at what temperature does this reaction become spontaneous (assuming $\Delta H°$ and $S°$ values do not change with temperature)?

23.15 Use the thermodynamic quantities given in Appendix C to calculate $\Delta G°$, $\Delta H°$, and $\Delta S°$ for the reaction corresponding to roasting of PbO in a CO atmosphere (Equation 23.5). Approximate the thermodynamic quantities of Pb(l) using the thermodynamic properties of Pb(s). Is this reaction spontaneous at 25 °C under standard conditions? Is it exothermic or endothermic?

23.16 Assess the feasibility of reducing TiO_2 to titanium metal by roasting in carbon monoxide. **(a)** Write a reaction for this process. **(b)** Use the thermodynamic quantities given in Appendix C to calculate $\Delta G°$, $\Delta H°$, and $\Delta S°$ for this reaction. Is this reaction spontaneous at 25 °C under standard conditions? **(c)** If we assume that $\Delta H°$ and $S°$ values do not change with temperature, at what temperature would this process become spontaneous? Do you think this process would be practical?

23.17 What is the major reducing agent in the reduction of iron ore in a blast furnace? Write a balanced chemical equation for the reduction process.

23.18 Write balanced chemical equations for the reduction of FeO and Fe_2O_3 by H_2 and by CO.

23.19 What role does each of the following materials play in the chemical processes that occur in a blast furnace: **(a)** air, **(b)** limestone, **(c)** coke, **(d)** water? Write balanced chemical equations to illustrate your answers.

23.20 **(a)** In the basic oxygen process for steel formation, what reactions cause the temperature in the converter to increase? **(b)** Write balanced chemical equations for the oxidation of carbon, sulfur, and silicon in the converter.

23.21 Describe how electrometallurgy could be employed to purify crude cobalt metal. Describe the compositions of the electrodes and electrolyte, and write out all electrode reactions.

23.22 The element tin is generally recovered from deposits of the ore cassiterite (SnO_2). The oxide is reduced with carbon, and the crude metal is purified by electrolysis. Write balanced chemical equations for the reduction process and for the electrode reactions in the electrolysis. (Assume that an acidic solution of $SnSO_4$ is employed as an electrolyte in the electrolysis.)

Metals and Alloys

23.23 Sodium is a highly malleable substance, whereas sodium chloride is not. Explain this difference in properties.

23.24 Germanium has the same crystal structure as diamond (Figure 11.41). Based on this fact, do you think germanium is likely to exhibit metallic properties? Explain your answer.

23.25 Explain how the electron-sea model accounts for the high electrical and thermal conductivity of metals.

23.26 **(a)** Compare the electronic structures of atomic chromium and atomic selenium. In what respects are they similar, and in what respects do they differ? **(b)** Chromium is a metal, and selenium is a nonmetal. What factors are important in determining this difference in properties?

23.27 The densities of the elements K, Ca, Sc, and Ti are 0.86, 1.5, 3.2, and 4.5 g/cm^3, respectively. What factors are likely to be of major importance in determining this variation? Which factor do you think will be the most important?

23.28 Explain this trend in melting points: Y 1522 °C, Zr 1852 °C, Nb 2468 °C, Mo 2617 °C.

23.29 Which would you expect to be the more ductile element, (a) Ag or Mo, (b) Zn or Si? In each case explain your reasoning.

23.30 How do you account for the observation that the alkali metals, like sodium and potassium, are soft enough to be cut with a knife?

23.31 Tin exists in two allotropic forms: Gray tin has a diamond structure, and white tin has a close-packed structure. One of these allotropic forms is a semiconductor with a small band gap while the other is a metal. Which one is which? Which form would you expect to have the longer Sn—Sn bond distance?

[23.32] The electrical conductivity of titanium is approximately 2500 times greater than that of silicon. Titanium has a hexagonal close-packed structure, and silicon has the diamond structure. Explain how the structures relate to the relative electrical conductivities of the elements.

23.33 Define the term *alloy*. Distinguish among solution alloys, heterogeneous alloys, and intermetallic compounds.

23.34 Distinguish between substitutional and interstitial alloys. What conditions favor formation of substitutional alloys?

23.35 For each of the following alloy compositions indicate whether you would expect it to be a substitutional alloy, an interstitial alloy, or an intermetallic compound: (a) $Fe_{0.97}Si_{0.03}$, (b) $Fe_{0.60}Ni_{0.40}$, (c) Cu_3Au.

23.36 For each of the following alloy compositions indicate whether you would expect it to be a substitutional alloy, an interstitial alloy, or an intermetallic compound: (a) $Cu_{0.66}Zn_{0.34}$, (b) Ag_3Sn, (c) $Ti_{0.99}O_{0.01}$.

Transition Metals

23.37 Which of the following properties are better considered characteristic of the free isolated atoms, and which are characteristic of the bulk metal: (a) electrical conductivity, (b) first ionization energy, (c) atomic radius, (d) melting point, (e) heat of vaporization, (f) electron affinity?

23.38 Which of the following species would you expect to possess metallic properties: (a) $TiCl_4$, (b) NiCo alloy, (c) W, (d) Ge, (e) Hg_2^{2+}? Explain in each case.

23.39 Zirconium and hafnium are the group 4B elements in the second and third transition series. The radii of these elements are virtually the same (Figure 23.20). Explain this similarity.

23.40 What is meant by the term lanthanide contraction? What properties of the transition elements are affected by the lanthanide contraction?

23.41 Write the formula for the fluoride corresponding to the highest expected oxidation state for (a) Sc, (b) Co, (c) Zn, (d) Mo.

23.42 Write the formula for the oxide corresponding to the highest expected oxidation state for (a) Cd, (b) V, (c) W, (d) Ru.

23.43 Why does chromium exhibit several oxidation states in its compounds, whereas aluminum exhibits only the +3 oxidation state?

23.44 The element vanadium exhibits multiple oxidation states in its compounds, including +2. The compound VCl_2 is known, whereas $ScCl_2$ is unknown. Use electron configurations and effective nuclear charges to account for this difference in behavior.

23.45 Write the expected electron configuration for (a) Cr^{3+}, (b) Au^{3+}, (c) Ru^{2+}, (d) Cu^+, (e) Mn^{4+}, (f) Ir^+.

23.46 What is the expected electron configuration for (a) Ti^{2+}, (b) Co^{3+}, (c) Pd^{2+}, (d) Mo^{3+}, (e) Ru^{3+}, (f) Ni^{4+}?

23.47 Which would you expect to be more easily oxidized, Ti^{2+} or Ni^{2+}?

23.48 Which would you expect to be the stronger reducing agent, Cr^{2+} or Fe^{2+}?

23.49 How does the presence of air affect the relative stabilities of ferrous and ferric ions in solution?

23.50 (a) Give the chemical formulas and colors of the chromate and dichromate ions. (b) Which of these ions is more stable in acidic solution? (c) What type of reaction is involved in their interconversion in solution?

23.51 Write balanced chemical equations for the reaction between iron and (a) hydrochloric acid, (b) nitric acid.

23.52 MnO_2 reacts with aqueous HCl to yield $MnCl_2(aq)$ and chlorine gas. (a) Write a balanced chemical equation for the reaction. (b) Is this an oxidation-reduction reaction? If yes, identify the oxidizing and reducing agents.

23.53 On the atomic level, what distinguishes a paramagnetic material from a diamagnetic one? How does each behave in a magnetic field?

23.54 On the atomic level, what distinguishes an antiferromagnetic material from a diamagnetic one?

23.55 (a) On the atomic level, what distinguishes ferromagnetic, ferrimagnetic, and antiferromagnetic materials from each other? (b) Which one of these types of magnetic materials *cannot* be used to make a permanent magnet?

23.56 The two most important iron oxide minerals are magnetite, Fe_3O_4, and hematite, Fe_2O_3. One is a ferrimagnetic material, while the other is an antiferromagnetic material. (a) Based on the oxidation states of iron, which one is more likely to be ferrimagnetic? (b) Would it be possible to use magnetic fields to separate these minerals?

ADDITIONAL EXERCISES

23.57 Write a chemical equation for the reaction that occurs when PbS is roasted in air. Why might a sulfuric acid plant be located near a plant that roasts sulfide ores?

23.58 Explain why aluminum, magnesium, and sodium metals are obtained by electrolysis instead of by reduction with chemical reducing agents.

23.59 Make a list of the chemical reducing agents employed in the production of metals, as described in this chapter. For each of them, identify a metal that can be formed using that reducing agent.

23.60 Write balanced chemical equations for each of the following verbal descriptions: **(a)** Vanadium oxytrichloride ($VOCl_3$) is formed by the reaction of vanadium(III) chloride with oxygen. **(b)** Niobium(V) oxide is reduced to the metal with hydrogen gas. **(c)** Iron(III) ion in aqueous solution is reduced to iron(II) ion in the presence of zinc dust. **(d)** Niobium(V) chloride reacts with water to yield crystals of niobic acid ($HNbO_3$).

23.61 Write a balanced chemical equation to correspond to each of the following verbal descriptions: **(a)** $NiO(s)$ can be solubilized by leaching with aqueous sulfuric acid. **(b)** After concentration, an ore containing the mineral carrollite ($CuCo_2S_4$) is leached with aqueous sulfuric acid to produce a solution containing copper ions and cobalt ions. **(c)** Titanium dioxide is treated with chlorine in the presence of carbon as a reducing agent to form $TiCl_4$. **(d)** Under oxygen pressure $ZnS(s)$ reacts at 150 °C with aqueous sulfuric acid to form soluble zinc sulfate, with deposition of elemental sulfur.

23.62 The crude copper that is subjected to electrorefining contains tellurium as an impurity. The standard reduction potential between tellurium and its lowest common oxidation state, Te^{4+}, is

$$Te^{4+}(aq) + 4\,e^- \longrightarrow Te(s) \qquad E°_{red} = 0.57\ V$$

Given this information, describe the probable fate of tellurium impurities during electrorefining.

23.63 Why is the +2 oxidation state common among the transition metals? Why do so many transition metals exhibit a variety of oxidation states?

[23.64] Write balanced chemical equations that correspond to the steps in the following brief account of the metallurgy of molybdenum: Molybdenum occurs primarily as the sulfide, MoS_2. On boiling with concentrated nitric acid, a white residue of MoO_3 is obtained. This is an acidic oxide; when it is dissolved in excess hot concentrated ammonia, ammonium molybdate crystallizes on cooling. On heating ammonium molybdate, white MoO_3 is obtained. On further heating to 1200 °C in hydrogen, a gray powder of metallic molybdenum is obtained.

23.65 Antimony and niobium both possess five valence-shell electrons and are metallic conductors. Which element would you expect to be a better conductor of electricity?

[23.66] Introduction of carbon into a metallic lattice generally results in a harder, less ductile substance with lower electrical and thermal conductivities. Explain why this might be so.

23.67 The thermodynamic stabilities of the three complexes $Zn(H_2O)_4^{2+}$, $Zn(NH_3)_4^{2+}$, and $Zn(CN)_4^{2-}$ increase from the H_2O to the NH_3 to the CN^- complex. How do you expect the reduction potentials of these three complexes to compare?

23.68 Indicate whether each of the following compounds is expected to be diamagnetic or paramagnetic, and give a reason for your answer in each case: **(a)** $NbCl_5$, **(b)** $CrCl_2$, **(c)** $CuCl$, **(d)** RuO_4, **(e)** $NiCl_2$.

[23.69] Associated with every ferromagnetic solid is a temperature known as its Curie temperature. When heated above its Curie temperature, the substance no longer exhibits ferromagnetism but rather becomes paramagnetic. Use the kinetic-molecular theory of solids to explain this observation.

23.70 Associated with every antiferromagnetic solid is a temperature known as its Néel temperature. When heated above its Néel temperature the magnetic behavior changes from antiferromagnetic to paramagnetic. In contrast diamagnetic substances do not generally become paramagnetic upon heating. How do you explain this difference in behavior?

23.71 Write balanced chemical equations for each of the following reactions characteristic of elemental manganese: **(a)** It reacts with aqueous HNO_3 to form a solution of manganese(II) nitrate. **(b)** When solid manganese(II) nitrate is heated to 450 K, it decomposes to MnO_2. **(c)** When MnO_2 is heated to 700 K, it decomposes to Mn_3O_4. **(d)** When solid $MnCl_2$ is reacted with $F_2(g)$, it forms MnF_3 (one of the products is ClF_3).

[23.72] Based on the chemistry described in this chapter and others, propose balanced chemical equations for the following sequence of reactions involving nickel: **(a)** The ore millerite, which contains NiS, is roasted in an atmosphere of oxygen to produce an oxide. **(b)** The oxide is reduced to the metal, using coke. **(c)** Dissolving the metal in hydrochloric acid produces a green solution. **(d)** Adding excess sodium hydroxide to the solution causes a gelatinous green material to precipitate. **(e)** Upon heating, the green material loses water and yields a green powder.

[23.73] Indicate whether each of the following solids is likely to be an insulator, a metallic conductor, or a semiconductor: **(a)** TiO_2, **(b)** Ge, **(c)** Cu_3Al, **(d)** Pd, **(e)** SiC, **(f)** Bi.

INTEGRATIVE EXERCISES

23.74 (a) A charge of 3.3×10^6 kg of material containing 27% Cu_2S and 13% FeS is added to a converter and oxidized. What mass of $SO_2(g)$ is formed? **(b)** What is the molar ratio of Cu to Fe in the resulting mixture of oxides?

(c) What are the likely formulas of the oxides formed in the oxidation reactions, assuming an excess of oxygen? **(d)** Write balanced equations representing each of the oxidation reactions.

23.75 Using the concepts discussed in Chapter 13, indicate why the molten metal and slag phases formed in the blast furnace shown in Figure 23.4 are immiscible.

23.76 In an electrolytic process nickel sulfide is oxidized in a two-step reaction:

$$Ni_3S_2(s) \longrightarrow Ni^{2+}(aq) + 2\,NiS(s) + 2\,e^-$$
$$NiS(s) \longrightarrow Ni^{2+}(aq) + S(s) + 2\,e^-$$

What mass of Ni^{2+} is produced in solution by passing a current of 67 A for a period of 11.0 hr, assuming the cell is 90% efficient?

[23.77] **(a)** Using the data in Appendix C, estimate the free-energy change for the following reaction at 1200 °C:

$$Si(s) + 2\,MnO(s) \longrightarrow SiO_2(s) + 2\,Mn(s)$$

(b) What does this value tell you about the feasibility of carrying out this reaction at 1200 °C?

[23.78] **(a)** In the converter employed in steel formation (Figure 23.6), oxygen gas is blown at high temperature directly into a container of molten iron. Iron is converted to rust on exposure to air at room temperature, but the iron is not extensively oxidized in the converter. Explain why this is so. **(b)** The oxygen introduced into the converter reacts with various impurities, particularly with carbon, phosphorus, sulfur, silicon, and impurity metals. What are the products of these reactions, and where do they end up in the process?

23.79 Copper(I) is an uncommon oxidation state in aqueous acidic solution because $Cu^+(aq)$ disproportionates into Cu^{2+} and Cu. Use data from Appendix E to calculate the equilibrium constant for the reaction

$$2\,Cu^+(aq) \rightleftharpoons Cu^{2+}(aq) + Cu(s)$$

23.80 The reduction of metal oxides is often accomplished using carbon monoxide as a reducing agent. Carbon (coke) and carbon dioxide are usually present, leading to the following reaction:

$$C(s) + CO_2(g) \rightleftharpoons 2\,CO(g)$$

Using data from Appendix C, calculate the equilibrium constant for this reaction at 298 K and at 2000 K, assuming that the enthalpies and entropies of formation do not depend upon temperature.

23.81 An important process in the metallurgy of titanium is the reaction between titanium dioxide and chlorine in the presence of carbon, which acts as a reducing agent, leading to the formation of gaseous $TiCl_4$. **(a)** Write a balanced chemical equation for this reaction, and use it with the values listed in Appendix C to calculate the standard enthalpy change of this reaction. Is this reaction exothermic or endothermic? **(b)** Write a reaction for the direct reaction between titanium dioxide and chlorine to form $TiCl_4$ and oxygen. Is this reaction exothermic or endothermic?

23.82 Magnesium is obtained by electrolysis of molten $MgCl_2$. **(a)** Why isn't an aqueous solution of $MgCl_2$ used in the electrolysis? **(b)** Several cells are connected in parallel by very large copper buses that convey current to the cells. Assuming that the cells are 96% efficient in producing the desired products in electrolysis, what mass of Mg is formed by passing a current of 97,000 A for a period of 24 hr?

23.83 Vanadium(V) fluoride is a colorless substance that melts at 19.5 °C and boils at 48.3 °C. Vanadium(III) fluoride, on the other hand, is yellow-green in color and melts at 800 °C. **(a)** Suggest a structure and bonding for VF_5 that accounts for its melting and boiling points. Can you identify a compound of a nonmetallic element that probably has the same structure? **(b)** VF_3 is prepared by the action of HF on heated VCl_3. Write a balanced equation for this reaction. **(c)** While VF_5 is a known compound, the other vanadium(V) halides are unknown. Suggest why these compounds might be unstable. (*Hint:* The reasons might have to do with both size and electronic factors.)

23.84 The galvanizing of iron sheet can be carried out electrolytically using a bath containing a zinc sulfate solution. The sheet is made the cathode, and a graphite anode is used. Calculate the cost of the electricity required to deposit a 0.49-mm layer of zinc on both sides of an iron sheet 2.0 m wide and 80 m long if the current is 30 A, the voltage is 3.5 V, and the energy efficiency of the process is 90%. Assume the cost of electricity is $0.082 per kilowatt hour. The density of zinc is 7.1 g/cm^3.

23.85 As mentioned in the text, Ni_3Al is used in the turbines of aircraft engines because of its strength and low density. Nickel metal has a cubic close-packed structure with a face-centered cubic unit cell, while Ni_3Al has the ordered cubic structure shown in Figure 23.17(b). The length of the cubic unit cell edge is 3.53 Å for nickel and 3.56 Å for Ni_3Al. Use this data to calculate and compare the densities of these two materials.

[23.86] Silver is found as Ag_2S in the ore argentite. **(a)** By using data in Table 17.1 and Appendix D.3, determine the equilibrium constant for the cyanidation of Ag_2S to $Ag(CN)_2^-$. **(b)** Based on your answer to part (a), would you consider cyanidation to be a practical means of leaching silver from argentite ore? **(c)** Silver is also found as AgCl in the mineral cerargyrite. Would it be feasible to use cyanidation as a leaching process for this ore?

[23.87] The heats of atomization, ΔH_{atom}, in kJ/mol, of the first transition series of elements are as follows:

Element	Ca	Sc	Ti	V	Cr	Mn	Fe	Co	Ni	Cu
ΔH_{atom}	178	378	471	515	397	281	415	426	431	338

(a) Write an equation for the process involved in atomization, and describe the electronic and structural changes that occur. **(b)** ΔH_{atom} varies irregularly in the series following V. How can this be accounted for, at least in part, using the electronic configurations of the gaseous atoms? (*Hint:* Recall the discussions of Sections 6.8 and 6.9.)

CHAPTER 24

CHEMISTRY OF COORDINATION COMPOUNDS

THE GRAND CENTRAL STAINED-GLASS SKYLIGHT IN THE CONCERT HALL OF THE PAUAU DE LA MUSICA CATALANA IN BARCELONA, SPAIN. The colors in stained glass are obtained by adding oxides of the transition metals to the molten glass before the glass sheets are formed.

24.1 Metal Complexes
We begin by introducing the concepts of *metal complexes* and *ligands* and by providing a brief history of the development of *coordination chemistry*.

24.2 Ligands with More than One Donor Atom
We examine some of the common geometries exhibited by coordination complexes for different *coordination numbers*.

24.3 Nomenclature of Coordination Chemistry
We introduce the *nomenclature* used to name coordination compounds.

24.4 Isomerism
We will see that coordination compounds exhibit *isomerism*, in which two compounds have the same composition but different structures. The types of isomerism exhibited by coordination compounds are described, including *structural isomers, geometric isomers,* and *optical isomers.*

24.5 Color and Magnetism
We discuss the basic notions of *color* and *magnetism* in coordination compounds.

24.6 Crystal-Field Theory
We explore how *crystal-field theory* allows us to explain some of the interesting spectral and magnetic properties of coordination compounds.

THE COLORS ASSOCIATED WITH CHEMISTRY ARE BOTH BEAUTIFUL AND INFORMATIVE—providing insights into the structure and bonding of matter. Compounds of the transition metals constitute an important group of colored substances. Some of them are used in paint pigments; others produce the colors

in glass and precious gems. For example, the colors in the stained-glass skylight shown in the chapter-opening photograph are due mainly to compounds of the transition metals. Why do these compounds have color, and why do these colors change as the ions or molecules bonded to the metal change? The chemistry we explore in this chapter will help us to answer these questions.

In earlier chapters we have seen that metal ions can function as Lewis acids, forming covalent bonds with a variety of molecules and ions that function as Lewis bases. ∞ (Section 16.11) We have encountered many ions and compounds that result from such interactions. We discussed $[Fe(H_2O)_6]^{3+}$ and $[Ag(NH_3)_2]^+$, for example, in our coverage of equilibria in Sections 16.11 and 17.5. Hemoglobin is an important iron compound responsible for the oxygen-carrying capacity of blood. ∞ (Sections 13.6 and 18.4) In Section 23.3 we saw that hydrometallurgy depends on the formation of species such as $[Au(CN)_2]^-$. In this chapter we will focus on the rich and important chemistry associated with such complex assemblies of metals surrounded by molecules and ions. Metal compounds of this kind are called *coordination compounds.*

24.1 METAL COMPLEXES

Species such as $[Ag(NH_3)_2]^+$ that are assemblies of a central metal ion bonded to a group of surrounding molecules or ions are called **metal complexes**, or merely *complexes*. If the complex carries a net charge, it is generally called a *complex ion*. ∞ (Section 17.5) Compounds that contain complexes are known as **coordination compounds**. Most of the coordination compounds that we will examine contain transition-metal ions, although ions of other metals can also form complexes.

The molecules or ions that bond to the metal ion in a complex are known as **ligands** (from the Latin word *ligare*, meaning "to bind"). There are two NH_3 ligands bonded to Ag^+ in $[Ag(NH_3)_2]^+$. Each ligand functions as a Lewis base, donating a pair of electrons to the metal to form a bond with the metal. ∞ (Section 16.11) Thus, every ligand has at least one unshared pair of valence electrons, as illustrated in the following examples:

$$\ddot{\underset{H}{\overset{}{O}}}-H \qquad \underset{H}{\overset{H}{\underset{|}{\overset{|}{N}}}}-H \qquad :\ddot{\underset{..}{Cl}}:^- \qquad :C\equiv N:^-$$

These examples illustrate that most ligands are either polar molecules or anions. In forming a complex, the ligands are said to *coordinate* to the metal.

The Development of Coordination Chemistry: Werner's Theory

Because compounds of the transition metals exhibit beautiful colors, the chemistry of these elements greatly fascinated chemists even before the periodic table was introduced. During the late 1700s through the 1800s, many coordination compounds were isolated and studied. These compounds showed properties that seemed puzzling in light of the bonding theories at the time. Table 24.1▼, for example, lists a series of compounds that result from the reaction of cobalt(III) chloride with ammonia. These compounds have strikingly different colors. Even the last two listed, which were both formulated as $CoCl_3 \cdot 4\,NH_3$, have different colors.

The modern formulations of these compounds are based on various lines of experimental evidence. For example, all the compounds in Table 24.1 are strong electrolytes (Section 4.1), but they yield different numbers of ions when dissolved in water. Dissolving $CoCl_3 \cdot 6\,NH_3$ in water yields four ions per formula unit (that is, the $[Co(NH_3)_6]^{3+}$ ion and three Cl^- ions), whereas $CoCl_3 \cdot 5\,NH_3$ yields only three ions per formula unit (that is, the $[Co(NH_3)_5Cl]^{2+}$ ion and two Cl^- ions). Furthermore, the reaction of the compounds with excess aqueous silver nitrate leads to the precipitation of variable amounts of $AgCl(s)$; the precipitation of $AgCl(s)$ in this way is often used to test for the number of "free" Cl^- ions in an ionic compound. When $CoCl_3 \cdot 6\,NH_3$ is treated with excess $AgNO_3(aq)$, 3 mol of $AgCl(s)$ are produced per mole of complex, so all three Cl^- ions in the formula can react to form $AgCl(s)$. By contrast, when $CoCl_3 \cdot 5\,NH_3$ is treated with $AgNO_3(aq)$ in an analogous fashion, only 2 mol of $AgCl(s)$ precipitate per mole of complex; one of the Cl^- ions in the compound does not react to form $AgCl(s)$. These results are summarized in Table 24.1.

TABLE 24.1 ■ Properties of Some Ammonia Complexes of Cobalt(III)

Original Formulation	Color	Ions per Formula Unit	"Free" Cl^- Ions per Formula Unit	Modern Formulation
$CoCl_3 \cdot 6\,NH_3$	Orange	4	3	$[Co(NH_3)_6]Cl_3$
$CoCl_3 \cdot 5\,NH_3$	Purple	3	2	$[Co(NH_3)_5Cl]Cl_2$
$CoCl_3 \cdot 4\,NH_3$	Green	2	1	*trans*-$[Co(NH_3)_4Cl_2]Cl$
$CoCl_3 \cdot 4\,NH_3$	Violet	2	1	*cis*-$[Co(NH_3)_4Cl_2]Cl$

In 1893 the Swiss chemist Alfred Werner (1866–1919) proposed a theory that successfully explained the observations in Table 24.1. This theory became the basis for understanding coordination chemistry. Werner proposed that metal ions exhibit both "primary" and "secondary" valences. The primary valence is the oxidation state of the metal, which for the complexes in Table 24.1 is +3. ∞ (Section 4.4) The secondary valence is the number of atoms directly bonded to the metal ion, which is also called the **coordination number**. For these cobalt complexes, Werner deduced a coordination number of 6 with the ligands in an octahedral arrangement (Figure 9.9) around the Co^{3+} ion.

Werner's theory provided a beautiful explanation for the results in Table 24.1. The NH_3 molecules in the complexes are ligands that are bonded to the Co^{3+} ion; if there are fewer than six NH_3 molecules, the remaining ligands are Cl^- ions. The central metal and the ligands bound to it constitute the **coordination sphere** of the complex. In writing the chemical formula for a coordination compound, Werner suggested using square brackets to set off the groups within the coordination sphere from other parts of the compound. He therefore proposed that $CoCl_3 \cdot 6\,NH_3$ and $CoCl_3 \cdot 5\,NH_3$ are better written as $[Co(NH_3)_6]Cl_3$ and $[Co(NH_3)_5Cl]Cl_2$, respectively. He further proposed that the chloride ions that are part of the coordination sphere are bound so tightly that they do not dissociate when the complex is dissolved in water. Thus, dissolving $[Co(NH_3)_5Cl]Cl_2$ in water produces a $[Co(NH_3)_5Cl]^{2+}$ ion and two Cl^- ions; only the two "free" Cl^- ions are able to react with $Ag^+(aq)$ to form $AgCl(s)$.

Werner's ideas also explained why there are two distinctly different forms of $CoCl_3 \cdot 4\,NH_3$. Using Werner's postulates, we formulate the compound as $[Co(NH_3)_4Cl_2]Cl$. As shown in Figure 24.1▶, there are two different ways to arrange the ligands in the $[Co(NH_3)_4Cl_2]^+$ complex, called the *cis* and *trans* forms. In *cis*-$[Co(NH_3)_4Cl_2]^+$ the two chloride ligands occupy adjacent vertices of the octahedral arrangement. In *trans*-$[Co(NH_3)_4Cl_2]^+$ the chlorides are opposite one another. As seen in Table 24.1, the difference in these arrangements causes the complexes to have different colors.

The insight that Werner provided into the bonding in coordination compounds is even more remarkable when we realize that his theory predated Lewis's ideas of covalent bonding by more than 20 years! Because of his tremendous contributions to coordination chemistry, Werner was awarded the 1913 Nobel Prize in Chemistry.

Violet

(a)

Green

(b)

▲ **Figure 24.1 The two forms (isomers) of [Co(NH₃)₄Cl₂]⁺.** In (a) *cis*-[Co(NH₃)₄Cl₂]⁺ the two Cl ligands occupy adjacent vertices of the octahedron, whereas in (b) *trans*-[Co(NH₃)₄Cl₂]⁺ they are opposite one another.

■■ **SAMPLE EXERCISE 24.1** | Identifying the Coordination Sphere of a Complex

Palladium(II) tends to form complexes with a coordination number of 4. One such compound was originally formulated as $PdCl_2 \cdot 3\,NH_3$. **(a)** Suggest the appropriate coordination compound formulation for this compound. **(b)** Suppose an aqueous solution of the compound is treated with excess $AgNO_3(aq)$. How many moles of $AgCl(s)$ are formed per mole of $PdCl_2 \cdot 3\,NH_3$?

SOLUTION

Analyze: We are given the coordination number of Pd(II) and a chemical formula indicating that NH_3 and Cl^- are the potential ligands. We are asked to determine **(a)** what ligands are attached to Pd(II) in the compound and **(b)** how the compound behaves toward $AgNO_3$ in aqueous solution.

Plan: (a) Because of their charge, the Cl^- ions can be either in the coordination sphere, where they are bonded directly to the metal, or outside the coordination sphere, where they are bonded ionically to the complex. Because the NH_3 ligands are neutral, they must be in the coordination sphere. **(b)** The chlorides that are in the coordination sphere will not be precipitated as AgCl.

Solve:

(a) By analogy to the ammonia complexes of cobalt(III), we predict that the three NH_3 groups of $PdCl_2 \cdot 3\,NH_3$ serve as ligands attached to the Pd(II) ion. The fourth ligand around Pd(II) is one of the chloride ions. The second chloride ion is not a ligand; it serves only as an anion in this ionic compound. We conclude that the correct formulation is $[Pd(NH_3)_3Cl]Cl$.

(a)

(b)

▲ **Figure 24.2 Reaction of Fe³⁺(aq) and SCN⁻(aq).** (a) An aqueous solution of NH₄SCN is in the pipette, and an aqueous solution of Fe³⁺ is in the flask. (b) The intensely colored [Fe(H₂O)₅NCS]²⁺ ion forms when the NH₄SCN is added to Fe³⁺(aq).

(b) The chloride ion that serves as a ligand will not be precipitated as AgCl(s) following the reaction with $AgNO_3(aq)$. Thus, only the single "free" Cl^- can react. We therefore expect to produce 1 mol of AgCl(s) per mole of complex. The balanced equation is $[Pd(NH_3)_3Cl]Cl(aq) + AgNO_3(aq) \longrightarrow [Pd(NH_3)_3Cl]NO_3(aq) + AgCl(s)$. This is a metathesis reaction (Section 4.2) in which one of the cations is the $[Pd(NH_3)_3Cl]^+$ complex ion.

■ **PRACTICE EXERCISE**

Predict the number of ions produced per formula unit in an aqueous solution of $CoCl_2 \cdot 6\,H_2O$.
Answer: three (the complex ion, $[Co(H_2O)_6]^{2+}$, and two chloride ions)

The Metal–Ligand Bond

The bond between a ligand and a metal ion is an example of an interaction between a Lewis base and a Lewis acid. ∞ (Section 16.11) Because the ligands have unshared pairs of electrons, they can function as Lewis bases (electron-pair donors). Metal ions (particularly transition-metal ions) have empty valence orbitals, so they can act as Lewis acids (electron-pair acceptors). We can picture the bond between the metal ion and ligand as the result of their sharing a pair of electrons that was initially on the ligand:

$$Ag^+(aq) + 2\ :\!\underset{\underset{\displaystyle H}{|}}{\overset{\overset{\displaystyle H}{|}}{N}}\!\!-\!\!H(aq) \longrightarrow \left[H\!-\!\underset{\underset{\displaystyle H}{|}}{\overset{\overset{\displaystyle H}{|}}{N}}\!:\!Ag\!:\!\underset{\underset{\displaystyle H}{|}}{\overset{\overset{\displaystyle H}{|}}{N}}\!-\!H \right]^+ (aq) \qquad [24.1]$$

The formation of metal–ligand bonds can markedly alter the properties we observe for the metal ion. A metal complex is a distinct chemical species that has physical and chemical properties different from the metal ion and the ligands from which it is formed. Complexes, for example, may have colors that differ dramatically from those of their component metal ions and ligands. Figure 24.2 ◀ shows the color change that occurs when aqueous solutions of SCN^- and Fe^{3+} are mixed, forming $[Fe(H_2O)_5NCS]^{2+}$.

Complex formation can also significantly change other properties of metal ions, such as their ease of oxidation or reduction. The silver ion, Ag^+, for example, is readily reduced in water:

$$Ag^+(aq) + e^- \longrightarrow Ag(s) \qquad E° = +0.799\ V \qquad [24.2]$$

In contrast, the $[Ag(CN)_2]^-$ ion is not so easily reduced because complexation by CN^- ions stabilizes silver in the +1 oxidation state:

$$[Ag(CN)_2]^-(aq) + e^- \longrightarrow Ag(s) + 2\,CN^-(aq) \qquad E° = -0.31\ V \quad [24.3]$$

Hydrated metal ions are actually complex ions in which the ligand is water. Thus, $Fe^{3+}(aq)$ consists largely of $[Fe(H_2O)_6]^{3+}$. ∞ (Section 16.11) Complex ions form in aqueous solutions from reactions in which ligands such as NH_3, SCN^-, and CN^- replace H_2O molecules in the coordination sphere of the metal ion.

▲ **GIVE IT SOME THOUGHT**

Write a balanced chemical equation for the reaction that causes the color change in Figure 24.2.

Charges, Coordination Numbers, and Geometries

The charge of a complex is the sum of the charges on the central metal and on its surrounding ligands. In $[Cu(NH_3)_4]SO_4$ we can deduce the charge on the complex ion if we first recognize that SO_4 represents the sulfate ion and therefore has a 2− charge. Because the compound is neutral, the complex ion must have a 2+ charge, $[Cu(NH_3)_4]^{2+}$. We can then use the charge of the complex ion to deduce the oxidation number of copper. Because the NH_3 ligands are neutral molecules, the oxidation number of copper must be +2.

$$+2 + 4(0) = +2$$
$$[Cu(NH_3)_4]^{2+}$$

■■ **SAMPLE EXERCISE 24.2** | Determining the Oxidation Number of a Metal in a Complex

What is the oxidation number of the central metal in $[Rh(NH_3)_5Cl](NO_3)_2$?

SOLUTION

Analyze: We are given the chemical formula of a coordination compound, and we are asked to determine the oxidation number of its metal atom.

Plan: To determine the oxidation number of the Rh atom, we need to figure out what charges are contributed by the other groups in the substance. The overall charge is zero, so the oxidation number of the metal must balance the charge that is due to the rest of the compound.

Solve: The NO_3 group is the nitrate anion, which has a 1− charge, NO_3^-. The NH_3 ligands are neutral, and the Cl is a coordinated chloride ion, which has a 1− charge, Cl^-. The sum of all the charges must be zero.

$$x + 5(0) + (-1) + 2(-1) = 0$$
$$[Rh(NH_3)_5Cl](NO_3)_2$$

The oxidation number of rhodium, x, must therefore be +3.

■ **PRACTICE EXERCISE**

What is the charge of the complex formed by a platinum(II) metal ion surrounded by two ammonia molecules and two bromide ions?
Answer: zero

■■ **SAMPLE EXERCISE 24.3** | Determining the Formula of a Complex Ion

A complex ion contains a chromium(III) bound to four water molecules and two chloride ions. What is its formula?

SOLUTION

Analyze: We are given a metal, its oxidation number, and the number of ligands of each kind in a complex ion containing the metal, and we are asked to write the chemical formula of the ion.

Plan: We write the metal first, then the ligands. We can use the charges of the metal ion and ligands to determine the charge of the complex ion. The oxidation state of the metal is +3, water is neutral, and chloride has a 1− charge.

Solve:

$$+3 + 4(0) + 2(-1) = +1$$
$$Cr(H_2O)_4Cl_2$$

The charge on the ion is 1+, $[Cr(H_2O)_4Cl_2]^+$.

■ **PRACTICE EXERCISE**

Write the formula for the complex described in the Practice Exercise accompanying Sample Exercise 24.2.
Answer: $[Pt(NH_3)_2Br_2]$

▶ **Figure 24.3 Four-coordinate complexes.** Structures of (a) $[Zn(NH_3)_4]^{2+}$ and (b) $[Pt(NH_3)_4]^{2+}$, illustrating the tetrahedral and square-planar geometries, respectively. These are the two common geometries for complexes in which the metal ion has a coordination number of 4.

(a)   (b)

Recall that the number of atoms directly bonded to the metal atom in a complex is called the *coordination number*. The atom of the ligand bound directly to the metal is called the **donor atom**. Nitrogen, for example, is the donor atom in the $[Ag(NH_3)_2]^+$ complex shown in Equation 24.1. The silver ion in $[Ag(NH_3)_2]^+$ has a coordination number of 2, whereas each cobalt ion in the Co(III) complexes in Table 24.1 has a coordination number of 6.

Some metal ions exhibit constant coordination numbers. The coordination number of chromium(III) and cobalt(III) is invariably 6, for example, and that of platinum(II) is always 4. The coordination numbers of most metal ions vary with the ligand, however. The most common coordination numbers are 4 and 6.

The coordination number of a metal ion is often influenced by the relative sizes of the metal ion and the surrounding ligands. As the ligand gets larger, fewer ligands can coordinate to the metal ion. Thus, iron(III) is able to coordinate to six fluorides in $[FeF_6]^{3-}$ but coordinates to only four chlorides in $[FeCl_4]^-$. Ligands that transfer substantial negative charge to the metal also produce reduced coordination numbers. For example, six neutral ammonia molecules can coordinate to nickel(II), forming $[Ni(NH_3)_6]^{2+}$, but only four negatively charged cyanide ions can coordinate, forming $[Ni(CN)_4]^{2-}$.

Four-coordinate complexes have two common geometries—tetrahedral and square planar—as shown in Figure 24.3▲. The tetrahedral geometry is the more common of the two and is especially common among nontransition metals. The square-planar geometry is characteristic of transition-metal ions with eight *d* electrons in the valence shell, such as platinum(II) and gold(III).

The vast majority of six-coordinate complexes have an octahedral geometry, as shown in Figure 24.4(a)▼. The octahedron is often represented as a planar square with ligands above and below the plane, as in Figure 24.4(b). Recall, however, that all positions on an octahedron are geometrically equivalent. ∞∞(Section 9.2)

▲ **GIVE IT SOME THOUGHT**

What are the geometries most commonly associated with **(a)** coordination number 4, **(b)** coordination number 6?

▶ **Figure 24.4 Six-coordinate complex.** Two representations of an octahedral coordination sphere, the common geometric arrangement for complexes in which the metal ion has a coordination number of 6. In (a) the light blue polyhedron is an octahedron. Representation (b) is easier to draw than (a) and is therefore a more common way to depict an octahedral complex.

(a)   (b)

24.2 LIGANDS WITH MORE THAN ONE DONOR ATOM

The ligands that we have discussed so far, such as NH_3 and Cl^-, are called **monodentate ligands** (from the Latin, meaning "one-toothed"). These ligands possess a single donor atom and are able to occupy only one site in a coordination sphere. Some ligands have two or more donor atoms that can simultaneously coordinate to a metal ion, thereby occupying two or more coordination sites. They are called **polydentate ligands** ("many-toothed"). Because they appear to grasp the metal between two or more donor atoms, polydentate ligands are also known as **chelating agents** (from the Greek word *chele*, "claw"). One such ligand is *ethylenediamine*:

$$CH_2{-}CH_2$$
$$H_2\ddot{N} \qquad \ddot{N}H_2$$

Ethylenediamine, which is abbreviated en, has two nitrogen atoms (shown in color) that have unshared pairs of electrons. These donor atoms are sufficiently far apart that the ligand can wrap around a metal ion with the two nitrogen atoms simultaneously bonding to the metal in adjacent positions. The $[Co(en)_3]^{3+}$ ion, which contains three ethylenediamine ligands in the octahedral coordination sphere of cobalt(III), is shown in Figure 24.5▶. Notice that the ethylenediamine has been written in a shorthand notation as two nitrogen atoms connected by an arc. Ethylenediamine is a **bidentate ligand** ("two-toothed" ligand) because it can occupy two coordination sites. Table 24.2▼ shows several common ligands.

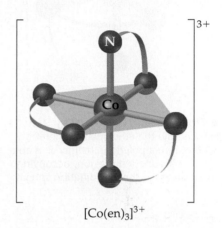

[Co(en)₃]³⁺

▲ **Figure 24.5 The [Co(en)₃]³⁺ ion.** Notice how each bidentate ethylenediamine ligand is able to occupy two positions in the coordination sphere.

TABLE 24.2 ■ Some Common Ligands

Ligand Type	Examples			
Monodentate	$H_2\ddot{O}$: Water	:$\ddot{\underset{..}{F}}$:⁻ Fluoride ion	[:C≡N:]⁻ Cyanide ion	[:$\ddot{\underset{..}{O}}$—H]⁻ Hydroxide ion
	:NH_3 Ammonia	:$\ddot{\underset{..}{Cl}}$:⁻ Chloride ion	[:$\ddot{S}$=C=$\ddot{N}$:]⁻ Thiocyanate ion ⌞—or—⌟	[:$\ddot{O}$—N=$\ddot{O}$:]⁻ Nitrite ion ⌞or⌟

Bidentate

$$H_2C{-}CH_2$$
$$H_2\ddot{N} \qquad \ddot{N}H_2$$
Ethylenediamine (en)

Bipyridine (bipy)

Ortho-phenanthroline (*o*-phen)

Oxalate ion

Carbonate ion

Polydentate

$$H_2C{-}CH_2 \quad CH_2{-}CH_2$$
$$H_2\ddot{N} \qquad \ddot{N}H \qquad \ddot{N}H_2$$
Diethylenetriamine

Triphosphate ion

Ethylenediaminetetraacetate ion (EDTA⁴⁻)

CoEDTA⁻

▲ **Figure 24.6 The [CoEDTA]⁻ ion.** Notice how the ethylenediaminetetraacetate ion, a polydentate ligand, is able to wrap around a metal ion, occupying six positions in the coordination sphere.

The ethylenediaminetetraacetate ion, abbreviated $[EDTA]^{4-}$, is an important polydentate ligand that has six donor atoms:

$[EDTA]^{4-}$

It can wrap around a metal ion using all six of these donor atoms, as shown in Figure 24.6 ◀, although it sometimes binds to a metal using only five of its six donor atoms.

In general, chelating ligands form more stable complexes than do related monodentate ligands. The formation constants for $[Ni(NH_3)_6]^{2+}$ and $[Ni(en)_3]^{2+}$, shown in Equations 24.4 and 24.5, illustrate this observation:

$$[Ni(H_2O)_6]^{2+}(aq) + 6\,NH_3(aq) \rightleftharpoons [Ni(NH_3)_6]^{2+}(aq) + 6\,H_2O(l)$$

$$K_f = 1.2 \times 10^9 \qquad [24.4]$$

$$[Ni(H_2O)_6]^{2+}(aq) + 3\,en(aq) \rightleftharpoons [Ni(en)_3]^{2+}(aq) + 6\,H_2O(l)$$

$$K_f = 6.8 \times 10^{17} \qquad [24.5]$$

Although the donor atom is nitrogen in both instances, $[Ni(en)_3]^{2+}$ has a formation constant that is more than 10^8 times larger than that of $[Ni(NH_3)_6]^{2+}$. The generally larger formation constants for polydentate ligands as compared with the corresponding monodentate ligands is known as the **chelate effect**. We examine the origin of this effect in greater detail in "A Closer Look: Entropy and the Chelate Effect" in this section.

Chelating agents are often used to prevent one or more of the customary reactions of a metal ion without actually removing it from solution. For example, a metal ion that interferes with a chemical analysis can often be complexed and its interference thereby removed. In a sense, the chelating agent hides the metal ion. For this reason, scientists sometimes refer to these ligands as sequestering agents. (The word *sequester* means to remove, set apart, or separate.)

Phosphates such as sodium tripolyphosphate, shown here, are used to complex or sequester metal ions such as Ca^{2+} and Mg^{2+} in hard water so these ions cannot interfere with the action of soap or detergents: ⚬⚬ (Section 18.6)

$$Na_5\left[O-\overset{\overset{O}{\|}}{\underset{\underset{O}{|}}{P}}-O-\overset{\overset{O}{\|}}{\underset{\underset{O}{|}}{P}}-O-\overset{\overset{O}{\|}}{\underset{\underset{O}{|}}{P}}-O \right]$$

▲ **Figure 24.7 Lichens growing on a rock surface.** Lichens obtain the nutrients needed for growth from a variety of sources. Using chelating agents, they are able to extract needed metallic elements from the rocks on which they grow.

Chelating agents such as EDTA are used in consumer products, including many prepared foods such as salad dressings and frozen desserts, to complex trace metal ions that catalyze decomposition reactions. Chelating agents are used in medicine to remove metal ions such as Hg^{2+}, Pb^{2+}, and Cd^{2+}, which are detrimental to health. One method of treating lead poisoning is to administer $Na_2[Ca(EDTA)]$. The EDTA chelates the lead, allowing it to be removed from the body via urine. Chelating agents are also quite common in nature. Mosses and lichens secrete chelating agents to capture metal ions from the rocks they inhabit (Figure 24.7 ◀).

▲ **GIVE IT SOME THOUGHT**

Cobalt(III) has a coordination number of 6 in all of its complexes. Is the carbonate ion acting as a monodentate or as a bidentate ligand in the $[Co(NH_3)_4CO_3]^+$ ion?

Metals and Chelates in Living Systems

Ten of the 29 elements known to be necessary for human life are transition metals. ∞ (Section 2.7, "Chemistry and Life: Elements Required by Living Organisms") These ten elements—V, Cr, Mn, Fe, Co, Ni, Cu, Zn, Mo, and Cd— owe their roles in living systems mainly to their ability to form complexes with a variety of donor groups present in biological systems. Metal ions are integral parts of many enzymes, which are the body's catalysts. ∞ (Section 14.7)

Although our bodies require only small quantities of metals, deficiencies can lead to serious illness. A deficiency of manganese, for example, can lead to convulsive disorders. Some epilepsy patients have been helped by the addition of manganese to their diets.

Among the most important chelating agents in nature are those derived from the *porphine* molecule, which is shown in Figure 24.8▶. This molecule can coordinate to a metal using the four nitrogen atoms as donors. Upon coordination to a metal, the two H atoms shown bonded to nitrogen are displaced. Complexes derived from porphine are called **porphyrins**. Different porphyrins contain different metal ions and have different substituent groups attached to the carbon atoms at the ligand's periphery. Two of the most important porphyrin or porphyrinlike compounds are *heme*, which contains Fe(II), and *chlorophyll*, which contains Mg(II).

▲ **Figure 24.8 The porphine molecule.** This molecule forms a tetradentate ligand with the loss of the two protons bound to nitrogen atoms. Porphine is the basic component of porphyrins, complexes that play a variety of important roles in nature.

◢ *A Closer Look*   ENTROPY AND THE CHELATE EFFECT

When we examined thermodynamics in Chapter 19, we learned that the spontaneity of chemical processes is favored by a positive change in the entropy of the system and by a negative change in its enthalpy. ∞ (Section 19.5) The special stability associated with the formation of chelates, called the *chelate effect*, can be explained by looking at the entropy changes that occur when polydentate ligands bind to a metal ion. To understand this effect better, let's look at some reactions in which two H_2O ligands of the square-planar Cu(II) complex $[Cu(H_2O)_4]^{2+}$ are replaced by other ligands. First, let's consider replacing the H_2O ligands with NH_3 ligands at 27 °C to form $[Cu(H_2O)_2(NH_3)_2]^{2+}$, the structure of which is shown in Figure 24.9(a)▶:

$$[Cu(H_2O)_4]^{2+}(aq) + 2\,NH_3(aq) \rightleftharpoons$$
$$[Cu(H_2O)_2(NH_3)_2]^{2+}(aq) + 2\,H_2O(l)$$

$$\Delta H° = -46\ kJ;\quad \Delta S° = -8.4\ J/K;\quad \Delta G° = -43\ kJ$$

The thermodynamic data provide us with information about the relative abilities of H_2O and NH_3 to serve as ligands in these systems. In general, NH_3 binds more tightly to metal ions than H_2O, so these kinds of substitution reactions are exothermic ($\Delta H < 0$). The stronger bonding of the NH_3 ligands also causes $[Cu(H_2O)_2(NH_3)_2]^{2+}$ to be more rigid, which

(a)   (b)

▲ **Figure 24.9 Ball-and-stick models of two related copper complexes.** The square-planar complexes (a) $[Cu(H_2O)_2(NH_3)_2]^{2+}$ and (b) $[Cu(H_2O)_2(en)]^{2+}$ have the same donor atoms, but (b) has a bidentate ligand.

is probably the reason that the entropy change for the reaction is slightly negative. By using Equation 19.17, we can use the value of $\Delta G°$ to calculate the equilibrium constant of the reaction at 27 °C. The resulting value, $K = 3.1 \times 10^7$, tells us that the equilibrium lies far to the right, favoring substitution of H_2O by NH_3. For this equilibrium, the change in the enthalpy is large and negative enough to overcome the negative change in the entropy.

continues on next page

continued

How does this situation change if instead of two NH_3 ligands we use a single bidentate ethylenediamine (en) ligand and form $[Cu(H_2O)_2(en)]^{2+}$ [Figure 24.9(b)]? The equilibrium reaction and thermodynamic data are

$$[Cu(H_2O)_4]^{2+}(aq) + en(aq) \rightleftharpoons$$
$$[Cu(H_2O)_2(en)]^{2+}(aq) + 2\,H_2O(l)$$

$$\Delta H^\circ = -54\text{ kJ}; \; \Delta S^\circ = +23\text{ J/K}; \; \Delta G^\circ = -61\text{ kJ}$$

The en ligand binds slightly more strongly to a Cu^{2+} ion than two NH_3 ligands, so the enthalpy change on forming $[Cu(H_2O)_2(en)]^{2+}$ is slightly more negative than for $[Cu(H_2O)_2(NH_3)_2]^{2+}$. There is a big difference in the entropy change, however. Whereas the entropy change for forming $[Cu(H_2O)_2(NH_3)_2]^{2+}$ is negative, the entropy change for forming $[Cu(H_2O)_2(en)]^{2+}$ is positive. We can explain this positive value by using the concepts we discussed in Section 19.3. Because a single en ligand occupies two coordination sites, two molecules of H_2O are released upon binding one en ligand. Thus, there are three molecules on the right side of the equation, whereas there are only two on the left side, all of which are part of the same aqueous solution. The greater number of molecules on the right leads to the positive entropy change for the equilibrium. The slightly more negative value of ΔH° coupled with the positive entropy change leads to a much more negative value of ΔG° and a correspondingly larger equilibrium constant: $K = 4.2 \times 10^{10}$.

We can combine the earlier equations to show that the formation of $[Cu(H_2O)_2(en)]^{2+}$ is thermodynamically preferred over the formation of $[Cu(H_2O)_2(NH_3)_2]^{2+}$. If we add the second reaction to the reverse of the first reaction, we obtain

$$[Cu(H_2O)_2(NH_3)_2]^{2+}(aq) + en(aq) \rightleftharpoons$$
$$[Cu(H_2O)_2(en)]^{2+}(aq) + 2\,NH_3(aq)$$

The thermochemical data for this equilibrium reaction can be obtained from those given earlier:

$$\Delta H^\circ = (-54\text{ kJ}) - (-46\text{ kJ}) = -8\text{ kJ}$$
$$\Delta S^\circ = (+23\text{ J/K}) - (-8.4\text{ J/K}) = +31\text{ J/K}$$
$$\Delta G^\circ = (-61\text{ kJ}) - (-43\text{ kJ}) = -18\text{ kJ}$$

Notice that at 27 °C (300 K), the entropic contribution $(-T\Delta S^\circ)$ to the free-energy change is negative and greater in magnitude than the enthalpic contribution (ΔH°). The resulting value of the equilibrium constant, K, for this reaction, 1.4×10^3, shows that the formation of the chelate complex is favored.

The chelate effect is important in biochemistry and molecular biology. The additional thermodynamic stabilization provided by entropic effects helps to stabilize biological metal-chelate complexes, such as porphyrins, and can allow changes in the oxidation state of the metal ion while retaining the structural integrity of the complex.

Related Exercises: 24.21, 24.22, and 24.79

▶ Figure 24.10 Myoglobin. Myoglobin is a protein that stores oxygen in cells. The molecule has a molecular weight of about 18,000 amu and contains one heme unit, shown in orange. The heme unit is bound to the protein through a nitrogen-containing ligand, as shown on the left side of the heme. In the oxygenated form an O_2 molecule is coordinated to the heme group, as shown on the right side of the heme. The continuous purple cylinder represents the three-dimensional structure of the protein chain. The dashed lines denote the helical sections. The protein wraps around to make a pocket for the heme group.

Figure 24.10▼ shows a schematic structure of myoglobin, a protein that contains one heme group. Myoglobin is a *globular protein*, one that folds into a compact, roughly spherical shape. Globular proteins are generally soluble in water and are mobile within cells. Myoglobin is found in the cells of skeletal muscle, particularly in seals, whales, and porpoises. It stores oxygen in cells until it is needed for metabolic activities. Hemoglobin, the protein that transports oxygen in human blood, is made up of four heme-containing subunits, each of which is very similar to myoglobin.

The coordination environment of the iron in myoglobin and hemoglobin is illustrated schematically in Figure 24.11▶. The iron is coordinated to the four nitrogen atoms of the porphyrin and to a nitrogen atom from the protein chain. The sixth position around the iron is occupied either by O_2 (in oxyhemoglobin, the bright red form) or by water (in deoxyhemoglobin, the purplish red form). The oxy form is shown in Figure 24.11. Some substances, such as CO, are poisonous because they bind to iron more strongly than does O_2. ∞ (Section 18.4)

The **chlorophylls**, which are porphyrins that contain Mg(II), are the key components in the conversion of solar energy into forms that can be used by living organisms. This process, called **photosynthesis**, occurs in the leaves of green plants. In photosynthesis, carbon dioxide and water are converted to carbohydrate, with the release of oxygen:

$$6\ CO_2(g) + 6\ H_2O(l) \longrightarrow C_6H_{12}O_6(aq) + 6\ O_2(g) \qquad [24.6]$$

The formation of one mole of glucose, $C_6H_{12}O_6$, requires the absorption of 48 moles of photons from sunlight or other sources of light. Chlorophyll-containing pigments in the leaves of plants absorb the photons. The structure of the most abundant chlorophyll, called chlorophyll *a*, is shown in Figure 24.12▶.

Chlorophylls contain a Mg^{2+} ion bound to four nitrogen atoms arranged around the metal in a planar array. The series of alternating, or *conjugated*, double bonds in the ring surrounding the metal ion is similar to ones found in many organic dyes. This system of conjugated double bonds makes it possible for chlorophyll to absorb light strongly in the visible region of the spectrum, which corresponds to wavelengths ranging from 400–700 nm. Figure 24.13▼ compares the absorption spectrum of chlorophyll to the distribution of visible solar energy at Earth's surface. Chlorophyll is green because it absorbs red light (maximum absorption at 655 nm) and blue light (maximum absorption at 430 nm) and transmits green light. Plant photosynthesis is nature's solar-energy–conversion machine. All living systems on Earth depend on photosynthesis for continued existence (Figure 24.14▼).

GIVE IT SOME THOUGHT

What property of the porphine ligand makes it possible for chlorophyll to play a role in plant photosynthesis?

▲ **Figure 24.11 Coordination sphere of oxymyoglobin and oxyhemoglobin.** The iron is bound to four nitrogen atoms of the porphyrin, to a nitrogen from the surrounding protein, and to an O_2 molecule.

▼ **Figure 24.12 Chlorophyll *a*.** All chlorophyll molecules are essentially alike; they differ only in details of the side chains.

◀ **Figure 24.13 The absorption of light by chlorophyll.** The absorption spectrum of chlorophyll (green curve) is compared with the distribution of solar radiation at ground level (red curve).

▶ **Figure 24.14 Photosynthesis.** The absorption and conversion of solar energy that occurs in leaves provides the energy necessary to drive all the living processes of the plant, including growth.

Although iron is the fourth most abundant element in Earth's crust, living systems have difficulty assimilating enough iron to satisfy their needs. Consequently, iron-deficiency anemia is a common problem in humans. Chlorosis, an iron deficiency in plants that results in yellowing of leaves, is also commonplace. Living systems have difficulty assimilating iron because most iron compounds found in nature are not very soluble in water. Microorganisms have adapted to this problem by secreting an iron-binding compound, called a *siderophore*, that forms an extremely stable water-soluble complex with iron(III). One such complex is called *ferrichrome*; its structure is shown in Figure 24.15▼. The iron-binding strength of a siderophore is so great that it can extract iron from Pyrex® glassware, and it readily solubilizes the iron in iron oxides.

The overall charge of ferrichrome is zero, which makes it possible for the complex to pass through the rather hydrophobic walls of cells. When a dilute solution of ferrichrome is added to a cell suspension, iron is found entirely within the cells in an hour. When ferrichrome enters the cell, the iron is removed through an enzyme-catalyzed reaction that reduces the iron(III) to iron(II). Iron in the lower oxidation state is not strongly complexed by the siderophore. Microorganisms thus acquire iron by excreting a siderophore into their immediate environment and then taking the resulting iron complex into the cell. The overall process is illustrated in Figure 24.16▶.

In humans, iron is assimilated from food in the intestine. A protein called *transferrin* binds iron and transports it across the intestinal wall to distribute it to other tissues in the body. The normal adult carries a total of about 4 g of iron. At any one time, about 3 g, or 75%, of this iron is in the blood, mostly in the form of hemoglobin. Most of the remainder is carried by transferrin.

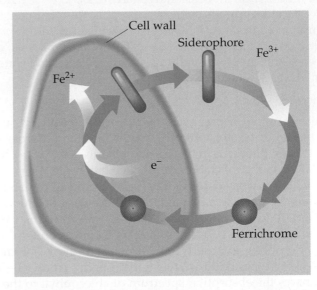

▲ **Figure 24.16 The iron-transport system of a bacterial cell.** The iron-binding ligand, called a siderophore, is synthesized inside the cell and excreted into the surrounding medium. It reacts with Fe^{3+} ion to form ferrichrome, which is then absorbed by the cell. Inside the cell the ferrichrome is reduced, forming Fe^{2+}, which is not tightly bound by the siderophore. Having released the iron for use in the cell, the siderophore may be recycled back into the medium.

A bacterium that infects the blood requires a source of iron if it is to grow and reproduce. The bacterium excretes a siderophore into the blood to compete with transferrin for the iron it holds. The formation constants for iron binding are about the same for transferrin and siderophores. The more iron available to the bacterium, the more rapidly it can reproduce, and, thus, the more harm it can do. Several years ago, New Zealand clinics regularly gave iron supplements to infants soon after birth. However, the incidence of certain bacterial infections was eight times higher in treated than in untreated infants. Presumably, the presence of more iron in the blood than absolutely necessary makes it easier for bacteria to obtain the iron needed for growth and reproduction.

In the United States it is common medical practice to supplement infant formula with iron sometime during the first year of life. However, iron supplements are not necessary for infants who breast-feed, because breast milk contains two specialized proteins, lactoferrin and transferrin, which provide sufficient iron while denying its availability to bacteria. Even for infants fed with infant formulas, supplementing with iron during the first several months of life may be ill advised.

For bacteria to continue to multiply in the bloodstream, they must synthesize new supplies of siderophores. Synthesis of siderophores in bacteria slows, however, as the temperature is increased above the normal body temperature of 37 °C, and it stops completely at 40 °C. This suggests that fever in the presence of an invading microbe is a mechanism used by the body to deprive bacteria of iron.

Related Exercise: 24.56

▲ **Figure 24.15 Ferrichrome.** In this complex six oxygen atoms coordinate an Fe^{3+} ion. The complex is very stable; it has a formation constant of about 10^{30}. The overall charge of the complex is zero.

24.3 NOMENCLATURE OF COORDINATION CHEMISTRY

When complexes were first discovered and few were known, they were named after the chemist who originally prepared them. A few of these names persist; for example, the dark red substance $NH_4[Cr(NH_3)_2(NCS)_4]$ is still known as Reinecke's salt. Once the structures of complexes were more fully understood, it became possible to name them in a more systematic manner. Let's consider two examples:

These examples illustrate how coordination compounds are named. The rules that govern naming of this class of substances are as follows:

1. *In naming salts, the name of the cation is given before the name of the anion.* Thus, in $[Co(NH_3)_5Cl]Cl_2$ we name the $[Co(NH_3)_5Cl]^{2+}$ cation and then Cl^-.

2. *Within a complex ion or molecule, the ligands are named before the metal. Ligands are listed in alphabetical order, regardless of charge on the ligand. Prefixes that give the number of ligands are not considered part of the ligand name in determining alphabetical order.* Thus, in the $[Co(NH_3)_5Cl]^{2+}$ ion we name the ammonia ligands first, then the chloride, then the metal: pentaamminechlorocobalt(III). In writing the formula, however, the metal is listed first.

3. *The names of the anionic ligands end in the letter* o, *whereas neutral ones ordinarily bear the name of the molecules.* Some common ligands and their names are listed in Table 24.3▼. Special names are given to H_2O (aqua), NH_3 (ammine), and CO (carbonyl). For example, $[Fe(CN)_2(NH_3)_2(H_2O)_2]^+$ would be named as diamminediaquadicyanoiron(III) ion.

TABLE 24.3 ■ Some Common Ligands

Ligand	Name in Complexes	Ligand	Name in Complexes
Azide, N_3^-	Azido	Oxalate, $C_2O_4^{2-}$	Oxalato
Bromide, Br^-	Bromo	Oxide, O^{2-}	Oxo
Chloride, Cl^-	Chloro	Ammonia, NH_3	Ammine
Cyanide, CN^-	Cyano	Carbon monoxide, CO	Carbonyl
Fluoride, F^-	Fluoro	Ethylenediamine, en	Ethylenediamine
Hydroxide, OH^-	Hydroxo	Pyridine, C_5H_5N	Pyridine
Carbonate, CO_3^{2-}	Carbonato	Water, H_2O	Aqua

4. *Greek prefixes (di-, tri-, tetra-, penta-, and hexa-) are used to indicate the number of each kind of ligand when more than one is present. If the ligand itself contains a prefix of this kind (for example, ethylenediamine), or is polydentate, alternate prefixes are used (bis-, tris-, tetrakis-, pentakis-, hexakis-) and the ligand name is placed in parentheses.* For example, the name for $[Co(en)_3]Br_3$ is tris(ethylenediamine)-cobalt(III) bromide.

5. *If the complex is an anion, its name ends in -ate.* The compound $K_4[Fe(CN)_6]$ is named potassium hexacyanoferrate(II), for example, and the ion $[CoCl_4]^{2-}$ is named tetrachlorocobaltate(II) ion.

6. *The oxidation number of the metal is given in parentheses in Roman numerals following the name of the metal.*

The following substances and their names demonstrate the application of these rules:

$[Ni(NH_3)_6]Br_2$	Hexaamminenickel(II) bromide
$[Co(en)_2(H_2O)(CN)]Cl_2$	Aquacyanobis(ethylenediamine)cobalt(III) chloride
$Na_2[MoOCl_4]$	Sodium tetrachlorooxomolybdate(IV)

■ SAMPLE EXERCISE 24.4 | Naming Coordination Compounds

Name the following compounds: **(a)** $[Cr(H_2O)_4Cl_2]Cl$, **(b)** $K_4[Ni(CN)_4]$.

SOLUTION

Analyze: We are given the chemical formulas for two coordination compounds, and we are assigned the task of naming them.

Plan: To name the complexes, we need to determine the ligands in the complexes, the names of the ligands, and the oxidation state of the metal ion. We then put the information together following the rules listed previously.

Solve: (a) As ligands, there are four water molecules, which are indicated as tetraaqua, and two chloride ions, indicated as dichloro. The oxidation state of Cr is +3.

$$+3 + 4(0) + 2(-1) + (-1) = 0$$

$$[Cr(H_2O)_4Cl_2]Cl$$

Thus, we have chromium(III). Finally, the anion is chloride. Putting these parts together, the name of the compound is

tetraaquadichlorochromium(III) chloride

(b) The complex has four cyanide ions, CN^-, as ligands, which we indicate as tetracyano. The oxidation state of the nickel is zero.

$$4(+1) + 0 + 4(-1) = 0$$

$$K_4[Ni(CN)_4]$$

Because the complex is an anion, the metal is indicated as nickelate(0). Putting these parts together and naming the cation first, we have

potassium tetracyanonickelate(0)

■ PRACTICE EXERCISE

Name the following compounds: **(a)** $[Mo(NH_3)_3Br_3]NO_3$, **(b)** $(NH_4)_2[CuBr_4]$. **(c)** Write the formula for sodium diaquabis(oxalato)ruthenate(III).

Answers: **(a)** triamminetribromomolybdenum(IV) nitrate, **(b)** ammonium tetrabromocuprate(II) **(c)** $Na[Ru(H_2O)_2(C_2O_4)_2]$

24.4 ISOMERISM

When two or more compounds have the same composition but a different arrangement of atoms, we call them **isomers**. Isomerism—the existence of isomers—is a characteristic feature of both organic and inorganic compounds. Although isomers are composed of the same collection of atoms, they usually differ in one or more physical or chemical properties such as color, solubility, or rate of reaction with some reagent. We will consider two main kinds of

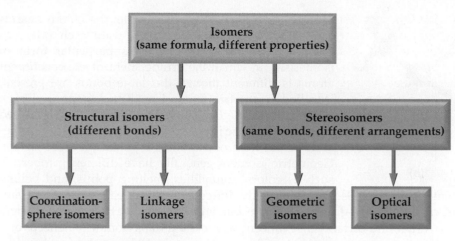

▲ **Figure 24.17 Forms of isomerism in coordination compounds.**

isomers in coordination compounds: **structural isomers** (which have different bonds) and **stereoisomers** (which have the same bonds but different ways in which the ligands occupy the space around the metal center). Each of these classes also has subclasses, as shown in Figure 24.17▲.

Structural Isomerism

Many different types of structural isomerism are known in coordination chemistry. Figure 24.17 gives two examples: linkage isomerism and coordination-sphere isomerism. **Linkage isomerism** is a relatively rare but interesting type that arises when a particular ligand is capable of coordinating to a metal in two different ways. The nitrite ion, NO_2^-, for example, can coordinate through either a nitrogen or an oxygen atom, as shown in Figure 24.18▶. When it coordinates through the nitrogen atom, the NO_2^- ligand is called *nitro*; when it coordinates through the oxygen atom, it is called *nitrito* and is generally written ONO^-. The isomers shown in Figure 24.18 have different properties. The N-bonded isomer is yellow, for example, whereas the O-bonded isomer is red. Another ligand capable of coordinating through either of two donor atoms is thiocyanate, SCN^-, whose potential donor atoms are N and S.

Coordination-sphere isomers differ in the ligands that are directly bonded to the metal, as opposed to being outside the coordination sphere in the solid lattice. For example, there are three compounds whose chemical formula is $CrCl_3(H_2O)_6$: $[Cr(H_2O)_6]Cl_3$ (a violet compound), $[Cr(H_2O)_5Cl]Cl_2 \cdot H_2O$ (a green compound), and $[Cr(H_2O)_4Cl_2]Cl \cdot 2\,H_2O$ (also a green compound). In the two green compounds the water has been displaced from the coordination sphere by chloride ions and occupies a site in the crystal lattice.

Stereoisomerism

Stereoisomerism is the most important form of isomerism. **Stereoisomers** have the same chemical bonds but different spatial arrangements. In the square-planar

Nitro isomer   Nitrito isomer

▲ **Figure 24.18 Linkage isomerism.** Yellow N-bound isomer (left) and red O-bound isomer (right) of $[Co(NH_3)_5NO_2]^{2+}$.

▲ **Figure 24.19 Geometric isomerism.**
(a) Cis and (b) trans geometric isomers
of the square-planar [Pt(NH₃)₂Cl₂].

complex $[Pt(NH_3)_2Cl_2]$, for example, the chloro ligands can be either adjacent to or opposite each other, as illustrated in Figure 24.19 ◄. This particular form of isomerism, in which the arrangement of the constituent atoms is different though the same bonds are present, is called **geometric isomerism**. The isomer in Figure 24.19(a), with like ligands in adjacent positions, is called the cis isomer. The isomer in Figure 24.19(b), with like ligands across from one another, is called the trans isomer. Geometric isomers generally have different properties, such as colors, solubilities, melting points, and boiling points. They may also have markedly different chemical reactivities. For example, *cis*-$[Pt(NH_3)_2Cl_2]$, also called *cisplatin*, is effective in the treatment of testicular, ovarian, and certain other cancers, whereas the trans isomer is ineffective.

Geometric isomerism is possible also in octahedral complexes when two or more different ligands are present. The cis and trans isomers of the tetraamminedichlorocobalt(III) ion were shown in Figure 24.1. As noted in Section 24.1 and Table 24.1, these two isomers have different colors. Their salts also possess different solubilities in water.

Because all the corners of a tetrahedron are adjacent to one another, cis-trans isomerism is not observed in tetrahedral complexes.

■■ SAMPLE EXERCISE 24.5 | Determining the Number of Geometric Isomers

The Lewis structure of the CO molecule indicates that the molecule has a lone pair of electrons on the C atom and one on the O atom ($:C\equiv O:$). When CO binds to a transition-metal atom, it nearly always does so by using the lone pair on the C atom. How many geometric isomers are there for tetracarbonyldichloroiron(II)?

SOLUTION

Analyze: We are given the name of a complex containing only monodentate ligands, and we must determine the number of isomers the complex can form.

Plan: We can count the number of ligands, thereby determining the coordination number of the Fe in the complex and then use the coordination number to predict the geometry of the complex. We can then either make a series of drawings with ligands in different positions to determine the number of isomers, or we can deduce the number of isomers by analogy to cases we have discussed.

Solve: The name indicates that the complex has four carbonyl (CO) ligands and two chloro (Cl^-) ligands, so its formula is $Fe(CO)_4Cl_2$. The complex therefore has a coordination number of 6, and we can assume that it has an octahedral geometry. Like $[Co(NH_3)_4Cl_2]^+$ (Figure 24.1), it has four ligands of one type and two of another. Consequently, it possesses two isomers: one with the Cl^- ligands across the metal from each other (*trans*-$Fe(CO)_4Cl_2$) and one with the Cl^- ligands adjacent (*cis*-$Fe(CO)_4Cl_2$).

In principle, the CO ligand could exhibit linkage isomerism by binding to a metal atom via the lone pair on the O atom. When bonded this way, a CO ligand is called an *isocarbonyl* ligand. Metal isocarbonyl complexes are extremely rare, and we do not normally have to consider the possibility that CO will bind in this way.

Comment: It is easy to overestimate the number of geometric isomers. Sometimes different orientations of a single isomer are incorrectly thought to be different isomers. If two structures can be rotated so that they are equivalent, they are not isomers of each other. The problem of identifying isomers is compounded by the difficulty we often have in visualizing three-dimensional molecules from their two-dimensional representations. It is sometimes easier to determine the number of isomers if we use three-dimensional models.

■■ PRACTICE EXERCISE

How many isomers exist for square-planar $[Pt(NH_3)_2ClBr]$?
Answer: two

A second type of stereoisomerism is known as **optical isomerism**. Optical isomers, called **enantiomers**, are mirror images that cannot be superimposed on each other. They bear the same resemblance to each other that your left hand bears to your right hand. If you look at your left hand in a mirror, the image is

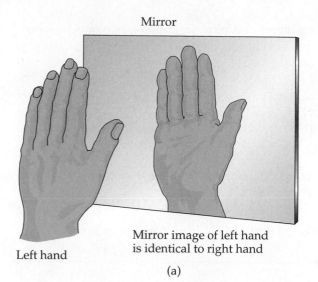

Mirror

Left hand

Mirror image of left hand
is identical to right hand

(a)

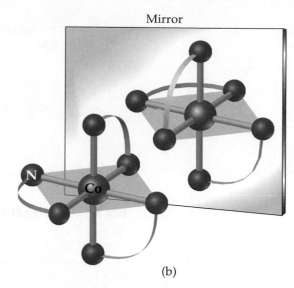

Mirror

N   Co

(b)

▲ **Figure 24.20 Optical isomerism.** Just as our hands are nonsuperimposable mirror images of each other (a), so too are optical isomers such as the two optical isomers of $[Co(en)_3]^{3+}$ (b).

identical to your right hand [Figure 24.20(a)▲]. No matter how hard you try, however, you cannot superimpose your two hands on one another. An example of a complex that exhibits this type of isomerism is the $[Co(en)_3]^{3+}$ ion. Figure 24.20(b) shows the two enantiomers of $[Co(en)_3]^{3+}$ and their mirror-image relationship to each other. Just as there is no way that we can twist or turn our right hand to make it look identical to our left, so also there is no way to rotate one of these enantiomers to make it identical to the other. Molecules or ions that are not superimposable on their mirror image are said to be **chiral** (pronounced KY-rul). Enzymes are among the most important chiral molecules and, as noted in Section 24.2, many enzymes contain complexed metal ions. A molecule need not contain a metal atom to be chiral, however; in Section 25.5, you will see that many organic molecules, including some of those that are important in biochemistry, are chiral.

■■ **SAMPLE EXERCISE 24.6** | Predicting Whether a Complex Has Optical Isomers

Does either *cis-* or *trans-*$[Co(en)_2Cl_2]^+$ have optical isomers?

SOLUTION

Analyze: We are given the chemical formula for two structural isomers, and we are asked to determine whether either one has optical isomers. The en ligand is a bidentate ligand, so the complexes are six-coordinate and octahedral.

Plan: We need to sketch the structures of the cis and trans isomers and their mirror images. We can draw the en ligand as two N atoms connected by a line, as is done in Figure 24.20. If the mirror image cannot be superimposed on the original structure, the complex and its mirror image are optical isomers.

Solve: The trans isomer of $[Co(en)_2Cl_2]^+$ and its mirror image is

where the dashed vertical line represents a mirror. Notice that the mirror image of the trans isomer is identical to the original. Consequently *trans-*$[Co(en)_2Cl_2]^+$ does not exhibit optical isomerism.

The mirror image of the cis isomer of $[Co(en)_2Cl_2]^+$, however, cannot be superimposed on the original:

Thus, the two cis structures are optical isomers (enantiomers): cis-$[Co(en)_2Cl_2]^+$ is a chiral complex.

■ PRACTICE EXERCISE

Does the square-planar complex ion $[Pt(NH_3)(N_3)ClBr]^-$ have optical isomers?
Answer: no

Most of the physical and chemical properties of optical isomers are identical. The properties of two optical isomers differ only if they are in a chiral environment—that is, one in which there is a sense of right- and left-handedness. In the presence of a chiral enzyme, for example, the reaction of one optical isomer might be catalyzed, whereas the other isomer would not react. Consequently, one optical isomer may produce a specific physiological effect within the body, whereas its mirror image produces a different effect or none at all. Chiral reactions are also extremely important in the synthesis of pharmaceuticals and other industrially important chemicals.

Optical isomers are usually distinguished from each other by their interaction with plane-polarized light. If light is polarized—for example, by passage through a sheet of polarizing film—the light waves are oscillating in a single plane, as shown in Figure 24.21 ▼. If the polarized light is passed through a solution containing one optical isomer, the plane of polarization is rotated either to the right (clockwise) or to the left (counterclockwise). The isomer that rotates the plane of polarization to the right is **dextrorotatory**; it is labeled the dextro, or d, isomer (Latin *dexter*, "right"). Its mirror image rotates the plane of polarization to the left; it is **levorotatory** and is labeled the levo, or l, isomer (Latin *laevus*, "left"). The isomer of $[Co(en)_3]^{3+}$ on the left in Figure 24.20(b) is found experimentally to be the l isomer of this ion. Its mirror image is the d isomer. Because of their effect on plane-polarized light, chiral molecules are said to be **optically active**.

▲ **Figure 24.21 Optical activity.** Effect of an optically active solution on the plane of polarization of plane-polarized light. The unpolarized light is passed through a polarizer. The resultant polarized light thereafter passes through a solution containing a dextrorotatory optical isomer. As a result, the plane of polarization of the light is rotated to the right relative to an observer looking toward the light source.

► GIVE IT SOME THOUGHT

> ▲ GIVE IT SOME THOUGHT
>
> What is the similarity and what is the difference between the *d* and *l* isomers of a compound?

When a substance with optical isomers is prepared in the laboratory, the chemical environment during the synthesis is not usually chiral. Consequently, equal amounts of the two isomers are obtained; the mixture is said to be **racemic**. A racemic mixture will not rotate polarized light because the rotatory effects of the two isomers cancel each other.

24.5 COLOR AND MAGNETISM

Studies of the colors and magnetic properties of transition-metal complexes have played an important role in the development of modern models for metal–ligand bonding. We have discussed the various types of magnetic behavior in Section 23.7, and we have discussed the interaction of radiant energy with matter in Section 6.3. Let's briefly examine the significance of these two properties for transition-metal complexes before we try to develop a model for metal–ligand bonding.

Color

In Figure 23.21 we saw the diverse range of colors exhibited by salts of transition-metal ions and their aqueous solutions. ∞∞ (Section 23.7) In these examples water molecules occupy the coordination sphere about the metal. In general, the color of a complex depends on the particular element, its oxidation state, and the ligands bound to the metal. Figure 24.22 ▼ shows how the pale blue color characteristic of $[Cu(H_2O)_4]^{2+}$ changes to a deep blue color as NH_3 ligands replace H_2O ligands to form $[Cu(NH_3)_4]^{2+}$.

For a compound to have color, it must absorb visible light. Visible light consists of electromagnetic radiation with wavelengths ranging from approximately 400 to 700 nm. ∞∞ (Section 6.1) White light contains all wavelengths in this

▲ **Figure 24.22 Effect of ligands on color.** An aqueous solution of $CuSO_4$ is pale blue because of $[Cu(H_2O)_4]^{2+}$ (left). When $NH_3(aq)$ is added (middle and right), the deep blue $[Cu(NH_3)_4]^{2+}$ ion forms.

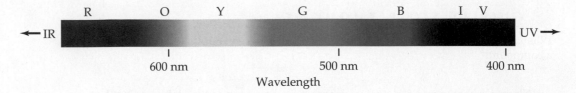

R   O   Y   G   B   I   V

←IR   UV→

600 nm   500 nm   400 nm

Wavelength

▲ **Figure 24.23 Visible spectrum.** The relationship between color and wavelength for visible light. (See also Figure 6.4.)

visible region. It can be dispersed into a spectrum of colors, each of which has a characteristic range of wavelengths, as shown in Figure 24.23▲. The energy of this or any other electromagnetic radiation is inversely proportional to its wavelength: ∞(Section 6.2)

$$E = h\nu = h(c/\lambda) \qquad [24.7]$$

A compound will absorb visible radiation when that radiation possesses the energy needed to move an electron from its lowest energy (ground) state to some excited state. ∞(Section 6.3) Thus, *the particular energies of radiation that a substance absorbs dictate the colors that it exhibits.*

When a sample absorbs visible light, the color we perceive is the sum of the remaining colors that are reflected or transmitted by an object and strike our eyes. An opaque object reflects light, whereas a transparent one transmits it. If an object absorbs all wavelengths of visible light, none reaches our eyes from that object. Consequently, it appears black. If it absorbs no visible light, it is white or colorless. If it absorbs all but orange light, the material appears orange. We also perceive an orange color, however, when visible light of all colors except blue strikes our eyes. Orange and blue are **complementary colors**; the removal of blue from white light makes the light look orange, and vice versa. Thus, an object has a particular color for one of two reasons: (1) It reflects or transmits light of that color; (2) it absorbs light of the complementary color. Complementary colors can be determined using an artist's color wheel, shown in Figure 24.24◄. The wheel shows the colors of the visible spectrum, from red to violet. Complementary colors, such as orange and blue, appear as wedges opposite each other on the wheel.

650 nm O 580 nm

R   Y

750 nm   560 nm
400 nm

V   G

430 nm B 490 nm

▲ **Figure 24.24 Artist's color wheel.** Colors with approximate wavelength ranges are shown as wedges. The colors that are complementary to each other lie opposite each other.

■ SAMPLE EXERCISE 24.7 | **Relating Color Absorbed to Color Observed**

The complex ion *trans*-$[Co(NH_3)_4Cl_2]^+$ absorbs light primarily in the red region of the visible spectrum (the most intense absorption is at 680 nm). What is the color of the complex?

SOLUTION

Analyze: We need to relate the color absorbed by a complex (red) to the color observed for the complex.

Plan: The color observed for a substance is complementary to the color it absorbs. We can use the color wheel of Figure 24.24 to determine the complementary color.

Solve: From Figure 24.24, we see that green is complementary to red, so the complex appears green.

Comment: As noted in Section 24.1, in the text discussing Table 24.1, this green complex was one of those that helped Werner establish his theory of coordination. The other geometric isomer of this complex, *cis*-$[Co(NH_3)_4Cl_2]^+$, absorbs yellow light and therefore appears violet.

■ PRACTICE EXERCISE

The $[Cr(H_2O)_6]^{2+}$ ion has an absorption band at about 630 nm. Which of the following colors—sky blue, yellow, green, or deep red—is most likely to describe this ion?
Answer: sky blue

▲ **Figure 24.25 Experimental determination of an absorption spectrum.** The prism is rotated so that different wavelengths of light pass through the sample. The detector measures the amount of light reaching it, and this information can be displayed as the absorption at each wavelength. The absorbance is a measure of the amount of light absorbed.

The amount of light absorbed by a sample as a function of wavelength is known as its **absorption spectrum**. The visible absorption spectrum of a transparent sample can be determined as shown in Figure 24.25▲. The spectrum of $[Ti(H_2O)_6]^{3+}$, which we will discuss in Section 24.6, is shown in Figure 24.26▶. The absorption maximum of $[Ti(H_2O)_6]^{3+}$ is at about 500 nm, but the band is broad. Because the sample absorbs most strongly in the green and yellow regions of the visible spectrum, it appears red-violet.

Magnetism

Many transition-metal complexes exhibit simple paramagnetism, as described in Sections 9.8 and 23.7. In such compounds the individual metal ions possess some number of unpaired electrons. It is possible to determine the number of unpaired electrons per metal ion from the degree of paramagnetism. The experiments reveal some interesting comparisons. Compounds of the complex ion $[Co(CN)_6]^{3-}$ have no unpaired electrons, for example, but compounds of the $[CoF_6]^{3-}$ ion have four unpaired electrons per metal ion. Both complexes contain Co(III) with a $3d^6$ electron configuration. ∞ (Section 7.4) Clearly, there is a major difference in the ways in which the electrons are arranged in the metal orbitals in these two cases. Any successful bonding theory must explain this difference, and we will present such a theory in Section 24.6.

▲ **Figure 24.26 The color of $[Ti(H_2O)_6]^{3+}$.** (a) A solution containing the $[Ti(H_2O)_6]^{3+}$ ion. (b) The visible absorption spectrum of the $[Ti(H_2O)_6]^{3+}$ ion.

▲ **GIVE IT SOME THOUGHT**

What is the electron configuration for (a) the Co atom and (b) the Co^{3+} ion? How many unpaired electrons does each possess? (See Section 7.4 to review electron configurations of ions.)

24.6 CRYSTAL-FIELD THEORY

Scientists have long recognized that many of the magnetic properties and colors of transition-metal complexes are related to the presence of d electrons in metal orbitals. In this section we will consider a model for bonding in transition-metal complexes, called the **crystal-field theory**, that accounts for many of the observed properties of these substances.*

The ability of a metal ion to attract ligands such as water around itself is a Lewis acid–base interaction. ∞ (Section 16.11) The base—that is, the ligand— donates a pair of electrons into a suitable empty orbital on the metal, as shown

---

*The name crystal field *arose because the theory was first developed to explain the properties of solid crystalline materials, such as ruby. The same theoretical model applies to complexes in solution.*

Bond dipole

Empty hybrid
orbital on metal

Filled sp^3
hybrid orbital
on ammonia

Resulting molecular orbital
of σ bond between metal
and ammonia

▲ **Figure 24.27 Metal–ligand bond formation.** The ligand, which acts as a Lewis base, donates charge to the metal via a metal hybrid orbital. The bond that results is strongly polar, with some covalent character. It is often sufficient to assume that the metal–ligand interaction is entirely electrostatic in character, as is done in the crystal-field model.

in Figure 24.27 ◄. Much of the attractive interaction between the metal ion and the surrounding ligands is due, however, to the electrostatic forces between the positive charge on the metal and negative charges on the ligands. If the ligand is ionic, as in the case of Cl^- or SCN^-, the electrostatic interaction occurs between the positive charge on the metal center and the negative charge on the ligand. When the ligand is neutral, as in the case of H_2O or NH_3, the negative ends of these polar molecules, which contain an unshared electron pair, are directed toward the metal. In this case the attractive interaction is of the ion–dipole type. ∞ (Section 11.2) In either case the result is the same: The ligands are attracted strongly toward the metal center. Because of the electrostatic attraction between the positive metal ion and the electrons of the ligands, the complex formed by the metal ion and the ligands has a lower energy than the fully separated metal and ligands.

Although the positive metal ion is attracted to the electrons in the ligands, the d electrons on the metal ion feel a repulsion from the ligands (negative charges repel one another). Let's examine this effect more closely, and particularly for the case in which the ligands form an octahedral array around the metal ion. For the crystal-field model, we consider the ligands to be negative points of charge that repel the electrons in the d orbitals. Figure 24.28 ◄ shows the effects of these point charges on the energies of the d orbitals in two steps. In the first step the *average* energy of the d orbitals is raised by the presence of the point charges. Hence, all five d orbitals are raised in energy by the same amount. In the second step we consider what happens to the energies of the individual d orbitals when the ligands sit in an octahedral arrangement.

In a six-coordinate octahedral complex we can envision the ligands approaching along the x-, y-, and z-axes, as shown in Figure 24.29(a)►; this arrangement is called an *octahedral crystal field*. Because the d orbitals on the metal ion have different shapes, they do not all have the same energy under the influence of the crystal field. To see why, we must consider the shapes of the d orbitals and how their lobes are oriented relative to the ligands.

Figure 24.29(b–f) shows the five d orbitals in an octahedral crystal field. Notice that the d_{z^2} and $d_{x^2-y^2}$ orbitals have lobes directed *along* the x-, y-, and z-axes pointing *toward* the point charges, whereas the d_{xy}, d_{xz}, and d_{yz} orbitals have lobes that are directed *between* the axes along which the charges approach. The high symmetry of the octahedral crystal field dictates that the d_{z^2} and $d_{x^2-y^2}$ orbitals experience the same amount of repulsion from the crystal field. Those two orbitals therefore have the same energy in the presence of the crystal field. Likewise, the d_{xy}, d_{xz}, and d_{yz} orbitals experience exactly the same repulsion, so those three orbitals remain together in energy. Because their lobes point directly at the negative ligand charges, electrons in the d_{z^2} and $d_{x^2-y^2}$ orbitals experience stronger repulsions than those in the d_{xy}, d_{xz}, d_{yz} orbitals. As a result, an energy separation, or splitting, occurs between the three lower-energy d orbitals (called the t_2 set of orbitals) and the two higher-energy ones (called the e set),* as shown on the right side of Figure 24.28. The energy gap between the two sets of d orbitals is labeled Δ, a quantity that is often called the *crystal-field splitting energy*.

▼ **Figure 24.28 Energies of d orbitals in an octahedral crystal field.** On the left the energies of the d orbitals of a free ion are shown. When negative charges are brought up to the ion, the average energy of the d orbitals increases (center). On the right the splitting of the d orbitals that is due to the octahedral field is shown. Because the repulsion felt by the d_{z^2} and $d_{x^2-y^2}$ orbitals is greater than that felt by the d_{xy}, d_{xz}, and d_{yz} orbitals, the five d orbitals split into a lower-energy set of three (the t_2 set) and a higher-energy set of two (the e set).

e set (d_{z^2}, $d_{x^2-y^2}$)

Δ

t_2 set (d_{xy}, d_{xz}, d_{yz})

Averaged repulsions
between ligands and
d-electrons

Energy

Free metal ion

Metal ion plus ligands
(negative point charges)

Splitting because of
octahedral crystal field

*The labels t_2 for the d_{xy}, d_{xz}, and d_{yz} orbitals and e for the d_{z^2} and $d_{x^2-y^2}$ orbitals come from the application of a branch of mathematics called group theory to crystal-field theory. Group theory can be used to analyze the effects of symmetry on molecular properties.

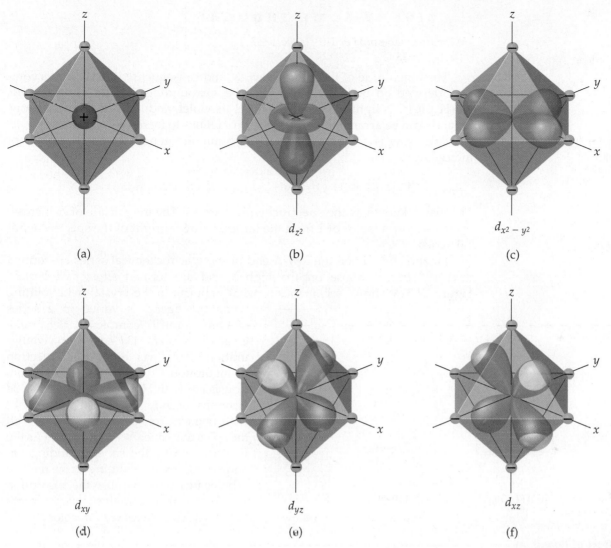

(a)

(b) d_{z^2}

(c) $d_{x^2-y^2}$

(d) d_{xy}

(e) d_{yz}

(f) d_{xz}

▲ **Figure 24.29 The five d orbitals in an octahedral crystal field.** (a) An octahedral array of negative charges approaching a metal ion. (b–f) The orientations of the d orbitals relative to the negative charges. Notice that the lobes of the d_{z^2} and $d_{x^2-y^2}$ orbitals (b and c) point toward the charges, whereas the lobes of the d_{xy}, d_{yz}, and d_{xz} orbitals (d–f) point between the charges.

GIVE IT SOME THOUGHT

Which *d* orbitals have lobes that point directly toward the negative charges in an octahedral crystal field?

The crystal-field model helps us account for the observed colors in transition-metal complexes. The energy gap between the *d* orbitals, Δ, is of the same order of magnitude as the energy of a photon of visible light. It is therefore possible for a transition-metal complex to absorb visible light, which excites an electron from the lower-energy *d* orbitals into the higher-energy ones. In the $[Ti(H_2O)_6]^{3+}$ ion, for example, the Ti(III) ion has an [Ar]$3d^1$ electron configuration (recall that when determining the electron configurations of transition-metal ions, we remove the *s* electrons first). ∞ (Section 7.4) Ti(III) is thus called a "d^1 ion." In the ground state of $[Ti(H_2O)_6]^{3+}$ the single 3*d* electron resides in one of the three lower-energy orbitals in the t_2 set. Absorption of light with a wavelength of 495 nm (242 kJ/mol) excites the 3*d* electron from the lower t_2 set to the upper *e* set of orbitals, as shown in Figure 24.30▶, generating the absorption spectrum shown in Figure 24.26. Because this transition involves exciting an electron from one set of *d* orbitals to the other, we call it a **d-d transition**. As noted earlier, the absorption of visible radiation that produces this *d-d* transition causes the $[Ti(H_2O)_6]^{3+}$ ion to appear red-violet.

▲ **Figure 24.30 Electronic transition accompanying absorption of light.** The 3*d* electron of $[Ti(H_2O)_6]^{3+}$ is excited from the lower-energy *d* orbitals to the higher-energy ones when irradiated with light of 495-nm wavelength.

Why are compounds of Ti(IV) colorless?

The magnitude of the energy gap, Δ, and consequently the color of a complex depend on both the metal and the surrounding ligands. For example, $[Fe(H_2O)_6]^{3+}$ is light violet, $[Cr(H_2O)_6]^{3+}$ is violet, and $[Cr(NH_3)_6]^{3+}$ is yellow. Ligands can be arranged in order of their abilities to increase the energy gap, Δ. The following is an abbreviated list of common ligands arranged in order of increasing Δ:

$$\xrightarrow{\quad \text{Increasing } \Delta \quad}$$
$$Cl^- < F^- < H_2O < NH_3 < en < NO_2^- \text{(N-bonded)} < CN^-$$

This list is known as the **spectrochemical series**. The magnitude of Δ increases by roughly a factor of 2 from the far left to the far right of the spectrochemical series.

Ligands that lie on the low-Δ end of the spectrochemical series are termed *weak-field ligands*; those on the high-Δ end are termed *strong-field ligands*. Figure 24.31 ▼ shows schematically what happens to the crystal-field splitting when the ligand is varied in a series of chromium(III) complexes. Because a Cr atom has an $[Ar]3d^54s^1$ electron configuration, Cr^{3+} has an $[Ar]3d^3$ electron configuration; Cr(III), therefore, is a d^3 ion. Consistent with Hund's rule, the three $3d$ electrons occupy the t_2 set of orbitals, with one electron in each orbital and all the spins the same. ∞ (Section 6.8) As the field exerted by the six surrounding ligands increases, the splitting of the metal d orbitals increases. Because the absorption spectrum is related to this energy separation, these complexes vary in color.

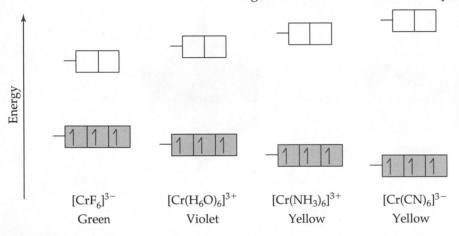

| $[CrF_6]^{3-}$ | $[Cr(H_6O)_6]^{3+}$ | $[Cr(NH_3)_6]^{3+}$ | $[Cr(CN)_6]^{3-}$ |
| Green | Violet | Yellow | Yellow |

▲ **Figure 24.31 Effect of ligand on crystal-field splitting.** This series of octahedral chromium(III) complexes illustrates how Δ, the energy gap between the t_2 and e orbitals, increases as the field strength of the ligand increases.

■■ **SAMPLE EXERCISE 24.8** | Using the Spectrochemical Series

Which of the following complexes of Ti^{3+} exhibits the shortest wavelength absorption in the visible spectrum: $[Ti(H_2O)_6]^{3+}$, $[Ti(en)_3]^{3+}$, or $[TiCl_6]^{3-}$?

SOLUTION

Analyze: We are given three octahedral complexes, each containing Ti in the +3 oxidation state. We need to predict which complex absorbs the shortest wavelength of visible light.

Plan: Ti(III) is a d^1 ion, so we anticipate that the absorption is due to a d-d transition in which the $3d$ electron is excited from the lower-energy t_2 set to the higher-energy e set. The wavelength of the light absorbed is determined by the magnitude of the energy difference, Δ. Thus, we use the position of the ligands in the spectrochemical series to predict the relative values of Δ. The larger the energy, the shorter the wavelength (Equation 24.7).

Solve: Of the three ligands involved—H_2O, en, and Cl^-—we see that ethylenediamine (en) is highest in the spectrochemical series and will therefore cause the largest splitting, Δ, of the t_2 and e sets of orbitals. The larger the splitting, the shorter the wavelength of the light absorbed. Thus, the complex that absorbs the shortest-wavelength light is $[Ti(en)_3]^{3+}$.

■■ **PRACTICE EXERCISE**

The absorption spectrum of $[Ti(NCS)_6]^{3-}$ shows a band that lies intermediate in wavelength between those for $[TiCl_6]^{3-}$ and $[TiF_6]^{3-}$. What can we conclude about the place of NCS^- in the spectrochemical series?
Answer: It lies between Cl^- and F^-; that is, $Cl^- < NCS^- < F^-$.

◄ Figure 24.32 Electron
configurations in octahedral
complexes. Orbital occupancies are
shown for complexes with one, two,
and three d electrons in an octahedral
crystal field.

Electron Configurations in Octahedral Complexes

The crystal-field model also helps us understand the magnetic properties and some important chemical properties of transition-metal ions. From Hund's rule, we expect that electrons will always occupy the lowest-energy vacant orbitals first and that they will occupy a set of degenerate orbitals one at a time with their spins parallel. ∞ (Section 6.8) Thus, if we have a d^1, d^2, or d^3 octahedral complex, the electrons will go into the lower-energy t_2 set of orbitals, with their spins parallel, as shown in Figure 24.32▲. When a fourth electron must be added, a problem arises. If the electron is added to a lower-energy t_2 orbital, an energy gain of magnitude Δ is realized, as compared with placing the electron in a higher-energy e orbital. There is a penalty for doing this, however, because the electron must now be paired up with the electron already occupying the orbital. The energy required to do this, relative to putting it in another orbital with parallel spin, is called the **spin-pairing energy**. The spin-pairing energy arises from the greater electrostatic repulsion of two electrons that share an orbital as compared with two that are in different orbitals with the same electron spin.

The ligands that surround the metal ion and the charge on the metal ion often play major roles in determining which of the two electronic arrangements arises. In both the $[CoF_6]^{3-}$ and $[Co(CN)_6]^{3-}$ ions, the ligands have a 1− charge. The F^- ion, however, is on the low end of the spectrochemical series so it is a weak-field ligand. The CN^- ion is on the high end of the spectrochemical series, so it is a strong-field ligand. It produces a larger energy gap than the F^- ion. The splittings of the d-orbital energies in these two complexes are compared in Figure 24.33►.

Cobalt(III) has an [Ar]$3d^6$ electron configuration, so these are both d^6 complexes. Let's imagine that we add these six electrons one at a time to the d orbitals of the CoF_6^{3-} ion. The first three will go into the lower-energy t_2 orbitals with their spins parallel. The fourth electron could go into one of the t_2 orbitals, pairing up with one of those already present. Doing so would result in an energy gain of Δ as compared with putting the electron in one of the higher-energy e orbitals. However, it would cost energy in an amount equal to the spin-pairing energy. Because F^- is a weak-field ligand, Δ is small, and the more stable arrangement is the one in which the electron is placed in one of the e orbitals. Similarly, the fifth electron we add goes into the other e orbital. With all of the orbitals containing at least one electron, the sixth must be paired up, and it goes into a lower-energy t_2 orbital; we end up with four of the electrons in the t_2 set of orbitals and two electrons in the e set. In the case of the $[Co(CN)_6]^{3-}$ complex, the crystal-field splitting is much larger. The spin-pairing energy is smaller than Δ, so all six electrons are paired in the t_2 orbitals, as illustrated in Figure 24.33.

The $[CoF_6]^{3-}$ complex is a **high-spin complex**; that is, the electrons are arranged so that they remain unpaired as much as possible. The $[Co(CN)_6]^{3-}$ ion, on the other hand, is a **low-spin complex**; that is, the electrons are arranged so that they remain paired as much as possible. These two different electronic arrangements can be readily distinguished by measuring the magnetic properties of the complex, as described earlier. The absorption spectrum also shows characteristic features that indicate the electronic arrangement.

Metal ions of the second and third transition series (those with $4d$ and $5d$ electrons) have larger d-orbitals than those from the first series ($3d$ electrons). Thus, these ions interact more strongly with ligands, resulting in a larger crystal-field splitting. Consequently, metal ions of the second and third transition series are invariably low spin in an octahedral crystal field.

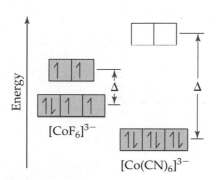

▲ Figure 24.33 High-spin and low-spin complexes. Populations of d orbitals are compared for the high-spin $[CoF_6]^{3-}$ ion (small Δ) and the low-spin $[Co(CN)_6]^{3-}$ ion (large Δ). Both complexes contain cobalt(III), which has six $3d$ electrons, $3d^6$.

▲ **Figure 24.34 Energies of the *d* orbitals in a tetrahedral crystal field.**

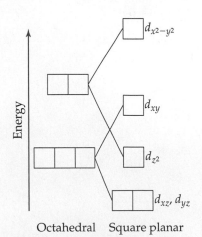

▲ **Figure 24.35 Energies of the *d* orbitals in a square-planar crystal field.** The boxes on the left show the energies of the *d* orbitals in an octahedral crystal field. As the two negative charges along the *z*-axis are removed, the energies of the *d* orbitals change, as indicated by the lines between the boxes. When the charges are completely removed, the square-planar geometry results, giving the splitting of *d*-orbital energies represented by the boxes on the right.

■ **SAMPLE EXERCISE 24.9** | Predicting the Number of Unpaired Electrons in an Octahedral Complex

Predict the number of unpaired electrons in six-coordinate high-spin and low-spin complexes of Fe^{3+}.

SOLUTION

Analyze: We must determine how many unpaired electrons there are in the high-spin and low-spin complexes of the metal ion Fe^{3+}.

Plan: We need to consider how the electrons populate *d* orbitals for Fe^{3+} when the metal is in an octahedral complex. There are two possibilities: one giving a high-spin complex and the other giving a low-spin complex. The electron configuration of Fe^{3+} gives us the number of *d* electrons. We then determine how these electrons populate the t_2 set and *e* set of *d* orbitals. In the high-spin case, the energy difference between the t_2 and *e* orbitals is small, and the complex has the maximum number of unpaired electrons. In the low-spin case, the energy difference between the t_2 and *e* orbitals is large, causing the t_2 orbitals to be filled before any electrons occupy the *e* orbitals.

Solve: Fe^{3+} is a d^5 ion. In a high-spin complex, all five of these electrons are unpaired, with three in the t_2 orbitals and two in the *e* orbitals. In a low-spin complex, all five electrons reside in the t_2 set of *d* orbitals, so there is one unpaired electron:

High spin   Low spin

■ **PRACTICE EXERCISE**

For which *d* electron configurations in octahedral complexes is it possible to distinguish between high-spin and low-spin arrangements?
Answer: d^4, d^5, d^6, d^7

Tetrahedral and Square-Planar Complexes

Thus far we have considered the crystal-field model only for complexes having an octahedral geometry. When there are only four ligands about the metal, the geometry is generally tetrahedral, except for the special case of metal ions with a d^8 electron configuration, which we will discuss in a moment. The crystal-field splitting of the metal *d* orbitals in tetrahedral complexes differs from that in octahedral complexes. Four equivalent ligands can interact with a central metal ion most effectively by approaching along the vertices of a tetrahedron. It turns out—and this is not easy to explain in just a few sentences—that the splitting of the metal *d* orbitals in a tetrahedral crystal is just the opposite of that for the octahedral case. That is, the three metal *d* orbitals in the t_2 set are raised in energy, and the two orbitals in the *e* set are lowered, as illustrated in Figure 24.34 ◄. Because there are only four ligands instead of six, as in the octahedral case, the crystal-field splitting is much smaller for tetrahedral complexes. Calculations show that for the same metal ion and ligand set, the crystal-field splitting for a tetrahedral complex is only four-ninths as large as for the octahedral complex. For this reason, all tetrahedral complexes are high spin; the crystal field is never large enough to overcome the spin-pairing energies.

Square-planar complexes, in which four ligands are arranged about the metal ion in a plane, can be envisioned as formed by removing two ligands from along the vertical *z*-axis of the octahedral complex. The changes that occur in the energy levels of the *d* orbitals are illustrated in Figure 24.35 ◄. Note in particular that the d_{z^2} orbital is now considerably lower in energy than the $d_{x^2-y^2}$ orbital because the ligands along the vertical *z*-axis have been removed.

Square-planar complexes are characteristic of metal ions with a d^8 electron configuration. They are nearly always low spin; that is, the eight d electrons are spin-paired to form a diamagnetic complex. Such an electronic arrangement is particularly common among the ions of heavier metals, such as Pd^{2+}, Pt^{2+}, Ir^+, and Au^{3+}.

GIVE IT SOME THOUGHT

Why is the energy of the d_{xz} and d_{yz} orbitals lower than that of the d_{xy} orbital in square-planar complexes?

■ SAMPLE EXERCISE 24.10 | Populating *d* Orbitals in Tetrahedral and Square-Planar Complexes

Four-coordinate nickel(II) complexes exhibit both square-planar and tetrahedral geometries. The tetrahedral ones, such as $[NiCl_4]^{2-}$, are paramagnetic; the square-planar ones, such as $[Ni(CN)_4]^{2-}$, are diamagnetic. Show how the d electrons of nickel(II) populate the d orbitals in the appropriate crystal-field splitting diagram in each case.

SOLUTION

Analyze: We are given two complexes containing Ni^{2+}, a tetrahedral one and a square-planar one. We are asked to use the appropriate crystal-field diagrams to describe how the d electrons populate the d orbitals in each case.

Plan: We need to first determine the number of d electrons possessed by Ni^{2+} and then use Figure 24.34 for the tetrahedral complex and Figure 24.35 for the square-planar complex.

Solve: Nickel(II) has an electron configuration of $[Ar]3d^8$. The population of the d electrons in the two geometries is

Tetrahedral   Square planar

Comment: Notice that the tetrahedral complex is paramagnetic with two unpaired electrons, whereas the square-planar complex is diamagnetic. Nickel(II) forms octahedral complexes more frequently than square-planar ones, whereas heavier d^8 metals tend to favor square planar coordination.

■ PRACTICE EXERCISE

How many unpaired electrons do you predict for the tetrahedral $[CoCl_4]^{2-}$ ion?
Answer: three

We have seen that the crystal-field model provides a basis for explaining many features of transition-metal complexes. In fact, it can be used to explain many observations in addition to those we have discussed. Many lines of evidence show, however, that the bonding between transition-metal ions and ligands must have some covalent character. Molecular-orbital theory ∞ (Sections 9.7 and 9.8) can also be used to describe the bonding in complexes, although the application of molecular-orbital theory to coordination compounds is beyond the scope of our discussion. The crystal-field model, although not entirely accurate in all details, provides an adequate and useful first description of the electronic structure of complexes.

A Closer Look | CHARGE-TRANSFER COLOR

▲ **Figure 24.36 Effect of charge-transfer transitions.** The compounds $KMnO_4$, K_2CrO_4, and $KClO_4$ are shown from left to right. The $KMnO_4$ and K_2CrO_4 are intensely colored because of ligand-to-metal charge transfer (LMCT) transitions in the MnO_4^- and CrO_4^{2-} anions. There are no valence d orbitals on Cl, so the charge-transfer transition for ClO_4^- requires ultraviolet light and $KClO_4$ is white.

In the laboratory portion of your course, you have probably seen many colorful compounds of the transition metals. Many of these exhibit color because of d-d transitions, in which visible light excites electrons from one d orbital to another. There are other colorful transition-metal complexes, however, that derive their color from a rather different type of excitation involving the d orbitals. Two such common substances are the deep violet permanganate ion (MnO_4^-) and the bright yellow chromate ion (CrO_4^{2-}), salts of which are shown in Figure 24.36 ▲. Both MnO_4^- and CrO_4^{2-} are tetrahedral complexes.

The permanganate ion strongly absorbs visible light with a maximum absorption at a wavelength of 565 nm. The strong absorption in the yellow portion of the visible spectrum is responsible for the violet appearance of salts and solutions of the ion (violet is the complementary color to yellow). What is happening during this absorption? The MnO_4^- ion is a complex of Mn(VII), which has a d^0 electron configuration. As such, the absorption in the complex cannot be due to a d-d transition because there are no d electrons to excite! That does not mean, however, that the d orbitals are not involved in the transition. The excitation in the MnO_4^- ion is due to a *charge-transfer transition*, in which an electron on one of the oxygen ligands is excited into a vacant d orbital on the Mn atom (Figure 24.37 ▶). In essence, an electron is transferred from a ligand to the metal, so this transition is called a *ligand-to-metal charge-transfer (LMCT) transition*. An LMCT transition is also responsible for the color of the CrO_4^{2-}, which is a d^0 Cr(VI) complex. Also shown in Figure 24.36 is a salt of the perchlorate ion (ClO_4^-). Like MnO_4^-, the ClO_4^- is tetrahedral and has

its central atom in the +7 oxidation state. However, because the Cl atom does not have low-lying d orbitals, exciting an electron requires a more energetic photon than for MnO_4^-. The first absorption for ClO_4^- is in the ultraviolet portion of the spectrum, so all the visible light is transmitted and the salt appears white.

Other complexes exhibit charge-transfer excitations in which an electron from the metal atom is excited to an empty orbital on a ligand. Such an excitation is called a *metal-to-ligand charge-transfer (MLCT) transition*.

Charge-transfer transitions are generally more intense than d-d transitions. Many metal-containing pigments used for oil painting, such as cadmium yellow (CdS), chrome yellow ($PbCrO_4$), and red ochre (Fe_2O_3) have intense colors because of charge-transfer transitions.

Related Exercise: 24.64

◀ **Figure 24.37 Ligand-to-metal charge transfer (LMCT) transition in MnO_4^-.** As shown by the blue arrow, an electron is excited from a nonbonding pair on O into one of the empty d orbitals on Mn.

■■ **SAMPLE INTEGRATIVE EXERCISE** | Putting Concepts Together

The oxalate ion has the Lewis structure shown in Table 24.2. **(a)** Show the geometrical structure of the complex formed by coordination of oxalate to cobalt(II), forming $[Co(C_2O_4)(H_2O)_4]$. **(b)** Write the formula for the salt formed upon coordination of three oxalate ions to Co(II), assuming that the charge-balancing cation is Na^+. **(c)** Sketch all the possible geometric isomers for the cobalt complex formed in part (b). Are any of these isomers chiral? Explain. **(d)** The equilibrium constant for the formation of the cobalt(II) complex produced by coordination of three oxalate anions, as in part (b), is 5.0×10^9. By comparison, the formation constant for formation of the cobalt(II) complex with three molecules of *ortho*-phenanthroline (Table 24.2) is 9×10^{19}.

From these results, what conclusions can you draw regarding the relative Lewis base properties of the two ligands toward cobalt(II)? **(e)** Using the approach described in Sample Exercise 17.14, calculate the concentration of free aqueous Co(II) ion in a solution initially containing 0.040 M oxalate ion and 0.0010 M $Co^{2+}(aq)$.

SOLUTION

(a) The complex formed by coordination of one oxalate ion is octahedral:

(b) Because the oxalate ion has a charge of 2−, the net charge of a complex with three oxalate anions and one Co^{2+} ion is 4−. Therefore, the coordination compound has the formula $Na_4[Co(C_2O_4)_3]$.

(c) There is only one geometric isomer. The complex is chiral, however, in the same way as the $[Co(en)_3]^{3+}$ complex, shown in Figure 24.20(b). These two mirror images are not superimposable, so there are two enantiomers:

(d) The *ortho*-phenanthroline ligand is bidentate, like the oxalate ligand, so they both exhibit the chelate effect. Thus, we can conclude that *ortho*-phenanthroline is a stronger Lewis base toward Co^{2+} than oxalate. This conclusion is consistent with what we learned about bases in Section 16.7, namely that nitrogen bases are generally stronger than oxygen bases. (Recall, for example, that NH_3 is a stronger base than H_2O.)

(e) The equilibrium we must consider involves three moles of oxalate ion (represented as Ox^{2-}).

$$Co^{2+}(aq) + 3\,Ox^{2-}(aq) \rightleftharpoons [Co(Ox)_3]^{4-}(aq)$$

The formation-constant expression is

$$K_f = \frac{[[Co(Ox)_3]^{4-}]}{[Co^{2+}][Ox^{2-}]^3}$$

Because K_f is so large, we can assume that essentially all of the Co^{2+} is converted to the oxalato complex. Under that assumption, the final concentration of $[Co(Ox)_3]^{3-}$ is 0.0010 M and that of oxalate ion is $[Ox^{2-}] = (0.040) - 3(0.0010) = 0.037\ M$ (three Ox^{2-} ions react with each Co^{2+} ion).

We then have

$$[Co^{2+}] = x\,M, [Ox^{2-}] \cong 0.037\ M, [[Co(Ox)_3]^{4-}] \cong 0.0010\ M$$

Inserting these values into the equilibrium-constant expression, we have

$$K_f = \frac{(0.0010)}{x(0.037)^3} = 5 \times 10^9$$

Solving for x, we obtain $4 \times 10^{-9}\ M$. From this, we can see that the oxalate has complexed all but a tiny fraction of the Co^{2+} present in solution.

CHAPTER REVIEW

SUMMARY AND KEY TERMS

Section 24.1 **Coordination compounds** are substances that contain **metal complexes**. Metal complexes contain metal ions bonded to several surrounding anions or molecules known as **ligands**. The metal ion and its ligands comprise the **coordination sphere** of the complex. The atom of the ligand that bonds to the metal ion is the **donor atom**. The number of donor atoms attached to the metal ion is the **coordination number** of the metal ion. The most common coordination numbers are 4 and 6; the most common coordination geometries are tetrahedral, square planar, and octahedral.

Sections 24.2 and 24.3 Ligands that occupy only one site in a coordination sphere are called **monodentate ligands**. If a ligand has several donor atoms that can coordinate simultaneously to the metal ion, it is a **polydentate ligand** and is referred to as a **chelating agent**. Two common examples are ethylenediamine, denoted en, which is a **bidentate ligand**, and the ethylenediaminetetraacetate ion, $[EDTA]^{4-}$, which has six potential donor atoms. In general, chelating agents form more stable complexes than do related monodentate ligands, an observation known as the **chelate effect**. Many biologically important molecules, such as the **porphyrins**, are complexes of chelating agents. A related group of plant pigments known as **chlorophylls** are important in **photosynthesis**, the process by which plants use solar energy to convert CO_2 and H_2O into carbohydrates.

As in the nomenclature of other inorganic compounds, specific rules are followed for naming coordination compounds. In general, the number and type of ligands attached to the metal ion are specified, as is the oxidation state of the metal ion.

Section 24.4 **Isomers** are compounds with the same composition but different arrangements of atoms and therefore different properties. **Structural isomers** differ in the bonding arrangements of the ligands. One simple form of structural isomerism, known as **linkage isomerism**, occurs when a ligand is capable of coordinating to a metal ion through either of two donor atoms. **Coordination-sphere isomers** contain different ligands in the coordination sphere.

Stereoisomers are isomers with the same chemical bonding arrangements but different spatial arrangements of ligands. The most common forms of stereoisomerism are **geometric isomerism** and **optical isomerism**. Geometric isomers differ from one another in the relative locations of donor atoms in the coordination sphere; the most common are cis-trans isomers. Optical isomers are nonsuperimposable mirror images of one another. Geometric isomers differ from one another in their chemical and physical properties; optical isomers or **enantiomers** are **chiral**, however, meaning that they have a specific "handedness" and they differ only in the presence of a chiral environment. Optical isomers can be distinguished from one another by their interactions with plane-polarized light; solutions of one isomer rotate the plane of polarization to the right (**dextrorotatory**), and solutions of its mirror image rotate the plane to the left (**levorotatory**). Chiral molecules, therefore, are **optically active**. A 50–50 mixture of two optical isomers does not rotate plane-polarized light and is said to be **racemic**.

Section 24.5 Studies of the colors and magnetic properties of transition-metal complexes have played important roles in the formulation of bonding theories for these compounds. A substance has a particular color because it either (1) reflects or transmits light of that color or (2) absorbs light of the **complementary color**. The amount of light absorbed by a sample as a function of wavelength is known as its **absorption spectrum**. The light absorbed provides the energy to excite electrons to higher-energy states.

It is possible to determine the number of unpaired electrons in a complex from its degree of paramagnetism. Compounds with no unpaired electrons are diamagnetic.

Section 24.6 The **crystal-field theory** successfully accounts for many properties of coordination compounds, including their color and magnetism. In this model the interaction between metal ion and ligand is viewed as electrostatic. Because some d orbitals are pointing right at the ligands whereas others point between them, the ligands split the energies of the metal d orbitals. For an octahedral complex, the d orbitals are split into a lower-energy set of three degenerate orbitals (the t_2 set) and a higher-energy set of two degenerate orbitals (the e set). Visible light can cause a **d-d transition**, in which an electron is excited from a lower-energy d orbital to a higher-energy d orbital. The **spectrochemical series** lists ligands in order of their ability to split the d-orbital energies in octahedral complexes.

Strong-field ligands create a splitting of d-orbital energies that is large enough to overcome the **spin-pairing energy**. The d electrons then preferentially pair up in the lower-energy orbitals, producing a **low-spin complex**. When the ligands exert a weak crystal field, the splitting of the d orbitals is small. The electrons then occupy the higher-energy d orbitals in preference to pairing up in the lower-energy set, producing a **high-spin complex**.

The crystal-field model also applies to tetrahedral and square-planar complexes, which leads to different d-orbital splitting patterns. In a tetrahedral crystal field, the splitting of the d orbitals is exactly opposite that of the octahedral case. The splitting by a tetrahedral crystal field is much smaller than that of an octahedral crystal field, so tetrahedral complexes are always high-spin complexes.

KEY SKILLS

- Determine the oxidation number and number of *d* electrons for metal ions in complexes.
- Name coordination compounds given their formula and write their formula given their name.
- Recognize and draw the geometric isomers of a complex.
- Recognize and draw the optical isomers of a complex.
- Use crystal-field theory to explain the colors and to determine the number of unpaired electrons in a complex.

VISUALIZING CONCEPTS

24.1 **(a)** Draw the structure for $Pt(en)Cl_2$. **(b)** What is the coordination number for platinum in this complex, and what is the coordination geometry? **(c)** What is the oxidation state of the platinum? [Section 24.1]

24.2 Draw the Lewis structure for the ligand shown below. **(a)** Which atoms can serve as donor atoms? Classify this ligand as monodentate, bidentate, or tridentate. **(b)** How many of these ligands are needed to fill the coordination sphere in an octahedral complex? [Section 24.2]

$NH_2CH_2CH_2NHCH_2CO_2^-$

24.3 The complex ion shown below has a 1− charge. Name the complex ion. [Section 24.3]

● = N
● = Cl
○ = H
● = Pt

24.4 There are two geometric isomers of octahedral complexes of the type MA_3X_3, where M is a metal and A and X are monodentate ligands. Of the complexes shown here, which are identical to (1) and which are the geometric isomers of (1)? [Section 24.4]

(1)   (2)   (3)   (4)   (5)

24.5 Which of the following complexes are chiral? Explain. [Section 24.4]

● = Cr   ◠ = $NH_2CH_2CH_2NH_2$ ● = Cl   ● = NH_3

(1)   (2)   (3)   (4)

24.6 Assume that the two solutions shown below have an absorption spectrum with a single absorption peak like that shown in Figure 24.26. What color does each solution absorb most strongly? [Section 24.5]

24.7 Consider the following crystal-field splitting diagrams. Select the one that fits each of the following descriptions: **(a)** a weak-field octahedral complex of Fe^{3+}, **(b)** a strong-field octahedral complex of Fe^{3+} **(c)** a tetrahedral complex of Fe^{3+}, **(d)** a tetrahedral complex of Ni^{2+}? (The diagrams do not indicate the relative magnitudes of Δ.) [Section 24.6]

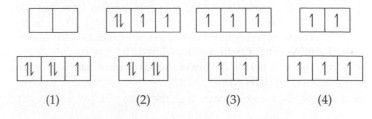

(1)   (2)   (3)   (4)

24.8 Consider the accompanying linear crystal field, in which the negative charges are on the z-axis. Using Figure 24.29 as a guide, predict which *d* orbital has lobes closest to the charges. Which two have lobes farthest from the charges? Predict the crystal-field splitting of the *d* orbitals in linear complexes. [Section 24.6]

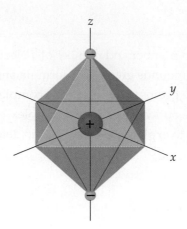

EXERCISES

Introduction to Metal Complexes

24.9 **(a)** What is the difference between Werner's concepts of *primary valence* and *secondary valence*? What terms do we now use for these concepts? **(b)** Why can the NH_3 molecule serve as a ligand but the BH_3 molecule cannot?

24.10 **(a)** What is the meaning of the term *coordination number* as it applies to metal complexes? **(b)** Generally speaking, what structural feature characterizes substances that can serve as ligands in metal complexes? Give an example of a ligand that is neutral and one that is negatively charged. **(c)** Would you expect ligands that are positively charged to be common? Explain. **(d)** What type of chemical bonding is characteristic of coordination compounds? Illustrate with the compound $Co(NH_3)_6Cl_3$.

24.11 A complex is written as $NiBr_2 \cdot 6 NH_3$ **(a)** What is the oxidation state of the Ni atom in this complex? **(b)** What is the likely coordination number for the complex? **(c)** If the complex is treated with excess $AgNO_3(aq)$, how many moles of AgBr will precipitate per mole of complex?

24.12 A certain complex of metal M is formulated as $MCl_3 \cdot 3 H_2O$. The coordination number of the complex is not known but is expected to be 4 or 6. **(a)** Would conductivity measurements provide information about the coordination number? **(b)** In using conductivity measurements to test which ligands are bound to the metal ion, what assumption is made about the rate at which ligands enter or leave the coordination sphere of the metal?

24.13 Indicate the coordination number of the metal and the oxidation number of the metal in each of the following complexes:
(a) $Na_2[CdCl_4]$
(b) $K_2[MoOCl_4]$
(c) $[Co(NH_3)_4Cl_2]Cl$
(d) $[Ni(CN)_5]^{3-}$
(e) $K_3[V(C_2O_4)_3]$
(f) $[Zn(en)_2]Br_2$

24.14 Indicate the coordination number of the metal and the oxidation number of the metal in each of the following complexes:
(a) $K_3[Co(CN)_6]$
(b) $Na_2[CdBr_4]$
(c) $[Pt(en)_3](ClO_4)_4$
(d) $[Co(en)_2(C_2O_4)]^+$
(e) $NH_4[Cr(NH_3)_2(NCS)_4]$
(f) $[Cu(bipy)_2I]I$

24.15 Determine the number and type of each donor atom in each of the complexes in Exercise 24.13.

24.16 What are the number and types of donor atoms in each of the complexes in Exercise 24.14?

Polydentate Ligands; Nomenclature

24.17 **(a)** What is the difference between a monodentate ligand and a bidentate ligand? **(b)** How many bidentate ligands are necessary to fill the coordination sphere of a six-coordinate complex? **(c)** You are told that a certain molecule can serve as a tridentate ligand. Based on this statement, what do you know about the molecule?

24.18 For each of the following polydentate ligands, determine (i) the maximum number of coordination sites that the ligand can occupy on a single metal ion and (ii) the number and type of donor atoms in the ligand: **(a)** ethylenediamine (en), **(b)** bipyridine (bipy), **(c)** the oxalate anion ($C_2O_4^{2-}$), **(d)** the 2− ion of the porphine molecule (Figure 24.8); **(e)** $[EDTA]^{4-}$.

24.19 Polydentate ligands can vary in the number of coordination positions they occupy. In each of the following, identify the polydentate ligand present and indicate the probable number of coordination positions it occupies:
(a) $[Co(NH_3)_4(o\text{-phen})]Cl_3$
(b) $[Cr(C_2O_4)(H_2O)_4]Br$
(c) $[Cr(EDTA)(H_2O)]^-$
(d) $[Zn(en)_2](ClO_4)_2$

24.20 Indicate the likely coordination number of the metal in each of the following complexes:
(a) $[Rh(bipy)_3](NO_3)_3$
(b) $Na_3[Co(C_2O_4)_2Cl_2]$
(c) $[Cr(o\text{-phen})_3](CH_3COO)_3$
(d) $Na_2[Co(EDTA)Br]$

24.21 (a) What is meant by the term *chelate effect*? (b) What thermodynamic factor is generally responsible for the chelate effect? (c) Why are polydentate ligands often called *sequestering agents*?

24.22 *Pyridine* (C_5H_5N), abbreviated py, is the following molecule:

(a) Why is pyridine referred to as a monodentate ligand?
(b) Consider the following equilibrium reaction:

$$[Ru(py)_4(bipy)]^{2+} + 2\ py \rightleftharpoons [Ru(py)_6]^{2+} + bipy$$

What would you predict for the magnitude of the equilibrium constant for this equilibrium? Explain the basis for your answer.

24.23 Write the formula for each of the following compounds, being sure to use brackets to indicate the coordination sphere:
(a) hexaamminechromium(III) nitrate
(b) tetraamminecarbonatocobalt(III) sulfate
(c) dichlorobis(ethylenediamine)platinum(IV) bromide
(d) potassium diaquatetrabromovanadate(III)
(e) bis(ethylenediamine)zinc(II) tetraiodomercurate(II)

Isomerism

24.27 By writing formulas or drawing structures related to any one of these three complexes,

$$[Co(NH_3)_4Br_2]Cl$$
$$[Pd(NH_3)_2(ONO)_2]$$
$$cis\text{-}[V(en)_2Cl_2]^+$$

illustrate (a) geometric isomerism, (b) linkage isomerism, (c) optical isomerism, (d) coordination-sphere isomerism.

24.28 (a) Draw the two linkage isomers of $[Co(NH_3)_5SCN]^{2+}$.
(b) Draw the two geometric isomers of $[Co(NH_3)_3Cl_3]^{2+}$.
(c) Two compounds with the formula $Co(NH_3)_5ClBr$ can be prepared. Use structural formulas to show how they differ. What kind of isomerism does this illustrate?

Color; Magnetism; Crystal-Field Theory

24.33 (a) To the closest 100 nm, what are the largest and smallest wavelengths of visible light? (b) What is meant by the term *complementary color*? (c) What is the significance of complementary colors in understanding the colors of metal complexes? (d) If a complex absorbs light at 610 nm, what is the energy of this absorption in kJ/mol?

24.34 (a) A complex absorbs light with wavelength of 530 nm. Do you expect it to have color? (b) A solution of a compound appears green. Does this observation necessarily mean that all colors of visible light other than green are absorbed by the solution? Explain. (c) What information is usually presented in a *visible absorption spectrum* of a compound? (d) What energy is associated with the absorption at 530 nm in kJ/mol?

24.35 In crystal-field theory, ligands are modeled as if they are point negative charges. What is the basis of this assumption, and how does it relate to the nature of metal–ligand bonds?

24.24 Write the formula for each of the following compounds, being sure to use brackets to indicate the coordination sphere:
(a) tetraaquadibromomanganese(III) perchlorate
(b) bis(bipyridyl)cadmium(II) chloride
(c) potassium tetrabromo(*ortho*-phenanthroline)cobaltate (III)
(d) cesium diamminetetracyanochromate(III)
(e) tris(ethylenediamine)rhodium(III) tris(oxalato)cobaltate(III)

24.25 Write the names of the following compounds, using the standard nomenclature rules for coordination complexes:
(a) $[Rh(NH_3)_4Cl_2]Cl$
(b) $K_2[TiCl_6]$
(c) $MoOCl_4$
(d) $[Pt(H_2O)_4(C_2O_4)]Br_2$

24.26 Write names for the following coordination compounds:
(a) $[Cd(en)Cl_2]$
(b) $K_4[Mn(CN)_6]$
(c) $[Cr(NH_3)_5CO_3]Cl$
(d) $[Ir(NH_3)_4(H_2O)_2](NO_3)_3$

24.29 A four-coordinate complex MA_2B_2 is prepared and found to have two different isomers. Is it possible to determine from this information whether the complex is square planar or tetrahedral? If so, which is it?

24.30 Consider an octahedral complex MA_3B_3. How many geometric isomers are expected for this compound? Will any of the isomers be optically active? If so, which ones?

24.31 Sketch all the possible stereoisomers of (a) tetrahedral $[Cd(H_2O)_2Cl_2]$, (b) square-planar $[IrCl_2(PH_3)_2]^-$, (c) octahedral $[Fe(o\text{-phen})_2Cl_2]^+$.

24.32 Sketch all the possible stereoisomers of (a) $[Rh(bipy)(o\text{-phen})_2]^{3+}$, (b) $[Co(NH_3)_3(bipy)Br]^{2+}$, (c) square-planar $[Pd(en)(CN)_2]$.

24.36 Explain why the d_{xy}, d_{xz}, and d_{yz} orbitals lie lower in energy than the d_{z^2} and $d_{x^2-y^2}$ orbitals in the presence of an octahedral arrangement of ligands about the central metal ion.

24.37 (a) Sketch a diagram that shows the definition of the *crystal-field splitting energy* (Δ) for an octahedral crystal field. (b) What is the relationship between the magnitude of Δ and the energy of the *d-d* transition for a d^1 complex? (c) Calculate Δ in kJ/mol if a d^1 complex has an absorption maximum at 590 nm.

24.38 As shown in Figure 24.26, the *d-d* transition of $[Ti(H_2O)_6]^{3+}$ produces an absorption maximum at a wavelength of about 500 nm. (a) What is the magnitude of Δ for $[Ti(H_2O)_6]^{3+}$ in kJ/mol? (b) What is the *spectrochemical series*? How would the magnitude of Δ change if the H_2O ligands in $[Ti(H_2O)_6]^{3+}$ were replaced with NH_3 ligands?

24.39 Explain why many cyano complexes of divalent transition-metal ions are yellow, whereas many aqua complexes of these ions are blue or green.

24.40 The $[Ni(H_2O)_6]^{2+}$ ion has an absorption maximum at about 725 nm, whereas the $[Ni(NH_3)_6]^{2+}$ ion absorbs at about 570 nm. Predict the color of each ion. **(b)** The $[Ni(en)_3]^{2+}$ ion absorption maximum occurs at about 545 nm, and that of the $[Ni(bipy)_3]^{2+}$ ion occurs at about 520 nm. From these data, indicate the relative strengths of the ligand fields created by the four ligands involved.

24.41 Give the number of d electrons associated with the central metal ion in each of the following complexes: **(a)** $K_3[TiCl_6]$, **(b)** $Na_3[Co(NO_2)_6]$, **(c)** $[Ru(en)_3]Br_3$, **(d)** $[Mo(EDTA)]ClO_4$, **(e)** $K_3[ReCl_6]$.

24.42 Give the number of d electrons associated with the central metal ion in each of the following complexes: **(a)** $K_3[Fe(CN)_6]$, **(b)** $[Mn(H_2O)_6](NO_3)_2$, **(c)** $Na[Ag(CN)_2]$, **(d)** $[Cr(NH_3)_4Br_2]ClO_4$, **(e)** $[Sr(EDTA)]^{2-}$.

24.43 For each of the following metals, write the electronic configuration of the atom and its 2+ ion: **(a)** Mn, **(b)** Ru, **(c)** Rh. Draw the crystal-field energy-level diagram for the d orbitals of an octahedral complex, and show the placement of the d electrons for each 2+ ion, assuming a strong-field complex. How many unpaired electrons are there in each case?

24.44 For each of the following metals, write the electronic configuration of the atom and its 3+ ion: **(a)** Ru, **(b)** Mo, **(c)** Co. Draw the crystal-field energy-level diagram for the d orbitals of an octahedral complex, and show the placement of the d electrons for each 3+ ion, assuming a weak-field complex. How many unpaired electrons are there in each case?

24.45 Draw the crystal-field energy-level diagrams and show the placement of d electrons for each of the following: **(a)** $[Cr(H_2O)_6]^{2+}$ (four unpaired electrons), **(b)** $[Mn(H_2O)_6]^{2+}$ (high spin), **(c)** $[Ru(NH_3)_5H_2O]^{2+}$ (low spin), **(d)** $[IrCl_6]^{2-}$ (low spin), **(e)** $[Cr(en)_3]^{3+}$, **(f)** $[NiF_6]^{4-}$.

24.46 Draw the crystal-field energy-level diagrams and show the placement of electrons for the following complexes: **(a)** $[VCl_6]^{3-}$, **(b)** $[FeF_6]^{3-}$ (a high-spin complex), **(c)** $[Ru(bipy)_3]^{3+}$ (a low-spin complex), **(d)** $[NiCl_4]^{2-}$ (tetrahedral), **(e)** $[PtBr_6]^{2-}$, **(f)** $[Ti(en)_3]^{2+}$.

24.47 The complex $[Mn(NH_3)_6]^{2+}$ contains five unpaired electrons. Sketch the energy-level diagram for the d orbitals, and indicate the placement of electrons for this complex ion. Is the ion a high-spin or a low-spin complex?

24.48 The ion $[Fe(CN)_6]^{3-}$ has one unpaired electron, whereas $[Fe(NCS)_6]^{3-}$ has five unpaired electrons. From these results, what can you conclude about whether each complex is high spin or low spin? What can you say about the placement of NCS⁻ in the spectrochemical series?

ADDITIONAL EXERCISES

24.49 Based on the molar conductance values listed here for the series of platinum(IV) complexes, write the formula for each complex so as to show which ligands are in the coordination sphere of the metal. By way of example, the molar conductances of 0.050 M NaCl and BaCl₂ are 107 ohm⁻¹ and 197 ohm⁻¹, respectively.

Complex	Molar Conductance (ohm⁻¹)* of 0.050 M Solution
$Pt(NH_3)_6Cl_4$	523
$Pt(NH_3)_4Cl_4$	228
$Pt(NH_3)_3Cl_4$	97
$Pt(NH_3)_2Cl_4$	0
$KPt(NH_3)Cl_5$	108

*The ohm is a unit of resistance; conductance is the inverse of resistance.

24.50 **(a)** A compound with formula $RuCl_3 \cdot 5\ H_2O$ is dissolved in water, forming a solution that is approximately the same color as the solid. Immediately after forming the solution, the addition of excess $AgNO_3(aq)$ forms 2 mol of solid AgCl per mole of complex. Write the formula for the compound, showing which ligands are likely to be present in the coordination sphere. **(b)** After a solution of $RuCl_3 \cdot 5\ H_2O$ has stood for about a year, addition of $AgNO_3(aq)$ precipitates 3 mol of AgCl per mole of complex. What has happened in the ensuing time?

24.51 Sketch the structure of the complex in each of the following compounds:
(a) cis-$[Co(NH_3)_4(H_2O)_2](NO_3)_2$
(b) $Na_2[Ru(H_2O)Cl_5]$

(c) trans-$NH_4[Co(C_2O_4)_2(H_2O)_2]$
(d) cis-$[Ru(en)_2Cl_2]$

24.52 **(a)** Give the full name for each of the compounds in Exercise 24.51. **(b)** Will any of the complexes be optically active? Explain.

24.53 The molecule *dimethylphosphinoethane* $[(CH_3)_2PCH_2\text{-}CH_2P(CH_3)_2$, which is abbreviated dmpe] is used as a ligand for some complexes that serve as catalysts. A complex that contains this ligand is $Mo(CO)_4(dmpe)$. **(a)** Draw the Lewis structure for dmpe, and compare it with ethylenediamine as a coordinating ligand. **(b)** What is the oxidation state of Mo in $Na_2[Mo(CN)_2(CO)_2(dmpe)]$? **(c)** Sketch the structure of the $[Mo(CN)_2(CO)_2(dmpe)]^{2-}$ ion, including all the possible isomers.

24.54 Although the cis configuration is known for $[Pt(en)Cl_2]$, no trans form is known. **(a)** Explain why the trans compound is not possible. **(b)** Suggest what type of ligand would be required to form a trans-bidentate coordination to a metal atom.

[24.55] The acetylacetone ion forms very stable complexes with many metallic ions. It acts as a bidentate ligand, coordinating to the metal at two adjacent positions. Suppose that one of the CH_3 groups of the ligand is replaced by a CF_3 group, as shown here,

Trifluoromethyl acetylacetonate (tfac)

Sketch all possible isomers for the complex with three tfac ligands on cobalt(III). (You can use the symbol ●○ to represent the ligand.)

24.56 Give brief statements about the relevance of the following complexes in living systems: **(a)** hemoglobin, **(b)** chlorophylls, **(c)** siderophores.

24.57 Write balanced chemical equations to represent the following observations. (In some instances the complex involved has been discussed previously in the text.) **(a)** Solid silver chloride dissolves in an excess of aqueous ammonia. **(b)** The green complex [Cr(en)$_2$Cl$_2$]Cl, on treatment with water over a long time, converts to a brown-orange complex. Reaction of AgNO$_3$ with a solution of the product precipitates 3 mol of AgCl per mole of Cr present. (Write two chemical equations.) **(c)** When an NaOH solution is added to a solution of Zn(NO$_3$)$_2$, a precipitate forms. Addition of excess NaOH solution causes the precipitate to dissolve. (Write two chemical equations.) **(d)** A pink solution of Co(NO$_3$)$_2$ turns deep blue on addition of concentrated hydrochloric acid.

24.58 Some metal complexes have a coordination number of 5. One such complex is Fe(CO)$_5$, which adopts a *trigonal bipyramidal* geometry (see Figure 9.8). **(a)** Write the name for Fe(CO)$_5$, using the nomenclature rules for coordination compounds. **(b)** What is the oxidation state of Fe in this compound? **(c)** Suppose one of the CO ligands is replaced with a CN$^-$ ligand, forming [Fe(CO)$_4$(CN)]$^-$. How many geometric isomers would you predict this complex could have?

24.59 Which of the following objects is chiral? **(a)** a left shoe, **(b)** a slice of bread, **(c)** a wood screw, **(d)** a molecular model of Zn(en)Cl$_2$, **(e)** a typical golf club.

24.60 The complexes [V(H$_2$O)$_6$]$^{3+}$ and [VF$_6$]$^{3-}$ are both known. **(a)** Draw the *d*-orbital energy-level diagram for V(III) octahedral complexes. **(b)** What gives rise to the colors of these complexes? **(c)** Which of the two complexes would you expect to absorb light of higher energy? Explain.

[24.61] One of the more famous species in coordination chemistry is the Creutz–Taube complex,

$$\left[(NH_3)_5RuN\bigcirc NRu(NH_3)_5 \right]^{5+}$$

It is named for the two scientists who discovered it and initially studied its properties. The central ligand is pyrazine, a planar six-membered ring with nitrogens at opposite sides. **(a)** How can you account for the fact that the complex, which has only neutral ligands, has an odd overall charge? **(b)** The metal is in a low-spin configuration in both cases. Assuming octahedral coordination, draw the *d*-orbital energy-level diagram for each metal. **(c)** In many experiments the two metal ions appear to be in exactly equivalent states. Can you think of a reason that this might appear to be so, recognizing that electrons move very rapidly compared to nuclei?

24.62 Solutions of [Co(NH$_3$)$_6$]$^{2+}$, [Co(H$_2$O)$_6$]$^{2+}$ (both octahedral), and [CoCl$_4$]$^{2-}$ (tetrahedral) are colored. One is pink, one is blue, and one is yellow. Based on the spectrochemical series and remembering that the energy splitting in tetrahedral complexes is normally much less than that in octahedral ones, assign a color to each complex.

24.63 Oxyhemoglobin, with an O$_2$ bound to iron, is a low-spin Fe(II) complex; deoxyhemoglobin, without the O$_2$ molecule, is a high-spin complex. **(a)** Assuming that the coordination environment about the metal is octahedral, how many unpaired electrons are centered on the metal ion in each case? **(b)** What ligand is coordinated to the iron in place of O$_2$ in deoxyhemoglobin? **(c)** Explain in a general way why the two forms of hemoglobin have different colors (hemoglobin is red, whereas deoxyhemoglobin has a bluish cast). **(d)** A 15-minute exposure to air containing 400 ppm of CO causes about 10% of the hemoglobin in the blood to be converted into the carbon monoxide complex, called carboxyhemoglobin. What does this suggest about the relative equilibrium constants for binding of carbon monoxide and O$_2$ to hemoglobin?

[24.64] Consider the tetrahedral anions VO$_4$$^{3-}$ (orthovanadate ion), CrO$_4$$^{2-}$ (chromate ion), and MnO$_4$$^-$ (permanganate ion). **(a)** These anions are *isoelectronic*. What does this statement mean? **(b)** Would you expect these anions to exhibit *d-d* transitions? Explain. **(c)** As mentioned in "A Closer Look" on charge-transfer color, the violet color of MnO$_4$$^-$, is due to a *ligand-to-metal charge transfer* (LMCT) transition. What is meant by this term? **(d)** The LMCT transition in MnO$_4$$^-$ occurs at a wavelength of 565 nm. The CrO$_4$$^{2-}$ ion is yellow. Is the wavelength of the LMCT transition for chromate larger or smaller than that for MnO$_4$$^-$? Explain. **(e)** The VO$_4$$^{3-}$ ion is colorless. Is this observation consistent with the wavelengths of the LMCT transitions in MnO$_4$$^-$ and CrO$_4$$^{2-}$?

[24.65] The red color of ruby is due to the presence of Cr(III) ions at octahedral sites in the close-packed oxide lattice of Al$_2$O$_3$. Draw the crystal-field-splitting diagram for Cr(III) in this environment. Suppose that the ruby crystal is subjected to high pressure. What do you predict for the variation in the wavelength of absorption of the ruby as a function of pressure? Explain.

24.66 In 2001, chemists at SUNY-Stony Brook succeeded in synthesizing the complex *trans*-[Fe(CN)$_4$(CO)$_2$]$^{2-}$, which could be a model of complexes that may have played a role in the origin of life. **(a)** Sketch the structure of the complex. **(b)** The complex is isolated as a sodium salt. Write the complete name of this salt. **(c)** What is the oxidation state of Fe in this complex? How many *d* electrons are associated with the Fe in this complex? **(d)** Would you expect this complex to be high spin or low spin? Explain.

[24.67] When Alfred Werner was developing the field of coordination chemistry, it was argued by some that the optical activity he observed in the chiral complexes he had prepared was because of the presence of carbon atoms in the molecule. To disprove this argument, Werner synthesized a chiral complex of cobalt that had no carbon atoms in it, and he was able to resolve it into its enantiomers. Design a cobalt(III) complex that would be chiral if it could be synthesized and that contains no carbon atoms. (It may not be possible to synthesize the complex you design, but we won't worry about that for now.)

[24.68] Generally speaking, for a given metal and ligand, the stability of a coordination compound is greater for the metal in the 3+ rather than in the 2+ oxidation state. Furthermore, for a given ligand the complexes of the bivalent metal ions of the first transition series tend to increase in stability in the order Mn(II) < Fe(II) < Co(II) < Ni(II) < Cu(II). Explain how these two observations are consistent with one another and also consistent with a crystal-field picture of coordination compounds.

24.69 Many trace metal ions exist in the bloodstream as complexes with amino acids or small peptides. The anion of the amino acid glycine, symbol gly, is capable of acting as a bidentate ligand, coordinating to the metal through nitrogen and oxygen atoms.

$$H_2NCH_2\overset{\displaystyle O}{\overset{\displaystyle \|}{C}}-O^-$$

How many isomers are possible for **(a)** [Zn(gly)$_2$] (tetrahedral), **(b)** [Pt(gly)$_2$] (square-planar), **(c)** [Co(gly)$_3$] (octahedral)? Sketch all possible isomers. Use N$\smile$O to represent the ligand.

[24.70] Suppose that a transition-metal ion was in a lattice in which it was in contact with just two nearby anions, located on opposite sides of the metal. Diagram the splitting of the metal d orbitals that would result from such a crystal field. Assuming a strong field, how many unpaired electrons would you expect for a metal ion with six d electrons? (*Hint:* Consider the linear axis to be the z-axis).

INTEGRATIVE EXERCISES

24.71 Metallic elements are essential components of many important enzymes operating within our bodies. *Carbonic anhydrase*, which contains Zn^{2+}, is responsible for rapidly interconverting dissolved CO_2 and bicarbonate ion, HCO_3^-. The zinc in carbonic anhydrase is coordinated by three nitrogen-containing groups and a water molecule. The enzyme's action depends on the fact that the coordinated water molecule is more acidic than the bulk solvent molecules. Explain this fact in terms of Lewis acid–base theory (Section 16.11).

24.72 Two different compounds have the formulation $CoBr(SO_4) \cdot 5\,NH_3$. Compound A is dark violet, and compound B is red-violet. When compound A is treated with $AgNO_3(aq)$, no reaction occurs, whereas compound B reacts with $AgNO_3(aq)$ to form a white precipitate. When compound A is treated with $BaCl_2(aq)$, a white precipitate is formed, whereas compound B has no reaction with $BaCl_2(aq)$. **(a)** Is Co in the same oxidation state in these complexes? **(b)** Explain the reactivity of compounds A and B with $AgNO_3(aq)$ and $BaCl_2(aq)$. **(c)** Are compounds A and B isomers of one another? If so, which category from Figure 24.17 best describes the isomerism observed for these complexes? **(d)** Would compounds A and B be expected to be strong electrolytes, weak electrolytes, or nonelectrolytes?

24.73 A manganese complex formed from a solution containing potassium bromide and oxalate ion is purified and analyzed. It contains 10.0% Mn, 28.6% potassium, 8.8% carbon, and 29.2% bromine by mass. The remainder of the compound is oxygen. An aqueous solution of the complex has about the same electrical conductivity as an equimolar solution of $K_4[Fe(CN)_6]$. Write the formula of the compound, using brackets to denote the manganese and its coordination sphere.

[24.74] The $E°$ values for two iron complexes in acidic solution are as follows:

$$[Fe(o\text{-}phen)_3]^{3+}(aq) + e^- \rightleftharpoons [Fe(o\text{-}phen)_3]^{2+}(aq) \quad E° = 1.12\ V$$

$$[Fe(CN)_6]^{3-}(aq) + e^- \rightleftharpoons [Fe(CN)_6]^{4-}(aq) \quad E° = 0.36\ V$$

(a) What do the relative $E°$ values tell about the relative stabilities of the Fe(II) and Fe(III) complexes in each case? **(b)** Account for the more positive $E°$ value for the

(*o-phen*) complex. Both of the Fe(II) complexes are low spin. (*Hint:* consider the charges carried by the ligands in the two cases.)

24.75 A palladium complex formed from a solution containing bromide ion and pyridine, C_5H_5N (a good electron-pair donor), is found on elemental analysis to contain 37.6% bromine, 28.3% carbon, 6.60% nitrogen, and 2.37% hydrogen by mass. The compound is slightly soluble in several organic solvents; its solutions in water or alcohol do not conduct electricity. It is found experimentally to have a zero dipole moment. Write the chemical formula, and indicate its probable structure.

24.76 **(a)** In early studies it was observed that when the complex [Co(NH$_3$)$_4$Br$_2$]Br was placed in water, the electrical conductivity of a 0.05 M solution changed from an initial value of 191 ohm^{-1} to a final value of 374 ohm^{-1} over a period of an hour or so. Suggest an explanation for the observed results. (See Exercise 24.49 for relevant comparison data.) **(b)** Write a balanced chemical equation to describe the reaction. **(c)** A 500-mL solution is made up by dissolving 3.87 g of the complex. As soon as the solution is formed, and before any change in conductivity has occurred, a 25.00-mL portion of the solution is titrated with 0.0100 M AgNO$_3$ solution. What volume of AgNO$_3$ solution do you expect to be required to precipitate the free $Br^-(aq)$? **(d)** Based on the response you gave to part (b), what volume of AgNO$_3$ solution would be required to titrate a fresh 25.00-mL sample of [Co(NH$_3$)$_4$Br$_2$]Br after all conductivity changes have occurred?

24.77 The total concentration of Ca^{2+} and Mg^{2+} in a sample of hard water was determined by titrating a 0.100-L sample of the water with a solution of EDTA^{4-}. The EDTA^{4-} chelates the two cations:

$$Mg^{2+} + [EDTA]^{4-} \longrightarrow [Mg(EDTA)]^{2-}$$

$$Ca^{2+} + [EDTA]^{4-} \longrightarrow [Ca(EDTA)]^{2-}$$

It requires 31.5 mL of 0.0104 M [EDTA]$^{4-}$ solution to reach the end point in the titration. A second 0.100-L sample was then treated with sulfate ion to precipitate Ca^{2+} as calcium sulfate. The Mg^{2+} was then titrated with 18.7 mL of 0.0104 M [EDTA]$^{4-}$. Calculate the concentrations of Mg^{2+} and Ca^{2+} in the hard water in mg/L.

24.78 Carbon monoxide is toxic because it binds more strongly to the iron in hemoglobin (Hb) than does O_2, as indicated by these approximate standard free-energy changes in blood:

$$Hb + O_2 \longrightarrow HbO_2 \qquad \Delta G° = -70 \text{ kJ}$$
$$Hb + CO \longrightarrow HbCO \qquad \Delta G° = -80 \text{ kJ}$$

Using these data, estimate the equilibrium constant at 298 K for the equilibrium

$$HbO_2 + CO \rightleftharpoons HbCO + O_2$$

[24.79] The molecule *methylamine* (CH_3NH_2) can act as a monodentate ligand. The following are equilibrium reactions and the thermochemical data at 298 K for reactions of methylamine and en with $Cd^{2+}(aq)$:

$$Cd^{2+}(aq) + 4\, CH_3NH_2(aq) \rightleftharpoons [Cd(CH_3NH_2)_4]^{2+}(aq)$$
$$\Delta H° = -57.3 \text{ kJ}; \quad \Delta S° = -67.3 \text{ J/K}; \quad \Delta G° = -37.2 \text{ kJ}$$

$$Cd^{2+}(aq) + 2\, en(aq) \rightleftharpoons [Cd(en)_2]^{2+}(aq)$$
$$\Delta H° = -56.5 \text{ kJ}; \quad \Delta S° = +14.1 \text{ J/K}; \quad \Delta G° = -60.7 \text{ kJ}$$

(a) Calculate $\Delta G°$ and the equilibrium constant K for the following *ligand exchange* reaction:

$$[Cd(CH_3NH_2)_4]^{2+}(aq) + 2\, en(aq) \rightleftharpoons$$
$$[Cd(en)_2]^{2+}(aq) + 4\, CH_3NH_2(aq)$$

(b) Based on the value of K in part (a), what would you conclude about this reaction? What concept is demonstrated? **(c)** Determine the magnitudes of the enthalpic ($\Delta H°$) and the entropic ($-T\Delta S°$) contributions to $\Delta G°$ for the ligand exchange reaction. Explain the relative magnitudes. **(d)** Based on information in this exercise and in the "A Closer Look" box on the chelate effect, predict the sign of $\Delta H°$ for the following hypothetical reaction:

$$[Cd(CH_3NH_2)_4]^{2+}(aq) + 4\, NH_3(aq) \rightleftharpoons$$
$$[Cd(NH_3)_4]^{2+}(aq) + 4\, CH_3NH_2(aq)$$

24.80 The value of Δ for the $[CrF_6]^{3-}$ complex is 182 kJ/mol. Calculate the expected wavelength of the absorption corresponding to promotion of an electron from the lower-energy to the higher-energy *d*-orbital set in this complex. Should the complex absorb in the visible range? (You may need to review Sample Exercise 6.3; remember to divide by Avogadro's number.)

[24.81] A Cu electrode is immersed in a solution that is 1.00 *M* in $[Cu(NH_3)_4]^{2+}$ and 1.00 *M* in NH_3. When the cathode is a standard hydrogen electrode, the emf of the cell is found to be +0.08 V. What is the formation constant for $[Cu(NH_3)_4]^{2+}$?

[24.82] The complex $[Ru(EDTA)(H_2O)]^-$ undergoes substitution reactions with several ligands, replacing the water molecule with the ligand.

$$[Ru(EDTA)(H_2O)]^- + L \longrightarrow [Ru(EDTA)L]^- + H_2O$$

The rate constants for several ligands are as follows:

Ligand, L	$k\ (M^{-1}s^{-1})$
Pyridine	6.3×10^3
SCN^-	2.7×10^2
CH_3CN	3.0×10

(a) One possible mechanism for this substitution reaction is that the water molecule dissociates from the complex in the rate-determining step, and then the ligand L fills the void in a rapid second step. A second possible mechanism is that L approaches the complex, begins to form a new bond to the metal, and displaces the water molecule, all in a single concerted step. Which of these two mechanisms is more consistent with the data? Explain. **(b)** What do the results suggest about the relative basicities of the three ligands toward Ru(III)? **(c)** Assuming that the complexes are all low spin, how many unpaired electrons are in each?

THE CHEMISTRY OF LIFE: ORGANIC AND BIOLOGICAL CHEMISTRY

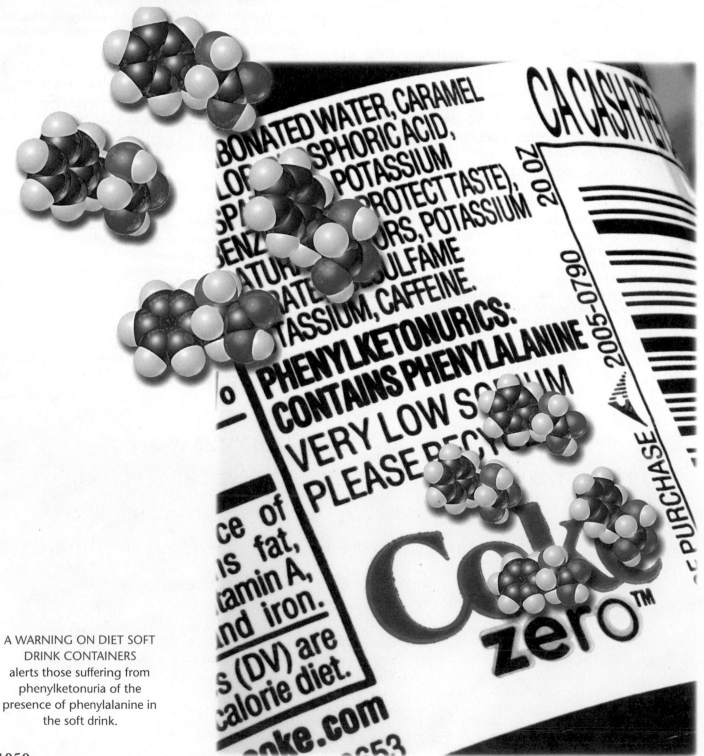

A WARNING ON DIET SOFT DRINK CONTAINERS alerts those suffering from phenylketonuria of the presence of phenylalanine in the soft drink.

WHAT'S AHEAD

25.1 Some General Characteristics of Organic Molecules
We begin with a brief review of the structures and reactivities of organic compounds.

25.2 Introduction to Hydrocarbons
We consider the *hydrocarbons*, compounds containing only C and H.

25.3 Alkanes, Alkenes, and Alkynes
We next explore three classes of hydrocarbons: those in which all bonds are single bonds called *alkanes*, those with one or more C≡C bonds called *alkenes*, and those with one or more C≡C bonds called *alkynes*. *Aromatic* hydrocarbons have at least one planar ring with delocalized π electrons. One reason for the huge number of organic compounds is the existence of *isomers*, compounds with identical compositions whose molecules have different structures.

25.4 Organic Functional Groups
We recognize that a central organizing principle of organic chemistry is the *functional group*, a group of atoms at which most of the compound's chemical reactions occur. Common organic functional groups that include oxygen and nitrogen atoms are described.

25.5 Chirality in Organic Chemistry
We learn that compounds with nonsuperimposable mirror images are *chiral*. Chirality plays very important roles in organic and biological chemistry.

25.6 Introduction to Biochemistry
We introduce the chemistry of living organisms, known as *biochemistry, biological chemistry,* or *chemical biology.* Important classes of compounds that occur in living systems are *proteins, carbohydrates, lipids,* and *nucleic acids.*

25.7 Proteins
We learn that proteins are polymers of *amino acids* linked with *amide* (also called *peptide*) bonds. Proteins are used for structural support (muscles, cartilage), molecular transporters (hemoglobin, which transports oxygen in blood), and enzymes (proteins that catalyze many metabolic reactions).

25.8 Carbohydrates
We observe that carbohydrates are sugars and polymers of sugars that are primarily used as fuel for organisms (glucose) or as structural supports in plants (cellulose).

25.9 Lipids
We recognize that lipids are a large class of molecules that are primarily used for energy storage in organisms (fats, oils).

25.10 Nucleic Acids
Finally, we learn that nucleic acids are polymers of nucleotides that comprise an organism's genetic information. *Deoxyribonucleic acid* (DNA) and *ribonucleic acid* (RNA) are polymers composed of *nucleotides.*

MANY CANS OF DIET SOFT DRINK CARRY THE WARNING "PHENYLKETONURICS: CONTAINS PHENYLALANINE." Phenylalanine is one of the essential amino acids, which means that we must have it in our diets to survive. However, in about one out of 10,000 to 20,000 Caucasian or Asian births, an enzyme

that converts phenylalanine to another amino acid, tyrosine, is completely or nearly completely deficient because of a genetic defect. The result is that phenylalanine accumulates in the blood and in body tissues. The disease that results is called phenylketonuria (PKU), which causes mental retardation and seizures. Newborns are now routinely tested for PKU at about 3 days of age. The disease can be managed by a diet that provides just enough phenylalanine for proper nutrition, without exceeding the limit. Hence, the government requires warnings on diet drink containers.

Although biological systems are almost unimaginably complex, they are nevertheless constructed of molecules of quite modest size, put together in nature to form a host of complex, interacting structures. The example of phenylalanine and PKU illustrates the point that to understand biology, we need to understand the chemical behaviors of molecules of low molar mass. This chapter is about the molecules, composed mainly of carbon, hydrogen, oxygen, and nitrogen, that form the basis of both organic and biological chemistry.

The element carbon forms a vast number of compounds. Over 16 million carbon-containing compounds are known. Chemists make thousands of new compounds every year, about 90% of which contain carbon. The study of carbon compounds constitutes a separate branch of chemistry known as **organic chemistry**. This term arose from the eighteenth-century belief that organic compounds could be formed only by living systems. This idea was disproved in 1828 by the German chemist Friedrich Wöhler when he synthesized urea (H_2NCONH_2), an organic substance found in the urine of mammals, by heating ammonium cyanate (NH_4OCN), an inorganic substance.

The study of the chemistry of living species is called *biological chemistry, chemical biology*, or **biochemistry**. In this final chapter we present a brief view of some of the elementary aspects of organic chemistry and biochemistry. Many of you will study these subjects in greater detail by taking additional courses devoted entirely to these topics. As you read the material that follows, you will notice that many of the concepts important for understanding the fundamentals of organic chemistry and biochemistry have been developed in earlier chapters.

25.1 SOME GENERAL CHARACTERISTICS OF ORGANIC MOLECULES

What is it about carbon that leads to the tremendous diversity in its compounds, and allows it to play such crucial roles in biology and society? Let's consider some general features of organic molecules, and as we do, let's review some principles that we learned in earlier chapters.

The Structures of Organic Molecules

The three-dimensional structures of organic and biochemical molecules play an essential role in determining their physical and chemical behaviors. Because carbon has four valence electrons ($[He]2s^22p^2$), it forms four bonds in virtually all its compounds. When all four bonds are single bonds, the electron pairs are disposed in a tetrahedral arrangement. ∞(Section 9.2) In the hybridization model the carbon 2s and 2p orbitals are then sp^3 hybridized. ∞(Section 9.5) When there is one double bond, the arrangement is trigonal planar (sp^2 hybridization). With a triple bond, it is linear (sp hybridization). Examples are shown in Figure 25.1▼.

▶ **Figure 25.1 Carbon geometries.** These molecular models show the three common geometries around carbon: (a) tetrahedral in methane (CH_4), where the carbon is bonded to four other atoms; (b) trigonal planar in formaldehyde (CH_2O), where the carbon is bonded to three other atoms; and (c) linear in acetonitrile (CH_3CN), where the top carbon is bonded to two atoms.

sp^3 Hybridized

sp^2 Hybridized

sp Hybridized

(a) Tetrahedral   (b) Trigonal planar   (c) Linear

C—H bonds occur in almost every organic molecule. Because the valence shell of H can hold only two electrons, hydrogen forms only one covalent bond. As a result, hydrogen atoms are always located on the surface of organic molecules, as in the propane molecule:

$$
\begin{array}{ccccccc}
 & H & & H & & H & \\
 & | & & | & & | & \\
H & - & C & - & C & - & C & - H \\
 & | & & | & & | & \\
 & H & & H & & H &
\end{array}
$$

The C—C bonds form the backbone, or skeleton, of the molecule, while the H atoms are on the surface, or "skin," of the molecule.

The bonding arrangements about individual atoms are important in determining overall molecular shape. In turn, the overall shapes of organic and biochemical molecules are important in determining how they will react with other molecules, and how rapidly. They also determine important physical properties.

The Stabilities of Organic Substances

In Section 8.8 we learned about the average strengths of various chemical bonds, including those that are characteristic of organic molecules, such as C—H, C—C, C—N, C—O, and C=O bonds. Carbon forms strong bonds with a variety of elements, especially H, O, N, and the halogens. Carbon also has an exceptional ability to bond to itself, forming a variety of molecules with chains or rings of carbon atoms. As we saw in Chapter 8, double bonds are generally stronger than single bonds, and triple bonds are stronger than double bonds. Increasing bond strength with bond order is accompanied by a shortening of the bond. Thus, carbon–carbon bond lengths decrease in the order C—C > C=C > C≡C.

We know from calorimetric measurements that reaction of a simple organic substance such as methane (CH_4) with oxygen is highly exothermic. ∞(Sections 5.6, 5.7, and 5.8) Indeed, the combustion of methane (natural gas) keeps many of our homes warm during the winter months! Although the reactions of most organic compounds with oxygen are exothermic, great numbers of them are stable indefinitely at room temperature in the presence of air because the activation energy required for combustion is large.

Most reactions with low or moderate activation barriers begin when a region of high electron density on one molecule encounters a region of low electron density on another molecule. The regions of high electron density may be due to the presence of a multiple bond or to the more electronegative atom in a polar bond. Because of their strength and lack of polarity, C—C single bonds are relatively unreactive. C—H bonds are also largely unreactive for the same reasons. The C—H bond is nearly nonpolar because the electronegativities of C (2.5) and H (2.1) are close. To better understand the implications of these facts, consider ethanol:

$$
\begin{array}{ccccccc}
 & H & & H & & & \\
 & | & & | & & & \\
H & - & C & - & C & - & O & - H \\
 & | & & | & & & \\
 & H & & H & & &
\end{array}
$$

The differences in the electronegativity values of C (2.5) and O (3.5) and those of O and H (2.1) indicate that the C—O and O—H bonds are quite polar. Thus, the chemical reactions of ethanol involve these bonds. A group of atoms such as the C—O—H group, which determines how an organic molecule functions or reacts, is called a **functional group**. The functional group is the center of reactivity in an organic molecule.

GIVE IT SOME THOUGHT

Which of these bond types is most likely to be the seat of a chemical reaction: C=N, C—C, or C—H?

Glucose ($C_6H_{12}O_6$)

Ascorbic acid ($HC_6H_7O_6$)

Stearate ($C_{17}H_{35}COO^-$)

▲ **Figure 25.2 Water-soluble organic molecules.** (a) Glucose ($C_6H_{12}O_6$), a simple sugar; (b) ascorbic acid ($HC_6H_7O_6$), known as vitamin C; (c) the stearate ion ($C_{17}H_{35}COO^-$), an ion that functions as a surfactant. (To fit this illustration in the allocated space, the scale at which the surfactant molecule is drawn is different from the scale used for the glucose and ascorbic acid molecules.)

Solubility and Acid–Base Properties of Organic Substances

In most organic substances, the most prevalent bonds are carbon–carbon and carbon–hydrogen, which have low polarity. For this reason, the overall polarity of organic molecules is often low. They are generally soluble in nonpolar solvents and not very soluble in water. ∞ (Section 13.3) Molecules that are soluble in polar solvents such as water are those that have polar groups on the surface of the molecule, such as found in glucose [Figure 25.2(a) ◄] or ascorbic acid [vitamin C, Figure 25.2(b)]. Surfactant organic molecules have a long, nonpolar part that extends into a nonpolar medium and a polar, ionic "head group" that extends into a polar medium, such as water [Figure 25.2(c)]. ∞ (Section 13.6) This type of structure is found in many biochemically important substances, as well as soaps and detergents.

Many organic substances contain acidic or basic groups. The most important acidic substances are the carboxylic acids, which bear the functional group —COOH. ∞ (Section 4.3 and Section 16.10) The most important basic substances are amines, which bear the —NH_2, —NHR, or —NR_2 groups, where R is an organic group consisting of some combination of C—C and C—H bonds, such as —CH_3 or —C_2H_5. ∞ (Section 16.7)

As you read this chapter, you will find many concept links to related materials in earlier chapters, many of them to the sections just discussed. *We strongly encourage you to follow these links and review the earlier materials.* Doing so will definitely enhance your understanding and appreciation of organic chemistry and biochemistry.

25.2 INTRODUCTION TO HYDROCARBONS

Because the compounds of carbon are so numerous, it is convenient to organize them into families that exhibit structural similarities. The simplest class of organic compounds is the *hydrocarbons*, compounds composed only of carbon and hydrogen. The key structural feature of hydrocarbons (and of most other organic substances) is the presence of stable carbon–carbon bonds. Carbon is the only element capable of forming stable, extended chains of atoms bonded through single, double, or triple bonds.

Hydrocarbons can be divided into four general types, depending on the kinds of carbon–carbon bonds in their molecules. Figure 25.3 ▶ shows an example of each of the four types: alkanes, alkenes, alkynes, and aromatic hydrocarbons. In these hydrocarbons, as well as in other organic compounds, each C atom invariably has four bonds (four single bonds, two single bonds and one double bond, or one single bond and one triple bond).

Alkanes are hydrocarbons that contain only single bonds, as in ethane (C_2H_6). Because alkanes contain the largest possible number of hydrogen atoms per carbon atom, they are called *saturated hydrocarbons*. **Alkenes**, also known as olefins, are hydrocarbons that contain at least one C=C double bond, as in ethylene (C_2H_4). **Alkynes** contain at least one C≡C triple bond, as in acetylene (C_2H_2). In **aromatic hydrocarbons** the carbon atoms are connected in a planar ring structure, joined by both σ and π bonds between carbon atoms. Benzene (C_6H_6) is the best-known example of an aromatic hydrocarbon. Alkenes, alkynes, and aromatic hydrocarbons are called *unsaturated hydrocarbons* because they contain less hydrogen than an alkane with the same number of carbon atoms.

The members of these different classes of hydrocarbons exhibit different chemical behaviors, as we will see shortly. Their physical properties, however, are similar in many ways. Because carbon and hydrogen do not differ greatly in electronegativity, hydrocarbon molecules are relatively nonpolar. Thus, they are

ALKANE
Ethane CH_3CH_3

(a)

ALKENE
Ethylene $CH_2{=}CH_2$

(b)

ALKYNE
Acetylene $CH{\equiv}CH$

(c)

AROMATIC
Benzene C_6H_6

(d)

▲ Figure 25.3 **The four hydrocarbon types.** Names, molecular formulas, and geometrical structures for examples of each type of hydrocarbon.

almost completely insoluble in water, but they dissolve readily in other nonpolar solvents. Their melting points and boiling points are determined by London dispersion forces. Hence, hydrocarbons tend to become less volatile with increasing molar mass. ∞ (Section 11.2) As a result, hydrocarbons of very low molecular weight, such as C_2H_6 (bp = −89 °C), are gases at room temperature; those of moderate molecular weight, such as C_6H_{14} (bp = 69 °C), are liquids; and those of high molecular weight, such as $C_{22}H_{46}$ (mp = 44 °C), are solids.

25.3 ALKANES, ALKENES, AND ALKYNES

Table 25.1 ▼ lists several of the simplest alkanes. Many of these substances are familiar because they are used so widely. Methane is a major component of natural gas and is used for home heating and in gas stoves and water heaters. Propane is the major component of bottled gas used for home heating and cooking in areas where natural gas is not available. Butane is used in disposable lighters and in fuel canisters for gas camping stoves and lanterns. Alkanes with from 5 to 12 carbon atoms per molecule are found in gasoline.

The formulas for the alkanes given in Table 25.1 are written in a notation called *condensed structural formulas*. This notation reveals the way in which atoms are bonded to one another but does not require drawing in all the bonds.

	TABLE 25.1 ■ First Several Members of the Straight-Chain Alkane Series		
Molecular Formula	Condensed Structural Formula	Name	Boiling Point (°C)
CH_4	CH_4	Methane	−161
C_2H_6	CH_3CH_3	Ethane	−89
C_3H_8	$CH_3CH_2CH_3$	Propane	−44
C_4H_{10}	$CH_3CH_2CH_2CH_3$	Butane	−0.5
C_5H_{12}	$CH_3CH_2CH_2CH_2CH_3$	Pentane	36
C_6H_{14}	$CH_3CH_2CH_2CH_2CH_2CH_3$	Hexane	68
C_7H_{16}	$CH_3CH_2CH_2CH_2CH_2CH_2CH_3$	Heptane	98
C_8H_{18}	$CH_3CH_2CH_2CH_2CH_2CH_2CH_2CH_3$	Octane	125
C_9H_{20}	$CH_3CH_2CH_2CH_2CH_2CH_2CH_2CH_2CH_3$	Nonane	151
$C_{10}H_{22}$	$CH_3CH_2CH_2CH_2CH_2CH_2CH_2CH_2CH_2CH_3$	Decane	174

For example, the Lewis structure and the condensed structural formulas for butane (C_4H_{10}) are

$$\begin{array}{ccccc} & H & H & H & H \\ & | & | & | & | \\ H- & C- & C- & C- & C-H \\ & | & | & | & | \\ & H & H & H & H \end{array}$$

$$H_3C-CH_2-CH_2-CH_3$$

or

$$CH_3CH_2CH_2CH_3$$

We will frequently use either Lewis structures or condensed structural formulas to represent organic compounds. Notice that each carbon atom in an alkane has four single bonds, whereas each hydrogen atom forms one single bond. Notice also that each succeeding compound in the series listed in Table 25.1 has an additional CH_2 unit.

GIVE IT SOME THOUGHT

How many C—H and C—C bonds are formed by the middle carbon atom of propane, Table 25.1?

Structures of Alkanes

The Lewis structures and condensed structural formulas for alkanes do not tell us anything about the three-dimensional structures of these substances. According to the VSEPR model, the geometry about each carbon atom in an alkane is tetrahedral; that is, the four groups attached to each carbon are located at the vertices of a tetrahedron. ∞ (Section 9.2) The three-dimensional structures can be represented as shown for methane in Figure 25.4 ▼. The bonding may be described as involving sp^3-hybridized orbitals on the carbon. ∞ (Section 9.5)

▲ **Figure 25.4 Representations of the three-dimensional arrangement of bonds about carbon in methane.**

Rotation about a carbon–carbon single bond is relatively easy, and it occurs very rapidly at room temperature. To visualize such rotation, imagine grasping either methyl group in Figure 25.5▶, which shows the structure of propane, and spinning it relative to the rest of the structure. Because motion of this sort occurs very rapidly in alkanes, a long-chain alkane molecule is constantly undergoing motions that cause it to change its shape, something like a length of chain that is being shaken.

Structural Isomers

The alkanes listed in Table 25.1 are called *straight-chain hydrocarbons* because all the carbon atoms are joined in a continuous chain. Alkanes consisting of four or more carbon atoms can also form *branched chains* called *branched-chain hydrocarbons*. Figure 25.6▼ shows space-filling models, full structural formulas, and

▲ **Figure 25.5 Rotation about a C—C bond.** Either representation of propane (C_3H_8) can help you visualize rotation about a C—C single bond.

$$CH_3CH_2CH_2CH_3$$
Butane
mp −138 °C
bp −0.5 °C

$$CH_3—CH—CH_3$$
$$\quad\quad\quad |$$
$$\quad\quad\quad CH_3$$
Isobutane
(2-methylpropane)
mp −159 °C
bp −12 °C

$$CH_3CH_2CH_2CH_2CH_3$$
Pentane
mp −130 °C
bp +36 °C

$$\quad\quad\quad CH_3$$
$$\quad\quad\quad |$$
$$CH_3—CH—CH_2—CH_3$$
Isopentane
(2-methylbutane)
mp −160 °C
bp +28 °C

$$\quad\quad\quad\quad CH_3$$
$$\quad\quad\quad\quad |$$
$$CH_3—C—CH_3$$
$$\quad\quad\quad\quad |$$
$$\quad\quad\quad\quad CH_3$$
Neopentane
(2,2-dimethylpropane)
mp −16 °C
bp +9 °C

▲ **Figure 25.6 Hydrocarbon isomers.** There are two structural isomers having the formula C_4H_{10}: butane and isobutane. For the formula C_5H_{12}, there are three structural isomers: pentane, isopentane, and neopentane. In general, the number of structural isomers increases as the number of carbons in the hydrocarbon increases.

condensed structural formulas for all the possible structures of alkanes containing four or five carbon atoms. There are two ways that four carbon atoms can be joined to give C_4H_{10}: as a straight chain (left) or a branched chain (right). For alkanes with five carbon atoms (C_5H_{12}), there are three different arrangements.

Compounds with the same molecular formula but with different bonding arrangements (and hence different structures) are called **structural isomers**. The structural isomers of a given alkane differ slightly from one another in physical properties. Note the melting and boiling points of the isomers of butane and pentane, given in Figure 25.6. The number of possible structural isomers increases rapidly with the number of carbon atoms in the alkane. There are 18 possible isomers having the molecular formula C_8H_{18}, for example, and 75 possible isomers with the molecular formula $C_{10}H_{22}$.

Nomenclature of Alkanes

The first names given to the structural isomers shown in Figure 25.6 are the so-called common names. The isomer in which one CH_3 group is branched off the major chain is labeled the *iso-* isomer (for example, isobutane). As the number of isomers grows, however, it becomes impossible to find a suitable prefix to denote each isomer. The need for a systematic means of naming organic compounds was recognized early in the history of organic chemistry. In 1892 an organization called the International Union of Chemistry met in Geneva, Switzerland, to formulate rules for systematic naming of organic substances. Since that time the task of updating the rules for naming compounds has fallen to the International Union of Pure and Applied Chemistry (IUPAC). Chemists everywhere, regardless of their nationality, subscribe to a common system for naming compounds.

The IUPAC names for the isomers of butane and pentane are the ones given in parentheses for each compound in Figure 25.6. These names as well as those of other organic compounds have three parts to them:

prefix	base	suffix

What substituents?   How many carbons?   What family?

The following steps summarize the procedures used to arrive at the names of alkanes, which all have names ending with the suffix *-ane*. We use a similar approach to write the names of other organic compounds.

1. *Find the longest continuous chain of carbon atoms, and use the name of this chain (Table 25.1) as the base name of the compound.* The longest chain may not always be written in a straight line, as seen in the following example:

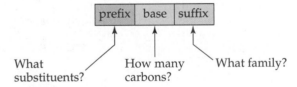

2-Methyl*hexane*

Because this compound has a chain of six C atoms, it is named as a substituted hexane. Groups attached to the main chain are called *substituents* because they are substituted in place of an H atom on the main chain.

2. *Number the carbon atoms in the longest chain, beginning with the end of the chain that is nearest to a substituent.* In our example we number the C atoms from the upper right because that places the CH_3 substituent on the second C atom of the chain; if we number from the lower right, the CH_3 would be on the fifth C atom. The chain is numbered from the end that gives the lowest number for the substituent position.

3. *Name and give the location of each substituent group.* A substituent group that is formed by removing an H atom from an alkane is called an **alkyl group**. Alkyl groups are named by replacing the *-ane* ending of the alkane name with *-yl*. The methyl group (CH_3), for example, is derived from methane (CH_4). Likewise, the ethyl group (C_2H_5) is derived from ethane (C_2H_6). Table 25.2▶ lists several common alkyl groups. The name 2-methylhexane indicates the presence of a methyl (CH_3) group on the second carbon atom of a hexane (six-carbon) chain.

▲   **GIVE IT SOME THOUGHT**

What is the chemical formula of the propyl group?

4. *When two or more substituents are present, list them in alphabetical order.* When there are two or more of the same substituent, the number of substituents of that type is indicated by a prefix: *di-* (two), *tri-* (three), *tetra-* (four), *penta-* (five), and so forth. Notice how the following example is named:

3-Ethyl-2,4,5-trimethylheptane

TABLE 25.2 ■ Condensed Structural Formulas and Common Names for Several Alkyl Groups

Group	Name
CH_3—	Methyl
CH_3CH_2—	Ethyl
$CH_3CH_2CH_2$—	Propyl
$CH_3CH_2CH_2CH_2$—	Butyl
	Isopropyl
	tert-Butyl

■ **SAMPLE EXERCISE 25.1** | Naming Alkanes

Give the systematic name for the following alkane:

SOLUTION

Analyze: We are given the structural formula of an alkane and asked to give its name.

Plan: Because the hydrocarbon is an alkane, its name ends in *-ane*. The name of the parent hydrocarbon is based on the longest continuous chain of carbon atoms, as summarized in Table 25.1. Branches are alkyl groups, named after the number of C atoms in the branch and located by counting C atoms along the longest continuous chain.

Solve: The longest continuous chain of C atoms extends from the upper left CH_3 group to the lower left CH_3 group and is seven C atoms long:

The parent compound is thus heptane. There are two CH_3 (methyl) groups that branch off the main chain. Hence, this compound is a dimethylheptane. To specify the location of the two methyl groups, we must number the C atoms from the end that gives the lowest number possible to the carbons bearing side chains. This means that we should start numbering with the upper left carbon. There is a methyl group on carbon 3, and one on carbon 4. The compound is thus 3,4-dimethylheptane.

■■ **PRACTICE EXERCISE**

Name the following alkane:

$$CH_3—CH—CH_3$$
$$CH_3—CH—CH_2$$
$$CH_3$$

Answer: 2,4-dimethylpentane

■■ **SAMPLE EXERCISE 25.2** | **Writing Condensed Structural Formulas**

Write the condensed structural formula for 3-ethyl-2-methylpentane.

SOLUTION

Analyze: We are given the systematic name for a hydrocarbon and asked to write its structural formula.

Plan: Because the compound's name ends in -*ane*, it is an alkane, meaning that all the carbon–carbon bonds are single bonds. The parent hydrocarbon is pentane, indicating five C atoms (Table 25.1). There are two alkyl groups specified, an ethyl group (two carbon atoms, C_2H_5) and a methyl group (one carbon atom, CH_3). Counting from left to right along the five-carbon chain, the ethyl group will be attached to the third C atom and the methyl group will be attached to the second C atom.

Solve: We begin by writing a string of five C atoms attached to each other by single bonds. These represent the backbone of the parent pentane chain:

$$C—C—C—C—C$$

We next place a methyl group on the second C and an ethyl group on the third C atom of the chain. We then add hydrogens to all the other C atoms to make the four bonds to each carbon, giving the following condensed structure:

$$CH_3$$
$$CH_3—CH—CH—CH_2—CH_3$$
$$CH_2CH_3$$

The formula can be written even more concisely in the following style:

$$CH_3CH(CH_3)CH(C_2H_5)CH_2CH_3$$

In this formula the branching alkyl groups are indicated in parentheses.

■■ **PRACTICE EXERCISE**

Write the condensed structural formula for 2,3-dimethylhexane.

Answer:
$$CH_3 \quad CH_3$$
$$CH_3CH—CHCH_2CH_2CH_3 \quad or \quad CH_3CH(CH_3)CH(CH_3)CH_2CH_2CH_3$$

Cycloalkanes

Alkanes can form rings or cycles. Alkanes with this form of structure are called **cycloalkanes**. Figure 25.7▶ illustrates a few cycloalkanes. Cycloalkane structures are sometimes drawn as simple polygons in which each corner of the polygon represents a CH_2 group. This method of representation is similar to that used for benzene rings. ∞ (Section 8.6) (Remember from our benzene discussion that in aromatic structures each vertex represents a CH group, not a CH_2 group.)

Carbon rings containing fewer than five carbon atoms are strained because the C—C—C bond angle in the smaller rings must be less than the 109.5° tetrahedral angle. The amount of strain increases as the rings get smaller. In cyclopropane, which has the shape of an equilateral triangle, the angle is only 60°; this molecule is therefore much more reactive than propane, its straight-chain analog.

Cyclohexane   Cyclopentane   Cyclopropane

Cycloalkanes, particularly the small-ring compounds, sometimes behave chemically like unsaturated hydrocarbons, which we will discuss shortly. The general formula for cycloalkanes, C_nH_{2n}, differs from the general formula for straight-chain alkanes, C_nH_{2n+2}.

Reactions of Alkanes

Because they contain only C—C and C—H bonds, most alkanes are relatively unreactive. At room temperature, for example, they do not react with acids, bases, or strong oxidizing agents, and they are not even attacked by boiling nitric acid. Their low chemical reactivity is due primarily to the strength and lack of polarity of C—C and C—H bonds.

Alkanes are not completely inert, however. One of their most commercially important reactions is *combustion* in air, which is the basis of their use as fuels. ∞ (Section 3.2) For example, the complete combustion of ethane proceeds as follows:

$$2\,C_2H_6(g) + 7\,O_2(g) \longrightarrow 4\,CO_2(g) + 6\,H_2O(l) \qquad \Delta H = -2855\ \text{kJ}$$

In the following sections we will see that hydrocarbons can be modified to impart greater reactivity by introducing unsaturation into the carbon–carbon framework and by attaching other reactive groups to the hydrocarbon backbone.

Chemistry Put to Work   GASOLINE

Petroleum, or crude oil, is a complex mixture of organic compounds, mainly hydrocarbons, with smaller quantities of other organic compounds containing nitrogen, oxygen, or sulfur. The tremendous demand for petroleum to meet the world's energy needs has led to the tapping of oil wells in such forbidding places as the North Sea and northern Alaska.

The usual first step in the *refining*, or processing, of petroleum is to separate it into fractions on the basis of boiling point. The fractions commonly taken are shown in Table 25.3 ▼.

Because gasoline is the most commercially important of these fractions, various processes are used to maximize its yield.

Gasoline is a mixture of volatile hydrocarbons containing varying amounts of aromatic hydrocarbons in addition to alkanes. In a traditional automobile engine a mixture of air and gasoline vapor is compressed by a piston and then ignited by a spark plug. The burning of the gasoline should create a strong, smooth expansion of gas, forcing the piston outward and imparting force along the drive shaft of the engine. If the gas

TABLE 25.3 ■ Hydrocarbon Fractions from Petroleum

Fraction	Size Range of Molecules	Boiling-Point Range (°C)	Uses
Gas	C_1 to C_5	−160 to 30	Gaseous fuel, production of H_2
Straight-run gasoline	C_5 to C_{12}	30 to 200	Motor fuel
Kerosene, fuel oil	C_{12} to C_{18}	180 to 400	Diesel fuel, furnace fuel, cracking
Lubricants	C_{16} and up	350 and up	Lubricants
Paraffins	C_{20} and up	Low-melting solids	Candles, matches
Asphalt	C_{36} and up	Gummy residues	Surfacing roads

continues on next page

continued

▲ **Figure 25.8 Octane rating.** The octane rating of gasoline measures its resistance to knocking when burned in an engine. The octane rating of this gasoline is 89, as shown on the face of the pump.

burns too rapidly, the piston receives a single hard slam rather than a strong, smooth push. The result is a "knocking" or "pinging" sound and a reduction in the efficiency with which energy produced by the combustion is converted to work.

The *octane number* of a gasoline is a measure of its resistance to knocking. Gasolines with high octane numbers burn more smoothly and are thus more effective fuels (Figure 25.8▲). Branched alkanes and aromatic hydrocarbons have higher octane numbers than the straight-chain alkanes. The octane number of gasoline is obtained by comparing its knocking characteristics with those of "isooctane" (2,2,4-trimethylpentane) and heptane. Isooctane is assigned an octane number of 100, whereas heptane is assigned 0. Gasoline with the same knocking characteristics as a mixture of 91% isooctane and 9% heptane is rated as 91 octane.

The gasoline obtained directly from fractionation of petroleum (called *straight-run* gasoline) contains mainly straight-chain hydrocarbons and has an octane number around 50. It is therefore subjected to a process called *reforming*, which converts the straight-chain alkanes into more desirable branched-chain ones (Figure 25.9▶). *Cracking* is used to produce aromatic hydrocarbons and to convert some of the less volatile kerosene and fuel-oil fraction into compounds with lower molecular weights that are suitable for use as automobile fuel. In the cracking process the hydrocarbons are mixed with a catalyst and heated to 400 °C to 500 °C. The catalysts used are naturally occurring clay minerals or synthetic Al_2O_3–SiO_2 mixtures. In addition to forming molecules more suitable for gasoline, cracking results in the formation of hydrocarbons of lower molecular weight, such as ethylene and propene. These substances are used in a variety of reactions to form plastics and other chemicals.

▲ **Figure 25.9 Fractional distillation.** Petroleum is separated into fractions by distillation and is subjected to catalytic cracking in a refinery, as shown here.

Adding certain compounds called *antiknock agents* or octane enhancers increases the octane rating of gasoline. Until the mid-1970s the principal antiknock agent was tetraethyl lead, $(C_2H_5)_4Pb$. It is no longer used, however, because of the environmental hazards of lead and because it poisons catalytic converters. ∞(Section 14.7: "Chemistry Put to Work: Catalytic Converters") Aromatic compounds such as toluene ($C_6H_5CH_3$) and oxygenated hydrocarbons such as ethanol (CH_3CH_2OH) and methyl *tert*-butyl ether (MTBE, shown below) are now generally used as antiknock agents.

$$CH_3-\underset{\underset{CH_3}{|}}{\overset{\overset{CH_3}{|}}{C}}-O-CH_3$$

The use of MTBE has been banned in several states, however, because it finds its way into drinking-water supplies from spills and leaking storage tanks, giving the water a bad smell and taste and perhaps producing adverse health effects. *Related Exercises: 25.29 and 25.30*

Alkenes

Alkenes are unsaturated hydrocarbons that contain at least one C=C bond. The simplest alkene is $CH_2=CH_2$, called ethene (IUPAC) or ethylene. Ethylene is a plant hormone. It plays important roles in seed germination and fruit ripening. The next member of the series is $CH_3-CH=CH_2$, called propene or propylene. For alkenes with four or more carbon atoms, several isomers exist for each molecular formula. For example, there are four isomers of C_4H_8, as shown in Figure 25.10▶. Notice both their structures and their names.

Methylpropene
bp −7 °C

1-Butene
bp −6 °C

cis-2-Butene
bp +4 °C

trans-2-Butene
bp +1 °C

▲ **Figure 25.10 C$_4$H$_8$ structural isomers.** Full structural formulas, names, and boiling points of alkenes having the molecular formula C$_4$H$_8$.

The names of alkenes are based on the longest continuous chain of carbon atoms that contains the double bond. The name given to the chain is obtained from the name of the corresponding alkane (Table 25.1) by changing the ending from -*ane* to -*ene*. The compound on the left in Figure 25.10, for example, has a double bond as part of a three-carbon chain; thus, the parent alkene is propene.

The location of the double bond along an alkene chain is indicated by a prefix number that designates the number of the carbon atom that is part of the double bond and is nearest an end of the chain. The chain is always numbered from the end that brings us to the double bond sooner and hence gives the smallest-numbered prefix. In propene the only possible location for the double bond is between the first and second carbons; thus, a prefix indicating its location is unnecessary. For butene (Figure 25.10) there are two possible positions for the double bond, either after the first carbon (1-butene) or after the second carbon (2-butene).

GIVE IT SOME THOUGHT

How many distinct locations are there for one double bond in a five-carbon linear chain?

If a substance contains two or more double bonds, each is located by a numerical prefix. The ending of the name is altered to identify the number of double bonds: diene (two), triene (three), and so forth. For example, CH$_2$=CH—CH$_2$—CH=CH$_2$ is 1,4-pentadiene.

The two isomers on the right in Figure 25.10 differ in the relative locations of their terminal methyl groups. These two compounds are **geometric isomers**, compounds that have the same molecular formula and the same groups bonded to one another, but differ in the spatial arrangement of these groups. ∞(Section 24.4) In the *cis* isomer the two methyl groups are on the same side of the double bond, whereas in the *trans* isomer they are on opposite sides. Geometric isomers possess distinct physical properties and often differ significantly in their chemical behavior.

Geometric isomerism in alkenes arises because, unlike the C—C bond, the C=C bond resists twisting. Recall that the double bond between two carbon atoms consists of a σ and a π bond. ∞(Section 9.6) Figure 25.11 ▶ shows a *cis* alkene. The carbon–carbon bond axis and the bonds to the hydrogen atoms and to the alkyl groups (designated R) are all in a plane. The *p* orbitals that overlap sideways to form the π bond are perpendicular to the molecular plane. As Figure 25.11 shows, rotation around the carbon–carbon double bond requires the π bond to be broken, a process that requires considerable energy (about 250 kJ/mol). While rotation about a double bond does not occur easily, it is a key process in the chemistry of vision. ∞(Section 9.6: "Chemistry and Life: Chemistry of Vision")

π bond

90° rotation
around
double bond

▲ **Figure 25.11 Rotation about a carbon–carbon double bond.** In an alkene, the overlap of the *p* orbitals that form the π bond is lost in the rotation. For this reason, rotation about carbon–carbon double bonds does not occur readily.

■ **SAMPLE EXERCISE 25.3** | Drawing Isomers

Draw all the structural and geometric isomers of pentene, C_5H_{10}, that have an un-branched hydrocarbon chain.

SOLUTION

Analyze: We are asked to draw all the isomers (both structural and geometric) for an alkene with a five-carbon chain.

Plan: Because the compound is named pentene and not pentadiene or pentatriene, we know that the five-carbon chain contains only one carbon–carbon double bond. Thus, we can begin by first placing the double bond in various locations along the chain, remembering that the chain can be numbered from either end. After finding the different distinct locations for the double bond, we can consider whether the molecule can have *cis* and *trans* isomers.

Solve: There can be a double bond after either the first carbon (1-pentene) or second carbon (2-pentene). These are the only two possibilities because the chain can be numbered from either end. (Thus, what we might erroneously call 4-pentene is actually 1-pentene, as seen by numbering the carbon chain from the other end.)

Because the first C atom in 1-pentene is bonded to two H atoms, there are no *cis-trans* isomers. On the other hand, there are *cis* and *trans* isomers for 2-pentene. Thus, the three isomers for pentene are

$$CH_3-CH_2-CH_2-CH=CH_2$$
1-Pentene

$$\underset{H}{\overset{CH_3-CH_2}{>}}C=C\underset{H}{\overset{CH_3}{<}}$$
cis-2-Pentene

$$\underset{H}{\overset{CH_3-CH_2}{>}}C=C\underset{CH_3}{\overset{H}{<}}$$
trans-2-Pentene

(You should convince yourself that *cis*- or *trans*-3-pentene is identical to *cis*- or *trans*-2-pentene, respectively.)

■ **PRACTICE EXERCISE**

How many straight-chain isomers are there of hexene, C_6H_{12}?
Answer: five (1-hexene, *cis*-2-hexene, *trans*-2-hexene, *cis*-3-hexene, *trans*-3-hexene)

Alkynes

Alkynes are unsaturated hydrocarbons containing one or more C≡C bonds. The simplest alkyne is acetylene (C_2H_2), a highly reactive molecule. When acetylene is burned in a stream of oxygen in an oxyacetylene torch, the flame reaches about 3200 K. The oxyacetylene torch is widely used in welding, which requires high temperatures. Alkynes in general are highly reactive molecules. Because of their higher reactivity, they are not as widely distributed in nature as alkenes; alkynes, however, are important intermediates in many industrial processes.

Alkynes are named by identifying the longest continuous chain in the molecule containing the triple bond and modifying the ending of the name as listed in Table 25.1 from *-ane* to *-yne*, as shown in Sample Exercise 25.4.

■ **SAMPLE EXERCISE 25.4** | Naming Unsaturated Hydrocarbons

Name the following compounds:

(a) $$CH_3CH_2CH_2-\underset{\underset{H}{\overset{\displaystyle |}{C}}}{\overset{\displaystyle CH_3}{\overset{\displaystyle |}{CH}}} C=C\underset{H}{\overset{CH_3}{<}}$$

(b) $$CH_3CH_2CH_2\underset{\overset{\displaystyle |}{CH_2CH_2CH_3}}{\overset{\displaystyle |}{CH}}-C≡CH$$

SOLUTION

Analyze: We are given the structural formulas for two compounds, the first an alkene and the second an alkyne, and asked to name the compounds.

Plan: In each case the name is based on the number of carbon atoms in the longest continuous carbon chain that contains the multiple bond. In the case of the alkene, care must be taken to indicate whether *cis-trans* isomerism is possible and, if so, which isomer is given.

Solve:

(a) The longest continuous chain of carbons that contains the double bond is seven in length. The parent compound is therefore heptene. Because the double bond begins at carbon 2 (numbering from the end closest to the double bond), the parent hydrocarbon chain is named 2-heptene. A methyl group is found at carbon atom 4. Thus, the compound is 4-methyl-2-heptene. The geometrical configuration at the double bond is *cis* (that is, the alkyl groups are bonded to the double bond on the same side). Thus, the full name is 4-methyl-*cis*-2-heptene.

(b) The longest continuous chain of carbon atoms containing the triple bond is six, so this compound is a derivative of hexyne. The triple bond comes after the first carbon (numbering from the right), making it a derivative of 1-hexyne. The branch from the hexyne chain contains three carbon atoms, making it a propyl group. Because it is located on the third carbon atom of the hexyne chain, the molecule is 3-propyl-1-hexyne.

■ PRACTICE EXERCISE

Draw the condensed structural formula for 4-methyl-2-pentyne.

Answer: $CH_3-C\equiv C-CH-CH_3$
$$\qquad\qquad\qquad\qquad |$$
$$\qquad\qquad\qquad\quad CH_3$$

Addition Reactions of Alkenes and Alkynes

The presence of carbon–carbon double or triple bonds in hydrocarbons markedly increases their chemical reactivity. The most characteristic reactions of alkenes and alkynes are **addition reactions**, in which a reactant is added to the two atoms that form the multiple bond. A simple example is the addition of a halogen such as Br_2 to ethylene:

$$H_2C{=}CH_2 + Br_2 \longrightarrow H_2C-CH_2 \qquad [25.1]$$
$$\qquad\qquad\qquad\qquad\quad |\quad\;\; |$$
$$\qquad\qquad\qquad\qquad\; Br\;\; Br$$

The pair of electrons that forms the π bond in ethylene is uncoupled and is used to form two new σ bonds to the two bromine atoms. The σ bond between the carbon atoms is retained.

Addition of H_2 to an alkene converts it to an alkane:

$$CH_3CH{=}CHCH_3 + H_2 \xrightarrow{\text{Ni, 500 °C}} CH_3CH_2CH_2CH_3 \qquad [25.2]$$

The reaction between an alkene and H_2, referred to as *hydrogenation*, does not occur readily under ordinary conditions of temperature and pressure. One reason for the lack of reactivity of H_2 toward alkenes is the high bond enthalpy of the H_2 bond. To promote the reaction, it is necessary to use a catalyst that assists in rupturing the H—H bond. The most widely used catalysts are finely divided metals on which H_2 is adsorbed. ∞ (Section 14.7)

Hydrogen halides and water can also add to the double bond of alkenes, as illustrated by the following reactions of ethylene:

$$CH_2{=}CH_2 + HBr \longrightarrow CH_3CH_2Br \qquad [25.3]$$
$$CH_2{=}CH_2 + H_2O \xrightarrow{H_2SO_4} CH_3CH_2OH \qquad [25.4]$$

The addition of water is catalyzed by a strong acid, such as H_2SO_4.

The addition reactions of alkynes resemble those of alkenes, as shown in the following examples:

$$CH_3C{\equiv}CCH_3 + Cl_2 \longrightarrow \underset{\underset{\displaystyle CH_3}{|}}{\overset{\overset{\displaystyle Cl}{|}}{C}}{=}\underset{\underset{\displaystyle Cl}{|}}{\overset{\overset{\displaystyle CH_3}{|}}{C}} \qquad [25.5]$$

2-Butyne   *trans*-2,3-Dichloro-2-butene

$$CH_3C{\equiv}CCH_3 + 2\,Cl_2 \longrightarrow CH_3{-}\underset{\underset{\displaystyle Cl}{|}}{\overset{\overset{\displaystyle Cl}{|}}{C}}{-}\underset{\underset{\displaystyle Cl}{|}}{\overset{\overset{\displaystyle Cl}{|}}{C}}{-}CH_3 \qquad [25.6]$$

2-Butyne   2,2,3,3-Tetrachlorobutane

■■ SAMPLE EXERCISE 25.5 | Identifying the Product of a Hydrogenation Reaction

Write the structural formula for the product of the hydrogenation of 3-methyl-1-pentene.

SOLUTION

Analyze: We are asked to predict the compound formed when a particular alkene undergoes hydrogenation (reaction with H_2).

Plan: To determine the structural formula of the reaction product, we must first write the structural formula or Lewis structure of the reactant. In the hydrogenation of the alkene, H_2 adds to the double bond, producing an alkane. (That is, each carbon atom of the double bond forms a bond to an H atom, and the double bond is converted to a single bond.)

Solve: The name of the starting compound tells us that we have a chain of five C atoms with a double bond at one end (position 1) and a methyl group on the third C from that end (position 3):

$$CH_2{=}CH{-}\underset{\underset{\displaystyle CH_3}{|}}{CH}{-}CH_2{-}CH_3$$

Hydrogenation—the addition of two H atoms to the carbons of the double bond—leads to the following alkane:

$$CH_3{-}CH_2{-}\underset{\underset{\displaystyle CH_3}{|}}{CH}{-}CH_2{-}CH_3$$

Comment: The longest chain in the product alkane has five carbon atoms; its name is therefore 3-methylpentane.

■■ PRACTICE EXERCISE

Addition of HCl to an alkene forms 2-chloropropane. What is the alkene?
Answer: propene

Mechanism of Addition Reactions

As the understanding of chemistry has grown, chemists have been able to advance from simply cataloging reactions known to occur to explaining *how* they occur. An explanation of how a reaction proceeds is called a *mechanism*. ∞ (Section 14.6)

In Equation 25.3 we considered the addition reaction between HBr and an alkene. This reaction is thought to proceed in two steps. In the first step, which is rate determining [∞ (Section 14.6)], the HBr molecule attacks the electron-rich double bond, transferring a proton to one of the two alkene carbons.

In the reaction of 2-butene with HBr, for example, the first step proceeds as follows:

$$CH_3CH{=}CHCH_3 + HBr \longrightarrow \left[\begin{array}{c} \overset{\delta+}{CH_3CH}{=}{=}{=}CHCH_3 \\ \vdots \\ H \\ \vdots \\ \overset{}{Br}\,^{\delta-} \end{array} \right] \longrightarrow CH_3\overset{+}{CH}{-}CH_2CH_3 + Br^- \quad [25.7]$$

The pair of electrons that formed the π bond between the carbon atoms in the alkane is used to form the new C—H bond.

The second step, involving the addition of Br^- to the positively charged carbon, is faster:

$$CH_3\overset{+}{CH}{-}CH_2CH_3 + Br^- \longrightarrow \left[\begin{array}{c} \overset{\delta+}{CH_3CH}{-}CH_2CH_3 \\ \vdots \\ Br\,^{\delta-} \end{array} \right] \longrightarrow \underset{Br}{CH_3CHCH_2CH_3} \quad [25.8]$$

In this reaction the bromide ion (Br^-) donates a pair of electrons to the carbon, forming the new C—Br bond.

Because the first rate-determining step in the reaction involves both the alkene and acid, the rate law for the reaction is second order, first order each in the alkene and HBr:

$$\text{Rate} = -\frac{\Delta[CH_3CH{=}CHCH_3]}{\Delta t} = k[CH_3CH{=}CHCH_3][HBr] \quad [25.9]$$

The energy profile for the reaction is shown in Figure 25.12▼. The first energy maximum represents the transition state in the first step of the mechanism. The second maximum represents the transition state for the second step. Notice that there is an energy minimum between the first and second steps of the reaction. This energy minimum corresponds to the energies of the intermediate species, $CH_3\overset{+}{CH}{-}CH_2CH_3$ and Br^-.

◀ **Figure 25.12 Energy profile for addition of HBr to 2-butene.** The chemical formulas given in brackets in Equations 25.7 and 25.8 are the transition states.

To show electron movement in reactions like these, chemists often use curved arrows, which point in the direction of the electron flow (from a negative charge toward a positive charge). For the addition of HBr to 2-butene, the shifts in electron positions are

$$CH_3CH{=}CHCH_3 + H{-}\ddot{B}r\!: \xrightarrow{\text{slow}} CH_3{-}\overset{+}{C}{-}\overset{H}{\underset{H}{C}}{-}CH_3 + \!:\!\ddot{B}r\!:^{-} \xrightarrow{\text{fast}} CH_3{-}\overset{H}{\underset{:\ddot{B}r:}{C}}{-}\overset{H}{\underset{H}{C}}{-}CH_3$$

Aromatic Hydrocarbons

Aromatic hydrocarbons are a large and important class of hydrocarbons. The simplest member of the series is benzene (C_6H_6), whose structure is shown in Figure 25.3. The planar, highly symmetrical structure of benzene, with its 120° bond angles, suggests a high degree of unsaturation. You might therefore expect benzene to resemble the unsaturated hydrocarbons and to be highly reactive. The chemical behavior of benzene, however, is unlike that of alkenes or alkynes. Benzene and the other aromatic hydrocarbons are much more stable than alkenes and alkynes because the π electrons are delocalized in the π orbitals. ∞ (Section 9.6)

We can estimate the stabilization of the π electrons in benzene by comparing the energy required to add hydrogen to benzene to form a saturated compound with the energy required to hydrogenate certain alkenes. The hydrogenation of benzene to form cyclohexane can be represented as

$$\Delta H° = -208 \text{ kJ/mol}$$

The enthalpy change in this reaction is −208 kJ/mol. The heat of hydrogenation of the cyclic alkene cyclohexene is −120 kJ/mol:

$$+ H_2 \longrightarrow \qquad \Delta H° = -120 \text{ kJ/mol}$$

Similarly, the heat released on hydrogenating 1,4-cyclohexadiene is −232 kJ/mol:

$$+ \ 2\,H_2 \longrightarrow \qquad \Delta H° = -232 \text{ kJ/mol}$$

From these last two reactions, it appears that the heat of hydrogenating each double bond is roughly 116 kJ/mol for each bond. Benzene contains the equivalent of three double bonds. We might expect, therefore, that the heat of hydrogenating benzene would be about 3 times −116, or −348 kJ/mol, if benzene behaved as though it were "cyclohexatriene"; that is, if it behaved as though it had three isolated double bonds in a ring. Instead, the heat released is 140 kJ less than this, indicating that benzene is more stable than would be expected for three double bonds. The difference of 140 kJ/mol between the "expected" heat of hydrogenation, −348 kJ/mol, and the observed heat of hydrogenation, −208 kJ/mol, is due to stabilization of the π electrons through delocalization in the π orbitals that extend around the ring in this aromatic compound.

Each aromatic ring system is given a common name as shown in Figure 25.13 ◀. Although aromatic hydrocarbons are unsaturated, *they do not readily undergo addition reactions*. The delocalized π bonding causes aromatic compounds

Benzene   Naphthalene

Anthracene   Toluene (Methylbenzene)

Pyrene

▲ **Figure 25.13 Condensed structural formulas and common names of several aromatic compounds.** The aromatic rings are represented by hexagons with a circle inscribed inside to denote aromatic character. Each corner represents a carbon atom. Each carbon is bound to three other atoms—either three carbons or two carbons and a hydrogen. The hydrogen atoms are not shown.

to behave quite differently from alkenes and alkynes. Benzene, for example, does not add Cl_2 or Br_2 to its double bonds under ordinary conditions. In contrast, aromatic hydrocarbons undergo **substitution reactions** relatively easily. In a substitution reaction one atom of a molecule is removed and replaced (substituted) by another atom or group of atoms. When benzene is warmed in a mixture of nitric and sulfuric acids, for example, hydrogen is replaced by the nitro group, NO_2:

$$[25.10]$$

More vigorous treatment results in substitution of a second nitro group into the molecule:

$$[25.11]$$

There are three possible isomers of benzene with two nitro groups attached. These three isomers are named *ortho-*, *meta-*, and *para-*dinitrobenzene:

ortho-Dinitrobenzene   *meta*-Dinitrobenzene   *para*-Dinitrobenzene
mp 118 °C   mp 90 °C   mp 174 °C

Mainly the *meta* isomer is formed in the reaction of nitric acid with nitrobenzene.

The bromination of benzene, which is carried out using $FeBr_3$ as a catalyst, is another substitution reaction:

$$[25.12]$$

In a similar reaction, called the *Friedel-Crafts reaction*, alkyl groups can be substituted onto an aromatic ring by reaction of an alkyl halide with an aromatic compound in the presence of $AlCl_3$ as a catalyst:

$$[25.13]$$

> ## GIVE IT SOME THOUGHT
>
> When naphthalene is reacted with nitric and sulfuric acids, two compounds containing one nitro group are formed. Draw the structures of these two compounds.

Chemistry and Life   POLYCYCLIC AROMATIC HYDROCARBONS

If you look up the U.S. Environmental Protection Agency's list of priority pollutants, you will find "PAHs" among the top 25. PAHs are "polycyclic aromatic hydrocarbons," a term which refers to an entire class of molecules whose structures are fused benzene rings, such as naphthalene and anthracene (Figure 25.13). Traditionally, molecules with three fused benzene rings or more are considered PAHs; some examples are shown below.

Tetracene

Chrysene

Pyrene

Benzo[a]pyrene

Pentacene

Coronene

You will notice that only one representative resonance structure is shown for each molecule, and that all these molecules resemble graphite. ∞ (Section 11.8) PAHs are found in coal and tars, and occur as side products of combustion of carbon-containing compounds. Therefore, PAHs can be found as particulate matter in air, water, and adsorbed to sediments and soil. They are relatively insoluble in water, as you would expect; the larger the molecule, the less water soluble, but more fat soluble, it is. Many PAHs are carcinogenic, meaning that they are known to cause cancer. Benzo[a]pyrene was the first chemical carcinogen discovered; it is one of the many components of cigarette smoke.

In spite of the toxicity of PAHs, there is great scientific interest in their electronic properties. Tetracene and pentacene are large enough that their delocalized π electrons provide a very high conductivity in the plane of the rings, but these molecules are also small enough that they are chemically easier to work with compared to larger PAHs. Tetracene is yellow, and pentacene is dark blue, due to their delocalized π electron systems. ∞ (Section 9.6) Pentacene is being explored as a component in organic thin-film transistors and solar energy conversion devices. ∞ (Section 12.3)

25.4 ORGANIC FUNCTIONAL GROUPS

The reactivity of organic compounds can be attributed to particular atoms or groups of atoms within the molecules. A site of reactivity in an organic molecule is called a *functional group* because it controls how the molecule behaves or functions. ∞ (Section 25.1) As we have seen, the presence of C=C double bonds or C≡C triple bonds in a hydrocarbon markedly increases its reactivity. Furthermore, these functional groups each undergo characteristic reactions. Each distinct kind of functional group often undergoes the same kinds of reactions in every molecule, regardless of the size and complexity of the molecule. Thus, the chemistry of an organic molecule is largely determined by the functional groups it contains.

Table 25.4 ▶ lists the most common functional groups and gives examples of each. Notice that in addition to C=C double bonds or C≡C triple bonds, there are also many functional groups that contain elements other than just C and H. Many of the functional groups contain other nonmetals such as O and N.

We can think of organic molecules as being composed of functional groups that are bonded to one or more alkyl groups. The alkyl groups, which are made of C—C and C—H single bonds, are the less reactive portions of the organic molecules. In describing general features of organic compounds, chemists often use the designation R to represent any alkyl group: methyl, ethyl, propyl, and so on. Alkanes, for example, which contain no functional group, are represented as R—H. Alcohols, which contain the —OH, or alcohol functional group, are represented as R—OH. If two or more different alkyl groups are present in a

TABLE 25.4 ■ Common Functional Groups in Organic Compounds

Functional Group	Type of Compound	Suffix or Prefix	Example	Systematic Name (common name)
$\diagdown$C$=$C$\diagup$	Alkene	*-ene*	(H)(H)C$=$C(H)(H)	Ethene (Ethylene)
—C≡C—	Alkyne	*-yne*	H—C≡C—H	Ethyne (Acetylene)
—C—Ö—H	Alcohol	*-ol*	H—C(H)(H)—Ö—H	Methanol (Methyl alcohol)
—C—Ö—C—	Ether	*ether*	H—C(H)(H)—Ö—C(H)(H)—H	Dimethyl ether
—C—Ẍ: (X = halogen)	Haloalkane	*halo-*	H—C(H)(H)—C̈l:	Chloromethane (Methyl chloride)
—C—N̈—	Amine	*-amine*	H—C(H)(H)—C(H)(H)—N̈(H)—H	Ethylamine
(:O:) —C—H	Aldehyde	*-al*	H—C(H)(H)—C(=O)—H	Ethanal (Acetaldehyde)
(:O:) —C—C—C—	Ketone	*-one*	H—C(H)(H)—C(=O)—C(H)(H)—H	Propanone (Acetone)
(:O:) —C—Ö—H	Carboxylic acid	*-oic acid*	H—C(H)(H)—C(=O)—Ö—H	Ethanoic acid (Acetic acid)
(:O:) —C—Ö—C—	Ester	*-oate*	H—C(H)(H)—C(=O)—Ö—C(H)(H)—H	Methyl ethanoate (Methyl acetate)
(:O:) —C—N̈—	Amide	*-amide*	H—C(H)(H)—C(=O)—N̈(H)—H	Ethanamide (Acetamide)

molecule, we will designate them as R, R′, R″, and so forth. In this section we examine the structure and chemical properties of two functional groups, alcohols and ethers. Later in this section we consider some additional functional groups that contain C$=$O bonds.

$$CH_3\!-\!\overset{\displaystyle |}{\underset{\displaystyle OH}{CH}}\!-\!CH_3 \qquad CH_3\!-\!\overset{\displaystyle CH_3}{\underset{\displaystyle OH}{\overset{\displaystyle |}{\underset{\displaystyle |}{C}}}}\!-\!CH_3 \qquad \overset{\displaystyle |}{\underset{\displaystyle OH}{CH_2}}\!-\!\overset{\displaystyle |}{\underset{\displaystyle OH}{CH_2}}$$

2-Propanol   2-Methyl-2-propanol   1,2-Ethanediol
Isopropyl alcohol;   *t*-Butyl alcohol   Ethylene glycol
rubbing alcohol

Phenol

$$\overset{\displaystyle OH}{\underset{}{\bigcirc\!\!\!\!\!\bigcirc}}$$

$$\overset{\displaystyle |}{\underset{\displaystyle OH}{CH_2}}\!-\!\overset{\displaystyle |}{\underset{\displaystyle OH}{CH}}\!-\!\overset{\displaystyle |}{\underset{\displaystyle OH}{CH_2}}$$

Phenol   1,2,3-Propanetriol   Cholesterol
           Glycerol; glycerin

▲ Figure 25.14 Structural formulas of six important alcohols. Common names are given in blue.

▲ Figure 25.15 Everyday alcohols. Many of the products we use every day—from rubbing alcohol and throat lozenges to hair spray and antifreeze—are composed either entirely or mainly of alcohols.

Alcohols

Alcohols are hydrocarbon derivatives in which one or more hydrogens of a parent hydrocarbon have been replaced by a *hydroxyl* or *alcohol* functional group, —OH. Figure 25.14 ◀ shows the structural formulas and names of several alcohols. Note that the name for an alcohol ends in *-ol*. The simple alcohols are named by changing the last letter in the name of the corresponding alkane to *-ol*—for example, etha*ne* becomes etha*nol*. Where necessary, the location of the OH group is designated by an appropriate numeral prefix that indicates the number of the carbon atom bearing the OH group, as shown in many of the examples in Figure 25.14.

The O—H bond is polar, so alcohols are much more soluble in polar solvents such as water than are hydrocarbons. The —OH functional group can also participate in hydrogen bonding. As a result, the boiling points of alcohols are much higher than those of their parent alkanes.

Figure 25.15 ◀ shows several familiar commercial products that consist entirely or in large part of an organic alcohol. Let's consider how some of the more important alcohols are formed and used.

The simplest alcohol, methanol (methyl alcohol), has many important industrial uses and is produced on a large scale. Carbon monoxide and hydrogen are heated together under pressure in the presence of a metal oxide catalyst:

$$CO(g) + 2\,H_2(g) \xrightarrow[400\,°C]{200-300\,\text{atm}} CH_3OH(g) \qquad [25.14]$$

Because methanol has a very high octane rating as an automobile fuel, it is used as a gasoline additive and as a fuel in its own right.

Ethanol (ethyl alcohol, C_2H_5OH) is a product of the fermentation of carbohydrates such as sugar and starch. In the absence of air, yeast cells convert carbohydrates into a mixture of ethanol and CO_2, as shown in Equation 25.15. In the process, yeast derives energy necessary for growth.

$$C_6H_{12}O_6(aq) \xrightarrow{\text{yeast}} 2\,C_2H_5OH(aq) + 2\,CO_2(g) \qquad [25.15]$$

This reaction is carried out under carefully controlled conditions to produce beer, wine, and other beverages in which ethanol is the active ingredient.

Many polyhydroxyl alcohols (those containing more than one OH group) are known. The simplest of these is 1,2-ethanediol (ethylene glycol, $HOCH_2CH_2OH$). This substance is the major ingredient in automobile antifreeze. Another common polyhydroxyl alcohol is 1,2,3-propanetriol [glycerol, $HOCH_2CH(OH)CH_2OH$]. It is a viscous liquid that dissolves readily in water and is widely used as a skin softener in cosmetic preparations. It is also used in foods and candies to keep them moist.

Phenol is the simplest compound with an OH group attached to an aromatic ring. One of the most striking effects of the aromatic group is the greatly increased acidity of the OH group. Phenol is about 1 million times more acidic in water than a typical nonaromatic alcohol such as ethanol. Even so, it is not a very strong acid ($K_a = 1.3 \times 10^{-10}$). Phenol is used industrially to make several kinds of plastics and dyes. It is also used as a topical anesthetic in many sore throat sprays.

Cholesterol, shown in Figure 25.14, is a biochemically important alcohol. The OH group forms only a small component of this rather large molecule, so cholesterol is only slightly soluble in water (0.26 g per 100 mL of H_2O). Cholesterol is a normal component of our bodies; when present in excessive amounts, however, it may precipitate from solution. It precipitates in the gallbladder to form crystalline lumps called *gallstones*. It may also precipitate against the walls of veins and arteries and thus contribute to high blood pressure and other cardiovascular problems. The amount of cholesterol in our blood is determined by how much cholesterol we eat and by total dietary intake. There is evidence that excessive caloric intake leads the body to synthesize excessive cholesterol.

Ethers

Compounds in which two hydrocarbon groups are bonded to one oxygen are called **ethers**. Ethers can be formed from two molecules of alcohol by splitting out a molecule of water. The reaction is catalyzed by sulfuric acid, which takes up water to remove it from the system:

$$CH_3CH_2-OH + H-OCH_2CH_3 \xrightarrow{H_2SO_4} CH_3CH_2-O-CH_2CH_3 + H_2O$$

$$[25.16]$$

A reaction in which water is split out from two substances is called a *condensation reaction*. ∞ (Sections 12.4 and 22.8)

Ethers are used as solvents. Both diethyl ether and the cyclic ether tetrahydrofuran are common solvents for organic reactions:

$$CH_3CH_2-O-CH_2CH_3$$

$$\begin{array}{c} CH_2-CH_2 \\ | \qquad | \\ CH_2 \quad CH_2 \\ \diagdown O \diagup \end{array}$$

Diethyl ether   Tetrahydrofuran (THF)

Aldehydes and Ketones

Several of the functional groups listed in Table 25.4 contain a $C=O$ double bond. This particular group of atoms is called a **carbonyl group**. The carbonyl group, together with the atoms that are attached to the carbon of the carbonyl group, defines several important functional groups that we consider in this section. In **aldehydes** the carbonyl group has at least one hydrogen atom attached, as in the following examples:

$$\begin{array}{cc} O & O \\ || & || \\ H-C-H & CH_3-C-H \end{array}$$

Methanal   Ethanal
Formaldehyde   Acetaldehyde

In **ketones** the carbonyl group occurs at the interior of a carbon chain and is therefore flanked by carbon atoms:

$$\begin{array}{cc} O & O \\ || & || \\ CH_3-C-CH_3 & CH_3-C-CH_2CH_3 \end{array}$$

Propanone   2-Butanone
Acetone   Methyl ethyl ketone

Aldehydes and ketones can be prepared by carefully controlled oxidation of alcohols. It is fairly easy to oxidize alcohols. Complete oxidation results in formation of CO_2 and H_2O, as in the burning of methanol:

$$CH_3OH(g) + \tfrac{3}{2}O_2(g) \longrightarrow CO_2(g) + 2\,H_2O(g)$$

Controlled partial oxidation to form other organic substances, such as aldehydes and ketones, is carried out by using various oxidizing agents such as air, hydrogen peroxide (H_2O_2), ozone (O_3), and potassium dichromate ($K_2Cr_2O_7$).

(a)

(b)

▲ **Figure 25.16 Everyday carboxylic acids and esters.** (a) Spinach and some cleaners contain oxalic acid; vinegar contains acetic acid; vitamin C is ascorbic acid; citrus fruits contain citric acid; and aspirin is acetylsalicylic acid (which is both an acid and an ester). (b) Many sunburn lotions contain the ester benzocaine; some nail polish remover is ethyl acetate; vegetable oils are also esters.

▲ **GIVE IT SOME THOUGHT**

Write the condensed formula for the ketone that would result from partial oxidation of this alcohol.

$$CH_2—CHOH$$
$$CH_2 \qquad CH_2$$
$$CH_2$$

Many compounds found in nature possess an aldehyde or ketone functional group. Vanilla and cinnamon flavorings are naturally occurring aldehydes. Two isomers of the ketone carvone impart the characteristic flavors of spearmint leaves or caraway seeds.

Ketones are less reactive than aldehydes and are used extensively as solvents. Acetone, which boils at 56 °C, is the most widely used ketone. The carbonyl functional group imparts polarity to the solvent. Acetone is completely miscible with water, yet it dissolves a wide range of organic substances. 2-Butanone ($CH_3COCH_2CH_3$), which boils at 80 °C, is also used industrially as a solvent.

Carboxylic Acids and Esters

We discussed carboxylic acids in Sections 4.3 and 16.10. **Carboxylic acids** contain the *carboxyl* functional group, which is often written as COOH. These weak acids are widely distributed in nature and are common in consumer products [Figure 25.16(a) ◀]. They are also important in the manufacture of polymers used to make fibers, films, and paints. Figure 25.17▼ shows the structural formulas of several carboxylic acids. Notice that oxalic acid and citric acid contain two and three carboxyl groups, respectively. The common names of many carboxylic acids are based on their historical origins. Formic acid, for example, was first prepared by extraction from ants; its name is derived from the Latin word *formica*, meaning "ant."

Carboxylic acids can be produced by oxidation of alcohols in which the OH group is attached to a CH_2 group, such as ethanol or propanol. Under appropriate conditions the corresponding aldehyde may be isolated as the first product of oxidation. These transformations are shown for ethanol in the following equations, in which (O) represents an oxidant that can provide oxygen atoms:

$$\underset{\text{Ethanol}}{CH_3CH_2OH} + (O) \longrightarrow \underset{\text{Acetaldehyde}}{CH_3\overset{O}{\overset{\|}{C}}H} + H_2O \qquad [25.17]$$

▶ **Figure 25.17 Structural formulas of common carboxylic acids.** The IUPAC names are given in black type for the monocarboxylic acids, but they are generally referred to by their common names (blue type).

$$CH_3—\overset{OH}{\underset{|}{CH}}—\overset{O}{\overset{\|}{C}}—OH$$
Lactic acid

$$H—\overset{O}{\overset{\|}{C}}—OH$$
Methanoic acid
Formic acid

$$HO—\overset{O}{\overset{\|}{C}}—CH_2—\overset{\overset{O}{\overset{\|}{C}}}{\underset{OH}{\overset{|}{C}}}—CH_2$$
Citric acid

Acetylsalicylic acid
Aspirin

$$CH_3—\overset{O}{\overset{\|}{C}}—OH$$
Ethanoic acid
Acetic acid

$$HO—\overset{O}{\overset{\|}{C}}—\overset{O}{\overset{\|}{C}}—OH$$
Oxalic acid

Phenyl methanoic acid
Benzoic acid

$$CH_3\overset{\overset{\displaystyle O}{\|}}{C}H \; + \; (O) \longrightarrow CH_3\overset{\overset{\displaystyle O}{\|}}{C}OH \qquad [25.18]$$

Acetaldehyde   Acetic acid

The air oxidation of ethanol to acetic acid is responsible for causing wines to turn sour, producing vinegar.

Acetic acid can also be produced by the reaction of methanol with carbon monoxide in the presence of a rhodium catalyst:

$$CH_3OH + CO \xrightarrow{\;catalyst\;} CH_3 \overset{\overset{\displaystyle O}{\|}}{-}C-OH \qquad [25.19]$$

This reaction involves, in effect, the insertion of a carbon monoxide molecule between the CH_3 and OH groups. A reaction of this kind is called *carbonylation*.

Carboxylic acids can undergo condensation reactions with alcohols to form **esters**:

$$CH_3-\overset{\overset{\displaystyle O}{\|}}{C}-OH \; + \; HO-CH_2CH_3 \longrightarrow CH_3-\overset{\overset{\displaystyle O}{\|}}{C}-O-CH_2CH_3 + H_2O \;\; [25.20]$$

Acetic acid   Ethanol   Ethyl acetate

Esters are compounds in which the H atom of a carboxylic acid is replaced by a carbon-containing group:

$$-\overset{\overset{\displaystyle O}{\|}}{C}-O-\overset{\overset{\displaystyle |}{}}{\underset{|}{C}}-$$

Figure 25.16(b) shows some common esters, which are named by using first the group from which the alcohol is derived and then the group from which the acid is derived. For example:

$$CH_3CH_2CH_2\overset{\overset{\displaystyle O}{\|}}{C}-OCH_2CH_3$$

Ethyl butyrate

Esters generally have very pleasant odors. They are largely responsible for the pleasant aromas of fruit. Pentyl acetate ($CH_3COOCH_2CH_2CH_2CH_2CH_3$), for example, is responsible for the odor of bananas.

When esters are treated with an acid or a base in aqueous solution, they are hydrolyzed; that is, the molecule is split into its alcohol and acid components:

$$CH_3CH_2-\overset{\overset{\displaystyle O}{\|}}{C}-O-CH_3 + Na^+ + OH^- \longrightarrow$$

Methyl propionate

$$CH_3CH_2-\overset{\overset{\displaystyle O}{\|}}{C}-O^- + Na^+ + CH_3OH \qquad [25.21]$$

Sodium propionate   Methanol

In this example the hydrolysis was carried out in basic medium. The products of the reaction are the sodium salt of the carboxylic acid and the alcohol.

Chemistry Put to Work   PORTRAIT OF AN ORGANIC CHEMICAL

Few other chemicals have been associated with human history for as long as acetic acid has. It is the primary ingredient of vinegar, formed by the oxidation of ethyl alcohol in dilute solutions formed from fruit juices. The oxidation of ethyl alcohol, which is the process responsible for the spoilage of wine and ciders, is aided by bacteria. When such "spoiled" solutions are distilled, the result is vinegar, a dilute (4–8%) solution of acetic acid.

A more complete fractional distillation of vinegar can yield nearly pure acetic acid—a clear, colorless liquid that boils at 118 °C and freezes at 16.7 °C. Because acetic acid solidifies when stored at temperatures not much below room temperature, the term "glacial acetic acid" came to be applied to pure or nearly pure acetic acid. Although vinegar is relatively harmless and is a common household solution used in cooking and cleaning, pure acetic acid is highly corrosive and capable of causing severe burns. You may be surprised to learn that the pure acid is also flammable.

Acetic acid is an important industrial chemical. More than 3×10^9 kg are produced in the United States each year, either by the process shown in Equation 25.19 or by a similar process. The most important uses are for the formation of acetate esters, as illustrated in Equation 25.20. For example, aspirin (acetylsalicylic acid) is formed by the reaction of acetic acid with salicylic acid, which also happens to have a hydroxyl group attached to the benzene ring:

▲ **Figure 25.18 Cellulose acetate fiber.** The blue garment, woven from cellulose acetate fiber, has a silky, vibrant texture compared with the white garment, woven from simple cotton fiber.

wood, is a polymeric material containing multiple hydroxyl groups. The condensation reaction of cellulose with acetic acid yields cellulose acetate, which has many uses. When spun into fibers, to name just one use, cellulose acetate forms rayon-like material (Figure 25.18 ▲).

Reaction of acetic acid with vinyl alcohol yields vinyl acetate. When vinyl acetate is polymerized, the product is polyvinyl acetate, which is used in water-based latex paints and in glues for paper and wood:

$$4n \ CH_2{=}CH$$

Vinyl acetate

Polyvinylacetate

Related Exercises: 25.49, 25.50

Notice that this is a condensation reaction, in which water is split out. The key to this reaction is that the OH group on the benzene ring undergoes the condensation reaction more readily than the OH group of the carboxylic acid. As we will learn later in this chapter, cellulose, which is found in cotton and

---

◀ **Figure 25.19 Saponification.** Saponification of fats and oils has long been used to make soap. This etching shows a step in the soap-making process during the mid-nineteenth century.

The hydrolysis of an ester in the presence of a base is called **saponification**, a term that comes from the Latin word for soap (*sapon*). Naturally occurring esters include fats and oils. In the soap-making process an animal fat or a vegetable oil is boiled with a strong base, usually NaOH. The resultant soap consists of a mixture of sodium salts of long-chain carboxylic acids (called fatty acids), which form during the saponification reaction (Figure 25.19 ◀).

■ **SAMPLE EXERCISE 25.6** | **Naming Esters and Predicting Hydrolysis Products**

In a basic aqueous solution, esters react with hydroxide ion to form the salt of the carboxylic acid and the alcohol from which the ester is constituted. Name each of the following esters, and indicate the products of their reaction with aqueous base.

(a) [benzene ring]—$\overset{\overset{\text{O}}{\|}}{\text{C}}$—$OCH_2CH_3$ **(b)** $CH_3CH_2CH_2$—$\overset{\overset{\text{O}}{\|}}{\text{C}}$—O—[benzene ring]

SOLUTION

Analyze: We are given two esters and asked to name them and to predict the products formed when they undergo hydrolysis (split into an alcohol and carboxylate ion) in basic solution.

Plan: Esters are formed by the condensation reaction between an alcohol and a carboxylic acid. To name an ester, we must analyze its structure and determine the identities of the alcohol and acid from which it is formed. We can identify the alcohol by adding an OH to the alkyl group attached to the O atom of the carboxyl (COO) group. We can identify the acid by adding an H group to the O atom of the carboxyl group. We have learned that the first part of an ester name indicates the alcohol portion and the second indicates the acid portion. The name conforms to how the ester undergoes hydrolysis in base, reacting with base to form an alcohol and a carboxylate anion.

Solve:

(a) This ester is derived from ethanol (CH_3CH_2OH) and benzoic acid (C_6H_5COOH). Its name is therefore ethyl benzoate. The net ionic equation for reaction of ethyl benzoate with hydroxide ion is

[benzene ring]—$\overset{\overset{\text{O}}{\|}}{\text{C}}$—$OCH_2CH_3(aq)$ + $OH^-(aq)$ $\longrightarrow$

[benzene ring]—$\overset{\overset{\text{O}}{\|}}{\text{C}}$—$O^-(aq)$ + $HOCH_2CH_3(aq)$

The products are benzoate ion and ethanol.

(b) This ester is derived from phenol (C_6H_5OH) and butanoic acid (commonly called butyric acid) ($CH_3CH_2CH_2COOH$). The residue from the phenol is called the phenyl group. The ester is therefore named phenyl butyrate. The net ionic equation for the reaction of phenyl butyrate with hydroxide ion is

$CH_3CH_2CH_2\overset{\overset{\text{O}}{\|}}{\text{C}}$—O—[benzene ring]$(aq)$ + $OH^-(aq)$ $\longrightarrow$

$CH_3CH_2CH_2\overset{\overset{\text{O}}{\|}}{\text{C}}$—$O^-(aq)$ + HO—[benzene ring](aq)

The products are butyrate ion and phenol.

■ **PRACTICE EXERCISE**

Write the structural formula for the ester formed from propyl alcohol and propionic acid.

Answer: $CH_3CH_2\overset{\overset{\text{O}}{\|}}{\text{C}}$—O—$CH_2CH_2CH_3$

Amines and Amides

Amines are organic bases. ∞ (Section 16.7) They have the general formula R_3N, where R may be H or a hydrocarbon group, as in the following examples:

$CH_3CH_2NH_2$ $(CH_3)_3N$ [benzene ring]—NH_2

Ethylamine   Trimethylamine   Phenylamine
                                                Aniline

Amines containing a hydrogen bonded to nitrogen can undergo condensation reactions with carboxylic acids to form **amides**:

$$CH_3\overset{\displaystyle O}{\overset{\|}{C}}-OH \ + \ H-N(CH_3)_2 \longrightarrow CH_3\overset{\displaystyle O}{\overset{\|}{C}}-N(CH_3)_2 \ + \ H_2O \qquad [25.22]$$

We may consider the amide functional group to be derived from a carboxylic acid with an NR_2 group replacing the OH of the acid, as in these additional examples:

$$CH_3\overset{\displaystyle O}{\overset{\|}{C}}-NH_2 \qquad \qquad \overset{\displaystyle O}{\overset{\|}{C}}-NH_2$$

Ethanamide   Phenylmethanamide
Acetamide   Benzamide

The amide linkage

$$R-\overset{\displaystyle O}{\overset{\|}{C}}-\underset{\underset{\displaystyle H}{|}}{N}-R'$$

where R and R′ are organic groups, is the key functional group in the structures of proteins, as we will see in Section 25.7.

25.5 CHIRALITY IN ORGANIC CHEMISTRY

A molecule possessing a nonsuperimposable mirror image is termed **chiral** (Greek *cheir*, "hand"). ∞ (Section 24.4) *Compounds containing carbon atoms with four different attached groups are inherently chiral.* A carbon atom with four different attached groups is called a *chiral center*. For example, the structural formula of 2-bromopentane is

$$CH_3-\underset{\underset{\displaystyle H}{|}}{\overset{\overset{\displaystyle Br}{|}}{C}}-CH_2CH_2CH_3$$

▲ **Figure 25.20 The two enantiomeric forms of 2-bromopentane.** The mirror-image isomers are not superimposable on each other.

All four groups attached to the second carbon are different, making that carbon a chiral center. Figure 25.20 ◄ illustrates the two nonsuperimposable mirror images of this molecule. If you imagine trying to move the left-hand molecule to the right and turning it in every possible way, you will conclude that it cannot be superimposed on the right-hand molecule. Nonsuperimposable mirror images are called optical isomers or *enantiomers*. ∞ (Section 24.4)

Organic chemists use the labels *R* and *S* to distinguish the two forms. We need not go into the rules for deciding on the labels. Enantiomers, such as the two forms of 2-bromopentane shown in Figure 25.20, have identical physical properties, such as melting and boiling points, and identical chemical properties when they react with nonchiral reagents. Only in a chiral environment do they exhibit different behaviors. One of the more interesting properties of chiral substances is that their solutions may rotate the plane of polarized light, as explained in Section 24.4.

Chirality is very common in organic substances. It is not often observed, however, because when a chiral substance is synthesized in a typical chemical reaction, the two enantiomeric species are formed in precisely the same quantity.

The resulting mixture of isomers is called a *racemic* mixture. A racemic mixture of enantiomers does not rotate the plane of polarized light because the two forms rotate the light to equal extents in opposite directions. ∞∞(Section 24.4)

Naturally occurring substances often are found as just one enantiomer. An example is tartaric acid, which has the structural formula*

$$
\begin{array}{c}
\text{COOH} \\
|\\
\text{H—C—OH} \\
|\\
\text{H—C—OH} \\
|\\
\text{COOH}
\end{array}
$$

This compound has not one, but two chiral centers. Both the inner carbon atoms are attached to four different groups. Tartaric acid is found in nature, either as the free acid or as a salt of calcium or potassium, in fruit extracts and especially as crystals deposited during wine fermentation. The naturally occurring form is optically active in solution. Tartaric acid is typical; many biologically important molecules are chiral. In Section 25.7 we will examine the amino acids, all of which (except for glycine) are chiral and found in nature as just one of the enantiomers.

Many drugs of importance in human medicine are chiral substances. When a drug is administered as a racemic mixture, it often turns out that only one of the enantiomers has beneficial results. The other is either inert or nearly so, or may even have a harmful effect. For this reason, a great deal of attention has been given in recent years to methods for synthesizing the desired enantiomer of chiral drugs. Synthesizing just one enantiomer of a chiral substance can be very difficult and costly, but the rewards are worth the effort. Worldwide sales of single-enantiomer drugs amount to over $125 billion annually! As an example of the different behavior of enantiomers, the drug *(R)*-albuterol (Figure 25.21▶) is an effective bronchodilator used to relieve the symptoms of asthma. The enantiomer *(S)*-albuterol is not only ineffective as a bronchodilator but actually counters the effects of *(R)*-albuterol. As another example, the nonsteroidal analgesic ibuprofen (marketed under the trade names Advil™, Motrin™, and Nuprin™) is a chiral molecule usually sold as the racemic mixture. However, a preparation consisting of just the more active enantiomer, *(S)*-ibuprofen (Figure 25.22▼), relieves pain and reduces inflammation more rapidly than the racemic mixture. For this reason, the chiral version of the drug may in time come to replace the racemic mixture.

▲ Figure 25.21 *(R)*-Albuterol. This compound, which acts as a bronchodilator in patients with asthma, is one member of a pair of optical isomers. The other member, *(S)*-albuterol, does not have the same physiological effect.

◢ GIVE IT SOME THOUGHT

What are the requirements on the four groups attached to a carbon atom in order that it be a chiral center?

▲ Figure 25.22 *(S)*-Ibuprofen. For relieving pain and reducing inflammation, the ability of this enantiomer far outweighs that of the (R) isomer.

Louis Pasteur discovered chirality while he was studying crystalline samples of salts of tartaric acid.

25.6 INTRODUCTION TO BIOCHEMISTRY

The several types of organic functional groups discussed in Section 25.4 generate a vast array of molecules with very specific chemical reactivities. Nowhere is this specificity more apparent than in *biochemistry*—the chemistry of living organisms.

Before we discuss specific biochemical molecules, we can make some general observations. As we will see, many biologically important molecules are quite large. The synthesis of these molecules is a remarkable aspect of biochemistry, one that places large demands on the chemical systems in living organisms. Organisms build biomolecules from much smaller and simpler substances that are readily available in the biosphere. The synthesis of large molecules requires energy because most of the reactions are endothermic. The ultimate source of this energy is the Sun. Mammals and other animals have essentially no capacity for using solar energy directly; rather, they depend on plant photosynthesis to supply the bulk of their energy needs. ∞ (Section 24.2)

In addition to requiring large amounts of energy, living organisms are highly organized. The complexity of all the substances that make up even the simplest single-cell organisms and the relationships among all the many chemical processes that occur are truly amazing. In thermodynamic terms this high degree of organization means that the entropy of living systems is much lower than the entropy of the random mix of raw materials from which the systems formed. Thus, living systems must continuously work against the spontaneous tendency toward increased entropy that is characteristic of all organized systems. Maintaining a high degree of order places additional energetic requirements on organisms. (You may want to review the second law of thermodynamics in Section 19.3 and the "Chemistry and Life" box there.)

We have introduced you to some important biochemical applications of fundamental chemical ideas in the "Chemistry and Life" essays that appear throughout this text. A complete listing of the topics covered and their locations in the text are included in the table of contents. The remainder of this chapter will serve as only a brief introduction to other aspects of biochemistry. Nevertheless, you will see some patterns emerging. Hydrogen bonding ∞ (Section 11.2), for example, is critical to the function of many biochemical systems, and the geometry of molecules can govern their biological importance and activity. Many of the large molecules in living systems are polymers (Section 12.6) of much smaller molecules. These **biopolymers** can be classified into three broad categories: proteins, polysaccharides (carbohydrates), and nucleic acids. Lipids are another common class of molecules in living systems, but they are usually large molecules, not biopolymers.

25.7 PROTEINS

Proteins are macromolecular substances present in all living cells. About 50% of your body's dry weight is protein. Proteins serve as the major structural components in animal tissues; they are a key part of skin, nails, cartilage, and muscles. Other proteins catalyze reactions, transport oxygen, serve as hormones to regulate specific body processes, and perform other tasks. Whatever their function, all proteins are chemically similar, being composed of the same basic building blocks, called **amino acids**.

Amino Acids

The building blocks of all proteins are α-amino acids, which are substances in which the amino group is located on the carbon atom immediately adjacent to the carboxylic acid group. Thus, there is always one carbon atom between the amino group and the carboxylic acid group. The general formula for an α-amino acid is represented in the following ways:

$$H_2N-\overset{\overset{\displaystyle H}{|}}{\underset{\underset{\displaystyle R}{|}}{C}}-\overset{\overset{\displaystyle O}{\parallel}}{C}-OH \quad \text{or} \quad {}^{+}H_3N-\overset{\overset{\displaystyle H}{|}}{\underset{\underset{\displaystyle R}{|}}{C}}-\overset{\overset{\displaystyle O}{\parallel}}{C}-O^{-}$$

The doubly ionized form, called the zwitterion, usually predominates at near-neutral values of pH. This form is a result of the transfer of a proton from the carboxylic acid group to the basic amine group. ∞ (Section 16.10: "Chemistry and Life: The Amphiprotic Behavior of Amino Acids")

Amino acids differ from one another in the nature of their R groups. Twenty-two amino acids have been identified in nature. Figure 25.23▼ shows the structural formulas of the 20 amino acids found in the proteins of our bodies. Those amino acids for which the side group is relatively nonpolar are shown in Figure 25.23(a). Those that contain a relatively polar or charged side group are shown in Figure 25.23(b)▼. Our bodies can synthesize ten of the amino acids in sufficient amounts for our needs. The other ten must be ingested and are called *essential amino acids* because they are necessary components of our diet.

(a)

▲ **Figure 25.23(a) The 9 amino acids found in the human body that contain relatively nonpolar, hydrophobic side groups.** The acids are shown in the zwitterionic form in which they exist in water at near-neutral pH. The names for each amino acid are given with their three-letter codes and one-letter codes in parentheses.

▶ **Figure 25.23(b)**
The 11 amino acids found in the human body that contain polar side groups. The acids are shown in the zwitterionic form in which they exist in water a near-neutral pH. The names for each amino acid are given with their three-letter codes and one-letter codes in parentheses.

$$CH_3-\underset{\underset{OH}{|}}{CH}-\underset{\underset{H}{|}}{\overset{\overset{H_3N^+}{|}}{C}}-\overset{\overset{O}{||}}{C}-O^-$$

Threonine (Thr; T)

$$HO-CH_2-\underset{\underset{H}{|}}{\overset{\overset{H_3N^+}{|}}{C}}-\overset{\overset{O}{||}}{C}-O^-$$

Serine (Ser; S)

$$HS-CH_2-\underset{\underset{H}{|}}{\overset{\overset{H_3N^+}{|}}{C}}-\overset{\overset{O}{||}}{C}-O^-$$

Cysteine (Cys; C)

$$HO-\bigcirc-CH_2-\underset{\underset{H}{|}}{\overset{\overset{H_3N^+}{|}}{C}}-\overset{\overset{O}{||}}{C}-O^-$$

Tyrosine (Tyr; Y)

$$\underset{O}{\overset{H_2N}{\diagdown}}C-CH_2-\underset{\underset{H}{|}}{\overset{\overset{H_3N^+}{|}}{C}}-\overset{\overset{O}{||}}{C}-O^-$$

Asparagine (Asn; N)

$$\underset{O}{\overset{H_2N}{\diagdown}}C-CH_2-CH_2-\underset{\underset{H}{|}}{\overset{\overset{H_3N^+}{|}}{C}}-\overset{\overset{O}{||}}{C}-O^-$$

Glutamine (Gln; Q)

$$NH_2-\overset{\overset{+NH_2}{||}}{C}-HN-(CH_2)_3-\underset{\underset{H}{|}}{\overset{\overset{H_3N^+}{|}}{C}}-\overset{\overset{O}{||}}{C}-O^-$$

Arginine (Arg; R)

$$H_3N^+-(CH_2)_4-\underset{\underset{H}{|}}{\overset{\overset{H_3N^+}{|}}{C}}-\overset{\overset{O}{||}}{C}-O^-$$

Lysine (Lys; K)

$$HN\underset{N}{\diagup}-CH_2-\underset{\underset{H}{|}}{\overset{\overset{H_3N^+}{|}}{C}}-\overset{\overset{O}{||}}{C}-O^-$$

Histidine (His; H)

$$^-O-\overset{\overset{O}{||}}{C}-CH_2-\underset{\underset{H}{|}}{\overset{\overset{H_3N^+}{|}}{C}}-\overset{\overset{O}{||}}{C}-O^-$$

Aspartic acid (Asp; D)

$$^-O-\overset{\overset{O}{||}}{C}-CH_2-CH_2-\underset{\underset{H}{|}}{\overset{\overset{H_3N^+}{|}}{C}}-\overset{\overset{O}{||}}{C}-O^-$$

Glutamic acid (Glu; E)

(b)

▲ **Figure 25.24 Alanine enantiomers.** The middle carbon of alanine is chiral, and therefore there are two enantiomers, which by definition are nonsuperimposable mirror images of each other.

The α-carbon atom of the amino acids, which bears both the ammonium and carboxylate groups, has four different groups attached to it. The amino acids are thus chiral (except for glycine, which has two hydrogens attached to the central carbon). Figure 25.24 ◀ shows the two enantiomers of the amino acid alanine. For historical reasons, the two enantiomeric forms of amino acids are often distinguished by the labels D (from the Latin *dexter*, "right") and L (from the Latin *laevus*, "left"). Nearly all amino acids normally found in living organisms are "left-handed"; that is, all have the L configuration at the chiral center (except for glycine, which is not chiral because its central carbon has two hydrogens attached). The principal exception to the dominance of L amino acids in nature are the proteins that make up the cell walls of bacteria, which can contain considerable quantities of the D isomers.

Polypeptides and Proteins

Amino acids are linked together into proteins by amide groups, one of the functional groups introduced in Section 25.4:

$$R-\overset{\overset{O}{||}}{C}-\underset{\underset{H}{|}}{N}-R \qquad\qquad [25.23]$$

Each of these amide groups is called a **peptide bond** when it is formed by amino acids. A peptide bond is formed by a condensation reaction between the carboxyl group of one amino acid and the amino group of another amino acid.

Alanine and glycine, for example, can react to form the dipeptide glycylalanine:

$$H_3\overset{+}{N}-\underset{\underset{H}{|}}{\overset{\overset{H}{|}}{C}}-\overset{\overset{O}{\|}}{C}-O^- + H-\underset{\underset{H}{|}}{\overset{\overset{H}{|}}{\overset{+}{N}}}-\underset{\underset{CH_3}{|}}{\overset{\overset{H}{|}}{C}}-\overset{\overset{O}{\|}}{C}-O^- \longrightarrow$$

Glycine (Gly; G) Alanine (Ala; A)

$$H_3\overset{+}{N}-\underset{\underset{H}{|}}{\overset{\overset{H}{|}}{C}}-\overset{\overset{O}{\|}}{C}-\underset{\underset{H}{|}}{N}-\underset{\underset{CH_3}{|}}{\overset{\overset{H}{|}}{C}}-\overset{\overset{O}{\|}}{C}-O^- + H_2O$$

Glycylalanine (Gly–Ala; GA)

The amino acid that furnishes the carboxyl group for peptide-bond formation is named first, with a *-yl* ending; then the amino acid furnishing the amino group is named. Based on the three-letter codes for the amino acids from Figure 25.23, glycylalanine can be abbreviated Gly-Ala, or as GA using the one-letter codes for the amino acids. In this notation it is understood that the unreacted amino group is on the left and the unreacted carboxyl group on the right. The artificial sweetener *aspartame* (Figure 25.25▶) is the methyl ester of the dipeptide of aspartic acid and phenylalanine:

$$H_2N-\underset{\underset{\underset{\underset{O}{\|}}{C-OH}}{\overset{|}{CH_2}}}{\overset{\overset{H}{|}}{C}}-\overset{\overset{O}{\|}}{C}-\underset{\underset{H}{|}}{N}-\underset{\underset{\underset{\hexagon}{CH_2}}{|}}{\overset{\overset{H}{|}}{C}}-\overset{\overset{O}{\|}}{C}-O-CH_3$$

▲ **Figure 25.25 Sweet stuff.**
The artificial sweetener aspartame is the methyl ester of a dipeptide.

■■ **SAMPLE EXERCISE 25.7** | **Drawing the Structural Formula of a Tripeptide**

Draw the full structural formula for alanylglycylserine, or AGS.

SOLUTION

Analyze: We are given the name of a substance with peptide bonds and asked to write its full structural formula.

Plan: The name of this substance suggests that three amino acids—alanine, glycine, and serine—have been linked together, forming a *tripeptide*. Note that the ending *-yl* has been added to each amino acid except for the last one, serine. By convention, the sequence of amino acids that make up peptides and proteins are written from the nitrogen end to the carbon end: the first-named amino acid (alanine, in this case) has a free amino group and the last-named one (serine) has a free carboxyl group. Thus, we can construct the structural formula of the tripeptide from its amino acid building blocks (Figure 25.23).

Solve: We first combine the carboxyl group of alanine with the amino group of glycine to form a peptide bond and then the carboxyl group of glycine with the amino group of serine to form another peptide group. The resulting tripeptide consists of three "building blocks" connected by peptide bonds:

Amino group ⟶ Carboxyl group

Ala Gly Ser
A   G   S

We can abbreviate this tripeptide as Ala-Gly-Ser, or AGS.

▲ **Figure 25.26 α-Helix structure for a protein.**
The symbol R represents any one of the several side
chains shown in Figure 25.23.

▲ **Figure 25.27 β-Sheet structure for a protein.**
The symbol R represents any one of the several side
chains shown in Figure 25.23.

Polypeptides are formed when a large number of amino acids are
linked together by peptide bonds. Proteins are linear (that is, unbranched) polypeptide molecules with molecular weights ranging
from about 6000 to over 50 million amu. Because up to 22 different
amino acids are linked together in proteins and because proteins consist of hundreds of amino acids, the number of possible arrangements
of amino acids within proteins is virtually limitless.

Protein Structure

The arrangement, or sequence, of amino acids along a protein chain is
called its **primary structure**. The primary structure gives the protein
its unique identity. A change in even one amino acid can alter the biochemical characteristics of the protein. For example, sickle-cell anemia is a genetic disorder resulting from a single replacement in a
protein chain in hemoglobin. The chain that is affected contains 146
amino acids. The substitution of a single amino acid with a hydrocarbon side chain for one that has an acidic functional group in the side
chain alters the solubility properties of the hemoglobin, and normal
blood flow is impeded. ∞(Section 13.6: "Chemistry and Life: Sickle-
Cell Anemia")

Proteins in living organisms are not simply long, flexible chains
with random shapes. Rather, the chains self-assemble into particular
structures based on the same kinds of intermolecular forces we learned
about in Chapter 11. The **secondary structure** of a protein refers to how
segments of the protein chain are oriented in a regular pattern.

One of the most important and common secondary structure
arrangements is the **α helix**, first proposed by Linus Pauling and
R. B. Corey. The helix arrangement is shown in schematic form in
Figure 25.26 ◄. Imagine winding a long protein chain in a helical fashion around a long cylinder. The helix is held in position by hydrogen-
bond interactions between N—H bonds and the oxygens of nearby
carbonyl groups in the amide backbone of the protein; the R groups
are not involved. The pitch of the helix and the diameter of the cylinder must be such that (1) no bond angles are strained and (2) the
N—H and C=O functional groups on adjacent turns are in proper
position for hydrogen bonding. An arrangement of this kind is possible for some amino acids along the chain, but not for others. Large
protein molecules may contain segments of the chain that have the α-
helical arrangement interspersed with sections in which the chain is
in a random coil.

Another common structural motif in proteins is the beta (β) sheet.
Beta sheets are made of two strands of peptides that hydrogen bond
their backbone amide groups more like a zipper than an alpha-helical
spiral (Figure 25.27 ◄). A single protein chain can form a β sheet if it
has a flexible loop connecting the two β sheet partners.

GIVE IT SOME THOUGHT

If you heat a protein to break the intramolecular hydrogen bonds, will you maintain the α helical or β sheet structure?

Proteins are not active biologically unless they are in a particular shape in solution. The process by which the protein adopts its biologically active shape is called **folding**. The overall shape of a protein in its folded form—determined by all the bends, kinks, and sections of rodlike α-helical, β-sheet, or flexible coil components—is called the **tertiary structure**. Figure 24.11 shows the tertiary structure of myoglobin, a protein with a molecular weight of about 18,000 amu and containing one heme group. ∞∞ (Section 24.2) Notice that some sections of the protein consist of α-helices.

Myoglobin is a *globular protein*, one that folds into a compact, roughly spherical shape. Globular proteins are generally soluble in water and are mobile within cells. They have nonstructural functions, such as combating the invasion of foreign objects, transporting and storing oxygen, and acting as catalysts. The *fibrous proteins* form a second class of proteins. In these substances the long coils align themselves in a more-or-less parallel fashion to form long, water-insoluble fibers. Fibrous proteins provide structural integrity and strength to many kinds of tissue and are the main components of muscle, tendons, and hair. The largest known proteins, in excess of 27,000 amino acids long, are muscle proteins called the titins.

The tertiary structure of a protein is maintained by many different interactions. Certain foldings of the protein chain lead to lower-energy (more stable) arrangements than do other folding patterns. For example, a globular protein dissolved in aqueous solution folds in such a way that the nonpolar hydrocarbon portions are tucked within the molecule, away from the polar water molecules. Most of the more polar acidic and basic side chains, however, project into the solution where they can interact with water molecules through ion–dipole, dipole–dipole, or hydrogen-bonding interactions.

One of the most important classes of proteins is *enzymes*, large protein molecules that serve as catalysts. ∞∞ (Section 14.7) Enzymes usually catalyze only very specific reactions. Their tertiary structure generally dictates that only certain substrate molecules can interact with the active site of the enzyme (Figure 25.28 ▶).

▼ **Figure 25.28 An enzyme and its substrate.** A computer-generated structure of an enzyme showing the carbon backbone as a green ribbon. The substrate (violet) is shown at the active site.

Chemistry and Life   THE ORIGINS OF CHIRALITY IN LIVING SYSTEMS

The presence of a "handedness" in the molecules that make up living systems is a key feature of life on Earth. The insistence of nature on just one chiral form in the molecules of life is called *homochirality*. How did the dominance of the L-amino acids arise? Why are the naturally occurring helices of protein and DNA, which we will discuss in Section 25.10, all right-turn helices? Homochirality could have arisen by chance in the course of evolution or because it was "seeded" in some way at the beginnings of life. One theory is that chirality was introduced early in Earth's evolutionary history, through seeding by chiral amino acids that fell on the planet from outer space.

Examination of the Murchison meteorite, which fell to Earth in 1969, has revealed the presence of amino acids. For some of the acids, there appears to be a surplus of the L form. One theory proposes that the chiral amino acids could have been synthesized in interstellar space by the action of circularly

polarized starlight.* Astronomers in Australia have recently observed circular polarization in the infrared light from a region of intense star birth in the Orion nebula. These workers calculate that a similar degree of circular polarization could be present in the visible and ultraviolet light from this source. Light that has the energy required to break chemical bonds, if circularly polarized, could give rise to new chiral molecules with a preference for one enantiomer over the other. Perhaps the homochirality we observe on Earth today arose, through a process of amplification and refinement in the course of evolutionary development, from molecules that were formed in interstellar space when the planet was young.

Circularly polarized light is like plane-polarized light, as shown in Figure 24.21, except that the plane continuously rotates either to the left or to the right. Thus, in a sense, circularly polarized light is chiral.

▲ **Figure 25.29 Linear structure of glucose and fructose.**

25.8 CARBOHYDRATES

Carbohydrates are an important class of naturally occurring substances found in both plant and animal matter. The name **carbohydrate** (hydrate of carbon) comes from the empirical formulas for most substances in this class, which can be written as $C_x(H_2O)_y$. For example, **glucose**, the most abundant carbohydrate, has the molecular formula $C_6H_{12}O_6$, or $C_6(H_2O)_6$. Carbohydrates are not really hydrates of carbon; rather, they are polyhydroxy aldehydes and ketones. Glucose, for example, is a six-carbon aldehyde sugar; whereas *fructose*, the sugar that occurs widely in fruit, is a six-carbon ketone sugar (Figure 25.29 ◄).

Glucose, having both alcohol and aldehyde functional groups and having a reasonably long and flexible backbone, can react with itself to form a six-member-ring structure, as shown in Figure 25.30 ▼. In fact, only a small percentage of the glucose molecules are in the open-chain form in aqueous solution. Although the ring is often drawn as if it were planar, the molecules are actually nonplanar because of the tetrahedral bond angles around the C and O atoms of the ring.

Figure 25.30 shows that the ring structure of glucose can have two relative orientations. In the α form the OH group on carbon 1 and the CH_2OH group on carbon 5 point in opposite directions. In the β form they point in the same direction. Although the difference between the α and β forms might seem small, it has enormous biological consequences. As we will soon see, this one small change in structure accounts for the vast difference in properties between starch and cellulose.

Fructose can cyclize to form either five- or six-member rings. The five-member ring forms when the OH group on carbon 5 reacts with the carbonyl group on carbon 2:

The six-member ring results from the reaction between the OH group on carbon 6 and the carbonyl group on carbon 2.

▶ **Figure 25.30 Cyclic glucose.** The carbonyl group of a glucose molecule can react with one of the hydroxyl groups to form either of two six-member-ring structures, designated α and β.

α-Glucose β-Glucose

SAMPLE EXERCISE 25.8 | Identifying Chiral Centers

How many chiral carbon atoms are there in the open-chain form of glucose (Figure 25.29)?

SOLUTION

Analyze: We are given the structure of glucose and asked to determine the number of chiral carbons in the molecule.

Plan: A chiral carbon has four different groups attached (Section 25.5). We need to identify those carbon atoms in glucose.

Solve: The carbon atoms numbered 2, 3, 4, and 5 each have four different groups attached to them, as indicated here:

Thus, there are four chiral carbon atoms in the glucose molecule.

PRACTICE EXERCISE

How many chiral carbon atoms are there in the open-chain form of fructose (Figure 25.29)?

Answer: three

Disaccharides

Both glucose and fructose are examples of **monosaccharides**, simple sugars that cannot be broken into smaller molecules by hydrolysis with aqueous acids. Two monosaccharide units can be linked together by a condensation reaction to form a *disaccharide*. The structures of two common disaccharides, *sucrose* (table sugar) and *lactose* (milk sugar), are shown in Figure 25.31 ▼.

The word *sugar* makes us think of sweetness. All sugars are sweet, but they differ in the degree of sweetness we perceive when we taste them. Sucrose is about six times sweeter than lactose, slightly sweeter than glucose, but only about half as sweet as fructose. Disaccharides can be reacted with water (hydrolyzed) in the presence of an acid catalyst to form monosaccharides.

▲ **Figure 25.31 Two disaccharides.** The structures of sucrose (left) and lactose (right).

When sucrose is hydrolyzed, the mixture of glucose and fructose that forms, called *invert sugar*,* is sweeter to the taste than the original sucrose. The sweet syrup present in canned fruits and candies is largely invert sugar formed from hydrolysis of added sucrose.

Polysaccharides

Polysaccharides are made up of many monosaccharide units joined together by a bonding arrangement similar to those shown for the disaccharides in Figure 25.31. The most important polysaccharides are starch, glycogen, and cellulose, which are formed from repeating glucose units.

Starch is not a pure substance. The term refers to a group of polysaccharides found in plants. Starches serve as a major method of food storage in plant seeds and tubers. Corn, potatoes, wheat, and rice all contain substantial amounts of starch. These plant products serve as major sources of needed food energy for humans. Enzymes within the digestive system catalyze the hydrolysis of starch to glucose.

▲ **Figure 25.32 Structure of a starch molecule.** The molecule consists of many units of the kind enclosed in brackets, joined by linkages of the α form. (That is, the C—O bonds on the linking carbons are on the opposite side of the ring from the CH_2OH groups.)

Some starch molecules are unbranched chains, whereas others are branched. Figure 25.32 ◄ illustrates an unbranched starch structure. Notice, in particular, that the glucose units are in the α form (that is, the bridging oxygen atom is opposite the CH_2OH group on the adjacent carbon).

Glycogen is a starchlike substance synthesized in the animal body. Glycogen molecules vary in molecular weight from about 5000 to more than 5 million amu. Glycogen acts as a kind of energy bank in the body. It is concentrated in the muscles and liver. In muscles it serves as an immediate source of energy; in the liver it serves as a storage place for glucose and helps to maintain a constant glucose level in the blood.

Cellulose forms the major structural unit of plants. Wood is about 50% cellulose; cotton fibers are almost entirely cellulose. Cellulose consists of an unbranched chain of glucose units, with molecular weights averaging more than 500,000 amu. The structure of cellulose is shown in Figure 25.33 ▼. At first glance this structure looks very similar to that of starch. In cellulose, however, the glucose units are in the β form (that is, the bridging oxygen atom is on the same side as the CH_2OH groups).

The distinction between starch and cellulose is made clearer when we examine their structures in a more realistic three-dimensional representation, as shown in Figure 25.34 ▶. The individual glucose units have different relationships to one another in the two structures. Because of this fundamental difference, enzymes

▲ **Figure 25.33 Structure of cellulose.** Like starch, cellulose is a polymer. The repeating unit is shown between brackets. The linkage in cellulose is of the β form, different from that in starch (see Figure 25.32).

The term invert sugar *comes from the fact that rotation of the plane of polarized light by the glucose-fructose mixture is in the opposite direction, or inverted, from that of the sucrose solution.*

that readily hydrolyze starches do not hydrolyze cellulose. Thus, you might eat a pound of cellulose and receive no caloric value from it whatsoever, even though the heat of combustion per unit weight is essentially the same for both cellulose and starch. A pound of starch, in contrast, would represent a substantial caloric intake. The difference is that the starch is hydrolyzed to glucose, which is eventually oxidized with release of energy. However, enzymes in the body do not readily hydrolyze cellulose, so it passes through the digestive system relatively unchanged. Many bacteria contain enzymes, called cellulases, that hydrolyze cellulose. These bacteria are present in the digestive systems of grazing animals, such as cattle, that use cellulose for food.

▲ **GIVE IT SOME THOUGHT**

Which type of linkage would you expect to join the sugar molecules of glycogen?

25.9 LIPIDS

Lipids are another important class of biological molecules. Lipids are nonpolar and are used by organisms for long-term energy storage (fats, oils) and as elements of biological structures (phospholipids, cell membranes, waxes).

In vertebrate organisms such as ourselves, excess glycogen is converted into fats for long-term storage. Plants will more typically store energy in the form of oils. Both fats and oils are chemically derived from glyercol and three long-chain fatty acids (Figure 25.35 ▼); fatty acids are carboxylic acids (RCOOH) in which R is a hydrocarbon chain with at least four carbons.

In saturated fatty acids (and the saturated fats derived from them) the R groups are alkanes. Saturated fatty acids, such as shortening, are commonly solids at room temperature and are less reactive than unsaturated fats. In unsaturated fatty acids (and the unsaturated fats derived from them) the R groups are alkenes. The *cis* and *trans* nomenclature we learned earlier for alkenes is the same here: *Trans* fats have Hs on the opposite sides of the C=C double bond, and *cis* fats have Hs on the same sides of the C=C double bond. Unsaturated fats (such as olive oil and peanut oil) are usually liquid at room temperature and are more often found in plants. For example, the major component (approximately 60–80%) of olive oil is oleic acid, *cis*-$CH_3(CH_2)_7CH=CH(CH_2)_7COOH$.

(a) Starch

(b) Cellulose

▲ **Figure 25.34 Structural differences in starch and cellulose.** These representations show the geometrical arrangements of bonds about each carbon atom. The glucose rings are oriented differently with respect to one another in the two structures.

▼ **Figure 25.35 Structures of selected fats.** Fats are derived from glycerol and fatty acid molecules.

Stearic acid

Oleic acid

Linoleic acid

Glycerol

Fatty acid

Fat molecule

$$RO-\overset{\displaystyle\overset{O}{\|}}{\underset{\displaystyle\underset{-O}{|}}{P}}-O-CH_2$$

$$H-\overset{|}{\underset{|}{C}}-O-\overset{O}{\overset{\|}{C}}-CH_2CH_2CH_2CH_2CH_2CH_2CH_2CH_2CH_2CH_2CH_2CH_2CH_2CH_2CH_3$$

$$H-\overset{|}{\underset{|}{C}}-O-\overset{}{\underset{\displaystyle\underset{O}{\|}}{C}}-CH_2CH_2CH_2CH_2CH_2CH_2CH_2CH=CHCH_2CH_2CH_2CH_2CH_2CH_2CH_3$$

$$\overset{|}{H}$$

▲ **Figure 25.36 Structure of phospholipids.** Phospholipids are closely related to fats. They form bilayers that are the key components of cell membranes.

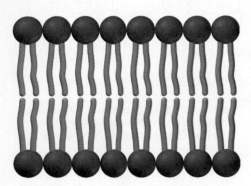

▲ **Figure 25.37 The cell membrane.** Living cells are encased in membranes that are typically made of phospholipid bilayers. The bilayer structure is stabilized by the favorable interactions of the hydrophobic tails of the phospholipids away from water, while the charged headgroups face the water that is both outside and inside the cell.

Oleic acid is also an example of a *monounsaturated* fatty acid, meaning that there is one carbon–carbon double bond in the chain. In contrast, *polyunsaturated* fatty acids have more than one carbon–carbon double bond in the chain.

For humans, *trans* fats are not nutritionally required, which is why some governments are moving to ban them in foods. How, then, did *trans* fats end up in our food? The process that converts unsaturated fats (such as oils) into saturated fats (such as shortening) is hydrogenation. ∞ (Section 25.3) The by-products of this hydrogenation process include *trans* fats. If we eat food that has been cooked in *trans* fats, we inevitably ingest some *trans* fats.

Some fatty acids that are essential for human health must be available in our diets, because our metabolism is incapable of synthesizing them from other substances. These essential fatty acids are ones that have the carbon–carbon double bonds either three carbons or six carbons away from the —CH_3 end of the chain. These are called omega-3 and omega-6 fatty acids, where omega refers to the last carbon in the chain (the carboxylic acid carbon is considered the first, or alpha, one).

Phospholipids are similar in chemical structure to fats and oils, with the important addition of a charged phosphate group to the glycerol (Figure 25.36 ▲). In water, phospholipids cluster together with their charged polar heads facing the water and their nonpolar tails facing inward. ∞ (Section 13.6) The phospholipids thus form a bilayer that is a key component of cell membranes (Figure 25.37 ◄).

25.10 NUCLEIC ACIDS

Nucleic acids are a class of biopolymers that are the chemical carriers of an organism's genetic information. **Deoxyribonucleic acids (DNA)** are huge molecules [Figure 1.2(c)] whose molecular weights may range from 6 million to 16 million amu. **Ribonucleic acids (RNA)** are smaller molecules, with molecular weights in the range of 20,000 to 40,000 amu. Whereas DNA is found primarily in the nucleus of the cell, RNA is found mostly outside the nucleus in the *cytoplasm*, the nonnuclear material enclosed within the cell membrane. The DNA stores the genetic information of the cell and specifies which proteins the cell can synthesize. The RNA carries the information stored by DNA out of the nucleus of the cell into the cytoplasm, where the information is used in protein synthesis.

The monomers of nucleic acids, called **nucleotides**, are formed from the following units:

1. A phosphoric acid molecule, H_3PO_4

2. A five-carbon sugar

3. A nitrogen-containing organic base

The sugar component of RNA is *ribose*, whereas that in DNA is *deoxyribose*.

Ribose   Deoxyribose

*Deoxy*ribose differs from ribose only in having one fewer oxygen atom at carbon 2.

The following nitrogen bases are found in DNA and RNA:

Adenine (A)	Guanine (G)	Cytosine (C)	Thymine (T)	Uracil (U)
DNA	DNA	DNA	DNA	RNA
RNA	RNA	RNA		

The base is attached to a ribose or deoxyribose molecule through a bond to the nitrogen atom shown in color above. An example of a nucleotide in which the base is adenine and the sugar is deoxyribose is shown in Figure 25.38▼.

Nucleic acids are polynucleotides formed by condensation reactions between an OH group of the phosphoric acid unit on one nucleotide and an OH group of the sugar of another nucleotide. Figure 25.39► shows a portion of the polymeric chain of a DNA molecule.

▲ **Figure 25.38 A nucleotide.** Structure of deoxyadenylic acid, a nucleotide formed from phosphoric acid, deoxyribose, and the organic base adenine.

◄ **Figure 25.39 Structure of a polynucleotide.** Because the sugar in each nucleotide is deoxyribose, this polynucleotide is of the form found in DNA.

▶ **Figure 25.40 Two views of DNA.**
(a) A computer-generated model of a DNA double helix. The dark blue and light blue atoms represent the sugar-phosphate chains that wrap around the outside. Inside the chains are the bases, shown in red and yellow-green. (b) A schematic illustration of the double helix showing the hydrogen-bond interactions between complementary base pairs.

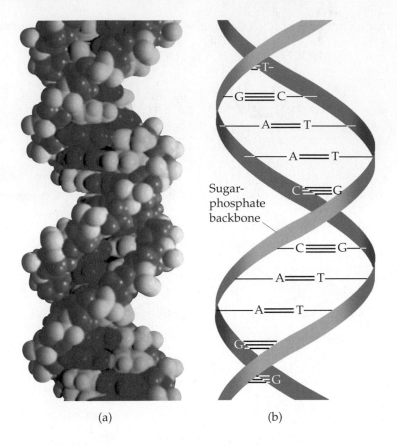

(a)   (b)

DNA molecules consist of two deoxyribonucleic acid chains or strands that are wound together in the form of a **double helix**, as shown in Figure 25.40 ▲. The drawing in Figure 25.40(b) has been simplified to show the essential features of the structure. The sugar and phosphate groups form the backbone of each strand. The bases (represented by the letters T, A, C, and G) are attached to the sugars. The two strands are held together by attractions between the bases in one strand and those in the other strand. These attractions involve London dispersion interactions, dipole-dipole and hydrogen bonds. ∞(Section 11.2) As shown in Figure 25.41 ▼, the structures of thymine (T) and adenine (A) make them perfect partners for hydrogen bonding. Likewise, cytosine (C) and guanine (G) form ideal hydrogen-bonding partners. In the double-helix structure, therefore, each thymine on one strand is opposite an adenine on the other strand. Likewise, each cytosine is opposite a guanine. The double-helix structure with complementary bases on the two strands is the key to understanding how DNA functions.

The two strands of DNA unwind during cell division, and new complementary strands are constructed on the unraveling strands (Figure 25.42 ▶).

Thymine   Adenine   Cytosine   Guanine

T=A   C≡G

▲ **Figure 25.41 Hydrogen bonding between complementary base pairs.** The hydrogen bonds are responsible for formation of the double-stranded helical structure of DNA.

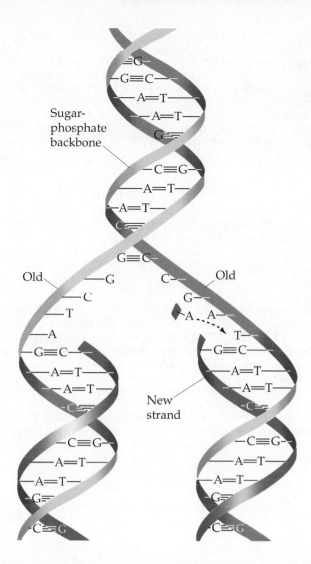

Sugar-phosphate backbone

Old

New strand

Old

◀ Figure 25.42 **DNA replication.** The original DNA double helix partially unwinds, and new nucleotides line up on each strand in complementary fashion. Hydrogen bonds help align the new nucleotides with the original DNA chain. When the new nucleotides are joined by condensation reactions, two identical double-helix DNA molecules result.

This process results in two identical double-helix DNA structures, each containing one strand from the original structure and one newly synthesized strand. This replication process allows genetic information to be transmitted when cells divide. The structure of DNA is also the key to understanding protein synthesis, the means by which viruses infect cells, and many other problems of central importance to modern biology. These themes are beyond the scope of this book. If you take courses in the life sciences, however, you will learn a good deal about such matters.

■■ **SAMPLE INTEGRATIVE EXERCISE** | Putting Concepts Together

Pyruvic acid has the following structure:

$$CH_3 - \overset{\overset{\displaystyle O}{\|}}{C} - \overset{\overset{\displaystyle O}{\|}}{C} - OH$$

   It is formed in the body from carbohydrate metabolism. In the muscle it is reduced to lactic acid in the course of exertion. The acid-dissociation constant for pyruvic acid is 3.2×10^{-3}. **(a)** Why does pyruvic acid have a higher acid-dissociation constant than acetic acid? **(b)** Would you expect pyruvic acid to exist primarily as the neutral acid or as dissociated ions in muscle tissue, assuming a pH of about 7.4 and an acid concentration of 2×10^{-4} M? **(c)** What would you predict for the solubility properties of pyruvic acid? Explain. **(d)** What is the hybridization of each carbon atom in pyruvic acid? **(e)** Assuming H atoms as the reducing agent, write a balanced chemical equation for the reduction of pyruvic acid to lactic acid (Figure 25.17). (Although H atoms do not exist as such in biochemical systems, biochemical reducing agents deliver hydrogen for such reductions.)

SOLUTION

(a) The acid ionization constant for pyruvic acid should be somewhat greater than that of acetic acid because the carbonyl function on the α-carbon atom exerts an electron-withdrawing effect on the carboxylic acid group. In the C—O—H bond system the electrons are shifted from hydrogen, facilitating loss of the hydrogen as a proton. ∞ (Section 16.10)

(b) To determine the extent of ionization, we first set up the ionization equilibrium and equilibrium-constant expression. Using HPv as the symbol for the acid, we have

$$HPv \rightleftharpoons H^+ + Pv^-$$

$$K_a = \frac{[H^+][Pv^-]}{[HPv]} = 3.2 \times 10^{-3}$$

Let $[Pv^-] = x$. Then the concentration of undissociated acid is $2 \times 10^{-4} - x$. The concentration of $[H^+]$ is fixed at 4.0×10^{-8} (the antilog of the pH value). Substituting, we obtain

$$3.2 \times 10^{-3} = \frac{[4.0 \times 10^{-8}][x]}{[2 \times 10^{-4} - x]}$$

Solving for x, we obtain $x[3.2 \times 10^{-3} + 4.0 \times 10^{-8}] = 6.4 \times 10^{-7}$.

The second term in the brackets is negligible compared to the first, so $x = [Pv^-] = 6.4 \times 10^{-7}/3.2 \times 10^{-3} = 2 \times 10^{-4}\ M$. This is the initial concentration of acid, which means that essentially all the acid has dissociated. We might have expected this result because the acid is quite dilute and the acid-dissociation constant is fairly high.

(c) Pyruvic acid should be quite soluble in water because it has polar functional groups and a small hydrocarbon component. It is miscible with water, ethanol, and diethyl ether.

(d) The methyl group carbon has sp^3 hybridization. The carbon carrying the carbonyl group has sp^2 hybridization because of the double bond to oxygen. Similarly, the carboxylic acid carbon is sp^2 hybridized.

(e) The balanced chemical equation for this reaction is

$$\begin{array}{ccc} \text{O} & & \text{OH} \\ \| & & | \\ CH_3CCOOH + 2\,(H) & \longrightarrow & CH_3CCOOH \\ & & | \\ & & \text{H} \end{array}$$

Essentially, the ketonic functional group has been reduced to an alcohol.

Strategies in Chemistry WHAT NOW?

If you are reading this box, you have made it to the end of our text. We congratulate you on the tenacity and dedication that you have exhibited to make it this far!

As an epilogue, we offer the ultimate study strategy in the form of a question: What do you plan to do with the knowledge of chemistry that you have gained thus far in your studies? Many of you will enroll in additional courses in chemistry as part of your required curriculum. For others, this will be the last formal course in chemistry that you will take. Regardless of the career path you plan to take—whether it is chemistry, one of the biomedical fields, engineering, the liberal arts, or another field—we hope that this text has increased your appreciation of the chemistry in the world around you. If you pay attention, you will be aware of encounters with chemistry on a daily basis, from food and pharmaceutical labels to gasoline pumps, sports equipment to news reports.

We have also tried to give you a sense that chemistry is a dynamic, continuously changing science. Research chemists synthesize new compounds, develop new reactions, uncover chemical properties that were previously unknown, find new applications for known compounds, and refine theories. The understanding of biological systems in terms of the underlying chemistry has become increasingly important as new levels of complexity are uncovered. You may wish to participate in the fascinating venture of chemical research by taking part in an undergraduate research program. Given all the answers that chemists seem to have, you may be surprised at the large number of questions that they still find to ask.

Finally, we hope you have enjoyed using this textbook. We certainly enjoyed putting so many of our thoughts about chemistry on paper. We truly believe it to be the central science, one that benefits all who learn about it and from it.

CHAPTER REVIEW

SUMMARY AND KEY TERMS

Introduction and Section 25.1 This chapter introduces **organic chemistry**, which is the study of carbon compounds (typically compounds containing carbon–carbon bonds), and **biochemistry**, which is the study of the chemistry of living organisms. We have encountered many aspects of organic chemistry in earlier chapters. Carbon forms four bonds in its stable compounds. The C—C single bonds and the C—H bonds tend to have low reactivity. Those bonds that have a high electron density (such as multiple bonds or bonds with an atom of high electronegativity) tend to be the sites of reactivity in an organic compound. These sites of reactivity are called **functional groups**.

Section 25.2 The simplest types of organic compounds are hydrocarbons, those composed of only carbon and hydrogen. There are four major kinds of hydrocarbons: alkanes, alkenes, alkynes, and aromatic hydrocarbons. **Alkanes** are composed of only C—H and C—C single bonds. **Alkenes** contain one or more carbon–carbon double bonds. **Alkynes** contain one or more carbon–carbon triple bonds. **Aromatic hydrocarbons** contain cyclic arrangements of carbon atoms bonded through both σ and delocalized π bonds. Alkanes are saturated hydrocarbons; the others are unsaturated.

Section 25.3 Alkanes may form straight-chain, branched-chain, and cyclic arrangements. Isomers are substances that possess the same molecular formula, but differ in the arrangements of atoms. In **structural isomers** the bonding arrangements of the atoms differ. Different isomers are given different systematic names. The naming of hydrocarbons is based on the longest continuous chain of carbon atoms in the structure. The locations of **alkyl groups**, which branch off the chain, are specified by numbering along the carbon chain. Alkanes with ring structures are called **cycloalkanes**. Alkanes are relatively unreactive. They do, however, undergo combustion in air, and their chief use is as sources of heat energy produced by combustion.

The names of alkenes and alkynes are based on the longest continuous chain of carbon atoms that contains the multiple bond, and the location of the multiple bond is specified by a numerical prefix. Alkenes exhibit not only structural isomerism but geometric (cis-trans) isomerism as well. In **geometric isomers** the bonds are the same, but the molecules have different geometries. Geometric isomerism is possible in alkenes because rotation about the C=C double bond is restricted.

Alkenes and alkynes readily undergo **addition reactions** to the carbon–carbon multiple bonds. Additions of acids, such as HBr, proceed via a rate-determining step in which a proton is transferred to one of the alkene or alkyne carbon atoms. Addition reactions are difficult to carry out with aromatic hydrocarbons, but **substitution reactions** are easily accomplished in the presence of catalysts.

Section 25.4 The chemistry of organic compounds is dominated by the nature of their functional groups. The functional groups we have considered are

R, R', and R" represent hydrocarbon groups—for example, methyl (CH_3) or phenyl (C_6H_5).

Alcohols are hydrocarbon derivatives containing one or more OH groups. **Ethers** are formed by a condensation reaction of two molecules of alcohol. Several functional groups contain the **carbonyl** (C=O) **group**, including **aldehydes, ketones, carboxylic acids, esters,** and **amides**. Aldehydes and ketones can be produced by oxidation of certain alcohols. Further oxidation of the aldehydes produces carboxylic acids. Carboxylic acids can form esters by a condensation reaction with alcohols, or they can form amides by a condensation reaction with amines. Esters undergo hydrolysis (**saponification**) in the presence of strong bases.

Section 25.5 Molecules that possess nonsuperimposable mirror images are termed **chiral**. The two nonsuperimposable forms of a chiral molecule are called *enantiomers*. In carbon compounds a chiral center is created when all four groups bonded to a central carbon atom are different, as in 2-bromobutane. Many of the molecules occurring in living systems, such as the amino acids, are chiral and exist in nature in only one enantiomeric form. Many drugs of importance in human medicine are chiral, and the enantiomers may produce very different biochemical effects. For this reason, synthesis of only the effective isomers of chiral drugs has become a high priority.

Sections 25.6 and 25.7 Many of the molecules that are essential for life are large natural polymers that are constructed from smaller molecules called monomers. Three of these **biopolymers** were considered in this chapter: proteins, polysaccharides (carbohydrates), and nucleic acids.

Proteins are polymers of **amino acids**. They are the major structural materials in animal systems. All naturally occurring proteins are formed from 22 amino acids are formed from 22 amino acids, although only 20 are common. The amino acids are linked by **peptide bonds**. A **polypeptide** is a polymer formed by linking many amino acids by peptide bonds.

Amino acids are chiral substances. Usually only one of the enantiomers is found to be biologically active. Protein structure is determined by the sequence of amino acids in the chain (its **primary structure**), the coiling or stretching of the chain (its **secondary structure**), and the overall shape of the complete molecule (its **tertiary structure**). Two important secondary structures are the α **helix** and the β **sheet**. The process by which a protein assumes its biologically active tertiary structure is called **folding**.

Sections 25.8 and 25.9 **Carbohydrates**, which are polyhydroxy aldehydes and ketones, are the major structural constituents of plants and are a source of energy in both plants and animals. **Glucose** is the most common

monosaccharide, or simple sugar. Two monosaccharides can be linked together by means of a condensation reaction to form a disaccharide. **Polysaccharides** are complex carbohydrates made up of many monosaccharide units joined together. The three most important polysaccharides are **starch**, which is found in plants; **glycogen**, which is found in mammals; and **cellulose**, which is also found in plants. **Lipids** are small molecules derived from glycerol and fatty acids that comprise fats, oils, and **phospholipids**. Fatty acids can be *saturated, unsaturated, cis,* or *trans* depending on their chemical formulas and structures.

Section 25.10 **Nucleic acids** are biopolymers that carry the genetic information necessary for cell reproduction; they also control cell development through control of protein synthesis. The building blocks of these biopolymers are **nucleotides**. There are two types of nucleic acids, the **ribonucleic acids (RNA)** and the **deoxyribonucleic acids (DNA)**. These substances consist of a polymeric backbone of alternating phosphate and ribose or deoxyribose sugar groups, with organic bases attached to the sugar molecules. The DNA polymer is a double-stranded helix (a **double helix**) held together by hydrogen bonding between matching organic bases situated across from one another on the two strands. The hydrogen bonding between specific base pairs is the key to gene replication and protein synthesis.

KEY CONCEPTS

- Draw hydrocarbon structures based on their names and name hydrocarbons based on their structures.
- Know the structures of the functional groups: alkene, alkyne, alcohol, carbonyl, ether, aldehyde, ketone, carboxylic acid, amine, amide.
- Distinguish between addition reactions and substitution reactions.
- Understand what makes a compound chiral and be able to recognize a chiral substance.
- Recognize the amino acids and understand how they form peptides and proteins via amide bond formation.
- Understand the difference between primary, secondary, and tertiary structure of proteins.
- Be able to explain the difference between α helix and β sheet peptide and protein structures.
- Understand the distinction between starch and cellulose structures.
- Classify molecules as saccharides or lipids based on their structures.
- Understand the difference between *cis* and *trans*, and saturated and unsaturated fatty acids.
- Understand the structure of nucleic acids.

VISUALIZING CONCEPTS

25.1 Which of the following molecules is unsaturated? [Section 25.2]

$$CH_3CH_2CH_2CH_3$$

(a)

$$\begin{array}{cc} CH_2 - CH_2 \\ CH_2 \quad\quad CH_2 \\ CH_2 \end{array}$$

(b)

$$\underset{\text{(c)}}{CH_3\overset{\displaystyle O}{\overset{\|}{C}}-OH} \qquad \underset{\text{(d)}}{CH_3CH=CHCH_3}$$

25.2 Which of the following molecules will most readily undergo an addition reaction? [Section 25.3]

(a)

$$\underset{\text{(b)}}{CH_3CH_2\overset{\displaystyle O}{\overset{\|}{C}}-OH}$$

$$\begin{array}{c} CH=CH \\ CH_2 \quad\quad CH_2 \\ CH_2 - CH_2 \end{array}$$

(c)

$$\underset{\underset{\text{(d)}}{\displaystyle NH_2}}{CH_3\overset{\displaystyle O}{\underset{|}{\overset{\|}{C}H\overset{\|}{C}}}-OH}$$

25.3 Which of the following compounds would you expect to have the highest boiling point? Explain. [Section 25.4]

(a) CH₃CH (O double bond) (b) CH₃CH₂OH (c) CH₃C≡CH (d) HCOCH₃ (O double bond)

25.4 Which of the following compounds is capable of possessing isomerism? In each case where isomerism is possible, identify the type or types of isomerism. [Sections 25.3 and 25.4]

(a) CH₃CHCHC—O⁻ (with CH₃, O, NH₃⁺) (b) benzene ring with C—OH and Cl

(c) CH₃CH₂CH=CHCH₃ (d) CH₃CH₂CH₃

25.5 From examination of the following ball-and-stick molecular models, choose the substance that **(a)** can be hydrolyzed to form a solution containing glucose, **(b)** is capable of forming a zwitterion, **(c)** is one of the four bases present in DNA, **(d)** reacts with an acid to form an ester, **(e)** is a lipid. [Sections 25.6–25.10]

(iii)

(iv)   (v)

(i)

(ii)

25.6 All the structures that follow have the same molecular formula, C₈H₁₈. Which structures are the same molecule? (*Hint:* One way to do this is to determine the chemical name for each.) [Section 25.3]

(a) CH₃CCH₂CHCH₃ (with CH₃, CH₃, CH₃)

(b) CH₃CHCHCH₂ (with CH₃, CH₃, CH₂, CH₃)

(c) CH₃CHCHCH₃ (with CH₃, CHCH₃, CH₃)

(d) CH₃CHCHCH₃ (with CH₃CHCH₃, CH₃)

EXERCISES

Introduction to Organic Compounds; Hydrocarbons

25.7 Predict the ideal values for the bond angles about each carbon atom in the following molecule. Indicate the hybridization of orbitals for each carbon.

CH₃CCCH₂COOH

25.8 Identify the carbon atom(s) in the structure shown that has (have) each of the following hybridizations: **(a)** sp^3, **(b)** sp, **(c)** sp^2.

N≡C—CH₂—CH₂—CH=CH—CHOH
                              |
                              C=O
                              |
                              H

25.9 Are carbon monoxide or ammonia considered organic molecules? Why or why not?

25.10 Organic compounds containing C—O and C—Cl bonds are more reactive than simple alkane hydrocarbons. Considering the comparative values of C—H, C—C, C—O, and C—Cl bond energies (Table 8.4), why is this so?

25.11 **(a)** What is the difference between a straight-chain and branched-chain alkane? **(b)** What is the difference between an alkane and an alkyl group? **(c)** Why are alkanes said to be saturated? **(d)** Give an example of an unsaturated molecule.

25.12 What structural features help us identify a compound as **(a)** an alkane, **(b)** a cycloalkane, **(c)** an alkene, **(d)** an alkyne, **(e)** a saturated hydrocarbon, **(f)** an aromatic hydrocarbon?

25.13 Give the molecular formula of a hydrocarbon containing five carbon atoms that is **(a)** an alkane, **(b)** a cycloalkane, **(c)** an alkene, **(d)** an alkyne. Which are saturated and which are unsaturated hydrocarbons?

25.14 Give the molecular formula of a cyclic alkane, a cyclic alkene, a linear alkyne, and an aromatic hydrocarbon that in each case contains six carbon atoms. Which are saturated and which are unsaturated hydrocarbons?

25.15 Enediynes are a class of compounds that include some antibiotic drugs. Draw the structure of an "enediyne" fragment that contains six carbons in a row. (*Hint:* "di" means "two".)

25.16 Give the general formula for any cyclic alkene, that is, a cyclic hydrocarbon with one double bond.

25.17 Write the condensed structural formulas for as many alkenes and alkynes as you can think of that have the molecular formula C_6H_{10}.

25.18 Draw all the possible noncyclic structural isomers of C_5H_{10}. Name each compound.

25.19 What are the characteristic hybrid orbitals employed by **(a)** carbon in an alkane, **(b)** carbon in a double bond in an alkene, **(c)** carbon in the benzene ring, **(d)** carbon in a triple bond in an alkyne?

25.20 What are the approximate bond angles **(a)** about carbon in an alkane, **(b)** about a doubly bonded carbon atom in an alkene, **(c)** about a triply bonded carbon atom in an alkyne?

25.21 Draw the structural formula or give the name, as appropriate, for the following:

(a)
$$\begin{array}{c} CH_3\,H\ \ \ H\ \ H\ \ H \\ |\ \ \ |\ \ \ \ |\ \ \ |\ \ \ | \\ H-C-C-C-C-C-H \\ |\ \ \ |\ \ \ \ |\ \ \ |\ \ \ | \\ H\ \ \ H\ \ \ H\ \ CH_3\,H \end{array}$$

(b)
$$CH_3CH_2CH_2CH_2CH_2CH_2CCH_2CHCH_3$$
with substituents CH_2 CH_3 and CH_3

(c) 2-methylheptane

(d) 4-ethyl-2,3-dimethyloctane
(e) 1,2-dimethylcyclohexane

25.22 Draw the structural formula or give the name, as appropriate, for the following:

(a)
$$\begin{array}{c} CH_3CH_2\ \ CH_2CH_3 \\ |\ \ \ \ \ \ \ \ | \\ CH_3CCH_2CH \\ |\ \ \ \ \ \ \ \ | \\ CH_3\ \ \ CH_3 \end{array}$$

(b)
$$\begin{array}{c} CH_3 \\ | \\ CH_3CH_2CH_2CCH_3 \\ | \\ CH_3CHCH_2CH_3 \end{array}$$

(c) 2,5,6-trimethylnonane
(d) 3-propyl-4,5-methyldecane
(e) 1-ethyl-3-methylcyclohexane

25.23 Name the following compounds:

(a)
$$\begin{array}{c} CH_3CHCH_3 \\ | \\ CHCH_2CH_2CH_2CH_3 \\ | \\ CH_3 \end{array}$$

(b)
$$\begin{array}{c} \ \ \ \ \ \ \ \ \ \ \ \ \ \ \ \ \ CH_3 \\ \ \ \ \ \ \ \ \ \ \ \ \ \ \ \ \ \ | \\ CH_3CH_2 \ \ \ \ \ \ CH_2CHCH_2CH_3 \\ \ \ \ \ \ \ \ \ C=C \\ \ \ \ \ \ H \ \ \ \ \ \ \ \ H \end{array}$$

(c) *(para-dibromobenzene structure with Br at top and bottom)*

(d)
$$\begin{array}{c} CH_2CH_3 \\ | \\ HC\equiv CCH_2CCH_3 \\ | \\ CH_3 \end{array}$$

(e) *(cyclobutane ring)*—CH_3

25.24 Name the following compounds:

(a)
$$\begin{array}{c} CH_2-CH_2 \\ /\ \ \ \ \ \ \ \ \ \backslash \\ Cl-CH\ \ \ \ \ \ HC-Cl \\ \backslash\ \ \ \ \ \ \ \ \ / \\ CH_2-CH_2 \end{array}$$

(b) $HC\equiv C-CH_2-Cl$

(c) *trans*-$CH_3CH=CHCH_2CH_2CH_3$

(d)
$$\begin{array}{c} CH_3 \\ | \\ CH_3CH_2-C-CH_2Cl \\ | \\ (benzene\ ring) \end{array}$$

(e) *cis*-$CH_2=CH-CH=CH-CH_2Cl$

25.25 Why is geometric isomerism possible for alkenes, but not for alkanes and alkynes?

25.26 Draw all structural and geometric isomers of butene and name them.

25.27 Indicate whether each of the following molecules is capable of geometrical (cis-trans) isomerism. For those that are, draw the structures: **(a)** 1,1-dichloro-1-butene, **(b)** 2,4-dichloro-2-butene, **(c)** 1,4-dichlorobenzene, **(d)** 4,5-dimethyl-2-pentyne.

25.28 Draw all the distinct geometric isomers of 2,4-hexadiene.

25.29 What is the octane number of a mixture of 35% heptane and 65% isooctane?

25.30 Describe two ways in which the octane number of a gasoline consisting of alkanes can be increased.

Reactions of Hydrocarbons

25.31 **(a)** What is the difference between a substitution reaction and an addition reaction? Which one is commonly observed with alkenes, and which one with aromatic hydrocarbons? **(b)** Using condensed structural formulas, write the balanced equation for the addition reaction of 2-pentene with Br_2 and name the resulting compound. **(c)** Write a balanced chemical equation for the substitution reaction of Cl_2 with benzene to make *para*-dichlorobenzene in the presence of $FeCl_3$ as a catalyst.

25.32 Using condensed structural formulas, write a balanced chemical equation for each of the following reactions: **(a)** hydrogenation of cyclohexene; **(b)** addition of H_2O to *trans*-2-pentene using H_2SO_4 as a catalyst (two products); **(c)** reaction of 2-chloropropane with benzene in the presence of $AlCl_3$.

25.33 **(a)** When cyclopropane is treated with HI, 1-iodopropane is formed. A similar type of reaction does not occur with cyclopentane or cyclohexane. How do you account for the reactivity of cyclopropane? **(b)** Suggest a method of preparing ethylbenzene, starting with benzene and ethylene as the only organic reagents.

25.34 **(a)** One test for the presence of an alkene is to add a small amount of bromine, a red-brown liquid, and look for the disappearance of the red-brown color. This test does not work for detecting the presence of an aromatic hydrocarbon. Explain. **(b)** Write a series of reactions leading to *para*-bromoethylbenzene, beginning with benzene and using other reagents as needed. What isomeric side products might also be formed?

25.35 The rate law for addition of Br_2 to an alkene is first order in Br_2 and first order in the alkene. Does this fact prove that the mechanism of addition of Br_2 to an alkene proceeds in the same manner as for addition of HBr? Explain.

25.36 Describe the intermediate that is thought to form in the addition of a hydrogen halide to an alkene, using cyclohexene as the alkene in your description.

25.37 The molar heat of combustion of gaseous cyclopropane is -2089 kJ/mol; that for gaseous cyclopentane is -3317 kJ/mol. Calculate the heat of combustion per CH_2 group in the two cases, and account for the difference.

25.38 The heat of combustion of decahydronaphthalene ($C_{10}H_{18}$) is -6286 kJ/mol. The heat of combustion of naphthalene ($C_{10}H_8$) is -5157 kJ/mol. (In both cases $CO_2(g)$ and $H_2O(l)$ are the products.) Using these data and data in Appendix C, calculate the heat of hydrogenation of naphthalene. Does this value provide any evidence for aromatic character in naphthalene?

Functional Groups and Chirality

25.39 Identify the functional groups in each of the following compounds:

(a) H_3C-CH_2-OH

(b)

(c) $H_3C-\overset{H}{\underset{}{N}}-CH_2CH-CH_2$

(d)

(e)

(f)
$$CH_3CH_2CH_2CH_2 \quad \overset{O}{\underset{}{}} \quad CH_2CH_2CH_2CH_3$$

(e) $CH_3CH_2CH_2CH_2CHO$

(f) CH_3CCCH_2COOH

25.40 Identify the functional groups in each of the following compounds:

(a)
$$H_3C-\overset{O}{\underset{}{C}}-O-CH_2CH_2CH_2CH_2CH_2CH_3$$

(b)

(c)

(d)

25.41 Give the structural formula for **(a)** an aldehyde that is an isomer of acetone, **(b)** an ether that is an isomer of 1-propanol.

25.42 **(a)** Give the empirical formula and structural formula for a cyclic ether containing four carbon atoms in the ring. **(b)** Write the structural formula for a straight-chain compound that is a structural isomer of your answer to part (a).

25.43 The IUPAC name for a carboxylic acid is based on the name of the hydrocarbon with the same number of carbon atoms. The ending *-oic* is appended, as in ethanoic acid, which is the IUPAC name for acetic acid. Draw the structure of the following acids: **(a)** methanoic acid, **(b)** pentanoic acid, and **(c)** 2-chloro-3-methyldecanoic acid.

25.44 Aldehydes and ketones can be named in a systematic way by counting the number of carbon atoms (including the carbonyl carbon) that they contain. The name of the aldehyde or ketone is based on the hydrocarbon with the same number of carbon atoms. The ending *-al* for aldehyde or *-one* for ketone is added as appropriate. Draw the structural formulas for the following aldehydes or ketones: **(a)** propanal, **(b)** 2-pentanone, **(c)** 3-methyl-2-butanone, **(d)** 2-methylbutanal.

25.45 Draw the condensed structure of the compounds formed by condensation reactions between **(a)** benzoic acid and ethanol, **(b)** ethanoic acid and methylamine, **(c)** acetic acid and phenol. Name the compound in each case.

25.46 Draw the condensed structures of the compounds formed from **(a)** butanoic acid and methanol, **(b)** benzoic acid and 2-propanol, **(c)** propanoic acid and dimethylamine. Name the compound in each case.

25.47 Write a balanced chemical equation using condensed structural formulas for the saponification (base hydrolysis) of **(a)** methyl propionate, **(b)** phenyl acetate.

25.48 Write a balanced chemical equation using condensed structural formulas for **(a)** the formation of butyl propionate from the appropriate acid and alcohol, **(b)** the saponification (base hydrolysis) of methyl benzoate.

25.49 Would you expect pure acetic acid to be a strongly hydrogen-bonded substance? How do the melting and boiling points of the substance (see the "Chemistry Put to Work" box on p. 1076 for data) support your answer?

25.50 Acetic anhydride is formed from acetic acid in a condensation reaction that involves the removal of a molecule of water from between two acetic acid molecules. Write the chemical equation for this process, and show the structure of acetic anhydride.

25.51 Write the condensed structural formula for each of the following compounds: **(a)** 2-pentanol, **(b)** 1,2-propanediol, **(c)** ethyl acetate, **(d)** diphenyl ketone, **(e)** methyl ethyl ether.

25.52 Write the condensed structural formula for each of the following compounds: **(a)** 3,3-dichlorobutyraldehyde, **(b)** methyl phenyl ketone, **(c)** *para*-bromobenzoic acid, **(d)** methyl-*trans*-2-butenyl ether, **(e)** *N, N*-dimethylbenzamide.

25.53 Draw the structure for 2-bromo-2-chloro-3-methylpentane, and indicate any chiral carbons in the molecule.

25.54 Does 3-chloro-3-methylhexane have optical isomers? Why or why not?

Proteins

25.55 **(a)** What is an α-amino acid? **(b)** How do amino acids react to form proteins? **(c)** Draw the bond that links amino acids together in proteins. What is this called?

25.56 What properties of the side chains (R groups) of amino acids are important in affecting the amino acid's overall biochemical behavior? Give examples to illustrate your reply.

25.57 Draw the two possible dipeptides formed by condensation reactions between leucine and tryptophan.

25.58 Write a chemical equation for the formation of methionyl glycine from the constituent amino acids.

25.59 **(a)** Draw the condensed structure of the tripeptide Gly-Gly-His. **(b)** How many different tripeptides can be made from the amino acids glycine and histidine? Give the abbreviations for each of these tripeptides, using the three-letter and one-letter codes for the amino acids.

25.60 **(a)** What amino acids would be obtained by hydrolysis of the following tripeptide?

$$
\underset{(CH_3)_2CH}{\overset{O}{\underset{|}{H_2NCHCNHCHCNHCHCOH}}} \quad \underset{H_2COH}{\overset{O}{\underset{|}{}}} \quad \underset{\underset{O}{\overset{}{\underset{}{}}}}{\overset{O}{\underset{|}{H_2CCH_2COH}}}
$$

(b) How many different tripeptides can be made from the amino acids glycine, serine, and glutamic acid? Give the abbreviation for each of these tripeptides, using the three-letter codes and one-letter codes for the amino acids.

[25.61] **(a)** Describe the primary, secondary, and tertiary structures of proteins. **(b)** *Quaternary structures* of proteins arise if two or more smaller polypeptides or proteins associate with each other to make an overall much larger protein structure. The association is due to the same hydrogen bonding, electrostatic, and dispersion forces we have seen before. Hemoglobin, the protein used to transport oxygen molecules in our blood, is an example of a protein that has quaternary structure. Hemoglobin is a tetramer; it is made of four smaller polypeptides, two "alphas" and two "betas." (These names do not imply anything about the number of alpha helices or beta sheets in the individual polypeptides.) What kind of experiments would provide sound evidence that hemoglobin exists as a tetramer and not as one enormous polypeptide chain? You may need to look into the chemical literature to discover techniques that chemists and biochemists use to make these decisions.

25.62 What is the difference between the α helix and β sheet secondary structures in proteins?

Carbohydrates and Lipids

25.63 In your own words, define the following terms: **(a)** carbohydrate, **(b)** monosaccharide, **(c)** disaccharide, **(d)** polysaccharide.

25.64 What is the difference between α-glucose and β-glucose? Show the condensation of two glucose molecules to form a disaccharide with an α linkage; with a β linkage.

25.65 What is the empirical formula of cellulose? What is the unit that forms the basis of the cellulose polymer? What form of linkage joins these monomeric units?

25.66 What is the empirical formula of glycogen? What is the unit that forms the basis of the glycogen polymer? What form of linkage joins these monomeric units?

25.67 The structural formula for the linear form of D-mannose is:

(a) How many chiral carbons are present in the molecule? **(b)** Draw the structure of the six-member-ring form of this sugar.

25.68 The structural formula for the linear form of galactose is:

(a) How many chiral carbons are present in the molecule? **(b)** Draw the structure of the six-member-ring form of this sugar.

25.69 Describe the chemical structures of lipids and phospholipids. Why can phospholipids form a bilayer in water?

25.70 Using data from Table 8.4 on bond energies, show that the more C—H bonds a molecule has compared to C—O and O—H bonds, the more energy it can store.

Nucleic Acids

25.71 Adenine and guanine are members of a class of molecules known as *purines*; they have two rings in their structure. Thymine and cytosine, on the other hand, are *pyrimidines*, and have only one ring in their structure. Predict which have larger dispersion forces in aqueous solution, the purines or the pyrimidines.

25.72 A nucleoside consists of an organic base of the kind shown in Section 25.10, bound to ribose or deoxyribose. Draw the structure for deoxyguanosine, formed from guanine and deoxyribose.

25.73 Just as the amino acids in a protein are listed in the order from the amine end to the carboxylic acid end (the *protein sequence*), the bases in nucleic acids are listed in the order 5′ to 3′, where the numbers refer to the position of the carbons in the sugars (shown here for deoxyribose):

The base is attached to the sugar at the 1′ carbon. The 5′ end of a DNA sequence is a phosphate of an OH group, and the 3′ end of a DNA sequence is the OH group. What is the DNA sequence for the molecule shown here?

25.74 When samples of double-stranded DNA are analyzed, the quantity of adenine present equals that of thymine. Similarly, the quantity of guanine equals that of cytosine. Explain the significance of these observations.

25.75 Imagine a single DNA strand containing a section with the following base sequence: 5′-GCATTGGC-3′. What is the base sequence of the complementary strand? (The two strands of DNA will come together in an *antiparallel* fashion: that is, 5′-TAG-3′ will bind to 3′-ATC-5′.)

25.76 Explain the chemical differences between DNA and RNA.

ADDITIONAL EXERCISES

25.77 Draw the condensed structural formulas for two different molecules with the formula C_3H_4O.

25.78 How many structural isomers are there for a five-member straight carbon chain with one double bond? For a six-member straight carbon chain with two double bonds?

25.79 There are no known stable cyclic compounds with ring sizes of seven or less that have an alkyne linkage in the ring. Why is this? Could a ring with a larger number of carbon atoms accommodate an alkyne linkage? Explain.

25.80 Draw the condensed structural formulas for the cis and trans isomers of 2-pentene. Can cyclopentene exhibit cis-trans isomerism? Explain.

25.81 Although there are silicon analogs of alkanes, silicon analogs of alkenes or alkynes are virtually unknown. Suggest an explanation.

25.82 If a molecule is an "ene-one," what functional groups must it have?

25.83 Write the structural formulas for as many alcohols as you can think of that have empirical formula C_3H_6O.

[25.84] Dinitromethane, $CH_2(NO_2)_2$, is a dangerously reactive substance that decomposes readily on warming. On the other hand, dichloromethane is relatively unreactive. Why is the nitro compound so reactive compared to the chloro compound? (*Hint:* Consider the oxidation numbers of the atoms involved and the possible products of decomposition.)

25.85 Identify each of the functional groups in the following molecules:

(a)

(Responsible for the odor of cucumbers)

(b)

(Quinine – an antimalarial drug)

(c)

(Indigo – a blue dye)

(d)

(Acetaminophen – aka Tylenol)

25.86 Write a condensed structural formula for each of the following: **(a)** an acid with the formula $C_4H_8O_2$, **(b)** a cyclic ketone with the formula C_5H_8O, **(c)** a dihydroxy compound with formula $C_3H_8O_2$, **(d)** a cyclic ester with formula $C_5H_8O_2$.

25.87 Although carboxylic acids and alcohols both contain an —OH group, one is acidic in water and the other is not. Explain the difference.

[25.88] Indole smells rather terrible in high concentrations but has a pleasant floral-like odor when highly diluted. It has the following structure:

Indole is a planar molecule. The nitrogen is a very weak base, with a K_b of 2×10^{-12}. Explain how this information indicates that the indole molecule is aromatic in character.

25.89 Locate the chiral carbon atoms, if any, in each of the following substances:

(a) $HOCH_2CH_2\overset{\displaystyle O}{\overset{\displaystyle \|}{C}}CH_2OH$

(b) $HOCH_2\overset{\displaystyle OH}{\overset{\displaystyle |}{C}}H\overset{\displaystyle O}{\overset{\displaystyle \|}{C}}CH_2OH$

(c) $HO\overset{\displaystyle O}{\overset{\displaystyle \|}{C}}\overset{\displaystyle CH_3}{\overset{\displaystyle |}{C}}H\overset{\displaystyle |}{\underset{\displaystyle NH_2}{C}}HC_2H_5$

25.90 Which of the following peptides have a net positive charge at pH 7? **(a)** Gly-Ser-Lys, **(b)** Pro-Leu-Ile, **(c)** Phe-Tyr-Asp.

25.91 Glutathione is a tripeptide found in most living cells. Partial hydrolysis yields Cys-Gly and Glu-Cys. What structures are possible for glutathione?

25.92 Starch, glycogen, and cellulose are all polymers of glucose. What are the structural differences among them?

25.93 Monosaccharides can be categorized in terms of the number of carbon atoms (pentoses have five carbons and hexoses have six carbons) and according to whether they contain an aldehyde (*aldo-* prefix, as in aldopentose) or ketone group (*keto-* prefix, as in ketopentose). Classify glucose and fructose in this way.

25.94 Can a DNA strand bind to a complementary RNA strand? Explain.

INTEGRATIVE EXERCISES

25.95 Explain why the boiling point of ethanol (78 °C) is much higher than that of its isomer, dimethyl ether (−25 °C) and why the boiling point of CH_2F_2 (−52 °C) is far above that of CH_4 (−128 °C).

[25.96] An unknown organic compound is found on elemental analysis to contain 68.1% carbon, 13.7% hydrogen, and 18.2% oxygen by mass. It is slightly soluble in water. Upon careful oxidation it is converted into a compound that behaves chemically like a ketone and contains 69.7% carbon, 11.7% hydrogen, and 18.6% oxygen by mass. Indicate two or more reasonable structures for the unknown.

25.97 An organic compound is analyzed and found to contain 66.7% carbon, 11.2% hydrogen, and 22.1% oxygen by mass. The compound boils at 79.6 °C. At 100 °C and 0.970 atm, the vapor has a density of 2.28 g/L. The compound has a carbonyl group and cannot be oxidized to a carboxylic acid. Suggest a structure for the compound.

[25.98] An unknown substance is found to contain only carbon and hydrogen. It is a liquid that boils at 49 °C at 1 atm pressure. Upon analysis it is found to contain 85.7% carbon and 14.3% hydrogen by mass. At 100 °C and 735 torr, the vapor of this unknown has a density of 2.21 g/L. When it is dissolved in hexane solution and bromine water is added, no reaction occurs. What is the identity of the unknown compound?

25.99 The standard free energy of formation of solid glycine is −369 kJ/mol, whereas that of solid glycylglycine is −488 kJ/mol. What is $\Delta G°$ for the condensation of glycine to form glycylglycine?

25.100 One of the most important molecules in biochemical systems is adenosine triphosphate (ATP), for which the structure is

ATP

ATP is the principal carrier of biochemical energy. It is considered an energy-rich compound because the hydrolysis of ATP to yield adenosine diphosphate (ADP) and inorganic phosphate is spontaneous under aqueous biochemical conditions. **(a)** Write a balanced equation for the reaction of ATP with water to yield ADP and inorganic phosphate ion. [*Hint:* Hydrolysis reactions are just the reverse of condensation reactions (Section 22.8).] **(b)** What would you expect for the sign of the free-energy change for this reaction? **(c)** ADP can undergo further hydrolysis. What would you expect for the product of that reaction?

25.101 A typical amino acid with one amino group and one carboxylic acid group, such as serine (Figure 25.23), can exist in water in several ionic forms. **(a)** Suggest the forms of the amino acid at low pH and at high pH. **(b)** Amino acids generally have two pK_a values, one in the range of 2 to 3 and the other in the range of 9 to 10. Serine, for example, has pK_a values of 2.19 and 9.21. Using species such as acetic acid and ammonia as models, suggest the origin of the two pK_a values. **(c)** Glutamic acid is an amino acid that has three pK_a's: 2.10, 4.07, and 9.47. Draw the structure of glutamic acid, and assign each pK_a to the appropriate part of the molecule.

[25.102] The protein ribonuclease A in its native, or most stable, form is folded into a compact globular shape. **(a)** Does the native form have a lower or higher free energy than the denatured form, in which the protein is an extended chain? **(b)** What is the sign of the entropy change in going from the denatured to the folded form? **(c)** In the folded form the ribonuclease A has four —S—S— bonds that bridge between parts of the chain, as shown in the accompanying figure. What effect do you predict that these four linkages have on the free energy and entropy of the folded form as compared with a hypothetical folded structure that does not possess the four —S—S— linkages? Explain. **(d)** A gentle reducing agent converts the four —S—S— linkages to eight —S—H bonds. What effect do you predict this would have on the tertiary structure and entropy of the protein? **(e)** Which amino acid must be present for —SH bonds to exist in ribonuclease A?

Native ribonuclease A

25.103 The monoanion of adenosine monophosphate (AMP) is an intermediate in phosphate metabolism:

$$A-O-\overset{\displaystyle O^-}{\underset{\displaystyle O}{\overset{|}{\underset{\|}{P}}}}-OH = AMP-OH^-$$

where A = adenosine. If the pK_a for this anion is 7.21, what is the ratio of $[AMP-OH^-]$ to $[AMP-O^{2-}]$ in blood at pH 7.4?

A Mathematical Operations

A.1 EXPONENTIAL NOTATION

The numbers used in chemistry are often either extremely large or extremely small. Such numbers are conveniently expressed in the form

$$N \times 10^n$$

where N is a number between 1 and 10, and n is the exponent. Some examples of this *exponential notation*, which is also called *scientific notation*, follow.

1,200,000 is 1.2×10^6 (read "one point two times ten to the sixth power")

0.000604 is 6.04×10^{-4} (read "six point zero four times ten to the negative fourth power")

A positive exponent, as in the first example, tells us how many times a number must be multiplied by 10 to give the long form of the number:

$$1.2 \times 10^6 = 1.2 \times 10 \times 10 \times 10 \times 10 \times 10 \times 10 \quad \text{(six tens)}$$

$$= 1,200,000$$

It is also convenient to think of the *positive exponent* as the number of places the decimal point must be moved to the *left* to obtain a number greater than 1 and less than 10: If we begin with 3450 and move the decimal point three places to the left, we end up with 3.45×10^3.

In a related fashion, a negative exponent tells us how many times we must divide a number by 10 to give the long form of the number.

$$6.04 \times 10^{-4} = \frac{6.04}{10 \times 10 \times 10 \times 10} = 0.000604$$

It is convenient to think of the *negative exponent* as the number of places the decimal point must be moved to the *right* to obtain a number greater than 1 but less than 10: If we begin with 0.0048 and move the decimal point three places to the right, we end up with 4.8×10^{-3}.

In the system of exponential notation, with each shift of the decimal point one place to the right, the exponent *decreases* by 1:

$$4.8 \times 10^{-3} = 48 \times 10^{-4}$$

Similarly, with each shift of the decimal point one place to the left, the exponent *increases* by 1:

$$4.8 \times 10^{-3} = 0.48 \times 10^{-2}$$

Many scientific calculators have a key labeled EXP or EE, which is used to enter numbers in exponential notation. To enter the number 5.8×10^3 on such a calculator, the key sequence is

$$\boxed{5} \; \boxed{\cdot} \; \boxed{8} \; \boxed{\text{EXP}} \; (\text{or} \; \boxed{\text{EE}}) \; \boxed{3}$$

On some calculators the display will show 5.8, then a space, followed by 03, the exponent. On other calculators, a small 10 is shown with an exponent 3.

To enter a negative exponent, use the key labeled $+/-$. For example, to enter the number 8.6×10^{-5}, the key sequence is

$$\boxed{8}\;\boxed{\cdot}\;\boxed{6}\;\boxed{\text{EXP}}\;\boxed{+/-}\;\boxed{5}$$

When entering a number in exponential notation, do not key in the 10 if you use the EXP or EE button.

In working with exponents, it is important to recall that $10^0 = 1$. The following rules are useful for carrying exponents through calculations.

1. *Addition and Subtraction* In order to add or subtract numbers expressed in exponential notation, the powers of 10 must be the same.

$$(5.22 \times 10^4) + (3.21 \times 10^2) = (522 \times 10^2) + (3.21 \times 10^2)$$

$$= 525 \times 10^2 \quad (3 \text{ significant figures})$$

$$= 5.25 \times 10^4$$

$$(6.25 \times 10^{-2}) - (5.77 \times 10^{-3}) = (6.25 \times 10^{-2}) - (0.577 \times 10^{-2})$$

$$= 5.67 \times 10^{-2} \quad (3 \text{ significant figures})$$

When you use a calculator to add or subtract, you need not be concerned with having numbers with the same exponents, because the calculator automatically takes care of this matter.

2. *Multiplication and Division* When numbers expressed in exponential notation are multiplied, the exponents are added; when numbers expressed in exponential notation are divided, the exponent of the denominator is subtracted from the exponent of the numerator.

$$(5.4 \times 10^2)(2.1 \times 10^3) = (5.4)(2.1) \times 10^{2+3}$$

$$= 11 \times 10^5$$

$$= 1.1 \times 10^6$$

$$(1.2 \times 10^5)(3.22 \times 10^{-3}) = (1.2)(3.22) \times 10^{5+(-3)} = 3.9 \times 10^2$$

$$\frac{3.2 \times 10^5}{6.5 \times 10^2} = \frac{3.2}{6.5} \times 10^{5-2} = 0.49 \times 10^3 = 4.9 \times 10^2$$

$$\frac{5.7 \times 10^7}{8.5 \times 10^{-2}} = \frac{5.7}{8.5} \times 10^{7-(-2)} = 0.67 \times 10^9 = 6.7 \times 10^8$$

3. *Powers and Roots* When numbers expressed in exponential notation are raised to a power, the exponents are multiplied by the power. When the roots of numbers expressed in exponential notation are taken, the exponents are divided by the root.

$$(1.2 \times 10^5)^3 = (1.2)^3 \times 10^{5 \times 3}$$

$$= 1.7 \times 10^{15}$$

$$\sqrt[3]{2.5 \times 10^6} = \sqrt[3]{2.5} \times 10^{6/3}$$

$$= 1.3 \times 10^2$$

Scientific calculators usually have keys labeled x^2 and $\sqrt{x}$ for squaring and taking the square root of a number, respectively. To take higher powers or roots, many calculators have y^x and $\sqrt[x]{y}$ (or INV y^x) keys. For example, to perform the operation $\sqrt[3]{7.5 \times 10^{-4}}$ on such a calculator, you would key in 7.5×10^{-4}, press the $\sqrt[x]{y}$ key (or the INV and then the y^x keys), enter the root, 3, and finally press =. The result is 9.1×10^{-2}.

■■ **SAMPLE EXERCISE 1** | Using Exponential Notation

Perform each of the following operations, using your calculator where possible:

(a) Write the number 0.0054 in standard exponential notation

(b) $(5.0 \times 10^{-2}) + (4.7 \times 10^{-3})$

(c) $(5.98 \times 10^{12})(2.77 \times 10^{-5})$

(d) $\sqrt[4]{1.75 \times 10^{-12}}$

SOLUTION

(a) Because we move the decimal point three places to the right to convert 0.0054 to 5.4, the exponent is −3:

$$5.4 \times 10^{-3}$$

Scientific calculators are generally able to convert numbers to exponential notation using one or two keystrokes. Consult your instruction manual to see how this operation is accomplished on your calculator.

(b) To add these numbers longhand, we must convert them to the same exponent.

$$(5.0 \times 10^{-2}) + (0.47 \times 10^{-2}) = (5.0 + 0.47) \times 10^{-2} = 5.5 \times 10^{-2}$$

(Note that the result has only two significant figures.) To perform this operation on a calculator, we enter the first number, strike the + key, then enter the second number and strike the = key.

(c) Performing this operation longhand, we have

$$(5.98 \times 2.77) \times 10^{12-5} = 16.6 \times 10^{7} = 1.66 \times 10^{8}$$

On a scientific calculator, we enter 5.98×10^{12}, press the × key, enter 2.77×10^{-5}, and press the = key.

(d) To perform this operation on a calculator, we enter the number, press the $\sqrt[x]{y}$ key (or the INV and y^x keys), enter 4, and press the = key. The result is 1.15×10^{-3}.

■■ **PRACTICE EXERCISE**

Perform the following operations:

(a) Write 67,000 in exponential notation, showing two significant figures

(b) $(3.378 \times 10^{-3}) - (4.97 \times 10^{-5})$

(c) $(1.84 \times 10^{15})(7.45 \times 10^{-2})$

(d) $(6.67 \times 10^{-8})^3$

Answers: **(a)** 6.7×10^4, **(b)** 3.328×10^{-3}, **(c)** 2.47×10^{16}, **(d)** 2.97×10^{-22}

A.2 LOGARITHMS

Common Logarithms

The common, or base-10, logarithm (abbreviated log) of any number is the power to which 10 must be raised to equal the number. For example, the common logarithm of 1000 (written log 1000) is 3 because raising 10 to the third power gives 1000.

$$10^3 = 1000, \text{ therefore, } \log 1000 = 3$$

Further examples are

$$\log 10^5 = 5$$

$$\log 1 = 0 \qquad \text{(Remember that } 10^0 = 1)$$

$$\log 10^{-2} = -2$$

In these examples the common logarithm can be obtained by inspection. However, it is not possible to obtain the logarithm of a number such as 31.25 by inspection. The logarithm of 31.25 is the number x that satisfies the following relationship:

$$10^x = 31.25$$

Most electronic calculators have a key labeled LOG that can be used to obtain logarithms. For example, on many calculators we obtain the value of log 31.25 by entering 31.25 and pressing the LOG key. We obtain the following result:

$$\log 31.25 = 1.4949$$

Notice that 31.25 is greater than 10 (10^1) and less than 100 (10^2). The value for log 31.25 is accordingly between log 10 and log 100, that is, between 1 and 2.

Significant Figures and Common Logarithms

For the common logarithm of a measured quantity, the number of digits after the decimal point equals the number of significant figures in the original number. For example, if 23.5 is a measured quantity (three significant figures), then log 23.5 = 1.371 (three significant figures after the decimal point).

Antilogarithms

The process of determining the number that corresponds to a certain logarithm is known as obtaining an *antilogarithm*. It is the reverse of taking a logarithm. For example, we saw above that log 23.5 = 1.371. This means that the antilogarithm of 1.371 equals 23.5.

$$\log 23.5 = 1.371$$

$$\text{antilog } 1.371 = 23.5$$

The process of taking the antilog of a number is the same as raising 10 to a power equal to that number.

$$\text{antilog } 1.371 = 10^{1.371} = 23.5$$

Many calculators have a key labeled 10^x that allows you to obtain antilogs directly. On others, it will be necessary to press a key labeled INV (for *inverse*), followed by the LOG key.

Natural Logarithms

Logarithms based on the number *e* are called natural, or base *e*, logarithms (abbreviated ln). The natural log of a number is the power to which *e* (which has the value 2.71828...) must be raised to equal the number. For example, the natural log of 10 equals 2.303.

$$e^{2.303} = 10, \text{ therefore ln } 10 = 2.303$$

Your calculator probably has a key labeled LN that allows you to obtain natural logarithms. For example, to obtain the natural log of 46.8, you enter 46.8 and press the LN key.

$$\ln 46.8 = 3.846$$

The natural antilog of a number is *e* raised to a power equal to that number. If your calculator can calculate natural logs, it will also be able to calculate natural antilogs. On some calculators there is a key labeled e^x that allows you to calculate natural antilogs directly; on others, it will be necessary to first press the INV key followed by the LN key. For example, the natural antilog of 1.679 is given by

$$\text{Natural antilog } 1.679 = e^{1.679} = 5.36$$

The relation between common and natural logarithms is as follows:

$$\ln a = 2.303 \log a$$

Notice that the factor relating the two, 2.303, is the natural log of 10, which we calculated above.

Mathematical Operations Using Logarithms

Because logarithms are exponents, mathematical operations involving logarithms follow the rules for the use of exponents. For example, the product of z^a and z^b (where z is any number) is given by

$$z^a \cdot z^b = z^{(a+b)}$$

Similarly, the logarithm (either common or natural) of a product equals the *sum* of the logs of the individual numbers.

$$\log ab = \log a + \log b \qquad \ln ab = \ln a + \ln b$$

For the log of a quotient,

$$\log(a/b) = \log a - \log b \qquad \ln(a/b) = \ln a - \ln b$$

Using the properties of exponents, we can also derive the rules for the logarithm of a number raised to a certain power.

$$\log a^n = n \log a \qquad \ln a^n = n \ln a$$
$$\log a^{1/n} = (1/n) \log a \qquad \ln a^{1/n} = (1/n) \ln a$$

pH Problems

One of the most frequent uses for common logarithms in general chemistry is in working pH problems. The pH is defined as $-\log[H^+]$, where $[H^+]$ is the hydrogen ion concentration of a solution (Section 16.4). The following sample exercise illustrates this application.

■ **SAMPLE EXERCISE 2** | Using Logarithms

(a) What is the pH of a solution whose hydrogen ion concentration is 0.015 M?
(b) If the pH of a solution is 3.80, what is its hydrogen ion concentration?

SOLUTION

1. We are given the value of $[H^+]$. We use the LOG key of our calculator to calculate the value of $\log[H^+]$. The pH is obtained by changing the sign of the value obtained. (Be sure to change the sign *after* taking the logarithm.)

$$[H^+] = 0.015$$
$$\log[H^+] = -1.82 \qquad \text{(2 significant figures)}$$
$$pH = -(-1.82) = 1.82$$

2. To obtain the hydrogen ion concentration when given the pH, we must take the antilog of $-pH$.

$$pH = -\log[H^+] = 3.80$$
$$\log[H^+] = -3.80$$
$$[H^+] = \text{antilog}(-3.80) = 10^{-3.80} = 1.6 \times 10^{-4}\ M$$

■ **PRACTICE EXERCISE**

Perform the following operations: **(a)** $\log(2.5 \times 10^{-5})$, **(b)** $\ln 32.7$, **(c)** antilog -3.47, **(d)** $e^{-1.89}$.
Answers: **(a)** -4.60, **(b)** 3.487, **(c)** 3.4×10^{-4}, **(d)** 1.5×10^{-1}

A.3 QUADRATIC EQUATIONS

An algebraic equation of the form $ax^2 + bx + c = 0$ is called a *quadratic equation*. The two solutions to such an equation are given by the quadratic formula:

$$x = \frac{-b \pm \sqrt{b^2 - 4ac}}{2a}$$

■ SAMPLE EXERCISE 3 | Using the Quadratic Formula

Find the values of x that satisfy the equation $2x^2 + 4x = 1$.

SOLUTION

To solve the given equation for x, we must first put it in the form

$$ax^2 + bx + c = 0$$

and then use the quadratic formula. If

$$2x^2 + 4x = 1$$

then

$$2x^2 + 4x - 1 = 0$$

Using the quadratic formula, where $a = 2$, $b = 4$, and $c = -1$, we have

$$x = \frac{-4 \pm \sqrt{(4)(4) - 4(2)(-1)}}{2(2)}$$

$$= \frac{-4 \pm \sqrt{16 + 8}}{4} = \frac{-4 \pm \sqrt{24}}{4} = \frac{-4 \pm 4.899}{4}$$

The two solutions are

$$x = \frac{0.899}{4} = 0.225 \quad \text{and} \quad x = \frac{-8.899}{4} = -2.225$$

Often in chemical problems the negative solution has no physical meaning, and only the positive answer is used.

A.4 GRAPHS

Often the clearest way to represent the interrelationship between two variables is to graph them. Usually, the variable that is being experimentally varied, called the *independent variable*, is shown along the horizontal axis (x-axis). The variable that responds to the change in the independent variable, called the *dependent variable*, is then shown along the vertical axis (y-axis). For example, consider an experiment in which we vary the temperature of an enclosed gas and measure its pressure. The independent variable is temperature, and the dependent variable is pressure. The data shown in Table A-1▶ can be obtained by means of this experiment. These data are shown graphically in Figure A.1▶. The relationship between temperature and pressure is linear. The equation for any straight-line graph has the form

$$y = mx + b$$

where m is the slope of the line and b is the intercept with the y-axis. In the case of Figure 1, we could say that the relationship between temperature and pressure takes the form

$$P = mT + b$$

where P is pressure in atm and T is temperature in °C. As shown in Figure 1, the slope is 4.10×10^{-4} atm/°C, and the intercept—the point where the line crosses the y-axis—is 0.112 atm. Therefore, the equation for the line is

$$P = \left(4.10 \times 10^{-4} \frac{\text{atm}}{°\text{C}} \right) T + 0.112 \text{ atm}$$

TABLE A-1 ■ Interrelation between Pressure and Temperature

Temperature (°C)	Pressure (atm)
20.0	0.120
30.0	0.124
40.0	0.128
50.0	0.132

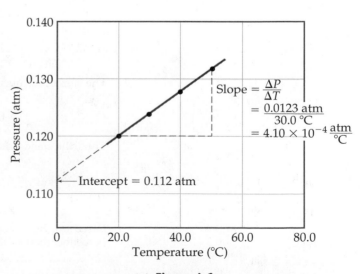

▲ Figure A.1

A.5 STANDARD DEVIATION

The standard deviation from the mean, s, is a common method from describing precision. We define the standard deviation as follows:

$$s = \sqrt{\frac{\sum_{i=1}^{N}(x_i - \bar{x})^2}{N - 1}}$$

where N is the number of measurements, $\bar{x}$ is the average (also called the mean), and x_i represents the individual measurements. Electronic calculators with built-in statistical functions can calculate s directly by inputting the individual measurements.

A smaller value of s indicates a higher precision, meaning that the data is more closely clustered around the average. The standard deviation has a statistical significance. Thus, if a large number of measurements is made, 68% of the measured values is expected to be within one standard deviation of the average, assuming only random errors are associated with the measurements.

■ SAMPLE EXERCISE 4 | Calculating an Average and Standard Deviation

The percent carbon in a sugar is measured four times: 42.01%, 42.28%, 41.79%, and 42.25%. Calculate **(a)** the average and **(b)** the standard deviation for these measurements.

SOLUTION

(a) The average is found by adding the quantities and dividing by the number of measurements:

$$\bar{x} = \frac{42.01 + 42.28 + 41.79 + 42.25}{4} = \frac{168.33}{4} = 42.08$$

(b) The standard deviation is found using the equation above:

$$s = \sqrt{\frac{\sum_{i=1}^{N}(x_i - \bar{x})^2}{N - 1}}$$

Let's tabulate the data so the calculation of $\sum_{i=1}^{N}(x_i - \bar{x})^2$ can be seen clearly.

Percent C	Difference between Measurement and Average, $(x_i - \bar{x})$	Square of Difference, $(x_i - \bar{x})^2$
42.01	$42.01 - 42.08 = -0.07$	$(-0.07)^2 = 0.005$
42.28	$42.28 - 42.08 = 0.20$	$(0.20)^2 = 0.040$
41.79	$41.79 - 42.08 = -0.29$	$(-0.29)^2 = 0.084$
42.25	$42.25 - 42.08 = 0.17$	$(0.17)^2 = 0.029$

The sum of the quantities in the last column is

$$\sum_{i=1}^{N}(x_i - \bar{x})^2 = 0.005 + 0.040 + 0.084 + 0.029 = 0.16$$

Thus, the standard deviation is

$$s = \sqrt{\frac{\sum_{i=1}^{N}(x_i - \bar{x})^2}{N - 1}} = \sqrt{\frac{0.16}{4 - 1}} = \sqrt{\frac{0.16}{3}} = \sqrt{0.053} = 0.23$$

Based on these measurements, it would be appropriate to represent the measured percent carbon as 42.08 ± 0.23.

B Properties of Water

Density:	0.99987 g/mL at 0 °C
	1.00000 g/mL at 4 °C
	0.99707 g/mL at 25 °C
	0.95838 g/mL at 100 °C
Heat of fusion:	6.008 kJ/mol at 0 °C
Heat of vaporization:	44.94 kJ/mol at 0 °C
	44.02 kJ/mol at 25 °C
	40.67 kJ/mol at 100 °C
Ion-product constant, K_w:	1.14×10^{-15} at 0 °C
	1.01×10^{-14} at 25 °C
	5.47×10^{-14} at 50 °C
Specific heat:	Ice (at −3 °C) 2.092 J/g-K
	Water (at 14.5 °C) 4.184 J/g-K
	Steam (at 100 °C) 1.841 J/g-K

Vapor Pressure (torr)

T (°C)	P	T (°C)	P	T (°C)	P	T (°C)	P
0	4.58	21	18.65	35	42.2	92	567.0
5	6.54	22	19.83	40	55.3	94	610.9
10	9.21	23	21.07	45	71.9	96	657.6
12	10.52	24	22.38	50	92.5	98	707.3
14	11.99	25	23.76	55	118.0	100	760.0
16	13.63	26	25.21	60	149.4	102	815.9
17	14.53	27	26.74	65	187.5	104	875.1
18	15.48	28	28.35	70	233.7	106	937.9
19	16.48	29	30.04	80	355.1	108	1004.4
20	17.54	30	31.82	90	525.8	110	1074.6

Substance	$\Delta H_f°$ (kJ/mol)	$\Delta G_f°$ (kJ/mol)	$S°$ (J/mol-K)	Substance	$\Delta H_f°$ (kJ/mol)	$\Delta G_f°$ (kJ/mol)	$S°$ (J/mol-K)
Aluminum				$C_2H_4(g)$	52.30	68.11	219.4
Al(s)	0	0	28.32	$C_2H_6(g)$	−84.68	−32.89	229.5
$AlCl_3(s)$	−705.6	−630.0	109.3	$C_3H_8(g)$	−103.85	−23.47	269.9
$Al_2O_3(s)$	−1669.8	−1576.5	51.00	$C_4H_{10}(g)$	−124.73	−15.71	310.0
				$C_4H_{10}(l)$	−147.6	−15.0	231.0
Barium				$C_6H_6(g)$	82.9	129.7	269.2
Ba(s)	0	0	63.2	$C_6H_6(l)$	49.0	124.5	172.8
$BaCO_3(s)$	−1216.3	−1137.6	112.1	$CH_3OH(g)$	−201.2	−161.9	237.6
BaO(s)	−553.5	−525.1	70.42	$CH_3OH(l)$	−238.6	−166.23	126.8
				$C_2H_5OH(g)$	−235.1	−168.5	282.7
Beryllium				$C_2H_5OH(l)$	−277.7	−174.76	160.7
Be(s)	0	0	9.44	$C_6H_{12}O_6(s)$	−1273.02	−910.4	212.1
BeO(s)	−608.4	−579.1	13.77	CO(g)	−110.5	−137.2	197.9
$Be(OH)_2(s)$	−905.8	−817.9	50.21	$CO_2(g)$	−393.5	−394.4	213.6
				$CH_3COOH(l)$	−487.0	−392.4	159.8
Bromine							
Br(g)	111.8	82.38	174.9	**Cesium**			
$Br^-(aq)$	−120.9	−102.8	80.71	Cs(g)	76.50	49.53	175.6
$Br_2(g)$	30.71	3.14	245.3	Cs(l)	2.09	0.03	92.07
$Br_2(l)$	0	0	152.3	Cs(s)	0	0	85.15
HBr(g)	−36.23	−53.22	198.49	CsCl(s)	−442.8	−414.4	101.2
Calcium				**Chlorine**			
Ca(g)	179.3	145.5	154.8	Cl(g)	121.7	105.7	165.2
Ca(s)	0	0	41.4	$Cl^-(aq)$	−167.2	−131.2	56.5
$CaCO_3(s, \text{calcite})$	−1207.1	−1128.76	92.88	$Cl_2(g)$	0	0	222.96
$CaCl_2(s)$	−795.8	−748.1	104.6	HCl(aq)	−167.2	−131.2	56.5
$CaF_2(s)$	−1219.6	−1167.3	68.87	HCl(g)	−92.30	−95.27	186.69
CaO(s)	−635.5	−604.17	39.75				
$Ca(OH)_2(s)$	−986.2	−898.5	83.4	**Chromium**			
$CaSO_4(s)$	−1434.0	−1321.8	106.7	Cr(g)	397.5	352.6	174.2
				Cr(s)	0	0	23.6
Carbon				$Cr_2O_3(s)$	−1139.7	−1058.1	81.2
C(g)	718.4	672.9	158.0				
C(s, diamond)	1.88	2.84	2.43	**Cobalt**			
C(s, graphite)	0	0	5.69	Co(g)	439	393	179
$CCl_4(g)$	−106.7	−64.0	309.4	Co(s)	0	0	28.4
$CCl_4(l)$	−139.3	−68.6	214.4				
$CF_4(g)$	−679.9	−635.1	262.3	**Copper**			
$CH_4(g)$	−74.8	−50.8	186.3	Cu(g)	338.4	298.6	166.3
$C_2H_2(g)$	226.77	209.2	200.8	Cu(s)	0	0	33.30

Substance	ΔH_f° (kJ/mol)	ΔG_f° (kJ/mol)	S° (J/mol-K)	Substance	ΔH_f° (kJ/mol)	ΔG_f° (kJ/mol)	S° (J/mol-K)
$CuCl_2(s)$	−205.9	−161.7	108.1	$MgO(s)$	−601.8	−569.6	26.8
$CuO(s)$	−156.1	−128.3	42.59	$Mg(OH)_2(s)$	−924.7	−833.7	63.24
$Cu_2O(s)$	−170.7	−147.9	92.36	**Manganese**			
Fluorine				$Mn(g)$	280.7	238.5	173.6
$F(g)$	80.0	61.9	158.7	$Mn(s)$	0	0	32.0
$F^-(aq)$	−332.6	−278.8	−13.8	$MnO(s)$	−385.2	−362.9	59.7
$F_2(g)$	0	0	202.7	$MnO_2(s)$	−519.6	−464.8	53.14
$HF(g)$	−268.61	−270.70	173.51	$MnO_4^-(aq)$	−541.4	−447.2	191.2
Hydrogen				**Mercury**			
$H(g)$	217.94	203.26	114.60	$Hg(g)$	60.83	31.76	174.89
$H^+(aq)$	0	0	0	$Hg(l)$	0	0	77.40
$H^+(g)$	1536.2	1517.0	108.9	$HgCl_2(s)$	−230.1	−184.0	144.5
$H_2(g)$	0	0	130.58	$Hg_2Cl_2(s)$	−264.9	−210.5	192.5
Iodine				**Nickel**			
$I(g)$	106.60	70.16	180.66	$Ni(g)$	429.7	384.5	182.1
$I^-(aq)$	−55.19	−51.57	111.3	$Ni(s)$	0	0	29.9
$I_2(g)$	62.25	19.37	260.57	$NiCl_2(s)$	−305.3	−259.0	97.65
$I_2(s)$	0	0	116.73	$NiO(s)$	−239.7	−211.7	37.99
$HI(g)$	25.94	1.30	206.3	**Nitrogen**			
Iron				$N(g)$	472.7	455.5	153.3
$Fe(g)$	415.5	369.8	180.5	$N_2(g)$	0	0	191.50
$Fe(s)$	0	0	27.15	$NH_3(aq)$	−80.29	−26.50	111.3
$Fe^{2+}(aq)$	−87.86	−84.93	113.4	$NH_3(g)$	−46.19	−16.66	192.5
$Fe^{3+}(aq)$	−47.69	−10.54	293.3	$NH_4^+(aq)$	−132.5	−79.31	113.4
$FeCl_2(s)$	−341.8	−302.3	117.9	$N_2H_4(g)$	95.40	159.4	238.5
$FeCl_3(s)$	−400	−334	142.3	$NH_4CN(s)$	0.0	—	—
$FeO(s)$	−271.9	−255.2	60.75	$NH_4Cl(s)$	−314.4	−203.0	94.6
$Fe_2O_3(s)$	−822.16	−740.98	89.96	$NH_4NO_3(s)$	−365.6	−184.0	151
$Fe_3O_4(s)$	−1117.1	−1014.2	146.4	$NO(g)$	90.37	86.71	210.62
$FeS_2(s)$	−171.5	−160.1	52.92	$NO_2(g)$	33.84	51.84	240.45
Lead				$N_2O(g)$	81.6	103.59	220.0
$Pb(s)$	0	0	68.85	$N_2O_4(g)$	9.66	98.28	304.3
$PbBr_2(s)$	−277.4	−260.7	161	$NOCl(g)$	52.6	66.3	264
$PbCO_3(s)$	−699.1	−625.5	131.0	$HNO_3(aq)$	−206.6	−110.5	146
$Pb(NO_3)_2(aq)$	−421.3	−246.9	303.3	$HNO_3(g)$	−134.3	−73.94	266.4
$Pb(NO_3)_2(s)$	−451.9	—	—	**Oxygen**			
$PbO(s)$	−217.3	−187.9	68.70	$O(g)$	247.5	230.1	161.0
Lithium				$O_2(g)$	0	0	205.0
$Li(g)$	159.3	126.6	138.8	$O_3(g)$	142.3	163.4	237.6
$Li(s)$	0	0	29.09	$OH^-(aq)$	−230.0	−157.3	−10.7
$Li^+(aq)$	−278.5	−273.4	12.2	$H_2O(g)$	−241.82	−228.57	188.83
$Li^+(g)$	685.7	648.5	133.0	$H_2O(l)$	−285.83	−237.13	69.91
$LiCl(s)$	−408.3	−384.0	59.30	$H_2O_2(g)$	−136.10	−105.48	232.9
Magnesium				$H_2O_2(g)$	−187.8	−120.4	109.6
$Mg(g)$	147.1	112.5	148.6	**Phosphorus**			
$Mg(s)$	0	0	32.51	$P(g)$	316.4	280.0	163.2
$MgCl_2(s)$	−641.6	−592.1	89.6	$P_2(g)$	144.3	103.7	218.1

Substance	ΔH_f° (kJ/mol)	ΔG_f° (kJ/mol)	S° (J/mol-K)	Substance	ΔH_f° (kJ/mol)	ΔG_f° (kJ/mol)	S° (J/mol-K)
Phosphorus (cont.)				$AgCl(s)$	−127.0	−109.70	96.11
$P_4(g)$	58.9	24.4	280	$Ag_2O(s)$	−31.05	−11.20	121.3
$P_4(s, \text{red})$	−17.46	−12.03	22.85	$AgNO_3(s)$	−124.4	−33.41	140.9
$P_4(s, \text{white})$	0	0	41.08	**Sodium**			
$PCl_3(g)$	−288.07	−269.6	311.7	$Na(g)$	107.7	77.3	153.7
$PCl_3(l)$	−319.6	−272.4	217	$Na(s)$	0	0	51.45
$PF_5(g)$	−1594.4	−1520.7	300.8	$Na^+(aq)$	−240.1	−261.9	59.0
$PH_3(g)$	5.4	13.4	210.2	$Na^+(g)$	609.3	574.3	148.0
$P_4O_6(s)$	−1640.1	—	—	$NaBr(aq)$	−360.6	−364.7	141.00
$P_4O_{10}(s)$	−2940.1	−2675.2	228.9	$NaBr(s)$	−361.4	−349.3	86.82
$POCl_3(g)$	−542.2	−502.5	325	$Na_2CO_3(s)$	−1130.9	−1047.7	136.0
$POCl_3(l)$	−597.0	−520.9	222	$NaCl(aq)$	−407.1	−393.0	115.5
$H_3PO_4(aq)$	−1288.3	−1142.6	158.2	$NaCl(g)$	−181.4	−201.3	229.8
				$NaCl(s)$	−410.9	−384.0	72.33
Potassium				$NaHCO_3(s)$	−947.7	−851.8	102.1
$K(g)$	89.99	61.17	160.2	$NaNO_3(aq)$	−446.2	−372.4	207
$K(s)$	0	0	64.67	$NaNO_3(s)$	−467.9	−367.0	116.5
$KCl(s)$	−435.9	−408.3	82.7	$NaOH(aq)$	−469.6	−419.2	49.8
$KClO_3(s)$	−391.2	−289.9	143.0	$NaOH(s)$	−425.6	−379.5	64.46
$KClO_3(aq)$	−349.5	−284.9	265.7	$Na_2OH_4(s)$	−1387.1	−1270.2	149.6
$K_2CO_3(s)$	−1150.18	−1064.58	155.44	**Strontium**			
$KNO_3(s)$	−492.70	−393.13	132.9	$SrO(s)$	−592.0	561.9	54.9
$K_2O(s)$	−363.2	−322.1	94.14	$Sr(g)$	164.4	110.0	164.6
$KO_2(s)$	−284.5	−240.6	122.5	**Sulfur**			
$K_2O_2(s)$	−495.8	−429.8	113.0	$S(s, \text{rhombic})$	0	0	31.88
$KOH(s)$	−424.7	−378.9	78.91	$S_8(g)$	102.3	49.7	430.9
$KOH(aq)$	−482.4	−440.5	91.6	$SO_2(g)$	−296.9	−300.4	248.5
Rubidium				$SO_3(g)$	−395.2	−370.4	256.2
$Rb(g)$	85.8	55.8	170.0	$SO_4^{2-}(aq)$	−909.3	−744.5	20.1
$Rb(s)$	0	0	76.78	$SOCl_2(l)$	−245.6	—	—
$RbCl(s)$	−430.5	−412.0	92	$H_2S(g)$	−20.17	−33.01	205.6
$RbClO_3(s)$	−392.4	−292.0	152	$H_2SO_4(aq)$	−909.3	−744.5	20.1
Scandium				$H_2SO_4(l)$	−814.0	−689.9	156.1
$Sc(g)$	377.8	336.1	174.7	**Titanium**			
$Sc(s)$	0	0	34.6	$Ti(g)$	468	422	180.3
Selenium				$Ti(s)$	0	0	30.76
$H_2Se(g)$	29.7	15.9	219.0	$TiCl_4(g)$	−763.2	−726.8	354.9
Silicon				$TiCl_4(l)$	−804.2	−728.1	221.9
$Si(g)$	368.2	323.9	167.8	$TiO_2(s)$	−944.7	−889.4	50.29
$Si(s)$	0	0	18.7	**Vanadium**			
$SiC(s)$	−73.22	−70.85	16.61	$V(g)$	514.2	453.1	182.2
$SiCl_4(l)$	−640.1	−572.8	239.3	$V(s)$	0	0	28.9
$SiO_2(s, \text{quartz})$	−910.9	−856.5	41.84	**Zinc**			
Silver				$Zn(g)$	130.7	95.2	160.9
$Ag(s)$	0	0	42.55	$Zn(s)$	0	0	41.63
$Ag^+(aq)$	105.90	77.11	73.93	$ZnCl_2(s)$	−415.1	−369.4	111.5
				$ZnO(s)$	−348.0	−318.2	43.9

TABLE D-1 ■ Dissociation Constants for Acids at 25 °C				
Name	**Formula**	K_{a1}	K_{a2}	K_{a3}
Acetic	CH_3COOH (or $HC_2H_3O_2$)	1.8×10^{-5}		
Arsenic	H_3AsO_4	5.6×10^{-3}	1.0×10^{-7}	3.0×10^{-12}
Arsenous	H_3AsO_3	5.1×10^{-10}		
Ascorbic	$H_2C_6H_6O_6$	8.0×10^{-5}	1.6×10^{-12}	
Benzoic	C_6H_5COOH (or $HC_7H_5O_2$)	6.3×10^{-5}		
Boric	H_3BO_3	5.8×10^{-10}		
Butanoic	C_3H_7COOH (or $HC_4H_7O_2$)	1.5×10^{-5}		
Carbonic	H_2CO_3	4.3×10^{-7}	5.6×10^{-11}	
Chloroacetic	$CH_2ClCOOH$ (or $HC_2H_2O_2Cl$)	1.4×10^{-3}		
Chlorous	$HClO_2$	1.1×10^{-2}		
Citric	$HOOCC(OH)(CH_2COOH)_2$ (or $H_3C_6H_5O_7$)	7.4×10^{-4}	1.7×10^{-5}	4.0×10^{-7}
Cyanic	$HCNO$	3.5×10^{-4}		
Formic	$HCOOH$ (or $HCHO_2$)	1.8×10^{-4}		
Hydroazoic	HN_3	1.9×10^{-5}		
Hydrocyanic	HCN	4.9×10^{-10}		
Hydrofluoric	HF	6.8×10^{-4}		
Hydrogen chromate ion	$HCrO_4^-$	3.0×10^{-7}		
Hydrogen peroxide	H_2O_2	2.4×10^{-12}		
Hydrogen selenate ion	$HSeO_4^-$	2.2×10^{-2}		
Hydrosulfuric	H_2S	9.5×10^{-8}	1×10^{-19}	
Hypobromous	$HBrO$	2.5×10^{-9}		
Hypochlorous	$HClO$	3.0×10^{-8}		
Hypoiodous	HIO	2.3×10^{-11}		
Iodic	HIO_3	1.7×10^{-1}		
Lactic	$CH_3CH(OH)COOH$ (or $HC_3H_5O_3$)	1.4×10^{-4}		
Malonic	$CH_2(COOH)_2$ (or $H_2C_3H_2O_4$)	1.5×10^{-3}	2.0×10^{-6}	
Nitrous	HNO_2	4.5×10^{-4}		
Oxalic	$(COOH)_2$ (or $H_2C_2O_4$)	5.9×10^{-2}	6.4×10^{-5}	
Paraperiodic	H_5IO_6	2.8×10^{-2}	5.3×10^{-9}	
Phenol	C_6H_5OH (or HC_6H_5O)	1.3×10^{-10}		
Phosphoric	H_3PO_4	7.5×10^{-3}	6.2×10^{-8}	4.2×10^{-13}
Propionic	C_2H_5COOH (or $HC_3H_5O_2$)	1.3×10^{-5}		
Pyrophosphoric	$H_4P_2O_7$	3.0×10^{-2}	4.4×10^{-3}	2.1×10^{-7}
Selenous	H_2SeO_3	2.3×10^{-3}	5.3×10^{-9}	
Sulfuric	H_2SO_4	Strong acid	1.2×10^{-2}	
Sulfurous	H_2SO_3	1.7×10^{-2}	6.4×10^{-8}	
Tartaric	$HOOC(CHOH)_2COOH$ (or $H_2C_4H_4O_6$)	1.0×10^{-3}	4.6×10^{-5}	

TABLE D-2 ▪ Dissociation Constants for Bases at 25 °C

Name	Formula	K_b
Ammonia	NH_3	1.8×10^{-5}
Aniline	$C_6H_5NH_2$	4.3×10^{-10}
Dimethylamine	$(CH_3)_2NH$	5.4×10^{-4}
Ethylamine	$C_2H_5NH_2$	6.4×10^{-4}
Hydrazine	H_2NNH_2	1.3×10^{-6}
Hydroxylamine	$HONH_2$	1.1×10^{-8}
Methylamine	CH_3NH_2	4.4×10^{-4}
Pyridine	C_5H_5N	1.7×10^{-9}
Trimethylamine	$(CH_3)_3N$	6.4×10^{-5}

TABLE D-3 ▪ Solubility-Product Constants for Compounds at 25 °C

Name	Formula	K_{sp}	Name	Formula	K_{sp}
Barium carbonate	$BaCO_3$	5.0×10^{-9}	Lead(II) fluoride	PbF_2	3.6×10^{-8}
Barium chromate	$BaCrO_4$	2.1×10^{-10}	Lead(II) sulfate	$PbSO_4$	6.3×10^{-7}
Barium fluoride	BaF_2	1.7×10^{-6}	Lead(II) sulfide*	PbS	3×10^{-28}
Barium oxalate	BaC_2O_4	1.6×10^{-6}	Magnesium hydroxide	$Mg(OH)_2$	1.8×10^{-11}
Barium sulfate	$BaSO_4$	1.1×10^{-10}	Magnesium carbonate	$MgCO_3$	3.5×10^{-8}
Cadmium carbonate	$CdCO_3$	1.8×10^{-14}	Magnesium oxalate	MgC_2O_4	8.6×10^{-5}
Cadmium hydroxide	$Cd(OH)_2$	2.5×10^{-14}	Manganese(II) carbonate	$MnCO_3$	5.0×10^{-10}
Cadmium sulfide*	CdS	8×10^{-28}	Manganese(II) hydroxide	$Mn(OH)_2$	1.6×10^{-13}
Calcium carbonate (calcite)	$CaCO_3$	4.5×10^{-9}	Manganese(II) sulfide*	MnS	2×10^{-53}
Calcium chromate	$CaCrO_4$	7.1×10^{-4}	Mercury(I) chloride	Hg_2Cl_2	1.2×10^{-18}
Calcium fluoride	CaF_2	3.9×10^{-11}	Mercury(I) iodide	Hg_2I_2	1.1×10^{-28}
Calcium hydroxide	$Ca(OH)_2$	6.5×10^{-6}	Mercury(II) sulfide*	HgS	2×10^{-53}
Calcium phosphate	$Ca_3(PO_4)_2$	2.0×10^{-29}	Nickel(II) carbonate	$NiCO_3$	1.3×10^{-7}
Calcium sulfate	$CaSO_4$	2.4×10^{-5}	Nickel(II) hydroxide	$Ni(OH)_2$	6.0×10^{-16}
Chromium(III) hydroxide	$Cr(OH)_3$	1.6×10^{-30}	Nickel(II) sulfide*	NiS	3×10^{-20}
Cobalt(II) carbonate	$CoCO_3$	1.0×10^{-10}	Silver bromate	$AgBrO_3$	5.5×10^{-5}
Cobalt(II) hydroxide	$Co(OH)_2$	1.3×10^{-15}	Silver bromide	$AgBr$	5.0×10^{-13}
Cobalt(II) sulfide*	CoS	5×10^{-22}	Silver carbonate	Ag_2CO_3	8.1×10^{-12}
Copper(I) bromide	$CuBr$	5.3×10^{-9}	Silver chloride	$AgCl$	1.8×10^{-10}
Copper(II) carbonate	$CuCO_3$	2.3×10^{-10}	Silver chromate	Ag_2CrO_4	1.2×10^{-12}
Copper(II) hydroxide	$Cu(OH)_2$	4.8×10^{-20}	Silver iodide	AgI	8.3×10^{-17}
Copper(II) sulfide*	CuS	6×10^{-37}	Silver sulfate	Ag_2SO_4	1.5×10^{-5}
Iron(II) carbonate	$FeCO_3$	2.1×10^{-11}	Silver sulfide*	Ag_2S	6×10^{-51}
Iron(II) hydroxide	$Fe(OH)_2$	7.9×10^{-16}	Strontium carbonate	$SrCO_3$	9.3×10^{-10}
Lanthanum fluoride	LaF_3	2×10^{-19}	Tin(II) sulfide*	SnS	1×10^{-26}
Lanthanum iodate	$La(IO_3)_3$	6.1×10^{-12}	Zinc carbonate	$ZnCO_3$	1.0×10^{-10}
Lead(II) carbonate	$PbCO_3$	7.4×10^{-14}	Zinc hydroxide	$Zn(OH)_2$	3.0×10^{-16}
Lead(II) chloride	$PbCl_2$	1.7×10^{-5}	Zinc oxalate	ZnC_2O_4	2.7×10^{-8}
Lead(II) chromate	$PbCrO_4$	2.8×10^{-13}	Zinc sulfide*	ZnS	2×10^{-25}

*For a solubility equilibrium of the type $MS(s) + H_2O(l) \rightleftharpoons M^{2+}(aq) + HS^-(aq) + OH^-(aq)$

E

Standard Reduction Potentials at 25 °C

Half-Reaction	$E°$(V)	Half-Reaction	$E°$(V)
$Ag^+(aq) + e^- \longrightarrow Ag(s)$	+0.799	$2\,H_2O(l) + 2\,e^- \longrightarrow H_2(g) + 2\,OH^-(aq)$	−0.83
$AgBr(s) + e^- \longrightarrow Ag(s) + Br^-(aq)$	+0.095	$HO_2^-(aq) + H_2O(l) + 2\,e^- \longrightarrow 3\,OH^-(aq)$	+0.88
$AgCl(s) + e^- \longrightarrow Ag(s) + Cl^-(aq)$	+0.222	$H_2O_2(aq) + 2\,H^+(aq) + 2\,e^- \longrightarrow 2\,H_2O(l)$	+1.776
$Ag(CN)_2^-(aq) + e^- \longrightarrow Ag(s) + 2\,CN^-(aq)$	−0.31	$Hg_2^{2+}(aq) + 2\,e^- \longrightarrow 2\,Hg(l)$	+0.789
$Ag_2CrO_4(s) + 2\,e^- \longrightarrow 2\,Ag(s) + CrO_4^{2-}(aq)$	+0.446	$2\,Hg^{2+}(aq) + 2\,e^- \longrightarrow Hg_2^{2+}(aq)$	+0.920
$AgI(s) + e^- \longrightarrow Ag(s) + I^-(aq)$	−0.151	$Hg^{2+}(aq) + 2\,e^- \longrightarrow Hg(l)$	+0.854
$Ag(S_2O_3)_2^{3-}(aq) + e^- \longrightarrow Ag(s) + 2\,S_2O_3^{2-}(aq)$	+0.01	$I_2(s) + 2\,e^- \longrightarrow 2\,I^-(aq)$	+0.536
$Al^{3+}(aq) + 3\,e^- \longrightarrow Al(s)$	−1.66	$2\,IO_3^-(aq) + 12\,H^+(aq) + 10\,e^- \longrightarrow$ $I_2(s) + 6\,H_2O(l)$	+1.195
$H_3AsO_4(aq) + 2\,H^+(aq) + 2\,e^- \longrightarrow$ $H_3AsO_3(aq) + H_2O(l)$	+0.559	$K^+(aq) + e^- \longrightarrow K(s)$	−2.925
$Ba^{2+}(aq) + 2\,e^- \longrightarrow Ba(s)$	−2.90	$Li^+(aq) + e^- \longrightarrow Li(s)$	−3.05
$BiO^+(aq) + 2\,H^+(aq) + 3\,e^- \longrightarrow Bi(s) + H_2O(l)$	+0.32	$Mg^{2+}(aq) + 2\,e^- \longrightarrow Mg(s)$	−2.37
$Br_2(l) + 2\,e^- \longrightarrow 2\,Br^-(aq)$	+1.065	$Mn^{2+}(aq) + 2\,e^- \longrightarrow Mn(s)$	−1.18
$2\,BrO_3^-(aq) + 12\,H^+(aq) + 10\,e^- \longrightarrow$ $Br_2(l) + 6\,H_2O(l)$	+1.52	$MnO_2(s) + 4\,H^+(aq) + 2\,e^- \longrightarrow$ $Mn^{2+}(aq) + 2\,H_2O(l)$	+1.23
$2\,CO_2(g) + 2\,H^+(aq) + 2\,e^- \longrightarrow H_2C_2O_4(aq)$	−0.49	$MnO_4^-(aq) + 8\,H^+(aq) + 5\,e^- \longrightarrow$ $Mn^{2+}(aq) + 4\,H_2O(l)$	+1.51
$Ca^{2+}(aq) + 2\,e^- \longrightarrow Ca(s)$	−2.87	$MnO_4^-(aq) + 2\,H_2O(l) + 3\,e^- \longrightarrow$ $MnO_2(s) + 4\,OH^-(aq)$	+0.59
$Cd^{2+}(aq) + 2\,e^- \longrightarrow Cd(s)$	−0.403	$HNO_2(aq) + H^+(aq) + e^- \longrightarrow NO(g) + H_2O(l)$	+1.00
$Ce^{4+}(aq) + e^- \longrightarrow Ce^{3+}(aq)$	+1.61	$N_2(g) + 4\,H_2O(l) + 4\,e^- \longrightarrow 4\,OH^-(aq) + N_2H_4(aq)$	−1.16
$Cl_2(g) + 2\,e^- \longrightarrow 2\,Cl^-(aq)$	+1.359	$N_2(g) + 5\,H^+(aq) + 4\,e^- \longrightarrow N_2H_5^+(aq)$	−0.23
$2\,HClO(aq) + 2\,H^+(aq) + 2\,e^- \longrightarrow$ $Cl_2(g) + 2\,H_2O(l)$	+1.63	$NO_3^-(aq) + 4\,H^+(aq) + 3\,e^- \longrightarrow NO(g) + 2\,H_2O(l)$	+0.96
$ClO^-(aq) + H_2O(l) + 2\,e^- \longrightarrow Cl^-(aq) + 2\,OH^-(aq)$	+0.89	$Na^+(aq) + e^- \longrightarrow Na(s)$	−2.71
$2\,ClO_3^-(aq) + 12\,H^+(aq) + 10\,e^- \longrightarrow$ $Cl_2(g) + 6\,H_2O(l)$	+1.47	$Ni^{2+}(aq) + 2\,e^- \longrightarrow Ni(s)$	−0.28
$Co^{2+}(aq) + 2\,e^- \longrightarrow Co(s)$	−0.277	$O_2(g) + 4\,H^+(aq) + 4\,e^- \longrightarrow 2\,H_2O(l)$	+1.23
$Co^{3+}(aq) + e^- \longrightarrow Co^{2+}(aq)$	+1.842	$O_2(g) + 2\,H_2O(l) + 4\,e^- \longrightarrow 4\,OH^-(aq)$	+0.40
$Cr^{3+}(aq) + 3\,e^- \longrightarrow Cr(s)$	−0.74	$O_2(g) + 2\,H^+(aq) + 2\,e^- \longrightarrow H_2O_2(aq)$	+0.68
$Cr^{3+}(aq) + e^- \longrightarrow Cr^{2+}(aq)$	−0.41	$O_3(g) + 2\,H^+(aq) + 2\,e^- \longrightarrow O_2(g) + H_2O(l)$	+2.07
$CrO_7^{2-}(aq) + 14\,H^+(aq) + 6\,e^- \longrightarrow$ $2\,Cr^{3+}(aq) + 7\,H_2O(l)$	+1.33	$Pb^{2+}(aq) + 2\,e^- \longrightarrow Pb(s)$	−0.126
$CrO_4^{2-}(aq) + 4\,H_2O(l) + 3\,e^- \longrightarrow$ $Cr(OH)_3(s) + 5\,OH^-(aq)$	−0.13	$PbO_2(s) + HSO_4^-(aq) + 3\,H^+(aq) + 2\,e^- \longrightarrow$ $PbSO_4(s) + 2\,H_2O(l)$	+1.685
$Cu^{2+}(aq) + 2\,e^- \longrightarrow Cu(s)$	+0.337	$PbSO_4(s) + H^+(aq) + 2\,e^- \longrightarrow Pb(s) + HSO_4^-(aq)$	−0.356
$Cu^{2+}(aq) + e^- \longrightarrow Cu^+(aq)$	+0.153	$PtCl_4^{2-}(aq) + 2\,e^- \longrightarrow Pt(s) + 4\,Cl^-(aq)$	+0.73
$Cu^+(aq) + e^- \longrightarrow Cu(s)$	+0.521	$S(s) + 2\,H^+(aq) + 2\,e^- \longrightarrow H_2S(g)$	+0.141
$CuI(s) + e^- \longrightarrow Cu(s) + I^-(aq)$	−0.185	$H_2SO_3(aq) + 4\,H^+(aq) + 4\,e^- \longrightarrow S(s) + 3\,H_2O(l)$	+0.45
$F_2(g) + 2\,e^- \longrightarrow 2\,F^-(aq)$	+2.87	$HSO_4^-(aq) + 3\,H^+(aq) + 2\,e^- \longrightarrow$ $H_2SO_3(aq) + H_2O(l)$	+0.17
$Fe^{2+}(aq) + 2\,e^- \longrightarrow Fe(s)$	−0.440	$Sn^{2+}(aq) + 2\,e^- \longrightarrow Sn(s)$	−0.136
$Fe^{3+}(aq) + e^- \longrightarrow Fe^{2+}(aq)$	+0.771	$Sn^{4+}(aq) + 2\,e^- \longrightarrow Sn^{2+}(aq)$	+0.154
$Fe(CN)_6^{3-}(aq) + e^- \longrightarrow Fe(CN)_6^{4-}(aq)$	+0.36	$VO_2^+(aq) + 2\,H^+(aq) + e^- \longrightarrow VO^{2+}(aq) + H_2O(l)$	+1.00
$2\,H^+(aq) + 2\,e^- \longrightarrow H_2(g)$	0.000	$Zn^{2+}(aq) + 2\,e^- \longrightarrow Zn(s)$	−0.763

Answers to Selected Exercises

Chapter 1

1.1 (a) Pure element: i, v (b) mixture of elements: vi (c) pure compound: iv (d) mixture of an element and a compound: ii, iii **1.4** The aluminum sphere (density = 2.70 g/cm^3) is lightest, then nickel (density = 8.90 g/cm^3), then silver (density = 10.49 g/cm^3). **1.6** (a) 7.5 cm, 2 significant figures (sig figs) (b) 140 °F, 2 sig figs **1.8** (a) 2 significant figures (sig figs) (b) 2 sig figs **1.11** (a) Heterogeneous mixture (b) homogeneous mixture (heterogeneous if there are undissolved particles) (c) pure substance (d) homogeneous mixture. **1.13** (a) S (b) Mg (c) K (d) Cl (e) Cu (f) fluorine (g) nickel (h) sodium (i) aluminum (j) silicon **1.15** C is a compound; it contains carbon and oxygen. A is a compound; it contains at least carbon and oxygen. B is not defined by the data given; it is probably a compound because few elements exist as white solids. **1.17** Physical properties: silvery white; lustrous; melting point = 649 °C; boiling point = 1105 °C; density at 20 °C = 1.738 g/cm^3; pounded into sheets; drawn into wires; good conductor. Chemical properties: burns in air; reacts with Cl$_2$. **1.19** (a) Chemical (b) physical (c) physical (d) chemical (e) chemical **1.21** (a) Add water to dissolve the sugar; filter this mixture, collecting the sand on filter paper and the sugar water in a flask. Evaporate water from the flask to recover solid sugar. (b) Heat the mixture until sulfur melts, then decant the liquid sulfur. **1.23** (a) 1 × 10^{-1} (b) 1 × 10^{-2} (c) 1 × 10^{-15} (d) 1 × 10^{-6} (e) 1 × 10^6 (f) 1 × 10^3 (g) 1 × 10^{-9} (h) 1 × 10^{-3} (i) 1 × 10^{-12} **1.25** (a) 17 °C (b) 422.1 °F (c) 506 K (d) 108 °F (e) 1644 K **1.27** (a) 1.59 g/ mL. Carbon tetrachloride, 1.59 g/mL, is more dense than water, 1.00 g/mL; carbon tetrachloride will sink rather than float on water. (b) 1.609 kg (c) 50.35 cm^3 **1.29** (a) Calculated density = 0.86 g/mL. The substance is probably toluene, density = 0.866 g/mL. (b) 40.4 mL ethylene glycol (c) 1.11 × 10^3 g nickel **1.31** 4.6 × 10^{-8} m; 46 nm **1.33** Exact: (c), (d), and (f) **1.35** (a) 3 (b) 2 (c) 5 (d) 3 (e) 5 **1.37** (a) 1.025 × 10^2 (b) 6.570 × 10^5 (c) 8.543 × 10^{-3} (d) 2.579 × 10^{-4} (e) −3.572 × 10^{-2} **1.39** (a) 21.11 (b) 237.4 (c) 652 (d) 7.66 × 10^{-2} **1.41** (a) 1 × 10^6 nm/mm (b) 1 × 10^{-6} kg/mg (c) 3.28 × 10^3 km/ft (d) 16.4 cm^3/in^3 **1.43** (a) 76 mL (b) 50. nm (c) 6.88 × 10^{-4} s (d) 2.3 × 10^2 g (e) 1.55 g/L (f) 6.151 × 10^{-3} L/s **1.45** (a) 4.32 × 10^5 s (b) 88.5 m (c) \$0.499/L (d) 46.6 km/hr (e) 1.420 L/s (f) 707.9 cm^3 **1.47** (a) 1.2 × 10^2 L (b) 4 × 10^2 mg (c) 9.64 km/L (d) 26 mL **1.49** 52 kg air **1.51** Use the cm as a unit for comparison (1 in. ≈ 2.5 cm). 57 cm = 57 cm; 14 in. ≈ 35 cm; 1.1 m = 110 cm. The order of length from shortest to longest is 14 in. < 57 cm < 1.1 m. **1.53** (a) \$0.91 (b) Since coins come in integer numbers, 3 coins are required. **1.55** Composition is the contents of a substance; structure is the arrangement of these contents. **1.58** 8.47 g O; the *law of constant composition* **1.61** (a) Volume (b) area (c) volume (d) density (e) time (f) length (g) temperature **1.64** (a) 1.13 × 10^5 quarters (b) 6.41 × 10^5 g (c) \$2.83 × 10^4 (d) 3.1 × 10^8 stacks (approximately 310 million stacks!) **1.67** The most dense liquid, Hg, will sink; the least dense, cyclohexane, will float; H$_2$O will be in the middle. **1.70** Density of solid = 1.63 g/mL **1.73** (a) Density of peat = 0.13 g/cm^3, density of topsoil = 2.5 g/cm^3. It is not correct to say that peat is "lighter" than topsoil. Volumes must be specified in order to compare masses. (b) Buy 7 bags of peat (6.9 are needed). **1.76** The inner diameter of the tube is 1.13 cm. **1.79** The separation is successful if two distinct spots are seen on the paper. To quantify the characteristics of the separation, calculate a reference value for each spot: distance travelled by spot/distance travelled by solvent. If the values for the two spots are fairly different, the separation is successful. **1.81** (a) Density = 1.04 g/mL (b) 5.2 × 10^3 g propylene glycol provide the same antifreeze protection as 1.00 gal of ethylene glycol. (c) 1.3 gal propylene glycol provide the same antifreeze protection as 1.00 gal of ethylene glycol.

Chapter 2

2.1 (a) The path of the charged particle bends because it is repelled by the negatively charged plate and attracted to the positively charged plate. (b) (−) (c) increase (d) decrease **2.4** The particle is an ion. $^{32}_{16}$S^{2-} **2.6** Formula: IF$_5$; name: iodine pentafluoride; the compound is molecular. **2.9** Postulate 4 of the atomic theory states that the relative number and kinds of atoms in a compound are constant, regardless of the source. Therefore, 1.0 g of pure water should always contain the same relative amounts of hydrogen and oxygen, no matter where or how the sample is obtained. **2.11** (a) 0.5711 g O/1 g N; 1.142 g O/1 g N; 2.284 g O/1 g N; 2.855 g O/1 g N (b) The numbers in part (a) obey the *law of multiple proportions*. Multiple proportions arise because atoms are the indivisible entities combining, as stated in Dalton's atomic theory. **2.13** (1) Electric and magnetic fields deflected the rays in the same way they would deflect negatively charged particles. (2) A metal plate exposed to cathode rays acquired a negative charge. **2.15** (a) If the positive plate were lower than the negative plate, the oil drops "coated" with negatively charged electrons would be attracted to the positively charged plate and would descend much more quickly. (b) The more times a measurement is repeated, the better the chance of detecting and compensating for experimental errors. Millikan wanted to demonstrate the validity of his result via its reproducibility. **2.17** (a) 0.19 nm; 1.9 × 10^2 or 190 pm (b) 2.6 × 10^6 Kr atoms (c) 2.9 × 10^{-23} cm^3 **2.19** (a) Proton, neutron, electron (b) proton = +1, neutron = 0, electron = −1 (c) The neutron is most massive. (The neutron and proton have very similar masses.) (d) The electron is least massive. **2.21** (a) Atomic number is the number of protons in the nucleus of an atom. Mass number is the total number of nuclear particles, protons plus neutrons, in an atom. (b) mass number **2.23** (a) ^{40}Ar: 18 p, 22 n, 18 e (b) ^{65}Zn: 30 p, 35 n, 30 e (c) ^{70}Ga: 31 p, 39 n, 31 e (d) ^{80}Br: 35 p, 45 n, 35 e (e) ^{184}W: 74 p, 110 n, 74 e (f) ^{243}Am: 95 p, 148 n, 95e

2.25

Symbol	^{52}Cr	^{55}Mn	^{112}Cd	^{222}Rn	^{207}Pb
Protons	24	25	48	86	82
Neutrons	28	30	64	136	125
Electrons	24	25	48	86	82
Mass no.	52	55	112	222	207

2.27 (a) $^{196}_{78}$Pt (b) $^{84}_{36}$Kr (c) $^{75}_{33}$As (d) $^{24}_{12}$Mg **2.29** (a) $^{12}_{6}$C (b) Atomic weights are average atomic masses, the sum of the mass of each naturally occurring isotope of an element times its fractional abundance. Each B atom will have the mass of one of the naturally occurring isotopes, while the "atomic weight" is an average value. **2.31** 63.55 amu

2.33 (a) In Thomson's cathode-ray experiments and in mass spectrometry, a stream of charged particles is passed through the poles of a magnet. The charged particles are deflected by the magnetic field according to their mass and charge. (b) The *x*-axis label is atomic weight, and the *y*-axis label is signal intensity. (c) Uncharged particles are not deflected in a magnetic field. The effect of the magnetic field on charged moving particles is the basis of their separation by mass.   **2.35** (a) average atomic mass = 24.31 amu

(b)

2.37 (a) Cr (metal) (b) He (nonmetal) (c) P (nonmetal) (d) Zn (metal) (e) Mg (metal) (f) Br (nonmetal) (g) As (metalloid) **2.39** (a) K, alkali metals (metal) (b) I, halogens (nonmetal) (c) Mg, alkaline earth metals (metal) (d) Ar, noble gases (nonmetal) (e) S, chalcogens (nonmetal)   **2.41** An empirical formula shows the simplest mole ratio of elements in a compound. A molecular formula shows the exact number and kinds of atoms in a molecule. A structural formula shows which atoms are attached to which.   **2.43** (a) $AlBr_3$ (b) C_4H_5 (c) C_2H_4O (d) P_2O_5 (e) C_3H_2Cl (f) BNH_2 **2.45** (a) 6 (b) 6 (c) 12

2.47

(a) C_2H_6O,   H—C—O—C—H

(b) C_2H_6O,   H—C—C—O—H

(c) CH_4O,   H—C—O—H

(d) PF_3,   F—P—F

2.49

Symbol	$^{59}Co^{3+}$	$^{80}Se^{2-}$	$^{192}Os^{2+}$	$^{200}Hg^{2+}$
Protons	27	34	76	80
Neutrons	32	46	116	120
Electrons	24	36	74	78
Net Charge	3+	2−	2+	2+

2.51 (a) Mg^{2+} (b) Al^{3+} (c) K^+ (d) S^{2-} (e) F^- **2.53** (a) GaF_3, gallium(III) fluoride (b) LiH, lithium hydride (c) AlI_3, aluminum iodide (d) K_2S, potassium sulfide   **2.55** (a) $CaBr_2$ (b) K_2CO_3 (c) $Al(CH_3COO)_3$ (d) $(NH_4)_2SO_4$ (e) $Mg_3(PO_4)_2$

2.57

Ion	K^+	NH_4^+	Mg^{2+}	Fe^{3+}
Cl^-	KCl	NH_4Cl	$MgCl_2$	$FeCl_3$
OH^-	KOH	NH_4OH	$Mg(OH)_2$	$Fe(OH)_3$
CO_3^{2-}	K_2CO_3	$(NH_4)_2CO_3$	$MgCO_3$	$Fe_2(CO_3)_3$
PO_4^{3-}	K_3PO_4	$(NH_4)_3PO_4$	Mg_3PO_4	$FePO_4$

2.59 Molecular: (a) B_2H_6 (b) CH_3OH (f) NOCl (g) NF_3. Ionic: (c) $LiNO_3$ (d) Sc_2O_3 (e) CsBr (h) Ag_2SO_4 **2.61** (a) ClO_2^- (b) Cl^- (c) ClO_3^- (d) ClO_4^- (e) ClO^- **2.63** (a) calcium, 2+; oxide, 2− (b) sodium, 1+; sulfate, 2− (c) potassium, 1+; perchlorate, 1− (d) iron, 2+, nitrate, 1− (e) chromium, 3+; hydroxide, 1− **2.65** (a) Magnesium oxide (b) aluminum chloride (c) lithium phosphate (d) barium perchlorate (e) copper(II) nitrate (cupric nitrate) (f) iron(II) hydroxide (ferrous hydroxide) (g) calcium acetate (h) chromium(III) carbonate (chromic carbonate) (i) potassium chromate (j) ammonium sulfate **2.67** (a) $Al(OH)_3$ (b) K_2SO_4 (c) Cu_2O (d) $Zn(NO_3)_2$ (e) $HgBr_2$ (f) $Fe_2(CO_3)_3$ (g) NaBrO   **2.69** (a) Bromic acid (b) hydrobromic acid (c) phosphoric acid (d) HClO (e) HIO_3 (f) H_2SO_3 **2.71** (a) Sulfur hexafluoride (b) iodine pentafluoride (c) xenon trioxide (d) N_2O_4 (e) HCN (f) P_4S_6 **2.73** (a) $ZnCO_3$, ZnO, CO_2 (b) HF, SiO_2, SiF_4, H_2O (c) SO_2, H_2O, H_2SO_3 (d) PH_3 (e) $HClO_4$, Cd, $Cd(ClO_4)_2$ (f) VBr_3 **2.75** (a) A hydrocarbon is a compound composed of the elements hydrogen and carbon only.

(b)

Molecular: C_4H_{10}
Empirical: C_2H_5

2.77 (a) Functional groups are groups of specific atoms that are constant from one molecule to the next. (b) —OH

(c)

2.79 (a) Dalton postulated the atomic theory, which is based on the indivisible atom as the smallest unit of an element that can combine with other elements. (b) Thomson measured the mass-to-charge ratio of the electron and proposed the "plum pudding" model of the atom. (c) Millikan measured the charge of an electron. (d) Rutherford postulated the nuclear atom, where most of the mass of the atom is concentrated in a small, dense nucleus and electrons move through space around the nucleus.   **2.82** (a) 2 protons, 1 neutron, 2 electrons (b) Tritium, ^3H, is more massive. (c) A precision of 1×10^{-27} g would be required to differentiate between ^3H and ^3He. **2.85** Arrangement A, 4.1×10^{14} atoms/cm² (b) Arrangement B, 4.7×10^{14} atoms/cm² (c) The ratio of atoms going from arrangement B to arrangement A is 1.2 to 1. In three dimensions, arrangement B leads to a greater density for Rb metal.   **2.88** (a) $^{16}_8O$, $^{17}_8O$, $^{18}_8O$ (b) All isotopes are atoms of the same element, oxygen, with the same atomic number, 8 protons in the nucleus and 8 electrons. We expect their electron arrangements to be the same and their chemical properties to be very similar. Each has a different number of neutrons, a different mass number, and a different atomic mass.   **2.91** (a) $^{69}_{31}Ga$, 31 protons, 38 neutrons; $^{71}_{31}Ga$, 31 protons, 40 neutrons (b) $^{69}_{31}Ga$, 60.3%, $^{71}_{31}Ga$, 39.7%.   **2.94** (a) 5 significant figures (b) An electron is 0.05444% of the mass of an ^1H atom.   **2.98** (a) Nickel(II) oxide, 2+ (b) manganese(IV) oxide, 4+ (c) chromium(III) oxide, 3+ (d) molybdenum(VI) oxide, 6+   **2.102** (a) Sodium chloride (b) sodium bicarbonate (or sodium hydrogen carbonate) (c) sodium hypochlorite (d) sodium hydroxide (e) ammonium carbonate (f) calcium sulfate   **2.104** (a) CaS, $Ca(HS)_2$ (b) HBr, $HBrO_3$ (c) AlN, $Al(NO_2)_3$ (d) FeO, Fe_2O_3 (e) NH_3, NH_4^+ (f) K_2SO_3, $KHSO_3$ (g) Hg_2Cl_2, $HgCl_2$ (h) $HClO_3$, $HClO_4$

Chapter 3

3.1 Equation (a) best fits the diagram.   **3.3** (a) NO_2 (b) No, because we have no way of knowing whether the empirical and molecular formulas are the same. NO_2 represents the simplest ratio of atoms in a molecule but not the only possible molecular formula.   **3.5** (a) $C_2H_5NO_2$ (b) 75.0 g/mol (c) 225 g (d) Mass %N in glycine is 18.7%.

3.7

$N_2 + 3 H_2 \longrightarrow 2 NH_3$. Eight N atoms (4 N_2 molecules) require 24 H atoms (12 H_2 molecules) for complete reaction. Only 9 H_2 molecules are available, so H_2 is the limiting reactant. Nine H_2 molecules (18 H atoms) determine that 6 NH_3 molecules are produced. One N_2 molecule is in excess.
3.9 (a) Conservation of mass (b) Subscripts in chemical formulas should not be changed when balancing equations, because changing the subscript changes the identity of the compound (*law of constant composition*). (c) $H_2O(l)$, $H_2O(g)$, NaCl(*aq*), NaCl(*s*)
3.11 (a) $2 CO(g) + O_2(g) \longrightarrow 2 CO_2(g)$
(b) $N_2O_5(g) + H_2O(l) \longrightarrow 2 HNO_3(aq)$
(c) $CH_4(g) + 4 Cl_2(g) \longrightarrow CCl_4(l) + 4 HCl(g)$
(d) $Al_4C_3(s) + 12 H_2O(l) \longrightarrow 4 Al(OH)_3(s) + 3 CH_4(g)$
(e) $2 C_5H_{10}O_2(l) + 13 O_2(g) \longrightarrow 10 CO_2(g) + 10 H_2O(l)$
(f) $2 Fe(OH)_3(s) + 3 H_2SO_4(aq) \longrightarrow Fe_2(SO_4)_3(aq) + 6 H_2O(l)$
(g) $Mg_3N_2(s) + 4 H_2SO_4(aq) \longrightarrow$
$$3 MgSO_4(aq) + (NH_4)_2SO_4(aq)$$
3.13 (a) $CaC_2(s) + 2 H_2O(l) \longrightarrow Ca(OH)_2(aq) + C_2H_2(g)$
(b) $2 KClO_3(s) \xrightarrow{\Delta} 2 KCl(s) + 3 O_2(g)$
(c) $Zn(s) + H_2SO_4(aq) \longrightarrow ZnSO_4(aq) + H_2(g)$
(d) $PCl_3(l) + 3 H_2O(l) \longrightarrow H_3PO_3(aq) + 3 HCl(aq)$
(e) $3 H_2S(g) + 2 Fe(OH)_3(s) \longrightarrow Fe_2S_3(s) + 6 H_2O(g)$
3.15 (a) Determine the formula by balancing the positive and negative charges in the ionic product. All ionic compounds are solids. $2 Na(s) + Br_2(l) \longrightarrow 2 NaBr(s)$ (b) The second reactant is $O_2(g)$. The products are $CO_2(g)$ and $H_2O(l)$.
$2 C_6H_6(l) + 15 O_2(g) \longrightarrow 12 CO_2(g) + 6 H_2O(l)$
3.17 (a) $Mg(s) + Cl_2(g) \longrightarrow MgCl_2(s)$
(b) $BaCO_3(s) \xrightarrow{\Delta} BaO(s) + CO_2(g)$
(c) $C_8H_8(l) + 10 O_2(g) \longrightarrow 8 CO_2(g) + 4 H_2O(l)$
(d) $C_2H_6O(l) + 3 O_2(g) \longrightarrow 2 CO_2(g) + 3 H_2O(l)$
3.19 (a) $2 Al(s) + 3 Cl_2(g) \longrightarrow 2 AlCl_3(s)$ combination
(b) $C_2H_4(g) + 3 O_2(g) \longrightarrow 2 CO_2(g) + 2 H_2O(l)$ combustion
(c) $6 Li(s) + N_2(g) \longrightarrow 2 Li_3N(s)$ combination
(d) $PbCO_3(s) \longrightarrow PbO(s) + CO_2(g)$ decomposition
(e) $C_7H_8O_2(l) + 8 O_2(g) \longrightarrow 7 CO_2(g) + 4 H_2O(l)$ combustion
3.21 (a) 63.0 amu (b) 158.0 amu (c) 310.3 amu (d) 60.1 amu (e) 235.7 amu (f) 392.3 amu (g) 137.5 amu   **3.23** (a) 16.8% (b) 16.1% (c) 21.1% (d) 28.8% (e) 27.2% (f) 26.5%   **3.25** (a) 79.2% (b) 63.2% (c) 64.6%   **3.27** (a) 6.022×10^{23} (b) The formula

weight of a substance in amu has the same numerical value as the molar mass expressed in grams.   **3.29** 23 g Na contains 1mol of atoms; 0.5 mol H_2O contains 1.5 mol atoms; 6.0×10^{23} molecules contains 2 mol of atoms.
3.31 4.37×10^{25} kg (assuming 160 lb has 3 significant figures). One mole of people weighs 7.31 times as much as Earth.   **3.33** (a) 35.9 g $C_{12}H_{22}O_{11}$ (b) 0.7577 mol $Zn(NO_3)_2$ (c) 6.0×10^{17} CH_3CH_2OH molecules (d) 2.47×10^{23} C atoms
3.35 (a) 0.373 g $(NH_4)_3PO_4$ (b) 5.737×10^{-3} mol Cl^- (c) 0.248 g $C_8H_{10}N_4O_2$ (d) 387 g cholesterol/mol
3.37 (a) molar mass = 162.3 g (b) 3.08×10^{-5} mol allicin (c) 1.86×10^{19} allicin molecules (d) 3.71×10^{19} S atoms
3.39 (a) 2.500×10^{21} H atoms (b) 2.083×10^{20} $C_6H_{12}O_6$ molecules (c) 3.460×10^{-4} mol $C_6H_{12}O_6$ (d) 0.06227 g $C_6H_{12}O_6$
3.41 3.2×10^{-8} mol C_2H_3Cl/L; 1.9×10^{16} molecules/L
3.43 (a) C_2H_6O (b) Fe_2O_3 (c) CH_2O **3.45** (a) $CSCl_2$ (b) C_3OF_6 (c) Na_3AlF_6 **3.47** (a) C_6H_{12} (b) NH_2Cl **3.49** (a) Empirical formula, CH; molecular formula, C_8H_8 (b) empirical formula, $C_4H_5N_2O$; molecular formula, $C_8H_{10}N_4O_2$ (c) empirical formula and molecular formula, $NaC_5H_8O_4N$ **3.51** (a) C_7H_8 (b) The empirical and molecular formulas are $C_{10}H_{20}O$.   **3.53** $x = 10$; $Na_2CO_3 \cdot 10 H_2O$ **3.55** If the equation is not balanced, the mole ratios derived from the coefficients will be incorrect and lead to erroneous calculated amounts of products.   **3.57** (a) 2.40 mol HF (b) 5.25 g NaF (c) 0.610 g Na_2SiO_3 **3.59** (a) $Al(OH)_3(s) + 3 HCl(aq) \longrightarrow$ $AlCl_3(aq) + 3 H_2O(l)$ (b) 0.701 g HCl (c) 0.855 g $AlCl_3$; 0.347 g H_2O (d) Mass of reactants = 0.500 g + 0.701 g = 1.201 g; mass of products = 0.855 g + 0.347 g = 1.202 g. Mass is conserved, within the precision of the data.   **3.61** (a) $Al_2S_3(s) + 6 H_2O(l) \longrightarrow 2 Al(OH)_3(s) + 3 H_2S(g)$ (b) 14.7 g $Al(OH)_3$ **3.63** (a) 2.25 mol N_2 (b) 15.5 g NaN_3 (c) 548 g NaN_3 **3.65** (a) 5.50×10^{-3} mol Al (b) 1.47 g $AlBr_3$ **3.67** (a) The *limiting reactant* determines the maximum number of product moles resulting from a chemical reaction; any other reactant is an *excess reactant*. (b) The limiting reactant regulates the amount of products because it is completely used up during the reaction; no more product can be made when one of the reactants is unavailable. (c) Combining ratios are molecule and mole ratios. Since different molecules have different masses, comparing initial masses of reactants will not provide a comparison of numbers of molecules or moles.
3.69 (a) 2255 bicycles (b) 50 frames left over, 305 wheels left over (c) the handlebars   **3.71** NaOH is the limiting reactant; 0.925 mol Na_2CO_3 can be produced; 0.075 mol CO_2 remain.
3.73 (a) $NaHCO_3$ is the limiting reactant. (b) 0.524 g CO_2 (c) 0.24 g citric acid remain   **3.75** 0.00 g $AgNO_3$ (limiting reactant), 1.94 g Na_2CO_3, 4.06 g Ag_2CO_3, 2.50 g $NaNO_3$
3.77 (a) The theoretical yield is 60.3 g C_6H_5Br. (b) 70.1% yield
3.79 28 g S_8 actual yield   **3.81** (a) $C_2H_4O_2(l) + 3 O_2(g) \longrightarrow$ $2 CO_2(g) + 2 H_2O(l)$ (b) $Ca(OH)_2(s) \longrightarrow CaO(s) + H_2O(l)$ (c) $Ni(s) + Cl_2(g) \longrightarrow NiCl_2(s)$ **3.83** (a) 1.25×10^{22} C atoms (b) 2.77×10^{-3} mol $HC_9H_7O_4$; 1.67×10^{21} $HC_9H_7O_4$ molecules   **3.85** (a) 8×10^{-20} g Si (b) 2×10^3 Si atoms (with 2 significant figures, 1700 Si atoms) (c) 1×10^3 Ge atoms (with 2 significant figures, 1500 Ge atoms)   **3.87** (a) $C_{10}H_{18}O$ (b) $C_{10}H_{18}O$. The empirical formula is also the molecular formula.   **3.89** C_6H_5Cl **3.91** (a) Atomic weight of X = 138.9 g/mol. (b) lanthanum, La   **3.93** (a) 1.19×10^{-5} mol NaI (b) 8.1×10^{-3} g NaI   **3.95** 1.1 kg H_2O **3.97** 7.5 mol H_2 and 4.5 mol N_2 present initially   **3.99** (a) $2 C_2H_2(g) + 5 O_2(g) \longrightarrow 4 CO_2(g) + 2 H_2O(g)$ (b) O_2 is the limiting reactant. (c) 6.7 g C_2H_2, 0 g O_2 (limiting reactant), 11.0 g CO_2, 2.25 g H_2O **3.101** 6.46×10^{24} O atoms   **3.103** (a) 88 kg CO_2 (b) 4×10^2 (400) kg CO_2 **3.105** (a) $S(s) + O_2(g) \longrightarrow SO_2(g)$; $SO_2(g) + CaO(s) \longrightarrow CaSO_3(s)$ (b) 7.9×10^7 g CaO (c) 1.7×10^8 g $CaSO_3$ **3.107** (a) 14 g HCN (b) 26 g NaCN (c) Fire produces 8.7×10^2 g HCN, much more than a lethal dose.

Chapter 4

4.1 Diagram (c) represents Li_2SO_4 **4.3** (a) AX is a nonelectrolyte. (b) AY is a weak electrolyte. (c) AZ is a strong electrolyte.   **4.5** Solid A is NaOH, solid B is AgBr, and solid C is glucose.   **4.7** (b) NO_3^- and (c) NH_4^+ will always be spectator ions.   **4.9** In a redox reaction, electrons are transferred from the oxidized substance to the reduced substance. In an acid–base reaction, protons are transferred from an acid to a base.   **4.11** No. Electrolyte solutions conduct electricity because the dissolved ions carry charge through the solution from one electrode to the other.   **4.13** Although H_2O molecules are electrically neutral, there is an unequal distribution of electrons throughout the molecule. The partially positive ends of H_2O molecules are attracted to anions in the solid, while the partially negative ends are attracted to cations. Thus, both cations and anions in an ionic solid are surrounded and separated (dissolved) by H_2O.
4.15 (a) $ZnCl_2(aq) \longrightarrow Zn^{2+}(aq) + 2\,Cl^-(aq)$
(b) $HNO_3(aq) \longrightarrow H^+(aq) + NO_3^-(aq)$
(c) $(NH_4)_2SO_4(aq) \longrightarrow 2\,NH_4^+(aq) + SO_4^{2-}(aq)$
(d) $Ca(OH)_2(aq) \longrightarrow Ca^{2+}(aq) + 2\,OH^-(aq)$
4.17 HCOOH molecules, H^+ ions, and $HCOO^-$ ions;
$HCOOH(aq) \longrightarrow H^+(aq) + HCOO^-(aq)$ **4.19** (a) Soluble (b) insoluble (c) soluble (d) insoluble (e) insoluble
4.21 (a) $Na_2CO_3(aq) + 2\,AgNO_3(aq) \longrightarrow Ag_2CO_3(s) +$ $2\,NaNO_3(aq)$ (b) No precipitate (c) $FeSO_4(aq) +$ $Pb(NO_3)_2(aq) \longrightarrow PbSO_4(s) + Fe(NO_3)_2(aq)$ **4.23** (a) Na^+, SO_4^{2-} (b) Na^+, NO_3^- (c) NH_4^+, Cl^- **4.25** The solution contains Pb^{2+}.
4.27

Compound	$Ba(NO_3)_2$ Result	NaCl Result
$AgNO_3(aq)$	No ppt	AgCl ppt
$CaCl_2(aq)$	No ppt	No ppt
$Al_2(SO_4)_3(aq)$	$BaSO_4$ ppt	No ppt

This sequence of tests would definitely identify the bottle contents.   **4.29** LiOH is a strong base, HI is a strong acid, and CH_3OH is a molecular compound and nonelectrolyte. The strong acid HI will have the greatest concentration of solvated protons.   **4.31** (a) A monoprotic acid has one ionizable (acidic) H, whereas a diprotic acid has two. (b) A strong acid is completely ionized in aqueous solution, whereas only a fraction of weak acid molecules are ionized. (c) An acid is an H^+ donor, and a base is an H^+ acceptor.   **4.33** As strong acids, HCl, HBr and HI are completely ionized in aqueous solution. They exist exclusively as $H^+(aq)$ and $Cl^-(aq)$, $H^+(aq)$ and $Br^-(aq)$, $H^+(aq)$ and $I^-(aq)$, respectively. As a weak acid, HF is only partially ionized. In solution it exists as a mixture of neutral $HF(aq)$ molecules, $H^+(aq)$ and $F^-(aq)$.   **4.35** (a) Acid, mixture of ions and molecules (weak electrolyte) (b) none of the above, entirely molecules (nonelectrolyte) (c) salt, entirely ions (strong electrolyte) (d) base, entirely ions (strong electrolyte)   **4.37** (a) H_2SO_3, weak electrolyte (b) C_2H_5OH, nonelectrolyte (c) NH_3, weak electrolyte (d) $KClO_3$, strong electrolyte (e) $Cu(NO_3)_2$, strong electrolyte
4.39 (a) $2\,HBr(aq) + Ca(OH)_2(aq) \longrightarrow CaBr_2(aq) + 2\,H_2O(l)$; $H^+(aq) + OH^-(aq) \longrightarrow H_2O(l)$ (b) $Cu(OH)_2(s) +$ $2\,HClO_4(aq) \longrightarrow Cu(ClO_4)_2(aq) + 2\,H_2O(l)$; $Cu(OH)_2(s) +$ $2\,H^+(aq) \longrightarrow 2\,H_2O(l) + Cu^{2+}(aq)$ (c) $Al(OH)_3(s) +$ $3\,HNO_3(aq) \longrightarrow Al(NO_3)_3(aq) + 3\,H_2O(l)$; $Al(OH)_3(s) +$ $3\,H^+(aq) \longrightarrow 3\,H_2O(l) + Al^{3+}(aq)$ **4.41** (a) $CdS(s) +$ $H_2SO_4(aq) \longrightarrow CdSO_4(aq) + H_2S(g)$; $Cd(s) + 2\,H^+(aq) \longrightarrow$ $H_2S(g) + Cd^{2+}(aq)$ (b) $MgCO_3(s) + 2\,HClO_4(aq) \longrightarrow$ $Mg(ClO_4)_2(aq) + H_2O(l) + CO_2(g)$; $MgCO_3(s) + 2\,H^+(aq) \longrightarrow$

$H_2O(l) + CO_2(g) + Mg^{2+}(aq)$ **4.43** (a) $CaCO_3(s) +$ $2\,HNO_3(aq) \longrightarrow Ca(NO_3)_2(aq) + H_2O(l) + CO_2(g)$; $2\,H^+(aq) +$ $CaCO_3(s) \longrightarrow Ca^{2+}(aq) + H_2O(l) + CO_2(g)$ (b) $FeS(s) +$ $2\,HBr(aq) \longrightarrow FeBr_2(aq) + H_2S(g)$; $FeS(s) + 2\,H^+(aq) \longrightarrow$ $Fe^{2+}(aq) + H_2S(g)$ **4.45** (a) In terms of electron transfer, oxidation is the loss of electrons by a substance and reduction is the gain of electrons (LEO says GER). (b) Relative to oxidation numbers, when a substance is oxidized, its oxidation number increases. When a substance is reduced, its oxidation number decreases.   **4.47** Metals in region A are most easily oxidized. Nonmetals in region D are least easily oxidized.   **4.49** (a) +4 (b) +4 (c) +7 (d) +1 (e) 0 (f) −1   **4.51** (a) $N_2 \longrightarrow 2\,NH_3$, N is reduced; $3\,H_2 \longrightarrow 2\,NH_3$, H is oxidized (b) $Fe^{2+} \longrightarrow$ Fe, Fe is reduced; $Al \longrightarrow Al^{3+}$, Al is oxidized (c) $Cl_2 \longrightarrow 2\,Cl^-$, Cl is reduced; $2\,I^- \longrightarrow I_2$ I is oxidized (d) $S^{2-} \longrightarrow SO_4^{2-}$, S is oxidized; $H_2O_2 \longrightarrow H_2O$; O is reduced
4.53 (a) $Mn(s) + H_2SO_4(aq) \longrightarrow MnSO_4(aq) + H_2(g)$; $Mn(s) + 2\,H^+(aq) \longrightarrow Mn^{2+}(aq) + H_2(g)$ (b) $2\,Cr(s) +$ $6\,HBr(aq) \longrightarrow 2\,CrBr_3(aq) + 3\,H_2(g)$; $2\,Cr(s) + 6\,H^+(aq) \longrightarrow$ $2\,Cr^{3+}(aq) + 3\,H_2(g)$ (c) $Sn(s) + 2\,HCl(aq) \longrightarrow SnCl_2(aq) +$ $H_2(g)$; $Sn(s) + 2\,H^+(aq) \longrightarrow Sn^{2+}(aq) + H_2(g)$ (d) $2\,Al(s) +$ $6\,HCOOH(aq) \longrightarrow 2\,Al(HCOO)_3(aq) + 3\,H_2(g)$; $2\,Al(s) +$ $6\,HCOOH(aq) \longrightarrow 2\,Al^{3+}(aq) + 6\,HCOO^-(aq) + 3\,H_2(g)$
4.55 (a) $Fe(s) + Cu(NO_3)_2(aq) \longrightarrow Fe(NO_3)_2(aq) + Cu(s)$ (b) NR (c) $Sn(s) + 2\,HBr(aq) \longrightarrow SnBr_2(aq) + H_2(g)$ (d) NR (e) $2\,Al(s) + 3\,CoSO_4(aq) \longrightarrow Al_2(SO_4)_3(aq) + 3\,Co(s)$
4.57 (a) i. $Zn(s) + Cd^{2+}(aq) \longrightarrow Cd(s) + Zn^{2+}(aq)$; ii. $Cd(s) + Ni^{2+}(aq) \longrightarrow Ni(s) + Cd^{2+}(aq)$ (b) Cd is between Zn and Ni on the activity series. (c) Place an iron strip in $CdCl_2(aq)$. If $Cd(s)$ is deposited, Cd is less active than Fe; if there is no reaction, Cd is more active than Fe. Do the same test with Co if Cd is less active than Fe or with Cr if Cd is more active than Fe.   **4.59** (a) Intensive; the ratio of amount of solute to total amount of solution is the same, regardless of how much solution is present. (b) The term 0.50 mol HCl defines an amount (~18 g) of the pure substance HCl. The term $0.50\,M$ HCl is a ratio; it indicates that there are 0.50 mol of HCl solute in 1.0 liter of solution.   **4.61** (a) $0.0500\,M$ NH_4Cl (b) 0.125 mol HNO_3 (c) 183 mL of $1.50\,M$ KOH
4.63 16 g $Na^+(aq)$ **4.65** BAC of $0.08 = 0.02\,M$ CH_3CH_2OH (alcohol)   **4.67** (a) 5.21 g KBr (b) $0.06537\,M$ $Ca(NO_3)_2$ (c) 10.2 mL of $1.50\,M$ Na_3PO_4 **4.69** (a) $0.15\,M$ K_2CrO_4 has the highest K^+ concentration. (b) 30.0 mL of $0.15\,M$ K_2CrO_4 has more K^+ ions.   **4.71** (a) $0.25\,M$ Na^+, $0.25\,M$ NO_3^-
(b) $1.3 \times 10^{-2}\,M$ Mg^{2+}, $1.3 \times 10^{-2}\,M$ SO_4^{2-} (c) $0.0150\,M$ $C_6H_{12}O_6$ (d) $0.111\,M$ Na^+, $0.111\,M$ Cl^-, $0.0292\,M$ NH_4^+, $0.0146\,M$ CO_3^{2-} **4.73** (a) 16.9 mL $14.8\,M$ NH_3 (b) $0.296\,M$ NH_3
4.75 (a) Add 21.4 g $C_{12}H_{22}O_{11}$ to a 250-mL volumetric flask, dissolve in a small volume of water, and add water to the mark on the neck of the flask. Agitate thoroughly to ensure total mixing. (b) Thoroughly rinse, clean, and fill a 50-mL buret with the $1.50\,M$ $C_{12}H_{22}O_{11}$. Dispense 23.3 mL of this solution into a 350-mL volumetric container, add water to the mark, and mix thoroughly.   **4.77** $1.398\,M$ $HC_2H_3O_2$ **4.79** 0.227 g KCl   **4.81** (a) 38.0 mL of $0.115\,M$ $HClO_4$ (b) 769 mL of $0.128\,M$ HCl (c) $0.408\,M$ $AgNO_3$ (d) 0.275 g KOH   **4.83** 27 g $NaHCO_3$ **4.85** $1.22 \times 10^{-2}\,M$ $Ca(OH)_2$ solution; the solubility of $Ca(OH)_2$ is 0.0904 g in 100 mL solution.   **4.87** (a) $NiSO_4(aq) +$ $2\,KOH(aq) \longrightarrow Ni(OH)_2(s) + K_2SO_4(aq)$ (b) $Ni(OH)_2$ (c) KOH is the limiting reactant. (d) 0.927 g $Ni(OH)_2$ (e) $0.0667\,M$ $Ni^{2+}(aq)$, $0.0667\,M$ $K^+(aq)$, $0.100\,M$ $SO_4^{2-}(aq)$ **4.89** 91.40% $Mg(OH)_2$
4.93 The first precipitate is $AgCl(s)$, the second is $SrSO_4(s)$. Since no precipitate forms on addition of $OH^-(aq)$ to the remaining solution, Ni^{2+} and Mn^{2+} must be absent from the original solution.   **4.94** (a, b) Expt. 1: NR; Expt. 2: $2\,Ag^+(aq) +$ $CrO_4^{2-}(aq) \longrightarrow Ag_2CrO_4(s)$ red precipitate; Expt. 3: NR;

Expt. 4: $2 Ag^+(aq) + C_2O_4^{2-}(aq) \longrightarrow Ag_2C_2O_4(s)$ white precipitate; Expt. 5: $Ca^{2+}(aq) + C_2O_4^{2-}(aq) \longrightarrow CaC_2O_4(s)$ white precipitate; Expt. 6: $Ag^+(aq) + Cl^-(aq) \longrightarrow AgCl(s)$ white precipitate. (c) Chromate salts appear to be more soluble than oxalate salts.   **4.100** $1.42\ M$ KBr   **4.104** $0.496\ M\ H_2O_2$ **4.106** (a) The molar mass of the acid is 136 g/mol. (b) The molecular formula is $C_8H_8O_2$.   **4.109** (a) $Mg(OH)_2(s) + 2 HNO_3(aq) \longrightarrow Mg(NO_3)_2(aq) + 2 H_2O(l)$ (b) HNO_3 is the limiting reactant. (c) 0.0923 mol $Mg(OH)_2$, 0 mol HNO_3, and 0.00250 mol $Mg(NO_3)_2$ are present.   **4.113** 1.766% Cl^- by mass   **4.115** 2.8×10^{-5} g Na_3AsO_4 in 1.00 L H_2O

Chapter 5

5.1 As the book falls, potential energy decreases and kinetic energy increases. At the instant before impact, all potential energy has been converted to kinetic energy, so the book's total kinetic energy is 85 J, assuming no transfer of energy as heat.
5.4 (a) No. The distance traveled to the top of a mountain depends on the path taken by the hiker. Distance is a path function, not a state function. (b) Yes. Change in elevation depends only on the location of the base camp and the height of the mountain, not on the path to the top. Change in elevation is a state function, not a path function.   **5.6** (a) The temperatures of the system and surroundings will equalize, so the temperature of the hotter system will decrease and the temperature of the colder surroundings will increase. The sign of q is ($-$); the process is exothermic. (b) If neither volume nor pressure of the system changes, $w = 0$ and $\Delta E = q = \Delta H$. The change in internal energy is equal to the change in enthalpy.
5.9 (a) $\Delta H_A = \Delta H_B + \Delta H_C$. The diagram and equation both show that the net enthalpy change for a process is independent of path, that ΔH is a state function. (b) $\Delta H_Z = \Delta H_X + \Delta H_Y$. (c) Hess's law states that the enthalpy change for net reaction Z is the sum of the enthalpy changes for steps X and Y, regardless of whether the reaction actually occurs via this path. The diagrams are a visual statement of Hess's law.   **5.11** An object can possess energy by virtue of its motion or position. Kinetic energy depends on the mass of the object and its velocity. Potential energy depends on the position of the object relative to the body with which it interacts.   **5.13** (a) 84 J (b) 20 cal (c) As the ball hits the sand, its speed (and hence its kinetic energy) drops to zero. Most of the kinetic energy is transferred to the sand, which deforms when the ball lands. Some energy is released as heat through friction between the ball and the sand.
5.15 1 Btu = 1054 J   **5.17** (a) The *system* is the well-defined part of the universe whose energy changes are being studied. (b) A *closed system* can exchange heat but not mass with its surroundings.   **5.19** (a) Work is a force applied over a distance. (b) The amount of work done is the magnitude of the force times the distance over which it is applied. $w = F \times d$.
5.21 (a) Gravity; work is done because the force of gravity is opposed and the pencil is lifted. (b) Mechanical force; work is done because the force of the coiled spring is opposed as the spring is compressed over a distance.   **5.23** (a) In any chemical or physical change, energy can be neither created nor destroyed; energy is conserved. (b) The *internal energy (E)* of a system is the sum of all the kinetic and potential energies of the system components. (c) Internal energy of a closed system increases when work is done on the system and when heat is transferred to the system.   **5.25** (a) $\Delta E = 76$ kJ, endothermic (b) $\Delta E = 0.84$ kJ, endothermic (c) $\Delta E = -80.0$ kJ, exothermic
5.27 (a) Since no work is done by the system in case (2), the gas will absorb most of the energy as heat; the case (2) gas will have the higher temperature. (b) In case (2) $w = 0$ and $q = 100$ J. In case (1) energy will be used to do work on the surroundings ($-w$), but some will be absorbed as heat ($+q$). (c) ΔE is greater for case (2) because the entire 100 J increases the internal energy of the system, rather than a part of the energy

doing work on the surroundings.   **5.29** (a) A *state function* is a property that depends only on the physical state (pressure, temperature, etc.) of the system, not on the route used to get to the current state. (b) Internal energy is a state function; heat is not a state function. (c) Work is not a state function. The amount of work required to move from one state to another depends of the specific series of processes or path used to accomplish the change.   **5.31** (a) ΔH is usually easier to measure than ΔE because at constant pressure $\Delta H = q$. The heat flow associated with a process at constant pressure can easily be measured as a change in temperature, while measuring ΔE requires a means to measure both q and w. (b) The process is exothermic.   **5.33** At constant pressure, $\Delta E = \Delta H - P \Delta V$. The values of either P and ΔV or T and Δn must be known to calculate ΔE from ΔH.   **5.35** $\Delta E = -97$ kJ; $\Delta H = -79$ kJ
5.37 (a) $CH_3COOH(l) + 2 O_2(g) \longrightarrow 2 H_2O(l) + 2 CO_2(g)$, $\Delta H = -871.7$ kJ

(b) $CH_3COOH(l) + 2 O_2(g)$

$$\Delta H = \boxed{-871.7\ kJ} \Big\downarrow$$

$$2 H_2O(l) + 2 CO_2(g)$$

5.39 The reactant, 2 Cl(g), has the higher enthalpy.
5.41 (a) Exothermic (b) -59 kJ heat transferred (c) 6.43 g MgO produced (d) 112 kJ heat absorbed   **5.43** (a) -13.1 kJ (b) -1.14 kJ (c) 9.83 J   **5.45** (a) $\Delta H = 726.5$ kJ (b) $\Delta H = -1453$ kJ (c) The exothermic forward reaction is more likely to be thermodynamically favored. (d) Vaporization is endothermic. If the product were $H_2O(g)$, the reaction would be more endothermic and would have a smaller negative ΔH.
5.47 (a) J/mol-°C or J/mol-K (b) J/g-°C or J/g-K (c) To calculate heat capacity from specific heat, the mass of the particular piece of copper pipe must be known.   **5.49** (a) 4.184 J/g-K (b) 75.40 J/mol-°C (c) 774 J/°C (d) 904 kJ   **5.51** 3.00×10^4 J
5.53 $\Delta H = -45.7$ kJ/mol NaOH   **5.55** $\Delta E_{rxn} = -25.5$ kJ/g $C_6H_4O_2$ or -2.75×10^3 kJ/mol $C_6H_4O_2$ **5.57** (a) Heat capacity of the complete calorimeter = 14.4 kJ/°C (b) 5.40 °C
5.59 Hess's law is a consequence of the fact that enthalpy is a state function. Since ΔH is independent of path, we can describe a process by any series of steps that add up to the overall process. ΔH for the process is the sum of ΔH values for the steps.   **5.61** $\Delta H = -1300.0$ kJ   **5.63** $\Delta H = -2.49 \times 10^3$ kJ
5.65 (a) *Standard conditions* for enthalpy changes are $P = 1$ atm and some common temperature, usually 298 K. (b) *Enthalpy of formation* is the enthalpy change that occurs when a compound is formed from its component elements. (c) *Standard enthalpy of formation* ΔH_f° is the enthalpy change that accompanies formation of one mole of a substance from elements in their standard states.   **5.67** (a) $\frac{1}{2} N_2(g) + \frac{3}{2} H_2(g) \longrightarrow NH_3(g)$, $\Delta H_f^\circ = -46.19$ kJ (b) $S(s) + O_2(g) \longrightarrow SO_2(g)$, $\Delta H_f^\circ = -296.9$ kJ (c) $Rb(s) + \frac{1}{2} Cl_2(g) + \frac{3}{2} O_2(g) \longrightarrow RbClO_3(s)$, $\Delta H_f^\circ = -392.4$ kJ (d) $N_2(g) + 2 H_2(g) + \frac{3}{2} O_2(g) \longrightarrow NH_4NO_3(s)$, $\Delta H_f^\circ = -365.6$ kJ   **5.69** $\Delta H_{rxn}^\circ = -847.6$ kJ
5.71 (a) $\Delta H_{rxn}^\circ = -196.6$ kJ (b) $\Delta H_{rxn}^\circ = 37.1$ kJ (c) $\Delta H_{rxn}^\circ = -976.94$ kJ (d) $\Delta H_{rxn}^\circ = -68.3$ kJ
5.73 $\Delta H_f^\circ = -248$ kJ   **5.75** (a) $C_8H_{18}(l) + \frac{25}{2} O_2(g) \longrightarrow 8 CO_2(g) + 9 H_2O(g)$, $\Delta H = -5064.9$ kJ (b) $8 C(s, g) + 9 H_2(g) \longrightarrow C_8H_{18}(l)$ (c) $\Delta H_f^\circ = -259.5$ kJ
5.77 (a) $C_2H_5OH(l) + 3 O_2(g) \longrightarrow 2 CO_2(g) + 3 H_2O(g)$ (b) $\Delta H_{rxn}^\circ = -1234.8$ kJ (c) 2.11×10^4 kJ/L heat produced (d) 0.071284 g CO_2/kJ heat emitted   **5.79** (a) *Fuel value* is the amount of heat produced when 1 g of a substance (fuel) is combusted. (b) 5 g of fat   **5.81** 104 or 1×10^2 Cal/serving

5.83 59.7 Cal   **5.85** (a) $\Delta H_{comb} = -1850$ kJ/mol C_3H_4, -1926 kJ/mol C_3H_6, -2044 kJ/mol C_3H_8 (b) $\Delta H_{comb} = -4.616 \times 10^4$ kJ/kg C_3H_4, -4.578×10^4 kJ/kg C_3H_6, -4.635×10^4 kJ/kg C_3H_8 (c) These three substances yield nearly identical quantities of heat per unit mass, but propane is marginally higher than the other two.   **5.87** (a) 469.4 m/s (b) 5.124×10^{-21} J (c) 3.086 kJ/mol   **5.90** The spontaneous air bag reaction is probably exothermic, with $-\Delta H$ and thus $-q$. When the bag inflates, work is done by the system, so the sign of w is also negative.   **5.93** $\Delta H = 38.95$ kJ; $\Delta E = 36.48$ kJ   **5.96** (a) 8.0×10^{10} kJ energy released (b) 1.9×10^4 ton dynamite   **5.100** 4.90 g CH_4 **5.103** (a) $\Delta H_f^\circ = 35.4$ kJ (b) We need to measure the heat of combustion of $B_5H_9(l)$.   **5.105** (a) $\Delta H^\circ = -631.3$ kJ (b) 3 mol of acetylene gas has greater enthalpy. (c) Fuel values are 50 kJ/g $C_2H_2(g)$, 42 kJ/g $C_6H_6(l)$.   **5.109** If all work is used to increase the man's potential energy, the stair-climbing uses 58 Cal and will not compensate for the extra order of 245 Cal fries. (More than 58 Cal will be required to climb the stairs, because some energy is used to move limbs and some will be lost as heat.)   **5.112** (a) 1.479×10^{-18} J/molecule (b) 1×10^{-15} J/photon. The X-ray has approximately 1000 times more energy than is produced by the combustion of 1 molecule of $CH_4(g)$.   **5.115** (a) ΔH° for neutralization of the acids is HNO_3, -55.8 kJ; HCl, -56.1 kJ; NH_4^+, -4.1 kJ. (b) $H^+(aq) + OH^-(aq) \longrightarrow H_2O(l)$ is the net ionic equation for the first two reactions. $NH_4^+(aq) + OH^-(aq) \longrightarrow NH_3(aq) + H_2O(l)$ (c) The ΔH° values for the first two reactions are nearly identical, -55.9 kJ and -56.2 kJ. Since spectator ions do not change during a reaction and these two reactions have the same net ionic equation, it is not surprising that they have the same ΔH°. (d) Strong acids are more likely than weak acids to donate H^+. Neutralization of the two strong acids is energetically favorable, while the third reaction is barely so. NH_4^+ is likely a weak acid.   **5.117** (a) $\Delta H^\circ = -65.7$ kJ (b) ΔH° for the complete molecular equation will be the same as ΔH° for the net ionic equation. Since the overall enthalpy change is the enthalpy of products minus the enthalpy of reactants, the contributions of spectator ions cancel. (c) ΔH_f° for $AgNO_3(aq)$ is -100.4 kJ/mol.

Chapter 6

6.2 (a) 0.1 m or 10 cm (b) No. Visible radiation has wavelengths much shorter than 0.1 m. (c) Energy and wavelength are inversely proportional. Photons of the longer 0.1 m radiation have less energy than visible photons. (d) Radiation with $\lambda = 0.1$ m is in the low energy portion of the microwave region. The appliance is probably a microwave oven.   **6.5** (a) $n = 1, n = 4$ (b) $n = 1, n = 2$ (c) In order of increasing wavelength and decreasing energy: (iii) < (iv) < (ii) < (i)   **6.8** (a) In the left-most box, the two electrons cannot have the same spin. (b) Flip one of the arrows in the left-most box, so that one points up and the other down. (c) Group 6A   **6.9** (a) Meters (b) 1/second (c) meters/second   **6.11** (a) True (b) False. The frequency of radiation decreases as the wavelength increases. (c) False. Ultraviolet light has shorter wavelengths than visible light. (d) False. X-rays travel at the same speed as microwaves. (e) False. Electromagnetic radiation and sound waves travel at different speeds.   **6.13** Wavelength of X-rays < ultraviolet < green light < red light < infrared < radio waves   **6.15** (a) 3.0×10^{13} s^{-1} (b) 5.45×10^{-7} m = 545 nm (c) The radiation in (b) is visible, the radiation in (a) is not. (d) 1.50×10^4 m   **6.17** 5.64×10^{14} s^{-1}; green.   **6.19** Quantization means that energy changes can happen only in certain allowed increments. If the human growth quantum is one foot, growth occurs instantaneously in one-foot increments. The child experiences growth spurts of one foot; her height can change only by one-foot increments.

6.21 (a) 4.61×10^{14} s^{-1} (b) 1.84×10^5 J (c) $\Delta E = 3.06 \times 10^{-19}$ J   **6.23** (a) $\lambda = 3.3 \, \mu m$, $E = 6.0 \times 10^{-20}$ J; $\lambda = 0.154$ nm, $E = 1.29 \times 10^{-15}$ J (b) The 3.3 μm photon is in the infrared and the 0.154 nm photon is in the X-ray region; the X-ray photon has the greater energy.   **6.25** (a) 6.11×10^{-19} J/photon (b) 368 kJ/mol (c) 1.64×10^{15} photons (d) 368 kJ/mol   **6.27** (a) The $\sim 1 \times 10^{-6}$ m radiation is in the infrared portion of the spectrum. (b) 8.1×10^{16} photons/s   **6.29** (a) $E_{min} = 7.22 \times 10^{-19}$ J (b) $\lambda = 275$ nm (c) $E_{120} = 1.66 \times 10^{-18}$ J. The excess energy of the 120 nm photon is converted into the kinetic energy of the emitted electron. $E_k = 9.3 \times 10^{-19}$ J/electron.   **6.31** When applied to atoms, the notion of quantized energies means that only certain values of ΔE are allowed. These are represented by the lines in the emission spectra of excited atoms.   **6.33** (a) Emitted (b) absorbed (c) emitted   **6.35** (a) $E_2 = -5.45 \times 10^{-19}$ J; $E_6 = -0.606 \times 10^{-19}$ J; $\Delta E = 4.84 \times 10^{-19}$ J; $\lambda = 410$ nm; visible, violet. (b) $E_1 = -2.18 \times 10^{-18}$ J; $E_\infty = 0$ J; $\Delta E = 2.18 \times 10^{-18}$ J/electron = 1.31×10^3 kJ/mol (c) The ionization energy of hydrogen calculated from the Bohr model agrees with the experimental result to three significant figures.   **6.37** (a) Only lines with $n_f = 2$ represent ΔE values and wavelengths that lie in the visible portion of the spectrum. Lines with $n_f = 1$ have shorter wavelengths and lines with $n_f > 2$ have longer wavelengths than visible radiation. (b) $n_i = 3, n_f = 2; \lambda = 6.56 \times 10^{-7}$ m; this is the red line at 656 nm. $n_i = 4, n_f = 2; \lambda = 4.86 \times 10^{-7}$ m; this is the blue line at 486 nm. $n_i = 5, n_f = 2; \lambda = 4.34 \times 10^{-7}$ m; this is the violet line at 434 nm.   **6.39** (a) Ultraviolet region (b) $n_i = 6, n_f = 1$ **6.41** (a) $\lambda = 5.6 \times 10^{-37}$ m (b) $\lambda = 2.65 \times 10^{-34}$ m (c) $\lambda = 2.3 \times 10^{-13}$ m (d) $\lambda = 1.51 \times 10^{-11}$ m   **6.43** 4.14×10^3 m/s   **6.45** (a) $\Delta x \geq 4 \times 10^{-27}$ m (b) $\Delta x \geq 3 \times 10^{-10}$ m   **6.47** (a) The uncertainty principle states that there is a limit to how precisely we can simultaneously know the position and momentum (a quantity related to energy) of an electron. The Bohr model states that electrons move about the nucleus in precisely circular orbits of known radius and energy. This violates the uncertainty principle. (b) De Broglie stated that electrons demonstrate the properties of both particles and waves, that each particle has a wave associated with it. A wave function is the mathematical description of the matter wave of an electron. (c) Although we cannot predict the exact location of an electron in an allowed energy state, we can determine the probability of finding an electron at a particular position. This statistical knowledge of electron location is the *probability density* and is a function of Ψ^2, the square of the wave function Ψ.   **6.49** (a) $n = 4, l = 3, 2, 1, 0$ (b) $l = 2, m_l = -2, -1, 0, 1, 2$ (c) $m_l = 2, l \geq 2$ or $l = 2, 3$ or 4 **6.51** (a) $3p$: $n = 3, l = 1$ (b) $2s$: $n = 2, l = 0$ (c) $4f$: $n = 4, l = 3$ (d) $5d$: $n = 5, l = 2$ **6.53** (a) impossible, $1p$ (b) possible (c) possible (d) impossible, $2d$

6.55 (a)   (b)   (c)

6.57 (a) The hydrogen atom $1s$ and $2s$ orbitals have the same overall spherical shape, but the $2s$ orbital has a larger radial extension and one more node than the $1s$ orbital. (b) A single $2p$ orbital is directional in that its electron density is concentrated along one of the three Cartesian axes of the atom. The $d_{x^2-y^2}$ orbital has electron density along both the x- and y-axes, while the p_x orbital has density only along the x-axis. (c) The average distance of an electron from the nucleus in a $3s$ orbital is greater than for an electron in a $2s$ orbital.

(d) $1s < 2p < 3d < 4f < 6s$ **6.59** (a) In the hydrogen atom, orbitals with the same principal quantum number, n, have the same energy. (b) In a many-electron atom, for a given n value, orbital energy increases with increasing l value: $s < p < d < f$. **6.61** (a) There are two main pieces of experimental evidence for electron "spin." The Stern-Gerlach experiment shows that atoms with a single unpaired electron interact differently with an inhomogeneous magnetic field. Examination of the fine details of emission line spectra of multi-electron atoms reveals that each line is really a close pair of lines. Both observations can be rationalized if electrons have the property of spin.

(b)   (c)

6.63 (a) 6 (b) 10 (c) 2 (d) 14   **6.65** (a) "Valence electrons" are those involved in chemical bonding. They are part or all of the outer-shell electrons listed after the core. (b) "Core electrons" are inner shell electrons that have the electron configuration of the nearest noble-gas element. (c) Each box represents an orbital. (d) Electron spin is represented by the direction of the half-arrows.   **6.67** (a) Cs, [Xe]$6s^1$ (b) Ni, [Ar]$4s^23d^8$ (c) Se, [Ar]$4s^23d^{10}4p^4$ (d) Cd, [Kr]$5s^24d^{10}$ (e) U, [Rn]$5f^36d^17s^2$ (f) Pb, [Xe]$6s^24f^{14}5d^{10}6p^2$ **6.69** (a) F$^-$, [He]$2s^22p^6$ or [Ne] (b) I$^-$, [Kr]$5s^25p^6$ or [Xe] (c) O^{2-}, [He]$2s^22p^6$ or [Ne] (d) K$^+$, [Ne]$3s^23p^6$ or [Ar] (e) Mg^{2+}, [He]$2s^22p^6$ or [Ne] (f) Al^{3+}, [He]$2s^22p^6$ or [Ne] **6.71** (a) He (b) O (c) Cr (d) Te (e) H   **6.73** (a) The fifth electron would fill the $2p$ subshell before the $3s$. (b) Either the core is [He], or the outer electron configuration should be $3s^23p^3$. (c) The $3p$ subshell would fill before the $3d$.   **6.75** (a) $\lambda_A = 3.6 \times 10^{-8}$ m, $\lambda_B = 8.0 \times 10^{-8}$ m (b) $\nu_A = 8.4 \times 10^{15}$ s^{-1}, $\nu_B = 3.7 \times 10^{15}$ s^{-1} (c) A, ultraviolet; B, ultraviolet   **6.77** 66.7 min **6.80** 430–490 nm   **6.84** (a) The Paschen series lies in the infrared. (b) $n_i = 4$, $\lambda = 1.87 \times 10^{-6}$ m; $n_i = 5$, $\lambda = 1.28 \times 10^{-6}$ m; $n_i = 6$, $\lambda = 1.09 \times 10^{-6}$ m   **6.87** $\nu = 1.02 \times 10^7$ m/s   **6.90** (a) l (b) n and l (c) m_s (d) m_l **6.96** If m_s had three allowed values instead of two, each orbital would hold three electrons instead of two. Assuming that there is no change in the n, l, and m_l values, the number of elements in each of the first four rows would be: 1st row, 3 elements; 2nd row, 12 elements; 3rd row, 12 elements; 4th row, 27 elements   **6.99** (a) 1.7×10^{28} photons (b) 34 s   **6.103** (a) Bohr's theory was based on the Rutherford nuclear model of the atom: a dense positive charge at the center and a diffuse negative charge surrounding it. Bohr's theory then specified the nature of the diffuse negative charge. The prevailing theory before the nuclear model was Thomson's plum pudding model: discrete electrons scattered about a diffuse positive charge cloud. Bohr's theory could not have been based on the Thomson model of the atom. (b) De Broglie's hypothesis is that electrons exhibit both particle and wave properties. Thomson's conclusion that electrons have mass is a particle property, while the nature of cathode rays is a wave property. De Broglie's hypothesis actually rationalizes these two seemingly contradictory observations about the properties of electrons.

Chapter 7

7.2 The billiard ball is a more appropriate analogy for the non-bonding atomic radius. The billiard ball has a definite "hard" boundary, and can be used to model nonbonding interactions in which there is no penetration of electron clouds. If we use the billiard ball to represent the bonding atomic radius of a fluorine atom, we overestimate this radius. When atoms bond, attractive interactions cause their electron clouds to penetrate

each other, bringing the nuclei closer together than during a nonbonding (billiard ball) collision.   **7.5** The energy change for the reaction is the ionization energy of A plus the electron affinity of A. The process is endothermic for both chlorine and sodium.   **7.7** Mendeleev placed elements with similar chemical and physical properties within a family or column of his table. For undiscovered elements, he left blanks. He predicted properties for the "blanks" based on properties of other elements in the family and on either side.   **7.9** Even though Si is the second most abundant element in Earth's crust, its abundant forms are compounds. In order to be "discovered," elemental Si had to be chemically removed from one of its compounds. Discovery awaited more sophisticated chemical techniques available in the nineteenth century.
7.11 (a) *Effective nuclear charge*, Z_{eff}, is a representation of the average electrical field experienced by a single electron. It is the average environment created by the nucleus and the other electrons in the molecule, expressed as a net positive charge at the nucleus. (b) Going from left to right across a period, effective nuclear charge increases.   **7.13** (a) For both Na and K, $Z_{eff} = 1$. (b) For both Na and K, $Z_{eff} = 2.2$. (c) Slater's rules give values closer to the detailed calculations: Na, 2.51; K, 3.49. (d) Both approximations give the same value of Z_{eff} for Na and K; neither accounts for the gradual increase in Z_{eff} moving down a group.   **7.15** The $n = 3$ electrons in Kr experience a greater effective nuclear charge and thus have a greater probability of being closer to the nucleus.   **7.17** (a) Atomic radii are determined by measuring distances between atoms in various situations. (b) Bonding radii are calculated from the internuclear separation of two atoms joined by a chemical bond. Nonbonding radii are calculated from the internuclear separation between two gaseous atoms that collide and move apart but do not bond. (c) For a given element, the nonbonding radius is always larger than the bonding radius.   **7.19** 1.37 Å   **7.21** From the sum of the atomic radii, As—I = 2.52 Å. This is very close to the experimental value of 2.55 Å.   **7.23** (a) Decrease (b) increase (c) F < S < P < As   **7.25** (a) Be < Mg < Ca (b) Br < Ge < Ga (c) Si < Al < Tl   **7.27** (a) Electrostatic repulsions are reduced by removing an electron from a neutral atom, effective nuclear charge increases, and the cation is smaller. (b) The additional electrostatic repulsion produced by adding an electron to a neutral atom decreases the effective nuclear charge experienced by the valence electrons, and increases the size of the anion. (c) Going down a column, valence electrons are further from the nucleus, and they experience greater shielding by core electrons. The greater radial extent of the valence electrons outweighs the increase in Z.   **7.29** The red sphere is a metal; its size decreases on reaction, characteristic of the change in radius when a metal atom forms a cation. The blue sphere is a nonmetal; its size increases on reaction, characteristic of the change in radius when a nonmetal atom forms an anion.   **7.31** (a) An isoelectronic series is a group of atoms or ions that have the same number of electrons and the same electron configuration. (b) (i) Al^{3+}:Ne (ii) Ti^{4+}:Ar (iii) Br$^-$:Kr (iv) Sn^{2+}:Cd   **7.33** (a) Na$^+$ (b) Assume a He core of 2 electrons. F$^-$:$Z_{eff} = 7$; Na$^+$:$Z_{eff} = 9$. (c) F$^-$:$Z_{eff} = 4.85$; Na$^+$:$Z_{eff} = 6.85$. (d) For isoelectronic ions, electron configurations and therefore shielding values (S) are the same. As nuclear charge (Z) increases, effective nuclear charge (Z_{eff}) increases and ionic radius decreases.   **7.35** Cl < S < K (b) K$^+$ < Cl$^-$ < S^{2-} (c) In the neutral atoms, the n-value of the outer electron in K is larger than the n-value of valence electrons in S and Cl, so K atoms are largest. When the $4s$ electron is removed, K$^+$ is isoelectronic with Cl$^-$ and S^{2-}, so the ion with the largest Z value, K$^+$, is smallest.   **7.37** (a) O^{2-} is larger than O because increased repulsions that accompany addition of electrons cause the electron cloud to expand. (b) S^{2-} is larger than O^{2-}, because for particles with like charges, size increases going down a family. (c) S^{2-} is larger than K$^+$ because the two

ions are isoelectronic and K^+ has the larger Z and Z_{eff}. (d) K^+ is larger than Ca^{2+}, because for isoelectronic particles, the ion with the smaller Z has the larger radius. **7.39** $B(g) \longrightarrow$ $B^+(g) + e^-$; $B^+(g) \longrightarrow B^{2+}(g) + e^-$; $B^{2+}(g) \longrightarrow B^{3+}(g) + e^-$ **7.41** (a) The electrons in any atom are bound to the atom by their electrostatic attraction to the nucleus. Therefore, energy must be added in order to remove an electron from an atom. The ionization energy, ΔE for this process, is thus positive. (b) F has a greater first ionization energy than O because F has a greater Z_{eff} and the outer electrons in both elements are approximately the same distance from the nucleus. (c) The second ionization energy of an element is greater than the first because more energy is required to overcome the larger Z_{eff} of the 1+ cation than that of the neutral atom. **7.43** (a) The smaller the atom, the larger its first ionization energy. (b) Of the nonradioactive elements, He has the largest, and Cs the smallest first ionization energy. **7.45** (a) Ar (b) Be (c) Co (d) S (e) Te **7.47** (a) In^{3+}, $[Kr]4d^{10}$ (b) Sb^{3+}, $[Kr]5s^24d^{10}$ (c) Te^{2-}, $[Kr]5s^24d^{10}5p^6$ or $[Xe]$ (d) Te^{6+}, $[Kr]4d^{10}$ (e) Hg^{2+}, $[Xe]4f^{14}5d^{10}$ (f) Rh^{3+}, $[Kr]4d^6$ **7.49** (a) Ni^{2+}, $[Ar]3d^8$, 2 unpaired electrons (b) Sn^{2+}, $[Kr]5s^24d^{10}$, 0 unpaired electrons **7.51** Positive, endothermic, values for ionization energy and electron affinity mean that energy is required to either remove or add electrons. Valence electrons in Ar experience the largest Z_{eff} of any element in the third row, resulting in a large, positive ionization energy. When an electron is added to Ar, the n = 3 electrons become core electrons which screen the extra electron so effectively that Ar^- has a higher energy than an Ar atom and a free electron. This results in a large positive electron affinity. **7.53** Electron affinity of Br: $Br(g) + 1 e^- \longrightarrow Br^-(g)$; $[Ar]4s^23d^{10}4p^5 \longrightarrow [Ar]4s^23d^{10}4p^6$; electron affinity of Kr: $Kr(g) + 1 e^- \longrightarrow Kr^-(g)$; $[Ar]4s^23d^{10}4p^6 \longrightarrow [Ar]4s^23d^{10}4p^65s^1$. Br^- adopts the stable electron configuration of Kr; the added electron experiences essentially the same Z_{eff} and stabilization as the other valence electrons and electron affinity is negative. In Kr^- ion, the added electron occupies the higher energy 5s orbital. A 5s electron is farther from the nucleus, effectively shielded by the spherical Kr core and not stabilized by the nucleus; electron affinity is positive. **7.55** (a) Ionization energy (I_1) of Ne: $Ne(g) \longrightarrow Ne^+(g) + 1 e^-$; $[He]2s^22p^6 \longrightarrow [He]2s^22p^5$; electron affinity ($E_1$) of F: $F(g) + 1 e^- \longrightarrow F^-(g)$; $[He]2s^22p^5 \longrightarrow [He]2s^22p^6$. (b) I_1 of Ne is positive; E_1 of F is negative. (c) One process is apparently the reverse of the other, with one important difference. Ne has a greater Z and Z_{eff}, so we expect I_1 for Ne to be somewhat greater in magnitude and opposite in sign to E_1 for F. **7.57** The smaller the first ionization energy of an element, the greater the metallic character of that element. **7.59** Based on ionization energies, the metallic character of Al is similar to that of Ca and Sr; it is clearly more metallic than the metalloids in groups 4A and 5A. **7.61** Ionic: MgO, Li_2O, Y_2O_3; molecular: SO_2, P_2O_5, N_2O, XeO_3. Ionic compounds are formed by combining a metal and a nonmetal; molecular compounds are formed by two or more nonmetals. **7.61** Ionic: MgO, Li_2O, Y_2O_3; molecular: SO_2, P_2O_5, N_2O, XeO_3. Ionic compounds are formed by combining a metal and a nonmetal; molecular compounds are formed by two or more nonmetals. **7.63** (a) An acidic oxide dissolved in water produces an acidic solution; a basic oxide dissolved in water produces a basic solution. (b) Oxides of nonmetals, such as SO_3, are acidic; oxides of metals, such as CaO, are basic. **7.65** (a) Dichlorine-septoxide (b) $2 Cl_2(g) + 7 O_2(g) \longrightarrow 2 Cl_2O_7(l)$ (c) While most nonmetal oxides we have seen, such as CO_2 or SO_2, are gases, a boiling point of 81 °C is not totally unexpected for a large molecule like Cl_2O_7. (d) Cl_2O_7 is an acidic oxide, so it will be more reactive to base, OH^-. **7.67** (a) $BaO(s) + H_2O(l) \longrightarrow$ $Ba(OH)_2(aq)$ (b) $FeO(s) + 2 HClO_4(aq) \longrightarrow Fe(ClO_4)_2(aq) + H_2O(l)$ (c) $SO_3(g) + H_2O(l) \longrightarrow H_2SO_4(aq)$ (d) $CO_2(g) + 2 NaOH(aq) \longrightarrow Na_2CO_3(aq) + H_2O(l)$ **7.69** (a) Na, $[Ne]3s^1$;

Mg, $[Ne]3s^2$ (b) When forming ions, both adopt the stable configuration of Ne; Na loses one electron and Mg two electrons to achieve this configuration. (c) The effective nuclear charge of Mg is greater, so its ionization energy is greater. (d) Mg is less reactive because it has a higher ionization energy. (e) The atomic radius of Mg is smaller because its effective nuclear charge is greater. **7.71** (a) Ca is more reactive because it has a lower ionization energy than Mg. (b) K is more reactive because it has a lower ionization energy than Ca. **7.73** (a) $2 K(s) + Cl_2(g) \longrightarrow 2 KCl(s)$ (b) $SrO(s) + H_2O(l) \longrightarrow Sr(OH)_2(aq)$ (c) $4 Li(s) + O_2(g) \longrightarrow 2 Li_2O(s)$ (d) $2 Na(s) + S(l) \longrightarrow Na_2S(s)$ **7.75** (a) The ionization energy of H fits between those of C and N. (b) The ionization energy of Li fits between those of Na and Mg. (c) These series are consistent with the assignment of H as a nonmetal and Li as a metal. **7.77** (a) F, $[He]2s^22p^5$; Cl, $[Ne]3s^23p^5$ (b) F and Cl are in the same group, and both adopt a 1− ionic charge. (c) The 2p valence electrons in F are closer to the nucleus and more tightly held than are the 3p electrons of Cl, so the ionization energy of F is greater. (d) The high ionization energy of F coupled with a relatively large exothermic electron affinity makes it more reactive than Cl toward H_2O. (e) While F has approximately the same effective nuclear charge as Cl, its small atomic radius gives rise to large repulsions when an extra electron is added, so the overall electron affinity of F is less exothermic than that of Cl. (f) The 2p valence electrons in F are closer to the nucleus so the atomic radius is smaller than that of Cl. **7.79** (a) The term "inert" was dropped because it no longer described all the Group 8A elements. (b) In the 1960s, scientists discovered that Xe would react with substances having a strong tendency to remove electrons, such as F_2. Thus, Xe could not be categorized as an "inert" gas. (c) The group is now called the noble gases. **7.81** (a) $2 O_3(g) \longrightarrow 3 O_2(g)$ (b) $Xe(g) + F_2(g) \longrightarrow XeF_2(g)$; $Xe(g) + 2 F_2(g) \longrightarrow XeF_4(s)$; $Xe(g) + 3 F_2(g) \longrightarrow XeF_6(s)$ (c) $S(s) + H_2(g) \longrightarrow H_2S(g)$ (d) $2 F_2(g) + 2 H_2O(l) \longrightarrow 4 HF(aq) + O_2(g)$ **7.83** Up to Z = 82, there are three instances where atomic weights are reversed relative to atomic numbers: Ar and K; Co and Ni; Te and I. In each case the most abundant isotope of the element with the larger atomic number has one more proton, but fewer neutrons than the element with the smaller atomic number. The smaller number of neutrons causes the element with the larger Z to have a smaller than expected atomic weight. **7.85** (a) 5+ (b) 4.8+ (c) Shielding is greater for 3p electrons, owing to penetration by 3s electrons, so Z_{eff} for 3p electrons is less than that for 3s electrons. (d) The first electron lost is a 3p electron, because it has a smaller Z_{eff} and experiences less attraction for the nucleus than a 3s electron does. **7.88** Moving across the representative elements, electrons added to ns or np valence orbitals do not effectively screen each other. The increase in Z is not accompanied by a similar increase in S. Z_{eff} increases and atomic size decreases. Moving across the transition elements, electrons are added to $(n - 1)d$ orbitals, which do significantly screen the ns valence electrons. The increase in Z is accompanied by a corresponding increase in S. Z_{eff} increases more slowly and atomic size decreases more slowly. **7.91** The completed 4f subshell in Hf leads to a much larger change in Z and Z_{eff} going from Zr to Hf than in going from Y to La. This larger increase in Z_{eff} going from Zr to Hf leads to a smaller increase in atomic radius than in going from Y to La. **7.94** I_1 through I_4 represent loss of the 2p and 2s electrons from the outer shell of the atom. Z is constant while removal of each electron reduces repulsion between the remaining electrons, so Z_{eff} increases and I increases. I_5 and I_6 represent loss of the 1s core electrons. These electrons are closer to the nucleus and experience the full nuclear charge, so the values of I_5 and I_6 are significantly greater than $I_1 - I_4$. I_6 is larger than I_5 because all electron–electron repulsion has been eliminated.

7.96 O, [He]$2s^2 2p^4$

⇅	⇅	↑	↑

2s   2p

O^{2-}, [He]$2s^2 2p^6$ = [Ne]

⇅	⇅	⇅	⇅

2s   2p

O^{3-}, [Ne]$3s^1$

The third electron would be added to the 3s orbital, which is farther from the nucleus and more strongly shielded by the [Ne] core. The overall attraction of this 3s electron for the oxygen nucleus is not large enough for O^{3-} to be a stable particle. **7.98** (a) The group 2B metals have complete $(n - 1)d$ subshells. An additional electron would occupy an np subshell and be substantially shielded by both ns and $(n - 1)d$ electrons. This is not a lower energy state than the neutral atom and a free electron. (b) Group 1B elements have the generic electron configuration $ns^1(n - 1)d^{10}$. An additional electron would complete the ns subshell and experience repulsion with the other ns electron. Going down the group, size of the ns subshell increases and repulsion effects decrease, so effective nuclear charge increases and electron affinities become more negative. **7.101** O$_2$ < Br$_2$ < K < Mg. O$_2$ and Br$_2$ are nonpolar nonmetals. O$_2$, with the much lower molar mass, should have the lower melting point. K and Mg are metallic solids with higher melting points than the two nonmetals. Since alkaline earth metals are typically harder, more dense, and higher melting than alkali metals, Mg should have the highest melting point of the group. This order of melting points is confirmed by data in Tables 7.4, 7.5, 7.6, and 7.7. **7.106** (a) Li, [He]$2s^1$; $Z_{eff} \approx 1+$. (b) $I_1 \approx 5.45 \times 10^{-19}$ J/mol ≈ 328 kJ/mol (c) The estimated value of 328 kJ/mol is less than the Table 7.4 value of 520 kJ/mol. Our estimate for Z_{eff} was a lower limit; the [He] core electrons do not perfectly shield the 2s electron from the nuclear charge. (d) Based on the experimental ionization energy, $Z_{eff} = 1.26$. This value is greater than the estimate from part (a), but agrees well with the "Slater" value of 1.3 and is consistent with the explanation in part (c). **7.109** (a) Mg$_3$N$_2$ (b) Mg$_3$N$_2$(s) + 3 H$_2$O(l) $\longrightarrow$ 3 MgO(s) + 2 NH$_3$(g); the driving force is the production of NH$_3$(g) (c) 17% Mg$_3$N$_2$ (d) 3 Mg(s) + 2 NH$_3$(g) $\longrightarrow$ Mg$_3$N$_2$(s) + 3 H$_2$(g). NH$_3$ is the limiting reactant and 0.46 g H$_2$ are formed. (e) $\Delta H^\circ_{rxn} = -368.70$ kJ

Chapter 8

8.2 (a) A^{2+} combines with Y^{2-} and B$^+$ combines with X$^-$ to form compounds with a 1:1 ratio of cations and anions. (b) AY. Lattice energy increases as ionic charge increases and decreases as inter-ionic distance increases. A and Y have the greater charges, while the A—Y and B—X separations are nearly equal. (c) BX has the smaller lattice energy. **8.5** (a) Moving from left to right along the molecule, the first C needs 2 H atoms, the second needs 1, the third needs none and the fourth needs 1. (b) In order of increasing bond length: 3 < 1 < 2 (c) In order of increasing bond enthalpy: 2 < 1 < 3 **8.7** (a) Valence electrons are those that take part in chemical bonding. This usually means the electrons beyond the core noble-gas configuration of the atom, although it is sometimes only the outer-shell electrons. (b) A nitrogen atom has 5 valence electrons. (c) The atom (Si) has 4 valence electrons. **8.9** P, $1s^2 2s^2 2p^6 3s^2 3p^3$. The $n = 3$ electrons are valence electrons; the others are nonvalence electrons. Valence electrons participate in chemical bonding, the others do not.

8.11 (a) ·Äl· (b) :B̈r: (c) :Är: (d) ·Sr

8.13 Mg⁀ + :Ö: ⟶ Mg^{2+} + $\left[:\ddot{O}:\right]^{2-}$ **8.15** (a) AlF$_3$ (b) K$_2$S

(c) Y$_2$O$_3$ (d) Mg$_3$N$_2$ **8.17** (a) Sr^{2+}, [Ar]$4s^2 3d^{10}4p^6$ = [Kr], noble-gas configuration (b) Ti^{2+}, [Ar]$3d^2$ (c) Se^{2-},

[Ar]$4s^2 3d^{10}4p^6$ = [Kr], noble-gas configuration (d) Ni^{2+}, [Ar]$3d^8$ (e) Br$^-$, [Ar]$4s^2 3d^{10}4p^6$ = [Kr], noble-gas configuration (f) Mn^{3+}, [Ar]$3d^4$ **8.19** (a) *Lattice energy* is the energy required to totally separate one mole of solid ionic compound into its gaseous ions. (b) The magnitude of the lattice energy depends on the magnitudes of the charges of the two ions, their radii and the arrangement of ions in the lattice. **8.21** KF, 808 kJ/mol; CaO, 3414 kJ/mol; ScN, 7547 kJ/mol. The interionic distances in the three compounds are similar. For compounds with similar ionic separations, the lattice energies should be related as the product of the charges of the ions. The lattice energies above are approximately related as 1:4:9. Slight variations are due to the small differences in ionic separations. **8.23** Since the ionic charges are the same in the two compounds, the KBr and CsCl separations must be approximately equal. **8.25** The large attractive energy between oppositely charged Ca^{2+} and O^{2-} more than compensates for the energy required to form Ca^{2+} and O^{2-} from the neutral atoms. **8.27** The lattice energy of RbCl(s) is +692 kJ/mol. This value is smaller than the lattice energy for NaCl because Rb$^+$ has a larger ionic radius than Na$^+$ and therefore cannot approach Cl$^-$ as closely as Na$^+$ can. **8.29** (a) A *covalent bond* is the bond formed when two atoms share one or more pairs of electrons. (b) Any simple compound whose component atoms are nonmetals, such as H$_2$, SO$_2$, and CCl$_4$, are molecular and have covalent bonds between atoms. (c) Covalent, because it is a gas at room temperature and below.

8.31 :Cl· + :Cl· + :Cl· + ·N: ⟶ Cl—N bonded with three Cl

8.33 (a) Ö=Ö (b) A double bond is required because there are not enough electrons to satisfy the octet rule with single bonds and unshared pairs. (c) The greater the number of shared electron pairs between two atoms, the shorter the distance between the atoms. An O=O double bond is shorter than an O—O single bond. **8.35** (a) Electronegativity is the ability of an atom in a molecule to attract electrons to itself. (b) The range of electronegativities on the Pauling scale is 0.7–4.0. (c) Fluorine is the most electronegative element. (d) Cesium is the least electronegative element that is not radioactive. **8.37** (a) O (b) C (c) P (d) Be **8.39** The bonds in (a), (c), and (d) are polar. The more electronegative element in each polar bond is: (a) F (c) O (d) I **8.41** (a) The calculated charge on H and O is 0.38e. (b) From Sample Exercise 8.5, the calculated charge on H and Cl in HCl is 0.178e. The O—H bond in OH is more polar than the H—Cl bond in HCl. (c) Yes. The electronegativity of O is greater than that of Cl, so the electronegativity difference between O and H is greater than that between Cl and H. Based on electronegativity, we expect OH to be more polar than HCl. **8.43** (a) SiCl$_4$, molecular, silicon tetrachloride; LaF$_3$, ionic, lanthanum(III) fluoride (b) FeCl$_2$, ionic, iron(II) chloride; ReCl$_6$, molecular (metal in high oxidation state), rhenium hexachloride. (c) PbCl$_4$, molecular (by contrast to the distinctly ionic RbCl), lead tetrachloride; RbCl, ionic, rubidium chloride

8.45 (a) H—Si—H (with H top and bottom) (b) :C≡O: (c) :F̈—S̈—F̈:

(d) :Ö—S—Ö—H (with Ö above and below S) (e) $\left[:\ddot{O}—\ddot{Cl}—\ddot{O}:\right]^-$ (f) H—N̈—Ö—H (with H below N)

8.47 (a) *Formal charge* is the charge on each atom in a molecule, assuming all atoms have the same electronegativity. **(b)** Formal charges are not actual charges. They are a bookkeeping system that assumes perfect covalency, one extreme for the possible electron distribution in a molecule. **(c)** Oxidation numbers are a bookkeeping system that assumes the more electronegative element holds all electrons in a bond. The true electron distribution is some composite of the two extremes. **8.49** Formal charges are shown on the Lewis structures; oxidation numbers are listed below each structure.

(a) $[:N\equiv O:]^+$
N, +3; O, −2

(b) $0:\ddot{Cl}-\overset{+1}{P}-\ddot{Cl}:0$ with $:\ddot{Cl}:$ below (0) and $:\ddot{O}:^{-1}$ above
P, +5; Cl, −1; O, −2

(c) $\left[-1:\ddot{O}-\overset{+3}{Cl}-\ddot{O}:-1\right]^-$ with $:\overset{-1}{\ddot{O}}:$ above and $:\overset{-1}{\ddot{O}}:$ below
Cl, +7; O, −2

(d) $-1:\ddot{O}-\overset{+2}{Cl}-\overset{0}{O}-H$ with $:\overset{-1}{\ddot{O}}:$ below
Cl, +5; H, +1; O, −2

8.51 (a) $[\ddot{O}=\dot{N}-\ddot{O}:]^- \longleftrightarrow [:\ddot{O}-\dot{N}=\ddot{O}]^-$

(b) O_3 is isoelectronic with NO_2^-; both have 18 valence electrons. **(c)** Since each N—O bond has partial double-bond character, the N—O bond length in NO_2^- should be shorter than an N—O single bond. **8.53** The more electron pairs shared by two atoms, the shorter the bond. Thus, the C—O bond lengths vary in the order $CO < CO_2 < CO_3^{2-}$. **8.55 (a)** Two equally valid Lewis structures can be drawn for benzene.

The concept of resonance dictates that the true description of bonding is some hybrid or blend of these two Lewis structures. The most obvious blend of these two resonance structures is a molecule with six equivalent C—C bonds with equal lengths. **(b)** In order for the six C—C bonds in benzene to be equivalent, each must have some double-bond character. That is, more than one pair but fewer than two pairs of electrons are involved in each C—C bond. This model predicts a uniform C—C bond length that is shorter than a single bond but longer than a double bond. **8.57 (a)** The octet rule states that atoms will gain, lose, or share electrons until they are surrounded by eight valence electrons. **(b)** The octet rule applies to the individual ions in an ionic compound. For example, in $MgCl_2$, Mg loses 2 e⁻ to become Mg^{2+} with the electron configuration of Ne. Each Cl atom gains one electron to form Cl⁻ with the electron configuration of Ar. **8.59** The most common exceptions to the octet rule are molecules with more than eight electrons around one or more atoms.

8.61 (a) $\left[:\ddot{O}-\ddot{S}-\ddot{O}:\right]^{2-}$ with $:\ddot{O}:$ below

Other resonance structures that violate the octet rule can be drawn, but the single best structure, the one shown here, obeys the octet rule

(b) H—Al—H with H below

6 electrons around Al

(c) $\left[:N\equiv N-\dot{N}:\right]^- \longleftrightarrow \left[:\dot{N}-N\equiv N:\right]^- \longleftrightarrow$

$\left[:\ddot{N}=N=\ddot{N}:\right]^-$

(d) $:\ddot{Cl}:$ / $:\ddot{Cl}-C-H$ / H (with another $:\ddot{Cl}:$)

(e) $:\ddot{F}-Sb$ bonded to five F atoms ($:\ddot{F}:$)

10 electrons around Sb

8.63 (a) $:\ddot{Cl}-\overset{0}{Be}-\ddot{Cl}:$ (with 0, 0, 0) This structure violates the octet rule.

(b) $\overset{+1}{\ddot{Cl}}=\overset{-2}{Be}=\overset{+1}{\ddot{Cl}} \longleftrightarrow :\overset{0}{\ddot{Cl}}-\overset{-2}{Be}\equiv\overset{+2}{Cl} \longleftrightarrow \overset{+2}{Cl}\equiv\overset{-2}{Be}-\overset{0}{\ddot{Cl}}:$

(c) Since formal charges are minimized on the structure that violates the octet rule, this form is probably most important. This is a different conclusion than for molecules that have resonance structures with expanded octets that minimize formal charge. **8.65 (a)** $\Delta H = -304$ kJ **(b)** $\Delta H = -82$ kJ **(c)** $\Delta H = -467$ kJ **8.67 (a)** $\Delta H = -321$ kJ **(b)** $\Delta H = -103$ kJ **(c)** $\Delta H = -203$ kJ **8.69 (a)** Exothermic **(b)** The ΔH calculated from bond enthalpies (−97 kJ) is slightly more exothermic (more negative) than that obtained using ΔH_f° values (−92.38 kJ). **8.71** The average Ti—Cl bond enthalpy is 430 kJ/mol. **8.73 (a)** Six (nonradioactive) elements. Yes, they are in the same family, assuming H is placed with the alkali metals. The Lewis symbol represents the number of valence electrons of an element, and all elements in the same family have the same number of valence electrons. By definition of a family, all elements with the same Lewis symbol must be in the same family. **8.75 (a)** 779 kJ/mol **(b)** 627 kJ/mol **(c)** 2195 kJ/mol **8.79 (a)** A polar molecule has a measurable dipole moment while a nonpolar molecule has a zero net dipole moment. **(b)** Yes. If X and Y have different electronegativities, the electron density around the more electronegative atom will be greater, producing a charge separation or dipole in the molecule. **(c)** $\mu = Qr$. The dipole moment, μ, is the product of the magnitude of the separated charges, Q, and the distance between them, r. **8.83 (a)** The empirical formula of Compound 1 is RuO_2. **(b)** The empirical formula of Compound 2 is RuO_4. **(c)** The yellow molecular compound is RuO_4, ruthenium tetroxide. The black ionic compound is RuO_2, ruthenium(IV) oxide.

8.85 (a)

The structures shown here minimize formal charges on the atoms. **(b)** Carbon–nitrogen bond distances should be in the range 1.40–1.41 Å, intermediate between C—N and C=N bonds. **8.88 (a)** +1 **(b)** −1 **(c)** +1 (assuming the odd electron is on N) **(d)** 0 **(e)** +3 **8.91** ΔH is +42 kJ for the first reaction and −200 kJ for the second. The latter is much more favorable because the formation of 2 mol of O—H bonds is more exothermic than the formation of 1 mol of H—H bonds.

8.94 (a)

(b) Each terminal O=N—O group has two possible placements for the N=O; this leads to 8 resonance structures with the carbon–nitrogen framework shown in (a). Structures with N=N can be drawn, but these do not minimize formal charge. There are 7 resonance structures with one or more N=N, for a total of 15 resonance structures.
(c) $C_3H_6N_6O_6(s) \longrightarrow 3\,CO(g) + 3\,N_2(g) + 3\,H_2O(g)$ (d) According to Table 8.4, N—N single bonds have the smallest bond enthalpy and are weakest. (e) The enthalpy of decomposition for the resonance structure in (a) is -1668 kJ/mol $C_3H_6N_6O_6$. For the resonance structure with 3 N=N and 6 N—O bonds, $\Delta H = -2121$ kJ/mol. The actual value is somewhere between these two values. The enthalpy change for 5.0 g of RDX is then in the range 38–48 kJ.   **8.96** (a) 1.77 Å
(b) 1.75–1.76 Å (c) The resonance structures for SO_2 indicate that the sulfur–oxygen bond is intermediate between a double and single bond, so the bond length of 1.43 Å should be significantly shorter than the single bond distance of 1.75 Å.

(d)

In order to rationalize the sulfur–oxygen bond distance of 1.48 Å, a resonance structure with S=O must contribute to the true structure. The sulfur atom attached to oxygen in this resonance structure has more than an octet of electrons and violates the octet rule.   **8.97** (a) Ti^{2+}, $[Ar]3d^2$; Ca, $[Ar]4s^2$. The 2 valence electrons in Ti^{2+} and Ca are in different principal quantum levels and different subshells. (b) In Ca the $4s$ is lower in energy than the $3d$, while in Ti^{2+} the $3d$ is lower in energy than the $4s$. (c) There is only one $4s$ orbital, so the 2 valence electrons in Ca are paired; there are 5 degenerate $3d$ orbitals, so the 2 valence electrons in Ti^{2+} are unpaired.   **8.100** (a) A, In_2S; B, InS; C, In_2S_3 (b) A, In(I); B, In(II); C, In(III) (c) In(I), $[Kr]5s^24d^{10}$; In(II), $[Kr]5s^14d^{10}$; In(III), $[Kr]4d^{10}$. None of these are noble-gas configurations. (d) In(III) in compound C has the smallest ionic radius. Removing successive electrons from an atom reduces electron repulsion, increases the effective nuclear charge experienced by the valence electrons and decreases the ionic radius. (e) Lattice energy increases as ionic charge increases and as inter-ionic distance decreases. Only the charge and size of In changes in the three compounds. In(I) in compound A has the lowest charge and largest ionic radius, so compound A has the smallest lattice energy and lowest melting point. Compound C has the largest lattice energy and highest melting point.
8.103 (a) Azide ion is N_3^-. (b) Resonance structures with formal charges are shown.

(c) The left structure minimizes formal charges and is probably the main contributor. (d) The N—N distances will be equal, and have the approximate length of a N—N double bond, 1.24 Å.

Chapter 9

9.1 Removing an atom from the equatorial plane of the trigonal bipyramid in Figure 9.3 creates a see-saw shape.   **9.3** (a) Square pyramidal (b) Yes, there is one nonbonding electron

domain on A. If there were only the 5 bonding domains, the shape would be trigonal bipyramidal. (c) Yes. If the B atoms are halogens, A is also a halogen.   **9.5** (a) Zero energy corresponds to two separate, non-interacting Cl atoms. This infinite Cl—Cl distance is beyond the right extreme of the horizontal axis on the diagram. (b) According to the valence bond model, valence orbitals on the approaching atoms overlap, allowing two electrons to mutually occupy space between the two nuclei and be stabilized by two nuclei rather than one. (c) The Cl—Cl distance at the energy minimum on the plot is the Cl—Cl bond length. (d) At interatomic separations shorter than the bond distance, the two nuclei begin to repel each other, increasing the overall energy of the system. (e) The y-coordinate of the minimum point on the plot is a good estimate of the Cl—Cl bond energy or bond strength.   **9.7** sp^3 hybrid orbitals
9.9 (a) i, Formed by two s atomic orbitals; ii, σ-type MO; iii, antibonding MO (b) i, formed by two p atomic orbitals overlapping end-to-end; ii, σ-type MO; iii, bonding MO (c) i, formed by two p atomic orbitals overlapping side-to-side; ii, π-type MO; iii, antibonding MO   **9.11** (a) Yes. The stated shape defines the bond angle and the bond length tells the size. (b) No. Atom A could have 2, 3 or 4 nonbonding electron pairs.
9.13 (a) An *electron domain* is a region in a molecule where electrons are most likely to be found. (b) Like the balloons in Figure 9.5, each electron domain occupies a finite volume of space, so they also adopt an arrangement where repulsions are minimized.   **9.15** (a) 2 (b) 1 (c) none (d) 3   **9.17** The electron-domain geometry indicated by VSEPR describes the arrangement of all bonding and nonbonding electron domains. The molecular geometry describes just the atomic positions. In H_2O there are 4 electron domains around oxygen, so the electron-domain geometry is tetrahedral. Because there are 2 bonding and 2 nonbonding domains, the molecular geometry is bent.
9.19 (a) Tetrahedral, tetrahedral (b) trigonal bipyramidal, T-shaped (c) octahedral, square pyramidal (d) octahedral, square planar   **9.21** (a) linear, linear (b) tetrahedral, trigonal pyramidal (c) trigonal bipyramidal, see-saw (d) octahedral, octahedral (e) tetrahedral, tetrahedral (f) linear, linear
9.23 (a) i, trigonal planar; ii, tetrahedral; iii, trigonal bipyramidal (b) i, 0; ii, 1; iii, 2 (c) N and P (d) Cl (or Br or I). This T-shaped molecular geometry arises from a trigonal bipyramidal electron-domain geometry with 2 nonbonding domains. Assuming each F atom has 3 nonbonding domains and forms only single bonds with A, A must have 7 valence electrons and be in or below the third row of the periodic table to produce these electron-domain and molecular geometries.   **9.25** (a) 1–109°, 2–109° (b) 3–109°, 4–109° (c) 5–180° (d) 6–120°, 7–109°, 8–109°   **9.27** $PF_5 > SF_4 > ClF_3$. As the number of nonbonding domains in the equatorial plane increases, they push back the axial A—F bonds, decreasing the F(axial)—A—F(equatorial) bond angles.   **9.29** (a) Although both ions have 4 bonding electron domains, the 6 total domains around Br require octahedral domain geometry and square-planar molecular geometry, while the 4 total domains about B lead to tetrahedral domain and molecular geometry. (b) The shape of the $(H_2O)^{4+}$ ion is linear.   **9.31** (a) Yes. The net dipole moment vector points along the Cl—S—Cl angle bisector. (b) No, $BeCl_2$ does not have a dipole moment.
9.33 (a) In Exercise 9.23, molecules (ii) and (iii) will have nonzero dipole moments. Molecule (i) has no nonbonding electron pairs on A, and the 3 A—F bond dipoles are oriented so that they cancel. Molecules (ii) and (iii) have nonbonding electron pairs on A and their bond dipoles do not cancel. (b) In Exercise 9.24, molecules (i) and (ii) have a zero dipole moment.
9.35 (a) IF (d) PCl_3 and (f) IF_5 are polar.

9.37 (a) Lewis structures

Molecular geometries

| Polar | Nonpolar | Polar |

(b) The middle isomer has a zero net dipole moment.
(c) C_2H_3Cl has only one isomer, and it has a dipole moment.
9.39 (a) *Orbital overlap* occurs when valence atomic orbitals on two adjacent atoms share the same region of space. (b) A chemical bond is a concentration of electron density between the nuclei of two atoms. This concentration can take place because orbitals on the two atoms overlap.   **9.41** (a) H—Mg—H, linear electron domain and molecular geometry (b) The linear electron domain geometry in MgH_2 requires sp hybridization.

(c)

9.43

Molecule	Molecular Structures	Electron Domain Geometry	Hybridization of Central Atom	Dipole Moment? Yes or No.
CO_2	O=C=O	Linear	sp	no
NH_3		Tetrahedral	sp^3	yes
CH_4		Tetrahedral	sp^3	no
BH_3		Trigonal planar	sp^2	no
SF_4		Trigonal bipyramidal	dsp^3	yes
SF_6		Octahedral	d^2sp^3	no
H_2CO		Trigonal planar	sp^2	yes
PF_5		Trigonal bipyramidal	dsp^3	no
XeF_2	F—Xe—F	Octahedral	d^2sp^3	no

9.45 (a) B, $[\text{He}]2s^2 2p^1$. One $2s$ electron is "promoted" to an empty $2p$ orbital. The $2s$ and two $2p$ orbitals that each contain one electron are hybridized to form three equivalent hybrid orbitals in a trigonal planar arrangement. (b) sp^2 (d) A single $2p$ orbital is unhybridized. It lies perpendicular to the trigonal plane of the sp^2 hybrid orbitals.   **9.47** (a) sp^2 (b) sp^3 (c) sp (d) sp^3d (e) sp^3d^2

9.49 (a)

(b)

σ

π

(c) A σ bond is generally stronger than a π bond because there is more extensive orbital overlap. (d) No. Overlap of two s orbitals results in electron density along the internuclear axis, while a π bond has none.   **9.51** (a)

H—C—C—H (with H above and below each C)   C=C (with H's)   H—C≡C—H

(b) sp^3, sp^2, sp (c) nonplanar, planar, planar (d) 7 σ, 0 π; 5 σ, 1 π; 3 σ, 2 π (e) The Si analogs would have the same hybridization as the C compounds given in part (b). That Si is in the row below C means it has a larger bonding atomic radius and atomic orbitals than C. The close approach of Si atoms required for π bond formation is unlikely; Si_2H_4 and Si_2H_2 probably do not exist under standard conditions.   **9.53** (a) 18 valence electrons (b) 16 valence electrons form σ bonds (c) 2 valence electrons form π bonds (d) no valence electrons are nonbonding (e) The left and central C atoms are sp^2 hybridized; the right C atom is sp^3 hybridized.   **9.55** (a) ~109° about the leftmost C, sp^3, ~120° about the right-hand C, sp^2 (b) The doubly bonded O can be viewed as sp^2, and the other as sp^3; the nitrogen is sp^3 with approximately 109° bond angles. (c) nine σ bonds, one π bond   **9.57** (a) In a localized π bond, the electron density is concentrated between the two atoms forming the bond. In a delocalized π bond, the electron density is spread over all the atoms that contribute p orbitals to the network. (b) The existence of more than one resonance form is a good indication that a molecule will have delocalized π bonding. (c) delocalized   **9.59** (a) Hybrid orbitals are mixtures of atomic orbitals from a single atom and remain localized on that atom. Molecular orbitals are combinations of atomic orbitals from two or more atoms and are delocalized over at least two atoms. (b) Each MO can hold a maximum of two electrons. (c) Antibonding molecular orbitals can have electrons in them.

9.61 (a)

σ_{1s}^*

$1s$ $1s$

σ_{1s}

H_2^+

(b) There is one electron in H_2^+. (c) $\sigma_{1s}^{\,1}$ (d) BO = $\frac{1}{2}$ (e) Fall apart. If the single electron in H_2^+ is excited to the σ_{1s}^* orbital, its energy is higher than the energy of an H $1s$ atomic orbital and H_2^+ will decompose into a hydrogen atom and a hydrogen ion.

9.63

(a) 1 σ bond (b) 2 π bonds (c) 1 σ^* and 2 π^* **9.65** (a) When comparing the same two bonded atoms, bond order and bond energy are directly related, while bond order and bond length

are inversely related. When comparing different bonded nuclei, there are no simple relationships. (b) Be_2 is not expected to exist; it has a bond order of zero and is not energetically favored over isolated Be atoms. Be_2^+ has a bond order of 0.5 and is slightly lower in energy than isolated Be atoms. It will probably exist under special experimental conditions.   **9.67** (a, b) Substances with no unpaired electrons are weakly repelled by a magnetic field. This property is called *diamagnetism*. (c) O_2^{2-}, Be_2^{2+} **9.69** (a) B_2^+, $\sigma_{2s}^2 \sigma_{2s}^{*2} \pi_{2p}^1$, increase (b) Li_2^+, $\sigma_{1s}^2 \sigma_{1s}^{*2} \sigma_{2s}^1$, increase (c) N_2^+, $\sigma_{2s}^2 \sigma_{2s}^{*2} \pi_{2p}^4 \sigma_{2p}^1$, increase (d) Ne_2^{2+}, $\sigma_{2s}^2 \sigma_{2s}^{*2} \sigma_{2p}^2 \pi_{2p}^4 \pi_{2p}^{*4}$, decrease   **9.71** CN, $\sigma_{2s}^2 \sigma_{2s}^{*2} \sigma_{2p}^2 \pi_{2p}^3$, bond order = 2.5; CN^+, $\sigma_{2s}^2 \sigma_{2s}^{*2} \sigma_{2p}^2 \pi_{2p}^2$, bond order = 2.0; CN^-, $\sigma_{2s}^2 \sigma_{2s}^{*2} \sigma_{2p}^2 \pi_{2p}^4$, bond order = 3.0. (a) CN^- (b) CN, CN^+ **9.73** (a) $3s$, $3p_x$, $3p_y$, $3p_z$ (b) π_{3p} (c) 2 (d) If the MO diagram for P_2 is similar to that of N_2, P_2 will have no unpaired electrons and be diamagnetic.   **9.76** SiF_4 is tetrahedral, SF_4 is seesaw, XeF_4 is square planar. The shapes are different because the number of nonbonding electron domains is different in each molecule, even though all have four bonding electron domains. Bond angles and thus molecular shape are determined by the total number of electron domains.   **9.79** (a) 2 σ bonds, 2 π bonds (b) 2 σ bonds, 2 π bonds (c) 3 σ bonds, 1 π bond (d) 4 σ bonds, 1 π bond   **9.81** BF_3 is trigonal planar, the B—F bond dipoles cancel and the molecule is nonpolar. PF_3 has a tetrahedral electron domain geometry with one position occupied by a nonbonding electron pair. The nonbonding electron pair ensures an asymmetrical electron distribution and the molecule is polar.

9.85

H   H

H   H

(a) The molecule is nonplanar. (b) Allene has no dipole moment. (c) The bonding in allene would not be described as delocalized. The π electron clouds of the two adjacent C=C are mutually perpendicular, so there is no overlap and no delocalization of π electrons.   **9.87** (a) All O atoms have sp^2 hybridization. (b) The two σ bonds are formed by overlap of sp^2 hybrid orbitals, the π bond is formed by overlap of atomic p orbitals, one nonbonded pair is in a p atomic orbital and the other five nonbonded pairs are in sp^2 hybrid orbitals. (c) unhybridized p atomic orbitals (d) four, two from the π bond and two from the nonbonded pair in the p atomic orbital   **9.91** $\sigma_{2s}^2 \sigma_{2s}^{*2} \pi_{2p}^4 \sigma_{2p}^1 \pi_{2p}^{*1}$ (a) Paramagnetic (b) The bond order of N_2 in the ground state is 3; in the first excited state it has a bond order of 2. Owing to the reduction in bond order, N_2 in the first excited state has a weaker N—N bond.   **9.97** (a) $2\,\text{SF}_4(g) + \text{O}_2(g) \longrightarrow 2\,\text{OSF}_4(g)$

(b)

:O:
‖
:F—S—F:
⁄   ⧵
:F: :F:

(c) $\Delta H = -551$ kJ, exothermic (d) The electron domain geometry is trigonal bipyramidal. The O atom can be either equatorial or axial. (e) Since F is more electronegative than O, the structure that minimizes 90° F—S—F angles, the one with O axial, is preferred.   **9.99** (a) sp^2 hybridization at the two central C atoms (b) Isomerization requires 180° rotation around the C=C double bond. A 90° rotation eliminates all overlap of the p orbitals that form the π bond and it is broken. (c) 4.42×10^{-19} J/molecule (d) $\lambda = 450$ nm (e) Yes, 450 nm light is in the visible portion of the spectrum. A cis-trans isomerization in the retinal portion of the large molecule rhodopsin is the first step in a sequence of molecular transformations in the eye that leads to vision. The sequence of events enables the eye to detect visible photons, in other words, to see.   **9.101** From bond enthalpies, $\Delta H = 5364$ kJ;

according to Hess's law, $\Delta H° = 5535$ kJ. The difference in the two results, 171 kJ, is due to the resonance stabilization in benzene. The amount of energy actually required to decompose 1 mol of $C_6H_6(g)$ is greater than the sum of the localized bond enthalpies.   **9.103** (a) $3d_{z^2}$
(b)

The "donuts" of the $3d_{z^2}$ orbitals have been omitted from the diagram for clarity. (c) A node is generated in σ^*_{3d} because antibonding MOs are formed when atomic orbital lobes with opposite phases interact. Electron density is excluded from the internuclear region and a node is formed in the MO.
(d)

(e) The bond order of Sc_2 is 1.0.

Chapter 10

10.2

(a) $V_2 = \frac{5}{3} V_1$ (b) $V_2 = \frac{1}{2} V_1$

300 K, V_1 500 K, V_2 1 atm, V_1 2 atm, V_2

10.5 Over time, the gases will mix perfectly. Each bulb will contain 4 blue and 3 red atoms. The "blue" gas has the greater partial pressure after mixing, because it has the greater number of particles at the same T and V as the "red" gas.   **10.8** (a) Curve A is $O_2(g)$ and curve B is $He(g)$. (b) For the same gas at different temperatures, curve A represents the lower temperature and curve B the higher temperature.   **10.11** (a) A gas is much less dense than a liquid. (b) A gas is much more compressible than a liquid. (c) All mixtures of gases are homogenous. Similar liquid molecules form homogeneous mixtures, while very dissimilar molecules form heterogeneous mixtures.   **10.13** 1.8×10^3 kPa   **10.15** (a) 10.3 m (b) 2.1 atm   **10.17** (a) The tube can have any cross-sectional area. (b) At equilibrium the force of gravity per unit area acting on the mercury column at the level of the outside mercury is not equal to the force of gravity per unit area acting on the atmosphere. (c) The column of mercury is held up by the pressure of the atmosphere applied to the exterior pool of mercury.   **10.19** (a) 0.349 atm (b) 265 mm Hg (c) 3.53×10^4 Pa (d) 0.353 bar   **10.21** (a) $P = 773.4$ torr (b) The pressure in Chicago is greater than standard atmospheric pressure, and so it makes sense to classify this weather system as a "high-pressure system."

10.23 (i) 0.30 atm (ii) 1.073 atm (iii) 0.136 atm   **10.23** (a) If V decreases by a factor of 4, P increases by a factor of 4. (b) If T decreases by a factor of 2, P decreases by a factor of 2. (c) If n decreases by a factor of 2, P decreases by a factor of 2. **10.27** (a) If equal volumes of gases at the same temperature and pressure contain equal numbers of molecules and molecules react in the ratios of small whole numbers, it follows that the volumes of reacting gases are in the ratios of small whole numbers. (b) Since the two gases are at the same temperature and pressure, the ratio of the numbers of atoms is the same as the ratio of volumes. There are 1.5 times as many Xe atoms as Ne atoms.   **10.29** (a) $PV = nRT$; P in atmospheres, V in liters, n in moles, T in kelvins. (b) An ideal gas exhibits pressure, volume, and temperature relationships described by the equation $PV = nRT$.   **10.31** Flask A contains the gas with $\mathcal{M} = 30$ g/mol, and flask B contains the gas with $\mathcal{M} = 60$ g/mol.

10.33

P	V	n	T
2.00 atm	1.00 L	0.500 mol	48.7 K
0.300 atm	0.250 L	3.05×10^{-3} mol	27 °C
650 torr	11.2 L	0.333 mol	350 K
10.3 atm	585 mL	0.250 mol	295 K

10.35 8.2×10^2 kg He   **10.37** 5.15×10^{22} molecules **10.39** (a) 91 atm (b) 2.3×10^2 L   **10.41** (a) 29.8 g Cl_2 (b) 9.42 L (c) 501 K (d) 1.90 atm   **10.43** (a) $n = 2 \times 10^{-4}$ mol O_2 (b) The roach needs 8×10^{-3} mol O_2 in 48 h, more than 100% of the O_2 in the jar.   **10.45** For gas samples at the same conditions, molar mass determines density. Of the three gases listed, (c) Cl_2 has the largest molar mass.   **10.47** (c) Because the helium atoms are of lower mass than the average air molecule, the helium gas is less dense than air. The balloon thus weighs less than the air displaced by its volume.   **10.49** (a) $d = 1.77$ g/L (b) $\mathcal{M} = 80.1$ g/mol   **10.51** $\mathcal{M} = 89.4$ g/mol **10.53** 3.5×10^{-9} g Mg   **10.55** 21.4 L CO_2 **10.57** 0.402 g Zn **10.59** (a) When the stopcock is opened, the volume occupied by $N_2(g)$ increases from 2.0 L to 5.0 L. $P_{N_2} = 0.40$ atm (b) When the gases mix, the volume of $O_2(g)$ increases from 3.0 L to 5.0 L. $P_{O_2} = 1.2$ atm (c) $P_t = 1.6$ atm   **10.61** (a) $P_{He} = 1.67$ atm, $P_{Ne} = 0.978$ atm, $P_{Ar} = 0.384$ atm, (b) $P_t = 3.03$ atm **10.63** $P_{CO_2} = 0.305$ atm, $P_t = 1.232$ atm   **10.65** $P_{N_2} = 0.98$ atm, $P_{O_2} = 0.39$ atm, $P_{CO_2} = 0.20$ atm   **10.67** 2.5 mole % O_2 **10.69** $P_t = 3.04$ atm   **10.71** (a) Increase in temperature at constant volume or decrease in volume or increase in pressure (b) decrease in temperature (c) increase in volume, decrease in pressure (d) increase in temperature   **10.73** The fact that gases are readily compressible supports the assumption that most of the volume of a gas sample is empty space.   **10.75** (a) Average kinetic energy of the molecules increases. (b) Average speed of the molecules increases. (c) Strength of an average impact with the container walls increases. (d) Total collisions of molecules with walls per second increases. **10.77** (a) In order of increasing speed and decreasing molar mass: HBr < NF_3 < SO_2 < CO < Ne (b) $u_{NF_3} = 324$ m/s **10.79** The order of increasing rate of effusion is $^2H^{37}Cl < {^1H^{37}Cl} < {^2H^{35}Cl} < {^1H^{35}Cl}$.   **10.81** As_4S_6 **10.83** (a) Non-ideal-gas behavior is observed at very high pressures and low temperatures. (b) The real volumes of gas molecules and attractive intermolecular forces between molecules cause gases to behave nonideally. (c) According to the ideal-gas law, the ratio PV/RT should be constant for a given gas sample at all combinations of pressure, volume, and temperature. If this ratio changes with increasing pressure, the gas sample is not behaving ideally.   **10.85** Ar ($a = 1.34, b = 0.0322$) will behave more like an ideal gas than CO_2 ($a = 3.59, b = 0.427$) at high pressures.   **10.87** (a) $P = 4.89$ atm (b) $P = 4.69$ atm

(c) Qualitatively, molecular attractions are more important as the amount of free space decreases and the number of molecular collisions increases. Molecular volume is a larger part of the total volume as the container volume decreases.   **10.89** A mercury barometer with water trapped in its tip will not read the correct pressure. Water at the top of the Hg column establishes a vapor pressure that exerts downward pressure in addition to gravity and partially counterbalances the pressure of the atmosphere.   **10.92** 742 balloons can be filled completely, with a bit of He left over.   **10.95** (a) 13.4 mol $C_3H_8(g)$ (b) 1.47×10^3 mol $C_3H_8(l)$ (c) The ratio of moles liquid to moles gas is 110. Many more molecules and moles of liquid fit in a container of fixed volume because there is much less space between molecules in the liquid phase.   **10.98** $P_t = 5.3 \times 10^2$ torr **10.101** 42.2 g O_2 **10.104** $X_{Ne} = 0.6482$ **10.107** (a) As a gas is compressed at constant temperature, the number of intermolecular collisions increases. Intermolecular attraction causes some of these collisions to be inelastic, which amplifies the deviation from ideal behavior. (b) As the temperature of a gas is increased at constant volume, a larger fraction of the molecules has sufficient kinetic energy to overcome intermolecular attractions and the effect of intermolecular attraction becomes less significant.   **10.110** (a) 0.036% of the total volume is occupied by Ar atoms. (b) 3.6% of the total volume is occupied by Ar atoms.   **10.112** (a) The molecular formula is C_3H_6. (b) Ar and C_3H_6 are nonpolar, have similar molar masses, and experience London disperson forces. We expect the effect of both volume and intermolecular attractions to be more significant for the structurally more complex C_3H_6. Cyclopropane will deviate more from ideal behavior at the specified conditions. **10.114** (a) 44.58% C, 6.596% H, 16.44% Cl, 32.38% N (b) $C_8H_{14}N_5Cl$ **10.117** $\Delta H = -1.1 \times 10^{14}$ kJ (assuming $H_2O(l)$ is a product)   **10.120** (a) 5.02×10^8 L $CH_3OH(l)$ (b) $CH_4(g) + 2 O_2(g) \longrightarrow CO_2(g) + 2 H_2O(l)$, $\Delta H° = -890.4$ kJ; ΔH for combustion of the methane is -1.10×10^{13} kJ. $CH_3OH(l) + 3/2 O_2(g) \longrightarrow CO_2(g) + 2 H_2O(l)$, $\Delta H° = -726.6$ kJ; ΔH for combustion of the methanol is -9.00×10^{12} kJ.

Chapter 11

11.1 The diagram best describes a liquid. The particles are close together, mostly touching, but there is no regular arrangement or order. This rules out a gaseous sample, where the particles are far apart, and a crystalline solid, which has a regular repeating structure in all three directions.   **11.4** (a) 385 mm Hg (b) 22 °C (c) 47 °C   **11.7** (a) 3 Nb atoms, 3 O atoms (b) NbO (c) This is primarily an ionic solid, because Nb is a metal and O is a nonmetal. There may be some covalent character to the Nb $\cdots$ O bonds.   **11.9** (a) Solid < liquid < gas (b) gas < liquid < solid (c) Matter in the gaseous state is most easily compressed, because particles are far apart and there is much empty space.   **11.11** (a) The density of olive oil is less than 1.00 g/cm^3. (b) At higher temperature, greater molecular motion and collisions cause the volume of the liquid to increase and the density to decrease. Olive oil is less dense at higher temperature.   **11.13** (a) London dispersion forces (b) dipole–dipole and London dispersion forces (c) dipole–dipole forces and in certain cases hydrogen bonding **11.15** (a) Nonpolar covalent molecule; London dispersion forces only (b) polar covalent molecule with O—H bonds; hydrogen bonding, dipole–dipole forces and London dispersion forces (c) polar covalent molecule; dipole–dipole and London dispersion forces (but not hydrogen bonding) **11.17** (a) Polarizability is the ease with which the charge distribution in a molecule can be distorted to produce a transient dipole. (b) Sb is most polarizable because its valence electrons are farthest from the nucleus and least tightly held. (c) in order of increasing polarizability: $CH_4 < SiH_4 < SiCl_4 < GeCl_4 < GeBr_4$ (d) The magnitudes of London dispersion forces and thus the boiling points of molecules increase as polarizability increases.

The order of increasing boiling points is the order of increasing polarizability given in (c).   **11.19** (a) H_2S (b) CO_2 (c) GeH_4 **11.21** Both rodlike butane molecules and spherical 2-methylpropane molecules experience dispersion forces. The larger contact surface between butane molecules facilitates stronger forces and produces a higher boiling point.   **11.23** (a) A molecule must contain H atoms, bound to either N, O or F atoms, in order to participate in hydrogen bonding with like molecules. (b) CH_3NH_2 and CH_3OH.   **11.25** (a) Replacing a hydroxyl hydrogen with a CH_3 group eliminates hydrogen bonding in that part of the molecule. This reduces the strength of intermolecular forces and leads to a lower boiling point. (b) $CH_3OCH_2CH_2OCH_3$ is a larger, more polarizable molecule with stronger London dispersion forces and thus a higher boiling point. **11.27**

Physical Property	H_2O	H_2S
Normal boiling point, °C	100.00	−60.7
Normal melting point, °C	0.00	−85.5

(a) Based on its much higher normal melting point and boiling point, H_2O has much stronger intermolcular forces. H_2O has hydrogen bonding, while H_2S has dipole–dipole forces. (b) H_2S is probably a typical compound with less empty space in the ordered solid than the liquid, so that the solid is denser than the liquid. For H_2O, maximizing the number of hydrogen bonds to each molecule in the solid requires more empty space than in the liquid, and the solid is less dense. (c) Specific heat is the energy required to raise the temperature of one gram of the substance one degree celsius. Hydrogen bonding in water is such a strong attractive interaction that the energy required to disrupt it and increase molecular motion is large.   **11.29** (a) As temperature increases, the number of molecules with sufficient kinetic energy to overcome intermolecular attractive forces increases, and viscosity and surface tension decrease. (b) The same attractive forces that cause surface molecules to be difficult to separate (high surface tension) cause molecules elsewhere in the sample to resist movement relative to each other (high viscosity).   **11.31** (a) $CHBr_3$ has a higher molar mass, is more polarizable, and has stronger dispersion forces, so the surface tension is greater. (b) As temperature increases, the viscosity of the oil decreases because the average kinetic energy of the molecules increases. (c) Adhesive forces between polar water and nonpolar car wax are weak, so the large surface tension of water draws the liquid into the shape with the smallest surface area, a sphere. (d) Adhesive forces between nonpolar oil and nonpolar car wax are similar to cohesive forces in oil, so the oil drops spread out on the waxed car hood.   **11.33** (a) Melting, endothermic (b) evaporation, endothermic (c) deposition, exothermic (d) condensation, exothermic   **11.35** Melting does not require separation of molecules, so the energy requirement is smaller than for vaporization, where molecules must be separated.   **11.37** 2.3×10^3 g H_2O **11.39** (a) 24.0 kJ (b) 5.57×10^3 kJ   **11.41** (a) The critical pressure is the pressure required to cause liquefaction at the critical temperature. (b) As the force of attraction between molecules increases, the critical temperature of the compound increases. (c) All the gases in Table 11.5 can be liquefied at the temperature of liquid nitrogen, given sufficient pressure.   **11.43** (a) No effect (b) no effect (c) Vapor pressure decreases with increasing intermolecular attractive forces because fewer molecules have sufficient kinetic energy to overcome attractive forces and escape to the vapor phase. (d) Vapor pressure increases with increasing temperature because average kinetic energies of molecules increase. (e) Vapor pressure decreases with increasing density because attractive intermolecular forces increase.   **11.45** (a) $CBr_4 < CHBr_3 < CH_2Br_2 < CH_2Cl_2 < CH_3Cl < CH_4$. The trend is dominated by dispersion forces even though four of the molecules are polar. The order of increasing volatility is the order of increasing vapor pressure, decreasing molar mass, and decreasing strength of dispersion forces. (b) Boiling point increases as the strength of intermolecular forces increases; this is the order of decreasing volatility and the reverse of the order in part (a). $CH_4 < CH_3Cl < CH_2Cl_2 < CH_2Br_2 < CHBr_3 < CBr_4$ **11.47** (a) The temperature of the water in the two pans is the same. (b) Vapor pressure does not depend on either volume or surface area of the liquid. At the same temperature, the vapor pressures of water in the two containers are the same. **11.49** (a) Approximately 48 °C (b) approximately 340 torr (c) approximately 16 °C (d) approximately 1000 torr **11.51** (a) The critical point is the temperature and pressure beyond which the gas and liquid phases are indistinguishable. (b) The line that separates the gas and liquid phases ends at the critical point because at conditions beyond the critical temperature and pressure, there is no distinction between gas and liquid. In experimental terms a gas cannot be liquefied at temperatures higher than the critical temperature, regardless of pressure. **11.53** (a) $H_2O(g)$ will condense to $H_2O(s)$ at approximately 4 torr; at a higher pressure, perhaps 5 atm or so, $H_2O(s)$ will melt to form $H_2O(l)$. (b) At 100 °C and 0.50 atm, water is in the vapor phase. As it cools, water vapor condenses to the liquid at approximately 82 °C, the temperature where the vapor pressure of liquid water is 0.50 atm. Further cooling results in freezing at approximately 0 °C. The freezing point of water increases with decreasing pressure, so at 0.50 atm the freezing temperature is very slightly above 0 °C.

11.55

11.57

Crystalline   Amorphous

11.59 (a) $SrTiO_3$ (b) Each Sr atom is coordinated to six O atoms, three in this unit cell and three in adjacent unit cells. **11.61** (a) $r = 1.355$ Å (b) density $= 22.67$ g/cm³ **11.63** Atomic weight $= 55.8$ g/mol **11.65** (a) $a = 4.70$ Å (b) 2.69 g/cm³ **11.67** $a = 6.13$ Å **11.69** (a) Intermolecular (really interparticle) forces among Ar atoms are dispersion forces. (b) Solid Ar is not a covalent network solid. Atoms in a covalent network solid are joined by strong covalent bonds, whereas atoms in Ar(s) are held in place by weak dispersion forces. **11.71** (a) Hydrogen bonding, dipole-dipole forces, London dispersion forces (b) covalent chemical bonds (c) ionic bonds (d) metallic bonds **11.73** In molecular solids relatively weak intermolecular forces bind the molecules in the lattice, so relatively little energy is required to disrupt these forces. In covalent-network solids, covalent bonds join atoms into an extended network. Melting or deforming a covalent-network solid means breaking covalent bonds, which requires a large amount of energy. **11.75** Because of its relatively high melting point and properties as a conducting solution, the solid must be ionic. **11.77** (a) Xe, greater atomic weight,

stronger dispersion forces (b) SiO_2, covalent-network lattice versus weak dispersion forces (c) KBr, strong ionic versus weak dispersion forces (d) C_6Cl_6, both are influenced by dispersion forces, C_6Cl_6 has the higher molar mass **11.79** (a) Decrease (b) increase (c) increase (d) increase (e) increase (f) increase (g) increase **11.83** When a halogen is substituted for H in benzene, molar mass, polarizability and strength of dispersion forces increase; the order of increasing molar mass is the order of increasing boiling points for the first three compounds. C_6H_5OH experiences hydrogen bonding, the strongest force between neutral molecules, so it has the highest boiling point. **11.88** (a) Evaporation is an endothermic process. The heat required to vaporize sweat is absorbed from your body, helping to keep it cool. (b) The vacuum pump reduces the pressure of the atmosphere above the water until atmospheric pressure equals the vapor pressure of water and the water boils. Boiling is an endothermic process, and the temperature drops if the system is not able to absorb heat from the surroundings fast enough. As the temperature of the water decreases, the water freezes. **11.92** The large difference in melting points is due to the very different forces imposing atomic order in the solid state. Much more kinetic energy is required to disrupt the delocalized metallic bonding in gold than to overcome the relatively weak London dispersion forces in Xe. **11.95** Diffraction, the phenomenon that enables us to measure interatomic distances in crystals, is most efficient when the wavelength of light is similar to or smaller than the size of the object doing the diffracting. Atom sizes are on the order of 1–10 Å, and the wavelengths of x-rays are also in this range. Visible light, 400–700 nm or 4000–7000 Å, is too long to be diffracted effectively by atoms (electrons) in crystals. **11.98** 16 Al atoms, 8 Mg atoms, 32 O atoms **11.101**

(i) $\mathcal{M} = 44$ (ii) $\mathcal{M} = 72$ (iii) $\mathcal{M} = 123$

(iv) $\mathcal{M} = 58$ (v) $\mathcal{M} = 123$ (vi) $\mathcal{M} = 60$

(a) Molar mass: Compounds (i) and (ii) have similar rodlike structures. The longer chain in (ii) leads to greater molar mass, stronger London-dispersion forces, and higher heat of vaporization. (b) Molecular shape: Compounds (iii) and (v) have the same chemical formula and molar mass but different molecular shapes. The more rodlike shape of (v) leads to more contact between molecules, stronger dispersion forces, and higher heat of vaporization. (c) Molecular polarity: Compound (iv) has a smaller molar mass than (ii) but a larger heat of vaporization, which must be due to the presence of dipole-dipole forces. (d) Hydrogen bonding interactions: Molecules (v) and (vi) have similar structures. Even though (v) has larger molar mass and dispersion forces, hydrogen bonding causes (vi) to have the higher heat of vaporization.
11.105 P(benzene vapor) $= 98.6$ torr.

Chapter 12

12.1 The band structure of material A is that of a metal. **12.3** Polymer (a), with ordered regions, is denser than branched polymer (b). The ordered regions in polymer (a) also indicate that it has stronger intermolecular forces and higher melting point than polymer (b). **12.5** (a) Molecule (i), with a terminal C=C, is the only monomer shown that is capable of addition polymerization. (b) Molecule (iii) contains both a carboxyl group and an amine group; it can "condense" with like monomers to form a polymer and NH_3. (c) Molecule (ii), with a —CN group, a rodlike 5 C chain and a planar benzenelike (phenyl) ring, has both dipole–dipole and dispersion forces that are likely to encourage the long range order required to form a liquid crystal   **12.7** (a) Semiconductor (b) insulator (c) semiconductor (d) metal

12.9

(a) Six AOs require six MOs (b) Zero nodes in the lowest energy orbital (c) Five nodes in highest energy orbital (d) Two nodes in the HOMO (e) Three nodes in the LUMO.   **12.11** (a) False. Semiconductors have a smaller band gap; they conduct some electricity, while insulators do not. (b) True. Dopants create either more electrons or more "holes," both of which increase the conductivity of the semiconductor. (c) True. Metals conduct electricity because delocalized electrons in the lattice provide a mechanism for charge mobility. (d) True. Metal oxides are ionic substances with essentially localized electrons. This barrier to charge mobility renders them insulators.
12.13 (a) CdS (b) GaN (c) GaAs   **12.15** Ge or Si (Ge is closer to Ga in bonding atomic radius.)   **12.17** As a semiconductor material for integrated circuits, silicon has the advantages that it is abundant, cheap, nontoxic; can be highly purified; and grows nearly perfect enormous crystals.   **12.19** Traditionally, the insulator silicon dioxide, SiO_2, is the material in a MOSFET gate. The next-generation material is HfO_2.   **12.21** (a) A 1.1 eV photon corresponds to a wavelength of 1.1×10^{-6} m. (b) According to the figure, Si can absorb a portion of the visible light that comes from the sun.   **12.23** $\lambda = 560$ nm   **12.25** The band gap is approximately 1.85 eV, which corresponds to a wavelength of 672 nm.   **12.27** Ceramics are not readily recyclable because of their extremely high melting points and rigid ionic or covalent-network structures.   **12.29** Very small, uniformly sized and shaped particles are required for the production of a strong ceramic object by sintering. Upon heating to initiate condensation reactions, the more uniform the particle size and the greater the total surface area of the solid, the more chemical bonds are formed and the stronger the ceramic object.
12.31 The ceramics are MgO, soda-lime glass, ZrB_2, Al_2O_3, and TaC. The criteria are a combination of chemical formula (with corresponding bonding characteristics) and Knoop values. Ceramics are ionic or covalent-network solids with fairly large hardness values. Hardness alone is not a sufficient criterion for classification as a ceramic. A metal, Cr, lies in the middle of the hardness range for ceramics. Bonding characteristics as well as hardness must be taken into account when classifying materials as ceramics.   **12.33** A superconducting material offers no resistance to the flow of electrical current. Superconductive materials could transmit electricity with much greater efficiency than current carriers.   **12.35** The sharp drop in resistivity of MgB_2 near 39 K is the superconducting transition temperature,

T_c.   **12.37** The phenomenon that superconductors exclude all magnetic fields from their volume is the Meisner effect. It can be used to levitate trains by having either the tracks or the train wheels made from a magnetic material, and the other from a superconductor, with the superconductor cooled below its transition temperature. It is more practical, and therefore more likely, that the train wheels are made of the superconductor and cooled.   **12.39** Monomers are small molecules with low molecular mass that are joined together to form polymers. Three monomers mentioned in this chapter are

Propylene (propene)   Styrene (phenyl ethene)   Isoprene (2-methyl-1,3-butadiene)

12.41

If a dicarboxylic acid and a dialcohol are combined, there is the potential for propagation of the polymer chain at both ends of both monomers.

12.43 (a)

(b)

(c)

12.45

12.47 Flexibility of molecular chains causes flexibility of the bulk polymer. Flexibility is enhanced by molecular features that inhibit order, such as branching, and diminished by features that encourage order, such as cross-linking or delocalized π electron density. Cross-linking, the formation of chemical bonds between polymer chains, reduces flexibility of the molecular chains, increases the hardness of the material, and decreases the chemical reactivity of the polymer.   **12.49** The function of the polymer determines whether high molecular mass and high degree of crystallinity are desirable properties. If the polymer will be used as a flexible wrapping or fiber, rigidity that is due to high molecular mass is an undesirable property.   **12.51** Is the neoprene biocompatible? Does it provoke inflammatory reactions? Does neoprene meet the physical requirements of a flexible lead? Will it remain resistant to degradation and maintain elasticity? Can neoprene be prepared in sufficiently pure form so that it can be classified as medical grade?   **12.53** Current vascular graft materials cannot be lined with cells similar to those in the native artery. The body detects that the graft is "foreign," and platelets attach to the inside surfaces, causing blood clots. The inside surfaces of future vascular implants need to accommodate a lining of cells that do not attract or attach to platelets.   **12.55** In order for skin cells in a culture medium to develop into synthetic skin, a mechanical matrix that holds the cells in contact with one another must be present. The matrix must be strong, biocompatible, and biodegradable. It probably has polar functional groups that form hydrogen bonds with biomolecules in the tissue cells. **12.57** Both an ordinary liquid and a nematic liquid crystal phase are fluids; they are converted directly to the solid phase upon cooling. The nematic phase is cloudy and more viscous than an ordinary liquid. Upon heating, the nematic phase is converted to an ordinary liquid.   **12.59** In the solid state the relative orientation of molecules is fixed and repeating in all three dimensions. When a substance changes to the nematic liquid crystalline phase, the molecules remain aligned in one dimension; translational motion is allowed, but rotational motion is restricted. Transformation to the isotropic liquid phase destroys the one-dimensional order, resulting in free translational and rotational motion.   **12.61** The presence of polar groups or nonbonded electron pairs leads to relatively strong dipole–dipole interactions between molecules. These are a significant part of the orienting forces necessary for liquid crystal formation.   **12.63** In the nematic phase there is one-dimensional order, while in a smectic phase there is two-dimensional order. In a smectic phase the long directions of the molecules and the ends of the molecules are aligned.   **12.65** A nematic phase is composed of sheets of molecules aligned along their lengths, with no additional order within the sheet or between sheets. A cholesteric phase also contains this kind of sheet, but with some ordering between sheets.   **12.67** If a solid has nanoscale dimensions of 1–10 nm, there may not be enough atoms contributing atomic orbitals to produce continuous energy bands of molecular orbitals.   **12.69** (a) False. As particle size decreases, the band gap increases. (b) False. As particle size decreases, wavelength decreases.   **12.71** 2.47×10^5 Au atoms **12.73** Semiconductors have a difference in energy, the band gap, between a filled valence band and an empty conduction bond. When a semiconductor is heated, more electrons have sufficient energy to jump the band gap, and conductivity increases. Metals have a partially-filled continuous energy band. Heating a metal increases the average kinetic energy, including vibrational energy, of the metal atoms. The greater vibrational energy of the atoms leads to imperfections in the lattice and discontinuities in the energy band. This creates barriers to electron delocalization and reduces the conductivity of the metal.

12.75

Teflon™ is formed by addition polymerization.   **12.79** At the temperature where a substance changes from the solid to the liquid crystalline phase, kinetic energy sufficient to overcome

most of the long-range order in the solid has been supplied. A relatively small increase in temperature is required to overcome the remaining aligning forces and produce an isotropic liquid. **12.83** $TiCl_4(g) + 2\,SiH_4(g) \longrightarrow TiSi_2(s) + 4\,HCl(g) + 2\,H_2(g)$. As a ceramic, $TiSi_2$ will have a three-dimensional network structure similar to that of Si. At the surface of the thin film, there will be Ti and Si atoms with incomplete valences that will chemically bond with Si atoms on the surface of the substrate. This kind of bonding would not be possible with a Cu thin film.   **12.84** (a) $\Delta H = -82$ kJ/mol (b) $\Delta H = -14$ kJ/mol (of either reactant) (c) $\Delta H = 0$ kJ   **12.87** (a) $x = 0.22$ (b) Hg and Cu both have more than one stable oxidation state. If different ions in the solid lattice have different charges, the average charge is a noninteger value. Ca and Ba are stable only in the +2 oxidation state and are unlikely to have a noninteger average charge. (c) Ba^{2+} is largest; Cu^{2+} is smallest.

Chapter 13

13.1 The ion-solvent interaction should be greater for Li^+. The smaller ionic radius of Li^+ means that the ion–dipole interactions with polar water molecules are stronger.   **13.5** Diagram (b) is the best representation of a saturated solution. There is some undissolved solid with particles that are close together and ordered, in contact with a solution containing mobile, separated solute particles. As much solute has dissolved as can dissolve, leaving some undissolved solid in contact with the saturated solution.   **13.9** (a) Yes, the *molarity* changes with a change in temperature. Molarity is defined as moles solute per unit volume of solution. A change of temperature changes solution volume and molarity. (b) No, *molality* does not change with change in temperature. Molality is defined as moles solute per kilogram of solvent. Temperature affects neither mass nor moles.   **13.13** If the magnitude of ΔH_3 is small relative to the magnitude of ΔH_1, ΔH_{soln} will be large and endothermic (energetically unfavorable) and not much solute will dissolve.   **13.15** (a) Dispersion (b) hydrogen bonding (c) ion–dipole (d) dipole–dipole   **13.17** (a) ΔH_1 (b) ΔH_3 **13.19** (a) Since the solute and solvent experience very similar London dispersion forces, the energy required to separate them individually and the energy released when they are mixed are approximately equal. $\Delta H_1 + \Delta H_2 \approx -\Delta H_3$. Thus, ΔH_{soln} is nearly zero. (b) Since no strong intermolecular forces prevent the molecules from mixing, they do so spontaneously because of the increase in randomness.   **13.21** (a) Supersaturated (b) Add a seed crystal. A seed crystal provides a nucleus of prealigned molecules, so that ordering of the dissolved particles (crystallization) is more facile.   **13.23** (a) Unsaturated (b) saturated (c) unsaturated (d) unsaturated **13.25** The liquids water and glycerol form homogenous mixtures (solutions) regardless of the relative amounts of the two components. The —OH groups of glycerol facilitate strong hydrogen bonding similar to that in water; like dissolves like. **13.27** (a) Dispersion interactions among nonpolar $CH_3(CH_2)_{16}$— chains dominate the properties of stearic acid, causing it to be more soluble in nonpolar CCl_4. (b) Dioxane can act as a hydrogen bond acceptor, so it will be more soluble than cyclohexane in water.   **13.29** (a) CCl_4 is more soluble because dispersion forces among nonpolar CCl_4 molecules are similar to dispersion forces in hexane. (b) C_6H_6 is a nonpolar hydrocarbon and will be more soluble in the similarly nonpolar hexane. (c) The long, rodlike hydrocarbon chain of octanoic acid forms strong dispersion interactions and causes it to be more soluble in hexane.   **13.31** (a) A sealed container is required to maintain a partial pressure of $CO_2(g)$ greater than 1 atm above the beverage. (b) Since the solubility of gases increases with decreasing temperature, some $CO_2(g)$ will remain dissolved in the beverage if it is kept cool.   **13.33** $S_{He} = 5.6 \times 10^{-4}\,M$, $S_{N_2} = 9.0 \times 10^{-4}\,M$ **13.35** (a) 2.15% Na_2SO_4

by mass (b) 3.15 ppm Ag   **13.37** (a) X_{CH_3OH} = 0.0427 (b) 7.35% CH_3OH by mass (c) 2.48 m CH_3OH **13.39** (a) 1.46 × 10^{-2} M $Mg(NO_3)_2$ (b) 1.12 M $LiClO_4 \cdot 3\ H_2O$ (c) 0.350 M HNO_3
13.41 (a) 4.70 m C_6H_6 (b) 0.235 m NaCl   **13.43** (a) 43.01% H_2SO_4 by mass (b) $X_{H_2SO_4}$ = 0.122 (c) 7.69 m H_2SO_4 (d) 5.827 M H_2SO_4
13.45 (a) X_{CH_3OH} = 0.227 (b) 7.16 m CH_3OH (c) 4.58 M CH_3OH
13.47 (a) 0.150 mol $SrBr_2$ (b) 1.56 × 10^{-2} mol KCl
(c) 4.44 × 10^{-2} mol $C_6H_{12}O_6$ **13.49** (a) Weigh out 1.3 g KBr, dissolve in water, dilute with stirring to 0.75 L. (b) Weigh out 2.62 g KBr, dissolve it in 122.38 g H_2O to make exactly 125 g of 0.180 m solution. (c) Weigh 244 g KBr and dissolve in enough H_2O to make 1.85 L solution. (d) Weigh 10.1 g KBr, dissolve it in a small amount of water, and dilute to 0.568 L.   **13.51** 71% HNO_3 by mass   **13.53** (a) 3.82 m Zn (b) 26.8 M Zn
13.55 1.8 × 10^{-3} M CO_2 **13.57** Freezing point depression, $\Delta T_f = K_f(m)$; boiling point elevation, $\Delta T_b = K_b(m)$; osmotic pressure, $\pi = MRT$; vapor pressure lowering, $P_A = X_A P_A^\circ$
13.59 (a) Sucrose has a greater molar mass than glucose, so the sucrose solution will contain fewer particles and have a higher vapor pressure.   **13.61** (a) P_{H_2O} = 186.4 torr (b) 78.9 g $C_3H_8O_2$ **13.61** (a) X_{Eth} = 0.2812 (b) P_{soln} = 238 torr (c) X_{Eth} in vapor = 0.472   **13.65** (a) Because NaCl is a strong electrolyte, one mole of NaCl produces twice as many dissolved particles as one mole of the molecular solute $C_6H_{12}O_6$. Boiling-point elevation is directly related to total moles of dissolved particles, so 0.10 m NaCl has the higher boiling point. (b) 0.10 m NaCl: ΔT_b = 0.101 °C, T_b = 100.1 °C; 0.10 m $C_6H_{12}O_6$: ΔT_b = 0.051 °C, T_b = 100.1 °C (c) Interactions between ions in solution result in non-ideal behavior.   **13.67** 0.050 m Li Br < 0.120 m glucose < 0.050 m $Zn(NO_3)_2$ **13.69** (a) T_f = −115.0 °C, T_b = 78.7 °C (b) T_f = −67.3 °C, T_b = 64.2 °C (c) T_f = −0.604 °C, T_b = 100.2 °C   **13.71** 167 g $C_2H_6O_2$ **13.73** π = 0.0168 atm = 12.7 torr   **13.75** Experimental molar mass of adrenaline is 1.8 × 10^2 g. The structure shows a molecular formula of $C_9H_{13}NO_3$, with a molar mass of 183 g. The two values agree to two significant figures, the precision of the experiment.
13.77 Molar mass of lysozyme = 1.39 × 10^4 g
13.79 (a) i = 2.76 (b) The more concentrated the solution, the greater the ion-pairing and the smaller the measured value of i.
13.81 (a) In the gaseous state, particles are far apart and inter-molecular attractive forces are small. When two gases combine, all terms in Equation 13.1 are essentially zero and the mixture is always homogeneous. (b) To determine whether Faraday's dispersion is a true solution or a colloid, shine a beam of light on it. If light is scattered, the dispersion is a colloid.   **13.83** (a) Hydrophobic (b) hydrophilic (c) hydrophobic (d) hydrophobic (but stabilized by adsorbed charges).   **13.85** When electrolytes are added to a suspension of proteins, dissolved ions form ion pairs with the protein surface charges, effectively neutralizing them. The protein's capacity for ion–dipole interactions with water is diminished and the colloid separates into a protein layer and a water layer.   **13.87** The periphery of the BHT molecule is mostly hydrocarbon-like groups, such as —CH_3. The one —OH group is rather buried inside and probably does little to enhance solubility in water. Thus, BHT is more likely to be soluble in the nonpolar hydrocarbon hexane, C_6H_{14}, than in polar water.   **13.90** (a) k_{Rn} = 7.27 × 10^{-3} mol/L-atm (b) P_{Rn} = 1.1 × 10^{-4} atm; S_{Rn} = 8.1 × 10^{-7} M
13.93 (a) 2.69 m LiBr (b) X_{LiBr} = 0.0994 (c) 81.1% LiBr by mass
13.96 X_{H_2O} = 0.808; 0.273 mol ions; 0.136 mol NaCl
13.99 (a) −0.6 °C (b) −0.4 °C   **13.102** (a) X_{KSCN} = 0.110; 6.86 m KSCN; 5.02 M KSCN (b) In this solution, there are 4.12 mol KSCN, 8.24 mol of ions and 33.3 mol H_2O, or approximately 4 water molecules for each ion. This is too few water molecules to completely hydrate the anions and cations in the solution. For a solution this concentrated, we expect significant ion-pairing, because the ions are not completely surrounded and separated by H_2O molecules. The effective number of

particles will be less than that indicated by m and M, so the observed colligative properties will be significantly different from those predicted by formulas for ideal solutions.   **13.104** (a) CF_4, 1.7 × 10^{-4} m; $CClF_3$, 9 × 10^{-4} m; CCl_2F_2, 2.3 × 10^{-2} m; $CHClF_2$, 3.5 × 10^{-2} m (b) Molality and molarity are numerically similar when kilograms solvent and liters solution are nearly equal. This is true when solutions are dilute and when the density of the solvent is nearly 1 g/mL, as in this exercise. (c) Water is a polar solvent; the solubility of solutes increases as their polarity increases. Nonpolar CF_4 has the lowest solubility and the most polar fluorocarbon, $CHClF_2$, has the greatest solubility in H_2O (d) The Henry's law constant for $CHClF_2$ is 3.5 × 10^{-2} mol/L-atm. This value is greater than the Henry's law constant for $N_2(g)$ because $N_2(g)$ is nonpolar and of lower molecular mass than $CHClF_2$.

13.108

(a) cation (g) + anion (g) + solvent

Ionic solid + solvent

(b) Lattice energy (U) is inversely related to the distance between ions, so salts with large cations like $(CH_3)_4$ N^+ have smaller lattice energies than salts with simple cations like Na^+. Also, the —CH_3 groups in the large cation are capable of dispersion interactions with nonpolar groups of the solvent molecules, resulting in a more negative solvation energy of the gaseous ions. Overall, for salts with larger cations, lattice energy is smaller (less positive), the solvation energy of the gaseous ions is more negative, and ΔH_{soln} is less endothermic. These salts are more soluble in polar nonaqueous solvents.
13.111 The process is spontaneous with no significant change in enthalpy, so we suspect that there is an increase in entropy. In general, dilute solutions of the same solute have greater entropy than concentrated ones, because the solute particles are more free to move about the solution. In the limiting case that the more dilute solution is pure solvent, there is a definite increase in entropy as the concentrated solution is diluted. In Figure 13.23, there may be some entropy decrease as the dilute solution loses solvent, but this is more than offset by the entropy increase that accompanies dilution. There is a net increase in entropy.

Chapter 14

14.1 (a) Product (b) The average rate of reaction is greater between points 1 and 2 because they are earlier in the reaction when the concentrations of reactants are greater.   **14.4** Container (1) has the fastest initial rate, only slightly faster than container (2).   **14.6** (a) The half-life is 15 min. (b) After four half-life periods, 1/16 of the reactant remains.   **14.10** There is one intermediate, B, and two transition states, A $\longrightarrow$ B and B $\longrightarrow$ C. The B $\longrightarrow$ C step is faster, and the overall reaction is exothermic.   **14.13** (a) *Reaction rate* is the change in the amount of products or reactants in a given amount of time. (b) Rates depend on concentration of reactants, surface area of reactants, temperature, and presence of catalyst. (c) The stoichiometry of the reaction (mole ratios of reactants and products) must be known to relate rate of disappearance of reactants to rate of appearance of products.

14.15

Time (min)	Mol A	(a) Mol B	[A] (mol/L)	Δ[A] (mol/L)	(b) Rate (M/s)
0	0.065	0.000	0.65		
10	0.051	0.014	0.51	−0.14	2.3×10^{-4}
20	0.042	0.023	0.42	−0.09	1.5×10^{-4}
30	0.036	0.029	0.36	−0.06	1.0×10^{-4}
40	0.031	0.034	0.31	−0.05	0.8×10^{-4}

(c) $\Delta[B]_{avg}/\Delta t = 1.3 \times 10^{-4} M/s$

14.17 (a)

Time (s)	Time Interval (s)	Concen- tration (M)	ΔM	Rate (M/s)
0		0.0165		
2,000	2,000	0.0110	−0.0055	28×10^{-7}
5,000	3,000	0.00591	−0.0051	17×10^{-7}
8,000	3,000	0.00314	−0.00277	9.23×10^{-7}
12,000	4,000	0.00137	−0.00177	4.43×10^{-7}
15,000	3,000	0.00074	−0.00063	2.1×10^{-7}

(b) From the slopes of the tangents to the graph, the rates are $12 \times 10^{-7} M/s$ at 5000 s, $5.8 \times 10^{-7} M/s$ at 8000 s.
14.19 (a) $-\Delta[H_2O_2]/\Delta t = \Delta[H_2]/\Delta t = \Delta[O_2]/\Delta t$
(b) $-\frac{1}{2}\Delta[N_2O]/\Delta t = \frac{1}{2}\Delta[N_2]/\Delta t = \Delta[O_2]/\Delta t$
(c) $-\Delta[N_2]/\Delta t = -\frac{1}{3}\Delta[H_2]/\Delta t = \frac{1}{2}\Delta[NH_3]\Delta t$
14.21 (a) $-\Delta[O_2]/\Delta t = 0.43$ mol/s; $\Delta[H_2O]/\Delta t = 0.85$ mol/s
(b) P_{total} decreases by 12 torr/min   **14.23** (a) If [A] doubles, there is no change in the rate or the rate constant. The overall rate is unchanged because [A] does not appear in the rate law; the rate constant changes only with a change in temperature. (b) The reaction is zero order in A, second order in B, and second order overall. (c) units of $k = M^{-1}s^{-1}$ **14.25** (a) Rate = $k[N_2O_5]$ (b) Rate = 1.16×10^{-4} M/s (c) When the concentration of N_2O_5 doubles, the rate of the reaction doubles. **14.27** (a, b) $k = 1.7 \times 10^2 M^{-1} s^{-1}$ (c) If $[OH^-]$ is tripled, the rate triples.   **14.29** (a) Rate = $k[OCl^-][I^-]$ (b) $k = 60 M^{-1} s^{-1}$ (c) Rate = $6.0 \times 10^{-5} M/s$ **14.31** (a) Rate = $k[BF_3][NH_3]$ (b) The reaction is second order overall. (c) $k_{avg} = 3.41 M^{-1} s^{-1}$ (d) $0.170 M/s$ **14.33** (a) Rate = $k[NO]^2[Br_2]$ (b) $k_{avg} = 1.2 \times 10^4 M^{-2} s^{-1}$ (c) $\frac{1}{2}\Delta[NOBr]/\Delta t = -\Delta[Br_2]/\Delta t$ (d) $-\Delta[Br_2]/\Delta t = 8.4 M/s$ **14.35** (a) $[A]_0$ is the molar concentration of reactant A at time 0. $[A]_t$ is the molar concentration of reactant A at time t. $t_{1/2}$ is the time required to reduce $[A]_0$ by a factor of 2. k is the rate constant for a particular reaction. (b) A graph of ln[A] versus time yields a straight line for a first-order reaction.   **14.37** (a) $k = 3.0 \times 10^{-6} s^{-1}$ (b) $t_{1/2} = 3.2 \times 10^4 s$ **14.39** (a) $P = 20$ torr (b) $t = 51$ s   **14.41** Plot $\ln P_{SO_2Cl_2}$ versus time, $k = -$slope $= 2.19 \times 10^{-5} s^{-1}$ **14.43** (a) The plot of $1/[A]$ versus time is linear, so the reaction is second order in [A]. (b) $k = 0.040 M^{-1} min^{-1}$ (c) $t_{1/2} = 38$ min   **14.45** (a) The plot of $1/[NO_2]$ versus time is linear, so the reaction is second order in NO_2. (b) $k =$ slope $= 10 M^{-1}s^{-1}$ **14.47** (a) The energy of the collision and the orientation of the molecules when they collide determine whether a reaction will occur. (b) At a higher temperature, there are more total collisions and each collision is more energetic.   **14.49** $f = 4.94 \times 10^{-2}$. At 400 K approximately 1 out of 20 molecules has this kinetic energy.

14.51 (a)

$E_a = 7$ kJ

$\Delta E = -66$ kJ

(b) E_a (reverse) = 73 kJ   **14.53** Reaction (b) is fastest and reaction (c) is slowest.   **14.55** (a) $k = 1.1 s^{-1}$ (b) $k = 13 s^{-1}$ **14.57** A plot of ln k versus $1/T$ has a slope of -5.71×10^3; $E_a = -R$(slope) $= 47.5$ kJ/mol.   **14.59** The reaction will occur 88 times faster at 50 °C, assuming equal initial concentrations. **14.61** (a) An *elementary reaction* is a process that occurs in a single event; the order is given by the coefficients in the balanced equation for the reaction. (b) A *unimolecular* elementary reaction involves only one reactant molecule; a *bimolecular* elementary reaction involves two reactant molecules. (c) A *reaction mechanism* is a series of elementary reactions that describe how an overall reaction occurs and explain the experimentally determined rate law.   **14.63** (a) Unimolecular, rate = $k[Cl_2]$ (b) bimolecular, rate = $k[OCl^-][H_2O]$ (c) bimolecular, rate = $k[NO][Cl_2]$ **14.65** (a) Two intermediates, B and C. (b) three transition states (c) C $\longrightarrow$ D is fastest. (d) endothermic   **14.67** (a) $H_2(g) + 2 ICl(g) \longrightarrow I_2(g) + 2 HCl(g)$ (b) HI is the intermediate. (c) first step: rate = $k_1[H_2][ICl]$, second step: rate = $k_2[HI][ICl]$ (d) If the first step is slow, the observed rate law is rate = $k[H_2][ICl]$ **14.69** (a) Rate = $k[NO][Cl_2]$ (b) The second step must be slow relative to the first step.   **14.71** (a) A catalyst increases the rate of reaction by decreasing the activation energy, E_a, or increasing the frequency factor, A. (b) A homogeneous catalyst is in the same phase as the reactants, while a heterogeneous catalyst is in a different phase.   **14.73** (a) Multiply the coefficients in the first reaction by 2 and sum. (b) $NO_2(g)$ is a catalyst because it is consumed and then reproduced in the reaction sequence. (c) This is a homogeneous catalysis.   **14.75** (a) Use of chemically stable supports makes it possible to obtain very large surface areas per unit mass of the precious metal catalyst because the metal can be deposited in a very thin, even monomolecular, layer on the surface of the support. (b) The greater the surface area of the catalyst, the more reaction sites, the greater the rate of the catalyzed reaction. **14.77** To put two D atoms on a single carbon, it is necessary that one of the already existing C—H bonds in ethylene be broken while the molecule is adsorbed, so that the H atom moves off as an adsorbed atom and is replaced by a D atom. This requires a larger activation energy than simply adsorbing C_2H_4 and adding one D atom to each carbon.   **14.79** (a) Living organisms operate efficiently in a very narrow temperature range; the role of enzymes as homogeneous catalysts that speed up desirable reactions, without heating and undesirable side effects, is crucial for biological systems. (b) catalase: $2 H_2O_2 \longrightarrow 2 H_2O + O_2$; nitrogenase: $N_2 \longrightarrow 2 NH_3$ (nitrogen fixation)   **14.81** (a) The catalyzed reaction is approximately 10,000,000 times faster at 25 °C. (b) The catalyzed reaction is 180,000 times faster at 125 °C.   **14.83** A balanced equation shows the overall, net change of a chemical reaction. Most reactions occur as a series of steps. The rate law contains only those reactants that form the activated complex (transition state) of the rate-determining step. If a reaction occurs in a single elementary step, the rate law can be written directly from the balanced equation for the step.   **14.86** (a) Rate = $k[HgCl_2][C_2O_4^{2-}]^2$ (b) $k = 8.7 \times 10^{-3} M^{-2} s^{-1}$

(c) rate $= 5.4 \times 10^{-5}$ M/s **14.89** (a) $t_{1/2} = 9.9 \times 10^2$ s
(b) $k = 2.05 \times 10^{-4}$ s^{-1} **14.92** (a) 1.08×10^{-4} M
(b) $k = 4.91 \times 10^{-4}$ s^{-1} (c) $t_{1/2} = 1.41 \times 10^3$ s $= 23.5$ min
(d) $t = 3.67 \times 10^3$ s $= 61.1$ min   **14.95** A plot of ln k versus
$1/T$ is linear with slope $= -1.751 \times 10^4$. $E_a = -$(slope)$R =$
1.5×10^2 kJ/mol.   **14.99** (a) $Cl_2(g) + CHCl_3(g) \longrightarrow$
$HCl(g) + CCl_4(g)$ (b) $Cl(g)$, $CCl_3(g)$ (c) reaction 1, unimolecular;
reaction 2, bimolecular; reaction 3, bimolecular (d) Reaction 2,
the slow step, is rate determining. (e) Rate $= k[CHCl_3][Cl_2]^{1/2}$
14.101 *Enzyme*: carbonic anhydrase; *substrate*: carbonic acid
(H_2CO_3); *turnover number*: 1×10^7 molecules/s.   **14.104** partial
pressure of $O_2 = 0.402$ atm   **14.106** (a) Use an open-end
manometer, a clock, a ruler, and a constant-temperature bath.
Load the flask with HCl(aq), and read the height of the Hg in
both arms of the manometer. Quickly add Zn(s) to the flask,
and record time $= 0$ when the Zn(s) contacts the acid. Record
the height of the Hg in one arm of the manometer at convenient time intervals such as 5 s. Calculate the pressure of $H_2(g)$
at each time. Since $P = (n/V) RT$, $\Delta P/\Delta t$ at constant temperature is an acceptable measure of reaction rate. (b) Keep the
amount of Zn(s) constant, and vary the concentration of
HCl(aq) to determine the reaction order for H$^+$ and Cl$^-$. Keep
the concentration of HCl(aq) constant, and vary the amount of
Zn(s) to determine the order for Zn(s). Combine this informa-
tion to write the rate law. (c) $-\Delta[H^+]/\Delta t = 2\Delta[H_2]/\Delta t$; the rate
of disappearance of H$^+$ is twice the rate of appearance of $H_2(g)$.
$[H_2] = P/RT$ (d) By changing the temperature of the constant-
temperature bath, measure the rate data at several temperatures and calculate the rate constant k at these temperatures.
Plot ln k versus $1/T$; the slope of the line is $-E_a/R$. (e) Measure
rate data at constant temperature, HCl concentration, and mass
of Zn(s), varying only the form of the Zn(s). Compare the rate
of reaction for metal strips and granules.   **14.109** Changes in
temperature change the kinetic energy of the various groups
on the enzyme and their tendency to form intermolecular associations or break free from them. At temperatures above the
temperature of maximum activity, sufficient kinetic energy has
been imparted so that the three-dimensional structure of the
enzyme is destroyed. This is the process of denaturation. In the
lock-and-key model of enzyme action, the active site is the specific location in the enzyme where the reaction takes place. The
precise geometry of the active site both accommodates and ac-
tivates the substrate. When an enzyme is denatured, its activity
is destroyed because the active site has collapsed.

Chapter 15

15.1 $k_f > k_r$ (b) The equilibrium constant is greater than 1.
15.4 (a) $A_2 + B \longrightarrow A_2B$ (b) $K_c = [A_2B]/[A_2][B]$ (c) $\Delta n = -1$
(d) $K_p = K_c (RT)^{\Delta n}$ **15.6** (a) Mixture (ii) is at equilibrium.
(b) Mixture (i) proceeds toward reactants; mixture (iii) proceeds toward products.   **15.9** K_c decreases as T increases, so
the reaction is exothermic.   **15.11** (a) $K_p = K_c = 0.12$ (b) Since
$k_f < k_r$, in order for the two rates to be equal, [A] must be
greater than [B], and the partial pressure of A is greater than the
partial pressure of B.   **15.13** (a) $K_c = [N_2O][NO_2]/[NO]^3$;
homogeneous (b) $K_c = [CS_2][H_2]^4/[CH_4][H_2S]^2$;
homogeneous (c) $K_c = [CO]^4/[Ni(CO)_4]$; heterogeneous (d) $K_c =$
$[H^+][F^-]/[HF]$; homogeneous (e) $K_c = [Ag^+]^2/[Zn^{2+}]$; heterogeneous   **15.15** (a) Mostly reactants (b) mostly products
15.17 $K_p = 1.0 \times 10^{-3}$ **15.19** (a) $K_c = 77$ (b) The equilibrium
favors NO and Br$_2$ at this temperature.   **15.21** (a) $K_p = 0.541$
(b) $K_p = 3.42$ (c) $K_c = 281$ **15.23** $K_c = 0.14$
15.25 (a) $K_p = P_{O_2}$ (b) The molar concentration, the ratio of
moles of a substance to volume occupied by the substance, is a
constant for pure solids and liquids.   **15.27** $K_c = 0.0184$
15.29 $K_p = 51$ **15.31** (a) $[H_2] = 0.012$ M, $[N_2] = 0.019$ M,
$[H_2O] = 0.138$ M (b) $K_c = 653.7 = 7 \times 10^2$
15.33 (a) $P_{CO_2} = 4.10$ atm, $P_{H_2} = 2.05$ atm, $P_{H_2O} = 3.28$ atm

(b) $P_{CO_2} = 3.87$ atm, $P_{H_2} = 1.82$ atm, $P_{H_2O} = 3.51$ atm
(c) $K_p = 0.11$ **15.35** (a) A reaction quotient is the result of a
general set of concentrations whereas the equilibrium constant
requires equilibrium concentrations. (b) to the right (c) The
concentrations used to calculate Q must be equilibrium concentrations.   **15.37** (a) $Q = 1.1 \times 10^{-8}$, the reaction will proceed to the left. (b) $Q = 5.5 \times 10^{-12}$, the reaction will proceed
to the right. (c) $Q = 2.2 \times 10^{-10}$, the mixture is at equilibrium.
15.39 $P_{Cl_2} = 4.96$ atm   **15.41** (a) $[Br_2] = 0.00767$ M,
$[Br] = 0.00282$ M, 0.0451 g Br(g) (b) $[H_2] = 0.014$ M,
$[I_2] = 0.00859$ M, $[HI] = 0.081$ M, 21 g HI
15.43 $[NO] = 0.002$ M, $[N_2] = [O_2] = 0.099$ M **15.45** The
equilibrium pressure of Br$_2(g)$ is 0.416 atm.   **15.47** (a)
$[Ca^{2+}] = [SO_4^{2-}] = 4.9 \times 10^{-3}$ M (b) A bit more than 2.0 g
CaSO$_4$ is needed in order to have some undissolved CaSO$_4(s)$
in equilibrium with 3.0 L of saturated solution.
15.49 $[IBr] = 0.447$ M, $[I_2] = [Br_2] = 0.0267$ M **15.51** (a) Shift
equilibrium to the right (b) decrease the value of K (c) shift
equilibrium to the left (d) no effect (e) no effect (f) shift equilibrium to the right   **15.53** (a) No effect (b) no effect (c) increase
equilibrium constant (d) no effect   **15.55** (a) $\Delta H° = -155.7$ kJ
(b) The reaction is exothermic, so the equilibrium constant will
decrease with increasing temperature. (c) Δn does not equal
zero, so a change in volume at constant temperature will affect
the fraction of products in the equilibrium mixture.   **15.57** (a)
$K_c = K_p = 1.5 \times 10^{-39}$ (b) Reactants are much more plentiful
than products at equilibrium.   **15.60** $K_c = 0.71$ **15.63** (a)
$P_t = 1.14$ atm (b) $K_p = 1.7$ **15.66** (a) $K_c = 7.0 \times 10^{-5}$
(b) $[PH_3] = 2.5 \times 10^{-3}$ M **15.69** (a) $Q = 1.7$; $Q > K_p$; the reaction will shift to the left. (b) $Q = 1.7$; $Q > K_p$; the reaction will
shift to the left. (c) $Q = 5.8 \times 10^{-3}$; $Q < K_p$; the reaction will
shift to the right.   **15.73** The maximum value of Q is 4.1, which
is less than K_p; reduction will occur.   **15.76** $P_{H_2} = P_{I_2} =$
1.48 atm; $P_{HI} = 10.22$ atm   **15.79** The patent claim is false.
A catalyst does not alter the position of equilibrium in a system, only the rate of approach to the equilibrium condition.
15.80 At equilibrium, $[H_6IO_4^-] = 0.0024$ M **15.83** (a) At
equilibrium, the forward and reverse reactions occur at equal
rates. (b) Reactants are favored at equilibrium. (c) A catalyst
lowers the activation energy for both the forward and reverse
reactions. (d) The ratio of the rate constants remains unchanged. (e) The value of K will increase with increasing
temperature.   **15.86** At 850 °C, $K_p = 14.1$; at 950 °C, $K_p = 73.8$;
at 1050 °C, $K_p = 2.7 \times 10^2$; at 1200 °C, $K_p = 1.7 \times 10^3$. Because
K increases with increasing temperature, the reaction is
endothermic.

Chapter 16

16.1 (a) HX, the H$^+$ donor, is the Brønsted–Lowry acid. NH$_3$,
the H$^+$ acceptor, is the Brønsted–Lowry base. (b) HX, the electron pair acceptor, is the Lewis acid. NH$_3$, the electron pair
donor, is the Lewis base.   **16.3** (a) HY is a strong acid. There
are no neutral HY molecules in solution, only H$^+$ cations and
Y$^-$ anions. (b) HX has the smallest K_a value. It has most neutral
acid molecules and fewest ions. (c) HX has fewest H$^+$ and
highest pH.   **16.6** (a) The order of base strength is the reverse
of the order of acid strength; the diagram of the acid with the
most free H$^+$ has the weakest conjugate base Y$^-$. The order of
increasing basicity is: Y$^- <$ Z$^- <$ X$^-$. (b) The strongest base,
X$^-$, has the largest K_b value.   **16.9** (a) Both molecules are oxyacids; the ionizable H atom is attached to O. The right molecule is a carboxylic acid; the ionizable H is part of a carboxyl
group. (b) Increasing the electronegativity of X increases the
strength of both acids. As X becomes more electronegative and
attracts more electron density, the O—H bond becomes weaker, more polar, and more likely to be ionized. An electronegative X group also stabilizes the anionic conjugate base, causing

the ionization equilibrium to favor products and the value of K_a to increase.   **16.13** Solutions of HCl and H_2SO_4 conduct electricity, taste sour, turn litmus paper red (are acidic), neutralize solutions of bases, and react with active metals to form $H_2(g)$. HCl and H_2SO_4 solutions have these properties in common because both compounds are strong acids. That is, they both ionize completely in H_2O to form $H^+(aq)$ and an anion. (HSO_4^- is not completely ionized, but the first ionization step for H_2SO_4 is complete.) The presence of ions enables the solutions to conduct electricity; the presence of $H^+(aq)$ in excess of 1×10^{-7} M accounts for all other properties listed.   **16.15** (a) The Arrhenius definition of an acid is confined to aqueous solution; the Brønsted–Lowry definition applies to any physical state. (b) HCl is the Brønsted–Lowry acid; NH_3 is the Brønsted–Lowry base.   **16.17** (a) (i) IO_3^- (ii) NH_3 (b) (i) OH^- (ii) H_3PO_4

16.19

Acid	+	Base	⇌	Conjugate Acid	+	Conjugate Base
(a) $NH_4^+(aq)$		$CN^-(aq)$		$HCN(aq)$		$NH_3(aq)$
(b) $H_2O(l)$		$(CH_3)_3N(aq)$		$(CH_3)_3NH^+(aq)$		$OH^-(aq)$
(c) $HCOOH(aq)$		$PO_4^{3-}(aq)$		$HPO_4^{2-}(aq)$		$HCOO^-(aq)$

16.21 (a)
Acid: $HC_2O_4^-(aq) + H_2O(l) \rightleftharpoons C_2O_4^{2-}(aq) + H_3O^+(aq)$;
Base: $HC_2O_4^-(aq) + H_2O(l) \rightleftharpoons H_2C_2O_4(aq) + OH^-(aq)$
(b) $H_2C_2O_4$ is the conjugate acid of $HC_2O_4^-$. $C_2O_4^{2-}$ is the conjugate base of $HC_2O_4^-$.   **16.23** (a) CH_3COO^-, weak base; CH_3COOH, weak acid (b) HCO_3^-, weak base; H_2CO_3, weak acid (c) O_2^-, strong base; OH^-, strong base (d) Cl^-, negligible base; HCl, strong acid (e) NH_3, weak base; NH_4^+, weak acid
16.25 (a) HBr. It is one of the seven strong acids. (b) F^-. HCl is a stronger acid than HF, so F^- is the stronger conjugate base.
16.27 (a) $OH^-(aq) + OH^-(aq)$, the equilibrium lies to the right. (b) $H_2S(aq) + CH_3COO^-(aq)$, the equilibrium lies to the right. (c) $HNO_3(aq) + OH^-(aq)$, the equilibrium lies to the left.
16.29 (a) *Autoionization* is the ionization of a neutral molecule into an anion and a cation. The equilibrium expression for the autoionization of water is $H_2O(l) \rightleftharpoons H^+(aq) + OH^-(aq)$. (b) Pure water is a poor conductor of electricity because it contains very few ions. (c) If a solution is acidic, it contains more H^+ than OH^-.   **16.31** (a) $[H^+] = 2.2 \times 10^{-11}$ M, basic (b) $[H^+] = 1.1 \times 10^{-6}$ M, acidic (c) $[H^+] = 1.0 \times 10^{-8}$ M, basic
16.33 $[H^+] = [OH^-] = 3.5 \times 10^{-8}$ M   **16.35** (a) $[H^+]$ changes by a factor of 100. (b) $[H^+]$ changes by a factor of 3.2   **16.37** (a) $[H^+]$ decreases, pH increases (b) The pH is between 3 and 4. By calculation, pH = 3.2; the solution is acidic. (c) pH = 5.2 is between pH 5 and pH 6, closer to pH = 5. A good estimate is 7×10^{-6} M H^+ and 3×10^{-9} M OH^-. By calculation, $[H^+] = 6 \times 10^{-6}$ M and $[OH^-] = 2 \times 10^{-9}$ M.
16.39

$[H^+]$	$[OH^-]$	pH	pOH	Acidic or Basic
7.5×10^{-3} M	1.3×10^{-12} M	2.12	11.88	acidic
2.8×10^{-5} M	3.6×10^{-10} M	4.56	9.44	acidic
5.6×10^{-9} M	1.8×10^{-6} M	8.25	5.75	basic
5.0×10^{-9} M	2.0×10^{-6} M	8.30	5.70	basic

16.41 $[H^+] = 4.0 \times 10^{-8}$ M, $[OH^-] = 6.0 \times 10^{-7}$ M, pOH = 6.22   **16.43** (a) A strong acid is completely ionized in aqueous solution. (b) $[H^+] = 0.500$ M (c) HCl, HBr, HI
16.45 (a) $[H^+] = 8.5 \times 10^{-3}$ M, pH = 2.07 (b) $[H^+] = 0.0419$ M, pH = 1.377 (c) $[H^+] = 0.0250$ M, pH = 1.602
(d) $[H^+] = 0.167$ M, pH = 0.778

16.47 (a) $[OH^-] = 3.0 \times 10^{-3}$ M, pH = 11.48 (b) $[OH^-] = 0.3758$ M, pH = 13.5750 (c) $[OH^-] = 8.75 \times 10^{-5}$ M, pH = 9.942 (d) $[OH^-] = 0.17$ M, pH = 13.23
16.49 3.2×10^{-3} M NaOH   **16.51** (a) $HBrO_2(aq) \rightleftharpoons H^+(aq) + BrO_2^-(aq)$, $K_a = [H^+][BrO_2^-]/[HBrO_2]$; $HBrO_2(aq) + H_2O(l) \rightleftharpoons H_3O^+(aq) + BrO_2^-(aq)$, $K_a = [H_3O^+][BrO_2^-]/[HBrO_2]$ (b) $C_2H_5COOH(aq) \rightleftharpoons H^+ + C_2H_5COO^-(aq)$, $K_a = [H^+][C_2H_5COO^-]/[C_2H_5COOH]$, $C_2H_5COOH(aq) + H_2O(l) \rightleftharpoons H^+ + C_2H_5COO^-(aq)$, $K_a = [H_3O^+][C_2H_5COO^-]/[C_2H_5COOH]$ **16.53** $K_a = 1.4 \times 10^{-4}$ **16.55** $[H^+] = [ClCH_2COO^-] = 0.0110$ M, $[ClCH_2COOH] = 0.089$ M, $K_a = 1.4 \times 10^{-3}$ **16.57** 0.089 M CH_3COOH **16.59** $[H^+] = [C_6H_5COO^-] = 1.8 \times 10^{-3}$ M, $[C_6H_5COOH] = 0.048$ M   **16.61** (a) $[H^+] = 1.1 \times 10^{-3}$ M, pH = 2.95 (b) $[H^+] = 1.7 \times 10^{-4}$ M, pH = 3.76 (c) $[OH^-] = 1.4 \times 10^{-5}$ M, pH = 9.15   **16.63** $[H^+] = 2.0 \times 10^{-2}$ M, pH = 1.71   **16.65** (a) $[H^+] = 2.8 \times 10^{-3}$ M, 0.69% ionization (b) $[H^+] = 1.4 \times 10^{-3}$ M, 1.4% ionization (c) $[H^+] = 8.7 \times 10^{-4}$ M, 2.2% ionization   **16.67** $HX(aq) \rightleftharpoons H^+(aq) + X^-(aq)$; $K_a = [H^+][X^-]/[HX]$. Assume that the percent of acid that ionizes is small. Let $[H^+] = [X^-] = y$, $K_a = y^2/[HX]$; $y = K_a^{1/2}[HX]^{1/2}$. Percent ionization $= y/[HX] \times 100$. Substituting for y, percent ionization $= 100\ K_a^{1/2}[HX]^{1/2}/[HX]$ or $100\ K_a^{1/2}/[HX]^{1/2}$. That is, percent ionization varies inversely as the square root of the concentration of HX.   **16.69** $[H^+] = 5.72 \times 10^{-3}$ M, pH = 2.24, $[C_6H_5O_7^{3-}] = 1.2 \times 10^{-9}$ M. The approximation that the first ionization is less than 5% of the total acid concentration is not valid; the quadratic equation must be solved. The $[H^+]$ produced from the second and third ionizations is small with respect to that present from the first step; the second and third ionizations can be neglected when calculating the $[H^+]$ and pH.   **16.71** All Brønsted–Lowry bases contain at least one unshared (lone) pair of electrons to attract H^+.
16.73 (a) $(CH_3)_2NH(aq) + H_2O(l) \rightleftharpoons (CH_3)_2NH_2^+(aq) + OH^-(aq)$; $K_b = [(CH_3)_2NH_2^+][OH^-]/[(CH_3)_2NH]$ (b) $CO_3^{2-}(aq) + H_2O(l) \rightleftharpoons HCO_3^-(aq) + OH^-(aq)$; $K_b = [HCO_3^-][OH^-]/[(CO_3^{2-})]$ (c) $HCOO^-(aq) + H_2O(l) \rightleftharpoons HCOOH(aq) + OH^-(aq)$ $K_b = [HCOOH][OH^-]/[HCOO^-]$
16.75 From the quadratic formula, $[OH^-] = 6.6 \times 10^{-3}$ M, pH = 11.82.   **16.77** (a) $[C_{10}H_{15}ON] = 0.033$ M, $[C_{10}H_{15}ONH^+] = [OH^-] = 2.1 \times 10^{-3}$ M (b) $K_b = 1.4 \times 10^{-4}$
16.79 (a) For a conjugate acid/conjugate base pair such as $C_6H_5OH/C_6H_5O^-$, K_b for the conjugate base can always be calculated from K_a for the conjugate acid, so a separate list of K_b values is not necessary. (b) $K_b = 7.7 \times 10^{-5}$ (c) Phenolate is a stronger base than NH_3.   **16.81** (a) Acetic acid is stronger. (b) Hypochlorite ion is the stronger base.   **16.83** (a) $[OH^-] = 1.4 \times 10^{-3}$ M, pH = 11.15 (b) $[OH^-] = 3.8 \times 10^{-3}$ M, pH = 11.58 (c) $[OH^-] = 3.3 \times 10^{-6}$ M, pH = 8.52   **16.85** (a) Acidic (b) acidic (c) basic (d) neutral (e) acidic   **16.87** K_b for the anion of the unknown salt is 1.4×10^{-11}; K_a for the conjugate acid is 7.1×10^{-4}. The conjugate acid is HF and the salt is NaF.
16.89 $[OH^-] = 5.0 \times 10^{-6}$ M, pH = 8.70   **16.91** (a) As the electronegativity of the central atom (X) increases, the strength of the oxyacid increases. (b) As the number of nonprotonated oxygen atoms in the molecule increases, the strength of the oxyacid increases.   **16.93** (a) HNO_3 is a stronger acid because it has one more nonprotonated oxygen atom and thus a higher oxidation number on N. (b) For binary hydrides, acid strength increases going down a family, so H_2S is a stronger acid than H_2O. (c) H_2SO_4 is a stronger acid because H^+ is much more tightly held by the anion HSO_4^-. (d) For oxyacids, the greater the electronegativity of the central atom, the stronger the acid, so H_2SO_4 is the stronger acid. (e) CCl_3COOH is stronger because the electronegative Cl atoms withdraw electron density from other parts of the molecule, which weakens the O—H bond and stabilizes the anionic conjugate base. Both effects favor increased ionization and acid strength.   **16.95** (a) BrO^- (b) BrO^- (c) HPO_4^{2-} **16.97** (a) True (b) False. In a series of

acids that have the same central atom, acid strength increases with the number of nonprotonated oxygen atoms bonded to the central atom. (c) False. H_2Te is a stronger acid than H_2S because the H—Te bond is longer, weaker, and more easily ionized than the H—S bond.   **16.99** Yes. The Arrhenius definition of a base, an $OH^-(aq)$ donor, is most restrictive; the Brønsted definition, an H^+ acceptor, is more general; and the Lewis definition, an electron-pair donor, is most general. Any substance that fits the narrow Arrhenius definition will fit the broader Brønsted and Lewis definitions.   **16.101** (a) Acid, $Fe(ClO_4)_3$ or Fe^{3+}; base, H_2O (b) Acid, H_2O; base, CN^- (c) Acid, BF_3; base, $(CH_3)_3N$ (d) Acid, HIO; base, NH_2^- **16.103** (a) Cu^{2+}, higher cation charge (b) Fe^{3+}, higher cation charge (c) Al^{3+}, smaller cation radius, same charge   **16.105** (a) K_w is the equilibrium constant for the reaction of two water molecules to form hydronium ion and hydroxide ion. (b) K_a is the equilibrium constant for the reaction of any acid, neutral or ionic, with water to form hydronium ion and the conjugate base of the acid. (c) pOH is the negative log of hydroxide ion concentration; pOH decreases as hydroxide ion concentration increases. (d) pK_b is the negative log of K_b; pK_b increases as base strength decreases.   **16.108** $K = 3.3 \times 10^7$ **16.111** (a) For solutions with equal concentrations, the weaker acid will have a lower $[H^+]$ and higher pH. (b) The acid with $K_a = 8 \times 10^{-5}$ is the weaker acid, so it has the higher pH. (c) The base with $pK_b = 4.5$ is the stronger base, has the greater $[OH^-]$ and smaller $[H^+]$, so higher pH.   **16.113** $K_a = 1.4 \times 10^{-5}$ **16.117** $[H^+] = 0.010\ M$, $[H_2PO_4^-] = 0.010\ M$, $[HPO_4^{2-}] = 6.2 \times 10^{-8}\ M$, $[PO_4^{3-}] = 2.6 \times 10^{-18}\ M$ **16.121** $6.0 \times 10^{13}\ H^+$ ions   **16.124** (a) To the precision of the reported data, the pH of rainwater 40 years ago was 5.4, no different from the pH today. With extra significant figures, $[H^+] = 3.61 \times 10^{-6}\ M$, pH = 5.443 (b) A 20.0-L bucket of today's rainwater contains 0.02 L (with extra significant figures, 0.0200 L) of dissolved CO_2.   **16.127** [cocaine] = $2.93 \times 10^{-3}\ M$, $[OH^-] = 3.4 \times 10^{-6}\ M$, $K_b = 3.9 \times 10^{-9}$ **16.129** (a) $K(i) = 5.6 \times 10^3$, $K(ii) = 10$ (b) Both (i) and (ii) have $K > 1$, so both could be written with a single arrow.

Chapter 17

17.1 The middle box has the highest pH. For equal amounts of acid HX, the greater the amount of conjugate base X^-, the smaller the amount of H^+ and the higher the pH.   **17.4** (a) Drawing 3 (b) Drawing 1 (c) Drawing 2   **17.7** (a) The red curve corresponds to the more concentrated acid solution. (b) On the titration curve of a weak acid, pH = pK_a at the volume half-way to the equivalence point. At this volume, the red curve has the smaller pK_a and the larger K_a.   **17.10** (a) The right diagram (b) the left diagram (c) the center diagram   **17.13** (a) The extent of ionization of a weak electrolyte is decreased when a strong electrolyte containing an ion in common with the weak electrolyte is added to it. (b) $NaNO_2$ **17.15** (a) $[H^+] = 1.8 \times 10^{-5}\ M$, pH = 4.73 (b) $[OH^-] = 4.8 \times 10^{-5}\ M$, pH = 9.68 (c) $[H^+] = 1.4 \times 10^{-5}\ M$, pH = 4.87   **17.17** (a) 4.5% ionization (b) 0.018% ionization   **17.19** In a mixture of CH_3COOH and CH_3COONa, CH_3COOH reacts with added base and CH_3COO^- combines with added acid, leaving $[H^+]$ relatively unchanged. Although HCl and Cl^- are a conjugate acid-base pair, Cl^- has no tendency to combine with added acid to form undissociated HCl. Any added acid simply increases $[H^+]$ in an HCl—NaCl mixture.   **17.21** (a) pH = 3.82 (b) pH = 3.96   **17.23** (a) pH = 4.60 (b) $Na^+(aq) + CH_3COO^-(aq) + H^+(aq) + Cl^-(aq) \longrightarrow CH_3COOH(aq) + Na^+(aq) + Cl^-(aq)$ (c) $CH_3COOH(aq) + Na^+(aq) + OH^-(aq) \longrightarrow CH_3COO^-(aq) + H_2O(l) + Na^+(aq)$ **17.25** 0.18 mol NaBrO   **17.27** (a) pH = 4.86 (b) pH = 5.0 (c) pH = 4.71   **17.29** (a) $[HCO_3^-]/[H_2CO_3] = 11$ (b) $[HCO_3^-]/[H_2CO_3] = 5.4$ **17.31** 360 mL of 0.10 M HCOONa, 640 mL of 0.10 M HCOOH   **17.33** (a) Curve B (b) pH at the approximate equivalence point of curve A = 8.0, pH at the approximate equivalence point of curve B = 7.0 (c) For equal

volumes of A and B, the concentration of acid B is greater, since it requires a larger volume of base to reach the equivalence point.   **17.35** (a) Above pH 7 (b) below pH 7 (c) at pH 7   **17.37** (a) HX is weaker. The higher the pH at the equivalence point, the stronger the conjugate base (X^-) and the weaker the conjugate acid HX. (b) Phenolphthalein, which changes color in the pH 8–10 range, is perfect for HX and probably appropriate for HY.   **17.39** (a) 42.4 mL NaOH soln (b) 35.0 mL NaOH soln (c) 29.8 mL NaOH soln   **17.41** (a) pH = 1.54 (b) pH = 3.30 (c) pH = 7.00 (d) pH = 10.69 (e) pH = 12.74   **17.43** (a) pH = 2.78 (b) pH = 4.74 (c) pH = 6.58 (d) pH = 8.81 (e) pH = 11.03 (f) pH = 12.42   **17.45** (a) pH = 7.00 (b) $[HONH_3^+] = 0.100\ M$, pH = 3.52 (c) $[C_6H_5NH_3^+] = 0.100\ M$, pH = 2.82   **17.47** (a) The concentration of undissolved solid does not appear in the solubility product expression because it is constant. (b) $K_{sp} = [Ag^+][I^-]$; $K_{sp} = [Sr^{2+}][SO_4^{2-}]$; $K_{sp} = [Fe^{2+}][OH^-]^2$; $K_{sp} = [Hg_2^{2+}][Br^-]^2$;   **17.49** (a) $K_{sp} = 7.63 \times 10^{-9}$ (b) $K_{sp} = 2.7 \times 10^{-9}$ (c) 5.3×10^{-4} mol $Ba(IO_3)_2/L$ **17.51** $K_{sp} = 2.3 \times 10^{-9}$ **17.53** (a) 7.1×10^{-7} mol AgBr/L (b) 1.7×10^{-11} mol AgBr/L (c) 5.0×10^{-12} mol AgBr/L   **17.55** (a) 1.4×10^3 g $Mn(OH)_2/L$ (b) 0.014 g/L (c) 3.6×10^{-7} g/L   **17.57** More soluble in acid: (a) $ZnCO_3$ (b) ZnS (d) AgCN (e) $Ba_3(PO_4)_2$ **17.59** $[Cu^{2+}] = 2 \times 10^{-12}\ M$ **17.61** $K = K_{sp} \times K_f = 8 \times 10^4$ **17.63** (a) $Q < K_{sp}$; no $Ca(OH)_2$ precipitates (b) $Q < K_{sp}$; no Ag_2SO_4 precipitates   **17.65** pH = 11.5   **17.67** AgI will precipitate first, at $[I^-] = 4.2 \times 10^{-13}\ M$.   **17.69** The first two experiments eliminate group 1 and 2 ions (Figure 17.22). The absence of insoluble carbonate precipitates in the filtrate from the third experiment rules out group 4 ions. The ions that might be in the sample are those from group 3, Al^{3+}, Fe^{3+}, Zn^{2+}, Cr^{3+}, Ni^{2+}, Co^{2+}, or Mn^{2+}, and from group 5, NH_4^+, Na^+, or K^+.   **17.71** (a) Make the solution acidic with 0.5 M HCl; saturate with H_2S. CdS will precipitate; ZnS will not. (b) Add excess base; $Fe(OH)_3(s)$ precipitates, but Cr^{3+} forms the soluble complex $Cr(OH)_4^-$. (c) Add $(NH_4)_2HPO_4$; Mg^{2+} precipitates as $MgNH_4PO_4$; K^+ remains soluble. (d) Add 6 M HCl; precipitate Ag^+ as AgCl(s); Mn^{2+} remains soluble.   **17.73** (a) Base is required to increase $[PO_4^{3-}]$ so that the solubility product of the metal phosphates of interest is exceeded and the phosphate salts precipitate. (b) K_{sp} for the cations in group 3 is much larger, and so to exceed K_{sp}, a higher $[S^{2-}]$ is required. (c) They should all redissolve in strongly acidic solution.   **17.75** $pOH = pK_b + \log\{[BH^+]/[B]\}$ **17.77** (a) pH = 3.025 (b) pH = 2.938 (c) pH = 12.862   **17.80** (a) pH of buffer A = pH of buffer B = 3.74. For buffers containing the same conjugate acid and base components, the pH is determined by the ratio of concentrations of conjugate acid and base. Buffers A and B have the same ratio of concentrations, so their pH values are equal. (b) Buffer capacity is determined by the absolute amount of buffer components available to absorb strong acid or strong base. Buffer A has the greater capacity because it contains the greater absolute concentrations of HCOOH and $HCOO^-$. (c) Buffer A: pH = 3.74, ΔpH = 0.00; buffer B: pH = 3.66, ΔpH = -0.12 (d) Buffer A: pH = 3.74, ΔpH = 0.00; buffer B: pH = 2.74, ΔpH = -1.00 (e) The results of parts (c) and (d) are quantitative confirmation that buffer A has a significantly greater capacity than buffer B.   **17.82** (a) molar mass = 82.2 g/mol (b) $K_a = 3.8 \times 10^{-7}$ **17.85** (a) $[OH^-] = 4.23 \times 10^{-3}\ M$, pH = 11.63 (b) It will require 40.00 mL of 0.100 M HCl to reach the first equivalence point; HCO_3^- is the predominant species at this point. (c) An additional 40.00 mL are required to form H_2CO_3, the predominant species at the second equivalence point. (d) $[H^+] = 1.2 \times 10^{-4}\ M$, pH = 3.92   **17.88** 1.6 L of 1.0 M NaOH   **17.91** (a) CdS (b) $BaCrO_4$ (c) $NiCO_3$ (d) Ag_2SO_4 **17.94** The solubility of $Mg(OH)_2$ in 0.50 M NH_4Cl is 0.057 mol/L   **17.96** $[KMnO_4] = [MnO_4^-] = 0.11\ M$ **17.99** $[OH^-] = 1.7 \times 10^{-11}\ M$, pH of the buffer = 3.22   **17.102** (a) $H^+(aq) + HCOO^-(aq) \longrightarrow HCOOH(aq)$ (b) $K = 5.6 \times 10^3$ (c) $[Na^+] = [Cl^-] = 0.075\ M$,

$[H^+] = [HCOO^-] = 3.7 \times 10^{-3}\ M$, $[HCOOH] = 0.071\ M$
17.105 At these conditions, the concentration of the anion is $1 \times 10^{-5}\ M$, the extent of ionization is 0.3%, and 99.7% of the aspirin is in the form of neutral molecules.   **17.108** $[Sr^{2+}] = [SO_4^{2-}] = 5.7 \times 10^{-4}\ M$, $K_{sp} = 3.2 \times 10^{-7}$

Chapter 18

18.1 A greater volume than 22.4 L (b) The gas will occupy more volume at 85 km than at 50 km.   **18.3** Ozone concentration varies with altitude because conditions favorable to ozone formation and unfavorable to its decomposition vary with altitude. Above 60 km, there are too few O_2 molecules. Below 30 km, there are too few O atoms. Between 30 km and 60 km, O_3 concentration varies depending on the concentrations of O, O_2 and M*.   **18.5** $CO_2(g)$ dissolves in seawater to form $H_2CO_3(aq)$. Carbon is removed from the ocean as $CaCO_3(s)$ in the form of sea shells, coral, and other carbonates. As carbon is removed, more $CO_2(g)$ dissolves to maintain the balance of complex and interacting acid-base and precipitation equilibria.   **18.7** The guiding principle of green chemistry is that processes should be designed to minimize or eliminate solvents and waste, generate nontoxic waste, be energy efficient, employ renewable starting materials, and take advantage of catalysts that enable the use of safe and common reagents.   **18.9** (a) Its temperature profile (b) troposphere, 0 to 12 km; stratosphere, 12 to 50 km; mesosphere, 50 to 85 km; thermosphere, 85 to 110 km   **18.11** The partial pressure of O_3 is 3.0×10^{-7} atm (2.2×10^{-4} torr).   **18.13** 8.4×10^{16} CO molecules   **18.15** 570 nm   **18.17** (a) *Photodissociation* is cleavage of a bond such that two neutral species are produced. *Photoionization* is absorption of a photon with sufficient energy to eject an electron, producing an ion and the ejected electron. (b) Photoionization of O_2 requires 1205 kJ/mol. Photodissociation requires only 495 kJ/mol. At lower elevations, high-energy short-wavelength solar radiation has already been absorbed. Below 90 km, the increased concentration of O_2 and the availability of longer wavelength radiation cause the photodissociation process to dominate.   **18.19** A hydrofluorocarbon is a compound that contains hydrogen, fluorine, and carbon; it contains hydrogen in place of chlorine. HFCs are potentially less harmful than CFCs because photodissociation does not produce Cl atoms, which catalyze the destruction of ozone.   **18.21** (a) The C—F bond requires more energy for dissociation than the C—Cl bond and is not readily cleaved by the available wavelengths of UV light. (b) Chlorine is present as chlorine atoms and chlorine oxide molecules, Cl and ClO, respectively.   **18.23** (a) Methane, CH_4, arises from decomposition of organic matter by certain microorganisms; it also escapes from underground gas deposits. (b) SO_2 is released in volcanic gases and also is produced by bacterial action on decomposing vegetable and animal matter. (c) Nitric oxide, NO, results from oxidation of decomposing organic matter and is formed in lightning flashes. (d) CO is a possible product of some vegetable matter decay.   **18.25** (a) $H_2SO_4(aq) + CaCO_3(s) \longrightarrow CaSO_4(s) + H_2O(l) + CO_2(g)$ (b) The $CaSO_4(s)$ would be much less reactive with acidic solution, since it would require a strongly acidic solution to shift the relevant equilibrium to the right. $CaSO_4(s) + 2 H^+(aq) \rightleftharpoons Ca^{2+}(aq) + 2 HSO_4^-(aq)$ **18.27** (a) Ultraviolet (b) 357 kJ/mol (c) The average C—H bond energy from Table 8.4 is 413 kJ/mol. The C—H bond energy in CH_2O, 357 kJ/mol, is less than the "average" C—H bond energy.

(d) H—C(=O)—H + $h\nu$ $\longrightarrow$ H—C(=O)· + H·

18.29 Incoming and outgoing energies are in different regions of the electromagnetic spectrum. CO_2 is transparent to incoming visible radiation but absorbs outgoing infrared radiation.   **18.31** 0.099 M Na^+ **18.33** 4.361×10^5 g CaO   **18.35** The minimum pressure required to initiate reverse osmosis is greater than 5.1 atm.   **18.37** (a) $CO_2(g)$, HCO_3^-, $H_2O(l)$, SO_4^{2-}, NO_3^-, HPO_4^{2-}, $H_2PO_4^-$ (b) $CH_4(g)$, $H_2S(g)$, $NH_3(g)$, $PH_3(g)$ **18.39** 2.5 g O_2 **18.41** $Mg^{2+}(aq) + Ca(OH)_2(s) \longrightarrow Mg(OH)_2(s) + Ca^{2+}(aq)$ **18.43** 0.42 mol $Ca(OH)_2$, 0.18 mol Na_2CO_3 **18.45** $4 FeSO_4(aq) + O_2(aq) + 2 H_2O(l) \longrightarrow 4 Fe^{3+}(aq) + 4 OH^- + 4 SO_4^{2-}(aq)$; $Fe^{3+}(aq) + 3 HCO_3^-(aq) \longrightarrow Fe(OH)_3(s) + 3 CO_2(g)$ **18.47** The fewer steps in a process, the less waste is generated. Processes with fewer steps require less energy at the site of the process, and for subsequent cleanup or disposal of waste.   **18.49** (a) H_2O (b) It is better to prevent waste than to treat it. Produce as little nontoxic waste as possible. Chemical processes should be efficient. Raw materials should be renewable.   **18.51** (a) Acid rain is rain with a larger $[H^+]$ and a lower pH than expected. The additional H^+ is produced by the dissolution of sulfur and nitrogen oxides in rain droplets to form sulfuric and nitric acid. (b) A greenhouse gas such as CO_2 or H_2O absorbs infrared or "heat" radiation emitted from Earth's surface and helps to maintain a relatively constant temperature at the surface. A significant build-up of greenhouse gases in the atmosphere could cause a corresponding increase in the average surface temperature and stimulate global climate change. (c) Photochemical smog is an unpleasant collection of atmospheric pollutants initiated by photochemical dissociation of NO_2 to form NO and O atoms. The major components are $NO(g)$, $NO_2(g)$, $CO(g)$, $O_3(g)$, and unburned hydrocarbons from gasoline. (d) Ozone depletion is the loss of stratospheric ozone due to reactions between O_3 and Cl. Insufficient stratospheric ozone allows damaging ultraviolet radiation to reach our ecosystem.   **18.56** (a) Photodissociation of $CBrF_3$ to form Br atoms requires less energy than the production of Cl atoms and should occur readily in the stratosphere. (b) $CBrF_3(g) \xrightarrow{h\nu} CF_3(g) + Br(g)$; $Br(g) + O_3(g) \longrightarrow BrO(g) + O_2(g)$ **18.59** The formation of $NO(g)$ is endothermic, so K increases with increasing temperature. The oxidation of $NO(g)$ to $NO_2(g)$ is exothermic, so the value of K decreases with increasing temperature.   **18.61** (a) $CH_4(g) + 2 O_2(g) \longrightarrow CO_2(g) + 2 H_2O(g)$ (b) $2 CH_4(g) + 3 O_2(g) \longrightarrow 2 CO(g) + 4 H_2O(g)$ (c) 9.5 L dry air   **18.63** Most of the 390 watts/m^2 radiated from Earth's surface is in the infrared region of the spectrum. Greenhouse gases absorb much of this radiation, which warms the atmosphere close to Earth's surface and makes the planet liveable.   **18.64** 7.1×10^8 m^2 **18.67** Because NO has an odd electron, like $Cl(g)$, it could act as a catalyst for decomposition of stratospheric ozone. This would lead to an increase in the amount of damaging ultraviolet radiation that reaches the troposphere. In the troposphere NO is oxidized to NO_2 by O_2. On dissolving in water, NO_2 disproportionates into $NO_3^-(aq)$ and $NO(g)$. Over time the NO in the troposphere will be converted into NO_3^-, which is in turn incorporated into soils.   **18.69** (a) $P_{NO_2} = 1.4 \times 10^{-5}$ torr (b) 2×10^{19} NO_2 molecules   **18.72** (a) H—Ö—H $\longrightarrow$ H· + ·Ö—H (b) 258 nm (c) The overall reaction is $O_3(g) + O(g) \longrightarrow 2 O_2(g)$. OH(g) is the catalyst in the overall reaction, because it is consumed and then reproduced.   **18.75** (a) The energy required is 9.8×10^3 kJ/gal H_2O. (b) The energy cost is \$0.23/gal H_2O. (c) 18% of the total cost is energy.   **18.78** (a) rate = $k[O_3][H]$ (b) $k_{avg} = 1.13 \times 10^{44}\ M^{-1}\ s^{-1}$ **18.81** 2×10^6 kg H_2SO_4

Chapter 19

19.1 (a)

(b) $\Delta H = 0$ for mixing ideal gases. ΔS is positive, because the disorder of the system increases. (c) The process is spontaneous and therefore irreversible. (d) Since $\Delta H = 0$, the process does not affect the entropy of the surroundings.   **19.4** (a) phase changes: 1 = melting (fusion), 2 = vaporization. (b) The larger volume and greater motional freedom of the gas phase causes ΔS for vaporization (2) to be larger than ΔS for fusion (1). **19.6** (a) At 250 K, $\Delta G = 0$ and the system is at equilibrium. (b) The reaction is spontaneous at temperatures above 250 K. (c) ΔH is the y-intercept of the graph, and is positive. (d) ΔG decreases as T increases, so ΔS is positive.   **19.9** Spontaneous: a, b, c, d; nonspontaneous: e   **19.11** (a) $NH_4NO_3(s)$ dissolves in water, as in a chemical cold pack. Naphthalene (moth balls) sublimes at room temperature. (b) Melting of a solid is spontaneous above its melting point but nonspontaneous below its melting point.   **19.13** (a) Endothermic (b) at or above 100 °C (c) below 100 °C (d) at 100 °C   **19.15** (a) For a *reversible* process, the forward and reverse changes occur by the same path. In a reversible process, both the system and the surroundings are restored to their original condition by exactly reversing the change. A reversible change produces the maximum amount of work. (b) There is no net change in the surroundings. (c) The vaporization of water to steam is reversible if it occurs at the boiling temperature of water for a specified external (atmospheric) pressure, and if the heat needed is added infinitely slowly. (d) No. Natural processes are spontaneous in the direction they occur and nonspontaneous in the opposite direction. By definition they are irreversible.   **19.17** (a) If the ideal gas is contained in a closed system at constant volume, a decrease in external temperature leads to a decrease in both temperature and pressure of the gas. (b) If the ideal gas is contained in a closed system at constant pressure, a decrease in external temperature leads to a decrease in both temperature and volume of the gas. (c) No. ΔE is a state function. $\Delta E = q + w$; q and w are not state functions. Their values do depend on path, but their sum, ΔE, does not.   **19.19** (a) An ice cube can melt reversibly at the conditions of temperature and pressure where the solid and liquid are in equilibrium. (b) We know that melting is a process that increases the energy of the system even though there is no change in temperature. ΔE is not zero for the process.   **19.21** (a) At constant temperature, $\Delta S = q_{rev}/T$, where q_{rev} is the heat that would be transferred if the process were reversible. (b) No. ΔS is a state function, so it is independent of path.   **19.23** (a) Entropy increases (b) 213 J/K **19.25** (a) For a spontaneous process, the entropy of the universe increases; for a reversible process, the entropy of the universe does not change. (b) For a reversible process, if the entropy of the system increases, the entropy of the surroundings must decrease by the same amount. (c) For a spontaneous process, the entropy of the universe must increase, so the entropy of the surroundings must decrease by less than 42 J/K. **19.27** $\Delta S = 0.762$ J/K   **19.29** (a) An increase in temperature produces more available microstates for a system. (b) A decrease in volume produces fewer available microstates for a system. (c) Going from liquid to gas, the number of available microstates increases.   **19.31** (a) ΔS is positive. (b) S of the system clearly increases in 19.9 (b) and (d).   **19.33** S increases in (a) and (c); S decreases in (b)   **19.35** (a) The entropy of a pure crystalline substance at absolute zero is zero. (b) In translational motion the entire molecule moves in a single direction; in rotational motion the molecule rotates or spins around a fixed axis. In vibrational motion the bonds within a molecule stretch and bend, but the average position of the atoms does not change.

19.37

19.39 (a) $Ar(g)$ (b) $He(g)$ at 1.5 atm (c) 1 mol of $Ne(g)$ in 15.0 L (d) $CO_2(g)$ **19.41** (a) $\Delta S < 0$ (b) $\Delta S > 0$ (c) $\Delta S < 0$ (d) $\Delta S > 0$ **19.43** (a) $C_2H_6(g)$ (b) $CO_2(g)$ **19.45** (a) $Sc(s)$, 34.6 J/mol-K; $Sc(g)$, 174.7 J/mol-K. In general, the gas phase of a substance has a larger $S°$ than the solid phase because of the greater volume and motional freedom of the molecules. (b) $NH_3(g)$, 192.5 J/mol-K; $NH_3(aq)$, 111.3 J/mol-K. Molecules in the gas phase have more motional freedom than molecules in solution. (c) 1 mol of $P_4(g)$, 280 J/K; 2 mol of $P_2(g)$, 2(218.1) = 436.2 J/K. More particles have a greater motional energy (more available microstates). (d) C (diamond), 2.43 J/mol-K; C (graphite), 5.69 J/mol-K. The internal entropy in graphite is greater because there is translational freedom among planar sheets of C atoms, while there is very little freedom within the covalent-network diamond lattice.   **19.47** For elements with similar structures, the heavier the atoms, the lower the vibrational frequencies at a given temperature. This means that more vibrations can be accessed at a particular temperature, resulting in greater absolute entropy for the heavier elements.   **19.49** (a) $\Delta S° = -120.5$ J/K. $\Delta S°$ is negative because there are fewer moles of gas in the products. (b) $\Delta S° = +176.6$ J/K. $\Delta S°$ is positive because there are more moles of gas in the products. (c) $\Delta S° = +152.39$ J/K. $\Delta S°$ is positive because the product contains more total particles and more moles of gas. (d) $\Delta S° = +92.3$ J/K. $\Delta S°$ is positive because there are more moles of gas in the products.   **19.51** (a) $\Delta G = \Delta H - T\Delta S$ (b) If ΔG is positive, the process is nonspontaneous, but the reverse process is spontaneous. (c) There is no relationship between ΔG and rate of reaction.   **19.53** (a) Exothermic (b) $\Delta S°$ is negative; the reaction leads to a decrease in disorder. (c) $\Delta G° = -9.9$ kJ (d) If all reactants and products are present in their standard states, the reaction is spontaneous in the forward direction at this temperature.   **19.55** (a) $\Delta H° = -537.22$ kJ, $\Delta S° = 13.7$ J/K, $\Delta G° = -541.40$ kJ, $\Delta G° = \Delta H° - T\Delta S° = -541.31$ kJ (b) $\Delta H° = -106.7$ kJ, $\Delta S° = -142.2$ J/K, $\Delta G° = -64.0$ kJ, $\Delta G° = \Delta H° - T\Delta S° = -64.3$ kJ (c) $\Delta H° = -508.3$ kJ, $\Delta S° = -178$ J/K, $\Delta G° = -465.8$ kJ, $\Delta G° = \Delta H° - T\Delta S° = -455.1$ kJ. The discrepancy in $\Delta G°$ values is due to experimental uncertainties in the tabulated thermodynamic data. (d) $\Delta H° = -165.9$ kJ, $\Delta S° = 1.4$ J/K, $\Delta G° = -166.2$ kJ, $\Delta G° = \Delta H° - T\Delta S° = -166.3$ kJ **19.57** (a) $\Delta G° = -140.0$ kJ, spontaneous (b) $\Delta G° = +104.70$ kJ, nonspontaneous (c) $\Delta G° = +146$ kJ, nonspontaneous

(d) $\Delta G° = -156.7$ kJ, spontaneous　**19.59** (a) $C_6H_{12}(l) +$ $9\,O_2(g) \longrightarrow 6\,CO_2(g) + 6\,H_2O(l)$ (b) Because $\Delta S°$ is negative, $\Delta G°$ is less negative than $\Delta H°$.　**19.61** (a) The forward reaction is spontaneous at low temperatures, but becomes nonspontaneous at higher temperatures. (b) The reaction is nonspontaneous in the forward direction at all temperatures. (c) The forward reaction is nonspontaneous at low temperatures, but becomes spontaneous at higher temperatures.　**19.63** $\Delta S > +76.7$ J/K　**19.65** (a) $T = 330$ K (b) nonspontaneous　**19.67** (a) $\Delta H° = 155.7$ kJ, $\Delta S° = 171.4$ J/K. Since $\Delta S°$ is positive, $\Delta G°$ becomes more negative with increasing temperature. (b) $\Delta G° = 19$ kJ. The reaction is not spontaneous under standard conditions at 800 K (c) $\Delta G° = -15.7$ kJ. The reaction is spontaneous under standard conditions at 1000 K.　**19.69** (a) $T_b = 79$ °C (b) From the *Handbook of Chemistry and Physics*, 74th Edition, $T_b = 80.1$ °C. The values are remarkably close; the small difference is due to deviation from ideal behavior by $C_6H_6(g)$ and experimental uncertainty in the boiling point measurement and the thermodynamic data.　**19.71** (a) $C_2H_2(g) + \frac{5}{2}O_2(g) \longrightarrow$ $2\,CO_2(g) + H_2O(l)$ (b) -1299.5 kJ of heat produced/mol C_2H_2 burned (c) $w_{max} = -1235.1$ kJ/mol C_2H_2　**19.73** (a) ΔG becomes more negative. (b) ΔG becomes more positive. (c) ΔG becomes more positive.　**19.75** (a) $\Delta G° = -5.40$ kJ (b) $\Delta G = 0.30$ kJ　**19.77** (a) $\Delta G° = -16.77$ kJ, $K = 870$ (b) $\Delta G° = 8.0$ kJ, $K = 0.04$ (c) $\Delta G° = -497.9$ kJ, $K = 2 \times 10^{87}$ **19.79** $\Delta H° = 269.3$ kJ, $\Delta S° = 0.1719$ kJ/K (a) $P_{CO_2} = 6.0 \times 10^{-39}$ atm (b) $P_{CO_2} = 1.6 \times 10^{-4}$ atm　**19.81** (a) $HNO_2(aq) \rightleftharpoons H^+(aq) + NO_2^-(aq)$ (b) $\Delta G° = 19.1$ kJ (c) $G = 0$ at equilibrium (d) $\Delta G = -2.72$ kJ　**19.84** (a) $\Delta H > 0$, $\Delta S > 0$ (b) $\Delta H < 0$, $\Delta S < 0$ (c) $\Delta H > 0$, $\Delta S > 0$ (d) $\Delta H > 0$, $\Delta S > 0$ (e) $\Delta H < 0$, $\Delta S > 0$　**19.88** (a) For all three compounds listed, there are fewer moles of gaseous products than reactants in the formation reaction, so we expect $\Delta S_f°$ to be negative. If $\Delta G_f° = \Delta H_f° - T\Delta S_f°$ and $\Delta S_f°$ is negative, $-T\Delta S_f°$ is positive and $\Delta G_f°$ is more positive than $\Delta H_f°$.　**19.92** (a) $K = 4 \times 10^{15}$ (b) An increase in temperature will decrease the mole fraction of CH_3COOH at equilibrium. Elevated temperatures must be used to increase the speed of the reaction. (c) $K = 1$ at 836 K or 563 °C.　**19.96** (a) $\Delta G = 8.77$ kJ (b) $w_{min} = 8.77$ kJ. In practice, a larger than minimum amount of work is required.　**19.100** (a) Acetone, $\Delta S_{vap}° = 88.4$ J/mol-K; dimethyl ether, $\Delta S_{vap}° = 86.6$ J/mol-K; ethanol, $\Delta S_{vap}° = 110$ J/mol-K; octane, $\Delta S_{vap}° = 86.3$ J/mol-K; pyridine, $\Delta S_{vap}° = 90.4$ J/mol-K. Ethanol does not obey Trouton's rule. (b) Hydrogen bonding (in ethanol and other liquids) leads to more ordering in the liquid state and a greater than usual increase in entropy upon vaporization. Liquids that experience hydrogen bonding are probably exceptions to Trouton's rule. (c) Owing to strong hydrogen bonding interactions, water probably does not obey Trouton's rule. $\Delta S_{vap}° = 109.0$ J/mol-K. (d) ΔH_{vap} for $C_6H_5Cl \approx 36$ kJ/mol　**19.103** (a) S increases (b) S decreases (c) S increases (d) S decreases　**19.106** (a) For any given total pressure, the condition of equal moles of the two gases can be achieved at some temperature. For individual gas pressures of 1 atm and a total pressure of 2 atm, the mixture is at equilibrium at 328.5 K or 55.5 °C. (b) 333.0 K or 60 °C (c) 374.2 K or 101.2 °C (d) The reaction is endothermic, so an increase in the value of K as calculated in parts (a)–(c) should be accompanied by an increase in T.

Chapter 20

20.1 In a Brønsted–Lowry acid–base reaction, H^+ is transferred from the acid to the base. In a redox reaction, one or more electrons are transferred from the reductant to the oxidant. The greater the tendency of an acid to donate H^+, the lesser the tendency of its conjugate base to accept H^+. The stronger the acid, the weaker its conjugate base. Similarly, the greater the tendency of a reduced species to donate electrons, the lesser the tendency of the corresponding oxidized species to accept electrons.

The stronger the reducing agent, the weaker the corresponding oxidizing agent.　**20.4** (a) Add 1 M $A^{2+}(aq)$ to the beaker with the A(s) electrode. Add 1 M $B^{2+}(aq)$ to the beaker with the B(s) electrode. Add a salt bridge to enable the flow of ions from one compartment to the other. (b) The A electrode functions as the cathode. (c) Electrons flow through the external circuit from the anode to the cathode, from B to A in this cell. (d) $E_{cell}° = 1.00$ V　**20.7** (a) Line 3 (b) $E_{red} = E_{red}° = 0.80$ V **20.11** (a) *Oxidation* is the loss of electrons. (b) Electrons appear on the products' side (right side). (c) The *oxidant* is the reactant that is reduced. (d) An *oxidizing agent* is the substance that promotes oxidation; it is the oxidant.　**20.13** (a) True (b) false (c) true　**20.15** (a) I, +5 to 0; C, +2 to +4 (b) Hg, +2 to 0; N, −2 to 0 (c) N, +5 to +2; S, −2 to 0 (d) Cl, +4 to +3; O, −1 to 0 **20.17** (a) $TiCl_4(g) + 2\,Mg(l) \longrightarrow Ti(s) + 2\,MgCl_2(l)$ (b) Mg(l) is oxidized; $TiCl_4(g)$ is reduced. (c) Mg(l) is the reductant; $TiCl_4(g)$ is the oxidant.　**20.19** (a) $Sn^{2+}(aq) \longrightarrow Sn^{4+}(aq) + 2e^-$, oxidation (b) $TiO_2(s) + 4\,H^+(aq) + 2e^- \longrightarrow Ti^{2+}(aq) + 2\,H_2O(l)$, reduction (c) $ClO_3^-(aq) + 6\,H^+(aq) + 6e^- \longrightarrow Cl^-(aq) + 3\,H_2O(l)$, reduction (d) $N_2(g) + 8\,H^+(aq) + 6e^- \longrightarrow 2\,NH_4^+(aq)$, reduction (e) $4\,OH^-(aq) \longrightarrow O_2(g) + 2\,H_2O(l) + 4e^-$, oxidation (f) $SO_3^{2-}(aq) + 2\,OH^-(aq) \longrightarrow SO_4^{2-}(aq) + H_2O(l) + 2e^-$, oxidation (g) $N_2(g) + 6\,H_2O(l) + 6e^- \longrightarrow 2\,NH_3(g) + 6\,OH^-(aq)$, reduction　**20.21** (a) $Cr_2O_7^{2-}(aq) + I^-(aq) + 8\,H^+(aq) \longrightarrow 2\,Cr^{3+}(aq) + IO_3^-(aq) + 4\,H_2O(l)$; oxidizing agent, $Cr_2O_7^{2-}$; reducing agent, I^- (b) $4\,MnO_4^-(aq) + 5\,CH_3(aq) + 12\,H^+(aq) \longrightarrow 4\,Mn^{2+}(aq) + 5\,HCO_2H(aq) + 11\,H_2O(l)$; oxidizing agent, MnO_4^-; reducing agent, CH_3OH (c) $I_2(s) + 5\,OCl^-(aq) + H_2O(l) \longrightarrow 2\,IO_3^-(aq) + 5\,Cl^-(aq) + 2\,H^+(aq)$; oxidizing agent, OCl^-; reducing agent, I_2 (d) $As_2O_3(s) + 2\,NO_3^-(aq) + 2\,H_2O(l) + 2\,H^+(aq) \longrightarrow 2\,H_3AsO_4(aq) + N_2O_3(aq)$; oxidizing agent, NO_3^-; reducing agent, As_2O_3 (e) $2\,MnO_4^-(aq) + Br^-(aq) + H_2O(l) \longrightarrow 2\,MnO_2(s) + BrO_3^-(aq) + 2\,OH^-(aq)$; oxidizing agent, MnO_4^-; reducing agent, Br^- (f) $Pb(OH)_4^{2-}(aq) + ClO^-(aq) \longrightarrow PbO_2(s) + Cl^-(aq) + 2\,OH^-(aq) + H_2O(l)$; oxidizing agent, ClO^-; reducing agent, $Pb(OH)_4^{2-}$　**20.23** (a) The reaction $Cu^{2+}(aq) + Zn(s) \longrightarrow Cu(s) + Zn^{2+}(aq)$ is occurring in both figures. In Figure 20.3 the reactants are in contact, while in Figure 20.4 the oxidation half-reaction and reduction half-reaction are occurring in separate compartments. In Figure 20.3 the flow of electrons cannot be isolated or utilized; in Figure 20.4 electrical current is isolated and flows through the voltmeter. (b) Na^+ cations are drawn into the cathode compartment to maintain charge balance as Cu^{2+} ions are removed. **20.25** (a) Fe(s) is oxidized, $Ag^+(aq)$ is reduced. (b) $Ag^+(aq) + 1e^- \longrightarrow Ag(s)$; $Fe(s) \longrightarrow Fe^{2+}(aq) + 2e^-$ (c) Fe(s) is the anode, Ag(s) is the cathode. (d) Fe(s) is negative; Ag(s) is positive. (e) Electrons flow from the Fe electrode (−) toward the Ag electrode (+). (f) Cations migrate toward the Ag(s) cathode; anions migrate toward the Fe(s) anode. **20.27** *Electromotive force*, emf, is the potential energy difference between an electron at the anode and an electron at the cathode of a voltaic cell. (b) One *volt* is the potential energy difference required to impart 1 J of energy to a charge of 1 coulomb. (c) *Cell potential*, E_{cell}, is the emf of an electrochemical cell. **20.29** (a) $2\,H^+(aq) + 2e^- \longrightarrow H_2(g)$ (b) A *standard* hydrogen electrode, SHE, has components that are at standard conditions, 1 M $H^+(aq)$ and $H_2(g)$ at 1 atm. (c) The platinum foil in a SHE serves as an inert electron carrier and a solid reaction surface.　**20.31** (a) A *standard reduction potential* is the relative potential of a reduction half-reaction measured at standard conditions. (b) $E_{red}° = 0$　**20.33** (a) $Cr^{2+}(aq) \longrightarrow Cr^{3+}(aq) + e^-$; $Tl^{3+}(aq) + 2e^- \longrightarrow Tl^+(aq)$ (b) $E_{red}° = 0.78$ V

(c)

Movement of cations →
← Movement of anions

20.35 (a) $E° = 0.823$ V (b) $E° = 1.89$ V (c) $E° = 1.211$ V
(d) $E° = 1.21$ V   **20.37** (a) $3 Ag^+(aq) + Cr(s) \longrightarrow 3 Ag(s) +$
$Cr^{3-}(aq)$, $E° = 1.54$ V (b) Two of the combinations have essentially equal $E°$ values: $2 Ag^+(aq) + Cu(s) \longrightarrow 2 Ag(s) +$
$Cu^{2+}(aq)$, $E° = 0.462$ V; $3 Ni^{2+}(aq) + 2 Cr(s) \longrightarrow 3 Ni(s) +$
$2 Cr^{3+}(aq)$, $E° = 0.46$ V   **20.39** (a) Anode, $Sn(s)$; cathode, $Cu(s)$.
(b) The copper electrode gains mass as Cu is plated out, and
the tin electrode loses mass as Sn is oxidized. (c) $Cu^{2+}(aq) +$
$Sn(s) \longrightarrow Cu(s) + Sn^{2+}(aq)$. (d) $E° = 0.473$ V.   **20.41** (a) $Cl_2(g)$
(b) $Ni^{2+}(aq)$ (c) $BrO_3^-(aq)$ (d) $O_3(g)$ **20.43** (a) $Cl_2(aq)$, strong
oxidant (b) $MnO_4^-(aq)$, acidic, strong oxidant (c) $Ba(s)$ strong
reductant (d) $Zn(s)$, reductant   **20.45** (a) $Cu^{2+}(aq) < O_2(g) <$
$Cr_2O_7^{2-}(aq) < Cl_2(g) < H_2O_2(aq)$ (b) $H_2O_2(aq) < I^-(aq) <$
$Sn^{2+}(aq) < Zn(s) < Al(s)$ **20.47** Al and $H_2C_2O_4$ **20.49** (a)
$2 Fe^{2+}(aq) + S_2O_6^{2-}(aq) + 4 H^+(aq) \longrightarrow 2 Fe^{3+}(aq) +$
$2 H_2SO_3(aq)$; $2 Fe^{2+}(aq) + N_2O(aq) + 2 H^+(aq) \longrightarrow$
$2 Fe^{3+}(aq) + N_2(g) + H_2O(l)$; $Fe^{2+}(aq) + VO_2^+(aq) +$
$2 H^+(aq) \longrightarrow Fe^{3+}(aq) + VO^{2+}(aq) + H_2O(l)$ (b) $E° = -0.17$ V,
$\Delta G° = 33$ kJ; $E° = -2.54$ V, $\Delta G° = 4.90 \times 10^2$ kJ; $E° = 0.23$ V,
$\Delta G° = -22$ kJ (c) $K = 1.8 \times 10^{-6} = 10^{-6}$; $K = 1.2 \times 10^{-86} =$
10^{-86}; $K = 7.8 \times 10^3 = 8 \times 10^3$ **20.51** $\Delta G° = 21.8$ kJ, $E°_{cell} =$
-0.113 V   **20.53** (a) $E° = 0.16$ V, $K = 2.54 \times 10^5 = 3 \times 10^5$
(b) $E° = 0.277$ V, $K = 2.3 \times 10^9$ (c) $E° = 0.45$ V, $K =$
$1.5 \times 10^{75} = 10^{75}$ **20.55** (a) $K = 9.8 \times 10^2$ (b) $K = 9.5 \times 10^5$
(c) $K = 9.3 \times 10^8$ **20.57** (a) The Nernst equation is applicable
when the components of an electrochemical cell are at nonstandard conditions. (b) $Q = 1$ (c) Q decreases and E increases
20.59 (a) E decreases (b) E decreases (c) E decreases (d) no
effect   **20.61** (a) $E° = 0.48$ V (b) $E = 0.53$ V (c) $E = 0.46$ V
20.63 (a) $E° = 0.46$ V (b) $E = 0.37$ V   **20.65** (a) The compartment with $[Zn^{2+}] = 1.00 \times 10^{-2}$ M is the anode. (b) $E° = 0$
(c) $E = 0.0668$ V (d) In the anode compartment $[Zn^{2+}]$ increases;
in the cathode compartment $[Zn^{2+}]$ decreases   **20.67** $E° =$
0.763 V, pH = 1.6   **20.69** (a) The emf of a battery decreases as
it is used. The concentrations of products increase and the
concentrations of reactants decrease, causing Q to increase and
E_{cell} to decrease. (b) A D-size battery contains more reactants
than a AA, enabling the D to provide power for a longer time.
20.71 464 g PbO_2 **20.73** (a) The anode (b) $E° = 3.50$ V (c) The
emf of the battery, 3.5 V, is exactly the standard cell potential
calculated in part (b). (d) At ambient conditions, $E \approx E°$, so
$\log Q \approx 1$. Assuming that the value of $E°$ is relatively constant
with temperature, the value of the second term in the Nernst
equation is approximately zero at 37 °C, and $E \approx 3.5$ V.
20.75 (a) The cell emf will have a smaller value. (b) NiMH batteries use an alloy such as $ZrNi_2$ as the anode material. This
eliminates the use and disposal problems associated with Cd,

a toxic heavy metal.   **20.77** The main advantage of a fuel cell is
that fuel is continuously supplied, so that it can produce electrical current for a time limited only by the amount of available
fuel. For the hydrogen-oxygen fuel cell, this is also a disadvantage because volatile and explosive hydrogen must be acquired
and stored. Alkaline batteries are convenient, but they have a
short lifetime, and the disposal of their zinc and manganese
solids is more problematic than disposal of water produced by
the hydrogen-oxygen fuel cell.   **20.79** (a) anode:
$Fe(s) \longrightarrow Fe^{2+}(aq) + 2e^-$; cathode: $O_2(g) + 4 H^+(aq) +$
$4e^- \longrightarrow 2 H_2O(l)$ (b) $2 Fe^{2+}(aq) + 3 H_2O(l) + 3 H_2O(l) \longrightarrow$
$Fe_2O_3 \cdot 3 H_2O(s) + 6 H^+(aq) + 2e^-$; $O_2(g) + 4 H^+(aq) +$
$4e^- \longrightarrow 2 H_2O(l)$ **20.81** (a) Mg is called a "sacrificial anode"
because it has a more negative $E°_{red}$ than the pipe metal and is
preferentially oxidized when the two are coupled. It is sacri-
ficed to preserve the pipe. (b) $E°_{red}$ for Mg^{2+} is -2.37 V, more
negative than most metals present in pipes, including Fe and
Zn.   **20.83** Under acidic conditions, air (O_2) oxidation of $Zn(s)$,
1.99 V, $Fe(s)$, 1.67 V, and $Cu(s)$, 0.893 V, are all spontaneous.
When the three metals are in contact, Zn will act as a sacrificial anode for both Fe and Cu, but after the Zn is depleted,
Fe will be oxidized (corroded).   **20.85** (a) *Electrolysis* is an
electrochemical process driven by an outside energy source.
(b) By definition, electrolysis reactions are nonspontaneous.
(c) $2 Cl^-(l) \longrightarrow Cl_2(g) + 2e^-$ **20.87** (a) 236 g $Cr(s)$ (b) 2.51 A
20.89 $w_{max} = -8.19 \times 10^4$ J   **20.91** (a) 4.0×10^5 g Li
(b) 0.24 kWh/mol Li   **20.93** (a) $2 Ni^+(aq) \longrightarrow Ni(s) + Ni^{2+}(aq)$
(b) $3 MnO_4^{2-}(aq) + 4 H^+(aq) \longrightarrow 2 MnO_4^-(aq) + MnO_2(s) +$
$2 H_2O(l)$ (c) $3 H_2SO_3(aq) \longrightarrow S(s) + 2 HSO_4^-(aq) +$
$2 H^+(aq) + H_2O(l)$ (d) $Cl_2(aq) + 2 OH^-(aq) \longrightarrow Cl^-(aq) +$
$ClO^-(aq) + H_2O(l)$ **20.96** (a) $E° = 0.672$ V, spontaneous
(b) $E° = -0.82$ V, nonspontaneous (c) $E° = 0.93$ V, spontaneous (d) $E° = 0.183$ V, spontaneous   **20.99** (a) $E° = 1.08$ V
(b) $[Ni^{2+}]$ increases as the cell operates. (c) Initial $[Ag^+] =$
$0.474 = 0.5$ M   **20.102** $K = 1.6 \times 10^6$ **20.105** The ship's hull
should be made negative. The ship, as a negatively charged
"electrode," becomes the site of reduction, rather than oxidation, in an electrolytic process.   **20.107** (a) 6.4×10^6 C
(b) 6.4×10^5 amp (c) 16 kWh   **20.110** A battery is a voltaic cell
in which the cathode compartment contains the positive terminal of the battery. In alkaline batteries, $OH^-(aq)$ is produced at
the cathode. The wire that turns the indicator pink is in contact
with $OH^-(aq)$, so the rightmost wire is connected to the positive terminal of the battery.   **20.113** (a) The oxidation potential
of A is equal in magnitude but opposite in sign to the reduction
potential of A^+. (b) $Li(s)$ has the highest oxidation potential,
$Au(s)$ the lowest. (c) The relationship is reasonable because
both oxidation potential and ionization energy describe removing electrons from a substance. Ionization energy is a property
of gas phase atoms or ions, while oxidation potential is a property of the bulk material.   **20.115** (a) $E° = 0.028$ V (b) anode:
$Fe^{2+}(aq) \longrightarrow Fe^{3+}(aq) + 1e^-$; cathode: $Ag^+(aq) + e^- \longrightarrow$
$Ag(s)$ (c) $\Delta S° = 148.5$ J. Since $\Delta S°$ is positive, $\Delta G°$ will become
more negative and $E°$ will become more positive as tempera-
ture is increased.   **20.118** K_{sp} for AgSCN is 1×10^{12}.

Chapter 21

21.1 (a) ^{24}Ne; outside; reduce neutron-to-proton ratio via
β decay (b) ^{32}Cl; outside; increase neutron-to-proton ratio
via positron emission or orbital electron capture (c) ^{108}Sn;
outside; increase neutron-to-proton ratio via positron emission
or orbital electron capture (d) ^{216}Po; outside; nuclei with $Z \geq 84$
usually decay via α emission.   **21.4** (a) 7 min (b) 0.1 min^{-1}
(c) 30% (3/10) of the sample remains after 12 min. (d) $^{88}_{41}Nb$
21.7 (a) 25 protons, 30 neutrons (b) 80 protons, 121 neutrons
(c) 19 protons, 20 neutrons   **21.9** (a) ^1_0n (b) ^4_2He or $^4_2\alpha$
(c) $^0_0\gamma$ **21.11** (a) $^{90}_{37}Rb \longrightarrow ^{90}_{38}Sr + ^{0}_{-1}e$ (b) $^{72}_{34}Se +$
$^{0}_{-1}e$ (orbital electron) $\longrightarrow ^{72}_{33}As$ (c) $^{76}_{36}Kr \longrightarrow ^{76}_{35}Br + ^{0}_{1}e$

(d) $^{226}_{88}\text{Ra} \longrightarrow ^{222}_{86}\text{Rn} + ^4_2\text{He}$ **21.13** (a) $^{211}_{82}\text{Pb} \longrightarrow ^{211}_{83}\text{Bi} + ^{\ 0}_{-1}\beta$
(b) $^{50}_{25}\text{Mn} \longrightarrow ^{50}_{42}\text{Cr} + ^0_1\text{e}$ (c) $^{179}_{74}\text{W} + ^{\ 0}_{-1}\text{e} \longrightarrow ^{179}_{73}\text{Ta}$
(d) $^{230}_{90}\text{Th} \longrightarrow ^{266}_{88}\text{Ra} + ^4_2\text{He}$ **21.15** 7 alpha emissions, 4 beta
emissions   **21.17** (a) Positron emission (for low atomic num-
bers, positron emission is more common than electron capture)
(b) beta emission (c) beta emission (d) beta emission   **21.19** (a)
Stable: $^{39}_{19}\text{K}$, 20 neutrons is a magic number (b) stable: $^{209}_{83}\text{Bi}$,
126 neutrons is a magic number (c) stable: $^{58}_{28}\text{Ni}$ even proton,
even neutron more likely to be stable; $^{65}_{28}\text{Ni}$ has high neutron/
proton ratio   **21.21** (a) ^4_2He (c) $^{40}_{20}\text{Ca}$ (e) $^{126}_{82}\text{Pb}$ **21.23** The alpha
particle, ^4_2He, has a magic number of both protons and neutrons,
while the proton is an odd proton, even neutron particle.
Alpha is a very stable emitted particle, which makes alpha
emission a favorable process. The proton is not a stable emitted
particle, and its formation does not encourage proton emission
as a process.   **21.25** Protons and alpha particles are positively
charged and must be moving very fast to overcome electrostatic forces which would repel them from the target nucleus. Neutrons are electrically neutral and not repelled by the nucleus.
21.27 (a) $^{252}_{98}\text{Cf} + ^{10}_5\text{B} \longrightarrow 3\,^1_0\text{n} + ^{259}_{103}\text{Lr}$ (b) $^2_1\text{H} + ^3_2\text{He} \longrightarrow$
$^4_2\text{He} + ^1_1\text{H}$ (c) $^1_1\text{H} + ^{11}_5\text{B} \longrightarrow 3\,^4_2\text{He}$ (d) $^{122}_{53}\text{I} \longrightarrow ^{122}_{54}\text{Xe} + ^{\ 0}_{-1}\text{e}$
(e) $^{59}_{26}\text{Fe} \longrightarrow ^{\ 0}_{-1}\text{e} + ^{59}_{27}\text{Co}$ **21.29** (a) $^{238}_{92}\text{U} + ^4_2\text{He} \longrightarrow$
$^{241}_{94}\text{Pu} + ^1_0\text{n}$ (b) $^{14}_7\text{N} + ^4_2\text{He} \longrightarrow ^{17}_8\text{O} + ^1_1\text{H}$ (c) $^{56}_{26}\text{Fe} + ^4_2\text{He} \longrightarrow$
$^{60}_{29}\text{Cu} + ^{\ 0}_{-1}\text{e}$ **21.31** (a) True. The decay rate constant and
half-life are inversely related. (b) False. If X is not radioactive,
its half-life is essentially infinity. (c) True. Changes in the
amount of A would be substantial and measurable over the
40-year time frame, while changes in the amount of X would
be very small and difficult to detect.   **21.33** 28.1 mg tritium
remain after 12.3 yr, 0.201 mg after 100 yr   **21.35** (a) The source
must be replaced after 2.18 yr or 26.2 months; this corresponds
to August of 2008. (b) Partially decayed cobalt-60 can be encased in a gamma-absorbing material such as lead and stored
in a safe, remote facility until it is no longer active.   **21.37** (a)
4.1×10^{-11} g $^{226}_{88}\text{Ra}$ decays in 5.0 min, 1.1×10^{11} alpha particles
emitted (b) 10 mCi   **21.39** $k = 1.21 \times 10^{-4}$ yr^{-1}; $t =$
4.3×10^3 yr   **21.41** $k = 5.46 \times 10^{-10}$ yr^{-1}; $t = 3.0 \times 10^9$ yr
21.43 $\Delta m = 9.474 \times 10^{-9}$ g per mole of Fe_2O_3 **21.45** $\Delta m =$
0.2414960 amu, $\Delta E = 3.604129 \times 10^{-11}$ J/^{27}Al nucleus re-
quired, 8.044234×10^{13} J/100 g ^{27}Al   **21.47** (a) mass defect =
0.098940 amu, binding energy/nucleon = 1.2305×10^{-12} J
(b) mass defect = 0.340423 amu, binding energy/nucleon =
1.37312×10^{-12} J (c) mass defect = 0.949189 amu,
binding energy/nucleon = 1.37533×10^{-12} J   **21.49** (a)
1.71×10^5 kg/d (b) 2.1×10^8 g ^{235}U   **21.51** (a) ^{59}Co; it has the
largest binding energy per nucleon, and binding energy gives
rise to mass defect.   **21.53** (a) NaI is a good source of iodine
because iodine is a large percentage of its mass, it is completely
dissociated into ions in aqueous solution and iodine in the form
of I$^-(aq)$ is mobile and immediately available for bio-uptake.
(b) A Geiger counter placed near the thyroid immediately after
injestion will register background, then gradually increase in
signal until the concentration of iodine in the thyroid reaches a
maximum. Over time, iodine-131 decays, and the signal decreases. (c) The radioactive iodine will decay to 0.01% of the
original amount in approximately 82 days.   **21.55** (a) In a fission chain reaction, one neutron initiates a nuclear transformation that produces more than one neutron. The product
neutrons initiate more transformations, so that the reaction is
self-sustaining. (b) Critical mass is the mass of fissionable material required to sustain a chain reaction so that only one
product neutron is effective at initiating a new transforma-
tion.   **21.57** (a) $^2_1\text{H} + ^2_1\text{H} \longrightarrow ^3_2\text{He} + 4\,^1_0\text{n}$ (b) $^{233}_{92}\text{U} + ^1_0\text{n} \longrightarrow$
$^{133}_{51}\text{Sb} + ^{98}_{41}\text{Nb} + 3\,^1_0\text{n}$ **21.59** (a) The extremely high temperature
is required to overcome electrostatic charge repulsions
between the nuclei so that they can come together to react.

(b) The sun is not solid. No element or compound is solid at
temperatures of 1,000,000 to 10,000,000 K.   **21.61** Hydrogen
abstraction: $\text{RCOOH} + \cdot\text{OH} \longrightarrow \text{RCOO}\cdot + \text{H}_2\text{O}$; deprotonation: $\text{RCOOH} + \text{OH}^- \longrightarrow \text{RCOO}^- + \text{H}_2\text{O}$. Hydroxyl
radical is more toxic to living systems, because it produces
other radicals when it reacts with molecules in the organism.
Hydroxide ion, OH$^-$, on the other hand, will be readily neutralized in the buffered cell environment. The acid-base reactions of OH$^-$ are usually much less disruptive to the organism
than the chain of redox reactions initiated by $\cdot$OH radical.
21.63 (a) 5.3×10^8 dis/s, 5.3×10^8 Bq (b) 6.1×10^2 mrad,
6.1×10^{-3} Gy (c) 5.8×10^3 mrem, 5.8×10^{-2} Sv   **21.65** $^{210}_{82}\text{Pb}$
21.69 (a) $^6_3\text{Li} \longrightarrow ^{56}_{28}\text{Ni} + ^{62}_{31}\text{Ga}$ (b) $^{40}_{20}\text{Ca} + ^{248}_{96}\text{Cm} \longrightarrow$
$^{147}_{62}\text{Sm} + ^{141}_{54}\text{Xe}$ (c) $^{88}_{38}\text{Sr} + ^{84}_{36}\text{Kr} \longrightarrow ^{116}_{46}\text{Pd} + ^{56}_{28}\text{Ni}$
(d) $^{40}_{20}\text{Ca} + ^{238}_{92}\text{U} \longrightarrow ^{70}_{30}\text{Zn} + 4\,^1_0\text{n} + 2\,^{102}_{41}\text{Nb}$ **21.73** The
C—OH bond of the acid and the O—H bond of the alcohol
break in this reaction. Initially, ^{18}O is present in the C—^{18}OH
group of the alcohol. In order for ^{18}O to end up in the ester,
the ^{18}O—H bond of the alcohol must break. This requires
that the C—OH bond in the acid also breaks. The unlabeled O
from the acid ends up in the H$_2$O product.   **21.75** ^{7}Be,
8.612×10^{-13} J/nucleon; ^{9}Be, 1.035×10^{-12} J/nucleon;
^{10}Be: 1.042×10^{-12} J/nucleon. The binding energies/nucleon
for ^{9}Be and ^{10}Be are very similar; that for ^{10}Be is slightly higher.
21.81 1.4×10^4 kg C_8H_{18}

Chapter 22

22.2 (a) Acid-base (Brønsted) (b) From left to right in the reac-
tion, the charges are: 0, 0, 1+, 1− (c) $\text{NH}_3(aq) + \text{H}_2\text{O}(l) \rightleftharpoons$
$\text{NH}_4^+(aq) + \text{OH}^-(aq)$ **22.5** (a) N_2O_5. Resonance structures
other than the ones shown are possible. Those with double
bonds to the central oxygen do not minimize formal charge
and make a smaller contribution to the net bonding model.

(b) N_2O_4. Resonance structures other than the ones shown are
possible.

(c) NO_2. We place the odd electron on N because of electronegativity arguments.

(d) N_2O_3.

(e) NO. We place the odd electron on N because of electronegativity arguments.

$\ddot{\text{N}}{=}\ddot{\text{O}}$

(f) N_2O. The rightmost structure does not minimize formal charge
and makes a smaller contribution to the net bonding model.

$:\text{N}{\equiv}\text{N}{-}\ddot{\ddot{\text{O}}}: \longleftrightarrow :\ddot{\text{N}}{=}\text{N}{=}\ddot{\text{O}}: \longleftrightarrow :\ddot{\ddot{\text{N}}}{-}\text{N}{\equiv}\text{O}:$

22.8 Only ionization energy fits the trend. Electronegativity
varies smoothly, and there is no value for Ar. Both atomic
radius and melting point vary in the direction opposite that
shown in the diagram. X—X single bond enthalpies show no
consistent trend and there should be no value for Ar.

22.11 Metals: (b) Sr, (c) Mn, (e) Rh; nonmetals: (a) P, (d) Se, (f) Kr; metalloids: none.   **22.13** (a) O (b) Br (c) Ba (d) O (e) Co
22.15 (a) N is too small a central atom to fit five fluorine atoms, and it does not have available d orbitals, which can help accommodate more than eight electrons. (b) Si does not readily form π bonds, which are necessary to satisfy the octet rule for both atoms in the molecule. (c) As has a lower electronegativity than N; that is, it more readily gives up electrons to an acceptor and is more easily oxidized.
22.17 (a) $Mg_3N_2(s) + 6 H_2O(l) \longrightarrow 2 NH_3(aq) + 3 Mg(OH)_2(s)$
(b) $2 C_3H_7OH(l) + 9 O_2(g) \longrightarrow 6 CO_2(g) + 8 H_2O(l)$
(c) $MnO_2(s) + C(s) \xrightarrow{\Delta} CO(g) + MnO(s)$ or $MnO_2(s) + 2 C(s) \xrightarrow{\Delta} 2 CO(g) + Mn(s)$ or $MnO_2(s) + C(s) \xrightarrow{\Delta} CO_2(g) + Mn(s)$ (d) $AlP(s) + 3 H_2O(l) \longrightarrow PH_3(g) + Al(OH)_3(s)$ (e) $Na_2S(s) + 2 HCl(aq) \longrightarrow H_2S(g) + 2 NaCl(aq)$
22.19 (a) ^1_1H, protium; ^2_1H, deuterium; ^3_1H, tritium (b) in order of decreasing natural abundance: protium > deuterium > tritium
22.21 Like other elements in group 1A, hydrogen has only one valence electron and its most common oxidation number is +1.
22.23 (a) $Mg(s) + 2 H^+(aq) \longrightarrow Mg^{2+}(aq) + H_2(g)$
(b) $C(s) + H_2O(g) \xrightarrow{1000\,°C} CO(g) + H_2(g)$
(c) $CH_4(g) + H_2O(g) \xrightarrow{1100\,°C} CO(g) + 3 H_2(g)$
22.25 (a) $NaH(s) + H_2O(l) \longrightarrow NaOH(aq) + H_2(g)$
(b) $Fe(s) + H_2SO_4(aq) \longrightarrow Fe^{2+}(aq) + H_2(g) + SO_4^{2-}(aq)$
(c) $H_2(g) + Br_2(g) \longrightarrow 2 HBr(g)$
(d) $2 Na(l) + H_2(g) \longrightarrow 2 NaH(s)$
(e) $PbO(s) + H_2(g) \xrightarrow{\Delta} Pb(s) + H_2O(g)$
22.27 (a) Ionic (b) molecular (c) metallic   **22.29** Vehicle fuels produce energy via combustion reactions. The combustion of hydrogen is very exothermic and its only product, H_2O, is a nonpollutant.   **22.31** Xenon has a lower ionization energy than argon; because the valence electrons are not as strongly attracted to the nucleus, they are more readily promoted to a state in which the atom can form bonds with fluorine. Also, Xe is larger and can more easily accommodate an expanded octet of electrons.   **22.33** (a) ClO_3^-, +5 (b) HI, −1 (c) ICl_3; I, +3, Cl, −1 (d) NaOCl, +1 (e) $HClO_4$, +7 (f) XeF_4; Xe, +4, F, −1   **22.35** (a) iron(III) chlorate (b) chlorous acid (c) xenon hexafluoride (d) bromine pentafluoride (e) xenon oxide tetrafluoride (f) iodic acid   **22.37** (a) van der Waals intermolecular attractive forces increase with increasing number of electrons in the atoms. (b) F_2 reacts with water. $F_2(g) + H_2O(l) \longrightarrow 2 HF(g) + O_2(g)$. That is, fluorine is too strong an oxidizing agent to exist in water. (c) HF has extensive hydrogen bonding. (d) Oxidizing power is related to electronegativity. Electronegativity and oxidizing power decrease in the order given.   **22.39** Perchlorate anion, ClO_4^-, must be extremely unreactive in aqueous solutions and aerobic environments. Although chlorine is in a very high oxidation state, it is not readily reduced, because the ion has a stable, symmetric structure that protects it against reactions.
22.41 (a) As an oxidizing agent in steelmaking; to bleach pulp and paper; in oxyacetylene torches; in medicine to assist in breathing (b) synthesis of pharmaceuticals, lubricants, and other organic compounds where C=C bonds are cleaved; in water treatment.
22.43 (a) $2 HgO(s) \xrightarrow{\Delta} 2 Hg(l) + O_2(g)$
(b) $2 Cu(NO_3)_2(s) \xrightarrow{\Delta} 2 CuO(s) + 4 NO_2(g) + O_2(g)$
(c) $PbS(s) + 4 O_3(g) \longrightarrow PbSO_4(s) + 4 O_2(g)$
(d) $2 ZnS(s) + 3 O_2(g) \longrightarrow 2 ZnO(s) + 2 SO_2(g)$
(e) $2 K_2O_2(s) + 2 CO_2(g) \longrightarrow 2 K_2CO_3(s) + O_2(g)$
22.45 (a) acidic (b) acidic (c) amphoteric (d) basic   **22.47** (a) H_2SeO_3, +4 (b) $KHSO_3$, +4 (c) H_2Te, −2 (d) CS_2, −2 (e) $CaSO_4$, +6
22.49 (a) $2 Fe^{3+}(aq) + H_2S(aq) \longrightarrow 2 Fe^{2+}(aq) + S(s) + 2 H^+(aq)$
(b) $Br_2(l) + H_2S(aq) \longrightarrow 2 Br^-(aq) + S(s) + 2 H^+(aq)$

(c) $2 MnO_4^-(aq) + 6 H^+(aq) + 5 H_2S(aq) \longrightarrow 2 Mn^{2+}(aq) + 5 S(s) + 8 H_2O(l)$
(d) $2 NO_3^-(aq) + H_2S(aq) + 2 H^+(aq) \longrightarrow 2 NO_2(aq) + S(s) + 2 H_2O(l)$
22.51 (a)

Trigonal pyramidal

(b)

Bent (free rotation around S–S bond)

(c)

Tetrahedral (around S)

22.53 (a) $SO_2(s) + H_2O(l) \rightleftharpoons H_2SO_3(aq) \rightleftharpoons H^+(aq) + HSO_3^-(aq)$
(b) $ZnS(s) + 2 HCl(aq) \longrightarrow ZnCl_2(aq) + H_2S(g)$
(c) $8 SO_3^{2-}(aq) + S_8(s) \longrightarrow 8 S_2O_3^{2-}(aq)$
(d) $SO_3(aq) + H_2SO_4(l) \longrightarrow H_2S_2O_7(l)$
22.55 (a) $NaNO_2$, +3 (b) NH_3, −3 (c) N_2O, +1 (d) NaCN, −3 (e) HNO_3, +5 (f) NO_2, +4
22.57 (a)

The molecule is bent around the central oxygen and nitrogen atoms; the four atoms need not be coplanar. The right-most form does not minimize formal charges and is less important in the actual bonding model.

(b)

The molecule is linear.

(c)

The geometry is tetrahedral around the left nitrogen, trigonal pyramidal around the right.

(d)

The ion is trigonal planar; it has three equivalent resonance forms.
22.59 (a) $Mg_3N_2(s) + 6 H_2O(l) \longrightarrow 3 Mg(OH)_2(s) + 2 NH_3(aq)$
(b) $2 NO(g) + O_2(g) \longrightarrow 2 NO_2(g)$
(c) $N_2O_5(g) + H_2O(l) \longrightarrow 2 H^+(aq) + 2 NO_3^-(aq)$
(d) $NH_3(aq) + H^+(aq) \longrightarrow NH_4^+(aq)$
(e) $N_2H_4(l) + O_2(g) \longrightarrow N_2(g) + 2 H_2O(g)$
22.61 (a) $HNO_2(aq) + H_2O(l) \longrightarrow NO_3^-(aq) + 2e^-$
(b) $N_2(g) + H_2O(l) \longrightarrow N_2O(aq) + 2 H^+(aq) + 2e^-$
22.63 (a) H_3PO_3, +3 (b) $H_4P_2O_7$, +5 (c) $SbCl_3$, +3 (d) Mg_3As_2, +5 (e) P_2O_5, +5   **22.65** (a) Phosphorus is a larger atom than nitrogen, and P has energetically available $3d$ orbitals, which participate in the bonding, but nitrogen does not. (b) Only one of the three hydrogens in H_3PO_2 is bonded to oxygen. The other two are bonded directly to phosphorus and are not easily ionized. (c) PH_3 is a weaker base than H_2O so any attempt to add H^+ to PH_3 in the presence of H_2O causes protonation of H_2O. (d) Because of the severely strained bond angles in P_4 molecules, white phosphorus is highly reactive.

22.67 (a) $2 Ca_3PO_4(s) + 6 SiO_2(s) + 10 C(s) \longrightarrow P_4(g) +$
$6 CaSiO_3(l) + 10 CO(g)$
(b) $PBr_3(l) + 3 H_2O(l) \longrightarrow H_3PO_3(aq) + 3 HBr(aq)$
(c) $4 PBr_3(g) + 6 H_2(g) \longrightarrow P_4 + 12 HBr(g)$
22.69 (a) HCN (b) $Ni(CO)_4$ (c) $Ba(HCO_3)_2$ (d) CaC_2
22.71 (a) $ZnCO_3(s) \xrightarrow{\Delta} ZnO(s) + CO_2(g)$
(b) $BaC_2(s) + 2 H_2O(l) \longrightarrow Ba^{2+}(aq) + 2 OH^-(aq) + C_2H_2(g)$
(c) $2 C_2H_2(g) + 5 O_2(g) \longrightarrow 4 CO_2(g) + 2 H_2O(g)$
(d) $CS_2(g) + 3 O_2(g) \longrightarrow CO_2(g) + 2 SO_2(g)$
(e) $Ca(CN)_2(s) + 2 HBr(aq) \longrightarrow CaBr_2(aq) + 2 HCN(aq)$
22.73 (a) $2 CH_4(g) + 2 NH_3(g) + 3 O_2(g) \xrightarrow{800\,°C}$
$2 HCN(g) + 6 H_2O(g)$
(b) $NaHCO_3(s) + H^+(aq) \longrightarrow CO_2(g) + H_2O(l) + Na^+(aq)$
(c) $2 BaCO_3(s) + O_2(g) + 2 SO_2(g) \longrightarrow 2 BaSO_4(s) + 2 CO_2(g)$
22.75 (a) H_3BO_3, +3 (b) $SiBr_4$, +4 (c) $PbCl_2$, +2
(d) $Na_2B_4O_7 \cdot 10 H_2O$, +3 (e) B_2O_3, +3   **22.77** (a) Lead (b) carbon, silicon, and germanium (c) silicon   **22.79** (a) Tetrahedral
(b) Metasilicic acid will probably adopt the single-strand silicate chain structure shown in Figure 22.46 (a). The Si to O ratio is correct and there are two terminal O atoms per Si that can accommodate the two H atoms associated with each Si atom of the acid.   **22.81** (a) Diborane has bridging H atoms linking the two B atoms. The structure of ethane has the C atoms bound directly, with no bridging atoms. (b) B_2H_6 is an electron-deficient molecule. The 6 valence electron pairs are all involved in B—H sigma bonding, so the only way to satisfy the octet rule at B is to have the bridging H atoms shown in Figure 22.48. (c) The term 'hydridic' indicates that the H atoms in B_2H_6 have more than the usual amount of electron density for a covalently bound H atom.
22.85 $BrO_3^-(aq) + XeF_2(aq) + H_2O(l) \longrightarrow Xe(g) +$
$2 HF(aq) + BrO_4^-(aq)$
22.87 (a) SO_3 (b) Cl_2O_5 (c) N_2O_3 (d) CO_2 (e) P_2O_5 **22.90** (a) PO_4^{3-}, +5; NO_3^-, +5, (b) The Lewis structure for NO_4^{3-} would be:

$$\left[\begin{array}{c} :\ddot{O}: \\ | \\ :\ddot{O}-N-\ddot{O}: \\ | \\ :\ddot{O}: \end{array} \right]^{3-}$$

The formal charge on N is +1 and on each O atom is −1. The four electronegative oxygen atoms withdraw electron density, leaving the nitrogen deficient. Since N can form a maximum of four bonds, it cannot form a π bond with one or more of the O atoms to regain electron density, as the P atom in PO_4^{3-} does. Also, the short N—O distance would lead to a tight tetrahedron of O atoms subject to steric repulsion.
22.93 $GeO_2(s) + C(s) \xrightarrow{\Delta} Ge(l) + CO_2(g)$
$Ge(l) + 2 Cl_2(g) \longrightarrow GeCl_4(l)$
$GeCl_4(l) + 2 H_2O(l) \longrightarrow GeO_2(s) + 4 HCl(g)$
$GeO_2(s) + 2 H_2(g) \longrightarrow Ge(s) + 2 H_2O(l)$
22.95 (a) $N_2H_4(aq) + 2 H_2O_2(aq) \longrightarrow N_2(g) + 4 H_2O(l)$
(b) $SO_2(g) + 2 OH^-(aq) + 2 H_2O_2(aq) \longrightarrow SO_4^{2-}(aq) + 2 H_2O(l)$
(c) $NO_2^-(aq) + H_2O_2(aq) \longrightarrow NO_3^-(aq) + H_2O(l)$
(d) $H_2S(g) + H_2O_2(aq) \longrightarrow S(s) + 2 H_2O(l)$
(e) $2 Fe^{2+}(aq) + H_2O_2(aq) + 2 H^+(aq) \longrightarrow 2 Fe^{3+}(aq) + 2 H_2O(l)$
22.97 (a) 88.2 g H_2 (b) 980 L   **22.100** [HClO] = 0.036 M
22.103 (a) $SO_2(g) + 2 H_2S(aq) \longrightarrow 3 S(s) + 2 H_2O(l)$ or
$8 SO_2(g) + 16 H_2S(aq) \longrightarrow 3 S_8(s) + 16 H_2O(l)$ (b) 2.0×10^3 mol $= 5.01 \times 10^4$ L $H_2S(g)$ (c) 9.5×10^4 g S = 210 lb S per ton of coal combusted.   **22.106** (a) covalent network
(b) $MnSi(s) + HF(aq) \longrightarrow SiH_4(g) + MnF_4(s)$. Reduction of Mn(IV) to Mn(II) is unlikely, because F^- is an extremely weak reducing agent.   **22.109** (a) $HOOC-CH_2-COOH \xrightarrow{P_2O_5}$
$C_3O_2 + 2 H_2O$ (b) 24 valence e^-, 12 e^- pair $\ddot{O}=C=C=C=\ddot{O}$

(c) C=O; about 1.23 Å; C=C, 1.34 Å or less. Consecutive C=C bonds require sp hybrid orbitals on C, so we expect the orbital overlap requirements of this bonding arrangement to require smaller than usual C=C distances.
(d) Two possible products are:

$$\ddot{O}=C-\overset{\displaystyle H}{\underset{\displaystyle H}{C}}-C=\ddot{O} \qquad H-\ddot{O}-C=C=C-\ddot{O}-H$$

Chapter 23

23.1 The diagram indicates that the roasting of ZnS is exothermic. The roasting reaction, once under way, will increase the temperature of the oven. The thermodynamic characteristics of the reaction do not affect its rate. Although the reaction produces heat when underway, heating is probably required so that the roasting reaction occurs at a practical rate.   **23.3** The stronger the metal-metal bonding of the element, the greater the magnitude of $\Delta H_f^\circ(g)$. Molybdenum, with valence orbitals that are exactly half-full, has a full bonding band and an empty valence band. It has stronger metal-metal bonding than zirconium and a larger $\Delta H_f^\circ(g)$. Cadmium atoms have filled valence orbitals, filled bonding and valence bands and weaker metal-metal bonding than zirconium. It has a smaller $\Delta H_f^\circ(g)$ than zirconium.   **23.5** Moving from left to right in a period, Z_{eff} increases. Increasing Z_{eff} leads to increasing ionization energy and electronegativity, but decreasing atomic radius. The chart shows a general decrease in magnitude of the property from K to Ge, so the property must be atomic radius.   **23.7** Iron: hematite, Fe_2O_3; magnetite, Fe_3O_4. Aluminum: bauxite, $Al_2O_3 \cdot xH_2O$. In ores, iron is present as the 3+ ion or as both the 2+ and 3+ ions as in magnetite. Aluminum is always present in the +3 oxidation state.   **23.9** An ore consists of a little bit of the stuff we want (chalcopyrite, $CuFeS_2$) and lots of other junk (gangue).
23.11 (a) $Cr_2O_3(s) + 6 Na(l) \longrightarrow 2 Cr(s) + 3 Na_2O(s)$
(b) $PbCO_3(s) \xrightarrow{\Delta} PbO(s) + CO_2(g)$
(c) $2 CdS(s) + 3 O_2(s) \xrightarrow{\Delta} 2 CdO(s) + 2 SO_2(g)$
(d) $ZnO(s) + CO(g) \xrightarrow{\Delta} Zn(l) + CO_2(g)$
23.13 (a) $SO_3(g)$ (b) $CO(g)$ provides a reducing environment for the transformation of Pb^{2+} to Pb. (c) $PbSO_4(s) \longrightarrow PbO(s) + SO_3(g)$; $PbO(s) + CO(g) \longrightarrow Pb(s) + CO_2(g)$
23.15 $\Delta H° = -65.7$ kJ; $\Delta G° = -325.9$ kJ; $\Delta S° = 15.9$ J/K. Both $\Delta H°$ and $\Delta G°$ are negative; the reaction is exothermic and spontaneous under standard conditions at 25 °C.
23.17 The major reducing agent is CO, formed by partial oxidation of the coke (C) with which the furnace is charged. $Fe_2O_3(s) + 3 CO(g) \longrightarrow 2 Fe(l) + 3 CO_2(g)$; $Fe_3O_4(s) + 4 CO(g) \longrightarrow 3 Fe(l) + CO_2(g)$ **23.19** (a) Air serves mainly to oxidize coke to CO; this exothermic reaction also provides heat for the furnace: $2 C(s) + O_2(g) \longrightarrow 2 CO(g)$, $\Delta H = -221$ kJ.
(b) Limestone, $CaCO_3$, is the source of basic oxide for slag formation: $CaCO_3(s) \xrightarrow{\Delta} CaO(s) + CO_2(g)$; $CaO(l) + SiO_2(l) \longrightarrow CaSiO_3(l)$. (c) Coke is the fuel for the blast furnace and the source of CO, the major reducing agent in the furnace. $2 C(s) + O_2(g) \longrightarrow 2 CO(g)$; $4 CO(g) + Fe_3O_4(s) \longrightarrow 4 CO_2(g) + 3 Fe(l)$ (d) Water acts as a source of hydrogen and as a means of controlling temperature. $C(s) + H_2O(g) \longrightarrow CO(g) + H_2(g)$, $\Delta H = +131$ kJ   **23.21** To purify crude cobalt electrochemically, use an electrolysis cell in which the crude metal is the anode, a thin sheet of pure cobalt is the cathode, and the electrolyte is an aqueous solution of a soluble cobalt salt such as $CoSO_4 \cdot 7 H_2O$. Reduction of water doesn't occur, because of kinetic effects. Anode reaction: $Co(s) \longrightarrow Co^{2+}(aq) + 2e^-$; cathode reaction: $Co^{2+}(aq) + 2e^- \longrightarrow Co(s)$.

23.23 Sodium is metallic; each atom is bonded to many others. When the metal lattice is distorted, many bonds remain intact. In NaCl the ionic forces are strong, and the ions are arranged in very regular arrays. The ionic forces tend to be broken along certain cleavage planes in the solid, and the substance does not tolerate much distortion before cleaving.   **23.25** In the electron-sea model, valence electrons move about the metallic lattice, while metal atoms remain more or less fixed in position. Under the influence of an applied potential, the electrons are free to move throughout the structure, giving rise to thermal and electrical conductivity.   **23.27** Moving left to right in the period, atomic mass and Z_{eff} increase. The increase in Z_{eff} leads to smaller bonding atomic radii and shorter metal-metal bond distances. It seems that the extent of metal-metal bonding increases in the series, which is consistent with greater occupancy of the bonding band as the number of valence electrons increases up to 6. The increasing metal-metal bond strength in the series is probably the most important factor influencing the increase in density.   **23.29** (a) Ag (b) Zn. Ductility decreases as the strength of metal-metal bonding increases, producing a stiffer lattice, less susceptible to distortion.   **23.31** White tin has a structure characteristic of a metal, while gray tin has the diamond structure characteristic of group 4A semiconductors. Metallic white tin has the longer bond distance because the valence electrons are shared with twelve nearest neighbors rather than being localized in four bonds as in gray tin. **23.33** An *alloy* contains atoms of more than one element and has the properties of a metal. In a solution alloy the components are randomly dispersed. In a heterogeneous alloy the components are not evenly dispersed and can be distinguished at a macroscopic level. In an intermetallic compound the components have interacted to form a compound substance, as in Cu_3As.   **23.35** (a) Interstitial alloy (b) solution alloy (c) intermetallic compound   **23.37** Isolated atoms: (b) and (f); bulk metal: (a), (c), (d), and (e). Although it seems that atomic radius is a property of isolated atoms, it can only be measured on a bulk sample.   **23.39** Hf has a completed $4f$ subshell, while Zr does not. The build-up in Z that accompanies the filling of the $4f$ orbitals, along with additional shielding of valence electrons, offsets the typical effect of the larger n value for the valence electrons of Hf. Thus, the atomic radius of Hf is about the same as that of Zr, the element above it in group 4B.   **23.41** (a) ScF_3 (b) CoF_3 (c) ZnF_2 (d) MoF_6 **23.43** Chromium, $[Ar]4s^13d^5$, has six valence electrons, some or all of which can be involved in bonding, leading to multiple stable oxidation states. Al, $[Ne]3s^23p^1$, has only three valence electrons, which are all lost or shared during bonding, producing the +3 state exclusively. **23.45** (a) Cr^{3+}, $[Ar]3d^3$ (b) Au^{3+}, $[Xe]4f^{14}5d^8$ (c) Ru^{2+}, $[Kr]4d^6$ (d) Cu^+, $[Ar]3d^{10}$ (e) Mn^{4+}, $[Ar]3d^3$ (f) Ir^+, $[Xe]4f^{14}5d^8$ **23.47** Ti^{2+} **23.49** Fe^{2+} is a reducing agent that is readily oxidized to Fe^{3+} in the presence of O_2 from air.   **23.51** (a) $Fe(s) + 2 HCl(aq) \longrightarrow$ $FeCl_2(aq) + H_2(g)$ (b) $Fe(s) + 4 HNO_3(g) \longrightarrow$ $Fe(NO_3)_3(aq) + NO(g) + 2 H_2O(l)$ **23.53** The unpaired electrons in a paramagnetic material cause it to be weakly attracted into a magnetic field. A diamagnetic material, where all electrons are paired, is very weakly repelled by a magnetic field. **23.55** (a) In ferromagnetic materials, coupled electron spins are aligned in the same direction. In antiferromagnetic materials, coupled spins are aligned in opposite directions and the opposing spins exactly cancel. In ferrimagnetic materials coupled spins are aligned in opposite directions but the opposing spins do not cancel. (b) Antiferromagnetic materials have no net electron spin and cannot be used to make permanent magnets. **23.57** $PbS(s) + O_2(g) \longrightarrow Pb(l) + SO_2(g)$. $SO_2(g)$ is a product of roasting sulfide ores. In an oxygen-rich environment, $SO_2(g)$ is oxidized to $SO_3(g)$, which dissolves in $H_2O(l)$ to form sulfuric acid, $H_2SO_4(aq)$. A sulfuric acid plant near a roasting plant would provide a means for disposing of hazardous $SO_2(g)$ that would also generate a profit.   **23.59** $CO(g)$: $Pb(s)$; $H_2(g)$: $Fe(s)$;

$Zn(s)$: $Au(s)$ **23.62** Based on its reduction potential, Te is less active and more difficult to oxidize than Cu. While Cu is oxidized from the crude anode during electrorefining, Te will not be oxidized. It is likely to accumulate, along with other impurities less active than Cu, in the so-called anode sludge. **23.67** $E°$ will become more negative as the stability (K_f value) of the complex increases.   **23.71** (a) $Mn(s) + 2 HNO_3(aq) \longrightarrow$ $Mn(NO_3)_2(aq) + H_2(g)$ (b) $Mn(NO_3)_2(s) \xrightarrow{\Delta} MnO_2(s) +$ $2 NO_2(g)$ (c) $MnO_2(s) \xrightarrow{\Delta} Mn_3O_4(s) + O_2(g)$ (d) $2 MnCl_2(s) + 9 F_2(g) \longrightarrow 2 MnF_3(s) + 4 ClF_3(g)$ **23.73** (a) Insulator (b) semiconductor (c) metallic conductor (d) metallic conductor (e) insulator (f) metallic conductor   **23.74** (a) 6.7×10^8 g SO_2 (b) 2.3 mol Cu/mol Fe (c) CuO and Fe_2O_3 (d) $Cu_2S(s) + 2 O_2(g) \longrightarrow 2 CuO(s) + SO_2(g)$; $4 FeS(s) +$ $7 O_2(g) \longrightarrow 2 Fe_2O_3(s) + 4 SO_2(g)$ **23.76** 7.3×10^2 g $Ni^{2+}(aq)$ **23.79** $K = 1.6 \times 10^6$ **23.82** (a) The standard reduction potential for $H_2O(l)$ is much greater than that of $Mg^{2+}(aq)$ (-0.83 V vs. -2.37 V). In aqueous solution $H_2O(l)$ would be preferentially reduced and no $Mg(s)$ would be obtained. (b) 1.0×10^3 kg Mg   **23.85** The density of pure Ni is 8.86 g/cm^3; the density of Ni_3Al is 7.47 g/cm^3. In Ni_3Al, one of every four Ni atoms is replaced with a lighter Al atom. The mass of the unit cell contents of Ni_3Al is ~85% of that of pure Ni, and the densities show the same relationship.

Chapter 24

24.1 (a)

(b) coordination number = 4, coordination geometry = square planar (c) oxidation number = +2   **24.4** Structures (3) and (4) are identical to (1); (2) and (5) are geometric isomers of (1).   **24.6** The yellow-orange solution absorbs blue-violet; the blue-green solution absorbs orange-red.   **24.9** (a) In Werner's theory, *primary valence* is the charge of the metal cation at the center of the complex. *Secondary valence* is the number of atoms bound or coordinated to the central metal ion. The modern terms for these concepts are oxidation state and coordination number, respectively. (b) Ligands are the Lewis base in metal-ligand interactions. As such, they must possess at least one unshared electron pair. NH_3 has an unshared electron pair but BH_3, with less than 8 electrons about B, has no unshared electron pair and cannot act as a ligand.   **24.11** (a) +2 (b) 6 (c) 2 mol $AgBr(s)$ will precipitate per mole of complex. **24.13** (a) Coordination number = 4, oxidation number = +2 (b) 5, +4 (c) 6, +3 (d) 5, +2 (e) 6, +3 (f) 4, +2   **24.15** (a) 4 Cl^- (b) 4 Cl^-, 1 O^{2-} (c) 4 N, 2 Cl^- (d) 5 C (e) 6 O (f) 4 N   **24.17** (a) A monodentate ligand binds to a metal via one atom, a bidentate ligand binds through two atoms. (b) Three bidentate ligands fill the coordination sphere of a six-coordinate complex. (c) A tridentate ligand has at least three atoms with unshared electron pairs in the correct orientation to simultaneously bind one or more metal ions.   **24.19** (a) *Ortho*-phenanthroline, o-phen, is bidentate (b) oxalate, $C_2O_4^{2-}$, is bidentate (c) ethylenediaminetetraacetate, EDTA, is pentadentate (d) ethylenediamine, en, is bidentate.   **24.21** (a) The term *chelate effect* refers to the special stability associated with formation of a metal complex containing a polydentate (chelate) ligand relative to a complex containing only monodentate ligands. (b) The increase in entropy, $+\Delta S$, associated with the substitution of a chelating ligand for two or more monodentate ligands generally gives rise to the *chelate effect*. Chemical reactions with $+\Delta S$ tend to be spontaneous, have negative ΔG and large values of K. (c) Polydentate ligands are used as *sequestering agents* to bind metal ions and prevent them from undergoing unwanted chemical reactions without removing them from solution. **24.23** (a) $[Cr(NH_3)_6](NO_3)_3$ (b) $[Co(NH_3)_4CO_3]_2SO_4$

(c) [Pt(en)$_2$Cl$_2$]Br$_2$ (d) K[V(H$_2$O)$_2$Br$_4$] (e) [Zn(en)$_2$][HgI$_4$]
24.25 (a) tetraamminedichlororhodium(III) chloride (b) potassium hexachlorotitanate(IV) (c) tetrachlorooxomolybdenum(VI) (d) tetraaqua(oxalato)platinum (IV) bromide

24.27 (a)

cis   trans

(b) [Pd(NH$_3$)$_2$(ONO)$_2$], [Pd(NH$_3$)$_2$(NO$_2$)$_2$]

(c)

(d) [Co(NH$_3$)$_4$Br$_2$]Cl, [Co(NH$_3$)$_4$BrCl]Br **24.29** Yes. No structural or stereoisomers are possible for a tetrahedral complex of the form MA$_2$B$_2$. The complex must be square planar with cis and trans geometric isomers. **24.31** (a) One isomer (b) trans and cis isomers with 180° and 90° Cl—Ir—Cl angles, respectively (c) trans and cis isomers with 180° and 90° Cl—Fe—Cl angles, respectively. The cis isomer is optically active.
24.33 (a) Visible light has wavelengths between 400 and 700 nm. (b) *Complementary* colors are opposite each other on an artist's color wheel. (c) A colored metal complex absorbs visible light of its complementary color. (d) 196 kJ/mol
24.35 Most of the attraction between a metal ion and a ligand is electrostatic. Whether the interaction is ion-ion or ion-dipole, the ligand is strongly attracted to the metal center and can be modeled as a point negative charge.

24.37 (a)

(b) The magnitude of Δ and the energy of the d-d transition for a d^1 complex are equal. (c) Δ = 203 kJ/mol **24.39** A yellow color is due to absorption of light around 400 to 430 nm, a blue color to absorption near 620 nm. The shorter wavelength corresponds to a higher-energy electron transition and larger Δ value. Cyanide is a stronger-field ligand, and its complexes are expected to have larger Δ values than aqua complexes.

24.41 (a) Ti^{3+}, d^1 (b) Co^{3+}, d^6 (c) Ru^{3+}, d^5 (d) Mo^{5+}, d^1, (e) Re^{3+}, d^4 **24.43** (a) Mn, [Ar]$4s^23d^5$; Mn^{2+}, [Ar]$3d^5$; 1 unpaired electron (b) Ru, [Kr]$5s^14d^7$; Ru^{2+}, [Kr]$4d^6$; 0 unpaired electrons (c) Rh, [Kr]$5s^14d^8$; Rh^{2+}, [Kr]$4d^7$; 1 unpaired electron **24.45** All complexes in this exercise are six-coordinate octahedral.

(a) d^4, high spin (b) d^5, high spin (c) d^6, low spin (d) d^5, low spin

(e) d^3 (f) d^8 **24.47** high spin

24.49 [Pt(NH$_3$)$_6$]Cl$_4$; [Pt(NH$_3$)$_4$Cl$_2$]Cl$_2$; [Pt(NH$_3$)$_3$Cl$_3$]Cl; [Pt(NH$_3$)$_2$Cl$_4$]; K[Pt(NH$_3$)Cl$_5$] **24.52** (a) [24.51(a)] *cis*-tetraamminediaquacobalt(II) nitrate; [24.51(b)] sodium aquapentachlororuthenate(III); [24.51(c)] ammonium *trans*-diaquabisoxalatocobaltate(III); [24.51(d)] *cis*-dichlorobisethylenediamineruthenium(II) (b) Only the complex in 24.51(d) is optically active. The mirror images of (a)–(c) can be superimposed on the original structure. The

chelating ligands in (d) prevent its enantiomers from being superimposable. **24.54** (a) In a square-planar complex, if one pair of ligands is trans, the remaining two coordination sites are also trans to each other. The bidentate ethylenediamine ligand is too short to occupy trans coordination sites, so the trans isomer of [Pt(en)Cl$_2$] is unknown. (b) The minimum steric requirement for a trans-bidentate ligand is a medium-length chain between the two coordinating atoms that will occupy the trans positions. A polydentate ligand such as EDTA is much more likely to occupy trans positions because it locks the metal ion in place with multiple coordination sites and shields the metal ion from competing ligands present in the solution.
24.57 (a) AgCl(s) + 2 NH$_3$(aq) $\longrightarrow$ [Ag(NH$_3$)$_2$]$^+$(aq) + Cl$^-$(aq)
(b) [Cr(en)$_2$Cl$_2$]Cl(aq) + 2 H$_2$O(l) $\longrightarrow$
[Cr(en)$_2$(H$_2$O)$_2$]$^{3+}$(aq) + 3 Cl$^-$(aq); 3 Ag$^+$(aq) + 3 Cl$^-$(aq) $\longrightarrow$
3 AgCl(s) (c) Zn(NO$_3$)$_2$(aq) + 2 NaOH(aq) $\longrightarrow$ Zn(OH)$_2$(s) +
2 NaNO$_3$(aq); Zn(OH)$_2$(s) + 2 NaOH(aq) $\longrightarrow$
[Zn(OH)$_4$]$^{2-}$(aq) + 2 Na$^+$(aq) (d) Co^{2+}(aq) + 4 Cl$^-$(aq) $\longrightarrow$
[CoCl$_4$]$^{2-}$(aq)

24.60 (a)

d^2

(b) Visible light with $\lambda = hc/\Delta$ is absorbed by the complex, promoting one of the d electrons into a higher energy d orbital. The remaining wavelengths are reflected or transmitted; the combination of these wavelengths is the color we see. (c) [V(H$_2$O)$_6$]$^{3+}$ will absorb light with higher energy because it has a larger Δ than [VF$_6$]$^{3-}$. H$_2$O is in the middle of the spectrochemical series and causes a larger Δ than F$^-$, a weak-field ligand.
24.62 [Co(NH$_3$)$_6$]$^{3+}$, yellow; [Co(H$_2$O)$_6$]$^{2+}$, pink; [CoCl$_4$]$^{2-}$, blue
24.64 (a) The term isoelectronic means that the three ions have the same number of valence electrons and the same electron configuration. (b) The three metal atoms are in their maximum oxidation states and have no d-electrons, so there should be no d-d transitions. (c) A ligand-to-metal charge transfer transition occurs when an electron in a filled ligand orbital is excited to an empty d-orbital of the metal. (d) Compounds appear the complementary color of the light they absorb. Permanganate appears purple because it is absorbing 565 nm yellow light. Chromate appears yellow because it is absorbing violet light of approximately 420 nm. The wavelength of the LMCT transition for chromate is smaller than that in permanganate. (e) Yes. A white compound indicates that no visible light is absorbed. Moving left on the periodic chart from Mn to Cr, the wavelength of the LMCT decreases. It is consistent that the ion containing V, further left on the chart, has a LMCT at an even shorter wavelength in the ultraviolet region of the spectrum.

24.66 (a)

(b) sodium dicarbonyltetracyanoferrate(II) (c) +2, 6d electrons (d) We expect the complex to be low spin. Cyanide (and carbonyl) are high on the spectrochemical series, which means the complex will have a large Δ splitting, characteristic of low-spin complexes. **24.71** In carbonic anhydrase, Zn^{2+} is the Lewis acid, withdrawing electron density from H$_2$O, the Lewis base. The O—H bond is polarized and H becomes more ionizable, more acidic than the bulk solvent. **24.73** K$_4$[Mn(ox)$_2$Br$_2$]
24.75 The chemical formula is [Pd(NC$_5$H$_5$)$_2$Br$_2$]. This is an

electrically neutral square-planar complex of Pd(II), a nonelectrolyte whose solutions do not conduct electricity. Because the dipole moment is zero, it must be the trans isomer.   **24.77** 47.3 mg Mg^{2+}/L, 53.4 mg Ca^{2+}/L **24.80** $\Delta E = 3.02 \times 10^{-19}$ J/photon, $\lambda = 657$ nm. The complex will absorb in the visible around 660 nm and appear blue-green.

Chapter 25

25.1 Molecules (c) and (d) are unsaturated.   **25.3** Compound (b), which has hydrogen bonding, has the highest boiling point.   **25.5** (a) Molecule (i), disaccharide (b) molecule (iv), amino acid (c) molecule (iii), organic base (d) molecule (v), alcohol (e) molecule (ii), fatty acid   **25.7** Numbering from the right on the condensed structural formula, C1 has trigonal planar electron domain geometry, 120° bond angles and sp^2 hybridization; C2 and C5 have tetrahedral electron-domain geometry, 109° bond angles, and sp^3 hybridization; C3 and C4 have linear electron domain geometry, 180° bond angles and sp hybridization.   **25.9** Neither NH_3 nor CO are typical organic molecules. NH_3 contains no carbon atoms. Carbon monoxide contains a C atom that does not form four bonds. **25.11** (a) A straight-chain hydrocarbon has all carbon atoms connected in a continuous chain. A branched-chain hydrocarbon has a branch; at least one carbon atom is bound to three or more carbon atoms. (b) An alkane is a complete molecule composed of carbon and hydrogen in which all bonds are σ bonds. An alkyl group is a substituent formed by removing a hydrogen atom from an alkane. (c) Alkanes are said to be saturated because they cannot undergo addition reactions, such as those characteristic of carbon-carbon double bonds. (d) Ethylene (or ethene), $CH_2{=}CH_2$, is unsaturated. **25.13** (a) C_5H_{12} (b) C_5H_{10} (c) C_5H_{10} (d) C_5H_8; saturated: (a), (b); unsaturated: (c), (d)   **25.15** One possible structure is $CH{\equiv}C{-}CH{=}CH{-}C{\equiv}CH$ **25.17** There are at least 46 structural isomers with the formula C_6H_{10}. A few of them are

$$CH_3CH_2CH_2CH_2C{\equiv}CH \qquad CH_3CH_2CH_2C{\equiv}CCH_3$$

25.19 (a) sp^3 (b) sp^2 (c) sp^2 (d) sp **25.21** (a) 2-methylhexane (b) 4-ethyl-2,4-dimethyldecane

25.23 (a) 2,3-dimethylheptane (b) cis-6-methyl-3-octene (c) para-dibromobenzene (d) 4,4-dimethyl-1-hexyne (e) methylcyclobutane   **25.25** Geometric isomerism in alkenes is the result of restricted rotation about the double bond. In alkanes bonding sites are interchangeable by free rotation about the C—C single bonds. In alkynes there is only one additional bonding site on a triply bound carbon, so no isomerism results. **25.27** (a) No

(b)

(c) no (d) no   **25.29** 65   **25.31** (a) An addition reaction is the addition of some reagent to the two atoms that form a multiple bond. In a substitution reaction one atom or group of atoms replaces another atom. Alkenes typically undergo addition, while aromatic hydrocarbons usually undergo substitution.

(b) $CH_3CH_2CH{=}CH{-}CH_3$ + Br_2 $\longrightarrow$
2-pentene
$$CH_3CH_2CH(Br)CH(Br)CH_3$$
2, 3-dibromopentane

(c)

$$C_6H_6 + Cl_2 \xrightarrow{FeCl_3} C_6H_4Cl_2$$

25.33 (a) The 60° C—C—C angles in the cyclopropane ring cause strain that provides a driving force for reactions that result in ring opening. There is no comparable strain in the five- or six-membered rings. (b) $C_2H_4(g)$ + HBr(g) $\longrightarrow$ $CH_3CH_2Br(l)$; $C_6H_6(l)$ + $CH_3CH_2Br(l)$ $\xrightarrow{AlCl_3}$ $C_6H_5CH_2CH_3(l)$ + HBr(g)   **25.35** Not necessarily. That the two rate laws are first order in both reactants and second order overall indicates that the activated complex in the rate-determining step in each mechanism is bimolecular and contains one molecule of each reactant. This is usually an indication that the mechanisms are the same, but it does not rule out the possibility of different fast steps, or a different order of elementary steps.   **25.37** ΔH_{comb}/mol CH_2 for cyclopropane = 696.3 kJ, for cyclopentane = 663.4 kJ. ΔH_{comb}/CH_2 group for cyclopropane is greater because C_3H_6 contains a strained ring. When combustion occurs, the strain is relieved and the stored energy is released.   **25.39** (a) Alcohol (b) amine, alkene (c) ether (d) ketone, alkene (e) aldehyde (f) carboxylic acid, alkyne   **25.41** (a) Propionaldehyde (or propanal):

(b) ethylmethyl ether:

25.43 (a)

(b) $CH_3CH_2CH_2CH_2\overset{\overset{\displaystyle O}{\|}}{C}-OH$

or

(c) $CH_3CH_2CH_2CH_2CH_2CH_2CH_2\overset{\overset{\displaystyle CH_3}{|}}{CH}-\overset{\overset{\displaystyle Cl}{|}}{C}-\overset{\overset{\displaystyle O}{\|}}{C}-OH$

25.45

(a) $CH_3CH_2O-\overset{\overset{\displaystyle O}{\|}}{C}-$

Ethylbenzoate

(b) $CH_3\overset{\overset{\displaystyle H}{|}}{N}-\overset{\overset{\displaystyle O}{\|}}{C}CH_3$

N-methylethanamide or
N-methylacetamide

(c) $-O-\overset{\overset{\displaystyle O}{\|}}{C}CH_3$

Phenylacetate

25.47

(a) $CH_3CH_2\overset{\overset{\displaystyle O}{\|}}{C}-O-CH_3 + NaOH \longrightarrow \left[CH_3CH_2C\overset{\diagup O}{\diagdown_O}\right]^-$

$+ Na^+ + CH_3OH$

(b) $CH_3\overset{\overset{\displaystyle O}{\|}}{C}-O-$ $+ NaOH \longrightarrow \left[CH_3C\overset{\diagup O}{\diagdown_O}\right]^- + Na^+$

$+$ OH

25.49 The presence of both —OH and —C=O groups in pure acetic acid leads us to conclude that it will be a strongly hydrogen-bonded substance. That the melting and boiling points of pure acetic acid are both higher than those of water, a substance we know to be strongly hydrogen-bonded, supports this conclusion.   **25.51** (a) $CH_3CH_2CH_2CH(OH)CH_3$
(b) $CH_3CH(OH)CH_2OH$

(c) $CH_3\overset{\overset{\displaystyle O}{\|}}{C}OCH_2CH_3$

(d)

(e) $CH_3OCH_2CH_3$

25.53 $H-\overset{\overset{\displaystyle H}{|}}{\underset{\underset{\displaystyle H}{|}}{C}}-\overset{\overset{\displaystyle H}{|}}{\underset{\underset{\displaystyle H}{|}}{C}}-\overset{\overset{\displaystyle CH_3}{|}}{\underset{\underset{\displaystyle H}{|}}{C^*}}-\overset{\overset{\displaystyle Br}{|}}{\underset{\underset{\displaystyle Cl}{|}}{C^*}}-\overset{\overset{\displaystyle H}{|}}{\underset{\underset{\displaystyle H}{|}}{C}}-H$ *chiral C atoms

25.55 (a) An α-amino acid contains an NH_2 group attached to the carbon adjacent to the carboxylic acid function. (b) In protein formation, amino acids undergo a condensation reaction between the amino group of one molecule and the carboxylic

acid group of another to form the amide linkage. (c) The bond that links amino acids in proteins is called the peptide bond.

25.57

25.59 (a)

(b) Three tripeptides ar possible: Gly-Gly-His, GGH; Gly-His-Gly, GHG; His-Gly-Gly, HGG   **25.61** The primary structure of a protein refers to the sequence of amino acids in the chain. The secondary structure is the configuration (helical, folded, open) of the protein chain. The tertiary structure is the overall shape of the protein determined by the way the segments fold together (b) X-ray crystallography is the primary and preferred technique for determining protein structure.   **25.63** (a) Carbohydrates, or sugars, are polyhydroxyaldehydes or ketones composed of carbon, hydrogen, and oxygen. They are derived primarily from plants and are a major food source for animals. (b) A monosaccharide is a simple sugar molecule that cannot be decomposed into smaller sugar molecules by hydrolysis. (c) A disaccharide is a carbohydrate composed of two simple sugar units. Hydrolysis breaks a disaccharide into two monosaccharides. (d) A polysaccharide is a polymer composed of many simple sugar units.   **25.65** The empirical formula of cellulose is $C_6H_{10}O_5$. As in glycogen, the six-membered ring form of glucose forms the monomer unit that is the basis of the polymer cellulose. In cellulose, glucose monomer units are joined by β linkages.   **25.67** (a) In the linear form of mannose, the aldehydic carbon is C1. Carbon atoms 2, 3, 4, and 5 are chiral because they each carry four different groups. (b) Both the α (left) and β (right) forms are possible.

25.69 Two important kinds of lipids are fats and fatty acids. Structurally, fatty acids are carboxylic acids with a hydrocarbon chain of more than four carbon atoms (typically 16–20 carbon atoms). Fats are esters formed by condensation of an alcohol, often glycerol, and a fatty acid. Phospholipids are glycerol esters formed from one phosphoric acid $[RPO(OH)_2]$ and two fatty acid (RCOOH) molecules. At body pH, the phosphate group is depronated and has a negative charge. The long, nonpolar hydrocarbon chains do not readily mix with water, but they do interact with the nonpolar chains of other phospholipid molecules to form the inside of a bilayer. The charged phosphate heads interact with polar water molecules on the outsides of the bilayer.   **25.71** *Purines*, with the larger electron cloud and molar mass, will have larger dispersion forces than *pyrimidines* in aqueous solution.
25.73 5′–TACG–3′   **25.75** The complimentary strand for 5′–GCATTGGC–3′ is 3′–CGTAACCG–5′.

25.77

25.80

Cyclopentene does not show cis-trans isomerism because the existence of the ring demands that the C—C bonds be cis to one another.

25.83

(Structures with the —OH group attached to an alkene carbon atom are called "vinyl alcohols" and are not the major form at equilibrium.)   **25.85** (a) Aldehyde, trans-alkene, cis-alkene (b) ether, alcohol, alkene, amine (two of these, one aliphatic and one aromatic) (c) ketone (two of these), amine (two of these) (d) amide, alcohol (aromatic)   **25.91** Glu-Cys-Gly is the only possible order. Glutamic acid has two carboxyl groups that can form a peptide bond with cysteine, so there are two possible structures for glutathione.

25.97

Answers to Give It Some Thought

Chapter 1

page 3 (a) about 100 elements, (b) atoms and molecules

page 6 oxygen, O

page 8 The water molecule contains atoms of two different elements, hydrogen and oxygen. A compound consists of two or more different elements.

page 11 (a) is a chemical change because a new substance is being formed. (b) is a physical change because the water merely changes its physical state and not its composition.

page 14 1 pg, which equals 10^{-12} g.

page 17 $2.5 \times 10^2 \text{ m}^3$. The volume of a rectangular object is length $\times$ width $\times$ height. The units for volume, based on the SI unit of length, m, are m^3. 5.77 L/s is a different derived unit because it contains time in the denominator.

page 20 (b) is inexact because it is a measured quantity. Both (a) and (c) are exact; (a) involves counting; and (c) is a defined value.

page 26 Whenever possible, we must avoid using a conversion factor that has fewer significant figures than the data whose units are being converted. It is best to use at least one more significant figure in the conversion factor than in the data, which is what was done in Sample Exercise 1.9.

Chapter 2

page 39 (a) the law of multiple proportions. (b) The second compound must contain two oxygen atoms for each carbon atom (that is, twice as many carbon atoms as the first compound).

page 43 (top) Most α particles pass through the foil without being deflected because most of the volume of the atoms that comprise the foil is empty space.

page 43 (bottom) (a) The atom has 15 electrons because atoms have equal numbers of electrons and protons. (b) The protons reside in the nucleus of the atom.

page 47 Any single atom of chromium must be one of the isotopes of that element. The isotope mentioned has a mass of 52.94 amu and is probably ^{53}Cr. The atomic weight differs from the mass of any particular atom because it is the average atomic mass of the naturally occurring isotopes of the element.

page 51 (a) Cl, (b) third period and group 7A, (c) 17, (d) nonmetal

page 54 (a) C_2H_6, (b) CH_3, (c) probably the ball-and-stick model because the angles between the sticks indicate the angles between the atoms.

page 58 We write the empirical formulas for ionic compounds. Thus, the formula is CaO.

page 60 The transition metals can form more than one type of cation, and the charges of these ions are therefore indicated explicitly with Roman numerals: Chromium(II) ion is Cr^{2+}. Calcium, on the other hand, always forms the Ca^{2+} ion, so there is no need to distinguish it from other calcium ions with different charges.

page 61 An *-ide* ending usually means a monatomic anion, although there are some anions with two atoms that are also named this way. An *-ate* ending indicates an oxyanion. The most common oxyanions have the *-ate* ending. An *-ite* ending also indicates an oxyanion, but one having less O than the anion whose name ends in *-ate*.

page 62 BO_3^{3-} and SiO_4^{4-}. The borate has three O atoms, like the other oxyanions of the second period in Figure 2.27, and its charge is 3−, following the trend of increasing negative charge as you move to the left in the period. The silicate has four O atoms, as do the other oxyanions in the third period in Figure 2.27, and its charge is 4−, also following the trend of increasing charge moving to the left.

Chapter 3

page 80 Each $Mg(OH)_2$ has 1 Mg, 2 O, and 2 H; thus, $3 Mg(OH)_2$ represents 3 Mg, 6 O, and 6 H.

page 84 The product is an ionic compound involving Na^+ and S^{2-}, and its chemical formula is therefore Na_2S.

page 91 (a) A mole of glucose. By inspection of their chemical formulas we find that glucose has more atoms of hydrogen and oxygen than water and in addition it also has carbon atoms; thus, a molecule of glucose has a greater mass than a molecule of water. (b) They both contain the same number of molecules because a mole of each substance contains 6.02×10^{23} molecules.

page 98 There are experimental uncertainties in the measurements.

page 99 3.14 mol because 2 mol $H_2 \stackrel{\wedge}{=} 1$ mol O_2 based on the stoichiometry of the chemical reaction.

page 100 The number of grams of product formed is the sum of the masses of the two reactants, 50 g. When two substances react in a combination reaction only one substance is formed as a product. According to the law of conservation of mass, the mass of the product must equal the masses of the two reactants.

Chapter 4

page 122 (a) $K^+(aq)$ and $CN^-(aq)$, (b) $Na^+(aq)$ and $ClO_4^-(aq)$

page 123 $MgBr_2$, because it leads to ions in solution.

page 127 Yes, the Na^+ ion. It appears both a reactant and a product in the same state and does change to form a new substance.

page 129 Three. Each COOH group will partially ionize in water to form $H^+(aq)$.

page 130 HBr

page 135 $SO_2(g)$

page 137 (a) Ne, (b) 0

page 142 $Ni^{2+}(aq)$. In Table 4.5 the ease of reduction of ions increases down the table.

page 144 1.00×10^{-2} M solution of sucrose is more concentrated; that is, it has the larger value of molarity. The smaller the value of x in 10^{-x}, the larger the value of 10^{-x}.

page 148 The molarity decreases to 0.25 M. Molarity is directly proportional to the number of moles of solute (not changed) and inversely proportional to the volume of the solution in liters. Doubling the volume changes moles/V to moles/$2V$ and the molarity is reduced by one half.

page 150 12.50 mL. The stoichiometry of the reaction between HBr and NaOH shows a 1:1 mole ratio. The given concentration of NaOH is twice that of HBr and, thus, a given volume of NaOH contains twice as many moles of solute than an equivalent volume of HBr. Therefore, the volume of NaOH required to reach the equivalence point is half of the original volume of HBr.

Chapter 5

page 167 (a) kinetic energy, (b) potential energy, (c) heat, (d) work

page 168 Open system. Humans exchange matter and energy with their surroundings.

page 171

page 172 Endothermic. In Figure 5.5, the final state has a higher internal energy than the initial state and this is a characteristic of an endothermic process. Also, the figure shows $\Delta E > 0$ which means energy flows into the system from the surroundings.

page 173 The balance (current state) does not depend on the ways the money may have been transferred into the account or on the particular expenditures made in withdrawing money from the account. It depends only on the net total of all the transactions.

page 175 (top) No. If ΔV is zero then the expression $w = -P\Delta V$ is also zero.

page 175 (bottom) It provides us with a state function that allows us to focus on heat flow, which is easier to measure than the work that accompanies a process.

page 178 The coefficients indicate the numbers of moles of reactants and products that give rise to the stated enthalpy change.

page 181 Hg(l). Rearranging equation 5.22 gives $\Delta T = \dfrac{q}{C_s \times m}$. When q and m are constant for a series of substances then $\Delta T = \dfrac{\text{constant}}{C_s}$. Therefore, the element with the smallest C_s in Table 5.2 has the largest ΔT, Hg(l).

page 182 (a) The energy lost by a system is gained by its surroundings. (b) $q_{\text{system}} = -q_{\text{surroundings}}$

page 186 (a) The sign of ΔH changes. (b) The magnitude of ΔH doubles.

page 189 $2\,C(s) + H_2(g) \longrightarrow C_2H_2(g)$ $\Delta H_f^{\circ} = 226.7$ kJ

page 194 Fats, because they have the largest fuel value of the three.

Chapter 6

page 215 No. Both visible light and X-rays are forms of electromagnetic radiation. They therefore both travel at the speed of light, c. The differing ability to penetrate skin is due to the different energies of visible light and X-rays, which we will discuss in the next section.

page 216 As temperature increases, the average energy of the emitted radiation increases. Blue-white light is at the short end of the visible spectrum (at about 400 nm), whereas red light is closer to the other end of the visible spectrum (about 700 nm). Thus, the blue-white light has a higher frequency, is more energetic, and is consistent with higher temperatures.

page 218 Ultraviolet. Figure 6.4 shows that a photon in the ultraviolet region of electromagnetic radiation has a higher frequency and greater energy than a photon in the infrared.

page 220 According to the third postulate, photons of only certain allowed frequencies can be absorbed or emitted as the electron changes energy state. The lines in the spectrum correspond to the allowed frequencies.

page 221 Absorb, because it is moving from a lower-energy state ($n = 3$) to a higher-energy state ($n = 7$).

page 223 Yes, all moving objects produce matter waves, but the wavelengths associated with macroscopic objects, such as the baseball, are too small to allow for any way of observing them.

page 224 The small size and mass of a subatomic particle. The term $h/4\pi$ in the uncertainty principle is a very small number that becomes important only when considering extremely small objects, such as electrons.

page 226 Yes, there is a difference. The first statement says that the electron's position is known exactly, which violates the uncertainty principle. The second statement says that there is a high probability of where the electron is, but there is still uncertainty in its position.

page 227 Bohr proposed that the electron in the hydrogen atom moves in a well-defined circular orbit about the nucleus, which violates the uncertainty principle. An orbital is a wave function that gives the probability of finding the electron at any point in space, in accord with the uncertainty principle.

page 228 The energy of an orbital is proportional to $-1/n^2$. The difference between $-1/(2)^2$ and $-1/(1)^2$ is much greater than the difference between $-1/(3)^2$ and $-1/(2)^2$.

page 230 The radial probability function for the 4s orbital will have four maxima and three nodes.

page 232 The probability of finding the electron when it is in this p orbital is greater in the interior of the lobe than on the edges, which corresponds to changes in intensity of the pink color.

page 233 No. We know that the 4s orbital is higher in energy than the 3s orbital. Likewise, we know that the 3d orbitals are higher in energy than the 3s orbital. But, without more information, we do not know whether the 4s orbital is higher or lower in energy than the 3d orbitals.

page 239 The 6s orbital, which starts to hold electrons at element 55, Cs.

page 243 We can't conclude anything! Each of the three elements has a different valence electron configuration for its nd and $(n + 1)s$ subshells: For Ni, $3d^8 4s^2$; for Pd, $4d^{10}$; and for Pt, $5d^9 6s^1$.

Chapter 7

page 257 The atomic number of an element depends on the number of protons in the nucleus, whereas the atomic weight depends (mainly) on the number of protons and the number of neutrons in the nucleus. Another example: Co has a lower atomic number (27) than that of Ni (28) but Co has a larger atomic weight (58.933 amu) than that of Ni (58.693 amu).

page 259 The 2p electron of an Ne atom experiences a greater Z_{eff}. The atomic number of Na is greater than that of Ne, but an electron in the 3s orbital of Na is farther from the nucleus and therefore more screened than is an electron in the 2p orbital of Ne.

page 262 The trends work against each other. Atomic radii tend to increase down a column in the periodic table. Atomic radius is determined by the distribution of the electrons in an atom and the volume they occupy in space. The effective nuclear charge experienced by outer electrons increases slightly down a column and this would tend to reduce orbital "size" and determines the trend in atomic radii of atoms in a column.

page 266 (top) The process in [7.4] requires energy with a shorter wavelength. The second ionization energy is associated with the process in [7.4] and it is a greater endothermic quantity than the first ionization energy associated with the process in [7.3]. An inverse relationship exists between energy of electromagnetic radiation and wavelength: the larger the energy, the smaller the wavelength.

page 266 (bottom) I_2 for a carbon atom. In each process an electron is being removed from an atom or ion with five electrons, either $B(g)$ or $C^+(g)$. The higher nuclear charge of the carbon nucleus makes I_2 for carbon greater than I_1 for boron.

page 269 The same: $[Ar]3d^3$. The $4s$ electrons are removed before the $3d$ electrons in forming transition metal ions.

page 271 The first ionization energy of $Cl^-(g)$ is the energy needed to remove an electron from Cl^-, forming $Cl(g) + e^-$. That is the reverse process of Equation 7.6, so the first ionization energy of $Cl^-(g)$ is +349 kJ/mol.

page 273 In general, increasing ionization energy correlates with decreasing metallic character.

page 275 Molecular, because it has a relatively low melting point. P, because it is a nonmetallic element.

page 278 Cs has the lowest ionization energy among the alkali metals.

page 281 Gastric fluids in the stomach are very acidic (see the "Chemistry at Work" box on antacids in Section 4.3). Metal carbonates are soluble in acidic solution, where they react with the acid to form $CO_2(g)$ and soluble salts, as in Equations 4.19 and 4.20.

page 284 We can extrapolate the data in the table to make intelligent guesses for these numbers. Notice that the atomic radii increase by 0.15 Å and 0.19 Å, respectively, from Cl to Br and from Br to I. We might therefore expect an increase of 0.15–0.2 Å from I to At, leading to an estimate of roughly 1.5 Å for the atomic radius of At. Similarly, we might expect I_1 for At to be about 900 kJ/mol.

Chapter 8

page 299 No. Cl has seven valence electrons. The first and second Lewis symbols are both correct—they both show seven valence electrons, and it doesn't matter which of the four sides has the single electron. The third symbol shows only five electrons and is incorrect.

page 301 (top) CaO is an ionic compound consisting of Ca^{2+} and O^{2-} ions. When Ca and O_2 react to form CaO, two Ca atoms lose two electrons each to form two Ca^{2+} ions and each oxygen atom in O_2 takes up two electrons to form two O^{2-} ions. Thus, we can say that each Ca atom transfers two electrons to each oxygen atom.

page 301 (bottom) No. Figure 7.11 shows that the alkali metal with the smallest first ionization energy is Cs with a value of +376 kJ/mol. Figure 7.12 shows that the halogen with the largest electron affinity is Cl with a value of −349 kJ/mol. The sum of the two energies gives a positive energy (endothermic). Therefore, all other combinations of alkali metals with halogens will also have positive values.

page 305 (top) Palladium, Pd

page 305 (bottom) Weaker. In both H_2 and H_2^+ the two H atoms are principally held together by the electrostatic attractions between the nuclei and the electron(s) concentrated between them. H_2^+ has only one electron between the nuclei whereas H_2 has two and this results in the H—H bond in H_2 being stronger.

page 307 Triple bond. CO_2 has two C—O double bonds. Because the C—O bond in carbon monoxide is shorter, it is likely to be a triple bond.

page 308 Electron affinity measures the energy released when an isolated atom gains an electron to form a 1− ion. The electronegativity measures the ability of the atom to hold on to its own electrons and attract electrons from other atoms in compounds.

page 309 Polar covalent. The difference in electronegativity between N and Si is 3.0 − 1.8 = 1.2. Based on the examples of F_2, HF, and LiF, the difference in electronegativity is great enough to introduce some polarity to the bond, but not sufficient to cause a complete electron transfer from one atom to the other.

page 311 IF. Because the difference in electronegativity between I and F is greater than that between Cl and F, the magnitude of Q should be greater for IF. In addition, because I has a larger atomic radius than Cl, the bond length in IF is longer than that in ClF. Thus, both Q and r are larger for IF, and therefore $\mu = Qr$ will be larger for IF.

page 312 Smaller dipole moment for C—H. The magnitude of Q should be similar for C—H and H—I bonds because the difference in electronegativity for each bond is 0.4. The C—H bond length is 1.1 Å and the H—I bond length is 1.6 Å. Therefore $\mu = Qr$ will be greater for H—I because it has a larger bond length (larger r).

page 314 OsO_4. The data suggests that the yellow substance is a molecular species with its low melting and boiling points. Os in OsO_4 has an oxidation number of +8 and Cr in Cr_2O_3 has an oxidation number of +3. In Section 8.4, we learn that a compound with a metal in a high oxidation state should show a high degree of covalence and OsO_4 fits this situation.

page 317 There is probably a better choice of Lewis structure than the one chosen. Because the formal charges must add up to 0 and the formal charge on the F atom is +1, there must be an atom that has a formal charge of −1. Because F is the most electronegative element, we don't expect it to carry a positive formal charge.

page 319 Yes. There are two resonance structures for ozone that each contribute equally to the overall description of the molecule. Each O—O bond is therefore an average of a single bond and a double bond, which is a "one-and-a-half" bond.

page 320 As "one-and-a-third" bonds. There are three resonance structures, and each of the three N—O bonds is single in two of those structures and double in the third. Each bond in the actual ion is an average of these: $(1 + 1 + 2)/3 = 1\frac{1}{3}$.

page 321 No, it will not have multiple resonance structures. We can't "move" the double bonds, as we did in benzene, because the positions of the hydrogen atoms dictate specific positions for the double bonds. We can't write any other reasonable Lewis structures for the molecule.

page 322 The formal charge of each atom is shown below:

$$\ddot{N}=\ddot{O} \qquad \ddot{N}=\ddot{O}$$
$$\text{F.C.} \quad 0 \qquad 0 \qquad\quad -1 \qquad +1$$

The first structure shows each atom with a zero formal charge and therefore it is the preferred Lewis structure. The second one shows a positive formal charge for an oxygen atom, which is a highly electronegative atom, and this is not a favorable situation.

page 325 The atomization of ethane produces $2\,C(g) + 6\,H(g)$. In this process, six C—H bonds and one C—C bond are broken. We can use $6D(C-H)$ to estimate the amount of enthalpy needed to break the six C—H bonds. The difference between that number and the enthalpy of atomization is an estimate of the bond enthalpy of the C—C bond, $D(C-C)$.

Chapter 9

page 344 (top) Octahedral. Removing two atoms that are opposite each other leads to a square planar geometry.

page 344 (bottom) The molecule does not follow the octet rule because it has ten electrons around the central A atom. There are four electron domains around A: two single bonds, one double bond, and one nonbonding pair.

page 349 Yes. Based on one resonance structure, we might expect the electron domain that is due to the double bond to "push" the domains that are due to the single bonds, leading to angles slightly different from 120°. However, we must remember that there are two other equivalent resonance structures—each of the three O atoms has a double bond to N in one of the three resonance structures (Section 8.6). Because of resonance, all three O atoms are equivalent, and they will experience the same amount of repulsion, which leads to bond angles equal to 120°.

page 351 A tetrahedral arrangement of electron domains is preferred because the bond angles are 109.5° compared to 90° bond angles in a square planar arrangement of electron domains. The larger bond angles result in smaller repulsions among electron domains and a more stable structure.

page 354 No. The C—O and C—S bond dipoles exactly oppose each other, like in CO_2, but because O and S have different electronegativities, the magnitudes of the bond dipoles will be different. As a consequence, the bond dipoles will not cancel each other and the OCS molecule has a nonzero dipole moment.

page 357 Decrease. Figure 9.15 shows the potential energy of the system increasing when the H—H bond length is shorter than the equilibrium H—H bond position. The equilibrium H—H bond length corresponds to the system in its lowest energy state and any other H—H bond lengths correspond to weaker H—H bond strengths.

page 358 (top) The three $2p$ orbitals are equivalent to one another; they differ only in their orientation. Thus, the two Be—F bonds would be equivalent to each other. Because p orbitals are perpendicular to one another, we would expect an F—Be—F bond angle of 90°. Experimentally, the bond angle is 180°.

page 358 (bottom) The unhybridized p orbital is oriented perpendicular to the plane defined by the three sp^2 hybrids (trigonal planar array of lobes), with one lobe on each side of the plane.

page 363 The molecule should not be linear. Because there are three electron domains around each N atom, we expect sp^2 hybridization and H—N—N angles of approximately 120°.

The molecule is expected to be planar; the unhybridized $2p$ orbitals on the N atoms can form a π bond only if all four atoms lie in the same plane. You might notice that there are two ways in which the H atoms can be arranged: They can be both on the same side of the N=N bond or on opposite sides of the N=N bond.

page 370 The molecule would fall apart. With one electron in the bonding MO and one in the antibonding MO, there is no net stabilization of the electrons relative to two separate H atoms.

page 372 Yes. In Be_2^+ there would be two electrons in the σ_{2s} MO but only one electron in the σ_{2s}^* MO, and therefore the ion is predicted to have a bond order of $\frac{1}{2}$. It should (and does) exist.

page 377 No. If the σ_{2p} MO were lower in energy than the π_{2p} MOs, we would expect the σ_{2p} MO to hold two electrons and the π_{2p} MOs to hold one electron each, with the same spin. The molecule would therefore be paramagnetic.

Chapter 10

page 395 The major reason is the relatively large distance between molecules of gases. Each molecule is acting almost independently from the other molecules.

page 396 The height of the column decreases because atmospheric pressure decreases with increasing altitude.

page 400 (top) As the pressure increases, the volume decreases. Doubling the pressure causes the volume to decrease to half its original value.

page 400 (bottom) The volume decreases, but it doesn't decrease to half because the volume is proportional to the temperature on the Kelvin scale but not on the Celsius scale.

page 403 Because 22.41 L is the volume of one mole of the gas at STP, it contains Avogadro's number of molecules, 6.022×10^{23}.

page 407 Because water has a lower molar mass (18.0 g/mol) than N_2 (28.0 g/mol), the water vapor is less dense. Note that density is proportional to the molar mass of a gas as shown in Equation 10.10.

page 410 According to Dalton's law of partial pressures, the pressure that is due to N_2 (its partial pressure) does not change. However, the total pressure that is due to the partial pressures of both N_2 and O_2 increases.

page 414 The average kinetic energies depend only on temperature and not on the identity of the gas. Thus, the trend in average kinetic energies is HCl (298 K) = H_2 (298 K) < O_2 (350 K).

page 415 Average speed: HCl < O_2 < H_2. The average kinetic energy (ε) of gas molecules in a sample is $\varepsilon = \frac{1}{2}m\mu^2$, where μ is the root mean square speed (rms) and m is the mass of a molecule. Rms is not the same as average speed, but it is very close. Thus, the rms of a sample of a gas is inversely proportional to the square root of m, $\mu = \sqrt{\dfrac{2}{m}\varepsilon}$.

The greater the mass of a gas particle, the smaller the rms, and average speed. At the same temperature, all gases have the same average kinetic energy and ε is a constant for the three gases. The gases are listed in order of increasing average speed (decreasing m).

page 420 (a) Mean free path decreases because the molecules are crowded closer together. (b) There is no effect. Although the molecules are moving faster at the higher temperature, they are not crowded any closer together.

page 421 (b) Gases deviate from ideality most at low temperatures and high pressures. Thus, the helium gas would deviate most from ideal behavior at 100 K (the lowest temperature listed) and 5 atm (the highest pressure listed).

page 422 The fact that real gases deviate from ideal behavior can be attributed to the finite sizes of the molecules and to the attractions that exist between molecules.

Chapter 11

page 439 (a) In a gas the energy of attraction is less than the average kinetic energy. (b) In a solid, the energy of attraction is greater than the average kinetic energy.

page 440 $Ca(NO_3)_2$ in water. CH_3OH is a molecular substance and a nonelectrolyte. When it dissolves in water, no ions are present. $Ca(NO_3)_2$ is an ionic compound and a strong electrolyte. When it dissolves in water, the Ca^{2+} ions and the NO_3^- ions interact with the polar water molecules by ion–dipole attractive forces.

page 441 The magnitude of the dipole–dipole force depends on both the magnitude of the dipoles and the distance between the dipoles. We cannot judge the distance from Table 11.2, but we see that acetonitrile has the largest dipole moment and the highest boiling point, suggesting that the dipole–dipole attractions are greatest for that substance.

page 442 (a) Polarizability increases in order of increasing molecular size and molecular weight: $CH_4 < CCl_4 < CBr_4$. (b) The strength of the dispersion forces follows the same order: $CH_4 < CCl_4 < CBr_4$.

page 446 For nearly all substances, the solid phase is more dense than the liquid phase. For water, however, the solid is less dense than the liquid.

page 448 (top) (a) Both viscosity and surface tension decrease with increasing temperature because of the increased molecular motion. (b) Both properties increase as the strength of intermolecular forces increases.

page 448 (bottom) The strengths of the cohesive and adhesive forces are equal.

page 450 Melting (or fusion), endothermic

page 455 CCl_4. Both compounds are nonpolar. Consequently, only dispersion forces exist between the molecules. Because dispersion forces are stronger for the larger, heavier CBr_4 than for CCl_4, CBr_4 has a lower vapor pressure than CCl_4. The substance with the larger vapor pressure at a given temperature is more volatile. Thus, CCl_4 is more volatile than CBr_4.

page 457 The line slopes to the right with increasing pressure because a liquid is more dense than a gas.

page 460 Crystalline solids melt at a specific temperature, whereas amorphous ones tend to melt over a temperature range.

page 461 Density is the mass of the unit cell in grams divided by the volume of the unit cell in cm^3. Calculate the total mass of the unit cell by summing the atomic masses (amu) of the net number of atoms contained within the unit cell. This mass is converted to grams using the equivalence of 1 gram equals 6.02×10^{23} amu. The volume of the unit cell is the length $\times$ width $\times$ height. Assuming a cubic unit cell, the length of an edge is cubed. Volume is converted to units of cm^3. Density is then calculated.

page 464 The higher the coordination number of the particles in a crystal, the greater the packing efficiency.

page 466 C_6H_6. Molecular solids are composed of molecules or nonmetal atoms. Because Co is a metal and K_2O is an ionic

substance, they do not form molecular solids. C_6H_6, however, is a molecular substance and forms a molecular solid.

Chapter 12

page 502 Yes, the molecule has both the $—NH_2$ and $—COOH$ groups in it, which can react as in nylon to make a polymer.

page 503 Vinyl acetate interferes with intermolecular interactions between adjacent ethylene chains, and so reduces crystallinity and melting point.

Chapter 13

page 529 The lattice energy of NaCl(s) must be overcome to separate Na^+ and Cl^- ions and disperse them into a solvent. C_6H_{14} is a nonpolar hydrocarbon and C_6H_{14} molecules are held together by London dispersion forces. An ion is normally not attracted to a nonpolar molecule. (In some situations a weak ion-induced dipole interaction can exist.) Thus, the energy required to separate the ions in NaCl is not recovered in the form of ion–C_6H_{14} interactions and prevents NaCl from dissolving in C_6H_{14}.

page 531 (a) exothermic, (b) endothermic

page 532 No, because the AgCl is not dispersed throughout the liquid phase.

page 535 No. The concentration of sodium acetate is higher than the stable equilibrium value, so some of the dissolved solute comes out of solution when a seed crystal initiates the process. The concentration of sodium acetate eventually reaches its equilibrium value.

page 538 Considerably lower because there would no longer be hydrogen bonding with water.

page 541 Dissolved gases are less soluble as temperature increases and they begin to escape below the boiling point of water. Also, adsorbed oxygen on the surface of the cooking pot begins to escape with increasing temperature.

page 542 230 ppm (1 ppm is 1 part in 10^6). 2.30×10^5 ppb (1 ppb is 1 part in 10^9).

page 544 For dilute solutions the two concentration units are approximately the same numerically. Molarity is the number of moles of solute per liter of solution. Molality is the number of moles of solute per kilogram of solvent. In a liter of solution there will generally be less than 1 kg of water because the solute takes up some of the volume. However, in dilute solutions the two ratios are essentially the same.

page 547 NaCl. Raoult's law is $P_A = X_A P_A^\circ$. From this relationship we see that as the mole fraction of solvent, X_A, decreases (increasing mole fraction of solute), the partial pressure exerted by the solvent vapor, P_A, decreases. The solute with the greater mole fraction will cause a larger reduction in P_A. When 1 mole of NaCl dissolves in water it forms one mole each of Na^+ and Cl^- ions whereas 1 mole of $C_6H_{12}O_6$ does not dissociate because it is a nonelectrolyte. Thus, the mole fraction of NaCl is twice that of $C_6H_{12}O_6$ and the NaCl solution has a greater vapor pressure lowering. A useful relationship can be derived from Raoult's law: $\Delta P_A = X_B P_A^\circ$ where ΔP_A is the vapor pressure lowering of the solvent and X_B is the mole fraction of solute.

page 548 P_A° is the vapor pressure of pure solvent and P_A is the vapor pressure of the solvent when a solute is present.

page 549 Not necessarily; if the solute dissociates into particles, it could have a lower molality and still cause an increase of 0.51 °C. The total molality of all the particles in the solution is $1\ m$.

page 553 The 0.20 *m* solution is hypotonic with respect to the 0.5 *m* solution. A hypotonic solution has the lower osmotic pressure of two solutions.

page 555 They would have the same osmotic pressure because they would have the same total concentration of dissolved particles.

page 558 The smaller droplets carry negative charges because of the embedded stearate ions and thus repel one another.

Chapter 14

page 575 Increasing the partial pressure of a gas increases the number of gas molecules in a given volume and therefore increases the concentration of the gas. Rates normally increase with increasing reactant concentrations.

page 577 The rate of a reaction is measured by the change in concentration of a reactant in a given unit of time. As a reaction proceeds, the concentration of a reactant decreases. As the concentration of a reactant decreases, the frequency with which the reactant particles collide with one another decreases, resulting in a smaller change in reactant concentration, and thus a decreased reaction rate.

page 578 The size of the triangle is mostly a matter of convenience. Drawing a larger triangle will give larger values of both $\Delta[C_4H_9Cl]$ and Δt, but their ratio, $\Delta[C_4H_9Cl]/\Delta t$, remains constant. Thus, the calculated slope is independent of the size of the triangle.

page 581 (a) The rate law for any chemical reaction is the equation that relates the concentrations of reactants to the rate of the reaction. The general form for a rate law is given in Equation 14.7. (b) The quantity *k* in any rate law is the rate constant.

page 582 (a) The rate law is second order in NO, first order in H_2, and third order overall. (b) No. Doubling [NO] will cause the rate to increase fourfold, whereas doubling [H_2] will merely double the rate.

page 587 In (a) the intercept is the initial partial pressure of the CH_3NC, 150 torr. In (b) it is the natural logarithm of this pressure, $\ln(150) = 5.01$.

page 590 As the reaction proceeds, the concentration of the reactant decreases. At the end of each half-life the substance loses half of its initial amount at the start of each half-life. At the end of the first half-life, the substance loses 5.0 g and 5.0 g remains. At the end of the second half-life, 2.5 g remains, and at the end of the third half-life, 1.3 g remains (rounded value). In general, the amount of a substance remaining

after *n* half-lives, is: (initial amount)$\left(\dfrac{1}{2^n}\right)$.

page 591 According to Equation 14.17 the half-life of a second order reaction depends inversely on the concentration of the reactant at the start of each half-life reaction. At the end of each half-life the concentration of the reactant is reduced and thus the half-life is larger for the next half-life reaction. That is, the half-life of a second-order reaction increases during the course of the reaction.

page 592 Molecules must collide in order to react.

page 594 The molecules must not merely collide in order to react, but they must collide in the proper orientation and with an energy greater than the activation energy for the reaction.

page 598 Because the elementary reaction involves two molecules, it is bimolecular.

page 601 The rate law depends not on the overall reaction, but on the slowest step in the mechanism.

page 606 Generally, a catalyst lowers the activation energy by providing a different, lower-energy pathway (different mechanism) for a reaction.

page 608 A heterogeneous catalyst is more easily removed. A heterogeneous catalyst is in a different phase from the reactants and therefore a heterogeneous mixture exists. Typically a heterogeneous catalyst is a solid and the reactants are either in a liquid or gas phase. A heterogeneous catalyst is identifiable as a separate phase in the mixture and an experimental method, such as filtration, can be designed to remove it. In a homogenous mixture, a one phase system, the catalyst is uniformly dispersed throughout the mixture, and its identification and removal is a more complex process.

page 610 (a) active site, (b) substrate

Chapter 15

page 630 (top) (a) The rates of opposing reactions are equal. (b) If the forward step is faster than the reverse step, $k_f > k_r$ and the constant, which equals k_f/k_r, will be greater than 1.

page 630 (bottom) The fact that concentrations no longer change with time indicates that equilibrium has been reached.

page 633 (top) It is independent of the starting concentrations of reactants and products.

page 633 (bottom) They represent equilibrium constants. K_c is obtained when equilibrium concentrations expressed in molarity are substituted into the equilibrium-constant expression. K_p is obtained when equilibrium partial pressures expressed in atmospheres are substituted into the expression.

page 635 $0.00140 \ M/1 \ M = 0.00140$ (no units)

page 638 Because the coefficients in the equation have been multiplied by 3, the exponents in K_p are also multiplied by 3, so that the magnitude of the equilibrium constant will be $(K_p)^3$.

page 639 Because the pure H_2O liquid is omitted from the equilibrium-constant expression, $K_p = P_{H_2O}$. Thus, at any particular temperature, the equilibrium vapor pressure is a constant.

page 641 $K_c = \dfrac{[NH_4^+][OH^-]}{[NH_3]}$. Water is the solvent and is excluded from the equilibrium-constant expression.

page 649 (a) The equilibrium shifts to the right, using up some of the added O_2 and forming NO_2. (b) The equilibrium shifts to the left, forming more NO to replace some of the NO that was removed.

page 650 Increasing the volume of the system decreases the total gas pressure. The system is no longer at equilibrium and it responds to this stress by favoring the side of the equilibrium which counteracts the effect of the change. The total gas pressure is directly related to the total number of gas molecules in the system. Because there are three gas molecules on the left in the equation and only two on the right, the equilibrium position shifts to the left. This change increases the total number of gas molecules by converting SO_3 into SO_2 and O_2 and also increases the total gas pressure until equilibrium is restored.

page 652 Evaporation is an endothermic process. Increasing the temperature of an endothermic process shifts the equilibrium to the right, forming more product. Because the product of the evaporation is the vapor, the vapor pressure increases.

page 655 No; catalysts have no effect on the position of an equilibrium, although they do affect how quickly equilibrium is reached.

Chapter 16

page 668 The H^+ ion for acids and the OH^- ion for bases.

page 670 NH_3 is the base because it accepts a H^+ from HSO_4^- as the reaction moves from the left side of the equation to the right side.

page 672 HNO_3 is a strong acid, which means that NO_3^- has negligible basicity. In most questions involving the conjugate base of a strong acid we assume it does affect the concentration of hydrogen or hydroxide ions of a solution.

page 676 (a) The solution is neutral, $[H^+] = [OH^-]$. (b) The pH increases as $[OH^-]$ increases.

page 678 (top) pH = 14.00 − 3.00 = 11.00. The solution is basic because pH > 7.

page 678 (bottom) From Figure 16.7, we see that the pH must be above about 8, meaning that the solution is basic.

page 681 Because CH_3^- is the conjugate base of a substance that has negligible acidity, CH_3^- must be a strong base. Bases stronger than OH^- abstract H^+ from water molecules: $CH_3^- + H_2O \longrightarrow CH_4 + OH^-$.

page 685 Because weak acids typically undergo very little ionization, often less than 1%. Normally we make the assumption and then check its validity based on the concentration of conjugate based formed in the calculation. If it is ≤5% of the initial concentration of the weak acid, we will use the assumption. If not, we have to do an exact calculation.

page 688 This is the acid-dissociation constant for the loss of the third and final proton from H_3PO_4: $HPO_4^{2-}(aq) \rightleftharpoons H^+(aq) + PO_4^{3-}(aq)$.

page 696 Because the NO_3^- ion is the conjugate base of a strong acid, it will not affect pH. (NO_3^- has negligible basicity.) Because the CO_3^{2-} ion is the conjugate base of a weak acid, it will affect pH, increasing the pH.

page 697 K^+, an alkali metal cation, does not affect pH. Most transition metal ions with a 2+ charge or higher form acidic solutions.

page 700 The increasing acidity down a column is due mainly to decreasing H—X bond strength. The trend across a period is due mainly to the increasing electronegativity of X, which weakens the H—X bond.

page 703 The carboxyl group, —COOH.

page 704 It must have an unshared pair of electrons which can be shared with another atom.

page 706 Fe^{3+}, because it has the highest charge. The ratio $\dfrac{Z_+}{r_+}$

is an indicator we can use to compare relative abilities of cations to form acidic solutions. The greater the ratio for a cation, the greater is its tendency to react with water and form an acidic solution. Note that a large radius and a small positive charge for an ion, such as for Na^+, results in a low value for the ratio. Na^+ exhibits no reaction with water to form an acidic solution. Fe^{3+} has a large charge and a relatively small ion size; thus, it will have a high value for the ratio.

Chapter 17

page 723 (top) (a) $[NH_4^+] = 0.10\ M$; $[Cl^-] = 0.10\ M$; $[NH_3] = 0.12\ M$. (b) Cl^-, (c) Equation 17.2.

page 723 (bottom) HNO_3 and NO_3^-. This is a strong acid and its conjugate base. Buffers are composed of weak acids and their conjugate bases. The NO_3^- ion is merely a spectator in any acid–base chemistry and is therefore ineffective in helping control the pH of a solution.

page 724 (a) The OH^- of NaOH (a strong base) reacts with the acid member of the buffer ($HC_2H_3O_2$), abstracting a proton. Thus, $[HC_2H_3O_2]$ decreases and $[C_2H_3O_2^-]$ increases. (b) The H^+ of HCl (a strong acid) reacts with the base member of the buffer ($C_2H_3O_2^-$). Thus, $[C_2H_3O_2^-]$ decreases and $[HC_2H_3O_2]$ increases.

page 727 The solution will most effectively resist a change in either direction when its pH = pK_a. Thus, the optimal pH is pH = pK_a = $-\log(1.8 \times 10^{-5})$ = 4.74, and the pH range of the buffer is about 4.7 ± 1.

page 730 pH increases. When NaOH, the titrant, is added to the HCl solution, a strong acid-strong base neutralization reaction occurs. The amount of hydrogen ion decreases and pH increases.

page 733 pH = 7. The neutralization of a strong acid and a strong base at the equivalence point gives a solution of a salt and water. The salt contains ions which do not change the pH of water.

page 737 (top) The nearly vertical part of the titration curve at the equivalence point is smaller for the weak acid–strong base titration, and fewer indicators undergo their color change within this narrow range.

page 737 (bottom) The following titration curve shows the titration of 25 mL of Na_2CO_3 with HCl, both with 0.1 M concentrations. The overall reaction between the two is

$$Na_2CO_3(aq) + 2\,HCl(aq) \longrightarrow 2\,NaCl(aq) + CO_2(g) + H_2O(l)$$

The initial pH [sodium carbonate in water only] is near 11 because CO_3^{2-} is a weak base in water. The graph shows two equivalence points, **A** and **B**. The first point, **A**, is reached at a pH of about 9:

$$Na_2CO_3(aq) + HCl(aq) \longrightarrow NaCl(aq) + NaHCO_3(aq)$$

HCO_3^- is weakly basic in water, and is a weaker base than the carbonate ion.

The second point, **B**, is reached at a pH of about 4:

$$NaHCO_3(aq) + HCl(aq) \longrightarrow NaCl(aq) + CO_2(g) + H_2O(l)$$

H_2CO_3, a weak acid, forms and decomposes to carbon dioxide and water.

page 739 Because all three compounds produce the same number of ions, their relative solubilities correspond directly to their K_{sp} values, with the compound with the largest K_{sp} value being most soluble, AgCl.

page 750 They are insoluble in water but dissolve in the presence of either an acid or a base.

page 753 A high concentration of H_2S and a low concentration of H^+ (that is, high pH) will shift the equilibrium to the right, reducing $[Cu^{2+}]$.

page 754 The solution must contain one or more of the cations in group 1 of the qualitative analysis scheme shown in Figure 17.22: Ag^+, Pb^{2+}, or Hg_2^{2+}.

Chapter 18

page 773 It looks like the area of the upper curve to the left of the visible portion is about twice as large as the area of the lower curve. The upper curve corresponds to radiation at the "top" of the atmosphere, and the bottom curve corresponds to radiation at sea level; therefore, we estimate about half of the ultraviolet light that arrives at Earth from the Sun gets absorbed by the upper atmosphere and does not make it to the ground.

page 775 Atomic Cl is the catalyst. In Equation 18.6, atomic Cl is produced by the reaction of CFCs with solar radiation with wavelengths in the range of 190 to 225 nm. Atomic Cl is a catalyst for the overall reaction shown in Equation 18.10. The sequence of reactions above Equation 18.10 shows atomic Cl reacting and reforming, which is characteristic of a catalyst.

page 778 SO_2 in the atmosphere reacts with oxygen to form SO_3. SO_3 reacts with water in the atmosphere to form H_2SO_4. Air containing sulfuric acid is referred to as "acid rain" because the presence of sulfuric acid lowers the pH to about 4.

page 780 Equation 18.13

page 783 The higher the humidity of air, the higher the quantity of water vapor in the atmosphere. Water vapor absorbs infrared radiation and at night it radiates part of the absorbed infrared energy back to the surface of the Earth. This warms the surface and helps to reduce cooling at night.

page 786 Biodegradable organic materials consume oxygen in water. A decrease in oxygen concentration in water stored in a closed container over five days shows a significant presence of these materials.

page 790 As with so many other things, there are trade-offs in selecting processes for green chemistry. Yes, increasing CO_2 is expected to lead to global warming; however, the main source of CO_2 is from combustion of fossil fuels. If we can reduce the amount of CO_2 entering the atmosphere from that source, perhaps we can afford to use supercritical or liquid CO_2 in industrial processes, which are safer for workers than other chemicals and form less harmful by-products. Supercritical (or just liquid) water would be a better choice environmentally if the industrial process you want to change still works with water as a solvent.

Chapter 19

page 804 No, nonspontaneous processes can occur so long as they receive some continuous outside assistance. Examples of nonspontaneous processes with which we may be familiar include the building of a brick wall and the electrolysis of water to form hydrogen gas and oxygen gas.

page 805 No. Just because the system is restored to its original condition doesn't mean that the surroundings have likewise been restored to their original condition.

page 807 ΔS depends not merely on q but on q_{rev}. Although there are many possible paths that could take a system from its initial to final state, there is always only one reversible isothermal path between two states. Thus, ΔS has only one particular value regardless of the path taken between states.

page 809 Because rusting is a spontaneous process, ΔS_{univ} must be positive. Therefore the entropy of the surroundings must increase, and that increase must be larger than the entropy decrease of the system.

page 810 A molecule can vibrate (atoms moving relative to one another) and rotate (tumble), whereas a single atom cannot undergo these motions.

page 811 $S = 0$, based on Equation 19.5 and the fact that $\ln 1 = 0$.

page 817 It must be a perfect crystal at 0 K (third law of thermodynamics), which means it has only a single accessible microstate.

page 819 Depends on whether the system is open, closed, or isolated. For an open or closed system, ΔS_{surr} always increases. The change in entropy of the surroundings in an isothermal process is $\Delta S_{surr} = \dfrac{-q_{sys}}{T}$. The energy evolved by an open or closed system is transferred to the surroundings and $-q_{sys}$ is a positive number. Thus ΔS_{surr} is a positive number and the entropy of the surroundings increases. If it is an isolated system, then energy is not transferred to the surroundings and the entropy of the surroundings does not change.

page 820 In any spontaneous process the entropy of the universe increases. In any spontaneous process operating at constant temperature, the free energy of the system decreases.

page 822 It indicates that the process to which the thermodynamic quantity refers has taken place under standard conditions, as summarized in Table 19.3.

page 825 Above the boiling point, vaporization is spontaneous, and $\Delta G < 0$. Therefore, $\Delta H - T\Delta S < 0$, and $\Delta H < T\Delta S$.

Chapter 20

page 845 Oxygen is first assigned a -2 oxidation number. Nitrogen must have a $+3$ oxidation number for the sum of oxidation numbers to equal -1, the charge of the ion.

page 848 No. Electrons should appear in the two half-reactions but cancel when the half-reactions are added properly. That is, the oxidation half-reaction and the reduction half-reaction must show the same number of electrons, but on different sides of the arrow. The electrons cancel when the two half-reactions are added. So, when you are balancing a redox equation and find yourself with e^-'s on either side of the reaction arrow when you are done, you know you should go back and check your work.

page 853 The anode is where oxidation takes place. Because electrons are being removed from the anode, negatively charged anions must migrate to the anode to maintain charge balance.

page 854 The zinc atoms on the surface lose two electrons, forming Zn^{2+} ions, which leave and go into the solution. The surface of the electrode becomes pitted and the size of the electrode diminishes as the reaction proceeds.

page 855 Yes. A redox reaction in a cell with a positive potential is spontaneous.

page 858 1 atm pressure of $Cl_2(g)$ and $1\ M$ solution of $Cl^-(aq)$.

page 859 True. $E^\circ_{cell} = E^\circ_{red}$ (cathode) $- E^\circ_{red}$ (anode) for a cell at standard-state conditions. E°_{cell} is a measure of the net driving force for the overall redox reaction; thus, the smaller the number, the smaller is the net driving force for the overall redox reaction.

page 863 Pb(s) is stronger because it lies above Hg(l) in the activity series, which means that it more readily loses electrons.

page 876 Al, Zn. Both are easier to oxidize than Fe.

Chapter 21

page 895 The mass number decreases by 4.

page 897 Only the neutron, as it is the only neutral particle listed.

page 898 From Figure 21.2, we see that the center of the belt of stability for a nucleus containing 70 protons lies at about 102 neutrons.

page 903 Particle accelerators are designed to accelerate particles with a charge. Normal particle accelerators cannot accelerate neutrons, which are not charged particles. Fast-moving neutrons are typically formed by high-energy protons crashing into a nucleus, resulting in the emission of a number of fast-moving particles, including neutrons.

page 906 (top) Spontaneous radioactive decay is a unimolecular process: A $\longrightarrow$ Products. The rate law that fits this observation is a first-order kinetic rate law, rate = k[A]. A second-order kinetic process has rate = k[A]^2 and the elementary reaction is bimolecular: A + A $\longrightarrow$ Products. A zero-order kinetic process has rate = k, and the rate does not change until the limiting reactant is entirely consumed. The latter two rate laws do not fit a unimolecular process.

page 906 (bottom) (a) Yes; doubling the mass would double the amount of radioactivity of the sample as shown in equation 21.18. (b) No; changing the mass would not change the half-life as shown in equation 21.20.

page 909 No. Alpha particles are more readily absorbed by matter than beta or gamma rays. Geiger counters must be calibrated for the radiation they are being used to detect.

page 913 No. Stable nuclei having mass numbers around 100 are the most stable nuclei. They could not form a still more stable nucleus with an accompanying release of energy.

Chapter 22

page 933 No. Phosphorus, being a third-row element, does not have a capacity for forming strong π bonds. Instead, it exists as a solid in which the phosphorus atoms are singly bonded to one another.

page 938 No. As shown in Figure 22.7, the free energy of formation of H$_2$Se(g) is positive, indicating that the equilibrium constant for the reaction will be small. (Section 19.7)

page 943 NaBrO$_3$ and NaClO$_3$ should be strong oxidizing agents because the halogen in each is in the +5 oxidation state and readily reducible to a lower oxidation state such as 0 or −1. In the Appendix, we find the standard reduction potential in acid solution for BrO$_3^-$ is +1.52 V and for ClO$_3^-$ it is +1.47 V. At standard state conditions, the bromate ion is a slightly better oxidizing agent than the chlorate ion.

page 946 HIO$_3$. Note that the iodine remains in the +5 oxidation state.

page 951 No; the oxidation number of sulfur is the same in both reactants and products.

page 956 The nitrogen atom in nitric acid has an oxidation number of +5 and that in nitrous acid has an oxidation number of +3.

page 963 MgCO$_3$(s). The bicarbonate ion of washing soda ionizes to form a proton and CO$_3^{2-}$ ion. Mg^{2+} reacts with CO$_3^{2-}$ to precipitate MgCO$_3$

Chapter 23

page 984 No. CaO contains a metal which is too active to be reduced with CO. Yes. Ag$_2$O should reduce readily with CO because Ag is not an active metal.

page 987 Gold is the reducing agent, oxidized from 0 to +1 oxidation state. Dissolved oxygen is the oxidizing agent, reduced from 0 to −2 oxidation state.

page 988 Al$_2$O$_3$

page 989 Na$^+$ and Cl$^-$

page 990 Ionic conductivity. Charles Hall needed an ionic salt that melted to form a conducting medium, dissolved aluminum oxide and did not interfere with electrolysis. Ions move toward the anode and cathode in the molten salt solution and electrons in the outer circuit of the cell.

page 991 Cu^{2+} and H$^+$ as cations, and SO$_4^{2-}$ as the anion

page 995 Hg. Hg has more valence electron per atom than W and also lies to the right of the elements in the middle of the transition columns. Thus, Hg has more antibonding orbitals occupied. W is to the left of the elements in middle and has no antibonding orbitals occupied. Hg, with antibonding orbitals occupied, should have weaker bonding between atoms compared to the bonding structure in W. Hg is a liquid at room temperature and has a melting point of −38.8 °C and W is a solid at room temperature and has a melting point of 3422 °C.

page 997 Interstitial, because B is so much smaller than Pd.

page 999 Cr has a smaller bonding atomic radius than Mn as shown in Table 23.4. From Figure 23.19, we see in the group 7B metals that the elements of the third series have greater bonding atomic radii than those of the first series. Therefore, Re, which is in the third series, is larger than either Cr or Mn. Cr has the smallest bonding atomic radius.

page 1002 Yes. From Section 6.9 we expect Cr to have an outer $4s^2 3d^4$ valence electron configuration and Cr^{3+} to have a $3d^3$ configuration. Cr^{3+} would possess three unpaired electrons. Note: even if we wrote $4s^2 3d^1$ instead of $3d^3$ it would still have one unpaired electron.

Chapter 24

page 1016 [Fe(H$_2$O)$_6$]$^{3+}$(aq) + SCN$^-$(aq) $\longrightarrow$
[Fe(H$_2$O)$_5$NCS]$^{2+}$(aq) + H$_2$O(l)

page 1018 (a) tetrahedral and square planar; (b) octahedral

page 1021 Each NH$_3$ has one donor atom. Consequently, the CO$_3^{2-}$ ion must have two donor atoms to give the cobalt atom a coordination number of 6. Thus, CO$_3^{2-}$ is acting as a bidentate ligand.

page 1023 The porphine ligand has conjugated double bonds which permit it to strongly absorb light in the visible region.

page 1031 The two compounds have the same composition and bonds, but they are optical isomers of each other (nonsuperimposable mirror images). The *d* isomer rotates plane-polarized light to the right (dextrorotatory), whereas the *l* isomer rotates plane-polarized light to the left (levorotatory).

page 1033 (a) [Ar]$4s^2 3d^7$, three unpaired electrons; (b) [Ar]$3d^6$, four unpaired electrons

page 1035 The d_{z^2} and $d_{x^2-y^2}$ orbitals

page 1036 The Ti(IV) ion has no *d* electrons, so there can be no *d-d* transitions, which are the ones usually responsible for the color of transition-metal compounds.

page 1039 The d_{xy} orbital, which has electron density in the *xy* plane, strongly interacts with the four ligands which are along the *x* and *y* axes. The d_{xz} and d_{yz} orbitals interact less strongly with the ligands in a tetrahedral complex compared to an octahedral complex because the two ligands along the vertical *z*-axis are removed. (Note that the two orbitals have a significant *z*-component of electron density.) These differences result in the d_{xy} orbital having a higher energy than that of the d_{xz} or d_{yz} orbitals.

Chapter 25

page 1053 C=N, because it is a polar double bond. C—H and C—C bonds are relatively unreactive.

page 1056 Two C—H bonds and two C—C bonds

page 1059 C_3H_7. The propyl group is formed by removing one hydrogen atom from propane, C_3H_8.

page 1063 Only two of the four possible C=C bond sites are distinctly different in a linear chain of five carbon atoms with one double bond.

page 1069

page 1074

page 1079 All four groups must be different from one another.

page 1085 No. Breaking the hydrogen bonds between N—H and O=C groups in a protein by heating causes the α-helix structure to unwind and the β-sheet structure to separate.

page 1089 The α form of the C—O—C linkage. Glycogen serves as a source of energy in the body, which means that the body's enzymes must be able to hydrolyze it to sugars. The enzymes work only on polysaccharides having the α linkage.

Glossary

absorption spectrum   A pattern of variation in the amount of light absorbed by a sample as a function of wavelength. (Section 24.5)

accuracy   A measure of how closely individual measurements agree with the correct value. (Section 1.5)

acid   A substance that is able to donate a H^+ ion (a proton) and hence increases the concentration of $H^+(aq)$ when it dissolves in water. (Section 4.3)

acid-dissociation constant (K_a)   An equilibrium constant that expresses the extent to which an acid transfers a proton to solvent water. (Section 16.6)

acidic anhydride (acidic oxide)   An oxide that forms an acid when added to water; soluble nonmetal oxides are acidic anhydrides. (Section 22.5)

acidic oxide (acidic anhydride)   An oxide that either reacts with a base to form a salt or with water to form an acid. (Section 22.5)

acid rain   Rainwater that has become excessively acidic because of absorption of pollutant oxides, notably SO_3, produced by human activities. (Section 18.4)

actinide element   Element in which the $5f$ orbitals are only partially occupied. (Section 6.8)

activated complex (transition state)   The particular arrangement of atoms found at the top of the potential-energy barrier as a reaction proceeds from reactants to products. (Section 14.5)

activation energy (E_a)   The minimum energy needed for reaction; the height of the energy barrier to formation of products. (Section 14.5)

active site   Specific site on a heterogeneous catalyst or an enzyme where catalysis occurs. (Section 14.7)

activity   The decay rate of a radioactive material, generally expressed as the number of disintegrations per unit time. (Section 21.4)

activity series   A list of metals in order of decreasing ease of oxidation. (Section 4.4)

addition polymerization   Polymerization that occurs through coupling of monomers with one another, with no other products formed in the reaction. (Section 12.6)

addition reaction   A reaction in which a reagent adds to the two carbon atoms of a carbon–carbon multiple bond. (Section 25.3)

adsorption   The binding of molecules to a surface. (Section 14.7)

alcohol   An organic compound obtained by substituting a hydroxyl group ($-OH$) for a hydrogen on a hydrocarbon. (Sections 2.9 and 25.4)

aldehyde   An organic compound that contains a carbonyl group ($C=O$) to which at least one hydrogen atom is attached. (Section 25.4)

alkali metals   Members of group 1A in the periodic table. (Section 7.7)

alkaline earth metals   Members of group 2A in the periodic table. (Section 7.7)

alkanes   Compounds of carbon and hydrogen containing only carbon–carbon single bonds. (Sections 2.9 and 25.2)

alkenes   Hydrocarbons containing one or more carbon–carbon double bonds. (Section 25.2)

alkyl group   A group that is formed by removing a hydrogen atom from an alkane. (Section 25.3)

alkynes   Hydrocarbons containing one or more carbon–carbon triple bonds. (Section 25.2)

alloy   A substance that has the characteristic properties of a metal and contains more than one element. Often there is one principal metallic component, with other elements present in smaller amounts. Alloys may be homogeneous or heterogeneous in nature. (Section 23.6)

alpha (α) helix   A protein structure in which the protein is coiled in the form of a helix, with hydrogen bonds between $C=O$ and $N-H$ groups on adjacent turns. (Section 25.7)

alpha particles   Particles that are identical to helium-4 nuclei, consisting of two protons and two neutrons, symbol $_2^4He$ or $_2^4\alpha$. (Section 21.1)

amide   An organic compound that has an NR_2 group attached to a carbonyl. (Section 25.4)

amine   A compound that has the general formula R_3N, where R may be H or a hydrocarbon group. (Section 16.7)

amino acid   A carboxylic acid that contains an amino ($-NH_2$) group attached to the carbon atom adjacent to the carboxylic acid ($-COOH$) functional group. (Section 25.7)

amorphous solid   A solid whose molecular arrangement lacks a regular, long-range pattern. (Section 11.7)

amphiprotic   Refers to the capacity of a substance to either add or lose a proton (H^+). (Section 16.2)

amphoteric   Capable of behaving as either an acid or a base. (Section 17.5)

amphoteric oxides and hydroxides   Oxides and hydroxides that are only slightly soluble in water, but which dissolve on addition of either acid or base. (Section 17.5)

angstrom   A common non-SI unit of length, denoted Å, that is used to measure atomic dimensions: $1 \text{ Å} = 10^{-10}$ m. (Section 2.3)

anion   A negatively charged ion. (Section 2.7)

anode   An electrode at which oxidation occurs. (Section 20.3)

antibonding molecular orbital   A molecular orbital in which electron density is concentrated outside the region between the two nuclei of bonded atoms. Such orbitals, designated as σ^* or π^*, are less stable (of higher energy) than bonding molecular orbitals. (Section 9.7)

antiferromagnetism   A form of magnetism in which unpaired electron spins on adjacent sites point in opposite directions and cancel each other's effects. (Section 23.8)

aqueous solution   A solution in which water is the solvent. (Chapter 4: Introduction)

aromatic hydrocarbons   Hydrocarbon compounds that contain a planar, cyclic arrangement of carbon atoms linked by both σ and delocalized π bonds. (Section 25.2)

Arrhenius equation   An equation that relates the rate constant for a reaction to the frequency factor, A, the activation energy, E_a, and the temperature, T: $k = Ae^{-E_a/RT}$. In its logarithmic form it is written $\ln k = -E_a/RT + \ln A$. (Section 14.5)

atmosphere (atm)   A unit of pressure equal to 760 torr; $1 \text{ atm} = 101.325$ kPa. (Section 10.2)

atom   The smallest representative particle of an element. (Sections 1.1 and 2.1)

atomic mass unit (amu)   A unit based on the value of exactly 12 amu for the mass of the isotope of carbon that has six protons and six neutrons in the nucleus. (Sections 2.3 and 3.3)

atomic number   The number of protons in the nucleus of an atom of an element. (Section 2.3)

atomic radius   An estimate of the size of an atom. See **bonding atomic radius**. (Section 7.3)

atomic weight   The average mass of the atoms of an element in atomic mass units (amu); it is numerically equal to the mass in grams of one mole of the element. (Section 2.4)

autoionization   The process whereby water spontaneously forms low concentrations of $H^+(aq)$ and $OH^-(aq)$ ions by proton transfer from one water molecule to another. (Section 16.3)

Avogadro's hypothesis   A statement that equal volumes of gases at the same temperature and pressure contain equal numbers of molecules. (Section 10.3)

Avogadro's law   A statement that the volume of a gas maintained at constant temperature and pressure is directly proportional to the number of moles of the gas. (Section 10.3)

Avogadro's number   The number of ^{12}C atoms in exactly 12 g of ^{12}C; it equals 6.022×10^{23}. (Section 3.4)

band   An array of closely spaced molecular orbitals occupying a discrete range of energy. (Section 12.2)

band gap   The energy gap between an occupied valence band and a vacant band called the conduction band. (Section 12.2)

bar   A unit of pressure equal to 10^5 Pa. (Section 10.2)

base   A substance that is an H^+ acceptor; a base produces an excess of $OH^-(aq)$ ions when it dissolves in water. (Section 4.3)

base-dissociation constant (K_b)   An equilibrium constant that expresses the extent to which a base reacts with solvent water, accepting a proton and forming $OH^-(aq)$. (Section 16.7)

basic anhydride (basic oxide)   An oxide that forms a base when added to water; soluble metal oxides are basic anhydrides. (Section 22.5)

basic oxide (basic anhydride)   An oxide that either reacts with water to form a base or reacts with an acid to form a salt and water. (Section 22.5)

battery   A self-contained electrochemical power source that contains one or more voltaic cells. (Section 20.7)

Bayer process   A hydrometallurgical procedure for purifying bauxite in the recovery of aluminum from bauxite-containing ores. (Section 23.3)

becquerel   The SI unit of radioactivity. It corresponds to one nuclear disintegration per second. (Section 21.4)

Beer's law   The light absorbed by a substance (A) equals the product of its molar absorptivity constant (a), the path length through which the light passes (b), and the molar concentration of the substance (c): $A = abc$. (Section 14.2)

beta particles   Energetic electrons emitted from the nucleus, symbol $_{-1}^{0}e$. (Section 21.1)

beta sheet   A structural form of protein in which two stands of amino acids are hydrogen-bonded together in a zipperlike configuration. (Section 25.7)

bidentate ligand   A ligand in which two linked coordinating atoms are bound to a metal. (Section 24.2)

bimolecular reaction   An elementary reaction that involves two molecules. (Section 14.6)

biochemistry   The study of the chemistry of living systems. (Chapter 25: Introduction)

biocompatible   Any substance or material that can be compatibly placed within living systems. (Section 12.7)

biodegradable   Organic material that bacteria are able to oxidize. (Section 18.6)

biomaterial   Any material that has a biomedical application. (Section 12.7)

biopolymer   A polymeric molecule of high molecular weight found in living systems. The three major classes of biopolymers are proteins, carbohydrates, and nucleic acids. (Section 25.6)

body-centered cubic cell   A cubic unit cell in which the lattice points occur at the corners and at the center. (Section 11.7)

bomb calorimeter   A device for measuring the heat evolved in the combustion of a substance under constant-volume conditions. (Section 5.5)

bond angles   The angles made by the lines joining the nuclei of the atoms in a molecule. (Section 9.1)

bond dipole   The dipole moment that is due to unequal electron sharing between two atoms in a covalent bond. (Section 9.3)

bond enthalpy   The enthalpy change, ΔH, required to break a particular bond when the substance is in the gas phase. (Section 8.8)

bonding atomic radius   The radius of an atom as defined by the distances separating it from other atoms to which it is chemically bonded. (Section 7.3)

bonding molecular orbital   A molecular orbital in which the electron density is concentrated in the internuclear region. The energy of a bonding molecular orbital is lower than the energy of the separate atomic orbitals from which it forms. (Section 9.7)

bonding pair   In a Lewis structure a pair of electrons that is shared by two atoms. (Section 9.2)

bond length   The distance between the centers of two bonded atoms. (Section 8.8)

bond order   The number of bonding electron pairs shared between two atoms, minus the number of antibonding electron pairs: bond order = (number of bonding electrons − number of antibonding electrons)/2. (Section 9.7)

bond polarity   A measure of the degree to which the electrons are shared unequally between two atoms in a chemical bond. (Section 8.4)

boranes   Covalent hydrides of boron. (Section 22.11)

Born–Haber cycle   A thermodynamic cycle based on Hess's law that relates the lattice energy of an ionic substance to its enthalpy of formation and to other measurable quantities. (Section 8.2)

Boyle's law   A law stating that at constant temperature, the product of the volume and pressure of a given amount of gas is a constant. (Section 10.3)

Brønsted–Lowry acid   A substance (molecule or ion) that acts as a proton donor. (Section 16.2)

Brønsted–Lowry base   A substance (molecule or ion) that acts as a proton acceptor. (Section 16.2)

buffer capacity   The amount of acid or base a buffer can neutralize before the pH begins to change appreciably. (Section 17.2)

buffered solution (buffer)   A solution that undergoes a limited change in pH upon addition of a small amount of acid or base. (Section 17.2)

calcination   The heating of an ore to bring about its decomposition and the elimination of a volatile product. For example, a carbonate ore might be calcined to drive off CO_2. (Section 23.2)

calorie   A unit of energy, it is the amount of energy needed to raise the temperature of 1 g of water by 1 °C from 14.5 °C to 15.5 °C. A related unit is the joule: 1 cal = 4.184 J. (Section 5.1)

calorimeter   An apparatus that measures the evolution of heat. (Section 5.5)

calorimetry   The experimental measurement of heat produced in chemical and physical processes. (Section 5.5)

capillary action   The process by which a liquid rises in a tube because of a combination of adhesion to the walls of the tube and cohesion between liquid particles. (Section 11.3)

carbide   A binary compound of carbon with a metal or metalloid. (Section 22.9)

carbohydrates   A class of substances formed from polyhydroxy aldehydes or ketones. (Section 25.8)

carbon black   A microcrystalline form of carbon. (Section 22.9)

carbonyl group   The $C{=}O$ double bond, a characteristic feature of several organic functional groups, such as ketones and aldehydes. (Section 25.4)

carboxylic acid   A compound that contains the —COOH functional group. (Sections 16.10 and 25.4)

catalyst   A substance that changes the speed of a chemical reaction without itself undergoing a permanent chemical change in the process. (Section 14.7)

cathode   An electrode at which reduction occurs. (Section 20.3)

cathode rays   Streams of electrons that are produced when a high voltage is applied to electrodes in an evacuated tube. (Section 2.2)

cathodic protection   A means of protecting a metal against corrosion by making it the cathode in a voltaic cell. This can be achieved by attaching a more easily oxidized metal, which serves as an anode, to the metal to be protected. (Section 20.8)

cation   A positively charged ion. (Section 2.7)

cell potential   A measure of the driving force, or "electrical pressure," for an electrochemical reaction; it is measured in volts: 1 V = 1 J/C. Also called electromotive force. (Section 20.4)

cellulose   A polysaccharide of glucose; it is the major structural element in plant matter. (Section 25.8)

Celsius scale   A temperature scale on which water freezes at 0° and boils at 100° at sea level. (Section 1.4)

ceramic   A solid inorganic material, either crystalline (oxides, carbides, silicates) or amorphous (glasses). Most ceramics melt at high temperatures. (Section 12.1)

chain reaction   A series of reactions in which one reaction initiates the next. (Section 21.7)

changes of state   Transformations of matter from one state to a different one, for example, from a gas to a liquid. (Section 1.3)

charcoal   A form of carbon produced when wood is heated strongly in a deficiency of air. (Section 22.9)

Charles's law   A law stating that at constant pressure, the volume of a given quantity of gas is proportional to absolute temperature. (Section 10.3)

chelate effect   The generally larger formation constants for polydentate ligands as compared with the corresponding *monodentate* ligands. (Section 24.2)

chelating agent   A polydentate ligand that is capable of occupying two or more sites in the coordination sphere. (Section 24.2)

chemical bond   A strong attractive force that exists between atoms in a molecule. (Section 8.1)

chemical changes   Processes in which one or more substances are converted into other substances; also called **chemical reactions**. (Section 1.3)

chemical equation A representation of a chemical reaction using the chemical formulas of the reactants and products; a balanced chemical equation contains equal numbers of atoms of each element on both sides of the equation. (Section 3.1)

chemical equilibrium A state of dynamic balance in which the rate of formation of the products of a reaction from the reactants equals the rate of formation of the reactants from the products; at equilibrium the concentrations of the reactants and products remain constant. (Section 4.1; Chapter 15: Introduction)

chemical formula A notation that uses chemical symbols with numerical subscripts to convey the relative proportions of atoms of the different elements in a substance. (Section 2.6)

chemical kinetics The area of chemistry concerned with the speeds, or rates, at which chemical reactions occur. (Chapter 14: Introduction)

chemical nomenclature The rules used in naming substances. (Section 2.8)

chemical properties Properties that describe a substance's composition and its reactivity; how the substance reacts or changes into other substances. (Section 1.3)

chemical reactions Processes in which one or more substances are converted into other substances; also called **chemical changes**. (Section 1.3)

chemistry The scientific discipline that treats the composition, properties, and transformations of matter. (Chapter 1: Introduction)

chiral A term describing a molecule or an ion that cannot be superimposed on its mirror image. (Sections 24.4 and 25.5)

chlorofluorocarbons Compounds composed entirely of chlorine, fluorine, and carbon. (Section 18.3)

chlorophyll A plant pigment that plays a major role in conversion of solar energy to chemical energy in photosynthesis. (Section 24.2)

cholesteric liquid crystalline phase A liquid crystal formed from flat, disc-shaped molecules that align through a stacking of the molecular discs. (Section 12.8)

coal A naturally occurring solid containing hydrocarbons of high molecular weight, as well as compounds containing sulfur, oxygen, and nitrogen. (Section 5.8)

coke An impure form of carbon, formed when coal is heated strongly in the absence of air. (Section 22.9)

colligative properties Those properties of a solvent (vapor-pressure lowering, freezing-point lowering, boiling-point elevation, osmotic pressure) that depend on the total concentration of solute particles present. (Section 13.5)

collision model A model of reaction rates based on the idea that molecules must collide to react; it explains the factors influencing reaction rates in terms of the frequency of collisions, the number of collisions with energies exceeding the activation energy, and the probability that the collisions occur with suitable orientations. (Section 14.5)

colloids (colloidal dispersions) Mixtures containing particles larger than normal solutes but small enough to remain suspended in the dispersing medium. (Section 13.6)

combination reaction A chemical reaction in which two or more substances combine to form a single product. (Section 3.2)

combustion reaction A chemical reaction that proceeds with evolution of heat and usually also a flame; most combustion involves reaction with oxygen, as in the burning of a match. (Section 3.2)

common-ion effect A shift of an equilibrium induced by an ion common to the equilibrium. For example, added Na_2SO_4 decreases the solubility of the slightly soluble salt $BaSO_4$, or added NaF decreases the percent ionization of HF. (Section 17.1)

complementary colors Colors that, when mixed in proper proportions, appear white or colorless. (Section 24.5)

complete ionic equation A chemical equation in which dissolved strong electrolytes (such as dissolved ionic compounds) are written as separate ions. (Section 4.2)

complex ion (complex) An assembly of a metal ion and the Lewis bases (ligands) bonded to it. (Sections 17.5 and 24.1)

compound A substance composed of two or more elements united chemically in definite proportions. (Section 1.2)

compound semiconductor A semiconducting material formed from two or more elements. (Section 12.0)

concentration The quantity of solute present in a given quantity of solvent or solution. (Section 4.5)

concentration cell A voltaic cell containing the same electrolyte and the same electrode materials in both the anode and cathode compartments. The emf of the cell is derived from a difference in the concentrations of the same electrolyte solutions in the compartments. (Section 20.6)

condensation polymerization Polymerization in which molecules are joined together through condensation reactions. (Section 12.6)

condensation reaction A chemical reaction in which a small molecule (such as a molecule of water) is split out from between two reacting molecules. (Sections 12.6 and 22.8)

conduction band A band of molecular orbitals lying higher in energy than the occupied valence band, and distinctly separated from it. (Section 12.2)

conjugate acid A substance formed by addition of a proton to a Brønsted–Lowry base. (Section 16.2)

conjugate acid–base pair An acid and a base, such as H_2O and OH^-, that differ only in the presence or absence of a proton. (Section 16.2)

conjugate base A substance formed by the loss of a proton from a Brønsted–Lowry acid. (Section 16.2)

continuous spectrum A spectrum that contains radiation distributed over all wavelengths. (Section 6.3)

conversion factor A ratio relating the same quantity in two systems of units that is used to convert the units of measurement. (Section 1.6)

coordination compound or complex A compound containing a metal ion bonded to a group of surrounding molecules or ions that act as ligands. (Section 24.1)

coordination number The number of adjacent atoms to which an atom is directly bonded. In a complex the coordination number of the metal ion is the number of donor atoms to which it is bonded. (Sections 11.7 and 24.1)

coordination sphere The metal ion and its surrounding ligands. (Section 24.1)

coordination-sphere isomers Structural isomers of coordination compounds in which the ligands within the coordination sphere differ. (Section 24.4)

copolymer A complex polymer resulting from the polymerization of two or more chemically different monomers. (Section 12.6)

core electrons The electrons that are not in the outermost shell of an atom. (Section 6.8)

corrosion The process by which a metal is oxidized by substances in its environment. (Section 20.8)

covalent bond A bond formed between two or more atoms by a sharing of electrons. (Section 8.1)

covalent-network solids Solids in which the units that make up the three-dimensional network are joined by covalent bonds. (Section 11.8)

critical mass The amount of fissionable material necessary to maintain a nuclear chain reaction. (Section 21.7)

critical pressure The pressure at which a gas at its critical temperature is converted to a liquid state. (Section 11.4)

critical temperature The highest temperature at which it is possible to convert the gaseous form of a substance to a liquid. The critical temperature increases with an increase in the magnitude of intermolecular forces. (Section 11.4)

cross-linking The formation of bonds between polymer chains. (Section 12.6)

crystal-field theory A theory that accounts for the colors and the magnetic and other properties of transition-metal complexes in terms of the splitting of the energies of metal ion d orbitals by the electrostatic interaction with the ligands. (Section 24.6)

crystal lattice An imaginary network of points on which the repeating unit of the structure of a solid (the contents of the unit cell) may be imagined to be laid down so that the structure of the crystal is obtained. Each point represents an identical environment in the crystal. (Section 11.7)

crystalline solid (crystal) A solid whose internal arrangement of atoms, molecules, or ions shows a regular repetition in any direction through the solid. (Section 11.7)

crystallinity A measure of the extent of crystalline character (order) in a polymer. (Section 12.6)

crystallization The process in which a dissolved solute comes out of solution and forms a crystalline solid. (Section 13.2)

cubic close packing A close-packing arrangement in which the atoms of the third layer of a solid are not directly over those in the first layer. (Section 11.7)

curie A measure of radioactivity: 1 curie = 3.7×10^{10} nuclear disintegrations per second. (Section 21.4)

cycloalkanes Saturated hydrocarbons of general formula C_nH_{2n} in which the carbon atoms form a closed ring. (Section 25.3)

Dalton's law of partial pressures A law stating that the total pressure of a mixture of gases is the sum of the pressures that each gas would exert if it were present alone. (Section 10.6)

***d-d* transition** The transition of an electron in a transition metal compound from a lower-energy *d* orbital to a higher-energy *d* orbital. (Section 24.6)

decomposition reaction A chemical reaction in which a single compound reacts to give two or more products. (Section 3.2)

degenerate A situation in which two or more orbitals have the same energy. (Section 6.7)

delocalized electrons Electrons that are spread over a number of atoms in a molecule rather than localized between a pair of atoms. (Section 9.6)

density The ratio of an object's mass to its volume. (Section 1.4)

deoxyribonucleic acid (DNA) A polynucleotide in which the sugar component is deoxyribose. (Section 25.10)

desalination The removal of salts from seawater, brine, or brackish water to make it fit for human consumption. (Section 18.5)

deuterium The isotope of hydrogen whose nucleus contains a proton and a neutron: ^2_1H. (Section 22.2)

dextrorotatory, or merely dextro or *d* A term used to label a chiral molecule that rotates the plane of polarization of plane-polarized light to the right (clockwise). (Section 24.4)

diamagnetism A type of magnetism that causes a substance with no unpaired electrons to be weakly repelled from a magnetic field. (Section 9.8)

diatomic molecule A molecule composed of only two atoms. (Section 2.6)

diffusion The spreading of one substance through a space occupied by one or more other substances. (Section 10.8)

dilution The process of preparing a less concentrated solution from a more concentrated one by adding solvent. (Section 4.5)

dimensional analysis A method of problem solving in which units are carried through all calculations. Dimensional analysis ensures that the final answer of a calculation has the desired units. (Section 1.6)

dipole A molecule with one end having a partial negative charge and the other end having a partial positive charge; a polar molecule. (Section 8.4)

dipole–dipole force The force that exists because of the interactions of dipoles on polar molecules in close contact. (Section 11.2)

dipole moment A measure of the separation and magnitude of the positive and negative charges in polar molecules. (Section 8.4)

displacement reaction A reaction in which an element reacts with a compound, displacing an element from it. (Section 4.4)

disproportionation A reaction in which a species undergoes simultaneous oxidation and reduction [as in $N_2O_3(g) \longrightarrow NO(g) + NO_2(g)$.] (Section 22.5)

donor atom The atom of a ligand that bonds to the metal. (Section 24.1)

doping Incorporation of a hetero atom into a solid to change its electrical properties. For example, incorporation of P into Si. (Section 12.3)

double bond A covalent bond involving two electron pairs. (Section 8.3)

double helix The structure for DNA that involves the winding of two DNA polynucleotide chains together in a helical arrangement. The two strands of the double helix are complementary in that the organic bases on the two strands are paired for optimal hydrogen bond interaction. (Section 25.10)

Downs cell A cell used to obtain sodium metal by electrolysis of molten NaCl. (Section 23.4)

dynamic equilibrium A state of balance in which opposing processes occur at the same rate. (Section 11.5)

effective nuclear charge The net positive charge experienced by an electron in a many-electron atom; this charge is not the full nuclear charge because there is some shielding of the nucleus by the other electrons in the atom. (Section 7.2)

effusion The escape of a gas through an orifice or hole. (Section 10.8)

elastomer A material that can undergo a substantial change in shape via stretching, bending, or compression and return to its original shape upon release of the distorting force. (Section 12.6)

electrochemistry The branch of chemistry that deals with the relationships between electricity and chemical reactions. (Chapter 20: Introduction)

electrolysis reaction A reaction in which a nonspontaneous redox reaction is brought about by the passage of current under a sufficient external electrical potential. The devices in which electrolysis reactions occur are called electrolytic cells. (Section 20.9)

electrolyte A solute that produces ions in solution; an electrolytic solution conducts an electric current. (Section 4.1)

electrolytic cell A device in which a nonspontaneous oxidation-reduction reaction is caused to occur by passage of current under a sufficient external electrical potential. (Section 20.9)

electromagnetic radiation (radiant energy) A form of energy that has wave characteristics and that propagates through a vacuum at the characteristic speed of 3.00×10^8 m/s. (Section 6.1)

electrometallurgy The use of electrolysis to reduce or refine metals. (Section 23.4)

electromotive force (emf) A measure of the driving force, or *electrical pressure*, for the completion of an electrochemical reaction. Electromotive force is measured in volts: 1 V = 1 J/C. Also called the cell potential. (Section 20.4)

electron A negatively charged subatomic particle found outside the atomic nucleus; it is a part of all atoms. An electron has a mass 1/1836 times that of a proton. (Section 2.3)

electron affinity The energy change that occurs when an electron is added to a gaseous atom or ion. (Section 7.5)

electron capture A mode of radioactive decay in which an inner-shell orbital electron is captured by the nucleus. (Section 21.1)

electron configuration A particular arrangement of electrons in the orbitals of an atom. (Section 6.8)

electron density The probability of finding an electron at any particular point in an atom; this probability is equal to ψ^2, the square of the wave function. (Section 6.5)

electron domain In the VSEPR model, a region about a central atom in which an electron pair is concentrated. (Section 9.2)

electron-domain geometry The three-dimensional arrangement of the electron domains around an atom according to the VSEPR model. (Section 9.2)

electronegativity A measure of the ability of an atom that is bonded to another atom to attract electrons to itself. (Section 8.4)

electronic charge The negative charge carried by an electron; it has a magnitude of 1.602×10^{-19} C. (Section 2.3)

electronic structure The arrangement of electrons in an atom or molecule. (Chapter 6: Introduction)

electron-sea model A model for the behavior of electrons in metals. (Section 23.5)

electron shell A collection of orbitals that have the same value of *n*. For example, the orbitals with $n = 3$ (the 3*s*, 3*p*, and 3*d* orbitals) comprise the third shell. (Section 6.5)

electron spin A property of the electron that makes it behave as though it were a tiny magnet. The electron behaves as if it were spinning on its axis; electron spin is quantized. (Section 6.7)

element A substance that cannot be separated into simpler substances by chemical means. (Sections 1.1 and 1.2)

elemental semiconductor A semiconducting material composed of just one element (Section 12.3)

elementary reaction A process in a chemical reaction that occurs in a single event or step. An overall chemical reaction consists of one or more elementary reactions or steps. (Section 14.6)

empirical formula (simplest formula) A chemical formula that shows the kinds of atoms and their relative numbers in a substance in the smallest possible whole-number ratios. (Section 2.6)

enantiomers Two mirror-image molecules of a chiral substance. The enantiomers are non-superimposable. (Section 24.4)

endothermic process A process in which a system absorbs heat from its surroundings. (Section 5.2)

energy The capacity to do work or to transfer heat. (Section 5.1)

energy band A band of allowed energy states of electrons in metals and semiconductors. (Section 23.5)

energy-level diagram A diagram that shows the energies of molecular orbitals relative to the atomic orbitals from which they are derived. Also called a **molecular-orbital diagram**. (Section 9.7)

enthalpy A quantity defined by the relationship $H = E + PV$; the enthalpy change, ΔH, for a reaction that occurs at constant pressure is the heat evolved or absorbed in the reaction: $\Delta H = q_p$. (Section 5.3)

enthalpy (heat) of formation The enthalpy change that accompanies the formation of a substance from the most stable forms of its component elements. (Section 5.7)

enthalpy of reaction The enthalpy change associated with a chemical reaction. (Section 5.4)

entropy A thermodynamic function associated with the number of different equivalent energy states or spatial arrangements in which a system may be found. It is a thermodynamic state function, which means that once we specify the conditions for a system—that is, the temperature, pressure, and so on—the entropy is defined. (Sections 13.1 and 19.2)

enzyme A protein molecule that acts to catalyze specific biochemical reactions. (Section 14.7)

equilibrium constant The numerical value of the equilibrium-constant expression for a system at equilibrium. The equilibrium constant is most usually denoted by K_p for gas-phase systems or K_c for solution-phase systems. (Section 15.2)

equilibrium-constant expression The expression that describes the relationship among the concentrations (or partial pressures) of the substances present in a system at equilibrium. The numerator is obtained by multiplying the concentrations of the substances on the product side of the equation, each raised to a power equal to its coefficient in the chemical equation. The denominator similarly contains the concentrations of the substances on the reactant side of the equation. (Section 15.2)

equivalence point The point in a titration at which the added solute reacts completely with the solute present in the solution. (Section 4.6)

ester An organic compound that has an OR group attached to a carbonyl; it is the product of a reaction between a carboxylic acid and an alcohol. (Section 25.4)

ether A compound in which two hydrocarbon groups are bonded to one oxygen. (Section 25.4)

exchange (metathesis) reaction A reaction between compounds that when written as a molecular equation appears to involve the exchange of ions between the two reactants. (Section 4.2)

excited state A higher energy state than the ground state. (Section 6.3)

exothermic process A process in which a system releases heat to its surroundings. (Section 5.2)

extensive property A property that depends on the amount of material considered; for example, mass or volume. (Section 1.3)

face-centered cubic cell A cubic unit cell that has lattice points at each corner as well as at the center of each face. (Section 11.7)

faraday A unit of charge that equals the total charge of one mole of electrons: $1 F = 96,500$ C. (Section 20.5)

f-block metals Lanthanide and actinide elements in which the $4f$ or $5f$ orbitals are partially occupied. (Section 6.9)

ferrimagnetism A form of magnetism in which electron spins on different kinds of sites point in opposite directions but do not fully cancel out. (Section 23.8)

ferromagnetism A form of magnetism in which unpaired electron spins at lattice sites are permanently aligned. (Section 23.7)

first law of thermodynamics A statement of our experience that energy is conserved in any process. We can express the law in many ways. One of the more useful expressions is that the change in internal energy, ΔE, of a system in any process is equal to the heat, q, added to the system, plus the work, w, done on the system by its surroundings: $\Delta E = q + w$. (Section 5.2)

first-order reaction A reaction in which the reaction rate is proportional to the concentration of a single reactant, raised to the first power. (Section 14.4)

fission The splitting of a large nucleus into two smaller ones. (Section 21.6)

folding The process by which a protein adopts its biologically active shape. (Section 25.7)

force A push or a pull. (Section 5.1)

formal charge The number of valence electrons in an isolated atom minus the number of electrons assigned to the atom in the Lewis structure. (Section 8.5)

formation constant For a metal ion complex, the equilibrium constant for formation of the complex from the metal ion and base species present in solution. It is a measure of the tendency of the complex to form. (Section 17.5)

formula weight The mass of the collection of atoms represented by a chemical formula. For example, the formula weight of NO_2 (46.0 amu) is the sum of the masses of one nitrogen atom and two oxygen atoms. (Section 3.3)

fossil fuels Coal, oil, and natural gas, which are presently our major sources of energy. (Section 5.8)

free energy (Gibbs free energy, G) A thermodynamic state function that gives a criterion for spontaneous change in terms of enthalpy and entropy: $G = H - TS$. (Section 19.5)

free radical A substance with one or more unpaired electrons. (Section 21.9)

frequency The number of times per second that one complete wavelength passes a given point. (Section 6.1)

frequency factor (A) A term in the Arrhenius equation that is related to the frequency of collision and the probability that the collisions are favorably oriented for reaction. (Section 14.5)

fuel cell A voltaic cell that utilizes the oxidation of a conventional fuel, such as H_2 or CH_4, in the cell reaction. (Section 20.7)

fuel value The energy released when 1 g of a substance is combusted. (Section 5.8)

functional group An atom or group of atoms that imparts characteristic chemical properties to an organic compound. (Section 25.1)

fusion The joining of two light nuclei to form a more massive one. (Section 21.6)

galvanic cell See **voltaic cell**. (Section 20.3)

gamma radiation Energetic electromagnetic radiation emanating from the nucleus of a radioactive atom. (Section 21.1)

gas Matter that has no fixed volume or shape; it conforms to the volume and shape of its container. (Section 1.2)

gas constant (R) The constant of proportionality in the ideal-gas equation. (Section 10.4)

Geiger counter A device that can detect and measure radioactivity. (Section 21.5)

geometric isomers Compounds with the same type and number of atoms and the same chemical bonds but different spatial arrangements of these atoms and bonds. (Sections 24.4 and 25.4)

Gibbs free energy A thermodynamic state function that combines enthalpy and entropy, in the form $G = H - TS$. For a change occurring at constant temperature and pressure, the change in free energy is $\Delta G = \Delta H - T\Delta S$. (Section 19.5)

glass An amorphous solid formed by fusion of SiO_2, CaO, and Na_2O. Other oxides may also be used to form glasses with differing characteristics. (Section 22.10)

glucose A polyhydroxy aldehyde whose formula is $CH_2OH(CHOH)_4CHO$; it is the most important of the monosaccharides. (Section 25.8)

glycogen The general name given to a group of polysaccharides of glucose that are synthesized in mammals and used to store energy from carbohydrates. (Section 25.8)

Graham's law A law stating that the rate of effusion of a gas is inversely proportional to the square root of its molecular weight. (Section 10.8)

gray (Gy) The SI unit for radiation dose corresponding to the absorption of 1 J of energy per kilogram of biological material; $1 Gy = 100$ rads. (Section 21.9)

green chemistry Chemistry that promotes the design and application of chemical products and processes that are compatible with human health and that preserve the environment. (Section 18.7)

ground state The lowest-energy, or most stable, state. (Section 6.3)

group Elements that are in the same column of the periodic table; elements within the same group or family exhibit similarities in their chemical behavior. (Section 2.5)

Haber process The catalyst system and conditions of temperature and pressure developed by Fritz Haber and coworkers for the formation of NH_3 from H_2 and N_2. (Section 15.1)

half-life The time required for the concentration of a reactant substance to decrease to half its initial value; the time required for half of a sample of a particular radioisotope to decay. (Sections 14.4 and 21.4)

half-reaction An equation for either an oxidation or a reduction that explicitly shows the electrons involved, for example, $Zn^{2+}(aq) + 2 e^- \longrightarrow Zn(s)$. (Section 20.2)

Hall process A process used to obtain aluminum by electrolysis of Al_2O_3 dissolved in molten cryolite, Na_3AlF_6. (Section 23.4)

halogens Members of group 7A in the periodic table. (Sections 7.8 and 22.4)

hard water Water that contains appreciable concentrations of Ca^{2+} and Mg^{2+}; these ions react with soaps to form an insoluble material. (Section 18.6)

heat The flow of energy from a body at higher temperature to one at lower temperature when they are placed in thermal contact. (Section 5.1)

heat capacity The quantity of heat required to raise the temperature of a sample of matter by 1 °C (or 1 K). (Section 5.5)

heat of fusion The enthalpy change, ΔH, for melting a solid. (Section 11.4)

heat of sublimation The enthalpy change, ΔH, for vaporization of a solid. (Section 11.4)

heat of vaporization The enthalpy change, ΔH, for vaporization of a liquid. (Section 11.4)

hemoglobin An iron-containing protein responsible for oxygen transport in the blood. (Section 18.4)

Henderson–Hasselbalch equation The relationship among the pH, pK_a, and the concentrations of acid and conjugate base in an aqueous solution: $pH = pK_a + \log \dfrac{[base]}{[acid]}$. (Section 17.2)

Henry's law A law stating that the concentration of a gas in a solution, C_g, is proportional to the pressure of gas over the solution: $C_g = kP_g$. (Section 13.3)

Hess's law The heat evolved in a given process can be expressed as the sum of the heats of several processes that, when added, yield the process of interest. (Section 5.6)

heterogeneous alloy An alloy in which the components are not distributed uniformly; instead, two or more distinct phases with characteristic compositions are present. (Section 23.6)

heterogeneous catalyst A catalyst that is in a different phase from that of the reactant substances. (Section 14.7)

heterogeneous equilibrium The equilibrium established between substances in two or more different phases, for example, between a gas and a solid or between a solid and a liquid. (Section 15.4)

hexagonal close packing A close-packing arrangement in which the atoms of the third layer of a solid lie directly over those in the first layer. (Section 11.7)

high-spin complex A complex whose electrons populate the d orbitals to give the maximum number of unpaired electrons. (Section 24.6)

high-temperature superconductivity The "frictionless" flow of electrical current (superconductivity) at temperatures above 30 K in certain complex metal oxides. (Section 12.1)

hole A vacancy in the valence band of a semiconductor, created by doping. (Section 12.3)

homogeneous catalyst A catalyst that is in the same phase as the reactant substances. (Section 14.7)

homogeneous equilibrium The equilibrium established between reactant and product substances that are all in the same phase. (Section 15.4)

Hund's rule A rule stating that electrons occupy degenerate orbitals in such a way as to maximize the number of electrons with the same spin. In other words, each orbital has one electron placed in it before pairing of electrons in orbitals occurs. Note that this rule applies only to orbitals that are degenerate, which means that they have the same energy. (Section 6.8)

hybridization The mixing of different types of atomic orbitals to produce a set of equivalent hybrid orbitals. (Section 9.5)

hybrid orbital An orbital that results from the mixing of different kinds of atomic orbitals on the same atom. For example, an sp^3 hybrid results from the mixing, or hybridizing, of one s orbital and three p orbitals. (Section 9.5)

hydration Solvation when the solvent is water. (Section 13.1)

hydride ion An ion formed by the addition of an electron to a hydrogen atom: H^-. (Section 7.7)

hydrocarbons Compounds composed of only carbon and hydrogen. (Section 2.9)

hydrogen bonding Bonding that results from intermolecular attractions between molecules containing hydrogen bonded to an electronegative element. The most important examples involve OH, NH, and HF. (Section 11.2)

hydrolysis A reaction with water. When a cation or anion reacts with water, it changes the pH. (Section 16.9)

hydrometallurgy Aqueous chemical processes for recovery of a metal from an ore. (Section 23.3)

hydronium ion (H_3O^+) The predominant form of the proton in aqueous solution. (Section 16.2)

hydrophilic Water-attracting. The term is often used to describe a type of colloid. (Section 13.6)

hydrophobic Water-repelling. The term is often used to describe a type of colloid. (Section 13.6)

hypothesis A tentative explanation of a series of observations or of a natural law. (Section 1.3)

ideal gas A hypothetical gas whose pressure, volume, and temperature behavior is completely described by the ideal-gas equation. (Section 10.4)

ideal-gas equation An equation of state for gases that embodies Boyle's law, Charles's law, and Avogadro's hypothesis in the form $PV = nRT$. (Section 10.4)

ideal solution A solution that obeys Raoult's law. (Section 13.5)

immiscible liquids Liquids that do not dissolve in one another to a significant extent. (Section 13.3)

indicator A substance added to a solution that changes color when the added solute has reacted with all the solute present in solution. (Section 4.6)

instantaneous rate The reaction rate at a particular time as opposed to the average rate over an interval of time. (Section 14.2)

insulator A solid with extremely low electrical conductivity. (Section 12.1)

intensive property A property that is independent of the amount of material considered, for example, density. (Section 1.3)

interhalogens Compounds formed between two different halogen elements. Examples include IBr and BrF_3. (Section 22.4)

intermediate A substance formed in one elementary step of a multistep mechanism and consumed in another; it is neither a reactant nor an ultimate product of the overall reaction. (Section 14.6)

intermetallic compound A homogeneous alloy with definite properties and composition. Intermetallic compounds are stoichiometric compounds, but their compositions are not readily explained in terms of ordinary chemical bonding theory. (Section 23.6)

intermolecular forces The short-range attractive forces operating between the particles that make up the units of a liquid or solid substance. These same forces also cause gases to liquefy or solidify at low temperatures and high pressures. (Chapter 11: Introduction)

internal energy The total energy possessed by a system. When a system undergoes a change, the change in internal energy, ΔE, is defined as the heat, q, added to the system, plus the work, w, done on the system by its surroundings: $\Delta E = q + w$. (Section 5.2)

ion Electrically charged atom or group of atoms (polyatomic ion); ions can be positively or negatively charged, depending on whether electrons are lost (positive) or gained (negative) by the atoms. (Section 2.7)

ion–dipole force The force that exists between an ion and a neutral polar molecule that possesses a permanent dipole moment. (Section 11.2)

ion exchange A process by which ions in solution are exchanged for other ions held on the surface of an ion-exchange resin; the exchange of a hard-water cation such as Ca^{2+} for a soft-water cation such as Na^+ is used to soften water. (Section 18.6)

ionic bond A bond between oppositely charged ions. The ions are formed from atoms by transfer of one or more electrons. (Section 8.1)

ionic compound A compound composed of cations and anions. (Section 2.7)

ionic hydrides Compounds formed when hydrogen reacts with alkali metals and also the heavier alkaline earths (Ca, Sr, and Ba); these compounds contain the hydride ion, H^-. (Section 22.2)

ionic solids Solids that are composed of ions. (Section 11.8)

ionization energy The energy required to remove an electron from a gaseous atom when the atom is in its ground state. (Section 7.4)

ionizing radiation Radiation that has sufficient energy to remove an electron from a molecule, thereby ionizing it. (Section 21.9)

ion-product constant For water, K_w is the product of the aquated hydrogen ion and hydroxide ion concentrations: $[H^+][OH^-] = K_w = 1.0 \times 10^{-14}$ at 25 °C. (Section 16.3)

irreversible process A process that cannot be reversed to restore both the system and its surroundings to their original states. Any spontaneous process is irreversible. (Section 19.1)

isoelectronic series A series of atoms, ions, or molecules having the same number of electrons. (Section 7.3)

isomers Compounds whose molecules have the same overall composition but different structures. (Section 24.4)

isothermal process One that occurs at constant temperature. (Section 19.1)

isotopes Atoms of the same element containing different numbers of neutrons and therefore having different masses. (Section 2.3)

joule (J) The SI unit of energy, $1 \text{ kg-m}^2/\text{s}^2$. A related unit is the calorie: $4.184 \text{ J} = 1 \text{ cal}$. (Section 5.1)

Kelvin scale The absolute temperature scale; the SI unit for temperature is the kelvin. Zero on the Kelvin scale corresponds to -273.15 °C; therefore, $K = °C + 273.15$. (Section 1.4)

ketone A compound in which the carbonyl group ($C=O$) occurs at the interior of a carbon chain and is therefore flanked by carbon atoms. (Section 25.4)

kinetic energy The energy that an object possesses by virtue of its motion. (Section 5.1)

kinetic-molecular theory A set of assumptions about the nature of gases. These assumptions, when translated into mathematical form, yield the ideal-gas equation. (Section 10.7)

lanthanide contraction The gradual decrease in atomic and ionic radii with increasing atomic number among the lanthanide elements, atomic numbers 57 through 70. The decrease arises because of a gradual increase in effective nuclear charge through the lanthanide series. (Section 23.7)

lanthanide (rare earth) element Element in which the $4f$ subshell is only partially occupied. (Sections 6.8 and 6.9)

lattice energy The energy required to separate completely the ions in an ionic solid. (Section 8.2)

law of conservation of mass A scientific law stating that the total mass of the products of a chemical reaction is the same as the total mass of the reactants, so that mass remains constant during the reaction. (Section 3.1)

law of constant composition A law that states that the elemental composition of a pure compound is always the same, regardless of its source; also called the **law of definite proportions**. (Section 1.2)

law of definite proportions A law that states that the elemental composition of a pure substance is always the same, regardless of its source; also called the **law of constant composition**. (Section 1.2)

law of mass action The rules by which the equilibrium constant is expressed in terms of the concentrations of reactants and products, in accordance with the balanced chemical equation for the reaction. (Section 15.2)

leaching The selective dissolution of a desired mineral by passing an aqueous reagent solution through an ore. (Section 23.3)

Le Châtelier's principle A principle stating that when we disturb a system at chemical equilibrium, the relative concentrations of reactants and products shift so as to partially undo the effects of the disturbance. (Section 15.7)

levorotatory, or merely levo or *l* A term used to label a chiral molecule that rotates the plane of polarization of plane-polarized light to the left (counterclockwise). (Section 24.4)

Lewis acid An electron-pair acceptor. (Section 16.11)

Lewis base An electron-pair donor. (Section 16.11)

Lewis structure A representation of covalent bonding in a molecule that is drawn using Lewis symbols. Shared electron pairs are shown as lines, and unshared electron pairs are shown as pairs of dots. Only the valence-shell electrons are shown. (Section 8.3)

Lewis symbol (electron-dot symbol) The chemical symbol for an element, with a dot for each valence electron. (Section 8.1)

ligand An ion or molecule that coordinates to a metal atom or to a metal ion to form a complex. (Section 24.1)

light-emitting diode A semiconductor device in which electrical energy can be converted into radiant energy, usually in the form of visible light. (Section 12.3)

lime-soda process A method used in large-scale water treatment to reduce water hardness by removing Mg^{2+} and Ca^{2+}. The substances added to the water are lime, CaO [or slaked lime, $Ca(OH)_2$], and soda ash, Na_2CO_3, in amounts determined by the concentrations of the undesired ions. (Section 18.6)

limiting reactant (limiting reagent) The reactant present in the smallest stoichiometric quantity in a mixture of reactants; the amount of product that can form is limited by the complete consumption of the limiting reactant. (Section 3.7)

line spectrum A spectrum that contains radiation at only certain specific wavelengths. (Section 6.3)

linkage isomers Structural isomers of coordination compounds in which a ligand differs in its mode of attachment to a metal ion. (Section 24.4)

lipid A nonpolar molecule derived from glycerol and fatty acids that is used by organisms for long-term energy storage. (Section 25.9)

liquid Matter that has a distinct volume but no specific shape. (Section 1.2)

liquid crystal A substance that exhibits one or more partially ordered liquid phases above the melting point of the solid form. By contrast, in nonliquid crystalline substances the liquid phase that forms upon melting is completely unordered. (Section 12.8)

lithosphere That portion of our environment consisting of the solid Earth. (Section 23.1)

lock-and-key model A model of enzyme action in which the substrate molecule is pictured as fitting rather specifically into the active site on the enzyme. It is assumed that in being bound to the active site, the substrate is somehow activated for reaction. (Section 14.7)

London dispersion forces Intermolecular forces resulting from attractions between induced dipoles. (Section 11.2)

low-spin complex A metal complex in which the electrons are paired in lower-energy orbitals. (Section 24.6)

magic numbers Numbers of protons and neutrons that result in very stable nuclei. (Section 21.2)

main-group elements Elements in the *s* and *p* blocks of the periodic table. (Section 6.9)

mass A measure of the amount of material in an object. It measures the resistance of an object to being moved. In SI units, mass is measured in kilograms. (Section 1.4)

mass defect The difference between the mass of a nucleus and the total masses of the individual nucleons that it contains. (Section 21.6)

mass number The sum of the number of protons and neutrons in the nucleus of a particular atom. (Section 2.3)

mass percentage The number of grams of solute in each 100 g of solution. (Section 13.4)

mass spectrometer An instrument used to measure the precise masses and relative amounts of atomic and molecular ions. (Section 2.4)

matter Anything that occupies space and has mass; the physical material of the universe. (Section 1.1)

matter waves The term used to describe the wave characteristics of a particle. (Section 6.4)

mean free path The average distance traveled by a gas molecule between collisions. (Section 10.8)

metal complex (complex ion or complex) An assembly of a metal ion and the Lewis bases bonded to it. (Section 24.1)

metallic bond Bonding, usually in solid metals, in which the bonding electrons are relatively free to move throughout the three-dimensional structure. (Section 8.1)

metallic character The extent to which an element exhibits the physical and chemical properties characteristic of metals, for example, luster, malleability, ductility, and good thermal and electrical conductivity. (Section 7.6)

metallic elements (metals) Elements that are usually solids at room temperature, exhibit high electrical and heat conductivity, and appear lustrous. Most of the elements in the periodic table are metals. (Sections 2.5 and 12.1)

metallic hydrides Compounds formed when hydrogen reacts with transition metals; these compounds contain the hydride ion, H^-. (Section 22.2)

metallic solids Solids that are composed of metal atoms. (Section 11.8)

metalloids Elements that lie along the diagonal line separating the metals from the nonmetals in the periodic table; the properties of metalloids are intermediate between those of metals and nonmetals. (Section 2.5)

metallurgy The science of extracting metals from their natural sources by a combination of chemical and physical processes. It is also concerned with the properties and structures of metals and alloys. (Section 23.1)

metathesis (exchange) reaction A reaction in which two substances react through an exchange of their component ions: $AX + BY \longrightarrow AY + BX$. Precipitation and acid–base neutralization reactions are examples of metathesis reactions. (Section 4.2)

metric system A system of measurement used in science and in most countries. The meter and the gram are examples of metric units. (Section 1.4)

microstate The state of a system at a particular instant; one of many possible energetically equivalent ways to arrange the components of a system to achieve a particular state. (Section 19.3)

mineral A solid, inorganic substance occurring in nature, such as calcium carbonate, which occurs as calcite. (Section 23.1)

miscible Liquids that mix in all proportions. (Section 13.3)

mixture A combination of two or more substances in which each substance retains its own chemical identity. (Section 1.2)

molal boiling-point-elevation constant (K_b) A constant characteristic of a particular solvent that gives the increase in boiling point as a function of solution molality: $\Delta T_b = K_b m$. (Section 13.5)

molal freezing-point-depression constant (K_f) A constant characteristic of a particular solvent that gives the decrease in freezing point as a function of solution molality: $\Delta T_f = K_f m$. (Section 13.5)

molality The concentration of a solution expressed as moles of solute per kilogram of solvent; abbreviated m. (Section 13.4)

molar heat capacity The heat required to raise the temperature of one mole of a substance by 1 °C. (Section 5.5)

molarity The concentration of a solution expressed as moles of solute per liter of solution; abbreviated M. (Section 4.5)

molar mass The mass of one mole of a substance in grams; it is numerically equal to the formula weight in atomic mass units. (Section 3.4)

mole A collection of Avogadro's number (6.022×10^{23}) of objects; for example, a mole of H_2O is 6.022×10^{23} H_2O molecules. (Section 3.4)

molecular compound A compound that consists of molecules. (Section 2.6)

molecular equation A chemical equation in which the formula for each substance is written without regard for whether it is an electrolyte or a nonelectrolyte. (Section 4.2)

molecular formula A chemical formula that indicates the actual number of atoms of each element in one molecule of a substance. (Section 2.6)

molecular geometry The arrangement in space of the atoms of a molecule. (Section 9.2)

molecular hydrides Compounds formed when hydrogen reacts with nonmetals and metalloids. (Section 22.2)

molecularity The number of molecules that participate as reactants in an elementary reaction. (Section 14.6)

molecular orbital (MO) An allowed state for an electron in a molecule. According to molecular-orbital theory, a molecular orbital is entirely analogous to an atomic orbital, which is an allowed state for an electron in an atom. Most bonding molecular orbitals can be classified as σ or π, depending on the disposition of electron density with respect to the internuclear axis. (Section 9.7)

molecular-orbital diagram A diagram that shows the energies of molecular orbitals relative to the atomic orbitals from which they are derived; also called an **energy-level diagram**. (Section 9.7)

molecular-orbital theory A theory that accounts for the allowed states for electrons in molecules. (Section 9.7)

molecular solids Solids that are composed of molecules. (Section 11.8)

molecular weight The mass of the collection of atoms represented by the chemical formula for a molecule. (Section 3.3)

molecule A chemical combination of two or more atoms. (Sections 1.1 and 2.6)

mole fraction The ratio of the number of moles of one component of a mixture to the total moles of all components; abbreviated X, with a subscript to identify the component. (Section 10.6)

momentum The product of the mass, m, and velocity, v, of an object. (Section 6.4)

monodentate ligand A ligand that binds to the metal ion via a single donor atom. It occupies one position in the coordination sphere. (Section 24.2)

monomers Molecules with low molecular weights, which can be joined together (polymerized) to form a polymer. (Section 12.6)

monosaccharide A simple sugar, most commonly containing six carbon atoms. The joining together of monosaccharide units by condensation reactions results in formation of polysaccharides. (Section 25.8)

multiple bonding Bonding involving two or more electron pairs. (Section 8.3)

nanomaterial A material whose useful characteristics are the result of dimensions in the range from 1 to 100 nm. (Section 12.9)

nanotechnology Technology that relies on the properties of matter at the nanoscale, that is, in the range from 1 to 100 nm. (Section 12.9)

natural gas A naturally occurring mixture of gaseous hydrocarbon compounds composed of hydrogen and carbon. (Section 5.8)

nematic liquid crystalline phase A liquid crystal in which the molecules are aligned in the same general direction, along their long axes, but in which the ends of the molecules are not aligned. (Section 12.8)

Nernst equation An equation that relates the cell emf, E, to the standard emf, $E°$, and the reaction quotient, Q: $E = E° - (RT/nF) \ln Q$. (Section 20.6)

net ionic equation A chemical equation for a solution reaction in which soluble strong electrolytes are written as ions and spectator ions are omitted. (Section 4.2)

neutralization reaction A reaction in which an acid and a base react in stoichiometrically equivalent amounts; the neutralization reaction between an acid and a metal hydroxide produces water and a salt. (Section 4.3)

neutron An electrically neutral particle found in the nucleus of an atom; it has approximately the same mass as a proton. (Section 2.3)

noble gases Members of group 8A in the periodic table. (Section 7.8)

node A locus of points in an atom at which the electron density is zero. For example, the node in a 2s orbital is a spherical surface. (Section 6.6)

nonbonding pair In a Lewis structure a pair of electrons assigned completely to one atom; also called a lone pair. (Section 9.2)

nonelectrolyte A substance that does not ionize in water and consequently gives a nonconducting solution. (Section 4.1)

nonionizing radiation Radiation that does not have sufficient energy to remove an electron from a molecule. (Section 21.9)

nonmetallic elements (nonmetals) Elements in the upper right corner of the periodic table; nonmetals differ from metals in their physical and chemical properties. (Section 2.5)

nonpolar covalent bond A covalent bond in which the electrons are shared equally. (Section 8.4)

normal boiling point The boiling point at 1 atm pressure. (Section 11.5)

normal melting point The melting point at 1 atm pressure. (Section 11.6)

nuclear binding energy The energy required to decompose an atomic nucleus into its component protons and neutrons. (Section 21.6)

nuclear disintegration series A series of nuclear reactions that begins with an unstable nucleus and terminates with a stable one; also called a **radioactive series**. (Section 21.2)

nuclear transmutation A conversion of one kind of nucleus to another. (Section 21.3)

nucleic acids Polymers of high molecular weight that carry genetic information and control protein synthesis. (Section 25.10)

nucleon A particle found in the nucleus of an atom. (Section 21.1)

nucleotide Compounds formed from a molecule of phosphoric acid, a sugar molecule, and an organic nitrogen base. Nucleotides form linear polymers called DNA and RNA, which are involved in protein synthesis and cell reproduction. (Section 25.11)

nucleus The very small, very dense, positively charged portion of an atom; it is composed of protons and neutrons. (Section 2.2)

nuclide A nucleus of a specific isotope of an element. (Section 2.3)

octet rule A rule stating that bonded atoms tend to possess or share a total of eight valence-shell electrons. (Section 8.1)

optical isomers Stereoisomers in which the two forms of the compound are nonsuperimposable mirror images. (Section 24.4)

optically active Possessing the ability to rotate the plane of polarized light. (Section 24.4)

orbital An allowed energy state of an electron in the quantum mechanical model of the atom; the term orbital is also used to describe the spatial distribution of the electron. An orbital is defined by the values of three quantum numbers: n, l, and m_l. (Section 6.5)

ore A source of a desired element or mineral, usually accompanied by large quantities of other materials such as sand and clay. (Section 23.1)

organic chemistry The study of carbon-containing compounds, typically containing carbon–carbon bonds. (Section 2.9; Chapter 25: Introduction)

osmosis The net movement of solvent through a semipermeable membrane toward the solution with greater solute concentration. (Section 13.5)

osmotic pressure The pressure that must be applied to a solution to stop osmosis from pure solvent into the solution. (Section 13.5)

Ostwald process An industrial process used to make nitric acid from ammonia. The NH_3 is catalytically oxidized by O_2 to form NO; NO in air is oxidized to NO_2; HNO_3 is formed in a disproportionation reaction when NO_2 dissolves in water. (Section 22.7)

overall reaction order The sum of the reaction orders of all the reactants appearing in the rate expression when the rate can be expressed as $rate = k[A]^a[B]^b\ldots$. (Section 14.3)

overlap The extent to which atomic orbitals on different atoms share the same region of space. When the overlap between two orbitals is large, a strong bond may be formed. (Section 9.4)

oxidation A process in which a substance loses one or more electrons. (Section 4.4)

oxidation number (oxidation state) A positive or negative whole number assigned to an element in a molecule or ion on the basis of a set of formal rules; to some degree it reflects the positive or negative character of that atom. (Section 4.4)

oxidation-reduction (redox) reaction A chemical reaction in which the oxidation states of certain atoms change. (Chapter 20: Introduction)

oxidizing agent, or oxidant The substance that is reduced and thereby causes the oxidation of some other substance in an oxidation-reduction reaction. (Section 20.1)

oxyacid A compound in which one or more OH groups, and possibly additional oxygen atoms, are bonded to a central atom. (Section 16.10)

oxyanion A polyatomic anion that contains one or more oxygen atoms. (Section 2.8)

ozone The name given to O_3, an allotrope of oxygen. (Section 7.8)

paramagnetism A property that a substance possesses if it contains one or more unpaired electrons. A paramagnetic substance is drawn into a magnetic field. (Section 9.8)

partial pressure The pressure exerted by a particular gas in a mixture. (Section 10.6)

particle accelerator A device that uses strong magnetic and electrostatic fields to accelerate charged particles. (Section 21.3)

parts per billion (ppb) The concentration of a solution in grams of solute per 10^9 (billion) grams of solution; equals micrograms of solute per liter of solution for aqueous solutions. (Section 13.4)

parts per million (ppm) The concentration of a solution in grams of solute per 10^6 (million) grams of solution; equals milligrams of solute per liter of solution for aqueous solutions. (Section 13.4)

pascal (Pa) The SI unit of pressure: $1\ Pa = 1\ N/m^2$. (Section 10.2)

Pauli exclusion principle A rule stating that no two electrons in an atom may have the same four quantum numbers (n, l, m_l, and m_s). As a reflection of this principle, there can be no more than two electrons in any one atomic orbital. (Section 6.7)

peptide bond A bond formed between two amino acids. (Section 25.7)

percent ionization The percent of a substance that undergoes ionization on dissolution in water. The term applies to solutions of weak acids and bases. (Section 16.6)

percent yield The ratio of the actual (experimental) yield of a product to its theoretical (calculated) yield, multiplied by 100. (Section 3.7)

period The row of elements that lie in a horizontal row in the periodic table. (Section 2.5)

periodic table The arrangement of elements in order of increasing atomic number, with elements having similar properties placed in vertical columns. (Section 2.5)

petroleum A naturally occurring combustible liquid composed of hundreds of hydrocarbons and other organic compounds. (Section 5.8)

pH The negative log in base 10 of the aquated hydrogen ion concentration: $pH = -log[H^+]$. (Section 16.4)

pH titration curve See **titration curve**. (Section 17.3)

phase change The conversion of a substance from one state of matter to another. The phase changes we consider are melting and freezing (solid $\rightleftharpoons$ liquid), sublimation and deposition (solid $\rightleftharpoons$ gas), and vaporization and condensation (liquid $\rightleftharpoons$ gas). (Section 11.4)

phase diagram A graphic representation of the equilibria among the solid, liquid, and gaseous phases of a substance as a function of temperature and pressure. (Section 11.6)

phospholipid A form of lipid molecule that contains charged phosphate groups. (Section 25.9)

photochemical smog A complex mixture of undesirable substances produced by the action of sunlight on an urban atmosphere polluted with automobile emissions. The major starting ingredients are nitrogen oxides and organic substances, notably olefins and aldehydes. (Section 18.4)

photodissociation The breaking of a molecule into two or more neutral fragments as a result of absorption of light. (Section 18.2)

photoelectric effect The emission of electrons from a metal surface induced by light. (Section 6.2)

photoionization The removal of an electron from an atom or molecule by absorption of light. (Section 18.2)

photon The smallest increment (a quantum) of radiant energy; a photon of light with frequency ν has an energy equal to $h\nu$. (Section 6.2)

photosynthesis The process that occurs in plant leaves by which light energy is used to convert carbon dioxide and water to carbohydrates and oxygen. (Section 24.2)

physical changes Changes (such as a phase change) that occur with no change in chemical composition. (Section 1.3)

physical properties Properties that can be measured without changing the composition of a substance, for example, color and freezing point. (Section 1.3)

pi (π) bond A covalent bond in which electron density is concentrated above and below the internuclear axis. (Section 9.6)

pi (π) molecular orbital A molecular orbital that concentrates the electron density on opposite sides of an imaginary line that passes through the nuclei. (Section 9.8)

Planck's constant (h) The constant that relates the energy and frequency of a photon, $E = h\nu$. Its value is 6.626×10^{-34} J-s. (Section 6.2)

plastic A material that can be formed into particular shapes by application of heat and pressure. (Section 12.6)

polar covalent bond A covalent bond in which the electrons are not shared equally. (Section 8.4)

polarizability The ease with which the electron cloud of an atom or a molecule is distorted by an outside influence, thereby inducing a dipole moment. (Section 11.2)

polar molecule A molecule that possesses a nonzero dipole moment. (Section 8.4)

polyatomic ion An electrically charged group of two or more atoms. (Section 2.7)

polydentate ligand A ligand in which two or more donor atoms can coordinate to the same metal ion. (Section 24.2)

polymer   A large molecule of high molecular mass, formed by the joining together, or polymerization, of a large number of molecules of low molecular mass. The individual molecules forming the polymer are called monomers. (Section 12.6)

polypeptide   A polymer of amino acids that has a molecular weight of less than 10,000. (Section 25.7)

polyprotic acid   A substance capable of dissociating more than one proton in water; H_2SO_4 is an example. (Section 16.6)

polysaccharide   A substance made up of many monosaccharide units joined together. (Section 25.8)

porphyrin   A complex derived from the porphine molecule. (Section 24.2)

positron   A particle with the same mass as an electron but with a positive charge, symbol ^0_1e. (Section 21.1)

potential energy   The energy that an object possesses as a result of its composition or its position with respect to another object. (Section 5.1)

precipitate   An insoluble substance that forms in, and separates from, a solution. (Section 4.2)

precipitation reaction   A reaction that occurs between substances in solution in which one of the products is insoluble. (Section 4.2)

precision   The closeness of agreement among several measurements of the same quantity; the reproducibility of a measurement. (Section 1.5)

pressure   A measure of the force exerted on a unit area. In chemistry, pressure is often expressed in units of atmospheres (atm) or torr: 760 torr = 1 atm; in SI units pressure is expressed in pascals (Pa). (Section 10.2)

pressure–volume (PV) work   Work performed by expansion of a gas against a resisting pressure. (Section 5.3)

primary structure   The sequence of amino acids along a protein chain. (Section 25.7)

primitive cubic cell   A cubic unit cell in which the lattice points are at the corners only. (Section 11.7)

probability density (ψ^2)   A value that represents the probability that an electron will be found at a given point in space. (Section 6.5)

product   A substance produced in a chemical reaction; it appears to the right of the arrow in a chemical equation. (Section 3.1)

property   A characteristic that gives a sample of matter its unique identity. (Section 1.1)

protein   A biopolymer formed from amino acids. (Section 25.7)

protium   The most common isotope of hydrogen. (Section 22.2)

proton   A positively charged subatomic particle found in the nucleus of an atom. (Section 2.3)

pure substance   Matter that has a fixed composition and distinct properties. (Section 1.2)

pyrometallurgy   A process in which heat converts a mineral in an ore from one chemical form to another and eventually to the free metal. (Section 23.2)

qualitative analysis   The determination of the presence or absence of a particular substance in a mixture. (Section 17.7)

quantitative analysis   The determination of the amount of a given substance that is present in a sample. (Section 17.7)

quantum   The smallest increment of radiant energy that may be absorbed or emitted; the magnitude of radiant energy is $h\nu$. (Section 6.2)

racemic mixture   A mixture of equal amounts of the dextrorotatory and levorotatory forms of a chiral molecule. A racemic mixture will not rotate the plane of polarized light. (Section 24.4)

rad   A measure of the energy absorbed from radiation by tissue or other biological material; 1 rad = transfer of 1×10^{-2} J of energy per kilogram of material. (Section 21.9)

radial probability function   The probability that the electron will be found at a certain distance from the nucleus. (Section 6.6)

radioactive   Possessing **radioactivity**, the spontaneous disintegration of an unstable atomic nucleus with accompanying emission of radiation. (Section 2.2; Chapter 21: Introduction)

radioactive series   A series of nuclear reactions that begins with an unstable nucleus and terminates with a stable one. Also called **nuclear disintegration series**. (Section 21.2)

radioisotope   An isotope that is radioactive; that is, it is undergoing nuclear changes with emission of radiation. (Section 21.1)

radionuclide   A radioactive nuclide. (Section 21.1)

radiotracer   A radioisotope that can be used to trace the path of an element in a chemical system. (Section 21.5)

Raoult's law   A law stating that the partial pressure of a solvent over a solution, P_A, is given by the vapor pressure of the pure solvent, P_A°, times the mole fraction of a solvent in the solution, X_A: $P_A = X_A P_A^\circ$ (Section 13.5)

rare earth element   See **lanthanide element**. (Sections 6.8 and 6.9)

rate constant   A constant of proportionality between the reaction rate and the concentrations of reactants that appear in the rate law. (Section 14.3)

rate-determining step   The slowest elementary step in a reaction mechanism. (Section 14.6)

rate law   An equation that relates the reaction rate to the concentrations of reactants (and sometimes of products also). (Section 14.3)

reactant   A starting substance in a chemical reaction; it appears to the left of the arrow in a chemical equation. (Section 3.1)

reaction mechanism   A detailed picture, or model, of how the reaction occurs; that is, the order in which bonds are broken and formed and the changes in relative positions of the atoms as the reaction proceeds. (Section 14.6)

reaction order   The power to which the concentration of a reactant is raised in a rate law. (Section 14.3)

reaction quotient (Q)   The value that is obtained when concentrations of reactants and products are inserted into the equilibrium expression. If the concentrations are equilibrium concentrations, $Q = K$; otherwise, $Q \neq K$. (Section 15.6)

reaction rate   A measure of the decrease in concentration of a reactant or the increase in concentration of a product with time. (Section 14.2)

redox (oxidation-reduction) reaction   A reaction in which certain atoms undergo changes in oxidation states. The substance increasing in oxidation state is oxidized; the substance decreasing in oxidation state is reduced. (Section 4.4 and Chapter 20: Introduction)

reducing agent, or reductant   The substance that is oxidized and thereby causes the reduction of some other substance in an oxidation-reduction reaction. (Section 20.1)

reduction   A process in which a substance gains one or more electrons. (Section 4.4)

refining   The process of converting an impure form of a metal into a more usable substance of well-defined composition. For example, crude pig iron from the blast furnace is refined in a converter to produce steels of desired compositions. (Section 23.2)

rem   A measure of the biological damage caused by radiation; rems = rads × RBE. (Section 21.9)

renewable energy   Energy such as solar energy, wind energy, and hydroelectric energy derived from essentially inexhaustible sources. (Section 5.8)

representative (main-group) element   An element from within the s and p blocks of the periodic table (Figure 6.29). (Section 6.9)

resonance structures (resonance forms)   Individual Lewis structures in cases where two or more Lewis structures are equally good descriptions of a single molecule. The resonance structures in such an instance are "averaged" to give a more accurate description of the real molecule. (Section 8.6)

reverse osmosis   The process by which water molecules move under high pressure through a semipermeable membrane from the more concentrated to the less concentrated solution. (Section 18.5)

reversible process   A process that can go back and forth between states along exactly the same path; a system at equilibrium is reversible if equilibrium can be shifted by an infinitesimal modification of a variable such as temperature. (Section 19.1)

ribonucleic acid (RNA)   A polynucleotide in which ribose is the sugar component. (Section 25.10)

roasting   Thermal treatment of an ore to bring about chemical reactions involving the furnace atmosphere. For example, a sulfide ore might be roasted in air to form a metal oxide and SO_2. (Section 23.2)

root-mean-square (rms) speed (μ)   The square root of the average of the squared speeds of the gas molecules in a gas sample. (Section 10.7)

rotational motion   Movement of a molecule as though it is spinning like a top. (Section 19.3)

salinity   A measure of the salt content of seawater, brine, or brackish water. It is equal to the mass in grams of dissolved salts present in 1 kg of seawater. (Section 18.5)

salt An ionic compound formed by replacing one or more hydrogens of an acid by other cations. (Section 4.3)

saponification Hydrolysis of an ester in the presence of a base. (Section 25.4)

saturated solution A solution in which undissolved solute and dissolved solute are in equilibrium. (Section 13.2)

scientific law A concise verbal statement or a mathematical equation that summarizes a wide range of observations and experiences. (Section 1.3)

scientific method The general process of advancing scientific knowledge by making experimental observations and by formulating hypotheses, theories, and laws. (Section 1.3)

scintillation counter An instrument that is used to detect and measure radiation by the fluorescence it produces in a fluorescing medium. (Section 21.5)

secondary structure The manner in which a protein is coiled or stretched. (Section 25.7)

second law of thermodynamics A statement of our experience that there is a direction to the way events occur in nature. When a process occurs spontaneously in one direction, it is nonspontaneous in the reverse direction. It is possible to state the second law in many different forms, but they all relate back to the same idea about spontaneity. One of the most common statements found in chemical contexts is that in any spontaneous process the entropy of the universe increases. (Section 19.2)

second-order reaction A reaction in which the overall reaction order (the sum of the concentration-term exponents) in the rate law is 2. (Section 14.4)

semiconductor A solid with limited electrical conductivity. (Section 12.1)

sigma (σ) bond A covalent bond in which electron density is concentrated along the internuclear axis. (Section 9.6)

sigma (σ) molecular orbital A molecular orbital that centers the electron density about an imaginary line passing through two nuclei. (Section 9.7)

significant figures The digits that indicate the precision with which a measurement is made; all digits of a measured quantity are significant, including the last digit, which is uncertain. (Section 1.5)

silicates Compounds containing silicon and oxygen, structurally based on SiO_4 tetrahedra. (Section 22.10)

single bond A covalent bond involving one electron pair. (Section 8.3)

SI units The preferred metric units for use in science. (Section 1.4)

slag A mixture of molten silicate minerals. Slags may be acidic or basic, according to the acidity or basicity of the oxide added to silica. (Section 23.2)

smectic liquid crystalline phase A liquid crystal in which the molecules are aligned along their long axes and arranged in sheets, with the ends of the molecules aligned. There are several different kinds of smectic phases. (Section 12.8)

smelting A melting process in which the materials formed in the course of the chemical reactions that occur separate into two or more layers. For example, the layers might be slag and molten metal. (Section 23.2)

solar cell An electronic device composed of doped semiconductors in which radiant energy can be converted into electrical energy. (Section 12.3)

sol-gel process A process in which extremely small particles (0.003 to 0.1 μm in diameter) of uniform size are produced in a series of chemical steps, followed by controlled heating. (Section 12.4)

solid Matter that has both a definite shape and a definite volume. (Section 1.2)

solubility The amount of a substance that dissolves in a given quantity of solvent at a given temperature to form a saturated solution. (Sections 4.2 and 13.2)

solubility-product constant (solubility product) (K_{sp}) An equilibrium constant related to the equilibrium between a solid salt and its ions in solution. It provides a quantitative measure of the solubility of a slightly soluble salt. (Section 17.4)

solute A substance dissolved in a solvent to form a solution; it is normally the component of a solution present in the smaller amount. (Section 4.1)

solution A mixture of substances that has a uniform composition; a homogeneous mixture. (Section 1.2)

solution alloy A homogeneous alloy, with the components distributed uniformly throughout. (Section 23.6)

solvation The clustering of solvent molecules around a solute particle. (Section 13.1)

solvent The dissolving medium of a solution; it is normally the component of a solution present in the greater amount. (Section 4.1)

specific heat (C_s) The heat capacity of 1 g of a substance; the heat required to raise the temperature of 1 g of a substance by 1 °C. (Section 5.5)

spectator ions Ions that go through a reaction unchanged and that appear on both sides of the complete ionic equation. (Section 4.2)

spectrochemical series A list of ligands arranged in order of their abilities to split the d-orbital energies (using the terminology of the crystal-field model). (Section 24.6)

spectrum The distribution among various wavelengths of the radiant energy emitted or absorbed by an object. (Sections 6.3 and 24.6)

spin magnetic quantum number (m_s) A quantum number associated with the electron spin; it may have values of $+\frac{1}{2}$ or $-\frac{1}{2}$. (Section 6.7)

spin-pairing energy The energy required to pair an electron with another electron occupying an orbital. (Section 24.6)

spontaneous process A process that is capable of proceeding in a given direction, as written or described, without needing to be driven by an outside source of energy. A process may be spontaneous even though it is very slow. (Section 19.1)

standard atmospheric pressure Defined as 760 torr or, in SI units, 101.325 kPa. (Section 10.2)

standard electrode potential See **standard reduction potential**. (Section 20.4)

standard emf, also called the standard cell potential ($E°$) The emf of a cell when all reagents are at standard conditions. (Section 20.4)

standard enthalpy change ($\Delta H°$) The change in enthalpy in a process when all reactants and products are in their stable forms at 1 atm pressure and a specified temperature, commonly 25 °C. (Section 5.7)

standard enthalpy of formation ($\Delta H_f°$) The change in enthalpy that accompanies the formation of one mole of a substance from its elements, with all substances in their standard states. (Section 5.7)

standard free energy of formation ($\Delta G_f°$) The change in free energy associated with the formation of a substance from its elements under standard conditions. (Section 19.5)

standard hydrogen electrode (SHE) An electrode based on the half-reaction $2 H^+(1\ M) + 2 e^- \longrightarrow H_2(1\ atm)$. The standard electrode potential of the standard hydrogen electrode is defined as 0 V. (Section 20.4)

standard molar entropy ($S°$) The entropy value for a mole of a substance in its standard state. (Section 19.4)

standard reduction potential ($E°_{red}$) The potential of a reduction half-reaction under standard conditions, measured relative to the standard hydrogen electrode. A standard reduction potential is also called a **standard electrode potential**. (Section 20.4)

standard solution A solution of known concentration. (Section 4.6)

standard temperature and pressure (STP) Defined as 0 °C and 1 atm pressure; frequently used as reference conditions for a gas. (Section 10.4)

starch The general name given to a group of polysaccharides that acts as energy-storage substances in plants. (Section 25.8)

state function A property of a system that is determined by the state or condition of the system and not by how it got to that state; its value is fixed when temperature, pressure, composition, and physical form are specified; P, V, T, E, and H are state functions. (Section 5.2)

states of matter The three forms that matter can assume: solid, liquid, and gas. (Section 1.2)

stereoisomers Compounds possessing the same formula and bonding arrangement but differing in the spatial arrangements of the atoms. (Section 24.4)

stoichiometry The relationships among the quantities of reactants and products involved in chemical reactions. (Chapter 3: Introduction)

stratosphere The region of the atmosphere directly above the troposphere. (Section 18.1)

strong acid An acid that ionizes completely in water. (Section 4.3)

strong base A base that ionizes completely in water. (Section 4.3)

strong electrolyte A substance (strong acids, strong bases, and most salts) that is completely ionized in solution. (Section 4.1)

structural formula A formula that shows not only the number and kinds of atoms in the molecule but also the arrangement (connections) of the atoms. (Section 2.6)

structural isomers Compounds possessing the same formula but differing in the bonding arrangements of the atoms. (Sections 24.4 and 25.3)

subatomic particles Particles such as protons, neutrons, and electrons that are smaller than an atom. (Section 2.2)

subshell One or more orbitals with the same set of quantum numbers n and l. For example, we speak of the $2p$ subshell ($n = 2$, $l = 1$), which is composed of three orbitals ($2p_x$, $2p_y$, and $2p_z$). (Section 6.5)

substitutional alloy A homogeneous (solution) alloy in which atoms of different elements randomly occupy sites in the lattice. (Section 23.6)

substitution reactions Reactions in which one atom (or group of atoms) replaces another atom (or group) within a molecule; substitution reactions are typical for alkanes and aromatic hydrocarbons. (Section 25.4)

substrate A substance that undergoes a reaction at the active site in an enzyme. (Section 14.7)

superconductor A material capable of carrying an electrical current without apparent resistance when cooled below a transition temperature, T_c. (Section 12.50)

superconducting ceramic A complex metal oxide that undergoes a transition to a superconducting state at a low temperature. (Section 12.1)

superconducting transition temperature (T_c) The temperature below which a substance exhibits superconductivity. (Section 12.1)

superconductivity The "frictionless" flow of electrons that occurs when a substance loses all resistance to the flow of electrical current. (Section 12.1)

supercritical mass An amount of fissionable material larger than the critical mass. (Section 21.7)

supersaturated solutions Solutions containing more solute than an equivalent saturated solution. (Section 13.2)

surface tension The intermolecular, cohesive attraction that causes a liquid to minimize its surface area. (Section 11.3)

surroundings In thermodynamics, everything that lies outside the system that we study. (Section 5.1)

system In thermodynamics, the portion of the universe that we single out for study. We must be careful to state exactly what the system contains and what transfers of energy it may have with its surroundings. (Section 5.1)

termolecular reaction An elementary reaction that involves three molecules. Termolecular reactions are rare. (Section 14.6)

tertiary structure The overall shape of a large protein, specifically, the manner in which sections of the protein fold back upon themselves or intertwine. (Section 25.9)

theoretical yield The quantity of product that is calculated to form when all of the limiting reagent reacts. (Section 3.7)

theory A tested model or explanation that satisfactorily accounts for a certain set of phenomena. (Section 1.3)

thermochemistry The relationship between chemical reactions and energy changes. (Chapter 5: Introduction)

thermodynamics The study of energy and its transformation. (Chapter 5: Introduction)

thermonuclear reaction Another name for fusion reactions; reactions in which two light nuclei are joined to form a more massive one. (Section 21.8)

thermoplastic A polymeric material that can be readily reshaped by application of heat and pressure. (Section 12.6)

thermosetting plastic A plastic that, once formed in a particular mold, is not readily reshaped by application of heat and pressure. (Section 12.2)

third law of thermodynamics A law stating that the entropy of a pure, crystalline solid at absolute zero temperature is zero: $S(0\text{ K}) = 0$. (Section 19.3)

titration The process of reacting a solution of unknown concentration with one of known concentration (a standard solution). (Section 4.6)

titration curve A graph of pH as a function of added titrant. (Section 17.3)

torr A unit of pressure (1 torr = 1 mm Hg). (Section 10.2)

transistor An electrical device that forms the heart of an integrated circuit. (Section 12.4)

transition elements (transition metals) Elements in which the d orbitals are partially occupied. (Section 6.8)

transition state (activated complex) The particular arrangement of reactant and product molecules at the point of maximum energy in the rate-determining step of a reaction. (Section 14.5)

translational motion Movement in which an entire molecule moves in a definite direction. (Section 19.3)

transuranium elements Elements that follow uranium in the periodic table. (Section 21.3)

triple bond A covalent bond involving three electron pairs. (Section 8.3)

triple point The temperature at which solid, liquid, and gas phases coexist in equilibrium. (Section 11.6)

tritium The isotope of hydrogen whose nucleus contains a proton and two neutrons. (Section 22.2)

troposphere The region of Earth's atmosphere extending from the surface to about 12 km altitude. (Section 18.1)

Tyndall effect The scattering of a beam of visible light by the particles in a colloidal dispersion. (Section 13.6)

uncertainty principle A principle stating there is an inherent uncertainty in the precision with which we can simultaneously specify the position and momentum of a particle. This uncertainty is significant only for particles of extremely small mass, such as electrons. (Section 6.4)

unimolecular reaction An elementary reaction that involves a single molecule. (Section 14.6)

unit cell The smallest portion of a crystal that reproduces the structure of the entire crystal when repeated in different directions in space.

It is the repeating unit or building block of the crystal lattice. (Section 11.7)

unsaturated solutions Solutions containing less solute than a saturated solution. (Section 13.2)

valence band A band of closely spaced molecular orbitals that is essentially fully occupied by electrons. (Section 12.2)

valence-bond theory A model of chemical bonding in which an electron-pair bond is formed between two atoms by the overlap of orbitals on the two atoms. (Section 9.4)

valence electrons The outermost electrons of an atom; those that occupy orbitals not occupied in the nearest noble-gas element of lower atomic number. The valence electrons are the ones the atom uses in bonding. (Section 6.8)

valence orbitals Orbitals that contain the outer-shell electrons of an atom. (Chapter 7: Introduction)

valence-shell electron-pair repulsion (VSEPR) model A model that accounts for the geometric arrangements of shared and unshared electron pairs around a central atom in terms of the repulsions between electron pairs. (Section 9.2)

van der Waals equation An equation of state for nonideal gases that is based on adding corrections to the ideal-gas equation. The correction terms account for intermolecular forces of attraction and for the volumes occupied by the gas molecules themselves. (Section 10.9)

vapor Gaseous state of any substance that normally exists as a liquid or solid. (Section 10.1)

vapor pressure The pressure exerted by a vapor in equilibrium with its liquid or solid phase. (Section 11.5)

vibrational motion Movement of the atoms within a molecule in which they move periodically toward and away from one another. (Section 19.3)

viscosity A measure of the resistance of fluids to flow. (Section 11.3)

volatile Tending to evaporate readily. (Section 11.5)

voltaic (galvanic) cell A device in which a spontaneous oxidation-reduction reaction occurs with the passage of electrons through an external circuit. (Section 20.3)

vulcanization The process of cross-linking polymer chains in rubber. (Section 12.6)

watt A unit of power; 1 W = 1 J/s. (Section 20.9)

wave function A mathematical description of an allowed energy state (an orbital) for an electron in the quantum mechanical model of the atom; it is usually symbolized by the Greek letter ψ. (Section 6.5)

wavelength The distance between identical points on successive waves. (Section 6.1)

weak acid An acid that only partly ionizes in water. (Section 4.3)

weak base A base that only partly ionizes in water. (Section 4.3)

weak electrolyte A substance that only partly ionizes in solution. (Section 4.1)

work The movement of an object against some force. (Section 5.1)

Photo/Art Credits

ger/Color-Pic, Inc. **13.27b** Gene Rhoden/Visuals Unlimited **13.32** Oliver Meckes & Nicole Ottawa/Photo Researchers, Inc.

Chapter 14: CO14 Hunter, Jeff/Getty Images Inc. - Image Bank **14.1a** Michael S. Yamashita/Corbis/Bettmann **14.1b** S.C. Fried/Photo Researchers, Inc. **14.1c** David N. Davis/Photo Researchers, Inc. **14.2a** Michael Dalton/Fundamental Photographs, NYC **14.2b** Richard Megna/Fundamental Photographs, NYC **14.11** Richard Megna/Fundamental Photographs, NYC **14.19a–c** Richard Megna/Fundamental Photographs, NYC **14.22** Delphi Energy & Chassis, Troy, Michigan **14.23** Richard Megna/Fundamental Photographs, NYC **14.27a** Science Photo Library/Photo Researchers, Inc.

Chapter 15: CO15 Mira.com/Ron Niebrugge **15.1a–c** Richard Megna/Fundamental Photographs, NYC **15.5** Ed Degginger/Color Pic, Inc. Courtesy Farmland Industries, Inc. **15.13a–c** Richard Megna/Fundamental Photographs, NYC **15.14a–c** Richard Megna/Fundamental Photographs, NYC **p. 631** Grant Heilman Photography, Inc.

Chapter 16: CO16 Rosemary Calvert/Getty Images Inc. - Photographer's Choice Royalty Free **16.3** Richard Megna/Fundamental Photographs, NYC **16.6** Yoav Levy/Phototake NYC **16.8a–b** Donald Clegg and Roxy Wilson/Pearson Education/PH College **16.10** Michael Dalton/Fundamental Photographs, NYC **16.11a–c** Richard Megna/Fundamental Photographs, NYC **16.14** Frank LaBua/Pearson Education/PH College **16.16** Tom Pantages

Chapter 17: CO17 Russ Bishop/Alamy Images **17.1** Donald Clegg & Roxy Wilson/Pearson Education/PH College **17.4** Profs. P. Motta and S. Correr/Science Photo Library/Photo Researchers, Inc. **17.7a–b** Richard Megna/Fundamental Photographs, NYC **17.16a–b** Richard Megna/Fundamental Photographs, NYC **17.17b** Gerry Davis/Phototake NYC **17.19a–c** Richard Megna/Fundamental Photographs, NYC **17.20a–c** Richard Megna/Fundamental Photographs, NYC **17.21a–c** Richard Megna/Fundamental Photographs, NYC **p. 757** Richard Megna/Fundamental Photographs, NYC

Chapter 18: CO18 SPL/Photo Researchers, Inc. **18.2** Jorma Luhta/Nature Picture Library **18.3** Courtesy Earth Observatory/NASA and Nicholas M. Short, Sr. **18.5** NASA Headquarters **18.5** National Atmospheric Deposition Program/National Trends Network (http://nadp.sws.uiuc.edu) **18.7a–b** Don and Pat Valenti **18.9a** Dennis Kunkel/Phototake NYC **18.10** Tom Pantages **18.11** Ulf E. Wallin/Ulf Wallin Photography **18.13** C.D. Keeling and T.P Whorf, Carbon Dioxide Research Group, Scripps Institution of Oceanography, University of California, La Jolla, CA 92093-0444. **18.14** Australia Picture Library/Corbis/Bettmann **18.15L-R** Windows to the Universe, of the University Corporation for Atmospheric Research. Copyright © 2004 University Corporation for Atmospheric Research. All rights reserved. **18.16a** DuPont **18.17** © Katadyn North America. Used by permission. **18.18** John Sohlden/Visuals Unlimited **18.20** LifeStraw courtesy of Vestergaards Frandsen, Inc. **18.21** Sheila Terry/Science Photo Library **18.22** Kim Fennema/Visuals Unlimited **18.23** PPG Industries Inc. **p. 797** Reprinted with permission from Walter Leitner, "Supercritical Carbon Dioxide as a Green Reaction Medium for Catalysis", *Acc. Chem Res.*, 35 (9), 746–756, 2002. Copyright 2002 American Chemical Society.

Chapter 19: CO19 NASA/Photo Researchers, Inc. **19.2a–b** Richard Megna/Fundamental Photographs, NYC **19.3.1–2** Michael Dalton/Fundamental Photographs, NYC **19.7** Austrian Central Library for Physics, Vienna, Austria **19.10(left)** Richard Megna/Fundamental Photographs, NYC **19.12a** Klaus Pavsan/Peter Arnold, Inc. **19.12b** Biophoto Associates/Photo Researchers, Inc. **19.16** Library of Congress

Chapter 20: CO20 Roy Langstaff Photography **20.1a–b** Richard Megna/Fundamental Photographs, NYC **20.2a–c** Richard Megna/Fundamental Photographs, NYC **20.3a–b** Richard Megna/Fundamental Photographs, NYC **20.4** Richard Megna/Fundamental Photographs, NYC **20.9(left)** Jeff Gnass/Corbis/Stock Market **20.15** The Burndy Library, Dibner Institute for the History of Science and Technology, Cambridge, Massachusetts **20.23** John Mead/Science Photo Library/Photo Researchers, Inc. **20.28a–b** Reed Barton/Tom Pantages **p. 855** Jeff Gnass/GEOLIGHT

Chapter 21: CO21 NOVASTOCK/PhotoEdit Inc. **21.1** CNRI/Photo Researchers, Inc. **21.7** CERN/European Organization for Nuclear Research **21.9** Photo Researchers, Inc. **21.10** Terence Kearey Photography **21.12** Kevin Schafer/Peter Arnold, Inc. **21.18** Archival Photofiles, [apf2-00502], Special Collections Research Center, University of Chicago Library. **21.18** Gary Sheahan "Birth of the Atomic Age" Chicago (Illinois); 1957. Chicago Historical Society, ICHi-33305. **21.20b** Ed Degginger/Color-Pic, Inc. **21.24** Figure 9 from P. W. Swarzenski, "U/Th Radionuclides as Coastal Groundwater Tracers," *Chem. Rev. 2007*, 107, 663–674. Courtesy of Peter W. Swarzenski, PhD, Coastal Marine Geology Program, U.S. Geological Survey **21.25** Richard Megna/Fundamental Photographs, NYC

Chapter 22: CO22 © Dan Forer/Beateworks/CORBIS. All Rights Reserved **22.5** © Russell Curtis/Photo Researchers, Inc. **22.6a–b** Richard Megna/Fundamental Photographs, NYC **22.9** Donald Clegg and Roxy Wilson/Pearson Education/PH College **22.11** Paul Silverman/Fundamental Photographs, NYC **22.12a–b** Richard Megna/Fundamental Photographs, NYC **22.13** Beth Plowes - Proteapix **22.14** NASA/Johnson Space Center **22.16** Joseph Priestley (1733–1804): colored English engraving, 19th Century. The Granger Collection, New York. **22.16** John Hill/Getty Images Inc. - Image Bank **22.18a–c** Richard Megna/Fundamental Photographs, NYC **22.19** Courtesy of DuPont Nomex® **22.21a** Jeffrey A. Scovil/Jeffrey A. Scovil **22.22** Lawrence Migdale/Science Source/Photo Researchers, Inc. **22.24** Dan McCoy/Rainbow **22.25a–c** Kristen Brochmann/Fundamental Photographs, NYC **22.29a–c** Donald Clegg and Roxy Wilson/Pearson Education/PH College **22.31** Michael Dalton/Fundamental Photographs, NYC **22.33** Richard Megna/Fundamental Photographs, NYC **22.37** © Mufty Munir/epa/CORBIS All Rights Reserved **22.38** General Electric Corporate Research & Development Center **22.40** AP Wide World Photos **22.41** Chip Clark **22.42** David Muench/Corbis/Bettmann **22.43** Photo Courtesy of Texas Instruments Incorporated **22.47** National Institute for Occupational Safety & Health

Chapter 23: CO23 Frank Micelotta/Getty Images, Inc - Liaison **23.1b** Brownie Harris/Corbis/Stock Market **23.2a–b** Karl Hartmann/Traudel Sachs/Phototake NYC **23.2c** Jeffrey A. Scovil/Jeffrey A. Scovil **23.3** The Cleveland-Cliffs Iron Company **23.5** Robin Smith/Getty Images Inc. - Stone Allstock **23.10** Oberlin College Archives, Oberlin, Ohio **23.12** Richard Megna/Fundamental Photographs, NYC **23.14a** Peter Christopher/Masterfile Corporation **23.14b** IT Stock International/Index Stock Imagery, Inc. **23.15** Magneti Ljubljana, d.d. **23.19** Donald Clegg and Roxy Wilson/Pearson Education/PH College **23.22** Michael Dalton/Fundamental Photographs, NYC **23.23** Donald Clegg and Roxy Wilson/Pearson Education/PH College **23.24** Richard Megna/Fundamental Photographs, NYC **23.25** Donald Clegg and Roxy Wilson/Pearson Education/PH College **23.26** Paul Silverman/Fundamental Photographs, NYC **23.27** Donald Clegg and Roxy Wilson/Pearson Education/PH College

Chapter 24: CO24 Max Alexander © Dorling Kindersley **24.2a–b** Richard Megna/Fundamental Photographs, NYC **24.7** Gary C. Will/Visuals Unlimited **24.14** Kim Heacox Photography/DRK Photo **24.18L–R** Richard Megna/Fundamental Photographs, NYC **24.22** Richard Megna/Fundamental Photographs, NYC **24.26c** Nigel Forrow Photography, photographersdirect.com **24.36a–c** Tom Pantages **p. 1043** Carey B. Van Loon

Chapter 25: CO25 Sergio Piumatti **25.8** Ed Degginger/Color-Pic, Inc. **25.9** Wes Thompson/Corbis/Stock Market **25.15** Richard Megna/Fundamental Photographs, NYC **25.16a** Richard Megna/Fundamental Photographs, NYC **25.16b** Beth Plowes - Proteapix **25.18** Richard Megna/Fundamental Photographs, NYC **25.19** Culver Pictures, Inc. **25.25** Richard Megna/Fundamental Photographs, NYC **25.27** Oxford Molecular Biophysics Laboratory/Science Photo Library/Photo Researchers, Inc.

Index

AB_n molecules, 342–43
Absolute entropy, 817, 818
Absolute temperature, 595
Absorption, 558, 607
Absorption spectrum, 1033
Accuracy, 21
Acetaldehyde (ethanal), 570, 1071, 1073, 1075
  properties of, 441
Acetaminophen, 1102
Acetic acid (ethanoic acid), 129, 682, 702–3, 716, 720, 1071, 1074, 1076
  decarbonylation of, 838
  glacial, 1076
  in green chemistry, 790
  ionization of, 122–23
  pH of, 684–85
  production of, 1075
  properties of, 682
  shape of, 352
  titration with caustic soda (NaOH), 733–37
Acetic anhydride, 1100
Acetone (propanone), 478, 536, 570, 571, 1071, 1073, 1074
Acetonitrile, 443, 566, 569, 1052
  methyl isonitrate converted to, 587
  properties of, 441
Acetylene, 838, 945, 964, 1055, 1064, 1071
  production of, 430
  triple bonds in, 364
Acetylide ion, 964
Acetylsalicylic acid (aspirin), 2, 132, 387, 710, 713, 764, 1074, 1076
Acid(s), 129
  adipic, 106, 501–2
  Arrhenius, 668
  binary, 699–70
  carboxylic, 702–3
  conjugate, 670
  defined, 64
  diprotic, 129, 679
  as electron acceptor, 704
  factors affecting strength of, 699
  household, 128
  ionic properties of, 129
  Lewis, 704–7
  metal oxides reacting with, 274
  monoprotic, 129, 679
  names and formulas of, 64–65
  oxidation of metals by, 138–40
  oxyacids, 700–2, 943
  properties of, 129
  reactions of. *See also* Acid–base reactions
    with magnesium, 138–39
    with magnesium hydroxide, 133
  relating to anions, 64–65
  relative strengths of, 672–73
  strong, 130–31, 672, 679–80
    in buffered solutions, 727–28
  titrating, 150–53
  weak, 130–31, 672, 681–90
    acid-dissociation constant (K_a), 681–83, 684–88, 693–95
    common-ion effect on, 720–23
    percent ionization of, 683–84
    polyprotic, 688–90, 737

Acid–base equilibria, 666–717. *See also* Aqueous equilibria
  acid-dissociation constant and base-dissociation constant relationship, 693–95
  Arrhenius definitions, 668
  autoionization of water, 673–75
  Brønsted–Lowry acids and bases, 668–73
    conjugate acid–base pairs, 670–71
    H^+ ions in water and, 669
    proton-transfer reactions, 669–70
    relative strengths of, 672–73
  chemical structure and, 699–703
    binary acids, 699–700
    carboxylic acids, 702–3
    factors affecting acid strength, 699
    oxyacids, 700–2
  ion product, 674–75
  Lewis acids and bases, 704–7
    electron-pair acceptor/donor concept, 704–5
    metal ions and, 705–7
  in organic substances, 703
  pH scale, 675–79
    measuring, 678–79
    other "p" scales, 678
    of salt solutions, 695–99
    anion reaction with water, 696
    cation reaction with water, 696–97
    combined cation-anion effect, 697–99
  solubility equilibria and, 741
  strong acids and bases, 679–81
    in buffered solutions, 727–28
  weak acids and bases, 681–93
    acid-dissociation constant, 681–83, 684–88, 693–95
    common-ion effect on, 720–23
    percent ionization of, 683–84
    polyprotic acids, 688–90
    types of, 692–93
Acid–base indicators, 150–51, 678, 679
Acid–base properties, of organic substances, 1054
Acid–base reactions, 128–35. *See also* Acid–base equilibria
  electrolytes, 131–32
  with gas formation, 134–35
  gas-phase, 669–70
  neutralization reactions and salts, 132–34
Acid–base titrations, 730–37
  of polyprotic acids, 737
  strong, 730–33
  weak, 733–37
Acid-dissociation constant (K_a), 681
  base-dissociation constant (K_b) and, 693–95
  calculating from pH, 682–83
  calculating pH from, 684–88
  for hydrolysis reactions, 705
  for polyprotic acids, 689
Acidic anhydrides (acidic oxides), 946
Acidic solutions, 697–98
Acid inhibitors, 135
Acid-insoluble sulfides, 754
Acidosis, 729
Acid rain, 275, 491, 776–77, 946

Acid salts, 694
Acid spills, 135
Actinides, electron configurations of, 239–40
Activated complex (transition state), 593
Activation, entropy of, 841
Activation energy, 592–94
  catalysis and, 610, 654
  determining, 595–97
Active metals, 140
Active site, 609
Active transport, 555
Activity, 906
Activity series, 140–42, 863
Actual yield, 105
Addition, significant figures in, 23
Addition polymerization, 499–500
Addition polymers, 500, 841
Addition reactions
  of alkenes and alkynes, 1065–66
  mechanism of, 1066–68
Adenine, 477, 1091, 1092
Adenosine diphosphate (ADP), 830, 960
Adenosine monophosphate (AMP), 1103
Adenosine triphosphate (ATP), 830, 840, 959, 1103
Adhesive forces, 448
Adipic acid, 106, 501–502
ADP (adenosine diphosphate), 830, 960
Adrenaline, 568
Adsorption, 558, 606–607
Air
  combustion in, 86–87
  composition of, 394
Air bags, automobile, 409
Alanine, 703, 1081, 1082, 1083
Alanylglycylserine, 1083
R-Albuterol, 1079
Alchemy, 143
Alcohol(s), 66–67, 1070–71
  condensation reactions with, 1075
  as functional groups, 1072–73
  oxidation of, 1074
  solubilities of, 536–37
Aldehydes, 1073–74
Alizarin yellow R, 679
Alkali (group 1A) metals, 50, 140, 240, 276–79
  group trends for, 276–79
  in ionic hydrides, 937
  ionic hydroxides of, 680–81
  oxidation number of, 137
  as reducing agent, 861
Alkali metal ions, 754
Alkaline battery, 872
Alkaline earth (group 2A) metals, 50, 140, 240, 279–81
  group trends for, 279–81
  in ionic hydrides, 937
  ionic hydroxides of, 680–81
  oxidation number of, 137
  as reducing agent, 861
Alkalosis, 729
Alkanes, 66–67, 1054, 1055–62
  cycloalkanes, 1060–61
  derivatives of, 66–67

  nomenclature of, 1058–60
  reactions of, 1061
  structural isomers of, 1057–58
  structures of, 1056–57
Alka-Seltzer, 135
Alkenes, 1054, 1055, 1062–64
  addition reactions of, 1065–66
Alkyl groups, 1059, 1070–71
  in Friedel-Crafts reaction, 1069
Alkynes, 1054, 1055, 1064–65
  addition reactions of, 1065–66
Allene, 389
Allotropes, 282
Alloys, 981, 983, 995–98
  defined, 995
  heterogeneous, 995, 996
  intermetallic compounds, 996–97
  interstitial, 995, 996
  shape-memory, 997–98
  solution, 995
  steels, 995–96
  substitutional, 995, 996
Alloy steels, 995
Al_2O_3, 484
α-helix, 1084
α radiation, 41, 895, 896, 897, 919–20
Aluminates, 493
Aluminum (Al), 6, 50, 992
  electrometallurgy of, 989–90
  electron configuration of, 240
  electronic properties of, 484
  hydrometallurgy of, 988
  Lewis symbol for, 299
  oxidation number of, 137
  oxidation of, 874
  purification of ore, 750
  recycling, 989–90
  specific heat of, 181
Aluminum hydroxide, 750
Aluminum oxide, 989
American Chemical Society, 51
Americium-241, 623, 928
Amethyst, 459
Amides, 1077–78
Amine hydrochlorides, 694
Amines, 692, 694, 1054, 1077–78
Amino acids, 1080–82
  amphiprotic behavior of, 703
  chiral, 1085
  essential, 1081
  side chain, 1081–82
Ammeter, 891
Ammonia, 129, 394, 452, 754, 953
  as Arrhenius and Brønsted–Lowry base, 670
  bond angles, 348
  bonding in, 360
  critical temperature and pressure of, 452
  in fertilizers, 631
  formation of, 818–19
  Haber (Haber-Bosch) process for synthesizing, 630–31, 890
  free energy changes in, 826, 828
  hydrogen and, 937
  nitrogen and, 953
  temperature effects on, 642
  volume and pressure effects on, 646

Ammonia, *(cont.)*
  molecular geometry of, 346
  oxidation of, 955
  properties of, 691
  reactions of
    with boron trifluoride, 323
    with water, 671
  synthesis of, 630–31, 641–42, 644,
    654–55, 825, 828
Ammonium cyanate, 1052
Ammonium ions, 123
  reaction with nitrite ions in water,
    580
Ammonium nitrate, 530, 531
Ammonium perchlorate, 943, 944, 975,
    978
Ammonium thiocyanate, 172
Amorphous solid, 459–60
AMP (adenosine monophosphate),
    1103
Amphetamine hydrochloride, 694
Amphiprotic substances, 670, 749
Amphoteric hydroxides, 748–50
Amphoteric oxides, 748–50
Amphoteric substances, 948
Amphoterism, 741, 748–50
amu (atomic mass unit), 43, 47
Anemia
  iron-deficiency, 1024
  sickle-cell, 559
-*ane* suffix, 66
Angina pectoris, 198, 956
Angstrom (Å), 43
Angular momentum quantum
    number, 226
Anhydrides
  acidic, 946
  basic, 947
Aniline (phenylamine), 1077
Anion(s), 54
  borane, 969
  carboxylate, 703
  chemical formulas and, 123
  combined effect with cation,
    697–99
  common, 62
  in freshwater, 786
  names and formulas of, 60–63
  oxygen, 283
  reaction with water, 696
  relating to acids, 64–65
  size of, 263–64
Anode, 852
  sacrificial, 875
Antacids, 135
Anthracene, 391, 1068, 1070
Antibonding molecular orbital, 369
Anticancer drugs, 340, 342, 388
Antiferromagnetism, 1001
Antifreeze, 550
Antiknock agents, 1062
Antimony, 957
Aqua regia, 143
Aqueous equilibria, 718–65
  acid–base titrations, 730–37
    of polyprotic acids, 737
    strong, 730–33
    weak, 733–37
  buffered solutions, 723–29
    blood as, 729
    buffer capacity and pH, 726–27
    calculating pH of buffer, 724–26
    composition and action of, 723–24
    strong acids or bases in, 727–78
  common-ion effect, 720–23
    solubility and, 741–50
  precipitation and separation of ions,
    750–53

in qualitative analysis for metallic
    elements, 753–55
  solubility equilibria, 737–41
    solubility-product constant (K_{sp}),
    738–40
Aqueous reactions. *See also* Solution
    stoichiometry
  acid–base reactions, 128–35
    electrolytes, 131–32
    with gas formation, 134–35
    neutralization reactions and salts,
    132–34
  of ions, 750–53
  molarity, 543–46
  oxidation-reduction, 135–42
    activity series and, 140–42
    defined, 135
    molecular and net ionic equations
    for, 140
    oxidation numbers (oxidation
    states), 137–38
    oxidation of metals by acids and
    salts, 138–40
  precipitation, 124–28
    exchange (metathesis) reactions,
    126–27
    ionic equations, 127–28
    solubility guidelines for ionic
    compounds, 124–26
  selective, 751–53
Aqueous solution(s), 528. *See also*
    Acid–base equilibria
  activity series of metals in, 141
  defined, 119
  electrolysis of, 876–77
  electrolytic properties of, 120, 121
  freezing-point depression in, 551
  ionic compounds in water, 120–22
  molecular compounds in water, 122
  strong and weak electrolytes, 122–23
Argentite, 1011
Arginine, 1082
Argon (Ar), 238, 452, 938
  in air, 394
  critical temperature and pressure of,
    452
  electron affinity of, 270
  Lewis symbol for, 299
  properties of, 284
Aristotle, 38
Aromatic hydrocarbons, 1054, 1055,
    1068–70
  polycyclic, 1070
Aromatic molecules, 321
Array detector, 465
Arrhenius, Svante, 592, 668
Arrhenius acids and bases, 668
Arrhenius equation, 594–95
  catalysis and, 606
Arsenic, 957
  in drinking water, 162–63, 542,
    960–61
Artificial tissue, 509–10
Asbestos, 967–68
Ascorbic acid (vitamin C), 132, 538,
    566, 667, 689, 1054, 1074
  acid-dissociation constant of, 689
Asparagine, 1082
Aspartame, 112, 1083
Aspartic acid, 1082
Aspirin (acetylsalicylic acid), 2, 132,
    387, 710, 713, 764, 1074, 1076
Astatine, 283
  isotopes of, 940
-*ate* suffix, 61, 62, 63, 64
Atmosphere, 768–70
  composition of, 769–70
  methyl bromide in, 590, 775

outer regions of, 770–72
  ozone in upper, 772–75
    depletion of, 774–75
  regions of, 768–69
  temperature of, 768
  troposphere, 768–69, 775–83
    carbon monoxide, 778–79
    nitrogen oxides and
    photochemical smog, 779–80
    sulfur compounds and acid rain,
    776–78
    water vapor, carbon dioxide and
    climate, 780–83
Atmosphere (atm), 396
Atmospheric pressure, 176, 395–98
  standard, 396
Atmospheric residence time, 590
Atom(s), 2
  Dalton's definition of, 38
  donor, 1018
  of element, 5
  elements and, 38
  estimating number of, 90
  many-electron, 232–34
  nuclear, 41–42
  sizes of, 259–64
Atomic bomb, 420, 915
Atomic emission, 219
Atomic Energy Commission, 51
Atomic masses, 47. *See also*
    Stoichiometry
Atomic mass scale, 46–47
Atomic mass unit (amu), 43, 47
Atomic number, 44–46, 257, 895
  first ionization energy *vs.*, 267
Atomic orbitals, 224–28
  crystal-field theory and, 1034–35
  d, 232
  energies of, 232–33
  f, 232
  p, 231–32
  phases in, 373–75
  quantum numbers and, 226–28
  representations of, 228–32
  s, 228–30
Atomic perspective, 2–3
Atomic radius/radii, 259–64
  bonding (covalent), 261
  bond lengths and, 261–62
  nonbonding (van der Waals), 261
  periodic trends in, 261, 262–63
Atomic structure. *See also* Electronic
    structure
  discovery of, 39–43
    cathode rays and electrons, 39–41
    nuclear atom, 41–42
    radioactivity, 41
  modern view of, 43–46
    atomic numbers, mass numbers,
    isotopes, 44–46
    "plum pudding" model of, 42
Atomic theory of matter, 38–39
Atomic weights, 46–48, 87–88, 257
Atomization, 325
ATP (adenosine triphosphate), 830,
    840, 959, 1103
Aurora borealis, 769
Autoionization of water, 673–75
  ion product, 674–75
Automobiles. *See* Cars
Autotrophs, 815
Average atomic masses, 47
Average speed, 415
Avogadro, Amedeo, 89, 400, 401
Avogadro's hypothesis, 401
Avogadro's law, 400–1, 402, 405
Avogadro's number, 89–95
Axial electron domains, 349

Azide ion, 389
Azobenzene, 389–90
Azo dyes, 389–90
para-Azoxyanisole, 525

Background radiation, 920
Bacteria
  cyanobacteria, 815
  iron metabolism in, 1024
Badge dosimeter, 909
Baeyer-Villiger reaction, 797
Baking soda, 963
Balanced equations, 80–83
  quantitative information from,
    98–102
Ball-and-stick model, 53, 342
Balloons
  hot-air, 400
  lighter-than-air, 401
  weather, 399
"Ballpark" estimate, 26
Balmer, Johann, 219
Band gap, 484, 486–87
Bands of energy states, 483–84
  of metals, 993–95
Bangladesh, arsenic in water in, 960–61
Bar, 396
Bardeen, John, 497
Barium (Ba), 281
  electron configuration of, 239
  properties of, 279
Barium chloride dihydrate, 533
Barium hydroxide, 947
Barium hydroxide octahydrate, 172
Barium oxide, 947
Barometer, 395–98
Bartlett, Neil, 285
Base(s), 129. *See also* Acid–base
    reactions
  amphoteric, 748–50
  Arrhenius, 668
  conjugate, 670
  defined, 129
  household, 128
  Lewis, 704–7
  relative strengths of, 672–73
  strong, 130–31, 680–81
    in buffered solutions, 727–28
  weak, 130–31, 690–93
    common-ion effect of, 720–23
    types of, 692–93
Base-dissociation constant (K_b), 690
  acid-dissociation constant (K_a) and,
    693–95
Base units, 14
Basic anhydrides (basic oxides), 947
Basic solutions, 697–98
  balancing equations for reactions in,
    849–50
Batteries, 842, 843, 870–74
  alkaline, 872
  defined, 870
  fuel cells, 873
  lead-acid, 871
  nickel-cadmium, nickel-metal
    hydride, and lithium-ion,
    872–73
  oxidation-reduction in, 870–71
  primary and secondary, 871
  in series, 871
Battery acid, 871
Bauxite, 750, 988
Bayer process, 988, 1004
BCS theory, 497
Be$_2$, 372
Becquerel, Henri, 41, 908
Becquerel (Bq), 906
Bednorz, J. G., 497

Bellamy, Hillary, 147
Belt of stability, 898–99
"Bends," the (decompression sickness), 540
Bent geometry, 342–43, 354
Benzene, 72, 548, 1054, 1055, 1068–69
    bonding in, 1068
    bromination of, 1069
    hydrogenation of, 1068
    isomers of, 1069
    molal boiling-point elevation and freezing-point depression constants of, 549
    properties of, 466
    resonance in, 321
    sigma and pi bond networks in, 365–66
    in styrene manufacture, 789
Benzenesulfonic acid, 763
Benzo[a]pyrene, 1070
Benzocaine, 1074
Benzoic acid (phenyl methanoic acid), 183, 702, 713
    properties of, 682
Bernoulli, Daniel, 416
Bertholet, Marcellin, 804
Beryllium (Be)
    electron configuration of, 236, 240
    ionization energy of, 268
    Lewis symbol for, 299
    properties of, 279
Berzelius, Jons Jakob, 499
β radiation, 41, 896–97, 919
Beta sheet, 1084
BF₃, 361
BHT (butylated hydroxytoluene), 569
Bicarbonate ion, 265, 729
Bidentate ligand, 1019
Bile, 558
Bimolecular reaction, 598, 600
Binary acids, 699–70
Binary hydrides, 700
Binary hydrogen compounds, 937–38
Binary molecular compounds, names and formulas of, 65
Biochemistry, 1052, 1080
    carbohydrates, 1086–89
        disaccharides, 1087–88
        monosaccharides, 1087
        polysaccharides, 1088–89
    entropy and, 1080
    ionic radii and, 265
    lipids, 1089–90
    nucleic acids, 1090–94
    proteins. See Protein(s)
Biocompatibility, 507
Biodegradable material, 786
Biofuels, 18
Biological chemistry. See Biochemistry
Biological effects of radiation, 919–23
    dosage and, 920
    radon, 921
    therapeutic, 894, 910, 922
Biological fluids, pH of, 677
Biological systems, phosphorus compounds in, 959–60. See also Biochemistry; Living systems
Biomass energy, 197
Biomaterials, 505–10
    characteristics of, 507
    examples
        artificial tissue, 509–10
        heart repair, 508
        vascular grafts, 509
    polymeric, 507–8
Biopolymers, 1080
Bioremediation, 798

Bipolar affective disorder (manic-depressive illness), 280
Bismuth, 241, 285
    properties of, 956, 957
Bismuth subsalicylate, 285
Bisulfates (hydrogen sulfates), 952
Bisulfites, 950–51
Blackbody radiation, 215
Blast furnace, 985
Blood
    as buffered solution, 723, 729
    pH range of, 677
Blood alcohol concentration, 159
Blood gases, deep-sea diving and, 540
Blood pressure, 398
Blood sugar, 92, 193. See also Glucose
Blue vitriol, 1004
Body-centered cubic unit cell, 460, 461
Body temperature, 185
Bohr, Niels, 218, 224, 253
Bohr's model, 219–22
    energy states of hydrogen atom, 220–22
    limitations of, 222
Boiling point(s)
    of halogens, 442
    intermolecular forces and, 439–40, 441
    of metals, 993
    molecular weight and, 443
    of noble gases, 442
    normal, 455, 827
    vapor pressure and, 455
Boiling-point elevation, 549
Boltzmann, Ludwig, 809, 811
Boltzmann's equation, 810–13
Bomb (constant volume) calorimetry, 183–84
Bond(s) and bonding, 296–99. See also Molecular orbitals
    carbon–carbon, 499, 504–5, 965, 1053, 1057, 1062, 1063, 1089. See under Carbon–carbon
    in ceramics, 493
    covalent, 298, 305–7, 312–14
        bond enthalpies and strength of, 326–28
        intermolecular forces vs., 439
        octet rule exceptions, 322–25
        orbital overlap and, 355–57
        polar vs. nonpolar, 307–8
        strengths of, 325–30
    dipole moments, 310–12
    double, 307
        length of, 329–30
        in retinal, 367
        rotation about, 1063
    in double bonds, 362
    electronegativity and, 307–14
    hydrogen, 443–46
        in water, 445–46
    ionic, 298, 299–305, 309, 312–14
        electron configurations of ions and, 302–3
        energetics of formation, 301–2
        polyatomic ions, 316
        transition-metal ions, 303–5
    Lewis structures, 305–6
        drawing, 314–18
        formal charge, 316–18
    Lewis symbols, 298–99
    metallic, 298, 468, 991–95
        electron-sea model for, 992–93
        molecular-orbital model for (band theory), 993–95
    metal-ligand, 1016
    multiple, 307
        bond angles and, 348–49

electron domains for, 348
    molecular geometry and, 362–68
octet rule, 299
    exceptions to, 322–25
    in oxygen, 944
peptide, 1082–83
pi (π), 362–64
    in chemistry of vision, 367
    delocalized, 365–66
    length of, 329–30
polarity, 307–14
resonance structures, 319–21
    in benzene, 321
    in nitrate ion, 320
    in ozone, 319
sigma (σ), 362, 364
single, 307
    length of, 329–30
    rotations about, 1057
    in solids, 464–69
    in transitional-metal complexes. See Crystal-field theory
triple, 307, 364
    hybrid orbitals and, 364
    length of, 329–30
    in triple bonds, 362
    valence electrons and, 322–25
Bond angles, 342
    ideal, 346
    nonbonding electrons and multiple bonds and, 348–49
    predicting, 352
Bond dipole, 353
Bond energy, 771
Bond enthalpies, 325–30
    bond length and, 329–30
    enthalpies of reactions and, 326–28
Bonding atomic radius (covalent radius), 261
Bonding electrons, 369
Bonding molecular orbital, 369
Bonding pair, 344, 348
Bond length(s), 307
    atomic radii and, 261, 262
    bond enthalpy and, 329–30
    electronegativity and, 312
Bond order, 370
Bond polarity, 307–14, 353–55
    electronegativity and, 308–10
Borane anions, 969
Boranes, 969
Borax, 970
Boric acid, 243
Boric oxide, 969
Born, Max, 304
Born-Haber cycle, 304
Boron (B), 268, 969–71
    electron configuration of, 240
    isotopes of, 243
    Lewis symbol for, 299
Boron nitride, 466
Boron trifluoride, 322–23
Bosch, Karl, 631
Boyle, Robert, 399
Boyle's law, 399–400, 402, 404, 415
Bragg, William and Lawrence, 465, 478
Branched-chain hydrocarbons, 1057
Brass, 566, 995
Brattain, Walter H., 497
British thermal unit (btu), 202
Bromcresol green, 763
Bromide ion, 606
Bromine (Br), 50, 298, 940
    in atmosphere, 775
    properties of, 283
    reaction of with nitric oxide, 603–4
2-Bromopentane enantiomers, 1078
Bromthymol blue, 132, 679, 697

Brønsted, Johannes, 668
Brønsted acids and bases, 668–73
    conjugate acid–base pairs, 668–73
    H⁺ ions in water and, 669
    proton-transfer reactions, 669–70
    relative strengths of, 672–73
Bronze, 995
btu (British thermal unit), 202
Buckminster Fuller, R., 468
Buckminsterfullerene (buckyball), 468, 498
Buechler, William J., 997
Buffer capacity, 726–27
Buffered solutions, 723–29
    blood as, 723, 729
    buffer capacity and pH, 726–27
    calculating pH of buffer, 724–26
    composition and action of, 723–24
    strong acids or bases in, 727–28
Bulk solids, 483
Burets, 16, 17
Burfoot, Amby, 147
Butadiene, 389
Butane, 479, 1055, 1057
    combustion of, 781
    Lewis structure and condensed structural formula for, 1056
    in natural gas, 196
Butanol, solubility of, 536
2-Butanone (methyl ethyl ketone), 1073, 1074
1-Butene, 1063
2-Butene, 390, 1067
cis-2-Butene, 1063
trans-2-Butene, 1063
Butylated hydroxytoluene (BHT), 569
Butyl chloride, 577
Butyl group, 1059
tert-Butyl group, 1059
2-Butyne, 1066
Butyric acid, 716

C₆₀, 498
Cadaverine, 694
Cade, John, 280
Cadmium ion, 265
Cadmium phosphide, 514
Cadmium telluride, 524
Cadmium yellow, 1040
CaF₂, 740–42
Caffeine, 567, 694
Calcination, 984
Calcite, 296, 741, 964
Calcium (Ca), 6
    electron configuration of, 240
    as essential nutrient, 281
    oxidation of, 136
    properties of, 279
    reaction with water, 279–80
Calcium carbide, 964
Calcium carbonate (limestone), 281, 296, 297–98, 403, 744, 985
    corrosion of by acid rain, 777
    decomposition of, 85, 640
    dissolution of, 744
    precipitation of, 737
    reaction with sulfur dioxide, 777
    specific heat of, 181
Calcium chloride, 988–89
Calcium hydride, 430, 937–38
Calcium ions, 57
Calcium oxide (lime or quicklime), 85, 777, 964
Calcium sulfate, 413
Calcium sulfite, 777
Calculations
    involving many variables, 404
    significant figures in, 23–24

Calculators, 513
Caloric content, 193
Calorie (Cal), 168
calorie (cal), 167–68
Calorimeter, 179
　coffee-cup, 182
Calorimetry, 179–84
　bomb (constant-volume), 183–84
　constant-pressure, 182–83
　heat capacity and specific heat,
　　179–81
Camphor, 1074
Cancer
　lung, 921
　from radiation, 920, 921
　radiation therapy for, 894, 922
Capillary action, 448
Caproic acid, 716
Carbides, 493, 964–65
Carbohydrates, 193, 194, 1086–89
　disaccharides, 1087–88
　monosaccharides, 1087
　polysaccharides, 1088–89
Carbon (C), 6, 50, 57, 960–65.
　　See also Carbon dioxide;
　　Carbon monoxide
　bonds about, 1056–57
　carbides, 964–65
　carbonic acid and carbonates,
　　963–64
　electron configuration of, 237
　elemental forms of, 961
　fullerenes, 468
　inorganic compounds of, 965
　isotopes of, 45
　Lewis symbol for, 299
　in living organisms, 57
　organic compounds of. See Organic
　　chemistry
　other group 4A elements vs., 965
　oxides of, 962–63
Carbon-11, 910
Carbon-14, 905, 910
Carbonated beverages, 540
Carbonate ion, 691
Carbonates, 963–64
Carbon black, 961
Carbon–carbon bonds, 965, 1053
　in alkenes, 1062
　in polymers, 499
　rotation about, 1057
Carbon–carbon double bonds
　in lipids, 1089
　rotation about, 1063
　vulcanization of rubber and, 504–5
Carbon–carbon pi bond, 932
Carbon–carbon triple bonds, 1064
Carbon composites, 962
Carbon dioxide, 52, 394, 433, 452, 963
　acid–base reaction and, 134
　atmospheric, 776
　in blood, 729
　bonding in, 932
　critical temperature and pressure of,
　　452
　in green dry cleaning, 790
　as greenhouse gas, 433, 781–82
　molecular model of, 2
　nonpolarity of, 353
　phase diagram of, 457–58
　reaction with water, 275
　ridding body of, 265
　specific heat of, 181
　supercritical, 453
　in troposphere, 780–83, 784
Carbon dioxide budget, 794

Carbon dioxide fire extinguisher, 407
Carbon disulfide, 570, 841, 965
Carbon fibers, 962
Carbon geometries, 1052
Carbon–hydrogen bond, 1053
Carbonic acid, 134, 705, 729, 777, 784,
　946, 963–64
　acid-dissociation constant of, 689
Carbonic acid–bicarbonate buffer
　system, 729
Carbonic anhydrase, 265, 624, 1048
Carbon monoxide, 52, 390, 394, 776,
　778–79, 962–63
　affinity for hemoglobin, 778–79
　in atmosphere, 776
　as pollutant, 778
　reduction of metal oxides using, 984
　toxicity of, 779
Carbon nanotubes, 515–17, 961
Carbon suboxide, 979
Carbon tetrachloride, 407, 539, 941
　molal boiling-point elevation of and
　　freezing-point depression
　　constants of, 549
　reaction with hexane, 531
Carbonylation, 1075
Carbonyl group, compounds with,
　1073
　aldehydes and ketones, 1073–74
　amines and amides, 1077–78
　carboxylic acids and esters, 1074–77
Carborundum (silicon carbide), 965
Carboxyhemoglobin, 779
Carboxylate anion, 703
Carboxyl group, 702, 703
Carboxylic acids, 702–3, 1074–77, 1089
Carboxylic groups, 1054
Carlsbad Caverns (New Mexico), 964
Carnot, Sadi, 804, 805, 840
Cars
　air bags, 85
　battery, 871
　body coating, 790, 791
　emission standards for, 779
　hybrid, 196–97
　plastic, 506
Carvone, 1074
Cassini Huygens satellite mission, 281
Catalase, 609, 948
Catalysis, 575, 605–12
　defined, 605
　enzymes, 608–12
　equilibria and, 654–56
　heterogeneous, 606–8
　homogeneous, 605–6
Catalytic converters, 608, 780
Cathode, 852
Cathode rays, electrons and, 39–41
Cathode-ray tubes, 39, 40
Cathodic protection, 875–76
Cation(s), 54
　acidity of hydrated, 705
　with anions, 697–99
　chemical formulas and, 123
　common, 60
　effect on pH, 705
　in freshwater, 786
　names and formulas of, 59–60
　qualitative analysis to group, 753–54
　reaction with water, 696–97
　size of, 263
Caustic soda (NaOH)
　acetic acid titration with, 733–37
　hydrochloric acid titration with,
　　730–33
Cavendish, Henry, 935
Caves, limestone, 964

Cavities, tooth, 747
CdSc, 486
CdTe, 486
Cell emf, 855–61
　concentration effects on, 865–70
　equilibrium and, 865–67
　free-energy change and, 863–64
　oxidizing and reducing agents,
　　860–61
　standard reduction (half-cell)
　　potentials, 856–60
Cell membrane, lipids in, 1090
Cell metabolism, free energy and, 830
Cell phone tower range, 498
Cell potential (E_{cell}), 855
Cellulose, 620, 1076, 1088–89
Cellulose acetate, 1076
Celsius scale, 15
Cementite, 996
Ceramics, 493–95
　applications of, 494
　making, 494
　properties of, 493
　superconducting, 497–98
Cerium (Ce), 239, 625
CERN (Conseil Européen pour la
　Recherche Nucléaire), 902
Cesium (Cs), 239, 272, 277, 835
CFCs (chlorofluorocarbons), 590,
　774–75, 782
Chadwick, James, 42
Chain reactions, 914
Chalcocite, 830
Chalcogens (group 6A), 50, 948–52
　group trends for, 282–83
Chalk, 296
Challenger disaster, 178
Changes, 10–11. See also Reaction(s)
Changes of state, 10
Channel, transistor, 490
Charcoal, 961
Charge(s)
　atomic, 39
　conservation of, 127
　formal, 316–18
　ionic, 55–56
　in metal complexes, 1016–18
　partial, 318
Charged particles, accelerating, 902
Charge-transfer colors, 1040
Charge-transfer transition, 1040
Charles, Jacques, 400, 430
Charles's law, 400, 402, 405
Chelate effect, 1020
Chelating agents (polydentate
　ligands), 1019–22
　in living systems, 1021–24
Chemical analysis, 149–54
Chemical biology. See Biochemistry
Chemical bond, 298. See also Bond(s)
　and bonding
Chemical changes, 10–11. See also
　Reaction(s)
Chemical energy, 167
Chemical equilibrium. See
　Equilibrium/equilibria
Chemical industry, 4
Chemical kinetics, 574. See also
　Reaction rates
Chemical properties, 9, 37
Chemical reactions. See Reaction(s)
Chemical requirements, of
　biomaterials, 507
Chemicals, top ten produced, 4
Chemical structure, acid–base behavior
　and, 699–703
Chemiluminescence, 591

Chemistry
　atomic and molecular perspective
　　of, 2–3
　as central science, 3
　chemical industry and, 4
　defined, 1
　descriptive, 932
　language of, 3–4
　study of, 2–4
　　reasons for, 3–4
Chile saltpeter, 952
Chiral, 1029–30
Chirality
　in amino acids, 1085
　homo-, 1085
　in living systems, 1085
　in organic chemistry, 1078–79
Chloramine, 953
Chlorate salts, 943
Chloride group, 754
Chloride ions, 669
Chlorides, insoluble, 753
Chlorination, 791
Chlorine (Cl), 283, 284, 940, 941
　added to municipal water supplies,
　　429, 787
　bonding in, 992
　electron affinity of, 270
　first ionization energy of, 266
　Lewis symbol for, 299
　mass spectrum of, 48
　nuclides of, 928
　polarity of, 355
　properties of, 283
　reactions of
　　with magnesium, 280
　　with methane, 327
　　with nitric oxide, 663
　　with ozone, 774–75
　　with phosphorus trichloride gas,
　　　957
　　with sodium, 299–305
　uses of, 284
　in water purification, 791
Chlorine-36, 927
Chlorine atoms, mass spectrum, 48
Chlorine dioxide, 435, 791
Chlorine monoxide (ClO), 774
Chlorofluorocarbons (CFCs), 590,
　774–75, 782
Chloroform, 549, 571
Chloromethane, 1071
Chlorophyll a, 1023
Chlorophylls, 1021, 1023
Chlorosis, 1024
Cholesteric liquid crystalline phase, 512
Cholesterol, 1072–73
Cholesteryl benzoate, 510
Chromate ion, 1040
Chromatography, 12
Chrome yellow, 1040
Chromite, 1004–5
Chromium (Cr), 243, 272, 995, 1002
　electron configuration of, 243
　melting point of, 469
Chromium(III), 1018
Chromium(III) complexes, 1036
Chromium(III) ion, 1002
Chromium oxides, 948
Chrysene, 1070
Chrysotile, 968
Cinnabar, 950, 983
Cinnamon, 1074
cis fats, 1089
cis isomers, 1015
Cisplatin, 388, 1028
Cis-trans isomerization, 390

Citric acid, 132, 135, 667, 689, 713, 1074
   acid-dissociation constant of, 689
Citrus fruits, 666, 667
Clausius, Rudolf, 414, 804
Clausius–Clapeyron equation, 456
Clean Air Act, 777
Climate, 780–83, 784
Closed system, 168
Close-packed arrangement, 75
C_{60} molecules, 468
Coagulation, 560
Coal, 196
   combustion of, 776
   gasification of, 196
Coal-fired power plants, 491
Cobalt, 982
Cobalt-60, 894, 903, 922, 926, 927
Cobalt glass, 968
Cobalt(II) chloride, 651
Cobalt(III), 1018, 1037
Cobalt(III) chloride, 1014
Cocaine, 716
Codeine, 694, 713
Coefficient, subscript vs., 81
Coffee-cup calorimeter, 182
Cohesive forces, 448
Coke, 655, 985
Colligative properties of solutions,
      546–56
   boiling-point elevation, 549
   of electrolyte solutions, 554
   freezing-point depression, 550–51
   molar mass determination through,
         555–56
   osmosis, 551–55
   vapor pressure reduction, 546–48
Collision model, 592
Collisions, molecular, 592–93
Colloidal dispersions, 556
Colloids, 556–61
   hydrophilic and hydrophobic, 557–60
   removal of colloidal particles, 560–61
   types of, 557
Colors, 1012, 1013
   complementary, 251, 1032
   in coordination compounds,
         1031–33
      charge-transfer, 1040
      crystal-field theory and, 1036, 1040
      ligands and, 1031
   as function of temperature, 215
Color wheel, 1032
Columbia Space Shuttle, 943
Combination reactions, 84–86
Combining volumes, law of, 401
Combustion, 1061
   enthalpies of, 188
   heat of, 188
Combustion analysis, 97–98
Combustion reactions, 86–87, 138
   of alkanes, 1061
   balanced equations for, 86–87
   with oxygen, 934
Common-ion effect, 720–23, 741–42
   solubility and, 741–50
Complementary base pairs, 1092
Complementary colors, 251, 1032
Complete ionic equation, 127
Complexes. See Metal complexes
Complex ions, 1014
   solubility and, 745–48
Components
   of mixture, 8
   of solution, 528
Composition, 2
Compound(s), 5–8. See also Organic
         chemistry

with carbonyl group, 1073
   aldehydes and ketones, 1073–74
   amines and amides, 1077–78
   carboxylic acids and esters,
         1074–77
   coordination. See Coordination
         compounds
   gases as, 394
   interhalogen, 388
   intermetallic, 996–97
   ionic. See Ionic compounds
   of metals with nonmetals, 273
   molecular, binary, 65
   naming. See Nomenclature
   organic, 59, 66–67. See also Organic
         chemistry
Compound semiconductors, 484–85, 486
Computer screens, 513
Concentrated, 542
Concentration(s), 142–49, 542–46
   acid–base titration to determine,
         150–54
   conversion of units of, 544–46
   defined, 142
   dilution, 146–49
   effects on cell emf, 865–70
      concentration cells, 867–70
      Nernst equation, 865–67
   of electrolyte, 145
   equilibrium, 645–47
   free energy change and, 828
   interconverting molarity, moles, and
         volume, 145–46
   in mass percentage, 542
   molality, 543–44
   molarity, 144–45, 543–54
   in mole fraction, 543–44
   in parts per billion (ppb), 542
   in parts per million (ppm), 542
   in reactant or product, changes in,
         648–49
   reaction rates and, 574, 575, 580–85
      change with time, 585–91
      rate laws, 581–85
   of solid substance, 639
Concentration cells, 867–70
Condensation, heat of, 450
Condensation polymerization, 501–2
Condensation polymers, 500, 841
Condensation reaction, 494–95, 959,
         1073
Condensed electron configurations,
         237–38
Condensed phases, 438
Condensed structural formulas, 1055–56
   for alkyl groups, 1059
   for cycloalkanes, 1061
Conducting polymer, 516
Conduction band, 484, 485
Conductivity, 121, 482
Conductors, 482
Cones, 367
Conjugate acid, 670
Conjugate acid–base pairs, 670–71,
         693–95
Conjugate base, 670
Conservation of mass, law of (law of
         conservation of matter), 38,
         79–80
Constant(s)
   acid-dissociation. See Acid-
         dissociation constant
   base-dissociation, 690, 693–95
   decay, 906
   equilibrium. See Equilibrium
         constant
Faraday, 863

formation, 747
   gas, 402
   Henry's law, 540
   ion-product, 674
   molal boiling-point-elevation, 549
   molal freezing-point-depression, 550
   Planck's, 215
   rate, 581, 583, 592
   Rydberg, 219, 220
   screening, 258
   solubility-product. See under
         Solubility equilibria
   van der Waals, 422–23
Constant composition, law of (law of
      definite proportions), 8, 38
Constant-pressure calorimetry, 182–83
Constant-volume (bomb) calorimetry,
      183–84
Constructive combination, 374
Consumer products, chelating agents
      and, 1020
Continuous spectrum, 219
Contour representations, 230, 232
Control rods, 915–16
Convection, 185
Conversion factors, 25
   involving volume, 27–29
   using two or more, 26–27
Converter, iron-refining, 986
Cooling, 451–52
   equilibria and, 653
   super-, 452
Cooling liquid, in nuclear reactor, 916
Cooper, Leon, 497
Coordination compounds, 1012–49
   colors in, 1013, 1031–33
      charge-transfer, 1040
      crystal-field theory, 1033–41
         colors and, 1036, 1040
         d orbitals and, 1034–35
         electron configurations in
            octahedral complexes, 1037–38
         for tetrahedral and square-planar
            complexes, 1038–41
   defined, 1014
   isomerism in, 1026–31
      stereoisomerism, 1027–31
      structural, 1027
   ligands, 1014, 1019–24
      bidentate, 1019
      in living systems, 1021–24
      monodentate, 1019
      polydentate (chelating agents),
         1019
      weak field and strong field, 1036
   magnetism in, 1033
   metal complexes, 1014–18
      charges, coordination numbers,
         and geometries, 1016–18
      metal-ligand bond, 1016
      Werner's theory and, 1014–16
   molecular-orbital theory of, 1039
   nomenclature of, 1025–26
Coordination number, 464, 1015
   charges, geometries, and, 1016–18
Coordination sphere, 1015
Coordination-sphere isomers, 1027
Copolymers, 501
Copper (Cu), 50, 298, 992, 1004–5
   in alloys, 995
   electron configuration of, 243
   electronic properties of, 484
   electrorefining of, 990–91
   extraction from chalcocite, 830
   oxidation of, 140–41
   reaction with nitric acid, 11
Copper(II) sulfate, 875

Copper(II) sulfate pentahydrate, 533,
      1004
Copper sulfate, 1004
Core electrons, 238
Corey, R. B., 1084
Coronene, 1070
Corrosion, 136, 874–76
   of iron, 136, 874–76
   as spontaneous process, 802–3
Corrosion-protecting cars, 790, 791
Coster, D., 253
Coulomb's law, 45
Covalent bond(s), 298, 305–7, 312–14
   bond enthalpies and strength of,
      326–28
   in hydrogen, 936
   intermolecular forces vs., 439
   octet rule exceptions, 322–25
   orbital overlap and, 355–57
   polar vs. nonpolar, 307–8
   strengths of, 325–30
Covalent carbides, 965
Covalent-network solids, 464,
      466–67
Covalent radius (bonding atomic
      radius), 261
Crab Nebula, xxxii
Cracking, 1062
Crenation, 553
Crick, Francis, 465
Critical mass, 914
Critical point (C), 456, 457
Critical pressure, 452
Critical temperature, 452
Cross-linking of polymers, 504, 505
Crutzen, Paul, 774
Cryolite, 940, 989, 990
Crystal-field splitting energy, 1034, 1036
Crystal-field theory, 1033–41
   colors and, 1036, 1040
   d orbitals and, 1034–35
   electron configurations in
      octahedral complexes, 1037–38
   for tetrahedral and square-planar
      complexes, 1038–41
Crystal lattice, 460
Crystalline solids, 438, 439, 458–60
   entropy of, 816
Crystallinity of polymers, 502
Crystallites, 997
Crystallization, 534
Crystals
   liquid, 510–13
      phases of, 511–12
   X-ray diffraction by, 465
Cubic centimeter, 16
Cubic close packing, 463
Curie, Marie and Pierre, 41
Curie (Ci), 906
Curie temperature, 1002
Curium-242, 903
Curl, Robert, 468
Cyalume light sticks, 591
Cyanide, gold recovery and, 987
Cyanide ion, 316
Cyanobacteria, 815
Cyanogen, 424
Cycles, 212
Cyclic glucose, 1086
Cycloalkanes, 1060–61
Cyclohexane, 77, 537, 565, 837, 1061
Cyclohexanol, 1061
Cyclohexatriene, 1068
Cyclopentadiene, 623
Cyclopentadienide ion, 389
Cyclopentane, 1061
Cyclopropane, 434, 570, 1061

Cyclotrimethylenetrinitramine (RDX), 338
Cyclotron, 902
Cysteine, 1082
Cytochrome, 891
Cytoplasm, 1090
Cytosine, 477, 1091, 1092

Dacron, 508, 509
Dalton, John, 38–39, 393, 410
Dalton's law of partial pressures, 410–11
Data, 13
Dating, radiocarbon, 905
d atomic orbitals
    crystal-field theory and, 1034–35
d-bands, 994
d-d transition, 1035, 1040
de Broglie, Louis, 222
Debyes (D), 311
Decaffeination, 453
Decarbonylation reaction, 838
Decay, radioactive, 895, 896–98
    rates of, 903–8
    types of, 896–98
Decay constant, 906
Decomposition reactions, 84–86
Decompression sickness (the "bends), 540
Deep-sea diving, blood gases and, 540
Definite proportions, law of (law of constant composition), 8
Degenerations, 233
Degrees of freedom, 814
Dehydration, 147, 951
Delocalization, 365–66
Democritus, 38
Density
    electron, 225
    of gas, molar mass related to, 406–8
    of liquid and solid phases, 445–46
    probability, 225, 230–31
    radial probability, 229
    SI unit of, 17
    unit cells and, 462–63
    weight vs., 17
Deoxyhemoglobin, 1023
Deoxyribonucleic acid (DNA), 1090–92
    hydrogen bonds in, 444
    replication of, 1093
    X-ray diffraction of, 465
Deoxyribose, 1091
Department of Energy, 917
Deposition, heat of, 450
Derived SI units, 16
Desalination, 784–85
Descriptive chemistry, 932
Desiccant, 413
Destructive combination, 374
Detergents, 959, 1054
Deuteration, 935
Deuterium, 935
Deuterium oxide, 935
Developer solutions, 861
Dextrorotatory, 1030
Dextrose. See Glucose
Diabetes, 102
Diacid, 501
Dialysis, 560
Diamagnetism, 376–77
Diamine, 501–2
Diamond
    band gap in, 486
    electronic properties of, 484
    structure of, 466, 484–85
    synthetic, 961
Diastolic pressure, 398
Diatomic gases, 394

Diatomic molecules, 52
    bond enthalpies in, 325
    dipole moments of, 311
    electron configurations for, 376–78
    heteronuclear, 379–81
    homonuclear, 371
        energy-level diagram for, 371, 375
        molecular orbitals for, 371–81
Diborane, 72, 969
Dichlorobenzene, 385
trans-2,3-Dichloro-2-butene, 1066
Dichloroethylene, 385
1,2-Dichloroethylene, 477
Dichloromethane, 477
Dicyclopentadiene, 623
Diethyl ether, 431, 1073
    vapor pressure of, 455
Differential rate law, 586
Diffraction, 465
Diffraction grating, 465
Diffusion, 417–20
    mean free path and, 419–20
Difluoromethane, 383
Dilute, 542
Dilution, 146–49
Dimensional analysis, 24–29, 145, 404
    conversion factors in, 26–27
    conversions involving volume, 27–29
Dimethyl ether, 441, 1071
Dimethylhydrazine, 979
Dimethylphosphinoethane (dmpe), 1046
2,2-Dimethylpropane, 442, 1057
Dimethyl sulfide, 950
Dinitrobenzene, isomers of, 1069
Dinitrogen pentoxide, 624, 954
Dinitrogen tetroxide, 628–29, 884
Dinitrogen textroxide-nitrogen dioxide equilibrium, 632–33, 637–38
Dinitrogen trioxide, 954
Dinitromethane, 1102
Dioxane, 565
Dioxygen, 282, 377, 944
Dipole, 310
Dipole–dipole forces, 440–41, 444, 446
Dipole moments, 310–12, 353–55
Diprotic acid, 129, 679
Disaccharides, 1087–88
Disc heart valve, 408
Disilicate ion, 966
Disorder. See also Entropy(ies)
    solution formation and, 531–32
Dispersion forces, 441–43, 446
    solution formation and, 528
Displacement reactions, 138–39
Disproportionate reaction, 889, 948
Dissociation energy, 771
Dissolution
    of ionic compounds, 737
    in water, 120, 121
Distillation, 12, 548, 784
    fractional, 548, 1062
Division, significant figures in, 23
DNA (deoxyribonucleic acid), 3, 1090–92
    hydrogen bonds in, 444
    replication of, 1093
    X-ray diffraction of, 465
DNA double helix, 477
Dolomite, 964
Donor atom, 1018
Dopants, 276
Doping, 488
d orbitals, 232, 360, 361
    crystal-field theory and, 1034–35
    phases in, 374
    in square-planar complexes, 1038–41
    in tetrahedral complexes, 1038–41

in transition metals, 1000
Doses, radiation, 920
Double bonds, 307, 932, 1052
    bond length of, 329–30
    in retinal, 367
    rotation about, 1063
Double helix, 1092–93
Downs cell, 988, 989
Drain, semiconductor, 490
Drierite, 413
Drugs
    amine, 694
    anticancer, 340, 342, 388
    chirality and, 1079
    cisplatin, 388
    lithium, 280
    Taxol, 340, 342
    transplatin, 388
Dry cleaning, green, 790
Dry ice, 963
Ductility, 992
Dumas-bulb technique, 430
Dyes, 132, 150–51, 668
    azo, 389–90
Dynamic equilibrium, 454, 534, 627
Dynamite, 329

Earth, 432, 766
    elements in crust, 6
    surface temperature of, 780–81
ECF (extracellular fluid), 868
Edema, 555
EDTA (ethylenediaminetetraacetate ion), 1019, 1020
Effective nuclear charge (Z_{eff}), 257–59, 260
Effusion, 417–20
    Graham's law of, 418–19
Einstein, Albert, 216, 217, 218, 915
Eka-aluminum, 257
Eka-silicon, 257
Elastomers, 499, 508
Electrical circuit, 852
Electrical work, 176, 879–81
Electricity
    from nuclear fission, 916–17
    from nuclear reactions, 894
Electrocardiography, 868–69
Electrochemistry, 842–91. See also Oxidation-reduction (redox) reactions
    batteries, 842, 843, 870–74
        alkaline, 872
        fuel cells, 873
        lead-acid, 871
        nickel-cadmium, nickel-metal hydride, and lithium-ion, 872–73
        primary and secondary, 871
    cell emf, 855–61
        concentration effects on, 865–70
        equilibrium and, 865–67
        oxidizing and reducing agents, 860–61
        standard reduction (half-cell) potentials, 856–60
    corrosion, 874–76
        of iron, 874–76
    defined, 844
    electrolysis, 876–81
        with active electrodes, 877–78
        of aqueous solutions, 876–77
        electrical work, 879–81
        quantitative aspects of, 878–79
        of water, 7
    free energy and redox reactions, 862–64

of metals (electrometallurgy), 988–91
    oxidation states, 844–46
    voltaic (galvanic cells), 851–54
        molecular view of electrode process, 854
        work done by, 879–81
Electrode(s), 852
    electrolysis with active, 877–78
    molecular view of, 854
    standard hydrogen, 856–57
Electrolysis. See under Electrochemistry
Electrolyte(s), 120
    colligative properties of, 554
    common-ion effect and, 720
    concentration of, 145
    strong and weak, 122–23
        identifying, 131–32
    in voltaic cell, 852
Electrolytic cells, 876
Electrolytic properties, of aqueous solution, 120, 121
Electromagnetic force, 45
Electromagnetic radiation, 212–13
Electromagnetic spectrum, 213
Electrometallurgy, 988–91
    of aluminum, 989–90
    copper refinement, 990–91
    of sodium, 988–89
Electromotive force (emf). See Cell emf
Electron(s), 43
    bonding, 369
    cathode rays and, 39–41
    core, 238
    discovery of, 40
    mass of, 40–41, 43
    nonbonding, bond angles, and, 348–49
    odd number of, 322
    outer-shell, 238
    paired, 235
    shielding, 258
    unpaired, 235
    valence, 238, 242
        bonding and, 298
        effective nuclear charge experienced by, 259
Electron affinity(ies), 270–71
    electronegativity and, 308
    of halogens, 271
    ionization energy vs., 270
    of nonmetals, 275
Electron capture, 897
Electron configuration(s), 234–40, 375–76
    of actinides, 239–40
    anomalous, 243
    condensed, 237–38
    homonuclear diatomic molecules, 371
    Hund's rule, 235–37
    of ions, 268–70, 302–3
    of lanthanides, 239–40
    molecular properties and, 376–78
    in octahedral complexes, 1037–38
    periodic table and, 240–43
    of transition metals, 238–39, 1000–2
Electron density, 221
    in molecules, 305
    in p orbitals, 231
    in s orbitals, 232
Electron-domain geometry, 345–46
Electron domains, 344–41
    axial, 349
    equatorial, 349
    for multiple bonds, 342
Electronegativity, 301
    bond polarity and, 308–10
    of oxyacids, 702

Electronic charge, 43
Electronic materials
   plastic, 506
   silicon chip, 489
   solar energy conversion, 491
Electronic structure, 210–53
   atomic orbitals, 224–32
      d, 232
      f, 232
      p, 231–32
      quantum numbers and, 226–28
      s, 228–30
   Bohr model and, 219–22
      energy states of hydrogen atom, 220–22
      limitations of, 222
   defined, 212
   line spectra and, 218–19
   of many-electron atoms, 232–34
   of materials, 482–84
   photons and, 215–18
   quantization of energy and, 215–18
   quantum (wave) mechanics, 211
   wave behavior of matter and, 222–24
   wave nature of light and, 212–15
Electron microscope, 223
Electron-pair acceptor/donor concept, 704–5
Electron-sea model for metallic bonding, 992–93
Electron shell, 227
Electron spin, 233–34
Electron spin resonance (ESR), 252–53
Electron transfer, 288, 299–300
Electroplating, 877–78
Electrorefining, of copper, 990–91
Electrostatic potential energy, 167
Element(s), 5, 6. See also specific elements and groups
   atoms and, 5
   common, 6
   defined, 2
   discovery of, 256–57
   electronegativities of, 308
   molecules of, 2
   periodic properties of. See Periodic properties of elements
   representative (main-group), 240
   symbols for, 6
   trace, 57
   transuranium, 51, 903
Elemental form, atom in, 137
Elemental semiconductors, 484, 486
Elementary reactions, 597–98
   rate laws for, 599–600
emf. See Cell emf
Emission spectra, 215
Emission standards, 779
Empirical formulas, 52–53, 95–98
   calculating, 95–96
   combustion analysis, 97–98
   for ionic compound, 58
   molecular formula from, 96–97
Emulsifying agent, 558
Enantiomers (optical isomers), 1028–31, 1078–79, 1082
Endothermic processes, 172, 175, 530–31, 651–52
End point of titration, 151, 732
Energetics, of ionic bond formation, 301–2
Energy barrier, 593–94
Energy changes
   accompanying phase changes, 449–50
   in nuclear reactions, 911–13
   solution formation and, 529–31

Energy conversion, 822
Energy diagram, 171
Energy(ies), 165–69. See also Thermodynamics: first law of
   activation, 592–94
      catalysis and, 610, 654
      determining, 595–97
   biomass, 197
   chemical, 167
   crystal-field splitting, 1034, 1036
   defined, 166
   dissociation, 771
   electrostatic, 167
   fossil fuels and, 196–97
   geothermal, 197
   hydroelectric, 197
   internal, 170–71
   ionization, 264–70, 308
      of alkaline earth metals, 279
      electron affinity vs., 270
      of metals vs. nonmetals, 272
      periodic trends in, 267–68
   kinetic, 166–67
   lattice, 301–2, 304
   molecular motion and, 809–10
   nonrenewable, 197
   nuclear, 197
   nuclear binding, 912–13
   potential, 166–67
   quantization of, 215–16
   radiant, 212
   renewable, 197
   solar, 197–98, 380
      conversion of, 489–91
   sources of, in U.S., 196
   spin-pairing, 1037
   Sun as source of, 892, 893
   system and surroundings, 168
   thermal, 167
   transferring, 168–69
   units of, 167–68
   wind, 197
Energy-level diagram (molecular orbital diagram), 369–70
Energy states
   bands of, 483
   of hydrogen, 220–22
Engineering polymers, 506
Enstatite, 967
Enthalpy(ies) (H), 174–77
   bond, 325–30
      bond length and, 329–30
      enthalpies of reactions and, 326–28
   of combustion, 188
   defined, 175
   of formation, 188–93
   free energy and, 822
   of fusion, 449
   of reaction, 177–79, 190–93, 326–28
   spontaneous processes and, 180
   as state function, 175
   of vaporization, 188
Enthalpy change
   Hess's law and, 184–88
   solution formation and, 530–31
Enthalpy diagram, 178
Entropy change, 806–7
   expansion and, 808
   for phase changes, 807
   qualitative predictions about, 813–15
   in reactions, 817–19
   in surroundings, 818–19
Entropy(ies), 806–17
   absolute, 817, 818

of activation, 841
   biochemistry and, 1080
   chelate effect and, 1021–22
   of expansion, 808
   free energy and, 822
   heat transfer and temperature related to, 808–9
   life and, 805
   microstates and, 811
   molecular interpretation of, 809–17
   probability and, 812
   in reactions, 817–19
   second law of thermodynamics, 806–9
   solution formation and, 531–32
   standard molar, 817
   temperature and, 816
Environment, chemistry of, 766–99
   atmosphere, 768–70
      composition of, 769–70
      methyl bromide in, 590, 775
      outer regions of, 770–72
      ozone in upper, 772–75
      regions of, 768–69
      temperature of, 768
      troposphere, 768–69, 775–83
   freshwater, 785–88
      dissolved oxygen and quality of, 786
      softening of, 788
      treatment of municipal supplies of, 786–87
   green chemistry, 788–92
      corrosion-protecting cars, 790, 791
      dry cleaning, 790
      principles of, 788–89
      solvents and reagents, 789–90
      water purification, 790–92
   oceans, 783–85
Environment, radiation in, 919–23
Environmental Protection Agency (EPA), 791, 921, 944, 960, 1070
Enzyme inhibitors, 610
Enzymes, 265, 575, 608–12, 1085
   efficiency of, 610
   inhibition of, 610
   in nitrogen fixation, 610–11
   specificity of, 609
Enzyme-substrate complex, 609
EPA (Environmental Protection Agency), 791, 921, 944, 960, 1070
EPCOT Center, 198
Epoxy systems, 962
EPROM (Erasable Programmable Read-Only Memory) silicon microchip, 3
Epsom salts, 113
Equation(s), 80–83
   balanced, 80–83
      quantitative information from, 98–102
   balancing, 80–83
   direction of, 636–37
   equilibrium constants and, 637–38
   Henderson–Hasselbach, 725, 727
   ionic, 127–28
   molecular, 127
   nuclear, 895–96
   states of reactants and products, 83
   thermochemical, 178
Equatorial electron domains, 349
Equilibrium constant, 628, 630–35, 632
   applications of, 644–47
   calculating, 641–43, 645–47
   direction of chemical equation and, 636–39
   evaluating, 632–33
   Gibbs free energy and, 826–31

   magnitude of, 635–63
   in terms of pressure, 633–34
   units of, 635
Equilibrium/equilibria, 123, 626–65
   acid–base. See Acid–base equilibria
   aqueous. See Aqueous equilibria
   cell emf and, 865–67
   chemical, 123
   concept of, 628–30
   dynamic, 454
   free energy and, 820–21
   heterogeneous, 639–41
   homogeneous, 639–41
   Le Châtelier's principle, 648–56
      catalyst effects, 654–56
      change in reactant or product concentrations, 648–49
      nitric oxide emission control and, 656
      temperature changes, 651–54
      volume and pressure changes, 649–51
   static, 627
Equivalence point of titration, 150, 730, 734
   pH and, 732–33
Erasable Programmable Read-Only Memory (EPROM) silicon microchip, 3
Eskalith, 280
ESR (electron spin resonance), 252–53
Essential amino acids, 1081
Essential fatty acids, 1090
Esters, 1074–77
Estimating answers, 26
Ethanal (acetaldehyde), 1071, 1073
Ethanamide (acetamide), 1071, 1078
Ethane, 66, 67, 1055
   bond enthalpies in, 325
   combustion of, 1061
   formation of, 607
   molar entropies of, 817
   in natural gas, 196
1,2-Ethanediol (ethylene glycol). See Ethylene glycol
Ethanoic acid. See Acetic acid (ethanoic acid)
Ethanol (ethyl alcohol), 18, 66, 67, 1062, 1072
   molar boiling-point elevation and freezing-point depression constants of, 549
   molecular model of, 2
   solubility of, 536, 537
   standard enthalpy of formation for, 189
   vapor pressure of, 453, 455, 456
Ethene (ethylene), 52, 607, 1062, 1071
Ethers, 1073
Ethyl acetate, 387, 1074
Ethyl alcohol. See Ethanol
Ethylamine, 1071, 1077
Ethyl benzene, in styrene manufacture, 789
Ethylene, 164, 394, 1055
   carbon–carbon double bond in, 367
   empirical formula for, 53
   hybridization in, 363
   molecular formula for, 52
   molecular geometry of, 363
   pi bonds in, 363
   reactions of
      with halogens, 1065
      with hydrogen, 607
Ethylenediamine (en), 1019–20, 1022
Ethylenediaminetetraacetate ion (EDTA), 1019, 1020

Ethylene (ethene), 52, 1062, 1071
  reaction with hydrogen gas, 607
Ethylene glycol, 97, 477, 546, 550, 1072
  molecular model of, 2
  vapor pressure of, 455
Ethylene oxide, 570
Ethyl group, 1059
Ethyne (acetylene), 364, 430, 838, 945, 964, 1055, 1064, 1071
Eutrophication, 786
Evaporation, 185
Exact numbers, 20
Excess reactants (reagents), 106
Exchange (metathesis) reactions, 126–27
Excited state, 220, 228
Exelus Inc., 789
Exothermic processes, 172, 175, 651–53
  solution formation and, 530–31
  spontaneity and, 530, 532
Explosive balance, 329
Explosives, 328–29, 955
Exponents, in rate law, 581–82
Extensive properties, 10
Extracellular fluid (ECF), 868

Face-centered cubic unit cell, 460, 461
Fahrenheit scale, 15, 16
Faraday, Michael, 515, 863
Faraday (F), 863
Faraday's constant, 863
Fats, metabolism of, 193, 194
Fat-soluble vitamins, 538
Fatty acids, 1076, 1089–90
f-block metals, 240
FDA, 280
Fe³, 704–5, 1016
FeMo-cofactor, 611
Fermi, Enrico, 915
Fermi level, 994
Ferrichrome, 1024
Ferrimagnetism, 1002
Ferrochrome, 996
Ferromagnetism, 1001
Ferrous sulfate, 796
Fertilizers, 277, 631
  nitrogen, 953
  phosphorus, 959
FET (field effect transistor), 516
FFVs (flexible-fuel vehicles), 18
Fibers, 962
Fibrous proteins, 1085
Field effect transistor (FET), 516
Film, 908–9
Film badges, 909
Filtration, 11–12
Fire extinguisher, carbon dioxide, 407
Fireflies, 591
Fireworks, 281, 572, 573
First ionization energy, 264–65
  of alkaline earth metals, 279
  atomic number vs., 267
  of metals vs. nonmetals, 272
  periodic trends in, 267–68
First law of thermodynamics, 170–73
  endothermic and exothermic processes, 172
  heat and work related to internal energy changes, 171–72
  internal energy, 170–71
  state functions, 172–73
First-order reactions, 582, 586–87, 590–91
  half-life of, 589–91
Fission, nuclear, 913–17
  reactors using, 915–17
  waste from, 917

Fixed nitrogen, 953
Flame tests, 278
Flexible-fuel vehicles (FFVs), 18
Fluoridation, 747
Fluorine (F), 283, 284, 939, 940–41
  electronegativity of, 308
  Lewis symbol for, 299
  nonpolarity of, 355
  oxidation number of, 137
  properties of, 283
Fluorine-18, 910
Fluorite, 459, 467
Fluoroapatite, 747, 764
Fluorocarbons, 570
Fluorapatite, 940
Fluorspar, 940
Folding, 1085
Food and Drug Administration (FDA), 280
Foods, thermochemistry of, 193–95
Fool's gold (pyrite), 459, 950
f orbitals, 232
Force(s), 45, 166
  adhesive, 448
  cohesive, 448
  dipole–dipole, 440–41, 444, 446
  electromagnetic, 45
  electromotive. See Cell emf
  gravitational, 45
  intermolecular. See Intermolecular forces
  ion–dipole, 440, 446
  nuclear, 45
Formal charge, 316–18
Formaldehyde (methanol), 364, 944, 1052, 1073
Formation constant (K_f), 747
Formic acid (methanoic acid), 702, 763, 1074
  acid-dissociation constant of, 682–83
  decomposition of, 611–12
  percent ionization of, 684
Formula(s), 52
  condensed structural, 1055–56, 1059, 1061
  empirical, 52–53, 95–98
    calculating, 95–96
    combustion analysis, 97–98
    for ionic compound, 58
    molecular formula from, 96–97
  molecular, 52–53
    from empirical formulas, 96–97
  percentage composition from, 95
  relating relative numbers of anions and cations to, 123
  structural, 53
  subscripts in, 81
Formula units, 88
Formula weights, 87–89
  molar mass and, 92
  percentage composition from, 88–89
Fossil fuels, 196, 380
  greenhouse effect and combustion of, 781
Fractional distillation, 548, 1062
Franklin, Rosalind, 465
Free energy, 819–24
  equilibrium constant and, 826–31
  redox reactions and, 862–64
  spontaneity and, 820
  standard free-energy changes, 822–24
  temperature and, 824–26
Free-energy change
  emf and, 863–64
  Gibbs free energy and, 822–24
Free radical, 919

Freezing, heat of, 450
Freezing point, 457
Freezing-point depression, 550
  molar mass from, 555
Frequency, 212, 214–15
Frequency factor, 595
Freshwater, 785–88
  dissolved oxygen and quality of, 786
  pH of, 777
  softening of, 788
  treatment of municipal supplies of, 786–87
Friedel-Crafts reaction, 1069
Frisch, Otto, 915
Fructose, 1086
Fruits, citrus, 666, 667
Fuel cells, 873, 874
Fuels, 167
  fossil, 196, 380, 781
  thermochemistry of, 195–97
Fuel values, 193, 194–95
Fuller, R. Buckminster, 468
Fullerenes, 468, 961
Functional groups, 66, 1053, 1070–78
  alcohols, 1072–73
  aldehydes and ketones, 1073–74
  amines and amides, 1077–78
  carboxylic acids and esters, 1074–77
  ethers, 1073
Furchgott, Robert F., 956
Furoic acid, 763
Fusion, 449
  heat (enthalpy) of, 449
  nuclear, 913, 918

Galactose, 1101
Galena, 950
Gallium, 257, 272
Gallium arsenide (GaAs), 485, 486, 487
Gallstones, 1073
Galvani, Luigi, 868
Galvanic cell, 851–54
Galvanic (voltaic) cells, 851–54. See also Batteries
  concentration cells, 867–70
  electromotive force (emf) in, 855–61
    concentration effects on, 865–70
    equilibrium and, 865–67
    oxidizing and reducing agents, 860–61
    standard reduction (half-cell) potentials, 856–60
  molecular view of electrode process, 854
  standard cell potential of, 859
  work done by, 879–81
Galvanized iron, 875
Gamma (γ) radiation, 41, 213, 896, 897, 919
  therapeutic, 922
Gangue, 983
GaP, 486
Gas constant, 402
Gas(es), 5, 392–435
  absolute temperature of, 414
  acid–base reactions with, 134–35
  blood, deep-sea diving and, 540
  characteristics of, 394–95
  collecting over water, 412–13
  diatomic, 394
  diffusion and mean free path, 419–20
  entropy change and isothermal expansion of, 805–6
  expansion of, 802, 808
  in freshwater, 786
  Graham's law of effusion, 418–19
  greenhouse, 781, 782

ideal, 402, 548
ideal-gas equation, 402–6
  gas density-molar mass relationship and, 406–8
  gas laws and, 404–6
  gas volume in reactions and, 408–10
  inert, 285
  isothermal expansion of, 805–6
  kinetic-molecular theory of, 414–16
    gas laws and, 415–16
  mixtures of, 394, 410–13
  monoatomic, 394
  natural, 196, 262, 409, 435, 798, 840. See also Methane
  noble, 394
    group trends for, 284–86
  pressure, 395–98
    partial, 410–13
    pressure-volume work and, 174–75, 176
  properties of, 394, 438
  real, 420–24
    van der Waals equation, 422–24
  separations of, 420
  solids and liquids compared to, 394
  solubility of, 536, 539–41
  state of, 398
  water, 936
Gas laws, 398–402
  Avogadro's law (quantity-volume relationship), 400–1, 402
  Boyle's law (pressure-volume relationship), 399–400, 402
  Charles's law (temperature-volume relationship), 400, 402
  ideal-gas equation, and, 402–6
  kinetic molecular theory and, 415–16
Gasoline, 536, 548, 1061–62
  blends, 18
  combustion of, 822
  synthetic, 196–97
Gas pipelines, 409
Gate, semiconductor, 490
Gatorade, 147
Gay-Lussac, Joseph Louis, 400–1
Geiger counter, 909
Gel, 495
Geometric isomerism, 1028, 1063
Geothermal energy, 197
Gerlach, Walter, 234
Germanium (Ge), 257, 466, 484, 487, 965
Gibbs, Josiah Willard, 819
Gibbs free-energy change, 822-824
  emf and, 863–64
Gibbs free energy (G), 819–24
  equilibrium constant and, 826–31
  spontaneity and, 820
  standard free-energy changes, 822–24
  temperature and, 824–26
Ginkgo leaf, 815
Glacial acetic acid, 1076
Glass, 459, 942, 968
  etched, 943
  quartz, 459
  silica, 459
Glassware, volumetric, 17
Global warming, 767, 782
Globular proteins, 1022, 1085
Glucose, 72, 92, 1054, 1086
  cyclic, 1086
  as food, 193
  molecular formula of, 1086
  oxidation of, 100–1, 830
  solubility of, 537
  structure of, 1086
Glucose monitoring, 102

Glutamic acid, 1082
Glutamine, 1082
Glycerin, 547
Glycerol (1,2,3-propanetriol), 1072, 1089, 1090
Glycine, 387, 703, 716, 1081, 1083
Glycogen, 1088
Glycoprotein, 19
Glycylalanine, 1083
Goiter, 942
Gold, 50, 143, 298, 796, 992
    discovery of, 256
    electronic properties of, 484
    hydrometallurgy of, 987
    on nanoscale, 515
    properties of, 143, 992
    reactions of, 143
Gold alloys, 995
Golden Gate Bridge, 626
Gold(III), 1018
Gold leaf, 992
Goodyear, Charles, 504
Goodyear blimp, 429
Goudsmit, Samuel, 233
Graduated cylinder, 16, 17
Grafts, vascular, 509
Graham, Thomas, 418
Graham's law of effusion, 418–19
Gram (g), 15
Graphite, 961
    in batteries, 873
    in carbon fibers, 962
    electronic properties of, 484
    structure of, 466–67
Gravitational forces, 45, 395
Gravity, 166–67
Gray (Gy), 920
Gray tin, 486
Greek prefixes, 65, 1026
Green chemistry, 788–92
    corrosion-protecting cars, 790, 791
    dry cleaning, 790
    principles of, 788–89
    solvents and reagents, 789–90
    water purification, 790–92
Greenhouse effect, 781–82
    carbon dioxide and, 433, 781–82
    methane and, 782
Grigg, C. L., 280
Ground state, 220, 227
Group 1A elements. See Alkali (group 1A) metals
Group 2A elements. See Alkaline earth (group 2A) metals
Group 4A elements, 965–69. See also Carbon (C); Silicon (Si)
    general characteristics of, 965–66
Group 5A elements, 956–60. See also Nitrogen (N); Phosphorus (P)
    general characteristics of, 956–57
Group 7A elements. See Halogens
Group 6A elements (chalcogens), 50, 948–52. See also Oxygen (O)
    general characteristics of, 949
    group trends for, 282–83
    occurrences and preparation of, 949
    properties and uses of, 949–50
Group 8A (noble gases), 938–40
Groups, 50
Guanine, 477, 1091, 1092
Guitar strings, 980, 981
Guldberg, Cato Maximilian, 631
Guy-Lussac, Joseph Louis, 430

Haber, Fritz, 304, 631, 641–42, 654
Haber (Haber-Bosch) process, 630–31, 890

free energy changes in, 826, 828
    hydrogen and, 937
    nitrogen and, 953
    temperature effects on, 642
    volume and pressure effects on, 646
Hafnium, 253, 999
Hahn, Otto, 915
Half-cell potentials, 856–60
Half-life of reactions, 589–91, 904
    calculations based on, 905–8
Half-reaction, 846, 878
Half-reaction method, 846–49
Halides
    halogens and, 284
    hydrogen, 942
    phosphorus, 957–58
Hall, Charles M., 989, 990
Hall-Héroult process, 989
Halogen-containing molecules, ozone layer destruction and, 590
Halogens, 50, 241, 940–44
    boiling points of, 442
    electron affinities of, 271
    elemental, 284
    group trends for, 283–84
    hydrogen halides, 942
    interhalogen compounds, 943
    oxidation number of, 137
    as oxidizing agents, 860
    oxyacids and oxyanions of, 943
    properties and preparation of, 940–41
    reaction with ethylene, 1065
    uses of, 941–42
Halons, 797
Handedness. See Chirality
Hard materials, 482
Hard water, 788
Hardystonite, 967
HDPE (high-density polyethylene), 503, 507
Heart
    biomaterials for repairing, 508
    ion concentration and, 868–69
    nitroglycerin and, 956
Heat, 166. See also Enthalpy(ies)
    of combustion, 188
    condensation, 450
    of deposition, 450
    of formation, 188
    of freezing, 450
    of fusion, 449
    internal energy change and, 171–72
    of reaction, 177
    reversible flow of, 805
    specific, 179–81
    of sublimation, 450
    transferring energy and, 168–69
    of vaporization, 449
Heat capacity, 179–81
    molar, 180
Heat exhaustion, 185
Heating curves, 450–52
Heat packs, 530
Heat stroke, 147, 185
Heavy water, 935
Heisenberg, Werner, 224
Helium (He), 938
    deep-sea diving and, 540
    electron configuration of, 235
    properties of, 284
Helium-4, 913
Hematite, 985
Heme, 1021–22
Hemoglobin, 559, 715, 729, 1022–23
    carbon monoxide affinity for, 778–79
Hemolysis, 508, 553

Henderson–Hasselbach equation, 725, 727
Henry's law, 540
Henry's law constant, 540
Heptane, 1062
Héroult, Paul, 989, 990
Hertz (Hz), 213
Hess's law, 184–88, 326, 638
Heterogeneous alloys, 995, 996
Heterogeneous catalysis, 606–8
Heterogeneous equilibria, 639–41
Heterogeneous mixture, 8
Heteronuclear diatomic molecules, 379–81
Heterotrophs, 815
Hevea brasiliensis, 504
Hexafluorobenzene, 477
Hexafluorosilicic acid, 942
Hexagonal close packing, 463
Hexane
    reaction with carbon tetrachloride, 531
    solubility of, 536
    viscosity of, 448
Hexanol, 536
Hexatriene, 321
HFCs (hydrofluorocarbons), 775, 782
High-carbon steels, 995
High-density polyethylene (HDPE), 503
Highest occupied molecular orbital (HOMO), 380
High-spin complex, 1037
High-temperature superconductivity, 497
Hindenburg disaster, 178
H⁺ ions
    concentrations of
        pH values and, 675–77
        rate law and, 677
    in water, 669
Histidine, 1082
Holes, 488
Homochirality, 1085
Homogeneous catalysis, 605–6
Homogeneous equilibria, 639–41
Homogeneous mixture, 8. See also Solution(s)
HOMO (highest occupied molecular orbital), 380
Homonuclear diatomic molecules, 371
Hot-air balloons, 400
Hot springs, 436, 437, 718, 719
Household chemicals, 4
Human body
    elements in, 6
    temperature regulation in, 185
Hund's rule, 235–37, 238
Hybrid car, 196–97
Hybridization, 357
Hybrid orbitals
    molecular geometry and, 357–62
        involving d orbitals, 360
        sp, 357–58
        sp² and sp³, 358–60
        triple bonds and, 364
Hydrates, 533, 875
Hydration, 529, 705
    water of, 533
Hydrazine, 884, 953
Hydrazobenzene, 390
Hydrazoic acid, 713
Hydride ions, 277, 936, 937
Hydrides, 277
    binary, 700
    interstitial, 938
    metallic, 938
    molecular, 938

Hydrocarbon(s), 66–67, 1054–55
    branched-chain, 1057
    combustion of, 86
    derivatives of, 66–67
    immiscibility of, 536
    as pollutants, 780, 1070
    saturated (alkanes), 66–67, 1054, 1055–62
        cycloalkanes, 1060–61
        nomenclature of, 1058–60
        reactions of, 1061
        structural isomers of, 1057–58
        structures of, 1056–57
    straight-chain, 1057
    unsaturated
        alkenes, 1062–66
        alkynes, 1064–66
        aromatic, 1054, 1068–70
    viscosities of, 448
Hydrocarbon fractions, from petroleum, 1061
Hydrochloric acid, 128, 133, 134, 243, 577
    ionization of solution of, 122
    reactions of
        with cobalt(II) chloride, 651
        with nickel, 533
        with sodium hydroxide, 132, 668
        with zinc, 844–45
    stock solutions of, 146
    titration with caustic soda (NaOH), 730–33
Hydrocyanic acid, 682, 965
Hydroelectric energy, 197
Hydrofluoric acid, 130, 943
    properties of, 681, 682
Hydrofluorocarbons (HFCs), 775, 782
Hydrogen (H), 6, 7, 52, 281–82, 430, 936–38
    binary compounds of, 937–38
    combustion of, 177
    covalent bonds in, 305, 356
    electron configuration of, 235
    energy states of, 220–22
    group trends for, 281–82
    ionization energy of, 281
    isotopes of, 432, 935
    line spectrum of, 219
    molecular, 52, 937
        attractions and repulsions in, 305
        covalent bonding in, 305
        molecular orbitals in, 368–70
        reaction with oxygen, 873
        as reducing agent, 861
    nuclear spin in, 235
    orbitals of, 227–28, 368–70
    oxidation number of, 137
    preparation of, 936–37
    properties of, 7, 936
    radial probability function of, 259
    reactions of, 281–82
        with ethylene, 607
        with nonmetals, 281
        with oxygen, 177
    in rocket fuel, 800, 801, 873
    Schrodinger's equation for, 224–25
    uses of, 937
Hydrogenation, 1065, 1090
Hydrogen bonding, 443–46
    aqueous solubility and, 537
    between complementary base pairs, 1092
    in DNA, 1092
    solution formation and, 529
    trends in, 446
    in water, 445–46
Hydrogen carbonate ion, 744

Hydrogen carbonates, 963
Hydrogen chloride, 327, 668
Hydrogen compounds of nitrogen, 953
Hydrogen cyanide, 394, 965
Hydrogen economy, 337, 937
Hydrogen fuel cells, 873
Hydrogen halides, 942
    bond lengths and dipole moments
        of, 312
    charge separation in, 312
Hydrogen ion, 129
Hydrogen peroxide, 52, 283, 790, 948,
    1073
    decomposition of, 609
    reaction with bromide ions, 606
    structural formula for, 53
Hydrogen sulfates, 952
Hydrogen sulfide, 134, 394, 622, 950
    critical temperature and pressure of,
        452
Hydrogen sulfites, 671, 950–51
Hydrolysis, 696
    acid-dissociation constants for, 705
    of metal ions, 705–7
Hydrometallurgy, 987–88
Hydronium ions, 137, 669, 696–97
Hydrophilic colloids, 557–60
Hydrophobic colloids, 557–60
hydro- prefix, 64
Hydrosulfide ion, 691
Hydrous oxide, 1004
Hydroxide ions, 129
Hydroxides
    of alkali metals, ionic, 680–81
    of alkaline earth metals, ionic,
        680–81
    amphoteric, 748–50
    base-insoluble, 754
    ionic, 680
    metal, 130
    solubility of, 748–50
Hydroxyapatite, 281, 764
Hydroxylamine, 691, 954
Hydroxyl group, 1072
Hydroxyl radical, 797, 919
Hydroxypatite, 747
Hypertension, 398
Hypertonic solutions, 553
Hypochlorite ion, 691
Hypochlorite salts, 943
Hypochlorous acid, 682, 692, 701, 713
Hyponatremia, 147
hypo- prefix, 64
Hypothalamus, 185
Hypothermia, 185
Hypothesis (tentative explanation), 13
Hypotonic solutions, 553

S-Ibuprofen, 1079
Ice, 5, 445–46
    melting of, 180, 803, 813
    structure of, 813
Ice packs, 530, 531
ICF (intracellular fluid), 868
-ic suffix, 59, 64
Ideal bond angles, 346
Ideal gas
    defined, 402
    entropy and, 810–11
    Raoult's law and, 548
Ideal-gas equation, 402–6, 416
    gas densities and molar mass, 406–8
    gas laws and, 404–6
    gas volumes in chemical reactions,
        408–10
Ideal solutions, 548
-ide suffix, 62, 64, 65

Ignarro, Louis J., 956
Imaging, medical, 236, 496–97
Immiscible liquids, 536
Indicators, acid–base, 150–51
Indigo, 1102
Indole, 1102
Inert gases, 285
Inexact numbers, 20
Information, tabulating, 404
Infrared radiation, 212, 213
Inorganic compounds, 59
    nomenclature of, 59–65
        acids, 64–65
            binary molecular compounds, 65
            ionic compounds, 59–64
Insoluble chlorides, 753
Insoluble phosphates, 754
Instantaneous reaction rate, 577–78
Insulators, 482
    band structure of, 484
Insulin, 102
Integrated rate law, 586, 588
Intensive properties, 10
Interhalogen compounds, 388, 943
Intermediates, 598–99
Intermetallic compounds, 996–97
Intermolecular forces, 437, 439–47
    attractive, 422
    covalent bond vs., 439
    effect on gas pressure, 422
    in gases, 438, 439
    ion-dipole, 440
    in liquids, 438, 439
    molecular geometry and, 438, 440,
        442
    pressure and, 439
    Raoult's law and, 548
    in smectic phases, 511
    in solids, 438, 439
    in solution formation, 528–29
    surface tension and, 448
    van der Waals forces
        comparison of, 446–47
        dipole–dipole, 440–41
        hydrogen bonding, 443–46
        London dispersion, 441–43
Internal energy (E), 170–71
    change in, 170
        relating to heat and work, 171–72
    as state function, 172–73
International Union of Chemistry, 1058
International Union of Pure and
        Applied Chemistry (IUPAC),
        50, 1058
Internuclear axis, 362
Interstitial alloys, 995, 996
Interstitial carbides, 965
Interstitial hydrides, 938
Intracellular fluid (ICF), 868
Invertase, 624
Invert sugar, 1088
Iodine (I), 50, 283–84, 940, 942
    bonding atomic radius of, 261
    properties of, 283
Iodine-131, 910, 922, 927
Iodine pentafluoride, 435
Iodized salt, 942
Ion(s), 54–56
    calcium, 57
    carbonate, 691
    charges of, 55–56, 302–3
    complex, 1014
    electron configurations of, 268–70
    formation of complex, 745–48
    in human heart, 869
    hydride, 277
    hydrogen, 129

hydroxide, 129
    isoelectric series of, 264
    monatomic, 137
    negative. See Anion(s)
    orientation of polar molecules
        toward, 440
    oxide, 282
    peroxide, 283
    polyatomic, 55, 61, 316
    positive. See Cation(s)
    precipitation and separation of,
        750–53
    properties of, 55
    required by organisms, 57
    sizes of, 259–64
    spectator, 127, 128
    sulfide, 752
    superoxide, 278, 283
    transition-metal, 303–5
Ion concentration, measuring using
        conductivity, 121
Ion–dipole forces, 440, 446
    solution formation and, 528
Ion–dipole interaction, 1034
Ion exchange for water softening, 788
Ionic bonds, 298, 299–305, 309, 312–14,
        446
    electron configurations of ions and,
        302–3
    energetics of formation, 301–2
    polyatomic ions, 309
    transition-metal ions, 303–5
Ionic carbides, 964
Ionic compounds, 56–59
    dissolution or precipitation of, 737
    electrolytic behavior of, 131
    electrolytic strength of, 123
    formation of, 299–300
    identifying, 123
    lattice energies for, 301–2
    names and formulas of, 59–64
    solubilities of, 124–26, 744
    in water, 120–22
Ionic conductivity, 494
Ionic equation(s), 127–28
    complete, 127
    net, 127
        writing, 128
Ionic hydrides, 937–38
Ionic radii, 259–64
    biological chemistry and, 265
    periodic trends in, 263–64
Ionic solids, 464, 467
    in water, 529, 813
Ionization
    percent, 683–84, 686–88
    photoionization, 772
    of weak base, 722
Ionization energy, 264–70, 308
    of alkaline earth elements, 279
    electron affinity vs., 270
    electron configurations of ions and,
        267–68
    electronegativity and, 308
    of metals, 272
    periodic trends in, 263–64
Ionizing radiation, 919
Ion pair, 554
Ion product, 694
Ion-product constant, 674
Ion product of water, 674–75
Iron (Fe), 6, 1003–4
    corrosion of, 136, 874–76
    galvanized, 875
    in myoglobin and hemoglobin, 1024
    oxidation of, 136, 802–3

pig, 985, 986
    pyrometallurgy of, 985–86
    as reducing agent, 861
    refining of, 985–86
    specific heat of, 181
Iron(II), 1003
Iron(II) fluoride, 880–81
Iron(III), 1004, 1018
Iron(III) chloride hexahydrate, 533
Iron(III) oxide, 875, 947, 988, 1003
Iron(II) sulfide, 950
Iron pyrite (fool's gold), 459, 950
Irreversible process, 804–6
Isobutane (2-methylpropane), 1057
Isoelectronic series of ions, 264
Isolated system, 168
Isoleucine, 1081
Isomerism, 1026–31
    alkane, 1057–58
    cis-trans, 390
    coordination-sphere, 1027
    geometric, 1028, 1063
    linkage, 1027
    optical, 1028–31
    stereoisomerism, 1027–31
    structural, 1027
Isooctane (2,2,4-trimethylpentane), 1062
Isopentane (2-methylbutane), 1057
Isoprene, 504
Isopropyl alcohol (2-propanol), 67,
        97–98, 1072
Isopropyl group, 1059
Isothermal process, 805–6, 808
Isotonic solutions, 553
Isotopes, 45–46, 895
    abundance of, 47
    of hydrogen, 935
    stable, with even and odd
        numbers of protons and
        neutrons, 901
    synthetic, 904
-ite suffix, 61, 64
-ium suffix, 60
IUPAC (International Union of Pure and
        Applied Chemistry), 50, 1058

Jet engine, composition of, 981, 982
Joint Institute for Nuclear Research
        (JINR), 18
Joule (J), 167
Joule, James, 167

Kelvin, Lord (William Thomson), 400
Kelvin scale, 15–16, 400
Kennedy, Joseph, 51
Ketones, 797, 1073–74
Kilogram (kg), 15
Kilojoules (kJ), 167
Kilowatt-hour (kWh), 879–80
Kimax, 968
Kinetic energy, 166
    temperature and, 594
Kinetic-isotope effect, 935
Kinetic-molecular theory, 414–16
    gas laws and, 415–16
Kinetics, chemical. See Reaction rates
Knocking characteristics of gasoline,
        1062
Kroto, Harry, 468
Krypton (Kr), 239, 284, 285
Krypton compound, 940
Kyoto Protocols, 767

Labels, food, 193
Lactic acid, 388, 717, 725
Lactic acid-glycolic acid copolymers, 509
Lactones, 797

Lactose, 1087
Lanthanide contraction, 999
Lanthanides, 239–40
Lanthanum (La), 239
La rue Montorgueil (painting), 254
LAS (linear alkylsulfonate) surfactant, 959
Lattice energy, 301–2
  calculation of, 304
Lattice point, 460
Lattice structure, crystal, 461–62
Lauryl alcohol, 568
Lauterbur, Paul, 236
Lavoisier, Antoine, 80
Lawrence Livermore National
  Laboratory, 18
Laws
  of combining volumes, 401
  of conservation of mass, 79–80
  of constant composition, 8, 38
  of definite proportions, 8
  of mass action, 631
  of multiple proportions, 39
  scientific, 13
LCDs (liquid crystal displays), 513
LDPE (low-density polyethylene), 503
Leaching, 987
Lead (Pb), 50, 790, 965
Lead-206, 905, 907
Lead-acid battery, 871
Lead crystal glass, 968
Lead(II) chloride, 639
Lead iodide, 124
Lead nitrate, 124
Lead poisoning, 1020
Le Châtelier, Henri-Louis, 648
Le Châtelier's principle, 648–56
  catalyst effects, 654–56
  change in reactant or product
    concentrations, 648–49
  nitric oxide emission control and, 656
  temperature changes, 651–54
  volume and pressure changes, 649–51
LEDs (light-emitting diodes), 19,
  492–93, 510
Length, SI unit of, 14–15
Leucine, 1081
Leukemia, 920
Leveling effect, 672
Levitation, magnetic, 496
Levorotary, 1030
Lewis, G. N., 298, 305, 704
Lewis acids and bases, 704–7
  crystal-field theory and, 1033–34
  metal-ligand bond and, 1016
Lewis structures, 305–6
  drawing, 314–18
  formal charge and, 316–18
Lewis symbols, 298–99
Lichens, 1020
Life, entropy and, 815
LifeStraw, 787
Ligands, 1014, 1019–24
  bidentate, 1019
  color effects of, 1031
  in living systems, 1021–24
  monodentate, 1019
  nomenclature of, 1025
  polydentate (chelating agents),
    1019–22
  weak field and strong field, 1036
Ligand-to-metal charge-transfer
  (LMCT) transition, 1040, 1047
Light
  electron excitation by, 380
  monochromatic, 218
  scattering by colloidal particles, 557

speed of, 212, 214, 220
  visible, 212, 213
  wave nature of, 212–15
Light absorption
  molecular orbitals and, 380
  at p-n junction, 491
Lightbulbs, energy-efficient, 19
Light-emitting diodes (LEDs), 19,
  492–93, 510
Lighter-than-air balloons, 401
"Like dissolves like," 538
Lime (calcium oxide), 85, 777, 964
Lime-soda process, 788
Limestone. *See* Calcium carbonate
  (limestone)
Limestone caves, 120
Limiting reactants (reagents), 102–7
  theoretical yields, 105–7
Linear alkylsulfonate (LAS) surfactant,
  959
Linear molecular geometry, 342–43,
  345, 347, 350, 354, 361, 1052
Line spectra, 218–19
Linkage isomerism, 1027
Linoleic acid, 1089
Lipids, 1089–90
Liquid(s), 5, 447–48
  cholesteric, 512
  intermolecular attractive forces in, 438
  molecular comparison of solids and,
    438
  nematic, 511, 513
  phase changes in, 449–53
  properties of, 438, 439
  smectic, 511
  surface tension of, 448
  vapor pressure and, 453–56
  viscosity of, 447–48
  volatile, 454
Liquid crystal displays (LCDs), 513
Liquid crystals, 510–13
  phases of, 511–12
Liter (L), 16
Lithium (Li)
  electron configuration of, 235–36, 238
  flame test, 278
  molecular orbitals for, 371–72
  reaction with water, 277
Lithium drugs, 280
Lithium ion, as oxidizing agent, 860–61
Lithium-ion battery, 872–73
Lithosphere, 982
Litmus, 132, 668, 678
Living systems. *See also* Biochemistry
  chirality in, 1085
  iron in, 1024
  metals and chelates in, 1021–24
  radiation in, 919–23
LMCT transition, 1040, 1047
Lobes, in orbitals, 231
Lock-and-key model, 609
London, Fritz, 441
London dispersion forces, 441–43, 446
  in DNA, 1092
Lone pairs, 344, 690
Low-density polyethylene (LDPE), 503
Lowest unoccupied molecular orbital
  (LUMO), 380
Lowry, Thomas, 668
Low-spin complex, 1037
LUMO (lowest unoccupied molecular
  orbital), 380
Lung cancer, 921
Lysine, 703, 1082
Lysozyme, 568, 609

Macromolecules, 502

Macroscopic realm, 2
Magic numbers, 901
Magnesite, 964
Magnesium (Mg), 280, 298, 430, 992
  combustion of, 84, 85
  electron affinity of, 271
  electron configuration of, 240
  as essential nutrient, 281
  oxidation of, 874
  properties of, 279
  reactions of
    with acid, 138–39
    with titanium tetrachloride, 884
Magnesium diboride, 498
Magnesium hydroxide, 133
Magnesium metal, combustion of, 84, 85
Magnesium oxide, 298
Magnesium sulfate, 530
Magnet, permanent, 1001
Magnetic levitation, 496
Magnetic quantum number, 226
Magnetic resonance imaging (MRI),
  496–97
  nuclear spin and, 236
Magnetism, 1001–2, 1033
  antiferromagnetism, 1001
  diamagnetism, 376–78
  ferrimagnetism, 1002
  ferromagnetism, 1001
  paramagnetism, 376–77, 1001, 1033
Magnetite, 983, 985
Magnitude, of equilibrium constants,
  635–36
Main-group (representative) elements,
  240
Malachite, 8, 983
Malic acid, 667
Malignant tumor, 922
Malleability, 992
Mammoth Cave (Kentucky), 964
Mammoth Hot Springs (Yellowstone
  National Park), 718
Manganese dioxide, 945
Manganese(II) oxide, 313
Manganese silicide, 979
Manganese(VII) oxide, 313
Manhattan Project, 915
Manic-depressive illness, 280
Manometer, 397–98
Mansfield, Peter, 236
Many-electron atoms, 232–34
Marble, corrosion of, 777
Marconi, Guglielmo, 772
Marley, Ziggy, 980
Marsden, Ernest, 42
Mass action, law of, 631
Mass defect, 912–13
Mass(es)
  atomic, 43. *See also* Stoichiometry
  average atomic. *See* Atomic weights
  conservation of, 79–80
  critical, 914
  electron, 43
  interconverting mole and, 93
  interconverting number of particles
    and, 94–95
  molar, 91–92
    calculating, 92
    defined, 91
    determining through colligative
      properties, 555–56
    effusion rate and, 417–19
    gas density and, 406–8
  neutralization reaction and, 150
  neutron, 43
  in nuclear reaction, change in,
    911–12

proton, 43
SI unit of, 15
supercritical, 914–15
Mass numbers, 45, 895
Mass percentage, 542
Mass spectrometer, 48
Mass spectrum, 48
Materials, modern, 480–525
  ceramics, 493–95
    applications, 494
    making, 494–95
    superconductors, 497–98
  classes of, 482
  for electronics
    silicon chip, 489
    solar energy conversion, 489–91
  electronic structure of, 482–84
  liquid crystals, 510–13
    types of liquid crystalline phases,
      511–12
  for medicine (biomaterials), 505–10
    characteristics of, 507
    examples of, 508–10
    polymeric, 507–8
  for nanotechnology, 513–17
    carbon nanotubes, 515–17
    metals on the nanoscale, 515
    semiconductors on the nanoscale,
      513–14
  for optics
    liquid crystals, 510–13
    semiconductor light-emitting
      diodes, 492–93
  polymers and plastics, 499–505
    engineering polymers, 506
    making polymers, 499–502
    recycling, 501
    structure and physical properties
      of polymers, 502–5
  semiconductors, 482, 484–93
    semiconductor doping, 488
    semiconductor light-emitting
      diodes, 492–93
    silicon chip, 489
    solar energy conversion, 489–91
  superconductors, 495–87
    ceramic, 497–98
Matter
  classifications of, 4–9
    compounds, 5–8
    elements, 2, 5, 6
    mixtures, 6, 8
    pure substances, 5–6
  conservation of, 38, 79–80
  defined, 2
  properties of, 9–12
    chemical, 9, 37
    extensive, 10
    intensive, 10
    physical, 9, 37
    physical and chemical changes,
      10–11
    separation of mixtures, 11–12
  states of, 4–5, 438–39
  wave behavior of, 222–24
Matter waves, 222–23
McMillan, Edwin, 51
Mean free path
  diffusion and, 419–20
  of metals, 515
Measurement, 13–29
  dimensional analysis, 24–29
    conversion factors in, 26–27
    conversions involving volume,
      27–29
  SI units of, 13–17
    base units, 14

Measurement, *(cont.)*
density, 17
derived, 16
length and mass, 14–15
for speed, 16
temperature, 15–16
volume, 16
uncertainty in, 20–24
precision and accuracy, 20–21
significant figures, 21–24
uncertainty principle and, 225
Mechanical work, 176
Mechanism, of addition reactions, 1066–68
Medical grade, 507
Medical imaging, 236, 496–97
Medicine
chelating agents in, 1020
chiral drugs in, 1079
modern materials for, 505–10
characteristics of, 507
examples of, 508–10
polymeric, 507–8
radiotracers used in, 910
Medium steels, 995
Meissner effect, 496
Meitner, Lise, 915
Melting points, 457
intermolecular forces and, 440
of metals, 993
normal, 457
of solids, 460
Mendeleev, Dmitri, 256–57
Meniscus, 448
Mercuric oxide dry-cell batteries, 888
Mercury (Hg), 432
meniscus, 448
roasting of, 984
surface tension of, 448
Mesitylene, 96–97
Mesosphere, 768
Metabolism
of glucose, 102, 193
nitric oxide and, 379
peroxide ion by-product of, 948
of proteins, 193
Meta isomer, 385
Metal(s), 50, 980–1011
active, 140
group trends for, 276–81
activity series of, 140–42
alkali. *See* Alkali (group 1A) metals
alkaline earth. *See* Alkaline earth (group 2A) metals
alloys, 981, 995–98
defined, 995
heterogeneous, 995, 996
intermetallic compounds, 996–97
interstitial, 995, 996
shape-memory, 997–98
solution, 995
steels, 995–96
substitutional, 995, 996
corrosion of, 874–76
iron, 874–76
deficiencies of, 1021
energy bands in, 484
f-block, 240
group trends for, 272–74
ionic compounds and, 56–57
ions of, 54
in living systems, 1021–24
on nanoscale, 515
noble, 140
occurrence and distribution of, 982–83
oxidation of, 874

by acids and salts, 138–40
periodic properties of, 271–74
physical properties of, 992
qualitative analysis of, 753–55
reactions of
with hydrogen, 281–82
with nonmetal, 84
with oxygen, 282
standard reduction potentials and activity series of, 863
transition. *See* Transition metals
Metal alkoxide, 494
Metal chlorides, 754
Metal complexes, 1014–18
charges, coordination numbers, and geometries, 1016–18
metal-ligand bond, 1016
Werner's theory of, 1014–16
Metal ions
acidic solutions and, 697
complex ion formation and, 745–48
coordination numbers of, 1015
hydrolysis of, 705–7
Lewis acids and bases and, 745–48
in water, 697
Metallic bonds, 298, 991–95
electron-sea model, 992–93
molecular-orbital model, 993–95
Metallic character, 272
Metallic elements, 50. *See also* Metal(s)
Metallic hydrides, 938
Metallic materials, properties of, 493
Metallic radius, 289
Metallic solids, 464, 467–69
Metal-ligand bond, 1016, 1034
Metalloenzymes, 611
Metalloids, 50
periodic properties of, 271, 276
Metallurgy, 983
defined, 983
electrometallurgy, 988–91
of aluminum, 989–90
copper refinement, 990–91
of sodium, 988–89
hydrometallurgy, 987–88
pyrometallurgy, 984–87
formation of steel, 986
of iron, 985–86
Metal oxides, 273–74
Metal-oxide semiconductor field-effect transistor (MOSFET), 490
Metal-to-ligand charge-transfer (MLCT) transition, 1040
meta- prefix, 1069
Metathesis reactions, 126–27
Meter (m), 14
Methanal (formaldehyde), 364, 944, 1052, 1073
Methane, 52, 66, 67, 327, 394, 409, 1052, 1055. *See also* Natural gas
in atmosphere, 776
balanced chemical equation for, 82
bond enthalpies in, 325
bonds in, 348, 1056
combustion of, 1053
as greenhouse gas, 782
hydrogen production and, 936
molar entropies of, 817
reactions of
with chlorine, 327
with oxygen, 81
representations of, 53
structural formula for, 53
Methanoic acid. *See* Formic acid
Methanol fuel cells, 874
Methanol (methyl alcohol), 66, 67, 661, 662, 838, 1071, 1072

combustion of, 1073
dissolution of, 121
hydrogen in manufacture of, 937
reaction with water, 121
solubility of, 536
solution of, 122
Methionine, 1081
Methyl acetate, 929
Methylamine, 691, 692, 715, 1049
Methylbenzene (toluene), 466, 548, 789, 1062, 1068
Methyl bromide, 590, 775
Methyl chloride, 327, 441, 775
Methyl ethanoate, 1071
Methyl ethyl ketone (2-butanone), 1073, 1074
Methyl group, 1059
Methyl groups, 1063
Methylhydrazine, 953, 979
Methyl iodide, 443
Methyl isonitrile, 597
first-order rearrangement of, 589
isomerization of, 594
transformation to acetonitrile, 587
Methyl mercaptan, 262
Methyl orange, 678, 679
3-Methyl-1-pentene, 1066
2-Methylpropane, 1057
2-Methyl-2-propanol, 1072
Methylpropene, 1063
Methyl propionate, 1075
Methyl red indicator, 679, 733, 734, 736
Methyl *tert*-butyl ether (MTBE), 1062
Methyl violet, 679
Metric system, 13–14
Meyer, Lothar, 256
Mg(OH)$_2$, 739, 742–44
Michelson, A. A., 214
Microscope, electron, 223
Microstates, 810–13
Mild steels, 995
Milk of magnesia, 133
Millikan, Robert, 40
Millikan's oil-drop experiments, 40–41
Milliliter, 16
Millimeter of mercury (mm Hg), 396
Millivoltmeter, 678
Minerals, 982–83
as metal source, 982–83
Miscible, 536
Mixture(s), 6, 8
components of, 8
equilibrium, 628
of gases, 394
heterogeneous, 8
homogeneous. *See* Solution(s)
racemic, 1031, 1079
separation of, 11–12
MLCT transition, 1040
Moderator, 916
Modern materials. *See* Materials, modern
Molal boiling-point-elevation constant, 549
Molal freezing-point-depression constant, 550
Molality, 543–44
conversion of, 544–45
Molar entropies, 817
Molar heat capacity, 180
Molarity (M), 144–45, 543–44
conversion of, 545–46
interconverting moles, volume, and, 145–46
Molar mass, 91–92
calculating, 92
defined, 91

determination of, 555–56
effusion rate and, 417–19
gas densities and, 406–8
Mole, 89–95
defined, 90
interconverting masses and, 93
interconverting masses and numbers of particles, 94–95
interconverting molarity, volume, and, 145–46
molar mass, 91–92
Molecular architecture, 341
Molecular (bond) polarity, 353–55
Molecular collisions, 592–93
Molecular compounds, 52–54
binary, 65
electrolytic behavior of, 131
nomenclature of, 312–13
in water, 122
Molecular diffusion, 417–20
Molecular effusion, 417–20
Molecular equations, 127
for redox reactions, 140
Molecular formulas, 52–53
from empirical formula, 96–97
of metal complexes, 1016–18
Molecular geometry, 340–91
bent, 342–43, 354
covalent bonding and, 355–57
defined, 345
dispersion forces and, 442
hybrid orbitals and, 357–62
involving *d* orbitals, 360
sp, 357–58
sp^2 and *sp*^{3}, 358–60
intermolecular attraction and, 438, 442
linear, 342–43, 345, 347, 350, 354, 361
of metal complexes, 1018
molecular (bond) polarity and, 353–55
molecular orbitals (MO) and, 368–70
bond order and, 370
in hydrogen molecule, 368–70
light absorption and, 380
from 2*p* atomic orbitals, 372
in second-row diatomic molecules, 371–81
multiple bonds and, 362–68
octahedral, 345, 349, 350, 351, 354, 361
seesaw, 350
square planar, 344, 350, 351, 354
square pyramidal, 350, 351
tetrahedral, 342, 345, 347, 354, 361
trigonal bipyramidal, 345, 349, 350, 354, 361
trigonal planar, 343, 345, 347, 354, 361
trigonal pyramidal, 343, 346
T-shaped, 343, 350
valance-shell electron-pair repulsion (VSEPR) model of
basis of, 344–45
for larger molecules, 352–53
for molecules with expanded valence shells, 349–51
nonbonding electrons and multiple bonds, 348–49
valence-bond theory and, 355, 360
valence-shell electron-pair repulsion (VSEPR) model of, 343–53
Molecular hydrides, 938
Molecularity, 598
Molecular orbital diagram. *See* Energy-level diagram (molecular orbital diagram)

Molecular-orbital model for metals, 993–95
Molecular orbitals (MO), 368–70
  antibonding, 369
  bonding, 369
  energy and, 380
  highest occupied (HOMO), 380
  lowest unoccupied (LUMO), 380
  modern materials and, 482–83
  molecular geometry and
    bond order and, 370
    in hydrogen molecule, 368–70
    light absorption and, 380
    from 2p atomic orbitals, 372
    in second-row diatomic molecules, 371–81
  phases in, 373–75
  pi (π), 372
  sigma (σ), 369
Molecular-orbital theory, 368
  of coordination compounds, 1038
  of metals, 1000
Molecular oxygen, 774–75
Molecular perspective, 2–3
Molecular solids, 464–66
Molecular speed, distributions of, 414–15
Molecular weights, 46, 87–88
  boiling point and, 443
Molecule(s), 2, 51–54
  aromatic, 321
  chemical formulas and, 52
  of compound, 5
  defined, 52
  degrees of freedom of, 814
  diatomic, 52
    heteronuclear, 379–81
    homonuclear, 371
  electron density in, 305
  of element, 5
  optically active, 1030
  organic, 1052–53
  picturing, 53–54
  polar, 310, 442
  properties of states and, 5
Mole fraction, 543–44
  conversion of, 544–45
  partial pressures and, 411–12
Molina, Mario, 774
Momentum, 222
Monatomic ions, 137
Monet, Claude, 254
Monoatomic gases, 394
Monochromatic light, 218
Monochromatic radiation, 218
Monochromatic wavelength, 218
Monodentate ligands, 1019
Monomers, 499
mono- prefix, 65
Monoprotic acids, 129, 679
Monosaccharides, 1087
Monounsaturated fatty acid, 1090
Montreal Protocol on Substances That Deplete the Ozone Layer, 775
Mortar, 964
Moseley, Henry, 257
MOSFET (metal-oxide semiconductor field-effect transistor), 490
Motion, molecular, 809–10
Motor oils, 447
Mount San Antonio, 214
Mount St. Helens, 429
Mount Wilson, 214
MRI (magnetic resonance imaging), 236, 496–97
MTBE (methyl tert-butyl ether), 1062
Müller, K. A., 497

Multiple bonds, 307
  bond angles and, 348–49
  bond enthalpies of, 326
  electron domains for, 349
  molecular geometry and, 362–68
Multiple-choice questions, 106
Multiple proportions, law of, 39
Multiplication, significant figures in, 23
Multistep mechanisms, 598–99
  rate-determining step for, 600–1
Multiwall carbon nanotubes, 516
Murad, Ferid, 956
Murchison meteorite, 1085
Music, rock, 980, 981
Myoglobin, 1022, 1085

Naming compounds. See Nomenclature
Nanomaterials, 513–17
  carbon nanotubes, 515–17
  metals, 515
  semiconductors, 513–14
Naphthalene, 391, 1068, 1070
Natural gas, 196, 262, 409, 435, 798, 840. See also Methane
  combustion of, 798
  conversion of, 840
nde, 1102
Néel temperature, 1002
Negligible acidity, 672
Nematic liquid crystalline phase, 511, 513
Neon (Ne)
  electron configuration of, 237–38
  Lewis symbol for, 299
  light emitted by, 210–11, 219, 429
  properties of, 284
Neon gas, 44
Neopentane (2,2-dimethylpropane), 442, 1057
Neoprene, 523
Neptunium, 903
Nernst, Walther, 865
Nernst equation, 865–67
Net ionic equations, 127
  for oxidation-reduction reactions, 140
  writing, 128
Neurotransmitter, nitric oxide as, 955
Neutralization reactions, 132–34
  using mass relations in, 150
  writing chemical equations for, 133–34
Neutral solutions, 674, 697–98
Neutrons, 42, 43, 894, 897, 903
  mass of, 43
Neutron-to-proton ratio, 898–90
Newton (N), 396
Newton, Isaac, 38, 214
Niacin, 683, 686
Nickel (Ni), 460
  electronic band structure for, 993
  electroplating with, 877
  oxidation of, 142
  reactions of
    with hydrochloric acid, 533
    with oxygen, 273
Nickel-acid reaction, 533
Nickel and titanium (NiTi), 997
Nickel-cadmium (nicad) battery, 845, 872
Nickel carbonyl, 433
Nickel(II), 1018
Nickel(II) oxide, 298
Nickel-metal-hydride battery, 872
Nitrate anion, 125
Nitrate ion
  delocalized bonding in, 366
  resonance structures in, 320

Nitric acid, 754, 955
  reactions of
    with copper, 11
    with gold, 143
    with iron, 1003–4
Nitric oxide(s), 663, 954–55
  in atmosphere, 776, 779
  emissions of, 656
  energy-level diagram for, 379
  reactions of
    with bromine, 603–4
    with chlorine, 663
    with oxygen gas, 814
Nitrides, 493
Nitrite ions, 580
Nitrocellulose, 955
Nitrogen (N), 452, 952–56
  in atmosphere, 394, 769
  deep-sea diving and, 540
  dissociation energy of, 771
  fixed, 953
  hydrogen compounds of, 953
  molecular, 377
    bonding in, 307
    critical temperature and pressure in, 452
    ionization of, 772
    Lewis structure of, 307
    photodissociation of, 771
    properties of, 769
    specific heat of, 181
  oxidation states of, 952
  oxides and oxyacids of, 954–56
  2p orbital filling in, 268
  preparation and uses of, 952–53
  properties of, 952, 956, 957
  reduction of, 611
  in troposphere, 775
Nitrogen-13, 910
Nitrogen-14, 901
Nitrogenase, 610–11
Nitrogen cycle, 610
Nitrogen dioxide, 11, 394, 628–29, 799, 814, 954, 955
  decomposition of, 588–89
  dinitrogen tetroxide-nitrogen dioxide equilibrium, 632–33, 637–38, 649–50
  in smog, 779–80
Nitrogen fixation, 610–11, 631
Nitrogen gas, in air bags, 409
Nitrogen oxides, 776, 779–80
Nitroglycerin, 198–99, 328–29, 955, 956
Nitro isomer, 1027
Nitroto isomer, 1027
Nitrous acid, 955
  properties of, 682
  reaction with water, 670
Nitrous oxide (laughing gas), 394, 954
  decomposition of, 602
NMR (nuclear magnetic resonance), 236
N_2O_4, 633
Nobel, Alfred, 198, 328–29, 956
Noble-gas compounds, 939–40
Noble-gas core, 238
Noble metals, 140
Noble (rare) gases (group 8A), 50, 394, 938–40
  boiling points of, 442
  electron affinities for, 271
  group trends for, 284–86
Nodal plane, 374
Nodes, 229
Nomenclature, 59
  of alkanes, 1058–60
  of alkenes, 1063
  of alkynes, 1064

amino acids, 1083
  chemical, 59
  of coordination compounds, 1025–26
  of inorganic compounds, 59–65
    acids, 64–65
    binary molecular compounds, 65
    ionic compounds, 59–64
  of molecular compounds, 312–13
Nomex, 523
Nonbonding atomic radius, 261
Nonbonding electrons, bond angles and, 348–49
Nonbonding pair, 344, 348
Nonelectrolyte, 120
Nonionizing radiation, 919
Nonmetallic elements, 50
Nonmetals, 50, 930–79
  boron, 969–71
    electron configuration of, 240
    Lewis symbol for, 299
  group 4A, 965–69. See also Carbon (C); Silicon (Si)
    general characteristics of, 965–66
  group 5A, 956–60. See also Nitrogen (N); Phosphorus (P)
    general characteristics of, 956–57
  group 6A, 948–52. See also Oxygen (O)
    general characteristics of, 949
    group trends for, 282–83
    occurrences and preparation of, 949
    oxides, oxyacids, and oxyanions of sulfur, 950–52
    properties and uses of, 949–50
    sulfides, 950
  group 7A. See Halogens
  group 8A (noble gases), 938–40
    compounds, 939–40
    group trends for, 284–86
  halogens (7A)
    group trends for, 283–84
  hydrogen. See Hydrogen (H)
  ions of, 54
  oxidation number of, 137
  oxygen group (6A)
    group trends for, 282–83
  periodic properties of, 271–72, 274–76, 932–33
  reactions of, 933–34
    with hydrogen, 281
    hydrogen with, 281–82
    with metals, 84
Nonpolar covalent bond, 307–8
Nonpolar molecules, 353
Nonrenewable energy, 197
Nonspontaneous reactions, 830
Nonvolatile substance, 546
Normal boiling point, 455, 827
Normal melting point, 457
Normal salts, 696
Noryl GTX, 506
n-type semiconductor, 488, 491
Nuclear age, 915
Nuclear atom, 41–42
Nuclear binding energies, 912–13
Nuclear charge
  effective, 257–59, 260
  ionic radii and, 262
Nuclear chemistry, 51, 892–929
  biological effects of radiation, 919–23
    dosage and, 920
    radon, 921
    therapeutic, 894, 910, 922
  cancer treatment with, 894, 922
  defined, 893

Nuclear chemistry, (cont.)
  elementary particles in, 894–95
  energy changes in nuclear reactions, 911–13
    nuclear binding energy, 912–13
  fission, 913–17
    reactors using, 915–17
    waste from, 917
  fusion, 913, 918
  nuclear stability patterns
    even vs. odd number of nucleons, 901
    magic numbers and, 901
    neutron-to-proton ratio, 898–900
    radioactive series (nuclear disintegration series), 900
  nuclear transmutations, 901–3
  radioactive decay, 895, 896–98
    rates of, 903–8
    types of, 896–98
  radioactivity, 894–98
    detection of, 908–10
Nuclear disintegration series, 900
Nuclear energy, 197
Nuclear equations, 895–96
Nuclear fission reactor, 915
Nuclear force, 898
Nuclear magnetic resonance (NMR), 236
Nuclear power plant design, 916–17
Nuclear reactions. See Nuclear chemistry
Nuclear reactors, 915–17
Nuclear spin, magnetic resonance imaging and, 236
Nuclear transmutations, 901–3
Nuclear waste, 917
Nucleic acids, 1090–94
Nucleons, 894
Nucleotides, 1090–91
Nucleus (atomic), 42
  shell model of, 901
Nuclide, 895
Numbers
  exact vs. inexact, 20
  rounding off, 23
Nylons, 500, 501–2, 841, 965

Oceans, 526–27, 783–85
Octahedral complexes, electron configurations in, 1037–38
Octahedral crystal field, 1034–35
Octahedral geometry, 345, 349, 350, 351, 354, 361
Octane, 66, 67, 206, 447
  solution formation and, 531
Octane numbers, 1062
Octet, 237
Octet rule, 299
  exceptions to, 322–25
n-Octylglucoside, 564
Odors, 694, 715
  esters and, 1075
OH group, acid–base equilibria and, 700–2
Oil
  combustion of, 776–77
  crude (petroleum), 196, 1061
Oil-drop experiments, Millikan's, 40–41
Oleic acid, 1089, 1090
Omega-3 and omega-6 fatty acids, 1090
Onnes, H. Kamerlingh, 495
Open system, 168
Opsin, 367
Optical isomers (enantiomers), 1028–31, 1078–79, 1082
Optically active molecules, 1030

Orbital diagram, 235
Orbital overlap, 355–57
  band gap and, 486
Orbitals
  atomic. See Atomic orbitals
  hybrid, molecular geometry, and, 357–62
  molecular. See Molecular orbitals (MO)
  valence, 255
Ores, 982
  worldwide mine production of, 985
Organic chemistry, 66, 1050–1103
  chirality in, 1078–79
  compounds with carbonyl group
    aldehydes and ketones, 1073–74
    amine and amides, 1077–78
    carboxylic acids, 1074–77
    esters, 1074–77
  functional groups, 1053, 1070–78
    alcohols, 1072–73
    aldehydes and ketones, 1073–74
    amine and amides, 1077–78
    carboxylic acids and esters, 1074–77
    ethers, 1073
  general characteristics of organic molecules, 1052–54
  hydrocarbons, 66–67, 1054–55
    alkenes, 1062–66
    alkynes, 1064–66
    aromatic, 1054, 1068–70
    branched-chain, 1057
    as pollutants, 1070
    saturated (alkanes), 66–67, 1054, 1055–62
    straight-chain, 1057
Organic compounds, 59, 66–67
Organic molecules, structures of, 1052–53
Organic substances, stability of, 1053
Orientation factor in reaction rates, 592, 593
Orlon, 965
Ortho isomer, 385
ortho- prefix, 1069
Osmosis, 551–55
  in living systems, 554–55
  reverse, 784–85
Osmotic pressure, 553
  molar mass from, 556
Ostwald process, 955
-ous suffix, 59, 64
Outer-shell electrons, 238
Overall reaction order, 582
Overhydration, 147
Overlap, orbital, 355–57, 486
Oxalate ion, 1040–41
Oxalic acid, 667, 1074
  acid-dissociation constant of, 689
Oxidation, 136
  of alcohols, 1074
  of glucose, 830
  of iron, 136, 802–3
  of metals, 138–40
Oxidation numbers (oxidation states), 137–38, 273, 844–46
  acidity and, 701
  formal charge and, 318
  of transition metals, 1000–1
Oxidation reactions, 86
Oxidation-reduction equations, 846–50
Oxidation-reduction (redox) reactions, 135–42, 843–46
  activity series and, 140–42
  balancing, 846–50
    in basic solution, 849–50

    half-reaction method, 846–49
  in batteries, 870–71
  corrosion, 874–76
    of iron, 874–76
  defined, 135
  disproportionation, 889, 948
  electron movement in, 844, 845, 846, 848, 854
  free energy and, 862–64
  molecular and net ionic equations for, 140
  of nitrogen, 611
  oxidation numbers (oxidation states), 137–38
  oxidation of metals by acids and salts, 138–40
  spontaneity of, 851, 862, 863
  in voltaic cells, 851–54
    concentration cells, 867–70
    emf in, 855–61
Oxide(s), 493, 946–48
  acidic, 946
  amphoteric, 748–50
  basic, 947
  of boron, 969
  of carbon, 962–63
  of nitrogen, 954–55
  sulfur, 950–52
Oxide ion, 282
Oxidizing agent (oxidant), 845
  strengths of, 860–61
Oxyacetylene, 1064
Oxyacids, 700–2, 943
  of halogens, 943
  of nitrogen, 954–55
  sulfur, 950–52
Oxyanions, 61–62, 943
  common, 61
  of halogens, 943
  as oxidizing agents, 860
  of sulfur, 950–52
Oxy compounds of phosphorus, 958–60
Oxygen (O), 6, 452, 939, 944–48
  allotropes of, 282. See also Ozone
  in atmosphere, 394, 769
  in blood, 729
  dissociation energy of, 771
  electron configuration of, 255–56
  in green chemistry, 790
  Lewis symbol for, 299
  methane reacting with, 81
  molecular, 2, 7, 52, 164
    bonding in, 377
    combustion reactions with, 934
    critical temperature and pressure in, 452
    ionization of, 772
    Lewis structure for, 377
    paramagnetism of, 377–78
    photodissociation of, 771
    properties of, 769
  oxidation number of, 137
  oxides of, 946–48
  as oxidizing agent, 860, 945
  ozone, 945–46
  paramagnetism of, 378
  peroxides, 948
  2p orbital filling in, 268
  preparation of, 944–45
  properties of, 7, 944
  reactions of
    with alkali metals, 278
    dissolution in water, 786
    with hydrogen, 177
    with metals, 282

    with methane, 81
    with nickel, 273
    with nitric oxide, 814
    with sulfur tetrafluoride, 390
  in rocket fuel, 800, 801, 873
  solubility of, 541
  superoxides, 948
  in troposphere, 775
  uses of, 945
Oxygen anions, 283
Oxygen atom, 7
Oxygen-demanding wastes, 786
Oxygen group. See Group 6A elements (chalcogens)
Oxyhemoglobin, 778–79, 1022–23, 1047
Oxymyoglobin, 1023
Ozonation, 791, 946
Ozone, 19, 52, 282, 389, 590, 944, 945–46, 1073
  in atmosphere, 776, 946
  electron-distribution diagram for, 319
  molecular structure for, 319
  reaction with chlorine, 774–75
  resonance structures in, 319
  in smog, 780
  structure of, 945
  in upper atmosphere, 772–75
    depletion of, 774–75
Ozone hole, 19, 774
Ozone layer, 772-75
  depletion of, 774–75
    halogen-containing molecules and, 590
    photodecomposition, 773

Pacemaker cells, 868
Pacific yew tree, 340, 341–42
Paint pigments, 254
Paired electron, 235
Palladium(II), 1015–16
Paper chromatography, 12
Paraffin, 445
Para isomer, 385
Parallelepipeds, 460
Parallel spins, 237
Paramagnetism, 376–78, 1001, 1033
para- prefix, 1069
Partial charges, 318
Partial pressures, 410–13
  mole fractions and, 411–12
  pressure-volume changes and, 651
Particle accelerators, 902
Parts per billion (ppb), 542
Parts per million (ppm), 542, 769
Pascal (Pa), 396
Pascal, Blaise, 396
Paschen series, 252
Pattern recognition, 58
Pauau de la musica catalana, 1012
Pauli, Wolfgang, 233
Pauli exclusion principle, 233–34
Pauling, Linus, 308, 1084
p-bands, 994
PCl₅, 323
Pearlite, 996
"Pebble-bed" reactor design, 916–17
p elements
  electron affinities for, 271
  oxidation states, 273
Pentaborane, 969
Pentacene, 1070
Pentane, 1057
n-Pentane, 442
Pentanol, 536
Pentene, isomers of, 1064
Pentium 4 processor, 490

Pentyl acetate, 1075
Peptide bonds, 1082–83
Pepto-Bismol, 285
Percentage composition, 88–89
    empirical formula from, 95
Percent ionization, 683–84, 686–88
Percent yield, 105
Perchlorate, 943, 944
Perchloric acid, 701, 943
Perchloroethylene, 453
Periodic properties of elements, 254–95
    atomic radii, 259–64
        periodic trends in, 262–63
    effective nuclear charge (Z_{eff}),
        257–59, 260
    electron affinities, 270–71
    group trends for active metals,
        276–81
        alkali metals (group 1A), 276–79
        alkaline earth metals (group 2A),
            279–81
    group trends for nonmetals, 281–86
        halogens (group 7A), 283–84
        hydrogen, 281–82
        noble gases (group 8A), 284–86
        oxygen group (group 6A), 282–83
    ionic radii, 259–64
        biological chemistry and, 265
        periodic trends in, 263–64
    ionization energy, 264–70
        electron configurations of ions
            and, 268–70
        periodic trends in, 267–68
        variation in successive, 266–67
    metalloids, 276
    metals, 272–74
    nonmetals, 274–76
Periodic table, 6, 48–51
    development of, 256–57
    electron configurations and, 240–43
    groups in, 50
    ionic charges and, 56
    metallic elements or metals, 50
    metalloids, 50
    nonmetallic elements or nonmetals,
        50
    periods of, 50
    transuranium elements, 51
Periodic trends, 932–33
    in atomic radii, 262–63
    in first ionization energies, 267–68
    in ionic radii, 263–64
Periods, 50
Permanent magnet, 1001
Permanganate half-reaction, 848–49
Permanganate ion, 1040
Peroxidase, 948
Peroxide ion, 283, 387, 948
Peroxides, 278, 948
per- prefix, 62, 64
Perspective drawing, 53
PES (photoelectron spectroscopy), 295
PET (polyethylene terephthalate), 500,
    508
PET (positron emission tomography),
    910
Petroleum, 196, 1061
pH, 675–79. See also Acid–base
    equilibria; Aqueous equilibria
    of buffer, 724–27
    calculating acid-dissociation
        constant from, 682–83
    calculating from acid-dissociation
        constant, 684–88
    of common substances, 676
    measuring, 678–79
    salt effects on, 695–99

solubility and, 742–45
titration curve, 730
Phase changes in liquids, 449–53
    critical temperature and pressure
        and, 452
    energy changes accompanying,
        449–50
    entropy change and, 807
    heating curves, 450–52
Phase diagrams, 456–58, 549
Phases
    in atomic and molecular orbitals,
        373–75
    condensed, 438
ortho-Phenanthroline, 1041
Phenol, 1072
    properties of, 466
Phenol acid, 681, 682
Phenolphthalein, 151, 678, 679, 733,
    734, 736
Phenylacetic acid, 712
Phenylalanine, 1050, 1051, 1081
Phenylamine, 1077
Phenylketonuria, 1050, 1051
Phenylmethanamide, 1078
Phenyl methanoic acid (benzoic acid),
    183, 682, 702, 713
pH meter, 678, 891
Phosgene, 330, 348–49, 635
Phosphate ion, Lewis structures for,
    324
Phosphates
    insoluble, 754
    as sequestering agents, 1020
Phosphine, 452
Phospholipids, 1090
Phosphoric acid, 958, 959
    acid-dissociation constant of, 689
Phosphorous acid, 706–7, 958
Phosphors, 909
Phosphorus (P), 992
    allotropes of, 957
    halides of, 957–58
    Lewis symbol for, 299
    occurrence, isolation, and properties
        of, 957
    oxy compounds of, 958–60
    properties of, 956, 957
    red, 957
    silicon doping with, 488
    white, 957
Phosphorus halides, 957–58
Phosphorus(III) oxide, 958
Phosphorus pentachloride, 661
Phosphorus trichloride, 957
Phosphorus trihalides, 390
Phosphorus(V) oxide, 958
Photochemical smog, 779–80
Photoconductivity, 491
Photocopiers, 950
Photodissociation, 770–71, 774
Photoelectric effect, 215, 216–18
Photoelectron spectroscopy (PES), 295
Photographic plates/film, 908–9
Photoionization, 772
Photolithography, 490
Photoluminescence, 514
Photons, 216–18
Photoreceptors, 367
Photosynthesis, 1023
Photovoltaics, 198
pH range, 727
pH titration curve, 730, 736
Physical changes, 10–11
Physical properties, 9, 37
Physical requirements, of biomaterials,
    507

Pig iron, 985, 986
pi (π) bonds, 362–64
    in alkenes, 1063
    in aromatic hydrocarbons, 1068–69
    in chemistry of vision, 367
    delocalized, 365–66
    in double bonds, 362–63
    in ozone, 945
    periodic trends and, 932
    strength of, 362
    in triple bonds, 364
Pipelines, gas, 409
Pipets, 16, 17
pi (π) molecular orbitals, 372
    in aromatic hydrocarbons, 1068
Planck, Max, 215–16
Planck's constant, 215, 219, 224
Plastic car, 506
Plasticizers, 504
Plastics, 499–505
    polycarbonate, 500
    recycling, 501
    thermoset, 499, 508
Platinum(II), 1018
Plato, 38
Plum-pudding model of the atom, 42
Plutonium (Pu), 51, 240, 903
Plutonium-239, 914, 917
p-n diode, 492
pOH, 678
Poker, as analogy for microstates, 812
Polar covalent bond, 308
Polarity
    bonding, 307–14
    molecular (bond), 353–55
    proton transfer reactions and,
        669–70
    solubility and, 536
Polarizability, 441
Polar liquids, solubility of, 536
Polar molecules, 310, 354–55, 442
Pollutants
    atmospheric, 776
    carbon monoxide, 778
    CFCs, 590, 774–75, 782
    hydrocarbons, 780, 1070
    nitrogen oxides, 779–80
    polycyclic aromatic hydrocarbons
        as, 1070
    sulfur dioxide, 776–78
    in urban atmosphere, 776
Pollution
    smog, 779–80
    thermal, 541
    water, 960–61
Polonium (Po), 282, 948
Polonium-218, 921
Polyacetylene, 516
Polyacrylonitrile, 522
Polyatomic ions, 55, 61, 316
Polycarbonate, 500
Polychloroprene, 522
Polycyclic aromatic hydrocarbons,
    1070
Polydentate ligands (chelating agents),
    1019–22
Polyethylene, 67, 499–500, 841
    high-density, 503
    low-density, 503
    properties of, 502
Polyethylene terephthalate (PET), 500,
    508
cis-Polyisoprene, 504
Polymer(s), 499–505, 841
    biopolymers, 1080
    co-, 501
    commercially important, 500

conducting, 516
cross-linking of, 504
elastomeric, 499
engineering, 506
making, 499–502
plastic cars, 506
structure and physical properties of,
    502–5
types of, 499
Polymeric biomaterials, 507–8
Polymerization, 499
    addition, 499–500
    condensation, 501–2
Polynucleotide, 1091
Polypeptides, 1082–84
Polyphenylene ether (PPE), 506
Polyphosphate, 514
Polypropylene, 500
Polyprotic acids, 688–90
    acid-dissociation constants of, 689
    titrations of, 737
Polysaccharides, 1088–89
Polystyrene, 500
Polytetrafluoroethylene (Teflon), 790,
    911
Polyunsaturated fatty acids, 1090
Polyurethane, 500
Polyvinyl acetate, 1076
Poly(vinyl alcohol), 352–53
Polyvinyl chloride (PVC), 500, 504,
    665, 941
p orbitals, 231–32
    energy-level diagrams/electron
        configurations, 375–76
    periodic trends and, 932–33
    phases in, 373–74
Porphine molecule, 1021
Porphyrins, 1021
Positron, 897
Positron emission tomography (PET),
    910
Potassium (K), 238, 277
    flame test, 278
    reaction with water, 277
Potassium-40, 919, 922–23, 927
Potassium chlorate, 412–13, 943, 945
Potassium chloride, 298
Potassium dichromate, 1002, 1073
Potassium ferricyanide, 59
Potassium ferrocyanide, 59
Potassium hydrogen phthalate (KHP),
    763
Potassium iodide, 124
Potassium nitrate, 124
Potassium superoxide, 948
Potential difference, 855
Potential energy, 166–67
    electrostatic, 167
    free energy and, 820
ppb (parts per billion), 542
PPE (polyphenylene ether), 506
ppm (parts per million), 542, 769
Praseodymium, 239
Precipitate, 124
Precipitation, 124–28
    exchange (metathesis) reactions,
        126–27
    of ionic compounds, 737
        solubility guidelines for, 124–26
    ionic equations, 127–28
    of ions, 750–53
    selective, 751–53
Precision, 20–21
Prefixes
    binary compound, 65
    Greek, 65, 1026
    metric system, 14

Pressure, 395–98
  atmospheric, 176, 395–98
  blood, 398
  critical, 452
  defined, 395
  equilibria and, 649–51
  equilibrium constants in terms of, 633–34
  gas, 395–98
  intermolecular forces and, 439
  osmotic, 553
  partial, 410–13
    mole fractions and, 411–12
  real *vs.* ideal gas behavior and, 421
  solubility and, 539–41
  spontaneous processes and, 803
  vapor, 453–56
Pressure-volume (*P-V*) work, 174–75, 176
Pressure-volume relationship, 399–400
Priestley, Joseph, 944
Primary cells (batteries), 871
Primary structure, of proteins, 1084
Primitive cubic unit cell, 460, 461
Principal quantum number (*n*), 220, 226
Probability, entropy and, 812
Probability density, 225, 230–31, 373
Problem solving, 89
  procedure, 149
Products, 80
  calculating amounts of, 100–2
    from limiting reactant, 104–5
  change in concentration of, 648–49
  enthalpy change and state of, 178–79
  states of, 83
Proline, 1081
Propagation, of waves, 212
Propane, 66, 67, 394, 433, 452, 1055
  combustion of, 86, 191, 823–24
  critical temperature and pressure of, 452
  molar entropies of, 817
  in natural gas, 196
  properties of, 441
  rotation about carbon–carbon single bonds of, 1057
  states of, 439
1,2,3-Propanetriol (glycerol), 1072, 1089, 1090
Propanol, 536, 836
1-Propanol, 66–67
2-Propanol (isopropyl alcohol), 67, 97–98, 1072
Propanone (acetone), 478, 536, 570, 571, 1071, 1073, 1074
Propene (propylene), 386, 1063
Property, 2
Propionic acid, 713
Propylene, 386, 1063
Propyl group, 1059
Protactinium-234, 900
Protein(s), 1080–85
  amino acids, 1080–82
    amphiprotic behavior of, 703
    side chain of, 1081–82
  defined, 1080
  DNA structure and synthesis of, 1093
  as enzymes, 609
  fibrous, 1085
  globular, 1022, 1085
  metabolism of, 193, 194
  polypeptides and, 1082–84
  structure of, 1084–85
Protein chain, structure of, 507–8
Protein sequence, 1101
Protium, 935

Proton(s), 42, 43, 894–95, 897
  mass of, 43
  neutron-to-proton ratio, 898–90
Proton-transfer reactions, 669–70
Proust, Joseph Louis, 8
"p" scales, 678
p-type semiconductor, 488, 491
Pure substances, 5–6
  crystalline, 816
Purines, 1101
Putrescine, 694
PVC (polyvinyl chloride), 500, 504, 665, 941
Pyrene, 1070
Pyrex, 968
Pyridine, 691, 1045
Pyrimidines, 1101
Pyrite (fool's gold), 459, 950
Pyrolytic carbon, 508
Pyrometallurgy, 984–87
  formation of steel, 986
  of iron, 985–86
Pyrosulfuric acid, 951
Pyruvic acid, 1093–94

Qualitative analysis for metallic elements, 753–55
Quantitative analysis, 753
Quantitative information, from balanced equations, 98–102
Quantity-volume relationship, 400–1
Quantized energy, 215–18
Quantum, 215
Quantum dots, 115, 514
Quantum mechanics, 211, 224–28
Quantum number
  angular momentum, 226
  magnetic (*m*), 226
  orbitals and, 226–28
  principal (*n*), 220, 226
  spin magnetic (*m*$_s$), 233
Quantum theory, 211, 218
Quantum wells, 514
Quantum wires, 514
Quartz, 459, 466, 968
Quartz glass (silica glass), 459, 968
Quaternary structure, of protein, 1100
Questions in tests, types of, 106
Quicklime (calcium oxide), 85, 777, 964
Quinine, 694, 1102

Racemic mixtures, 1031, 1079
rad, 920
Radial probability density, 229
Radial probability function, 229–31, 259
Radiant energy, 212
Radiation, 185
  alpha (α), 41, 895, 896, 897, 919–20
  background, 920
  beta (β), 41, 896–97, 919
  biological effects of, 919–23
    dosage and, 920
    radon, 921
    therapeutic, 894, 910, 922
  blackbody, 215
  electromagnetic, 212–13
  gamma (γ), 41, 213, 896, 897, 919, 922
  infrared, 212, 213
  ionizing, 919
  monochromatic, 218
  nonionizing, 919
Radiation therapy, 894, 910, 922
Radicals, free, 919
Radioactive decay, 895, 896–98
  rates of, 903–8
  types of, 896–98

Radioactive series (nuclear disintegration series), 900
Radioactivity, 41, 894–98
  detection of, 908–10
Radiocarbon dating, 905
Radioisotopes, 895, 921
Radioisotope scanning, 894
Radiometric dating, 905
Radionuclides, 895
Radiotracers, 910
  medical applications of, 910
Radio waves, 772
Radium-226, 896, 926
Radius. *See* Atomic radius/radii; Ionic radii
Radon, 239, 284, 921, 938
Radon-222, 921
Radon-testing kits, 921
Randomness. *See* Entropy(ies)
Raoult's law, 547–48
Rare earth elements, 239
Rate constants, 581
  temperature and, 592
  units of, 583
Rate-determining (rate-limiting) step, 601
Rate laws, 581–85
  differential, 586
  for elementary steps, 599–600
  exponents in, 581–82
  H^+ concentration and, 677
  initial rates to determine, 584–85
  integrated, 586, 588
  for multistep mechanisms, 600–1
  units of rate constant, 583
RBE (relative biological effectiveness), 920
RDX (cyclotrimethylenetrinitramine), 338
Reactants (reagents), 80
  calculating amounts of, 100–2
  change in concentration of, 648–49
  enthalpy change and state of, 178–79
  environmentally friendly, 789–90
  excess, 106
  limiting (limiting reagents), 102–7
    theoretical yields, 105–7
  physical state of, 574
  states of, 83
Reaction(s), 10, 79, 573
  acid–base, 128–35. *See also*
    Acid–base equilibria
    electrolytes, 131–32
    with gas formation, 134–35
    gas-phase, 669–70
    neutralization reactions and salts, 132–34
  addition
    of alkenes and alkynes, 1065–66
    mechanism of, 1066–68
    of alkanes, 1061
  analyzing, 143
  aqueous. *See* Aqueous equilibria;
    Aqueous reactions; Aqueous solution(s)
  bimolecular, 598, 600
  carbonylation, 1075
  chain, 914
  chemiluminescent, 591
  combustion, 86–87
    balanced equations for, 86–87
    with oxygen, 934
  condensation, 959, 1073
    with alcohol, 1075
  decarbonylation, 838

  displacement, 138–39
  disproportionate, 948
  elementary, 597–98, 599–600
  endothermic, 172, 175
  enthalpies of, 177–79, 326–28
  entropy changes in, 813–14, 817–19
  exothermic, 172, 175
  first-order, 586–87, 590–91
  Friedel-Crafts, 1069
  gas volumes in, 408–10
  half-life of, 589–91
  heat of, 177
  involving nonmetals, 933–34
  mechanisms of. *See* Reaction mechanisms
  nonspontaneous, 830
  nuclear. *See* Nuclear chemistry
  periodic trends and, 933–34
  predicting direction of, 644–45
  proton-transfer, 669–70
  rates of. *See* Reaction rates
  redox. *See* Oxidation-reduction (redox) reactions
  second-order, 588–89, 591
  solution formation and, 533–34
  spontaneity of. *See* Spontaneous processes
  substitution, 1069
  termolecular, 598
  thermite, 172, 206
  thermonuclear, 918
  unimolecular, 598
  water and, 119
Reaction mechanisms, 597–604
  defined, 597
  elementary steps, 597–98
    rate laws for, 599–600
  with fast initial step, 602–4
  multistep, 598–99
    rate-determining step for, 600–1
  with slow initial step, 601–2
Reaction orders, 581–82
  overall, 582
Reaction quotient (*Q*), 644, 750
Reaction rates, 574–625
  catalysis and, 575, 605–12
    enzymes, 575, 608–12
    heterogeneous, 606–8
    homogeneous, 605–6
  concentration and, 574, 575
    change with time, 585–91
    rate laws, 580–85
  defined, 575
  factors affecting, 574–75
  instantaneous (initial rate), 577–78
  spectroscopic methods to measure, 580
  stoichiometry and, 578–80
  temperature and, 574, 591–97
    activation energy, 592–94, 595–97
    Arrhenius equation for, 594–95
    collision model of, 592
    orientation factor in, 592, 593
  time and, 577, 585–91
  for weak and strong acids, 686
Reactivity, patterns of, 84–87
  combinations and decomposition reactions, 84–86
  combustion reactions, 86–87
Reactors, nuclear, 915–17
Reagents. *See* Reactants (reagents)
Real gases, 420–24
  van der Waals equation, 422–24
Recycling symbols, 501
Red blood cells, 729
  osmosis and, 553
  sickled and normal, 559

Red ochre, 1040
Redox reactions, 843–44. *See* Oxidation-reduction (redox) reactions
Red phosphorus, 957
Reducing agent (reductant), 845
 strengths of, 861
Reduction. *See* Oxidation-reduction (redox) reactions
Refining, 984–85, 1061
Refrigerant, carbon dioxide as, 963
Reinitzer, Frederick, 510
Relative biological effectiveness (RBE), 920
rem (roentgen equivalent for man), 920
Remsen, Ira, 10
Renewable energy, 197
Representative (main-group) elements, 240
Resonance structures, 319–21, 365–66
 in benzene, 321
 in nitrate ion, 320
 in ozone, 319
Retinal, 367
Reverse osmosis, 784–85
Reversible process, 804–6
R groups, 703, 1081
Rhodopsin, 367
Rhombic sulfur, 949
Ribonuclease A, 1103
Ribonucleic acid (RNA), 1090–91
Ribose, 1091
Ring structure, of glucose, 1086
rms speed, 415, 417
RNA (ribonucleic acid), 1090–91
Roasting, 984
Rocket fuel, 800, 801, 873, 943, 944, 953, 979
Rock music, 980, 981
Rods, 367
Rømer, Ole, 214
Roosevelt, Franklin D., 915
Root-mean-square (rms) speed, 415, 417
Rotational motion, 810
Rounding off numbers, 23
Rowland, F. Sherwood, 774
Royal Institution of Great Britain's Faraday Museum, 515
Rubber, 459
 vulcanization of, 504–5, 950
Rubidium, 239, 277
Rubidium-87, 919
Rusting. *See* Corrosion
Rutherford, Daniel, 952
Rutherford, Ernest, 41–42, 218, 219, 257, 901
Rutile, 475–76
Rydberg constant, 219, 220
Rydberg equation, 219, 221

Sacrificial anode, 875
SAE (Society of Automotive Engineers), 447
St. Jude valve, 508
Salicylic acid, 1076
Salinity, of seawater, 783
Salt(s). *See also* Sodium chloride
 acid, 694
 acid–base reactions and, 132–34
 acid hydrolysis of, 705
 chlorate, 943
 defined, 132
 dissolving of, 813
 electrolysis of molten, 877
 hypochlorite, 943
 iodized, 942

naming, 1025
 normal, 696
 oxidation of metals by, 138–40
 solubility-pH relationship in, 746, 748
 of transition metals, 1000
Salt bridge, 852, 853, 868
Saltpeter, 952
Salt solutions
 acid–base properties of, 695–99
 anion reaction with water, 696
 cation reaction with water, 696–97
 combined cation-anion effect, 697–99
 conductivity of, 121
Saponification, 1076
Saturated hydrocarbons. *See* Alkanes
Saturated solution, 534–35, 738
Saturn, 281
Scale, 788
Scandium, 238
Scanning tunneling microscopy (STM), 36
Schrieffer, Robert, 497
Schrödinger, Erwin, 224
Schrödinger's wave equation, 224–25
Scientific law, 13
Scientific method, 13
Scintillation counter, 909
Screen, 258
Screening constant (*S*), 258
Seaborg, Glenn, 51
Seaborgium, 51
Seawater, 783–84
 desalination of, 784–85
Secondary cells (batteries), 871
Secondary structure, of proteins, 1084
Second ionization energy, 265
Second law of thermodynamics, 808–9
Second order overall, 582
Second-order reactions, 588–89, 591
Second-row diatomic molecules, 371–81
Seesaw geometry, 350
Selective precipitation of ions, 751–53
s elements, electron affinities for, 271
Selenium (Se), 948–50
Semiconductor GaAs (gallium arsenide), 39
Semiconductors, 276, 482, 484–93
 band structure of, 484
 compound, 484–85, 486
 doping of, 488
 elemental, 484, 486
 light-emitting diodes, 492–93
 on nanoscale, 513–14
 n-type, 488, 491
 p-type, 488, 491
 silicon chip, 489
 silicon in, 966
 solar energy conversion, 489–91
 transistor, 490
Semipermeable membranes, 551–52, 553, 560, 868
Separation, of ions, 750–53
Serine, 1082
Serpentine asbestos, 968
Seven Up, 280
Shape, molecular. *See* Molecular geometry
Shape-memory alloys, 997–98
Shell model of nucleus, 901
SHE (standard hydrogen electrode), 856–57
Shield, 258
Shiver, 185
Shockley, William, 497

Shock sensitive, 329
Sickle-cell anemia, 559
Side chain, amino acid, 1081–82
Siderite, 964
Siderophore, 1024
Sigma (σ) bonds, 362, 364
Sigma (σ) molecular orbitals, 369
Sigma (σ) symbol, 191
Significant figures, 21–23
 in calculations, 23–24
Silica
 amorphous, 495
 reaction with hydrofluoric acid, 942
Silica glass (quartz glass), 459, 968
Silicate minerals, 982
Silicates, 326, 493, 966–68
Silicon (Si), 6, 276, 466, 480, 491, 525, 874, 965, 992
 doping of, 488
 electronic properties of, 484
 Lewis symbol for, 299
 occurrence and preparation of, 966
 semiconductor, 276, 489
 structure of, 485
Silicon carbide, 466, 494, 965
Silicon chip, 489
Silicon dioxide, 931, 932, 966
Silicones, 969
Silicon nitride, 494
Silicon tetrachloride, 72
Silver (Ag), 50, 303, 992
 electronic properties of, 484
 on nanoscale, 515
 oxidation of, 140
 reaction with copper, 140–41
 sterling, 527–28
Silver chloride, 664
Silver chromate, 739
Simple cubic (primitive cubic), 460
Single bonds, 307, 932
 bond enthalpies of, 326
 length of, 329–330
 rotations about, 1052
Single-walled form, 516
Sinkholes, 744
Sintered, 494
SiO_2, 484, 968
SI units, 13–17
 base units, 14
 density, 17
 derived, 16
 length and mass, 14–15
 for speed, 16
 temperature, 15–16
 volume, 16
Slag, 984, 986
Slater's rules, 260
Sleight, A. W., 497
Smalley, Richard, 468
Smectic liquid crystalline phases, 511
Smelting, 984
Smog, 779
Soap, 132, 668, 1054, 1076
Soap scum, 788
Society of Automotive Engineers (SAE), 447
Soda-lime glass, 968
Sodium (Na), 277, 460
 condensed electron configuration of, 258
 effective nuclear charge of, 258
 electrometallurgy of, 988–89
 electron configuration of, 237
 flame test, 278
 ions of, 54
 Lewis symbol for, 299
 melting point of, 469

reactions of
 with chlorine, 299–305
 with water, 277
Sodium acetate, 535, 720
Sodium azide, 85, 409
Sodium bicarbonate, 134, 135
Sodium borohydride, 969
Sodium bromide, 942
Sodium cation, 54
Sodium chloride, 56–57, 861
 conductivity of solution of, 120
 crystal structure of, 461–63
 dissolution in water, 529
 electrolysis of
  aqueous, 876–77
  molten, 876
 formation of, 299, 300, 304
 states of, 439
Sodium chlorite, 943
Sodium chromate, 1002
Sodium fluoride, 747
Sodium formate, 763
Sodium hydroxide, 132, 668
Sodium hypochlorite, 692–93, 942
Sodium iodide, 942
Sodium lactate, 725
Sodium lithium hydroxide, 101–2
Sodium monofluorophosphate, 747
Sodium perchlorate, 929
Sodium propionate, 1075
Sodium stearate, 558–59
Sodium sulfate, dissociation of, 122
Sodium tripolyphosphate, 959, 1020
Sodium vapor lights, 219, 278
Softening of water, 788
Soft materials, 482
Sol, 494
Solar cells (photovoltaic devices), 198, 491
Solar energy, 197–98, 380
 conversion of, 489–91
Solar panels, 491
Solar spectrum, 770
Sol-gel process, 494
Solid(s), 5
 amorphous, 459–60
 bonding in, 464–69
 bulk, 483
 concentration of, 639
 covalent-network, 464, 466–67
 crystalline, 438, 439, 458–60
 intermolecular attractive forces in, 438
 ionic, 464, 467
 metallic, 464, 467–69
 molecular, 464–66
 molecular comparison of liquids and, 438
 properties of, 438
 structures of, 458–64
  close packing of spheres, 463–64
  sodium chloride, 461–63
  unit cells, 460–61
Solid solutes, 541
Solid solutions, 527–28
Solubility, 124, 534–35, 741–50
 amphoterism and, 748–50
 common-ion effect and, 741–42
 complex ion formation and, 745–48
 molar, 738
 of organic substances, 1054
 pH and, 742–45
 pressure effects on, 539–41
 solubility-product constant and, 738–40
 solubility-product constant *vs.*, 738–40

Solubility, (cont.)
    solute-solvent interactions and, 536–39
    temperature effects on, 541
Solubility equilibria, 737–41
    solubility-product constant, 738
        limitations of, 741
        reaction quotient and, 750
        solubility vs., 738–40
Solubility guidelines, for ionic compounds, 124–26
Solubility-product constant (K_{sp}, solubility product), 738
    limitations of, 741
Solutes, 120, 528
    molarity to calculate grams of, 146
    solid, 541
    titration to determine quantity of, 152
Solute-solvent interactions, 536–39
Solution(s), 8, 120, 526–71
    acidic, 697–98
    aqueous. See Aqueous solution(s)
    basic, 697–98
        balancing equations for reactions in, 849–50
    buffered, 723–29
        blood as, 723, 729
        buffer capacity and pH, 726–27
        calculating pH of buffer, 724–26
        composition and action of, 723–24
        strong acids or bases in, 727–28
    colligative properties, 546–56
        boiling-point elevation, 549
        of electrolyte solutions, 554
        freezing-point depression, 550–51
        molar mass determination through, 555–56
        osmosis, 551–55
        vapor pressure reduction, 546–48
    colloids, 556–61
        hydrophilic and hydrophobic, 557–60
        removal of colloidal particles, 560–61
        types of, 557
    concentration of, 142–49, 542–46
        conversion of units of, 544–46
        dilution, 146–49
        of electrolyte, 145
        interconverting molarity, moles, and volume, 145–46
        in mass percentage, 542
        molality, 543–44
        molarity, 144–45, 543–44
        in mole fraction, 543–44
        in parts per billion (ppb), 542
        in parts per million (ppm), 542
    defined, 120
    examples of, 528
    formation of, 528–34
        energy changes and, 529–31
        intermolecular forces and, 528–29
        reactions and, 533–34
        spontaneity, entropy, and, 531–32
    hypertonic, 553
    hypotonic, 553
    ideal, 548
    isotonic, 553
    neutral, 674, 697–98
    preparing by dilution, 148–49
    saturated, 534–35, 738
    solid, 527–28
    standard, 150
    stock, 146
    supersaturated, 535
    unsaturated, 535

Solution alloys, 995
Solution stoichiometry, 149–54
    titrations, 150–54
        acid–base, 730–37
Solvation, 122, 529
Solvent(s), 120, 528
    environmentally friendly, 789–90
    ethers as, 1073
    ketones as, 1074
    supercritical fluid, 789–90
    water as, 719
Sorbic acid, 714
s orbitals, 228–30
    energy-level diagrams/electron configurations, 375–76
    phases in, 373
Source, semiconductor, 490
Space-filling models, 53, 54, 342
Space shuttle, 800, 801, 873, 943
Spearmint, 1074
Specific heat, 179–81
Spectator ions, 127, 128
Spectrochemical series, 1036
Spectroscopy, measuring reaction rates with, 580
Spectrum, 218
    continuous, 219
Speed
    of light, 212, 214, 220
    root-mean-square (rms) vs. average, 415
Spheres, close packing of, 463–64
sp hybrid orbitals, 357–58, 361
sp^2 hybrid orbitals, 358–60, 361
sp^3 hybrid orbitals, 358–60, 361
Spinel, 478
Spin magnetic quantum number (m_s), 233
Spin-pairing energy, 1037
Spontaneous processes, 180, 802–6
    criterion for, 804
    exothermic processes and, 530, 532
    free energy and, 820
    oxidation-reduction reactions, 851, 862, 863
    pressure and, 803
    reversible and irreversible, 804–6
    solution formation and, 531–32
    temperature and, 803
Square-planar complexes, 1038–41
Square planar geometry, 344, 350, 351, 354, 1018, 1021
Square pyramidal geometry, 350, 351
Stability(ies)
    belt of, 898–99
    nuclear
        even vs. odd number of nucleons, 901
        magic numbers and, 901
        neutron-to-proton ratio, 898–900
        radioactive series (nuclear disintegration series), 900
    of organic substances, 1053
Stained glass, 515, 1012, 1013
Stainless steel, 874, 996
Standard atmospheric pressure, 396
Standard cell potential, 855–60
Standard deviation, 21
Standard emf, 855–60
Standard enthalpy change, 188
Standard enthalpy of formation, 189
Standard free-energy change, 822–23
Standard free energy of formation, 822–24
Standard hydrogen electrode (SHE), 856–57
Standard molar entropies, 817

Standard reduction (half-cell) potentials, 856–60
Standard solution, 150
Standard state, 188
Standard temperature and pressure (STP), 403
Stannous fluoride, 747
Starch, 193, 1088–89
State(s)
    changes of, 10
    of gas, 398
    of matter, 4–5
    of reactants and products, 83
State function(s), 172–73, 804
    enthalpy as, 175
Static equilibrium, 627
Statistical thermodynamics, 810
Steam turbines, 840
Stearate, 1054
Stearic acid, 1089
Steel, 995–96
    formation of, 986
    iron for, 986
    mild, 995
    stainless, 874, 996
Stereoisomers, 1027–31
Sterling silver, 527–28
Stern, Otto, 234
Stern-Gerlach experiment, 234
STM (scanning tunneling microscopy), 36
Stock solutions, 146
Stoichiometrically equivalent quantities, 99
Stoichiometry, 79–117
    Avogadro's number and the mole, 89–95
        interconverting masses and moles, 93
        interconverting masses and numbers of particles, 94–95
        molar mass, 91–92
    calculating equilibrium concentrations and, 646–47
    calculation, 727–28
    chemical equations, 80–83
        balancing, 80–83
        states of reactants and products, 83
    defined, 79
    empirical formulas from analyses, 95–98
        combustion analysis, 97–98
        molecular formula from, 96–97
    formula weights, 87–89
        percentage composition from formulas, 88–89
    of half-reaction, 878
    limiting reactions (limiting reagent), 102–7
        theoretical yields, 105–7
    patterns of chemical reactivity, 84–87
        combination and decomposition reactions, 84–86
        combustion reactions, 86–87
    problem-solving procedure for, 149
    quantitative information from balanced equations, 98–102
    reaction rates and, 578–80
    solution, 149–54
        titrations, 150–54
STP (standard temperature and pressure), 403
Straight-chain alkanes, 1056
Straight-chain hydrocarbons, 1057
Straight-run gasoline, 1062

Stratosphere, 768, 769
Strong acids, 130–31, 672, 679–80
    added to buffers, 727–28
Strong acid-strong base titrations, 730–33
Strong bases, 130–31, 680–81
    added to buffers, 727–28
Strong electrolytes, 122–23
    identifying, 131–32
Strong-field ligands, 1036
Strong nuclear forces, 45
Strontium (Sr), 279, 281
Strontium-90, 904, 907–8, 926
Strontium oxide, 838
Structural formulas, 53
    condensed, 1055–56, 1059, 1061
Structural isomers, 1027
    of alkanes, 1057–58
    of butene, 1063
Structure
    atomic. See Atomic structure; Electronic structure
Structure, chemical, 2
    acid–base equilibria and, 699–703
        binary acids, 699–700
        carboxylic acids, 702–3
        factors affecting acid strength, 699
        oxyacids, 700–2
Styrene, 789
Subatomic particles, 39
Sublimation, 449, 458
    heat of, 450
Submicroscopic realm, 3
Subscript, coefficient vs., 81
Subshell, 227–28
Substance, 5–6
Substitutional alloys, 995, 996
Substitution reactions, 1069
Substrates, 609
Subtraction, significant figures and, 23
Sucrose, 78, 298, 1087–88
    conductivity of solution of, 120, 122
    dehydration of, 951
    reaction with sulfuric acid, 951
    van't Hoff factor for, 554
Sugar, 1087
    invert, 1088
Sulfate ion, 952
Sulfates, 952
Sulfide ions, 752
Sulfides, 277, 950
    acid-insoluble, 754
    base-insoluble, 754
Sulfites, 950–51
Sulfur (S), 50, 282, 283, 948–50, 992
    electron configuration of, 256
    elemental, 283, 381
    Lewis symbol for, 298, 299
    occurrences and preparation of, 949
    oxides, oxyacids, and oxyanions of, 950–52
    properties and uses of, 949–50
Sulfur compounds, in troposphere, 776–78
Sulfur dioxide, 283, 394, 776, 838, 946, 950
    in atmosphere, 776–78
    dissolved in water, 950
    reaction with calcium carbonate, 777
Sulfuric acid, 688–89, 700, 776, 942, 950
    acid-dissociation constant of, 689
    commercial, 951–52
    formula weight of, 87
    reaction with sucrose, 951
Sulfurous acid, 688, 689
Sulfur oxides, 949
Sulfur tetrafluoride, 390

Sulfur trioxide, 951
Sun, as energy source, 892, 893
Superconducting ceramics, 497–98
Superconducting transition
    temperature, 495
Superconductivity, 495-98
    high-temperature, 497
Superconductors, 495–98
Supercooling, 452
Supercritical fluid, 453, 456, 457
Supercritical fluid extraction, 453
Supercritical fluid solvents, 789–90
Supercritical mass, 914–15
Supercritical water, 790
Superoxide ion, 278, 283, 387
Superoxides, 948
Supersaturated solutions, 535
Surface tension, 448
Surroundings, 168
    entropy changes in, 818–19
Sutures, dissolving, 510
Symbols, chemical, 6
Syngas (synthetic gas), 196–97
Synthetic diamond, 961
Syringes, 16, 17
System, 168
Système International d'Unités.
    See SI units
Systolic pressure, 398
Szilard, Leo, 915

Table salt. See Sodium chloride
Taconite, 985
Tailpipe emission standards, 779
Talc, 967
Tartaric acid, 713, 1079
    acid-dissociation constant of, 689
Tausonite, 475
Taxol (tamoxifen), 340, 342
Technetium, 256
Technetium-99, 910
Teflon (polytetrafluoroethylene), 790,
    941
Tellurium (Te), 282, 948–50
Temperature
    absolute, 595
    of atmosphere, 768
    body, 185
    color as function of, 215
    critical, 452
    Curie, 1002
    of Earth's surface, 780–81
    entropy and, 816
    equilibria and, 651–54
    fusion and, 918
    Gibbs free energy and, 824–26
    kinetic energy and, 594
    molecular speeds and, 414–15
    Néel, 1002
    reaction rates and, 574, 591–97
        activation energy, 592–94, 595–97
        Arrhenius equation for, 594–95
        collision model of, 592
        orientation factor in, 592, 593
    real vs. ideal gas behavior and, 421
    regulation in humans, 185
    SI units of, 15–16
    solubility and, 541
    spontaneity of reaction and, 825–26
    spontaneous processes and, 803
    standard enthalpy change,
        equilibrium constant, and,
        824–26
    superconducting transition, 495
    vapor pressure and, 454–55
    volume and, 400
Tentative explanation (hypothesis), 13

Terephthalic acid, 790
Termolecular reaction, 598
Tertiary structure, of proteins, 1085
Test-taking strategies, 106–7
Tetraboric acid, 970
Tetracene, 391, 1070
2,2,3,3-Tetrachlorobutane, 1066
Tetrachloroethylene, 790
Tetraethyl lead, 1062
Tetrahedral complexes, 1038–41
Tetrahedral geometry, 342, 345, 347,
    354, 361, 1018, 1052
Tetrahydrofuran (THF), 1073
Thallium (Tl), 625
Thallium-201, 894, 910
Thallium "stress test," 894
Theoretical yields, 105–7
Theory, 13
Thermal energy, 167
Thermal pollution, 541
Thermite reaction, 172, 206, 927
Thermochemical equations, 178
Thermochemistry, 164–209
    calorimetry, 179–84
        bomb (constant-volume), 183–84
        constant-pressure, 182–83
        heat capacity and specific heat,
            179–81
    defined, 166
    energy and, 166–69
        fossil fuels, 196–97
        kinetic and potential, 166–67
        nuclear, 197
        solar, 197–98
        system and surroundings, 168
        transferring, 168–69
        units of, 167–68
    enthalpy(ies), 174–77
        defined, 175
        of formation, 188–93
        of reaction, 177–79, 190–93
        spontaneous processes and, 180
    first law of thermodynamics, 170–73
        endothermic and exothermic
            processes, 172
        heat and work related to internal
            energy changes, 171–72
        internal energy, 170–71
        state functions, 172–73
    of foods, 193–95
    of fuels, 195–97
    Hess's law, 184–88
Thermodynamic equilibrium constant,
    635
Thermodynamics, 166, 800–41
    defined, 802
    entropy, 806–17
        absolute, 817, 818
        of expansion, 808
        heat transfer and temperature
            related to, 808–9
        life and, 815
        microstates and, 811
        molecular interpretation of,
            809–17
        probability and, 812
        in reactions, 817–19
        temperature and, 816
    first law of, 170–73
        endothermic and exothermic
            processes, 172
        heat and work related to internal
            energy changes, 171–72
        internal energy, 170–71
        state functions, 172–73
    Gibbs free energy, 819–31
        equilibrium constant and, 826–31

spontaneity and, 820
standard free-energy changes,
    822–24
temperature and, 824–26
    second law of, 808–9
    spontaneous processes, 802–6
        pressure and, 803
        reversible and irreversible, 804–6
        temperature and, 803
    statistical, 810
    third law of, 816–17
Thermodynamic sign convention, 270
Thermonuclear reactions, 918
Thermoplastics, 499, 508
Thermosetting plastics, 499, 508
Thermosphere, 768
THF (tetrahydrofuran), 1073
Thiosulfate ion, 952, 987
Third law of thermodynamics, 816–17
THMs (trihalomethanes), 791
Thomson, J. J., 40, 41, 42, 218
Thomson, William (Lord Kelvin), 400
Thorium-232, 919
Thorium-234, 895, 900
Thortveitite, 966
Threonine, 1082
Thymine, 477, 1091, 1092
Thymol blue, 679
Time, reaction rates and, 577, 585–91
Tin (Sn), 965
    gray, 486
    white, 486
Tire gauges, 397
Tissue, artificial, 509–10
Titanic, raising the, 890
Titanium, 992
    nickel and, 997
Titanium dioxide, 380
Titanium dioxide solar cell, 515
Titanium tetrachloride, 884
Titan (moon), 281, 428
Titins, 1085
Titration(s), 150–54
    acid–base, 730–37
        of polyprotic acids, 737
        strong, 730–33
        weak, 733–37
    end point of, 151
    equivalence point of, 150
TNT (trinitrotoluene), 328, 955
Tokamak, 918
Toluene (methylbenzene), 466, 548,
    789, 1062, 1068
Tooth decay, 747
Tornados, 392, 393
Torr, 396
Torricelli, Evangelista, 396
Trace elements, 57
Trains, high-speed, 496
trans fats, 1089–90
Transferrin, 1024
trans isomers, 1015
Transistor, 489, 490
Transition-metal ions, 303–5
Transition metals, 59, 998–1002
    chromium, 1002
    compounds of. See Coordination
        compounds
    copper. See Copper (Cu)
    electron configurations of, 238–39,
        1000–1
    iron. See Iron (Fe)
    magnetism, 1001–2
        paramagnetism, 1001
    oxidation states of, 1000–1
    physical properties of, 998–99
    radii of, 999

Transition state (activated complex),
    593
Translational motion, 810
Transplatin, 388
Transuranium elements, 51, 903
Trifluoromethyl acetylacetonate (tfac),
    1046
Trigonal bipyramidal geometry, 345,
    349, 350, 354, 361, 1047
Trigonal planar geometry, 343, 345,
    347, 354, 361, 1052
Trigonal pyramidal geometry, 343, 346
Trihalomethanes (THMs), 791
Trimethylamine, 1077
2,2,4-Trimethylpentane, 1062
Trinitroglycerin, 198–99
Trinitrotoluene (TNT), 328, 955
Triple bonds, 307, 1052
    hybrid orbitals and, 364, 369
    length of, 329–30
Triple point, 457
Trisodium phosphate, 714
Tristearin, 193
Tritium, 75, 935
Tropopause, 768
Troposphere, 768, 775–83
    carbon monoxide, 778–79
    nitrogen oxides and photochemical
        smog, 779–80
    sulfur compounds and acid rain,
        776–78
    water vapor, carbon dioxide, and
        climate, 780–83
Trouton's rule, 839
Tryptophan, 1081
T-shaped geometry, 343, 350
Tumor, malignant, 922
Tungsten carbide, 965
Turnover number, 610
Tutankhamun, 143
20-kiloton bomb, 915
Twisted nematic liquid crystal display,
    513
2p orbitals, molecular orbitals from,
    372
Tyndall effect, 557
Tyrosine, 1051, 1082

Uhlenbeck, George, 233
Uncertainty principle, 223–24
    measurement and, 225
UNICEF, 960
Unimolecular reactions, 598
Unit cells, 460–61
United Nations Children's Fund
    (UNICEF), 960
United Nations Intergovernmental
    Panel on Climate Change, 767
Unpaired electrons, 235
Unsaturated hydrocarbons, 1054
    alkenes, 1062–66
    alkynes, 1064–66
    aromatic, 1054, 1068–70
Unsaturated solutions, 535
Uracil, 1091
Uranium (U), 240
    isotopes of, 420, 895
Uranium-233, 914
Uranium-234, 895
Uranium-235, 420, 895, 913–14, 915
Uranium-238, 420, 895, 900, 905, 913,
    917
    abundance of, 919
    rate of decay, 903
Urbain, G., 253
Urea, 193, 623, 1052
Uric acid, 280

Valence band, 484, 485
Valence-bond theory, 355–57
  hybrid orbitals and, 357–62
    involving *d* orbitals, 360
    pi bonding and, 362–68
    sp, 357–58
    sp^2 and *sp*^{3}, 358–60
    VSEPR model and, 355, 357, 360
Valence electron configurations, 242
Valence electrons, 238, 242
  bonding and, 298
  effective nuclear charge of, 259
  less than an octet of, 322–23
  more than an octet of, 323–25
Valence orbitals, 255
Valence-shell electron-pair repulsion
    (VSEPR) model, 343–53
  basis of, 344–45
  for larger molecules, 352–53
  for molecules with expanded
    valence shells, 349–51
  nonbonding electrons and multiple
    bonds, 348–49
  valence-bond theory and, 355, 357,
    360
Valence shells, molecules with
    expanded, 349–51
Valine, 1081
Valve replacement, 508
Vanadium, 995
Vanadium(III) fluoride, 1011
Vanadium(V) fluoride, 1011
Vancomycin, 18–19
van der Waals, Johannes, 422, 440
van der Waals constants, 422–23
van der Waals equation, 422–24
van der Waals forces
  comparison of, 446–47
  dipole–dipole, 440–41
  hydrogen bonding, 443–46
    trends in, 443–45
    in water, 445–46
  ion–dipole, 440
  London dispersion, 441–43
van der Waals radius (nonbonding
    atomic radius), 261
Vanilla, 1074
van't Hoff factor (*i*), 554
Vapor, 5, 394. *See also* Gas(es)
Vaporization
  enthalpies of, 188
  heat of, 449
Vapor pressure, 453–56
  boiling point and, 455
  lowering, 546–48
  molecular-level explanation of, 454
  volatility, temperature, and, 454–55
Variables, calculations involving many,
    404
Vascular grafts, 509
Vector quantities, 353
Vibrational motion, 810
Vinegar, 1076
Vinyl acetate, 1076
Vinyl alcohol, 352–53
Vinyl chloride, 941
Viscosity, 447–48
Visible light, 212, 213
  color and, 1031–32
Visible spectrum, 1032
Vision, chemistry of, 367
Vitamins
  B, 538
  B$_6$, 564
  C (ascorbic acid), 132, 538, 566, 667,
    689, 1054, 1074

D, 538
E, 538, 564
fat- and water-soluble, 538
K, 538
A (retinol), 538
Volatile, 454
Volatile components, separating, 548
Volatile substance, 546
Volatility, 454–55
Volcanoes, 775
Volta, Alessandro, 868
Voltaic (galvanic) cells, 851–54. *See also*
    Batteries
  concentration cells, 867–70
  electromotive force (emf) in, 855–61
    concentration effects on, 865–70
    equilibrium and, 865–67
    oxidizing and reducing agents,
      860–61
    standard reduction (half-cell)
      potentials, 856–60
  molecular view of electrode process,
    854
  standard cell potential of, 859
  work done by, 879–81
Volume(s)
  conversions involving, 27–29
  equilibria and, 649–51
  of gas, 408–10
  interconverting molarity, moles,
    and, 145–46
  law of combining, 401
  molar, 403
  pressure-volume relationship,
    399–400
  quantity-volume relationship, 400–1
  real *vs.* ideal gas behavior and, 421
  SI unit of, 16
  temperature-volume relationship,
    400
Volumetric flasks, 16, 17
von Hevesy, G., 253
Vulcanization, 504–5, 950

Waage, Peter, 631
Wahl, Arthur, 51
Washing soda, 963
Wastes
  nuclear, 917
  oxygen-demanding, 786
Watches, 513
Water, 7, 52. *See also* Aqueous
    equilibria; Aqueous solution(s)
  as analogy for electron flow, 855
  arsenic in drinking, 162–63, 542,
    960–61
  autoionization of, 673–75
  boiling-point elevation of, 549
  bonding in, 348
    hydrogen bonding, 445–46
  chlorine dissolved in, 791
  collecting gases over, 412–13
  critical temperature and pressure of,
    452
  density of, 17
  desalination of, 784–85
  dissolution in, 120, 121
    of ionic solid, 813
    of oxygen, 786
    of sodium chloride, 529
  electrolysis of, 7
  evaporation of, 10
  excess consumption of, 147
  forms of, 394
  fresh, 785–88
  hard, 788

heating curve for, 450
heavy, 935
H$^+$ ions in, 669
of hydration, 533
hydrogonium ion in, 669
ionic compounds in, 120–22
ion product of, 674–75, 694
meniscus, 448
molar mass of, 91
molecular compounds in, 122
molecular model of, 2
from oxidation of glucose, 100–1
perchlorate in drinking, 944
phase diagram of, 457–58
phases of, 437
physical states of, 5
polarity of, 354
properties of, 7
reactions of, 119
  with alkali metals, 277
  with ammonia, 671
  with anions, 696
  with barium oxide, 947
  with butyl chloride, 577
  with calcium, 279–80
  with calcium hydride, 937–38
  with carbon dioxide, 275
  with cations, 696–97
  with hydrogen chloride, 668
  with nitrous acid, 670
sea, 783–84
softening of, 788
solubility of, 119
  guidelines for ionic compounds,
    125
solubility of gases in, 536, 540–41
as solvent, 122, 641, 719
specific heat of, 180, 181
standard reduction potentials in, 857
structural formula for, 53
supercritical, 790
surface tension of, 448
treatment of municipal supplies,
    786–87
valence bonding in, 360
vapor pressure of, 455, 547
vibrational and rotational motion in,
    810
wave characteristics of, 212
Water gas, 936
Water purification, 786–87, 790–92
Water softening, 788
Water-soluble vitamins, 538
Water vapor, 437
  climate and, 780–83
Watson, James, 465
Watt (W), 202
Wave behavior of matter, 222–24
Wave functions, 225
Wavelength, 212, 214–15
Wave mechanics, 224
Wave nature of light, 212–15
Waves, 222–23
Weak acid(s), 130–31, 672, 681–90
  acid-dissociation constant, 681–83,
    684–88, 693–95
  common-ion effect on, 720–23
  percent ionization of, 683–84
  polyprotic acids, 688–90
Weak acid-strong base titrations,
    733–37
Weak base(s), 130–31, 690–93
  common-ion effect on, 720–23
  types of, 692–93
Weak electrolytes, 122–23
  identifying, 131–32

Weak-field ligands, 1036
Weak nuclear force, 45
Weather
  air densities and, 407
  gases and, 392, 393
Weather balloon, 399
Weight
  atomic, 46–48, 87–88, 257
  density *vs.*, 17
  molecular, 46
Welding, 945
Werner, Alfred, 1015, 1047
Werner's theory, 1014–16
White cliffs of Dover, 296, 297
White light, 1031–32
White phosphorus, 957
White tin, 486
Wilkins, Maurice, 465
Wind energy, 197
Wind power, 937
Wires, quantum, 514
Wöhler, Friedrich, 1052
Work, 166
  electrical, 176
  internal energy change and, 171–72
  mechanical, 176
  pressure-volume (mechanical work),
    174–75, 176
  reversibility and, 168–69
Work function, 217
World Health Organization, 791
World Meteorological
    Organization/United Nations
    Environment Programme
    Scientific Assessment of Ozone
    Depletion, 19
World ocean, 783–85
Wrap wire, 981

Xenon, 284, 285
Xenon compounds, 939
Xenon hexafluoride, 389
Xenon tetrafluoride, 939
X-ray crystallography, 465
X-ray diffraction, 465
X-ray diffractometers, 465
X-rays, 212, 213, 218, 236
para-Xylene, 790

Yield
  actual, 105
  percent, 105
  theoretical, 105–7
Young Girl Reading (painting), 3
Yttrium-barium-copper oxide, 497
Yttrium hydroxide, 790, 791
Yucca Mountain, 917

Zeros, significant digits and, 22
Zinc (Zn), 238
  in cathodic protection, 875
  in galvanized iron, 875
  oxidation of, 140
  reaction with hydrochloric acid,
    844–45
  as reducing agent, 861
  in solution of Cu^{2+}, 851–52, 854
Zinc blende, 467, 485
Zinc ion, 265
Zinc sulfide, 909
Zinn, Walter, 915
Zircon, 967
Zirconia, 494
Zirconium, 253, 999
Zone refining, 966
Zwitterion, 703, 1081

Common Ions

Positive Ions (Cations)

1+

Ammonium (NH_4^+)
Cesium (Cs^+)
Copper(I) or cuprous (Cu^+)
Hydrogen (H^+)
Lithium (Li^+)
Potassium (K^+)
Silver (Ag^+)
Sodium (Na^+)

2+

Barium (Ba^{2+})
Cadmium (Cd^{2+})
Calcium (Ca^{2+})
Chromium(II) or chromous (Cr^{2+})
Cobalt(II) or cobaltous (Co^{2+})
Copper(II) or cupric (Cu^{2+})
Iron(II) or ferrous (Fe^{2+})
Lead(II) or plumbous (Pb^{2+})
Magnesium (Mg^{2+})
Manganese(II) or manganous (Mn^{2+})
Mercury(I) or mercurous (Hg_2^{2+})

Mercury(II) or mercuric (Hg^{2+})
Strontium (Sr^{2+})
Nickel(II) (Ni^{2+})
Tin(II) or stannous (Sn^{2+})
Zinc (Zn^{2+})

3+

Aluminum (Al^{3+})
Chromium(III) or chromic (Cr^{3+})
Iron(III) or ferric (Fe^{3+})

Negative Ions (Anions)

1−

Acetate (CH_3COO^- or $C_2H_3O_2^-$)
Bromide (Br^-)
Chlorate (ClO_3^-)
Chloride (Cl^-)
Cyanide (CN^-)
Dihydrogen phosphate ($H_2PO_4^-$)
Fluoride (F^-)
Hydride (H^-)
Hydrogen carbonate or
  bicarbonate (HCO_3^-)

Hydrogen sulfite or bisulfite (HSO_3^-)
Hydroxide (OH^-)
Iodide (I^-)
Nitrate (NO_3^-)
Nitrite (NO_2^-)
Perchlorate (ClO_4^-)
Permanganate (MnO_4^-)
Thiocyanate (SCN^-)

2−

Carbonate (CO_3^{2-})
Chromate (CrO_4^{2-})
Dichromate ($Cr_2O_7^{2-}$)
Hydrogen phosphate (HPO_4^{2-})
Oxide (O^{2-})
Peroxide (O_2^{2-})
Sulfate (SO_4^{2-})
Sulfide (S^{2-})
Sulfite (SO_3^{2-})

3−

Arsenate (AsO_4^{3-})
Phosphate (PO_4^{3-})

Fundamental Constants*

Atomic mass unit	1 amu	$= 1.660538782 \times 10^{-27}$ kg
	1 g	$= 6.02214179 \times 10^{23}$ amu
Avogadro's number	N	$= 6.02214179 \times 10^{23}$/mol
Boltzmann's constant	k	$= 1.3806504 \times 10^{-23}$ J/K
Electron charge	e	$= 1.602176487 \times 10^{-19}$ C
Faraday's constant	F	$= 9.64853399 \times 10^4$ C/mol
Gas constant	R	$= 0.082058205$ L-atm/mol-K
		$= 8.314472$ J/mol-K
Mass of electron	m_e	$= 5.48579909 \times 10^{-4}$ amu
		$= 9.10938215 \times 10^{-31}$ kg
Mass of neutron	m_n	$= 1.008664916$ amu
		$= 1.674927211 \times 10^{-27}$ kg
Mass of proton	m_p	$= 1.007276467$ amu
		$= 1.672621637 \times 10^{-27}$ kg
Pi	π	$= 3.1415927$
Planck's constant	h	$= 6.62606896 \times 10^{-34}$ J-s
Speed of light	c	$= 2.99792458 \times 10^8$ m/s

*Fundamental constants are listed at the National Institute of Standards and Technology Web site:
http://physics.nist.gov/PhysRefData/contents.html

Useful Conversion Factors and Relationships

Length

SI unit: meter (m)

$$1 \text{ km} = 0.62137 \text{ mi}$$
$$1 \text{ mi} = 5280 \text{ ft}$$
$$= 1.6093 \text{ km}$$
$$1 \text{ m} = 1.0936 \text{ yd}$$
$$1 \text{ in.} = 2.54 \text{ cm (exactly)}$$
$$1 \text{ cm} = 0.39370 \text{ in.}$$
$$1 \text{ Å} = 10^{-10} \text{ m}$$

Mass

SI unit: kilogram (kg)

$$1 \text{ kg} = 2.2046 \text{ lb}$$
$$1 \text{ lb} = 453.59 \text{ g}$$
$$= 16 \text{ oz}$$
$$1 \text{ amu} = 1.660538782 \times 10^{-24} \text{ g}$$

Temperature

SI unit: Kelvin (K)

$$0 \text{ K} = -273.15 \text{ °C}$$
$$= -459.67 \text{ °F}$$
$$\text{K} = \text{°C} + 273.15$$
$$\text{°C} = \tfrac{5}{9}(\text{°F} - 32°)$$
$$\text{°F} = \tfrac{9}{5}\text{°C} + 32°$$

Energy (derived)

SI unit: Joule (J)

$$1 \text{ J} = 1 \text{ kg-m}^2/\text{s}^2$$
$$1 \text{ J} = 0.2390 \text{ cal}$$
$$= 1 \text{ C} \times 1 \text{ V}$$
$$1 \text{ cal} = 4.184 \text{ J}$$
$$1 \text{ eV} = 1.602 \times 10^{-19} \text{ J}$$

Pressure (derived)

SI unit: Pascal (Pa)

$$1 \text{ Pa} = 1 \text{ N/m}^2$$
$$= 1 \text{ kg/m-s}^2$$
$$1 \text{ atm} = 101{,}325 \text{ Pa}$$
$$= 760 \text{ torr}$$
$$= 14.70 \text{ lb/in}^2$$
$$1 \text{ bar} = 10^5 \text{ Pa}$$
$$1 \text{ torr} = 1 \text{ mm Hg}$$

Volume (derived)

SI unit: cubic meter (m³)

$$1 \text{ L} = 10^{-3} \text{ m}^3$$
$$= 1 \text{ dm}^3$$
$$= 10^3 \text{ cm}^3$$
$$= 1.0567 \text{ qt}$$
$$1 \text{ gal} = 4 \text{ qt}$$
$$= 3.7854 \text{ L}$$
$$1 \text{ cm}^3 = 1 \text{ mL}$$
$$1 \text{ in}^3 = 16.4 \text{ cm}^3$$

Index of Useful Tables and Figures

Activity Series of the Metals	Table 4.5
Atomic and Ionic Radii	Figure 7.8
Atomic Radii	Figure 7.7
Average Bond Enthalpies	Table 8.4
Average Bond Lengths	Table 8.5
Common Strong Acids and Bases	Table 4.2
Electromagnetic Spectrum	Figure 6.4
Electron Configurations of the Elements	Figure 6.31
Electronegativities of the Elements	Figure 8.6
Summary of Electrolytic Behavior of Common Soluble Ionic and Molecular Compounds	Table 4.3
Common Functional Groups in Organic Compounds	Table 25.4
SI Base Units	Table 1.4
Selected Prefixes used in the Metric System	Table 1.5
Solubility Guidelines for Common Ionic Compounds in Water	Table 4.1